GENERAL AVIATION AIRCRAFT DESIGN

Dedication

I dedicate this book to my Father and Mother, to whom I am forever indebted, and my wonderful wife Linda for her unconditional support and patience.

GENERAL AVIATION AIRCRAFT DESIGN: APPLIED METHODS AND PROCEDURES

SNORRI GUDMUNDSSON BScAE, MScAE, FAA DER(ret.)
Assistant Professor of Aerospace Engineering,
Embry-Riddle Aeronautical University

AMSTERDAM • BOSTON • HEIDELBERG • LONDON
NEW YORK • OXFORD • PARIS • SAN DIEGO
SAN FRANCISO • SINGAPORE • SYDNEY • TOKYO

Butterworth-Heinemann is an imprint of Elsevier

Butterworth-Heinemann is an imprint of Elsevier
The Boulevard, Langford Lane, Kidlington, Oxford OX5 1GB, UK
225 Wyman Street, Waltham, MA 02451, USA

First edition 2014

Copyright © 2014 Elsevier Inc. All rights reserved.

No part of this publication may be reproduced, stored in a retrieval system or transmitted in any form or by any means electronic, mechanical, photocopying, recording or otherwise without the prior written permission of the publisher

Permissions may be sought directly from Elsevier's Science & Technology Rights Department in Oxford, UK: phone (+44) (0) 1865 843830; fax (+44) (0) 1865 853333; email: permissions@elsevier.com. Alternatively you can submit your request online by visiting the Elsevier web site at http://elsevier.com/locate/permissions, and selecting Obtaining permission to use Elsevier material

Notice
No responsibility is assumed by the publisher for any injury and/or damage to persons or property as a matter of products liability, negligence or otherwise, or from any use or operation of any methods, products, instructions or ideas contained in the material herein. Because of rapid advances in the medical sciences, in particular, independent verification of diagnoses and drug dosages should be made

Library of Congress Cataloging-in-Publication Data
A catalog record for this book is available from the Library of Congress

British Library Cataloguing-in-Publication Data
A catalogue record for this book is available from the British Library

ISBN: 978-0-12-809998-8

For information on all Butterworth-Heinemann publications visit our website at elsevierdirect.com

Transferred to Digital Printing in 2016

Contents

Preface xiii
Acknowledgments xv
Helpful Notes xvi
   Helpful Websites for the Aircraft Designer xvi
   The Greek Alphabet xvi
   List of Abbreviations and Common Terms xvi
   Handy Conversion Factors xix
   A Note about Mass and Force xix
   A Note about Format xx
   Aircraft Design DOs and DON'Ts xxi
   Professor Gudmundsson's Cake Philosophy xxi

1. The Aircraft Design Process

1.1 Introduction 2
   1.1.1 The Content of this Chapter 5
   1.1.2 Important Elements of a New Aircraft Design 5
1.2 General Process of Aircraft Design 11
   1.2.1 Common Description of the Design Process 11
   1.2.2 Important Regulatory Concepts 13
1.3 Aircraft Design Algorithm 15
   1.3.1 Conceptual Design Algorithm for a GA Aircraft 16
   1.3.2 Implementation of the Conceptual Design Algorithm 16
1.4 Elements of Project Engineering 19
   1.4.1 Gantt Diagrams 19
   1.4.2 Fishbone Diagram for Preliminary Airplane Design 19
   1.4.3 Managing Compliance with Project Requirements 21
   1.4.4 Project Plan and Task Management 21
   1.4.5 Quality Function Deployment and a House of Quality 21
1.5 Presenting the Design Project 27
   Variables 32
   References 32

2. Aircraft Cost Analysis

2.1 Introduction 33
   2.1.1 The Content of this Chapter 34
   2.1.2 A Review of the State of the GA Industry 34
2.2 Estimating Project Development Costs 36
   2.2.1 Quantity Discount Factor 36
   2.2.2 Development Cost of a GA Aircraft – the Eastlake Model 37
   2.2.3 Development Cost of a Business Aircraft – the Eastlake Model 44
2.3 Estimating Aircraft Operational Costs 46
   2.3.1 Operational Cost of a GA Aircraft 46
   2.3.2 Operational Cost of a Business Aircraft 49
   Exercises 51
   Variables 52
   References 53

3. Initial Sizing

3.1 Introduction 55
   3.1.1 The Content of this Chapter 56
   3.1.2 Fundamental Concepts 56
   3.1.3 Software Tools 57
3.2 Constraint Analysis 57
   3.2.1 General Methodology 58
   3.2.2 Introduction of Stall Speed Limits into the Constraint Diagram 65
3.3 Introduction to Trade Studies 66
   3.3.1 Step-by-step: Stall Speed – Cruise Speed Carpet Plot 67
   3.3.2 Design of Experiments 69
   3.3.3 Cost Functions 72
   Exercises 74
   Variables 75

4. Aircraft Conceptual Layout

4.1 Introduction 77
   4.1.1 The Content of this Chapter 78
   4.1.2 Requirements, Mission, and Applicable Regulations 78
   4.1.3 Past and Present Directions in Aircraft Design 79
   4.1.4 Aircraft Component Recognition 79
4.2 The Fundamentals of the Configuration Layout 82
   4.2.1 Vertical Wing Location 82
   4.2.2 Wing Configuration 86
   4.2.3 Wing Dihedral 86
   4.2.4 Wing Structural Configuration 87
   4.2.5 Cabin Configurations 88
   4.2.6 Propeller Configuration 89
   4.2.7 Engine Placement 89
   4.2.8 Landing Gear Configurations 91
   4.2.9 Tail Configurations 92

    4.2.10 Configuration Selection Matrix 92
    Variables 93
    References 95

5. Aircraft Structural Layout

5.1 Introduction 97
    5.1.1 The Content of this Chapter 98
    5.1.2 Notes on Aircraft Loads 98
5.2 Aircraft Fabrication and Materials 98
    5.2.1 Various Fabrication Methods 100
    5.2.2 Aluminum Alloys 103
    5.2.3 Steel Alloys 106
    5.2.4 Titanium Alloys 107
    5.2.5 Composite Materials 108
5.3 Airframe Structural Layout 116
    5.3.1 Important Structural Concepts 116
    5.3.2 Fundamental Layout of the Wing Structure 120
    5.3.3 Fundamental Layout of the Horizontal and Vertical Tail Structures 126
    5.3.4 Fundamental Layout of the Fuselage Structure 128
    Variables 131
    References 131

6. Aircraft Weight Analysis

6.1 Introduction 134
    6.1.1 The Content of this Section 135
    6.1.2 Definitions 135
    6.1.3 Fundamental Weight Relations 137
    6.1.4 Mission Analysis 137
6.2 Initial Weight Analysis Methods 138
    6.2.1 Method 1: Initial Gross Weight Estimation Using Historical Relations 138
    6.2.2 Method 2: Historical Empty Weight Fractions 140
6.3 Detailed Weight Analysis Methods 141
6.4 Statistical Weight Estimation Methods 142
    6.4.1 Method 3: Statistical Aircraft Component Methods 142
    6.4.2 Statistical Methods to Estimate Engine Weight 145
6.5 Direct Weight Estimation Methods 147
    6.5.1 Direct Weight Estimation for a Wing 147
    6.5.2 Variation of Weight with AR 154
6.6 Inertia Properties 161
    6.6.1 Fundamentals 161
    6.6.2 Reference Locations 162
    6.6.3 Total Weight 162
    6.6.4 Moment about (X_0, Y_0, Z_0) 162
    6.6.5 Center of Mass, Center of Gravity 164
    6.6.6 Determination of CG Location by Aircraft Weighing 165
    6.6.7 Mass Moments and Products of Inertia 165
    6.6.8 Moment of Inertia of a System of Discrete Point Loads 167
    6.6.9 Product of Inertia of a System of Discrete Point Loads 168
    6.6.10 Inertia matrix 168
    6.6.11 Center of Gravity Envelope 168
    6.6.12 Creating the CG Envelope 169
    6.6.13 In-flight Movement of the CG 173
    6.6.14 Weight Budgeting 173
    6.6.15 Weight Tolerancing 174
    Exercises 176
    Variables 178
    References 180

7. Selecting the Power Plant

7.1 Introduction 182
    7.1.1 The Content of this Section 183
    7.1.2 Power Plant Options for Aviation 183
    7.1.3 The Basics of Energy, Work, and Power 183
    7.1.4 Thermodynamics of the Power Plant 183
    7.1.5 General Theory of Thrust Generation 184
    7.1.6 Fundamental Definitions 185
    7.1.7 Fuel Basics 187
7.2 The Properties of Selected Engine Types 190
    7.2.1 Piston Engines 190
    7.2.2 Turboprops 196
    7.2.3 Turbojets 199
    7.2.4 Turbofans 200
    7.2.5 Electric Motors 203
    7.2.6 Computer code: Thrust as a Function of Altitude and Mach Number 207
7.3 Aircraft Power Plant Installation 209
    7.3.1 Piston Engine Installation 210
    7.3.2 Piston Engine Inlet and Exit Sizing 213
    7.3.3 Installation of Gas Turbines 222
    7.3.4 Jet Engine Inlet Sizing 223
7.4 Special Topics 227
    7.4.1 The Use of Gearboxes 227
    7.4.2 Step-by-step: Extracting Piston Power from Engine Performance Charts 228
    7.4.3 Extracting Piston Power Using the Petty Equation 229
    Exercises 232
    Variables 232
    References 234

8. The Anatomy of the Airfoil

8.1 Introduction 236
    8.1.1 The Content of this Section 237
    8.1.2 Dimensional Analysis — Buckingham's Π Theorem 237
    8.1.3 Representation of Forces and Moments 238
    8.1.4 Properties of Typical Airfoils 239
    8.1.5 The Pressure Coefficient 241
    8.1.6 Chordwise Pressure Distribution 242
    8.1.7 Center of Pressure and Aerodynamic Center 243
    8.1.8 The Generation of Lift 245
    8.1.9 Boundary Layer and Flow Separation 247
    8.1.10 Estimation of Boundary Layer Thickness 250
    8.1.11 Airfoil Stall Characteristics 251
    8.1.12 Analysis of Ice-accretion on Airfoils 252
    8.1.13 Designations of Common Airfoils 253
    8.1.14 Airfoil Design 254

8.2 The Geometry of the Airfoil 256
    8.2.1 Airfoil Terminology 256
    8.2.2 NACA Four-digit Airfoils 257
    8.2.3 NACA Five-digit Airfoils 261
    8.2.4 NACA 1-series Airfoils 263
    8.2.5 NACA 6-series Airfoils 264
    8.2.6 NACA 7-series Airfoils 266
    8.2.7 NACA 8-series Airfoils 267
    8.2.8 NACA Airfoils in Summary – Pros and Cons and Comparison of Characteristics 267
    8.2.9 Properties of Selected NACA Airfoils 267
    8.2.10 Famous Airfoils 267
8.3 The Force and Moment Characteristics of the Airfoil 275
    8.3.1 The Effect of Camber 276
    8.3.2 The Two-dimensional Lift Curve 276
    8.3.3 The Maximum Lift Coefficient, C_{lmax} 276
    8.3.4 The Effect of Reynolds Number 277
    8.3.5 Compressibility Effects 278
    8.3.6 Compressibility Modeling 280
    8.3.7 The Critical Mach Number, M_{crit} 281
    8.3.8 The Effect of Early Flow Separation 282
    8.3.9 The Effect of Addition of a Slot or Slats 283
    8.3.10 The Effect of Deflecting a Flap 284
    8.3.11 The Effect of Cruise Flaps 284
    8.3.12 The Effect of Deploying a Spoiler 285
    8.3.13 The Effect of Leading Edge Roughness and Surface Smoothness 286
    8.3.14 Drag Models for Airfoils Using Standard Airfoil Data 287
    8.3.15 Airfoil Selection How-to 289
    Exercises 294
    Variables 294
    References 296

9. The Anatomy of the Wing

9.1 Introduction 301
    9.1.1 The Content of this Chapter 301
    9.1.2 Definition of Reference Area 302
    9.1.3 The Process of Wing Sizing 302
9.2 The Trapezoidal Wing Planform 303
    9.2.1 Definitions 303
    9.2.2 Poor Man's Determination of the MGC 306
    9.2.3 Planform Dimensions in Terms of b, λ, and AR 307
    9.2.4 Approximation of an Airfoil Cross-sectional Area 308
9.3 The Geometric Layout of the Wing 309
    9.3.1 Wing Aspect Ratio, AR 309
    9.3.2 Wing Taper Ratio, TR or λ 315
    9.3.3 Leading Edge and Quarter-chord Sweep Angles, Λ_{LE} and $\Lambda_{C/4}$ 317
    9.3.4 Dihedral or Anhedral, Γ 318
    9.3.5 Wing Twist – Washout and Washin, ϕ 319
    9.3.6 Wing Angle-of-incidence, i_W 325
    9.3.7 Wing Layout Properties of Selected Aircraft 329
9.4 Planform Selection 329
    9.4.1 The Optimum Lift Distribution 329
    9.4.2 Methods to Present Spanwise Lift Distribution 331
    9.4.3 Constant-chord ("Hershey Bar") Planform 332
    9.4.4 Elliptic Planforms 333
    9.4.5 Straight Tapered Planforms 334
    9.4.6 Swept Planforms 336
    9.4.7 Cranked Planforms 338
    9.4.8 Delta Planforms 340
    9.4.9 Some Exotic Planform Shapes 341
9.5 Lift and Moment Characteristics of a 3D Wing 342
    9.5.1 Properties of the Three-dimensional Lift Curve 342
    9.5.2 The Lift Coefficient 344
    9.5.3 Determination of Lift Curve Slope, $C_{L\alpha}$, for a 3D Lifting Surface 345
    9.5.4 The Lift Curve Slope of a Complete Aircraft 347
    9.5.5 Step-by-step: Transforming the Lift Curve from 2D to 3D 348
    9.5.6 The Law of Effectiveness 349
    9.5.7 Flexible Wings 349
    9.5.8 Ground Effect 350
    9.5.9 Impact of C_{Lmax} and Wing Loading on Stalling Speed 353
    9.5.10 Step-by-step: Rapid C_{Lmax} Estimation [5] 353
    9.5.11 Step-by-step: C_{Lmax} Estimation per USAF DATCOM Method 1 355
    9.5.12 Step-by-step: C_{Lmax} Estimation per USAF DATCOM Method 2 356
    9.5.13 C_{Lmax} for Selected Aircraft 360
    9.5.14 Estimation of Oswald's Span Efficiency 363
9.6 Wing Stall Characteristics 366
    9.6.1 Growth of Flow Separation on an Aircraft 367
    9.6.2 General Stall Progression on Selected Wing Planform Shapes 369
    9.6.3 Deviation from Generic Stall Patterns 369
    9.6.4 Tailoring the Stall Progression 369
    9.6.5 Cause of Spanwise Flow for a Swept-back Wing Planform 374
    9.6.6 Pitch-up Stall Boundary for a Swept-back Wing Planform 375
    9.6.7 Influence of Manufacturing Tolerances on Stall Characteristics 378
9.7 Numerical Analysis of the Wing 379
    9.7.1 Prandtl's Lifting Line Theory 379
    9.7.2 Prandtl's Lifting Line Method – Special Case: The Elliptical Wing 384
    9.7.3 Prandtl's Lifting Line Method – Special Case: Arbitrary Wings 386
    9.7.4 Accounting for a Fuselage in Prandtl's Lifting Line Method 390
    9.7.5 Computer code: Prandtl's Lifting Line Method 393
    Exercises 396
    Variables 397
    References 398

10. The Anatomy of Lift Enhancement

10.1 Introduction 402
    10.1.1 The Content of this Chapter 403
10.2 Leading Edge High-lift Devices 403
    10.2.1 Hinged Leading Edge (Droop Nose) 403
    10.2.2 Variable-camber Leading Edge 405

10.2.3 Fixed Slot 406
  10.2.4 The Krüger Flap 408
  10.2.5 The Leading Edge Slat 412
  10.2.6 Summary of Leading Edge Device Data 416
10.3 Trailing Edge High-lift Devices 417
  10.3.1 Plain Flap 417
  10.3.2 Split Flap 420
  10.3.3 Junkers Flap or External Flap 423
  10.3.4 The Single-slotted Flap 425
  10.3.5 Double-slotted Flaps 427
  10.3.6 Fowler Flaps 430
  10.3.7 Gurney Flap 432
  10.3.8 Summary of Trailing Edge Device Data 434
10.4 Effect of Deploying High-lift Devices on Wings 436
  10.4.1 Lift Distribution on Wings with Flaps Deflected 436
  10.4.2 Wing Partition Method 437
10.5 Wingtip Design 441
  10.5.1 The Round Wingtip 443
  10.5.2 The Spherical Wingtip 443
  10.5.3 The Square Wingtip 444
  10.5.4 Booster Wingtips 444
  10.5.5 Hoerner Wingtip 446
  10.5.6 Raked Wingtip 446
  10.5.7 Endplate Wingtip 448
  10.5.8 The Winglet 448
  10.5.9 The Polyhedral Wing(tip) 452
  10.5.10 Comparison Based on Potential Flow Theory 453
  Variables 455
  References 456

11. The Anatomy of the Tail

11.1 Introduction 460
  11.1.1 The Content of this Chapter 461
  11.1.2 The Process of Tail Sizing 461
11.2 Fundamentals of Static Stability and Control 462
  11.2.1 Fundamentals of Static Longitudinal Stability 463
  11.2.2 Modeling the Pitching Moment for a Simple Wing-HT System 466
  11.2.3 Horizontal Tail Downwash Angle 466
  11.2.4 Historical Values of $C_{m\alpha}$ 468
  11.2.5 Longitudinal Equilibrium for Any Configuration 468
  11.2.6 The Stick-fixed and Stick-free Neutral Points 472
  11.2.7 Fundamentals of Static Directional and Lateral Stability 475
  11.2.8 Requirements for Static Directional Stability 476
  11.2.9 Requirements for Lateral Stability 477
  11.2.10 Historical Values of $C_{n\beta}$ and $C_{l\beta}$ 477
  11.2.11 The Dorsal Fin 477
  11.2.12 The Ventral Fin 480
  11.2.13 Tail Design and Spin Recovery 482
11.3 On the Pros and Cons of Tail Configurations 483
  11.3.1 Conventional Tail 483
  11.3.2 Cruciform Tail 486
  11.3.3 T-tail 486
  11.3.4 V-tail or Butterfly Tail 489
  11.3.5 Inverted V-tail 493
  11.3.6 Y-tail 493
  11.3.7 Inverted Y-tail 494
  11.3.8 H-tail 494
  11.3.9 Three-surface Configuration 495
  11.3.10 A-tail 495
  11.3.11 Twin Tail-boom or U-tail Configuration 496
  11.3.12 Canard Configuration 496
  11.3.13 Design Guidelines when Positioning the HT for an Aft Tail Configuration 497
11.4 The Geometry of the Tail 499
  11.4.1 Definition of Reference Geometry 499
  11.4.2 Horizontal and Vertical Tail Volumes 500
  11.4.3 Design Guidelines for HT Sizing for Stick-fixed Neutral Point 501
  11.4.4 Recommended Initial Values for V_{HT} and V_{VT} 502
11.5 Initial Tail Sizing Methods 502
  11.5.1 Method 1: Initial Tail Sizing Optimization Considering the Horizontal Tail Only 503
  11.5.2 Method 2: Initial Tail Sizing Optimization Considering the Vertical Tail Only 510
  11.5.3 Method 3: Initial Tail Sizing Optimization Considering Horizontal and Vertical Tail 513
  Exercises 517
  Variables 517
  References 519

12. The Anatomy of the Fuselage

12.1 Introduction 521
  12.1.1 The Content of this Chapter 521
  12.1.2 The Function of the Fuselage 522
12.2 Fundamentals of Fuselage Shapes 523
  12.2.1 The Frustum-shaped Fuselage 523
  12.2.2 The Pressure Tube Fuselage 523
  12.2.3 The Tadpole Fuselage 524
12.3 Sizing the Fuselage 526
  12.3.1 Initial Design of the External Shape of the Fuselage 526
  12.3.2 Refining the External Shape of the Fuselage 529
  12.3.3 Internal Dimensions of the Fuselage 531
  12.3.4 Cockpit Layout 532
12.4 Estimating the Geometric Properties of the Fuselage 535
  12.4.1 Simple Estimation of the Surface Area of a Body of Revolution 536
  12.4.2 Fundamental Properties of Selected Solids 536
  12.4.3 Surface Areas and Volumes of a Typical Tubular Fuselage 538
  12.4.4 Surface Areas and Volumes of a Tadpole Fuselage 539
  12.4.5 Surface Areas and Volumes of a Pod-style Fuselage 544
12.5 Additional Information 544
  Variables 545
  References 545

13. The Anatomy of the Landing Gear

13.1 Introduction 548
  13.1.1 The Content of this Chapter 548
  13.1.2 Landing Gear Arrangement 548
  13.1.3 Landing Gear Design Checklist 549

- 13.2 Tires, Wheels, and Brakes 551
  - 13.2.1 Important Dimensions and Concepts for Landing Gear Design 551
  - 13.2.2 Retractable Landing Gear 553
  - 13.2.3 Types and Sizes of Tires, Wheels and Brakes 555
  - 13.2.4 Types of Landing Gear Legs 560
  - 13.2.5 Reaction of Landing Gear Forces 565
  - 13.2.6 Comparing the Ground Characteristics of Taildragger and Tricycle Landing Gear 565
- 13.3 Geometric Layout of the Landing Gear 567
  - 13.3.1 Geometric Layout of the Tricycle Landing Gear 567
  - 13.3.2 Geometric Layout of the Taildragger Landing Gear 569
  - 13.3.3 Geometric Layout of the Monowheel Landing Gear with Outriggers 571
  - 13.3.4 Tricycle Landing Gear Reaction Loads 571
  - 13.3.5 Taildragger Landing Gear Reaction Loads 576
  - Variables 579
  - References 580

14. The Anatomy of the Propeller

- 14.1 Introduction 582
  - 14.1.1 The Content of this Chapter 584
  - 14.1.2 Propeller Configurations 584
  - 14.1.3 Important Nomenclature 586
  - 14.1.4 Propeller Geometry 587
  - 14.1.5 Geometric Propeller Pitch 588
  - 14.1.6 Windmilling Propellers 593
  - 14.1.7 Fixed Versus Constant-speed Propellers 594
  - 14.1.8 Propulsive or Thrust Power 595
- 14.2 Propeller Effects 595
  - 14.2.1 Angular Momentum and Gyroscopic Effects 596
  - 14.2.2 Slipstream Effects 597
  - 14.2.3 Propeller Normal and Side Force 598
  - 14.2.4 Asymmetric Yaw Effect 599
  - 14.2.5 Asymmetric Yaw Effect for a Twin-engine Aircraft 599
  - 14.2.6 Blockage Effects 602
  - 14.2.7 Hub and Tip Effects 604
  - 14.2.8 Effects of High Tip Speed 605
  - 14.2.9 Skewed Wake Effects – $A \cdot q$ Loads 605
  - 14.2.10 Propeller Noise 606
- 14.3 Properties and Selection of the Propeller 607
  - 14.3.1 Tips for Selecting a Suitable Propeller 607
  - 14.3.2 Rapid Estimation of Required Prop Diameter 608
  - 14.3.3 Rapid Estimation of Required Propeller Pitch 610
  - 14.3.4 Estimation of Required Propeller Efficiency 610
  - 14.3.5 Advance Ratio 611
  - 14.3.6 Definition of Activity Factor 613
  - 14.3.7 Definition of Power- and Thrust-related Coefficients 614
  - 14.3.8 Effect of Number of Blades on Power 615
  - 14.3.9 Propulsive Efficiency 618
  - 14.3.10 Moment of Inertia of the Propeller 619
- 14.4 Determination of Propeller Thrust 620
  - 14.4.1 Converting Piston BHP to Thrust 620
  - 14.4.2 Propeller Thrust at Low Airspeeds 621
  - 14.4.3 Step-by-step: Determining Thrust Using a Propeller Efficiency Table 630
  - 14.4.4 Estimating Thrust From Manufacturer's Data 631
  - 14.4.5 Other Analytical Methods 632
- 14.5 Rankine-Froude Momentum Theory 632
  - 14.5.1 Formulation 633
  - 14.5.2 Ideal Efficiency 635
  - 14.5.3 Maximum Static Thrust 636
  - 14.5.4 Computer code: Estimation of Propeller Efficiency Using the Momentum Theorem 638
- 14.6 Blade Element Theory 640
  - 14.6.1 Formulation 641
  - 14.6.2 Determination of α_i Using the Momentum Theory 650
  - 14.6.3 Compressibility Corrections 654
  - 14.6.4 Step-by-step: Prandtl's Tip and Hub Loss Corrections 655
  - 14.6.5 Computer code: Determination of the Propeller Induced Velocity 656
  - Variables 657
  - References 658

15. Aircraft Drag Analysis

- 15.1 Introduction 663
  - 15.1.1 The Content of this Chapter 665
- 15.2 The Drag Model 665
  - 15.2.1 Basic Drag Modeling 666
  - 15.2.2 Quadratic Drag Modeling 666
  - 15.2.3 Approximating the Drag Coefficient at High Lift Coefficients 670
  - 15.2.4 Non-quadratic Drag Modeling 672
  - 15.2.5 Lift-induced Drag Correction Factors 672
  - 15.2.6 Graphical Determination of L/D_{max} 673
  - 15.2.7 Comparing the Accuracy of the Simplified and Adjusted Drag Models 673
- 15.3 Deconstructing the Drag Model: the Drag Coefficients 674
  - 15.3.1 Basic Drag Coefficient: C_{D0} 674
  - 15.3.2 The Skin Friction Drag Coefficient: C_{Df} 675
  - 15.3.3 Step-by-step: Calculating the Skin Friction Drag Coefficient 680
  - 15.3.4 The Lift-induced Drag Coefficient: C_{Di} 686
  - 15.3.5 Total Drag Coefficient: C_D 691
  - 15.3.6 Various Means to Reduce Drag 691
- 15.4 The Drag Characteristics of the Airplane as a Whole 693
  - 15.4.1 The Effect of Aspect Ratio on a Three-dimensional Wing 693
  - 15.4.2 The Effect of Mach Number 695
  - 15.4.3 The Effect of Yaw Angle β 695
  - 15.4.4 The Effect of Control Surface Deflection – Trim Drag 695
  - 15.4.5 The Rapid Drag Estimation Method 696
  - 15.4.6 The Component Drag Build-up Method 697
  - 15.4.7 Component Interference Factors 700
  - 15.4.8 Form Factors for Wing, HT, VT, Struts, Pylons 700
  - 15.4.9 Form Factors for a Fuselage and a Smooth Canopy 701

15.5 Miscellaneous or Additive Drag 708
    15.5.1 Cumulative Result of Undesirable Drag (CRUD) 709
    15.5.2 Trim Drag 710
    15.5.3 Cooling Drag 714
    15.5.4 Drag of Simple Wing-like Surfaces 715
    15.5.5 Drag of Streamlined Struts and Landing Gear Pant Fairings 715
    15.5.6 Drag of Landing Gear 718
    15.5.7 Drag of Floats 724
    15.5.8 Drag of Deployed Flaps 725
    15.5.9 Drag Correction for Cockpit Windows 726
    15.5.10 Drag of Canopies 727
    15.5.11 Drag of Blisters 728
    15.5.12 Drag Due to Compressibility Effects 730
    15.5.13 Drag of Windmilling and Stopped Propellers 731
    15.5.14 Drag of Antennas 731
    15.5.15 Drag of Various Geometry 732
    15.5.16 Drag of Parachutes 735
    15.5.17 Drag of Various External Sources 736
    15.5.18 Corrections of the Lift-induced Drag 737
15.6 Special Topics Involving Drag 740
    15.6.1 Step-by-step: Extracting Drag from L/D_{max} 740
    15.6.2 Step-by-step: Extracting Drag from a Flight Polar Using the Quadratic Spline Method 741
    15.6.3 Step-by-step: Extracting Drag Coefficient for a Piston-powered Propeller Aircraft 744
    15.6.4 Computer code 15-1: Extracting Drag Coefficient from Published Data for Piston Aircraft 748
    15.6.5 Step-by-step: Extracting Drag Coefficient for a Jet Aircraft 749
    15.6.6 Determining Drag Characteristics from Wind Tunnel Data 750
15.7 Additional Information – Drag of Selected Aircraft 752
    15.7.1 General Range of Subsonic Minimum Drag Coefficients 752
    15.7.2 Drag of Various Aircraft by Class 752
    Exercises 756
    Variables 757
    References 759

16. Performance – Introduction

16.1 Introduction 761
    16.1.1 The Content of this Chapter 762
    16.1.2 Performance Padding Policy 762
16.2 Atmospheric Modeling 763
    16.2.1 Atmospheric Ambient Temperature 763
    16.2.2 Atmospheric Pressure and Density for Altitudes below 36,089 ft (11,000 m) 764
    16.2.3 Atmospheric Property Ratios 764
    16.2.4 Pressure and Density Altitudes below 36,089 ft (11,000 m) 765
    16.2.5 Density of Air Deviations from a Standard Atmosphere 765
    16.2.6 Frequently Used Formulas for a Standard Atmosphere 767
16.3 The Nature of Airspeed 768
    16.3.1 Airspeed Indication Systems 768
    16.3.2 Airspeeds: Instrument, Calibrated, Equivalent, True, and Ground 769
    16.3.3 Important Airspeeds for Aircraft Design and Operation 771
16.4 The Flight Envelope 774
    16.4.1 Step-by-step: Maneuvering Loads and Design Airspeeds 775
    16.4.2 Step-by-step: Gust Loads 778
    16.4.3 Step-by-step: Completing the Flight Envelope 782
    16.4.4 Flight Envelopes for Various GA Aircraft 782
16.5 Sample Aircraft 783
    16.5.1 Cirrus SR22 783
    16.5.2 Learjet 45XR 785
    Exercises 786
    Variables 787
    References 789

17. Performance – Take-Off

17.1 Introduction 791
    17.1.1 The Content of this Chapter 792
    17.1.2 Important Segments of the T-O Phase 792
    17.1.3 Definition of a Balanced Field Length 795
17.2 Fundamental Relations for the Take-Off Run 797
    17.2.1 General Free-body Diagram of the T-O Ground Run 797
    17.2.2 The Equation of Motion for a T-O Ground Run 798
    17.2.3 Review of Kinematics 799
    17.2.4 Formulation of Required Aerodynamic Forces 800
    17.2.5 Ground Roll Friction Coefficients 800
    17.2.6 Determination of the Lift-off Speed 800
    17.2.7 Determination of Time to Lift-off 802
17.3 Solving the Equation of Motion of the T-O 802
    17.3.1 Method 1: General Solution of the Equation of Motion 802
    17.3.2 Method 2: Rapid T-O Distance Estimation for a Piston-powered Airplane [5] 805
    17.3.3 Method 3: Solution Using Numerical Integration Method 807
    17.3.4 Determination of Distance During Rotation 813
    17.3.5 Determination of Distance for Transition 813
    17.3.6 Determination of Distance for Climb Over an Obstacle 814
    17.3.7 Treatment of T-O Run for a Taildragger 815
    17.3.8 Take-off Sensitivity Studies 816
17.4 Database – T-O Performance of Selected Aircraft 817
    Exercises 817
    Variables 819
    References 820

18. Performance – Climb

18.1 Introduction 821
    18.1.1 The Content of this Chapter 822
18.2 Fundamental Relations for the Climb Maneuver 822

18.2.1 General Two-dimensional Free-body Diagram for an Aircraft 822
18.2.2 General Planar Equations of Motion of an Airplane 823
18.2.3 Equations of Motion for Climbing Flight 823
18.2.4 Horizontal and Vertical Airspeed 824
18.2.5 Power Available, Power Required, and Excess Power 824
18.2.6 Vertical Airspeed in Terms of Thrust or Power 824
18.2.7 Rate-of-climb 825
18.3 General Climb Analysis Methods 825
18.3.1 General Rate-of-climb 825
18.3.2 General Climb Angle 827
18.3.3 Max Climb Angle for a Jet 827
18.3.4 Airspeed for θ_{max} for a Jet (Best Angle-of-climb Speed) 828
18.3.5 ROC for θ_{max} for a Jet 829
18.3.6 Airspeed for Best ROC for a Jet 829
18.3.7 Best ROC for a Jet 831
18.3.8 Airspeed for θ_{max} for a Propeller-powered Airplane 832
18.3.9 Airspeed for Best ROC for a Propeller-powered Airplane 833
18.3.10 Best Rate-of-climb for a Propeller-powered Airplane 834
18.3.11 Time to Altitude 836
18.3.12 Absolute/Service Ceiling Altitude 838
18.3.13 Numerical Analysis of the Climb Maneuver — Sensitivity Studies 840
18.4 Aircraft Database — Rate-of-climb of Selected Aircraft 842
Variables 844
References 845

19. Performance — Cruise

19.1 Introduction 848
19.1.1 The Content of this Chapter 849
19.1.2 General Free-body Diagram for Steady Level Flight 849
19.1.3 Planar Equations of Motion (Assumes No Rotation About y-axis) 849
19.1.4 Important Airspeeds for Propeller Aircraft 850
19.1.5 Important Airspeeds for Subsonic Jet Aircraft 851
19.2 General Cruise Analysis Methods for Steady Flight 851
19.2.1 Plotting the Drag Polar 852
19.2.2 Drag Breakdown 852
19.2.3 Required versus Available Thrust 855
19.2.4 Airspeed in Terms of Thrust 857
19.2.5 Minimum Airspeed, V_{min} 860
19.2.6 Stalling Speed, V_S 860
19.2.7 Airspeed of Minimum Power Required, V_{PRmin} 864
19.2.8 Airspeed of Minimum Thrust Required, V_{TRmin}, or Best Glide Speed, V_{BG}, V_{LDmax} 867
19.2.9 Best Range Airspeed for a Jet, V_{Rmax} 875
19.2.10 Maximum Level Airspeed, V_{max} 877
19.2.11 Flight Envelope 880
19.2.12 Power Required 882
19.2.13 Power Available for a Piston-powered Aircraft 883
19.2.14 Computer code: Determining Maximum Level Airspeed, V_{max}, for a Propeller Aircraft 883
19.2.15 Computer code: Determining Maximum Level Airspeed, V_{max}, for a Jet 884
19.3 General Analysis Methods for Accelerated Flight 885
19.3.1 Analysis of a General Level Constant-velocity Turn 885
19.3.2 Extremes of Constant-velocity Turns 889
19.3.3 Energy State 891
Variables 893
References 894

20. Performance — Range Analysis

20.1 Introduction 896
20.1.1 The Content of this Chapter 896
20.1.2 Basic Cruise Segment for Range Analysis 896
20.1.3 Basic Cruise Segment in Terms of Range Versus Weight 896
20.1.4 The "Breguet" Range Equation 897
20.1.5 Basic Cruise Segment for Endurance Analysis 897
20.1.6 The "Breguet" Endurance Equation 898
20.1.7 Notes on SFC and TSFC 898
20.2 Range Analysis 899
20.2.1 Mission Profiles 899
20.2.2 Range Profile 1: Constant Airspeed/Constant Altitude Cruise 899
20.2.3 Range Profile 2: Constant Altitude/Constant Attitude Cruise 901
20.2.4 Range Profile 3: Constant Airspeed/Constant Attitude Cruise 902
20.2.5 Range Profile 4: Cruise Range in the Absence of Weight Change 903
20.2.6 Determining Fuel Required for a Mission 907
20.2.7 Range Sensitivity Studies 908
20.3 Specific Range 909
20.3.1 Definitions 909
20.3.2 CAFE Foundation Challenge 910
20.4 Fundamental Relations for Endurance Analysis 911
20.4.1 Endurance Profile 1: Constant Airspeed/Constant Altitude Cruise 911
20.4.2 Endurance Profile 2: Constant Attitude/Altitude Cruise 912
20.4.3 Endurance Profile 3: Constant Airspeed/Attitude Cruise 913
20.5 Analysis of Mission Profile 914
20.5.1 Basics of Mission Profile Analysis 915
20.5.2 Methodology for Mission Analysis 916
20.5.3 Special Range Mission 1: IFR Cruise Mission 918
20.5.4 Special Range Mission 2: NBAA Cruise Mission 919
20.5.5 Payload-Range Sensitivity Study 919
Exercises 921
Variables 922
References 923

21. Performance — Descent

21.1 Introduction 925
    21.1.1 The Content of this Chapter 926
21.2 Fundamental Relations for the Descent Maneuver 926
    21.2.1 General Two-dimensional Free-body Diagram for an Aircraft 926
    21.2.2 Planar Equations of Motion (Assumes no Rotation about y-axis) 927
21.3 General Descent Analysis Methods 927
    21.3.1 General Angle-of-descent 927
    21.3.2 General Rate-of-descent 927
    21.3.3 Equilibrium Glide Speed 929
    21.3.4 Sink Rate 930
    21.3.5 Airspeed of Minimum Sink Rate, V_{BA} 931
    21.3.6 Minimum Angle-of-descent 931
    21.3.7 Best Glide Speed, V_{BG} 931
    21.3.8 Glide Distance 932
    Variables 933
    References 934

22. Performance — Landing

22.1 Introduction 935
    22.1.1 The Content of this Chapter 936
    22.1.2 Important Segments of the Landing Phase 936
22.2 Fundamental Relations for the Landing Phase 938
    22.2.1 General Free-body Diagram of the Landing Roll 938
    22.2.2 The Equation of Motion for the Landing Roll 938
    22.2.3 Formulation of Required Aerodynamic Forces 938
    22.2.4 Ground Roll Friction Coefficients 939
    22.2.5 Determination of the Approach Distance, S_A 939
    22.2.6 Determination of the Flare Distance, S_F 940
    22.2.7 Determination of the Free-roll Distance, S_{FR} 940
    22.2.8 Determination of the Braking Distance, S_{BR} 940
    22.2.9 Landing Distance Sensitivity Studies 942
    22.2.10 Computer code: Estimation of Landing Performance 942
22.3 Database — Landing Performance of Selected Aircraft 944
    Variables 945
    References 946

23. Miscellaneous Design Notes

23.1 Introduction 948
    23.1.1 The Content of this Chapter 948
23.2 Control Surface Sizing 948
    23.2.1 Introduction to Control Surface Hinge Moments 948
    23.2.2 Fundamentals of Roll Control 949
    23.2.3 Aileron Sizing 960
    23.2.4 Fundamentals of Pitch Control 962
    23.2.5 Fundamentals of Yaw Control 964
23.3 General Aviation Aircraft Design Checklist 964
    23.3.1 Crosswind Capability at Touch-Down 965
    23.3.2 Balked Landing Capability 965
    23.3.3 Take-off Rotation Capability 966
    23.3.4 Trim at Stall and Flare at Landing Capability 966
    23.3.5 Stall Handling Capability 966
    23.3.6 Stall Margin for Horizontal Tail 967
    23.3.7 Roll Authority 967
    23.3.8 Control System Harmony 968
    23.3.9 Climb Capability 968
    23.3.10 One-engine-inoperative Trim and Climb Capability 969
    23.3.11 Natural Damping Capability 969
    23.3.12 Fuel Tank Selector 969
    23.3.13 Control System Stretching 969
    23.3.14 Control System Jamming 970
    23.3.15 Ground Impact Resistance 970
    23.3.16 Reliance Upon Analysis Technology 971
    23.3.17 Weight Estimation Pitfalls 972
    23.3.18 Drag Estimation Pitfalls 972
    23.3.19 Center of Gravity Travel During Flight 972
    23.3.20 Wing/Fuselage Juncture Flow Separation 972
23.4 Faults and Fixes 972
    23.4.1 Stability and Control — Dorsal Fin and Rudder Locking 973
    23.4.2 Stability and Control — Ventral Fin and Deep Stall 973
    23.4.3 Stability and Control — Ventral Fin and Dutch Roll 973
    23.4.4 Stability and Control — Forebody Strakes 973
    23.4.5 Stability and Control — Taillets and Stabilons 974
    23.4.6 Stability and Control — Control Horns 975
    23.4.7 Stall Handling — Stall Strips 975
    23.4.8 Stall Handling — Wing Fence 976
    23.4.9 Stall Handling — Wing Pylons 977
    23.4.10 Stall Handling — Vortilons 978
    23.4.11 Stall Handling — Wing Droop (Cuffs, Leading Edge Droop) 978
    23.4.12 Flow Improvement — Vortex Generators 978
    23.4.13 Trailing Edge Tabs for Multi-element Airfoils 980
    23.4.14 Flow Improvement — Nacelle Strakes 981
    23.4.15 Flow Improvement — Bubble-drag, Turbulators, and Transition Ramps 981
    Variables 982
    References 983

Appendix A 985
Appendix B 997
Index 1007

Preface

The purpose of this book is to gather, in a single place, a diverse set of information and procedures that are particularly helpful to the designer of General Aviation aircraft. Additionally, it provides step-by-step derivations of many mathematical methods, as well as easy-to-follow examples that help illustrate their application. The procedures range from useful project management tools to practical geometric layout methods, as well as sophisticated aerodynamics, performance, and stability and control analysis methods.

The design of an airplane generally begins with the introduction of specific requirements: how fast, how far, how many, what amenities, what mission? Once introduced to such requirements, the entry level designer often asks: "What's next? Where do I even begin?" This book provides step-by-step procedures that lead the reader through the entire process; from a clean-sheet-of-paper to the proof-of-concept aircraft. They were selected and developed by the author's many years of experience in the aircraft industry; initially as a flight-test engineer, then structural engineer, aerodynamicist, and eventually an aircraft designer. Subsequent years of experience in academia have allowed the presentation methods to be polished, based on student feedback. In the author's own design experience, a book such as this would have been extremely helpful in the form presented here, both as a resource and guide. This book is written with that in mind.

An effective design process not only answers whether the proposed design will meet the desired requirements, but also what remedies are viable in case it does not. During this phase, the speed of analysis is almost always of the utmost importance, and the competent designer should be able to predict differences between variations of the desired vehicle. However, the design process is multifaceted — it is more than just solving equations — managing the process is also imperative. It is not only necessary to wield the proper tools, but also to know when to apply them. This is particularly important for the manager of the design team; they should always know what steps follow the current one and what tools and resources are required.

This book is intended to provide the experienced, as well as the aspiring, designer with clear and effective analysis procedures. There is already a good collection of well written college textbooks available on aerodynamics, structures, flight dynamics, and airplane design available for the engineering student. Many are written solely with the student of aerospace engineering in mind and, consequently, often present simple problems inspired more by mathematical convenience than practical situations. Such conveniences are usually absent in the industrial environment, where problems involve natural processes that do not always accommodate "equation friendly" shortcuts. This book also offers a large chapter on propellers, a topic many textbooks, sadly, ignore. The propeller is here to stay for the foreseeable future and this warrants the large space dedicated to it.

This book differs from such textbooks as it is solely written with the analysis of real airplanes in mind. Most of the examples presented involve actual production aircraft, allowing results to be directly compared to published data. This gives the reader a great sense for the accuracy of the various analysis methods. It also provides a number of numerical methodologies that take advantage of the power of the modern desktop or laptop computer. This comes in the form of powerful program snippets and spreadsheet setups intended for analysis work with Microsoft Excel. This book offers the student a thorough introduction to practical and industry proven methods, and the practicing engineer with a great go-to text. I am certain you will find it a very helpful book and that it will increase your productivity.

Snorri Gudmundsson
Assistant Professor of Aerospace Engineering
Embry-Riddle Aeronautical University
Daytona Beach, Florida.

For supporting materials please visit:

http://booksite.elsevier.com/9780123973085

Acknowledgments

A large book like this is a substantial undertaking. It can only become a reality with contributions from many individuals and companies who, in one way or another, participated in its making. I want to use this opportunity and thank these individuals and companies for their help in providing various information and support so that I would be able to provide you, the reader, with material of greater depth than otherwise possible.

I want to begin by thanking my editor: Mr. Joe Hayton, and Project Managers Ms. Chelsea Johnston and Ms. Pauline Wilkinson of Elsevier Publishing, for invaluable guidance during the development and production of the book. I'd also like to thank Dr. Howard Curtis, my fellow Professor of Aerospace Engineering at Embry-Riddle Aeronautical University, who believed strongly enough in the project to suggest it to Joe.

The following individuals and companies deserve an expression of my gratitude. I want to thank Mr. Don Pointer of the Dassault Falcon Jet Corporation for providing information about Dassault business jets. I also want to extend my thanks to Flightglobal.com, Williams International, Price Induction, Hirth Engines, and Electraflyer for material provided by them. I want to thank Mr. Raymond Ore for providing cutaways of the Spitfire and Mosquito aircraft and the Ed Coates collection. I am indebted to my former student, Mr. Phil Rademacher, for the large number of photographs he supplied for the project. Mr. Rademacher is an expert in aircraft recognition and, as such, has won a number of intercollegiate competitions. Phil provided me with an enormous pool of aircraft photos, of which many can be found throughout this book. Another student of mine, Mr. Nick Candrella, also provided selected pictures. A former colleague of mine, Mr. Jake Turnquist provided selected pictures as well and also deserves thanks. I also want to thank Nirmit Prabahkar, Manthan Joshi, Thomas Ford, Brian Smith, Teddy Li, Matthew Clark, and Fabio An for data collection. I also want to thank Dr. Laksh Naraynaswami for proofreading Section 7, *The Selection of the Powerplant*, and providing priceless guidance regarding turbo-machinery and inlet design. I also want to thank Mr. Brian Meyer of Hartzell Propellers Inc. for his contribution to the book. Mr. Meyer provided priceless guidance and help in proofreading Section 14, *The Anatomy of the Propeller*, supplied material and suggestions that made the section much better. I want to further extend thanks to Hartzell Propellers for their permission to use selected material on propellers. I want to thank Mr. Dale Klapmeier of Cirrus Aircraft for permitting detailed information about the SR20 and SR22 aircraft to be presented in the book. I also want to thank Mr. Paul Johnston, Cirrus' chief engineer for initial proofreading and helpful suggestions. I want to thank Mr. Bruce Barrett for several anecdotal nuggets from his colorful career as a flight-test pilot. Finally, I want to express my gratitude to Professor Emeritus Charles Eastlake who provided most of the material on the development cost analysis of Section 2 in this book, in addition for his proofreading effort and insightful comments.

Snorri Gudmundsson

DISCLAIMER

Every effort has been made to trace and acknowledge copyright. The author would welcome any information from people who believe their photos have been used without due credit.

Helpful Notes

HELPFUL WEBSITES FOR THE AIRCRAFT DESIGNER

FAA regulations:	http://www.faa.gov/
NACA/NASA Report Server	http://ntrs.nasa.gov/search.jsp
Aircraft three-view drawing database:	http://richard.ferriere.free.fr/3vues/3vues.html
Aircraft picture database:	http://www.airliners.net/search/index.main
Airfoil databases:	http://www.worldofkrauss.com/
	http://www.ae.illinois.edu/m-selig/ads/coord_database.html

THE GREEK ALPHABET

Αα	Alpha	Νν	Nu
Ββ	Beta	Ξξ	Xi
Γγ	Gamma	Οο	Omicron
Δδ	Delta	Ππ	Pi
Εε	Epsilon	Ρρ	Rho
Ζζ	Zeta	Σσς	Sigma
Ηη	Eta	Ττ	Tau
Θθ	Theta	Υυ	Upsilon
Ιι	Iota	Φφ	Phi
Κκ	Kappa	Χχ	Chi
Λλ	Lambda	Ψψ	Psi
Μμ	Mu	Ωω	Omega

LIST OF ABBREVIATIONS AND COMMON TERMS

Abbreviation	Description	Units (UK and SI)
AC	Standard Airworthiness Certificate (context dependent)	
AC	Advisory Circular (context dependent)	
AD	Airworthiness Directives	
AF	Activity Factor	
AIAA	American Institute of Aeronautics and Astronautics	
AISI	American Iron and Steel Institute	
ALF	Artificial Laminar Flow	
AMM	Aircraft Maintenance Manual	
AOA	Angle of Attack	Degrees or radians
AOD	Angle of Descent	Degrees or radians
AOG	Angle of Glide	Degrees or radians
AOI	Angle of Incidence	Degrees or radians
AOL	Aircraft Operating Limitations	
AOY	Angle of Yaw	Degrees or radians
AR	Aspect Ratio	
AR_{HT}	Horizontal Tail Aspect Ratio	
AR_{VT}	Vertical Tail Aspect Ratio	
AR_W	Wing Aspect Ratio	

Abbreviation	Description	Units (UK and SI)
ASTM	American Society for Testing and Materials	
BET	Blade Element Theory	
BFL	Balanced Field Length	ft or m
BHP	Brake Horse Power	HP
BL	Boundary Layer	
CAA	Civil Aviation Authority	
CAD	Computer Aided Design	
CAR	Civil Aviation Regulation	
CAT	Clear Air Turbulence	
C_{bhp}	SFC of a piston engine in terms of BHP	
CDBM	Component Drag Build-up Method	
CER	Cost Estimating Relationship	
CFD	Computational Fluid Dynamics	
CFR	Code of Federal Regulations	
CG	Center of Gravity	
C_{jet}	SFC of a turbojet, turbofan, or a pulsejet engine	
CPI	Consumer Price Index	
C_r	Wing root chord	
CRP	Carbon Reinforced Plastics	
CRUD	Cumulative Result of Undesirable Drag	
CS	Certification Specification	
C_t	Wing tip chord	
C_{ws}	SFC of a piston engine in terms of WattSeconds	
DAPCA	Development and Procurement Cost of Aircraft	
EASA	European Aviation Safety Agency	
ELOS	Equivalent Level of Safety	
ESDU	Engineering Sciences Data Unit (formerly)	

Abbreviation	Description	Units (UK and SI)
FAA	Federal Aviation Administration	
FAR	Federal Aviation Regulations	
FF	Form Factor	
FOD	Foreign Object Damage	
FRP	Fiberglass Reinforced Plastics	
GA	General Aviation	
GAMA	General Aviation Manufacturers Association	
GDT	Geometric Dimensioning and Tolerancing	
GRP	Graphite Reinforced Plastic	
HLFC	Hybrid Laminar Flow Control	
HT	Horizontal Tail	
IF	Interference Factor	
IFR	Instrument Flight Rules	
IGE	In Ground Effects	
IPT	Integrated Product Teams	
ISA	International Standard Atmosphere	
JAA	Joint Aviation Authorities	
JAR	Joint Aviation Regulations	
KCAS	Knots, Calibrated Airspeed	Knots
KEAS	Equivalent airspeed	Knots
KGS	Ground speed	Knots
KIAS	Knots indicated airspeed	Knots
KTAS	Knots, True Airspeed	Knots
LCO	Life Cycle Oscillations	
LE	Leading Edge	
LFC	Laminar Flow Control	
LIFTOFF	The event when an airplane's landing gear is no longer in contact with the ground	

(Continued)

Abbreviation	Description	Units (UK and SI)
LSA	Light sport aircraft	
MAC	Mean Aerodynamic Chord	
MAP	Manifold Pressure	
MAV	Micro Air Vehicle	
MCP	Maximum Continuous Power	
MFTS	Master Flight Test Schedule	
MGC	Mean Geometric Chord	ft or m
MLG	Main Landing Gear	
MMPDS	*Metallic Materials Properties Development and Standardization*	
NACA	National Advisory Committee for Aeronautics	
NASA	National Aeronautics and Space Administration	
NBAA	National Business Aviation Association	
NLF	Natural Laminar Flow	
NLG	Nose Landing Gear	
NSCFD	Navier-Stokes Computation Fluid Dynamics	
OAT	Outside Air Temperature	
OEI	One Engine Inoperative	
OGE	Out of Ground Effects	
OML	Outside Mold Line	
PFM	Pilots Flight Manual	
PFT	Potential Flow Theory	
PIO	Pilot Induced Oscillation	
PMA	Parts Manufacturer Approval	
POC	Proof-of-Concept	
POH	Pilot's Operating Handbook	
QFD	Quality Function Deployment	
RFP	Request for Proposal	
ROC	Rate of Climb	
ROD	Rate of Descent	
RPM	Revolutions per Minute	
RTM	Resin Transfer Molding	
S-AC	Special Airworthiness Certificate	
SAE	Society of Automotive Engineers	
SAS	Stability Augmentation System	
SB	Service Bulletin	
SCS	Stability Coordinate System	
SFC	Specific Fuel Consumption	
SHP	Shaft Horse Power	HP
S-L	Sea Level	
Stall	The minimum airspeed at which an airplane can fly level	
STC	Supplemental Type Certificate	
TC	Type Certificate	
TCDS	Type Certificate Data Sheet	
TE	Trailing Edge	
TED	Trailing Edge Down	
TEL	Trailing Edge Left	
TER	Trailing Edge Right	
TEU	Trailing Edge Up	
T-O	Take-off	
TR	Taper Ratio	
TRA	Tire and Rim Association	
TSFC	Thrust specific fuel consumption	
TSO	Technical Standard Order	
TSOA	Technical Standard Order Authorization	
UAV	Unmanned Aerial Vehicle	
USAF	United States Air Force	
VBA	Visual Basic for Applications	
VG	Vortex Generator	
VLM	Vortex lattice method	
VT	Vertical Tail	

HANDY CONVERSION FACTORS

1 ft	=	0.3048 m
1 m	=	3.28084 ft
1 mi (statute mile)	=	5280 ft
1 nm (nautical mile)	=	6076 ft
1 BHP	=	0.746 kW
1 BHP	=	746 W
1 BHP	=	33000 ftlb$_f$/min
1 BHP	=	550 ftlb$_f$/sec
1 kW	=	1.340483 BHP
1 W	=	0.001340483 BHP
1 ft/s	=	0.59242 knots
1 ft/s	=	0.3048 m/s
1 mph	=	1.467 ft/s
1 knot	=	1.688 ft/s
1 US gallon of Avgas	=	6.0 lb$_f$ (2.718 kg)
1 US gallon of Jet A	=	6.7 lb$_f$ (3.035 kg)
1 US gallon	=	3.785412 liters
Fuel tank volume: 1 in^3	=	0.004328704 US gal
Fuel tank volume: 1 US Gal	=	231.02 in^3
1 GPa (giga-pascal)	=	145037.73773 psi
1 MPa (mega-pascal)	=	145.03773773 psi

A NOTE ABOUT MASS AND FORCE

Often several forms of units of force are presented in the UK-system. Examples include lbs, lb$_f$, lb$_{st}$ (engine thrust), lb$_t$ (engine thrust), and so on. Usually this is done to distinguish between mechanical and other kinds of forces. In this book, though, the intention is to keep everything simple and straight-forward and for that reason, the same unit will be used at all times:

If we are talking about a pound mass we will use: lb$_m$
If we are talking about a pound force we will use: lb$_f$

This will be done regardless of the source of the force.

A NOTE ABOUT FORMAT

This book has been designed in a fashion intended to be particularly useful to the reader (Note that QED is Latin for *Quod Erat Demonstrandum* means "Now it has been demonstrated"):

ANALYSIS 7: AIRSPEED FOR MAXIMUM L/D RATIO — Article's title bar

Knowing the airspeed at which the maximum L/D ratio is achieved is imperative, not only from a standpoint of safety but also as the airspeed of minimum thrust required (also see Equation (15-21). Pilots of single aircraft are trained to establish this airspeed as soon as possible in case of engine failure as it will result in a maximum glide distance (see ANALYSIS 8 in Section 17). It is also known as the *Airspeed for Minimum Thrust Required*. — Main article (white background)

$$V_{LD_{max}} = \sqrt{\frac{2}{\rho}\left(\frac{W}{S}\right)}\sqrt{\frac{k}{C_{D_{min}}}} \qquad (15\text{-}24)$$

Derivation: — Derivation (gray background)

We showed that C_L for LD_{max} was given by:
$C_L = \sqrt{\frac{C_{D_{min}}}{k}}$

Insert this into the lift equation and solve for V:

$$V = \sqrt{\frac{2W}{\rho S C_L}} = \sqrt{\frac{2W}{\rho S \sqrt{\frac{C_{D_{min}}}{k}}}} = \sqrt{\frac{2}{\rho}\left(\frac{W}{S}\right)}\sqrt{\frac{k}{C_{D_{min}}}}$$

QED

EXAMPLE 15-5 — Example (boxed text)

Determine the airspeed the pilot should maintain in order to achieve maximum lift-to-drag ratio for the sample aircraft at 30000 ft and 20000 lb:

$$V_{LD_{max}} = \sqrt{\frac{2}{\rho}\left(\frac{W}{S}\right)}\sqrt{\frac{k}{C_{D_{min}}}}$$

$$= \sqrt{\frac{2}{0.0008897}\left(\frac{20000}{311.6}\right)}\sqrt{\frac{0.05236}{0.020}} = 487.5 \text{ ft/s}$$

$$V_{LD_{max}} = \left(\frac{487.5}{1.688}\right)\sqrt{\frac{0.0008897}{0.002378}} \approx 177 \text{ KCAS}$$

Determine the airspeed at 30000 ft and 15000 lb:

$$V_{LD_{max}} = \sqrt{\frac{2}{\rho}\left(\frac{W}{S}\right)}\sqrt{\frac{k}{C_{D_{min}}}}$$

$$= \sqrt{\frac{2}{0.0008897}\left(\frac{15000}{311.6}\right)}\sqrt{\frac{0.05236}{0.020}} = 418.4 \text{ ft/s}$$

$$V_{LD_{max}} = \left(\frac{418.4}{1.688}\right)\sqrt{\frac{0.0008897}{0.002378}} \approx 152 \text{ KCAS}$$

AIRCRAFT DESIGN DOS AND DON'TS

DO DEFINE THE MISSION!
Airplanes are designed for specific missions.
Define your mission clearly.

DON'T FALL IN LOVE WITH YOUR DESIGN!
Care for it.
There are two kinds of designers;
Designers who love their design like a 5 year old kid loves his mother: *"My mother is perfect and how dare you say anything bad about her!"*
Designers who love their design like a mother loves her 5 year old kid. *"My child has flaws and I have to nurture it to help it become the best it can be."*

DO WELCOME CRITICISM OF YOUR DESIGN!
Don't take it personally. Use it to improve your design or to articulate why a certain feature is not needed or is a bad idea.
There are four kinds of criticism: Criticism stemming from malicious intent - IGNORE IT!
Criticism stemming from ignorance – IGNORE IT, but educate!
Attempt at constructive criticism, but poorly delivered – CONSIDER IT!
Genuinely constructive criticism – USE IT TO IMPROVE YOUR DESIGN!

DON'T CRITICIZE YOUR PEER'S DESIGN TACTLESSLY!
Do you want your design to be judged unfairly? If not, don't judge others unfairly. Realize there may be reasons for the inclusion of a feature – ask for its purpose. If you're not sure, ask before dishing out criticism.
If your criticism is tactful, it will be well received.

(Continued)

DO DESIGN FOR SAFETY!
You are responsible for other people's lives. Do you really know whether or not your suggested "feature" of fancy is detrimental to safety? What have other people done? How did they do it? Was it abandoned? Was it safe?

DO DESIGN FOR MINIMUM DRAG!
Drag is easy to increase, hard to reduce. Apply conservatism to your drag estimates. Nature is never as optimistic as you!
Slow on paper, but fast in reality is good.
Fast on paper, but slow in reality is BAD!

DO SELF-REFLECT
Are you sure your invention is greatest thing since sliced bread?
If so, why hasn't it already been invented? Is it possible it was already invented, but rejected[1]?

PROFESSOR GUDMUNDSSON'S CAKE PHILOSOPHY

All events are analogous to a cake recipe:

To make a cake requires a multitude of ingredients. Each must be added to the mix in the proper amount and order. Attributing the texture and taste of an entire cake to a single ingredient is misleading at best and dishonest at worst.

---

[1] ...because it was terrible! Is it possible the inventor can't see it because he is infatuated with it?

CHAPTER 1

The Aircraft Design Process

OUTLINE

1.1 Introduction — 2
 1.1.1 The Content of this Chapter — 5
 1.1.2 Important Elements of a New Aircraft Design — 5
  Definition of the Mission — 5
  Performance Requirements and Sensitivity — 5
  Handling Requirements (Stability and Control) — 5
  Ease of Manufacturing — 6
  Certifiability — 6
  Features and Upgradability (Growth) — 7
  Aesthetics — 7
  Maintainability — 8
  Lean Engineering and Lean Manufacturing — 8
  Integrated Product Teams (IPT) — 9
  Fundamental Phases of the Aircraft Design Process — 9
  Development Program Phase — 11
  Post-development Programs — 11

1.2 General Process of Aircraft Design — 11
 1.2.1 Common Description of the Design Process — 11
  Elementary Outline of the Design Process — 11
  Design Process per Torenbeek — 12
  Typical Design Process for GA Aircraft — 13
 1.2.2 Important Regulatory Concepts — 13
  Type Certificate (TC) — 13
  Supplemental Type Certificate (STC) — 14
  Standard Airworthiness Certificate (AC) — 14
  Special Airworthiness Certificate (SAC) — 14
  Maintenance Requirements — 14
  Airworthiness Directives (AD) — 14
  Service Bulletin (SB) — 15
  Advisory Circular (AC) — 15
  Technical Standard Order (TSO) — 15
  Technical Standard Order Authorization (TSOA) — 15
  Parts Manufacturer Approval (PMA) — 15

1.3 Aircraft Design Algorithm — 15
 1.3.1 Conceptual Design Algorithm for a GA Aircraft — 16
 1.3.2 Implementation of the Conceptual Design Algorithm — 16

1.4 Elements of Project Engineering — 19
 1.4.1 Gantt Diagrams — 19
 1.4.2 Fishbone Diagram for Preliminary Airplane Design — 19
 1.4.3 Managing Compliance with Project Requirements — 21
 1.4.4 Project Plan and Task Management — 21
 1.4.5 Quality Function Deployment and a House of Quality — 21
  Step 1: Customer Requirements — 22
  Step 2: Technical Requirements — 24
  Step 3: Roof — 25
  Step 4: Interrelationship Matrix — 26
  Step 5: Targets — 27
  Step 6: Comparison Matrix — 27

1.5 Presenting the Design Project — 27
  Three-view Drawings — 28
  Images Using Solid Modelers — 28
  Images Using Finite Element Modelers — 29
  Images Using Computational Fluid Dynamics Software — 29
  Cutaway Drawings — 30
  Engineering Reports — 30
  Engineering Drawings — 32

Variables — 32

References — 32

1.1 INTRODUCTION

What is a *design*? It is probably appropriate to begin a book on design by discussing the term itself, especially considering the concept is often erroneously defined and sometimes even characterized through zeal rather than a true understanding of its meaning. The author recalls a past interview with a renowned designer who, during a TV interview, was asked to define the term. The show that ensued was a disappointing mixture of superficial self-importance and an embarrassing unpreparedness for the question. Following an artful tiptoeing around the issue the concluding response could be summarized as; "well, everything is designed." No, nothing could be further from the truth: not everything is designed. Some things are designed while other things are not. When self-proclaimed designers have a hard time defining the term properly it should not be surprising when laypeople misuse the word and apply it to things that are clearly not designed. The least we can expect of any designer is to accurately define the concept to laypeople, some of whom have openly demonstrated an inability to distinguish a *regular pattern* from a *design*.

Any attempt at defining the word properly requires an insight into how the brain perceives the geometry that surrounds us. It is a question of great intrigue. How can the brain tell apart a clamshell and a cloud, or a raccoon and a road? This is achieved using the brain's innate ability called *rapid pattern recognition*, common to all animals. It is one of the most important biological traits in any species that relies on optics as a primary sensory organ. In fact, this ability, a consequence of a biological evolution lasting over hundreds of millions of years, is imperative to the survival of the species. Its most important strength is that it allows animals to make a distinction between the facial features of a mother and the silhouette of a dangerous predator (for instance, see [1]).

If you see a face when you look at the front end of a typical car, or the silhouette of people or animals when looking at rock formations or clouds, you should know that this is your brain's pattern recognition subroutine working overtime. It is desperately trying to construct a recognizable image from any pattern that hits the retina to help you quickly identify friends from foes. The faster a member of a species can accomplish this feat, the greater is the chance it may escape a dangerous predator or identify a concealed prey, providing a clear evolutionary advantage. It helps a falcon see a rodent from great heights as much as it helps an antelope identify a lurking lion. However, just as rapid pattern recognition is capable of discerning predator and prey; it can also play tricks with the brain and cause it to assemble random patterns into images of easily recognizable things that are simply not there. This condition is called *apophenia*. The lack of public education on this elementary biological function is stunning and renders some laypeople altogether incapable of realizing that the interplay of dark and light areas on their toast or potato chip that looks like their favorite celebrity is not a design but only a random pattern the brain has managed to assemble into a recognizable shape. Deprived of knowledge to know any better, many yield to wishful thinking and allow the imagination to run wild.

In short, a *regular pattern* is a combination of geometrical, physical, or mathematical features that may or may not be random, but "appears" either repetitious or regular through some characterization, such as learning. In fact, our environment is jam-packed with regular patterns. The repetition (or regularity) of a pattern allows the brain to separate it from the truly random background. People, familiar with the term "design," erroneously deduce that since a pattern appears to be regular it must be designed, when in fact it is not. A *design* is a pattern of geometrical, physical, or mathematical features that is the consequence of an intent and purpose. A design requires an originator who intended for the pattern to look a certain way so it could serve its proposed purpose. This way, *a design is a subset of regular patterns and one that has a preconceived goal, requires planned actions to prepare, and serves a specific purpose.*

Consider the natural shapes in Figure 1-1. The mountain range to the left was not designed but formed by the mindless forces of nature. There was no preconceived plan that the range should look this way and not some other way. It just formed this way over a long time — it is a random but repetitious pattern. The contrails that criss-cross the sky over the Yosemite National Park, in the right image, were not planned either. They are a consequence of random departure times of different airplanes in different parts of the USA, headed in different directions at different altitudes. While the arrangement of the airway system is truly designed it was not conceived with the contrails in mind but for a different purpose altogether. No one planned the airway system so this pattern would form over El Capitan in this fashion and not some other. No one was ever tasked with figuring out that this particular day the winds aloft would allow the pattern to stay so regular. Its appearance is nothing but a coincidence. The geometry of the contrails, just like the mountain range in the left image, is the consequence of random events that were not designed. Claiming these are designs, automatically inflicts a burden-of-proof obligation on the petitioner: Show the plans, the originator, and explain the purpose and, if unable to, simply call it by its proper name until such plans surface: a pattern is a pattern until it can be shown to be a design.

FIGURE 1-1 Examples of random patterns. The mountain range to the left is shaped by the random forces of nature. The pattern of contrails over the Yosemite National Park is the consequence of random departure times of the aircraft involved that are influenced by random decisions of air traffic controllers. There is absolutely no intentional intelligence that forms these shapes. They just appear that way. *(Photos by* Snævarr Guðmundsson*)*

With the philosophy of design behind us, we can now focus on the primary topic of this book – the design of aircraft, in particular General Aviation aircraft. According to the Federal Aviation Administration, the term *General Aviation aircraft* (from here on called GA aircraft) refers to *all aircraft other than airlines and military operations* [2]. This includes a large body of aircraft, ranging from sailplanes and airships to turbofan jets. Most aircraft are designed to comply with strict regulatory standards. In the USA these are managed and maintained by the Federal Aviation Administration (FAA). In Europe the standards are set by the European Aviation Safety Agency (EASA). These standards are similar in most ways, which results from an effort between the two agencies to harmonize them. Table 1-1 lists a number of standards for selected classes of aircraft.

In the USA, a light sport aircraft (LSA) is treated differently from an aircraft certified to 14 CFR Part 23 or 25. Instead of a direct involvement in the certification process, the FAA accepts compliance based on so-called consensus standards. These standards are neither established nor maintained by the agency itself but by some other organization. Some of these are really "watered down" FAA rules that are far less burdensome to comply with than the originals. This can partially be justified on the basis that the airplanes they apply to are much simpler than regular aircraft.

The acceptance of consensus standards (LSA) is effectively based on the "honor system." In other words, a manufacturer tells the FAA its product complies with the applicable standards and, in return, receives an airworthiness certificate. This is done as long as no "issues" surface. The system is a form of "self-regulation" and is designed to keep the FAA out of the loop. The LSA industry recognizes that responsible compliance is the only way to avoid more burdensome regulations. According to FAA officials in 2012, this system has been more or less problem free, excluding one instance [3].

Currently, the American Society for Testing and Materials (ASTM) is the primary organization that establishes and maintains consensus standards for LSA. ASTM has developed a number of standards that apply to different types of aircraft. The FAA accepts some of these in lieu of 14 CFR. Which standard ultimately depends on the subclass of aircraft (aircraft, glider, gyrocopter, lighter-than-air, powered parachutes, and weight-shift control)

TABLE 1-1 Certification Basis for Several Classes of Aircraft

Class	Regulations	Comments
General Aviation	14 CFR Part 23 (USA) CS-23 (Europe)	
Commercial Aviation	14 CFR Part 25 (USA) CS-25 (Europe)	
Sailplanes	14 CFR 21.17(b) (USA) CS-22 (Europe)	14 CFR 21.17(b) allows the FAA to tailor the certification on a need-to basis to sailplanes. Then, by referring to AC 21.17-2A, the FAA accepts the former JAR-22 as a certification basis, which have now been superceded by CS-22.
Airships	14 CFR 21.17(b) (USA) CS-30 and CS-31HA	14 CFR 21.17(b) allows the FAA to tailor the certification on a need-to basis to airships.
Non-conventional Aircraft	14 CFR 21.17(b) (USA) CS-22 (Europe)	14 CFR 21.17(b) allows the FAA to tailor the certification on a need-to basis to non-conventional aircraft.
Light Sport Aircraft (LSA)	Consensus (USA) CS-LSA (Europe)	See discussion below regarding LSA acceptance in the USA.

and on a number of specific fields (design and performance; required equipment; quality assurance; and many others). For instance, for the subclass *aircraft*, design and performance is accepted if it complies with ASTM F2245, required equipment must also comply with ASTM F2245, but quality assurance must comply with ASTM F2279, maintenance and inspection with ASTM F2483, and so on. Gliders, gyroplanes, and other light aircraft must comply with different ASTM standards. The matrix of requirements can be obtained from the FAA website [4].

While this book will mostly focus on the design of new GA aircraft, other classes of aircraft will be discussed when needed. The designer of GA aircraft should be well rounded in other types of aircraft as well, a point that will be made repeatedly throughout this book.

GA aircraft certified under 14 CFR Part 23 are subject to a number of limitations as stipulated under 14 CFR Part 23.3-Airplane categories. The regulations place aircraft to be certified into four categories; *normal*, *utility*, *aerobatic*, and *commuter*. These categories must abide by the restrictions listed in Table 1-2. With the exception of the commuter category, an aircraft may be certified in more than one category provided all requirements of each are met.

New aircraft are designed for a variety of reasons, but most are designed to fulfill a specific role or a mission as dictated by prospective customers. For economic reasons, some aircraft (primarily military aircraft) are designed to satisfy more than one mission; these are *multi-role* aircraft. Others, for instance homebuilt aircraft, are designed for much less demanding reasons and are often solely based on what appeals to the designer.

No matter the type of aircraft or the reason for its design, specific tasks must be completed before it can be built and flown. The order of these tasks is called the *design process*. This process is necessitated by the fact that it costs a lot of money to develop a new aircraft. Organizations that develop new aircraft do not invest large amounts of funds in a design project until convinced it can perform what it is intended to. A design process makes this possible by systematically evaluating critical aspects of the design. This is primarily done using mathematical procedures, as well as specific testing of structural configuration, materials, avionics, control system layout, and many more.

The order of the tasks that constitute the design process may vary depending on the company involved. Usually there is an overlap of tasks. For instance, it is possible that the design of the fuselage structure is already in progress before the sizing of the wing or stabilizing surfaces is fully completed. Generally, the actual process will depend on the size and maturity of the company in which it takes place and the order of tasks often varies. However, there are certain steps that must be completed in all of them; for instance, the estimation of weight; sizing of lifting surfaces and the fuselage; estimation of performance; and other essential tasks.

In mature companies, the design process is managed by individuals who understand the big picture. They understand the scope of the project and are aware of the many pitfalls in scheduling, hiring, design, and other

TABLE 1-2 Restrictions for Aircraft Classes Certified under 14 CFR Part 23

Restriction	Commuter	Normal	Utility	Aerobatic
Number of pilots	1 or 2	1	1	1
Max number of occupants	19	9	9	9
Max T-O weight	19,000 lb$_f$	12,500 lb$_f$	12,500 lb$_f$	12,500 lb$_f$
Aerobatics allowed?	No	No	Limited	Yes
Non-aerobatic operations permitted	Normal flying Stalls (no whip stalls) Steep turns ($\phi < 60°$)	Normal flying Stalls (no whip stalls) Lazy eights Chandelles Steep turns ($\phi < 60°$)	Normal flying Stalls (no whip stalls) Lazy eights Chandelles Steep turns ($\phi < 90°$) Spins (if approved)	N/A
Max maneuvering g-loading, n_+	$2.1 + \dfrac{24,000}{W + 10,000} < n_+ \leq 3.8$		4.4	6.0
Min maneuvering g-loading, n_-	$-0.4 n_+ < n_- \leq -1.52$		-1.76	-3.0

W = maximum T-O weight. Maneuvering loads are based on 14 CFR 23.337.
A whip stall may occur when the airplane is stalled while in a slip. This can cause the outer wing to stall first and drop abruptly [5].

tasks, that many engineers consider less than glamorous. These people must be well-rounded in a number of disciplines: aerodynamics; performance analysis; stability and control; handling; power plants; weight analysis; structural layout; environmental restrictions; aviation regulations; history of aviation; and aircraft recognition, to name a few. Although not required to be an expert in any of these fields, their understanding must be deep enough to penetrate the surface. Knowing what to do, how to do it, and when to do it, is the key to a successful aircraft development program.

1.1.1 The Content of this Chapter

- **Section 1.2** presents a general description of the aircraft design process.
- **Section 1.3** presents two specific algorithms intended to guide the aircraft designer through the conceptual design process. If you are unsure of "what to do next," refer to these. They are based on actual industry experience and are not academic "cookbook" approaches.
- **Section 1.4** presents project management tools. Many beginning project leaders are often at a loss as to how to manage a project. If this is your predicament you need to study these tools. Project management revolves around knowing what to do and when to do it. Thus, the manager must construct a chronological order of the tasks that need to be completed.
- **Section 1.5** presents helpful approaches to describing engineering ideas using graphics ranging from three-view drawings to composite photo images. These are extremely helpful when trying to "sell" an idea.

1.1.2 Important Elements of a New Aircraft Design

Before going further, some specific topics must be brought up that the lead airplane designer must introduce and discuss thoroughly with the design team. Among those are:

Definition of the Mission

It is imperative that the mission of the new aircraft is very clearly defined. Is it primarily intended to serve as a cruiser? If so, what airspeed and cruising altitude is it most likely to see during its operation? Is it a cargo transport aircraft? How much weight must it carry? How fast, far, and high shall it fly? Is it a fighter? What energy state or loitering capabilities are required? The mission must be clearly defined because the airplane will be sized to meet that particular mission. An aircraft designed in this fashion will be most efficient when performing that mission. Clarity of this nature also has an unexpected redeeming power for the designer: It is very common during the development of aircraft that modifications to capabilities are suggested by outside agencies. In spite of being well meant, some such suggestions are often detrimental to the mission. A clearly defined mission allows the designer to turn down a disadvantageous suggestion on the basis that it compromises the primary mission.

Performance Requirements and Sensitivity

Performance requirements must be clearly defined and are usually a part of the mission definition. It is imperative to quantify characteristics such as the take-off distance, time to cruise altitude, cruise range, and even environmental noise for some types of aircraft. But it is also important to understand how deviations from the design conditions affect the performance. This is referred to as *performance sensitivity*. How does high altitude and a hot day affect the take-off distance? How about the upward slope of the runway? How does having to cruise, say, some 5000 ft below the design altitude affect the range? How about if the airplane is designed for a cruising speed higher than would be permitted by air traffic considerations and, therefore, is consistently operated at a lower cruising speed? How will that affect the range? Clearly, there are many angles to designing an aircraft, but rather than regarding it as a nuisance the designer should turn it into strength by making people in management and marketing aware of the deficiencies. And who knows — perhaps the new aircraft is less sensitive than the competition and this could be turned into a marketing advantage.

Handling Requirements (Stability and Control)

How important is the handling of the aircraft? Is this a small aircraft that is operated manually, rendering stick forces and responsiveness imperative? Is it a heavy aircraft with hydraulic or electric actuators, so stick forces are fed back to the pilot electronically and, thus, can be adjusted to be whatever is considered good? How about unsuspected responses to, say, thrust forces?

The Lockheed SA-3 Viking, an anti-submarine warfare aircraft, features a high wing with two powerful turbofan engines mounted on pylons. When spooling up, the aircraft experiences a powerful nose pitch-up tendency that is captured by a stability augmentation system (SAS) that was not originally designed into the prototype. The Boeing B-52 Stratofortress uses spoilers for banking. When banking hard, the spoiler on the down-moving wing is deployed and this reduces lift on the outboard part of that wing. This, in turn, means the center of lift moves forward, causing a nose pitch-up tendency, which the pilot must react to by pushing the yoke forward (for nose pitch-down). Handling issues of this nature must be anticipated and their severity resolved.

Ease of Manufacturing

Is it imperative that the aircraft will be easy to manufacture? Ease of manufacture will have a profound impact on the engineering of the product and its cost to the customer. A straight constant-chord wing can be manufactured at a lower cost than one that has tapered planform and compound surfaces, but it will be less efficient aerodynamically. Which is more important? The designer must have means to demonstrate why a particular geometry or raw material is required for the project. The concept of ease of marketing always looks good on paper, but this does not guarantee its success. For instance, it is simple to select composites for a new aircraft design on the grounds that this will make it easier to manufacture compound surfaces. But are they really needed? For some aircraft, the answer is a resounding *yes*, but for others the answer is simply *no*.

As an example, consider the de Havilland of Canada DHC-2 Beaver (see Figure 1-2). Designing this otherwise sturdy airplane from composites would be an unwise economic proposition. In the current environment it would simply be more expensive to build using composites and sell at the same or lower price than the aluminum version. To begin with, it is not easy to justify the manufacturing of an aerodynamically inefficient frustum-style fuselage[1] and constant-chord wing featuring a non-laminar flow airfoil with composites. Composites are primarily justifiable when compound surfaces or laminar flow wings must be manufactured. They require expensive molds to be built and maintained, and, if the aircraft ends up being produced in large numbers, the molds have to be manufactured as well; each may only last for perhaps 30–50 units.

The interested reader is encouraged to jump to Section 2.2, *Estimating project development costs*, for further information about manufacturing costs (in particular see Example 2-3, which compares development and manufacturing costs for a composite and aluminum aircraft). Cost analysis methods, such as the widely used DAPCA-IV, predict man-hours for the engineering development of composite aircraft to be around two times greater than that of comparable aluminum aircraft. They also predict tooling hours to double and manufacturing hours to be 25% greater than for aluminum aircraft. Labor and material are required not only to manufacture the airplane, but also to

FIGURE 1-2 The de Havilland of Canada DHC-2 Beaver. *(Photo from Wikipedia Commons)*

manufacture and maintain expensive tooling. As a result, composite aircraft are more expensive to manufacture in spite of substantial reduction in part count.

This inflicts an important and serious constraint on the scope of production. Composites require heating rooms to ensure the resin cures properly so it can provide maximum strength. Additionally, vacuum bagging or autoclaves are often required[2] to force tiny air bubbles out of the resin during cure to guarantee that the certified strength is achieved. The manufacturer must demonstrate to the authorities that material strength is maintained by a constant production of coupons for strength testing. Special provisions must be made to keep down moisture and prevent dust from entering the production area, not to mention supply protective clothing and respirators to all technicians who work with the material. All of this adds more cost and constraints to the production and all of it could have been avoided if the designer had realized that requiring composites was more a marketing ploy than a necessity. This is not to say that composites do not have their place – they certainly do – but just because composites are right for one application, does not mean they are appropriate for another one.

Certifiability

Will the aircraft be certified? If the answer is yes, then the designer must explore all the stipulations this is likely to inflict. If no, the designer bears a moral obligation to ensure the airplane is as safe to operate as possible. Since non-certified airplanes are destined to be small, this can be accomplished by designing it to prevailing certification standards, for instance, something like 14 CFR Part 23 or ASTM F2245 (LSA aircraft).

Regulations often get a bad rap through demagoguery by politicians and ideologues, most of whom

---

[1] A frustum-style fuselage is a tapered structure that does not feature compound surfaces. It is discussed in Chapter 12, *The Anatomy of the Fuselage*.

[2] Note that some manufacturers of composite structures claim that curing composites using vacuum "bagging" is equally effective as using an autoclave – it is certainly more economical. For instance see: http://www.gmtcomposites.com/why/autoclave.

sound like they have less than no understanding of their value. In fact, regulations are to be thanked for the current level of safety in commercial aviation; *commercial air travel is the safest mode of transportation because of regulations and this would be unachievable in their absence.* As an example, commercial aviation in the USA, which operates under 14 CFR Part 121, operated with fewer than 1.5 accidents per million departures and no fatalities over the years 2007−2009 [6]. Up until May 2012, there had only been one fatal accident in commercial aviation since 2007, the ill-fated Colgan Air Flight 3407 [7]. Unfortunately, GA suffers from approximately 6 accidents and a little over 1 fatal accident per 100,000 flight hours − a statistic that has remained relatively constant since 2000.

Of course it is possible to make regulations so strict they smother industry; however, this is neither the intent nor does it benefit anyone. The intent is public safety. The modern aircraft is a very complicated machine, whose failure may have catastrophic effects on people and property. The early history of aviation is wrought with losses of life that highlight this fact. For this reason the flying public has the right to know the risk involved before embarking on a flight. Although admittedly extreme, a number of intriguing questions can be posed: would we board an aircraft knowing there was a 50% chance it would crash? No? What percentage would we accept? And generally, how do we know commercial aviation is safe enough to accept the risk?

The fact is that the flying public is completely oblivious to the risk they take when boarding an airplane. While most have heard that aviation is the safest mode of transportation, how do people really know? Aviation is far too complex for anyone but experts to evaluate the level of safety. So, what has convinced people that the risk is indeed very low? The answer is statistics; statistics driven by standards not intended to guarantee profits, but the manufacturing of safe aircraft; standards that all players must adhere to by law.

There is only one way to promote such adherence and it is to employ a body powerful enough to enforce the standards and prevent negligence that otherwise would be rampant. This body is the government. History is wrought with examples of industries that let operational and product safety take the back-seat to profits. After all, this is what spurred the so-called "strict-liability" clause of 1963 (see Section 2.1.2, A Review of the State of the General Aviation Industry). While, at the time of this writing, LSA is a form of self-regulation that seems to work, this cannot be extrapolated to other industries. It works for LSA because compliance is less expensive than refusing to comply. But this does not hold for all other industries. Rather than focusing a futile effort on getting rid of regulations, the focus should be on streamlining them to make sure they work for everybody.

Regulations are akin to a computer code − they contain wrinkles that need to be ironed out. There is a golden medium between regulations that are too strict and no regulations at all. In the experience of this author, this is exactly what industry and the authorities are trying to accomplish. Aviation authorities are well aware of the effects burdensome regulations have on businesses, and for that reason, on a regular basis, review regulatory paragraphs for the purpose of streamlining them. Such reviews always include representatives from industry, who wield a deep understanding of the topic, and also benefit from making aviation regulations the safest and least burdensome possible. That aside, it is in fact very beneficial for industry that everybody must comply with the same set of rules − nothing is worse for industry than rules that favor one company over another. Other aspects of regulations are discussed in more detail in Section 1.2.2, *Important Regulatory Concepts*.

Features and Upgradability (Growth)

The weight of most civilian and military aircraft increases with time. It is not a question of if, but when and how much. There is not an example of the opposite, to the knowledge of this author. Requirements for added capabilities and systems raise the weight and often require substantial changes such as the introduction of a more powerful engine, and even wing enlargement. Additionally, it is often discovered during prototyping that the selected material and production methodology leads to a heavier aircraft than initially thought. The careful designer sizes the aircraft for a weight that is 5−10% higher than the projected gross weight.

Aesthetics

In light of the above topics, looks may seem like a secondary concern. But it is one that should never be underestimated. While beauty is in the eye of the beholder, it is a fact of business that aircraft that have a certain look appeal to a larger population of potential buyers and, therefore, sell better, even if their performance is less than that of the competition. The so-called Joint Strike Fighter program is a great example of such appeal (even though difference in performance is not the issue). The purpose of the program was to introduce an aircraft for the US armed forces that simultaneously replaced the F-16, A-10, F/A-18, and AV-8B tactical fighter aircraft. Three versions of the aircraft were planned and in order to keep development, production, and operating costs down, a common shape was proposed for which 80% of parts were interchangeable. There were two participants in the contract bid; Lockheed Martin and Boeing. Lockheed's entry was the X-35 and Boeing's X-32 (see Figure 1-3). Both aircraft were thought to be worthy candidates, but on October 26, 2001, Lockheed's design

FIGURE 1-3 Does the Boeing X-32 or the Lockheed X-35 look better? *(Photos: (left) by Jake Turnquist; (right) by Damien A. Guarnieri, Lockheed Martin)*

was announced as the winner. The reason cited by the Department of Defense, according to The Federation of American Scientists, an independent, nonpartisan think tank, was:

> "The Lockheed Martin X-35 was chosen over the competing Boeing X-32 primarily because of Lockheed's lift-fan STOVL design, which proved superior to the Boeing vectored-thrust approach." [8]

Apparently, in hover, the X-32's engine exhaust would return to the intake, and lower its thrust. However, soon thereafter rumors began running rampant that the real reason was in fact the looks of the two proposals, a claim denied by James Roche, the then secretary of the Air Force [9]. Rumor held that military pilots did not like the looks of the Boeing proposal, some allegedly referring to it as 'the flounder.' This rumor cannot be confirmed, but perhaps the reader has an opinion on whether the looks of the two aircraft in Figure 1-3 could have had an impact on its acceptance.

Another case in point is the Transavia PL-12 Airtruk, shown in Figure 1-4. It was originally developed in New Zealand as the Bennett Airtruck (later Waitomo Airtruk). It is a single-engine agricultural sesquiplane of all-metal construction. Among many unusual features is a cockpit mounted on top of the engine, twin tailbooms that are only connected at the wing, designed to allow a fertilizer truck to back up and refill the airplane's hopper, and the sesquiplane configuration generates four wingtip vortices that help better spread fertilizer. It is a capable aircraft, with a 2000 lb fertilizer capacity, and can be used as a cargo, ambulance or aerial survey aircraft, as well. But a strange-looking beast it is, at least to this author.

Maintainability

Maintainability is the ease at which an airplane can be kept airworthy by the operator. Maintainability is directly related to ease of manufacturing. Complicated manufacturing processes can result in an aircraft that is both hard and costly to maintain. One of the advantages of aluminum is how relatively easy it is to repair. Composites on the other hand can be very hard to maintain. Maintainability also extends to the ergonomics of repairing. Are expensive tools required? Will the mechanic need to contort like an acrobat to replace that part? Will it take 10 hours of labor to access a part that will take 5 minutes to replace? These are all issues that affect maintainability. It cannot be emphasized enough that a novice engineer should consult with A&P mechanics and try to understand their perspective. Many valuable lessons are learned from the technicians who actually have to do the work of fabrication, assembly, and maintenance.

Lean Engineering and Lean Manufacturing

The concepts *lean engineering* and *lean manufacturing* refer to design and production practices whose target is to minimize waste and unnecessary production steps. For instance, consider the production of a hypothetical wooden kitchen chair. Assume that pride has the manufacturer attach a gold-plated metal plaque to the lower surface of the seat that reads: 'World's finest kitchen chairs, since 1889.' Assume it takes five

FIGURE 1-4 The Transavia PL-12 Airtruk agricultural aircraft. *(Photo by Geoff Goodall, Ed Coates' Collection)*

separate steps to attach the plaque: two drilling operations of pilot holes, one alignment operation, and two operations in which the plaque is screwed to the seat's lower surface. Not only would the plaque have to be attached, but an overhead labor is required to order it from an outside vendor, transport it to the manufacturer, keep it in stock, and so on. Strictly speaking, the purpose of a chair is to allow someone to sit on it and, then, the said plaque is not visible. It can be argued the plaque serves no other purpose than to brag about the manufacturer and, as such, it brings no added value to the customer. In fact, it only brings up the cost of production; it certainly does not make the seating experience any more enjoyable. The plaque is therefore simply wasteful and from the standpoint of a lean production should be eliminated from the process.

The purpose of lean manufacturing is to refine the production process to ensure minimum waste. This increases the profitability of a business through efficiency. Since production processes either add value or waste to the end product, the purpose of lean engineering is to refine the design of a product so that simple, effective, and non-wasteful production processes may be employed. The scope of lean manufacturing is large and can entail topics such as optimizing the layout of templates for cutting material for clothing to minimize the amount of material that goes to waste; to the operation of the stock room, where parts are ordered from vendors just before they need to be assembled (so-called just-in-time philosophy), so the assembler won't have to keep capital in parts in an inventory. The overall consequence of such practices is a far more efficient production, and therefore, less expensive products, both to the customer and Mother Earth.

The philosophy behind lean manufacturing is usually attributed to Toyota the car manufacturer, which is renowned for its adherence to it in its production processes. For this reason, it is also known as *Toyotism*, and the success of the company's philosophy has afforded it plenty of attention. An imperative step in Toyota's approach is to identify what is called the Seven Wastes [10]. The approach was originally developed by Toyota's chief engineer Taiichi Ohno, who identified seven common sources of waste inside companies: (1) *overproduction*, as in the manufacturing of products before they are needed; (2) *waiting*, which occurs when parts are not moving smoothly in the production flow; (3) *transporting*, as in moving a product in between processes; (4) *unnecessary processing*, when expensive, high-precision methods are used when simpler methods would suffice; (5) *unnecessary inventory*, which is the accumulation of vendor parts and components in stockrooms; (6) *excessive or unnecessary motion*, which is the lack of ergonomics on the production floor, which may increase production time; and (7) *production defects*, which are inflicted on the production floor and are costly due to the inspection and storage requirements.

The above discussion barely scratches the surface of lean manufacturing, but is intended to whet the appetite of the reader.

Integrated Product Teams (IPT)

An *integrated product team* is a group of people with a wide range of skills who are responsible for the development of a product or some feature. The formation of IPTs are very common in the aviation industry, as the modern airplane is a compromise of a number of disciplines. To better understand how IPTs work, consider the development of a pressurization system for an aircraft. An example IPT could consist of the following members:

(1) A structural analyst, whose task is to determine pressurization stresses in the airframe and suggest airframe modifications if necessary.
(2) A performance analyst, whose responsibility is the evaluation of the benefits of the higher cruise altitude and airspeed the pressurization will permit.
(3) A power plant expert, who solves engine-side problems, such as those associated with bleed air, heat exchangers, and liaison duties between the engine and airframe manufacturers.
(4) An interior expert, who evaluates the impact of the pressurization system on the interior decoration, such as those that stem from the requirement of sealing and condensation.
(5) An electrical expert, who evaluates the electrical work required to allow the pilot to operate the pressurization system.
(6) A systems expert, to work on the pressurization system ducting layout, interface issues with heat exchangers, cabin pressure relief valves, cabin sealing, and so on.

Such a group would meet, perhaps once a week, to discuss issues and come up with resolutions, often with the inclusion of representatives of the manufacturers of the various systems.

Fundamental Phases of the Aircraft Design Process

In general, the aircraft design process involves several distinct phases. These are referred to as:

(1) Requirements phase
(2) Conceptual design phase
(3) Preliminary design phase

(4) Detail design phase, and
(5) Proof-of-concept aircraft construction and testing phase.

These will now be discussed in greater depth. It should be stressed that these differ in detail from company to company. Some tasks that here are presented in the conceptual design phase may be a part of the preliminary design phase in one company and a part of a different phase in another. The exact order of task is not important — its completion is.

(1) REQUIREMENTS PHASE

From a certain point of view, *requirements* are akin to a wish list. It is a list of expectations that the new design must meet. It specifies the aircraft capabilities, such as how fast, how far, how high, how many occupants, what payload, and so on (in other words, its mission). The requirements may be as simple as a few lines of expected capabilities (e.g. range, cruising speed, cruising altitude, and number of occupants) or as complex as documents containing thousands of pages, stipulating environmental impact, operating costs, maintainability, hardware, and avionics, just to name a few. It is the responsibility of the design lead to ensure the airplane has a fair chance of meeting the requirements and this is usually demonstrated during the next phase, the conceptual design phase.

(2) CONCEPTUAL DESIGN PHASE

The conceptual design phase formally establishes the initial idea. It absorbs just enough engineering to provide management with a reliable assessment of likely performance, possible looks, basic understanding of the scope of the development effort, including marketability, labor requirements, and expected costs. Typically, the following characteristics are defined during this phase:

- Type of aircraft (piston, turboprop, turbojet/fan, helicopter)
- Mission (the purpose of the design)
- Technology (avionics, materials, engines)
- Aesthetics (the importance of "good looks")
- Requirements for occupant comfort (pressurization, galleys, lavatories)
- Ergonomics (occupants and occupant ergonomics)
- Special aerodynamic features (flaps/slats, wing sweep, etc.)
- Certification basis (LSA, Part 23, Part 25, Military)
- Ease of manufacturing (how will it be produced)
- Maintainability (tools, labor, and methods required to maintain the aircraft)
- Initial cost estimation
- Evaluation of marketability.

The conclusion of this phase is an *initial loft* and a *conceptual design evaluation*, which allows management to make a well-reasoned call as to whether to proceed with the design by entering the preliminary design phase.

(3) PRELIMINARY DESIGN PHASE

The preliminary design ultimately answers whether the idea is viable. It not only exposes potential problems, as well as possible solutions to those problems, but yields a polished loft that will allow a flying prototype to be built. Some of the specific tasks that are accomplished during this phase are:

- Detailed geometry development
- Layout of major load paths
- Weight estimation
- Details of mission
- Performance
- Stability and control
- Evaluation of special aerodynamic features
- Evaluation of certifiability
- Evaluation of mission capability
- Refinement of producibility
- Maintainability is defined
- Preliminary production cost estimation.

Ideally, the conclusion of this phase is a *drawing package* and a *preliminary design evaluation*. If this evaluation is negative, this usually spells a major change to, if not a cancellation of, the program. If positive, a decision to go ahead with the fabrication of a *proof-of-concept* (POC) aircraft is usually taken.

(4) DETAIL DESIGN PHASE

The detail design process primarily involves the conversion of the loft from the preliminary design into something that can be built and ultimately flown. Of course, it is far more complicated than that, and a limited description of the work that takes place is listed below:

- Detail design work (structures, systems, avionics, etc.)
- Study of technologies (vendors, company cooperation, etc.)
- Subcontractor and vendor negotiations
- Design of limited (one-time use) tooling (fixtures and jigs)
- Structural detail design
- Mechanical detail design
- Avionics and electronics detail design
- Ergonomics detail design
- Mock-up fabrication
- Maintenance procedures planning
- Material and equipment logistics.

The conclusion of this phase is the final *outside mold line* and *internal structure for the POC*. This is generally the beginning of the construction planning, although it almost always begins long before the detail design phase is completed.

(5) PROOF-OF-CONCEPT AIRCRAFT CONSTRUCTION AND TESTING PHASE

The construction of the POC aircraft or *prototype*, begins during the detail design phase. For established companies that intend to produce the design, this is a very involved process, as the production process, with all its paperwork and quality assurance protocols, is being prepared at the same time. Some of the tasks that are accomplished are listed below:

- Detail design revisions (structures, systems, avionics, etc.)
- Application of selected technologies
- Tooling design and fabrication
- Fabrication and assembly
- Structural testing
- Aeroelastic testing (ground vibration testing)
- Mechanical testing
- Avionics testing
- Maintenance procedures and refinement of maintainability.

The culmination of this phase is *maiden flight of the POC*. This is followed by the development flight testing as discussed below.

Development Program Phase

A development program follows a successful completion of the preliminary design. The development of this phase usually begins long before the maiden flight and is usually handled by flight test engineers, flight test pilots, and management.

- Establish aircraft operating limitations (AOL)
- Establish pilots' operating handbook (POH)
- Prepare master flight test schedule (MFTS)
- Envelope expansion schedule (or Matrix)
- Test equipment acquisition
- Flight support crew training
- Group roles must be trained prior to flight – not on the job
- Establish emergency procedures
- Establish group responsibilities
- Revision of AOL, POH, MFTS
- Flight readiness review.

The conclusion of this phase is a *certifiable aircraft*. This means the organization understands the risks and scope of the required certification effort and is convinced the certification program can be successfully completed.

Post-development Programs

A lot of work still remains, even though the development program comes to an end in a successful manner. A viable aircraft design continues in development when customers begin its operation and discover features that "would greatly benefit the design". There is the advancement of avionics. New equipment must be installed and these must be engineered. A broad scope of various post-development programs is listed below:

- Development flight test/structural/systems/ avionics program
- Certification flight test/structural/systems/avionics program
- Aircraft is awarded with a type certificate
- Production process design
- Production tooling design and fabrication
- Delivery of produced aircraft
- Eventual reception of production certificate.

1.2 GENERAL PROCESS OF AIRCRAFT DESIGN

This section presents several views of the general process of airplane design, from the introduction of a request for proposal (RFP) to a certified product. Since great effort is exerted in designing the airplane to comply with civil aviation regulations, a brief introduction to a number of very important regulatory concepts is given as well.

1.2.1 Common Description of the Design Process

Elementary Outline of the Design Process

A general description of the aircraft design process is provided in several aircraft design textbooks intended for university students of aerospace engineering. Understanding this process is of great importance to the aircraft designer, in particular design team leaders. An elementary depiction of the design process is presented in Figure 1-5. While the diagram correctly describes the chronological order of steps that must be accomplished before the POC is built, it is somewhat misleading as the overlap between phases is not presented. In a real industry environment there really is not a set date at which conceptual design ends and preliminary design begins. Instead there is substantial overlap between the phases, as this permits a more efficient use of the workforce. Instead, the conceptual design stage is slowly and surely phased out.

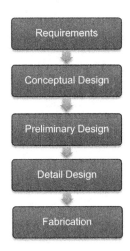

FIGURE 1-5 An elementary outline of the aircraft design process.

As an example, in the form presented in Figure 1-5, the engineers responsible for the detail design would effectively be idle until the design reached the detail design stage. This would present a costly situation for any business. Instead, the preliminary design takes place in various stages that are parallel to the detail design stage. This way, the preliminary design of the fuselage might take place after the wings, whilst the detail design of the wing takes place at the same time.

Design Process per Torenbeek

In his classic text, Torenbeek [11] discusses the process in detail and presents a depiction, reproduced in Figure 1-6. This diagram demonstrates the process in a realistic manner, by showing overlapping activities. There is really not a specific date beyond which the previous phase ends and a new one begins. It also shows important milestones, such as a configuration freeze, go-ahead approval, and acceptance of type

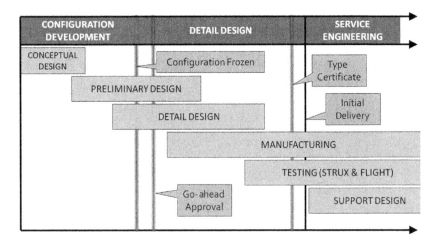

FIGURE 1-6 Aircraft design process per Torenbeek (1986).

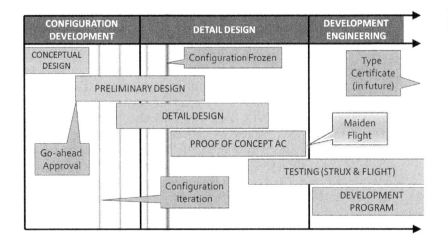

FIGURE 1-7 Aircraft design process for a typical GA aircraft.

certificate. A *configuration freeze* is a set date after which no changes are allowed to the external geometry or the outside mold line (OML), even if a better geometric shape is discovered. It marks the date for the aerodynamics group to cease geometric optimization, as the "frozen" configuration is adequate to meet the requirements. The *go-ahead approval* is the date at which upper management gives the green light for the design team to proceed with the selected configuration and develop an actual prototype. In other words, it marks the readiness of the organization to fund the project. A *type certificate* is described in Section 1.2.2.

The diagram correctly shows that the preliminary design begins before the conceptual design is completed. Of course, most major geometric features (wings, fuselage, tail, etc.) have already been sized by then, but many others remain as work in progress. Evaluation of the effectiveness of winglets, control surface sizing and hinge moment estimation, landing gear retraction mechanism, and many others, are examples of such tasks.

Torenbeek's diagram also shows that detail design begins during the preliminary design phase, and even manufacturing overlaps the other two. The manufacturing phase includes the design and construction of production tooling, establishment of vendor relations and other preparatory tasks.

Typical Design Process for GA Aircraft

The following process is based on the author's experience and explains the development of typical GA aircraft. It parallels Torenbeek's depiction in most ways, but accounts for iteration cycles often required during the preliminary design phase. This reflects the fact that during the preliminary design phase, issues will arise that may require the OML of the configuration to be modified; in particular if the design is somewhat unorthodox. Such an issue might be a higher engine weight than expected, requiring it to be moved to a new location to maintain the original empty-weight CG position. This, in turn, calls for a reshaping of the engine cowling or nacelle, calling for other modifications. Such changes are handled by numbering each version of the OML as if it were the final version, because, eventually, the one with the highest number will be the configuration that gets frozen. This allows the design team to proceed with work on structures and other internal features of the aircraft, rather than waiting for a configuration freeze.

1.2.2 Important Regulatory Concepts

It is the duty of the government of the country in which the aircraft is designed and manufactured to ensure it is built to standards that provide sufficient safety. If the aircraft is built in one country and is being certified in another, it is still the duty of the government of the latter to ensure it complies with those same standards. This task requires the standards (the law of the land) to be enforced — and may call for a denial of an airworthiness certificate (see below) should the manufacturer refuse to comply with those standards. This is why this important task should remain in the hands of a government — for it has the power to force the compliance.

From the standpoint of manufacturers, certifying aircraft in a different country can become contentious when additional, seemingly local, requirements must be complied with as well. Such scenarios can often inflict great frustration, if not controversy, on the certification process. International harmonization of aircraft certification standards remains an important topic. In effect, such harmonization would allow the demonstrated compliance in one country to be accepted in another, upon review of certification documents. This would be helpful to already cash-strained businesses that have clearly demonstrated they already side with safety. It underlines a fair complaint often leveled by industry that a serious review and justification for the presence of selected paragraphs in the regulatory code is in order.

The standards that aircraft are designed and built to have names like the Civil Aviation Regulations (CAR, now obsolete), Federal Aviation Regulations (FAR, current in the United States), Joint Aviation Regulations (JAR, European, obsolete as of September 28, 2003), or Certification Specifications [12] (CS, current in Europe). The current government agencies that enforce adherence to these standards have names like the Civil Aviation Authority (CAA, now obsolete), Federal Aviation Administration (FAA, current in the United States), Joint Aviation Authorities (JAA, European, obsolete as of June 30, 2009), or European Aviation Safety Agency (EASA, current in Europe). With respect to FAR, the convention is to refer to them as Title 14 of the Code of Federal Regulations, or simply 14 CFR. This way, a particular section of the regulations is cited by adding it to that code. For instance Part 23 would be written as 14 CFR Part 23, and so on.

Type Certificate (TC)

Once the manufacturer of a civilian (i.e. non-military) aircraft, engine, or propeller has demonstrated that their product meets or exceeds the current airworthiness standards set by its regulatory agency, it is awarded a *TC* by publishing a type certificate data sheet (TCDS). The TCDS is a document that contains important information about operating limitations, applicable regulations, and other restrictions. This means the aircraft is now "officially defined" by the TC. TCDS for all civilian aircraft can be viewed on the FAA website [13].

Obtaining a TC is a very costly proposition for the manufacturer, but it is also very valuable in securing marketability of the product. It can be stated with a high level of certainty that a specific product without a TC (and thus considered experimental) is unlikely to sell in the same quantity or at the same price it would if it had a TC. The reason being that the TC guarantees product quality, which is imperative to the customer: it makes the product "trustworthy." The reason why a TC is so costly is that it requires the product to undergo strenuous demonstration of its safe operation and quality of material and construction. Additionally, the TC serves as a basis for producing the aircraft.

Supplemental Type Certificate (STC)

Many owners of airplanes want to add features to the model. A replacement of a piston engine with a gas turbine is an example of a very common change made to existing certified aircraft. Another example is the conversion of an airplane to allow it to transport patients, something it was very unlikely to have been originally designed for. Such changes are possible, but require the aviation authorities to approve the installation or change. Once convinced the change does not compromise the continued airworthiness of the aircraft, a *supplemental type certificate* is issued. The STC specifies what change was made to the aircraft, details how it affects the TC, specifies new or revised operational limitations, and lists what serial numbers are affected. The list of serial numbers is called *effectivity*.

Standard Airworthiness Certificate (AC)

Once the TC has been obtained, each unit of the now mass-produced aircraft will receive a *standard airworthiness certificate*. This is only issued once each aircraft has been demonstrated to conform to the TC and be assembled in accordance with industry practice; is ready for safe operation; and has been registered (giving it a tail number). Each aircraft produced is tracked using serial numbers. The AC allows the aircraft to be operated, as long as its maintenance is performed in accordance with regulations.

Special Airworthiness Certificate (SAC)

A special airworthiness certificate can be issued for airplanes that, for some reason, must be operated in a specialized fashion (e.g. ferry flying, agricultural use, experimental, marketing, etc.), but precludes it from being used for commercial transportation of people or freight. An S-AC is issued in accordance with 14 CFR 21.175 in the following subclasses: primary, restricted, limited, light-sport, provisional, special flight permits, and experimental. Of these, the prototypes of new aircraft designs typically receive an experimental permit while they are being flight tested or for market surveys.

Once the manufacturer is nearing the completion of the certification process and it is apparent it will comply with the remaining regulations, the authorities often allow the manufacturer to begin delivering aircraft by issuing provisional permits. This helps the manufacturer begin to recover the extreme costs of developing the aircraft. The provisional permit inflicts limitations to the operation of the aircraft that are lifted once the manufacturer finally receives the TC. An example of this could be a GA airplane designed for an airframe lifetime of, say, 12,000 hours.[3] Since fatigue testing is one of the last compliances to be demonstrated, it is possible the aircraft would receive a provisional S-AC with a 2000 hr airframe limitation. Since GA aircraft usually operate some 300–400 flight hours per year, the 2000 hr limitation will not affect the operator for several years, allowing the manufacturer to complete the certification while being able to deliver aircraft. Once the 12,000 hr lifetime is demonstrated, the 2000 hr limitation on already delivered aircraft is lifted, provided their airframe is deemed to qualify.

Maintenance Requirements

The use of an aircraft subjects it to wear and tear that eventually will call for repairs. Such repairs can be of a *preventive* type, such as the replacement of a component expected to fail within a given period of time, or the *restorative* type, such as the addition of a doubler to improve the integrity of a structural part beginning to show signs of fatigue. The aviation authorities require manufacturers to stipulate frequency and severity of preventive maintenance by instructing what tasks must be accomplished and when, in a maintenance program. If the owner or operator of the aircraft does not comply with this satisfactorily, the aircraft may lose its AC and is then said to be "grounded."

Airworthiness Directives (AD)

Sometimes the operation of a specific aircraft type develops unanticipated issues that may compromise its safety. If such issues arise, the manufacturer is obligated to notify the aviation authorities. The authorities will issue an Airworthiness Directive (AD) to the manufacturer and to all operators worldwide. The AD is a document that stipulates redesign effort or maintenance action that must be accomplished to prevent the issue from developing into a catastrophic event. Compliance with the AD is required or the AC for the specific aircraft may be cancelled. ADs for different aircraft types can be viewed on the FAA website [14].

---

[3]GA aircraft often specify airframe lifetime in terms of flight hours rather than cycles because they are operated in a much less rigorous environment than commercial aircraft.

Service Bulletin (SB)

In due course the manufacturer inevitably gains experience from the operation of the aircraft. This experience results from dealing with individual customers as well as from the manufacturer's sustaining engineering effort. This experience usually results in the improvement of the aircraft or its operation and is therefore very valuable. Consequently, it is important to share it with other operators. This is done by publishing service bulletins (SB). Although the recommendations in a SB are most often discretionary (i.e. it is up to the customer to comply), they will sometimes relay information required to comply with an AD.

Advisory Circular (AC)

An *advisory circular* is a means for the FAA to share information with the aviation community regarding specific regulations and recommended operational practices. This information is sometimes detailed enough to be presented in the form of a textbook (e.g. AC36-3H – Estimated Airplane Noise Levels in A-Weighted Decibels) or as simple as a few pages (e.g. AC 11-2A – Notice of Proposed Rulemaking Distribution System). A complete list of ACs is provided on the FAA website www.faa.gov.

Technical Standard Order (TSO)

A *technical standard order* is a minimum performance standard that particular materials, parts, processes, and appliances used on civil aircraft are subjected to. Effectively, a TSO is a letter to the manufacturers of a given product that states that if they (the manufacturers) wish to get their products TSOd, they will have to meet the performance requirements and submit a list of engineering documentations (drawings, specifications, diagrams, etc.) that are specified in the letter. Effectively, a TSO is an official certificate that confirms the part is safe for use in a specific aircraft. In other words: it is airworthy. This puts the manufacturer at a significant advantage over another one whose product is not TSOd. It is also essential for pilots to know that the equipment they are using is airworthy.

Technical Standard Order Authorization (TSOA)

A *technical standard order authorization* is a document that authorizes the manufacturer to produce parts and components in accordance with a particular TSO. As an example, consider a battery manufacturer who wants to produce a battery for use in a particular type of aircraft. The TSO tells the manufacturer what the battery must be capable of (e.g. amp-hours, temperature tolerance, etc.). The TSOA tells the manufacturer that in the eyes of the FAA the product is qualified and can now be produced.

Parts Manufacturer Approval (PMA)

Parts manufacturer approval authorizes a manufacturer to produce and sell replacement or modification parts for a given aircraft. This way, the manufacturer can produce airworthy parts even if they were not the original manufacturer.

1.3 AIRCRAFT DESIGN ALGORITHM

This section presents a step-by-step method intended to help the novice designer begin the conceptual design of a GA aircraft and bring it to the preliminary phase. As stated earlier, the *conceptual design phase* formally establishes the initial specifications of and defines the external geometry of the aircraft. It is imperative that proper analyses are selected during this phase, as this is an opportunity to design as many problems out of the airplane as possible.

The conceptual design process is one of iteration. The designer should realize this from the start of the project. A well-organized project is one that allows analyses to be conducive to iteration. This means that as the design of the aircraft progresses, it is inevitable that new things will be discovered that call for many of the previous calculations to be redone. For instance, it might be discovered that the wingspan needs to be increased, something that will affect all parts of the wing geometry. Aspect ratio, wing area, and so on, will need to be recalculated – often many times during this process. We wish to prepare the design process so the calculations of this nature are swift. Ideally, if we change the wingspan, all parameters that depend on it should be updated automatically, from the most elementary geometry to the most complex weight, drag, performance analyses.

The modern spreadsheet is ideal for such an analysis approach. This book is written to provide the designer with methods to make the implementation of this process easier for spreadsheet analysis – which is why many of the graphs also feature the exact equations used to generate them. When necessary, the author has painstakingly digitized a great number of graphs for which there are no data available other than the graphs themselves. Additionally, many methods are presented using computer codes. All of this is done to help the designer to more easily answer fundamental questions about his or her creation.

The design algorithm presented below assumes the implementation of the conceptual approach using a spreadsheet. Of course, it is not the only way to take care of business. However, it is based on a careful evaluation of what went well and not so well in the industry design experience of this author. Another word for algorithm is process; it is simply a list of tasks that are arranged in a logical order. In addition to the algorithms, the designer should keep handy the information from Section 23.3, *General Aviation Aircraft Design Checklist*, which describes a number of pitfalls to avoid.

1.3.1 Conceptual Design Algorithm for a GA Aircraft

This algorithm treats the design process almost as if it were a computer program (see Table 1-3). First, a number of initialization tasks are performed, followed by a set of iterative tasks. The table provides a complete conceptual design process and several tasks to help bring the design into the preliminary design phase. Where appropriate, the reader is directed toward a section in this book that will provide solutions and analysis methods.

It should be pointed out that sketching the airplane is not suggested until Step 10. This may appear strange to some readers, however, the reason is simple: not enough information exists for an effective sketch until Step 10. This of course does not mean a sketch cannot be or should not be drawn before that — just that an accurate depiction of the airplane is not possible. For one, the wing and tail geometry are determined in Steps 8 and 9, so an earlier sketch is unlikely to represent those with any precision. For this reason, and in the humble view of this author, an earlier sketch is a bit like a shot in the dark. Of course, adhering to this algorithm is not the law of the land. It merely represents how this author does things. The reader is welcome to make modifications to the algorithm and bend it to his or her own style. What works best for the reader is of greater importance.

The implementation of this algorithm is best accomplished through the use of a spreadsheet. Organize the spreadsheet in a manner that is *conducive to iteration*. What this means is that if (or more precisely, *when*) any parameter changes, all parameters that depend on it will be automatically updated.

1.3.2 Implementation of the Conceptual Design Algorithm

Consider the ways of the past when engineers didn't have the powerful tool that the modern spreadsheet is. Months were spent on estimating performance; stability and control; structural analyses, etc., and any change to the external geometry of the aircraft would call for a major recalculation effort. So, let's entertain the scenario in which the wing area has to be increased by 5% to reduce stalling speed. This would call for an update in drag analysis, because the change in area increases the drag, which in turn changes the performance. The geometric change also modifies the airplane's stability. Additionally, greater area changes the distribution of the lift and the magnitude of the bending moments. So, all wing structural calculations will have to be revised as well. And this effort takes a lot of time, perhaps many weeks. Fortunately for the modern engineer, those days are gone, because, if properly prepared, a spreadsheet will automatically re-calculate in a heartbeat everything that depends on the value that is changed.

The spreadsheet is best prepared in the manner shown in Figure 1-8. Modern spreadsheet software such as Microsoft Excel or the free-of-charge Open Office Calc are three-dimensional, which means they allow multiple *worksheets*. Each cell in a worksheet can contain formulas that refer to any other cell in the spreadsheet, which means each worksheet can link to another worksheet within the same spreadsheet. It is ideal to use this capability to organize the spreadsheet in a manner particularly useful to aircraft design. This requires specific worksheets to act either as a hub or a spoke in a hub-and-spoke hierarchy. The hub, which is called the general worksheet, acts primarily as a data entry page, where, at best, only relatively simple calculations take place. For instance, the user would enter empty and gross weight and the useful load might be calculated by a simple subtraction. However, all the sophisticated analyses take place in subsequent worksheets, which should be considered the spokes. This is shown as the systematic hierarchy in Figure 1-8.

As an example of how this would work, consider the *tail sizing* worksheet. It requires the wingspan of the aircraft as a part of that analysis (see Chapter 11, *The Anatomy of the Tail*). However, rather than entering that parameter on the worksheet itself, the worksheet would fetch it from the *general* worksheet. The same holds for the *stability and control* worksheet. It also requires the wingspan in its calculations. That worksheet would also fetch its value from the *general* tab. This way, if for some reason the wingspan has to be changed, the designer can simply enter the new value on the *general* worksheet, and the *tail sizing*, and *stability and control* worksheets will be automatically updated. The only thing for the designer to be mindful of is to be rigorous in the application of this philosophy from the start of the project. It will save him countless hours of review work.

To give the reader a better insight into how this is implemented in a real spreadsheet, consider Figure 1-9, which shows how the hierarchy appears in practice. Note that two easily identifiable colors have been chosen for cells to indicate where the user shall enter information and where a formula has been entered. This reduces the risk of the user accidentally deleting important formulas and helps with making the spreadsheet appear better organized and more professional.

1.3 AIRCRAFT DESIGN ALGORITHM

TABLE 1-3 Conceptual Design Algorithm for a GA Aircraft

Step	Task	Chapter/Section
1	Understand requirements, mission definition, and the implications of the regulations to which the airplane will be certified.	—
2	Study aircraft that fall into the same class as the one to be designed. These may present you with great design ideas and solutions. They can also show you what to steer away from — which is priceless! • Evaluate what configuration layout may best suit the mission. • Decide on a propulsion methodology (propeller, turbofan, others?).	—
3	If the target weight and maximum level airspeed are known, estimate the development and manufacturing costs for a projected 5-year production run. If the target weight is not known perform this task once it is known (see Step 6 or 12). Evaluate how many units must be produced to break even and the required retail prices. Evaluate operational costs as well. How do these compare to the competition?	2.2 2.3
4	Create a *constraint diagram* based on the requirements of Step 1 (target performance).	3.2
5	Select critical performance parameters (T/W or BHP/W and W/S) from the constraint diagram. Once T/W and W/S are known, the next step is to estimate the gross weight so that wing area and required engine thrust (or power) can be extracted.	3.2
6	Estimate initial empty and gross weight using W-ratios and/or historical relations.	6.2
7	Using the results from the constraint diagram of Step 4 and the initial gross weight of Step 6, estimate the initial wing area. Keep in mind the requirements for stall speeds (e.g. LSA limit is 45 KCAS, 14 CFR Part 23 is 61 KCAS, etc.) to ensure the selected W/S and T/W (or BHP/W) will allow the design to simultaneously meet all performance requirements and stall speeds.	3.2
8	Estimate initial tail surface area and special position using V_{HT} and V_{VT} methodology.	11.4 11.5
9	Propose a wing layout that suits the mission by establishing an initial AR, TR, airfoils, planform shape, dihedral, washout, etc. Note that many of these parameters are likely to change in the next iteration. For the airfoil selection use a method like the one shown in Section 8.3.15, *Airfoil selection how-to*.	8, 9
10	If not already done, sketch several initial configurations and methodically evaluate their pros and cons. Select a candidate configuration.	4
11	Based on the selected propulsion methodology (see Step 2), select the engine type and layout (number of, types, properties of, location of) to be evaluated.	7
12	Using the candidate configuration, estimate empty, gross, and fuel weight using the appropriate combination of *statistical*, *direct*, and/or *known weights* methods.	6.3 6.4 6.5
13	Determine the empty weight CG, CG loading combination cloud, gross weight CG, movement due to fuel burn, and inertia properties (I_{xx}, I_{yy}, ...).	6.6
14	Determine a candidate CG envelope based on results from Step 13. Expect this to change once Step 16 is completed.	6.6
15	Layout fuselage (space claims, occupant location, baggage, cargo) using a method similar to that of Section 12.3, *Sizing the fuselage*.	12.3
16	Perform a detailed static and dynamic stability analysis of the candidate configuration.	Various
17	Modify the tail surface geometry in accordance with the results from the static and dynamic stability analysis of Step 13. Note that dynamic stability modes should be converging and the geometry will likely have to be "morphed" to eliminate any diverging dynamic modes.	11
18	Evaluate the following layout design modifications as needed, based on the above analyses: • Structural load paths (wing, HT, VT, fuselage, etc.) • Control system layout (manual, hydraulic, fly-by-wire/light) • Flight control layout (geometry, aerodynamic balancing, trim tabs) • High-lift systems and layout (flap types, LE devices) • Landing gear layout (tri-cycle, tail-dragger, fixed, retractable, etc.)	Various

TABLE 1-3 Conceptual Design Algorithm for a GA Aircraft—cont'd

Step	Task	Chapter/Section
19	Modify the design for benign stall characteristics (via washout, airfoils, slats, flaps).	9, 10
20	Perform a detailed drag analysis of the candidate configuration. Design for minimum drag by polishing the geometry for elimination of separation areas, including the addition of wing fairings.	15
21	Perform a detailed performance analysis (T-O, climb, cruise, range, descent, landing). Perform sensitivity analyses of T-O, climb, cruise, range, and landing. Create a payload-versus-range plot.	16–22
22	Optimize and refine where possible.	Various
23	Perform a regulatory evaluation and answer the following questions: (1) Will the candidate configuration meet the applicable aviation regulations? (2) Does it meet all requirements of Step 1? (3) Does it satisfy the mission described in Step 1? If the answer to any of these questions is NO, then go back to Step 10 and modify the candidate configuration. If all can be answered with a YES, then continue to the next step.	14 CFR Part 23
24	Freeze OML. Do this by the release of a document controlled electronic solid model of the vehicle.	—
25	Create a V-n diagram.	16.4
26	Detailed load analysis.	—
27	Move into the preliminary design phase.	—

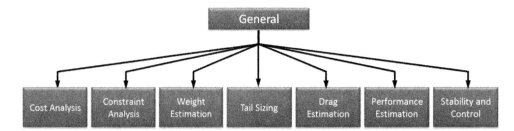

FIGURE 1-8 Organizational hierarchy of a spreadsheet (see Section 1.3.2 for explanation).

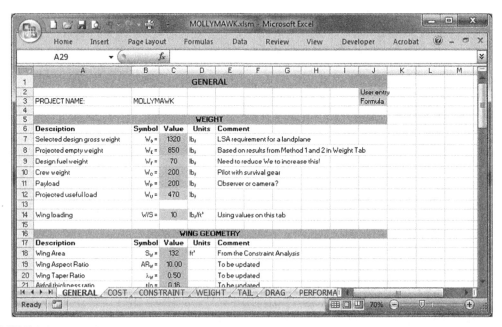

FIGURE 1-9 Organizational hierarchy implemented in an actual spreadsheet (see Section 1.3.2 for explanation).

1.4 ELEMENTS OF PROJECT ENGINEERING

The purpose of this section is to present a few tools that are at the disposal of the project engineer. The reader is reminded that this is not a complete listing; there is a multitude of ways to conduct business. Experienced engineers may not find anything helpful in this section, but that is all right, this section is not intended for them, but rather the novice engineer who is not sure where to begin or how to do things.

Any serious engineering project requires many activities to be managed simultaneously. Scheduling, communication, hiring, conflict resolution, coordination and interaction between groups of specialists are but a few tasks that are required to move the project along. Such projects usually require someone, frequently an experienced engineer, to perform these duties; this person is the *project engineer*.

In addition to the aforementioned tasks, there are multiple others that the project engineer is responsible for and some of those seem to have very little in common with the engineering the person was trained in. He or she serves as a liason between management of the company and the engineering workforce ensuring scheduled deadlines are met and, as such, plan overtime and effective delegation of tasks. He or she may have to evaluate, negotiate with and work with vendors, as well as organize training of the workforce and frequently help select between design alternatives and facilitate the resolution of development problems. He or she may even be required to help resolve personal problems between individuals in the workforce, through mediation or other means. Some of these are tasks the project engineer has never even heard about as a student.

Nowadays, project engineering is offered as an elective course in many universities. Typically, such courses emphasize various project engineering skills, such as time management, how to make meetings more effective, scheduling, leadership, effective communication, delegation styles, lean engineering, and engineering economic analyses.

People often ask what skills a project engineer must master in order to be a good project manager. Six important skills are often cited; communication, organization, team building, leadership, coping, and technological skills. *Communication skills* involve the ability to listen to people and being able to persuade them to act in a manner that favors the goals of the project. Having organization skills means you have the ability to plan, set goals, and analyze difficult situations. *Team-building skills* involve having the ability to empathize with other team members, demonstrating loyalty to the team, and the motivation to work as part of the team in order to ensure that the project is successful. *Leadership skills* involve setting a good example and the exercise of professionalism. Having good leadership skills means being enthusiastic, having a positive outlook, and involves being able to effectively delegate tasks. A good leader sees the "big picture" and can communicate it to the team members. *Coping skills* involve flexibility, patience, persistence, and openness to suggestions from others. It makes the leader resolute and able to adjust to changing conditions. *Technological skills* involve the utilization of prior experience, knowledge of the project, and the exercise of good judgment.

1.4.1 Gantt Diagrams

A Gantt diagram is a graphical method of displaying the chronological flow of a project and is a standard method used by the industry. The diagram breaks the project down into individual major and subtasks, allowing the manager to assign a start and end date to each, as well as multiple other information, such as human resources, and equipment. These appear as horizontal bars, as can be seen in Figure 1-10. It is possible to buy software that allows this to be done effectively once it has been mastered. Gantt diagrams also generally feature important project completion dates or milestones.

1.4.2 Fishbone Diagram for Preliminary Airplane Design

The *Fishbone diagram*, more formally known as an *Ishikawa diagram* or a *cause-and-effect diagram*, is named after Kaoru Ishikawa (1915–1989), a Japanese quality control statistician. At its core, the diagram focuses on effects and their causes by drawing them in a special manner. The causes are written along the perimeter of a page, with arrows pointing towards the effect or consequence, which is the horizontal arrow in the middle of the diagram. The resulting graph is reminiscent of a fish skeleton, which explains its name.

This diagram is used in a slightly modified fashion for aircraft design (see Figure 1-11). In this application the horizontal arrow is actually a timeline. It starts at the initiation of the project and terminates at its completion. This format can be applied to the entire development program, or sub-projects. The causes can be thought of as major tasks that are broken down into sub tasks. The causal arrows points to a milestone or a representative time location on the horizontal line, as shown in the figure. The advantage of this diagram is that it allows the project manager to (1) demonstrate the status of the project to upper management (for instance compare Figure 1-11 and Figure 1-12), (2) to anticipate when to ramp up for specific sub-projects, and (3) understand the "big-picture" of the project.

20　1. THE AIRCRAFT DESIGN PROCESS

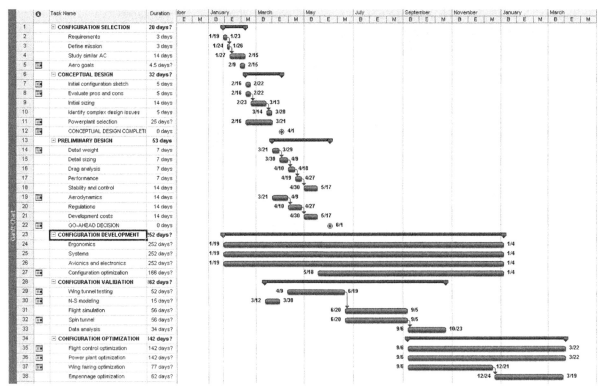

FIGURE 1-10　A Gantt diagram showing a hypothetical conceptual and preliminary design of a simple aircraft.

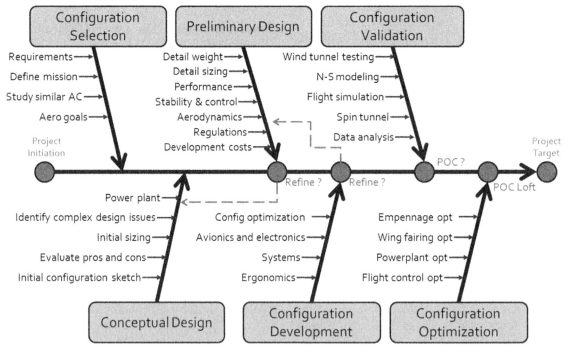

FIGURE 1-11　A typical fishbone diagram.

A fishbone diagram for a preliminary airplane design typically consists of the following categories:

- Configuration selection
- Conceptual design
- Preliminary design
- Configuration development
- Configuration validation
- Configuration optimization.

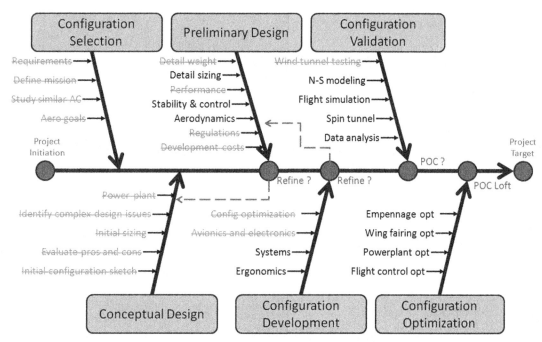

FIGURE 1-12 The same fishbone diagram at a later time during the design process. Completed tasks have been struck through. The diagram displays progress made at a glance and gives the manager a great tool to reallocate resources.

Proper documentation (reports) should always be created along the way in a manner stipulated by the corresponding company. In small companies the designer may choose to use a system based on limited reporting; however, a more comprehensive reporting system may be employed to generate certification-style reports along the way. This may help free up resources during the actual certification program.

1.4.3 Managing Compliance with Project Requirements

For project management purposes, it is imperative to list all design requirements clearly and then regularly evaluate the status of each, in order to indicate compliance. An example of such a list is shown in Figure 1-13. Some requirements are not as much "required" as they are desired, and these should be indicated in the 'required' column, as such. Note that some customer requirement may "invoke" other requirements the customer may or may not have a deep understanding of. An obvious example of this is federal regulatory requirements. Figure 1-13 lists one of the customer's requirements as Certification in the Light Sport Aircraft (LSA) category. This invokes a set of complex new requirements, so important that they may compromise the viability of the entire project.

For a complex project with a large number of milestones, a milestone list should be prepared. It is helpful if it is based on the fishbone diagram as shown in Figure 1-11 and Figure 1-12. This will allow the manager to keep a close tally on how each set of tasks is progressing.

1.4.4 Project Plan and Task Management

Figure 1-14 shows an example of a project milestone list, which is used to monitor the progress of a complicated project and help the manager relay to the workforce when certain tasks have to be completed. This way, target days can be set and proper scheduling of resources can take place.

Figure 1-15 shows an example of a project design plan. The design plan breaks the entire project into subparts and tasks, allowing the manager to keep track of the progress of various tasks. The hardest part of the plan is to keep the plan suitably detailed. Planning too many details will result in a plan that's impossible to achieve on deadline (see rightmost column in the figure) and will absorb the manager's time by requiring constant revisions. Remember, the plan is not the boss, the boss is. The plan is simply a tool to help run things smoothly. By the same token, a plan with too little detail is useless. This can be seen by considering the comparison in Table 1-4. As is so often required, a golden balance should be struck between the two extremes.

1.4.5 Quality Function Deployment and a House of Quality

An important requirement of sophisticated products is that they must simultaneously satisfy a large number

PROJECT REQUIREMENTS

	REQUIRED	MET (Y/N)
Customer Requirements		
Light single engine aircraft that comfortably seats two people	Y	
Cruising speed >100 KIAS	Y	
Service ceiling >10000 ft	Y	
Certifiable in LSA category	Y	
Target range: 200 pax·mi/gallon	Y	
Engine: Rotax 912	Y	
Regulatory Requirements for LSA		
Max T-O weight is 1320 lbf or less	Y	
Max level speed (VH) with MCP is 120 KCAS or less	Y	
Max stalling speed (VS1) is 45 KCAS or less	Y	
Max seating capacity of two persons	Y	
A single, reciprocating engine	Y	
A fixed or ground adjustable propeller	Y	
A non-pressurized cabin	Y	
Fixed landing gear	Y	
Additional Requirements		

FIGURE 1-13 An example of a (simple) project requirement list.

of requirements. Among them are customer and engineering requirements. In order to enhance the likelihood that a product will satisfy the needs of the customer, it is necessary to survey what it is they know or think they need. Unfortunately, survey responses can often be vague and it is, thus, necessary to convert them to statements that allow them to be measured. For instance, a statement like "I don't want to pay a lot of money for maintenance" can be translated to "reliability." This, in turn, can be measured in terms of how frequently parts fail and require repairs. It is inevitable that some of these requirements conflict with each other, in addition to depending on each other. For instance, the weight of an aircraft will have a great impact on its rate-of-climb, but none on its reliability.

Quality function deployment (QFD) is a method intended to help in the design of complex products, by taking various customer wishes into account. This is accomplished using a sophisticated selection matrix that helps evaluate the impact of the various customer wishes on areas such as the engineering development. The output can be used to highlight which customer wishes to focus on. The primary drawback is that it can take considerable effort to develop and it suffers from being highly dependent on the perspectives of the design team members. It was developed by the Japanese specialists Dr. Yoji Akao and Shigero Mizuno and is widely used in disparate industries. One of the method's best-known tools is known as a *house of quality* (also called a *quality functional deployment matrix*), a specialized matrix, resembling a sketch of a house, designed to convert customer requirements into a numeric score that helps define areas for the designer to focus on.

The preparation of a house of quality (HQ) is best explained through an example. Generally, the HQ consists of several matrixes that focus on different aspects of the development of a product (see Figure 1-16). This way, the impact of desired (or customer) requisites on the technical requirements and their interrelation is identified. Ultimately, the purpose is to understand which requirements are of greater importance than others and how this complicates the development of the product.

Below, a much simplified version of the HQ, tackling the development of a small GA airplane, is presented. The reader is reminded that the HQ can be implemented in a number of ways — and a form that suits, say, the textile industry does not necessarily apply directly to the aviation industry.

Step 1: Customer Requirements

Assume that customer surveys have been collected for the design of a simple aircraft and the desired

DESIGN PROJECT MILESTONES - Revision A

PROJECT:	YOUR PROJECT NAME			
NAME:			SECTION:	

		REQUIRED	STATUS	REPORT ID
A	**Configuration Selection**			
A1	Requirement study	Y	0	
A2	Define mission	Y	0	DN-001
A3	Study aircraft in same/similar class	Y	0	
A4	Study aerodynamic goals	Y	0	
			DUE DATE:	TBD
B	**Conceptual Design**			
B1	Initial configuration sketch	Y	0	
B2	Evaluate pros and cons	Y	0	
B3	Initial sizing	Y	0	DN-002
B4	Identify complex design issues	Y	0	
B5	Obtain design information for power plant	Y	0	
			DUE DATE:	TBD
C	**Preliminary Design**			
C1	Weight analysis	Y	0	
C2	Airfoil selection	Y	0	
C3	Refined sizing	Y	0	
C4	Drag analysis	Y	0	
C5	Performance analysis	Y	0	
C6	Stability and control analysis	Y	0	DN-002
C7	Handling characteristics	Y	0	
C8	V-n diagram, Flight envelope, Aeroelastic envelope	Y	0	
C9	Airloads	Y	0	
C10	Compliance with regulations (LSA)	Y	0	
C11	Project cost estimation	Y	0	
			DUE DATE:	TBD
D	**Configuration Development**			
D1	Ergonomics	Opt	0	N/R
D2	Systems	Opt	0	N/R
D3	Avionics and electronics	N	0	N/R
D4	Configuration optimization using analytical methods	Opt	0	N/R
			DUE DATE:	N/R
E	**Configuration Validation**			
E1	Wind tunnel testing	Opt	0	N/R
E2	Spin tunnel testing	N	0	N/R
E3	CFD analysis (Navier-Stokes modeling)	Opt	0	N/R
E4	Flight simulation	Opt	0	N/R
E5	Data analysis	N	0	N/R
E6	Other	N	0	N/R
			DUE DATE:	N/R
F	**Configuration Optimization**			
F1	Optimization of flight control system	Y	0	N/R
F2	Optimization of power plant system	N	0	N/R
F3	Optimization of wing fairing	N	0	N/R
F4	Optimization of empennage	Opt	0	N/R
F5	Other	N	0	N/R
			DUE DATE:	N/R

FIGURE 1-14 An example of a project milestone list.

DESIGN PLAN - Revision A				
	STEP	TASK	STATUS	DEADLINE
Selection DN-001	1	Create a fishbone diagram of approach	0	
	2	Study and understand requirements	0	
	3	Define the mission	0	
	4	Create a database of similar aircraft	0	
	5	Study and understand what aerodynamic goals must be met	0	
Conceptual DN-002	6	Perform first weight estimate of the airplane	0	
	7	Perform first estimate of critical performance parameters by retrieving them from the database. (C_{Lmax}, L/D_{max}, T/W, W/S)	0	
	8	Use critical performance parameters to estimate fundamental geometry (wing area, max fuel qty, and others)	0	
	9	Sketch the configuration	0	
Preliminary DN-003	10	Perform detailed weight estimate of the airplane	0	
	11	Perform static stability analysis to evaluate CG limitations	0	
	12	Move components in the airplane to ensure CG location will be achievable	0	
	13	Complete dynamic stability analysis	0	
	14	Complete handling analysis	0	
	15	Complete performance analysis	0	
	16	Complete regulatory evaluation	0	
	17	Complete project cost estimation	0	
	18	Optimize as necessary	0	
	19	Incorporate needed changes into the design and, if necessary, go back to Step 10	0	
	20	Repeat until requirements are met and airplane is certifiable	0	
	21	Create a detailed three-view of the aircraft	0	

FIGURE 1-15 An example of a project design plan.

requirements are *fast, efficient, reliable, spacious*, and *inexpensive* (see Figure 1-17). An actual HQ would almost certainly have more than five requirements, but, again, this demonstration will be kept simple.

The survey has requested that potential customers rate the requirements using values between 1 (not important) and 5 (very important). This is placed in a matrix as shown in the figure [1-17]. This way, 'fast' has a rating of 3.0 (moderately important), 'efficient' has a rating of 5.0 (very important), and so on. Then, the ratings are added and the sum (18.5) is entered as shown. The column to the right shows the percentages of the ratings. For instance, the percentage associated with the requirement 'fast' is $100 \times 3.0/18.5 = 16.2\%$.

Step 2: Technical Requirements

The next step requires the designing team to list a number of engineering challenges that relate to the customer requirements. For instance, the requirement for 'efficiency' calls for special attention to the lift and drag characteristics of the aircraft. These have been listed in Figure 1-18 with some other engineering challenges, such as 'size of aircraft', 'drag', 'weight', and so on. These will be revisited in Step 4.

TABLE 1-4 Examples of Plans of Different Detail

A Simple (Useless) Plan	A Clear and Effective Plan	A Complex and Ineffective Plan
Buy a new car.	List what I want in my new car.	List what I want in my new car.
	Look for a car that satisfies my list.	List what my spouse wants in my new car.
	Get a bank loan.	List what my kids want in my new car.
	Purchase my new car.	Compile a comprehensive list of every ones' needs and wishes.
		Start up computer.
		Start Internet Explorer and go to Google.
		Search for cars online.
		Look for cars at dealerships.
		Refill my current car with gas after driving to dealerships.
		Go to three banks and talk to loan officers about specifics.
		Evaluate which bank to choose.
		Discuss the choice with spouse.
		Go to the bank of choice and get loan.
		Celebrate by taking spouse to dinner.
		Get gas again after running all the errands.
		Order car online or buy from dealership.
		Celebrate again by opening a bottle of champagne.

Step 3: Roof

The roof (see Figure 1-18) is used to indicate interrelationships between the various engineering challenges. It must be kept in mind that the roof sits on top of the technical requirements matrix and the diagonals enclose the columns of engineering challenges. This arrangement must be kept in mind for the following discussion.

The roof consists of two parts: the roof itself and, for a lack of a better term, the fascia. The fascia is used to indicate whether the challenge listed below (e.g. 'drag' or 'weight') has a favorable effect on the product. This way, more 'power' has a favorable effect (more power is good) and this is indicated using the arrowhead that points up. 'Production cost', on the other hand, has negative effects on it, so the arrow points down.

The other challenges have been identified in a similar manner, except the last one ('size of aircraft'). It is not clear whether or not a larger or a smaller version of the aircraft is beneficial to the product, so it is left without an arrow. Naturally, this may change if the team decides this is important; all parts of the HQ are ultimately decided by the design team and its consensus may differ from what is being shown here.

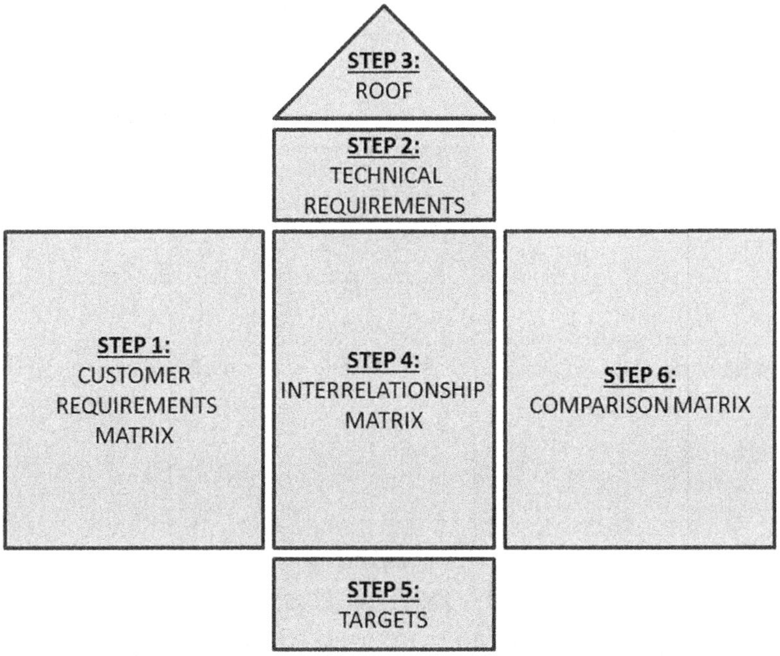

FIGURE 1-16 A basic house of quality.

From customer surveys

Customer Requirements	Importance 1 (not) – 5 (very)	%
FAST	3.0	16.2%
EFFICIENT	5.0	27.0%
RELIABLE	4.5	24.3%
SPACIOUS	4.0	21.6%
INEXPENSIVE	2.0	10.8%
SUM =	18.5	100%

FIGURE 1-17 Customer requirements matrix.

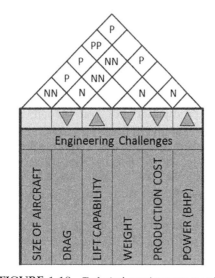

FIGURE 1-18 Technical requirements matrix.

Next consider the roof itself, shown as the diagonal lines in Figure 1-18. It is used to indicate positive and negative relationships between the challenges. These are typically denoted with symbols (e.g. + for positive and − for negative), but here the following letters are used:

NN — means there is a strong negative relationship between the two engineering challenges.
N — means there is a negative relationship.
P — means there is a positive.
PP — means there is a strong positive relationship between the two engineering challenges.

Consider the columns containing 'size of aircraft' and 'drag'. It can be argued that there is a strong negative relationship between the 'size of aircraft' and 'drag' (large aircraft = high drag). This is indicated by entering NN at the intersection of their diagonals. Similarly, there is a positive relation between 'size of aircraft' and 'lift', indicated by the P at the intersection. Some might argue there should be a strong positive relationship; however, if the size refers to the volume of the fuselage rather than the wings, then the relationship is arguably only positive. This shows that the build-up of these relationships is highly dependent on interpretation, requiring the design team to reach a consensus. Once complete, the example letter combinations are entered as shown in Figure 1-18.

Step 4: Interrelationship Matrix

The next step is to try to place weight on the engineering challenges as they relate to the customer requirements. This is accomplished using the interrelationship matrix (see Figure 1-19). In other words, consider the customer requirement 'fast'. It will have a strong influence on the engineering challenge 'drag'. However, 'lift' will be less affected by it. Similarly, the customer requirement 'reliable' will not have any effect on the 'weight', and so on.

The design team must come up with a scale that can be used to indicate the severity of such associations. It is not uncommon to use a scale such as the one shown below:

9 — means the customer requirement has great influence.
3 — means the customer requirement has moderate influence.
1 — means the customer requirement has weak influence.

For clarity, omit entering numbers in cells where no influence exists. Some people enter special symbols in the cells that mean the same, but in this author's view it only adds an extra layer of confusion. Note that these numbers will be used as multipliers in the next step, which makes it much simpler to enter them directly.

Engineering Challenges					
SIZE OF AIRCRAFT	DRAG	LIFT CAPABILITY	WEIGHT	PRODUCTION COST	POWER (BHP)
9	9	3	9	9	9
1	9	9		9	3
1				1	3
9	3			3	1
3	3	3	9	9	9
4.24	4.86	3.24	2.43	5.76	4.19
17.2	19.7	13.1	9.8	23.3	16.9

FIGURE 1-19 The interrelationship matrix.

FIGURE 1-20 The target matrix.

Customer Requirements	Importance 1 (not) – 5 (very)		SIZE OF AIRCRAFT	DRAG	LIFT CAPABILITY	WEIGHT	PRODUCTION COST	POWER (BHP)
FAST	3.0	16.2%	9	9	3	9	9	9
EFFICIENT	5.0	27.0%	1	9	9		9	3
RELIABLE	4.5	24.3%	1				1	3
SPACIOUS	4.0	21.6%	9	3			3	1
INEXPENSIVE	2.0	10.8%	3	3	3	9	9	9
SUM =	18.5	100%	4.24	4.86	3.24	2.43	5.76	4.19
			17.2	19.7	13.1	9.8	23.3	16.9

(From customer surveys; Engineering Challenges)

Step 5: Targets

The target matrix (see Figure 1-20) represents the results of a cross-multiplication and summation that is used to determine where to place the most effort during the development of the product. The operation takes place as follows:

Consider the percentage column of the customer requirements matrix (16.2%, 27.0%, etc.) and the first column of the technical requirements column ('size of aircraft', 9, 1, 1, etc.). These are multiplied and summed as follows:

$$0.162 \times 9 + 0.270 \times 1 + 0.243 \times 1 + 0.216 \times 9 + 0.108 \times 3 = 4.24$$

The remaining columns are multiplied in this fashion, always using the percentage column, yielding 4.86, 3.24, 2.43, and so on.

The next step is to convert the results into percentages. First, add all the results (4.24 + 4.86 + ...) to get 24.73. Second, for the first column, the percentage of the total is $100 \times 4.24/24.73 = 17.2\%$, $100 \times 4.86/24.73 = 19.7\%$ for the second one, and so forth. These numbers are the most important part of the HQ, as the highest one indicates where most of the development effort should be spent. The results and the entire HQ can be seen in Figure 1-21. It can be seen that, in this case, the 'production cost' and 'drag' are the two areas that should receive the greatest attention.

Step 6: Comparison Matrix

It is often helpful to create a matrix to compare, perhaps, an existing company product to that of the competition. This helps to identify shortcomings in the company products and to improve them. A comparison matrix is shown in Figure 1-21, where they have been "graded" in light of the customer requirements, allowing differences to be highlighted. This way, while the customer requirement 'fast' has a score of 3.0, it is possible the design team values it a tad lower, or at 2.5. However, the team may also conclude that competitor aircraft 1 and 2 emphasize it even less. Such a conclusion should be based on hard numbers, such as drag coefficients or cruising speed, and not subjective opinions.

The purpose of this section was to introduce the reader to the HQ as a possible tool to help with the development of a new product (or the redesign of an existing one). The interested reader is directed toward the multitude of online resources that add further depth to this topic.

1.5 PRESENTING THE DESIGN PROJECT

A picture is worth a thousand words. This old adage is particularly true in the world of engineering, where detailed information about complicated mechanisms, machinery, and vehicles, must be communicated clearly and effectively. While the topic of geometric dimensioning and tolerancing (GDT) and industry standards in technical drafting is beyond the scope of this book, saying a few words about the presentation of images is not. The practicing engineer will participate in many meetings and design reviews, where often a large number of experts in various fields gather and try to constructively criticize a new design. The process is often both exhausting and humbling, but is invaluable as a character-builder. At such moments, being able to adequately describe the functionality of

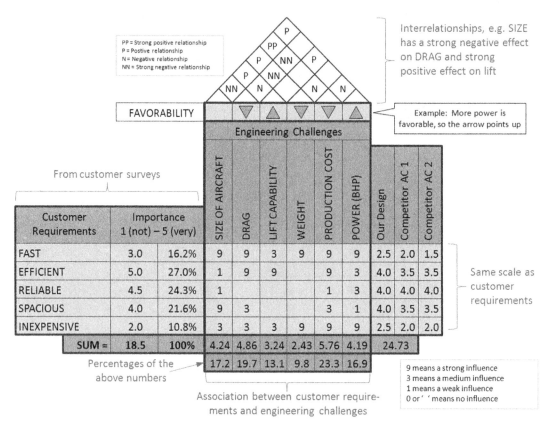

FIGURE 1-21   The completed house of quality.

one's design is priceless and no tool is better for that than a figure, an image, or a schematic. Three-dimensional depictions are particularly effective. The modern aircraft designer benefits greatly from computer-aided design (CAD) tools such as solid modelers (Pro/E, Solidworks, CATIA, and others), which allow very complicated three-dimensional geometry to be depicted with a photo-realistic quality. Highly specialized software, for instance, finite element analysis (FEA) and computational fluid dynamics (CFD) programs, allow the engineer to describe the pros and cons of very complicated structural concepts and three-dimensional flow fields, and even add a fourth dimension by performing time-dependent analyses. It cannot be over-emphasized to the entry-level engineer to get up to speed on this technology. It not only helps with communication, but also develops a strong three- and four-dimensional insight into engineering problems.

Three-view Drawings

The three-view drawing is a fundamental presentation tool the engineer should never omit. Airplane types are commonly displayed using three-view drawings, showing their top, side, and frontal views. Such drawings are an essential part of the complete submittal proposal package for any aircraft. Although such presentation images date back to the beginning of aviation, they are as important to any proposal as wings are to flying. Figure 1-22 shows a typical drawing, with the added modern flare in the form of a three-dimensional perspective rendering.

Images Using Solid Modelers

The modern solid modeler software (CAD) has revolutionized aircraft design. Long gone are the sloped drafting tables the technical drafter used to work with, as are the special architectural pens that deliver uniform line thicknesses and other tools of the past. These began to disappear in the late 1980s and early 1990s. Now, drafters are equipped with personal computers or workstations and model complicated parts and assemblies in virtual space. At the time of writing, programs such as Pro/E (Pro-Engineer), Solidworks, and CATIA are the most common packages used and pack an enormous sophistication in their geometric engines. They can be used on any desk- or laptop computer and allow mechanical linkages to be animated, photorealistic renderings of the design to be made, and some even offer limited finite element and computational fluid dynamics capabilities. These methodologies allow

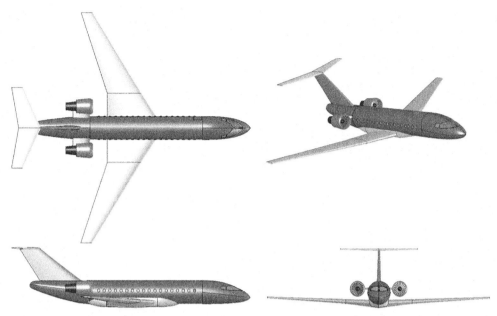

FIGURE 1-22 A non-standard three-view drawing, made using modern solid modeling software.

highly mathematical surfaces, referred to as NURBS, to be defined and modified on a whim. Such tools provide perfect mathematical definitions of complicated compound surfaces and, therefore, allow curvature-perfect OML to be created. Images from such programs can be quite persuasive and informative. Figure 1-23 shows an image of a twin-engine regional jet design from one such package, superimposed on a background image taken at some 18,000 ft. The resulting image can be of great help in engineering and marketing meetings.

FIGURE 1-23 A solid model of a modern regional jet superimposed on a photographic background, showing the capability of modern computer-aided design software.

Images Using Finite Element Modelers

The modern structural analysis often includes very sophisticated finite element analyzers, which are capable of producing very compelling images. While such images should be used with care (as their compelling nature tricks many into thinking they actually represent reality, which they may not), they can give even a novice an excellent understanding of load paths as well as where stress concentrations reside. While such images are usually available only after detailed design work has begun, images from previous design exercises can sometimes be helpful in making a point about possible structural concepts.

Figure 1-24 shows stress concentrations in a forward shear-web of the wing attachment/spar carry-through for a small GA aircraft, subject to an asymmetric ultimate load. The elongated diamonds in the center of the spar carry-through are corrugations intended to stiffen the shear-web, but these cause high stress concentrations on their own.

Images Using Computational Fluid Dynamics Software

Computational fluid dynamics is a vibrant field within the science of fluid mechanics. Spurred by a need to predict and investigate aerodynamic flow around three-dimensional bodies, this computational technology has become the stalwart of the modern aerodynamics group. Similar advice as above should be given to the entry-level aircraft designer. The images generated by the modern CFD packages are often mind-boggling in their sophistication (Figure 1-25). It is

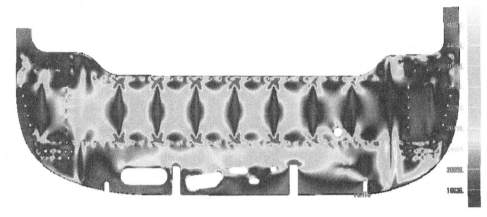

FIGURE 1-24   An image of a stress field in the spar carry-through of a small GA aircraft due to asymmetric wing loads, generated by a popular finite element analysis software.

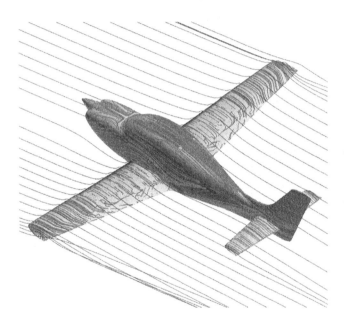

FIGURE 1-25   Streamlines and oilflow plots speak volumes about the nature of airflow around this SR22, showing the strength of Navier-Stokes CFD software. *(Courtesy of Cirrus Aircraft)*

therefore easy to be lulled into trusting them blindly — but they may not necessarily show what happens in real flow. This is not to say they never resemble reality, only that they do not always.

Cutaway Drawings

Few visual representations are as capable of illustrating the complexity of an airplane as the cutaway drawing. Such images are normally extremely detailed and require a great depth of knowledge of the internal structure of an airplane to correctly prepare. A case in point is Figure 1.26, which shows a cutaway of the Dassault Falcon 7X business jet. While certified to 14 CFR Part 25 (Commercial Aviation) rather than 14 CFR Part 23 (General Aviation) the figure depicts the state of the art in civil aircraft design in the early twenty-first century.

Engineering Reports

The work of the engineer is primarily of the "mental" kind; it largely involves the process of thinking. This poses an interesting challenge for anyone hiring an engineer — how can this intangible product be captured so it does not have to be recreated over and over again? The answer is the engineering report and engineering drawing.

An engineering report is a document that describes the details of a particular idea. Engineering reports encompass a very large scope of activities. It can be a mathematical derivation of a particular formula, listing of test setup or analysis of test results, justification for a particular way to fabricate a given product, evaluation of manufacturing cost or geometric optimization. The list goes on. Engineering reporting can also be the completion of proposals or even the writing of scientific papers. Regardless of its purpose, the report must always be written with the emphasis on completeness and detail. Such technical reports are how a company retains the thinking of the engineer so it does not have to be 'rethought' next time around — it turns the intangible into something physical.

The organization and format of reports vary greatly. It is not practical to present any particular method here on how to write a report. However, what all reports have in common is that they should be objective, concise, and detailed. One of the primary mistakes made by entry-level engineers is to ignore the documentation of what appears trivial. The author is certainly guilty of making such mistakes. While working on a specific assignment, one effectively becomes an expert on that topic. Grueling work on such

1.5 PRESENTING THE DESIGN PROJECT

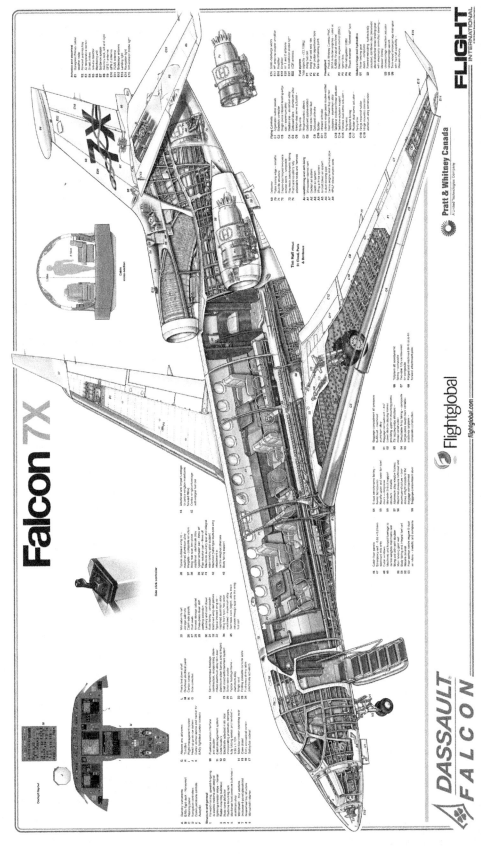

FIGURE 1-26 A cutaway of the Dassault Falcon 7X reveals details about the structure, systems, aerodynamic features, and accommodation, etc. (*Courtesy of Flightglobal and Dassault Falcon Jet Corporation*)

a project for a number of weeks or months can blur the senses to what needs to be included in the engineering report. The expertise, surprisingly, skews one's perspective; complex concepts become so trivial in the mind of the engineer that their definitions or other related details get omitted from the documentation. Then, several months or even a few years later, one has become an expert on a different topic. The previous work is a distant memory, securely archived in the digital vaults of the organization. Then, something happens that calls for a review of that past work. It is then that one realizes how many important concepts one left out of the original report and these, now, call for an extra effort and time for reacquaintance.

Additionally, detailed and careful documentation is priceless when you have to defend your work in a deposition. It is what US companies use every day to defend themselves against accusations of negligence, saving themselves billions of dollars.

Engineering Drawings

The modern engineering drawing has become a very sophisticated method of relaying information about the geometry of parts and assemblies. The details of what is called an "industry standard drawing" will not be discussed here, other than mentioning that such drawings must explain tolerance stack-ups and feature a bill of materials and parts to be employed. Today, engineering drawings are almost exclusively created using computers by a specialized and important member of the engineering team – the drafter. A competent drafter knows the ins and outs of the drafting standards and ensures these do not become a burden to the engineer.

Engineering drawings are typically of two kinds: part drawing and assembly drawing. The part drawing shows the dimensions of individual parts (a bracket, an extrusion, a tube, a bent aluminum sheet, etc.), while the assembly drawing shows how these are to be attached in relation to each other. A homebuilt kitplane may require 100-200 drawings, a GA aircraft may require 10,000, and a fighter or a commercial jetliner 50,000 to over 100,000 drawings. For this reason, a logical numbering system that allows parts and assemblies to be quickly located is strongly recommended. This way, all drawings pertaining to the left wing aileron could start with WL-A-drw number, while the right wing flap system would be WR-F-drw number. Such systems increase productivity by speeding up drawing searches – which are very frequent.

VARIABLES

Symbol	Description	Units (UK and SI)
AR	Wing aspect ratio	
BHP/W	Brake horse power-to-weight ratio	BHP/lb_f or BHP/N
CG	Center of gravity	ft, m, or %MAC
KCAS	Knots calibrated airspeed	ft/s or m/s
MAC	Mean aerodynamic chord	ft or m
n	Load factor	
T	Rated thrust	lb_f
T/W	Thrust-to-weight ratio	
TR	Wing taper ratio	
V_{HT}	Horizontal tail volume coefficient	
V_{VT}	Vertical tail volume coefficient	
W	Weight	lb_f or N
W/S	Wing loading	lb_f/ft^2 or N/m^2

References

[1] Boyer Pascal. Religion Explained. Basic Books 2001.
[2] http://faa.custhelp.com/app/answers/detail/a_id/154/kw/%22general%20aviation%22/session/L3RpbWUvMTMzNTgwOTk4MS9zaWQvSkxqTW9ZV2s%3D.
[3] Zodiac CH. 601 XL Airplane, Special Review Team Report, January 2010.
[4] http://www.faa.gov/aircraft/gen_av/light_sport/media/StandardsChart.pdf.
[5] Anonymous. Airplane Flying Handbook. FAA-H-8083–3A 2004.
[6] NTSB/ARA-11/01. Review of U.S. Civil Aviation Accidents 2007–2009. NTSB 2011.
[7] Anonymous. Aircraft Accident Report: Loss of Control on Approach; Colgan Air, Inc.; Operating as Continental Connection Flight 3407; Bombardier DHC-8-400, N200WQ; Clarence Center, New York; February 12, 2009. NTSB/AAR-10/01 2010.
[8] http://www.fas.org/programs/ssp/man/uswpns/air/fighter/f35.html.
[9] U.S. Department of Defense News Transcript. Briefing on the Joint Strike Fighter Contract Announcement. 4:30 pm EDT, http://www.defense.gov/transcripts/transcript.aspx?transcriptid=2186; October 26, 2001.
[10] Source. http://www.emsstrategies.com/dm090203article2.html.
[11] Torenbeek Egbert. Synthesis of Subsonic Aircraft Design. 3rd ed. Delft University Press; 1986. p. 499.
[12] http://www.easa.europa.eu/agency-measures/certification-specifications.php.
[13] http://www.airweb.faa.gov/Regulatory_and_Guidance_Library/rgMakeModel.nsf/MainFrame.
[14] http://rgl.faa.gov/Regulatory_and_Guidance_Library/rgAD.nsf/Frameset.

CHAPTER 2

Aircraft Cost Analysis

OUTLINE

2.1 Introduction — 34
  2.1.1 *The Content of this Chapter* — 34
  2.1.2 *A Review of the State of the GA Industry* — 34

2.2 **Estimating Project Development Costs** — 36
  2.2.1 *Quantity Discount Factor* — 36
  2.2.2 *Development Cost of a GA Aircraft — the Eastlake Model* — 37
    Product Liability Costs — 37
    Number of Engineering Man-hours — 37
    Number of Tooling Man-hours — 38
    Number of Manufacturing Labor Man-hours — 38
    Cost Analysis — 39
    Total Cost of Engineering — 39
    Total Cost of Development Support — 39
    Total Cost of Flight Test Operations — 40
    Total Cost of Tooling — 40
    Total Cost of Manufacturing — 40
    Total Cost of Quality Control — 40
    Total Cost of Materials — 40
    Total Cost to Certify — 40
    Cost of Retractable Landing Gear per Airplane — 41
    Cost of Avionics — 41
    Cost of Power Plant (engines, propellers) — 41
    Break-even Analysis — 43
    Derivation of Equation (2-19) — 43
  2.2.3 *Development Cost of a Business Aircraft — the Eastlake Model* — 44
    Number of Engineering Man-hours — 44
    Number of Tooling Man-hours — 45
    Number of Manufacturing Labor Man-hours — 45
    Total Cost of Engineering — 45
    Total Cost of Development Support — 45
    Total Cost of Flight Test Operations — 45
    Total Cost of Tooling — 45
    Total Cost of Manufacturing — 45
    Total Cost of Quality Control — 45
    Total Cost of Materials — 45
    Total Cost to Certify — 46
    Cost of Retractable Landing Gear per Airplane — 46
    Cost of Avionics — 46
    Cost of Power Plant (engines, propellers) — 46

2.3 **Estimating Aircraft Operational Costs** — 46
  2.3.1 *Operational Cost of a GA Aircraft* — 46
    Maintenance Cost ($ per year) — 46
    Maintenance to Flight Hour Ratio — 46
    Storage Cost ($ per year) — 47
    Annual Fuel Cost ($ per year) — 47
    Annual insurance cost ($ per year) — 47
    Annual Inspection Cost ($ per year) — 47
    Engine Overhaul Fund ($ per year) — 47
    Monthly Loan Payment — 47
    Annual Loan Payment ($ per year) — 47
    Total Yearly Cost — 48
    Cost per Flight Hour — 48
  2.3.2 *Operational Cost of a Business Aircraft* — 49
    Maintenance Cost ($ per year) — 49
    Maintenance to Flight Hour Ratio — 49
    Storage Cost ($ per year) — 49
    Annual Fuel Cost ($ per year) — 49
    Annual Insurance Cost ($ per year) — 49
    Annual Insurance Cost ($ per year) — 49
    Engine Overhaul Fund ($ per year) — 49
    Crew Cost — 49
    Hourly Crew — 49
    Annual Loan Payment ($ per year) — 51
    Total Yearly Cost — 51

Exercises — 51

Variables — 52

References — 53

2.1 INTRODUCTION

The estimation of the cost of developing an aircraft is an essential part of the design process. We may have conceived of the world's most interesting airplane, but is it worth the effort and cost to manufacture? If we are convinced it is, how many airplanes do we intend to manufacture? What will be the cost of each to the customer? How many will we need to manufacture before we break-even? How many engineers or technicians are needed to develop the aircraft? All of these are very important questions and this chapter is intended to answer some of them.

The development and procurement costs of aircraft (DAPCA) is a method used to estimate the development cost of new military aircraft. It is developed by the RAND Corporation, and is described in a report [1] available on its website [2]. The method, which is commonly referred to as DAPCA-IV,[1] establishes special cost estimating relationships (CERs), which are a set of statistical equations that predict aircraft acquisition costs using only basic information like empty weight and maximum airspeed. The DAPCA-IV can be used to estimate cost for research, development, testing, and evaluation (RDT&E) and even allows workforce estimation to take place. In short, the CERs estimate the cost of (1) engineering, tooling, manufacturing labor, and quality control; (2) manufacturing material, development support, and flight testing; and (3) total program cost.

The CERs are presented as a set of exponential equations that were developed by applying multi-variable least-squares regression analysis to practically all US military aircraft in service and production at the time of its inception. Consequently, the model is highly biased toward the price structure adapted by the Pentagon, which does not apply to the GA aircraft industry. This can be seen in the grossly overestimated development costs for GA aircraft predicted using the unmodified DAPCA-IV method. Professor Emeritus Charles Eastlake of Embry-Riddle Aeronautical University has adapted the original DAPCA-IV formulation to GA aircraft to better reflect the development and operational cost of such airplanes. The method presented in Ref. [3] explains and makes justification for the modifications made to the original DAPCA-IV model. This method has been extended to include executive-class aircraft as well.

2.1.1 The Content of this Chapter

- **Section 2.2** presents methods to estimate the costs involved in developing new GA aircraft. The method, which is based on the DAPCA-IV aircraft procurement cost analysis method, has been especially tailored for GA aircraft.
- **Section 2.3** presents methods to help estimate the cost of operating GA aircraft. Such methods are essential when trying to demonstrate whether the new aircraft will be more or less expensive to operate than competitor aircraft.

2.1.2 A Review of the State of the GA Industry

A word of caution; the reader should apply realism to any cost analysis. The numbers returned by such methods are not exact, but rather in the ballpark at best and inaccurate at worst. It is easy to estimate how much an airplane will cost if 100, 1000, or 10,000 units are produced. However, this has no bearing on how many will actually sell. It is impossible to provide a mathematical expression to estimate that part of the equation. It is best to consult manufacturer's data, such as Ref. [4], an annual compilation by the General Aviation Manufacturers Association (GAMA), available for download from the organization's website [5]. The reference lists deliveries of all GA aircraft over a number of years and gives an important glimpse into the state of the industry, some of which is reflected in Figure 2-1. It further breaks down deliveries by airplane make and model. The likelihood of grabbing substantial market share from established players, who have worldwide networks of support structure such as spares and service stations, should not be overestimated.

General Aviation: Status of the Industry, Related Infrastructure, and Safety Issues (Anonymous 2001) provides a healthy dose of realism on the nature of the industry since people began tracking this information in 1946. Figure 2-2 shows that since 1946, overall, there has been an enormous drop in aircraft deliveries, albeit with periods of growth. The first drop takes place immediately after WWII, when aircraft production plummeted from about 35,000 in 1946, over a 5 year period, to 2302 in 1951. The thought at the time was that owning a private airplane would become the norm after the war, not unlike what happened to the ownership of the automobile after WWI. This view did not materialize and a large surplus of aircraft was generated that took more than 5 years to dispose of. This was followed by a period of steady growth that peaked in 1966, when 15,768 units were delivered. In 1963, the California Supreme Court made a decision that adopted the rule of "strict liability" with respect to negligence [6]. This meant that companies could (and can) be held liable for harm caused by their products even if there is no evidence of negligence.

---

[1] The DAPCA-IV is preceded by the now obsolete DAPCA-III (R-1854-PR from 1976) and so on.

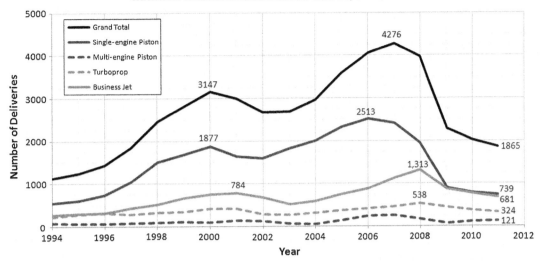

FIGURE 2-1 Sales prospects for GA aircraft from 1994 to 2011 (based on *General Aviation Statistical Databook and Industry Outlook 2011*).

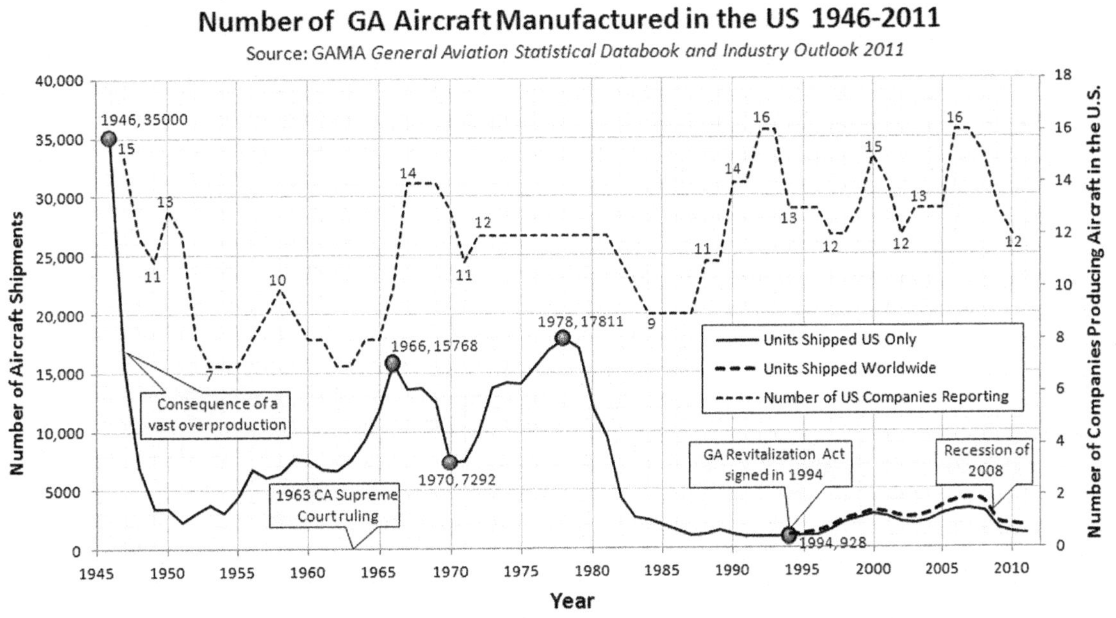

FIGURE 2-2 Number of aircraft produced in the USA has been dropping since 1946, with intermittent periods of growth (based on *General Aviation Statistical Databook and Industry Outlook 2011*).

Other states soon followed suit, shifting the liability burden from the customer onto the industry. This caused a sharp rise in liability suits against industry. The response of the aviation industry was to purchase protection in the form of liability insurance and add this to the price tag of new aircraft. This, in turn, increased the price of new aircraft, causing demand to fall. To cut a long story short, this explains the drop in aircraft deliveries between 1978, when production reached a high since 1946, with 17,811 deliveries; and the low of 928 units in 1994, which is the year that then-President Bill Clinton signed into law the General Aviation Revitalization Act (sometimes called the *tort reform*), whose purpose was to limit the extent of liabilities. This has led to a modest growth in the industry, with a subsequent drop when the economic recession of 2008 began.

The point of this discussion is to emphasize that even though financial models, like the DAPCA-IV, make reasonable predictions, the reader must be mindful that it is the economy that is unpredictable. What may seem like a viable business model today may not be so tomorrow — and vice versa.

2.2 ESTIMATING PROJECT DEVELOPMENT COSTS

The method presented below uses a highly modified version of the DAPCA-IV model to estimate the development cost of light GA aircraft based on expected weight of the bare airframe (without engines, tires, controls, and so on) and maximum level airspeed. Special correction factors[2] are used to account for aircraft that require more complicated manufacturing technologies such as the fabrication of tapered wings, complex flap systems,[3] and pressurization.

The first step in the application of the method is to estimate man-hours for three important areas of the project: engineering, tooling, and manufacturing. The next step uses these to estimate the actual cost in dollars, but also introduces additional cost-related issues that must be accounted for. Once this is completed, it is possible to estimate the price per unit, number of units to break-even, and other factors of interest.

2.2.1 Quantity Discount Factor

Since the DAPCA-IV method neither accounts for propulsive devices nor avionics, these are added (or purchased) later. The cost is adjusted using a special quantity discount factor (QDF) whose value depends on the quantity purchased and the application of a "learning curve" or, more appropriately, *experience effectiveness adjustment factor*.[4] In short, increased experience improves the productivity of the technician. This way, 80% experience effectiveness means that if it takes a technician 100 hrs to put together, say, a batch of 10 assemblies, the next batch will only take 80% of that time, or 80 hrs, and the next batch will take 64 hrs, and so on. Effectively this means that each time the total number of units produced is doubled, the price per unit drops to 80% of the previous price. This spurs the creation of the QDF as a tool to adjust the cost of doing business. Figure 2-3 shows the QDF for four values of experience effectiveness — 80%, 85%, 90%, and 95%. The application of the QDF to the engine, propeller, and avionics is justified assuming that buying in bulk from vendors will allow prices to be negotiated. The QDF is calculated using the following expression:

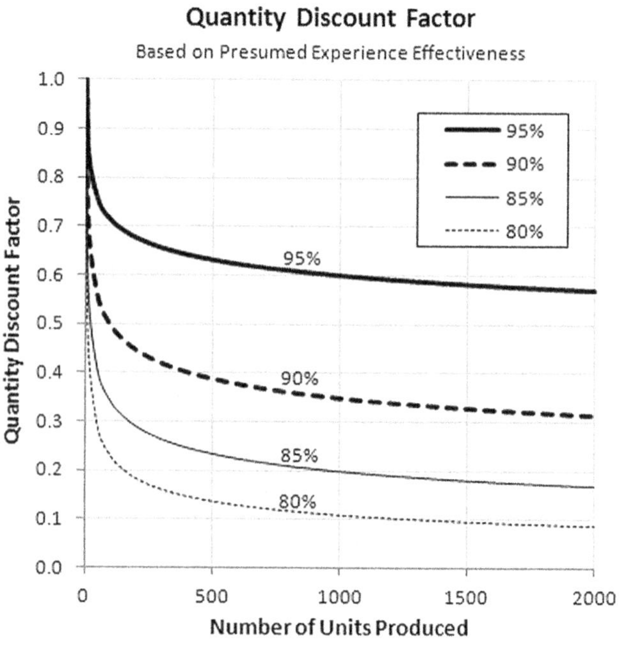

FIGURE 2-3 QDF depends on presumed experience effectiveness.

$$\text{Quantity Discount Factor}: \quad QDF = (F_{\text{EXP}})^{1.4427 \cdot \ln N} \quad (2\text{-}1)$$

where

F_{EXP} = experience effectiveness (= 0.8 for 80%, 0.9 for 90%, and so on),
N = number of units produced.

It is common to assume an F_{EXP} of 80%, but some people with direct experience of a production environment contend this is too optimistic. Manufacturing companies often suffer from a large labor turnover rate, which adds costs through the recruitment of new replacements,[5] something that inevitably is detrimental to the overall experience effectiveness. Just consider the curve for the 80% effectiveness in Figure 2-3. It implies that an engine manufacturer who sells a single engine for $300,000 will lower the price to $30,000 per engine if 1000 units are purchased — something we can safely generalize is unlikely to happen. Here, the Eastlake

---

[2]Ok, fudge factors.

[3]A complex flap system is somewhat subject to engineering judgment. In this context fixed hinge flaps are considered simple, whereas translating hinges are complex. An exception to this distinction would be the flap system on the typical Cessna aircraft, which is considered simple (albeit clever).

[4]Comment: it is recognized that many people simply call it learning curve and the author takes no issue with that, other than considering the term "experience effectiveness" more appropriate because once a person has "learned" to insert and tighten a fastener, there really is not much else to be "learned." It is the experience, on the other hand, that allows the person to perform the task faster and faster.

[5]Recruiting requires interviewing time, associated administrative work, training, supervisory time, and overtime that is paid to employees who temporarily have to take on additional tasks until a new replacement is found.

model uses a more realistic F_{EXP} of 95% and this yields results that better match actual aircraft production. Also, various associated costs, hourly employee rates, fuel costs, and others, must be used and these vary from location to location.

2.2.2 Development Cost of a GA Aircraft — the Eastlake Model

This model was originally developed in 1986. However, as presented here, the costs are calculated assuming the cost of living in the year 2012. All appropriate constants (excluding exponentials) have been updated to reflect this. This means that for later (or earlier) years, the costs must be adjusted to reflect current values. This is usually done through the application of the consumer price index (CPI), known informally as the cost of living index. This means that if the reader is applying this method, say, in 2022, the CPI (denoted by the term CPI_{2012} in the following formulation) must be updated relative to the year 2012. This information can be obtained from the website of the Bureau of Labor Statistics.[6]

Product Liability Costs

An important calculation modification is the estimation of the manufacturer's product liability costs. While it could be incorporated in some of the statistical formulas, this cost is added directly in order to, as Eastlake and Blackwell (2000) put it, "force students to think about this reality of being in business in the US." According to information from the insurance industry, the product liability cost for any particular manufacturer depends on the number of aircraft sold and their accident rate. It is next to impossible to predict how a particular product will fare once in production. Therefore, account for this assuming 12–17% above total cost to produce (for instance see Table 2-1).

Number of Engineering Man-hours

The number of engineering man-hours required to design the aircraft and perform the necessary RDT&E can be estimated from the following expression:

$$H_{ENG} = 0.0396 \cdot W_{airframe}^{0.791} \cdot V_H^{1.526} \cdot N^{0.183} \cdot F_{CERT} \cdot F_{CF} \cdot F_{COMP} \cdot F_{PRESS}$$

(2-2)

where

$W_{airframe}$ = weight of the structural skeleton
V_H = maximum level airspeed in KTAS
N = number of planned aircraft to be produced over a 5-year period
$F_{CERT} = 0.67$ if certified as LSA, $= 1$ if certified as a 14 CFR Part 23 aircraft
$F_{CF} = 1.03$ for a complex flap system, $= 1$ if a simple flap system
$F_{COMP} = 1 + f_{comp}$, a factor to account for the use of composites in the airframe

TABLE 2-1 Project Cost Analysis

	Man-hours	Rate, $/hr	Total Cost	Cost per Unit	
Engineering	205,670	90	$38,814,240	$38,814	
Development support			$1,363,399	$1,363	
Flight test operations			$247,569	$248	
Tooling	190,300	60	$23,942,391	$23,942	
Certification Cost			$64,367,598		
Manufacturing labor	1,366,628	50	$143,284,146	$143,284	
Quality control			$27,940,409	$27,940	
Materials/equipment			$19,159,003	$19,159	
Units produced in 5 years				1000	
Quantity Discount Factor				0.5998	
				Without QDF	With QDF
Fixed landing gear discount				-$7,500	-$4,498
Engine(s)				$53,953	$32,360
Propeller(s)				$14,087	$8,449
Avionics				$15,000	$8,997
TOTAL COST TO PRODUCE				$330,291	$300,059
Manufacturer's liability insurance					$50,000
MINIMUM SELLING PRICE					$350,059

---

[6] http://www.bls.gov. In particular, see http://www.bls.gov/data/inflation_calculator.htm, which is a calculator that returns the index using simple user inputs. Also, explanations on how the CPI is calculated can be seen at http://www.bls.gov/cpi/cpifaq.htm#Question_11.

f_{comp} = fraction of airframe made from composites (= 1 for a complete composite aircraft)

F_{PRESS} = 1.03 for a pressurized aircraft, = 1 if unpressurized.

Note that the structural skeleton weighs far less than the empty weight of the aircraft. This weight can be approximated by considering the empty weight less engines, avionics, seats, furnishing, control system, and other components. In the absence of such information, assume it is about 65% of empty weight.

Number of Tooling Man-hours

This is the number of man-hours required to design and build tools, fixtures, jigs, molds, and so on. Note that some recurring variables (e.g. F_{PRESS}) have new values.

$$H_{TOOL} = 1.0032 \cdot W_{airframe}^{0.764} \cdot V_H^{0.899} \cdot N^{0.178} \\ \cdot Q_m^{0.066} \cdot F_{TAPER} \cdot F_{CF} \cdot F_{COMP} \cdot F_{PRESS} \quad (2\text{-}3)$$

where

Q_m = estimated production rate in number of aircraft per month (= $N/60$ for 60 months/5 years)

F_{TAPER} = 0.95 for a constant-chord wing, = 1 for a tapered wing

F_{CF} = 1.02 for a complex flap system, = 1 if a simple flap system

$F_{COMP} = 1 + f_{comp}$, a factor to account for the use of composites in the airframe

F_{PRESS} = 1.01 for a pressurized aircraft, = 1 if unpressurized

Number of Manufacturing Labor Man-hours

The number of man-hours required to build the aircraft.

$$H_{MFG} = 9.6613 \cdot W_{airframe}^{0.74} \cdot V_H^{0.543} \cdot N^{0.524} \\ \cdot F_{CERT} \cdot F_{CF} \cdot F_{COMP} \quad (2\text{-}4)$$

where

F_{CERT} = 0.75 if certified as LSA, = 1 if certified as a 14 CFR Part 23 aircraft

F_{CF} = 1.01 for a complex flap system, = 1 if a simple flap system

$F_{COMP} = 1 + 0.25 \cdot f_{comp}$ a factor to account for the use of composites in the airframe

---

EXAMPLE 2-1

(a) Estimate the man-hours required to produce a single-engine, piston-powered composite aircraft if its airframe is expected to weigh 1100 lb$_f$ ($W_{airframe}$) and it is designed for a maximum level airspeed of 185 KTAS (V_H). It is expected that 1000 aircraft (N) will be produced in the first 5 years (Q_m = 1000 units/60 months ≈ 17 units per month). The unpressurized aircraft will be certified under 14 CFR Part 23 and will feature a tapered wing with a simple flap system.

(b) If it is assumed that the engineering staff will work 40 hrs a week for 48 weeks a year, how many engineers are required to accomplish the development over a period of 3 years?

(c) What is the average time it will take to manufacture a single unit?

(d) Determine and compare the corresponding values if the airplane is made from aluminum (i.e. only change the factor F_{COMP}).

Solution

Refer to the descriptions of equations for variables.

(a) Number of engineering man-hours required:

$$H_{ENG} = 0.0396 \cdot W_{airframe}^{0.791} \cdot V_H^{1.526} \cdot N^{0.183} \cdot F_{CERT} \\ \cdot F_{CF} \cdot F_{COMP} \cdot F_{PRESS} \\ = 0.0396 \cdot (1100)^{0.791} \cdot (185)^{1.526} \\ \cdot (1000)^{0.183} \cdot 1 \cdot 1 \cdot 2 \cdot 1 = 205\,670 \text{ hrs}$$

Number of man-hours for constructing tooling:

$$H_{TOOL} = 1.0032 \cdot W_{airframe}^{0.764} \cdot V_H^{0.899} \cdot N^{0.178} \cdot Q_m^{0.066} \\ \cdot F_{TAPER} \cdot F_{CF} \cdot F_{COMP} \cdot F_{PRESS} \\ = 1.0032 \cdot (1100)^{0.764} \cdot (185)^{0.899} \cdot (1000)^{0.178} \\ \cdot (17)^{0.066} \cdot 1 \cdot 1 \cdot 2 \cdot 1 \\ = 190{,}300 \text{ hrs}$$

Number of man-hours required to produce airplane:

$$H_{MFG} = 9.6613 \cdot W_{airframe}^{0.74} \cdot V_H^{0.543} \cdot N^{0.524} \cdot F_{CERT} \\ \cdot F_{CF} \cdot F_{COMP} \\ = 9.6613 \cdot (1100)^{0.74} \cdot (185)^{0.543} \cdot (1000)^{0.524} \\ \cdot 1 \cdot 1 \cdot (1.25) \\ = 1{,}366{,}628 \text{ hrs}$$

2.2 ESTIMATING PROJECT DEVELOPMENT COSTS

EXAMPLE 2-1 (cont'd)

(b) Number of engineers needed to develop the aircraft over a period of 3 years:

$$N_{ENG} = \frac{205{,}670 \text{ hrs}}{(3 \text{ years})(48 \text{ weeks})(40 \text{ hrs/week})}$$

$$\approx 36 \text{ engineers}$$

(c) Average time to manufacture a single unit:

$$t_{AC} = \frac{1{,}366{,}628 \text{ hrs}}{1000 \text{ units}} = 1{,}367 \text{ hrs}$$

(d) Performing the same calculations for the production of aluminum aircraft and comparing to the composite aircraft yielded the following results:

	Man-hours	
	Composite AC	Aluminum AC
Engineering man-hours	205,670	102,835
Tooling	190,300	95,150
Manufacturing labor	1,366,628	1,093,303
	Other	
Number of engineers	36	18
Time to manufacture each unit	1367	1093

The results from parts (b) and (c) in Example 2-1 above need further explanation. The number of engineers indicates the average over the development period. Most projects have few engineers at first and then, as the project moves into the preliminary design phase, additional engineers are hired. There might be six engineers working on the project at first and 60 toward the end.

The average number of hours to build each unit appears reasonable considering a fully optimized manufacturing process for a small airplane, but it takes a long time to polish the process to get it to that level. The reader should be careful in trusting such numbers as they may mislead. It may take 5000–6000 hrs/aircraft to assemble the first few aircraft. Some businesses do not possess the financial capacity to pass through that hurdle.

Cost Analysis

Once the number of hours has been determined, the next step is to estimate costs by multiplying these with rates in currency per hour. This is precisely what is done below, although some of the other equations determine costs based on the weight and speed, as done above. In 2012, a typical rate for engineering was $92 per hour, tooling labor was $61 per hour, and manufacturing labor was $53 per hour. It should be stated that these figures include overheads – according to www.engineersalary.com, an engineer with a B.Sc. and M.Sc. and 10 years of experience on the West Coast of the United States should be making about $100,000 a year. This amounts to about $48 an hour. A technician in a typical aircraft plant could make anywhere from $12 to $20 an hour.

Total Cost of Engineering

Total cost of engineering the aircraft:

$$C_{ENG} = 2.0969 \cdot H_{ENG} \cdot R_{ENG} \cdot CPI_{2012} \quad (2\text{-}5)$$

where

R_{ENG} = rate of engineering labor in $ per hour (e.g. $92/hour)

CPI_{2012} = consumer price index relative to the year 2012

By definition, CPI_{2012} for the year 2012 relative to the same year is 1. The constant 2.0969 is the CPI for the years 1986 to 2012, which is when the CER models were developed. The following equations have been corrected as well.

Total Cost of Development Support

The cost of overheads, administration, logistics, human resources, facilities maintenance personnel and similar entities required to support the development effort; calculate and pay salaries; and other necessary tasks.

$$C_{DEV} = 0.06458 \cdot W_{airframe}^{0.873} \cdot V_H^{1.89} \cdot N_P^{0.346} \cdot CPI_{2012}$$
$$\cdot F_{CERT} \cdot F_{CF} \cdot F_{COMP} \cdot F_{PRESS}$$

(2-6)

where

N_P = number of prototypes

F_{CERT} = 0.5 if certified as LSA, = 1 if certified as a 14 CFR Part 23 aircraft

$F_{CF} = 1.01$ for a complex flap system, $= 1$ if a simple flap system

$F_{COMP} = 1 + 0.5 \cdot f_{comp}$, a factor to account for the use of composites in the airframe

$F_{PRESS} = 1.03$ for a pressurized aircraft, $= 1$ if unpressurized

Total Cost of Flight Test Operations

Total cost of completing the development and certification flight-test program:

$$C_{FT} = 0.009646 \cdot W_{airframe}^{1.16} \cdot V_H^{1.3718} \cdot N_P^{1.281} \cdot CPI_{2012} \cdot F_{CERT} \quad (2\text{-}7)$$

where

$F_{CERT} = 10$ if certified as LSA, $= 1$ if a 14 CFR Part 23 aircraft

Total Cost of Tooling

This entails the cost of designing, fabricating, and maintaining jigs, fixtures, molds, and other tools required to build the airplane. The tooling requires industrial and manufacturing engineers for the design work and technicians to fabricate and maintain.

$$C_{TOOL} = 2.0969 \cdot H_{TOOL} \cdot R_{TOOL} \cdot CPI_{2012} \quad (2\text{-}8)$$

where

$R_{TOOL} = $ rate of tooling labor in \$ per hour (e.g. \$61/hour)

Total Cost of Manufacturing

This entails the cost of manufacturing labor required to produce the aircraft.

$$C_{MFG} = 2.0969 \cdot H_{MFG} \cdot R_{MFG} \cdot CPI_{2012} \quad (2\text{-}9)$$

where

$R_{MFG} = $ rate of manufacturing labor in \$ per hour (e.g. \$53/hour)

Total Cost of Quality Control

This entails the cost of technicians and the equipment required to demonstrate that the product being manufactured is indeed the airplane shown in the drawing package.

$$C_{QC} = 0.13 \cdot C_{MFG} \cdot F_{CERT} \cdot F_{COMP} \quad (2\text{-}10)$$

where

$F_{CERT} = 0.5$ if certified as LSA, $= 1$ if certified as a 14 CFR Part 23 aircraft

$F_{COMP} = 1 + 0.5 \cdot f_{comp}$, a factor to account for use of composites in the airframe

Total Cost of Materials

This is the cost of raw material (aluminum sheets, pre-impregnated composites, landing gear, avionics, etc.) required to fabricate the airplane.

$$C_{MAT} = 24.896 \cdot W_{airframe}^{0.689} \cdot V_H^{0.624} \cdot N^{0.792} \cdot CPI_{2012} \cdot F_{CERT} \cdot F_{CF} \cdot F_{PRESS} \quad (2\text{-}11)$$

where

$F_{CERT} = 0.75$ if certified as LSA, $= 1$ if certified as a 14 CFR Part 23 aircraft

$F_{CF} = 1.02$ for a complex flap system, $= 1$ if a simple flap system

$F_{PRESS} = 1.01$ for a pressurized aircraft, $= 1$ if unpressurized

Total Cost to Certify

The total cost to certify is the cost of engineering, development support, flight test, and tooling (assuming production tooling is used to produce at least some of the prototypes).

$$C_{CERT} = C_{ENG} + C_{DEV} + C_{FT} + C_{TOOL} \quad (2\text{-}12)$$

EXAMPLE 2-2

Estimate the total cost to certify the airplane of Example 2-1, assuming engineering, tooling, and manufacturing rates are \$90, \$60, and \$50 per hour, respectively. The planned number of prototypes is four. In the year 2012, the $CPI_{2012} = 1$.

Solution

Total cost of engineering:

$$C_{ENG} = 2.0969 \cdot H_{ENG} \cdot R_{ENG} \cdot CPI_{2012}$$

$$= 2.0969 \cdot (205670) \cdot (90) \cdot (1)$$

$$= \$38{,}814{,}248$$

EXAMPLE 2-2 (cont'd)

Total cost of development support:

$$C_{DEV} = 0.06458 \cdot W_{airframe}^{0.873} \cdot V_H^{1.89} \cdot N_P^{0.346} \cdot CPI_{2012}$$
$$\cdot F_{CERT} \cdot F_{CF} \cdot F_{COMP} \cdot F_{PRESS}$$
$$= 0.06458 \cdot (1100)^{0.873} \cdot (185)^{1.89} \cdot (4)^{0.346}$$
$$\cdot (1) \cdot 1 \cdot 1 \cdot (1.5) \cdot 1$$
$$= \$1,363,399$$

Total cost of flight test operations:

$$C_{FT} = 0.009646 \cdot W_{airframe}^{1.16} \cdot V_H^{1.3718} \cdot N_P^{1.281} \cdot CPI_{2012} \cdot F_{CERT}$$
$$= 0.009646 \cdot (1100)^{1.16} \cdot (185)^{1.3718} \cdot (4)^{1.281} \cdot (1) \cdot 1$$
$$= \$247,576$$

Total cost of tooling:

$$C_{TOOL} = 2.0969 \cdot H_{TOOL} \cdot R_{TOOL} \cdot CPI_{2012}$$
$$= 2.0969 \cdot (190,300) \cdot (60) \cdot (1) = \$23,942,404$$

Total cost of manufacturing:

$$C_{MFG} = 2.0969 \cdot H_{MFG} \cdot R_{MFG} \cdot CPI_{2012}$$
$$= 2.0969 \cdot (1,366,628) \cdot (50) \cdot (1) = \$143,284,113$$

Total cost of quality control:

$$C_{QC} = 0.13 \cdot C_{MFG} \cdot F_{CERT} \cdot F_{COMP}$$
$$= 0.13 \cdot (143,284,113) \cdot 1 \cdot (1.5) = \$27,940,402$$

Total cost of materials:

$$C_{MAT} = 24.896 \cdot W_{airframe}^{0.689} \cdot V_H^{0.624} \cdot N^{0.792} \cdot CPI_{2012}$$
$$\cdot F_{CERT} \cdot F_{CF} \cdot F_{PRESS}$$
$$= 24.896 \cdot (1100)^{0.689} \cdot (185)^{0.624} \cdot (1000)^{0.792}$$
$$\cdot (1) \cdot 1 \cdot 1 \cdot 1$$
$$= \$19,158,623$$

The total cost to certify:

$$C_{CERT} = C_{ENG} + C_{DEV} + C_{FT} + C_{TOOL}$$
$$= \$38,814,248 + \$13,53,399 + \$247,576$$
$$+ \$23,942,404$$
$$= \$64,357,627$$

With the total costs determined, additional costs of various components and items can be estimated as follows:

Cost of Retractable Landing Gear per Airplane

The cost of retractable landing gear is already assumed in the DAPCA-IV formulation, so an adjustment is only made if the airplane has fixed landing gear. If so, subtract $7500 per airplane.

Cost of Avionics

In the absence of more accurate information, in 2012 US dollars add $15,000 per airplane if it is certified to 14 CFR Part 23. Add $4500 per airplane if it is certified as an LSA.

Cost of Power Plant (engines, propellers)

The cost of the engine depends on the number of (N_{PP}) and type of engine (piston, turboprop, turbojet, or turbofan). For piston and turboprop engines the cost depends on the rated brake-horsepower (P_{BHP}) or shaft-horsepower (P_{SHP}). For turbojets and turbofans it is based on the rated thrust (T).

Piston engines:

$$C_{PP} = 174.0 \cdot N_{PP} \cdot P_{BHP} \cdot CPI_{2012} \quad (2\text{-}13)$$

Turboprop engines:

$$C_{PP} = 377.4 \cdot N_{PP} \cdot P_{SHP} \cdot CPI_{2012} \quad (2\text{-}14)$$

Turbojet engines:

$$C_{PP} = 868.1 \cdot N_{PP} \cdot T^{0.8356} \cdot CPI_{2012} \quad (2\text{-}15)$$

Turbofan engines:

$$C_{PP} = 1035.9 \cdot N_{PP} \cdot T^{0.8356} \cdot CPI_{2012} \quad (2\text{-}16)$$

Since piston and turboprop engines also require propellers, this cost must be determined as well. The two most common types are the fixed-pitch and the constant-speed propellers. The typical fixed-pitch propeller cost around $3145 in 2012. However, constant-speed propellers are more expensive and an expression that takes into account the diameter of the propeller (D_P, in feet) and P_{SHP} has been derived.

Fixed-pitch propellers:

$$C_{FIXPROP} = 3145 \cdot N_{PP} \cdot CPI_{2012} \quad (2\text{-}17)$$

Constant-speed propellers:

$$C_{CSTPROP} = 209.69 \cdot N_{PP} \cdot CPI_{2012} \cdot D_P^2 \cdot \left(\frac{P_{SHP}}{D_P}\right)^{0.12} \quad (2\text{-}18)$$

EXAMPLE 2-3

(a) Estimate the cost of the engine and propeller for the airplane of Example 2-1, if its single piston engine is 310 BHP, swings a 6.5 ft constant-speed propeller, and the average cost per avionics suite is $15,000. $CPI_{2012} = 1$.

(b) Using the information from this and Example 2-1 and Example 2-2, tabulate the cost to produce 1000 units of the airplane over a 5-year period, and estimate the necessary manufacturer's selling price. Assume no profits and include a manufacturer's liability insurance of $50,000 per airplane. This value represents the minimum selling price.

(c) Plot how the number of units produced affects the minimum selling price.

(d) Perform the preceding analysis for an aluminum aircraft and compare to the composite airplane, by only changing the factor F_{COMP}. Solution

(a) Engine cost:

$$C_{PP} = 174.0 \cdot N_{PP} \cdot P_{BHP} \cdot CPI_{2012}$$
$$= 174.0 \cdot (1) \cdot (310) \cdot (1) = \$53,940$$

Propeller cost:

$$C_{CSTPROP} = 209.69 \cdot N_{PP} \cdot CPI_{2012} \cdot D_P^2 \cdot \left(\frac{P_{SHP}}{D_P}\right)^{0.12}$$
$$= 209.69 \cdot (1) \cdot (1) \cdot (6.5)^2 \cdot \left(\frac{310}{6.5}\right)^{0.12}$$
$$= \$14,087$$

(b) The entire cost estimation is tabulated in Table 2-1, indicating a minimum selling price of $350,059. Note that there may be slight numerical discrepancies between the table and the above calculations, but this can be attributed to their implementation in a spreadsheet, which retains 16 significant digits.

In short, in order to pay the development and manufacturing costs at this price, 1000 airplanes must be sold (remember this was the premise — 1000 units over 5 years). The price shown does not generate any profits and, it probably does not have to be said, would hardly keep the company in business for long. Next, we will assume profit margins and evaluate how this affects the bottom line.

This evaluation is based on airplanes like the Cirrus SR22 and Cessna 350 Corvalis. According to information from www.cirrusaircraft.com, in 2012 the base price for a brand new Cirrus SR22, featuring its famed Cirrus Airframe Parachute System, seatbelts with airbags, three-bladed propeller, and air conditioning is $449,900. The Cessna 350 is no longer produced, so its price is not known. However, for comparison reasons, its more powerful brother, the Cessna 400 TTX, costs about $733,950 [7], while a comparable Cirrus SR22T costs about $633,600 [8]. Both airplanes have turbo-normalized engines and other features that increase the price.

(c) The graph shown in Figure 2-4 was created by evaluating the minimum selling price considering a

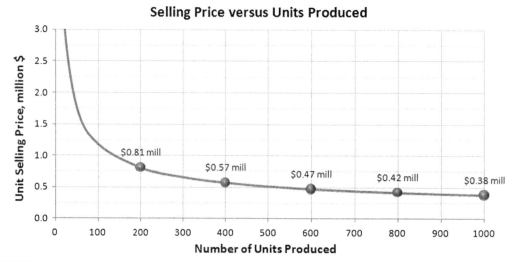

FIGURE 2-4 The selling price in millions of $ as a function of units produced shows a rapid drop in price at first.

> # EXAMPLE 2-3 (cont'd)
>
> number of production scenarios with differing numbers of units produced. It shows how the price drops rapidly with the number of units produced and then becomes more horizontal with higher production rates.
>
> (d) A comparison of the cost of development and manufacturing between a composite and aluminum aircraft is shown in Table 2-2. It reveals that the DAPCA-IV type statistical analyses predict composite aircraft to be of the order of 25–30% more expensive to manufacture than a comparable aluminum aircraft.

TABLE 2-2 Project Cost Comparison between a Composite and Aluminum Aircraft

	Costs per Unit		
	Composite AC	Aluminum AC	Ratio Comp/Al
Engineering	$38,814	$19,407	2.00
Development support	$1,363	$909	1.50
Flight test operations	$248	$248	1.00
Tooling	$23,942	$11,971	2.00
Certification Cost	**$64,368**	**$32,535**	**1.98**
Manufacturing labor	$143,284	$114,627	1.25
Quality control	$27,940	$14,902	1.88
Materials/equipment	$19,159	$19,159	1.00
Fixed landing gear discount	-$4,498	-$4,498	1.00
Engine(s)	$32,360	$32,360	1.00
Propeller(s)	$8,449	$8,449	1.00
Avionics	$8,997	$8,997	1.00
Manufacturer's liability insurance	$50,000	$50,000	1.00
TOTAL COST TO PRODUCE	**$300,059**	**$226,531**	**1.32**
MINIMUM SELLING PRICE	**$350,059**	**$276,531**	**1.27**

Break-even Analysis

Break-even analysis is used to determine how many units must be produced before revenue equals the cost incurred. Using the standard cost-volume-profit-analysis the following expression is used to determine this:

Number of units to break-even:

$$N_{BE} = \frac{\text{total fixed cost}}{\text{unit sales price } - \text{ unit variable cost}} \quad (2\text{-}19)$$

In this context the certification cost can be considered the total fixed cost, while the sum of manufacturing labor, quality control, materials/equipment, landing gear, engines, propellers, avionics, and manufacturer's liability insurance, divided by the number of units produced, constitutes unit variable cost. Example 2-4 shows the application of this approach.

Derivation of Equation (2-19)

The total cost of developing N units is given by:

$$(\text{total fixed cost}) + (\text{unit variable cost}) \times N$$

The total revenue from selling N units is:

$$(\text{unit sales price}) \times N$$

When the two are equal, we have broken even, i.e.,

$$(\text{total fixed cost}) + (\text{unit variable cost}) \times N$$
$$= (\text{unit sales price}) \times N$$

If we designate the number of units to break-even by the variable N_{BE}, we can write:

$$N_{BE} = \frac{\text{total fixed cost}}{\text{unit sales price } - \text{ unit variable cost}}$$

QED

EXAMPLE 2-4

Estimate how many airplanes must be produced before the manufacturer can expect to break-even, if the price is set at $400,000. Plot the production cost and revenue versus number of units produced, assuming a retail price of $350,000, $400,000, and $450,000 (see Figure 2-5). Plot total production cost and revenue versus number of units produced. Indicate break-even points on the plot.

Solution

Total fixed costs:

$$C_{CERT} = \$64{,}357{,}627$$

Unit variable cost is obtained by adding the cost for one unit for:

$$\begin{aligned}C_{CERT} &= \$143{,}284 + \$27{,}940 + \$19{,}159 - \$4{,}498 \\ &\quad + \$32{,}360 + \$8{,}449 + \$8{,}997 + \$50{,}000 \\ &= \$285{,}691\end{aligned}$$

Break-even point:

$$N_{BE} = \frac{\text{total fixed cost}}{\text{unit sales price } - \text{ unit variable cost}}$$

$$= \frac{\$64{,}357{,}627}{\$400{,}000 - 285{,}691} = 563 \text{ units}$$

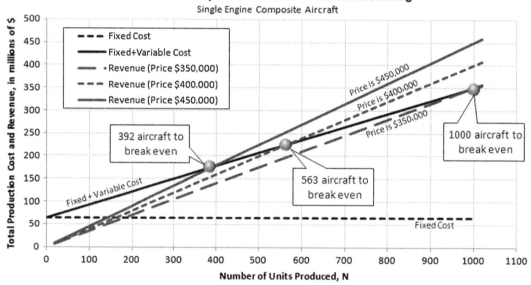

FIGURE 2-5 Break-even analysis assuming three different prices.

2.2.3 Development Cost of a Business Aircraft – the Eastlake Model

As stated before, the Eastlake model has also been adapted to the development of a business or an executive aircraft. This model is much closer to the DAPCA-IV model. Only applicable equations will be presented – the calculations are effectively identical to those for GA aircraft. Since such business aircraft would always be certified under either 14 CFR Part 23 or Part 25, provisions are made to account for this. Certification under the latter category will be more expensive due to the difference in the regulations. The factors denoted by the common variable F_{CERT} are best guesses for the cost difference – the reader can modify those values as suitable by corresponding experience.

Number of Engineering Man-hours

The number of man-hours of engineering time required to design the aircraft and perform the necessary RDT&E can be estimated from the following expression:

$$H_{ENG} = 4.86 \cdot W_{airframe}^{0.777} \cdot V_H^{0.894} \cdot N^{0.163} \cdot F_{CERT} \\ \cdot F_{CF} \cdot F_{COMP} \cdot F_{PRESS} \quad (2\text{-}20)$$

where

$W_{airframe}$ = weight of the structural skeleton
V_H = maximum level airspeed in KTAS
N = number of planned aircraft to be produced over a 5 year period
$F_{CERT} = 1$ if certified as a 14 CFR Part 23, $= 1.15$ if certified as a 14 CFR Part 25
$F_{CF} = 1.03$ for a complex flap system, $= 1$ if a simple flap system
$F_{COMP} = 1 + f_{comp}$, a factor to account for the use of composites in the airframe
f_{comp} = fraction of airframe made from composites ($= 1$ for a complete composite aircraft)
$F_{PRESS} = 1.03$ for a pressurized aircraft, $= 1$ if unpressurized

Number of Tooling Man-hours

The number of man-hours required to design and build tools, fixtures, jigs, molds, and so on.

$$H_{TOOL} = 5.99 \cdot W_{airframe}^{0.777} \cdot V_H^{0.696} \cdot N^{0.263} \cdot F_{CERT} \\ \cdot F_{TAPER} \cdot F_{CF} \cdot F_{COMP} \cdot F_{PRESS} \quad (2\text{-}21)$$

where

Q_m = estimated production rate in number of aircraft per month ($= N/60$ for 60 weeks/5 years)
$F_{CERT} = 1$ if certified as a 14 CFR Part 23, $= 1.05$ if certified as a 14 CFR Part 25
$F_{TAPER} = 0.95$ for a constant-chord wing, $= 1$ for a tapered wing
$F_{CF} = 1.02$ for a complex flap system, $= 1$ if a simple flap system
$F_{COMP} = 1 + f_{comp}$, a factor to account for the use of composites in the airframe
$F_{PRESS} = 1.01$ for a pressurized aircraft, $= 1$ if unpressurized

Number of Manufacturing Labor Man-hours

The number of man-hours required to build the aircraft.

$$H_{MFG} = 7.37 \cdot W_{airframe}^{0.82} \cdot V_H^{0.484} \cdot N^{0.641} \\ \cdot F_{CERT} \cdot F_{CF} \cdot F_{COMP} \quad (2\text{-}22)$$

where

$F_{CF} = 1.01$ for a complex flap system, $= 1$ if a simple flap system
$F_{CERT} = 1$ if certified as a 14 CFR Part 23, $= 1.05$ if certified as a 14 CFR Part 25
$F_{COMP} = 1 + 0.25 \cdot f_{comp}$, a factor to account for the use of composites in the airframe

Total Cost of Engineering

Use Equation (2-5).

Total Cost of Development Support

$$C_{DEV} = 95.24 \cdot W_{airframe}^{0.63} \cdot V_H^{1.3} \cdot CPI_{2012} \cdot F_{CERT} \\ \cdot F_{CF} \cdot F_{COMP} \cdot F_{PRESS} \quad (2\text{-}23)$$

where

$F_{CERT} = 1$ if certified as a 14 CFR Part 23, $= 1.10$ if certified as a 14 CFR Part 25
$F_{CF} = 1.01$ for a complex flap system, $= 1$ if a simple flap system
$F_{COMP} = 1 + 0.5 \cdot f_{comp}$, a factor to account for the use of composites in the airframe
$F_{PRESS} = 1.03$ for a pressurized aircraft, $= 1$ if unpressurized

Total Cost of Flight Test Operations

Total cost of completing the development and certification flight-test program:

$$C_{FT} = 2606.51 \cdot W_{airframe}^{0.325} \cdot V_H^{0.822} \cdot N_P^{1.21} \cdot CPI_{2012} \cdot F_{CERT} \quad (2\text{-}24)$$

where

$F_{CERT} = 1$ if certified as a 14 CFR Part 23, $= 1.50$ if certified as a 14 CFR Part 25

Total Cost of Tooling

Use Equation (2-8).

Total Cost of Manufacturing

Use Equation (2-9).

Total Cost of Quality Control

$$C_{QC} = 0.133 \cdot C_{MFG} \cdot F_{CERT} \cdot F_{COMP} \quad (2\text{-}25)$$

where

$F_{CERT} = 1$ if certified as a 14 CFR Part 23, $= 1.50$ if certified as a 14 CFR Part 25
$F_{COMP} = 1 + 0.5 \cdot f_{comp}$, a factor to account for use of composites in the airframe

Total Cost of Materials

$$C_{MAT} = 23.066 \cdot W_{airframe}^{0.921} \cdot V_H^{0.621} \cdot N^{0.799} \\ \cdot CPI_{2012} \cdot F_{CERT} \cdot F_{CF} \cdot F_{PRESS} \quad (2\text{-}26)$$

where

$F_{CERT} = 1$ if certified as a 14 CFR Part 23, $= 1.15$ if certified as a 14 CFR Part 25

$F_{CF} = 1.02$ for a complex flap system, $= 1$ if a simple flap system

$F_{PRESS} = 1.01$ for a pressurized aircraft, $= 1$ if unpressurized

Total Cost to Certify

Use Equation (2-12).

Cost of Retractable Landing Gear per Airplane

The cost of retractable landing gear is already assumed in the DAPCA-IV formulation, so an adjustment is only made if the airplane has fixed landing gear. If so, subtract $7500 per airplane.

Cost of Avionics

In the absence of more accurate information, in 2012 US dollars add $60,000 per airplane if it is certified to 14 CFR Part 23. Add $100,000 per airplane if it is certified to 14 CFR Part 25. This is in part to account for the installation of passenger entertainment systems.

Cost of Power Plant (engines, propellers)

Use Equations (2-13) through (2-18) as appropriate.

Follow the same procedures as presented in Section 2.2.2, *Development cost of a GA aircraft — the Eastlake model*.

2.3 ESTIMATING AIRCRAFT OPERATIONAL COSTS

A part of manufacturing and selling airplanes is to persuade potential customers to purchase your airplane rather than someone else's. In order to bring forth a convincing argument the manufacturer runs a sales department whose purpose is to provide a true comparison of the cost of ownership between comparable aircraft. One of the most important figures of merit used for this is the cost of ownership; the amount of money required to own and operate the aircraft per hour flown. This section focuses on the operation of a GA aircraft, estimating operational cost per flight hour for a privately owned and flown GA aircraft.

The following model was generated from scratch and is based on actual experience of aircraft ownership. It comprises basic book-keeping and tracking of several years of costs associated with privately owned aircraft. The primary inputs are flight hours per year, cost of fuel, amount of money borrowed to purchase the aircraft (to include loan payments in the model), and the amount of insurance coverage. The model assumes 0.3 maintenance man-hours required per flight hour (denoted by the term F_{MF}) for a single-engine, fixed-gear, fixed-pitch prop aircraft. This number is adjusted for characteristics that affect the maintenance effort. It is increased to account for factors such as difficult engine access, retractable landing gear, wet wings, complex avionics equipment, and complex high-lift devices. Negative increments are given for the cost savings achievable from maintenance performed by the owner to the extent of what is allowed by FAA regulations and for the simpler craft intended to be certified as LSA.

2.3.1 Operational Cost of a GA Aircraft

As stated above, it is assumed that the owner is the pilot and, thus, does not incur any costs for a flight crew. Storage cost, annual inspections, and contributing to an engine overhaul bank are also included in the model. The cost is ultimately presented in dollars per flight hour, allowing convenient comparison with rental cost for a similar aircraft. The number of flight hours per year (Q_{FLGT}) for normal GA aircraft varies greatly, from around 100 hours a year for an underutilized aircraft, to 1000 hours[7] or more for a student trainer aircraft. Personal aircraft are flown in the ballpark of 100–500 hours per year, with 300 hours being a reasonable average.

Maintenance Cost ($ per year)

$$C_{AP} = F_{MF} \cdot R_{AP} \cdot Q_{FLGT} \qquad (2\text{-}27)$$

where

F_{MF} = ratio of maintenance man-hours to flight hours (see below)

R_{AP} = an hourly rate for a certified Airframe and Powerplant (A&P) mechanic (typ. $53–67 per hr)

Q_{FLGT} = number of flight hours per year.

Maintenance to Flight Hour Ratio

$$F_{MF} = 0.30 + F_1 + F_2 + F_3 + F_4 + F_5 + F_6 + F_7 + F_8$$
$$(2\text{-}28)$$

where

$F_1 = -0.15$ if maintenance is performed by owner and 0 if performed by an A&P mechanic

$F_2 = 0$ for an easy engine access, $= 0.02$ for a difficult access

$F_3 = 0$ for a fixed landing gear, $= 0.02$ for a retractable landing gear

---

[7] A primary trainer airplane flown on average 4 hours, five days a week, flies $4 \times 5 \times 52 = 1040$ hrs per year.

$F_4 = 0$ if no VFR radios are installed, $= 0.02$ if VFR radios are installed

$F_5 = 0$ if no IFR radios are installed, $= 0.04$ if IFR radios are installed

$F_6 = 0$ if no integral fuel tanks are installed, $= 0.01$ if such tanks are installed

$F_7 = 0$ for a simple flap system, $= 0.02$ for a complex flap system

$F_8 = 0$ for 14 CFR Part 23 certification, $= -0.10$ for LSA certification

Airplane owners usually have to pay for storage at a main base. Assume the rate per month is $250.

Storage Cost ($ per year)

$$C_{STOR} = 12 \cdot R_{STOR} \qquad (2\text{-}29)$$

where

R_{STOR} = storage rate ($\approx$ $250 per month)

Annual Fuel Cost ($ per year)

$$C_{FUEL} = \frac{BHP_{CRUISE} \cdot SFC_{CRUISE} \cdot Q_{FLGT} \cdot R_{FUEL}}{6.5}$$

$$= FF_{CRUISE} \cdot Q_{FLGT} \cdot R_{FUEL}$$

$$(2\text{-}30)$$

where

BHP_{CRUISE} = Typical horsepower during cruise
SFC_{CRUISE} = typical specific fuel consumption during cruise
FF_{CRUISE} = total *fuel flow* in gallons per hour
R_{FUEL} = price of fuel in $/gallon

The insurance cost is a nebulous value that is disclosed by insurance companies on an individual basis. It takes into account factors such as pilot experience; use of aircraft; price and type of aircraft; and so on. Low-time pilots generally have to pay a higher premium than their high-time contemporaries. Agricultural aircraft engage in high-risk operations and this increases the premium as well. For instance, in the year 2012, the premium for a Cessna 172 might have been around $1000–1500 a year. Included in the policy is a hull value of $50,000 with standard liability of $100,000 per passenger and maximum liability of $1,000,000. On the other hand, a modern Cirrus SR22 aircraft, valued closer to $600,000, owned and operated by a low-time pilot, might cost $20,000 a year to insure. At the same time, a high-time pilot owning a less expensive Cirrus might only have to pay $3000 annually. The operational cost model being presented does not account for these variations and it is up to the reader to obtain the appropriate figure. Here, a simple model is employed that gives a reasonable estimate for many instances, but may not be applicable to others.

Annual insurance cost ($ per year)

$$C_{INS} = 500 + 0.015 \cdot C_{AC} \qquad (2\text{-}31)$$

where

C_{AC} = insured value of the aircraft.

If estimating the operational cost of a new design, the C_{AC} amounts to the purchase price of the aircraft.

Annual Inspection Cost ($ per year)

$$C_{INSP} = \$500 \qquad (2\text{-}32)$$

The airplane's engine(s) will require regular overhaul, as stipulated by the engine's required time between overhaul (TBO). It is prudent to assume this maintenance requirement is amortized over the total flight hours of the airplane. One way of doing this is to obtain an estimate for the cost of this major maintenance event and divide by the TBO of the engine. For instance, Lycoming and Continental engines usually have a TBO of around 2000 hrs. If the cost of the overhaul is expected to be $10,000, it follows that it is reasonable to charge $5 per flight hour. This is reflected in the expression below:

Engine Overhaul Fund ($ per year)

$$C_{OVER} = 5 \cdot N_{PP} \cdot Q_{FLGT} \qquad (2\text{-}33)$$

where

N_{PP} = number of engines

If the airplane was fully or partially funded through financial institutions, the annual cost of paying back those loans should be included as well. This is accounted for as shown below, using the standard mortgage formula:

Monthly Loan Payment

$$C_{month} = \frac{Pi}{1 - 1/(1+i)^n} \qquad (2\text{-}34)$$

where

P = the principal or amount of money originally borrowed
i = monthly interest rate
n = number of pay periods in months. This way 15 years would be $12 \cdot 15 = 180$ pay periods.

Annual Loan Payment ($ per year)

$$C_{LOAN} = \frac{12 \cdot Pi}{1 - 1/(1+i)^n} \qquad (2\text{-}35)$$

Total Yearly Cost

$$C_{YEAR} = C_{AP} + C_{STOR} + C_{FUEL} + C_{INS} + C_{INSP} + C_{OVER} + C_{LOAN}$$

(2-36)

And finally, the cost per each hour flown should be:

Cost per Flight Hour

$$C_{HR} = \frac{C_{YEAR}}{Q_{FLGT}}$$

(2-37)

EXAMPLE 2-5

Estimate the operational for the airplane in Example 2-1, assuming the following scenario:

(1) The airplane is certified to 14 CFR Part 23.
(2) It is maintained by an A&P mechanic who charges $60 per hour.
(3) It has easy engine access, fixed landing gear, IFR radios only, integral fuel tanks, and a simple flap system.
(4) It is flown 300 hours per year. Its 310 BHP engine consumes 16 gal/hr of fuel on average at $5/gallon.
(5) Storage cost is $250 per year.
(6) Use the given insurance model and the price of the airplane in Example 2-3, or $350,059.
(7) The airplane is fully paid for with a 15-year loan that has an APR of 9%.

Solution

Start by estimating the maintenance to flight hour ratio:

$$F_{MF} = 0.30 + F_1 + F_2 + F_3 + F_4 + F_5 + F_6 + F_7 + F_8$$
$$= 0.30 + 0 + 0 + 0 + 0 + 0.04 + 0.01 + 0 + 0$$
$$= 0.35$$

Annual maintenance cost:

$$C_{AP} = F_{MF} \cdot R_{AP} \cdot Q_{FLGT} = 0.35 \cdot 60 \cdot 300 = \$6300$$

Storage cost:

$$C_{STOR} = 12 \cdot R_{STOR} = 12 \cdot 250 = \$3000$$

Annual fuel cost:

$$C_{FUEL} = FF_{CRUISE} \cdot Q_{FLGT} \cdot R_{FUEL} = 16 \cdot 300 \cdot 5$$
$$= \$24,000$$

Annual insurance cost:

$$C_{INS} = 500 + 0.015 \cdot C_{AC} = 500 + 0.015 \cdot (380,065)$$
$$= \$6201$$

Annual inspection cost:

$$C_{INSP} = \$500$$

Engine overhaul fund:

$$C_{OVER} = 5 \cdot 1 \cdot 300 = \$1500$$

Annual loan payment:

$$C_{LOAN} = \frac{12 \cdot Pi}{1 - 1/(1+i)^n} = \frac{12 \cdot (380,065)(0.09/12)}{1 - 1/(1 + (0.09/12))^{(12 \times 15)}}$$
$$= \$46\ 258$$

The monthly payment is $3855. Therefore, the total annual cost of owning and operating the airplane amounts to the sum of these, or:

$$C_{YEAR} = C_{AP} + C_{STOR} + C_{FUEL} + C_{INS} + C_{INSP} + C_{OVER} + C_{LOAN}$$
$$= \$6300 + \$3000 + \$24,000 + \$6201 + \$500 + \$1500 + \$46,258$$
$$= \$87,759$$

If operated 300 hours a year, the cost per flight hour is:

$$C_{HR} = \frac{\$87,759}{300 \text{ hrs}} = \$293 \text{ per hour}$$

In comparison, a cost of $235.60 per flight hour for the $734,000 Cessna 400 TTX Corvalis is cited (http://se.cessna.com/single-engine/cessna-400/cessna-400-pricing.html#). It does not appear to include the cost of financing.

2.3.2 Operational Cost of a Business Aircraft

This estimation assumes a professionally flown aircraft with high quality maintenance and the other costs that have already been detailed in Section 2.3.1, *Operational cost of a GA aircraft*. This presentation is intended to give an idea of the costs associated with GA business aircraft. For business jets, certified to 14 CFR Part 25, the reader can seek more precise information from companies such as Conklin and de Decker [9], which collects it in great detail for all aircraft currently in service. A listing of cost-related items for such aircraft is provided in Table 2-3 and is based on the Conklin and de Decker approach.

The number of flight hours per year (Q_{FLGT}) for normal business aircraft varies greatly, from around 100 hours a year for an underutilized aircraft, to 1000 hours or more for a student trainer aircraft.

Maintenance Cost ($ per year)

$$C_{AP} = F_{MF} \cdot R_{AP} \cdot Q_{FLGT} \quad (2\text{-}38)$$

where

F_{MF} = ratio of maintenance man-hours to flight hours (see below)
R_{AP} = an hourly rate for a certified airframe and power plant (A&P) mechanic (typ. $53–67 per hr)
Q_{FLGT} = number of flight hours per year.

Maintenance to Flight Hour Ratio

$$F_{MF} = 2.00 + F_1 + F_2 + F_3 + F_4 + F_5 + F_6 \quad (2\text{-}39)$$

where

$F_1 = 0$ for an easy engine access, $= 0.2$ for difficult access
$F_2 = 0$ for fixed landing gear, $= 0.2$ for retractable landing gear
$F_3 = 0$ if simple avionics are installed, $= 0.2$ if complex avionics are installed
$F_4 = 0$ if no integral fuel tanks are installed, $= 0.1$ if such tanks are installed
$F_5 = 0$ for a simple flap system, $= 0.2$ for a complex flap system
$F_6 = 0$ for 14 CFR Part 23 certification, $= 0.5$ for 14 CFR Part 25 certification

Storage Cost ($ per year)

Use Equation (2-29), but assume R_{STOR} = storage rate $\approx$ $250–$1500 per month, depending on size of hangar space needed.

Annual Fuel Cost ($ per year)

Use Equation (2-30).

Annual Insurance Cost ($ per year)

Use Equation (2-31) in the absence of better information.

Annual Insurance Cost ($ per year)

$$C_{INSP} = \$1000 - \$15{,}000 \quad (2\text{-}40)$$

The airplane's engine(s) will require regular overhaul, as stipulated by the engine's required time between overhaul (TBO). It is prudent to assume this maintenance requirement is amortized over the total flight hours of the airplane. One way of doing this is to obtain an estimate for the cost of this major maintenance event and divide by the TBO of the engine. For instance, Williams International FJ44 engines usually have TBO around 4000 hrs, Pratt & Whitney PW306 are around 6000 hrs. If the cost of the overhaul is expected to be $30,000–40,000, it follows that it is reasonable to charge $6.7 to $7.5 per flight hour per engine. The higher value is reflected in the expression below:

Engine Overhaul Fund ($ per year)

$$C_{OVER} = 7.5 \cdot N_{PP} \cdot Q_{FLGT} \quad (2\text{-}41)$$

where

N_{PP} = number of engines

Crew Cost

Some business aircraft are operated by flight hours only. The associated crew cost is then based on the number of hours flown annually. In the absence of better information the following expression can be used to estimate this cost:

Hourly Crew

$$C_{CREW} = N_{CREW} \cdot R_{CREW} \cdot Q_{FLGT} \quad (2\text{-}42)$$

where

N_{CREW} = number of crew members required to operated the airplane.
R_{CREW} = hourly rate of crew per hour – business-dependent.

The term R_{CREW} ultimately depends on the business involved and can range from $50 to $150 per hour. Other business aircraft have full-time pilots and even a flight attendant, with the associated annual salary and benefit costs (see Table 2-3). Yet other businesses may keep only one full-time pilot on board, and hire a co-pilot and a flight attendant on a need-to basis. In this case Equation (2-42) may be used to account for the additional crew and its value added to that of the full-time pilot.

TABLE 2-3 A Variable and Fixed-cost Analysis for a Typical Business Jet Aircraft

	Small Jet	Medium Jet	Large Jet
ESTIMATED HOURLY VARIABLE COSTS			
FUEL COSTS			
Fuel (typical jet fuel costs $6.90 in 2012)	$1,650.00	$1,950.00	$2,650.00
Fuel additives	-	-	-
Lubricants	-	-	-
MAINTENANCE LABOR COSTS			
Maintenanc labor cost per hour	$93	$93	$93
Maintenance hours/Flight hour	5.11	1.55	0.7
Maintenance labor	$475.23	$144.15	$65.10
Parts airframe/Eng/Avionics	$321.03	$140.60	$123.23
POWERPLANT OVERHAUL FUND			
Engine restoration	$350.00	$400.00	$600.00
Thrust reverser allowance	$14.00	$26.00	$45.00
APU allowance	$14.00	$60.00	$60.00
MISCELLANEOUS EXPENSES			
Landing/Parking	$20.00	$45.00	$75.00
Crew expenses	$70.00	$280.00	$280.00
Supplies/Catering	$35.00	$150.00	$150.00
Carbon offset	-	-	-
Other	-	-	-
TOTALS			
Total Variable Cost/Hour	$2,949.26	$3,195.75	$4,048.33
ANNUAL FIXED COST			
CREW AND OPERATIONAL COSTS			
Crew salaries - Captain (NBAA rates)	$95,000	$135,000	$161,000
- Co Pilot	$60,000	$90,000	$94,000
- Flight attendant	-	$84,000	$89,000
- Flight Eng/Other	-	-	-
- Benefits	$47,000	$93,000	$104,000
Hangar - Typical	$25,700	$61,300	$82,500
INSURANCE COSTS			
Insurance - Hull	$7,735	$32,300	$41,760
Single limit liability	$10,500	$16,500	$16,500
OTHER COSTS			
Recurrent training	$27,200	$58,200	$85,600
Aircraft modernization (?)	$45,000	$33,333	$33,333
Navigation chart service	$4,800	$18,500	$18,500
Refurbishing (?)	$16,800	$78,500	$117,500
Computer Mx. Program (?)	$9,750	$12,000	$12,000
Weather service	$700	$700	$700
Other fixed costs	-	-	-
Mgmt fee/Yr	-	-	-
TOTALS			
Total Fixed Cost/Year	$350,185	$713,333	$856,393

Annual Loan Payment ($ per year)

use Equations (2-34) and (2-35).

Total Yearly Cost

$$C_{YEAR} = C_{AP} + C_{STOR} + C_{FUEL} + C_{INS} + C_{INSP} + C_{OVER} + C_{CREW} + C_{INS} + C_{LOAN} \quad (2\text{-}43)$$

Follow the same procedures as presented in Section 2.3.1, *Operational cost of a GA aircraft*.

EXERCISES

(1) An LSA aircraft is being designed by a startup business and you have been hired to evaluate the business case. It is planned that the lifting surfaces of the new aircraft will be composite, but the fuselage will be made from aluminum. This way, it is estimated that 50% of the aircraft will be composite and 50% aluminum. The estimated airframe weight is 530 lb$_f$ and the maximum level airspeed is 120 KTAS (V_H). It is estimated that 250 airplanes will be manufactured over a 5-year period. The airplane features a tapered wing with a simple flap system and, as required for LSA aircraft, the fuselage is unpressurized and it has a 69 inch diameter fixed-pitch propeller driven by a piston engine. With this in mind, estimate the following:

(a) Number of man-hours of engineering time.
(b) Number of man-hours for construction tooling.
(c) Number of man-hours to produce 250 airplanes.
(d) Estimate manpower required for each of the above, assuming 40 hrs a week for 48 weeks a year and production run over 5 years (as stated above). In other words, how many engineers, tooling, and technicians will be required over the period of time?
(e) Estimate the average number of hours required to produce each airframe.

(2) Using the airplane from Exercise (1), estimate the total cost to certify and manufacture 250 units over the 5-year period assuming 95% experience effectiveness, engineering, tooling, and manufacturing rates are $95, $65, and $55 per hour, respectively. Assume 15% product liability cost. The planned number of prototypes is two. Use the consumer price index for the year 2012 (i.e. $CPI_{2012} = 1$). Solve the problem using spreadsheet software and prepare an estimate similar to that in Table 2-1 and validate using standard hand calculations. Determine:

(a) Cost to certify.
(b) Total cost per unit to produce.
(c) Break-even analysis for retail prices at $15,000, $30,000, and $45,000 above total cost per unit, assuming the sales agent is paid $7000 for each airplane sold (i.e. add $7000 to the three retail prices).
(d) Determine the price of three LSA aircraft by researching manufacturers, websites (for instance go to: http://www.lightsportaircrafthq.com/ for a listing of manufacturers).

(3) Estimate the hourly operational cost for the airplane in Exercise (1) for the three retail price options from Exercise (2), assuming it is maintained by an A&P mechanic who charges $50 per hour. It has easy engine access, fixed landing gear, IFR radios only, integral fuel tanks, and a simple flap system. It is flown 150 hours per year. Its 100 BHP engine consumes 6 gal/hr of fuel on average, at $5/gallon. Storage cost is $50 per month. The engine time between overhaul (TBO) is 1500 hours and the cost to overhaul is $4500. Include the acquisition cost for the airplane by assuming it is purchased using a 20% down-payment with the remainder borrowed at 9% APR for 15 years. Note that C_{AC} is the sum of the total cost per unit, the markup, and the sales commission, i.e. the total paid by the customer as a fly-away price.

(4) (a) The total cost of developing a brand new airplane can be expressed as the sum of the fixed cost (constant), denoted by FC, and the variable cost, which can be expressed as $U \cdot N$, where U is the unit variable cost and N is the number of units produced. Consider a scenario in which the retail price of the product is variable rather than constant in order to help initially market the airplane. As an example of such a variable retail price structure, consider a situation where the unit sales price (call it P_1) is low at first to help market the airplane, but is then raised to P_2 after a specific number of units, N_1, have been produced. Derive an expression for the break-even point, i.e. the total number of units, N, required to break-even.

(b) Calculate the number of units that must be produced to break-even for a scenario in which $FC = \$50$ million, $U = \$0.285$ million/unit, $P_1 = \$0.350$ million, $P_2 = \$0.450$ million, and $N_1 = 300$. How many units does it take if the price is not increased and it is offered a P_1?

VARIABLES

Symbol	Description	Units (UK and SI)
AR	Wing aspect ratio	
BHP/W	Brake horse power-to-weight ratio	BHP/lb$_f$ or BHP/N
C_{AC}	Insured valued of aircraft	$
C_{AP}	Yearly maintenance cost	$/yr
C_{CERT}	Total cost for certification	$
C_{CREW}	Crew cost	$/hr
$C_{CSTPROP}$	Cost of constant-speed propellers	$
C_{DEV}	Total development support cost	$
C_{Dmin}	Coefficient of minimum drag	
C_{ENG}	Total cost of engineer	$
$C_{FIXPROP}$	Cost of fixed-pitch propellers	$
C_{FT}	Total cost for flight test operations	$
C_{FUEL}	Annual fuel cost	$/yr
CG	Center of gravity	ft, m, or %MAC
C_{HR}	Cost per flight hour	$/hr
C_{INS}	Annual cost for insurance	$/yr
C_{INSP}	Annual inspection cost	$/yr
C_{Lmax}	Maximum lift coefficient	
C_{LOAN}	Monthly loan payment	$/yr
$C_{L\alpha}$	Aircraft lift curve slope	deg or radians
C_{MAT}	Total material cost	$
C_{MFG}	Total manufacturing cost	$
C_{OVER}	Engine overhaul fund	$/yr
CPI_{2012}	Consumer price index relative to the year 2012	
C_{PP}	Cost of engine	$
C_{QC}	Total cost of quality control	$
C_{STOR}	Cost for storage	$
C_{TOOL}	Total tooling cost	$
C_{YEAR}	Yearly operational cost	$/yr
D_P	Propeller diameter	ft or m
F_{CERT}	Certification factor	
F_{CF}	Complex flap system factor	
F_{COMP}	A factor to account for the use of composites in the airframe	
f_{comp}	Fraction of airframe made from composites	
F_{EXP}	Experience effectiveness adjustment factor	
FF_{CRUISE}	Total fuel flow	gal/hr
F_{MF}	Required maintenance man-hours for every flight hour	
F_{PRESS}	Pressurization factor	
F_{TAPER}	Chord taper factor	
H_{ENG}	Number of engineering man-hours	hrs
H_{MFG}	Number of manufacturing labor hours	hrs
H_{TOOL}	Number of tooling man-hours	hrs
k	Coefficient for lift-induced drag	
KCAS	Knots calibrated airspeed	ft/s or m/s
KIAS	Knots indicated airspeed	ft/s or m/s
LE Sweep	Wing leading edge sweep	deg or radians
MAC	Mean aerodynamic chord	ft or m
n	Load factor	
N	Number of units produced (context-dependent)	
N	Number of planned aircraft to be produced (context-dependent)	
N_{BE}	Number of sold units to break-even	
N_{CREW}	Number of crew members to operate aircraft	
N_{ENG}	Number of engineers	
N_P	Number of prototypes	
N_{PP}	Number of engines	
P_{BHP}	Rated brake horsepower	ft·lb$_f$/s or N·m/s
P_{SHP}	Rated shaft power	ft·lb$_f$/s or N·m/s
QDF	Quality discount factor	
Q_{FLGT}	Flight hours per year	hrs/yr
Q_m	Aircraft production rate	number of aircraft/month
R_{AP}	Rate for certified Airframe and Power plant mechanic	$/hr
R_{CREW}	Rate for crew	$/hr
Re	Reynolds number	
R_{ENG}	Rate of engineering labor	$/hr
R_{FUEL}	Cost of fuel	$/gal
R_{MFG}	Rate of manufacturing labor	$/hr
R_{STOR}	Rate for storage	$/yr
R_{TOOL}	Rate of tooling labor	$/hr
SFC_{CRUISE}	Specific fuel consumption at cruise condition	lb$_f$/hr/BHP

Symbol	Description	Units (UK and SI)
T	Rated thrust	lb_f
t/c	Thickness-to-chord ratio	
T/W	Thrust-to-weight ratio	
t_{AC}	Average to to manufacture a single unit	hrs
TR	Wing taper ratio	
V_H	Maximum level airspeed in KTAS	ft/s
V_{HT}	Horizontal tail volume coefficient	
V_{VT}	Vertical tail volume coefficient	
W	Weight	lb_f or N
W/S	Wing loading	lb_f/ft^2 or N/m^2
$W_{airframe}$	Weight of structural skeleton	lb_f

References

[1] Hess, R. W., and H. P. Romanoff. *Aircraft Airframe Cost Estimating Relationships*. R-3255-AF, RAND Corporation, December 1987.
[2] http://www.rand.org.
[3] Eastlake CN, Blackwell HW. Cost Estimating Software for General Aviation Aircraft Design. St. Louis, MO: Proceedings of the ASEE National Conference; 2000.
[4] Anonymous. *General Aviation Statistical Databook and Industry Outlook 2011*. General Aviation Manufacturers Association, 2012.
[5] http://www.gama.aero.
[6] Anonymous. *General Aviation: Status of the Industry, Related Infrastructure, and Safety Issues*. Report to Congressional Requesters, GAO-01-916, U.S. General Accounting Office, August 2001, p. 18.
[7] http://se.cessna.com/single-engine/cessna-400/cessna-400-pricing.html# (June 2012).
[8] http://cirrusaircraft.com/media/pricesheets/sr22t.pdf (June 2012).
[9] http://www.conklindd.com.

CHAPTER 3

Initial Sizing

OUTLINE

3.1 Introduction 55
  3.1.1 The Content of this Chapter 56
  3.1.2 Fundamental Concepts 56
    Cost 56
    Effectiveness 56
    Cost-effectiveness 56
  3.1.3 Software Tools 57
    Software for Statistical Analysis and Data Visualization 57
    Software for Multi-disciplinary Optimization 57

3.2 Constraint Analysis 57
  3.2.1 General Methodology 58
    T/W for a Level Constant-velocity Turn 58
    T/W for a Desired Specific Energy Level 58
    T/W for a Desired Rate of Climb 58
    T/W for a Desired T-O Distance 59
    T/W for a Desired Cruise Airspeed 59
    T/W for a Service Ceiling (ROC = 100 fpm or 0.508 m/s) 59
    Additional Notes 59
    Derivation of Equations (3-1) and (3-2) 61
    Derivation of Equation (3-3) 62
    Derivation of Equation (3-4) 62
    Derivation of Equation (3-5) 63
    Derivation of Equation (3-6) 63
  3.2.2 Introduction of Stall Speed Limits into the Constraint Diagram 65
    C_{Lmax} for a Desired Stalling Speed 66
    Derivation of Equation (3-7) 66

3.3 Introduction to Trade Studies 66
  3.3.1 Step-by-step: Stall Speed − Cruise Speed Carpet Plot 67
    Step 1: Preliminary Data 67
    Step 2: Decide Plot Limits 67
    Step 3: Calculate Aerodynamic Properties 67
    Step 4: Tabulate Stall Speeds as a Function of C_{Lmax} and Wing Area 67
    Step 5: Calculate and Tabulate Maximum Airspeeds as a Function of Wing Area 67
    Step 6: Tabulate Maximum Airspeeds in Preparation for Plotting 68
    Step 7: Create Carpet Plot 68
  3.3.2 Design of Experiments 69
  3.3.3 Cost Functions 72

Exercises 74

Variables 75

3.1 INTRODUCTION

A successful aircraft development program is the consequence of a satisfactory solution to a large number of dissimilar problems. Ideally, we want our airplane to have a low empty weight, good performance, easy handling, a stout light structure, be inexpensive to manufacture and operate, and so on. All of these present different problems to the design process and each require a specific solution. For this reason, the best solution to each individual problem is not necessarily the best solution from a synergistic standpoint.

In fact, the application of such methods is integral to the design process. Designers conduct various trade studies to find the best solution to problems that simultaneously involve a large range of engineering disciplines and project economics. Ultimately, the purpose is to bring to market a useful product that reduces acquisition and operational costs, while improving the performance of the previous technology. Achieving this requires *compromise* and *balance*. Airplanes designed with only one discipline in mind will unlikely satisfy other requirements.

A classic example of this is the sizing of the wing area. A large wing area leads to a lower stalling speed; if the design problem is simply "low stalling speed" or a "lot of fuel capacity," then a large wing area is an obvious solution. However, if speed and efficiency are a goal too, a large wing area is detrimental. It increases drag and structural weight and ignoring these will result in a poor design, no matter how low the stalling speed.

A correct sizing of an airplane depends on a large number of very important variables such as those discussed in Section 1.1.2, *Important elements of a new aircraft design*. A clearly stated mission plays a paramount role in this respect and allows the sizing to be accomplished using mathematical tools. This section presents a few optimization methods that focus on the external geometry of the airplane. Of these, the most prominent is the so-called *constraint analysis*. The method is used to determine the appropriate wing area and engine power (or thrust) for a new design, based on a number of performance requirements. The section will also introduce a method called design of experiments (DOE), carpet plots, and a simple discussion of cost functions.

3.1.1 The Content of this Chapter

- **Section 3.2** presents a powerful method, called constraint analysis, that helps the designer determine the wing sizing (W/S) and thrust-to-weight ratio (T/W or P/W) for the new design, so that it will meet all prescribed performance requirements.
- **Section 3.3** presents several trade study methods, which are powerful tools for the solution of various engineering problems.

3.1.2 Fundamental Concepts

Among a number of concepts frequently brought up in trade studies are *cost*, *effectiveness*, and *cost-effectiveness*. While these concepts apply to a large number of topics, in aviation they usually apply to operational and performance efficiency of the aircraft. In this context, *operational efficiency* refers to the cost associated with acquiring, maintaining, and using (operating) the vehicle. The operational efficiency puts a serious constraint on unconventional aircraft configurations. Historically, unconventional aircraft do not enjoy great market acceptance (they are unconventional after all). As demonstrated in Section 2.2.2, *Development cost of a GA aircraft — the Eastlake model*, the cost of acquisition is directly related to the number of units delivered. It is a serious drawback for such designs that they lack the operational experience of their "conventional" competitor aircraft.

In its simplest terms, *performance efficiency* refers to the magnitude of the maximum lift-to-drag ratio. As is evident from the formulation of the performance chapters of this book, this ratio has a profound effect on key performance parameters such as range and endurance. High fuel costs make this figure of merit even more important than before. This implies that the aircraft designer should strive to develop an aircraft that is aerodynamically sleek and one that requires low engine power for cruise.

Cost

The cost of an aircraft includes the cost of the resources needed to design, manufacture, and operate it. Resources include engineers, technicians, and administrative staff; as well as contractors, materials, office and manufacturing facilities; and specialized equipment such as wind tunnels. Cost is most often represented using monetary units such as dollars. Trade studies always try to minimize cost.

Effectiveness

The effectiveness of an aircraft refers to a quantitative measure of how well the aircraft achieves its mission. This is highly dependent upon its performance. For instance, consider an aircraft designed for a long-distance flight. Its effectiveness can be defined as the ratio of the actual range to the intended range. Effectiveness sells. Trade studies always try to maximize effectiveness.

Cost-effectiveness

The cost-effectiveness of a system combines both the cost and the effectiveness of the system in the context of its objectives. While it may be necessary to measure either or both of those in terms of several numbers, it is sometimes possible to combine the components into a meaningful, single-valued objective function for use in design optimization. Even without knowing how to relate effectiveness to cost, designs that have lower cost and higher effectiveness are always preferred.

Usually the hardest part of any trade study is to formulate the various effects. Some examples of how this can be done will be shown in this text. Three types of trade studies are discussed: constraint analysis, graphical trade study, and Multi-Disciplinary Optimization (MDO). The reader should know that such studies are a field of specialization within engineering and, therefore, only an elementary introduction is given here.

3.1.3 Software Tools

A large number of software packages have emerged that provide substantial help in harnessing the computational power of the modern computer. A number of sophisticated data visualization packages are available, as well as software that make running MDO easy. Detailed description of such software, other than a brief explanation of what such packages do, is beyond the scope of this text.

Software for Statistical Analysis and Data Visualization

A large number of software packages capable of presenting statistical analyses results using powerful visualization are available, both as commercial packages and as freeware.[1] Some software is ideal for work involving design of experiment (see Section 3.3.2, *Design of experiments*), for instance JMP Software (www.jmp.com). Others include statistical analyses with other capabilities, such as Mathematica (www.wolfram.com/mathematica), Maple (www.maplesoft.com), and MATLAB (www.mathworks.com).

Software for Multi-disciplinary Optimization

A number of powerful packages intended for MDO are also available.[2] Many such software packages allow the user to connect multiple unrelated external software solutions together and run them from a central hub (e.g. AIMMS PRO from www.aimms.com, PHX ModelCenter from www.phoenix-int.com, and HEEDS MDO from www.redcedartech.com/products/heeds_mdo). For instance, it is possible to connect Microsoft Excel analysis to some unrelated analyses (e.g. software and dynamic link libraries written using FORTRAN, C, C++, Visual Basic, etc.) that could be running as an MS-DOS or Windows process. This is done by "telling" the program what the input and output variables mean and how they relate to one another. Such software will then often rewrite the input files, run the associated external software through a batch operation, read the resulting output file, and use the data as a part of the overall optimization. The power of such software is realized when companies can put existing "in-house" programs to use with very limited effort.

3.2 CONSTRAINT ANALYSIS

One of the first tasks in any new aircraft design is to perform a constraint analysis using a special graph called a constraint analysis graph. The primary advantage of this graph is that it can be used to assess the required wing area and power plant for the design, such that it will meet all performance requirements.

Constraint analysis is used to assess the relative significance of performance constraints on the design. This is done by plotting the constraints on a special two-dimensional graph called the *design space* (see Figure 3-1). Commonly the two axes represent characteristics such as (y-axis) thrust-to-weight ratio (T/W) and (x-axis) wing loading (W/S). The graph is then read by noting that any combinations of W/S and T/W that are above the constraint curves will result in a design

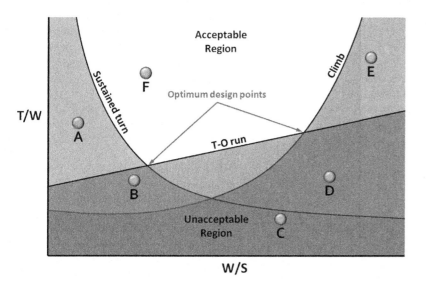

FIGURE 3-1 Typical design space. Only a combination of T/W and W/S that lie in the white (acceptable) region constitute viable design. Here F is the only viable design, albeit not optimal.

---

[1] For a list of packages see: http://en.wikipedia.org/wiki/List_of_statistical_packages.

[2] For a list of packages see: http://en.wikipedia.org/wiki/List_of_optimization_software.

that meets those requirements. The white-colored region in the figure is the domain of acceptable solutions. The shaded region represents unacceptable solutions. For instance, Design A would meet the T-O run and climb requirements, design C would meet none, and design E all but the climb. Design F meets all requirements. The graph allows the designer to see at a glance the combination of W/S and T/W that allows the requirements to be met.

The graph also features *optimum design points*, at which the least amount of power or thrust is required to meet all applicable requirements. Of the two points shown, the true optimum is the one that offers the lower W/S and T/W. This point results in minimum power required and this usually means a power plant that is less expensive to acquire and operate. The designer should also consider the impact of the W/S on the stalling speed of the aircraft. For instance, the 14 CFR Part 23 stall speed maximum is 61 KCAS[3] or 45 KCAS for Light Sport Aircraft (LSA).[4] Section 3.2.2, *Introduction of stall speed limits into the constraint diagram*, will introduce how to incorporate this important limit into the constraint diagram.

The key to creating a constraint graph is the conversion of applicable formulas into a form such that T/W is a function of W/S. Transcendental functions have to be solved iteratively. The optimum points indicate where the design can be accomplished at a minimum cost (less power required usually means smaller, less expensive power plant).

3.2.1 General Methodology

The general methodology of constraint analysis requires some performance characteristics of interests to be described using mathematical expressions. In order to be useful, the expressions must be converted into the proper format that allows them to be evaluated. This section focuses on converting the performance characteristics into the form $T/W = f(W/S)$. However, the reader should be mindful that other criteria besides T/W and W/S may be considered. For the design of GA aircraft, the following basic formulation is very practical. They are all based on the simplified drag model (See Chapter 15, *Aircraft drag analysis*), so the designer should expect some deviations from a more sophisticated analysis. This is not important, however, because little is known about the design when they are used anyway — they represent some of the earliest steps in the design process.

T/W for a Level Constant-velocity Turn

The following expression is used to determine the T/W ratio required to maintain a specific banking load factor (n) at a specific airspeed and altitude, without losing altitude. For instance, consider a project where the design is required to maintain a 45° bank angle at a given airspeed. The first step would be to convert the angle into a load factor, n, using Equation (19-36). Then, the expression would be used to determine the required T/W as a function of W/S.

$$\frac{T}{W} = q\left[\frac{C_{Dmin}}{(W/S)} + k\left(\frac{n}{q}\right)^2\left(\frac{W}{S}\right)\right] \quad (3\text{-}1)$$

where

C_{Dmin} = minimum drag coefficient
k = lift-induced drag constant
q = dynamic pressure at the selected airspeed and altitude (lb$_f$/ft^2 or N/m^2)
S = wing area (ft^2 or m^2)
T = thrust (lb$_f$ or N)
W = weight (lb$_f$ or N)
n = load factor = $1/\cos\phi$

Note that Equation (3-1) corresponds to specific energy density $P_S = 0$ (see Section 19.3.3, *Energy State*).

T/W for a Desired Specific Energy Level

Sometimes it is of importance to evaluate the T/W for a specific energy level other than $P_S = 0$, as was done above. The following expression is used for this purpose. For instance, consider a project where the design is required to possess a specific energy level amounting to 20 ft/s at a given load factor, airspeed, and altitude. Such an evaluation could be used for the design of an aerobatic airplane, for which the capability of a rival aircraft might be known and used as a baseline.

$$\frac{T}{W} = q\left[\frac{C_{Dmin}}{(W/S)} + k\left(\frac{n}{q}\right)^2\left(\frac{W}{S}\right)\right] + \frac{P_S}{V} \quad (3\text{-}2)$$

where

q = dynamic pressure at the selected airspeed and altitude
P_S = specific energy level at the condition
V = airspeed

T/W for a Desired Rate of Climb

The following expression is used to determine the T/W required to achieve a given rate of climb. An

---

[3] Per 14 CFR 23.49(d), *Stalling period*.
[4] Per 14 CFR 1.1, *Definitions*.

example of its use would be the extraction of T/W for a design required to climb at 2000 fpm at S-L or 1000 fpm at 10,000 ft.

$$\frac{T}{W} = \frac{V_V}{V} + \frac{q}{(W/S)} C_{Dmin} + \frac{k}{q} \cdot \left(\frac{W}{S}\right) \quad (3\text{-}3)$$

where

q = dynamic pressure at the selected airspeed and altitude
V = airspeed
V_V = vertical speed

Note that ideally the airspeed, V, should be an estimate of the best rate-of-climb airspeed (V_Y — see Section 18.3, *General climb analysis methods*). Since this requires far more information than typically available when this tool is used, resort to historical data by using V_Y for comparable aircraft. However, it may still be possible to estimate a reasonable V_Y for propeller aircraft using Equation (18-27).

T/W for a Desired T-O Distance

The following expression is used to determine the T/W required to achieve a given ground run distance during T-O. An example of its use would be the extraction of T/W for a design required to have a ground run no longer than 1000 ft.

$$\frac{T}{W} = \frac{V_{LOF}^2}{2g \cdot S_G} + \frac{q \cdot C_{D\ TO}}{W/S} + \mu\left(1 - \frac{q \cdot C_{L\ TO}}{W/S}\right) \quad (3\text{-}4)$$

where

$C_{L\ TO}$ = lift coefficient during T-O run
$C_{D\ TO}$ = drag coefficient during T-O run
q = dynamic pressure at $V_{LOF}/\sqrt{2}$ and selected altitude
S_G = ground run
V_{LOF} = liftoff speed
μ = ground friction constant
g = acceleration due to gravity

T/W for a Desired Cruise Airspeed

The following expression is used to determine the T/W required to achieve a given cruising speed at a desired altitude. An example of its use would be the extraction of T/W for a design required to cruise at 250 KTAS at 8000 ft.

$$\frac{T}{W} = q C_{Dmin}\left(\frac{1}{W/S}\right) + k\left(\frac{1}{q}\right)\left(\frac{W}{S}\right) \quad (3\text{-}5)$$

where

q = dynamic pressure at the selected airspeed and altitude
S = wing area

T/W for a Service Ceiling (ROC = 100 fpm or 0.508 m/s)

The following expression is used to determine the T/W required to achieve a given service ceiling, assuming it is where the best rate-of-climb of the airplane has dropped to 100 fpm. An example of its use would be the extraction of T/W for a design required to have a service ceiling of 25,000 ft.

$$\frac{T}{W} = \frac{V_V}{\sqrt{\frac{2}{\rho}\left(\frac{W}{S}\right)}\sqrt{\frac{k}{3 \cdot C_{Dmin}}}} + 4\sqrt{\frac{k \cdot C_{Dmin}}{3}} \quad (3\text{-}6)$$

where

ρ = air density at the desired altitude.
V_V = rate-of-climb = 1.667 ft/s if using the UK-system and 0.508 m/s if using the SI-system

Note that service ceiling implies V_Y (the best rate-of-climb airspeed), as this yields the highest possible value. This is particularly important to keep in mind when converting the T/W to thrust and then to power for propeller aircraft (and as is demonstrated later). For this reason, V_Y should be estimated, for instance using Equation (18-27) or other suitable techniques.

Additional Notes

(1) Note that when constructing constraint diagrams for propeller aircraft, the analysis is complicated by having to convert the thrust-to-weight ratio to P/W. Such diagrams are far more convenient for propeller-powered aircraft because conventional piston and turboprop engines are rated in terms of horsepower. The conversion is accomplished using Equation (14.38), repeated here for convenience:

$$T = \frac{\eta_p \times 550 \times P_{BHP}}{V} \Leftrightarrow P_{BHP} = \frac{TV}{\eta_p \times 550} \quad (14.38)$$

(2) Normalization of thrust and power: Since engine thrust or power depends on altitude, a proper comparison requires a transfer of all altitude characteristics to S-L. This can be accomplished for piston-engine power using the Gagg-Ferrar model, presented in Equation (7-16). For gas turbines the method of Mattingly et al., presented in Section 7.2, *The properties of selected engine types*, can be used. Alternatively, if the airplane is powered with a gas turbine, the designer may have access to an "engine deck" which is software written by the engine manufacturer that allows the thrust (or SHP) at a given altitude, airspeed, and other operational parameters to be extracted. Using such software, the designer should make sure the engine can generate the required thrust or SHP in the corresponding conditions.

A case in point is the constraint diagrams seen in Figure 3-2 and Figure 3-3 for Example 3-1, developed in conjunction with the constraint curves superimposed on the design space. The fact is that some are plotted at S-L conditions, while others represent higher altitudes. Consider the curve representing the service ceiling in Figure 3-3. At $W/S = 30$ lb_f/ft^2 this requirement calls for approximately

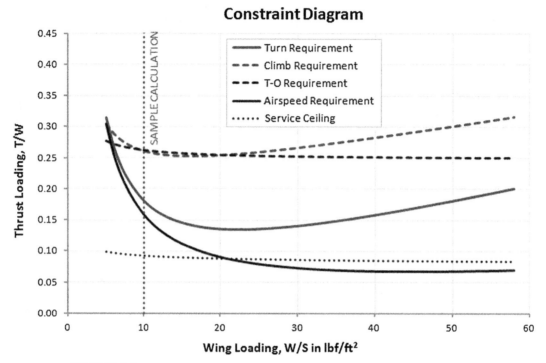

FIGURE 3-2 Constraint diagram. This graph only shows T/W versus wing loading.

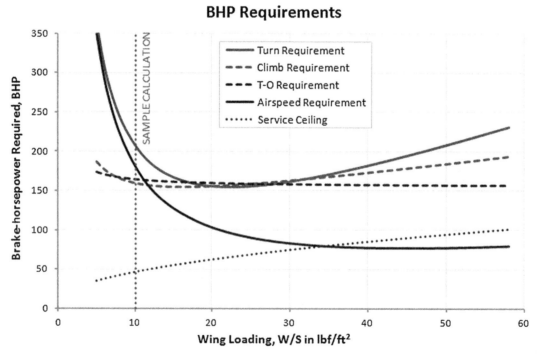

FIGURE 3-3 Plotting required BHP for the various requirements. This graph displays the T/W ratios of Figure 3-2 converted to corresponding power in BHP at each altitude.

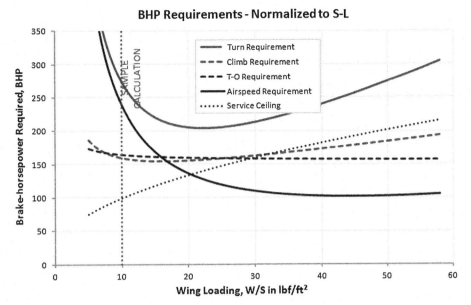

FIGURE 3-4 Required power in BHP for the various requirements normalized to S-L conditions. This graph makes it possible to realistically select an engine for the aircraft.

160 BHP at 20000 ft. The point is that, using the Gagg and Ferrar model, this means the engine must be capable of producing at least 340 BHP at S-L at that wing loading. Therefore, we say that the higher horsepower number (340 BHP) has been *normalized* to S-L conditions. Note that Figure 3-4 shows all the requirements presented in the example normalized in this fashion.

(3) Note that the dynamic pressure, q, is always calculated at the specific condition for which it refers. This way, the following rules apply to q:

Equation (3-1): q is calculated at the turning airspeed and the associated altitude.
Equation (3-3): q is calculated at the climb airspeed and the associated altitude.
Equation (3-4): q is calculated at $V_{LOF}/\sqrt{2}$ in accordance with Section 17.3.1, METHOD 1: *General solution of the equation of motion* and the associated altitude.
Equation (3-5): q is calculated at the desired cruising speed and the associated altitude.
Equation (3-6): ρ is at the desired service ceiling.

(4) One of the problems encountered at this stage of the design is that the geometry of the airplane is largely unknown. For that reason, one does not know important parameters such as C_{Dmin}, C_{DTO}, C_{LTO}, and k. To resolve this issue, the designer must look to existing aircraft in the same class as the one being designed and try to estimate them. Table 3-1 is intended to give the designer a range of typical values, in lieu of such a study. Also, consider Table 15-18, which lists C_{Dmin} for a number of selected aircraft.

Derivation of Equations (3-1) and (3-2)

Consider Equation (19-40) for thrust required in a sustained turn at a load factor n, here repeated for convenience (ignoring the trim drag):

$$T_{REQ} = qS\left[C_{Dmin} + k(nW/qS)^2\right]$$

This equation can be put into the desired form by dividing both sides by W (simplifying T_{REQ} by writing T):

$$T = qS\left[C_{Dmin} + k\left(\frac{nW}{qS}\right)^2\right] \Leftrightarrow$$

$$\frac{T}{W} = \frac{qS}{W}\left[C_{Dmin} + k\left(\frac{nW}{qS}\right)^2\right]$$

$$= \frac{q}{(W/S)}\left[C_{Dmin} + k\left(\frac{n}{q}\right)^2\left(\frac{W}{S}\right)^2\right]$$

Dividing through by W/S yields Equation (3-1). Note that this formulation features the simplified drag model (see Chapter 15, *Aircraft Drag Analysis*, in particular Equation (15-5)).

To derive Equation (3-2) we consider the case where there is more thrust available than the required thrust, T_{REQ}. Calling this thrust T_{AVAIL}, we can write this as the thrust required plus additional thrust, ΔT, i.e.:

$$T_{AVAIL} = T_{REQ} + \Delta T$$

TABLE 3-1 Typical Aerodynamic Characteristics of Selected Classes of Aircraft

Class	C_{Dmin}	C_{DTO}	C_{LTO}	Comment
Amphibious	0.040–0.055	0.050–0.065	≈ 0.7	Assumes flaps in T-O position.
Agricultural	0.035–0.045	0.045–0.055	≈ 0.7	Assumes flaps in T-O position.
Biplane	0.045–0.050	0.045–0.050	≈ 0.4	Assumes no flaps.
GA trainer	0.030–0.035	0.040–0.045	≈ 0.7	Assumes flaps in T-O position.
GA high-performance single	0.025–0.027	0.035–0.037	≈ 0.7	Assumes flaps in T-O position.
GA typical single, fixed gear	0.028–0.035	0.038–0.045	≈ 0.7	Assumes flaps in T-O position.
Turboprop commuter	0.025–0.035	0.035–0.045	≈ 0.8	Assumes flaps in T-O position.
Turboprop military trainer	0.022–0.027	0.032–0.037	≈ 0.7	Assumes flaps in T-O position.
Turbofan business jet	0.020–0.025	0.030–0.035	≈ 0.8	Assumes flaps in T-O position.
Modern passenger jetliner	0.020–0.028	0.030–0.038	≈ 0.8	Assumes flaps in T-O position.
1960s–70s passenger jetliner	0.022–0.027	0.032–0.037	≈ 0.6	Assumes flaps in T-O position.
World War II bomber	0.035–0.045	0.045–0.055	≈ 0.7	Assumes flaps in T-O position.
World War II fighter	0.020–0.025	0.030–0.035	≈ 0.5	Assumes flaps in T-O position.

Then, insert Equation (3-1) and note that since the power associated with ΔT is $\Delta P = \Delta T \cdot V$, we can write:

$$T_{AVAIL} = T_{REQ} + \Delta T = qS\left[C_{Dmin} + k\left(\frac{nW}{qS}\right)^2\right] + \Delta T$$

$$= qS\left[C_{Dmin} + k\left(\frac{nW}{qS}\right)^2\right] + \frac{\Delta P}{V}$$

Then, multiply the additional term by 1 or W/W and note that $P_S = \Delta P/W$. This comes from the fact that ROC is power divided by weight (see Chapter 18, *Performance – Climb*). Therefore, we get:

$$T_{AVAIL} = qS\left[C_{Dmin} + k\left(\frac{nW}{qS}\right)^2\right] + \frac{\Delta P}{V}\frac{W}{W}$$

$$= qS\left[C_{Dmin} + k\left(\frac{nW}{qS}\right)^2\right] + \frac{P_S}{V}W$$

Then dividing both sides of the equal sign by W similar to what was done for Equation (3-1) yields Equation (3-2).

QED

Derivation of Equation (3-3)

Consider Equation (18-18) for Rate-of-climb, also repeated for convenience:

$$V_V = V\left(\frac{T}{W} - q\left(\frac{S}{W}\right)C_{Dmin} - k\cdot\left(\frac{W}{S}\right)\frac{\cos^2\theta}{q}\right)$$

Assuming that small angle relations hold, $\cos\theta \sim 1$, we write:

$$V_V = V\left(\frac{T}{W} - q\left(\frac{S}{W}\right)C_{Dmin} - k\cdot\left(\frac{W}{S}\right)\frac{\cos^2\theta}{q}\right)$$

$$= V\left(\frac{T}{W} - \frac{q}{(W/S)}C_{Dmin} - \frac{k}{q}\cdot\left(\frac{W}{S}\right)\right)$$

Solving for T/W yields Equation (3-3).

QED

Derivation of Equation (3-4)

Assuming the ground run to start from rest, the kinematic relations between acceleration, speed, and distance can be written as:

$$S - S_0 = \frac{V^2 - V_0^2}{2a} \quad \Leftrightarrow \quad S_G = \frac{V_{LOF}^2}{2a}$$

Where a is the average acceleration during the T-O run, calculated at $V_{LOF}/\sqrt{2}$ using Equation (17-4):

$$a = g\left[\frac{T}{W} - \frac{D}{W} - \mu\left(1 - \frac{L}{W}\right)\right]$$

$$= g\left[\frac{T}{W} - \frac{qSC_{D\ TO}}{W} - \mu\left(1 - \frac{qSC_{L\ TO}}{W}\right)\right]$$

Inserting into the expression for S_G this leads to:

$$S_G = \frac{V_{LOF}^2}{2a} = \frac{V_{LOF}^2}{2g\left[\frac{T}{W} - \frac{qSC_{D\ TO}}{W} - \mu\left(1 - \frac{qSC_{L\ TO}}{W}\right)\right]}$$

Solving for T/W by algebraic manipulations, this leads to:

$$\frac{T}{W} - \frac{qSC_{D\ TO}}{W} - \mu\left(1 - \frac{qSC_{L\ TO}}{W}\right) = \frac{V_{LOF}^2}{2g \cdot S_G}$$

Writing this in terms of T/W to yield a convenient form for use in the design space will give Equation (3-4).

Note that the argument for the equation is effectively $V_{LOF}/\sqrt{2}$ (and not V_{LOF}), as this is used to calculate the acceleretion. Therefore, when extracting power for a propeller powered aircraft, use $P_{BHP} = T \cdot (V_{LOF}/\sqrt{2})/(\eta_p \cdot 550)$, where η_p is the propeller efficiency at $V_{LOF}/\sqrt{2}$.

QED

Derivation of Equation (3-5)

During cruise we may assume thrust to equal drag. Therefore, we can write:

$$T = D = \frac{1}{2}\rho V^2 S C_D = \frac{1}{2}\rho V^2 S \left(C_{Dmin} + \frac{C_L^2}{\pi \cdot AR \cdot e}\right)$$

Expand the above expression and insert the definition for the lift coefficient, $C_L = \dfrac{2W}{\rho V^2 S}$

$$T = \frac{1}{2}\rho V^2 S C_{Dmin} + \frac{\frac{1}{2}\rho V^2 S}{\pi \cdot AR \cdot e} C_L^2$$

$$= \frac{1}{2}\rho V^2 S C_{Dmin} + \frac{\frac{1}{2}\rho V^2 S}{\pi \cdot AR \cdot e}\left(\frac{2W}{\rho V^2 S}\right)^2$$

Manipulate algebraically by isolating the term W/S:

$$T = \frac{1}{2}\rho V^2 S C_{Dmin} + \frac{2\rho V^2 S}{\pi \cdot AR \cdot e \cdot \rho^2 \cdot V^4}\left(\frac{W}{S}\right)^2$$

$$= \frac{1}{2}\rho V^2 S C_{Dmin} + \frac{2S}{\pi \cdot AR \cdot e \cdot \rho \cdot V^2}\left(\frac{W}{S}\right)^2$$

Divide through by the weight W and manipulate:

$$\frac{T}{W} = \frac{\frac{1}{2}\rho V^2 S C_{Dmin} + \frac{2S}{\pi \cdot AR \cdot e \cdot \rho \cdot V^2}\left(\frac{W}{S}\right)^2}{W}$$

$$= \frac{\frac{1}{2}\rho V^2 S C_{Dmin}}{W} + \frac{2S}{\pi \cdot AR \cdot e \cdot \rho \cdot V^2 \cdot W}\left(\frac{W}{S}\right)^2$$

Rearrange in terms of W/S:

$$\frac{T}{W} = \frac{1}{2}\rho V^2 C_{Dmin}\left(\frac{1}{W/S}\right) + \frac{2}{\pi \cdot AR \cdot e \cdot \rho \cdot V^2}\left(\frac{W}{S}\right)$$

Since we define the dynamic pressure as $q = \frac{1}{2}\rho V^2$ and the lift-induced drag constant as $k = 1/(\pi \cdot AR \cdot e)$, these can be inserted into the above expression to yield Equation (3-5).

QED

Derivation of Equation (3-6)

It is assumed that the service ceiling is where the best rate of climb has dropped to 100 fpm. Therefore, using Equation (18-17), it is possible to write:

$$\frac{ROC_{fpm}}{60} = \frac{100}{60} = \frac{TV - DV}{W}$$

$$= \left(\frac{T}{W} - \frac{D}{W}\right)V \Leftrightarrow \frac{T}{W} = \frac{1.667}{V} + \frac{D}{W}$$

Assuming the simplified drag model, this can be rewritten as follows:

$$\frac{T}{W} = \frac{1.667}{V} + \frac{q}{W/S}\left(C_{Dmin} + k\left(\frac{W}{qS}\right)^2\right) \quad (i)$$

Noting that using the simplified drag model the best rate-of-climb airspeed for a propeller aircraft is given by:

$$V_Y = \sqrt{\frac{2}{\rho}\left(\frac{W}{S}\right)}\sqrt{\frac{k}{3 \cdot C_{Dmin}}}$$

Therefore, the dynamic pressure is given by:

$$q = \frac{1}{2}\rho V_Y^2 = \frac{1}{2}\rho\left[\frac{2}{\rho}\left(\frac{W}{S}\right)\sqrt{\frac{k}{3 \cdot C_{Dmin}}}\right]$$

$$= \left(\frac{W}{S}\right)\sqrt{\frac{k}{3 \cdot C_{Dmin}}}$$

Inserting both of these into Equation (i), yields:

$$\frac{T}{W} = \frac{1.667}{\sqrt{\frac{2}{\rho}\left(\frac{W}{S}\right)\sqrt{\frac{k}{3 \cdot C_{Dmin}}}}} + \frac{\left(\frac{W}{S}\right)\sqrt{\frac{k}{3 \cdot C_{Dmin}}}}{W/S}$$

$$\times \left(C_{Dmin} + k\left(\frac{W}{\left(\frac{W}{S}\right)\sqrt{\frac{k}{3 \cdot C_{Dmin}}}S}\right)^2\right)$$

Finally this yields:

$$\frac{T}{W} = \frac{1.667}{\sqrt{\frac{2}{\rho}\left(\frac{W}{S}\right)\sqrt{\frac{k}{3 \cdot C_{Dmin}}}}} + \sqrt{\frac{k}{3 \cdot C_{Dmin}}}(C_{Dmin} + 3 \cdot C_{Dmin})$$

$$= \frac{1.667}{\sqrt{\frac{2}{\rho}\left(\frac{W}{S}\right)\sqrt{\frac{k}{3 \cdot C_{Dmin}}}}} + 4\sqrt{\frac{k \cdot C_{Dmin}}{3}}$$

QED

EXAMPLE 3-1

A single piston-engine propeller airplane is being designed to meet the following requirements:

(1) Design gross weight shall be 2000 lb_f.
(2) It must sustain a 2g constant velocity turn while cruising at 150 KTAS at 8000 ft.
(3) It must be capable of climbing at least 1500 fpm at 80 KCAS at S-L.
(4) It must be capable of operating from short runways in which the ground run is no greater than 900 ft and liftoff speed of 65 KCAS at design gross weight.
(5) It must be capable of a cruising speed of at least 150 KTAS at 8000 ft.
(6) It must be capable of a service ceiling of at least 20,000 ft.

The designer's initial target is a minimum drag coefficient of 0.025 and an aspect ratio of 9. Furthermore, it is assumed the ground friction coefficient for the T-O requirement is 0.04, the T-O lift and drag coefficients are $C_{L\,TO} = 0.5$ and $C_{D\,TO} = 0.04$, respectively. Plot a constraint diagram for the design for these requirements in terms of W/S and T/W for values of W/S ranging from 5 to 58 lb_f/ft^2. Then determine the required wing area and horsepower for the airplane if its propeller efficiency is 0.80 (assume it is for all flight conditions, even though this would not hold for real propeller aircraft).

Solution

Step 1: Calculate span efficiency per Section 9.5.14, *Estimation of Oswald's Span Efficiency*, here using Method 1:

$$e = 1.78(1 - 0.045 AR^{0.68}) - 0.64$$
$$= 1.78(1 - 0.045(9)^{0.68}) - 0.64 = 0.7831$$

Step 2: Calculate the lift-induced drag constant, k, per Equation (15-7):

$$k = \frac{1}{\pi \cdot AR \cdot e} = \frac{1}{\pi \cdot (9) \cdot (0.783124)} = 0.04516$$

Step 3: Calculate the T/W per Equation (3-1). Here, only a sample value of the T/W will be calculated, using $W/S = 10$ lb_f/ft^2. Then a number of values were calculated using Microsoft Excel and plotted in Figure 3-2. Begin by computing the dynamic pressure at 8000 ft:

$$\rho = 0.002378(1 - 0.0000068756 \times 8000)^{4.2561}$$
$$= 0.001869 \text{ slugs/ft}^3$$

Then calculate the dynamic pressure:

$$q = \frac{1}{2}\rho V^2 = \frac{1}{2}(0.001869)(1.688 \times 150)^2 = 59.9 \text{ lb}_f/ft^2$$

Then calculate the T/W, say for a sample value of $W/S = 10$ lb_f/ft^2:

$$\frac{T}{W} = q\left[\frac{C_{Dmin}}{(W/S)} + k\left(\frac{n}{q}\right)^2\left(\frac{W}{S}\right)\right]$$
$$= (59.9)\left[\frac{0.025}{(10)} + (0.04516)\left(\frac{2}{59.9}\right)^2(10)\right] = 0.1799$$

Step 4: Calculate the T/W per Equation (3-2). Here, again, only a sample value will be calculated. Begin by noting that the ROC amounts to 1500 fpm/(60 s/min) = 25 ft/s and 80 KCAS at S-L amounts to 135.0 ft/s. First, calculate the dynamic pressure:

$$q = \frac{1}{2}\rho V^2 = \frac{1}{2}(0.002378)(135.0)^2 = 21.7 \text{ lb}_f/ft^2$$

Again, let's use a sample value of $W/S = 10$ lb_f/ft^2 to get the following value of T/W:

$$\frac{T}{W} = \frac{V_V}{V} + \frac{q}{(W/S)}C_{Dmin} + \frac{k}{q}\cdot\left(\frac{W}{S}\right)$$
$$= \frac{25}{135.0} + \frac{21.7}{(10)}(0.025) + \frac{0.04516}{21.7}\cdot(10) = 0.2602$$

Step 5: Calculate the T/W per Equation (3-4). Here we must first determine realistic values for the lift and drag coefficients during the T-O run. Start by calculating the dynamic pressure at $V = V_{LOF}/\sqrt{2}$:

$$q = \frac{1}{2}\rho V^2 = \frac{1}{2}(0.002378)\left(\frac{65 \times 1.688}{\sqrt{2}}\right)^2 = 7.16 \text{ lb}_f/ft^2$$

Therefore, for the sample value of $W/S = 10$ lb_f/ft^2, recalling that $\mu = 0.04$, $C_{L\,TO} = 0.5$, and $C_{D\,TO} = 0.04$, we get the following value of T/W:

$$\frac{T}{W} = \frac{V_{LOF}^2}{2g \cdot S_G} + \frac{q \cdot C_{D\,TO}}{W/S} + \mu\left(1 - \frac{q \cdot C_{L\,TO}}{W/S}\right)$$
$$= \frac{(65 \times 1.688)^2}{2(32.174)(900)} + \frac{(7.16)(0.04)}{10} + (0.04)\left(1 - \frac{(7.16)(0.5)}{10}\right)$$
$$= 0.2622$$

Step 6: Calculate the T/W per Equation (3-5), using the ρ and q calculated in **Step 3**. Then calculate the T/W for the sample value of $W/S = 10$ lb_f/ft^2 as follows:

$$\frac{T}{W} = qC_{Dmin}\left(\frac{1}{W/S}\right) + k\left(\frac{1}{q}\right)\left(\frac{W}{S}\right)$$
$$= (59.9)(0.025)\left(\frac{1}{10}\right) + 0.04516\left(\frac{1}{59.9}\right)(10) = 0.1573$$

EXAMPLE 3-1 (cont'd)

Step 7: Begin by computing the dynamic pressure at 20,000 ft:

$$\rho = 0.002378(1 - 0.0000068756 \times 20000)^{4.2561}$$
$$= 0.001267 \text{ slugs/ft}^3$$

Calculate T/W for the service ceiling using Equation (3-6):

$$\frac{T}{W} = \frac{1.667}{\sqrt{\frac{2}{\rho}\left(\frac{W}{S}\right)}\sqrt{\frac{k}{3 \cdot C_{Dmin}}}} + 4\sqrt{\frac{k \cdot C_{Dmin}}{3}}$$

$$= \frac{1.667}{\sqrt{\frac{2(10)}{0.001267}}\sqrt{\frac{0.4516}{3 \times 0.025}}} + 4\sqrt{\frac{0.4516 \times 0.025}{3}} = 0.09266$$

Step 8: By computing the values of T/W for a number of values of W/S we can plot the results as shown in Figure 3-2, where this is done for W/S ranging from 5 to 58. Note that the region above the curves is the acceptable region.

Step 9: Since T/W varies with W/S, pick a token value of W/S to use in a sample calculation in which BHP and wing area are extracted. First, considering the sample $W/S = 10$ lb$_f$/ft^2 the weight requirement of 2000 lb$_f$ will call for a propeller thrust of $2000 \times 0.1799 = 360$ lb$_f$ to be generated while turning with a 2g loading at 150 KTAS at 8000 ft. Using Equation (14.38) we determine that the engine horsepower at that altitude must be at least:

$$P_{BHP} = \frac{TV}{\eta \cdot 550} = \frac{(360)(1.688 \times 150)}{(0.80) \cdot 550} = 207 \text{ BHP}$$

This, of course, means that the engine must have a higher rating than 207 BHP, because this is what is needed at 8000 ft. Using the Gagg and Ferrar model (see Equation (7-16)), we can estimate what the S-L rating must be:

$$P_{BHP \ S-L} = \frac{P_{BHP}}{(1.132\sigma - 0.132)} = \frac{207}{\left(1.132\frac{0.001869}{0.002378} - 0.132\right)}$$
$$= 273 \text{ BHP}$$

This shows that in order to generate 207 BHP at 8000 ft, the engine will have to be rated at least 273 BHP at S-L.

Similarly, the required wing area can now be obtained from the wing loading. For the above sample of $W/S = 10$ lb$_f$/ft^2, the wing area must be $S = W/(W/S) = 2000/(10) = 200$ ft^2. Compared to modern airplanes both BHP and S are much larger than the norm for a gross weight of 2000 lb$_f$. Clearly we picked a value of W/S that is not representative of aircraft currently in production. Let us calculate the value of BHP for the range of W/S for all the requirements and find out if anything else can be learned. This has been done in the graph in Figure 3-3.

Figure 3-4 shows this graph normalized. It can be seen that the curves for the turn, airspeed, and service ceiling requirements are shifted upward when compared to Figure 3-3. It is evident that the minimum T/W for the turn requirement occurs when $W/S \approx 22.5$ lb$_f$/ft^2. The resulting airplane will only require around 205 BHP to meet all requirements. This wing loading can be used to establish the required wing area. Doing so, we get the following requirement for the wing area: $S = 2000/(22.5) = 89$ ft^2. Should this aircraft be certified under FAR 23 (which requires a maximum stall speed of 61 KCAS) the design will require a maximum lift coefficient of:

$$C_{Lmax} = 2W/(\rho \cdot V^2 \cdot S)$$
$$= 2(2000)/(0.002378 \cdot (61 \times 1.688)^2 \cdot 89) = 1.78$$

This calls for a simple high lift system to meet the 61 KCAS stall requirement. It will also allow the pilot to perform some steep approaches for landing.

3.2.2 Introduction of Stall Speed Limits into the Constraint Diagram

It is very important to consider stall speed limitations imposed by the regulatory authorities when constructing the constraint diagram. Otherwise, it is possible to inadvertently select a combination of W/S and T/W that, while meeting all other requirements, results in a stalling speed that is too high. It is strongly recommended that the stalling speed is incorporated into the constraint diagram as shown in Figure 3-5. This is accomplished by plotting isobars for selected stalling speeds using a second vertical axis. To do this, the maximum lift coefficient

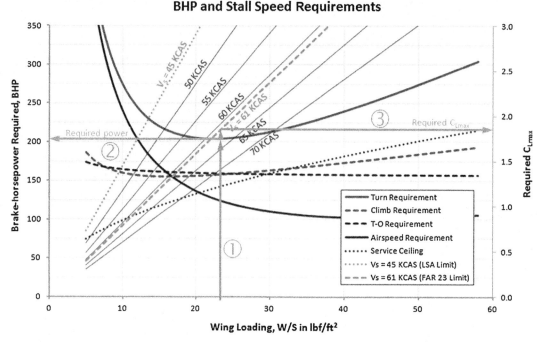

FIGURE 3-5 Constraint diagram with stall speed limits superimposed.

for a given stalling speed is plotted as a function of the wing loading as explained below.

C_{Lmax} for a Desired Stalling Speed

For this purpose, the maximum lift coefficient can be considered a function of the wing loading, W/S, for a constant dynamic pressure, q_{stall}, using the expression below:

$$C_{Lmax} = \frac{1}{q_{stall}}\left(\frac{W}{S}\right) \quad (3\text{-}7)$$

In using this technique, a target stalling speed is selected. Then, the dynamic pressure, q_{stall}, is determined, after which the maximum lift coefficient is calculated for a range of wing loadings, W/S. It is preferable to repeat this for a number of stalling speeds, say, 5 KCAS apart. Then, these are superimposed on the constraint diagram as isobars using a secondary vertical axis for the C_{Lmax}. Figure 3-5 shows this added to Figure 3-3 of Example 3-1. In the example, $W/S = 22.5$ lb$_f$/ft² was shown to be an optimum. To use this information, consider that we are interested in evaluating the C_{Lmax} required for the airplane to stall at $V_S = 61$ KCAS. Following arrow ① we move from the horizontal axis to the optimum point and then continue along arrow ② to read about 205 BHP from the left vertical axis. Then, we go back to the optimum point and move to the diagonal isobar labeled '$V_S = 61$ KCAS (FAR 23 Limit).' Then move horizontally along arrow ③ until the right vertical axis is intersected. There we can read that a C_{Lmax} of about 1.78 is required to meet that restriction — something easily accomplished with a simple high-lift system.

Derivation of Equation (3-7)

The expression is simply obtained from the standard equation for lift, $L = \frac{1}{2}\rho V^2 S C_L$. At stall we may write:

$$W \approx L = \frac{1}{2}\rho V_S^2 S C_{Lmax} = q_{stall} S C_{Lmax} \Leftrightarrow C_{Lmax} = \frac{1}{q_{stall}}\left(\frac{W}{S}\right)$$

QED

3.3 INTRODUCTION TO TRADE STUDIES

The term *trade study* (aka *trade-off*) refers to various methods whose purpose is to identify the most balanced technical solution among a set of proposed viable solutions. A good understanding of such methods is essential for the design engineer and they have a lot of power to offer. For instance, consider the problem of determining a suitable wing area for an aircraft. A large wing area will generally bring down the stalling speed, which is good. However, it will also increase drag and structural weight, which makes the airplane less efficient, which is not good. The three characteristics (lift, drag, and weight) are counterproductive. The designer must eventually weigh the importance of each and make a suitable compromise, which is reflected in the selected wing area. Such tasks are most easily accomplished using trade studies of the nature presented below.

3.3.1 Step-by-step: Stall Speed — Cruise Speed Carpet Plot

The first method is the preparation of a *stall speed — cruise speed* carpet plot, which is intended to help the designer select a wing area such that both desired stalling and cruising speed targets can be met. It requires a number of key parameters to be known, unlike the constraint diagram, which requires much less initial knowledge. This renders the method a tool to use <u>after</u> the constraint diagram has been prepared. As such, it is ideal when considering the modification (growth) of existing airplane types.

The creation of the carpet plot appears complicated at first, but in fact it is not, as long as the following steps are followed. The presentation below applies to a propeller powered aircraft and for clarity it will use actual numbers that are similar to the Cirrus SR22, as we pretend we are designing a similar aircraft. This is done because the consequent comparison gives a deeper insight into the usefulness of this method. Finally, note that the method presented uses Microsoft's Excel spreadsheet program for the analysis.

Step 1: Preliminary Data

Establish design parameters similar to what is shown in Table 3-2.

Note that it is prudent to aim at a stalling speed approximately 2 knots below the 14 CFR Part 23 minimum of 61 KCAS. This is why 59 KCAS is entered as a "desired" number. Also, looking at the competition, it is prudent to aim for at least 180 KTAS at 8000 ft.

Step 2: Decide Plot Limits

Decide the range and steps of maximum lift coefficient, C_{Lmax}, to analyze. Here we will pick $1.7 \leq C_{Lmax} \leq 2.3$ with seven steps, of which the magnitude of each is 0.1.

Decide the range of wing areas, S, using the same number of steps as for the C_{Lmax}. Similarly, we will pick $120 \leq S \leq 180$ ft^2, also with seven steps, each being 10 ft^2 in magnitude.

Step 3: Calculate Aerodynamic Properties

Calculate the parameters shown in Table 3-3. Note that the density is calculated using Equation (16-18) or (16-19), Oswald's span efficiency is calculated using Equations (9-89), (9-91), or (9-92), and power at altitude is calculated using Equation (7-16):

Step 4: Tabulate Stall Speeds as a Function of C_{Lmax} and Wing Area

Prepare a table of stalling speeds similar to the one shown in Table 3-4 below. The stalling speed in the body of the table is calculated using Equation (19-7) at S-L. Note that the calculated stalling speeds are in terms of ft/s and are converted to KCAS by dividing by the factor 1.688 knots/(ft/s). For instance, using $S = 150$ ft^2 and $C_{Lmax} = 2.0$ gives a stalling speed of 97.9 ft/s, which when converted equals 58 KCAS.

Step 5: Calculate and Tabulate Maximum Airspeeds as a Function of Wing Area

Having completed tabulating the stalling speeds, next calculate the maximum airspeed that corresponds to the range of wing areas selected (here 120 to 180 ft^2). It is imperative to realize that this can be accomplished in several ways. Here, methods that use the Visual Basic for Applications functions that have already been developed in this book are picked. For this exercise, two of them will be used; Section 14.5.4, *COMPUTER CODE: Estimation of propeller efficiency using the momentum theorem* and Section 19.2.14, *COMPUTER CODE: Determining maximum level airspeed, V_{max}, for a propeller aircraft*. They are used together to create a table similar to Table 3-5. Note that the two arrows display important relationships that must be kept in mind when preparing the spreadsheet in order to prevent a circular error from occurring in Excel. The rightmost column labeled V

TABLE 3-2 Example of Preliminary Data

Parameter	Symbol	Value	Units
Desired stalling speed	V_S	59	KCAS
Desired cruising speed	V_C	180	KTAS
Desired cruise altitude	H	8000	ft
Target gross weight	W	3400	lb$_f$
Expected minimum drag coefficient	C_{Dmin}	0.025	
Expected aspect ratio (to estimate e)	AR	10	
Expected taper ratio (to estimate e)	λ	0.5	
If propeller, rated power	P_0	310	BHP

TABLE 3-3 Determination of Selected Required Parameters

Parameter	Symbol	Value	Units
Density	ρ	0.001869	slugs/ft^3
Oswald's span efficiency	e	0.7566	
Max horsepower at altitude	P	235	BHP

TABLE 3-4 Stall Speeds as a Function of C_{Lmax} and Wing Area

Stalling Speed KCAS		WING AREA, ft²						
		120	130	140	150	160	170	180
Maximum Lift Coefficient	1.7	70	67	65	63	61	59	57
	1.8	68	65	63	61	59	57	56
	1.9	66	64	61	59	57	56	54
	2.0	65	62	60	**58**	56	54	53
	2.1	63	61	58	56	55	53	52
	2.2	62	59	57	55	53	52	50
	2.3	60	58	56	54	52	51	49

TABLE 3-5 Maximum Airspeeds as a Function of Wing Area

S	η_p	Vmax	V token
ft²		KTAS	KTAS
120	0.828	191	191
130	0.826	186	186
140	0.825	182	182
150	0.823	178	177
160	0.822	174	174
170	0.820	171	171
180	0.819	168	168

TABLE 3-6 Maximum Airspeeds as a Function of Wing Area to be Used for Plotting

S	Vmax						
ft²	KTAS	KTAS	KTAS	KTAS	KTAS	KTAS	KTAS
120	191	186	182	178	174	171	168
130	191	186	182	178	174	171	168
140	191	186	182	178	174	171	168
150	191	186	182	178	174	171	168
160	191	186	182	178	174	171	168
170	191	186	182	178	174	171	168
180	191	186	182	178	174	171	168

token contains airspeeds that are only used with the function PropEfficiency(BHP, V, H, Dp, Nv). These are actually entered manually and iteratively until they are as close as possible to V_{max} — a nuisance step, but necessary to avoid circular error. The PropEfficiency function is used to calculate the propeller efficiency in the column labeled η_p. This is done assuming a viscous profile efficiency, N_v, of 0.85 (see the computer code of Section 14.5.4 for details). Then, the airspeed in the column labeled V_{max} uses this value as an argument when using the function Vmax_Prop(S, k, CDmin, W, rho, BHP, eta). If PropEfficiency refers to V_{max} directly, a circular error will occur.

Step 6: Tabulate Maximum Airspeeds in Preparation for Plotting

Next prepare a table similar to Table 3-6. This table and those from previous steps will be used to prepare the carpet plot in the next step. Note that in the table below, the airspeeds are repeated, here seven times. This is necessary to create the carpet plot, as will become evident in the next step.

Step 7: Create Carpet Plot

In this step, the carpet plot shown in Figure 3-6 will be created. This involves using the specific rows and columns of Tables 3-4 and 3-5 as shown in Table 3-7 below. To plot the curves with the shallow slope, plot the rows and columns indicated by the shading below. Note how the X and Y values are selected. The table from Step 5 corresponds to X-values and the one from Step 4 is used for Y-values.

Similarly, the vertical lines are plotted using the scheme shown in Table 3-8. The table in Step 6 is used for the X-values and the one in Step 4 is used for the Y-values.

Once the carpet plot is complete (see Figure 3-6), it is time to interpret it. Considering the vertical lines first, it can be seen that the greater the wing area the slower will

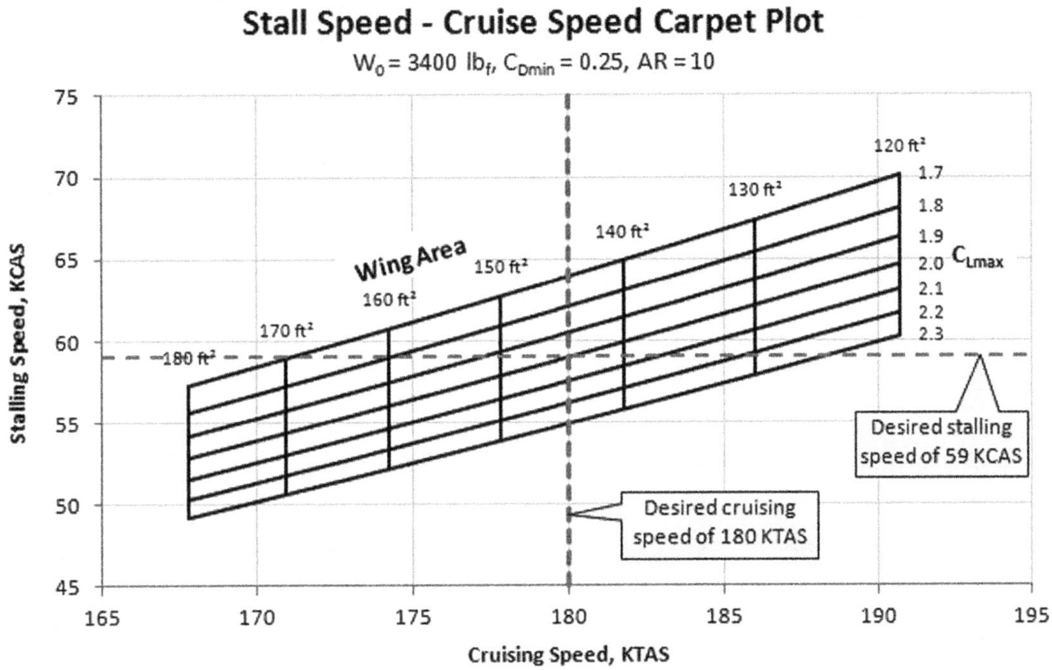

Stall Speed - Cruise Speed Carpet Plot
$W_0 = 3400 \text{ lb}_f, C_{Dmin} = 0.25, AR = 10$

FIGURE 3-6 The resulting carpet plot.

be the cruising speed. This way, should the desired C_{Dmin} of 0.025 indeed be realized, a 180 ft² wing would result in a cruising speed of 167 KTAS, while a 120 ft² wing would yield 191 KTAS. The analysis performed here does not account for a larger C_{Dmin} for the larger wing, but this can easily be included by performing drag analysis.

Continuing with the 120 ft² wing, it can be seen that if it featured a high-lift system capable of generating a $C_{Lmax} = 2.3$, it would have a stalling speed of 60 KCAS. However, if a very simple flap system was employed and it only achieved a $C_{Lmax} = 1.7$, it would stall at 70 KCAS. In either case, it would bust the desired stalling speed of 59 KCAS.

The desired stalling and cruising speeds have been superimposed on the graph. Where the two lines intersect it can be seen that a wing area close to 145 ft² and a high-lift system that generates a $C_{Lmax} \approx 2.0$ should achieve both target airspeeds. Interestingly, this is precisely what the SR22 features, except that its POH cruising speed at 8000 ft and 78% power is 183 KTAS. This lends great support to the value of this method.

3.3.2 Design of Experiments

Design of experiments (DOE) is a method used to determine which variable(s) from a collection from variables is the most effective contributor to some process. The method is best explained using an example.

Consider the development of a vertical tail (VT) for some airplane and that we are interested in understanding what properties contribute to its directional stability derivative, $C_{n\beta}$. To keep things manageable we will analyze the simple constant-chord VT configuration shown in Figure 3-7. The tail is mounted to the hinged tail arm, which allows it to rotate freely, and we can assume the hinge represents the location of the airplane's center of gravity. Aerodynamic theory dictates that the derivative is affected by the tail arm (l_{VT}), tail planform area (S_{VT}), tail span (b_{VT}), and leading edge sweep (Λ_{VT}). Other contributions, such as that of taper ratio or airfoil type, will be ignored.

The following question can now be asked: Which of the above variables affect the value of the $C_{n\beta}$ the most? For instance, if any of the variables change by, say, 10% from its initial value, which will change $C_{n\beta}$ the most? The answer is important because if we want to change $C_{n\beta}$, the result simply tells us where to focus our effort.

Before these questions can be answered, the appropriate formulation must be developed. First, the yawing moment, C_N, is the product of the lift force acting on the tail, L_{VT}, and its distance from the hinge or tail arm, denoted by l_{VT}:

Directional moment:

$$C_N = \frac{N_{VT}}{qSb} = \frac{L_{VT} \times l_{VT}}{qSb} = \frac{qS(C_{L_{\beta VT}} \times \beta) \times l_{VT}}{qSb}$$
$$= \left(\frac{l_{VT}}{b}\right)(C_{L_{\beta VT}} \times \beta) \quad (3\text{-}8)$$

TABLE 3-7 Selecting Data for Plotting the Curves with the Shallow Slope

S ft²	η_p	Vmax KTAS	V token KTAS
120	0.828	191	191
130	0.826	186	186
		182	182
	0.823	178	177
160	0.822	174	174
170	0.820	171	171
180	0.819	168	168

This column is the 'X'

Stalling Speed KCAS	Maximum Lift Coefficient	WING AREA, ft²						
		120	130	140	150	160	170	180
	1.7	70	67	65	63	61	59	57
	1.8	68	65	63	61	59	57	56
	1.9	66	64	61	59	57	56	54
	2.0	65	62	60	**58**	56	53	53
	2.1	63	61	58	56	55	53	52
	2.2	62	59	57	55	53	52	50
	2.3	60	58	56	54	52	51	49

This row is the 'Y'

TABLE 3-8 Selecting Data for Plotting the Vertical Curves

Vmax						
KTAS	KTAS	KTAS	KTAS	KTAS	KTAS	KTAS
191	186	182	178	174	171	168
191	186	182	178	174	171	168
191	186	182	178	174	171	168
191	186	182	178	174	171	168
191	186	182	178	174	171	168
191	186	182	178	174	171	168
191	186	182	178	174	171	168

This column is the 'X'

Stalling Speed KCAS	Maximum Lift Coefficient	WING AREA, ft²						
		120	130	140	150	160	170	180
	1.7	70	67	65	63	61	59	57
	1.8	68	65	63	61	59	57	56
	1.9	66	64	61	59	57	56	54
	2.0	65	62	60	**58**	56	54	53
	2.1	63	59			55	53	52
	2.2	62				53	52	50
	2.3	60	58	56	54	52	51	49

This row is the 'Y'

3.3 INTRODUCTION TO TRADE STUDIES

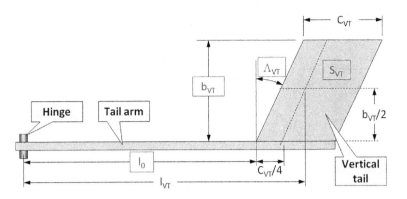

FIGURE 3-7 The geometric relations of a simple tail. The boxed variables will be evaluated.

where

N_{VT} = yawing moment
L_{VT} = lift force generated by the tail
l_{VT} = tail arm
β = yaw angle
q = dynamic pressure
S = wing reference area
b = wing reference span
$C_{L_{\beta VT}}$ = three-dimensional lift curve slope of the tail

Then, the directional stability, $C_{n\beta}$, at low yaw angles can be approximated from:

$$C_{n_\beta} = \frac{\partial C_N}{\partial \beta} \approx \left(\frac{l_{VT}}{b}\right) C_{L_{\beta VT}} \qquad (3\text{-}9)$$

The tail arm, l_{VT}, is based on the dimensions in Figure 3-7 and is calculated from (see the dimensions in the figure):

$$l_{VT} = l_0 + \frac{C_{VT}}{4} + \frac{b_{VT}}{2} \tan \Lambda_{VT} \qquad (3\text{-}10)$$

where

l_0 = a tail arm basic length (to the leading edge of the root)
C_{VT} = average chord of the VT
b_{VT} = the span of the VT

In this analysis, it is better to use the term l_0 to control the length of the tail arm. The lift curve slope of the tail, $C_{L_{\beta VT}}$, can be calculated using Equation (9-57), but this is directly dependent on Λ_{VT} since the configuration features a constant-chord. By varying each of the four variables (l_0, b_{VT}, S_{VT}, and Λ_{VT}) over a range of 10%, using some representative numbers for the variables q, S, and b, the graphs of Figure 3-8 were created. The results will now be discussed.

When a specific variable is varied over a range of 10% of its baseline magnitude this simply means that lower and upper bound values are determined and the $C_{n\beta}$ is calculated for each. For instance, consider the VT span, b_{VT}. The lower bound would be calculated as $0.9 \cdot b_{VT}$ and the upper as $1.1 \cdot b_{VT}$. This way, Figure 3-8 reveals the impact of such a variation on the $C_{n\beta}$. It can be seen it is significantly affected by the variables l_0, b_{VT}, and S_{VT}, while Λ_{VT} practically has no effect on it. The effect of l_0, b_{VT}, and S_{VT} appears mostly equal. If we discovered our airplane had insufficient directional stability, it would be wise to focus on those and ignore modifying the leading edge sweep.

Results like the ones presented above can help with keeping project research costs down. For instance,

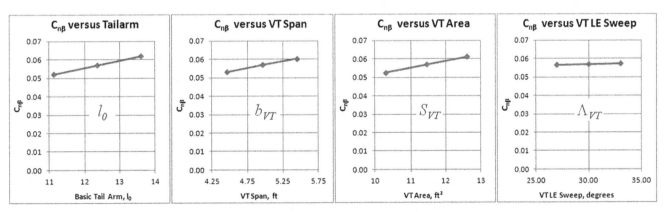

FIGURE 3-8 The results from a DOE analysis.

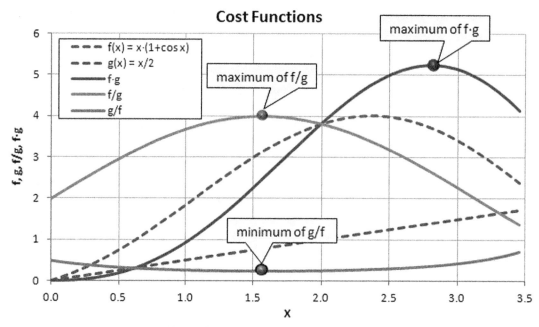

FIGURE 3-9 Definition of cost functions.

when planning a wind-tunnel test, the analysis indicates that investigating variations in the leading edge sweep is not necessary. On the other hand, focusing on investigating the effect of changes in the tail arm, span, and area of the VT is warranted. This way, the number of research variables is reduced from four to three and the time required to complete the wind-tunnel testing should be reduced as well.

3.3.3 Cost Functions

The viability of proposed solutions in a trade study can be judged using so-called *cost functions*. A cost function is a product of, a sum of, or some mathematical combination of two or more parameters that yields a value that can be used to evaluate the quality of the combination.

Cost functions are essential to many kinds of trade studies and can be priceless as figures of merit when evaluating multiple characteristics. A typical use of a cost function involves the maximization or minimization of a particular parameter or a combination of parameters. Consider two functions $f = x \cdot (1 + \cos x)$ and $g = x/2$, that represent some properties of interest in a trade study. Figure 3-9 shows these functions plotted as dashed curves over the range $0 < x < 3.5$. It is possible that f and g describe favorable characteristics, or f a favorable one and g an unfavorable one, or vice versa. We may now be interested in three optimized solutions:

(1) The maximum of the product of both functions, which requires $f \cdot g$ to be calculated.
(2) We want to maximize f while minimizing g, which requires f to be divided by g.
(3) We want to maximize g while minimizing f, which requires g to be divided by f.

Note that points (2) and (3) appear to effectively describe the same scenario (we would only have to swap the functions); however, this is incorrect because the two functions are indeed dissimilar. We now want to plot these cases as shown in Figure 3-9. It can be seen that the maximum of $f \cdot g$ is near $x = 2.8$. For f/g it is near $x = 1.6$ and for g/f the maximums are near $x = 0$ and $x = 3.5$. The minimum is near $x = 1.75$.

EXAMPLE 3-2

Five engine models are being considered for an airplane (see table below). It has been determined that all five will function well in the airplane, but we want to find out which engine is the best choice based on weight (W), cost (C), power (P), and fuel consumption (SFC) as listed below. Suggest cost functions that can be used to indicate the most suitable engine type.

EXAMPLE 3-2 (cont'd)

Type	Weight lb$_f$	Cost $	Power BHP	SFC lb$_f$/(hr·BHP)
Engine 1	125	3500	97	0.48
Engine 2	115	4700	120	0.46
Engine 3	135	4250	115	0.47
Engine 4	142	4260	105	0.48
Engine 5	126	3950	117	0.47

Solution

The approach is to maximize favorable properties and minimize unfavorable ones. For instance, we want the engine to have a low weight (W), be inexpensive (C), and have low fuel consumption (SFC). Assuming we put equal emphasis on all three, the engine with the lowest value of $W \times C \times SFC$ is a potential winner. However, we also want the highest power possible. Therefore, an appropriate cost function would be:

$$Cost = \frac{P}{W \times C \times SFC}$$

This ratio is highest for an engine with high power, low weight, low price, and low SFC. It may not have the highest power or the lightest weight, but the most favorable combination of the selected characteristics. Calculated for the example engines, this would result in the following costs, which indicates Engine 5 is the best option:

Type	Weight lb$_f$	Cost $	Power BHP	SFC lb$_f$/(hr·BHP)	Cost hr·BHP^2/(lb$_f$·$)
Engine 1	125	3500	97	0.48	0.000462
Engine 2	115	4700	120	0.46	0.000483
Engine 3	135	4250	115	0.47	0.000426
Engine 4	142	4260	105	0.48	0.000362
Engine 5	126	3950	117	0.47	0.000500

Cost functions can be defined in other ways too. For instance, we could evaluate ratios such as power/weight (BHP/lb$_f$), power/cost (BHP/$), and power/SFC (lb$_f$·BHP^2/hr). Since the high values of P/W, P/C, and P/SFC are desirable, a suitable cost function could be defined as the sum of these, i.e.

$$Cost = \frac{P}{W} + \frac{P}{C} + \frac{P}{SFC}$$

This would result in the following costs:

Type	P/W BHP/lb$_f$	P/C BHP/$	P/SFC hr·BHP^2/lb$_f$	Cost hr·BHP^3/(lb$_f^2$·$)
Engine 1	0.776	0.028	202	202.9
Engine 2	1.043	0.026	261	261.9
Engine 3	0.852	0.027	245	245.6
Engine 4	0.739	0.025	219	219.5
Engine 5	0.929	0.030	249	249.9

It can be seen that Engine 2 has the highest power-to-weight ratio, but the power-to-cost ratio of Engine 5 is the best. Overall, according to this scheme, Engine 2 is the best choice.

Sometimes it is desirable to emphasize one ratio above others. In other words, it is possible the power/weight ratio is of greater importance to the designer than, say, the power/cost ratio. This can be handled by introducing weighing fractions in a variety of ways. As an example, if the importance of P/W is considered 4 times more important than P/C and 10 times more important than P/SFC, we could introduce this as shown below:

$$Cost = \frac{P}{W} + \frac{1}{4}\frac{P}{C} + \frac{1}{10}\frac{P}{SFC}$$

Implementing these yields the following table:

Type	P/W BHP/lb$_f$	0.25·P/C BHP/$	0.1·P/SFC hr·BHP^2/lb$_f$	Cost hr·BHP^3/(lb$_f^2$·$)
Engine 1	0.776	0.006929	20.21	20.991
Engine 2	1.043	0.006383	26.09	27.137
Engine 3	0.852	0.006765	24.47	25.327
Engine 4	0.739	0.006162	21.88	22.621
Engine 5	0.929	0.007405	24.89	25.830

Again, for this example, Engine 2 comes out as the best option.

EXERCISES

(1) A single engine piston-engine propeller airplane is being designed to meet the following requirements:
  (a) The design shall comply with LSA requirements as stipulated by ASTM F2245.
  (b) Design gross weight shall be 1320 lb$_f$ in accordance with LSA requirements.
  (c) It must sustain a 1.5g constant velocity turn while cruising at 100 KCAS.
  (d) It must be capable of climbing at least 1000 fpm at 70 KCAS at S-L.
  (e) It must be capable of operating from short runways in which the ground run is no greater than 500 ft and liftoff speed of 55 KCAS at design gross weight.
  (f) It must be capable of a cruising speed of at least 110 KTAS at 8000 ft.
  (g) It must be capable of a service ceiling of at least 14,000 ft.

  The designer's initial target is a minimum drag coefficient of 0.035 and an aspect ratio of 7. Furthermore, it is assumed the ground friction coefficient for the T-O requirement is 0.04, the T-O lift and drag coefficients are $C_{L\ TO} = 0.5$ and $C_{D\ TO} = 0.04$, respectively. Plot a constraint diagram for these requirements in terms of W/S and T/W for values of W/S ranging from 10 to 40 lb$_f$/ft^2. Then, determine the required wing area and horsepower for the airplane if its propeller efficiency at cruise is 0.80, 0.7 during climb, and 0.6 at other low-speed operations.

(2) A twin piston-engine propeller airplane is being designed to meet the following requirements:
  (a) Design gross weight shall be 5000 lb$_f$.
  (b) It must sustain a 1.5g constant velocity turn while cruising at 180 KTAS at 12,000 ft.
  (c) It must be capable of climbing at least 1800 fpm at 100 KCAS at S-L.
  (d) It must be capable of operating from short runways in which the ground run is no greater than 1200 ft and liftoff speed of 75 KCAS at design gross weight.
  (e) It must be capable of a cruising speed of at least 180 KTAS at 12,000 ft.
  (f) It must be capable of a service ceiling of at least 25,000 ft.

  The designer's initial target is a minimum drag coefficient of 0.035 and an aspect ratio of 7. Furthermore, it is assumed the ground friction coefficient for the T-O requirement is 0.04, the T-O lift and drag coefficients are $C_{L\ TO} = 0.5$ and $C_{D\ TO} = 0.04$, respectively. Plot a constraint diagram for these requirements in terms of W/S and T/W for values of W/S ranging from 10 to 40 lb$_f$/ft^2. Then, determine the required wing area and horsepower for the airplane if its propeller efficiency at cruise is 0.80, 0.7 during climb, and 0.6 at other low speed operations.

(3) Prepare a stall speed–cruise speed carpet plot for a small twin-engine jet aircraft for which the following parameters are given:

Parameter	Symbol	Value	Units
Desired stalling speed	V_S	90	KCAS
Desired cruising speed	V_C	475	KTAS
Desired cruise altitude	H	41,000	ft
Target gross weight	W	8000	lb$_f$
Expected minimum drag coefficient	C_{Dmin}	0.020	
Expected LE sweep angle (to estimate e)	Λ_{LE}	20°	
Expected aspect ratio (to estimate e)	AR	8	
Expected taper ratio (to estimate e)	λ	0.5	
Max thrust at S-L	T_0	2000	lb$_f$

(4) Four avionics suites are being considered for a new small airplane (see table below) and you have been tasked with recommending one over the others. Using weight (W), cost (C), voltage (V), IFR rating (R), number of software features (F), screen width (S), and screen resolution area ($w \cdot h$) as variables, suggest a cost function that can be used to indicate the most suitable avionics suite. (Hint: use the min or max of each column as a reference value, noting that low weight, cost, and voltage; IFR rating; high number of software features; large screen width and resolution are favorable).

Type	Weight lb$_f$	Cost $	Voltage Amp	IFR Rated	Software Features	Screen Width, in	Resolution w × h
Avionics Suite 1	9.8	8900	12	1	35	10	800 × 6000
Avionics Suite 2	11.2	13,500	12	1	82	12	1024 × 726
Avionics Suite 3	10.9	9100	18	1	36	10	800 × 600
Avionics Suite 4	8.8	6000	12	0	24	7	480 × 386

VARIABLES

Symbol	Description	Units (UK and SI)
AR	Aspect ratio	
b	Wingspan	ft or m
b_{VT}	Vertical tail span	ft or m
$C_{D\,TO}$	Drag coefficient during T-O run	
C_{Dmin}	Minimum drag coefficient	
$C_{L\,TO}$	Lift coefficient during T-O run	
C_{Lmax}	Maximum lift coefficient	
$C_{L\beta VT}$	Vertical tail lift curve slope	per deg or per rad
C_N	Yawing moment	$lb_f \cdot ft$ or $N \cdot m$
$C_{n\beta}$	Directional stability derivative	deg^{-1} or rad^{-1}
D_P	Propeller diameter	ft or m
e	Oswald efficiency	
g	Acceleration due to gravity	ft/s^2 or m/s^2
H	Cruise altitude	ft or m
k	Coefficient for lift-induced drag	
KCAS	Knots calibrated airspeed	knots
KTAS	Knots true airspeed	knots
l_0	Basic length of a tail arm (to leading edge of tail root)	ft or m
l_{VT}	Vertical tail moment arm	ft or m
L_{VT}	Vertical tail lift force	lb_f or N
n	Load factor	
P_0	Rated power (for propellers)	BHP
P_{BHP}	Power of a piston engine	BHP
P_S	Specific energy level	ft/s or m/s
q	Dynamic pressure	lb_f/ft^2 or N/m^2
q_{stall}	Dynamic pressure @ stall condition	lb_f/ft^2 or N/m^2
ROC	Rate of climb	ft/min of m/min
S	Surface area	ft^2 of m^2
SFC	Specific fuel consumption	$lb_f/(hr \cdot BHP)$
S_G	Ground run	ft of m
S_{VT}	Vertical tail surface area	ft^2 or m^2
T	Thrust	lb_f or N
T/W	Thrust-to-weight ratio	
T_0	Thrust at sea level	lb_f or N
T_{REQ}	Required thrust for specified condition	lb_f or N
V	Airspeed	ft/s or m/s
V_C	Cruise speed	ft/s or m/s
V_{LOF}	Liftoff speed	ft/s or m/s
V_s	Stall speed	ft/s or m/s
V_V	Vertical speed	ft/s m/s
W	Weight	lb_f or N
W/S	Wing loading	lb_f/ft^2 or N/m^2
W_0	Gross weight	lb_f or kg
μ	Ground friction constant	
η	Propeller efficiency factor	
λ	Taper ratio	
Λ_{VT}	Vertical tail leading edge sweep	deg or rad
ρ	Density	$slugs/ft^3$ or kg/m^3

CHAPTER 4

Aircraft Conceptual Layout

OUTLINE

4.1 Introduction 77
   4.1.1 The Content of this Chapter 78
   4.1.2 Requirements, Mission, and Applicable Regulations 78
   4.1.3 Past and Present Directions in Aircraft Design 79
   4.1.4 Aircraft Component Recognition 79

4.2 The Fundamentals of the Configuration Layout 82
   4.2.1 Vertical Wing Location 82
      Field-of-view 84
      Impact on Airframe Design 84
      Impact on Flight and Operational Characteristics 85
      Parasol Wings 85

   4.2.2 Wing Configuration 86
   4.2.3 Wing Dihedral 86
   4.2.4 Wing Structural Configuration 87
   4.2.5 Cabin Configurations 88
   4.2.6 Propeller Configuration 89
   4.2.7 Engine Placement 89
   4.2.8 Landing Gear Configurations 91
   4.2.9 Tail Configurations 92
   4.2.10 Configuration Selection Matrix 92

Variables 93

References 95

4.1 INTRODUCTION

Of the seemingly countless tasks confronting the aircraft designer, one of the most important is the determination of a suitable configuration. Should it be a monoplane or a biplane; a single-engine or a multiengine; an unmanned aircraft or one with 800 seats? Should it be driven by a propeller or a turbofan? If propeller-powered, should it be a pusher or a tractor? What layout will truly best serve the mission of the airplane? The answers to such questions have a profound impact on all other tasks and, thus, are of substantial importance to the entire design process. Today's aircraft designer has access to an enormous database of possible configurations. Many of those have a long operational history that allows the designer to realistically evaluate important pros and cons, and predict their capabilities more accurately than possible before. Some even argue that not many configurations remain to be invented. It is no exaggeration that when it comes to issues like the positioning of wings, landing gear, and engines; or the shape and size of stabilizing surfaces; and even aesthetics, the modern designer can practically go window-shopping for ideas using this vast database.

Aesthetics (also discussed in Section 1.1.2) is a sensitive topic for many. It turns out that the looks of an airplane play a very important role in its marketability. There really is no true method of defining good looks; beauty is in the eyes of the beholder, as mentioned before. It is tempting to propose that the number of produced and sold units of a particular aircraft versus another might serve as an indicator, but even this is unreliable because there are so many other factors that affect sales; for instance the cost of acquiring, maintaining, and operating the airplane can easily offset what most people would consider "good looks." In spite of that, the designer must always keep in mind that the customer, who spends an exorbitant amount of money on a brand new aircraft, wants the purchase to look good. He or she wants it to look like the million dollars just spent. The designer of new aircraft should be mindful of the psychology of the process of purchasing. The cost of an expensive consumer product is sometimes justified by its looks. With respect to aircraft, the competent designer not only knows what the most efficient aerodynamic shape looks like, but also that some of that efficiency may have to be sacrificed in favor of improved looks. An ugly but efficient aircraft is less likely to sell than a good-looking airplane that is slightly less efficient. Of course, sometimes efficiency and good looks go hand-in-hand. For instance, winglets have a wide appeal and they tend to improve the efficiency of aircraft. There are plenty of examples of the opposite, i.e. a feature that improves efficiency but looks awkward, although none will be brought up here to prevent bruised feelings.

Then, there are some designers of aircraft who appear to try their best to stay away from "conventional" configurations. Some strive to put their mark on aviation through original and unorthodox geometry. While the originality of such thinking is to be respected and encouraged, the current state of aircraft design is also to be respected. Originality should be the consequence of mission requirements and not for the sake of being original. There is a good reason why most airplanes look the way they do: evolution. The airplane, as we know it today, is the consequence of an evolution that shares many parallels with biological evolution. A multitude of configurations has been "selected" by the market due to cost-effectiveness (purchase and operational price) and safety statistics, while many others have been weeded out of the pool of options. One hundred years of aircraft evolution have yielded a number of geometries that have proven to be safe and reliable. There are plenty of configurations that simply haven't cut it. Reliability is a very important property if you have to pay for the operation of an aircraft. Safety is a very important property if you have to fly in one. And safety is more important than reliability. The designer who pursues an unconventional aircraft configuration should be concerned about operational safety and reliability. He or she must have the big picture in mind. Safety and reliability are far more important than a signature shape. The selection of the configuration will affect not only areas such as performance and handling, but also areas often less considered, such as maintenance and operation in the field. While the capability of our design methods is deep enough to allow for the design of unconventional aircraft operating in attached flow, the same does not hold for separated flow conditions. And that's where things tend to go wrong.

A part of the conceptual layout is the determination of the class of aircraft the new design belongs to. This is important for several reasons. First, it greatly reduces the number of aircraft one needs to evaluate for comparison purposes. Second, it defines the class of aviation regulations that must be considered in the design of the aircraft. Aircraft are typically placed in order based on characteristics, such as number of engines, mission, and performance. And, third, it allows one to include and exclude specific problems the various types of aircraft have experienced. It is also of importance that the aircraft designer is able not only to identify aircraft components, but also knows their specific purpose on the aircraft. This is discussed in Section 4.1.4, *Aircraft component recognition*.

In this chapter, we will present various examples of aircraft, many of which are being operated on a daily basis. This will be followed by the introduction of a number of common and not so common aircraft configurations for the purpose of giving the aspiring aircraft designer some ideas as to what shape to select for a particular mission. The advantages and disadvantages of each configuration will be discussed, as this is an imperative part of the selection process. Also note that a more detailed discussion of wings, tails, and high-lift devices is also presented elsewhere in this book. The purpose of the chapter is to help the designer ponder the implications of the selected configuration on the scope of the design process.

Note that more details on the conceptual design of specific airplane types are provided in Appendix C, provided online.

4.1.1 The Content of this Chapter

Section 4.2 presents a number of important design considerations and discusses their advantages and disadvantages. This is intended to help the designer to develop a keener eye for the implications of selecting a particular configuration. This awareness can avoid costly mistakes for any company designing a new aircraft.

4.1.2 Requirements, Mission, and Applicable Regulations

As stated in Step 1 of the *Conceptual Design Algorithm for a GA Aircraft* presented in Section 1.3.1, the design process begins by the execution of the statement: "Understand requirements, mission definition, and the implications of the regulations to which the airplane will be certified." This means:

(1) The plane's requirements simply mean: how far, how fast, how high, how heavy, how long a take-off and landing distance etc, must the airplane be capable of.

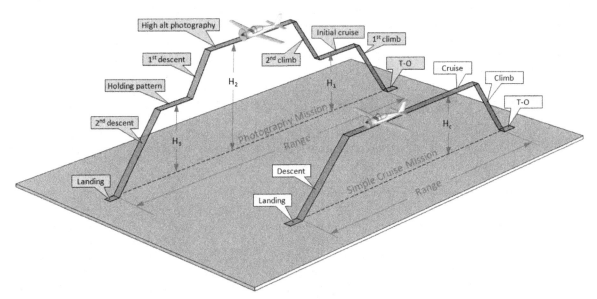

FIGURE 4-1 Two example missions: a simple cruise, and a high-altitude photography mission.

(2) The airplane's mission simply means what it is the airplane is supposed to do. In other words, is it a passenger transport plane that takes off, climbs to the cruise altitude, and cruises for a while before descending for landing? Or is the mission more complicated (see Figure 4-1)? Whatever the mission, its details should be clearly defined for the reasons stated in Section 1.1.2, *Important elements of a new aircraft design.*

(3) Regulations refers to the airplane's designer clearly knowing which regulations the airplane will be designed to.

These three have a profound impact on the airplane configuration.

4.1.3 Past and Present Directions in Aircraft Design

It may strike many as a surprise that aircraft design would be affected by fashion. It would seem that something as vain as style would be beyond engineering, but a review of the history of aviation reveals this is not the case. It is actually vibrant with shapes and components that were popular at one time, but later became a part of history, although others stuck around and became the norm. Table 4-1 lists a few fads that are clearly visible by observing the evolution of the aircraft from early times to modernity.

4.1.4 Aircraft Component Recognition

In the discussion that follows, it is imperative the reader recognizes the terminology used. Figures 4-2 through 4-5 are intended to familiarize the reader with

TABLE 4-1 Fads in Aircraft Design

Era	Fashion
1910s	Rotary engines, biplanes, engine-synchronized machine guns (necessity more than fashion).
1920s	Corrugated aluminum aircraft, wheel fairings for fixed landing gear, open cockpits–closed passenger cabins.
1930s	Engines inside the wing, the birth of scheduled passenger transportation, closed cockpits–closed cabins, retractable landing gear, round wingtips, taildraggers, seaplanes.
1940s	Elliptical wings, engine supercharging, sliding canopies for fighters, tricycle landing gear.
1950s	Passenger turboprops, Jetsons'-style jet geometry,[1] supersonic aircraft.
1960s	VTOL aircraft, supersonic passenger transport, Yehudi flaps for commercial jet aircraft, multi-slotted Fowler flaps for jetliners, low-bypass-ratio jet engines, delta wings for fighters.
1970s	Reduced field-of-view (FOV) cockpits in fighters, "walk-about-cabin" for business jets, STOL aircraft.
1980s	Composites, NLF airfoils, wide-body jets, increased FOV fighter cockpits, and T-tails, more simplified high-lift system for jetliners resulting from reduction in LE-sweep angles, which was a consequence of the development of airfoils using computers.
1990s	Propfans, joined wing design, high-bypass-ratio turbofans, ETOPS-certified commercial jetliners, glass cockpits for commercial jetliners, hush-kits for older jetliners.
2000s	Winglets, glass cockpits for GA aircraft, LSA aircraft.
2010s	Chevrons for jet engines, raked wingtips, electrical aircraft.

[1] *Generally, this means fuselages with a bullet-shaped nose with a Pitot sticking out of it. The term is really the author's preference and is admittedly used to give name to something particularly difficult to describe.*

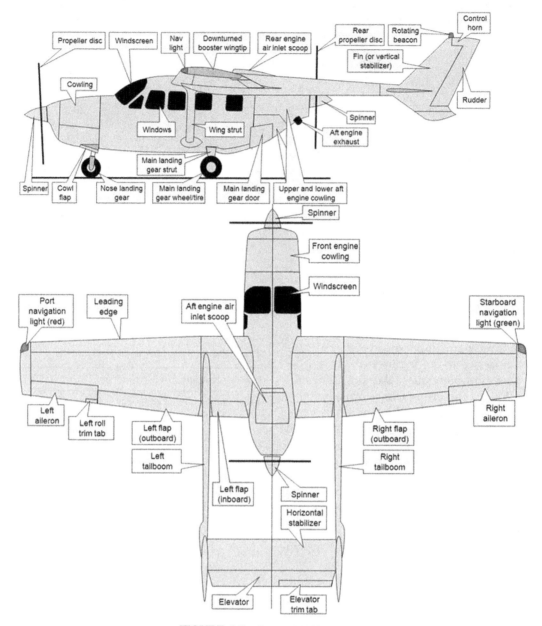

FIGURE 4-2 Cessna 337 Skymaster.

the external parts of typical aircraft. All aircraft feature many parts and components that are directly exposed to airflow and affect not only the performance and operation of the aircraft, but also the cost of manufacturing and maintenance. The location of most of these components (e.g. Pitot tubes, static sources, antennas, and others) is usually the consequence of hard work which involves the various design groups coming to an agreement on the most suitable location. For instance, a static port must be installed in an area where surface pressure remains relatively constant with angle-of-attack. This area, on the other hand, may be prime real-estate for an antenna, which requires an unobstructed signal path in order to work well, or the installation of a NACA duct for an inlet of cooling air for the avionics or other components. Having an understanding of where specific components may have to be placed on the aircraft will help the designer anticipate and avoid possible detail design conflicts.

The main components of an aircraft are: wings; fuselage; nacelle; empennage; horizontal and vertical tail; power plant; and landing gear, to name a few. These components can be broken down further into subcomponents. For instance, it is possible to break the wing into a main element, flap, aileron, spoiler, wingtip, and so on. Sometimes it is convenient, if not necessary, to

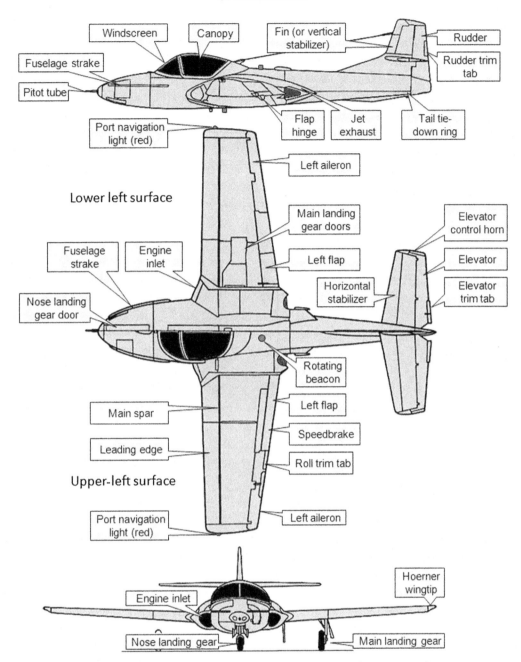

FIGURE 4-3 Cessna T-37 Dragonfly (also known as the Tweety Bird).

break these down further. For instance, the flap consists of a spar, ribs, skin, attachment brackets, access panels, and so on. Of the above primary components, three need some further definition:

A *fuselage* is a structural body not intended to generate lift (although it may) whose purpose is to contain engine, fuel, occupants, baggage, and mission-related equipment, although not always simultaneously. A fuselage is always mounted to lifting and stabilizing surfaces, if not directly, then through structural members.

An *empennage* refers to the horizontal and vertical tail of a conventional aircraft configuration. The word is of French origin, where it refers to the tail feathers of an arrow. Sometimes it is taken to mean the general region or assembly of the fuselage that contains both the horizontal and vertical tail.

A *nacelle* is a fuselage that does not carry an empennage. Nacelles usually carry an engine, but may or may not house occupants. Nacelles can be mounted to a lifting surface, such as a wing, or to a non-lifting geometry like a fuselage.

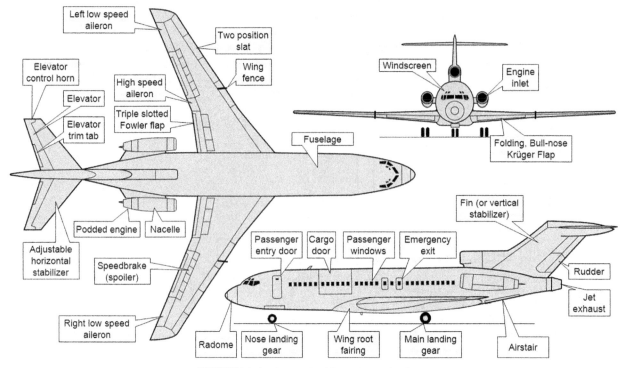

FIGURE 4-4 Boeing B-727 commercial jetliner.

4.2 THE FUNDAMENTALS OF THE CONFIGURATION LAYOUT

This section presents important concepts to keep in mind when selecting a particular configuration, as well as arguments for and against their selection.

Before starting the design of an airplane the novice designer should familiarize him/herself with Table 4-2, which shows typical dimensions for some selected classes of aircraft. Students of aircraft design who have yet to develop a keen sense for dimensions and weights of airplanes are encouraged to study the table in detail. This is not to say that a new design cannot be outside the shown limits, but rather that most aircraft ever built fall somewhere between the extremes cited. If the specifications of the new airplane fall outside these limits, inadvertently, the table may encourage the designer to take a second look at the numbers.

4.2.1 Vertical Wing Location

One of the most prominent features of any aircraft is the vertical placement of the wing. This section presents five common such configurations and presents arguments for and against each. The vertical wing location may end up being based on a number of factors such as:

Accessibility (freight, passengers, fuel)
Length of landing gear legs
Stability and control
Protection of occupants
Operation (amphibians, land only)
Aesthetics

Field-of-view
Manufacturing issues
Structural issues
Interference with passenger cabin
Aerodynamic drag
Manufacturer's (or designer's) preference

Ideally, the designer weighs each factor and decides the best position for the wing. It is important to avoid succumbing to biases such as "low-wing airplanes are always faster" or "high-wing airplanes have better stall characteristics." There is no law of nature that says that one or the other is superior. It all depends on other details, such as overall drag, airfoil selection, geometry of the airplane, and so on. It is the interaction of the components constituting the complete aircraft that matters. In fact, a case can be made that a high wing will be faster as it has less destructive effect on wing lift over the fuselage. There are fast and slow examples of either configuration. Examples include the low-wing P-51 Mustang (fast) versus Evans VP-1 Volksplane (slow) and the high-wing Mitsubishi MU-2 (fast) versus the Piper J-3 Cub (slow). The most common vertical wing placements are shown in Figure 4-6. The designer is urged to consider the consequences of the selection that are detailed below.

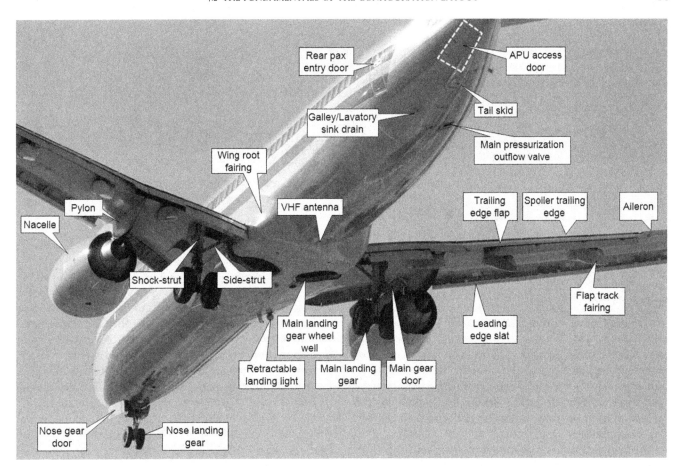

FIGURE 4-5 A Boeing 737-800 in landing configuration. *(Photo by Phil Rademacher)*

TABLE 4-2 Typical Properties of Aircraft Based on Class [1]

	LSA[a]	GA Aircraft[b]	Sailplanes[c]	Commuter Propliners[d]	Bizjets	Commercial Jetliners[e]
Wingspan, ft	17–35	30–45	35–101	45–100	44–70	90–290
Wing area, ft^2	75–160	150–400	120–250	300–860	200–1400	900–10,000
Wing aspect ratio	5–12	6–11	10–51	6–13	5–12.8	7–10
Wing taper ratio	0.5–1.0	0.3–1.0	0.4–0.5	0.35–1.0	0.3–0.5	0.20–0.5
HT aspect ratio	3–5	3–5	5–7.7	3–6	4.5–6.5	3–4
HT taper ratio	0.5–1.0	0.5–1.0	0.5–1.0	0.5–1.0	0.4–0.7	0.3–0.7
VT aspect ratio	0.7–3	1–2	1–3	1–3	1–3	1–3
VT taper ratio	0.3–1.0	0.5–1.0	0.5–1.0	0.5–0.9	0.4–0.9	0.5–1.0
Empty weight, lb$_f$	200–880	800–3000	100–1100	7000–26,000	7000–50,000	40,000–550,000
Gross weight, lb$_f$	400–1430	1500–12,500	280–1700	12,000–55,000	20,000–100,000	75,000–1,300,000

[a] *Includes typical homebuilt and other experimental category aircraft.*
[b] *Refers to 14 CFR Part 23 or EASA CS-23 certified aircraft.*
[c] *Includes motorgliders.*
[d] *Refers to the typical turboprop powered domestic aircraft and a handful of piston aircraft.*
[e] *Refers to 14 CFR Part 25 passenger jetliners used both domestically and internationally.*

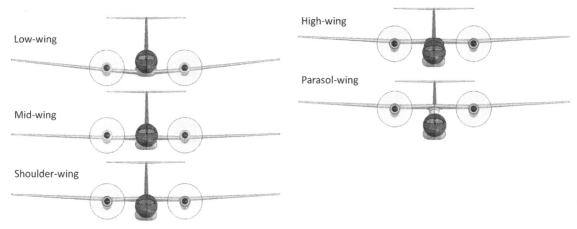

FIGURE 4-6 Vertical wing location nomenclature.

Field-of-view

The high-wing configuration offers a great field-of-view downward, whereas it may obstruct the pilot's view when banking (turning). This is an important issue for small airplanes (when the pilot sits below the wing) as it arguably increases the risk of mid-air collision for such aircraft. Therefore, the designer should consider the installation of transparencies on the roof to remedy this shortcoming.

The opposite holds true for a low-wing configuration. There is less downward visibility, but superior field-of-view in the direction of the turn. Of course, this argument does not hold for large commuter aircraft because the cockpit is placed far ahead of the wing. On small airplanes, the shoulder wing configuration improves visibility upward and downward, but requires the wing to be swept forward in order to ensure the center of gravity (CG) is placed properly on the mean geometric chord (MGC).

Impact on Airframe Design

The high-wing configuration, for a light aircraft, might rely on a gravity-fed fuel system, whereas a low-wing configuration may require a fuel pump (an added system). Fueling a high-wing aircraft is a drawback, because the fuel tanks are generally in the wing with the fuel caps on top of the wing. This requires a step ladder, which may not be available at all airfields. Larger airplanes solve this issue by featuring fueling points in the fuselage, where fuel is pumped under pressure. That option is impractical for GA aircraft that operate off airfields that do not have such equipment.

Entry into a high-wing configuration is often as simple as opening a door and stepping into the cabin. Small aircraft with low wings require a reinforced walkway on the wing and an external step that usually remains exposed to the airstream.[2] This usually means a walkway with sandpaper texture that is known to detrimentally affect flight characteristics of some aircraft.[3]

Many high-wing airplanes use wing-struts, which substantially reduce the shear and bending moments (see Section 4.2.4, *Wing structural configuration*), rendering the wing structure lighter than if built using cantilevered beam principles. Such struts are subjected to tension forces in normal flight, whereas struts on low-wing aircraft would be in compression, exposing them to a buckling failure.

The structure of the low wing can be used to attach landing gear, resulting in shorter and lighter landing gear. In small aircraft, the low configuration also allows the occupant seats to be attached to the main spar and the fuselage structure necessitated by the aft spar (or shear web). Both result in a more efficient structure.

The low, high, and parasol wing configurations open up the passenger volume as the wing structure does not pass through the cabin. This is very important in the design of passenger aircraft. In contrast, the shoulder and mid-wing configurations both require the wing spar to be accommodated inside the cabin. Frequently, in single- and two-seat variants of aerobatic aircraft, this is solved by placing the main spar well ahead of the occupant, typically in the area of the instrument panel. This allows the legs of the pilot to pass comfortably below the structure.

---

[2]The Cessna 310 is an example of an aircraft that features a retractable step.

[3]For instance see AD/DA42/4, *Wing Stub Safety Walkway*, issued in June 2008 by the Australian CASA, which applies to selected Diamond DA-42 aircraft.

The mid-wing configuration was widely used in aircraft design during the Second World War. Bombers from the era featured wings with hard points for bombs in the wing center structure and yet providing underbelly volume. The configuration is never used in large passenger aircraft as the wing structure would penetrate and occupy a part of the cabin, in addition to making traffic between forward and aft cabin impractical. In spite of this, the configuration has been used in a few aircraft; for instance, the 10-passenger IAI-1124 Westwind business jet and the 6-seat Piper Aerostar (formerly Ted Smith Aerostar). A possible remedy to the structural detriment could be special hoop frames that would allow the wing loads to be reacted "around" the enclosed volume. However, such a structure would be very inefficient and, thus, heavier. The Hamburger Flugzeugbau HFB-320 Hansa Jet, manufactured in the late 1960s and early 1970s, solved the problem with a forward-swept mid-wing whose primary structural element was behind the cabin. That configuration is not practical for the typical commuter aircraft. Although not a GA aircraft, the General Dynamics F-16 is an example of a mid-wing aircraft that solves the problem with stout machined hoop-frames around its single engine. These are justified by the engine placement.

Impact on Flight and Operational Characteristics

High-wing aircraft are less affected by ground effect and, thus, float less than low-wing aircraft when landing. This may be an important consideration in the design of bush-planes, where accuracy in making a landing spot on a short unprepared runway is imperative. Additionally, a low position of the wing may increase the risk of an accidental ground strike when operating off unprepared fields. More bush planes are of the high-wing configuration than the low-wing.

The configuration increases roll stability (or dihedral effect $-C_{l\beta}$), although this may present a disadvantage for heavy transport (e.g. cargo) aircraft, requiring anhedral to remedy. In small aircraft, the mid-wing configuration is frequently chosen for aerobatic airplanes to ensure neutral roll stability. This allows rapid roll maneuvers with minimum yaw coupling, something very desirable for precision aerobatic maneuvers. A low wing position has limited lateral stability, requiring the wing to feature dihedral angle to make up for it.

The high and low wing configurations often present some challenges in the geometry of the wing/fuselage fairing. Mid wings usually need smaller wing root fairings, although this may not hold for the aft part of the wing. It is a good configuration for aerobatic aircraft and a number of such designs are popular. Among those are the Slick series of aerobatic aircraft (Slick Evolution, Slick 360, etc.), Laser Z-300, Sukhoi Su-31, and Extra 300.

Snow bank collision is an interesting and surprising consideration to keep in mind for the design of aircraft primarily slated for operation off unimproved strips in cold climates (e.g. Alaska). Many isolated cold-climate communities rely on aviation to transport supplies in winter. Unimproved strips are then prepared by pushing snow on frozen lakes to form a runway, leaving snow banks several feet high along the perimeter of a very narrow runway. These can pose challenges to pilots of low-wing aircraft attempting to land in crosswinds.

Parasol Wings

Parasol wings are not common in modern aircraft design, although a few examples exist. The configuration consists of the wing separate from and placed high above the fuselage. From a certain point of view, the fuselage is hung from below the wing. The best-known aircraft to feature such a wing is undoubtedly the Consolidated PBY-5 Catalina (Figure 4-7) which was designed in the 1930s. Among others that sport the configuration are a series of aircraft built by Dornier, such as the Do J Wal (designed in the 1920s), Dornier Libelle (1920s), Do-18 (1930s), Do-24 (1930s), Dornier Seastar (1980s), and the Dornier S-Ray 007 (2000s), an amphibious sport aircraft. The configuration

FIGURE 4-7 The Consolidated PBY-5 Catalina is an example of an airplane featuring a parasol wing. *(Photo by Phil Rademacher)*

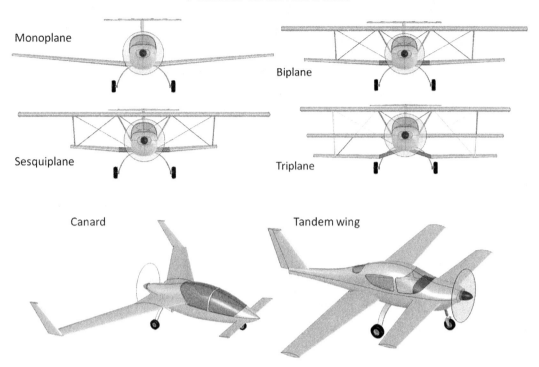

FIGURE 4-8 Typical wing configurations.

is beneficial for propeller-powered amphibious aircraft if the engine is mounted on the wing, helping to keep the propeller out of the spray of water.

This placement arguably results in an aerodynamically "cleaner" and thus a more efficient wing. The absence of a fuselage restores the lift potential of the wing, yielding a lower lift-induced drag. However, it also results in two sources of interference drag, one at the fuselage side and the other at the wing side.

For wing-mounted engines, the configuration may have a lower flutter speed due to the engine mass being mounted on a relatively flexible wing structure. This is compounded by how the fuselage is separated from the wing. An additional drawback is the high thrust line of the configuration, which makes power effects noticeable to the pilot.

Dihedral effect may be excessive and may require added vertical tail area to increase directional stability to counteract its effect on dynamic stability modes such as Dutch roll.

4.2.2 Wing Configuration

Airplanes are also categorized based on the number of wings they have. The monoplane is by far the most common configuration as it is without a doubt the easiest to make aerodynamically efficient. The primary advantage of the biplane or triplane configuration is the large wing area that can be packed in a small wingspan. This allows for very maneuverable airplanes with relatively low stalling speed without flaps. The drawback of the configuration is the aerodynamic inefficiency that stems from placing the low-pressure region of the lower wing close to the high pressure region of the upper wing. This reduces the production of lift, requiring higher AOA to generate the same lift coefficient and, consequently, higher lift-induced drag.

The difference between a sesquiplane and a biplane is the shorter span of the lower wing (Figure 4-8). This improves the efficiency of the outboard part of the upper wing by enabling higher pressure to be generated on its lower surface. It also results in a phenomenon that makes the configuration ideal for agricultural aircraft: the generation of four distinct wingtip vortices that help spread fertilizer or insecticide more effectively. This book primarily focuses on monoplanes, but details of biplane design are provided in Appendix C1.1.3, Conceptual Design of Small Biplanes.

The difference between a canard and a tandem plane is in the size of the forward wing. Generally, the elevator is installed in the forward lifting surface. Both lifting surfaces generate upward-pointing lift vectors in level flight and both forward surfaces are highly destabilizing, longitudinally.

4.2.3 Wing Dihedral

The dihedral angle is the angle the wing plane makes with the horizontal. It allows the aircraft designer to provide the airplane with roll stability and a way to affect the severity of dynamic modes such as Dutch roll. Its primary effect is on the stability derivative $C_{l\beta}$ (dihedral

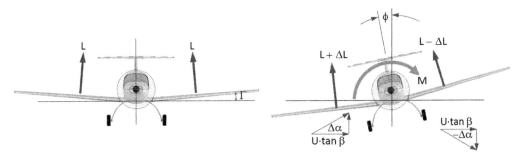

FIGURE 4-9 Dihedral effect explained.

effect). In addition to the dihedral angle, the magnitude of the dihedral effect depends on the vertical location of the wing and sweep angle. Ultimately, the designer must predict the dynamic stability characteristics of the airplane design and evaluate the appropriate dihedral angle.

Figure 4-9 shows an airplane at airspeed U, banking through an angle ϕ. The banking results in a side-slip, which effectively results in a yaw angle β. The yaw angle results in side flow component amounting to $U \cdot \tan \beta$, which when combined with the change in vertical flow due to the roll causes a net change in angle-of-attack, $\Delta \alpha$, on each wing. The subsequent change in lift (ΔL) on each wing is shown in the figure, and it generates a restoring rolling moment (one that tends to rotate the aircraft back to level flight), here denoted by the letter M.

Common dihedral configurations are shown in Figure 4-10. Of these the three on the left are most commonly used. The cranked dihedral is used extensively on the French Jodel and selected Robin aircraft, as well as on some sailplanes. However, it is also featured on the Argentinian FMA IA-58 Pucará twin turboprop ground attack aircraft and of course the McDonnell-Douglas F-4 Phantom.

The gull-wing configuration is fairly rare, being most famously used on the Vought F4U Corsair, where its purpose was to increase the propeller clearance for carrier operations. It was also used for various reasons on the Blohm & Voss BV-137, Caproni Ca-331b Raffica, Dewoitine HD-780, Fairey AS-1 Gannet, Heinkel He-112 B, and Junkers Ju-87 Stuka.

The inverted gull-wing configuration is often used for twin engine seaplanes, where it helps take wing-mounted engines and propellers away from the spray of water. It is featured on the Beriev Be-6, Be-12, Chyetverikov MDR-6, and Moskalev 16 amphibians and seaplanes. It is also used on the Göppingen Gö-3 Minimoa sailplane, and the PZL P-1, PZL P-11, Piaggio P-166, and Supermarine 224 landplanes. The Stinson SR-10 is an example of an aircraft that could fit into this class, featuring a wing whose upper surface has a distinct gull-wing break. However, the lower surface forms a straight line and the spar does not have a break, rendering it more of a transitional form.

4.2.4 Wing Structural Configuration

Nowadays, the wing's structural layout is based on either a cantilevered or a strut-braced methodology (see Figure 4-11). Both have their pros and cons, with strut-braced having a higher drag than the cantilevered configuration. However, the maximum shear and

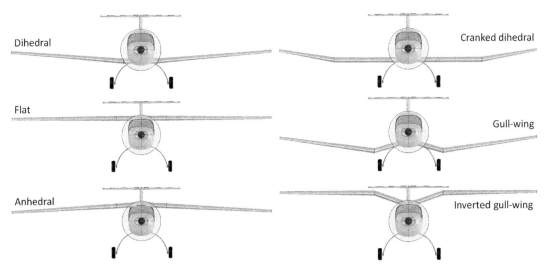

FIGURE 4-10 Wing dihedral nomenclature.

FIGURE 4-11 Wings are typically either cantilever or braced with struts.

bending loads of the strut-braced wing are much less than those of the cantilevered wing, resulting in a lighter wing structure.

As an example, consider Figure 4-12, which shows a strut-braced (top) and cantilevered wing (bottom) subject to an equal aerodynamic load, represented by the simplified trapezoidal lift distribution. The lift distribution of real wings is not trapezoidal, but the accuracy of its shape is not important to the point being made here.

The upper part of the figure shows where the maximum shear and bending moment occurs on the strut-braced configuration. Their corresponding magnitudes are given by V_{max} and M_{max}, respectively.

The lower image shows shear and moment diagrams for the cantilever configuration. It shows the maximum shear is 2.3× greater than that of the strut-braced wing and the moment is 4× greater. Conversely, although not shown, there is a substantial compression load that must be reacted between the wing-to-fuselage and strut-to-fuselage attachment points. It follows that the structural weight of the strut-braced wing will be much less and it should be given a serious consideration if aerodynamic efficiency is not a factor.

4.2.5 Cabin Configurations

Here, the discussion of cabin configuration will be limited to light aircraft only, as cabins for passenger aircraft are presented in more detail in Chapter 12, *The anatomy of the fuselage*. Typically there are two kinds of cabin styles: canopy and roofed (see Figure 4-13). An advantage of the roofed cabin is increased protection in the case of a turnover accident. This configuration requires an entry door to be added, preferably one on each side. These may present some fit and function issues in production, although similar arguments can be made against the canopy. The roof also limits the field-of-view.

A canopy offers exceptional field-of-view to the pilot, which is very desirable for many travelers, in addition to reducing the risk of a mid-air collision. However, turnover mishaps are of considerable concern for such aircraft. This is reflected in the regulation 14 CFR 23.561, *General*. The applicant must demonstrate compliance by reinforcing the window frame to which the windscreen is attached to prevent a collapse that would harm the occupants. Often called the A-pillar, this frame effectively becomes a rollover cage, increasing its girth and weight.

Excessively high cabin temperatures due to green-house effects are a drawback of the canopy. The configuration should allow for a canopy that can be left open during ground operations (while taxiing) for cabin cooling. This is particularly important if the airplane does not have air conditioning (such systems are not common in small aircraft). Reduction in green-house effect makes the roofed cabin configuration a viable candidate.

The acrylic canopy must be installed and operated with care (if flexible), as cracks may develop around fastener holes. The configurations with the canopy should feature an appropriate mechanism to prevent it from opening in flight due to aerodynamic forces. For instance, should the latching mechanism fail in flight, an aft-hinged or side-hinged canopy is at risk of opening due to aerodynamic forces. If the canopy departed the aircraft, it might damage the HT or VT, possibly rendering the vehicle uncontrollable. If the canopy opened up and stayed with the vehicle, a

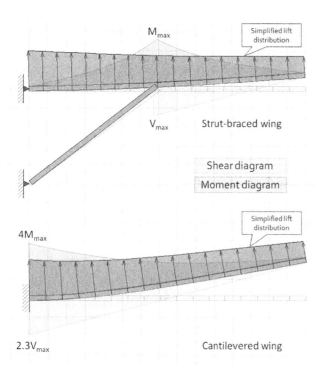

FIGURE 4-12 Shear and moment diagrams expose the structural implications of selecting a strut-braced versus cantilevered wing configuration.

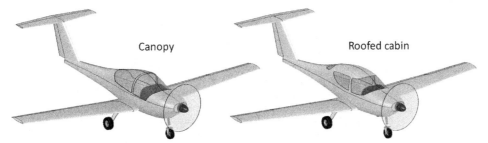

FIGURE 4-13 Typical cabin configurations for small aircraft.

substantial asymmetry in loads could result rendering the aircraft uncontrollable as well. Of course, either canopy-mounting technique makes it easier to board the airplane than a forward-hinged canopy.

4.2.6 Propeller Configuration

There are two fundamental ways to mount a propeller to an engine; as a *tractor* or as a *pusher* (see Figure 4-14). Either configuration is practical for piston engine, gas turbines, and electric motors.

The pros and cons of these configurations are discussed in detail in Section 14.1.2, *Propeller arrangement*.

The tractor configuration is a proven arrangement that is generally suitable for most applications. It provides undisturbed air for the propeller although the higher airspeed and lower quality air increase the drag of the body immersed in the propwash.

The pusher propeller is a good solution to some specialized mission requirements, for instance for single-engine reconnaissance or observation missions. This will remove the propeller from the field of view and allow a high-visibility cockpit to be designed. There are a number of issues the designer should be aware of. Propeller manufacturers are generally apprehensive about the arrangement as it introduces problems not always anticipated by the designer. Some of those are detailed in Section 14.1.2, *Propeller configurations*. However, the aspiring designer should not let this aversion influence the introduction of a particular configuration, as propeller manufacturers are happy to work on any such project. They only want the designer to recognize the shortcomings.

4.2.7 Engine Placement

As intuition would hold, the placement of any significant source of force on an aircraft is of great concern. Engine thrust is an example of such a source and, in magnitude, is second only to that of the wing lift. The moment generated by this force must be arrested by the stabilizing surfaces. If the engine thrust is placed above the CG of the airplane, the consequence will be a nose pitch-down moment that, on conventional aircraft, must be trimmed out using elevator trailing edge up (TEU) deflection (see Figure 4-15). The opposite holds for an engine whose thrustline is below the CG. The higher the thrust, the greater is the deflection required, although a better method is to enlarge the size of the elevator or increase the planform area of the horizontal tail, or a combination thereof.

In short, the larger the value of Δz is in Figure 4-15, the larger must be the elevator authority. The moment generated by the thrust must be taken into account during the design phase to prevent the dangerous potential of an undersized stabilizing surface. The possibility is serious enough to warrant a discussion in Section 23.3, *GA aircraft design checklist*.

Noticeable power effects are another consequence of engine placement. These can be quite complicated for propeller-powered aircraft, as is discussed in detail in Section 14.2, *Propeller effects*, and some of those are shared by the jet engine. While not necessarily dangerous, they can be particularly annoying to the pilot. If each power change calls for swift pilot reaction, the airplane will simply be less pleasant to fly. In spite of such nuisances, some aircraft require the thrustline to be high, such as seaplanes, as this protects the propeller

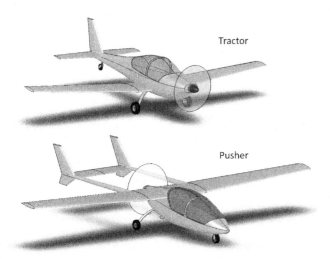

FIGURE 4-14 The two propeller configurations.

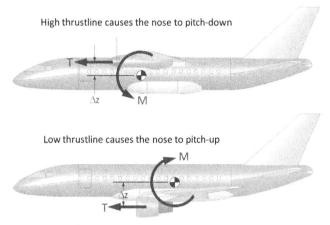

FIGURE 4-15 The effect of high or low thrustlines is a nose pitch-down or pitch-up tendency.

from water spray. For such airplanes, pitch changes with power settings are accepted because it saves the propeller. For the operators, it is just something the pilot has to get used to.

Another important consideration for the layout of a propeller is the effect of propwash. If it flows over a control surface like the horizontal and vertical tail, it will improve the control authority at high power settings, but the higher airspeed also increases the drag of the surface, albeit modestly.

Figure 4-16 shows a number of common engine placements. **Configuration A** features the jet engines in pods (or nacelles) mounted to the aft part of the fuselage. This configuration was first introduced in the 1960s in the French Sud-Est Caravelle passenger jetliner. The placement results in modest, if any, pitch effects and is intended to reduce engine noise in the cabin, although noise in the aft-most part of the cabin is increased, if anything.

Configuration B mounts the engines below the wing, using pylons. This configuration will result in substantial pitch effects, although some kind of stability augmentation system (SAS) can be employed to reduce the effect. The configuration is vulnerable to foreign object damage (FOD) but, in spite of that, is the most common engine placement found on passenger jetliners. The placement is beneficial from a structural standpoint as the weight of the engines introduces bending moment relief, which ultimately reduces airframe weight. Additionally, its forward position has a favorable effect on the flutter characteristics of the wings.

Configuration C features the engine above the wing and will generate nose pitch-down moment at high thrust settings. The configuration was first introduced in the 1970s on the German WFV-Fokker 614 jet, but later adopted on the Hondajet, where the intent is to avoid the ground clearance problem of under-wing nacelles. However, it can introduce peculiar aerodynamic and flutter issues [2,3].

Configuration D is a twin-engine turboprop aircraft with propellers mounted in nacelles on the wing. This is the most common method to install turboprops in such aircraft. However, it can lead to serious asymmetric thrust condition in a one engine inoperative (OEI) situation. Also, the rotating propellers can cause a so-called whirl-flutter or subject the airframe to fatigue through

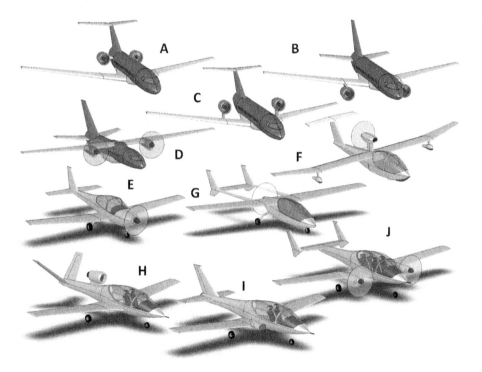

FIGURE 4-16 Common engine placements.

life-cycle oscillations (LCO) of the engine-wing combination.

Configuration E is a *tractor* propeller configuration, which has limited thrust effects due to engine placement, but more due to the physics of propeller thrust generation.

Configuration F is a seaplane, with the engine placed on top to protect the propeller from sea spray. It is a pusher configuration. It suffers from substantial pitch effects but, as stated earlier, this detriment is accepted due to the protection the prop enjoys.

Configuration G is a pusher configuration, which has the thrustline sitting somewhat high, although this is not as much of a problem as one might first think. The propeller will help keep flow attached on the aft part of the fuselage. It is subjected to some drawbacks of the pusher propeller configuration, although the particular configuration shown improves safety by making it hard to accidentally walk into a rotating propeller.

Configuration H is a single-engine jet that features a turbofan engine on a pylon on top of the fuselage. It would suffer from substantial pitch effects, although this can be partly remedied by deflecting the nozzle a few degrees up. The advantage of the engine placement is that it places the inlet in the airstream so it has a high pressure recovery, even at high angles-of-attack (AOA).

Configuration I features a buried engine, which results in minimal pitch effects with thrust, if any. The drawback is that the bifurcated inlet reduces pressure recovery at the front face of the compressor, reducing maximum available thrust. The bifurcated duct is also problematic if the airplane is operated in icy conditions as ice will accrete in the bend of the inlet.

Configuration J is a small four-seat twin-engine propeller-powered aircraft, suitable as a light VIP transport or reconnaissance aircraft. Its piston engines are mounted on the wings and must sit high enough to prevent damage to the propellers due to small objects that might be thrown from the operation of the nose landing gear on unimproved runways, or if it is subject to a flat tire on any of the landing gear. The nacelles are designed to accommodate the retractable landing gear. It features an H-tail to help generate restoring yawing moment in the case of an OEI situation.

4.2.8 Landing Gear Configurations

A large number of different landing gear configurations have been developed for use in aircraft. Six examples are shown in Figure 4-17. It should be stressed it does not show all the options, only those that are used on 99.99% of all GA aircraft. The most widely used configuration is the tricycle, followed by the taildragger. One of the advantages of a taildragger is less drag than the tricycle. An example of improvements attained by a small aircraft is that of the Cessna 150. It is claimed that converting a tricycle version of the aircraft to a taildragger gave it an increased cruising speed of nearly 8 knots. [4]. The benefit is always a function of the airplane and its overall drag, but a 4–10 knot increase in cruising speed is reasonable.

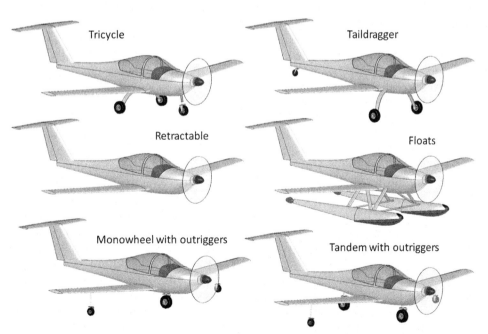

FIGURE 4-17 Selected landing gear configurations.

The monowheel with outriggers is a popular design for sailplanes and motor gliders like the British-designed Europa XS and the German Scheibe Tandem-Falke. The monowheel configuration reduces the drag of the landing gear. The same is true of the tandem wheel configuration, although it is rarely used in GA aircraft. The British Hawker Harrier is the best-known example of tandem wheel configurations.

Tricycle, taildraggers, mono- and tandem wheel configurations may all be retractable. Fixed landing gear will increase the drag of the airplane and, if this is the case, the designer should strongly consider wheel fairings for drag reduction. The floats increase drag substantially, but allow operation on land and water. They are still popular among many pilots who are more interested in access to obscure wilderness retreats than high airspeeds. Floats and amphibious airplanes are dealt with in Appendix C3, *Design of seaplanes*.

The tricycle landing gear makes the vehicle dynamically stable on the ground and reduces the risk of a ground loop. For this reason, it is better for inexperienced pilots and, thus, better suited for trainer aircraft. Similarly, the taildragger configuration is dynamically unstable and more prone to ground looping. It is better for operation off unimproved runways. The pros and cons of and the conceptual design of the landing gear are presented in far more detail in Chapter 13, *The anatomy of the landing gear*.

Taildragger aircraft have a number of advantages that make them more attractive for "bush-plane" operations. The primary advantage is the high AOA that can be generated at a low airspeed (in fact at zero airspeed). This way the airplane can be allowed to accelerate and then the pilot can quickly "drop" the tail, allowing the airplane to lift off in ground effect. Such techniques give the configuration markedly shorter runway requirements and, thus, make it better suited as a bush-plane. Interestingly, this is not always reflected in the data. For instance, the Cessna Model 180 (taildragger) and 182 (tricycle) are effectively identical excluding the landing gear. Jane's All the World's Aircraft 1970−71 [5] reports each having identical T-O and landing distances at the same weight, but the 180 is favored as the bush-plane option, providing evidence that pilot technique is imperative. The configuration is better suited for operation off unimproved landing strips because two wheels on the ground reduce the possibility of the landing gear hitting ground obstructions (large rocks or other obstructions) during the T-O run (two wheels versus three).

The taildragger configuration is generally thought to be harder to land and maneuver on the ground than a tricycle gear due to a high deck angle, which makes it harder to see over the nose of the airplane. This configuration is usually used on small aircraft, although it has been featured on large aircraft. The largest taildragger ever built is the eight-engine Soviet Tupolev ANT-20 Maxim Gorky, with a gross weight of 116,600 lb_f. The Curtiss C-46 Commando is another large taildragger, although its gross weight of 48,000 lb_f is dwarfed by the ANT-20.

The total structural weight of the float is the highest, but least for the monowheel. The structure required to react the main landing gear impact load will weigh less than the structure required to react the impact loads of both tricycle and taildragger. The configuration is also the least expensive to manufacture. It is a drawback that it is vulnerable to crosswinds and too much taxiing on the ground. The same holds for the tandem wheel.

4.2.9 Tail Configurations

A number of possible tail configurations are shown in Figure 4-18. **Configuration A** is a conventional tail, **B** is a cruciform tail, **C** is a T-tail, **D** is a V-tail, **E** is an H-tail, **G** a Y-tail, **H** an inverted Y-tail, and **Configuration I** is an inverted V-tail. The pros and cons of these tails are detailed in Chapter 11, *The anatomy of the tail*, and will not be further addressed here.

4.2.10 Configuration Selection Matrix

Many manufacturers of aircraft know up front what configuration is to be designed. Regardless of the internal debate that may take place, and to which we are not privy, the history of aviation shows there are certain themes that present themselves in these designs. All single-engine Cessna aircraft are high-wing (except the Ag Wagon and Ag Cat agricultural aircraft). All Piper and Beech are low-wing (except the Piper Cub and Tripacer), as are all Mooneys, which also feature their signature straight LE horizontal and vertical tails. However, this is not always the case and the resulting configuration is a consequence of internal debate. In this section, a method to help with the configuration selection is presented (see Table 4-3). Often, the selection is compounded by the fact that all candidate designs can be shown to meet the performance and operational requirements. An observation of modern-day regional jets reveals that a number of dissimilar configurations can clearly perform the design missions effectively. Aircraft as disparate as the Dornier 328 (high wing, engines on wing), Bombardier CRJ200 (low wing, aft podded engines), and Embraer 175 (low wing, engines on wing) are examples of this. Granted there is a difference in fuel efficiency and the economics of each, however, what are the reasons behind these configurations? Why would these three manufacturers come to such different conclusions about their particular configuration?

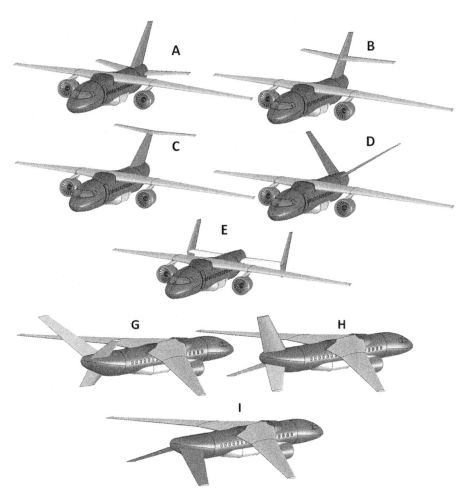

FIGURE 4-18 Eight tail configurations.

If many different candidate configurations are being considered, it is prudent to evaluate each based on a list of desirable and undesirable characteristics. The selection of a configuration is something that, in particular, baffles the student of aircraft design and the less experienced (some would say less-opinionated) designer. The weighted tabulation of Table 4-3, which considers the development of a two-seat trainer aircraft, can be very helpful when down-selecting a configuration. The trick is to phrase features such that a beneficial one receives the high score and an unfavorable one a low score and to allow the same score to be given more than once.

The number and selection of "questions" should be the result of an internal evaluation, where everyone who has a stake in the outcome has an opportunity to incorporate their concerns and goals. Note that in this case, configuration B beats the other configurations although all score well, indicating they are all plausible candidates for the particular mission. The greatest drawback of this approach is that, in addition to the nature of questions included, it is very susceptible to the weighting factors, and an honest debate about each should take place prior to the evaluation. The reader should keep in mind it is presented here as *one*, and not the *only* approach.

VARIABLES

Symbol	Description	Units (UK and SI)
α	Angle of attack	Deg or rad
β	Yaw angle	Deg or rad
$C_{l\beta}$	Dihedral effect	
ϕ	Bank angle	Deg or rad
Γ	Dihedral angle	Deg or rad
L	Lift	lb_f or N
M	Moment (context-dependent)	$ft \cdot lb_f$ or $N \cdot m$
M_{max}	Maximum bending moment	$ft \cdot lb_f$ or $N \cdot m$
T	Thrust	lb_f or N
U	Velocity along the body x-axis	ft/s or m/s
V_{max}	Maximum shear force	lb_f or N

TABLE 4-3 Example of the Down-selection of a Two-seat, Single-engine GA Aircraft Configuration by Configuration Selection Matrix

Feature	Weight	Configuration Score 1 − Good, 2 − Better, 3 − Best		
		E	C	A
CONFIGURATIONS TO BE EVALUATED →				
Does the design allow for the desired side-by-side seating arrangement?	1	3	3	3
Is 360° field-of-view (practically) possible?	1	3	3	1
Suitability for bush-plane operation	2	1	3	3
Expected cost of developing canopy and windscreen (less cost is best)	1	1	1	2
Cabin susceptibility to greenhouse effects and uncomfortable sun exposure (low is best)	0.5	1	1	3
Does design offer the desired tricycle landing gear configuration?	1	3	1	1
Can retractable landing gear be accommodated?	0.25	1	1	1
Propeller configuration (tractor is less costly)	1	3	3	3
Propeller configuration (enclosed prop is safest)	2	1	1	1
Low drag − is $LD_{max} > 14$ achievable?	0.75	1	2	1
Low drag − can NLF lifting surfaces be employed?	0.5	3	3	3
Low drag − are winglets necessary? (no is best)	0.25	3	3	3
Low drag − does configuration allow the inclusion of smooth fillets and clean wing?	0.25	3	3	2
Fuel system − is large fuel volume in wing possible?	0.5	2	2	2
Fuel system − is simple fuel system possible?	0.5	1	1	3
Can the design handle a large CG envelope?	1	3	3	3
Potentially the best spin recovery	3	3	3	2
Aesthetics − does design look unconventional?	0.5	1	2	3
Aesthetics − stylish fin	0.5	2	3	1
Aesthetics − stylish wingtips	0.25	1	1	1
Manufacturing − composite fuselage	1	3	3	3
Manufacturing − composite wing	1	3	3	3
Maintenance − composite wing	1	1	1	1
Maintenance − engine access	1	1	1	1
Operation − cabin entry (easy is best)	2	1	1	3
Operation − refueling (low tank fill point is best)	2	3	3	1
Operation − preflight oil check	1	3	3	3
SUM OF WEIGHT × SCORE:		55.75	59.5	56.25

References

[1] Taylor, John WR. Jane's All the World's Aircraft. Jane's Yearbooks, various years.

[2] Fujino M, Kawamura Y. Wave-drag characteristics of an over-the-wing nacelle business-jet configuration. Journal of Aircraft Nov–Dec 2003;40(6).

[3] Fujino M, Oyama H, Omotani H. Flutter Characteristics of an Over-the-Wing Engine Mount Business-Jet Configuration. AIAA-2003-1942 2003.

[4] Clarke B. The Cessna 150 & 152. TAB Books. 2nd ed. 1993. p. 180–181.

[5] Taylor JWR, editor. Jane's All the World's Aircraft 1970-71. Jane's Yearbooks, 1971.

CHAPTER 5

Aircraft Structural Layout

OUTLINE

5.1 Introduction	97
5.1.1 The Content of this Chapter	98
5.1.2 Notes on Aircraft Loads	98
5.2 Aircraft Fabrication and Materials	98
5.2.1 Various Fabrication Methods	100
Casting	100
Molding	100
Sheet Metal Forming	100
Extrusions	101
Forging	101
Machining	101
Welding	101
Joining	102
5.2.2 Aluminum Alloys	103
5.2.3 Steel Alloys	106
5.2.4 Titanium Alloys	107
5.2.5 Composite Materials	108
Types of Composite	108
Structural Analysis of Composite Materials	109
Pros and Cons of Composite Materials	110
Fibers	111
Resin	112
Sandwich Core Materials	113
Glass Transition Temperature	113
Gelcoat	113
Pre-cure	114
Aircraft Construction Methodologies	114
Fabrication Methods	114
5.3 Airframe Structural Layout	116
5.3.1 Important Structural Concepts	116
Monocoque and Semi-monocoque Structure	117
Wood Construction	117
Steel Truss Covered with Fabric	119
Aluminum Construction	119
Composite Sandwich Construction	119
5.3.2 Fundamental Layout of the Wing Structure	120
5.3.3 Fundamental Layout of the Horizontal and Vertical Tail Structures	126
Fabrication and Installation of Control Surfaces	126
Unconventional Tails: T-tail, V-tail, and H-tail	127
5.3.4 Fundamental Layout of the Fuselage Structure	128
Fuselage Structural Assembly — Conventional Aluminum Construction	128
Fuselage Structural Assembly — Composite Construction	128
Special Considerations: Pressurization	129
Variables	131
References	131

5.1 INTRODUCTION

It should not come as a surprise that the layout of the airframe is one of the most important elements of the entire aircraft development process. The structural layout dictates whether empty weight targets will be met and, thus, whether other design requirements can be achieved. Interestingly, a poorly laid out structure may cause problems that, in a worst-case scenario, can potentially lead to the termination of an otherwise viable aircraft development program. For instance, poorly conceived load paths in a pressurized fuselage may result in detrimental structural deformation that can make it impossible to maintain an advertised pressure differential. How such a flaw would affect the development program would ultimately depend on how far along it had progressed when discovered. The required fix could be a major redesign of the fuselage structure and, depending on the program status, its financial stability could be compromised. On the other hand, even the ideal airframe layout will not guarantee the development program becomes a success. An aircraft can be structurally optimized while simultaneously suffering from aerodynamic, power, or systems inadequacies that, ultimately, may bring about its demise. The important point is that while the structural layout cannot *make*, it can certainly *break* the viability of the program.

In this section, we will look at some general layouts of aircraft structures, albeit without too much structural analysis, as the focus of this book is primarily conceptual and preliminary design and not detail design. The purpose of the section is to help the designer visualize the implications of the various configuration choices on the resulting structure.

Note that the material properties presented are in the UK system. Use the following factors to convert to the SI-system.

To convert psi to GPa (giga-pascal), multiply by 145037.73773
To convert psi to MPa (mega-pascal), multiply by 145.03773773
To convert lb_f/in^3 to specific density, multiply by 27.7334934
To convert lb_f/in^3 to g/cm^3, multiply by 27.7334934
1 ksi equals 1000 psi

5.1.1 The Content of this Chapter

- **Section 5.2** presents characteristics and properties of typical materials used for the construction of the modern GA aircraft.
- **Section 5.3** presents a description of the fabrication and installation of various structural components of an aircraft.

5.1.2 Notes on Aircraft Loads

Aircraft are designed to react several types of loads as discussed below:

(1) *Aerodynamic loads* (or *airloads* for short) refer to forces and moments caused by the dynamic pressure to which the aircraft is subjected. Airloads include forces, such as wing lift and drag, and moments, like wing torsion and bending. Their magnitude depends on the weight of the aircraft, the load factor, its geometry, and, again, dynamic pressure. The total magnitude is defined based on requirements set forth by the aviation authorities — for instance, 14 CFR Part 23 and 25. However, the local values depend on the geometry. Thus, consider two aircraft, A and B, of equal weight and wing area that differ only by the wing aspect ratio (AR) and taper ratio (TR). Assume aircraft A has the higher AR and lower TR. For reasons that will be detailed in Chapter 9, *The anatomy of the wing*, it will generate substantially higher bending moment than aircraft B.

(2) *Inertia loads* refer to forces and moments caused by subjecting aircraft components to acceleration. An example is the battery, which does not experience any aerodynamic loads. Its support structure must be capable of reacting the forces that result from the applied load factors. Other components, such as a piston engine, are simultaneously subjected to both aerodynamic and inertia loads.

(3) *Operational loads* refer to loads other than aerodynamics and inertia that are simply caused by the fact that the airplane is being used. Examples of such loads include door hinge and locking loads, floor loading loads, wing step-on loads and other similar loads. Such loads are often tricky to define, but are usually small compared to, say, the wing loads. Operational loads usually lead to wear and tear.

In addition to the primary role of the airframe, the location and shape of all the major load paths has a major influence on weight. From a certain point of view, it is the responsibility of the structural engineer to design the structure so it will only carry the loads it is likely to encounter in operation. This is imperative to the success of the design. An airplane whose strength is greater than the operational loads is in fact *overdesigned*; it is stronger and heavier than it needs to be. As a consequence, during each flight, it will carry around a lot of material whose weight would better be a part of the useful load. Additionally, the wings, stabilizing surfaces, engines, and landing gear have major effects on the weight and location of its center of gravity. This can bring about loading problems that may have to be solved using heavy ballast, again, whose weight would better be a part of the useful load.

5.2 AIRCRAFT FABRICATION AND MATERIALS

The selection of structural material for a new aircraft can be an involved process that requires a number of very important considerations. If an airplane is mostly to be constructed from a single source of material, its selection will clearly have a profound impact on a couple of important areas: manufacturing and maintenance. Established companies tend to stick with the material and fabrication processes they know best from past projects; they are unlikely to change a manufacturing process that may have taken decades and a substantial amount of investment to develop. For this reason, manufacturers of aluminum aircraft are unlikely to invest in the development of a composite aircraft, and vice versa. This does not preclude the introduction of a new material to an airplane, although this will happen on a smallscale at first. Then, if the material is promising, the manufacturer might increase its use in a process of evolution. This approach has been very evident among manufacturers of jet commercial aircraft, such as Boeing and Airbus. The introduction of new materials should not be done unless its characteristics have been carefully evaluated. The following listing provides some areas the designer should understand before new material is selected:

Availability
Compatibility with other materials
Corrosion and embrittlement
Cost of certification
Electrical characteristics
Environmental stability
Erosion and abrasion
Fabrication characteristics
Fatigue
Fracture toughness and crack growth
Material costs
Producibility
Static strength/weight
Thermal characteristics
Wear characteristics.

At the time of writing, the most common material used for aircraft remains aluminum. However, the use of composites has gained great popularity and is even seeing extensive use in the fabrication of new commercial aircraft, such as the Boeing 787. Already, several all-composite aircraft, such as the Cirrus SR20 and SR22; Cessna Corvalis (formerly Columbia); and Diamond DA40 Katana and DA42 Twinstar, are certified under 14 CFR Part 23. Aluminum has a number of very important properties that lend themselves well to the construction of vehicles that must be light and yet stiff. Composites are a somewhat recent introduction to the aircraft industry, although their history and use dates back to the early 1950s.

Arguably the best source for material properties data for aerospace vehicles is a document structural engineers know as the *MIL-HDBK-5*. It contains design information on the strength properties of metallic materials and elements for aerospace vehicle structures. The data in the document are published based on a collaboration effort of the US armed forces, the Federal Aviation Administration (FAA), and the industry, and until recently, were maintained as a joint effort of the FAA and the Department of Defense [1]. MIL-HDBK-5 has now been superseded by the *Metallic Materials Properties Development and Standardization* (or the MMPDS). The MMPDS is the FAA's effort to maintain the MIL-HDBK-5 handbook, which is recognized worldwide as the most reliable source available for statistically based allowables in the design of aircraft, as well as for repairs, alterations and modifications.

When using MIL-HDBK-5, one must be aware of the limitations of the statistical methods used to present material properties. The reader must be mindful that the production of aircraft requires uniformity and repeatability. The aircraft produced today must be equally strong as the airplane produced last month, within some statistical limits. This can only be accomplished by uniformity and repeatability in the manufacturing process of material. Each batch of material transported to the aircraft manufacturer is either tested by the manufacturer himself or has a certificate of testing from a third-party test lab that demonstrates its strength is no lower than some specific value. The handbook refers to this as *data basis*[1] and cites four types of room-temperature mechanical properties. These are listed below based on the least to the highest statistical confidence.

- *Typical Basis* — a typical average value of the material property (e.g. yield stress in tension) and has no statistical assurance associated with it.
- *S-Basis* — means that the value of the material property is based on industry specifications or federal or military standards. As an example, industry specifications can be those of the SAE or ASTM.
- *B-Basis* — means that at least 90% of the test coupons are expected to equal or exceed a statistically calculated mechanical property value with a statistical confidence of 95%. For instance, consider the ultimate tensile strength of 2024-T3 sheet, which for a specific sheet thickness might be 64,000 psi. If we test the ultimate strength of 10 coupons of this material, at least nine must equal or exceed 64,000 psi, with 95% confidence.
- *A-Basis* — means that at least 99% of the test coupons are expected to equal or exceed a statistically calculated mechanical property value with a statistical confidence of 95%.

Typically, structural analysis uses A-Basis allowables for structural members whose failure is considered catastrophic. B-Basis allowables are used for redundant structural members whose failure would result in the redistribution of loads without compromising safety of flight. The reader is directed toward MIL-HDBK-5 for more details.

---

[1] See Section 9.1.6, MIL-HDBK-5J.

5.2.1 Various Fabrication Methods

There are a number of manufacturing techniques the aircraft designer must keep in mind during the design stage. Cognizance of manufacturing difficulties that are not always obvious when a particular feature or geometry is suggested, is particularly important. The following are common manufacturing methodologies. Note that introducing these in detail is beyond the scope of this introductory text and the aspiring engineer should acquire as much knowledge of general assembly and construction methods as possible.

Casting

Casting is one of the oldest manufacturing methods known to man, dating back to at least 4000 BCE [2, Table 1, p. 6]. The process entails the following steps:

(1) A mold is created from an already existing part, for instance, by making an imprint of the part in granular material such as sand.
(2) The material for the part is heated until it becomes liquefied at which time it is poured into the mold. An example of this is molten aluminum.
(3) The part is then allowed to cool ("freeze") for a specific time, during which it solidifies.
(4) Once sufficiently cool, the part is removed from the mold, which is typically destroyed in the process. This gives rise to the saying 'one part, one mold,' making the casting process very labor-intensive.

The advantage is that the original model of the part can be shaped from material less strong than the material used in the casting. For instance, it is possible to make the original part from wood, whereas the copies are made from some metal. There are a number of different casting methods in existence that depend on the material used or the desired shape. For this reason, casting takes considerable expertise to do well. Casting of aircraft metals (aluminum or steel) will leave the material fully annealed and thus lacking strength. For this reason casting should never be used for critical aircraft structure.

Molding

The difference between casting and molding is that molding involves the construction of a heat-tolerant mold that is used to make multiple part copies, whereas casting involves one mold per part. Molding has become a very sophisticated manufacturing process that requires considerable expertise. An example of such processes is injection molding, in which material in a liquid form is injected under high pressure into the mold — an operation intended to eliminate air bubbles from the material, which are a source of stress concentrations in the material that can render it far less durable than otherwise.

Sheet Metal Forming

The concept *forming* refers to the process of forcing the material into a particular shape. Industry has developed a large number of methods to force metals into particular shapes. Presenting all of them is beyond the scope of this book, however, when it comes to aircraft, sheet metal forming and forging are the best known. Forging is presented separately as it is considered by many to be in a class on its own.

Sheet metal is most often formed to introduce flanges to stiffen the material so it can be used for stringers and spars or for joining with other sheet metal parts (see paragraph about joining below). It is sometimes formed to provide the skin curvatures of lifting surfaces, although this is only required when the skin thickness becomes too large for the panel to flex freely. Thicknesses of that magnitude are common in the inboard wing skin panels of commercial jetliners or military aircraft. The thickness of aluminum sheets used for GA aircraft is usually small enough to allow it to flex with ease. The material is usually cut to shape using hand- or hydraulically actuated *shears*, depending on sheet thickness and cut length. Then, the forming takes place using a special tool called a *sheet metal brake*. The bending operation requires some planning, as there are limits to how tight the bend radius can be. As a consequence, allowances have to be made for extra material for the bend itself. Another phenomenon, *springback*, must be considered when working with sheet metals. It requires the operator to bend the sheet to a predetermined angle which is slightly greater than the desired angle. Once removed from the metal brake, the sheet will spring back to the desired angle.

There are two kinds of surface flexing the engineer must be aware of: *simple* and *compound* (see Figure 5-1). All metals will readily undergo a simple surface flex (or deformation), which in effect is a simple plate

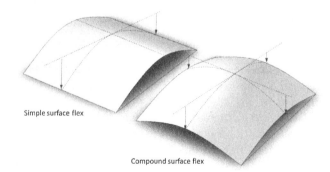

FIGURE 5-1 The difference between a simple and compound surface flex.

bending. Compound flex, on the other hand, is accompanied by internal twist (shearing) of the material molecules, in addition to bending deformation about two axes. Metals resist this type of deformation so it is practically impossible to form the compound flex unless its internal molecular structure is stretched using specialized forming methods such as hydraulic pressing. This fact is imperative when selecting material for aircraft components. The compound surface is where composite materials shine — but the manufacturing of composites is usually more expensive than for aluminum. If it is imperative that the surface features a compound flex, as is the case for very low-drag aircraft such as sailplanes, consider composites. If it suffices to use a simple surface flex, such as frustum fuselages and simply tapered wings, consider sheet metal.

Extrusions

An *extrusion* is the process of forcing an ingot of near-molten metal through a die with a specific geometric pattern. This is a common process for aluminum alloys intended for use in airframes, although it is also used to produce structural steel for buildings. The process converts the half-molten ingot into a long and straight column of structural material featuring a constant cross-sectional shape. When made from aluminum alloys it is ideal for use as longerons or stringers in airframes. It is common to find extrusions whose cross-section resembles letters such as 'H,' 'L,' 'T,' 'U' (also called 'C') and 'Z,' and far more complicated shapes are available. Of these, the L-extrusion, usually called an *angle extrusion*, is of great use as a stringer or a spar cap in aluminum spars. The C-extrusion, usually called a *C-channel*, is of great use for various brackets and hinges designed to react high structural loads. The use of extrusions in aircraft is extensive and includes not only stringers, but seat-tracks, brackets, wing attachment fittings, and countless other applications. Extrusions have higher material strengths than plates as the formation of it compresses the grain structure.

Forging

The best-known and probably the oldest forming operation is *forging*, dating back perhaps as far as 8000 BCE [2, p. 384]. Forging is when a metal is locally subjected to large compressive forces in the form of "hammering" using various dies and tools. This hammering can be done to either cold or hot parts, but it usually increases the strength, toughness, and durability of the material through the process of work hardening. For this reason, it is not unusual to see aircraft components that must react large forces, such as landing gear struts, made from forged metals [2, p. 384]. Work hardening is a consequence of the deformation of the grain structure of the material. Since most forged parts are subjected to secondary machining operations to improve appearance, the work hardening tends to complicate the manufacturing process by making it harder to finish. Forging metals such as iron and steel at elevated temperatures will reduce work hardening and make them easier to post-process. With respect to the economics of forging, the designer must choose carefully, as the cost of forging a non-critical part may be much higher than, say, molding.

Machining

Machining is the fabrication of a part through the removal of excess material. There are a number of ways machining takes place, the most common being sawing, cutting, turning, and milling. Like the other methods above, machining takes a lot of expertise and experience to do well, but an understanding of what can or cannot be machined can open the door of success and close the door of failure. Machining aluminum and low-carbon steels (e.g. AISI 1025) is relatively easy, but this becomes gradually more difficult when the carbon content increases. Hardened steels are very difficult to machine and require sophisticated tools to accomplish this. Machining is much easier to accomplish when the material is in its annealed state. Afterwards, it is necessary to heat-treat the part to acquire adequate strength, even though many parts will undergo further machining post heat-treating.

Welding

Welding is the joining of parts made from identical metals by heating them to a point of surface melting and then bringing them together to allow their molecules to *coalesce*. A filler material is often used to create a stronger joint. Welding is one of the most common ways to join parts and forms a very strong and durable bond between the parts. This contrasts joining of parts using *soldering* or *brazing*, both of which do not melt the working parts. A large number of methods can be used to perform the welding: most notable are a gas flame, an electric arc, a laser, and an electron beam. Low-carbon-grade steels are easily welded or brazed by all techniques and the filler material should be comparable in strength to the base metals. Steels with higher carbon levels will often require stress-relieving after the welding has been completed and sometimes even subsequent heat treating. Welding is commonly used to join parts making up engine mounts, landing gear, and fuselages, demonstrating the method can take a beating if properly done. It is a drawback that the process often leads to warping that may change the intended geometry. The welding of critical structural aircraft parts should always be done by a certified welder. Critical structural parts should not be made from welded aluminum due to a reduction in fatigue life.

Joining

Joining is typically used to assemble a large part from many smaller ones. It includes operations such as riveting, using threaded fasteners, welding, and so on. For instance, consider the fabrication of an engine mount, which requires a number of typically tubular parts to be joined through the process of welding (see above). Also, the fabrication of an aluminum wing is accomplished through the joining of ribs, spars, stringers, and skin through the process of riveting sheet metal. The aspiring aircraft designer should acquire a deep understanding of the two most common riveting techniques used in the industry: *bucking* and *blind riveting*.

Of the two, bucking is the primary method used and is employed when two (or more) aluminum sheets are to be joined (or to join a sheet to an extrusion) and there is ample access to both sides of the parts to be joined, for the technician. The standard procedure is shown in Figure 5-2 in four steps. First the sheets are aligned using carefully placed clamps (not shown). Then, holes are drilled at specific intervals depending on the shear stress to be transferred from one sheet to the next through the rivets. Since the drilling operation typically forms sharp edges (or burrs) on the opposite side, these must be removed prior to the insertion of the rivets. Otherwise, the joining will not develop full strength.

The technician usually and temporarily inserts a special tool through selected holes called a Cleco®. This prevents the sheets from slipping during further drilling or bucking operations. The third step involves inserting the proper rivets into the hole, and the fourth is the actual bucking operation. It often requires two technicians to accomplish, in particular if large sheets are being joined. The technician on the head side of the rivet places an air hammer against the rivet, while the other places a heavy metal block called a *bucking-bar* against the opposite side of the rivet. When both are ready, the operator of the air hammer presses a trigger on the air hammer to generate a short burst of hammering to the rivet. The inertia of the bucking bar will then deform the rivet such that a solid and strong attachment is formed, as the hammering will cold work the rivet. Bucking takes practice and careless handling of the tools may damage the sheets around the rivet.

Blind-riveting is only used when lack of access to the back side of the sheets prevents the use of an air hammer and bucking bar. It is also used for non-critical structural assembly. Driving a blind rivet is a very simple two-step operation (see Figure 5-3) in which a special tool, a *rivet gun*, is used to pull out the stem (or spindle) until it snaps at its weak spot, where its diameter has been deliberately reduced. This allows the stem to be pulled up just enough to compress the opposite end of the rivet and lock it in place. Blind rivets are also available as structural rivets and, as stated earlier, are sometimes the only choice. Cherrymax® is the best known brand for such rivets. Blind riveting, while far easier to perform than bucking, still requires care in installation in order to avoid tilting of the stem, which might misalign the rivet. It is also considerably more expensive that conventional bucking rivet installation.

Finally, there are a number of different types of rivet heads, but as presenting all of them is not appropriate in this text, only the two most common types will be cited. These are: *universal* and *counter-sunk* rivet heads (Figure 5-4). The universal head (as shown in Figure 5-2 and Figure 5-3) is typically used for low-performance aircraft (in terms of airspeed), whereas the counter-sunk rivet head will be flush with respect to the surface of the sheet. This reduces the drag of the airplane and is thus implemented on high-speed aircraft. However, this brings additional complication to the table that increases the cost of the riveting: counter-sunk rivets require an indentation to be made for the rivet head, either by a special drilling operation or forming of a

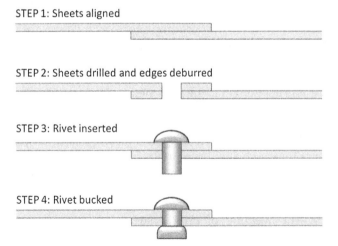

FIGURE 5-2 Standard procedure to join two aluminum sheets by bucking a rivet.

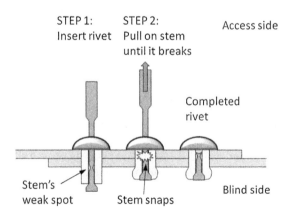

FIGURE 5-3 Standard procedure to join two aluminum sheets using a blind rivet.

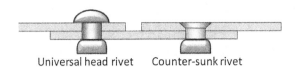

FIGURE 5-4 The two most common rivet head types: *universal* and *counter-sunk*.

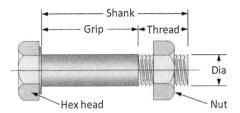

FIGURE 5-5 The nomenclature for a basic threaded fastener.

dimple using a special tool. Either one increases production costs.

Next to rivets, threaded fasteners (or bolts) are the most commonly used fasteners in aircraft (see Figure 5-5). Such fasteners have superior tensile (and shear) strengths compared to rivets (which are only intended for use in shear), but are far more expensive to use. Like all aircraft hardware, threaded fasteners must be traceable to an approved manufacturing process. Most bolts used for aircraft applications are general-purpose (e.g. AN-3 through AN-20 bolts), internal-wrenching (e.g. MS-20004 through MS-20024), and close tolerance (e.g. the hex-headed AN-173 through AN-186 or NAS-80 through NAS-86).

The shank of these bolts features a smooth section, called the grip, and a threaded section onto which the nut is mounted. The length of the grip must be equal to or slightly exceed the thickness of the material it is intended to hold. The nut must be tightened or torqued to the right amount to preload the fastener. This ensures the joined parts do not slip during service, ensures a more uniform transfer of loads, and increases the fatigue life of the fastener. Nuts are usually self-locking or non-self-locking. Castellated nuts are a type of the latter that are locked in place using special safety-pins called cotter-pins. Such nuts are required for all structurally critical parts, such as engine mounts, landing gear, and wing attachments. The installation of threaded fasteners should always use flat washers (e.g. AN960) so that torquing the nut will not damage the surface of the joining materials.

5.2.2 Aluminum Alloys

Aluminum is a lightweight and corrosion-resistant structural material that can be strengthened further by chemical and mechanical means. Chemically, the strength is increased by adding specific elements to it (see Table 5-1). It is this process that turns the aluminum into an alloy. Mechanically, the strength is increased via cold working and heat treatment. The primary advantages of aluminum alloys are low density, high strength-to-weight ratio, good corrosion resistance (Alclad), ease of fabrication, diversity of form, electrical conductivity, isotropy, abundance, and generally repeatable properties.

One of the most important properties of aluminum is that it is mostly isotropic. Isotropic materials offer strength and stiffness regardless of the orientation of the force being applied. Aluminum sheets used for aircraft construction are mostly isotropic as there is a slight difference between the "rolled" and "transverse" directions. Aluminum sheets are produced by first casting molten aluminum into a thick sheet, which is then hot rolled (at 500 °F to 650 °F) until a specific thickness is achieved. Then the hot-rolled sheet is annealed and cold rolled until a desired "retail-ready" thickness is produced. This process gives the sheet bi-directional properties, although the structure featuring it is analyzed is if it were isotropic. Repairing aluminum is much easier than most other materials used for aircraft construction. This is another very important property as it makes field repairs practical. Aluminum has been the primary material for aircraft construction since before World War II, although the use of composites has begun to threaten its stature.

At this time, aluminum accounts for about 75–80% of commercial and military aircraft. According to data from the General Aviation Manufacturers Association (GAMA) from 2005,[2] some 65–70% of GA aircraft were made from aluminum. Of the number of different aluminum alloys available, generally three types are used more than others: 2024, 6061, and 7075. Table 5-1 lists the major alloying element for the different types of aluminum. *Wrought alloys* are rolled from an ingot or extruded into specific shapes. The word "wrought" is the archaic past tense of the verb "to work." "Wrought alloy" literally means "worked alloy." *Cast alloys* are melted and poured in a liquid form into molds where they are allowed to cool. These two methods lead to two very different classes of alloys, in which wrought alloys are stronger as a consequence of special post-processes such as cold working, heat treatment and precipitation hardening.

Ultimately, the properties of aluminum alloys are determined by the alloy content and method of fabrication. Besides strength, the designer must be aware of some specific characteristics of aluminum, such as grain

---

[2]Of 3580 aircraft delivered in 2005, some 2535 were made from conventional materials, of which aluminum was by far the most common material. Some 1045 were composite aircraft. Source: http://www.gama.aero/files/documents/2005ShipmentReport.pdf.

TABLE 5-1 Basic Designation for Wrought and Cast Aluminum Alloys

Wrought Alloys		Cast Alloys	
Alloy Group	Major Alloying Elements	Alloy Group	Major Alloying Elements
1XXX	99.00% minimum aluminum	1XX.0	99.00% minimum aluminum
2XXX	Copper	2XX.0	Copper
3XXX	Manganese	3XX.0	Silicon with added copper and/or magnesium
4XXX	Silicon	4XX.0	Silicon
5XXX	Magnesium	5XX.0	Magnesium
6XXX	Magnesium and silicon	6XX.0	Unused series
7XXX	Zinc	7XX.0	Zinc
8XXX	Other elements	8XX.0	Tin
9XXX	Unused series	9XX.0	Other elements

Reproduced from Table 3.1 of MIL-HDBK-5J [2].

TABLE 5-2 Basic Temper Designation System for Aluminum Alloys

Temper	Temper Description
F	Fabricated. Indicates that no special control over thermal conditions or strain-hardening is employed.
O	Annealed. Used with wrought products that are annealed to obtain the lowest strength temper, and to cast products which are annealed to improve ductility and dimensional stability. The O may be followed by a digit other than zero.
H	Strain-hardened (wrought products only). Applies to products which have their strength increased by strain-hardening, with or without supplementary thermal treatments to produce some reduction in strength. The H is always followed by two or more digits.
W	Solution heat-treated. An unstable temper applicable only to alloys which spontaneously age at room temperature after solution heat treatment. This designation is specific only when the period of natural aging is indicated: for example, W ½ hr.
T	Thermally treated to produce stable tempers other than F, O, or H. Applies to products which are thermally treated, with or without supplementary strain-hardening, to produce stable tempers. The T is always followed by one or more digits.

Reproduced from Table 3.1.2 of MIL-HDBK-5J [2].

direction, dependence of strength on plate thickness, corrosion properties, and fatigue. These are beyond the scope of this discussion, but the designer should refer to MIL-HDBK-5 or the MMPDS.

Wrought and cast aluminum and aluminum alloys are identified by a special 4-digit numerical designation. First consider the wrought alloys shown in the left part of Table 5-1. The first digit '2' indicates the alloy group. An example is the widely used 2024-T3 alloy. It indicates that 2024 contains copper as the major alloying element. The second digit '0' indicates the kind of modifications made to the original alloy or impurity limits. This value is usually '0' for structural alloys used for GA aircraft (e.g. 2024, 6061, 7075). Then consider the cast alloys in the right part of Table 5-1. The second and third digits identify the aluminum alloy, while the digit right of the decimal point indicates the product: XXX.0 means casting; XXX.1 and XXX.2 mean the metal is in ingot form.

The designation of both wrought and cast aluminum alloys uses special suffixes to identify their temper properties and is based on the sequences of basic treatments used to produce the various tempers. Thus, 2024-T3 means the aluminum is solution heat-treated, cold worked, and naturally aged to a substantially stable condition. The basic temper designation system is listed in Table 5-2. The designation of the numerical codes, e.g. '3' in '-T3,' is beyond the scope of this introduction, but an interested reader is encouraged to review MIL-HDBK-5 for more details.

Aluminum alloys have at least three important flaws the aircraft designer must be aware of. First is the absence of an *endurance limit*, the second is *stress corrosion*, and the third is *galvanic corrosion*.

The **endurance limit** (also called *fatigue limit*) is a property of many metals, for instance steel, which allows them to resist cyclic stress loading. This means that if the maximum amplitude of the cyclic stress during cyclic loading is below a certain value the material can react the loading forever. If the stress levels are higher than that limit, the material will eventually succumb to fatigue and fail. Some metals have very definite endurance limits, for instance steels. Aluminum, on the other hand, does not have a definite endurance limit [3, p. 81]. This means that for even very low stresses, if the number of cycles is large enough it will fail (see Figure 5-6 for an example life-cycle plot for 2024-T3 aluminum from MIL-HDBK-5J). Some engineers analyze aluminum structures assuming an endurance limit of some 10,000–12,000 psi, but such structures should still be subject to periodic inspection of crack growth.

Consider a structural member made from a 2024 aluminum alloy, whose limit tensile stress is 47,000 psi. Further assume it reacts a cyclic load ranging from −30,000 to 30,000 psi. Clearly the stress is well below the yield limit and, therefore, at first glance, intuition would hold that since the load is lower than the limit tensile stress, the cyclic loading could be applied indefinitely. However, common sense is sometimes a poor measure of reality. It turns out that the structure can

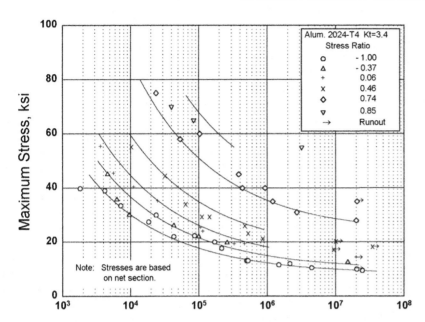

FIGURE 5-6 Figure 3.2.3.1.8(d) in MIL-HDBK-5J displays one of aluminum's primary flaws — no clear endurance limit. *(Figure from MIL-HDBK-5J)*

only be exposed to this load a finite number of times, perhaps some 100,000 times. If such a part belonged to an engine that rotated at a rate of 2500 RPM, it should be expected to fail in only (100,000 cycles/2500 cycles per minute) or some 40 minutes. The implication of this law is that structures made from such material need far more material than required to react the maximum loads — the structure must be heavier than a static stress analysis would indicate. The lifetime limitation of a critical aluminum structure requires such cyclic loads to be taken into account during detail design. For this reason, all aluminum aircraft structures have limited operational life, no matter how low the stress level, further requiring regular inspection of the structure.

The greatest challenge in evaluating the fatigue life of a structure is to define the loads that act on the aircraft. The problem is compounded by the fact that the load varies rapidly during each flight. On a calm day, the airplane will experience less load excursion due to gusts than on a bumpy day. Additionally, the frequency and magnitude of the loads will depend on how the airplane is used. For instance, a trainer will experience hard landings far more frequently than a professionally flown transport aircraft. In order to account for this variety, airplane fatigue loads are based on so-called load spectra. A load spectrum diagram is used to indicate the probability of a particular airplane experiencing given load levels during its lifetime. For instance, an ordinary normal category airplane (see Table 1-2), designed to operate for a 12,000 hour life, might be expected to reach 3.8 g once or twice in its lifetime. It may experience 1.5 g several thousands of times.

The FAA published the report AFS-120-73-2, *Fatigue Evaluation of Wing and Associated Structure on Small Airplanes* [3], in 1973. The purpose of the report is to provide methods for estimating the *safe life* of an aircraft structure. It is often the first step toward determining the life expectancy of the airplane. It provides scatter factors and load spectra for various types of aircraft and operation (e.g. taxi loads, landing impact loads, gust and maneuver load, etc.). These factors and load spectra are used to develop the probability that structural components, such as wing or tail, could reach the end of their design life (in terms of ground-air-ground cycles) without developing detectable fatigue cracks.

Stress corrosion is a phenomenon that occurs to ductile alloys that are exposed to high tensile stresses in a corrosive environment. Examples of corrosive environments include water vapor, aqueous solutions, organic liquids, and liquid metals. The corrosion manifests itself as cracking along grain boundaries in the material. Research shows that aluminum alloys that contain substantial amounts of soluble alloying elements, primarily copper, magnesium, silicon, and zinc, are particularly susceptible to stress-corrosion cracking. Examples of such alloys include 7079-T6, 7075-T6, and 2024-T3, which comprise more than 90% of the in-service failures of all high-strength aluminum alloys [4].

Galvanic corrosion occurs when two electrochemically dissimilar metals are in close proximity to one another in a structure, for instance, when aluminum is joined to steel. Besides the electrochemical dissimilarity, an electrically conductive path between the two metals must exist to allow metal ions to move from the metal that acts as the anode to the one that acts as a cathode. While this is primarily an issue during detail design, the aircraft designer must be aware of potential implications stemming from insisting on dissimilar metals

TABLE 5-3 Typical Applications of Aluminum Alloys in GA Aircraft

Aluminum Alloy	Typical Application
2024-T3, 2024-T4	Used for high-strength tension application such as wing, fuselage, and tail structure. Has good fracture toughness[a], slow crack growth, and good fatigue life compared to other aluminum alloys [5, p. 102].
6061-T6	Used for resilient secondary structures such as access panels, piston engine baffles, cockpit instrument panels, etc.
7075-T6, T651	Used for high-stress applications similar to those of the 2024. It is stronger than 2024, but has lower fracture toughness and fatigue resistance.

[a]*MIL-HDBK-5J defines fracture toughness as "The fracture toughness of a material is literally a measure of its resistance to fracture. As with other mechanical properties, fracture toughness is dependent upon alloy type, processing variables, product form, geometry, temperature, loading rate, and other environmental factors" [2].*

being joined in the airframe. Of course, joining dissimilar metals is frequently done in the aviation industry, but this should be avoided when possible. The galvanic corrosion problem can be remedied by applying special plating or finishing to the metals as a protection.

Table 5-3 lists several aluminum alloys commonly used in GA aircraft structures in the aviation industry. The designer should regard these as alloys for primary and secondary structures.

Table 5-4 shows selected properties for a few aluminum alloys that are frequently used in GA aircraft. Table 5-5 shows common sheet thicknesses of commercially available aluminum alloys. Note that to save space, the sheet thicknesses are stacked in two columns for each unit.

TABLE 5-5 Common Sheet Metal Thicknesses for Aluminum Alloys

Inches		mm	
0.012	0.072	0.30	1.83
0.016	0.081	0.41	2.06
0.020	0.091	0.51	2.31
0.025	0.102	0.64	2.59
0.032	0.125	0.81	3.18
0.040	0.156	1.02	3.96
0.051	0.188	1.30	4.78
0.064	0.250	1.63	6.35

5.2.3 Steel Alloys

The superior strength of steel often renders it the only material fit for use in highly stressed regions of the airplane. Among frequent use are the landing gear, engine mounts, high-strength fasteners, and many other mechanical parts for which durability and strength are essential.

By definition, steel is iron (Fe) that has been modified through the introduction of alloying elements, such as nickel (Ni), vanadium (V), cobalt (Co), chromium (Cr), magnesium (Mg), molybdenum (Mo), carbon (C), and other elements. The introduction of these elements has a profound and very desirable effect on the properties of the iron and practically converts it into a new material that is in all very strong, stiff, and durable. These properties can be further enhanced through the introduction of processes such as annealing, quenching, cold working, and heat treating. The branch of materials science that deals with such processes is called metallurgy and

TABLE 5-4 Selected Properties of Common Aluminum Alloys (A-Basis and Longitudinal Direction)

Description	Density	Tensile Modulus	Shear Modulus	Poisson Ratio	Yield Tensile	Ultimate Tensile	Ultimate Shear	Ultimate Bearing $e/D = 1.5$
Symbol	ρ	E	G	μ	F_{ty}	F_{tu}	F_{su}	F_{bru}
Units	lb_f/in^3	ksi	ksi		ksi	ksi	ksi	ksi
2024-T3 Sheet 0.01–0.125"	0.100	10.5×10^3	4.0×10^3	0.33	47	64	39	104
2024-T4 Sheet 0.01–0.249"	0.100	10.5×10^3	4.0×10^3	0.33	40	62	37	93
—	0.098	9.9×10^3	3.8×10^3	0.33	16	30	20	48
6061-T6 Sheet 0.01–0.125"	0.098	9.9×10^3	3.8×10^3	0.33	36	42	36	67
7075-T6 Sheet 0.040–0.125"	0.101	10.3×10^3	3.9×10^3	0.33	70	78	47	121

Reproduced from MIL-HDBK-5J.

since it is beyond the scope of this text, only an elementary introduction will be given (see, for example, Ref. [6] for more detail).

In general, metallurgical processes allow specific properties, such as hardness, ductility, toughness, and so on, to be modified. For instance, annealing is a process in which the metal is heated to a specific temperature, where it is kept for a given time, after which it is cooled at a specific rate. This process relieves stresses that may be in the material and "softens" it (makes it more ductile and less hard) so it is easier to cut, stamp, or grind. Quenching is the rapid cooling of steel and produces grain structure that is particularly hard. It is used for a class of steels called low-carbon steels and the austenitic stainless steels. This improves the durability of the steel and makes it ideal for use as highly loaded precision parts. Cold working is used to increase the yield strength of a metal. This can be done by methods such as cold rolling, cold extrusion, and cold drawing, to name a few. Heat treating is a process in which material is heated and cooled at specific rates in order to modify the arrangement of their molecular structure. It is the primary way steels other than the low-carbon and austenitic stainless steels are strengthened.

The properties of a selection of commonly used steels are presented in Table 5-6. Of these, AISI 1025 is a general-purpose steel used for various shop projects, such as to make jigs, fixtures, mock-ups, and similar. Generally, the steel is not used for operational aircraft, although it is possible to get it in an aircraft quality. Steels such as AISI 4130 and 4340 are also known as "chromoly," as they contain traces of both chromium and molybdenum. Because of the reliable heat-treating practices and processing techniques for these steels they are very common in aircraft construction, where they are used for engine mounts, landing gear, truss fuselages, and other high-stress components. They are readily available as sheet, plate, and tubing stock.

5.2.4 Titanium Alloys

Titanium is a great choice for applications that require high strength and light weight in a demanding environment. It is a relatively lightweight metal that has good strength-to-weight ratio, low coefficient of thermal expansion, good toughness, and good oxidation resistance. It also has a higher melting point than steel (1660 °C versus 1650 °, respectively). The metal was discovered in 1791 by a British chemist, William Gregor (1761–1817), and then independently again in 1793 by the German chemist Martin Heinrich Klaproth (1743–1817). The metal is one of the most abundant elements in nature, although it is expensive to extract and isolate. Today, it is found in a large array of applications, ranging from engine components and airframes to various biomedical implants, as well as golf clubs [7].

The properties of a selection of commonly used titanium alloys are presented in Table 5-7. Titanium is almost always alloyed with aluminum for use in aircraft structures. Among several common titanium alloys are Ti-6Al-4V and Ti-4Al-4Mo-2Sn-0.5Si, of which the former is thought to be the most widely used [5, p. 109]. In addition to titanium (Ti), it contains 6% aluminum (Al), 4% vanadium (V), and a trace of

TABLE 5-6 Selected Properties of Common Steels

Description	Density	Tensile Modulus	Shear Modulus	Poisson Ratio	Yield Tensile	Ultimate Tensile	Ultimate Shear	Ultimate Bearing $e/D = 1.5$
Symbol	ρ	E	G	μ	F_{ty}	F_{tu}	F_{su}	F_{bru}
Units	lb_f/in^3	ksi	ksi		ksi	ksi	ksi	ksi
AISI 1025 sheet, strip, and plate	0.284	29.0×10^3	11.0×10^3	0.32	36	55	35	90*
AISI 4130 ($t \leq 0.188''$) sheet Normalized, stress-relieved	0.283	29.0×10^3	11.0×10^3	0.32	75	95	57	200*
AISI 4130 ($t > 0.188''$) sheet Normalized, stress-relieved	0.283	29.0×10^3	11.0×10^3	0.32	70	90	54	190*
AISI 4130 ($t \leq 0.188''$) tubing Quenched and tempered	0.283	29.0×10^3	11.0×10^3	0.32	100	125	75	146 175*
AISI 4340 Bar, forging, tubing	0.283	29.0×10^3	11.0×10^3	0.32	217	260	156	347 440*
300M	0.283	29.0×10^3	11.0×10^3	0.32	220	270	162	414 506*

* For $e/D = 2.0$
Reproduced from MIL-HDBK-5J.

TABLE 5-7 Selected Properties of Titanium

Description	Density	Tensile Modulus	Shear Modulus	Poisson Ratio	Yield Tensile	Ultimate Tensile	Ultimate Shear	Ultimate Bearing $e/D = 1.5$
Symbol	ρ	E	G	μ	F_{ty}	F_{tu}	F_{su}	F_{bru}
Units	lb_f/in^3	ksi	ksi		ksi	ksi	ksi	ksi
Pure Ti (sheet, plate) CP-1 (AMS 4901)	0.165	15.5×10^3	6.5×10^3	—	70	80	42	120
Ti-6Al-4V, $t \leq 0.1875$, B-Basis	0.160	16.0×10^3	6.2×10^3	0.31	131	139	90	221
Ti-6Al-4V, $0.1875 < t \leq 2.000$, B-Basis	0.160	16.0×10^3	6.2×10^3	0.31	125	135	84	214

Reproduced from MIL-HDBK-5J.

iron (Fe) and oxygen (O). The first large-scale use of this material was in the production of the famous Lockheed SR-71 Blackbird reconnaissance aircraft. The development of the aircraft, as well as that of the now infamous North American XB-70, solved many of the production problems accompanying its use and made it a suitable alternative to aluminum alloys that offers greater strength, stiffness, competitive weight, and high heat resistance. Of course, all this comes at a higher price. At the time of writing, the price of pure aluminum was in the $2.5 per kg range, but titanium remained just shy of $9 per kg [8]. This price renders its competitiveness limited in the GA industry and makes it a material resorted to for special requirements.

5.2.5 Composite Materials

In the aircraft industry, the term *composite* applies to structures that consist of more than one constituent material so the combination yields properties that are superior to those of the constituent materials. Composites are a large and disparate class of materials, ranging from steel-reinforced concrete used for buildings to stiffened plywood-balsa-plywood sandwich panels used in airplanes. Nowadays, when it comes to aircraft, composites almost exclusively refer to various fiber-reinforced plastics that are used as both primary and secondary structures. It is essential that the aircraft designer is familiar with the numerous terms that are used in industry. This article defines and explains most of the common terminology used by engineers and technicians.

Types of Composite

There are three common forms of composite used for industrial applications:

1. *Fibrous composites*, which consist of fibers embedded in a matrix (resin). FRPs are examples of this.
2. *Laminated composites*, which consist of layers of various materials. Composite sandwich panels are the best examples of a laminated composite. Such composites are simply referred to as *laminates* and the constituent layers are called *plies*.
3. *Particulate composites*, which are composed of particles in a matrix. Steel-reinforced concrete is an example of this. At the time of writing, particulate composites are not used to construct airplanes and will thus be omitted from further discussion.

In its most basic form, composites consist of layers of *fibers* in the form of a cloth that are impregnated with some type of *plastic matrix* (or resin) and then *cured* to form a rigid structure. An example of this is fiberglass cloth embedded in epoxy resin. This is how the so-called *fiberglass-reinforced plastics* (FRP) and *carbon-reinforced plastics* (CRP) are prepared. Sometimes a third constituent material, called a *core*, is added to fabricate the so-called *composite sandwich*. The purpose of the core is to separate the plies by a given thickness and that way increase the stiffness of the structure. The resulting panels are light, stiff, and strong and are ideal for use as skin for wing, HT, VT, or fuselage structures. Panels so stiffened allow multiple ribs and frames to be eliminated from the structure, simplifying the airframe.

To better understand the potential the core has in the stiffness of composites, consider Figure 5-7, which shows three $10''$ long cantilevered composite beams. The top one is a simple 4-ply laminate consisting of typical aircraft-grade fiberglass laid up using a $[+45°/-45°]_S$ layup (S stands for symmetrical). The center and bottom ones feature the same fiberglass layup, with the addition of a $0.375''$ and $0.75''$ core, respectively. The resulting thicknesses and normalized densities (the density of the bottom beam is $1.69 \times$ that of the top one) can be seen in the figure. Then, some load is applied to the tip (right end) of the laminates such that the top one deflects $1''$. Applying the

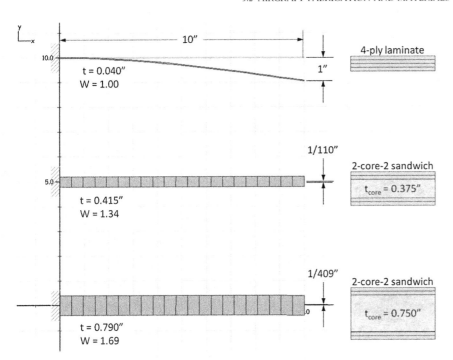

FIGURE 5-7 The effect of deflection of cantilevered beams under identical load is used here to compare the stiffness of a laminate and sandwich composite materials.

same load to the center and bottom beams would cause them to deflect 1/110″ and 1/409″ respectively. This means that the bending stiffness of the center beam is 110× greater and the bottom one is 409× stiffer than the top laminate. The huge increase in stiffness only costs a very modest increase in weight.

Structural Analysis of Composite Materials

There are two approaches used to perform structural analysis of composite materials; micro- and macromechanics. *Micromechanics* examines the interaction of the constituent materials (i.e. of the fibers and matrix) on a microscopic level. One of the outputs is the predicted "average" properties (such as strength and stiffness) of a composite laminate in terms of the properties and behavior of the constituent materials. Another one is the prediction of the distributions of stresses and strains in the laminate on a ply-to-ply basis.

An example of the capability of micromechanics is shown in Figure 5-8. A 6-core-6 sandwich with an unsymmetrical layup $[+45°/+45°/0°/+45°/0°/+45°]_S$ with bi-directional cloth is subjected to pure bending. The three left columns show strains in the composite as a whole, whereas the three right columns show strains in each of the plies making up the composite. The darker shaded region on the top and bottom of each column is the fiberglass plies and the lighter and thicker (taller) center region represents the thickness of the core (0.375″). Considering the composite strains it can be seen that the moment creates strains identical to those predicted by classic solid mechanics theory.

The moment (applied about the *x*-axis) creates strain about both *x*- and *y*-axes through Poisson's ratio. However, when considering the ply strains, it can be seen that the largest strains are picked up by the four 0° plies and the core. The core has a very low modulus of elasticity (Young's modulus) so it can stretch quite a bit without the formation of large stresses. The plies, on the other hand, have a very high modulus of elasticity, so the four 0° plies will generate substantially larger stresses than the +45° ones. As a consequence, if the applied moment becomes large enough, they are the first plies to fail. The application of micromechanics further allows the structural analyst to evaluate whether, if this happens, the remaining six plies will be capable of reacting the moment, or will fail subsequently. That sort of analysis is called *residual strength analysis* and is a standard procedure in the development of composite aircraft.

Macromechanics is the study of composite materials assuming they can be approximated as if they were homogeneous and the effects of the constituent materials are detected only as averaged apparent properties of the composite. This way, composite structural members are treated almost as if they were isotropic (except with different properties along each material axis), yielding a convenient analysis workaround for use in finite element analysis software.

The structural analysis of composites is performed using the *classic laminate theory*, which allows the prediction of stresses and strains in composite laminates. Among others, Tsai [9] and Jones [10] provide a good

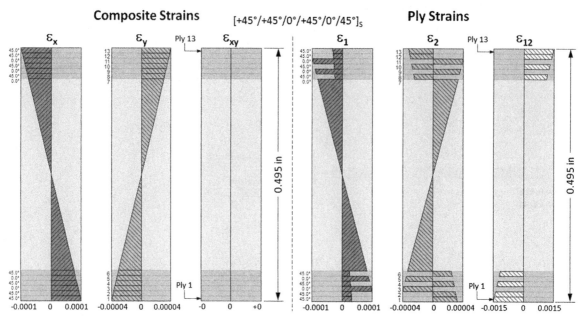

FIGURE 5-8 The effect of a pure bending moment on the strains (and therefore stresses) in a 6-core-6 laminate. The thicknesses of the core and plies is proportionally accurate.

treatise of the theory. It is based on the following assumptions:

- The material is orthotropic.
- Ply properties are linearly elastic.
- There is no coupling between the normal and shear strains, ε and γ, or the normal and shear stresses, σ and τ. In the case of a unidirectional composite, where the stress/strain coordinate axes are referred to as the principal material directions, the assumption is justified on the basis of material symmetries.

General directions of the stress and strain axes are denoted as shown in Figure 5-9. Then, material properties in the principal material directions 1, 2, and 3 are as follows:

E_1, E_2, and E_3 = Young's (elastic) moduli in the principal material directions
G_{23}, G_{31}, and G_{21} = shear moduli
v_{ij} = Poisson's ratio for transverse strain in the j-direction, when stressed in the i-direction.

The examples presented in Figure 5-7 and Figure 5-8 are prepared using the theory.

Pros and Cons of Composite Materials

FRPs offer many benefits over traditional materials, among which are high strength, light weight, flexibility in design, ease in the fabrication of compound surfaces, part consolidation, high dielectric strength, dimensional stability, and corrosion resistance. At this time, composites are being used in a seemingly endless number of applications, ranging from recreational boats, where they have practically replaced other traditional building methods, to aircraft. In the aircraft industry, composites are exceptional because of a favorable strength-to-weight ratio and the ease with which compound surfaces can be fabricated. Such surfaces are essential for drag reduction in aircraft. Due to good electrical insulating properties composites are also ideal for use in appliances, tools, and other machinery. Furthermore, they are corrosion-resistant and offer extended service life over metals. The author has had

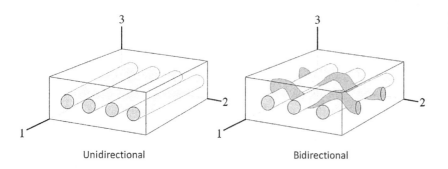

FIGURE 5-9 Difference between a uni- and bi-directional fiberglass cloth.

an experimental composite structure exposed to the elements since 1989 to evaluate this aspect of composites. When last inspected, degradation was impossible to detect visually, although admittedly no coupons have been pulled off this structure to validate whether the strength properties have changed. Overall, in addition to high tensile strength, glass fibers offer excellent thermal properties, as well as great impact- and chemical-resistance characteristics.

FRPs also come with disadvantages. To begin with, the resin is highly toxic; if not handled with care, it can easily result in serious dermatitis. It is also subject to storage limitations, strength variability, and impact sensitivity, all of which may cause serious strength degradation. Impact may also cause delamination, which is a separation of plies that results in strength, stiffness, and buckling issues. It is a serious flaw that composites tend to fail with limited warning. Metals, in contrast, fail only after an inelastic elongation. The strength of composite structures is vulnerable to fabrication flaws such as wrinkling, bridging, and dry fibers which will compromise its strength. In the professional manufacturing environment, the structure is carefully inspected against these, which adds cost to the production. Composites are notoriously poor in bearing and require careful attention to cleanliness during the construction process. Additionally, they often require specific surface finish requirements. For instance, FRPs and CRPs require light colors, preferably white, on surfaces exposed to sunlight to minimize heat absorption. Heat can be very detrimental to the strength of the resin being used. As a consequence, their operational temperature limits are well below that of aluminum, which is not that great to begin with. Additionally, the fact that FRPs are good electrical insulators makes them very vulnerable to catastrophic failure if struck by lightning. This is particularly critical to airplanes and calls for specific methods to carry electrical current by the introduction of metal conductors that have to be co-cured with the composite. To add insult to injury, these conductors are typically a "one-shot deal." They have to be replaced upon landing, unless of course a second flash of lightning strikes first.

Fibers

While it is this combination of matrix, fiber and manufacturing process that gives the composite its superior performance, it is helpful to consider these elements separately.

It is primarily the glass fibers that are embedded in the resin that account for the strength advantage FRPs have over unreinforced plastics. Fibers typically come in three forms: uni-directional, bi-directional, and as fiber mats. The first two are shown in Figure 5-9. Fiber mats are chopped strands of fibers that are randomly assembled into a cloth. They are not to be used for primary structures in aircraft as their strength and stiffness properties are unacceptably poor. On the other hand they are acceptable as secondary structures provided the ply thickness is low enough. Fiber mats are commonly used for boats, swimming pools, and jacuzzis.

The fibers play an imperative role in a composite structure reacting loads. Since the fibers are much stiffer than the matrix, the load inevitably is reacted by the fibers. The resin matrix, in contrast, serves to distribute the load among the fibers, besides retaining the intended shape of the structure. Several types of fibers are available commercially, of which the most common are introduced below:

Aramid fibers: a class of very strong, lightweight, and heat-resistant multifilament fibers used for a myriad of applications ranging from bulletproof vests and helmets to parachute tethers. Introduced in 1961 by the DuPont Company. They are widely used in the aerospace industry, for instance, under the name Nomex.

Boat glass: a name commonly used to identify fiberglass used for boat construction. It is also called fiberglass mat or, simply, glass mat. Boat glass consists of fiberglass chopped into short strands that are then pressed together to form a mat. The mat offers far more uniform properties than uni- or bi-directional fiberglass, only much poorer. The glass mat requires approximately 1.5 to 2 times its own weight in resin to be fully saturated.

Boron fibers: boron is a class of sophisticated fibers that are high-strength and lightweight. They are widely used in various advanced aerospace structures, for instance in aircraft such as the F-14, F-15, B-1 Lancer, and even the Space Shuttle. They are also found in bicycle frames, golf shafts, and fishing rods.

Carbon fibers: another advanced high-strength, high-stiffness, and lightweight fiber used in a variety of applications, ranging from baseball bats and bicycle frames to automotive and aerospace vehicles. They are used in micro air vehicles (MAVs) as well as the fuselage of the new Boeing 787 Dreamliner. Carbon fibers are also known under the name graphite. The primary drawback of laminates made from carbon fibers is their vulnerability to damage, which is compounded by the difficulty in detecting damage visually.

C-glass: specially developed to provide good corrosion resistance to hydrochloric and sulfuric acid. It gets its name for this property, which is short for corrosion-resistant fiber.

E-glass: the most popular type of fiberglass and typically the baseline when comparing composites.

In particular, E-glass offers good strength properties at a low cost; and it accounts for more than 90% of all glass fiber reinforcements. Named for its good electrical resistance, E-glass is particularly well-suited to applications where radio-signal transparency is desired, as in aircraft radomes and antennae. E-glass is also used extensively in computer circuit boards to provide stiffness and electrical resistance. Along with more than 50% silica oxide, this fiber also contains oxides of aluminum, boron and calcium, as well as other compounds.

Graphite fibers: see carbon fibers.

Kevlar®: the registered trademark of a version of aramid fibers developed by DuPont in 1965. The resulting fibers are extremely strong and resilient and are probably best known for their use in body armor and military helmets. It sees wide use in the civilian aviation industry as well as, for instance, as rotor-burst protection in jet engines, and even as the risers in the Cirrus Airframe Parachute System (CAPS) in the Cirrus SR20 and SR22 aircraft.

R-glass (AKAS-glass or T-glass): a type of fiberglass that offers greater strength (~30%) and better temperature tolerance than E-glass. It is primarily used for aerospace applications. Also called high-strength glass fiber. When greater strength and lower weight are desired, S-glass is a possible candidate instead of other advanced fibers, such as carbon. High-strength glass is generally known as S-type glass in the USA; it is often called R-glass in Europe and T-glass in Japan. Originally developed for military applications in the 1960s, a lower-cost version, S-2 glass, was later developed for commercial applications. High-strength glass has appreciably higher silica oxide, aluminum oxide, and magnesium oxide content than E-glass. Typically, S-2 glass is approximately 40% to 70% stronger than E-glass.

S-2 glass: can be used as a substitute for E-glass. In comparison, it has higher tensile and compressive strength, is stiffer, and exhibits improved impact resistance and toughness. In the aviation industry, S-2 glass is used for helicopter blades, aircraft flooring and interiors, but it can also be found in applications well beyond aviation. Like C-glass, it has good corrosion resistance to hydrochloric and sulfuric acid.

Resin

The purpose of the resin is to bind the fibers together into a single structural unit and, in the process, distribute strains among them while protecting them from the elements. Generally, there are two kinds of resin: *thermosets* and *thermoplastics*. The difference depends on the chemistry of the polymers, both of which contain highly complex molecular chains. In the case of thermosets, as the resin cures, molecular chains *crosslink* to form a rigid structure that cannot be changed through the further application of heat; the final product is *irreversible*. Thermoplastics, on the other hand, can be processed at higher temperatures; they can be reheated and reshaped more than once; the final product is *reversible*.

THERMOSETS

This is the resin used for aircraft structural applications. They are relatively inexpensive, simple to use, and offer good mechanical and electrical properties, as well as resistance to the elements. They are best known as a plastic that once cured cannot be converted to it original state (contrasting thermoplastics) The most common resin thermosets are listed below. It is a drawback that they ususally cure during an exothermic chemical process. They have a stable shelf life of several months, but when mixed with the proper catalyst ("hardener"), cure within minutes.

Epoxies: the most common resin used for aerospace applications. The nickname "epoxy" comes from its chemical name "ployepoxide." Epoxies are more expensive than the polyesters but offer greater strength and stiffness, as well as less shrinkage. They are highly resistant to solvents and alkalis and even some acids. They are easily incorporated into most composite manufacturing processes and allow specific properties, such as chemical or electrical, to be modified through the proper catalyst. Some common types of epoxy resins for aircraft use are: Safe-T-Poxy, which was especially developed to reduce the development of dermatitis, a common allergic reaction. It is no longer produced, and has been replaced by a new resin called E-Z Poxy, which offers the same handling and physical properties. MGS Epoxy, is used for certified aircraft applications; AlphaPoxy, is used for secondary structures; and Aeropoxy, is used for primary structures. Also well known are Rutan Aircraft Epoxy (RAE) systems.

Phenolic resins: used for a multitude of applications, some of which take advantage of their high temperature tolerance (brakes, rocket nozzles). Used to impregnate Nomex honeycomb floors and interior cabin liners in some aircraft, where it meets smoke, combustion, and toxicity requirements.

Polybutadienes: have great electrical properties and chemical resistance and as such are used for radomes as an alternative to E-glass/epoxy laminates. High resilience renders them a popular choice in the production of tires.

Polyesters: used for a multitude of applications, such as boats, bathtubs, and auto body parts. Polyester resins are solvents for many types of synthetic foams (see below), so the user must make sure the proper core is used if making composite sandwiches.

Polyurethanes: can be formed into either thermoset or thermoplastic resin. As a thermoset, it is primarily used for applications involving automotive bumpers.

Vinylesters: are used for many of the same applications as polyesters, but are more expensive. They are better than polyesters in applications exposed to high moisture environment, such as for boat manufacturing.

THERMOPLASTICS

They are less widely used for aviation applications than thermosets. Their best-known property is that when heated they become liquid, but then return back to a solid state when cooled. The property renders the material highly practical for all sorts of applications, ranging from soda bottles, nylon garments, monofilament fishing lines, to engine fuel lines. Thermoplastics can be melted and frozen repeatedly, rendering them recyclable.

Sandwich Core Materials

The sandwich core can be made from a multitude of materials, although with some constraints. First, the resin must not be a solvent for the core; and second, it has to be resilient enough to not fail before the fiberglass. The following materials are well suited for use in aircraft composite sandwiches, although some are not used for certified aircraft:

Urethane foam: costly, but easy to work with. It is impervious to most solvents and can thus be used with less expensive polyester resin. It can easily be cut and carved to shape, and then sanded to shape with bits of itself [12]. It is useful for making wingtips and fairings in homebuilt aircraft, as well as compound surfaces. Readily available in sheets that are $24'' \times 48''$, in thicknesses from $\frac{1}{2}''$ to $2''$, at $2-4$ lb_f/ft^3. It gives off toxic fumes when it melts and should not be used to hot-wire (see later). Not used for certified aircraft.

Clark foam: more expensive and dense (4.5 lb_f/ft^3) variety of urethane foam. Renowned for versatility and famous for use as core in surfboards. Not made since 2005. Not used for certified aircraft.

Styrofoam: blue-colored styrofoam is the most popular material for use as core in wings of homebuilt aircraft and is also used for insulation in homes. Readily available in sheets that are as large as $48'' \times 96''$, in thicknesses from $\frac{3}{4}''$ to $4''$, at 2 lb_f/ft^3. Not used for certified aircraft.

Polystyrene: commonly used for marine applications, it is also used as core in the wings of several homebuilt aircraft. Well known for its use as insulation in homes and as packing material. Easily recognizable as the aggregate of small foam balls. It is very susceptible to solvents and will be 'eaten' by polyester resin. Available in blocks that are as large as $14'' \times 109''$, and $7''$ thick, at 1.6 to 2.0 lb_f/ft^3. Not used for certified aircraft.

Klegecell®: registered trademark for a PVC foam that meets all FAA regulations for fireproof aviation materials. Has been in production for over 50 years. Unaffected by UV rays and very stable with respect to resins. Has extremely high strength-to-weight ratio, excellent thermal and acoustic insulation properties, low water absorption and good chemical resistance. Available in sheets that are as large as $48'' \times 96''$, in thicknesses from $\frac{1}{4}''$ to $2''$, at $3-6.25$ lb_f/ft^3. Used for certified aircraft.

Divinycell®: registered trade mark for a PVC foam that also meets all FAA regulations for fireproof aviation materials. Unaffected by UV rays and very stable with respect to resins. Available in sheets that are as large as $48'' \times 96''$, in thicknesses from $\frac{1}{4}''$ to $2''$, at $3-6$ lb_f/ft^3. Used for certified aircraft.

Honeycomb: Honeycomb refers to a class of materials used as sandwich cores in which thin material, ranging from paper to alloys, is formed into hexagonal cells to use as core. Honeycomb can be used for both flat and curved panels, however, bonding fibers to the comb is more difficult. There are three relatively well known types of honeycomb: (1) aluminum honeycomb, which has one of the highest strength-to-weight ratios of any structural material; (2) Nomex honeycomb, which is made from Nomex paper dipped in phenolic resin and is widely used in the aviation industry; and (3) thermoplastic honeycomb, which is used in a multitude of transport applications. Used for certified aircraft.

Glass Transition Temperature

In terms of FRPs and GRPs, the glass transition temperature, T_G, refers to the temperature at which the resin transitions from a hard and relatively brittle state into a molten (or soft) state. Reaching this temperature in operation could be catastrophic to a primary structure as it renders the laminate incapable of reacting the applied loads. Most FRPs and GRPs used for aviation applications have a TG in excess of 180 °F.

Gelcoat

Gelcoat is what provides the glossy, high-quality finish on the exposed surface of FRPs and GRPs. It is a polyester or epoxy resin specifically prepared with chemicals to control viscosity and cure-time, as well as pigment with the desired color. Gelcoat is sprayed into the mold ahead of the plies being laid up.

Pre-cure

Pre-cure is a term used for flat laminated plates that are cured prior to being used as a supplemental structural material. Think of it as a flat sheet of aluminum alloy, except it is made from FRP. Having these at one's disposal is priceless, as one can cut them to a desired shape, and then co-cure them with a laminate layup. Pre-cures are frequently used to place hard points in a sandwich laminate, through which metal fasteners may be used. Their thickness is then equal to the thickness of the core. This will form a kind of island of solid laminate in the sandwich panel, which, as stated earlier, is ideal to provide bearing strength and transfer fastener load into the sandwich.

Aircraft Construction Methodologies

There are primarily two methods used to build composite airplanes; *moldless composite sandwich construction*, and *molded composite construction*. The former is typically used for homebuilt or kit aircraft and is a method generally thought to have been pioneered by the well-known Burt Rutan to permit customers to fabricate the experimental Rutan VariEze and LongEze kit planes [11]. The method is explained in detail by Lambie [12] and Clarke [13]. The first step in the application of the method is to "preform" the sandwich core using a multitude of methods. Once the core has been prepared, it is covered with fiberglass cloth and subsequently impregnated with resin (or "wetlay"). The impregnation takes place by pouring resin over the cloth and then paintbrushes and squeegees are used to spread it and to wet the entire cloth. This step requires careful attention to prevent too much resin from being used. If more than one ply is required, the second ply is laid on top of the first while it is still wet, and more resin is added, and so on. The part is then allowed to cure.

The pre-forming of the core is done by a multitude of methods. It can be something as simple (but crude) as carving or sanding the foam to shape, although for parts that require greater accuracy (note that accuracy is a relative term) the foam is cut to shape using an electrically heated wire (or "hot-wiring"). While satisfactory for homebuilt aircraft, this method is never used for certified aircraft because of its inherent flaw of quality irregularity.

Instead, molded composite construction is used for certified aircraft. The method uses "female" or cavity molds that have been accurately shaped to form the outside mold line (OML) of the part. Then fiberglass cloth called "pre-preg" (because it is already impregnated with resin) is laid inside the mold. If more than one ply is required, another pre-preg is laid on top of the previous one, and so on. Once the layup has been completed, some strands of sticky putty are laid around the part and then a plastic sheet is draped over it and tacked to the putty. This encloses the part in a hermetically sealed environment ("vacuum-bagging"). Then a vacuum pump is connected to the plastic and turned on to form a vacuum under the plastic veil. This is a part of a production process to be explained in a moment. Then, the part sitting in the mold is rolled into a warming room (perhaps some 150–180 °F) where it is allowed to cure for a specific number of hours.

So, why is all of this preparation necessary?

The pre-preg is a special fiberglass (or graphite) cloth impregnated with resin under controlled circumstances. This ensures repeatability. In order to certify an airplane, the material qualities have to be repeatable. This means that the strength of the composite laminate should not vary from day to day – it should be the same no matter the time of month or position of the moon. The warming room ensures the resin cures at an optimum temperature, but this maximizes the strength of the laminate. It also lowers the viscosity of the impregnated resin, something taken advantage of through the application of pressure. With the formation of a vacuum on the part side of the plastic veil, atmospheric pressure squeezes air-bubbles out of the pre-preg and helps spread the resin uniformly throughout the laminate. Both improve the quality of the laminate, ensuring the proper fiber/resin ratio. Sometimes, rather than using vacuum, which only applies a 1 atmospheric pressure to the laminate, the laminate is brought into a pressurized container, called an "autoclave." There it is subjected to as much as 5–10 times the atmospheric pressure.

Fabrication Methods

There are a few fabrication methods used to manufacture FRPs and GRPs that are worth presenting in this context:

Hand layup and spray up: the simplest and least expensive method to manufacture FRP or GRP parts. Plies of fibers are placed into a mold, after which it is impregnated with resin, unless the cloth is a pre-preg. The impregnation takes place either by simply pouring the resin onto the cloth and spreading it out using squeegees and paintbrushes, or it is sprayed on using special spray-guns.

Resin transfer molding – RTM: consists of a rigid heated mold that contains gelcoat, surfacing veil, and the fiberglass cloth, into which resin is pumped under pressure. The mold is typically 100–120 °F (40–50 °C).

The warm and pressurized resin flows through the tool and uniformly impregnates the laminate. The primary advantage of this method is the superior surface quality of parts, as well as dimensional tolerances and consistency of parts.

Compression molding: consists of placing the material to be molded (a thermoset), preheated, in a heated open male-female mold. Then the mold halves are brought together and the material is compressed, which forcefully spreads it uniformly over the entire mold surface. Compression molding is the oldest manufacturing method used by the plastics industry.

Injection molding: the most common means of producing parts out of plastic material. Melted plastic is forced under pressure into a mold of the desired part and is allowed to cool and solidify. The method is very versatile and most plastic parts commonly found in one's environment are made using this process.

Filament winding: filament winding is a process in which resin-wet fibers are threaded through a roving delivery device called a feedeye. The feedeye moves back and forth along a rotating mandrel with the desired shape — a body of revolution. The fibers are wound helically in this fashion until a desired thickness is achieved. The method is used to create pipes, tanks (e.g. external fuel tanks), and even airplane fuselages. The fiber angle is controlled with the rotation speed of the mandrel and typically varies between 7° and 90°. The process compacts the laminate, making vacuum bagging unnecessary.

Pultrusion: pultrusion consists of strands of fiber that are pulled through a die to form a column of some specific cross section. The operation involves pre-wetting the strands in liquid resin before they are pulled through a heated steel die. The process is analogous to forming an extrusion from aluminum alloys, except the fibers are being pulled out of the die rather than being pressed through it. The fibers are pulled through the machine using two powerful pulling clamps, of which only one pulls at a time. When the active clamp reaches the end of its track, the second clamp picks up the slack, allowing the first to get back to its initial position where it resides until it takes over when the second reaches the end of its travel.

The above, which barely scratches the surface, should demonstrate that the field of composites and plastics is booming. Selected properties of typical composite materials are shown in Table 5-8. Note that there is a large variation in properties between fiber brands, fiber volume, resin system, layup process, and other factors. The table should not be used for structural analyses — it is only presented to give ballpark values. Figure 5-10 compares the density, cost, strength, and stiffness of several composite materials (and polyethylene plastic), using E-glass as a baseline. Such a comparison matrix is helpful when selecting material for an application.

The reader wanting to learn more about composite materials and their use and certification in the aviation industry is directed to MIL-HDBK-17 [14], AC-20-107B [15], and AC-21-26 [16].

TABLE 5-8 Selected Properties of Typical FRPs and CRPs

Description	Density	Tensile Modulus	Shear Modulus	Poisson Ratio	Yield Tensile	Ultimate Tensile	Ultimate Shear
Symbol	ρ	E	G	μ	F_{ty}	F_{tu}	F_{su}
Units	lb_f/in^3	$\times 10^3$ ksi	$\times 10^3$ ksi		ksi	ksi	ksi
Epoxy (resin)	0.046	≈0.6	≈0.23	≈0.34	—	—	—
Polyester (resin)	0.042	≈0.47	≈0.17	≈0.38	—	—	—
Vinylester (resin)	0.046	≈0.5	≈0.17	≈0.38	—	—	—
E-glass	0.094	≈10	≈4	≈0.2	No yield	≈27	—
S-glass	0.092	≈7	≈0.6	≈0.26	No yield	≈50 to 90	10 to 12
High-modulus carbon	0.072	≈53	≈2.7	≈0.2	No yield	≈190	—
High-strength carbon	0.065	≈35	≈3.6	≈0.3	No yield	≈320	—
Boron	0.090	≈170	≈26	≈0.35	No yield	—	—
Aramid (Kevlar)	0.052	≈18	≈4	≈0.36	No yield	≈40	—

(Based on Ref. [17] and other sources)

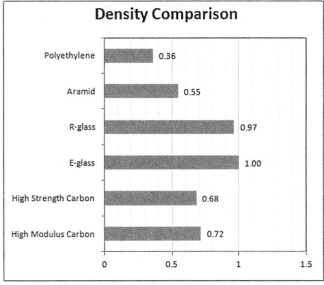
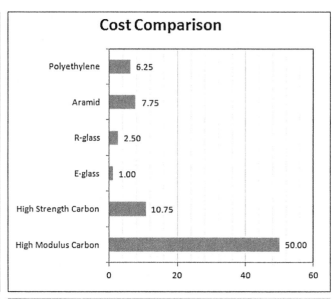
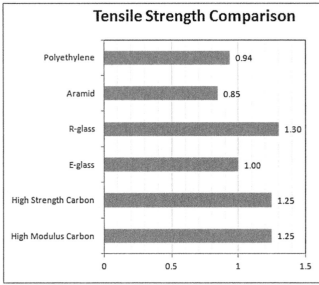
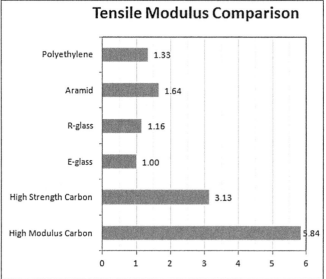

FIGURE 5-10 A comparison of several composite materials, normalized to E-glass. *(Based on http://www.hexcel.com)*

5.3 AIRFRAME STRUCTURAL LAYOUT

Detailed topics in structural design are beyond the scope of this book, however, it is important to present a brief overview of structural layout to help the designer select the appropriate fabrication methodology. In modern times, there are generally four distinct construction techniques used to fabricate aircraft: wood, welded steel trusses, stiffened skin construction, and composites. The last two are most widely used, however, wood and welded trusses, while infrequent, may be the right fit for a specific design project. It is the duty of the designer to fully understand the strengths and shortcomings of the available fabrication methodologies and select the proper one for the job. This section presents the application of these methods to real aircraft and introduces both important structural concepts and challenges that are experienced in their development.

5.3.1 Important Structural Concepts

A number of structural terms are introduced in the discussion below, necessitating their brief definition [18]:

- A *stiffener* is a longitudinal or transverse member intended to reinforce a structure by increasing its stiffness.
- A *flange* is a longitudinal stiffener that runs along the edge of a flexible shell (or sheet), whose purpose is to increase the stiffness of the shell.

- A *stringer* is a longitudinal stiffener that is not a flange.
- A *boom* is a beam in the shape of a shell.
- A *rib* is a transverse stiffener in an open shell, or the end of a closed shell.
- A *bulkhead* is a transverse member in a closed shell. Strictly speaking the term includes members such as wing ribs and fuselage hoop-frames, although a bulkhead is often used for transverse members that are more substantial than frames and to which other structures attach, for instance, wing, horizontal tail, and engines.
- A *primary structure* is one that reacts primary loads applied to a structure. Airworthiness Circular 25.1529-1 defines it as structure that significantly contributes to the carrying of flight, ground, or pressure loads.
- A secondary structure carries significantly lower operational loads than a primary structure and is usually used for fairings and doors for unpressurized containers, for example.
- An *allowable* is a maximum allowable stress value of some specific material property. For instance, the ultimate tensile stress allowable for a 2024-T3 aluminum sheet of 0.125 inch thickness is 64,000 psi.
- A *notched allowable* is an allowable assuming it has notches in it. This always results in a reduction in the allowable, sometimes as much as 50%.
- *Fail-safe* means that should the primary load path in a structure fail during operation, an alternative load path exists that prevents a catastrophic failure of the structure as a whole.

Monocoque and Semi-monocoque Structure

The word monocoque comes from the Greek word *mono* (single) and the French word *coque* (shell). Monocoque is a structural technique in which stresses are reacted by a thin membrane or a shell of material, rather than a collection of beams. Such structures are stiff in bending, and light, and are therefore ideal for weight-sensitive vehicles such as airplanes. A good way to visualize a monocoque structure is to fold a sheet of paper into a cylinder and tape the free edge using something like Scotch tape. Although the resulting structure is stiff in bending this will reveal its greatest weakness — structural instability. A monocoque structure has a great tendency to fail in buckling or crippling, something easily demonstrated by a person standing on top of an empty aluminum beverage can. The empty container can support a grown man, but push the side in with a pencil with the person standing on it and the can will be compressed in the blink of an eye. This instability necessitates the addition of an internal support structure that resists these failure modes. Such a support structure is generally an assembly of frames, bulkheads, stringers, and longerons (Figure 5-11). The combination is referred to as *semi-monocoque* construction. Although the addition of the support structure adds to the overall weight of the configuration, it retains its light and stiff characteristics while reacting the applied loads. The advent of the monocoque structure was a major breakthrough in the development of aircraft structures. As stated earlier, such structures react the applied loads in the skin (hence the name "stressed-skin" construction).

Wood Construction

Not too many aircraft are built from wood in modern times, although a few are still being operated today. The most prominent is arguably the de Havilland DH-98 Mosquito, a twin-engine, multi-role combat aircraft, made famous during World War II (see Figure 5-12). The fuselage of the Mosquito was made from a composite consisting of sheets of balsawood bonded to sheets of birch. The wing was a one-piece, all-wood construction. It featured two spars made from spruce and plywood, and the skin was a plywood sheet. A cutaway of the Mosquito can be seen in Figure 5-13, showing details of how ribs, spars, bulkheads, and skins were assembled to make this historic airplane.

The largest flying boat ever built, the Hughes H-4 Hercules, also known as the Spruce Goose, was built from plywood and to this day boasts the largest wingspan of any aircraft in the history of aviation. Its wingspan of 97.5 m (320 ft) is larger even that of the world's current largest aircraft, the Antonov An-225 Mriya (with 88.4 m or 290-ft wingspan). Well-known examples of GA aircraft made from wood are the Bellanca Viking (designed in 1960) and various types of aircraft made by Jodel (originating in 1946) and the

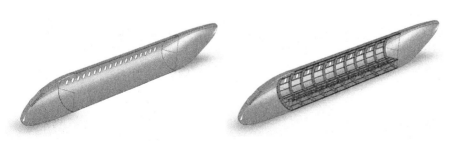

FIGURE 5-11 The difference between a pure monocoque (left) and semi-monocoque (right) fuselage structure.

118   5. AIRCRAFT STRUCTURAL LAYOUT

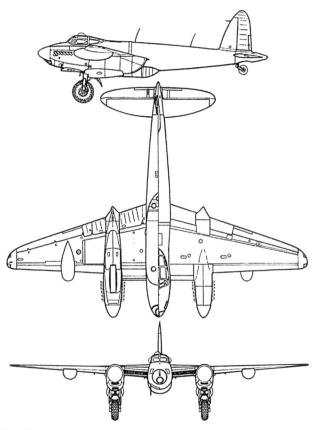

FIGURE 5-12 The de Havilland DH-98 Mosquito is arguably the most sophisticated wooden aircraft in the history of aviation.

Robin DR400. Additionally, there are a number of kit planes for amateur builders that feature wooden construction.

As with everything, constructing aircraft out of wood has pros and cons. Among the advantages of wood are availability, affordability, comparable strength and good impact resistance, and it is usually easy to work with. Among the disadvantages are inconsistent material properties; crack growth (splitting); low Young's moduli; possibility of rotting and even termite infestation; flammability; water absorbability (moisture variation); reduction in strength if moisture content exceeds fiber-saturation; and sensitivity to grain direction (anisotropy).

Plywood is an excellent structural material for use in wooden aircraft. It is usually made from an odd number of thin sheets (plies), each oriented at a 90° angle to the adjacent ply. The primary advantage is it offers bi-directional material properties, in addition to offering greater resistance to splitting and much-improved dimensional stability with moisture content. Plywood used in aircraft must comply with standards set by MIL-P-6070, which requires it to be tested for dimensional discrepancies, glue strength, strength properties, and others. Plywood is typically used for wing skin, fuselage skin, ribs, and frames. Common types of plywood for use in aircraft are made from birch, poplar, fir, maple, and mahogany.

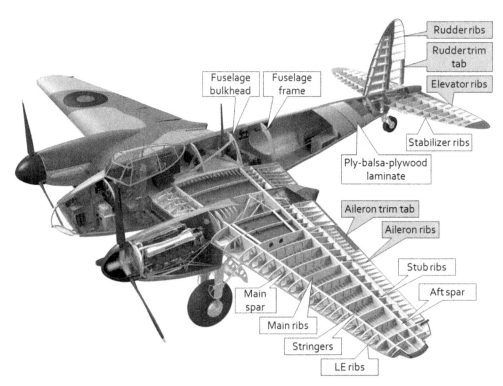

FIGURE 5-13 A cutaway of the de Havilland DH-98 Mosquito, showing important elements of its wooden construction. Dark labels indicate aluminum and light indicate wood. *(Courtesy of Raymond Ore, www.raymondore.co.uk)*

Parts made from wood are primarily joined by two means: bonding and mechanical joints. The use of joining shapes (such as lap-joints; tongue and grooves; tenon and mortise, etc.) is not recommended as these invariably lead to stress concentrations that may ultimately lead to failure. Milling or routing parts is acceptable if corners are rounded. Wooden parts require special protection internally, as well as externally.

The adhesive typically used is called Aerodux-500 Resorcinol Adhesive. It is a water- and boil-proof resorcinol/formaldehyde adhesive designed for use in structural wood beams. It requires a hardener to cure, mixed in the ratio 1:1, and can cure at temperatures as low as 7 °C (45 °F).

The reader interested in designing an aircraft constructed from wood is directed toward the documents ANC-18 *Design Of Wood Aircraft Structures* [19], a classic text on best practices and structural analysis of wooden structures, and NACA R-354 [20], a 34-page report with tips regarding selection and the properties of wood.

Steel Truss Covered with Fabric

Many aircraft feature a fuselage consisting of a truss structure made from steel tubes welded to form a stiff, strong, and light structure. Normally, the truss (see Figure 5-14) is then covered with fabric and dope. Such fuselages are usually made from straight sections of steel tubing (nowadays from 4130 chromium-molybdenum steel) and, more often than not, are rectangular in shape. While robust, such a structure is not exactly ideal for low-drag fuselages. This is not to say that there are not low-drag aircraft around that feature such a structure. As an example, the German Scheibe SF-25 Motor Falke motor glider features a steel truss fuselage structure, although it is an exception to the rule. The method is commonly used for aerobatic and agricultural airplanes, which take a severe beating operationally and for which the truss structure serves well.

Aluminum Construction

Aluminum remains the most common aircraft construction method at the time of writing. Stressed-skin construction has revealed itself as a very efficient means of producing aircraft, thanks to a sheet metal skin riveted to sheet metal frames and bulkheads. Such a structure is very light and stiff, and the industry has developed a large number of impressive tools and techniques to put together aircraft in a short time. The stressed skin means that shear, torsion, and bending loads are reacted by the skin, which is stabilized (made less susceptible to buckling and crippling) using frames and stringers. Of course, a part of the load is reacted by this extra structure, but the difference is that the skins actually transfer loads in the form of stresses, unlike the truss-and-fabric construction of the past.

Aluminum also offers an important benefit – damage repair is relatively straightforward. As such it is surprisingly forgiving of damage, although some would argue it is relatively susceptible to damage. However, the number of parts required to assemble such aircraft is usually significant. This is due to the fact that the aluminum is in the form of thin sheets that are folded and cut to specific shapes before being riveted together. The thickness of the aluminum sheets renders it very flexible. Therefore, it has very limited buckling and crippling resistance, until it is stabilized using stiffeners — which are long sections of aluminum made either by folding other aluminum sheets or from extrusions.

Figure 5-15 shows a cutaway of the famous Supermarine Spitfire and reveals a large number of parts required to make this typical high-performance aluminum aircraft.

Composite Sandwich Construction

Composites and composite sandwich construction have already been discussed. The advent of FAA-certified aircraft such as the Cirrus SR20, SR22, Cessna Corvalis, and a series of aircraft produced by Diamond Aircraft reveals the advantage of such constructions. All feature modern tadpole fuselages and NLF airfoils, making them very efficient. For instance, the SR22 and Corvalis, both of which have fixed landing gear and wide fuselages (50 inches), offer cruising speeds that are right up there with that of rival aircraft such as the aluminum Mooney Ovation, which has retractable landing gear and a narrow fuselage (43.5-inch internal width).

The most obvious difference between composite aircraft construction and conventional wood or aluminum construction is the number of parts. For instance, the wing spar of a composite aircraft is typically a one-piece component, tip-to-tip. An aluminum structure, in contrast, consists of multiple parts: spar caps, shear webs, stiffeners, all of which are assembled using rivets. A composite wing also contains far fewer ribs, as the skins are stiffened and, thus, do not need

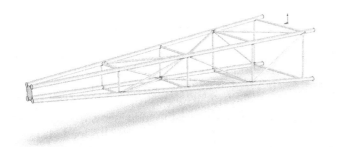

FIGURE 5-14 Example truss structure intended for the empennage of an airplane.

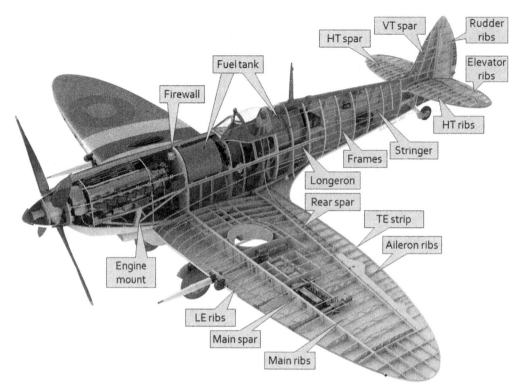

FIGURE 5-15 A cutaway of the Supermarine Spitfire, showing important elements of its aluminum construction. *(Courtesy of Raymond Ore, www.raymondore.co.uk)*

the same number of ribs, and they are devoid of stringers. The typical composite airplane is bonded together, using an actual adhesive, something it has in common with wooden aircraft. From a certain point of view, assembling a modern composite (certified) aircraft is not unlike putting together a plastic model. The shells of components are bonded together, effectively using "glue." Of course the analogy ends there, but the process requires far fewer parts than do aluminum aircraft.

5.3.2 Fundamental Layout of the Wing Structure

It can be argued that the wing is the most important structure of the airplane. It generates the largest aerodynamic load of all components and often features complex mechanical systems that themselves are subject to substantial loads. The wing must provide volume to store fuel, as well as to accommodate the control system required to actuate ailerons and the high-lift system. Furthermore, if the landing gear is retractable and mounted to the wing structure, the proper cutouts and reinforcement must be made to the wing to allow that function.

The wing structure is designed to react shear forces and moments that result from lift and drag. They are almost always reacted as three mutually orthogonal shear forces (lift, drag, and an inboard force if the wing has dihedral), and three mutually orthogonal moments (bending moment, drag moment, and wing torsion). To react these loads, the wing features a number of load-carrying members that have to be carefully assembled so that that the wing will be (1) geometrically symmetrical (i.e. left and right wing halves are identical mirror images of each other); and (2) as close to the intended geometry as possible. A typical wing structure is shown in Figure 5-16. It consists of the following parts (note the labeling of A through N for easier identification):

The *main spar* (**A**) is the primary load path in the wing and is intended to react wing bending and shear loads. The idealized spar consists of a relatively thin sheet of vertical structure called the *main spar shear web* (**B**). Two thicker members, called the *main spar caps* (**C**), are attached to the shear web, one along the bottom edge and the other along the top edge. The shear load reacted by the main wing spar is simply the wing lift and, by convention, the shear web is designed to react a significant part of the total wing lift. This load is distributed from the tip to the root. This load also generates bending moment and this is reacted by the spar caps. Both the shear and moment are equal to zero at the wingtip and reach a maximum value at the root.

5.3 AIRFRAME STRUCTURAL LAYOUT

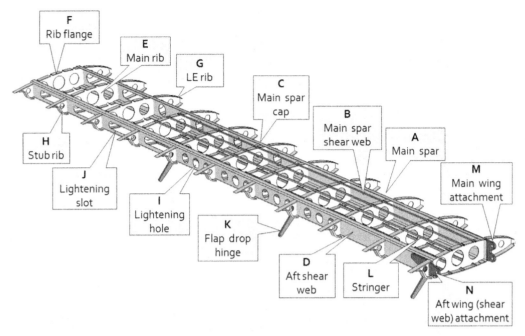

FIGURE 5-16 A simple schematic of a typical structural layout of a wing for GA aircraft.

For this reason, the ideal wing spar would allow the thickness to vary from tip to root, although this is hard to do in practice, unless the spar is machined. If the spar is made from aluminum alloy, it is designed to have a stepwise reduction in shear web thickness moving from root to tip, with an increasing frequency and size of lightening holes.

Several typical main spar cross sections are shown in Figure 5-17. Note that some of the spars feature dissimilar spar cap thicknesses on the top and bottom of the spar. This is indicative of the careful nature of aircraft structural analysis — material is used only where absolutely necessary. Aircraft are generally designed to react larger forces up than down. For this reason, the upper spar cap sees higher compressive loads than the lower cap. Therefore, it must provide more meat to resist column buckling or crippling failure and is made thicker.

The *aft shear web* (D) is also a primary load path; however, it is only intended to react a fraction of the lift, with the remainder being reacted by the main spar. It also reacts the wing torsion, which is generated by the airfoil's pitching moments, and the moment generated by the wing drag. The main spar and aft shear

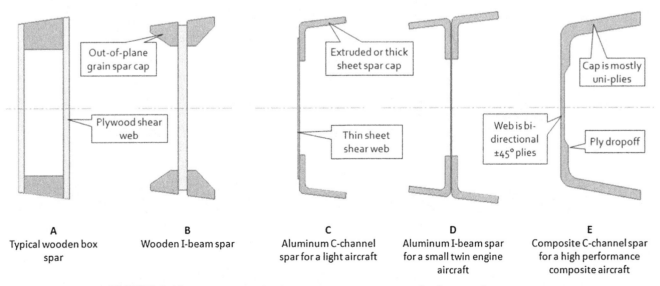

FIGURE 5-17 A schematic of typical main spar cross sections for GA aircraft.

web split the entire lift load generated by the wing. The amount of the split depends on factors like airfoil geometry, AOA, and control surface deflection, but generally around 60—70% is reacted by the main spar and 30—40% by the aft shear web. This structural member also transfers wing torsion to the ribs and the aft attachment bracket (see later), where it is reacted as a shear force. Additionally, it reacts the fore-aft chordwise force that results from the projection of the lift and drag on the chord plane with the main spar. This force is peculiar in the sense that at high airspeed it is mostly drag that places the aft wing attachment into compression. At low airspeed and high AOA the projection of the wing lift onto the chord plane becomes larger than that of the drag and this would force the wing forward if this were not prevented by the aft attachment, which then reacts this force in tension.

This structure is also called *aft spar*, but this nomenclature is erroneous unless the aft attachment is designed to transfer bending moments in the form of a couple. For smaller aircraft this is very rare. The rule is that if the aft attachment has a single fastener hole it will act as a simple supported joint that only resists a shear force. In this case, the structural member is a shear web. If the aft attachment has two fastener holes it will transfer bending moments and only now is it appropriate to call it an aft spar. In this situation the aft spar (like the front one) is designed to react the entire lift force, making the structure *fail-safe*. A fail-safe wing structure can react the entire flight loads using either the main or aft spars. This means extra safety in the case of structural failure. Such a structural philosophy is commonly employed in passenger aircraft and is based on the fact that a statically indeterminate structure is inherently safer after being subject to damage. The Rockwell 114 and 115 aircraft are examples of light GA aircraft that feature a fail-safe wing structure.

The above discussion should show that it is very practical to take advantage of the wing structure to pick up landing gear loads for low-wing aircraft. The designer should always try to take advantage of the major load paths as this will render alternative load paths unnecessary, in turn, reducing the weight of the aircraft.

The *main ribs* **(E)** are primary structural members that extend between the main spar and the aft shear web and tie them together. As such, the ribs serve several purposes [5, p. 278]. First, they stabilize the wing skin and prevent it from buckling while reacting wing torsion. This is an imperative function for two reasons: it helps maintain the intended aerodynamic shape and the skin's ability to transfer torsional loads. Second, the ribs shorten the effective column length of the stringers (see later), making them more resistant to column buckling. Third, they transfer wing torsion to the spars and eventually to the wing attachment hard points. Fourth, they react crushing loads due to wing bending. Fifth, they redistribute concentrated loads, such as those due to the landing gear, flap deployment, and engine pylons. And sixth, they react diagonal tension loads from the skin if subjected to skin wrinkling.

Each rib is formed so it has a *rib flange* **(F)**, but these are used to rivet (aluminum wing) or bond (composite wing) the rib to the skins and spars, forming a solid structure. The rib spacing is a task accomplished during the detail design phase. Selecting the largest possible rib spacing will save weight and simplify assembly.

An important question often asked during the layout of swept-back wings is whether the ribs should be mounted normal to the main spar or parallel to the direction of flight as shown in Figure 5-18. At first glance, it would seem the latter (configuration B) is more

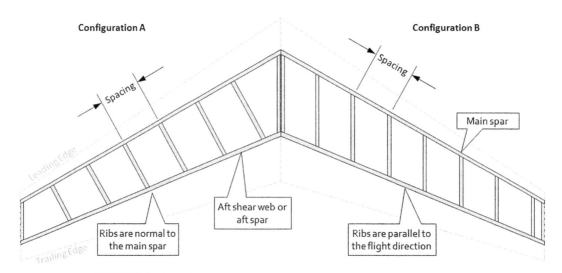

FIGURE 5-18 A schematic of two possible rib layouts for a swept-back wing.

reasonable because the rows of rivets along the skin (assuming aluminum construction) will cause less disruption to the boundary layer. However, that argument can be deflated by noting that the rows of stringers (discussed later) that extend spanwise from root to tip, and are necessary structural elements as well, will also be riveted to the skin — disrupting the boundary layer.

An important drawback of configuration B is that since the rib spacing is the same as that of Configuration A, the length of the ribs will be greater. Therefore, one should expect the arrangement to be heavier. Another complication is that it is actually harder to install the ribs for configuration B because they are not at a 90° angle with respect to the spar. On the other hand, the ribs belonging to configuration A are normal to the main spar (although their aft parts are not) and this offers production advantages.

In practice, both configurations are used. One reason is that configuration A is not practical next to the fuselage — it is simpler to mount the ribs parallel to it. Since the rib orientation is not usually changed immediately from B to A, it follows that a number of parallel ribs are installed before the orientation is changed. Another reason is that it might be convenient to use configuration B rib in an airplane where the engine is mounted to the wing, although this is not always the case either.

An inspection of aircraft with swept-back wings reveals that the vast majority of commercial jetliners feature all ribs that are normal to the main spar (configuration A). However, there are many exceptions where a combination of the two approaches is used. For instance, all inboard ribs of the Bombardier CRJ 1000 are parallel to the flight direction, while outside the flaps they are normal to the main spar. The Gulfstream G650 has the first six ribs parallel to the flight direction and the remaining normal to the main spar.

Another important question has to do with how the rib spacing is selected. The answer does include structural analysis, which belongs to the detail design phase and, thus, is beyond the scope of this book. However, only an elementary explanation of the process will be given. The procedure begins by assuming specific rib spacing. Then, a structural analysis that determines the material thicknesses required to react the air loads is performed. This allows the weight of the ribs and skin to be estimated. This is repeated for a few other candidate rib spacings. Eventually, a high enough number of rib spacings (a minimum of three is required to approximate the weight using a quadratic polynomial) has been evaluated to create a graph similar to the one shown in Figure 5-19. The graph shows the weight of the skin, ribs, and their combination as a function of rib spacing. If a minimum exists, as shown in the figure, it is selected for use in the wing design.

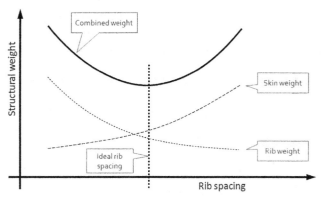

FIGURE 5-19 A schematic demonstrating the selection of rib spacing.

There are two additional ribs mounted to the wing that are of importance. The *leading edge rib* (G) is extremely important as it provides the forward shape of the airfoil. It also transfers large loads to the main spar. This happens when the wing generates lift at high AOAs. This will front-load the wing, but this means that the substantial low-pressure peaks form on the LE of the wing. This, in turn, means that most of the lift is generated by the forward part of the wing and this requires the structure to be very stout in this region.

The other rib type of interest is the *stub rib* (H), which is attached to the aft shear web. Stub ribs maintain the aft shape of the intended wing airfoil, while allowing control cables and pulleys to be threaded through various openings in the aileron. They also provide support to various control-system brackets and components. For instance, aileron hinge brackets are typically attached to stub ribs, which then transfer the air load to the aft shearweb. Aileron hinges are always mounted to stub ribs that have main ribs in front of them. This ensures the aileron loads are transferred more uniformly into the wing structure.

In order to keep the weight of the wing structure to a minimum, *lightening holes* (I) and *lightening slots* (J) are made in the structure. By doing this, a substantial weight of material that otherwise would simply be there for the ride is removed. Such holes and slots are far more common in aluminum and wooden structures than in composite structures, as the orthotropic nature of composites does not always lend itself well to such removal of material. Instead, composite spars allow for a more practical ply drop-off in the spanwise direction, resulting in more uniform thickness reduction, as each ply amounts to some 0.01 inch thickness. Aluminum spars for light airplanes often resort to lightening slots at the wing outboard region, as the shear and torsion have reduced significantly. Examples of such wing structures are found in many Cessna propeller aircraft.

The deployment of flaps can inflict very large loads on the wing structure that can result in a substantial increase in wing torsion. The flap load is transferred directly to the aft shear web and main spar through the flap hinges, of which the *dropped flap hinge* (**K**) is an example. Such hinges are almost always mounted right to the aft shear web and always have a stout main rib in front of them to ensure the main spar also picks up this load.

Stringers (**L**) are long columns of comparatively small cross section that are used to stiffen the skins and, thus, prevent them from buckling under load. For small aircraft the stringers are usually made from a folded strip of aluminum alloy sheet, whereas larger aircraft have stringers made from extrusions. And even larger (and expensive) aircraft often have integral stringers, in which the skin is machined from a thick plate of aluminum alloy so the stringers and skin are a single unit. This produces the lightest possible skin panel thanks to the efficient means by which stress is distributed throughout the panel. Additionally, the machining allows for stringers to be tapered smoothly along the span, as well as around holes and ribs [5, p. 258].

The *main wing attachment* (**M**) bracket is a primary load path and the most important hardpoint in the entire airplane. Generally, there are two kinds of attachment found in aircraft: *fixed* and *rotary*. The latter are primarily used for military aircraft with swiveling (F-14, F-111, Tornado, etc.) or folding wings (A-7, F-4, F-18, etc.) and are omitted from this text. A good discussion of those can be found in Ref. 5.

Figure 5-20 shows several methods to mount the main spar to the fuselage of the fixed type. The detailed appearance of the layouts shown varies greatly in practice and the figure should be regarded more from a stylistic perspective than precise. Configurations A, B, and C are used for high-wing aircraft. A and B are used in some commuter aircraft, e.g. Fokker F-27. The wing bending moments are fully reacted by the wing and the fuselage is effectively hanging down below it, using pinned joints. This way the fuselage does not have to be reinforced to react the wing bending moments, although it must react internal moments due to the difference in the reaction forces between the two attachment points.

Configuration C is used by many Cessna aircraft that feature wing struts. The hard points on each wing form a triangle that is structurally rigid, although it is not failsafe. This configuration, too, has pinned supports so bending moments are not reacted by the fuselage, although a substantial compression load has to be reacted along a line going through the lower strut pin and the wing root pin. Configuration D is used in many airplanes, e.g. the Beech Bonanza and Eclipse 500. This configuration uses a so-called *spar carry-through* to react the wing loads. The carry-through is by far the stiffest single structural member in the aircraft. This means that it will pick up the wing bending loads, largely bypassing the fuselage to which it is attached.

Configuration E is almost exclusively used on mid-wing fighter aircraft (e.g. F-104, F-16). The load-carrying frame requires sophisticated and costly machining from a solid ingot of alloy, making it a very

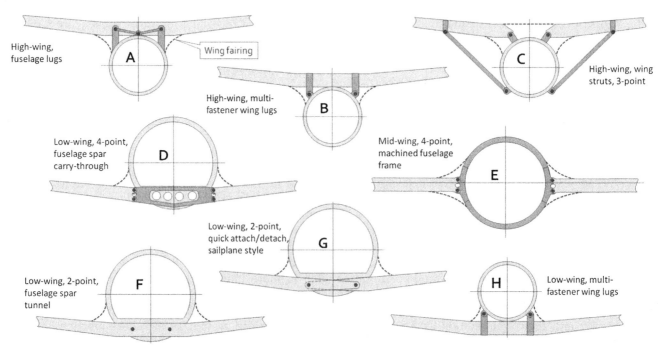

FIGURE 5-20 A schematic of common and possible wing attachment methods.

expensive component to make. Additionally, a number of such frames are installed for the typical multi-spar fighter wing. Such aircraft feature the engine in the cavity between the wings, which is the reason for the selected configuration. The orientation of the fastener in the wing attachment is often normal to, rather than parallel to, the fuselage, as shown here. Configuration F is effectively configuration B inverted. It is used on the Cirrus SR20 and SR22 aircraft.

Configuration G is used in aircraft that have to be quickly assembled and dis-assembled for transportation purposes. Such airplanes include sailplanes and some homebuilt aircraft. Finally, configuration H is used in many business jets, where the fuselage sits on top of the wing. Naturally, such airplanes feature much greater structural complexity than depicted, as their attachments are statically indeterminate (fail-safe). The attachments transfer loads in a variety of ways. Some react bending moments as a force couple, others do not react them at all.

It is important to be aware of the tremendous load transferred by the wing to the fuselage structure. Figure 5-21 shows a simple example of a beam structure intended to resemble a typical cantilevered wing configuration. It is assumed this airplane weighs 2000 lb_f and features a 40-ft wingspan. Furthermore, assume it is certified to the normal category of 14 CFR Part 23, which requires the airframe to be designed for a limit load of 3.8 g. The picture shows that when the airplane reacts this load symmetrically (both wings generate equal load), each attachment point transfers a shear force of 3807 lb_f or 1.9 times the gross weight. When reacting an asymmetric load per 14 CFR 23.349(a)(2), which presumes one wing is generating the full half-span load and the other merely 75% (60% for aerobatic airplanes), the attachment point on the fully loaded side will transfer a shear force of 5520 lb_f or 2.76 times the gross weight. This is a surprising result for those seeing it for the first time, even though it is derived using elementary statics. The purpose of the paragraph is to ensure this fact is not overlooked when designing aircraft.

Considering the wing layout in Figure 5-16, the entire bending moment and most of the shear force is transferred to the fuselage by the main spar. The main wing attachment must provide ample surface area for reacting wing loads in bearing and be designed such that the loads to be transferred rise gradually, rather than sharply. This means that there should be a number of fasteners present in the skin and ribs to pick up the wing load and transfer it to the main spar and then through the main spar attachment. This hardpoint reacts substantial loads, even during normal flight, rendering it very susceptible to fatigue. Consequently, the wing attachments must be analyzed and their material thickness increased beyond what is required for operational loads. There is also a concern about the dissimilarity of

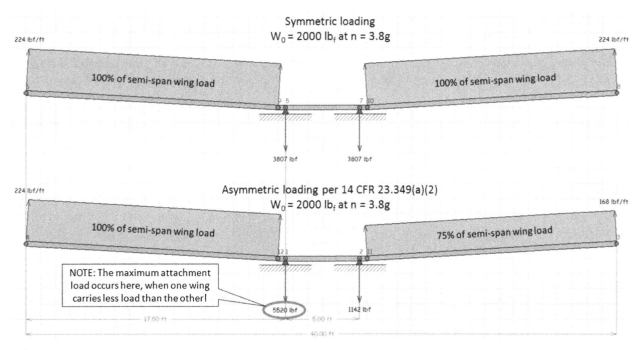

FIGURE 5-21 Reaction loads due to the aerodynamic lift generated by the left and right wing-halves of a 2000 lb_f hypothetical aircraft (erroneously assuming a uniform distribution of lift). Two load conditions are shown: symmetrical (upper) and asymmetrical (lower). Surprisingly, the maximum wing attachment load is achieved when the wing carries the asymmetric 100/75 load and not 100/100 (normal and utility category).

metals that may be used. For instance, using steel fasteners with aluminum brackets or sheets will cause galvanic corrosion and is a recipe for disaster unless proper precautions are taken [21].

The *aft wing attachment* **(N)** is a primary load path and the second most important hardpoint in the airplane. For the wing layout shown in Figure 5-16, it transfers a part of the shear force and reacts the wing torsion.

5.3.3 Fundamental Layout of the Horizontal and Vertical Tail Structures

The horizontal tail (HT) and vertical tail (VT) are aptly described as simplified versions of the accompanying wing. Compared to the wing, the loads of the HT and VT are modest. In small aircraft the structure often consists of a single spar to which a number of ribs are riveted (or bonded), and then covered with skin. Sometimes, the loads are so modest it suffices to stiffen the skin using corrugations. Examples of this can be seen in many small GA aircraft, for instance the V-tailed Beech Bonanza and many single-engine Cessna and Piper aircraft, which use corrugated skins to stiffen their elevators, flaps, and ailerons.

The main spar of light aircraft is usually placed at approximately 65–70% of the chord, selected specifically for structural efficiency. This allows the elevator and rudder hinges to be mounted directly to the spar. The Piper series of single- and twin-engine aircraft feature very simple all-movable stabilators consisting of two-spar corrugated-skin construction, designed with the main spar along the hinge line and the auxiliary spar (or stiffener) mounted closer to the trailing edge, allowing an anti-servo tab to be mounted to it.

Larger airplanes feature a single main spar and a smaller auxiliary spar. Heavy aircraft, in contrast, have an HT and VT whose load paths are structurally superior to the wings of small aircraft. Such stabilizing surfaces usually have two spars and statically indeterminate fail-safe structure.

Furthermore, the incidence of the stabilizer section of heavy aircraft is designed to be adjusted in flight, providing a powerful means to trim the airplane over a wide CG range. The incidence of the stabilizer is typically changed using a powerful jackscrew drive. The stabilizer of the Boeing 727 commercial jetliner, shown in Figure 5-22, is an example of this. The picture reveals many of the complexities inherent in advanced aircraft: redundant rudders (rudder A and B) and trim tab; vortex generators; redundant elevator trim tabs; hydraulic systems, and so on.

Fabrication and Installation of Control Surfaces

The HT is usually attached to the fuselage using specific hardpoints, which are fundamentally analogous to wing hardpoints, albeit simpler. Many composite aircraft bond the skin of the HT to a wide flange designed into the fuselage. This ensures the horizontal loads are transferred through a wide bond area. The VT is often an integral part of the fuselage airframe – in particular in composite aircraft. In aluminum aircraft the VT spar is sometimes integral to the aft-most bulkhead, whose purpose may be twofold; to provide a load path for the stabilizing surfaces and a means to anchor the control-system pulleys or bellcranks, which often react substantial loads.

The stabilizing surfaces typically feature symmetric airfoils, such as NACA 0008 through 0012, whose thickness ranges from 8% to 12% of the chord (as indicated by

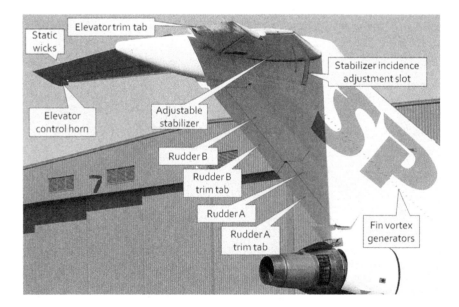

FIGURE 5-22 The T-tail assembly of a Boeing 727 commercial jetliner. *(Photo by Phil Rademacher)*

their designations). Such airfoils are low drag and yet offer volume to accommodate control system cables, pushrods, pulleys, and bellcranks. This thickness also results in a stiff structure that is free of flutter within the operating envelope of the aircraft. However, selecting a symmetric airfoil is not a rule. Some propeller-powered aircraft use either a cambered airfoil or a symmetrical airfoil at an angle-of-incidence with respect to the fuselage to reduce propeller effects (see Section 14.2, *Propeller effects*). Many aircraft feature symmetric airfoils such as NACA 63-008 through NACA 63-012, whose maximum thickness is farther aft than that of the double-o series, allowing the stabilizing surfaces to sustain NLF more extensively. When possible, the designer of efficient aircraft should consider such NLF airfoils for stabilizing surfaces. However, it must be emphasized this is not always practical.

It is important to bring up two issues that may lead to difficulties in the production of control surfaces. The first is the selection of airfoils that have very narrow trailing edges. Often, this is an unfortunate consequence of selecting NLF airfoils. Such airfoils often feature a cusp that makes the space between the upper and lower surfaces very small at the trailing edge. This calls for ribs so small they cannot be installed without being shortened and this may lead to a partially unsupported trailing edge, susceptible to flexing under loads. Changing from one airfoil to another on a tapered wing may require the wing skin to be stretched. If built from aluminum, this may call for costly manufacturing methods in a mass-produced aircraft. A solution often utilized is to simply ignore the cusp and replace it with a flat section (see Figure 5-23). Of course the resulting airfoil is not the one the designer initially intended. It is important to be aware of such problems and (1) avoid design solutions that are impractical from a manufacturing standpoint; and (2) ensure the production airfoils are those that are represented in the design analysis work and not the theoretical ones that are impossible to fabricate.

The other issue also has to do with cusped NLF airfoils. The high pressure generated by the lower surface of the cusp results in hinge moments that deflect the surface trailing edge up at higher AOAs, for instance, during climb (see Figure 5-24). This invalidates the drag coefficient, modifies the airfoil and reduces the performance of the airfoil. The designer should insist that the engineers designing the flight control system are aware of such detrimental tendency.

Unconventional Tails: T-tail, V-tail, and H-tail

In addition to the normal loads that a specific stabilizing assembly must react, the T-tail must react higher asymmetric loads than a conventional tail. When an aircraft featuring a T-tail yaws at a high AOA, each side of the HT generates very different normal loads. The windward side will see a high-pressure region form between the fin and the HT. The leeward side of the HT will generate much less load. This creates asymmetry in the loading on the HT that the fuselage must react in addition to the moment generated by the VT. As a consequence, the fuselage and the VT both end up being heavier than a conventional tail. Additionally, aeroelastically, the mass of the HT placed at the tip of the VT reduces its natural frequency, which, in turn, lowers its flutter airspeed. The remedy is to stiffen the VT, which again adds weight to the airframe.

Similar concerns can be raised about a V-tail. In order to achieve static and dynamic stability in a V-tail aircraft, relatively large surfaces are required. The rudder functionality of the tail requires the ruddervators on the left and right sides to deflect opposite to each other, similar to ailerons. The deflection can result in substantial forces on the two tail surfaces — one acting up and

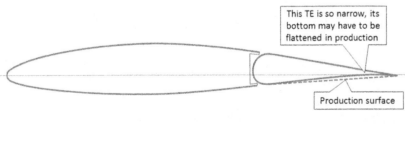

FIGURE 5-23 The manufacture of thin trailing edges is often solved by ignoring it and replacing it with a flat, rather than curved surface. Does this change invalidate the design analysis work?

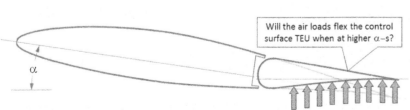

FIGURE 5-24 Cusped trailing edges may deflect at higher AOAs, unless the control system is very stiff, which it usually isn't unless it is hydraulically actuated. The actual aircraft will be likely to suffer from a reduced climb performance and reduced lift at stall.

the other down. As a consequence, a large torsion is generated that the fuselage must react. The designer of V-tail aircraft must be cognizant of this effect and ensure the cross-sectional area of the fuselage in the area of the V-tail is large enough to bring the shear flow down to acceptable levels.

The H-tail brings similar complications to the empennage loads as do the T- and V-tails. Additionally, as in the case of the T-tail, the two fins placed at the tip of the HT can be considered point masses at the end of a cantilevered beam, which brings down its natural frequency and, therefore, the flutter speed. The Lockheed Electra, a twin-engine, 10-passenger military transport, designed in the 1930s, suffered catastrophic flutter of its H-tail. The designer must select thick enough airfoils to reduce the likelihood of such events.

The reader is directed to Section 11.3, *On the pros and cons of tail configurations*, for additional information about these and other tail configurations.

5.3.4 Fundamental Layout of the Fuselage Structure

Fuselage Structural Assembly – Conventional Aluminum Construction

A conventional aluminum semi-monocoque fuselage structure is shown in Figure 5-25 and Figure 5-26. This structure typically consists of a row of hoop-frames that are joined with stringers and floor beams and floor frames by rivets or metal bonding. A schematic of a typical cylindrical center structure is shown in Figure 5-26. The aluminum sheets that form the fuselage skin are then riveted to the frames and stringers, forming a very stiff but light structure. In this structure, cutouts must have a generous radius rather than sharp corners, as these will generate substantial stress concentrations that will significantly reduce the life of the structure.

Figure 5-25 shows a very important feature of such a structure: all cutouts feature generously rounded corners or are elliptical or circular in some airplanes. The purpose of this is to reduce stress concentrations that result from the removal of material and, thus, increase the durability (and safety) the structure.

Fuselage Structural Assembly – Composite Construction

A semi-monocoque fuselage made from composites features a significantly different philosophy. The skin is stiffened using a composite sandwich that consists of a layer of material like honeycomb or foam sandwiched between layers of fiberglass or graphite or similar. This way, the total number of frames can be reduced substantially. Such fuselages are typically made by first bonding the internal structure together and allowing it to cure. Then, this assembly is bonded as a single piece to the stiffened fuselage skin halves. The fuselage skins have joggles to which the adhesive is applied. Joining the skins' halves together properly is easier said, than done.

A common difficulty in the assembly of semi-monocoque composite fuselages is to control the thickness of the bonds (adhesive). Certified composite aircraft have to demonstrate the repeatability of the bondline strengths. This is done by the manufacturer, which assesses a range of bondline thicknesses expected

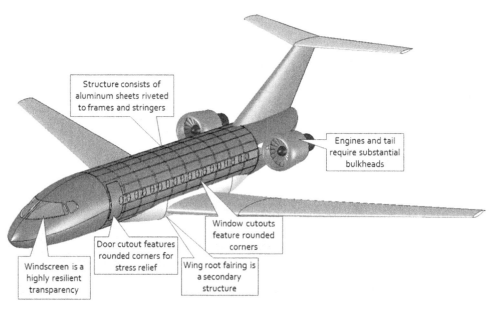

FIGURE 5-25 Typical fuselage for a passenger transportation aircraft consists of aluminum sheets riveted to an underlying rigid structure.

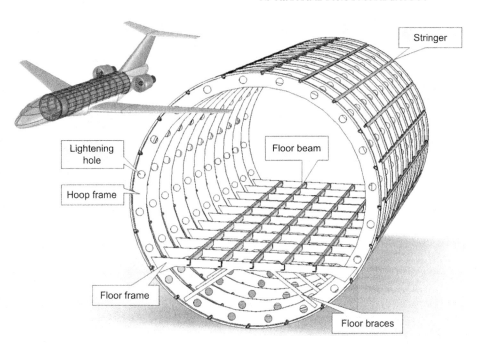

FIGURE 5-26 The underlying fuselage structure consists of hoop frames, stringers, floor frames, and floor beams (nose and tail structure is omitted).

to be seen during production (e.g. 0.040″ to 0.125″). Then, repeated strength tests of specimens using those thicknesses are performed, and this allows the strength of the bondline thicknesses to be established. Then, during production, the bondline thickness of all bonds is inspected and if found outside these limits a repair must be designed, basically using the assumption that the bondline strength is zero. Then, a technician performs the repair. It can be seen that poor production quality can cost the corresponding manufacturer a fortune. For this reason, established manufacturers perform the bonding operations using special machinery that substantially cuts down on such deviations (or non-conformances as they are called in industry).

Special Considerations: Pressurization

Tens of thousands of passenger and business aircraft operate every day at altitudes ranging from 25,000 to 51,000 ft. At those altitudes, especially altitudes above 40,000 ft, people quickly die if exposed to the outside atmosphere. For this reason, any aircraft designed to routinely operate at such altitudes must provide oxygen to the crew and passengers. Effectively, such aircraft are *pressure vessels* and are capable of maintaining higher pressure inside the cabin than that of the ambient atmosphere. Generally, people will begin to suffer from oxygen deficiency at altitudes as low as 14,000 ft. The individual capability varies, with some individuals capable of climbing mountains as high as 28,600 ft,[3] and as low as 6000−8000 ft for people who suffer from heart problems. Requirements in 14 CFR Part 121 to supply oxygen to the occupants when operating an aircraft are stipulated in the following paragraphs:

> 121.327 — *Supplemental oxygen: Reciprocating engine powered airplanes.*
> 121.329 — *Supplemental oxygen for sustenance: Turbine engine powered airplanes.*
> 121.331 — *Supplemental oxygen requirements for pressurized cabin airplanes: Reciprocating engine powered airplanes.*
> 121.333 — *Supplemental oxygen for emergency descent and for first aid; turbine engine powered airplanes with pressurized cabins.*

However, it is paragraphs 14 CFR 23.841, *Pressurized cabins*, and 14 CFR 25.841, *Pressurized cabins*, that stipulate what capability the airframe must possess in order to sustain cabin pressure in the case of a system failure. Design guidelines are also given in SAE ARP1270, *Aircraft Cabin Pressurization Control Criteria*.

With respect to aircraft, requiring pressurizations inflicts serious design, manufacturing, maintenance, and operational limitations on the aircraft. It is essential that the designer becomes aware of some of these requirements as they expose a number of challenges, ranging from fuselage deformation to system and equipment installation.

Figure 5-27 shows a cabin pressurization schedule for a typical commercial jetliner. As the airplane climbs, the pressurization will immediately begin to delay the

---

[3]For instance, in 1979, Reinhold Messner and Michl Dacher ascended K2 without supplemental oxygen. This feat is well out of the norm of human capability and took immense training and preparation.

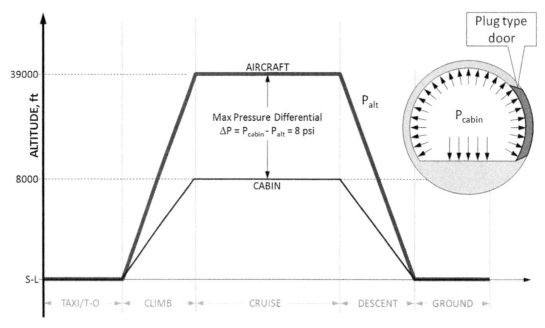

FIGURE 5-27 Cabin air pressure scheduling for a typical passenger jetliner.

altitude reduction in the cabin. As the airplane reaches its intended cruise altitude (here 39,000 ft) the pressure difference between the atmosphere at that altitude and the cabin is about 8 psi (atmospheric pressure at S-L is 14.7 psi), but this is equivalent to an atmospheric pressure at 8000 ft, enough for all except the weakest of us to survive.

From a structural standpoint, the most efficient pressure vessel is one that reacts lateral or *out-of-plane* stresses as tensile stresses [22]. This implies that for a given volume, a spherical pressure vessel is the most efficient one (see Figure 5-28). However, when it comes to transporting passengers and maintaining acceptable performance, and stability and control, this geometry is not practical. The next best shape is a cylinder, which is of course a sphere that has been split along a meridional and the two halves attached to a cylinder (again see

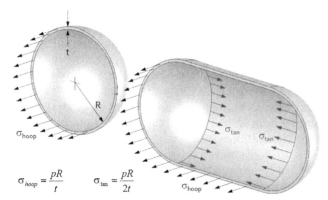

FIGURE 5-28 The difference between hoop and tangential stresses. The term p is the internal-external pressure difference. R and t are radius and wall thickness, respectively.

Figure 5-28). From a structural standpoint, this is the ideal shape for a pressurized aircraft, and this explains why this form prevails in the aviation industry.

As shown in Figure 5-28, the sphere reacts the out-of-plane pressure load as hoop stresses only. However, the cylinder reacts it as hoop and tangential stresses. When used for fuselages, this requires an especially reinforced structure to be placed at either end of the fuselage. This structure is called a *pressure bulkhead*, and it must react a substantial pressure force. The layperson is often oblivious to the forces the fuselage must support which are solely attributed to the pressure differential. For instance, the typical passenger door in a commercial aircraft is 42 × 72 inches (type A door). When exposed to an 8 psi pressure differential (see Figure 5-27) the total out-of-plane force acting on it amounts to 42 × 72 × 8 = 24,192 lb$_f$. This explains why it is so amusing when angry passengers threaten to open the doors in flight.

In all seriousness, this number depicts the robustness of the reinforcement required to hold the doors in place. A common method is to use doors that are shaped like a plug (see Figure 5-27). Such doors help distribute the pressure load around the door frame. It is sad but true that doors in pressurized aircraft are overlooked by almost anyone except the engineers who designed them. It is sad because these doors are truly a marvel of modern engineering. Not only do they have to react some 12 tons of load, when opened, most types swing to the outside of the airplane by a simply operated door handle. Think about that next time you board an airplane. How would you get a plug type door to do this feat? In addition to an immense pressurization load, cutouts for windows and doors will further cause stress concentration requiring a yet greater amount of material to sustain. The aspiring designer insisting on super-large entry doors and windows for a pressurized airplane should be very cognizant of the structural challenges this may cause. While large openings can be implemented technically, it will most certainly reduce the useful load of the aircraft.

VARIABLES

Symbol	Description	Units (UK and SI)
E	Young's (elastic) modulus	ksi or MPa
F_{bru}	Ultimate bearing stess (per MIL-HDBK)	ksi or MPa
F_{su}	Ultimate shear stress (per MIL-HDBK)	ksi or MPa
F_{tu}	Ultimate tensile stress (per MIL-HDBK)	ksi or MPa
F_{ty}	Yield tensile stress (per MIL-HDBK)	ksi or MPa
G	Shear modulus	ksi or MPa
P	Pressure	psi or Pa
T_G	Glass transition temperature	°R or K
ε	Normal strain	in/in or mm/mm
γ	Shear strain	in/in or mm/mm
μ	Poisson ratio (per MIL-HDBK)	
ν	Poisson ratio	
ρ	Density	lb_f/in^3 or kg/m^3
σ	Normal stress	psi or Pa
τ	Shear stress	psi or Pa

References

[1] MIL-HDBK-5J. Metallic Materials and Elements for Aerospace Vehicle Structures. Department of Defence; 2003.
[2] Kalpakjian S. Manufacturing Engineering and Technology. Addison-Wesley; 1989.
[3] AFS-120-73-2. Fatigue Evaluation of Wing and Associated Structure on Small Airplanes. FAA Engineering and Manufacturing Division. Airframe Branch; May, 1973.
[4] Article 17. Stress Corrosion Cracking of Aluminum Alloys. http://www.keytometals.com/Article17.htm.
[5] Niu, Chung-Yung M. Airframe Structural Design. Conmilit Press; 1988.
[6] Flinn RA, Trojen PK. Engineering Materials and Their Applications. 3rd ed. Houghton Mifflin; 1986.
[7] http://asm.matweb.com/search/SpecificMaterial.asp?bassnum=MTP641.
[8] http://www.steelonthenet.com/charts.html.
[9] Tsai SW. Composites Design. 4th ed. Think Composites; 1987.
[10] Jones RM. Mechanics of Composite Materials. Hemisphere; 1975.
[11] Rutan B. Moldless Composite Sandwich Aircraft Construction. Rutan Aircraft; 2005.
[12] Lambie J. Composite Construction for Homebuilt Aircraft. Aviation Publishers; 1984.
[13] Clarke B. Building, Owning, and Flying a Composite Homebuilt. TAB Books; 1985.
[14] MIL-HDBK-17F. Composite Materials Handbook. Department of Defence; 2002.
[15] AC-20-107B. Composite Aircraft Structures. FAA 09/08/2009.
[16] AC-21-26. Quality Control for the Manufacture of Composite Structures. FAA 06/26/1989.
[17] Hoskin BC, Baker AA. Composite Materials for Aircraft Structures. AIAA Education Series, 1986.
[18] Kuhn P. Stresses in Aircraft and Shell Structures. McGraw-Hill; 1956.
[19] ANC-18. Design Of Wood Aircraft Structures. Army-Navy-Commerce Committee; 1944.
[20] NACA R-354. Aircraft Woods: Their Properties, Selection and Characteristics. Markwardt, L. J. 1930.
[21] Advisory Circular AC 43.13-1B. Acceptable Methods, Techniques, and Practices – Aircraft Inspection and Repair. FAA 1998.
[22] Bruhn EF. An Analysis and Design of Flight Vehicle Structures. Jacobs 1973, p. A16.1.

CHAPTER 6

Aircraft Weight Analysis

OUTLINE

6.1 Introduction 134
    Initial Weight Estimation 134
    Detailed Weight Estimation 134
    Weight Estimation Advice 135
    A Comment About Units of Weight 135
  6.1.1 *The Content of this Chapter* 135
  6.1.2 *Definitions* 135
    Empty Weight – $W_e = W_{empty}$ 135
    Design Gross Weight – W_0 135
    Useful Load – $W_u = W_{useful\ load}$ 135
    Payload – $W_p = W_{payload}$ 135
    Crew Weight – $W_c = W_{crew}$ 135
    Fuel Weight – $W_f = W_{fuel}$ 135
    Ramp Weight – $W_R = W_{ramp}$ 135
    Maximum Landing Weight – W_{LDG} 135
    Maximum Zero Fuel Weight – W_{MZF} 135
  6.1.3 *Fundamental Weight Relations* 137
  6.1.4 *Mission Analysis* 137

6.2 Initial Weight Analysis Methods 138
  6.2.1 *Method 1: Initial Gross Weight Estimation Using Historical Relations* 138
  6.2.2 *Method 2: Historical Empty Weight Fractions* 140

6.3 Detailed Weight Analysis Methods 141

6.4 Statistical Weight Estimation Methods 142
  6.4.1 *Method 3: Statistical Aircraft Component Methods* 142
    Wing Weight 142
    Horizontal Tail (HT) Weight 143
    Vertical Tail (VT) weight 143
    Fuselage Weight 143
    Main Landing Gear Weight 143
    Nose Landing Gear Weight 143
    Installed Engine Weight 144
    Fuel System Weight 144
    Flight Control-system Weight 144
    Hydraulic System Weight 144
    Avionics Systems Weight 144
    Electrical System 144
    Air-conditioning and Anti-icing 144
    Furnishings 144
  6.4.2 *Statistical Methods to Estimate Engine Weight* 145
    Weight of Piston Engines 145
    Weight of Turboprop Engines 146
    Weight of Turbofan Engines 146

6.5 Direct Weight Estimation Methods 147
  6.5.1 *Direct Weight Estimation for a Wing* 147
  6.5.2 *Variation of Weight with AR* 154
    Baseline Definitions for a Trapezoidal Wing 155
    Derivation of Equation (6-49) 156
    Derivation of Equation (6-50) 156
    Derivation of Equation (6-51) 156
    Derivation of Equation (6-52) 156
    Derivation of Equation (6-53) 156
    Derivation of Equation (6-54) 157
    Derivation of Equation (6-55) 157
    Method of Fractions 158
    Derivation of Equation (6-56) 159
    Derivation of Equation (6-57) 159
    Derivation of Equation (6-58) 159
    Derivation of Equation (6-59) 159
    Derivation of Equation (6-60) 159
    Derivation of Equation (6-61) 160
    Derivation of Equation (6-62) 160
    Derivation of Equation (6-63) 160

6.6 Inertia Properties 161
  6.6.1 *Fundamentals* 162
  6.6.2 *Reference Locations* 162
  6.6.3 *Total Weight* 162
  6.6.4 *Moment About (X_0, Y_0, Z_0)* 162
  6.6.5 *Center of Mass, Center of Gravity* 164
  6.6.6 *Determination of CG Location by Aircraft Weighing* 165
  6.6.7 *Mass Moments and Products of Inertia* 165
    Fundamental Relationships 165

Parallel-axis Theorem for Moments of Inertia	166	Determination of the Aft CG Limit	170
Derivation of Equation (6-78)	167	Determination of the Forward CG Limit	170
Parallel-plane Theorem for Products of Inertia	167	Loading Cloud	171
6.6.8 Moment of Inertia of a System of Discrete Point Loads	167	6.6.13 In-flight Movement of the CG	173
		6.6.14 Weight Budgeting	173
		6.6.15 Weight Tolerancing	174
6.6.9 Product of Inertia of a System of Discrete Point Loads	168	Exercises	176
6.6.10 Inertia Matrix	168	Variables	178
6.6.11 Center of Gravity Envelope	168	References	180
Design Guidelines	168		
6.6.12 Creating the CG Envelope	169		

6.1 INTRODUCTION

One of the most important steps in the aircraft design process is the estimation of the weight of the vehicle. While usually not requiring complicated mathematical tools, this task can bring considerable challenges for the cognizant engineer. One of the challenges is that excessive under- or overestimation of an airplane's empty weight can bring dire consequences to a development program. Weight-related issues in aircraft development programs have been brought to the forefront recently by the Lockheed-Martin F-35, which allegedly was nearly 3000 lb_f over target. On April 7, 2004, Lockheed declared a stand-down day, notifying its employees that further development of the aircraft would halt until the weight was lost [1]. A number of problems were introduced — if its gross weight was allowed to stand, it would suffer a reduction in useful load. If it was increased, the reduced thrust-to-weight ratio would lead to detriments in performance and render it unable to accomplish its advertised short take-off vertical landing (STOVL) capability.

The history of aviation is wrought with overweight aircraft. In civilian circles the two latest are reports of the Boeing 787 Dreamliner [2] and Airbus 380 [3]. If these established manufacturers can make mistakes in their estimations, then certainly we can too. This section introduces several methods to help the designer estimate the weight of the airplane. The complexity and accuracy of these methods vary. The simplest ones are intended for initial estimation only and their accuracy will only land you in the ballpark. As one would expect, the more complex methods are more accurate, but also depend on some specific knowledge about the geometry and other parameters, such as wing area, quantity of fuel, or length of landing gear. Therefore, they can only be used once the primary dimensions of the aircraft have been established.

As detailed in Section 1.3, *Aircraft design algorithm*, a new estimate of the weight is required during each design iteration. The first iteration calls for a simple weight estimate to be performed and this can be accomplished using either of the two methods presented here. Both are based on historical data and assume the class of the aircraft is known. These methods will only give a rudimentary idea as to how heavy the airplane may become and are intended to allow basic sizing to take place. The second set of methods are more detailed estimations that usually include a mixture of known weights (such as for engines, landing gear components, etc.), statistical weights, and direct weight estimation (based on the geometry and density of materials chosen).

Once the design iterations begin, the designer should update all calculations that refer to the previous weight, as stipulated in Section 1.3.

Initial Weight Estimation

Initial weight analysis methods are only intended to be used during the first iteration of the airplane design. The estimation of the weight is equally important to that of the drag. Overly optimistic weight estimation has killed a number of design programs in the history of aviation and is an engineering problem that is not likely to become a "thing of the past" for the foreseeable future. This section will present some helpful methods to get you started with your new design; however, you must advance into more sophisticated methods during subsequent design iterations.

Detailed Weight Estimation

Detailed weight analysis is performed during the second and any subsequent iteration. It is one of the most important tasks in the design process and must be approached with great care and seriousness. A flawed analysis can contribute to a program cancellation. The most common pitfall is underestimating the empty weight, as stated under maximum zero fuel weight — W_{MZF} below. This section discusses methods used to correctly estimate a likely empty weight of an aircraft, but the reader should be aware that their proper use demands great engineering judgment.

Weight Estimation Advice

Be realistic: do not expect your design to weigh less than airplanes in a similar class — at least not for the first iteration. The aircraft that you're comparing with have often gone through costly weight-reduction programs and many can be considered largely weight-optimized. Remember that the people who worked on these aircraft were smart people as well and probably spent a lot of time trying to get unnecessary weight out. Your design will not be optimized at first.

Be careful: if your airplane ends up weighing less than planned, then great — it will have greater utility or growth capacity than planned and your boss may even pat you on the back. If your airplane ends up heavier than planned, don't be surprised if the project is cancelled. That's really how simple it is, although you would most likely be asked to sharpen your pencil first.

A Comment About Units of Weight

Weight is a force. Its unit in the UK system is lb_f, pronounced "pounds" or "pounds-force". In the SI system, the correct unit is N, pronounced "newtons." Most metric countries, while technically incorrect, specify weight using the unit of mass — kg (kilograms), rather than newtons. This way, an airplane is specified to "weigh," say 906 kilograms (or kilos) and not 19,614 N (which corresponds to 2000 lb_f). For this reason, when referring to the weight of aircraft using the metric system, this convention will be adhered to. The reader is reminded to keep this convention in mind when coming across weight stated as a unit of mass.

6.1.1 The Content of this Chapter

- **Section 6.2** presents two methods intended to assess the first estimate of the airplane's weight.
- **Section 6.3** discusses detailed weight analysis methods, a precursor to the statistical and direct weight estimation methods.
- **Section 6.4** presents a method to estimate the weight of GA aircraft.
- **Section 6.5** presents direct weight estimation methods.
- **Section 6.6** discusses the various inertia properties, including numerical estimation of moments and products of inertia. The importance of weight budgeting in aircraft design is presented, as well as methods to evaluate uncertainty in the prediction of the CG and other inertia properties.

6.1.2 Definitions

The following are standard definitions for weight used in the aircraft industry.

Empty Weight — $W_e = W_{empty}$

Weight of an aircraft without useful load. Includes oil, unusable fuel, and hydraulic fluids.

Design Gross Weight — W_0

The maximum T-O weight for the mission the airplane is designed for.

Useful Load — $W_u = W_{useful\ load}$

Useful load is defined as the difference between the design gross weight and the empty weight. It is the weight of everything the aircraft will carry besides its own weight. This typically includes occupants, fuel, freight, etc.

Payload — $W_p = W_{payload}$

The part of the useful load that yields revenue for the operator. Typically it is the passengers and freight.

Crew Weight — $W_c = W_{crew}$

Weight of occupants required to operate the aircraft.

Fuel Weight — $W_f = W_{fuel}$

Weight of the fuel needed to complete the design mission.

Ramp Weight — $W_R = W_{ramp}$

Design gross weight + small amount of fuel to accommodate warm-up and taxi into T-O position.

Maximum Landing Weight — W_{LDG}

Maximum weight at which the aircraft may land without compromising airframe strength.

Maximum Zero Fuel Weight — W_{MZF}

Max zero fuel weight is the maximum weight the airplane can carry with no fuel on board. Note that the maximum zero fuel weight implies that all weight above W_{MZF} lbs must be fuel.

Of the above definitions, only last of these requires elaboration. It is a relatively common occurrence in the aviation industry that gross weight must be increased. There are typically two underlying reasons for this: (1) the design team underestimated the gross weight, and (2) it is desired that the airplane is capable of being operated at a greater weight than initially anticipated. Note that from a certain point of view, the second reason is a different version of the first reason.

So, consider the predicament confronting a hypothetical design team after initially estimating the empty and gross weight of a new airplane to be 3000 lb$_f$ and 5000 lb$_f$, respectively. This means a useful load of 2000 lb$_f$. Then one day, during the detail design phase, as the design team is meticulously trying to comply with aviation regulations and accommodate management and customer requests for added features, the weights engineer drops a bomb during a design review meeting announcing that there is no way the gross weight target can be met! This airplane will weigh 3500 lb$_f$ empty! Based on the numbers above it is easy to see that this sudden and unwelcome injection of reality means the projected useful load just dropped from 2000 to 1500 lb$_f$ — a whopping one-quarter, eradicating the design's competitive edge! So, what can be done? Should the wing area be increased? Since the wings may already have been designed, this implies a major redesign effort with associated delay in delivery and costs. Perhaps the weights engineer should be fired? Well, a new one will have to be hired and he or she will probably only repeat the bad news, so this won't help. Should a major structural optimization be initiated in the hope that 500 lb$_f$ of lard can be found and designed out of the airplane? If the airframe design is being accomplished by an experienced engineering team it is unlikely there will be more than 50—100 lb$_f$ of extra weight at this stage — and it will be costly and also delay delivery. It is then that a knowledgeable engineer in the group saves the day by suggesting the concept of W_{MZF}. This change is mostly a matter of paperwork and is contingent on the approval of the aviation authorities. If approved, it works as follows and as illustrated in Figure 6-1:

(1) In order to maintain the same competitive useful load of 2000 lb$_f$, the best choice is to increase the gross weight from 5000 to 5500 lb$_f$.

(2) If the fuel is carried in the wings it will act as a bending moment relief and, therefore, not require a change in its internal structure.

(3) A large number of loading combinations are studied during the design in the form of so-called *load cases* (or *load conditions*). As an example, one of the load cases would consist of the airplane with fuel tanks filled to the brim and a few passengers on board; in another load case, all of that fuel has been consumed; and in a third, the airplane is filled with passengers and has very little fuel, and so forth. And it would be shown for all the load combinations that the airframe does indeed comply with aviation regulations. For this reason, it can be argued that this airplane may be loaded indiscriminately up to 5000 lb$_f$; however, once there, all remaining weight (500 lb$_f$) must be fuel in the wing. If this is done, then it can be argued that the weight of the fuel will counter the added aerodynamic load that the wing must generate and react. To see this, assume we add 250 lb$_f$ of fuel to each wing. This weight calls for a 250 lb$_f$ increase in lift, canceling the impact on the shear force that must be reacted. In other words, once airborne, the stress in the spar caps, skin, and shear webs is effectively unchanged.

This elementary physics is often surprisingly hard for many to grasp. So consider the following thought experiment. Imagine an extremely light wing model inside a wind tunnel and at a given airspeed and AOA, some 100 lb$_f$ of lift is measured. This means that the wing attachment is reacting 100 lb$_f$ in shear, right? Now, consider a new model of this wing is installed in the tunnel, except this one weighs 100 lb$_f$. When run at the same airspeed and AOA as the previous model, what

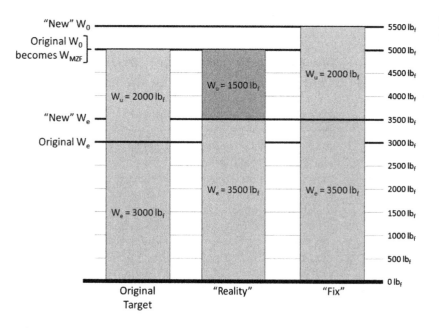

FIGURE 6-1 Justification for maximum zero fuel weight.

is the shear force reacted by the wing attachment? 100 lb$_f$? 0 lb$_f$? If you said 0 lb$_f$, you are correct. By the same token, if 25 lb$_f$ of weight is now added to the wing (so it weighs 125 lb$_f$) and it is made to generate 125 lb$_f$ of lift, the same thing happens; again the net shear being reacted by the attachment is 0 lb$_f$. From the standpoint of the material in the attachment, it sees exactly the same magnitude of stresses in either case.

Of course it's not so simple in practice. There are other consequences of the maximum zero fuel weight. One is an increase in the stalling speed, which, as a consequence of this approach, may call for a redesign of the high-lift system (which is easier than redesigning the entire wing). The other is reputation. The very concept implies some mistakes in weight estimation may have been made along the way, whether this is true or not. Regardless, the concept of maximum zero fuel weight is something the aircraft designer should be aware of and understand. This can come in very handy when least expected.

6.1.3 Fundamental Weight Relations

The following relations are fundamental expressions for an aircraft weight. In comparing Equations (6-1) and (6-3), the reader should be mindful that Equation (6-1) is a primary governing equation and (6-3) is an expression of what the useful load might consist of.

Design gross weight:

$$W_0 = W_e + W_u \qquad (6\text{-}1)$$

Useful load:

$$W_u = W_c + W_f + W_p \qquad (6\text{-}2)$$

Design gross weight:

$$W_0 = W_e + W_c + W_f + W_p \qquad (6\text{-}3)$$

Note that Equation (6-1) gives the maximum or "official" W_u, while Equation (6-2) gives the "current" W_u. It may or may not be equal to the "official" W_u. Weight ratios are imperative in estimating the weight during the first design iteration, but are also necessary for mission analyses. The fundamental weight ratios are the empty weight and fuel weight ratios:

Empty weight ratio:

$$\frac{W_e}{W_0} \qquad (6\text{-}4)$$

Fuel weight ratio:

$$\frac{W_f}{W_0} \qquad (6\text{-}5)$$

EXAMPLE 6-1

Determine the useful load, empty weight ratio, and fuel weight ratio for a Cessna 150 aircraft, using the following information:

Gross weight:
$$W_0 = 1600 \text{ lb}_f$$

Empty weight:
$$W_e = 1100 \text{ lb}_f$$

Fuel weight:
$$W_f = (35 \text{ gal})\left(6 \text{ lb}_f/\text{gal}\right) = 210 \text{ lb}_f$$

Solution

Useful load:
$$W_u = W_0 - W_e = 1600 - 1100 = 500 \text{ lb}_f$$

Empty weight ratio:
$$\frac{W_e}{W_0} = \frac{1100}{1600} = 0.6875$$

Fuel weight ratio:
$$\frac{W_f}{W_0} = \frac{210}{1600} = 0.13125$$

6.1.4 Mission Analysis

When designing aircraft to specific missions, where range or endurance requirements are clearly spelled out, it becomes imperative to assess the amount of fuel required to complete the mission. Mission analysis is a tool intended to help assess the amount of fuel for this purpose. This tool is imperative during the conceptual design stage and is closely related to the estimation of range and endurance. For this reason, further discussion is presented in Section 20.5, *Analysis of mission profile*.

6.2 INITIAL WEIGHT ANALYSIS METHODS

Before beginning serious development on a new aircraft it is absolutely imperative to review the weight of aircraft that belong to the same class as the proposed design. Since such work focuses on historical aircraft, the formulations that result are referred to as *historical relations*. These are typically ratios, for instance, empty and fuel weight ratios. They are based on the assumption that the new airplane will be certified using similar regulations as the historical reference aircraft and, therefore, must react identical load factors. Consequently, as long as the new aircraft carries a similar payload, is similarly configured, and made from similar materials as the reference aircraft, their weights are likely to be similar. The accuracy of these methods depends on the number of reference aircraft and how closely they resemble the one being drafted.

6.2.1 Method 1: Initial Gross Weight Estimation Using Historical Relations

Guidance: Use this method if the gross weight *is not known* beforehand. Be careful — it can yield an over-estimation. Ensure that the reference aircraft database consists of aircraft in the same class and is not a mix of properties. For instance, if designing a piston-propeller aircraft, do not mix turboprop or turbofan aircraft in the database. Also, do not mix small two-place and large 19-place aircraft, or low-performance VFR and high-performance IFR aircraft, and so on.

Consider the design of a new 6-seat, twin-engine, piston-powered aircraft. It fits perfectly into the class of twin-engine GA aircraft — or does it? This class includes other twin-engine aircraft such as the Piper PA-23 Apache, Beech Model 76 Duchess, Beech Model 58 Baron, Piper PA-31 Navajo, Cessna Model 303, and Cessna Model 421. The empty weight of these aircraft ranges from about 3200 to 4500 lb$_f$ and the gross weight from 5200 to 7500 lb$_f$. They are all powered by a number of different piston engine types, and carry different amounts of payload, from six to eight occupants or so. It is prudent to *include* these aircraft in the statistical analysis. However, the Piper PA-42 Cheyenne or the Beech Model 100 King Air, not to mention its larger relatives, the Model 200 and 300 Super King Air, should be *excluded*. These aircraft weigh some 6900 to 7800 lb$_f$ empty; 11,200 to 12,500 lb$_f$ loaded; and are powered by gas turbines, are pressurized, and of higher performance than the aforementioned aircraft. The point being made is that the selection of candidate aircraft must be refined enough to exclude aircraft that either have or lack features that could skew the results. In the case of the above aircraft we have major differences in properties such as pressurized versus unpressurized, piston engine versus turbine, and so forth.

Besides serving as a "sanity check" and, thus, possibly preventing disastrous consequences of under- or overestimating the weight, selective inclusion and exclusion of candidate aircraft improves the reliability and allows a realistic "first stab" estimation of the airplane's gross weight to take place. First, however, the fuel and empty weight ratios must be defined.

Fuel weight ratio:

$$W_f = \left(\frac{W_f}{W_0}\right) W_0 \qquad (6\text{-}6)$$

Empty weight ratio:

$$W_e = \left(\frac{W_e}{W_0}\right) W_0 \qquad (6\text{-}7)$$

Once these have been established, we can rewrite the expression for the *design gross weight* as the sum of the empty weight, crew weight, fuel weight, and payload:

$$W_0 = \left(\frac{W_e}{W_0}\right) W_0 + W_c + \left(\frac{W_f}{W_0}\right) W_0 + W_p \qquad (6\text{-}8)$$

This can be solved for W_0, yielding an expression that can be used to estimate gross weight in terms of the weight ratios.

$$W_0 = \frac{W_c + W_p}{\left[1 - \left(\frac{W_e}{W_0}\right) - \left(\frac{W_f}{W_0}\right)\right]} \qquad (6\text{-}9)$$

Then, the gross weight is estimated as follows: (1) Establish the desired payload, W_p, and crew weight, W_c, for the new design. (2) Determine historical values for fuel and empty weight ratios of similar aircraft. (3) Calculate the proposed gross weight using Equation (6-9).

The ratios W_e/W_0 and W_f/W_0 can be obtained from historical data, providing a possible solution to Equation (6-9). Relationships for W_e/W_0 are provided by Raymer [4], Torenbeek [5], and Nicolai [6]. The ratio

TABLE 6-1 Establishing Weight Ratios for Light Sport Aircraft

	Weight of Selected LSA Aircraft						
	W_e lb_f	W_o lb_f	Q_f gals	W_f lb_f	W_e/W_o	W_u/W_o	W_f/W_o
Aerostar Festival LSA	848	1318	21	127	0.643	0.357	0.096
Allegro 2007	622	1320	13	79	0.471	0.529	0.060
Cessna C-162 Skycatcher	834	1320	24	144	0.632	0.368	0.109
ECO1 EXP (medium)	655	1322	13	79	0.495	0.505	0.060
ECO1 EXP (short)	655	1322	13	79	0.495	0.505	0.060
ECO1 LSA	655	1318	13	79	0.497	0.503	0.060
ECO1 UL	655	1318	13	79	0.497	0.503	0.060
ECO1 VLA (medium)	655	1322	13	79	0.495	0.505	0.060
ECO1 VLA (short)	655	1322	13	79	0.495	0.505	0.060
Flight Design CTLS	717	1320	34	204	0.543	0.457	0.155
Flight Design CTSW	693	1320	34	206	0.525	0.475	0.156
Flyfabriken LN-3 Seagull	639	1323	26	158	0.483	0.517	0.120
RANS S-19 Venterra	820	1320	24	144	0.621	0.379	0.109
RANS S-6 Coyote II	700	1320	18	108	0.530	0.470	0.082
RANS S-7 Courier	732	1320	26	156	0.555	0.445	0.118
Remos GX	639	1320	22	132	0.484	0.516	0.100
Sky Arrow 600 Sport	840	1320	26	155	0.636	0.364	0.117
Sky Arrow 650 Sport	877	1433	17	103	0.612	0.388	0.072
AVERAGES	716	1327	20	122	0.540	0.460	0.092
STANDARD DEVIATIONS	87	27	7	43	0.061	0.061	0.033

Sources of data are various manufacturers' websites. Data may contain erroneous weights.

W_f/W_0 is best obtained by research into similar aircraft. Additionally, aircraft specifications in the public domain, such as those found in *Jane's All the World's Aircraft* or on type certificate data sheets, can be used to build such relationships. Table 6-1 shows an example of such analysis for a number of light sport aircraft. Note that W_e stands for empty weight, W_u is useful load, W_f is fuel load, and W_0 is gross weight.

The statistical values in Table 6-1 suggest that the empty weights of the selected aircraft are 716 ± 87 lb_f and the gross weights 1327 ± 27 lb_f. Furthermore, for the designer of an LSA aircraft, it suggests an empty weight ratio, useful load ratio, and fuel weight ratio of around 0.540, 0.460, and 0.092, respectively. This information is vital for the initial sizing, for instance when performing constraint analysis. Statistical equations for a number of classes of aircraft are presented below in Section 6.2.2, *Method 2: Historical empty weight fractions.*

NB: Many real airplanes exceed their gross weight with all seats loaded and full fuel tanks. This way, an airplane may be capable of a fuel weight ratio of 0.20, but only 0.10 if all seats are occupied. Using the former value with Equation (6-9) would therefore yield gross weight that is unrealistically high. For this reason, the reader is urged to exercise caution and select the value of W_f/W_0 accordingly. For instance, adjust the historical fuel weight ratio by simply calculating $W_f = W_0 - W_e - W_p - W_c$ and compute an adjusted W_f/W_0. Alternatively, the designer is at liberty to decide that when all seats are occupied, the airplane can hold some specific quantity of fuel that may or may not refer to full fuel tanks. This, by the way, is not done in Example 6-2 below, and explains why the gross weight is so unrealistically high.

EXAMPLE 6-2

A four-seat trainer is being designed and it is required to carry 300 lb_f of baggage in addition to the occupants. Assume a crew of one, 200 lb_f/person, and use the ratios W_e/W_0 and W_f/W_0 obtained from the analysis of the Cessna 150 in Example 6-1. For the conceptual design estimate initial values for:

(1) Gross weight.
(2) Empty weight.
(3) Fuel weight.

Solution

Crew weight: $W_c = 200$ lb_f

Payload:

$$W_p = (3 \text{ persons})(200 \text{ lb}_f/\text{person}) + 300 = 900 \text{ lb}_f$$

Empty weight and fuel weight ratios are:

$$W_e/W_0 = 0.6875 \quad \text{and} \quad W_f/W_0 = 0.13125$$

(1) An initial gross weight is:

$$W_0 = \frac{W_c + W_p}{[1 - (W_e/W_0) - (W_f/W_0)]}$$
$$= \frac{200 + 900}{[1 - (0.6875) - (0.13125)]} = 6069 \text{ lb}_f$$

(2) An initial empty weight is:

$$W_e = \left(\frac{W_e}{W_0}\right) W_0 = (0.6875)(6069) = 4172 \text{ lb}_f$$

(3) An initial fuel weight is:

$$W_f = \left(\frac{W_f}{W_0}\right) W_0 = (0.13125)(6069) = 797 \text{ lb}_f$$

6.2.2 Method 2: Historical Empty Weight Fractions

Guidance: Use this method if the gross weight *is known* beforehand. This is the case for many types of aircraft, e.g. LSA, which should not weigh more than 1320 lb_f or 1430 lb_f if amphibious. Do not "back out" W_0 from a desired empty weight ratio using this method.

The modern aerospace engineer is in the enviable position of having access to a large collection of different kinds of airplanes to compare the design to (Figure 6-2). This is priceless during the conceptual design phase, when one or more airplanes can typically be found that

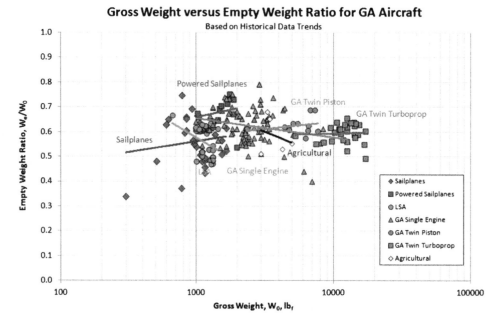

FIGURE 6-2 Historical empty weight fraction trends based on aircraft class.

compare well to the one being designed. In particular, this is helpful when it comes to estimating weight. Empty weight fractions are an important early step in the design process and people have devised statistical formulas based on historical aircraft that allow an initial empty or gross weight to be estimated by a class of aircraft.

Sailplanes:

$$\frac{W_e}{W_0} = \begin{cases} 0.2950 + 0.0386 \cdot \ln W_0 & \text{if } W_0 \text{ is in lb}_f \\ 0.3255 + 0.0386 \cdot \ln W_0 & \text{if } W_0 \text{ is in kg} \end{cases}$$
(6-10)

Powered sailplanes:

$$\frac{W_e}{W_0} = \begin{cases} 0.3068 + 0.0510 \cdot \ln W_0 & \text{if } W_0 \text{ is in lb}_f \\ 0.3471 + 0.0510 \cdot \ln W_0 & \text{if } W_0 \text{ is in kg} \end{cases}$$
(6-11)

Light sport aircraft (land):

$$\frac{W_e}{W_0} = \begin{cases} 1.5451 - 0.1402 \cdot \ln W_0 & \text{if } W_0 \text{ is in lb}_f \\ 1.4343 - 0.1402 \cdot \ln W_0 & \text{if } W_0 \text{ is in kg} \end{cases}$$
(6-12)

Light sport aircraft (amphib):

$$\frac{W_e}{W_0} = \begin{cases} 1.6351 - 0.1402 \cdot \ln W_0 & \text{if } W_0 \text{ is in lb}_f \\ 1.5243 - 0.1402 \cdot \ln W_0 & \text{if } W_0 \text{ is in kg} \end{cases}$$
(6-13)

GA single-engine:

$$\frac{W_e}{W_0} = \begin{cases} 0.8841 - 0.0333 \cdot \ln W_0 & \text{if } W_0 \text{ is in lb}_f \\ 0.8578 - 0.0333 \cdot \ln W_0 & \text{if } W_0 \text{ is in kg} \end{cases}$$
(6-14)

GA twinpiston:

$$\frac{W_e}{W_0} = \begin{cases} 0.4074 + 0.0253 \cdot \ln W_0 & \text{if } W_0 \text{ is in lb}_f \\ 0.4274 + 0.0253 \cdot \ln W_0 & \text{if } W_0 \text{ is in kg} \end{cases}$$
(6-15)

GA twin turboprop:

$$\frac{W_e}{W_0} = \begin{cases} 0.5319 + 0.0066 \cdot \ln W_0 & \text{if } W_0 \text{ is in lb}_f \\ 0.5371 + 0.0066 \cdot \ln W_0 & \text{if } W_0 \text{ is in kg} \end{cases}$$
(6-16)

Agricultural:

$$\frac{W_e}{W_0} = \begin{cases} 1.4029 - 0.0995 \cdot \ln W_0 & \text{if } W_0 \text{ is in lb}_f \\ 1.3242 - 0.0995 \cdot \ln W_0 & \text{if } W_0 \text{ is in kg} \end{cases}$$
(6-17)

6.3 DETAILED WEIGHT ANALYSIS METHODS

Upon completion of the first design iteration[1], which yields the initial weight estimate, it is now time for the design team to sharpen their pencils and begin the estimation of a more accurate empty weight for the aircraft. The initial weight estimate should only be considered a value that gives an "idea of the weight" until the detailed weight analysis is completed.

During the second (and even subsequent) design iterations, things change frequently and this can pose problems for any trade study. An effective weight analyst will therefore try to create relationships between size and weight of airplane parts, as this will be of great help to those conducting trade studies. As an example, such relationships should allow the user to estimate the weight of the proposed design as a function of parameters such as wing area or number of occupants, and so on.

Another important result of weight analysis is information that allows weight budgeting to be prepared. The process will yield weight for components such as the wing, HT, VT, fuselage, etc., which can then be used to establish target weights for the components. It is of utmost importance that the structural design team is aware of the weight target of the structure as this will help in directing the design toward a lighter one. Weight targeting via weight budgeting is discussed in some detail in Section 6.6.14, *Weight budgeting*.

Typically, detailed weight estimation methods include:

Known weights	—this section
Statistical weight estimation	—see Section 6.4
Direct weight estimation	—see Section 6.5

The concept "known weights" refers to parts and components that can either be weighed with reasonable accuracy or whose manufacturer (if the component is obtained from an outside vendor) can disclose with reasonable confidence. Most of the time the weight analyst uses all three methods simultaneously, but known weights always supersede both the statistical and direct weight estimations. The following parts can be expected to have published weights:

Engines
Propellers
Wheels, tires, brakes, etc.
Landing gear
Standard parts (electronics, avionics, antennas, instruments, fasteners, etc.)

---

[1] See for instance Section 1.3.1, *Conceptual design algorithm for a general aviation aircraft*.

Statistical and direct weight estimations will now be treated in some detail.

6.4 STATISTICAL WEIGHT ESTIMATION METHODS

Statistical weight estimation methods are based on historical data derived from existing airplanes. For instance, if we know the weight of the wing structure for a population of aircraft that fall into a specific class (e.g. twin-engine propeller aircraft), it is possible to derive relationships that could be based on geometric parameters such as wing area, aspect ratio, and taper ratio, as well as limit or ultimate load factors. The assumption is that the wing weight of two types of aircraft in the same class that are certified to the same set of regulations and whose gross weight is similar should be similar, even if made by different manufacturers. The statistical relationship established by the entire class of aircraft can thus be used to estimate the wing weight of any aircraft of the same class as long as it falls between the extremes of the aircraft in that class. Such estimation methods usually require some dimensions to have been established beforehand (e.g. *AR*, *TR*, sweep, *S*, etc.). Such methods are often developed in industry or in academia. Since many airplanes feature aluminum and composites alike, the user must use such statistical methods with care, as these may be solely based on aluminum aircraft.

Statistical weight estimation methods are always based on a specific class of aircraft, for instance, GA aircraft, commercial aircraft, fighters, and so on. Such classes share commonalities that improve the accuracy of the formulation. However, be mindful that some classes of aircraft have seen advances, such as an increased use of composites, that may skew the resulting weights.

6.4.1 Method 3: Statistical Aircraft Component Methods

Guidance: Only use this method once you have more information about the geometry of the aircraft. This is not an initial weight estimation method like methods 1 and 2, presented earlier; it requires a large amount of data that results from analysis that follows the use of those methods. *It is to be used after!* **Also note that this method yields the empty weight, W_e, of the aircraft. To get the gross weight, W_0, occupants, freight, and fuel must be added.**

The following equations are presented in Raymer [4] and Nicolai [6] and are intended for conventional GA aircraft only. Both references cite their primary sources for these methods and both provide estimation methods for other aircraft types besides GA aircraft. The interested reader is directed to those sources for further details.

Note that the equations must be used with care as they are unit-sensitive; e.g. some arguments are in ft, while others are in inches. Also, the reader may ask: "Which method should I select, and why?" The short answer is that both methods should be used and, when possible, engineering judgment should be used to narrow down the two options. In some cases, the average of the two values should be pondered. In other cases, the two methods present identical equations (because they come from the same original source). The reader is strongly urged to apply it to a number of aircraft in the same class and evaluate how "close" it matches their empty weight. If the results do not match well, the results should be used to develop scaling factors. For instance, if the predicted weight is, say, 25% lighter than the actual empty weight of the reference aircraft, the results for the new aircraft should be multiplied by a factor of 1.25.

Wing Weight
Raymer:

$$W_W = 0.036 \cdot S_W^{0.758} W_{FW}^{0.0035} \left(\frac{AR_W}{\cos^2 \Lambda_{C/4}}\right)^{0.6} \\ \times q^{0.006} \lambda^{0.04} \left(\frac{100 \cdot t/c}{\cos \Lambda_{C/4}}\right)^{-0.3} (n_z W_O)^{0.49} \quad (6\text{-}18)$$

Nicolai:

$$W_W = 96.948 \cdot \left[\left(\frac{n_z W_O}{10^5}\right)^{0.65} \left(\frac{AR_W}{\cos^2 \Lambda_{C/4}}\right)^{0.57} \\ \times \left(\frac{S_W}{100}\right)^{0.61} \left(\frac{1+\lambda}{2(t/c)}\right)^{0.36} \sqrt{1+\frac{V_H}{500}}\right]^{0.993} \quad (6\text{-}19)$$

where

W_W = predicted weight of wing in lb$_f$
S_W = trapezoidal wing area in ft^2
W_{FW} = weight of fuel in wing in lb$_f$ (If $W_{FW} = 0$ then let $W_{FW}^{0.0035} = 1$)
AR = Aspect Ratio of wing, HT, or VT, per the appropriate subscripts
$\Lambda_{C/4}$ = wing sweep at 25% MGC
q = dynamic pressure at cruise
λ = wing taper ratio
t/c = wing thickness to chord ratio
n_Z = ultimate load factor
W_0 = design gross weight in lb$_f$
V_H = maximum level airspeed at S-L in KEAS

Horizontal Tail (HT) Weight

Raymer:

$$W_{HT} = 0.016(n_z W_O)^{0.414} q^{0.168} S_{HT}^{0.896} \left(\frac{100 \cdot t/c}{\cos \Lambda_{HT}}\right)^{-0.12}$$

$$\cdot \left(\frac{AR_{HT}}{\cos^2 \Lambda_{HT}}\right)^{0.043} \lambda_{HT}^{-0.02} \qquad (6\text{-}20)$$

Nicolai:

$$W_{HT} = 127 \left[\left(\frac{n_z W_O}{10^5}\right)^{0.87} \left(\frac{S_{HT}}{100}\right)^{1.2} \left(\frac{l_{HT}}{10}\right)^{0.483} \sqrt{\frac{b_{HT}}{t_{HT\max}}}\right]^{0.458}$$

$$(6\text{-}21)$$

where

W_{HT} = predicted weight of HT in lb$_f$
S_{HT} = trapezoidal HT area in ft^2
Λ_{HT} = HT sweep at 25% MGC
λ_{HT} = HT taper ratio
l_{HT} = horizontal tail arm, from wing C/4 to HT C/4 in ft
b_{HT} = HT span in ft
$t_{HT\,max}$ = max root chord thickness of HT in inches

Vertical Tail (VT) Weight

Raymer:

$$W_{VT} = 0.073(1 + 0.2 F_{tail})(n_z W_O)^{0.376} q^{0.122}$$

$$S_{VT}^{0.873} \left(\frac{100 \cdot t/c}{\cos \Lambda_{VT}}\right)^{-0.49} \cdot \left(\frac{AR_{VT}}{\cos^2 \Lambda_{VT}}\right)^{0.357} \lambda_{VT}^{0.039}$$

$$(6\text{-}22)$$

Nicolai:

$$W_{VT} = 98.5 \left[\left(\frac{n_z W_O}{10^5}\right)^{0.87} \left(\frac{S_{VT}}{100}\right)^{1.2} \sqrt{\frac{b_{VT}}{t_{VT\max}}}\right] \qquad (6\text{-}23)$$

where

W_{VT} = predicted weight of VT in lb$_f$
F_{tail} = 0 for conventional tail, = 1 for T-tail
S_{VT} = trapezoidal VT area in ft^2
Λ_{VT} = VT sweep at 25% MGC
λ_{VT} = VT taper ratio
b_{VT} = HT span in ft
$T_{VT\,max}$ = max root chord thickness of VT in inches

Fuselage Weight

Raymer:

$$W_{FUS} = 0.052 \cdot S_{FUS}^{1.086} (n_z W_O)^{0.177} l_{HT}^{-0.051} \left(\frac{l_{FS}}{d_{FS}}\right)^{-0.072} q^{0.241}$$

$$+ 11.9(V_P \Delta P)^{0.271} \qquad (6\text{-}24)$$

Nicolai:

$$W_{FUS} = 200 \left[\left(\frac{n_z W_O}{10^5}\right)^{0.286} \left(\frac{l_F}{10}\right)^{0.857} \right.$$

$$\left. \left(\frac{w_F + d_F}{10}\right) \left(\frac{V_H}{100}\right)^{0.338} \right]^{1.1} \qquad (6\text{-}25)$$

where

W_{FUS} = predicted weight of the fuselage in lb$_f$
S_{FUS} = fuselage wetted area in ft^2
l_{FS} = length of fuselage structure (forward bulkhead to aft frame) in ft
d_{FS} = depth of fuselage structure in ft
V_P = volume of pressurized cabin section in ft^3
ΔP = cabin pressure differential, in psi (typically 8 psi)
l_F = fuselage length in ft
w_F = fuselage max width in ft
d_F = fuselage max depth in ft

Main Landing Gear Weight

Raymer:

$$W_{MLG} = 0.095(n_l W_l)^{0.768}(L_m/12)^{0.409} \qquad (6\text{-}26)$$

Nicolai:

$$W_{MNLG} = 0.054(n_l W_l)^{0.684}(L_m/12)^{0.601} \qquad (6\text{-}27)$$

where

W_{MLG} = predicted weight of the main landing gear in lb$_f$
n_l = ultimate landing load factor
W_l = design landing weight in lb$_f$
L_m = length of the main landing gear strut in inches
W_{MNLG} = predicted weight of the entire landing gear in lb$_f$

Nose Landing Gear Weight

Raymer:

$$W_{NLG} = 0.125(n_l W_l)^{0.566}(L_n/12)^{0.845} \qquad (6\text{-}28)$$

where

W_{NLG} = predicted weight of the nose landing gear in lb$_f$
n_l = ultimate landing load factor
W_l = design landing weight in lb$_f$
L_n = length of the nose landing gear strut in inches

Installed Engine Weight

Raymer:
$$W_{EI} = 2.575 W_{ENG}^{0.922} N_{ENG} \quad (6\text{-}29)$$

Nicolai:
$$W_{EI} = 2.575 W_{ENG}^{0.922} N_{ENG} \quad (6\text{-}30)$$

where

W_{EI} = predicted weight of the engine installed in lb_f
W_{ENG} = uninstalled engine weight in lb_f
N_{ENG} = number of engines

Fuel System Weight

Raymer:
$$W_{FS} = 2.49 Q_{tot}^{0.726} \left(\frac{Q_{tot}}{Q_{tot} + Q_{int}}\right)^{0.363} N_{TANK}^{0.242} N_{ENG}^{0.157} \quad (6\text{-}31)$$

Nicolai:
$$W_{FS} = 2.49 \left[Q_{tot}^{0.6} \left(\frac{Q_{tot}}{Q_{tot} + Q_{int}}\right)^{0.3} N_{TANK}^{0.2} N_{ENG}^{0.13} \right]^{1.21} \quad (6\text{-}32)$$

where

W_{FS} = predicted weight of the fuels system in lb_f
Q_{tot} = total fuel quantity in gallons
Q_{int} = fuel quantity in integral fuel tanks in gallons
N_{TANK} = number of fuel tanks

Flight Control-system Weight

Raymer:
$$W_{CTRL} = 0.053 l_{FS}^{1.536} b^{0.371} \left(n_z W_O \times 10^{-4}\right)^{0.80} \quad (6\text{-}33)$$

Nicolai:
$$W_{CTRL} = 1.08 W_O^{0.7} \quad \text{(Powered control system)}$$
$$W_{CTRL} = 1.066 W_O^{0.626} \quad \text{(Manual control system)} \quad (6\text{-}34)$$

where

W_{CTRL} = predicted weight of the flight control system in lb_f
b = wingspan in ft

Hydraulic System Weight

Raymer:
$$W_{HYD} = 0.001 W_O \quad (6\text{-}35)$$

where

W_{HYD} = predicted weight of the hydraulics system in lb_f

Avionics Systems Weight

Raymer:
$$W_{AV} = 2.117 W_{UAV}^{0.933} \quad (6\text{-}36)$$

Nicolai:
$$W_{AV} = 2.117 W_{UAV}^{0.933} \quad (6\text{-}37)$$

where

W_{AV} = predicted weight of the avionics installation in lb_f
W_{UAV} = weight of the uninstalled avionics in lb_f

Electrical System

Raymer:
$$W_{EL} = 12.57 (W_{FS} + W_{AV})^{0.51} \quad (6\text{-}38)$$

Nicolai:
$$W_{EL} = 12.57 (W_{FS} + W_{AV})^{0.51} \quad (6\text{-}39)$$

Air-conditioning and Anti-icing

Raymer:
$$W_{AC} = 0.265 W_O^{0.52} N_{OCC}^{0.68} W_{AV}^{0.17} M^{0.08} \quad (6\text{-}40)$$

Nicolai:
$$W_{AC} = 0.265 W_O^{0.52} N_{OCC}^{0.68} W_{AV}^{0.17} M^{0.08} \quad (6\text{-}41)$$

where

W_{AC} = predicted weight of the AC and anti installation in lb_f
N_{OCC} = number of occupants (crew and passengers)
M = Mach number

Furnishings

Raymer:
$$W_{FURN} = 0.0582 W_O - 65 \quad (6\text{-}42)$$

Nicolai:
$$W_{FURN} = 34.5 N_{CREW} q_H^{0.25} \quad (6\text{-}43)$$

where

W_{FURN} = predicted weight of furnishings in lb_f
N_{CREW} = number of crew.
q_H = dynamic pressure at max level airspeed, lb_f/ft^2

EXAMPLE 6-3

Determine the wing weight for a light airplane with the following specifications:

- A = wing aspect ratio = 16
- N_Z = ultimate load factor = 1.5 × 4.0 = 6.0g
- Q = dynamic pressure at cruise = 0.5 × 0.002378 × $(120 \times 1.688)^2 = 48.8$ lb$_f$/ft^2
- S_W = trapezoidal wing area = 130 ft^2
- t/c = wing thickness to chord ratio = 0.16
- W_{dg} = design gross weight = 1320 lb$_f$
- W_{fw} = weight of fuel in wing = 100 lb$_f$
- λ = wing taper ratio = 0.5
- Λ = wing sweep at 25% MAC = 0°

Solution

Wing weight:

$$W_{wing} = 0.036 \cdot S_w^{0.758} W_{fw}^{0.0035} \left(\frac{A}{\cos^2 \Lambda}\right)^{0.6}$$
$$\times Q^{0.006} \lambda^{0.04} \left(\frac{100 \cdot t/c}{\cos \Lambda}\right)^{-0.3} \left(N_z W_{dg}\right)^{0.49}$$
$$= 0.036 \cdot (130)^{0.758} (100)^{0.0035} \left(\frac{16}{1}\right)^{0.6} (48.8)^{0.006}$$
$$\times (0.5)^{0.04} \left(\frac{16}{1}\right)^{-0.3} (6 \times 1320)^{0.49}$$
$$= 273 \text{ lb}_f$$

6.4.2 Statistical Methods to Estimate Engine Weight

The following methods have been derived based on a large number of piston, turboprops, and turbofans. These weights correspond to W_{ENG} in Equations (6-29) and (6-30). They are based on manufacturers' data.

Weight of Piston Engines

Figure 6-3 shows the correlation between the uninstalled weight and rated brake horsepower of a variety of current piston engines. It is helpful for the designer when estimating the weight of a new engine, for instance, during the weight estimation phase or for multi-disciplinary optimization. Numerical analysis of the data reveals that if some specific rated BHP is being sought, the weight of the resulting engine can be estimated from:

Piston engines:

$$W_{ENG} = \frac{P_{rated} - 21.55}{0.5515} \qquad (6\text{-}44)$$

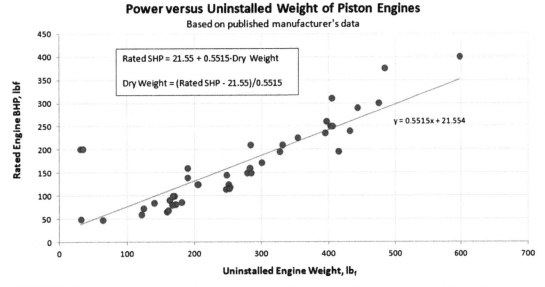

FIGURE 6-3 There is a correlation between the uninstalled weight of piston engines and their rated power.

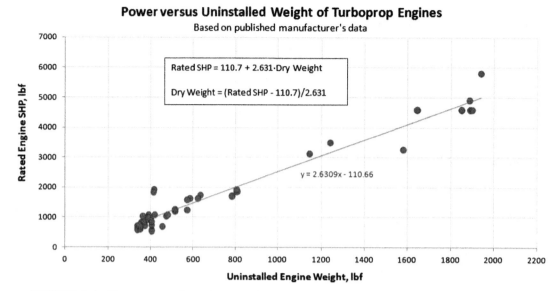

FIGURE 6-4   There is a correlation between the uninstalled weight of turboprops and their rated power.

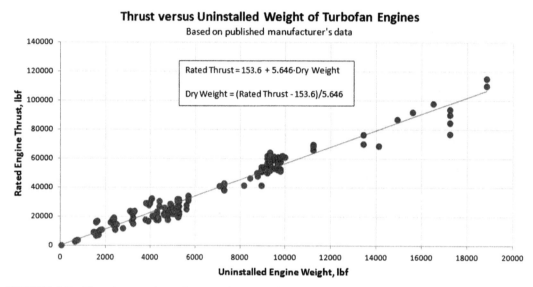

FIGURE 6-5   There is a correlation between the uninstalled weight of turbofan engines and their rated thrust.

Weight of Turboprop Engines

Figure 6-4 shows the correlation between the uninstalled weight and rated shaft-horsepower of a variety of current turboprop engines. This is a helpful tool for the designer, when a new engine is being developed for a specific application whose weight remains to be established. Numerical analysis of the data reveals that if some specific rated SHP is being sought, the weight of the resulting engine can be estimated from:

Turboprop engines:

$$W_{ENG} = \frac{P_{rated} - 110.7}{2.631} \qquad (6\text{-}45)$$

where P_{rated} = rated power of the engine in BHP (pistons) or SHP (turboprops)

Of course, and as is evident from Figure 6-4, there are outliers on the graph that the designer should be mindful of.

Weight of Turbofan Engines

Figure 6-5 shows the correlation between the uninstalled weight and rated thrust of a variety of current turbofan engines. This is a helpful tool for the designer, when a new engine is being developed for a specific application whose weight remains to be established. Statistical analysis of the data reveals that if some specific rated thrust is being sought, the weight of the resulting engine can be estimated from:

Turbofan engines:

$$W_{ENG} = \frac{T_{rated} - 153.6}{5.646} \qquad (6\text{-}46)$$

6.5 DIRECT WEIGHT ESTIMATION METHODS

Components such as wings, fuselage, HT, VT, and control surfaces frequently require direct weight estimation. Nowadays, access to solid modeling software simplifies the effort considerably and can become fairly accurate (but remember: garbage in − garbage out!). However, if one doesn't have access to such software, one must resort to weight modeling via geometric analysis. The quality of effort is dependent on the analyst. An example of how such estimation is detailed in this section, using a simplified representation of a wing, whose total lift is denoted by L. It can be easily extended to quickly assess material requirements for other lifting surfaces as well.

6.5.1 Direct Weight Estimation for a Wing

It is important to recognize that this method is just intended to get you "in the ballpark." It is not a substitute for a detailed load and structural analysis. It will not consider other failure modes (buckling, crippling, etc.) and loading (asymmetric, deflected controls, etc.). Also, a single-cell torsion box structure is assumed, but a more refined version of this method could assume more than one spar (multi-cell).

Note that for initial design purposes it is common to break the wing structure into categories based on structural role:

Skin only reacts the wing torsion.
Spar shear web only reacts the vertical shear force.
Spar caps only react the bending loads.

Consider the simple wing of Figure 6-6, whose top view is shown in Figure 6-7. Such a wing, if made from aluminum, would typically feature a main spar, aft spar (or shear web), ribs, and skin riveted together to form a stiff but light structure. Note that although the following discusses wings in particular, the method applies for any lifting surface that features spars, ribs, and skins, such as horizontal and vertical tails.

Figure 6-8 shows an arbitrary cross section of the wing. The upper image shows the extent of the control surface (e.g. flaps or aileron), while the lower shows structural detail inside it.

In order to allow a rapid estimate of the weight of the wing, the structure is idealized. This is shown in Figure 6-9. The entire cross-sectional area of all spar caps[2] is concentrated in the upper and lower spar caps. If multiple spars are used, the average height of the spars should be used to avoid overestimation of the structural depth. Similarly, the entire thickness of all shear webs is concentrated in the idealized shear web. A further idealization takes place by assuming the skin and airfoil to be represented by a parabolic D-cell section as shown in the figure. It will be assumed that spacing between ribs is one-half the average cell length and that their thickness equals that of the skin. There are limitations to this

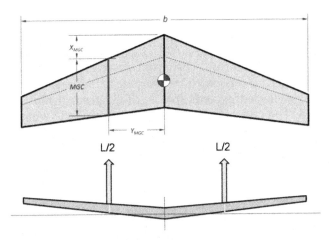

FIGURE 6-6 Lift is applied as a point load at the mean geometric chord.

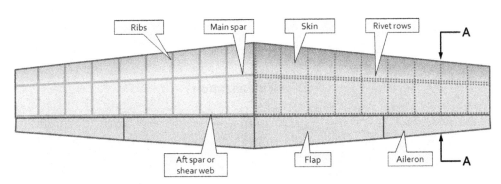

FIGURE 6-7 Simple straight tapered wing.

---

[2] The term "all spars" refers to wing structures that feature multiple spars.

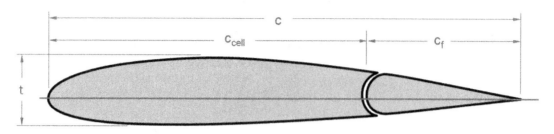

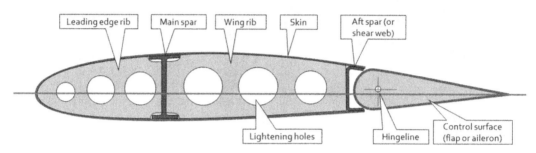

FIGURE 6-8 Section A-A showing structural detail.

idealization that the designer should be aware of. These include the omission of electrical harnesses; fuel and control system; and hard-points for landing gear or external ordnance.

The critical loads reacted by the structure must be identified before the weight can be calculated. The designer must know the expected *ultimate* load factor the airplane will be designed to, as well as a representative candidate airfoil. The airfoil is necessary so we can calculate the torsion to be reacted. If the torsion with flaps deflected exceeds that at dive speed, it must be used. The critical loads are then applied to the wing as shown in Figure 6-10. If the design gross weight of the aircraft is denoted by W_0, the maximum lift of the

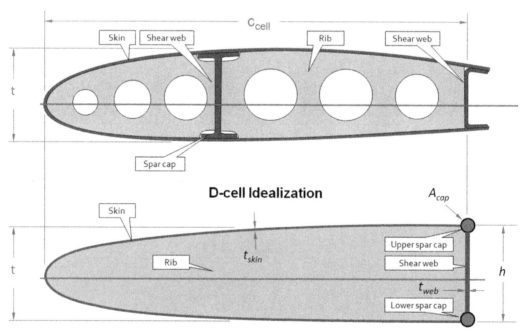

FIGURE 6-9 Idealization of the structural detail.

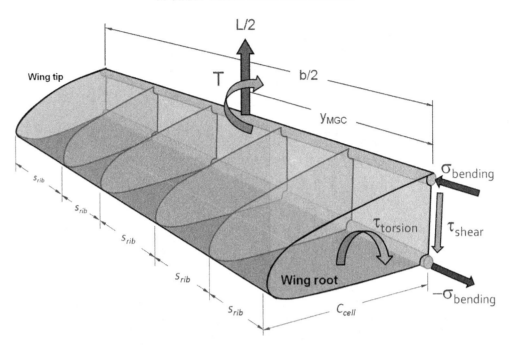

FIGURE 6-10 Loads reacted by the idealized wing segment.

airplane, L, will depend on the ultimate load factor as follows:

$$L = n_{ult} W_0 \qquad (6\text{-}47)$$

If we ignore the fact that a part of this lift will be carried by the fuselage and horizontal tail, this assumes the wing reacts the entire lift. Each wing, thus, reacts one-half of that total force and torsion, T, both of which are calculated as follows:

$$\begin{aligned} \frac{L}{2} &= \frac{n_{ult} W_0}{2} \\ T &= \frac{1}{2}\rho V^2 \frac{S}{2} \cdot c_{MGC} \cdot C_m = \frac{1}{4}\rho V^2 S \cdot c_{MGC} \cdot C_m \end{aligned} \qquad (6\text{-}48)$$

The method below assumes these to be applied to the wing, as shown in Figure 6-10. The lift force must be applied at the spanwise location of the mean geometric chord to ensure the bending moment is accounted for and reacted in the spar caps. Since the lift and torsion are really distributed loads and not a point load and moment as indicated in Figure 6-10, both are zero at the wingtip and reach the maximum at the wing root, as approximated by Equation (6-48). For this reason we will allow the geometry of the structure to taper to a certain minimum value, as it would in a real wing. Otherwise, we will greatly overestimate the weight of the structure. A suitable spar cap area, A_{cap}, at the tip can be assumed to be 5–10% of the area at the root. The wing skin and shear web can be assumed to gradually reduce to the minimum aluminum sheet thickness 0.020″.

Note that compression strength is picked rather than tensile strength to be conservative. See the list of variables for definitions of terms.

Step 1: Weight of the Wing Skin

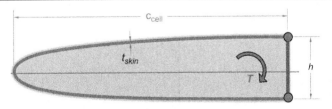

Assumption:	The torsional moment of the wing half is reacted entirely by the wing skin.
Cell area at root:	$A_{cell} = \dfrac{4C_{cell}h/2}{3} = \dfrac{2C_{cell}h}{3}$

Step 1: Weight of the Wing Skin

Cell area at tip:
$$A_{cell_T} = \frac{2(\lambda C_{cell})(\lambda h)}{3} = \lambda^2 A_{cell}$$

Cell arc length at root:
$$s_{cell} = \sqrt{(h/2)^2 + 4C_{cell}^2} + \frac{(h/2)^2}{4C_{cell}}\sinh^{-1}\left(\frac{2C_{cell}}{h/2}\right) = \sqrt{\frac{h^2}{4} + 4C_{cell}^2} + \frac{h^2}{16C_{cell}}\sinh^{-1}\left(\frac{4C_{cell}}{h}\right)$$

Cell arc length at tip:
$$s_{cell_T} = \lambda s_{cell}$$

Torsion at root:
$$T = \frac{1}{2}\rho V^2 \frac{S}{2} \cdot c_{MGC} \cdot C_m = \frac{1}{4}\rho V^2 S \cdot c_{MGC} \cdot C_m$$

Skin shear stress at root:
$$\tau_{skin} = \frac{|T|}{2At} = \frac{\frac{1}{4}\rho V^2 S \cdot c_{MGC} \cdot |C_m|}{2A_{cell}t_{skin}} = \frac{\rho V^2 S \cdot c_{MGC} \cdot |C_m|}{8A_{cell}t_{skin}}$$

Required minimum skin thickness at root:
$$\tau_{max} > \frac{\rho V^2 S \cdot c_{MGC} \cdot |C_m|}{8A_{cell}t_{skin}} \Rightarrow t_{skin} > \frac{\rho V^2 S \cdot c_{MGC} \cdot |C_m|}{8A_{cell}\tau_{max}}$$

Don't select skin thickness less than 0.020".

Required minimum skin thickness at tip:
$$t_{skin_T} > 0.15 t_{skin}$$

Don't select skin thickness less than 0.020".

Weight of skin:
$$W_{skin} = \rho_{skin}\frac{(t_{skin} + t_{skin_T})}{2}\frac{b}{2}\frac{(s_{cell} + s_{cell_T})}{2} = \rho_{skin}\frac{b(t_{skin} + t_{skin_T})s_{cell}(1+\lambda)}{8}$$

Step 2: Weight of Wing Shear Web

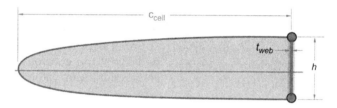

Assumption: The shear force of the wing half is reacted entirely by the shear web.

Moment of inertia at root:
$$I = \frac{t_{web}h^3}{12}$$

First area moment at root:
$$Q = A_{halfweb} \cdot \bar{y} = \left(\frac{h}{2}\right)t_{web}\left(\frac{h}{4}\right) = \frac{h^2 t_{web}}{8}$$

Shear force at root:
$$V = \frac{n_{ult}W}{2} = \frac{3n_{lim}W}{4}$$

Shear web stress at root:
$$\tau_{web} = \frac{VQ}{It} = \frac{\left(\frac{n_{ult}W}{2}\right)\left(\frac{h^2 t_{web}}{8}\right)}{\left(\frac{t_{web}h^3}{12}\right)t_{web}} = \frac{12(n_{ult}W)(h^2 t_{web})}{16(t_{web}h^3)t_{web}} = \frac{3(n_{ult}W)}{4t_{web}h}$$

Shear web thickness at root:
$$\tau_{max} > \frac{3(n_{ult}W)}{4t_{web}h} \Rightarrow t_{web} > \frac{3(n_{ult}W)}{4h\tau_{max}}$$

Don't select web thickness less than 0.020".

Shear web thickness at tip:
$$t_{web_T} > 0.15 t_{web}$$

Don't select web thickness less than 0.020".

Weight of shear web:
$$W_{web} = \rho_{web}\frac{(t_{web} + t_{web_T})}{2}\frac{b}{2}\frac{h(1+\lambda)}{2} = \rho_{web}\frac{b(t_{web} + t_{web_T})h(1+\lambda)}{8}$$

Step 3: Weight of Wing Spar Caps

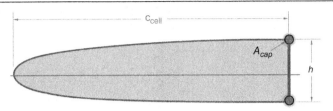

Assumption:	The bending moment of the wing half is reacted entirely by the spar caps.
Bending moment at root:	$M = F_{bend} \times h = \dfrac{n_{ult} W}{2} \times Y_{MGC}$
"Bending force" at root (the couple that reacts M):	$F_{bend} = \dfrac{n_{ult} W}{2} \dfrac{Y_{MGC}}{h}$
Bending stress at root:	$\sigma_{bending} = \dfrac{F_{bend}}{A_{cap}} = \dfrac{\frac{n_{ult} W}{2} \frac{Y_{MGC}}{h}}{A_{cap}} = \dfrac{n_{ult} \cdot W \cdot Y_{MGC}}{2 \cdot h \cdot A_{cap}}$
Spar cap area at root:	$\sigma_{max} > \dfrac{n_{ult} \cdot W \cdot Y_{MGC}}{2 \cdot h \cdot A_{cap}} \;\Rightarrow\; A_{cap} > \dfrac{n_{ult} \cdot W \cdot Y_{MGC}}{2 \cdot h \cdot \sigma_{max}}$
Spar cap area at tip:	$A_{cap_T} > 0.05 A_{cap}$ Don't select cap area less than 0.010 in^2.
Weight of spar caps:	$W_{caps} = 2 \times \rho_{caps} \dfrac{(A_{cap} + A_{cap_T})}{2} \dfrac{b}{2} = \rho_{caps} \dfrac{b(A_{cap} + A_{cap_T})}{2}$ Note that there are two spar caps (upper and lower) and that is why the weight is multiplied by 2.

Step 4: Weight of Ribs

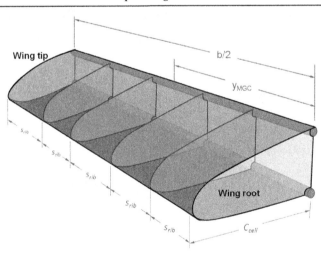

Assumption:	The spacing of ribs is approximately one-half of c_{cell}.
Number of ribs:	$N_{rib} \approx INT\left(\dfrac{b}{C_{MGC}}\right) + 1 \approx INT\left(\dfrac{b}{C_{avg}}\right) + 1$ Where INT stands for the integer value of the ratio.
Thickness of ribs:	$t_{rib} = t_{skin}$ Don't select rib thickness less than 0.020″
Weight of ribs:	$W_{ribs} = \rho_{ribs} \dfrac{(A_{cell} + A_{cell_T})}{2} \dfrac{(t_{skin} + t_{skin_T})}{2} N_{ribs} = N_{ribs} \rho_{ribs} \dfrac{A_{cell}(1 + \lambda^2)(t_{skin} + t_{skin_T})}{4}$

EXAMPLE 6-4

Estimate the wing weight for a light airplane of Example (6-3) with the following specifications:

AR = wing aspect ratio = 16
C_m = average airfoil pitching moment = -0.1
L = wing sweep at 25% MAC = 0°
N_Z = ultimate load factor = 1.5 × 4.0 = 6.0g
S = trapezoidal wing area = 130 ft²
t/c = wing thickness to chord ratio = 0.16
V_D = dive speed = 150 KCAS
W_{dg} = design gross weight = 1320 lb$_f$
λ = wing taper ratio = 0.5

Aluminum sheets are available in the following thicknesses:

0.016″, 0.020″, 0.025″, 0.032″, 0.040″, 0.050″, 0.063″

Compare this with the result of Example (6-3). Assume the wing main element chord is 70% of wing chord. Assume an ultimate shear strength of 38,000 psi and ultimate compression strength of 44,000 psi. Assume constant material thickness for the entire wing and select thicknesses from the Available Thickness table. ρ_{2024} = 0.1 lb$_f$/in³.

Solution

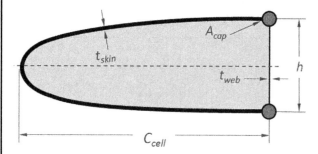

FIGURE 6-11 Example wing structure.

Step 1: Determine Basic Wing Geometry

Wing geometry:

$$AR = \frac{b^2}{S} \Leftrightarrow b = \sqrt{AR \times S} = \sqrt{16 \times 130} = 45.6 \text{ ft}$$

Average chord:

$$AR = \frac{b^2}{S} = \frac{b}{C_{avg}} \Leftrightarrow C_{avg} = \frac{b}{AR} = \frac{45.6}{16} = 2.850 \text{ ft}$$

Root chord:

$$C_{avg} = \frac{C_R}{2}(1+\lambda) \Leftrightarrow C_R = \frac{2C_{avg}}{1+\lambda} = \frac{2(2.850)}{1+0.5}$$
$$= 3.80 \text{ ft}$$

Tip chord:

$$C_T = \lambda C_R = 0.5 \times 3.80 = 1.90 \text{ ft}$$

Step 2: Determine Mean Geometric Properties

MGC:

$$MGC = \left(\frac{2}{3}\right)C_r\left(\frac{1+\lambda+\lambda^2}{1+\lambda}\right)$$
$$= \left(\frac{2}{3}\right)(3.80)\left(\frac{1+0.5+0.25}{1+0.5}\right) = 2.956 \text{ ft}$$

Y-location of MGC$_{LE}$:

$$y_{MGC} = \left(\frac{b}{6}\right)\left(\frac{1+2\lambda}{1+\lambda}\right) = \left(\frac{45.6}{6}\right)\left(\frac{1+1}{1+0.5}\right) = 10.135 \text{ ft}$$

Step 3: Determine Structural Geometry at the Root and Tip

Note that the subscript "R" stands for root and "T" for tip.

Cell length at root:

$$C_{cell} = 0.7C_R = 0.7(3.80) = 2.660 \text{ ft}$$

Structural depth at root:

$$h = 0.16C_R = 0.16(3.80) = 0.6081 \text{ ft}$$

Cell area:

$$A_{cell} = \frac{2C_{cell}h}{3} = \frac{2(0.7C_R)(0.16C_R)}{3} = 0.07467(3.80)^2$$
$$= 1.079 \text{ ft}^2$$

Arc length at root:

$$s_{cell} = \sqrt{(h/2)^2 + 4C_{cell}^2} + \frac{(h/2)^2}{4C_{cell}}\sinh^{-1}\left(\frac{2C_{cell}}{h/2}\right)$$
$$= \sqrt{\frac{0.6081^2}{4} + 4(2.660)^2} + \frac{0.6081^2}{16 \times 2.660}\sinh^{-1}\left(\frac{4 \times 2.660}{0.6081}\right)$$
$$= 5.360 \text{ ft}$$

Arc length at tip:

$$s_{cell_T} = \lambda s_{cell} = (0.5)(5.360) = 2.680 \text{ ft}$$

EXAMPLE 6-4 (cont'd)

Step 4: Determine Skin Thickness

This is the skin thickness required at root.

Torsion:

$$T = \frac{1}{4}\rho V^2 S \bar{c} C_m$$
$$= \frac{1}{4}(0.002378)(150 \times 1.688)^2(130)(2.956)(-0.1)$$
$$= -1465 \text{ ft·lb}_f$$

Skin thickness:

$$t_{skin} > \frac{|T|}{2A_{cell_R}\tau_{max}} = \frac{1465 \times 12}{2(1.079 \times 144)(38000)} = 0.0012 \text{ in}$$

Using the Available Thickness table the minimum sheet thickness larger than this value is 0.016″. However, we will select the larger recommended minimum of 0.020″ thickness and this is also the thickness at the tip of the wing half.

Step 5: Determine the Weight of the Skin

$$W_{skin} = \rho_{skin}\frac{b(t_{skin} + t_{skin_T})s_{cell}(1+\lambda)}{8}$$
$$= (0.1)\frac{(45.6 \times 12)(0.020 + 0.020)(5.360 \times 12)(1+0.5)}{8}$$
$$= 26.4 \text{ lb}_f$$

Step 6: Determine Shear Web Thickness

Shear force:

$$V = \frac{n_{ult}W}{2} = \frac{6.0(1320)}{2} = 3960 \text{ lb}_f$$

Shear web thickness at root:

$$t_{web} > \frac{3(n_{ult}W)}{4h\tau_{max}} = \frac{3(6.0 \times 1320)}{4(0.6081 \times 12)(38000)} = 0.0214 \text{ in}$$

Shear web thickness at tip:

$$t_{web_T} > 0.15 t_{web} = 0.15 \times 0.0214 = 0.0032 \text{ in}$$

We will thus select a web thickness of 0.025″ for the root area and 0.020″ for the tip area.

Step 7: Determine the Weight of the Shear Web

$$W_{web} = \rho_{web}\frac{b(t_{web} + t_{web_T})h(1+\lambda)}{8}$$
$$= (0.1)\frac{(45.6 \times 12)(0.025 + 0.020)(0.6081 \times 12)(1+0.5)}{8}$$
$$= 3.37 \text{ lb}_f$$

Step 8: Determine Spar Cap Area

"Bending force":

$$F_{bend} = \frac{n_{ult}W}{2}\frac{Y_{MGC}}{h} = (3960)\frac{10.135}{0.6081} = 66000 \text{ lb}_f$$

This is the axial force (or force couple) in the spar caps at the root of the wing.

Spar cap area at root:

$$A_{cap} > \frac{n_{ult} \cdot W \cdot Y_{MGC}}{2 \cdot h \cdot \sigma_{max}} = \frac{66000}{44000} = 1.50 \text{ in}^2$$

Spar cap area at tip:

$$A_{cap_T} > 0.05 A_{cap} = 0.05 \cdot 1.50 = 0.075 \text{ in}^2$$

Step 9: Determine Weight of Spar Caps

$$W_{caps} = \rho_{caps}\frac{b(A_{cap} + A_{cap_T})}{2}$$
$$= (0.1)\frac{(45.6 \times 12)(1.50 + 0.075)}{2} = 43.1 \text{ lb}_f$$

Step 10: Determine Number of Ribs

Number of ribs:

$$N_{rib} \approx INT\left(\frac{b}{C_{avg}}\right) + 1 = INT\left(\frac{45.6}{2.850}\right) + 1 = 17$$

Step 11: Determine Weight of Ribs

$$W_{ribs} = N_{ribs}\rho_{ribs}\frac{A_{cell}(1+\lambda^2)(t_{skin} + t_{skin_T})}{4}$$
$$= (17)(0.1)\frac{(1.079 \times 144)(1+0.5^2)(0.020 + 0.020)}{4}$$
$$= 3.30 \text{ lb}_f$$

Step 12: Determine Wing Weight

Left or right wing:

$$W_{skin} + W_{web} + W_{caps} + W_{ribs} = 26.4 + 3.37 + 43.1 + 3.30$$
$$= 76.2 \text{ lb}_f$$

Total wing:

$$2 \times 76.2 = 152 \text{ lb}_f$$

The weight of this wing is significantly less than the result of Example 6-3. The reason is that this only represents the weight of the main wing element. It omits control surfaces, their attachment hard points, wing attachments, control system, fuel system, electric system, wingtip fairing and

(Continued)

EXAMPLE 6-4 (cont'd)

so on. The ratio between the two is 273/152 = 1.80. There is no guarantee this ratio is maintained for other configurations.

Note

It is of interest to consider how the weight of these components varies with wing area. Such a plot is shown in Figure 6-12. Such graphs come in handy when doing trade studies to evaluate an airplane's weight as a function of wing area (see Chapter 5).

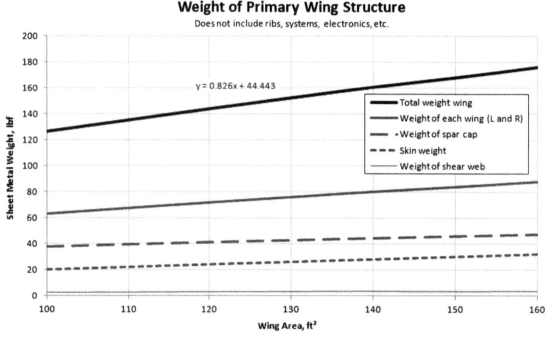

FIGURE 6-12 The variation of various structural components as a function of wing area.

6.5.2 Variation of Weight with AR

The impact of the aspect ratio on the weight of the wing is of utmost importance in the design of the aircraft. Many designs must meet strict range or endurance requirements that call for a high-AR wing. However, it is easy to overlook that the cost of such wings is extra weight required by the high AR. At other times, the designer may want to evaluate the impact of an AR change of a mature design, or even an existing aircraft. This section presents an approximation for the evaluation of such changes.

Consider the special case for which the wing area, S, and taper ratio, λ, are constant, but the AR is allowed to vary (see Figure 6-14). Assume we have a baseline wing and want to compare it to a modified wing with the same S and λ; the only change is in the AR (and therefore the dimensions of the root and tip chord). The weight of the modified AR wing can be approximated by the following assumption:

(1) There is no change in wing's airfoil. This means the thickness ratio is constant. As a consequence, given a constant S, a higher AR results in a "thinner" wing whose chords are also shortened.
(2) It is assumed that changes in geometry are "small" enough so that change in the wing skin shear stress can be ignored. The skin shear stresses are caused by the wing torsion due to the airfoil's pitching moment and torsion due to forward- or aft-swept wings. A large wing chord will have a greater cross-sectional area to react this torsion than a smaller wing chord, but the smaller chord wing will also generate lower pitching moments. It is prudent for the designer to evaluate whether this assumption is valid for the particular wing, but here it will be

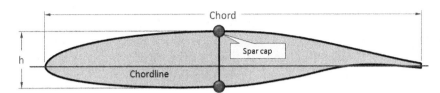

FIGURE 6-13 Structural depth, h, of an airfoil.

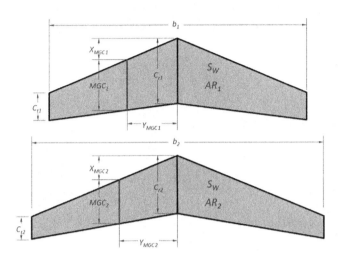

FIGURE 6-14 Two wings of equal area but different aspect ratios.

assumed the change in shear stresses are small enough to permit the same skin thickness to be maintained.

(3) It is assumed the change in AR does not require a change in other geometry that would cause components other than the wing weight to change (e.g. empennage geometry, etc.).
(4) It is assumed that there is no change in vertical shear. Therefore, the shear web thickness does not change. This is justified on the basis that changing the AR will not alter the airplane's gross weight, only its empty weight.
(5) The maximum bending moment at the root is directly related to the location of the center of lift, which is assumed to act at the spanwise station for the MGC.
(6) The change in bending stresses is equal to the change in bending moments. If the bending moments change by 25%, then so will the bending stresses.
(7) The change in material geometry required to react the bending moment is directly related to the change in stress levels — and therefore it is assumed the goal is to maintain similar stress levels in the spar caps before and after change.
(8) The material allowable, σ_{max}, is assumed the same for both wing geometries.
(9) Assume the structural depth of an airfoil to be based on its maximum thickness (see Figure 6-13).
(10) Assume the spar caps have a circular cross section, separated by the structural depth (see Figure 6-13).
(11) Assume the cross-sectional area of the spar cap at tip to be 10% of that of the root.

From these assumptions it can be seen that the only change is in the dimensions of the spar caps. This implies that the only change in the weight of the structure will be associated with the change in the spar caps geometry. To estimate the magnitude of this change we begin by establishing relationships between geometry and stress.

Baseline Definitions for a Trapezoidal Wing

The following expressions are needed to begin the weight estimation and are all based on the parameters S, AR, and λ. For instance, they can be used to calculate the properties of the baseline wing.

Wingspan:
$$b = \sqrt{AR \times S} \qquad (9\text{-}15)$$

Mean geometric chord:
$$MGC = \frac{4}{3}\sqrt{\frac{S}{AR}}\left(\frac{1+\lambda+\lambda^2}{1+2\lambda+\lambda^2}\right) \qquad (6\text{-}49)$$

Spanwise location of the center of lift:
$$y_{MGC} = \frac{\sqrt{AR \times S}}{6}\left(\frac{1+2\lambda}{1+\lambda}\right) \qquad (6\text{-}50)$$

Structural depth at the MGC:
$$h = \frac{4}{3}\sqrt{\frac{S}{AR}}\left(\frac{1+\lambda+\lambda^2}{1+2\lambda+\lambda^2}\right)\left(\frac{t}{c}\right) \qquad (6\text{-}51)$$

Maximum bending moment:
$$M_{max} = \frac{n_{ult} W \sqrt{AR \times S}}{12}\left(\frac{1+2\lambda}{1+\lambda}\right) \qquad (6\text{-}52)$$

Moment of inertia:
$$I_{XX} = \frac{16A}{18}\left(\frac{S}{AR}\right)\left(\frac{1+\lambda+\lambda^2}{1+2\lambda+\lambda^2}\right)^2\left(\frac{t}{c}\right)^2 \qquad (6\text{-}53)$$

Required spar cap area:

$$A_{cap} > \frac{n_{ult} W \times AR}{16\sigma_{max}\left(\frac{t}{c}\right)} \left(\frac{1+3\lambda+2\lambda^2}{1+\lambda+\lambda^2}\right) \quad (6\text{-}54)$$

Required spar cap weight:

$$W_{cap} = 1.1 \times \rho_{cap} \times A_{cap} \sqrt{AR \times S} \quad (6\text{-}55)$$

where

A_{cap} = cross-sectional area of the upper or lower spar cap in m^2 or ft^2
AR = aspect ratio
b = wingspan in m or ft
n_{ult} = ultimate flight load in gs.
S = wing area in m^2 or ft^2
W = airplane design gross weight in N or lb$_f$
λ = wing taper ratio
ρ_{cap} = weight density of spar cap material in N/m^3 or lb$_f$/ft^3
σ_{max} = tensile stress allowable of spar cap material in Pa or lb$_f$/ft^2

With respect to the ultimate flight load, n_{ult}, the load must be the maneuvering or gust load, whichever is larger. Note that many of the derivations below refer to equations in Chapter 7, *The wing planform*.

Derivation of Equation (6-49)

Insert Equation (6-6) into Equation (9-20) and manipulate:

$$MGC = \left(\frac{4b}{3AR}\right)\left(\frac{1+\lambda+\lambda^2}{1+2\lambda+\lambda^2}\right)$$

$$= \left(\frac{4\sqrt{AR \times S}}{3AR}\right)\left(\frac{1+\lambda+\lambda^2}{1+2\lambda+\lambda^2}\right)$$

$$= \frac{4}{3}\sqrt{\frac{S}{AR}}\left(\frac{1+\lambda+\lambda^2}{1+2\lambda+\lambda^2}\right)$$

QED

Derivation of Equation (6-50)

Insert Equation (6-6) into Equation (9-8) to determine the location of the center of lift:

$$y_{MGC} = \left(\frac{b}{6}\right)\left(\frac{1+2\lambda}{1+\lambda}\right) = \frac{\sqrt{AR \times S}}{6}\left(\frac{1+2\lambda}{1+\lambda}\right)$$

QED

Derivation of Equation (6-51)

Consider Figure 6-13, which defines the structural depth of the airfoil, h. At the MGC this depth is given by:

$$h = MGC\left(\frac{t}{c}\right)$$

Then, insert Equation (9-20) and manipulate algebraically:

$$h = \left(\frac{4b}{3AR}\right)\left(\frac{1+\lambda+\lambda^2}{1+2\lambda+\lambda^2}\right)\left(\frac{t}{c}\right)$$

$$= \left(\frac{4\sqrt{AR \times S}}{3AR}\right)\left(\frac{1+\lambda+\lambda^2}{1+2\lambda+\lambda^2}\right)\left(\frac{t}{c}\right)$$

$$= \frac{4}{3}\sqrt{\frac{S}{AR}}\left(\frac{1+\lambda+\lambda^2}{1+2\lambda+\lambda^2}\right)\left(\frac{t}{c}\right)$$

QED

Derivation of Equation (6-52)

The maximum bending moment is given by[3]:

$$M_{max} \approx \frac{L}{2} \times y_{MGC} = \frac{n_{ult} W}{2} \times y_{MGC}$$

where L is the lift, n_{ult} is the ultimate load factor, and W is the weight of the airplane. Inserting Equation (6-50) for y_{MGC} yields:

$$M_{max} \approx \frac{n_{ult} W}{2} \times y_{MGC} = \frac{n_{ult} W}{2} \times \frac{\sqrt{AR \times S}}{6}\left(\frac{1+2\lambda}{1+\lambda}\right)$$

$$= \frac{n_{ult} W \sqrt{AR \times S}}{12}\left(\frac{1+2\lambda}{1+\lambda}\right)$$

QED

Derivation of Equation (6-53)

The moment of inertia can be calculated from the parallel-axis theorem, assuming the spar caps have an area A_{cap} and are separated by the structural depth h:

$$I_{XX} = 2 \times A_{cap} \times \left(\frac{h}{2}\right)^2 = \frac{A_{cap} h^2}{2}$$

Inserting Equation (6-51) for structural height yields:

$$I_{XX} = \frac{A_{cap} h^2}{2} = \frac{A_{cap}}{2} \times \left(\frac{4b}{3AR}\right)^2 \left(\frac{1+\lambda+\lambda^2}{1+2\lambda+\lambda^2}\right)^2 \left(\frac{t}{c}\right)^2$$

---

[3] Although the maximum bending moment is generally determined at the location of the wing attachments, for this method this is assumed at the plane of symmetry.

Finally, yielding:

$$I_{XX} = \frac{16A_{cap}}{18}\left(\frac{\sqrt{AR \times S}}{AR}\right)^2\left(\frac{1+\lambda+\lambda^2}{1+2\lambda+\lambda^2}\right)^2\left(\frac{t}{c}\right)^2$$

$$= \frac{16A_{cap}}{18}\left(\frac{S}{AR}\right)\left(\frac{1+\lambda+\lambda^2}{1+2\lambda+\lambda^2}\right)^2\left(\frac{t}{c}\right)^2$$

QED

Derivation of Equation (6-54)

Maximum stress at the outer fibers may not exceed:

$$\sigma_{max} > \frac{M_{max} \times (h/2)}{I_{XX}} = \frac{M_{max} \times h}{2I_{XX}} = \frac{M_{max} \times h}{2\left(\frac{Ah^2}{2}\right)}$$

$$= \frac{M_{max}}{Ah}$$

We can use this expression to determine the minimum area A_{cap} required for the spar caps.

$$A_{cap} > \frac{M_{max}}{\sigma_{max}h}$$

Inserting the proper relations for M_{max} and h:

$$A_{cap} > \frac{M_{max}}{\sigma_{max}h} = \frac{\frac{n_{ult}W\sqrt{AR \times S}}{12}\left(\frac{1+2\lambda}{1+\lambda}\right)}{\sigma_{max}\left(\frac{4b}{3AR}\right)\left(\frac{1+\lambda+\lambda^2}{1+2\lambda+\lambda^2}\right)\left(\frac{t}{c}\right)}$$

$$= \frac{n_{ult}W\sqrt{AR^3 \times S}}{16b\sigma_{max}\left(\frac{t}{c}\right)}\left(\frac{1+2\lambda}{1+\lambda}\right)\frac{(1+\lambda)^2}{1+\lambda+\lambda^2}$$

$$= \frac{n_{ult}W\sqrt{AR^3 \times S}}{16\sqrt{AR \times S}\sigma_{max}\left(\frac{t}{c}\right)}\frac{(1+2\lambda)(1+\lambda)}{1+\lambda+\lambda^2}$$

$$= \frac{n_{ult}W \times AR}{16\sigma_{max}\left(\frac{t}{c}\right)}\left(\frac{1+3\lambda+2\lambda^2}{1+\lambda+\lambda^2}\right)$$

QED

Derivation of Equation (6-55)

The total volume of spar caps, assuming the thickness at the tip is 10% of that at the root:

$$V_{cap} = 2\frac{A_{cap}(1+0.1)}{2}b = 1.1A_{cap}\sqrt{AR \times S}$$

The spar cap weight is thus:

$$W_{cap} = \rho_{cap}V_{cap} = 2\frac{A_{cap}(1+0.1)}{2}b$$

$$= 1.1 \times \rho_{cap} \times A_{cap}\sqrt{AR \times S}$$

QED

EXAMPLE 6-5

Let's evaluate how accurate these expressions are by comparing them to an existing aircraft, the Beech Bonanza A36. The Bonanza's design gross weight is 3600 lb$_f$, wing area 181 ft^2, AR is 6.2, and λ is 0.538. The Bonanza's airfoils are the NACA 23016.5 at the root (t/c = 0.165) and 23012 at the tip (t/c = 0.12). Use the root thickness ratio, 0.165, for the variable t/c. The airplane is certified under 14 CFR, Part 23, in the utility category. This means the ultimate load factor is $4.4g \times 1.5 = 6.6g$. Assume the spar caps are fabricated from 2024-T3 extrusion, whose density is 0.1 lb$_f$/in^3 and σ_{max} = 65,000 psi (or 9,360,000 psf). Evaluate the above parameters based on these numbers and compare to values that are in the public domain.

Solution

Wingspan is:

$$b = \sqrt{AR \times S} = \sqrt{6.2 \times 181} = 33.5 \text{ ft}$$

Mean geometric chord:

$$MGC = \frac{4}{3}\sqrt{\frac{S}{AR}}\left(\frac{1+\lambda+\lambda^2}{1+2\lambda+\lambda^2}\right)$$

$$= \frac{4}{3}\sqrt{\frac{181}{6.2}}\left(\frac{1+0.538+0.538^2}{1+2 \times 0.538+0.538^2}\right) = 5.566 \text{ ft}$$

Spanwise location of the center of lift:

$$y_{MGC} = \frac{\sqrt{AR \times S}}{6}\left(\frac{1+2\lambda}{1+\lambda}\right)$$

$$= \frac{\sqrt{6.2 \times 181}}{6}\left(\frac{1+2 \times 0.538}{1+0.538}\right) = 7.536 \text{ ft}$$

Structural depth at the MGC:

$$h = \frac{4}{3}\sqrt{\frac{S}{AR}}\left(\frac{1+\lambda+\lambda^2}{1+2\lambda+\lambda^2}\right)\left(\frac{t}{c}\right)$$

$$= \frac{4}{3}\sqrt{\frac{181}{6.2}}\left(\frac{1+0.538+0.538^2}{1+2 \times 0.538+0.538^2}\right)(0.165) = 0.918 \text{ ft}$$

EXAMPLE 6-5 (cont'd)

Maximum bending moment at the plane of symmetry:

$$M_{max} = \frac{n_{ult}W\sqrt{AR \times S}}{12}\left(\frac{1+2\lambda}{1+\lambda}\right)$$

$$= \frac{6.6 \times 3600\sqrt{6.2 \times 181}}{12}\left(\frac{1+2 \times 0.538}{1+0.538}\right)$$

$$= 89{,}530 \text{ ft·lb}_f$$

Required maximum spar cap area at the plane of symmetry:

$$A_{cap} > \frac{n_{ult}W \times AR}{16\sigma_{max}\left(\frac{t}{c}\right)}\left(\frac{1+3\lambda+2\lambda^2}{1+\lambda+\lambda^2}\right)$$

$$= \frac{6.6 \times 3600 \times 6.2}{16 \times 9{,}360{,}000(0.165)}\left(\frac{1+3 \times 0.538 + 2 \times 0.538^2}{1+0.538+0.538^2}\right)$$

$$= 0.01042 \text{ ft}^2 = 1.500 \text{ in}^2$$

Moment of inertia at the plane of symmetry:

$$I_{XX} = \frac{16A_{cap}}{18}\left(\frac{S}{AR}\right)\left(\frac{1+\lambda+\lambda^2}{1+2\lambda+\lambda^2}\right)^2\left(\frac{t}{c}\right)^2$$

$$= \frac{16 \times 0.01042}{18}\left(\frac{181}{6.2}\right)\left(\frac{1+0.538+0.538^2}{1+2\times 0.538+0.538^2}\right)^2(0.165)^2$$

$$= 0.004394 \text{ ft}^4 = 91.1 \text{ in}^4$$

Spar cap weight:

$$W_{cap} = 1.1 \times \rho_{cap} \times A_{cap}\sqrt{AR \times S}$$

$$= 1.1 \times (0.1 \times 12^3) \times 0.01042\sqrt{6.2 \times 181} = 66.3 \text{ lb}_f$$

Comparison of the approximated to official numbers are shown in Table 6-2 below:

These results show this method is in good agreement with the "official" values, and lends support to its validity.

TABLE 6-2 Comparison of "Official" to Analysis for the Beech Bonanza

Property	Symbol	"Official" Value	Analysis	Comment
Wingspan	b	33.5 ft	33.5 ft	Analysis values based on published data.
Mean geometric chord	C_{MGC}	5.441 ft	5.566 ft	Official value obtained from analysis of a 3-view drawing.
Spanwise location of center of lift	y_{MGC}	7.445 ft	7.536 ft	Official value obtained from a standard estimate based on a 3-view drawing.
Structural depth	h	0.941 ft	0.918 ft	Official value measured by author on the actual airplane.
Maximum bending moment	M_{max}	79,358 ft·lb$_f$	89,530 ft·lb$_f$	Official value based on vortex-lattice analysis of the aircraft, which accounts for lift on fuselage and horizontal tail, whereas this analysis assumes all lift is generated by the wings.
Required spar cap area	A	1.490 in^2	1.500 in^2	Official value measured by author on actual airplane.
Moment of inertia	I_{XX}	95.567 in^4	91.1 in^4	Official value calculated using parallel-axis theorem with A and h.
Weight of spar caps	W_{cap}	Unknown	66.3 lb$_f$	Official value is not known, but analysis value is considered reasonable.

Method of Fractions

Once the baseline properties are known, we can now estimate the properties of a modified wing whose only geometric change is the AR (S and λ are assumed constant for both). Assume we have defined a baseline configuration, denoted by the subscript 1, and a comparison configuration, denoted by the subscript 2. Then, the following ratios hold between the two wings.

Wingspan:

$$b_2 = b_1 \sqrt{\frac{AR_2}{AR_1}} \quad (6\text{-}56)$$

Mean geometric chord:

$$MGC_2 = MGC_1 \sqrt{\frac{AR_1}{AR_2}} \quad (6\text{-}57)$$

Spanwise location of the center of lift:

$$y_{MGC\ 2} = y_{MGC\ 1} \sqrt{\frac{AR_2}{AR_1}} \quad (6\text{-}58)$$

Structural depth:

$$h_2 = h_1 \sqrt{\frac{AR_1}{AR_2}} \quad (6\text{-}59)$$

Maximum bending moment:

$$M_{max2} = M_{max1} \sqrt{\frac{AR_2}{AR_1}} \quad (6\text{-}60)$$

Spar cap areas:

$$A_{cap2} = A_{cap1} \frac{AR_2}{AR_1} \quad (6\text{-}61)$$

Moment of inertia:

$$I_{XX_2} = I_{XX_1} \quad (6\text{-}62)$$

Required spar cap weight:

$$W_{cap2} = W_{cap1} \left(\frac{AR_2}{AR_1}\right)^{3/2} \quad (6\text{-}63)$$

Change in spar cap weight:

$$\Delta W_{cap} = W_{cap2} - W_{cap1} \quad (6\text{-}64)$$

Derivation of Equation (6-56)

Using Equation (9-15) with the two subscripts:

$$\frac{b_2}{b_1} = \sqrt{\frac{AR_2 \times S}{AR_1 \times S}} \Rightarrow b_2 = b_1 \sqrt{\frac{AR_2}{AR_1}}$$

QED

Derivation of Equation (6-57)

Using Equation (6-49) and applying the proper subscripts then dividing one MGC by the other leads to:

$$\frac{MGC_2}{MGC_1} = \frac{\frac{4}{3}\sqrt{\frac{S}{AR_2}}\left(\frac{1+\lambda+\lambda^2}{1+2\lambda+\lambda^2}\right)}{\frac{4}{3}\sqrt{\frac{S}{AR_1}}\left(\frac{1+\lambda+\lambda^2}{1+2\lambda+\lambda^2}\right)} = \frac{\sqrt{AR_1}}{\sqrt{AR_2}}$$

$$\Rightarrow MGC_2 = MGC_1 \sqrt{\frac{AR_1}{AR_2}}$$

QED

Derivation of Equation (6-58)

Using Equation (6-50) and applying the proper subscripts and dividing one y_{MGC} by the other leads to:

$$\frac{y_{MGC\ 2}}{y_{MGC\ 1}} = \frac{\frac{\sqrt{AR_2 \times S}}{6}\left(\frac{1+2\lambda}{1+\lambda}\right)}{\frac{\sqrt{AR_1 \times S}}{6}\left(\frac{1+2\lambda}{1+\lambda}\right)} = \frac{\sqrt{AR_2}}{\sqrt{AR_1}}$$

$$\Rightarrow y_{MGC\ 2} = y_{MGC\ 1} \sqrt{\frac{AR_2}{AR_1}}$$

QED

Derivation of Equation (6-59)

Using Equation (6-51) and applying the proper subscripts and dividing one h by the other leads to:

$$\frac{h_2}{h_1} = \frac{\frac{4}{3}\sqrt{\frac{S}{AR_2}}\left(\frac{1+\lambda+\lambda^2}{1+2\lambda+\lambda^2}\right)\left(\frac{t}{c}\right)}{\frac{4}{3}\sqrt{\frac{S}{AR_1}}\left(\frac{1+\lambda+\lambda^2}{1+2\lambda+\lambda^2}\right)\left(\frac{t}{c}\right)} = \frac{\sqrt{AR_1}}{\sqrt{AR_2}} \Rightarrow h_2 = h_1 \sqrt{\frac{AR_1}{AR_2}}$$

QED

Derivation of Equation (6-60)

Using Equation (6-52) and applying the proper subscripts and dividing one M_{max} by the other leads to:

$$\frac{M_{max2}}{M_{max1}} = \frac{\frac{n_{ult} W \sqrt{AR_2 \times S}}{12}\left(\frac{1+2\lambda}{1+\lambda}\right)}{\frac{n_{ult} W \sqrt{AR_1 \times S}}{12}\left(\frac{1+2\lambda}{1+\lambda}\right)} = \frac{\sqrt{AR_2}}{\sqrt{AR_1}}$$

$$\Rightarrow M_{max2} = M_{max1} \sqrt{\frac{AR_2}{AR_1}}$$

QED

Derivation of Equation (6-61)

Using Equation (6-54) and applying the proper subscripts and dividing one I_{XX} by the other leads to:

$$\frac{A_{cap2}}{A_{cap1}} = \frac{\frac{n_{ult} W \times AR_2}{16 \sigma_{max}(\frac{t}{c})} \left(\frac{1+3\lambda+2\lambda^2}{1+\lambda+\lambda^2}\right)}{\frac{n_{ult} W \times AR_1}{16 \sigma_{max}(\frac{t}{c})} \left(\frac{1+3\lambda+2\lambda^2}{1+\lambda+\lambda^2}\right)} = \frac{AR_2}{AR_1}$$

$$\Rightarrow A_{cap2} = A_{cap1} \frac{AR_2}{AR_1}$$

QED

Derivation of Equation (6-62)

Using Equation (6-18) and applying the proper subscripts and dividing one I_{XX} by the other leads to:

$$\frac{I_{XX_2}}{I_{XX_1}} = \frac{\frac{16 A_{cap2}}{18} \left(\frac{S}{AR_2}\right) \left(\frac{1+\lambda+\lambda^2}{1+2\lambda+\lambda^2}\right)^2 \left(\frac{t}{c}\right)^2}{\frac{16 A_{cap1}}{18} \left(\frac{S}{AR_1}\right) \left(\frac{1+\lambda+\lambda^2}{1+2\lambda+\lambda^2}\right)^2 \left(\frac{t}{c}\right)^2}$$

$$= \left(\frac{A_{cap2}}{A_{cap1}}\right) \left(\frac{AR_1}{AR_2}\right)$$

$$\Rightarrow I_{XX_2} = I_{XX_1} \left(\frac{A_{cap2}}{A_{cap1}}\right) \left(\frac{AR_1}{AR_2}\right)$$

Now, let's insert Equation (6-61) and simplify:

$$I_{XX_2} = I_{XX_1} \left(\frac{A_{cap2}}{A_{cap1}}\right) \left(\frac{AR_1}{AR_2}\right) = I_{XX_1} \left(\frac{AR_2}{AR_1}\right) \left(\frac{AR_1}{AR_2}\right)$$

$$= I_{XX_1}$$

QED

Derivation of Equation (6-63)

Maximum stress at the outer fibers may not exceed:

$$\frac{W_{cap2}}{W_{cap1}} = \frac{1.1 \times \rho_{cap} \times A_{cap2} \sqrt{AR_2 \times S}}{1.1 \times \rho_{cap} \times A_{cap1} \sqrt{AR_1 \times S}}$$

$$= \left(\frac{A_{cap2}}{A_{cap1}}\right) \frac{\sqrt{AR_2}}{\sqrt{AR_1}}$$

$$\Rightarrow W_{cap2} = W_{cap1} \left(\frac{A_{cap2}}{A_{cap1}}\right) \sqrt{\frac{AR_2}{AR_1}}$$

Now, let's insert Equation (6-61) and simplify:

$$W_{cap2} = W_{cap1} \left(\frac{A_{cap2}}{A_{cap1}}\right) \sqrt{\frac{AR_2}{AR_1}} = W_{cap1} \left(\frac{AR_2}{AR_1}\right) \sqrt{\frac{AR_2}{AR_1}}$$

$$= W_{cap1} \left(\frac{AR_2}{AR_1}\right)^{3/2}$$

QED

EXAMPLE 6-6

Use the method of fraction to estimate the change in empty weight for the Beech Bonanza A36 aircraft of Example 6-5, for AR increasing from 6.2 to 14. This assumes the only change is in the weight of the spar caps. The standard empty weight is 2247 lb_f. Plot the change in empty weight and maximum bending moments.

Solution

All baseline values are calculated in Example 6-5, including the estimated baseline weight for the spar caps of 66.3 lb_f and maximum baseline bending moment is 89530 $ft \cdot lb_f$. The baseline AR_1 is 6.2. Let's calculate a sample value using $AR_2 = 10$.

Maximum bending moment for $AR_2 = 10$:

$$M_{max2} = M_{max1} \sqrt{\frac{AR_2}{AR_1}} = 89,530 \sqrt{\frac{10}{6.2}} = 113,703 \text{ ft} \cdot lb_f$$

In order to estimate the empty weight we first compute the spar cap weight for $AR_2 = 10$ and then use Equation (6-29) to determine the difference between the two. We then add this difference to the baseline empty weight.

$$W_{cap2} = W_{cap1} \left(\frac{AR_2}{AR_1}\right)^{3/2} = 66.3 \left(\frac{10}{6.2}\right)^{3/2} = 135.9 \text{ lb}_f$$

Therefore, the difference is:

$$\Delta W_{cap} = W_{cap2} - W_{cap1} = 135.9 - 66.3 = 69.6 \text{ lb}_f$$

The empty weight is therefore:

$$W_{e\ AR=10} = W_e + \Delta W_{cap} = 2247 + 69.6 = 2317 \text{ lb}_f$$

EXAMPLE 6-6 (cont'd)

The remaining values are plotted in Figure 6-15, which shows how AR can affect aircraft weight.

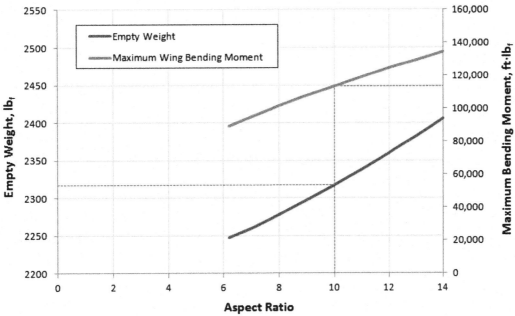

FIGURE 6-15 Predicted empty weight and maximum bending moments versus aspect ratio. The dotted lines indicate the sample values calculated in the example.

6.6 INERTIA PROPERTIES

The determination of various inertia properties is imperative during the design process. Properties such as moments and products of inertia are required to predict dynamic stability, and, thus, play a major role in the development of the OML. Table 6-3 lists a number of required inertia properties need for such analyses. In this book, the term "*at a specific condition*" refers to some particular atmospheric conditions or a specific event during a flight, for instance the end of cruise or the start of descent. During the design stage, inertia properties are considered at many such specific events. They differ from the properties at take-off, as fuel would have been consumed, or external stores dropped (for military aircraft). Fuel for jets usually constitutes as much as 20% of their T-O weight, and some 10–15% for piston engines. Therefore, there can be a large change

TABLE 6-3 Important Inertia Properties

Property	Symbol	Section
Weight at a specific condition	W_{tot}	6.6.3
Center of gravity (CG) in terms of location	X_{CG}, Y_{CG}, Z_{CG}	6.6.5
CG is also given in terms of %MAC (in particular X_{CG})	X_{CG}	6.6.5
Moment of inertia about the x-, y-, and z-axes	I_{XX}, I_{YY}, I_{ZZ}	6.6.7
Product of inertia in the xy-, xz-, and yz-planes	I_{XY}, I_{XZ}, I_{YZ}	6.6.7

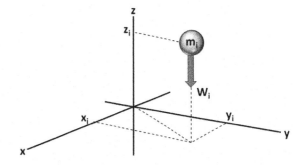

FIGURE 6-16 The definition of a point load in three-dimensional space.

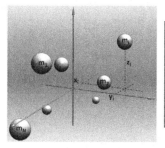

Item	Weight	X	Y	Z
Item 1	W_1	X_1	Y_1	Z_1
Item 2	W_2	X_2	Y_2	Z_2
Item 3	W_3	X_3	Y_3	Z_3
...	...	...	...	...
Item i	W_i	X_i	Y_i	Z_i
...	...	...	...	...
Item N	W_N	X_N	Y_N	Z_N

FIGURE 6-17 A collection of point loads in three-dimensional space (left) and a tabular representation (right).

in the inertia properties between T-O and landing, and this can have a profound effect on dynamic stability.

6.6.1 Fundamentals

The process of determining inertia properties typically involves treating components as a collection of point loads; i.e. as a weight and position in space (see Figure 6-16). This allows properties such as the weight, or moments and products of inertia of the vehicle to be estimated using the methods to be introduced shortly. Then, the inertia properties for the entire collection are determined by a simple summation. The moments and products of inertia of large objects, such as wings, and heavy objects, such as engines, should be included in the total for further accuracy and this is indicated in the formulation that follows. Depending on the location and shape of such components, the moments and products of inertia of the parts themselves can easily add 25%, and even higher values, to the total amount calculated by the parallel-axis theorem.

In this section, the inertia properties shown in Table 6-3 will be calculated.

The formulation presented in this section assumes the airplane can be represented by a collection of point loads in three-dimensional space, as shown in Figure 6-17. Note that each arbitrary point is denoted by the subscript i.

6.6.2 Reference Locations

The aerospace engineer should use terminology commonly used in the aviation industry when referring to a point in space at which a specific weight is located near an aircraft. For instance, consider the location of avionics equipment or an occupant. The physical location is referred to using terms such as:

FS – fuselage station	BL – butt (or buttock) line
WL – water line	WS – wing station
HS – horizontal station	VS – vertical station

FS, BL, and WL, are depicted in Figure 6-18. When an airplane features swept wings or tail, it is convenient to represent locations using a wing, horizontal, or vertical station. These are effectively a BL aligned to something like the quarter-chord line or another conveniently selected datum.

EXAMPLE 6-7

The CG of an airplane is reported to be at FS191.3 (in other words: at fuselage station 191.3 inches). If the LE of the MGC airfoil is at FS185 and the MGC is 6.32 ft, where is the CG with respect to the MGC?

Solution

$$\%MAC = \%MGC = 100\left(\frac{X_{CG} - X_{MGC}}{MGC}\right)$$

$$= 100\left(\frac{191.3 - 185}{6.32 \times 12}\right) = 8.31\ \%MGC$$

6.6.3 Total Weight

As applied to an airplane, we break it into a finite number of subcomponents: engine, propeller, left wing, right wing, horizontal tail, fuselage, left main landing gear, right main gear, and so on, whose weights we can assess. These are denoted by W_i, where the index i is assigned to each component. Then, the total weight of the collection of components is calculated as follows:

Total weight:

$$W_{tot} = \sum_{i=1}^{N} W_i \qquad (6\text{-}65)$$

6.6.4 Moment About (X_0, Y_0, Z_0)

Moments about an arbitrary reference point (X_0, Y_0, Z_0) are calculated using the expressions below. This is

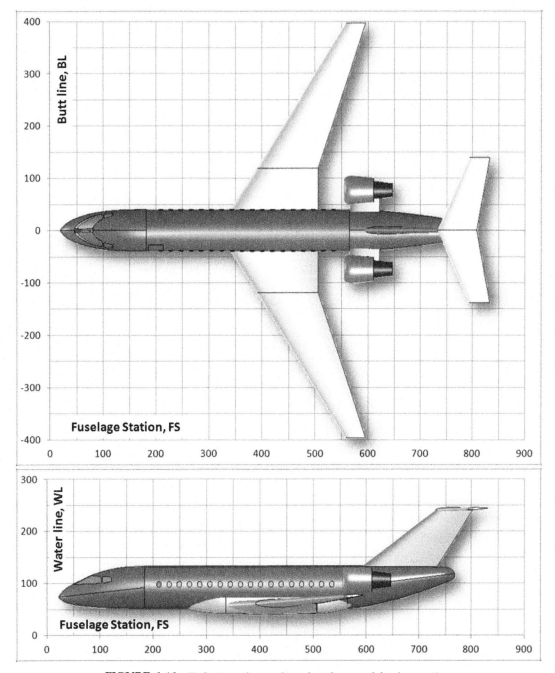

FIGURE 6-18 Definition of water lines, butt lines, and fuselage station.

a necessary intermediary step before the center of gravity (CG) can be calculated:

$$M_X = \sum_{i=1}^{N} W_i \times (X_i - X_0)$$
$$M_Y = \sum_{i=1}^{N} W_i \times (Y_i - Y_0) \qquad (6\text{-}66)$$
$$M_Z = \sum_{i=1}^{N} W_i \times (Z_i - Z_0)$$

Unless otherwise specified, our reference point is always (0, 0, 0) and this is assumed in the following formulation. We rewrite Equation (6-66) by writing the moments about the point (0, 0, 0):

$$M_X = \sum_{i=1}^{N} W_i \times X_i \quad M_Y = \sum_{i=1}^{N} W_i \times Y_i \quad M_Z$$
$$= \sum_{i=1}^{N} W_i \times Z_i \qquad (6\text{-}67)$$

6.6.5 Center of Mass, Center of Gravity

Consider a system of matter (this could be a collection of solid objects, liquids, gases, or any combination thereof) distributed in three-dimensional space. Then, we define *the center of mass (CM) of the system as the point in space at which uniform force acting on the whole system is equivalent to that force acting at just that point.*

In a uniform gravitational field, the force acting on the system can be considered to act at the CM, in which case we refer to it as the center of gravity (CG). While the CM and CG are often used interchangeably, this does not hold true in non-uniform acceleration fields. However, in the context of this book, it is always assumed the airplane is operated in a uniform acceleration field and, therefore, the CM and CG are always the same point. The position of the CM, $\mathbf{R} = (X_{CM}, Y_{CM}, Z_{CM})$, can be calculated from:

$$\mathbf{R} = \frac{\sum m_i r_i}{\sum m_i} \quad (6\text{-}68)$$

where

m_i = mass of a specific object within the collection of objects

$r_i = (x_{CM}, y_{CM}, z_{CM})_i$ = location of mass m_i

Similarly, the location of the center of gravity of the collection of components is estimated using the following expressions:

$$X_{CG} = \frac{M_X}{W_{tot}} = \frac{\sum_{i=1}^{N} W_i \times X_i}{W_{tot}}$$

$$Y_{CG} = \frac{M_Y}{W_{tot}} = \frac{\sum_{i=1}^{N} W_i \times Y_i}{W_{tot}} \quad (6\text{-}69)$$

$$Z_{CG} = \frac{M_Z}{W_{tot}} = \frac{\sum_{i=1}^{N} W_i \times Z_i}{W_{tot}}$$

It is very common to present the X-location of the CG in terms of %MAC (although it really refers to MGC). This would be calculated as follows, where X_{MGC} is a reference distance to the leading edge of the MGC, although this is shown to reference the apex of the swept back wing in Figure 6-19:

$$X_{CG_{MGC}} = 100 \times \left(\frac{X_{CG} - X_{MGC}}{MGC} \right) \quad (6\text{-}70)$$

EXAMPLE 6-8

The following collection of point loads is given in Table 6-4. Determine the total weight, moments, and the location of the CG in 3-dimensional space.

Solution

Total weight:

$$W_{tot} = \sum_{i=1}^{N} W_i = 3.25 + 7.50 + 2.50 + 1.25 \\ + 5.00 + 2.50 + 2.50 = 24.50 \text{ lb}_f$$

TABLE 6-4 Collection of Point Loads

i	W lb_f	X ft	Y ft	Z ft
1	3.25	5.0	3.0	1.0
2	7.50	3.5	-2.0	3.0
3	2.50	6.5	-2.5	-2.0
4	1.25	10.0	3.0	-2.5
5	5.00	12.0	-3.0	1.5
6	2.50	8.0	-1.0	-3.5
7	2.50	7.5	4.0	-4.0

Moments about the origin of the coordinate system:

$$M_X = \sum_{i=1}^{N} W_i \times X_i = 3.25 \times 5.0 + 7.50 \times 3.5 + \cdots \\ + 2.50 \times 7.5 = 170.0 \text{ ft} \cdot \text{lb}_f$$

$$M_Y = \sum_{i=1}^{N} W_i \times Y_i = 3.25 \times 3.0 + 7.50 \times (-2.5) + \cdots \\ + 2.50 \times 4.0 = -15.25 \text{ ft} \cdot \text{lb}_f$$

$$M_Z = \sum_{i=1}^{N} W_i \times Z_i = 3.25 \times 1.0 + 7.50 \times 3.0 + \cdots \\ + 2.50 \times (-4.0) = 6.38 \text{ ft} \cdot \text{lb}_f$$

CG with respect to the origin of the coordinate system:

$$X_{CG} = \frac{M_X}{W_{tot}} = \frac{170.0 \text{ ft} \cdot \text{lb}_f}{24.50 \text{ lb}_f} = 6.939 \text{ ft}$$

$$Y_{CG} = \frac{M_Y}{W_{tot}} = \frac{-15.25 \text{ ft} \cdot \text{lb}_f}{24.50 \text{ lb}_f} = -0.622 \text{ ft}$$

$$Z_{CG} = \frac{M_Z}{W_{tot}} = \frac{6.38 \text{ ft} \cdot \text{lb}_f}{24.50 \text{ lb}_f} = 0.260 \text{ ft}$$

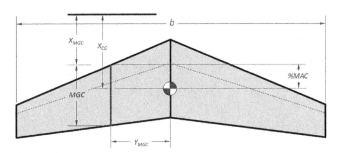

FIGURE 6-19 The location of the CG is commonly referred to in terms of the %MGC.

6.6.6 Determination of CG Location by Aircraft Weighing

The CG location of actual aircraft is always determined by direct weighing. Small aircraft are parked on specially designed weighing kits, which consist of three separate electronic scales; one for the nose gear and two for the main gear (see Figure 6-20). A special device simultaneously connects to all three, allowing the weight on each wheel and the total to be read. Larger aircraft are often equipped with special jacking points used for the same purpose. The advantage of such hardpoints is that their spatial location is known. This contrasts with many fixed landing gear configurations, whose measurements are affected by structural flex, which introduces inaccuracy. Then, once the distance between the weighing points is known, the measured weights can be used to calculate the location of the CG as shown below:

Location of CG from nose gear:

$$x_N = \left(\frac{R_M}{R_M + R_N}\right) x_{NM} = \left(\frac{R_M}{W}\right) x_{NM} \quad (6\text{-}71)$$

Location of CG from main gear:

$$x_M = \left(1 - \frac{R_M}{W}\right) x_{NM} \quad (6\text{-}72)$$

where

R_M = main gear reaction, the sum of both main gear scales
R_N = nose gear reaction
W = total aircraft weight = $R_N + R_M$
x_N, x_M, and x_{NM} = distances defined in Figure 6-20

Note that the CG location is usually determined with respect to some datum. However, this differs among airplane types. The above expression is generic and by locating the nose landing gear with respect to such a datum the CG can be represented in terms convenient to the designer or operator. Additional methodologies are provided by D'Estout [7]. Note that when weighing an aircraft in this fashion, it is imperative that it is leveled as accurately as possible and that no wind conditions prevail where the weighing takes place.

6.6.7 Mass Moments and Products of Inertia

Fundamental Relationships

Any object that rotates about some axis has a tendency to continue that motion, just like an object moving along a straight path has a tendency to move along that path. The former is an example of rotational momentum (think of a flywheel) and the latter of linear momentum. Unless acted on by some force, both will continue this motion indefinitely. The tendency of a rotating body to continue its motion depends on two properties: its mass and the distance of its CG from the axis of rotation. This leads to the definition of *mass moment of inertia* as the property of an object that is to rotational momentum what mass is to linear momentum.

The moment of inertia of a point mass, m, with respect to such an axis is defined as the object's mass times its distance, r, from the axis squared (see Figure 6-21).

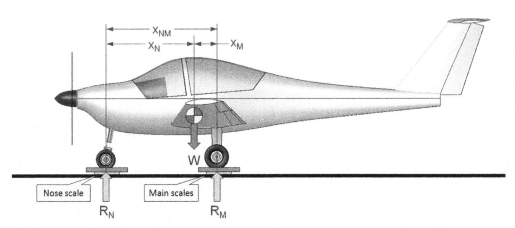

FIGURE 6-20 Typical setup of scales when weighing aircraft.

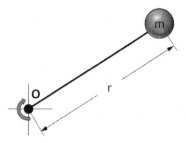

FIGURE 6-21 The definition of mass moment of inertia.

Mathematically, the moment of inertia of the mass about point O is given by:

$$I = mr^2 = \frac{W}{g} r^2 \qquad (6\text{-}73)$$

where W is the weight of the object and g is acceleration due to gravity. In aircraft stability and control theory we are primarily interested in evaluating the moment of inertia about the x-, y-, and z-axes. Therefore, Equation (6-70) must be evaluated for each axis and the results would be presented as follows:

$$I_i = mr_i^2 \quad \text{where} \quad i = x, y, z \qquad (6\text{-}74)$$

The mass moment of inertia of an arbitrary body of constant density and continuous mass distribution can be determined by integrating the contribution of the infinitesimal mass, dm, over the volume of the body about an arbitrary axis of rotation, O (see Figure 6-22):

$$I = \int_B r^2 dm \qquad (6\text{-}75)$$

where B is used to indicate the integration is performed over the entire body. This way, the moment of inertia about an axis is a measure of the *distribution of matter about that axis*. As stated earlier, aircraft stability and control theory requires the moments of inertia to be determined about three mutually orthogonal axes that go through the CG. Equation (6-75) is then rewritten accordingly for each axis by noting that r^2 about the x-axis is given by $(y^2 + z^2)$,

r^2 about the y-axis is given by $(x^2 + z^2)$, and so on. Therefore, the moment of inertia about the point O is given by:

$$I_{xx} = \int_B (y^2 + z^2) dm \quad I_{yy} = \int_B (x^2 + z^2) dm$$
$$I_{zz} = \int_B (x^2 + y^2) dm \qquad (6\text{-}76)$$

where the double-subscripts refer to a rotation about the axis indicated. The value of the moment of inertia is always positive. Note that the three orthogonal vectors that form the coordinate system about which the body rotates necessarily form three separate and mutually orthogonal planes; the xy-, xz-, and yz-planes. Rotational motion in three-dimensional space is strongly affected by how uniformly (or symmetrically) the body is distributed on each side of these planes. This inertia property is called the *products of inertia* and is determined as follows:

$$I_{xy} = I_{yx} = \int_B (xy) dm \quad I_{xz} = I_{zx} = \int_B (xz) dm$$
$$I_{yz} = I_{zy} = \int_B (yz) dm \qquad (6\text{-}77)$$

The value of the product of inertia can be negative or positive. The products of inertia are a measure of the *symmetry of the object*. An airplane is often symmetrical about one of the planes — the xz-plane for a standard coordinate system. For this reason, this plane is often called the *plane of symmetry*. The product of inertia about the xz-plane is often taken to be 0; however, this is false if the airplane has asymmetric mass loading, such as unbalanced fuel in the wing tanks, or a single pilot in a two-seat, side-by-side, cabin. Considering $I_{xz} = 0$ for such an asymmetric loading can only be justified if its magnitude is negligible compared to the other moments and products of inertia. It is not justifiable if one wing fuel tank is full and the other is empty.

Parallel-axis Theorem for Moments of Inertia

Consider Figure 6-23, which shows an arbitrary body rotating about a point other than its CG. The distance

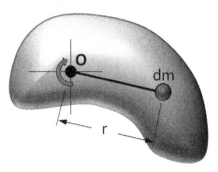

FIGURE 6-22 The mass moment of inertia of an arbitrary body about some arbitrary axis of rotation.

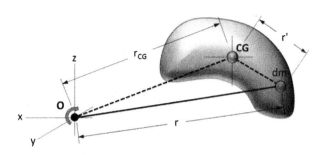

FIGURE 6-23 The parallel-axis theorem explained.

between the CG and the axis of rotation, O, is denoted by r_{CG}. The distance from the axis of rotation to the infinitesimal mass dm is given by r and the distance between it and the CG is given by r'. If the moment of inertia must be evaluated for this situation, evaluating Equation (6-76) leads to a convenient theorem in which the moment of inertia is given by the following expression

$$I_O = mr_{CG}^2 + \int_B r'dm = mr_{CG}^2 + I_{CG} \qquad (6\text{-}78)$$

where

I_{CG} = moment of inertia of the body about its own CG.
r' = distance from CG to an infinitesimal mass dm
r_{CG} = distance from the reference point O to the CG of the body
m = the mass of the body

Equation (6-78) is referred to as the *parallel-axis theorem*. A more practical form of it is shown below:

$$\begin{aligned} I_{xx} &= I_{xx_{CG}} + m(y_{CG}^2 + z_{CG}^2) \\ I_{yy} &= I_{yy_{CG}} + m(x_{CG}^2 + z_{CG}^2) \\ I_{zz} &= I_{zz_{CG}} + m(x_{CG}^2 + y_{CG}^2) \end{aligned} \qquad (6\text{-}79)$$

where

$I_{xx_{CG}}, I_{yy_{CG}}, I_{zz_{CG}}$ = moment of inertia of the body about its own CG
x_{CG}, y_{CG}, z_{CG} = distance from the reference point O to the CG of the body
m = the mass of the body

Derivation of Equation (6-78)

The moment of inertia about the arbitrary point in Figure 6-23:

$$\begin{aligned} I_O &= \int_B r^2 dm = \int_B (r_{CG} + r')^2 dm \\ &= \int_B (r_{CG}^2 + 2r_{CG}r' + r'^2) dm \\ &= \int_B r_{CG}^2 dm + \int_B 2r_{CG}r' dm + \int_B r'^2 dm \end{aligned}$$

Since r_{CG} is a constant, we can simplify this and write:

$$I_O = r_{CG}^2 \int_B dm + 2r_{CG} \int_B r' dm + \int_B r'^2 dm$$

By inspection, the first term represents the moment of inertia of the mass, acting as a point mass, as it rotates about point O and this is given by (remember that r_{CG} is constant):

$$I_P = r_{CG}^2 \int_B dm = mr_{CG}^2$$

where the subscript P denotes the *parallel axis term*. The second term is zero, because the origin of r' is at the CG, but the mass is distributed about the CG. To better see this, consider the coordinate system superimposed on the CG in Figure 6-23. The contribution of the mass lying above the x-axis will be cancelled by equal mass that lies below it. In fact, the integral is effectively a moment integral analogous to Equation (6-66), where X_0 is the x_{CG}. Finally, the third term is the moment of inertia of the body about its own CG and is given by:

$$I_{CG} = \int_B r'^2 dm$$

From which we can write:

$$I_O = I_P + I_{CG} = mr_{CG}^2 + \int_B r'^2 dm$$

QED

Parallel-plane Theorem for Products of Inertia

The parallel-axis theorem can be extended to the product of inertia in a similar fashion, in which case it is referred to as the *parallel-plane theorem*.

$$\begin{aligned} I_{xy} &= I_{xy_{CG}} + m(x_{CG}^2 + y_{CG}^2) \\ I_{xz} &= I_{xz_{CG}} + m(x_{CG}^2 + z_{CG}^2) \\ I_{yz} &= I_{yz_{CG}} + m(y_{CG}^2 + z_{CG}^2) \end{aligned} \qquad (6\text{-}80)$$

where

$I_{xx_{CG}}, I_{yy_{CG}}, I_{zz_{CG}}$ = moment of inertia of the body about its own CG
x_{CG}, y_{CG}, z_{CG} = distance from the reference point O to the CG of the body
m = the mass of the body

The derivation is similar to that of the parallel-axis theorem.

6.6.8 Moment of Inertia of a System of Discrete Point Loads

The form of the moment of inertia most helpful for our analysis of the airplane is when they are written in terms of a collection of discrete mass points, similar to the treatment in Section 6.6.5, *Center of mass, center of gravity*. Thus we write the moments of inertia of a system of discrete point loads as follows:

$$I_{XX} = \frac{1}{g} \sum_{i=1}^{N} W_i(X_i - X_{CG})^2 + I_{XX_i}$$

$$I_{YY} = \frac{1}{g} \sum_{i=1}^{N} W_i(Y_i - Y_{CG})^2 + I_{YY_i} \quad (6\text{-}81)$$

$$I_{ZZ} = \frac{1}{g} \sum_{i=1}^{N} W_i(Z_i - Z_{CG})^2 + I_{ZZ_i}$$

Note that the terms involving the product of the weight and distance from the CG represent the application of the parallel-axis theorem. The last term of each equation is the moment of inertia of the body itself. For instance, a typical piston engine can have a significant moment of inertia about its own CG and this should be included in the estimation. However, the moment of inertia of some particular piece of avionics about its own CG is usually negligible and may be omitted.

6.6.9 Product of Inertia of a System of Discrete Point Loads

The product of inertia is estimated in a similar fashion as the moments of inertia, although added care must be exercised, as the position of a component on either side of each plane must be included with the proper sign.

$$I_{XY} = \frac{1}{g} \sum_{i=1}^{N} W_i(X_i - X_{CG})(Y_i - Y_{CG})_i + I_{XY_i}$$

$$I_{XZ} = \frac{1}{g} \sum_{i=1}^{N} W_i(X_i - X_{CG})(Z_i - Z_{CG}) + I_{XZ_i}$$

$$I_{YZ} = \frac{1}{g} \sum_{i=1}^{N} W_i(Y_i - Y_{CG})(Z_i - Z_{CG})_i + I_{YZ_i}$$

$$(6\text{-}82)$$

where

W_i = weight of item i
X_i = X-location of item i
Y_i = Y-location of item i
Z_i = Z-location of item i

6.6.10 Inertia Matrix

The moments and products of inertia about a specific point are often represented in a matrix format, as this lends itself conveniently for various dynamic stability analyses. This matrix is called the *inertia matrix* and it is always symmetric. Here, it is shown in a format that assumes the axes of interest of go through the airplane's CG.

$$[I_{CG}] = \begin{bmatrix} I_{xx} & -I_{xy} & -I_{xz} \\ -I_{xy} & I_{yy} & -I_{yz} \\ -I_{xz} & -I_{yz} & I_{zz} \end{bmatrix}_{CG} \quad (6\text{-}83)$$

The inertia matrix is dependent on the orientation of the axes going through the CG. One particular orientation results in the matrix becoming diagonal. The axes that cause this special form are called the *principal axes* of the matrix. The principal axes are dependent on the location of the reference point; shifting the reference point to a new location will change the orientation of the principal axes.

6.6.11 Center of Gravity Envelope

Currently, aviation regulations dictate that all aircraft certified as GA aircraft (14 CFR Part 23) must be statically stable and, dynamically, the Dutch roll mode must be stable, while spiral stability and phugoid modes may be slightly divergent. This implies the CG of the aircraft must remain in the neighborhood of a specific location in space, with respect to the airplane, throughout the duration of any flight. This neighborhood is called the *center of gravity envelope*, or simply, a *CG envelope* (see Figure 6-24). The location of the CG is of paramount importance to the pilot of an airplane and it is his or her responsibility to ensure the aircraft is loaded such that the CG remains inside the allowable envelope throughout the entire duration of the flight.

Design Guidelines

In the interests of safety, the reference point (0, 0, 0) to which the CG refers is usually placed far in front of and below the nose of the aircraft. This ensures that when moments due to the positions of discrete weights are being calculated, all the spatial locations only have a positive sign and not a combination of positive and negative signs. This reduces the chance of mathematical error creeping into calculations, and, therefore, the chance of erroneous results that might indicate the CG was inside the CG envelope, when in fact it was outside it.

There are two common ways to indicate the location of the CG for GA aircraft. The first presents it in terms of percentage of the mean geometric chord. The second expresses it more explicitly in terms of the fuselage station (FS). The use of each differs between aircraft manufacturers. The design engineer should be familiar with both methods.

The CG envelope in Figure 6-24 is based on the Type Certification Data Sheet (TCDS 3A15 [8]) for the Beech F33C Bonanza and shows the location of the CG may

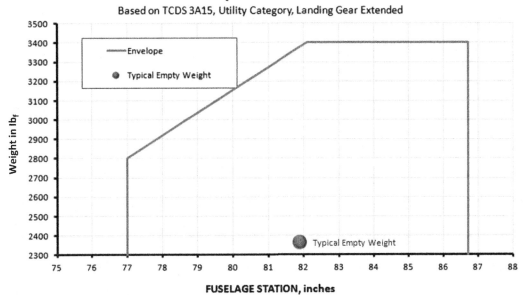

FIGURE 6-24 A typical CG envelope for a light GA aircraft.

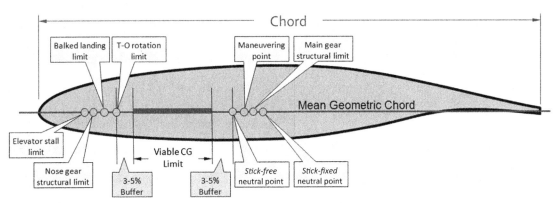

FIGURE 6-25 Factors that affect the viable CG limit.

vary from fuselage station 77, or FS77 to FS86.7 up to 2800 lb$_f$, but from FS82.1 to FS86.7 at 3400 lb$_f$. The FS are in units of inches. Additionally, the plot shows the CG location and weight for a typical empty F33C Bonanza aircraft.

6.6.12 Creating the CG Envelope

The aforementioned section shows that the creation of the CG envelope is a crucial and necessary step. Its goal is to determine how far forward and aft the CG may travel without compromising the safe operation of the aircraft. Unfortunately, this is not as simple as that, as these limits may depend on aerodynamic (i.e. stability and control) and structural issues. Figure 6-25 shows parameters that are typically considered when determining a viable CG envelope. Note that the order of the critical CG locations may be transposed. Also, other considerations may apply to your particular aircraft design that are not reflected here. Ultimately, it is the responsibility of the design leader to ensure that a viable CG envelope has been established. The safe operation of the airplane depends on this task being accomplished correctly.

In the opinion of this author, the designers of selected kit planes have done a less than acceptable job at this and designed airplanes that operate with the CG too far aft. This manifests itself in airplanes with very light stick forces and requires piloting more like what one would expect from a marginally stable aircraft. This may have been done under the pretense that it makes the airplane more "responsive" and "fun to fly," but a more plausible explanation is that it stems from the designer's lack of understanding of stability and control theory. For instance, an airplane that the pilot is advised to handle with care because "it is so easy to start a PIO"[4] is an airplane with unacceptable longitudinal characteristics. Such airplanes are dangerous, no matter how "fun to fly." The customers of such airplanes are often unaware of the risks involved in flying such airplanes and sometimes pay dearly. Make sure you don't fall into this trap.

Determination of the Aft CG Limit

The first step in determining a viable CG envelope lies in the determination of the so-called *stick-fixed neutral point*. The name comes from the fact that it assumes the elevator is immovable, typically in the neutral position. This point is generally a good indicator of how far aft the CG can go, but is not the final answer. There are at least two other points that must also be determined and have to do with stability and control; the *stick-free neutral point* and *maneuvering point*. The former requires knowledge of the elevator hinge moment and the latter about the pitch damping characteristics of the airplane. Each of these points will yield a maximum value for the CG location beyond which the aft CG may not cross. Furthermore, for a conventional tricycle landing gear, the structural capabilities of the main landing gear should also be accounted for. If the CG moves too far aft, the main landing gear reaction loads increase and can eventually lead to structural failure (the same holds for the nose landing gear if the CG is too far forward). The landing gear or airframe design teams typically furnish the aerodynamicist with this information.

Determination of the Forward CG Limit

Characteristics that affect the forward points are also shown in Figure 6-25 and Figure 6-26. The *elevator stall limit* indicates where the maximum deflection of the elevator has been reached and the lift coefficient required to trim the airplane would cause one of the following two scenarios: the elevator stalls, which means there is no additional elevator authority remaining; or the lift coefficient required by the horizontal tail (with the elevator fully deflected trailing edge up) is so high it implies it has exceeded its stall AOA. Generally, rather than allowing the controls to reach such extremes, the experienced aerodynamicist places a limit on each (see Section 23.3, *GA aircraft design checklist*).

A *balked landing limit* represents a condition in which the pilot of an airplane, in the landing configuration, aborts the landing procedure for a go-around. This is a very demanding situation for the airplane because it is flying slowly, with flaps fully deployed, and at full power. This limit is similar to the elevator stall limit in the sense it represents a demand for too large an elevator deflection. The *T-O rotation limit* results from the elevator being unable to lift the nose gear off the ground during the T-O run.

These limiting CG locations cited above fall around the viable CG envelope and dictate the usable limit. Some limits, such as the landing gear structural limits or the elevator limits, are typically presented as isobars (see Figure 6-26). The structural isobars are determined during the detailed design phase. The elevator limit isobar is determined using stability and control theory.

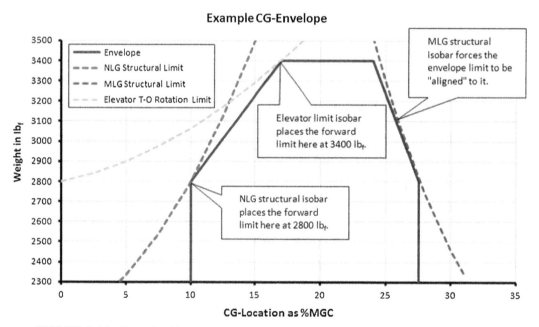

FIGURE 6-26 Example of how nose or main landing gear structural limits may affect CG limits.

---

[4]PIO stands for pilot-induced oscillation.

The figure shows how the three isobars affect the shape of the envelope in different ways. The NLG isobar dictates the forward light limit (here 10% MGC at 2800 lb$_f$). The elevator limit isobar forces the forward limit at 3400 lb$_f$ at 17% MGC) and the MLG isobar forces the aerodynamicist to move the aft envelope line farther forward (sloping it in the process), even though the airplane might be quite capable of being more aft. With respect to Figure 6-25, the viable envelope should generally offer between 3% and 5% MAC buffer, forward and aft of the corresponding limiting points, in case there are analysis inaccuracies.

Loading Cloud

The generation of a *loading cloud* is an important step that should be completed for any aircraft that carries more than one occupant. It is a graph showing the CG envelope and as many combinations of occupants, baggage, and fuel as is practical. An example of this is shown in Figure 6-27, constructed using the data of Table 6-5. The plot gives a lot of clues about the range of weights and CG locations the airplane must operate within and whether it is likely to be operated outside the proposed CG limits. An airplane that requires the operator to constantly worry about whether it is loaded outside the CG limits is one that is likely to accrue criticism. As such, the loading cloud is helpful during both the preliminary stage and development of production vehicles.

As stated above, the graph is created by calculating the CG location and weight for selected combinations of occupants, baggage, and fuel weight. The basic graph of Figure 6-27 is based on the single-engine, four-seat Beechcraft F33C Bonanza aircraft and uses the calculation methodology of Section 6.6.11, *Center of gravity envelope*. The data calculated in Table 6-5 are superimposed to help one visualize how well the CG envelope contains the loading combinations. As can be seen, some of the combinations consist of various amounts of fuel, occupant, and baggage weights. This way, each row has its own total weight and CG location, which is plotted in Figure 6-27. The figure shows that the F33C hardly has a forward loading problem, but is easily loaded outside the aft CG limit. This results from the relative aft position of even the front seats. In defense of the F33C, some of the load combinations presented are arguably "preposterous" or "unfair." However, they are presented to emphasize the importance of reviewing the rationale behind particular loading scenarios. For instance, considering Combo IDs 11 and 12, it is not plausible that a competent pilot would store 270 lb$_f$ in the baggage area, with an empty seat farther forward, not to mention that 270 lb$_f$ might exceed the allowable load in the baggage area.

The loading cloud reveals primarily two properties of the load combination: the most forward and aft CG location the airplane is likely to ever see in practice; and the maximum and minimum weight margin expected

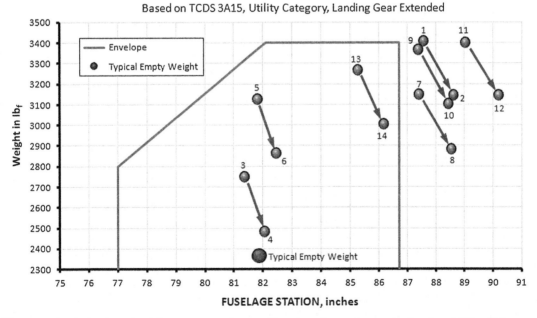

FIGURE 6-27 Example of a loading cloud. It exposes a serious problem, here destined to render the aircraft "illegal" for a careless pilot. If this happens to your aircraft and the load combinations are "practical" (unlike the ones shown here), then do something about it! — move the wing, move heavy parts around to place the empty weight CG in a better location, so the envelope can accommodate most of the cloud. Just do something.

TABLE 6-5 Tabulated Loading Combinations Used to Create the Loading Cloud of Figure 6-27. Note that weights are in lb_f and arms in inches

Combo ID	Empty Weight		Front Left Seat		Front Right Seat		Aft Left Seat		Aft Right Seat		Baggage Area		Fuel		Total	
	W_e	X_{CGe}	W	Arm	W	Arm	W	Arm	W	Arm	W	Arm	W	Arm	W	X_{CG}
1	2363	81.9	170	85.0	170	85.0	170	121.0	170	121.0	100	150.50	264	75.0	3407	87.6
2	2363	81.9	170	85.0	170	85.0	170	121.0	170	121.0	100	150.50	0	75.0	3143	88.6
3	2363	81.9	120	85.0	0	85.0	0	121.0	0	121.0	0	150.50	264	75.0	2747	81.4
4	2363	81.9	120	85.0	0	85.0	0	121.0	0	121.0	0	150.50	0	75.0	2483	82.0
5	2363	81.9	250	85.0	250	85.0	0	121.0	0	121.0	0	150.50	264	75.0	3127	81.8
6	2363	81.9	250	85.0	250	85.0	0	121.0	0	121.0	0	150.50	0	75.0	2863	82.4
7	2363	81.9	250	85.0	0	85.0	0	121.0	0	121.0	270	150.50	264	75.0	3147	87.4
8	2363	81.9	250	85.0	0	85.0	0	121.0	0	121.0	270	150.50	0	75.0	2883	88.5
9	2363	81.9	120	85.0	120	85.0	250	121.0	250	121.0	0	150.50	264	75.0	3367	87.4
10	2363	81.9	120	85.0	120	85.0	250	121.0	250	121.0	0	150.50	0	75.0	3103	88.4
11	2363	81.9	170	85.0	170	85.0	0	121.0	170	121.0	270	150.50	257	75.0	3400	89.1
12	2363	81.9	170	85.0	170	85.0	0	121.0	170	121.0	270	150.50	0	75.0	3143	90.2
13	2363	81.9	170	85.0	250	85.0	0	121.0	120	121.0	100	150.50	264	75.0	3267	85.3
14	2363	81.9	170	85.0	250	85.0	0	121.0	120	121.0	100	150.50	0	75.0	3003	86.2

during operation. The importance of performing this sort of analysis cannot be overemphasized. For instance, when certifying a new airplane, the authorities will require the manufacturer to demonstrate compliance to the regulations at each extreme of the CG envelope. This means the airplane must be loaded to these extremes, and demonstrated to satisfy applicable regulations. This can lead to situations that are better avoided,

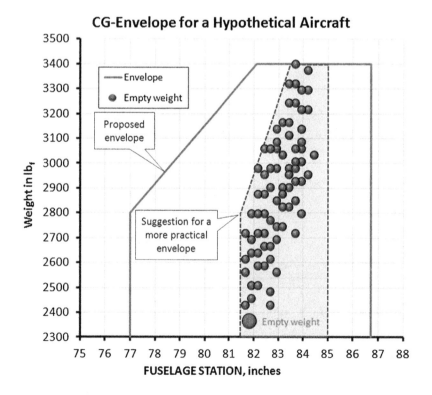

FIGURE 6-28 Sometimes the CG envelope proposed turns out to be larger than actually needed.

as explained in the next paragraph and highlighted in Figure 6-28.

Consider a scenario that sometimes comes up in practice; too large a CG envelope. The initial CG envelope may have been based on analysis and even preliminary flight testing. However, it may be found to exceed what the airplane will ever be exposed to in operation. This may be the consequence of the loading possibilities of the aircraft; for instance due to seating arrangement (e.g. side-by-side seating), limited travel with fuel consumption, and other factors. An unnecessarily large CG envelope may bring about a serious headache for the flight test team because the airworthiness of the vehicle must be demonstrated at the extremes of the envelope. Our hypothetical aircraft would require heavy ballast to be mounted in awkward places to allow it to be flown at the extremes of the envelope. It might even require special temporary hardpoints to be designed, fabricated, and installed to carry the ballast. It is really a wasted effort, rendering it far more sensible to simply redraw a narrower envelope to resemble that of the shaded area in Figure 6-28. This will shave off certification cost and effort and the resulting airplane will be equally useful as the one featuring the wider envelope.

6.6.13 In-flight Movement of the CG

All aircraft that burn fossil fuels will reduce their weight during flight. Some aircraft will reduce their weight by other means, such as airplanes dropping parachutists, or military aircraft jettisoning empty fuel tanks or ordnance (missiles, bombs, etc.). This movement must be considered during the design phase of the aircraft, and may not cause the CG to move out of the CG envelope. Often this requires components to be relocated to ensure the airplane can contain the CG location inside the CG envelope regardless of such weight changes. Figure 6-29 shows an example of the Beech F33C Bonanza with two 200 lb_f-people and full tanks of fuel (74 gals usable) as the fuel is completely consumed. This example is based on specifications given by TCDS 3A15 for the airplane. It shows that the CG moves back an inch or so but, more importantly, that the CG stays inside the airplane's envelope.

6.6.14 Weight Budgeting

The purpose of weight budgeting is to provide constraints and impetus for airframe designers to design parts with strong emphasis on weight. It is a common problem in the aviation industry that components are overdesigned, which means that parts are unnecessarily strong and, thus, too heavy. This is one of the primary reasons why empty weight targets get busted during the development of aircraft. Weight budgeting helps the weight reduction effort to stay where it belongs: with the cognizant designer (airframe, avionics, power plant, and so on).

Another important purpose is that it helps the engineer understand where effort toward weight reduction is most likely to bear fruit. This is important if empty weight

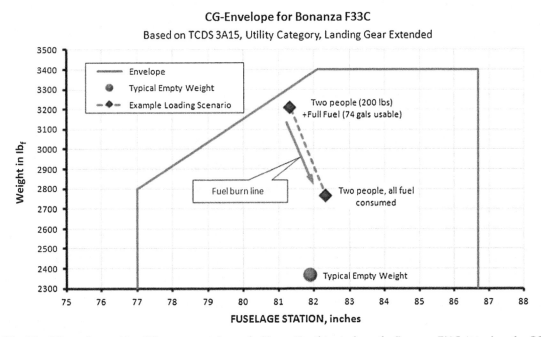

FIGURE 6-29 The CG envelope with a CG movement due to fuel burn. For this airplane, the Bonanza F33C, it is clear the CG will remain inside the CG envelope even if all the usable fuel were consumed.

TABLE 6-6 Example of a Weight Budget Being Compared to Actual Weights

Category	Weight for Category (Budget) lb$_f$	Projected Weight (Actual) lb$_f$	Source
Wings	600	555	Direct
Horizontal/vertical tail	100	131	Direct
Fuselage	500	580	Direct
Weight penalty for pressurization	Included in Fuselage	35	Measured
Main landing gear	180	250	Measured
Nose landing gear	95	75	Measured
Nacelle	100	86	Direct
Fuel system	40	126	Statistical
Power plant	450	485	Measured
Flight control system	60	91	Statistical
Hydraulic system	40	24	Statistical
Electrical systems	120	260	Statistical
HVAC	60	125	Statistical
Bleed air system	15		Statistical
Pressurization system	15		Statistical
De-icing system	65		Statistical
Oxygen system	30		Direct
Furnishings	200	216	Statistical
Other	100	0	Direct
TOTAL	2770	3039	

TABLE 6-7 Point Loads with Tolerances

Item	Weight	X	Y	Z
Item 1	$W_1 \pm \Delta W_1$	$X_1 \pm \Delta X_1$	$Y_1 \pm \Delta Y_1$	$Z_1 \pm \Delta Z_1$
Item 2	$W_2 \pm \Delta W_2$	$X_2 \pm \Delta X_2$	$Y_2 \pm \Delta Y_2$	$Z_2 \pm \Delta Z_2$
Item 3	$W_3 \pm \Delta W_3$	$X_3 \pm \Delta X_3$	$Y_3 \pm \Delta Y_3$	$Z_3 \pm \Delta Z_3$
...				
Item i	$W_i \pm \Delta W_i$	$X_i \pm \Delta X_i$	$Y_i \pm \Delta Y_i$	$Z_i \pm \Delta Z_i$
...				
Item N	$W_N \pm \Delta W_N$	$X_N \pm \Delta X_N$	$Y_N \pm \Delta Y_N$	$Z_N \pm \Delta Z_N$

targets have not been met. Remember the adage: "reducing 1000 lbs by 1% is better than reducing 1 lb by 50%."

The weight budgeting process begins by breaking the complete aircraft into categories, such as Wings, Horizontal Tail, Vertical Tail, Fuselage, Landing Gear, Powerplant, etc. Sometimes such categories are broken down further, e.g. Wings-Left Main for the main element of the left wing, Wings-Ailerons, Wings-Flaps, Wings-Electrics, and so on. An example of a weight budget breakdown as used for weight management in the development of a prototype is shown in Table 6-6.

This example shows that the projected weight of this aircraft is some 269 lb$_f$ higher than planned. This might be acceptable for a test vehicle, although it could cause complications in a flight test program, in particular if the weight of available fuel for test flying is compromised. But it is unacceptable for the production airplane as it amounts to 10% over target.

6.6.15 Weight Tolerancing

During the preliminary design phase it is impossible to pinpoint a precise location for the CG or magnitudes of moments and products of inertia, as final weights and CG locations of individual components are constantly changing. For instance, the location of the engine's CG may be specified as some value $X_{engine} \pm \Delta x$. It may be an engine in development and its weight may be given as $W_{engine} \pm \Delta W$. As an example, the engine manufacturer might specify the weight of the engine to be 356 ± 15 lb$_f$. Consequently, the moment contribution of this engine to the total moment about the reference point (0, 0, 0) would be a range along the x-axis, rather than a specific point, computed as follows:

$$\begin{aligned} M_{engine} &= \left(X_{engine} \pm \Delta x\right)\left(W_{engine} \pm \Delta W\right) \\ &= X_{engine} W_{engine} \pm \left(X_{engine} \Delta W + W_{engine} \Delta x \right. \\ &\quad \left. + \Delta x \Delta W\right) \end{aligned}$$

(6-84)

Under such circumstances it is better to consider these as a "sphere" of possible values, rather than a single specific point. In the process, the designer can assess the probability of the CG being outside the allowable limits. This section develops equations that allow the aircraft designer to keep track of these important parameters by assigning tolerances to them.

Consider a collection of point loads positioned in three-dimensional space whose weight and position are known to a certain level of accuracy (tolerance) only (see Table 6-7). Then, the inertia properties for such a collection are defined as follows:

Total weight:

$$W_{tot} = \sum_{i=1}^{N} W_i \pm \Delta W_i = \sum_{i=1}^{N} W_i \pm \sum_{i=1}^{N} \Delta W_i$$

$$= \begin{cases} W_{TOT_{min}} \\ W_{TOT_{nom}} \\ W_{TOT_{max}} \end{cases} \quad (6\text{-}85)$$

X-moments about the point (0, 0, 0):

$$M_X = \begin{cases} \sum_{i=1}^{N}(W_i - \Delta W_i)(X_i - \Delta X_i) = M_{X_{min}} \\ \sum_{i=1}^{N} W_i \times X_i = M_X \\ \sum_{i=1}^{N}(W_i + \Delta W_i)(X_i + \Delta X_i) = M_{X_{max}} \end{cases} \quad (6\text{-}86)$$

Y-moments about the point (0, 0, 0):

$$M_Y = \begin{cases} \sum_{i=1}^{N}(W_i - \Delta W_i)(Y_i - \Delta Y_i) = M_{Y_{min}} \\ \sum_{i=1}^{N} W_i \times Y_i = M_Y \\ \sum_{i=1}^{N}(W_i + \Delta W_i)(Y_i + \Delta Y_i) = M_{Y_{max}} \end{cases} \quad (6\text{-}87)$$

Z-moments about the point (0, 0, 0):

$$M_Z = \begin{cases} \sum_{i=1}^{N}(W_i - \Delta W_i)(Z_i - \Delta Z_i) = M_{Z_{min}} \\ \sum_{i=1}^{N} W_i \times Z_i = M_Z \\ \sum_{i=1}^{N}(W_i + \Delta W_i)(Z_i + \Delta Z_i) = M_{Z_{max}} \end{cases} \quad (6\text{-}88)$$

Location of CG:

$$X_{CG_{min}}, Y_{CG_{min}}, Z_{CG_{min}} = \frac{M_{X_{min}}}{W_{TOT_{min}}}, \frac{M_{Y_{min}}}{W_{TOT_{min}}}, \frac{M_{Z_{min}}}{W_{TOT_{min}}}$$

$$X_{CG}, Y_{CG}, Z_{CG} = \frac{M_X}{W_{TOT}}, \frac{M_Y}{W_{TOT}}, \frac{M_Z}{W_{TOT}}$$

$$X_{CG_{min}}, Y_{CG_{max}}, Z_{CG_{max}} = \frac{M_{X_{max}}}{W_{TOT_{max}}}, \frac{M_{Y_{max}}}{W_{TOT_{max}}}, \frac{M_{Z_{max}}}{W_{TOT_{max}}}$$

(6-89)

EXAMPLE 6-9

A collection of point loads is given in Table 6-8. Determine a probable location of the CG along the x-axis by accounting for tolerances.

Solution

Total weight:

$$W_{tot} = \sum_{i=1}^{N} W_i \pm \sum_{i=1}^{N} \Delta W_i = 60.0 \pm 5.0$$

$$= \begin{cases} W_{TOT_{min}} = 55.0 \text{ lb}_f \\ W_{TOT_{nom}} = 60.0 \text{ lb}_f \\ W_{TOT_{max}} = 65.0 \text{ lb}_f \end{cases}$$

X-moments about the point (0, 0, 0):

$$M_{X_{min}} = \sum_{i=1}^{N}(W_i - \Delta W_i)(X_i - \Delta X_i)$$

$$= (10.0 - 2.0)(5.0 - 0.5) + (20.0 - 1.0)(3.5 - 0.5)$$
$$+ (30.0 - 2.0)(8.5 - 0.5)$$

$$= 317.0 \text{ ft} \cdot \text{lb}_f$$

$$M_X = \sum_{i=1}^{N} W_i \times X_i$$

$$= 10.0 \times 5.0 + 20.0 \times 3.5 + 30.0 \times 8.5 = 375.0 \text{ ft} \cdot \text{lb}_f$$

(Continued)

EXAMPLE 6-9 (cont'd)

$$M_{X_{max}} = \sum_{i=1}^{N}(W_i + \Delta W_i)(X_i + \Delta X_i)$$
$$= (10.0+2.0)(5.0+0.5) + (20.0+1.0)(3.5+0.5)$$
$$+ (30.0+2.0)(8.5+0.5)$$
$$= 438.0 \text{ ft} \cdot \text{lb}_f$$

CG range:

$$X_{CG_{min}} = \frac{M_{X_{min}}}{W_{TOT_{min}}} = \frac{317.0 \text{ ft} \cdot \text{lb}_f}{55.0 \text{ lb}_f} = 5.764 \text{ ft}$$

$$X_{CG} = \frac{M_X}{W_{TOT}} = \frac{375.0 \text{ ft} \cdot \text{lb}_f}{60.0 \text{ lb}_f} = 6.250 \text{ ft}$$

$$X_{CG_{max}} = \frac{M_{X_{max}}}{W_{TOT_{max}}} = \frac{438.0 \text{ ft} \cdot \text{lb}_f}{65.0 \text{ lb}_f} = 6.738 \text{ ft}$$

Another way of presenting this would be:

$$X_{CG} = 6.250 \begin{cases} -0.483 \\ +0.486 \end{cases} \text{ ft} \approx 6.25 \pm 0.48 \text{ ft}$$

TABLE 6-8 Collection of Point Loads

i	W lb$_f$	ΔW ft	X ft	ΔX ft
1	10.0	2.0	5.0	0.5
2	20.0	1.0	3.5	0.5
3	30.0	2.0	8.5	0.5

where

W_i = weight of item i
X_i = X-location of item i
Y_i = Y-location of item i
Z_i = Z-location of item i
ΔW_i = Tolerance assigned to the weight of item i
ΔX_i = Tolerance assigned to the X-location of item i
ΔY_i = Tolerance assigned to the Y-location of item i
ΔZ_i = Tolerance assigned to the Z-location of item i

EXERCISES

(1) Determine the useful load, empty weight ratio, and fuel weight ratio for an airplane whose gross weight is 1650 lb$_f$, empty weight is 950 lb$_f$, and which can carry 33 gallons of avgas.

(2) You have been asked to design a twin-engine, piston-powered GA aircraft that requires only one pilot for operation. The customer wants the airplane capable of taking off at the design gross weight with a useful load that consists of eight 200 lb$_f$-individuals (which includes the pilot), 200 gallons of avgas and 350 lb$_f$ of baggage on board. Additionally, he or she wants the design's empty weight to amount to no more than

TABLE 6-9 Weight Data for a Hypothetical Amphibious LSA

Empty Weight Componnts				
Description	Weight, lb$_f$	X, ft	Y, ft	Z, ft
Wing	207	2.16	0.00	0.34
Horizontal tail	17	15.82	0.00	0.00
Vertical tail	24	16.83	0.00	0.00
Fuselage	84	-1.91	0.00	-0.77
Main gear	91	2.50	0.00	-2.50
Nose gear	29	-5.50	0.00	-2.50
Rotax 912	189	3.50	0.00	-0.50
Tail arm	20	9.52	0.00	0.00
Propeller	35	4.85	0.00	-0.25
Ballast	75	-3.04	0.00	-1.00
Buffer weight (5%)	40	0.00	0.00	0.00
Operational Items				
Description	Weight, lb$_f$	X, ft	Y, ft	Z, ft
Observer	200	-1.04	1.25	-1.00
Pilot	200	-1.04	-1.25	-1.00
Fuel tank	95	2.34	2.43	0.00
Fuel tank	95	2.34	-2.43	0.00
Baggage	25	0.00	0.00	0.00

65% of gross weight. Determine the empty weight (W_e), gross weight (W_o), useful load (W_u), payload (W_p), crew weight (W_c), fuel weight ratio (W_f/W_o), and empty weight ratio (W_e/W_o). Compare the empty weight ratio to the one obtained using the formulation of Section 6.2.2, *Method 2: Historical empty weight fractions*, for the same class of airplanes.

(3) (a) Estimate the empty and gross weight and the corresponding CG positions using the data for the amphibious LSA aircraft depicted in Table 6-9. (b) Also estimate the airplane's moment and products of inertia.

(4) This problem is intended to demonstrate a "typical" operational scenario for an airplane design using an actual aircraft. A CG envelope for the Beech F33C Bonanza is shown in Figure 6-24. The empty weight of the airplane is 2363 lb_f and the empty weight CG location is at fuselage station (FS) 81.9 inches. The gross weight of the airplane is 3400 lb_f. The airplane is to be loaded for a flight trip in accordance with the data of Table 6-10.

Determine the following:

(a) The maximum fuel the pilot may take-off with, in lb_f and US gallons.
(b) Determine the weight and FS with T-O fuel on board.
(c) Determine the weight and FS with all fuel consumed.
(d) Plot the points representing the empty weight CG, as well as those from (b) and (c).

TABLE 6-10 Weight Data For a Hypothetical Amphibious LSA

Item	FS, inches	Weight, lb_f
Empty weight	81.9	2363
Pilot (front left seat)	85.0	180
Pax 1 (front right seat)	85.0	140
Pax 2 (aft left seat)	121	200
Pax 3 (aft right seat)	121	120
Baggage	150	50
Fuel	75.0	?

(e) Is there a problem with the pilot's planned loading of this airplane? If so, what is it? Is there a simple remedy he or she can apply to solve the problem? What is is? (Support with numbers where appropriate.)

(5) Consider the side view of the Cessna 177 RG Cardinal shown in Figure 6-30, as it is being weighed. This is done by placing scales under its nose wheel and the two main wheels. The airplane contains only unusable fuel and is otherwise empty. Determine the following for the airplane if the nose scale (R_N) reads 275 lb_f, the left main (R_{LEFT}) reads 725 lb_f, and the right main (R_{RIGHT}) reads 695 lb_f:

(a) Empty weight (Ans: 1695 lb_f).
(b) Fuselage station of the CG (Ans: FS107.7).

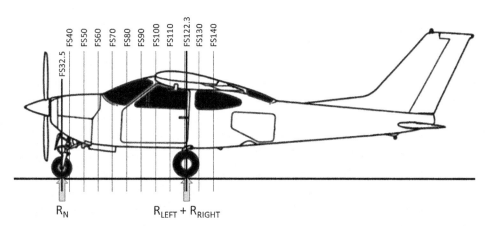

FIGURE 6-30 A side view of the Cessna 177 RG Cardinal.

VARIABLES

Symbol	Description	Units (UK and SI)
$\dot{w}_j$	Average fuel flow during segment j	lb_f/unit time or N/unit time
A_{cap}	Spar cap area	ft^2 or m^2
A_{cell}	Area of idealized cell	ft^2 or m^2
$A_{halfweb}$	Area of half of the spar web	ft^2 or m^2
AR	Aspect ratio	
b	Wingspan	ft or m (context dependent)
b_{HT}	Horizontal tail span	ft or m (context dependent)
b_{VT}	Vertical tail span	ft or m (context dependent)
C_{avg}	Average chord length	ft or m
C_{cell}	Chord length of idealized cell	ft or m
C_m	Coefficient of moment	
c_{MGC}	Mean geometric chord	ft or m
d_F	Maximum fuselage depth	ft or m
F_{bend}	Bending force	lb_f or N
F_{tail}	Vertical tail weight factor	
g	Acceleration due to gravity	ft/s^2 or m/s^2
h	Structural depth at MGC	ft or m
i	Node index for mission segment	
I	Moment of inertia	ft^4 or m^4
I_{CG}	Moment of inertia of a body about its own CG	$slugs \cdot ft^2$ or $kg \cdot m^2$
I_{XX}	Moment of inertia about the X-axis	$slugs \cdot ft^2$ or $kg \cdot m^2$
I_{XY}	Product of inertia in the XY-plane	ft^4 or m^4
I_{XZ}	Product of inertia in the XZ-plane	ft^4 or m^4
I_{yy}	Moment of inertia about the Y-axis	$slugs \cdot ft^2$ or $kg \cdot m^2$
I_{YZ}	Product of inertia in the YZ-plane	ft^4 or m^4
I_{zz}	Moment of inertia about the Z-axis	$slugs \cdot ft^2$ or $kg \cdot m^2$
j	Mission segment	
l_{FS}	Length of fuselage structure (forward bulkhead to aft frame)	ft or m (context dependent)
l_{HT}	Horizontal tail arm	ft or m (context dependent)
L_m	Length of main landing gear strut	inch
L_n	Length of nose landing gear strut	inch
M	Mach number	
M	Bending moment	$lb_f \cdot ft$ or $N \cdot m$
M_{max}	Maximum bending moment	$lb_f \cdot ft$ or $N \cdot m$
M_X	Moment about the x-axis	$lb_f \cdot ft$ or $N \cdot m$
M_Y	Moment about the y-axis	$lb_f \cdot ft$ or $N \cdot m$
M_Z	Moment about the z-axis	$lb_f \cdot ft$ or $N \cdot m$
N_{ENG}	Number of engines	
n_l	Ultimate landing load factor	
n_{lim}	Structural limit load	
N_{OCC}	Number of occupants (crew and passengers)	
N_{rib}	Number of ribs	
N_{TANK}	Number of fuel tanks	
n_{ult}	Ultimate structural load	
n_z	Ultimate load factor	
P_{rated}	Rated power of engine	BHP
q	Dynamic pressure	lb_f/ft^2 or N/m^2 (context dependent)
Q	First area moment	ft^3 or m^3
Q_f	Fuel quantity	gal or L
Q_{int}	Quantity of fuel in integral tanks	gal or L (context dependent)
Q_{TOT}	Total fuel quantity	gal or L (context dependent)
r'	Distance from CG to an infinitesimal mass	ft or m
r_{CG}	Distance from CG to reference point	ft or m
R_m	Main gear reaction force	lb_f or N
R_n	Nose gear reaction force	lb_f or N
S_{cell}	Surface area of idealized cell	ft^2 or m^2
S_{FUS}	Fuselage wetted area	ft^2 or m^2 (context dependent)
S_{HT}	Horizontal tail area	ft^2 or m^2 (context dependent)
S_{VT}	Vertical tail area	ft^2 or m^2 (context dependent)
S_w	Wing planform area	ft^2 or m^2
t/c	Thickness to chord ratio	
$t_{HT\ max}$	Maximum root chord thickness	in
t_i	Time at node index	lb_f or N
t_{i-1}	time at previous node index	lb_f or N

Symbol	Description	Units (UK and SI)
T_{rated}	Rated thrust of engine	lb$_f$ or N
t_{rib}	Thickness of rib	ft or m
t_{skin}	Skin thickness	ft or m
$t_{VT\ max}$	Maximum root chord thickness	in
t_{web}	Thickness of spar web	ft or m
V	Shear force (context dependent)	lb$_f$ or N
V_{cap}	Shear force in cap	lb$_f$ or N
V_H	Maximum level airspeed at S-L	KEAS
W_0	Design gross weight of aircraft	lb$_f$ or N
W_{AC}	Predicted weight of air-conditioning and anti-icing	lb$_f$ or N (context dependent)
W_{AV}	Predicted weight of avionic systems	lb$_f$ or N (context dependent)
W_c	Weight of crew	lb$_f$ or N
W_{caps}	Weight of spar caps	lb$_f$ or N
W_{CTRL}	Predicted weight of flight control systems	lb$_f$ or N (context dependent)
W_e	Empty weight of aircraft	lb$_f$ or N
W_e/W_0	Empty weight ratio	
W_{EI}	Predicted weight of installed engine	lb$_f$ or N (context dependent)
W_{EL}	Predicted weight of electrical systems	lb$_f$ or N (context dependent)
W_{ENG}	Uninstalled engine weight	lb$_f$ or N (context dependent)
W_f	Weight of fuel	lb$_f$ or N
W_F	Maximum fuselage width	ft or m (context dependent)
W_f/W_0	Fuel weight ratio	
W_{FS}	Predicted weight of fuel system	lb$_f$ or N (context dependent)
W_{FURN}	Predicted weight of furnishing	lb$_f$ or N (context dependent)
W_{FUS}	Predicted fuel weight of fuselage	lb$_f$ or N (context dependent)
W_{FW}	Weight of fuel in wing	lb$_f$ or N (context dependent)
W_{HT}	Predicted weight of horizontal tail	lb$_f$ or N (context dependent)
W_{HYD}	Predicted weight of hydraulic system	lb$_f$ or N (context dependent)
W_i	Aircraft weight at node index	lb$_f$ or N
W_{i-1}	Aircraft weight at previous node index	lb$_f$ or N
W_j/W_0	Segment fuel ratio	lb$_f$ or N
W_l	Design landing weight	lb$_f$ or N (context dependent)
W_{LDG}	Maximum landing weight	lb$_f$ or N
W_{MLG}	Predicted weight of main landing gear	lb$_f$ or N (context dependent)
W_{MNLG}	Predicted weight of entire landing gear	lb$_f$ or N (context dependent)
W_{MZF}	Maximum zero fuel weight	lb$_f$ or N
W_{NLG}	Predicted weight of nose landing gear	lb$_f$ or N (context dependent)
W_p	Weight of payload	lb$_f$ or N
W_R	Ramp weight	lb$_f$ or N
W_{skin}	Wight of skin	lb$_f$ or N
W_{total}	Total weight of specified conditions	lb$_f$ or N
W_u	Weight of useful load	lbf or N
W_{UAV}	Weight of uninstalled avionic systems	lb$_f$ or N (context dependent)
W_{VT}	Predicted weight of vertical tail	lb$_f$ or N (context dependent)
W_W	Predicted weight of wing	lb$_f$ or N
W_{web}	Weight of shear web	lb$_f$ or N
X_{CG}	Length-wise location of CG	ft or m or %MGC
x_M	Location of CG from main gear	ft or m
x_N	Location of CG from nose gear	ft or m
x_{NM}	Distance between nose gear and main gear	ft or m
Y_{CG}	Span-wise location of CG	ft or m
y_{MGC}	Span-wise location of MGC	ft or m
Z_{CG}	Height-wise location of CG	ft or m
λ	Wing taper ratio	
$\Lambda_{C/4}$	Wing sweep at 25% MGC	deg or rad (context dependent)
Λ_{HT}	Horizontal tail sweep at 25% MGC	deg or rad (context dependent)
λ_{HT}	Horizontal tail taper ratio	
Λ_{VT}	Vertical tail sweep at 25% MGC	deg or rad (context dependent)
λ_{VT}	Vertical tail taper ratio	
ρ_{caps}	Density of spar cap material	slugs/ft^3 or kg/m^3
ρ_{skin}	Density of skin material	slugs/ft^3 or kg/m^3
$\sigma_{bending}$	Normal stress due to bending	psi or Pa
τ_{max}	Maximum shear stress in structure of interest	psi or Pa
τ_{skin}	Shear stress in skin	psi or Pa
$\tau_{torsion}$	Shear stress due to torsion	psi or Pa

References

[1] Pappalardo Joe. Weight watchers. How a team of engineers and a crash diet saved the joint strike fighter. Air and Space Magazine October/November, 2006:66—73.

[2] Thomas Geoffrey. Following Quantas cuts, Attention turns to 787. Air Transport World April 15, 2009.

[3] Moores Victoria. Emirates seeks A380 and 747-8 Weight Control. Flightglobal October 24, 2007.

[4] Raymer Daniel. Aircraft Design: A Conceptual Approach. AIAA Education Series 1996.

[5] Torenbeek Egbert. Synthesis of Subsonic Aircraft Design. 3rd ed. Delft University Press; 1986.

[6] Nicolai Leland. Fundamentals of Aircraft and Airship Design. Volume I. AIAA Education Series 2010.

[7] D'Estout Henri G. Aircraft Weight and Balance Control. 4th ed. Aero Publishers; 1967.

[8] TCDS 3A15. Hawker Beechcraft Corporation, Revision 94, 02/25/2007. FAA.

CHAPTER 7

Selecting the Power Plant

OUTLINE

7.1 Introduction 182
   7.1.1 The Content of this Chapter 183
   7.1.2 Power Plant Options for Aviation 183
   7.1.3 The Basics of Energy, Work, and Power 183
   7.1.4 Thermodynamics of the Power Plant 183
      Piston Engines 183
      Gas Turbines 183
   7.1.5 General Theory of Thrust Generation 184
      Propeller and Gas Turbines 185
      Rockets 185
      Thrust-to-weight Ratio 185
   7.1.6 Fundamental Definitions 185
      Brake Horsepower (BHP) 185
      Shaft Horsepower (SHP) 185
      Equivalent Horsepower (EHP) 185
      Thrust Horsepower (THP) 185
      Engine Torque and Turboprop Engines 186
      Engine Power and Thrust Ratings 186
   7.1.7 Fuel Basics 187
      Density of Aviation Gasoline (Avgas) 187
      Energy Content of Fuel for Piston Engines 187
      Fuel Octane Rating and Fuel Grades for Piston Engines 187
      Fuel Grades for Jet Fuel 188
      Specific Fuel Consumption (SFC) 188
      SFC for Jets 189
      SFC for Pistons 189

7.2 The Properties of Selected Engine Types 190
   7.2.1 Piston Engines 190
      Two-stroke versus Four-stroke Engines 190
      Operation of the Four-stroke Engine 191
      Air-to-fuel Ratio 192
      Compression and Pressure Ratios 192
      Displacement 192
      Typical Specific Fuel Consumption for Piston Engines 192
      Effect of Airspeed on Engine Power 192
      Effect of Altitude on Engine Power 192
      Effect of Temperature on Engine Power 195
      Effect of Manifold Pressure and RPM on Engine Power 195
   7.2.2 Turboprops 196
      Typical Fuel Consumption 196
      Turboprop Inertial Separators 196
      Throttle Ratio (TR) 196
      Step-by-step: Effect of Altitude and Airspeed on Turboprop Engine Thrust 198
   7.2.3 Turbojets 199
      Typical Fuel Consumption 199
      Step-by-step: Effect of Altitude and Airspeed on Turbojet Engine Thrust 200
   7.2.4 Turbofans 200
      Typical Fuel Consumption 201
      Step-by-step: Effect of Altitude and Airspeed on Turbofan Engine Thrust 201
   7.2.5 Electric Motors 203
      Energy Density 204
      Batteries 204
      Fuel Cells 205
      Hybrid Electric Aircraft 206
      The Pure Electric Aircraft 206
      Formulation 206
   7.2.6 Computer code: Thrust as a Function of Altitude and Mach Number 207

7.3 Aircraft Power Plant Installation 209
      Fireproofing 209
      The Firewall 210
      Danger Zones Around Propeller Aircraft 210
   7.3.1 Piston Engine Installation 210
      Systems Integration 212
      Types of Engine Mounts 212
      Fuel System 212
   7.3.2 Piston Engine Inlet and Exit Sizing 213
      Method 1: Inlet-exit-dependent Heat Transfer 215
      Exit Area and Cowl Flaps 216
      Method 2: Inlet-radiator-exit Method 217
   7.3.3 Installation of Gas Turbines 222

Installation of Jet Engines	222
Installation of Turboprop Engines	222
Turbo Machinery and Rotor-burst	222
7.3.4 *Jet Engine Inlet Sizing*	223
Inlet Types for Jet Engines	223
The Diffuser Inlet	224
Step 1: Required Mass Flow Rate	225
Step 2: Determine Airspeed Limitations at the Inlet Lip and Compressor	225
Step 3: Establish Known and Unknown Flow Conditions	225
Step 4: Determine Conditions at Station ①	226
Step 5: Determine Conditions at Station ②	226
Pressure Recovery	226
Inlet Lip Radius	226
Diffuser Length	226
Stagger or Rake Angle	227
Derivation of Equation (7-48)	227
7.4 Special Topics	**227**
7.4.1 *The Use of Gearboxes*	227
7.4.2 *Step-by-step: Extracting Piston Power from Engine Performance Charts*	228
7.4.3 *Extracting Piston Power Using the Petty Equation*	229
Determination of the Polynomials Describing $P_{BHP\ max}$ and P_{FHP}	230
Derivation of Equation (7-59)	231
Exercises	**232**
Variables	**232**
References	**234**

7.1 INTRODUCTION

Clearly, the power plant is what makes powered flight possible through the generation of a force used to produce forward velocity, which then is used to create wing lift. As such, the engine selection is fundamental to the success of the design. Sometimes airplanes are designed around a particular engine, although often the engine is selected once performance requirements are realized. There is a range of power plant options available for the modern aircraft: piston engines, gas turbines, jet engines (turbojets and turbofans), to name a few.

Piston engines, gas turbines, and jet engines are the most commonly used in aircraft, while the pulsejet and rocket motors are rarely used. The most famous use of a pulsejet is in the German V1 drone used during the Second World War. While rocket engines offer high thrust in small packages, they have seen very limited use in just a few experimental aircraft, but are mostly used to improve T-O performance in military transport by supplementing the normal source of thrust. An example of this is the use of the RATO[1] packs for military transport aircraft such as the Lockheed C-130 Hercules.

Both the pulsejet and rocket types suffer from abysmal thermal efficiency, in particular the latter. Electric motors, on the other hand, represent the most efficient source of power currently available. At this time, there is an ongoing surge in their use in light experimental aircraft. Unfortunately, an efficient source of power does not translate into a "lot of" power, only that the conversion of electric to mechanical energy is very efficient (~95%). Currently, the greatest drawback of electric power is a lack of power due to a low "energy-per-mass" ratio. Nevertheless, battery technology is rapidly developing, making such engines more and more attractive.

At the present, batteries for electric engines have a low endurance, perhaps some 40–60 minutes, although strides toward improvements are rapidly being made. Fuel cell technology is advancing at a swift pace and this offers a potential for vastly greater endurance for such aircraft. In 2011, in a competition organized by NASA and the CAFE Foundation (Comparative Aircraft Flight Efficiency), the Pipistrel G4, a four-seat, twin-fuselage electric sailplane, made history by flying nearly 196 miles non-stop, in 1 hour 49 minutes [1]. This translates to an airspeed a tad over 107 mph (93 KTAS). Some of the technology demonstrators that participated ran on electric power for almost two hours!

The aircraft designer must keep a number of properties in mind when selecting a power plant: for instance, power output; type; cost, and availability of fuel; fuel efficiency; weight; mechanical complexity; price; maintainability; cooling requirements; inlet and exhaust requirements; and many others. For instance, jet engines require high fuel flow at low altitude, and, therefore, are very inefficient and costly to operate near sea level. However, as the altitude is increased, the maintenance of a constant fuel-to-air ratio guarantees a drop in fuel consumption (albeit at the cost of thrust). For this reason, jet engines become very attractive for high airspeeds at high altitudes. Consequently, if these aircraft transport people, which most of them do, the cabin must be

---

[1]RATO stands for rocket-assisted take-off.

pressurized, but this has a major impact on the structure and systems required. In short, engine selection has a major impact on the overall design of the aircraft.

In this section, we will present a formulation applicable to these engines that helps the designer *select* rather than *design* engines. Engine design is a field of specialization and will not be discussed in great detail here. On the other hand, a formulation will be presented that helps select an engine and to make fundamental sizing of the installation and to extract thrust to support performance and stability and control analyses.

7.1.1 The Content of this Chapter

- **Section 7.2** reviews the most common types of power plants available for or used in modern aircraft. Design information regarding piston engines, turboprops, turbojets, turbofans, and electric motors is presented. In addition, methods to estimate power and thrust as a function of atmospheric conditions and airspeed are provided.
- **Section 7.3** discusses topics that affect the installation of engines. Topics ranging from fireproofing to inlet sizing are presented.
- **Section 7.4** presents specialized topics related to engines, ranging from how to account for gearboxes, to the extraction of power information from typical engine performance charts, and the use of the 'petty equation' for the prediction of the same.

7.1.2 Power Plant Options for Aviation

Modern airplanes are typically propelled by any of the following classes of power plant:

(1) Piston engines (piston propellers)
(2) Gas turbines (turboprops)
(3) Jet engines (turbojets and turbofans)
(4) Pulsejets (very rare)
(5) Rockets (very rare, sometimes supplements other engines for T-O)
(6) Electric motors (propellers).

Piston engines and turboprops are by far the most common means of generating thrust in GA aircraft (FAR Part 23 aircraft), followed by turbofan engines. Turboprops and turbofans are the most common choices for commercial aircraft (FAR Part 25). Turbojets and low-bypass-ratio turbofans are most common for military applications. Small microturbo turbojets have been used for aircraft such as the homebuilt Bede BD-5J (the world's smallest jet) and the Caproni C22J (a candidate for a military trainer that has GA written all over it).

7.1.3 The Basics of Energy, Work, and Power

A review of the fundamentals of energy, power, and torque is presented here as a familiarity with various energy and power concepts is essential for the discussion that follows. The basics are shown in Table 7-1.

7.1.4 Thermodynamics of the Power Plant

The mechanism by which chemical energy is converted into mechanical energy often requires some medium, such as air. Such processes will change the pressure and volume of the gas, allowing the change in the thermodynamic state of the medium to be used to move a mechanical device, such as a piston or a turbine blade. Consequently, the thermodynamic operation of such devices is often described using *thermodynamic cycles*, which are represented as a graph on which the pressure and volume of the gas is plotted. Such graphs are plotted in Figure 7-1.

Piston Engines

Piston engines typically operate in accordance with a thermodynamic theory known as the four-stroke Otto cycle. In this cycle, the first step entails a compression of a mixture of air and fuel inside the combustion chamber. This process is shown in Figure 7-1 as a reduction in volume (horizontal axis) and an associated increase in pressure (side 1-2). This is accomplished by moving the piston inside the cylinder so the volume is decreased. The next step is the combustion, in which chemical energy obtained by the rapid burn of the fuel-air mixture increases the pressure without additional change in volume. This is depicted as the vertical increase in pressure (side 2-3). This pressure forces the piston in the opposite direction, increasing the volume as shown by side 3-4. Finally, once the piston reaches the position of maximum volume, a valve is opened, allowing the gases to escape (exhaust). This will drop the pressure inside the cylinder without additional change in volume (side 4-1). This operation is then repeated in the engine.

Gas Turbines

A similar thermodynamic cycle for gas turbines is called the Brayton cycle (see Figure 7-1). In this cycle, air enters the intake to the engine at a specific pressure. It is then compressed using a multi-bladed compressor and forced through ducting that reduces the volume of the air (side 1-2). The air is then directed into a combustion chamber, at which point it mixes with fuel, and the mixture is ignited. The geometry of the chamber forces volumetric expansion without change in pressure (side 2-3). The fuel-air mixture rushes through an opening in the combustion chamber, and impinges on a turbine wheel(s) with the associated conversion to mechanical

TABLE 7-1 The Basics of Energy, Work, and Power

Concept	Formulation	Units SI system	UK system
ENERGY			
The conservation of mass-energy is one of the fundamental conservation laws of physics. It basically says that energy can neither be created nor destroyed, but it changes form. The form of energy refers to potential, kinetic, electrical, nuclear, chemical, and other forms of energy.	Kinetic energy: $KE = \frac{1}{2}mV^2$ Potential energy: $PE = mgh$	Joules (J) kWh 1 kWh = 3.6×10^6 J	BTU
WORK			
Work is defined as the product of force applied to move an object a given distance. Work is also the same as *torque*.	Work ≡ Force × Distance	Joules N·m	ft·lb$_f$
POWER			
Power is defined as the amount of work done in a given time. It is also possible to define it as shown.	Power ≡ $\frac{\text{Work}}{\text{Time}}$ ≡ $\frac{\text{Force} \times \text{Distance}}{\text{Time}}$ ≡ Force × Speed ≡ $\frac{\text{Torque}}{\text{Time}}$	W J/sec N·m/s	hp ft·lb$_f$/s
One "horsepower"		746 W 0.746 kW	33000 ft·lb$_f$/min 550 ft·lb$_f$/s

energy. As this happens, the pressure of the mixture drops and its volume increases (side 3-4). Side 4-1 represents the completion of this cycle as fresh air again enters the inlet to the engine.

7.1.5 General Theory of Thrust Generation

All methods that convert mechanical energy to thrust effectively use the energy to accelerate a mass of air (or water in the case of ships), utilizing Newton's third law of motion, which states that action requires an equal and opposite reaction. Consider the stream tube displayed in Figure 7-2. The parameters A, P and ρ are the cross-sectional area, pressure, and density, where the inlet is denoted by the subscript 'i' and the exit by 'e'. The mass flow rate through the inlet is thus:

$$\dot{m}_i = \rho_i \cdot V_i \cdot A_i \quad (7\text{-}1)$$

Then consider the introduction of fuel to the system, somewhere between the inlet and exit, denoted by $\dot{m}_{fuel}$. This will increase the mass flow rate, so at the exit it will amount to:

$$\dot{m}_e = \dot{m}_i + \dot{m}_{fuel} \quad (7\text{-}2)$$

Therefore, the *net force* acting on the stream tube will be the difference between the pressure at the inlet, p_i,

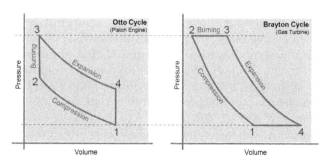

FIGURE 7-1 Thermodynamic cycles for pistons and gas turbine engines.

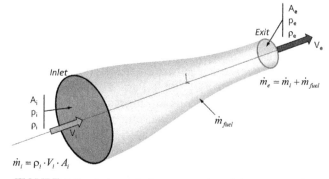

FIGURE 7-2 A theoretical representation of thrust generation.

and exit, p_e, times the area over which it acts, plus the rate of change of momentum of the mixture, i.e.:

$$T_{net} = p_e A_e - p_i A_i + (\dot{m}_i + \dot{m}_{fuel})V_e - \dot{m}_i V_i \quad (7\text{-}3)$$

Equation (7-3) is modified based on the type of thrust device it is applied to, as shown in the discussion that follows.

Propeller and Gas Turbines

The magnitude of $\dot{m}_{fuel}$ is generally so small that it can be ignored. Therefore, we rewrite Equation (7-3) as follows:

$$T_{net} = p_e A_e - p_i A_i + \dot{m}_i(V_e - V_i) \quad (7\text{-}4)$$

The only issue here is the determination of the inlet and exit airspeed and pressures. It is possible to extend Equation (7-4) to include propellers by assuming that $\dot{m}_i = 0$; and that the distance between A_e and A_i approaches zero, so $A_e = A_i$. This is the result of what is called the *actuator-disc* or *Rankine-Froude momentum theory*, presented in Chapter 14, *The anatomy of the propeller*.

Rockets

For rockets, the values of $\dot{m}_i$ and V_i are zero, since no air is introduced (it is effectively replaced by the oxidizing chemical), so $\dot{m}_{fuel}$ takes on a governing role and is more appropriately referred to as $\dot{m}_{propellant}$ where $\dot{m}_{propellant} = \dot{m}_{fuel} + \dot{m}_{oxidizer}$. Additionally, A_e and A_i both refer to the area at the nozzle exit plane and p_i is the ambient pressure. Therefore, we rewrite Equation (7-3) as follows:

$$T_{net} = (p_e - p_i)A_e + \dot{m}_{propellant}V_e \quad (7\text{-}5)$$

Thrust-to-weight Ratio

The thrust-to-weight ratio, denoted by T/W, is a very important figure of merit for aircraft. It ranges from a value of 0 for un-powered aircraft, to values in the ballpark of 1.1 for many fighter aircraft (see Table 7-2). Of course, thrust changes with airspeed, and weight with time aloft, so the true value of T/W changes in flight. Therefore, the stated T/W usually refers to a specific weight, such as maximum gross weight or some other specific weight, while the thrust refers to the maximum static thrust.

The importance of high enough T/W cannot be overemphasized. As an example, the de Havilland DH-106 Comet 1 suffered two hull losses, one with fatalities, both directly attributed to its low T/W of 0.18. The first incident occurred on October 26, 1952, and the second on March 3, 1953 [2]. In general, due to the low T/W, the airplane had to be rotated with utmost care to avoid a sharp rise in drag and a subsequent reduction of acceleration. In comparison, the modern jetliner has a T/W of the order of 0.3–0.35.

7.1.6 Fundamental Definitions

Brake Horsepower (BHP)

BHP usually refers to the amount of power delivered at the engine output shaft of a piston engine. It is measured using an instrument called a *dynamometer*, which is a mechanical or electric braking device. In the UK system, the horsepower corresponds to the work required to raise a weight of 33,000 lb_f, 1 ft in 1 minute. This also corresponds to the work required to raise a weight of 550 lb_f, 1 ft in 1 second. This is written as follows: 1 hp = 33,000 ft·lb_f/min = 550 ft·lb_f/s. Horsepower can be converted to watts (J/s) in the SI system by multiplying by a factor of 746. In other words:

$$1 \text{ hp} = 746 \text{ W} = 0.746 \text{ kW}$$

Shaft Horsepower (SHP)

SHP refers to the amount of power delivered at the propeller shaft of a gas turbine engine. If the engine features a gear reduction drive (to reduce the RPM to keep the propeller tip airspeed below the speed of sound), sometimes called a *propeller speed reduction unit* (PSRU), this value may be somewhat less than the horsepower generated by the engine due to losses in the mechanism.

Equivalent Horsepower (EHP)

EHP is a term used only for turboprops and refers to the combination of the SHP and the residual thrust available from its jet exhaust. The EHP is usually about 5% higher than the SHP.

Thrust Horsepower (THP)

THP refers to the amount of power used to propel an aircraft through the air in terms of horsepower. The thrust of turbojets, turbofans, pulsejets, and rockets is generated by accelerating the fluid directly. For that reason, such engines are always rated in terms of the thrust they generate. This contrasts with piston engines

TABLE 7-2 Typical T/W for Selected Classes of Aircraft

Type of aircraft	T/W range
Unpowered (sailplanes, hang gliders, etc.)	0
General aviation	0.30–0.45
Commercial jetliners of the 1950s–1960s	0.18–0.26
Commercial jetliners of the 1970s-present	0.20–0.35
Modern fighter aircraft	0.55–1.15

and turboprops, whose mechanical work is used to rotate a propeller, which then creates the thrust. This way, it is not the engine per se that generates the thrust, but the propeller, and this is why it is more appropriate to rate such engines in terms of their power. This is easier to see by noting that the thrust generated by such engines depends on the propeller — for instance, one can mount two different types of propeller on the same engine and generate two different levels of thrust at the same power level. This is not the case for a turbofan or a turbojet. One of the nuisances with this difference is that sometimes it is helpful to convert the thrust into horsepower, for instance, to compare the effective power of a piston or a turboprop to a turbofan. This must be done by multiplying the thrust T of the turbofan (or turbojet or pulsejet, etc.) with the airspeed V at which it is flying. If working in the UK system, the thrust is given in lb_f and the airspeed in ft/s. The unit for the power is thus in terms of $ft \cdot lb_f/s$, which can be converted to horsepower (i.e. THP) by dividing the product by 550, using the expression below:

UK system (T in lb_f, V in ft/s):

$$THP = \frac{TV}{550} \quad (7\text{-}6)$$

SI system (T in N, V in m/s):

$$THP = \frac{TV}{746} \quad (7\text{-}7)$$

Engine Torque and Turboprop Engines

The power output of turboprop aircraft is often represented using torque and RPM, rather than horsepower. Consequently, when considering performance data for turboprop aircraft, it is often helpful to convert the torque and RPM into horsepower.

Torque is equivalent to work (force × distance) and not power (force × speed). Thus, torque is power × time, or power is torque divided by time. Consider a rotating arm of length d rotating at rate RPM due to the force F. Then the work done each minute by the force F is the product of it and the distance over which it is applied. For rotary motion, this can be written as $F \cdot 2\pi d \cdot RPM$. If we use the UK system of units, this will be in terms of $ft \cdot lb_f/min$. It can, thus, easily be converted to horsepower by dividing by 33,000 $ft \cdot lb_f/min$, yielding the following relationship:

To convert torque and RPM to SHP:

$$SHP = \frac{torque \times 2\pi \times RPM}{33,000} = \frac{torque \times RPM}{5252} \quad (7\text{-}8)$$

Thus, a turboprop operating at a torque of 1500 $ft \cdot lb_f$ and 2000 RPM is delivering 571 SHP of power.

Engine Power and Thrust Ratings

The following concepts are frequently used when specifying engine power and thrust ratings:

(1) *Take-off, wet* either refers to thrust generation of a jet engine utilizing afterburner or the maximum available T-O thrust of an engine that uses water injection. The latter is more representative of GA aircraft. Water injection dates back to early gas turbine technology, where it was used on turboprop engines such as the Rolls-Royce Dart, which powered aircraft such as the Fokker F-27 and Vickers Viscount. It was also used in jet engines such as the Pratt & Whitney JT3D, which powered the Boeing 707 jetliner. In short, water injection is the mixing of atomized water with air before it enters the combustion chamber. This reduces the temperature of the combustor inlet air and consequently, the turbine entry temperature (TET). The same effect can be achieved by direct injection into the combustion chamber. Since the TET may not exceed material limitations, the injection of water is really a trick to add more fuel (to get more thrust or power) without exceeding engine temperature limits. Injecting water in this fashion is usually limited to a short time, perhaps some 5 minutes. Its use is constrained by altitude and ambient air and water temperature. While water injection is no longer commonly used, Daggett et al. [3] discuss it as a means to reduce NOx emissions and engine operating costs by lowering hot section temperatures and, thus, prolonging its operational life.

Water injection is related to throttle ratio, discussed in a Section 7.2.2, *Turboprops*. If the throttle ratio is 1, the TET will be at its maximum for engine operation at standard S-L temperature. If the engine operates at a higher than standard temperature, the engine power must be decreased since the combustor entry temperature would be too high and the limiting TET must not be exceeded. Newer engines have a throttle ratio greater than 1, and, therefore, at standard temperature operation, the TET is less than the maximum allowable value. Water injection is thus no longer necessary at higher ambient temperatures and nowadays is often associated with turbo-machinery of yesteryear.

(2) *Take-off, dry* either refers to thrust generation of a jet engine without the use of an afterburner, or the maximum thrust available for T-O without the use of water injection. The latter is more representative of GA aircraft. The thrust setting is usually limited to 5 minutes. The setting is also permitted to provide reverse thrust during landing.

(3) *Maximum continuous power* (MCP) or *thrust* (MCT) refers to the maximum power (or thrust) setting that can be used continuously, although it is usually

intended to be used in an emergency (eg. one engine inoperative) situation. In piston-engine aircraft, maximum level airspeed, V_H, is obtained at this power setting.

(4) *Maximum climb power* or *thrust* refers to the power or thrust setting used during normal climb operations. For piston engines, this is often the same as the MCP, or close to it.

(5) *Maximum cruise power* or *thrust* refers to the power setting used for cruise.

(6) *Flat rating*. An engine is said to be flat-rated if it generates the same thrust or power over a range of temperatures. The thrust or power of an engine operating at wide-open throttle varies inversely with temperature. Thus, its output will be higher on a cold day than a hot one. This is unfortunate for the operation of aircraft as there will be less power (or thrust) available when taking off on a hot day, when other characteristics of the airplane are deficient as well.

Engine manufacturers have responded to this by offering engines with "reported" power ratings that are lower than what the engine is actually capable of generating. Then the fuel control system is used to control the maximum power available by metering the ambient temperature. In this way, as far as the pilot is concerned, the same engine power is available whether operating on a hot or a cold day. This is called flat rating and may be explained by a simple example: consider a piston engine capable of 300 BHP on a standard day (OAT = 518.67 °R or 273.15 K). According to Equation (7-20), the engine power on a day that is 40 °C above a standard day (often written ISA + 40 °C) will only generate some 280 BHP. Therefore, the manufacturer may market it as a 280 BHP flat-rated engine. Then, as far as the pilot is concerned, the airplane is equipped with a 280 and not a 300 BHP engine. The extra 20 BHP are then used to make up for the power lost on a hot day, giving the pilot a fixed (flat) 280 BHP for T-O even on a day that is 40 °C above a standard day.

7.1.7 Fuel Basics

Density of Aviation Gasoline (Avgas)

The density of avgas is 0.71 kg/liter. In the UK system, its weight is 5.9–6.0 lb$_f$/gallon. For analysis work in this text, a weight of 6.0 lb$_f$/gallon is always assumed.

Energy Content of Fuel for Piston Engines

Piston engine power is directly related to the amount of air mass flow into the intake manifold.

In the UK system: 1 hp ~ 620 × mass flow (in lb$_m$/s).

In the SI system: 1 kW ~ 1019 × mass flow (in kg/s).

The theoretically ideal (stochiometric) ratio for piston engines is 1 kg of fuel to 14.7 kg of air. When achieved, this ratio results in the highest temperature during combustion and is, therefore, of concern when it comes to engine durability. If the air-to-fuel mixture is less than 14.7:1 (e.g. 14:1), it is called *rich*. If greater (e.g. 15:1), it is called *lean*. These two concepts are of great importance to pilots.

The energy content of a 1 lb$_f$ of avgas is 14,800,000 ft·lb$_f$ (20.07 MJ). Burning 1 lb$_f$ of avgas in 1 minute with 100% efficiency would generate:

$$\frac{14{,}800{,}000 \text{ ft} \cdot \text{lb}_f}{33{,}000 \text{ ft} \cdot \text{lb}_f/\text{BHP}} = 448 \text{ BHP}$$

A typical, modern medium-sized piston engine, such as the Continental IO-360, delivers 200 BHP at maximum power, while consuming 16 gallons of avgas per hour; 16 gal/h amounts to 16 gal/h × 6 lb$_f$/gal = 96 lb$_f$ of fuel per hour, which amounts to 96 lb$_f$/60 = 1.6 lb$_f$/min. The equivalent energy content of this fuel is:

$$1.6 \times \frac{14{,}800{,}000 \text{ ft} \cdot \text{lb}_f}{33{,}000 \text{ ft} \cdot \text{lb}_f/\text{BHP}} = 716.8 \text{ BHP}$$

However, only some 200 BHP is delivered as mechanical energy. The resulting efficiency is thus 200/716.8 or 27.9%. In a long-range cruise mode, the same engine delivers 55% of its rated power (110 BHP), while consuming some 8.4 gal/h. Applying the same calculation method we find the efficiency amounts to 29.2%.

Fuel Octane Rating and Fuel Grades for Piston Engines

Fuel octane rating is a measure of the capability of a fuel to resist compression before it spontaneously self-ignites. Thus, fuel with a higher octane number can withstand greater pressure inside the cylinder before igniting in this fashion. For this reason high-octane-number fuel must be used in high-compression (high-performance) engines or they will suffer from engine knocking. That aside, the concept is one that is often misunderstood as there are a number of different octane ratings (e.g. Research Octane Number, RON; Motor Octane Number, MON, etc.). These definitions are outside the scope of this book.

Fuel used to power piston engines aircraft is commonly known as avgas, or aviation gasoline. This contrasts with mogas, or motor gasoline, which is used in cars and some experimental and a few GA aircraft. The difference between the two is that avgas contains a toxic chemical called tetraethyl lead (TEL), which is used to improve the combustion properties of the fuel. The fuel octane rating is used to differentiate between a few grades of fuel, which are offered in different colors to prevent incorrect selection (see Table 7-3).

TABLE 7-3 Common Fuel Grades for Piston Engine Use

Fuel grades	Color	Comment
80/87	Red	The first number (80) is the octane rating assuming a lean mixture. The second number (87) indicates the rating at a rich mixture. Used for aircraft engines with low compression ratios. No longer produced.
82UL	Purple	UL stands for unleaded. Similar to mogas, but without automotive additives. Intended for low-compression engines such as those common in experimental aircraft and aircraft that have STCs permitting the use of mogas. No longer produced.
91/96 91/96UL	Brown	UL stands for unleaded. Avgas is often intended for military use (e.g. UAVs). Produced today by the Swedish fuel manufacturer Hjelmco in a clear color.
100LL	Blue	LL stands for low lead. The most common avgas in use today. The fuel can be used with engines designed for 80/87.
100/130	Green	Also called avgas 100. Has been superseded by 100LL, although still available in limited quantities.
115/145	Purple	Leaded fuel produced for warbirds and the supercharged radial engines used to power the passenger planes of the 1940s–1960s. Now produced in limited quantities for air races. This fuel is necessary in order to obtain rated power in such engines.

Fuel Grades for Jet Fuel

There is also a wide range of fuel grades intended for jet engines. Table 7-4 lists those most common for civilian aircraft. A range of jet fuel with specifications for different countries is available to, but not presented here. More details are available from Ref. [4]. For analysis work in this text, a density of 6.7 lb_f/gallon is always assumed.

Specific Fuel Consumption (SFC)

Specific fuel consumption (*SFC*) is one of the most important metrics employed in aviation. It is important not only in aircraft design but also in the operation of the aircraft. First and foremost, *SFC* indicates how efficiently a power plant converts chemical into mechanical energy. While there is usually not a great variation in *SFC* between engines within a specific class of power plants, there is a huge variation between the classes. Thus, piston engines are generally more efficient than turbo machinery, which is far more efficient than, say, rockets. This is important when ensuring the selection of power plant matches the mission of the airplane, although most of the time this is not a problem. The primary importance is when estimating range and endurance of the aircraft.

As discussed in Section 7.1.5, *General theory of thrust generation*, the operation of any mechanical engine requires chemical energy to be converted into mechanical energy, typically by an intermediary conversion to thermal energy. In conventional piston and jet engines, the chemical energy stored in the fuel is consumed in an exothermic chemical reaction (via combustion) that adds considerable thermal energy to the air, rapidly increasing pressure or volume. The change in these states is then used to move the mechanical elements of the machine.

Consider two engines, call them A and B. If engine A requires less fuel to generate a given power (or thrust for jet engines) than engine B, then we say it is the more efficient of the two. This would place engine B at a competitive disadvantage. So, efficiency clearly plays an imperative role in the marketability of an engine (just like weight).

It is crucial for the engineer to be able to compare the efficiency of different power plants for design purposes and this is accomplished by the definition of *fuel consumption* as the quantity of fuel burned in a unit time (lbs/hr, kg/min, etc.). This is sometimes referred to as *fuel flow* (FF). We then define *specific fuel consumption* (*SFC*) as the quantity of fuel burned in unit time required to produce a given engine

TABLE 7-4 Common Fuel Grades for Civilian Jet Engine Use (based on Refs [4] and [5])

	Fuel Grades			
Property	Jet A	Jet A-1	Jet B	TS-1 (Regular)
Flash point	100 °F (38 °C)	100 °F (38 °C)	—	82.4 °F (28 °C)
Freeze point	−40 °F (−40 °C)	−52.6 °F (−47 °C)	−59.8 °F (−51 °C)	< −76 °F (−60 °C)
Density at 15 °C	6.48–7.02 lb_f/gal (0.775–0.840 kg/liter)	6.48–7.02 lb_f/gal (0.775–0.840 kg/liter)	6.27–6.69 lb_f/gal (0.750–0.801 kg/liter)	6.48 lb_f/gal (0.775 kg/liter)
Comment	Suitable for most gas turbines. Primarily available in the USA.	Suitable for most gas turbines. Widely available.	An alternative to Jet A-1 but more flammable. A cold-climate jet fuel.	Primarily used in Russia and the CIS states.

output. *SFC is a technical figure of merit that indicates how efficiently the engine converts fuel into power.*

We will now consider *SFC* for jet engines and piston engines separately.

SFC *for Jets*

The fuel consumption of jet engines is always measured in terms of mass or weight of fuel flow, per unit time, per unit thrust force. For instance, a small jet engine may burn some 1600 lb_f (725 kg) of fuel in a matter of an hour at a T-O thrust setting. If the thrust of the engine is known, the *SFC* can be computed as shown below:

UK system:

$$SFC = c_{jet} \equiv \frac{\text{weight of fuel in lbs/hour}}{\text{thrust in pounds force}} = \frac{lb_f/hr}{lb_f} = \frac{1}{hr} \quad (7\text{-}9)$$

SI system:

$$SFC = c_{jet} \equiv \frac{\text{mass of fuel in grams/sec}}{\text{thrust in Newtons}} = \frac{g}{N \cdot sec} \quad (7\text{-}10)$$

The pure turbojet, of the kind used in the early days of the jet propulsion, has very poor efficiency. The high-bypass turbofan, in comparison, has superb efficiency, something that explains their wide use in modern passenger transport aircraft.

EXAMPLE 7-1

A Williams FJ33 turbofan is found to consume 1000 lb_f of fuel per hour while generating 1500 lb_f of thrust. Determine c_{jet} in the UK and SI systems:

Solution

$$c_{jet} = \frac{1000 \, lb_f/hr}{1500 \, lb_f} = 0.667 \frac{lb_f/hr}{lb_f}$$

$$c_{jet} = \frac{(453 \, kg/hr)}{6664 \, N} = \frac{(0.1258 \, kg/s)}{6664 \, N}$$
$$= \frac{(125.8 \, g/s)}{6664 \, N} = \frac{(125833 \, mg/s)}{6664 \, N}$$
$$= 18.9 \frac{mg}{Ns}$$

SFC *for Pistons*

The fuel consumption of piston engines is always measured in terms of mass or weight of fuel flow per unit time, per unit of power. For instance, a small piston engine may burn some 100 lb_f (45 kg) of fuel in a matter of an hour at a T-O power setting. If the power of the engine is known in BHP or kW, the *SFC* can be computed as shown below:

UK system:

$$SFC = c_{bhp} \equiv \frac{\text{weight of fuel in lbs/hour}}{\text{power in brake horsepower}} = \frac{lb_f/hr}{BHP} \quad (7\text{-}11)$$

SI system:

$$SFC = c_{ws} \equiv \frac{\text{mass of fuel in grams/sec}}{\text{power in watts}}$$
$$= \frac{g}{W \times sec} = \frac{g}{J} \quad (7\text{-}12)$$

EXAMPLE 7-2

A piston engine is found to consume 10 gallons of fuel per hour while generating 150 BHP. Determine c_{bhp} and c_{ws} (in the UK and SI systems, respectively). Fuel weighs 6 lbs per gallon:

Solution

$$c_{bhp} = \frac{10 \, \text{gallons/hr}}{150 \, BHP} = \frac{60 \, lb_f/hr}{150 \, BHP} = 0.400 \frac{lb_f/hr}{BHP}$$

$$c_{ws} = \frac{(27.18 \, kg/hr)}{150 \, BHP} = \frac{(0.00755 \, kg/s)}{111.9 \, kW} = \frac{(7.55 \, g/s)}{111.9 \, kW}$$
$$= \frac{(7550 \, mg/s)}{111,900 \, W} = 0.06747 \frac{mg}{W \times s}$$

7.2 THE PROPERTIES OF SELECTED ENGINE TYPES

A large number of devices have been invented that convert the chemical energy available in fuel into mechanical energy and presenting them all would take volumes. Therefore, only the three main engine types that are commonly used in GA aircraft are presented in this section: the piston engine, gas turbine, and electric motor. The piston engine and gas turbine are used for the vast majority of all aircraft and will thus take precedence. The electric motor is an emerging technology that has been used for a long time in radio-controlled aircraft, but advances in battery technology are already allowing their use in light aircraft and are therefore also discussed.

There is a secondary aim of this section, which is to provide tools to help the aircraft designer estimate power and thrust for use in performance and stability analyses.

7.2.1 Piston Engines

The piston engine has been a stalwart of the aviation industry since the Wright brothers flew the first airplane on December 17, 1903. The modern piston engine is very reliable, affordable, common, and relatively light, which has made it ideal for use in aircraft. For airplanes, piston engines come in a variety of sizes, ranging from tiny single-piston, glow-plug engines such as the Cox Tee Dee 0.010 in^3 engine, used to power small radio-controlled aircraft, to the giant 5000 BHP Lycoming XR-7755 36-cylinder [6] radial piston engine, intended to power huge airplanes like the early Cold-war era Convair B-36 'Peacemaker.' Some modern manufacturers of piston engines for UAV and GA aircraft are listed in Table 7-5. Note that at the time of writing, all the manufacturers were still in business. Out-of-business piston engine manufacturers are not included.

The piston engines considered in this chapter all share the commonality of consisting of one or more pistons that attach to a crankshaft and operate inside a cylinder. There are many variations of the concept, of which the so-called Wankel engine is the best known, but the engines dealt with here are all conventional pistons as described above, of which there are two types: two- and four-stroke. Several common types of piston engines are shown in Table 7-6.

Two-stroke versus Four-stroke Engines

Two-stroke engines are valve-less so they are simpler, lighter, and less expensive to manufacture. They are less durable than four-stroke engines because they lack a dedicated lubrication system. Instead, they require oil to be mixed in with the fuel (about 4 oz per gallon of fuel). For this reason they burn a considerable amount of oil when compared to the four-stroke engine.

A two-stroke engine will exhaust combustion gases and draw in a fresh fuel/air mixture on the down-stroke. It will then compress and ignite the mixture on the up-stroke. A four-stroke engine will ignite with a subsequent down-stroke. On the following up-stroke the combustion gases are forced out of the cylinder. As the piston's next down-stroke begins, the fuel/air mixture will be drawn into the cylinder and be compressed and ignited on the subsequent up-stroke. Thus, ignition occurs once every revolution in a two-stroke engine, but once every other revolution in a four-stroke. This gives the two-stroke engine a significant power boost and allows it, potentially, to double the power for the same-displacement engine. A two-stroke engine manufactured by Hirth Engines is shown in Figure 7-4.

TABLE 7-5 Selected Manufacturers of Piston Engines for GA and UAV Aircraft

Maker	Country	Application	Horsepower range (BHP)	Website
Continental Motors	USA	GA, LSA	75–360	www.genuinecontinental.aero
JPX	France	LSA, UAV	14–90	www.jpx.fr
Limbach Engines	Germany	GA, LSA	20–167	www.limflug.de
Hirth Engines	Germany	GA, LSA	14.6–102	www.hirth-motoren.de
Rotax Engines	Austria	GA, LSA	40–115	http://www.rotax-aircraft-engines.com/
SMA Engines	France	GA	227	www.smaengines.com
Textron Lycoming	USA	GA, LSA, UAV	115–400	www.lycoming.com
ULPower Aero Engines	Belgium	GA, LSA, UAV	97–130	www.ulpower.com
Zenoah	Japan	UAV, RC	1.68–5.82	www.zenoah.net
Jabiru Engines	Australia	GA, LSA, UAV	85–120	www.jabiru.net.au

GA = general aviation aircraft, LSA = light sport aircraft, UAV = unmanned aerial vehicles, RC = radio-controlled aircraft.

7.2 THE PROPERTIES OF SELECTED ENGINE TYPES

TABLE 7-6 Power and Weight of Selected Piston Engines for GA and Experimental Aircraft

Manufacturer	Type	Cylinders	Dsplcmnt in^3	TBO hours	Weight lb$_f$	RPM	Rated power BHP	SFC lb$_f$/hr/BHP
Lycoming	O-235	4	235	2400	243–255	2800	115–125	0.6
	O-320	4	320	2000	268–299	2700	150–160	0.6
	O-360	4	360	2000	280–301	2700	168–180	0.6
	IO-390	4	390	2000	308	2700	210	0.6
	IO-580	6	580	–	444	2700	315	0.6
	IO-720	8	720	–	593–607	2650	400	0.6
Continental Motors	IO-360	6	360	–	327–331	2800	200	0.6
	IO-550	6	550	–	467–470	2700	300–310	0.6
Hirth Motoren	3003[a,b]	4	63.6	1000	93	6500	102	0.83–1.80
	3501[a,b]	2	38.1	1000	78	5500	60	
	3701[a,b]	3	57.3	1000	100	6000	100	
Rotax	447UL[a]	2	26.6	300	72	6800	40	–
	503UL[a]	2	30.3	300	85	6800	49	–
	582UL[a]	2	35.4	300	79	6800	65	–
	912UL[a]	4	73.9	1500	122	5800	81	0.47
	912ULS[a]	4	73.9	1500	125	5800	100	0.47
	914UL[a]	4	73.9	1200	154	5800	115	–

[a]Non-certified.
[b]Two-stroke.
TBO = time between overhauls, RPM = revolutions per minute, SFC = specific fuel consumption.

The operation of a two-stroke engine is less efficient than that of a four-stroke engine. This results from the use of cleaner gasoline in a four-stroke engine, which is not mixed with oil as is the fuel for two-stroke engines. As a consequence the combustion burns the fuel more completely and at a higher temperature than possible in a two-stroke engine and both lead to higher efficiency. Additionally, the two-stroke approach leaves remnants of combusted gases inside the cylinder during compression and ignition and forces unburned gas into the exhaust, resulting in greater emission of environmentally harmful chemicals.

Operation of the Four-stroke Engine

Figure 7-3 shows a schematic of a four-cylinder internal combustion engine. The piston and cylinders have been labeled 1 through 4 and are shown during the four stages of the Otto cycle (refer to Figure 7-1).

(1) **Injection:** piston is at the end of its down-stroke and has drawn in a fresh fuel/air mixture.
(2) **Compression:** piston is at the end of its up-stroke and has compressed the fuel/air mixture just prior to ignition.
(3) **Combustion:** mixture has been ignited and the down-stroke of the piston is beginning.
(4) **Exhaust:** piston is beginning its up-stroke, forcing the combustion gases out and into the exhaust tube.

There are a number of concepts concerning piston engines that one must be aware of.

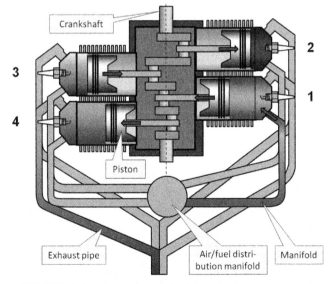

FIGURE 7-3 The workings of a four-cylinder piston engine.

FIGURE 7-4 A Hirth 3702, 84 BHP (62 kW) three cylinder, two-stroke, water-cooled piston engine for light experimental (home-built) aircraft. *(Courtesy of Hirth Engines. www.hirth-motoren.de)*

Air-to-fuel Ratio

As stated in Section 7.1.7, *Fuel basics*, the ideal stochiometric (air-to-fuel) ratio is 14.7:1 and it yields the least amount of carbon monoxide emissions. However, the engine's maximum power is generally achieved at about 12:1 to 13:1 (rich mixture). On the other hand, minimum fuel consumption is achieved at approximately 16:1 (lean mixture).

Compression and Pressure Ratios

The *compression ratio* is defined as the ratio between the volume of the cylinder with the piston in the bottom position, V_{bottom} (largest volume), and in the top position, V_{top} (smallest volume). The higher this ratio, the greater will be the power output from a given engine. It is generally in the 6–10 range.

Similarly, the *pressure ratio* is defined as the ratio of the pressures inside the cylinder with the piston in the top and bottom positions, denoted by p_{top} and p_{bottom}, respectively. Assuming adiabatic compression inside the cylinder (no heat energy is added when compressing the gas), the relation between the pressure and volume can be shown to comply with the following expression:

$$p_{bottom} V_{bottom}^{\gamma} = p_{top} V_{top}^{\gamma} \Leftrightarrow \frac{p_{top}}{p_{bottom}} = \left(\frac{V_{bottom}}{V_{top}}\right)^{\gamma}$$

(7-13)

TABLE 7-7 Specific Fuel Consumption of Typical Normally Aspirated Piston Engines for Aircraft

	lb$_f$/hr/BHP	gr/kW/hr
Two-stroke	0.83–1.80	280–600
Four-stroke	0.42–0.60	140–205

Displacement

Displacement is the total volume of the combustion chamber in the piston engine. The diameter of each cylinder is called a *bore*. The total distance a piston moves is called a *stroke*. The displacement of an engine with N cylinders is defined as follows:

$$V_{disp} = \frac{\pi}{4} N (bore^2 \times stroke)$$

(7-14)

The displacement represents the maximum volume of the combustion chambers of all the cylinders assuming the piston is simultaneously at the bottom of the stroke for all (an impossible scenario).

Typical Specific Fuel Consumption for Piston Engines

The fuel consumption of piston engines varies by type, as shown in Table 7-7.

A typical breakdown of how energy is wasted in piston engines can be seen in Table 7-8.

Effect of Airspeed on Engine Power

Generally, the power generated by a piston engine is assumed constant with airspeed, making it an *airspeed-independent power plant*. Effectively, this means that if a piston engine produces 100 BHP at a specific power setting, say, at stalling speed, it will also generate 100 BHP at the same power setting at its maximum airspeed. In real applications, however, the power output from the engine depends on the pressure recovery at the manifold. If the pressure recovery is airspeed-dependent, for instance as a consequence of the changes in the attitude of the aircraft, then the engine power will become airspeed-dependent, albeit only slightly so. However, for all intents and purposes, during design work, piston engine power output can be considered independent of airspeed.

Effect of Altitude on Engine Power

The power output of normally aspirated engines depends on how efficiently the mixture of air and fuel burns inside the cylinder during combustion, a process

TABLE 7-8 Energy Wasted in a Piston Engine (Based on Ref. [7])

Cause	Percentage
Available in fuel	100
Heat lost to oil	−2
Heat lost to cooling air	−11
Heat lost to radiation	−5
Heat lost to exhaust	−52
Mechanical losses	−5
SUM	≈25%

that sharply increases the pressure inside the cylinder and pushes the piston in the opposite direction. This, in turn, depends on the total quantity of oxygen molecules (O_2) initially inside the cylinder as the piston begins the compression stroke. The quantity of molecules inside the enclosed volume of the cylinder, of course, is the density and it is directly related to the initial pressure in the cylinder, as realized through the equation of state. For this reason, pressure and density are fundamental variables in the operation of piston engines. This, of course, implies that such engines are highly dependent on the density of air, which is a function of altitude.

Initial pressure in the cylinders can be increased by two means: by recovering as much ram air pressure as possible in the engine manifold (pertains to normally aspirated engines); and by artificially increasing the pressure in the manifold. The former is achieved by ensuring the intake is not blocked and is located in an area where air is allowed to stagnate with minimum losses. The latter can be done through the process of *turbo-charging* or *turbo-normalizing*.

To estimate the impact of altitude on the power output of an engine some specialized models are applied. The simplest one, presented below, assumes that the engine power is directly dependent on the density ratio:

Simple altitude-dependency model:

$$P = P_{SL}\left(\frac{\rho}{\rho_{SL}}\right) = P_{SL}\sigma \qquad (7\text{-}15)$$

where P, ρ and σ are power, density, and density ratio at altitude, respectively; and P_{SL} and ρ_{SL} correspond to S-L values. Another more accurate altitude model for piston engines is the Gagg and Ferrar model [8], here presented in its three most frequently encountered forms:

Gagg and Ferrar model:

$$P = P_{SL}\left(\sigma - \frac{(1-\sigma)}{7.55}\right) = P_{SL}(1.132\sigma - 0.132)$$
$$= P_{SL}\frac{(\sigma - 0.117)}{0.883} \qquad (7\text{-}16)$$

where

P_{SL} = power in terms of BHP at S-L
σ = density ratio

The two expressions above are used with normally aspirated engines only. Figure 7-5 shows a comparison between the simplified and Gagg and Ferrar models. The horizontal axis shows the percentage power and the vertical altitude in ft. One way of reading the graph is to ask how much power a piston engine delivers at a given altitude. For instance, how much power does an engine rated as 200 BHP at S-L deliver at full throttle at 15,000 ft? By tracing the horizontal line extending from 15,000 ft to the point where it intersects the thick curve of the Gagg and Ferrar model, it can be seen it delivers approximately 57.5% power, or $0.575 \times 200 = 115$ BHP. By the same token, consider the same engine, at S-L, at some arbitrary throttle setting that generates 100 BHP. At the same altitude it will deliver a mere 57.5 BHP at the same throttle setting. The Gagg and Ferrar model matches manufacturers' data far better than the former and is recommended for design work.

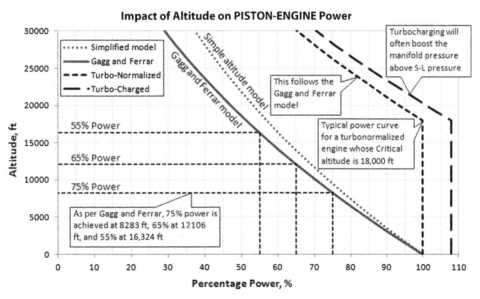

FIGURE 7-5 A comparison showing the difference between several models used to describe how piston-engine power is affected by change in altitude.

Straight lines representing 55%, 65%, and 75% power have been plotted in Figure 7-5, but these represent typical power settings reported by manufacturers or widely known publications such as Jane's *All the World's Aircraft*. It can be seen that once the engine is at 8283 ft, even at full throttle the maximum engine power will not exceed 75% of its rated S-L value as long as it is normally aspirated. Corresponding altitudes for 65% and 55% are shown as 12,106 ft and 16,324 ft, respectively. In order to prevent this sort of power loss with altitude, the air flowing into the manifold must be introduced at a higher pressure. It must be *pre-pressurized*.

Ideal manifold pressure for a normally aspirated engine is around 29.92 in Hg, depending on the ambient pressure, so ideally such pre-pressurization should maintain this pressure to as high an altitude as possible, and even boost it a tad. There are three common ways of doing this: by supercharging, turbo-charging, or turbo-normalizing. Discussing these devices in detail is beyond the scope of this book, so only a few introductory facts will be included here.

Supercharging is the oldest method known to pre-compress air prior to entering the cylinder of a piston engine. A *supercharger* is a compressor, often of a centrifugal design, that is directly or indirectly connected to the engine. Such devices may operate at rotation rates as high as 120,000 RPM. And while requiring additional energy from the engine to drive it, the increase in power far overweighs the cost of its production. A supercharger will often generate manifold pressure in the 50−60 in Hg range. For instance, Ref. [9] specifies 52 inHg MAP for the C-46 Commando, a twin-engine heavy transport aircraft from the Second World War era.

A *turbocharger* is a centrifugal compressor that features a turbine that is driven by the hot exhaust gases from the engine. For this reason, when operating at optimal conditions it puts no additional load on the engine. Rather it utilizes the thermal energy that otherwise would go unharnessed into the environment. This causes it to inflict reduced load on the engine, making it more efficient than a supercharger. However, its presence causes back-pressure to mount in the exhaust manifold, reducing the efficiency of the engine. The magnitude of the back-pressure is largely dependent on a complex interaction between important parameters such as RPM, throttle setting, ambient pressure, and geometry of the inlet [10]. Back-pressure is the obstruction to free flow of the exhaust gases through the tailpipe. Additionally, an exhaust-driven compressor is less efficient than a mechanically driven one.

The third type is *turbo-normalization*, which really is a turbocharger designed to limit (or normalize) the manifold pressure to that at S-L up to its critical altitude, hence its name. The critical altitude is one at which the ambient pressure has dropped so much that the compressor can no longer deliver air at S-L pressure into the manifold. The turbo-normalizing differs from turbocharging in that it limits the maximum manifold pressure to that of the S-L pressure up to the critical altitude. The turbocharger, on the other hand, increases the manifold pressure above what is possible at S-L. Thus, super- and turbochargers may maintain the manifold pressure to between 32 and 38 in Hg, in contrast to 29.92 inHg average pressure at S-L. The increased pressure implies a greater number of air molecules that must combine with fuel to ensure proper combustion and, therefore, the process increases power, but fuel consumption and engine wear. Turbo-normalizing, in theory, maintains rated S-L power and fuel consumption with altitude. Thus, its operation at altitude is not unlike what it is at S-L.

The impact of altitude on turbocharged and turbo-normalized engines is shown in Figure 7-5. The turbocharged engine is shown to yield power higher than that possible at S-L. The turbo-normalized engine, on the other hand, provides S-L power up to its critical altitude. The critical altitude differs from installation to installation. Some typical installations have a critical altitude of 18,000 ft, while some GA aircraft are capable of even higher altitudes. For instance, the Cirrus SR22 is offered with a turbo-normalized engine capable of S-L power up to 25,000 ft. Such a feature has great customer value because it effectively means its maximum available power is independent of altitude.

Engines that are pressurized in this fashion can be analyzed by assuming power to decrease after the critical altitude has been achieved in accordance with the Gagg and Ferrar model. Mathematically, if the S-L power is given by P_{SL} and critical altitude by h_{crit}, then power can be estimated using the following expression:

If $h \leq h_{crit}$ then

$$P = P_{SL} \qquad (7\text{-}17)$$

If $h > h_{crit}$ then

$$P = P_{SL}\left(1.132(1 - 0.0000068756(h - h_{crit}))^{4.2561} - 0.132\right) \qquad (7\text{-}18)$$

It is inevitable that the compression of air greatly increases its temperature. Therefore, some such mechanical systems use intercoolers to cool the air exiting the centrifugal compressor. The temperature rise can be approximated by assuming a thermodynamic process called an isentropic process (an isotropic process is one which is both adiabatic and reversible). Using this process, the change in temperature can be estimated from:

$$p_2 = p_1(T_2/T_1)^{\gamma/(\gamma-1)} \Rightarrow T_2 = T_1\left(\frac{p_2}{p_1}\right)^{(\gamma-1)/\gamma} \quad (7\text{-}19)$$

Effect of Temperature on Engine Power

Since power is affected by density and pressure, it follows suit that it is also influenced by temperature. The following expression is used to correct power at non-standard temperature conditions:

$$\frac{P}{P_{std}} = \sqrt{\frac{T_{std}}{T_{OAT}}} = \sqrt{\frac{518.67(1-\kappa h)}{T_{OAT}\,°R}} = \sqrt{\frac{273.15(1-\kappa h)}{T_{OAT}\,K}} \quad (7\text{-}20)$$

where

P_{std} = standard power at altitude and ISA κ = lapse rate constant (Table 16-2)
T_{std} = standard day temperature
T_{OAT} = outside air temperature at condition
h = pressure altitude at condition

Note that Equation (7-20) is recommended by several piston engine manufacturers without specifically presenting derivations for it. As will be demonstrated in the following example, it overestimates the engine power when compared to the Gagg and Ferrar model with density ratio based on the ideal gas law. The author recommends the latter method as it yields more conservative performance estimations.

EXAMPLE 7-3

Estimate the power of a piston engine rated at 100 BHP while being operated at full power at 10,000 ft on a day on which the OAT is 30 °F (or 30 °R) higher than ISA.

Solution

Method 1: Use of Equation (7-20)

Lapse rate factor:

$$(1 - \kappa h) = (1 - 0.0000068756 \times 10,000) = 0.9312$$

Standard day temperature at 10,000 ft:

$$T_{std} = 518.67 \times 0.9312 = 483.0 \,°R$$

Density ratio at 10,000 ft (standard day):

$$\sigma = 0.9312^{4.2561} = 0.7385$$

Maximum power at 10,000 ft per Gagg and Ferrar:

$$P = P_{SL}(1.132\sigma - 0.132)$$
$$= 100(1.132 \times 0.7385 - 0.132) = 70.4 \text{ BHP}$$

This is further reduced by the warmer-than-normal day using Equation (7-20) as follows:

$$P = P_{std}\sqrt{\frac{T_{std}}{T_{OAT}}} = 70.4\sqrt{\frac{483.0}{483.0 + 30}} = 68.3 \text{ BHP}$$

Method 2: Ideal gas law with Gagg and Ferrar

The answer can also be estimated using the ideal gas equation as follows:

Pressure at 10,000 ft:

$$p = 2116(1 - kh)^{5.2561} = 2116 \times 0.9312^{5.2561} = 1455 \text{ psf}$$

Density at 10,000 ft:

$$\rho = \frac{p}{RT} = \frac{1455}{1716 \times (483 + 30)} = 0.001653 \text{ slugs/ft}^3$$

Density ratio at 10,000 ft:

$$\sigma = \frac{\rho}{\rho_0} = \frac{0.001653}{0.002378} = 0.6951$$

Gagg and Ferrar:

$$P = 100(1.132 \times 0.6951 - 0.132) = 65.5 \text{ BHP}$$

Effect of Manifold Pressure and RPM on Engine Power

The relationship between the manifold pressure and RPM is complex and is usually presented by the piston engine manufacturer in the form of a special graph called an *engine performance chart*. An example of such a chart is shown in Section 7.4.2, *Step-by-step: Extracting piston power from engine performance charts*. The primary

FIGURE 7-6 A typical installation of a turboprop on an agricultural aircraft (Air Tractor AT-802). The comparative light weight of a gas turbine requires it to be mounted far ahead of the wing, something that also increases propeller effects. *(Photo by Phil Rademacher)*

drawback of this plot is that it lends itself very poorly to modern utilization in spreadsheets. The solution to this remedy is found in a specialized formula, called the Petty equation. This powerful tool is presented in Section 7.4.3, *Extracting piston power using the Petty equation*.

7.2.2 Turboprops

A turboprop is a gas turbine engine designed to primarily drive a propeller as the thrust generator. It differs from a jet engine in that the thrust generated by the engine itself is very small compared to the contribution of the propeller. The turboprop engine packs far more power per unit weight than the piston engine. As an example, the power-to-weight ratio of turboprops is generally in the 2.3 to 2.7 SHP/lb_f range for the modern gas turbine (1.5–1.7 SHP/lb_f for gas turbines from the 1950s to 1970s). This compares very favorably to the 0.5 to 1.0 BHP/lb_f range for the typical piston engine. Figure 7-6 shows a typical turboprop installation in an agricultural aircraft.

Since the gas turbine rotates at a high rate (often in the 20,000–40,000 RPM range), a gear reduction drive must be used to reduce the rotation rate of the propeller. Turboprops have proven themselves a very reliable powerplant for aircraft, not to mention generating ample power at high altitudes (where they become more efficient). These qualities have resulted in great popularity among domestic and utility aircraft, as they allow aircraft to fly onto small and sometimes unimproved strips and then take-off and climb to altitudes that allow them to fly above weather.

One of the most popular such engines is the Pratt & Whitney Canada PT6 turbine, currently used by more than 6500 operators over the entire planet. It is a family of engines that offer power ranging from 500 SHP to 2000 SHP. It was developed by a company that traces its origins to the 1930s, to a small organization that specialized in the maintenance of Pratt & Whitney Wasp radial engines [11]. The development of the PT6 began in the 1950s and it was introduced to the market in the 1960s. To date, some 36,000 units have been delivered, making it the world's most widely used gas turbine engine in its class.

Typical Fuel Consumption

One of the primary drawbacks of turbo machinery is high fuel consumption. This makes gas turbines primarily useful for high-altitude operation, where low-density-reduced fuel consumption (as a consequence of maintaining a constant fuel-to-air ratio) and high true airspeed make their use practical. Table 7-9 shows the typical specific fuel consumption of selected turboprop engines.

Turboprop Inertial Separators

The proximity of the tip of a rotating propeller tends to throw up a cloud of dust and particles that can get sucked into the inlet of gas turbine engines. For this reason many are equipped with an inertial separator that the pilot can activate while maneuvering at low speeds on the ground. The inertial separator refers to the geometry of the inlet, which forces air to make a sharp turn before entering the engine inlet. The sharp turn cannot be made by heavy particles, which get separated from the stream of air and are thrown out of the engine inlet. Inertial separators save a lot of engines from great harm.

Throttle Ratio (TR)

Ambient air is the propellant in gas turbine engines. For this reason, it should not be surprising that atmospheric properties significantly influence their performance. The pressure of air affects engine thrust through its effect on mass flow rate and, since both pressure and density are lower at higher altitudes, engine thrust will also reduce with altitude. However, the total temperature, T_{t0}, of which static temperature, T_0, is a part per the expression $T_{t0} = T_0[1 + (\gamma - 1)M^2/2]$, affects engine behavior in more profound ways than both pressure and density.

As the flight Mach number and altitude change, so does the total temperature. Engine materials and cooling methodologies place limitations on *compressor exit temperature* (CET) and *turbine entry temperature* (TET). The former is the temperature of air as it leaves the last compressor stage and the latter is the temperature after combustion immediately before it enters the first turbine stage. In order to keep compressor efficiencies at a practical level, the engine may operate at its maximum TET and maximum compressor pressure ratio only at one specific value of T_{t0}, say, T_{t0des} (call it design total

TABLE 7-9 Typical T-O Power and SFC of Selected Turboprop Engines. From Ref. [12]

Engine type	Weight, dry	Prop RPM	T-O Power rating[a]	SFC (T-O)
Garrett TPE331-10	380 lb$_f$ 172 kg	—	1000 SHP 746 kW	0.560 lb$_f$/hr/SHP 94.6 μg/J
Garrett TPE331-5/6	360 lb$_f$ 163 kg	—	840 SHP 626 kW	0.626 lb$_f$/hr/SHP 105.8 μg/J
Motorlet Walter M 601B	425.5 lb$_f$ 193 kg	2450	691 SHP 515 kW	0.656 lb$_f$/hr/EHP 110.8 μg/J
Motorlet Walter M 601E	425.5 lb$_f$ 193 kg	2450	751 SHP 515 kW	0.649 lb$_f$/hr/EHP 109.7 μg/J
Pratt & Whitney Canada P&WC PT6A-11	314 lb$_f$ 142.4 kg	2200	528 EHP 373 kW	0.647 lb$_f$/hr/EHP 109.4 μg/J
Pratt & Whitney Canada P&WC PT6A-21	316 lb$_f$ 143.3 kg	2200	580 EHP 410 kW	0.630 lb$_f$/hr/EHP 106.5 μg/J
Pratt & Whitney Canada P&WC PT6A-34	320 lb$_f$ 145.1 kg	2200	783 EHP 559 kW	0.595 lb$_f$/hr/EHP 100.6 μg/J
Pratt & Whitney Canada P&WC PT6A-41	391 lb$_f$ 177.3 kg	2000	903 EHP 634 kW	0.591 lb$_f$/hr/EHP 99.9 μg/J
Pratt & Whitney Canada P&WC PW118	861 lb$_f$ 391 kg	1300	1892 EHP 1342 kW	0.498 lb$_f$/hr/EHP 84.2 μg/J
Pratt & Whitney Canada P&WC PW120	921 lb$_f$ 417.8 kg	1200	2100 EHP 1491 kW	0.485 lb$_f$/hr/EHP 82.0 μg/J
Pratt & Whitney Canada P&WC PW123	992 lb$_f$ 450 kg	1200	2502 EHP 1775 kW	0.470 lb$_f$/hr/EHP 79.4 μg/J
Pratt & Whitney Canada P&WC PW127	1060 lb$_f$ 480 kg	1200	2880 EHP 2051 kW	0.459 lb$_f$/hr/EHP 77.6 μg/J
WSK-PZL TVD-10B	507 lb$_f$ 230 kg	—	1011 SHP 754 kW	0.570 lb$_f$/hr/SHP 96.4 μg/J
Rolls-Royce Dart 535	1340 lb$_f$ 607 kg	1395	2080 SHP 1551 kW	0.615 lb$_f$/hr/SHP 104 μg/J
Rolls-Royce Dart 536	1257 lb$_f$ 569 kg	1395	2120 SHP 1580 kW	0.615 lb$_f$/hr/SHP 104 μg/J

[a] $EHP \approx 1.05 \times SHP$.

temperature). The ratio of this total temperature to standard S-L temperature (518.67 °R) is known as *theta-break*. If the actual T_{t0} is less than T_{t0des} (which could be caused by lower flight Mach number or higher altitude) then TET must be decreased to maintain efficiency. If the actual T_{t0} is greater than T_{t0des} (due to operation at higher Mach or lower altitude) then the power generated by the turbine is reduced, causing the compressor pressure ratio to decline.

High compressor pressure ratio results in good *thrust specific fuel consumption* (TSFC) and high TET yields high specific thrust $F/\dot{m}$, where F is the thrust generated and $\dot{m}$ is the mass flow rate through the engine. These imply that it is best to operate the engine always at its theta-break; however, this is never possible since an engine is required to satisfactorily operate at various Mach numbers and altitudes. As a compromise, an engine may be designed so that it has a theta-break value roundabout the most desired combination of altitude and Mach number. The engine is then said to have *throttle ratio* (TR) equal to the theta-break value. For instance, if the desired combination is, say, $M = 0.6$, sea level, then the theta-break will be $(T_{t0}/T_{std}) = 1.072$, and the engine has $TR = 1.072$. On the other hand, if $M = 0$ and $h = 0$ (takeoff condition) is chosen as the desired combination, then the theta-break (and TR) will be 1.

To the extent of the author's knowledge, the concept of throttle ratio was first introduced by Mattingly, Heiser, and Pratt [13], in their aircraft engine design textbook. The book presents a very convenient method to estimate the thrust of gas turbine engines (turboprops, turbojets, and turbofans) as a function of flight conditions. The fundamentals of the method were developed by Mattingly [14] and utilizes simple empirical algebraic equations dependent on Mach number and altitude to determine the change in thrust from some baseline thrust value. The method is ideal for use in spreadsheets or software routines during the conceptual and preliminary design stage of aircraft featuring such engines (for instance see the Microsoft Excel Visual Basic routine presented in the Section 7.2.6, *Computer code: Thrust as a function of altitude and Mach number*).

The *TR* is selected by the engine manufacturer and is required for an accurate prediction of engine thrust and, thus, for aircraft performance analysis. An important consequence of the selection of *TR* can be seen from the following: as stated above, an engine with $TR = 1$ will be at its theta-break point at $M = 0$, $h = 0$, and, consequently, its TET will be at its maximum (assuming standard day conditions – ISA). Therefore, no further increase in the turbine entry temperature is possible. This means that when the aircraft is operating at an elevated temperature, say for instance 20 °C above ISA, the turbine power will be reduced. This, in turn, will detrimentally affect the T-O distance and rate of climb. It is possible to restore some of this loss using technology such as water injection.

By the same token, an engine selected so its $TR > 1$, say $TR = 1.072$, will operate at less than its maximum TET at $M = 0, h = 0$. In response to elevated ambient temperature the TET may be raised to compensate, and this allows constant thrust to be maintained up to some specific altitude. In this case, the thrust may be maintained up to ambient temperature of $518.67 \times 1.072 = 556$ °R (87 °F or 35 °C). From the standpoint of aircraft design, this is very favorable as it permits a far more impressive performance on hot days.

The name "theta-break" can be inferred from gas turbine performance graphs (e.g. see Figure 6.E8 of Ref. 13), which show a decisive break in each curve. For instance, in the cited figure, the break occurs at $M = 0.6$ at S-L, but at $M = 1.45$ at 36,000 ft. The particular engine has a unique *TR* (1.07 in this case), but the break shifts to higher Mach numbers with increasing altitude due to decline in T_0. Beyond 36,000 ft, however, the break remains at the same Mach number (stratosphere).

Step-by-step: Effect of Altitude and Airspeed on Turboprop Engine Thrust

The effect of altitude and airspeed on the thrust of turboprop engines can be modeled using the Mattingly method of Ref. [13].

Step 1: Determine the baseline thrust to use at S-L, F_{SL}, for instance the maximum static thrust at ISA.

Step 2: Calculate temperature ratio:

$$\theta_0 = \frac{T_{tot}}{T_0} = \frac{T}{T_0}\left(1 + \frac{\gamma-1}{2}M^2\right) \quad (7\text{-}21)$$

Step 3: Calculate pressure ratio

$$\delta_0 = \frac{p_{tot}}{p_0} = \frac{p}{p_0}\left(1 + \frac{\gamma-1}{2}M^2\right)^{\frac{\gamma}{\gamma-1}} \quad (7\text{-}22)$$

Step 4: If $M \leq 0.1$ then

$$F = F_{SL}\delta_0 \quad (7\text{-}23)$$

If $M > 0.1$ and $\theta_0 \leq TR$ then

$$F = F_{SL}\delta_0\left[1 - 0.96(M - 0.1)^{0.25}\right] \quad (7\text{-}24)$$

If $M > 0.1$ and $\theta_0 > TR$ then

$$F = F_{SL}\delta_0\left[1 - 0.96(M - 0.1)^{0.25} - \frac{3(\theta_0 - TR)}{8.13(M - 0.1)}\right]$$
$$(7\text{-}25)$$

where
F = thrust at (the atmospheric) condition
F_{SL} = thrust the engine would be producing at a given power lever setting at S-L
p = pressure at condition
p_0 = standard S-L pressure
p_{tot} = total pressure at condition
T = temperature at condition
T_0 = standard S-L temperature
T_{tot} = total temperature at condition

The thrust of a generic turboprop engine is plotted for several altitudes as a thrust ratio (i.e. as F/F_{SL}) in Figure 7-7 using the above formulation and a $TR = 1.072$. The sharp drop around $M = 0.1$ occurs when the TET reaches a maximum because of the increase in total air temperature. Until that point, it is possible to maintain constant thrust – something that offers great performance improvements in the operation of turboprop aircraft. Of course, this comes at the cost of a reduction in compression ratio at elevated temperatures in order to keep the TET below its maximum permissible value. In other words, the engine is flat-rated to a given thrust output at standard atmospheric conditions (ISA), even though, theoretically, it could generate thrust greater than what its flat-rating indicates. Plotting the thrust ratio, as is done in Figure 7-7, rather than for a specific rated S-L thrust, allows the curves to be applied to nonspecific turboprop engines. It can be shown using the formulation that at an altitude of approximately 7800 ft the maximum thrust has fallen to 75% of the rated S-L value.

Austyn-Mair and Birdsall [15] present the effect of altitude and airspeed on turboprop power rather than thrust using the following expression:

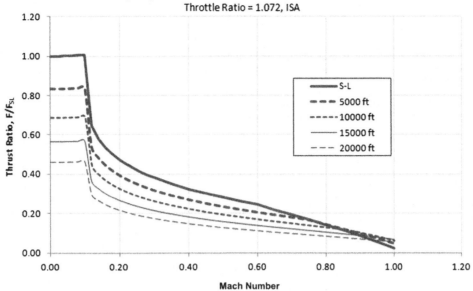

FIGURE 7-7 Thrust ratio of a turboprop as a function of Mach number and altitude.

$$\frac{P}{P_{SL}} = AM^n \qquad (7\text{-}26)$$

where

P = power at (the atmospheric) condition
P_{SL} = power setting at S-L
A = engine-dependent constant
M = Mach number
n = engine-dependent constant

The engine-dependent constants A and n must be selected based on engine data provided by the engine manufacturer. The designer should request power output at specific altitudes and Mach numbers and use this to determine both constants. The constant n is a fraction between 0 and 1, and is often close to 0.5. It reflects the fact that the available shaft power increases with ram pressure in the engine intake.

7.2.3 Turbojets

While the turbojet is rarely used for certified GA aircraft, there are a number of such installations on experimental aircraft and UAVs. One of the best-known uses of a jet engine in such an aircraft was the installation of a Microturbo TRS-18 turbojet in a Bede BD-5J single-seat kit plane. And as stated earlier, a pair of those was used to power the Caproni C22J trainer aircraft. Microturbo (www.microturbo.com) is a well-known manufacturer of small turbojets for drones. PBS VB (www.pbsvb.com) is a company in the Czech Republic that manufactures small turbojets for experimental aircraft and even self-launching sailplanes. Its engines have been installed on a Sonex experimental aircraft and Blanik, Salto, and TST-14J BonusJet sailplanes. The Blanik has been operated at altitudes as high as 32,000 ft using the engine.

Typical Fuel Consumption

Table 7-10 shows the typical specific fuel consumption of selected turbojet engines. Note that the reported values are all at a maximum T-O thrust.

TABLE 7-10 Typical T-O Thrust and SFC of Selected Turbojet Engines. From Ref. [12]

Engine type	T-O Thrust rating	SFC (T-O)
Instytut Lotnictwa IL K-15	3305 lb_f 14.7 kN	1.006 $lb_f/hr/lb_f$ 28.49 mg/Ns
LM WP6	6614 lb_f 29.42 kN	0.980 $lb_f/hr/lb_f$ 27.76 mg/Ns
Microturbo TRS 18-046	202 lb_f 0.898 kN	1.27 $lb_f/hr/lb_f$ 35.97 mg/Ns
Microturbo TRS 18-056	221 lb_f 0.982 kN	1.27 $lb_f/hr/lb_f$ 35.97 mg/Ns
Microturbo TRI 60	772 lb_f 3.430 kN	1.25 $lb_f/hr/lb_f$ 35.40 mg/Ns
PBS VB TJ100 A	247 lb_f 1.097 kN	1.090 $lb_f/hr/lb_f$ 30.87 mg/Ns
PBS VB TJ100 C	225 lb_f 1.000 kN	1.177 $lb_f/hr/lb_f$ 33.33 mg/Ns

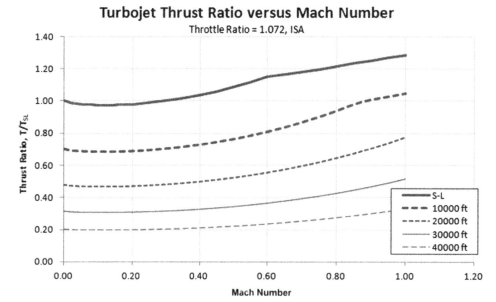

FIGURE 7-8 Thrust ratio of a turbojet as a function of Mach number and altitude.

Step-by-step: Effect of Altitude and Airspeed on Turbojet Engine Thrust

Reference [14] also presents a method to estimate the effect of altitude and airspeed on the thrust of turbojet engines, similar to the method of Section 7.2.2, *Turboprops*. Turbojets are typically used in military applications and, thus, feature formulation for maximum and military thrust (with afterburner used).

Step 1: Calculate temperature ratio:

$$\theta_0 = \frac{T_{tot}}{T_0} = \frac{T}{T_0}\left(1 + \frac{\gamma-1}{2}M^2\right) \quad (7\text{-}21)$$

Step 2: Calculate pressure ratio

$$\delta_0 = \frac{p_{tot}}{p_0} = \frac{p}{p_0}\left(1 + \frac{\gamma-1}{2}M^2\right)^{\frac{\gamma}{\gamma-1}} \quad (7\text{-}22)$$

Step 3a: Maximum thrust (means afterburner is on)

If $\theta_0 \leq TR$ then

$$F = F_{SL}\delta_0\left[1 - 0.3(\theta_0 - 1) - 0.1\sqrt{M}\right] \quad (7\text{-}27)$$

If $\theta_0 > TR$ then

$$F = F_{SL}\delta_0\left[1 - 0.3(\theta_0 - 1) - 0.1\sqrt{M} - \frac{1.5(\theta_0 - TR)}{\theta_0}\right] \quad (7\text{-}28)$$

Step 3b: Military thrust (means afterburner is off)

If $\theta_0 \leq TR$ then

$$F = 0.8F_{SL}\delta_0\left[1 - 0.16\sqrt{M}\right] \quad (7\text{-}29)$$

If $\theta_0 > TR$ then

$$F = 0.8F_{SL}\delta_0\left[1 - 0.16\sqrt{M} - \frac{24(\theta_0 - TR)}{(9+M)\theta_0}\right] \quad (7\text{-}30)$$

where

F = thrust at (the atmospheric) condition
F_{SL} = thrust the engine would be producing at the power setting at S-L

The thrust of a generic turbojet engine is plotted for several altitudes as a thrust ratio (i.e. as T/T_{SL}) in Figure 7-8 using the above formulation and a $TR = 1.072$.

7.2.4 Turbofans

Turbofan engines are used in a number of GA aircraft. While most are business jets, recently a few new jets have emerged that are powered by turbofans; these are known as *Personal*, Jets, and are designed to be owner flown and operated, analogous to the operation of single-engine, piston-powered aircraft. This trend has led to the development of small turbofans, such as the DGEN 380 in Figure 7-9. Examples of such jets are the Cirrus SF50 Vision and the Diamond Jet. These airplanes are made possible by the development of low-thrust certified turbofans manufactured by Williams International and Pratt & Whitney Canada. One such engine, the Williams International FJ44, is shown in Figure 7-12.

The turbofan differs from the turbojet in that the diameter of the forward compressor wheel is larger

FIGURE 7-9 A cutaway of the Price Induction DGEN 380 730 lb$_f$ thrust turbofan engine specifically designed for low and slow GA aircraft. (*Courtesy of Price Induction*)

than that of the engine behind it (the *core*). The core contains the remaining compressor wheels (the fan), the combustion chamber and the turbine wheels. A part of the air passing through the fan flows through the core and a part over it. The ratio of the air bypassing the core to the air that flows through it is called the *bypass ratio* (BPR). As a rule of thumb, the higher the bypass ratio the more fuel-efficient is the engine, although other details come into play as well. A turbofan whose BPR is less than 2 is considered a low-bypass-ratio engine.

Typical Fuel Consumption

Table 7-11 shows the typical specific fuel consumption of selected turbofan engines.

Step-by-step: Effect of Altitude and Airspeed on Turbofan Engine Thrust

Reference [13] also presents a method to estimate the effect of altitude and airspeed on the thrust of turbofan engines, similar to those presented above. There are generally two classes of turbofans: those with low and those with high bypass ratios. A low-bypass-ratio turbofan is one whose bypass ratio is less than 1. Low-bypass-ratio turbofans are usually intended for military applications whereas high-bypass-ratio are typically used in civilian applications. For this reason, two types of formulation are presented, one for each type.

Step 1: Calculate temperature ratio

$$\theta_0 = \frac{T_{tot}}{T_0} = \frac{T}{T_0}\left(1 + \frac{\gamma - 1}{2}M^2\right) \quad (7\text{-}21)$$

Step 2: Calculate pressure ratio

$$\delta_0 = \frac{p_{tot}}{p_0} = \frac{p}{p_0}\left(1 + \frac{\gamma - 1}{2}M^2\right)^{\frac{\gamma}{\gamma-1}} \quad (7\text{-}22)$$

Step 3a: Maximum thrust of low-bypass-ratio turbofans

If $\theta_0 \leq TR$ then

$$F = F_{SL}\delta_0 \quad (7\text{-}31)$$

If $\theta_0 > TR$ then

$$F = F_{SL}\delta_0\left[1 - \frac{3.5(\theta_0 - TR)}{\theta_0}\right] \quad (7\text{-}32)$$

Step 3b: Military thrust of low-bypass-ratio turbofans

If $\theta_0 \leq TR$ then

$$F = 0.6F_{SL}\delta_0 \quad (7\text{-}33)$$

If $\theta_0 > TR$ then

$$F = 0.6F_{SL}\delta_0\left[1 - \frac{3.8(\theta_0 - TR)}{\theta_0}\right] \quad (7\text{-}34)$$

Step 3c: High-bypass-ratio turbofans

If $\theta_0 \leq TR$ then

$$F = F_{SL}\delta_0\left[1 - 0.49\sqrt{M}\right] \quad (7\text{-}35)$$

If $\theta_0 > TR$ then

$$F = F_{SL}\delta_0\left[1 - 0.49\sqrt{M} - \frac{3(\theta_0 - TR)}{1.5 + M}\right] \quad (7\text{-}36)$$

where

F = thrust at (the atmospheric) condition
F_{SL} = thrust the engine would be producing at the power lever setting at S-L

TABLE 7-11 Typical T-O Thrust and SFC of Selected Turbojet Engines. From Ref. [12]

Engine type	Bypass ratio	T-O Thrust rating	SFC (T-O)
Pratt & Whitney Canada P&WC JT15D-4B	3.3	2500 lb$_f$ 11.12 kN	0.562 lb$_f$/hr/lb$_f$ 15.92 mg/Ns
Pratt & Whitney Canada P&WC JT15D-5A	3.3	2900 lb$_f$ 12.9 kN	0.551 lb$_f$/hr/lb$_f$ 15.61 mg/Ns
Pratt & Whitney Canada P&WC530A	–	2887 lb$_f$ 12.8 kN	–
Pratt & Whitney Canada P&WC910F	–	950 lb$_f$ 4.22 kN	–
Turbomeca-SNECMA Larzac 04-C6	1.13	2966 lb$_f$ 13.19 kN	0.71 lb$_f$/hr/lb$_f$ 20.1 mg/Ns
CFM56-3B2	5.0	22000 lb$_f$ 97.90 kN	0.655 lb$_f$/hr/lb$_f$ 18.55 mg/Ns
CFM56-5C2	6.6	31200 lb$_f$ 138.8 kN	0.567 lb$_f$/hr/lb$_f$ 16.06 mg/Ns
Williams International FJ33	–	1000–1900 lb$_f$ 4.44–8.44 kN	–
Williams International FJ44-2A	4.1	2300 lb$_f$ 10.23 kN	0.460 lb$_f$/hr/lb$_f$ 13.03 mg/Ns
Price Induction DGEN 380[a]	7.6	560 lb$_f$ 2.49 kN	–
Price Induction DGEN 390[a]	6.9	730 lb$_f$ 3.24 kN	–

[a]Still in development.

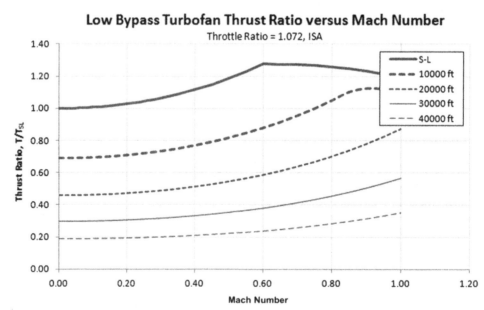

FIGURE 7-10 Thrust ratio of a low-bypass-ratio turbofan as a function of Mach number and altitude.

The thrust of a generic-low-bypass ratio turbofan engine is plotted for several altitudes as a thrust ratio (i.e. as T/T_{SL}) in Figure 7-10 using the above formulation and a $TR = 1.072$. The thrust of a generic high-bypass-ratio turbofan engine is also plotted for several altitudes in Figure 7-11 as a thrust ratio in using the above formulation and a $TR = 1.072$.

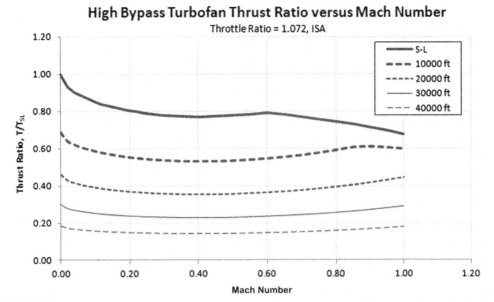

FIGURE 7-11 Thrust ratio of a high-bypass-ratio turbofan as a function of Mach number and altitude.

7.2.5 Electric Motors

Electric motors are no longer only for parasailing and ultra-lights but are quickly becoming an important trend in the light aircraft industry, particularly in light sports aircraft (LSA). Electric motors, capable of delivering power in excess of 80 HP (60 kW), are now being used to power aircraft that can carry as many as four people. At the time of writing, aircraft such as the two-seat e-Genius and the four-seat, two-hull Pipistrel Taurus G4 prototype aircraft demonstrate what the future of electric aviation beholds.

Two recent electric aircraft types are of great interest because they reveal some of the challenges of this emerging technology. The first is the two-seat Yuneec e430, acclaimed as the "world's first commercially

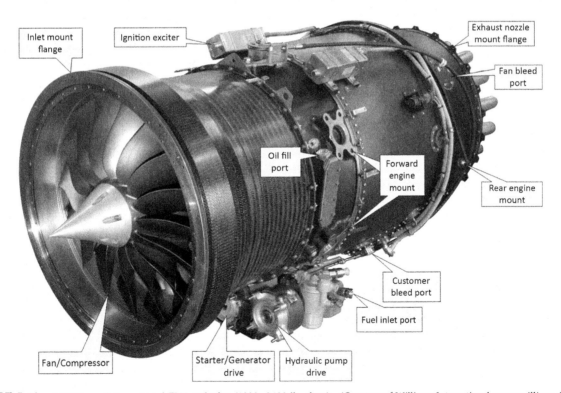

FIGURE 7-12 A Williams International FJ44 turbofan (1900–3600 lb$_f$ class). (*Courtesy of Williams International, www.williams-int.com*)

FIGURE 7-13 An experimental Electraflyer C (a modified Moni motor glider), powered by a 13.5 kW (18 hp) direct-drive electric motor. The photo to the right shows that the installation of electric motors is particularly clean and simple when compared to other engines. *(Courtesy of www.electraflyer.com)*

produced electric Aircraft" [16]. It is a two-seat, powered sailplane that has the appearance of a high-wing trainer. It has a glide ratio of 24, almost two times that of a typical avgas powered aircraft. According to product information, it has an empty weight without batteries of 377 lb_f (172 kg), empty weight with batteries of 561 lb_f (255 kg), and maximum T-O weight of 946 lb_f (430 kg). The battery pack weighs 184 lb_f (83.5 kg) and is a lithium-ion polymer (LiPo) that takes 3–4 hours to recharge at 220 V (double or triple that for 110 V) and provides about 2 hrs of flying time. The e430 stalls at 35 KCAS, cruises at 52 KTAS, and has a top speed of 80 KTAS.

The second aircraft is the Electraflyer-C (see Figure 7-13). It is a modified single-seat Monnett Moni motorglider that has been equipped with a 13.5 kW (18 hp) electric motor. Its empty weight with a battery pack is 380 lb_f (172 kg) and maximum T-O weight is 625 lb_f (283 kg). It cruises at 61 KTAS and has a maximum level airspeed of 78 KTAS. It can stay aloft for 1.5 hours using a 5.6 kWh LiPo battery pack. Such a pack weighs about 78 lb_f (36 kg).

Two observations can be made from the above aircraft. First, both aircraft feature high-AR wings. In order for the electric aircraft to be viable, it must be more efficient than the conventional GA aircraft and require far less power for operation. For this reason, aircraft must have a higher-AR wing. Second, the weight of the battery packs is a major part of the gross weight, indicative of the low energy density of electric power when compared to fossil fuels.

Energy Density

Energy density is the amount of energy stored in a unit weight of a battery. It is denoted by E_{BATT} and is typically given in terms of watt·hours/kg or Wh/kg for short. The energy density of even the best batteries (LiPo) is substantially lower than that of fossil fuels, about 60 times less [17], leaving electric aviation at a significant disadvantage. The energy density for typical aircraft battery packs is given in terms of kWh (kilowatt-hours). A 5.6 kWh battery can deliver 5600 W over a period of an hour. This corresponds to the energy required to keep a 100-watt light-bulb lit for 56 hours (2.3 days) or a 1500 W microwave running for 3.7 hours.

Batteries

A battery is a device that converts chemical energy into electrical energy (or the other way around). It can also be thought of as a device that allows electrical energy to be carried around. The invention, according to some, is ancient and dates back to the beginning of the Common Era, some 2000 years ago, with the so-called Parthian batteries. While this is debated, it is known that the battery in the modern form dates back to the 1800s with the invention by Alessandro Volta (1745–1827). The battery has come a long way since then and currently, spurred by growing interest in environmental protection, there is considerable research going on. Some of what is said here will possibly be obsolete in a few years time and the reader is encouraged to do their own research to ensure up-to-date information.

There are several issues that concern batteries and must be kept in mind:

(1) The energy density of current battery technology is very low compared to that of gasoline. Therefore, a lot of battery power requires a lot of battery — i.e. a lot of weight.

(2) Energy durability — shelf life. There is more to battery capability than just energy density. For instance, $LiCO_2$ batteries have a higher initial energy density than, say, $LiFePO_4$ batteries. However, after a year of frequent recharge-discharge cycles, the $LiFePO_4$ battery has similar residual energy density to the $LiCO_2$ battery. In 2 years it does better. This is battery durability.

(3) Discharge voltage depends on the remaining charge and battery temperature. The initial discharge voltage is usually high, but diminishes with the energy used. This means that initially after a battery recharge is completed, the battery appears to "contain a lot of power." However, this drops rapidly. For aircraft this means that a fully charged battery yields a reported T-O distance, but the first touch-n-go requires a much longer runway. This is not acceptable for aircraft transporting people for commercial purposes.

(4) The current battery technology poses fire hazards.

The ideal battery for use in airplanes should be light, rechargeable, have a long durability, and with the highest energy density possible. The current state of technology has been largely driven by demand for laptop computers and cell phones, where bright screens and long endurance is of the essence. Batteries of the kind people are mostly familiar with, such as those used for flashlights (D, C, AA, and AAA style) or conventional car batteries, are not suitable at all, due to low energy density and high weight. The modern lithium-ion battery is a huge improvement over the batteries of the past, although it is in fact marginal. Currently, two types of batteries are suitable for use in aircraft: LiFePO$_4$ and LiCoO$_2$. Both have their pros and cons.

The current battery technology consists of the types of batteries listed in Table 7-12.

Fuel Cells

A fuel cell is a device that produces electricity by combining hydrogen and oxygen, forming water and heat as byproducts (see Figure 7-14). The fuel cell is superior to batteries in many ways. It has the potential of being a zero-emission device (if the electrical energy is generated using renewable energy) and it overcomes a serious drawback of chemical batteries

TABLE 7-12 Common Battery Types

Battery type	Comment
Lead-acid	Best known as the car battery. Low energy option not suited for use in aircraft.
NiCad	Used to be popular for radio-controlled aircraft. Largely obsolete.
NiMH	Popular as rechargeable batteries for robots.
Li-Ion	Lithium-ion battery, best known as battery packs for laptop computers.
LiPo	Current power packages for electric aircraft.

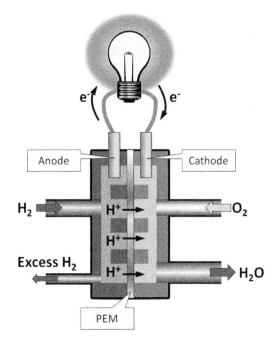

FIGURE 7-14 The workings of a fuel cell (see text).

in which voltage reduces as a function of the battery charge. For batteries, this effectively means that as its charge drains it is no longer possible to get maximum power obtained as when it was fully charged. For airplanes, this means reduced and "variable" top performance, one which depends on charge remaining. In order for electric aircraft to be truly compatible with a conventional fuel-powered aircraft, it is necessary that maximum power can be drawn regardless of the state of charge. This is not an issue with fuel cells and renders them particularly attractive for use in electric aircraft.

While there are several types of fuel cells, this text only considers those that consist of a thin membrane, called a proton exchange membrane (PEM). One side of it is exposed to pure hydrogen gas (H_2) and the other to oxygen (O_2). The PEM catalytically strips the electrons off the hydrogen, converting it into hydrogen ions (H^+). Furthermore, it ensures the ions can only pass through it in one direction: to the side that is bathed in oxygen. The electrons that were stripped off take a different path and flow through the anode to the cathode, generating electric current in the process. At the same time, the hydrogen ions that pass through the PEM encounter oxygen and the electrons flowing through the cathode react to form H_2O, completing the process.

At the anode: $H_2 \rightarrow 2H^+ + 2e^-$
At the cathode: $\frac{1}{2}O_2 + 2H^+ + 2e^- \rightarrow H_2O$

Hybrid Electric Aircraft

The low energy density of electric motor installations has led to the introduction of engine configurations that bridge the gap between gasoline engines and electric motors through hybrid functionality. A hybrid electric aircraft is one that features a combination of electric motor and some other type of power plant. Strictly speaking, such a power plant can be any of the other types discussed in this section. However, in this text the use of the term will be limited to aircraft driven by a propeller powered by a combination of an electric motor and a piston engine or a gas turbine. In this context we further classify hybrid electric aircraft into the following types:

(1) A *parallel hybrid* has an electric motor and a gasoline engine connected to the same drive-train. In this way, the same propeller can be driven using either electrical or gasoline power or it can use both simultaneously.
(2) In a *series hybrid* the propeller is driven using the electric motor only. However, it has a gasoline engine that runs a generator that charges the battery when it runs low.

In 2011, Embry-Riddle Aeronautical University flew a modified Stemme S-10 (parallel) hybrid-powered sailplane using a Rotax 912 and an electric motor coupled to the same driveshaft — arguably the first hybrid light plane in the history of aviation. In this configuration, the airplane uses conventional piston power for T-O and climb, but once reaching cruise altitude the gasoline engine is turned off and the electric power is switched on for cruise.

The Pure Electric Aircraft

Electric propulsion has a number of advantages over propulsion that depends on fossil fuel. Among those is the very quiet operation of such engines. This is compounded using large-diameter, low-RPM propellers that also generate relatively low noise. This contrasts with noisy piston or gas turbine engines and propellers associated with conventional installations. The operation of electric motors is practically without vibration as long as the propeller is well-balanced. The other usual nuisances of gasoline engines are absent as well. There are no residues, odors, or stains associated with the operation of such motors — they are extremely clean. Additional advantages include a very simple and reliable start and operation of the motor. The motor itself is a very reliable device that requires minimal maintenance when compared to a piston engine. Tune-ups and expensive overhauls are not required. Another important consideration is pilot and passenger safety; since no fossil fuels are consumed there is no chance of carbon monoxide poisoning. Another advantage is that batteries can be recharged by simply plugging them into a household outlet and, at this time, recharging batteries is inexpensive. Electric airplanes are also environmentally friendly and don't emit greenhouse gases, although this is offset by the fact that in many places the production of electricity releases harmful greenhouse chemicals into the environment. This holds for electricity produced by oil or coal plants. Renewable energy is of course the answer, giving the electric airplane a unique potential as an environmentally friendly transportation vehicle.

Unfortunately, the production of electricity is not the only drawback. One of the most important downsides is the storage of the energy on board an aircraft. There are primarily two ways electrical energy is provided to run the motor: via batteries or via fuel cells. At this time, both options are very heavy and a high toll must be paid in terms of reduction in useful load. In fact, as will be discussed later, an airplane really must be specifically designed to effectively use electric propulsion; it is very impractical to convert existing gasoline-powered aircraft into electric ones. One of the issues with batteries is their relatively low energy content. For longer flights this calls for a large amount of matter (i.e. mass) to be carried around, which reduces the useful load of the airplane. Fuel cells, on the other hand, require large quantities of hydrogen to be carried in highly pressurized bottles, sometimes as high as 5000–10,000 psi. In comparison, the pressure inside the combustion chamber of the Space Shuttle main engine is in the 3000 psi range. This means that if the hydrogen bottles burst, they pose a serious threat to the airplane and its occupants.

Formulation

The following expressions are helpful when solving various problems that involve electrical power.

$$\text{Voltage:} \quad V = \begin{cases} I \times R \\ P/I \\ \sqrt{P \times R} \end{cases} \text{Volts}$$

$$\text{Current:} \quad I = \begin{cases} \sqrt{P/R} \\ P/V \\ V/R \end{cases} \text{Amps}$$

$$\text{Resistance:} \quad R = \begin{cases} V/I \\ V^2/P \\ P/I^2 \end{cases} \text{Ohms}$$

$$\text{Power:} \quad P = \begin{cases} V^2/R \\ R \times I^2 \\ V \times I \end{cases} \text{Watts}$$

where

I = current (amps)
P = power (watts)
R = resistance (ohms)
V = voltage (volts)

EXAMPLE 7-4

Powered paragliding is currently a popular sport. Some paragliders are powered by a compact electric propeller power-pack the pilot straps to his or her back. The data below are given for one manufacturer of such power packs. Determine an equivalent horsepower rating for the electric motor. Also compute the current to the motor and its internal resistance. If the battery pack stores 30 Ah (Amp-hours) of energy, how long can the motor be run at peak force (assuming no overheating takes place and Peukert's law is not applied)?

Voltage: 66.6 V
Output: 10 kW

Solution

Power:

$$P = 10 \text{ kW} = 10{,}000 \text{ W} = \frac{10{,}000 \text{ W}}{746 \text{ W/hp}} = 13.4 \text{ hp}$$

Current:

$$I = \frac{P}{V} = \frac{10{,}000 \text{ W}}{66.6 \text{ V}} \approx 150 \text{ Amps}$$

Resistance:

$$R = \frac{V}{I} = \frac{66.6 \text{ V}}{150 \text{ Amps}} \approx 0.444 \text{ Ohms}$$

Duration:

$$\text{Duration} = \frac{\text{Energy content}}{I} = \frac{30 \text{ Ah}}{150 \text{ Amps}}$$

$$= \frac{30 \text{ Amps} \cdot \text{hr}}{150 \text{ Amps}} \approx 0.2 \text{ hr} = 12 \text{ min}$$

7.2.6 Computer code: Thrust as a Function of Altitude and Mach Number

One of the most important capabilities of the designer is the ability to predict the thrust of the selected engine as a function of altitude and airspeed. This is imperative for performance and stability and control theories. The preceding methods all lend themselves to incorporation in a function that can be called from analysis spreadsheets such as Microsoft Excel. The following function incorporates all the preceding methods and allows piston engines as well as high-bypass-ratio turbofans to be evaluated. The input arguments are thrust (F0), throttle ratio (TR), altitude (H), Mach Number (M), and deviation from ISA (deltaOAT). Note that the routine treats piston engines in a different manner than the other engines, as it assumes the there is a constant-power device.

```
Function Engine_Thrust(F0 As Single, TR As Single, H As Single, M As Single, deltaOAT_C As Single, Mode As
Byte) As Single
'This routine can be used to estimate the change in thrust depending on
'flight conditions for the following engines:
'
'   1. Piston engines (Mode = 0)
'   2. Turboprops (Mode = 1)
'   3. Turbojets, maximum power (Mode = 2)
'   4. Turbojets, military power (Mode = 3)
'   5. Low Bypass Ratio Turbofans (Mode = 4)
'   6. High Bypass Ratio Turbofans (Mode = 5)
'
'Variables: F0         = Engine thrust, in lbf
'           TR         = Throttle ratio
'           H          = Altitude at condition, in ft
'           M          = Mach Number
'           deltaOAT   = Deviation from ISA in °C
'
'NOTE1: For piston engines, the function only treats the power, so F0 is
'       the rated engine power at S-L and ISA.
```

```
'NOTE2: The function calls the AtmosProperty function, so it must be present.
'
'Initialize
    'Atmospheric properties
    Dim P As Single, OAT As Single, rho As Single
    'Property ratios
    Dim sigma As Single, delta As Single, theta As Single
'Presets
    P = AtmosProperty(H, 11)        'Pressure at H in lbf/ft²
    OAT = AtmosProperty(H, 10)      'Standard OAT at H in °R
    OAT = OAT + deltaOAT_C * 1.8    'Include temperature deviation. Note T °R = 1.8T K
    rho = P / (1716 * OAT)          'Density in slugs/ft3
    'Pressure ratio
    delta = P / 2116 * (1 + 0.2 * M ^ 2) ^ 3.5
    'Temperature ratio
    theta = OAT / 518.67 * (1 + 0.2 * M ^ 2)
    'Density ratio
    sigma = rho / 0.002377427
'Process
    Select Case Mode
    Case 0 'Piston per Gagg and Ferrar
        Engine_Thrust = F0 * (1.132 * sigma - 0.132)
    Case 1 'Turboprop per Mattingly, et al.
        If M <= 0.1 Then
            Engine_Thrust = F0 * delta
        Else
            If theta <= TR Then
                Engine_Thrust = F0 * delta * (1 - 0.96 * (M - 0.1) ^ 0.25)
            Else
                Engine_Thrust = F0 * delta * (1 - 0.96 * (M - 0.1) ^ 0.25 - 3 * (theta - TR) / (8.13 * (M - 0.1)))
            End If
        End If
    Case 2, 3 'Turbojet per Mattingly, et al.
        If Mode = 2 Then     'Max thrust
            If theta <= TR Then
                Engine_Thrust = F0 * delta * (1 - 0.3 * (theta - 1) - 0.1 * Sqr(M))
            Else
                Engine_Thrust = F0 * delta * (1 - 0.3 * (theta - 1) - 0.1 * Sqr(M) - 1.5 * (theta - TR) / theta)
            End If
        ElseIf Mode = 3 Then    'Military thrust (afterburner)
            If theta <= TR Then
                Engine_Thrust = 0.8 * F0 * delta * (1 - 0.16 * Sqr(M))
            Else
                Engine_Thrust = 0.8 * F0 * delta * (1 - 0.16 * Sqr(M) - 24 * (theta - TR) / ((9 + M) * theta))
            End If
        End If
    Case 4 'LBR Turbofan per Mattingly, et al.
        If theta <= TR Then
            Engine_Thrust = F0 * delta
        Else
            Engine_Thrust = F0 * delta * (1 - 3.5 * (theta - TR) / theta)
        End If
    Case 5 'HBR Turbofan per Mattingly, et al.
        If theta <= TR Then
            Engine_Thrust = F0 * delta * (1 - 0.49 * Sqr(M))
        Else
            Engine_Thrust = F0 * delta * (1 - 0.49 * Sqr(M) - 3 * (theta - TR) / (1.5 + M))
        End If
    End Select
End Function
```

7.3 AIRCRAFT POWER PLANT INSTALLATION

The type of engines selected will largely dictate the layout of the installation. Propeller engines (pistons or turboprops) require stout bulkheads to which the engine mounts are attached. Such bulkheads are called *firewalls*. For practical reasons, most of the firewall is normal to the flight direction, calling for measures that reduce drag, most effectively in the shape of an aerodynamically shaped cowling. Jet engines require inlets and exhausts that ideally should be short and without excessive bends. If mounted inside the fuselage, a jet engine requires fireproofing and careful structural design in case of a rotor-burst. If outside the fuselage, bulkheads to mount the engine are still required, but trusty fireproofing is replaced by having to perforate the outside mold line (OML) of the fuselage (or wing) to route the structure, wiring, fuel, and hydraulic lines. Subsequently, this calls for an aerodynamic refinement in the form of a pylon and nacelle. In all cases, the loads generated by the thrust and inertia of the engine must be reacted by a highly reinforced structure in the region of the installation.

In this section, typical engine installations for piston and jet engines will be presented from the standpoint of impact on aesthetics, aerodynamics, and, to a limited extent, structures. While the structural design belongs to the detail design phase, the implications of a particular installation should be understood by the designer. In particular, the regulatory aspect of the installation should be pondered, but for GA aircraft these can be found under 14 CFR 23.901, *Installation*, which is best recited for convenience:

---

§23.901 INSTALLATION

(a) For the purpose of this part, the airplane powerplant installation includes each component that—
  (1) Is necessary for propulsion; and
  (2) Affects the safety of the major propulsive units.
(b) Each powerplant installation must be constructed and arranged to—
  (1) Ensure safe operation to the maximum altitude for which approval is requested.
  (2) Be accessible for necessary inspections and maintenance.
(c) Engine cowls and nacelles must be easily removable or openable by the pilot to provide adequate access to and exposure of the engine compartment for preflight checks.
(d) Each turbine engine installation must be constructed and arranged to—
  (1) Result in carcass vibration characteristics that do not exceed those established during the type certification of the engine.
  (2) Ensure that the capability of the installed engine to withstand the ingestion of rain, hail, ice, and birds into the engine inlet is not less than the capability established for the engine itself under §23.903(a)(2).
(e) The installation must comply with—
  (1) The instructions provided under the engine type certificate and the propeller type certificate.
  (2) The applicable provisions of this subpart.
(f) Each auxiliary power unit installation must meet the applicable portions of this part.

Note that the term 'part' in subparagraph (f) refers to the subpart of 14 CFR Part 23 that treats power plant, aptly called *Subpart E — Powerplant*. The subchapter consists of the paragraphs 23.901 through 23.1203. The designer should be familiar with those that pertain to the particular engine type being selected. Note that many of the paragraphs specifically refer to other systems that are necessary to operate the engine, such as the fuel and oil systems.

Fireproofing

On the 2nd of June 1983, a DC-9 commercial jetliner en route from Dallas/Fort Worth to Toronto, Canada, suffered a cabin fire thought to have originated behind in the aft lavatory, perhaps by a careless smoker who threw a cigarette, not yet extinguished, into a paper towel container.[2] The fire spread between the outside wall of the aircraft and the inside decorative walls. The cabin filled with toxic smoke, causing it to be diverted to an alternative airport, the Greater Cincinnati International Airport, Covington, Kentucky. Once on the ground, some 60 to 90 seconds after the cabin doors were opened, fresh oxygen fueled the fire now extensively burning, although hiding under the decorative wall, causing it to quickly engulf

---

[2] Note that this is a speculation and not an established fact.

the cabin. The fire killed 23 of the 46 occupants on board [18]. Consequently, the installation of smoke detectors and fire extinguishers as well as proper training of crew members found their way into current Federal Aviation Regulations. This tragedy serves as a reminder that in many ways the aviation regulations are written in the blood of victims; they make flying safer for the rest of us.

Fire is a real threat in all aircraft. Unlike a car, the inability to stop the vehicle on a moment's notice makes fire in airplanes a very serious issue. *Fireproofing* is the addition of fire-resisting or retarding material and the installation of fire-suppression chemical dispensers to aircraft intended to make it more fire-resistant and, thus, safer in the case of such an emergency. There are three areas in an aircraft that are more susceptible to fire than others: the engine compartment, the cabin area, and anywhere electrical wiring is placed. Designing the fireproofing does not require much mathematics, but rather should demonstrate compliance with the applicable federal regulations. GA aircraft must comply with the regulation of 14 CFR Part 23:

23.865 — Fire protection of flight controls, engine mounts, and other flight structure,	23.1193 — Cowling and nacelle,
23.1181 — Designated fire zones; regions included,	23.1195 — Fire extinguishing systems,
23.1182 — Nacelle areas behind firewalls,	23.1197 — Fire extinguishing agents,
23.1183 — Lines, fittings, and components,	23.1199 — Extinguishing agent containers,
23.1189 — Shutoff means,	23.1201 — Fire extinguishing systems materials, and
23.1191 — Firewalls,	23.1203 — Fire detector system.

The Firewall

Piston and turboprop engines are typically mounted on a truss structure made from welded chrome-molybdenum steel (SAE 4130), which is then mounted to the airframe and separated from it via the *firewall*. Its purpose is to prevent fire from spreading beyond the engine compartment. The firewall is usually made from stainless steel or other heat-resistant material. Per 14 CFR 23.1191, some materials are exempt from fire retardation testing and these are listed below:

0.015 inch thick stainless steel sheet,
0.018 inch thick mild steel sheet (coated with aluminum or otherwise protected against corrosion)
0.018 inch thick terne plate
0.018 inch thick Monel metal
0.016 inch thick titanium sheet

Additionally, steel or copper base alloy firewall fittings are exempt as well. Other materials must be demonstrated to provide fire resistance — for instance, they must be subjected to a flame that is 2000 ± 150 °F for at least 15 minutes without penetration. Furthermore, the material must be protected against corrosion.

External jet engine installations also feature firewalls on the pylon, as will be discussed later.

Danger Zones Around Propeller Aircraft

Awareness of danger zones around the aircraft that stem from an engine selection is important. The following applies to both piston and turboprop engines. These introduce two kinds of danger zones: those while operating on the ground and those while while airborne. These are due to propwash, turbine exhaust (applies to turboprops only), prop strike (a person walking into the prop), and blade separation. These are depicted in Figure 7-15.

Prop wash can be dangerous if it picks up and blows heavy objects that may harm a person standing behind it. The exhaust zone contains dangerous fumes from the engine that can be harmful if inhaled and can burn a person standing close to the exhaust. Sadly, people walk into rotating propellers several times each year. Not much needs to be said about the outcome of that battle. Blades separating from propeller hubs are, fortunately, very rare, but can happen both in flight and also for reasons that may be easily overlooked. For instance, propeller blades often break off during emergency landings, when landing gear has failed to extend or lock. One of the precautions enforced in such situations in commuter aircraft is to move the passengers inside the danger zone to a different part of the airplane.

7.3.1 Piston Engine Installation

The piston engine installation must meet several requirements:

(1) Be structurally sound enough to react all loads generated by the engine.
(2) Allow for easy access for maintenance.
(3) Allow engine controls to be easily routed to and from the engine. This includes the electrical system, fuel plumbing, and engine controls (throttle, mixture, pitch control).
(4) The installation must be resistant to fire.
(5) The propeller must have a type certificate and be free of vibration.

The loads generated by the engine installation are primarily inertia loads due to gravitation and maneuvering loads and loads generated by the engine itself, such as thrust and gyroscopic moments. For GA aircraft, the loads the installation must be designed to react are stipulated in 14 CFR 23.901(b) (thrust) and 23.361, *Engine*

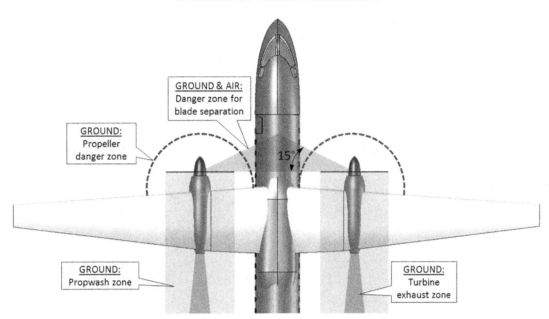

FIGURE 7-15 Danger zones around a typical turboprop.

torque, and 23.363, *Side load on engine mount*. In summary, these paragraphs state that the engine installation must react the worst of the following loads:

(1) Simultaneous application of max T-O thrust, torque, and 75% of the limit load factor (see Table 1-2 for load factors).
(2) Simultaneous application of max continuous thrust, torque, and 100% of the limit load factor.
(3) For turboprops, to account for a sudden malfunction (e.g. quick feathering), the simultaneous application of max T-O thrust, torque, and 1*g* load, multiplied by 1.6.
(4) For turbine engine installations, torque due to sudden engine stoppage (such as compressor jamming) or maximum acceleration of the engine.
(5) For all engine types, account for a lateral loading by multiplying engine weight by $n_1/3$, where n_1 is the limit load factor, with a minimum value of 1.33.

When determining the torque in (1) and (2), the appropriate mean torque must be multiplied by the following factors of safety:

- Turboprops multiply by 1.25.
- For piston engines with five or more cylinders, multiply by 1.33.
- For piston engines with fewer than five cylinders, multiply by $(6 - N_{cylinder})$.

These loads are applied at the CG of the engine, except propeller thrust, which is applied at the hub, as is shown in Figure 7-16.

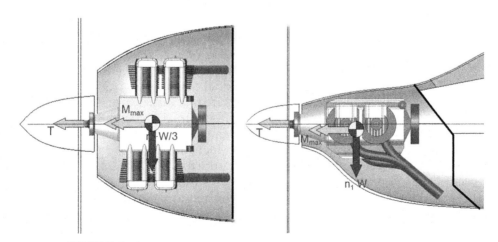

FIGURE 7-16 Application of engine loads to the CG and the propeller hub.

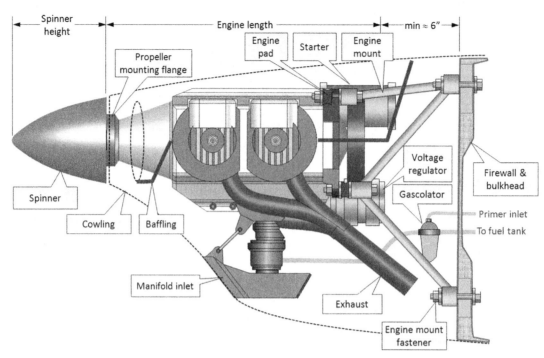

FIGURE 7-17 A typical piston engine installation.

Systems Integration

Figure 7-17 shows a typical installation of a small piston engine and identifies a number of different systems required for proper operation of the engine. This means that a number of perforations have to be made through the firewall. For typical piston engine installations, provisions have to be made for the following instruments:

(1) Oil pressure gage.
(2) Oil temperature gage.
(3) Tachometer (RPM indicator).
(4) Manifold pressure gage (MAP – often omitted for low-powered engines).
(5) Fuel tank quantity gages.
(6) Fuel flow indicator (omitted for low-performance aircraft).
(7) Hobbs indicator (shows the number of hours on the engine).

These call for electrical connectors and instruments to be mounted on either side of the firewall. Additionally, the following electrical and fuel related equipment must be accounted for:

(1) Starter and ignition switch wiring.
(2) Battery, which is often inside the engine compartment, unless it serves a secondary purpose as ballast.
(3) Voltage regulator.
(4) Primer inlet and fuel lines.
(5) Mixture control.
(6) Throttle control.

(7) Carburetor heat control, unless the engine features fuel-injection technology.

Types of Engine Mounts

There are three common means of mounting a piston engine to the airplane. *Dynafocal mounts* arrange the fastener pattern such that the fasteners point toward the CG of the engine. This reduces engine vibration, but requires the engine mount and motor pads to be welded at an angle, making their fabrication harder. *Conical mounts* align the fasteners parallel to the crankshaft, while *bed mounts* align the fasteners perpendicular to the crankshaft (see Figure 7-18).

Fuel System

A typical fuel system layout for a low-wing high-performance piston-engine airplane is shown in Figure 7-19. Normally, there are two fuel tanks, one in each wing. The fuel is gravity-fed from each tank into special tanks called *collector tanks*. Their purpose is to ensure that maneuvering the airplane will not result in a drop in fuel pressure that might interrupt the engine operation. Float-type sensors in each tank detect remaining fuel quantity.

Since the collector tank is below the engine, it must be pumped to the engine's injector manifold, where it is delivered to individual cylinders. Two fuel pumps are used for this purpose – one is driven directly by the engine and the other is electric and is used when starting the engine and during critical operations, such as T-O

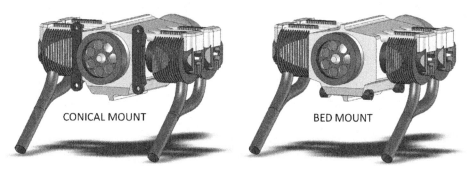

FIGURE 7-18 Conical and bed mounts.

and landing. It is also used for vapor suppression during climb and is left on for up to 30 minutes once the plane levels off for cruise. The pumps draw fuel from the collector tank selected by the pilot. Excess fuel not used in the manifold is returned to the selected fuel tank.

The operation of the fuel system depends on proper venting. If venting were not provided, the pressure in the tank would reduce as the fuel is consumed. Eventually this would lead to decreased flow of fuel and an inevitable engine fuel starvation and stoppage. The fuel vents ensure ambient pressure is provided no matter the remaining fuel quantity. This is often done by exposing a vent line to a stagnation pressure port. This will pressurize the tank and help with suitable fuel flow.

7.3.2 Piston Engine Inlet and Exit Sizing

The size of the inlet for a piston engine is primarily based on the cooling requirements of the engine and relies on allowing air into and out of the engine compartment. The nature of this flow is very complex. Not only is the airflow highly turbulent, it is compounded by imperfect pressure recovery (i.e. how much of the dynamic pressure gets converted to stagnation pressure), heat transfer, and pressure drop as it is forced through the radiator (or cylinder cooling fins). For this reason, the designer should be realistic about the accuracy of the sizing methods presented in here; they are not accurate. However, they offer (1) a way to establish a baseline inlet and (2) an understanding of what parameters affect the cooling. The final inlet and exit dimensions will have to be polished in flight testing. Examples of typical inlets are shown in Figure 7-20.

Adequate cooling requires a steady flow of air through the engine compartment. For this reason, in addition to an inlet, an adequately sized exit must be provided to ensure a steady and continuous supply of low-temperature air through the compartment. But there also has to be a balance between the inlet and exit areas. If the exit size is smaller than the inlet, the volume of air flowing through the compartment will also be low — most of it will simply flow around the nacelle and cooling capacity will suffer. If the exit is larger than the inlet, only

air that amounts to the air coming through the inlet will exit, but there may also be backflow of warm air into the engine compartment. This can cause uneven heating and even overheating in areas.

Common ways to cool engines in both tractor and pusher aircraft configurations are shown in Figure 7-21 and Figure 7-22. The cooling method introduces air through a front-facing inlet and it is then directed around the engine cylinders using thin panels called *baffles*. The purpose of baffles is to ensure that air entering the engine compartment flows around the exposed cylinder cooling fins by blocking other possible paths.

The cooling air is introduced to the engine using either what is called an *updraft* or a *downdraft* methodology. The difference between the two is displayed in Figure 7-21 and Figure 7-22. Note that there is a serious drawback to the updraft cooling method for single-engine tractor configurations. Engine failure is often accompanied by oil leaks and, considering the configuration shown, oil could easily cover the windscreen, challenging the safe operation of the aircraft. However, it may be the right configuration for multiengine aircraft with the engines in on-wing nacelles.

The cooling of pusher configurations can be problematic, as the cooling air must often be introduced through curved ducts besides having to ingest air with reduced energy. The designer should anticipate this possibility up front and consider scoop-type inlets for maximum pressure recovery.

The cooling takes place as air flows across the surface and around the fins of the cylinders or the heat exchanger (radiator). This carries away thermal energy in the thermodynamic process of convection. The greater the mass flow the greater is the cooling. Proper cooling depends on airspeed, ambient temperature, density, and engine power (power output, fuel consumption, and operating temperature).

Consider the three inlets in Figure 7-23. The top figure shows the engine without a cowling, which from a certain point of view can be regarded as the largest possible "inlet". This inlet is way too big and results in high drag without any additional cooling effectiveness. The center inlet is tiny and would result in engine

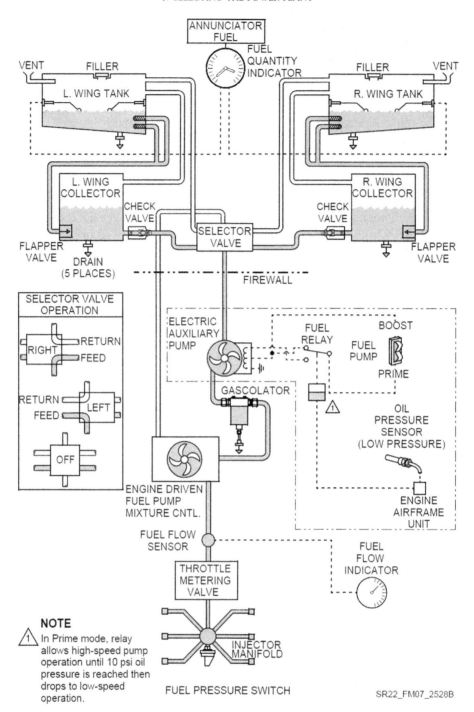

FIGURE 7-19 The fuel system for a single-piston-engine high-performance aircraft. *(Courtesy of Cirrus Aircraft)*

overheating because of the poor pressure recovery and momentum losses inside. The bottom inlet is of the correct (and compromised) size. It provides adequate cooling at all operating conditions and while it generates more drag than the center inlet this is acceptable because of its cooling capability.

This shows that the size of the inlet can have a wide range without impacting the cooling capability of the heat exchanger. This is because the cooling is dependent on the area of the radiator, and this does not change with the larger inlet size. A large inlet simply increases the drag of the airplane and for that reason the smallest inlet that provides adequate cooling over the operational range of the aircraft is the right one. The range is limited at the lower end by momentum losses inside the diffuser.

Note that a proper sizing of the cooling air exits should account for the presence of other heat exchangers, such as oil coolers, intercoolers, and cabin air heaters.

7.3 AIRCRAFT POWER PLANT INSTALLATION 215

FIGURE 7-20 Engine inlets for selected aircraft. *(Photos by Phil Rademacher)*

Method 1: Inlet-exit-dependent Heat Transfer

This method assumes that the inlet area must be equal to the exit area required to carry the engine heat away. The heat carried away by the air flowing through the engine compartment can be found from:

$$\dot{Q} = \dot{m} \cdot C_p \cdot \Delta T \tag{7-37}$$

where

$\dot{m}$ = mass flow of air flowing through the engine compartment

C_p = specific heat of pressure for air and

ΔT = rise in temperature as it flows through the radiator or the cylinder fins

For air, the value of C_p is as follows: $C_p = 1000 \, \frac{J}{kg \cdot K}$
The mass flow rate is given by

$$\dot{Q} = \dot{m} \cdot C_p \cdot \Delta T = \rho V_E A_E \cdot C_p \cdot \Delta T \tag{7-38}$$

where

V_E = airspeed at the exit of the engine compartment
A_E = exit area

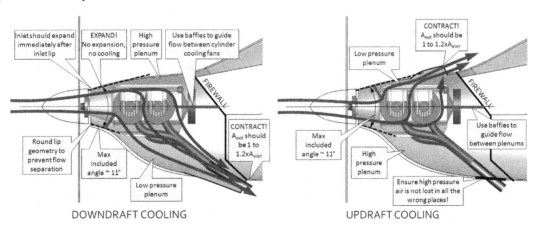

FIGURE 7-21 Airflow through a conventional tractor engine installation.

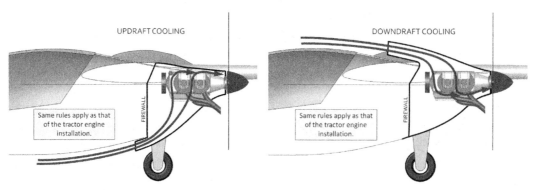

FIGURE 7-22 Airflow through a conventional pusher engine installation.

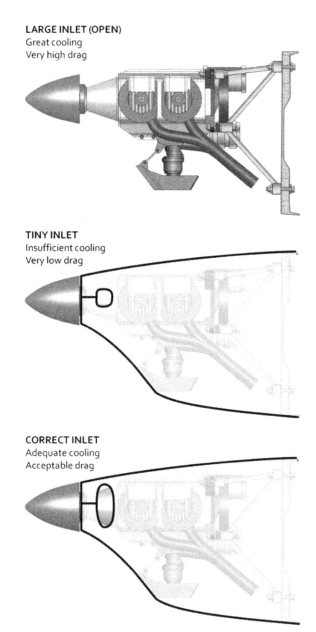

FIGURE 7-23 Extreme inlet sizing.

This can be used to solve for A_E, which then can be used as the inlet area, A_I:

$$A_E = A_I = \frac{\dot{Q}}{\rho V_E \cdot C_p \cdot \Delta T} \qquad (7\text{-}39)$$

The resulting area will be highly dependent on the selection of the exit airspeed, V_E. Therefore, the engineer should evaluate area requirements during climb (low speed, full power), approach (low speed, low power), and cruise (high speed, moderate power) on hot days and use these to evaluate the area. The fact that it takes the engine some time to overheat is sometimes used to get by with smaller inlets and outlets. Some airplanes have insufficient cooling capability during climb and depend on the airplane reaching higher (and cooler) altitudes quickly so cruise can commence, where cooling is usually sufficient.

Exit Area and Cowl Flaps

Equation (7-39) shows the exit area is a function of the airspeed. When the airspeed is high, such as at cruise, the required exit area is small. When the airspeed is low, a larger area is required. It is to be expected that a small area that generates relatively low cooling drag at cruise may lead to cooling problems at lower airspeeds, such as before landing. This makes it tricky to determine an area that works for both conditions. Sometimes this is impossible to accomplish and potential for over-heating must be solved using a so-called *cowl-flap*, a mechanical device that is used to adjust the exit area on a need-to basis.

Many high-performance piston-engine aircraft feature such flaps. Normally, the cowl-flap is open for low-speed operations (T-O, climb, landing) and closed during cruise. Cowl-flaps are also useful when operating the engine in very cold climate, when over-cooling the engine is a realistic possibility. In such conditions the cowl flap is kept closed, ensuring the engine runs at ideal cylinder-head temperature.

EXAMPLE 7-5

The installation manual for a Rotax 912 four-stroke aircraft piston engine recommends a radiator capable of transferring thermal energy that amounts to 28 kW. If air warms up by 50 K as it flows through the radiator, size the exit area as a function of the airspeed through it in m/s. How large must the exit area be for a cruising speed of 50 m/s and during climb at 30 m/s?

Solution

The heat carried away by air can be found from:

$$\dot{Q} = \dot{m} \cdot C_p \cdot \Delta T = 28{,}000 \ W$$

Therefore we can write:

$$\dot{Q} = \dot{m} \cdot C_p \cdot \Delta T = \rho V_E A_E \cdot C_p \cdot \Delta T$$

Solve for A_E:

$$A_E = \frac{\dot{Q}}{\rho V_E \cdot C_p \cdot \Delta T} = \frac{28{,}000}{(1.223) V_E (1000)(50)} = \frac{0.4579}{V_E} \ m^2$$

At airspeed of 50 m/s the exit area must be 0.00916 m² (0.0986 ft² or 14.2 in²)

At airspeed of 30 m/s the exit area must be 0.0153 m² (0.164 ft² or 23.7 in²)

Method 2: Inlet-radiator-exit Method

McCormick [19] presents the following method, derived from recommendations by Lycoming for estimating cooling drag for piston engines. The method is also helpful when sizing their inlet and exit area as well. Its primary drawback is that it requires data from the engine manufacturer, which, unfortunately are not always available. The piston engine (or a heat exchanger, such as a radiator) can be analyzed based on the idealized configuration shown in Figure 7-24.

The method idealizes the engine by considering it as a system of *inlet-radiator-exit*. Then, it is assumed that adiabatic compression and expansion of air takes place inside the inlet and exit, shown in Figure 7-25, which themselves are idealized as a diffuser and nozzle, respectively.

The method assumes that air entering the diffuser will slow down gradually to a minimum value at the forward face of the radiator. The effectiveness of the cooling requires a pressure differential to exist across the radiator (i.e. between its forward and aft faces). This pressure differential is referred to as a "pressure drop" and is imperative as it forces air to flow through the radiator and carry away thermal energy. The process can be improved by greater slowing of air in the inlet, i.e. by a greater recovery of the far-field dynamic pressure. A common pressure recovery in airplane piston engine cowling inlets is of the order of 60–80%, based on airspeed. For carefully designed scoops, it can be as large as 80–90% depending on airspeed.

Figure 7-26 shows the model with the flow properties identified at each important station. The flow conditions can be estimated to the front face of the radiator using the flow conditions upstream, assuming an adiabatic compression. Similarly, it is possible to estimate the conditions up to the aft face of the radiator using the downstream conditions, assuming an adiabatic expansion. It remains to tie the two together, something that requires the change across the radiator to be known (information the engine manufacturer should be able to provide). The temperature and pressure in an adiabatic compression or expansion can be estimated using the isentropic flow relations:

$$\frac{p}{p_0} = \left(\frac{T}{T_0}\right)^{\frac{\gamma}{\gamma-1}} \Rightarrow \frac{p}{p_0} = \frac{p_0 + kq}{p_0} = \left(\frac{T}{T_0}\right)^{\frac{\gamma}{\gamma-1}} \quad (7\text{-}40)$$

where

p and p_0 = pressure at condition and reference, respectively

T and T_0 = temperature at condition and reference, respectively

k = pressure recovery coefficient (1 = complete recovery, 0.5 = 50% recovery, etc.)

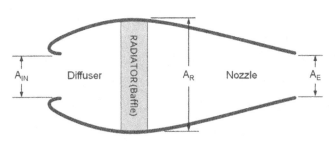

FIGURE 7-24 Idealization of a conventional engine installation.

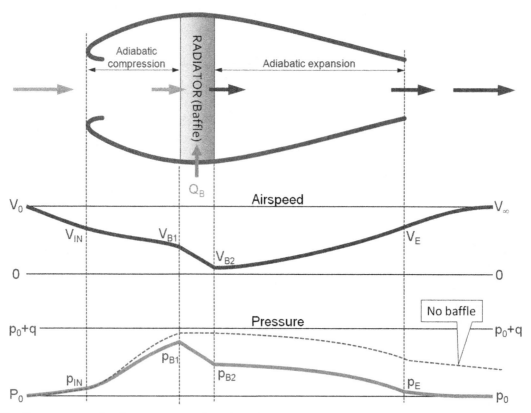

FIGURE 7-25 Changes in speed and pressure as air flows through engine installation (drop and rise in airspeed and pressure are not to scale).

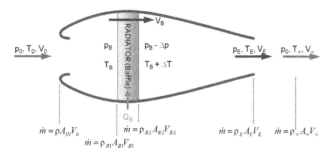

FIGURE 7-26 Flow requirements that must be met.

The factor k indicates how much of the dynamic pressure is preserved as the speed of the airflow is slowed down and is an indicator of the efficiency of the diffuser. If $k = 0$, there is no recovery and the total pressure remains that of the ambient pressure in the far-field. If $k = 1$, there is 100% recovery and all the dynamic pressure is converted into total air pressure without any losses. This is generally highly desirable.

A_{IN} = inlet area	V_0 = far-field airspeed
A_B = reference area of the baffle (radiator)	V_{B1} = airspeed in front of the baffle
A_E = exit area	V_{B2} = airspeed aft of the baffle
$\dot{m}$ = mass (or weight) flow rate	V_E = airspeed at the exit
ρ = density of air	$V\infty$ = airspeed in the streamtube behind the nozzle
T_0 = far-field temperature	p_0 = far-field pressure
T_{B1} = temperature at the baffle forward face	p_{B1} = pressure at the baffle (radiator) forward face
T_{B2} = temperature at the baffle aft face	p_{B2} = pressure at the baffle aft face
T_E = temperature at the exit	p_E = pressure at the exit
ΔT = temperature increase through the baffle	Δp = pressure drop through the baffle
T_∞ = temperature in the streamtube behind the nozzle	p_0 = far-field pressure
Q_B = heat flow into heat exchanger	

With the pressure known, the airspeed at condition can be determined using the compressible Bernoulli equation:

$$V = \sqrt{V_0^2 + \frac{2\gamma}{\gamma - 1}\left(\frac{p_0}{\rho_0} - \frac{p}{\rho}\right)} \quad (7\text{-}41)$$

Determining the pressure drop through the radiator is not a simple task. Usually this is done by empirical methods (read as trial and error). However, the pressure drop across baffle depends on:

$$\Delta p \propto \frac{1}{2}\rho V_{B1}^2 \quad (7\text{-}42)$$

where

V_{B1} = airspeed at the forward face of the radiator.

The mass flow through the radiator is given by:

$$\dot{m} = \rho A_B V_B \quad (7\text{-}43)$$

This implies that the drop in pressure is related to the mass flow rate and density as described by the following relationship, which is derived by inserting Equation (7-43) into (7-42):

$$\Delta p \propto \frac{\dot{m}^2}{\rho} \quad (7\text{-}44)$$

EXAMPLE 7-6

A piston engine is being operated at 10,000 ft and airspeed of 185 KTAS where it delivers 230 BHP. OAT is 30 °F above standard temperature. The manufacturer recommends a constant cylinder-head temperature (CHT) of 450 °F for maximum engine life. Size the inlet and exit area assuming 75% pressure recovery at the radiator and that air temperature rises by 150 °F across the cylinders. Estimate how much engine power is lost to cooling.

Solution

This problem assumes a typical piston engine installation to estimate the required exit area based on hypothetical flight condition. The pressure drop through the baffle is based on experimental measurements, and ultimately requires pressure, temperature, and airspeed to be evaluated at four stations through the inlet, radiator, and nozzle. These are denoted as stations 0, B_1, B_2, and E in Figure 7-27. Also, note that the problem was solved using a calculator with double-floating-point accuracy. Therefore, following along with a calculator and rounding off will yield slightly different numbers in the third or fourth significant digit.

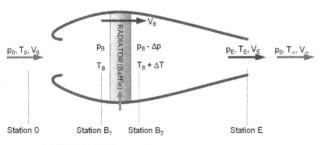

FIGURE 7-27 Definition of stations of interest.

Step 1: Determine Conditions at Station 0

Calculate the far-field temperature using the information given in the problem:

$$T_0 = 518.69(1 - 0.0000068756 \times 10,000) + 30 = 513.0°R$$

Calculate pressure at altitude using the hydrostatic gas equation:

$$p_0 = 2116(1 - 0.0000068756 \times 10,000)^{5.2561} = 1455 \text{ psf}$$

Calculate density using the ideal gas equation:

$$\rho = \frac{p_0}{RT_0} = \frac{1455}{(1716)(513.0)} = 0.001653 \text{ slugs/ft}^3$$

Airspeed:

$$V_0 = 185 \times 1.688 = 312.3 \text{ ft/s}$$

We have thus completely defined p, T, and V in the far-field (station 0).

Step 2: Determine Conditions at Station B_1

Determine the temperature at the radiator front face using the adiabatic gas relation, assuming that adiabatic expansion takes place inside the diffuser. This allows us to determine the flow characteristics at the forward face based on the flight conditions already determined at station 0 using the adiabatic relations of Equation (7-40). Therefore, the temperature at the baffle is:

$$\frac{p_0 + kq}{p_0} = \left(\frac{T_{B1}}{T_0}\right)^{\frac{\gamma}{\gamma-1}} \Rightarrow T_{B1} = T_0\left(\frac{p_0 + kq}{p_0}\right)^{\frac{\gamma-1}{\gamma}}$$

EXAMPLE 7-6 (cont'd)

Where k is the pressure recovery factor. With that said, let's calculate the dynamic pressure in the far-field to evaluate the impact of the pressure recovery:

$$q = \frac{1}{2}(0.001653)(312.3)^2 = 80.60 \text{ psf}$$

Inserting values (where k = 0.75 for 75% pressure recovery):

$$T_{B1} = T_0\left(\frac{p_0 + kq}{p_0}\right)^{\frac{\gamma-1}{\gamma}} = (513.0)\left(\frac{1455 + 0.75(80.60)}{1455}\right)^{\frac{0.4}{1.4}}$$
$$= 519.0°R$$

This corresponds to approximately 59.4 °F. Pressure at the baffle assuming 75% pressure recovery:

$$p_{B1} = p_0 + kq = 1455 + 0.75 \times 80.60 = 1516 \text{ psf}$$

Calculate density using the ideal gas equation:

$$\rho = \frac{p_{B1}}{RT_{B1}} = \frac{1516}{(1716)(519.0)} = 0.001702 \text{ slugs/ft}^3$$

The airspeed at the baffle can now be calculated from the compressible Bernoulli equation:

$$V_{B1} = \sqrt{V_0^2 + \frac{2\gamma}{\gamma-1}\left(\frac{p_0}{\rho_0} - \frac{p_{B1}}{\rho_{B1}}\right)} = 159.5 \text{ ft/s}$$

We have thus completely defined p, T, and V in the far-field (station B1).

Step 3: Determine Conditions at Station B2

This step relies on information that must be obtained from the engine manufacturer and this typically consists of two graphs similar to the ones shown in Figure 7-28. It is used to determine the conditions on the downstream side of the radiator. Since the temperature and pressure on the upstream side of the radiator have been determined (Step 2), the pressure altitude must be determined because the graphs of Figure 7-28 depend on this value. The pressure altitude this corresponds to can be estimated using Equation (16-8):

$$H_P = 145442\left(1 - \left(\frac{p}{p_0}\right)^{0.19026}\right)$$
$$= 145442\left(1 - \left(\frac{1516}{2116}\right)^{0.19026}\right) = 8947 \text{ ft}$$

Next, determine the pressure loss through the baffle using the left graph of Figure 7-28. Locate 59.4 °F on the horizontal axis and move along Arrow ① to the isobar designated for a CHT = 450 °F. Then move along Arrow ② to read approximately 2.7 lb_f/s of required cooling airflow at this condition. This means that 2.7 lb_f of air must flow through the radiator every second to adequately cool the engine. Then, extend the arrow to locate 8947 ft on the right graph (between the 5000 and 10,000 ft curves). Finally, follow Arrow ③ to locate 35 lb_f/ft² as the pressure drop across the radiator.

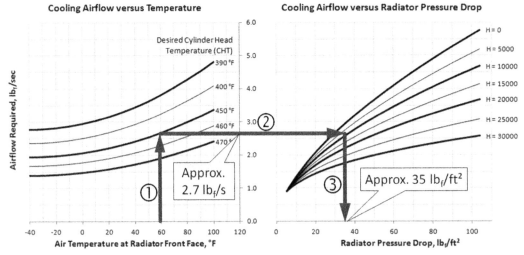

FIGURE 7-28 Special graphs supplied by the engine manufacturer are used to extract the required cooling airflow for the engine. The graphs do not represent any particular engine type.

EXAMPLE 7-6 (cont'd)

Required cooling airflow:

$$\dot{m} = 2.7 \text{ lb}_f/s$$

Resulting pressure drop:

$$\Delta p = 35 \text{ psf}$$

Pressure downstream of the baffle:

$$p_{B2} = p_{B1} - \Delta p = 1516 - 35 = 1481 \text{ psf}$$

Temperature (+150 °F) rise downstream of the baffle:

$$T_{B2} = T_{B1} + \Delta T = 519.0 + 150.0 = 669.0 \text{ °R}$$

It is assumed the speed of air is near zero as it exits the aft face of the radiator. This is based on the speed already being accounted for in the value of the pressure loss, which was empirically determined (by the engine manufacturer). Therefore we say that $V_{B2} = 0$ ft/s and, thus, claim we have completely defined p, T, and V in the far-field (station B1).

Step 4: Determine Conditions at Station E

We will determine some properties at the exit assuming an adiabatic expansion. This will require us to know an additional property at the downstream face of the baffle, namely the density:

$$\rho_{B2} = \frac{p_{B2}}{RT_{B2}} = \frac{1481}{(1716)(669.0)} = 0.001290 \text{ slugs/ft}^3$$

The pressure at the exit is assumed to be the atmospheric pressure in the far-field. The reader should be aware that this is not necessarily true. The pressure depends also on whether or not there is flow separation occurring at the exit, which would lower the pressure and affect our results. However, this is a reasonable first stab assumption as long as we are cognizant of its limitations and until we are able to actually measure it:

$$p_E = p_0$$

The density at the exit can be found from the adiabatic relation:

$$\frac{p_E}{p_{B2}} = \left(\frac{\rho_E}{\rho_{B2}}\right)^\gamma \Rightarrow \rho_E = \rho_{B2}\left(\frac{p_E}{p_{B2}}\right)^{\frac{1}{\gamma}} \Rightarrow \rho_E$$

$$= 0.001290\left(\frac{1455}{1481}\right)^{\frac{1}{\gamma}} = 0.001274 \text{ slugs/ft}^3$$

And the airspeed at the exit can be found using the compressible Bernoulli equation:

$$\frac{V^2}{2} + \frac{\gamma}{\gamma - 1}\frac{p}{\rho} = \text{constant}$$

Inserting the variables in this example leads to:

$$\frac{V_{B2}^2}{2} + \frac{\gamma}{\gamma - 1}\frac{p_{B2}}{\rho_{B2}} = \frac{V_E^2}{2} + \frac{\gamma}{\gamma - 1}\frac{p_E}{\rho_E}$$

Assuming speed V_{B2} through the baffle to be small:

$$0 + \frac{\gamma}{\gamma - 1}\frac{p_{B2}}{\rho_{B2}} = \frac{V_E^2}{2} + \frac{\gamma}{\gamma - 1}\frac{p_E}{\rho_E}$$

$$\Rightarrow V_E = \sqrt{\frac{2\gamma}{\gamma - 1}\left(\frac{p_{B2}}{\rho_{B2}} - \frac{p_E}{\rho_E}\right)}$$

The airspeed at the exit comes to:

$$V_E = \sqrt{\frac{2\gamma}{\gamma - 1}\left(\frac{p_{B2}}{\rho_{B2}} - \frac{p_E}{\rho_E}\right)}$$

$$= \sqrt{\frac{2(1.4)}{1.4 - 1}\left(\frac{1487}{0.001295} - \frac{1455}{0.001275}\right)} = 199.3 \text{ ft/s}$$

Temperature at the exit:

$$T_E = T_{B2}\left(\frac{p_E}{p_{B2}}\right)^{\frac{\gamma-1}{\gamma}} = (669.0)\left(\frac{1455}{1481}\right)^{\frac{0.4}{1.4}} = 665.7 \text{ °R}$$

Now we have enough information to size the inlet area:

$$\dot{m} = \rho_0 V_0 A_{IN} \Rightarrow A_{IN} = \frac{\dot{m}}{\rho_0 V_0} \Rightarrow A_{IN}$$

$$= \frac{2.7/32.174}{(0.001653)(312.3)} = 0.163 \text{ ft}^2$$

And the outlet area:

$$\dot{m} = \rho_E V_E A_E \Rightarrow A_E = \frac{\dot{m}}{\rho_E V_E}$$

$$\Rightarrow A_E = \frac{2.7/32.174}{(0.001274)(199.3)} = 0.331 \text{ ft}^2$$

An area corresponding to about 5×5 inches^2 will suffice for an inlet and 6×6 inches^2 for the exit at this airspeed. It can be seen that this particular flight condition will not dictate the final size of the inlet or exit and other flight conditions, in particular ones involving low airspeed, must be evaluated in this fashion as well. The results are shown graphically in Figure 7-29.

In this example it is assumed air temperature rises by 150 °F across the cylinders. Let's see how

EXAMPLE 7-6 (cont'd)

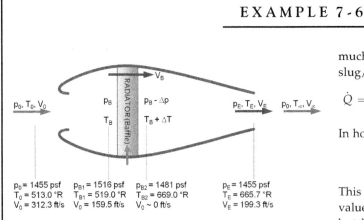

FIGURE 7-29 All properties are defined at the four stations of interest.

much power this corresponds to (C_p units are ft/lb$_f$/slug/°R):

$$\dot{Q} = \dot{m}c_p(\Delta T) = \left(\frac{2.7}{32.174}\right)(6000)(150) = 75527 \text{ ft·lb}_f/\text{s}$$

In horsepower:

$$\dot{Q} = \frac{75527}{550} = 137.3 \text{ BHP}$$

This is about 57% of the engine power. Generally, this value should range between 40% and 50% (per Ref. 19), but here it is actually based on graphs that are simple examples that do not represent an actual engine type.

7.3.3 Installation of Gas Turbines

Due to the complexity of gas turbine installation, many design details are accomplished with direct help from the engine manufacturer. Polishing the inlet and exhaust design, fuel system layout, and bleed air and other systems requires expertise and experience only they wield. The following discussion should therefore only be considered introductory.

The relatively small diameter of gas turbines allows them to be mounted on the aft fuselage or on pylons on the fuselage or wings in a podded configuration. However, they are also buried inside the wing or in the fuselage. There are usually good reasons to justify each type of installation, but the buried installation poses more serious challenges in the case of a fire. A fire in a podded engine is likely to burn itself out without causing damage to the engine mounts or the nearby airframe. A buried engine, on the other hand, may present serious risks to the surrounding airframe and, thus, requires reliable fireproofing. Even for small aircraft the fireproofing can weigh in excess of 100 lb$_f$, something easy to overlook during the design phase, but can easily shorten range by 50 to 200 nautical miles (nm).

Installation of Jet Engines

Installation of jet engines must comply with largely the same requirements as piston engines, with some exceptions. Jet engines are installed either internally (inside a fuselage) or externally (on a pylon). However, in either case, identical connectors and lines are required. Figure 7-30 shows a schematic of a typical jet engine installation. Internal and external engine installations are also equipped with fire-suppressant systems, which increase the weight further. Additionally, if the aircraft is to be certified for flight into known icing (FIKI), the leading edge of the inlet (usually called inlet lip) must feature anti-ice capability, compounding the complexity.

Installation of Turboprop Engines

Turboprops typically follow a similar process to the piston engine. They are substantially lighter and, while longer than piston engines of comparable power, their girth (diameter) is more compact. For instance, the 400 BHP Lycoming IO-720 weighs about 600 lb$_f$, whereas the 600–1100 SHP Pratt & Whitney Canada PT6 weighs about 350–400 lb$_f$. For this reason, if a piston engine is to be replaced with a turboprop, the installation will require the propeller to be mounted farther forward, giving the airplane a very distinct appearance. Consequently, all forces acting on the propeller result in higher moments, which call for greater control surface authority to react.

Turbo Machinery and Rotor-burst

The compressor and turbine in a typical gas turbine rotate at very high rates when compared to a piston engine. The typical gas turbine rotates at 20–40 thousand RPM. This means the compressor or turbine blades react a substantial centripetal force and, if one breaks off, two things will happen:

(1) The fragments turn into lethal projectiles and can cut through a stout structure like a hot knife through butter (except a tad faster).
(2) The support structure will have to react substantial oscillatory loading due to the imbalanced compressor or turbine rotors. Albeit rare, such events happen from time to time in aircraft and are called *rotor-bursts*.

It is an extremely dangerous event that must be accounted for in the detail design of the airplane.

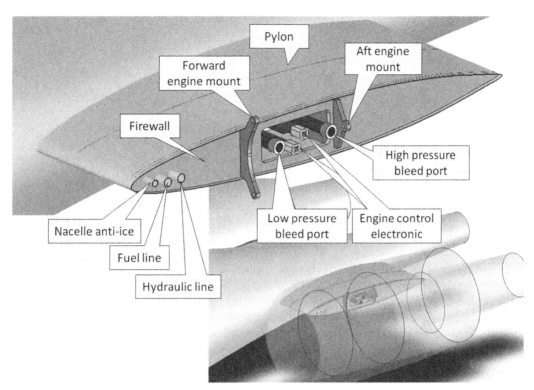

FIGURE 7-30 A typical external jet engine installation.

7.3.4 Jet Engine Inlet Sizing

Since this book is aimed at GA aircraft design, only subsonic inlets will be presented. Subsonic inlets are generally relatively short compared to their supersonic counterparts and their geometry is that of a diffuser (the cross-section at the inlet lip is smaller than at the front face of the compressor). The purpose of the inlet is to allow air to be brought smoothly to the compressor and to slow its speed to bring up the pressure inside the inlet. At rest, the inlet is designed to allow the required mass flow of air to enter the engine to ensure the maximum rated thrust can be generated. In flight, the inlet is designed to slow air from the far-field airspeed to the most efficient inlet airspeed for the selected jet engine. For instance, a jet aircraft cruising at a high subsonic Mach number (around Mach 0.8 or so) may require the inlet airspeed to be slowed down to the Mach 0.4 to 0.5 range at the front of the compressor.

It is paramount that pressure losses are kept to a minimum in the inlet because of the magnifying effect of the compressor. The pressure in the combustor is reduced by an amount that corresponds to the inlet pressure loss multiplied by the compressor pressure ratio. Thus, a pressure drop of 1 psia in the inlet can cause up to a 25 psia drop in the combustor [20, p. 209].

Losses can stem from several sources, external and internal. For instance, the formation of shocks outside the inlet can affect the total pressure upstream, in addition to increasing the aerodynamic drag of the engine. This can happen at high subsonic airspeeds if the nacelle has a sharp leading edge or when air flows through too narrow a gap between the fuselage and nacelle on a podded engine configuration. Flow inside the inlet can experience unfavorable pressure gradients that can cause flow separation on the inside walls, which can have the following effects:

(1) Flow no longer slows down isentropically, which leads to lower total pressure inside the inlet.
(2) The separated region along the inside wall narrows the effective cross-sectional area, which leads to higher than desired airspeed (and thus lower total pressure) at the front face of the compressor.
(3) The separated flow is inherently unstable and looses "smoothness" as it enters the compressor, reducing its efficiency and stability of operation.

Inlet Types for Jet Engines

Two types of installation methodologies are used for jet engines; *external* and *internal* (see Figure 7-31). The external inlet is also called a *pitot* inlet. For internal inlets, there are predominantly two main subtypes: the *NACA inlet* and the *bifurcated* inlet. These inlets have the following pros and cons.

From an operational standpoint, the external (or pitot) inlet offers great pressure recovery for the entire range of *AOA* and *AOY* the aircraft is likely to see in practice. It is easier to access for maintenance and engine

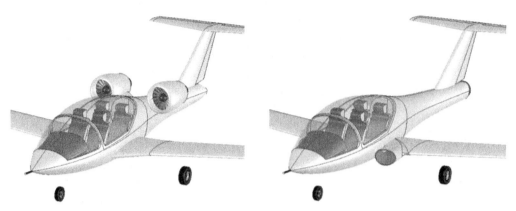

FIGURE 7-31 An external (left) and integrated (right) jet engine installation. The external one is also called a *pitot* inlet. The integrated inlet is of a *bifurcated* type.

removal and replacement. However, it also generates more aerodynamic drag due to the nacelle and pylon.

The internal bifurcated inlet is a good solution for single-engine installations, allowing the engine to be placed behind the cabin and be positioned so thrust effects on stability and control are minimized. This tends to result in less aerodynamic drag since the engine installation adds much less wetted area than the external one. Of course placing the fuselage in front of the inlet will cause boundary layer growth to be ingested, unless steps are taken to avoid it, such as the use of boundary layer diverters. This can be problematic if the airplane yaws and separated air enters the inlet on one side. Other problems are pressure losses due to bending of the internal inlet, as well as problematic ice accretion that may occur on the forward-facing curves of the inlet, which requires anti-ice remedies for aircraft certified for FIKI.

The internal NACA inlet generates even less aerodynamic drag than the bifurcated inlet and can be the right solution for specific single-engine installations. It more often sees use as a generic low-drag inlet for engine or component cooling than as a jet engine inlet, because it suffers from much greater pressure losses than its bifurcated counterpart. Since turbofan engines are more sensitive to inlet pressure losses, it can be argued that the NACA inlet is better suited for turbojet applications and should be avoided for turbofans.

The Diffuser Inlet

The process of slowing air down with minimal losses in pressure recovery is most effectively accomplished using adiabatic expansion. As stated earlier, this is imperative, as any reduction in the total pressure at the front face of the compressor will lead to reduced thrust. The most efficient inlet geometry for this purpose is the diffuser (see Figure 7-32). A diffuser has a capture area slightly smaller than the compressor disc area.

The airspeed of an airplane has a profound effect on the nature of how a jet engine ingests air. Two extremes are shown in Figure 7-33. At rest, the jet engine will

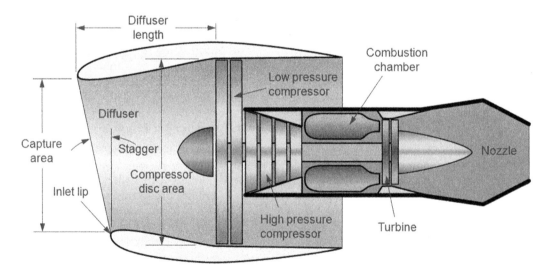

FIGURE 7-32 A typical external jet engine installation features an inlet designed to slow airspeed from the far-field airspeed to some target airspeed.

7.3 AIRCRAFT POWER PLANT INSTALLATION

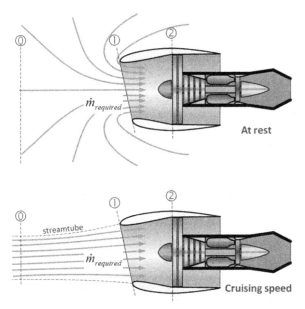

FIGURE 7-33 Shape of the flow field entering a jet engine at rest (top) and at cruise (bottom).

effectively ingest air from around and in front of the inlet, a very important fact for technicians to keep in mind while working on the ground near such machines. The inflow can be tremendously powerful and, sadly, every year an unfortunate mishap takes place in which a person is sucked into a large jet engine with fatal consequences. It also shows the engine can easily draw in FOD off the ground and sustain considerable damage.

As the speed of the airplane increases, the streamlines shown in the upper diagram become more and more aligned with the engine axis and begin to form a distinct streamtube. A distinct streamtube means that only air inside this volume will be ingested by the engine. The lower diagram of Figure 7-33 shows this at cruising speed, during which much more air than required by the engine is available.

While the design of the inlet is best implemented with the participation of the engine manufacturer, the airframe designer may want to perform preliminary sizing of the inlet to expedite the layout of the airplane. The following Step-by-step is provided for that purpose. Note that Flack [20]; Seddon and Goldsmith [21]; and Mattingly [14] provide good guidance for the general inlet design and should be referenced as well.

The inlet design involves a number of important geometric parameters to be considered. For analysis reasons, three stations of interest have been superimposed on the engines of Figure 7-33, denoted by ⓪, ①, and ②. The inlet sizing requires the flow conditions and geometry to be known at each station and these are pressure, density, temperature, airspeed, and cross-sectional area. Station ⓪ represents the far-field. In this way, at rest, the far-field area is infinite in size

and the Mach number is thus assumed zero. However, at cruise, the area in the far-field is substantially smaller and is less than the area at the inlet lip, denoted by ①. Generally, the following procedure can be used to determine the inlet lip geometry.

Step 1: Required Mass Flow Rate

Obtain the maximum mass flow rate required by the engine and call it $\dot{m}_{required}$. This information is usually provided by the engine manufacturer and is usually dictated by the engine performance at static T-O thrust, making it the critical flight condition for the design. This contrasts with the cruise condition, in which the airspeed is so high that the inlet stream tube necks down in the far-field as shown in Figure 7-33, indicating plenty of air can enter the engine. The remaining steps assume this to be the case.

Step 2: Determine Airspeed Limitations at the Inlet Lip and Compressor

Generally, the airspeed at the inlet lip (station ①) should not exceed Mach 0.8. Similarly, the airspeed at the front face of the compressor (station ②) is limited to Mach 0.4 to 0.5 and is ultimately specified by the engine manufacturer. This is intended to prevent the airspeed at the tip of the fan becoming too high, resulting in decreased efficiency.

Step 3: Establish Known and Unknown Flow Conditions

The knowns and unknowns can be established in a format similar to the one shown below. Question marks indicate the parameter is initially unknown. Recall that the total pressure is the sum of the ambient and dynamic pressures. Thus, if the engine is moving at some airspeed V, the total pressure is given by $p_{total} = p_{ambient} + p_{dynamic}$. However, as stated above, for this particular condition the engine is assumed to be at rest, so $p_{dynamic} = 0$. For this reason, the total pressure in the far-field is simply the ambient pressure.

Flow condition	Station ⓪ (far-field)	Station ① (inlet)	Station ② (compressor)
Area	$A_0 = \infty$	$A_1 = ?$	$A_2 = $ given
Mach number	$M_0 = 0$	$M_1 = 0.8$	$M_2 = 0.4$ to 0.5
Total pressure	$p_{T0} = p_0 = $ given	$p_{T1} = ?$	$p_{T2} = ?$
Temperature	$T_0 = $ given	$T_1 = ?$	$T_2 = ?$
Density	$\rho_0 = $ given	$\rho_1 = ?$	$\rho_2 = ?$
Pressure recovery ratio	—	—	$\pi_2 = p_{T2}/p_{T0}$

Step 4: Determine Conditions at Station ①

Even though the airspeed in the far-field is zero as the engine spools up to T-O thrust, air can easily accelerate to very high airspeeds at the inlet lip. Normally, the inlet lip area is sized so the airspeed does not exceed Mach 0.8. If it is assumed that isentropic flow relations hold between stations ⓪ and ① and that the ratio of specific heats of air is $\gamma = 1.4$, it is possible to determine the flow variables P_T, T, and ρ, to be determined as follows (using $M_1 = 0.8$):

Total pressure:

$$p_{T1} = \frac{p_{T0}}{\left(1 + \frac{\gamma-1}{2}M_1^2\right)^{\frac{\gamma}{\gamma-1}}} = \frac{p_0}{\left(1 + \frac{\gamma-1}{2}M_1^2\right)^{\frac{\gamma}{\gamma-1}}} = 0.65602 p_0 \quad (7\text{-}45)$$

Temperature:

$$T_1 = \frac{T_0}{\left(1 + \frac{\gamma-1}{2}M_1^2\right)} = 0.88652 T_0 \quad (7\text{-}46)$$

Density:

$$\rho_1 = \frac{p_{T1}}{RT_1} = \frac{0.65602 p_0}{(1716)(0.88652 T_0)} = 0.00043123 \frac{p_0}{T_0} \quad (7\text{-}47)$$

Inlet area (capture area):

$$A_1 = \frac{\dot{m}_{required}}{M_1}\sqrt{\frac{1}{\gamma p_0}\left(\frac{\rho_0}{\rho_1}\right)} = 1.25 \dot{m}_{required}\sqrt{\frac{1}{\gamma p_0}\left(\frac{\rho_0}{\rho_1}\right)} \quad (7\text{-}48)$$

Step 5: Determine Conditions at Station ②

Assuming the airspeed (M_2) at the front face of the compressor to be in the Mach 0.4 to 0.5 range, the remaining parameters are again determined using isentropic flow relations between stations ① and ②:

Total pressure:

$$p_{T2} = \frac{p_{T0}}{\left(1 + \frac{\gamma-1}{2}M_2^2\right)^{\frac{\gamma}{\gamma-1}}} = \frac{p_0}{\left(1 + \frac{\gamma-1}{2}M_2^2\right)^{\frac{\gamma}{\gamma-1}}} \quad (7\text{-}49)$$

Temperature:

$$T_2 = \frac{T_0}{\left(1 + \frac{\gamma-1}{2}M_2^2\right)} \quad (7\text{-}50)$$

Density:

$$\rho_2 = \frac{p_{T2}}{RT_2} \quad (7\text{-}51)$$

Again, it is important to remember that the above equations reflect the assumption the engine is at rest.

Pressure Recovery

For perfect inlets, the pressure recovery ratio, π_2, is 1. This will yield the maximum thrust for a given flight condition. Short inlets common to podded jet engines of the type shown in Figure 7-33 usually have π_2 close to 1. Integrated inlets (see below) have π_2 values that are often well below 1.

Inlet Lip Radius

It is imperative that the lip radius is carefully designed. A small radius can cause flow separation at high *AOA* or *AOY*. A large lip radius tends to reduce pressure distortion at high *AOA* or *AOY* and it also results in higher nacelle drag. A pressure distortion is localized deviation of the expected average pressure at the front face of the compressor, usually as less than the average pressure. This is very undesirable as it may cause oscillation in the airloads of the fan blades. Practical lip radii for subsonic aircraft range from about 6% to 10% of the inlet diameter.

Diffuser Length

The length of the diffuser is of great importance as well. A schematic of a pitot inlet is shown in Figure 7-34. The parameters of importance are the included angle, θ, the diffuser length, L, and the inlet lip diameter, D_1. The ratio L/D_1 is called the *inlet aspect ratio*. For a given aspect ratio, L/D_1, too large an included angle θ indicates the diffuser is expanding too rapidly. This promotes detrimental flow separation on the inside wall. Conversely, for a given θ, too long a diffuser also promotes flow separation. The phenomenon is detailed by Schlichting [22, pp. 222–224], who demonstrates that no matter the included angle, if the diffuser length exceeds a certain distance, separation is inevitable. Additionally, such a diffuser is bound to have a larger nacelle wetted area and, therefore, increases the aerodynamic drag of the installation in addition to being

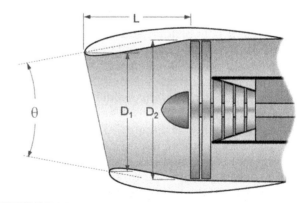

FIGURE 7-34 Dimensions for empirical diffuser length evaluation.

heavier. There is also a range of dimensions for which the formation of the flow separation is transitory; i.e. flow separation may fluctuate. The result may be compressor blade flutter and onset of early fatigue due to oscillatory loading of the blades.

As indicated by Flack [20], the optimum pressure difference between stations ① and ② should comply with:

Ideal pressure coefficient:

$$C_{p_{1\to 2}} = \frac{2(p_{T2} - p_{T1})}{\rho V_1^2} < 0.6 \qquad (7\text{-}52)$$

The pressure between the stations must be allowed to rise over a suitable distance and this usually requires more sophisticated analysis methods that can account for the intricacies of the desired geometry. However, in the absence of such schemes, a convenient empirical method for simple diffusers is provided by Flack [20] to evaluate whether the geometry will be prone to separation. Separation is unlikely if the included angle θ is less than the minimum value obtained from the following expression:

Minimum angle:

$$\ln\theta_{\min} = 3.28 - 0.46\ln\left(\frac{L}{D_1}\right) - 0.031\left(\ln\left(\frac{L}{D_1}\right)\right)^2 \qquad (7\text{-}53)$$

Conversely, separation is all but guaranteed if the included angle θ is greater than the maximum value obtained from the expression:

Maximum angle:

$$\ln\theta_{\max} = 3.39 - 0.38\ln\left(\frac{L}{D_1}\right) - 0.020\left(\ln\left(\frac{L}{D_1}\right)\right)^2 \qquad (7\text{-}54)$$

where θ is in degrees. In between the two values is the transitory separation, in which there may or may not be separation. The trends based on this formulation are plotted in Figure 7-35. As an example consider an inlet with an aspect ratio of 3. The graph shows that keeping the included angle less than 15° will prevent flow separation inside the inlet.

Stagger or Rake Angle

Stagger refers to lip geometry in which the upper lip is forward of the lower one (see Figure 7-32). The arrangement improves flow quality at higher AOAs, by reducing the airspeed at the lower lip and, thus, reduces tendency for flow separation [21]. Staggering is usually very modest in subsonic aircraft and ranges from 0° to perhaps 5°.

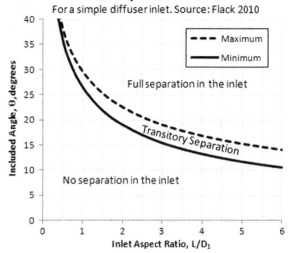

FIGURE 7-35 Flow separation trends for simple diffusers.

Derivation of Equation (7-48)

Of the set of equations, only Equation (7-48) needs to be derived. First, the required mass flow rate is given by:

$$\dot{m}_{required} = \rho_1 A_1 V_1 \qquad (i)$$

The airspeed, V_1, is Mach number multiplied by the speed of sound, i.e. $V_1 = M_1 \cdot a$, where a is the speed of sound at the inlet lip. The speed of sound can be calculated from the ideal gas expression:

$$a = \sqrt{\frac{\gamma p_0}{\rho_0}}$$

Inserting this into Equation (i) and expanding and subsequently solving for the inlet area yields:

$$\dot{m}_{required} = \rho_1 A_1 V_1 = \rho_1 A_1 (M_1 a)$$

$$= \rho_1 A_1\left(M_1\sqrt{\frac{\gamma p_0}{\rho_0}}\right) \Leftrightarrow A_1 = \frac{\dot{m}_{required}}{M_1}\sqrt{\frac{\rho_0}{\gamma p_0 \rho_1}}$$

QED

7.4 SPECIAL TOPICS

This section introduces several topics related to engine power that are of interest to the aircraft designer.

7.4.1 The Use of Gearboxes

Aircraft engines are designed to deliver maximum power at a relatively low RPM, when compared to car or snowmobile engines. For instance, Continental and Lycoming aircraft engines deliver their maximum power at approximately 2700–2800 RPM, whereas a car engine may deliver this at 5000–6000 RPM. Many airplane designers, in particular those designing homebuilt aircraft, often adapt automobile engines to their designs. These engines rotate at a very high rate so if connected directly to the selected propeller its tip speed would be supersonic, with the associated noise and propulsive losses. For this reason gearboxes are attached to such engines to bring down the rotation rate. Turboprops usually rotate at some 20,000–40,000 RPM so a gearbox is a standard part of the engine unit.

Attaching a gearbox to an engine will reduce the power delivered slightly due to its internal friction. However, in the following discussion it is assumed such losses are negligible. Consider the gears of Figure 7-36. The radius of gearwheel 1 is R_1 and that of gearwheel 2 is R_2. Assume gearwheel 1 rotates at a constant rate Ω_1 and delivers torque T_1. Then the following holds for gearwheel 2, where V_1 and V_2 are the linear speeds of a point on the perimeter of the wheels.

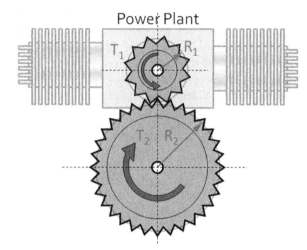

FIGURE 7-36 A schematic of a gearbox. R_1 is the gear wheel connected to the crankshaft. R_2 is the gear wheel connected to the output axle, e.g. the propeller axis.

RPM of gearwheel 2:

$$V_1 = V_2 \Rightarrow \Omega_1 R_1 = \Omega_2 R_2 \Leftrightarrow \Omega_2 = \Omega_1 \left(\frac{R_1}{R_2}\right) \quad (7\text{-}55)$$

Torque of gearwheel 2:

$$\left.\begin{array}{c} T_1 = F_1 \times R_1 \\ T_2 = F_2 \times R_2 \end{array}\right\} \Rightarrow F_1 = F_2 \Leftrightarrow T_2 = T_1\left(\frac{R_2}{R_1}\right) \quad (7\text{-}56)$$

Power of gearwheel 1:

$$P_1 = F_1 \times V_1 = \left(\frac{T_1}{R_1}\right) \times (\Omega_1 R_1) = T_1 \Omega_1 \quad (7\text{-}57)$$

Power of gearwheel 2:

$$P_2 = T_2 \Omega_2 = T_1\left(\frac{R_2}{R_1}\right) \times \Omega_1\left(\frac{R_1}{R_2}\right) = T_1 \Omega_1 = P_1 \quad (7\text{-}58)$$

From which we see that the gearwheels will change the RPM and Torque, but power remains unchanged.

EXAMPLE 7-7

A four-cylinder Rotax 912ULS engine is capable of generating some 100 BHP of power at 5800 RPM. If fitted with a 2.43:1 reduction drive, determine the reduced RPM and horsepower.

Solution

Rotation rate after gear reduction: $\quad \Omega_2 = \Omega_1\left(\frac{R_1}{R_2}\right)$

$$= 5800\left(\frac{1}{2.43}\right) = 2387 \text{ RPM}$$

From Equation (7-58), horsepower after the reduction drive is the same as before, or 100 BHP.

7.4.2 Step-by-step: Extracting Piston Power from Engine Performance Charts

Manufacturers of piston engines usually provide aircraft designers with engine performance charts similar to the one in Figure 7-37. Such charts are used to extract BHP for an engine based on its RPM and manifold pressure (*MAP*), which are parameters obtained from easily visible instruments in most piston-engine aircraft. Then, further corrections are made by accounting for the outside air temperature (OAT) at the condition. Note that the MAP is usually given in terms of inches mercury (in Hg).

Such charts are read as explained below. It is easier to follow by example, so consider the performance chart in Figure 7-37 and assume the engine is being operated at

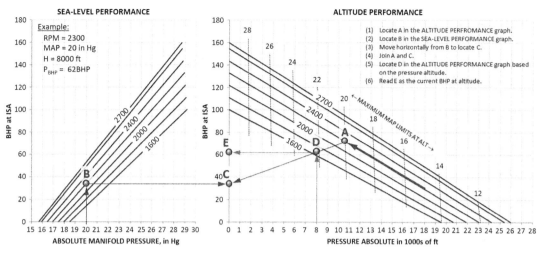

FIGURE 7-37 An example of a piston-engine performance chart for a typical certified 160 BHP aircraft engine. The chart is used to extract P_{BHP} for an engine based on its RPM and manifold pressure (*MAP*). See text for details of how to read the chart.

8000 ft at 2300 RPM and MAP of 20 inHg. Then the following steps are performed:

Step 1: Locate A in the altitude performance section of the graph by moving along the curve that indicates 2300 RPM.
Step 2: Locate B in the sea level performance graph.
Step 3: Move horizontally from B to locate C.
Step 4: Join A and C.

Step-by-step: Extracting piston power from engine performance charts, that reading such charts is not a suitable approach for analyses conducive to iteration. An analytical equation has been developed by Dr. James S. Petty of the US Air Force Wright Aeronautical Laboratories (AFWAL), which converts the performance chart into a handy equation that is easy to implement in a spreadsheet or a computer code. This equation is recognized as the *Petty equation*.

$$P_{BHP} = P_{BHP\ max} \sqrt{\frac{T_{std}}{T_{OAT}}} \frac{\left[R_m - R_f(1 - R_m)\right]\left(\sigma_{std} - R_m^{0.8097}\right) + \left(\frac{R_m^{0.8097} - 0.117}{0.883}\right)(1 - \sigma_{std})}{1 - R_m^{0.8097}} \quad (7\text{-}59)$$

Step 5: Locate D in the ALTITUDE PERFORMANCE graph based on the pressure altitude.
Step 6: Read E as the current BHP at altitude.
Step 7: Correct for temperature deviation using Equation (7-20):

$$P_{BHP} = P_{BHP\ E} \sqrt{\frac{T_{std}}{T_{OAT}}}$$

where

T_{std} = standard day temperature at altitude
T_{OAT} = outside air temperature at condition

7.4.3 Extracting Piston Power Using the Petty Equation

The extraction of engine power from the engine performance chart is a rather cumbersome and time-consuming effort. It is clear from Section 7.4.2,

where

P_{BHP} = horsepower at condition (specified *MAP*, RPM, and altitude)
$P_{BHP\ max}$ = maximum S-L horsepower as a function of RPM (typically a polynomial)
P_{FHP} = friction horsepower as a function of RPM (typically a polynomial — PFHP = PBHP at MAP = 0)
h = pressure altitude in feet
MAP = manifold pressure in inches Hg
MAP_{max} = maximum manifold pressure as a function of RPM (typically a polynomial)
R_m = manifold pressure ratio = MAP/MAP_{max}
R_f = friction horsepower ratio = $|PFHP|/PBHP_{max}$ (always a positive value)
σ = density ratio for standard atmosphere = $(1 - 0.0000068756h)^{4.2561}$

TABLE 7-13 Determining $P_{BHP\,max}$ and P_{FHP} Polynomials

① RPM	② $P_{BHP\,max}$	③ MAP	④ m	⑤ P_{FHP}
1600	100	18.50	9.524	−76.19
1800	110	18.00	10.000	−70.00
2000	122	17.50	10.609	−63.65
2200	133	17.00	11.083	−55.42
2400	145	16.50	11.600	−46.40
2600	156	16.00	12.000	−36.00
2700	163	15.75	12.302	−30.75

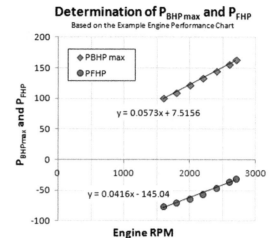

FIGURE 7-38 A simple curve fit is used to represent P_{BHPmax} and P_{FHP}.

The equation is a very valuable tool at the disposal of the designer of piston-powered aircraft as it effectively allows the power of a piston engine to be modeled using a spreadsheet or other computer software. It is priceless and more than worth the effort of setting it up. The primary drawback in its use is the preparation of the two polynomials, for the $P_{BHP\,max}$ and P_{FHP}. These must be determined using the original performance chart obtained from the engine manufacturer. Both are a function of RPM and can be determined as follows.

Determination of the Polynomials Describing $P_{BHP\,max}$ and P_{FHP}

Consider the engine performance chart of Figure 7-37, in particular the sea level performance side. The trick to creating these polynomials is to tabulate the endpoints and then fit a curve to these (see Table 7-13). For instance, consider the generation of the polynomial for the values of $P_{BHP\,max}$. Column ① contains the selected RPM values. Column ② contains the corresponding values of $P_{BHP\,max}$, which have been obtained by extending all the curves to a MAP of 29.

The next three columns pertain to the determination of P_{FHP}. The easiest way to determine the values of P_{FHP} is to read the value of the MAP for $P_{BHP} = 0$, and then use the equation of a line to determine the values of P_{FHP} when $MAP = 0$. These values can be seen in column ③. Column ④ contains the slope of the lines calculated from:

$$m = \frac{P_{BHPmax}}{29 - MAP_{P_{BHP}=0}}$$

Then, the P_{FHP}, which is contained in column ⑤, can be calculated as follows:

$$P_{BHP} = P_{FHP} + m \cdot P_{BHP\,max}$$
$$\Leftrightarrow P_{FHP} = P_{BHP} - m \cdot P_{BHPmax}$$

The points from Table 7-13 have been plotted in Figure 7-38. Then, it is an easy task to determine the best fit curve. Here, an equation of a line turned out to provide an acceptable fit, but this is not guaranteed. Commonly one must resort to quadratic and even cubic polynomials.

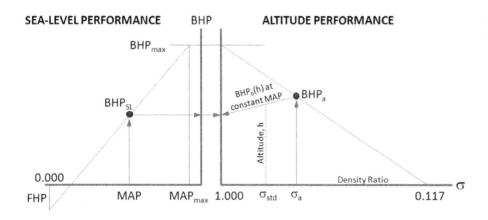

FIGURE 7-39 A simplified piston-engine performance chart that depicts only one value of the RPM.

Derivation of Equation (7-59)

Consider the simplified performance chart in Figure 7-39, on which only a single RPM is shown. The sea level curve, in the sea level performance side, extends from a negative friction power (P_{FHP}), which is the norm at $MAP = 0$, to a maximum power, $P_{BHP\ max}$, at the maximum manifold pressure, MAP_{max}. The representative P_{BHP} versus Altitude curve is shown in the altitude performance side. It depends on σ and extends from the S-L P_{BHP} value to where MAP can no longer be maintained. The locus of these limit points results in a specific P_{BHP} versus Altitude curve for each RPM, starting at σ = 1 to σ = 0.117.

The sea level performance curve can be easily represented using the following parametric expression (noting that the MAP axis extends from 0 to MAP_{max} so the parameter can be defined as $R_m = MAP/MAP_{max}$):

$$P_{BHP\ SL} = P_{FHP}(1 - R_m) + P_{BHP\ max} R_m \quad \text{(i)}$$

Then, define the ratio $R_f = P_{FHP}/P_{BHP\ max}$. Then, rewrite Equation (i) and insert this definition and simplify.

$$P_{BHP\ SL} = \frac{P_{BHP\ max}}{P_{BHP\ max}} P_{FHP}(1 - R_m) + P_{BHP\ max} R_m$$
$$= P_{BHP\ max} R_f (1 - R_m) + P_{BHP\ max} R_m$$
$$= P_{BHP\ max} \left(R_f(1 - R_m) + R_m \right)$$

Note that in order to prevent sign errors from occurring, the ratio R_f will from hereon be considered positive. However, since P_{FHP} is always negative, the above result is rewritten as follows, as this will preserve the sign:

$$P_{BHP\ SL} = P_{BHP\ max}\left[R_m - R_f(1 - R_m) \right] \quad \text{(ii)}$$

The altitude performance curve is based on the Gagg and Ferrar piston engine power correction of Equation (7-16), here repeated for convenience.

$$P_{BHP\ a} = P_{BHP\ max}\left(\frac{\sigma_a - 0.117}{0.883} \right) \quad \text{(iii)}$$

where the subscript "a" indicates this is taken from the altitude side. The manifold pressure for each RPM varies linearly with the pressure ratio and, thus using Equation (16-6), can be written as follows:

$$\frac{MAP_a}{MAP_{max}} = \frac{P}{P_{SL}} = \sigma_a^{1.235} \quad \text{(iv)}$$

Or, conversely:

$$\sigma_a = \left(\frac{MAP_a}{MAP_{max}} \right)^{0.8097} = R_m^{0.8097} \quad \text{(v)}$$

Then, substitute Equation (v) into Equation (iii) to yield:

$$P_{BHP\ a} = P_{BHP\ max}\left(\frac{\sigma_a - 0.117}{0.883} \right)$$
$$= P_{BHP\ max}\left(\frac{R_m^{0.8097} - 0.117}{0.883} \right) \quad \text{(vi)}$$

$P_{BHP\ a}$ is the power corrected for altitude effects only and this remains to be adjusted to standard day and corrected for temperature at the flight altitude, h. The adjustment takes place by locating the uncorrected power, denoted by $P_{BHP\ b}$, along the constant manifold pressure line on the altitude side of the performance chart using the equation:

$$P_{BHPb} = P_{BHP\ SL} + m(1 - \sigma_{std}) \quad \text{(vii)}$$

where

σ_{std} = standard day density ratio
$= (1 - 0.0000068756h)^{4.2561}$
h = altitude in ft
m = slope of the constant manifold pressure line =
$(P_{BHP\ a} - P_{BHP\ SL})/(1 - \sigma_a)$

Inserting this into Equation (vii) yields:

$$P_{BHP\ b} = P_{BHP\ SL} + m(1 - \sigma_{std})$$
$$= P_{BHP\ SL} + \frac{P_{BHP\ a} - P_{BHP\ SL}}{1 - \sigma_a}(1 - \sigma_{std}) \quad \text{(viii)}$$

Further algebraic manipulations:

$$P_{BHP\ b} = P_{BHP\ SL} + (P_{BHP\ a} - P_{BHP\ SL})\frac{1 - \sigma_{std}}{1 - \sigma_a}$$
$$= \frac{P_{BHP\ SL}(\sigma_{std} - \sigma_a) + P_{BHP\ a}(1 - \sigma_{std})}{1 - \sigma_a} \quad \text{(ix)}$$

The uncorrected power must then be corrected for temperature and this is done using Equation (7-20):

$$P_{BHP} = P_{BHP\ b}\sqrt{\frac{T_{std}}{T_{OAT}}} \quad \text{(x)}$$

And this is the power that one would used in performance calculations. Combining Equations (ii), (iii), (v), and (ix) with (x) leads to the following expression:

$$P_{BHP} = \sqrt{\frac{T_{std}}{T_{OAT}}} \frac{P_{BHP\,max}\left[R_m - R_f(1-R_m)\right]\left(\sigma_{std} - R_m^{0.8097}\right) + P_{BHP\,max}\left(\frac{R_m^{0.8097} - 0.117}{0.883}\right)(1 - \sigma_{std})}{1 - R_m^{0.8097}}$$

Further simplification yields:

$$P_{BHP} = P_{BHP\,max} \sqrt{\frac{T_{std}}{T_{OAT}}} \frac{\left[R_m - R_f(1-R_m)\right]\left(\sigma_{std} - R_m^{0.8097}\right) + \left(\frac{R_m^{0.8097} - 0.117}{0.883}\right)(1 - \sigma_{std})}{1 - R_m^{0.8097}} \quad \text{(xi)}$$

This is the Petty equation.

QED

EXERCISES

(1) Determine the *SFC* for the following scenarios in the UK and SI systems. (a) For a jet engine consuming 660 lb$_f$ of fuel per hour while generating 990 lb$_f$ of thrust. (b) For a piston engine consuming 21 gallons while producing 275 BHP.

(2) Using the Gagg and Ferrar model tabulate the altitudes at which a normally aspirated piston engine generates 100%, 95%, 90%, ..., 45% power for ISA, ISA−30°C, ISA, and ISA+30°C.

(3) Using the Mattingly method estimate the maximum power of a Garrett TPE331-10 at 30,000 ft and Mach 0.45, using the basic T-O data of Table 7-9 and assuming a throttle ratio of 1.072. Estimate the fuel consumption in gallons of Jet-A per hour. Assuming a cruise segment at that altitude, how far can the airplane fly on 500 gallons if equipped with two TPE331s?

(4) Using the Mattingly method to estimate the thrust of a Microturbo TRS 18-056 at 10,000 ft and Mach 0.39, using the basic T-O data of Table 7-10 and assuming a throttle ratio of 1.000. Estimate the fuel consumption in gallons of Jet-A per hour. How much fuel must this airplane carry to cover a 300 nm cruise segment at that altitude if equipped with only one TRS 18-056?

(5) Compare the thrust and fuel consumption of the two CFM56 engines in Table 7-11, at 35,000 ft and Mach 0.8 in terms of lb$_f$ of Jet-A per hour, assuming it can be based on the T-O data.

(6) A single-engine electric aircraft is equipped with a 40 kW motor and battery packs that store 150 Ah of energy. Estimate its total endurance if it uses full power for 5 minutes, 65% power for 5 minutes, and 30% for the remaining charge (assuming no overheating and ignore Peukert's law).

(7) A simple diffuser inlet is being designed for a specialty turbofan engine that requires the cruise speed Mach number of 0.65 to be reduced to Mach 0.3 at the front face of the compressor. The diameter of the compressor is 30 inches. (a) Determine the required inlet lip diameter so the Mach number at the lip at T-O will not exceed 0.8. (b) If the included angle is expected to be 17°, how long must the diffuser be in order to prevent flow separation?

(8) Estimate the power of the example engine in Section 7.4.3, *Extracting piston power using the Petty equation* at altitudes of 5000 ft, 10,000 ft, and 15,000 ft.

VARIABLES

Symbol	Description	Units (UK and SI)
a	Speed of sound	ft/s or m/s
A	Cross-sectional area	ft^2 or m^2
A	Engine-dependent constant (for piston engines)	
A_0	Cross-sectional area at station 0 (far-field)	ft^2 or m^2
A_1	Cross-sectional area at station 1 (inlet)	ft^2 or m^2
A_2	Cross-sectional area at station 2 (compressor)	ft^2 or m^2
A_B	Reference area of the baffle (radiator)	ft^2 or m^2
A_e	Exit cross-sectional area	ft^2 or m^2
A_i	Inlet cross-sectional area	ft^2 or m^2
Bore	Diameter of each cylinder	ft or m
c_{bhp}	Specific fuel consumption for piston engines	(lb$_f$/hr)/BHP or g/J
c_{jet}	Specific fuel consumption for jet engines	1/hr or g/(N·s)
C_p	Specific heat of pressure	BTU/(sl·°R) or J/(kg·K)

VARIABLES

Symbol	Description	Units (UK and SI)
$C_{P1 \to 2}$	Ideal pressure coefficient (for turbine inlet design)	
D_1	Inlet lip diameter	ft or m
E_{batt}	Energy density (of a battery)	(W·hr)/kg (SI only)
F	Force or thrust at condition	lb_f or N
F_1	Force produced by gearwheel 1	lb_f or N
F_2	Force produced by gearwheel 2	lb_f or N
F_{SL}	Thrust produced by engine at given power setting at S-L	lb_f or N
g	Gravity	ft/s² or m/s²
h	Altitude	ft or m
h_{crit}	Critical altitude	ft or m
I	Current	A (amperes) (SI only)
k	Pressure recovery coefficient	
KE	Kinetic energy	BTU or J
L	Length	ft or m
m	Mass	slugs or kg
m	Slope of the constant manifold pressure line	
M	Mach number	
$\dot{m}$	Mass flow rate of air through the engine compartment	sl/s or kg/s
M_0	Mach number at station 0 (far-field)	
M_1	Mach number at station 1 (inlet)	
M_2	Mach number at station 2 (compressor)	
MAP	Manifold pressure	inHg
MAP_{max}	Maximum manifold pressure as a function of RPM	inHg
$\dot{m}_e$	Exit mass flow rate	sl/s or kg/s
$\dot{m}_{fuel}$	Fuel mass flow rate	sl/s or kg/s
$\dot{m}_i$	Inlet mass flow rate	sl/s or kg/s
$\dot{m}_{propellant}$	Propellant mass flow rate	sl/s or kg/s
$\dot{m}_{required}$	Required maximum mass flow rate (for turbine inlet)	sl/s or kg/s
n	Engine-dependent constant (for piston engines)	
p	Pressure at condition	lb_f/ft^2 or N/m^2
P	Power at condition	hp or W
p_0	Standard S-L pressure or far-field pressure	lb_f/ft^2 or N/m^2
P_1	Power of gearwheel 1	hp or W
P_2	Power of gearwheel 2	hp or W
p_{B1}	Pressure at the baffle forward face	lb_f/ft^2 or N/m^2
p_{B2}	Pressure at the baffle aft face	lb_f/ft^2 or N/m^2
P_{BHP}	Horsepower at condition	hp (or W after conversion to SI)
$P_{BHP\,max}$	Maximum sea level horsepower as a function of RPM	hp (or W after conversion to SI)
p_{bottom}	Pressure with piston in bottom position	lb_f/ft^2 or N/m^2
p_e	Exit pressure	lb_f/ft^2 or N/m^2
PE	Potential energy	BTU or J
P_{FHP}	Friction horsepower as a function of RPM	hp
P_{SL}	Brake horsepower at sea-level	hp (or W after conversion to SI)
P_{std}	Standard power at altitude and ISA	hp (or W after conversion to SI)
p_{T0}	Total pressure at station 0 (far-field)	lb_f/ft^2 or N/m^2
p_{T1}	Total pressure at station 1 (inlet)	lb_f/ft^2 or N/m^2
p_{T2}	Total pressure at station 2 (compressor)	lb_f/ft^2 or N/m^2
p_{top}	Pressure with piston in top position	lb_f/ft^2 or N/m^2
p_{tot}	Total pressure at condition	lb_f/ft^2 or N/m^2
q	Dynamic pressure	lb_f/ft^2 or N/m^2
$\dot{Q}$	Heat transfer rate	BTU/s or J/s
Q_B	Heat flow into heat exchanger	BTU or J
R	Resistance	Ω (ohms) (SI only)
R	Ideal gas constant	ft·lb_f/(slug·°R) or J/(kg·K)
R_1	Radius of gearwheel 1	ft or m
R_2	Radius of gearwheel 2	ft or m
R_f	Friction horsepower ratio	
R_m	Manifold pressure ratio	
SFC	Specific fuel consumption	See c_{bhp} or c_{jet}
$Stroke$	Total distance piston moves	ft or m
T	Temperature	°R or K
T	Thrust	lb_f or N
T	Torque	in·lb_f or N·m
T/W	Thrust-to-weight ratio	
T_0	Far-field temperature	°R or K

(Continued)

Symbol	Description	Units (UK and SI)
T_0	Standard S-L temperature	°R or K
T_0	Static temperature	°R or K
T_0	Temperature at station 0 (far-field)	°R or K
T_{0des}	Design total temperature	°R or K
T_1	Temperature at station 1 (inlet)	°R or K
T_1	Torque of gearwheel 1	in·lb$_f$ or N·m
T_2	Temperature at station 2 (compressor)	°R or K
T_2	Torque of gearwheel 2	in·lb$_f$ or N·m
T_∞	Temperature in the streamtube behind the nozzle	°R or K
T_{B1}	Temperature at the baffle forward face	°R or K
T_{B2}	Temperature at the baffle aft face	°R or K
THP	Total horsepower	hp (or W after conversion to SI)
T_{net}	Net thrust	lb$_f$ or N
T_{OAT}	Outside air temperature at condition	°R or K
TR	Throttle ratio	
T_{std}	Standard day temperature	°R or K
T_{t0}	Total temperature	°R or K
T_{tot}	Total temperature at condition	°R or K
V	Velocity	ft/s or m/s
V	Voltage	Volts (SI only)
V	Volume	ft^3 or m^3
V_∞	Airspeed in the streamtube behind the nozzle	ft/s or m/s
V_0	Far-field airspeed	ft/s or m/s
V_{B1}	Airspeed in front of the baffle	ft/s or m/s
V_{B2}	Airspeed aft of the baffle	ft/s or m/s
V_{bottom}	Volume of the cylinder with piston at bottom position	ft^3 or m^3
V_{disp}	Displacement of the piston engine	ft^3 or m^3
V_e	Exit velocity	ft/s or m/s
V_i	Inlet velocity	ft/s or m/s
V_{top}	Volume of the cylinder with piston at top position	ft^3 or m^3
W	Work	ft·lb$_f$ or J
Ω_1	Angular speed of gearwheel 1	RPM
Ω_2	Angular speed of gearwheel 2	RPM

Symbol	Description	Units (UK and SI)
δ_0	Pressure ratio	
γ	Ratio of specific heats	
κ	Lapse rate constant	1/ft or 1/m
π_2	Pressure recovery ratio (turbines only)	
θ	Included angle	deg or rad
θ_0	Temperature ratio	
ρ	Density	sl/ft^3 or kg/m^3
ρ_0	Density at station 0 (far-field)	sl/ft^3 or kg/m^3
ρ_1	Density at station 1 (inlet)	sl/ft^3 or kg/m^3
ρ_2	Density at station 2 (compressor)	sl/ft^3 or kg/m^3
ρ_{SL}	sea level density	sl/ft^3 or kg/m^3
σ	Density ratio	
σ_{std}	Standard day density ratio	

References

[1] http://cafefoundation.org/v2/gfc_2011_results.html.
[2] http://aviation-safety.net/index.php.
[3] Daggett David L, Silvio Ortanderl, David Eames, Jeffrey J. Berton, and Christopher A. Snyder. *Revisiting Water Injection for Commercial Aircraft*. SAE 2004-01-3108, 2004.
[4] Anonymous. *ExxonMobil: World Jet Fuel Specifications with Avgas Supplement*. 2005 Edition.
[5] http://www.shell.com/home/content/aviation/products/fuels/types/civil_jet_fuel_grades/.
[6] http://www.lycoming.com/engines/legacy/.
[7] Stinton D, The Design of the Aeroplane. Collins.
[8] Avco Lycoming Corporation Report 2268. *Horsepower Correction Factors and Operating Techniques for Engine Development and Calibration*. August 1960.
[9] Anonymous. *Pilot Training Manual for the C-46 Commando*. AAF Manual No. 50–16.
[10] Benser WA, Moses JJ, An Investigation of Backflow Phenomenon in Centrifugal Compressors. NACA R-806.
[11] http://www.pwc.ca/en/about/history.
[12] Taylor, John WR, and Mark Lambert, Eds. *Jane's All the World's Aircraft*. Jane's Information Group, various years.
[13] Mattingly JD, Heiser WH, Pratt DT, Aircraft Engine Design, 2nd edn. AIAA Education Series.
[14] Mattingly JD, Elements of Gas Turbine Propulsion. McGraw-Hill.
[15] Austyn-Mair W, Birdsall DL, Aircraft Performance. Cambridge University Press, 102–106.
[16] http://yuneeccouk.site.securepod.com/Aircraft.html.
[17] Eberhardt JJ, Fuels of the Future for Cars and Trucks. U.S. Department of Energy.
[18] Accident report NTSB/AAR-86/02, January 31, 1986.
[19] McCormick BW, Aerodynamics, Aeronautics, and Flight Mechanics. John Wiley & Sons. p. 420.
[20] Flack RD, Fundamentals of Jet Propulsion with Applications. Cambridge Aerospace Series. Cambridge University Press.
[21] Seddon J, Goldsmith EL, Intake Aerodynamics. AIAA Education Series.
[22] Schlichting H, Boundary Layer Theory. Pergamon Press Ltd.

CHAPTER 8

The Anatomy of the Airfoil

OUTLINE

8.1 Introduction 236
  8.1.1 The Content of this Chapter 237
  8.1.2 Dimensional Analysis — Buckingham's Π Theorem 237
  8.1.3 Representation of Forces and Moments 238
    The Smeaton Lift Equation (Obsolete) 239
  8.1.4 Properties of Typical Airfoils 239
    Section Lift Coefficient, C_l 239
    Maximum and Minimum Lift Coefficients, C_{lmax} and C_{lmin} 239
    Lift Curve Slope, $C_{l\alpha}$ 239
    Angle-of-attack at Zero Lift, α_{ZL} 240
    Linear Range 240
    C_l at Zero AOA, C_{lo} 241
    Minimum Drag Coefficient, C_{dmin} 241
    Lift Coefficient of Minimum Drag, C_{lmind} 241
  8.1.5 The Pressure Coefficient 241
    Derivation of Equation (8-13) 241
    Derivation of Equation (8-14) 241
    The Canonical Pressure Coefficient 241
  8.1.6 Chordwise Pressure Distribution 242
    "Conventional" Lift Distribution 242
    Stratford Distribution 243
  8.1.7 Center of Pressure and Aerodynamic Center 243
    Center of Pressure 243
    Aerodynamic Center and Quarter Chord Moment 244
    Derivation of Equation (8-16) 244
  8.1.8 The Generation of Lift 245
    Momentum Theorem 245
    Bernoulli Theorem 246
    Kutta-Joukowski Circulation Theorem 246
  8.1.9 Boundary Layer and Flow Separation 247
    Reynolds Number 247
    The Effect of Flow Separation 247
    Boundary Layer Transition 248
    Flow Separation 249
    Factors Affecting Laminar Flow 250
  8.1.10 Estimation of Boundary Layer Thickness 250
  8.1.11 Airfoil Stall Characteristics 251
    Trailing Edge (TE) Stall 251
    Leading Edge (LE) Stall 251
  8.1.12 Analysis of Ice-accretion on Airfoils 252
  8.1.13 Designations of Common Airfoils 253
  8.1.14 Airfoil Design 254
    Xfoil and XFLR5 255
    PROFILE ("The Eppler Code") 255
    AeroFoil 255
    JavaFoil 256
    Design Process 256
    Direct Analysis Method 256
    Inverse Airfoil Design Method 256

8.2 The Geometry of the Airfoil 256
  8.2.1 Airfoil Terminology 256
    Thickness, Mean-line and Camber 256
    LE Radius 256
    Square TE 257
    Geometric Description 257
  8.2.2 NACA Four-digit Airfoils 257
    Applications 258
    Numbering System 258
    Computation of Airfoil Ordinates 258
    Step 1: Preliminary Values 259
    Step 2: Airfoil Resolution 259
    Step 3: Prepare Ordinate Table 259
    Step 4: Calculate Thickness 259
    Step 5: Compute the y-value for the Mean-line 259
    Step 6: Calculate the Slope of the Mean-line 259
    Step 7: Calculate the Ordinate Rotation Angle 260
    Step 8: Calculate the Upper and Lower Ordinates 260
    Generation of the NACA 4415 — an Example Implementation 260
  8.2.3 NACA Five-digit Airfoils 261
    Applications 261
    Numbering System 262
    Computation of Airfoil Ordinates 262

Step 1: Preliminary Values	262
Step 2: Airfoil Resolution	262
Step 3: Prepare Ordinate Table	262
Step 4: Calculate Thickness	262
Step 5: Compute the y-value for the Mean-line	262
Step 6: Calculate the Slope of the Mean-line	263
Step 7: Calculate the Ordinate Rotation Angle	263
Step 8: Calculate the Upper and Lower Ordinates	263
8.2.4 *NACA 1-series Airfoils*	263
Applications	263
Numbering System	263
8.2.5 *NACA 6-series Airfoils*	264
Applications	265
Numbering System	265
8.2.6 *NACA 7-series Airfoils*	266
Applications	266
Numbering System	266
8.2.7 *NACA 8-series Airfoils*	267
Applications	267
Numbering System	267
8.2.8 *NACA Airfoils in Summary — Pros and Cons and Comparison of Characteristics*	267
8.2.9 *Properties of Selected NACA Airfoils*	267
8.2.10 *Famous Airfoils*	267
CLARK Y	267
USA-35B	268
NACA 23012	268
GA(W)-1 or LS(1)-0417	271
Davis Wing Airfoil	272
"Peaky" Airfoils	273
Supercritical Airfoils	274
Joukowski Airfoils	274
Liebeck Airfoils	275
8.3 **The Force and Moment Characteristics of the Airfoil**	275
8.3.1 *The Effect of Camber*	276
8.3.2 *The Two-dimensional Lift Curve*	276
8.3.3 *The Maximum Lift Coefficient, C_{lmax}*	276
Maximum Theoretical Lift Coefficient	276
8.3.4 *The effect of Reynolds Number*	277
8.3.5 *Compressibility Effects*	278
Compressibility Effect on Lift	278
Compressibility Effect on Drag	278
Compressibility Effect on Lift and Drag Exemplified	279
Compressibility Effect on Pitching Moment	279
8.3.6 *Compressibility Modeling*	280
8.3.7 *The Critical Mach Number, M_{crit}*	281
Step-by-step: Determining M_{crit} for a Body	281
Step 1: Establish the Minimum Pressure Coefficient	281
Step 2: Select Compressibility Correction Method	281
Step 3: Solve to Determine M_{crit}	281
Derivation of Equation (8-62)	281
8.3.8 *The Effect of Early Flow Separation*	282
8.3.9 *The Effect of Addition of a Slot or Slats*	283
8.3.10 *The Effect of Deflecting a Flap*	284
8.3.11 *The Effect of Cruise Flaps*	284
8.3.12 *The Effect of Deploying a Spoiler*	285
8.3.13 *The Effect of Leading Edge Roughness and Surface Smoothness*	286
8.3.14 *Drag Models for Airfoils Using Standard Airfoil Data*	287
8.3.15 *Airfoil Selection How-to*	289
Impact on Drag	289
Impact on Flow Separation	289
Impact on Maximum Lift and Stall Handling	289
Impact on Longitudinal Trim	289
Critical Mach Number	289
Impact on Wing-fuselage Juncture	289
Impact on Structural Depth	289
Evaluation of a Target Zero-lift AOA	290
Selection of an Airfoil	290
NACA Recommended Criteria	294
Exercises	294
Variables	294
References	296

8.1 INTRODUCTION

Any object that moves through a fluid induces a pressure field in its vicinity. The pressure field changes the pressure on its surface and induces a resultant pressure force, R, which acts on the object. Then, we define *lift*, L, as the component of this force that is normal to the trajectory (or flight path). Similarly, *drag*, D, is defined as the component of this force tangent to the trajectory. In addition to the pressure force, viscous friction adds to the total drag force. The lift generated by three-dimensional objects is treated in Chapter 9, *The anatomy of the wing* and drag is treated in Chapter 15, *Aircraft drag analysis*. However, the purpose of this section is to focus

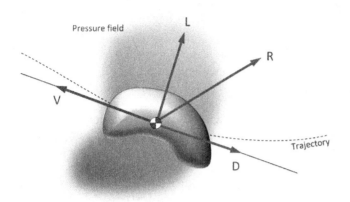

FIGURE 8-1 An object moving in fluid induces pressure a field.

specifically on the important geometric shape used for lifting surfaces: the airfoil.

What sets the airfoil shape apart from other geometry, like, say, the kidney in Figure 8-1, is that its resultant force approaches being normal to the tangent to the trajectory. This results in a lift force component substantially larger than the drag component. Consequently, such geometry generates lift far more effectively than other shapes. The treatment of the airfoil in this section is purely two-dimensional, but this will be expanded to three dimensions in Chapters 9 and 15.

Note that it is a convention in the literature to denote forces and moments for two-dimensional geometry by a lower-case letter but with capitalization when referring to three-dimensional geometry. Thus, lift, drag, and moment for an airfoil would be denoted by l, d, and m, respectively, but using L, D, and M for a three-dimensional wing. The difference between the two, fundamentally, is that a wing has a finite aspect ratio (AR), whereas an airfoil can be considered like a wing of infinite span and, thus, infinite AR. This convention will be adhered to in this text. Consequently, the lift, drag, and moment coefficients for an airfoil are written using lower-case identifiers; C_l, C_d, C_m. Capitalized identifiers are used for 3D wings or an aircraft as a whole: C_L, C_D, C_M.

Many areas of the aircraft design process rely on accurate estimation of these forces and moments. This includes performance analysis, determination of useful load, and structural analysis, to name a few. As has already been alluded to, any competitive aircraft design requires the useful load to be maximized. This implies the empty weight must be minimized and this can only be done if the distribution of pressure loads on the vehicle can be accurately estimated.

When it comes to airplanes, we are mostly interested in bodies whose geometry results in a lift force (L) that is substantially larger than the drag (D). Airfoils are examples of such bodies and, at low angles-of-attack their lift is significantly greater than their drag. As an example, the lift component for modern-day airfoils can be in the excess of 200 times the drag force at some specific orientation in the flow.

8.1.1 The Content of this Chapter

- **Section 8.1** presents fundamental concepts and theories regarding airfoil lift and drag generation. It contains very important definitions. Additionally, it introduces how pressure is distributed along the upper and lower surfaces of the airfoil and how it affects the growth of the boundary layer and, eventually, flow separation.
- **Section 8.2** defines important geometric properties of airfoils. It also presents information intended to make the aircraft designer better rounded when comes to identifying various airfoil types, such as NACA airfoils, and understanding of their background. For this purpose, the section introduces a number of airfoils that have gained fame or notoriety in the history of aviation.
- **Section 8.3** discusses the generation of forces and moments on the airfoil. It details how various outside agents, such as very high airspeeds, high angle-of-attack, deflection of control surfaces, and even contamination, affects their aerodynamic properties. Finally, it presents a method to help the designer select the proper airfoil for a new aircraft design.

8.1.2 Dimensional Analysis — Buckingham's Π Theorem

Dimensional analysis is a tool used by scientists to confirm derived equations that describe physical phenomena, by enforcing unit consistency. In physics, the fundamental units are mass (m), length (L), time (t), ampère (A), and kelvin (K). All other physical quantities have units that are based on these. As an example, consider pressure, which is defined as force per area. Force is defined as mass times acceleration or $m \cdot L/t^2$. For this reason, it is possible to write pressure as $(m \cdot L/t^2)/L^2 = m/(L \cdot t^2)$. In aerodynamics, forces are denoted as follows:

Pressure force:

$$F_{\text{press}} = pA = \frac{m}{L \cdot t^2}L^2 = \frac{mL}{t^2} \qquad (8\text{-}1)$$

This matches the units for the force, showing it is dimensionally consistent.

The Buckinham Π theorem, named after Edgar Buckingham (1867–1940), is used to derive the proper form for aerodynamic forces. In general, observation shows that the force generated in a fluid flowing over a body depends on the density of the fluid (more density, larger force), the relative speed of the fluid with respect to the body (more speed, larger force), and the size of the body (larger body, larger force). It is possible to relate these using the following expression:

Force due to fluid flow:

$$F = k\rho^a V^b l^c \quad (8\text{-}2)$$

where

ρ = density, m/L^3
V = airspeed, L/t
l = characteristic length, L
k = unknown constant of proportionality
a, b, c = exponents to be determined

Inserting the dimensions into Equation (8-2) yields:

$$\frac{mL}{t^2} = k\left(\frac{m}{L^3}\right)^a \left(\frac{L}{t}\right)^b L^c \quad (8\text{-}3)$$

Simplification on the right side leads to:

$$\frac{mL}{t^2} = k \frac{m^a}{L^{3a}} \frac{L^b}{t^b} L^c = k \times \frac{m^a \times L^{b+c-3a}}{t^b}$$

Since the dimensions on the left- and right-hand sides must be consistent, we can determine $a, b,$ and c as follows:

$m^1 = m^a \Leftrightarrow a = 1$
$t^{-2} = t^{-b} \Leftrightarrow b = 2$
$L^1 = L^{b+c-3a} \Rightarrow 1 = b + c - 3a = 2 + c - 3 \Leftrightarrow c = 2$

Therefore, we can rewrite Equation (8-3) as follows:

Force due to fluid flow:

$$F = k\rho V^2 l^2 \quad (8\text{-}4)$$

This formulation serves as the basis for all forces and moments used in aerodynamic theory.

8.1.3 Representation of Forces and Moments

The total force (or resultant force) generated by a wing can be found to depend on several parameters: the wing's geometry, density of air, airspeed, and the angle the chord line of the wing's airfoils make to the flow of air, the angle-of-attack (from here on also referred to as *AOA*). While the wing is a three-dimensional, it is usually treated as a set of two two-dimensional geometric features; the airfoil (*x-z* plane as shown in Figure 8-2) and planform (*x-y* plane, see Section 9.4, *Planform Selection*). It was shown by dimensional analysis (i.e. via the Buckingham's Π theorem) that the equation describing this resultant force, r, is given by:

$$r = \frac{1}{2}\rho V^2 \cdot S \cdot C_r \quad (8\text{-}5)$$

where

ρ = density of air, in kg/m^3 or slugs/ft^3
V = airspeed, in m/s or ft/s
S = reference area, typically wing area, in m^2 or ft^2
C_r = non-dimensional coefficient that relates *AOA* to the force

Figure 8-2 shows that the lift (force normal to the airspeed), drag (force parallel to the airspeed), and pitching moment (which all are assumed to act at the quarter chord) can be defined as follows:

$$l = \frac{1}{2}\rho V^2 \cdot S \cdot C_l = \frac{1}{2}\rho V^2 \cdot S \cdot C_r \cos\alpha$$
$$d = \frac{1}{2}\rho V^2 \cdot S \cdot C_d = \frac{1}{2}\rho V^2 \cdot S \cdot C_r \sin\alpha \quad (8\text{-}6)$$
$$m = \frac{1}{2}\rho V^2 \cdot S \cdot c \cdot C_m$$

Lift and drag are less important to the structural engineer than the normal and chordwise forces, f_n and

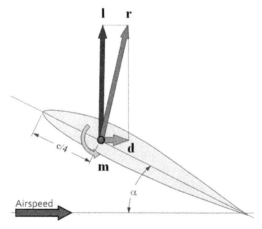
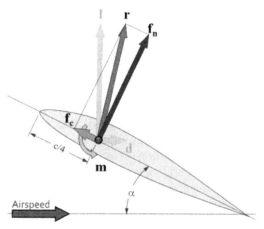

FIGURE 8-2 Forces and moments acting on an airfoil (left) and the definition of normal and chordwise force on an airfoil at a high *AOA* (right).

f_c, respectively (see Figure 8-2). The normal force is perpendicular (as the name implies) to the wing plane (the hypothetical plane formed by the span- and chordwise vectors) and generates the bending moment. The chordwise force, on the other hand, is parallel to the chord plane. At low angles-of-attack the magnitude of f_c is close to the drag force and points toward the trailing edge of the airfoil. However, at high angles-of-attack f_c actually points forward, toward the leading edge. The effect tends to move the wing in a forward direction! This effect must be taken into account in structural analysis as it places the aft spar attachment in tension, whereas at low angles-of-attack it places the aft attachment in compression. Figure 8-2 shows that the normal and chordwise forces can be defined as follows:

$$f_n = \frac{1}{2}\rho V^2 \cdot S \cdot (C_l \cos\alpha + C_d \sin\alpha)$$
$$f_c = \frac{1}{2}\rho V^2 \cdot S \cdot (C_d \cos\alpha - C_l \sin\alpha)$$
(8-7)

For three-dimensional objects such as aircraft, the representation of forces and moments that correspond to Equation (8-6) is given by:

$$L = \frac{1}{2}\rho V^2 \cdot S \cdot C_L$$
$$D = \frac{1}{2}\rho V^2 \cdot S \cdot C_D \qquad (8\text{-}8)$$
$$M = \frac{1}{2}\rho V^2 \cdot S \cdot C_{MGC} \cdot C_M$$

where L, D, M refer to the three-dimensional lift, drag, and pitching moment, respectively. Thus, C_L is the three-dimensional lift coefficient, C_D the drag coefficient, and C_M the pitching moment coefficient of the complete aircraft. These will be treated in more detail in Chapters 9, 11, and 15. C_{MGC} is the wing's mean geometric chord and S is the reference wing area. Both are presented in detail in Chapter 9, *The anatomy of the wing*.

The Smeaton Lift Equation (Obsolete)

From a historical standpoint, it is of interest to consider the now-obsolete Smeaton lift equation (which was how the Wright brothers determined the wing area required for their Flyer.) It was attributed to the English civil engineer John Smeaton (1724–1792), who is often referred to as the father of civil engineering, and was in use up until the beginning of the twentieth century. Smeaton's lift equation is given as follows:

$$L = \text{pressure factor} \times \text{velocity}^2 \times \text{wing area}$$
$$\times \text{lift factor} = kV^2 A c_l \qquad (8\text{-}9)$$

where

L = lift force in lb$_f$
k = Smeaton's coefficient
A = wing area in ft^2
V = airspeed in ft/s
c_l = lift coefficient

Engineers at the time considered the lift coefficient as the ratio of the object's lift force to its drag force, where the drag was for a flat plate of area A mounted perpendicular to the airstream. Smeaton's coefficient, k, is the drag of a 1 ft^2 flat plate at 1 mph. At the turn of the century (~1900) the accepted value for this coefficient was 0.005 and this had been the value used by Otto Lilienthal in the design of his gliders. In fact, it was Smeaton himself who came up with this particular value; however, other sources claimed it ranged from 0.0027 to 0.005. The Wright brothers concluded the coefficient was wrong and experimentally determined it to be closer to 0.0033. The modern value is 0.00326 [1].

8.1.4 Properties of Typical Airfoils

A typical presentation of change in lift coefficients with *AOA* is shown in Figure 8-3, with important properties labeled. The graph shows true wind tunnel test results for the NACA 4415 airfoil. A number of important observations can be made based on the figure.

Section Lift Coefficient, C_l

The two-dimensional lift coefficient is commonly called the *section lift coefficient*. This concept is of great importance to the airplane designer and will be discussed at length later. One of this concept's most useful properties is that it can indicate both the effective angle-of-attack of the airfoil and how close to stalling it is.

Maximum and Minimum Lift Coefficients, C_{lmax} and C_{lmin}

The largest and smallest magnitudes of the lift coefficient are denoted by C_{lmax} and C_{lmin}, respectively. The former always has a positive magnitude and the latter a negative one. These values are extremely important because they dictate the stalling speed (at positive and negative loading) of the aircraft (and therefore wing size and airplane weight), as well as impacting other important characteristics such as the maneuvering loads and spin behavior. Generally, the stall is defined as the flow conditions that follow the first lift curve peak, which is where the C_{lmax} (or C_{lmin}) occurs [2, p.1].

Lift Curve Slope, $C_{l\alpha}$

$C_{l\alpha}$ indicates how rapidly lift changes with angle-of-attack. The maximum value of the slope is predicted

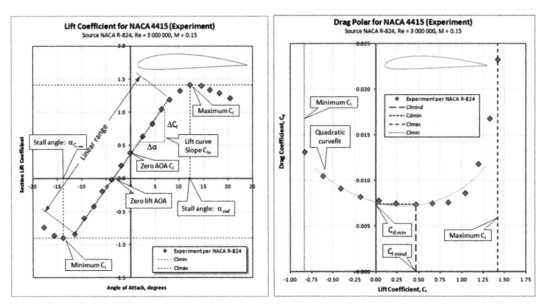

FIGURE 8-3 A typical two-dimensional lift curve and drag polar for an airfoil.

by linear thin airfoil theory for incompressible flow to be 2π (approximately 6.283) and most airfoils indeed achieve a value close to that. For instance, the slope of the red dotted line displaying the linear range in Figure 8-3 is approximately 5.90. The lift curve slope is usually linear at low *AOA*s; however, it is nonlinear outside this range and can have a negative slope. A nonlinear lift curve slope is indicative of flow separation prevailing on the body.

Angle-of-attack at Zero Lift, α_{ZL}

This refers to the angle-of-attack at which the airfoil produces no lift. It is important when considering wing washout (see Section 9.3.5, *Wing twist — washout and washin*, ϕ) and when converting the two-dimensional lift curve to three dimensions, as the change can be approximated by, effectively, rotating the lift curve around this point, toward a shallower slope (see Section 9.5.5, *Step-by-step: Transforming the lift curve from 2D to 3D*). The α_{ZL} is shown in Figure 8-4.

Linear Range

The linear range (here shown ranging from $AOA = -12°$ through 8°) is called so because, within it, one can estimate the lift coefficient for any *AOA* using a simple linear expression, such as the one below:

$$C_l = C_{l_0} + C_{l_\alpha} \alpha \qquad (8\text{-}10)$$

The extent of this region ultimately depends on the geometry and the operational airspeeds (via Reynolds numbers as discussed in Section 8.3.4, *The effect of Reynolds number*).

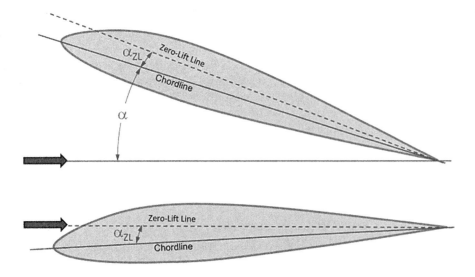

FIGURE 8-4 An airfoil at an angle-of-attack, α. The upper figure depicts the zero-lift angle-of-attack, α_{ZL}. The lower figure shows the airfoil at α_{ZL}, as no lift is generated.

C_l at Zero AOA, C_{lo}

C_{lo} is the value of the lift coefficient at zero *AOA*. This is of great importance in the selection of the airfoil as it will affect the angle-of-incidence at which the wing must be mounted. Generally, this value ranges from 0.0 (for symmetric airfoils) to 0.6 (for highly cambered airfoils). It is negative for under-cambered airfoils (e.g. airfoils used near the root of high subsonic jet aircraft). If the *AOA* of zero lift, α_{ZL}, and lift curve slope, $C_{l\alpha}$, are known, the value of C_{lo} can be calculated from:

$$C_{l_0} = -C_{l_\alpha} \alpha_{ZL} \qquad (8\text{-}11)$$

Minimum Drag Coefficient, C_{dmin}

C_{dmin} is the lowest value of the drag coefficient that can be found on the drag polar. Its magnitude is vital to the selection of the airfoil. Ideally, C_{dmin} should be a low as possible, but it also has to be low where it counts; in the region of intended lift coefficient of cruise.

Lift Coefficient of Minimum Drag, C_{lmind}

C_{lmind} is the lift coefficient where the minimum drag coefficient occurs on the drag polar. This location impacts the selection of the airfoil for the same reason as C_{dmin}.

8.1.5 The Pressure Coefficient

The *pressure coefficient* is of considerable importance in the discussion that follows, so a brief review is warranted. It is very useful to represent pressure in terms of a dimensionless quantity, similar to that of lift and drag. The incompressible pressure coefficient is defined as follows:

$$C_p \equiv \frac{p - p_\infty}{\frac{1}{2}\rho_\infty V_\infty^2} = \frac{\Delta p}{q} \qquad (8\text{-}12)$$

where

q = dynamic pressure, in lb_f/ft^2 or N/m^2
p = pressure, in lb_f/ft^2 or N/m^2
p_∞ = far-field pressure, in lb_f/ft^2 or N/m^2
V_∞ = far-field airspeed, in ft/s or m/s
ρ_∞ = far-field density, in $slugs/ft^3$ or kg/m^3

The incompressible pressure coefficient can also be written as follows:

$$C_p = 1 - \left(\frac{V}{V_\infty}\right)^2 \qquad (8\text{-}13)$$

The maximum possible value of the C_p at the stagnation point in incompressible flow is 1. The C_p in compressible flow can become larger than 1 if the flow is supersonic. The compressible C_p is given by:

$$C_p = \frac{2}{\gamma M_\infty^2}\left(\frac{p}{p_\infty} - 1\right) \qquad (8\text{-}14)$$

where

M_∞ = far-field Mach number
γ = ratio of specific heats = 1.4 at altitudes where aircraft typically operate

Derivation of Equation (8-13)

From Bernoulli's equation:

$$p_\infty + \frac{1}{2}\rho_\infty V_\infty^2 = p + \frac{1}{2}\rho_\infty V^2 \Rightarrow p - p_\infty$$
$$= \frac{1}{2}\rho_\infty(V_\infty^2 - V^2)$$

Inserting this into Equation (8-12):

$$C_p \equiv \frac{p - p_\infty}{\frac{1}{2}\rho_\infty V_\infty^2} = \frac{\frac{1}{2}\rho_\infty(V_\infty^2 - V^2)}{\frac{1}{2}\rho_\infty V_\infty^2} = \frac{V_\infty^2 - V^2}{V_\infty^2}$$
$$= 1 - \left(\frac{V}{V_\infty}\right)^2$$

QED

Derivation of Equation (8-14)

Using the equation of state ($p = \rho RT$), the speed of sound can be found from:

$$a_0 = \sqrt{\gamma R T_\infty} = \sqrt{\gamma p_\infty / \rho_\infty}$$

Therefore, the dynamic pressure can be written as follows:

$$q_\infty = \frac{1}{2}\rho_\infty V_\infty^2 = \frac{1}{2}\frac{\gamma p_\infty}{\gamma p_\infty}\rho_\infty V_\infty^2 = \frac{\gamma p_\infty}{2}\left(\frac{\rho_\infty}{\gamma p_\infty}\right)V_\infty^2$$
$$= \frac{\gamma p_\infty}{2}\left(\frac{1}{a_0^2}\right)V_\infty^2 = \frac{\gamma p_\infty}{2}M_\infty^2$$

Inserting this into Equation (8-12) yields:

$$C_p = \frac{p - p_\infty}{\frac{1}{2}\rho_\infty V_\infty^2} = \frac{p - p_\infty}{q_\infty} = \frac{p - p_\infty}{\frac{\gamma p_\infty}{2}M_\infty^2} = \frac{2}{\gamma M_\infty^2}\left(\frac{p}{p_\infty} - 1\right)$$

QED

The Canonical Pressure Coefficient

The *canonical pressure coefficient* is regarded by many as a better way to represent airfoil pressure distribution. The concept was introduced by A. M. O. Smith [3] to evaluate the adverse pressure gradient and help determine the onset of flow separation. The approach scales the pressure coefficient, so it varies between 0 and 1. This is done by selecting the peak pressure (at the start of the adverse pressure gradient, i.e. where pressure

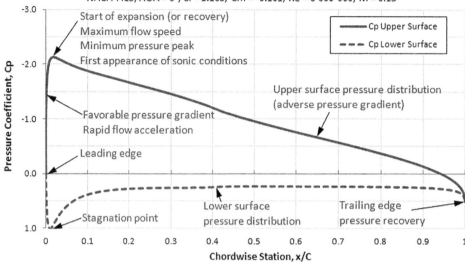

FIGURE 8-5 Important properties of a pressure distribution curves for a typical airfoil.

begins to increase). The canonical pressure coefficient is defined as follows:

$$\overline{C}_p = 1 - \left(\frac{V}{V_m}\right)^2 \quad (8\text{-}15)$$

8.1.6 Chordwise Pressure Distribution

The distribution of pressure along the surface of an airfoil is of fundamental importance for a number of reasons. These range from the determination of structural loads to the magnitudes of drag, lift, pitching moment, shock formation, laminar-to-turbulent boundary layer transition, hinge moments, and many other characteristics.

The pressure distribution is usually shown using the pressure coefficient (see Section 8.1.5, *The pressure coefficient*) plotted separately for the upper and lower surfaces (for instance, see the solid and dashed curves in Figure 8-5, respectively). For clarity, the pressure distribution for the upper surface is plotted above the one for the lower surfaces. Since the pressure coefficients are normally negative for the upper surface, the vertical axis is inverted, with the negative values above the positive ones.

"Conventional" Lift Distribution

Figure 8-6 shows a pressure distribution generated by a conventional cambered airfoil (the NACA 4415) at a subsonic condition and an AOA of $2°$. The thick solid line shows the pressure along the upper surface and the dashed one along the lower surface. It is evident that, at this AOA, the pressure on the upper surface reaches its lowest value relatively close to the leading edge, or around 20% of the chord length. This means that a favorable pressure gradient extends only as far aft as

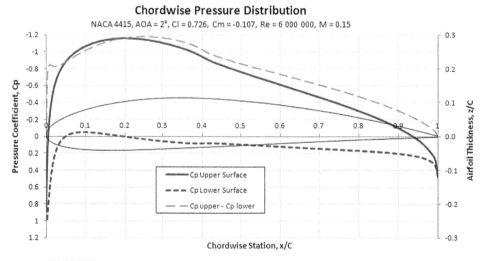

FIGURE 8-6 The chordwise distribution for a conventional airfoil at $AOA = 2°$.

20%. In other words, assuming a smooth surface, the laminar boundary layer is promoted only over the first 20% of the chord length. This does not mean a transition will occur immediately thereafter, but this is very likely unless the surface of the airfoil is super-smooth.

Also note that the pressure distribution along the lower surface starts with stagnation condition ($C_p = 1$ at $x = 0$), which then develops into a low-pressure dip near the 10% of the chord. This is caused by the airflow accelerating around the curved geometry in the area of the leading edge. It highlights that if curvature is present in a fluid flow, the local pressure will always be reduced — an important fact to keep in mind when trying to reduce hinge moments of control surfaces.

Figure 8-6 also shows the pressure differential (difference between the upper and lower surface pressure). It is negative along the entire chord, indicating all segments of the chord are contributing to the lift, albeit in different capacities. For instance, it is evident that the first 50% of the chord contributes substantially more to the overall lift of the airfoil than does the remaining 50%. This fact begs the question: is it possible to design an airfoil whose chordwise distribution would be uniform? Such an airfoil would have to be the most efficient possible, because all chordwise stations would contribute equally, right? Well, strictly speaking the answer is yes, although this is not achievable in reality due to the tendency of air to separate. However, airfoils that attempt this have been designed and are already in use in many aircraft, in particular, sailplanes and high-performance composite GA aircraft. A discussion of their pressure distribution follows.

Stratford Distribution

In contrast, consider the chordwise distribution of the NLF(1)-0414F airfoil shown in Figure 8-7 at the same AOA of $2°$. This distribution forms a distinct flat pressure contour on the upper surface. This contour is commonly referred to as a "rooftop" or Stratford pressure distribution. Such a distribution promotes an extensive laminar boundary layer, provided the surface is sufficiently smooth.

Consider a hypothetical pressure distribution that is uniform along the chord of the airfoil, from the LE to the TE. This would yield the most efficient generation of lift, as each chordwise segment is contributing equally to the total lift of the airfoil. Such a pressure distribution, on the other hand, is physically impossible because this would require the air pressure to change from a low to ambient pressure instantaneously. In real applications, the air pressure must be allowed to rise over a given distance. If the distance is too short, a separation will occur. If the distance is too long, full advantage is not taken of the benefits of the laminar boundary layer. If the distance is just right, the flow will be on the verge of separation and the flat roof will extend as far aft as possible and skin friction drag will be optimized for that airfoil. Modern airfoils are specifically designed to generate a chordwise pressure distribution along the upper surface that is uniform across much of the chord. However, this can only be achieved for a small range of specific AOAs (say $1°-2°$). For this reason, it is imperative to define the mission of the aircraft clearly so the AOA at which it will be operated can be determined and used to design the airfoils.

8.1.7 Center of Pressure and Aerodynamic Center

Center of Pressure

As has already been stated, a body immersed in fluid flow induces a pressure field that, in turn, generates a resultant force and moment. The magnitude of the moment depends on the position on the body about

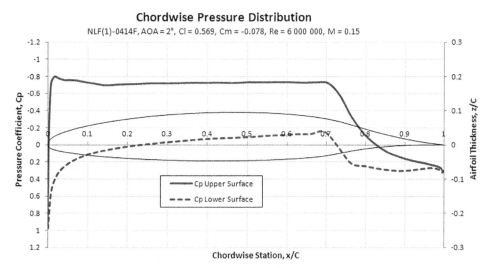

FIGURE 8-7 The chordwise distribution for an NLF airfoil at $AOA = 2°$. The solid line shows the Stratford distribution.

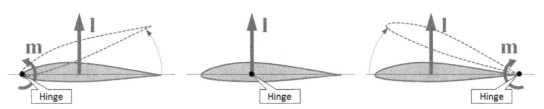

FIGURE 8-8 Center of pressure explained.

which it is measured. Then, the *center of pressure* is defined as the point where the magnitude of the moment equals zero.

To better explain the concept, consider the three identical airfoils shown in Figure 8-8, all of which feature hinges in different places; one on the leading edge (left), the next at the center of pressure (center), and the trailing edge (right). Considering the leftmost airfoil, the lift force generates a moment that will rotate it leading edge down (counter-clockwise). By the same token, the rightmost airfoil, which is hinged at the trailing edge, would be rotated in the opposite direction: leading edge up (clockwise). It follows there must be a point between the leading and trailing edge about which no rotation would take place. This point is the center of pressure. Its location depends on the distribution of pressure along the airfoil and will change with *AOA*.

Aerodynamic Center and Quarter Chord Moment

The *aerodynamic center* is the point on a body about which the aerodynamic moment is independent of the *AOA*. Its presence is plotted in all standard NACA wind tunnel test graphs, where its consequence can be seen as the pitching moment curve that is constant over almost the entire α-sweep range. If the slope of the lift curve, C_{l_α}, and the pitching moment about the quarter chord, $C_{m_{\alpha/4}}$, are known, the aerodynamic center can be computed from:

$$x_{ac} = 0.25 - \frac{C_{m_{\alpha/4}}}{C_{l_\alpha}} \qquad (8\text{-}16)$$

The pitching moment of airfoils is often reported at the aerodynamic center, although it is needed at the quarter-chord for many stability and control problems. The following expression can be used to transfer the moment from the aerodynamic center to the quarter chord:

$$C_{m_{c/4}} = C_{m_{ac}} + C_l(0.25 - x_{ac}) = C_{m_{ac}} + C_{l_\alpha}\alpha(0.25 - x_{ac}) \qquad (8\text{-}17)$$

Derivation of Equation (8-16)

Based on Figure 8-9, the moment about the aerodynamic center can be obtained by summing the forces (l) and moments (m) as follows:

$$m_{ac} = m_{c/4} + l(cx_{ac} - c/4) = m_{c/4} + l(x_{ac} - 0.25)c$$

Dividing through with $q \cdot S \cdot c$, where $q = \frac{1}{2}\rho V^2$ and by referencing Equation (8-6), this can be written in a coefficient form as shown below:

$$\frac{m_{ac}}{qSc} = \frac{m_{c/4}}{qSc} + \frac{l}{qSc}(x_{ac} - 0.25)c \Rightarrow C_{m_{ac}}$$
$$= C_{m_{c/4}} + C_l(x_{ac} - 0.25)$$

This can easily be solved for $C_{m_{c/4}}$ to give Equation (8-17). Then, differentiating with respect to α and recalling that the definition of the aerodynamic center means that the change in moment with change in α is zero, this above is rewritten as shown:

$$\frac{dC_{m_{ac}}}{d\alpha} = \frac{dC_{m_{c/4}}}{d\alpha} + \frac{dC_l}{d\alpha}(x_{ac} - 0.25) = 0$$

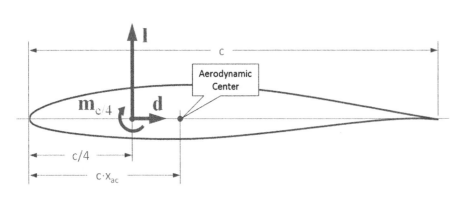

FIGURE 8-9 Location of the aerodynamic center.

Using convention in writing the slopes and solving for x_{ac} results in the following:

$$x_{ac} = 0.25 - \frac{C_{m_{ac/4}}}{C_{l_\alpha}}$$

QED

8.1.8 The Generation of Lift

Generally, the generation of lift can be explained in at least three ways. These are known by their casual names *momentum theorem*, *Bernoulli theorem*, and the *circulation theorem*. We will briefly introduce each in order to provide clarity as each is occasionally referred to in other places in the text. As an interesting side note, some people outside industry and academia have engaged in debates about the importance of the first two methods. There are some who claim that Bernoulli's theorem is wrong and the momentum theorem is the definitive explanation, whilst others claim the opposite. While it can be argued that the momentum theory is easier for laypeople to relate to (see below), such debates are silly and only reveal the participant's limited understanding of aerodynamics. The fact is that all three are excellent ways to describe the generation of lift and all have their pros and cons.

Momentum Theorem

The *momentum theorem* explains lift as the consequence of a wing moving through a mass of air and giving it a downward motion (see Figure 8-10). Since the mass of air is initially at rest, the downward motion means the vertical speed of the air changes from zero to some finite value in a given amount of time. This, in turn, means that a force will be generated in the opposite direction in accordance with *Newton's third law of motion*. The magnitude of this force can be estimated using *Newton's second law of motion*. The second law of motion states it is the rate of change of momentum of the mass of air that generates the force. The third law states an equal force that acts in the opposite direction of the motion of the mass is also generated. It is this force that we call *lift*. A common analogy used to describe this phenomenon is the recoil of a firearm. As is well known, the change in the momentum of a bullet generates a force that acts in the opposite direction of its motion. Thus, the sudden down flow imparted on the air effectively generates "continuous recoil" — the lift.

The downward motion of the air is called *downwash*, here denoted by the letter w, and it represents the vertical speed of air behind the wing. It contrasts with the horizontal speed; a result of the wing moving through air. If we know the downwash and the mass flow of air being deflected, the magnitude of the lift can be

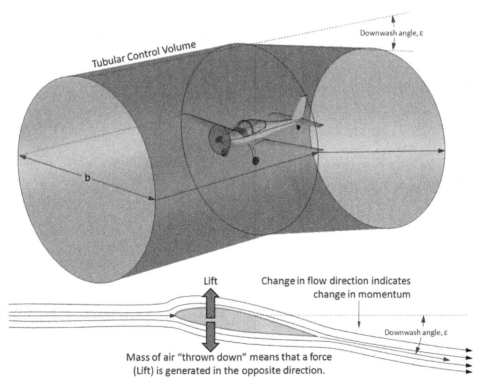

FIGURE 8-10 An airplane's motion causes a downward deflection of a tube of air; in turn, in accordance with Newton's second law of motion, its rate of change of momentum generates a force (lift) in the opposite direction.

estimated, as stated earlier, using Newton's second law of motion:

$$L = \dot{m}w \quad (8\text{-}18)$$

where $\dot{m}$ = mass flow rate inside the cylinder = $\rho V \cdot \pi b^2 / 4$.

The mass flow rate in the stream tube is given by $\dot{m} = \rho A_{tube} V$, where A_{tube} is the cross-sectional area of the stream tube. If it is assumed that the diameter of the stream tube in Figure 8-10 equals that of the wingspan (denoted by b) then the rate of change of momentum (lift force) can be estimated from:

$$L = \dot{m}w = \rho A_{tube} V w = \rho \left(\frac{\pi}{4}\right) b^2 V w \quad (8\text{-}19)$$

Equating this with the standard expression for lift (Equation (8-6)) allows the magnitude of the downwash to be estimated:

$$L = \rho\left(\frac{\pi}{4}\right)b^2 Vw = \frac{1}{2}\rho V^2 S C_L \Leftrightarrow w = 2V\frac{S}{\pi b^2}C_L \quad (8\text{-}20)$$

Noting that $S = b \cdot C_{avg}$ and $AR = b/C_{avg}$ this can be rewritten as follows:

$$w = 2V\frac{b \cdot C_{avg}}{\pi b^2}C_L = 2V\frac{C_{avg}}{\pi b}C_L = \frac{2C_L}{\pi AR}V \quad (8\text{-}21)$$

Since the downwash can approximated by $w = \varepsilon V$ (see Figure 8-10), we can write:

$$\varepsilon = \frac{2C_L}{\pi AR} \quad (8\text{-}22)$$

This equation leads to another result, which as will be shown later, is very helpful in stability and control theory:

$$\frac{d\varepsilon}{d\alpha} = \frac{2}{\pi AR}\frac{d}{d\alpha}(C_{L_0} + C_{L_\alpha}\alpha) = \frac{2C_{L_\alpha}}{\pi AR} \quad (8\text{-}23)$$

Bernoulli Theorem

The *Bernoulli theorem* has been used for decades to explain the formation of lift to generations of engineers and pilots. It postulates that lift is the consequence of the difference in pressure between the upper and lower surfaces of an airfoil, although this would more properly be explained as the resultant of integrating the pressure over the entire surface of a body. Named after the Swiss mathematician Daniel Bernoulli (1700–1782), the theorem stipulates there is a relationship between the pressure and speed of the fluid at a point and along a streamline that goes through that point. This is expressed as follows:

Incompressible Bernoulli equation

$$\frac{p}{\rho} + \frac{V^2}{2} + gz = \text{constant} \quad (8\text{-}24)$$

Compressible Bernoulli equation

$$\left(\frac{\gamma}{\gamma - 1}\right)\frac{p}{\rho} + \frac{V^2}{2} + gz = \text{constant} \quad (8\text{-}25)$$

where

g = acceleration due to gravity, in ft/s^2 or m/s^2
p = pressure, in lb$_f$/ft^2 or N/m^2
V = fluid speed, ft/s or m/s
z = elevation above or below some reference plane, in ft or m
ρ = fluid density, in slugs/ft^3 or kg/m^3
γ = ratio of specific heat for the fluid (1.4 for air at altitudes below 100 km).

Bernoulli's theorem is used daily by thousands of people in industry and academia to estimate the aerodynamic forces acting on a body. Computational fluid dynamics (CFD) software uses the theorem to estimate aerodynamic forces and moments acting on a body with great success.

Kutta-Joukowski Circulation Theorem

The *Kutta-Joukowski circulation theorem* is much more a mathematical method than it is an explanation. Named after the German mathematician Martin Wilhelm Kutta (1867–1944) and the Russian scientist Nikolay Yegorovich Joukowski[1] (1847–1921), the theorem postulates that the lift generated by an airfoil can be considered the product of density (ρ), forward airspeed (V), and a mathematical concept called *circulation* (Γ). In short, the Kutta-Joukowski theorem states that the airfoil's lift can be expressed as follows:

$$L = \rho V \Gamma b \quad (8\text{-}26)$$

where the circulation is calculated using the expression [4]:

$$\Gamma = -\oint_C \vec{V} \cdot d\vec{s} \quad (8\text{-}27)$$

where

$\vec{V}$ = velocity (see Figure 8-11)
C = closed curve in a flow field (see Figure 8-11)

---

[1] This is often spelled Zhukovsky.

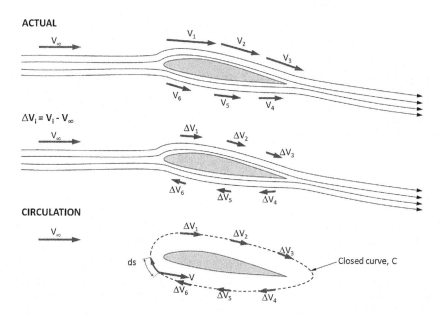

FIGURE 8-11 The top image shows the actual airflow over a wing and that the airspeed along the upper surface the wing is faster than along the lower surface. Assuming the speed at each point is known, the total value can be subtracted from the far-field velocity. This is shown in the center image and reveals that there is a general flow backwards along the upper surface and forward along the lower. This motion is reminiscent of a circulation around the airfoil, shown in the bottom image.

To better understand the concept of circulation, consider Figure 8-11, which shows the airflow around an airfoil. The top figure shows the far-field airspeed, V_∞, and some representative airspeed V_1 through V_6, positioned at selected locations in the flow-field, above and below the airfoil. Then, it is possible to calculate the difference between each of those airspeeds and the far-field airspeed. This is shown as the airspeeds ΔV_1 through ΔV_6 in the center figure. The important observation is that the airspeed differences along the upper surface are positive and point in the flow direction. Conversely, they are negative along the lower surface and point in a direction opposite that of the general airflow. As can be seen in the bottom figure, these differentials effectively form a path around the airfoil. This path is the circulation.

The Kutta-Joukowski theorem is very useful in a number of computational fluid dynamics (CFD) methods, such as the lifting line method (see Section 9.7.1, *Prandtl's lifting line theory*), and the vortex-lattice methods, where it is directly used to estimate the lift and induced drag force generated by a lifting surface.

8.1.9 Boundary Layer and Flow Separation

An understanding of the nature of fluid as it moves over a surface is imperative in aircraft design. This is not only true to explain a number of phenomena that occur on airplanes in flight, but also when wind tunnel testing a scaled model of a larger aircraft and interpreting the results. Much of this understanding is borrowed from *boundary layer theory* (BLT), which describes the nature of viscous flow near the surface of a body through expressions developed using conservation laws and the Navier-Stokes equations. This theory is extensive and while essential to treatment of viscous fluid flow, it is too large in scope to be suitable for this book. For this reason, for theoretical derivations refer to texts such as those of Schlichting [5] and Young [6]. Only important results will be presented in this text.

Reynolds Number

The Reynolds number (R_e) is a measure of the ratio of inertial forces to viscous forces in a fluid flow. It is of great importance in the analysis of the boundary layer. It is defined as follows:

Reynolds number:

$$R_e = \frac{\rho V L}{\mu} \tag{8-28}$$

where

L = reference length (e.g. wing chord being analyzed), in ft or m
V = reference airspeed, in ft/s or m/s
ρ = air density, in slugs/ft^3 or kg/m^3
μ = air viscosity, in lb$_f$·s/ft^2 or N·s/m^2

A simple expression, valid for the UK system at sea-level conditions only, is (V and L are in ft/s and ft, respectively):

$$R_e \approx 6400 V C \tag{8-29}$$

For the SI system at sea-level conditions only, the expression becomes (V and L are in m/s and m, respectively):

$$R_e \approx 68500 V C \tag{8-30}$$

The Effect of Flow Separation

Consider two identical aircraft whose only difference is size or scale. Imagine these are immersed in

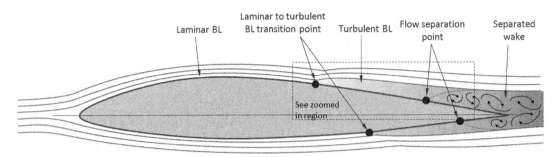

FIGURE 8-12 Three types of fluid flow: laminar, turbulent, and separated on an airfoil. See Figure 8-13 for the zoomed-in region.

fluid flow at an identical airspeed such that, as far as the aircraft is concerned, it differs in *Reynolds numbers* only. Now consider them rotated slowly through an alpha-sweep, from 0° to 90°, while we collect their force and moment coefficients. At first, while the flow is mostly attached, both aircraft will effectively generate identical and linear force and moment coefficients. However, after some *AOA* (perhaps at $\alpha = 8°$) has been reached, the flow begins to separate on the smaller body, while it remains attached on the larger one. The resulting force and moment coefficients for the small body now turn nonlinear, while remaining linear for the larger one. Eventually, the larger body too begins to experience the same effect (perhaps at $\alpha = 12°$). The flow separation causes a large increase in drag and reduction in lift. The pitching moment may increase or reduce depending on the overall geometry. This calls for a distinction in the nature of the flow over the two bodies. Generally, three such types are identified, based on the character of the boundary layer that envelopes the body. They are called *laminar*, *turbulent*, and *separated flow* (see Figure 8-12).

Boundary Layer Transition

A body moving in a fluid will generally be exposed to three distinct types of fluid flow; one is characterized by the presence of a *laminar boundary layer*, another by a *turbulent boundary layer*, and the third by *separated flow* which contains molecules that flow in all possible directions, even upstream (see Figure 8-13). A *laminar boundary layer* occurs when the streamlines inside the boundary layer flow smoothly. In contrast, streamlines inside a *turbulent boundary layer* are chaotic. An important phenomenon takes place as the initially laminar boundary layer changes to a turbulent boundary layer, due to a process called *transition*. When the airflow strikes a smooth body a laminar boundary layer will form immediately, but will later degenerate when it transitions into a turbulent boundary layer. This generally occurs when the local Reynolds number equals approximately 5×10^5. This is referred to as the *transition Reynolds number*.

Unfortunately, the transition is complicated by factors that may change the transition Reynolds number and may cause the transition to happen at a different (lower or higher) Reynolds number [7]. Note that the term *early*

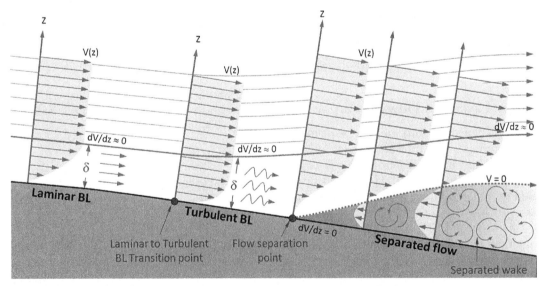

FIGURE 8-13 The nature of fluid flow inside laminar and turbulent boundary layers, and separated flow.

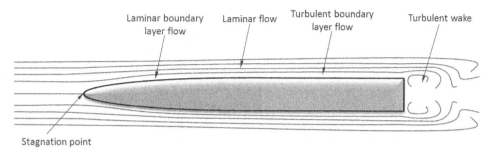

FIGURE 8-14 Flow over an object with a small leading edge radius results in a turbulent wake only.

transition means that the transition takes place earlier than indicated by the value of the Reynolds number. By the same token, *delayed transition* refers to the opposite. Factors that may change the Reynolds numbers:

(1) *Surface roughness.* The transition process is dependent on flow disturbances that take place inside the boundary layer. For this reason, the presence of small surface imperfections (roughness) will excite these disturbances and expedite the transition — even to very low values of the Reynolds number. Also see Section 8.3.13, *The effect of leading edge roughness and surface smoothness.*

(2) *Surface temperature.* The thickness of the boundary layer increases with temperature, as does the energy contained within it. The increase in temperature and thickness is associated with an earlier transition. A cold surface tends to delay transition.

(3) *Pressure gradient.* A *favorable pressure gradient* (i.e. a reduction in static pressure along the flow direction, which is caused by flow acceleration) will stabilize the laminar boundary layer and delay transition. The opposite holds for an *adverse pressure gradient* (i.e. an increase in static pressure associated with a flow deceleration). Also see Section 8.1.6, *Chordwise pressure distribution* and Figure 15-14.

(4) *Mach number.* The transition Reynolds number increases with higher Mach number (i.e. in compressible flow).

Flow Separation

In a *separated flow* the streamlines are separated from the surface. Such flow may actually flow upstream, against the direction of the free stream airflow. Flow separates when the gradient dV/dz equals zero on the surface ($z = 0$). **The external geometry of an airplane should be shaped so the areas of flow separation at the mission condition are minimized, if not eliminated.**

All three types usually manifest themselves in fluid flow in the order shown in Figure 8-13. This typically occurs on bodies as shown in Figure 8-14 or Figure 8-15. The former shows flow over an object that features a rounded, small-radius leading edge (LE). The flow forms a laminar boundary layer on the forward part of the object (the bow), followed by a transition to a turbulent boundary layer, and, finally, by the flow separating into a turbulent wake at the stern. Note that the streamlines outside the boundary layer are smooth and can be treated as if they belonged to an inviscid fluid. In fact, for computational efficiency, this is how most Navier-Stokes solvers treat fluid flow: viscid inside the boundary layer, but inviscid outside of it.

Figure 8-15 shows a different scenario, in which an object with a very blunt LE causes flow separation on the LE that results in (highly) turbulent, circulatory flow inside the flow separation region. The laminar streamlines flowing over this region attach aft of this region before separating into the turbulent wake at the stern.

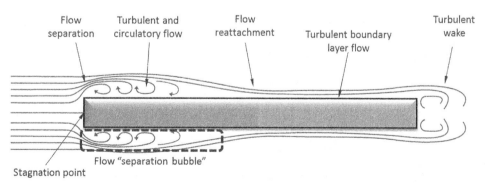

FIGURE 8-15 Flow over a blunt LE object results in an immediate forward flow separation, subsequent reattachment, and finally, a turbulent wake. The forward flow separation region is called a "separation bubble."

The forward flow separation region is usually referred to as a "separation bubble." Such bubbles can also form on objects that are considerably less blunt than the one in Figure 8-15. They can even occur on airfoils if their geometry is conducive to such formation. This may be caused by sudden change in curvature. Besides an increase in drag, separation bubbles on airfoils may lead to very detrimental stall characteristics.

Separation bubbles can occur on any aircraft, in particular at the wing/fuselage juncture. However, the phenomenon also occurs on lifting surfaces, in particular on small chords (or low Reynolds numbers). As discussed in Section 8.2.10, *Famous airfoils*, The NACA 23012 airfoil is used for a number of GA aircraft in spite of its abrupt stall characteristics, which are attributed to the formation of a separation bubble. But they are a prevalent problem for even smaller Reynolds numbers, such as those in which radio-controlled aircraft operate (60,000−500,000). In this region a stable bubble may form in the laminar boundary layer along the leading edge of the wing, increasing the drag of the vehicle. This is referred to by many as "bubble drag" [8]. One way of reducing this drag is to design the airfoil such that its transition ramp[2] reduces the chance of a bubble formation (refer to Ref. [8] for more details). Another way is to place a transition strip (often called a "trip strip") along the leading edge to force the laminar boundary layer to transition into a turbulent one without forming the separation bubble. Currently this is a trial-and-error approach and a trip strip that is ideal for one Reynolds number may be detrimental for another one. Research shows their effect is contingent upon the size of the bubble, its intensity, *AOA*, and geometry of the airfoil.

Factors Affecting Laminar Flow

The laminar boundary layer is very sensitive to a number of factors. Some are out of the control of the designer, whereas others are not and, effectively, depend on his awareness. Among those are (list partially based on Bertin [9]):

1. Geometry (e.g. sweep and surface curvature).
2. Surface smoothness (or lack thereof).
3. Surface temperature.
4. Compressibility effects (Mach number, Reynolds number).
5. Atmospheric conditions (ice crystals, rain).
6. Manufacturing quality (waviness, smoothness, steps and gaps in surface joints).
7. Leading edge quality (insects, dirt, erosion, icing).
8. Suction or blowing at the surface (surface openings, distribution of boundary layer control).
9. Noise (engine, propwash).

8.1.10 Estimation of Boundary Layer Thickness

The estimation of the thickness of the boundary layer has many applications. As an example, it is essential in the design of boundary layer diverters for jet engine installation. It is also used in many computational fluid dynamics applications, when fluid flow is being modeled as inviscid and the geometry of the immersed body must be modified (enlarged) to account for the thickness of the viscous layer.

The thickness of the boundary layer depends on whether it is laminar or turbulent. A turbulent boundary layer grows at a faster rate and is thicker than a laminar boundary layer at the same conditions. Their thicknesses can be estimated from (based on Ref. [9]):

Laminar BL:

$$\delta_{lam} = \frac{5x}{\sqrt{R_e(x)}} = 5\sqrt{\frac{\mu}{\rho V}}\sqrt{x} \qquad (8\text{-}31)$$

Turbulent BL:

$$\delta_{turb} = \frac{0.3747 \cdot x}{(R_e(x))^{0.2}} = \frac{0.3747 \cdot x^{0.8}}{(\rho V/\mu)^{0.2}} \qquad (8\text{-}32)$$

The dimensions δ_{lam} and δ_{turb} are defined in Figure 8-16.

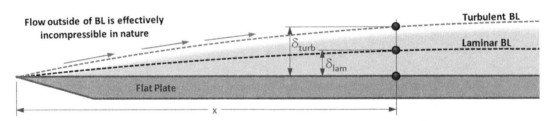

FIGURE 8-16 The thickness of the boundary layer depends on whether it is laminar and turbulent.

---

[2]A transition ramp is used to describe the shape of the pressure coefficient curve aft of where it peaks. The ramp can be tailored to modify the adverse pressure gradient to control the transition of the laminar into the turbulent boundary layer without promoting the formation of a large separation bubble.

8.1.11 Airfoil Stall Characteristics

In this text, *stall* refers to the flow condition that follows the first peak of the lift curve. It is a consequence of the formation of a large separation located between the leading and trailing edges of the airfoil. The thickness of the airfoil largely dictates how flow separation develops on the airfoil. If the airfoil is thick, the separation tends to begin at the trailing edge and move forward as the *AOA* increases. On the other hand, if the airfoil is thin, the separation tends to begin at the leading edge in the form of a separation bubble. This has a profound effect on the maximum subsonic section lift coefficient as well as drag. However, other parameters besides airfoil thickness affect the maximum lift as well. Among those are the location of the airfoil's maximum thickness, camber and its chordwise location, Mach number, Reynolds number, free-stream turbulence, and the surface condition (roughness) [2]. All affect the nature of the flow separation that ultimately places an upper limit on the maximum lift. Reference 10 classifies the nature of the flow separation in terms of *trailing edge* and *leading edge* stalls, as described below.

Trailing Edge (TE) Stall

This is undoubtedly the best known among the three types of stalls and typically occurs on thick airfoils, whose thickness-to-chord ratio is around 12% or greater [10]. Such airfoils are characterized by a smooth change in C_l and C_m between the negative and positive lift peaks (see C_{lmax} and C_{lmin} in Figure 8-3 and Section 8.1.4, *Properties of typical airfoils*) and often well beyond those. The shape of the peak of the lift curve is rounded with a mild drop in C_l. Growth in post-stall drag polar is gradual, and the pitching moment curve is without sharp breaks. The flow stays mostly attached to an *AOA* of about 10°, beyond which it progressively moves forward. At maximum lift the flow is separated to approximately mid-chord the rear half of the airfoil.

Leading Edge (LE) Stall

Leading edge stalls are much less familiar to the General Aviation community than TE stalls. They are associated with sharp leading edge radii of thin airfoils (which are almost never used for GA aircraft — excluding the NACA 23012 airfoil, which is used on a number of GA aircraft). The sharp LE results in a larger pressure peak than airfoils with larger radius LE. This is evident in Figure 8-17, which shows the thinner NACA 4408 airfoil peaks at $C_p \approx -16.5$ versus $C_p \approx -6.0$ for the thicker NACA 4415 airfoil. As a consequence, the pressure recovery, behind the peak for the thinner airfoil, is much steeper than for the thicker one. This is a problem nature solves by separating the laminar boundary layer. This separation can occur well below the *AOA* where the airfoil actually stalls. It forces the laminar boundary layer to transition into a turbulent one, forming a bubble of trapped low-energy air between the surface of the airfoil and the boundary layer. An example of such a bubble is shown in Figure 8-18. On a wing the bubble is a spanwise vortex.

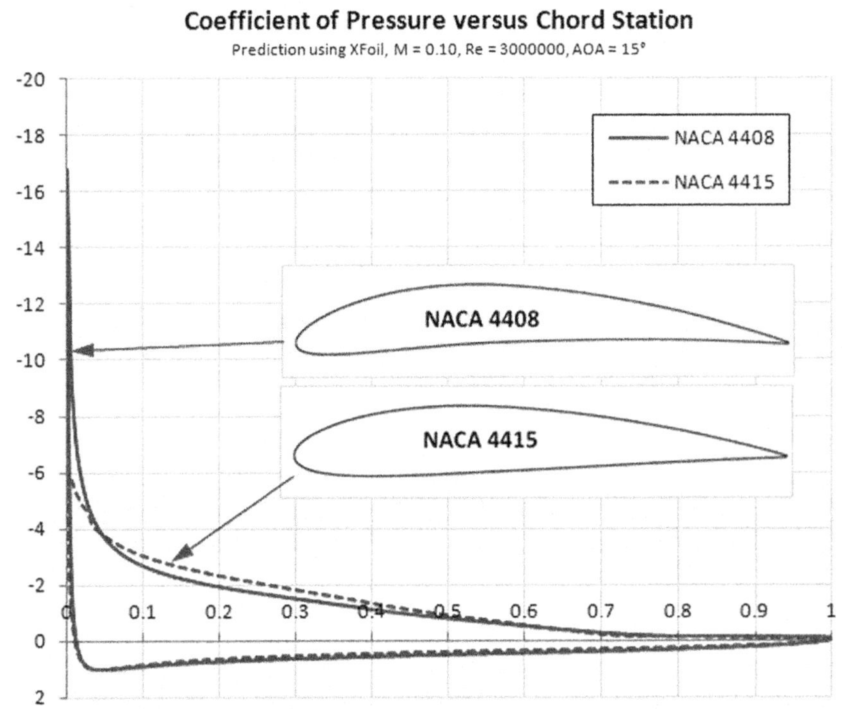

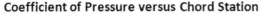

FIGURE 8-17 Chordwise pressure distribution for a thin (NACA 4408) and a thick (NACA 4415) airfoil at $AOA = 15°$, showing the thinner airfoil results in a much greater pressure peak than the thicker one.

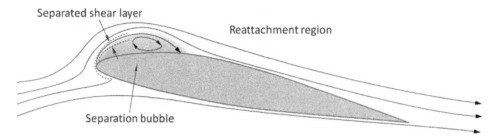

FIGURE 8-18 Formation of a separation bubble on a thin airfoil (based on Ref. [11]).

Research has shown that the bubbles come in two distinct forms, short or long, and each predicates different scenarios [11]. The difference between the two is determined using Owen's criterion, which computes the Reynolds number of the displacement thickness of the boundary layer using the following expression:

$$R_{\delta_1} = \frac{V}{\nu} \delta_1 \quad (8\text{-}33)$$

where V is the velocity at the edge of the boundary layer, ν is the kinematic viscosity and δ_1 is the displacement thickness. In Ref. [12] this value was always found to be greater than 500 for a short bubble and less than 500 for a long one. Note that the size of the bubble, in effect, depends on the free stream Reynolds number, so it is possible the stall behavior of the airfoil changes from one to the other. If the Reynolds number is high enough, chances are no bubble will form. The difference between the two is as follows:

(1) **Short-bubble leading edge stall:** the length of the bubble is about 1% of the chord at low $AOAs$, but reduces in size with increased AOA. The bubble has limited effect on the pressure distribution and high peak suction can continue to rise despite the bubble's presence up until some specific AOA, when the flow abruptly and finally separates from the airfoil's surface. This results in a violent stall, accompanied by large change in lift and pitching moment [10].

(2) **Long-bubble leading edge stall:** the length of the bubble is about 2—3% of the chord at a low AOA. However, this grows rapidly with AOA until a reattachment fails to take place, causing the bubble to combine with the full flow separation over the airfoil. A long bubble will affect the pressure distribution over the airfoil in profound ways and will cause the peak suction to collapse. The maximum lift for the long bubble is less than that for the short one, but the stall is less abrupt [10].

The maximum lift coefficient of thin airfoils that stall because of flow separation at the leading edge can be determined based on the leading edge geometry. A so-called *leading-edge parameter*, which is defined as the difference between the upper-surface ordinates of the airfoil at the 0.15% and 6% chord stations, has been used for this purpose with good results [13] (for instance see Section 9.5.12, *Step-by-step: C_{Lmax} estimation per USAF DATCOM method* 2). Additional correction is required for thicker airfoils. Compressibility effects are important on thick airfoils as they cause reduction in the maximum lift coefficient starting at $M \approx 0.2$. Reference [13] also presents a method to estimate the maximum lift coefficient of several airfoils. The three types of airfoil stalls are illustrated in Figure 8-19.

8.1.12 Analysis of Ice-accretion on Airfoils

Flight into inclement weather is commonplace nowadays, with aircraft being certified to fly into known icing conditions that arguably represent the greatest challenge any aircraft can be exposed to. While the engineering of ice removal is beyond the scope of this section, it is helpful for the aircraft designer to understand the fundamentals of ice-accretion on airfoils.

NASA's Glenn Research Center has been a pioneer in the development of computational methods that estimate the collection of ice (or ice accretion) on an airfoil's leading edge. This development included the implementation of these methods in a computer code called LEWICE (after the research center's former name Lewis Research Center). The code accurately predicts the growth of ice under a range of meteorological conditions and, because it has been so extensively validated by NASA, is considered very reliable in industry.

Generally, such codes work as follows. First, a table containing the geometric description of the airfoil (i.e. its x- and y-coordinates) is read and analyzed using a panel-code solver. This will allow the code to determine the flow field around the airfoil, including the stagnation points. Then, the accumulation of ice at the stagnation points over a specific time (which the user may specify directly or have the program determine) is estimated. Clearly this will modify (or grow) the initial geometry and this is used to define a new geometry (i.e. the airfoil

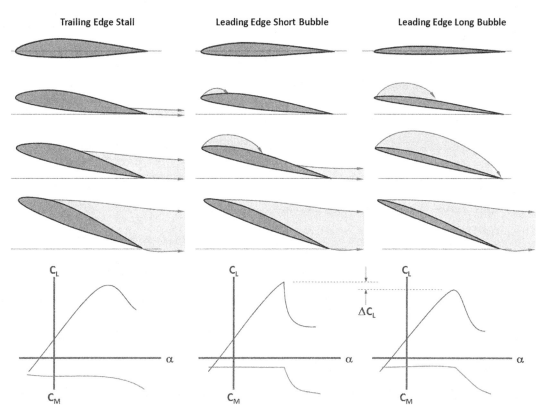

FIGURE 8-19 Types of stall and its effect on the lift and pitching moment curves (based on Refs [11] and [12]).

with the small amount of ice growth or *accretion*), which is then fed again into the solver as the input "airfoil" for the next iteration. The process is then repeated for a specified time. The user will have to specify various properties of the air during the flight condition, such as its relative humidity, liquid-water content, droplet size, temperature, airspeed, and other parameters. Figure 8-20 shows an example output from LEWICE for a common GA airfoil, the NACA 64-215. It shows that ice accretion is a formidable foe for aircraft wings.

The purpose of such research is to provide aircraft manufacturers with a reliable tool to estimate what is called the *impingement limits*, which are points on the upper and lower surfaces beyond which limited ice accretion takes place. It must be defined to determine how far aft of the leading edge ice protection must wrap. Impingement limits are determined for a variety of flight conditions and atmospheric conditions and are ultimately based on the collectively aft-most limits.

8.1.13 Designations of Common Airfoils

A large number of different airfoils have been designed since the dawn of flight. These offer a range of properties, some ideal, while others are less desirable.

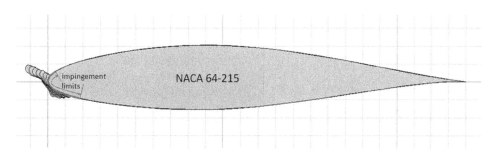

FIGURE 8-20 A LEWICE prediction showing ice accretion on an unprotected NACA 64-215 airfoil after 45 minutes exposure to supercooled liquid water at $-4.75\ °C$ (23.4 °F) at an airspeed of 90 m/s (295 ft/s) and *AOA* of 4°. The chord is 1.0 m (about 40 inches).

As a consequence, the sheer number sometimes makes the airfoil selection process a bit daunting. The literature[3] presents a number of airfoil designations that appear perplexing at first, although eventually one discovers the same airfoil designations appear repeatedly. Table 8-1 lists designations of airfoils that are found in use on various airplanes.

TABLE 8-1 Designations of Common Airfoils

AG	Ashok Gopalarathnam, an independent airfoil designer
ARA	The Aircraft Research Association Ltd, Britain
Clark	Col. Virginius Clark of the NACA
Davis	David Davis, an independent airfoil designer
DESA	Douglas El Segundo Airfoil
DLBA	Douglas Long Beach Airfoil
Do	Dornier
DSMA	Douglas Santa Monica Airfoil
DFVLR	The German Research and Development Establishment for Air and Space Travel
DLR	The German Aerospace Center
Drela	Dr. Mark Drela of MIT
EC	The National Physical Laboratories, Britain
Eiffel	Gustave Eiffel, an early French aeronautical researcher
Eppler	Dr. Richard Eppler of the University of Stuttgart
FX	Dr. F.X. Wortmann of the University of Stuttgart
GU	University of Glasgow in Scotland
Gilchrist	Ian Gilchrist of Analytical Methods Inc.
Gottingen	The AV Gottingen aerodynamics research center in Germany
Joukowsky	Nicolay Egorovich Joukowsky, an early Russian aeronautical researcher
K	Dr. Yasuzu Naito of Nakajima
LB	Dr. Ichiro Tani of Tokyo University
Liebeck	Dr. Robert Liebeck of McDonnell Douglas, now Boeing
Lissaman	Dr. Peter Lissaman of AeroVironment Inc.
MAC	Airfoils designed at Mitsubishi. During the 1940s, the designer was Tsutomu Fujino
McWilliams	Rick McWilliams, an independent airfoil designer
Narramore	Jim Narramore of Bell Helicopter Textron
NACA	The US National Advisory Committee for Aeronautics

TABLE 8-1 Designations of Common Airfoils—cont'd

NASA	The US National Aeronautics and Space Administration
NN	Dr. Hideki Itokawa of Nakajima
NPL	The National Physical Laboratories, Britain
Navy	The US Navy, Philadelphia Navy Yard
Onera	The French National Aerospace Research Establishment
RAE	The Royal Aeronautical Establishment, Britain
RAF	The National Physical Laboratories, Britain
Riblett	Harry Riblett, an independent airfoil designer
Roncz	John Roncz, an independent airfoil designer
Selig	Dr. Michael Selig of the University of Illinois, Urbana-Champaign
Somers	Dan Somers of Airfoils Inc.
TH	Dr. Tatsuo Hasegawa of Tachikawa
TsAGI	The Russian Central Aerodynamics and Hydrodynamics Institute
USA	The US Army
Viken	Jeff Viken of NASA Langley Research Center

Primarily based on http://www.public.iastate.edu/~akmitra/aero361/design_web/airfoil_usage.htm.

8.1.14 Airfoil Design

During the history of aviation, thousands of different airfoils have been designed for applications ranging from aircraft, turbo machinery, wind turbines, propellers, and even ships (hydrofoils). It is likely that a useful airfoil for a new design may be found in that database. However, the modern manufacturer of aircraft will instead often opt for airfoils specifically designed for new aircraft. This way, the new airfoil may have a higher maximum lift than the older airfoil, but this also guarantees the aircraft can truly operate at a minimum drag during its cruise mission. The latter can have a profound impact on the bottom line for the customer and result in substantial savings in fuel costs in the process, making the airplane more marketable. The cost of designing an airfoil is usually a minor expenditure of the complete development program, in particular for commercial aircraft, and is therefore something that should be seriously considered at the beginning of the design process.

Airfoils are typically designed by two means; *direct analysis* and *inverse design*. Nowadays, this is always done using computer software. Two-dimensional computer programs such as Xfoil [14]; XFLR5 (this program

---

[3]This means documents such as *Jane's All the World's Aircraft*, technical data sheets for aircraft, aerospace engineering books, scientific publications, etc.

is actually a user-friendly interface for Xfoil, which it runs in the background) [15]; the Eppler Code [16]; or AeroFoil [17], are widely used and run on any personal computer (PC). Two of these, Xfoil and XFLR5, are even distributed free of charge. All of the following programs allow polars (C_L versus α, C_D versus C_L, etc.) to be plotted and airfoils to be designed using an *inverse design* method.

Xfoil and XFLR5

Xfoil is probably the best known of the above codes. It dates back to 1986 and was written by Dr. Mark Drela, an aerodynamics professor at the Massachusetts Institute of Technology. It uses a high-order panel method and a fully-coupled viscous/inviscid interaction method to evaluate drag, boundary layer transition and separation. Xfoil is widely used in the aircraft industry and generally speaking is a reliable tool, although, in the view of this author, it suffers from a poor user interface when compared to many other codes. The user of modern computer operating systems is averse to the unfriendly interfaces of the bygone MS-DOS era. This has been solved in a program called XFLR5 [18], developed by Mr. André Deperrois, which makes Xfoil analyses much easier to perform (see Figure 8-21).

Xfoil allows the user to perform viscous and inviscid analysis of existing airfoils. The user can specify where a laminar boundary layer transitions into a turbulent one, or have the program predict the movement of the transition point with *AOA*. The viscous analysis can be used to predict C_l, C_d, and C_m to just beyond C_{lmax}, and uses Karman-Tsien compressibility correction at high subsonic airspeeds (see Section 8.3.6). The program allows the user to simulate control surface deflection by specifying hinge point and deflection angle.

PROFILE ("The Eppler Code")

The software PROFILE was written by Dr. Richard Eppler of the University of Stuttgart and Dan Somers, a consulting aerodynamicist. The program uses a conformal-mapping method for the design of airfoils for low-speed applications with prescribed velocity-distribution characteristics. The program is claimed to be user-friendly by the distributer of the program, www.airfoils.com/.

AeroFoil

The software AeroFoil was developed by Mr. Donald Reid, a professional nuclear engineer who has a background in aerospace engineering, and "is intended to be the most 'user-friendly' of its type," as stated on its website, aerofoilengineering.com/index.htm. The software uses a vortex-panel method coupled with integral boundary layer equations to calculate the aerodynamic properties of airfoils. It allows up to three airfoils to be compared simultaneously. Validation examples are provided on the website and show the predictions made by the program are in good agreement with experiment.

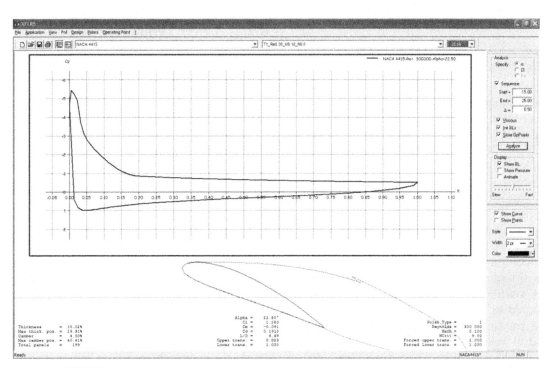

FIGURE 8-21 The XFLR5 user interface makes it easier to access the capabilities of Xfoil.

JavaFoil

JavaFoil is simple and easy-to-use software developed by the German aerodynamicist Dr. Martin Hepperle. The program performs a potential flow analysis using a higher-order panel method, in which the vorticity varies linearly along each panel representing the airfoil. Then, an integral boundary layer method is applied, using a separate boundary layer analysis module. Beginning at the stagnation point, the method solves a set of differential equations to help evaluate the characteristics of the boundary layer. According to information on the developer's website, the equations and criteria for transition and separation were developed by Dr. Eppler. The program is free and available from www.mh-aerotools.de/airfoils/javafoil.htm. It provides a routine that allows a large number of airfoils to be generated.

Design Process

Once a decision has been made to design an airfoil, the first step in the design is to list the desired characteristics. Such a list can consist of a range of operational lift coefficients and conditions (Mach number, Reynolds number) at which specific properties such as maximum value of the lift should occur, desirable entry into the post-stall region, or where minimum drag should occur, extent of laminar flow, the magnitude of the pitching moment coefficient, as well as desirable geometric characteristics such as thickness and its location along the chord. The next step is to decide on a methodology: direct or inverse method (see below). Some designers use existing airfoils as a baseline and modify them, typically using both methods and a trial-and-error approach until most of the desired characteristics have been achieved.

Direct Analysis Method

A *direct analysis* is the evaluation of the pressure field around an airfoil that has already been defined. In direct analysis the ordinates of the airfoil are entered into software such as Xfoil, PROFILE, or AeroFoil and then the software will predict lift, drag, and pitching moment of the airfoil at the angles-of-attack specified. The most important feature of such software is an accurate prediction of flow separation growth with AOA, subsequent stall, and width and depth of the drag bucket at lower AOA. The software packages cited above are all capable of such predictions, although the accuracy of the results is not being commented on. The software cited sometimes yields different predictions for an identical airfoil at identical conditions. It is always the responsibility of the user to learn to properly use the software and this requires the predicted results to be compared with reliable wind tunnel tests.

The direct analysis method is not an ideal way to design an airfoil for a desired pressure distribution, as this would require substantial trial and error to accomplish. The *inverse airfoil design method* discussed below is a far better approach.

Inverse Airfoil Design Method

The inverse method allows the airfoil designer to specify a specific velocity distribution along the surface, which is then used to calculate the geometry that will generate such a distribution. The knowledgeable designer will know the consequences of a specific speed distribution, where it will promote laminar flow or cause early separation, and so on. For this reason, airfoil design is a field of specialization and calls for the evaluation of large numbers of airfoils to build a database of behavior based on experience. Inverse methods were responsible for significant advances in airfoil design in the 1950s, when enough computational power was realized to allow integral boundary layer methods to be coupled with potential-flow solutions. The computational prowess of the modern computer is now used to implement such methods using the full Navier-Stokes equations.

8.2 THE GEOMETRY OF THE AIRFOIL

This section presents important properties of the geometry of the airfoil and a number of famous airfoils the aircraft designer should be aware of, as some offer interesting possibilities, while others should be avoided.

8.2.1 Airfoil Terminology

The discussion of airfoils in this text assumes the designer is familiar with the concepts defined below in Figure 8-22 and Figure 8-23.

Thickness, Mean-line and Camber

NACA defines airfoils based on a specific thickness distribution and mean-line. The camber is defined as the maximum distance between the mean-line and the chordline. Camber strongly affects the downwash behind the airfoil and, thus, how much lift is generated. The rule-of-thumb is that the larger the camber, the greater the maximum lift of the airfoil and greater the thickness the greater the stall angle-of-attack and drag. Of course there are exceptions. Generally, the greater the camber the greater is the drag as well.

LE Radius

The leading edge radius impacts the maximum lift of the airfoil, as well as its drag in cruise. Generally, the larger the radius the more lift will be generated at high AOA, as this delays flow separation near the LE. This

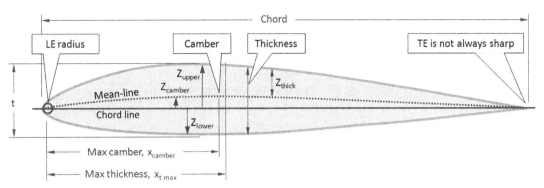

FIGURE 8-22 Airfoil nomenclature for a conventional airfoil.

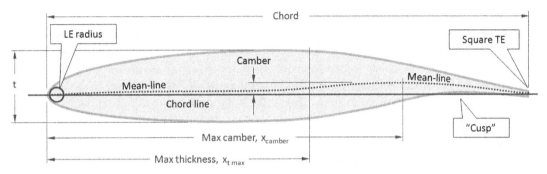

FIGURE 8-23 Airfoil nomenclature for a modern natural laminar flow airfoil.

often manifests itself as a less abrupt reduction in lift at stall ("stall break"). Large radius can also increase the airfoil's drag, although this is also dependent on the geometry of the airfoil downstream.

Square TE

A square trailing edge is sometimes employed to decrease adverse pressure gradients[4] on airfoils. This is important for NLF airfoils (which feature the maximum camber way back along the chord) to help stabilize the boundary layer on the aft part of the airfoil. This way, the formation of a separation bubble is prevented and, consequently, both lift and drag characteristics are improved. It is of importance how the TE is squared. A sharp trailing edge cannot just be made blunt, as this will not increase the thickness of the airfoil upstream. Rather, the TE must be deliberately thickened to improve adverse pressure gradients. Blunt TE airfoils are sometimes called "flatbacks."

Geometric Description

The geometric description of airfoils is generally given in so-called *ordinate tables*. Such tables often list the airfoils upper and lower surfaces separately in terms of x and z coordinates. The x-axis is called the chord line and the x-ordinate represents the chordwise station and the z-ordinate the height above or below the x-axis. Then we define the mean-line and thickness as follows:

$$\text{mean} - \text{line} = m = \frac{1}{2}\left(z_{\text{upper}} + z_{\text{lower}}\right) \quad (8\text{-}34)$$

$$\text{thickness} = t = z_{\text{upper}} - z_{\text{lower}} \quad (8\text{-}35)$$

If the mean-line and thickness are tabulated rather than the ordinates, then these can be obtained from:

$$z_{\text{upper}} = m + \frac{1}{2}t \quad (8\text{-}36)$$

$$z_{\text{lower}} = m - \frac{1}{2}t \quad (8\text{-}37)$$

8.2.2 NACA Four-digit Airfoils

The NACA four-digit airfoils were the product of Eastman N. Jacobs and his colleagues at the NACA Variable-Density Wind Tunnel, who around 1929 demonstrated that the characteristics of an airfoil are largely dependent on its thickness and mean-line [19].

---

[4]An adverse pressure gradient is the rate of rise in pressure from the low pressure peak on the airfoil to the atmospheric pressure aft of the trailing edge. If this pressure rise occurs too rapidly, air will separate and a thick wake will form.

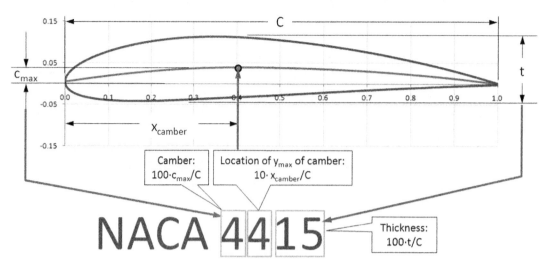

FIGURE 8-24 Interpretation of NACA four-digit airfoil designation.

This allowed the airfoils to be described using a mathematical formulation and a designation system that reflects the airfoil's geometric properties. These airfoils have designations like 2412, 3308, or 4415 (see Figure 8-24). The figure shows how the digits are interpreted. Further development of these airfoils for propellers was done by Albert von Doenhoff [20]. The development NACA 4-digit series airfoils are detailed in Refs [19,20].

Applications

The airfoils are widely used in GA aircraft, with the best-known aircraft being a family of Cessna airplanes. Cambered versions are used for wings, while symmetric ones are used for HT and VT. Symmetric airfoils are also used for helicopter rotors, antennas, and even for some supersonic aircraft and missile fins.

Numbering System

As shown in Figure 8-24, the numbering system is based on the geometry of the airfoil. The first digit indicates the camber in percentage of chord. The second digit indicates the distance from the leading edge to the maximum value of the camber in tenths of the chord. The last two digits indicate the thickness of the airfoil in percent of the chord. Thus, the NACA 4415 airfoil has a 4% camber located at 40% of the chord and is 15% thick. Also, NACA 0009 is a symmetrical airfoil as indicated by the first two digits 00. The airfoil is 9% thick.

Computation of Airfoil Ordinates

The geometry of the airfoil can be determined by computing the x- and y-values in accordance with the Step-by-step that follows. The algorithm is based on the report NACA TN-460 and is set up for analysis using a spreadsheet. Note how the ordinates of the upper and lower surfaces are calculated. These are rotated with respect to the slope of the mean-line. Thus, the x-value of the point on the upper surface is not the same as that of the point on the lower surface. This necessitates the evaluation of the slope of the mean-line and the rotation about the point (x, y_c) as shown in Figure 8-25.

The definition of airfoil is always based on the assumption that the leading edge is located at

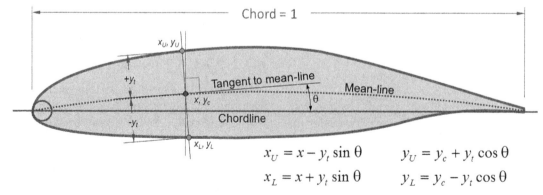

$$x_U = x - y_t \sin \theta \qquad y_U = y_c + y_t \cos \theta$$
$$x_L = x + y_t \sin \theta \qquad y_L = y_c - y_t \cos \theta$$

FIGURE 8-25 Determination of the ordinates for the upper and lower surfaces of a NACA 4-series airfoil.

$x = 0$ and the trailing edge at $x = 1$. The resulting chord is of a unit length. The advantage of defining the airfoil in terms of a unit chord is that it can be easily scaled up by a direct multiplication of the ordinates by the desired chord. The geometric definition is in the form of a table containing the x- and y-coordinates of the airfoil. As stated in the previous section, such a table is called an *ordinate table*, where the word ordinate refers to the y-value of a specific co-ordinate.

The ordinate table lists the x-values for the airfoil by distributing them along the x-axis. While the x-values are sometimes distributed uniformly, a more desirable distribution is based on a cosine scheme as shown in Figure 8-26. Doing so will guarantee a better definition of the leading edge, where curvature is greater. This scheme consists of a unit circle, which is uniformly sectored at an angle $\Delta\phi$. The value of $\Delta\phi$ is determined from the number of points as follows:

$$\Delta\phi = \frac{90°}{N-1} \quad \text{or} \quad \Delta\phi = \frac{\pi}{2(N-1)} \quad (8\text{-}38)$$

Now consider the thick dark line extending from $x = -1$ to 0 in Figure 8-26 (QII). The x-values of the intersection of the sector lines and the circle are projected vertically on to this line, revealing a tight separation of points close to $x = -1$. The spacing of the vertical line gradually gets more sparse as we approach $x = 0$. If we shift this pattern by adding 1 to each x so obtained, the resulting set of x-values results in a tight separation close to $x = 0$ (the leading edge of our airfoil) and more sparse as we approach the trailing edge at $x = 1$.

To prepare this scheme we number each x using indexes ranging from 1 through N, where N is the number of points. Mathematically, we can write:

$$\begin{aligned} x_1 &= 0 \\ x_2 &= 1 - \cos(\Delta\phi) \\ x_3 &= 1 - \cos(2\Delta\phi) \\ &\vdots \\ x_i &= 1 - \cos((i-1)\Delta\phi) \\ &\vdots \\ x_N &= 1 - \cos((N-1)\Delta\phi) = 1 \end{aligned} \quad (8\text{-}39)$$

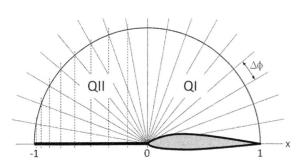

FIGURE 8-26 A preparation of a cosine scheme.

These definitions allow us to prepare the following Step-by-step to calculate the geometry of any 4-digit airfoil.

Step 1: Preliminary Values

Decide the thickness ratio, camber, and location of the camber using the variables t, C, and x_{camber}, respectively. If the thickness ratio is 15%, then $t = 0.15$. If the camber is 4%, then $C = 0.04$. If the location of the camber is 40%, then $x_{camber} = 0.4$. (These numbers represent the NACA 4415 airfoil.)

Step 2: Airfoil Resolution

Decide how many x-values (and thus y-values) to include in the analysis. Here, let's call that value N (e.g. $N = 100$ for 100 points).

Step 3: Prepare Ordinate Table

Tabulate the x-ordinates along the unit chord using the cosine scheme where $\Delta\phi$ is calculated using Equation (8-38). These values should range from 0 to 1, where $x = 0$ represents the leading edge and $x = 1$ the trailing edge.

Step 4: Calculate Thickness

Calculate the thickness of the upper and lower surfaces of the airfoil for each x-value, from:

$$\pm y_t = 5t\big(0.29690\sqrt{x} - 0.12600x - 0.35160x^2 \\ + 0.28430x^3 - 0.10150x^4\big) \quad (8\text{-}40)$$

Step 5: Compute the y-value for the Mean-line

The next step involves computing the y-value of the mean-line for each x, which we call y_c. This depends on whether x is larger or smaller than the location of the camber and is given by:

If $x \leq x_{camber}$ then

$$y_c(x) = C\frac{(2x_{camber} - x)x}{x_{camber}^2} \quad (8\text{-}41)$$

If $x > x_{camber}$ then

$$y_c(x) = C\frac{(1 - 2x_{camber}) + 2x_{camber}x - x^2}{(1 - x_{camber})^2} \quad (8\text{-}42)$$

Step 6: Calculate the Slope of the Mean-line

Calculate the slope of the mean-line at the point:

If $x \leq x_{camber}$ then

$$\frac{dy_c}{dx} = \frac{2C}{x_{camber}}\left(1 - \frac{x}{x_{camber}}\right) \quad (8\text{-}43)$$

If $x > x_{camber}$ then

$$\frac{dy_c}{dx} = \frac{2C(x_{camber} - x)}{(1 - x_{camber})^2} \quad (8\text{-}44)$$

Step 7: Calculate the Ordinate Rotation Angle

Calculate the angle θ as follows:

$$\theta = \tan^{-1}\left(\frac{dy_c}{dx}\right) \quad (8\text{-}45)$$

Step 8: Calculate the Upper and Lower Ordinates

Calculate the upper and lower surface ordinates as follows:

$$\begin{array}{ll} x_U = x - y_t \sin\theta & y_U = y_c + y_t \cos\theta \\ x_L = x + y_t \sin\theta & y_L = y_c - y_t \cos\theta \end{array} \quad (8\text{-}46)$$

Generation of the NACA 4415 – an Example Implementation

This procedure has been implemented as shown in Figure 8-27, for a NACA 4415 airfoil, using $N = 30$. More points should be used for a serious design project. Let's perform a sample calculation for the row with ID number 10 or $i = 10$ (see the highlighted row in the figure). Beginning with column 2, we calculate the cumulative sector angle ϕ based on Equation (8-38) as follows:

$$\phi = (i-1)\Delta\phi = (i-1)\frac{90°}{N-1} = (10-1)\frac{90°}{30-1}$$
$$= 27.93°$$

where $\Delta\phi = 3.1034°$ and i is the index in the first column (titled "ID").

NACA 4-DIGIT AIRFOILS

Number of points	N =	30
Angular spacing	Δφ =	3.1034
Thickness	t =	0.15
Camber	C =	0.04
Location of max C	x_{camber} =	0.4

1	2	3	4	5	6	7	8	9	10	11
ID	φ	x	$y_t(x)$	$y_c(x)$	$y_c'(x)$	θ	x_U	y_U	x_L	y_L
1	0.00	0.0000	0.0000	0.0000	0.2000	11.31	0.0000	0.0000	0.0000	0.0000
2	3.10	0.0015	0.0084	0.0003	0.1993	11.27	-0.0002	0.0085	0.0031	-0.0079
3	6.21	0.0059	0.0165	0.0012	0.1971	11.15	0.0027	0.0173	0.0090	-0.0150
4	9.31	0.0132	0.0243	0.0026	0.1934	10.95	0.0086	0.0264	0.0178	-0.0212
5	12.41	0.0234	0.0317	0.0045	0.1883	10.66	0.0175	0.0357	0.0292	-0.0266
6	15.52	0.0365	0.0387	0.0070	0.1818	10.30	0.0295	0.0451	0.0434	-0.0311
7	18.62	0.0523	0.0453	0.0098	0.1738	9.86	0.0446	0.0544	0.0601	-0.0349
8	21.72	0.0710	0.0514	0.0129	0.1645	9.34	0.0627	0.0636	0.0794	-0.0378
9	24.83	0.0924	0.0569	0.0163	0.1538	8.74	0.0838	0.0726	0.1011	-0.0399
10	27.93	0.1165	0.0617	0.0199	0.1418	8.07	0.1078	0.0810	0.1252	-0.0412
11	31.03	0.1431	0.0659	0.0235	0.1284	7.32	0.1347	0.0889	0.1515	-0.0419
12	34.14	0.1723	0.0693	0.0270	0.1138	6.49	0.1645	0.0959	0.1802	-0.0419
13	37.24	0.2039	0.0720	0.0304	0.0980	5.60	0.1969	0.1020	0.2109	-0.0413
14	40.34	0.2378	0.0738	0.0334	0.0811	4.64	0.2319	0.1070	0.2438	-0.0402
15	43.45	0.2740	0.0748	0.0360	0.0630	3.60	0.2693	0.1107	0.2787	-0.0386
16	46.55	0.3123	0.0750	0.0381	0.0438	2.51	0.3090	0.1130	0.3156	-0.0368
17	49.66	0.3526	0.0743	0.0394	0.0237	1.36	0.3509	0.1137	0.3544	-0.0348
18	52.76	0.3948	0.0728	0.0400	0.0026	0.15	0.3946	0.1128	0.3950	-0.0328
19	55.86	0.4388	0.0705	0.0398	-0.0086	-0.49	0.4394	0.1103	0.4382	-0.0306
20	58.97	0.4844	0.0674	0.0392	-0.0188	-1.08	0.4857	0.1066	0.4832	-0.0281
21	62.07	0.5316	0.0636	0.0381	-0.0292	-1.67	0.5334	0.1016	0.5297	-0.0254
22	65.17	0.5801	0.0590	0.0364	-0.0400	-2.29	0.5825	0.0954	0.5777	-0.0226
23	68.28	0.6299	0.0539	0.0341	-0.0511	-2.92	0.6326	0.0879	0.6271	-0.0197
24	71.38	0.6807	0.0481	0.0312	-0.0624	-3.57	0.6837	0.0793	0.6777	-0.0168
25	74.48	0.7325	0.0418	0.0277	-0.0739	-4.23	0.7355	0.0694	0.7294	-0.0139
26	77.59	0.7850	0.0348	0.0235	-0.0856	-4.89	0.7880	0.0582	0.7821	-0.0112
27	80.69	0.8382	0.0274	0.0187	-0.0974	-5.56	0.8409	0.0459	0.8356	-0.0086
28	83.79	0.8919	0.0194	0.0131	-0.1093	-6.24	0.8940	0.0324	0.8898	-0.0061
29	86.90	0.9459	0.0108	0.0069	-0.1213	-6.92	0.9472	0.0176	0.9446	-0.0038
30	90.00	1.0000	0.0016	0.0000	-0.1333	-7.59	1.0002	0.0016	0.9998	-0.0016

FIGURE 8-27 Implementation of the Step-by-step for a NACA four-digit airfoil in a spreadsheet.

Column 3 is the x-value using the cosine scheme, calculated using Equation (8-39):

$$x_{10} = 1 - \cos((i-1)\Delta\phi) = 1 - \cos((10-1)3.1034°)$$
$$= 0.1165$$

Column 4 is the thickness, calculated using Equation (8-40):

$$\pm y_t = 5t\big(0.29690\sqrt{x_{10}} - 0.12600 x_{10} - 0.35160 x_{10}^2$$
$$+ 0.28430 x_{10}^3 - 0.10150 x_{10}^4\big)$$
$$= \pm 0.0617$$

Column 5 is the y-value of the mean-line, calculated using Equation (8-41) since $x_{10} < x_{\text{camber}}$:

$$y_c(x_{10}) = C\frac{(2x_{\text{camber}} - x_{10})x_{10}}{x_{\text{camber}}^2}$$
$$= 0.04 \frac{(2 \times 0.4 - 0.1165)0.1165}{0.4^2} = 0.0199$$

Column 6 is the slope of the mean-line, calculated using Equation (8-43) since $x_{10} < x_{\text{camber}}$:

$$\frac{dy_c}{dx} = \frac{2C}{x_{\text{camber}}}\left(1 - \frac{x_{10}}{x_{\text{camber}}}\right) = \frac{2 \times 0.04}{0.4}\left(1 - \frac{0.1165}{0.4}\right)$$
$$= 0.1418$$

Column 7 is the slope of the mean-line in degrees, calculated using Equation (8-45):

$$\theta = \tan^{-1}\left(\frac{dy_c}{dx}\right) = \tan^{-1}(0.1418) = 8.07°$$

The remaining columns 8–12 are calculated using Equation (8-46). The spreadsheet can be used to determine the geometry of any NACA four-digit airfoil by simply changing the thickness fraction, camber, and the location of the camber.

8.2.3 NACA Five-digit Airfoils

The NACA five-digit airfoils can be traced to the work of Jacobs and followed the development of the four-digit airfoils. The thickness distribution was identical to that of the four-digit series; however, the mean-line was modified to place the "maximum camber unusually far forward," to quote the title of NACA-TR-537 [21], which details their investigation. The investigation of these airfoils followed the revelation that the forward position of the maximum camber resulted in an increase of the maximum lift [22]. The 5-series airfoils were designed to generate a high maximum lift coefficient, and low drag and pitching moment coefficients. A family of the five-digit series features a reflexed camber to provide a zero C_m but have seen limited use. A typical NACA 5-digit airfoil is shown in Figure 8-28.

The reader is directed to Section 8.2.10, *Famous airfoils*, which discusses a few issues affecting a member of this family of airfoils, the NACA 23012 airfoil. As shown in Refs [21,22], these airfoils generally have poor stall characteristics because of a sharp loss in lift immediately after stall and, although generating impressive maximum lift and minimum drag, this is a very serious drawback.

Applications

The airfoils are widely used in GA aircraft, commuters, and business jets, where they are used for wings. Among aircraft using five-digit airfoils are a number of models manufactured by Beechcraft.

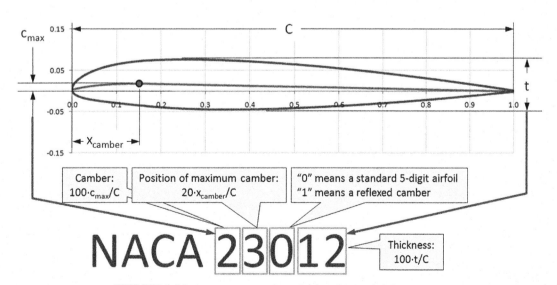

FIGURE 8-28 Interpretation of NACA five-digit airfoil designation.

Numbering System

Per Ref. [21], "the first digit is used to designate the relative magnitude of the camber." NACA R-824 [23] adds that "the first digit indicates the amount of camber in terms of the relative magnitude of the design lift coefficient; the design lift coefficient in tenths is thus three-halves of the first integer." Thus, considering the airfoil NACA 23012, the 2 means that the maximum camber height is 2%/100 = 0.02 and the design lift coefficient is $(2/10)\cdot(3/2) = 0.3$. The second digit, when divided by 20, would give the chordwise location of the maximum camber (0.15). The third digit is '0' for normal camber and '1' for reflexed airfoils like those used for flying wings. The last two digits represent the thickness of the airfoil (0.12 or 12%). Using this nomenclature, the various members of the family of five-digit airfoils would be represented as shown in Table 8-2.

Computation of Airfoil Ordinates

This information can now be used to mathematically compute the shape of any 5-digit airfoil as shown in the Step-by-step below:

Step 1: Preliminary Values

Decide the thickness ratio, camber, and location of the maximum camber using the variables t, C, and x_{camber}, respectively. If the thickness ratio is 12%, then $t = 0.12$. If the camber is 2%, then $C = 0.02$. If the location of the maximum camber is 15%, then $x_{camber} = 0.15$. (These numbers represent the NACA 23012 airfoil.)

Step 2: Airfoil Resolution

Decide how many x-values (and thus y-values) to include in the analysis. Here, let's call that value N (e.g. $N = 100$ for 100 points).

Step 3: Prepare Ordinate Table

Tabulate the x-ordinates along the unit chord using the cosine scheme where $\Delta\phi$ is calculated using Equation (8-38). These values should range from 0 to 1, where $x = 0$ represents the leading edge and $x = 1$ the trailing edge.

Step 4: Calculate Thickness

Calculate the thickness of the upper and lower surfaces of the airfoil for each x-value, using Equation (8-40), repeated here for convenience:

$$\pm y_t = 5t\left(0.29690\sqrt{x} - 0.12600x - 0.35160x^2 + 0.28430x^3 - 0.10150x^4\right) \quad (8\text{-}40)$$

Step 5: Compute the y-value for the Mean-line

The next step involves computing the y-value of the mean-line for each x, which we call y_c. This depends on whether x is larger or smaller than the location of the camber and is given by:

If $x \leq x_{camber}$ then

$$y_c = \frac{k_1}{6}\left[x^3 - 3mx^2 + m^2(3-m)x\right] \quad (8\text{-}47)$$

If $x > x_{camber}$ then

$$y_c = \frac{k_1 m^3}{6}(1 - x) \quad (8\text{-}48)$$

where the constants m and k_1 are obtained from Table 8-3.

TABLE 8-3 Mean-line Designations for NACA Five-digit Airfoils

Mean-line Designation	x_{camber}	m	k_1
210	0.05	0.0580	361.400
220	0.10	0.1260	51.640
230	0.15	0.2025	15.957
240	0.20	0.2900	6.643
250	0.25	0.3910	3.230

TABLE 8-2 Various Members of the NACA Five-digit Airfoil Series

Location of Maximum Camber	Camber Position, x_{camber}	Conventional airfoil	Example	Reflexed airfoil	Example
5% or 0.05c	0.05	10	NACA 21012	11	NACA 21112
10% or 0.10c	0.10	20	NACA 22012	21	NACA 22112
15% or 0.15c	0.15	30	NACA 23012	31	NACA 23112
20% or 0.20c	0.20	40	NACA 24012	41	NACA 24112
25% or 0.25c	0.25	50	NACA 25012	51	NACA 25112

Step 6: Calculate the Slope of the Mean-line

Calculate the slope of the mean-line at the point using Equations (8-43) and (8-44), also repeated for convenience:

If $x \leq x_{camber}$ then

$$\frac{dy_c}{dx} = \frac{2C}{x_{camber}}\left(1 - \frac{x}{x_{camber}}\right) \qquad (8\text{-}43)$$

If $x > x_{camber}$ then

$$\frac{dy_c}{dx} = \frac{2C(x_{camber} - x)}{(1 - x_{camber})^2} \qquad (8\text{-}44)$$

Step 7: Calculate the Ordinate Rotation Angle

Calculate the angle θ using Equation (8-45):

$$\theta = \tan^{-1}\left(\frac{dy_c}{dx}\right) \qquad (8\text{-}45)$$

Step 8: Calculate the Upper and Lower Ordinates

Calculate the upper and lower surface ordinates using Equations (8-46):

$$\begin{array}{ll} x_U = x - y_t \sin\theta & y_U = y_c + y_t \cos\theta \\ x_L = x + y_t \sin\theta & y_L = y_c - y_t \cos\theta \end{array} \qquad (8\text{-}46)$$

It is of interest to note that it is possible to calculate the constant m and k_1 for any position of the x_{camber} using the following expressions, which are obtained through simple curvefit to the data in the table in Step 5:

$$\begin{array}{l} m = 0.003 + x_{camber} + 2.2x_{camber}^2 \\ k_1 = 0.05991 x_{camber}^{-2.919} \end{array} \qquad (8\text{-}49)$$

Note that the equation for k_1 did not have a perfect correlation like the one for m. For this reason, values of k_1 calculated using Equation (8-49) will differ slightly from those in the table, although the accuracy gives acceptable results. This allows the evaluation of alternative versions of the five-digit series airfoils. Corresponding values for m and k_1 for reflexed airfoils (211, 221, etc.) are given in Ref. [21].

This procedure has been implemented as shown in Figure 8-29, for a NACA 23012 airfoil, using $N = 30$. More points should be used for a serious design project. In fact, the primary difference between this scheme and the one for the four-digit airfoil is the determination of m, k_1, and then the calculation of the mean-line. Here $m = 0.2025$, $k_1 = 15.957$, although it is calculated using Equation (8-49) A sample calculation for the row with ID number 10 or $i = 10$ (see the highlighted row in the figure) is provided. First, columns 2 and 3 are identical to the previous approach. The thickness calculated in column 4 also uses Equation (8-40), but this time it yields 0.0494.

Column 5 is the y-value of the mean-line, calculated using Equation (8-47) since $x_{10} < x_{camber}$:

$$y_c(x_{10}) = \frac{15.957}{6}\left[x_{10}^3 - 3(0.2025)x_{10}^2 \right. \\ \left. + (0.2025)^2(3 - 0.2025)x_{10}\right] = 0.0170$$

Column 6 is the slope of the mean-line, calculated using Equation (8-43) since $x_{10} < x_{camber}$:

$$\frac{dy_c}{dx} = \frac{2C}{x_{camber}}\left(1 - \frac{x_{10}}{x_{camber}}\right) = \frac{2 \times 0.02}{0.15}\left(1 - \frac{0.1165}{0.15}\right) \\ = 0.0596$$

Column 7 is the slope of the mean-line in degrees, calculated using Equation (8-45):

$$\theta = \tan^{-1}\left(\frac{dy_c}{dx}\right) = \tan^{-1}(0.0596) = 3.41°$$

The remaining columns 8–12 are calculated using Equation (8-46). The spreadsheet can be used to determine the geometry of any NACA 5-digit airfoil by simply changing the appropriate parameters.

8.2.4 NACA 1-series Airfoils

The 1-series airfoils were designed in the late 1930s after the four- and five-digit series (which explains the order of the airfoils in this presentation). The geometry was based on thin airfoil theory rather than on specifying the location of the max camber and then applying the thickness distribution of the preceding families of airfoils. They marked the first time inverse airfoil design was put to use. Theses airfoils are primarily used for propellers, as they prevent the formation of large pressure peaks that are detrimental to airfoils near supersonic airspeeds. It is primarily the 16-version of the 1-series airfoils that is mostly used, so these are sometimes referred to as the 16-series rather than the latter. Reference [24] presents a computer code to help develop ordinates for NACA 16-series airfoils. A typical NACA 1-series airfoil is shown in Figure 8-30.

Applications

Widely used for aircraft and ship propellers.

Numbering System

Typical 1-series airfoils are designated by a five-digit number such as NACA 16-212. The first integer '1'

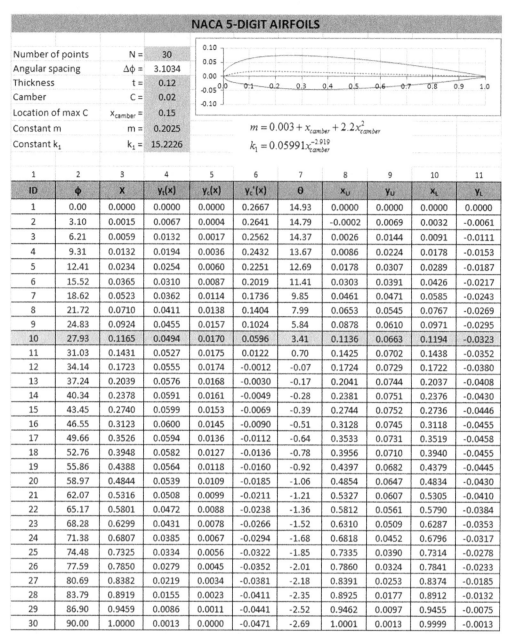

FIGURE 8-29 Implementation of the step-by-step for a NACA five-digit airfoil in a spreadsheet.

represents the series. The second digit '6' denotes the distance in tenths to the chordwise location of the minimum pressure when the symmetrical airfoil is at zero lift (60%). The first number following the dash '2' is the amount of camber in terms of the design lift coefficient in tenths (0.2). The final two digits '12' represent the thickness of the airfoil (0.12 or 12%).

8.2.5 NACA 6-series Airfoils

The 6-series airfoils represent a family of airfoils that were designed to sustain the laminar boundary layer over a larger portion of the chord than the four- or five-digit series. This was accomplished by delaying the development of an adverse pressure gradient by bringing the thickness of the airfoil as far aft as possible before an inevitable separation would ensue. The thickness of these airfoils was developed using exact airfoil theory and for that reason no simple formulation is available in the literature like that for the four- and five-digit airfoils. However, their camber, which is based on thin-airfoil theory, is described in NACA R-824 [23].

A sub-family of the NACA 6-series is called the 6A-series airfoils. They were designed to eliminate the trailing edge cusp associated with the former, which posed great difficulties in their fabrication [25]. For

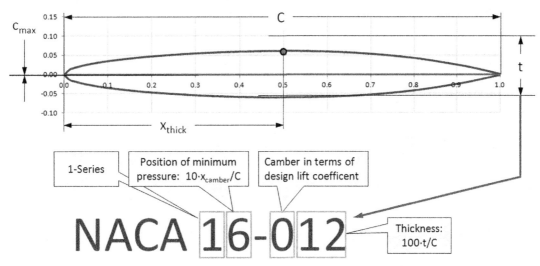

FIGURE 8-30 Interpretation of NACA 1-series airfoil designation.

instance, conventional construction methods that require folding aluminum sheet to form such a trailing edge results in geometry far too tight to accommodate supporting ribs in the trailing edge. As a consequence, the trailing edge is unsupported and at higher angles-of-attack it can flex, effectively modifying the airfoil. Reference [26] presents a Fortran IV code to develop the ordinates of NACA 6-series airfoils. Note that Ref. [27] presents an updated version of the code that is more portable between machines.

Applications

The airfoils are widely used for aircraft ranging from WWII era fighters to high-performance GA aircraft, business jets, and military trainers.

Numbering System

The NACA 6-series airfoils (Figure 8-31) feature a six-digit designation with an indicator of the mean-line used. For instance, the designation for the airfoil **NACA 65$_3$-415, a = 0.5**, the '6' refers to the 6-series airfoils, '5' denotes the chordwise location of the maximum camber in tenths of the chord (50%). The subscript '3' gives the range of lift coefficients below and above the design lift coefficient in which favorable pressure gradients exists on both surfaces ($\pm$ 0.3). The '4' following the dash indicates the design lift coefficient in tenths (0.4). In this way, the particular airfoil is expected to sustain laminar flow for lift coefficients ranging from 0.1 to 0.7 (i.e. 0.4 $\pm$ 0.3). The last two digits indicate the airfoil thickness as percent of the chord (0.15 or 15%). The

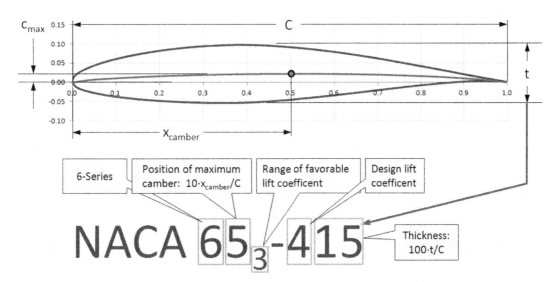

FIGURE 8-31 Interpretation of NACA 6-series airfoil designation.

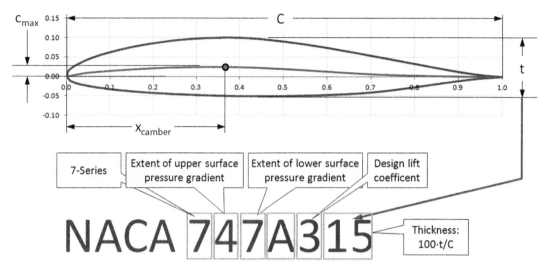

FIGURE 8-32 Interpretation of NACA 7-series airfoil designation.

designation "a = 0.5" refers to the mean-line used, but this is used in the derivation of the ordinates. When a mean-line designation is omitted the default value of a = 1.0 is used.

The airfoil designation has a number of variations. For instance, the above airfoil (NACA 65$_3$-415) is sometimes represented as NACA 65(3)-415 or 65,3-415.

The camber line can be a combination of more than one such line and then the design lift coefficient is the sum of all the camber lines used; for instance, the "2" in 218 in the following airfoil designation:

$$\text{NACA } 65_3 - 218 \begin{cases} a = 0.5, \ C_{lmind} = 0.3 \\ a = 1.0, \ C_{lmind} = -0.1 \end{cases}$$

If the thickness distribution of an airfoil is obtained by linearly increasing or decreasing the ordinates of one of the original thickness distributions, then the airfoil is denoted as shown in the example below:

$$\text{NACA } 65(318) - 217, \ a = 0.5$$

Where the number '3' in the parentheses denotes the low-drag range and the '18' denotes the thickness ratio of the original thickness distribution. NACA 6-series airfoils whose thickness ratio is less than 0.12 usually do not have a subscript as their low-drag range is less than 0.1 (meaning the subscript should be a fraction). In this case, the subscript is omitted, as shown below:

$$\text{NACA } 65 - 210$$

There are other deviations, but an interested reader is directed to Ref. [23] for more details.

8.2.6 NACA 7-series Airfoils

The NACA 7-series airfoils were designed to maximize the extent of laminar flow on the upper and lower surfaces, recognizing the two can be different from each other. A typical NACA 1-series airfoil is shown in Figure 8-32.

Applications

Not widely used. The University of Illinois at Urbana-Champaign (UIUC) airfoil database indicates five airplane types that use such airfoils; the Brändli BX-2 Cherry, DRS Sentry HP (UAV), Jameson RJJ-1 Gypsy Hawk, MacDonald S-20, and the RFB RW.3-A3 Multoplane.

Numbering System

The NACA 7-series have their own numbering system best explained by considering a typical type: **NACA 747A315**. The '7' indicates the series number. The '4' indicates the extent of favorable pressure gradient over the upper surface of the airfoil in tenths of the chord length (40%). The next, '7', similarly, indicates the extent of favorable pressure region over the lower surface in tenths of the chord length (70%). This means that as long as the surface qualities are smooth, the laminar boundary layer should extend at least that far aft. The series of three numbers '315' following the letter 'A' is identical to that of the 6-series, i.e. '3' is the design lift coefficient (0.3) and '15' is the thickness ratio in hundredths (0.15 or 15%). The intent of the 'A' is to distinguish between airfoils that have properties that would lead to an identical digit, but differ in camber or thickness distribution. As an example, another airfoil of the series with an equal coverage of favorable pressure gradients, but with a different camber-line or thickness distribution, would be distinguished from the first one using the serial letter 'B'. As with the 6-series airfoils, the 7-series also feature camber lines that are the combination of two or more lines.

8.2.7 NACA 8-series Airfoils

In 1949, NACA developed an additional family of airfoils, called the 8-series [28]. These airfoils were developed to prevent the abrupt loss of lift (exemplified in Figure 8-51) near the critical Mach number (see Section 8.3.7, *The critical Mach number, M_{crit}*). These airfoils have not seen much use and were abandoned in favor of Peaky airfoils (see Section 8.2.10, *Famous airfoils*), which were designed some 10–15 years later. They are really presented here in the interests of completeness.

Applications
No known application.

Numbering System

The numbering system for typical NACA 8-series airfoils is best explained by considering a representative type, e.g. **NACA 835A216**. The first digit, '8', identifies the series. The second one, '3', denotes the position of the minimum pressure on the upper surface in tenths (0.30). Similarly, the third digit, '5', is the position of the minimum pressure on the lower surface. The letter 'A' has an identical function as in the 7-series airfoils, as do the remaining three digits, where the first '2' is the design lift coefficient in tenths (0.2) and the two remaining characters are the airfoil thickness in hundredths (0.16 or 16%).

8.2.8 NACA Airfoils in Summary — Pros and Cons and Comparison of Characteristics

The general advantages and disadvantages of NACA airfoils are summarized in Table 8-4. Also note Figure 8-33 and Figure 8-34, which show typical differences in lift, pitching moment, and drag for the different classes of airfoils.

8.2.9 Properties of Selected NACA Airfoils

A substantial amount of aerodynamic data exists on NACA airfoils, as has already been pointed out in the preceding sections. Table 8-5 shows a number of properties for selected airfoils. Note that the widths of the shaded bars in the columns denoted by C_{lmax}, $C_{l\alpha}$, C_{dmin}, and $C_{m\ ac}$ denote the most favorable values. Thus the widest bar in the C_{lmax} column shows the highest C_{lmax}. The shortest bar in the C_{dmin} column denotes the lowest values of the drag coefficient, and so on.

This allows the table to be further manipulated and sorted by the highest C_{lmax} or the lowest C_{dmin}. This is shown in Tables 8-6 and 8-7. The reader should be reminded that these only apply to the airfoils shown in Table 8-5.

8.2.10 Famous Airfoils

CLARK Y

The Clark Y airfoil (see Figure 8-35) is famous for being one of the most widely used airfoils in the history of aviation, although its use is mostly in airplanes designed before World War II. It is named after Colonel Virginius E. Clark (1886–1948), who designed many airfoils around the time of World War I. The Clark Y airfoil, in

TABLE 8-4 Pros and Cons of the NACA Airfoils Presented

	Pros	Cons
Four-digit	Generally thick airfoils with benign stall characteristics. Insensitive to non-smooth surfaces. Center of pressure has limited movement over a wider range of AOA.	Relatively low C_{lmax} and high C_{dmin} and $|C_m|$.
Five-digit	High C_{lmax} and relatively low C_{dmin} and $|C_m|$. Insensitive to non-smooth surfaces.	Abrupt stall characteristics.
16-Series	Prevents high pressure peaks that lead to detrimental performance near $M = 1$.	Low C_{lmax}
6-Series	Relatively high C_{lmax} and low C_{dmin}. Sustains extensive laminar flow if surface is smooth and forms a drag bucket. Relatively thick.	Sensitive to non-smooth surfaces. High $|C_m|$. Higher drag outside the drag bucket than that of the 4- and 5-digit airfoils. Some have poor stall characteristics.
7-Series	Low C_{dmin}. Sustains extensive laminar flow if surface is smooth and forms a drag bucket. Lower $|C_m|$ than the 6-series. Some display good stall characteristics.	Sensitive to non-smooth surfaces. Low C_{lmax}.
8-Series	N/A	N/A

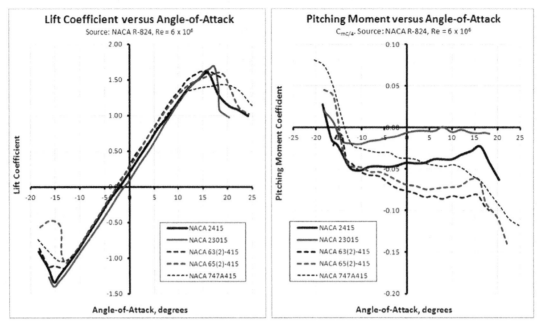

FIGURE 8-33 Lift curve and pitching moment characteristics of selected NACA series airfoils (from Ref. [23]).

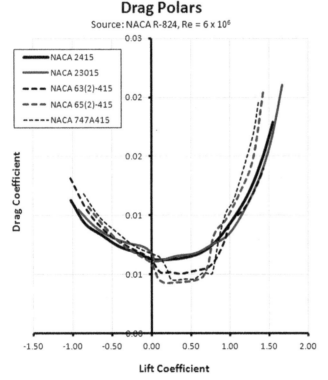

FIGURE 8-34 Drag polars of selected NACA series airfoils (from Ref. [23]).

particular, was designed in 1922. One of its distinguishing features is the flat lower surface, which extends from 30% chord to the trailing edge. The aerodynamic properties of the Clark Y airfoil can be found in NACA R-502 [29]. Among famous aircraft types that feature this airfoil are the Ryan NYP Spirit of St. Louis, flown by Charles Lindbergh across the Atlantic in 1927; the Lockheed Vega, flown by Amelia Earhart across the Atlantic; and Wiley Post (which he named "Winnie Mae") twice around the globe. The University of Illinois at Urbana-Champaign (UIUC) airfoil database cites this airfoil 493 times for the 7420 aircraft type instances.

USA-35B

The USA series of airfoils was designed by engineers of the United States Army (USA) in the era before 1920. The USA35B (see Figure 8-36) is the best known. It is a flat-bottomed airfoil used in many well-known aircraft, especially in a line of aircraft made by Piper Aircraft, such as the Piper J-3 Cub, PA-25 Pawnee, PA-23 Apache and Aztec twin-engine aircraft. Like the Clark Y airfoil, it features a substantial camber and generates a high maximum lift coefficient and gentle entry into the post-stall region, but also high drag.

NACA 23012

The NACA four-digit airfoils were developed in the 1930s. This work led to the development of the 23-thousand series airfoils. One of the best known of those airfoils is the NACA 23012 (see Figure 8-37), which is used in a number of aircraft. The UIUC airfoil database cites the airfoil 397 times for the 7420 aircraft type instances, including very common aircraft such as Beech's Bonanza and B1900 Airliner, just to name a few. It is also featured on the twin-engine Britten-Norman BN-2 Islander, Cessna 208 Caravan and even the Citation jet.

8.2 THE GEOMETRY OF THE AIRFOIL

TABLE 8-5 Properties of Selected NACA Airfoils (More shading indicates more desirable characteristics)

Airfoil	C_{lmax}	α_{STALL}	α_{ZL}	$C_{l\alpha}$	$C_{l\,min\,d}$	C_d C_{dmin}	$C_l = 0$	$C_l = 0.4$	$C_l = 0.8$	$C_{m\,ac}$	x_{ac}	M_{crit}	Reference
0006	0.88	13.0	0.0	0.102	0.00	0.0052	0.0052	0.0058	0.0090	-0.002	0.250	0.805	A and B
0009	1.33	14.0	0.0	0.101	0.00	0.0056	0.0056	0.0060	0.0084	-0.003	0.250	0.766	B and C
0012	1.53	17.0	0.0	0.101	0.00	0.0083				-0.002	0.250		B
0015	1.53	17.0	0.0	0.100	0.00	0.0093				0.000	0.240		B
0018	1.49	17.0	0.0	0.098	0.00	0.0108				-0.002	0.240		B
0021	1.38	17.0	0.0	0.094	0.00	0.0120				-0.001	0.230		B
0025	1.20	16.0	0.0	0.089	0.00	0.0143				-0.003	0.230		B
1408	1.35	14.0	-1.0	0.105	0.20	0.0052	0.0057	0.0057	0.0070	-0.024	0.250		A
1410	1.52	14.0	-1.0	0.109	0.20	0.0055	0.0057	0.0570	0.0070	-0.020	0.250		A
1412	1.57	15.0	-1.2	0.103	0.20	0.0057	0.0058	0.0061	0.0076	-0.023	0.252	0.720	A and C
23012	1.78	18.0	-1.2	0.104	0.10	0.0060	0.0068	0.0060	0.0068	-0.015	0.247	0.672	A and C
23015	1.72	18.0	-1.0	0.103	0.10	0.0062	0.0072	0.0064	0.0072	-0.008	0.243	0.663	A and C
23018	1.62	16.0	-1.2	0.104	0.30	0.0069	0.0069	0.0068	0.0078	-0.005	0.243	0.655	A and C
23021	1.51	15.5	-1.2	0.100	0.10	0.0070	0.0070	0.0073	0.0089	-0.004	0.238	0.623	A and C
2412	1.62	16.0	-1.8	0.101	0.20	0.0085	0.0060	0.0060	0.0072	-0.047	0.247	0.690	A and C
2415	1.65	16.0	-2.0	0.106	0.20	0.0063	0.0064	0.0064	0.0077	-0.045	0.246	0.677	A and C
2418	1.47	14.0	-2.1	0.099	0.25	0.0067	0.0068	0.0071	0.0081	-0.045	0.241	0.650	A and C
2421	1.47	16.0	-1.9	0.099	0.20	0.0071	0.0071	0.0072	0.0088	-0.040	0.241	0.630	A and C
2424	1.30	16.0	-1.9	0.093	0.10	0.0075	0.0075	0.0078	0.0099	-0.040	0.231	0.606	A and C
4412	1.66	14.0	-4.0	0.106	0.40	0.0060	0.0062	0.0060	0.0064	-0.092	0.247	0.647	A and C
4415	1.64	14.0	-4.0	0.106	0.40	0.0064	0.0066	0.0064	0.0070	-0.093	0.245	0.635	A and C
4418	1.54	14.0	-4.0	0.100	0.40	0.0066	0.0070	0.0066	0.0076	-0.088	0.242	0.620	A and C
4421	1.46	16.0	-4.0	0.096	0.27	0.0072	0.0075	0.0073	0.0081	-0.086	0.238	0.602	A and C
63_1-412	1.77	15.0	-3.0	0.110	0.35	0.0044	0.0062	0.0044	0.0075	-0.075	0.271		A
64_1-006	0.83	8.0-10.0	0.0	0.104	0.00	0.0038	0.0038	0.0057	-	0.000	0.256	0.836	A and C
64_1-009	1.17	11.0	0.0	0.108	0.00	0.0040	0.0040	0.0061	0.0082	0.000	0.262	0.785	A and C
64_1-012	1.44	15.0	0.0	0.110	0.00	0.0042	0.0042	0.0062	0.0081	0.000	0.262	0.744	A and C
64_1-206	1.03	12.0	-1.2	0.104	0.18	0.0038	0.0050	0.0057	0.0062	-0.040	0.253	0.793	A and C
64_1-209	1.40	13.0	-1.4	0.104	0.20	0.0040	0.0053	0.0060	0.0075	-0.040	0.261	0.760	A and C
64_1-212	1.55	15.0	-1.2	0.108	0.19	0.0042	0.0043	0.0050	0.0077	-0.028	0.262	0.728	A and C
64_1-412	1.67	15.0	-2.6	0.112	0.36	0.0044	0.0058	0.0046	0.0076	-0.070	0.267	0.700	A and C
64_2-215	1.57	15.0	-1.5	0.111	0.22	0.0045	0.0046	0.0048	0.0081	-0.030	0.265	0.700	A and C
64_2-415	1.66	16.0	-2.8	0.114	0.40	0.0046	0.0060	0.0049	0.0080	-0.070	0.264	0.678	A and C
64_3-418	1.56	14.0-20.0	-3.0	0.117	0.40	0.0050	0.0060	0.0050	0.0062	-0.065	0.273	0.655	A and C
64_3-421	1.48	18.0-24.0	-2.8	0.119	0.40	0.0050	0.0056	0.0050	0.0058	-0.065	0.276	0.642	A and C
65_1-006	0.93	12.0	0.0	0.105	0.00	0.0035	0.0035	0.0058	-	0.000	0.258	0.838	A and C
65_1-009	1.08	11.0	0.0	0.106	0.00	0.0041	0.0041	0.0060	0.0081	0.000	0.264	0.790	A and C
65_1-012	1.36	14.0	0.0	0.106	0.00	0.0038	0.0038	0.0061	0.0081	0.000	0.261	0.750	A and C
65_1-206	1.06	10.0-16.0	-1.2	0.106	0.20	0.0036	0.0050	0.0055	0.0071	-0.030	0.257	0.791	A and C
65_1-209	1.30	12.0	-1.2	0.108	0.20	0.0038	0.0052	0.0058	0.0075	-0.032	0.259	0.755	A and C
65_1-212	1.48	14.0	-1.2	0.108	0.20	0.0040	0.0047	0.0057	0.0082	-0.032	0.261	0.726	A and C
65_1-410	1.52	14.0	-2.6	0.110	0.36	0.0037	0.0058	0.0038	0.0072	-0.068	0.262	0.714	A and C
65_2-412	1.64	16.0	-3.0	0.109	0.38	0.0038	0.0056	0.0038	0.0075	-0.070	0.265	0.697	A and C
65_2-415	1.62	16.0	-2.6	0.107	0.38	0.0042	0.0060	0.0042	0.0084	-0.060	0.268	0.675	A and C
65_3-418	1.55	18.0	-2.6	0.106	0.40	0.0043	0.0062	0.0043	0.0072	-0.060	0.265	0.656	A and C
65_4-421	1.55	22.0	-3.2	0.111	0.42	0.0044	0.0048	0.0044	0.0053	-0.065	0.272	0.637	A and C
66_3-418	1.57	18.0	-3.0	0.100	0.40	0.0037	0.0065	0.0037	0.0110	-0.075	0.266		A

References: A is NACA R-824, B is NACA TN-460, C is Perkins & Hage

Research on the lift and drag properties of the symmetrical four-digit NACA airfoils showed that while the drag was quite small, the same could be said of the maximum lift. It was proposed that the maximum lift could be improved by introducing a small camber to the airfoil. This was accomplished by simply deflecting the forward part of the leading edge. As can be seen in Figure 8-38, the 23012 airfoil is effectively a NACA 0012 airfoil whose forward 15% has been deflected leading edge down. This new airfoil showed great promise in wind tunnel testing, offering a low C_{dmin}, very high C_{lmax}, and low C_m, all are great properties that make

TABLE 8-6 The Top Ten NACA Airfoils with the Highest C_{lmax} (More shading indicates more desirable characteristics)

Airfoil	C_{lmax}	α_{STALL}	α_{ZL}	$C_{l\alpha}$	$C_{l\,min\,d}$	C_{dmin}	C_d $C_l=0$	$C_l=0.4$	$C_l=0.8$	$C_{m\,ac}$	x_{ac}	M_{crit}	Reference
23012	1.78	18.0	-1.2	0.104	0.10	0.0060	0.0068	0.0060	0.0068	-0.015	0.247	0.672	A and C
63_1-412	1.77	15.0	-3.0	0.110	0.35	0.0044	0.0062	0.0044	0.0075	-0.075	0.271		A
23015	1.72	18.0	-1.0	0.103	0.10	0.0062	0.0072	0.0064	0.0072	-0.008	0.243	0.663	A and C
64_1-412	1.67	15.0	-2.6	0.112	0.36	0.0044	0.0058	0.0046	0.0076	-0.070	0.267	0.700	A and C
64_2-415	1.66	16.0	-2.8	0.114	0.40	0.0046	0.0060	0.0049	0.0080	-0.070	0.264	0.678	A and C
4412	1.66	14.0	-4.0	0.106	0.40	0.0060	0.0062	0.0060	0.0064	-0.092	0.247	0.647	A and C
2415	1.65	16.0	-2.0	0.106	0.20	0.0063	0.0064	0.0064	0.0077	-0.045	0.246	0.677	A and C
65_1-412	1.64	16.0	-3.0	0.109	0.38	0.0038	0.0056	0.0038	0.0075	-0.070	0.265	0.697	A and C
4415	1.64	14.0	-4.0	0.106	0.40	0.0064	0.0066	0.0064	0.0070	-0.093	0.245	0.635	A and C
65_2-415	1.62	16.0	-2.6	0.107	0.38	0.0042	0.0060	0.0042	0.0084	-0.060	0.268	0.675	A and C

TABLE 8-7 The Top Ten NACA Airfoils with the Lowest C_{dmin} (More shading indicates more desirable characteristics)

Airfoil	C_{lmax}	α_{STALL}	α_{ZL}	$C_{l\alpha}$	$C_{l\,min\,d}$	C_{dmin}	C_d $C_l=0$	$C_l=0.4$	$C_l=0.8$	$C_{m\,ac}$	x_{ac}	M_{crit}	Reference
65_1-006	0.93	12.0	0.0	0.105	0.00	0.0035	0.0035	0.0058	-	0.000	0.258	0.838	A and C
65_1-206	1.06	10.0-16.0	-1.2	0.106	0.20	0.0036	0.0050	0.0055	0.0071	-0.030	0.257	0.791	A and C
66_3-418	1.57	18.0	-3.0	0.100	0.40	0.0037	0.0065	0.0037	0.0110	-0.075	0.266		A
65_1-410	1.52	14.0	-2.6	0.110	0.36	0.0037	0.0058	0.0038	0.0072	-0.068	0.262	0.714	A and C
65_1-412	1.64	16.0	-3.0	0.109	0.38	0.0038	0.0056	0.0038	0.0075	-0.070	0.265	0.697	A and C
65_1-012	1.36	14.0	0.0	0.106	0.00	0.0038	0.0038	0.0061	0.0081	0.000	0.261	0.750	A and C
65_1-209	1.30	12.0	-1.2	0.108	0.20	0.0038	0.0052	0.0058	0.0075	-0.032	0.259	0.755	A and C
64_1-206	1.03	12.0	-1.2	0.104	0.18	0.0038	0.0050	0.0057	0.0062	-0.040	0.253	0.793	A and C
64_1-006	0.83	8.0-10.0	0.0	0.104	0.00	0.0038	0.0038	0.0057	-	0.000	0.256	0.836	A and C
65_1-212	1.48	14.0	-1.2	0.108	0.20	0.0040	0.0047	0.0057	0.0082	-0.032	0.261	0.726	A and C

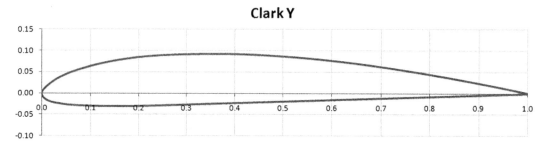

FIGURE 8-35 The Clark Y airfoil.

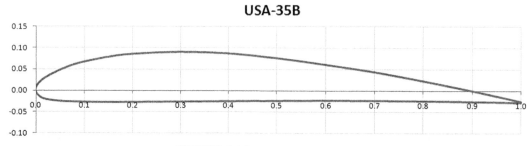

FIGURE 8-36 The USA-35B airfoil.

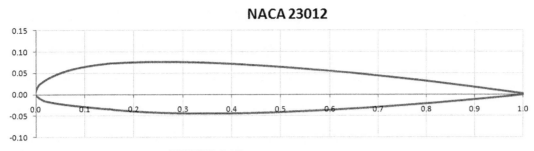

FIGURE 8-37 The NACA 23012 airfoil.

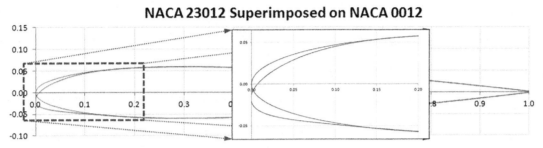

FIGURE 8-38 The NACA 23012 superimposed on the NACA 0012 airfoil reveals the only difference between the two is from the leading edge to 0.15C. A closer look at the leading edge of the two airfoils shows how the leading edge of the NACA 23012 airfoil.

the airfoil very tempting for the aircraft designer. However, what was less touted in the NACA literature was the abruptness of its stall. And this is a serious problem. In fact, it is not just the NACA 23012 that poses such problems, but a host of airplanes that feature a root/tip combination of 23015/23009 and similar.

As stated in Section 8.1.11, *Airfoil stall characteristics*, a leading-edge separation bubble may form on airfoils whose thickness-to-chord ratio is 12% or less. Compounding this issue on the NACA 230XX series airfoil is the nature of its geometry: the leading edge deflection at 15% chord. This introduces a sharp discontinuity in the curvature of the airfoils that may also contribute to the formation of the separation bubble [30]. As the *AOA* of the airfoil is increased, the airflow will reattach behind the bubble (see schematic to the left in Figure 8-39) and separate again at the trailing edge. With further increase of *AOA* the second separation point will continue to move forward, eventually and suddenly combining with the leading-edge separation bubble. When this happens, there will be an abrupt and large drop in the lift coefficient. This is clearly visible as the sharp drop in Figure 8-39, from the wind tunnel test data presented in NACA TR-537 [31].

Once featured on a wing, fabrication differences between the two wing halves on either side of the plane of symmetry will result in a high probability that one wing will stall (abruptly) before the other. In other words, an uncontrollable roll-off at stall is a very plausible scenario the designer should be strongly aware of. And while it is certainly possible to "beat" this detrimental stall behavior into submission with adequate wing washout and stall strips, this is likely to take extra development time and cost, making the series of airfoils something the aircraft designer should stay away from. There are multiple other options at the disposal to the designer which offer similar favorable properties but with far more benign stall characteristics.

GA(W)-1 or LS(1)-0417

The GA(W)-1 (see Figure 8-40) was designed in 1972 specifically for GA aircraft using computational fluid dynamics by Robert T. Whitcomb and associates at the NASA Langley Research Center [32]. Details of its characteristics can be found in the report NASA TN D-7428 [33]. It is also known as NASA/Langley/Whitcomb LS(1)-0417. The report tested the airfoils at Reynolds numbers ranging from 2×10^6 to 20×10^6 and Mach numbers ranging from 0.15 to 0.28 and demonstrated C_{Lmax} ranging from 1.6 to 2.0. A selection of lift-to-drag ratios between 65–85 were obtained at a climb lift coefficient of 1.0. These characteristics appeared very promising and the airfoil was featured on at least two aircraft designs from the 1970s: the Piper PA-38 Tomahawk and Beechcraft 77 Skipper. Other aircraft using the airfoil are the cancelled American Aviation Hustler, the Edgley Optica, and the cancelled military trainer Fairchild T-46 Eaglet.

Among other favorable characteristics is the structural depth offered by its thickness-to-chord ratio of

272　　　　　　　　　　　　　　　　　　　　　　8. THE ANATOMY OF THE AIRFOIL

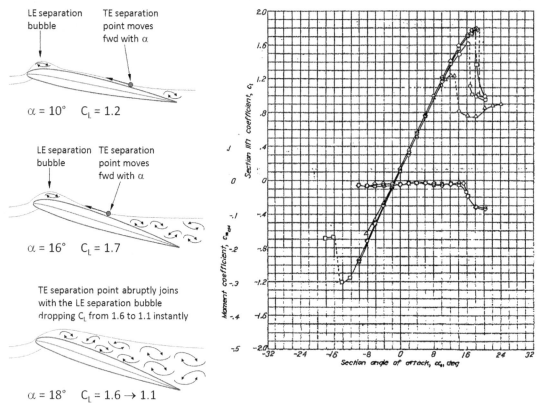

FIGURE 8-39　Data from NACA R-537 for the NACA 23012 airfoil. It clearly displays the sharp drop in the lift coefficient at stall.

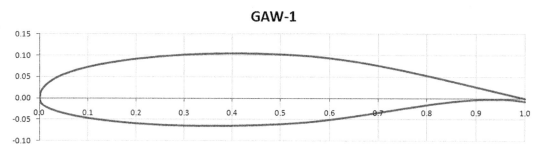

FIGURE 8-40　The GA(W)-1 airfoil.

17%. This provides the aircraft with ample volume for fuel and should result in a lighter wing structure. The cusp (reflexed curvature) near the trailing edge of the airfoil usually presents an issue when using NLF airfoils, as it results in a thin trailing thickness that makes it practically impossible to fit ribs that extend to the trailing edge. The last 30% of the chord is usually allotted to control surfaces and usually these are constructed using conventional stressed skin aluminum construction. The cusp will then require ribs whose flanges must be curved, which requires costly tooling to fabricate. If the control surfaces are built using composite materials the cusp can be accommodated, although such surfaces usually end up being heavier than if they were made from aluminum. Overall, the cusp presents a hard detail in real production environment and sometimes forces the manufacturer to modify the geometry by making the control surfaces flat-bottomed.

Davis Wing Airfoil

The Davis wing is the topic of discussion in online Appendix C1.5.1, *The Davis wing*, but it stood for a wing design philosophy used for many military aircraft, of which the Consolidated B-24 Liberator is probably the best known. The airfoil, which was inspired by what, at the time, was thought to be the shape of a teardrop (see Figure 8-41), had a number of interesting characteristics, most prominent of which was low drag and high lift. The airfoil was evaluated in NACA WR-L-677 [34]. An Xfoil analysis of the airfoil at a Reynolds number

8.2 THE GEOMETRY OF THE AIRFOIL

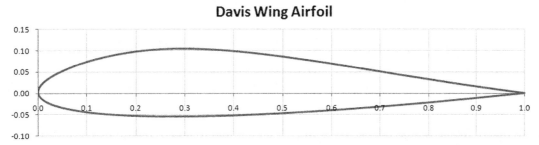

FIGURE 8-41 The basic Davis wing airfoil for the B-24 Liberator.

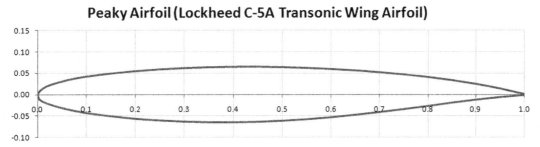

FIGURE 8-42 A typical "peaky" airfoil. This one is the basic C-5A Galaxy airfoil.

of 9×10^6 matches the experiment reasonably well and indicates it has many interesting properties. Among those are gentle stall characteristics, LD_{max} in excess of 160, and a C_{lmind} of 0.65 (suitable for a lumbering heavy transport aircraft). Xfoil under-predicts the C_{lmax} of about 1.35 versus 1.4 from experiment, but the C_{lmind} between prediction and experiment match well.

"Peaky" Airfoils

A *peaky airfoil* is a name used for airfoils that were designed in the 1960s for use in aircraft operating in the transonic range (see Figure 8-42). When conventional airfoils accelerate to high subsonic airspeeds a shockwave begins to form well below Mach 1. This takes place as the speed of air flowing around the airfoil locally reaches and exceeds the speed of sound on the airfoil. As soon as shocks begin to form in the flow there will be a sharp rise in the drag of the airfoils, as shown in Figure 8-50. The airspeed at which this drag rise begins is the *critical Mach number* (see Section 8.3.7, *The critical Mach number, M_{crit}*). M_{crit} for typical NACA four-digit or five-digit airfoils may be as low as a hair over Mach 0.6, depending on thickness. This renders them impractical for use in high-speed aircraft. A strong shock wave will form on such airfoils, greatly increasing their drag and reducing their lift, and may even introduce aeroelastic problems to the vehicle [35]. It is possible to delay these effects by increasing the sweep of the wing, but eventually they are inescapable.

As soon as the shock forms, the center of lift of the airfoil moves from approximately the quarter-chord to the mid-chord, further compounding the issue by substantially increasing stability. Additionally, the shock will induce separation, which thickens the wake aft of the shock (see Figure 8-43), increasing the drag further [36]. What follows is a complex interaction of the movement of the upper and lower shockwaves that ultimately results in a drastic drop in the lift coefficient being produced, causing a *shock-stall*.

Research on such high-speed effects began early, or in the 1940s, but serious work on modifying airfoils began

FIGURE 8-43 A Schlieren photograph of the flow past a 10% RAE 102 airfoil. Visible is the severe separation at the foot of the upper-surface shock, as well as an lower-surface shock. *AOA* is 2° and $M = 0.88$. *(Photo from Ref. [36])*

in 1955 with the pioneering work of H. H. Pearcey, who tried to experimentally obtain "an essentially shock-free flow" [37]. Pearcey showed that by reshaping the leading edge of an airfoil it was possible to weaken the shock. The process was to allow the flow to expand rapidly from the stagnation point and become supersonic in the leading edge area. This would form a series of compression shock waves that slowed the local Mach number, so the final shock wave was substantially weakened [38]. Less drag was one important benefit from this change. It was thought the modified airfoils would be capable of weakening the shock for maximum local Mach numbers as high as 1.4. The resulting pressure distribution has a prominent pressure peak near the leading edge and was described as being *peaky*. The airfoil has a relatively flat upper surface, which necessitates a cusp in the trailing edge to improve its lift generation.

Experience in using peaky airfoils on combat aircraft has revealed that high-speed, high-g maneuvers (which call for high *AOA*) result in the formation of significant shock strength that may cause sudden loss of lift. When introduced to the Hawker-Siddeley Kestrel FGA.1, the prototype of the production Hawker-Siddeley Harrier, this shock would occur above Mach 0.8, causing a serious wing-rocking at an *AOA* where a gentle buffet would be expected [38].

Supercritical Airfoils

Supercritical airfoils are often referred to as one of the three major contributions to aviation made by the famous aerodynamicist Richard T. Whitcomb (1921–2009), whom many historians of aviation call the most distinguished alumnus of the NASA Langley Research Center. His other two contributions being the area-rule and the winglet.

Like the peaky airfoils, supercritical airfoils are intended for high-speed aircraft. They feature a large radius (blunt) leading edge, considerably flatter upper surface than the peaky airfoils, and a significant cusp near the trailing edge (see Figure 8-44). The blunt leading edge softens the suction peak of the smaller radius peaky airfoil. The flatter top surface helps keep down the local Mach number and keeps down adverse pressure gradients. The cusp was introduced to help generate lift without lowering M_{crit} by forming a high-pressure region under the airfoil.

The characteristics of supercritical airfoils can be summarized as follows: (1) the drag rise Mach number, M_{crit}, is substantially higher than for more conventional airfoils. As an example, at a section lift coefficient of 0.65, the M_{crit} of one such airfoil is 0.79, which compares very favorably to the 0.67 of a NACA 64A-series airfoil of an equal thickness ratio [39]. (2) The section pitching moment coefficient for supercritical airfoils is substantially more negative than for conventional airfoils. This is caused by the high-pressure region that forms at the lower surface due to the cusp. (3) The supercritical airfoil also increases M_{crit} at off-design section lift coefficients. (4) The airfoil also increases the maximum lift coefficient at high subsonic Mach numbers.

Wing design that uses supercritical airfoils offers many advantages to a high-speed aircraft. First, and as has already been stated, is the increase in M_{crit} for a given thickness and wing sweep. Second, their thickness offers structural depth and volume for fuel. Third, they allow the designer to reduce the wing sweep, which offers a host of benefits. Among those are improvements in low-speed characteristics and performance capabilities.

Joukowski Airfoils

Joukowski airfoils are a much more clever mathematical tool used to demonstrate aerodynamic theory than a practical solution to aircraft design. The airfoils are named after the Russian scientist Nikolay Yegorovich Joukowsky (sometimes spelt Zhukovsky, 1847–1921), who is also the originator of the circulation theorem (see Section 8.1.8, *The generation of lift*). The airfoils are generated in conformance with the transformation of a circle on to the complex plane, using the complex number ζ:

$$z = x + \mathbf{i}y = \zeta + \frac{1}{\zeta} \qquad (8\text{-}50)$$

Consider the complex number ζ to be given by $\zeta = a + \mathbf{i}b$. Inserting this into Equation (8-50) and

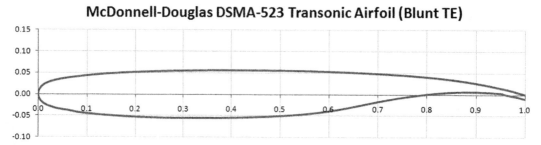

FIGURE 8-44 A transonic blunt TE airfoil developed by McDonnell-Douglas under the designation DSMA-523.

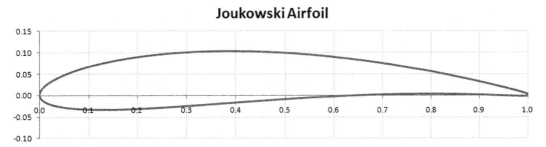

FIGURE 8-45 A Joukowski airfoil.

manipulating yields the two spatial variables x and y that are given by the following relation:

$$x = \frac{a(a^2 + b^2 + 1)}{(a^2 + b^2)} \quad \text{and} \quad y = \frac{b(a^2 + b^2 - 1)}{(a^2 + b^2)} \quad (8\text{-}51)$$

By varying the values of a and b between -1 and 1 (e.g. try $a = -0.1$ and $b = 0.1$) the shape of the airfoil can be modified (see Figure 8-45). Then, using elementary flow concepts, such as uniform flow, sources, sinks, and vortex flow, airflow around the resulting airfoil can be simulated. Joukowski airfoils are a subject in most texts on theoretical aerodynamics, although they are rarely used on actual aircraft. In fact, the UIUC airfoil database indicates five instances, all of them sailplanes (the Schempp-Hirth Gö-4 III Goevier, Schleicher ASK 14, ASK 16, ASK 18, and Schleicher Ka-6 Rhönsegler).

Liebeck Airfoils

The unconventionally shaped airfoil shown in Figure 8-46 is named after Dr. Robert Liebeck, who was one of the first to try and find out what sort of "airfoil-shaped" geometry would yield the highest possible C_{lmax}. The answer is the shape devised by Dr. Liebeck, which is remarkable for achieving the highest lift-to-drag ratio of any airfoil. The airfoils are designed to utilize Stratford distribution (see Section 8.1.6, *Chordwise pressure distribution*) and achieve an L/D ratio in excess of 600. The airfoils are intended for low Reynolds number applications. There is a large amount of scientific literature available that investigates the properties of these airfoils.

8.3 THE FORCE AND MOMENT CHARACTERISTICS OF THE AIRFOIL

As discussed in Section 8.1.3, *Representation of forces and moments*, the lift, drag, and pitching moment generated by the airfoil is almost always converted into a dimensionless coefficient form. The primary advantage of this is transferability. Transferability means that if the coefficients are known for a particular airfoil geometry, they can be used to extract forces and moments for any other size of the airfoil, airspeed and atmospheric conditions. These coefficients are defined as shown below

$$l = \frac{1}{2}\rho V^2 \cdot S \cdot C_l$$
$$d = \frac{1}{2}\rho V^2 \cdot S \cdot C_d \quad (8\text{-}52)$$
$$m = \frac{1}{2}\rho V^2 \cdot S \cdot c \cdot C_m$$

where

V = airspeed, in m/s or ft/s
l = lift force, in N or lb$_f$
d = drag force, in N or lb$_f$
m = pitching moment, in N·m or ft·lb$_f$
S = reference area, typically wing area, in m^2 or ft^2
C_l = non-dimensional lift coefficient that relates *AOA* to the lift force
C_d = non-dimensional drag coefficient that relates *AOA* to the drag force
C_m = non-dimensional pitching moment coefficient that relates *AOA* to the pitching moment
ρ = density of air, in kg/m^3 or slugs/ft^3

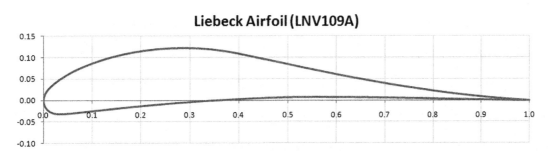

FIGURE 8-46 A Liebeck airfoil.

8.3.1 The Effect of Camber

Introducing positive (or negative) camber to an airfoil will have a profound effect on its lift curve and drag polar. These are illustrated in Figure 8-47. A positive camber will shift the lift curve to the left and up, resulting in a higher C_{lmax}, and introduce a negative zero lift angle (α_{ZL}) and a positive value to the lift coefficient at $\alpha = 0°$ (C_{lo}). The opposite happens if a negative camber is introduced.

The camber will also change the drag polar. If the camber is modest, as is usual for most airfoils, the change will primarily be in the shifting of the drag polar (see Figure 8-47). This change introduces a new variable, C_{lmind}, which is the value of the lift coefficient where the airfoil's minimum drag coefficient, C_{dmin}, occurs. C_{dmin} for a symmetrical airfoil is at $C_l = 0$. If the change in the camber is large (as happens when flaps are deflected) the shape of the drag polar will change and be shifted to a higher C_{dmin} as well.

8.3.2 The Two-dimensional Lift Curve

The lift coefficient relates the *AOA* to the lift force. If the lift force is known at a specific airspeed the lift coefficient can be calculated from:

$$C_l = \frac{2l}{\rho V^2 S} \quad (8-53)$$

In the linear region, at low *AOA*, the lift coefficient can be written as a function of *AOA* as shown below:

$$C_l = C_{l_0} + C_{l_\alpha}\alpha \quad (8-54)$$

Equation (8-54) allows the *AOA* corresponding to a specific lift coefficient to be determined, provided the lift curve slope known:

$$\alpha = \frac{C_l - C_{l_0}}{C_{l_\alpha}} \quad (8-55)$$

Equation (8-55) is only applicable in the linear region, like the one shown in Figure 8-3.

8.3.3 The Maximum Lift Coefficient, C_{lmax}

Standard lift curve graphs, which plot the variation of the lift coefficient with *AOA*, usually show a very distinct peak (e.g. see Figure 8-39). This peak is the *maximum lift coefficient* and both it and the *AOA* at which it occurs are of profound importance in the development of any mechanical device that depends on aerodynamic forces.

The shape of the lift curve in this region is of importance as well. Airfoils that display a sharp drop in the lift curve at stall should be avoided for use in aircraft, in particular if benign stall characteristics are being sought.

Maximum Theoretical Lift Coefficient

Questions such as "what is the maximum value of the lift coefficient?" and "what kind of shape yields that value?" are often asked. Answers to these questions are addressed in a classic paper by Smith [3]. Knowing the theoretical limits is important for comparison reasons. It allows efficiency and potential for improvements to be quantified. Figure 8-48 shows a cylinder of diameter *d*, rotating in a uniform inviscid flow of velocity *V*, exposed to three values of the

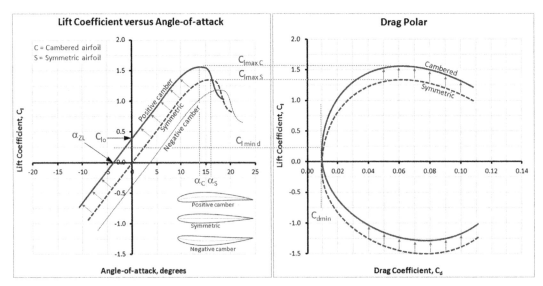

FIGURE 8-47 The effect of Reynolds number on the lift curve and drag polar.

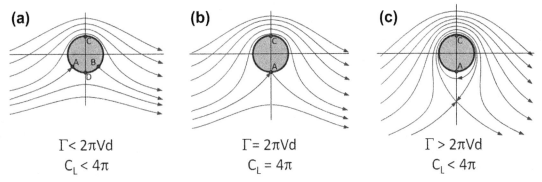

$\Gamma < 2\pi Vd$ $\Gamma = 2\pi Vd$ $\Gamma > 2\pi Vd$
$C_L < 4\pi$ $C_L = 4\pi$ $C_L < 4\pi$

FIGURE 8-48 Potential flow past a circular cylinder drawn for three values of the circulation Γ. *(Based on Ref. [46])*

circulation, Γ. Since the flow is inviscid it will not separate, but rather flow around the cylinder as shown. While such a flow exists only in the imagination, it is still of considerable importance in aerodynamics. Not only does it show what could be were it not for viscosity, it also allows the aerodynamicist to determine theoretical limits.

Using *potential flow theory* (PFT), it is possible to simulate airflow around a rotating cylinder by adding the elementary flow functions of uniform and vortex flow. For instance, consider the Figure 8-48(a), which shows two stagnation points, A and B. In Figure 8-48(b), the circulation is strong enough to bring the two stagnation points together, and in (c) it is strong enough to move the stagnation point off the surface. It can be shown using PFT that the circulation required to create the flow in Figure 8-48(b) amounts to $2\pi Vd$. For instance see Anderson [40, pp 249–251]. Using Equation (8-26), we can write:

$$l = \rho V \Gamma = \rho V (2\pi Vd) = 2\pi \rho V^2 d.$$

But it is also possible to write lift using Equation (8-52). Therefore, the lift coefficient, C_l, can be found to equal:

$$l = 2\pi \rho V^2 d = \frac{1}{2}\rho V^2 d C_l \Rightarrow C_l = 4\pi$$

The corresponding lift coefficients for values of the circulation on either side of $2\pi Vd$ result in lift coefficients that are less than 4π. A geometry that yields a higher value of the lift coefficients is not known by this author. The value $C_{lmax} = 4\pi$ can therefore be considered the *maximum theoretical lift coefficient*.

8.3.4 The effect of Reynolds Number

Figure 8-49 shows how the Reynolds number (R_e) affects the airfoil's lift curve and drag polar. The left graph shows that the slope and intersection of the linear region to the horizontal and vertical axes are unaffected by the R_e. However, the higher the R_e, the higher is the stall *AOA* (denoted by α_1, α_2, and α_3), maximum lift

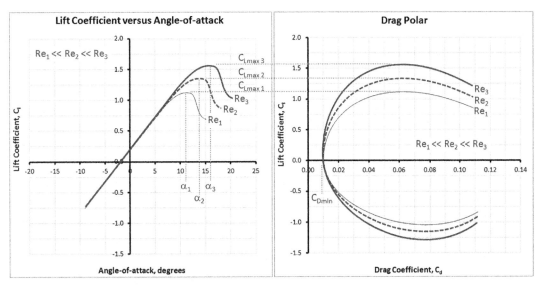

FIGURE 8-49 The effect of Reynolds number on the lift curve and drag polar.

coefficient (denoted by $C_{Lmax\ 1}$, $C_{Lmax\ 2}$, and $C_{Lmax\ 3}$), and the point where the slope becomes non-linear. This effect is purely viscous so it is usually estimated through wind-tunnel testing or using viscous computational fluid dynamics flow solvers.

A typical change in the drag polar is shown in the right graph of Figure 8-49. Generally, an increase in R_e will not have a large effect on the minimum drag, unless it causes a significant change in the nature of the flow, such as if a laminar boundary layer transitions early into a turbulent one, or as a result of poor surface qualities (e.g. see Section 8.3.13, *The effect of leading edge roughness and surface smoothness*). On the other hand, it will reduce the C_d elsewhere, thanks to the delayed tendency for flow to separate. For thick airfoils, the high R_e delays the forward movement of the trailing edge separation point to a higher *AOA*, ultimately delaying stall. For this reason, as the R_e increases, the drag polar typically gets wider.

8.3.5 Compressibility Effects

Generally, flow can be considered incompressible up to Mach 0.3 and even as high as Mach 0.5. This fact is very convenient to the designer of slow aircraft, as he or she will not have to deal with the added complexity of compressibility. Compressibility requires the introduction of equivalent airspeed (see Section 16.3.2, *Airspeeds: instrument, calibrated, equivalent, true, and ground*) in addition to modifications that have to be made to the lift, drag and pitching moment coefficients.

Compressibility Effect on Lift

Figure 8-50 shows the effect compressibility has on the airfoil's lift curve and drag. Compressibility increases the slope of the lift curve. In practice, this means that a smaller *AOA* is required to generate a given lift coefficient, as is evident from the left graph in Figure 8-50. The designer of GA aircraft is generally only interested in the stall portion of the lift curve at low Mach numbers because such airplanes always stall at low subsonic airspeeds. However, the stall portion is important in fighter aircraft design as such aircraft pull high *g*s at high airspeeds at *AOA*s near or exceeding stall.

Compressibility Effect on Drag

Compressibility also has important effect on the drag polar (also discussed in Section 15.4.2, *The effect of Mach number*). The drag coefficient at a given *AOA* (or C_l) remains relatively constant until the airfoil approaches Mach numbers in the range 0.6 to 0.8. These are called the *critical Mach number* (see Section 8.3.7, *The critical Mach number*, M_{crit}). Increasing the airspeed further will result in a sudden rise in drag. This is a consequence of a shock forming on the airfoil, as the local velocities exceed the speed of sound. The effect is shown as the sharp rise in the compressible drag coefficient, C_{DC}, in the right graph of Figure 8-50. These are but a part of the intriguing (if not surprising) behavior of fluids near this critical airspeed ($M = 1$). The so-called *drag divergence Mach number*, denoted by M_{DD}, is defined after the coefficient has clearly begun to diverge from its incompressible value. Generally, there are two common definitions by two prestigious manufacturers; Boeing and Douglas (which is now a part of Boeing). The Boeing approach defines M_{DD} as the Mach number where the drag coefficient exceeds its incompressible value ($C_{Dincompressible}$ in Figure 8-50) by 0.0020. The Douglas approach defined it as the Mach number where the slope dC_{Dc}/dM first reaches 0.10 [41]. Section 15.5.12, *Drag due*

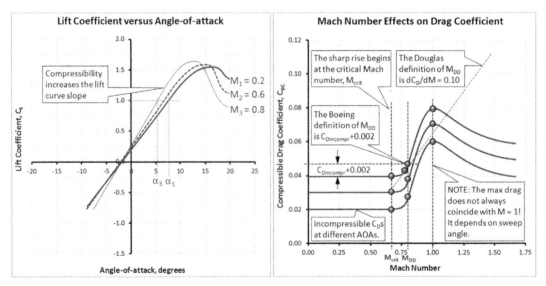

FIGURE 8-50 The effect of Mach number on lift and drag.

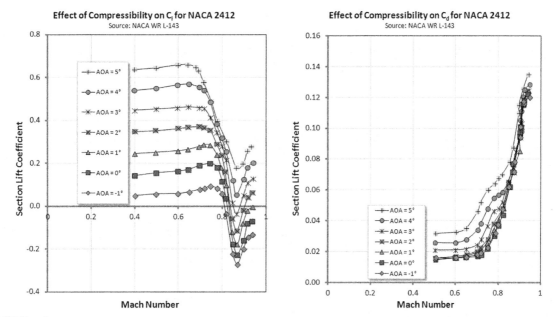

FIGURE 8-51 The effect of Mach number on the lift and drag coefficient of a NACA 2412 airfoil. *(Reproduced from Ref. [47].)*

to compressibility effects, presents a method to account for this drag.

Compressibility Effect on Lift and Drag Exemplified

Compressibility affects both the section lift coefficient and the section drag coefficient detrimentally at high Mach numbers, as shown in Figure 8-51. These show that at a fixed *AOA*, there is an abrupt reduction in lift and a sharp rise in drag when the airspeed approaches sonic conditions. This effect happens at the *critical Mach number* (see Section 8.3.7, *The critical Mach number*, M_{crit}). It is indicative of local airspeeds over the airfoil having exceeded Mach one, forming a normal shock-wave that alters the pressure distribution, causing the large deviations seen.

Figure 8-52 shows the effect the M_{crit} has on the lift curve slope and the zero *AOA* lift coefficient. The first observation is that the lift curves at $M = 0.4$ and 0.6 do not show a large change. However, there is a sharp reduction in the lift curve slope at $M = 0.8$ (note that the values of $C_{l\alpha}$ and C_{lo} shown are based on the four points at $1°$ through $4°$). This is followed by a slight recovery at the minimum value of the lift coefficients, which occurs at a Mach number of 0.87; however, the zero *AOA* lift is now negative and almost doubled in magnitude. Any airplane featuring this airfoil (NACA 2412) operating at these Mach numbers would suffer serious stability issues.

Compressibility Effect on Pitching Moment

Compressibility has a major impact on the behavior of airfoils and wings at high Mach numbers. It is appropriate to discuss both at the same time. This is shown schematically in Figure 8-53 for a hypothetical aircraft trimmed at $M = 0.5$. It is of interest to consider the effect in terms of the pitch stability, denoted by $C_{m\alpha}$, and stick force, which is the force the pilot must exert in order to control the airplane longitudinally. Of course the graph would be different for other aircraft, CG locations, control system, airfoil types, trim points, and so on. However, considering these at the moment is not necessary to make the following point. As will be discussed in Chapter 11, *The anatomy of the tail*, static stability of aircraft requires $C_{m\alpha}$ to be negative (here, for the example aircraft, it is near -1.0 at low Mach numbers. This means that at low airspeeds, the pilot must pull on the yoke (or stick or column) in order to slow down further. Alternatively, he or she must push in order to fly at Mach numbers higher than 0.5.

With respect to the airfoil, as the far-field airspeed increases, a shock-wave begins to form and change the pressure distribution over the wing. The airplane has entered the transonic region. The effect of the shock-wave depends on where on the wing it forms first (inboard, midspan, outboard) and grows with further increase in the airspeed. It is possible to experience a number of different scenarios, besides the one shown in Figure 8-53. The bottom line is that once a given airspeed is exceeded, the pitch stability and stick force will rise rapidly. This is caused by the center of pressure moving from approximately the quarter-chord to the mid-chord. The effect is akin to the CG being moved forward by that amount at subsonic speeds. If the airplane has not been designed for this, it is possible that the magnitude of the stick force will render the airplane

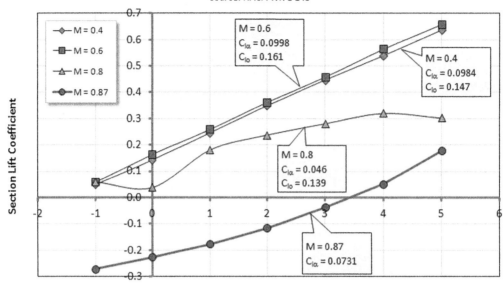

FIGURE 8-52 The effect of Mach number on the lift curve of the NACA 2412 airfoil. The lift curve slope and the lift at zero *AOA* are shown in the bullets. *(Based on analysis using data from Ref. [47])*

uncontrollable — it will enter a dive (recall the moment tends to point the nose down), which will increase the airspeed further, and from which it may not recover.

8.3.6 Compressibility Modeling

The following methods are commonly used to account for the impact of compressibility on pressure coefficients:

Prandtl-Glauert [42]:

$$C_p = \frac{C_{po}}{\sqrt{1-M^2}} \qquad (8\text{-}56)$$

Kármán-Tsien:

$$C_p = \frac{C_{po}}{\sqrt{1-M^2} + \left(\frac{M^2}{1+\sqrt{1-M^2}}\right)\frac{C_{po}}{2}} \qquad (8\text{-}57)$$

Laitone:

$$C_p = \frac{C_{po}}{\sqrt{1-M^2} + \left(\frac{M^2(1+0.2M^2)}{2\sqrt{1-M^2}}\right)C_{po}} \qquad (8\text{-}58)$$

Frankl-Voishel correction for skin friction drag:

$$C_{Df} = C_{Dfo}(0.000162M^5 - 0.00383M^4 + 0.0332M^3 \\ - 0.118M^2 + 0.0204M + 0.996) \qquad (8\text{-}59)$$

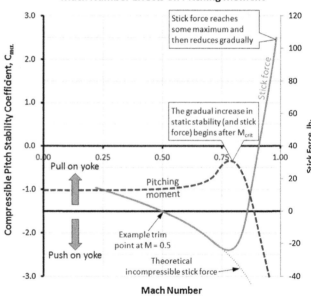

FIGURE 8-53 The effect of Mach number on pitching moment.

while compressibility effects can be included with the lift and moment coefficients as follows (here using the Prandtl-Glauert model):

$$C_l = C_{l_0}/\sqrt{1-M^2} \qquad (8\text{-}60)$$

$$C_m = C_{m_0}/\sqrt{1-M^2} \qquad (8\text{-}61)$$

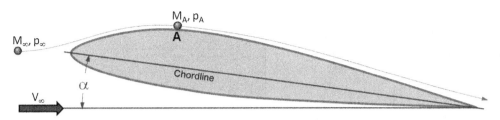

FIGURE 8-54 An airfoil immersed in airflow with the point of minimum pressure identified as point A.

where the subscripts '0' refer to the incompressible values. The same correction does not hold for the drag coefficient, as it incorporates skin friction as well as pressure drag and will give erroneous results.

8.3.7 The Critical Mach Number, M_{crit}

Consider the airfoil (or some arbitrary body) immersed in fluid flow as shown in Figure 8-54. The presence of the body will cause the fluid to accelerate over it, resulting in local speeds that are higher (and lower) in places compared to the far-field speed. In the case of the fluid being air; if the far-field Mach number is denoted by M_∞, there will be a different Mach number detected at some arbitrary point A on the body. Then, we define the *critical Mach number*, M_{crit}, as the far-field speed at which the local velocity, anywhere on the body, first achieves $M = 1$. As has already been discussed, this manifests itself as a sudden rise in the drag of the body.

Typical airfoils begin to experience this phenomenon when the far-field Mach numbers exceed 0.60 to 0.80. The actual value is highly dependent on the geometry of the airfoil, primarily the thickness: M_{crit} is much lower for thick airfoils than thin ones. Examples of M_{crit} for selected NACA airfoils can be seen in Table 8-5. The M_{crit} of airfoils used for high-speed aircraft is normally increased by sweeping the wing forward or aft (see Chapter 9, *The anatomy of the wing*). Additionally, the effect depends on the *AOA*. The higher the *AOA*, the higher are the local airspeeds over the airfoil and this will expedite the onset of the drag rise.

Step-by-step: Determining M_{crit} for a Body

The M_{crit} can be determined for a body, such as an airfoil or a fuselage, by the application of the following method, based on Anderson [40, pp. 674–679] and Perkins and Hage [43].

Step 1: Establish the Minimum Pressure Coefficient

Using either a theoretical or experimental method determine the low-speed, incompressible minimum pressure coefficient, C_{po}, on the body. For airfoils this can be done by any of the freely (and commercially) available computer codes cited in Section 8.1.14, *Airfoil design*. References [23,44] also list airspeed ratios for a number of NACA airfoils.

Step 2: Select Compressibility Correction Method

Using any of the methods presented with Equations (8-56), (8-57), or (8-58), select the appropriate compressibility correction. Note that Equations (8-57) and (8-58) are generally considered more precise than the Prandtl-Glauert method.

Step 3: Solve to Determine M_{crit}

Depending on the compressibility correction method selected, the predicted M_{crit} can be found by setting it equal to the following expression, which determines the pressure coefficient required for sonic conditions to prevail at the point of minimum pressure:

$$C_{p_{crit}} = \frac{2}{\gamma M_{crit}^2}\left\{\left[\frac{2}{1+\gamma}\left(1+\frac{(\gamma-1)}{2}M_{crit}^2\right)\right]^{\frac{\gamma}{\gamma-1}} - 1\right\} \quad (8\text{-}62)$$

An example of this procedure is shown in Figure 8-55, as it is applied to the NASA NLF(1)-0414F airfoil in Figure 8-7.

Derivation of Equation (8-62)

As the airspeed (Mach number) increases, the local velocities over an airfoil increase as well. Consider point A on the airfoil in Figure 8-54. Denote the static pressure in the far field with p_∞ and at A using p_A. Assuming the flow to be isentropic (adiabatic and irreversible) the pressure in the far-field and at A can be related to the total pressure as follows:

Total-to-far-field pressure ratio:

$$\frac{p_{tot}}{p_\infty} = \left[1 + \frac{\gamma-1}{2}M_\infty^2\right]^{\frac{\gamma}{\gamma-1}} \quad \text{(i)}$$

Total-to-point A pressure ratio:

$$\frac{p_{tot}}{p_A} = \left[1 + \frac{\gamma-1}{2}M_A^2\right]^{\frac{\gamma}{\gamma-1}} \quad \text{(ii)}$$

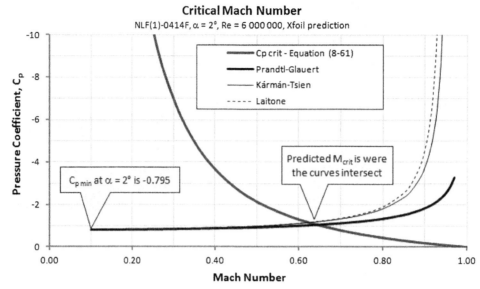

FIGURE 8-55 The critical Mach number of the NASA NLF(1)-0414F airfoil at an *AOA* of 2° is approximately at $M = 0.63$ to 0.65, depending on compressibility model.

Therefore, the pressure at point A can be related to that in the far-field by dividing Equation (i) by (ii) as follows:

$$\frac{p_A}{p_\infty} = \frac{p_{tot}/p_\infty}{p_{tot}/p_A} = \frac{\left[1 + \frac{\gamma-1}{2}M_\infty^2\right]^{\frac{\gamma}{\gamma-1}}}{\left[1 + \frac{\gamma-1}{2}M_A^2\right]^{\frac{\gamma}{\gamma-1}}}$$

$$= \left[\frac{1 + \frac{\gamma-1}{2}M_\infty^2}{1 + \frac{\gamma-1}{2}M_A^2}\right]^{\frac{\gamma}{\gamma-1}} = \left[\frac{2 + (\gamma-1)M_\infty^2}{2 + (\gamma-1)M_A^2}\right]^{\frac{\gamma}{\gamma-1}} \quad \text{(iii)}$$

Of special interest is the case when the airspeed at point A becomes sonic, i.e. $M_A = 1$. Then Equation (iii) can be rewritten as shown below:

$$\frac{p_A}{p_\infty} = \left[\frac{2 + (\gamma-1)M_\infty^2}{2 + (\gamma-1)M_A^2}\right]^{\frac{\gamma}{\gamma-1}} = \left[\frac{2 + (\gamma-1)M_\infty^2}{1 + \gamma}\right]^{\frac{\gamma}{\gamma-1}}$$

$$= \left[\frac{2}{1+\gamma}\left(1 + \frac{(\gamma-1)}{2}M_\infty^2\right)\right]^{\frac{\gamma}{\gamma-1}} \quad \text{(iv)}$$

Inserting Equation (iii) into Equation (8-14) and renaming M_∞ as M_{crit} yields Equation (8-62).

QED

8.3.8 The Effect of Early Flow Separation

Figure 8-56 and Figure 8-57 show the effect early flow separation has on the airfoil's lift curve and drag polar. This is in fact easy to detect from the lift curve and, thus, the designer reviewing wind tunnel test results should be aware of its nature and not be caught off guard. The effect appears in a similar fashion for two-dimensional airfoils and three-dimensional aircraft as a reduction in the lift curve slope. It can be seen in Figure 8-56 that the slope change is followed by a stall *AOA* that may be slightly less than what would be expected, and the maximum lift coefficient may be substantially lower. This can cause severe detriments to the low-speed performance of an aircraft, and in the most severe case make a new design impossible to certify. In addition to a smaller C_{Lmax}, the slope change will require the airplane to make up for the shallow slope by operating at a higher *AOA*, with the associated increase in lift-induced drag.

The early separation of two-dimensional bodies, such as airfoils, can be the consequence of the pressure recovery region of the airfoil being too short, the trailing edge region of the airfoil being too steep, or the discontinuity of a surface caused by the presence of a control surface. For efficiency the lift curve should be linear to as high a lift coefficient as possible.

It follows from the coupling of lift and drag that the effect of flow separation impacts the drag polar as well, although it is harder to make out. As shown in Figure 8-57, the effect will narrow the drag polar and deviate from a quadratic approximation (in the absence of a drag bucket) much earlier than what one would expect. As stated above, this means that there will be an unexpected rise in the drag of the aircraft, even at climb *AOA*, that will lower rate of climb, and reduce optimum operating points, such as endurance and range, which are often achieved at relatively high *AOA*s.

8.3 THE FORCE AND MOMENT CHARACTERISTICS OF THE AIRFOIL

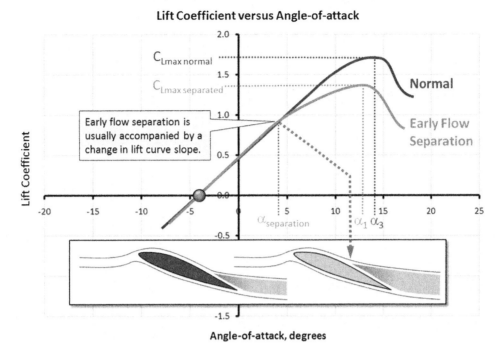

FIGURE 8-56 The effect of early flow separation on the lift curve.

8.3.9 The Effect of Addition of a Slot or Slats

Figure 8-58 shows the effect of introducing a leading edge slot or slat to an airfoil. The airfoil's lift curve is modified in an important way as as shown in the figure. Usually, the literature depicts the effect of deploying the slat by showing that the point of stall on the lift curve is extended to a higher AOA and C_{lmax} (or C_{Lmax} for three-dimensional geometry). However, two additional effects must be considered as well. The first one is that this is only accurate as long as the deployment changes the overall camber in minute ways only. The second one is that the deployment does not significantly change the chord length of the airfoil. If the shape of the camber

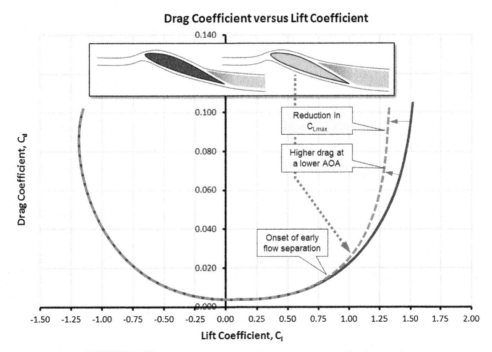

FIGURE 8-57 The effect of early flow separation on the drag polar.

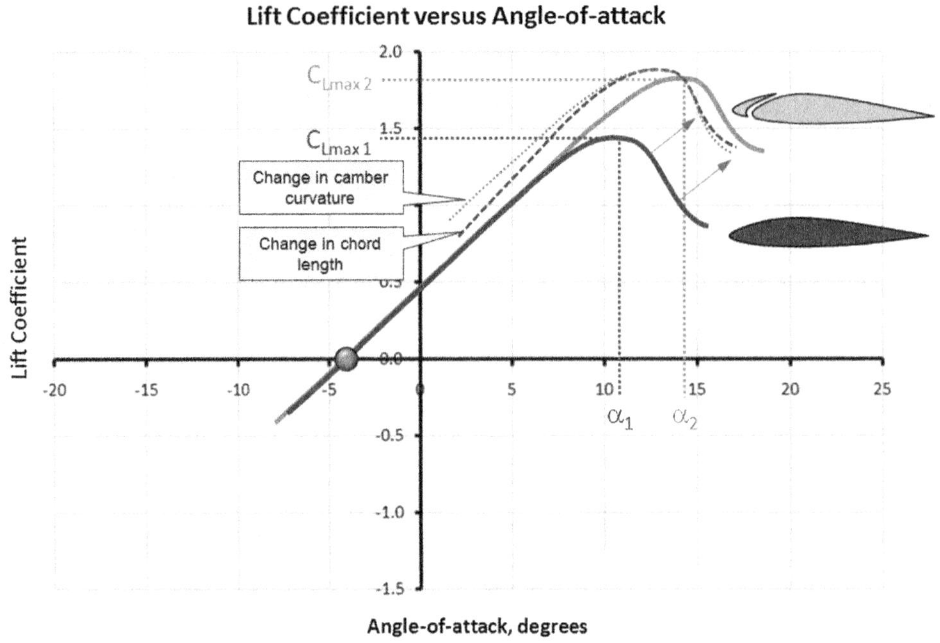

Lift Coefficient versus Angle-of-attack

FIGURE 8-58 The effect of introducing a slot or slat to the airfoil on the lift curve.

changes such that its curvature increases, the lift curve will shift up and left, similar to that of a flap deflection (see Section 8.3.10, *The effect of deflecting a flap*). If the chord of the airfoil increases as a consequence of deployment, the slope of the lift curve will change as well. In this case, it is most clear to envision the slope change being caused by a larger lift force being converted to a lift coefficient using the original chord-length (or wing area in the case of a three-dimensional wing). If both parameters change noticeably, the resulting lift curve will display a combination of the two effects and not that typically displayed in the literature.

Various leading edge devices are presented in Section 10.2, *Leading edge high-lift devices*.

8.3.10 The Effect of Deflecting a Flap

Flaps have a significant impact on the aerodynamic characteristics of airfoils. At a given airspeed and *AOA*, the magnitudes of lift, drag, and pitching moment of an airfoil increase. Figure 8-59 shows the effect of deflecting a flap on a wing's lift curve. Figure 8-60 show the effect on the drag polar. It is important to keep in mind that deflecting the flap trailing edge down (TED) will move the lift curve up and sideways to the left. This is a consequence of an increase in the airfoil's camber, but this increases its lift at a given *AOA*. The opposite takes place if the flap is deflected trailing edge up (TEU). This decambers the airfoil, effectively shifting the lift curve downward and to the right. This means that the stall *AOA* is shifted as well, to the left or right, depending on the deflection.

The deflection also affects the drag and increases it regardless of whether the flap is deflected TEU or TED. However, the magnitude and sign of the pitching moment change significantly. Deflecting a flap TEU allows wings to be made statically stable without the use of stabilizing surfaces such as horizontal tails or canards. The magnitude of the change in lift largely depends on the design of the flap, but this is discussed in more detail in Section 10.3, *Trailing edge high-lift devices*.

8.3.11 The Effect of Cruise Flaps

A *cruise flap* is a wing control surface whose purpose is to shift the drag polar to the left when deflected (see Figure 8-61). To achieve this effect, the flap must be deflected trailing edge up. As a consequence, the drag characteristics of the aircraft will be significantly affected. The primary effect is that the maximum L/D will now move to a lower value of the lift coefficient. This means that the aircraft will be more efficient at a higher airspeed — and this is favorable because people generally want to travel from A to B fast and efficiently.

Cruise flaps are a very common feature on sailplanes, where they allow the pilot to fly from thermal to thermal faster to minimize sinking in-between. The total drag coefficient will increase a bit as the polar is shifted to the left and up. However, for low deflection angles, the transverse movement is much larger than the upward one, so there is not a large increase in drag. Cruise flaps will also reduce the C_m as the up-deflected trailing edge will effectively transform the airfoil into a reflexed one. In this way, cruise flaps can reduce trim drag.

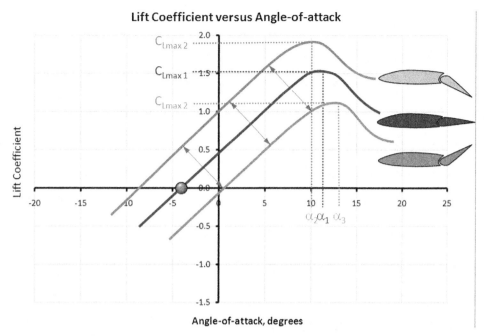

FIGURE 8-59 The effect of introducing a flap to the airfoil on the lift curve.

Cruise flaps are almost exclusively an extension of the functionality of regular flaps. In other words, the flap control system is designed such that the flap can deflect, say, 5° to 10° trailing edge up in addition to its normal downward deflection. As a consequence, they inflict very limited complexity on its design, although the use of slotted flap types may cause issues due to the upper wing surface extending farther aft (called *slot lip*).

8.3.12 The Effect of Deploying a Spoiler

A *spoiler* is a device whose purpose is to increase the drag of the airplane. This is necessary for low-drag airplanes such as sailplanes and jets, to allow them to descend at relatively steep flight path angles. Without such a device, a sleek aircraft would have to descend at a very shallow flight path angle to keep down the airspeed. The resulting landing would be challenging as

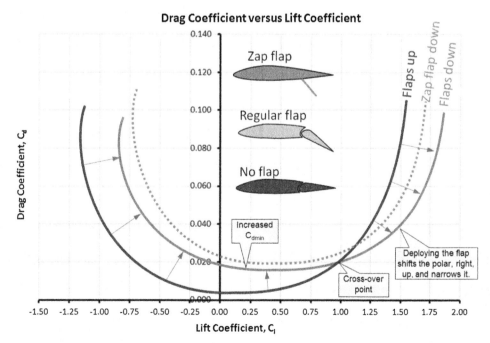

FIGURE 8-60 The effect of introducing a flap to the airfoil on the drag polar.

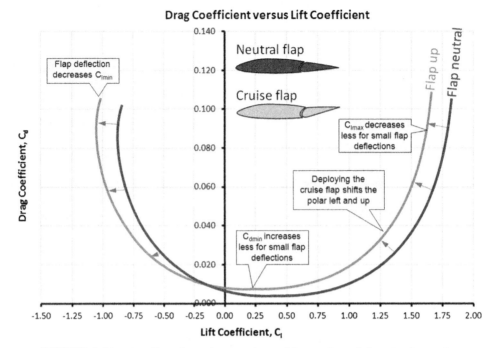

FIGURE 8-61 The effect of introducing a "cruise" flap to the airfoil on the drag polar.

the airplane would float long distances down the runway before touching down. A spoiler remedies this by allowing for a steeper descent angle at a given airspeed. Devices of this kind are also referred to as an *airbrake* or *speed-brake*. Spoilers are very useful for sailplanes, but serve two purposes on jets: to allow a rapid descent from some initial altitude to a lower one; and to ensure the airplane does not bounce after the first touch-down during landing, by "killing" lift. A jet that bounces during landing gets airborne and consumes a lot of runway.

Some people make a distinction between a spoiler and a speed brake, using the term spoiler if the device is deployed from the wing (it "spoils" the lift and increases drag), but using the term speed brake if it is deployed from another part of the aircraft, for instance the fuselage. In this case it only generates drag. The advantage of speed-brakes is that they increase the drag without changing the *AOA*.

In most cases, the effect of deploying a speed-brake can be approximated by shifting the drag polar upward to a higher C_{dmin}, while reducing the C_{lmax}. This is shown in Figure 8-62. Note that while the C_{lmax} of the airfoil is decreased significantly due to a "forced" flow separation, the magnitude of the C_{lmin} increases significantly as well (i.e. it becomes even more negative). The effect is similar to using a split flap on an inverted airfoil.

8.3.13 The Effect of Leading Edge Roughness and Surface Smoothness

The quality of lifting surfaces is of great importance and is an effect that was investigated by NACA as early as the late 1930s [45] and showed that surface roughness caused a large increase in drag. It was also shown that smooth surfaces are important even when extensive laminar flow is not to be expected [23]. However, it was also shown that surfaces do not have to be superbly smooth or polished. On the other hand, polishing a surface that is not aerodynamically smooth will reduce its drag. The experimenting showed that 320-grit sandpaper produced acceptable surface smoothness for all aerodynamically smooth surfaces. It also showed that surface particles are more detrimental than surface scratches when it comes to the transition from laminar to turbulent boundary layer.

Figure 8-63 shows the effect of surface finish on the two-dimensional section coefficient of a NACA 64-420 airfoil with two kinds of surface finish; a smooth one and unpolished camouflage paint. It can be seen how the camouflage paint affects the drag by nearly doubling in magnitude near a R_e of 20×10^6. It is essential for the designer to be aware of such trends in the performance estimation of an aircraft intended, perhaps, as a military trainer. Ignoring the effect could make an airplane that on the drawing board appears competitive far less viable in practice. The magnitude of the drag increase is as follows:

If $R_e < 20 \times 10^6$: $\Delta C_d = 0.000453$
If $R_e \geq 20 \times 10^6$: $\Delta C_d = 0.00308$

The effect of a contaminated leading edge (LE) is also presented in Ref. [23] and is reflected in the drag polar in Figure 8-64. The drag polar, initially smooth, is shifted

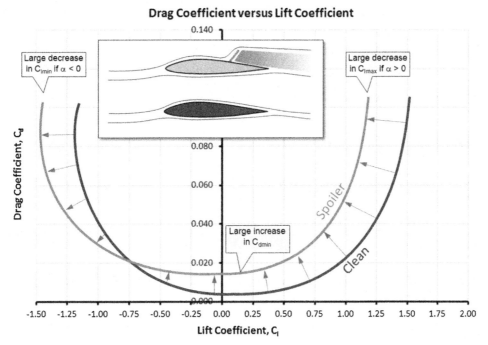

FIGURE 8-62 Increase in C_{Dmin} due to change in the deployment of a spoiler. Generally, the shape of the drag polar changes through a vertical upward shift and contraction, possibly causing a change in the location of C_{lminD} for the shifted drag polar.

sharply upward as a consequence of grain contamination as small as 0.004 inches. The polar is shifted farther upward when the grain size grows to 0.011 inches; however, even though this is an increase by a factor of three, the drag polar only shifts up by a fraction — the damage is already done. Reference [23] also shows that LE roughness reduces section lift-curve slope and maximum lift coefficient.

8.3.14 Drag Models for Airfoils Using Standard Airfoil Data

Figure 8-65 shows a standard presentation of airfoil drag and pitching moment data from technical documents that all aircraft designers should be intimately familiar with: *NACA R-824* [23] and *Theory of Wing Sections* by Abbott and Doenhoff [44]. The data of interest in this section is the drag polar, which describes the

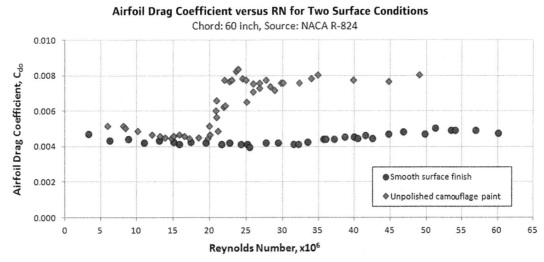

FIGURE 8-63 The effect of surface finish on the minimum drag of an airfoil. The smooth airfoil is only slightly affected by the change in R_e, whereas the unpolished camouflage paint job experiences a doubling in magnitude at R_e around 20×10^6. *(Based on Ref. [23])*

288   8. THE ANATOMY OF THE AIRFOIL

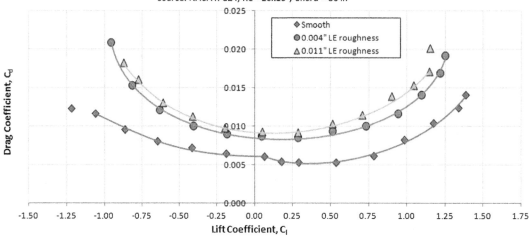

FIGURE 8-64 The effect of introducing surface roughness on the drag polar. *(Based on Ref. [23])*

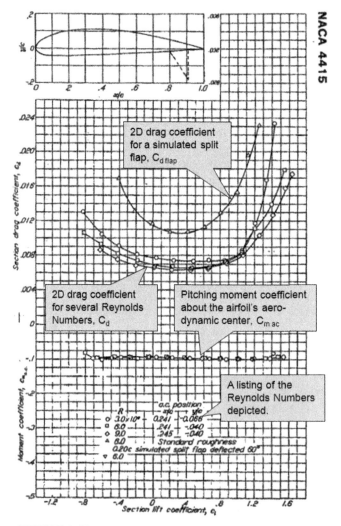

FIGURE 8-65 Typical representation of airfoil data (Ref. [23]).

drag coefficient as a function of the lift coefficient. For standard aircraft drag analysis the airfoil is of course only a part of the total aircraft drag; the remaining drag comes from the fuselage, engines, and other parts, as well as the interference of these with each other. However, the airfoil drag is relatively manageable, mathematically, provided a number of rules and realizations are adhered to.

A considerable amount of information on airfoil drag is at the disposal of the aircraft designer, from the literature, such as the aforementioned NACA publications. Additionally, as stated in Section 8.1.14, *Airfoil design*, it is even possible to obtain airfoil analysis software, free of charge, from the internet.

8.3.15 Airfoil Selection How-to

The selection of airfoils is often a challenging task for the aircraft designer. The airfoil can have a profound effect on the performance and handling of the aircraft, as well as the structure and weight. If the designer is forced to resort to selecting an airfoil from sources such as the NACA R-824 report or the Stuttgarter Profilkatalog, the following properties should be kept in mind:

Impact on Drag

The geometry of the airfoil will affect where the laminar boundary layer transitions into a turbulent one. This can have a major effect on the magnitude of the skin friction drag and, therefore, the performance of the airplane. The geometry of the airfoil dictates whether or not an airfoil can develop extensive laminar flow. In order to take advantage of NLF airfoils the designer must ensure the mission C_L resides inside the drag bucket. If the width of the drag bucket envelopes the C_L for best rate of climb, this will also improve climb performance. This can be done by the selection of the appropriate wing area.

Impact on Flow Separation

The geometry also affects at what *AOA* flow separation begins and how rapidly it grows with additional increase in *AOA*. Modern NLF airfoils are quite sensitive to imperfections and this may pose a problem if it causes asymmetry between the left and right wings at stall. This may lead to undesirable stall handling, such as an uncontrollable roll-off to the left or right. Additionally, leading edge radius and smoothness for the first 15–20% of the chord can have a significant effect on the stall characteristics.

Impact on Maximum Lift and Stall Handling

The geometry of the airfoil dictates what its maximum lift coefficient will be. The more prominent the camber the higher is the maximum lift coefficient. The smaller the leading edge radius, the lower will be the maximum lift coefficient. A small leading edge radius may also promote a sudden drop in lift at stall, with the associated detrimental effect on stall handling. The shape of the airfoil may also encourage the formation of a separation bubble that may have a profound detrimental effect on stall handling (e.g. see discussion of the NACA 23012 series airfoils in Section 8.2.10, *Famous airfoils*).

Impact on Longitudinal Trim

The geometry of the airfoil dictates its pitching moment coefficient, C_m. This may have important implications for the sizing of the horizontal tail and trim drag in cruise. Generally, the larger the camber the greater is the magnitude of the pitching moment coefficient. A symmetrical airfoil will have a zero C_m about its aerodynamic center. Cambered airfoils will have negative C_m. Deploying flaps will make it even more negative (stable), but requires a larger horizontal tail area or greater elevator deflection to balance.

Critical Mach Number

The geometry of the airfoil is instrumental in how early the shockwaves begin to form at high subsonic Mach numbers. A thick airfoil will cause a greater local acceleration of air, which will form a shockwave at a lower M than will a less thick airfoil. An airplane designed to cruise at, say, $M = 0.5$ should not feature airfoils much thicker than 15% without modifying the thickness distribution to prevent shocks at dive speed. This is particularly important in the juncture between the fuselage and wing, where the local airspeeds are even greater.

Impact on Wing-fuselage Juncture

The wing-fuselage juncture can present particularly challenging effects on airflow, where it accelerates to even greater airspeeds than elsewhere on the wing. The effect is generally twofold: acceleration to a higher airspeed means a lower critical Mach number; and acceleration to a higher airspeed means that air must decelerate that much more. If this deceleration (called pressure recovery) takes place over a short distance a flow separation will occur. Such a separation may even take place at a low *AOA*. The consequence is higher drag and impaired performance.

Impact on Structural Depth

The geometry of the airfoil dictates entirely the structural depth of the wing. A thick airfoil provides a large structural depth that accommodates a taller spar. A taller spar, in turn, brings down bending stresses in the spar caps and results in a lighter wing structure that offers greater fuel volume. For this reason, selecting a thick rather than thin airfoil is essential, as long as it does not increase section drag too much. Modern NLF airfoils are thicker and generate less drag than previous airfoils, which is one of the reasons why established manufacturers design mission-based airfoils.

Evaluation of a Target Zero-lift AOA

It is of importance to identify what zero lift AOA, α_{ZL}, to aim for in the airfoil selection process. Using the data of Ref. 23 (Equation [16]), Ref. 13 presents a helpful expression that can be used to estimate target α_{ZL} for the selection process. Assuming the designer knows the design AOA, denoted by α_i (in degrees), and the design lift coefficient, denoted by C_{li}, the zero-lift AOA is given by:

Zero-lift AOA:

$$\alpha_{ZL} = K\left(\alpha_i - \frac{90}{\pi^2}C_{l_i}\right) \qquad (8\text{-}63)$$

where

$K = 0.93$ for NACA four-digit airfoils
$K = 1.08$ for NACA five-digit airfoils
$K = 0.74$ for NACA 6-series airfoils

Consider an aircraft intended to cruise so its NACA 6-series airfoil is at an AOA near $1°$ and its section lift coefficient is 0.40. Using Equation (8-63) we readily find the target α_{ZL} should be near $-2°$. Generally, this method is accurate, although it is recommended that test data is used when possible.

Selection of an Airfoil

Assuming the designer is aware of the above limitations and issues, the following guidelines can be kept in mind when selecting an airfoil.

1. If an NLF airfoil is selected for the airplane, look for one with the widest possible drag bucket to help take advantage of its properties (see Figure 8-66). A wide drag bucket makes it likelier that the section lift coefficients ($C_{l\ climb}$ and $C_{l\ cruise}$) generated by the wing (or other lifting surfaces) will lie inside it during climb and cruise, regardless of the combination of weight and CG location. The wing area and planform of the airplane must be properly sized to make this possible, otherwise it the section lift coefficients may simply lie outside the drag bucket (see Figure 8-67).

   A large wing area results in small values of cruise C_l but an increased skin friction drag. Even though inside the drag bucket the reduction in drag may be less than the rise due to skin friction. Too small a wing area requires a higher climb and cruise C_l and may cause the airplane to operate outside the drag bucket (see Chapter 9, *The anatomy of the wing*, for guidance). Clearly, the wing must be sized and shaped such that at cruise (for most applications cruise has priority over climb) the values of $C_{l\ cruise}$ place it inside the drag bucket. Note that if the reader cannot specifically generate section lift coefficients C_l as a function of spanwise station, the C_L may be used, although the limitations of doing this must be understood. The inaccuracy comes into the picture when considering that the magnitude of a section lift coefficient at a specific wing station may be very different from the actual C_L. The difference may be large toward the tip or root and this would cause its drag to be greater than the configuration with a larger wing area. Of course, a larger wing area will lower the stalling speed, which is desirable. A wide drag bucket will also improve rate-of-climb and L/D ratio.

2. Select an airfoil with the largest C_{lmax} possible. However, *avoid airfoils with sharp drop in C_l immediately after stall* (see Figure 8-68). The sharp drop indicates the presence of a separation bubble near the leading edge on the upper surface of the

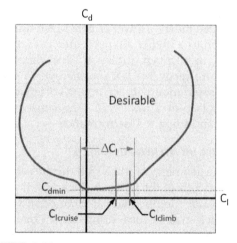
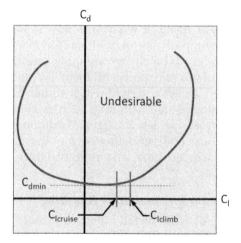

FIGURE 8-66 A drag polar featuring a wide drag bucket (left) is always more desirable than without one (right).

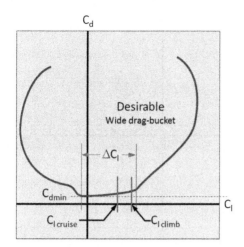
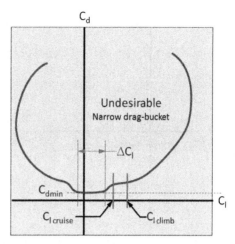

FIGURE 8-67 A drag polar featuring a wide drag bucket (left) is always more desirable than a narrow one (right).

airfoil, which suddenly combines with the flow separation point near the stall *AOA*, resulting in a sudden drop in the C_l. This can have very serious consequences for real airplanes. Typically, it causes one wing to stall before the other, resulting in a powerful roll-off that can have profound impact on production flight testing. It is very costly to manufacturers to have to fix the stall characteristics of production airplanes. Always place a great emphasis on good stall characteristics even if it means a loss of a few knots in cruise.

NACA reports often present the ratio C_{lmax}/C_{dmin} as a figure of merit for designers evaluating airfoils. This implies that the greater the ratio, the farther apart are the two arguments and the better the airfoil. A high C_{lmax} and low C_{dmin}, what more can one ask for? Well, one *can* ask for something more — good stall characteristics. The ratio C_{lmax}/C_{dmin} alone does not tell the whole story. It does not say anything about stall characteristics. Therefore, the ratio should be prefaced with: *the greatest ratio C_{lmax}/C_{dmin} that has benign stall characteristics.*

3. After having identified airfoils with low enough C_{dmin} and high enough C_{lmax} select the one with the lowest $C_{mC/4}$ possible at the cruise C_l (see Figure 8-69). However, this coefficient should have less weight in the selection process than the other two. It should be used to pick between two airfoils that happen to offer similar values of C_{dmin} and C_{lmax}. While the lower $C_{m\,C/4}$ will help reduce trim drag, it can always be accommodated by a proper sizing of the stabilizer.

4. Ideally, the maximum *L/D* ratio (LD_{max}) should be achieved close to the cruise *AOA*. Unfortunately, in practice, the two *AOA*s are separated by a wide margin. Therefore, when considering two airfoils that offer the same LD_{max}, place a higher weight on the one whose *AOA* of LD_{max} is closer to the cruise *AOA*. A higher *LD* at cruise yields a more efficient aircraft. As an example consider the graphs of Figure 8-70.

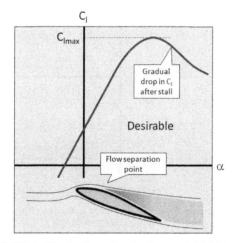
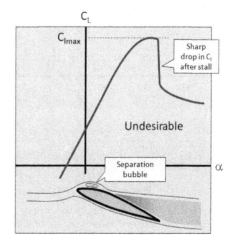

FIGURE 8-68 A sharp drop in C_l at stall (right) is undesirable, as it usually leads to poor stall characteristics.

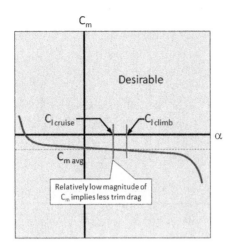
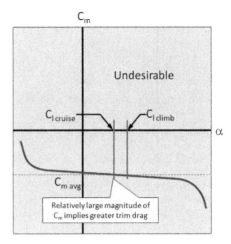

FIGURE 8-69 The larger the magnitude of C_m (right) the greater will be the trim drag.

These show a convenient method to check how the applicable lift and drag curves match the intended performance characteristics. The two plots compare the aerodynamic properties of NACA 23015 and 65_2-415 airfoils. They assume some hypothetical targets for the cruise and climb lift coefficients. For instance, the cruising speed requires the lift coefficient to vary between $0.25 < C_{LC} < 0.32$ (depending on weight) and the climb lift coefficient is expected to be close to $C_{L\ CLIMB} \approx 0.7$.

Figure 8-70 also shows that not only does the cruise range reside inside the drag bucket of the NACA 65_2-415 airfoil, the climb lift coefficient is much closer to the its maximum L/D than that of the NACA 23015 airfoil. An airplane using the NACA 65_2-415 airfoil will be more efficient than one featuring the 23015 airfoil. It can also be seen that the wing with the NACA 23015 must be installed at a higher angle of incidence than the NACA 65_2-415 airfoil (these depend on the 3D geometry of the wing).

Additionally, while the NACA 65_2-415 results in a slightly higher stalling speed, its stall characteristics are far more benign.

To exemplify the selection process, consider the airfoil selection for a new airplane design. To help with the process, Table 8-8 is prepared. It contains three example airfoils whose pertinent properties have been tabulated to allow them to be scored against one another. The data for the candidate airfoils was obtained by digitizing the appropriate

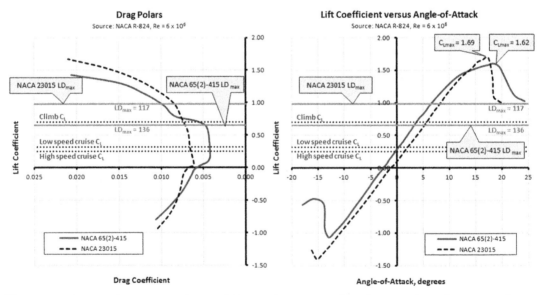

FIGURE 8-70 The aerodynamic characteristics of two airfoils plotted to compare their characteristics to the target performance. Note how well (or poorly) the desired C_L for climb and cruise coincide with the airfoil properties. For instance, notice how the LD_{max} of the NACA 65-series airfoil falls close to the target climb C_L (this is very desirable). Also notice how the low- and high-speed cruise C_Ls are inside the drag bucket for both airfoils, although the 65-series airfoil yields a substantially lower C_d (as long as laminar flow can be achieved) than the five-digit airfoil.

TABLE 8-8 Table Used to Down-select Candidate Airfoils*

ID	Parameter	Airfoil 1	Airfoil 2	Airfoil 3	Score		
1	R_e for stall	≈ 2.4×10^6			Score the airfoils by entering 1 to indicate the airfoil with the best characteristics. It is possible all the airfoils deserve a score.		
2	R_e for best ROC	≈ 4.1×10^6					
3	R_e for cruise	≈ 7.5×10^6					
4	Target C_{Lmax}	1.45					
5	C_L for best ROC, $C_{L\,ROCmax}$	≈ 0.70					
6	C_L for target cruise, C_{LC}	≈ 0.25–0.32					
7	Name	65$_2$-415	23015	747A415	1	2	3
8	Thickness ratio (high is best)	0.15	0.15	0.15	1	1	1
9	Sensitive to surface quality?	Y/**N**	**Y**/N	Y/**N**		1	
10	R_e for data below	6×10^6	6×10^6	6×10^6			
11	C_l for AOA = 0° (close to C_{LC} is best)	0.3	0.1	0.2	1		
12	AOA for C_l = 0	−2.6°	−1.0°	−2.0°			
13	C_{lmax} (highest is best)	1.58	1.69	1.43		1	
14	AOA of C_{lmax} (highest is best)	16.5°	17.4°	18.8°			1
15	Stall characteristics (A, B, C)	A	C	A	1		1
16	C_{dmin} (lowest is best)	0.00424	0.0062	0.00453	1		
17	C_l of C_{dmin} (close to cruise is best)	0.4	0.1	0.5	1		
18	$(C_l/C_d)_{max}$ (highest is best)	136	117	150			1
19	C_l of $(C_l/C_d)_{max}$ (low C_l is best)	0.65	0.98	0.77	1		
20	Cruise C_m	−0.084	−0.004	−0.038		1	
21	Drag bucket start at C_l	0.2	0.0	0.2			
22	Drag bucket ends at C_l	0.6	0.1	0.77			
23	Is $C_{L\,ROCmax}$ inside drag bucket?	Y/**N**	Y/**N**	**Y**/N			1
24	Is $C_{L\,CRUISE}$ inside drag bucket?	**Y**/N	Y/**N**	**Y**/N	1		1
				Sum:	7	4	6

Stall Characteristics: A = gentle, B = between A and C, C = abrupt.   *Shaded cells contain information not directly used for scoring.

graphs in Ref. [23] and, then, after simple analyses, their properties were entered as depicted in the graphs.

While established aircraft manufacturers often design airfoils that are specifically tailored to the mission of the aircraft, the comparison performed in Table 8-8 is still helpful when evaluating custom airfoils.

In this example the airfoils are scored by simply entering a '1' for a winning airfoil in the columns on the right-hand side. The numbers 1, 2, and 3 refer to the three airfoils. Then, the score for each airfoil is summed and the total is entered at the bottom of the table. The airfoil with the highest score is obviously the one to consider, although the results may be more ambiguous than that. Sometimes more than one airfoil simultaneously meets a given criterion — in which case all the appropriate airfoils receive a '1'. An example of this can be seen in the table, in lines 8, 15, and 24. Other times, two or more airfoils end up with a total score that is identical. In this case the process must be polished some more. For instance, there could be a weighting factor associated with some of the criteria. For instance, a score of 2 can be given for the lowest drag coefficient, 0.5 for the highest pitching moment coefficient, and so on. The designer can also add other criteria to the table to help with the selection process as well.

While most of the rows in Table 8-8 are self-explanatory, a few need to be elaborated on.

Rows 11 and 12: as a rule of thumb, we expect the three-dimensional lift coefficient to be close to 85–90% of the two-dimensional one. The score here is based on the fact that this places Airfoil 1 inside the cruise range of the C_L, but the others outside it.

Row 16: note that Airfoil 1 has the most favorable C_l of C_{dmin} for cruise. Airfoil 2 (NACA 23015) has it closer; however, it is at a lower value than the cruise C_L. For this reason, the airplane would have to be diving to benefit from it. This is further supported by the data in rows 21 and 22, which show that the drag bucket for the 23015 airfoil (if we want to call it that) ranges from $C_l = 0$ to 0.1, which is beyond where the airplane will be operated. For Airfoils 1 and 3, on the other hand, the cruise C_l is guaranteed to be inside the drag bucket.

NACA Recommended Criteria

The conclusion section of Ref. [23] lists a number of things to keep in mind when selecting airfoils. These are paraphrased below.

Airfoils permitting extensive laminar flow, such as the NACA 6- and 7- series, have substantially less drag at typical cruise lift coefficients than other kinds of airfoils. However, these characteristics are realized only if the quality of the lifting surface is smooth. Wind tunnel tests have shown that extensive laminar flow is possible on smooth three-dimensional wings if the surface quality is smooth and like that provided by sanding in the chordwise direction with No. 320 carborundum sand paper.

Wings of moderate thickness ratios with such surface qualities can achieve a minimum drag coefficient of the order of 0.0080. In fact, the C_{Dmin} depends more on the surface quality than on the chosen airfoil. Thus, at high Reynolds numbers where laminar flow is no longer achievable, drag can be kept low by ensuring smooth surface qualities.

The maximum lift coefficient for moderately cambered 6-series airfoils are as high as those achieved using NACA 24- and 44-series airfoils. The NACA 230-series airfoils with thickness ratio less than 20% generally achieve the highest maximum lift coefficients. The maximum lift coefficient with flaps is about the same for moderately thick 6-series airfoils as it is for the NACA 23012 with flaps. However, the thinner 6-series airfoils have substantially lower C_{lmax} with flaps. The lift curve slope for smooth 6-series airfoils is slightly steeper than that of the 24-, 44-, and 230-series airfoils. It exceeds the theoretical value (2π) for thin airfoils.

Leading edge contamination (roughness) causes large reductions in C_{lmax} for plain and flapped airfoils. The magnitude of the reduction is similar for both. The leading edge contamination also reduces the lift curve slope, especially for thicker airfoils that have the location of minimum pressure farther aft than do thinner ones. This lends support to the importance of understanding the impact of poor surface quality and leading edge contamination on the overall characteristics of the airfoils and, thus, the performance of the aircraft when designing wings.

The NACA 6-series airfoils have higher critical Mach numbers than the earlier airfoil types at typical cruise lift coefficients. However, their critical Mach numbers are lower at higher lift coefficients than are the same types of airfoils. The 6-series airfoils also offer better lift coefficients at higher Mach numbers than the earlier airfoils.

EXERCISES

(1) The purpose of this exercise is to train the student in retrieving information from the available literature. Here, information from NACA R-824 will be used. It is available for download (free of charge) from the NASA Technical Report Server (http://ntrs.nasa.gov/search.jsp). Perform the following tasks for the NACA 23012 airfoil:

(a) Locate the *Station and Ordinate* table (p. 359) and plot the airfoil assuming a wing chord of 39.4 inches (100 cm). This task can easily be accomplished using spreadsheet software such as Microsoft Excel. More specifically, plot the upper, lower, and mean-lines. Also determine the *x*-locations of the airfoil's maximum thickness and camber.

(b) Using the graph of p. 375, determine the critical Mach number (M_{crit}) of the airfoil at a lift coefficient $C_l = 0.30$.

(c) Using the graph of p. 404, determine the following characteristics for Reynolds numbers of 3 and 6 million. For the R_e of 6 million, use both the "clean" and "standard roughness" data. Determine C_{lmax}, $C_{l\alpha}$ of the linear range, "average" C_m for $\alpha < 14°$, C_{dmin}, and C_l of C_{dmin}.
Answers: (b) 0.6, (c) for R_e of 3×10^6, $C_{lmax} = 1.62$, $C_{l\alpha} = 0.104$, $C_m = -0.05$, $C_{dmin} = 0.006$ and C_l of $C_{dmin} = 0.1$.

(2) An airfoil is subjected to airspeed of 315 ft/s on a standard day at S-L, when the pressure at a specific point on it is found to be 13.3 psi. Determine the incompressible and compressible pressure coefficient, C_p, using all three compressibility models of Section 8.3.6, *Compressibility modeling*.
Answer: -1.702, -1.774, -1.840, and -1.918.

(3) Using the information about the critical Mach number obtained in Exercise (1), determine the critical pressure coefficient, $C_{p\ crit}$, of the NACA 23012 airfoil at the lift coefficient of 0.3.
Answer: -1.294.

VARIABLES

Symbol	Description	Units (UK and SI)
$\overline{C}_p$	Canonical pressure ratio	
$\dot{m}$	Mass flow rate	slugs/s or kg/s
$\frac{dy_c}{dx}$	Slope of camber line at point	
A	Area	ft² or m²
a	Mean-line designation for NACA 6-series airfoils	
A_{tube}	Cross-sectional area of stream tube	ft² or m²
c	Airfoil chord length	ft or m

VARIABLES

Symbol	Description	Units (UK and SI)
C	Camber	
C_{avg}	Average chord length	ft or m
C_d	2D coefficient of drag	
C_D	Wing drag coefficient	
C_{Df}	Skin friction drag coefficient	
C_{Dfo}	Incompressible skin friction drag coefficient	
$C_{Dincompressible}$	Incompressible drag coefficient at some condition	
C_{dmin}	Minimum 2D drag coefficient	
C_l	2D lift coefficient	
C_L	3D lift coefficient	
C_{lmind}	2D lift coefficient at minimum drag	
$C_{L\ ROCmax}$	3D lift coefficient at maximum rate of climb	
C_{lclimb}	2D lift coefficient at climb	
$C_{lcruise}$	2D lift coefficient at cruise	
C_{lmax}	Maximum 2D lift coefficient	
C_{lmin}	Minimum 2D lift coefficient	
C_{lo}	2D lift coefficient at $\alpha = 0$ (context dependent)	
C_{lo}	Incompressible 2D lift coefficient (context dependent)	
$C_{l\alpha}$	2D lift curve slope	deg or rad
C_m	2D coefficient of moment	
C_M	3D coefficient of moment	
$C_{m\ ac}$	2D coefficient of moment about aerodynamic center	
$C_{m\ avg}$	Average 2D coefficient of moment	
$C_{mc/4}$	Airfoil pitching moment coefficient about the quarter chord	
C_{mo}	2D coefficient of moment at $\alpha = 0$ (context dependent)	
C_{mo}	Incompressible coefficient of moment (context dependent)	
C_p	Pressure coefficient	
$C_{p\ crit}$	Pressure coefficient at critical Mach	
$C_{p\ o}$	Reference pressure coefficient	
C_r	Non-dimensional coefficient that relates AOA to force	
d	Airfoil drag force (context dependent)	lb_f or N
d	Diameter of circular cylinder (context dependent)	ft or m
F	Force due to fluid flow	lb_f or N
f_c	Airfoil chord-wise force	lb_f or N
f_n	Airfoil normal force	lb_f or N
F_{press}	Pressure force	lb_f or N
g	Acceleration due to gravity	ft/s^2 or m/s^2
k	Unknown constant of proportionality (context dependent)	
k	Smeaton's coefficient (context dependent)	
k_1	Constant used with NACA 5-digit airfoils	
L	Reference length (context dependent)	ft or m
L	Lift force (context dependent)	lb_f or N
l	Characteristic length (context dependent)	ft or m
l	Airfoil lift force (context dependent)	lb_f or N
LD_{max}	Maximum lift-to-drag ratio	
m	Mass (context dependent)	slugs or kg
m	Airfoil pitching moment (context dependent)	$ft \cdot lb_f$ or $N \cdot m$
m	Constant for airfoil design (context dependent)	
M_∞	Far-field Mach number	
M_A	Mach number at some point A on an airfoil	$ft \cdot lb_f$ or $N \cdot m$
$m_{c/4}$	Pitching moment about quarter chord	$ft \cdot lb_f$ or $N \cdot m$
M_{crit}	2D critical Mach number	
M_{DD}	Drag divergence Mach number	
N	Number of airfoil points	
p	Pressure	lb_f/ft^2 or Pa
p_∞	Far-field pressure	lb_f/ft^2 or Pa
p_A	Pressure at some point A on an airfoil	lb_f/ft^2 or Pa
p_{tot}	Total pressure	lb_f/ft^2 or Pa
q	Dynamic pressure	lb_f/ft^2 or N/m^2
r	Resultant aerodynamic force	lb_f or N
R_e	Reynold number	
$R_{\delta 1}$	Reynolds number of boundary layer	
S	Reference area	ft^2 or m^2
t	time	s
t/c	Thickness to chord ratio	
V	Airspeed	ft/s or m/s
V_∞	Far-field airspeed	ft/s or m/s

(Continued)

Symbol	Description	Units (UK and SI)
x	Location of camber line along x-axis (context dependent)	% chord
x_{ac}	Location of aerodynamic center along x-axis	% chord
x_{camber}	Location of maximum camber	% chord
x_L	Location of lower airfoil point along the x-axis	% chord
x_U	Location of upper airfoil point along the x-axis	% chord
y_c	Location of camber line along y-axis	% chord
y_L	Location of lower airfoil point along the y-axis	% chord
y_U	Location of upper airfoil point along the y-axis	% chord
z	Elevation above or below a reference point	ft or m
Z_{camber}	Distance to camber line from the chordline	ft or m
Z_{lower}	Distance to lower surface of airfoil from the chordline	ft or m
Z_{thick}	Distance to airfoil surface from camber line	ft or m
Z_{upper}	Distance to upper surface of airfoil from the chordline	ft or m
ΔC_L	Change in lift coefficient	
Δp	Difference in pressure	lb_f/ft^2 or Pa
$\Delta \phi$	Angular spacing	deg or rad
Γ	2D circulation	$ft^2 \cdot rad/s$ or $m^2 \cdot rad/s$
α	Angle-of-attack	deg or rad
α_{stall}	2D stall angle-of-attack (context dependent)	deg or rad
α_{ZL}	Angle-of-attack at zero lift	deg or rad
δ	Thickness of boundary layer	in or mm
δ_{lam}	Thickness of laminar boundary layer	in or mm
δ_{turb}	Thickness of turbulent boundary layer	in or mm
ε	Downwash	
γ	Specific heat ratio	
μ	Air viscosity	$lbf \cdot s/ft^2$ or $N \cdot s/m^2$
θ	Angle between line tangent to camber line and chord line	deg or rad
ρ	Density	$slugs/ft^3$ or kg/m^3
ρ_∞	Far-field density	$slugs/ft^3$ or kg/m^3

References

[1] http://wright.nasa.gov/airplane/liftold.html.
[2] NACA-TN-2502. Examples of Three Representative Types of Airfoil-Section Stall at Low Speed. McCullough, Goerge B; Gault, Donald E; 1951.
[3] Smith AMO. High-lift aerodynamics. Journal of Aircraft 1975;12(6).
[4] Anderson Jr John D. Introduction to Flight. 3rd ed. McGraw-Hill; 1989. Section 5.19.
[5] Schlichting Hermann. Boundary Layer Theory. Pergamon Press; 1955.
[6] Young AD. Boundary Layers. AIAA Education Series; 1989.
[7] Bertin John, J, Smith Michael L. Aerodynamics for Engineers. Prentice-Hall; 1979. p. 95.
[8] Gopalarathnam Ashok, et al. Design of low Reynolds number airfoils with trips. Journal of Aircraft 2003;40(4).
[9] Bertin John J. Engineering Fluid Mechanics. Prentice-Hall; 1984.
[10] Hoak DE. USAF Stability and Control DATCOM. Flight Control Division, Air Force Flight Dynamics Laboratory; 1978. Section 4.1.1.3.
[11] R.&M. No. 3122 The Formation of Regions of Separated Flow on Wing Surfaces. Aeronautical Research Council, L. F. Crabtree; 1959.
[12] Kundu, Kumar Ajoy. Aircraft Design. Cambridge University Press; 2010.
[13] Hoak DE. USAF Stability and Control DATCOM. Flight Control Division, Air Force Flight Dynamics Laboratory; 1978. Section 4.1.1.4.
[14] http://web.mit.edu/drela/Public/web/xfoil/
[15] http://www.xflr5.com/xflr5.htm
[16] http://www.airfoils.com/index.htm
[17] http://aerofoilengineering.com/
[18] Meschia Francesco. Model analysis with XFLR5. Radio Controlled Soaring Digest 2008;25(2).
[19] NACA TR-460 The Characteristics of 78 Related Airfoil Sections from Tests in the Variable-Density Wind Tunnel. Jacobs, Eastman N, Kenneth E. Ward and Robert M. Pinkerton; 1933.
[20] NACA TR-492 Tests of 16 Related Airfoils at High Speed. Stack, John, and Albert E. von Doenhoff; 1935.
[21] NACA TR-537 Tests in the Variable-Density Wind Tunnel of Related Airfoils having the Maximum Camber Unusually Far Forward. Jacobs, Eastman N., and Robert M. Pinkerton; 1936.
[22] NACA TR-610 Tests of Related Forward-camber Airfoils in the Variable-density Wind Tunnel. Jacobs, Eastman N; Robert M. Pinkerton, and Harry Greenberg; 1937.
[23] NACA R-824 Summary of Airfoil Data. Abbott, Ira H., Albert E. von Doenhoff and Louis S. Stivers Jr; 1945.
[24] NASA TM-X-3284 Development of a Computer Program to Obtain Ordinates for NACA 4-Digit, 4-Digit Modified, 5-Digit, and 16 Series Airfoils. Ladson, C. L., and Cuyler W. Brooks Jr; 1975.
[25] NACA TR-903 Theoretical and Experimental Data for a Number of NACA 6A-Series Airfoil Sections. Loftin, Laurence K., Jr; 1948.
[26] NASA TM-X-3069 Development of a Computer Program to Obtain Ordinates for NACA-6 and 6A-Series Airfoils. Ladson, C. L., and Cuyler W. Brooks Jr; 1974.
[27] NASA TM-4741 Computer Program to Obtain Ordinates for NACA Airfoils. Ladson, Charles L., Cuyler W. Brooks Jr., Acquilla S. Hill, and Darrell W. Sproles; 1996.

REFERENCES

[28] NACA TR-947 The Development of Cambered Airfoil Sections having Favorable Lift Characteristics at Supercritical Mach Numbers. Graham, Donald J; 1949.

[29] NACA R-502 Scale Effect on Clark Y Airfoil Characteristics from NACA Full Scale Wind-Tunnel Tests. Silverstein, Abs, NACA; 1934.

[30] Riblett Harry. GA Airfoils – A Catalog of Airfoils for General Aviation Use. 5th ed. Self-published; 1988.

[31] NACA TR-537 Tests in the Variable-Density Wind Tunnel of Related Airfoils having the Maximum Camber Unusually Far Forward. Jacobs, Eastman N., and Robert M. Pinkerton; 1936.

[32] NASA CR-2443 Development of a Fowler Flap System for a High Performance General Aviation Airfoil. Wentz, W. H., Jr., and H. C. Seetharam; 1974.

[33] NASA TN D-7428 Low-Speed Aerodynamic Characteristics of a 17-Percent-Thick Airfoil Section Designed for General Aviation Applications. McGhee, Robert J., and William D. Beasley; 1973.

[34] NACA WR-L-677 Lift and Drag Tests of Three Airfoil Models with Fowler Flaps. submitted by Consolidated Aircraft Corporation. Turner, Harold R., Jr; 1941.

[35] NASA TT-F-749 On Peaky Airfoil Sections. Sato, Junzo; 1973.

[36] R.&M. No. 3108 Some Effects of Shock-induced Separation of Turbulent Boundary Layers in Transonic Flow past Aerofoils. Aeronautical Research Council, H. H. Pearcey; 1955.

[37] Pearcey HH. The aerodynamic design of section shapes for swept wings. Advances Aero. Sci. 1962;3:277–322. Proceedings of the 2nd International Congress on Aeronautical Science.

[38] Whitford Ray. Design for Air Combat. Jane's; 1987.

[39] NASA TM-X-1109 An Airfoil Shape for Efficient Flight at Supercritical Mach Numbers. Whitcomb, R. T., and L. R. Clark; 1965.

[40] Anderson John D. Fundamentals of Aerodynamics. 4th ed. McGraw-Hill; 2007.

[41] Roskam Jan, Lan Chuan-Tau Edward. Airplane Aerodynamics & Performance. DARcorporation; 2000.

[42] Glauert H. The effect of compressibility on the lift of an aerofoil. Proceedings of the Royal Society, London, CXVIII; 1928. 113–119.

[43] Perkins Courtland D, Hage Robert E. Airplane Performance, Stability, and Control. John Wiley & Sons; 1949. 59–63.

[44] Abbott Ira H, Von Doenhoff Albert E. Theory of Wing Sections. Dover; 1959.

[45] NACA TN-695 The Effects of Some Surface Irregularities on Wing Drag. Hood, Manley J; 1939.

[46] Thwaites Bryan. Incompressible Aerodynamics. Dover; 1960.

[47] NACA WR L-143 Completed Tabulation in the United States Of Tests of 24 Airfoils at High Mach Numbers. Ferri, Antonio; 1945.

CHAPTER 9

The Anatomy of the Wing

OUTLINE

9.1 Introduction 301
   9.1.1 *The Content of this Chapter* 301
   9.1.2 *Definition of Reference Area* 302
   9.1.3 *The Process of Wing Sizing* 302
      The Establishment of Basic Datum 302

9.2 **The Trapezoidal Wing Planform** 303
   9.2.1 *Definitions* 303
      Aspect Ratio for HT and VT 305
      Aspect Ratio for Multi-wing Configurations 305
      Derivation of Equations (9-6), (9-8), and (9-10) 305
   9.2.2 *Poor Man's Determination of the MGC* 306
   9.2.3 *Planform Dimensions in Terms of S, λ, and AR* 307
      Derivation of Equation (9-20) 307
      Derivation of Equation (9-21) 308
   9.2.4 *Approximation of an Airfoil Cross-sectional Area* 308
      Derivation of Equations (9-22) and (9-23) 308

9.3 **The Geometric Layout of the Wing** 309
   9.3.1 *Wing Aspect Ratio, AR* 309
      Illustrating the Difference Between Low and High AR Wings 310
      Determining AR Based on a Desired Range 312
      Derivation of Equation (9-26) 312
      Determining AR Based on Desired Endurance 313
      Derivation of Equation (9-28) 313
      Estimating AR Based on Minimum Drag 314
      Estimating AR for Sailplane Class Aircraft 315
   9.3.2 *Wing Taper Ratio, TR or λ* 315
   9.3.3 *Leading Edge and Quarter-chord Sweep Angles, Λ_{LE} and $\Lambda_{C/4}$* 317
      Impact of Sweep Angle on the Critical Mach Number 317
      Impact of Sweep Angle on the C_{Lmax} 318
      Impact of Sweep Angle on Structural Loads 318
      Impact of Sweep Angle on Stall Characteristics 318

   9.3.4 *Dihedral or Anhedral, Γ* 318
   9.3.5 *Wing Twist — Washout and Washin, ϕ* 319
      Geometric Washout 319
      Aerodynamic Washout 319
      A Combined Geometric and Aerodynamic Washout 323
      Panknin and Culver Wing Twist Formulas 324
   9.3.6 *Wing Angle-of-incidence, i_W* 325
      Step 1: Determine the Optimum AOA for the Fuselage 326
      Step 2: Determine the Representative AOA at Cruise 326
      Step 3: Determine the Recommended Wing Angle-of-incidence 326
      Step 4: Determine the Recommended HT Angle-of-incidence 327
      Derivation of Equation (9-43) 327
      Derivation of Equation (9-45) 327
      Decalage Angle for a Monoplane 328
   9.3.7 *Wing Layout Properties of Selected Aircraft* 329

9.4 **Planform Selection** 329
   9.4.1 *The Optimum Lift Distribution* 329
   9.4.2 *Methods to Present Spanwise Lift Distribution* 331
   9.4.3 *Constant-chord ("Hershey-bar") Planform* 332
      Pros 333
      Cons 333
   9.4.4 *Elliptic Planforms* 333
      Pros 334
      Cons 334
   9.4.5 *Straight Tapered Planforms* 334
      Pros 334
      Cons 334
      Straight Leading or Trailing Edges 335
      Pros 335
      Cons 335
      The Compound Tapered Planform 335
   9.4.6 *Swept Planforms* 336
      The Aft-swept Planform 336
      Pros 336

	Cons	336	**Step 3:** Determine the AOA at which Local C_l Intersects C_{lmax}	355
	The Forward-swept Planform	337	**Step 4:** Compute C_{Lmax}	355
	Variable Swept Planform	338	9.5.12 *Step-by-step: C_{Lmax} Estimation per USAF DATCOM Method 2*	356
9.4.7	*Cranked Planforms*	338	Step 1: Determine the Taper Ratio Correction Factor	357
	Semi-tapered Planform	339	Step 2: Determine if Wing Qualifies as "High AR"	357
	Crescent Planform	339	Step 3: Determine the Leading Edge Parameter	357
	Schuemann Planform	339	Step 4: Determine the Max Lift Ratio	357
9.4.8	*Delta Planforms*	340	Step 5: Determine the Mach Number Correction Factor	357
	The Delta Planform Shape	340	Step 6: Calculate the C_{Lmax}	358
	The Double-delta Planform Shape	341	Step 7: Determine Zero Lift Angle and Lift Curve Slope	358
9.4.9	*Some Exotic Planform Shapes*	341	Step 8: Determine a Correction for the Stall Angle-of-attack	358
	Disc- or Circular-shaped Planform	341	Step 9: Determine the Wing Stall Angle-of-attack	359
	Other Configurations	342	9.5.13 C_{Lmax} *for Selected Aircraft*	360
9.5	**Lift and Moment Characteristics of a 3D Wing**	**342**	9.5.14 *Estimation of Oswald's Span Efficiency*	363
9.5.1	*Properties of the Three-dimensional Lift Curve*	342	Basic Definition	363
	Lift Curve Slope, $C_{L\alpha}$	343	Method 1: Empirical Estimation for Straight Wings	363
	Maximum and Minimum Lift Coefficients, C_{Lmax} and C_{Lmin}	343	Method 2: Empirical Estimation for Swept Wings	363
	C_L at Zero AOA, C_{Lo}	343	Method 3: Douglas Method	364
	Angle-of-attack at Zero Lift, α_{ZL}	343	Method 4: Lifting Line Theory	364
	Linear Range	343	Method 5: USAF DATCOM Method for Swept Wings — Step-by-step	364
	Angle-of-attack where Lift Curve Becomes Non-linear, α_{NL}	343	Step 1: Calculate the Lift Curve Slope	365
	Angle-of-attack for Maximum Lift Coefficient, α_{stall}	343	Step 2: Calculate the Leading Edge Suction Parameter	365
	Design Lift Coefficient, C_{LC}	343	Step 3: Calculate Special Parameter 1	365
	Angle-of-attack for Design Lift Coefficient, α_C	343	Step 4: Calculate Special Parameter 2	365
	Derivation of Equation (9-49)	343	Step 5: Read or Calculate the Leading Edge Suction Parameter	365
9.5.2	*The Lift Coefficient*	344	**9.6** **Wing Stall Characteristics**	**366**
	The Relationship between Airspeed, Lift Coefficient, and Angle-of-attack	344	9.6.1 *Growth of Flow Separation on an Aircraft*	367
	Wide-range Lift Curve	344	9.6.2 *General Stall Progression on Selected Wing Planform Shapes*	369
9.5.3	*Determination of Lift Curve Slope, $C_{L\alpha}$, for a 3D Lifting Surface*	345	9.6.3 *Deviation from Generic Stall Patterns*	369
	Derivation of Equation (9-53)	346	9.6.4 *Tailoring the Stall Progression*	369
	Derivation of Equation (9-54)	347	Design Guidelines	369
9.5.4	*The Lift Curve Slope of a Complete Aircraft*	347	Tailoring Stall Characteristics of Wings with Multiple Airfoils	371
	Derivation of Equations (9-59) and (9-60)	347	Other Issues Associated with Wings with Multiple Airfoils	373
9.5.5	*Step-by-step: Transforming the Lift Curve from 2D to 3D*	348	9.6.5 *Cause of Spanwise Flow for a Swept-back Wing Planform*	374
9.5.6	*The Law of Effectiveness*	349	9.6.6 *Pitch-up Stall Boundary for a Swept-back Wing Planform*	375
9.5.7	*Flexible Wings*	349		
9.5.8	*Ground Effect*	350		
9.5.9	*Impact of C_{Lmax} and Wing Loading on Stalling Speed*	353		
9.5.10	*Step-by-step: Rapid C_{Lmax} Estimation*	353		
9.5.11	*Step-by-step: C_{Lmax} Estimation per USAF DATCOM Method 1*	355		
	Step 1: Determine the Two-dimensional Maximum Lift Coefficient	355		
	Step 2: Plot the Distribution of Section Lift Coefficients	355		

Derivation of Equations (9-94) and (9-95)	377	Change in Induced Angle-of-attack	387
9.6.7 *Influence of Manufacturing Tolerances on Stall Characteristics*	378	Derivation of Equations (9-124) through (9-128)	387
9.7 Numerical Analysis of the Wing	**379**	Derivation of Equation (9-130)	388
9.7.1 *Prandtl's Lifting Line Theory*	379	9.7.4 *Accounting for a Fuselage in Prandtl's Lifting Line Method*	390
The Vortex Filament and the Biot-Savart Law	380	9.7.5 *Computer code: Prandtl's Lifting Line Method*	393
Helmholz's Vortex Theorems	381	**Exercises**	**396**
Lifting Line Formulation	381	**Variables**	**397**
9.7.2 *Prandtl's Lifting Line Method — Special Case: The Elliptical Wing*	384	**References**	**398**
Derivation of Equations (9-113) through (9-119)	384		
9.7.3 *Prandtl's Lifting Line Method — Special Case: Arbitrary Wings*	386		

9.1 INTRODUCTION

Now that the characteristics of airfoils have been discussed, it is time to evaluate their use in lifting surfaces. A lifting surface can be defined as a three-dimensional body whose primary purpose is to generate aerodynamic loads; primarily lift. A wing is a lifting surface capable of structurally reacting the load it generates, up to certain airspeeds of course.

The geometry of lifting surfaces is usually defined in terms of two two-dimensional shapes; the *airfoil* and *planform*. The former has already been treated in Chapter 8, *The anatomy of the airfoil*. However, the combination of the two is the topic of this section. The history of aviation shows that a great many combinations of airfoils and planform shapes have been used, some with great success, others less so. It is the responsibility of the design team, usually the aerodynamics group, to select airfoils and a planform shape that not only fulfill the aerodynamic and performance goals, but also that allow the requirements of other design groups to be met. A part of this responsibility calls for the geometry of the wing and stabilizing surfaces, and their relative spatial location, to be defined so the combination yields an aircraft that is both safe and easy to handle. At the same time, the wing must feature enough internal volume to accommodate systems, landing gear, fuel, and so on. The aero group must also recognize that any systems required to enhance lift will increase the cost of manufacturing and maintenance, not to mention increase risk when flight-tested. The goal should always be to select the simplest system that does the job.

We have seen in Chapter 5, *Aircraft structural layout*, how the wing and stabilizing surfaces are typically constructed. A spar-rib-stringers-skin is always used for aluminum structures and spar-rib-stiffened-skin construction for composite structures. A planform shape that deviates from the shapes shown in Section 9.4, *Planform selection*, can present serious structural challenges. If a challenging planform is chosen, it would be wise to lay out the structure early in the process and work out possible manufacturing problems and not just aerodynamic ones. Such problems may range from limited fuel volume, space claims for retractable landing gear, the high-lift system and primary control system, curved spars, to unexpected aeroelastic issues.

In short, this section assumes the wing area and vertical position of the wing (low, mid, high, parasol) have already been determined and what remains is:

(1) determine the planform shape (general geometry, taper ratio, and leading edge sweep)
(2) determine wing twist (washout)
(3) determine the lift and pitching moment characteristics ($C_{L\alpha}$, C_{Lmax}, C_{Lmin}, α_{stall}, C_M, Oswald's span efficiency, and others).

Note that weight and drag characteristics of the wing are treated in Chapter 6, *Aircraft weight analysis*, and Chapter 15, *Aircraft drag analysis*, respectively.

9.1.1 The Content of this Chapter

- **Section 9.2** presents a handy formulation to calculate various properties of trapezoidal surfaces, which is the most common planform shape for lifting surfaces such as wings, and horizontal and vertical tails.
- **Section 9.3** presents various concepts, topics, and methods for laying out the wing planform. These include aspect ratio, taper ratio, washout, and wing incidence angle.
- **Section 9.4** is intended to help with the selection of a suitable planform shape. It introduces a number of

different geometries and presents information about their pros and cons, in addition to giving ideas of the distribution of section lift coefficients.
- **Section 9.5** presents a number of methods to evaluate the lift and pitching moment characteristics of the wing (methods to evaluate drag are presented in Chapter 15, *Aircraft drag analysis*).
- **Section 9.6** presents a number of issues that have to do with wing stall characteristics and how to improve them.
- **Section 9.7** presents Prandtl's Lifting Line Theory, which is a numerical method used to estimate the aerodynamic properties of the wing. A computer function, written in Visual Basic for Applications, intended for use with Microsoft Excel is also presented.

9.1.2 Definition of Reference Area

One of the most important concepts in aircraft design (as well as in the entire discipline of aerodynamics) is the *reference area*. Characteristics such as lift, drag, and moment coefficients all require this area to be specified, as does performance; stability and control; and load analyses, just to name a few. The shaded areas in Figure 9-1 show how aircraft designers typically define the reference area. However, this is not to say it couldn't be defined in some other way; in fact, the shaded regions may not be at all what the manufacturers did for these particular aircraft, just what would be a plausible approach.

Since the definition of this area takes place in the early stages of the airplane's design, it is almost always defined in a simple manner. As a consequence, when the geometry gets modified at a later time (for instance, a wingtip is reshaped) the reference area is not changed (as this would be likely to require a large number of documents to be revised, even certification documents), but the change will be manifested through a change in maximum lift coefficient, modified drag polar and similar definitions. An added wingtip shape, strakes, or leading edge extensions are typically omitted because they are often afterthoughts or added as a consequence of a wind tunnel or a flight test program (stall and spin recovery) or added features (wingtip tanks for greater range or modifications to the initial wing/fuselage fairing).

9.1.3 The Process of Wing Sizing

The concept of *Wing Sizing* refers to the process required to determine the size, shape, and three-dimensional positioning of the wing surfaces. This procedure is listed in Table 9-1.

The Establishment of Basic Datum

The positioning of the wing is the determination of where, with respect to some reference point in space, the wing shall be located. During the design of the aircraft it is usually decided that the reference datum for the airplane may be a point on the fuselage or the wing or a point in space (typically ahead of the nose of the aircraft as discussed in Section 6.6.11, *Center of gravity envelope*).

Among those are evaluation of dynamic stability and compliance with the design checklist of Section 23.3, *GA aircraft design checklist*. Typically, this will consider the following:

- Stability derivatives such as $C_{m\alpha}$, $C_{m\delta e}$, C_{mq}, $C_{y\beta}$, $C_{n\beta}$, $C_{n\delta r}$, C_{nr}, and others.

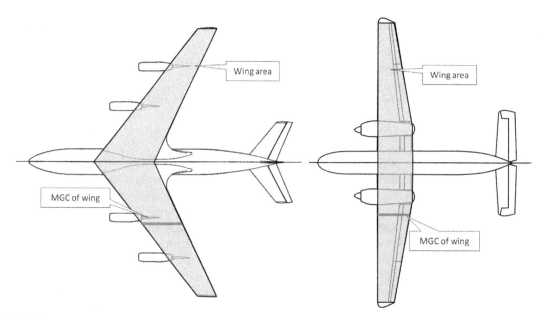

FIGURE 9-1 Reference area for the Boeing 707 and De Havilland of Canada DHC-7 Caribou (drawing is not to scale).

TABLE 9-1 Initial Sizing and Layout Algorithm for the Wing of a GA Aircraft

Step	Task	Section
1	It is assumed that the required wing area has already been determined using the constraint analysis method of Section 3.2, *Constraint analysis*. Verify the predicted stalling speed is within regulatory margins (e.g. 45 KCAS if LSA, 61 KCAS is 14 CFR Part 23, etc.) before proceeding.	3.2
2	Select the wing planform geometry (see Section 9.4, *Planform selection*). This may or may not be the final shape.	9.4
3	As a starting point, select a candidate aspect ratio and taper ratio. For *AR* consider the use of Equations (9-26) or (9-28). In the absence of specific target values, initial values can be selected by considering the class of aircraft being designed. For instance, refer to Tables 4-2 regarding the specific class of airplanes and then Tables 9-3 and 9-4 for additional help. These are initial values that are likely to change, but are a reasonable starting point.	4.2 9.3.1 9.3.2
4	Select candidate airfoils for the root and tip. Do not assign a wing washout at this time.	8
5	Evaluate the need for wing sweep. Is this a high-subsonic aircraft? Is there a possible problem with CG location that calls for a wing sweep to solve?	9.3.3
6	Once the airplane takes shape, a more sophisticated reshaping is required, in accordance with the design algorithm of Section 1.3.1, *Conceptual design algorithm for a GA aircraft*. This reshaping takes into account performance, stability and control, structural, and systems considerations.	1.3.2

- Impact on the non-linear behavior of the above stability derivatives.
- Structures (structural weight, aeroelasticity, etc.).
- Control authority in extreme flight conditions.
- Control system complexity.
- Operation (e.g. tail length may interfere with T-O rotation, require excessive control surface deflection, and subject the structure to high stresses, etc.).
- Stall tailoring; i.e. the design of the wing for benign stall characteristics.

9.2 THE TRAPEZOIDAL WING PLANFORM

The trapezoidal planform represents the simplest geometric shape selected for aircraft lifting surfaces. Although basic, it is important enough to warrant a specific discussion. One of the primary advantages of this geometry is the simple mathematics that can be used to describe it. Absence of curvature or leading and trailing edge discontinuities results in a simple mathematical representation, in turn, offering convenience when sizing the airplane and extremely helpful when performing trade studies or during optimization.

This class of planform shapes includes a range of geometric shapes ranging from elementary forms, such as the constant-chord planform ("Hershey-bar") to tapered planform shapes that may be swept forward or aft. It is important to know well the geometric relationships between span, aspect ratio, taper ratio, and chord. These will come in handy in a multitude of analyses, ranging from weight estimation to multi-disciplinary optimization.

Methods to treat planform shapes that are either cranked or described with piecewise continuous mathematical functions are given in Appendix D: *Geometry of lifting surfaces*.

9.2.1 Definitions

Figure 9-2 shows the general definitions for the geometry of the trapezoidal planform and indicates important details of such sections. A *leading edge* is the part of a lifting surface that faces the direction of intended movement. The *trailing edge* is opposite to that side. The *root* is the inboard side of the planform and, as can be seen in the figure, is where the two trapezoids join. This would typically be the centerline of the aircraft or the plane of symmetry. Although the term root sometimes refers to where the wing intersects the fuselage, that definition is not used in this text because the geometry of the fuselage may be changing and thus defining a wing chord there would lead to some undesirable issues. The *tip* is the side opposite of the root — it is the outboard side. The distance between the left and right tip is called the *span*, denoted by the variable b. The transverse dimension is called *chord*. Thus we talk about a *root-chord* or *tip-chord* as the length of the corresponding sides, denoted by the variables C_r and C_t, respectively. We also call the ratio of the tip-chord to the root-chord the *taper ratio*, denoted by λ. Another important ratio is the *aspect ratio*, AR, which indicates the slenderness of the wing. Both are expressed mathematically below.

The *quarter-chord line* is drawn from a point one-fourth of the distance from the leading edge of the root-chord to a point one-fourth of the distance from the leading edge of the tip chord. It is important because

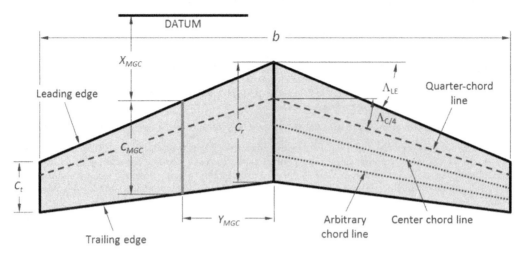

FIGURE 9-2 Fundamental definitions of a trapezoidal wing planform.

two-dimensional aerodynamic data frequently uses the quarter-chord point as a reference when presenting pitching moment data. Scientific literature, for instance the USAF DATCOM, regularly uses the quarter-chord as a reference point for its graphs and computational techniques. Additionally, this is often the location of the main spar in lifting surfaces and it is very convenient to the structural analysts to not have to perform moment transformation during the design of the structure.

The *center chord line* is obtained in a similar fashion to the quarter-chord line and is sometimes used as a reference in scientific literature, although not as frequently. Other important parameters are the sweep angles of the leading edge and quarter-chord line, denoted by Λ_{LE} and $\Lambda_{C/4}$, respectively.

The *mean geometric chord* (MGC) of the planform is often (and erroneously) referred to as the *mean aerodynamic chord* (MAC), which is the chord at the location on the planform at which the center of pressure is presumed to act. The problem is that this location is dependent on three-dimensional influences, such as that of the airfoils, twist, sweep and other factors, not to mention angle-of-attack, in particular when flow separation begins. Authors who refer to the MGC as the MAC typically acknowledge the shortcoming, present the geometric formulation presented here to calculate it, before continuing to call it the MAC. In this text, we will break from this convention and simply call it by its appropriate title — the MGC. The importance of the MGC is that it can be considered a reference location on a wing, to which the location of the center of gravity is referenced and even for a quick preliminary estimation of wing bending moments inside it. Therefore, it is important to also estimate what its spanwise station, y_{MGC}, is.

Considering Figure 9-2, we will now derive mathematical expressions for the aforementioned important properties of the wing:

Wing area:
$$S = b\left(\frac{C_r + C_t}{2}\right) \tag{9-1}$$

Aspect ratio — general:
$$AR = \frac{b^2}{S} \tag{9-2}$$

Aspect ratio — constant-chord:
$$AR = \frac{b}{C_{avg}} \tag{9-3}$$

Taper ratio:
$$\lambda = \frac{C_t}{C_r} \tag{9-4}$$

Average chord:
$$C_{avg} = \frac{C_r + C_t}{2} = \frac{C_r}{2}(1 + \lambda) \tag{9-5}$$

Mean geometric chord:
$$C_{MGC} = \left(\frac{2}{3}\right)C_r\left(\frac{1 + \lambda + \lambda^2}{1 + \lambda}\right) \tag{9-6}$$

Mean aerodynamic chord:
$$C_{MAC} \approx C_{MGC} \tag{9-7}$$

Y-location of MGC_{LE}:
$$y_{MGC} = \left(\frac{b}{6}\right)\left(\frac{1 + 2\lambda}{1 + \lambda}\right) \tag{9-8}$$

X-location of MGC_{LE}:
$$x_{MGC} = y_{MGC} \tan \Lambda_{LE} \tag{9-9}$$

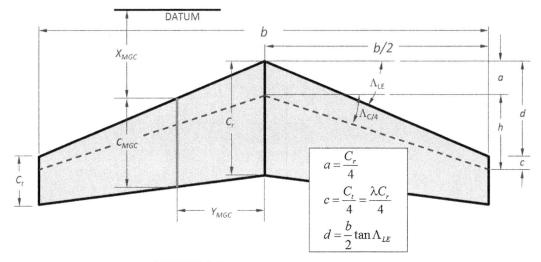

FIGURE 9-3 Deriving the quarter-chord angle.

Angle of quarter-chord line:

$$\tan\Lambda_{C/4} = \tan\Lambda_{LE} + \frac{C_r}{2b}(\lambda - 1) \qquad (9\text{-}10)$$

Angle of an arbitrary chord line:

$$\tan\Lambda_n = \tan\Lambda_m - \frac{4}{AR}\left[(n - m)\frac{1 - \lambda}{1 + \lambda}\right] \qquad (9\text{-}11)$$

Equation (9-11) is obtained from USAF DATCOM [1], where m and n are chordwise fractions of the chord line (0.25 for quarter-chord, 0.5 for the center chord line, etc.) and m is the fraction for a known angle, and n for the unknown angle (see Example 9-1 for example of use).

Aspect Ratio for HT and VT

The aspect ratio for a horizontal tail (HT) is calculated exactly as it is for the wing. The aspect ratio for a vertical tail (VT) is calculated using the following expression:

Aspect ratio for a VT:

$$AR_{VT} = \frac{b_{VT}^2}{S_{VT}} \qquad (9\text{-}12)$$

where

b_{VT} = height of the VT from its base to tip, in ft or m
S_{VT} = area of the VT, in ft² or m²

The aspect ratio for a twin tail is calculated by applying Equation (9-12) to *one tail only*, where S_{VT} refers to one half the total area of the VT.

Aspect Ratio for Multi-wing Configurations

The aspect ratio for multi-wing aircraft, such as biplanes and triplanes, is given below:

Aspect ratio for a biplane:

$$AR_{biplane} = \frac{2b_{\text{larger wing}}^2}{S} \qquad (9\text{-}13)$$

Aspect ratio for a triplane:

$$AR_{triplane} = \frac{3b_{\text{largest wing}}^2}{S} \qquad (9\text{-}14)$$

Derivation of Equations (9-6), (9-8), and (9-10)

Equation (9-6): see Section D.5.3, *Derivation of MGC*.
Equation (9-8): see Section D.5.2, *Spanwise location of the MGC*.
Equation (9-10): refer to the Figure 9-3 to define the dimensions a, $b/2$, c, d, and h.

$$\left.\begin{array}{l} a = \dfrac{C_r}{4} \\[4pt] d = \dfrac{b}{2}\tan\Lambda_{LE} \\[4pt] c = \dfrac{C_t}{4} = \dfrac{\lambda C_r}{4} \end{array}\right\} \Rightarrow h = d + c - a$$

$$= \frac{b}{2}\tan\Lambda_{LE} + \frac{\lambda C_r}{4} - \frac{C_r}{4}$$

$$h = \frac{1}{2}\left[b\tan\Lambda_{LE} + \frac{\lambda C_r}{2} - \frac{C_r}{2}\right] = \frac{1}{2}\left[b\tan\Lambda_{LE} + \frac{C_r}{2}(\lambda - 1)\right]$$

$$\tan\Lambda_{C/4} = \frac{h}{b/2} = \frac{\left[b\tan\Lambda_{LE} + \frac{C_r}{2}(\lambda - 1)\right]}{b}$$

$$= \tan\Lambda_{LE} + \frac{C_r}{2b}(\lambda - 1)$$

QED

EXAMPLE 9-1: TRAPEZOIDAL WING PLANFORM

Determine the primary characteristics of the wing shown in Figure 9-4. Also calculate the angle of the center chord line using Equation (9-11).

Solution

Wing area:

$$S = b\left(\frac{C_r + C_t}{2}\right) = (10)\left(\frac{2+1}{2}\right) = 15 \text{ ft}^2$$

Aspect ratio – general:

$$AR = \frac{b^2}{S} = \frac{10^2}{15} = 6\,\tfrac{2}{3} \approx 6.667$$

Taper ratio:

$$\lambda = C_t/C_r = 0.5$$

Mean geometric chord:

$$C_{MGC} = \left(\frac{2}{3}\right)C_r\left(\frac{1+\lambda+\lambda^2}{1+\lambda}\right) = \left(\frac{2}{3}\right)(2)\left(\frac{1+0.5+0.5^2}{1+0.5}\right)$$
$$= 1\,\tfrac{5}{9} \text{ ft} \approx 1.556 \text{ ft}$$

Mean aerodynamic chord:

$$C_{MAC} \approx C_{MGC}$$

Y-location of MGC_{LE}:

$$y_{MGC} = \left(\frac{b}{6}\right)\left(\frac{1+2\lambda}{1+\lambda}\right) = \left(\frac{10}{6}\right)\left(\frac{1+2\times 0.5}{1+0.5}\right)$$
$$= 2\,\tfrac{2}{9} \text{ ft} \approx 2.222 \text{ ft}$$

X-location of MGC_{LE}:

$$x_{MGC} = y_{MGC}\tan\Lambda_{LE} = (2\,\tfrac{2}{9})\tan(30°) \approx 1.283 \text{ ft}$$

Angle of quarter-chord line:

$$\tan\Lambda_{C/4} = \tan\Lambda_{LE} + \frac{C_r}{2b}(\lambda - 1) = \tan(30)$$
$$+ \frac{2}{2(10)}(0.5 - 1) \approx 0.5274 \Rightarrow \Lambda_{C/4} \approx 27.8°$$

Angle of the center chord line using the quarter-chord line (m = 0.25, n = 0.50) is obtained from Equation (9-11):

$$\tan\Lambda_n = \tan\Lambda_m - \frac{4}{AR}\left[(n-m)\frac{1-\lambda}{1+\lambda}\right]$$

Inserting the given values, we get:

$$\tan\Lambda_{0.50C} = \tan\Lambda_{0.25C} - \frac{4}{6.667}\left[(0.50 - 0.25)\frac{1-0.5}{1+0.5}\right]$$
$$= 0.4772 \Rightarrow \Lambda_{0.50C} = 25.5°$$

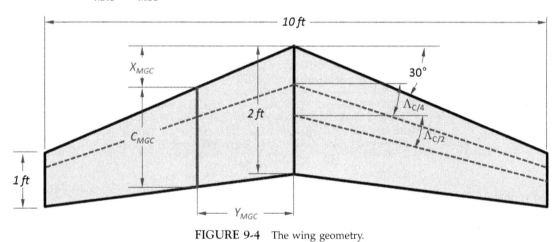

FIGURE 9-4 The wing geometry.

9.2.2 Poor Man's Determination of the MGC

In the absence of computational tools the designer can determine the location of the MGC using the graphical scheme in Figure 9-5. It is important to remember that such graphical tools are the relics of a bygone era and at best are what the slide-rule is to the modern calculator — results from a digital calculator should always take precedence.

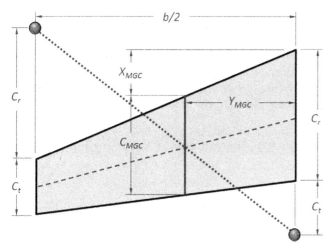

FIGURE 9-5 Graphical scheme to determine the MGC for a simple trapezoidal planform.

9.2.3 Planform Dimensions in Terms of S, λ, and AR

During the design phase, the basic dimensions of the wing (span and chord at root and tip) are sometimes defined by wing area, taper ratio, and aspect ratio as governing variables. This is convenient because these parameters can tell the designer a lot about the likely aerodynamic properties of the wing. With the wing parameters defined in this fashion use the following formulation to determine the required wingspan, root chord, and tip chord of a simple tapered planform like the one in Figure 9-6. These are based on Equations (9-2), (9-3), and (9-10):

From

$$AR = \frac{b^2}{S} \Rightarrow b = \sqrt{AR \cdot S} \qquad (9\text{-}15)$$

From

$$AR = \frac{b}{C_{avg}} \Rightarrow C_{avg} = \left(\frac{C_r + C_t}{2}\right) = \frac{b}{AR} \qquad (9\text{-}16)$$

Average chord:

$$C_{avg} = \frac{C_r + C_t}{2} = \frac{C_r}{2}(1+\lambda) = \frac{S}{b} \qquad (9\text{-}17)$$

From

$$2C_{avg} = \frac{2b}{AR} = C_r(1+\lambda) \Rightarrow C_r = \frac{2b}{(1+\lambda)AR} \qquad (9\text{-}18)$$

From

$$2C_{avg} = \frac{2S}{b} = C_r(1+\lambda) \Rightarrow C_r = \frac{2S}{(1+\lambda)b} \qquad (9\text{-}19)$$

Mean geometric chord:

$$C_{MGC} = \left(\frac{4b}{3AR}\right)\left(\frac{1+\lambda+\lambda^2}{1+2\lambda+\lambda^2}\right) \qquad (9\text{-}20)$$

Chord for a straight tapered wing:

$$c(y) = C_r\left(1 + \frac{2(\lambda-1)}{b}y\right) \qquad (9\text{-}21)$$

> **Derivation of Equation (9-20)**
>
> Using Equation (9-18), we can calculate the root chord, C_r, using any set of values for the wingspan, b, and aspect ratio, AR:
>
> $$C_r = \frac{2b}{(1+\lambda)AR}$$

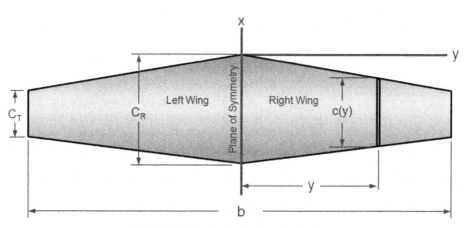

FIGURE 9-6 A simple tapered planform.

We then insert this into Equation (9-6) and manipulate as follows:

$$C_{MGC} = \left(\frac{2}{3}\right)C_r\left(\frac{1+\lambda+\lambda^2}{1+\lambda}\right)$$
$$= \left(\frac{2}{3}\right)\left(\frac{2b}{(1+\lambda)AR}\right)\left(\frac{1+\lambda+\lambda^2}{1+\lambda}\right)$$
$$= \left(\frac{4b}{3AR}\right)\left(\frac{1+\lambda+\lambda^2}{1+2\lambda+\lambda^2}\right)$$

QED

Derivation of Equation (9-21)

Parametric formulation of the chord c as a function of spanwise station y for a straight tapered wing whose taper ratio is λ, root chord C_r, and span is b is obtained by inserting the geometric parameters as follows (where the parameter t is $2y/b$ and ranges from 0 to 1):

$$c(y) = C_r(1-t) + \lambda C_r t = C_r\left(1 - \frac{2y}{b}\right) + \lambda C_r\frac{2y}{b}$$
$$= C_r\left[1 - \frac{2y}{b} + \lambda\frac{2y}{b}\right] = C_r\left(1 + \frac{2(\lambda-1)}{b}y\right)$$

QED

EXAMPLE 9-2: TRAPEZOIDAL WING PLANFORM

A wing is to be designed such that $S = 140$ ft^2, $\lambda = 0.5$, and $AR = 10$. Determine the wingspan, root chord, and tip chord required for the wing. Also determine the chord at $y = 10$ ft.

Solution

Wingspan:

$$AR = \frac{b^2}{S} \Rightarrow b = \sqrt{AR \cdot S} = \sqrt{(10)\cdot(140)} = 37.42 \text{ ft}$$

Wing root chord:

$$C_r = \frac{2S}{b(1+\lambda)} = \frac{2(140)}{(37.42)(1+0.5)} = 4.99 \text{ ft}$$

Wing tip chord:

$$C_t = \lambda C_r = 0.5(4.99) = 2.49 \text{ ft}$$

Chord at $y = 10$ ft:

$$c(10) = C_R\left(1 + \frac{2(\lambda-1)}{b}y\right)$$
$$= (4.99)\left(1 + \frac{2(0.5-1)}{37.42}(10)\right) = 3.66 \text{ ft}$$

9.2.4 Approximation of an Airfoil Cross-sectional Area

Often, the cross-sectional area (internal area) of an airfoil must be evaluated as it yields useful clues about the internal volume of a wing available for fuel storage. In the absence of precise airfoil data the following approximation is used to estimate the internal area of geometry resembling that of an airfoil:

Total area:

$$A_{airfoil} = \frac{(k+3)C \cdot t}{6} \quad (9\text{-}22)$$

where

C = airfoil chord, in ft or m
k = location of the airfoil's maximum thickness as a fraction of C
t = airfoil thickness, in ft or m

Sometimes it is more convenient to present the thickness using the thickness-to-chord ratio, denoted by (t/c).

This way, the thickness is expressed using the product $(t/c) \cdot C$. Thus, Equation (9-22) is written as follows:

Total area:

$$A_{airfoil} = \frac{(k+3)}{6}\left(\frac{t}{c}\right)C^2 \quad (9\text{-}23)$$

Derivation of Equations (9-22) and (9-23)

Consider Figure 9-7, which shows an airfoil of chord C approximated by a parabolic D-cell and a triangular section. It is assumed the two sections join at the chord station of maximum thickness, t, whose location is given by $k \cdot C$, where k is the location of the airfoil's maximum thickness as a fraction of C. The cross-sectional areas of the two sections and the total area are given by the following expressions:

Parabolic section:

$$A_1 = \frac{2}{3}k \cdot C \cdot t \quad \text{(i)}$$

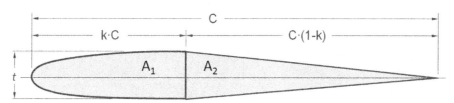

FIGURE 9-7 An approximation of an airfoil using elementary geometry.

Triangular section:

$$A_2 = \frac{1}{2}C \cdot (1-k)t \qquad \text{(ii)}$$

Therefore, the internal area of the airfoil can be approximated by adding the two as shown below:

$$A_{airfoil} = A_1 + A_2 = \frac{2k \cdot C \cdot t}{3} + \frac{C \cdot (1-k)t}{2} = \frac{(k+3)C \cdot t}{6}$$

Then, multiply by C/C (=1) to get Equation (9-23).

QED

9.3 THE GEOMETRIC LAYOUT OF THE WING

Once the wing area and wingspan have been determined, it is possible to begin to establish the remaining geometric properties. The geometric layout of the wing refers to properties such as the aspect ratio, taper ratio, wing sweep, dihedral, and so on. These constitute the geometric layout of the wing. The layout has profound influence on the entire design process, affecting a large number of other areas in the development. These are not just aerodynamics, performance, and stability and control, but also structures and systems layout, just to name a few. The layout process assesses the topics listed below. The corresponding sections are listed as well.

The layout constitutes the geometric parameters presented in Table 9-2.

Of these, the AR, TR, and the LE sweep give the designer fundamental control over the aerodynamic characteristics of the wing. This is not to say the others are not important too, but rather that they can be considered more as dials to "fine-tune" the wing design. The determination of the AR, TR, and sweep may be the consequence of a complicated optimization; however, this is not always the case.

9.3.1 Wing Aspect Ratio, AR

The wing aspect ratio (AR) is a fundamental property that simultaneously affects the magnitude of the lift-induced drag (C_{Di}) and the slope of the lift curve ($C_{L\alpha}$). As such, it also directly influences both performance and stability and control. Table 9-3 shows typical values of the AR for several classes of aircraft.

Figure 9-8 shows two planform shapes of equal planform area, but different AR. The stubby configuration ($AR = 4$) has a shallower $C_{L\alpha}$ and lower C_{Lmax}, but higher stall AOA (α_{stall}) than the slender one ($AR = 16$). The stubby planform has less roll damping than the high AR and, for that reason, is better suited for airplanes that require roll responsiveness, such as aerobatic aircraft. Additionally, it generates higher lift-induced drag (C_{Di}) than the slender one. A high

TABLE 9-2 Typical Values of AR for Various Classes of Aircraft

Geometric property	Section
Planform shape	9.4
Aspect ratio (AR)	9.3.1
Taper ratio (TR or λ)	9.3.2
Leading edge (LE) or quarter chord (C/4) sweep angle	9.3.3
Dihedral angle	9.3.4
Washout	9.3.5
Angle-of-incidence	9.3.6
Positioning of the wing on the fuselage	4.2.1
Partitioning of the wing into a roll control region (ailerons), high lift region (flaps, slats), and lift suppression region (spoilers)	10

TABLE 9-3 Typical Values of AR for Various Classes of Aircraft

Type of Vehicle or Vehicle Component	AR Range
Missiles	0.5–1
Military fighters	2.5–4.0
GA aircraft	6–11
Aerobatic airplanes	5–6
Twin-engine commuters	10–14
Commercial jetliners	7–10
Sailplanes	10–51

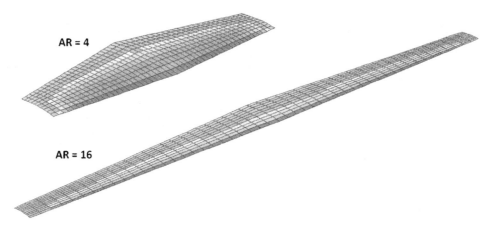

FIGURE 9-8 Two planform shapes of equal areas, but different aspect ratios.

AR wing is required for sailplanes and airplanes that are required to have long-range or endurance. They have high roll damping and will be structurally heavier because of larger bending moments.

Table 9-4 gives a general rule-of-thumb about the impact the AR has on selected aerodynamic properties.

The aspect ratio is defined as follows for monoplane and biplane configurations:

Monoplane:
$$AR = b^2/S \qquad (9\text{-}24)$$

Biplane:
$$AR = 2b^2/S \qquad (9\text{-}25)$$

where
b = wingspan
S = wing area

Illustrating the Difference Between Low and High AR Wings

The difference between a low and high AR is illustrated in Figure 9-9, which shows two wings of equal wing areas and taper ratios. The upper wing has a relatively low AR (6.67) while the lower one has a relatively large one (18.57). The airspeed in both cases is 100 KTAS and both are positioned at an AOA such they generate 2000 lb_f of lift. The figure shows that the low AR wing requires an AOA of 1.55° but the high AR one 0.66°. In accordance with Section 15.3.4, *The lift-induced drag coefficient: C_{Di}*, the higher AOA means the low AR wing generates higher induced drag. The greater flow field disruption is even visible by the larger wingtip vortex generated by the low AR wing in Figure 9-9. Here, a lift-induced drag value of 71 lb_f was estimated for the low AR wing and 45 lb_f for the high AR one. Additionally, the lift curve slope, $C_{L\alpha}$, for the low AR wing

TABLE 9-4 Typical Impact of Aspect Ratio on Aerodynamic Properties of the Wing

Aspect Ratio, AR	Pros	Cons
1.0	High stall angle-of-attack. High flutter speed. Low roll damping. Low structural weight. Great gust penetration capability (because of the shallow $C_{L\alpha}$).	Inefficient because of high induced drag. Shallow $C_{L\alpha}$ requires large changes in AOA with airspeed. Low C_{Lmax} (high stalling speed). Low LD_{max}.
5–7	Good roll response. Relatively high flutter speed. Limited adverse yaw. Reasonable gust penetration.	Inefficient to marginally efficient for long-range design. Relatively high induced drag.
7–12	Good balance between low induced drag and roll response. Good glide characteristics for powered aircraft.	Some adverse yaw may be noticed toward the higher extreme of AR. Resonable to marginal gust penetration capability (steep $C_{L\alpha}$).
20+	Low induced drag. Great glide characteristics. Steep $C_{L\alpha}$ (large change in lift with small changes in α). High maximum lift coefficient.	High structural weight. Low flutter speed. High roll damping. High adverse yaw. Steep $C_{L\alpha}$ results in higher gust loads.

9.3 THE GEOMETRIC LAYOUT OF THE WING 311

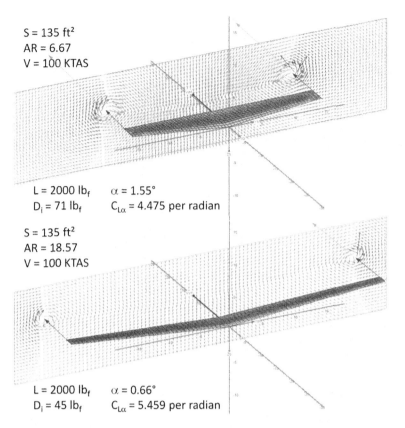

S = 135 ft²
AR = 6.67
V = 100 KTAS

L = 2000 lb$_f$ α = 1.55°
D$_i$ = 71 lb$_f$ $C_{L\alpha}$ = 4.475 per radian

S = 135 ft²
AR = 18.57
V = 100 KTAS

L = 2000 lb$_f$ α = 0.66°
D$_i$ = 45 lb$_f$ $C_{L\alpha}$ = 5.459 per radian

FIGURE 9-9 The difference between a low *AR* (top) and high *AR* (bottom) is illustrated. Both wings are of equal area and each generates 2000 lb$_f$ of lift at 100 KTAS. The flow solution is shown in a plane positioned aft of the wing trailing edge.

(4.475 per radian) is less than that of the high *AR* wing (5.459 per radian). Therefore, the low *AR* wing will stall at a higher *AOA* and airspeed than the high *AR* wing. However, it is far more responsive as its roll damping coefficient, C_{lp}, is lower.

The difference between the two configurations is further illustrated in Figure 9-10. The graphs were generated for three wings using the potential flow theory. The wings all have a constant *TR* (0.5), but varying *AR*. The left graph shows the spanwise distribution of section lift coefficients at an *AOA* of 10°. The tip loading that results from the *TR* (and will be discussed next) is evident. It can be seen that the high *AR* of 16 generates the highest C_l at the give *AOA*, as is to be expected since its $C_{l\alpha}$ is the

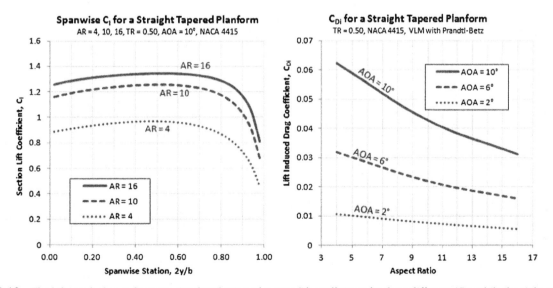

FIGURE 9-10 The left graph shows the spanwise distribution of section lift coefficients for three different *AR*s, while the right graph shows the effect of *AR* on the lift induced drag coefficient.

highest. Other than the magnitude of the section lift coefficients, it is also evident that the *AR* does not have major changes on the general shape of the distribution; in other words, the *AR* has great effects on the magnitude of section lift coefficients at a given *AOA*, and relatively small influence on the shape of the distribution.

The left graph in Figure 9-10 shows how the *AR* affects the lift induced drag. This is illustrated for three *AOA*s; 2°, 6°, and 10°. The graph shows substantial reduction in lift-induced drag is to be had with larger *AR*.

Determining AR Based on a Desired Range

It is possible to evaluate and recommend *AR* for some special design cases. One such is the determination of *AR* based on a desired range, *R*, at some desired cruising airspeed, V_C. If the aircraft being designed belongs to a class of aircraft for which C_{Dmin} can be estimated with reasonable accuracy and the expected initial and final cruise weights are known, denoted by W_{ini} and W_{fin}, respectively, then the average cruise lift coefficient can be used with Equation (20-12) to extract the effective aspect ratio, AR_e, for the vehicle using the following expression:

$$AR_e = \frac{C_{LC}^2}{\pi} \frac{1}{\left(\frac{V_C}{R}\frac{C_{LC}}{c_t}\ln\left(\frac{W_{ini}}{W_{fin}}\right) - C_{Dmin}\right)} \quad (9\text{-}26)$$

where

C_{LC} = average cruise lift coefficient, corresponding to the average of W_{ini} and W_{fin}.
R = range in ft
V_C = average cruising speed in ft/s
c_t = thrust specific fuel consumption in 1/s

The denominator in Equation (9-26) results in a singularity when $C_{Dmin} = \frac{V_C}{R}\frac{C_{LC}}{c_t}\ln\left(\frac{W_{ini}}{W_{fin}}\right)$. In fact, it returns negative values if the airspeed is lower than obtained by the following expression:

$$V_C > \frac{R \cdot c_t \cdot C_{Dmin}}{C_{LC}\ln\left(W_{ini}/W_{fin}\right)} \quad (9\text{-}27)$$

For this reason, Equation (9-26) should only be applied to airspeeds that are greater than the value obtained by the above expression. It should be used to guide the selection of the *AR* rather than obtain it directly, for instance by plotting isobars.

Note that AR_e is the product of the *AR* and Oswald's span efficiency, $AR \cdot e$. Once this product has been determined, it is the responsibility of the designer to devise a proper combination of *AR* and planform geometry to achieve this value. Also note that if the airplane features a swept back planform that the *AR* limits of Equation (9-94) must be considered as well.

Although the low *AR* surface is less efficient than the high *AR*, it is much better suited as a control surface for tail-aft configurations (i.e. as a horizontal and vertical tail). This results from the higher stall *AOA*, which is a consequence of the low *AR*. Delayed stall introduces a certain level of safety to the operation of tail-aft airplanes because it will require them to reach very high *AOA* or *AOY* before the stabilizing moments begin to drop. This is the reason why the horizontal and vertical tail surfaces on airplanes typically feature low aspect ratios.

Derivation of Equation (9-26)

Assuming the simplified drag model, Equation (20-12) can be solved for the drag coefficient as follows:

$$R = \frac{V}{c_t}\frac{C_L}{C_D}\ln\left(\frac{W_{ini}}{W_{fin}}\right) \Leftrightarrow C_D = C_{Dmin} + \frac{C_L^2}{\pi \cdot AR_e}$$

$$= \frac{V}{R}\frac{C_L}{c_t}\ln\left(\frac{W_{ini}}{W_{fin}}\right)$$

This can then be solved for the effective *AR* as follows:

$$AR_e = \frac{C_L^2}{\pi} \frac{1}{\left(\frac{V}{R}\frac{C_L}{c_t}\ln\left(\frac{W_{ini}}{W_{fin}}\right) - C_{Dmin}\right)}$$

Note that this is a generic case and the subscript C is used to denote some desired cruise conditions.

QED

EXAMPLE 9-3

The Cirrus SR22 pilot operating handbook (POH) reveals that the airplane has an 899 nm range at 65% power at 10,000 ft. The POH states the cruising speed at this condition is 174 KTAS and fuel consumption is 15.4 gal/hr. This implies an SFC = 15.4 gal/hr × (6 lb$_f$/gal)/ (0.65 × 310 BHP) = 0.4586 lb$_f$/hr/BHP. Evaluate the reliability of Equation (9-26) by considering the hypothetical design of a SR22 class aircraft, which "happens" to share a

EXAMPLE 9-3 (cont'd)

number of parameters. This hypothetical airplane is designed for a range of 900 nm precisely at the same condition. Estimate the effective AR for this airplane using Equation (9-26) and compare to that of the SR22. Assume the weight at the beginning of cruise is 3379 lb$_f$ and 2910 lb$_f$ at end of cruise. Assume the wing area is 145 ft^2 and $C_{Dmin} = 0.02541$ (as calculated per Example 15-18).

Solution

Average weight during cruise:

$$W_{avg} = \frac{W_{ini} + W_{fin}}{2} = \frac{3379 + 2910}{2} = 3145 \text{ lb}_f$$

Density at 10,000 ft:

$$\rho = 0.002378(1 - 0.0000068756 \times 10,000)^{4.2561}$$
$$= 0.001756 \text{ slugs/ft}^3$$

The airspeed in ft/s is $174 \times 1.688 = 293.7$ ft/s, so the lift coefficient at cruise is:

$$C_{LC} = \frac{2W_{avg}}{\rho V_C^2 S} = \frac{2 \times 3145}{(0.001756)(293.7)^2(145)} = 0.2864$$

The thrust specific fuel consumption is found using Equation (20-9):

$$c_t = \frac{c_{bhp} V}{1,980,000 \eta_p} = \frac{0.4586 \times 293.7}{1,980,000(0.85)} = 8.0 \times 10^{-5} \text{ 1/s}$$

The desired range of 900 nm amounts to 5,468,400 ft, yielding the following effective AR:

$$AR_e = AR \cdot e = \frac{C_{LC}^2}{\pi} \frac{1}{\left(\frac{V_C}{R}\frac{C_{LC}}{c_t}\ln\left(\frac{W_{ini}}{W_{fin}}\right) - C_{Dmin}\right)}$$

$$= \frac{(0.2864)^2}{\pi} \frac{1}{\left(\frac{(293.7)}{5,468,400}\frac{(0.2864)}{8.0 \times 10^{-5}}\ln\left(\frac{3379}{2910}\right) - 0.2541\right)}$$

$$= 7.86$$

Note that the actual AR of the SR22 is 10. The answer therefore implies its Oswald efficiency is 0.786. This compares favorably to Equation (9-89) as follows:

$$e = 1.78(1 - 0.045 \times 10^{0.68}) - 0.64 = 0.7566$$

Determining AR Based on Desired Endurance

The AR can also be determined based on a desired endurance, E, at some desired cruising airspeed, V_C, similar to what was done for range and using the same basic variables. This time the method uses Equation (20-22) to extract the effective aspect ratio, AR_e, for the vehicle using the following expression:

$$AR_e = \frac{C_{LC}^2}{\pi} \frac{1}{\left(\frac{1}{E}\frac{C_{LC}}{c_t}\ln\left(\frac{W_{ini}}{W_{fin}}\right) - C_{Dmin}\right)} \quad (9\text{-}28)$$

where E = endurance in seconds.

Like Equation (9-26), the denominator in Equation (9-28) has similar limitations and should be used with cruise lift coefficients greater than those obtained with the following expression:

$$C_{LC} > \frac{E \cdot c_t \cdot C_{Dmin}}{\ln\left(W_{ini}/W_{fin}\right)} \quad (9\text{-}29)$$

Derivation of Equation (9-28)

Assuming the simplified drag model, Equation (20-22) can be solved for the drag coefficient as follows:

$$E = \frac{1}{c_t}\frac{C_L}{C_D}\ln\left(\frac{W_{ini}}{W_{fin}}\right) \Leftrightarrow C_D = C_{Dmin} + \frac{C_L^2}{\pi \cdot AR_e}$$

$$= \frac{1}{E}\frac{C_L}{c_t}\ln\left(\frac{W_{ini}}{W_{fin}}\right)$$

This can then be solved for the effective AR as follows:

$$AR_e = \frac{C_L^2}{\pi} \frac{1}{\left(\frac{1}{E}\frac{C_L}{c_t}\ln\left(\frac{W_{ini}}{W_{fin}}\right) - C_{Dmin}\right)}$$

Note that this is a generic case and the subscript C is used to denote some desired cruise conditions.

QED

> # EXAMPLE 9-4
>
> Use Equation (9-28) to estimate the effective AR for the SR22 class aircraft of Example 9-3, using the same parameters, if it is designed to cruise 900 nm at 174 KTAS.
>
> $$AR_e = \frac{(0.2864)^2}{\pi} \frac{1}{\left(\frac{1}{(5.172 \times 3600)} \frac{(0.2864)}{8.0 \times 10^{-5}} \ln\left(\frac{3379}{2910}\right) - 0.2541\right)}$$
>
> $$= 7.87$$
>
> ## Solution
>
> Using Equation (9-28), the desired range of 900 nm at 174 KTAS amounts to 5.172 hrs, yielding the following effective AR:
>
> effectively the same result as in the previous example.

Figure 9-11 shows the effect aspect ratio has on a wing's lift curve and drag polar. It compares the properties of some airfoil (for which AR_1 is considered ∞) to the same airfoil being used in a three-dimensional wing (whose AR_2 is much, much smaller). It can be seen that as the AR reduces, so does the lift curve slope (denoted by $C_{l\alpha}$ for the airfoil and $C_{L\alpha}$ for the wing) and maximum lift coefficient (denoted by C_{lmax} for the airfoil and C_{Lmax} for the wing). If the airfoil generates a specific lift coefficient at α_1, it must be placed at a higher AOA, α_2, once used on a three-dimensional wing. The cost of this is an increase in drag, which changes from C_d for the airfoil to C_D for the wing, as is shown in Figure 9-11. The difference between the two is the three-dimensional lift induced drag coefficient and the rise in the airfoil drag due to growth in the flow separation region.

Estimating AR Based on Minimum Drag

The AR can also be set based on the minimum drag of the airplane. This approach requires the design lift coefficient to be determined (e.g. see Equation [9-49]). Then, assuming constant wing area, a relatively sophisticated estimation of the total drag coefficient of the airplane is accomplished. The estimation takes into account changes in the wing geometry as a function of AR (wing chords change since S is constant) and evaluates skin friction assuming either a fully turbulent or a mixed boundary layer. Then, using the appropriate form and interference factors, the minimum drag coefficient is estimated (see Chapter 15, *Aircraft drag analysis*). The lift-induced drag coefficient is also calculated, using an appropriate model of the Oswald's span efficiency. Adding the two coefficients comprises the total drag coefficient, C_D. This can then be plotted versus the AR, in a carpet plot, similar to that of Figure 9-12. The carpet plot is usually prepared using a range of lift coefficients. Such a plot reveals the location of the optimum AR. The upper and lower bounds in the figure are simply the optimum $AR \pm 2$, but the designer can widen or narrow this range depending on the project. Figure 9-12

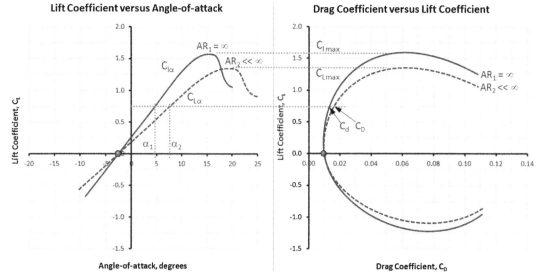

FIGURE 9-11 The effect of aspect ratio on a 3D lift curve and drag polar.

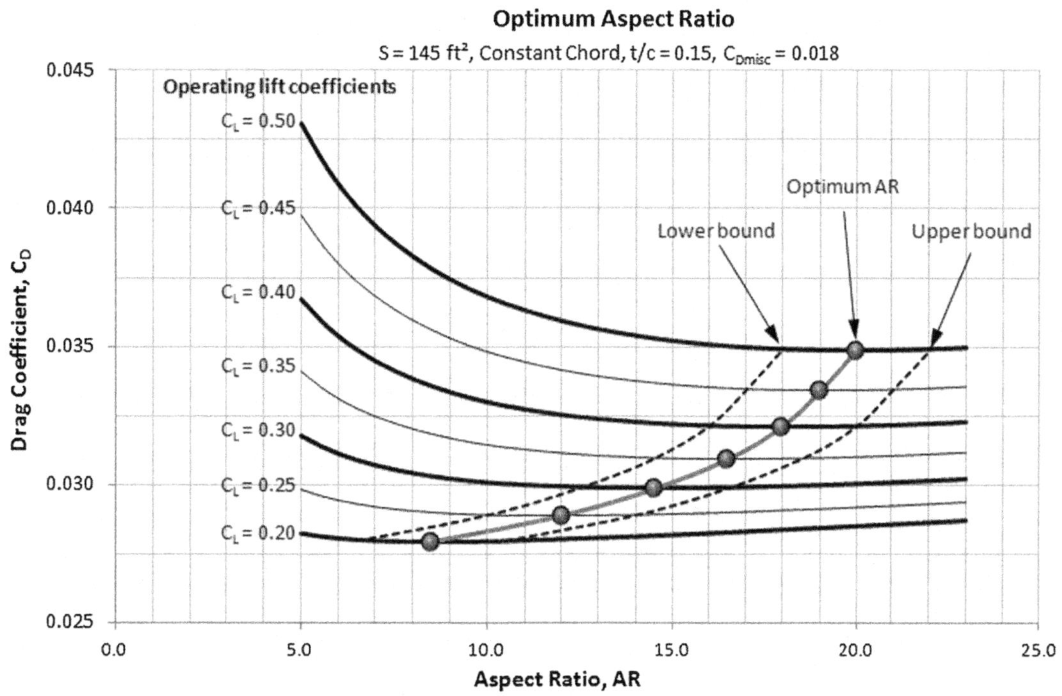

Optimum Aspect Ratio
S = 145 ft², Constant Chord, t/c = 0.15, C_{Dmisc} = 0.018

FIGURE 9-12 The effect of aspect ratio on the drag coefficient of a hypothetical aircraft.

shows that for the hypothetical aircraft being presented, practically any AR between the two limits will yield the lowest value of the total drag coefficient. This gives the designer room to accommodate other concerns such as structural weight.

Estimating AR for Sailplane Class Aircraft

A plausible AR for powered and unpowered sailplanes can be established using the historical data in Figure 9-13. The data points represent contemporary manufacturers' information collected from Ref. [2]. The maximum lift-to-drag ratios (LD_{max}) presented includes the drag of the fuselage and stabilizing surfaces and demonstrates the sophistication of the modern sailplane. The graph shows trends for both regular and powered sailplanes and presents accompanying least-squares curves. Note that these estimates do not replace analyses of the kind presented by Ref. [3], but rather supply initial estimates based on historical sailplanes.

Also plotted are theoretical predictions of the LD_{max} using the *simplified drag model* (see Chapter 15, *Aircraft drag analysis*) and C_{Dmin} of 0.01 (typical of many sailplanes). This model is often the first choice of the novice airplane designer, but it is inaccurate for sailplanes designed for extensive NLF as it does not model the drag bucket associated with such aircraft. These curves are merely presented here to demonstrate why the simplified drag model is a poor predictor.

The theoretical LD_{max} was calculated using Equation (19-18), where the Oswald's span efficiency was calculated per Equations (9-91) and (9-89) and. The curves shown assume a minimum drag coefficient of 0.01, which is a typical value for clean sailplanes. The graph shows clearly the limitation of the simplified drag model.

The following empirical formulation can be used to estimate the AR for the conceptual design of unpowered and powered sailplanes (as long as $AR < 36$):

Sailplanes:

$$AR \approx 44.482 - \sqrt{1672.2 - 28.41 LD_{max}} \quad (9\text{-}30)$$

Powered sailplanes:

$$AR \approx \frac{LD_{max} + 0.443}{1.7405} \quad (9\text{-}31)$$

Naturally, the calculated AR will not guarantee that a desired LD_{max} will be achieved. Rather it indicates that, historically, sailplanes with that AR have achieved the said LD_{max}. Achieving it will require the utmost attention to anything on the airplane that generates drag.

9.3.2 Wing Taper Ratio, TR or λ

The *taper ratio* is the second of the three most important geometric properties of a wing. It has a profound impact on how lift is distributed along the wingspan. For instance, consider Figure 9-14, which shows how it changes the spanwise distribution of section lift

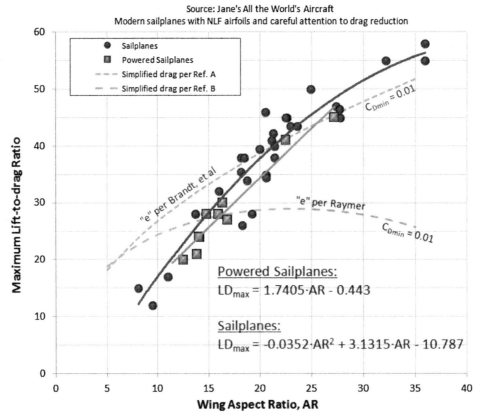

FIGURE 9-13 Maximum lift-to-drag ratio for modern sailplanes and powered sailplanes as a function of AR. Ref. A is Ref. [4] and Ref. B is Ref. [5].

coefficients (left graph) and lift force per unit length (right graph). It shows clearly how a highly tapered wing becomes "tip-loaded," whereas a constant-chord wing is "root-loaded," terms that refers to the distribution of section lift coefficients. This is of great importance when tailoring stall characteristics and controlling the effectiveness of the wing. The right graph shows how the actual lift force of the highly tapered wing moves inboard. This reduces the bending moments and can help reduce the structural weight of the wing.

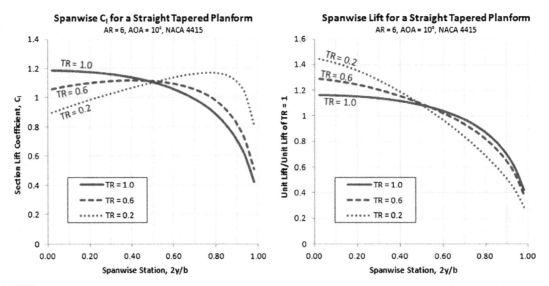

FIGURE 9-14 The effect of taper ratio on the spanwise distribution of section lift coefficients (left); and lift force (right).

TABLE 9-5 Typical Impact of Taper Ratio on Aerodynamic Properties of the Wing

Taper Ratio, λ	Pros	Cons
0.3	Induced drag is close to that of an elliptical wing planform, but is much simpler to manufacture.	Poor stall characteristics. Tip-loaded planform requires a large washout to delay tip stall.
0.5	Good balance between low induced drag and good stall characteristics.	Stall begins mid-span and spreads to tip and root. Usually requires moderate washout.
1.0	Good stall progression. Washout generally not required. Simple to manufacture.	High induced drag.

Too much tip-loading of a wing can have serious consequences for its stall characteristics. It is evident from the shape of the spanwise C_l that the wing whose $TR = 0.2$ will stall near the wingtip, as it peaks there. This would make the wing very susceptible to roll instability at stall, which could have dire consequences if ignored. The constant-chord wing, on the other hand, has its C_l peak inboard, near the root, which retains roll stability and results in benign stall characteristics. Such characteristics are very desirable for trainers and, frankly, should be one of the primary goals of any passenger-carrying aircraft.

Additionally, the left graph of Figure 9-14 shows that a *TR* in the neighborhood of 0.5–0.6 is a good compromise between efficiency and stall behavior, even though some wing twist (to be discussed later) will be required to ensure stall progression that protects roll stability at stall. The right graph of Figure 9-14 shows that even though the outboard airfoils of the highly tapered wing are working hard to generate lift, the actual force is less than that of the constant-chord wing because the chord is so much shorter. This is very beneficial from a structural standpoint as it brings the center of force inboard and reduces the wing bending moments, although torsional rigidity on the outboard wing suffers. Table 9-5 gives a general rule-of-thumb about the impact the *TR* has on selected aerodynamic properties.

9.3.3 Leading Edge and Quarter-chord Sweep Angles, Λ_{LE} and $\Lambda_{C/4}$

The purpose of sweeping the wing forward or aft is primarily twofold: (1) to fix a CG problem and (2) to delay the onset of shockwaves. The latter is the reason for using swept wings for high-speed aircraft (high subsonic and supersonic). However, the former is surprisingly common as well. Strictly speaking, sweep should be avoided unless necessary. It not only makes the wing less efficient aerodynamically, it is also detrimental to stall characteristics if swept back. Additionally, for slow-flying GA aircraft, it implies the spar has discontinuous breaks in it that make it less efficient structurally.

Of course, this is not to say that sweep is all bad. It is very helpful for the design of high-speed aircraft, where it enables the use of thicker airfoils, which lightens the airframe and makes up for some of the lost efficiency. For low-speed airplanes, it is a tool that allows a project to be salvaged if it is discovered that the CG is farther forward or aft than anticipated. This is the reason behind the aft-swept wings featured on the venerable Douglas DC-3 Dakota (C-47) [6]. The same holds for forward-swept configurations. The military scouting and trainer airplane SAAB MFI-15 Safari was designed with an improved field-of-view in mind, which is why it features a shoulder-mounted wing with the main spar behind the two occupants. The wing is swept forward to solve the CG issue that results from the engine and the two occupants sitting in front of the main spar.

Impact of Sweep Angle on the Critical Mach Number

The greatest benefit of wing sweep is a reduction in the strength of and delay in the onset of shock formation. The shock formation will not only cause a sharp increase in drag; it also changes the chordwise pressure distribution on the airfoil, causing the center of lift to move from approximately the airfoil's quarter-chord to mid-chord. The consequence of this is called "Mach-tuck," a severe increase in nose-down pitching moment. The change in drag due to this effect is detailed in Section 15.4.2, *The effect of Mach number*. Also, Figure 15-28 shows how the leading edge sweep increases the critical Mach number, M_{crit}, and delays the onset of the peak of the compressibility drag coefficient. This is helpful as it allows thicker and more structurally efficient airfoils to be used in the wing.

Consider the swept wing in Figure 9-15 and the deconstruction of the far-field airspeed into two components: one parallel to and the other perpendicular to the leading edge. In its simplest form the wing sweep theory contends that it is the airspeed component normal to the leading edge that dictates when shockwaves begin to form. This allows the following correction to the critical Mach number to be made (see Section 8.3.7, *The critical Mach number, M_{crit}*).

$$M_{crit} = \frac{(M_{crit})_{\Lambda_{LE}=0}}{\cos \Lambda_{LE}} \quad (9\text{-}32)$$

In practice, only half of this reduction is experienced.

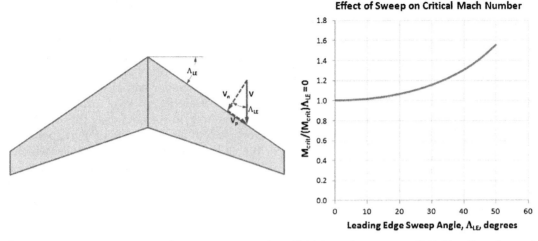

FIGURE 9-15 The deconstruction of the far-field into normal (n) and parallel (p) speed components (left). The effect of sweeping the leading edge is shown to increase the critical Mach number (right).

Impact of Sweep Angle on the C_{Lmax}

The maximum lift coefficient is reduced with an increase in wing sweep angle. This is detailed in Section 9.5.10, *Step-by-step: Rapid C_{Lmax} estimation*.

Impact of Sweep Angle on Structural Loads

The sweep has very important effect on structural loads. As discussed in Chapter 8, *The anatomy of the airfoil*, cambered airfoils inherently generate a pitching moment about their quarter chords whose tendency is to rotate the airfoil LE down. All flapped airfoils, cambered or not, generate extra pitching moment (torsion), which is added to the baseline moment. The total moment, regardless of composition, is reacted as shear flow in the wing structure. If the wing is swept aft, an additional and usually much larger pitching moment is generated because of the center of lift being moved a large distance aft. Consequently, the swept wing structure will be heavier than a straight wing designed for the same flight condition. Similarly, a forward-swept wing may also introduce a large LE up torsion, if the sweep angle is large enough, although this is reduced by an amount corresponding to the innate torsion of the airfoils of the wing.

Impact of Sweep Angle on Stall Characteristics

Sweep angle has very detrimental effects on the stall characteristics for reasons detailed in Sections 9.6.5, *Cause of spanwise flow for a swept-back wing planform* and 9.6.6, *Pitch-up stall boundary for a swept-back wing planform*.

9.3.4 Dihedral or Anhedral, Γ

The *dihedral* (or *anhedral*) is the angle the wing makes with respect to the ground plane or the x-y plane (as it is normally called) when viewing the airplane from the front (or back). A dihedral refers to the wingtip being higher (with respect to the ground) than the wing root. The opposite holds true for anhedral. The dihedral affects the lift of the configurations in two ways: due to the tilting of the lift force and how the dihedral changes the AOA of the wing. Consider a wing that has dihedral angle Γ subjected to an AOA given by α. The geometry of the configuration reveals that the AOA seen by the airfoil, and denoted by α_N, is reduced by the factor $\cos \Gamma$ (for instance, if $\Gamma = 0°$, then $\alpha_N = \alpha$ and if $\Gamma = 90°$, then $\alpha_N = 0°$, no matter the magnitude of α). Therefore, we can write:

Impact on *AOA*:

$$\alpha_N = \alpha \cdot \cos \Gamma \qquad (9\text{-}33)$$

Similarly, the lift generated in the plane of symmetry is the product of the lift normal to the wing surface, denoted by L_N, reduced by the same factor, $\cos \Gamma$.

Tilting of lift force:

$$L = L_N \cdot \cos \Gamma \qquad (9\text{-}34)$$

Since L_N can be written as $L_N = qSC_{LN} = qSC_{L\alpha}\alpha_N = qSC_{L\alpha}\alpha\cos \Gamma$, the lift in the plane of symmetry can be presented as follows:

Lift of a wing with dihedral:

$$L = qSC_{L\alpha}\alpha \cos^2 \Gamma \qquad (9\text{-}35)$$

where

$C_{L\alpha}$ = lift curve slope of the wing assuming $\Gamma = 0°$
q = dynamic pressure
S = reference wing area

This result has been experimentally confirmed, for instance see Ref. [7]. It allows wing lift to be determined in terms of variables that refer to a wing with no dihedral. Furthermore, see discussion of V-tails in Section 11.3.4, *V-tail or butterfly tail*. Typical wings feature a dihedral of $4°-7°$. The term $\cos^2 \Gamma$ amounts to 0.995 to 0.985 and, therefore, is usually ignored in the estimation of stability derivatives.

Dihedral plays an essential role in the generation of *dihedral effect*. It is discussed further in Section 4.2.3, *Wing dihedral*. Values for several classes of aircraft are given in Table 9-6.

9.3.5 Wing Twist — Washout and Washin, ϕ

Many aircraft feature wings that are twisted so the tip airfoil is at a different angle of incidence with respect to the fuselage than the root airfoil. This is called wing *washout* if the incidence of the tip is less than that of the root, and *washin* if it is larger. Washin is rarely used, but is mostly used to describe the aeroelastic effects high *AOA* has on forward-swept wings, where aerodynamic loads tend to twist the wing so the *AOA* at the tip is increased even further. This article, however, deals with the intentional twisting of the wing, where its purpose is generally twofold:

1. **To prevent the wingtip region from stalling before the wing root.** If the wingtip stalls before the root, there is an increased risk the aircraft will roll abruptly and uncontrollably at stall, a condition that may lead to incipient stall. Roll tendency at stall contributes significantly to fatal accidents[1] as it will often take place at low altitudes when an airplane banks to establish final approach before landing. An aircraft with such a roll-off tendency may stall if the pilot banks steeply and it may, ultimately, result in a crash as lack of altitude will prevent successful recovery. In the real world, manufacturing tolerances inevitably lead to airplanes not being perfectly symmetrical. This promotes one side of the airplane to stall before the other one, generating a rolling moment at stall. A wing twist can be used to build a "buffer" so the wingtips remain un-stalled when the rest of the wing stalls, greatly improving roll stability.
2. **To modify the spanwise lift distribution to help achieve minimum drag at mission condition.** The wing is most efficient when the section lift coefficients are constant along the span, as it is for an elliptical wing planform. Twisting the wing will allow the spanwise distribution of section lift coefficients to be modified for a tip-loaded wing, bringing the general distribution closer to that of an elliptical wing. A byproduct of this is a reduction in wing bending moments as the center of lift is brought closer to the plane of symmetry.

Geometric Washout

Geometric washout usually refers to the difference in angles of incidence of the root and tip airfoils (see Figure 9-16). It is denoted by ϕ_G. Generally the wing twist ranges from $0°$ to $-4°$, where the negative sign indicates that the leading edge of the tip is lower than that of the root. Thus, if we say the "washout is $3°$" we mean that $\phi_G = -3°$ and the LE of the tip is lower than that of the root. If we say the "washin is $3°$" we mean that $\phi_G = +3°$ and the LE of the tip is higher than that of the root. For instance, the inserted image of the root and tip airfoils in Figure 9-16 displays a washout. However, sometimes twist is highly complicated along the wingspan as in the case of the McDonnell-Douglas AV-8B, whose twist varies in a segmented fashion to a maximum of $-8°$ at the tip [8].

Note that if the twist is linear, the relative angle of twist at any spanwise station can be determined using the following expression:

$$\phi(y) = \frac{2y}{b}\phi_G \qquad (9\text{-}36)$$

The expression assumes the reference angle is 0 when $y = 0$ (the plane of symmetry) and becomes ϕ_G at the wingtip (where $y = b/2$).

Aerodynamic Washout

Aerodynamic washout is another way of designing roll stability at stall into the wing. In this case, two different airfoils are selected for the root and tip that are specifically based on one of two parameters: (1) their individual zero-lift *AOA* or (2) their two-dimensional stall *AOA* or maximum section lift coefficient. The former is favored when analyzing wings using methods such as the lifting line theory. Typically the idea is to provide roll stability at high *AOA* by ensuring the tip airfoil remains un-stalled before the inboard airfoil.

The primary reason for selecting an aerodynamic washout is that it allows the main wing spar caps to be straight, something that is important for the construction of composite aircraft. A spar cap in a composite aircraft mostly consists of unidirectional fibers, whose strength is very sensitive to fiber alignment. Once two airfoils have been selected, the aerodynamic washout is defined as follows (see Figure 9-17):

---

[1]For instance, see the *"Nall Report,"* an annual publication that evaluates aircraft accidents.

TABLE 9-6 Various Wing Layout Properties for Selected Aircraft (Based on Ref. [9])

Aircraft	Aspect Ratio AR	Taper Ratio λ	Dihedral Γ	Washout ϕ_G	Incidence (Root) i_{root}	Leading Edge Sweep Λ_{LE}
COMMERCIAL JETLINERS						
Airbus A300-600	7.7	-	-	-	-	28° (0.250C)
Airbus A310	8.8	0.260	11.13° (IB) 4.05° (OB)	-	5.05°	28° (0.250C)
Airbus A320-200	9.4	-	5.2°	-	-	25° (0.250C)
Boeing 707	7.056	0.275	7°	-	2°	35° (0.250C)
Boeing 727	7.2	0.304	3°	-	2°	32° (0.250C)
Boeing 737-100, -200	8.83	0.340	6°	-	1°	25° (0.250C)
Boeing 747-100, -200	6.96	0.245	7°	-	2°	37.5° (0.250C)
Boeing 757	7.77	0.211	5°	-	3.2°	25° (0.250C)
Boeing 767	7.9	0.267	6°	-	4.25°	31.5° (0.250C)
Fokker F28 Fellowship	8.4	0.239	2.5°	-	-	17.4° (0.250C)
McDonnell-Douglas DC-9/MD87	9.62	0.156	3°	-	1.25°	24.5° (0.250C)
McDonnell-Douglas DC-10-30/40	7.5	0.252	5.24° (IB) 3.03° (OB)	-	-	35° (0.250C)
MILITARY JET AIRCRAFT						
Aermacchi MB-339	6.1	-	-	-	-	11.3°
Aero L-39 Albatros	4.4	-	2.5°	-	2°	6.43°
BAC 167 Strikemaster	5.84	0.545	6°	3°	3°	-
Cessna A-37 Dragonfly	6.2	0.682	3°	2.63°	3.63°	0° (0.225C)
DBD Alpha Jet	4.8	-	−6°	-	-	28° (0.250C)
Hawker Siddeley Harrier	3.175	0.354	−12°	−8°	1.75°	34° (0.250C)
Microjet 200B	9.3	-	5.03°	-	3°	0° (0.300C)
TURBOPROPS, COMMUTER AND MILITARY						
ATR 42	11.08	0.549	2.5°	-	2°	3.1° (0.250C)
ATR 72	12.0	0.549	2.5°	-	2°	2.8° (0.250C)
De Havilland DHC-6 Twin Otter	10.1	1	-	-	-	-
De Havilland DHC-7 Dash 7	10.1	0.441	4.5°	-	3°	3.2° (0.250C)
De Havilland DHC-8 Dash 8	12.4	-	2.5°	-	3°	3.03° (0.250C)
Dornier Do 28 Skyservant	8.3	1	1.5°	-	4°	0°
Fokker F27 Friendship	12.0	0.404	2.5°	2°	3°	0° (0.250C)
IAI-101, 201 Arava	10	1	1.5°	-	0.45°	0°
LET L-410	10.79	-	1.75°	2.5°	2°	0° (0.250C)
Lockheed C-130 Hercules	10.09	0.852	2.5°	3°	3°	0° (0.250C)
Mitsubishi MU-2	7.71	-	0°	3°	2°	0.35° (0.250C)
SAAB 340	11.0	0.375	7°	-	2°	3.6° (0.250C)
Shorts SD3-30	12.3	1	3°	-	-	0°
Shorts Skyvan Series 3M	11	1	2.03°	-	2.5°	0°

TABLE 9-6 Various Wing Layout Properties for Selected Aircraft (Based on Ref. [9])—cont'd

Aircraft	Aspect Ratio AR	Taper Ratio λ	Dihedral Γ	Washout ϕ_G	Incidence (Root) i_{root}	Leading Edge Sweep Λ_{LE}
GA SINGLE-ENGINE, PISTON AND TURBOPROP						
Beechcraft Sierra/Sundowner	7.5	1	6.5°	2°	3°	0°
Beechcraft T-34C Mentor	6.22	0.412	7°	3°	4°	0° (0.250C)
Beechcraft V35 Bonanza	6.2	0.5	6°	3°	4°	0° (0.250C)
Cessna 150/152	6.7	0.687	1°	1°	1°	-
Cessna 172 Skyhawk	7.52	0.687	2.73°	3°	1.5°	-
Cessna 177 Cardinal	7.31	0.726	1.5°	3°	3.5°	0°
Cessna 182 Skylane	7.37	0.669	1.73°	3.62°	0.78°	0°
Cessna 208 Caravan	9.61	0.616	3°	3.22°	2.62°	-
Cessna 210 Centurion	7.66	0.726	1.5°	3°	1.5°	-
Cirrus SR20/22	10	0.5	4.5°	0°	0.25°	0° (0.250C)
Pilatus PC-6 Turbo-Porter	8.4	1	1°	-	2°	0°
Piper PA-28 Cherokee Arrow II	6.11	1	7°	-	2°	0°
Piper PA-28 Cherokee Warrior	7.24	0.669	7°	3°	2°	-
Piper PA-46 Malibu	10.57	-	4°	-	-	-
SIAI-Marchetti SF260	6.3	0.49	6.33°	2.75°	2.75°	0° (0.250C)
Valmet L-70 Miltrainer	6.62	-	-	-	-	-
Valmet L-90 Redigo	7.25	0.6	6°	3°	3°	-
Zlin 142	6.4	1	6°	-	-	−4.33°
GA TWIN-ENGINE						
Beechcraft B58 Baron	7.16	0.42	6°	4°	4°	-
Beechcraft B60 Duke	7.243	0.32	6°	4°	4°	-
Beechcraft B99 Airliner	7.51	0.5	7°	4.8°	4.8°	-
Britten-Norman BN-2A (standard)	7.4	1	0°	-	2°	0°
Cessna 310	7.3	0.674	5°	3°	2.5°	-
Cessna 337 Skymaster	7.18	0.667	3°	2°	4.5°	-
Partenavia P.68 Victor	7.7	1	1°	-	1.5°	0°
Piaggio P.180 Avanti	11.8	0.352	2°	-	0°	0° (0.150C)
Piper PA-31 Cheyenne	7.22	-	5°	2.5°	1.5°	0° (0.300C)
Piper PA-31-310 Navajo	7.22	-	-	1.0° (aero) 2.5° (geo)	-	-
BUSINESS JETS						
Cessna M550 Citation II	7.8	-	4°	3°	2.5°	-
Cessna M650 Citation III	8.94	-	3°	-	-	25° (0.250C)
Dassault Falcon 20	6.4	-	2°	-	1.5°	30° (0.250C)
Dassault Falcon 100	6.5	-	2°	-	1.5°	30° (0.250C)
Gates Learjet 55	6.72	0.391	-	-	-	13° (0.250C)

(Continued)

TABLE 9-6 Various Wing Layout Properties for Selected Aircraft (Based on Ref. [9])—cont'd

Aircraft	Aspect Ratio AR	Taper Ratio λ	Dihedral Γ	Washout ϕ_G	Incidence (Root) i_{root}	Leading Edge Sweep Λ_{LE}
EXPERIMENTAL AND HOMEBUILT						
Colomban MC 15 Cri Cri	7.75	1	4°	1.5°	1°	0°
Corby Starlet	5.00	1	6°	3.5°	2.5°	0°
Rutan Varieze	9.20	-	-	−3°	-	-
Verilite Model 100 Sunbird	7.56	1	2°	2°	2°	0°
SAILPLANES						
Rolladen-Schneider LS4	21.4	-	4°	-	-	0°
Schempp-Hirth Nimbus-3D	35.9	-	3°	-	1.5°	−2°
Schleicher ASK 21	16.1	-	4°	-	-	−1.5° (0.250C)
Schweizer SGM 2-37	18.1	-	3.5°	1°	1°	0°

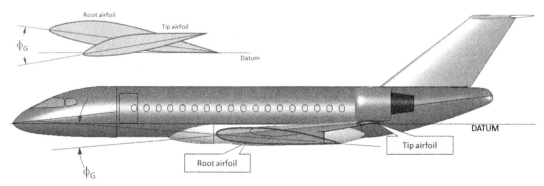

FIGURE 9-16 Definition of geometric washout.

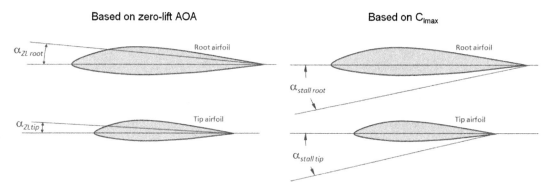

FIGURE 9-17 Definition of aerodynamic washout based on the stall AOA.

Twist based on zero lift AOA:

$$\phi_A = \alpha_{ZL_{root}} - \alpha_{ZL_{tip}} \quad (9\text{-}37)$$

Twist based on stall AOA:

$$\phi_A = \alpha_{stall_{root}} - \alpha_{stall_{tip}} \quad (9\text{-}38)$$

where

$\alpha_{ZL_{root}}$ = two-dimensional zero-lift AOA for the root airfoil

$\alpha_{ZL_{tip}}$ = two-dimensional zero-lift AOA for the tip airfoil

$\alpha_{stall_{root}}$ = two-dimensional stall AOA for the root airfoil

$\alpha_{stall_{tip}}$ = two-dimensional stall AOA for the tip airfoil

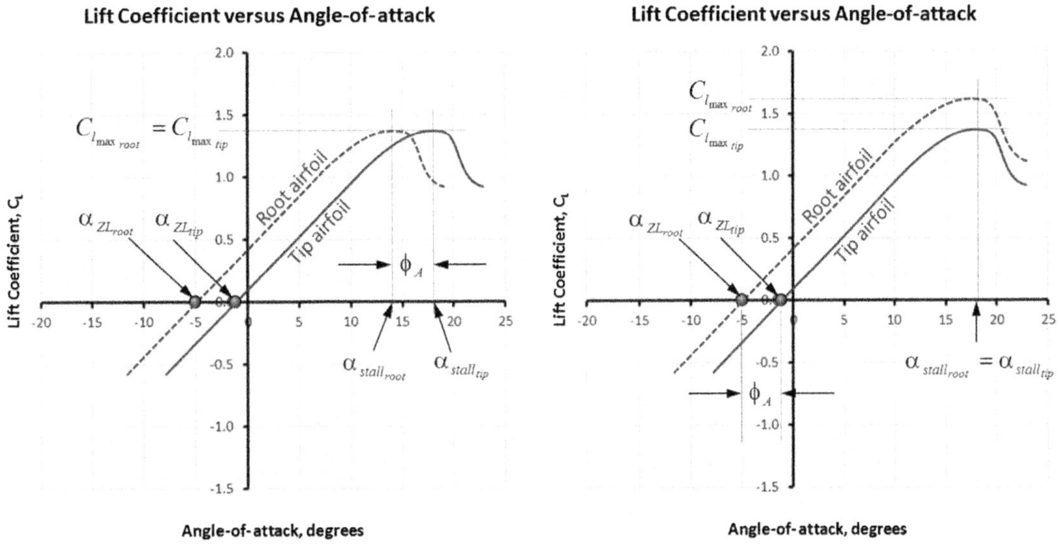

FIGURE 9-18 Idealized effect of an aerodynamic washout. In the left graph, the root airfoil stalls at a lower AOA than the tip providing roll stability at stall. A problem for the zero-lift definition of the aerodynamic washout is shown to the right. The root airfoil has a higher C_{lmax} than the tip airfoil but stalls at the same AOA; the effective washout is zero. This renders the definition of washout based on zero-lift angles misleading.

Note that there is an inherent problem with the definition based on Equation (9-37) and this is displayed in Figure 9-18. To begin with, the left graph of Figure 9-18 depicts the idea behind the aerodynamic washout: that the root airfoil stalls at an AOA less than the tip. It assumes that (1) $\alpha_{ZL\,tip} - \alpha_{ZL\,root} = \alpha_{stall\,root} - \alpha_{stall\,tip}$, which is not necessarily achieved in practice, and (2) requires $C_{Lmax\,tip} = C_{Lmax\,root}$ (otherwise assumption (1) will not hold). This allows the aerodynamic washout can effectively be represented as the difference in the zero-lift AOAs.

One of the problems with this definition is shown in the right graph of Figure 9-18, which depicts the common case of the root airfoil being thicker than the tip airfoil. Consequently, its C_{lmax} may be larger than that of the tip airfoil and its stall AOA is greater. Furthermore, if the wing is tapered, the difference in Reynolds number between the two can make the $C_{Lmax\,tip}$ and $\alpha_{stall\,tip}$ substantially smaller than that of the root. The figure effectively depicts a scenario in which the true washout is zero and possibly a washin, whereas relying on the zero-lift AOAs might indicate ample washout and therefore is misleading. Additional complexity must be considered in the magnitude of the section lift coefficients, which vary along the span. The above definition, thus, only gives a part of the whole picture; the remainder calls for analysis of the distribution of section lift coefficients along the span.

A Combined Geometric and Aerodynamic Washout

If two different airfoils are selected for the root and tip, in addition to a geometric washout, the combined effect can be calculated from:

$$\phi_A = \alpha_{stall_{root}} - \alpha_{stall_{tip}} + \phi_G \quad (9\text{-}39)$$

EXAMPLE 9-5

An airplane features two dissimilar airfoils at the root and tip whose $\alpha_{stall\,root} = 16.5°$ and $\alpha_{stall\,tip} = 15.0°$. What is the combined washout for a geometric washout of 0° and 3°?

Solution

Combined washout for $\phi_G = 0°$:

$$\phi_C = \alpha_{stall_{root}} - \alpha_{stall_{tip}} + \phi_G = 16.5° - 15.0° + 0°$$
$$= +1.5° \text{ (washin)}$$

Combined washout for $\phi_G = 3°$:

$$\phi_C = \alpha_{stall_{root}} - \alpha_{stall_{tip}} + \phi_G = 16.5° - 15.0° + (-3°)$$
$$= -2.5°$$

Recall that a negative angle means the root is at a greater AOA than that of the tip.

Panknin and Culver Wing Twist Formulas

The wing twist of a flying wing is such a fundamental parameter that it should be determined early on in the design phase — effectively as soon as the design lift coefficient has been determined. The designer of tailless aircraft can determine the proper washout of the wing using one of two special formulas specifically devised to help with the layout of such configurations. They are called the Panknin and Culver twist formulas. They are presented here without derivation for completeness of the discussion in this article.

The *Panknin twist formula* is attributed to Dr. Walter Panknin, who in 1989 presented a method to the designers of radio-controlled (RC) aircraft intended to help them properly lay out the washout of flying wings so the elevons are in trail at cruise. The formulation has been applied with great success to many RC flying wing designs [2]. It determines the geometric twist ϕ_G between the inboard and outboard airfoils of the wing (see Figure 9-19) using the following expression:

$$\phi_G = \frac{\left(K_1 C_{m_{root}} + K_2 C_{m_{tip}}\right) - C_{LC} \cdot K_{SM}}{1.4 \times 10^{-5} \times AR^{1.43} \times \Lambda_{C/4}} - \left(\alpha_{ZL_{root}} - \alpha_{ZL_{tip}}\right) \quad (9\text{-}40)$$

where

AR = wing aspect ratio
$C_{m_{tip}}$ = pitching moment coefficient of the tip airfoil
$C_{m_{tip}}$ = pitching moment coefficient of the tip airfoil
C_{LC} = lift coefficient at cruise (for which the aircraft is designed)
K_{SM} = fraction design static margin (e.g. if SM = 10% then K_{SM} = 0.1)

$$K_1 = \frac{3 + 2\lambda + \lambda^2}{4(1 + \lambda + \lambda^2)}$$

$K_2 = 1 - K_1$
$\alpha_{ZL_{root}}$ = zero-lift AOA for the root airfoil, in degrees
$\alpha_{ZL_{tip}}$ = zero-lift AOA for the tip airfoil, in degrees
$\Lambda_{C/4}$ = quarter-chord sweep angle, in degrees

The formula determines the geometric twist angle ϕ_G based on the selected static margin, but allows the designer to select airfoils, AR, and quarter-chord sweep angle. It is applicable to both forward- and aft-swept wings. Note that one must exercise care in the application of units for the quarter-chord sweep. Its units must be in terms of degrees.

The *Culver twist formula* is attributed to Mr. Irv Culver, a former engineer at Lockheed Skunk Works. Like the Panknin formula, his formula is widely used by designers of small RC tailless aircraft. It is intended for flying wings of moderate sweep angles (typically $\approx 20°$) and design lift coefficients of 0.9–1.2, where the lower value indicates a high-speed glider and the higher is for high-performance sailplanes [8]. Note that the presentation here differs slightly from that of Ref. [8] in the simplification of terms.

$$\phi_G = \pi \cdot \Lambda_{C/2} \left(\frac{AR}{AR+1}\right) \frac{C_{LC}}{C_{L_\alpha}} \quad (9\text{-}41)$$

where

AR = wing aspect ratio
C_{L_α} = lift curve slope
$\Lambda_{C/2}$ = center-chord sweep angle, in degrees or radians

Once the washout is known its distribution along the span can be found using the following expression:

$$\phi(y) = \phi_G (1-y)^{1 + AR/2\pi} \quad (9\text{-}42)$$

EXAMPLE 9-6

A flying wing is being designed for operation at a cruise C_L of 0.5 (C_{LC} = 0.5) at a static margin of 10% (K_{SM} = 0.1). It has been determined that its AR shall be 10, taper ratio (λ) shall be 0.5, and quarter-chord sweep angle, $\Gamma_{C/4}$, is 30°. During the airfoil selection process, the designers compare the use of a NACA 4415 airfoil (a conventional highly cambered airfoil, whose C_m is -0.1 and $\alpha_{ZL} = -4°$) and a NACA 0015 airfoil (whose C_m is 0.0 and $\alpha_{ZL} = 0°$). What is the required washout using each airfoil, in accordance with the Panknin formula?

Solution

The solution is effectively "plug and chug" into Equation (9-40). First, the parameters K_1 and K_2 must be determined.

$$K_1 = \frac{3 + 2\lambda + \lambda^2}{4(1 + \lambda + \lambda^2)} = \frac{3 + 2(0.5) + (0.5)^2}{4\left(1 + (0.5) + (0.5)^2\right)}$$
$$= 0.6071 \Rightarrow K_2 = 1 - K_1 = 0.3929$$

Using these with Equation (9-40) for the NACA 4415 airfoil leads to:

EXAMPLE 9-6 (cont'd)

$$\phi_G = \frac{\left(K_1 C_{m_{root}} + K_2 C_{m_{tip}}\right) - C_{L_{cruise}} \cdot K_{SM}}{1.4 \times 10^{-5} \times AR^{1.43} \times \Lambda_{C/4}} - \left(\alpha_{ZL_{root}} - \alpha_{ZL_{tip}}\right)$$

$$= \frac{(0.6071 \times -0.1 + 0.3929 \times -0.1) - 0.5 \times 0.1}{1.4 \times 10^{-5} \times 10^{1.43} \times 30°}$$

$$-(-4° + 4°) = -13.3°$$

Using these with Equation (9-40) for the NACA 0015 airfoil leads to:

$$\phi_G = \frac{\left(K_1 C_{m_{root}} + K_2 C_{m_{tip}}\right) - C_{L_{cruise}} \cdot K_{SM}}{1.4 \times 10^{-5} \times AR^{1.43} \times \Lambda_{C/4}} - \left(\alpha_{ZL_{root}} - \alpha_{ZL_{tip}}\right)$$

$$= \frac{(0 + 0) - 0.5 \times 0.1}{1.4 \times 10^{-5} \times 10^{1.43} \times 30°} - (0° + 0°) = -4.42°$$

The analysis shows that using a conventional cambered airfoil like the NACA 4415 is a bad choice, as it would require a 13.3° washout. Using a symmetrical airfoil like the NACA 0015 would bring this down to 4.42°.

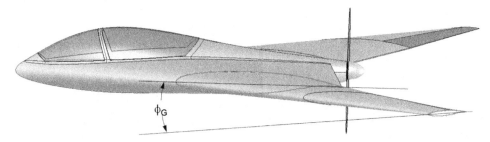

FIGURE 9-19 The Panknin and Culver formulas apply to the geometric washout of tailless aircraft.

9.3.6 Wing Angle-of-incidence, i_W

Once the general shape of the fuselage, wing, and horizontal tail has been selected, it is time to decide on the relative orientation of the three. Consider a commercial jet designed to cruise at a specified airspeed, V_C, and altitude (e.g. $M = 0.8$ at 35,000 ft). Once the desired cruise altitude has been reached, we note the weight of the airplane at the beginning of its cruise segment as W_1. We also note its weight at the end of the cruise segment as W_2. It should be evident that if the airplane is operated using conventional fossil fuel, then W_1 will be greater than W_2. This is particularly true for jet aircraft as large quantities of fuel are consumed en route. Consequently, the airplane will initially be cruising at a higher AOA than at the end of its mission. Keeping the change in AOA in mind it is prudent to determine the AOA at mid-cruise and use this as a representative AOA for the entire cruise segment. This AOA will be closer to both the initial and final AOAs than if either the initial or final AOA were selected.

Now consider Figure 9-20, which shows three possible orientations of a fuselage during the cruise. The top fuselage has its nose too low, the center too high, and the bottom one shows the correct orientation. Note that all three fuselage placements show a representative root airfoil mounted at the same cruise AOA (or α_C) – it is only the fuselage that is mounted differently. Note that it is more convenient to refer the wing incidence to the root airfoil rather than the MGC airfoil, because this is usually directly referred to in the fuselage lofting.

The two top schematics in Figure 9-20 show fuselage and wing configurations that result in a higher total airplane drag than the bottom one. It is the responsibility of the designer to determine the optimum AOA for the fuselage and make sure this is the orientation of the fuselage at the selected cruise mission point. This AOA may be based on the minimum drag position of the fuselage, or its maximum lift-to-drag ratio, or its maximum lift contribution when combined with the wing. In other words, the optimum AOA of the fuselage should be determined as the one that maximizes the efficiency of the airplane as a whole. For instance, a minimum drag AOA for a typical commercial jetliner fuselage would be close to being parallel to the flight path during the design cruise condition and then the wing should be mounted at its required representative cruise AOA at that condition. The determination of the wing AOI can be considered a four-step process. The first step is to determine the fuselage minimum drag AOA, the second is to determine the required cruise AOA, the third is to add the two together, and the fourth is to determine the horizontal incidence angle that minimizes trim drag.

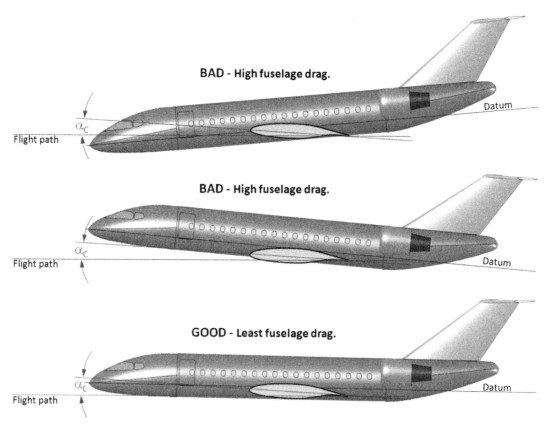

FIGURE 9-20 Determination of the fuselage minimum drag position. The two top configurations will generate greater drag at cruise than the bottom one, which places the fuselage in its minimum drag orientation at cruise. For a conventional pressurized aircraft the centerline of the fuselage will be close to parallel to the flight path during cruise.

Step 1: Determine the Optimum AOA for the Fuselage

The first step will be highly dependent on the geometry of the aircraft, although when considering fuselages for conventional commercial aircraft it is good to assume it is parallel to the centerline of the tubular structure that forms the passenger cabin (see the datum in Figure 9-20). A wind tunnel testing or a CFD analysis should be used to determine the optimum orientation of the fuselage as its influence on the three-dimensional flow field may be quite complicated. The result from this step would be denoted by α_{Fopt}, for AOA of optimum fuselage orientation, using the following stipulation:

(1) If the optimum AOA results in the fuselage having a nose-down attitude, then let $\alpha_{Fopt} > 0$.
(2) If the optimum AOA results in the fuselage having a nose-up attitude, then let $\alpha_{Fopt} < 0$.

Step 2: Determine the Representative AOA at Cruise

The value of the midpoint cruise AOA, α_C, can be determined from Equation (9-43) below:

$$\alpha_C = \left(\frac{1}{C_{L_\alpha}}\right) \frac{(W_1 + W_2)}{\rho V_C^2 S} + \alpha_{ZL} \qquad (9\text{-}43)$$

where

$C_{L\alpha}$ = lift curve slope
S = reference wing area
α_{ZL} = zero lift angle-of-attack
ρ = density at cruise altitude

Step 3: Determine the Recommended Wing Angle-of-incidence

The angle-of-incidence (see Figure 9-21) at the root airfoil, denoted by i_W, can now be determined from:

$$i_W = \alpha_C + \alpha_{Fopt} - \Delta\phi_{MGC} \qquad (9\text{-}44)$$

where

i_W = recommended AOI
α_C = wing angle-of-attack at cruise midpoint
α_{Fopt} = fuselage minimum drag angle-of-attack
$\Delta\phi_{MGC}$ = correction to account for wing twist (see below)

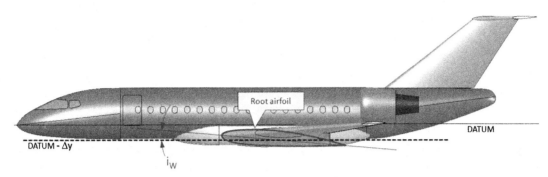

FIGURE 9-21 Definition of a wing incidence angle.

Note that since the first two terms of the above expression really return the AOI at the MGC rather than at the root, a correction must be introduced if the wing features a washout. This is denoted by the term $\Delta\phi_{MGC}$, which is the difference in angular incidences between the root and MGC airfoils. If the wing has a linear twist the value of $\Delta\phi_{MGC}$ can be determined as follows by inserting the expression for the spanwise location of the MGC, y_{MGC}, of Equation (9-8) into Equation (9-36). Note that the minus sign in Equation (9-44) is required to ensure that a washout will be added to determine:

$$\Delta\phi_{MGC} = \left(\frac{1+2\lambda}{3+3\lambda}\right)\phi_G \quad (9\text{-}45)$$

Step 4: Determine the Recommended HT Angle-of-incidence

Once a recommended wing AOI has been determined, the incidence of the HT can be determined. Again using the representative cruise mission weight $(W_1 + W_2)/2$ the horizontal tail should be mounted so the elevator deflection at this point will be neutral. This is called flying the horizontal in trail and simply means neutral deflection. This will result in a minimum trim drag.

Derivation of Equation (9-43)

Assuming the general lift properties of the wing are known, the lift coefficient can be written as follows combining either Equation (9-43) or (9-50):

$$C_L = C_{L0} + C_{L\alpha}\alpha = \frac{2W}{\rho V^2 S} \quad (i)$$

The weight of the airplane at the midpoint during the cruise is the average of the initial cruise weight, W_1, and the weight at the end of cruise, W_2:

$$C_L = C_{L0} + C_{L\alpha}\alpha_C = \frac{2\left(\frac{W_1+W_2}{2}\right)}{\rho V^2 S} = \frac{(W_1+W_2)}{\rho V^2 S} \quad (ii)$$

Where α_C is the midpoint cruise AOA. Solving for α_C leads to:

$$C_{L\alpha}\alpha_C = \frac{(W_1+W_2)}{\rho V^2 S} - C_{L0} \Leftrightarrow \alpha_C = \frac{(W_1+W_2)}{\rho V^2 S C_{L\alpha}} - \frac{C_{L0}}{C_{L\alpha}}$$

The last term is simply the zero lift AOA, α_{ZL}, which can be seen from:

$$C_{L0} + C_{L\alpha}\alpha = 0 \Rightarrow C_{L\alpha}\alpha = -C_{L0} \Rightarrow \alpha_{ZL} = -\frac{C_{L0}}{C_{L\alpha}}$$

Therefore, we can write:

$$\alpha_C = \frac{(W_1+W_2)}{\rho V^2 S C_{L\alpha}} - \frac{C_{L0}}{C_{L\alpha}} = \left(\frac{1}{C_{L\alpha}}\right)\frac{(W_1+W_2)}{\rho V^2 S} + \alpha_{ZL}$$

QED

Derivation of Equation (9-45)

Assuming the general lift properties of the wing are known, the lift coefficient can either be written as follows combining Equations (9-43) or (9-50):

$$\Delta\phi_{MGC} = \phi(y_{MGC}) = \frac{2y_{MGC}}{b}\phi_G$$
$$= \left(\frac{2}{b}\right)\left(\frac{b}{6}\right)\left(\frac{1+2\lambda}{1+\lambda}\right)\phi_G = \left(\frac{1}{3}\right)\left(\frac{1+2\lambda}{1+\lambda}\right)\phi_G$$
$$= \left(\frac{1+2\lambda}{3+3\lambda}\right)\phi_G$$

QED

EXAMPLE 9-7

A reconnaissance aircraft is being designed and is expected to weigh 20,000 lb$_f$ at the start of its design mission cruise segment at a cruising speed V_C = 250 KTAS at 25,000 ft, and 15,500 lb$_f$ at the end. Determine a suitable angle-of-incidence for its wing if the following parameters have been determined (ignore compressibility effects): where

$C_{L\alpha}$ = lift curve slope = 4.2 per radian
b = reference wingspan = 52.9 ft
S = reference wing area = 350 ft^2
α_{ZL} = zero lift angle-of-attack = $-2.5°$
α_{Fopt} = fuselage optimum angle-of-attack = 1.2° nose-up
λ = Wing taper ratio = 0.5
ϕ_G = wing washout = $-3.0°$

Solution

Step 1: Establish the optimum *AOA* for the fuselage. Since the fuselage minimum drag position is 1.2° nose up, we set $\alpha_{Fopt} = -1.2°$.

Step 2: Determine the representative *AOA* at the midpoint of the cruise segment (note that density at 25,000 ft is 0.001066 slugs/ft^3).

$$\alpha_C = \left(\frac{1}{C_{L\alpha}}\right)\frac{(W_1 + W_2)}{\rho V_C^2 S} + \alpha_{ZL}$$
$$= \left(\frac{1}{4.2 \times \pi/180}\right)\frac{(20000 + 15500)}{(0.001066)(250 \times 1.688)^2(350)}$$
$$- 2.5° = 4.789°$$

In other words, the MGC should be approximately at an *AOA* of 4.8° at the midpoint of the cruise. On a completely different note, at an initial wing loading of 20,000 lb$_f$/350 ft^2 = 57.1 lb$_f$/ft^2, the designer may want to reconsider whether this airplane should have a greater wing area. Let us not digress.

Step 3: Determine the recommended AOI based on the minimum drag configuration of the fuselage at the cruise midpoint. First, calculate the correction angle since the wing has a 3° washout.

$$\Delta\phi_{MGC} = \phi(y_{MGC}) = \left(\frac{1+2\lambda}{3+3\lambda}\right)\phi_G$$
$$= \left(\frac{1+2(0.5)}{3+3(0.5)}\right)(-3°) = -1.333°$$

Then, finish by calculating the recommended AOI for the wing root chord as follows:

$$i_W = \alpha_C + \alpha_{Fopt} - \Delta\phi_{MGC} = 4.789° - 1.2° - (-1.333°)$$
$$= 4.922°$$

It can be seen that the *AOA* of the MGC is given by α_C and is 4.789° and if the fuselage optimum *AOA* were 0°, rather than $-1.2°$, the recommended AOI would be 4.789° $-$ ($-1.333°$) = 6.122°. The resulting geometry is shown in Figure 9-22.

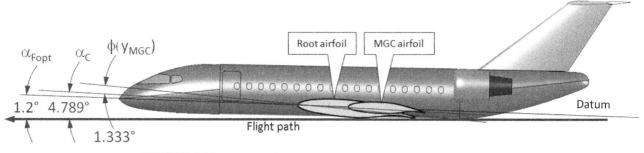

FIGURE 9-22 Wing incidence angle determined for the example aircraft.

Decalage Angle for a Monoplane

On a monoplane, a *decalage angle* is the difference between the incidence angles of the wing and horizontal tail: Figure 9-23. The angle is an important indicator of the stability of the aircraft and generally requires substantial stability analyses to determine. Strictly speaking, the AOI of the HT, i_{HT}, is determined based on the wing AOI, i_W, and

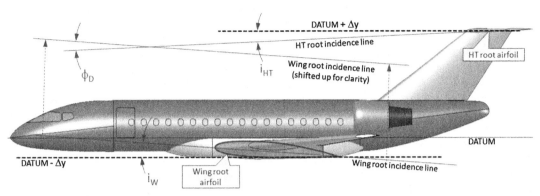

FIGURE 9-23 Definition of a decalage angle for a monoplane.

then the decalage angle can be calculated as shown below:

$$\phi_D = i_W - i_{HT} \qquad (9\text{-}46)$$

9.3.7 Wing Layout Properties of Selected Aircraft

The purpose of Table 9-6 is to present the reader with typical values of some of the wing parameters presented above to help assessing appropriate values.

9.4 PLANFORM SELECTION

As stated in the introduction to this section, a glance at the history of aviation reveals a large number of planform shapes have been used. The success of some is glaring (e.g. constant-chord, tapered, swept back), while others have been a disappointment (e.g. triplanes, disc-shaped, circular, channel wings). This section evaluates the impact of varying the wing planform shape on the wing's capability to generate lift and to lose it at stall.

For convenience, the distribution of the section lift coefficients for the planform shapes will be compared to that of a rectangular constant-chord wing. This is done to help the reader realize the impact of selecting a particular geometry. All the planform shapes feature the same reference area (10 ft^2), the same airfoil (NACA 4416), and are exposed to airspeed of 100 KCAS at a 10° angle-of-attack. They should be thought of as a wing planform study for a generic airplane, so the physical geometry should be ignored and they should rather be evaluated in terms of parameters such as their aspect ratio (AR), taper ratio (TR), and leading edge sweep angles. Wing dihedral and washout is 0° for all examples. The AR and TR are varied to ensure all the examples have an equal area, while featuring values that are representative of typical airplanes. This gives an excellent insight into the lifting capabilities of the planform. Examples of aircraft that use the said planform shapes are cited as well. The reader should be aware that the true maximum lift coefficient of each wing planform shape will ultimately depend on the airfoil selection and various viscous phenomena. Such effects are too complicated to deal with in general terms in this text and may have a profound impact on the actual stall characteristics.

9.4.1 The Optimum Lift Distribution

Generally, the objective during planform selection is to select geometry that (1) generates lift through an effective use of the available span, (2) does not generate excessive bending moment, (3) results in docile stall characteristics, and (4) offers acceptable roll responsiveness. Let's consider these in more detail.

Figure 9-24 shows the distribution of section lift coefficients, C_l, along the span of some arbitrary wing planform. The figure shows the frontal view of a cantilevered wing (the left wing is shown, looking toward the leading edge). The plane of symmetry (left) is where the wing root would be located and the right side is the left wingtip. The figure shows two kinds of distribution of C_l. The

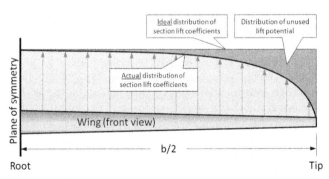

FIGURE 9-24 Ideal, actual, and wasted lift distributions.

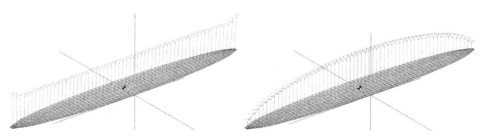

FIGURE 9-25 The difference between the distribution of section lift coefficients (left) and lift force (right). The distribution of section lift coefficients is indicative of stall tendency and induced drag coefficient. The lift force distribution is important for structural issues.

first can be considered an *ideal* spanwise distribution, which would be achieved if the laws of physics didn't require the lift to gradually go to zero at the wingtip. This distribution would result in each spanwise station contributing uniformly to the total lift coefficient. As a consequence, it would require the least amount of AOA at any given airspeed to maintain altitude. And as is shown in Chapter 15, the less the AOA, the less is the generation of lift-induced drag.

Figure 9-24 also shows the *actual* distribution of C_l. It represents the true section lift coefficients generated along the wing and it differs from the ideal distribution. Thus, the area between the ideal and actual distributions represents the distribution of *unused* lift potential. The smaller this area, the more efficient is the wing. The larger this area, the higher will be the stalling speed of the aircraft, because the missing lift must be made up for with dynamic pressure (i.e. airspeed) at stall. Similarly, the larger this area, the greater will be the induced drag at cruise, because the missing lift, again, must be made up by a larger AOA (which implies an increase in pressure and induced drag). In short, the design goal of the lifting surfaces should always be to minimize the wasted lift distribution.

Unfortunately, things are a bit more complicated than reflected by this. It turns out that wings designed to generate as uniform a distribution of C_ls as possible have a serious side effect: stall. Theoretically, if such a wing uses an identical airfoil (ignoring the viscous effects), every spanwise station stalls at the same instant. In real airplanes this causes the wing to roll to one side or the other, depending on factors like yaw angle, propeller rotation, deviations from the ideal loft, and others. A wing roll-off at stall can initiate a spin — a very dangerous scenario if the airplane is close to the ground, such as when maneuvering (banking) to establish final approach before landing. For this reason, some of the efficiency must be sacrificed for added safety in the handling the airplane, using techniques such as wing washout, dissimilar airfoils, discontinuous leading edges, and many others.

It turns out that there actually is a wing planform that, at least in theory, achieves the ideal lift distribution; the elliptic planform (see Section 9.4.4, *Elliptic planforms*). The planform will generate section lift coefficients that are uniform along the entire span (see Figure 9-25). Unfortunately, as always, there is a catch — a very serious catch. The uniform distribution achieved by the elliptic planform is great for cruise, but awful near stall. As the elliptic planform reaches higher and higher AOA, the wing will stall fully — instantly, rather than progressing gradually from the inboard to the outboard wing, something that helps to maintain roll stability. This means that if there are manufacturing discrepancies — and there always are manufacturing discrepancies — the left or right wing may stall suddenly before the opposite wing. The consequence is a powerful wing roll to the left or right. This tendency can be remedied by a decisive wing washout; however, the resulting distribution of section lift coefficients will no longer be uniform. It is also possible to help stall progression with stall strips; however, whatever the remedy, the project will possibly be hampered by a troublesome stall improvement development during flight testing.

Another important observation can also be made from Figure 9-25. This is the difference between the distribution of section lift coefficients and that of lift force. While the distribution of section lift coefficients is a very important indicator of stall tendency and induced drag coefficient, the lift force distribution is imperative for structural reasons. The vertical shear force can be determined by integrating the distributed lift force along the span. Bending moments are obtained by integrating the shear along the span. With this in mind, the right graph of Figure 9-14 effectively shows how low taper ratio reduces the wing bending moments by moving the center of the lift inboard. For this reason tapered wings are essential to help reduce structural weight.

All the following images assume the leading edge faces up and the trailing edge down, and simple expressions of wing area and AR are provided for convenience. Always be mindful of the difference in Reynolds numbers for tapered wing planform shapes (root R_e is different from tip R_e). This comparison is implemented

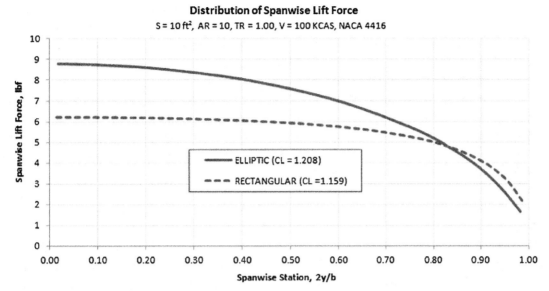

FIGURE 9-26 Comparison of spanwise force generated by a rectangular and elliptic planforms.

via the vortex-lattice method (VLM) using the commercially available code SURFACES [10]. All the models have 720 panels and are aligned along the camber line of the NACA 4416 airfoil.

9.4.2 Methods to Present Spanwise Lift Distribution

The lift distribution is usually presented in three forms: force distribution, section lift distribution, and a distribution of the fraction of section lift coefficients to the total lift coefficient generated by the wing. All have their advantages and shortcomings. These are compared in Figure 9-26, Figure 9-27, and Figure 9-28 using the rectangular and elliptic wing planform shapes of the previous section.

Figure 9-26 shows the distribution of lift as a force along the span. The area under the graph is the shear force along the span and, assuming a cantilevered wing, the progression of integration from tip inboard to root will provide the shear diagram. Then, integrating the shear diagram from tip to root will generate the moment diagram. Note that sometimes this graph is presented using the physical dimensions of the wingspan (e.g. $b = 35$ ft or similar). In this case, the units for the y-axis of the graph are more correctly represented as

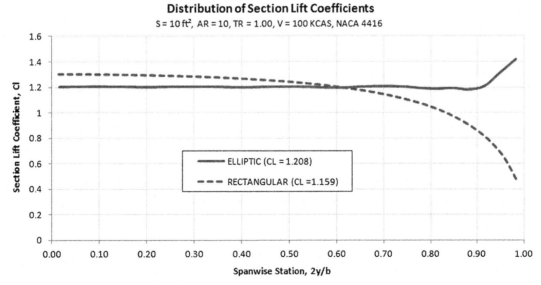

FIGURE 9-27 Comparison of section lift coefficient.

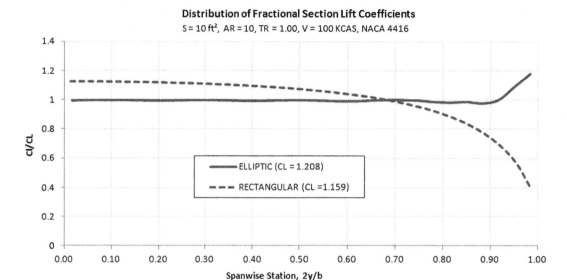

FIGURE 9-28 Comparison of fractional section lift coefficient.

lb$_f$/ft (assuming the UK system is being used). Another important observation is the difference in the force generated by the elliptical wing planform when compared to the rectangular form.

Figure 9-27 shows the distribution of section lift coefficients along the span. The advantage of this presentation is that it can be used to evaluate the magnitude of geometric (or aerodynamic) twist required to help control stall progression or even bending moments. Since an airfoil will stall at a specific two-dimensional lift coefficient and this format presents the current two-dimensional lift coefficient (i.e. the section lift coefficient) this can be used to design good stall characteristics into the airplane (see Section 9.6.4, *Tailoring the stall progression*).

Figure 9-28 shows the distribution of the contribution of the section lift coefficient to the total lift generated by the geometry. It helps demonstrate regions that contribute a lot or little to the overall lift.

9.4.3 Constant-chord ("Hershey-bar") Planform

Nicknamed for its simple constant-width geometry, this planform is widely used for many different kinds of aircraft, small and large. Light planes that feature Hershey-bar wings include the Beech Model 77 Skipper, Piper PA-38 Tomahawk, Piper PA-28 Cherokee, and Piper J-3 Cub. Among larger aircraft are the de Havilland of Canada DHC-6 Twin Otter and the Shorts 360.

The distribution of spanwise section lift coefficients, C_l, for a Hershey-bar wing is shown in Figure 9-29. This is done for two *AOAs*; 0° and 10°. For comparison purposes, the latter curve will be superimposed on all subsequent planform shapes so the reader can get a quick glimpse of how the lift distribution of other wing planform shapes differs from that of the constant-chord wing.

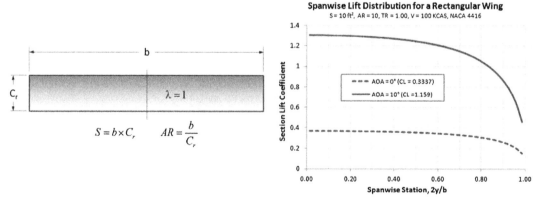

FIGURE 9-29 Basic geometry and lifting characteristics of the *constant-chord* wing planform.

9.4 PLANFORM SELECTION

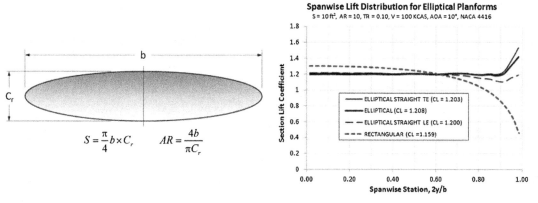

FIGURE 9-30 Basic geometry and lifting characteristics of the *elliptical* wing planform.

Pros

The configuration has two primary advantages. The first is forgiving stall characteristics, attributed to the reduction in section lift coefficients toward the wingtip (see Figure 9-29). This promotes a very favorable stall progression (growth of flow separation) that begins at the root and moves outboard to the tip, rendering it the last part of the wing to stall. This effect is very desirable as it gives the configuration important roll stability at stall.

The other important advantage is the lower manufacturing costs because all ribs have the same basic geometry and the spar is a constant-thickness beam. This simplicity affects not only the wing ribs and spar, but also the control surfaces.

This configuration is ideal for trainer aircraft or airplanes for which manufacturing cost is imperative.

Cons

Since the wingtip contributes less to the lift generation than the root (in terms of section lift coefficients), the planform is particularly inefficient. For this reason, the planform should never be used for efficient aircraft, such as sailplanes or long-range aircraft. This includes wings and stabilizing surfaces.

9.4.4 Elliptic Planforms

The elliptical wing planform (see Figure 9-30) is most famous for its use in one of the most formidable fighters of the Second World War, the British Supermarine Spitfire. Other aircraft that feature this configuration include: the American Republic P-47 Thunderbolt; and the German Heinkel He-70 He-112B, Mudry CAP-10, and CAP-20. Interestingly, the aerodynamic properties of the planform are widely publicized, even among some laypeople.

The planform shown in Figure 9-30 is a pure ellipse, which means that a straight line extending from tip-to-tip is located at the 50% chord. However, there is no requirement that an elliptic planform has to comply with that geometry. For instance, it would be structurally practical to design it so the quarter-chord was a straight line. This would simplify the design by allowing a straight main spar (albeit with a curved spar height) to be positioned at the quarter-chord.

Elliptical planforms with a straight leading edge or trailing edge (see Figure 9-31) are common on radio-controlled aircraft and sailplanes and are sometimes referred to as crescent-shaped wings. The graph of Figure 9-31 shows that, fundamentally, the distribution of section lift coefficients for the three planform shapes is the same, except at the tip. While all three planform shapes generate a similar total C_L at 10°, the one with the straight LE generates a lower section lift coefficient at the tip than the others, implying a less severe washout is required to improve stall characteristics. However, one must be careful in the interpretation of such linear curves because the sharp tip introduces important viscous effects at higher *AOA*s that render the linear

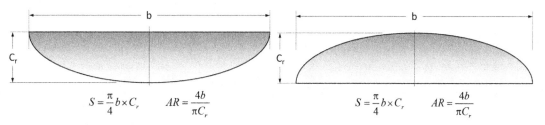

FIGURE 9-31 Basic geometry of *elliptical* planform with a *straight LE* and *TE*.

predictions invalid. These are discussed in Section 9.6.3, *Deviation from generic stall patterns*.

Pros

Its primary advantage is the uniform distribution of section lift coefficients (see Figure 9-30). This distribution makes the planform very efficient when it comes to utilizing the entire span and results in the least amount of lift-induced drag of any planform.

Cons

The planform's greatest drawback is its producibility, which is severely impaired by the complex compound surfaces. It is very difficult to manufacture using aluminum, as this would call for sheets to be stretched through hydroforming (or similar). However, it is much easier to produce using modern composites.

Another serious drawback is that the section lift coefficients are mostly uniform along the span, which causes the entire wing to stall at once (again assuming a constant airfoil and no washout). This can pose a serious problem for low-speed (or high *AOA*) operations and requires a decisive washout or airfoil selection to remedy.

9.4.5 Straight Tapered Planforms

A straight tapered wing planform is one for which the root chord is different from the tip chord, such that the smaller chord resides inside two lines that are perpendicular to the plane of symmetry and that are drawn at the LE and TE of the larger chord (see Figure 9-32).

Some authors define straight wings as those that have zero sweep on any spanwise line between 25% and 70% chord. The drawback with that definition is that it excludes common GA aircraft that feature a straight LE or TE.[2]

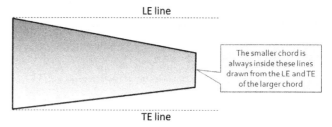

FIGURE 9-32 Definition of a straight tapered wing planform.

The list of airplane types that use this wing planform shape is very long. A generic tapered planform and its spanwise distribution of section lift coefficient is shown in Figure 9-33.

Pros

The primary advantage of tapered wings is reduction in bending moments and lift-induced drag. Straight tapered wings offer improved efficiency over the Hershey-bar wing as the section lift coefficients are higher toward the tip. Consequently, the wingtips contribute more to the total lift coefficient. This, combined with the relative geometric simplicity of the linear taper, which is easier to manufacture than the elliptical planform, renders the planform ideal for most airplanes. The improved efficiency of the configuration usually warrants the increased production complications.

Cons

The taper compromises the stall characteristics and requires a geometric or aerodynamic wing washout, or a combination of both, to be employed. An additional solution might be to select an airfoil for the tip that has a higher stall *AOA* than the root. It is common to locate root and tip airfoils such that their quarter-chords are on a line perpendicular to the plane of symmetry.

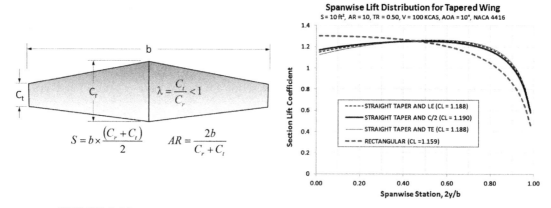

FIGURE 9-33 Basic geometry and lifting characteristics of *straight tapered* wing planforms.

---

[2]For example, it excludes a great many aircraft; a family of Zlin and Mooney aircraft, to name a few.

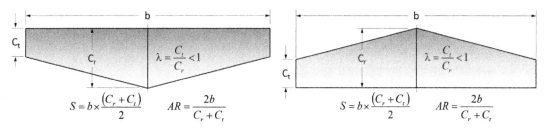

FIGURE 9-34 Basic geometry of the *straight-LE tapered* wing planform.

This allows a straight main spar to be placed at the quarter-chord of the wing, but this is an effective location as it results in good structural depth that provides the potential for reasonably large fuel tanks behind the spar. The configuration causes manufacturing complications due to the different geometry of each rib.

Straight Leading or Trailing Edges

Other airplanes feature straight tapered wings whose leading or trailing edge is perpendicular to the plane of symmetry (see Figure 9-34). This means that a spanwise line going through the quarter-chord of the root and tip is swept forward or back.

Pros

Improved structural and aerodynamic efficiency over the constant-chord configuration. The straight LE planform may be advantageous if it is foreseen that the operational CG will be too far forward. The opposite holds for an operational CG that turns out to be too far aft.

Cons

It is a drawback that if the wing features a spar that is perpendicular to the plane of symmetry and extends from tip to tip, its caps will be curved. This can be eliminated by sweeping the spar forward, which makes the use of a single-piece spar more challenging.

The hinge line of control surfaces can have a significant forward sweep, which makes them less efficient aerodynamically and can complicate the design of the control system. Among a number of examples of aircraft that feature the straight LE are the Arado Ar-79, Cessna 177 Cardinal, Commonwealth Ca-12 Boomerang, and Fairey Barracuda, and a series of Mooney aircraft feature the planform shape for both the wing and stabilizing surfaces.

On the Mooney aircraft, the straight LE of the horizontal and vertical tails arguably renders them more efficient aerodynamically than the more popular aft-swept tails and, in theory, this allows the tail be a tad smaller, with less wetted area. This is an argument often heard among laypeople. However, an investigation reveals this is not true. Some of that efficiency is simply lost in lowered control authority of the less effective control surfaces whose hinge lines are inevitably swept forward. Of course, the same argument can be made for aft-swept stabilizers – they also have hinge lines that are swept aft. The difference is that the aft sweep adds arm that makes up for the loss in aerodynamic efficiency. A VLM analysis of the Mooney M-20R revealed that its directional stability derivative, $C_{n\beta}$, is in the range of 0.054 per radian. Introducing a 45° aft sweep in the LE of the VT (no change in surface or wetted area and root chord precisely the same place) increased it to 0.060 per radian, directly contradicting any claims of greater effectiveness of the straight LE tail. This simply means that the reduction in $C_{L\alpha}$ of the VT reduces less than the tail arm increases. The important point is that in the case of the Mooney, the tail is a signature feature for the brand. One can always recognize a Mooney tail and for the brand this is important. But more effective it is not.

Another version of the straight tapered wing is the straight trailing edge planform in Figure 9-34. Many of the same arguments for or against hold for this planform. It is featured on a number of aircraft, among others the Aermacchi MB-326 and MB-339, Bücker Bü-180 Student, Zlin Z-526, 726, and Z-26.

The Compound Tapered Planform

The compound tapered wing planform consists of two tapered sections; the inboard section is an inverse taper and the outboard section has standard taper (see Figure 9-35). The lift distribution for this planform is also shown. It can be seen that section lift coefficients peak inboard, indicating this is where stall would initiate. This is significantly reduced midspan and then increases toward the tip. The analysis indicates the planform is not particularly efficient although the discontinuity at the leading edge may provide some viscous benefits. Among well-known aircraft featuring this planform is the Westland Lysander, a British single-engine observation aircraft with short take-off and landing (STOL) designed in the 1930s. The reason for this choice was pilot visibility – its STOL capability can be attributed to the high-lift system it was equipped with and not the planform itself.

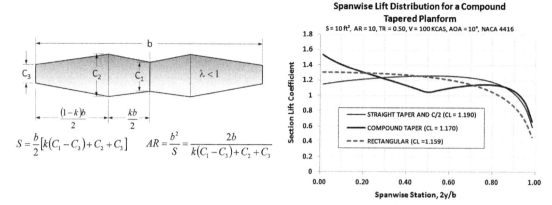

FIGURE 9-35 Basic geometry and lifting characteristics of the *compound tapered* wing planform.

9.4.6 Swept Planforms

Like the tapered wing planform, the swept planform is one of the most common types of geometry found in aviation. Practically all commercial aircraft use swept wings, and even aircraft that operate at low subsonic speeds commonly use the planform shape for their stabilizing surfaces. For this reason, this geometry must be elaborated on in some detail.

The Aft-swept Planform

The invention of the planform is generally thought to date back to the 1930s. However, it was not possible without the work of scientists who preceded the era. Meier [11] takes this history back to Isaac Newton (1643−1727), Pierre-Simon, Marquis de Laplace (1749−1827), Ernst Mach (1838−1916), and many others, up to Adolf Busemann (1901−1986), who is generally credited with the invention of the swept-back wing. Busemann belonged to a group of very famous German scientists who were led by Ludwig Prandtl (1875−1953). The best known were Theodore von Kármán (1881−1963), Max M. Munk (1890−1986), and Jakob Ackeret (1898−1981) [12]. The result of Busemann's work was first published at the 5th Volta conference in Rome in 1935 [13]. The Volta conference was an annual conference of physics, mathematics, history, and philosophy, named to honor the great Italian physicist Alessandro Volta (1745−1827). In the USA, Robert T. Jones tested swept-back wings toward the end of 1944. It was the first such work in the USA. It was published in 1945 in NACA TR-863 [14]. In it, Jones cites Busemann's research from 1935 as one of his references.

Pros

One of the primary advantages of the configuration is that when used with high-speed aircraft, it delays the formation of shockwaves to a higher Mach number (see Section 9.3.3, *Leading edge and quarter chord sweep Angles*, Λ_{LE} *and* $\Lambda_{C/4}$).

The aft-swept configuration is less susceptible to flutter than either the straight or the forward-swept configurations. This results from the tendency of a positive (vertical upward) lift force to reduce the *AOA* of the tip through aeroelastic effects.

The aft-swept (or forward-swept) configuration is also a possible solution when the CG is expected to end up too far aft or forward on a proposed configuration. The Messerschmitt Me-262 Schwalbe was the first jet aircraft to feature a swept-back wing. However, the modest sweep angle of 18.5° was insufficient to greatly impact divergence drag number and was a consequence of its Junkers Jumo jet engines being heavier than expected. For this reason its originally straight wings were swept aft to solve an issue with the location of its center of gravity [15].

Cons

Among drawbacks is the deterioration of airflow close to the tip of the planform with an increase in *AOA*. The reason for this deterioration is explained in Section 9.6.5, *Cause of spanwise flow for a swept-back wing planform*. Two important consequences of this phenomenon is a powerful nose pitch-up moment that develops as the *AOA* approaches stall and an accompanying deterioration in roll stability and aileron effectiveness.

Since the center of lift is positioned behind the wing root attachment there is an increase in wing torsion, often substantial, that increases the weight of the airframe.

The configuration is more susceptible to control reversal as a result of the tendency of the lift force to reduce the *AOA* of the tip through aeroelastic effects. The same effect moves the center of lift forward with *AOA* if the wing is flexible and regardless of flexibility

9.4 PLANFORM SELECTION

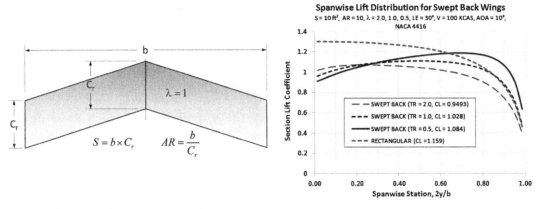

FIGURE 9-36 Basic geometry of the constant-chord swept-back wing planform.

when the wing stalls. It is the cause of the powerful and dangerous nose pitch-up moment at stall.

Aircraft with swept wings often experience issues with the wing fuel tanks. For instance, as the airplane rotates to take-off and begins to climb, fuel will flow toward the outboard and aft side of the tank. This may cause the CG to shift aft. This is solved using so-called *baffle check valves*, which are one-way flapper valves that allow fuel only to flow inboard [16]. Large transport aircraft, some carrying as much as 40% of their gross weight in fuel, can experience large changes in CG location as fuel is consumed. This often calls for a computer-controlled fuel management system that monitors and controls how fuel from the wing tanks is used.

In order to understand the advantages and disadvantages of the swept aft configuration, it is helpful to break it down depending on taper ratio, λ. This way it is possible to consider three classes; $\lambda < 1$, $\lambda = 1$, $\lambda > 1$ (see Figure 9-36 and Figure 9-37). Most aircraft with aft-swept wings fall into the first category. In fact, there as so many different types of aircraft in this class that it would be unfair to name any particular one. Almost all high-speed military and passenger transport aircraft feature the configuration. These are particularly easy to locate.

The other two classes contain considerably fewer members. When it comes to the second class, for which $\lambda = 1$, there is a handful of examples. In current times, the Boeing B-2 Spirit is probably the best known.

However, there are others. Among those are the Dunne Monoplane and Biplane, designed in the 1910s by John W. Dunne (1875–1949). The English Electric Lightning, MiG-8 Outcka, MiG I-320, MiG-17, Sukhoï Su-15P, and Yakovlev Yak-25 (Flashlight) have a λ close to 1. And in the third category, the Republic XF-91 Thunderceptor is the only contender. Designed with the tip chord larger than the root chord, the idea was to reduce the section lift coefficients at the tip in order to improve low-speed handling.

The Forward-swept Planform

One of the aerodynamic advantages of the forward-swept wing can be seen in Figure 9-38: the reduced section lift coefficients at the tip. The distribution ensures the inboard wing stalls first and gives the configuration great roll stability at stall (remember this also means while pulling high gs), making it almost impossible to tip stall. The reduced tip loading also means that the ailerons retain far more authority at high *AOA* than on an aft-swept configuration. This improves controllability during critical phases of the flight, such as during landing. The shape of the lift distribution places the center of lift closer to the plane of symmetry, reducing bending moments (although it does not prevent divergent torsional tendency). Additionally, spanwise flow is directed inboard rather than outboard, where the fuselage more or less acts like an endplate or a fence that prevents early separation.

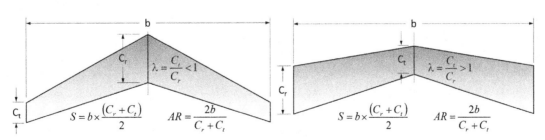

FIGURE 9-37 Basic geometry of the *tapered* and *inverse tapered swept-back* wing planform.

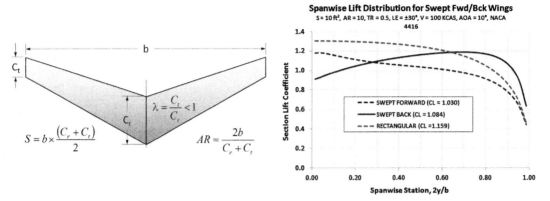

FIGURE 9-38 Basic geometry and basic lifting characteristics of the forward-swept wing, compared to the constant-chord and the tapered swept-back planform shapes.

The primary disadvantage of the geometry is divergent aeroelastic deformation. As the *AOA* of the wing is increased, the elastic torsional deformation twists the wing, increasing the *AOA* further. Forward-swept wings must be especially reinforced to keep the divergent deformation to a minimum, inevitably increasing their weight.

The first major development of a forward-swept wing configuration was the German Junker Ju-287. An interesting discussion of its development is given in Ref. [11]. The configuration has been used in a few other designs, including the Grumman X-29, the HFB-340 Hansajet, and the Sukhoi Su-47. A modest forward sweep is relatively common, for instance in sailplanes and a number of GA aircraft. However, it is almost always a solution to a CG problem and not for compressibility fixes.

Variable Swept Planform

The variable sweep planform (see schematic in Figure 9-39) dates back to the German Messerschmitt P.1101, whose incomplete prototype was discovered at the end of WWII. It featured a swiveling wing design whose sweep angle was to be selected and set manually before each flight. Later, this innovative design led to fighter and bomber models, such as the General Dynamics F-111, Grumman F-14, Sukhoi Su-17, Panavia Tornado, and others. The configuration is not used in any GA aircraft, as it inevitably leads to a heavier airframe, control system complexity, and other complications. It is really a configuration suited for supersonic aircraft, where it solves problems at the high- and low-speed extremes of the flight envelope. It is included here for the sake of completion.

9.4.7 Cranked Planforms

The term "cranked" refers to a break in the leading or trailing edge of a wing that changes the leading (or trailing) edge sweep angle. It turns out that such planform shapes are surprisingly common — for instance a large number of Cessna single-engine aircraft feature cranked wings, as do other popular aircraft such as the Piper PA-28 Cherokee Archer or the Beechcraft Bonanza.

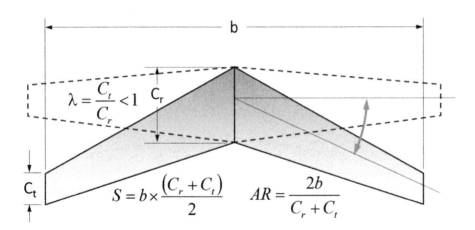

FIGURE 9-39 Basic geometry of the *variable sweep* wing planform.

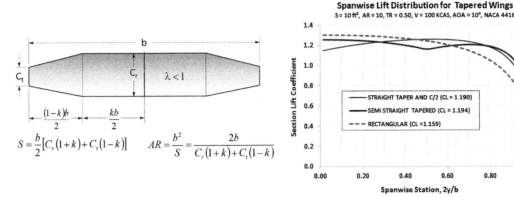

FIGURE 9-40 Basic geometry and lifting characteristics of the *semi-tapered* wing, compared to the constant-chord and the tapered swept-back planform.

Although included in a separate class, the double delta configuration is also technically a cranked planform. In fact, any aircraft that features a leading edge extension falls into this category. However, by convention, double deltas are considered a member of delta wings. Regardless, this class is inclusive and includes a number of wing planform shapes that otherwise might be considered unrelated.

Semi-tapered Planform

The semi-tapered planform is one that has a constant-chord inboard section and a tapered outboard section (see Figure 9-40). The planform increases the section lift coefficients on the outboard wing and improves its efficiency over that of the Hershey-bar wing. A side benefit of this is improvement in aileron effectiveness. The drawback is the added complexity of construction, and reduction in the aileron effectiveness if its hingeline becomes highly swept.

The configuration is best known on a family of single-engine Cessna aircraft: the 150, 152, 172, 182, 206, and many others. It is also used on a number of Piper PA-28 Warrior and Cherokee Archer aircraft.

Crescent Planform

The *crescent* wing planform (Figure 9-41) has seen very limited use. It is best known for its use on the British Handley-Page Victor, where the highly swept inboard section of the wing allowed the use of a thick airfoil without introducing early shock formation at high Mach numbers. The thick airfoil was needed to accommodate its four jet engines, which were buried in the wing root. The wing of the Victor varied in quarter-chord sweep and thickness, progressively from 22° and 4% t/c at the tip to 53° and 16% at the root. This allowed a constant M_{crit} to be maintained along the wing and resulted in more efficient reaction of bending moments. A consequence of this geometry was improved aileron control authority and reduced tendency for tip stall and subsequent nose pitch-up.

Schuemann Planform

The Schuemann planform shape has already been thoroughly introduced in the Appendix C1.5.2, *The Schuemann wing*. It is featured on a number of sailplanes, for instance, the Stemme S-10 and the DG-1000. The wing style has also been introduced on commuter

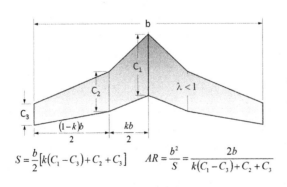
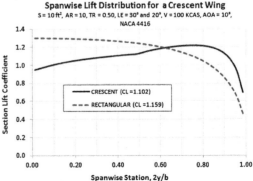

FIGURE 9-41 Basic geometry of the *crescent* wing planform.

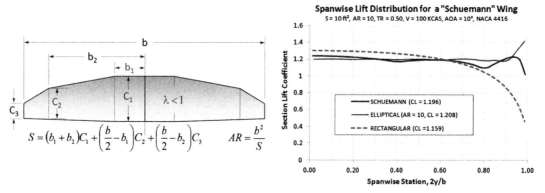

FIGURE 9-42 Basic geometry and lifting characteristics of the *Schuemann* wing planform.

aircraft such as the Dornier Do-228, Dornier 328, and Fairchild-Dornier 328JET.

Figure 9-42 shows the fundamental dimensions of the planform shape and the distribution of the section lift coefficients compared to those of a Hershey-bar and elliptical planform shapes. It can be seen that the distribution approximates that of the elliptical planform, leading to a reduction in lift-induced drag. This explains the wing's popularity for sailplanes. The figure also reveals that a potential problem with the wing is an early wing tip stall, not unlike the elliptical wing and is caused by the sharp outboard taper. This is of course complicated by viscous effects at the wingtip, as highly swept wingtips may position the tip vortex so it suppresses early tip stall tendency, not unlike delta wings. Sailplane wingtips at a high AOA will flex substantially due to aeroelastic effects. This effectively unloads the wingtip and loads up the center portion of the wing, which then stalls before the wingtip. For this reason, sailplanes usually have good stall characteristics. A short and stiff wing, likelier to be used for a GA aircraft, will not flex nearly as much as the high AR wing and should thus be expected to feature decisive wing washout at the tip, unless the tip segment of the wing features a high enough sweep.

9.4.8 Delta Planforms

The Delta Planform Shape[3]

A discussion of the pros and cons of delta wings (see schematic in Figure 9-43) is given in Appendix C1.5.5, *The delta wing*, and will not be elaborated on here, other than that they are planform shapes intended for high-subsonic or supersonic aircraft and not low-subsonic airplanes. Even though it is certainly possible to use delta wings for low-speed airplanes (as evident by the Dyke Delta) it is a choice that is hard to justify for reasons other than fun flying; with reduced storage space (thanks to a short wingspan), being another drawback.

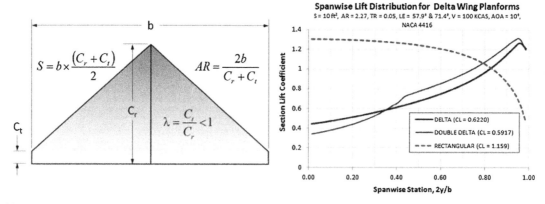

FIGURE 9-43 Basic geometry and lifting characteristics of the *delta* wing planform. Note that the results for the lift distribution are erroneous at the given AOA, as they do not reflect change in lift due to vortex lift.

---

[3]Note that the linear vortex-lattice method does not predict the viscous phenomenon of the leading edge vortex system experienced by delta wings that would already have begun to form at the AOA of 10°. The LE vortex would affect both the C_L and the lift distribution.

9.4 PLANFORM SELECTION

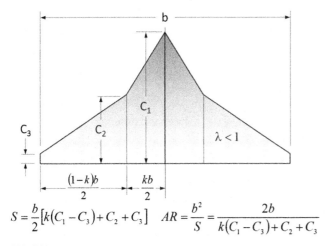

FIGURE 9-44 Basic geometry of the *double delta* wing planform.

Deltas stall at very high angles of attack and generate low C_{Lmax} compared to straight un-swept wings; they will therefore require high approach speed and deck angles. As an example, the 2000 lb$_f$ Dyke JD-2 Delta reportedly stalls at about 61–65 KCAS [17]. This means that the 173 ft^2 wing area generates a C_{Lmax} around 0.8–0.9. That is 50–65% less than conventional straight wings. They also have poor LD_{max}, which is of concern for engine-out emergencies. The best-known aircraft that uses the configuration is the Dassault Mirage III and its various derivative aircraft (e.g. Mirage IV, 2000, Rafale).

The Double-delta Planform Shape

The double delta (see Figure 9-44) is also known as the *compound delta*. They have an advantage over conventional delta wings in that they produce a vortex pair, rather than a single vortex over each wing that mutually interfere with each other. The resulting system increases the lift of the double delta over that of the conventional delta, rendering supersonic fighter aircraft far more maneuverable [18]. When it comes to subsonic aircraft, the configuration suffers from similar limitations to the single delta. The configuration has been used on military aircraft like the SAAB J-35 Draken and General Dynamics F-16E cranked arrow.

9.4.9 Some Exotic Planform Shapes

Disc- or Circular-shaped Planform

In short, the disc-shaped planform (see Figure 9-45) is something no self-respecting aircraft designer should propose as a primary lifting surface. Perhaps inspired by the fad of UFO sightings in the 1950s, the well-respected but long-defunct Avro Canada Ltd, a company that at one time employed some 50,000 people, dared to develop the AVRO Canada VZ-9A, nicknamed the AVRO car. A secret military project, the Avrocar employed the Coandă effect to generate lift and thrust. The project was cancelled in 1961 when it became clear that insufficient thrust and stability issues would hamper its success.

Figure 9-45 shows a comparison of spanwise lift distribution between the disc planform and an elliptical and a Hershey-bar wing. In spite of the uniform distribution of section lift coefficients, the graph clearly shows how inefficient this planform shape is when compared to the Hershey-bar or elliptical wings; it barely ekes out one-third of the lift at the selected *AOA* (10°). From an efficiency standpoint, the disc-shaped wing is simply beyond objectionable. However, it is presented here as it is an ideal shape for a radar disc of the kind installed on a reconnaissance or early warning and control system (AWACS) military aircraft such as the Boeing E-3 Sentry (a modified Boeing 707) or Grumman E-2 Hawkeye. For such applications a shallow lift curve slope is essential for stability and control reasons.

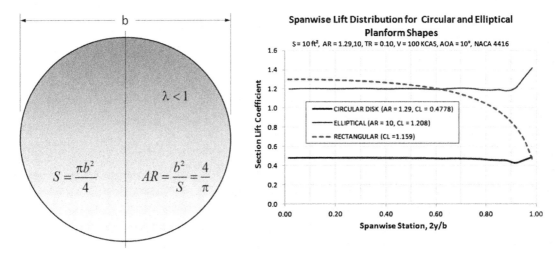

FIGURE 9-45 Basic geometry and lifting characteristics of the *disc or circular* wing planform.

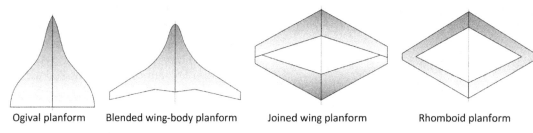

Ogival planform Blended wing-body planform Joined wing planform Rhomboid planform

FIGURE 9-46 Selected unorthodox wing planform shapes.

Other Configurations

Four other configurations are shown in Figure 9-46: an ogival, joined wing, blended wing-body, and a rhomboid planform. Lift distribution curves for these shapes will not be presented here, since these planform shapes are highly specialized or experimental.

9.5 LIFT AND MOMENT CHARACTERISTICS OF A 3D WING

As has already been thoroughly discussed, lift, drag, and pitching moment are almost always converted into a dimensionless coefficient form as this allows transferability (see Equation [8-8]). The lift, drag, and pitching moment coefficients are defined as shown below (note the capitalization of the forces and moments for the three-dimensional object):

$$L = \frac{1}{2}\rho V^2 \cdot S \cdot C_L$$
$$D = \frac{1}{2}\rho V^2 \cdot S \cdot C_D \qquad (9\text{-}47)$$
$$M = \frac{1}{2}\rho V^2 \cdot S \cdot c \cdot C_M$$

where the variables have already been defined elsewhere (e.g. see Variables at the end of this chapter).

In this section, the three-dimensional lift and moment characteristics of the wing will be evaluated, whereas drag is treated in Chapter 15, *Aircraft drag analysis*. Here, a number of important characteristics will be discussed, including construction of the three-dimensional lift curve, estimation of span efficiency and the maximum three-dimensional lift coefficient, stall behavior, tailoring of stall characteristics, and others.

9.5.1 Properties of the Three-dimensional Lift Curve

Figure 9-47 shows a typical three-dimensional lift curve. Overall, the curve displays identical characteristics to that of the two-dimensional curve of Section 8.1.4, *Properties of typical airfoils*, except it represents the behavior of the entire aircraft (wings, fuselage, HT, etc.). The contribution of the individual components of the aircraft to the overall non-linear shape of the curve may make it very different from that of the selected airfoils. As before, the aircraft designer is most interested in the following characteristics:

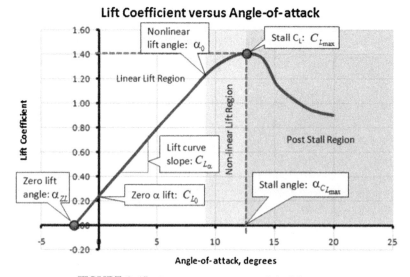

FIGURE 9-47 Important properties of the lift curve.

Lift Curve Slope, $C_{L\alpha}$

The lift curve slope is a measure of how rapidly the wing generates lift with change in *AOA*. As stated in Section 8.1.4, the theoretical maximum is 2π, although real airfoils deviate from it. The lift curve slope of a three-dimensional wing is *always less* than that of the airfoils it features (see Section 9.5.3, *Determination of lift curve slope, $C_{L\alpha}$, for a 3D lifting surface*). Once a certain *AOA* has been achieved the wing will display a pronounced reduction in the lift curve slope (see Figure 9-47). This point is called *stall* and, although not shown in the figure, occurs both at a positive and negative angle-of-attack. The lift at stall dictates how much wing area the aircraft must feature for a desired stalling speed.

Maximum and Minimum Lift Coefficients, C_{Lmax} and C_{Lmin}

The largest and smallest magnitudes of the lift coefficient are denoted by C_{Lmax} and C_{Lmin}, respectively. It indicates at what angle-of-attack the airplane will achieve its minimum airspeed (stalling speed), or what wing area is required for a desired stalling speed. As for the airfoil, the stall is defined as the flow conditions that follow the first lift curve peak, which is where the C_{Lmax} (or C_{Lmin}) occur (see, for instance, Ref. [19]). Both values are required when generating aerodynamic loads for the structures group.

C_L at Zero AOA, C_{Lo}

C_{Lo} is the value of the lift coefficient of the wing at zero *AOA*. It is of great importance in the scheme of things, because it affects the angle-of-incidence at which the wing must be mounted. Generally this value ranges from 0.0 (for symmetric airfoils) to 0.6 (for highly cambered airfoils). It is negative for under-cambered airfoils (e.g. airfoils used near the root of high subsonic jet aircraft).

Angle-of-attack at Zero Lift, α_{ZL}

This is the angle at which the wing generates no lift. For positively cambered airfoils this angle is always negative, unless some specific components (e.g. cambered fuselage) affect it greatly. For symmetrical airfoils it is always 0°.

Linear Range

The linear range is analogous to that of the airfoil, except it applies to the entire aircraft. In this range, the following equation of a line can be used to describe how lift varies with *AOA*.

$$C_L = C_{L_0} + C_{L_\alpha}\alpha \qquad (9\text{-}48)$$

Angle-of-attack where Lift Curve Becomes Non-linear, α_{NL}

Once a certain *AOA* is reached the wing begins to display a pronounced reduction in the lift curve slope. This always happens before the stall *AOA* is reached.

Angle-of-attack for Maximum Lift Coefficient, α_{stall}

Once a certain *AOA* is reached, a pronounced reduction in lift curve takes place; this is the stall.

Design Lift Coefficient, C_{LC}

The design lift coefficient is the C_L at which the aircraft is expected to operate during the mission for which it is designed. This is usually the lift coefficient during cruise. It is important to select an airfoil that has been designed to generate the least amount of drag at that lift coefficient (see Section 8.3.15, *Airfoil selection how-to*), as well as ensure flow separation areas are completely suppressed for minimum drag. In the case of a constant cruising speed (or loiter) mission, if the target airspeed is known, it is possible to estimate this design cruising speed using the weight of the airplane at the beginning and end of the mission using the following expression:

$$C_{L_C} = C_{L0} + \frac{(W_1 + W_2)}{\rho V_C^2 S} + C_{L\alpha}\alpha_{ZL} \qquad (9\text{-}49)$$

where

$C_{L\alpha}$ = lift curve slope
S = reference wing area
α_{ZL} = zero lift angle-of-attack
ρ = density at cruise altitude
W_1, W_2 = aircraft weight at the beginning (1) and end (2) of the design mission

Angle-of-attack for Design Lift Coefficient, α_C

Ideally, the airplane should be flying at an *AOA* that generates the least amount of drag during the intended mission for which it is designed. For instance, this could be a cruise or long-range performance point. In the case of the design mission being a cruise or loiter mission, this *AOA* can be calculated from Equation (9-43).

Derivation of Equation (9-49)

We start with Equation (9-43) for the design cruise *AOA*, repeated here for convenience:

$$\alpha_C = \left(\frac{1}{C_{L\alpha}}\right)\frac{(W_1 + W_2)}{\rho V_C^2 S} + \alpha_{0L}$$

Then we insert it into the Equation (9-51) and manipulate algebraically:

$$C_{L_C} = C_{L0} + C_{L\alpha}\alpha_C$$
$$= C_{L0} + C_{L\alpha}\left(\left(\frac{1}{C_{L\alpha}}\right)\frac{(W_1 + W_2)}{\rho V_C^2 S} + \alpha_{0L}\right)$$
$$= C_{L0} + \frac{(W_1 + W_2)}{\rho V_C^2 S} + C_{L\alpha}\alpha_{0L}$$

QED

9.5.2 The Lift Coefficient

The lift coefficient relates the AOA to the lift force. If the lift force is known at a specific airspeed the lift coefficient is obtained from Equation (9-47) and can be calculated from:

$$C_L = \frac{2L}{\rho V^2 S} \qquad (9\text{-}50)$$

In the linear region of the lift curve, at low AOA, the lift coefficient can be written as a function of AOA as shown below:

$$C_L = C_{L_0} + C_{L_\alpha}\alpha \qquad (9\text{-}51)$$

Equation (9-51) allows the AOA corresponding to a specific lift coefficient to be determined provided the lift curve slope known:

$$\alpha = \frac{C_L - C_{L_0}}{C_{L_\alpha}} \qquad (9\text{-}52)$$

The Relationship between Airspeed, Lift Coefficient, and Angle-of-attack

Consider Figure 9-48, which shows an airplane being operated in horizontal flight at different airspeeds, denoted by the ratio V/V_S, where V_S is its stalling speed. Starting with the upper left image, the aircraft is at a high airspeed (e.g. if $V_S = 50$ KCAS, the figure shows it at $V = 3 \times 50 = 150$ KCAS). This results in a very low lift coefficient, C_L, and an attitude that is slightly nose-down. As the aircraft slows down, it can only maintain altitude by exchanging less airspeed for a higher C_L, which calls for a higher α. As it slows down further, a higher and higher nose-up attitude is required to generate a larger and larger C_L. Eventually, a maximum value of the C_L is achieved, C_{Lmax}, after which the airplane can no longer maintain horizontal flight. This is followed by an immediate and forceful drop of the nose caused by the sudden loss of lift. The airplane begins a dive toward the ground, which increases its airspeed, making stall recovery possible. This is shown as the left bottom image, which shows the aircraft recovering and, while in a nose-down attitude, its higher airspeed has already lowered the α.

Wide-range Lift Curve

A typical change in the lift coefficient with AOA ranging from 0° to 90° is shown in Figure 9-49. The graph is based on true wind tunnel test data, although the actual values have been normalized to the maximum lift coefficient. Two important observations can be made. The first is the linear range at low AOA (here shown ranging from $AOA = 0°$ through $10°$). Note that the extent of this linear region depends on the geometry and operational airspeeds (via Reynolds numbers). The second observation is the relatively large value of the C_L at an AOA around 45°–50°, which, while large, is inefficient because of the high drag associated with it.

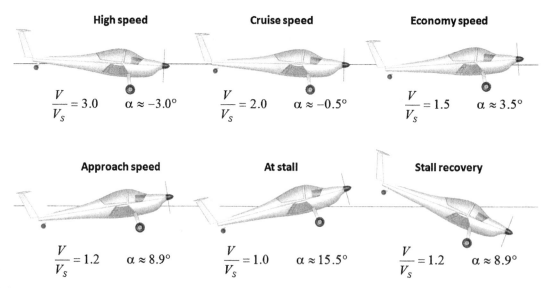

FIGURE 9-48 A schematic showing the attitude of an aircraft at different airspeeds, V. Stalling speed is denoted by V_S. The AOA are typical values for light planes.

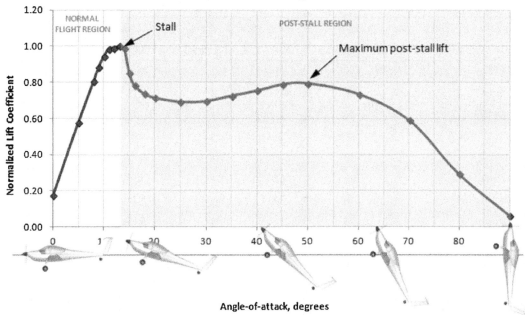

FIGURE 9-49 Example of change in lift coefficient with *AOA* ranging from 0° to 90° for a complete aircraft.

9.5.3 Determination of Lift Curve Slope, $C_{L\alpha}$, for a 3D Lifting Surface

Consider the lift curve slope of an airfoil used for some specific lifting surface (which could be a wing, an HT, or a VT). For reasons that become clear in Section 9.7, *Numerical analysis of the wing*, the surface induces larger upwash in the flow field than the airfoil alone. Consequently, its effective *AOA* is less than that of the airfoil (because the induced *AOA* is larger). Therefore, the wing must operate at a larger *AOA* to generate the same lift coefficient as the airfoil. The lift curve slope for the wing is less steep than for the airfoil. This fact is used to derive the following expressions that allow the two-dimensional lift curve slope of an airfoil ($C_{l\alpha}$) to be converted to three-dimensions for a wing ($C_{L\alpha}$).

The transformation is usually derived using Prandtl's Lifting Line Theory (see Section 9.7.2, *Prandtl's lifting line method — special case: the elliptical wing*). The following expression is used with elliptical wings only:

Lift curve slope for an elliptical wing:

$$C_{L\alpha} = \frac{C_{l\alpha}}{1 + \frac{C_{l\alpha}}{\pi \cdot AR}} \qquad (9\text{-}53)$$

A common (but not necessarily correct) assumption is that the lift curve slope of an airfoil is 2π. This yields the following expression:

Elliptical wing with $C_{l\alpha} = 2\pi$:

$$C_{L\alpha} = 2\pi \frac{AR}{AR + 2} \qquad (9\text{-}54)$$

NACA TN-817 [20] and TN-1175 [21] present methods to make Equation (9-54) suited for $AR < 4$, but these are generally unwieldy. The following expression is an attempt to extend Equation (9-53) to more arbitrary wing shapes and requires the correction factor, τ, to be determined:

Lift curve slope for an arbitrary wing:

$$C_{L\alpha} = \frac{C_{l\alpha}}{1 + \frac{C_{l\alpha}}{\pi \cdot AR}(1 + \tau)} \qquad (9\text{-}55)$$

The factor τ is a function of the Fourier coefficients determined using the lifting line method and represents the following correction to the induced *AOA*, as shown in Dommasch [22]. The actual value of τ is calculated and provided by Glauert [23].

$$\alpha_i = \frac{C_L(1 + \tau)}{\pi \cdot AR}$$

By making some approximations and determining the downwash at the $^3/_4$ chord station, rather than the $^1/_4$ station, Helmbold [24] derived the expression shown below:

TABLE 9-7 Comparing $C_{L\alpha}$ Calculated Using Three Selected Methods and the Vortex-lattice Method (VLM)

Wing Planform	AR	$\Lambda_{C/2}$	$C_{L\alpha}$ per Radian Eq. (9-54)	Eq. (9-56)	Eq. (9-57)	VLM
Elliptical, straight C/2	10	0.00	5.24	5.15	4.90	5.02
Elliptical, straight LE	10	−6.60	5.24	5.15	4.87	4.99
Elliptical, straight TE	10	6.60	5.24	5.15	4.87	4.99
Schuemann	10	5.04	5.24	5.15	4.88	4.99
Semi-straight taper	10	0.00	5.24	5.15	4.90	4.98
Straight taper, straight C/2	10	0.00	5.24	5.15	4.90	4.96
Straight taper, straight LE	10	−3.81	5.24	5.15	4.89	4.95
Straight taper, straight TE	10	3.81	5.24	5.15	4.89	4.95
Compound taper	10	0.00	5.24	5.15	4.90	4.88
Rectangular	10	0.00	5.24	5.15	4.90	4.82
Crescent	10	22.48	5.24	5.15	4.59	4.60
Swept back, $TR = 0.5$, LE sweep 30°	10	28.55	5.24	5.15	4.40	4.52
Swept forward, $TR = 2.0$, LE sweep 30°	10	−32.78	5.24	5.15	4.24	4.32
Swept back, $TR = 1.0$, LE sweep 30°	10	30.00	5.24	5.15	4.35	4.29
Swept back, $TR = 2.0$, LE sweep 30°	10	27.05	5.24	5.15	4.45	3.98
Delta	2.27	38.59	3.34	2.84	2.51	2.44
Double delta	2.26	45.08	3.33	2.83	2.39	2.30
Disk	1.29	0.00	2.46	1.85	1.83	1.84

General lift curve slope:

$$C_{L\alpha} = \frac{2\pi \cdot AR}{2 + \sqrt{AR^2 + 4}} \quad (9\text{-}56)$$

Finally, the following expression, referred to as the Polhamus equation, is derived from NACA TR-3911 [25] and is based on a modification made to Helmbold's equation. It is also presented in USAF DATCOM [26]. The expression accounts for compressibility, deviation from the 2π airfoil lift curve slope, and taper ratio. While TR does not explicitly appear in the equation, Reference [25] demonstrates that if the mid-chord sweep angle ($\Lambda_{C/2}$) is used, the TR can be eliminated. The resulting expression is only valid for non-curved planform shapes and $M \leq 0.8$:

General lift curve slope:

$$C_{L\alpha} = \frac{2\pi \cdot AR}{2 + \sqrt{\left(\frac{AR \cdot \beta}{\kappa}\right)^2 \left(1 + \frac{\tan^2 \Lambda_{C/2}}{\beta^2}\right) + 4}} \quad (9\text{-}57)$$

where

AR = wing aspect ratio
β = Mach number parameter (Prandtl-Glauert) = $(1 - M^2)^{0.5}$
κ = ratio of two-dimensional lift curve slope to 2π
$\Lambda_{C/2}$ = sweepback of mid-chord

Of the above methods, Equation (9-57) compares well with experiment (and this author's experience). This can be assessed indirectly by comparing results using it to that of the vortex-lattice method (VLM), which, as has been shown before, compares well with experiment. Such a comparison is shown in Table 9-7. The general trend is that Equations (9-54) and (9-56) (intended for elliptical planform shapes) predict steeper lift curve slopes than Equation (9-57) and the VLM. Also, note the insensitivity of Equations (9-54) and (9-51) to other characteristics, such as sweep and general planform shape of of the wing.

Derivation of Equation (9-53)

The graph in Figure 9-50 shows the lift curve for two ARs; $AR = \infty$ (an airfoil) and an elliptical wing of an arbitrary AR. The lift coefficient for the airfoil can be written as follows:

$$C_L = \text{constant} + C_{l\alpha} \cdot \alpha$$

The wing induces upwash that reduces the α by an amount denoted by α_i (induced AOA). Therefore, the lift coefficient for the wing is given by:

$$C_L = \text{constant} + C_{l\alpha} \cdot (\alpha - \alpha_i)$$

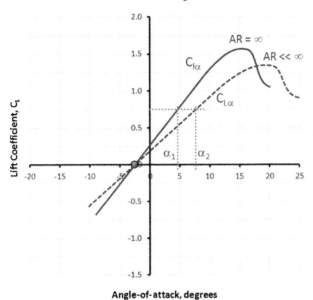

FIGURE 9-50 Lift curves for an airfoil and an elliptical wing.

The value of α_i is given by Equation (9-87). Inserting this yields:

$$C_L = \text{constant} + C_{l\alpha} \cdot \left(\alpha - \frac{C_L}{\pi AR}\right)$$

The lift curve slope can now be found by differentiating with respect to α:

$$\frac{dC_L}{d\alpha} = C_{l\alpha} - C_{l\alpha} \cdot \frac{1}{\pi AR}\frac{dC_L}{d\alpha}$$

$$= C_{l\alpha} - C_{l\alpha} \cdot \frac{1}{\pi AR}\frac{dC_L}{d\alpha} \Rightarrow \frac{dC_L}{d\alpha} = C_{L\alpha}$$

$$= \frac{C_{l\alpha}}{\left(1 + \frac{C_{l\alpha}}{\pi AR}\right)}$$

QED

Derivation of Equation (9-54)

We begin with Equation (9-53) and replace the airfoil lift-curve slope with 2π:

$$C_{L\alpha} = \frac{C_{l\alpha}}{\left(1 + \frac{C_{l\alpha}}{\pi AR}\right)} = \frac{2\pi}{\left(1 + \frac{2\pi}{\pi AR}\right)}$$

Then manipulate algebraically:

$$C_{L\alpha} = \frac{2\pi}{\left(1 + \frac{2\pi}{\pi AR}\right)} = \frac{2\pi}{\left(\frac{\pi AR}{\pi AR} + \frac{2\pi}{\pi AR}\right)} = 2\pi\frac{\pi AR}{\pi AR + 2\pi}$$

$$= 2\pi\frac{AR}{AR + 2}$$

QED

9.5.4 The Lift Curve Slope of a Complete Aircraft

A complete aircraft typically consists of a wing, a stabilizing surface, such as a horizontal tail or a canard (or both), a fuselage, and, sometimes, engine nacelles and external stores. All of these components contribute to the total lift developed by the aircraft and, often, their contribution causes the lift curve slope of the aircraft to differ from that of the wing alone. It can be seen that, for instance, the HT produces lift that adds to the wing lift (assuming a fixed neutral elevator). If the combination is attributed to the reference wing area alone, it would "appear" the wing is generating greater lift than its actual contribution. This is important to keep in mind when considering gust loads for airframe loads and in some stability and control analyses. The following method can be used to estimate the total lift curve slope of a wing and HT, approximating that of the complete aircraft. Effectively, we want to write the total lift coefficient of the airplane, $C_{L_{tot}}$, as follows:

Lift coefficient:

$$C_{L_{tot}} = C_{L_{0tot}} + C_{L_{\alpha tot}}\alpha \qquad (9\text{-}58)$$

where the zero AOA lift and lift curve slope are computed from:

Zero AOA lift:

$$C_{L_{0tot}} = C_{L_0} + \frac{S_{HT}}{S}C_{L_{0HT}} \qquad (9\text{-}59)$$

Lift curve slope:

$$C_{L_{\alpha tot}} = C_{L_{\alpha W}} + \frac{S_{HT}}{S}C_{L_{\alpha HT}}\left(1 - \frac{\partial\varepsilon}{\partial\alpha}\right) \qquad (9\text{-}60)$$

where

S = wing reference area
S_{HT} = HT planform area
C_{L_0} = zero AOA lift of the wing
$C_{L_{\alpha W}}$ = wing lift curve slope
$C_{L_{0HT}}$ = zero AOA lift of the HT (0 for symmetrical airfoils)
$C_{L_{\alpha HT}}$ = HT lift curve slope
$\frac{\partial\varepsilon}{\partial\alpha}$ = wing downwash angle $\approx \frac{2C_{L_{\alpha W}}}{\pi \cdot AR}$ for elliptical wings

Derivation of Equations (9-59) and (9-60)

We can write the total lift of the wing and HT as follows:

$$qSC_{L_{tot}} = qSC_{L_W} + qS_{HT}C_{L_{HT}}$$

Divide through by qS to get the lift coefficient form:

$$C_{L_{tot}} = C_{L_{0tot}} + C_{L_{\alpha tot}}\alpha_W = C_{L_W} + \frac{S_{HT}}{S}C_{L_{HT}}$$

Expand in terms of component properties:

$$C_{L_{tot}} = \left(C_{L_0} + C_{L_{\alpha_W}}\alpha_W\right) + \frac{S_{HT}}{S}\left(C_{L_{0_{HT}}} + C_{L_{\alpha_{HT}}}\alpha_{HT}\right)$$

Insert the AOA the HT is subjected to:

$$C_{L_{tot}} = \left(C_{L_0} + C_{L_{\alpha_W}}\alpha_W\right) + \frac{S_{HT}}{S}$$
$$\times \left(C_{L_{0_{HT}}} + C_{L_{\alpha_{HT}}}\alpha_w\left(1 - \frac{\partial \varepsilon}{\partial \alpha}\right)\right)$$

And finally, gather like terms to yield Equations (9-59) and (9-60):

$$C_{L_{tot}} = \left(C_{L_0} + \frac{S_{HT}}{S}C_{L_{0_{HT}}}\right)$$
$$+ \left[C_{L_{\alpha_W}} + \frac{S_{HT}}{S}C_{L_{\alpha_{HT}}}\left(1 - \frac{\partial \varepsilon}{\partial \alpha}\right)\right]\alpha_w$$

QED

9.5.5 Step-by-step: Transforming the Lift Curve from 2D to 3D

An important part of working with airfoils is the realization that the lift capabilities of a two-dimensional airfoil are superior to that of a three-dimensional wing. Figure 9-11 reveals how a two-dimensional lift curve changes once it is introduced to a wing of finite aspect ratio. Among noticeable effects is a reduction in the lift curve slope and lift at zero AOA. The maximum lift coefficient is reduced although the stall AOA increases. This section presents a method that allows the transformation of the two-dimensional lift curve into a three-dimensional one. The method assumes the same airfoil along the wing. If more than one airfoil is used, the properties of the airfoil at the MGC can be assumed.

Step 1: Compute a three-dimensionl lift curve slope using Equation (9-57).

Step 2: Compute the zero-lift angle for the two-dimensional airfoil using the following expression,

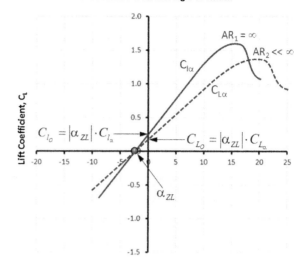

FIGURE 9-51 Determination of the lift coefficient at zero AOA for a three-dimensional lift curve.

which is obtained by inspection of the curves in Figure 9-51:

$$C_{lo} = -\alpha_{ZL}\cdot C_{l\alpha} \Leftrightarrow \alpha_{ZL} = -\frac{C_{lo}}{C_{l\alpha}}$$

Step 3: Compute lift at $AOA = 0°$ for the three-dimensional wing using:

Zero α lift:

$$C_{L0} = |\alpha_{ZL}|\cdot C_{L\alpha} \quad (9\text{-}61)$$

Step 4: Compute pitching moment for three-dimensional wing, denoted by $C_{m\alpha}$. Note that since:

$$C_{m\alpha} = C_{l\alpha}\cdot \Delta x \Leftrightarrow \Delta x = C_{m\alpha}/C_{l\alpha}$$

It follows that:

$$C_{m\alpha} = C_{L\alpha}\cdot \Delta x = C_{L\alpha}\cdot\left(\frac{C_{m\alpha}}{C_{l\alpha}}\right) \quad (9\text{-}62)$$

EXAMPLE 9-8

A Hershey-bar wing ($\lambda = 1$) with an aspect ratio of 20 is being evaluated for use in a low-speed vehicle ($M \approx 0$). One of the airfoils being considered is a NACA 23012.

Convert the following two-dimensional data for the airfoil (which has been extracted from the *Theory of Wing Sections*, by Abbott and Doenhoff) for the three-dimensional:

EXAMPLE 9-8 (cont'd)

Lift curve slope:

$$C_{l\alpha} = 0.1051 \text{ per deg}$$

Lift at zero *AOA*:

$$C_{lo} = 0.1233$$

Slope of the pitching moment curve:

$$C_{m\alpha} = 0.00020 \cdot \alpha - 0.01198$$

The last expression is obtained from interpolation.

Solution

Step 1: Compute a 3D lift curve slope using Equation (9-57):

$$C_{L\alpha} = \frac{AR \cdot C_{l_\alpha}}{2 + \sqrt{\frac{AR^2 \beta^2}{\kappa^2}\left(1 + \frac{\tan^2 \Lambda_{C/2}}{\beta^2}\right) + 4}}$$

where

AR = wing aspect ratio = 20
β = Mach number parameter (Prandtl-Glauert)
    = $(1 - M^2)^{0.5} \approx 1$

κ = ratio of 2D lift curve slope to 2π = 0.1051 × $(180/\pi)/(2\pi)$ = 0.95840
$\Lambda_{c/2}$ = sweepback of mid-chord = 0°

$$C_{L\alpha} = \frac{2\pi \cdot 20}{2 + \sqrt{\frac{400}{0.95840^2}\left(1 + \frac{0}{1^2}\right) + 4}} = 5.472 \text{ per rad}$$

$$= 0.09551 \text{ per deg}$$

Step 2: Compute zero lift angle for the two-dimensional airfoil:

$$\alpha_0 = -\frac{C_{lo}}{C_{l\alpha}} = -\frac{0.1233}{0.1051} = -1.173°$$

Step 3: Compute lift at zero angle for the 3D wing using:

$$C_{L0} = -\alpha_0 \cdot C_{L\alpha} = -(-1.173°) \cdot 0.09551 = 0.1121$$

Step 4: Compute pitching moment for three-dimensional wing:

$$(C_{m_\alpha})_{3D} = C_{L_\alpha} \cdot \left(\frac{C_{m_\alpha}}{C_{l_\alpha}}\right) = (0.09551) \cdot \left(\frac{-0.01198}{0.1051}\right)$$

$$= -0.01089$$

9.5.6 The Law of Effectiveness

Consider the estimation of the three-dimensional lift curve slope of a wing that has two distinct airfoils at the root and tip. As we have seen, the estimation of the wing's lift curve slope requires a representative two-dimensional lift curve slope of an airfoil, but which should be used? The *law of effectiveness* is a handy rule of thumb that helps solve this problem. This law contends that any representative two-dimensional aerodynamic property of a multi-airfoil wing can be approximated by the value at its area centroid, or the mean geometric chord (MGC). A mathematical representation is given by the following equation, obtained using a linear parametric equation:

The law of effectiveness:

$$P_{MGC} = P_{root} + \frac{2y_{MGC}}{b}\left(P_{tip} - P_{root}\right) \qquad (9\text{-}63)$$

where *P* stands for *property*. The property can be the lift curve slope, maximum lift coefficient, drag coefficient, pitching moment, and similar. When applied to a wing for which we intend to determine a representative lift curve slope, Equation (9-63) becomes:

$$\left(C_{l_\alpha}\right)_{MGC} = \left(C_{l_\alpha}\right)_{root} + \frac{2y_{MGC}}{b}\left[\left(C_{l_\alpha}\right)_{tip} - \left(C_{l_\alpha}\right)_{root}\right]$$

$$(9\text{-}64)$$

9.5.7 Flexible Wings

The high-aspect-ratio wings of sailplanes tend to flex excessively during maneuvers and even in normal flight. Flex as large as 6 ft (2 m) is not unheard of in some cases. The same phenomenon occurs on many commercial aircraft (e.g. Boeing 747 and Airbus A380). Regardless of aircraft class, if excessive wing flex is anticipated, it is important to consider its effects on the lift capability of such aircraft.

Figure 9-52 shows the effect of wing flex on the distribution of section lift coefficients. In this example, a wing with a 50-ft wingspan is deflected so its tip is 5 ft higher than that of the unflexed wing. The figure shows that for a given *AOA*, the center of lift moves inboard, reducing the bending moment. However, less lift is also being

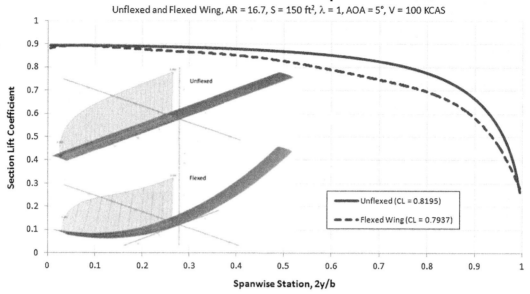

FIGURE 9-52 Lift distribution of the un-flexed and flexed wings compared.

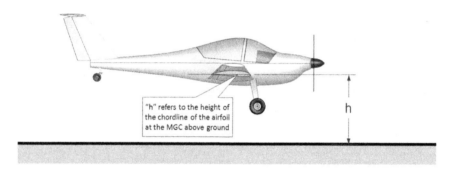

FIGURE 9-53 An airplane in ground effect.

produced (in this particular case, some 3% less), so the resulting aircraft will have to operate at a higher *AOA*. This will lead to a slightly higher operational lift-induced drag and will diminish the wing's long-range efficiency. Additionally, the lift curve slope reduces by about 3%. While this may sound detrimental, there are actually two sides to the topic; the other is discussed in Section 10.5.9, *The polyhedral wing(tip)*.

9.5.8 Ground Effect

Ground effect is the change in the aerodynamic forces as a consequence of the body being in close proximity to the ground. As the aircraft nears the ground, the ground will get in the way of the downwash, effectively preventing it from fully developing. This modifies the entire flow field around the aircraft and, thus, affects a number of its characteristics. Some formulation indicates the aerodynamic properties of the airplane begin to change when it is as much as 2 wingspans from the ground. However, the changes are negligible at that height and it is more reasonable to include ground effect once the airplane is about 1 wingspan from the ground or less. Pilots begin to detect those effects at an even lower height, typically around half a wingspan.

The problem of ground effect was studied as early as 1912 by Albert Betz [27] (1885–1968). Using Prandtl's lifting line theory, Wieselberger [28] developed a formulation to estimate the reduction in the lift-induced drag near the ground. His work is translated in Ref. [29]. It does this by calculating a special *ground influence coefficient*, denoted by Φ. The presentation here resembles that of Ref. [30], in which h stands for height of the wing above the ground and b is the wingspan. It differs only slightly from the presentation of Wieselberger, who used $h/2$ for the height above the ground. Additionally, in order to correspond to the other two presentations shown below, Wieselberger's coefficient (denoted by σ) is subtracted from 1, yielding:

Φ per Wieselberger:

$$\Phi = 1 - \frac{1 - 1.32(h/b)}{1.05 + 7.4(h/b)} \quad (9\text{-}65)$$

This approximation appears in good agreement with experiment [29] for values of h/b between 0.033 and 0.25. Using the Biot-Savart law applied to a horseshoe vortex whose span is $\pi b/4$, McCormick [31] shows that the ground influence coefficient can be estimated from:

Φ per McCormick:

$$\Phi = \frac{(16 \cdot h/b)^2}{1 + (16 \cdot h/b)^2} \quad (9\text{-}66)$$

Assuming an elliptical lift distribution of a straight wing of $AR \approx 5$ and using the lifting line theory, Asselin [32] estimates the following value of the ground influence coefficient:

Φ per Asselin:

$$\Phi = 1 - \frac{2}{\pi^2} \ln\left(1 + \frac{\pi}{8 \cdot h/b}\right) \quad (9\text{-}67)$$

These approximations are compared in Figure 9-54. The ground influence coefficient is then used to adjust the following characteristics of the airplane:

Lift-induced drag:

$$(C_{Di})_{IGE} = \Phi \times (C_{Di})_{OGE} \quad (9\text{-}68)$$

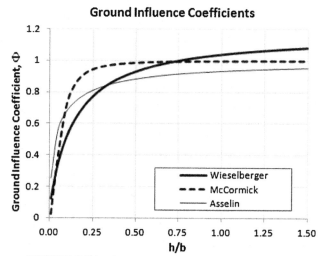

FIGURE 9-54 Comparing ground influence coefficients.

Maximum lift-to-drag ratio:

$$(LD_{\max})_{IGE} = \frac{(LD_{\max})_{OGE}}{\sqrt{\Phi}} \quad (9\text{-}69)$$

where IGE stands for *in ground effect* and OGE stands for *out of ground effect*.

Modeling the ground effect can be accomplished using vortex theories such as Prandtl's lifting line theory or Weissinger's vortex-lattice theory. By creating an inverted mirror image of the wing (or airfoil) with bound vortices of equal strength but rotating in opposite directions (see Figure 9-55) the resulting flow field will

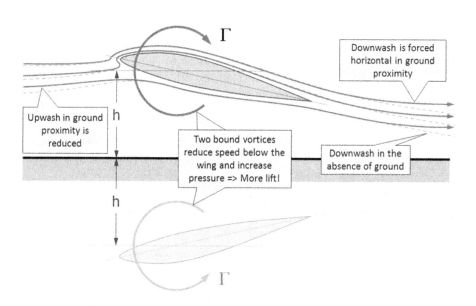

FIGURE 9-55 Modeling the ground effect is done using a mirror image airfoil (or wing).

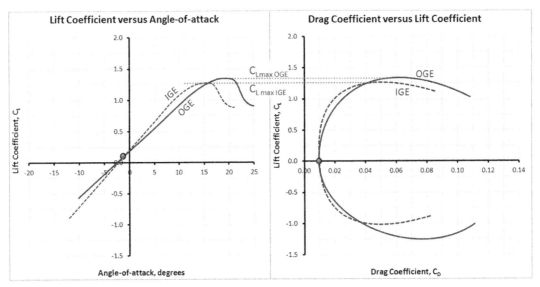

FIGURE 9-56 The impact of ground effect on the lift curve and drag polar.

feature a horizontal streamline along the ground plane. The streamlines above and below will be realigned when compared to the flow field in the absence of the ground as is shown in Figure 9-55. The impact on the lift curve, drag polar, and pitching moment curve is shown in Figure 9-56 and Figure 9-57. The following changes can be noted (these are in part based on Ref. 31). Note that IGE stands for in ground effect and OGE stands for out of ground effect.

(1) Up- and downwash in the proximity of the ground is reduced compared to that in the far-field.
(2) This reduction lowers the induced AOA, resulting in less aft tilting of the lift vector and, thus, less induced drag (see Section 15.3.4, *The lift-induced drag coefficient: C_{Di}*). The minimum drag, however, is not reduced.
(3) The reduced up- and downwash also reduces the lift. However, the two bound vortices (on either side of the ground plane) will cause a greater reduction in airspeed under the airfoil, increasing the pressure along the lower surface above that in the absence of the ground. This increase is greater than the reduction due to the diminished downwash, yielding an overall increase in lift at a given AOA.
(4) The lift increase causes an increase in the lift curve slope. Based on Ref. [29] this tends to result in a small reduction in the zero-AOA lift.
(5) The steepening of the lift curve slope increases the pitching moment of the wing and shifts it downward.
(6) The effective AR is increased because of the reduction in lift-induced drag. This, and the accompanying increase in lift at a given AOA, increases the L/D ratio, causing a "floating" tendency.
(7) The trim AOA is reduced, which means that the airplane has a tendency to go to a lower AOA.
(8) Elevator effectiveness of a conventional tail-aft configuration is reduced as the low-pressure region on the lower surface is counteracted by the

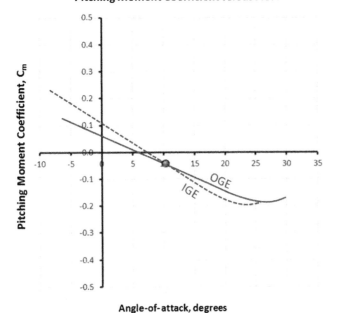

FIGURE 9-57 The impact of ground effect on the pitching moment.

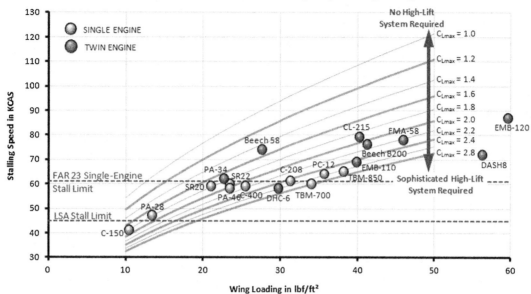
FIGURE 9-58 A carpet plot showing stalling speed as a function of wing loading and maximum lift coefficient.

formation of a high-pressure region similar to that of the wing. By the same token, the elevator effectiveness of a canard configuration will increase.

9.5.9 Impact of C_{Lmax} and Wing Loading on Stalling Speed

During the design stage the stalling speed and required wing area must be determined. Figure 9-58 highlights how wing loading (W/S) and maximum lift coefficient (C_{Lmax}) affect the stalling speed (here shown in KCAS) by displaying it as a carpet plot. The figure shows the regulatory FAR 23 stall speed limit of 61 KCAS for single-engine aircraft and the 45 KCAS limit set forth in the Light Sport Aircraft (LSA) category. The graph has selected aircraft superimposed. Note that single-engine aircraft, such as the PC-12 and TBM-850, are turboprop aircraft that were granted exemption from the 61 KCAS rule on the grounds of envelope protection equipment they feature.

As an example of use, consider an airplane slated for FAR 23 certification characterized by a wing loading of some 25 lb_f/ft^2. It can be seen it must feature a high-lift system capable of at least $C_{Lmax} = 2.0$ in order to meet the 61 KCAS requirement.

Alternatively, consider another example in which the designer of an airplane slated for FAR 23 certification wants to feature a simple high-lift system capable of $C_{Lmax} = 1.8$. Figure 9-58 reveals that as long as the wing loading is less than 22 lb_f/ft^2 the FAR 23 single-engine stall speed limit (61 KCAS) will be complied with. The following expression is used to plot the curves in Figure 9-58, of which the right hand approximation is only valid at S-L. The constants (1/1.688 = 0.592 and 29/1.688 = 17.18) convert the value in ft/s to knots.

$$V = 0.592\sqrt{\frac{2W}{\rho S C_{Lmax}}} \approx 17.18\sqrt{\frac{(W/S)}{C_{Lmax}}} \quad (9\text{-}70)$$

9.5.10 Step-by-step: Rapid C_{Lmax} Estimation [5]

The primary purpose of this method is to allow a fast estimation of a likely stalling speed. This method is very simple and, consequently, accuracy is suspect. It is presented here because it is acknowledged that, during the conceptual design stage, the designer needs a fast and simple method that has a "fair chance" of providing reasonably accurate results. It is acceptable only during the conceptual design phase and should be replaced with more accurate methods, once the design progresses.

Step 1: Calculate a representative C_{lmax} (two-dimensional) for the airfoil at the MGC using the *law of effectiveness*:

$$C_{lmax} = (C_{l_{max}})_{root} + \frac{2y_{MGC}}{b}\left[(C_{l_{max}})_{tip} - (C_{l_{max}})_{root}\right]$$

$$(9\text{-}71)$$

Step 2: Calculate the three-dimensional C_{Lmax} from the following expression (this is the straight wing C_{Lmax}):

$$C_{L_{max0}} = 0.9 \times C_{l_{max}} \qquad (9\text{-}72)$$

Step 3: Correct for wing sweep angle. This reduces the maximum lift over that of a straight wing, highlighting yet another challenge for the operation of aircraft with swept wings. Generally, the reduction in the maximum lift can be estimated from the following expression:

$$C_{L_{max}} = C_{L_{max0}} \times K_\Lambda \qquad (9\text{-}73)$$

where

C_{Lmax0} = max lift coefficient of the unswept wing
K_Λ = Sweep correction factor (see Figure 9-59)

Raymer [5], Jenkinson [33], and others define K_Λ as follows, where it is based on the sweep of the quarter-chord:

$$K_\Lambda = \cos\Lambda_{C/4}$$

Young [34], on the other hand, defines it in terms of the leading edge sweep. It matches historical data well:

$$K_\Lambda = \cos^3\Lambda_{LE}$$

For instance, a wing sweep of 35° will reduce the C_{Lmax} by about 15% per Figure 9-59. Note that these methods are also extended to evaluate the reduction in effectiveness of control surfaces with swept hingelines. In this case, replace $\Lambda_{C/4}$ by the hingeline sweep angle.

EXAMPLE 9-9

Compare the calculated maximum lift coefficient for the Cirrus SR22 using the Rapid C_{Lmax} Method and compare to the "known" value of C_{Lmax} of the airplane, which can be calculated using published information in the aircraft's POH (S = 144.9 ft², W = 3400 lb$_f$, and V_S = 70 KCAS [M = 0.10]) and the calculated values of Table 16-6. Assume the airfoil for the airplane is NACA 65$_2$-415 for both the root and tip.

Solution

Start by estimating the maximum three-dimensional lift coefficient to compare to, based on the POH information:

$$C_{L_{max}} = \frac{2W}{\rho V^2 S} = \frac{2 \cdot 3400}{0.002378 \cdot (70 \cdot 1.688)^2 \cdot 144.9} = 1.41$$

Next, let's figure out the Reynolds number at stall, using the average chord for the airplane (in lieu of the MGC). The average chord can be found from Equation (9-17):

$$C_{avg} = \frac{S}{b} = \frac{144.9 \text{ ft}^2}{38.3 \text{ ft}} = 3.783 \text{ ft}$$

Therefore, using Equation (8-28) the Reynolds number equals:

$$R_e \approx 6400VL = 6400 \times (70 \times 1.688) \times (3.783) \approx 2\,860\,000$$

We use this information to extract the maximum lift coefficient for the NACA 65-415 airfoil using the wind tunnel data in NACA R-824 [35]. Using the plot for R_e = 3.0 million (which is closest to 2.86 million), displayed in Figure 9-60, the C_{lmax} = 1.45 and this will be used at the root.

Since the airplane's taper ratio is 0.5, the R_e at the root is two times that at the tip. By inspecting the graph in Figure 9-60 it is estimated that the tip C_{lmax} is approximately 1.35. Therefore, the 3D C_{Lmax} for the SR22, assuming NACA 65-415 for root and tip, can be estimated as follows:

$$C_{lmax} = (C_{l_{max}})_{root} + \frac{2y_{MGC}}{b}\left[(C_{l_{max}})_{tip} - (C_{l_{max}})_{root}\right]$$

$$= 1.45 + \frac{2 \times 8.51}{38.3}[1.35 - 1.45] = 1.406$$

By observation, the quarter-chord sweep angle is 0°. Therefore, the three-dimensional maximum lift coefficient can be found to equal:

$$C_{L_{max}} = 0.9 \times C_{l_{max}} \times \cos\Lambda_{C/4}$$

$$= 0.9 \times 1.406 \times \cos 0° = 1.265$$

The difference between the POH and estimated value using the POH is some 10% — in this case, underestimating the capability of the airplane. That level of accuracy could adversely affect the wing sizing of a brand-new aircraft and must be kept in mind, although it will give some idea of the maximum lift coefficient. However, the method does not account for the lift generated by the fuselage and HT at this high *AOA*.

9.5 LIFT AND MOMENT CHARACTERISTICS OF A 3D WING

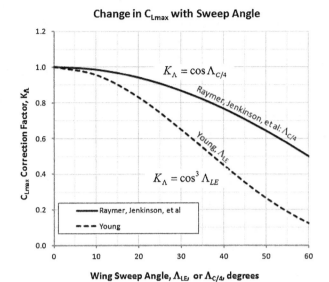

FIGURE 9-59 Sweep correction factor for C_{Lmax}.

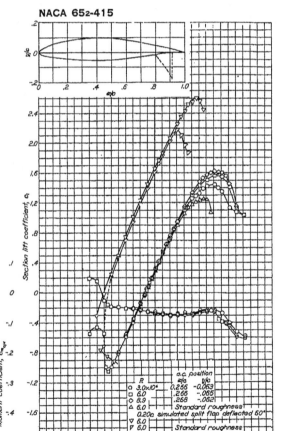

FIGURE 9-60 Lifting properties of the NACA 65-415 airfoils (from Ref. [34]).

9.5.11 Step-by-step: C_{Lmax} Estimation per USAF DATCOM Method 1

The wing's maximum lift coefficient can also be calculated using the following method from the USAF DATCOM (where it is referred to as Method 1). This method requires access to an accurate wingspan-wise loading analysis program, such as the vortex-lattice or doublet-lattice methods. The method is limited to moderately swept wing planforms, where LE vortex effects are not yet significant. Delta wings are excluded, unless the LE vortex can be estimated. Furthermore, the spanwise location where stall is first detected should be limited within a band extending from one local chord-length away from the wing root and tip. The DATCOM considers this method superior to Method 2 (see next section).

Step 1: Determine the Two-dimensional Maximum Lift Coefficient

Determine the 2D C_{lmax} along the span for the wing, based on the appropriate Mach number and Reynolds number. Experimental data should always be used if available.

Step 2: Plot the Distribution of Section Lift Coefficients

Plot the C_{lmax} along a normalized spanwise station ($2y/b$, ranging from 0 to 1), as shown by the dashed line in Figure 9-61.

Step 3: Determine the AOA at which Local C_l Intersects C_{lmax}

Then plot the distribution of section lift coefficients for an *AOA* until the maximum of a local C_l intersects the airfoil's C_{lmax}. This is where the stall will first occur. An approximate value of where this occurs can be estimated from (Ref. [1], Article 4.1.3.4, Wing Maximum Lift) (note that $\lambda =$ taper ratio):

$$\eta_{stall} = 1 - \lambda \qquad (9\text{-}74)$$

Step 4: Compute C_{Lmax}

Calculate the three-dimensional C_{Lmax} by integrating along the span:

$$C_{Lmax} = \frac{1}{S} \int_0^{1.0} b \cdot C_l(\eta) \cdot C(\eta) d\eta \qquad (9\text{-}75)$$

where

$C_l(\eta)$ = section lift coefficient as a function of spanwise station
$C(\eta)$ = wing chord as a function of spanwise station
η = spanwise station = $2y/b$; ranges from 0 to 1 (for $b/2$)

EXAMPLE 9-10

Predict the maximum lift coefficient for the Cirrus SR22 using the DATCOM method and compare to the "known" value of C_{Lmax3D} (= 1.41), calculated in Example 9-9. Assume the airplane has an airfoil in the NACA 65_2-415 class and account for the effect of Reynolds number on the C_{lmax} at the root and tip.

Solution

The results in Figure 9-61 were obtained using the vortex-lattice solver SURFACES, but similar data should be obtainable from other solvers as well. The model is based on measurements taken from Figure 16-15 and included the wing's leading edge extensions (or cuffs), the fuselage, and horizontal tail. Note that the program calculates the total lift coefficient, C_L (see values in parentheses in the legend) and these represent values that would be returned by Equation (9-75) if the shape of the lift distribution were to be presented as a function of the spanwise station.

The stall limit (blue straight dashed line) indicates the C_{lmax} = 1.45 at the root of the NACA 65_2-415 airfoil and 1.35 at tip, determined in Example 9-9. This allows the stall *AOA* and the maximum lift coefficient to be estimated using the DATCOM method as follows:

$$C_{Lmax} \approx 1.468 \quad \text{and} \quad \alpha_{stall} \approx 14°$$

The difference between the POH (C_{Lmax} = 1.41) and the estimated value (C_{Lmax} = 1.468) is some 4.1% and, indeed, shows a very good agreement. This method generally agrees well with experiment and differs by approximately ±6%.

Finally this; it is acknowledged that not all readers have access to codes like the one used in this example and this may present frustration to some. While the author empathizes with such emotions, ultimately, engineers working in industry have access to such codes and this example is intended for them. It is possible to use the method with the lifting line theory presented in Section 9.7, *Numerical analysis of the wing*.

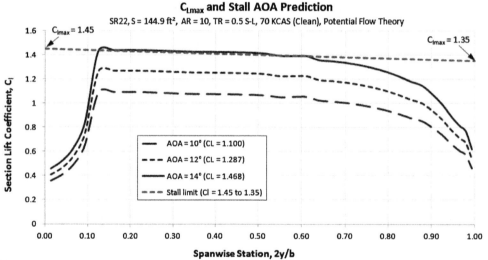

FIGURE 9-61 Lift distribution near stall as predicted by the vortex-lattice method. The waviness in places is due to the discontinuity in the segment of the wing, which features a leading edge extension.

9.5.12 Step-by-step: C_{Lmax} Estimation per USAF DATCOM Method 2

The wing's maximum lift coefficient can be calculated using the following method from the USAF DATCOM (where it is referred to as Method 2). While somewhat involved, the method allows the estimation of the maximum lift and angle-of-attack for maximum lift at subsonic speeds. The method is empirically derived and is based on experimental data for predicting the subsonic maximum lift and the angle-of-attack for maximum lift of high aspect ratio, untwisted, constant section wings.

Generally, the maximum lift of high-aspect-ratio wings at subsonic speeds is directly related to the maximum lift of the wing section or airfoil. According to the DATCOM, the wing planform shape influences the maximum lift obtainable, although its effect is less important than that of the airfoil's section characteristics.

Step 1: Determine the Taper Ratio Correction Factor

First determine if the wing in question complies with the DATCOM's "definition" of a high-aspect-ratio wing. Do this by determining the *taper ratio correction factor* (TRCF), C_1, from Figure 9-62. Alternatively, the TRCF can be approximated using the following empirical expression, based on the curve of Figure 9-62.

$$C_1 = \frac{1}{2}\sin\left\{\pi(1-\lambda)^{1.5+0.8\sin^{0.4}\left(\pi(1-\lambda)^2\right)}\right\} \quad (9\text{-}76)$$

Step 2: Determine if Wing Qualifies as "High AR"

Then determine whether the wing complies with the DATCOM's "definition" of a high-aspect-ratio wing:

$$AR > \frac{4}{(C_1 + 1)\cos\Lambda_{LE}} \quad (9\text{-}77)$$

If the airplane's AR is larger than the ratio of Equation (9-77) then the procedure is applicable to it.

Step 3: Determine the Leading Edge Parameter

Determine the leading edge parameter (LEP), denoted by Δy, which is used several steps later. The parameter Δy is the difference between airfoil ordinate at 6% chord and ordinate at 0.15% chord and is represented in terms of %. Thus, a value of 0.03 would be written as 3.00. Since this method assumes a single airfoil wing, it is appropriate to approximate the leading edge parameter based on the geometry of the airfoil at the mean geometric chord. Figure 9-63 illustrates the process. *Note that this is based on an airfoil ordinate table that has been normalized to a chord of unity (C = 1).*

The listing below contains expressions for the LEP, Δy, for selected types of airfoils, based on their thickness-to-chord ratios (t/c). The explicit expressions in the list may make it easier to determine Δy without having to perform calculations based on the ordinate table. The list is based on Figure 2.2.1-8 of the USAF DATCOM [1].

NACA 4- and 5-digit Series airfoils:

$$\Delta y = 25(t/c) \quad (9\text{-}78)$$

NACA 63-Series:

$$\Delta y = 22.132(t/c) \quad (9\text{-}79)$$

NACA 64-Series:

$$\Delta y = 20.411(t/c) \quad (9\text{-}80)$$

NACA 65-Series:

$$\Delta y = 19.091(t/c) \quad (9\text{-}81)$$

NACA 66-Series:

$$\Delta y = 18.182(t/c) \quad (9\text{-}82)$$

Biconvex:

$$\Delta y = 11.667(t/c) \quad (9\text{-}83)$$

Double wedge:

$$\Delta y = 5.882(t/c) \quad (9\text{-}84)$$

Step 4: Determine the Max Lift Ratio

Determine the ratio $C_{L\max}/C_{l\max}$ using Figure 9-64. The figure illustrates the variation of the ratio between the wing's maximum lift coefficient and the section maximum lift coefficient as a function of the leading edge sweep and the LEP Δy:

where

$C_{l\max}$ = section maximum lift coefficient
$C_{L\max}$ = maximum three-dimensional lift coefficient

Step 5: Determine the Mach Number Correction Factor

Determine the Mach number correction factor (MNCF), $\Delta C_{L\max}$, from Figure 9-65, using the LEP, the wing's leading edge sweep (Λ_{LE}), and Mach number evaluated at the stalling speed. Note that the reference

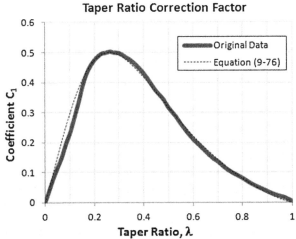

FIGURE 9-62 Taper ratio correction factor *(Based on Ref. [1])*.

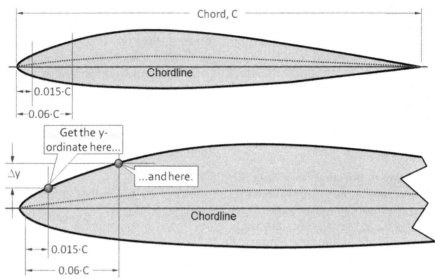

FIGURE 9-63 Determination of the LEP *(Based on Ref. [1])*.

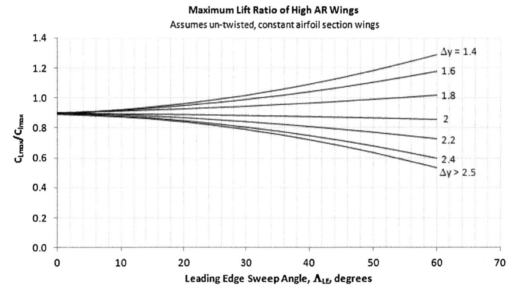

FIGURE 9-64 C_{Lmax}/C_{lmax} ratio data plot *(Based on Ref. [1])*.

document also presents similar graphs for $\Lambda_{LE} = 40°$ and $60°$, but such sweeps are rarely used on GA aircraft.

Step 6: Calculate the C_{Lmax}

Calculate the wing's maximum lift coefficient C_{Lmax} using the expression below:

$$C_{Lmax} = \left(\frac{C_{Lmax}}{C_{lmax}}\right) C_{lmax} + \Delta C_{Lmax} \qquad (9\text{-}85)$$

where

C_{Lmax}/C_{lmax} = ratio obtained from Figure 9-64

C_{lmax} = section maximum lift coefficient
ΔC_{Lmax} = Mach number correction factor obtained from Figure 9-65

Step 7: Determine Zero Lift Angle and Lift Curve Slope

Determine the wing's zero lift angle, α_{ZL}, and lift curve slope, $C_{L\alpha}$. Both have to be in terms of degrees.

Step 8: Determine a Correction for the Stall Angle-of-attack

Determine the correction angle, $\Delta\alpha_{stall}$, for the nonlinear effects of vortex flow from Figure 9-66.

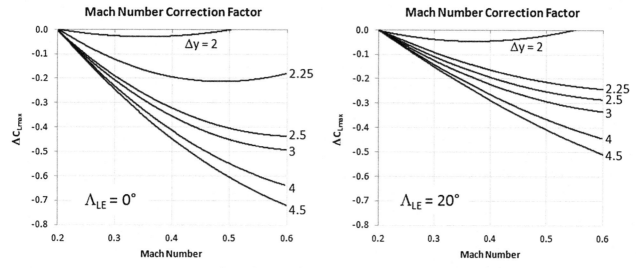

FIGURE 9-65 Determine ΔC_{Lmax} using the above graphs *(Based on Ref. [1])*.

Step 9: Determine the Wing Stall Angle-of-attack

The angle-of-attack of the wing a stall can be calculated from the following equation:

$$\alpha_{stall} = \frac{C_{Lmax}}{C_{L\alpha}} + \alpha_{ZL} + \Delta\alpha_{stall} \qquad (9\text{-}86)$$

where

$C_{L\alpha}$ = wing lift curve slope

α_{ZL} = wing zero lift angle

$\Delta\alpha_{stall}$ = correction factor from Figure 9-66

EXAMPLE 9-11

Compare the calculated maximum lift coefficient for the Cirrus SR22 using the DATCOM method to the known value of C_{Lmax} for the airplane, calculated as shown in Example 9-9.

Solution

All values in this solution were obtained by scaling the airplane in the 3-view (see Figure 16-15) based on the indicated wingspan. A reader trying to reproduce the solution may come up with slightly different dimensions.

Based on the 3-view, the leading edge sweep angle was found to equal: $\Lambda_{LE} = 1.93°$

The aspect ratio is found using the wingspan and area information in the 3-view:

$$AR = \frac{b^2}{S} = \frac{38.3^2}{144.9} = 10.12$$

Using Figure 16-15 the taper ratio is estimated to be 0.5. Additionally, the leading edge parameter Δy for the NACA 65-415 airfoil is approximately 2.86%, obtained using Equation (9-81).

The wing was checked to see if it complies with the DATCOM definition of a high AR wing. To do this, we calculate the taper ratio correction factor using Equation (9-76), which turns out to be $C_1 = 0.3064$.

Therefore, from Equation (9-77):

$$\frac{4}{(C_1 + 1)\cos\Lambda_{LE}} = \frac{4}{(0.3064 + 1)\cos(1.93°)} = 3.064 < 9.32$$

This simply means that the method is applicable to the SR22. Next read Figure 9-64 to determine the ratio C_{Lmax}/C_{lmax}. The resulting ratio is approximately:

$$C_{L_{max}}/C_{l_{max}} = 0.89$$

Note that a maximum section lift coefficient is estimated at $C_{lmax} = 1.40$, using Figure 9-60 (and as shown in Example 9-9). Next, let's estimate the Mach number correction factor (MNCF), ΔC_{Lmax}. We do this using Figure 9-65 with $\Lambda_{LE} = 0$ (since the LE sweep is only 1.93°), $M = 0.10$ and $\Delta y = 3.5$:

$$\Delta C_{Lmax} = 0$$

EXAMPLE 9-11 (cont'd)

We can now estimate the three-dimensional maximum lift coefficient for the wing using Equation (9-85):

$$C_{L_{max}} = \left(\frac{C_{L_{max}}}{C_{l_{max}}}\right)C_{l_{max}} + \Delta C_{L_{max}} = (0.89)1.40 + 0 = 1.246$$

The difference between the POH and estimated value using the POH is some 11.6%. Next, estimate the stall AOA. We do this by first estimating the lift curve slope for the SR22 using Equation (9-57), repeated for convenience.

$$C_{L_\alpha} = \frac{2\pi \cdot AR}{2 + \sqrt{\frac{AR^2 \beta^2}{\kappa^2}\left(1 + \frac{\tan^2 \Lambda_{C/2}}{\beta^2}\right) + 4}}$$

where
AR = wing aspect ratio = 10.12
β = Mach number parameter (Prandtl-Glauert) = $(1 - M^2)^{0.5} \approx 0.995$. Call it 1
κ = ratio of 2D lift curve slope to $2\pi = 0.107 \times (180/\pi)/(2\pi) = 0.9757$

$\Lambda_{c/2}$ = sweepback of mid-chord is close $0°$

$$C_{L_\alpha} = \frac{2\pi \cdot 10.12}{2 + \sqrt{\frac{10.12^2}{0.9757^2}(1) + 4}} = 5.061 \text{ per rad}$$
$$= 0.08834 \text{ per deg}$$

From Table 8-5, the zero-lift angle for the 65_2-415 airfoil is $\alpha_{ZL} = -2.6°$. Next, let's determine the nonlinear vortex flow-correction angle from Figure 9-66. Using the leading edge sweep angle $\Lambda_{LE} = 1.93°$ and the leading edge parameter $\Delta y = 2.86\%$, the correction angle is approximately $\Delta \alpha_{stall} = 1.3°$. This allows us to calculate the stall angle-of-attack using Equation (9-16):

$$\alpha_{C_{L_{max}}} = \frac{C_{L_{max}}}{C_{L_\alpha}} + \alpha_{ZL} + \Delta\alpha_{stall} = \frac{1.291}{0.08834} - 2.6° + 1.3°$$
$$= 13.3°$$

The actual stall angle for the SR22 is not published, so a comparison cannot take place. However, it is thought that this angle is probably 1° to 3° too low.

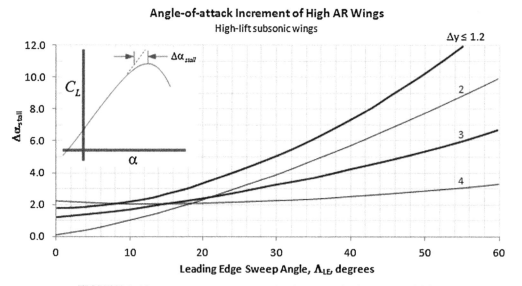

FIGURE 9-66 Determine $\Delta\alpha_{stall}$ using the above graphs *(Based on Ref. [1])*.

9.5.13 C_{Lmax} for Selected Aircraft

Examples of the maximum lift coefficient for selected aircraft are shown in Table 9-8. The aspiring designer is encouraged to be realistic and careful when estimating this value. It has a profound impact on the capability of the aircraft; an overestimation inevitably results in an undersized wing area, which could have a major impact on the stall and low speed characteristics of the

TABLE 9-8 Maximum Lift Coefficients for Selected Aircraft

Name	Gross Weight lb$_f$	Wing Area ft^2	Stalling Speed, KCAS		Maximum Lift Coefficient		Source
			V_0 (flaps)	V_1 (clean)	C_{Lmax0}	C_{Lmax1}	
SINGLE-ENGINE							
Aerotec A-122A	1825	145	39	57	2.44	1.16	1
Bede BD-5A Micro	660	47	48	54	1.78	1.41	1
Cirrus SR20	3050	145	59	67	1.79	1.38	2
Cirrus SR22	3400	145	60	70	1.92	1.41	2
Cessna 162 Skycatcher	1320	120	40	44	2.03	1.68	1
Cessna 172 Skyhawk	2450	174	47	51	1.88	1.60	1
Cessna 182 Skylane	2950	174	50	56	2.00	1.60	2
Cessna 208B Grand Caravan	8750	279	61	78	2.48	1.52	2
Embraer EMB-201	3417	194	51	57	2.00	1.60	1
Let Z-37 Čmelák (Bumblebee)	3855	256	45	49	2.19	1.85	1
Neiva N621A	3306	185	57	64	1.65	1.31	1
Piper PA-46-350 Malibu	4340	175	58	69	2.18	1.56	1
Taylor J.T.1 Monoplane	610	76	33	40	2.18	1.48	1
Transavia PL-12 Airtruk	3800	256	52	55	1.62	1.45	1
TWIN-ENGINE PROPELLER							
Beechcraft Baron 55	5100	199	73	88	1.42	0.98	1
Beechcraft Duke B60	6775	213	76	86	1.63	1.27	3
Beechcraft Queenair B80	8800	294	71	85	1.75	1.22	1
Cessna 421 Golden Eagle	4501	215	74	83	1.13	0.90	6
Cessna 337 Skymaster	4630	201	61	70	1.83	1.39	1
Partenavia P.68	1960	200	56	64	0.92	0.71	1
Pilatus Britten-Norman BN-2B Islander	2993	325	40	50	1.70	1.09	1
Piper Seminole	3800	184	55	57	2.02	1.88	5
Rockwell Commander 112A	2650	152	54	61	1.77	1.38	4
Vulcanair P.68 Observer	4594	200	57	68	2.09	1.47	1
COMMUTER TURBOPROPS							
Beechcraft Kingair C90	9650	294	72	80	1.87	1.51	1
Beechcraft Kingair 100	10,600	298	76	92	1.82	1.24	1
CASA C-212 Aviocar	13,889	431	62	72	2.48	1.84	1
Frakes conversion Turbo-Mallard	14,000	444	66	76	2.14	1.61	1
Let L-140 Turbolet	11,905	354	64	83	2.43	1.44	1
Lockheed Model 35 Orion P-3	135,000	1300	112	133	2.44	1.73	1
Nomad N22	8000	324	47	65	3.30	1.73	1

TABLE 9-8 Maximum Lift Coefficients for Selected Aircraft—cont'd

Name	Gross Weight lb$_f$	Wing Area ft^2	Stalling Speed, KCAS		Maximum Lift Coefficient		Source
			V$_0$ (flaps)	V$_1$ (clean)	C$_{Lmax0}$	C$_{Lmax1}$	
Piper PA-31P Pressurized Navajo	7800	229	72	80	1.94	1.57	1
Rockwell Commander 630A	10,250	266	77	82	1.92	1.69	1
Shorts SD3-30	22,000	453	74	92	2.62	1.69	1
BUSINESS JETS							
Beechjet 400A	16,100	241	82	87	2.93	2.60	1
Cessna Citation CJ1	10,700	240	77	82	2.22	1.96	1
Cessna Citation Mustang	8645	210	73	91	2.23	1.47	7
Dassault Falcon 900	45,500	527	85	106	3.53	2.27	1
Dassault-Breguet Mystère-Falcon 900	20,640	528	82	104	1.72	1.07	1
Dassault Falcon 2000X	41,000	527	84	98	3.26	2.39	1
Embraer Phenom 100	10,472	202	77	100	2.58	1.53	1
Gates Learjet 24D	13,500	232	99	126	1.75	1.08	1
Gulfstream Aerospace IV	71,700	950	108	120	1.91	1.55	1
Safire S-26	5130	143	69	92	2.23	1.26	1
COMMERCIAL JETLINERS							
A320-200	170,000	1320	121	179	2.60	1.19	1
A320-B4	360,000	2800	120	150	2.64	1.69	8
A330	520,000	3892	120	137	2.74	2.10	8
A340-200	610,000	3892	135	145	2.54	2.20	8
Boeing 727-200	172,000	1700	106	171	2.66	1.02	1
B737-400	150,000	1135	146	206	1.83	0.92	8
B757-200	255,000	1994	120	145	2.62	1.80	8
B777-200A	545,000	4005	170	150	2.43	1.55	8
Fokker 100	95,000	1006	109	160	2.35	1.09	1
Lockheed L-1011-1 Tristar	430,000	3456	125	166	2.35	1.33	1

Sources:
1. Jane's All the World's Aircraft
2. Type POH
3. http://www.classg.com/aircraft_specs.i?cmd=compare&manid1=56&model1=60+Duke
4. http://www.classg.com/aircraft_specs.i?cmd=compare&acids=568%2C520%2C172%2C922&replaceid=520&manid2=89&model2=phenom+100
5. http://www.classg.com/aircraft_specs.i?cmd=compare&acids=568%2C520%2C172%2C922&replaceid=520&manid2=89&model2=phenom+100
6. http://www.aeroresourcesinc.com/store_/images/classifieds/119-1.pdf
7. http://viewer.zmags.com/publication/f8e9ba38#/f8e9ba38/1
8. http://webpages.charter.net/anw/ANW/performance.html

TABLE 9-9 Examples of Oswald's Span Efficiencies for Selected Aircraft *(Based on Ref. [39])*

Single-engine Propeller						Twin-engine Propeller					
Manufacturer	Model	S,ft²	b, ft	AR	e	Manufacturer	Model	S, ft²	b, ft	Ag	e
Beechcraft	35	184	33.5	6.10	0.82	Beechcraft	AT-7	349	47.7	6.51	0.74
Boeing-Stearman	PT-l8	298	32.2	6.94	0.75	Cessna	AT-8	295	41.9	5.96	0.61
Cessna	OE-2	174	36.0	7.45	0.7	Douglas	A-26B	540	70.0	9.07	0.79
Cessna	180	174	35.8	7.38	0.75	Curtiss	C-46A	1360	108.1	8.59	0.88
Cessna	150	160	33.3	6.94	0.77	Douglas	C-47B	987	95.5	9.24	0.89
Cessna	172	174	36.1	7.48	0.77	North American	B-25D	610	67.6	7.49	0.78
Cessna	182	174	36.0	7.45	0.84	Martin	B-26F	658	71.0	7.66	0.76
Cessna	185	174	35.8	7.38	0.86	Cessna	310	175	35.0	7.00	0.73
Cessna	177	174	35.5	7.24	0.57	Gulfstream	G-I	610.3	76.5	9.59	0.78
Douglas	O-46A	332	45.8	6.30	0.8	SAAB	SF340	450	70.3	10.99	0.8
Stinson	L-5	155	34.0	7.46	1.02						

new design. Always compare your maximum lift coefficient to that of similar aircraft to ensure unrealistic overestimation is avoided.

9.5.14 Estimation of Oswald's Span Efficiency

The *Oswald span efficiency* is a vital parameter required to predict the lift-induced drag of an airplane. It is named after W. Bailey Oswald, who first defined it in a NACA report published in 1933 [36]. Interestingly, Oswald called it the *airplane efficiency factor*. It is not always easy to estimate, but here several methods will be demonstrated. Note that examples of the Oswald efficiency for selected single- and twin-engine aircraft are shown in Table 9-9.

Basic Definition

The span efficiency can be defined as the resultant of the lift and side force, divided by the product of π, AR, and the lift-induced drag coefficient, as shown below. The expression assumes that C_{Di} is already known, for instance through flight or wind tunnel testing.

Definition of the span efficiency:

$$e = \frac{C_L^2 + C_Y^2}{\pi \cdot AR \cdot C_{Di}} \quad (9\text{-}87)$$

where

C_L = lift coefficient
C_Y = side force coefficient
C_{Di} = induced drag coefficient
AR = reference aspect ratio

If the wing has winglets the aspect ratio should be corrected by modifying the AR using the following expression [32]:

$$AR_{corr} = AR\left(1 + \frac{1.9h}{b}\right) \quad (9\text{-}88)$$

where

AR = original "clean wing" AR
AR_{corr} = boosted AR
b = wingspan
h = height of winglets

Method 1: Empirical Estimation for Straight Wings

Raymer [5] presents the following statistical expression to estimate the Oswald efficiency of straight wings. Note that it omits dependency on taper ratio, but is still handy for conceptual design work. The expression is limited to lower AR only:

$$e = 1.78(1 - 0.045AR^{0.68}) - 0.64 \quad (9\text{-}89)$$

Method 2: Empirical Estimation for Swept Wings

Raymer [5] also presents the following statistical expression to estimate the Oswald efficiency of swept wings. It has limitations similar to Equation (9-89):

$$e = 4.61(1 - 0.045AR^{0.68})(\cos \Lambda_{LE})^{0.15} - 3.1 \quad (9\text{-}90)$$

Brandt et al. [4] present the following expression to estimate the factor:

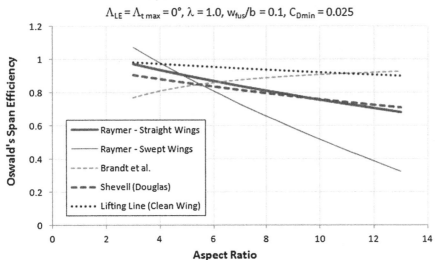

FIGURE 9-67 A comparison of four methods for estimating the Oswald span efficiency factor.

$$e = \frac{2}{2 - AR + \sqrt{4 + AR^2(1 + \tan^2\Lambda_{tmax})}} \quad (9\text{-}91)$$

where

Λ_{LE} = leading edge sweep angle
$\Lambda_{t\,max}$ = sweep angle of the maximum wing thickness line

Method 3: Douglas Method

Shevell [37] presents the following expression to calculate the Oswald efficiency, which is in part based on unpublished studies by the Douglas aircraft company. Its presentation has been modified slightly to better fit the discussion here. Other than that, it returns exactly the same values.

$$e = \frac{1}{\pi \cdot AR \cdot r \cdot C_{Dmin} + 1/((1 + 0.03t - 2t^2)u)} \quad (9\text{-}92)$$

where

t = fuselage width to wingspan ratio = w_{fus}/b
w_{fus} = maximum width of the fuselage and b is the wingspan
u = correction factor for non-elliptical wing planform, typically 0.98 to 1.00.
r = parasitic correction factor and $r = 0.38 - \Lambda_{LE}/3000 + \Lambda_{LE}^2/15000$ and Λ_{LE} is in degrees

Method 4: Lifting Line Theory

The *lifting line theory* is presented in Section 9.7, *Numerical analysis of the wing* and Section 15.3.4, *The lift-induced drag coefficient*: C_{Di}. The Oswald span efficiency can be calculated using the method shown in Section 9.7.5, *Computer code: Prandtl's lifting line method*.

Note that Methods 1 through 4 are compared in Figure 9-67 with a straight Hershey-bar wing. The results for the lifting line theory are theoretical results for a clean wing. Also note that Methods 1 and 2 do not reflect dependency on the taper ratio, λ.

Method 5: USAF DATCOM Method for Swept Wings – Step-by-step

The USAF DATCOM [1, p. 2.2.1-8] is based on a paper by Frost and Rutherford [38] published in 1963. The paper suggests that the Oswald efficiency depends on the factor R, which is the ratio between the actual drag force of a wing and that of an elliptical wing. The idea assumes that both utilize a symmetrical airfoil. Using statistical analysis of a large number of NACA reports, the authors devised a method to estimate the span efficiency for a larger range of AR. Per Ref. [37] the method compared to test data that included following planform and flight conditions:

Aspect ratio:

$$2 \leq AR \leq 10.7$$

Taper ratio:

$$0 \leq \lambda \leq 0.713$$

Leading edge sweep angle:

$$19.1° \leq \Lambda_{LE} \leq 63.4°$$

Leading edge suction parameter:

$$2 \leq R_{l_{LER}} \leq 10$$

Mach number:

$$0.13 \leq M \leq 0.81$$

This does not mean it is not applicable to other planform shapes. The method is used to estimate a special factor, called the *leading edge suction parameter, R*, which is used with the following expression to estimate the span efficiency:

$$e = \frac{1.1(C_{L_\alpha}/AR)}{R(C_{L_\alpha}/AR) + (1-R)\pi} \quad (9\text{-}93)$$

where R = leading edge suction parameter (must be read from Figure 4.7 in the reference document).

This method requires several parameters to be determined, which are then used to extract the LE suction parameter, R, for use with Equation (9-93).

Step 1: Calculate the Lift Curve Slope

Lift curve slope can be calculated from Equation (9-57) (where the variables are explained):

$$C_{L_\alpha} = \frac{\partial C_L}{\partial \alpha} = \frac{2\pi \cdot AR}{2 + \sqrt{\left(\frac{AR \cdot \beta}{\kappa}\right)^2 \left(1 + \frac{\tan^2 \Lambda_{C/2}}{\beta^2}\right) + 4}}$$

Step 2: Calculate the Leading Edge Suction Parameter

Leading edge suction parameter:

$$R_{l_{LER}} = \frac{\rho V l_{LER}}{\mu}$$

where
l_{LER} = leading edge radius (from airfoil data)
V = airspeed
μ = air viscosity, in $lb_f \cdot s/ft^2$
ρ = air density, slugs/ft^3

Step 3: Calculate Special Parameter 1

Special parameter 1:

$$P_1 = \frac{AR \cdot \lambda}{\cos \Lambda_{LE}}$$

where
λ = taper ratio
Λ_{LE} = sweep of the LE

Step 4: Calculate Special Parameter 2

Special parameter 2:

$$P_2 = R_{l_{LER}} \times \cot \Lambda_{LE} \sqrt{1 - M^2 \cos^2 \Lambda_{LE}}$$

where
M = Mach number

Step 5: Read or Calculate the Leading Edge Suction Parameter

If $P_2 \leq 1.3 \times 10^5$ determine R from Figure 9-68 or calculate from the following expression:

$$R = \left(-4.728 + 2.185 \cdot \log_{10}(P_2) - 0.2131 \cdot \log_{10}(P_2)^2\right) + 0.095 \cdot \sin\left(\frac{\pi P_1}{20}\right)$$

If $P_2 > 1.3 \times 10^5$ determine R from Figure 9-68 or calculate from the following expression:

$$R = 0.86 + 0.1119 \cdot \left(\frac{P_1}{10}\right)^{\frac{1}{1.8 + P_1}}$$

Neither equation is based on theoretical analysis, but rather derived using a curve fit methodology that results in acceptable fit to the graphs.

EXAMPLE 9-12

Determine the Oswald's span efficiency for the Learjet 45XR, whose $AR = 7.33$ and $TR = 0.391$. Compare methods 2 and 4. Assume the airfoil has a section lift curve slope of 2π, a LE sweep of 17°, mid-chord sweep of 10.5°, and a LE radius of 0.1 ft. Assume the maximum thickness is at the mid-chord as well. Assume an airspeed of $M = 0.3$ at S-L on a standard day ($V = 335$ ft/s) and ignore the fact the airplane has winglets.

Solution

Method 2

Using Equation (9-90):

$$e = 4.61(1 - 0.045 AR^{0.68})(\cos \Lambda_{LE})^{0.15} - 3.1$$
$$= 4.61\left(1 - 0.045(7.33)^{0.68}\right)(\cos(17°))^{0.15} - 3.1$$
$$= 0.6807$$

> **EXAMPLE 9-12** *(cont'd)*
>
> Using Equation (9-91):
>
> $$e = \frac{2}{2 - AR + \sqrt{4 + AR^2(1 + \tan^2 \Lambda_{tmax})}}$$
>
> $$= \frac{2}{2 - 7.33 + \sqrt{4 + 7.33^2(1 + \tan^2(10.5°))}} = 0.8374$$
>
> **Method 4**
>
> **Step 1:** Mach number parameter:
>
> $$\beta = \sqrt{1 - M^2} = \sqrt{1 - 0.3^2} = 0.9539$$
>
> $$C_{L_\alpha} = \frac{2\pi \cdot AR}{2 + \sqrt{\left(\frac{AR \cdot \beta}{\kappa}\right)^2 \left(1 + \frac{\tan^2 \Lambda_{C/2}}{\beta^2}\right) + 4}}$$
>
> $$= \frac{2\pi \cdot 7.33}{2 + \sqrt{\left(\frac{7.33 \times 0.9539}{1}\right)^2 \left(1 + \frac{\tan^2 10.5°}{0.9539^2}\right) + 4}}$$
>
> $$= 4.90$$
>
> **Step 2:**
>
> $$R_{l_{LER}} = \frac{\rho V l_{LER}}{\mu} = \frac{(0.002378)(335)(0.1)}{3.745 \times 10^{-7}} = 212718$$
>
> **Step 3:**
>
> $$P_1 = \frac{AR \cdot \lambda}{\cos \Lambda_{LE}} = \frac{7.33 \times 0.391}{\cos 17°} = 2.997$$
>
> **Step 4:**
>
> $$P_2 = R_{l_{LER}} \times \cot \Lambda_{LE} \sqrt{1 - M^2 \cos^2 \Lambda_{LE}}$$
>
> $$= 212718 \times \cot(17°)\sqrt{1 - 0.3^2 \cos^2(17°)} = 6.67 \times 10^5$$
>
> **Step 5:** Since $P_2 > 1.3 \times 10^5$, R can be determined as follows:
>
> $$R = 0.86 + 0.1119 \cdot \left(\frac{P_1}{10}\right)^{\frac{1}{1.8 + P_1}}$$
>
> $$= 0.86 + 0.1119 \cdot \left(\frac{2.997}{10}\right)^{\frac{1}{1.8 + 2.997}} = 0.947$$
>
> **Step 6:**
>
> $$e = \frac{1.1 (C_{L_\alpha}/AR)}{R(C_{L_\alpha}/AR) + (1 - R)\pi}$$
>
> $$= \frac{1.1(4.90/7.33)}{0.947(4.90/7.33) + (1 - 0.947)\pi} = 0.919$$
>
> Clearly, there is a range of values to consider, begging the question: which one should we pick? Unless there is confidence in a particular method, this author would take the average of the three (0.812) until a better number is determined.

9.6 WING STALL CHARACTERISTICS

All normal airplanes need to exceed a certain minimum airspeed before they can become airborne and maintain level flight. The minimum airspeed required for level flight is the stalling speed. What transpires at this speed arguably inflicts one of the most important design challenges for the wing design, or for that matter, the entire airplane. This section is dedicated to the stall and intended to provide important information about this well-known phenomenon.

It is helpful to visualize the stall maneuver in terms of altitude as well as airspeed. A pilot-controlled stall maneuver is depicted in Figure 9-69. It begins with engine power being cut and a subsequent deceleration. For compliance with 14 CFR Part 23.201, *Wings level stall*, this deceleration should be as close to 1 KCAS per second as possible.

As the airplane slows down, a larger and larger α is required to maintain altitude and the *approach to stall* phase culminates in the stall itself, as the airplane reaches α_{STALL} or α_{CLmax} and breaks the stall by the sudden drop of the nose. The drop leads to a dive, which, in turn, results in altitude being lost, as shown in the figure. This altitude loss depends on the size of the aircraft and can be as small as 25–50 ft for a small and light homebuilt or ultralight aircraft; 200–400 ft for a two- to four-seat single-engine GA aircraft; to 2000 ft or more for a large commercial jetliner aircraft. The *stall recovery* phase ends with power being added and a subsequent climb to altitude.

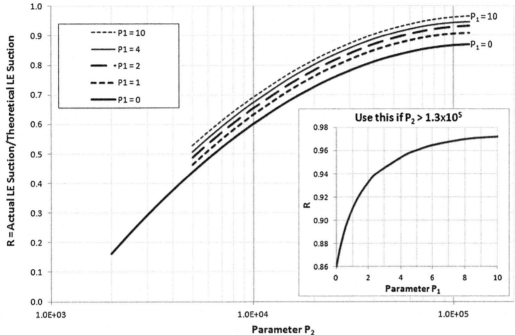

FIGURE 9-68 Leading edge suction parameter *(Based on Ref. [4])*.

9.6.1 Growth of Flow Separation on an Aircraft

Flow separation can be a serious issue for aircraft and one that the designer must be fully aware of in order to minimize it. Consider the series of images in Figure 9-70, which shows an airplane seen from the same perspective at different airspeeds. Note that the airspeed ratios and *AOA* are approximate and correspond to the design in the figure only, although they would be applicable to many aircraft types. If the aircraft is well designed there should not be any flow separation regions at the cruising speed. This ensures it will be as efficient as possible at cruise because the flow separation is a source of increased pressure drag. Once slowing down from cruising speed the *AOA* begins to rise and it is inevitable that separation regions begin to form and increase as well.

As soon as the airplane has decelerated to approximately its economy cruising speed (or best rate-of-climb airspeed), a separation region has already begun to form in two places: mid-span and at the wing/fuselage juncture. This is highly undesirable,

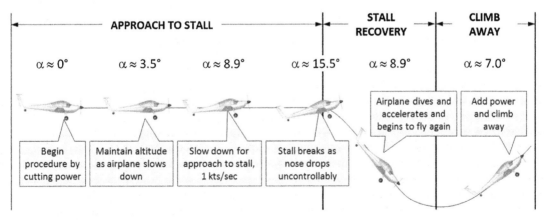

FIGURE 9-69 A schematic showing an approach to stall, stall recovery and subsequent climb away.

368 9. THE ANATOMY OF THE WING

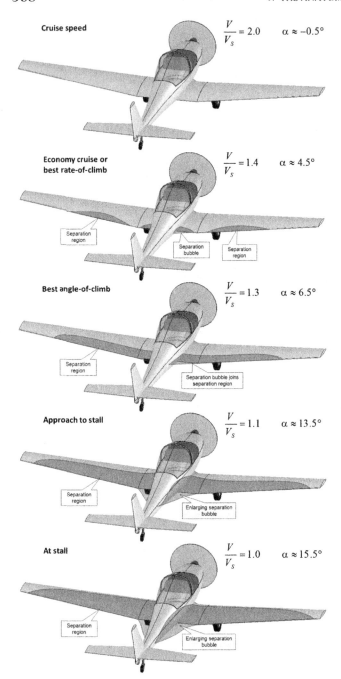

FIGURE 9-70 Growth of flow separation region on an aircraft as seen from a fixed point from the aircraft.

but unavoidable. The separation on the wing depends solely on its geometry. For instance, the aircraft shown in Figure 9-70 has a tapered wing planform whose section lift coefficients are highest around the mid span (see Section 9.4.5, *Straight tapered planforms*), although an attempt has been made here to suppress them at the tip with a geometric wing washout. This causes the flow to begin to separate mid-span rather than elsewhere on the wing. Of course, we seek a progression of the separation that begins at the root and moves toward the tip, but achieving this would require a large wing washout that would be detrimental to the wing's efficiency, which, by the way, is the reason why a tapered planform was chosen in the first place.

With respect to the flow separation at the wing/fuselage juncture, we must distinguish between separation caused by the high *AOA* on the wing, and that caused by poor geometry between the two. The former is unavoidable with an increase in *AOA*, whereas the latter forms prematurely at relatively low *AOA*, sometimes even before any separation is visible on the wing itself. It is the responsibility of the aerodynamicist to suppress this formation as long as possible and this can be achieved by a careful tailoring of the wing root fairing. The shape of this separation is best described as one having a distinct volume that extends as far as a chord length or more into the flow field behind the juncture. For this reason it is appropriately called a *separation bubble*. The airplane in Figure 9-70 does not have a wing root fairing in this area, but the formation of the separation bubble should encourage the aspiring aircraft designer to design a fairing to suppress it. The pressure inside it is less than in the surrounding area and, therefore, it increases the drag of the aircraft. Additionally, it reduces the airplane's lift curve slope, requiring it to fly at a higher *AOA* than otherwise, with the associated increase in induced drag. A manifestation of such a slope increase is shown in Figure 8-56.

Separation bubbles form easily because of the large rise in airspeed in the channel formed by the wing and the fuselage juncture. To visualize why, one must keep in mind that as a volume of air approaches the airplane, the pressure within it changes from the static or atmospheric pressure it had. As the volume approaches and passes the vehicle, it undergoes a rapid rise and reduction in pressure and a subsequent rise back to atmospheric pressure. The pressure change is associated with the change in the speed of the molecules within the volume, but this deceleration or acceleration is ultimately caused by the geometry over which the volume flows. When the pressure drops, as it does when the airspeed increases, we call the rate at which this takes place a *favorable pressure gradient*. When the pressure rises, as happens when the speed of the molecules slows down, we call it an *adverse pressure gradient*.

These concepts are fundamental to understanding the nature of flow separation. For instance, the volume of air flowing over the upper surfaces of the wing accelerates to a maximum airspeed on the highest part of the airfoil (and depends on the *AOA*). There is also a similar acceleration in airspeed along the fuselage side. At the juncture of the wing and fuselage the effects of the two combine, to make the resulting airspeed greater than elsewhere along the wing (or fuselage). This higher airspeed results in a lower pressure in that region than elsewhere. Clearly, this pressure must rise back to the atmospheric pressure aft of the wing, however, because it was lower to begin with it must do so more rapidly. This results in a large adverse pressure gradient, which generates a flow deceleration problem that nature solves by separating the flow from the surface. Depending on the geometry of the airplane, this separation can easily begin to form at moderate *AOA*s, even during a high-speed climb.

Further deceleration of the aircraft to, say, its best angle-of-climb airspeed, causes the two separation regions to join into a single one that extends from the fuselage to a specific span station. The separation bubble at the wing/fuselage juncture continues to grow into the flow field and this is represented as the volume aft of the wing root trailing edge. As the aircraft approaches stall, a larger and larger area of the wing is covered with flow separation. The direction of the progression of the separation should be away from the fuselage toward the wingtips. Eventually, at stall, the wing is mostly separated, but if well-designed the wingtip should still be un-stalled for roll stability.

General progression of stall on selected wing planform shapes is illustrated in the next section.

9.6.2 General Stall Progression on Selected Wing Planform Shapes

9.6.3 Deviation from Generic Stall Patterns

While Figures 9-71 and 9-72 provide a fundamental understanding of the impact planform has on stall progression, this is further complicated by the selection of airfoils and wing washout that may be employed for those planform shapes. As an example of this consider Figure 9-73, which shows the stall progression over three separate straight tapered planform shapes [40–42]. The figure shows that once different airfoils, *AR*, λ, and even surface roughness are accounted for the impact of the stall progression will be modified. This introduces asymmetry in the stall progression, as well as regions of initial flow separation, and intermittent and complete stalling. It indicates that each wing style must be evaluated based on its own specific geometry.

Another example is depicted in Figure 9-74, which compares the progression of the separation region on untwisted elliptical and crescent-shaped wing planform shapes, based on a paper written by van Dam, Vijgent, and Holmes [43], showing how the stall progression can be highly affected by vortical flow forming along a highly swept outboard leading edge. In the paper, the authors indicate that the separation-induced vortex flow over the highly swept tips of the crescent wing improved its stall characteristics when compared to the unmodified elliptical wing (whose stall characteristics were shown to be abrupt and unsteady). This viscous phenomenon delayed full stall of the experimental wing to a higher stall *AOA* (14.5° versus 13.0° for the elliptical wing) and yielded a higher C_{Lmax} (1.06 versus 0.98, respectively). At lower *AOA*, the lift characteristics were found to be practically identical, something important to keep in mind for cruise operations. The complex surface flow depicted in Figure 9-80 shows that at 14° the outboard portion of the elliptical wing is fully separated, while it is mostly attached on the crescent wing. The crescent-shaped wing planform is reminiscent of that of numerous species of birds.

9.6.4 Tailoring the Stall Progression

Good stall characteristics are simply a question of safety. An airplane that constantly rolls left or right at stall is at an increased risk of entering spin. If the airplane stalls close to the ground, perhaps as a consequence of the pilot banking hard to turn on final approach, there simply is no time (or altitude) to recover, no matter how good spin recovery characteristics the airplane has or how proficient the pilot. The consequence is usually a fatal crash.

There is no good reason to develop an airplane without good stall characteristics, especially considering these can be tailored into the airplane from its inception. It is not being claimed that this is easy, although nowadays it is easier using CFD solvers. CFD methods such as vortex-lattice, doublet-lattice, and other panel codes can be used to determine the distribution of section lift coefficients along the span of the aircraft, even though such solvers ignore viscosity. Navier-Stokes solvers can be of even greater use, as long as the selected turbulence model does not mislead the user in the extent and shape of the flow separation (for instance, see Section 23.3.16, *Reliance upon analysis technology*). Then, armed with an understanding of how the stall progresses along the wing, the designer can select a combination of airfoil types and wing washout to control the stall progression along span of the wing.

Design Guidelines

The target stall pattern should always begin at the root and progress toward the tip as the *AOA* is increased. This ensures the wingtips will be the last part of the wing to stall, providing vital roll stability and control throughout the maneuver. If inviscid design methodology is used to tailor the stall progression, the goal should be to ensure the section lift coefficient (C_l) at the 70% span station is no higher than the maximum lift coefficient (C_{lmax}) of the airfoil at that station. Furthermore, from 70% to 100%, C_l should gradually

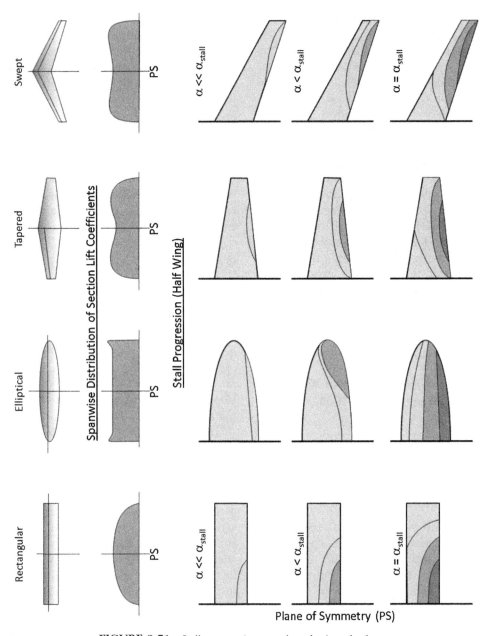

FIGURE 9-71 Stall progression on selected wing planforms.

fall to zero. Some authors (e.g. Torenbeek [44]) recommend (C_{lmax} − 0.1), but this may be hard to achieve in practice without excessive washout. At any rate, the idea is to promote roll stability at stall and washout is a powerful tool to provide this function. If a viscous analysis method is used (e.g. Navier-Stokes solvers), then a more pinpointed tailoring can be accomplished, but only if the flow separation prediction is deemed trustworthy. For something as serious as stall tailoring, flow visualization of the separated region obtained from wind tunnel testing should always be used to validate the CFD model.

Figure 9-75 shows an example of typical linear analysis for a tapered wing ($\lambda = 0.5$) that features the same airfoil (NACA 65_2-415) throughout the wing. Since the Reynolds number at the tip is only one-half of the root value, the C_{lmax} is less at the tip and this should be taken into account for airplanes that feature tapered planform shapes. The baseline wing design (solid curve) has no washout, whereas the other three have a 2°, 4°, and 6° washout, respectively. The graph shows the distribution of C_l for these four models at AOA of 16°. The thick dashed line shows the distribution of the C_{lmax} from root to tip.

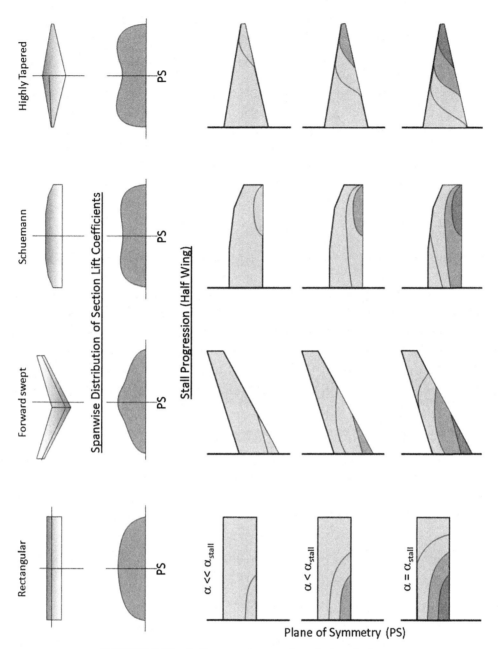

FIGURE 9-72 Stall progression on selected wing planforms.

The graphs in Figure 9-75 show that the C_l for the baseline wing exceeds the C_{lmax} between Spanwise Station 0 to about 0.87. This means the wing should be expected to be fully stalled to that point, something that would result in poor stall characteristics, as barely 10% of the span (at the tip) is un-stalled. The proposals for a 2°, 4°, and 6° washout all lead to improvements, especially the last one (6°), which brings the stall to Spanwise Station 0.65. However, as can be seen in Table 9-6, washout higher than 4° is rare. Excessive washout can lead to an increase in the lift-induced drag as the wing must be operated at a higher AOA to generate the same airplane lift coefficient (C_L). A better solution would be to feature less washout, say somewhere between 2° and 4°, and feature a tip airfoil that has a higher C_{lmax} than the root (an aerodynamic washout). Of course it is assumed such an airfoil would offer gentle stall characteristics (not those of the NACA 23012 airfoil presented in Section 8.2.10, *Famous airfoils*).

Tailoring Stall Characteristics of Wings with Multiple Airfoils

A possible stall tailoring remedy is proposed in the graph of Figure 9-75. It consists of increasing the C_{lmax}

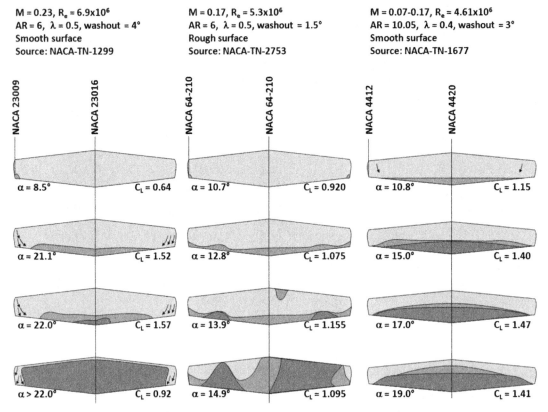

FIGURE 9-73 Stall progression on straight tapered wing planforms differing in AR, λ, washout, and surface roughness *(Based on Refs [39–41])*.

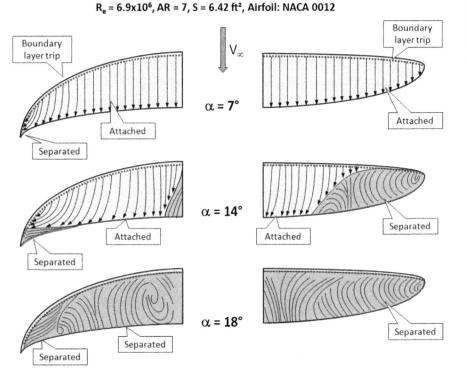

FIGURE 9-74 Stall progression on a crescent-shaped and an elliptical wing planform of comparable geometry shows the complex flow inside a separation region *(Based on Ref. [42])*.

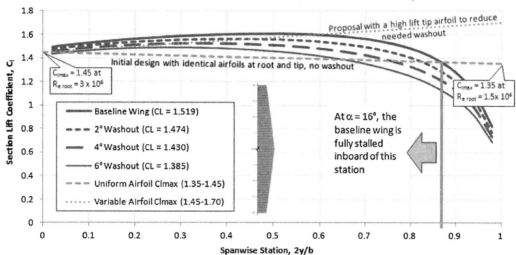

FIGURE 9-75 The effect of washout on probable stall progression. The baseline wing is more tip-loaded than the ones with washout and this will cause it to stall closer to the wingtip, which may cause roll-off problems. Washout is a powerful means to control stall progression and can be enhanced by selecting a high-lift airfoil at the tip.

of the tip airfoil from 1.57 to about 1.7, by defining a new tip airfoil. Of course, this may pose some challenges; such an airfoil might possess some undesirable characteristics too. However, assuming this is achievable; it may be possible to manufacture the wing without a geometric washout. This can be an advantage for some composite wing designs, as it allows uni-directional plies in the spar to be laid up in a more manufacturing-friendly fashion than a twisted spar. In practice, however, tailoring the wing for good stall progression is solved using a combination of both geometric and aerodynamic washouts.

Other Issues Associated with Wings with Multiple Airfoils

Multi-airfoil wings are the norm for high-performance aircraft, but are also common in smaller and simpler GA aircraft. High-performance aircraft require wings that allow the airplane to operate at low airspeeds while avoiding compressibility effects associated with high-speed flight. It takes sophistication in manufacturing to produce such wings. This is particularly challenging in the production of wings made from alloys, as the difference in geometry inevitably requires the wing skin to be stretched to conform to the resulting compound surface. This will become clearer in a moment.

A hypothetical multi-airfoil wing is shown in Figure 9-76, Figure 9-77, and Figure 9-78, with the layout presented in Figure 9-76. It should be stressed there is no rhyme or reason why the particular airfoils have been selected other than to demonstrate aerodynamic, structural, and manufacturing complexities that may arise in such a wing design. The wing is defined using three airfoils: at the plane of symmetry (the "root"); at the intersection of the flap and aileron; and at the tip. Structurally, the intersection of the flap and aileron is a good location to anchor a new airfoil as a rib is required there to mount the hard-points for the control surfaces.

Figure 9-76 also shows that the airfoil selection will affect the extent of the laminar boundary layer on the wing's upper surface (assuming this is achievable). This change must be accounted for in the drag estimation for the wing. Achieving a laminar boundary layer is difficult, as has already been discussed in Chapter 8, *The anatomy of the airfoil*. Attempting this on the above wing will require careful and more expensive manufacturing tooling.

Figure 9-77 shows the distribution of section lift coefficients along both wing halves at $\alpha = 16.5°$ and is based on potential flow analysis. It also shows how the maximum lift varies along the span because of the three airfoils and that the progression of stall begins just outside the 50% span station. Also plotted is the distribution of C_l for the same wing with a 3° washout, showing improvements, although the wing is still highly tip-loaded. The reader is also reminded that the linear method used does not correctly predict flow separation due to chord- and spanwise flow (recall this is a swept wing design). Therefore, although the linear method is helpful in understanding the airflow around the wing, ultimately the graph represents an ideal flow scenario that is not present in the real flow.

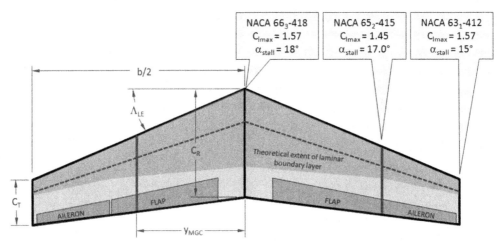

FIGURE 9-76 An example wing layout, showing the theoretical extent of laminar boundary layer and variation in its maximum lift capability.

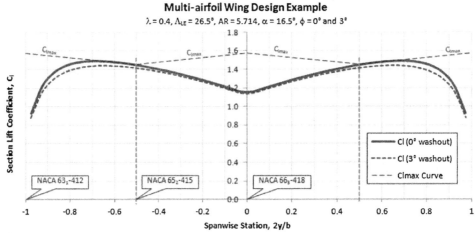

FIGURE 9-77 An example of a potential flow lift analysis of the multi-airfoil wing layout that features 0° and 3° washout.

Figure 9-78 shows some structural and manufacturing issues that present themselves in wings with multiple airfoils. The geometry that results often comes as a surprise to novice engineers designing the structure of such wings. Since the three airfoils used have dissimilar geometry, the spar extending from root to tip will be subject to a geometric non-linearity; the skin will form a compound surface. In the figure, the wing is cut along a proposed spar-plane. The view along the cut shows two situations are present: first, there is a discontinuity in the spar OML at the mid-span station. Second, mathematically, the spar cap must also feature a slightly curved surface, extending from root to mid-span and then to the tip. If it is required that the skin must adhere perfectly to the compound surface and it will be made from aluminum, the sheet metal for the spar and skin will have to be pressed to shape using hydraulic presses. This will greatly increase the cost of production. In real aircraft, especially less expensive ones, the discontinuity is usually solved by sheets terminating along the discontinuity. The curvature may have to be solved by straightening the spar cap and accepting that the wing will not be what the aerodynamics group really wants or by inserting shims between the spar and the skin. This may also make it more challenging to achieve a laminar boundary layer, as the resulting shape may no longer present the intended theoretical shape.

9.6.5 Cause of Spanwise Flow for a Swept-back Wing Planform

Swept-back wings can experience a significant and uncontrollable pitch-up moment at high angles-of-attack. The reason for this is twofold:

(1) The aft-swept planform induces local upwash near the tip which increases the local section lift coefficients. This means that the tip airfoils reach

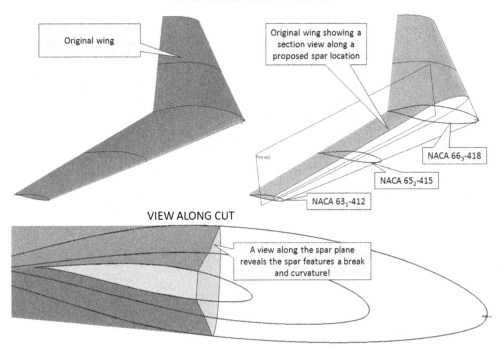

FIGURE 9-78 An example of geometric complexities arising from the multi-airfoil wing layout.

their stall section lift coefficients sooner than the inboard airfoils.

(2) Air begins to flow in a spanwise direction near the tip, but this leads to an early flow separation.

The first cause is somewhat hard to explain in layman's terms, but we will try. Imagine the wing is cut into a finite number of small sections along the span of the wing. The sections extend from the root to the tip such that the inboard section is always upstream of the adjacent outboard section. For this reason, the inboard section will begin to disturb the flowfield before the section outboard of it. A part of this disturbance is an upwash ahead of that section which extends spanwise into the flowfield. Then, when the outboard section begins to disturb the flowfield ahead of it, there is already an upwash component in it, induced by the section inboard of it. This section will then impart it own influence on the flowfield, which manifests itself as a slightly greater upwash for the section outboard of it, and so on. The upwash implies a greater local *AOA*. A greater local *AOA* implies a higher section lift coefficient.

The cause of the spanwise flow, on the other hand, and which is the topic of this section, can be explained as follows. Consider the swept-back wing of Figure 9-79, which shows an aircraft with a swept-back wing at some *AOA*. Two sample chordwise pressure distributions are drawn on the right wing. Also, the locus of the peak spanwise pressure distribution along the wing is shown as the dotted curve drawn at the pressure peaks. It can be seen that the pressure peak on the outboard wing is higher than inboard. Now, consider a line perpendicular to the centerline of the fuselage at some arbitrary chord station. It cuts through the chordwise pressure distributions as indicated by the vertical arrows. The inboard arrow is shorter than the outboard one, indicating higher pressure than on the outboard one (remember these distributions represent low and not high pressures). For this reason the higher pressure on the inboard station forces air to flow from the inboard to the outboard station, giving the flow field an overall outboard spanwise speed component. This is shown as the streamlines on the left wing.

9.6.6 Pitch-up Stall Boundary for a Swept-back Wing Planform

As stated in the previous section, swept-back wings suffer from significant pitch-up moment near and at stall. The effect depends on the quarter-chord sweep angle, $\Lambda_{C/4}$, and aspect ratio, *AR*. The effect is investigated in NACA TR-1339 [45] and NACA TN-1093 [46]. Figure 9-80 is reproduced from those references and summarizes the effect. It shows that the higher the *AR*, the less is the $\Lambda_{C/4}$ at which the pitch-up is experienced. This is very important in the development of long-range high-subsonic aircraft as a high *AR* favors long-range but high $\Lambda_{C/4}$ favors high airspeed. These properties are therefore mutually detrimental. As contemporary airliners demonstrate, the development of the modern airfoil has remedied this limitation with its greater critical Mach number (at which shocks begin to form). This has allowed aircraft with lower $\Lambda_{C/4}$ ($\approx 20°-27°$)

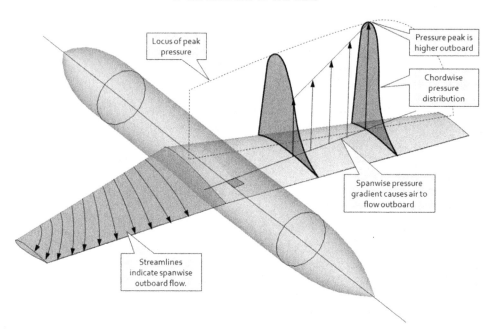

FIGURE 9-79 The spanwise pressure gradient on an aft-swept wing results in an outboard spanwise flow. The opposite holds true for a forward swept wing *(Based on Ref. [17])*.

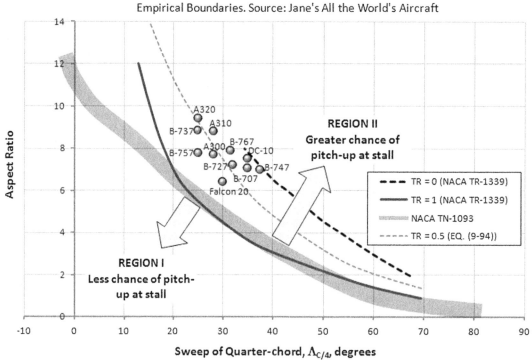

FIGURE 9-80 Empirical pitch-up boundary for a swept-back wing. *Reproduced based on Refs [44] and [45]*.

to utilize higher AR and, thus, operate more efficiently at the usual cruising speeds of such aircraft ($M \approx 0.78-0.82$).

An empirical equation based on the data in Ref. [44] can now be developed. It relates the taper ratio, λ, and, sweep angle of the quarter chord in degrees, $\Lambda_{C/4}$, to calculate an AR limit for swept-back wings. For a given $\Lambda_{C/4}$ the selected AR should be less than this limit:

$$AR_{\lim} \leq 17.714(2 - \lambda)e^{-0.04267\Lambda_{C/4}} \qquad (9\text{-}94)$$

Alternatively, given a target AR, the $\Lambda_{C/4}$ in degrees should not exceed the value below:

$$\Lambda_{C/4_{lim}} \leq 23.436[\ln(17.714(2-\lambda)) - \ln(AR)] \quad (9\text{-}95)$$

As shown in Figure 9-81, the combination of AR, λ, and $\Lambda_{C/4}$ can lead to desirable or undesirable results in terms of nose pitch-down or pitch-up characteristics at stall. Aircraft that feature aft-swept wings should always be wind tunnel tested for stability at stall, but the awareness of the data used to produce Figure 9-80 can go a long way in preventing serious stall problems from presenting themselves.

Finally, NACA RM-L8D29 [47] presents various results that are helpful to the designers of swept-back wings. It investigated the effect of a number of high-lift devices and fences on the stall characteristics on stall characteristics (see Figure 9-82). It concluded that the half-span, leading-edge slats eliminated the tip stall and prevented the nose-up pitching moment. The flaps complicated the stall characteristics and formed a loop in the pitching moment curve (in the figure), although it was suggested it could be brought under control using an appropriately stabilizing surface.

Derivation of Equations (9-94) and (9-95)

The derivation is based on the data obtained using Ref. [44], which presents two curves that are functions of the quarter-chord sweep angle, $\Lambda_{C/4}$; one represents a taper ratio $\lambda = 0$ and the other $\lambda = 1$. The idea is to derive an empirical expression that is a function of λ and $\Lambda_{C/4}$ that closely fits both expressions. This can be done with some numerical analysis. A least-squares exponential fit to the two curves yields the two following expressions:

For $\lambda = 0$:
$$AR = 35.885 \cdot e^{-0.04206\Lambda_{C/4}}$$

For $\lambda = 1$:
$$AR = 18.171 \cdot e^{-0.04329\Lambda_{C/4}}$$

Since the values of the two exponents are relatively close to each other it can be deduced that the value of λ has minimal impact on it and primarily affects the constant. For this reason the average of the two (−0.04267) can justifiably be used for the empirical expression. Assuming a function of the form:

$$AR = f(\lambda) \cdot e^{-0.04267\Lambda_{C/4}} \quad (i)$$

The function $f(\lambda)$ can then be approximated by noting that the constant changes from 35.885 when $\lambda = 0$ to 18.171 when $\lambda = 1$. It is convenient to employ a parametric representation for this variation, in particular considering that λ can be used unmodified as the parameter. Thus, Equation (i) can be rewritten as follows:

$$\begin{aligned} AR &= (35.885(1-\lambda) + 18.171\lambda)e^{-0.04267\Lambda_{C/4}} \\ &= (35.885 - 35.885\lambda + 18.171\lambda)e^{-0.04267\Lambda_{C/4}} \\ &= 17.714(2-\lambda)e^{-0.04267\Lambda_{C/4}} \quad (ii) \end{aligned}$$

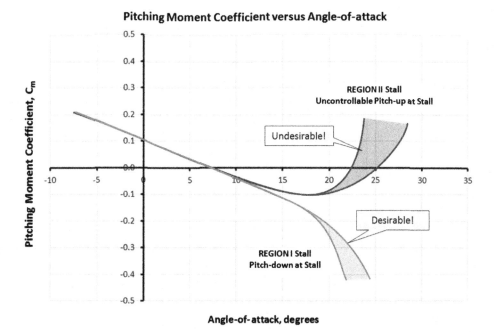

FIGURE 9-81 The combination of AR, TR, and quarter-chord sweep can lead to desirable or undesirable pitch characteristics at stall.

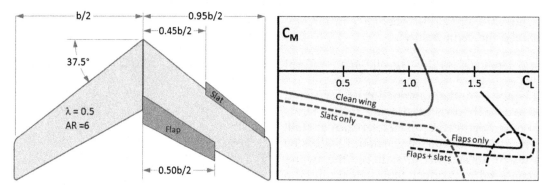

FIGURE 9-82 The effect of various combinations of a double-slotted flaps and slats on the pitching moment of a 37° swept wing. The loop is attributed to the section characteristics of the double-slotted flap *(Based on Ref. [46])*.

And this is Equation (9-94). Equation (9-95) is simply obtained by solving for $\Lambda_{C/4}$.

$$AR = 17.714(2 - \lambda)e^{-0.04267\Lambda_{C/4}}$$
$$\Rightarrow \ln(AR) = \ln(17.714(2-\lambda)) - 0.04267\Lambda_{C/4}$$
$$\Rightarrow \Lambda_{C/4} = 23.436[\ln(17.714(2-\lambda)) - \ln(AR)]$$

QED

9.6.7 Influence of Manufacturing Tolerances on Stall Characteristics

Inherent roll instability at stall is one of the most common handling deficiencies affecting aircraft. In fact, most aircraft ever built have displayed the condition to some extent, requiring the introduction of "fixes" to remedy it. The manifestation of this condition occurs at stall as the aircraft does not drop the nose with wings level, but rather rolls uncontrollably to the left or right side (see Figure 9-83). Its root causes are complicated and one should not assume there is a single cause, but rather a combination of factors.

It may sound strange, but all aircraft are inherently asymmetric, even though this is usually impossible to discern with the naked eye. Nevertheless, every single serial number differs slightly from the previous or the following one in its deviation from the intended outside mold line (OML). Not even the left wing of any given airplane is a perfect mathematical mirror image of the

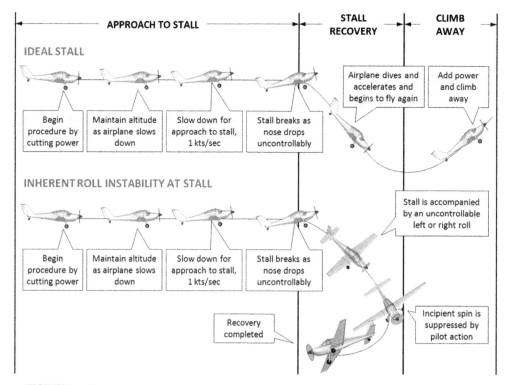

FIGURE 9-83 Comparing an "ideal stall" (top) to one with an inherent roll instability (bottom).

right one, although it is completely invisible to the casual observer. There are always subtle differences in washout, thickness, waviness, and their corresponding locations on each wing, all of which may combine to promote roll instability at stall.

A typical deviation from the OML in the aviation industry amounts to ±0.125 inches, although most wings must usually meet tolerances ranging from ±0.050 to ±0.100 inches over the lifting surfaces. Manufacturers of GA aircraft that feature NLF lifting surfaces often take it one step further and maintain even tighter tolerances; sometimes as tight as ±0.005 inches along the leading edge. While maintaining tolerances is imperative in the manufacture of aircraft, overly tight tolerances are of detrimental value. They call for expensive and robust inspection procedures and additional manpower to demonstrate such wings meet the set specifications. Very tight tolerances should always be justified with research.

One of the main concerns manufacturers have with NLF wings is their sensitivity to deviations from the OML, as this may promote early transition of a laminar boundary layer into a turbulent one, which, in turn, increases drag. This often requires expensive repairs to be made to wings that arguably are just fine. The author of this book once investigated the some 300 production aircraft with NLF wings to determine whether there was a correlation between the magnitude of such small manufacturing deviations and stall characteristics reported by the production flight test pilots. No correlation could be found. Tight tolerances had subjected the production to costly inspection procedures without measurable benefits.

While this is not intended to reject the application of tight tolerances in the manufacturing of NLF wings, there clearly is a point of diminishing returns. It turns out that near stall *AOA*, the flow is altogether separated anyway and, therefore, insensitive to minor deviation from the OML. The source of roll instability should be looked for elsewhere, such as in deviation in thickness, twist, dihedral, and objects that are asymmetrically exposed to the airflow.

It would seem from the above discussion that if too tight tolerances are a problem, then perhaps loose tolerances may be desirable. However, this is not the case. Loose tolerances allow an airplane to be excessively asymmetric, not just in surface qualities, but in large deviation from the OML of the surface that almost certainly would cause an inherent roll instability. It is best to design sensitivity to small deviations out of the OML by providing assertive aerodynamic roll stability at high *AOA*. This can be done by increasing the camber of the wingtip airfoil (by specifically selecting a high-lift airfoil), or by reducing its *AOA* using washout, or by providing a leading-edge extension (see Section 23.4.11, *Stall handling — wing droop (cuffs, leading-edge droop)*, or by introducing slats.

9.7 NUMERICAL ANALYSIS OF THE WING

The advent of the digital computer has clearly revolutionized science and engineering. As far as aircraft design is concerned, the technology allows far more complicated and realistic analyses to be performed than previously possible. In its most sophisticated form, the use of computer technology allows the full Navier-Stokes equations (NSE) to be solved for extremely complex situations. This ranges from airflow resembling that experienced by an insect to that of a tumbling meteorite entering Earth's atmosphere at hypersonic speeds. At the present time, however, the application of this method is both time-consuming and expensive as it requires a detailed digital model of the geometry to be prepared and flow solution obtained using a cluster of interconnected computers. Meshing the model is often more art than science and requires an experienced individual to complete. Improperly designed mesh simply yields erroneous results. This is further compounded by the amount of time required to solve the problem. It can easily take a couple hours to a couple of days to accomplish, depending on the complexity of the body and the computational prowess of the individual (or company) involved. For this reason, conceptual design using NSE solvers is very impractical and should not even be considered. Currently, NSE is really a *post-conceptual* design tool; it should be used *after* the concept has been designed and not to design it.

There are a number of practical numerical methods available to the designer that, in comparison, are lightning fast and, for attached flow, are equally as accurate as the solution obtained using the NSE. Among those are the *lifting line method* and panel methods such as the vortex-lattice or doublet-lattice. While programmatically considerably older than the modern Navier-Stokes solver, they remain far more practical during the conceptual design phase and are simple enough to implement using a desktop computer. Due to space limitations, only the most basic of these methods will be demonstrated; *Prandtl's lifting line method*. For other methods, the reader is directed toward excellent texts such as those by Katz and Plotkin [48]; Pope [49]; Bertin and Smith [50]; and Moran [51].

9.7.1 Prandtl's Lifting Line Theory

Developed by Ludwig Prandtl (1875–1953) and his colleagues at the University of Göttingen between

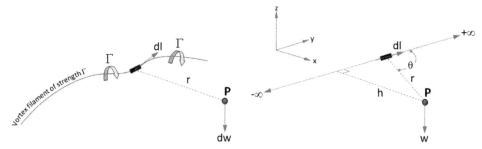

FIGURE 9-84 A curved vortex filament (left) and an infinite, straight vortex filament (right).

1911–1918, the lifting line theory can be used to determine the aerodynamic characteristics of straight three-dimensional wings. The method does not treat wings with dihedral or sweep, but can account for wing twist and varying airfoils and chords along the span of the wing. It is reliable for wings whose AR is no smaller than about 4. The method mathematically replaces the wing with a number of constant-strength vortices, here denoted with the Greek letter Γ. The problem, effectively, revolves around determining their strength; however, once this has been accomplished, it is possible to estimate a number of characteristics, such as lift and drag, downwash, and distribution of lift along the wing. It is thus useful not only for aerodynamic properties, but also for structures and stability and control. A derivation of the method will now be presented, but first the following mathematical constructs must be introduced.

The Vortex Filament and the Biot-Savart Law

A vortex filament is an imaginary spatial curve that induces a rotary flow in the space through which it passes (see Figure 9-84). The best analogy is to think of it as the center of a tornado with the associated circulatory flow around its core. The ability of the filament to induce circulation around it depends on its strength, denoted by Γ. Consider an infinitely small vector segment **dl** along the filament and some arbitrary point **P** in space. The small segment will induce a velocity **dw** at the point **P**, whose magnitude can be determined using the *Biot-Savart*[4] *law*:

Biot-Savart law:

$$dw = \frac{\Gamma}{4\pi} \frac{\mathbf{dl} \times \mathbf{r}}{|\mathbf{r}|^3} \qquad (9\text{-}96)$$

The Biot-Savart law is named after the French mathematician and physicist Jean-Baptiste Biot (1774–1862) and Félix Savart (1791–1841), a fellow Frenchman who trained for a career in medicine, although his mind was absorbed by natural philosophy.

The derivation of the law is beyond the scope of this book, but interested readers can refer to almost any textbook on electrical engineering. The law was actually derived to relate a magnetic field induced by an electric current in a wire, but it can also be used to estimate the circulation of flow around a wing. This is because the Biot-Savart law is one of numerous solutions of *Laplace's equation*, which is the governing equation for irrotational, incompressible fluid flow. It is shown below for convenience:

Laplace's equation

$$\nabla^2 \phi = \frac{\partial^2 \phi}{\partial x^2} + \frac{\partial^2 \phi}{\partial y^2} + \frac{\partial^2 \phi}{\partial z^2} = 0 \qquad (9\text{-}97)$$

where ϕ is the *velocity potential*. Note that the direction of the velocity is imperative. In accordance with a right-handed coordinate system, assume the thumb of the right hand to point in a direction indicated by the filament **dl** in Figure 9-84. Then the other four fingers curl around the filament as if holding a rope. The direction of the velocity is always in the direction the four fingers make. This is an important concept to keep in mind for what follows.

In the development of the lifting-line theory, we will be applying the Biot-Savart law to a number of straight (versus curved) vortex filaments that are infinitely long. Such a straight segment is shown in Figure 9-84. It stretches from $-\infty$ to $+\infty$. Knowing the contribution of the tiny segment **dl**, the total velocity induced at the point **P** can be determined by integrating the contribution along the entire filament, i.e.:

$$w = \int_{-\infty}^{+\infty} \frac{\Gamma}{4\pi} \frac{\mathbf{dl} \times \mathbf{r}}{|\mathbf{r}|^3} \qquad (9\text{-}98)$$

In order to constrain the discussion here to the bare essentials, the solution of the integral is omitted and only the final result presented. The interested reader

---

[4] Pronounced *bee-yo-suh-var*.

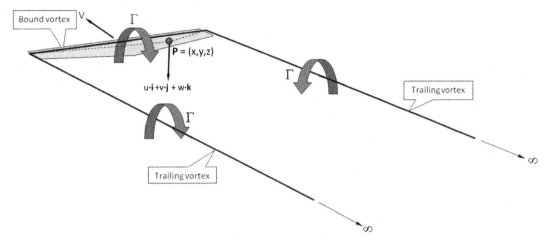

FIGURE 9-85 The flow field around a wing approximated by three connected constant-strength straight vortex filaments. On the wing plane, $u = v = 0$.

can, for instance, refer to Anderson [52] for the evaluation. The solution involves relating the parameters h, r, and θ in Figure 9-84, and inserting into the integral prior to its evaluation. Thus, the fluid speed at the point **P**, induced by the straight vortex filament of strength Γ, is given by:

Velocity induced a point **P**:

$$w = \frac{\Gamma}{2\pi h} \qquad (9\text{-}99)$$

Helmholz's Vortex Theorems

In 1858, the German scientist Hermann von Helmholz (1821–1894) made the next step by using the vortex filament to analyze inviscid, incompressible fluid flow. In doing so he established what has become known as *Helmholz's vortex theorems*. These state that:

(1) The strength of the vortex filament is constant along its entire length; and
(2) A vortex filament cannot end in a fluid, but must either extend to infinity or form a closed path.

These theorems are used to evaluate a special kind of vortex called a horseshoe vortex. The horseshoe vortex has important properties that will be discussed shortly and that allow it to be used to represent the lift of a finite wing.

Lifting Line Formulation

Now that we have established that a straight vortex induces a circulation around the vortex filament, it is possible to extend the idea to a three-dimensional wing. An observation of real finite wings reveals two important facts:

(1) The wing induces circulation that extends from tip to tip; and

(2) Each wingtip sheds a vortex that extends far into the flow field behind the wing.

This fact can be approximated using three separate vortex filaments as shown in Figure 9-85. First, a *trailing vortex* extends from infinity to the left wingtip. Then, a *bound vortex* extends from the left wingtip to the right wingtip. And finally, a third *trailing vortex* extends from the right wingtip back to infinity. This system of vortices not only satisfies Helmholz's vortex theorems, it also induces a flow field that resembles that of real three-dimensional wings. Additionally, the three vortices, which are of a constant strength Γ, all have the same sign, as can be observed using the right-handed rule (the thumb of the right hand should point forward for the first vortex, to the right for the second one, and backward for the third one).

While promising, there is a problem with this system in that it leads to a spanwise lift distribution that is constant. However, this implies that the lift at the wingtips is non-zero, but this is physically impossible. In fact, a single vortex like this will cause the downwash at the wingtip to go to infinity. The solution to this dilemma is to add more horseshoe vortices to the system, each of lesser span than the next, and with its own constant circulation (see Figure 9-86). The inboard vortices usually have greater strength than the outboard ones, but the interaction of all of them causes the resulting downwash to resemble that shown by experiment. This way, the strength of each individual vortex will be constant, but their total interaction along the span will result in a variable spanwise load distribution.

The lifting line method is based on the assumption that the vortex strength along the span is known. Consider that, at some spanwise wing station y, the vortex strength is given by the value $\Gamma(y)$. This means that if we move a distance Δy to another spanwise station, the

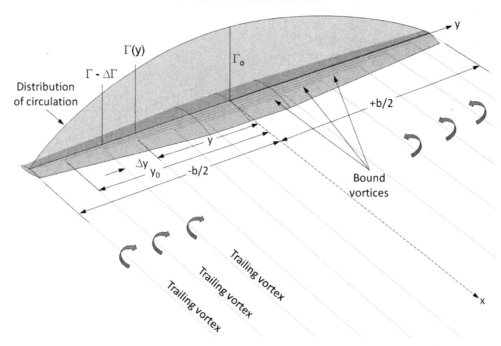

FIGURE 9-86 A wing simulated with a system of vortices.

change in the strength of the vortex, $\Delta\Gamma$, can be found from:

$$\Delta\Gamma = \frac{d\Gamma}{dy}\Delta y \qquad (9\text{-}100)$$

Now consider the spanwise station at y_0 and the vortex at y, shown in Figure 9-86. It is possible to determine the downwash (i.e. the vertical speed) induced at y_0 by the vortex at y (see depiction in Figure 9-87).

Let's denote the contribution to the total downwash at y_0 by δw_{y_0}. Since the trailing vortex extends from infinity to $x = 0$ where it stops (because that is where we have planted our vortex system, so it does not extend to $-\infty$ also), the influence will be half of that shown by Equation (9-99), or:

$$\delta w_{y_0} = \frac{1}{2}\left[+\frac{d\Gamma}{dy}dy\frac{1}{2\pi(y-y_0)}\right] \qquad (9\text{-}101)$$

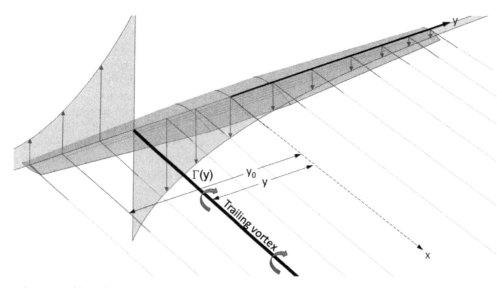

FIGURE 9-87 A depiction of how the single trailing vortex at y influences the vertical flow at the other stations. The other trailing vortices affect the one at y in a similar fashion.

Note that as depicted in Figure 9-86 and Figure 9-87, the vortex at y will induce an upward pointing contribution at y_0. This is emphasized by the $+$ sign in the equation. In order to calculate the total downwash at y_0 due to all the vortices distributed along the span, Equation (9-101) is integrated from the left wingtip ($y = -b/2$) to the right one ($y = +b/2$):

$$w_{y_0} = +\frac{1}{4\pi} \int_{-b/2}^{b/2} \frac{d\Gamma/dy}{(y - y_0)} dy \quad (9\text{-}102)$$

As can be seen, the downwash will ultimately depend on the strength of the circulation along the span. The contributions of other vortices in the system eventually yield a downward-pointed velocity at y_0. With the downwash at y_0 known, the downwash angle $\alpha_{i_{y_0}}$ at y_0 can now be calculated as follows:

$$\alpha_{i_{y_0}} = \tan^{-1}\left(-\frac{w_{y_0}}{V}\right) \approx -\frac{w_{y_0}}{V} \quad (9\text{-}103)$$

where V denotes the far-field airspeed. As stated in Section 15.3.4, *The lift-induced drag coefficient: C_{Di}*, the downwash behind the wing and corresponding upwash in front of it effectively "tilt" the undisturbed air through an *induced AOA*, denoted by α_i, reducing the geometric angle-of-attack, α, at the quarter-chord by a magnitude, to what is called the *effective AOA*, α_e (see Figure 9-88):

Effective *AOA*:

$$\alpha_e = \alpha - \alpha_i \quad (9\text{-}104)$$

Note that if the wing features washout, α_i and thus α_e become a function of the spanwise station. Similarly, cambered airfoils are treated by subtracting the angle of zero lift from the geometric *AOA*. The presence of the induced *AOA* tilts the lift force back by the angle ε (see Figure 15-20) and must therefore be resolved into two force components: one normal to the flight path (lift) and the other parallel to the flight path (lift-induced drag). The Kutta-Joukowski theorem (see Section 8.1.8, *The generation of lift*) makes it possible to calculate these two forces per unit span using the following expressions:

Lift per unit span:

$$l(y) = \rho V \cdot \Gamma(y) \quad (9\text{-}105)$$

Lift-induced drag per unit span:

$$d_i(y) = -\rho w(y) \cdot \Gamma(y) \quad (9\text{-}106)$$

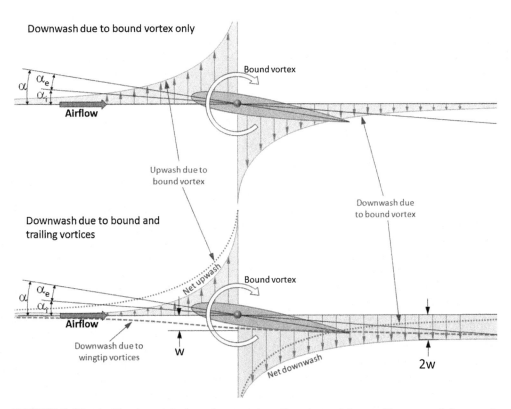

FIGURE 9-88 A side view of the bound vortex along the wing and the resulting up- and downwash.

The minus sign is necessary to ensure the negative value of the downwash produces a positive drag force. The total lift and lift-induced drag forces can now be determined as shown below:

Lift:
$$L = \rho V \int_{-b/2}^{b/2} \Gamma(y)dy \quad (9\text{-}107)$$

Drag:
$$D_i = -\rho \int_{-b/2}^{b/2} w(y)\Gamma(y)dy \quad (9\text{-}108)$$

Consequently, the lift and lift-induced drag coefficient are given by:

Lift coefficient:
$$C_L = \frac{2L}{\rho V^2 S} = \frac{2}{VS} \int_{-b/2}^{b/2} \Gamma(y)dy \quad (9\text{-}109)$$

Lift-induced drag coefficient:
$$C_{Di} = \frac{2D_i}{\rho V^2 S} = -\frac{2}{V^2 S} \int_{-b/2}^{b/2} w(y)\Gamma(y)dy \quad (9\text{-}110)$$

If there is a representative zero-lift *AOA* for the wing, the lift curve slope can be calculated from:

Lift curve slope:
$$C_{L_\alpha} = \frac{C_L}{\alpha - \alpha_{ZL}} \quad (9\text{-}111)$$

The subsequent articles present several applications of this method.

9.7.2 Prandtl's Lifting Line Method — Special Case: The Elliptical Wing

As discussed in Section 9.4.4, *Elliptic planforms*, from a standpoint of efficiency the elliptical wing planform is very practical in aircraft design and is therefore of great interest to the aircraft designer. In this article, the lifting line method is applied to an elliptical planform and several useful closed-form solutions of selected aerodynamic characteristics are derived. These characteristics are helpful even if they are only used for comparison reasons. The solution assumes that the distribution of circulation is known and is given by:

Elliptical lift distribution:
$$\Gamma(y) = \Gamma_0 \sqrt{1 - \left(\frac{2y}{b}\right)^2} \quad (9\text{-}112)$$

Using this distribution, the following aerodynamic characteristics for the elliptical planform can be derived (see derivation section that follows):

Vortex strength (constant):
$$\Gamma_0 = \frac{2C_L S}{\pi b} V \quad (9\text{-}113)$$

Downwash (constant):
$$w = -\frac{\Gamma_0}{2b} \quad (9\text{-}114)$$

Induced *AOA*:
$$\alpha_i \approx \frac{\Gamma_0}{2bV} = \frac{C_L}{\pi AR} \quad (9\text{-}115)$$

Lift force:
$$L = \rho V \cdot \Gamma_0 \frac{\pi b}{4} \quad (9\text{-}116)$$

Lift-induced drag force:
$$D_i = \rho \Gamma_0^2 \frac{\pi}{8} \quad (9\text{-}117)$$

Lift coefficient:
$$C_L = \frac{\pi b \Gamma_0}{2VS} \quad (9\text{-}118)$$

Lift-induced drag coefficient:
$$C_{Di} = \frac{C_L^2}{\pi AR} \quad (9\text{-}119)$$

Derivation of Equations (9-113) through (9-119)

First evaluate the derivative $d\Gamma/dy$. This can be done using the chain rule of differentiation:

$$\frac{d\Gamma}{dy} = \Gamma_0 \frac{d}{dy}\left(1 - \left(\frac{2}{b}\right)^2 y^2\right)^{\frac{1}{2}}$$

$$= \Gamma_0 \frac{1}{2}\left(1 - \left(\frac{2}{b}\right)^2 y^2\right)^{-\frac{1}{2}}\left(-\frac{8}{b^2}y\right)$$

$$= -\frac{4\Gamma_0}{b^2}\frac{y}{\sqrt{1 - (2y/b)^2}}$$

This can be used with Equation (9-102) to calculate the total downwash at any spanwise station, say y_0 (for the sake of consistency):

$$w_{y_0} = \frac{1}{4\pi} \int_{-b/2}^{b/2} \frac{d\Gamma/dy}{(y-y_0)} dy = \frac{1}{4\pi} \int_{-b/2}^{b/2} \frac{-\frac{4\Gamma_0}{b^2}\frac{y}{\sqrt{1-(2y/b)^2}}}{(y-y_0)} dy$$

$$= -\frac{\Gamma_0}{\pi b^2} \int_{-b/2}^{b/2} \frac{y}{\sqrt{1-(2y/b)^2}(y-y_0)} dy$$

It is helpful to evaluate the integral by transforming spanwise stations into an angular form as follows:

$$y = (b/2)\cos\phi \Rightarrow dy = -(b/2)\sin\phi d\phi \quad (9\text{-}120)$$

Using this transformation, we find the lower limit (-b/2) becomes 0 and the upper one (+b/2) becomes π. If we define $y_0 = (b/2)\cos\phi_0$ the above integral can be rewritten as follows:

$$w_{y_0} = -\frac{\Gamma_0}{\pi b^2}$$
$$\times \int_{\pi}^{0} \frac{(b/2)\cos\phi}{\sqrt{(b/2)^2 - (b/2)^2\cos^2\phi} \times ((b/2)\cos\phi - (b/2)\cos\phi_0)}$$
$$\times (-(b/2)\sin\phi d\phi)$$

Simplifying further yields:

$$w_{y_0} = \frac{\Gamma_0}{2\pi b} \int_{\pi}^{0} \frac{\cos\phi \sin\phi d\phi}{\sqrt{1-\cos^2\phi} \times (\cos\phi - \cos\phi_0)}$$

$$= \frac{\Gamma_0}{2\pi b} \int_{\pi}^{0} \frac{\cos\phi \sin\phi d\phi}{\sin\phi \times (\cos\phi - \cos\phi_0)}$$

Finally, this leads to:

$$w_{y_0} = -\frac{\Gamma_0}{2\pi b} \int_{0}^{\pi} \frac{\cos\phi}{\cos\phi - \cos\phi_0} d\phi \quad \text{(i)}$$

The evaluation of this integral is beyond the scope of this text, but a solution is provided in Karamcheti [53]. An alternative solution method based on the original expression omitting the transformation is presented by Bertin and Smith [50]. In either case, the resulting expression reduces to:

$$w_{y_0} = -\frac{\Gamma_0}{2b} \quad \text{(ii)}$$

This is Equation (9-114). Note that the distribution of downwash is constant along the span. The downwash angle can thus be obtained from Equation (9-103):

$$\alpha_{i y_0} \approx -\frac{w_{y_0}}{V} = \frac{\Gamma_0}{2bV} \quad \text{(iii)}$$

This is Equation (9-115). Note that this angle is constant along the span as well. Inserting Equation (9-112) into (9-107) yields the total lift generated by the wing:

$$L = \rho V \int_{-b/2}^{b/2} \Gamma(y) dy$$

$$= \rho V_\infty \int_{-b/2}^{b/2} \Gamma_0 \sqrt{1 - \left(\frac{2y}{b}\right)^2} dy \quad \text{(iv)}$$

Using the above angular transformations, Equation (iv) can be rewritten and simplified as follows:

Lift:

$$L = \rho V \cdot \Gamma_0 \int_{0}^{\pi} \sqrt{1 - \cos^2\phi} (b/2) \sin\phi d\phi$$

$$= \rho V_\infty \Gamma_0 \frac{\pi b}{4} \quad \text{(v)}$$

This is Equation (9-116). The lift coefficient can now be determined from:

$$C_L = \frac{2L}{\rho V_\infty^2 S} = \frac{\pi b \Gamma_0}{2VS} \quad \text{(vi)}$$

This is Equation (9-118). Alternatively, the vortex strength Γ_0 is readily obtained from Equation (vi) as follows:

$$\Gamma_0 = \frac{2SC_L}{\pi b} V \quad \text{(vii)}$$

This is Equation (9-113). Similarly, the lift-induced drag can be computed from Equation (9-108):

$$D_i = -\rho \int_{-b/2}^{b/2} w(y) \Gamma(y) dy$$

$$= -\rho \int_{-b/2}^{b/2} w(y) \Gamma_0 \sqrt{1 - \left(\frac{2y}{b}\right)^2} dy \quad \text{(viii)}$$

Inserting Equation (9-114) and using the spanwise angular transformation of Equation (9-120), we get Equation (9-117):

$$D_i = -\rho \int_{-b/2}^{b/2} \frac{-\Gamma_0}{2b}\Gamma_0\sqrt{1-\left(\frac{2y}{b}\right)^2}\,dy$$

$$= \frac{\rho\Gamma_0^2}{2b}\int_0^\pi \sqrt{1-\cos^2\phi}\,(b/2)\sin\phi\,d\phi = \rho\Gamma_0^2\frac{\pi}{8} \quad \text{(ix)}$$

The lift-induced drag coefficient can now be determined as follows, using Equation (vii):

$$C_{Di} = \frac{2D_i}{\rho V^2 S} = \frac{2\left(\rho\Gamma_0^2\frac{\pi}{8}\right)}{\rho V^2 S} = \frac{\pi}{4V^2 S}\Gamma_0^2$$

$$= \frac{\pi}{4V^2 S}\left(\frac{2SC_L}{\pi b}V\right)^2 = \frac{SC_L^2}{\pi b^2} = \frac{C_L^2}{\pi AR} \quad \text{(x)}$$

This is Equation (9-119).

QED

9.7.3 Prandtl's Lifting Line Method — Special Case: Arbitrary Wings

The history of aviation reveals that most aircraft do not feature elliptical wing planform shapes. It follows that it is desirable to extend the lifting line method to non-elliptical wings. One way of accomplishing this is to represent the spanwise distribution of vortex strengths using a Fourier sine series, consisting of N terms:

Fourier series lift distribution:

$$\Gamma(\phi) = 2bV\sum_{n=1}^{N} A_n \sin n\phi \quad (9\text{-}121)$$

Note that if the lift distribution is symmetrical, all even values of n are omitted. The first equation of interest is the one that allows the coefficients $A_1, A_2, A_3, \ldots, A_N$ to be solved.

Monoplane equation:

$$\sum_{n=1}^{N} A_n \sin n\phi (\mu n + \sin\phi) = \mu(\alpha - \alpha_{ZL})\sin\phi \quad (9\text{-}122)$$

The monoplane equation is used to construct a system of simultaneous equations, as shown below. The left-hand side of this equation is called the *aerodynamic influence matrix*:

where

α_{ZLi} = zero lift angle for the airfoil at station ϕ
ϕ_i = angles (in radians) that correspond to the spanwise angular stations obtained from $y_i = (b/2)\cos\phi_i$ (i.e. Equation [9-120]), in radians
$\mu = C \cdot C_{l_\alpha}/4b$, dimensionless; the airfoil's lift curve slope, C_{l_α}, may be a function of ϕ
b = wingspan, in ft or m
C = chord at station ϕ_i, in ft or m

Solving the system leads to a number of interesting results, some of which are presented below.

Lift force:

$$L = \frac{1}{2}\rho V \cdot S \cdot (\pi \cdot AR \cdot A_1) = \frac{1}{2}\rho V \cdot b^2 \cdot \pi \cdot A_1 \quad (9\text{-}124)$$

Lift-induced drag:

$$D_i = \frac{\rho V^2}{2}b^2\pi \sum n A_n^2 \quad (9\text{-}125)$$

Lift coefficient: $\quad C_L = \pi \cdot AR \cdot A_1 \quad (9\text{-}126)$

The lift curve slope can thus be calculated from:
Lift curve slope:

$$C_{L_\alpha} = \frac{C_L}{\alpha - \alpha_{ZL}} = \frac{\pi \cdot AR \cdot A_1}{\alpha - \alpha_{ZL}} \quad (9\text{-}127)$$

And lift-induced drag coefficient:

$$C_{Di} = \frac{C_L^2}{\pi \cdot AR}(1 + \delta) = \frac{C_L^2}{\pi \cdot AR \cdot e} \quad (9\text{-}128)$$

where $\delta = \sum_{n=2}^{N} n\left(\frac{A_n}{A_1}\right)^2$ and $e = \frac{1}{1+\delta}$.

As stated above, if the spanwise distribution of lift is symmetrical, all the even indexed constants ($A_2, A_4, \ldots$) of the summation are set to zero. This can be written as follows:

$$\delta = \sum_{n=2}^{N} n\left(\frac{A_n}{A_1}\right)^2 = 3\left(\frac{A_3}{A_1}\right)^2 + 5\left(\frac{A_5}{A_1}\right)^2 + 7\left(\frac{A_7}{A_1}\right)^2 + \ldots$$

(9-129)

See Figure 15-22 for a graph plotting δ as a function of taper ratio and aspect ratio for straight tapered wings.

$$\begin{bmatrix} \sin\phi_1(\mu+\sin\phi_1) & \sin 2\phi_1(2\mu+\sin\phi_1) & \cdots & \sin N\phi_1(N\mu+\sin\phi_1) \\ \sin\phi_2(\mu+\sin\phi_2) & \sin 2\phi_2(2\mu+\sin\phi_2) & \cdots & \sin N\phi_2(N\mu+\sin\phi_2) \\ \vdots & \vdots & \ddots & \vdots \\ \sin\phi_N(\mu+\sin\phi_N) & \sin 2\phi_N(2\mu+\sin\phi_N) & \cdots & \sin N\phi_N(N\mu+\sin\phi_N) \end{bmatrix} \begin{Bmatrix} A_1 \\ A_2 \\ \vdots \\ A_N \end{Bmatrix} = \begin{Bmatrix} \mu(\alpha-\alpha_{ZL_1})\sin\phi_1 \\ \mu(\alpha-\alpha_{ZL_2})\sin\phi_2 \\ \vdots \\ \mu(\alpha-\alpha_{ZL_N})\sin\phi_N \end{Bmatrix} \quad (9\text{-}123)$$

Change in Induced Angle-of-attack

The induced *AOA* on an arbitrary planform differs from that of an elliptical planform of an equal *AR* according to the following expression:

$$\alpha_{wing} - \alpha_{airfoil} = \frac{C_L}{\pi AR}(1 + \tau) \qquad (9\text{-}130)$$

where τ is given by:

$$(1 + \tau) = \frac{\alpha_{wing}}{A_1} - \frac{\pi AR}{C_{l_\alpha}} \qquad (9\text{-}131)$$

Derivation of Equations (9-124) through (9-128)

In order to determine the coefficients $A_1, A_2, A_3, \ldots, A_N$, the circulation for the N vortices must be determined. The procedure begins by relating the section lift coefficient at any angular station ϕ as follows:

$$C_l(\phi) = \frac{\text{lift per unit span}}{\frac{1}{2}\rho V^2 C} = \frac{\rho V \cdot \Gamma(\phi)}{\frac{1}{2}\rho V^2 C(\phi)} = \frac{2\Gamma(\phi)}{VC(\phi)} \qquad \text{(i)}$$

where $C(\phi)$ is the chord at a specific angular station ϕ. However, if the lift curve slope, $C_{l_\alpha} = dC_l/d\alpha$, of the airfoil at the angular station ϕ is known, the section lift coefficient can also be determined from:

$$C_l = \left(\frac{dC_l}{d\alpha}\right)(\alpha_e - \alpha_{ZL}) = C_{l_\alpha}(\alpha - \alpha_i - \alpha_{ZL}) \qquad \text{(ii)}$$

where α_{ZL} is the zero-lift *AOA* of the airfoil at the angular station ϕ. Note that α, α_{ZL}, and α_i are all functions of the angular station ϕ. By equating Equations (i) and (ii) we get:

$$\frac{2\Gamma(\phi)}{VC(\phi)} = C_{l_\alpha}(\alpha - \alpha_i - \alpha_{ZL})$$

$$\Rightarrow \frac{2\Gamma(\phi)}{C_{l_\alpha}C(\phi)} = V(\alpha - \alpha_{ZL}) - V\alpha_i \qquad \text{(iii)}$$

The term $V\alpha_i$ is given by Equation (9-102):

$$V\alpha_i = -w = -\frac{1}{4\pi} \int_{-b/2}^{+b/2} \frac{d\Gamma/dy}{y - y_0} dy \qquad \text{(iv)}$$

where the minus sign indicates a downward flow of air. Using the definition of $\Gamma(\phi)$ given by Equation (9-121) this can be rewritten as follows:

$$-w(\phi) = -V \frac{\sum n A_n \sin n\phi}{\sin \phi} \qquad \text{(v)}$$

Therefore, Equation (iii) can be rewritten as follows:

$$\frac{2\Gamma(\phi)}{C(\phi) \cdot C_{l_\alpha}} = V(\alpha - \alpha_{ZL}) - V\frac{\sum n A_n \sin n\phi}{\sin \phi} \qquad \text{(vi)}$$

Expanding by inserting $\Gamma(\phi)$ per Equation (9-121) yields:

$$\frac{4b}{C \cdot C_{l_\alpha}}\sum A_n \sin n\phi = (\alpha - \alpha_{ZL}) - \frac{\sum n A_n \sin n\phi}{\sin \phi} \qquad \text{(vii)}$$

Let $\mu = C \cdot C_{l_\alpha}/4b$ and insert into Equation (vii). This leads to:

$$\sin\phi \sum A_n \sin n\phi = \mu(\alpha(\phi) - \alpha_{ZL}(\phi))\sin\phi - \mu\sum n A_n \sin n\phi$$

$$\Rightarrow \sin\phi \sum A_n \sin n\phi + \mu\sum n A_n \sin n\phi$$
$$= \mu(\alpha(\phi) - \alpha_{ZL}(\phi))\sin\phi$$

Simplify terms in the summation sign to yields a governing equation that is called the *monoplane equation*.

$$\sum_{n=1}^{N} A_n \sin n\phi(\mu n + \sin\phi) = \mu(\alpha - \alpha_{ZL})\sin\phi \qquad (9\text{-}122)$$

This equation yields a set of equations that must be solved simultaneously to obtain the constants $A_1, A_2, A_3, \ldots, A_N$. The angle ϕ refers to a spanwise angle transformation of Equation (9-120). If the problem deals with symmetrical load distribution, all even terms ($n = 2, 4$, etc.) are omitted from the solution.

Next consider the lift coefficient generated by the wing, given by:

$$C_L = \frac{2L}{\rho V^2 S} = \frac{2}{VS} \int_{-b/2}^{b/2} \Gamma(y) dy \qquad \text{(viii)}$$

Again using the definition of $\Gamma(\phi)$ given by Equation (9-121), Equation (viii) becomes:

$$C_L = \frac{2}{VS} \int_{-b/2}^{b/2} \Gamma(y) dy$$

$$= \frac{2}{VS} \int_{0}^{\pi} \left[2bV \sum_{n=1}^{N} A_n \sin n\phi\right]\left(\left(\frac{b}{2}\right)\sin\phi d\phi\right)$$

$$= \frac{2b^2}{S} \int_{0}^{\pi} \left[\sum_{n=1}^{N} A_n \sin n\phi\right]\sin\phi d\phi$$

The integral can be solved noting that:

$$\int_{0}^{\pi} \sin n\phi \sin\phi d\phi = \begin{cases} \pi/2 & \text{if } n = 1 \\ 0 & \text{if } n \neq 1 \end{cases}$$

This allows Equation (ix) to be rewritten as follows:

$$C_L = \frac{2b^2}{S}\int_0^\pi \left[\sum_{n=1}^N A_n \sin n\phi\right]\sin\phi\,d\phi = \frac{2b^2}{S}\frac{\pi}{2}A_1$$

$$= \frac{\pi b^2 A_1}{S} = \pi \cdot AR \cdot A_1 \qquad (ix)$$

This is Equation (9-126). Then consider the lift-induced drag coefficient generated by the wing. Equation (9-108) defines the lift-induced drag, D_i. Inserting Equation (v) for $w(y)$ and Equation (9-121) for $\Gamma(\phi)$ we get:

$$D_i = -\rho \int_{-b/2}^{b/2} w(y)\Gamma(y)\,dy$$

$$= -\rho \int_{-b/2}^{b/2} \left[-V\frac{\sum nA_n \sin n\phi}{\sin\phi}\right]$$

$$\times \left[2bV \sum A_n \sin n\phi\right]\left(\left(\frac{b}{2}\right)\sin\phi\,d\phi\right)$$

$$= \frac{\rho V 2bbV}{2}\int_0^\pi \left[\frac{\sum nA_n \sin n\phi}{\sin\phi}\right]$$

$$\times \left[\sum A_n \sin n\phi\right](\sin\phi\,d\phi)$$

This reduces to:

$$D_i = \rho V^2 b^2 \int_0^\pi \sum nA_n \sin n\phi \sum A_n \sin n\phi\,d\phi$$

Where the evaluation of the integral yields:

$$\int_0^\pi \sum nA_n \sin n\phi \sum A_n \sin n\phi\,d\phi = \frac{\pi}{2}\sum nA_n^2$$

Therefore, the lift-induced drag is given by:

$$D_i = \frac{\rho V^2}{2}b^2 \pi \sum nA_n^2 \qquad (x)$$

Using Equation (9-110) it is now possible to determine the lift-induced drag coefficient:

$$C_{Di} = \frac{2D_i}{\rho V^2 S} = \frac{2\left[\frac{\rho V^2}{2}b^2 \pi \sum nA_n^2\right]}{\rho V^2 S} = \frac{b^2 \pi}{S}\sum nA_n^2$$

$$= \pi \cdot AR \cdot \sum nA_n^2 \qquad (xi)$$

Since $C_L = \pi \cdot AR \cdot A_1 \Leftrightarrow A_1 = \frac{C_L}{\pi \cdot AR}$ it is possible to rewrite this as follows:

$$C_{Di} = \pi \cdot AR \cdot \sum nA_n^2 = \pi \cdot AR \cdot \left(A_1^2 + \sum_{n=2}^N nA_n^2\right)$$

$$= \frac{C_L^2}{\pi \cdot AR}\left(1 + \sum_{n=2}^N n\left(\frac{A_n}{A_1}\right)^2\right) \qquad (xii)$$

This is a classic form of this equation, typically written in the following form, using the term δ to denote the lift-induced drag factor.

$$C_{Di} = \frac{C_L^2}{\pi \cdot AR}(1 + \delta) \qquad (xiii)$$

It is evident that the minimum C_{Di} is obtained when $\delta = 0$, but this represents an elliptic lift distribution.
QED

Derivation of Equation (9-130)

Consider a two-dimensional airfoil and three-dimensional wing whose cross-sectional geometry is that of the said airfoil. The airfoil is akin to a three-dimensional wing of infinite wingspan. Consider these at specific AOA such that they both generate an equal lift coefficient, C_L. This allows the aerodynamic AOA[5] for each "wing" to be determined as follows:

Airfoil:

$$\alpha_{airfoil} = C_L/C_{l_\alpha}$$

Wing:

$$\alpha_{wing} = C_L/C_{L_\alpha}$$

Subtracting one from the other leads to:

$$\alpha_{wing} - \alpha_{airfoil} = \frac{C_L}{C_{L_\alpha}} - \frac{C_L}{C_{l_\alpha}} = C_L\left(\frac{1}{C_{L_\alpha}} - \frac{1}{C_{l_\alpha}}\right)$$

This represents the difference in AOA between the two geometries. Next, insert $A_1 = C_L/\pi AR$ into the expression for the three-dimensional wing,

---

[5] The aerodynamic AOA is measured with respect to zero lift; in other words: $\alpha - \alpha_{ZL}$.

9.7 NUMERICAL ANALYSIS OF THE WING 389

i.e. $C_{L_\alpha} = C_L/\alpha_{wing}$. This yields the following relationship between $C_{L\alpha}$ and A_1:

$$A_1 = C_L/\pi AR \Rightarrow C_L = A_1 \pi AR \Rightarrow C_{L_\alpha} = C_L/\alpha_{wing}$$
$$= A_1 \pi AR/\alpha_{wing}$$

Replacing the corresponding term in the above equation for the difference in AOA leads to:

$$\alpha_{wing} - \alpha_{airfoil} = C_L\left(\frac{1}{C_{L_\alpha}} - \frac{1}{C_{l_\alpha}}\right)$$
$$= C_L\left(1\bigg/\left(\frac{C_L}{\alpha_{wing}}\right) - \frac{1}{C_{l_\alpha}}\right)$$
$$= \frac{C_L}{\pi AR}\left(\frac{\alpha_{wing}}{A_1} - \frac{\pi AR}{C_{l_\alpha}}\right)$$

QED

EXAMPLE 9-13

Estimate the aerodynamic characteristics of the SR22 wing planform (ignoring the presence of the fuselage) using the lifting line method. Determine this using four terms in the monoplane equation, assuming symmetrical lift distribution. Assume the aircraft is flying at an AOA of $5°$ (0.08727 radians) and that the wing uses the NACA 65-415 airfoil (see Figure 9-60). Assume the airfoil's zero-lift AOA, α_{ZL}, is approximately $-2.7°$ (-0.04712 radians) and its lift curve slope, $C_{l\alpha}$, is 2π. Note that the dimensions in Figure 9-89 are calculated based on the geometry shown and are not the "official" numbers (e.g. official $S = 144.9$ ft^2 and $AR = 10$). Determine the airspeed that the airplane must fly at S-L if it weighs 3400 lb$_f$.

Solution

Begin by creating the relationship between the physical and angular span stations (see Figure 9-90). Note that the physical stations are given by $y_i = (b/2)\cos\phi_i$ (Equation [9-120]). For this reason, the cosine of ϕ is in effect a parameter that varies between 1 and 0 that can be used with a parametric representation for the chord. Thus, the chord at any spanwise station can be calculated from the following parametric equation:

$$C_i = C_t\cos\phi_i + C_r(1 - \cos\phi_i)$$

where $C_r = 5.18$ ft and $C_t = 2.59$ ft.

Next tabulate the following values:

Column ① and ②: contain the spanwise angular stations from Figure 9-90 both in degrees and radians.

Column ③: contains the cosine of the angular station.

Column ④: contains the wing chord at the angular station. Calculated using $C_i = C_t\cos\phi_i + C_r(1 - \cos\phi_i)$.

Using the root and tip chords, $C_r = 5.18$ ft and $C_t = 2.59$ ft, respectively, this expression becomes:

$$C_i = 2.59\cos\phi_i + 5.18(1 - \cos\phi_i)$$

Columns ⑤ through ⑥: contain the various sines of the angular stations. These are needed to properly setup the matrix as required by the monoplane equation.

	①	②	③	④	⑤	⑥	⑦	⑧	⑨
	Angular Station			C(φ)					
ID	Deg	Rad	cos φ	ft	sin φ	sin 3φ	sin 5φ	sin 7φ	μ
1	22.5	0.3927	0.9239	2.79	0.3827	0.9239	0.9239	0.3827	0.1143
2	45.0	0.7854	0.7071	3.35	0.7071	0.7071	−0.7071	−0.7071	0.1373
3	67.5	1.1781	0.3827	4.19	0.9239	−0.3827	−0.3827	0.9239	0.1718
4	90.0	1.5708	0.0000	5.18	1.0000	−1.0000	1.0000	−1.0000	0.2124

EXAMPLE 9-13 (cont'd)

Column ⑨: is calculated as follows, using values from the first row (ID = 1):

$$\mu_1 = C_1 \cdot C_{l_\alpha}/4b = (2.79) \cdot (2\pi)/4(38.3) = 0.1143$$

Next prepare the aerodynamic influence matrix per Equation (9-123). Note that since the lift distribution is symmetrical, we are only concerned with the constants A_1, A_3, A_5, and A_7. This leads to the following setup.

$$\begin{bmatrix} 0.1902 & 0.6704 & 0.8816 & 0.4527 \\ 0.5971 & 0.7913 & -0.9856 & -1.1798 \\ 1.0123 & -0.5508 & -0.6823 & 1.9646 \\ 1.2124 & -1.6373 & 2.0622 & -2.4871 \end{bmatrix} \begin{Bmatrix} A_1 \\ A_3 \\ A_5 \\ A_7 \end{Bmatrix}$$

$$= \begin{Bmatrix} 0.005879 \\ 0.001305 \\ 0.002133 \\ 0.028551 \end{Bmatrix}$$

For instance, the first row in the above matrix is calculated as follows, using the monoplane equation:

$$A_{11} = \sin\phi_1(\mu + \sin\phi_1)$$
$$= (0.3827)(0.1143 + 0.3827) = 0.1902$$

$$A_{12} = \sin 3\phi_1(3\mu + \sin\phi_1)$$
$$= (0.9239)(3 \times 0.1143 + 0.3827) = 0.6704$$

$$A_{13} = \sin 5\phi_1(5\mu + \sin\phi_1)$$
$$= (0.9239)(5 \times 0.1143 + 0.3827) = 0.8816$$

$$A_{14} = \sin 7\phi_1(7\mu + \sin\phi_1)$$
$$= (0.3827)(7 \times 0.1143 + 0.3827) = 0.4527$$

$$B_1 = \mu(\alpha - \alpha_{ZL_1})\sin\phi_1$$
$$= (0.1143)(0.08727 - (-0.04712))(0.3827) = 0.005879$$

Solving the equations requires the square matrix to be inverted and then multiplied to the column matrix on the right-hand side, yielding the following values for the constants:

$$\begin{Bmatrix} A_1 \\ A_3 \\ A_5 \\ A_7 \end{Bmatrix} = \begin{Bmatrix} 2.251 \times 10^{-2} \\ 8.674 \times 10^{-4} \\ 1.195 \times 10^{-3} \\ -8.441 \times 10^{-5} \end{Bmatrix}$$

This allows the following aerodynamic parameters to be calculated. First, the lift coefficient can be calculated from Equation (9-126):

$$C_L = \pi \cdot AR \cdot A_1 = \pi \cdot (9.858) \cdot 2.251 \times 10^{-2} = 0.6971$$

The lift-induced drag coefficient can be found by first determining the factor δ:

$$\delta = \sum_{n=2}^{N} n\left(\frac{A_n}{A_1}\right)^2 = 3\left(\frac{A_3}{A_1}\right)^2 + 5\left(\frac{A_5}{A_1}\right)^2 + 7\left(\frac{A_7}{A_1}\right)^2$$
$$= 0.01865$$

This yields the following drag coefficient per Equation (9-128):

$$C_{Di} = \frac{C_L^2}{\pi \cdot AR}(1 + \delta) = \frac{(0.6971)^2}{\pi \cdot (9.858)}(1 + 0.01865) = 0.01599$$

Oswald's span efficiency is thus:

$$e = \frac{1}{1+\delta} = \frac{1}{1+0.01865} = 0.9817$$

Note that this value is for the wing without the detrimental effects of the fuselage. The lifting line method will always return efficiency that is too high for this reason. Finally, the wing's lift curve slope can be determined as follows:

$$C_{L_\alpha} = \frac{C_L}{(\alpha - \alpha_{ZL})} = \frac{0.6971}{(0.08727 + 0.04712)}$$
$$= 5.187 \text{ per radian}$$

Finally, the airspeed the airplane must fly at in order to generate 3400 lb$_f$ of lift at 5° AOA can be determined as follows:

$$V = \sqrt{\frac{2W}{\rho S C_L}} = \sqrt{\frac{2(3400)}{(0.002378)(148.8)(0.6971)}} = 166 \text{ ft/s}$$
$$= 98.4 \text{ KCAS}$$

9.7.4 Accounting for a Fuselage in Prandtl's Lifting Line Method

The lifting line method is seriously limited in that it applies directly to a "clean" wing; that is, the presence of a fuselage is not included. Consequently, the prediction will tend to return inflated span efficiencies, which can be detrimental to performance analysis. One way of correcting for this shortcoming is to consider only the exposed part of the wing. This is shown in Figure 9-91, which shows three wing planform shapes of equal spans, reference areas, and (therefore) AR. The top one is a "clean" elliptical wing, to which all other wings

9.7 NUMERICAL ANALYSIS OF THE WING   391

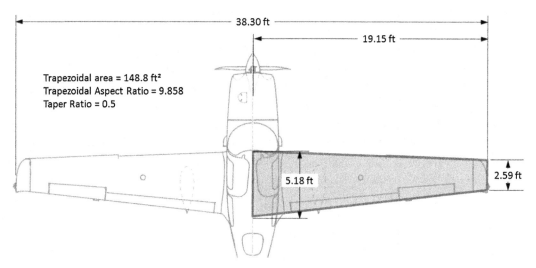

FIGURE 9-89 A top view of the SR-22 showing dimensions used for this example.

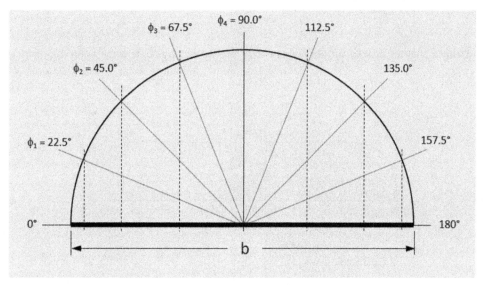

FIGURE 9-90 Relationship between physical and angular stations along the wingspan.

are compared when using the lifting line method. The wing in the center is a "clean" trapezoidal wing, whereas the bottom one is the same wing, except with a fuselage present. A schematic of the corresponding distribution of section lift coefficients is shown in the right portion of the figure.

It is evident from the preceding analysis that it compares the top (elliptical) and center wing (trapezoidal), whereas it would be more appropriate to compare the top and the bottom one. The fact that the fuselage so reduces the lift generated by the wing means that the airplane will have to fly at a higher AOA to make up for it. This is largely why the wing with the fuselage generates higher lift-induced drag than the clean wing and, when compared to the clean elliptical wing, reduces the Oswald span efficiency. The following method allows this to be taken into account when using the lifting line method, but first the following assumptions are introduced.

(1) Assume that the lift is entirely generated by the exposed wing panels and not by the region over the fuselage. This means that the section lift coefficients over the fuselage region are assumed to be zero.

(2) Assume that the fuselage acts as a wall. This allows the lifting line method to be applied to the exposed part of

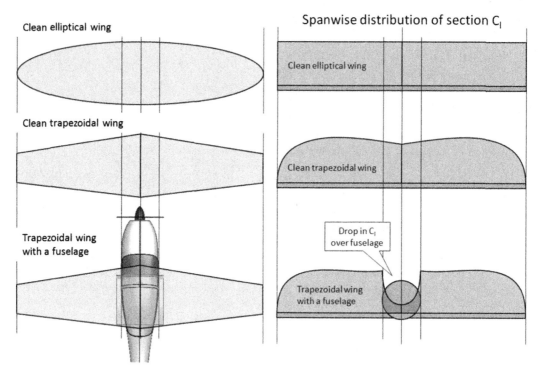

FIGURE 9-91 A schematic showing how the presence of a fuselage reduces the magnitude of the section lift coefficient over the fuselage.

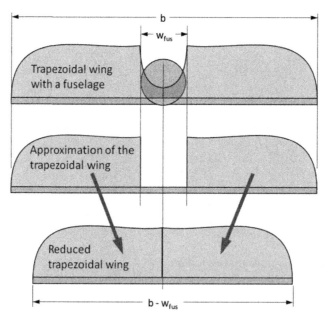

FIGURE 9-92 A schematic showing how the presence of a fuselage can be accounted for using the lifting line method.

the wing by simply reducing the wing geometry as shown in Figure 9-92. For instance, the wingspan, b, will be reduced by the width of the fuselage, w_{fus}.
Reduced wingspan, b_R:

$$b_R = b - w_{fus} \qquad (9\text{-}132)$$

Let the planform area corresponding to the wing inside the fuselage be given by $C_r \times w_{fus}$. Therefore, the reduced wing area S_R is given by:
Reduced wing area, S_R:

$$S_R = S - C_r w_{fus} \qquad (9\text{-}133)$$

Reduced aspect ratio, AR_R:

$$AR_R = \frac{b_R^2}{S_R} \quad (9\text{-}134)$$

Furthermore, reducing the wingspan will modify the taper ratio because the root chord changes. The reduced taper ratio can be estimated using a parametric representation for the chord:

$$C_{r_R} = C_r(1-t) + tC_t = \left(1 - \frac{w_{fus}}{b}\right)C_r + \left(\frac{w_{fus}}{b}\right)C_t$$

This can be used to determine the reduced taper ratio, λ_R:

$$\lambda_R = \frac{C_t}{C_{r_R}} = \frac{bC_t}{C_r\left(b - w_{fus}\right) + w_{fus}C_t}$$

$$= \frac{bC_t}{C_r\left(1 - w_{fus}\right) + w_{fus}C_t} \quad (9\text{-}135)$$

So, rather than analyzing the complete wing, its wing span and area should be reduced using the above expressions. Therefore, the wing will now require a higher AOA in order to generate the C_L required for a given flight condition.

EXAMPLE 9-14

Estimate the aerodynamic characteristics of the SR22 wing planform (including the presence of the fuselage) using the corrected lifting line method just discussed. Use all the information from Example 9-9 and assume a fuselage width of 50 inches (4.167 ft). Determine the C_L required for the airplane to generate 3400 lb$_f$ at 98.4 KCAS and use this to estimate the aerodynamic characteristics. Then, compare it to that of Example 9-9 (the AOA and C_L of Example 9-13 are 5° and 0.6971, respectively).

Solution

The procedure is identical to that of Example 9-13, so only results will be cited. First, compute the reduced wing parameters:

Reduced wingspan, b_R:

$$b_R = b - w_{fus} = 38.30 - 4.167 = 34.13 \text{ ft}$$

Reduced wing area, S_R:

$$S_R = S - C_r w_{fus} = 148.8 - 5.18 \times 4.167$$
$$= 127.2 \text{ ft}^2$$

Reduced aspect ratio, AR_R:

$$AR_R = \frac{b_R^2}{S_R} = \frac{34.13^2}{127.2} = 9.16$$

Reduced taper ratio, λ_R:

$$\lambda_R = \frac{bC_t}{C_r\left(b - w_{fus}\right) + w_{fus}C_t} = 0.5288$$

Then, estimate the C_L required by the smaller wing:

$$C_L = \frac{2W}{\rho V^2 S} = \frac{2(3400)}{(0.002378)(98.5 \times 1.688)^2(127.2)} = 0.8132$$

Proceeding with the analysis method presented before, the results presented in Table 9-10 were obtained. It can be seen that while the "clean" wing requires 5° to lift 3400 lb$_f$ at 98.4 KCAS, the reduced wing (more realistic) requires 6.43° (see Table 9-10 below, where 'original' refers to the clean wing and 'reduced' refers to the wing with the fuselage). Consequently, its induced drag coefficient is 47% greater than the "clean" wing.

9.7.5 Computer code: Prandtl's Lifting Line Method

The following code implements the lifting line method for an arbitrary wing using the above formulation. It is written using Visual Basic for Applications (VBA) and can be used as is in Microsoft Excel. It allows as many vortices as system resources allow to be used for the analysis, represented by the variable N (see the commented variable definitions in the code). A value of N larger than 50 is impractical; N in the ballpark of 10–14 is adequate in most instances.

In order to use the code and assuming Microsoft Excel 2007 is being used, the reader must select the *Developer* tab and then click on the *Visual Basic* icon to open a window containing the programming

TABLE 9-10 Comparing Results for the Original and Reduced Wing

		Elliptical	Original	Reduced	%Change
Taper ratio	$\lambda =$	N/A	0.5000	0.5288	5.8
Aspect ratio	$AR =$	9.86	9.86	9.16	−7.1
Wing area	$S =$	148.8	148.8	127.2	−14.5
Angle-of-attack	$\alpha =$	5.00	5.00	6.43	28.6
Lift coefficient	$C_L =$	0.6971	0.6971	0.8148	16.9
Induced drag factor	$\delta =$	0.0000	0.01865	0.01894	1.6
Lift induced drag coefficient	$C_{Di} =$	0.01569	0.01599	0.02351	47.0
Oswald's efficiency	$e =$	1.0000	0.9817	0.9814	0.0

environment. This window will feature a title like *Microsoft Visual Basic*, followed by the filename, visible in the upper-left corner. In the project pane, typically docked at the left-hand side of this window, right-click to reveal a pop-up menu. One of the commands is the *Insert* command and it has a submenu indicated by a dark triangle. Hover with the mouse cursor over this triangle until the submenu appears. Select the *Module* command. This will open a VBA editor inside the main window. The program below must be entered there. The function is then called from the spreadsheet itself as shown in Figure 9-93. It can be seen that simple cell references are used to pass arguments to the routine. Here 50 vortices are being used, and the Mode is 2, which means that the term δ is returned.

```
Function LiftingLine(AR As Single, TR As Single, S As Single, Cla As Single, AOA As Single, AOA_ZL As Single, N As Integer, Mode As Byte) As Single
'This function implements Prandtl's Lifting Line theory
'
'Variables: AR   = Aspect Ratio
'    TR   = Taper Ratio
'    S    = Wing area (ft^2)
'    Cla  = Average airfoil lift curve slope (per rad)
'    AOA  = Angle of attack (rad)
'    AOA_ZL = Average zero-lift AOA of airfoils (rad)
'    N    = Number of vortices
'
'Mode:   =0 Return CL
'    =1 Return CDi
'    =2 Return del
'    =3 Return e
'
'Initialize
   Dim i As Integer, j As Integer, m As Single
   Dim Pi As Single, Sum As Single
   Dim dPhi As Single, sinPhi As Single, t As Single, C_of_Phi As Single
   Dim Cr As Single, Ct As Single, b As Single
   Dim CL As Single, CDi As Single, CLalpha As Single, del As Single
'Dimension arrays
   ReDim Phi(N) As Single
   ReDim mu(N) As Single
```

```
    ReDim A(N, 1) As Double
    ReDim MatB(N, N) As Double
    ReDim MatC(N, 1) As Double
'Presets
    Pi = 3.14159265
    b = Sqr(S * AR)       'Wing span
    Cr = 2 * S / (b + b * TR) 'Root chord
    Ct = TR * Cr    'Tip chord
'Create stations
    dPhi = 0.5 * Pi / CSng(N) 'in Radians
    For i = 1 To N
    'Calculate station angle
    Phi(i) = Phi(i - 1) + dPhi

    'Calculate mu
    t = Cos(Phi(i))      'Parameter t
    C_of_Phi = (1 - t) * Cr + t * Ct    'Chord determined using parametric formulation
    mu(i) = 0.25 * C_of_Phi * Cla / b
    Next i
'Prepare aerodynamic influence coefficients
    For i = 1 To N
    'Calculate the sine of the angle Phi(i) so it won't have to be calculated over and over
    sinPhi = Sin(Phi(i))
    'Calculate the vortex influence matrix
    For j = 1 To N
    m = 2 * j - 1
    MatB(i, j) = Sin(m * Phi(i)) * (m * mu(i) + sinPhi)
    Next j
    'Calculate the boundary conditions
    MatC(i, 1) = mu(i) * (AOA - AOA_ZL) * sinPhi
    Next i
'Invert matrix
    i = MAT_GaussP(MatB(), MatC(), A())
'Calculate lift coefficient
    CL = A(1, 1) * Pi * AR
    CLalpha = CL / (AOA - AOA_ZL)
'Calculate lift induced drag coefficient
    Sum = 0
    For i = 2 To N
    m = 2 * i - 1
    Sum = Sum + m * A(i, 1) ↑ 2
    Next i
    del = Sum / (A(1, 1) ↑ 2)
    e = 1 / (1 + del)
    CDi = CL ↑ 2 * (1 + del) / (Pi * AR)
'Return
    If Mode = 0 Then LiftingLine = CL
    If Mode = 1 Then LiftingLine = CDi
    If Mode = 2 Then LiftingLine = del
    If Mode = 3 Then LiftingLine = e
    If Mode = 4 Then LiftingLine = CLalpha
End Function
```

λ	AR					
	4	6	8	10	12	14
0.000	=LiftingLine(S$76,$R77,B78, C10,D11,F16,50,2)					
0.025	0.065281	0.094001	0.1168008	0.135233	0.15044	0.163215
0.050	0.048375	0.071701	0.0907342	0.106382	0.119439	0.130495
0.075	0.036884	0.055819	0.0716159	0.084796	0.09591	0.105397
0.100	0.028594	0.044003	0.0571099	0.068196	0.077639	0.085766
0.125	0.022418	0.035006	0.0459075	0.055252	0.063296	0.070277
0.150	0.017731	0.028068	0.0371802	0.045099	0.051992	0.058032
0.200	0.011383	0.018526	0.0250679	0.030929	0.036164	0.040852
0.250	0.007684	0.012912	0.0179135	0.022556	0.026827	0.030751
0.300	0.005684	0.009896	0.0141123	0.018168	0.02201	0.025628
0.350	0.004839	0.008689	0.0126825	0.016632	0.020457	0.024124
0.400	0.004818	0.008789	0.0129866	0.017198	0.021323	0.025317
0.450	0.005404	0.009864	0.0145909	0.019347	0.024019	0.028555
0.500	0.00645	0.011681	0.0171895	0.022711	0.028121	0.033366
0.600	0.009535	0.016921	0.0245386	0.032065	0.039365	0.046386
0.700	0.013535	0.023628	0.0338427	0.043794	0.053344	0.062452
0.800	0.018141	0.031289	0.0443947	0.057018	0.069022	0.08039
0.900	0.023164	0.03958	0.0557491	0.071178	0.085743	0.099453
1.000	0.028477	0.04829	0.0676113	0.085907	0.103072	0.119149

FIGURE 9-93 An example of how calls can be made to the *Lifting Line* function from within Microsoft Excel.

EXERCISES

(1) Determine all the variables of the wing in Figure 9-94, if its planform area, S, is 200 ft^2, $AR = 7.5$, and $\lambda = 0.5$. Approximate its internal volume if it features a constant 15% thick airfoil, whose maximum thickness is at 50% chord. If 30% of this volume is to be used for the fuel tanks, how much fuel can the wing hold and how much will it weigh assuming Jet-A fuel (1 US gal = 231.02 in^3)?

Answer: $b = 38.73$ ft, $C_r = 6.885$ ft, MGC = 5.355 ft, $Y_{MGC} = 8.607$ ft, $X_{MGC} = 4.969$ ft, $\Lambda_{C/4} = 28.05°$, $\Lambda_{C/2} = 26.03°$, wing volume is approximately 100 ft^3; it can hold 225 US gallons, which weigh 1510 lb$_f$.

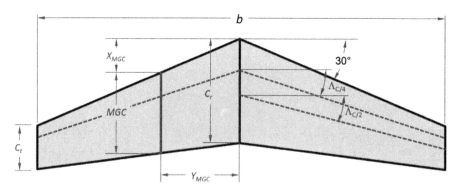

FIGURE 9-94 Wing used in Exercise (1).

VARIABLES

Symbol	Description	Units (UK and SI)
AOA	Angle-of-attack	Degrees or radians
AR	Aspect ratio	
AR_{corr}	Boosted aspect ratio	
AR_e	Effective aspect ratio ($AR \cdot e$)	
AR_{lim}	Aspect ratio limit	
AR_R	Reduced aspect ratio	
b	Wingspan	ft or m
b_R	Reduced wingspan	ft or m
c	Chord length	ft or m
c_{avg}	Average chord length	ft or m
c_{bhp}	Specific fuel consumption of a piston engine	(lb$_f$/hr)/BHP or g/J
C_d	Two-dimensional drag coefficient	
C_D	Three-dimensional drag coefficient	
C_{Di}	Lift-induced drag coefficient	
C_{DM}	Compressibility drag coefficient	
C_{Dmin}	Minimum drag coefficient	
C_l	Two-dimensional lift coefficient	
C_L	Three-dimensional lift coefficient	
C_{L0}	Zero AOA lift coefficient	
C_{LC}	Average cruise lift coefficient	
C_{Lmax}	Maximum 3D lift coefficient	
C_{Lmin}	Minimum coefficient of lift	
C_{lp}	Roll damping coefficient	
$C_{l\alpha}$	Two-dimensional lift curve slope	Per degree or per radian
$C_{L\alpha}$	Three-dimensional lift curve slope	Per degree or per radian
C_m	Two-dimensional pitching moment coefficient	
C_M	Pitching moment coefficient	
C_{mROOT}	Pitching moment coefficient of the root airfoil	
C_{mTIP}	Pitching moment coefficient of the tip airfoil	
$C_{m\alpha}$	Two-dimenstional pitching moment curve slope	Per degree or per radian
$C_{M\alpha}$	Three-dimensional pitching moment curve slope	Per degree or per radian
c_r	Root chord length	ft or m
c_{rR}	Reduced root chord length	ft or m
c_t	Tip chord length	ft or m
c_t	Thrust specific fuel consumption	1/s
C_Y	Side force coefficient	
D	Drag	lb$_f$ or N
$d_i(y)$	Lift-induced drag per unit span	lb$_f$/ft or N/m
dl	Infinitesimally small vector length	ft or m
dw	Velocity induced at arbitrary point P by dl	ft/s or m/s
e	Oswald's span efficiency	
E	Endurance	s
h	Component of distance to point P perpendicular to velocity	ft or m
h	Height of winglets	ft or m
i_{HT}	Horizontal tail angle-of-incidence	Degrees or radians
i_{root}	Wing root airfoil angle-of-incidence	Degrees or radians
i_W	Wing angle-of-incidence	Degrees or radians
K_{SM}	Fraction design static margin	
L	Lift	lb$_f$ or N
LD_{max}	Maximum lift-to-drag ratio	
l_{LER}	Leading edge radius	ft
M	Mach number	
M	Pitching moment	lb$_f$·ft or N·m
MAC	Mean aerodynamic chord	ft or m
M_{crit}	Critical Mach number	
MGC	Mean geometric chord	ft or m
P_1	Special parameter 1	
P_2	Special parameter 2	
r	Distance to arbitrary point P	
R	Leading edge suction parameter	
R	Range	ft
R_e	Reynolds number	
S	Planform area	ft^2 or m^2
S_R	Reduced wing area	
t	Airfoil thickness	ft or m
TR	Taper ratio	

(Continued)

Symbol	Description	Units (UK and SI)
u	x-component of total velocity vector	ft/s or m/s
v	y-component of total velocity vector	ft/s or m/s
V	Velocity	ft/s or m/s
V_C	Average cruising speed	ft/s
V_n	Normal component of velocity	ft/s or m/s
V_p	Parallel component of velocity	ft/s or m/s
V_S	Stall velocity	ft/s or m/s
w	Total velocity induced at arbitrary point P	ft/s or m/s
w	z-component of total velocity vector	ft/s or m/s
W_1	Aircraft weight at the beginning of the design mission	lb_f or N
W_2	Aircraft weight at the end of the design mission	lb_f or N
W_{avg}	Average weight during cruise	lb_f
W_{fin}	Final weight at cruise	lb_f
w_{fus}	Fuselage width	ft or m
W_{ini}	Initial weight at cruise	lb_f
w_{y0}	Downwash velocity induced by vortices	ft/s or m/s
X_{MGC}	x-distance to the leading edge of the MGC	ft or m
y	Spanwise station	ft or m
Y_{MGC}	y-distance from the root chord to the MGC	ft or m
ΔC_{Lmax}	Mach number correction factor	
Δy	Leading edge parameter	
$\Delta \alpha_{CLmax}$	Correction angle for stall AOA	Degrees or radians
$\Delta \phi_{MGC}$	Correction to account for wing twist	Degrees or radians
Γ	Dihedral/anhedral angle	Degrees or radians
Γ	Vortex filament strength	
$\Lambda_{c/2}$	Sweep of the mid-chord line	Degrees or radians
$\Lambda_{c/4}$	Sweep of the quarter-chord line	Degrees or radians
$\Lambda_{c/4\ lim}$	Limit sweep of the quarter-chord line	Degrees or radians
Λ_{LE}	Sweep of the leading edge	Degrees or radians
$\Lambda_{t\ max}$	Sweep of the maximum wing thickness line	Degrees or radians
α	Angle-of-attack	Degrees or radians
α_0	Nonlinear lift angle-of-attack	Degrees or radians
α_C	Cruise angle-of-attack	Degrees or radians
α_{stall}	Stall AOA	Degrees or radians
$\alpha_{stall\ root}$	Two-dimensional stall AOA for the root airfoil	Degrees or radians
$\alpha_{stall\ tip}$	Two-dimensional stall AOA for the tip airfoil	Degrees or radians
α_e	Effective AOA	Degrees or radians
$\alpha_{F\ opt}$	Optimum AOA for the fuselage	Degrees or radians
α_i	Induced AOA	Degrees or radians
α_{stall}	Stall angle-of-attack	Degrees or radians
α_{ZL}	Zero lift angle-of-attack	Degrees or radians
α_{ZLroot}	Two-dimensional zero-lift AOA for the root airfoil	Degrees or radians
α_{ZLtip}	Two-dimensional zero-lift AOA for the tip airfoil	Degrees or radians
β	Prandtl-Glauert Mach number parameter	
δ	Lift-induced drag factor	
ϕ	Velocity potential	
ϕ	Washin/washout angle	Degrees or radians
ϕ_A	Aerodynamic washout	Degrees or radians
ϕ_D	Decalage angle	Degrees or radians
ϕ_G	Geometric washout	Degrees or radians
η	Spanwise station (for $b/2$)	
η_P	Propeller efficiency	
κ	Ratio of 2D lift curve slope to 2π	
λ	Taper ratio	
λ_R	Reduced taper ratio	
μ	Air viscosity	$lb_f \cdot s/ft^2$
ρ	Density of air at altitude	$slugs/ft^3$ or kg/m^3
τ	Lift curve slope correction factor	

References

[1] Hoak DE. USAF Stability and Control DATCOM. Flight Control Division, Air Force Flight Dynamics Laboratory; 1978.
[2] Kuhlman Bill, Bunny. On the Wing…. R/C Soaring Digest October 2003.
[3] Thomas Fred. Fundamentals of Sailplane Design. College Park Press; 1999.

REFERENCES

[4] Brandt Steven A, Stiles Randall J, Bertin John J, Whitford Ray. Introduction to Aeronautics: A Design Perspective. AIAA Education Series; 1997.

[5] Raymer DP. Aircraft Design: A Conceptual Approach. AIAA Education Series; 1996.

[6] Loftin, Laurence Jr K. Quest for Performance; The Evolution of Modern Aircraft. NASA SP-468 1985:93.

[7] NACA R-823. Experimental Verification of a Simplified Vee-Tail Theory and Analysis for Available Data on Complete Model with Vee Tails. Purser, Paul E., and John P. Campell; 1944.

[8] AIAA-1977-607-721. The AV-8B Wing: Aerodynamic Concept and Design. Lacey, I.R., and K. Miller, AIAA; 1977.

[9] Taylor, John W. R. Jane's All the World's Aircraft. Jane's Yearbooks, various years.

[10] http://www.flightlevelengineering.com.

[11] Meier Hans-Ulrich. German Development of the Swept Wing 1935-1945. AIAA; 2006.

[12] Jones Robert T. Adolf Busemann, 1901-1986, vol. 3. Memorial Tributes. National Academy of Engineering; 1989.

[13] Busemann, Adolf. Aerodynamischer Auftrieb bei Überschallgeschwindigkeit. Luftfahrtforschung, Bd. 12, Nr. 6, Oct.

[14] NACA TR-863. Wing Plan Forms for High-Speed Flight. Jones, Robert T; 1945.

[15] Boyne Walter J. Messerschmitt Me 262: Arrow to the Future. Schiffer Publishing, Ltd; 1992.

[16] Langton Roy, et al. Aircraft Fuel Systems. John Wiley and Sons, Ltd; 2009.

[17] Sargent, Barnes Sparky. A Dyke Delta Reborn. EAA Sport Aviation Magazine; December, 2008.

[18] Whitford Ray. Design for Air Combat. Jane's; 1987.

[19] NACA-TN-2502. Examples of Three Representative Types of Airfoil-Section Stall at Low Speed. McCullough, George B., Gault, Donald E; 1951. p. 1.

[20] NACA TN-817. Correction of the Lifting Line Theory for the Effect of the Chord. Jones, Robert T; 1941.

[21] NACA TN-1175. Lifting-Surface-Theory Aspect-Ratio Corrections to the Lift and Hinge-Moment Parameters for Full-Span Elevators on Horizontal Tail Surfaces. Swanson, Robert S., and Stewart M. Crandall; 1947.

[22] Dommasch Daniel O, Sherby Sydney S, Connolly Thomas F. Airplane Aerodynamics. 4th ed. Pittman; 1967. pp. 154–160.

[23] Glauert Hermann. The Elements of Aerofoil and Airscrew Theory. Cambridge University Press; 1926.

[24] Helmbold, H. B. Der unverwundene Ellipsenflügel als tragende Flache. Jahrbuch 1942 der Deutschen Luftfahrtforschung, R. Oldenbourg (Munich), pp. I 111–I 113.

[25] NACA TR-3911. A Method for Predicting Lift Increments due to Flap Deflection at Low Angles of Attack in Incompressible Flow. Lowry, John G., and Edward C. Polhamus; 1957.

[26] Hoak DE. USAF Stability and Control DATCOM. Flight Control Division, Air Force Flight Dynamics Laboratory; 1978. pp. 4.1.3.2–49.

[27] Betz Albert. Lift and Drag of a Wing Near a Horizontal Surface. Zeitschrift für Flugtechnik und Motorluftschiffahrt 1912:212.

[28] Wieselberger C. Wing Resistance Near the Ground. Zeitschrift für Flugtechnik und Motorluftschiffahrt 1921;(No. 10).

[29] NACA TM-77. Wing Resistance Near the Ground. Wieselberger, C; 1922.

[30] NASA TN-D-970. Effect of Ground Proximity on the Aerodynamic Characteristics of Aspect-Ratio-1 Airfoils With and Without End Plates. Carter, Arthur W; 1961.

[31] McCormick Barnes W. Aerodynamics, Aeronautics, and Flight Mechanics. John Wiley & Sons; 1979. p. 420.

[32] Asselin Mario. An Introduction to Aircraft Performance. AIAA Education Series; 1997.

[33] Jenkinson Lloyd R, Simpkin Paul, Rhodes Darren. Civil Jet Aircraft Design. AIAA Education Series; 1999.

[34] Young AD. The Aerodynamic Characteristics of Flaps. R.&M. No. 2622, British A. R. C; 1947.

[35] NACA R-824. Summary of Airfoil Data. Abbott, Ira H., Albert E. von Doenhoff and Louis S. Stivers Jr; 1945.

[36] NACA TR-408. General Formulas and Charts for the Calculation of Airplane Performance. Oswald, Bailey W; 1933.

[37] Shevell Richard S. Fundamentals of Flight. Prentice Hall; 1983. 181–183.

[38] Frost Richard C, Rutherford Robbie. Subsonic Wing Span Efficiency. AIAA Journal April 1963;vol. 1(No. 4).

[39] Roskam Jan. Airplane Design, Part VI. DARcorporation; 2000.

[40] NACA TN-1299. Effects of Mach Number and Reynolds Number on the Maximum Lift Coefficient of a Wing of NACA 230-Series Airfoil Sections. Furlong, G. Chester, and James E. Fitzpatrick; 1947.

[41] NACA TN-2753. Effects of Mach Number Variation between 0.07 and 0.34 and Reynolds Number Variation between 0.97x106 and 8.10x106 on the Maximum Lift Coefficient of a Wing of NACA 64-210 Airfoil Sections. Fitzpatrick, James E., and William C. Schneider; 1952.

[42] NACA TN-1677. Experimental and Calculated Characteristics of Several High-Aspect-Ratio Tapered Wings Incorporating NACA 44-series, 230-series, and Low-Drag 64- Series Airfoil Sections. Bollech, Thomas V; 1948.

[43] van Dam CP, Vijgent PMHW, Holmes BJ. Aerodynamic Characteristics of Crescent and Elliptic Wings at High Angles of Attack. Journal Aircraft April 1991;vol. 28(No. 4).

[44] Torenbeek Egbert. Synthesis of Subsonic Aircraft Design. Delft University Press; 1986.

[45] NACA TR-1339. A Summary and Analysis of the Low-Speed Longitudinal Characteristics of Swept Wings at High Reynolds Numbers. Furlong GC, and McHugh JG; 1957.

[46] NACA TN-1093. Effect of Sweepback and Aspect Ratio on Longitudinal Stability Characteristics of Wings at Low Speeds. Shortal JA, and Maggin B; 1946.

[47] NACA RM-L8D29. Wind-Tunnel Investigation of High-Lift and Stall-Control Devices on a 37 degree Sweptback Wing of Aspect Ratio 6 at High Reynolds Numbers. Koven, William, and Robert R. Graham; 1948.

[48] Katz Joseph, Plotkin Allen. Low-Speed Aerodynamics. Cambridge University Press; 2001.

[49] Pope Alan. Basic Wing and Airfoil Theory 2009. Dover.

[50] Bertin John J, Smith Michael L. Aerodynamics for Engineers. Prentice-Hall; 1979. p. 171.

[51] Moran Jack. An Introduction to Theoretical and Computational Aerodynamics. John Wiley and Sons; 1984.

[52] Anderson, John Jr D. Fundamentals of Aerodynamics. 4th ed. McGraw-Hill; 2007.

[53] Karamcheti Krishnamurty. Principles of Ideal-Fluid Aerodynamics. John Wiley & Sons; 1966. Appendix E, p. 624.

CHAPTER 10

The Anatomy of Lift Enhancement

OUTLINE

10.1 Introduction — 402
   10.1.1 The Content of this Chapter — 403

10.2 Leading-Edge High-lift Devices — 403
   10.2.1 Hinged Leading Edge (Droop Nose) — 403
      General Design Guidelines — 406
      Aerodynamic Properties — 406
      Summary — 406
   10.2.2 Variable-camber Leading Edge — 406
      Aerodynamic Properties — 407
   10.2.3 Fixed Slot — 407
      General Design Guidelines — 408
      Aerodynamic Properties — 408
      Summary — 408
   10.2.4 The Krüger Flap — 408
      Simple Krüger Flap — 409
      Design Guidelines — 409
      Aerodynamic Properties — 410
      Summary — 410
      Folding, Bull-nose Krüger Flap — 411
      Aerodynamic Properties — 411
      Variable-camber Krüger Flap — 411
      Aerodynamic Properties — 411
   10.2.5 The Leading-Edge Slat — 412
      The Airload-actuated Slat (or the Automated Handley-Page Slat) — 413
      The Maxwell Leading Edge Slot — 414
      Three-position Slats — 415
      General Design Guidelines — 415
      Aerodynamic Properties — 416
      Summary for Slats — 416
      Summary for Maxwell Flap — 416
   10.2.6 Summary of Leading Edge Device Data — 416
      How to Use the Tables — 416

10.3 Trailing Edge High-lift Devices — 417
   10.3.1 Plain Flap — 417
      General Design Guidelines — 418
      Aerodynamic Properties — 419
   10.3.2 Split Flap — 420
      Split Flap — 420
      Zap Flap — 420
      General Design Guidelines — 422
      Aerodynamic Properties — 422
   10.3.3 Junkers Flap or External Flap — 423
      General Design Guidelines — 425
      Aerodynamic Properties — 425
   10.3.4 The single-slotted Flap — 425
      General Design Guidelines — 426
      Aerodynamic Properties — 426
   10.3.5 Double-slotted Flaps — 427
      Fixed-vane Double-slotted Flap — 428
      Articulating-vane Double-slotted Flap — 428
      Main/aft Double-slotted Flap — 428
      Triple-slotted Flap — 428
      General Design Guidelines — 430
   10.3.6 Fowler Flaps — 430
      Single-slotted Fowler Flap — 430
      General Design Guidelines — 431
      Aerodynamic Properties — 432
   10.3.7 Gurney Flap — 432
      General Design Guidelines — 433
      Aerodynamic Properties — 433
   10.3.8 Summary of Trailing Edge Device Data — 434
      How to Use the Tables — 434

10.4 Effect of Deploying High-lift Devices on Wings — 436
   10.4.1 Lift Distribution on Wings with Flaps Deflected — 436
      Partial Span Flaps — 436
      Full Span Flaps — Flaperons — 437
   10.4.2 Wing Partition Method — 437
      Estimation of the Maximum Lift Coefficient — 438
      Estimation of the Pitching Moment Coefficient — 439

10.5 Wingtip Design — 440
      Aerodynamic Effectiveness of Wingtips — 442
      Derivation of Equation (10-5) — 442
   10.5.1 The Round Wingtip — 443

10.5.2 *The Spherical Wingtip*	443	The Whitcomb Winglet	450
10.5.3 *The Square Wingtip*	444	The Blended Winglet	451
10.5.4 *Booster Wingtips*	444	10.5.9 *The Polyhedral Wing(tip)*	452
The Upturned Booster Wingtip	445	10.5.10 *Comparison Based on Potential*	
The Downturned Booster Wingtip	445	*Flow Theory*	453
10.5.5 *Hoerner Wingtip*	446	**Variables**	**455**
10.5.6 *Raked Wingtip*	446		
Rakelet	448	**References**	**456**
10.5.7 *Endplate Wingtip*	448		
10.5.8 *The Winglet*	448		

10.1 INTRODUCTION

It was discovered early on in aviation that in order to achieve high airspeeds, wings of small areas were needed. This is clearly evident from the drag equation, $D = \frac{1}{2}\rho V^2 \cdot S \cdot C_D$, which shows that the wing area affects the drag proportionally — halve the area, halve the drag. However, a large wing area is desirable for low-speed operations, take-off and landing, and this is apparent from the lift equation, $L = \frac{1}{2}\rho V^2 \cdot S \cdot C_L$. The solution to the conflicting problem of high lift and low drag has been to keep S as low as possible, and then try to increase lift capability by increasing C_{Lmax}. This is most effectively accomplished using special mechanical devices on the wing that enhance its lifting capability; these were the high-lift devices. They allowed the transformation of the wing from a shape useful for low-speed flight into one conducive to high airspeeds and back again.

In short, the purpose of lift enhancement is to (1) allow the airplane to operate at lower airspeeds, which translates into shorter runway requirements; (2) provide improved L/D in the T-O configuration to help complying with noise requirements during departure climb; (3) increase drag during landing, so that the approach glide angle can be made steeper, making the aircraft easier to land; (4) increase drag during landing in order to reduce floating in ground effect, and (5) reduce *AOA* near maximum lift so the airplane is easier to land at low airspeeds.

A great example of the use of high-lift devices is shown in one of the most interesting passenger jetliners ever produced, the Boeing B-727 tri-jet (see Figure 10-1). It was developed in the 1960s to allow operation from much shorter runways than the jetliners of the day, which all required long runways for operation. This was achieved using an impressive assortment of leading and trailing edge high-lift devices, giving it one of the highest C_{Lmax} of any aircraft. This chapter will present leading edge devices first, followed by trailing edge devices. Finally, methods to account for the addition of such devices to the airplane as a whole will be presented.

The capability of high-lift devices will be the focus of this chapter. These devices usually serve two

FIGURE 10-1 A Boeing B-727 passenger airliner taxiing into T-O position. Visible are parts of its sophisticated wing high-lift system; the *folding, bull-nose Krüger Flap* can be seen on the inboard part of the wing, and the *three-position slats* on the outboard part of the wing. *(Photo by Phil Rademacher)*

purposes: to change the camber of the wing's airfoils, and delay flow separation. As such, there are two kinds of high-lift devices: *passive* and *active*. Passive high-lift devices do not require additional energy to provide lift enhancement, whereas active ones do. Examples of active lift enhancement devices include the jet-blown flap and vectored thrust. Since they add substantial cost to the aircraft, in terms of both hardware and operation, they are never used on GA aircraft, but rather in specialized military aircraft. For this reason, only the former kind will be dealt with here.

There are also two kinds of passive high-lift devices: those that are mounted to the leading edge and those mounted to the trailing edge. As the reader will see shortly, a large number of such devices have been invented — in both groups. As usual, all have their pros and cons, but as a rule of thumb, the more complicated such a system is the more lift it supplies, but it also adds more weight, cost, and operational penalties.

10.1.1 The Content of this Chapter

- **Section 10.2** presents helpful design information for the selection of leading edge high-lift systems.
- **Section 10.3** presents helpful design information for the selection of trailing edge high-lift systems or flaps.
- **Section 10.4** presents methods to estimate the impact of adding partial span high-lift devices.
- **Section 10.5** introduces a number of different wingtip options and discusses their advantages and disadvantages.

10.2 LEADING-EDGE HIGH-LIFT DEVICES

Generally, the purpose of the leading edge high-lift device is to increase the stall angle and maximum lift coefficient of the airfoil without a significant shift in the lift curve, as happens with trailing edge devices. Thus, deploying a leading edge device will have much less effect on trim than do flaps. The effect is described in Section 8.3.9, *The effect of addition of a slot or slats*, in addition to the text here. A summary of the aerodynamic properties of many of the leading edge devices discussed are given in Table 10-1. This data is helpful during the conceptual and preliminary design phases. The aerodynamic data contains information on change in maximum lift coefficient, lift coefficient at zero *AOA*, minimum drag, and pitching moment.

It is important for the designer to keep in mind that, in general, leading edge devices increase the stall *AOA* so much that alone they are surprisingly impractical. For instance, a fixed slot will increase the stall *AOA* by some 9°. This means that an airplane normally stalling at an *AOA* of about 15° would stall around 24° with them deployed. This means that in order to realize the benefits of the higher C_{Lmax}, the airplane would have a deck angle that would be very impractical for the operation of the airplane. The remedy is to mix them with trailing edge flaps. These always reduce the stall *AOA*. Thus, the 9° of stall *AOA* added by the leading edge device is reduced to perhaps an overall 2° increase.

As stated earlier, the maximum lift capability of an airfoil can be improved by two means: by increasing curvature and delaying flow separation. It is also possible to improve stall by increasing the leading edge radius, although this is arguably an airfoil design topic. This section presents a description of a number of such devices, some of which only increase curvature while others do both. Some of those devices are extremely simple and are ideal for simple and slow-flying aircraft, for instance bushplanes. Others are far more complex and require sophisticated four-bar linkages to deliver them into a proper position and shape. They are intended for commercial jetliners and business jets. Movable mechanical leading edge devices are subjected to many challenges. They are fundamental to the operational safety of the airplane, so they have to be reliable — while reacting substantial loads. Jamming is not an option.

The effectiveness of the leading edge devices is best described by the resulting change in C_{lmax} and α_{stall}. This is a direct measure of their ability to postpone flow separation. This is a very important capability, because while it is possible to greatly increase the airfoil curvature geometrically, lift will not increase unless the flow can be made to follow that curvature. The effectiveness is thus highly dependent on the geometry of the airfoil and the leading edge device. There are other characteristics of importance too — impact on drag and pitching moment. These characteristics will be presented based on the availability of experimental data in this section.

10.2.1 Hinged Leading Edge (Droop Nose)

The *hinged leading edge* or *droop nose* or *leading edge flap* is a mechanical device that increases the leading edge camber and therefore C_{Lmax} and α_{stall} of the wing (see Figure 10-2). The device reduces the stalling speed and can also reduce roll instability at stall. Mechanically it is a very simple device and has limited impact on drag when retracted when compared to other such devices, as it seals the slot, preventing air from "leaking" from the lower to the upper surface. However, the increase in C_{Lmax} is limited due to the small-radius curvature on the upper surface, which may induce flow separation. This is caused by the absence of a slot and discontinuity in the curvature on the upper surface, which may spur a separation bubble. A typical deflection range is 15–40°. The main strengths of the device are its low cost, relative ease of manufacturing,

TABLE 10-1 Summary of the Aerodynamic Properties of Leading Edge Devices

Description	Section	Basic Airfoil	Re x10⁶	Flap Type	C_s/C	$(C_f/C)_1$	$(C_f/C)_2$	δ_s	$(\delta_f)_1$	$(\delta_f)_2$	C_{lmax}	α_{max}	C_{lo}	C_{dmin}	C_{mo}	ΔC_{lmax}	$\Delta\alpha_{max}$	ΔC_{lo}	ΔC_{dmin}	ΔC_m	Reference
HINGED LEADING EDGE																					
Airfoil only			6	-	0.15	-	-	0°	-	-	1.10	11.0	0.01	-	0.000	-	-	-	-	-	
Hinged LE			6	-	0.15	-	-	30°	-	-	1.66	18.8	-0.14	-	-0.048	0.56	7.8	-0.15	-	-0.048	NACA TN-3007
Hinged LE + TE Flap	10.2.1	NACA 64A010	6	Split	0.15	0.0557	0.1723	30°	60°	-	2.58	15.8	1.27	-	-0.215	1.48	4.8	1.26	-	-0.215	
Hinged LE + TE Flap			6	Dbl-Slotted	0.15	0.25	-	30°	30°	55°	3.09	10.5	2.23	-	-0.565	1.99	-0.5	2.22	-	-0.565	
VARIABLE CAMBER LEADING EDGE																					
Variable Camber LE	10.2.2	-	-	-	-	-	-	-	-	-	-	-	-	-	-	-	-	-	-	-	-
FIXED SLOT																					
Airfoil only	10.2.3	Clark Y	0.63	-	0.1475	-	-	-	-	-	1.297	15.0	0.36	0.0150	-	-	-	-	-	-	NACA TR-407
Fixed Slot			0.63	-	0.1475	-	-	-	-	-	1.751	24.0	0.28	0.0229	-	0.45	9.0	-0.08	0.0079	-	
SIMPLE KRÜGER																					
Airfoil only			2.14	-	0.1	-	-	0°	-	-	1.13	12.3	0.09	0.0082	-0.017	-	-	-	-	-	
Simple Krüger Flap	10.2.4	Mustang 2	2.14	-	0.1	-	-	120°	-	-	1.43	17.3	0.00	0.0230	-0.044	0.30	5.0	-0.09	0.0149	-0.027	NACA TM-1177
Simple Krüger Flap + TE Flap			2.14	Split	0.1	0.2	-	120°	60°	-	2.25	13.0	1.15	0.1656	-0.194	1.12	0.7	1.06	0.1575	-0.177	
FOLDING BULL-NOSE KRÜGER																					
Folding, Bull-nose Krüger Flap	10.2.5	-	-	-	-	-	-	-	-	-	-	-	-	-	-	-	-	-	-	-	-
VARIABLE CAMBER KRÜGER																					
Variable Camber Krüger Flap	10.2.6	-	-	-	-	-	-	-	-	-	-	-	-	-	-	-	-	-	-	-	-
SLATS																					
Airfoil only			6	-	0.17	-	-	0°	-	-	1.10	11.0	0.01	-	0.000	-	-	-	-	-	
Slat (optimum deflection)			6	-	0.17	-	-	25.6°	-	-	1.94	22.0	-0.20	-	-0.115	0.84	11.0	-0.21	-	-0.115	
Slat (opt. deflection) + TE Flap	10.2.7	NACA 64A010	6	Split	0.17	0.0557	0.1723	25.6°	60°	-	2.72	17.3	1.25	-	-0.283	1.62	6.3	1.24	-	-0.283	NACA TN-3007
Slat (opt. deflection) + TE Flap			6	Dbl-Slotted	0.17	0.25	-	25.6°	30°	55°	3.00	9.0	2.26	-	-0.551	1.90	-2.0	2.25	-	-0.551	
MAXWELL FLAP																					
Airfoil only			0.61	-	0.175	-	-	-	-	-	1.26	14.0	0.38	0.0201	-0.077	-	-	-	-	-	
Maxwell Flap 1			0.61	-	0.175	-	-	Open	-	-	1.81	23.0	0.25	0.0357	-0.076	0.55	9.0	-0.13	0.0156	0.0008	
Maxwell Flap 1 + TE Flap	10.2.7	Clark Y	0.61	Split	0.3	0.211	-	Open	60°	-	2.31	16.7	1.31	0.2835	-0.177	1.05	2.7	0.93	0.2634	-0.1004	NACA TN-598
Maxwell Flap 2			0.61	-	0.3	-	-	Open	-	-	2.07	25.8	0.21	0.0477	-0.113	0.81	11.8	-0.17	0.0276	-0.0365	
Maxwell Flap 2 + TE Flap			0.61	Split	0.3	0.211	-	Open	60°	-	2.53	21.6	1.27	0.2834	-0.219	1.27	7.6	0.89	0.2633	-0.1424	
Handley Page slat			0.61	-	0.147	0.211	-	Open	-	-	1.83	25.0	1.27	0.2834	-	0.57	11.0	0.89	0.2633	-	

Abbreviations:
LE = leading edge
TE = trailing edge
$R_{e\ test}$ = Reynolds number during test
C_S = chord of LE device
C = airfoil chord
Dbl = double
δ_S = deflection angle of LE device
$(\delta_f)_1$ = deflection of element 1 of a TE device
$(\delta_f)_1$ = deflection of element 2 of a TE device
C_{lmax} = max section lift coefficient
α_{max} = stall AOA
C_{lo} = lift coefficient at $\alpha = 0°$
C_{dmin} = minimum section drag coefficient
C_{mo} = pitching moment coefficient at $\alpha = 0°$

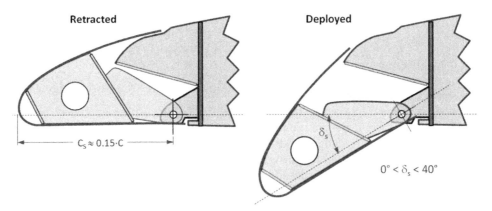

FIGURE 10-2 A schematic of the *hinged leading edge* (*droop nose*).

and low impact on weight and drag. An aircraft that uses the droop nose leading edge is the Lockheed F-104 Starfighter; the increase in maximum lift is far too small to make it practical for use in passenger aircraft.

Figure 10-3 shows flow visualization made of the hinged leading edge (and plain flap) and reveals several issues that are detrimental to the effectiveness of the device. The designer should be aware of such issues, as this will aid in the selection and justification for alternative leading edge devices. In all three pictures, the leading and trailing edge flaps, denoted by δ_s and δ_f, respectively, are deflected 33°. The top, center, and bottom

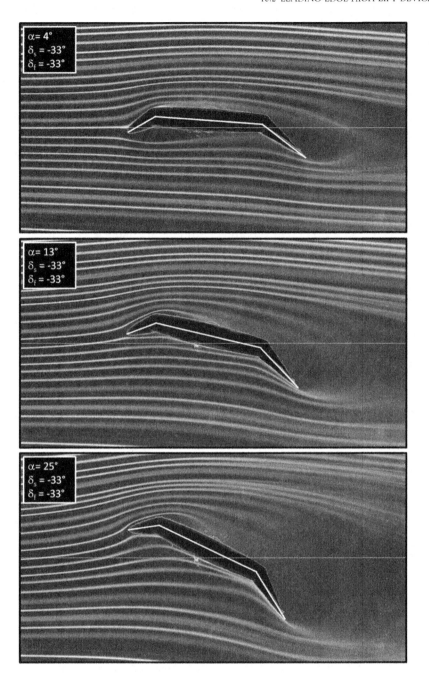

FIGURE 10-3 Flow visualization of the *hinged leading edge*. (Photos by Phil Rademacher)

pictures show the airfoil at a α of 4°, 13°, and 25°, respectively.

The top picture reveals that the stagnation point is located just above the nose radius of the leading edge and this requires the fluid to flow downward around the nose. However, the boundary layer lacks the energy to fully flow around the nose radius and instead separates from the surface and creates a separation bubble on the lower surface. This results in a diminished circulation around the airfoil and would cause additional and, possibly, unexpected drag and reduction in lift of the configuration. Of course, the solution is reduced angle of deflection of the nose droop at that AOA – but at least the picture shows the consequence of too much deflection. Also note the flow separation behind the flap. The flow in this experiment is at a low Reynolds number, but at a flap deflection of 33° even large airplanes featuring a plain flap (see Section 10.3.1, *Plain flap*) also suffer such massive separation. The effect of flow separation behind flaps is discussed in more detail in Section 10.3, *Trailing edge high-lift devices*.

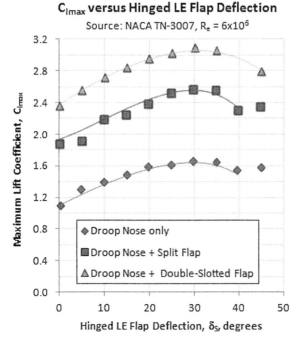

FIGURE 10-4 A schematic of the *hinged leading edge*.

The center picture in Figure 10-3 shows the airfoil at a higher *AOA* of 13°. The stagnation point has moved closer to the tip of the nose, causing the elimination of the separation bubble seen in the top picture, behind and below the drooped nose. However, another one has begun to form on the upper surface, right behind the discontinuity in the upper curve. This indicates the sensitivity of such surfaces to discontinuity in surfaces, and this is the primary drawback of the device. The bottom picture shows the airfoil deep in the post-stall region, at an *AOA* of 25°. The stagnation point is now below the nose tip and the separation bubble on the trailing edge of the drooped nose has increased substantially. It has, effectively, become a part of the massive separation region behind the stalled airfoil.

General Design Guidelines

NACA TN-3007 [1] provides helpful guidelines for the initial design of a fixed slot. Using a symmetrical NACA 64A010 airfoil, the results of an investigation of combinations of a leading edge slat and flap, and split flap and a double-slotted flap are presented. With respect to the leading edge flap, the investigation showed that for $R_e \approx 2 \times 10^6$ a droop deflection $\delta_s = 40°$ resulted in the highest C_{lmax}. For higher R_e, $\delta_s = 30°$ was found to result in the highest C_{lmax}. The reference did not investigate drag.

Aerodynamic Properties

Results for the hinged leading edge using a NACA 64A010 airfoil are given in Table 10-1 and are obtained from Ref. [1], for a Reynolds number of 6×10^6. Results for other airfoils and conditions are likely to be different. A maximum change in maximum section lift coefficient to be expected is $\Delta C_{lmax} \approx 0.56$. This will increase the stall *AOA* by about $\Delta \alpha_{max} \approx 7.8°$, and the change in pitching moment coefficient is approximately $\Delta C_m \approx -0.048$.

As intuition would hold, the maximum lift coefficient increases with the deflection of the hinged leading edge, up to a maximum around 30°. The variation of C_{lmax} is plotted against the deflection in Figure 10-4. Higher deflections reduce this maximum and can be attributed to separation of effects caused by the sharp curvature around the hingeline.

Summary

$$\Delta C_{l_{max}} \approx 0.56 \qquad \Delta C_{d_{min}} \approx$$
$$\Delta \alpha_{max} \approx 7° - 8° \qquad \Delta C_m \approx -0.048$$

10.2.2 Variable-camber Leading Edge

The *variable-camber leading edge* is a device designed to increase the airfoil camber at the leading edge (see Figure 10-5) while minimizing impact on cruise drag. Like its droop nose sibling, the device increases the low-pressure peak at higher *AOA*s and therefore C_{Lmax}. However, it offers improved continuity in the upper surface curvature, making it smoother than the hinged leading edge, discussed above. This reduces the

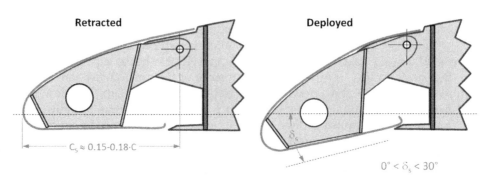

FIGURE 10-5 A schematic of the *variable-camber leading edge*.

possibility of the formation of a separation bubble on the top surface. The skin extending from the main element of the wing onto the leading edge prevents leakage drag, rendering the device more efficient (less drag). When used with an active flight control system the device can be controlled to offer mission adaptability. This makes it primarily suitable for aircraft that must maneuver at higher airspeeds, such as fighter aircraft.

The device uses a complicated mechanical linkage (not shown in schematic) to ensure the compound motion necessary to ensure a smooth flexible skin curvature. This mechanism adds weight and complexity to the structure and increases design and development costs, not to mention maintenance. Additionally, the device calls for the use of flexible skin and movable linkages in the mechanical system, yielding a leading edge that no longer contributes to the wing stiffness and flutter resistance. The device does not yield a large increase in C_{Lmax}, so it is not used for jetliners.

A typical deflection range is 15–30°. The main strength of the device is a smooth upper surface and, thus, reduced impact on drag at cruise. Among its drawbacks are high cost, complex manufacturing of mechanical operation, a moderate impact on weight, and reduced maximum lift due to the absence of slot flow. The F-111 AFTI (Advanced Fighter Technology Integration) [2] Mission Adaptive Wing test program is really the only example of this concept. Boeing also performed studies on a mission adaptive wing for a commercial jetliner [3]. The study assumed both the leading and trailing edge devices featured a smooth variable-camber mechanism and it was shown such a design would permit less sweep and improved aerodynamic efficiency below the design Mach number.

Aerodynamic Properties

Not available at this time. In the absence of better data, aerodynamic properties are expected to be similar to those of the droop nose leading edge flap, which are presented in Table 10-1.

10.2.3 Fixed Slot

The *fixed slot* is a wing design (or airfoil selection) philosophy and not a mechanical device, per se, as it is immovable (see Figure 10-8). Its invention is usually attributed to Gustav Lachmann and Sir Handley Page (see Section 10.2.5, The leading edge slat). It has been used with great success in many short take-off and landing (STOL) aircraft, among which are the Fieseler Fi-156 Storch (see Figure 10-7); Zenith STOL series CH 701, CH 750, CH 801, and Heintz Zenith; and Westland Lysander. The device is ideal for very simple slow-flying aircraft whose primary capability is getting into and out of small patches of land most people would not dignify with the noun "airfield." However, a more sophisticated variation of the fixed slot exists in high-speed aircraft too. The Douglas DC-8 commercial jetliner widely used in the 1970s–1990s features fixed slots near the engine pylons. These are closed during cruise, but before landing these are opened to improve low speed handling.

The slot works through what is called the *slat-effect*. Conventional wisdom has it that the slot provides an access for high-pressure air from the lower surface to flow to the upper surface. Additionally, its shape accelerates this flow and somehow energizes the airflow over the upper surface of the airfoil, yielding a higher C_{lmax} and stall AOA. On the other hand, in a famous paper, A. M. O. Smith [4] disputes this explanation and points out that the airflow through the slot is actually reduced. In fact, the slot works more like a small wing placed in front of the larger wing rather than a device that accelerates slot flow.

As shown in a paper by Liebeck [5], Smith's explanation can be realized by replacing the leading edge element with a vortex (see Figure 10-6). The circulation caused by this vortex induces velocity on the larger airfoil that reduces the airspeed over its leading edge and through the slot and this drops the pressure peak on the leading edge of the airfoil, delaying stall. This description is supported by both wind tunnel tests and analyses performed using computational fluid dynamics [4].

This way, the slat-effect acts to reduce the pressure peak and this delays the flow separation and allows the airfoil to reach a higher AOA and, thus, generate higher C_{lmax} than possible without it.

The fixed slot is very simple in construction compared to other leading edge devices. It is a very effective and inexpensive device that allows aircraft like the Fieseler Storch (see Figure 10-7) to achieve a C_{Lmax} in excess of 4.2.[1]

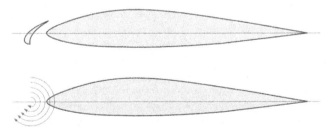

FIGURE 10-6 The workings of the slat explained using vortex flow (based on Ref. [4]).

---

[1] This assumes a gross weight of 2920 lb$_f$, wing area of 280 ft^2, and stalling speed of 27 KCAS.

FIGURE 10-7 A German Fieseler Fi-156 Storch STOL aircraft boasting its *fixed slot*. (Photo by Nick Candrella)

As stated above, the fixed slot is ideal for aircraft intended for STOL operation, as long as aerodynamic efficiency is not an issue. The slot prevents laminar flow beyond its trailing edge so, even at low *AOA*s, it greatly increases the drag of the airfoil when compared to the "clean" airfoil, rendering it impractical for high-speed aircraft. A fixed slot is used on the horizontal stabilator of the Cessna 177 Cardinal to prevent it from stalling when flaps are deployed and CG is in a forward position, something requiring a large stabilator deflection.

General Design Guidelines

NACA TR-407 [6] provides helpful guidelines for the initial design of a fixed slot. Spurred by the popularity of the Handley-Page slat, which would increase drag by a factor of more than 3 when deployed, the report details the investigation of a series of fixed slots in an attempt to reduce the minimum drag coefficient without detrimentally affecting the maximum lift coefficient and the stall *AOA* of a Clark Y airfoil. Of the configurations tested, the one shown in Figure 10-8 was found to yield the best results. It was found that reducing the drag coefficient of the auxiliary airfoil does not necessarily reduce the drag of the combination and may even cause it to increase and the maximum lift to decrease. Rounding the nose of the main airfoil was determined to be the most promising way to reduce the drag of the combination. It was also shown that moving the depth of the slot aft (to the right in Figure 10-8) did not have appreciable effect on the aerodynamic characteristics of the combination.

Aerodynamic Properties

Results for the fixed slot using a Clark Y airfoil are given in Figure 10-9 and Table 10-1 and are obtained from Ref. [6]. The test Reynolds number was about 0.63×10^6. A maximum change in maximum section lift coefficient to be expected is $\Delta C_{lmax} \approx 0.45$, about 0.11 less than that of the hinged leading edge. This will increase the stall *AOA* by about $\Delta \alpha_{max} \approx 9°$, the zero *AOA* lift coefficient will drop by about $\Delta C_{lo} \approx -0.074$, and the minimum drag coefficient will increase by about $\Delta C_{dmin} \approx 0.0079$ (79 drag counts). Ultimately, the gain depends on the shape and the experimental setup does not represent an optimized geometry.

Summary

$$\Delta C_{l_{max}} \approx 0.45 \quad \Delta C_{d_{min}} \approx 0.0079$$
$$\Delta \alpha_{max} \approx 9° \quad \Delta C_m \approx$$

10.2.4 The Krüger Flap

The *Krüger flap* was invented in 1943 by the German Werner Krüger (1910—), an aerodynamicist for Dornier, where he worked with the noted aerodynamicist

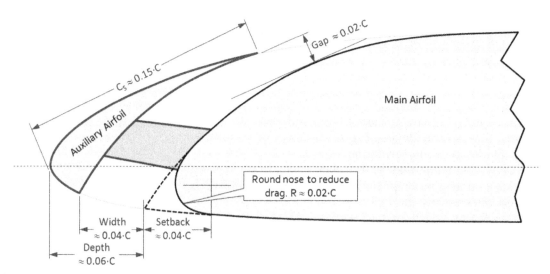

FIGURE 10-8 A schematic of a *fixed slot*. Airfoil (Clark Y) and dimensions are based on the optimum configuration as determined by Ref. [6].

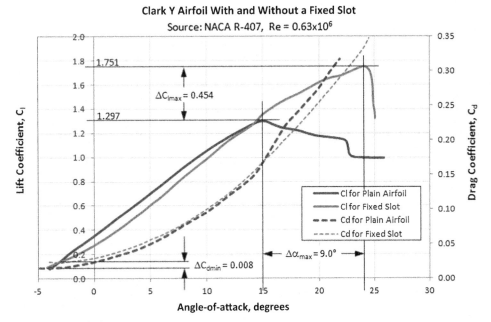

FIGURE 10-9 Effect of the fixed slot on the lift and drag characteristics of the Clark Y airfoil. *Reproduced from Ref. [6].*

Hermann Schlichting (1907–1982) [7, p. 200]. Krüger's career goal was to improve flight characteristics by boundary-layer control. He patented his invention in 1944 and it was first used in the Boeing 367-80, the prototype of the venerable Boeing 707. The flap is a two-position device (retracted-deployed) and, as such, at first glance is mechanically a simple device. The deployed position is generally biased toward landing (C_{Lmax}) and not T-O [8]. Optimizing it to other flight conditions would require other deployment angles and this would call for a more complex mechanism or improved geometry (e.g. bull-nose Krüger). Krüger flaps are ideal for the inboard wing as they delay flow separation to a lesser angle than the outboard slats and, therefore, improve roll stability. Of course the same effect can be achieved by other means too, rendering them largely obsolete.

Simple Krüger Flap

The *simple Krüger flap* is a high-lift device intended to increase the curvature of the camber of the airfoil to which it is mounted. It is used in some commercial jetliners to improve the high-lift capability of the undercambered airfoil near the root of the fuselage[2] and delay flow separation. As such, it is primarily used on the inboard part of the wing. The outside mold line (OML) of the device is essentially shaped like the lower surface of the leading edge of the airfoil. Its geometry is highly dependent on the geometry of the leading edge and this inflicts an aerodynamic limitation on its shape.

The effectiveness of the simple Krüger flap with variations in *AOA* is generally considered poor. It is not used on any modern airplane. The flap rotates into a position that is approximately 110° to 140° with respect to the chord line (see Figure 10-10). Examples of aircraft that use the simple Krüger flap are (as previously mentioned) the Boeing 707, Convair 880, and Convair 990, which featured it from root to tip.

Design Guidelines

The function of the Krüger flap is explained by noting that at the proper deflection angle the stagnation point of the incoming airflow is positioned near the leading edge of the flap. This reduces the excessive airspeed over the leading edge of the airfoil by allowing the airflow to accelerate over a longer distance. The reduction in local airspeeds, in turn, means the associated pressure peaks are reduced and this delays the formation of flow separation (see schematic in Figure 10-11). Consequently, the airfoil stalls at a higher *AOA* [7, pp. 170–171]. In short, the general design requires a trial and error approach through extensive wind tunnel testing, although some basic design guidelines can be presented.

Krüger investigated the device in NACA TM-1177 [9], with and without the presence of a split flap. He found that the chord length of the flap, C_S, should be 0.07C to 0.10C, where C is the airfoil chord length. He tested two chord lengths, one which was 0.05C and the other 0.10C, and found the shorter one actually reduced the maximum lift. From this he concluded that a minimum

---

[2] The purpose of which is to help prevent shock formation between the upper surface of the wing and the fuselage.

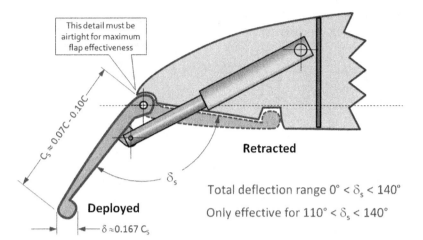

FIGURE 10-10 A schematic of the *simple Krüger flap*.

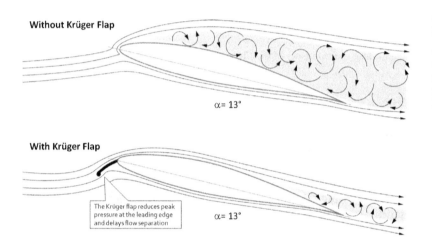

FIGURE 10-11 The *simple Krüger flap* in action. The upper figure shows an airfoil stalled at the given α. By repositioning the stagnation point, air is now allowed to accelerate over greater distance, but this reduces the pressure peaks, delaying the flow separation and pushing the stall to a higher α (based on Ref. [7]).

chord length should be $0.07C$, as shown in Figure 10-10. He suggested the leading edge of the flap should feature a ball at the leading edge with a radius around $0.167\,C_S$, although increasing it to $0.334\,C_S$ showed no detriment. However, allowing air to flow through a gap between the flap and the leading edge of the wing reduced the increase in ΔC_{lmax}, cited below, by about 50%–70%.

Aerodynamic Properties

Results for a $0.01 \cdot C$ Krüger flap using a Mustang 2 airfoil are given in Table 10-1. The test Reynolds number was about 2.14×10^6. Per Ref. [9] the maximum increase in ΔC_{lmax} for the Krüger alone without trailing edge flaps amounted to $\Delta C_{lmax} \approx 0.30$, at a deflection angle $\delta_s = 130°$. In fact, it must swing through approximately $90°$ before it even begins to increase the C_{lmax}. This gain is negatively affected by the presence of a deflected trailing edge flap. The reference also tested a $0.2 \cdot C$ split flap at $60°$ deflection and found that with the flap ΔC_{lmax} increased only by 0.15, albeit for δ_s ranging from $100°$ to $120°$. It is also stated that there was a substantial reduction in the magnitude of the low pressure region on the leading edge.

The Krüger flap will increase the stall AOA by about $\Delta \alpha_{max} \approx 5°$, zero AOA lift coefficient will drop by about $\Delta C_{lo} \approx -0.093$, and the minimum drag coefficient will increase by about $\Delta C_{dmin} \approx 0.0149$ (149 drag counts), so it add substantial drag to the airplane. Change in pitching moment coefficient is approximately $\Delta C_m \approx -0.027$. The combination of the Krüger flap and the $0.2 \cdot C$ split flap increased the stall AOA by some $\Delta \alpha_{max} \approx 0.7°$. While this may not seem like much, the reader is reminded that the primary purpose of the high-lift devices is to increase the lift without changing the stall AOA excessively (i.e. making it too large or too small).

Summary

$$\Delta C_{l_{max}} \approx 0.30 \quad \Delta C_{d_{min}} \approx 0.0149$$
$$\Delta \alpha_{max} \approx 5° \quad \Delta C_m \approx -0.027$$

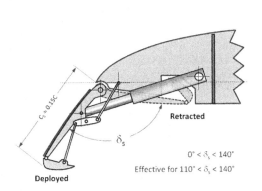

FIGURE 10-12 A schematic of the *folding, bull-nose Krüger flap* (based on Ref. [8]). The photo shows its implementation in a Boeing 727 commercial jetliner. *(Photo by Nick Candrella)*

Folding, Bull-nose Krüger Flap

The *folding, bull-nose Krüger flap* is an improvement over the simple Krüger flap, as it increases the curvature of the airfoil camber. The round bull-nose (see Figure 10-12) improves the effectiveness of the flap over a larger range of *AOA* by providing a larger area in which to "capture" the stagnation point. Just like the simple Krüger flap, the flap is a two-position device (retracted-deployed) whose deployed position is generally biased toward landing (C_{Lmax}) and not T-O. The folding, bull-nose Krüger flap is clearly more complicated than the simple one and often requires a slaved mechanical linkage that folds the bull-nose into the deployed and retracted positions. The flap is used on the Boeing 727 (see Figure 10-1 and Figure 10-12), 737, and 747 (see Figure 10-14) jetliners.

As with the simple Krüger flap, the drawback is a limited increase in C_{Lmax}. As with the simple Krüger flap, the rigid flap is highly dependent on the lower surface of the leading edge and this may inflict an aerodynamic detriment to the geometry. However, it can be the ideal surface to control stall progression along the wing.

Aerodynamic Properties

In the absence of design data assume the properties discussed above for the simple Krüger flap. Results for the simple Krüger flap are given in Table 10-1.

Variable-camber Krüger Flap

The *variable-camber Krüger flap* is intended to improve the shape of the simple and bull-nose Krügers. It is capable of developing far superior curvature, greatly improving its aerodynamic properties when compared to the others. It is a two-position device (retracted-deployed). The deployed position is generally biased toward landing (C_{Lmax}) and not T-O. Each flap panel must be flexible and is thus usually made from fiberglass and in short sections (spanwise speaking). Two hat sections parallel to the leading edge are used to stiffen it in the spanwise direction and ensure chordwise flex only.

The flap deflects through approximately 120° angle from the stowed position (depending on leading edge geometry), during which it transforms from a more or less flat panel into a highly curved one. This feat requires a complex four-bar linkage to make certain it acquires the right shape and position (see schematic in Figure 10-13). This mechanism also explains why the device is so far only a two-position one. Adding a third position (say a T-O setting) that does not violate the quality of the other positions is destined to be challenging. The flexible panels act like bug shields and protect the leading edge from contamination. Figure 10-14 shows the combination of a bull-nose and variable-camber Krüger flaps on a Boeing 747-400 commercial jetliner.

The sophistication of the actuation mechanism of the flap system is impressive — it is truly a marvel of engineering. However, in order to make the system work reliably, each part has to be made with tight tolerances. As a consequence, it is expensive to manufacture and maintain. The flap must be carefully rigged to prevent undesirable distortion of the panels under high air loads and be preloaded to avoid panel bulging (and the associated drag increase) when retracted.

In spite of the limitations, Krüger flaps should not be ignored as a potential candidate configuration for aircraft designed to sustain a laminar boundary layer over the wing's upper surface. Since the flap is stowed on the lower surface of the airfoil, an opportunity exists for a smooth upper surface. To the author's best knowledge, the Boeing 747 is the only aircraft currently in production to use the variable-camber Krüger flap. It was also featured on the proof-of-concept Boeing YC-14, a participant in the USAF Advanced Medium STOL Transport (AMST) project in the late 1970s.

Aerodynamic Properties

In the absence of design data assume the properties discussed above for the simple Krüger flap, although it is undoubtedly much better. Results for the simple Krüger flap are given in Table 10-1.

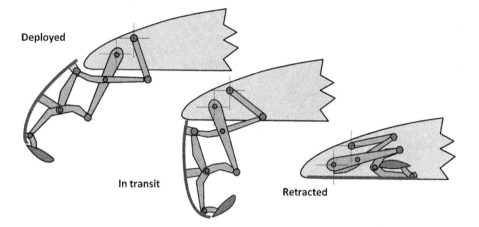

FIGURE 10-13 A schematic of the *variable-camber Krüger flap* (based on Ref. [8]).

FIGURE 10-14 A Boeing 747-400 taxiing into take-off position, boasting a combination of *bull-nose* (inboard) and *variable-camber Krüger* flaps. *(Photo by Phil Rademacher)*

10.2.5 The Leading-Edge Slat

The idea of the slat dates back to 1918 to work by the German aeronautical engineer Gustav Lachmann (1896–1966). Lachmann's original work has been translated into English in NACA TN-71 [10].

The slat was also developed independently by the British industrialist and aircraft manufacturer Sir Frederick Handley Page (1885–1962). In 1919, in order to avoid a patent conflict Page came to an agreement with Lachmann, who ended up working for Page until the end of his life [57]. The device was described publicly in a lecture given by Page before the British Royal Aeronautical Society in 1921 [4]. Handley Page seems to have been among the first to try it out in practice, put to use on a de Havilland D.H. 9. The device later became known as the *Handley Page slot*. Initially, it was fixed (as discussed in Section 10.2.3, Fixed slot), but later, the drag associated with it encouraged Handley-Page's chief engineer George Volkert and his assistant S. G. Ebel to develop a slot that was closed at low AOAs and automatically opened with increasing AOA, when the airplane slowed down [57] — the *leading edge slat*[3] was born.

The *two-position slat* is a device that increases C_{lmax} through the *slat-effect*, explained in Section 10.2.3, Fixed slot. Mechanically, it is moderately complex (Figure 10-15).

Important aerodynamic properties of slats are described in terms of the dimensions shown in Figure 10-16, but the motion is described in terms of:

(1) Extension forward of the leading edge (Δx or a).
(2) Downward drop below the leading edge (Δy or b).
(3) Size of the gap at the outlet of the slot (c).
(4) The rotation of the element (δ_S).

Generally, aerodynamic data is provided for the slat by referencing these parameters.

---

[3]An article on Wikipedia states the design was so successful that licensing fees to other companies are acclaimed as its main source of income in the early 1920s. This claim has not been verified and the author is reluctant to take it at face value. However, it is presented here as an anecdote, as it is interesting if true.

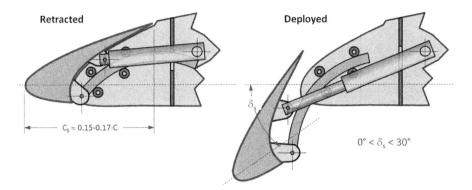

FIGURE 10-15 A schematic of the mechanical aspect of the *two-position slat* (based on Ref. [8]).

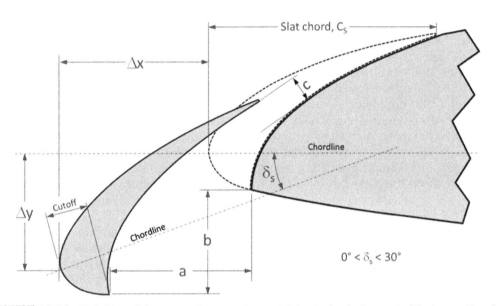

FIGURE 10-16 Definition of the geometric parameters pertaining to the deployment of the *two-position slat*.

The Airload-actuated Slat (or the Automated Handley-Page Slat)

As stated earlier, the *airload-actuated slat* is the automated version of Handley-Page's fixed slot. It is arguably an ingenious device designed to be extracted and retracted by the magnitude of the pressure acting at the leading edge of the wing (see Figure 10-17). This way its operation is passive; when the airplane is at a low speed it is automatically deployed and when at high speed the slat is stowed, again automatically. This happens solely due to the action of pressure forces, although some designs feature a spring to help retract the slat. As the airplane must operate at higher *AOAs* at low speeds, a low-pressure region forms around the leading edge that pulls the element forward out of its stowed position. By the same token, when the airplane's airspeed increases, its *AOA* reduces. The stagnation pressure now impinges on the element and forces it back into the stowed position. This system of passive leading edge slats is employed on aircraft like the Messerschmitt Me-262 Schwalbe, SOCATA Rallye 100 and McDonnell Douglas A-4 Skyhawk (see Figure 10-18).

The slat travels in or out from the leading edge, but the actuation rods may be slightly curved, leading to the slat deploying out and rotating through a slight angle downward. In the deployed position, the device

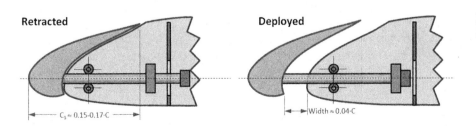

FIGURE 10-17 A schematic of the *airload-actuated slat*.

FIGURE 10-18 The airload actuated slats on the McDonnell-Douglas A-4 Skyhawk do not need a mechanical actuation system. *(Photos by Phil Rademacher)*

increases the pressure drop around the leading edge, increasing the stall *AOA* and C_{Lmax} of the wing. The fact that the slat retracts when the airplane accelerates to higher airspeeds reduces its drag significantly compared to the fixed slot. This renders it ideal for aircraft intended for STOL operations, but for which higher cruising speed is also important. Airplanes rolling rapidly at high speeds (fighter aircraft) have been known to occasionally have the slat on the down-travelling wing forced out, while the one on the up-travelling wing remains retracted. The resulting asymmetry may produce a violent roll departure.

The Maxwell Leading Edge Slot

The *Maxwell slot* is best described as an in-between the fixed slot and slat. Rather than translating and rotating like the standard (or Handley Page) slat, the device operates through rotation only about the leading edge, as is shown in Figure 10-19. From that perspective, the slat is mechanically simple, although the addition of the hinged door adds complexity. However, the cruise configuration of the airfoil is relatively smooth, although it would almost certainly trip the laminar boundary layer on both the upper and lower surfaces if used with an NLF airfoil. The position of the slat is referred to in terms of the gap opening (e.g. 0.01C or 0.0175C, etc.) rather than rotation in degrees, as this will vary based on the airfoil geometry.

The Maxwell slot can also be used without the hinged door shown in Figure 10-19. This will make its installation mechanically simpler and at first glance it will appear more like a fixed slot, however, with one important difference — in cruise, the gap size is zero, whereas it remains constant for the fixed slot. The device is investigated in Refs [11,12]. Reference [11] concludes that the increase in C_{lmax} of a Maxwell slot is similar to that achievable with

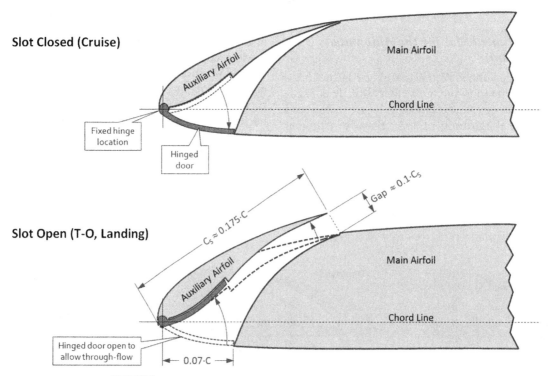

FIGURE 10-19 A schematic of the *Maxwell slot* (based on Ref. [11]).

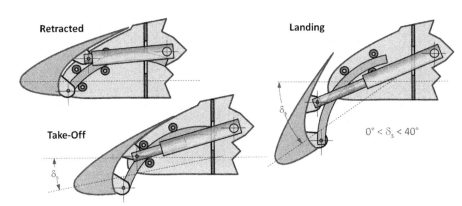

FIGURE 10-20 A schematic of the *three-position slat*.

a Handley Page slat. Reference [12] found the optimum gap width to be $0.0175 \cdot C$. The investigation shows that the drag of the slat increases with gap size.

Three-position Slats

The *three-position slat* is a version of the leading edge slat that features three positions (see Figure 10-20): retracted, take-off, and landing. In the take-off position, the slat either rests against the leading edge of the main airfoil, preventing air from flowing from the lower to the upper surface, or a very narrow slot opens. In this configuration, it works more to increase the camber of the basic airfoil, as the closed slot does not permit circulation to develop around the slat. The ideal motion of the slat is one of forward and downward motion, with limited rotation. The resulting airfoil geometry offers higher L/D than the open slot configuration and this improves T-O and climb performance. In the landing position, the slat rotates to open a slot, offering the maximum improvement in C_{Lmax}.

The track and mechanism required to move the slat in the above fashion is more complicated than shown in Figure 10-20, as such a system is biased toward translation at first and rotation later. However, many modern aircraft simply feature a circular arc design and live with a narrow slot for the T-O setting. Other aircraft, such as the B-777, have the main airfoil leading edge shaped such that it forms a seal with the trailing edge of the slat. Figure 10-21 shows the three-position slat in action on the MD-80 jetliner. The difference in slot sizes between the second and third setting is clearly visible in the center and right pictures, respectively. The three-position slat is the most common leading edge high-lift device in use today. Practically all commercial jetliners, except the B-747, use the device.

General Design Guidelines

Besides the geometry as noted, the effectiveness of the slat depends on its position and deflection as defined in Figure 10-16. The key design parameters include the chord length, C_S, gap, slot width, a (which is equal to the forward translation, Δx), vertical translation, Δy, and deflection angle, δ_S.

Generally, the chord length should be of the order of 0.15 to $0.175C$ and the gap around 0.001-$0.002 \cdot C$. The optimum deflection is usually in the neighborhood of $25°$. Figure 10-22 shows how the slat deflection angle affects the increase in C_{lmax}. It is actually a combination of the deflection angle and the gap opening that yields the highest increase.

FIGURE 10-21 Three-position slat in action on the MD-80 commercial jet liner. Cruise position (left), T-O and early descent position; (center), with slot closed; and landing position (right) with slot open. *(Photos by author)*

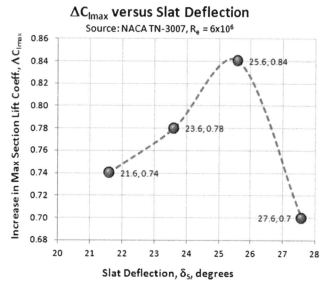

FIGURE 10-22 Change in C_{lmax} with a slat deflection angle (based on Ref. [1]).

Note that there is a discrepancy in the literature as to how the chord length is defined. The two options can be seen in Figure 10-15 and Figure 10-19. The interested reader should refer to the cited sources to determine which applies to the published data.

Aerodynamic Properties

Results for slats and Maxwell slots using a NACA 64A010 and Clark Y airfoils are given in Table 10-1. They are obtained from Refs [1,11]. Slats yield a maximum change in maximum section lift coefficient as high as $\Delta C_{lmax} \approx 0.84$ for an optimized configuration. For preliminary design, numbers ranging from 0.5 to 0.7 can be used. The slats increase the stall AOA by about $\Delta \alpha_{max} \approx 11°$ while the zero AOA lift coefficient drops by about $\Delta C_{lo} \approx -0.21$. This ignores the shift of the lift and assumes the slat will only extend the lift curve upward as shown in Figure 8-58. The minimum drag coefficient increases by about $\Delta C_{dmin} \approx 0.008$ (80 drag counts).

Summary for Slats

$$\Delta C_{l_{max}} \approx 0.84 \quad \Delta C_{d_{min}} \approx \text{not specified}$$
$$\Delta \alpha_{max} \approx 11° \quad \Delta C_m \approx -0.115$$

Summary for Maxwell Flap

$$\Delta C_{l_{max}} \approx 0.55 \text{ to } 0.81 \quad \Delta C_{d_{min}} \approx 0.0156 \text{ to } 0.0276$$
$$\Delta \alpha_{max} \approx 9 \text{ to } 11.8° \quad \Delta C_m \approx 0.0008 \text{ to } -0.03656$$

10.2.6 Summary of Leading Edge Device Data

Table 10-1 lists important aerodynamic characteristics for a number of leading edge high-lift devices. The purpose is to provide the designer with answers to questions like — what is the change in lift, drag, and pitching moment when such devices are added to the aircraft. All data presented below is based on wind tunnel testing.

How to Use the Tables

Remember that the tables are intended to get you "in the ballpark" during the conceptual design phase. Begin by locating the part of the table containing the type of leading edge flap. Make note of the airfoil and Reynolds number used during the wind tunnel testing, as well as the flap chord and deflection. These are important values to keep in mind if the geometry and operational conditions of the target flap deviate significantly from these values. For values relatively close, it is acceptable to prorate the characteristics of interests (see Example 10-1). The values to be extracted are ΔC_{lmax}, $\Delta \alpha_{max}$, ΔC_{dmin}, and ΔC_m.

EXAMPLE 10-1

A NACA 65_2-415 airfoil is to be used in a wing design and will feature a slat whose chord is 15% of the chord length and will be deflected to 22°. Determine the airfoil's C_{lmax}, α_{max}, C_{dmin}, and C_m at a Reynolds number of 6 million.

Solution

Using Table 8-5, the baseline airfoil characteristics are given by (generally, the data in the table is for $R_e = 6 \times 10^6$) $C_{lmax\ b} = 1.62$, $\alpha_{max\ b} = 16°$, $C_{dmin\ b} = 0.0042$, and $C_{m\ b} = -0.060$.

Using Table 10-1 the following values are obtained for the slat, as tested on a NACA 64A010 airfoil.

$\Delta C_{lmax} = 0.84$, $\Delta \alpha_{max} = 11°$, ΔC_{dmin} = not available, and $\Delta C_m = -0.115$

Unfortunately, we will not be able to determine the airfoil's minimum drag. Using the available data, we prorate the other properties as follows to account for differences in deflection and chordlength:

$$C_{lmax} = C_{lmaxb} + \Delta C_{lmax} \left(\frac{\delta_{Strue}}{\delta_S}\right)\left(\frac{C_{Strue}}{C_S}\right)$$
$$= 1.62 + 0.84 \left(\frac{22°}{25.6°}\right)\left(\frac{0.15}{0.17}\right) = 2.26$$

EXAMPLE 10-1 (cont'd)

$$\alpha_{\max} = \alpha_{\max b} + \Delta\alpha_{\max}\left(\frac{\delta_{Strue}}{\delta_S}\right)\left(\frac{C_{Strue}}{C_S}\right)$$
$$= 16° + 11°\left(\frac{22°}{25.6°}\right)\left(\frac{0.15}{0.17}\right) = 24.3°$$

$$C_m = C_{mb} + \Delta C_m\left(\frac{\delta_{Strue}}{\delta_S}\right)\left(\frac{C_{Strue}}{C_S}\right)$$
$$= -0.06 - 115\left(\frac{22°}{25.6°}\right)\left(\frac{0.15}{0.17}\right) = -0.147$$

10.3 TRAILING EDGE HIGH-LIFT DEVICES

In this section we will look at a number of trailing edge high-lift devices. Such devices are essential in reducing T-O and landing distances for aircraft. Not only do they make it possible to operate aircraft over a wide range of airspeeds; for commercial aircraft, they ultimately make or break their business case as they dictate the kind of airports the airplane can be operated from.

The purpose of the trailing edge high-lift device is to increase the maximum lift coefficient of the airfoil. Since this is accomplished by increasing the camber of the airfoil, this usually reduces the stall AOA as well. The effect is described in Section 8.3.10, *The effect of deflecting a flap*. Deploying a flap will affect not only C_{lmax} and α_{stall} but also the pitching moment and is, thus, of great importance when sizing control surfaces. The pitching moment increases substantially and this must be arrested by an adequately sized stabilizer and elevator. From an aircraft handling standpoint, deploying flaps has a major effect on the stall characteristics of the aircraft. It can also be problematic in the development of aircraft — an airplane may stall impeccably without flaps, and terribly with them deployed.

Flaps increase drag and require higher thrust to maintain straight and level flight. Naturally, this also means that the maximum level airspeed is reduced. Trailing edge devices greatly influence the distribution of lift over the wing and have a profound effect on the structural design. They develop large aerodynamic forces that must be reacted through the appropriate hard points in the wing. This also requires an increase in skin thickness to react increased wing torsion. A few small aircraft feature manually operated flaps. For such aircraft, the deflection loads are of concern and must be kept low enough to enable pilots with limited upper body strength to actuate them.

The investigation of trailing edge high-lift devices is made harder by the complexity of analysis methods. Their operation is affected by viscosity in major ways and this renders the wind tunnel as the primary methodology. Computational fluid dynamic codes utilizing Navier-Stokes solvers are making such efforts easier, although CFD should always be validated for each case using wind tunnel testing. Before the advent of such computational methods, flaps were analyzed using Glauert's extension of the thin airfoil theory. The theory allows lift, pitching moment and hinge moments to be estimated using closed-form solutions [13].

The invention of the flap dates back to the early days of aviation. One of the earliest pieces of research can be found in Ref. [14], published in 1914. In spite of this, the flap did not become widely used until the early 1930s, when aerodynamic efficiency and wing loading had reached a state where take-off and landing speeds presented serious challenges to the operation of aircraft [15]. The trailing edge flap was a clear solution and it allowed fast aircraft to both reduce airspeed and increase the glide slope. Since then a substantial number of configurations have been devised and utilized in aircraft. The most common ones are presented here, but the reader should be aware of that more exist.

10.3.1 Plain Flap

The *plain flap* is a simple high-lift surface that only moves through rotation without translation. It really is the simplest solution for use as a control surface, rather than a high-lift device. The control surface is effectively a semi-circle joined at the base of a triangle (see Figure 10-23). The hingeline is placed at the center of the circle, but this ensures that there will be no change in the gap between the control surface and the trailing edge of the skin of the main lifting surface element ahead of the control. This allows for a very simple and reliable control system to control the deflection of the flap, although as discussed under General Design Guidelines, the overhang (the part of the flap ahead of the hingeline) sometimes causes important complications. As such it is an effective and inexpensive means of increasing (and decreasing)

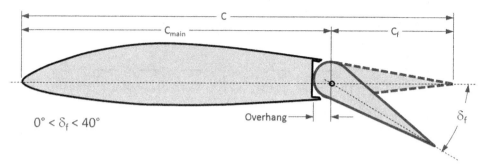

FIGURE 10-23 A schematic of the *plain flap*.

lift, provided large deflections are not needed. For this reason, the vast majority of aircraft, big and small, fast and slow, feature the type for use as an aileron, elevator, and rudder control surface.

Among drawbacks is a relatively low increase in C_{lmax}, when compared to more sophisticated flap types. As a control surface, plain flap surfaces with the hingeline at the center of the circle are difficult to mass balance and require the addition of structural arms to carry a block of mass balance. Also, a plain flap with large overhangs result in a large movement of its leading edge up and down, which requires the skin on the upper and lower surfaces to be removed to allow for this motion.

Note that the flap area is defined as the area behind the hingeline. The overhang is not a part of that area, even though it belongs to the same control surface.

The plain flap results in a relatively low increase in drag with deflection, especially for deflections in the $\pm 10°$ range. It increases the airfoil camber and circulation around the airfoil, yielding higher C_{lmax} and reduced α_{stall}. If deflected beyond $\pm 15°$, depending on the Reynolds number, separation begins to form on the flap reducing its effectiveness (see Figure 10-24). At lower Reynolds numbers (small UAVs and RC aircraft) such separation may begin at deflection as low as $\pm 10°$. For this reason, the maximum deflection of the flap should not be greater than $\pm 30°$, because any further deflection will not yield much, if any, improvement in C_{lmax}. Recommended maximum deflection angle is something closer to $\pm 25°$.

General Design Guidelines

The primary design variables for the plain flap are flap chord, flap deflection, hingeline location, and flap airfoil. The maximum lift of the plain flap is dependent on the flap chord, which should be of the order of 20%—30%. Some airplanes feature plain flaps for rudders with flap chords as much as 50%. The flap should be over-sized by approximately $0.0003 \cdot C$ to $0.0030 \cdot C$. Thus, if the chord is 3.28 ft (1.00 m), the over-sizing should amount to approximately 0.012″ to 0.120″ (0.3–3.0 mm) per side (see Figure 10-25). The purpose of the over-sizing is to re-energize the boundary layer by placing a small obstruction in its way. This helps air stay attached farther aft, improving the effectiveness of the surface at low deflections by preventing it from operating in dead air.

The overhang (see Figure 10-26) plays an important role in reducing hinge moments of the flap, which is imperative if it is to be used as a manually actuated

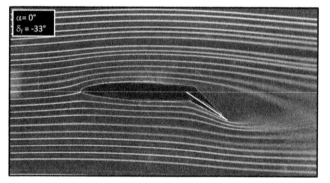

FIGURE 10-24 Streamlines show one of the drawbacks of the plain flap — massive flow separation that forms at deflection angles over 15°. This flap is deflected at 33°. *(Photo by Phil Rademacher)*

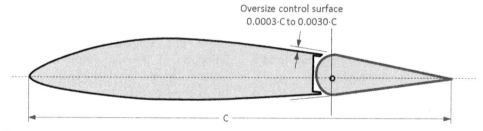

FIGURE 10-25 The plain flap should be slightly over-sized to help re-energize the airflow and help it stay attached.

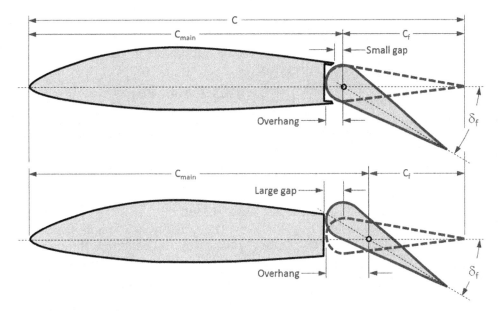

FIGURE 10-26 Enlarged overhang may cause geometric issues that may turn into aerodynamic issues.

control surface (aileron, elevator, or rudder). However, this may introduce unexpected aerodynamic complexities. Let's first consider the upper image in the figure. It shows the flap with an overhang that equals the leading edge radius of the flap. In other words, the hingeline is precisely the center of the leading edge arc. This allows the overlapping skin (or trailing edge lip as it is commonly called) to be extended so it practically touches the skin of the flap. Of course, the overlapping skin should never be in direct contact with the flap, as it flexes in flight, which can potentially lead to jamming. For many applications, the small gap may range from 0.050″ to 0.125″ (1.3 to 3.2 mm). A careful design of this detail will include an estimation of the most adverse structural deformation of the flap and size the gap to accommodate it. At any rate, the presence of the lip is an important advantage as this prevents a jet of air from the lower surface of the airfoil from streaming through the slot and detrimentally affecting the flow on the upper surface. Its absence can be a serious detriment to the stall characteristics of the airfoil, while its presence, in cruise, acts like a seal that reduces drag.

A large overhang requires the overlapping skin (trailing edge lip) to be removed. Otherwise, the leading edge of the flap will strike it and prevent further deflection. On the other hand, if the overhang is larger than the hingeline radius of the flap, the leading edge will translate with deflection as shown in Figure 10-26, and the associated (and inevitable) large gap. The presence of this gap is remedied on the Cirrus SR22 by a flexible plastic strip that is bonded to the upper surface and greatly reduces the gap size.

Aerodynamic Properties

One of the design variables is flap chord ratio. A comparison for three flap chords is provided in Ref. [13]. These are shown in Figure 10-27. Equation (10-1) is least squares fit to the data for the three flaps.

$$C_{lmax0.1C} = 1.262 + 0.01492\delta_f - 8.315 \times 10^{-5}\delta_f^2$$
$$C_{lmax0.2C} = 1.234 + 0.02216\delta_f - 1.611 \times 10^{-4}\delta_f^2 \quad (10\text{-}1)$$
$$C_{lmax0.3C} = 1.250 + 0.02418\delta_f - 2.093 \times 10^{-4}\delta_f^2$$

where δ_f is in degrees. The magnitude of the C_{lmax} using these equations is listed in the table below. It can be seen

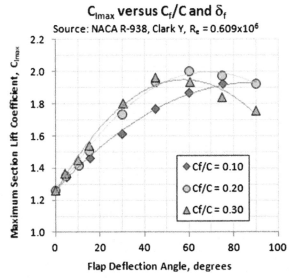

FIGURE 10-27 The maximum lift coefficient as a function of flap deflection angle and flap chord ratio.

that a flap chord ratio of $0.20 \cdot C$ to $0.30 \cdot C$ will yield the highest value of the maximum lift coefficient. Reference [13] recommends $0.25 \cdot C$ as the optimum flap chord. The highest C_{lmax} of the flap is obtained when it is deflected 60°. The magnitude of the C_{lmax} was not found to vary much within the range of Reynolds numbers tested (0.6×10^6 to 3.5×10^6). Drag coefficients were shown to increase rapidly once the lift coefficient exceeds 1.2, which can be attributed to the increased flow separation behind the flap.

Section C_{lmax} for Clark Y Airfoil			
	Flap Chord Ratio C_f/C		
δ_f	0.10	0.20	0.30
0	1.26	1.26	1.26
10	1.40	1.47	1.48
20	1.53	1.64	1.66
30	1.63	1.78	1.80
40	1.73	1.89	1.90
50	1.80	1.97	1.95
60	1.86	2.01	1.96
70	1.90	2.02	1.93

Another aerodynamic issue is the influence of a gap between the flap and the main wing. Such a gap allows air to flow from the higher pressure on the lower surface to the upper one. This reduces the effectiveness of the flap, as is clearly shown in Figure 10-28. At a deflection of 60°, the maximum lift coefficient drops from about 2.02 to 1.65 in the presence of a $0.0032 \cdot C$ gap. Other data can be seen in Table 10-2.

10.3.2 Split Flap

The *split flap* is really a two-member family of trailing edge flaps that consist of deflecting a plate on the lower surface without any change in the geometry of the upper surface. The two family members are called the *split flap* and the *Zap flap*.

Split Flap

The *split flap* is a simple flap concept that, like the plain flap, is deflected through rotation only (see Figure 10-29). It was invented in part by Orville Wright [16] and James M. H. Jacobs, who patented it jointly in 1924 under US Patent 1,504,663. Typical deflection angles range from 0° to 70°. The flap is very simple mechanically. When deployed it enlarges and magnifies the high-pressure region on the lower surface of the wing ahead of the flap, while generating a large separation region behind it. It provides great attitude and glide-slope control due to the high increase in drag without too much increase in lift or pitching moment. When used as a speed brake, it is superior to a spoiler (which is operated on the upper surface) because it increases drag without the lift reduction — it actually increases lift, but sharply reduces the L/D ratio, making the approach easier to control. The increase in C_{lmax} is low compared to more sophisticated flap types (although arguably it is surprisingly high) and for that reason the flap is not used on any modern airliner. However, it is used on the Douglas DC-3 (C-47) aircraft and on many fighter and bomber aircraft operated during WWII: for instance, the Curtiss SB2C Helldiver and Douglas SBD Dauntless dive bombers, both of which featured split flaps with a large number of holes to reduce buffeting effects [17]. Among GA aircraft featuring the flap is the Cessna 310 and the Yakovlev Yak-18T.

Zap Flap

The Zap flap is a variation of the split flap that introduces translation in addition to the rotation (see Figure 10-30). It gets its name from inventor Edward F. Zaparka, who patented it in 1933 as US Patent 2,147,360. In its simplest implementations, an actuator or a pushrod will force the leading edge of the flap backward. The presence of a special linkage forces the device to simultaneously rotate into position, increasing the chord. On a wing, it increases the wing area. As a consequence, the Zap flap generates higher lift than the split flap. The flap has not seen commercial use, but was tested in the late 1930s and early 1940s by NACA using

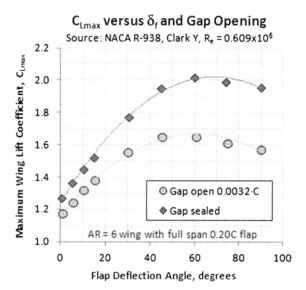

FIGURE 10-28 The effect of gap on the magnitude of the maximum lift coefficient.

TABLE 10-2 Summary of the Aerodynamic Properties of Trailing Edge Devices

PLAIN FLAPS

Description	Section	Basic Airfoil	Re_{test}	Flap Type	C_{main}/C	$(C_f/C)_1$	$(C_f/C)_2$	δ_s	$(\delta_f)_1$	$(\delta_f)_2$	C_{lmax}	α_{max}	C_{l0}	C_{dmin}	C_m	ΔC_{lmax}	$\Delta \alpha_{max}$	ΔC_{l0}	ΔC_{dmin}	ΔC_m	Reference
Airfoil only			0.63	-	1	-	-	-	-	-	1.297	15.0	0.36	0.0150	-	-	-	-	-	-	NACA TR-407
Plain flap	10.3.1	Clark Y	0.609	Plain	0.1	0.9	-	-	15	-	1.467	-	-	-	-	0.170	-	-	-	-	NACA TR-938
									30	-	1.635	-	-	-	-	0.338	-	-	-	-	
									40	-	1.726	-	-	-	-	0.429	-	-	-	-	
					0.2	0.8	-	-	15	-	1.558	-	-	-	-	0.261	-	-	-	-	
									30	-	1.782	-	-	-	-	0.485	-	-	-	-	
									40	-	1.891	-	-	-	-	0.594	-	-	-	-	
					0.3	0.7	-	-	15	-	1.578	-	-	-	-	0.281	-	-	-	-	
									30	-	1.799	-	-	-	-	0.502	-	-	-	-	
									40	-	1.894	-	-	-	-	0.597	-	-	-	-	

SPLIT FLAPS

Description	Section	Basic Airfoil	Re_{test}	Flap Type	C_{main}/C	$(C_f/C)_1$	$(C_f/C)_2$	δ_s	$(\delta_f)_1$	$(\delta_f)_2$	C_{lmax}	α_{max}	C_{l0}	C_{dmin}	C_m	ΔC_{lmax}	$\Delta \alpha_{max}$	ΔC_{l0}	ΔC_{dmin}	ΔC_m	Reference
Airfoil only		NACA 64A010	6x10^6	-	1	-	-	0	-	-	1.10	11.0	0.01	-	0.00	-	-	-	-	-	NACA TN-3007
Airfoil+Split Flap			6x10^6	Split		0.2	-	0	60	-	1.88	5.3	-	-	-	0.78	-5.7	-	-	-	
Airfoil only		Mustang 2	2.14x10^7	-		0.1	-	0	-	-	1.13	12.3	0.09	0.0082	-0.017	-	-	-	-	-	NACA TM-1177
Airfoil+Split Flap			2.14x10^7	Split		0.1	-	0	-	-	2.11	11.1	1.29	0.1594	-0.204	0.98	-1.2	1.20	0.1512	-0.19	
Airfoil only	10.3.2	NACA 66(215)-216		-		-	-	-	-	-	1.56	17.1	0.20	-	-0.045	-	-	-	-	-	
Airfoil+Split Flap				Split		0.2	-	-	70	-	2.61	11.0	1.73	-	-0.218	1.05	-6.10	1.53	-	-0.17	
Airfoil only		NACA 66$_1$-212	6x10^6	-	1	-	-	-	-	-	1.41	14.1	0.14	-	-0.039	-	-	-	-	-	NACA WR-L-140
Airfoil+Split Flap				Split		0.2	-	-	70	-	2.17	6.9	1.56	-	-0.212	0.76	-7.20	1.42	-	-0.17	
Airfoil only		NACA 65$_1$-212		-		-	-	-	-	-	1.49	14.7	0.15	-	-0.035	-	-	-	-	-	
Airfoil+Split Flap				Split		0.2	-	-	60	-	2.15	7.65	1.49	-	-0.199	0.66	-7.05	1.34	-	-0.16	

ZAP FLAPS

Description	Section	Basic Airfoil	Re_{test}	Flap Type	C_{main}/C	$(C_f/C)_1$	$(C_f/C)_2$	δ_s	$(\delta_f)_1$	$(\delta_f)_2$	C_{lmax}	α_{max}	C_{l0}	C_{dmin}	C_m	ΔC_{lmax}	$\Delta \alpha_{max}$	ΔC_{l0}	ΔC_{dmin}	ΔC_m	Reference
Wing only (3D airplane)				-		-	-	-	-	-	1.29	13	0.4	0.053	-0.025	-	-	-	-	-	
Wing + 0.947b Zap Flap	10.3.2	N-22	3.12x10^6	Zap	1	0.376	-	43	-	-	2.37	12	1.6	0.165	-0.25	1.08	-1.0	1.20	0.11	-0.225	NACA WR-L-437
Wing + 0.947b Zap Flap + Gap Increase				Zap		0.376	-	43	-	-	2.57	13	1.72	-	-	1.28	0.0	1.32	-	-	

JUNKERS FLAPS (ALL POSITIONS ARE OPTIMAL, BUT THE POSITION OF C_{lmax} DOES NOT NECESSARILY COINCIDE WITH THAT OF C_{Dmin})

Description	Section	Basic Airfoil	Re_{test}	Flap Type	C_{main}/C	$(C_f/C)_1$	$(C_f/C)_2$	δ_s	$(\delta_f)_1$	$(\delta_f)_2$	C_{lmax}	α_{max}	C_{l0}	C_{dmin}	C_m	ΔC_{lmax}	$\Delta \alpha_{max}$	ΔC_{l0}	ΔC_{dmin}	ΔC_m	Reference
Airfoil only		Clark Y (main) NACA 0012 (flap)	0.609x10^6	-	1.000	-	-	-	-	-	1.250	18	-	0.0155	-	-	-	-	-	-	NACA TN-524
Airfoil + Junkers Flap				Junkers		0.15	-	-	45	-	1.810	12	-	0.0146	-	0.56	-6.0	-	-0.0009	-	
Airfoil only		Clark Y (main) Clark Y (flap)	0.609x10^6	-		-	-	-	-	-	1.300	-	-	0.0145	-	-	-	-	-	-	
Airfoil + Junkers Flap				Junkers	1.000	0.2	-	-	30	-	1.96	-	-	0.0123	-	0.66	-	-	-0.0022	-	
Airfoil + Junkers Flap				Junkers		0.2	-	-	40	-	2.00	-	-	0.0123	-	0.70	-	-	-0.0022	-	
Airfoil only	10.3.3			-		-	-	-	-	-	1.145	-	-	0.0105	-	-	-	-	-	-	
Airfoil + Junkers Flap		NACA 23012 (main) Clark Y (flap)	0.609x10^6	Junkers	1.000	0.2	-	-	20	-	1.85	-	-	0.0101	-	0.71	-	-	-0.0004	-	NACA R-541
Airfoil + Junkers Flap				Junkers		0.2	-	-	30	-	1.98	-	-	0.0101	-	0.84	-	-	-0.0004	-	
Airfoil + Junkers Flap				Junkers		0.2	-	-	40	-	1.79	-	-	0.0101	-	0.65	-	-	-0.0004	-	
Airfoil only				-		-	-	-	-	-	1.145	-	-	0.0105	-	-	-	-	-	-	
Airfoil + Junkers Flap		NACA 23012 (main) Clark Y (flap)	0.609x10^6	Junkers	1.000	0.3	-	-	20	-	1.961	-	-	0.0098	-	0.82	-	-	-0.0007	-	
Airfoil + Junkers Flap				Junkers		0.3	-	-	30	-	2.108	-	-	0.0098	-	0.96	-	-	-0.0007	-	
Airfoil + Junkers Flap				Junkers		0.3	-	-	40	-	2.140	-	-	0.0098	-	1.00	-	-	-0.0007	-	
Airfoil + Junkers Flap				Junkers		1.3	-	-	60	-	1.908	-	-	0.0098	-	0.76	-	-	-0.0007	-	

SINGLE-SLOTTED FLAPS

Airfoil Section	C_f/C	C_{main}/C	Slot Entry Config.	Flap Nose Shape	δ_f (deg)	C_{lmax}	ΔC_{lmax}	x_f	y_f	Optimum position	Re x10^6	Reference
Clark Y (airfoil only)	-	-	-	-	-	1.297	-	0	-0.025	YES		NACA TR-407
Clark Y + Flap	0.200	1.000	b	A	30	2.44	1.143	0	-0.025	YES	0.61	NACA R-534
	0.300		b	A	40	2.83	1.533	0	-0.025	YES		
	0.400		b	A	40	3.10	1.803	0	-0.025	YES		
NACA 23012 (airfoil only)	-	1.000	-	-	-	1.55	-	-	-	-		NACA R-679
NACA 23012 + Flap	0.100	0.930	a	A	50	2.25	0.70	0.004	0.005	YES		NACA R-679
	0.150	1.000	b	A	30	2.68	1.13	0	0.015	YES		NACA ACR, Nov 1940
	0.250	1.000	b	A	40	3.22	1.67	0	0.015	YES		
	0.257	0.800	a	B	50	2.76	1.21	0.005	0.018	YES		NACA R-664
	0.257	0.830	a	A	50	2.81	1.26	0.005	0.016	YES		
	0.257	0.830	a	A	40	2.83	1.28	0.013	0.024	YES		NACA R-679
	0.257	1.000	b	A	30	2.90	1.35	0	0.025	NO	3.5	NACA R-664
	0.300	0.900	c	A	50	2.92	1.37	0.002	0.010	NO		
	0.300	0.900	c	A	40	2.92	1.37	0.002	0.020	NO		
	0.300	0.900	c	A	30	2.93	1.38	0.002	0.030	NO		NACA TN 808
	0.300	0.900	b	A	40	2.88	1.33	0.002	0.020	NO		
	0.300	1.000	b	A	40	3.29	1.74	0	0.015	NO		
	0.400	0.715	b	A	50	2.87	1.32	0.015	0.015	YES		NACA TN 715
	0.400	0.715	a	A	50	2.90	1.35	0.015	0.015	YES		
NACA 23021 (airfoil only)	-	1.000	-	-	-	1.35	-	-	-	-		NACA R-677
NACA 23021	0.150	1.000	b	A	60	2.59	1.24	0	0.015	YES		NACA TN 782
	0.150	1.000	b	A	60	2.66	1.31	0.050	0.030	YES		NACA ACR, Feb 1941
	0.250	1.000	b	A	40	3.17	1.82	0.025	0.015	YES		
	0.257	0.827	b	B	60	2.69	1.34	0	0.015	YES	3.5	NACA R-677
	0.257	0.827	a	B	60	2.74	1.39	0	0.015	YES		
	0.257	0.827	b	A	60	2.71	1.36	0.005	0.020	YES		
	0.257	0.827	a	A	50	2.82	1.47	0	0.025	YES		
	0.400	0.715	b	A	50	2.79	1.44	0.015	0.025	YES		NACA TN 728
	0.400	0.715	a	A	50	2.88	1.53	0.015	0.045	YES		

(Continued)

TABLE 10-2 Summary of the Aerodynamic Properties of Trailing Edge Devices—cont'd

SINGLE-SLOTTED FLAPS-Continued											
Airfoil Section	c_f/C	C_{main}/C	Slot Entry Config.	Flap Nose Shape	δ_f (deg)	C_{lmax}	ΔC_{lmax}	x_f	y_f	Optimum position	Re x10^6	Reference
NACA 23030	-	1.000	-	-	-	1.01	-	-	-	-		
	0.257	0.860	b	B	60	2.59	1.58	0.025	0.040	YES		
	0.257	0.860	a	B	60	2.68	1.67	-0.005	0.040	YES	3.5	NACA TN 755
	0.400	0.775	b	B	50	2.82	1.81	0.025	0.060	YES		
	0.400	0.775	a	B	50	2.90	1.89	0.025	0.060	YES		
63_4-420	0.250	0.880	b	B	35	3.00	-	0.018	0.045	NO	6	NACA R-824
63_4-421(approx)	0.243	0.835	a	A	40	3.21	-	0	0.027	YES	9	NACA MR, Jan 7, 1943
	0.250	0.840	c	A	45	2.47	-	0.009	0.010	YES		
65-210	0.250	0.900	c	A	41.3	2.48	-	0.014	0.009	YES	6	NACA TN 1191
	0.250	0.975	c	A	35	2.45	-	0.004	0.020	YES		
$65_{(112)}$-111 (approx.)	0.350	0.839	c	A	35	2.69	-	-0.020	0.032	YES	9	NACA TN 1463
65_1-213 (approx.)	0.336	0.889	c	A	40	2.63	1.45	0.019	0.046	NO	9	NACA MR L5L11
$65_{(215)}$-114	0.259	0.915	c	A	40	2.80	1.65	0.019	0.038	NO	9	
66(215)-216	0.250	-	-	-	40	2.34	0.92	-	-	-	6	NACA R-938
65_2-221	0.263	0.832	a	B	30	2.83	-	0.025	0.046	YES	9.95	NACA MR, June 4, 1942
$66_{(215)}$-116, a = 0.6	0.250	0.824	c	B	55	2.70	-	0	0.028	NO	6	NACA MR L4K21
66_2-116, a = 0.6	0.250	0.827	a	A	45	2.69	-	0.017	0.038	NO	6	NACA MR, May 23, 1942
	0.300	0.900	c	A	37	2.92	-	0	0.016	NO	6	NACA R-824
66,2-216	0.250	0.824	a	A	40	2.89	-	0.023	0.040	YES	5.1	NACA MR A4L28
	0.250	0.834	c	A	45	2.88	-	0.011	0.031	YES	5.1	
66_2-118	0.250	0.900	-	-	32.5	2.68	-	-	-	NO	6	NACA MR, Dec 2,1941
Davis (t/c = 0.18)	0.300	1.000	b	A	40	3.45	-	-	-	NO	6.2	NACA MR, Dec 29,1941
Davis (modified) (t/c = 0.18)	0.300	1.000	b	A	40	3.14	-	-	-	NO	6	
66-Series (approx) (t/c = 0.18)	0.300	1.000	b	A	30	3.34	-	-	-	NO	6.2	

SINGLE-SLOTTED FOWLER FLAPS											
Airfoil Section	c_f/C	C_{main}/C	Slot Entry Config.	Flap Nose Shape	δ_f (deg)	C_{lmax}	ΔC_{lmax}	x_f	y_f	Optimum position	Re x10^6	Reference
Clark Y (airfoil only)	-	-	-	-	-	1.27	-	-	-	-	6	NACA TN-419
Clark Y	0.400	1.000	-	-	40	3.17	1.90	0	0.025	YES	6	
GA(W)-1 (airfoil only)	-	1.000	-	-	-	1.56	-	-	-	-		
	0.290		-	-	15	2.72	1.16	0.900*	-0.055	YES		NASA CR-2443
GA(W)-1	0.290	0.960	-	-	30	3.35	1.79	0.940*	-0.043	YES	2.2	*Flap x_f and y_f are absolute values that reference leading edge of airfoil
	0.290		-	-	40	3.48	1.92	0.950*	-0.040	YES		
	0.300	1.000	-	-	20	3.17	1.61	0.953*	-0.025	YES		
GA(W)-1	0.300		-	-	30	3.55	1.99	0.978*	-0.025	YES		
	0.300		-	-	40	3.80	2.24	1.007*	-0.027	YES		
GA(W)-1	1.290	1.000	-	-	40	3.71	2.15	-	-	YES	2.2	NASA CR-3687

Fairchild 22 [18] and Fairchild XR2X-1 [19] aircraft. Some results are presented below.

General Design Guidelines

Split flaps of either kind may cause severe buffeting if deployed at high airspeed, for instance when used as a dive brake. The solution is to fabricate the flap with perforations. The hinge moment of the Zap flap may reverse if there is an enlargement of the gap between the wing and the leading edge of the flap as a consequence of the flap's translation and rotation.

Aerodynamic Properties

The split flap increases C_{lmax} by a value ranging from 0.66 to 1.05, depending on airfoil, chord, and deflection. Generally it is deflected to an angle of 60°. The references cited showed the stall AOA to change from $-1.2°$ to $-7.2°$. The change in C_m is -0.16 to -0.19. The zero AOA drag coefficient changed by some 0.1512.

The author has been unable to locate wind tunnel test data for specific airfoils featuring a Zap flap. The Zap flap was tested on a full-scale aircraft in 1942. A deployment of both the split and Zap flap leads to a relatively small shift in the stall AOA, in particular the Zap flap, which had a marginal increase in the stall AOA (which is very unusual for trailing edge high-lift devices) according to Refs [18,19].

Reference [19] found the Zap flap to increase maximum lift coefficient by 1.08 for a configuration that had a gap of 0.010C between the wing and the flap leading edge. The same flap with a gap of 0.037C increased this by 0.20. The minimum drag coefficient increased by

TABLE 10-2 Summary of the Aerodynamic Properties of Trailing Edge Devices—cont'd

	DOUBLE-SLOTTED FLAPS											
Airfoil Section	C_f/C	C_v/C	C_{main}/C	C_{lmax}	δ_f (deg)	δ_v (deg)	x_f	y_f	x_v	y_v	Optimum position	Re x10^6	Reference
NACA 1410	0.250	0.075	0.840	3.06	50	25	0.026	0.016	0.012	0.019	YES	6.0	NACA TN-1545
NACA 23012	0.100	0.189	0.830	2.99	70	40	0.009	0.009	0.014	0.024	YES	3.5	NACA R-679
	0.257	0.227	0.715	3.47	70	30	0.014	0.012	0.015	0.035	NO	3.5	NACA R-723
	0.257	0.117	0.826	3.30	60	25	-0.016	0.010	-0.004	0.017	YES	3.5	NACA ARR 3L10
NACA 23021	0.257	0.227	0.715	3.56	60	30	0.019	0.024	0.025	0.065	NO	3.5	NACA R-723
	0.257	0.147	0.827	3.32	70	30	0.017	0.027	0.007	0.024	YES	3.5	NACA ARR L4J05
NACA 23030	0.257	0.260	0.715	3.71	80	40	0.040	0.050	0.045	0.040	NO	3.5	NACA R-723
NACA 63_4-421(approx)	0.195	0.083	0.870	3.30	55	14	0.038	0.012	-0.009	0.016	NO	6.0	-
NACA 63-210	0.250	0.075	0.840	2.91	50	25	0.022	0.024	0.024	0.018	YES	6.0	NACA TN-1545
NACA 64-208	0.250	0.075	0.840	2.51	45	30	0.015	0.015	0.015	0.019	YES	6.0	NACA TN-1545
	0.250	0.056	0.840	2.40	50	25	0.018	0.014	0.015	0.024	YES	6.0	
NACA 64-210	0.250	0.075	0.840	2.82	55	30	0.023	0.006	0.012	0.018	YES	6.0	NACA TN-1545
NACA 64-212	0.250	0.075	0.840	3.03	50	30	0.021	0.020	0.010	0.019	YES	6.0	NACA TN-1545
NACA 64-A212	0.229	0.083	0.833	2.83	55	26	0.044	0.005	0.004	0.014	YES	6.0	NACA TN-1293
NACA $65_{(216)}$-215, a=0.8	0.248	0.096	0.820	3.38	70	12	0.024	0.010	0.025	0.032	NO	6.3	NACA RM L7A30
NACA 65-118	0.244	0.100	0.864	3.35	65	23	0.038	0.007	0.009	0.025	YES	6.0	NACA ACR 3120, R-824
NACA 65-121	0.236	0.109	0.850	3.08	51	20	0.029	0.017	0.012	0.024	YES	2.2	NACA TN-1395
NACA 65-210	0.250	0.075	0.840	2.72	50	25	0.025	0.011	0.009	0.024	YES	6.0	NACA TN-1545
NACA 65-418	0.236	0.106	0.851	3.50	65	21	0.027	0.007	0.012	0.028	YES	6.0	NACA TN-1071
NACA 66-210	0.250	0.075	0.840	2.64	55	25	0.029	0.023	0.012	0.022	YES	6.0	NACA TN-1545
	0.250	0.100	0.840	2.72	60	25	0.027	0.039	0.024	0.021	YES	6.0	
NACA 66-214(approx)	0.227	0.085	0.854	3.00	55	20	0.044	0.009	0.004	0.025	YES	9.0	NACA TN-1110
Douglas 7-series type t/c=0.154	0.250	0.056	0.820	3.15	50	19	0.017	0.016	0.012	0.024	NO	6.0	NACA MR L5C2a
Republic 6-series type t/c=0.17	0.238	0.092	0.880	3.55	60	25	0.015	0.020	-0.005	0.020	YES	8.5	NACA MR L6A08a
Republic 6-series type t/c=0.17	0.238	0.092	0.880	3.43	60	25	0.015	0.020	-0.005	0.020	YES	14.0	NACA MR L6A08a
Airfoil NACA 64A010	0.25	0.075	0.823	2.36	52.7	30	-	-	-	-	-	6	NACA TN-3007

Abbreviations:

LE = leading edge
TE = trailing edge
$R_{e\ test}$ = Reynolds number during test
C_S = chord of LE device
C = airfoil chord
Dbl = double
δ_S = deflection angle of LE device
$(\delta_f)_1$ = deflection of element 1 of a TE device
$(\delta_f)_1$ = deflection of element 2 of a TE device
C_{lmax} = max section lift coefficient
α_{max} = stall AOA
C_{lo} = lift coefficient at $\alpha = 0°$
C_{dmin} = minimum section drag coefficient
C_{mo} = pitching moment coefficient at $\alpha = 0°$

0.11 and the pitching moment coefficient decreased by −0.225. Other data can be seen in Table 10-2.

The data for the Zap flap pertains to a three-dimensional wing, rather than an airfoil, like the other devices in this chapter. The wing of the test vehicle had a constant chord of 4.34 ft, wingspan of 33.02 ft, and wing area of 141.5 ft^2, giving it an AR of 7.705.

10.3.3 Junkers Flap or External Flap

The *Junkers flap* (also called the *external flap*) is an unusual high-lift device in the sense that it resides entirely outside the wing. Mechanically, it is a simple design (see Figure 10-31) that results in a modest increase in C_{lmax} compared to more modern devices. However, when neutrally deflected, it even reduces the airfoil drag slightly.

The external flap is most notable for its use on the German Junkers 52-3 tri-motor military (and commercial) transport aircraft designed around 1930 (see Figure 10-32), hence its name. Additionally, it was used on the Junkers Ju-87 Stuka dive bomber. A similar device was patented by Charles E. Wragg in the USA in 1930 (U.S. Patent 1,756,272), although it is almost certain that it was conceived earlier in Germany. The device was referred to as the Wragg compound wing. Among other notable aircraft featuring the flap is the Colomban MC-15 Cri-Cri (often cited as the world's smallest twin-engine aircraft), the Miles Gemini, and Zenith CH 701, 7050, and 801 STOL kitplanes.

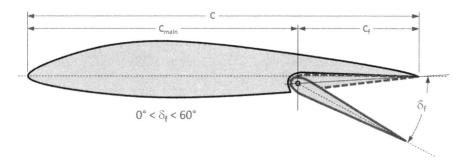

FIGURE 10-29 A schematic of the *split flap*.

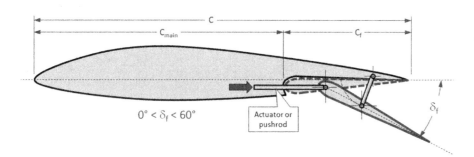

FIGURE 10-30 A schematic of the *Zap flap*.

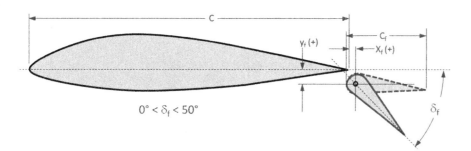

FIGURE 10-31 A schematic of the *Junkers flap*.

FIGURE 10-32 An airworthy example of a German Junkers Ju-52 military transport aircraft featuring *Junkers flap*. *(Photo by Nick Candrella)*

The flap lends itself well for use as a flaperon, in which the outboard part of the flap combines the functionality of a flap and aileron, although as such it may suffer from excessive adverse yaw. One of the flap's primary advantages is that it adds practically no drag to the installation and can even reduce it by a few drag counts if properly located near the trailing edge. Of course some of this is offset by the fact that the flap

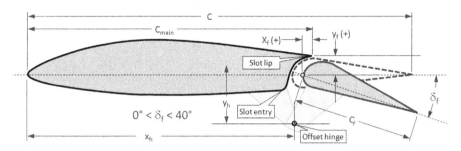

FIGURE 10-33 A schematic of the *single-element slotted flap*.

loads must be reacted by external hardpoints that increase the drag of the configuration. The flap does not generate a large enough C_{lmax} on its own to be suited to the modern GA aircraft (see below), but it allows for a simple installation which may be the right choice for selected applications, as the flap resides outside the boundary layer, making it more effective at low deflection angles than identically sized conventional ailerons that are fully submerged.

General Design Guidelines

Loads are reacted by flap hinges that are mounted externally. These should be aerodynamically faired, even though this appears not to have been a concern in the design of the Ju-52. The flap was investigated by NACA in 1935 and 1936. NACA TN-524 [20] found the ideal hinge location to be $0.0125 \cdot C$ (1.25%) behind the trailing edge and $0.025 \cdot C$ (2.5%) below the chord line. These dimensions are referred to as x_f and y_f in Figure 10-31. NACA R-541 [21] also presents similar results, where it was remarked that this position is quite critical. The flap must be neutrally balanced, or slightly nose-heavy about the hinge line to prevent flutter. If the hingeline of the flap is close to its leading edge, an external installation of flap balance will be required and this will further increase the drag of the design.

Aerodynamic Properties

Reference [20] wind tunnel tested a Junkers flap configuration featuring a Clark Y airfoil for the main wing element and NACA 0012 for the flap element. The C_{lmax} of the flap in the optimum location was found to equal 1.81, based on the total area, which included the combined area of the test wing element and the flap. This compares to 1.250 for the basic Clark Y at the test R_e of 0.609×10^6. Therefore, the increase in C_{lmax} amounts to merely 0.56. The minimum drag of the baseline airfoil presented in the report was 0.0155, but this was shown to be reduced to 0.0146 by deflecting the flap $-5°$ (TEU). Furthermore, a deflection angle of $+5°$ (TED) is suggested for use during climb.

Reference [21] investigated the use of a Junkers flap using a NACA 23012, 23021, and Clark Y basic airfoils. The flap chord tested was $0.20 \cdot C$ and $0.30 \cdot C$. The resulting aerodynamic effects can be found in Table 10-2. As one would expect, it found the optimum flap location to vary based on the airfoil, although not far from the position determined by Ref. [21].

10.3.4 The single-slotted Flap

The *single-slotted flap* (see Figure 10-33) is an improved version of the plain flap. The flap consists of an airfoil mounted to a hinge that is offset from the main airfoil. The resulting motion combines rotation and translation and increases the airfoil chord length by some 5–10%. The motion opens a slot along the trailing edge that is imperative to the functionality of the flap. Such a translation is usually referred to as a *Fowler motion*, although the single-slotted flap is not regarded as a Fowler flap (see Section 10.3.6, *Fowler flaps*).

The general wisdom explains the increase in lift over that of the plain flaps as a consequence of the energized airflow over the plain flap. This delays separation to a higher deflection angle than is possible with the plain flap. The author prefers A. M. O. Smith's vortex analogy [4], which also applies to leading edge devices such as slats (for instance see Figure 10-6). With respect to trailing edge flaps, the analogy considers the formation of two vortices: one on the main wing element and the other on the flap. The analogy helps explain the observed airspeed over the leading edge of the flap and the resulting low pressure peak. This fact has been well demonstrated experimentally in many of the references presented in this book. This not only explains well the increase in the C_{lmax} but also shows that the presence of the slot is essential to the formation of the two vortices. This is important to keep in mind in the design of variable-cambered airfoils (e.g. see Section 10.2.2, *Variable-camber leading edge*) – such airfoils are usually designed without slots – there is no separation of the leading or trailing edge elements and, therefore, they suffer from lower C_{lmax} that their separated counterparts.

There are a number of versions of the simple slotted flap, although only two will be discussed here: the *single-element slotted flap* and the *single-slotted Fowler flap*.

The single-element slotted flap is the most common type of trailing-edge high-lift devices used for GA aircraft. The flap features a single airfoil element of simple mechanical operation and it offers higher C_{lmax} than the plain flap and less drag than either the split or Zap flaps. The flap is allowed to rotate to an angle as large as 50°, although such deflections result in mostly separated flow over the flap. Figure 10-35 shows the capability of a slotted flap deflected to 40°. The ΔC_{lmax} gained is 0.926.

General Design Guidelines

Some forethought must be exercised when laying out the hinge geometry. If the highest point on the flap is ahead of the hingeline when stowed, it will travel to an even higher point as it is being deployed. This may cause the flap to bind or jam in the cove, either during deployment or during retraction. This problem is compounded by the fact that the flap will flex upward when loaded. This may cause the flap to jam in flight, even if no such problems are evident on the ground. Therefore, the flight test team should try to emulate the flap flexing on the ground to avoid the problem in the air. A jammed flap may be a nuisance that requires the airplane to return to base for a repair. This is the most likely scenario. However, it is also possible it would result in a more serious flap asymmetry if the control system was fabricated such the flap on one side of the plane of symmetry could deploy fully with the other one partially deployed.

Since the presence of the slot entry (or cove entry) increases drag, there are variations of this flap that feature a cove panel to open and close it via special slave linkage. An example is shown in Figure 1.10 of Ref. [8]. The flap system should be capable of deflecting to approximately 40°, as shown in Figure 10-34. Exceeding this has limited value and may actually aggravate stall characteristics. For smaller airplanes such added complexity should be avoided. However, sealing the cove is recommended, as shown in Figure 10-36. It reduces drag considerably at typical cruise lift coefficients.

Another issue frequently overlooked is one of actuation binding, which can be caused by very high friction in the flap actuation system. Actuation binding is a serious issue as the actuation motor may not have the power to overcome the friction. It also flexes the structure and is a fatigue issue waiting to happen. It is usually a result of multiple hinges, whose hinge points are not perfectly concentric or mounted on poor foundations (e.g. flexible brackets mounted to plates rather than stiff frames and bulkheads).

Recall that at least two hinges must be used for this flap. Often, in order to minimize flap flex, three or more hinges are necessary. Mathematically, the hingeline must be a straight line that goes through all the hinges — one must be able to simultaneously look through all the hinge-holes. This means that a tight

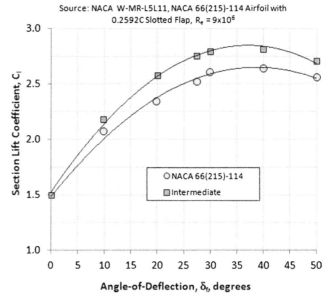

FIGURE 10-34 Typical change in the maximum lift of a *single-element slotted flap*, showing that C_{lmax} occurs near a 40° flap deflection.

tolerance must be associated with their location — and often the only way to mount the flap is through a high-precision match drilling after the flap hinges have been installed. It is not recommended to pre-drill the hinges as their installation will not guarantee the hinge points are on the same mathematical line.

Aerodynamic Properties

The presence of the slot entry is required on the lower surface of the airfoil. This increases the drag of the airfoil with the flap stowed (see Figure 10-36) by approximately 10 drag counts (i.e. 0.0010). This should be considered the cost of doing business. The expected increase in lift, ΔC_{lmax}, should be a fairly good 0.6 to 1.0, depending on geometry and Reynolds number.

A round slot entry is recommended, although it is slightly harder to manufacture. The cost of doing business can be kept down by providing a gap seal. It does not have to be like the rubber seal indicated in Figure 10-36; it can also be a carefully sculpted wing trailing edge that practically contacts the upper surface of the flap. The drawbacks of this configuration are that the paint on the flap surface may be scraped off (can be solved with a ultra-high-molecular-weight polyethylene tape, which has a high abrasion tolerance); and that the tight fit of the upper surface of the flap to the wing on the ground may become an extremely tight fit in the air, making retraction or extension all but impossible.

The airflow through the slot is not necessarily smooth, although one would think so at first glance — this depends on the slot geometry. However, an appropriate location for a rubber seal should not be an obstruction to the flow.

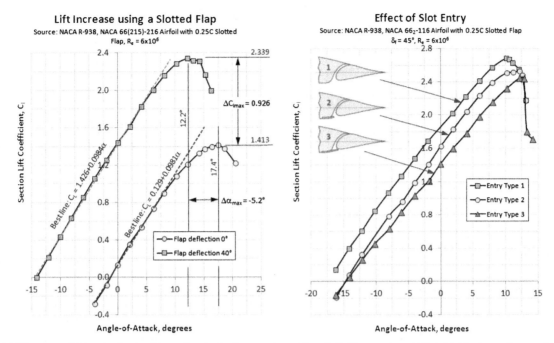

FIGURE 10-35 Typical lift gain obtained from the single-element slotted flap (left). The maximum lift is highly dependent on the cove shape (right). Rounded slot entry yields the highest maximum lift coefficient (based on Ref. [13]).

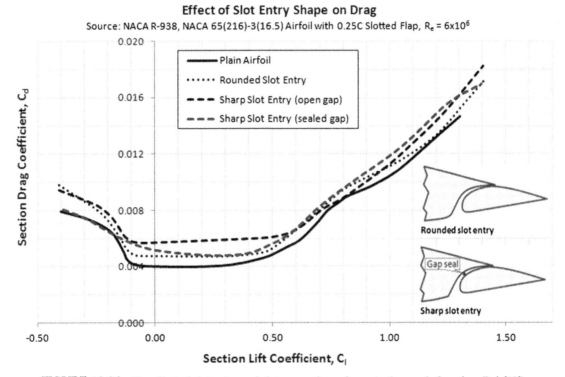

FIGURE 10-36 The effect of slot entry and slot gap seal are shown in the graph (based on Ref. [13]).

10.3.5 Double-slotted Flaps

The *double-slotted flap* increases the maximum lift coefficient in the same way the single-slotted flap does — by delaying flow separation over the flap element. The key difference is that the double-slotted flap adds a level of boundary-layer control not possible with the single-slotted one. The extra slot allows the flap to be deflected to an even higher angle before flow becomes excessively separated. This gives a great boost

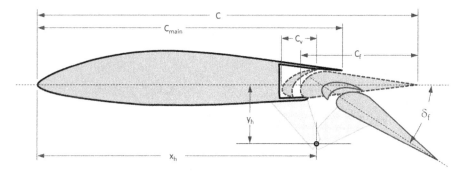

FIGURE 10-37 A schematic of the fixed-vane double-slotted flap. It can be seen that the position of the vane relative to the flap element does not change with the actuation.

to the C_{lmax}. From a certain point of view, the flap can be considered a single-slotted flap to which a turning vane has been added to help guide the air over it. However, the vane also generates a large lift on its own. The forward of the two flap elements is called the *vane* and the other is simply the *flap*.

The term double-slotted refers to a family of two-element flaps, which not unlike the family of Krüger flaps comprises more than one type of flap. The primary difference between members is how the two elements interact. Generally, the most important design parameters are flap deflection, flap size, flap extension, and the slot geometry, which dictates how efficiently air flows through the slot. The following listing cites the most common types of double-slotted flaps.

Fixed-vane Double-slotted Flap

The *fixed vane double-slotted flap* (see Figure 10-37) is a version of the slotted flap in which the flow separation over the flap element is delayed by a fixed vane placed in front it. The explanation for why the fixed slot prevents the aft (or main) flap element from stalling is provided by the vortex analogy of Section 10.2.5, *The leading edge slat*. A vortex placed at the leading edge of each segment reduces the pressure peak, delaying the flow separation.

The configuration increases the maximum lift of the slotted flap, but it is heavier and costlier to manufacture. However, it is lighter and less expensive than its articulating cousin, to be discussed next. The vane allows the main flap to be deflected to an angle as high as 55° before suffering reduced improvement in C_{Lmax}. The configuration increases drag over that of the single-slotted flap, even at lower deflection. This is remedied by the mechanically more sophisticated articulated vane, which is closed at low flap deflections to reduce drag and improve T-O and climb performance. The fixed vane is used on the McDonnell Douglas DC-9/MD-80 commercial jetliner, the Hondajet and the Cirrus SF50 Vision prototypes.

Articulating-vane Double-slotted Flap

There are two primary advantages of using an articulating vane for a double-slotted flap (see Figure 10-38). First, the actuation of the flap can be such that the vane closes the slot between it and the flap's main element. This reduces the drag of the configuration, helping to improve T-O and climb performance. Second, the articulation allows an increase in the total chord over that of the fixed vane with no increase in stowed space requirements. This way the Fowler motion of the flap can be increased, which increases the maximum lift coefficient of the configuration.

The primary drawback of the articulating vane is its mechanical complexity and weight. The vanes are usually spring-loaded and rest up against a stop. Once the main element has transited backward a certain distance, the vane slot will have opened fully and the vane now begins to transit aft with the main element of the flap. The vane is attached to the flap using straight or curved tracks that are hidden in the main element.

Main/aft Double-slotted Flap

This version of the double-slotted flap is mechanically more complicated than the fixed or articulating vane flaps discussed above. This is because both elements translate and rotate (see Figure 10-39). The chord increase due to these is also greater than that of the articulating vane, yielding a slightly higher maximum lift coefficient. The forward element is the larger of the two and is now referred to as the main element. The wing overlaps the forward element, which overlaps the aft element. The forward flap typically deflects 30° to 35° (δ_{f1}) whilst the aft flap deflects 28° to 30°, or through a total deflection of 58° to 65° (δ_{f2}).

Triple-slotted Flap

It is debatable whether to consider the *triple-slotted flap* a variation of the slotted flap or a Fowler flap. This is due to the large translation the flap is subjected to during transit (see Figure 10-40). Typical deflection of the front element is 30°, center element 45°, and aft

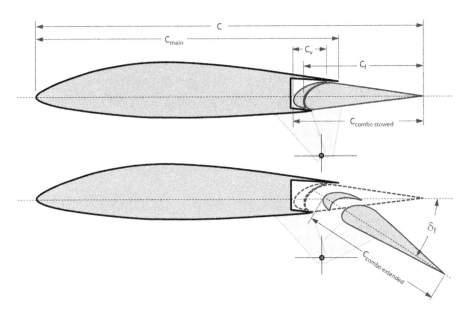

FIGURE 10-38 A schematic of the articulating-vane double-slotted flap.

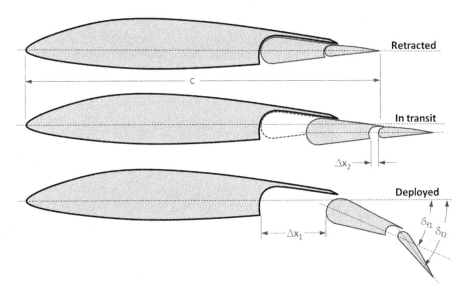

FIGURE 10-39 A schematic of the main/aft double-slotted flap.

element 80° (approximately). The flap increases airfoil chord length and camber, yielding a large increase in C_{Lmax}, C_D, and C_M. A very large increase in C_{Lmax} results in substantial reduction of the stalling speed of heavy aircraft. Using A. M. O. Smith's vortex analogy [4], there is a very complex interaction of gap jet airspeeds that combine to reduce flow separation over the elements, resulting in a high C_{Lmax}. The drag of the configuration is high, which is helpful to the pilot during the landing phase.

It is an important drawback that a very complicated and heavy mechanical system is required to deliver each flap element to its desired position. The elements must move in harmony on either side of the plane of symmetry to avoid flap asymmetry. The flaps generate very high loads that require substantial structure to support and they subject the wing to a large torsion. The flaps cause a large increase in drag, which may require high engine thrust to overcome. This can cause handling complaints if the engine spool-up time is slow. Even though the triple-slotted flap generates higher section lift coefficients than the double-slotted flap, once in three-dimensional flow it suffers from greater losses due to the flap tip vortices. The flaps result in a very large increase in pitching moment, which must be arrested by powerful stabilizing surfaces. From an operational standpoint, the mechanical system requires increased labor hours for maintenance.

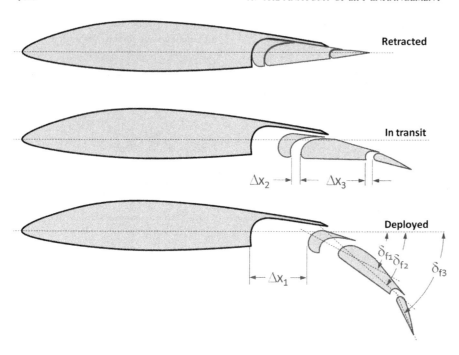

FIGURE 10-40 A schematic of the triple-slotted flap.

Triple-slotted flaps are primarily used for commercial jetliners. Among aircraft featuring the flap are the Boeing B-727, 737, and 747. Recent advances in computational fluid dynamics (CFD) have resulted in a reduction in wing sweep that, in turn, has reduced the need for such complex flaps. Jetliners such as the 757, 767, and those made by Airbus feature the main/aft double-slotted flaps.

General Design Guidelines

Designing a multi-element flap usually takes scores of engineers. The complexity of such flaps should not be underestimated. Mechanically sophisticated flaps, which require translation and rotation, can pose a number of development issues in addition to those that have already been brought up, such as excessive play in the actuation mechanism that leads to a reduced deflection' Not to mention the possibility of a flutter problem. Issues often arise with the overlap and gap between the flap and the wing, which may change the maximum lift. There are issues with the positioning of the flap with respect to the wing, when deployed. The optimum location of the flap for maximum lift as predicted by Navier-Stokes solvers usually differs from that obtained in wind tunnel testing. If tight tolerances are not maintained, asymmetric flap deflection between the left and right sides may result and this can lead to roll-off problems when stalling in either the T-O or landing configuration. The motion of the flap must be such that the vane remains in contact with the upper wing skin (or upper cove or spoiler) for the first 15°, making it a less draggy single-element flap for the T-O configuration. Finally, the more sophisticated the flap, the more internal wing volume it tends to absorb. This will affect available fuel volume for selected applications and should be considered during the flap selection process.

Ideally, the flaps should actuate reliably and repeatedly for many years without jamming, requiring only routine maintenance, such as regular lubrication. As usual, when it comes to mechanical design, the simplest configuration that does the job is always the best one.

10.3.6 Fowler Flaps

The term Fowler flap usually refers to a high-lift device that significantly increases the chord of an airfoil in addition to rotation. They are named after the American inventor and aeronautical engineer Harlan D. Fowler (1895–1982) [22]. Reference [23] offers a fascinating insight into the development of this flap. It was invented in 1924 and was first used by the Glenn L. Martin Company, who hired Fowler to design flaps. In 1937, the flaps were introduced on the Lockheed L-14 and later on the Boeing B-29 Superfortress.

The most notable feature of Fowler flaps is the large increase in chord it provides. Since the section lift coefficient is based on the shorter stowed chord, this greatly boosts its value. It is regarded by many as the first modern high-lift mechanical flap (e.g. see Ref. [4]).

Single-slotted Fowler Flap

The *single-slotted Fowler flap* is generally deployed by a large translation first, followed by a rotation to as much

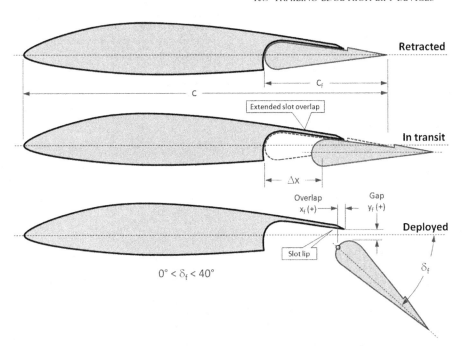

FIGURE 10-41 A schematic of the single-slotted Fowler flap.

as 40°–45° (see Figure 10-41). The translation requires the upper skin to extend much farther aft than the single-slotted flap. This means the upper aft wing skin must be stiffened in order to prevent it from bulging outward, creating a structural challenge. A single-slotted Fowler flap is used on the Boeing B-52 bomber and 747-SP commercial jetliner. A derivative single-slotted Fowler flap is used on the Cessna 152, 172, and other models under the name *para-lift* flaps. These translate far enough aft to justifiably be considered single-slotted Fowler flaps.

Figure 10-42 shows one way to acquire a Fowler motion for a flap design. The flap has two guide pins that move inside specially shaped flap track slots. The motion requires an actuator to move it forward and aft. That detail is left out, but can be accomplished using a linear actuator or jack-screws. The figure illustrates how the flap travels linearly about 60% of the total translation distance with merely 5° of deflection. The remaining 35° are accomplished over the remainder of the translation, resulting in substantial increase in the airfoil chord.

General Design Guidelines

A Fowler flap may present unexpected challenges. As stated earlier the upper skin that covers the flap must be stiffened to ensure it will not bulge outward due to the pressure differential between the upper and lower surfaces. This can be challenging if the flap is thick, as it will leave less thickness for structural reinforcement of the slot overlap. Play in the system can be a serious threat to the success of such flaps, as can the bending stiffness of the flap itself. These are important pitfalls. If a single actuator is used per flap side, it should be

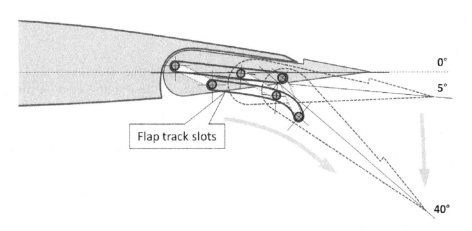

FIGURE 10-42 One way of acquiring Fowler motion. Double flap track slots force the flap to translate far aft before rotation begins.

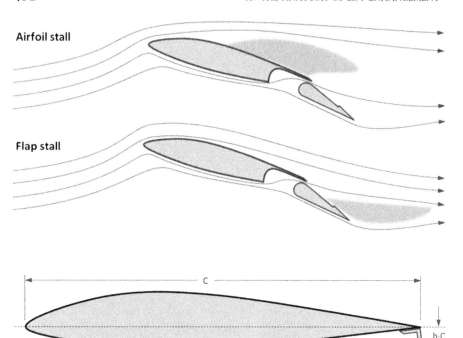

FIGURE 10-43 A flapped airfoil can stall in two ways; airfoil and flap stall (based on Ref. [25]).

FIGURE 10-44 A schematic of the *Gurney flap*.

placed near the mid-span station of the flap (ideally near the center of lift). This will minimize the flap moment that could twist (or yaw) the flap around its vertical axis. Such a moment can jam the flap in the guide slots as one end moves forward (or aft), in the opposite direction to the other end. Lack of bending stiffness can flex the flap between the flap tracks so it puts undue load on the tracks or causes the flap to strike the upper surface of the wing between the tracks.

Aerodynamic Properties

Reference [24] tested two types of Fowler flaps intended for use on a GA(W)-1 airfoil; a 29% chord flap designed using a computer method developed by Lockheed Aircraft, and a 30% chord flap designed by Robertson Aircraft, a company that specialized in the manufacture and installation of STOL STCs. The smaller flap achieved a C_{lmax} of 3.48 on the airfoil, while a C_{lmax} of 3.80 was achieved by the 30% chord flap. A range of optimized flap positions based on deflection angle is presented in the reference. The performance gain is highly dependent on the gap and the overlap, with an optimum gap width of approximately 2.5% to 3.0% chord for the GA(W)-1 airfoil tested.

The flap increases the pitching moment by a significant value, or $\Delta C_m = -0.75$. The increase in the minimum drag coefficient amounted to $\Delta C_{dmin} = 0.040$, and shifted it to a higher C_l.

Among other observations made in Ref. [24], the GA(W)-1 airfoil has a prominent cusp on the aft portion of the lower surface. It was found that straightening this, as is often necessary in a production environment, led to a performance penalty at $C_l > 0.8$. It was also found that adding vortex generators increased C_{lmax} by 0.2. The vortex generators increased drag at low AOAs, but decreased it at higher AOAs. Additionally, there is a negligible change in the stall AOA.

Observation indicates that a multi-element airfoil experiences two kinds of post-stall patterns referred to as an *airfoil stall* or a *flap stall* [25] (see Figure 10-43). The airfoil stall consists of the separation located mostly on the main element, whereas the flap stall resides mostly on the flap. The appearance of these is dependent on the gap between the cove and the flap, with a wide gap leading to a flap stall. The optimum gap will lead to an airfoil stall with a flow fully attached over the flap. This yields the highest maximum lift coefficient.

10.3.7 Gurney Flap

A Gurney flap (see Figure 10-44) is a small angle added to the pressure side of the trailing edge of an airfoil and helps increase its maximum lift. The device is extremely simple to fabricate, consisting of a strip of an extruded aluminum angle or folded sheet metal. It forms a circulation pool that modifies the shape of the overall circulation over the airfoil. Considering its small size, the device increases the maximum lift substantially. Similarly it increases zero-alpha lift, or alternatively, decreases the zero-lift AOA. It also increases the wing drag.

The magnitude of these increases depends on its height, denoted by the fraction h (see Figure 10-44). The flap further increases the nose-down pitching moment about its quarter-chord, offering a stabilizing effect to the aircraft if mounted to its wing.

The invention of the Gurney flap is attributed to Dan Sexton Gurney (1931–) an American race car legend, who in the early 1970s used it to help increase the down-force on race cars [26]. Since then, the flap has found a variety of applications ranging from wings on race cars, wind-turbine blades, and helicopter stabilators. It can also be seen on the trailing edge of the Cessna 208 Caravan single-engine turboprop.

General Design Guidelines

Select a suitable flap height using data from the literature, for instance Ref. [27]. Note that the flap height has great impact on the drag of the Gurney flap. Refer to Figure 10-45 and Figure 10-46 for more details.

Aerodynamic Properties

A large number of papers have been published on the Gurney flap. Liebeck [26] was probably the first to present information on the Gurney flap and to attempt to explain its effect. In the paper he cites early testing performed by Douglas Aircraft that seemed to indicate drag reduction, primarily for thick airfoils. However, many

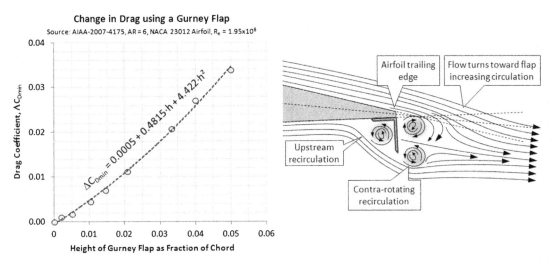

FIGURE 10-45 Drag characteristics of the Gurney flap as a function of its height (left) on a wing of AR = 6. A schematic of the flow field around a Gurney flap (right). (Both based on Ref. [27]).

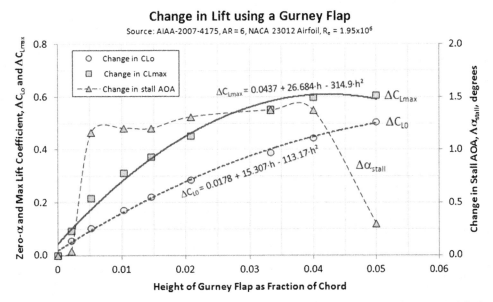

FIGURE 10-46 Lift characteristics of the Gurney flap as a function of its height (based on Ref. [27]).

later publications indicate there is in fact a drag increase, at least for airfoil geometry, useful in aircraft. For example publications by Cavanaugh et al. [27] and Myose et al. [28], both demonstrate increase in C_{Lmax} and C_{Dmin}. Neuhart [29] presents an example of a Gurney flap that increased the airfoil drag at low values of C_L three-fold. Similar results can be found in other papers.

Reference [27] investigated the impact of varying the height of a Gurney flap on a three-dimensional wing of $AR = 6$ in a wind-tunnel test whose results are presented below. The impact on lift, drag, and pitching moment observed gives a good insight into the effectiveness and shortcomings of the flap. In short, the results show that Gurney flaps increase C_{Lmax} and C_{Lo} (or decreases α_{ZL}). Furthermore, they increase C_{Dmin}, C_{mo}, and $C_{m\alpha}$. Reference [29] presented pressure measurements on the upper and lower surfaces and found it to decrease on the upper surface and increase on the lower surface. This is consistent with an increase in flow circulation.

The impact on lift and drag is shown in Figure 10-46 and Figure 10-45. The latter figure also shows a schematic of the flow field around the trailing edge and how the presence of the flap changes the circulation around the airfoil by shifting the wake downward. The formulas shown in the graphs are obtained using quadratic least squares fit to the published data. They can be used to get an idea about the effect of the Gurney flap for new aircraft design. Note that these results are valid for wings of $AR = 6$ at a R_e of 1.95×10^6 and should be used with care for wing shapes and flight conditions that deviate greatly from those.

An effort has been made to reduce the drag of the Gurney flap. For instance, Vijgen et al. [30] patented a serrated trailing edge for this purpose. Mayda and van Dam [31] proposed a Gurney flap with slits. The common denominator in these designs is that the reduction in minimum drag is accompanied by a drop in the maximum lift.

10.3.8 Summary of Trailing Edge Device Data

Table 10-2 lists important aerodynamic characteristics for a number of trailing edge high-lift devices. The purpose is to provide the designer with answers to questions like: what is the change in lift, drag, and pitching moment when such devices are added to the aircraft. All data presented below is based on wind tunnel testing.

How to Use the Tables

Remember that the tables are intended to get you "in the ballpark" during the conceptual design phase. Airfoil; flap type and chord; and the Reynolds number at the flight condition must be known. Begin by locating the part of the table containing the type of trailing edge flap. Make note of the airfoil and Reynolds number used during the wind tunnel testing, as well as the flap chord and deflection. These are important values to keep in mind if the geometry and operational conditions of the target flap deviate significantly from these values. For values relatively close, it is acceptable to prorate the characteristics of interests (e.g. see Example 10-1 or 10-2). The values to be extracted are ΔC_{lmax}, $\Delta \alpha_{max}$, ΔC_{dmin}, and ΔC_m. Also, make note of the geometry details shown in Figure 10-47 and Figure 10-48, which must be defined for proper referencing in Table 10-2.

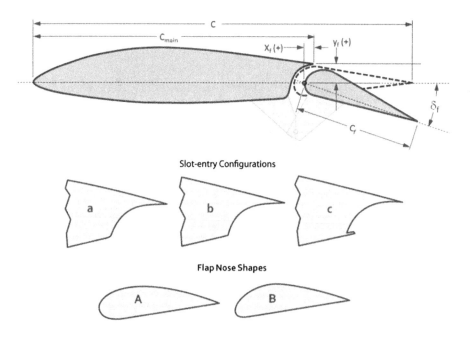

FIGURE 10-47 Reference geometry schematics for slotted flaps for Table 10-2.

10.3 TRAILING EDGE HIGH-LIFT DEVICES

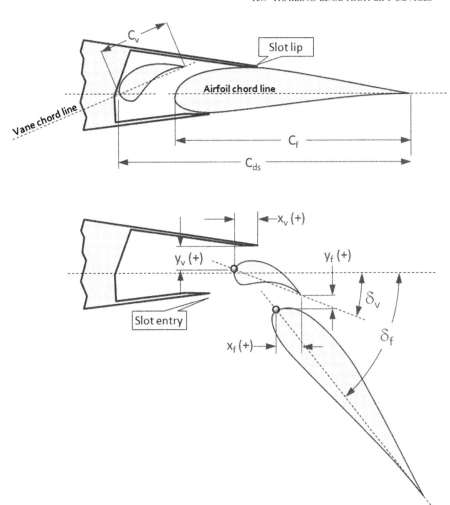

FIGURE 10-48 Reference geometry schematics for double-slotted flaps for Table 10-2.

EXAMPLE 10-2

A NACA 65_2-415 airfoil is to be used in a wing design and will feature a simple slotted flap whose chord is 30% of the chord length and will be deflected to 32°. Determine the airfoil's C_{lmax} at a Reynolds number of 6 million.

Solution

Using Table 8-5, the baseline airfoil characteristics are given by (generally the data in the table is for $R_e = 6 \times 10^6$) $C_{lmax\ b} = 1.62$, $\alpha_{max\ b} = 16°$, $C_{dmin\ b} = 0.0042$, and $C_{m\ b} = -0.060$. We must use Table 10-2 to get data for the flap type selected. Unfortunately, in this case, there is only datum available to determine $C_{l\ max}$. The other parameters will have to be determined by other means, e.g. the methods of Section 15.5.8, *Drag of deployed flaps*.

Looking through the data of Table 10-2 it can be seen there are two 65-series airfoils that feature a single-slotted flap and are "reasonably" similar to 65_2-415 airfoil (NACA 65_1-213 and $65_{(215)}$-114). Here we will use the latter. However, the increase in the maximum section lift coefficient is based on testing at $R_e = 9 \times 10^6$, which is one-third greater than our target R_e of 6×10^6. This exposes one of the problems with using test data for dissimilar airfoils exposed to dissimilar flight conditions: applicability. However, in the absence of more direct data, Table 10-2 is our best bet but calls for a sound engineering judgment. If possible, adjust the C_{lmax} to the appropriate

EXAMPLE 10-2 (cont'd)

Reynolds number. Here assume the C_{lmax} at 6 million to be 90% of that at 9 million. Using Table 10-2 the following values are obtained for the $65_{(215)}$-114 airfoil:

$\Delta C_{lmax} = 1.65$ for $\delta_f = 40°$ and $C_f/C = 0.256$

Therefore, we can estimate the maximum section lift coefficient as follows:

$$C_{lmax} = C_{lmaxb} + \Delta C_{lmax}\left(\frac{\delta_{f\ true}}{\delta_f}\right)\left(\frac{C_{f\ true}}{C_f}\right)$$
$$= 1.62 + (0.90 \times 1.65)\left(\frac{32°}{40°}\right)\left(\frac{0.30}{0.256}\right) = 3.01$$

10.4 EFFECT OF DEPLOYING HIGH-LIFT DEVICES ON WINGS

The purpose of this section is to provide methods that help the designer estimate the impact of deploying high-lift devices on a three-dimensional wing. The characteristics of interest usually include the maximum lift coefficient, lift curve slope, stall AOA, minimum drag, and pitching moment. In this section, a method to estimate the lift and pitching moment of a wing with deployed high-lift devices will be presented. A method to estimate the drag is presented in Section 15.5.8, *Drag of deployed flaps*.

The analysis methods presented use the dimensioning scheme shown in Figure 10-49. The subscript 's' stands for slat and refers to any kind of a leading edge high-lift device. The subscript 'f' stands for flap and can refer to any kind of a trailing edge high-lift device.

10.4.1 Lift Distribution on Wings with Flaps Deflected

Partial Span Flaps

Figure 10-50 shows the spanwise distribution of section lift coefficients with partial span flaps deflected 0°, 10°, 20°, and 30°. It is based on potential flow theory. The figure shows that the spanwise change in C_l over the wing is not instantaneous, but gradual. The coefficients begin to rise well outside the flap's root and tip.

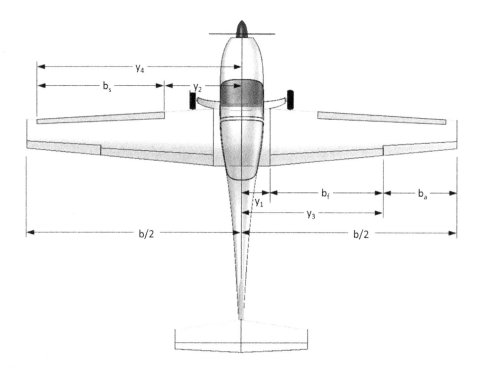

FIGURE 10-49 Layout and important dimensions of a typical aircraft with partial span flaps and slats.

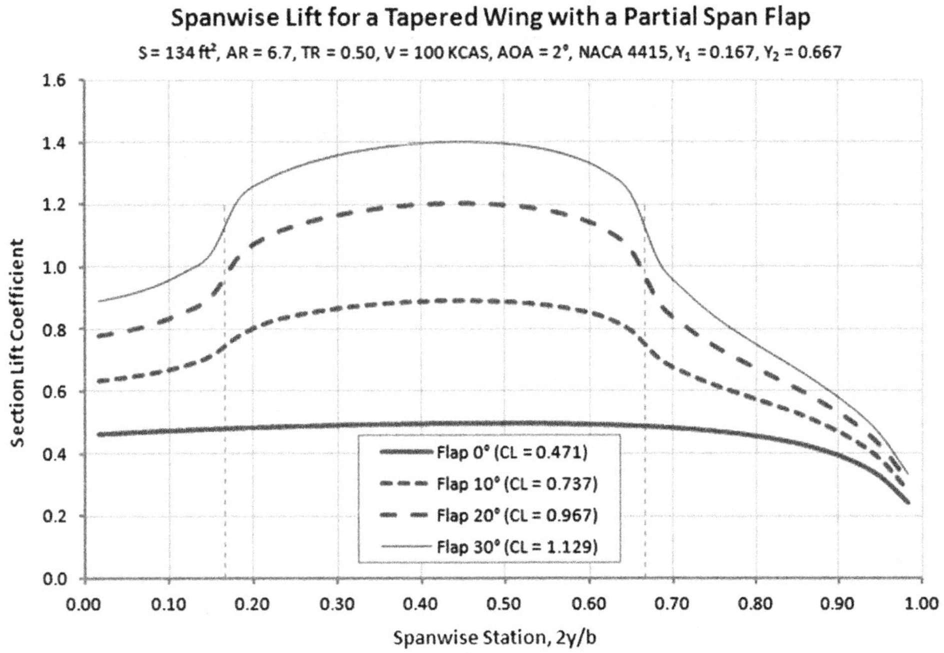
Spanwise Lift for a Tapered Wing with a Partial Span Flap
S = 134 ft², AR = 6.7, TR = 0.50, V = 100 KCAS, AOA = 2°, NACA 4415, Y1 = 0.167, Y2 = 0.667
Section Lift Coefficient
Spanwise Station, 2y/b
Flap 0° (CL = 0.471)
Flap 10° (CL = 0.737)
Flap 20° (CL = 0.967)
Flap 30° (CL = 1.129)
0.0
0.2
0.4
0.6
0.8
1.0
1.2
1.4
1.6
0.00
0.10
0.20
0.30
0.40
0.50
0.60
0.70
0.80
0.90
1.00

FIGURE 10-50 Spanwise distribution of section lift coefficients for partial span flaps ($C_f/C = 0.30$).

This is caused by the increased circulation around the flap, which increases the upwash in the flow field outside the flap ends. The figure also shows how the upwash complicates stall tailoring efforts, as the increase in section lift coefficients on the outboard wing will cause early flow separation in that area of the wing.

Full Span Flaps – Flaperons

Full span flaps, as the name implies, extend from the inboard wing to the wingtip. Such devices substantially increase the total lift capability of the wing, although they may introduce new problems not present in the unflapped wing. First, in order to maintain roll control, aileron functionality must be introduced to the flap control system. This is usually solved by splitting the flap into inboard and outboard segments. Then, the outboard segment is designed to function as both a flap and an aileron. Such a control system configuration is referred to as *flaperons*.

A second drawback stems from the fact that, when deployed, the overall spanwise distribution of section lift coefficients is modified (see Figure 10-51). This may cause roll instability at stall for the flapped wing, even when there are none present for the unflapped configuration. The problem is compounded by the fact that roll control requires one flaperon to be deflected to an even greater angle than the rest of the flap. This can lead to premature aileron stall at low speeds and high *AOA*, just when they are needed the most. The de Havilland of Canada DHC-6 Twin Otter is an example of an aircraft that features such flaps and solves this by varying the deflection of the flaps. This is readily evident from its TCDS [32], the deflection of its double-slotted flaps varies as shown in Table 10-3.

Thus, while the aft inboard trailing element deflects to 62.5°, the aileron (which is the trailing element of the outboard flap) is limited to 26° + 17.5° = 43.5°. Helped by the presence of a slot in the flap, flow separation is kept to a minimum. A third drawback is adverse yaw. The down-deflected aileron element, of course, is associated with the wing that rolls up and this adds substantial drag to the outboard wing. The opposite happens on the wing that rolls down; comparatively substantial reduction in drag. The combination may cause a noticeable adverse yaw that must be eliminated by assertive rudder deflection.

10.4.2 Wing Partition Method

In order to better estimate lift and pitching moments resulting from the deployment of various high-lift devices, it is common to use the *wing partition method*. The method is more or less a derivative of the blade element theory (BET) presented in Section 14.6. The method partitions the wing into the segments shown in Figure 10-52 and, then, the aerodynamic properties of the individual segments are estimated and used to calculate the total of the wing. The partitioning makes it is much easier to manage the change in aerodynamic properties. The scheme presented here is only for demonstration. Your airplane may be more, or less

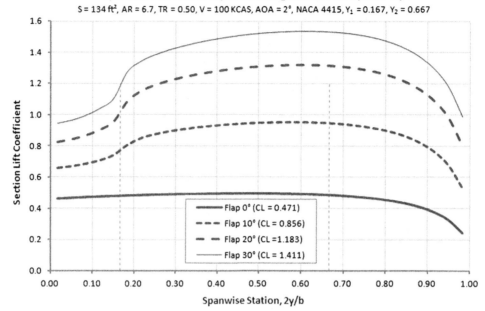

FIGURE 10-51 Spanwise distribution of section lift coefficients for full span flaps (e.g. flaperons) ($C_f/C = 0.30$).

TABLE 10-3 Deflection Characteristics of the DHC-6 Twin Otter Flaperons

Flap	Inboard element	Outboard element
Forward element	0° to 40°	0° to 26°
Aft (trailing) element	0° to 62.5°	Aileron (flaps up): Up 17.5°/Down 16.0° Aileron (flaps dn): Up 25.0°/Down 17.5°

complicated. A demonstration of the method is presented in Example 10-3.

The first step is to tabulate the geometric properties of each segment, using a table similar to Table 10-4. These are then summed in accordance with the formulation presented below (here shown listing four segments). An example of its use is shown in Example 10-3.

Estimation of the Maximum Lift Coefficient

The following expression, in effect, is the BET variation used to estimate the maximum lift coefficient:

$$C_{L\,\text{maxi}} \approx \frac{0.9}{S_{ref}} \left[\sum_{i=1}^{N_f} C_{l\,\text{maxi}} \times S_i \times \cos\left(\Lambda_{hingeline}\right)_i + \sum_{i=1}^{N_{uf}} C_{li} \times S_i \right]$$

(10-2)

where

N_f = number of flapped wing segments
N_{uf} = number of unflapped wing segments
$C_{l\text{max }i}$ = maximum lift coefficient of the flapped wing segment

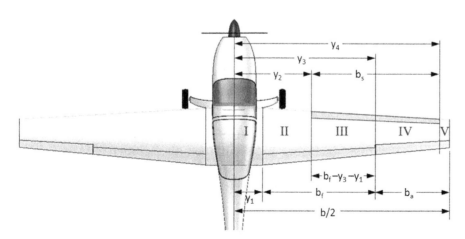

FIGURE 10-52 Partitioning of the wing into manageable segments helps with analysis.

10.4 EFFECT OF DEPLOYING HIGH-LIFT DEVICES ON WINGS

TABLE 10-4 Wing Geometric Properties Based on Segment

Segment	Y_a	Y_b	Area, S_i	$\Lambda_{hingeline}$	$C_{lmax\ a}$	$C_{lmax\ b}$	ΔC_{lmax}	$C_{lmax\ avg}$	$C_{lmax\ avg} \cdot S_i$
I									
II									
III									
IV									
								Sum =	

where Y_a = distance of the inboard end of the segment from the plane of symmetry
Y_b = distance of the outboard end of the segment from the plane of symmetry
S_i = trapezoidal planform area of the segment
$\Lambda_{hingeline}$ = hingeline angle (with respect to a vector normal to the plane of symmetry)
$C_{lmax\ a}$ = maximum lift coefficient at the inboard end of the segment
$C_{lmax\ b}$ = maximum lift coefficient at the outboard end of the segment
ΔC_{lmax} = increase in maximum lift coefficient due to a high-lift device
$C_{lmax\ avg}$ = average of $C_{lmax\ a}$ and $C_{lmax\ b}$

C_{li} = average lift coefficient of the unflapped wing segments at the stall AOA of the flapped one
$\Lambda_{hingeline}$ = angle of the hingeline to a normal to the plane of symmetry

The factor 0.9 is used to account for three-dimensional effect. It is imperative to account for the detrimental effects of hingeline sweep. Note that even straight-tapered wings will be subject to it, unless of course the planform is laid out such that the hingeline is perpendicular to the plane of symmetry.

Estimation of the Pitching Moment Coefficient

Since the pitching moment of an airfoil, effectively, is the product of the lift it generates and its location on the airfoil, it is possible to approximate the total pitching moment of the wing in a manner similar to that presented in Equation (10-2):

$$C_m = \frac{0.9}{S_{ref}} \sum_{i=1}^{N} C_{mi} \times S_i \quad (10\text{-}3)$$

where

N = total number of wing segments
$C_{m\ i}$ = average pitching moment coefficient of each wing segment, flapped and unflapped

---

EXAMPLE 10-3

Estimate the maximum lift coefficient for the SR22 using the wing partition method. The aircraft features single slotted flaps that deflect to 32°. Its reference area is 144.9 ft². Assume a NACA 65_2-415 airfoil for the entire wing (although this is not the case for the real aircraft). Use data from Table 8-5 for the airfoil. The pertinent dimensions for the wing are shown in Figure 10-53, where:

$y_1 = 3.00$ ft $y_2 = 11.75$ ft $y_3 = 16.42$ ft $y_4 = b/2 = 19.15$ ft
$b_f = 8.75$ ft $b_a = 4.67$ ft

Use the equation of the chord given for the SR22 in Section 16.5.1, *Cirrus SR22*, given by $5.160 - 0.1342 \cdot y$ (ft). Compare the value to the POH value, which is shown to be 1.99 in Table 16-6.

Solution

It is a tad tricky to find data specifically using the NACA 65_2-415 airfoil. Ultimately, we will have to use hybrid data, i.e. splice together a data set using airfoils that "resemble" the 65_2-415. First retrieve information from Table 8-5, from which the stall AOA, zero lift AOA, and lift curve slope for the airfoil are found to be as follows:

$\alpha_{stall} = 16.0°$ $\alpha_{ZL} = -2.6°$ $C_{l\alpha} = 0.107$ $C_{lmax} = 1.62$

Next, resort to Table 10-2 for a single slotted flap on airfoils similar to the 65_2-415. One of the airfoils is the NACA $65_{(215)}$-114, detailed in NACA MR L5L11. We will consider it "close enough" to use in this example. Researching the sources gives the following information. Figure 9 of the reference (i.e. NACA MR L5L11) shows a C_{lmax} of about 2.80 for 32° of flap deflection at a $R_e = 9 \times 10^6$. Figure 10 shows that (fortunately) there is not a large reduction in C_{lmax} by lowering the R_e to 3×10^6 (although there is a huge drop when one goes to 1×10^6). We will, thus, use the value of 2.80, however, to be conservative, we multiply it by 0.95. Therefore, consider $C_{lmax} = 0.95 \times 2.80 = 2.66$. Figure 8 of the reference document also shows the stall AOA to change by 4°, from 14° for the unflapped airfoil, to

EXAMPLE 10-3 (cont'd)

10° for the flapped airfoil. Using this data we now rewrite the properties as follows:

Unflapped airfoil:

$\alpha_{stall} = 16.0°$ $\alpha_{ZL} = -2.6°$
$C_{l\alpha} = 0.107$ $C_{lmax} = 1.62$

Flapped airfoil:

$\alpha_{stall} = 12.0°$ $\alpha_{ZL} = -2.6°$
$C_{l\alpha} = 0.107$ $C_{lmax} = 2.66$

We will not need all of the above information, but it is good to have nevertheless. Next determine the *AOA* of the unflapped airfoil when the flapped one stalls. To do this we resort to Equations (8-10) and (8-11), i.e.:

$$C_l = C_{l_0} + C_{l_\alpha}\alpha = -C_{l_\alpha}\alpha_{ZL} + C_{l_\alpha}\alpha$$
$$= -(0.107)(-2.6°) + (0.107)(12.0°) = 1.562$$

We now have enough information to begin to populate Table 10-4. First, take care of the initial geometry as shown in Table 10-5. Columns ① and ② are the spanwise stations in units of ft (i.e. distance from the plane of symmetry) that enclose the corresponding wing segment. Columns ③ and

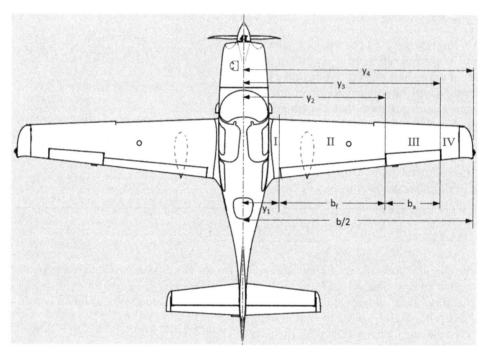

FIGURE 10-53 Partitioning of the wing of the SR22 for this example.

TABLE 10-5 Geometry of the Wing Segments

Segment	① Y_a	② Y_b	③ $C(Y_a)$	④ $C(Y_b)$	⑤ C_{avg}	⑥ Area, S_i
I	0	3	5.16	4.76	4.959	14.88
II	3	11.75	4.76	3.58	4.170	36.49
III	11.75	16.42	3.58	2.96	3.270	15.27
IV	16.42	19.15	2.96	2.59	2.773	7.57

EXAMPLE 10-3 (cont'd)

④ are the chords at those stations in ft. Column ⑤ is the average of the two chords; and Column ⑥ is the planform area of the segment.

Table 10-6 presents the analysis necessary to determine the maximum lift coefficient. It is ideal to review Equation (10-2) to recall the variables that must be included. The first three columns correspond to Columns ①, ②, and ⑥ of Table 10-5 and are merely here for reference. Column ④ is the hingeline angle (assumed 4° for the flapped section). Columns ⑤ and ⑥ are the maximum lift coefficients estimated at the two spanwise stations Y_a and Y_b. It is simply a linear interpolation of the C_{lmax} for the unflapped airfoils at the root and tip. Column ⑦ is the increase in the C_{lmax} of the segment due to the deflected flap. Here, it can be seen that only Segment II has an associated increase, as this is where the flap is. Column ⑦ is the average of the maximum lift coefficient of the airfoils at stations Y_a and Y_b. Finally, Column ⑧ is the segment weighted value of the average maximum lift coefficient of the segment; i.e. the larger the segment area, the greater is its contribution to the total C_{Lmax}.

Then, the maximum lift for the airplane can be determined using Equation (10-2) as shown below:

$$C_{Lmax} \approx \frac{0.9}{S_{ref}} \left[\sum_{i=1}^{N_f} C_{lmaxi} \times S_i \times \cos\left(\Lambda_{hingeline}\right)_i + \sum_{i=1}^{N_{uf}} C_{li} \times S_i \right]$$

$$= \frac{0.9 \times 2 \times 155.81}{144.9} = 1.936$$

The factor "2" is there to account for both wing halves, but the formulation in Table 10-6 only treats one wing half. The result is about 97.3% of the POH value of 1.99 (see Table 16-6) — not too a bad an agreement at all.

TABLE 10-6 Estimating the Contribution of the Wing Segments to the Total C_{Lmax}

	①	②	③	④	⑤	⑥	⑦	⑧	⑨
Segment	Y_a	Y_b	Area, S_i	$\Lambda_{hingeline}$	$C_{lmax\,a}$	$C_{lmax\,b}$	ΔC_{lmax}	$C_{lmax\,avg}$	$C_{lmax\,avg} \cdot S_i$
I	0	3	14.88	0	1.562	1.562	0.000	1.562	23.24
II	3	11.75	36.49	-4	1.562	1.562	1.100	2.662	96.90
III	11.75	16.42	15.27	0	1.562	1.562	0.000	1.562	23.85
IV	16.42	19.15	7.57	0	1.562	1.562	0.000	1.562	11.83
								Sum =	155.81

10.5 WINGTIP DESIGN

One of the most noticeable features of aircraft is the variety in their wingtip shapes. Wingtips come in all shapes and sizes, begging the question — do wingtips have an important role to play when it comes to aerodynamics, or are they primarily another expression of aesthetics? The answer to both questions is yes. Wingtips can offer fundamental improvements in efficiency and handling, but they are sometimes also selected for appearance. A large number of wingtip styles have been devised — we will only present a few and the most common ones here and primarily discuss their effectiveness from an aerodynamic standpoint. The looks are left to the reader to judge.

The literature shows that the shape of the wingtip affects both the minimum and lift-induced drag [33]. For instance, a square wingtip (see Section 10.5.3, *The square wingtip*) has a higher minimum drag than a spherical tip (see types below). However, the square tip results in less lift-induced drag than the spherical tip. This reduction is attributed to a distortion in the flow field that brings the wingtip vortices either closer or farther from the plane-of-symmetry. The greater the separation, the greater is the "apparent" AR of the wing and the less the lift-induced drag. Hoerner [34] states that the separation of the wingtip vortices "does not coincide with the geometrical span," but depends on the shape of the wingtips.

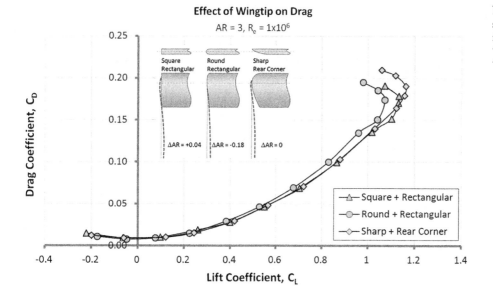

FIGURE 10-54 The effect of wingtip shape on the drag polar. The effect holds for higher AR wings as well. Note that the "Sharp + Rear Corner" wingtip is a "Hoerner" wingtip (based on Ref. [35]).

An example of this is shown in Figure 10-54. It can be seen that the round-shaped wingtip (as seen when looking at the front) results in the highest drag. Although the difference at low lift coefficients is small, it grows rapidly enough to yield a measurable reduction in rate of climb and maximum lift-to-drag ratio.

Aerodynamic Effectiveness of Wingtips

An important element in the installation of wingtip devices is an evaluation of their effectiveness. Low-speed characteristics, such as claims of improvements in stall handling, are difficult to measure. Any anecdotal claims of this must be regarded with great skepticism, in particular if they are associated with the marketing of aftermarket products. High-speed effectiveness is easier to measure and this is usually done by measuring the impact on lift-induced drag. This allows effectiveness to be stated in terms of something relatively easy to relate to, such as an effective change in AR, denoted by ΔAR. Thus, the value of a specific wingtip device can be evaluated by accounting for it when estimating drag in performance analyses. This can be done by simply adding ΔAR to the reference AR in the standard aircraft drag model (see Chapter 15, *Aircraft drag analysis*).

In real applications, the addition of a wingtip device affects more than just the lift-induced drag. For one, it always adds surface area and, thus, it increases wetted area. Additionally, some wingtips, such as endplates, also increase interference drag. An example of the combined effect of installing a typical winglet is presented in Figure 10-71. The cost of reduction in the lift-induced drag usually comes at a cost of higher parasitic drag. Since the parasitic addition is so highly dependent on the aircraft itself, we will only consider the effect on the lift-induced drag here. The goal is to modify the lift-induced drag as follows:

$$C_{Di} = \frac{C_L^2}{\pi \cdot (AR + \Delta AR) \cdot e} \quad (10\text{-}4)$$

where AR is the aspect ratio of the original (or baseline) wing, C_L is the lift coefficient at some reference flight condition, and e is the Oswald span efficiency of the original wing. Assume further that we modify the baseline wing by mounting a wingtip device, whose addition to the wingspan is given as Δb_t at each wingtip. While this increases both the wingspan and wing area, we will consider both the reference span and area to be unchanged. Additionally, we consider the difference in the Oswald efficiency of the two wing styles to stem solely from the addition of some finite aspect ratio, ΔAR, to the baseline AR. This can be justified based on the observation that as far as lift-induced drag is concerned, it is the product $AR \cdot e$ that matters and not AR or e on its own. Therefore, we can either modify AR, or e, or the product of the two. Here, however, for convenience, we will consider this change in terms of the AR as it is easier to relate to, as stated above. The gain in AR is calculated from:

$$\Delta AR = \left(\frac{C_L^2}{\pi \cdot AR \cdot e \cdot \Delta C_{Di} + C_L^2} - 1 \right) AR \quad (10\text{-}5)$$

Derivation of Equation (10-5)

The purpose of the formulation is to apply it to an airplane being operated at some desired C_L so the wingtip style being considered can be compared to the baseline wing. For this reason, we can specify the two following condition at some given lift coefficient, C_L:

Original wing:
$$(C_{Di})_1 = \frac{C_L^2}{\pi \cdot AR \cdot e}$$

Modified wing:
$$(C_{Di})_2 = \frac{C_L^2}{\pi \cdot (AR + \Delta AR) \cdot e}$$

This shows that the difference in lift-induced drag is given by:
$$\Delta C_{Di} = (C_{Di})_2 - (C_{Di})_1 = \frac{C_L^2}{\pi \cdot (AR + \Delta AR) \cdot e} - \frac{C_L^2}{\pi \cdot AR \cdot e}$$
$$= \frac{C_L^2}{\pi \cdot e}\left(\frac{1}{AR + \Delta AR} - \frac{1}{AR}\right)$$

Solve for ΔAR:
$$\Delta AR = \frac{C_L^2 AR}{\pi \cdot AR \cdot e \cdot \Delta C_{Di} + C_L^2} - AR$$
$$= \left(\frac{C_L^2}{\pi \cdot AR \cdot e \cdot \Delta C_{Di} + C_L^2} - 1\right) AR$$

QED

10.5.1 The Round Wingtip

The *round wingtip*, shown in Figures 10-55 and 10-56, is one of the oldest and simplest ways to terminate the wing at the tip. It consists of a semi-circular edge that extends from the leading to the trailing edge. The wingtip

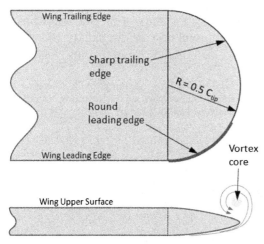

FIGURE 10-55 General shape of the spherical wingtip.

FIGURE 10-56 A Beechcraft D-18 boasting its round wingtips. The twin vertical tail surfaces, effectively, are endplates for the horizontal tail. *(Photo by Phil Rademacher)*

has been around since the 1920s. Hoerner [34] demonstrated it is less effective than once thought, as the wingtip vortex tends to move onto the upper surface. Consequently, the effective AR is in fact reduced. This means that when working with a specific AR in drag or stability and control analyses, the AR must be reduced by as much as 0.20.

The effectiveness of the tip is overestimated by potential flow theory, which renders it more effective than either a square or the spherical wingtip (see Section 10.5.10, *Comparison based on potential flow theory*). This can be explained by increasing section lift coefficients near the rapidly tapering tip, which at cruise AOA makes the configuration more efficient. However, this is contradicted by the results of Ref. [34], which are based on wind tunnel testing.

Gain in effectiveness: $\Delta AR = -0.19$

10.5.2 The Spherical Wingtip

The *spherical wingtip*, shown in Figure 10-57, is a simple and commonly used termination of the wing at the tip. It is simple to fabricate using composites and poses minimal challenges in the layout of its loft. Its planform view of the curvature is usually governed by the airfoil, as the geometry is a semi-circle along the tip chord.

Zimmer [36] refers to the wingtip under the name *Goettinger* wingtip, but the author has been unable to confirm that name.

Aerodynamically, the spherical tip is an inefficient shape, as it causes the wingtip vortex to roll up and inboard, as shown in Figure 10-58. This reduces the separation between the two wingtip vortices, which leads to a reduction in the effective AR (see Figure 15-69) and increased lift-induced drag.

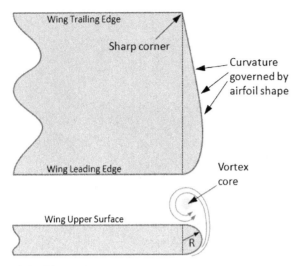

FIGURE 10-57 General shape of the spherical wingtip.

The geometry has no influence on dihedral effect. However, it is thought to have a detrimental effect on spin characteristics due to the formation of a low-pressure region around the wingtip, which helps drive the autorotation and, thus, can render spin recovery more difficult.

Gain in effectiveness: $\Delta AR = -0.18$

10.5.3 The Square Wingtip

The *square wingtip* (see Figure 10-59) is the simplest geometry possible for a wingtip. The sharp termination forces the wingtip vortex to form and reside on the outboard side of the tip (rather than above, as for the round tip). This effectively pushes the wingtip vortices farther apart from the centerline of the aircraft and increases the effective AR, albeit by a small amount. This, at least theoretically, reduces lift-induced drag by a small amount (see Section 15.5.18, *Corrections of the lift-induced drag*). The tip has no

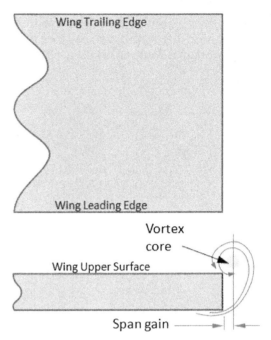

FIGURE 10-59 General shape of the square wingtip.

influence on dihedral effect. An example of a square wingtip is shown in Figure 10-60.

Gain in effectiveness: $\Delta AR = +0.004$

10.5.4 Booster Wingtips

Booster wingtips are noticeable by their distinct trailing edge curvature, which comes in two forms; upturned and downturned (see Figure 10-62 and Figure 10-63). Both are intended to bring the wingtip vortices farther outboard, increasing effective AR and, in that way, reducing the induced drag. The true effect on drag is not well known, although some manufacturers of aftermarket booster wingtips claim benefits such as low-speed roll stability and reduction in stalling speed. The author has been unable to locate any published papers that support such claims. The characteristics of the booster wingtip will now be discussed.

FIGURE 10-58 A Piper PA-28 Cherokee four-seat, single-engine sport aircraft featuring round wing tips. *(Photo by Phil Rademacher)*

FIGURE 10-60 An Extra 300 single-seat, single-engine, aerobatic aircraft featuring square tips for all lifting surfaces. *(Photo by Phil Rademacher)*

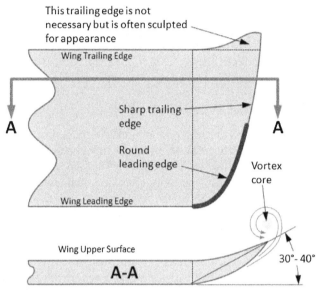

FIGURE 10-61 General shape of the upturned booster wingtip.

The Upturned Booster Wingtip

The upturned booster wingtip consists of geometry in which the lower and upper surfaces join in a sharp termination, or edge, that flares upward at the trailing edge, as shown in Figure 10-61 and Figure 10-62. The wingtip style is most easily made from composite materials, although the sharp edge sometimes requires fabrication techniques that minimize the formation of production nuisances such as dry fibers and "bridging."

Theoretically, the wingtip increases the separation of the wingtip vortices and, therefore, increases the effective *AR*. In practice, the increase is probably negligible. It also *increases* dihedral effect by a small amount.

Gain in effectiveness: $\Delta AR = 0.00$

The Downturned Booster Wingtip

The downturned booster wingtip is effectively an inversion of the upturned style. Just like the upturned booster tip, there is not much to be found in the literature regarding improvements in effectiveness (e.g. see Rokhsaz [37]). Naik and Ostowari [38] present an experimental investigation of several non-planar wingtip configurations, including a downturned droop. The results provided show limited to no benefit over the planar reference wing. In accordance with standard lateral stability theory, the downturned booster wingtip *decreases* dihedral effect, i.e. roll stability. However, there is a perception they improve handling near stall and, thus, have been marketed as a part of short t-o and landing (STOL) STCs for low-speed aircraft. Other schools of thought claim they have a detrimental effect on spin characteristics as a low-pressure region is formed around the wingtip, which will help drive

FIGURE 10-62 A Cirrus SR22 four-seat, single-engine, high-performance aircraft, featuring upturned booster wingtips. *(Photo by Phil Rademacher)*

FIGURE 10-63 A Cessna 206 six-seat, single-engine, high-performance aircraft featuring downturned booster wing tips. *(Photo by Phil Rademacher)*

autorotation and, thus, can render spin recovery more difficult.

Gain in effectiveness: $\Delta AR = 0.00$

10.5.5 Hoerner Wingtip

The Hoerner wingtip is named after the German aerodynamicist Sighard Hoerner (?–1975) who showed its effectiveness to be superior to that of the round wingtip. In fact, its primary advantage is that its impact on the effective AR is neutral, rather than negative like that of the round wingtip (see Section 15.5.18, *Corrections of the lift-induced drag*). Hoerner wingtips are used on a number of aircraft, most notably the Grumman American AA-5 series of aircraft, produced between 1971 and 2005 (see Figure 10-64).

The wingtip increases dihedral effect. Theoretically, it is intended to shift the wingtip vortex farther outboard and this requires a sharp edge separating the upper and lower surfaces along the aft trailing edge as shown in Figure 10-65. Normally, the first part of the leading edge features a round geometry to prevent the upper surface from stalling at low $AOAs$.

Gain in effectiveness: $\Delta AR = 0.00$

10.5.6 Raked Wingtip

The raked wingtip is a very efficient means to increase the effective AR. Its primary drawback is that it comes at the cost of increased wingspan and wing bending moments. The wingspan increase may spur parking and hangar space challenges, whereas the increase in bending moments will raise the airframe weight. It is an advantage that the device does not increase parasitic interference drag, although it will increase skin friction if it is an addition to an existing wing baseline. An example of a raked wingtip is shown on the Boeing 777 commercial transport aircraft in Figure 10-66. Such wingtips have also been used on the Boeing 767 and 787 Dreamliner aircraft. Their use in GA aircraft, at this time, is rare. The best example is the Coast Guard surveillance version of the Schweizer SGM 2-37 motorglider, called the RG-8A.

Contrary to popular belief, raked wing tips have a long history, having been first evaluated in 1921 in NACA TN-69 [39]. Both positive (forward sweep) and negative (aft sweep) rakes were studied and it was found that the maximum L/D ratio at an AOA of $4°$ increased by 7%. Interestingly, the best result was that of a $20°$ positive rake. The report concluded, "the effects of rake on the lift and drag are so small that considerations of strength and aileron efficiency should govern the wing tip form." This interesting conclusion highlights the difference in the airplane designs of the 1920s and 2010s — today, 7% translates into substantial fuel savings — which was much less of a problem in the early twentieth century.

The effectiveness of the raked wingtip depends on the *rake angle*, φ, but also on its own dihedral angle, Γ, and span, Δb_t (see definitions in Figure 10-67). Based on potential flow theory, the dihedral angle should be around $4-8°$, depending on wing geometry. A very high or low dihedral ($\pm 10°$ and greater) will actually increase the

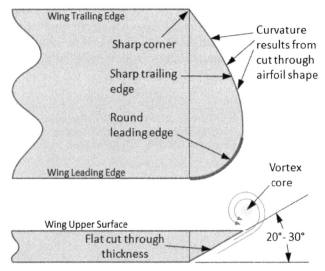

FIGURE 10-64 Typical geometric considerations for the layout of a Hoerner tip.

FIGURE 10-65 A Grumman American AA-5 four-seat, single-engine sport aircraft featuring a Hoerner wing tip. *(Photo by Phil Rademacher)*

FIGURE 10-66 A Boeing 777 commercial transport aircraft, operated by Emirates, featuring a raked wing tip. *(Photo by Phil Rademacher)*

lift-induced drag. An example of this is shown in Figure 10-68. The graph compares three raked wingtip styles to a baseline Hersey-bar wing of $AR = 6$. All the configurations have the same total wingspan and the raked tip span is $\Delta b_t = 0.125b$. For this reason, the wing and wetted area of the baseline wing are larger than those of the other configurations, called Tip A, Tip B, and Tip C. Tip A has a $5.71°$ dihedral, Tip B has a $26.57°$ dihedral, and Tip C has a $26.57°$ anhedral. It can be seen that Tip A (sometimes referred to as the Trilander wingtip, as in Britten Norman Trilander) generates the least lift-induced drag of the four wing styles,

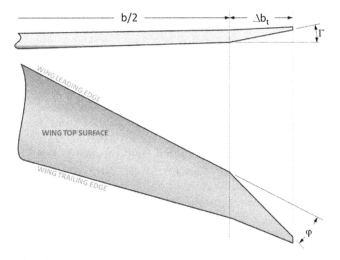

FIGURE 10-67 Definition of basic geometry of the raked wingtip.

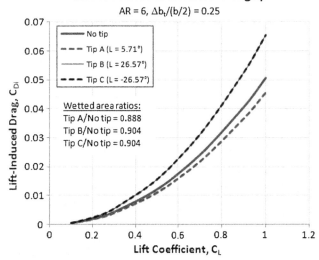

FIGURE 10-68 An example of a raked wingtip shows that a low dihedral decreases lift-induced drag, while a larger dihedral increases it.

in addition to having the least wetted area. This indicates it is the most efficient wing style of the four. Tips B and C generate precisely equal lift-induced drag and both are substantially less efficient in cruise than either the baseline or Tip A. This result also shows that designing a raked wingtip requires careful evaluation of the geometry.

Gain in effectiveness depends on the details of the geometry, but expect $\Delta AR = +0.5$ and possibly higher than $+1.5$ for well optimized wingtips. For instance, Tip A in Figure 10-68 achieves a ΔAR of around 0.88 without any optimization.

Rakelet

The *Rakelet* is a combination of the raked wingtip and a winglet, and is considered a new wingtip device. Investigated as a proposal for the KC-135 Stratotanker, by Diaz et al. [40], it is claimed to improve its range over that of the baseline wing by as much as 7.5%.

10.5.7 Endplate Wingtip

It was recognized early on that drag did not just depend on the geometry of the body, but also increased rapidly with *AOA*. The discovery of the endplate is attributed to the British scientist Frederick W. Lanchester (1868–1946), who discovered that vertical plates positioned at the tip of a lifting surface greatly improved its lift. Other serious studies date back to the early 1920s with the work of Reid [41]. The endplate has a similar effect as placing a wing section from wall-to-wall in a wind tunnel — it makes it behave more like a two-dimensional airfoil. The consequence is an increase in section lift coefficients, lift curve slope, and maximum lift coefficient [42]. It also reduces the lift-induced drag. Examples of theoretical treatment of lift and drag endplates are given in Refs [43, 44, 45].

Endplates have a number of drawbacks. First, the surface adds parasitic drag through increase in wetted area and interference drag where the plate joins the lifting surface. The combination cuts into the aerodynamic benefits and may even exceed the reduction in lift-induced drag. Second, the inertia of the endplates not only increases the weight, it also reduces the natural frequency of the lifting surface, which makes it more susceptible to flutter. To solve this issue, the lifting surface must be stiffened, which further adds to the weight. Third, if the endplates are used as a vertical tail and include two rudders, a more complicated control system is required. Fourth, in a production environment, using two vertical surfaces for a VT, rather than one, increases manufacturing complexity.

In spite of these shortcomings, endplates have been widely used in the aviation industry. Airplanes dating from to the 1930s and later, frequently featured the configuration as horizontal and vertical tail surfaces.

FIGURE 10-69 The familiar Airbus winglet is actually an endplate. *(Photo by Phil Rademacher)*

During the Second World War, a large number of multi-engine bombers and transport aircraft were designed with twin tails, which was helpful in an environment in which 'losing' an engine was a likely scenario. Immersing the two tails in the propwash of operational engines improved not only controllability, but survivability. The most recent example of such use is the tail surfaces of the Russian military transport aircraft Antonov An-225 Mriya, currently the world's heaviest airplane with a 1.41 million lb_f gross weight. This design solution helps to keep the overall height of the aircraft down, allowing it to be stored in existing hangars for maintenance. Endplates are also used in place of a winglet on the Airbus 320 family of commercial jetliners (see Figure 10-69).

The magnitude of the effect is highly dependent on the height of the endplates, as shown in Figure 10-70. When used on a horizontal tail, the endplate allows for a smaller horizontal tail surface, improved elevator effectiveness and the endplates serve well as vertical tails.

Gain in effectiveness (see Equation (9-88)):

$$\Delta AR \approx 1.9(h/b)AR \text{ for } h/b \text{ up to } 0.4.$$

10.5.8 The Winglet

From a certain point of view, the winglet is a sophisticated mutation of the endplate discussed above. As stated earlier, there are several drawbacks to the use of endplates. These detriments are lessened using winglets. Since winglets are often not completely vertical, and largely above the wing plane, they can increase the lateral stability derivative ($C_{l\beta}$ or dihedral effect) over that of the endplate. This may aggravate the Dutch roll damping, but can also affect the cross-wind capability of the aircraft. Of course, the endplate and winglet both increase the side force derivatives, $C_{y\beta}$. Winglets also pose structural ramifications if added to airplanes

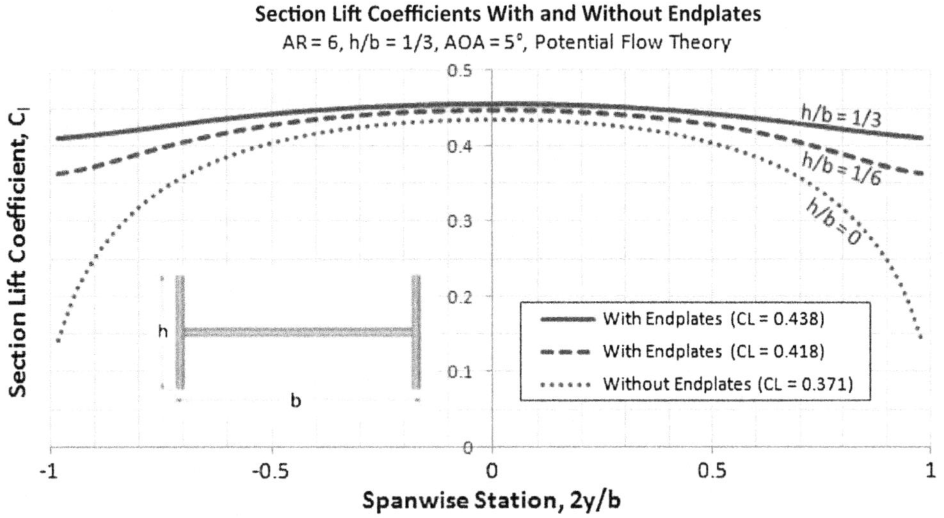

Section Lift Coefficients With and Without Endplates

FIGURE 10-70 Comparing the distribution of section lift coefficients with and without endplates shows great improvement in surface lift effectiveness depending on endplate height.

not originally designed with them. They increase the wing bending moments and lower the flutter speed (albeit less than endplates). But one of the most important differences between a 'clean' wing and one with winglets is in the nature of drag, as will now be explained.

Adding a winglet to an airplane increases two types of drag — skin friction and interference drag. They only reduce the lift-induced drag — when properly designed. In fact, they can easily increase the total drag. They are aerodynamically viable only when the reduction in the lift-induced drag is larger than the increase in the skin friction and interference drag. This is illustrated in Figure 10-71, which compares the potential installation of winglets on a Cirrus SR22 style aircraft to the original clean wing. While the numbers are "pulled out of thin air" (except the C_{Dmin} of the original configuration, which is calculated in Example 15-18), they still represent a realistic scenario. The presence of the winglet is treated as an increase in the original AR and C_{Dmin}. Figure 10-71 shows that when operating at $C_L <$ 0.55 (e.g. at cruise), the total C_D of the original airplane

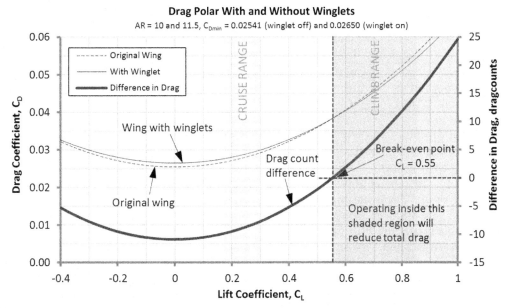

FIGURE 10-71 A comparison of drag polars before and after a winglet has been installed, helps explain their benefits. The thin dashed curve is the drag polar for the original wing, the solid thin curve represents the wing with the winglet, and the thick solid line is the difference between the two.

is actually lower than with the winglet. If operating at C_L > 0.55 (e.g. during climb) the total drag will be reduced over the original wing. It is easy to determine from its POH that the SR22 cruises at C_L ranging from 0.17 to 0.30. Therefore, a winglet installation would have to be justified on other merits, such as cosmetics or improved climb and T-O performance. Of course, this does not deny that certain winglet geometry may exist that moves the break-even point below what is shown in the graph.

Another interesting lesson evident in Figure 10-71 is that the winglet only shaves off a few drag counts. For instance, at C_L = 0.8 it reduces the drag by 12 drag counts. Compare that to 254 drag counts of the original aircraft (4.7%). At C_L = 0.3 the opposite holds true; the drag is increased by 8 drag counts. Another way of interpreting this difference is that the airplane may have to climb to a high altitude (where it would have to cruise at a higher AOA) in order for the winglets to be justified based on drag reduction. If the same airplane is cruising in denser air, it will do so at a lower AOA and, thus, with drag penalty.

Like the generation of lift, the workings of a winglet can be explained in a number of ways. The author's favorite explanation is that the magnitude of the section lift coefficients near the wingtip increases due to an enlarged low-pressure region generated inboard of the winglet. The resulting lift distribution resembles that of a wing with a larger AR than indicated by the plain planform geometry.

This, in effect, reduces C_{Di} for the same C_L. This view explains why the winglet has to be 'loaded up' in order to function. "Loaded up" simply means it must be placed at a relative AOA large enough to generate lift and, that way, boost the low-pressure region near the wingtip. For this reason the angle-of-incidence of the winglet must be placed at a specific "toe-in" or "toe-out" angle (Figure 10-72), depending on geometry. An unloaded winglet will not reduce the lift-induced drag because its contribution to the low-pressure field is negligible.

Another explanation attributes the reduction in the lift-induced drag to the distribution of the sheet of trailing vortices over greater distance, making it better resemble that of an elliptic planar wing (which has the minimum lift-induced drag). An insightful reasoning is provided by Hoerner [35] and McCormick [46], who explain it as the consequence of the lift force acting on the winglet surface, which is tilted forward due to the local AOA on the winglet. This way, the lift of the winglet, effectively, generates a forward force component — thrust — that can be subtracted from the drag. This perspective allows formulation to be derived to estimate the drag reduction. Regardless, all these explanations are different sides of the same box.

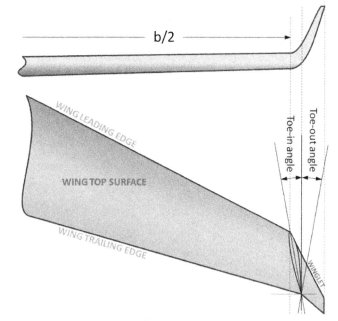

FIGURE 10-72 Definition of toe-in and toe-out angles.

Gain in effectiveness can be approximated using the following expression, by assuming the winglet resembles an endplate (where h is the total height of the winglet):

$$\Delta AR \approx 1.9(h/b)AR$$

The gain in AR is unlikely to be higher than about +1.5.

The Whitcomb Winglet

The development of the Whitcomb winglet (see Figure 10-73) is attributed to Richard T. Whitcomb (1921−2009), who in the early 1970s conducted research with his team of scientists in the Langley 8-foot Transonic Pressure Tunnel [47]. Part of this work was published in 1976, in a NASA report he authored [48]. It marks the first time a winglet was seriously considered for a large and heavy aircraft. Once this pioneering work demonstrated winglets were indeed effective, NASA began to conduct a substantial amount of research on the topic. Reference [49] was the first in a long line of technical papers evaluating its capabilities resulting from this effort. In the late 1970s, Whitcomb developed a winglet for use on a Boeing KC-135 tanker aircraft. The research showed the winglet increased its cruise range by as much as 7% [50]. This work encouraged many manufacturers of commercial, business, and GA aircraft to produce aircraft with winglets. The Gates Learjet Model 28/29 Longhorn was the first production airplane to feature winglets (although these had the lower winglet omitted). A Whitcomb winglet was also featured on the Rutan VariEze designed in 1974. The

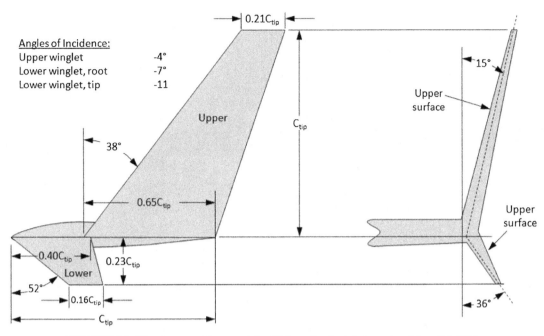

FIGURE 10-73 A general layout of the Whitcomb winglet *(reproduced from Ref. [48]).*

Boeing 747-400 was the first commercial jetliner produced with winglets.

A flight test evaluation of a Whitcomb winglet on the McDonnell-Douglas DC-10 commercial jetliner in the early 1980s revealed that the most efficient configuration consisted of a combination of a slight aileron droop with a reduced span winglet. The combination reduced fuel burn at cruise by 3%, increased range by 2%, and reduced T-O distance by 5% at maximum gross weight [51]. In 1990, McDonnell-Douglas began deliveries of MD-11s with such a winglet installed. The winglet differed from the basic Whitcomb type in that it had a lower AR and started much closer to the leading edge of the wing (see Figure 10-74).

Reference [48] compared lift, drag, and pitching moment of a wind-tunnel model of the KC-135 with three wingtip devices attached: a wingtip extension of 38% of the height of the upper winglet, the upper winglet, and the upper and lower winglets combined. The results for drag are reproduced in Figure 10-75. It shows that the winglet generates less drag than simply extending the wingtip. It also shows that the drag reduction obtained by the upper and lower combination is no better than with only the upper winglet. This has resulted in a modification to the general Whitcomb design, to be discussed next. Strictly speaking, the smaller winglet does not have to be smaller — it is just so to prevent it from striking the ground on low-wing aircraft [48, p. 8].

The Blended Winglet

The blended winglet differs from the Whitcomb winglet in the absence of the lower winglet. As can be seen in Figure 10-75, the drag benefit of the Whitcomb

FIGURE 10-74 A Whitcomb winglet on a McDonnell-Douglas MD-11 commercial jetliner. *(Photo by Phil Rademacher)*

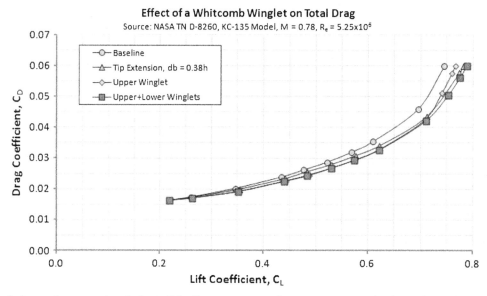

FIGURE 10-75 A drag polar comparing the basic KC-135 wing to one with a tip extension and Whitcomb winglet, one of which had the lower surface removed *(reproduced from Ref. [48])*.

winglet is marginal at low (cruise) lift coefficients to say the least. Comparing this contribution to the manufacturing complexity and weight, it is reasonable to simply remove it. Another difference between the two is the generous arc of the intersection between the wing and the winglet. This allows the wing airfoils to transition smoothly to the winglet airfoils. It also replaces the sharp intersection of the Whitcomb winglet, reducing the pressure peak and, therefore, local airspeeds. Keeping those low is important in preventing flow separation due to low-pressure peaks (interference drag) and the formation of normal shock at high Mach numbers (wave drag).

The blended winglet can be seen on many modern airliners. The most visible aircraft include aircraft such as the Boeing 737, Boeing 757 (see Figure 10-76), and many others.

The blended winglet was designed and patented by Louis B. Gratzer as U.S. Patent 5,348,253. The patent states the theoretical background behind the design, where it is intended to solve a common deficiency of winglet installation — sharp discontinuities of intersections that result in departure from optimum loading. This is resolved using the gradual transition of the blended geometry, which allows the wingtip airfoil geometry to change smoothly to that of the winglet. It was specifically selected by Boeing due to fewer changes being required to the existing airframe [52]. Some pros and cons of winglets are presented in Table 10-7.

10.5.9 The Polyhedral Wing(tip)

The *polyhedral wingtip* is really a special wing shape and is also referred to as the *cambered-span wing* or *non-planar wing* (see Figure 10-77 and Figure 10-78). It consists of a gradually increasing dihedral at several wing stations along the wing. The idea originated with Cone [53], who showed that cambered-span wings can increase the effective *AR* by as much as 50% compared to a flat elliptical wing of equal span, producing equal lift. Cone [54] also presented a method and an illustrative example to help with the design of cambered-spans and evaluate the reduction in lift-induced drag. A good description of Cone's highly theoretical work is presented by Jones [55].

FIGURE 10-76 A modern winglet on a Boeing 757 commercial jetliner. *(Photo by Phil Rademacher)*

TABLE 10-7 Summary of Pros and Cons of Winglets

Pros	Cons
• Increases the efficiency of the wing by reducing lift-induced drag. • Improved fuel consumption as a consequence of reduced lift-induced drag. • Reduces T-O distance and increases rate of climb due to the same (the higher the clean wing lift-induced drag, the greater is the reduction brought on by the winglet). • Thought by many to bring ramp appeal to aircraft. • Increases effective wing AR with minimum increase in wingspan.	• Increases dihedral effect. • Moves center of lift outboard, increasing wing bending moments. • Increases wing weight due to the higher bending moments. • Reduction in the natural frequency of the wing, which lowers the flutter speed. • Adds wetted area and interference, whose drag must be overcome by the reduction in lift-induced drag. For this reason, the winglets usually increase drag at low C_L (even though this is not apparent from Ref. 48). • The support structure for the winglet adds manufacturing complexity and cost.

An incorporation of the polyhedral wing increases manufacturing complexity due to the break in the wing-spar and -skin. It also increases the dihedral effect. The polyhedral wingtip has been used on at least two sailplanes, the Schempp-Hirth Discus-2 and the Glaser-Dirks DG-1000 (see Figure 10-77).

From a conceptual design standpoint, non-planar wings can be considered from at least two perspectives, each of which leads to very different results. As explained below, it is very important to keep these results in mind when designing such aircraft. For a better understanding, refer to Figure 10-79.

First, consider a stiff straight-wing design, denoted as Wing A in Figure 10-79. Assume that this wing has a baseline wingspan b, wing area S, and is generating some lift coefficient, C_L. Then consider Wing B, which unloaded is geometrically identical to Wing A. However, assume that it is so flexible structurally that in level 1g flight the static aero-elastic effects deform it so forms the shape of Wing B. As can be seen in Figure 10-79, the resulting wingspan will be shortened by the amount Δb and its effective wing area will be reduced by an amount ΔS. Therefore, Wing b will have to be flown at a slightly higher AOA in order to generate the same lift coefficient as Wing a. For this reason, Wing b will generate higher lift-induced drag than Wing a. An example of such a scenario is shown in Figure 9-52, which compares the lift of a flexed and straight wing at the same AOA.

Alternatively, consider the non-planar Wing c in Figure 10-79. It is also designed with the same wingspan and wing area as Wing a, in addition to featuring the same planform shape. Assume this wing to be stiff, so, in contrast to Wing b, it does not flex in flight, but maintains its original span and area. This wing will generate the same lift coefficient as Wing a at a slightly lower AOA. Consequently, its lift-induced drag will be less.

This peculiar result can be attributed to a more efficient reduction of pressure on the upper surface due to the fact that the spanwise bound vortex is longer than that of the straight wing. This can be readily demonstrated using potential flow theory (e.g. vortex-lattice). Of course, Wing C has a greater wetted area than Wing A, it will be slightly heavier and this may offset the benefits some. A comparison between the three wing styles as a function of flex height is shown in Figure 10-80.

10.5.10 Comparison Based on Potential Flow Theory

Potential flow theory can be used to evaluate the effectiveness of various wingtip devices, primarily when they are designed for operation at C_L for which flow separation is still limited. This section compares a few such designs for lift, drag, and contribution to lateral stability (see Table 10-8). It can be used for guidance when selecting the appropriate wingtip geometry. Note that the table does not feature the winglet or the polyhedral wing.

Of the wingtip types compared, it can be seen that the raked wingtip and the Trilander type yield the most efficient wings for cruising flight. Generally, the more efficient the wing style is, the steeper is its lift curve slope. The explanation for this can be seen in

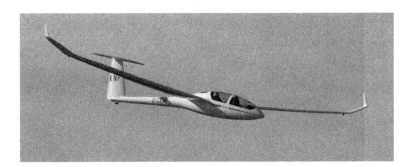

FIGURE 10-77 Glaser-Dirks DG-1000 sailplane features a polyhedral wing. *(Photo by Paul Hailday through Wikimedia Commons)*

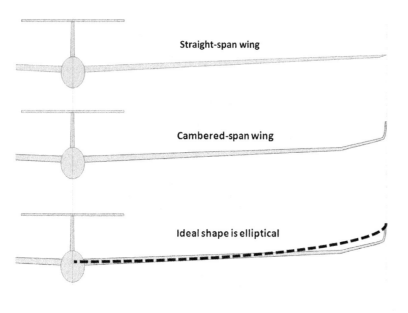

FIGURE 10-78 Comparison of the polyhedral wing, showing a straight wing, the polyhedral camber, and the ideal elliptical shape (based on Ref. [56]).

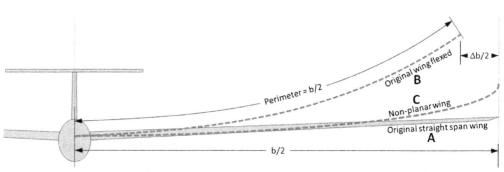

FIGURE 10-79 Comparison of a polyhedral wing to an actual wing flex. Wing A is the original straight wing of wingspan b and area S. Wing B is that same wing flexed due to airloads; it will suffer a reduction in its effective wing span and area. Wing C is a non-planar wing designed with wingspan b and S.

Figure 10-81, which shows that the section lift coefficients for the more efficient wing styles remain high along the tip. While favorable for cruise, this should also be kept in mind for stall characteristics. This means that the wingtip will be the first part of the wing to stall at higher *AOA*s. If the span of the wingtip is substantial, this can have a detrimental effect on the roll stability of the wing at stall. This would be less of a concern for a relatively short-span wingtip, such as the Trilander type, as it is subject to the formation of a leading edge vortex, similar to that of a deltawing. However, it may pose challenges for a long-span raked wingtip.

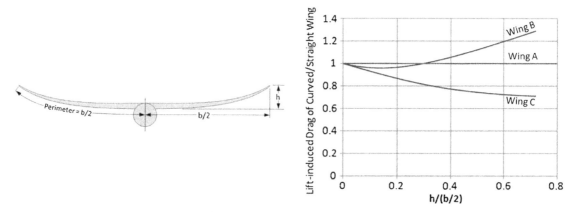

FIGURE 10-80 A comparison of the three wing styles discussed in this section (based on Ref. [55]).

TABLE 10-8 Summary of Characteristics of Several Wingtip Designs

Wingtip	AR	$S_{wet}/S_{wet\,o}$	α degrees	$C_{L\alpha}$ per radian	C_{Di}	e	AR.e	LD_{max}	h_n	$\Delta C_{l\beta}$ per degree
Spherical (baseline)	8.000	1.0000	8.00	4.564	0.0200	0.8099	6.48	22.6	0.250	0.00000
Upturned	8.076	1.0004	7.90.	4.620	0.0198	0.8095	6.54	22.7	0.254	-0.00029
Downturned	8.076	1.0004	7.90	4.620	0.0198	0.8095	6.54	22.7	0.254	0.00031
Round tip	8.240	0.9708	7.82	4.671	0.0194	0.8078	6.66	23.2	0.254	0.00000
Trilander type	8.427	0.9509	7.79	4.685	0.0191	0.8012	6.75	23.6	0.270	-0.00032
Raked	9.014	0.9509	7.59	4.813	0.0180	0.7974	7.19	24.4	0.298	-0.00042

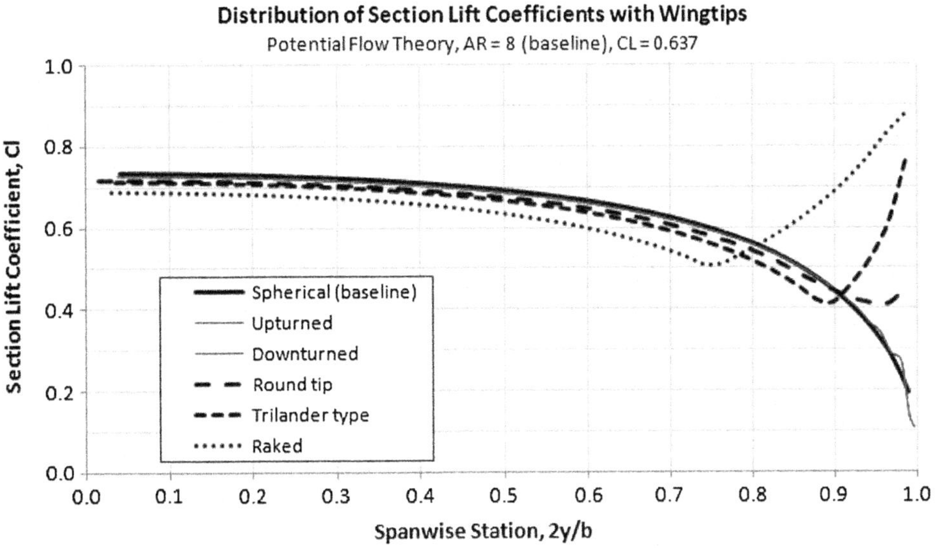

FIGURE 10-81 Distribution of section lift coefficients reveals how the Trilander type and raked wingtips boost lift at the wingtip, which helps explain how they improve the overall wing efficiency by allowing operation at a lower *AOA* to generate the same C_L.

VARIABLES

Symbol	Description	Units (UK and SI)
AOA	Angle-of-attack	Degrees or radians
AR	Aspect ratio	
b	Wing span	ft or m
b_a	Aileron span	ft or m
b_f	Flap span	ft or m
b_s	Slat span	ft or m
c	Size of the gap at outlet of the slot	ft or m
C	Airfoil chord length	ft or m
$C_{combo\,extended}$	Combined flap chord length when extended	ft or m
$C_{combo\,stowed}$	Combined flap chord length when stowed	ft or m

Symbol	Description	Units (UK and SI)
C_D	Drag coefficient	
C_{Di}	Induced drag coefficient	
C_L	Lift coefficient	
C_l	Section lift coefficient	
C_{L0}	3D zero *AOA* lift coefficient	
C_{li}	Average lift coefficient of the unflapped wing segments at stall *AOA* of flapped segments	
C_{lmax}	Maximum 2D lift coefficient	
C_{Lmax}	Maximum 3D lift coefficient	
$C_{lmax\,a}$	Maximum lift coefficient at the inboard end of the segment	
$C_{lmax\,avg}$	Average of $C_{lmax\,a}$ and $C_{lmax\,b}$	
$C_{lmax\,b}$	Maximum lift coefficient at the outboard end of the segment	

(Continued)

Symbol	Description	Units (UK and SI)
$C_{lmax\ i}$	Maximum lift coefficient of the flapped wing segment	
C_{Lmin}	Minimum coefficient of lift	
$C_{L\alpha}$	3D lift curve slope	Per degree or per radian
C_M	Pitching moment coefficient	
$C_{m\ i}$	Average pitching moment coefficient of each wing segment	
C_{M0}	3D zero AOA pitching moment coefficient	
C_{main}	Chord length of airfoil without flap	ft or m
$C_{M\alpha}$	3D pitching moment curve slope	Per degree or per radian
C_s	Slat chord length	ft or m
C_v	Vane chord length	ft or m
e	Oswald's span efficiency	
h	Flap height fraction (Gurney flap)	
h	Winglet height	ft or m
M	Mach number	
N	Number of wing segments	
N_f	Number of flapped wing segments	
N_{uf}	Number of unflapped wing segments	
R_e	Reynolds number	
S	Planform area	ft^2 or m^2
S_i	Trapezoidal planform area of segment	ft^2 or m^2
V	Airspeed	ft/s or m/s
x_f	x-displacement of flap hinge line from TE of chord airfoil (Junkers flaps)	ft or m
x_f	x-displacement of flap hinge line from slot lip (Fowler flaps)	ft or m
x_h	x-distance from leading edge of airfoil to flap hinge	ft or m
y_1	y-distance from aircraft centerline to inboard flap chord	ft or m
y_2	y-distance from aircraft centerline to inboard slat chord	ft or m
y_3	y-distance from aircraft centerline to outboard flap chord	ft or m
y_4	y-distance from aircraft centerline to outboard slat chord	ft or m
Y_a	Distance of inboard end of segment from plane of symmetry	ft or m
Y_b	Distance of ooutboard end of segment from plane of symmetry	ft or m
y_f	y-displacement of flap hinge from chord centerline (Junkers flap)	ft or m
y_f	y-displacement of flap hinge from slot lip (Fowler flap)	ft or m
y_h	y-distance from chord centerline to hinge	ft or m
ΔC_{dmin}	Change in 2D drag coefficient	
ΔC_{l0}	Change in 2D zero AOA lift coefficient	
ΔC_{lmax}	Change in maximum 2-D lift coefficient	
ΔC_m	Change in 2D pitching moment coefficient	
Δx	Extension forward of slat from leading edge	ft or m
Δx_1	Extension of first flap	ft or m
Δx_2	Extension of second flap	ft or m
Δx_3	Extension of third flap	ft or m
Δy	Downward drop of slat below the leading edge	ft or m
$\Delta \alpha_{max}$	Change in stall angle-of-attack	Degrees or radians
$\Lambda_{c/4}$	Quarter-chord sweep angle	Degrees or radians
$\Lambda_{hingeline}$	Hingeline angle	Degrees or radians
α_{stall}	Stall angle-of-attack	Degrees or radians
α_{ZL}	Zero-lift angle-of-attack	Degrees or radians
δ_f	Flap deflection angle	Degrees or radians
δ_s	Slat deflection angle	Degrees or radians
δ_v	Vane deflection angle	Degrees or radians
ρ	Density of air	slugs/ft^3 or kg/m^3

References

[1] NACA TN-3007. Lift and Pitching Moment at Low Speeds of The NACA 64A010 Airfoil Section Equipped with Various Combinations of a Leading-Edge Slat, Leading-Edge Flap, Split Flap, and Double-Slotted Flap. Hayter: Nora-Lee F., and John A. Kelly; 1953.

[2] NASA TM-4370. Variable-Camber Systems Integration and Operational Performance of the AFTI/F-111 Mission Adaptive Wing. Smith, John W., Wilton P. Lock, and Gordon A. Payne; 1992.

REFERENCES

[3] NASA CR-158930. Assessment of Variable Camber for Application to Transport Aircraft 1980.

[4] Smith AMO. High-Lift Aerodynamics. Journal of Aircraft 1975;Vol. 12(No. 6).

[5] Liebeck RH, Smyth DN. Study of Slat-Airfoil Combinations Using Computer Graphics. Journal of Aircraft 1973;Vol. 20(No. 4).

[6] NACA TR-407. The Characteristics of a Clark Y Wing Model Equipped with Several Forms of Low-Drag Fixed Slots. Weick: Fred E., and Carl J. Wenzinger; 1933.

[7] Meier Hans-Ulrich. German Development of the Swept Wing 1935-1945. AIAA 2010.

[8] NASA CR-4746. High-Lift System on Commercial Subsonic Airliners. Rudolph: Peter K. C; 1996.

[9] NACA TM-1177. Wind-Tunnel Investigations on a Changed Mustang Profile with Nose Flap Force and Pressure-Distribution Measurements. Krüger: Werner; 1947.

[10] NACA TN-71. Experiments with Slotted Wings. Lachmann: Gustav; 1921.

[11] NACA TN-598. Wind Tunnel Tests of a Clark Y Wing with "Maxwell" Leading Edge Slots. Gauvain: William E; 1937.

[12] NACA WR-L-574. Wind-Tunnel Investigation of an NACA 23012 Airfoil with an 18.05-Percent-Chord Maxwell Slat and with Trailing Edge Flap. Gillis: Clarence L., and John W. McKee; 1941.

[13] NACA R-938. Summary of Section Data on Trailing-Edge High-Lift Devices. Cahill: Jones F; 1949.

[14] Nayler, et al. Experiments on Models of Aeroplane Wings at the National Physical Laboratory. A.R.C. R.&M; March 1914. No. 110.

[15] Young AD. The Aerodynamic Characteristics of Flaps. R.&M. No. 2622, British A. R. C; 1947.

[16] Freedman Russell. The Wright Brothers: How They Invented the Airplane. Holiday House 1991:116.

[17] NACA WR-L-373. Wind-Tunnel Investigation of Perforated Split Flaps for Use as Dive Brakes on a Tapered NACA 23012 Airfoil. Purser: Paul E., and Thomas R. Turner; 1941.

[18] NACA TN-596. Full-Scale Wind-Tunnel and Flight Test of a Fairchild 22 Airplane Equipped with a Zap Flap and Zap Ailerons. Dearborn: C. H. and H. A. Soule; 1937.

[19] NACA WR-L-437. Full-Scale Wind-Tunnel and Flight Tests of a Fairchild XR2K-1 Airplane with a Zap Flap and Upper-Surface Aileron-Wing Installation. O'Sullivan: William J; 1942.

[20] NACA TN-524. Wind-tunnel Tests of a Wing with a Trailing-Edge Auxiliary Airfoil used as a Flap. Noyes: Richard W; 1935.

[21] NACA R-541. Aerodynamic Characteristics of Wings with Cambered External Airfoil Flaps, Including Lateral Control, with a Full-Span Flap. Platt: Robert C; 1935.

[22] Lott, Chabino Sarah. Guide to the Harlan D. Fowler Papers 1920-1980. Collection number: MSS-1995-04. San José State University Library; 2009.

[23] NASA SP-2007-4409. The Wind and Beyond. Hansen: James R; 2007.

[24] NASA CR-2443. Development of a Fowler Flap System for a High Performance General Aviation Airfoil. Wentz: W. H., Jr. and H. C. Seetharam; 1974.

[25] NASA-CR-3687. Additional Flow Field Studies of the GA(W)-1 Airfoil with 30-percent Chord Fowler Flap Including Slot-Gap Variations and Cove Shape Modifications. Wentz: W. H., Jr. and C. Ostowari; 1983.

[26] Liebeck Robert H. Design of Subsonic Airfoils for High Lift. Journal of Aircraft September 1978;Vol. 15(No. 9).

[27] Cavanaugh Michael A, Robertson Paul, Mason William H. Wind Tunnel Test of Gurney Flaps and T-Strips on an NACA 23012 Wing. AIAA 2007-4175 2007.

[28] Myose Roy, Papadakis Michael, Heron Ismael. Gurney Flap Experiments on Airfoils, Wings, and Reflection Plane Model. Journal of Aircraft March–April 1998;Vol. 35(No. 2).

[29] NASA-TM-4071. A Water Tunnel Study of Gurney Flaps. Neuhart: Dan H., and Odis C. Pendergraft, Jr; 1988.

[30] Vijgen Paul MH W, Howard Floyd G, Bushnell Dennis M, Holmes Bruce J. Serrated Trailing Edges for Improving Lift and Drag Characteristics of Lifting Surfaces. US-PATENT-5,088,665 1992.

[31] Mayda EA, van Dam CP, Nakafuji D. Computational Investigation of Finite Width Microtabs for Aerodynamic Load Control. AIAA-2005-1185 2005.

[32] TCDS A9EA. Viking Air Limited, Revision 15, 06/11/2012, FAA

[33] Hoerner Sighard F. Fluid-Dynamic Lift. L. Hoerner; 1965. 3–6.

[34] Hoerner Sighard. Aerodynamic Shape of the Wing Tips. Technical Report 5752, USAF, Air Material Command, Wright-Patterson Air Force Base 1949.

[35] Hoerner Sighard F. Fluid-Dynamic Drag. L. Hoerner; 1965.

[36] NASA TM-88534. The Aerodynamic Optimization of Wings at Subsonic Speeds and the Influence of Wingtip Design. Zimmer H; 1987. Dissertation 1983.

[37] Rokhsaz Kamran. A Brief Survey of Wing Tip Devices for Drag Reduction. SAE 1993;932574.

[38] Naik DA, Ostowari C. Effects of Nonplanar Outboard Wing Forms on a Wing. Journal of Aircraft 1990;Vol. 27:117–22.

[39] NACA TN-69. An Investigation on the Effect of Raked Wing Tips. Norton: F. H.; 1921.

[40] Diaz Mario R Avila, Yechout Thomas R, Bryant Elaine M. 'The Rakelet' - A Wingtip Modification Approach to Improve Endurance. Range and Fuel Savings, AIAA 2012-0669 2012.

[41] NACA TR-201. The Effects of Shielding the Tips of Airfoils. Reid: Elliott G; 1925.

[42] NACA TN-2440. Wind-Tunnel Investigation and Analysis of the Effects of End Plates on the Aerodynamic Characteristics of an Unswept Wing. Riley: Donald R; 1951.

[43] NACA TM-856. The Lift Distribution of Wings with End Plates. Mangler: W; 1938.

[44] NACA R-267. Drag of Wings with Endplates. Hemke: Paul E; 1927.

[45] Weber J. Theoretical Load Distribution on a Wing with Vertical Plates. A.R.C. R.&M. No. March 1956;2960.

[46] McCormick Barnes W. Aerodynamics, Aeronautics, and Flight Mechanics. John Wiley & Sons; 1979.

[47] NASA SP-2003-4529. Concept to Reality. Chambers: Joseph R.; 2003.

[48] NASA TN D-8260. A Design Approach and Selected Wind-Tunnel Results at High Subsonic Speeds for Wing-Tip Mounted Winglets. Whitcomb: Richard T; 1976.

[49] NASA TN D-8264. A High Subsonic Speed Wind Tunnel Investigation of Winglets on a Representative Second-Generation Jet Transport Wing. Whitcomb: Richard T., S. G. Flechner, and P. F. Jacobs; 1976.

[50] NASA SP-4303. On the Frontier - Flight Research at Dryden, 1946-1981. Hallion: Richard P., NASA History Series; 1984.

[51] NASA CR-3748. DC-10 Winglet Flight Evaluation – Summary Report. Taylor: A. B.; 1983.

[52] Anonymous. Blended Winglets for Improved Airplane Performance. Boeing Publication.

[53] NASA TR-R-139. The Theory of Induced Lift and Minimum Induced Drag of Nonplanar Lifting Systems. Cone: Clarence D. Jr; 1962.

[54] NASA TR-R-152. The Aerodynamic Design of Wings with Cambered Span having Minimum Induced Drag. Cone: Clarence D. Jr; 1960.

[55] Jones Robert T. Minimizing Induced Drag. Soaring and Motorgliding October 1979;Vol. 43(Number 10).

[56] Thomas Fred. Fundamentals of Sailplane Design. College Park Press; 1999.

[57] Barnes Chris H. Handley Page Aircraft since 1907, Putnam, 1976.

CHAPTER 11

The Anatomy of the Tail

OUTLINE

11.1 Introduction — 460
   11.1.1 The Content of this Chapter — 461
   11.1.2 The Process of Tail Sizing — 461

11.2 Fundamentals of Static Stability and Control — 462
   11.2.1 Fundamentals of Static Longitudinal Stability — 463
      Requirements for Static Longitudinal Stability — 463
      Forces and Moments for Longitudinal Equilibrium — 466
      Common Expressions for the Aerodynamic Coefficients — 466
   11.2.2 Modeling the Pitching Moment for a Simple Wing-HT System — 466
   11.2.3 Horizontal Tail Downwash Angle — 467
      Downwash per the Momentum Theory — 467
   11.2.4 Historical Values of $C_{m\alpha}$ — 468
   11.2.5 Longitudinal Equilibrium for Any Configuration — 468
   11.2.6 The Stick-fixed and Stick-free Neutral Points — 472
      Derivation of Equation (11-26) — 473
   11.2.7 Fundamentals of Static Directional and Lateral Stability — 475
      Roll or Bank — 475
      Yaw — 475
      Slipping or Sideslip — 475
   11.2.8 Requirements for Static Directional Stability — 476
   11.2.9 Requirements for Lateral Stability — 477
   11.2.10 Historical Values of $C_{n\beta}$ and $C_{l\beta}$ — 477
   11.2.11 The Dorsal Fin — 477
   11.2.12 The Ventral Fin — 480
   11.2.13 Tail Design and Spin Recovery — 482

11.3 On the Pros and Cons of Tail Configurations — 483
   11.3.1 Conventional Tail — 483
   11.3.2 Cruciform Tail — 486
   11.3.3 T-tail — 486
   11.3.4 V-tail or Butterfly Tail — 489
      Simplified Theory of V-tails — 493
   11.3.5 Inverted V-tail — 493
   11.3.6 Y-tail — 493
   11.3.7 Inverted Y-tail — 494
   11.3.8 H-tail — 494
   11.3.9 Three-surface Configuration — 495
   11.3.10 A-tail — 495
   11.3.11 Twin Tail-boom or U-tail Configuration — 496
      Inverted U-tail — 496
   11.3.12 Canard Configuration — 496
   11.3.13 Design Guidelines when Positioning the HT for an Aft Tail Configuration — 497

11.4 The Geometry of the Tail — 499
   11.4.1 Definition of Reference Geometry — 499
   11.4.2 Horizontal and Vertical Tail Volumes — 500
   11.4.3 Design Guidelines for HT Sizing for Stick-fixed Neutral Point — 501
   11.4.4 Recommended Initial Values for V_{HT} and V_{VT} — 502

11.5 Initial Tail Sizing Methods — 502
   11.5.1 Method 1: Initial Tail Sizing Optimization Considering the Horizontal Tail Only — 503
      Assumptions — 503
      Determination of Tail Arm for a Desired V_{HT} Such that Wetted Area is Minimized — 503
      Derivation of Equation (11-40) — 507

   11.5.2 Method 2: Initial Tail Sizing Optimization Considering the Vertical Tail Only — 510
      Assumptions — 510
      Determination of Tail Arm for a Desired V_{VT} such that Wetted Area is Minimized — 511
      Derivation of Equation (11-48) — 511

11.5.3 Method 3: Initial Tail Sizing Optimization Considering Horizontal and Vertical Tail	513	Exercises	517
Assumptions	513	Variables	517
Determination of Tail Arm for the Desired V_{HT} and V_{VT}	513	References	519
Derivation of Equation (11-56)	514		

11.1 INTRODUCTION

The purpose of the tail is to provide the aircraft with a means of stability and control. As such, it is one of the most important components of the entire airplane. The aircraft designer must determine not only its size, location and configuration, but also the type of controls it will feature. Should the controls be a deflectable flap or an all-movable lifting surface? If the choice is a flap, then what should be its dimensions? If all-movable, where should its hingeline be placed? In this text, the word "tail" refers to any configuration used to balance an airplane, and may be used with a conventional tail aft configuration, a canard, a three-surface configuration, and any other found suitable for that purpose, although an effort will be made to make the discussion clear. The word includes both the horizontal and vertical stabilizing surfaces, however, a *horizontal tail* (HT) refers to a surface intended to control the pitch of the aircraft, and *vertical tail* (VT) refers to one intended to control the yaw (and sometimes roll).

Stability and control theory shows that for some suitably small *AOA* and *AOY*, the pitch can be decoupled from the roll and yaw; in other words: the pitch can be considered independent of roll and yaw. This offers a great convenience to the stability analyst. However, roll and yaw are coupled and have to be treated as such. Yaw will generate a roll and vice versa. Generally, the roll is controlled using *ailerons*, pitch is controlled using an *elevator*, and yaw is controlled using a *rudder*. In this section, we will only focus on the elevator and the rudder and the control surfaces to which they connect: the *horizontal stabilizer* and the *fin*. Controls are detailed in Chapter 23, *Miscellaneous design notes*.

Consider the airplane in Figure 11-1. A conventional stability coordinate-system consisting of x-, y-, and z-axes have been superimposed on the figure. It should be mentioned that in stability and control theory the positive direction of the z-axis always points down, rather than up. The rotation about the x-axis is called *roll*, the rotation about the y-axis is called *pitch*, and the rotation about the z-axis is called *yaw*. Using the right-hand rule of rotation, a positive roll angle is one in which the right wing moves down and the left one up. Similarly, a positive pitch angle is nose-up and negative is nose pitched down. A positive yaw angle is one that moves the nose to the right and negative to the left. These positive rotations are indicated in Figure 11-1. A positive yaw angle is one which would rotate the nose to the right. A positive rudder deflection will generate a positive side force (in the direction of positive y). This means the rudder trailing edge will deflect to the left and a negative (nose left) yawing moment will be generated. A positive elevator deflection (trailing edge down) will produce an increase in lift and pitch the nose down. It is important to keep these conventions in mind for the discussion that follows.

It is important to realize that pitch- and yaw-control can be achieved by other means than just using an elevator and a rudder. For instance, many military fighters combine ailerons and elevator in an all-movable control surface that is located behind the wing, called an *elevon*. Flying wings often combine the rudder and aileron in a clamshell like control surface on the outboard wing. During flight the upper clamshell is deflected a few degrees trailing edge up (TEU), and the lower one trailing edge down (TED). This generates drag at the wing tips that creates directional stability (this is evident from pictures of the B-2 Spirit Bomber). Such devices are beyond the current discussion. Here we will only consider the more conventional shape, which can be extended to canards and V-tails, although care must be exercised when considering those.

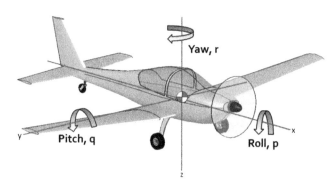

FIGURE 11-1 Definition of the x-, y-, and z-axis for an aircraft.

Before any stability and control analysis can begin, the designer must select the type of tail configuration. In other words, will the airplane feature a conventional tail, or T-tail, or other kind of a tail design? Refer to Section 11.3, *On the pros and cons of tail configurations*, for a discussion on the different kinds of tail surfaces. An imperative element of that decision involves determining how far from the center of gravity (or any other datum point) the tail surfaces should be placed and how large their lifting areas should be. Initial sizing schemes are introduced in this section. The reader should be mindful that ultimately, dynamic stability and handling of the aircraft (spin recovery) will be the final arbiters, but one must begin somewhere and the methods presented herein generally yield a good initial geometry. However, ultimately the aircraft designer should perform dynamic stability analysis. It turns out that an aircraft may be statically stable, but dynamically unstable. A proper dynamic stability analysis will reveal shortcomings and enable the designer to adjust the size or location of the HT and VT (among some other geometric features) such that the aircraft is dynamically stable as well.

11.1.1 The Content of this Chapter

- **Section 11.2** presents a general discussion of static stability and control. In addition to defining fundamental stability concepts, it discusses trends in longitudinal and lateral/directional stability derivatives.
- **Section 11.3** presents a general discussion of the pros and cons of different tail configurations.
- **Section 11.4** defines general horizontal and vertical tail geometry formulation.
- **Section 11.5** presents three methods to tailor the tail arm and the vertical tail surface based on the minimum wetted area of a fuselage that has the shape of a frustum. The three methods use the horizontal tail volume, the vertical tail volume, and a combination of the two.

11.1.2 The Process of Tail Sizing

The concept *tail sizing* refers to the process required to determine the size, shape, and three-dimensional positioning of the stabilizing surfaces. The process of defining the horizontal and vertical tail geometry is accomplished in the following steps:

Step 1: Determine which of the tail configurations in Section 11.3, *On the pros and cons of tail configurations*, suit the project. Ensure there is a deep and thorough (possibly non-mathematical) evaluation of possible pros and cons, in accordance with the discussion in this chapter, realizing that there may be additional benefits and flaws of each design, not mentioned here. Additionally, there may be other configurations that should be considered besides those presented here. Aesthetics should be seriously considered in this design step. All of the tail configurations presented here will work and have been used on actual aircraft, albeit in different capacities. The primary difference is in the cost of implementation, such as weight, mechanical complexity, and efficiency, to name a few. Unless there is a specific reason for choosing one tail configuration over the others, it will be very helpful to the decision process to draw the proposed vehicle with different tails to help with the tail options, as shown in the examples of Section 11.5, *Initial tail sizing methods*.

Step 2: Estimate the geometry based on historical data, such as shown in this chapter. This involves estimating the appropriate horizontal and vertical tail volume per Section 11.4.2, *Horizontal and vertical tail volumes*, and uses these to perform an initial tailoring of the stabilizing surfaces. It must be made clear that this is *only* an initial estimate based on historical trends. The geometry and characteristics of the airplane, such as the power plant and handling characteristics (both treated elsewhere), will ultimately dictate modification to this initial estimate and this is discussed in Step 3.

Step 3: Once the airplane takes shape, a far more sophisticated resizing, or at least a revision, of the geometry obtained in Step 2 must take place. Such reshaping will depend on a number of factors that do not strictly belong in this chapter. The following bullets constitute design guidelines that all new aircraft should be capable of demonstrating:

- The HT must be capable of trimming the airplane at low airspeeds at a forward CG location. This means airspeed at least as low as 1.2 V_S in the landing configuration.
- The HT must be capable of trimming the airplane at high airspeeds at an aft CG location. This means the pilot should be able to trim the aircraft for *cruise let-down*, i.e. to begin and maintain descent at airspeed as high as cruising speed plus 5 to 10 KTAS.
- The HT and VT must be of a low AR to reduce the risk of tail stall and yet be suitably large AR to make the stabilizing surfaces responsive to AOA or AOY changes. Typically this means that the AR of the HT should be $3 < AR_{HT} < 5$ and the VT should be $0.9 < AR_{VT} < 2$.
- The VT must provide means to prevent rudder locking. This usually means the addition of a dorsal fin (see Section 11.2.11, *The dorsal fin*).

- The HT and VT control surfaces must provide enough authority to allow the airplane to be controlled during demanding maneuvers such as balked landing and crosswind landing without excessive control surface deflections.
- The HT must allow the airplane to be stalled. This is imperative because the stalling speed (say +1 knot) is truly the slowest speed the airplane is capable of. If the airplane cannot stall because of limited elevator authority, its minimum speed will be higher than the stalling speed and this can result in higher approach speeds, demand longer runways, and even compromise the certifiability of the aircraft. This may happen if the minimum speed is higher than the regulatory limits (i.e. 61 KCAS for 14 CFR Part 23, or 45 KCAS for LSA).
- The HT and VT must result in stability derivatives such as $C_{m\alpha}$, C_{mde}, C_{mq}, C_{yb}, C_{nb}, C_{ndr}, C_{nr}, and others that ensure the airplane is naturally statically and dynamically stable. All 14 CFR Part 23 aircraft are required to be statically stable and demonstrate stable longitudinal short period and lateral directional oscillation (also known as Dutch roll) modes. However, the designer should strive to make the remaining dynamic modes convergent. Strictly speaking, GA aircraft do not need fly-by-wire control systems because by law they *have to be stable*. This, however, does not preclude the development of such systems for GA aircraft as these may offer supplemental benefits.
- The HT and VT should not have detrimental impact on the non-linear behavior of the above stability derivatives at high *AOAs* and *AOYs*. For instance, the derivatives should not acquire values that render the airplane unrecoverable in spins or deep-stall.
- The HT and VT must be designed for minimum structural weight, with the designer being cognizant of the manufacturability and aeroelastic consequences of a particular design.
- The control system should be simple and reliable and it should not require excessive control surface deflection to maneuver the airplane, even in demanding maneuvers. Large control surface deflection will cause the surface to stall, sharply lowering the control authority.
- If the control system is manual, it should not require excessive control forces to deflect the surfaces throughout the operating envelope of the aircraft. Consult 14 CFR Part 23.155, Elevator control force in maneuvers and 23.397, Limit control forces and torques, regarding regulatory limits.
- The designer must be cognizant of other operational limitations. For instance, excessive length of the tail arm may interfere with T-O rotation, subject the structure to high stresses, and lower the flutter speed of the airplane. Excessively short tail arms will require high deflection angles of control surfaces and may result in poor handling characteristics due to unacceptably low pitch damping.

Many of the above bullets are not treated directly in this section. Among those are the evaluation of dynamic stability and compliance with the design checklist of Section 23.3, *Preliminary aircraft design checklist*. At any rate, provided the designer does not violate other requirements, drag should always be minimized. Sections 11.4 and 11.5 develop a few techniques intended as a "first stab" tail sizing methods that minimize the wetted area of the empennage in order to reduce skin friction drag.

11.2 FUNDAMENTALS OF STATIC STABILITY AND CONTROL

The design of the tail is essential to the safe operation of the airplane. The design is highly dependent on the scientific discipline called *stability and control theory*. In this context, the science of *mechanics* is usually broken into two fields: *statics* and *dynamics*. Statics considers the equilibrium of matter for which linear and angular acceleration is zero, while dynamics studies the equilibrium of matter that undergoes linear and angular accelerations. Aircraft stability and control is a field within mechanics that applies specifically to vehicles subject to six degrees of freedom motion (three linear and three rotational).

The analysis of the total stability of an airplane is performed by considering the contribution of a number of components. Thus, there is a contribution due to the wings, HT, fuselage, landing gear, and power plant. There can be a further breakdown based on specific components — for instance, the contribution of the wing is broken down into that of the main wing element, flap, leading edge devices, and so forth. The magnitude of these is then summed along and about the three axes, which indicates the instantaneous stability and motion of the aircraft. An airplane can maintain steady unaccelerated flight only when the sum of all forces and moments about its CG vanishes.

Static and dynamic stability analyses revolve around developing the equations of motion, evaluating the component contributions, and using these to evaluate a number of static and dynamic stability characteristics. The standard stability coordinate system is defined so the x-axis points toward the nose, the y-axis points to the right wingtip, and the z-axis points downward. Positive rotations are defined according to the right-hand rule.

11.2.1 Fundamentals of Static Longitudinal Stability

Requirements for Static Longitudinal Stability

Consider the airplane in Figure 11-2. The image to the left shows it at a high *AOA* and the right one at a low *AOA*. The figure helps build an understanding of what is meant by longitudinal stability. In the left image, the horizontal tail (HT) generates a lift force, L_{HT}, which points upward and, thus, tends to reduce the *AOA* by lowering the nose. Using the standard stability coordinate system the resulting moment has a negative magnitude. This means that grabbing around the y-axis with the right hand to generate this nose-down rotation requires the thumb to point in the negative y-direction. To pitch the nose up requires the thumb to point in the positive y-direction.

The right image of Figure 11-2 shows the opposite. Due to the low *AOA*, the HT is generating lift in the downward direction causing a tendency to increase the *AOA*. This requires the moment to have a positive value. This means that somewhere between the two extremes is an *AOA* for which there is no tendency for the HT to increase or decrease the *AOA*. This is the *trim AOA*. An airplane whose stabilizing surface (here the HT) generates enough lift force to force the aircraft to a specific trim *AOA* is called a *stable aircraft*.

These two conditions in Figure 11-2 have been plotted in Figure 11-3. The conditions consist of $\alpha > 0$ and $M < 0$ in the left image and $\alpha < 0$ and $M > 0$ in the right image. The graph shows clearly that in order for the aircraft to be stable, the pitching moment curve must necessarily have a negative slope. This slope is denoted by the symbol $C_{m\alpha}$. Additionally, in order to be able to trim the airplane

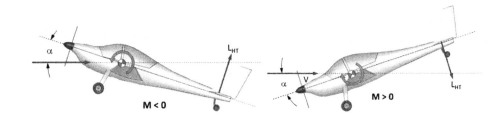

FIGURE 11-2 The generation of longitudinal stability. The left image has $\alpha > 0$ and $M < 0$ (the moment due to L_{HT} tends to pitch the nose down). The right image has $\alpha < 0$ and $M > 0$ (the moment due to L_{HT} tends to pitch the nose up). This airplane is statically stable.

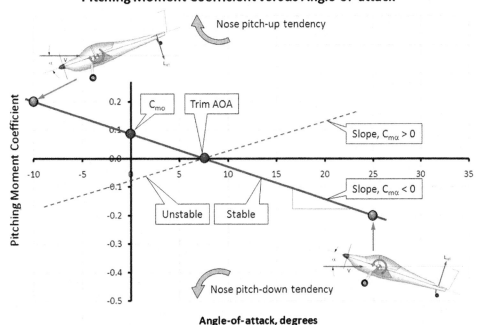

FIGURE 11-3 The pitching moment curve must have a negative slope for the airplane to be stable and intersect the y-axis (C_m-axis) at a positive value (C_{mo}) in order to be trimmable at a positive *AOA* (necessary condition to generate lift that opposes weight).

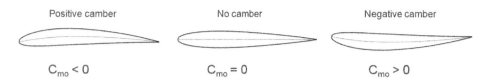

FIGURE 11-4 The requirement for trimmability requires an airfoil to feature a negative camber.

at an *AOA* that generates positive lift, the intersection to the *y*-axis (C_m-axis), denoted by C_{m0}, must be larger than zero. Mathematically, this is written as follows:

Requirement for static stability:

$$C_{m_\alpha} = \frac{\partial C_m}{\partial \alpha} < 0 \qquad (11\text{-}1)$$

Requirement for static trimmability:

$$C_{m_o} > 0 \qquad (11\text{-}2)$$

If these conditions are satisfied, then there exists an *AOA* > 0 for which the C_m is equal to zero. The importance of *AOA* > 0 is that the vehicle can generate lift in the opposite direction of the weight and simultaneously be statically stable — a necessary condition for flying in the absence of stability augmentation systems (SAS). As stated earlier, the former condition is the slope of the pitching moment curve, and is short-hand for:

$$C_{m_\alpha} = \frac{\partial C_m}{\partial \alpha} = \frac{\partial C_m}{\partial C_L}\frac{\partial C_L}{\partial \alpha} = \frac{\partial C_m}{\partial C_L} C_{L_\alpha} \qquad (11\text{-}3)$$

Note that the subscripts for the moment coefficients differ from those of the force coefficients, where capitalization is used to distinguish between two- and three-dimensional force coefficients. Here, the lowercase subscript of C_m can refer to the pitching moment of both a two-dimensional geometry, such as an airfoil, and a three-dimensional geometry, such as a wing. The distinction must be made by context. While it can be confusing, this is done by convention, as the purpose here is not to introduce a new format, but useful formulation. Thus, C_m can refer to the pitching moment of the airfoil or the airplane, separable by context only.

The condition of trimmability for an airfoil can be achieved by selecting the proper camber (see Figure 11-4). The positive camber has a negative C_{mo} (see Example 11-1), while the opposite is true for one that has a negative camber. In the case of airplanes, a positive C_{mo} is created by the addition of a stabilizing surface like a HT and by equipping it with an elevator. The elevator allows the moment curve to be moved up or down at will.

To understand why these conditions are necessary consider again Figure 11-3. The solid and dashed lines intersect the horizontal axis approximately at the α_{trim} = 7.5°. Focusing on the solid line first, consider the aircraft being perturbed from α_{trim} to, say, 5°. This implies the nose is lower than before and results in a nose-up tendency that will bring the airplane back to α_{trim}. Similarly, should the perturbation result in a slightly higher *AOA*, say 10°, the opposite happens and this brings the nose back down to α_{trim}. Alternatively, if the airplane is statically unstable, as represented by the dashed line, a perturbation that lowers the nose will be accompanied by a tendency to reduce the *AOA* further. A perturbation resulting in a higher *AOA* will similarly result in a tendency to increase it further.

EXAMPLE 11-1

Determine the pitching moment curve for the NACA 4412 airfoil in Figure 11-5, if its lift and pitching moment coefficients about the quarter chord are given by the following cubic approximations:

$$C_l(\alpha) = 0.40575 + 0.11885 \cdot \alpha + 0.00028215 \cdot \alpha^2 - 0.00017205 \cdot \alpha^3$$

$$C_m(\alpha) = -0.10076 + 0.001121 \cdot \alpha + 0.00025507 \cdot \alpha^2 - 0.000012469 \cdot \alpha^3$$

The weight, *W*, is placed at a distance *h* from the leading edge. Note that these approximations are shown in the left graph of Figure 11-6. Plot the resulting C_m curve of the airfoil for the following CG locations; $h = 0, 0.1, 0.25,$ and 0.3.

Solution

First we note the lift of the airfoil is given by:

$$L = qC \cdot C_l$$

Similarly, the moment is given by:

$$M_0 = qC^2 \cdot C_l$$

EXAMPLE 11-1 (cont'd)

where q is the dynamic pressure and C is the chord length. The sum of the moments about the CG (assuming clockwise rotation about the CG is positive) is therefore:

$$M = M_0 - L \cdot (C/4 - h)$$

Expanding leads to:

$$\begin{aligned} M &= M_0 - L \cdot (C/4 - h) \\ &= qC^2 \cdot C_m(\alpha) - qC \cdot C_l(\alpha) \cdot (C/4 - h) \end{aligned}$$

Convert to coefficient form by dividing through by qC:

$$C_m = C \cdot C_m(\alpha) - C_l(\alpha) \cdot (C/4 - h)$$

Insert the cubic splines:

$$\begin{aligned} C_m = C \cdot \big[&-0.10076 + 0.001121 \cdot \alpha + 0.00025507 \cdot \alpha^2 \\ &- 0.000012469 \cdot \alpha^3 \big] + \vec{c} \\ - \big[&0.40575 + 0.11885 \cdot \alpha + 0.00028215 \cdot \alpha^2 \\ &- 0.00017205 \cdot \alpha^3 \big] \cdot (C/4 - h) \end{aligned}$$

The resulting plot is shown in the right graph of Figure 11-6. It can be seen that when the weight is at the leading edge ($h = 0$) the airfoil has the greatest stability of the positions plotted (steepest negative slope). When the weight is at or behind the quarter chord ($h = 0.25$ and 0.3), the airfoil is neutrally stable or unstable, respectively.

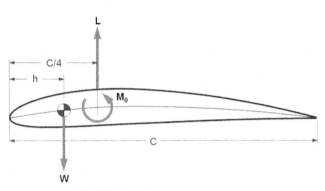

FIGURE 11-5 NACA 4412 airfoil.

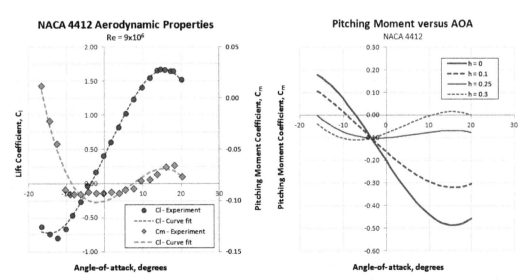

FIGURE 11-6 Experimental and curvefit lift and pitching moment coefficients for the NACA 4412 airfoil (left). Resulting pitching moment curves for the four positions of the CG (right).

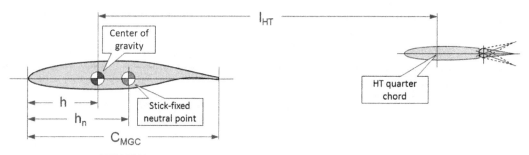

FIGURE 11-7 Dimensions pertinent to the pitching moment curve.

Forces and Moments for Longitudinal Equilibrium

The longitudinal equilibrium requires lift to be equal to weight, drag to be equal to thrust, and pitching moment to be equal to zero. Therefore, the following must hold for longitudinal flight conditions:

Along x-axis:

$$D_W + D_{HT} + D_{VT} + D_{FUS} + D_{LDG} + \cdots - T\cos\varepsilon = 0 \quad (11\text{-}4)$$

Along z-axis:

$$L_W + L_{HT} + L_{VT} + L_{FUS} + L_{LDG} + \cdots + T\sin\varepsilon = W \quad (11\text{-}5)$$

Moment about y-axis (pitch):

$$M_W + M_{HT} + M_{VT} + M_{FUS} + M_{LDG} + M_T + \cdots = 0 \quad (11\text{-}6)$$

where the subscripts refer to the wing, HT, VT, fuselage, landing gear, and so on, respectively.

Common Expressions for the Aerodynamic Coefficients

In the formulation that follows, the lift, drag, and moment for the entire aircraft are expressed in terms of coefficients that are linear combinations of various contributions. Note that the moment is always taken about the CG (refer to the variable list at the end of the section for the descriptions):

$$C_L = C_{L0} + C_{L\alpha}\alpha + C_{L\beta}\beta + C_{L\delta_e}\delta_e + C_{L\delta_f}\delta_f$$
$$+ C_{L\delta_{spoiler}}\delta_{spoiler} + \cdots \quad (11\text{-}7)$$

Similarly, the lift coefficient of the tail is given by:

$$C_{LHT} = C_{L0_{HT}} + C_{L\alpha_{HT}} \cdot \alpha_{HT} + C_{L_{\delta_e}} \cdot \delta_e + \cdots \quad (11\text{-}8)$$

The drag coefficient can be represented as follows:

$$C_D = C_{Dmin} + C_{D\alpha}\alpha + C_{D\beta}\beta + C_{D\delta_e}\delta_e + C_{D\delta_f}\delta_f$$
$$+ C_{D\delta_{spoiler}}\delta_{spoiler} + \cdots \quad (11\text{-}9)$$

The pitching moment coefficient can be represented as shown below:

$$C_m = C_{mo} + C_{m\alpha}\cdot\alpha + C_{m\beta}\cdot\beta + C_{m\delta_f}\delta_f + \cdots \quad (11\text{-}10)$$

11.2.2 Modeling the Pitching Moment for a Simple Wing-HT System

The solid line in Figure 11-3 is the *pitching moment curve* and is denoted by C_m. It is a function of the AOA and the location of the center of gravity. Consider Figure 11-7, which shows a simple three-dimensional system consisting of a wing and a HT. Shown are two airfoils that represent the mean geometric chord (MGC) of the wing and the HT. Furthermore, it positions the center of gravity (CG) and the stick-fixed neutral point (to be defined later), denoting the former using the term h and the latter h_n. It is given by the follow expression:

$$C_m = C_{mo} + C_{m\alpha}\cdot\alpha = C_{mo} + C_{L\alpha}\left(\frac{h - h_n}{C_{MGC}}\right)\alpha \quad (11\text{-}11)$$

where

h = physical location of the CG
h_n = physical location of the stick-fixed neutral point

In stability and control theory, the quantity $(h - h_n)/C_{MGC}$ is termed *Static Margin*. It is of utmost importance in the discussion that follows. It can be seen that if the location of the CG moves behind the neutral point, the quantity $(h - h_n)$ of Equation (11-11) will acquire a positive value. This means the airplane is unstable — at high AOAs, it will tend to increase its AOA further rather than reducing it and the opposite at low AOAs. This renders the aircraft uncontrollable for human pilots, although it can be controlled by a computer-controlled flight control system (fly-by-wire or fly-by-light). GA aircraft must be designed so they are naturally statically and dynamically stable, although the authorities have proven flexible in the certification of airplanes whose phugoid and spiral stability modes are divergent.

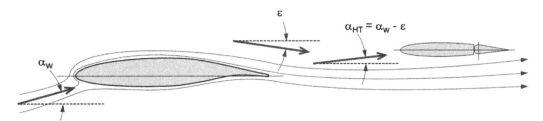

FIGURE 11-8 A schematic showing how the downwash angle, ε, affects the AOA on the horizontal tail.

11.2.3 Horizontal Tail Downwash Angle

If the wing is placed ahead of the horizontal tail (as is common for most airplanes) it will be subject to downwash that changes the AOA on the tail. For instance, consider the main wing at an AOA of 10°. At first glance, one might assume the horizontal tail is also at an AOA of 10°, but this is not the case. The downwash from the main wing will reduce the AOA on the HT, so it might only be 5°. Clearly this will affect the stability of the airplane and, thus, must be accounted for.

Downwash per the Momentum Theory

The momentum theory represents the simplest method to predict the downwash behind the wing. Its primary shortcoming is it assumes an elliptical planform and returns an average value whereas in real flow the magnitude varies with position in space. Its primary advantage is ease of estimation. Its results are generally acceptable for conceptual and preliminary design; however, using the method to position the height of the HT with respect to the wing is not reliable. The downwash is expressed using the following linear relation:

$$\varepsilon = \varepsilon_0 + \frac{d\varepsilon}{d\alpha}\alpha \quad (11\text{-}12)$$

where ε_0 is the residual downwash (when $\alpha = 0$) which is only present if the wing features cambered airfoils. The derivative $d\varepsilon/d\alpha$ indicates the change in downwash with AOA. In the absence of more sophisticated analysis, the downwash can be estimated using the momentum theorem of Section 8.1.8, *The generation of lift*. This allows the downwash angle to be presented in terms of the $C_{L\alpha}$ of the wing per Equation (8-22), repeated below for convenience:

$$\varepsilon \approx \frac{2C_{LW}}{\pi AR} \quad (8\text{-}21)$$

where C_{LW} is the lift coefficient of the wing and not the entire airplane, but the downwash is caused by the wing primarily. The units for the angle are radians. Once the downwash angle is known, it is easy to determine the AOA on the horizontal tail, using the schematic of Figure 11-8.

A more common way of presenting the downwash angle is to write:

$$\varepsilon = \varepsilon_0 + \frac{d\varepsilon}{d\alpha}\alpha = \frac{2C_{LW}}{\pi AR} = \frac{2}{\pi AR}\left(C_{L0_W} + C_{L\alpha_W}\alpha\right) \quad (11\text{-}13)$$

where, similarly, $C_{L\alpha W}$ is the lift curve slope of the wing. Therefore, we can define the residual and AOA-dependent downwash as follows:

$$\varepsilon_0 = \frac{2C_{L0_W}}{\pi AR} \text{ and } \frac{d\varepsilon}{d\alpha} = \frac{2C_{L\alpha W}}{\pi AR} \quad (11\text{-}14)$$

It is evident from the derivation of the above expressions that they represent the average for the entire column of air being deflected downward to generate lift. In real flow, the size and position of the HT in space will cause the downwash angle to vary along its span. Inserting Equation (11-14) into (11-13) yields:

$$\alpha_{HT} = \alpha_W - \varepsilon = \alpha_W\left(1 - \frac{2C_{L\alpha W}}{\pi AR}\right) - \frac{2C_{L0_W}}{\pi AR} \quad (11\text{-}15)$$

However, once an angle-of-incidence (AOI) is introduced to the wing and HT, things get a bit more complicated (see Figure 11-9). Note that a positive i_W and negative i_{HT} are shown as most aircraft feature that geometry. Increasing the wing AOI is akin to increasing the overall AOA on the wing. When adding i_{HT} to the sum (as shown), it will reduce the AOA on the HT. Therefore, the AOA of the HT can be summarized as follows:

$$\alpha_{HT} = \alpha_W - i_W - \varepsilon + i_{HT} \approx \alpha_W\left(1 - \frac{2C_{L\alpha W}}{\pi AR}\right) - \frac{2C_{L0_W}}{\pi AR} \\ - i_W + i_{HT} \quad (11\text{-}16)$$

It is convenient to define the AOA of the *wing-body* combination as shown in Figure 11-9 and write Equation (11-16) as follows:

$$\alpha_{HT} = \alpha_{WB} - \varepsilon + i_{HT} \quad (11\text{-}17)$$

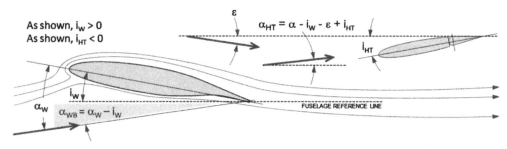

FIGURE 11-9 A schematic showing how introducing an *AOI* of the wing and HT the downwash angle, ε, affects the *AOA* on the horizontal tail.

11.2.4 Historical Values of $C_{m\alpha}$

As stated earlier, the purpose of the HT and VT is to make the airplane controllable and provide it with acceptable static stability. The term "acceptable static stability" is somewhat nebulous and depends on the class of airplanes being considered. It refers to the magnitude of the slope of the pitch stability, $C_{m\alpha}$. Thus, transport aircraft usually have "high" pitch stability, which is reflected in a low value of $C_{m\alpha}$, often in the -1.2 to -1.7 per radian range. GA aircraft are typically in the -0.6 to -1.0 per radian range. On the other hand, the modern fighter aircraft is purposely unstable ($C_{m\alpha} > 0$).

During the design of the aircraft it is important to know what longitudinal stability derivatives to aim for. As usual, the history of aviation presents us with a large number of candidate values. Table 11-1 lists a number of longitudinal stability derivatives, gathered from Refs 1, 2, 3, and 4, and the author's personal notes. A selection of the airplanes is shown plotted in Figure 11-10.

A high level of stability is a good quality in a transport-class aircraft, as the pilot (or the autopilot) will not have to work hard to maintain steady level flight. This is also a requirement by 14 CFR Part 23.173 (see introduction to this section) for GA aircraft and 25.173 for commercial aircraft. The primary drawback is reduced maneuverability (the rate at which the aircraft's orientation can be changed), but this is a secondary priority for transport aircraft — they only have to be maneuverable enough to allow for safe flying. Fighter aircraft, on the other hand, represent the other extreme. A fighter must be highly maneuverable as this, in addition to its energy level, largely dictates whether it can beat an opponent in a dogfight. Such aircraft have very high wing-loading and a high moment of inertia. If made too stable, the resulting aircraft would be unacceptably sluggish. The solution is a reduction in the static stability as this increases the maneuverability. In fact, the modern fighter is longitudinally unstable for this reason. The drawback is that it requires a computer-managed control system to control such an aircraft (fly-by-wire or the more modern fly-by-light). One important consequence of supersonic flight is that the center of lift moves aft, and this increases static stability. Thus, the modern fighter aircraft is longitudinally unstable while subsonic, but stable when supersonic. The static stability of GA aircraft fits snugly between these two extremes.

Note that although this book focuses on GA aircraft, it is still of interest to see how the derivatives are affected by high Mach numbers. For this reason, high speed values are included. It is important to realize that the $C_{m\alpha}$ is not truly a constant. It changes with Mach number because of compressibility effects, as well as when the airplane's configuration changes as a result of retracting landing gear and flaps. Compressibility effects are shown for the F-104 Starfighter, a third-generation fighter, as the solid line in Figure 11-10, which explains why its stability derivative, $C_{m\alpha}$, is less than zero over its Mach range. Marginally stable and unstable fighters emerged first with fourth-generation fighter aircraft. Also, note the range of stability of the F-4 Phantom fighter (also third-generation) which is marginal at $M = 0.2$ (landing configuration), but becomes noticeably stable at $M = 2$. Also notice how the Mach number changes the longitudinal stability of commercial jetliners, like the Boeing 727, 747, and the Convair CV-880.

11.2.5 Longitudinal Equilibrium for Any Configuration

One of the most important benefits of the longitudinal stability equations (see Equations (11-4) through (11-6)) is their use for determining the static stability. The following formulation is used to determine the *AOA* (α), elevator deflection (δ_e), and thrust (T) required for steady level flight. It is applicable to almost any configuration, as long as the stability derivatives are determined correctly. Converting these equations to a coefficient form yields:

Along x-axis:

$$C_{D_{min}} + C_{D_\alpha}\alpha + C_{D_{\delta_e}}\delta_e + C_{D_{\delta_f}}\delta_f - \frac{T\cos\varepsilon_T}{qS} = 0 \quad (11\text{-}18)$$

Along z-axis:

$$C_{L_0} + C_{L_\alpha}\alpha + C_{L_{\delta_e}}\delta_e + C_{L_{\delta_f}}\delta_f + \frac{T\sin\varepsilon_T}{qS} = \frac{W}{qS} \quad (11\text{-}19)$$

TABLE 11-1 Pitch Stability Derivatives for Selected Aircraft

		Longitudinal Stability for Selected Aircraft						
Make	Model	M	AR	C_{MGC}, ft	b, ft	S, ft^2	$C_{m\alpha}$	Reference
Bede	BD-5A*	0.27	6.76	2.60	17	42.78	−0.830	Author
Beech	D-18*	0.23	6.51	8.89	47.7	349	−1.748	Author
	V-35 Bonanza*	0.23	6.10	5.49	33.5	184	−0.455	Author
Boeing	727-100*	0.76	7.07	15.28	108	1650	−1.500	6
		0.60	7.07	15.28	108	1650	−1.672	Author
	747(Pwr Appr)	0.25	6.96	27.31	195.7	5500	−1.450	2
	747	0.30	6.96	27.31	195.7	5500	−1.216	2
		0.60	6.96	27.31	195.7	5500	−0.685	2
		0.80	6.96	27.31	195.7	5500	−0.630	2
Cessna	182 Skylane	0.14	7.45	4.83	36	174	−0.900	1
Convair	CV-880	0.60	7.20	18.94	120.0	2000	−0.522	2
		0.80	7.20	18.94	120.0	2000	−0.650	2
Convair	CV-880 (Pwr Appr)	0.25	7.20	18.94	120.0	2000	−0.903	2
De Havilland Canada	DHC-6	0.22	10.00	6.5	65.0	423	−1.200	4
Lockheed	A-4 Skyhawk	0.40	2.71	−	26.5	259	−0.380	3
		0.80	2.71	−	26.5	259	−0.410	3
	C-5 (Pwr Appr)	0.22	7.75	30.1	219.2	6200	−0.827	2
	F-104 Starfighter	0.80	2.45	9.55	21.9	196	−0.901	2
		2.00	2.45	9.55	21.9	196	−0.736	2
	Jetstar	0.60	5.33	10.93	53.8	543	−0.663	2
		0.80	5.33	10.93	53.8	543	−0.717	2
	Jetstar (Pwr Appr)	0.20	5.33	10.93	53.8	543	−0.800	2
	NT-34 (Pwr Appr)	0.20	6.00	6.72	37.5	235	−0.401	2
McDonnell Douglas	F-4	0.60	2.82	16.04	38.7	530	−0.279	2
		2.00	2.82	16.04	38.7	530	−0.606	2
	F-4 (Pwr Appr)	0.21	2.82	16.04	38.7	530	−0.098	2
Ryan	Navion	0.16	6.06	5.7	33.4	184	−0.683	3
Start & Flug	H-101 Salto*	0.06	21.51	2.07	44.6	92.4	−1.206	Author

* CG at quarter-chord of the MGC.

Moment about y-axis (pitch):

$$C_{m_0} + C_{m_\alpha}\alpha + C_{m_{\delta_e}}\delta_e + C_{m_{\delta_f}}\delta_f + C_{m_T} + C_{m_{TN}} = 0 \quad (11\text{-}20)$$

where ε_T is the thrust force angle. The coefficients are the sums of the coefficients determined in the previous section. For instance, $C_{m\alpha}$ is the sum of the contribution of the wing, HT, fuselage, landing gear, etc. The same holds for the other coefficients. Therefore, it is first necessary to rearrange the above equations:

Along x-axis:

$$C_{D_\alpha}\alpha + C_{D_{\delta_e}}\delta_e - \frac{\cos\varepsilon_T}{qS}T = -C_{D_{\min}} - C_{D_{\delta_f}}\delta_f \quad (11\text{-}21)$$

Along z-axis:

$$C_{L_\alpha}\alpha + C_{L_{\delta_e}}\delta_e + \frac{\sin\varepsilon_T}{qS}T = \frac{W}{qS} - C_{L_0} - C_{L_{\delta_f}}\delta_f \quad (11\text{-}22)$$

The moment equation poses a bit of a problem when dealing with propeller normal force, reflected in the

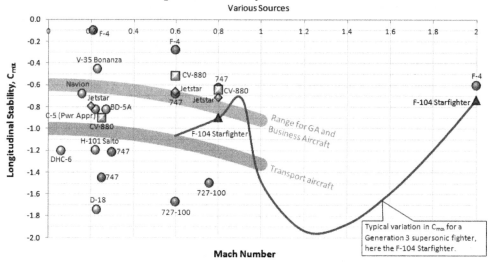

FIGURE 11-10 Trends of longitudinal stability derivatives for selected aircraft.

coefficient $C_{m_{TN}}$. The propeller normal force depends on thrust, but it is one of the unknowns. Therefore, the moment portion would have to be solved iteratively, which is not always convenient. The usual remedy is to omit thrust from the solution and solve only for α and δ_e. However, here, we also want to estimate thrust requirements as this gives important clues about the capability of the aircraft, such as, does it have enough power or thrust to sustain level flight at the selected airspeed? The moment equation is written as shown in Equation (11-23):

case, Equation (11-23) can be written as shown in Equation (11-24):

$$C_{m_\alpha}\alpha + C_{m_{\delta_e}}\delta_e + \frac{z_T}{qSC_{MGC}}T = -\frac{T_N \cdot x_T}{qSC_{MGC}} - C_{m_0} - C_{m_{\delta_f}}\delta_f \tag{11-24}$$

The equations can now be rearranged in the following matrix form of Equation (11-25):

$$\begin{bmatrix} C_{D_\alpha} & C_{D_{\delta_e}} & -\dfrac{\cos\varepsilon_T}{qS} \\ C_{L_\alpha} & C_{L_{\delta_e}} & \dfrac{\sin\varepsilon_T}{qS} \\ C_{m_\alpha} & C_{m_{\delta_e}} & \dfrac{z_T}{qSC_{MGC}} \end{bmatrix} \begin{Bmatrix} \alpha \\ \delta_e \\ T \end{Bmatrix} = \begin{Bmatrix} -C_{D_{min}} - C_{D_{\delta_f}}\delta_f \\ \dfrac{W}{qS} - C_{L_0} - C_{L_{\delta_f}}\delta_f \\ -\dfrac{T_N \cdot x_T}{qSC_{MGC}} - C_{m_0} - C_{m_{\delta_f}}\delta_f \end{Bmatrix} \tag{11-25}$$

Moment about y-axis (pitch):

$$C_{m_\alpha}\alpha + C_{m_{\delta_e}}\delta_e + C_{m_T} + C_{m_{TN}} = -C_{m_0} - C_{m_{\delta_f}}\delta_f \tag{11-23}$$

The workaround for the normal force problem is to treat it as a constant and calculate an average around an expected thrust value. This will reduce the error in the solution. The other option, as stated above, is to solve for the thrust and use that to calculate the normal force. Using that value the thrust is calculated again and used to get a new value of the normal force, and so on, until both approach a fixed value. In either

When solved, Equation (11-25) yields the *AOA*, elevator deflection, and thrust required for a longitudinally stable flight at a given airspeed and atmospheric conditions. The solution can be implemented using a simple method such as the 3 × 3 Cramer's Rule. Once implemented, the formulation can be used to estimate suitable elevator deflection range (by varying the CG location) and power requirements at various conditions, including high altitudes and with flaps (or landing gear) deployed.

Note that typical values of the derivatives for small GA aircraft fall within the following limits (note that all the derivatives are in terms of radians; note that all angles, α, δ_e, and δ_f are in radians):

$$0.015 < C_{D\min} < 0.06 \qquad 0 < C_{L_0} < 0.6 \qquad -0.2 < C_{m_o} < 0.2$$

$$C_{D_\alpha} = 2kC_{L_\alpha}C_L \qquad 3.0 < C_{L_\alpha} < 6.0 \qquad C_{m_\alpha} = C_{L_\alpha}(h_{CG} - h_{AC})$$

$$C_{D_{\delta_e}} \approx \frac{0.001}{\delta_{e\max}(in\ radians)} \qquad C_{L_{\delta_e}} \approx \frac{0.2}{\delta_{e\max}(in\ radians)} \qquad -0.5 < C_{m_{\delta_e}} < -1.0$$

$$C_{D_{\delta_f}} \approx \frac{0.01}{\delta_{f\max}(in\ radians)} \qquad C_{L_{\delta_f}} \approx \frac{0.9}{\delta_{f\max}(in\ radians)} \qquad -0.7 < C_{m_{\delta_f}} < -0.2$$

Typical $\delta_{e\max}$ is 20°−25° (0.349 − 0.436 radians) and $\delta_{f\max}$ is 30°−45° (0.524 − 0.785 radians). These are only intended to give "ballpark" numbers and are not a substitute for analysis. The derivatives of your airplane may deviated greatly from these. Also see Appendix C1, *Design of Conventional Aircraft* for additional tail sizing tools.

EXAMPLE 11-2

Determine the α, δ_e, and T required for a stable level flight with flaps retracted at S-L and (a) 180 KCAS (q = 109.8 lb$_f$/ft^2) and (b) 75 KCAS (q = 19.06 lb$_f$/ft^2). The aircraft weight amounts to 3400 lb$_f$, S = 144.9 ft^2, C_{MGC} = 3.78 ft, x_T = 13.06 ft, z_T = −0.5 ft, ε = 0°, k = 0.04207 and the pertinent coefficients are given by:

$C_{D_\alpha} = 0.0863 \quad C_{D_{\delta_e}} = 0.000175 \quad C_{D\min} = 0.02541$
$C_{L_\alpha} = 4.8 \quad C_{L_{\delta_e}} = 0.355 \quad C_{L_0} = 0.2$
$C_{m_\alpha} = -0.72 \quad C_{m_{\delta_e}} = -0.923 \quad C_{m_0} = 0 \quad T_N = 50\ \text{lb}_f$

Note that all the stability derivatives are in terms of radians. If we assume propeller efficiency, η_p, of 0.85 for the high-speed case and 0.60 for the low-speed one, how much engine power is required to achieve this? Use Equation (14-38).

Solution

(a) Begin by populating the matrices of Equation (11-25):

$$\begin{bmatrix} 0.0863 & 0.000175 & -\frac{1}{(109.8)(144.9)} \\ 4.8 & 0.355 & 0 \\ -0.72 & -0.923 & \frac{-0.5}{(109.8)(144.9)(3.78)} \end{bmatrix} \begin{Bmatrix} \alpha \\ \delta_e \\ T \end{Bmatrix}$$

$$= \begin{Bmatrix} -0.02541 \\ \frac{3400}{(109.8)(144.9)} - 0.2 \\ -\frac{(50)(13.06)}{(109.8)(144.9)(3.78)} \end{Bmatrix} = \begin{Bmatrix} -0.02541 \\ 0.01370 \\ -0.01086 \end{Bmatrix}$$

Then, solve for the three arguments α, δ_e, and T, using any matrix method. Here the solution yields:

$$\begin{Bmatrix} \alpha \\ \delta_e \\ T \end{Bmatrix} = \begin{Bmatrix} 0.002409\ rad \\ 0.006217\ rad \\ 407.5\ \text{lb}_f \end{Bmatrix} = \begin{Bmatrix} 0.138\ \text{deg} \\ 0.356\ \text{deg} \\ 407.5\ \text{lb}_f \end{Bmatrix}$$

The power requirements are thus:

$$P = \frac{TV}{\eta_p \times 550} = \frac{407.5 \times (180 \times 1.688)}{0.85 \times 550}$$

$$= 265\ \text{BHP} \quad (\text{or about 85\% Power})$$

(b) Proceed in a similar manner, except note that $C_{D\alpha} = 2kC_{L\alpha}C_L$ changes to 0.4973. The results are:

$$\begin{Bmatrix} \alpha \\ \delta_e \\ T \end{Bmatrix} = \begin{Bmatrix} 12.9\ \text{deg} \\ -7.26\ \text{deg} \\ 378\ \text{lb}_f \end{Bmatrix}$$

The power requirements are thus 145 BHP or about 47% power. In reality this power is underestimated as it is based on the simplified drag model (see Chapter 15, *Aircraft drag analysis*), which under-predicts the drag at high lift coefficients.

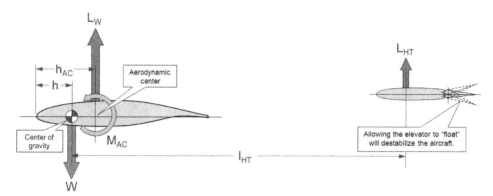

FIGURE 11-11 Wing-HT system used to derive Equation (11-26).

11.2.6 The Stick-fixed and Stick-free Neutral Points

As the CG of an aircraft is moved from a far forward to a far aft position (e.g. by moving useful load around), its longitudinal stability derivative, $C_{m\alpha}$, is modified greatly, from a large negative number to a large positive number (relatively speaking). This is reflected in Figure 11-3, which shows $C_m(\alpha)$ with both a positive and negative slope. As has already been discussed, the positive slope means the vehicle is statically unstable. By law, GA aircraft must be stable. For this reason, it is vital to be able to determine the CG location at which the slope becomes zero. This important point is called the *neutral point*.

There are two types of neutral points: *stick-fixed* and *stick-free*. The former refers to the stability with the elevator fixed in its neutral position (0° deflection angle), while the latter refers to the elevator being free to move. This distinction is of considerable importance, because at a given AOA (assuming $\alpha > 0$), a conventional elevator tends to float trailing edge up (as if to "help" the airplane getting to an even larger AOA). Therefore, the airplane is less stable than if the elevator is fixed. It should be evident that for typical aircraft, *the stick-free neutral point should be expected to be farther forward than the stick-fixed one*.

An important note should be made here regarding the stick-free neutral point. It is indeed possible for it to be aft of the stick-fixed neutral point. This depends on the magnitude of hinge moments due to deflection and AOA on the HT. However, during the conceptual design phase, a forward-lying stick-free neutral point is more critical as it narrows the CG envelope for conventional aircraft. And therein lies the problem with its determination; it depends on the elevator hinge moment. This, in turn, depends on the geometry of the horizontal tail, the size of the elevator, hinge line location, airfoil, the presence and geometry of a control horn, deflection of a trim tab, and other factors (see Section 23.2.1, *Introduction to control surface hinge moments*). Such details are simply not known during the conceptual design phase and this calls for some assumptions to be made to allow the HT to be sized.

On the other hand, the stick-fixed neutral point is less hard to determine, although it is by no means simple. The following method allows the first stab at the stick-fixed neutral point to be made. Then, the stick-free neutral point may be assumed to lie approximately 5% MGC ahead of the stick-fixed, allowing a preliminary aft CG limit to be established. *This will have to be revisited and estimated more accurately before the first flight of the prototype, when the geometry of the HT is known in detail.* For more information on the generation of the CG-envelope refer to Section 6.6.12, *Creating the CG envelope*.

The stick-fixed neutral point can be determined using the following expression, where the physical characteristics refer to those in Figure 11-11:

$$\frac{h_n}{C_{MGC}} = \frac{h_{AC}}{C_{MGC}} + \eta_{HT} \cdot V_{HT} \cdot \frac{C_{L_{\alpha HT}}}{C_{L_\alpha}} \cdot \left(1 - \frac{2C_{L_\alpha}}{\pi \cdot AR}\right) - \frac{C_{m_{\alpha AC}}}{C_{L_\alpha}} \quad (11\text{-}26)$$

where

AR = wing aspect ratio
h_n = physical location of the CG at which $C_{m\alpha} = 0$; i.e. the stick-fixed neutral point
h_{AC} = physical location of the aerodynamic center
η_{HT} = tail efficiency (see discussion in the derivation)
V_{HT} = horizontal tail volume
C_{L_α} = wing lift curve slope

$C_{L_{\alpha HT}}$ = HT lift curve slope
$C_{m_{\alpha AC}}$ = longitudinal stability contribution of components other than the wing

Note that the term $C_{m_{\alpha AC}}$ refers to the stabilizing effects of components such as the fuselage, nacelles, landing gear, the wing itself, and so on, as a function of the AOA. If the sum acts to rotate the LE down, then $M_{AC} < 0$ (has a negative sign and is stabilizing). If it acts to rotate LE up, then $M_{AC} > 0$ (has a positive sign and is destabilizing). The sign ultimately depends on the aircraft configuration. Note that the destabilizing effects of fuselages and nacelles can be estimated using the so-called Munk-Multhopp method, which is presented in Appendix C1.6, *Additional Tools for Tail Sizing*.

Derivation of Equation (11-26)

It is imperative to keep the orientation of the M_{AC} in mind in the following derivation. First, determine the sum of moments about the CG. For static stability, this must equal zero:

$$\sum M_{CG} = 0 \Rightarrow -L_W(h_{AC} - h) - L_{HT} \cdot l_{HT} + M_{AC} = 0 \quad \text{(i)}$$

Note that the sign for M_{AC} here is "+". Therefore, if M_{AC} is stabilizing ($M_{AC} < 0$) we will get $+(-|M_{AC}|) = -M_{AC}$, where |.| stands for the absolute value. Insert the definitions of $L_W = q \cdot S \cdot C_{L_W}$, the lift of the HT, given by $L_{HT} = q \cdot \eta_{HT} \cdot S_{HT} \cdot C_{L_{HT}}$, and $M_{AC} = q \cdot S \cdot C_{MGC} \cdot C_{m_{AC}}$ and simplify. Also note that the pitching moment coefficient about the aerodynamic center is given by: $C_{m_{AC}} = C_{m_{0AC}} + C_{m_{\alpha AC}} \cdot \alpha$. This of course implies its dependency on the attitude (or AOA) of the airplane.

$$-q \cdot S \cdot C_{L_W}(h_{AC} - h) - q \cdot \eta_{HT} \cdot S_{HT} \cdot C_{L_{HT}} \cdot l_{HT}$$
$$+ q \cdot S \cdot C_{MGC} \cdot C_{m_{AC}} = 0$$
$$S \cdot C_{L_W}(h_{AC} - h) + \eta_{HT} \cdot S_{HT} \cdot C_{L_{HT}} \cdot l_{HT}$$
$$- q \cdot S \cdot C_{MGC} \cdot C_{m_{AC}} = 0$$

The term η_{HT} is the *tail efficiency factor*. It ranges from 0.8 to 1.2, where numbers larger than 1 represents situations where a part of the HT is inside a propwash. Note that the sign for M_{AC} is now "−". Therefore, if M_{AC} is stabilizing ($C_{m_{AC}} < 0$) we will get $-(-|C_{m_{AC}}|) = +C_{m_{AC}}$.

Next, insert the definitions for C_{L_W} and $C_{L_{HT}}$ and divide through by q·S, as shown below:

$$(C_{L_0} + C_{L_\alpha}\alpha)(h_{AC} - h) + \eta_{HT} \cdot \frac{S_{HT}}{S} \cdot (C_{L_{0HT}} + C_{L_{\alpha HT}}\alpha_{HT})$$
$$\cdot l_{HT} - C_{MGC} \cdot C_{m_{AC}} = 0$$

Next, assume the HT features a symmetrical airfoil, i.e. $C_{L_{0HT}} = 0$:

$$(C_{L_0} + C_{L_\alpha}\alpha)(h_{AC} - h) + \eta_{HT} \cdot \frac{S_{HT}}{S} \cdot C_{L_{\alpha HT}} \alpha_{HT} \cdot l_{HT} - C_{MGC}$$
$$\cdot C_{m_{AC}} = 0 \quad \text{(ii)}$$

The AOA of the HT is affected by downwash from the wing upstream and can be approximated using the following expression (here assuming no angle-of-incidence). Note that accounting for the AOI is not necessary, as it will only affect the trim point (i.e. shift the C_m curve up or down) and not modify the slope of $C_{m\alpha}$:

$$\alpha_{HT} = \alpha - \varepsilon = \alpha - \alpha\frac{\partial\varepsilon}{\partial\alpha} = \alpha\left(1 - \frac{\partial\varepsilon}{\partial\alpha}\right)$$

A simple approximation for the rate of change of downwash with AOA is given by Equation (8-22) and is only valid for elliptical wings. However, it will give a reasonable prediction for other wing styles and, since we are eager to find out the approximate location of our stick-fixed neutral point, we employ it: $\frac{\partial\varepsilon}{\partial\alpha} = \frac{2C_{L_\alpha}}{\pi \cdot AR}$. Inserting this into Equation (ii) yields:

$$(C_{L_0} + C_{L_\alpha}\alpha)(h_{AC} - h) + \eta_{HT} \cdot \frac{S_{HT}}{S} \cdot C_{L_{\alpha HT}} \alpha\left(1 - \frac{\partial\varepsilon}{\partial\alpha}\right)$$
$$\cdot l_{HT} - C_{MGC} \cdot C_{m_{AC}} = 0$$
$$\Rightarrow (C_{L_0} + C_{L_\alpha}\alpha)(h_{AC} - h) + \eta_{HT} \cdot \frac{S_{HT}}{S} \cdot C_{L_{\alpha HT}} \alpha\left(1 - \frac{2C_{L_\alpha}}{\pi \cdot AR}\right)$$
$$\cdot l_{HT} - C_{MGC} \cdot C_{m_{AC}} = 0$$

Now, let $C_{m_0} = C_{L_0}(h_{AC} - h)$ and recall that $C_{m_{AC}} = C_{m_{0AC}} + C_{m_{\alpha AC}} \cdot \alpha$, simplify by gathering like terms:

$$C_{m_0} - C_{MGC} \cdot C_{m_{0AC}} + C_{L_\alpha}\alpha(h_{AC} - h) + \eta_{HT} \cdot \frac{S_{HT}}{S}$$

$$\cdot C_{L_{\alpha HT}}\alpha\left(1 - \frac{2C_{L_\alpha}}{\pi \cdot AR}\right) \cdot l_{HT} - C_{MGC} \cdot C_{m_{\alpha AC}} \cdot \alpha = 0$$

$$C_{m_0} + \left[C_{L_\alpha}(h_{AC} - h) + \eta_{HT} \cdot \frac{S_{HT}}{S} \cdot C_{L_{\alpha HT}}\left(1 - \frac{2C_{L_\alpha}}{\pi \cdot AR}\right)\right.$$

$$\left. \cdot l_{HT} - C_{MGC} \cdot C_{m_{\alpha AC}}\right]\alpha = 0$$

To determine the neutral point, the term inside the bracket must equal zero, i.e.:

$$C_{L_\alpha}(h_{AC} - h) + \eta_{HT} \cdot \frac{S_{HT}}{S} \cdot C_{L_{\alpha HT}}\left(1 - \frac{2C_{L_\alpha}}{\pi \cdot AR}\right) \cdot l_{HT} - C_{MGC}$$

$$\cdot C_{m_{\alpha AC}} = 0 \tag{iii}$$

This depends primarily on the location of the CG, denoted by h. Rearranging Equation (iii) in terms of h yields:

$$C_{L_\alpha}(h_{AC} - h) + \eta_{HT} \cdot \frac{S_{HT}}{S} \cdot C_{L_{\alpha HT}}\left(1 - \frac{2C_{L_\alpha}}{\pi \cdot AR}\right)$$

$$\cdot l_{HT} - C_{MGC} \cdot C_{m_{\alpha AC}} = 0$$

$$\Rightarrow C_{L_\alpha}(h_{AC} - h) = -\eta_{HT} \cdot \frac{S_{HT}}{S} \cdot C_{L_{\alpha HT}}\left(1 - \frac{2C_{L_\alpha}}{\pi \cdot AR}\right)$$

$$\cdot l_{HT} + C_{MGC} \cdot C_{m_{\alpha AC}}$$

This allows the location of the stick-fixed neutral point to be determined from:

$$h = h_{AC}$$

$$+ \frac{\eta_{HT} \cdot \frac{S_{HT}}{S} \cdot C_{L_{\alpha HT}}\left(1 - \frac{2C_{L_\alpha}}{\pi \cdot AR}\right) \cdot l_{HT} - C_{MGC} \cdot C_{m_{\alpha AC}}}{C_{L_\alpha}}$$

Since this point is typically denoted by h_n, we replace h by this symbol below. It is more convenient to be able to express the dimensions h and h_{AC} in terms of the fraction of chord length at the MGC, here denoted by C_{MGC}. For this reason we divide through by C_{MGC}:

$$\frac{h_n}{C_{MGC}} = \frac{h_{AC}}{C_{MGC}} + \left(\frac{1}{C_{MGC}}\right)$$

$$\times \frac{\eta_{HT} \cdot \frac{S_{HT}}{S} \cdot C_{L_{\alpha HT}}\left(1 - \frac{2C_{L_\alpha}}{\pi \cdot AR}\right) \cdot l_{HT} - C_{MGC} \cdot C_{m_{\alpha AC}}}{C_{L_\alpha}}$$

Rearranging further yields:

$$\frac{h_n}{C_{MGC}} = \frac{h_{AC}}{C_{MGC}} + \eta_{HT} \cdot \frac{S_{HT} \cdot l_{HT}}{S \cdot C_{MGC}} \cdot \frac{C_{L_{\alpha HT}}}{C_{L_\alpha}} \cdot \left(1 - \frac{2C_{L_\alpha}}{\pi \cdot AR}\right)$$

$$- \frac{C_{m_{\alpha AC}}}{C_{L_\alpha}}$$

Finally, define $V_{HT} = \dfrac{S_{HT} \cdot l_{HT}}{S \cdot C_{MGC}}$ as the *horizontal tail volume* and insert to obtain Equation (11-26).

QED

EXAMPLE 11-3

Determine the stick-fixed neutral point of the Cirrus SR22, using the following data, in part using geometry obtained from Figure 16-15 and Table 16-6, and in part by applying the methods of Chapter 9, *The anatomy of the wing*, to both the wing and HT geometry. First do this by ignoring the presence of the fuselage and, then, by assuming the token value given below for $C_{m_{\alpha AC}}$.

AR = wing aspect ratio = 10.21
C_{MGC} = 4.017 ft
$h_{AC} = 0.25 \cdot C_{MGC} = 1.004$ ft
$l_{HT} \approx 13.3$ ft
$S_{HT} = 29.75$ ft^2
V_{HT} = horizontal tail volume = 0.6782

C_{L_α} = wing lift curve slope = 5.299 per radian
$C_{L_{\alpha HT}}$ = HT lift curve slope = 4.516 per radian
$C_{m_{\alpha AC}}$ = 0.0 and 1.0
η_{HT} = tail efficiency = 1.05

EXAMPLE 11-3 (cont'd)

Solution

By plugging and chugging Equation (11-26) for *no fuselage*, we get:

$$\frac{h_n}{C_{MGC}} = \frac{h_{AC}}{C_{MGC}} + \eta_{HT} \cdot V_{HT} \cdot \frac{C_{L_{\alpha_{HT}}}}{C_{L_\alpha}} \cdot \left(1 - \frac{2C_{L_\alpha}}{\pi \cdot AR}\right) - \frac{C_{m_{\alpha AC}}}{C_{L_\alpha}}$$

$$= 0.25 + (1.05)(0.6782) \cdot \frac{4.516}{5.299} \cdot \left(1 - \frac{2 \times 5.299}{\pi \times 10.21}\right)$$

$$- \frac{0}{5.299} = 0.6564$$

By plugging and chugging Equation (11-26) for a *token fuselage*, we get:

$$\frac{h_n}{C_{MGC}} = \frac{h_{AC}}{C_{MGC}} + \eta_{HT} \cdot V_{HT} \cdot \frac{C_{L_{\alpha_{HT}}}}{C_{L_\alpha}} \cdot \left(1 - \frac{2C_{L_\alpha}}{\pi \cdot AR}\right) - \frac{C_{m_{\alpha AC}}}{C_{L_\alpha}}$$

$$= 0.25 + (1.05)(0.6782) \cdot \frac{4.516}{5.299} \cdot \left(1 - \frac{2 \times 5.299}{\pi \times 10.21}\right)$$

$$- \frac{1.0}{5.299} = 0.4677$$

The presence of a fuselage is seen to move the stick-fixed neutral point far forward.

11.2.7 Fundamentals of Static Directional and Lateral Stability

Before embarking on the analysis of lateral and directional stability, a few terms must be defined.

Roll or Bank

An airplane is said to be rolling or banking if a line drawn from wingtip to wingtip (assuming a symmetrical airplane) or some other normally horizontal reference line is sloped with respect to the y-z-axis as defined in Figure 11-1. This implies a rotation about its longitudinal axis — the x-axis or the roll-axis. Roll is the primary method used to change heading (direction of flight) and is controlled using the ailerons. The rudder is merely used to "fine tune" the heading change through coordination — in other words prevent skidding or slipping (discussed below).

The reason why the rudder is far less effective in changing heading than the bank maneuver can be explained using mechanics. A heading change results from acceleration in the horizontal plane that changes the original flight direction. To accomplish this rapidly, substantial force is required. The force generated by the VT through the deflection of a rudder is not a force large enough to change the heading fast enough for safe flight — for this a side force obtained using wing lift is required.

Yaw

An airplane is said to be yawed if its centerline is not parallel to the x-z-plane. This implies a rotation about its vertical axis — the z-axis or the yaw-axis. Based on the assumption that most airplanes are designed to be symmetrical about the x-z-plane, this rotation makes it un-symmetrical with respect to the airflow, which inevitably generates a side force and moment about the yaw axis.

Slipping or Sideslip

If the airplane is "yawed out of a turn", i.e. if the nose points outside of the trajectory of the turn, it is said to be *slipping* (see Figure 11-12). If banking left,

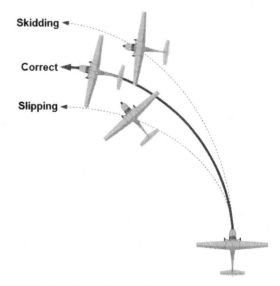

FIGURE 11-12 The definition of slipping (sideslip) and skidding.

this means the nose points to the right. Effectively, the bank angle of the airplane is steeper than the rate of turn would indicate. Slips primarily happen in two ways: as a consequence of uncoordinated deflection of ailerons and rudder; or the consequence of the intentional application of opposite rudder during a bank to increase drag or align the ground track while landing in a crosswind.

Slip is a trick sometimes used by pilots for altitude or airspeed control because the yaw that results increases the drag of the airplane.

If the airplane is "yawed into a turn", i.e. if the nose points to the inside of the trajectory of the turn, it is said to be *skidding* (see Figure 11-12). If banking left, this means the nose points further left than the rate of turn indicates; the bank angle of the airplane is shallower than indicated by the rate of turn. Skids primarily happen in two ways: (1) as a consequence of uncoordinated deflection of ailerons and rudder or (2) the consequence of the intentional and excessive application of pro-bank rudder.

All aircraft certified to 14 CFR Part 23 must be demonstrably laterally and directionally stable per 14 CFR 23.177 *Static Directional and Lateral Stability*. For this reason, slipping and skidding are pilot-induced maneuvers, as the airplane must be designed to suppress them. However, they may occur because of an airplane operating with asymmetric thrust, such as a multiengine aircraft with one engine inoperable (OEI).

11.2.8 Requirements for Static Directional Stability

Directional stability is the capability of the vehicle to weather vane. Imagine standing behind an actual weather vane with the wind directly in your face. If the vane is rotated so its nose points, say, right (and the tail points left) intuition tells us its tail will generate lift that points to the right, in the positive y-direction (see depiction in Figure 11-13). This, in turn, generates a moment whose tendency is to rotate the nose left and align it (and the tail) with the wind. Since the moment corrects the alignment, it is said to be *restoring*.

If the above weather vane is yawed nose right, then, using the stability coordinate system (SCS), the angle $\beta < 0°$. This means that if looking along the centerline of the vane, the wind would strike the left cheek. The restoring moment is negative because per the right-hand rule, the resulting rotation is analogous to grabbing around the z-axis with the right hand to rotate it with the right thumb pointing upward — in the negative z-direction. The opposite holds true if the weather vane is rotated nose left — a positive moment (thumb pointing down) is then required to bring the nose right to the initial position.

Figure 11-13 shows how this establishes requirements for static directional stability. It turns out that in order for this correcting tendency to be realized, the slope of

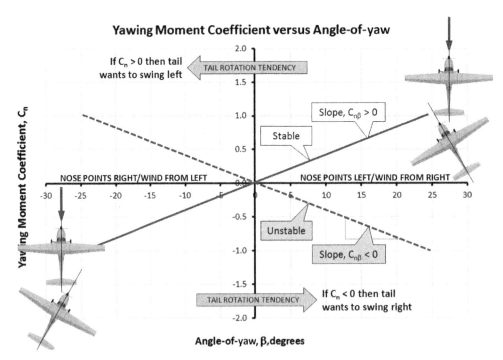

FIGURE 11-13 The requirement for directional stability.

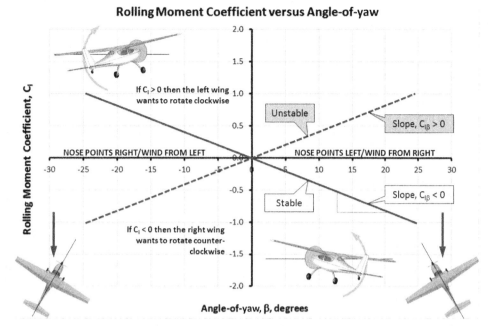

FIGURE 11-14 The requirement for lateral stability.

the yawing moment curve must have a positive slope. Mathematically this is written as follows:

Requirement for static directional stability:

$$C_{n_\beta} = \frac{\partial C_n}{\partial \beta} > 0 \quad \text{and} \quad C_n = 0 \quad \text{if} \quad \beta = 0 \quad (11\text{-}27)$$

11.2.9 Requirements for Lateral Stability

Next consider lateral stability (see Figure 11-14). It differs from both longitudinal and directional stability in that it requires sideslip (or yaw), and not roll itself, to be corrected (ignoring the application of devices like ailerons). This is the aforementioned *dihedral effect*. The geometric features of airplanes are such that when flying asymmetrically a restoring rolling as well as directional moments are created. It is the responsibility of the designer to decide how to manipulate the geometry to make these moments restoring or convergent (and not diverging). Dihedral effect has many sources as will become evident shortly.

Consider the airplane in the upper left part of Figure 11-14, whose nose points to the right of the wind direction (which is normal to the plane). For now consider only the contribution of the wing to the rolling moment. It can be seen from the top view in the lower left corner that the left wing leads the right one. This causes asymmetric loading on the wing that generates more lift on the left wing than the right one. The difference creates a rolling moment that tends to lift the left wing and bring it back to level. The rolling moment is positive because, according to the right-hand rule, the resulting moment vector points forward (the thumb would point forward) along the positive x-axis. The opposite holds if the nose is yawed to the left so the right wing leads the left one; a negative moment is created. By plotting a line between those two conditions, as is done in Figure 11-14, it can be seen that the rolling moment must have a negative slope in order to be stable. Mathematically:

Requirement for lateral stability:

$$C_{l_\beta} = \frac{\partial C_l}{\partial \beta} < 0 \quad \text{and} \quad C_l = 0 \quad \text{if} \quad \beta = 0 \quad (11\text{-}28)$$

11.2.10 Historical Values of $C_{n\beta}$ and $C_{l\beta}$

During the design of the aircraft it is important to know what directional and roll stability derivatives to aim for. As usual, the history of aviation presents us with a large number of candidate values. Table 11-2 lists a number of directional and lateral stability derivatives, gathered from References [1, 2, 3, and 4], and the author's own estimations. A selection of the airplanes are shown plotted in Figure 11-15 and Figure 11-16.

Note that although this book focuses on GA aircraft, it is still of interest to see how the derivatives are affected by high Mach numbers. For this reason, high-speed values are included.

11.2.11 The Dorsal Fin

A *dorsal fin* is a small surface extension installed at the leading edge of the root of the vertical tail (see

TABLE 11-2 Directional and Roll Stability Derivatives for Selected Aircraft

		Directional and Lateral Stability for Selected Aircraft								
Name	Model	M	AR	C_{MGC}, ft	b, ft	S, ft^2	$C_y\beta$	$C_n\beta$	$C_l\beta$	Reference
Bede	BD-SA	0.27	6.76	2.60	17	42.78	−1.428	0.105	−0.107	Author
Beech	ID-18	0.23	6.51	8.89	47.7	349	−0.927	0.162	−0.095	Author
	V-35 Bonanza	0.23	6.10	5.49	33.5	184	−0.331	0.060	−0.045	Author
Boeing	727-100	0.76	7.07	15.28	108	1650	—	0.086	—	1
		0.60	7.07	15.28	108	1650	−0.898	0.090	−0.098	Author
	747 (Pwr Appr)	0.25	6.96	27.31	195.7	5500	−1.080	0.184	−0.281	2
	747	0.30	6.96	27.31	195.7	5500	−0.886	0.140	−0.177	2
		0.60	6.96	27.31	195.7	5500	−0.893	0.159	−0.179	2
		0.80	6.96	27.31	195.7	5500	−0.889	0.196	−0.280	2
British Aerospace	Hawk	0.37	5.92	5.51	32.6	179.64	—	0.153	—	1
Cessna	182 Skylane	0.14	7.45	4.83	36	174	—	0.07	—	1
Convair	CV-880	0.60	7.20	18.94	120.0	2000	−0.788	0.128	−0.163	2
		0.80	7.20	18.94	120.0	2000	−0.813	0.129	−0.177	2
Convair	CV-880 (Pwr Appr)	0.25	7.20	18.94	120.0	2000	−0.877	0.139	−0.196	2
De Havilland Canada	DHC-6	0.22	10.00	6.5	65.0	423	−0.866	0.087	−0.119	4
Grumman	OV-1 Mohawk	0.22	6.40	7.5	48	360	—	0.267	—	1
Lockheed	A-4 Skyhawk	0.40	2.71	—	26.5	259	—	0.25	−0.12	3
		0.80	2.71	—	26.5	259	—	0.27	−0.14	3
	C-5 (Pwr Appr)	0.22	7.75	30.1	219.2	6200	−0.770	0.075	−0.123	2
	F-104 Starfighter	0.30	2.45	9.55	21.9	196	−1.170	0.335	−0.164	2
		0.80	2.45	9.55	21.9	196	—	0.285	−0.130	2
		2.00	2.45	9.55	21.9	196	—	0.158	−0.056	2
	Jetster	0.60	5.33	10.93	53.8	543	−0.718	0.1307	−0.077	2
		0.80	5.33	10.93	53.8	543	−0.742	0.12948	−0.055	2
	Jetster (Pwr Appr)	0.20	5.33	10.93	53.8	543	−0.720	0.137	−0.103	2
	NT-34 (Pwr Appr)	0.20	6.00	6.72	37.5	235	−0.720	0.049	−0.127	2
McDonnell Douglas	F-4	0.60	2.82	16.04	38.7	530	−0.642	0.114	−0.0895	2
		2.00	2.82	16.04	38.7	530	−0.647	0.065	−0.0156	2
	F-4 (Pwr Appr)	0.21	2.82	16.04	38.7	530	−0.655	0.199	−0.156	2
Northrop	T-38	0.79	3.77	6.72	25.3	170	—	0.279	—	1
Ryan	Navion	0.16	6.06	5.7	33.4	184	−0.564	0.071	−0.074	3
Start & Flug	H-101 Salto	0.06	21.51	2.07	44.6	92.4	−0.216	0.024	−0.063	Author

Figure 11-17). Its purpose is to add directional stability to the aircraft and that way prevent a serious condition known as *rudder-lock*. The dorsal fin, or dorsal as it is often referred to, can be as simple as a thin flat plate or as complicated as a curved compound surface stamped aluminum fairing riveted to an existing fin. In any case, its presence causes the leading edge to feature a discontinuity and this is imperative to its functionality. Older aircraft, such as the Douglas DC-4 (see Figure 11-18), DC-6, and many others) have curved

11.2 FUNDAMENTALS OF STATIC STABILITY AND CONTROL

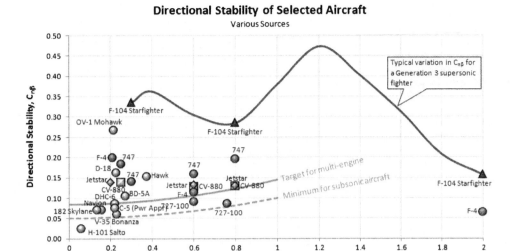

FIGURE 11-15 Trends of directional stability derivatives among selected aircraft.

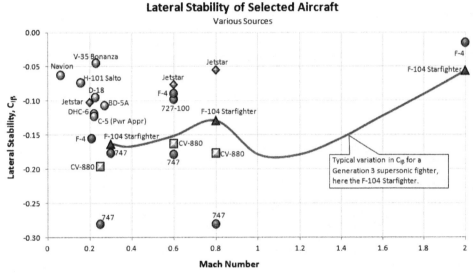

FIGURE 11-16 Trends of lateral stability derivatives among selected aircraft.

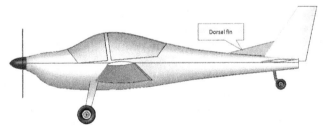

FIGURE 11-17 The dorsal fin is a surface extension between the VT and the fuselage, whose purpose is to improve directional stability at high angles of yaw.

dorsal fins, whose leading edges appear continuous, mathematically, and this makes it hard to perceive a specific discontinuity. However, this in no way prevents the formation of a vortex over the vertical tail.

The phenomenon of rudder locking is largely caused by insufficient directional stability of the airplane at high AOYs. As the airplane is yawed to a high AOY (for instance in a side-slip maneuver) two things may happen:

(1) If the vertical tail is poorly designed (e.g. features excessively large AR and small area) it may stall. A consequence of this will be a sharp drop in directional stability. The drop may be enough for the restoring moment generated by the vertical tail to be less than the destabilizing moment generated by the fuselage.

(2) The side loading resulting from the yaw will tend to force the rudder to the leeward side (if it is not there already as it may have been used to drive the

FIGURE 11-18 A dorsal fin on the Douglas DC-4 is typical of aircraft of that era, featuring a mathematically continuous and smooth curved leading edge. *(Photo by Phil Rademacher)*

airplane to this condition in the first place). This reduces the yaw contribution of the vertical tail further. If the hinge moments of the rudder are high, this compounds the problem by making it impossible for the pilot to step on the opposite rudder to recover from the situation. This way, the rudder is effectively "locked," hence the name.

A classic example of the severity of the rudder-locking problem was recorded on March 18, 1939, when an early production version of the Boeing 307 Stratoliner crashed during an evaluation test flight, killing all 10 people on board [5]. Officials of the Dutch airline KLM had expressed interest in the aircraft and sent a test pilot to evaluate it. The airplane, slated to be delivered to Pan Am, was at 10,000 ft when the KLM pilot shut down two of its engines on the same wing to evaluate its asymmetric thrust characteristics. He brought it to maneuvering speed, where it stalled and entered spin. The rudder locked during the spin, making recovery attempts by its two pilots impossible [6]. It is thought these attempts caused the wing and empennage to separate from the rest of the airplane. Wind tunnel testing after the mishap showed that the original tail on the airplane was too small and rudder area too large. It demonstrated that adding a dorsal would remedy the problem [7].

One solution for a rudder lock condition is to mount a dorsal to the fin. This will partition the vertical into two low-aspect-ratio segments. The smaller one will remain unstalled to an even higher yaw angle (because small AR surfaces stall at a higher AOA than high AR surfaces) and this helps maintain sufficient directional stability up to the higher AOY than without it.

Another solution, often featured on aircraft designed in the 1930—1940s, is an H-tail configuration, with small low-aspect-ratio ($AR \sim 1$) tail surfaces. Such surfaces stall at very high $AOAs$, as high as $30° - 40°$.

In the absence of the dorsal, the entire tail would be stalled. Its presence introduces a discontinuity in the leading edge of the VT, which at non-zero AOY generates a vortex as shown in Figure 11-19. The vortex effectively splits the fin into upper and lower halves. The upper half has a higher AR than the lower one and, thus, stalls at a lower AOY. And that is the important thing. The fact that the lower half is only partially stalled renders the $C_{n\beta}$ greater than if the dorsal was absent. This allows the $C_{n\beta}$ to be maintained to a higher AOY. Not only is the $C_{n\beta}$ increased due to the added area of the dorsal, it also guarantees it stays higher to greater AOY.

The recovery from a rudder lock requires the airspeed to be reduced by a roll or a pull-up maneuver to be performed or so the hinge moment drops to a magnitude that allows the pilot to bring the rudder to neutral with force [7, p. 221].

11.2.12 The Ventral Fin

The ventral fin is a lifting surface used for two purposes: to improve stall, or Dutch roll characteristics. Dorsal fins are rarely installed for other purposes, although the author of this book was involved in a project that featured one purely for looks (misguided one might add). The installation of a ventral fin typically means two things: there is insufficient nose pitch-down moment at stall; and there is insufficient Dutch roll damping. It is relatively easy to identify the reason for installation, as ventral fins installed for the former are usually substantially more horizontal than the latter (see Figure 11-20).

Ventral fins generally work as follows. At cruise $AOAs$ the AOA of the ideal ventral fin will be very low. This is necessary as it is essentially an aerodynamic fix for a high AOA condition and, thus, at cruise its drag should be as low as possible. Then, as the pitch angle (AOA) of the airplane increases the ventral becomes gradually more effective and it will begin to contribute to the longitudinal stability. This is depicted in Figure 11-21. The reader should also review Section 11.3.3, *T-tail*, for further insight.

The solid curve in Figure 11-21 represents the pitching moment curve for a hypothetical airplane. It can be seen that the nonlinearity in the original aircraft result in a trim point around $32°$. If this airplane for some

11.2 FUNDAMENTALS OF STATIC STABILITY AND CONTROL

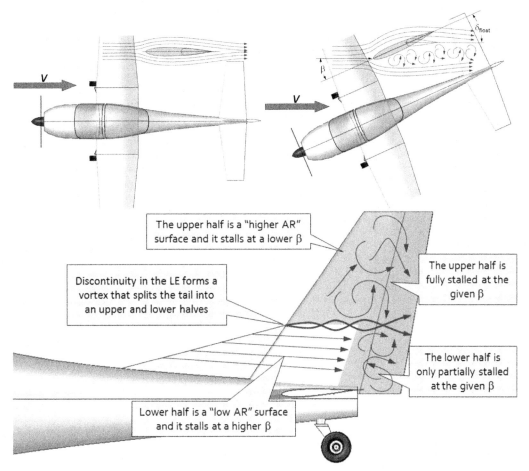

FIGURE 11-19 How a dorsal fin prevents rudder lock.

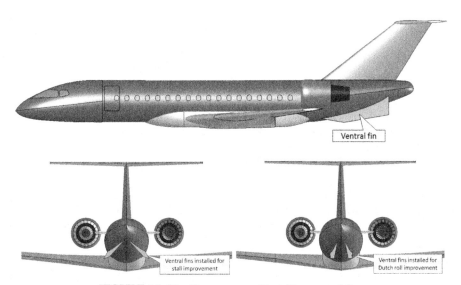

FIGURE 11-20 The purpose of installing ventral fins.

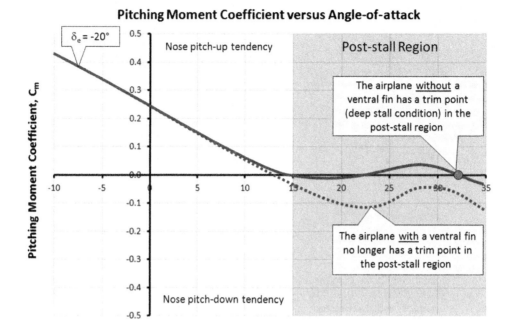

FIGURE 11-21 A ventral fin installed to prevent deep-stall changes the shape of the pitching moment curve such that a trim point no longer exists in the post-stall region.

reason reached that *AOA*, it would not have the elevator authority to recover (note that the graph represents the condition at full elevator aft condition; also refer to Figure 11-33 for further insight). The installation of the ventral will increase the nose pitch-down moment at this condition, effectively deleting the trim point. This would allow the modified aircraft to recover from the condition — it would not be affected by a deep-stall.

The installation of ventral fins must take into account the three-dimensional shape of the flow field around the aft end of the airplane, often rendering the orientation of the ventral fin somewhat "unintuitive."

11.2.13 Tail Design and Spin Recovery

Spin entry and recovery is of major concern in GA aircraft. Spin is a consequence of asymmetric wing stall that typically results in an inadvertent wing roll to the left or right, initiating a dynamic mode called *autorotation*. The motion is a rapid vertical descent along a helical path. The autorotation is caused as the lower wing, i.e. the wing on the inside of the yawing motion, has a higher *AOA* than the higher wing (the outside wing). The difference in *AOA* results in a difference in wing lift between the two wing halves, such that the outside wing lift is greater and this drives the rotation.

Spin regulations for certified GA aircraft are presented in 14 CFR 23.221, Spinning. The paragraph states that a normal, utility, and aerobatic category aircraft for which spin certification is sought *"must be able to recover from a one-turn spin or a three-second spin, whichever takes longer, in not more than one additional turn after initiation of the first control action for recovery, or demonstrate compliance with the optional spin resistant requirements of this section."*

The arrangement of the HT with respect to the VT is of great concern in aircraft design. This is because the wake generated by the HT may "blanket" the rudder, rendering it far less effective than otherwise. Blanketing means that the rudder is largely covered by separated airflow, but this may reduce its effectiveness so it becomes incapable of providing adequate counter-rotational moment during spin. This can cause a serious problem when attempting spin recovery, at times making it altogether impossible. Figure 11-22 shows common HT/VT arrangements and idealized wakes emanating from the HT. Some of the configurations are void of separated flow (e.g. V-tail, T-tail, and H-tail), whereas the conventional configuration is exposed to different levels of blanketing.

It must be said that the flow field around a spinning airplane is very complex and the spin is driven by a difference in the lift generated by the left and right wings. The flow field is further compounded by contributions from the fuselage, so what is said here really represents rules of thumb. It is generally recommended that at least 1/3 of the rudder should remain in undistorted airflow. The figure shows three conventional configurations, of which only the bottom one has that much rudder area

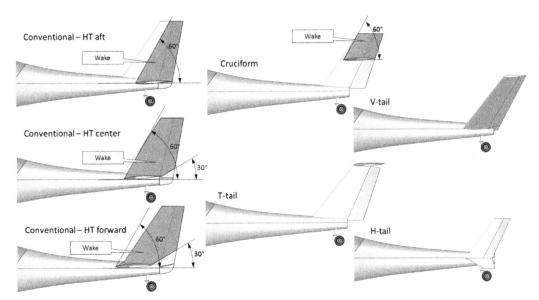

FIGURE 11-22   The blanketing of the rudder during spin is largely dependent on the tail configuration.

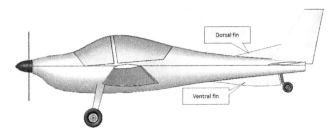

FIGURE 11-23   Potential solution to a spin recovery problem.

in "clean" air. Figure 11-23 shows a possible solution to a spin recovery problem — increase in vertical surface area using dorsal and ventral fins. There is no guarantee such a solution will work, but if it does, it usually affects the aesthetics of the airplane.

11.3 ON THE PROS AND CONS OF TAIL CONFIGURATIONS

The purpose of this section is to introduce a number of the various tail configurations that have seen use throughout the history of aviation. The selection itself can only be made easier if the designer is aware of the many advantages and disadvantages of each configuration, in terms of aerodynamics; stability and control; structural layout; aeroelasticity; and operational characteristics.

11.3.1 Conventional Tail

The *conventional tail* is the most common of all tail configurations. Possibly some 80% of aircraft ever built feature this tail configuration, ranging from ultralight aircraft to the world's largest passenger transport aircraft, the Airbus A380 and Boeing 747. A large number of GA aircraft also feature the tail, such as the Cessna single-engine series aircraft (140, 150, 152, 172, and so on) as well as the latest entries, the all-composite Cirrus SR20, SR22, and Cessna Corvalis high-performance aircraft. The configuration is often a "default" configuration unless some special requirements call for alternative solutions. For the discussion that follows, this configuration is considered the "baseline" configuration (see Figure 11-24).

The advantages of this configuration are many, beginning with it being a safe-and-tried configuration as evidenced by the fact that the majority of airplanes feature it. For single-engine propeller aircraft the configuration takes advantage of the propwash, which gives it a boost during the T-O run. This makes it possible to rotate at a lower airspeed, reducing the ground run (see the discussion about drawbacks below).

The horizontal tail is sometimes positioned so the wing wake at high AOA hits the tail, causing it to oscillate (often violently), providing a natural feedback to the pilot of an imminent stall, and then exiting the wake in the post-stall to allow for ready recovery (although often it remains inside the wake). This phenomenon is illustrated in Figure 11-25. Achieving this reliably and by design (as compared to "luck") requires a thorough understanding of the flow field behind the wing and calls for wind tunnel testing or computational fluid dynamics (CFD) analysis. It is also possible the tail can be positioned such that as the flaps are deployed, the increased nose pitch-down moment may be cancelled by an increase in downwash on the HT, which provides the

FIGURE 11-24 A conventional tail configuration.

necessary nose pitch-up moment. This would result in an aircraft that experiences modest, if any, change in trim settings when the flaps are deployed. However, the phenomenon is airspeed-dependent and, thus, cannot always be taken advantage of.

Another advantage is that the HT and VT are joined in the fuselage, where its girth is still somewhat large and, thus, is still effective in reacting torsional loads. This torsional rigidity also helps resisting empennage flutter.

Figure 11-25 shows an airplane at three specific elevator deflections and AOA. The top figure (A) shows it at a low AOA during cruise. The wing wake lies below the HT and its thickness is relatively small. The center figure (B) shows it at a high AOA, just before stall. The thickness of the wake has grown substantially and now the HT is inside it. The turbulent airflow would cause the tail to oscillate, and the pilot would sense this in the control stick (or yoke) in a mechanical control system, as is the norm for smaller aircraft.

However, the pilot might detect structural oscillations even if equipped with a digital control system. The bottom figure (C) shows the aircraft at a post-stall condition, with a fully stalled wake. Depending on its thickness, the HT may or may not reside inside the wake. If inside the wake, the elevator authority is noticeably reduced.

Figure 11-26 shows the C_m versus AOA for the aircraft during the above events (labeled A through C). Point A is trim point for a condition where the elevator deflection might be 3° TED. Point B represents an elevator deflection of 20° TEU and the aircraft is close to stall. Point C represents the aircraft, still with the elevator deflected 20° TEU, at an AOA of 25°, inside the post-stall region. Point D represents the condition of Figure 11-27, deep inside the post-stall region. Some aircraft, for instance the Cirrus SR20 and SR22, are designed to be laterally controllable in this condition, as a part of their Equivalent Level of Safety (ELOS) certification. However, many others are not. Some of those would experience uncontrollable roll tendency to the left or right.

Note that the upper half of the graph is labeled "Nose pitch-up tendency" and the lower half is labeled "Nose pitch-down tendency." This refers to the corrective tendency displayed by a statically stable airplane. For instance, consider an airplane trimmed at Point A using the center curve. If for some reason it is displaced to a lower AOA (i.e. nose down), it would effectively be moving from the trim point (A) along the curve to some positive value of C_m. This positive C_m would force the nose back up to the AOA at Point A. This would be reversed if it were displaced to a higher AOA (i.e. nose

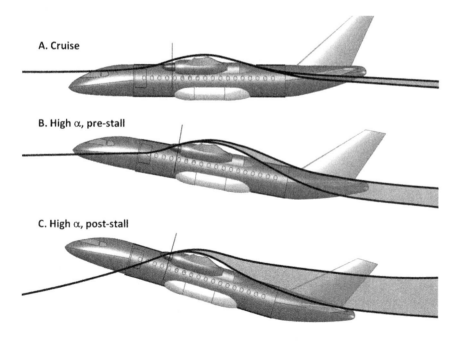

FIGURE 11-25 An aircraft at three different conditions. At cruise (A), the wing wake lies below the HT, but near stall it may strike the HT (B), causing it to oscillate, warning the pilot of an imminent stall. At post-stall (C), the HT may exceed the lower boundary of the wake, although this ultimately depends on the geometry of the airplane.

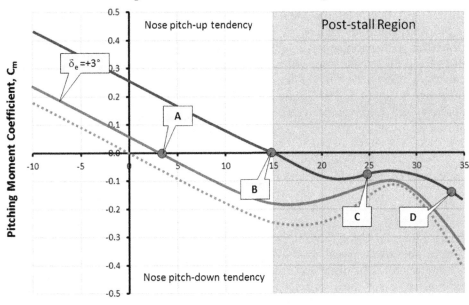

FIGURE 11-26 The conditions for the aircraft in Figure 11-25 and Figure 11-27.

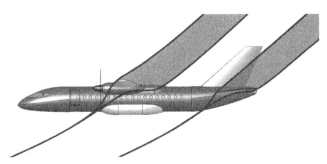

FIGURE 11-27 The same aircraft at a high *AOA* in a post-stall condition. The shaded areas represent the highly separated wakes emanating from the wing and HT. The HT wake covers a substantial portion of the rudder, reducing the rudder authority. This is Point D in Figure 11-26.

up) – this would generate a tendency to bring the nose back down to Point A.

The primary drawback of the conventional tail is that the wake from the HT may "blanket" the rudder at *AOA* in the post-stall range (which may easily involve in very high *AOA*s), although many such aircraft are specifically designed to ensure that adequate rudder area is outside the separated region. Aerobatic aircraft such as the Yakolev Yak-55 and the Extra 300 are examples of this. A blanketed rudder is very undesirable in abused flight conditions such as during a spin (see Figure 11-27) and makes recovery more challenging. The separated flow in the wake reduces the rudder authority substantially and may render the aircraft unrecoverable. However, one should not overlook the fact that a large number of aircraft known for excellent spin recoverability, such as the aforementioned aerobatic aircraft, feature conventional tails. Spin is a very complicated phenomenon and recoverability is affected by a great many features of the airplane besides the blanketing of the rudder. It is only a part of the puzzle. Also, the position of the HT prevents an "aft podded" engine configuration, in which the engines are mounted on the fuselage aft of the cabin to bring down engine noise in the cabin. Additionally, for propeller aircraft, if the tail is mounted inside the propwash, its drag contribution increases.

Again, in spite of the above discussion the designer should always evaluate specific drawbacks with a rational attitude and should ask, "How bad is the flaw really?" Is it possible to assign a number to it, for instance, weight or drag count? It is inevitable that some of the pros and cons cited throughout the book sound as if their impact is much greater than it actually may be in reality. The goal of such discussion is to point out possible flaws. However, be mindful of the real impact. For instance, consider an aircraft configuration that requires 350 lb_f of thrust at cruise. It is possible the impact of placing the HT in the propwash might add 1 lb_f of drag to the total drag of the aircraft, but it might reduce the T-O run by 150 ft. Which is of greater importance? Well, it is up to the designer to decide, but while it is imperative to point out everything that might be considered a pro or a con, putting figures of merit next to each is the only way to really understand the overall impact they have on the design.

11.3.2 Cruciform Tail

The cruciform tail configuration is often resorted to if it is desirable to feature aft podded engines. A wide range of aircraft use this tail arrangement, ranging from bombers such as the B-1B Lancer, the twin turboprop passenger transport aircraft Handley Page HP-137 Jetstream, or the now obsolete Sud-Est Caravelle, a classic early commercial jet transport aircraft. The Dassault Falcon family of business jets also features this tail configuration. NACA research on cruciform tails can be found in NACA reports TN-2175 [8] and RM-L54I089, among others. An example of a cruciform tail is shown in Figure 11-28.

The cruciform tail is a practical solution for an aft podded engine configuration and one that will result in lower structural loads than the T-tail (see later). It is also practical for freight aircraft designed to allow the empennage to be swung open for loading and unloading through the rear fuselage (as is the T-tail). Depending on the position of the break in the fuselage, a conventional HT might hit the wings when swung open and limit the allowable swing angle. Of course, there are conventional tail airplanes such as the four-engine Canadair CL-44 that show this is not necessary as it depends on the location of the opening and length of the fuselage. For such a configuration, the HT sits above the exhaust from the jet engines, as shown in Figure 11-29. The lower part of VT is exposed to "clean" airflow at high, post-stall AOA, improving spin recoverability. The position of the HT creates limited endplate effectiveness.

Among drawbacks are higher torsional loads than for the baseline configuration (the conventional), which results in a heavier structure (albeit lighter than T-tail). Also, the VT must feature structural reinforcement to allow the HT to be mounted to it. As a consequence, this configuration will be heavier than the baseline tail, assuming identical size and shape of the tail surfaces. Another drawback of this configuration is that the upper part of the rudder is not fully exposed to "clean" airflow during spin recovery. Additionally, the configuration requires a sectored elevator (i.e. a sector of the root of the elevator is removed) to allow the rudder to deflect freely, or the rudder must be split in the area of the HT for the same reason. This requires additional manufacturing steps that could increase the cost of production of the rudder (or elevator). Furthermore, the configuration has higher interference drag.

11.3.3 T-tail

T-tails have been common on aircraft since the mid-1960s, when they appeared for the first time on a commercial passenger transport aircraft – the Boeing B-727. The concept was tested by NACA as early as 1954 [9,10], but has since been studied by NASA and other scientists in multiple publications. The configuration is considered by many to be stylish, a feature of great importance to the marketability of aircraft. Besides commercial transports, T-tails are very common on business jets, where they are more common than the conventional tail. Business jets ranging from the early Gates Learjet 25 through to the Gulfstream G650 feature

FIGURE 11-28 A cruciform tail configuration.

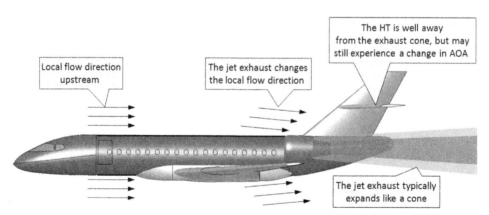

FIGURE 11-29 The cruciform tail configuration brings the HT out of the jet exhaust, although it may still be affected by a change in AOA on the tail with thrust setting.

FIGURE 11-30   A T-tail configuration.

the configuration. In the GA industry several aircraft feature the configuration. The best-known among smaller such aircraft are the twin-seat, single-engine trainers Piper PA-38 Tomahawk and Beech Model 77 Skipper. T-tails can also be found on larger twin-engine turboprops, such as the Beech 1900. Additionally, T-tails can be found on a range of sailplanes. An example of a T-tail is shown in Figure 11-30.

Among the advantages is that the tail leaves the rudder un-blanketed at post-stall *AOA*, giving the rudder a greater spin recovery potential (see Figure 11-31). The reader must be mindful of the shortcomings of the figure, which is a simple representation of an otherwise highly complex flow phenomenon. The flow field near the tail is dependent on the actual *AOA* and geometric features such as separation of wing and HT and shape of the fuselage. Figure 11-31 shows the aircraft at an *AOA* beyond deep-stall, discussed below. Additionally, as stated earlier, spin and spin recovery are complicated and an un-blanketed rudder is only part of what makes an airplane recoverable. Another advantage of the T-tail configuration is that it lends itself perfectly to aft podded engines, as are so common among business jets. If the VT is swept aft, which is also the norm for high-speed aircraft, an additional tail arm can be gained by placing the HT on top of the VT. As a consequence the tail can be slightly smaller than the baseline tail. Additionally, the endplate effect the HT gives to the VT can be utilized to reduce the size of the vertical tail, which allows for drag reduction (albeit modest). The high position of the HT effectively places it in a relatively undisturbed air, although this primarily holds true at low *AOA*. On sailplanes, where low drag is of utmost importance, placing the tail in undisturbed air is essential because this helps maintain NLF on their surfaces.

There are several disadvantages of this configuration. The high-mounted HT, coupled with the presence of the VT, can generate a substantial asymmetric lift in yaw. This results in high torsional loads that are reacted at the top of the fin and eventually by the fuselage. This torsion is compounded by the torsion from the VT that

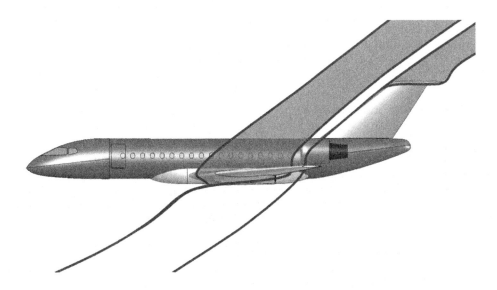

FIGURE 11-31   A schematic of a T-tail aircraft post-stall at a high *AOA*, showing the turbulent wake from the wing does not blanket the VT (see text for more details).

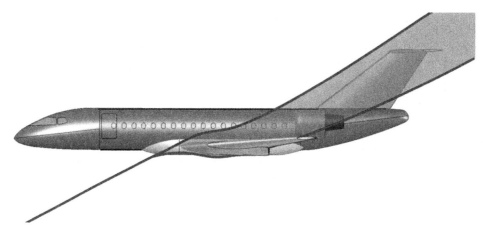

FIGURE 11-32 A T-tail aircraft at a high AOA, showing the HT inside the wing wake, causing the "deep stall" phenomenon.

simultaneously results from the yaw, making the loads much higher. This impacts the structure of both the VT and the fuselage, causing an overall increase in its weight. The high mass of the HT mounted at the tip of the VT reduces the tail's natural frequency of oscillation, effectively lowering its flutter speed. This may have to be remedied by additional structure to increase the tail's stiffness, further increasing its weight. Maintenance is impacted as repair stations will have to accommodate technicians working on the tail high above the ground.

One of the configuration's primary disadvantages is that the tail may end up inside the wing wake at stall or post-stall (see Figure 11-32). The figure shows how the HT may ultimately come to rest inside the highly separated wake from the wing. Therefore, the elevator authority is severely diminished and may even disappear. Note that the figure does not necessarily depict the only attitude possible — some aircraft have been reported descending mostly horizontal at a high AOA, others in a more nose-down attitude. This is a very serious condition that in the extreme case can render the aircraft incapable of dropping the nose and recovering from the stall. It is recognized as a *"deep-stall"* and has caused multiple development and operational vehicles to crash with fatal consequences. It is one of two important reasons why commercial transport aircraft with T-tails always feature "stick-shakers" or "stick-limiters" for the elevator control yoke in the cockpit. That way, the pilot cannot stall the airplane inadvertently.

To understand how the deep-stall phenomenon can affect an airplane with a T-tail, consider Figure 11-33, which shows a typical T-tail aircraft. Again, the three curves represent the C_m for the complete aircraft with different elevator deflections. The top curve represents C_m with the elevator deflected 20° TEU, the center curve depicts the elevator in a neutral position, and the bottom depicts the elevator deflected 15° TED. The three trim points in the pre-stall region are similar to those of Figure 23-19.

Now consider the right side of the graph of Figure 11-33, which is shaded and titled the "Post-stall Region." The airplane being represented stalls around 15°. If it is driven to stall abruptly, this would be accompanied by angular momentum about the pitch axis, which might rotate it to an α greater than 15°. If the angular momentum is high enough, the pitch angle would be driven farther to the right to an even higher α, where some strange things begin to happen.

The first thing to happen is that the negative slope of the aircraft becomes positive (i.e. unstable). This positive slope would be realized as a tendency to increase the α further. This is caused as the HT enters the wing wake, which suppresses its effectiveness. Consequently, the natural instability of the fuselage (as it is a body of revolution) will destabilize the vehicle, causing the curves to approach the horizontal axis (around 16°−18°). Eventually, they intersect it and ascend into the positive part of the graph before descending back down as the tail moves out of the wake due to the excessively high AOA. This tendency is common among T-tail aircraft. The consequence of this behavior (which is caused by the tail entering the highly turbulent wing wake as shown in Figure 11-32) is that each curve now has a corresponding trim point around an AOA of 30°−32°. These points pretty much cover the entire travel of the elevator. The problem with this is that the reduction in elevator authority at those AOAs, which are noticeable by the closeness of the three curves, renders the aircraft practically uncontrollable in pitch. It won't matter how much the pilot moves the yoke forward or back, the elevator deflection on its own cannot bring the airplane out of this equilibrium condition. Consequently, it will stay there at this high AOA, fully stalled and uncontrollable as it descends rapidly, eventually crashing. For some aircraft, applying full engine power might help nudge

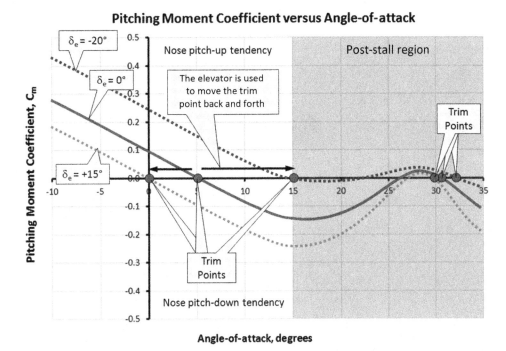

FIGURE 11-33 The physics of the "deep-stall" phenomenon explained.

it out of this stable post-stall *AOA*, although aircraft with jet engines may experience a flame-out to further compound the situation. The prototype of a new design that features a T-tail should not be stalled in flight testing without a spin recovery chute, in case it is susceptible to this phenomenon.

The history of aviation is wrought with deep-stall accidents in T-tail aircraft. A Gloster Javelin was lost on June 11, 1953, due to this phenomenon [11], killing the pilot. On March 23, 1962, a Handley Page Victor crashed killing two of the crew of five and two people on the ground [12]. On October 22, 1963, a BAC 1-11 crashed on a test flight due to deep-stall killing all seven occupants [13]. Coincidentally, the very same day a Tupolev Tu-134 crashed due to the same cause [14]. On June 3, 1966, a Hawker Siddeley HS-121 Trident crashed on a test flight, killing the crew of four [15]. These are but a few of a long list of such accidents in which a T-tail was the contributing factor.

11.3.4 V-tail or Butterfly Tail

Research on V-tail configurations dates back to 1931 with the publication of an anonymous paper that discussed the so-called Rudlicki V-tail.[1] The configuration combined the three surfaces of the conventional tail into two surfaces that form a distinct V (see Figure 11-34). This way the functionality of a typical single rudder surface and two elevator surfaces is combined in two surfaces that are called *ruddervators*. The ruddervators are deflected in a manner that effectively recreates the functionality of the rudder and elevator (see Figure 11-35).

V-tails can be found on a wide range of aircraft, ranging from sailplanes, to propeller and jet aircraft. Some of these are shown in Table 11-3. Among sailplanes is the Glasflügel H-101 Salto aerobatic sailplane. The best-known GA aircraft featuring a V-tail is the four-seat, single-engine Beech Model 35 Bonanza, which was first introduced in 1947. Two brand-new aircraft designs appeared in 2007, the Cirrus SF50 Vision and the Eclipse-jet, both being single-piloted personal jets, which feature a single Williams FJ33 turbofan mounted in a pod on top of the fuselage. The Northrop Grumman RQ-4 Global Hawk is a well-known unmanned aerial vehicle (UAV) that features V-tails. Additionally, the Fouga CM-170 Magister, a twin-engine, two-seat military trainer, introduced in 1948, features a V-tail.

Considerable research was done on such tails by NACA in the 1940s and multiple reports are available from the online NASA Technical Report Server [16].

---

[1] Jerzy Stanislaw Rudlicki (1893–1977) was a Polish aeronautical engineer who is considered the inventor of the V-tail. He acquired a Polish patent for it in 1930 (patent# 15938). The paper that investigated the tail is "The Rudlicki Vee Tail," *Aircraft Engineering*, vol. IV, no. 37, March 1932, pp. 63–64.

FIGURE 11-34 A V-tail configuration.

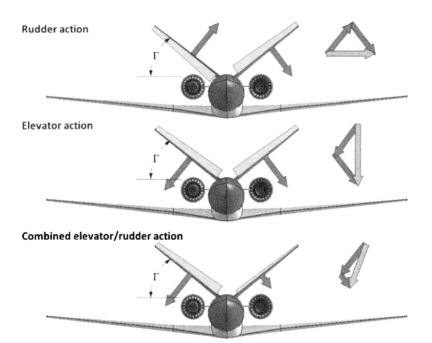

FIGURE 11-35 How ruddervators are used to generate the primary functionality of the conventional rudder, elevator, and combined action.

For instance, the NACA reports TN-1369 [17], TN-1478 [18] and R-823 [19] should be very helpful to the designers of V-tail aircraft. The last one (R-823) was particularly helpful to the author of this book during such a development program.

Let's consider the advantages of V-tails as stated by NACA R-823, where comparison is made to the conventional tail. First, there is less interference drag because the V-tail has fewer fuselage-tail junctures. Second, there is less tendency for "rudder locking," a phenomenon discussed in Section 11.2.11, *The dorsal fin*. Third, the higher location of the tail surfaces reduces required elevator deflection as the center of pressure is above the fuselage and therefore the drag component of the deflected ailerons will increase the nose pitch-up moment. This is depicted in Figure 11-36.

The pitch-up moment M can be written as: $M = F_V \cdot a + F_H \cdot b$, where F_V is vertical component of the resultant force acting on the V-tail, F_H is horizontal component, and a and b are the arms shown in the figure. The corresponding moment for a conventional tail would be closer to $F_V \cdot a$. Additionally; the surfaces extend farther out of ground effect than the conventional tail, making the tail more effective during T-O and landing. Fourth, for amphibious aircraft the tail will most likely be out of the spray during the T-O, making it more effective. Fifth, the tail is out of the wake of the wing and canopy in high-speed flight and therefore there is less chance of tail buffeting. Sixth, there are fewer tail surfaces to manufacture.

A V-tail is a great configuration for aircraft equipped with single jet engines and allows them to be mounted in

TABLE 11-3 Selected Aircraft Featuring V-tails

Aircraft type	Comment
Beechcraft V35 Bonanza	Single-engine, four-seat GA aircraft. Produced in large numbers.
Fouga CM-170 Magister	Twin-engine, twin-seat military trainer (195X). Produced in large numbers.
Potez-Heinkel C.M. 191	Based on the Fouga Magister. Prototypes only.
The Zipper	Ill-fated aircraft.
F-117 Nighthawk	Military fighter, stealth configuration, fly-by-wire.
Global Hawk	UAV.
Microjet	Twin-engine, twin-seat jet trainer candidate. Only prototypes built.
Dale AE Weejet 800	Prototype only.
P-63 Kingcobra	A Kingcobra fitted with a V-tail for testing purposes. Prototype only.
Eclipse V-jet	Prototype GA personal jet. Single-engine jet. Prototype only.
V-jet	Developed in participation with Williams International. Prototype only.
Cirrus SF50 Vision	Prototype GA personal jet. Single-engine Williams FJ33 jet.
Hütter 30 GFK	Sailplane.
Musters HKS 1	Sailplane.
Schleicher Ka 3	Sailplane.
Greif 1a (FK 1)	Sailplane.
Greif 2a (FK 2)	Sailplane.
Schreder HP-12	Sailplane.
Schreder HP-18	Sailplane.
Start & Flug H101 Salto	Sailplane, aerobatic.

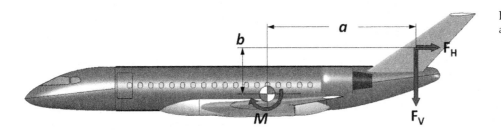

FIGURE 11-36 Increase in pitch authority explained.

a pod on top of the fuselage, on the aircraft's centerline. Two recent examples of such a design solution are the Cirrus SF50 Vision and the Eclipse-jet.

The following disadvantages are cited in NACA R-823 [19]. First, there is a possible interaction of elevator and rudder control forces. For instance, during a crosswind landing the pilot might use the rudder pedals to side-slip the airplane on final approach. Since this would cause a TEU deflection in one of the ruddervators and a TED in the other one, the elevator stick-forces would be affected when compared to those with the rudder neutral. Second, there is a possible interaction of elevator and rudder trimming, when trim tabs are deflected to large angles. Third, the control system requires a special treatment of the control inputs. This is done using a device called a *control system mixer* (see Figure 11-37). Fourth, it results in greater loads on the tail and fuselage, which tends to increase structural weight. This is caused by the fact that the V-tail must serve as the HT and the VT. This requires its size to be larger than if it only served one purpose. Consequently, during maneuvers, such as yawing, a larger force will be generated, which of course requires stouter and, thus, heavier structure.

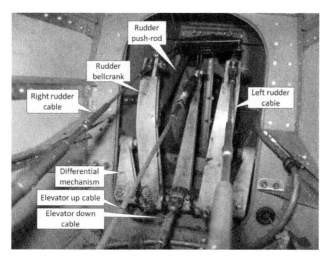

FIGURE 11-37 Rudder/elevator control mixer in a Beech Bonanza. *(Photo from author's collection)*

But there are other shortcomings not mentioned by R-823 [19]. First is that the application of rudder will cause the airplane to roll in a direction opposite to the intended bank. To understand what this means, consider a pilot wanting to execute a turn to the left. To accomplish this, the pilot will roll the left wing down. This causes the right wing to be raised and, consequently, it will move through an arc with a greater radius than the left wing. This makes it move faster through air than the left wing. It follows it generates higher drag than the left wing, and this causes a tendency for the nose to point to the right, even though the heading is changing to the left. This phenomenon is known as an *adverse yaw* and the pilot simply cancels it by stepping on the left rudder pedal. If we observe the aircraft from behind as this happens, we would see the pilot action forcing the tail to the right, in the process moving the nose to the left (the intended direction). For some aircraft the pilot actively corrects the turn in this fashion, while for others there is less need to. The preceding description is not a consequence of a V-tail, per se, but needs to be kept in mind when considering the following effects.

Now consider Figure 11-38, which compares a rudder input for a V-tail and an inverted V-tail. The upper image shows the ruddervators of a V-tail deflected to produce a nose-left yaw (move the tail to the right). The left ruddervator is deflected TED and the right one is deflected TEU. The resulting force vectors on the tail surfaces give rise to a tendency for the aircraft to roll to the right, which is the opposite of the intended turn. Such tendency is called an *adverse roll*.

Then, review the lower image in Figure 11-38. It shows how the ruddervators on an inverted V-tail would have to deflect in order to move the tail to the right. The left ruddervator is deflected TEU and the right one TED. The resulting force vectors on the tail cause a tendency that would roll the aircraft to the left — in the intended direction. In short, the inverted V-tail helps the pilot execute the turn. The short version of the above is: a V-tail rolls the airplane *out of* a turn. An inverted V-tail rolls the airplane *into* the turn. The latter is more desirable. Such tendency is called a *proverse roll*.

It is an additional drawback that the airplane's rotation about the pitch axis is due to the vertical component of the lift acting on the V-tail. For this reason, a larger tail is needed. There is a control system interaction, such that

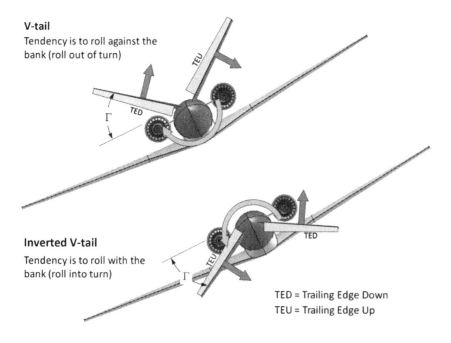

FIGURE 11-38 The difference in yaw response of a V-tail and an inverted V-tail. The intent of the pilot is to bank to the left and coordinate the turn with the rudder. If high rudder deflection is required, a conventional V-tail will tend to rotate the plane out of the turn, whereas an inverted V-tail will rotate it into the turn.

stepping on the rudder pedals may cause a nose pitch-up or -down moment, depending on the geometry of the tail and the airfoil. This may call for ruddervator deflection differentials to be built into the control system. For example, such a differential control may result in deflections in which TED ruddervator deflects $-6°$ while the TEU ruddervator deflects $+10°$. And finally, an important attribute of a V-tail is that it increases the dihedral effect of the aircraft. Consequently, a reduced Dutch roll damping should be expected as well as more of a "rolling" type of Dutch roll. Such aircraft, in particular high-flying jets, may have to feature a yaw damper (which of course all high-flying jets feature anyway).

Simplified Theory of V-tails

Reference 19 lists a number of expressions that are handy in the design of V-tails. The reference presents the following set of experimentally verified formulations to help with the design of the configuration (derivation is provided in the reference document):

Lift curve slope of the V-tail:

$$C_{L_{\alpha_{HT}}} = C_{L_{\alpha_N}} \cos^2 \Gamma \qquad (11\text{-}29)$$

Elevator authority of the V-tail:

$$C_{L_{\delta_e}} = C_{L_{\alpha_N}} \tau \cos \Gamma \qquad (11\text{-}30)$$

Side force slope of the V-tail:

$$C_{Y_{\beta_r}} = -K C_{L_{\alpha_N}} \sin^2 \Gamma \qquad (11\text{-}31)$$

Rudder authority of the V-tail:

$$C_{Y_{\delta_r}} = K C_{L_{\alpha_N}} \tau \sin \Gamma \qquad (11\text{-}32)$$

where
Γ = Dihedral of the V-tail
$C_{L_{\alpha_N}}$ = lift curve slope of the planform at a zero dihedral
τ = control effectiveness parameter
K = the ratio of sum of lifts obtained by equal and opposite changes in angle-of-attack of two semi-spans of tail to lift obtained by an equal change in angle-of-attack for complete tail

Formulas for the control effectiveness parameter, τ, are given in the reference document, but it is a function of the tail and wing geometry, the V-tail dihedral, and elevator authority, to name a few.

11.3.5 Inverted V-tail

The inverted V-tail (Figure 11-39) shares many of the pros and cons of the V-tail, but has seen far less use in the industry. The best-known example of its use is the

FIGURE 11-39 An inverted V-tail configuration.

homebuilt single-seat monoplane Aerocar Mini IMP, although some versions featured a vertical tail addition resulting in an inverted Y-tail (see later). The General Atomics Predator, an unmanned military aircraft, also has an inverted V-tail, although some would object to that definition as it also has a vertical fin between the two outboard tails — sort of like a "collapsed" inverted Y-tail. Additionally, the Elbit Hermes 90 unmanned aerial vehicle (UAV) features an interesting inverted V-tail on an elevated tailboom.

A great advantage of the inverted V-tail configuration is the proverse roll tendency discussed in the previous section. Its greatest drawback is that its tips reduce the T-O rotation and flare potential of the aircraft. For this reason, the tail must be mounted high on the fuselage and long main landing gear legs are likely to be required to improve the rotation capability of the airplane.

11.3.6 Y-tail

From a certain point of view, the Y-tail (Figure 11-40) is a variation of the V-tail. One might say it is a V-tail with an additional vertical surface. Like the V-tail it requires a ruddervator control mixer (otherwise it is just an inverted conventional tail with the horizontal featuring a very high dihedral). However, sometimes a rudder is also mounted on the vertical ventral fin on the bottom, giving the aircraft increased rudder authority. The most notable aircraft that feature this tail are the General Atomics Predator B and the cancelled LearAvia Lear Fan 2100 aircraft.

FIGURE 11-40 A Y-tail configuration.

Among the tail's advantages is an improvement to the directional stability and Dutch roll damping provided by the vertical element of the tail and this presents an improvement over the conventional V-tail configuration. This allows for a shorter-span V-tail. Even a relatively short-span vertical element can provide substantial improvement in the Dutch roll damping. A major drawback with the configuration is that it presents a serious challenge to rotation during T-O and flare during landing. This can put a significant limit on the permissible span of the vertical surface, reducing its ability to improve directional stability and Dutch roll damping. However, as usual, the practicality of a particular tail configuration ultimately depends on the general geometry and, as the aforementioned aircraft indicate, a Y-tail can be successfully featured.

11.3.7 Inverted Y-tail

Many of the pros and cons of the Y-tail also apply to the inverted Y-tail. However, the tail would bring about great improvements in stall recovery as the down-pointing tail feathers will pitch into unseparated flow, which contrasts the Y-tail's tendency to rotate into the wing wake. By far the best-known aircraft that features an inverted Y-tail is the cold-era fighter McDonnell Douglas F-4 Phantom, although some would argue it is a conventional tail configuration whose HT has a large anhedral. The greatest detriment for the configuration is shown in Figure 11-41, in the case of a conventional jetliner in which it would severely limit the ability of the aircraft to rotate for T-O or landing.

11.3.8 H-tail

The H-tail (Figure 11-42) is a good solution when the overall height of an airplane is an issue and a single VT is simply too tall. It is a good configuration for twin-engine propeller aircraft as well, providing a means to supplement the restoring yawing moment required to balance the asymmetric moment caused in a one engine inoperative (OEI) situation. By placing the two VT surfaces inside the propwash of each engine, should one engine be lost only one of the two surfaces experiences a reduction in dynamic pressure. Sometimes, the VT is installed with a slight incidence to further supplement the restoring yawing moment (e.g. Breguet Br-695), but this may not be required, as the propwash tends to follow the shape of the fuselage and be bent, placing the VT at an AOA that promotes the generation of this restoring yawing moment.

FIGURE 11-41 An inverted Y-tail configuration.

FIGURE 11-42 An H-tail configuration.

Military aircraft of the past often featured an H-tail, where the vertical surfaces were of a very small AR, frequently in the 1.0 range. The benefit of this is a VT surface that stalls at a very high AOY, which helps preventing the airplane from experiencing rudder lock (see Section 11.2.11, *The dorsal fin*). For this configuration a dorsal fin is not necessary. The configuration also increases the efficiency of the horizontal tail through endplate effects. This effect promotes a uniform lift distribution by eliminating lift drop-off at the tip and allows the size (and thus weight and drag) of the HT to be somewhat reduced (see Section 10.5.8, *Endplate wingtip*).

An important detriment of the configuration is reduced flutter speed. This is caused by the reduction in the natural frequency of the HT because of the mass of the VT being placed at its tip. This reduction in flutter speed must be improved by stiffening the HT, which will increase the overall weight of the configuration. The Lockheed Model 14 Electra, a twin-engine, 10-passenger, transport aircraft developed in the 1930s, is a classic example of an aircraft featuring an H-tail that suffered fatal accidents due to tail flutter.[2] It is not clear whether this was due to the rudder fluttering (which can happen on all aircraft) or the HT/VT combination itself, although this does not change the above statement. An additional flaw is that the tail experiences peculiar air loading scenarios at high AOY that may also call for an increase in structural weight. The configuration will also

---

[2]Source: www.airdisasters.co.uk/. The airplane was being operated by Northwest Airlines on October 1, 1938. The problem was solved by increasing the tail balancing – most likely that of the rudder.

require the rudder control system to feature a split so both rudders can move in harmony, which increases control system complexity. And finally, there is higher interference drag and the lift curve slope of the tail is small due to the small AR.

11.3.9 Three-surface Configuration

A three-surface configuration consists of a main wing with a foreplane and a horizontal tail. The foreplane is so called as it is positioned ahead of the wing. The horizontal tail is positioned aft of the main wing identical to that of the conventional configuration. The idea with the configuration is to place the CG such that all lifting surfaces can contribute to lift, making the configuration more efficient through a reduction in lift-induced drag. Of course, this is offset by an increase in interference drag.

The main lifting surface can consist of one or more separate wings. In this context, a biplane with two stabilizing surfaces (effectively a four-surface aircraft) should be considered a three-surface aircraft, because the intent of the two main wings is to generate lift and not stabilizing moments. Modern times have only seen a few examples of such aircraft making it to the production stage. However, a number of three-surface biplane configurations were developed in the early part of the twentieth century. Among the first was the Voisin 1907 biplane, also called the Voisin-Farman No. 1 after the French aviator and aircraft designer Henri Farman (1874−1958). Farman designed and built the Henri Farman biplane, which first flew in 1909. It is also known as the Farman III. Other airplanes include the Bristol Boxkite, which first flew in 1910, and the Cody Biplane No. 3, which first flew in 1911. At least three aircraft designed by Joseph Albessard featured monoplanes. These were the Albessard 1911, Albessard 1912, and the Albessard Triavion [20], which flew for the first time in 1926. All are peculiar aircraft with the main wing in front, a smaller wing in the middle of the fuselage, and a smaller yet horizontal tail located aft.

The best-known modern three-surface aircraft is the Piaggio P180 Avanti, an Italian twin-turboprop business-class transport aircraft that carries up to 10 occupants. The airplane is a truly unique design intended to compete with business jet aircraft and is touted as the fastest twin-turboprop ever built, capable of 402 KTAS at 38,000 ft ($M \approx 0.70$).[3] That aside, the manufacturer claims the three-surface configuration results in 34% less drag than comparable turbofan aircraft and offers 40% improvement in fuel consumption. The

FIGURE 11-43 A three-surface configuration.

aircraft has a flap on the forewing that deflects trailing edge down (TED) when the flaps are deployed to counter the resulting pitch-down moment. The elevator is located on the aft surface. This arrangement helps keep control system complexity manageable.

The configuration calls for more localized structural reinforcements and more interference drag, offsetting the reduction in lift-induced drag somewhat. The foreplane also places the inboard wing in downwash on, reducing the effectiveness of the wing further. Furthermore, considering the aircraft in Figure 11-43, the foreplane tip vortices might cause a distortion in flow into engine and cause detrimental asymmetry in the compressor loading.

11.3.10 A-tail

The A-tail (Figure 11-44) is a variant of the inverted V-tail, so the reader is referred to Section 11.3.5, *Inverted V-tail*, for further discussion of pros and cons. The configuration is already used in UAVs such as the AAI RQ-7 Shadow and UAV Factory Penguin B UAV, although it has seen very limited use in aircraft that carry people. The Vogt LO-120 ultralight is an example of a small experimental aircraft, as is the Advanced Aero LLC's Inverted V-tail Technology Demonstrator. The former is exceedingly rare and the latter has not

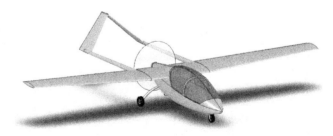

FIGURE 11-44 An A-tail configuration.

---

[3]Compare that to the four-engine Tupolev Tu-114 that holds the ultimate record with a maximum speed of 473.66 KTAS ($M \approx 0.82$), set on April 9, 1960, and would routinely cruise at 415 KTAS ($M \approx 0.72$).

flown yet, although the configuration experienced limited flight testing on a heavily modified Sadler Vampire experimental light plane in 2010.

11.3.11 Twin Tail-boom or U-tail Configuration

The twin tail-boom configuration has seen extensive use in the history of aviation, where it has been used for both large and small aircraft. Examples of GA aircraft include the Cessna 337, of which some 3000 aircraft were built; Adam A500 and A700, and others. Among famous military aircraft are the Northrop P-61 Black Widow, Lockheed P-38 Lightning, and the North American Rockwell OV-10 Bronco. The latest entry of such a configuration is the South African AHRLAC twin-seat reconnaissance aircraft. A typical GA implementation of this configuration is shown in Figure 11-45. Although the version shows extension on the lower VT, the tail effectively forms an arrangement that can be described with the letter U (hence the U-tail name).

For propeller configurations, the HT is often placed in the propwash, giving it boost for soft-field T-O maneuver and good responsiveness. However, this is not always the case. A great advantage of the configuration for smaller aircraft is that the tail-booms form an obstruction making it hard for people to accidentally walk into the prop. Pros and cons of the configuration are detailed in Section 11.3.8, *H-tail*.

Inverted U-tail

An inverted U-tail refers to a configuration in which the HT is placed on the top of the VT (see Figure 11-46). This configuration is used on the North American OV-10, Adam A500, and Adam A700. On propeller aircraft, it places the HT outside the propwash, reducing its drag, but also eliminating a propwash boost over it.

11.3.12 Canard Configuration

A canard configuration is one in which the horizontal stabilizing surface is placed in front of the main wing (see Figure 11-47). This placement generates a destabilizing pitching moment, which contrasts the stabilizing moment generated by tail-aft configurations. The configuration actually dates back to the early days of aviation. As an example, the Wright Flyer, the first heavier-than-air airplane, was a canard. The configuration sporadically appeared in various experimental designs throughout history and got its greatest boost in the early 1970s with advent of the Rutan VariEze and LongEze homebuilt aircraft. Canard aircraft are discussed in more detail in Appendix C2: *Design of canard aircraft*, so the discussion will be limited here.

One of the most touted advantages of the canard configuration is *stall-proofing* which refers to a characteristic in which the canard stalls before the main wing. This lowers the nose and prevents the main wing from stalling. This is often presented as an inability of the airplane to stall. In fact this is an inaccurate description; it *is* indeed possible to stall the main wing. For instance, by an abused stall entry. For instance, it is possible to drive the airplane to a high enough *AOA* to stall the main wing using high rotational momentum. However, the term stall-proofing is generally taken to mean a normal 1 knot/sec deceleration approach to stall. If this deceleration is maintained, such airplanes are indeed designed to gently drop the nose before the main wing stalls. Under those circumstances, the canard can be considered stall-proof.

Another advantage of the configuration is that in stable level flight the lift of the canard points upward, in contrast to a downward force generated by the conventional tail-aft configuration. This means the canard contributes to the total lift of the aircraft, rather than the weight (i.e. the downward-acting force). Consequently, it is commonly maintained the vehicle can be flown at a lower *AOA* in cruise, which means less lift-induced drag, making the concept more efficient than conventional aircraft.

FIGURE 11-45 A twin tail-boom (U-tail) configuration.

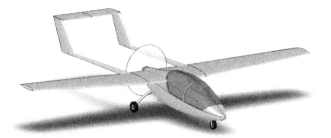

FIGURE 11-46 An inverted U-tail configuration.

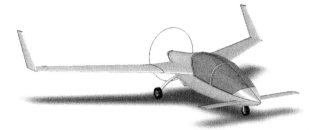

FIGURE 11-47 A canard configuration.

One of the drawbacks of the configuration is that it requires higher T-O airspeed than a tail-aft configuration and, thus, has longer runway requirements. This is a direct consequence of the stall-proofing. It leads to a configuration whose minimum speed is higher than if the main wing were indeed allowed to stall. In simple terms this means the airplane must accelerate to a higher rotation speed during which it covers greater distance. Its landing distance is longer for the same reason.

Another drawback is that the generally compact size of canards results in a tail arm that tends to be smaller than similarly sized conventional aircraft. This must be made up for by a larger stabilizing surface or, more commonly, by sweeping the main wing aft to move the MGC farther away from the canard — thus increasing the arm. Generally, while the popular canard configurations (LongEze, Cozy, and others) display good flight characteristics, their directional stability is often weak (for instance see Ref. [21]). This is due to the short tail arm of the winglets at the tip of the wing. Here, the aft-swept wing is one way to improve this.

While stall-proofing has come to mean an automatic stall recovery, the stall may result in a large nose-drop, depending on how aggressively it is performed. This is due to the fact, and as is pointed out by Larrabee and Abzug [7], that the angular pitch rotation following the stall *increases* the *AOA* on the canard and can delay stall recovery until a steep nose pitch-down attitude is achieved. They point out that an argument can be made that conventional tail-aft aircraft can also recover automatically with the appearance of sudden upwash on the HT as the wings' downwash disappears at stall.

Other drawbacks of the configuration are a reduction in forward and downward visibility caused by the canard which also reduces the effectiveness of the main wing. This must be made up by flying the vehicle at a higher *AOA* and, thus, a greater lift-induced drag than if the downwash were not present. This cuts into any advantage offered by the positive lift of the canard. Additionally, the reduction in the section lift coefficients of the main wing (that result from the downwash) increases the margin between the canard stall and the subsequent wing stall — something that may further increase the minimum speed of the vehicle. The downwash also means the outboard wing generates higher lift than it would in its absence — so the wing bending moments are increased.

11.3.13 Design Guidelines when Positioning the HT for an Aft Tail Configuration

It has been known for a long time that the location of the horizontal tail with respect to the wing of conventional aircraft configuration plays an essential role in the handling characteristics near or at stall. Such characteristics are only problematic if the airplane inherits a tendency to pitch-up uncontrollably near stall. Such problems can be caused by the distortion of the flow field because of the highly separated flow emanating from the wing and fuselage at high *AOA*s. The separated flow field reduces the tail effectiveness, in the worst-case scenario, enough to make recovery impossible. This has already been discussed in terms of T-tails (see Section 11.3.3, *T-tail*). It has also been discussed that T-tails are considered by many to offer an appealing appearance, making them an attractive option to the aircraft designer. However, it is imperative that the designer is aware of both this short-coming and its possible solution, for instance the ventral fin (see Section 11.2.12, *The ventral fin*).

The pitch-up problem was studied in NACA R-1339 [22] and showed that the wing sweep can have a large impact on the nonlinearity in the pitching moment curve near stall. However, this may be complicated by the presence and location of the HT (see Figure 11-48). In the absence of the tail the figure shows the swept wing on its own becomes unstable. However, with a T-tail, the same wing will become stable, only after crossing the horizontal axis. This, in turn, creates another trim point (here at approximately $AOA = 29°$), from which the configuration may not recover (deep-stall).

Straight wings with T-tails, on the other hand, may cause nonlinearities that are restricted to a small range of *AOA*s near stall. For this reason, such aircraft with the horizontal tails mounted high above the main wing may experience reduced stability but not detrimental instability. And they are less likely to have a trim point in the post-stall region. Therefore, the horizontal tail of such aircraft can be sized with trim capability throughout its operational range. In spite of this, this author recommends the designer considers the possibility of deep-stall tendency, either by wind tunnel testing or by a stall evaluation of a radio-controlled model. The former should be required for all aircraft intended to carry people for commercial purposes, the latter can be considered an option for smaller manufacturers or homebuilders. As stated in Ref. 23, the effect of the fuselage may be hard to predict otherwise and can easily devolve into a major design problem, in particular for low-*AR* wings with thin airfoils.

Swept-back wings suffer flow separation at lift coefficients well below that C_{Lmax} and are either unstable or marginally stable through stall. Therefore, configurations with swept-back wings must be treated in a much different and more stringent way than the straight wing and should be wind tunnel tested at all cost. Reference 23 is a compilation of many sources and provides the high-*AOA* properties of a large array of configurations, with straight and swept

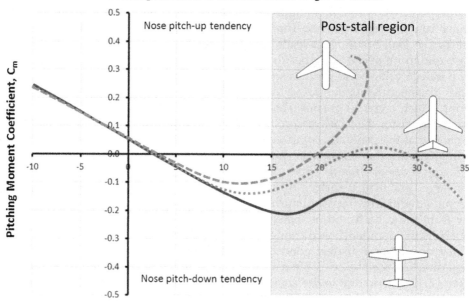

FIGURE 11-48 Pitching moment curves, for example swept-back and straight wing configurations. The curve for the swept-back wing is seen to wander into the upper part of the graph at higher *AOA*s. This is indicative of a nose pitch-up tendency characteristic of swept-back wings.

wings, with and without fuselages and various high-lift devices.

The pitch-up problem was also studied in NACA TN-1093 [23]. While focusing on the wing planform, the investigation, like that of Ref. [23], reaffirms that while the wing sweep can have a large impact on the nonlinearity in the pitching moment curve near stall, the impact of taper ratio and aspect ratio are imperative. In order to prevent instability a large *AR* should have reduced leading edge sweep. A low taper ratio aggravates tip stall and is thus detrimental to the stability. All of this impacts the selection of the HT location because it must remedy the instability issues. The combined effect of *AR* and leading edge sweep from Refs 23 and 24 are depicted in Figure 11-49.

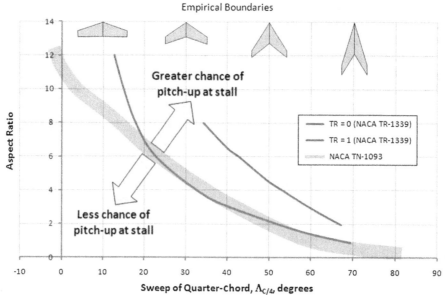

FIGURE 11-49 Empirical pitch-up boundary for a swept-back wing. *Reproduction based on Refs [22] and [23].*

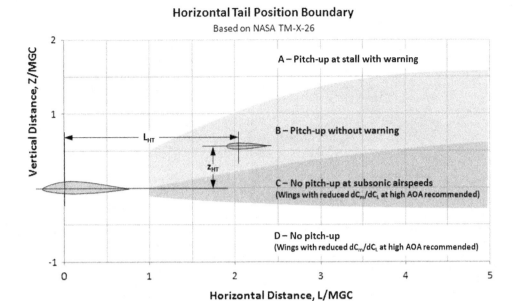

FIGURE 11-50 Recommended placement of the HT behind the wing. *Reproduction based on Ref. [24].*

The pitch-up problem was studied in NASA TM-X-26 [24] using low-aspect-ratio wings (AR ranged from 1.55 to 3.0) with substantial leading edge sweep, such as those more commonly associated with fighter aircraft. While the results, in general, apply only to aircraft with AR ranging from 2 to 5, they give important clues to the trends when comes to the positioning of the HT behind the wing. This is shown in Figure 11-50, which is a reproduction of Figure 3(b) in the reference document. The designer of tail-aft aircraft should compare the intended location of the tail in order to evaluate the probability of an uncontrollable pitch-up at stall. A horizontal tail positioned in Region A or B will likely call for aerodynamic fixes in the form of ventral fins.

11.4 THE GEOMETRY OF THE TAIL

In short, the purpose of the tail is to supply the vehicle with adequate pitch and yaw stability, which, of course, begs the question: how big should the tail surfaces be and where should they be located in order to provide "adequate" stability? In this section we will begin to answer the question by establishing means to evaluate tail geometry and then present some historical reference values of selected aircraft. In the next section we will close the loop by presenting methods that allow the size and location of the tail to be estimated.

Consider the simplified airplane shown in Figure 11-51. The fuselage extending from the wings' quarter-chord plane to the quarter chord of the horizontal or vertical tail (denoted by the dashed lines) has been approximated by a frustum. The stabilizing surfaces have been approximated by flat plates. We will now use the abbreviations HT and VT for the horizontal and vertical tails, respectively. The area of the HT, S_{HT}, is defined as the product of its average chord, C_{HT}, and span, b_{HT}. Similarly, the area of the VT, S_{VT}, is defined as the product of its average chord, C_{VT}, and span, b_{VT}. Then we define the horizontal tail arm, l_{HT}, as the distance from the quarter chord of the wings' MGC to the quarter chord of the MGC of the HT. The vertical tail arm, l_{VT}, is defined in a similar manner.

Note that the combination of the horizontal and vertical tail with its supporting structure is commonly referred to as the "empennage."

11.4.1 Definition of Reference Geometry

Just like for the wing reference area, the reference area for both the horizontal and vertical lifting surfaces must be defined. Figures 11-52 to 11-55 show how aircraft designers typically define the reference area; however, and this needs to be emphasized: the shaded regions may not be what the actual manufacturers did for these particular aircraft, just what would be a likely approach.

The distances l_{HT} and l_{VT} are defined to extend from the quarter chord of the MGC of the wing to the quarter chord of the MGC of the horizontal or vertical tails, respectively.

The aspect ratio of the HT is determined precisely like that of the wing; however, determining the AR of

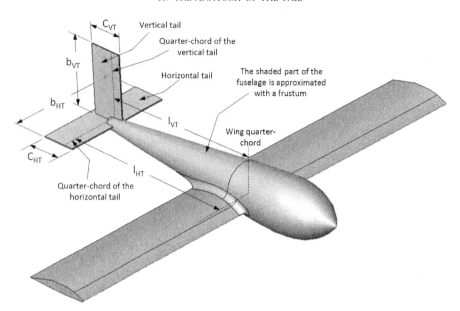

FIGURE 11-51 A simplified airplane layout.

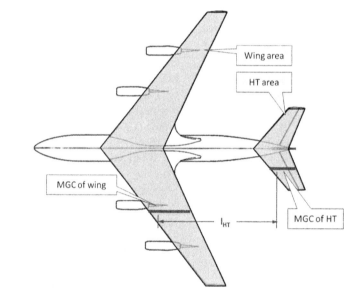

FIGURE 11-52 Definition of HT area and horizontal tail arm for the Boeing 707.

the VT is often confusing because the surface is in many ways a half a wing. This is not a problem though. The definition of the *AR* for the VT is straight forward as shown in Section 11.4.2, below.

11.4.2 Horizontal and Vertical Tail Volumes

The formulation of the static pitch and directional stability, using stability and control theory, leads to the occurrence of terms that have come to be called *horizontal* and *vertical tail volumes*. These non-dimensional ratios compare the product of the stabilizing planform area and its distance from the wing to the basic wing geometry, which is the product of either the reference wing chord and wing area or the reference wing span and wing area. It turns out that the value of this ratio can be correlated between classes of aircraft (see Section 11.4.4, *Recommended initial values for V_{HT}*

11.4 THE GEOMETRY OF THE TAIL

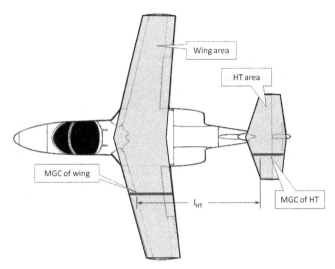

FIGURE 11-53 Definition of HT area and horizontal tail arm for the SAAB 105.

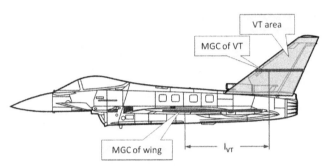

FIGURE 11-54 Definition of VT area and tail arm for the Eurofighter EF-2000 Typhoon. *(Source: www.wikipedia.org)*

and V_{VT}). We define the horizontal and vertical tail volumes as follows:

Horizontal tail volume:

$$V_{HT} = \frac{l_{HT} \cdot S_{HT}}{c_{REF} \cdot S_{REF}} \quad (11\text{-}33)$$

Vertical tail volume:

$$V_{VT} = \frac{l_{VT} \cdot S_{VT}}{b_{REF} \cdot S_{REF}} \quad (11\text{-}34)$$

where

c_{REF} = mean geometric chord, ft or m
b_{REF} = reference span (e.g. wingspan)
S_{REF} = reference area (e.g. wing area)
l_{HT} = horizontal tail arm
l_{VT} = vertical tail arm
S_{HT} = horizontal tail area
S_{VT} = vertical tail area
V_{HT} = horizontal tail volume
V_{VT} = vertical tail volume

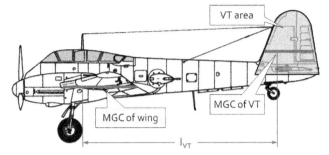

FIGURE 11-55 Definition of VT area and tail arm for the Messerschmitt Me-210 Hornisse. *(Source: www.wikipedia.org)*

If a desired value for the horizontal or vertical tail volume is known, Equations (11-33) and (11-34) can be used to extract a number of information, as will be shown in Section 11.5.

11.4.3 Design Guidelines for HT Sizing for Stick-fixed Neutral Point

It is the purpose of stability and control theory to determine the location of the stick-fixed neutral point and other stability-related limitations. Here, however, it is prudent to give some design guidelines, albeit only for the stick-fixed neutral point. The stick-free neutral point depends on the hinge moments of the elevator and, therefore, is a design concern later in the preliminary design.

It can be shown that the stick-fixed neutral point for aspect ratios in excess of 4 is mostly independent of its magnitude, but instead is mostly a function of the horizontal tail volume, V_{HT}, defined by Equation (11-33). This is shown in Figure 11-56, which depicts how the V_{HT} affects the location of the neutral point. The graph assumes the wing and HT only and symmetric airfoils for both.

As an example of how the graph in Figure 11-56 is read, assume an airplane whose $V_{HT} = 1.0$. The airplane should be expected to have a stick-fixed neutral point of approximately 0.68. This may drop to 0.45 or so once the fuselage is added. In fact, the graph allows the V_{HT} for a desired stick-fixed neutral point to be estimated. This is done using the following expression based on the curve fit in Figure 11-56:

$$V_{HT} = 3.193 - 3.426\sqrt{1.084 - h_n} \quad (11\text{-}35)$$

This expression can be used as a "first stab" estimation of the necessary V_{HT} assuming no fuselage. The presence of a fuselage might move the h_n some 0.1–0.15 forward, assuming the target V_{HT} ranges from 0.4 to 1.0. The reader must be mindful that this estimation says nothing about handling — some aircraft may need smaller or greater V_{HT} depending on other operational requirements.

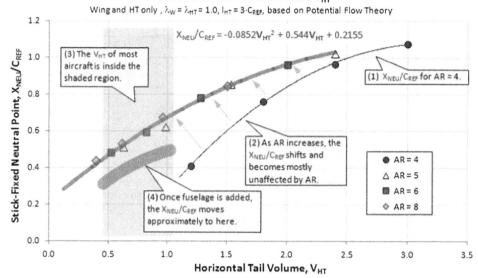

FIGURE 11-56 Impact of horizontal tail volume on the location of the stick-fixed neutral point. It can be seen that above an $AR = 4$, the AR has minimal influence on the location of the neutral point.

EXAMPLE 11-4

Estimate a plausible V_{HT} required for an airplane whose h_n is expected to be close to 0.4 C_{REF} including the fuselage.

Solution

Using Equation (11-35) and assuming that the stick-fixed neutral point, h_n, of the wing-HT system will be close to $0.4 + 0.1 \sim 0.5$; therefore, the designer should initially shoot for a V_{HT} close to:

$$V_{HT} = 3.193 - 3.426\sqrt{1.084 - h_n}$$
$$= 3.193 - 3.426\sqrt{1.084 - 0.5}$$
$$= 0.575$$

11.4.4 Recommended Initial Values for V_{HT} and V_{VT}

One of the first tasks in the initial sizing of the aircraft is to determine candidate dimensions for the HT and VT. This can be easily accomplished by investigating the horizontal and vertical tail volumes of existing classes of aircraft. Raymer [1] presents the V_{HT} and V_{VT} for different classes of aircraft shown in Table 11-4.

Table 11-5 presents helpful reference data for the taper ratio and volumes of the horizontal and vertical tail for a number of classes of aircraft. These are based on measurements of three-views. The designer is strongly urged to compare the tail volume of own design to historical data of the kind presented below.

11.5 INITIAL TAIL SIZING METHODS

Unless other factors are considered of greater importance, the designer should always try to minimize drag. Since the bulk of the airplane's drag results from skin friction, it follows that reducing the wetted area of the empennage is an important goal. This section shows how to size the horizontal and vertical tail such that the combination of its wetted area and that of the tail arm is minimized, while ensuring the desired tail volume is met. The section considers four different methods, each focusing on a selected lifting surface. Thus, the first method only sizes the tail arm assuming a specific horizontal tail volume, V_{HT}, is desired. A proper use of this method would then size

TABLE 11-4 Candidate Horizontal and Vertical Tail Volumes per Class of Aircraft (Based on Ref. [1])

	V_{HT}	V_{VT}
Sailplanes	0.50	0.02
Homebuilt	0.50	0.04
GA – single-engine	0.70	0.04
GA – twin-engine	0.80	0.07
Agricultural aircraft	0.50	0.04
Twin turboprop	0.90	0.08
Flying boat	0.70	0.06
Jet trainer	0.70	0.06
Jet fighter	0.40	0.07
Military cargo/bomber	1.00	0.08
Jet transport	1.00	0.09

the VT based on where it will be placed on the tail arm. The second method turns this around and sizes the VT first, followed by the HT. The third method sizes the HT and VT simultaneously, but this method is most appropriate for conventional airplanes. Finally, the fourth method attempts to minimize the wetted areas for an arbitrary fuselage, whose surface area can be described mathematically. The designer is expected to select one method only and stick with it and not use multiple methods.

11.5.1 Method 1: Initial Tail Sizing Optimization Considering the Horizontal Tail Only

The method assumes a fixed horizontal tail volume and returns the tail arm length that results in the minimum wetted area of the design. The method shown assumes a simplified geometry, but it can be adapted for a more complicated geometry than used here.

Guidance: Note that this method does not consider the presence of the vertical tail, only the horizontal tail. It is assumed that either an alternative method will be used for the VT or it will be placed using the tail arm calculated in this section. If using this method, the designer will have to determine the size of the VT based on the tail arm calculated here. See Section 11.5.3 for a method that considers both the horizontal and vertical tail volumes simultaneously.

Assumptions

The method assumes the fuselage aft of the wings' quarter chord can be approximated by a frustum and the horizontal tail (HT) as a constant-chord lifting surface (see Figure 11-57). If the airplane features a tapered rather than a constant-chord surface, the quarter chord of its MGC should be placed on the quarter-chord line of the constant-chord surface. The purpose of this section is to develop a formulation that allows the designer to determine: (1) the required horizontal tail arm (l_{HT}); (2) horizontal tail area (S_{HT}); and (3) horizontal tail span (b_{HT}) and average chord (c_{HT}) of the tail based on a desired horizontal tail volume, V_{HT}. Note that the method does not account for wetted area that should be removed at the wing and tailplane junctures. In order to prepare the appropriate formulation, some fundamental geometry must be defined. For this reason, consider the dimensions of the frustum and the HT shown in Figure 11-57 and the following definitions:

Horizontal tail area:

$$S_{HT} = b_{HT} \cdot c_{HT} \quad (11\text{-}36)$$

Horizontal tail volume:

$$V_{HT} = \frac{S_{HT} \cdot l_{HT}}{S_{REF} \cdot c_{REF}} \quad (11\text{-}37)$$

Surface area of frustum:

$$S_F = 2\pi \frac{(R_1 + R_2)}{2} \cdot l_{HT} = \pi(R_1 + R_2) \cdot l_{HT} \quad (11\text{-}38)$$

Total wetted area:

$$S_{wet} \approx \pi(R_1 + R_2) \cdot l_{HT} + 2S_{HT} \quad (11\text{-}39)$$

Determination of Tail Arm for a Desired V_{HT} Such that Wetted Area is Minimized

The distance between the wings' MGC quarter chord and the quarter chord of the horizontal tail is calculated using the following expression:

$$l_{HT} = \sqrt{\frac{2 \cdot V_{HT} \cdot S_{REF} \cdot c_{REF}}{\pi(R_1 + R_2)}} \quad (11\text{-}40)$$

Once the l_{HT} is known, the required area, span, and average chord for the horizontal tail can be computed from:

Horizontal tail area:

$$S_{HT} = \frac{V_{HT} \cdot S_{REF} \cdot C_{REF}}{l_{HT}} \quad (11\text{-}41)$$

HT span:

$$b_{HT} = \sqrt{AR_{HT} \times S_{HT}} \quad (11\text{-}42)$$

11. THE ANATOMY OF THE TAIL

TABLE 11-5 Geometric Properties of the Horizontal and Vertical Tails of Selected Aircraft

		GA Single-engine Piston				
			Taper Ratios		Tail Volumes	
Aircraft	AR_{WING}	λ_W	λ_{HT}	λ_{VT}	V_{HT}	V_{VT}
Aermacchi Al-60	7.17	0.660	0.684	0.368	0.695	0.0413
Aeronca Champion	7.27	1.000	0.471	0.339	0.419	0.0240
Auster Autocrat	7.01	1.000	0.328	0.350	0.372	0.0224
Aviamilano F.14 Nibbio	6.99	0.520	0.475	0.377	0.668	0.0418
Beagle Pup	8.04	0.565	1.000	0.909	0.723	0.0523
Beechcraft Bonanza G36	6.13	0.500	0.667	0.467	0.628	0.0492
Beechcraft Sierra Be-24	7.35	1.000	1.000	0.357	0.581	0.0348
Bölkow 208 Junior	6.37	1.000	1.000	0.444	0.493	0.0378
Cessna 152	6.81	0.693	0.604	0.461	0.557	0.0337
Cessna 172	7.45	0.638	0.563	0.443	0.761	0.0419
Cessna 210	8.13	0.690	0.543	0.484	0.766	0.0378
Cessna 400	9.19	0.714	0.694	0.274	0.774	0.0418
Christen A-1 Huskey	6.89	1.010	0.111	0.304	0.362	0.0190
DA20	10.18	0.689	0.361	0.556	0.820	0.0351
DA40 (XLS)	10.76	0.622	0.554	0.660	0.721	0.0423
Lancair Evolution	9.25	0.546	0.619	0.409	0.957	0.0472
Lancair IV-P	6.46	0.495	0.630	0.373	1.000	0.0601
Lancair Legacy	7.88	0.533	0.727	0.445	0.568	0.0563
Mooney Ovation2 GX	8.05	0.476	0.540	0.483	0.606	0.0407
Piper J-3 Cub	6.96	1.000	0.636	0.272	0.479	0.0180
PZL Koliber 160	7.48	1.000	1.000	0.720	0.926	0.0520
SR-20	10.14	0.549	0.633	0.500	0.686	0.0323
SR-22	10.14	0.498	0.613	0.502	0.637	0.0333

		GA Single-engine Piston Aerobatic				
			Taper Ratios		Tail Volumes	
Aircraft	AR_{WING}	λ_W	λ_{HT}	λ_{VT}	V_{HT}	V_{VT}
CAP 10	6.01	0.342	0.503	0.358	0.743	0.0710
CAP 232	5.38	0.479	0.522	0.327	0.499	0.0580
Fournier RF-6b	8.85	0.494	0.562	0.318	0.694	0.0392

		GA Single-engine Turboprop				
			Taper Ratios		Tail Volumes	
Aircraft	AR_{WING}	λ_W	λ_{HT}	λ_{VT}	V_{HT}	V_{VT}
Cessna Caravan	9.72	0.615	0.603	0.397	0.982	0.0560
Pilatus PC-6	8.36	1.000	1.000	1.000	0.592	0.0330
Pilatus PC-12	10.24	0.520	0.694	0.644	0.992	0.0495

TABLE 11-5 Geometric Properties of the Horizontal and Vertical Tails of Selected Aircraft—Cont'd

GA Multi-engine Piston

Aircraft	AR_{WING}	Taper Ratios			Tail Volumes	
		λ_W	λ_{HT}	λ_{VT}	V_{HT}	V_{VT}
Aero 45	8.78	0.394	0.580	0.338	0.369	0.0295
Beechcraft Baron G58	7.19	0.423	0.667	0.429	0.791	0.0464
Beechcraft Duchess	7.98	0.796	1.000	0.615	0.666	0.0477
Beechcraft Duke	7.24	0.327	0.545	0.386	0.821	0.0573
Britten-Norman BN-2 Islander	7.39	1.000	1.000	0.333	0.635	0.0489
Cessna 421	7.87	0.657	0.630	0.505	1.280	0.0782
Diamond DA42	10.11	0.545	0.481	0.511	0.500	0.0338
Partenavia P68	7.74	1.000	1.000	0.449	0.657	0.0488

Multi-engine Turboprop

Aircraft	AR_{WING}	Taper Ratios			Tail Volumes	
		λ_W	λ_{HT}	λ_{VT}	V_{HT}	V_{VT}
Antonov An-10	11.86	0.345	0.354	0.272	1.060	0.0639
Antonov An-24	11.20	0.268	0.438	0.348	1.097	0.0867
Beechcraft 1900	9.80	0.420	0.500	0.571	1.289	0.0796
Bombardier Dash 8 Q100	12.35	0.500	0.769	0.709	1.522	0.1157
Bombardier Dash 8 Q400	12.81	0.418	0.771	0.743	1.769	0.1252
Dassault MD-415 Communauté	7.50	0.361	0.629	0.344	0.579	0.0414
DHC-6 Twin Otter	10.06	0.988	0.990	0.519	0.918	0.0783
DHC-7 Dash 7	10.06	0.427	0.725	0.496	1.099	0.0888
Grumman G-159	10.10	0.379	0.470	0.275	0.869	0.0697
Ilyushin Il-114	10.99	0.375	0.417	0.398	1.140	0.0769
Ilyushin Il-14	12.73	0.368	0.484	0.539	0.847	0.0694
Ilyushin Il-18	9.99	0.333	0.467	0.342	0.869	0.0670
Let L-410 Turbolet	10.78	0.451	0.568	0.452	0.924	0.0773
Lockheed L-188 Electra	7.54	0.392	0.333	0.221	0.884	0.0694
NORD 262	9.15	0.580	0.517	0.543	1.182	0.0604

Business Jet Aircraft

Aircraft	AR_{WING}	Taper Ratios			Tail Volumes	
		λ_W	λ_{HT}	λ_{VT}	V_{HT}	V_{VT}
Bombardier Learjet 23	5.46	0.509	0.500	0.512	0.639	0.0743
Bombardier Learjet 29	7.25	0.381	0.462	0.600	0.613	0.0549
Bombardier Learjet 45	7.32	0.376	0.435	0.735	0.652	0.0519
IAI Westwind	6.04	0.335	0.429	0.333	0.698	0.0768
Lockheed JetStar	5.44	0.373	0.286	0.323	0.559	0.0597

(Continued)

TABLE 11-5 Geometric Properties of the Horizontal and Vertical Tails of Selected Aircraft—Cont'd

Commercial Jet Transport Aircraft						
		Taper Ratios			Tail Volumes	
Aircraft	AR_{WING}	λ_W	λ_{HT}	λ_{VT}	V_{HT}	V_{VT}
Airbus A320	9.49	0.317	0.272	0.263	1.409	0.1148
Airbus A340	13.53	0.263	0.424	0.284	1.106	0.0810
Airbus A380	7.52	0.249	0.386	0.369	0.719	0.0518
BAe BAC-111 Series 475	8.48	0.322	0.641	0.737	0.851	0.0425
BAe-146	8.89	0.382	0.435	0.679	1.537	0.1128
Boeing 707-320B	6.96	0.275	0.412	0.346	0.545	0.0431
Boeing 727-200	6.86	0.304	0.342	0.607	0.688	0.0667
Boeing 737-100	7.88	0.344	0.391	0.264	1.379	0.0746
ERJ-145	7.90	0.272	0.643	0.505	0.908	0.0709
Tupolev Tu-154	7.00	0.252	0.492	0.533	0.552	0.0589
Sailplanes						
		Taper Ratios			Tail Volumes	
Aircraft	AR_{WING}	λ_W	λ_{HT}	λ_{VT}	λ_{HT}	λ_{VT}
PW-5	17.83	0.184	1.000	0.414	0.766	0.0192
Akaflieg D-36	24.73	0.264	0.514	0.537	0.456	0.0166
Avialsa A-60	20.56	0.409	0.527	0.441	0.515	0.0182
Caproni A-21	25.67	0.270	0.432	0.451	0.464	0.0095
Centrair 101 Pegase	21.47	0.413	0.664	0.590	0.625	0.0188
Centrair 201 Marianne	20.01	0.357	0.581	0.675	0.490	0.0157
EIRI PIK-20E	22.46	0.374	0.680	0.684	0.674	0.0177
Fournier RF-4D	11.22	0.420	0.437	0.312	0.549	0.0185
Grob Astir CS	18.24	0.392	0.820	0.594	0.594	0.0135
Grob G-109	15.94	0.443	0.617	0.574	0.517	0.0139

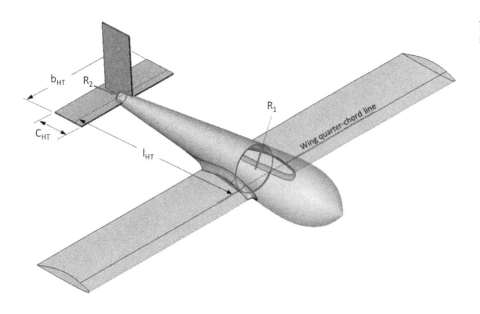

FIGURE 11-57 Definition of dimensions for the HT.

Average chord of the HT:

$$C_{avg} = \frac{b_{HT}}{AR_{HT}} \quad (11\text{-}43)$$

As we have seen before, Equation (11-41) is obtained from Equation (11-37), and the two remaining ones, Equations (11-42) and (11-43), are obtained from the formulation in Section 9.2, *The trapezoidal wing planform*. The formulation requires a prior knowledge of the aspect ratio of the HT, but this typically ranges from 3 to 5. The reader should evaluate the AR_{HT} for existing aircraft in the same class as the new design to use as an initial value.

Derivation of Equation (11-40)

First, we want to minimize wetted area, so let's solve Equation (11-37) for S_{HT}:

Tail area:

$$S_{HT} = \frac{V_{HT} \cdot S_{REF} \cdot c_{REF}}{l_{HT}} \quad (i)$$

Then, we rewrite Equation (11-39) for S_{WET} by inserting Equation (i) into it:

$$S_{wet} \approx \pi(R_1 + R_2) \cdot l_{HT} + 2S_{HT}$$
$$= \pi(R_1 + R_2) \cdot l_{HT} + 2\frac{V_{HT} \cdot S_{REF} \cdot c_{REF}}{l_{HT}} \quad (ii)$$

Let's differentiate Equation (ii) with respect to l_{HT}:

$$\frac{dS_{wet}}{dl_{HT}} = \frac{d}{dl_{HT}}\left(\pi(R_1 + R_2) \cdot l_{HT} + 2\frac{V_{HT} \cdot S_{REF} \cdot c_{REF}}{l_{HT}}\right)$$
$$= \pi(R_1 + R_2) - \frac{2 \cdot V_{HT} \cdot S_{REF} \cdot c_{REF}}{l_{HT}^2}$$

Set to zero to determine the optimum tail arm:

$$\frac{dS_{wet}}{dl_{HT}} = \pi(R_1 + R_2) - \frac{2 \cdot V_{HT} \cdot S_{REF} \cdot c_{REF}}{l_{HT}^2} = 0$$

$$\Rightarrow l_{HT}^2 = \frac{2 \cdot V_{HT} \cdot S_{REF} \cdot c_{REF}}{\pi(R_1 + R_2)}$$

$$\Rightarrow l_{HT} = \sqrt{\frac{2 \cdot V_{HT} \cdot S_{REF} \cdot c_{REF}}{\pi(R_1 + R_2)}} \quad (11\text{-}40)$$

QED

EXAMPLE 11-5

A small airplane has a reference area $S_{REF} = 110$ ft^2 and a reference chord $c_{REF} = 3.5$ ft. If the tail cone has the dimensions shown in Figure 11-58, determine l_{HT} for $V_{HT} = 0.75$.

Solution

Determine the optimum tail arm using Equation (11-44):

$$l_{HT} = \sqrt{\frac{2 \cdot V_{HT} \cdot S_{REF} \cdot c_{REF}}{\pi(R_1 + R_2)}} = \sqrt{\frac{2 \cdot (0.75) \cdot (110) \cdot (3.5)}{\pi(2.0 + 0.3)}}$$
$$= 8.94 \, ft$$

The resulting wetted area amounts to:

$$S_{wet} \approx \pi(R_1 + R_2) \cdot l_{HT} + 2\frac{V_{HT} \cdot S_{REF} \cdot c_{REF}}{l_{HT}}$$

Plugging in the numbers:

$$S_{wet} \approx \pi(2 + 0.3) \cdot (8.94) + 2\frac{(0.75) \cdot (110) \cdot (3.5)}{(8.94)}$$
$$= 129.2 \, ft^2$$

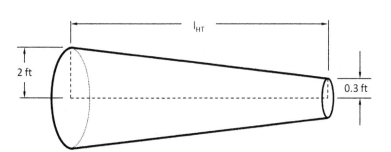

FIGURE 11-58 A definition of the frustum geometry.

EXAMPLE 11-5 (cont'd)

Plot the tail cone area as a function of tail arm (see Figure 11-59).

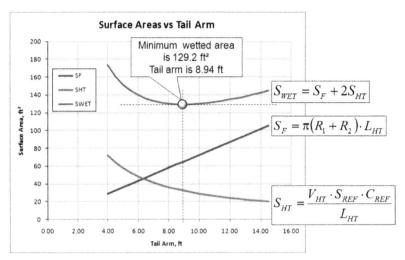

FIGURE 11-59 Graphical determination of the area of minimum drag.

EXAMPLE 11-6

For the airplane of Example 11-5 determine l_{HT} for V_{HT} ranging from 0.50 to 1.00.

Solution

The solution is plotted in Figure 11-60 for horizontal tail volumes (V_{HT}) ranging from 0.5 to 1.0. The left graph shows that as the surface area of the frustum remains constant it is the area of the HT that varies with the V_{HT}.

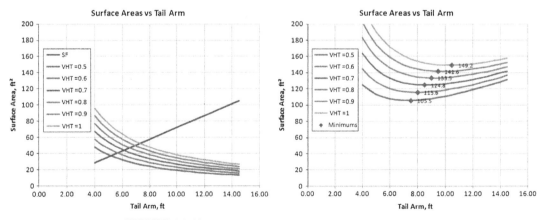

FIGURE 11-60 Change in area of minimum drag with V_{HT}.

EXAMPLE 11-7

A small airplane has a reference area $S_{REF} = 130$ ft², an AR of 16, and TR of 0.5. If the tail cone has the dimensions shown in Figure 11-61, determine l_{HT} for $V_{HT} = 0.75$. Size the HT for an AR of 4 assuming a rectangular planform. Use MGC for c_{ref}.

Solution

Wingspan:

$$AR = \frac{b^2}{S} = \frac{b}{c_{avg}} \Leftrightarrow b = \sqrt{AR \times S}$$
$$= \sqrt{16 \times 130} = 45.6 \text{ ft}$$

Average chord:

$$AR = \frac{b}{c_{avg}} \Leftrightarrow c_{avg} = \frac{b}{AR} = \frac{45.6}{16} = 2.85 \text{ ft}$$

Root chord:

$$c_R = \frac{2c_{avg}}{(1 + \lambda)} = \frac{2(2.85)}{(1 + 0.5)} = 3.80 \text{ ft}$$

Tip chord:

$$c_T = \lambda c_R = 0.5 \times 3.80 = 1.90 \text{ ft}$$

MGC:

$$MGC = \left(\frac{2}{3}\right) c_R \left(\frac{1 + \lambda + \lambda^2}{1 + \lambda}\right)$$
$$= \left(\frac{2}{3}\right)(3.80)\left(\frac{1 + 0.5 + 0.25}{1 + 0.5}\right)$$
$$= 2.956 \text{ ft}$$

Determine the optimum tail arm using Equation (11-7):

$$l_{HT} = \sqrt{\frac{2 \cdot V_{HT} \cdot S_{REF} \cdot c_{REF}}{\pi(R_1 + R_2)}}$$
$$= \sqrt{\frac{2 \cdot (0.75) \cdot (130) \cdot (2.956)}{\pi(1.25 + 0.15)}} = 11.45 ft$$

The resulting wetted area amounts to:

$$S_{wet} \approx \pi(R_1 + R_2) \cdot l_{HT} + 2\frac{V_{HT} \cdot S_{REF} \cdot C_{REF}}{l_{HT}}$$
$$= \pi(1.25 + 0.15) \cdot (11.45) + 2\frac{(0.75) \cdot (130) \cdot (2.956)}{(11.45)}$$
$$= 100.7 \text{ ft}^2$$

Determine tail area using Equation (11-5):

$$S_{HT} = \frac{V_{HT} \cdot S_{REF} \cdot C_{REF}}{l_{HT}} = \frac{0.75 \cdot 130 \cdot 2.956}{11.45} = 25.17 \text{ ft}^2$$

HT span:

$$b_{HT} = \sqrt{AR_{HT} \times S_{HT}} = \sqrt{4 \times 25.17} = 10.03 \text{ ft}$$

Average chord:

$$C_{avg} = \frac{b_{HT}}{AR_{HT}} = \frac{10.03}{4} = 2.51 \text{ ft}$$

When drawn to scale, this airplane looks as shown in Figure 11-62. There is no dihedral or twist yet applied to wings, or taper to the horizontal tail, and this might represent the first "sketch" of a new aircraft.

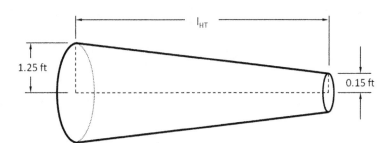

FIGURE 11-61 A definition of the frustum geometry.

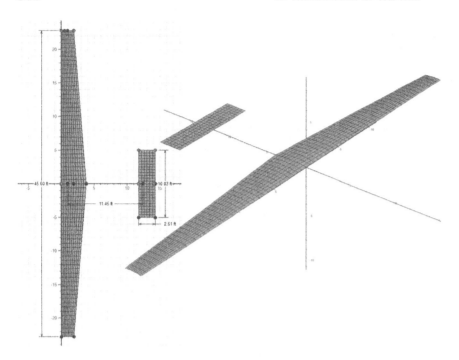

FIGURE 11-62 A simple model based on the analysis in Section 11.5.1.

11.5.2 Method 2: Initial Tail Sizing Optimization Considering the Vertical Tail Only

This method differs from the previous one in that it shows how to size the vertical tail (rather than the horizontal tail) such that the combination of its wetted area and that of the tail arm is minimized, while ensuring the desired vertical tail volume is met. As before, the method assumes a fixed (vertical) tail volume and returns the tail arm length that results in the minimum wetted area of the design. As for the horizontal tail sizing, the method shown assumes a simplified geometry, but it can be adapted for a more complicated one.

Guidance: Note that this method does not consider the presence of the horizontal tail, only the vertical tail. It is assumed that either an alternative method will be used for the HT or it will be placed using the tail arm calculated in this section. If using this method, the designer will have to determine the size of the HT based on the tail arm calculated here. See Section 11.5.3 for a method that considers both the horizontal and vertical tail volumes simultaneously.

Assumptions

Just as before, the method assumes the fuselage aft of the wings' quarter chord can be approximated by a frustum and the vertical tail (VT) as a constant-chord lifting surface (see Figure 11-63). If the airplane features a tapered surface rather than a constant-chord one, the quarter chord of its MGC should be placed on the quarter-chord line of the constant-chord surface. This section also develops a formulation that allows the designer to determine: (1) the required vertical tail arm (l_{VT}); (2) vertical tail area (S_{VT}); and (3) vertical tail span (b_{VT}), and average chord (c_{VT}) of the tail based on a desired vertical tail volume, V_{VT}. In order to prepare the appropriate formulation, some fundamental geometry must be defined. For this reason, consider the dimensions of the frustum and the VT shown in Figure 11-63 and the following definitions:

Vertical tail area:

$$S_{VT} = b_{VT} \cdot c_{VT} \qquad (11\text{-}44)$$

Vertical tail volume:

$$V_{VT} = \frac{S_{VT} \cdot l_{VT}}{S_{REF} \cdot b_{REF}} \qquad (11\text{-}45)$$

Surface area of frustum:

$$S_F = \pi(R_1 + R_2) \cdot l_{VT} \qquad (11\text{-}46)$$

Total wetted area:

$$S_{wet} \approx \pi(R_1 + R_2) \cdot l_{VT} + 2S_{VT} \qquad (11\text{-}47)$$

11.5 INITIAL TAIL SIZING METHODS

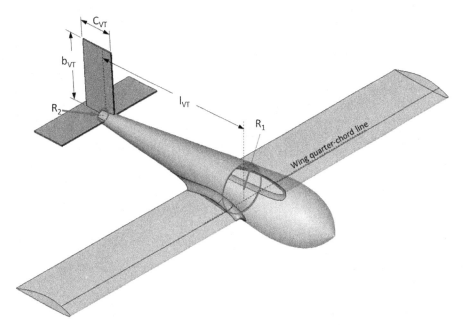

FIGURE 11-63 Definition of dimensions for the VT.

Determination of Tail Arm for a Desired V_{VT} such that Wetted Area is Minimized

The distance between the wings' MGC quarter chord and the quarter chord of the vertical tail is calculated using the following expression:

$$l_{VT} = \sqrt{\frac{2 \cdot V_{VT} \cdot S_{REF} \cdot b_{REF}}{\pi(R_1 + R_2)}} \quad (11\text{-}48)$$

Once the l_{VT} is known, the required area, span, and average chord for the vertical tail can be computed from:

Vertical tail area:

$$S_{VT} = \frac{V_{VT} \cdot S_{REF} \cdot b_{REF}}{l_{VT}} \quad (11\text{-}49)$$

VT span:

$$b_{VT} = \sqrt{AR_{VT} \times S_{VT}} \quad (11\text{-}50)$$

Average chord of the VT:

$$C_{avg} = \frac{b_{VT}}{AR_{VT}} \quad (11\text{-}51)$$

As we have seen before, Equation (11-49) is obtained from Equation (11-45), and the two remaining ones, Equations (11-50) and (11-51), are obtained from the formulation in Chapter 7, *Geometry of lifting surfaces*. The formulation requires a prior knowledge of the aspect ratio of the VT, but this typically ranges from 1.5 to 2.5. The reader should evaluate the AR_{VT} for existing aircraft in the same class as the new design to use as an initial value.

Derivation of Equation (11-48)

First, we want to minimize wetted area, so let's solve Equation (11-45) for S_{VT}:

Tail area:

$$S_{VT} = \frac{V_{VT} \cdot S_{REF} \cdot b_{REF}}{l_{VT}} \quad (i)$$

Then, we rewrite Equation (11-47) for S_{WET} by inserting Equation (i) into it:

$$S_{wet} \approx \pi(R_1 + R_2) \cdot l_{VT} + 2S_{VT}$$

$$= \pi(R_1 + R_2) \cdot l_{VT} + 2\frac{V_{VT} \cdot S_{REF} \cdot b_{REF}}{l_{VT}} \quad (ii)$$

Let's differentiate this with respect to l_{VT}:

$$\frac{dS_{wet}}{dl_{VT}} = \frac{d}{dl_{VT}}\left(\pi(R_1 + R_2) \cdot l_{VT} + 2\frac{V_{VT} \cdot S_{REF} \cdot b_{REF}}{l_{VT}}\right)$$

$$= \pi(R_1 + R_2) - \frac{2 \cdot V_{VT} \cdot S_{REF} \cdot b_{REF}}{l_{VT}^2}$$

Set to zero to determine the optimum tail arm:

$$\frac{dS_{wet}}{dl_{VT}} = \pi(R_1 + R_2) - \frac{2 \cdot V_{VT} \cdot S_{REF} \cdot b_{REF}}{l_{VT}^2} = 0$$

$$\Rightarrow l_{VT}^2 = \frac{2 \cdot V_{VT} \cdot S_{REF} \cdot b_{REF}}{\pi(R_1 + R_2)}$$

$$\Rightarrow l_{VT} = \sqrt{\frac{2 \cdot V_{VT} \cdot S_{REF} \cdot b_{REF}}{\pi(R_1 + R_2)}} \quad (11\text{-}48)$$

QED

EXAMPLE 11-8

The small airplane of Example 11-7 has a reference area $S_{REF} = 130$ ft^2, an AR of 16, TR of 0.5, and a tail arm of 11.45 ft. If the tail cone has the dimensions shown in Figure 11-64, determine l_{VT} for $V_{VT} = 0.020$. Select the longer tail arm of the two and size the VT for an AR of 2 assuming a rectangular planform.

Solution

Determine the optimum tail arm using Equation (11-15):

$$l_{VT} = \sqrt{\frac{2 \cdot V_{VT} \cdot S_{REF} \cdot b_{REF}}{\pi(R_1 + R_2)}}$$

$$= \sqrt{\frac{2 \cdot (0.020) \cdot (130) \cdot (45.6)}{\pi(1.25 + 0.15)}}$$

$$= 7.34 \text{ ft}$$

Since l_{HT} is larger than l_{VT} we will us the former because this will ensure the HT has adequately long arm. For this reason, we will have to determine the tail area based on using l_{HT} with Equation (11-13) rather than l_{VT}:

$$S_{VT} = \frac{V_{VT} \cdot S_{REF} \cdot b_{REF}}{l_{HT}} = \frac{0.020 \cdot 130 \cdot 45.6}{11.45} = 10.35 \text{ ft}^2$$

VT span:

$$b_{VT} = \sqrt{AR_{VT} \times S_{VT}} = \sqrt{2 \times 10.35} = 4.55 \text{ ft}$$

Average chord of the VT:

$$c_{avg} = \frac{b_{VT}}{AR_{VT}} = \frac{4.55}{2} = 2.28 \text{ ft}$$

So, we have defined the primary dimensions of the airplane in Example 11-7. The second iteration of the concept aircraft is shown in Figure 11-65.

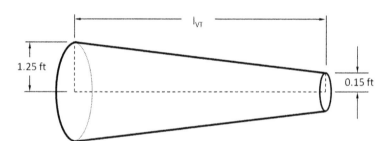

FIGURE 11-64 Dimensions of the frustum tail section.

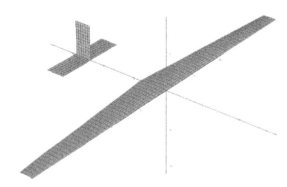

FIGURE 11-65 The simple airplane of Examples 11-3 and 11-4 with a VT added.

FIGURE 11-66 Method 3 applies to tail feathers whose geometric centroids are pretty much equi-distance from the CG, like that of the Aeronca 7AC Champion (left). On the other hand, it is not applicable to tail surfaces arranged like that of the Aero L-39C Albatros (right). *(Photos by Phil Rademacher)*

11.5.3 Method 3: Initial Tail Sizing Optimization Considering Horizontal and Vertical Tail

Many airplane configurations feature a HT and VT whose centers of lift are in close enough proximity with one another (along the longitudinal axis) to consider a simultaneous optimization for both. Figure 11-66 shows two example aircraft of many whose HT and VT can be considered close enough to optimize the tail dimensions based on both simultaneously. This section will formulate such an optimization.

Guidance: Use this method if the location of the centroids of the HT and VT are expected to be relatively close to one another along the x-axis.

Assumptions

In accordance with the previous methods, the fuselage aft of the wings' quarter chord is approximated by a frustum and the HT and VT both as constant-chord lifting surfaces (see Figure 11-67). If the airplane features a tapered surface rather than a constant-chord one, the quarter chord of its MGC should be placed on the quarter-chord line of the constant-chord surface. The purpose of this section is to develop a formulation that allows the designer to determine: (1) the required horizontal tail arm (l_T); (2) horizontal and vertical tail areas (S_{HT} and S_{VT}); and (3) horizontal and vertical tail span (b_{HT} and b_{VT}), and average chord (c_{HT} and c_{VT}) of the tail based on desired horizontal and vertical tail volumes, V_{HT} and V_{VT}. Wetted area that should be removed at the wing and tailplane junctures is not accounted for.

Formulate expressions for the S_{HT} and S_{VT} in terms of the tail arm L_T. The planform areas of the HT and VT are defined using Equations (11-36) and (11-44).

Horizontal tail volume:

$$V_{HT} = \frac{S_{HT} \cdot l_T}{S_{REF} \cdot c_{REF}} \quad (11\text{-}52)$$

Vertical tail volume:

$$V_{VT} = \frac{S_{VT} \cdot l_T}{S_{REF} \cdot b_{REF}} \quad (11\text{-}53)$$

Surface area of frustum:

$$S_F = \pi(R_1 + R_2) \cdot l_T \quad (11\text{-}54)$$

Wetted area:

$$S_{wet} \approx \pi(R_1 + R_2) \cdot l_T + 2 S_{HT} + 2 S_{VT} \quad (11\text{-}55)$$

Determination of Tail Arm for the Desired V_{HT} and V_{VT}

Calculate the vertical tail arm from the following expression:

$$l_T = \sqrt{\frac{2 \cdot S_{REF}(V_{HT} \cdot c_{REF} + V_{VT} \cdot b_{REF})}{\pi(R_1 + R_2)}} \quad (11\text{-}56)$$

Once the l_T is known, the required area, span, and average chord for the HT and VT can be computed from:

Horizontal tail area:

$$S_{HT} = \frac{V_{HT} \cdot S_{REF} \cdot c_{REF}}{l_T} \quad (11\text{-}57)$$

Vertical tail area:

$$S_{VT} = \frac{V_{VT} \cdot S_{REF} \cdot b_{REF}}{l_T} \quad (11\text{-}58)$$

HT span:

$$b_{HT} = \sqrt{AR_{HT} \times S_{HT}} \quad (11\text{-}59)$$

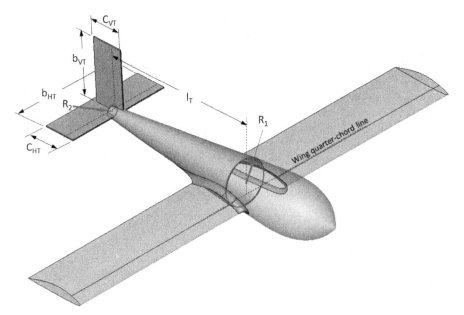

FIGURE 11-67 Definition of dimensions for the HT and VT.

VT span:

$$b_{VT} = \sqrt{AR_{VT} \times S_{VT}} \quad (11\text{-}60)$$

Average chord of HT:

$$(C_{avg})_{HT} = \frac{b_{HT}}{AR_{HT}} \quad (11\text{-}61)$$

Average chord of VT:

$$(C_{avg})_{VT} = \frac{b_{VT}}{AR_{VT}} \quad (11\text{-}62)$$

As we have seen before, Equation (11-49) is obtained from Equation (11-45), and the two remaining ones, Equations (11-50) and (11-51), are obtained from the formulation in Chapter 7, *Geometry of lifting surfaces*. The formulation requires a prior knowledge of the aspect ratio of the VT, but this typically ranges from 1.5 to 2.5. The reader should evaluate the AR_{VT} for existing aircraft in the same class as the new design to use as an initial value.

Derivation of Equation (11-56)

First, we want to minimize wetted area, so let's write Equation (11-15) explicitly for l_T:

$$S_{wet} = \pi(R_1 + R_2) \cdot l_T + 2\frac{V_{HT} \cdot S_{REF} \cdot c_{REF}}{l_T}$$
$$+ 2\frac{V_{VT} \cdot S_{REF} \cdot b_{REF}}{l_T} \quad (i)$$

Let's differentiate Equation (i) with respect to l_T:

$$\frac{dS_{wet}}{dl_T} = \frac{d}{dl_T}\left(\pi(R_1 + R_2) \cdot l_T + 2\frac{V_{HT} \cdot S_{REF} \cdot c_{REF}}{l_T}\right.$$
$$\left. + 2\frac{V_{VT} \cdot S_{REF} \cdot b_{REF}}{l_T}\right)$$

$$= \pi(R_1 + R_2) - \frac{2 \cdot S_{REF}}{l_T^2}(V_{HT} \cdot c_{REF} + V_{VT} \cdot b_{REF})$$

Set to zero to determine the optimum tail arm:

$$\frac{dS_{wet}}{dl_T} = \pi(R_1 + R_2) - \frac{2 \cdot S_{REF}}{l_T^2}(V_{HT} \cdot c_{REF} + V_{VT} \cdot b_{REF}) = 0$$

$$\Rightarrow l_T^2 = \frac{2 \cdot S_{REF}(V_{HT} \cdot c_{REF} + V_{VT} \cdot b_{REF})}{\pi(R_1 + R_2)}$$

$$\Leftrightarrow l_T = \sqrt{\frac{2 \cdot S_{REF}(V_{HT} \cdot c_{REF} + V_{VT} \cdot b_{REF})}{\pi(R_1 + R_2)}} \quad (11\text{-}56)$$

QED

EXAMPLE 11-9

A small airplane has a reference area $S_{REF} = 130$ ft^2, an AR of 16, and TR of 0.5. If the tail cone has the dimensions shown in Figure 11-68, determine l_T for $V_{HT} = 0.75$ and $V_{VT} = 0.04$. Size the HT for an AR of 4 and the VT for an AR of 2, assuming a rectangular planform. Here use MGC for c_{ref}.

Solution

Wingspan:

$$AR = \frac{b^2}{S} = \frac{b}{c_{avg}} \Leftrightarrow b = \sqrt{AR \times S} = \sqrt{16 \times 130}$$
$$= 45.6 \text{ ft}$$

Average chord:

$$AR = \frac{b}{c_{avg}} \Leftrightarrow c_{avg} = \frac{b}{AR} = \frac{45.6}{16} = 2.85 \text{ ft}$$

Root chord:

$$c_R = \frac{2c_{avg}}{(1+\lambda)} = \frac{2(2.85)}{(1+0.5)} = 3.80 \text{ ft}$$

Tip chord:

$$c_T = \lambda c_R = 0.5 \times 3.80 = 1.90 \text{ ft}$$

MGC:

$$MGC = \left(\frac{2}{3}\right) c_R \left(\frac{1+\lambda+\lambda^2}{1+\lambda}\right)$$
$$= \left(\frac{2}{3}\right)(3.80)\left(\frac{1+0.5+0.25}{1+0.5}\right) = 2.956 \text{ ft}$$

Determine the optimum tail arm using Equation (11-56):

$$l_T = \sqrt{\frac{2 \cdot S_{REF}(V_{HT} \cdot c_{REF} + V_{VT} \cdot b_{REF})}{\pi(R_1 + R_2)}}$$
$$= \sqrt{\frac{2 \cdot 130(0.75 \times 2.956 + 0.02 \times 45.6)}{\pi(1.25 + 0.15)}} = 13.60 \text{ ft}$$

Determine HT area using Equation (11-57):

$$S_{HT} = \frac{V_{HT} \cdot S_{REF} \cdot c_{REF}}{l_T} = \frac{0.75 \cdot 130 \cdot 2.956}{13.60} = 21.19 \text{ ft}^2$$

HT span:

$$b_{HT} = \sqrt{AR_{HT} \times S_{HT}} = \sqrt{4 \times 21.19} = 9.74 \text{ ft}$$

Average chord:

$$c_{avg} = \frac{b_{HT}}{AR_{HT}} = \frac{9.74}{4} = 2.44 \text{ ft}$$

Determine VT area using Equation (11-58):

$$S_{VT} = \frac{V_{VT} \cdot S_{REF} \cdot b_{REF}}{l_T} = \frac{0.020 \cdot 130 \cdot 45.6}{13.60} = 8.72 \text{ ft}^2$$

VT span:

$$b_{VT} = \sqrt{AR_{VT} \times S_{VT}} = \sqrt{2 \times 8.72} = 4.18 \text{ ft}$$

Average chord:

$$c_{avg} = \frac{b_{VT}}{AR_{VT}} = \frac{4.18}{2} = 2.09 \text{ ft}$$

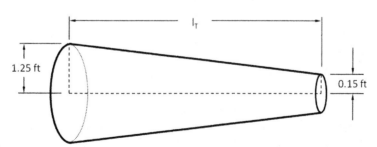

FIGURE 11-68 A definition of the frustum geometry.

EXAMPLE 11-9 (cont'd)

The resulting geometry is shown in Figure 11-69.

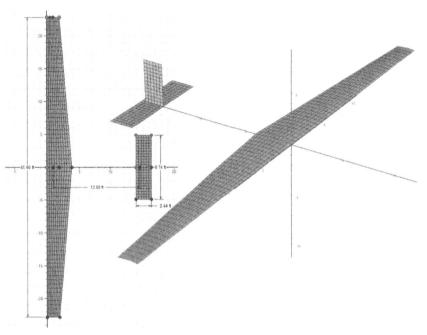

FIGURE 11-69 The simple aircraft of Examples 11-3 through 11-5 after applying the combined optimization method.

EXAMPLE 11-10

A small airplane is being designed with a twin tail-boom configuration (see Figure 11-70). Derive an expression for the drag optimized tail arm, similar to Equation (11-16). Use R_1 and R_2 to represent the base and top radii of each boom, and assume S_{VT} to represent to total VT area.

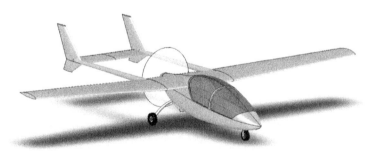

FIGURE 11-70 A small twin tail-boom configuration.

EXAMPLE 11-10 (cont'd)

Solution

Following the same procedure we define the total wetted area as follows, noting that we have two tailbooms and not one as before:

$$S_{wet} = 2\pi(R_1 + R_2) \cdot l_T + 2\frac{V_{HT} \cdot S_{REF} \cdot c_{REF}}{l_T}$$
$$+ 2\frac{V_{VT} \cdot S_{REF} \cdot b_{REF}}{l_T} \quad \text{(i)}$$

Let's differentiate Equation (i) with respect to l_T:

$$\frac{dS_{wet}}{dl_T} = \frac{d}{dl_T}\left(2\pi(R_1 + R_2) \cdot l_T + 2\frac{V_{HT} \cdot S_{REF} \cdot c_{REF}}{l_T} + 2\frac{V_{VT} \cdot S_{REF} \cdot b_{REF}}{l_T}\right)$$
$$= 2\pi(R_1 + R_2) - \frac{2 \cdot S_{REF}}{l_T^2}(V_{HT} \cdot c_{REF} + V_{VT} \cdot b_{REF})$$

Set to zero to determine the optimum tail arm:

$$\frac{dS_{wet}}{dl_T} = 2\pi(R_1 + R_2) - \frac{2 \cdot S_{REF}}{l_T^2}(V_{HT} \cdot c_{REF} + V_{VT} \cdot b_{REF})$$
$$= 0$$

$$\Rightarrow l_T^2 = \frac{S_{REF}(V_{HT} \cdot c_{REF} + V_{VT} \cdot b_{REF})}{\pi(R_1 + R_2)}$$

$$l_T = \sqrt{\frac{S_{REF}(V_{HT} \cdot c_{REF} + V_{VT} \cdot b_{REF})}{\pi(R_1 + R_2)}} \quad \text{(ii)}$$

EXERCISES

(1) The Cessna 172 Skyhawk is a very popular single-engine GA aircraft that has a reference area $S_{REF} = 174$ ft^2 and a wingspan of 36 ft. If its double-tapered wing has a reference TR of 0.7, determine a minimum drag L_T using historical data from Table 11-4. Size the HT for an AR of 4 and the VT for an AR of 2, assuming a rectangular planform. Here use MGC for c_{ref} and assume its wing is simply tapered. Compare the resulting HT and VT to that of the original airplane. If different, explain why it differs.

VARIABLES

Symbol	Description	Units (UK and SI)
$\frac{d\varepsilon}{d\alpha}$	Downwash gradient	/deg or /rad
AR	Wing aspect ratio	
AR_{HT}	Horizontal tail aspect ratio	
AR_{VT}	Vertical tail aspect ratio	
b	Wingspan	ft or m
b_{HT}	Span of horizontal tail	ft or m
b_{REF}	Wingspan	ft or m
b_{VT}	Span of vertical tail	ft or m
C	Chord length	ft or m
$C/4$	Quarter chord	ft or m
C_{avg}	Average chord length	ft or m
C_D	Drag coefficient	
C_{Dmin}	Minimum drag coefficient	
$C_{D\alpha}$	Change in drag coefficient due to AOA	/deg or /rad
$C_{D\beta}$	Change in drag coefficient due to sideslip angle	/deg or /rad
$C_{D\delta e}$	Change in drag coefficient due to elevator deflection	/deg or /rad
$C_{D\delta f}$	Change in drag coefficient due to flap deflection	/deg or /rad
$C_{D\delta spoiler}$	Change in drag coefficient due to spoiler deflection	/deg or /rad
C_{HT}	Chord length of horizontal tail	ft or m
C_L	Lift coefficient	
C_l	2D lift coefficient	
C_{LHT}	Horizontal tail lift coefficient	
C_{Lo}	Lift coefficient at zero AOA	
C_{LoHT}	Horizontal tail lift coefficient at zero AOA	
C_{LoW}	Wing lift coefficient	

Symbol	Description	Units (UK and SI)
C_{LW}	Wing lift coefficient	
$C_{L\alpha}$	Change in lift coefficient due to AOA	/deg or /rad
$C_{L\alpha HT}$	Change in horizontal tail lift coefficient due to AOA	/deg or /rad
$C_{l\beta}$	Change in coefficient of rolling moment due to sideslip angle	/deg or /rad
$C_{L\beta}$	Change in lift coefficient due to sideslip angle	/deg or /rad
$C_{L\delta e}$	Change in lift coefficient due to elevator deflection	/deg or /rad
$C_{L\delta f}$	Change in lift coefficient due to flap deflection	/deg or /rad
$C_{L\delta spoiler}$	Change in lift coefficient due to spoiler deflection	/deg or /rad
C_m	2D coefficient of moment	
C_{MGC}	Chord length of MGC	ft or m
C_{mo}	Coefficient of moment at zero AOA	/deg or /rad
C_{mq}	Change in coefficient of pitching moment due to pitch rate	
$C_{m\alpha}$	Change in coefficient of pitching moment due to AOA	/deg or /rad
$C_{m\beta}$	Change in coefficient of pitching moment due to sideslip angle	/deg or /rad
$C_{m\delta e}$	Change in coefficient of pitching moment due to elevator deflection	/deg or /rad
$C_{m\delta f}$	Change in drag coefficient of pitching moment due to flap deflection	/deg or /rad
C_{nr}	Change in coefficient of yawing moment due to yaw rate	
$C_{n\beta}$	Change in coefficient of yawing moment due to sideslip angle	/deg or /rad
$C_{n\delta r}$	Change in coefficient of yawing moment due to rudder deflection	
C_{REF}	Length of wing MGC	ft or m
C_{VT}	Chord length of vertical tail	ft or m
$C_{y\beta}$	Change in coefficient of side force due to sideslip angle	/deg or /rad
D_{FUS}	Drag due to fuselage	lb_f or N
D_{HT}	Drag due to horizontal tail	lb_f or N
D_{LDG}	Drag due to landing gear	lb_f or N
D_{VT}	Drag due to vertical tail	lb_f or N
D_W	Drag due to the wing	lb_f or N
F_H	Horizontal component of lift on a V-tail	lb_f or N
F_V	Vertical component of lift on a V-tail	lb_f or N
h	Distance from airfoil leading edge to CG	ft or m

Symbol	Description	Units (UK and SI)
i_{HT}	Incidence of horizontal tail	deg or rad
i_W	Incidence of wing	deg or rad
L	Lift (context dependent)	lb_f or N
L_{FUS}	Lift due to fuselage	lb_f or N
L_{HT}	Lift due to horizontal tail (context dependent)	lb_f or N
l_{HT}	Distance of horizontal tail $MGC_{C/4}$ to wing $MGC_{C/4}$ along the x-axis	ft or m
L_{LDG}	Lift due to landing gear	lb_f or N
l_T	Distance of both tail $MGC_{C/4}$s to wing $MGC_{C/4}$ along the x-axis	ft or m
L_{VT}	Lift due to vertical tail (context dependent)	lb_f or N
l_{VT}	Distance of vertical tail $MGC_{C/4}$ to wing $MGC_{C/4}$ along the x-axis	ft or m
L_W	Lift due to wing	lb_f or N
M	Moment (context dependent)	$ft \cdot lb_f$ or $N \cdot m$
M	Mach (context dependent)	
M_{FUS}	Pitching moment due to fuselage	$ft \cdot lb_f$ or $N \cdot m$
M_{HT}	Pitching moment due to horizontal tail	$ft \cdot lb_f$ or $N \cdot m$
M_{LDG}	Pitching moment due to landing gear	$ft \cdot lb_f$ or $N \cdot m$
M_T	Pitching moment due to thrust	$ft \cdot lb_f$ or $N \cdot m$
M_{VT}	Pitching moment due to vertical tail	$ft \cdot lb_f$ or $N \cdot m$
M_W	Pitching moment due to wing	$ft \cdot lb_f$ or $N \cdot m$
q	Dynamic pressure	lb_f/ft^2 or Pa
R_1	Radius of frustum base	ft or m
R_2	Radius of frustum tip	ft or m
S	Wing surface area	ft^2 or m^2
S_F	Frustum surface area	ft^2 or m^2
S_{HT}	Horizontal tail surface area	ft^2 or m^2
S_{REF}	Wing surface area	ft^2 or m^2
S_{VT}	Vertical tail surface area	ft^2 or m^2
S_{wet}	Total wetted surface area	ft^2 or m^2
T	Thrust	lb_f or N
V_{HT}	Horizontal tail volume	
V_s	Stalling velocity	ft/s or m/s
V_{VT}	Vertical tail volume	
W	Weight of aircraft	lb_f or N
X_{NEU}	Location of neutral along MGC	%MGC
x_T	Distance of thrustline from CG along x-axis	ft or m

Symbol	Description	Units (UK and SI)
y_T	Distance of thrustline from CG along y-axis	ft or m
z_{HT}	Distance of HT chordline to wing chordline along the z-axis	ft or m
z_T	Distance of thrustline from CG along z-axis	ft or m
Γ	Circulation	$ft^2 \cdot rad/s$ or $m^2 \cdot rad/s$
α	*AOA*	deg or rad
α_{trim}	Trim *AOA*	deg or rad
α_{WB}	Wing-body *AOA*	deg or rad
α_{HT}	Horizontal tail *AOA*	deg or rad
β	Sideslip angle	deg or rad
δ_e	Elevator deflection angle	deg or rad
δ_f	Flap deflection angle	deg or rad
$\delta_{spoiler}$	spoiler deflection angle	deg or rad
ε	Thrust angle (context dependent)	deg or rad
ε	Downwash angle (context dependent)	deg or rad
ε_o	Residual downwash angle	deg or rad
ε_T	Thrust angle	deg or rad
λ_{HT}	Horizontal tail taper ratio	
λ_{VT}	Vertical tail taper ratio	
λ_W	Wing taper ratio	
τ	Control effectiveness parameter	

References

[1] Raymer Daniel. Aircraft Design: A Conceptual Approach. 4th ed. AIAA Education Series; 2006.

[2] NASA CR-2144. Aircraft Handling Quality Data. Heffley, Robert K., and Wayne F. Jewell; Dec. 1972.

[3] Nelson Robert C. Flight Stability and Automatic Control. 2nd ed. McGraw-Hill; 1998.

[4] NASA TM-105977. Icing Effects on Aircraft Stability and Control Determined From Flight Data - Preliminary Results. Ratvasky, T.P., and R.J. Ranaudo; 1993.

[5] http://www.historylink.org/index.cfm?DisplayPage=output.cfm&file_id=2230.

[6] Cook William H. Road to the 707: The Inside Story of Designing the 707. William H. Cook; 1991.

[7] Abzug Malcolm J, Eugene Larrabee E. Airplane Stability and Control: A History of the Technologies that Made Aviation Possible. 2nd ed. Cambridge Aerospace Series; 2002.

[8] NACA TN-2175. Effect of an Unswept Wing on the Contribution of Unswept-tail Configurations to the Low-Speed Static- and Rolling-Stability Derivatives of a Midwing Airplane Model. Letko. William, and Donald Riley; 1950.

[9] NACA RM-L54I08. Investigation at High Subsonic Speeds of the Pressure Distribution on a 45° Sweptback Vertical Tail in Sideslip with a 45° Sweptback Horizontal Tail Mounted at 50-Percent and 100-Percent Vertical Tail Span. Wiley, Harleth G., and William C. Moseley, Jr; 1954.

[10] NACA TR-1171. Effect of Horizontal-Tail Span and Vertical Location on the Aerodynamic Characteristics of an Unswept Tail Assembly in Sideslip. Riley, Donald R 1954.

[11] http://en.wikipedia.org/wiki/Gloster_Javelin. http://www.ejection-history.org.uk/Aircraft_by_Type/Gloster_Javelin.htm. http://aviation-safety.net/wikibase/wiki.php?id=20519.

[12] http://www.thevictorassociation.org.uk/?p=491.

[13] http://aviation-safety.net/database/record.php?id=19631022-0.

[14] Duffy Paul, Kandalov Andrei. Tupolev: The Man and His Aircraft. Airlife Publishing; 1996. 142–143.

[15] http://aviation-safety.net/database/record.php?id=19660603-1.

[16] http://ntrs.nasa.gov/search.jsp.

[17] NACA TN-1369. Effect of Geometric Dihedral on the Aerodynamic Characteristics of Two Isolated Vee-tail Surfaces. Schade, Robert O; 1947.

[18] NACA TN-1478. Wind Tunnel Investigation of the Stability and Control Characteristics of a Complete Model Equipped with a Vee Tail. Polhamus, Edward C., and Robert J. Moss; 1947.

[19] NACA R-823. Experimental Verification of a Simplified Vee-tail Theory and Analysis fo Available Data on Complete Model with Vee Tails. Purser, Paul E., and John P. Campell; 1944.

[20] NACA AC-89. Albessard Triavion Aircraft. J. Serryer; 1929.

[21] NASA TP-2623. Wind-Tunnel Investigation of the Flight Characteristics of a Canard General-Aviation Airplane Configuration. Satran, Dale R; 1986.

[22] NACA TR-1339. A Summary and Analysis of the Low-Speed Longitudinal Characteristics of Swept Wings at High Reynolds Numbers. Furlong, G. C., and J. G. McHugh; 1957.

[23] NACA TN-1093. Effect of Sweepback and Aspect Ratio on Longitudinal Stability Characteristics of Wings at Low Speeds. Shortal, J. A., and B. Maggin; 1946.

[24] NASA TM-X-26. Design Guide for Pitch-Up Evaluation and Investigation at High Subsonic Speeds of Possible Limitations due to Wing-Aspect-Ratio Variations. Spreemann, Kenneth P; 1959.

CHAPTER 12

The Anatomy of the Fuselage

OUTLINE

12.1 Introduction 521
    12.1.1 The Content of this Chapter 521
    12.1.2 The Function of the Fuselage 522

12.2 Fundamentals of Fuselage Shapes 523
    12.2.1 The Frustum-shaped Fuselage 523
    12.2.2 The Pressure Tube Fuselage 523
    12.2.3 The Tadpole Fuselage 524

12.3 Sizing the Fuselage 526
    12.3.1 Initial Design of the External Shape of the Fuselage 526
    12.3.2 Refining the External Shape of the Fuselage 529
    12.3.3 Internal Dimensions of the Fuselage 531
    12.3.4 Cockpit Layout 532
        Pilot Visibility 533
        Large Aircraft 534
        General Passenger Cabin Layout 534
        Special Considerations: Galley Design 534
        Special Considerations: Lavatory Design 534

12.4 Estimating the Geometric Properties of the Fuselage 535
    12.4.1 Simple Estimation of the Surface Area of a Body of Revolution 536
    12.4.2 Fundamental Properties of Selected Solids 536
        Elliptic Cylinder (includes base and top) 537
        Uniform Cylinder ($D = D_1 = D_2$) (includes base and top) 537
        Paraboloid (includes base) 537
        Cone (includes base) 537
        Paraboloid (includes base) 538
        Test Values 538
    12.4.3 Surface Areas and Volumes of a Typical Tubular Fuselage 538
        Test Values 539
        Derivation of Equation (12-13) 539
        Derivation of Equation (12-14) 539
    12.4.4 Surface Areas and Volumes of a Tadpole Fuselage 539
        Derivation of Equation (12-16) 540
        Derivation of Equation (12-17) 542
        Test Values 542
    12.4.5 Surface Areas and Volumes of a Pod-style Fuselage 544

12.5 Additional Information 544

Variables 545

References 545

12.1 INTRODUCTION

The purpose of the fuselage is multifaceted. It does much more than just provide volume for occupants and freight to be transported by the aircraft or an aerodynamically efficient protection from the elements. Its design must be treated with a great deal of respect so it can truly serve the role for which it is intended. This section focuses on the design of the fuselage.

12.1.1 The Content of this Chapter

- **Section 12.2** discusses three elementary fuselage shapes commonly used in aircraft design, the frustum, tubular, and tadpole. The section presents a range of information on the pros and cons of each, that will help with the selection although the tubular shape is usually default geometry for pressurized and non-pressurized passenger transport aircraft.

- **Section 12.3** presents a simple method to sketch a fuselage for a GA aircraft. While there are many ways to accomplish initial fuselage sizing, this method forces the CG range and wing position to be considered from the beginning, an approach that quickly reveals unexpected problems with airplane passenger and payload combinations.
- **Section 12.4** presents simple methods to estimate geometric properties of typical fuselage shapes.
- **Section 12.5** presents typical dimensions for humans (male and female) that are helpful for the internal design of fuselages. It also tabulates the internal dimensions of a number of aircraft.

12.1.2 The Function of the Fuselage

The fuselage structure must allow components such as lifting surfaces, engines, and landing gear to be mounted and offer adequate load paths to react the large loads these generate. Among amenities that complicate the fuselage design are the various openings that are required for easy access into and out of the volume. The openings must be carefully laid out in order to keep the number of highly stressed regions to a minimum. Since doors are usually not intended to transfer axial and shear loads (except in the case of pressurized vessels, where doors must be capable of transferring the out-of-plane pressurization loads) the openings must be reinforced to relieve stress concentrations with minimum amount of deformation of the structure. It is inevitable that each such opening (door or window) will increase stress concentration, which calls for localized reinforcement. These, in turn, increase the empty weight of the vehicle. For this reason, the designer should evaluate objectively whether a given opening into the fuselage is justifiable: is it necessary or is it just desirable?

Some factors that will affect the design of the fuselage are:

(1) If the airplane transports people, sufficient internal space must be given to each person. Larger transport aircraft should offer ample space for the passengers and cabin crew members to move around (for instance, to go to a lavatory, or exit in case of an emergency).
(2) If the airplane is large, amenities (lavatories and galleys) must be provided for the occupants. Large passenger transport aircraft should have at least one lavatory per 50 passengers and one galley per 100 passengers. For instance, a typical 150-passenger Boeing 737 has two galleys (one in the front, the other in the back of the cabin) and three lavatories (one in the front, two in the back).
(3) The cockpit should be ergonomically laid out, regardless of airplane size. This means primary instruments and controls should all be within reach of the pilot and not require him or her to lean in order to access them.
(4) Windscreen shape and strength requirements will dictate the design of the forward part of the airplane and depend on airplane geometry and operational requirements (e.g. pressurization, bird strike, etc.).
(5) Layout of emergency exits: for instance 14 CFR Part 121.291 requires all operators of passenger aircraft with seating capacity greater than 44 to demonstrate it can be completely evacuated in less than 90 seconds.
(6) The layout of control, electrical, and other important systems. The fuselage structure should be expected to accommodate control cables, pushrods, pulleys, and wiring harnesses so they go around critical structural members and do not penetrate them.
(7) The fuselage should be designed with compartments intended to carry baggage and freight that are easily accessible. If the aircraft is large, such compartments must be accessible from the outside. The fuselage must provide structure to allow baggage to be tied down so it will not shift in flight, possibly altering the CG. This structure should be stout enough to react emergency landing loads as well.

If landing gear loads are reacted by the fuselage (in contrast to the wing) this will require hoop frames in the area of the landing gear to be substantially reinforced. Typically, the main landing gear will then retract into special aerodynamically shaped housings on the bottom of the fuselage. An opening should be provided in the front part of the airplane to house the nose landing gear. The author is not aware of any instance that features a nose landing gear that retracts into a separate housing unit and not the fuselage itself. It is good practice to examine existing aircraft of similar configuration and study how the landing gear housing is designed when evaluating the pros and cons of a design direction.

The fuselage must also provide structure to attach it to the wing. Commuters and similar passenger aircraft usually feature high or low wing configurations. Mid-wing commuters are practically unknown in modern times — the most recent one was the HFB-340 Hansajet, designed in the 1960s and operated until the early 1970s. There really is no good reason to mount the wing in the middle of the passenger cabin of such aircraft. High-wing aircraft typically feature hardpoints on the main and rear spars that allow the fuselage to be hung below them. Examples of such aircraft are the Shorts SD3-30; and Fokker F-27 and F50 twin-turboprop commuter aircraft. The second two have a part of the spar penetrate the ceiling inside the cabin; the protective spar cover in the cabin ceiling is clearly noticeable upon entry.

Tall people must lower their heads to avoid hitting the ceiling material that covers the spar.

Aircraft with low wings may have the fuselage mounted on top of the spar in a similar manner, although they more often feature a reinforced spar box under the cabin floor. In the case of low-wing aircraft, there is never a need for the spar to penetrate the floor – it would invite unacceptable risk to passengers, to have them maneuver around an elevation in a floor when moving about the cabin. Regardless of the wing attachment method employed, such aircraft usually feature external wing root fairings to prevent separation of air flowing through the juncture. Depending on the aircraft, wing root fairings can be massive and require an internal support structure on their own.

Sometimes the fuselage is designed to carry engines. This is very common for small single-engine aircraft, but also for selected twin-engine aircraft. Most small propeller aircraft feature the engine in the front, while there are also a few pusher configurations as well. Among such aircraft are the Cessna 337 Skymaster and Adam A500, both of which feature two engines; a tractor and a pusher. Neither one is being produced at this time. Regardless of the location of the engine, such configurations always require a special fire protection to be placed between the engine and the cabin. This protective wall is called a *firewall*. Requirements for firewalls in GA aircraft are stipulated in 14 CFR Part 23.1191, *Firewalls*. Among requirements is that the firewall must withstand a flame temperature of 2000 ± 150°F for at least 15 minutes.

Jet engines mounted to fuselages are attached either internally or externally. An internally mounted engine must provide fireproofing and a fail-safe structure around the engine. The fireproofing must be capable of protecting the primary structure should the engine catch on fire – the heat of such a fire cannot compromise the structural integrity of the aircraft. Requirements for this fireproofing are given in 14 CFR Part 23.1195, *Fire extinguishing systems*. The failsafe structure is necessary in case fragments from a possible rotor-burst penetrate a primary structural member. Such an emergency may not cause the aircraft itself to disintegrate. Primary systems in the aircraft must also be redundant – for instance a rotor-burst cannot take out the primary flight control system and if it does, there should be a secondary control system to allow the aircraft to continue flying.

As stated earlier, the fuselage must often feature various openings, ranging from access to an avionics bay to jet engine inlets for buried engine configurations. There must also be access to baggage compartments. Some pressurized aircraft feature special pressurized baggage compartments that allow animals to be transported safely, whereas small aircraft always have unpressurized baggage compartments. Some fuselages allow the main landing gear to retract inside it, requiring cutouts in the fuselage.

The fuselage is equally as important as the other components of the aircraft. Serving to contain occupants, freight, and important systems, it has to be carefully designed to provide spaciousness and yet be light and not generate too much drag. Just as the wing must be properly sized to carry the airframe and useful load, the fuselage must be sized to carry payload and shield it from the elements. This chapter is intended to discuss a number of topics that are important to the design of the fuselage.

12.2 FUNDAMENTALS OF FUSELAGE SHAPES

This section discusses three fundamental shapes of the fuselage; frustum, tubular, and tadpole. Of these, tubular fuselage was brought up in the introduction to the structural layout of the fuselage in Section 5.3, *Airframe structural layout*. It is imperative for the designer to be aware of the pros and cons of each configuration.

12.2.1 The Frustum-shaped Fuselage

The *frustum fuselage* is used to describe a fuselage whose empennage is effectively shaped like a frustum or a *trapezoidal prism*. An example of such a fuselage is shown in Figure 12-1. It is common to manufacturers like Beechcraft, Cessna, and Piper. Such fuselages are easily recognizable by a tapered boxlike appearance, although the term frustum refers to a tapered cylinder (see Figure 12-23). They are inexpensive to produce because they can be made from folded sheet metal riveted to frames, which produces a light and stiff structure. It is a drawback that they generate far more drag than tadpole fuselages. The configuration is the right choice for roomy, inexpensive, stiff, and strong fuselages, where drag is not an issue, but internal volume is. The frustum fuselage is ideal for utility transport aircraft, for instance, feeder aircraft for package services. However, if the goal is an aerodynamically efficient aircraft, frustum-shaped fuselages are the wrong choice. Such fuselages are indicative of the aircraft design philosophy of yesteryear and, today, are primarily justified by reduced production costs.

12.2.2 The Pressure Tube Fuselage

The *pressure tube fuselage* is used to describe a fuselage effectively shaped like a cigar, with a tubular main section and capped ends (see Figure 12-2). This fuselage shape is ideal for passenger aircraft, small and large,

FIGURE 12-1 A Piper PA-28 Cherokee featuring a conventional frustum style fuselage. *(Photo by Phil Rademacher)*

pressurized or not, of any airspeed range. If the airplane is pressurized, the cross-section is circular, as no geometry carries pressure loads more efficiently. The reaction of pressure loads is discussed in more detail in Section 5.3.4, *Fundamental layout of the fuselage structure*, so the presentation will be limited here.

The forward section typically ranges from 1.45 to 1.75 times the diameter of the fuselage. The length of the empennage ranges from 3 to 3.35 times the diameter for most airplanes. This fineness ratio has the least drag, as shown in Section 15.4.9, *Form factors for a fuselage and a smooth canopy*.

12.2.3 The Tadpole Fuselage

The tadpole fuselage is used to describe a fuselage whose empennage and forward portion resembles the shape of a *tadpole*, the larval stage of a frog. An example of such a fuselage is shown in Figure 12-3. Tadpole fuselages are more expensive to produce than frustum fuselages, in particular if made from aluminum as the geometry features compound surfaces that would call for expensive metal-forming processes. Their production can be achieved more economically using composites and this remains the primary method used for this purpose. All modern sailplanes feature a tadpole fuselage shape, as well as a number of modern propeller aircraft. Among those are the Cirrus SR20 and SR22, the Diamond DA-20 Katana, DA-40 Diamond Star, and DA-42 TwinStar. All are composite aircraft.

Galvao [1] pointed out the advantages of fuselages shaped like fish by citing examples from nature and used the superposition of sources and sinks to represent and evaluate the aerodynamic properties of such fuselages. He used what is called the *three halves power law* to determine the ratio between width and height of a fuselage and discussed means of converting an airfoil silhouette into a three-dimensional fuselage shape. While not a tadpole surface, the resulting fuselage resembles a short tadpole.

Althaus [2] was one of the first to investigate the properties of the tadpole fuselage. Dodbele et al. [3] used a surface singularity analysis method as a tool to help design such fuselages. Radespiel [4] presents the

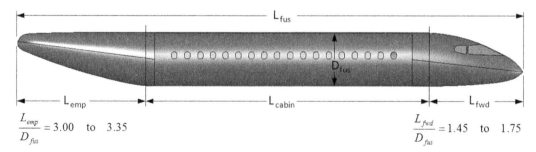

FIGURE 12-2 A schematic of a pressure tube fuselage, showing typical lengths of the forward and aft sections in terms of the fuselage diameter.

12.2 FUNDAMENTALS OF FUSELAGE SHAPES

FIGURE 12-3 A Rolladen-Schneider LS4 sailplane boasts a tadpole-style fuselage. *(Photo by Phil Rademacher)*

effect of proper contraction geometry (called *waisting*) on the transition region and, ultimately, the drag of the fuselage, supported by wind tunnel testing. The interested reader should be made aware that there is a plethora of literature on tadpole fuselages, presenting it in depth is beyond the scope of this book.

Tadpole fuselages generate far less drag than the frustum kind for two primary reasons: (1) their forward portion is shaped to sustain laminar boundary layer; and (2) their empennage shape results in as much as 30–40% less wetted area. In Ref. [2], Althaus says:

1. Similar to laminar airfoils, the front part of the body should produce favorable pressure gradients in all meridians even at incidences of about ±10°. At the same time, the whole surface should be smooth and leak-free in order to avoid any disturbances to the laminar flow.

2. Behind the transition front it is favorable to contract the cross-section. On one hand, this reduces the wetted surface; on the other, it shifts the unavoidable pressure rise to the thinner parts of the turbulent boundary layer, which is a well-known principle of favorable boundary layer control.

Althaus compared three tadpole-style fuselages to a frustum fuselage of equal length and diameter, whose fineness ratio (l/d) was 10. Both shapes were representative of those used for sailplanes (see Figure 12-4). The

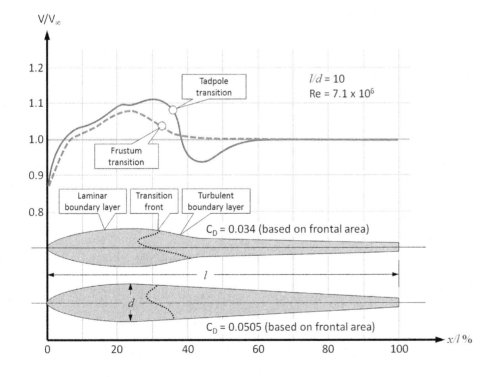

FIGURE 12-4 Difference in transition and total fuselage drag of a tadpole and frustum fuselage. The drag of the frustum is almost 49% that of the tadpole fuselage. *(Based on Ref. [2])*

study revealed an important difference in the transition front (the curve along which laminar-to-turbulent transition takes place) and their drag characteristics. As shown in the figure, the minimum drag coefficient (based on the frontal area) of the fuselage at a Reynolds number of 7.1×10^6 is 0.034 for the tadpole and 0.0505 for the frustum (at an AOA of $0°$). This means the drag of the frustum is 48.5% greater than that of the tadpole, explaining why such fuselages have become the norm in modern sailplane design.

The tadpole design requires careful attention to the curvature of the geometry aft of the maximum thickness. Too sharp a contraction will result in a flow separation that will increase the overall drag of the fuselage. Too small a contraction will not reduce the wetted area rapidly enough to make a dent in the total drag. As can be seen in Figure 12-4, the contraction also reduces the local airspeeds slowly (relatively speaking) in the area where the turbulent boundary layer is still relatively thin. This helps prevent an early separation. This area is akin to the pressure-recovery region of an NLF airfoil. The goal is to maintain the laminar boundary layer as far aft along the fuselage as possible, but, once the maximum thickness has been exceeded, give air enough distance to slow down without flow separation.

Another aspect of tadpole fuselage design is the downward tilt of the fuselage, shown in Figure 12-5. This is a response to the upwash caused in the airflow ahead of the wing. A straight fuselage, as shown in the upper figure, will be at a higher AOA and this will increase its drag. To reduce this, the forward portion of the fuselage is tilted downward to align it with the oncoming airflow, reducing its drag. Some even tilt the tailboom down or reshape it to better match the flow field behind the wing as well.

A flow-adaption design of this nature should only be performed using CFD technology or wind tunnel testing.

12.3 SIZING THE FUSELAGE

The initial design of the fuselage should be implemented once the designer has determined the fundamental dimensions of the wing and stabilizing surfaces and their representative locations in space. These should then be drawn to scale at the correct location. This is followed by the positioning of the most important components (such as engines and landing gear) and payload (passengers and freight), taking into account its dimensional restrictions. Finally, the outside mold line (OML) of the fuselage can be drawn such that it encompasses the entire layout. This process allows for modifications and shifting of components to ensure the CG will remain within reasonable margins.

12.3.1 Initial Design of the External Shape of the Fuselage

This author refrains from using statistical methodologies to estimate the physical dimensions of the fuselage design, as these are limited to specific shapes and do not prevent conflicts that may arise, for instance, due to the space claims of internal components. For example, what impact will the width and height of an engine have on the girth of the fuselage? What about the waterline adjacent to the head of the occupants? Will these be entirely on the inside of the airplane, or will parts thereof be exposed to the airstream? It is important to tackle space claim problems from the start and eliminate them head on. This can be done using the following

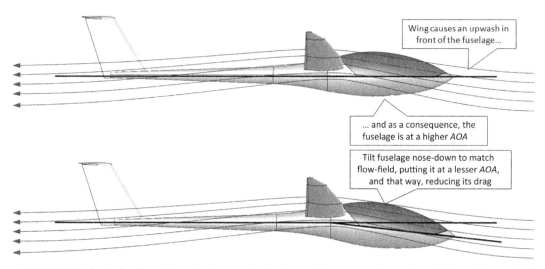

FIGURE 12-5 Reduction of drag by aligning the fuselage with the oncoming airflow. *(Adapted from Ref. [5])*

process, which is outlined in more detail below using images where possible.

Step 1: Determine the general shape of the wing, as well as the HT and VT of the aircraft, using information gathered from Steps 1 through 9 in Table 1-3. This data should be detailed enough to provide the span and chords of the wing and stabilizing surfaces and also the proper location of the stabilizing surfaces with respect to the wing's MGC/4 (see Figure 12-6). Enter this rudimentary data into a design drafting program or simply draw on a piece of paper. This image represents the aerodynamic requirement of the design.

Step 2: Indicate the desired CG envelope on the MGC (see Figure 12-7). For instance, the CG envelope for conventional aircraft may extend from 15% to 35% MGC. At this point in the game the final CG envelope remains unknown because far more detailed analysis remains to be done. However, we have to start somewhere and this is a reasonable first step.

Step 3: Estimate the weight of all known components making up the airplane. For instance, consider those presented in Section 6.4, *Statistical weight estimation methods* (omitting components that do not belong to the particular design, i.e. ignore pressurization or hydraulic systems for a small fixed-gear airplane). Tabulate these weights, ideally in a spreadsheet application, by entering them in a column format. The next column should be the x-location of the expected CG location of each of the components. For instance, estimate the CG location of the wing, HT, VT, and so on and enter in the second column. Summing these weights effectively yields the predicted empty weight of the airplane. Using these weights, estimate the empty weight CG location and indicate in the figure, as shown in Figure 12-8. Note that at this time we do not know if this CG location is viable — we will evaluate that in Step 4. Later, in Step 5, we may have to change the CG locations of selected components until the total CG can be "nudged" into the proper location.

Step 4: This step will require some trial and error, consisting of moving selected components and the pilot, passengers, and baggage around until the forward and aft CG limits mostly land inside the desired CG limits (see Figure 12-9). This process requires the designer to develop a *loading cloud* similar to the one described in Section 6.6.12, *Creating the CG envelope*, ideally using the spreadsheet tabulation from the previous step.

Using the components that make up the empty weight of the airplane, evaluate the CG locations of the empty aircraft plus as many combinations of occupants, fuel, and baggage as considered practical. Plot all the resulting CGs on the figure, as shown in Figure 12-9. If the bulk of the CG locations fall outside the limits, try to modify the CG locations of selected components, such as the landing gear, engine, avionics, and even the occupants. Remember that the location of the wing, HT, and VT should not be changed unless necessary.

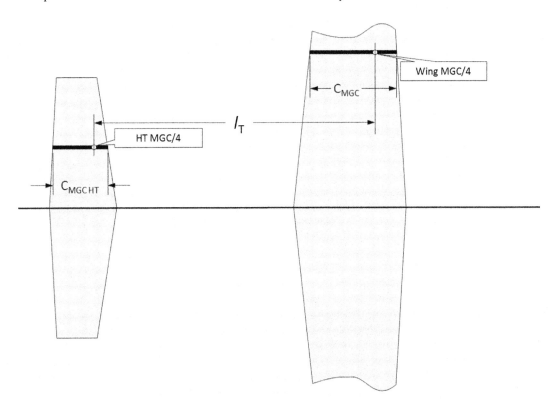

FIGURE 12-6 In Step 1, the lifting surfaces are drawn to scale and separated by the correct tail arm l_T.

528 12. THE ANATOMY OF THE FUSELAGE

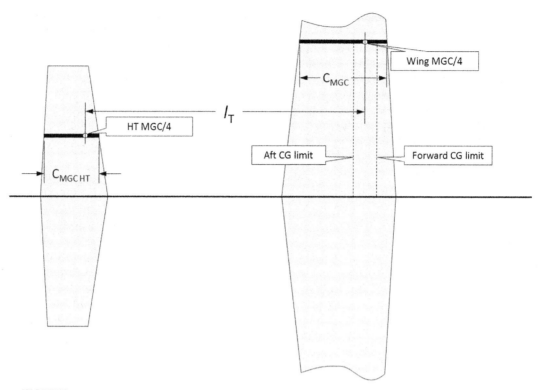

FIGURE 12-7 In Step 2, the expected CG limit around the quarter-chord of the wing's MGC is labeled.

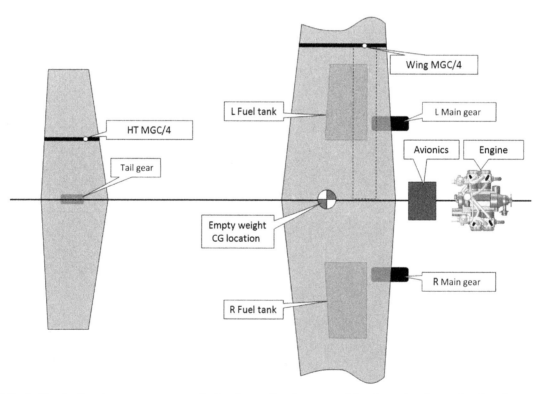

FIGURE 12-8 In Step 3, all major components constituting the predicted empty weight of the aircraft are placed in their proper location. The empty weight CG is indicated on the drawing. Note that at this time we do not know if this CG is in the proper location.

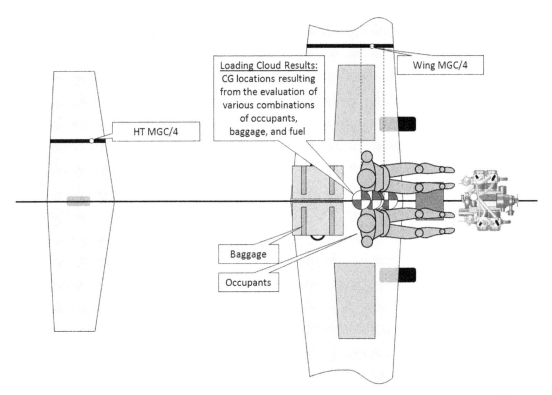

FIGURE 12-9 In Step 4, occupants, baggage, the engine, and other known "heavy" components are moved (or nudged) around until the CG loading cloud is mostly placed within the desired CG limits.

Note that if the HT or VT are moved to a different location, the resulting tail arm (l_T) will differ from the original one. This means that the horizontal and vertical tail volumes (denoted by V_{HT} and V_{VT}) have now changed. Therefore, the geometry of the HT and VT should be updated to ensure the selected V_{HT} and V_{VT} remain constant. This, in turn, will affect the weight of the said components, which affects the CG again. Welcome to the world of aircraft design. The process can be made much easier using a spreadsheet that automatically updates the necessary variables.

Some components will only require a small amount of nudging, while other may require a large relocation. While tedious, this process is necessary to prevent future "surprises" once the airplane has been developed further and making changes has become much harder. It is to be expected that components end up in "interesting" or "surprising" locations necessitated by placing the CG loading cloud in its proper location.

Step 5: Once the CG appears to be in a satisfactory location, indicate the occupants and component dimensions on the diagram. Trace a rudimentary fuselage that encloses occupants and all internal components, keeping in mind its aerodynamic and aesthetic appearance (see Figure 12-10). This effectively concludes the initial definition of the fuselage and it is now possible to begin refining it by considering other requirements it must satisfy.

12.3.2 Refining the External Shape of the Fuselage

A number of factors must be considered when refining the fuselage.

(1) The fuselage should be made as streamlined as possible. It should feature the appropriate forward shape that will not result in an excessively large stagnation region. It should feature an aft shape that takes into account the effect of fineness ratio (empennage length/diameter) on drag. This usually means that the empennage should have a fineness ratio between 3 and 3.35 (see Figure 12-2).

(2) The tail upsweep angle must be considered to guarantee the airplane can rotate for T-O and flare for landing with the selected landing gear layout. Assuming compressed landing gear at gross weight, the upsweep should allow the airplane to be rotated to its clean stall *AOA* (see Figure 12-11). Also, designers of passenger aircraft always think about growth potential: is it possible this aircraft will require a larger passenger volume in future? If fuselage "plugs" are used to accomplish such a development, will additional engineering resources be required to modify the empennage for ground clearance? The value of such forethought is shown in Figure 12-11.

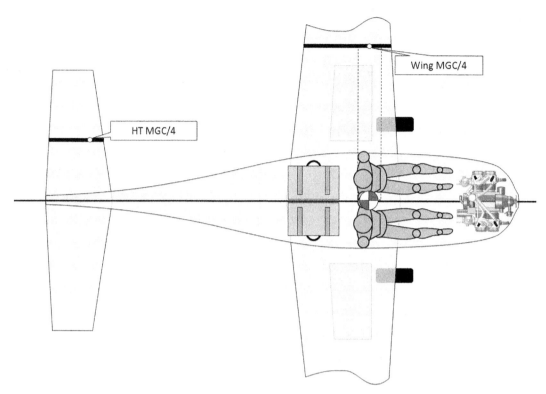

FIGURE 12-10 Step 5 is the final step, taken once the proper range of the CG has been determined. With the approximate dimensions of internal parts known, trace the outline of the fuselage, preferably using a tadpole or other low-drag fuselage shape.

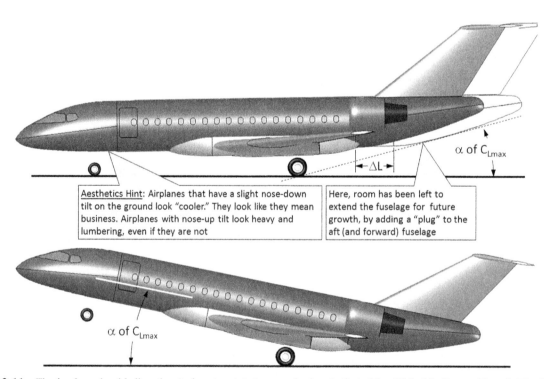

FIGURE 12-11 The fuselage should allow the airplane to rotate to an angle close to that of the AOA of its C_{Lmax}, although it is also prudent to think about growth.

(3) Designers of small composite airplanes should consider a tadpole shape as they generate less drag than conventional frustum-type fuselages. Airplanes made from aluminum almost always use the frustum shape, as a tadpole fuselage requires compound surfaces. Larger aircraft normally do not feature tadpole shapes as it reduces the available internal volume.

(4) The external dimensions of the cabin area of the fuselage must allow for the required internal dimensions plus the airframe around it. If the airplane is pressurized the designer should strongly consider a circular fuselage shape as this is structurally the most efficient configuration for pressure loads. An unpressurized fuselage is not limited by that shape.

(5) The designer should be cognizant of structural load paths when laying out openings for doors and windows. For instance, consider the aircraft of Figure 12-13. The frame (often called the A-pillar) to which the windscreen mounts should be placed as directly as possible where it is expected the wing's main spar will be placed. This way, the A-pillar can be attached to the structure already required to react the wing loads. Window openings in the fuselage, besides having to provide convenient viewing angles for the passengers, should be placed so they fall between hoop frames. It is to be expected that hoop frames will be placed at even intervals — so, place the passenger windows at even intervals between them.

(6) The location of the wing (high, mid, low, parasol) must be taken into account when refining the external fuselage, as well as the possible shape of the wing root fairing. Typically, the maximum fuselage width is where the wing joins the fuselage, although this is not always the case. However, think about how the wing root fairing design will be affected by the region near the trailing edge of the root.

(7) Jet engines buried in the fuselage will have a great impact on the external geometry of the fuselage, and inlet and exhaust paths must be clearly defined for space-claim reasons. The volume reserved for the engine is out of reach for other systems. It is surprisingly easy to locate equipment near such a volume, later to discover it penetrates the engine volume.

(8) Placement of necessary equipment: avionics, battery, hydraulic system, control system, should take into account access for maintenance crews. Also, many aircraft use collector tanks that are placed inside the fuselage, typically between the main spar and aft shear web.

(9) And finally, this simple food-for-thought advice: many people find airplanes that "lean a little" on the nose-gear in the ground attitude have a more appealing look than if level or nose-up. Of course this is only in the eye of the beholder — it is akin to people's views on the cosmetics of T-tails or winglets. However, even assuming the designing team agrees with this view, it is simply not always practical, as it is for many tractor propeller aircraft. Propeller aircraft must comply with prop-strike requirements and designing around this aesthetic view could call for excessively long main landing gear legs. However, if possible, consider giving the aircraft an approximately 1° nose-down ground attitude for that extra it-means-business look. Examples of aircraft that have an improved look (admittedly in the author's opinion) are the Rockwell OV-10 Bronco, the Lockheed P-3C Orion, and the Grumman Gulfstream II. Examples of the opposite are the Piper Cherokee Six and the Beech A36 Bonanza. Not surprisingly, both are tractor propeller aircraft.

12.3.3 Internal Dimensions of the Fuselage

If the aircraft is designed to carry passengers, the designer must be aware of the constraints the human body places on the external shape. For instance, Figure 12-12 shows three cross sections intended for a side-by-side seating. It can be seen that for occupant comfort, the two on the left and right provide ample

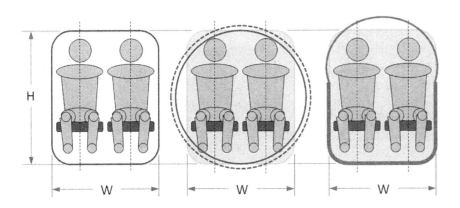

FIGURE 12-12 Cross sections of a fuselage for a small airplane reveal that different shapes result in different problems.

headroom, whereas the one in the center does not. For this reason, such a fuselage must be made larger, perhaps at least what is indicated by the dashed circle.

For GA aircraft a number of regulations apply to the design of the crew and passenger areas. These are:

PERSONNEL AND CARGO ACCOMMODATIONS

23.771 − Pilot compartment.
23.773 − Pilot compartment view.
23.775 − Windshields and windows.
23.777 − Cockpit controls.
23.779 − Motion and effect of cockpit controls.
23.781 − Cockpit control knob shape.
23.783 − Doors.
23.785 − Seats, berths, litters, safety belts, and shoulder harnesses.
23.787 − Baggage and cargo compartments.

23.791 − Passenger information signs.
23.803 − Emergency evacuation.
23.805 − Flightcrew emergency exits.
23.807 − Emergency exits.
23.811 − Emergency exit marking.
23.812 − Emergency lighting.
23.813 − Emergency exit access.
23.815 − Width of aisle.
23.831 − Ventilation.

Pressurization
23.841 − Pressurized cabins.
23.843 − Pressurization tests.

Fire protection
23.851 − Fire extinguishers.
23.853 − Passenger and crew compartment interiors.
23.855 − Cargo and baggage compartment fire protection.

Among factors that should be considered during the design of the fuselage are:

(1) Space amounting to 22 inches wide, 43 inches deep, and 65 inches high is typically given to each passenger.
(2) If the design features an aisle, it should be 76 inches high and 20 inches wide (if a stand-up cabin), to allow passengers to move around freely and for beverage carts to be able to fit comfortably. Note that the aisle widths given by 14 CFR 23.815, *Width of aisle*, are minimums.
(3) For weight estimation, the average weight of the modern passenger should be considered to be 200 lb_f, even though federal regulations assume 170 lb_f. The average weight of a person has increased in the last few decades and 170 lb_f is simply too light.
(4) If the duration of a typical design flight mission exceeds one hour, a lavatory should be provided. The number of lavatories and galleys depends on the number of passengers. Typically, there is one lavatory per 1 to 50 passengers and one galley per 50−100 passengers. Thus a 50 passenger Fokker F-50 has a single lavatory and a (½-a) galley. A 150-passenger Boeing 737 typically has three lavatories (one front, two in the back) and two galleys (one front and one back).
(5) Size of the cockpit depends on the aircraft. Primary controls should be within reach, as should secondary, although these should be placed away from the primary. Details of equipment layout in cockpit design are beyond the scope of this text.
(6) Passenger and crew emergency exits must be provided in accordance with 14 CFR 23.

12.3.4 Cockpit Layout

The term *accommodation* refers to both the passenger cabin and the cockpit. In this section we will consider the human aspect of aircraft design − i.e. how to best fit people into a given volume and the dimensions of a typical fuselage. General dimensions for the cabin of a small, side-by-side (or single-) seat aircraft are shown in Figure 12-13. This is based on the general internal dimensions of the Cessna 150/152, which is reasonably sized for most people, although its width is a minimum by modern standards. Figure 12-14 shows such a cabin with a 95th percentile mannequin. It can be seen that for this design (which differs from that of the 152) the proximity of the head to the ceiling is not satisfactory. A design review should improve that detail.

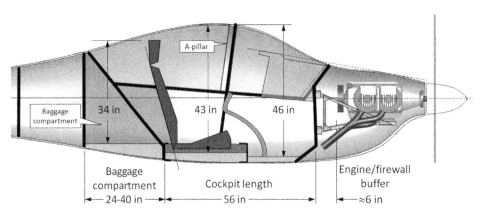

FIGURE 12-13 General cabin layout for a small airplane. The dimensions are typical.

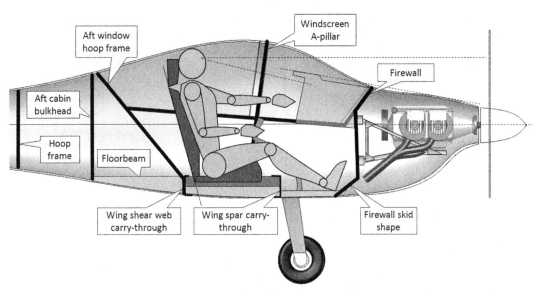

FIGURE 12-14 General accommodation of a large person (95th percentile) in the cabin of a small aircraft. A good fit that also accommodates smaller pilots is a challenge and should be studied in detail before the final loft for a proof-of-concept aircraft is released.

Minimum recommended width and height for a typical small aircraft are shown in Figure 12-15. The height assumes upright sitting. Sailplanes are exempted from the height requirement, as upright sitting results in a greater frontal area in a sailplane and, thus, greater drag. The cabin should allow at least 20″ (50 cm) per person. Anything else will likely make a part of the population claustrophobic, and, frankly, will be too narrow for the stouter pilot. Cabin dimensions for a number of aircraft are provided in Table 12-5.

Pilot Visibility

It should not come as a surprise that pilot visibility is of paramount importance in manned aircraft design. In fact, in certified GA aircraft, pilot visibility is subject to compliance to 14 CFR 23.773, *Pilot compartment view*. The recommended field-of-view (FOV) for a typical GA aircraft is shown in Figure 12-16. It is a drawback that a good FOV usually comes at the cost of reduced structural integrity, as transparencies do not provide acceptable structural integrity.

Transparencies are usually made from two kinds of plastics: acrylics or polycarbonate. Acrylic is a thermoplastic polymer and is both light and clear to look through, allowing more than 90% of external light to get through the material. It is the choice for non-pressurized aircraft. It allows complicate geometries to be formed with ease, by heating the material in molds. The primary drawback is cracking, which may start due to some imperfection and, if left to its own devices, will continue to propagate.

Stretched acrylics are a solution, in which the plastic is physically stretched using special machinery. This reorients the long polymer chains, giving the plastic much improved properties. Among those are improves tensile stress (of the order of 10,000 psi) and reduced crack propagation. This makes the material suitable for birdproof windshields and enables its use in pressurized aircraft.

Polycarbonate is a specialized plastic considered the strongest transparent plastic available. It offers outstanding toughness, temperature tolerance, and impact resistance, making it ideal for windscreens subjected to birdstrikes at high airspeeds, such as low-flying fighter aircraft. Its primary drawback is crazing (the formation of a network of small cracks), requiring it to have protective coating. Polycarbonate is generally not used for GA aircraft. Typical dimensions for various cabin layouts are shown in Figure 12-17.

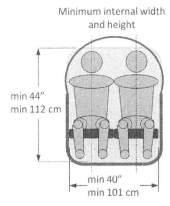

FIGURE 12-15 Minimum cabin/cockpit dimensions for upright seating.

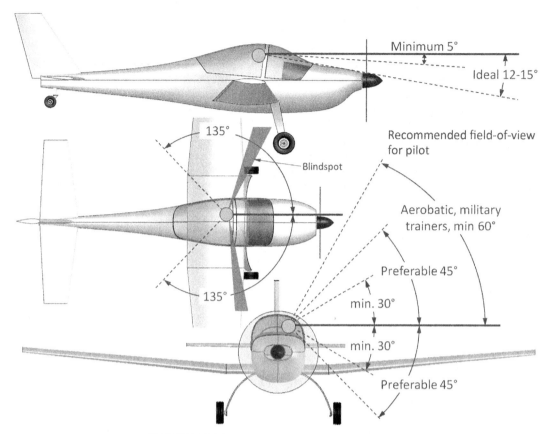

FIGURE 12-16 Recommended pilot's field-of-view.

Large Aircraft

It should go without saying that the design of the cockpit of a large aircraft is a monumental undertaking that would require volumes to present. While most of the cockpit design is really a task taken on during the detail design phase, it is imperative that the external shape of the aircraft allows for an internal geometry that will enable an ergonomically efficient cockpit to be designed. An example of a spacious cockpit is shown in Figure 12-18.

General Passenger Cabin Layout

Jenkinson et al. [6] cite common distances (or pitch) between seat rows in the passenger cabin of typical commuter or commercial aircraft. These are shown in Figure 12-19. An example of the luxurious spacing between passenger seats in a business jet is shown in Figure 12-20. Such comfort is of course out of the reach of most people, but is presented here to contrast the other extreme presented in Figure 12-15.

Special Considerations: Galley Design

A galley is effectively a kitchen in an aircraft. It is where food and beverage carts are stored, where hot beverages are brewed and ice cubes are stored, as well as containing ovens that are used to reheat chilled meals for passengers. The galley contains cabinets for dry goods and duty-free goods. The aircraft manufacturer does not produce equipment for the galley, but rather purchases it from a vendor. The production of equipment for the galley is an industry in its own right and a number of engineering companies specialize in the manufacturing and installation of such equipment. A number of provisions must be made in the region of the fuselage where a galley is to be placed. First, the galley should always be next to a service door. Such doors allow a galley service truck to park outside the galley for a rapid removal and replenishing of food and beverage carts, as well as trash. Second, there should be plumbing for a sink, so that liquid waste can be disposed of. Third, there should be electrical outlets that are capable of powering a number of ovens simultaneously. Fourth, there should be structural provisions to anchor the galley equipment, but these have to be capable of reacting substantial loads. These loads are typically obtained per 14 CFR 23.561(b)(3) and amount to $3.0g$ upward, $18.0g$ forward, and $4.5g$ sideward. A galley is only offered in larger multiengine passenger aircraft.

Special Considerations: Lavatory Design

Just like the galley, lavatory equipment is engineered and produced by specialized vendors. Such equipment

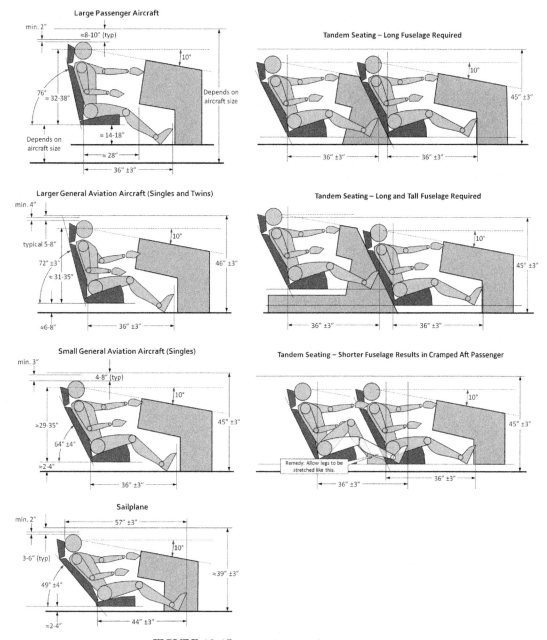

FIGURE 12-17 Typical seating for cockpit layout.

consists of a sink and a faucet, a counter, mirror, lighting, soap dispensers, amenity racks, and of course the toilet itself (Figure 12-21). The lavatory is a fully functional module that can be installed anywhere in the aircraft, as long as some preparations have been made. It consists of a centralized waste tank that connects to the toilets using small pipes. This allows the waste tank to be emptied in a single place, rather than for each individual toilet. Modern toilets use *vacuum flush* technology that does away with the recirculation of the blue-chemical flush liquid of yesteryear. Such toilets are substantially lighter and environmentally friendlier than the old-generation toilet.

12.4 ESTIMATING THE GEOMETRIC PROPERTIES OF THE FUSELAGE

During the early stages of the design process, there simply is not yet enough detail available to calculate the surface areas and volumes with a high degree of accuracy. Often the design consists of nothing more than some preliminary dimensions. For instance, we may have a reasonable idea about how long the fuselage might be, or its average diameter. Then, it is often convenient to estimate the geometric properties of the airplane using some generic shapes that resemble the proposed form. This section will present a few such shapes and

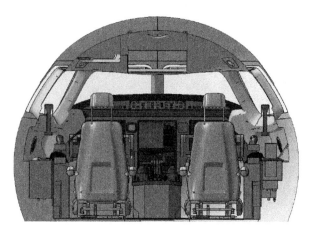

FIGURE 12-18 Cockpit layout in a large business jet, here the Dassault 7X. *(Courtesy of Dassault Falcon Jet Corporation)*

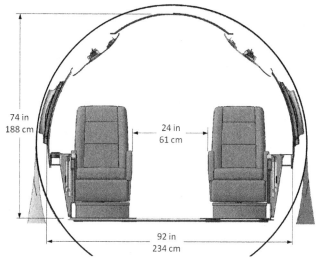

FIGURE 12-20 Cabin layout in a Dassault 7X business jet aircraft. *(Courtesy of Dassault Falcon Jet Corporation)*

simple and handy formulas to estimate the areas of the surface and volume. We will start with the geometry of a few fundamental shapes that can be combined to form shapes that resemble that of a fuselage.

12.4.1 Simple Estimation of the Surface Area of a Body of Revolution

The fastest method to estimate the wetted area of a fuselage is to evaluate the areas of the side and top silhouettes (Figure 12-22). This is based on the notion that the surface area of an elliptic cylinder can be estimated using the average of the major and minor widths times its length.π

Surface area:

$$S_{\text{FUS}} = \pi \left[\frac{A_{side} + A_{top}}{2} \right] \quad (12\text{-}1)$$

12.4.2 Fundamental Properties of Selected Solids

A more complex formulation for typical aircraft fuselage is based on segmenting it into simple geometric shapes such as the ones shown in Figure 12-23. This section presents formulas to determine the properties of various fuselage geometric types.

The advantage of this method is that it provides reasonable accuracy for more fuselages that are more complicated than the typical pressure tube fuselage. However, it is a drawback that the formulation is somewhat complicated, but this is the cost of their accuracy.

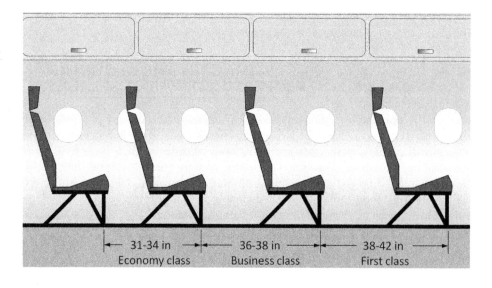

FIGURE 12-19 Typical pitch of seats in a commuter or commercial aircraft.*(Based Ref. [6])*

12.4 ESTIMATING THE GEOMETRIC PROPERTIES OF THE FUSELAGE   537

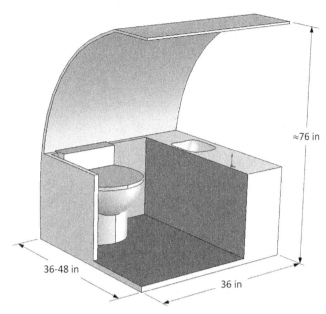

FIGURE 12-21 Space claims of a typical aircraft lavatory.

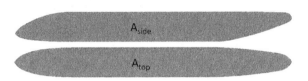

FIGURE 12-22 The least complicated method to estimate the wetted area of a body of revolution is to use the areas of its side and top silhouettes.

Elliptic Cylinder (includes base and top)

Surface area:

$$S_{EC} = \frac{\pi}{2} \cdot \left[D_1 D_2 + \left((D_1 + D_2) + \frac{(D_1 - D_2)^2}{4(D_1 + D_2)} \right) \cdot L \right] \quad (12\text{-}2)$$

Volume:
$$V_{EC} = \frac{\pi \cdot D_1 \cdot D_2 \cdot L}{4} \quad (12\text{-}3)$$

Uniform Cylinder ($D = D_1 = D_2$) (includes base and top)

Surface area:
$$S_{UC} = \frac{\pi \cdot D}{2} \cdot [D + 2L] \quad (12\text{-}4)$$

Volume:
$$V_{UC} = \frac{\pi \cdot D^2 \cdot L}{4} \quad (12\text{-}5)$$

Paraboloid (includes base)

Surface area:
$$S_P = \frac{\pi \cdot D}{12L^2} \cdot \left[\left(4L^2 + \frac{D^2}{4} \right)^{1.5} - \frac{D^3}{8} \right] \quad (12\text{-}6)$$

Volume:
$$V_P = \frac{\pi \cdot D^2 \cdot L}{8} \quad (12\text{-}7)$$

Cone (includes base)

Surface area:
$$S_C = \frac{\pi \cdot D}{2} \left(\frac{D}{2} + \sqrt{L^2 + \frac{D^2}{4}} \right) \quad (12\text{-}8)$$

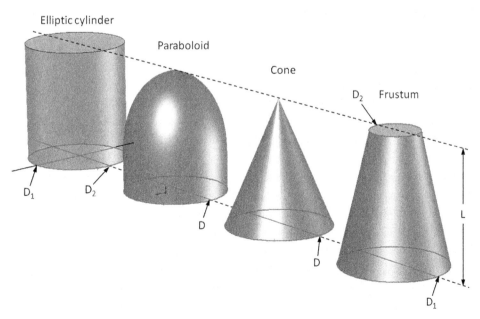

FIGURE 12-23 Elementary solids of equal height L.

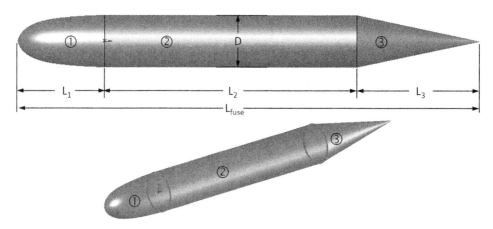

FIGURE 12-24 An approximation of a jet transport fuselage using elementary solids.

Volume:
$$V_C = \frac{\pi \cdot D^2 \cdot L}{12} \quad (12\text{-}9)$$

Paraboloid (includes base)

Surface area:
$$S_F = \pi \left(\frac{D_1^2 + D_2^2}{4} + \frac{(D_1 + D_2)}{2} \sqrt{L^2 + \frac{D_1^2 - D_2^2}{4}} \right) \quad (12\text{-}10)$$

Volume:
$$V_F = \frac{\pi \cdot L}{12} \left(D_1^2 + D_1 D_2 + D_2^2 \right) \quad (12\text{-}11)$$

Test Values

In order to minimize the risk of an error when entering these in a spreadsheet or other computer codes, the validation values in Table 12-1 are provided for the reader to compare own calculations to.

12.4.3 Surface Areas and Volumes of a Typical Tubular Fuselage

The following geometry can be used to estimate the geometric properties of a generic passenger transport aircraft. The geometry consists of a paraboloid, cylinder, and a cone, featuring the dimensions shown in Figure 12-24. Each section height, which here is conveniently described using the component length, is denoted by an appropriate subscript. The fundamental dimension D (diameter) is the same for all sections.

Total length:
$$L_{\text{fuse}} = L_1 + L_2 + L_3 \quad (12\text{-}12)$$

Surface area:
$$S_{FUSE} = \frac{\pi \cdot D}{4} \left(\frac{1}{3L_1^2} \cdot \left[\left(4L_1^2 + \frac{D^2}{4} \right)^{1.5} - \frac{D^3}{8} \right] \right.$$
$$\left. - D + 4L_2 + 2\sqrt{L_3^2 + \frac{D^2}{4}} \right) \quad (12\text{-}13)$$

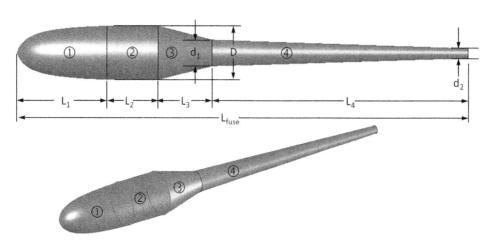

FIGURE 12-25 An approximation of a tadpole fuselage using elementary solids.

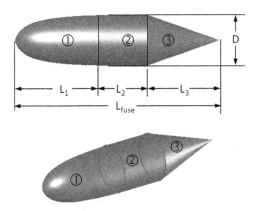

FIGURE 12-26 An approximation of a pod-style fuselage using elementary solids.

Volume:

$$V_{\text{PAX}} = \frac{\pi \cdot D^2}{4} \cdot \left(\frac{L_1}{2} + L_2 + \frac{L_3}{3} \right) \quad (12\text{-}14)$$

Test Values

The following validation values have been tabulated for the reader. Consider a fuselage whose dimensions are given by: $D = 2$, $L_1 = L_2 = L_3 = 2$. This way, we can simply add the values in Table 12-1 and compare them to the outcome of the above expressions. It is important to remember that the surface areas include the base and top, so these must be subtracted from the total surface area when combining solids, as is done below:

$$S_{FUSE} = \frac{\pi 2}{4} \left(\frac{1}{3(2)^2} \left[\left(4(2)^2 + \frac{2^2}{4} \right)^{1.5} - \frac{2^3}{8} \right] \right.$$
$$\left. - 2 + 4(2) + \sqrt[2]{(2)^2 + \frac{2^2}{4}} \right)$$
$$\approx 25.4938$$

$$V_{FUSE} = \frac{\pi \cdot D^2}{4} \cdot \left(\frac{L_1}{2} + L_2 + \frac{L_3}{3} \right) = \frac{\pi \cdot (2)^2}{4} \left(\frac{2}{2} + 2 + \frac{2}{3} \right)$$
$$= 11.5192$$

Derivation of Equation (12-13)

$$S_{FUSE} = \overbrace{\left\{ \frac{\pi D}{12 L_1^2} \left[\left(4 L_1^2 + \frac{D^2}{4} \right)^{1.5} - \frac{D^3}{8} \right] - \frac{\pi D^2}{4} \right\}}^{\text{Paraboloid}}$$

$$+ \overbrace{\{\pi D L_2\}}^{\text{Cylinder}} + \overbrace{\left\{ \frac{\pi D}{2} \sqrt{L_3^2 + \frac{D^2}{4}} \right\}}^{\text{Cone}}$$

$$= \frac{\pi D}{4} \left(\frac{1}{3 L_1^2} \left[\left(4 L_1^2 + \frac{D^2}{4} \right)^{1.5} - \frac{D^3}{8} \right] \right.$$
$$\left. - D + 4 L_2 + \sqrt[2]{L_3^2 + \frac{D^2}{4}} \right)$$

QED

Derivation of Equation (12-14)

$$V_{FUSE} = \frac{\pi \cdot D^2 \cdot L_1}{8} + \frac{\pi \cdot D^2 \cdot L_2}{4} + \frac{\pi \cdot D^2 \cdot L_3}{12}$$
$$= \frac{\pi \cdot D^2}{4} \cdot \left(\frac{L_1}{2} + L_2 + \frac{L_3}{3} \right)$$

QED

12.4.4 Surface Areas and Volumes of a Tadpole Fuselage

The following geometry can be used to estimate the geometric properties of the generic tadpole-shaped fuselage shown in Figure 12-25. This is handy when estimating the wetted area of sailplanes and the modern GA aircraft, such as the Cirrus SR20 and SR22. The geometry consists of a paraboloid, cylinder, and two sets

TABLE 12-1 Validation Sample

THESE ENTRY VALUES SHOULD YIELD				SURFACE	VOLUME
Base diameter	$D =$	2	Elliptic Cylinder	12.82817	3.14159
Diameter 1	$D_1 =$	2	Uniform Cylinder	18.84956	6.28319
Diameter 2	$D_2 =$	1	Paraboloid	9.04423	3.14159
Height	$L =$	2	Cone	10.16641	2.09440
			Frustum	14.19740	3.66519

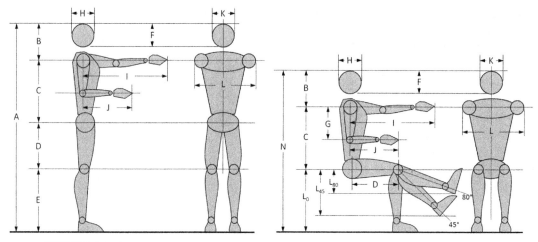

FIGURE 12-27 A standard erect and sitting mannequins used for space claims by occupants.

of frustum, featuring the dimensions shown. Each section height, which here is conveniently described using the component length, is denoted by an appropriate subscript. The fundamental dimension D (diameter) is the same for all sections.

Total length:

$$L_{\text{fuse}} = L_1 + L_2 + L_3 + L_4 \quad (12\text{-}15)$$

Surface area:

$$S_{TAD} = \frac{\pi D}{4}\left(\frac{1}{3L_1^2}\left[\left(4L_1^2 + \frac{D^2}{4}\right)^{1.5} - \frac{D^3}{8}\right]\right.$$

$$- D + 4L_2 + \frac{2(D+d_1)}{D}\sqrt{L_3^2 + \frac{D^2 - d_1^2}{4}}$$

$$\left. + \frac{2(d_1+d_2)}{D}\sqrt{L_4^2 + \frac{d_1^2 - d_2^2}{4}}\right) \quad (12\text{-}16)$$

Volume:

$$V_{TAD} = \frac{\pi}{4}\left[D^2 \cdot \left(\frac{L_1}{2} + L_2 + \frac{L_3}{3}\right) + \frac{L_3}{3}\left(Dd_1 + d_1^2\right)\right.$$

$$\left. + \frac{L_4}{3}\left(d_1^2 + d_1 d_2 + d_2^2\right)\right] \quad (12\text{-}17)$$

Derivation of Equation (12-16)

Begin by adding the contribution of all the elementary solids used, with tops and bases removed:

$$S_{TAD} = \overbrace{\left\{\frac{\pi D}{12L_1^2}\left[\left(4L_1^2 + \frac{D^2}{4}\right)^{1.5} - \frac{D^3}{8}\right] - \frac{\pi D^2}{4}\right\}}^{Paraboloid}$$

$$+ \overbrace{\{\pi D L_2\}}^{Cylinder} + \overbrace{\left\{\frac{\pi(D+d_1)}{2}\sqrt{L_3^2 + \frac{D^2 - d_1^2}{4}}\right\}}^{Frustum}$$

$$+ \underbrace{\left\{\frac{\pi(d_1+d_2)}{2}\sqrt{L_4^2 + \frac{d_1^2 - d_2^2}{4}}\right\}}_{Frustum}$$

Then simplify:

$$S_{TAD} = \frac{\pi D}{4}\left(\frac{1}{3L_1^2}\left[\left(4L_1^2 + \frac{D^2}{4}\right)^{1.5} - \frac{D^3}{8}\right]\right.$$

$$- D + 4L_2 + \frac{2(D+d_1)}{D}\sqrt{L_3^2 + \frac{D^2 - d_1^2}{4}}$$

$$\left. + \frac{2(d_1+d_2)}{D}\sqrt{L_4^2 + \frac{d_1^2 - d_2^2}{4}}\right)$$

QED

12.4 ESTIMATING THE GEOMETRIC PROPERTIES OF THE FUSELAGE

TABLE 12-2 Dimensions of the Erect Mannequin in Figure 12-27

			MALE AND FEMALE PHYSICAL CHARACTERISTICS					
			1st	5th	50th	95th	99th	
Stature	A	Male	63.1	64.8	69.1	73.5	75.2	in
			160.3	164.6	175.5	186.7	191.0	cm
		Female	58.4	60.2	64.1	68.4	70.1	in
			148.3	152.9	162.8	173.7	178.1	cm
Shoulder-head	B	Male	11.8	11.9	12.5	13.0	13.2	in
			30.0	30.2	31.8	33.0	33.5	cm
		Female	10.9	11.3	11.8	12.3	12.5	in
			27.7	28.7	30.0	31.2	31.8	cm
Torso	C	Male	19	19.3	20.1	20.8	21.1	in
			48.3	49.0	51.1	52.8	53.6	cm
		Female	17.5	17.8	18.4	19.2	19.2	in
			44.5	45.2	46.7	48.8	48.8	cm
Hip height	D	Male	14.9	15.4	16.7	18.0	18.5	in
			37.8	39.1	42.4	45.7	47.0	cm
		Female	14.3	14.7	15.9	17.1	17.8	in
			36.3	37.3	40.4	43.4	45.2	cm
Knee height	E	Male	17.4	18.2	19.8	21.7	22.4	in
			44.2	46.2	50.3	55.1	56.9	cm
		Female	15.7	16.4	18.0	19.8	20.6	in
			39.9	41.7	45.7	50.3	52.3	cm
Head height	F	Male	8.4	8.6	8.6	9.1	9.4	in
			21.3	21.8	21.8	23.1	23.9	cm
		Female	7.8	8.3	8.6	9.1	9.4	in
			19.8	21.1	21.8	23.1	23.9	cm
Upper arm	G	Male	11.3	11.5	12.3	13.5	14	in
			28.7	29.2	31.2	34.3	35.6	cm
		Female	10.5	10.7	11.5	12.4	12.8	in
			26.7	27.2	29.2	31.5	32.5	cm
Head length	H	Male	7.1	7.3	7.8	8.2	8.4	in
			18.0	18.5	19.8	20.8	21.3	cm
		Female	6.8	7.0	7.4	7.8	8.0	in
			17.3	17.8	18.8	19.8	20.3	cm
Functional reach	I	Male	28.4	29.1	31.5	34.1	35.3	in
			72.1	73.9	80.0	86.6	89.7	cm
		Female	25.9	26.7	28.9	31.4	32.4	in
			65.8	67.8	73.4	79.8	82.3	cm
Elbow-fingertip	J	Male	17.1	17.6	19.2	20.6	21.3	in
			43.4	44.7	48.8	52.3	54.1	cm
		Female	15.4	16.0	17.4	19.0	19.6	in
			39.1	40.6	44.2	48.3	49.8	cm
Head width	K	Male	5.1	5.6	6.0	6.3	6.5	in
			13.0	14.2	15.2	16.0	16.5	cm
		Female	5.2	5.4	5.7	6.0	6.1	in
			13.2	13.7	14.5	15.2	15.5	cm
Shoulder breadth	L	Male	17.1	17.7	19.3	21.1	21.7	in
			43.4	45.0	49.0	53.6	55.1	cm
		Female	15	15.6	17.0	18.6	19.4	in
			38.1	39.6	43.2	47.2	49.3	cm

Based on Ref. [7].

TABLE 12-3 Dimensions of the Sitting Mannequin in Figure 12-27

			MALE AND FEMALE PHYSICAL CHARACTERISTICS					
			1st	5th	50th	95th	99th	
Sitting stature	N	Male	48.2	49.4	52.4	55.5	56.7	in
			122.4	125.5	133.1	141.0	144.0	cm
		Female	44.1	45.5	48.2	51.3	52.3	in
			112.0	115.6	122.4	130.3	132.8	cm
Leg at 0°	L_0	Male	17.4	18.2	19.8	21.7	22.4	in
			44.2	46.2	50.3	55.1	56.9	cm
		Female	15.7	16.4	18	19.8	20.6	in
			39.9	41.7	45.7	50.3	52.3	cm
Leg at 45°	L_{45}	Male	12.3	12.9	14.0	15.3	15.8	in
			31.3	32.7	35.6	39.0	40.2	cm
		Female	11.1	11.6	12.7	14.0	14.6	in
			28.2	29.5	32.3	35.6	37.0	cm
Leg at 80°	L_{80}	Male	3.0	3.2	3.4	3.8	3.9	in
			7.7	8.0	8.7	9.6	9.9	cm
		Female	2.7	2.8	3.1	3.4	3.6	in
			6.9	7.2	7.9	8.7	9.1	cm

Based on Ref. [7].

Derivation of Equation (12-17)

$$V_{TAD} = \frac{\pi \cdot D^2 \cdot L_1}{8} + \frac{\pi \cdot D^2 \cdot L_2}{4} + \frac{\pi \cdot L_3}{12}\left(D^2 + Dd_1 + d_1^2\right) + \frac{\pi \cdot L_4}{12}\left(d_1^2 + d_1 d_2 + d_2^2\right)$$

$$= \frac{\pi}{4}\left[D^2 \cdot \left(\frac{L_1}{2} + L_2 + \frac{L_3}{3}\right) + \frac{L_3}{3}\left(Dd_1 + d_1^2\right) + \frac{L_4}{3}\left(d_1^2 + d_1 d_2 + d_2^2\right)\right]$$

QED

Test Values

In order to minimize the risk of an error when entering these in a spreadsheet or other computer codes, the following validation values have been tabulated for the reader to compare their own calculations to. Consider a fuselage whose dimensions are given by: $D = 2$, $d_1 = 1$, $d_2 = 0.5$, $L_1 = L_2 = L_3 = L_4 = 2$. This way we can simply add the individual values for each elementary component and compare the sum to the outcome of the above expressions. Thus, $S_{TAD} = 5.9026 + 12.5664 + 10.2704 + 4.8216 = 33.5610$ square units and $V_{TAD} = 3.1416 + 6.2832 + 3.6652 + 0.9163 = 14.0063$ cubic units. It can be seen that both match well.

TABLE 12-4 Typical Occupant Weights Used in Aircraft Design

	Standard Occupant Weight	
	lb$_f$	kg
Standard weight per 14 CFR 23.25	170	77
Recommended for GA design	200	91
Typical zero/zero ejection seat	250–300	113–136

TABLE 12-5 Cabin Dimension for Selected Aircraft (Based on Manufacturers' Information)

Turboprop Aircraft					
Aircraft	Passengers	Length, in	Width, in	Height, in	Baggage, ft^3
Beechcraft 1900D	19	303	54	71	192
Beechcraft King Air 200	10	200	54	57	54
Beechcraft King Air 90	6	155	57	54	—
Dornier 328	32	407	87	75	222
Embraer Brasilia 120	30	378	64	68	226
Jetstream 31	18	291	73	71	52–65
Metroliner 23	19	305	64	58	143.5
Piaggio Avanti P180	8	170	74	69	51

Single-engine Piston					
Aircraft	Passengers	Length, in	Width, in	Height, in	Baggage, ft^3
Beechcraft V35 Bonanza	4	—	42	50	—
Cessna 152	2	—	39.75	42	—
Cessna 172 Skyhawk	4	142	39.5	48	—
Cessna 182 Skyline	4	—	42	48	—
Cessna 206 Skywagon	6	145	44	49.5	—
Cessna Corvalis	4	139.5	49	49	—
Cirrus SR20 and SR22	4	—	49	50	—
Mooney 201	4	—	43.5	44.5	—
Piper Arrow	4	—	41	45	—
Piper Cherokee PA-28-140	4	—	40	44	—

Regional and Business Jets					
Aircraft	Passengers	Length, in	Width, in	Height, in	Baggage, ft^3
Canadair Regional Jet	30	545	98	73	250
Cessna 525C Citation CJ4	—	268	58	57	—
Cessna 680 Citation Sovereign	—	369	66	68	—
Cessna Citation Bravo	8	192	58	56	74
Cessna Citation EXCEL	8	223	67	68	87
Cessna Citation Ultra	8	208	59	58	67
Cessna Citation X	9	282	67	68	87
Cessna Citation jet/CJI	6	136	59	57	58
Challenger 300	16	343	86	73	131
Challenger 604	16	306	99	74	115
Challenger 800	30	581	98	73	250
Dassault Falcon 2000	13	315	92	74	134
Dassault Falcon 900	17	398	92	75	127
Dornier 328 Jet	32	408	88	75	277

(Continued)

TABLE 12-5 Cabin Dimension for Selected Aircraft (Based on Manufacturers' Information)—Cont'd

	Regional and Business Jets				
Aircraft	Passengers	Length, in	Width, in	Height, in	Baggage, ft^3
Galaxy	10	293	87	76	150
Global Express	16	540	98	76	325
Gulfstream 100	7	206	58	67	51
Gulfstream IV/SP	16	488	88	74	169
Gulfstream V	16	527	88	74	226
Hawker 125-700	9	256	71	69	40
Hawker 125-800	10	256	72	69	48
Hawker 450	–	250	59	57	–
Honda HA-420 Hondajet	–	214	60	59	–
LearJet 45	8	238	62	59	65
LearJet 60	7	212	71	68	59
Legacy	14	517	82	67	242

$$S_{TAD} = \frac{\pi 2}{4}\left(\frac{1}{3(2)^2}\left[\left(4(2)^2 + \frac{2^2}{4}\right)^{1.5} - \frac{2^3}{8}\right] - 2 + 4(2) + \frac{2(2+1)}{2}\sqrt{(2)^2 + \frac{2^2 - (1)^2}{4}}\right.$$

$$\left. + \frac{2(1+0.5)}{2}\sqrt{(2)^2 + \frac{(1)^2 - (0.5)^2}{4}}\right) = 33.5610$$

$$V_{TAD} = \frac{\pi}{4}\left[D^2 \cdot \left(\frac{L_1}{2} + L_2 + \frac{L_3}{3}\right) + \frac{L_3}{3}\left(Dd_1 + d_1^2\right) + \frac{L_4}{3}\left(d_1^2 + d_1 d_2 + d_2^2\right)\right]$$

$$= \frac{\pi}{4}\left[2^2 \cdot \left(\frac{2}{2} + 2 + \frac{2}{3}\right) + \frac{2}{3}\left(2(1) + (1)^2\right) + \frac{2}{3}\left((1)^2 + (1)(0.5) + (0.5)^2\right)\right]$$

$$= \frac{\pi}{4}\left[4 \cdot \left(3 + \frac{2}{3}\right) + \frac{6}{3} + \frac{2}{3}(1.75)\right] = 14.0063$$

12.4.5 Surface Areas and Volumes of a Pod-style Fuselage

The following geometry can be used to estimate the geometric properties of a generic pod-style fuselage shown in Figure 12-26, such as that of airplanes like the P-38 Lightning. The geometry consists of a paraboloid, cylinder, and a cone, featuring the dimensions shown. This is effectively identical to the passenger transport fuselage, except the cylindrical segment is substantially shorter. Use Equations (12-12), (12-13), and (12-14).

12.5 ADDITIONAL INFORMATION

A general mannequin is essential when trying to establish space claims of occupants during the layout of the cabin and cockpit. A simple one can be made using elementary geometry such as circles for joints and head, and sticks for torso, stomach, arms, and legs. Such a figure should feature joints to allow limbs to bend to help assess space claims when standing, sitting down, or reaching and so on. It should give a realistic representation of the size of buttocks and thighs so that when folded into a sitting position the height from the head to the buttock will be accurate. A common mistake is to size the geometry of smaller aircraft so a portion of the pilot's head ends up penetrating the fuselage side. An example of such a mannequin is shown in Figure 12-27. The dimensions indicated by the letters A through L can be obtained using documents such as *The Human Factors Design Guide* (HFDG) [7], which provides a collection of dimensions of human physical data and is a great resource for this purpose. The dimensions have been listed in Tables 12-2 and 12-3 for both males and females, ranging from the 1st percentile to the 99th. The concept of percentile refers to the percentage of people who are included in that particular class. For instance, the 50th percentile means that 50% of the general population does not surpass the cited physical dimensions, and so on. For instance, the stature (A) of the 95th percentile is 186.7 cm. This means that 95% of all people are equal to or less tall than that. The modern aircraft should be designed to accommodate the 95th percentile male.

Table 12-4 details typical occupant weights used in the design of GA aircraft. Table 12-5 shows the cabin dimensions for selected aircraft.

VARIABLES

Symbol	Description	Units (UK and SI)
A_{side}	Side area of fuselage	ft² or m²
A_{top}	Top area of fuselage	ft² or m²
C_D	Drag coefficient	
C_{Lmax}	Maximum lift coefficient	
C_{MGC}	Mean geometric chord length	ft or m
D	Diameter of geometric shape	ft or m
d_1, d_2	Diameters of frustum ends	ft or m
D_1, D_2	Major and minor diameters of an ellipsis	ft or m
D_{fus}	Maximum diameter of the fuselage	ft or m
H	Height of the fuselage	ft or m
L	Length of geometric shape	ft or m
l/d	Fineness ratio	
L_1, L_2, L_3, L_4	Fuselage segment lengths	ft or m
L_{cabin}	Length of the cabin	ft or m
L_{emp}	Length of the empennage	ft or m
L_{fus}	Total length of the fuselage	ft or m
L_{fwd}	Length of the forward section	ft or m
l_T	x-distance from wing MGC/4 to HT MGC/4	ft or m
Re	Reynolds number	
S_C	Surface area of a cone	ft² or m²
S_{EC}	Surface area of an elliptic cylinder	ft² or m²
S_F	Surface area of a frustum	ft² or m²
S_{FUS}	Fuselage surface area	ft² or m²
S_P	Surface area of a paraboloid	ft² or m²
S_{TAD}	Surface area of a tadpole fuselage	ft² or m²
S_{UC}	Surface area of a uniform cylinder	ft² or m²
V	Local airspeed	ft/s or m/s
V_∞	Freestream airspeed	ft/s or m/s
V_C	Volume of a cone	ft³ or m³
V_{EC}	Volume of an elliptic cylinder	ft³ or m³
V_F	Volume of a frustum	ft³ or m³
V_{HT}	Horizontal tail volume term	
V_P	Volume of a paraboloid	ft³ or m³
V_{PAX}	Volume of aircraft (passenger cabin)	ft³ or m³
V_{TAD}	Volume of a tadpole fuselage	ft³ or m³
V_{UC}	Volume of a uniform cylinder	ft³ or m³
V_{VT}	Vertical tail volume term	
W	Width of the fuselage	ft or m

References

[1] Galvao, Leme F. A Note on Low Drag Bodies. Poland: Paper presented at the XI OSTIV Congress, in Lesyno; 1968.
[2] Althaus D. NASA CR-2315, Motorless Flight Research. In: James L, editor. Nash-Webber 1972.
[3] Dodbele SS, van Dam CP, Vijgent PMHW, Holmes BJ. Shaping of Airplane Fuselages for Minimum Drag. AIAA Journal of Aircraft, May 1987;24(5).
[4] NASA TM-77014. Wind Tunnel Investigations of Glider Fuselages with Different Waisting and Wing Arrangements. Radespiel, R; 1983.
[5] Thomas F. Fundamentals of Sailplane Design. College Park Press; 1999.
[6] Jenkinson LR, Simpkin P, Rhodes D. Civil Jet Aircraft Design. Arnold; 1999.
[7] Human Factors Design Guide (HFDG). DOT/FAA/CT-96/1.

CHAPTER 13

The Anatomy of the Landing Gear

OUTLINE

13.1 Introduction — 548
   13.1.1 The Content of this Chapter — 548
   13.1.2 Landing Gear Arrangement — 548
   13.1.3 Landing Gear Design Checklist — 549
      Positioning — 549
      Ground Clearance — 549
      Landing Gear Component Vendors — 549
      Tires and Tire Inflation Pressure — 549
      Landing Impact and Braking — 549

13.2 Tires, Wheels, and Brakes — 551
   13.2.1 Important Dimensions and Concepts for Landing Gear Design — 551
      Turning Radius — 552
      Tire Footprint — 552
      Castering Nose and Tail Wheels — 553
   13.2.2 Retractable Landing Gear — 553
   13.2.3 Types and Sizes of Tires, Wheels and Brakes — 555
      Tire Types — 555
      Inflation Pressure — 555
      Tire Geometry — 555
      Tire Sizes — 555
      Selection of Tire Sizes — 555
      Wheels — 558
      Brakes — 559
      Derivation of Equation (13-6) — 560
      Other Useful Information — 560
   13.2.4 Types of Landing Gear Legs — 560
      Leaf-spring Landing Gear (A) — 563
      Oleo-pneumatic Landing Gear (B, C, D) — 564
      Oleo-pneumatic Trailing-link Landing Gear (C) — 564
      Various Landing Gear Issues — Shimmy — 564
      Various Landing Gear Issues — Whistling — 565
      Various Landing Gear Issues — Non-linear Loads During Retraction and Deployment — 565

   13.2.5 Reaction of Landing Gear Forces — 565
   13.2.6 Comparing the Ground Characteristics of Taildragger and Tricycle Landing Gear — 565

13.3 Geometric Layout of the Landing Gear — 567
   13.3.1 Geometric Layout of the Tricycle Landing Gear — 567
      Step 1 — 567
      Step 2 — 567
      Step 3 — 567
      Step 4 — 568
      Step 5 — 568
      Step 6 — 568
      Step 7 — 568
      Step 8 — 568
      Step 9 — 568
      Overturn Angle — 569
      Step 10 — 569
      Step 11 — 569
      Step 12 — 569
      Step 13 — 569
      Step 14 — 569
   13.3.2 Geometric Layout of the Taildragger Landing Gear — 569
      Step 3 — 570
      Step 4 — 570
      Step 5 — 570
      Step 6 — 570
      Step 7 — 570
      Step 8 — 570
      Step 9 — 570
      Step 10 — 570
   13.3.3 Geometric Layout of the Monowheel Landing Gear with Outriggers — 571
   13.3.4 Tricycle Landing Gear Reaction Loads — 571
      Special Case — Static Loads — 572
      Definition of Aerodynamic Loads — 572

Derivation of Equation (13-7)	573	Variables	579
13.3.5 Taildragger Landing Gear Reaction Loads	576	References	580
Special Case — Static Loads	576		
Derivation of Equation (13-10)	576		

13.1 INTRODUCTION

The purpose of the landing gear is obvious — to allow the aircraft to return to the ground without causing damage to the structure. To accomplish this, the landing gear must not only react substantial forces and moments, but also provide a means to deliver the load safely into the airframe. In addition to this primary function, the landing gear must also permit the secondary functions listed below:

(1) It must allow the airplane to maneuver easily on the ground, and during taxi, take-off, and landing (steering).
(2) It must provide a means to slow down after touchdown or when maneuvering on the ground (braking).
(3) Large aircraft must feature a means for being towed and pushed by airport vehicles.
(4) It must feature enough wheels to allow the weight of the airplane to be distributed so that damage will not be inflicted on taxi- and runways.

Designing the landing gear is not a trivial task and this is further compounded when the gear must be retractable as well. The challenges are in terms of both the kinematics of the landing gear mechanism and space-claims inside the affected airframe (wing or fuselage).

Generally, the following issues must be resolved before it is possible to begin the layout of the landing gear.

(1) Desired landing gear configuration:
   a. What is the landing gear arrangement (tricycle, taildragger, etc.)?
   b. What is the location of the wheels and tires?
   c. What is the wheelbase and wheel track?
   d. What is the number and size of wheels and tires?
(2) Should the landing gear be fixed or retractable?
   a. If retractable, should there be hydraulic or electric actuation?
   b. If retractable, will doors cover landing gear fully or partially?
(3) Landing gear details:
   a. What kind of shock absorption (leaf-spring, oleo-pneumatic, trailing link, etc.)?
   b. What kind of braking system (calipers, drogue chute)?
   c. What kind of floatation (ability of the airplane to operate on soft airfields)?

13.1.1 The Content of this Chapter

- **Section 13.2** presents important information about wheels, tires, and brakes used for aircraft, allowing the reader to size and select these landing gear components. Additionally, a number of important concepts regarding fixed and retractable landing gear are presented. It is recommended that the reader is familiar with these before beginning the layout of the landing gear.
- **Section 13.3** presents cookbook methods to place the nose, main, and tail wheels of the airplane, for conventional, taildragger, and monowheel configurations. Additionally, methods of calculating the reaction loads for the landing gear and to evaluate the airspeed required to rotate the airplane during T-O are presented.

13.1.2 Landing Gear Arrangement

The first decision of the designer regarding the development of the landing gear is the arrangement. A large number of different types of landing gear have been employed in the history of aviation. Figure 13-1 shows the four most common arrangements of a large number of possible options. Of those, the most common are the tricycle and taildragger by a large margin. While it may be necessary to consider alternative options for specific projects, the designer is well advised to consider these four carefully — before seeking alternative solutions. There is a reason why the tricycle and taildragger configurations are used in 99.5% of all aircraft, and the four arrangements cover perhaps 99.9% of all aircraft ever built.

To help with the decision let's consider the advantages and disadvantages of the above landing gear arrangements. Pazmany [1, p. 21] presents a number of pros and cons that apply to the configurations, some of which are presented in Table 13-1.

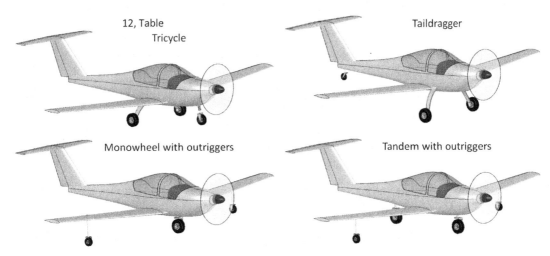

FIGURE 13-1 Examples of landing gear arrangement.

13.1.3 Landing Gear Design Checklist

Since the function of the landing gear is one of the most important in the entire aircraft, it is reasonable to list a number of features that the designer should be aware of and that influence a well-designed system.

Positioning

The main and auxiliary landing gears (nosewheel, tail wheel, etc.) must be properly positioned with respect to the CG to reduce the risk of ground looping, overturning, crosswind canting, tail angle, and to allow the airplane to be maneuvered as needed during the T-O and landing ground run. Methods that address this are presented in the following sections:

Section 13.3.1, *Geometric layout of the tricycle landing gear*
Section 13.3.2, *Geometric layout of the taildragger landing gear*
Section 13.3.3, *Geometric layout of the monowheel landing gear with outriggers*

The designer should keep in mind that an efficient structure places stout structural elements near the hardpoints for the landing gear so extra support structure will not be called for.

Ground Clearance

A well-designed landing gear provides adequate clearance between the ground and all other parts of the airplane, even in the worst combination of flat tire and deflated oleo-pneumatic shock absorber. This means propeller tips, nacelles, wingtips, deflected flaps, and so on. A tailskid or bumper should be installed to prevent damage to the airplane from a tail strike.

Landing Gear Component Vendors

Table 13-2 lists a few companies that manufacture and sell components for landing gear.

Tires and Tire Inflation Pressure

The inflation pressure should be selected based on the weight of the aircraft, number of tires, and the bearing capability of the airfield from which the airplane will be operated.

Nosewheel tire pressure should be based on allowable dynamic loads as follows:

Type III tires: $1.4\times$ static max vertical force
Type VII tires: $1.35\times$ static max vertical force

Main gear tire should allow for 25% growth in aircraft gross weight. The main gear tire load rating should be based on maximum gross weight and adverse CG location. Forged aluminum-alloy wheels are recommended.

Landing Impact and Braking

The landing gear is required by applicable regulations to react a number of landing scenarios. For GA aircraft, the applicable landing gear loads are stipulated by 14 CFR 23.471 to 23.511. For seaplanes, the hull landing loads are stipulated in 14 CFR 23.521 to 23.562. For land planes, there are individual requirements for landing on the main gear only, three-point landing, and side impact loads. The magnitude of these loads is very high and they are reacted as point loads in the airframe. In addition to the impact forces, the landing gear should provide good damping characteristics and should permit the aircraft to taxi over uneven ground without transmitting excessive shocks to the airframe. The airplane should possess a good, reliable, and safe braking system so that it can handle all braking conditions. It should offer a parking brake that can hold the airplane at gross weight on a 1:10 gradient slope or on a level runway with maximum T-O power applied on one engine. Anti-skid brakes are recommended.

TABLE 13-1 Pros and Cons of Several Landing Gear Arrangements

	Advantages	Disadvantages
Tricycle	• Dynamically stable on the ground so it is easier to maneuver. • Good ground control in crosswinds. • Good forward visibility because of low deck angle. • Floor deck angle on the ground is closer to being horizontal, making passenger entry and exit easier. • Propeller better protected from ground strike. • Hard braking on main wheels cannot cause the airplane to nose over. • Airplane pitches nose-down upon main gear touch-down, reducing lift. • Less bounce after touch-down. • Good acceleration during T-O due to lower *AOA*. • Shorter wheelbase permits tight turning radius. • Easier to land and, thus, more forgiving for inexperienced pilots.	• Requires a minimum airspeed before the airplane can be rotated for T-O. • Higher structural weight due to three highly loaded landing gear legs. • More costly for the same reason. • Higher cruise drag for the same reason. • Three landing gear legs make ground ride worse on uneven surfaces. • Subject to nosewheel shimmy that can be very damaging. • Nosewheel shimmy more likely a design problem than on a tailwheel. • Higher dynamic ground loads due to heavier load on nose gear than on a tail wheel. • More complex steering mechanism. • Nosewheel retraction can be challenging due to often limited space. • Low aero-drag due to the low *AOA* attitude during landing requires more braking effort. • Heavier braking unloads main wheels and may cause skidding.
Taildragger	• Allows the pilot to rotate at low speeds, reducing soft-field T-O distance. • Less structural weight due to only two highly loaded landing gear legs. • Less costly for the same reason. • Less cruise drag for the same reason. • Two landing gear legs improve ground ride on uneven surfaces. • Better adapted for rough fields (favored by bush pilots). • Structurally simple fuselage or wing attachment. • High *AOA* for landing results in increase aero-drag and lessens demand on brakes. • Light weight of tail wheel due to relatively low ground loads. • Simple steering mechanism. • Greater propeller clearance during taxi operations increases safety on unprepared runways.	• Dynamically unstable on the ground so it is harder to maneuver on the ground. • Tendency to ground-loop. • Poor directional control with strong crosswinds. • Poor forward visibility because of high deck angle. • Harder to land for inexperienced pilots. • Prop strikes are possible as a consequence of heavy braking that may force nose down. • High *AOA* during landing may cause the airplane to bounce. • Slower initial acceleration due to the higher deck angle.
Monowheel	• Eliminates reinforcement for landing gear hard points from the wing. • Low weight due to a single main landing gear versus multiple. • Less costly for the same reason. • Less cruise drag for the same reason. • Ground loads transferred directly to fuselage through a short load path.	• Requires outriggers or wingtip skids for ground stability. • Outriggers may require wider taxiways. • Poor ground control, especially in crosswind. • Reduced propeller ground clearance. • Reduced *AOA* for flare. • Retraction of main gear forward of CG is bound to be challenging because of space constraints. • Higher T-O speed to make up for the limited ability to rotate. • Not intended for extensive taxiing.
Tandem Wheel	• Eliminates reinforcement for landing gear hard points from the wing. • Low weight due to a single main landing gear versus multiple. • Less costly for the same reason. • Less cruise drag for the same reason. • Ground loads transferred directly to fuselage through a short load path.	• Requires outriggers or wingtip skids for ground stability. • Outriggers may require wider taxiways. • Poor ground control, especially in crosswind. • Reduced propeller ground clearance. • Reduced *AOA* for flare. • Retraction of main gear forward of CG is challenging because of space constraints. • Higher T-O speed to make up for the limited ability to rotate. • Not intended for extensive taxiing.

13.2 TIRES, WHEELS, AND BRAKES

TABLE 13-2 Selected Vendors of Landing Gear Components

Company	Website	Tires	Wheels	Brakes	Landing Gear	Comment
Aircraft Spruce	www.aircraftspruce.com	X	X	X	X	Retailer of gear for kitplanes; no manufacturing
Azusa	www.azusaeng.com	X	X	X		Non-certified products for kitplanes and ultralights
Bridgestone	www.bridgestone.com	X				Tires for most types of aircraft
Goodrich	www.goodrich.com		X	X	X	Large, business, and military aircraft
Goodyear	www.goodyearaviation.com	X				Tires for most types of aircraft
Grove	www.groveaircraft.com	X	X	X	X	Aluminum leaf-spring and fixed, for certified and experimental aircraft
Messier-Bugatti-Dowty	www.safranmbd.com		X	X	X	Large commercial, military, helicopter, business and regional aircraft
Michelin	www.airmichelin.com	X				Tires for most types of aircraft
Parker	www.parker.com		X	X		Business and GA aircraft
Triumph	www.triumphgroup.com				X	Contract design of landing gear
Wicks Aircraft	www.wicksaircraft.com	X	X	X	X	Retailer of gear for kitplanes; no manufacturing
Desser Tire & Rubber Company	www.desser.com	X				Manufacturer of Aero Classic tires; offers various other parts for landing gear
Dunlop Aircraft Tyres	www.dunlopaircrafttyres.com	X				Tires for most types of aircraft
Specialty Tires of America	www.stausaonline.com	X				Formerly McCreary Tire

13.2 TIRES, WHEELS, AND BRAKES

This section presents some important concepts to keep in mind when designing the layout of the landing gear. Concepts to be cognizant of include landing gear configuration, wheelbase, wheel track, and turning radius.

13.2.1 Important Dimensions and Concepts for Landing Gear Design

Figure 13-2 shows important concepts and dimensions used when discussing landing gear arrangement. The tail wheel spindle axis angle, ϕ, should

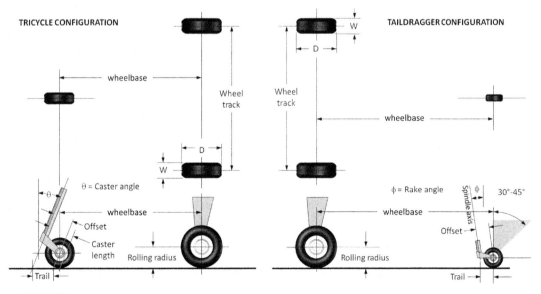

FIGURE 13-2 Important geometric definitions for a tricycle landing gear arrangement. *(Based on Ref. [1])*

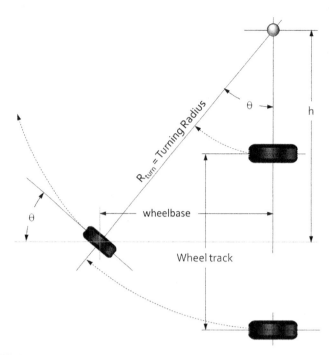

FIGURE 13-3 Important geometric definitions for determining turning radius.

be inclined forward by some 5°. The tail wheel trail should be 1/10th of the wheel's diameter. When deflected, the tail wheel shock absorber should place the wheel axle inside the shaded area.

Turning Radius

For maneuvering on the ground, in particular when turning into position on a narrow runway, the *turning radius* is a very important feature in aircraft ground operation. Figure 13-3 shows how the turning radius for a given rotation of the nose landing gear can be determined. The distance h denotes the location of the center of turn and can be calculated using:

Distance to turning center:

$$h = \frac{\text{Wheelbase}}{\tan \theta} \quad (13\text{-}1)$$

Equation (13-2) below allows the turning radius to be calculated, knowing only the wheelbase and the turning angle of the nose gear.

Turning radius:

$$R_{turn} = \sqrt{\text{Wheelbase}^2 + h^2}$$
$$= \text{Wheelbase}\sqrt{1 + \frac{1}{\tan \theta^2}} \quad (13\text{-}2)$$

Some aircraft are capable of turning on a dime, literally, by enabling a large enough nose gear turning angle. This requires the nose gear to be capable of turning at least:

$$h = \frac{\text{Wheelbase}}{\tan \theta} = \frac{\text{Wheel track}}{2}$$
$$\Rightarrow \tan \theta = \frac{2 \times \text{Wheelbase}}{\text{Wheel track}} \quad (13\text{-}3)$$

Tire Footprint

A tire in contact with the ground will flatten slightly on the bottom so its contact area resembles that of an ellipse (see Figure 13-4). The weight, F, supported by the tire is distributed over this area, generating an average pressure, P, on the area which amounts to:

$$P = F/S_e = \frac{4F}{\pi \cdot a \cdot b} \quad (13\text{-}4)$$

The higher the pressure inside the tire, the less are the values of a and b. High-pressure tires supporting large loads may cause rutting damage to the runway. On soft enough ground, the tire may sink, which explains why high-floatation tires (so-called "Tundra tires") have both a large diameter and low internal pressure. Note that the dimension a is usually taken to be 85% of the distance c.

Note that the compression of the tire increases the internal pressure due to the reduction of internal volume. The unloaded pressure is therefore always lower than the loaded one. Inflation pressure reported by manufacturers refers to the unloaded tire at standard S-L temperature. Typically, when the tire is loaded to the rated load,

13.2 TIRES, WHEELS, AND BRAKES

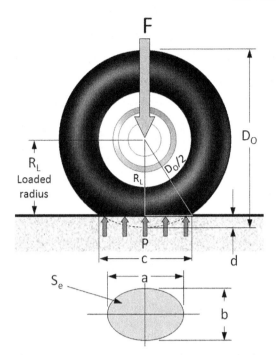

FIGURE 13-4 Tire footprint. *(Based on Ref. [1])*

the pressure will be approximately 4% higher than the rated pressure. Tires should always be allowed to cool for at least an hour before checking the pressure.

Equation (13-4) can be used to estimate the required internal pressure for a tire if the allowable ground pressure limits are known for a particular runway or taxiways. Effectively, P equals the internal tire pressure.

Manufacturers' data, such as that of Ref. 2, tabulates the *loaded radius*, R_L, enabling the estimation of the dimension a in Figure 13-4. The first step is relating it to the distance c as shown below:

$$\frac{D_0}{2} = \sqrt{R_L^2 + \frac{c^2}{4}} \Leftrightarrow c = \sqrt{D_0^2 - 4R_L^2}$$

The value of a is usually about 85% of c. Therefore, the dimension can be calculated as follows:

$$\Rightarrow a = 0.85c = 0.85\sqrt{D_0^2 - 4R_L^2} \quad (13\text{-}5)$$

Castering Nose and Tail Wheels

The use of castering nose landing gear and tail wheels in taildragger aircraft is a common solution for steerable landing gear in many light GA aircraft. The steering is usually implemented using differential braking, i.e. the pilot steps on the left or right rudder pedals to generate braking in the corresponding main wheel, which turns the aircraft on the ground.

The most common problem with castering landing gear is shimmy, which is a violent oscillation that is a function of speed and the inertia characteristics of the landing gear. Shimmy can begin when the aircraft moves over uneven ground, and even due to worn tires or landing gear parts. It ranges from a simple wobble to an oscillation so violent it literally breaks the landing gear off the airplane. For this reason, castering landing gear is sometimes equipped with a small device called a *shimmy damper*, which most often is a small cylinder filled with hydraulic fluid that has a piston inside that connects to the movable (rotational) part of the landing gear.

Several configurations are shown in Figure 13-5. Their properties are shown in Table 13-3.

13.2.2 Retractable Landing Gear

The advent of retractable landing gear was an important element in the evolution of aircraft. It allowed

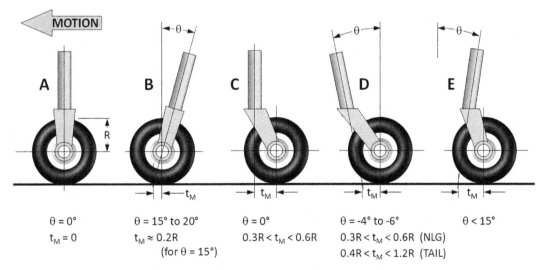

FIGURE 13-5 Five common styles of castering-wheel configurations used for nose and tail wheels. NLG = nose landing gear, TAIL = taildragger, t_M = mechanical trail. *(Based on Ref. [1])*

TABLE 13-3 Types of Castering Nose and Tail Wheels

Style	Used for Nosewheel	Used for Tail Wheel	Statically Stable?	Dynamically Stable?	Shimmy Damper Required?	Comment
A	YES	NO	Neutral	Neutral	–	Requires steered mounting.
B	YES	NO	NO	YES	–	Shimmy damper may be required.
C	YES	YES	Neutral	YES	YES	Shimmy damper not required if twin-wheeled or one wheel with double treads.
D	NO	YES	YES	YES	YES	Shimmy damper may not be required when $t_M > R$.
E	YES		NO	YES	YES	When sideways, the nose of the aircraft is lowered, so it is statically unstable.

Based on Ref. [1].

substantial reduction in drag and increase in cruising speed, making it a very attractive option for increasingly efficient aircraft. However, there were drawbacks, although not compelling enough to suppress its use. Not only were there challenges in functionality and failure mechanism, a new type of accident reared its ugly head: landings with perfectly operational landing gear retracted.

There are many elements that make the retraction and extension of retractable landing gear the most critical phase in its operation. Among them is the reliability of locking and unlocking of the extended gear mechanism, as well as the up-locking and release when retracted. This is compounded by the fact that the landing gear is often operated in severe environments, where it gets exposed to mud and ice. It is generally recommended that the actuation mechanism, doors, and support structure should be designed to allow actuation at gross weight and $1.6V_{S1}$ or higher. Some aircraft, like the high-flying Learjet business jets, allow the landing gear to be partially extended at even high airspeeds to provide additional drag in an emergency descent [3]. Actuation air loads remain one of the most serious challenges to retraction or extension and can easily exceed the capacity of the actuation mechanism. As mentioned before, partially extended retractable landing gear for fast and high-flying aircraft should be designed to also act as an emergency speed brake. If not possible, the landing gear mechanism should be designed to withstand at least $0.67V_C$. The time to retract and extend should be less than 10 and 15 seconds respectively [4].

In order to reap the benefit of retractable landing gear it should be stowed under aerodynamically smooth fairings. It also requires an emergency actuation system to allow the pilot to deploy it manually, should the normal actuation mechanism fail. The design is further complicated by the installation of a braking system on the main landing gear or steering mechanism for the nose gear. The design of retractable landing gear should be left to experts, as there are a number of pitfalls for the novice designer. For instance, retraction and deployment loads during transit are hard to predict accurately. These can be surprisingly large, making it impossible for the specified hydraulic mechanism to perform its duty. One possible consequence is the inability to extend the landing gear fully after a successful retraction. This explains why maiden flights are generally made with the landing gear locked in its extended position. Not much needs to be said about the consequences for a development program if the landing gear cannot be extended and the pilot must opt for a belly landing during the first flight.

The kinematics of landing gear retraction and extension is a challenging task, which is best understood by considering the massive landing loads that must be reacted by a compact frame that must repeatedly and effortlessly fold into and out of a small container (the wheel well). The complexity of the system becomes evident in the design of some aircraft, which feature landing gear doors that open temporarily, just to allow the landing gear to pass through, and then close again – not once, but tens of thousands of times. It takes foresight and engineering experience to design a reliable system like that. While this section presents methods to allow the most important geometric details of the landing gear to be designed, the various retraction kinematics are beyond the scope of this text. References 1, 4, and 5 are good resources for general design of landing gear.

Additionally, the size of the internal volume required to house the retracted gear must be kept in mind when sizing the OML. A potential mistake is to find out *after* substantial analysis has been completed that the wheel wells penetrate the cabin or engine compartment, requiring time-consuming revisions.[1] The retraction system should feature an emergency extension system that is independent of the primary system and should not use hand-pumping or cranking by the pilot. The up-locking of retractable landing gear should not use gear door locks and allow an override emergency release. Failure of electric locking systems should never prevent the emergency deployment of the landing gear. Mechanical sequencing (e.g. special landing gear door opening before the landing gear is retracted and the closing after retraction) should be minimized. Niu [5] provides a number of other requirements essential for the detailed design of retractable landing gear.

Figure 13-6 shows a number of so-called 'stick' diagrams. These are helpful when visualizing the kinematics of retractable landing gear. The images show the landing gear in its unloaded configuration (in the air) and do not reveal ground deflection. Black nodes represent *immovable joints*, while white ones are *movable*. An immovable joint only allows rotation about a fixed point. A movable joint allows rotation and translation. Note that an immovable joint that appears on top of a retracted wheel simply implies an offset hard point.

13.2.3 Types and Sizes of Tires, Wheels and Brakes

The selection of tires for the airplane involves the determination of the *type of tire*, *tire size*, *wheel*, and *brakes*. This section presents information to help the designer in this matter.

Tire Types

Of the number of standard tire designations used for aircraft, the ones presented in this book are based on definition by the Tire and Rim Association (TRA), which is the technical standardizing body of tires and rims in the USA. TRA designations for common tires are shown in Table 13-4. Note that the Three Part Type designation is the modern system. The designer of certified GA aircraft will pick the Three Part Type designation, whereas the designer of homebuilt aircraft can resort to other types of tires.

Inflation Pressure

The tire maintains its shape as a consequence of the inflation pressure. Tires are very flexible in bending but have very limited extension (or stretch). As the tire reacts loads its cross-section flexes and a side force will be generated on the flange of the rim. The tire is designed to operate at an approximately 32–35% deflection under static load, but this means it will bottom out at around 3× the static load. Shock absorbers are designed to bottom out at 3× the static load as well, and are intended to do so before the tire[1]. Table 13-5 shows ranges of pressure for various applications.

Tire Geometry

The following parameters are important when discussing tires (refer to Figure 13-7):

Tire aspect ratio: $AR = \dfrac{H}{W}$

Ensure there is no confusion with the AR of a wing. A low tire AR means it is wider than it is high. Such tires are intended for high speeds on smooth surfaces. High-AR tires are meant for rough field operations that call for improved floatation.

Type III have AR ranging from:	0.85 to 1.00
Type VII have AR ranging from:	0.77 to 0.90
Type VIII have AR ranging from:	0.65–0.77
Three Part Type have AR ranging from:	0.73 to 0.92

Lift ratio: $LR = \dfrac{D}{d}$

An LR of 1.5 to 2.0 is considered low, while 2.0 to 2.5 is desirable [1].

Tire Sizes

Unfortunately, when it comes to specifying tire sizes there is no one standard, but rather several that have evolved since the inception of tire standardization in 1903. This results in several choices at the disposal of the designer, although focusing on the Three Part Type is recommended for modern aircraft design. The TRA type tires shown in Table 13-4 have the sizing designation shown in Table 13-6. Figure 13-8 shows schematics of the three most common types, Type III, VII, and the Three Part Type.

Selection of Tire Sizes

The first step is to download a document like Ref. [2], which features an excellent selection guide in *Section 4 – Data Section – Tires*. The document lists tires by type and rated speed, load, inflation pressure, braking load, and bottoming load. It is an ideal resource to pick a potential tire during the conceptual and preliminary design phase. It can be downloaded free of charge from: www.goodyearaviation.com/resources/. Also, note that for the certification of

---

[1] The same actually holds when designing the cabin or cockpit. Heads or limbs penetrating a critical structure or the OML are often the result of a didn't-think-through attitude.

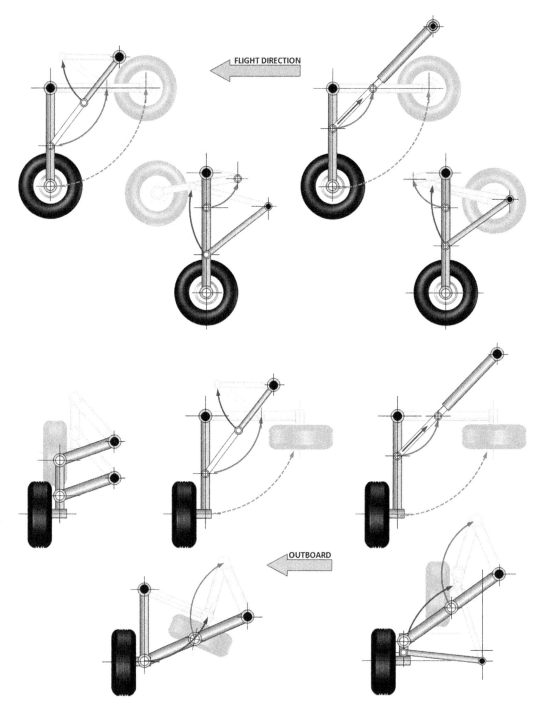

FIGURE 13-6 Some retraction methodologies for configurations that retract forward or aft with respect to flight direction (upper half) and some that retract inboard and extend outboard. (*Based on Refs [1 and 4]*)

GA aircraft, the tire selection is bound by 14 CFR 23.733 *Tires*.

It is problematic that Type III do not explicitly specify the nominal diameter of the tire. For specific dimensions refer to Ref. 2.

Main wheel tires: select main landing gear tires in accordance with the following guidelines:

(1) Account for a growth in the weight of the airplane (assuming it is intended to become a successful production). It is customary to assume a weight 25% greater than the initial gross weight. History shows that airplanes almost always increase their weight as new models are introduced in response to the desire for added capabilities. For this reason,

TABLE 13-4 Contemporary Types of Tires for Aircraft [2]

Type	Pressure	Used for	Speed Limit [1], KGS	Comment
I	Low	Piston aircraft	Low	This is the oldest size designation for tires and is based only on the outside diameter of the tire.
III	Low	Piston aircraft	<140	This is one of the earliest size designations and was used for early piston-prop type aircraft. It is a low-pressure tire intended to provide improved cushioning and floatation.
VII	100–250 psi	Jet aircraft	140–195	A more recent type than Type III; this type covers most of the older sizes and was designed for today's jet aircraft with its higher load capacity.
Three Part Type	–	Piston and jet aircraft	183–204	All new tires being developed follow this classification. This group was developed to meet the higher speeds and loads of today's aircraft. NOTE: Some sizes have a letter (such as 'H') in front of the diameter. This is to identify a tire that is designed for a higher percent deflection.
Radial	–	Piston and jet aircraft	183–204	Radial size nomenclature is the same as Three Part Type, except an 'R' replaces the '–' (dash) before the wheel/rim diameter to indicate a radial tire.
Metric	–	Piston and jet aircraft	183–204	This size designation is the same as Three Part Type, except the diameter and section width dimensions are in millimeters, and the wheel/rim diameter is in inches.

KGS stands for knots ground speed. See Chapter 16, *Performance – introduction* for more details. Note that the ground speeds shown refer to the maximums – the tire can always be used at a lower ground speed.

selecting a larger tire now can avoid costly re-engineering later.
(2) Selection should be based on the load caused by the aft-most CG location for tricycle, and most-forward CG location for taildraggers.
(3) Assume the most severe load-speed-time history during normal operation. This means determining the most adverse combination of high elevation and high ambient temperature take-off or landing. Low-density atmospheric conditions lead to higher ground speeds and this is critical to the tire selection process.

TABLE 13-5 Range of Inflation Pressures for Typical Aircraft

Pressure, psi	Comment
<40	Recommended for unimproved, low-strength and uneven runways. A high-*AR* tire and low pressure offers great impact cushioning, minimizes surface rutting, and improves the durability of the treads and tire in general. This is the pressure range for the typical automobile tire.
40–75	Used for tires expected to operate off improved runways.
75–125	Intended for high-performance aircraft operating off asphalt and concrete runways only. The higher pressure allows for a smaller and lighter tire, which is helpful for reduced aerodynamic drag and stowing volume.
>125	Generally not used for GA aircraft.

Nosewheel tires: select nose landing gear tires in accordance with the following guidelines:

(1) As with the main landing gear tire, account for a growth in the weight of the airplane by assuming a 25% weight growth.
(2) Estimate static and dynamic braking load on the tire. Consider 14 CFR 23.493 *Braked Roll Conditions* in this matter.
(3) Base selection on the most forward CG location.

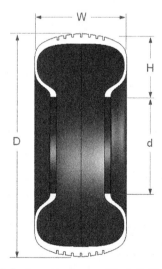

FIGURE 13-7 Important geometric definitions for a tire.

TABLE 13-6 Examples of Dimension Designation for Various Types of Tires for Aircraft

Type	Tire Size Example	Nominal Diameter	Nominal Section Width	Nominal Wheel/Rim Diameter
I	27	27.0 in	—	—
III	8.50–10	—	8.50 in	10.0 in
VII	49 x 17	49.0 in	17.0 in	—
Three Part Type	49 x 19.0–20	49.0 in	19.0 in	20.0 in
Radial	32 x 8.8R16	31.8 in	9.25 in	16.0 in
Metric	670 x 210–12	670 mm	210 mm	12.0 in

(Based on Ref. [2].)

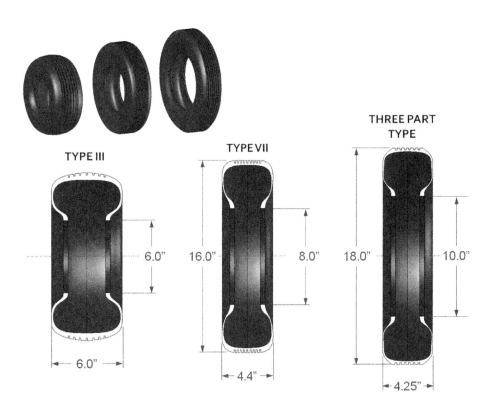

FIGURE 13-8 The three most common aircraft tire types. Left is Type III 6.00-6; center is Type VII 16×4.4 (with an 8-inch rim); and right is Three Part Type 18×4.25-10. Images are to scale.

Tail wheel and outrigger wheels: for smaller aircraft, the tail wheel and outrigger wheels are solid rubber. For larger aircraft, inflated tires are used and can be selected using a similar approach to the above. In the absence of rational analysis, if the outrigger wheels are intended to support landing impact, the attachment structure should be capable of supporting a vertical and side load amounting to $1/2\,g$. The wheel can then be selected based on the resulting load. For example, an airplane with an outrigger and 1000 lb$_f$ gross weight would be reinforced to react 500 lb$_f$ acting at the tire. This side load requirement is in accordance with 14 CFR 23.485 *Side Load Conditions*, although the regulations do not specifically address outrigger configuration. There is no set rule in outrigger structure. Aircraft like the Fournier RF-1 and RF-3 use outriggers not capable of reacting the above load, albeit sufficient to support taxi loads. The Fournier RF-2 does not even use wheels, but rather a wire loop, each end of which is mounted to the wing structure.

Wheels

The modern wheel is usually made from forged aluminum or magnesium alloy. Aluminum wheels are heavier, but are less expensive and have better corrosion

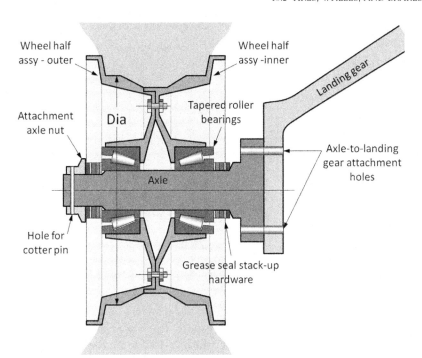

FIGURE 13-9 A schematic of a typical modern wheel made from joined halves.

resistance. Both are equal in robustness and have the same performance rating.

Wheels usually consist of two halves that are joined together by bolts (see Figure 13-9). Some modern designs are of a single-piece configuration that is easier to manufacture and maintain. They feature special lock-rings to provide the necessary lip for the tire.[2] The wheel contains tapered roller bearings that are a press fit into a space called the bearing housing. They are kept in place by spring-fit snap rings. Brake discs are often attached to the wheels using the fasteners that hold the two halves together.

There are three common sizes of wheels for small aircraft: 5″, 6″, and 7″ diameter. The diameter (Dia) is measured between the wheel shelves as shown in Figure 13-9.

Brakes

The brakes must provide the airplane with a means to: stop the aircraft after landing or an aborted take-off; keep the airplane at rest at full power (this means the critical engine only for a multi-engine aircraft); allow the airplane to be steered via differential braking or appropriate slow-downs during taxiing; and park. The brakes are not expected to retain the aircraft with the wheels locked should there be enough thrust to move the aircraft. A simplified schematic of the brake system in a light aircraft is shown in Figure 13-10.

There are two common kinds of brakes: *drum brakes* and *disc brakes*. Drum brakes usually reside inside the wheel, which reduces their frontal area and, if mounted on retractable landing gear, makes them a possible solution for space-claim issues in thin wings. Disc brakes have superior performance, are lighter, and easier to maintain.

The application of brakes generates considerable heat through friction. The amount of heat depends on factors such as the weight of the vehicle being stopped, the speed at which braking is applied, the effectiveness of the conversion of friction into heat, and the friction area of the brakes. This is where the difference between drums and discs become apparent. Drum brakes dissipate heat far less effectively than disc brakes. Since the brake is contained inside the wheel the result is inferior heat transfer from the drum and if too much heat is generated the braking capacity is lost ("fading") just when it is needed the most. In contrast, disc brakes reside on the outside of the wheel and are directly exposed to air, in addition to featuring calipers with two braking pads on each side of the disc (and sometimes multiple discs). Direct exposure to air leads to superior cooling and more effective braking under demanding conditions.

Manufacturers recommend that wheels and brakes be selected together. These should be selected in accordance with the following guidelines:

(1) Determine the static load on each wheel. Select wheels whose rating matches or exceeds this value.
(2) The selection of brakes for GA aircraft is bound by 14 CFR 23.735 *Brakes*. This requires the brakes to have

---

[2] For instance see selected Goodrich products.

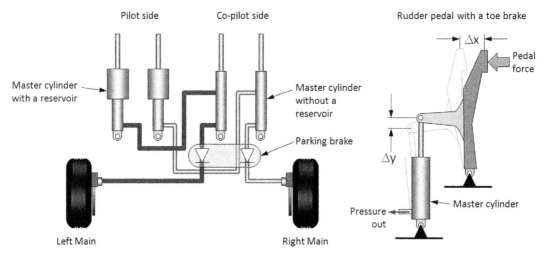

FIGURE 13-10 A schematic of a brake system for a small aircraft. It is recommended that $\Delta y = \tfrac{1}{2} \Delta x$. *(Based on Ref. [6])*

the capacity to absorb the kinetic energy of the aircraft based on one of two methods: conservative rational analysis and, in the absence of such analysis, on the following formula:

$$KE = \frac{0.0433 W V_{S0}^2}{N} \quad (13\text{-}6)$$

where

KE = kinetic energy in ft·lb$_f$
V_{S0} = stalling speed in the landing configuration in KTAS
N = number of main wheels featuring brakes.

Equation (13-6) yields the kinetic energy per brake and this value must be specified when contacting brake manufacturers.

Derivation of Equation (13-6)

For V_{S0} in ft/s or m/s and m in slugs or kg, the kinetic energy is given by:

$$KE = \frac{1}{2} m V_{S0}^2$$

The above regulatory paragraph assumed V_{S0} in KTAS and W in lb$_f$. Therefore,

$$KE = \frac{1}{2} m V_{S0}^2 \rightarrow \frac{KE}{N} = \frac{\tfrac{1}{2}\tfrac{W}{g}(1.688 V_{S0})^2}{N} = \frac{0.0433 W V_{S0}^2}{N}$$

QED

Other Useful Information

Aircraft tires are designed such that the internal tensile forces in each fabric layer are uniform when unloaded. When deflected, this force will vary from higher in the outer plies to lower in the inner ones. This variation generates shear stresses in the tire that must not be exceeded. Exceeding this is possible if the tire is under-inflated or over-loaded. This will cause a large deflection that leads to heat build-up which can cause ply separation and rapidly decrease the durability of the tire. It will also cause an uneven wearing of the tread and deterioration of the shoulder, in addition to possible damage to the sidewall of the tire during high impacts (landing). Excessive inflation pressure can also be detrimental due to uneven tread wear and reduced braking effectiveness, as well as making the tire more susceptible to damage by foreign objects.

Tires can withstand reasonably high temperatures, with typical maximum operational limits being a surface temperature that exceeds 225 °F (107 °C) or brake heat reaching a temperature of 300 °F (149 °C). Such a temperature can result from excessive braking after a high-speed landing.

Finally, when selecting wheels and tires, it can be helpful to know the types of tires used for existing fleets of aircraft. Tables 13-7, 13-8, and 13-9 provide such information for selected aircraft of various classes. The reader is directed toward Ref. 2 for more details.

13.2.4 Types of Landing Gear Legs

One of the most important capabilities of any landing gear is how it absorbs the landing impact load. For GA aircraft the most common way to accomplish this is through the use of leaf-springs, rubber doughnuts, rubber bungees, coiled steel springs, and oleo-pneumatic shock absorbers. The first three are depicted in Figure 13-11, showing typical deflection at impact (denoted by 'd'). The primary advantage of such landing

TABLE 13-7 Examples of Tires Used for Various Business Jet Aircraft

BUSINESS AIRCRAFT					
Aircraft Maker	Model	Main Tire		Nose or Tail Tire	
		Size	Ply Rating	Size	Ply Rating
Aérospatiale	SN-600 Corvette	26x6.6	10 TL	17.5x6.25-6DDT	8 TL
BAe-Hawker	HS-125	23x7.00-12BR	12 TL	18x4.25-10 DT	6 TL
Bombardier	CL-600/601-Challenger	26x6.6	14TL	18x4.4 DT	12 TL
Cessna	500 Citation I	22x8.00-10	10 TL	18x4.4 DDT	10 TL
	550 Citation II	22x8.00-10	12 TL	18x4.4 DDT	10 TL
	650 Citation III	22x5.75-12	10 TL	18x4.4 DDT	10 TL
	560 Citation V	22x8.00-10	12 TL	18x4.4 DDT	10 TL
	650 Citation VI	22x5.75-12	10 TL PR	18x4.4 DDT	10 TL
	650 Citation VII	22x5.7512	10 TL PR	18x4.4 DDT	10 TL
	750 Citation X	26x6.6R14	14 TL PR	18x4.4 DT	10 TL
Dassault	10 Falcon	22x5.75-12	10TL	18x5.75-8	8TL
	20 Falcon	26x6.6 / 26x6.6R14	10TL	13x5.0-4 DT	14TL

Based on Ref. [2]

TABLE 13-8 Examples of Tires Used for Various Single-engine Propeller Aircraft

SINGLE-ENGINE PROPELLER					
Aircraft Maker	Model	Main Tire		Nose or Tail Tire	
		Size	Ply Rating	Size	Ply Rating
American Champion	Decathlon/Super Decathlon	6.00-6/8.00-6	4 or 6	Scott	-
Commander	112/114 Commander	6.00-6/7.00-6	6	5.00-5	4 or 6
Beechcraft	C-17R Staggerwing	7.50-10	6	10.00SC	-
	BE-23 Musketeer, BE-A24 Sundowner	6.00-6	4	6.00-6	4
	BE-33A Bonanza	7.00-6	6	5.00-5	4
	BE-36 Bonanza	7.00-6	6	5.00-5	4
Bellanca	17-30A Viking, 17-31A Super-Viking	6.00-6	6	15x6.00-6	6/8
Cessna	150,-152 Commuter, Aerobat	6.00-6	4	5.00-5	4
	172 Skyhawk	6.00-6	4	5.00-5	4
	177RG Cardinal RG	15x6.00-6	6	5.00-5	4
	180 Skywagon	6.00-6	6	8.00" SC	6
	182RG, T182RG Skylane	15x6.00-6	6	5.00-5	4
	185 Skywagon	6.00-6	6	8.00" SC	8
	188 AG-Wagon	22x8.00-8	6	10.00" SC	6
	206 StationAir	6.00-6	6	5.00-5	6
	207 Skywagon	6.00-6	8	6.00-6	8
	208 (FLO) Caravan	8.50-10	8	22x8.00-8	6
	210 Centurion	6.00-6	8	5.00-5	6
Cirrus	SR20, SR22	15x6.00-6	6	5.00-5	6
Embraer	EMB-200 Ipanema	6.50-8	8TT	2.80/2.50-4	4TT
Mooney	MO-20 Ranger, MO-21C Super	6.00-6	6	5.00-5	4/6
	MO-22 Mustang	6.00-6	6	15x6.00-6	6
	MO-20E Chapparal	6.00-6	6	5.00-5	4
	MO-20F Executive	6.00-6	6	5.00-5	6
	MO-20J 201	6.00-6	6	5.00-5	6
	MO-20K Turbo 231	6.00-6	6	5.00-5	6
	M-20M Bravo, M-20S Eagle, M-20R Ovation	6.00-6	6	5.00-5	6
Piper	J3 Cub	8.00-4	4	8.00"SC	Solid
	PA18-135 Super Cub, PA18-150 Super Cub	6.00-6	4	8.00"SC	Solid
	PA25-150 Pawnee	7.00-6	4	8x3.00-4	4
	PA28-160 Cherokee	6.00-6	4	6.00-6	4
	PA28-161 Warrior II	6.00-6	4	5.00-5	4
	PA28-181 Archer II	6.00-6	4/6	6.00-6	4
	PA28R-201 Arrow	6.00-6	6	5.00-5	4
	PA32 6-300	6.00-6	6	6.00-6	6
	PA32R-301T Saratoga II TC	6.00-6	8	5.00-5	6
	PA36 Brave	8.50-10	6	10x3.5-4	6
	PA38 Tomahawk	6.00-6	6	5.00-5	6
	PA46-350P Malibu Mirage	6.00-6	8	5.00-5	6
Lake Aircraft	LA-4 Buccaneer, LA-250 Renegade/Seafury	6.00-6	4	5.00-4	4

Based on Ref. [2]

TABLE 13-9 Examples of Tires Used for Various Twin-engine Propeller Aircraft

Aircraft Maker	Model	Main Tire Size	Main Tire Ply Rating	Nose or Tail Tire Size	Nose or Tail Tire Ply Rating
Beechcraft	BE-18	11.00-12	8 TL	14.50SC	8 TL
	BE-50 Twin Bonanza	8.50-10	6	6.50-10	4
	BE-B55 Baron	6.50-8	8	5.00-5	6
	BE-60 Duke	19.5x6.75-8	10 TL	15x6.00-6	4
	BE-80 Queen Air	8.50-10	8 TL	6.50-10	6 TL
	BE-C90B King Air	8.50-10	8 TL	6.50-10	6 TL
	BE-200T Super King Air	22x6.75-10	8 TL	22x6.75-10	8 TL
	BE-99	18x5.5	8 TL	6.50-10	6 TL
	BE-300/350 Super King Air	22x6.75-10	8 TL	19.5x6.75-8	10 TL
	BE-1900 Airliner	22x6.75-10	8 TL	19.5x6.75-8	10 TL
Cessna	T303-Crusader	6.00-6	8	6.00-6	6
	310,-T310	6.50-10	6	6.00-6	4
	320,-340	6.50-10	6	6.00-6	4
	Model 336	6.00-6	6	15x6.00-6	6
	337-Super-Skymaster	18x5.5	8	15x6.00-6	6
	401,-402	6.50-10	8	6.00-6	6
	404-Titan	22x7.75-10	10 TL	6.00-6	6
	Model 411	6.50-10	8	6.00-6	6
	414-Chancellor	6.50-10	8	6.00-6	6
	421-Golden Eagle	6.50-10	8	6.00-6	6
	425-Conquest I	6.50-10	10	6.00-6	6
	441-Conquest II	22x7.75-10	10 TL	6.00-6	6
Piper	PA23,x160-235xApache	7.00-6	6	6.00-6	4
	PA23-250xAztec	7.00-6	8	6.00-6	4
	PA34-270T Seneca V	6.00-6	8	6.00-6	8
	PA44-180 Seminole	6.00-6	8	5.00-5	6
	PA42xCheyenne IIIA	6.50-10	12TL	17.5x6.25-6	10
	PA44-180xSeminole	6.00-6	6	6.00-6	6
	600/601Bx(PA60)xAerostar	6.50-8	TL	6.00-6	6
	PA30 Twin Commanche	6.00-6	4	6.00-6	6
	PA31 Chieftain	6.50-10	8	6.00-6	6
	PA31T Navajo	6.50-10	10	6.00-6	6
	PA34-220 Seneca	6.00-6	8	6.00-6	6

Based on Ref. [2]

gear is its simplicity and relative light weight. However, their shock absorption efficiency is less than ideal. To understand why, consider the graphs of Figure 13-12. Then a few words about shock absorption are necessary.

The purpose of a shock absorber is to help bring some initial speed of some mass to zero through deceleration over a given distance. Ideally, this deceleration is achieved through the force acting on the absorber, which changes from an initial zero force, to some maximum load just as the speed reaches zero. In airplanes, it serves to bring the rate-of-descent to zero. Intuition holds that if this deceleration takes place over a long distance, then, from the standpoint of the occupants, the landing will be soft and void of a "shock." Unfortunately, we do not have the luxury of a long distance. The vertical speed, which ranges from 10 to 20 ft/s, must be brought to zero in only a few inches. If a very flexible spring is used, then it will bottom out before the vertical speed reaches zero — and we are in for a very hard shock. If the landing gear is rigid — well, we are also in for a hard shock. The solution to this is to provide some spring action that is neither too flexible nor to stiff. Figure 13-12 shows this in action.

By definition, the ideal shock absorber has 100% efficiency because it achieves the maximum deceleration in the minimum distance — the comfort of the ride has no bearing on this value. This is denoted by the constant K_S in Figure 13-12, which shows that if $K_S = 1$, then the entire landing load will be reacted without deflection. A steel spring obeys Hooke's law and can be sized to react the impact over the available distance with 50% efficiency ($K_S = 0.5$). It will improve the ride, but demands much greater available distance for the deceleration to take place. A tire is even less efficient, with efficiency $K_S = 0.39$ to 0.47 (per Ref. [1]). However, the oleo-pneumatic strut is one of the most efficient means to slow down the aircraft. The graph shows it will react most of the landing load over a relatively short distance, and then the remaining load will be reacted while providing a much more comfortable ride for the occupants.

13.2 TIRES, WHEELS, AND BRAKES   563

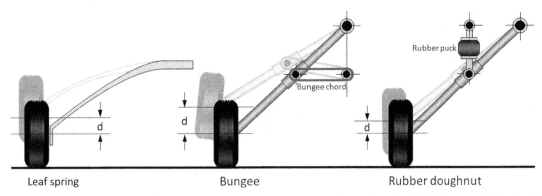

FIGURE 13-11 A schematic of typical leaf-spring, bungee, and rubber doughnut shock absorbers. The dimension 'd' is the deflection of the gear, which is indicative of the efficiency of the landing gear shock absorption.

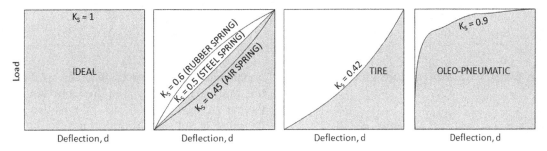

FIGURE 13-12 Load versus deflection graphs showing what is meant by shock absorption efficiency. (*Based on Ref. [1]*)

The landing gear comes in a wide variety of shapes and sizes; volumes could be written about the pros and cons of each. Here, however, in the interest of space, only the four common types shown in Figure 13-13 will be presented. These are the leaf-spring steel (or composite) main landing gear, (A); oleo-pneumatic strut retractable main landing gear, (B); oleo-pneumatic retractable trailing-link main landing gear, (C); and a steerable oleo-pneumatic strut nose landing gear, (D).

It should be stressed that there are many other ways to absorb the landing impact. Many are presented in Refs 1, 4, and 5, not to mention Ref. [7], which presents, not surprisingly, incredibly original landing gear designs devised by German engineers during World War II. It is a good source for anyone looking for interesting ideas for landing gear. However, those presented in Figure 13-13 represent the most common ways this is accomplished for the modern GA aircraft.

Leaf-spring Landing Gear (A)

The leaf-spring landing gear, as the name implies, consists of a relatively flat but stiff cantilever beam that reacts landing loads in bending. The primary advantage of such landing gear is that it is inexpensive, stout, durable, and is relatively easy to mount to an airplane. It really represents the simplest form of the landing gear. The leaf-spring landing gear is generally used as the main landing gear. Its relatively low thickness-to-chord ratio renders it a relatively low-drag external structure, although the wheels, tires, and braking calipers generate substantial drag and should be covered using a wheel fairing. The primary drawbacks are high reaction loads in the airframe, as the spring beam tends to have a large moment arm. Also, the landing gear does not lend itself well to a retractable configuration. Some Cessna aircraft feature retractable cantilevered landing gear that resembles leaf-spring gear, but really consists of tubular geometry. The leaf-spring has limited structural damping, but works well because of the damping provided by the scrubbing motion of the tires. It has a poor efficiency as a shock absorber, something remedied by the scrubbing motion as well. To the best knowledge of the author, the largest aircraft to currently use a leaf-spring landing gear is the de Havilland of Canada

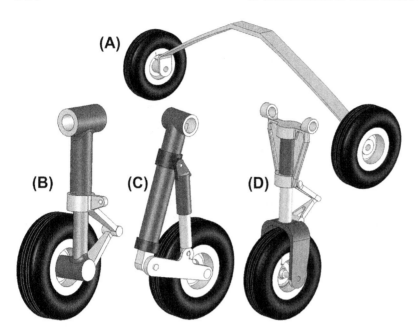

FIGURE 13-13 Schematics of common types of landing gear for small aircraft. (A) a leaf-spring steel (or composite) main landing gear; (B) an oleo-strut retractable main landing gear; (C) an oleo-strut retractable trailing-link main landing gear; and (D) a steerable oleo-strut nose landing gear.

DHC-6 Twin Otter, with a maximum gross weight of 12,500 lb_f.

Oleo-pneumatic Landing Gear (B, C, D)

(B), (C) and (D) in Figure 13-13 are types of landing gear that use a mechanical oleo-pneumatic shock absorber of the type shown in Figure 13-14. Such a shock absorber initially presses the oil against air or nitrogen, while dissipating the impact energy by forcing the oil to flow through a number of recoil orifices. This allows the shock from the landing impact to be reacted initially with a short deflection, and then a larger one as the oil seeps back into its chamber. Reference 4 provides methods to size oleo-pneumatic struts.

Oleo-pneumatic Trailing-link Landing Gear (C)

The oleo-pneumatic trailing-link landing gear has become a popular option in many aircraft due to the superior ride quality it offers on uneven ground. This is possible as the mechanism allows for much larger deflection of the wheel and tire assembly than is possible when mounted directly to the strut. The drawback of the trailing-link landing gear is its greater weight and cost. It also requires a larger internal volume to stow.

Various Landing Gear Issues – Shimmy

Shimmy is a violent dynamic instability that occurs, primarily in free-castering landing gear, due to unbalancing of the forces that act on the members of the trailing link. This is a frequent issue that can be violent enough to break off the nose landing gear. It is therefore of substantial importance to the manufacturer

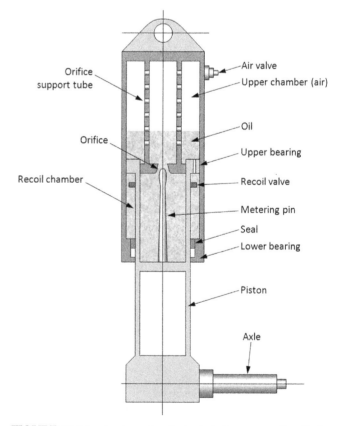

FIGURE 13-14 A schematic of a typical oleo-pneumatic cylinder.

and operator alike. The aircraft designer should be aware of it and possible solutions (shimmy dampers, high-friction castering, geometric relations, mass balancing, and so on).

Various Landing Gear Issues – Whistling

Whistling is a phenomenon in which the complex geometry of a landing gear exposed to the airflow in flight generates a high-pitched audible sound. One aircraft that repeatedly generates this sound is the Piper PA-28 Arrow with its retractable landing gear extended. There is not much that can be done to eliminate the whistling. The author is not aware of any particular studies that have been performed to evaluate the source either of the sound or of potential solutions.

Various Landing Gear Issues – Non-linear Loads During Retraction and Deployment

During the retraction or deployment of landing gear, the actuation system (linkages and hydraulic actuator) undergo variable loading. An actuator capable of brisk retraction or extension during ground tests (when airspeed in zero) may turn out to be grossly undersized when air loads compound the inertia loads. For instance, consider the retraction of a main landing gear into a wing, exemplified in Figure 13-15. As the gear rotates into the wheel well, the air loads acting on the landing gear door (whose effect is attenuated by the complex flow field around the landing gear leg and wheel) may overload the actuator unexpectedly (well, only prior to you reading this) so the gear is stuck in a partially retracted position. It might get stuck at a ϕ ranging from 30° to 60°. This may cause a serious problem in a flight test if the actuator is undersized – it should always be oversized so it has adequate power to power the gear. The test pilot should be ready with a contingency plan – for instance, try retraction just above stalling speed.

13.2.5 Reaction of Landing Gear Forces

The landing gear force matrix consists of three force components: a vertical force, F_{land}, drag force, F_{drag}, and side force, F_{side}. The determination of the magnitude of these loads is done by two means: regulatory or rational analysis. The regulatory loads comprise a "cookbook" style calculation that is designed to be very conservative. Its primary advantage (some would claim the only advantage) is that is saves a lot of analysis time required to complete a rational analysis. However, it can be argued that the resulting aircraft is heavier than it need be. For GA aircraft, 14 CFR 23.471 through 23.511 provide the necessary guidance to estimate all necessary loads for land-based aircraft to show compliance with the regulations.

Figure 13-16 shows a balanced free-body diagram depicting how the landing loads are reacted in a typical landing gear design. The landing gear reaction loads (R_1 through R_4 and M_1 and M_2) must be reacted at specific hard points, in either the wing or the fuselage. These, in turn, require a local reinforcement to transfer the reaction loads into the airframe. Local reinforcement is another way of saying "added structural weight." Of course, Nature leaves no choice to the designer – the only countermeasure is to try and minimize this weight by designing effective load paths.

13.2.6 Comparing the Ground Characteristics of Taildragger and Tricycle Landing Gear

One of the most important differences between a taildragger and tricycle landing gear is their handling on the ground. The taildragger is dynamically unstable, while the tricycle is dynamically stable. To see why, consider Figure 13-17. It shows a taildragger configuration, with a wheel track w_t in a straight and yawed

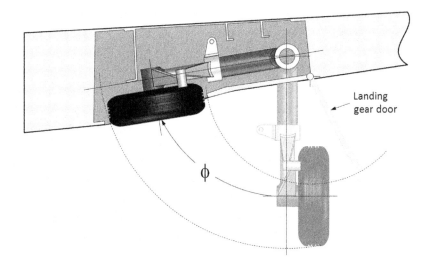

FIGURE 13-15 The retraction of the landing gear often causes unsuspected air loads to overload the actuator, with unforeseeable results.

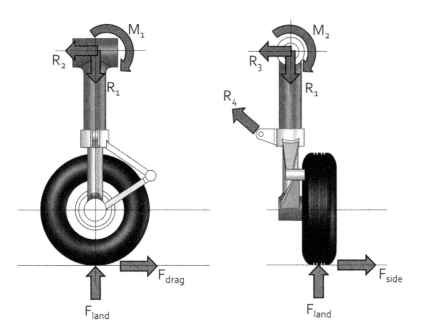

FIGURE 13-16 A free-body diagram showing reaction forces on a typical landing gear strut.

configuration. Note that the figure shows the bottom of the airplane, so the left wing is to the right and vice versa. In this situation, the ground friction of each main gear (F_R and F_L) tire creates an equal and opposite moment about the CG.

In the right image the aircraft is shown yawing to the right. This changes the moment arms of the two ground reactions. The right one (F_R) is increased by a distance Δw_R, resulting in an increased *destabilizing moment*. The left reaction (F_L) is decreased by a distance Δw_L, yielding a reduced *stabilizing moment*. The sum of the two is a destabilizing moment that promotes an even greater yawing tendency. Consequently, the configuration will be inherently unstable.

The opposite holds for a tricycle landing gear and this is shown in Figure 13-18. In the left image, the ground friction of the main gear creates an equal and opposite moment about the CG, just like it did for the taildragger. Note the addition of the reaction of the nose landing gear, F_N. In the right image, the aircraft is yawing to the right. As before, the ground friction from the right main and nose tires generates a destabilizing moment. However, this time they act along a decreasing moment arm. The ground friction of the left tire, on the other hand, is stabilizing and it acts along a moment arm that increases with the yaw angle. Consequently, the stabilizing moment increases with yaw angle, rendering the configuration inherently stable.

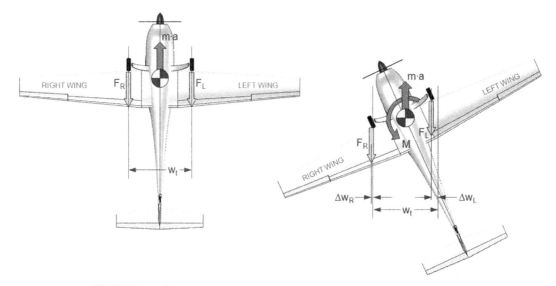

FIGURE 13-17 Ground instability of the taildragger configuration explained.

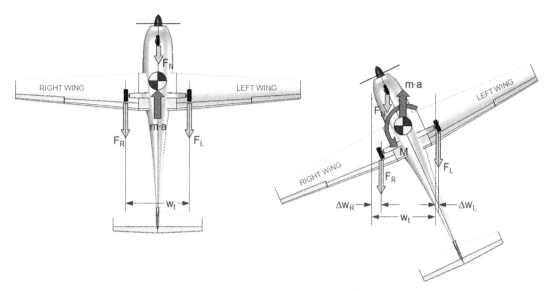

FIGURE 13-18 Ground stability of tricycle configuration explained.

In conclusion, the tricycle configuration is safer than the taildragger and, thus, is selected for most airplanes. In particular, primary trainers and commercial aircraft should always feature this configuration. However, as has already been discussed in Table 13-1, the taildragger configuration is a better choice for light aircraft that takeoff and land on poorly prepared runways (bushplanes).

13.3 GEOMETRIC LAYOUT OF THE LANDING GEAR

This section presents methods that help to accurately lay out the landing gear geometry for tricycle, taildragger, and monowheel configurations. Additionally, methods to evaluate loads on the landing gear will be developed.

13.3.1 Geometric Layout of the Tricycle Landing Gear

If a decision is made to proceed with a tricycle landing gear, the next step is to define the layout of its geometry.

The tricycle landing gear can be laid out as presented below; however, first the following must be completed:

(1) A side view of the design in a cruise attitude. Draw the mean geometric chord (MGC) on the side view.
(2) A top or bottom view of the airplane.
(3) A completed CG envelope.

Refer to Figure 13-19 for Steps 1 and 2.

Step 1

Draw the forward CG limit.

Step 2

Draw the aft CG limit.
Refer to Figure 13-20 for Steps 3 through 7.

Step 3

Determine the *highest* vertical CG location at the aft CG limit and plot on the diagram as shown. This position is the most critical as it is the first point to cross the vertical line drawn in Step 7. If that happens, the plane will fall on its tail.

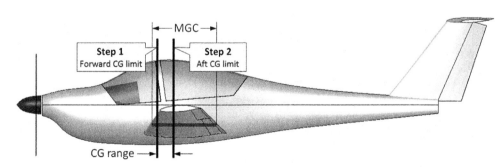

FIGURE 13-19 Steps 1 and 2 for the location of the tricycle landing gear.

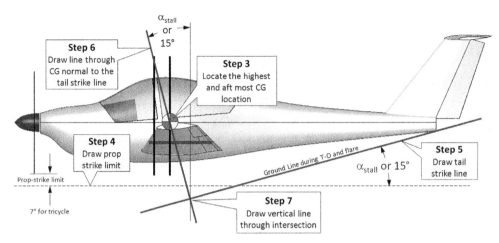

FIGURE 13-20 Steps 3 through 7 for the location of the tricycle landing gear.

Step 4

Draw the prop-strike limit as shown, ensuring it is parallel to the ground plane. This is as close to the ground the propeller should get under most adverse conditions, such as with a flat nose gear tire. The limit per 14 CFR 23.925(a) is 7 inches.

Step 5

Draw the tail-strike line as shown, ensuring it is at an angle of α_{stall} in the clean configuration (because this is the highest angle) up to a maximum value (recommended) of 15°. *The tail strike line is also the ground line.*

Step 6

Draw a line through the CG perpendicular to the tail-strike line. This is where the contact point of the tire should be, at static deflection of $1\,g$ load. IMPORTANT: This is the static deflection of $1\,g$ load. The landing gear will drop below the tail strike line once it is off the ground.

Step 7

Draw a vertical line through the intersection of the tail-strike line and the normal to it. It should form the same angle as the tail-strike line does to the horizontal. If done correctly, the three lines should all intersect at the same point.

Refer to Figure 13-21 for Steps 8 and 9.

Step 8

Position the main landing gear such the contact point of the wheel is at the intersection of the three lines as shown in Figure 13-21. If practical, align the main landing gear leg along the vertical line.

Step 9

Position the nose landing gear so it carries no more than 20% of the aircraft weight when the CG is at the forward limit, and no less than 10% when the CG is at the aft limit. Too much nose gear load will simply make it harder to rotate the airplane for lift-off. Too light a load will make steering the aircraft harder because of reduced ground friction. A light nose gear load can also promote "porpoising" on the ground.

A few more steps are required to properly position the landing gear.

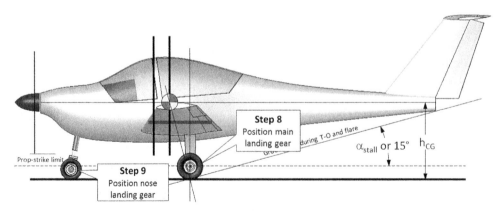

FIGURE 13-21 Steps 8 and 9 for the location of the tricycle landing gear.

Overturn Angle

The following steps are intended to verify that the wheel track is wide enough to provide lateral stability when taxiing and turning a corner. Since the CG is located distance h_{CG} above the ground, the centrifugal force it generates as the airplane turns will result in an overturning moment that can roll the airplane on the wing to the outside of the trajectory. To minimize the risk of this, the designer should check that the combination of wheel track and h_{CG} will not cause this to take place. As a rule of thumb, the closer to the ground the CG is, the better will be the lateral stability of the airplane. The farther forward the CG, the less is the lateral stability.

Refer to Figure 13-22 for Steps 10 through 14.

Step 10

Draw a line through the center of the contact points of the nose and the left or right main landing gear as shown in Figure 13-22. Here, the right landing gear has been chosen (note that we are looking up on the bottom of the airplane, so the right wing appears to the left).

Step 11

Draw a line through the CG that is parallel to the line drawn in Step 10. For clarity, draw both lines far enough from the airplane to allow additional lines to be marked with ease.

Step 12

Draw a line that is perpendicular to the lines drawn in Step 10 and Step 11. This line represents the ground.

Step 13

Place the CG distance h_{CG} above the ground line, along the line drawn in Step 11, as shown in Figure 13-22. Also refer to the definition of the h_{CG} in Figure 13-21.

Step 14

Draw a line that extends from the intersection of the line drawn in Step 10 and the ground line drawn in Step 12, to the CG. This line represents the *overturn angle*, θ. This angle should be less than 63° for general land-based aircraft and 54° for carrier-based aircraft. If it is larger, then any of the following measures can be taken (provided preceding requirements are not violated):

(1) The wheel track can be increased.
(2) The height of the landing gear can be reduced.
(3) The spacing between the nose and main landing gear can be increased.
(4) A combination of the above three.

13.3.2 Geometric Layout of the Taildragger Landing Gear

If a taildragger configuration is to be designed, the next step is the layout of its geometry. The procedure

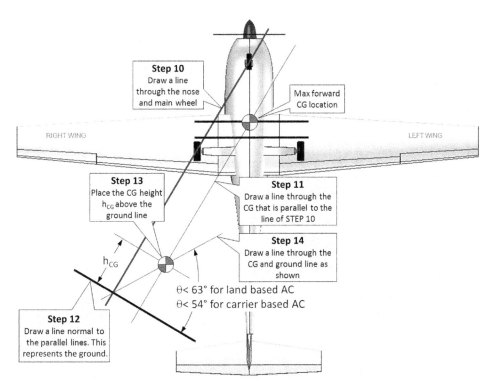

FIGURE 13-22 Steps 10 through 14 for the location of the landing gear.

is fundamentally similar, with a couple of exceptions. As before, the following must be completed:

(1) A side view of the design in a cruise (or tail-off-ground T-O) attitude. Draw the mean geometric chord (MGC) on the side view.
(2) A top or bottom view of the airplane.
(3) A completed CG envelope.

Refer to Figure 13-19 for Steps 1 and 2.

Step 1 and 2 are identical to those for the tricycle landing gear.

Refer to Figure 13-23 for Steps 3 through 7.

Step 3

Determine the *highest* vertical CG location at the *forward and aft CG limits* and plot on the diagram as shown. These two positions are critical to the proper positioning of the main landing gear.

Step 4

Draw the prop-strike limit as shown, ensuring it is parallel to the ground plane. This is as close to the ground as the propeller should get under most adverse conditions, such as with a flat nose gear tire. The limit per 14 CFR 23.925(a) is 9 inches.

Step 5

Draw a line at an incline of 15° that goes through the forward CG location. Note that some references cite 16°, but either one will work just fine.

Step 6

Draw a line at an incline of 25° that goes through the aft CG location. For most aircraft the CG envelope is enclosed between these two lines.

Step 7

Locate the intersection of the lines in Step 5 and Step 6. This is where the contact point of the tire should be, at a static deflection of $1g$ load. Draw a line at an incline of 12° to 15° that goes through this intersection, as shown in Figure 13-23. This line represents the ground and should be used to place the tail wheel. Note that the incline of the line shown is 12°. Note that, ideally, the angle of incidence made by the MGC with respect to the ground line should be that of α_{stall} up to a maximum value (recommended) of 15°.

Refer to Figure 13-24 for Steps 8 and 9.

Step 8

Position the main landing gear so the contact point of the wheel is at the intersection of the three lines as shown in Figure 13-24.

Step 9

Position the tail wheel landing gear such it carries no more than 5% of the aircraft weight when the CG is at the forward limit, and no greater than 10% when the CG is at the aft limit. Too much load on the tail wheel will make it harder to raise the airplane during the ground run. Too light a load will make steering and controlling the aircraft in a crosswind harder because of reduced ground friction.

Step 10

It is imperative that the main landing gear wheel track is wide enough to guarantee the airplane is stable while it corners on the ground. Ensure the width between the tires, W_t, is large enough so the overturn angle is 25° or greater (see Figure 13-25).

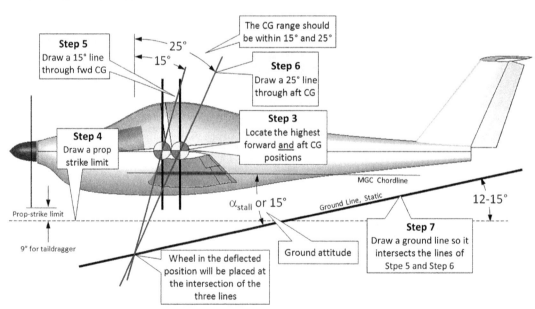

FIGURE 13-23 Steps 3 through 7 for the location of the taildragger landing gear.

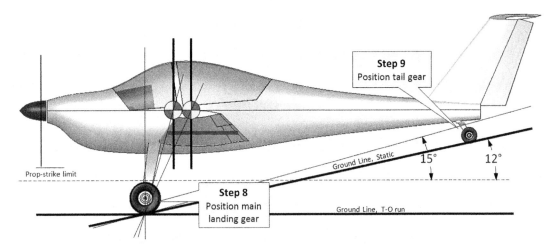

FIGURE 13-24 Steps 8 and 9 for the location of the taildragger landing gear.

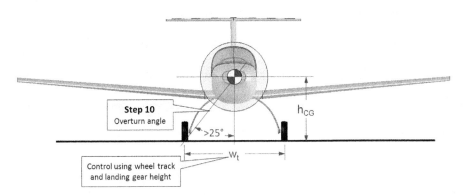

FIGURE 13-25 Step 10 ensures overturn angle is 25° or greater.

13.3.3 Geometric Layout of the Monowheel Landing Gear with Outriggers

The monowheel landing gear with outriggers is primarily used in motor gliders. Examples of such aircraft include the Scheibe SF-25 Falke; SF-28 Tandem Falke; Fournier RF-1, RF-2, RF-3, RF-4, RF-5; and the Europa XS. The procedure for positioning the main wheel and the outriggers is fundamentally similar to the previous methods, with a couple of exceptions. Important design guidelines are shown in Figure 13-26.

Stinton [8] indicates the ideal spanwise position of the outrigger is at the rolling radius of gyration[3] as this will result in a minimum outrigger load in an "outrigger first" landing. Exploring a database of such aircraft shows this methodology is not used in outrigger design in general.

Pazmany [1, p. 20] shows that the outrigger distance (a) from the plane of symmetry generally ranges from 33% to 100% of the wing half-span ($b/2$) (Figure 13-27). Placing the outriggers at the wingtip is not unheard of. For instance, the Rutan Solitaire and the Monnett Monerai both feature outriggers at the wingtip.

13.3.4 Tricycle Landing Gear Reaction Loads

Figure 13-28 shows loads acting on a tricycle landing gear configuration as it moves at a *constant* speed during the ground run. We want to determine the reaction loads on the nose and main landing gear, denoted by R_N and R_M, respectively.

Nose gear reaction, R_N:

$$R_N = \frac{T \cdot y_T + (W + L_{HT} - L_W) \cdot (x_M - \mu \cdot h_{CG}) + L_W \cdot x_W - L_{HT} \cdot l_{HT} - M_W}{x_N + x_M - 2\mu \cdot h_{CG}} \qquad (13\text{-}7)$$

---

[3]Which would be calculated using $R_{gyration} = \sqrt{I_{xx}/m}$ where I_{xx} is the mass moment of inertia about the roll axis and m is the mass of the airplane.

572   13. THE ANATOMY OF THE LANDING GEAR

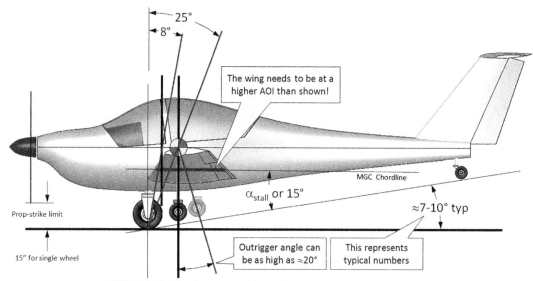

FIGURE 13-26 Design guidelines for a monowheel with outriggers.

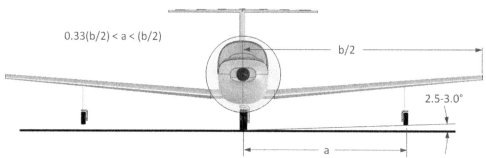

FIGURE 13-27 Design guidelines for positioning the outriggers.

Main gear reaction, R_M:

$$R_M = W + L_{HT} - L_W - R_N \quad (13\text{-}8)$$

where

L_W = wing lift, in lb_f or N
L_{HT} = horizontal tail lift, in lb_f or N
M_W = pitching moment, in $ft \cdot lb_f$ or $N \cdot m$
T = thrust, in lb_f or N

Special Case – Static Loads

For this case $V = 0$ and $T = 0$ (note that this makes friction irrelevant as it only affects the aircraft in motion), leading to the following results:

$$R_N = \frac{W \cdot x_M}{x_N + x_M} \quad \text{and} \quad R_M = W - R_N \quad (13\text{-}9)$$

Note that since the location of the CG is used as a reference in this formulation, a question of sign convention might be raised. For instance, in Figure 13-28, should x_M have a positive sign and x_N a negative sign? What about the forces?

The formulation setup here assumes absolute values of all dimensions and forces. Signs are taken care of in the formulation of both forces and moments. The only exception is the sign of M_W, which is typically negative (nose pitch-down). Its sign is the only one that must be accounted for when using the formulation.

Definition of Aerodynamic Loads

The following formulation of the lift and thrust is convenient, as it allows the designer to evaluate the impact of various aerodynamic characteristics on the landing gear reaction forces. For instance, if a spreadsheet has been prepared, the designer can evaluate the airspeed when rotation to lift-off can be achieved, and evaluate design changes to modify it if necessary.

Thrust for a propeller:

$$T = \frac{\eta_p \cdot 550 \cdot BHP}{V}$$

13.3 GEOMETRIC LAYOUT OF THE LANDING GEAR

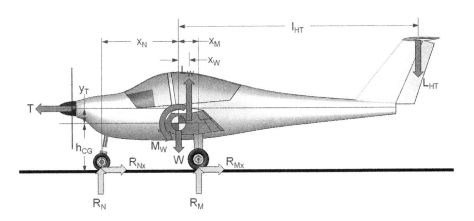

FIGURE 13-28 Balanced force diagram for a typical tricycle landing gear configuration.

Wing lift:

$$L_W = \frac{1}{2}\rho V^2 S C_L = \frac{1}{2}\rho V^2 S(C_{L_0} + C_{L_\alpha} \cdot \alpha_{TO})$$

HT lift:

$$L_{HT} = \frac{1}{2}\rho V^2 S_{HT} C_{L_{HT}} = \frac{1}{2}\rho V^2 S_{HT}(C_{L_{\alpha HT}} \cdot i_{HT} + C_{L_{\delta_e}} \cdot \delta_e)$$

Derivation of Equation (13-7)

Begin by defining the horizontal ground friction force:

$$R_{Nx} + R_{Mx} = \mu(R_N + R_M) \quad \text{(i)}$$

Sum forces in the vertical (z) direction, such that positive forces act upward:

$$\sum F_Z = 0 \;\Rightarrow\; R_N + R_M + L_W - W - L_{HT} = 0 \quad \text{(ii)}$$

This yields the main landing gear reaction force R_M:

$$\Rightarrow R_M = W + L_{HT} - L_W - R_N \quad \text{(iii)}$$

Sum moments about the center of gravity (CG), such that nose pitch-up moments are positive (note that the ground friction R_{Mx} and R_{Nx} are written explicitly as $\mu(R_N + R_M) \cdot h_{CG}$):

$$\sum M_{CG} = 0 \;\Rightarrow\; -T \cdot y_T + R_N \cdot x_N - R_M \cdot x_M$$
$$- L_W \cdot x_W + L_{HT} \cdot l_{HT}$$
$$- \mu(R_N + R_M) \cdot h_{CG} + M_W = 0 \quad \text{(iv)}$$

Note that typically the value of the wing pitching moment, M_W, is negative. Expand and rearrange the R_N and R_M terms:

$$-T \cdot y_T + R_N \cdot x_N - \mu \cdot R_N \cdot h_{CG} - R_M \cdot x_M$$
$$- \mu \cdot R_M \cdot h_{CG} - L_W \cdot x_W + L_{HT} \cdot l_{HT} + M_W = 0$$

Simplify in terms of R_N and R_M:

$$-T \cdot y_T + R_N \cdot (x_N - \mu \cdot h_{CG}) - R_M \cdot (x_M - \mu \cdot h_{CG})$$
$$- L_W \cdot x_W + L_{HT} \cdot l_{HT} + M_W = 0$$

Inserting Equation (iii) for R_M yields:

$$-T \cdot y_T + R_N \cdot (x_N - \mu \cdot h_{CG}) - (W + L_{HT} - L_W - R_N)$$
$$\cdot (x_M - \mu \cdot h_{CG}) - L_W \cdot x_W + L_{HT} \cdot l_{HT} + M_W = 0$$

Rearrange to combine the term R_N yields:

$$-T \cdot y_T + R_N \cdot (x_N - \mu \cdot h_{CG}) + R_N \cdot (x_M - \mu \cdot h_{CG})$$
$$- (W + L_{HT} - L_W) \cdot (x_M - \mu \cdot h_{CG})$$
$$- L_W \cdot x_W + L_{HT} \cdot l_{HT} + M_W = 0$$

Simplify further to yield:

$$-T \cdot y_T + R_N \cdot (x_N + x_M - 2\mu \cdot h_{CG}) - (W + L_{HT} - L_W)$$
$$\cdot (x_M - \mu \cdot h_{CG}) - L_W \cdot x_W + L_{HT} \cdot l_{HT} + M_W = 0$$

Finally, solving for the nose gear reaction force, R_N, yields:

$$\Rightarrow R_N = \frac{T \cdot y_T + (W + L_{HT} - L_W) \cdot (x_M - \mu \cdot h_{CG}) + L_W \cdot x_W - L_{HT} \cdot x_{HT} - M_W}{x_N + x_M - 2\mu \cdot h_{CG}} \quad \text{(iv)}$$

QED

EXAMPLE 13-1

Consider the dimensions of the aircraft in Figure 13-29 and whose pertinent characteristics are listed below. Determine the following:

(a) the static ground reaction loads without engine power;
(b) dynamic ground reaction loads on the NLG and MLG, assuming no rotation at $V = 40$ KTAS; and
(c) the airspeed at which the pilot can (slowly) rotate the nose of the aircraft to lift off, assuming the constant airspeed formulation.

For convenience, assume the propeller efficiency can be described using the function shown in Figure 14-45.

Wing area, $S_W = 110$ ft^2	Propeller diameter, $D = 5.75$ ft
Wing mean geometric chord, $C_{MGC} = 3.25$ ft	Propeller T-O rotation, RPM = 2500
Horizontal tail area, $S_{HT} = 25$ ft^2	Lift coefficient of HT during T-O run, $C_{L\ HT} = -0.825$
T-O weight, $W = 1320$ lb$_f$	Propeller efficiency, $\eta_p =$ 0.096574+1.703049 J−0.952281 J^2
Lift coefficient during T-O run, $C_{L\ TO} = 0.5$	
Engine T-O power, $P_{TO} = 100$ BHP	Pitching moment coefficient about the CG, $C_M = -0.05$

Solution

(a) **Static ground reaction loads without power:**
Nose reaction from Equation (13-9):

$$R_N = \frac{W \cdot x_M}{x_N + x_M} = \frac{1320 \times 1.3}{3.5 + 1.3} = 357.5 \text{ lb}_f$$

Main landing gear reaction:

$$R_M = W - R_N = 1320 - 357.5 = 962.5 \text{ lb}_f$$

(b) **Dynamic ground reaction loads at $V = 40$ KTAS:**
Airspeed in ft/s:

$$V = 40 \times 1.688 = 67.5 \text{ ft/s}$$

Dynamic pressure:

$$q = \frac{1}{2}\rho V^2 = \frac{1}{2}(0.002378)(67.5)^2 = 5.42 \text{ lb}_f/\text{ft}^2$$

Advance ratio:

$$J = \frac{60V}{RPM \times D} = \frac{60 \times 67.5}{2500(5.75)} = 0.2818$$

Propeller efficiency:

$$\eta_p = 0.096574 + 1.703049J - 0.952281J^2 = 0.5009$$

Thrust:

$$T = \frac{\eta_p \times 550 \times BHP}{V} = \frac{0.5009 \times 550 \times 100}{67.5} = 408 \text{ lb}_f$$

Wing lift:

$$L_W = qS_W C_{L\ TO} = (5.42)(110)(0.5) = 298.1 \text{ lb}_f$$

HT lift:

$$L_{HT} = qS_{HT}C_{HT} = (5.42)(25)|-0.825| = 111.8 \text{ lb}_f$$

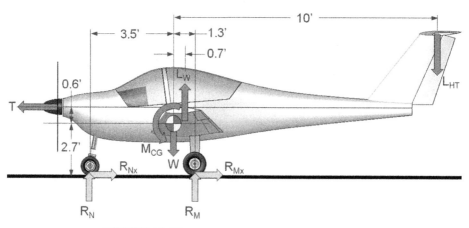

FIGURE 13-29 Example landing gear configuration.

EXAMPLE 13-1 (cont'd)

Pitching moment:

$$M_W = qS_W C_{MGC} C_M = (5.42)(110)(3.25)(-0.05)$$
$$= -96.9 \text{ ft·lb}_f$$

Nose landing gear reaction load:

$$R_N = \frac{T \cdot y_T + (W + L_{HT} - L_W) \cdot (x_M - \mu \cdot h_{CG}) + L_W \cdot x_W - L_{HT} \cdot l_{HT} - M_W}{x_N + x_M - 2\mu \cdot h_{CG}}$$

$$= \frac{408 \times 0.6 + (1320 + 111.8 - 298.1) \cdot (1.3 - 0.04 \times 2.7) + 298.1 \times 0.7 - 111.8 \times 10 + 96.9}{3.5 + 1.3 - 2(0.04) \cdot 2.7}$$

$$= \frac{244.8 + 1133.7 \times 1.192 + 208.67 - 1118 + 96.9}{4.584} = 171.0 \text{ lb}_f$$

Main landing gear reaction loads (total for both wheels):

$$R_M = W + L_{HT} - L_W - R_N$$
$$= 1320 + 111.8 - 298.1 - 171.0 = 962.7 \text{ lb}_f$$

(c) Determine the rotation airspeed:

The rotation airspeed can be determined by calculating the reaction loads for the nose landing gear for a range of airspeeds and then identifying where the NLG reaction load becomes zero. The calculations are easily implemented in a spreadsheet. The calculation shown in Part (b) was implemented for airspeeds ranging from 10 through 55 KTAS using Microsoft Excel. The results have been plotted in Figure 13-30. It is evident from the figure that the airplane will be capable of lifting the nosewheel at around 52.5 KTAS.

It should be pointed out that as soon as the nose landing gear lifts off the runway, the model represented here is no longer valid, because the rotation inevitably means change in AOA. This brings about a significant modification of the flow field, changing the lift coefficients of the wing and HT. However, the method presented here returns a reliable value below and up to the rotation speed.

It is absolutely imperative that the aircraft designer checks the capability of a new design to rotate below V_{LOF}, in particular if the design features a high thrust line. The analysis presented above will reveal whether or not an adequate elevator authority is available.

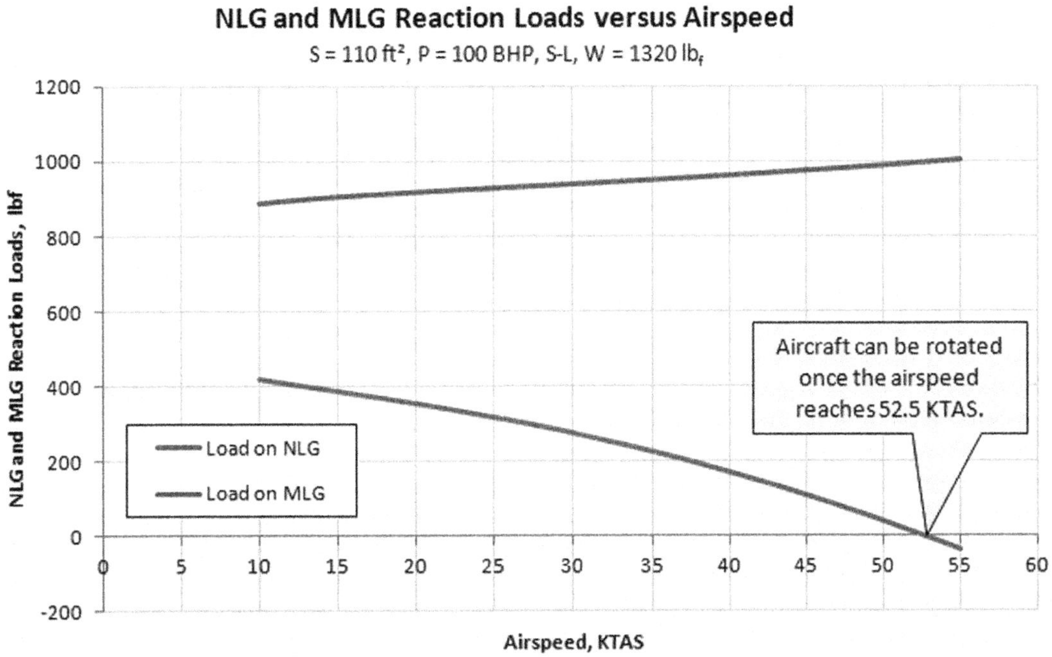

FIGURE 13-30 Nose and main landing gear reaction loads. Note that the NLG reaction load goes to zero at 52.5 KTAS. This means the aircraft must be accelerated to that airspeed before rotating the airplane is possible.

13.3.5 Taildragger Landing Gear Reaction Loads

Figure 13-31 shows loads acting on a taildragger landing gear configuration at a *constant* speed during the ground run, while the tail is still on the ground. We want to determine the reaction loads on the main and tail landing gear, denoted by R_M and R_T, respectively.

Tail wheel reaction, R_T:

$$R_T = \frac{-T \cdot y_T + (W - L_{HT} - L_W) \cdot (x_M - \mu \cdot h_{CG}) - L_W \cdot x_W - L_{HT} \cdot l_{HT} + M_W}{x_M + x_T} \quad (13\text{-}10)$$

Main gear reaction, R_M:

$$R_M = W - L_{HT} - L_W - R_T \quad (13\text{-}11)$$

where
L_W = wing lift, in lb$_f$ or N.
L_{HT} = horizontal tail lift, in lb$_f$ or N
M_W = pitching moment, in ft·lb$_f$ or N·m
T = thrust, in lb$_f$ or N

Special Case – Static Loads

For this case, $V = 0$ and $T = 0$, leading to the following results:

$$R_T = \frac{W \cdot x_M}{x_M + x_T} \quad \text{and} \quad R_M = W - R_T \quad (13\text{-}12)$$

As already discussed in Section 13.3.4, *Tricycle landing gear reaction loads*, the formulation is set up assuming absolute values of all dimensions and forces. With the exception of M_W, the proper signs are taken care of in the formulation of both forces and moments.

Another confusion that might arise is the direction of LHT, shown as pointing downward in Figure 13-31. Strictly speaking the direction of this force, as setup for the derivation below, does not matter. What matters is that the force direction is already defined as positive upward. If the elevator is neutral (as is shown in the figure) the force would actually point upward (in the positive direction). This would cause a tendency to lift the tail off the ground.

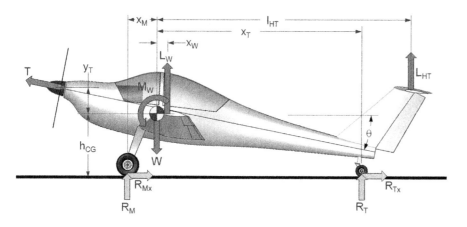

FIGURE 13-31 Balanced force diagram for a typical taildragger landing gear configuration.

Derivation of Equation (13-10)

Begin by defining the horizontal ground friction force:

$$R_{Mx} + R_{Tx} = \mu(R_M + R_T) \quad (i)$$

Sum forces in the vertical (z) direction, such that positive forces act upward:

$$\sum F_Z = 0 \quad \Rightarrow \quad R_M + R_T + L_W - W + L_{HT} = 0 \quad (ii)$$

This yields the main landing gear reaction force R_M:

$$\Rightarrow R_M = W - L_{HT} - L_W - R_T \quad (iii)$$

For simplicity, assume thrust is largely horizontal. Sum moments about the center of gravity (CG), such that nose pitch-up moments are positive:

$$\sum M_{CG} = 0 \quad \Rightarrow \quad -T \cdot y_T + R_M \cdot x_M - L_W \cdot x_W - R_T \cdot x_T$$
$$- L_{HT} \cdot l_{HT} - \mu(R_M + R_T) \cdot h_{CG} + M_W = 0 \quad (iv)$$

Note that typically the value of the wing pitching moment, M_W, is negative. Expand and rearrange the R_T and R_M terms:

$$-T \cdot y_T + R_M \cdot x_M - L_W \cdot x_W - R_T \cdot x_T - L_{HT} \cdot l_{HT} - \mu R_M \cdot h_{CG} - \mu R_T \cdot h_{CG} + M_W = 0$$

Simplify in terms of R_T and R_M:

$$-T \cdot y_T + R_M \cdot (x_M - \mu \cdot h_{CG}) - L_W \cdot x_W - R_T \cdot (x_T + \mu \cdot h_{CG}) - L_{HT} \cdot l_{HT} + M_W = 0$$

Inserting Equation (iii) for R_M yields:

$$-T \cdot y_T + (W - L_{HT} - L_W - R_T) \cdot (x_M - \mu \cdot h_{CG}) - L_W \cdot x_W - R_T \cdot (x_T + \mu \cdot h_{CG}) - L_{HT} \cdot l_{HT} + M_W = 0$$

Rearrange to combine the term R_T yields:

$$-T \cdot y_T + (W - L_{HT} - L_W) \cdot (x_M - \mu \cdot h_{CG}) - L_W \cdot x_W - R_T \cdot (x_M + x_T) - L_{HT} \cdot l_{HT} + M_W = 0$$

Finally, solving for the nose gear reaction force, R_T, yields

$$\Rightarrow R_T = \frac{-T \cdot y_T + (W - L_{HT} - L_W) \cdot (x_M - \mu \cdot h_{CG}) - L_W \cdot x_W - L_{HT} \cdot l_{HT} + M_W}{x_M + x_T} \quad \text{(v)}$$

QED

EXAMPLE 13-2

Consider the dimensions of the aircraft in Figure 13-29 and whose pertinent characteristics are listed below. Determine the following:

(a) the static ground reaction loads without engine power;
(b) the airspeed at which the tail wheel lifts off the ground for elevator deflection $\delta_e = -10°, 0°, +10°$, assuming the constant airspeed formulation.

For convenience, assume the propeller efficiency can be described using the function shown in Figure 14-45.

Wing area, $S_W = 110$ ft²	Propeller diameter, $D = 5.75$ ft
Wing mean geometric chord, $C_{MGC} = 3.25$ ft	Propeller T-O rotation, RPM = 2500
Horizontal tail area, $S_{HT} = 25$ ft²	Lift coefficient of HT during T-O run, $C_{L\ HT} = 0.72 + 0.055 \cdot \delta_e$
T-O weight, $W = 1320$ lb$_f$	Propeller efficiency, $\eta_p = 0.096574 + 1.703049\ J - 0.952281\ J^2$
Lift coefficient during T-O run, $C_{L\ TO} = 1.4$	Pitching moment coefficient about the CG, $C_M = -0.05$
Engine T-O power, $P_{TO} = 100$ BHP	

Solution

(a) Static ground reaction loads without power:

Nose reaction from Equation (13-12):

$$R_T = \frac{W \cdot x_M}{x_M + x_{TW}} = \frac{1320 \times 1.3}{1.3 + 10} = 151.9\ \text{lb}_f$$

Main landing gear reaction:

$$R_M = W - R_N = 1320 - 151.9 = 1168\ \text{lb}_f$$

This way, the tail wheel carries about 11.5% of the gross weight of the aircraft in the static configuration. This is the upper limit of what is advisable for static loads and the designer should consider moving the tail wheel a tad aft, unless other factors make it impossible.

(b) Determine the airspeed at which the tail wheel lifts off the ground for elevator deflection $\delta_e = -10°, 0°, +10°$:

Note that the HT lift coefficient is given by $C_{L\ HT} = 0.72 + 0.055 \cdot \delta_e$, where the value 0.72 reflects the fact that while the tail is on the ground the HT will be generating a positive lift, i.e. lift that points upward. The elevator deflection, reflected by δ_e, increases the lift coefficient when the elevator is deflected trailing edge down (TED) and decreases when it is deflected trailing edge up (TEU).

Validation calculation for $V = 40$ KTAS and $\delta_e = 0°$ (or $C_{L\ HT} = 0.72$):

Airspeed in ft/s:

$$V = 40 \times 1.688 = 67.5\ \text{ft/s}$$

Dynamic pressure:

$$q = \frac{1}{2}\rho V^2 = \frac{1}{2}(0.002378)(67.5)^2 = 5.42\ \text{lb}_f/\text{ft}^2$$

EXAMPLE 13-2 (cont'd)

Advance ratio:

$$J = \frac{60V}{RPM \times D} = \frac{60 \times 67.5}{2500(5.75)} = 0.2818$$

Propeller efficiency:

$$\eta_p = 0.096574 + 1.703049J - 0.952281J^2 = 0.5009$$

Thrust:

$$T = \frac{\eta_p \times 550 \times BHP}{V} = \frac{0.5009 \times 550 \times 100}{67.5} = 408 \text{ lb}_f$$

Wing lift:

$$L_W = qS_W C_{L\ TO} = (5.42)(110)(1.4) = 834.7 \text{ lb}_f$$

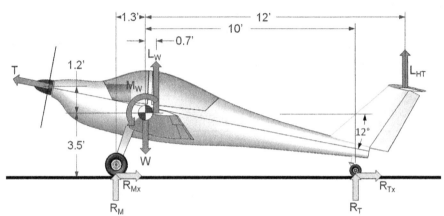

FIGURE 13-32 Example configuration.

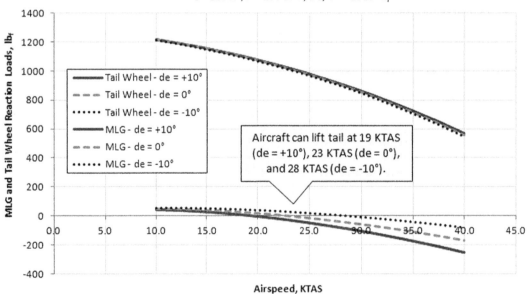

FIGURE 13-33 Main landing gear and tail wheel reaction loads. Note that the tail wheel reaction load goes to zero at 20, 24, and 29 KTAS, depending on elevator deflection (de = +10°, 0°, −10°). This means the aircraft must be accelerated to those airspeeds before it is possible to lift the tail wheel off the ground.

EXAMPLE 13-2 (cont'd)

HT lift:

$$L_{HT} = qS_{HT}C_{HT} = (5.42)(25)(0.72) = 97.6 \text{ lb}_f$$

Pitching moment:

$$M_W = qS_W C_{MGC} C_M = (5.42)(110)(3.25)(-0.05)$$
$$= -96.9 \text{ ft·lb}_f$$

Nose landing gear reaction load:

$$R_T = \frac{-T \cdot y_T + (W - L_{HT} - L_W) \cdot (x_M - \mu \cdot h_{CG}) - L_W \cdot x_W - L_{HT} \cdot l_{HT} + M_W}{x_M + x_T}$$

$$= \frac{-(408)(1.2) + (1320 - 97.6 - 834.7) \cdot (1.3 - 0.04 \cdot 3.5) - (834.7)(0.7) - (97.6)(12) - 96.9}{1.3 + 10}$$

$$= -167.5 \text{ lb}_f$$

A negative value means the tail wheel is already off the ground (which really means the calculation is no longer valid). Calculating the main landing gear reaction loads (total for both wheels) will still serve as a validation (e.g. in the preparation of automated calculations like that in a spreadsheet):

$$R_M = W - L_{HT} - L_W - R_N$$
$$= 1320 - 97.6 - 834.7 - (-167.5) = 555.2 \text{ lb}_f$$

Completing the same calculation for other airspeeds and plotting the data on a graph such as the one in Figure 13-33, allows one to determine what the airspeed must be.

VARIABLES

Symbol	Description	Units (UK and SI)
a	Elliptical major axis length	ft or m
AOA	Angle-of-attack	Degrees or radians
AR	Aspect ratio	
b	Elliptical minor axis length	ft or m
BHP	Brake horsepower of piston engine	HP
c	Length of tire footprint	ft or m
C_D	Drag coefficient	
C_L	3D lift coefficient of the wing	
$C_{L\ HT}$	3D lift coefficient of the horizontal tail	
C_{L0}	3D zero AOA lift coefficient	
$C_{L\alpha}$	3D lift curve slope of the wing	Per degree or per radian
$C_{L\alpha HT}$	3D lift curve slope of the horizontal tail	Per degree or per radian
$C_{L\delta e}$	3D coefficient of lift generated by elevator deflection	
C_M	Pitching moment coefficient	
C_{MGC}	Mean geometric chord length	ft or m
c_r	Root chord length	ft or m
c_t	Tip chord length	ft or m
d	Deflection of gear	ft or m
d	Diameter of wheel	ft or m
d	Difference between unloaded tire radius and loaded tire radius	ft or m
D	Diameter of propeller	ft or m
D	Diameter of tire	ft or m
F	Force	lb$_f$ or N
F_{drag}	Drag landing force	lb$_f$ or N
F_L	Left landing gear friction force	lb$_f$ or N
F_{land}	Vertical landing force	lb$_f$ or N
F_N	Nose gear friction force	lb$_f$ or N
F_R	Right landing gear friction force	lb$_f$ or N
F_{side}	Side landing force	lb$_f$ or N
h	Distance to turning center	ft or m
H	Radial distance from outside of wheel to outside of tire	ft or m

Symbol	Description	Units (UK and SI)
h_{CG}	Height of center of gravity above ground	ft or m
i_{HT}	Incidence of the horizontal tail	Degrees or radians
J	Advance ratio	
KE	Kinetic energy	ft·lb$_f$
K_S	Spring constant	lb$_f$/ft or N/m
L_{HT}	Lift of horizontal tail	lb$_f$ or N
l_{HT}	x-distance from CG to AC of horizontal tail	ft or m
LR	Lift ratio	
L_W	Lift force of wing	lb$_f$ or N
M_1, M_2	Landing gear reaction moments	lb$_f$·ft or N·m
M_W	Moment of wing about the CG	lb$_f$·ft or N·m
N	Number of main wheels featuring brakes	
P	Average pressure on tire	lb$_f$/ft^2 or N/m^2
q	Dynamic pressure	lb$_f$/ft^2 or N/m^2
R_1, R_2, R_3, R_4	Landing gear reaction forces	lb$_f$ or N
R_L	Loaded radius	ft or m
R_M	Main gear reaction force	lb$_f$ or N
R_N	Nose gear reaction force	lb$_f$ or N
RPM	Revolutions per minute of propeller	RPM
R_T	Tail wheel reaction force	lb$_f$ or N
R_{turn}	Turning radius	ft or m
S	Wing planform area	ft^2 or m^2
S_e	Elliptical area (of tire)	ft^2 or m^2
S_{HT}	Horizontal tail planform area	ft^2 or m^2
T	Thrust	lb$_f$ or N
t_M	Mechanical trail	ft or m
V	Airspeed	ft/s or m/s
V_C	Design cruising speed	ft/s or m/s
V_{LOF}	Liftoff speed	ft/s or m/s
V_{S0}	Stalling speed in landing configuration	KTAS
V_{S1}	Stalling speed with flaps retracted	ft/s or m/s
W	Weight of aircraft	lb$_f$ or N
W	Width of tire	ft or m
Wheel base	x-distance between main gear and nose/tail gear	ft or m
Wheel track	y-distance between wheels of main gear	ft or m
w_t	Wheel track	ft or m
x_M	x-distance from CG to main gear	ft or m
x_N	x-distance from CG to nose gear	ft or m
x_T	x-distance from CG to tail wheel	ft or m
x_W	x-distance from CG to AC of wing	ft or m
y_T	y-distance between CG and thrust	ft or m
Δw_L	Change in left wheel moment arm	ft or m
Δw_R	Change in right wheel moment arm	ft or m
Δx	Displacement of rudder pedal	ft or m
Δy	Displacement of brake cylinder piston	ft or m
α_{stall}	Stall angle-of-attack	Degrees or radians
α_{TO}	Take-off angle-of-attack	Degrees or radians
δ_e	Elevator deflection angle	Degrees or radians
ϕ	Landing gear retraction angle	Degrees or radians
ϕ	Tail wheel spindle axis angle/rake angle	Degrees or radians
η_P	Propeller efficiency	
μ	Ground friction coefficient	
θ	Caster angle	Degrees or radians
θ	Overturn angle	Degrees or radians
θ	Tipback angle	Degrees or radians
θ	Turning angle of nose gear	Degrees or radians
ρ	Density of air	lb$_f$/ft^2 or N/m^2

References

[1] Pazmany L. Landing Gear Design for Light Aircraft, vol. 1. Ladislao Pazmany; 1986.
[2] Goodyear Aircraft Tire Data Book. Goodyear Tire and Rubber Company; 2002.
[3] Anonymous. Gates Learjet 35A/36A FAA Approved Airplane Flight Manual. Gates Learjet Corporation; 1983.
[4] Curry NS. Aircraft Landing Gear Design: Principles and Practices. AIAA Education Series; 1988.
[5] Niu Michael Chun-Yung. Airframe Structural Design. Conmilit Press; 1988. p. 439–441.
[6] http://groveaircraft.com/brakedesign.html.
[7] Sengfelder G. German Aircraft Landing Gear. Schiffer; 1993.
[8] Stinton D. The Design of the Aeroplane. Collins; 1983.

CHAPTER 14

The Anatomy of the Propeller

OUTLINE

14.1 Introduction — 582
  14.1.1 The Content of this Chapter — 584
  14.1.2 Propeller Configurations — 584
  14.1.3 Important Nomenclature — 586
  14.1.4 Propeller Geometry — 587
  14.1.5 Geometric Propeller Pitch — 588
    Fundamental Formulation of
    Pitch Angle or Geometric Pitch — 589, 590
    Constant-Pitch Propeller — 590
    Variable-Pitch Propeller — 590
    Fundamental Relationships of Propeller Rotation — 591
    Determination of the Desired Pitch for Fixed-Pitch Propellers — 592
    Derivation of Equation (14-12) — 592
  14.1.6 Windmilling Propellers — 593
  14.1.7 Fixed Versus Constant-Speed Propellers — 594
  14.1.8 Propulsive or Thrust Power — 595

14.2 Propeller Effects — 595
  14.2.1 Angular Momentum and Gyroscopic Effects — 596
  14.2.2 Slipstream Effects — 597
  14.2.3 Propeller Normal and Side Force — 598
    Normal Force — 598
    Side Force — 598
  14.2.4 Asymmetric Yaw Effect — 599
  14.2.5 Asymmetric Yaw Effect for a Twin-Engine Aircraft — 599
  14.2.6 Blockage Effects — 602
  14.2.7 Hub and Tip Effects — 604
  14.2.8 Effects of High Tip Speed — 605
  14.2.9 Skewed Wake Effects — $A \cdot q$ Loads — 605
  14.2.10 Propeller Noise — 606

14.3 Properties and Selection of the Propeller — 607
  14.3.1 Tips for Selecting a Suitable Propeller — 607
  14.3.2 Rapid Estimation of Required Prop Diameter — 608
    Derivation of Equation (14-18) — 608
  14.3.3 Rapid Estimation of Required Propeller Pitch — 610
  14.3.4 Estimation of Required Propeller Efficiency — 610
  14.3.5 Advance Ratio — 611
  14.3.6 Definition of Activity Factor — 613
  14.3.7 Definition of Power- and Thrust-Related Coefficients — 614
  14.3.8 Effect of Number of Blades on Power — 615
    Derivation of Equation (14-30) — 616
  14.3.9 Propulsive Efficiency — 618
    Derivation of Equation (14-35) — 618
    Propeller Efficiency Map — 618
  14.3.10 Moment of Inertia of the Propeller — 619
    Derivation of Equation (14-37) — 619

14.4 Determination of Propeller Thrust — 620
  14.4.1 Converting Piston BHP to Thrust — 620
    Derivation of Equation (14-38) — 621
  14.4.2 Propeller Thrust at Low Airspeeds — 621
    Method 1: Quadratic Interpolation — 621
    Derivation of Equation (14-40) — 622
    Method 2: Cubic Spline Method for Fixed-Pitch Propellers — 623
    Derivation of Equation (14-42) — 624
    Method 3: Cubic Spline Method for Constant-Speed Propellers — 628
  14.4.3 Step-by-step: Determining Thrust Using a Propeller Efficiency Table — 629
    Step 1: Determine Advance Ratio — 629
    Step 2: Determine Power Coefficient — 631
    Step 3: Extract Propeller Efficiency — 631
    Step 4: Calculate Thrust — 631
  14.4.4 Estimating Thrust From Manufacturer's Data — 631
  14.4.5 Other Analytical Methods — 632

14.5 Rankine-Froude Momentum Theory — 632
  14.5.1 Formulation — 633
  14.5.2 Ideal Efficiency — 635
  14.5.3 Maximum Static Thrust — 636

    14.5.4 *Computer code: Estimation of Propeller Efficiency Using the Momentum Theorem*   638
        Step 1: Preliminaries   638
        Step 2: Set Initial Values   638
        Step 3: Determine Thrust   638
        Step 4: Determine Induced Velocity   638
        Step 5: Determine Ideal Efficiency   638
        Step 6: Determine the Next Propeller Efficiency   638
        Step 7: Determine the Change in Propeller Efficiency   638
        Step 8: Plan the Next Step   638
14.6 Blade Element Theory   640
    14.6.1 *Formulation*   641
        Observation 1   641
        Step 1: Table Columns 2–9   645
        Step 2: Table Columns 11–17   645
        Step 3: Table Columns 18–25   646
    14.6.2 *Determination of α_i Using the Momentum Theory*   650
        Step 1: Initial Value   652
        Step 2: Determine the Next w 652
        Step 3: Determine the Difference   652
    14.6.3 *Compressibility Corrections*   653
        Correction of Lift   653
        Correction of Drag   655
    14.6.4 *Step-by-step: Prandtl's Tip and Hub Loss Corrections*   655
        Step 1: Tip Correction Parameter   655
        Step 2: Tip Correction   655
        Step 3: Hub Correction Parameter   655
        Step 4: Hub Correction   655
        Step 5: Prandtl Correction   655
    14.6.5 *Computer code: Determination of the Propeller Induced Velocity*   656
Variables   657
References   658

14.1 INTRODUCTION

A propeller is a device that converts mechanical energy into a force, which we call *thrust*, and is used to propel the vehicle to which it is attached. The propeller features one or more lifting surfaces called *propeller blades*[1] that are rotated rapidly using an engine. The thrust is the aerodynamic[2] lift force produced by the blades and is identical to the force produced by a wing. Propellers are, by far, the most common means of generating thrust for General Aviation aircraft (14 CFR Part 23). Among Light Sport Aircraft (LSA) it is practically the only means. Several types of commercial aircraft (14 CFR Part 25) also use propellers for propulsion. And a number of military aircraft do so as well; primarily trainers and several multiengine military transport aircraft. In airplanes, propellers are usually driven by piston or gas turbine engines (which are called turboprops). Additionally, electrical power is also gaining popularity (albeit mostly for light airplanes at this time), and it is converted to propulsive power using propellers.

Propellers provide a very efficient means of generating thrust by giving a relatively large mass of matter a modest acceleration (see Section 7.1.5, *General theory of thrust generation*). In contrast, rockets give a relatively small mass of matter a large acceleration. Generally, the larger the acceleration, the greater is the amount of chemical energy (fuel) that must be converted into mechanical energy. Thus, the generation of thrust using a propeller consumes far less fuel than any other method, making it the most efficient propulsive option currently available for airplanes. In addition, manufacturing and maintaining a propeller is far less expensive than, say, a jet engine. Therefore, propellers are really the only option for low-cost aircraft. A drawback to propellers it they are usually limited to low subsonic applications. However, the history of aviation reveals there are a number of aircraft that have been capable of relatively high subsonic airspeeds, although the cost is always considerable engine power and sophistication in propeller design.

The McDonnell XF-88B is often cited as the fastest aircraft ever equipped with a propeller. Designed in the early 1950s, the airplane was a modified prototype of the XF-88 Voodoo fighter and was actually powered by two Westinghouse J-34 turbojets. It also featured an 1800 SHP Allison T38 turboprop engine in the nose that had been installed for research purposes. The engine rotated a three bladed wide-chord propeller

---

[1] Note that a one-bladed propeller usually features a single helical coil shaped like a wood screw. An example of this is David Bushnell's Turtle submarine from 1776.

[2] Or hydrodynamic in the case of watercraft.

and was capable of high-speed flight (just exceeding Mach 1), although only thanks to the thrust generated by its jet engines. The jet version of the XF-88 later saw service with the USAF in the 1960s and became better known as the F-101 Voodoo fighter. Another aircraft, the Republic XF-84H Thunderscreech, is often called the world's fastest true propeller aircraft.[3] The aircraft was designed in the early 1950s and was equipped with a 5850 SHP Allison XT40 turboprop engine. It is claimed it achieved a speed of Mach 0.83, although this is disputed by other sources, which say it achieved only Mach 0.70. The aircraft is considered by many to have been the loudest aircraft ever built, as the outer half of the propeller blades would travel at airspeed greater than Mach 1. Allegedly, the noise could be heard 25 miles (40 km) away and was the reason for the airplane's nickname.

The above aircraft were experimental projects that never made it to production. The fastest mass-produced propeller-powered aircraft is the four-engine Tupolev Tu-114 (Figure 14-1), which was capable of carrying over 220 passengers in a pressurized cabin at altitudes up to 39,000 ft. Its maximum speed was about 470 KTAS ($M \approx 0.82$), although its cruising speed was closer to 415 KTAS ($M \approx 0.72$). It featured four of the world's most powerful turboprop engines, the Kuznetsov NK-12, each capable of developing some 14800 SHP,[4] while swinging contra-rotating propellers. Although obsolete a long time ago, it was arguably way ahead of its time when designed in the mid-1950s, with its characteristic swept-back wings and lower deck galleys and deck crew rest areas. The airplane was in operation by Aeroflot from 1961 until 1976.

The propeller often poses serious analytical challenges for the aircraft designer. Ideally, the designer wants to determine thrust generated by a propeller using a simple variable; for instance, using the engine's power setting. Unfortunately, reality is far more complicated. The thrust generated by the propeller is the consequence of a complex interaction between the forward motion of the propeller, its rotational speed, and geometry. This requires the designer to understand a number of fundamentals not always covered in standard undergraduate engineering university coursework. This chapter will detail two important theories that can be used to calculate propeller thrust; the *Froude-Rankine momentum theory* and the so-called *blade-element theory*. The latter, while much more complicated to use, allows the power required to rotate the propeller to be estimated. There is a third theory of propulsive thrust generation that will not be covered in this text. This is the *propeller vortex theory*, which, effectively, is a version of the so-called Prandtl's lifting line theory applied to propellers. It uses potential flow theory to evaluate the flow circulation around the blades of a propeller and automatically includes the effect of the hub and propeller tip. Additionally, many authors (including the author of this book) have used the vortex-lattice method (VLM) for the same purpose, but both methods are highly mathematical and require computational proficiency well in excess of what is practical for aircraft design. These are really tools best left for propeller manufacturers to wield.

FIGURE 14-1 Tupolev Tu-114 passenger aircraft. Arguably the fastest propeller aircraft ever produced, with a cruising speed in excess of Mach 0.71. *(Photo by Dmitry Avdeev — from Wikipedia Commons)*

---

[3] For instance see NACA RM-SL53G10, Low-Speed Investigation of the Static Lateral Stability and Control Characteristics of a 1/6-Scale Model of the Republic XF-84H Airplane with the Propeller Operating, for investigation of the type.

[4] SHP stands for shaft horse-power, a unit of power used for turboprops.

In short, the propeller is here to stay. At the time of this writing, the propeller is by far the most efficient means of turning engine power into propulsive power currently available to the designer. When it comes to efficiency, it remains superior to the jet engine. The modern turbojet and -fan have undergone substantial advances in the last 70 years, rendering it vastly superior to the turbomachinery of yesteryear. However, its efficiency is lackluster when compared to a piston engine or gas turbine that rotates a propeller. The propeller can be thought of as a bridge between design requirements and available power plants. As an example consider the 310 BHP Continental IO-550 engine. This engine is used to power aircraft ranging from the amphibious Seawind 300C, whose cruising speed is 145 KTAS, to the high-performance 185 KTAS Cirrus SR22. The dissimilarity of such aircraft shows that the interplay of mission requirements and the available power plant is bridged only by the design of the propeller. It shows the design of the propeller needs to be considered early in the design phase and should not be treated as an afterthought. Propeller design is a field of specialization requiring much more space than can be allotted here. The designer of new aircraft is well advised to establish a good working relationship with propeller manufacturers by committing to using their products early in the design process; in fact, as soon as the engine has been selected. Not only will this will lead to more accurate performance predictions, it will also reduce a number of problematic design issues that are likely to arise and lie outside the field of expertise of the aircraft designer.

14.1.1 The Content of this Chapter

- **Section 14.2** discusses a number of effects the propeller has on an aircraft and which the aircraft designer must be aware of.
- **Section 14.3** presents tips to help the designer select the right propeller, ranging from initial estimation of diameter and pitch.
- **Section 14.4** presents methods to estimate propeller thrust.
- **Section 14.5** presents the so-called Rankine-Froude momentum theory, which can be used to estimate thrust, induced airspeed in the streamtube going through the propeller, and ideal propeller efficiency.
- **Section 14.6** presents the so-called blade element theory, which can be used to estimate thrust, propeller efficiency, and power required to rotate the propeller.

14.1.2 Propeller Configurations

Propellers can be mounted in a number of ways to an airplane. Three common methods are shown in Figure 14-2; a *tractor* (A), *pusher* (B), and a configuration featuring the engine and propeller mounted on the wing in *nacelles* (C). Configuration (C) is a variation of (A) or (B). The advantages and disadvantages of these installation methods are listed in Table 14-1. Note that so-called "inline" configurations, which consist of a tractor and pusher, can simply be treated as a combination of configurations (A) and (B).

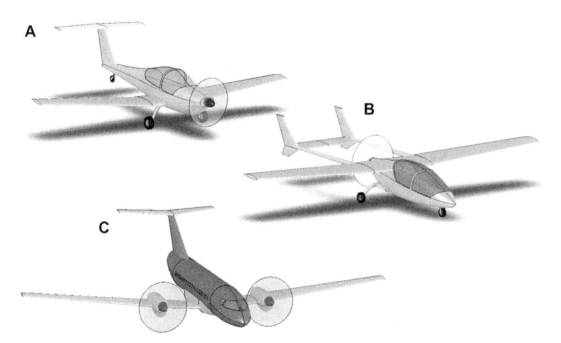

FIGURE 14-2 Three common methods to mount propellers (and engines). A is a *tractor* configuration, B is a *pusher* configuration, and C is a *tractor mounted on a nacelle* configuration.

TABLE 14-1 Pros and Cons of Tractor, Pusher, and Nacelle Configurations

Configuration	Advantages	Disadvantages
Tractor (A)	The incoming air is undisturbed. Ground clearance is not an issue during rotation at T-O, or flare before touch-down.[a] Placing the fuselage behind the propeller allows for a reduced streamtube inflow distortion and less asymmetric disc loading that would increase blade stresses. There is less chance that the propeller suffers damage due to FOD, in particular when the aircraft is moving. Propeller is not subject to excessive heat from the exhaust. Propwash can help with T-O rotation during soft- or short-field take-offs, by increasing the dynamic pressure at the horizontal tail.	Obstructed forward view, which usually requires the back side of the prop to be painted black. Increased cabin noise. Normal force forward of CG decreases stability. The turbulent high-speed wake flows over a fuselage or a nacelle and (at least theoretically) increases drag. When placed in the front with nose landing gear, there is a risk of ground-strike in a hard landing. Frequency of propwash pressure pulses may excite structural frequencies at some specific RPMs, resulting is structural vibrations.
Pusher (B)	Unobstructed forward view. Reduced cabin noise. Normal force aft of the CG increases stability. Turbulent high-speed wake does not flow over a fuselage or a nacelle, and, at least theoretically, will result in less drag. The streamtube will energize the flow in front of propeller and suppress flow separation on the body, even at high *AOA*.	Ground clearance may be an issue during rotation at T-O, or flare before landing. Fuselage ahead of the propeller may distort flow inside the streamtube, causing asymmetric disc loading and increased blade stresses. This distortion may affect the propeller's performance. Propeller may suffer FOD, in terms of both pebbles shot by tires and ingestion by ice shedding off a fuselage. Propeller may be subject to excessive heat from the exhaust. Special regulatory requirements for pushers are stipulated in 14 CFR 23.905. Results in a higher fly-over noise and possible propeller corrosion issues (explained below).
Nacelle (C)	The nacelle has many of the same advantages as the tractor configuration, but is more efficient since the blockage of the nacelle is less than that of a fuselage.	The nacelle has many of the same disadvantages as the tractor configuration; however, since the nacelle usually extends far forward from the wing's structural support, the airframe may be subject to structural oscillations that, if ignored, may cause premature structural failure, and even whirl flutter phenomena.

[a]*Per 14 CFR Part 23.925 (Propeller clearance), there must be a minimum 7" ground clearance for a tricycle landing gear and 9" for taildraggers, at the most adverse combination of CG, weight, the most adverse pitch position of the propeller, and at static ground deflection of the landing gear. Aircraft with leaf spring struts as landing gear must comply with 1.5 times its most adverse weight.*

A few additional explanations are needed for the pusher propeller. As stated in Table 14-1 special regulatory requirements affect the certification of pusher aircraft and are stipulated in 14 CFR 23.905(e) through (f). Paragraph 23.905(e) states that:

"All areas of the airplane forward of the pusher propeller that are likely to accumulate and shed ice into the propeller disc during any operating condition must be suitably protected to prevent ice formation, or it must be shown that any ice shed into the propeller disc will not create a hazardous condition."

Paragraph 23.905(f) states that:

"Each pusher propeller must be marked so that the disc is conspicuous under normal daylight ground conditions."

Paragraph 23.905(g) states that:

"If the engine exhaust gases are discharged into the pusher propeller disc, it must be shown by tests, or analysis supported by tests, that the propeller is capable of continuous safe operation."

And, finally, paragraph 23.905(h) states that:

"All engine cowling, access doors, and other removable items must be designed to ensure that they will not separate from the airplane and contact the pusher propeller."

There are further requirements made in paragraph 23.925(b) Propeller clearance that state that:

"(b) Aft-mounted propellers. In addition to the clearances specified in paragraph (a) of this section, an airplane with an aft-mounted propeller must be designed such that the propeller will not contact the runway surface when the airplane is in the maximum pitch attitude attainable during normal takeoffs and landings."

Excluding the last paragraph, these are extra requirements not demanded by the certification of tractor aircraft. Other drawbacks of the pusher propeller that need further explanation are the higher flyover noise and propeller corrosion. Experience has demonstrated that the ingestion of fuselage wake produces additional broadband noise,[5] often adding several dB(A) to the airplane's noise. Consequently, all other things being equal, pushers tend to be noisier than tractor configurations. A pusher propeller placed behind the engine exhaust will collect soot on the blades, forming acids that attack the propeller structure. Maintenance protocols for many pushers require the operator to wash the propeller after the last flight of the day as preventive maintenance. The competent aircraft designer keeps such drawbacks in mind when choosing a pusher over a tractor. As stated in Chapter 2, *Aircraft conceptual layout*, there are important consequences to a configuration selection, some of which have environmental, operational, and maintenance impacts. It is not wise to base the choice on looks only — operational practicality should lead such decisions.

14.1.3 Important Nomenclature

Any discussion about propellers involves jargon that the designer must be familiar with. For instance, it is likely that in a discussion about modern aircraft, one will hear pilots and mechanics alike describing a propeller as "an advanced composite five-blade, constant-speed, reversible, and fully featherable." What does it mean? The following definitions help clarify these terms and must be kept in mind for the discussion that follows:

A **fixed-pitch propeller** is a propeller whose pitch angle (the incidence of the blades with respect to the plane of rotation) cannot be changed. Such propellers are comparatively inexpensive, light, and require very little maintenance. The lack of a mechanical system renders them pretty much failsafe; however, they will windmill in the case of an engine failure and, thus, increase drag when a high glide ratio is sorely needed. Generally, such propellers are designed for best efficiency at either climb ("climb prop") or cruise ("cruise prop").

A **ground-adjustable propeller** is a propeller whose pitch angle can be adjusted using simple tools while stopped on the ground only. Thus, the operator can change the pitch from, say, a "climb" to a "cruise" style propeller (see Section 14.1.7, *Fixed versus constant-speed propellers*) between flights. Such propellers have become popular among homebuilders and LSA aircraft.

A **two-position propeller** is one that allows the pilot to change the pitch of the blades mechanically in flight, but only allows two settings (versus multiple ones for the ground-adjustable propeller). This way, the pilot could select a climb setting for the take-off and subsequent climb, and then adjust it to a coarser cruise setting once cruise altitude is reached.

A **controllable-pitch propeller** is one that allows the pilot to change the pitch over a range of possible pitch angles during flight. This is almost always accomplished through the use of hydraulic boost built into the hub of the propeller. Such a system allows the pilot to adjust the propeller pitch for the most efficient thrust output possible.

A **constant-speed propeller** is a propeller that will automatically adjust its pitch to maintain a preset RPM, which otherwise is highly affected by airspeed. It does this through the use of a controlling mechanism attached to the engine, called a *governor*, which balances centripetal and hydraulic forces.

A **feathering propeller** is a controllable-pitch or constant-speed propeller that allows the pilot to align the blades perfectly with the forward speed of the aircraft in the case of an emergency, such as an engine failure. This prevents the propeller from windmilling and, thus, reduces the drag impact of the inoperative engine.

A **reversing propeller** is a controllable or a constant-speed propeller that allows the pilot to align the blades to angles beyond what is possible with a feathering propeller. Effectively, it allows the thrust force to be pointed in a direction opposite to the forward motion of the aircraft. This feature is used during landing to decelerate the airplane and even to allow the aircraft to "back out" out from an air terminal using its own power. It is commonly found on turboprop engine installations.

An idealized power lever quadrant console is shown in Figure 14-3. Common reverse operation of propellers involves settings called *ground fine*, *reverse*, and *beta*. **Ground fine** is a low pitch setting used when taxiing the airplane with the engine running at a "reasonably high" rotation rate. Maintaining such RPMs on some turboprops is recommended to prevent damaging oscillation that may occur at lower RPMs. The ground fine setting prevents too high thrust from being generated at those rotation rates and this is most helpful for low-speed ground operations.

The **reverse** setting, as stated earlier, generates thrust in the opposite direction and is helpful immediately after landing. The setting labeled "Taxi" in the figure is where the pilot would operate the power lever while taxiing. In this range, the RPM is constant, but the blade pitch angles are adjusted. **Beta** refers to the range of power lever positions that are used on the ground that include the reverse and taxi power settings.

---

[5]Broadband noise refers to sound that, by definition, extends over a wide range of frequencies, perhaps even the entire range of audible frequencies.

14.1 INTRODUCTION   587

integral extension of the propeller hub (most common with constant-speed propellers). The purpose of the spacer or hub extension is to increase the spacing between the engine and propeller so that the geometry of the engine cover (cowling) may be made more aerodynamically efficient and, thus, reduce the drag penalty of the engine installation. This often leads to improved cooling. While heavier, the overall reduction in drag may easily surpass the penalty of the weight addition. Propeller extensions must be carefully installed because if the propeller becomes misaligned, a severe vibration may result. Ideally, it should have a centering boss to make the installation easier.

14.1.4 Propeller Geometry

A three-bladed propeller is shown in Figure 14-4, rotating about an axis we call the *axis of rotation*. The *spinner* is an aerodynamically shaped cover, whose purpose is to reduce the drag of the hub of the propeller and to protect it from the elements. The propeller blades are what generate the thrust of the device, denoted by T. The pressure differential between the front and aft face of the propeller blade results in a vortex that is shed from the tip of the blade and is carried back by the airflow going through the propeller. This forms the typical helical shape shown in the image. Note that only one vortex is shown, but two others are also formed by the other blades and are hidden for clarity.

A frontal projection of the three-bladed propeller is shown in Figure 14-5, where R is the blade radius, r is the radius to an arbitrary *blade station*, and Ω is the rotation rate, typically in radians per second or minutes.

The blade of a propeller is really a cantilevered wing that moves in a circular path rather than along a straight one. Just like an airplane's wing, the planform of the

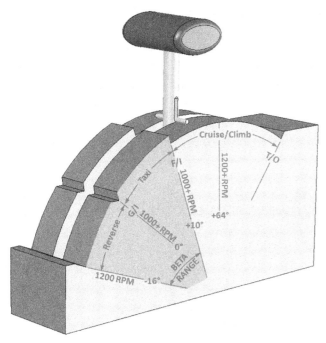

FIGURE 14-3 Idealized power lever quadrant for a typical turboprop. G/I is ground idle, F/I is flight idle, T/O is take-off.

A **beta control prop** is a propeller specifically designed to allow operation of the propeller as described above.

Material for propellers: the modern propeller is made from three different materials: wood, forged aluminum, or composites. All have their pros and cons, but only aluminum and composites should be used for high-performance aircraft. Wooden propellers are lighter and should not be overlooked as a solution for light airplanes that may have the CG too close to a forward limit (tractor) or aft limit (pusher).

A **propeller extension** can consist of a separate spacer piece (most commonly for fixed-pitch propellers) or an

FIGURE 14-4 Propeller helix.

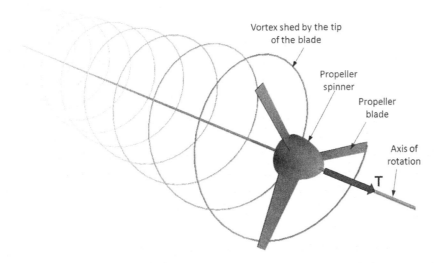

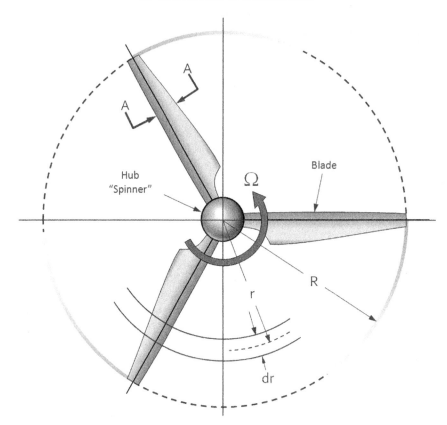

FIGURE 14-5 Propeller geometry.

propeller blade has a profound impact on the magnitude of the thrust force created, as well as at what "cost." What constitutes "cost" is the amount of power required to rotate it, as well as side effects such as noise. Figure 14-6 shows a picture of a typical blade planform and its cross-sectional airfoils, taken from NACA TR-339 [1]. The blade planform, along with geometric properties such as twist and airfoil camber, is of crucial importance to optimize a propeller. These affect not only the capability of the blade to generate thrust, but also its strength and natural frequencies.

14.1.5 Geometric Propeller Pitch

Consider the propeller in Figure 14-7, whose diameter is D and radius is R. As the propeller rotates through a

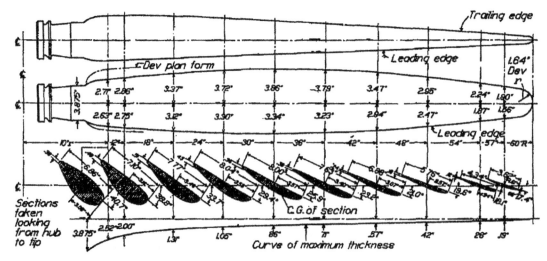

FIGURE 14-6 Geometry of a typical metal propeller blade. *(from Ref. [1])*

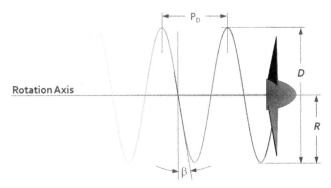

FIGURE 14-7 Schematic showing some propeller properties (note that wake contraction not shown).

Fundamental Formulation of

Considering the geometry shown in Figure 14-7 we can now define the following characteristics of the propeller:

Pitch angle:

$$\tan\beta = \frac{P_D}{2\pi r_{ref}} \quad (14\text{-}1)$$

where

r_{ref} = reference radius, usually 75% of the propeller radius R
P_D = pitch distance of the propeller

Generally, the value of P_D ranges from 60% to 85% of the diameter of the propeller. Propeller manufacturers specify the value of the pitch using the geometric pitch at $0.75R$. An example of this is the Sensenich 69CK propeller used as a replacement for the Cessna 152 two-seat, single engine trainer. Its diameter is 69 inches (1.753 m) and pitch ranges from 42 to 58 inches (1.067–1.473 m).[6] The *pitch-to-diameter ratio* is also used to identify propellers:

Pitch-to-diameter ratio:

$$\frac{P_D}{D} = \frac{\pi \cdot r}{R}\tan\beta \quad (14\text{-}2)$$

A propeller moving through a low-viscosity fluid like air will cover less distance per revolution than the geometric pitch would indicate (see Figure 14-8). Therefore, the angle formed between the rotation plane and a

full circle, its tip rotates through an arc length (circumference) of $C = \pi D = 2\pi R$. As the propeller rotates it "screws" itself forward a certain distance P for each full rotation. Now consider the abstract case where the propeller is assumed to rotate slowly through some highly viscous fluid. If this were possible, the propeller would move through this fluid like a metal screw moves into a piece of wood. The distance it would cover in one full revolution is called the *geometric pitch* or *pitch distance*, P_D, of the propeller. It is commonly specified in terms of inches of pitch. Thus a propeller designated as a 42-inch pitch prop would move 42 inches forward in one revolution (using the metal screw through wood analogy). The angle the helix makes to the rotation plane is called the *geometric pitch angle* and is denoted by β.

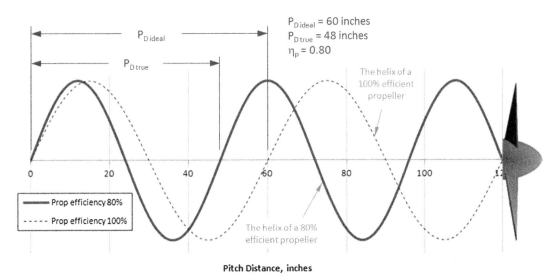

FIGURE 14-8 The propeller will advance a shorter distance (pitch distance) in a low-viscosity fluid than the geometric pitch indicates.

---

[6]The propeller's Type Certificate Data Sheet P904 69CK, 72CK, 72CC.

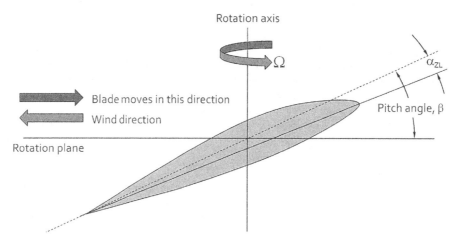

FIGURE 14-9 Definition of a propeller pitch angle.

tangent to the blade tip helix at each blade station is less than the geometric pitch angle. This angle is called the *helix angle* and is denoted by ϕ. It can be estimated if the forward speed of the propeller is known using the following expression:

Helix angle:

$$\tan \phi = \frac{2\pi \, r \, n}{V_0} = \frac{\pi \, r \, RPM}{30 \cdot V_0} \qquad (14\text{-}3)$$

Pitch Angle or Geometric Pitch

Consider the cross-section A-A of the propeller blade in Figure 14-5 at some arbitrary blade station at radius r. An example of such a cross-section is the airfoil shown in Figure 14-9. In order for the propeller to generate thrust in a forward direction its chord line must form a positive angle-of-attack to the relative wind as it moves about its rotation axis. Since it simultaneously moves along the rotation axis (upward in the figure), the propeller blade airfoil must be set at a specific geometric angle with respect to the rotation plane.

As stated before, this angle is the *pitch angle* or *geometric pitch*, here denoted by the Greek letter β (also shown in Figure 14-7). Since propeller blades usually feature cambered airfoils, their associated zero-lift AOA, α_{ZL}, is a negative number (negative AOA). Note that its reference datum is the zero-lift line of the airfoil and *not* the chord line. The angle between the two is α_{ZL}, as we expect from the airfoil geometric definition of Chapter 8.

Constant-Pitch Propeller

As the propeller moves in the direction of the rotation axis (i.e. in the direction of the thrust) a blade radial station near the hub experiences a relatively small airspeed component in the rotation plane, whereas the airspeed near the tip is comparatively large. The AOA experienced at a given radial station depends on this radial airspeed and the forward airspeed. Since the entire blade experiences the same forward airspeed but varying radial speed, the AOA "seen" by the airfoil at a specific radial station varies along the blade's span. The lift generated by each radial station of an untwisted propeller would be highly uneven and, at high forward speeds, might even produce reversed thrust. Propeller manufacturers solve this by twisting the blade so that all radial stations experience the optimum AOA at some specific mission conditions, e.g. climb or cruise. This AOA is typically the one where the airfoil's lift-to-drag ratio peaks. Therefore, the resulting propeller thrust is maximized.

The most common way of accomplishing this is to design the propeller blade such that each radial station moves a uniform pitch distance, P_D, in one full turn (again, imagine it rotating in a highly viscous fluid). Since the arc length that each radial station makes varies along the propeller blade, the pitch angle closer to the hub must be at a higher pitch angle than at another one farther from the hub. Since each radial station moves a constant pitch distance, such propellers are called *constant-pitch propellers* (see Figure 14-10). The concept should not be confused with *fixed-* or *variable-pitch* propellers, which are discussed in Section 14.1.7, *Fixed versus constant-speed propellers*. Propellers designed using the constant-pitch scheme allow for a simple mathematical representation of the pitch angle β as a function of any spanwise station r, based on Equation (14-1):

Constant-pitch propeller:

$$\tan \beta = \frac{P}{2\pi r} \qquad (14\text{-}4)$$

where; β = pitch angle and r = arbitrary blade station along a propeller blade.

Variable-Pitch Propeller

A *variable-pitch propeller* is one in which the pitch distance is not fixed, but changes gradually along the

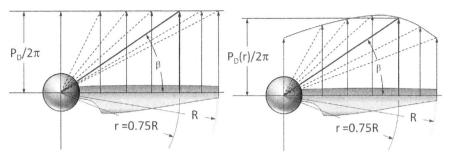

FIGURE 14-10 Constant-pitch (left) and variable-pitch (right) propellers.

span of the propeller (see Figure 14-10). The purpose of this is to load the propeller disc plane in some desirable way. For instance, it is possible the propeller designer is trying to prevent the inboard section of the propeller from stalling at some specific condition, or to prevent the section lift coefficients at the blade tip from exceeding a given maximum value at the airplane's mission operating conditions. Or the purpose may be to modify the distribution of section lift coefficients in an attempt to make the propeller more efficient. Section lift coefficients that are too high are caused by chordwise pressure that is too low, which in turn is associated with higher airspeeds that may even exceed the speed of sound, resulting in a noisy or less efficient propeller.

Fundamental Relationships of Propeller Rotation

Consider Figure 14-11, which shows two propeller blade sections. The upper one shows the blade airfoil at a zero forward airspeed (static condition, e.g. airplane sitting on ground at rest prior to T-O). The thick vector indicates the oncoming airflow seen by an imaginary observer on the blade. The angle β is the pitch angle, whereas the *AOA* is represented by α as usual. The magnitude of the airspeed at that blade station in this static condition is purely dependent on the rotation rate times the spanwise distance of the blade station from the rotation axis, or:

Propeller rotational speed:

$$V_{ROT} = \Omega \cdot r \qquad (14\text{-}5)$$

As is evident from Figure 14-11, the *AOA* is the angle between the blade's chord line and the rotation plane. Also note how the *AOA* changes with forward airspeed, V_0, as shown in the lower portion of the figure. Note that RPM can be converted to angular velocity as follows:

$$\Omega = 2\pi \left(\frac{RPM}{60}\right) \text{ rad/s} \qquad (14\text{-}6)$$

Additionally, we see that since the angular velocity Ω is measured in radians/second we can define the period and frequency of the rotational motion as follows:

Period of rotation:

$$\Omega = \frac{2\pi}{P_\Omega} \quad \Leftrightarrow \quad P_\Omega = \frac{2\pi}{\Omega} \text{ seconds} \qquad (14\text{-}7)$$

Frequency of rotation:

$$\Omega = 2\pi f_\Omega \quad \Leftrightarrow \quad f_\Omega = \frac{\Omega}{2\pi} \text{ Hertz} \qquad (14\text{-}8)$$

Now consider the lower portion of Figure 14-11, which shows the propeller as it moves at some airspeed V_0 in the direction of the rotation axis. Now our imaginary observer sees the airflow coming from a different direction, at an angle ϕ, i.e. the helix *angle*. It can be seen that α is now much smaller than in the static case and the resultant airspeed can be obtained from the Pythagorean relation:

$$V_R = \sqrt{V_0^2 + V_{ROT}^2} = \sqrt{V_0^2 + (\Omega \cdot r)^2} \qquad (14\text{-}9)$$

The dynamic pressure acting on the airfoil at radial station r is based on this airspeed.

The following expressions are useful to quickly estimate the helical tip speed and Mach number of the propeller.

Propeller helical tip speed:

$$V_{tip} = \sqrt{V_0^2 + (\Omega R_{prop})^2} = \sqrt{V_0^2 + (\pi n D)^2}$$
$$= \sqrt{V_0^2 + \left(\frac{\pi \cdot RPM \cdot D}{60}\right)^2} \qquad (14\text{-}10)$$

Tip Mach number:

$$M_{tip} = \frac{V_{tip}}{a_0} = \frac{V_{tip}}{\sqrt{\gamma R T}} \approx \frac{V_{tip}}{1116\sqrt{1 - 0.0000068753\,H}} \qquad (14\text{-}11)$$

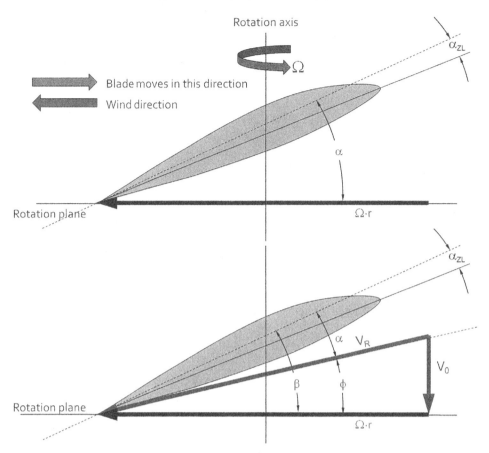

FIGURE 14-11 The upper propeller blade is rotating at static conditions ($V = 0$) and the lower one at some airspeed $V = V_0$.

Determination of the Desired Pitch for Fixed-Pitch Propellers

The selection of a propeller requires its diameter and pitch to be specified. While it is recommended the designer consults with a propeller manufacturer on the appropriate dimensions, the least he can do is to be armed with a "ballpark value." The following expression can be used to estimate the pitch of a fixed-pitch propeller whose rotation rate is denoted by RPM while operating at a desired cruising speed, denoted by V_{KTAS}, with efficiency denoted by η_P.

Propeller pitch distance:

$$P_D \approx 1251 \left(\frac{V_{KTAS}}{RPM}\right)\left(\frac{1}{\eta_P}\right) \qquad (14\text{-}12)$$

Derivation of Equation (14-12)

Consider a propeller rotating at a rate indicated by RPM or RPS = RPM/60. If we assume it can operate with efficiency denoted by η_P and we know the pitch distance, P_D, it follows the propeller screwing motion will move it forward at the rate shown below:

$$V = \left(\frac{RPM}{60}\right) P_D \times \eta_P$$

Generally, we want to refer to P_D in inches and V in KTAS (and call it V_{KTAS}). If P_D is in inches it follows that V will be in inches/second. Additionally, noting that 1 KTAS = 1.688 ft/s, we introduce the following conversions:

$$V_{KTAS} = V\left(\frac{1 \text{ ft}}{12 \text{ in}}\right)\left(\frac{1 \text{ KTAS}}{1.688 \text{ ft/s}}\right) = \frac{V}{20.256 \text{ in/s} \cdot \text{KTAS}}$$

$$= \frac{\left(\frac{RPM}{60}\right) P_D \times \eta_P}{20.256 \text{ in/s} \cdot \text{KTAS}} = \frac{RPM \times P_D \times \eta_P}{1251.36 \text{ in/KTAS}}$$

Solving for P_D leads to:

$$V_{KTAS} = \frac{RPM \times P_D \times \eta_P}{1251.36}$$

$$\Rightarrow P_D = 1251.36 \left(\frac{V_{KTAS}}{RPM}\right)\left(\frac{1}{\eta_P}\right)$$

QED

EXAMPLE 14-1

Calculate the required pitch of a fixed pitch propeller rotating at 2400 RPM, assuming 75% efficiency at 110 KTAS.

Solution

$$P_D \approx 1251 \left(\frac{V_{KTAS}}{RPM}\right)\left(\frac{1}{\eta_P}\right) = 1251\left(\frac{110}{2400}\right)\left(\frac{1}{0.75}\right)$$
$$= 76.5 \text{ inches}$$

14.1.6 Windmilling Propellers

Windmilling occurs in airplanes when power to the propeller is cut and it rotates solely due to the aerodynamic lift generated by the blades. A generalized schematic of this phenomenon is shown in Figure 14-12. First consider the left image. In normal operation the prop is driven by the engine at an angular velocity of Ω and is moving at airspeed V_0 through air. At a particular blade station r, the speed of the blade element is $\Omega \cdot r$, which yields the resulting airspeed V_R and α shown. The lift, L, generated by the entire blade is shown in blue and its component parallel to the rotation axis is the thrust, T. The term w is the induced velocity, which depends on the shape of the streamtube that goes through the propeller disc.

In the right schematic, the propeller is rotating at an angular velocity Ω and some forward airspeed V_0, both of which may be different from that of the left figure. This time, the speed components yield an α from the opposite side, producing lift, L, that points in the opposite direction. Again, thrust is generated, albeit in the opposite direction (and thus it should really be called drag).

Typical recovery procedures in airplanes following a non-catastrophic engine failure[7] require its nose to be lowered until a specified forward speed (here denoted by V_0) is achieved. This is typically the airplane's best glide speed (see Section 19.2.8, *Airspeed of minimum thrust required*, V_{TRmin}). As the angular velocity of the propeller gradually reduces, the *AOA* eventually becomes negative, as shown in the right image of Figure 14-12. This starts the propeller windmilling (or auto-rotating) and at this point drag begins to increase and can have detrimental effects on the

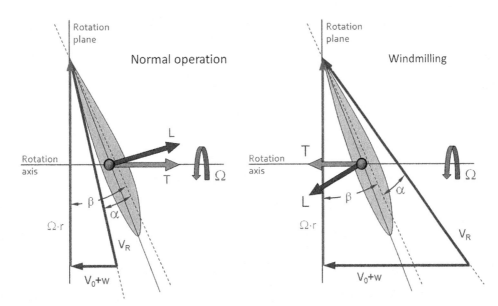

FIGURE 14-12 Normal operation of a propeller compared to windmilling.

---

[7]In which the engine components can still rotate.

L/D ratio. This effect must be considered in the aircraft design and the publication of a Pilots Operating Handbook (POH).

The reversed thrust (drag) generated by the windmilling propeller is higher than for the stationary propeller. The greater the airspeed, the greater is the propeller drag. Some pilots attempt to improve the glide ratio by stopping the autorotation. This requires the airplane to be slowed down so the propeller rotation is halted by the internal friction of the engine. More often than not this will require airspeeds in the neighborhood of the airplane's stalling speed and, consequently, such maneuvers should not be attempted at low altitudes. Propeller-powered multiengine aircraft are always equipped with feathering propellers, which circumvent this problem by changing the pitch of the propellers such that windmilling is no longer possible and the pitch places the blades in a low-drag position, greatly reducing the impact on drag. A method to estimate the drag of windmilling (and stopped) propellers is presented in Section 15.5.13 *Drag of windmilling and stopped propellers*.

14.1.7 Fixed Versus Constant-Speed Propellers

Propeller efficiency is discussed in greater length in Section 14.3.9, *Propulsive efficiency*, but it is an indicator as to how much engine power is being converted into propulsive power (thrust × airspeed). Thus, a particular propeller may be 0.80 efficient at a specific condition. This means that 80% of the engine power (BHP) is being converted into propulsive power. As we have seen before, the AOA of the blade varies with airspeed (as shown in the lower part of Figure 14-11) and the propeller's thrust will do so as well. It follows that propeller efficiency is a function of airspeed.

Generally, propellers for aircraft are designed for a particular airspeed or range of airspeeds specified by the airframe manufacturer. In this case the following rules of thumb apply:

For low-speed operations use a low pitch.
For high-speed operations use a large (coarse) pitch.[8]

A fixed-pitch propeller is one in which the blade pitch angles are permanently fixed. Such propellers are simple, light, and inexpensive. Their drawback is that their best efficiency is achieved at a particular airspeed only, so the designer must decide whether to emphasize climb or cruise performance and select a prop that favors either. Usually, two types of fixed-pitch propellers are available. The *fixed-pitch "climb"* propeller is designed to reach it maximum propeller efficiency at a relatively low airspeed. This makes it very suitable for use in airplanes where climb performance is of importance, such as trainers and bush-planes. The *fixed-pitch "cruise"* propeller is designed to reach its maximum efficiency at a higher airspeed, making it suitable for airplanes where higher cruising speed is of greater importance than climb performance.

Pilots who fly aircraft with fixed-pitch propellers notice that, if initially at cruise, the RPM of the engine will reduce if climb is initiated and increase if descent is initiated. In the former case, the climb inevitably slows down the aircraft and the propeller blades begin to experience a higher AOA. Thus they generate more drag, which, in turn, means higher torque is being generated and this slows the engine RPM. If we assume the pilot does not change the power setting, the RPM will simply drop as well and the prop will now be operating at a changed efficiency. In the latter case the opposite happens. The blades will experience lower AOA as the aircraft speeds up and generate less drag requiring less torque. The engine experiences this as reduced load so its RPM increases, again, changing the efficiency. If the aircraft is using a cruise prop, the change will be toward less efficiency, since it may very well have been operating at peak performance. If it is a climb prop, it would become more efficient as the airplane slows down, but less if it increases the airspeed.

An airplane specifically designed and operated as a cruiser will always feature a cruise-style propeller. Airspeed variations do therefore detrimentally affect its performance. This begs the question, is it possible to feature a prop that will not be so subject to such off-peak deviations? In other words: is it possible to design a prop so it tends to maintain the RPM and therefore its peak efficiency at the condition? The answer is yes. In fact, such propellers have existed for a long time and are called *constant-speed propellers*. However, such propellers went through an evolutionary phase, as in the 1920s manufactures, such as Pierre Levasseur, introduced manual control to adjust the blade pitch in flight [2] and that way "convert" a T-O prop into a climb prop, and a climb prop into a cruise one.

The variation in RPM with airspeed still occurred and this led to the development of an automatic controller, called a *propeller governor*, or simply a governor. It is designed to use accurate deviations from the setting selected by the pilot to maintain peak efficiency at all times. Figure 14-13 shows how the propeller efficiency typically varies with airspeed for three different but commercially available propeller types. The green and blue curves represent the efficiency of propellers whose pitch is fixed.

---

[8]If it helps, remember low pitch for low speed and high pitch for high speed.

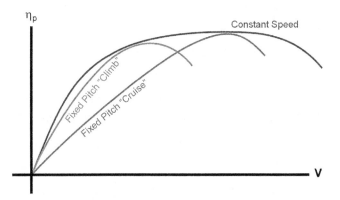

FIGURE 14-13 Two kinds of "fixed" pitch propellers versus a "constant-speed" propeller.

A detailed description of how the governor works is beyond the scope of this section; however, a brief introduction will be given. Once the desired RPM (using the cockpit RPM control) has been established the governor will automatically adjust the pitch of the propeller blades as needed to maintain that RPM. This is accomplished by means of throttle setting and flyweights. The governor mechanically detects RPM via the flyweights and uses it in conjunction with the throttle setting to maintain the required oil pressure inside a pressurized oil reservoir. Consider an airplane whose pitch angle deviates from a level flight such that an increase in RPM results (for instance if the nose pitches down). This implies an increase in airspeed and would cause the governor to increase the propeller pitch which would reduce the RPM. By the same token, a decrease in RPM would indicate a reduction in airspeed (for instance if the nose pitches up). The governor would then reduce the propeller pitch angle until the original RPM was again established. A section view of a hub of a constant-speed propeller is shown in Figure 14-14.

14.1.8 Propulsive or Thrust Power

Propulsive power is the power required to move a vehicle at a specific speed using specific force or thrust. Power is defined by:

$$\text{Power} = \frac{\text{Work}}{\text{Time}}$$

Since work is defined as the application of a force over a specific distance (force × distance) we get:

$$\text{Power} = \frac{\text{Work}}{\text{Time}} = \frac{\text{Force} \times \text{Distance}}{\text{Time}}$$
$$= \text{Force} \times \frac{\text{Distance}}{\text{Time}} = \text{Force} \times \text{Speed}$$

We therefore define propulsive or thrust power as follows:

$$P_R = T \cdot V \qquad (14\text{-}13)$$

where the subscript R stands for *required* as this term represents the power required in the performance analysis of later sections.

14.2 PROPELLER EFFECTS

This section considers important side-effects that are caused by the presence of the propeller. These effects are so significant that the designer must be aware of them

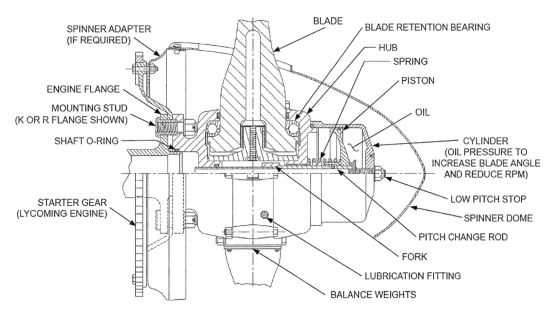

FIGURE 14-14 A section view of the inside of a constant-speed propeller. *(courtesy of Hartzell Propeller Inc.)*

and consider them in the development of the airplane. Pilots of propeller aircraft recognize these effects, although not all understand the physics behind them, and call them the *P-factor* (for prop-factor). While some sources specifically call out the asymmetric yaw as the P-factor (see Section 14.2.4, *Asymmetric yaw effect*) others, such as Stinton [3], call the combination of the angular momentum effects (Section 14.2.1, *Angular momentum and gyroscopic effects*), slipstream effects (Section 14.2.2, *Slipstream effects*), the normal force (Section 14.2.3, *Propeller normal and side force*), and the asymmetric yaw effect (Section 14.2.4, *Asymmetric yaw effect*) the P-factor. This author agrees with this classification for the simple reason it is impossible for the pilot to distinguish between gyroscopic, slipstream, and other effects resulting from the operation of the propeller. They only feel that running the propeller at high RPM introduces effects not present when idle. In this text, the P-factor encompasses all the cited effects.

However, there are other effects to bring up that are of importance in the development of the aircraft; for instance, the asymmetric yaw effect for multiengine aircraft (Section 14.2.5, *Asymmetric yaw effect for a twin-engine aircraft*) that dictates which engine is critical in the case of an engine failure, blockage effects (Section 14.2.6, *Blockage effects*), which reduce the thrust the propeller can develop, hub and tip effects (Section 14.2.7, *Hub and tip effects*), the effects of high tip speed (Section 14.2.8, *Effects of high tip speed*), and skewed wake effects (Section 14.2.9, *Skewed wake effects — A·q loads*) and the resulting $A \cdot q$ loads that are critical for turboprops and electrically powered aircraft.

14.2.1 Angular Momentum and Gyroscopic Effects

Angular momentum and gyroscopic effects play an important role in stability and control theory and, thus, must be taken into account in the design process. Consider the propeller to the left in Figure 14-15, which rotates at a constant angular velocity Ω_x (computed using Equation (14-6)). As it rotates about the *x*-axis (the axis of rotation), an angular momentum, h_{SRx}, (SR stand for spinning rotor) is being generated. In conventional single-engine aircraft this momentum must be reacted by the airplane, otherwise it would itself begin to spin in the opposite direction. For instance, the angular momentum of a propeller that rotates clockwise, as seen from behind (right blade is moving down) will tend to rotate the airplane in the opposite direction — to the left. Generally, in such airplanes, this is easily suppressed by a very slight aileron deflection, since the change in lift due to aileron deflection acts through a long arm. The magnitude of the angular momentum is given by:

$$h_{SRx} = (I_{XX})_{prop} \cdot \Omega_x \qquad (14\text{-}14)$$

where

$(I_{XX})_{prop}$ = moment of inertia of the propeller about its axis of rotation

Ω_x = angular velocity of the propeller; the subscript x denotes rotation about the *x*-axis

Some multiengine aircraft feature an even number of engines (two, four, six, etc.) and propellers that rotate in opposite directions about the plane of symmetry. As long as the engines all rotate at the same RPM, this cancels the angular moments.

Now consider the same propeller to the right in Figure 14-15, which shows the original propeller in the process of rotating to a new orientation. This change in orientation is called *gyroscopic precession*. As the motion takes place, additional gyroscopic couples, M_y and M_z, are generated. These moments are restoring since their action is to keep the propeller in the original orientation.

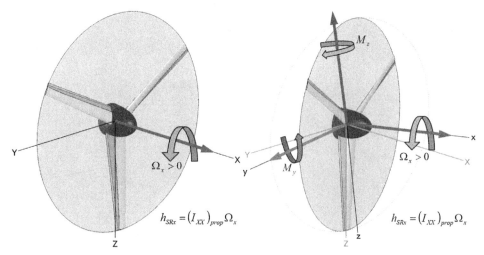

FIGURE 14-15 The left propeller rotates only along the *x*-axis. In the right image, it is in the process of changing orientation, which induces the restoring moments M_y and M_z.

14.2 PROPELLER EFFECTS

Note that as soon as the rotation ceases, so will those moments. If the angular velocity of the precession is given by $\vec{\omega}_P = p\mathbf{i} + q\mathbf{j} + r\mathbf{k}$, then the gyroscopic moments can be found from:

$$\mathbf{M}_P = \begin{bmatrix} M_x \\ M_y \\ M_z \end{bmatrix} = \vec{\omega}_P \times \mathbf{h}_{SR}$$

$$= \begin{bmatrix} i & j & k \\ p & q & r \\ h_{SRx} & h_{SRy} & h_{SRz} \end{bmatrix} \quad (14\text{-}15)$$

The components p, q, and r can be interpreted as the rotation rates of an airplane about its x-, y-, and z-axes, respectively. Note that for the propeller in Figure 14-15, the gyroscopic moments would be given by:

$$\begin{bmatrix} M_x \\ M_y \\ M_z \end{bmatrix} = \begin{bmatrix} \mathbf{i} & \mathbf{j} & \mathbf{k} \\ p & q & r \\ h_{SRx} & 0 & 0 \end{bmatrix}$$

$$= (0)\mathbf{i} - (0 - rh_{SRx})\mathbf{j} + (0 - qh_{SRx})\mathbf{k}$$

$$= rh_{SRx}\mathbf{j} - qh_{SRx}\mathbf{k} \quad (14\text{-}16)$$

14.2.2 Slipstream Effects

A typical single-engine aircraft is shown in Figure 14-16. It features a standard propeller whose rotation is clockwise as seen from the pilot, who sits behind it (clockwise motion of this nature is common among American propeller manufacturers, whereas counter-clockwise is common in Europe). The corkscrew-shaped blade tip vortex is sketched as it encloses the fuselage of the airplane. The tip vortex is caused by the pressure differential between the forward (low pressure) and rearward (high pressure) sides of the propeller. This pressure differential causes air on the back side of the propeller to flow toward the tip, while the opposite happens on the forward face. The opposite radial flow directions on the two sides cause the formation of this vortex at the tip, exactly as it does on a regular wing. The vortex is indicative of the downwash left by the propeller blade. The lower image of Figure 14-16 shows the vortex in the area of the vertical tail, subject to this downwash. It shows that the downwash component, which is perpendicular to the wake left by the propeller blade, changes its angle of attack, generating additional lift on it in the process. The

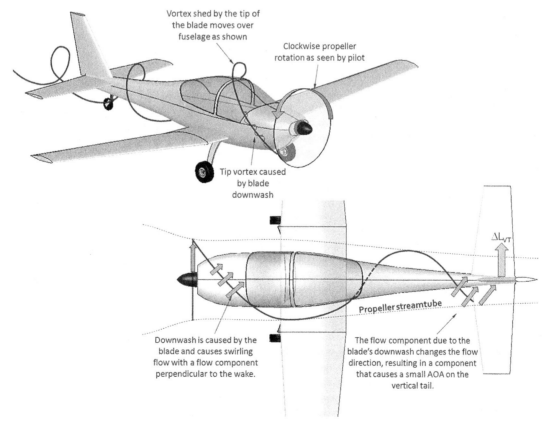

FIGURE 14-16 The swirling motion of air inside the propeller streamtube will induce a slight change in the *AOA* of the vertical tail; for a clockwise-turning propeller, this will introduce a left yaw tendency.

resulting yawing moment for this aircraft (assuming a clockwise propeller rotation) tends to move the nose to the left. It can be suppressed by a rudder deflected trailing edge right (step on the right rudder), or by a cambered airfoil (camber on the left side), or by a small angle-of-incidence adjustment (leading edge left). All of these suppress the yaw by generating an opposing lift on the vertical tail. The magnitude of the yawing moment depends on engine power, RPM, and the airspeed of the aircraft.

14.2.3 Propeller Normal and Side Force

Normal Force

When an aircraft is operated at an *AOA* that results in the incoming flow being at an angle other than normal to the propeller disc plane, its blades will load up asymmetrically. Consider the aircraft in Figure 14-17, which shows such a condition. The lower part of the image shows the left and right halves of the propeller with an asymmetric lift distribution. This is a consequence of the right propeller blade moving in a downward motion (here, if we assume a clock-wise rotation of a two-bladed propeller, the pilot would see the right blade moving down and the left one moving up). This motion is toward an upward component of the airspeed, which, in turn, increases its lift. At the same time, the blade moving upward (the left blade) moves away from the upward component, reducing its lift. It should be kept in mind that although most of the lift of the propeller points in the forward direction, there is a vertical component to it as well. The resulting increase in lift of the right blade is larger than the reduction of the left one (due to the V^2 of the dynamic pressure), causing a net force component acting in the vertical direction that adds to the normal component of the thrust. Unfortunately, the determination of the normal force is not a simple task, as it depends on the blade geometry, RPM, airspeed, and angle-of-attack of the airplane.

The CG of the tractor configuration in Figure 14-17 is aft of the propeller. Consequently, the normal force will destabilize the aircraft (introduce a nose pitch-up contribution). However, if the propeller is aft of the CG, as is the case for most pushers, the effect is opposite or stabilizing (nose pitch-down contribution). The moment about the vertical axis (z-axis) generated by the asymmetric disc loading also causes another effect, what pilots refer to as the "P-factor," to be discussed next.

Side Force

A force is generated in an identical manner to the normal force when the airplane is yawed, except it points sideways. At high power, a tractor configuration will experience a destabilizing moment (see Figure 14-18) that will tend to increase the yaw angle, requiring rudder input to correct and making it easier to exacerbate the

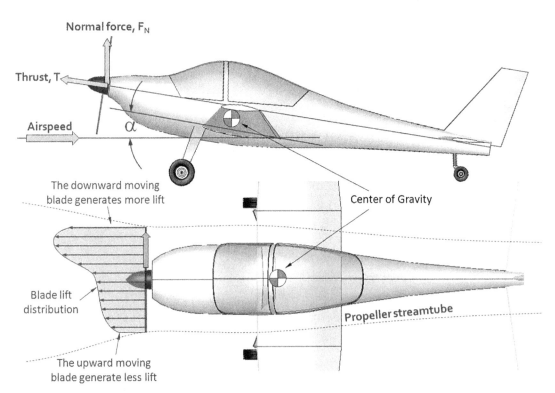

FIGURE 14-17 The generation of asymmetric disc loading.

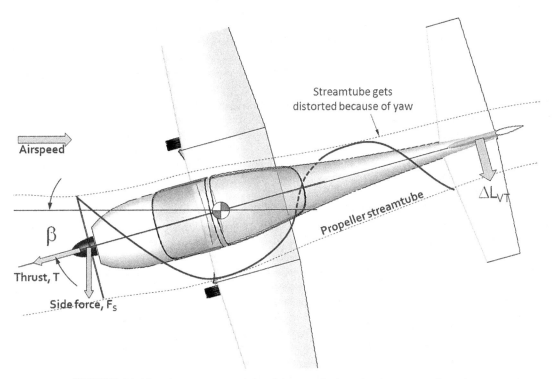

FIGURE 14-18   The generation of destabilizing side force for a tractor configuration.

yawed condition if care is not exercised. The designer should keep this effect in mind and ensure the directional stability ($C_{n\beta}$) is large enough to prevent this from becoming a problem.

14.2.4 Asymmetric Yaw Effect

Another effect can be understood from Figure 14-17. The lower image shows the lift distribution over the propeller disc is highly asymmetric. Consequently, the right area of the disc generates more thrust than the left one. This effect generates a couple that tends to turn the nose to the left (again assuming a clockwise propeller). The pilot must step on the right rudder to suppress it. The magnitude of this couple depends on the engine power, RPM, and forward airspeed. On the other hand, the suppressing moment, which is generated by a rudder deflection, is dependent on the airspeed of the aircraft. And when the airspeed is low, the suppressing moment is low. However, the asymmetric moment may not necessarily be as low since it is highly dependent on the radial velocity of the propeller, which depends on the RPM. This highlights a potential critical flight condition for the aircraft — high power, low airspeed, high AOA, which is typical for an initial climb after take-off or during a balked landing maneuver. It is imperative the aircraft designer is aware of this effect and sizes the vertical tail to handle it.

14.2.5 Asymmetric Yaw Effect for a Twin-Engine Aircraft

The asymmetric yaw effect has serious implications for multiengine propeller aircraft. To explain this effect, consider the twin-engine aircraft shown in Figure 14-19. The left image shows the aircraft with both engines operating normally and both propellers rotating clockwise as seen from behind (note that everything said here is reversed for propellers that rotate counter-clockwise).[9] As is unavoidable, both propellers are subject to asymmetric thrust loading that places the center of thrust (denoted by T) to the right of the rotation axis. The thrust of the left engine acts over a distance d_L and the thrust of the right engine acts over a distance d_R, which, as can be seen, is larger than d_L. Therefore, a relatively small moment M is generated about the CG and this must be reacted by introducing a small rudder deflection, $\delta_{r\ trim}$. For the clockwise propeller rotation shown, the moment M would result in a nose-left tendency. And while relatively small, the pilot will detect it by observing that the "ball" on the Turn and Bank indicator swings to the right of center. The pilot

---

[9]The De Havilland DHC-8 DASH 8 is an example of a twin-engine commuter aircraft that has both propellers rotating clockwise as seen from behind. The Fokker F-27 Friendship is an example of a twin-engine commuter that has both propellers rotating counter-clockwise.

600   14. THE ANATOMY OF THE PROPELLER

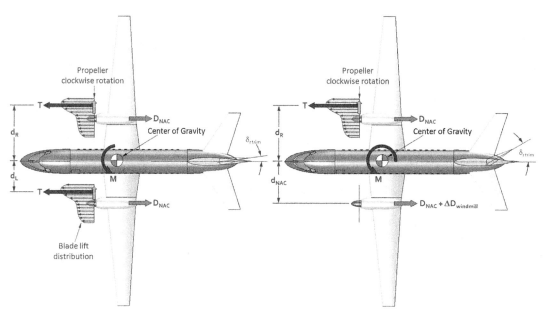

FIGURE 14-19 The consequence of asymmetric disc loading on a twin-engine aircraft in normal cruise (left) and in the case of an engine failure (right). Note that the rudder must be trimmed at a significantly higher deflection than for the normal operation, in addition to the recommended practice of "banking into the good engine" (see Figure 14-20 and text).

will respond by pressing the right rudder pedal[10] or, more appropriately, by trimming the airplane nose-right using the rudder trim in an attempt to center the Turn and Bank indicator ball. Some aircraft feature a yaw-string attached in front of the windscreen and this provides the pilot with additional help by allowing him to visually assess the severity of the yaw. Note that the drag of each nacelle is denoted by D_{NAC}, and both act over a distance d_{NAC}.

Now consider the right image in Figure 14-19, which shows a scenario in which the left (critical) engine has failed. This situation is referred to as flying with *one engine inoperative* (OEI) and can present a very serious situation to the pilot. An inspection of Figure 14-19 shows that a failure of the left engine results in a substantially greater moment than if the right engine fails. For this reason, the left engine is called the *critical engine*. In other words, an inoperative left engine presents a greater problem to the pilot than a failed right engine. The asymmetric thrust will cause a powerful moment, M, that forces the aircraft to yaw toward the dead engine. This may be compounded by the fact that the propeller may now be windmilling with the associated increase in drag ($\Delta D_{windmill}$) of the dead engine. The drag may be reduced by feathering the propeller (see Section 14.1.3, *Important nomenclature*), which is a mechanical feature offered on most twin-engine aircraft.

The standard procedure in this situation is for the pilot to eliminate the yaw by stepping on the rudder pedal on the same side as the functional engine. This is followed by banking the airplane between 2° and 5° toward the functional engine (see Figure 14-20). Once this has been established, the pilot will "feather" the dead engine.

Let's consider these actions in more detail. Pilots are trained to identify which side the dead engine is by stepping on the rudder pedals. It takes considerably less force to press the pedal on the same side as the dead engine compared to the one on the side of the functional engine.[11] This is caused by the floating tendency of the rudder, but without control inputs the condition causes the airplane to yaw (and bank) toward the dead engine. For the aircraft of Figure 14-19, before the pilots makes corrective control inputs, it will yaw so its right side is windward. The rudder will thus float trailing edge left and this is why the left rudder pedal feels dead.

Banking toward the functional engine establishes a turning tendency that reduces the rudder deflection required to maintain heading. Increasing banking may even eliminate the need for rudder input, but usually at the cost of the already impaired performance. There is a limit to how much banking should be applied. If the bank angle is too great the rate-of-climb could easily be detrimentally affected and could prevent the aircraft from maintaining altitude. As a rule of thumb, in order

---

[10] Pilots use the mnemonic: "step on the rudder that's on the same side as the ball."

[11] Pilots use the mnemonic: *"dead foot, dead engine."*

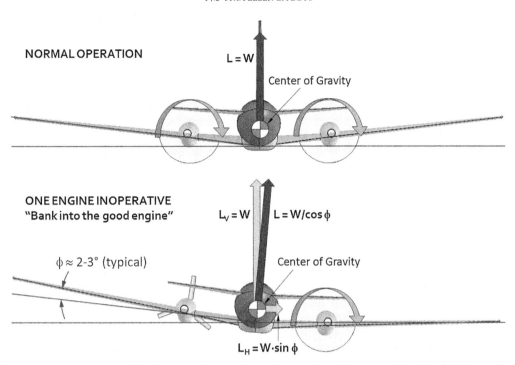

FIGURE 14-20 The standard practice of "banking into the good engine" is used to reduce required rudder deflection when flying with one engine inoperative. However, when slowing down for landing it becomes harder and harder to maintain altitude in the configuration.

to achieve the maximum rate-of-climb in the OEI configuration (V_{YSE}) a bank angle of 2° is recommended if the non-critical engine fails, and 3° if the critical engine fails. A bank angle in excess of 5° almost certainly results in rate-of-descent (loss of altitude) for underpowered twins.

Another situation may confront the pilot operating a twin in an OEI configuration. If the airplane is slowed down below a certain airspeed, control authority will be lost and the rudder will be incapable of opposing the yaw. This airspeed is called *minimum control airspeed*, denoted by V_{MC}. If the airspeed falls below this airspeed, the aircraft will roll upside down, possibly causing an inverted spin, if not impacting the ground first. At any rate, it is a likely fatal scenario for all the occupants. For this reason it is imperative the pilot maintains high enough airspeed at all times, and slows down only after reducing power and pointing the nose down. The minimum control airspeed (V_{MC}) is usually higher than the airplane's stalling speed (V_S) and rotating to take-off should not be performed until after it is reached and exceeded.

This scenario is remedied by installing a right engine/propeller combination that rotates counter-clockwise (or opposite to that of the left engine). Figure 14-21 shows such a configuration. Interestingly, an examination of aviation history reveals this is in fact rarely done. This is called a *counter-rotating* configuration.[12] Albeit safer, the primary drawback of that configuration is that it is more expensive to produce, as logistically it must feature two engines and propellers that are dissimilar because they rotate in opposite directions. An operation of such an aircraft results in greater expenses as engine parts are no longer interchangeable. This can pose a peculiar situation for an operator, which might have parts for the left engine on hand, while none are in stock for a failed right engine. Such scenarios should not be overlooked when selecting an engine configuration as this could severely impact the finances of an operator. Instead, the designer should size the control system and surfaces to handle the critical engine. For this reason it is imperative the aircraft designer is fully aware of this condition.

It is of interest to note that the Wright Flyer featured two counter-rotating propellers, although this was due to the nature of the single engine drive train. It is almost certain that the notion of asymmetric thrust would not have entered the minds of the pioneers of aviation at the time. Among other aircraft that feature this option are the De Havilland DH-103 Hornet, which was a derivative of the DH-98 Mosquito. The Lockheed P-38 Lightning featured counter-rotating propellers, although, surprisingly, both propellers rotated inversely of what is shown in Figure 14-21. This suggests that a

---

[12]Note that it differs from *contra-rotating* propellers, which consist of two propellers mounted to the same engine and rotating in the opposite direction. Such propellers are beyond the scope of this book.

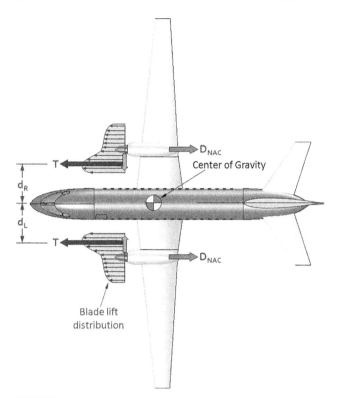

FIGURE 14-21 An aircraft with counter-rotating propellers as shown here does not have a critical engine and is a safer design, albeit operationally more expensive.

cancellation of propeller torque was driving that design decision and not asymmetric yaw. An identical arrangement is found on the Heinkel He-177 Greif. Among other aircraft is a series of Piper aircraft, in order of models, the Piper PA-31 Navajo Chieftain, PA-34 Seneca, PA-39 Twin Commanche, the cancelled PA-40 Arapaho, and the PA-44 Seminole. Other aircraft include the Cessna T303 Crusader, Beech Model 76 Duchess, and Diamond DA-42 Twinstar, all of which feature counter-rotating piston engines. All of these are designed with the correct propeller rotation shown in Figure 14-21.

It is interesting to note that large airplanes typically don't feature propellers that rotate in opposite direction. Even large aircraft, such as the four-engine Lockheed P-3C Orion and C-130 Hercules, feature propellers that all rotate in a clockwise direction. This is an example of a design decision that focuses on reducing maintenance costs in lieu of safety. Of course, for the above aircraft, an easy counter-argument is to be made that safety is in fact not compromised, but maintained by requiring rigorous pilot training, as well as providing hydraulically actuated control surfaces, and very powerful engines — an argument that is hard to dispute. Many larger aircraft provide a *rudder boost system*; a pneumatically powered actuators that force the rudder to the proper deflection angle to counter the asymmetry [4]. Such systems detect which engine has failed and react accordingly — providing substantial safety to the operation of the aircraft. In addition, the operational history of these aircraft supports this notion — there are not many accidents in current times that can be attributed to asymmetric thrust on larger propeller-powered aircraft. However, for small twins this is not the case, as will now be demonstrated.

It is of great importance that the designer understands the implications of selecting a twin-engine configuration, in particular if the engine power is relatively low. In addition to increased acquisition and maintenance cost, the selection has important safety and performance implications associated with it. From a development standpoint, such configurations generate more drag than a single-engine aircraft of the same power (assuming a conventional on-wing engine layout). This results from the presence of two nacelles on the wing in addition to the fuselage. This impact on drag can be reduced by a twin-engine configuration such as that of the Cessna 337 Skymaster and Adam A500 aircraft. However, what is more serious is the performance degradation that results if one engine fails. If a spacious cabin and high cruising speed are desired, a solution such as is offered by recent small GA aircraft should be considered. Aircraft such as the French SOCATA TBM-700 and TBM-850 and the Swiss Pilatus PC-12 are powered by a single Pratt & Whitney PT-6 turbine engine and are capable of cruising speeds in excess of 300 KTAS at 30,000 ft. This beats any of the above twin-engine piston aircraft (excluding the P-38).

To discuss the effect of engine power on the performance of a generic twin-engine piston aircraft consider Figure 14-22. The aircraft portrayed is in the small Piper Apache size class of aircraft. The figure shows that in normal operation, the aircraft climbs 1478 fpm at S-L and offers a maximum speed of 173 KCAS. Upon losing an engine the best rate of climb will drop to 239 fpm and its maximum speed to 97 KCAS. The analysis assumes a drag degradation associated with maintaining a straight heading with OEI by adding 50 drag counts ($\Delta C_{D\ OEI}$) to the original minimum drag. This is reasonable based on the airplane being flown cross-controlled in a yawed configuration. The problem with this turn of events is that the low rate of climb does not allow a lot of room for mistakes. In light of the low-power sensitivity, an inexperienced or frantic pilot may easily maneuver the airplane off the peak conditions, causing it to lose altitude.

14.2.6 Blockage Effects

There are two kinds of blockage effect. First is a body placed in the streamtube ahead of the propeller disc (a pusher configuration). The other is a body placed in the streamtube behind the disc (tractor configuration).

14.2 PROPELLER EFFECTS

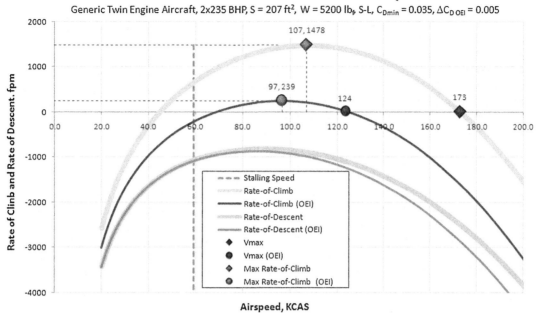

FIGURE 14-22 A performance degradation resulting from an engine-out scenario on a Piper Apache class aircraft.

An idealized streamtube going through a propeller disc is shown in Figure 14-46. Placing an obstruction such as a fuselage (or a nacelle) into the streamtube will distort it and prevent the formation of a proper vena contracta. This will reduce the acceleration of the flow in the streamtube and, thus, the resulting thrust. Figure 14-23 shows an ideal ("Unblocked") streamtube superimposed on what the actual ("Blocked") streamtube might look like.

An estimation of the impact of blockage effects requires the flow around the obstruction to be incorporated into the wake model (for instance, see the wake of Section 14.5, *Rankine-Froude momentum theory*). The most reliable approach (excluding empirical methods) is to use the method of superposition depicted in Figure 14-24. A version of this approach is detailed in NACA TM-492, *Influence of Fuselage on Propeller Design*, and NASA CR-4199, *An Analysis for High Speed*

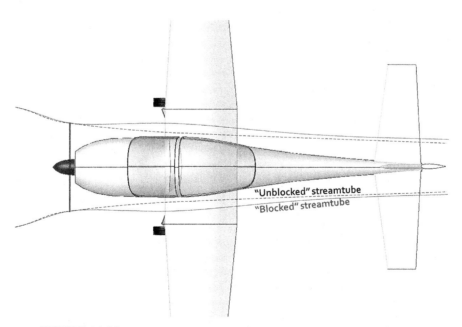

FIGURE 14-23 The difference between "blocked" and "unblocked" propwash.

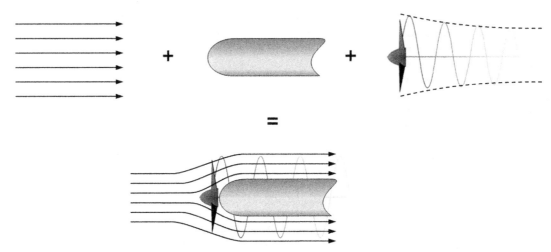

FIGURE 14-24 The philosophy of modeling blockage effects (based on NASA CR-4199).

Propeller-Nacelle Aerodynamic Performance Prediction. The mathematics of these methods are beyond the scope of this book. The manufacturers of propellers do consider blockage effects in the design of their propellers, although the methods used are usually proprietary. The aircraft designer can help mitigate blockage problems by considering a hub extension for piston-engine aircraft and select the largest spinner possible to cover the inboard and least efficient portion of the blade. This allows the cowling or nacelle to be sculpted in a manner that provides less obstruction to the flow. Ideally, the cowling or nacelle should be designed to be as axi-symmetric as possible, as this will yield higher installed propeller efficiency.

14.2.7 Hub and Tip Effects

The presence of the hub (root) and the tip of the blade will have a large effect on the distribution of section lift coefficients, as this is where lift must necessarily go to zero. The blade of a wind-turbine is shown in Figure 14-25 (hub is omitted from the view), and shows how the lift diminishes near the root (left) and tip (right). The effect is less noticeable near the root, because the

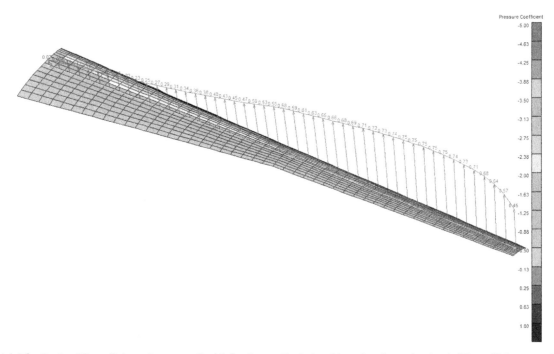

FIGURE 14-25 Section lift coefficients for a propeller blade of a small wind turbine, showing reduction in lift coefficient near the root (left) and the tip (right).

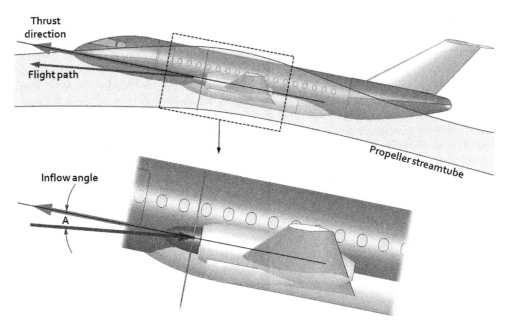

FIGURE 14-26 A flight condition (climb) that generates a skewed wake (upper image) and the definition of the inflow angle A (lower image).

radial speed of the air is less there. However, the tip shows how the section lift coefficients drop rapidly. Naturally, the twist of the blade and camber of the airfoil will influence this as well, but the figure illustrates that the phenomenon will affect thrust. For this reason, hub and tip effects must be taken into account when analyzing the thrust of the propeller. The so-called *Prandtl's tip and hub loss correction* is presented in Section 14.6.4, *Step-by-step: Prandtl's tip and hub loss corrections*, where it is used to correct analysis made by the *blade element method*. The correction method uses the geometric features of the propeller to approximate the reduction of the section lift coefficients and is relatively simple to apply.

14.2.8 Effects of High Tip Speed

The primary effect of a high tip speed is the risk of the formation of shock waves on the blade, which leads to a reduction in the propeller efficiency, increase in torque (requires more power to turn it), and an increase in the noise emitted by the propeller. The pitch, thickness, and curvature of the blade airfoil accelerate air well beyond what the tip speed might indicate. These effects are best avoided by selecting rotational tip speeds in the $M \approx 0.6$ range for wooden propellers and in the $M \approx 0.75-0.8$ range for metal and composite propellers.

14.2.9 Skewed Wake Effects — $A \cdot q$ Loads

A skewed wake is one for which the streamtube inflow through the propeller disc is not normal to its plane (see Figure 14-26). Such a wake is encountered any time a propeller-powered aircraft climbs or yaws. In general, there is no significant effect on the propeller performance, and usually the effect is most severe during transient maneuvers. However, the skewed wake has important effects on noise and propeller loads, especially for turboprops. Let's consider these effects in some detail.

Consider the airplane in Figure 14-26, which presents a condition in which the propeller thrust line is not aligned with the flight path. Consequently, the streamtube is at an angle A to the thrust line (or normal to the propeller disc plane). As the propeller rotates at an angular speed Ω, each down-turning blade will encounter an increase in its *AOA* and each up-turning blade will see a reduction in its *AOA*, as explained in the discussion about the normal force. This will load up each blade and subject it to cyclic loading at the frequency described by Equation (14-8). This is commonly referred to as 1P cyclic loading, where 1P stands for *once per revolution*. A byproduct of this is increased noise.

As stated above, the phenomena increases the propeller's 1P loads, which is the main source of loads for a turboprop. These loads are directly proportional to the product of the angle A and the dynamic pressure of the free-stream, q. For this reason they are often referred to as "$A \cdot q$" loads. For piston-propeller configurations, the primary load of the propeller is caused by the sudden acceleration resulting from the shock of the cylinders firing and the subsequent torsional deformation of the crankshaft. These loads are much greater than the $A \cdot q$ loads.

However, for turboprops (and for electric aircraft) the $A \cdot q$ loads are critical. For this reason, it is imperative the designer of turboprop and electric aircraft performs careful mission performance analyses to determine the *AOA* of the vehicle at a representative cruise condition and then align the thrustline so it is mostly parallel to the flight path (the inflow angle $A = 0°$). Clearly, this cannot be achieved over the entire airspeed range, so propeller manufacturers recommend a "balanced" approach. That is, the product $A \cdot q$ is determined at the maximum and minimum operating weights (see Figure 14-27). Then, the thrustline is aligned for zero inflow angle between the two minimums, guaranteeing the least $A \cdot q$ loads possible. Of course low load is desirable: the less the load, the lighter the propeller. Failure to follow these guidelines may result in the design failing article *14 CFR 23.907 Propeller vibration and fatigue*, which may cause a major redesign effort, with the associated development costs and program delays.

14.2.10 Propeller Noise

Noise associated with the operation of aircraft remains a topic of (often) hotly contested debate in society. One consequence of this debate has been a substantial reduction in aircraft noise over the last 25 years or so, in particular for jet aircraft. This has been achieved by a combination of new regulatory policies and increased research effort. Noise pollution generated by propellers has also been brought to more tolerable levels, although many aircraft types, primarily high-performance piston aircraft, still feature engines that operate at high RPMs at full power, resulting in objectionable noise pollution. Like all other aspects of aircraft design, noise level is a tradeoff with cost, weight, performance and other parameters. What remains is that noise pollution from propellers can be significant and should be seriously considered by the designer from the start of the design project. It is best to work with a propeller manufacturer and use their expertise to design and select propellers that are both quiet and efficient. Current manufacturers of propellers have responded with propeller blade designs that feature airfoils of lower thickness-to-chord ratio and swept planform geometry near the tip. An example of such a design is the scimitar-shaped propeller in Figure 14-28. Such blades are intended to reduce the magnitude of low-pressure peaks and delay the onset of normal shocks that increases the noise. Additionally, the trend has been toward an increased number of blades, as this allows the propeller diameter to be reduced, which, in turn, reduces the tip airspeed. Engine operating RPMs have also been reduced.

Propeller aircraft must comply with the requirements of 14 CFR Part 36 Appendix G [5]. Among a number of stipulations, this requires the aircraft to fly over a microphone placed some 8200 ft (2500 m) from brake release and the noise to be measured. The measured noise level must not exceed that shown in Figure 14-29.

The aircraft designer should not forget that two aircraft that use exactly the same engine and propeller combination can have two different measured noise

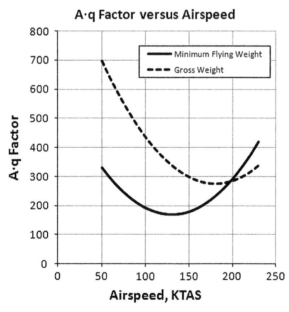

FIGURE 14-27 Skewed wakes at extreme weights reveal two distinct minimums — a compromise between the two should be made.

FIGURE 14-28 A modern three-bladed scimitar shaped propeller. *(courtesy of Hartzell Propeller Inc.)*

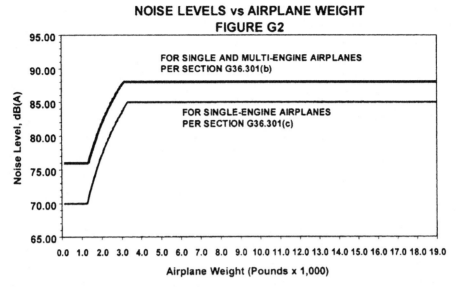

FIGURE 14-29 A graph from www.faa.gov showing the current noise level limitations for propeller aircraft (G36.301 Aircraft noise limits).

levels. All other things being equal, the aircraft with the larger wing area and wing span will lift off using a shorter ground run, and it will climb faster at a lower airspeed (V_Y; see Chapter 14, *Performance - climb*), resulting in a steeper climb gradient. In short, it will climb to a higher altitude in the given distance of 8200 ft, resulting in a lower measured noise level. If there is a concern the aircraft may not or marginally comply with 14 CFR Part 36, the designer should perform a sensitivity study of wing geometry and noise levels.

Reference [6] presents a simple method to theoretically predict the near-field and far-field noise of propellers.

14.3 PROPERTIES AND SELECTION OF THE PROPELLER

A number of issues must be considered during the selection of a propeller for a new aircraft design. The designer should consult the propeller manufacturer early on to help with the selection. Usually this process involves the designer providing the propeller manufacturer with detailed information about the new (or modified) aircraft. The propeller manufacturers generally have some form of design specification that the aircraft designer can complete to efficiently transmit all of the relevant information to the propeller manufacturer. This information is used by the propeller manufacturer to select an existing design, or in some cases a new blade design is warranted.

14.3.1 Tips for Selecting a Suitable Propeller

Although the aircraft designer should benefit from the experience of the propeller manufacturer, he should also be armed with a number of parameters. Among those are:

(1) An estimate of the propeller diameter. Section 14.3.2, *Rapid estimation of required prop diameter*, provides methods that help estimate the diameter of several types of propellers.

(2) An estimate of the propeller pitch. Section 14.3.3, *Rapid estimation of required propeller pitch*, provides methods that help estimate the required pitch for a fixed-pitch propeller.

(3) The airplane's target performance: requirements for take-off and climb performance, mission cruising speed and altitude strongly affect the selection of the propeller type. Are these best met using a constant-speed propeller, or will a fixed-pitch design do? Traditionally, if the engine power is greater than 180—200 BHP, and the cruising speed is in excess of 130—140 KTAS, a constant-speed propeller is recommended.

(4) Engine characteristics, such as the rated BHP, operational RPM, and torsional vibration play an important role in the selection of the propeller diameter and number of blades. The installation of the engine on the airframe may cause detrimental interferences between components of the airplane and the propeller blades. The shape of the nacelle influences the installation as well and sometimes requires a hub extension so that a more aerodynamically "friendly" external geometry can be designed. Additionally, the presence of a body in front or behind the propeller distorts the flow in the streamtube going through the propeller and will affect its efficiency. Therefore, the entire configuration of the aircraft must be considered. Is the propeller for a single or a multiengine aircraft? Shall it rotate clockwise or counter-clockwise? It is a tractor or a pusher? Moreover, a number of systems are required to operate the propeller. For instance, a de-ice system may be required for flight into known icing, a propeller pitch control, and possibly a gear-reduction drive. And finally, there is always a concern for cost, weight, noise, and efficiency. In the addition to the above, the selection process should also consider the following considerations:

- Tip (refers to rotational and not helical) speed should be as high as possible, but there is an upper limit. Too high a tip speed increases noise and reduces efficiency (via shock formation).
- Rotational tip speeds for wooden propellers[13] are in the $M \approx 0.6$ range.
- Rotational tip speeds for metal propellers[14] are in the $M \approx 0.75-0.8$ range.
- Rotational tip speeds for advance composite[15] propellers are in the $M \approx 0.75-0.8$ range.
- If the engine RPM is too high a gear-reduction drive may be required to slow it down.
- High rotational speed requires fewer blades or a smaller blade area to generate thrust.
- High-powered engines whose rotational rate limits the propeller radius need more blades to help convert engine power into thrust.
- Too large a diameter may result in high tip speeds and noise, but can also present some ground proximity problems.
- Weight — a three-bladed prop can be of a smaller diameter, but three blades weigh more than two blades.

When considering thrust per number of blades, the fewer the blades the more efficient is the propeller. However, this should not be taken to mean that a prop with a large number of blades can't have propeller efficiency equal to or higher than a comparable two-bladed propeller.

14.3.2 Rapid Estimation of Required Prop Diameter

The conceptual design of propeller-powered aircraft often requires the prop diameter to be quickly estimated. This is crucial in determining various characteristics, ranging from thrust for performance analysis to propeller noise and geometric constraints, such as the likelihood of a ground-strike. This section considers a couple of such rapid estimation methods. Stinton [3] presents a method to calculate the diameter of *wooden propellers in inches*, based on the intended cruising speed in KTAS, engine power, and RPM.

Two-bladed wooden propeller:

$$D = 10000 \cdot \sqrt[4]{\frac{P_{BHP}}{53.5 \times RPM^2 \times V_{TAS}}} \text{ (in inches)} \quad (14\text{-}17)$$

Three-bladed wooden propeller:

$$D = 10000 \cdot \sqrt[4]{\frac{P_{BHP}}{75.8 \times RPM^2 \times V_{TAS}}} \text{ (in inches)} \quad (14\text{-}18)$$

Four-bladed wooden propeller:

$$D = 10000 \cdot \sqrt[4]{\frac{P_{BHP}}{111 \times RPM^2 \times V_{TAS}}} \text{ (in inches)} \quad (14\text{-}19)$$

Derivation of Equation (14-18)

The source document does not derive Equations (14-17) or (14-19), but states that values for a three-bladed propeller can be obtained by interpolating between the two expressions. This leads to the following closed-form equation for the closed three-bladed propeller:

$$D = \frac{10000 \cdot \sqrt[4]{\frac{P_{BHP}}{53.5 \times RPM^2 \times V_{TAS}}} + 10000 \cdot \sqrt[4]{\frac{P_{BHP}}{111 \times RPM^2 \times V_{TAS}}}}{2}$$

Manipulating algebraically leads to:

$$D = 10000 \cdot \left[\left(\frac{P_{BHP}}{53.5 \times RPM^2 \times V_{TAS} \times 16} \right)^{0.25} + \left(\frac{P_{BHP}}{111 \times RPM^2 \times V_{TAS} \times 16} \right)^{0.25} \right]$$

---

[13] Refers to propellers whose structural core is wood. It is possible to buy wood propellers that have a thin, protective composite (usually fiberglass) coating and sheet-metal leading edges and are marketed as "composite." While true in the literal sense, the structural core of the blades is still wood and, thus, they should not be misclassified as an advanced composite propeller. These wood-composite propellers have thicker tips and trailing edges just like the classic varnished wood blades. Just think: if the composite shield is removed you'll still have a solid propeller behind it capable of reacting all the applied aerodynamic and inertia loads. If the wooden core is removed you'll have a thin layer of composite protection, unable to react the same.

[14] Refers to traditional forged-aluminum blades.

[15] Refers to typical carbon/epoxy blades with a metal leading edge erosion shield.

$$D = 10000 \cdot \frac{(P_{BHP})^{0.25}}{(RPM^2 \times V_{TAS})^{0.25}}$$

$$\times \left[\frac{1}{(53.5 \times 16)^{0.25}} + \frac{1}{(111 \times 16)^{0.25}} \right]$$

$$= 10000 \cdot \frac{(P_{BHP})^{0.25}}{(RPM^2 \times V_{TAS})^{0.25}} \left[\frac{1}{75.8^{0.25}} \right]$$

$$= 10000 \cdot \sqrt[4]{\frac{P_{BHP}}{75.8 \times RPM^2 \times V_{TAS}}}$$

QED

EXAMPLE 14-2

The Aerotrek A240 Light-Sport Aircraft is equipped with a 62.2 inch diameter three-bladed composite propeller. When powered by an 80 BHP Rotax 912UL engine it is capable of flying at 104 KTAS at 75% power. Calculate the recommended diameter of two-, three-, and four-bladed wooden propellers for this airplane, intended to rotate at 2300 RPM, and compare to the actual propeller diameter. What are the corresponding tip speeds?

Solution

Two-bladed propeller:

$$D = 10000 \cdot \sqrt[4]{\frac{P_{BHP}}{53.5 \times RPM^2 \times V_{TAS}}}$$

$$= 10000 \cdot \sqrt[4]{\frac{0.75 \times 80}{53.5 \times 2300^2 \times 104}} = 67.2 \text{ inches}$$

Rotation speed in terms of radians per second:

$$\Omega = 2\pi \left(\frac{RPM}{60} \right) = 2\pi \left(\frac{2300}{60} \right) = 240.9 \text{ rad/s}$$

Tip speed:

$$V_{tip} = \Omega \times R = 240.9 \times \frac{67.2/12}{2} = 675 \text{ ft/s}$$

Three-bladed propeller:

$$D = 10000 \cdot \sqrt[4]{\frac{P_{BHP}}{75.8 \times RPM^2 \times V_{TAS}}}$$

$$= 10000 \cdot \sqrt[4]{\frac{0.75 \times 80}{75.8 \times 2300^2 \times 104}} = 61.6 \text{ inches}$$

Tip speed:

$$V_{tip} = \Omega \times R = 240.9 \times \frac{61.6/12}{2} = 618 \text{ ft/s}$$

Four-bladed propeller:

$$D = 10000 \cdot \sqrt[4]{\frac{P_{BHP}}{111 \times RPM^2 \times V_{TAS}}}$$

$$= 10000 \cdot \sqrt[4]{\frac{0.75 \times 80}{111 \times 2300^2 \times 104}} = 56.0 \text{ inches}$$

Tip speed:

$$V_{tip} = \Omega \times R = 240.9 \times \frac{56.0/12}{2} = 562 \text{ ft/s}$$

Note that the result for the three-bladed propeller can also be obtained from $(67.2 + 56.0)/2 = 61.6$ inches.

The same source states that the diameter of three-bladed metal propellers in inches can be obtained from the following expressions.

Two-bladed metal propellers:

$$D = 22 \sqrt[4]{P_{BHP}} \text{ (in inches)} \quad (14\text{-}20)$$

Three-bladed metal propellers:

$$D = 18 \sqrt[4]{P_{BHP}} \text{ (in inches)} \quad (14\text{-}21)$$

Raymer [7] presents the following expression, independent of propeller material, which he states are modified from those of Stinton. The corresponding propeller diameter in ft or meters depending on K_p:

Propeller diameter:

$$D = K_p \sqrt[4]{P_{BHP}} \quad (14\text{-}22)$$

where the factor K_p is obtained from Table 14-2.

TABLE 14-2 Factor K_p for Typical Propeller Types

Type of Propeller	K_p for P in BHP and D in Inches	K_p for P in kW and D in m
Two-bladed	20.4	0.56
Three-bladed	19.2	0.52
Four or more blades	18.0	0.49

14.3.3 Rapid Estimation of Required Propeller Pitch

For a fixed-pitched propeller, the required pitch can be determined from Tables 14-3, 14-4, and 14-5. The first table represents the ideal, but physically impossible propeller with an efficiency of 100%, while the other two show pitch for 90% and 80% efficient respectively. The table is assembled using Equation (14-12) and is read as shown by the shaded cells. A 100% efficient propeller rotating at 2700 RPM must feature a 56 inch pitch in order to achieve 120 KTAS. The tables are intended to give the designer an idea of the propeller pitch required.

14.3.4 Estimation of Required Propeller Efficiency

Once a trustworthy drag model of an aircraft has been prepared, it is appropriate to evaluate the type of propeller that will be required based on its required propeller efficiency. This section shows how to do this in an effective manner. This involves the preparation of airspeed-thrust (see Figure 14-30) or airspeed-power (see Figure 14-31) maps. It does not matter which one is prepared, as long as one of them is. The maps show thrust and power in terms of propeller efficiency isobars. Then, important desirable performance characteristics can be added on a need-to basis. The map shows at a glance the propeller efficiency required to achieve those characteristics. It is easiest to explain such maps using an example. Here the SR22 is the obvious choice.

First consider Figure 14-30. The isobars are plotted using Equation (14-38) for constant values of η_p. The static thrust was calculated using Equation (14-63). Then the drag force is plotted based on the available drag model. The drag model for the SR22 was extracted using POH data in Example 15-18. It is then used to plot the drag force curve shown in the graph. The ROC curve is based on solving Equation (18-17) for thrust required to achieve a given ROC:

$$ROC = 60\left(\frac{TV - DV}{W}\right) \quad \Rightarrow \quad T_{REQ} = D + \frac{ROC}{60V}W$$

Here 1300 fpm was selected, as this is close to the capability of the SR22 at gross weight at S-L at $V = 91$ KCAS (recall that at S-L KCAS and KTAS are equal). This way, it is possible to explore the capability of its propeller. The graph of Figure 14-30 shows that a prop efficiency of around 0.85 is required to achieve a top speed of 186 KTAS at S-L. Similarly, in order to achieve the ROC at V_Y, the prop efficiency must be around 0.64.

TABLE 14-3 Propeller Pitch Table for 100% Efficient Propeller

		INTENDED CRUISING AIRSPEED in KTAS ($\eta_p = 1.00$)													
		50	60	70	80	90	100	110	120	130	140	150	160	170	180	190
RPM	2000	31	38	44	50	56	63	69	75	81	88	94	100	106	113	119
	2100	30	36	42	48	54	60	66	72	77	83	89	95	101	107	113
	2200	28	34	40	46	51	57	63	68	74	80	85	91	97	102	108
	2300	27	33	38	44	49	54	60	65	71	76	82	87	92	98	103
	2400	26	31	36	42	47	52	57	63	68	73	78	83	89	94	99
	2500	25	30	35	40	45	50	55	60	65	70	75	80	85	90	95
	2600	24	29	34	39	43	48	53	58	63	67	72	77	82	87	91
	2700	23	28	32	37	42	46	51	56	60	65	70	74	79	83	88
	2800	22	27	31	36	40	45	49	54	58	63	67	72	76	80	85
	2900	22	26	30	35	39	43	47	52	56	60	65	69	73	78	82
	3000	21	25	29	33	38	42	46	50	54	58	63	67	71	75	79
	3100	20	24	28	32	36	40	44	48	52	57	61	65	69	73	77
	3200	20	23	27	31	35	39	43	47	51	55	59	63	66	70	74
	3300	19	23	27	30	34	38	42	46	49	53	57	61	64	68	72
	3400	18	22	26	29	33	37	40	44	48	52	55	59	63	66	70
	3500	18	21	25	29	32	36	39	43	46	50	54	57	61	64	68

TABLE 14-4 Propeller Pitch Table for 90% Efficient Propeller

		\multicolumn{15}{c}{INTENDED CRUISING AIRSPEED in KTAS ($\eta_p = 0.90$)}
		50	60	70	80	90	100	110	120	130	140	150	160	170	180	190
RPM	2000	35	42	49	56	63	70	76	83	90	97	104	111	118	125	119
	2100	33	40	46	53	60	66	73	79	86	93	99	106	113	119	113
	2200	32	38	44	51	57	63	70	76	82	88	95	101	107	114	108
	2300	30	36	42	48	54	60	66	73	79	85	91	97	103	109	103
	2400	29	35	41	46	52	58	64	70	75	81	87	93	98	104	99
	2500	28	33	39	44	50	56	61	67	72	78	83	89	95	100	95
	2600	27	32	37	43	48	53	59	64	70	75	80	86	91	96	91
	2700	26	31	36	41	46	51	57	62	67	72	77	82	88	93	88
	2800	25	30	35	40	45	50	55	60	65	70	74	79	84	89	85
	2900	24	29	34	38	43	48	53	58	62	67	72	77	82	86	82
	3000	23	28	32	37	42	46	51	56	60	65	70	74	79	83	79
	3100	22	27	31	36	40	45	49	54	58	63	67	72	76	81	77
	3200	22	26	30	35	39	43	48	52	56	61	65	70	74	78	74
	3300	21	25	29	34	38	42	46	51	55	59	63	67	72	76	72
	3400	20	25	29	33	37	41	45	49	53	57	61	65	70	74	70
	3500	20	24	28	32	36	40	44	48	52	56	60	64	68	72	68

This sort of information can be very useful to the propeller manufacturer when helping to select or design a practical propeller.

Figure 14-31 is identical to Figure 14-30, except it uses power rather than force to relay the same information. Therefore, the propeller efficiency isobars are horizontal lines, which may make it a tad easier to evaluate the propeller efficiencies. Other than that, it gives exactly the same answers as the airspeed-thrust map.

14.3.5 Advance Ratio

The *advance ratio* is a measure of how far the propeller travels in unit time in terms of its diameter. The distance

TABLE 14-5 Propeller Pitch Table for 80% Efficient Propeller

		\multicolumn{15}{c}{INTENDED CRUISING AIRSPEED in KTAS ($\eta_p = 0.80$)}
		50	60	70	80	90	100	110	120	130	140	150	160	170	180	190
RPM	2000	39	47	55	63	70	78	86	94	102	109	117	125	133	141	149
	2100	37	45	52	60	67	74	82	89	97	104	112	119	127	134	142
	2200	36	43	50	57	64	71	78	85	92	100	107	114	121	128	135
	2300	34	41	48	54	61	68	75	82	88	95	102	109	116	122	129
	2400	33	39	46	52	59	65	72	78	85	91	98	104	111	117	124
	2500	31	38	44	50	56	63	69	75	81	88	94	100	106	113	119
	2600	30	36	42	48	54	60	66	72	78	84	90	96	102	108	114
	2700	29	35	41	46	52	58	64	70	75	81	87	93	98	104	110
	2800	28	34	39	45	50	56	61	67	73	78	84	89	95	101	106
	2900	27	32	38	43	49	54	59	65	70	76	81	86	92	97	102
	3000	26	31	36	42	47	52	57	63	68	73	78	83	89	94	99
	3100	25	30	35	40	45	50	56	61	66	71	76	81	86	91	96
	3200	24	29	34	39	44	49	54	59	64	68	73	78	83	88	93
	3300	24	28	33	38	43	47	52	57	62	66	71	76	81	85	90
	3400	23	28	32	37	41	46	51	55	60	64	69	74	78	83	87
	3500	22	27	31	36	40	45	49	54	58	63	67	72	76	80	85

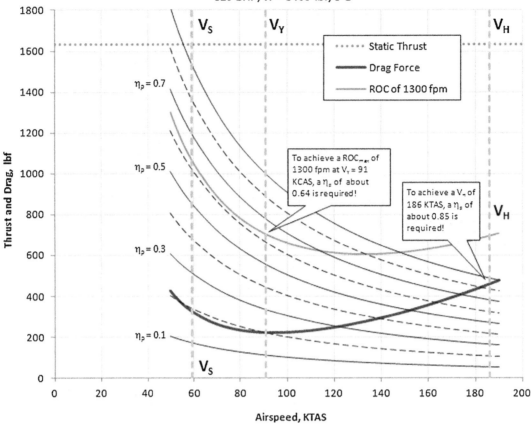

FIGURE 14-30 An airspeed-thrust map shows the propeller efficiency required to meet given performance characteristics in terms of thrust.

the prop travels per second is the vehicle's speed through the medium and is given by V_0. For instance, if we measure an advance ratio of 0.5 for a 10 ft diameter propeller rotating at 600 RPM it means the propeller (i.e. the vehicle it is propelling) is moving at a forward speed of 50 ft/s.

Advance ratio:

$$J = \frac{V_0}{nD} = \frac{60 \cdot V_0}{RPM \cdot D} \quad (14\text{-}23)$$

EXAMPLE 14-3

The SR22 is equipped with a 76 inch diameter propeller. If the airplane is observed to fly at 185 KTAS at 8000 ft with the propeller rotating at 2700 RPM, what are the helix angle, blade rotational and helical tip speeds, Mach number, and advance ratio?

Solution

Tip radius:

$$R = \left(\frac{76/2}{12}\right) = 3.167 \text{ ft}$$

Angular velocity:

$$\Omega = 2\pi \left(\frac{RPM}{60}\right) = 2\pi \left(\frac{2700}{60}\right) = 282.7 \text{ rad/s}$$

Helix angle:

$$\tan \phi = \frac{\pi R \; RPM}{30 \cdot V} = \frac{\pi (3.167)\;(2700)}{30 \cdot (185 \times 1.688)} = 2.867$$

$$\Rightarrow \phi = \tan^{-1}(2.867) = 70.8°$$

Propeller rotational tip speed calculated using Equation (14-5):

$$V_{ROT} = \Omega \cdot R = (2827) \cdot (3.167) = 895.3 \text{ ft/s}$$

EXAMPLE 14-3 (cont'd)

Propeller tip speed:

$$V_{tip} = \sqrt{V_0^2 + \left(\frac{\pi \cdot RPM \cdot D}{60}\right)^2}$$

$$= \sqrt{(185 \times 1.688)^2 + \left(\frac{\pi \cdot 2700 \cdot (76/12)}{60}\right)^2}$$

$$= 948 \text{ ft/s}$$

Tip Mach number:

$$M_{tip} = \frac{948}{1116\sqrt{1 - 0.0000068753 \times 8000}} = 0.874$$

Advance ratio:

$$J = \frac{V_0}{nD} = \frac{60 \cdot V_0}{RPM \cdot D} = \frac{60 \cdot (185 \times 1.688)}{2700 \cdot (76/12)} = 1.096$$

14.3.6 Definition of Activity Factor

The *activity factor* (*AF*) represents the power absorption capability of all blade elements. Small aircraft have an *AF* ranging from 80-105, while larger aircraft have an *AF* around 130-170. The *AF* is used when reading propeller efficiency maps, which are special contour maps prepared by the propeller manufacturer that allow propeller efficiency to be read as a function of the advance ratio and power coefficient (see Section 14.3.7, *Definition of power- and thrust-related coefficients*):

Blade activity factor:

$$AF = \frac{100\,000}{16} \int_{0.15}^{1.0} \left(\frac{c}{D}\right)\left(\frac{r}{R}\right)^3 d\left(\frac{r}{R}\right) \qquad (14\text{-}24)$$

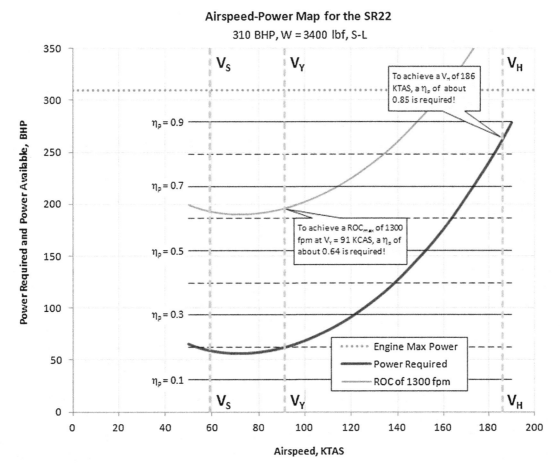

FIGURE 14-31 An airspeed-power map shows the propeller efficiency required to meet given performance characteristics in terms of power.

614　　14. THE ANATOMY OF THE PROPELLER

Total *AF*:

$$AF_{TOT} = N_B \times AF \quad (14\text{-}25)$$

where

R = blade radius.
N_B = number of blades

See the list of variables at the end of this chapter for other variable details.

EXAMPLE 14-4

What is the activity factor of each blade of the airplane of Example 14-3, whose diameter is 76 inches and chord is a constant 5 inches.

Solution

$$AF = \frac{100\ 000}{16} \int_{0.15}^{1.0} \left(\frac{c}{D}\right)\left(\frac{r}{R}\right)^3 d\left(\frac{r}{R}\right)$$

$$= \frac{100\ 000}{16}\left(\frac{5}{76}\right) \int_{0.15}^{1.0} x^3 dx = 411.2 \left[\frac{x^4}{4}\right]_{0.15}^{1.0}$$

$$AF = 411.2 \left[\frac{1 - 0.15^4}{4}\right] = 102.7$$

14.3.7 Definition of Power- and Thrust-Related Coefficients

The following concepts are definitions:

Power coefficient:

$$C_P = \frac{P}{\rho n^3 D^5} = \frac{550 \times P_{BHP}}{\rho \left(\frac{RPM}{60}\right)^3 D^5} = \frac{118\ 800\ 000 \times P_{BHP}}{\rho \cdot RPM^3 \cdot D^5}$$

$$(14\text{-}26)$$

Thrust coefficient:

$$C_T = \frac{T}{\rho n^2 D^4} = \frac{3600 \cdot T}{\rho \cdot RPM^2 D^4} \quad (14\text{-}27)$$

Torque coefficient:

$$C_Q = \frac{Q}{\rho n^2 D^5} = \frac{3600 \cdot Q}{\rho \cdot RPM^2 \cdot D^5} = \frac{C_P}{2\pi} \quad (14\text{-}28)$$

By combining Equations (14-26) and (14-28) we get:

Power-Torque relation:

$$C_Q = \frac{Q}{\rho n^2 D^5} = \frac{C_P}{2\pi} = \frac{P/\rho n^3 D^5}{2\pi} \Rightarrow P = 2\pi n Q$$

$$(14\text{-}29)$$

See the list of variables at the end of this chapter for variable details.

EXAMPLE 14-5

Compute C_P and C_Q for the airplane in Example 14-3 for the same condition, if it is known its engine is generating 225 BHP.

Solution

Power coefficient:

$$C_P = \frac{118\ 800\ 000 \times P_{BHP}}{\rho \cdot RPM^3 \cdot D^5}$$

$$= \frac{118\ 800\ 000 \times 225}{(0.002378)(2700)^3(76/12)^5} = 0.05604$$

Torque coefficient:

$$C_Q = \frac{Q}{\rho n^2 D^5} = \frac{C_P}{2\pi} = \frac{0.05604}{2\pi} = 0.008919$$

EXAMPLE 14-6

Estimate the required propeller diameter for a light airplane designed to operate around 120 KTAS, assuming:

(a) A two-bladed wooden prop, if the maximum expected engine power is 80 BHP and max engine RPM is 5800 RPM with a 2.27:1 gear reduction ratio.
(b) A three-bladed metal prop.
(c) Assume a max tip speed of $M = 0.75$ at S-L.

Solution

(a) First determine RPM delivered to propeller:

$$RPM = \frac{RPM_{engine}}{GR} = \frac{5800}{2.27} = 2555 \text{ RPM}$$

Then, calculate the diameter of a two-bladed wooden propeller:

$$D = 10000 \cdot \sqrt[4]{\frac{P_{BHP}}{53.5 \times RPM^2 \times V_{TAS}}}$$

$$= 10000 \cdot \sqrt[4]{\frac{80}{53.5 \times 2555^2 \times 120}} = 66 \text{ inches}$$

(b) A three-bladed metal propeller:

$$D = 18 \sqrt[4]{P_{BHP}} = 18 \sqrt[4]{80} = 54 \text{ inches}$$

(c) Maintaining max tip speed of $M = 0.75$:

$$M_{tip} = \frac{V_{tip}}{1116\sqrt{1 - 0.0000068753H}} < 0.75$$

$$\Rightarrow \frac{V_{tip}}{1116} < 0.75$$

$$\Rightarrow V_{tip} < 0.75 \times 1116 = 837 \text{ ft/s}$$

Therefore, the maximum diameter can be determined as follows:

$$V_{tip} = \sqrt{V^2 + \left(\frac{\pi \cdot RPM \cdot D}{60}\right)^2} < 837 \text{ ft/s}$$

$$\Rightarrow D = \frac{60\sqrt{V_{tip}^2 - V^2}}{\pi \cdot RPM}$$

$$= \frac{60\sqrt{(837)^2 - (120 \times 1.688)^2}}{\pi \cdot 2555}$$

$$= 6.07 \text{ ft (73 inches)}$$

14.3.8 Effect of Number of Blades on Power

The number of blades is fundamental to how the engine power is converted into propulsive power. Replacing propellers is a common occurrence in the aviation industry. Sometimes such changes involve a propeller that has a different number of blades than the original prop. For instance, consider a four-bladed propeller as a replacement for a two-bladed one. Assuming such a modification only involves a change in the number of blades, the addition will increase the torque required to swing the new propeller by a factor of two. This can be seen, for instance, in the formulation developed in Section 14.6, *Blade element theory*. However, since this propeller will be mounted on the same engine, the power available per blade will diminish. This will manifest itself in the engine being unable to rotate the propeller at the original RPM. With everything else being equal (e.g. propeller diameter and blade planform geometry being the same) less thrust will be generated and the performance of the airplane will suffer. The remedy is to reduce the torque generated per blade, as this will allow the propeller to be rotated at RPM closer to the original one. The easiest way to accomplish this is to reduce the diameter of the propeller. An advantage of such a modification is less propeller noise and weight. Also, the ground clearance of the prop will increase (assuming the number of blades is increased), rendering this an important option at the discretion of the designer struggling with potential prop strike issues.

It is of interest to estimate what propeller diameter is appropriate when changing the number of blades, regardless of whether the intent is to increase or reduce the number of blades. This can be done by solving the following expression for the new diameter, D_{new}, which is the diameter of the replacement prop:

$$V_0^2 D_{new} + (\pi n k)^2 D_{new}^3$$
$$= \left(V_0^2 D_{old} + (\pi n k)^2 D_{old}^3\right)\left(\frac{N_{B_{old}}}{N_{B_{new}}}\right)\left(\frac{S_{old}}{S_{new}}\right) \quad (14\text{-}30)$$

where

D_{old} = diameter of the old and new propellers, in ft or m
V_0 = intended reference airspeed for the airplane, in ft/s or m/s
n = propeller rotation rate in revolutions per second (rps)
N_{Bold}, N_{Bnew} = number of blades on the old and new propellers, respectively

S_{old}, S_{new} = blade planform area of the old and new propeller, respectively, in ft² or m²
k = fraction spanwise location of the center of pressure on the blade (typically 0.65–0.75)

The expression requires an iterative scheme for the solution. As an example, consider a 78 inch diameter two-bladed propeller used with some versions of the Cirrus SR22. Assume this prop is to be replaced with a three- or four-bladed propeller, whose blade areas are 90% of the original ones. Solving for D_{new} using the above method suggests prop diameters that are 90% and 82% of the original diameter, respectively. The recommended propeller diameter ratios are plotted in the two graphs shown in Figure 14-32. First consider the left graph, which shows how the ratio varies with the intended cruising speed, V_0. It shows that the dependency on the forward speed of the airplane is indeed negligible. The right graphs shows dependency on the blade area ratio and depicts how it, on the other hand, plays a fundamental role in the diameter requirements. Thus, a three-bladed propeller with a blade area ratio of 1.0 should have a diameter about 87% of the original one. A four-bladed propeller should be about 78% of the original diameter. The conclusion is that Equation (14-30) can be easily simplified by ignoring the contribution of V_0, yielding:

$$D_{new} = D_{old} \sqrt[3]{\left(\frac{N_{B_{old}}}{N_{B_{new}}}\right)\left(\frac{S_{old}}{S_{new}}\right)} \quad (14\text{-}31)$$

This expression allows simple arithmetic to take place with acceptable accuracy, as can be seen in Figure 14-32.

Finally, consider the special case in which the propeller chord, c, remains constant. In other words, the blades are reduced in length only. In this case it is possible to relate the blade area ratio to the diameter as shown below:

$$\left.\begin{array}{l} S_{old} = cD_{old}/2 \\ S_{new} = cD_{new}/2 \end{array}\right\} \Rightarrow \frac{S_{old}}{S_{new}} = \frac{cD_{old}/2}{cD_{new}/2} = \frac{D_{old}}{D_{new}}$$

In this case, Equation (14-30) reduces to (note the change in power of the diameters):

$$V_0^2 D_{new}^2 + (\pi n k)^2 D_{new}^4 = \left(V_0^2 D_{old}^2 + (\pi n k)^2 D_{old}^4\right)\left(\frac{N_{B_{old}}}{N_{B_{new}}}\right) \quad (14\text{-}32)$$

The reader should not rely blindly on this analysis method. It is only presented here to give an idea as to how much the prop diameter could change. Propeller replacement should always involve the manufacturer as other details, not included here, may play an important role. Among these are planform shape, modified airfoils, and others. For instance, the TCDS for the Cirrus SR22 [8] reveals it can feature a 78 inch two-bladed or a 76 inch three-bladed prop, a diameter ratio closer to 0.974. An inspection of the propeller blades reveals a dissimilar planform shape.

Derivation of Equation (14-30)

We begin by observing that there is a linear relationship between power absorption and number of blades. If the number of blades is doubled (without changing their dimensions), then the power, P, to rotate the prop at the original condition must be doubled as well (assuming the original propeller diameter and blade geometry). Here, however, we are considering replacing

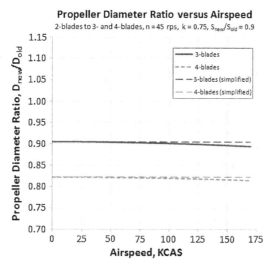
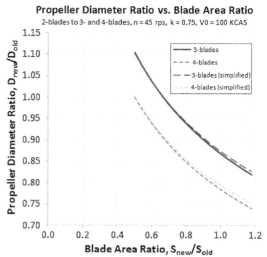

FIGURE 14-32 Influence of airspeed (left) and blade area ratio (right) on the size of a three- and four-bladed propellers intended to replace a two-bladed one.

the propeller on an existing engine. For this reason the power available will be the same before and after the change. Since power and torque are related through Equation (14-29), this implies that the torque generated by the new propeller cannot exceed that of the old prop. The following assumptions are made:

(1) The planform shape of the old and new propellers is the same. This means that the center of pressure of the blades occurs the same distance from the hub.
(2) Same engine, so power before and after is equal.
(3) The spanwise loading on each blade can be replaced by a single point force acting at the center of pressure.

The torque per blade can be considered as the aerodynamic drag force acting on the blade at some distance r from the hub, where r is the center of pressure on the blade. This is shown in Figure 14-33. Assume its location is known and r is therefore given by:

$$r_{old} = kR_{old} = kD_{old}/2 \qquad (i)$$

where k is some constant. The value of r_{new} is obtained in an identical manner. For instance, if $k = 0.5$ the center of pressure is assumed to be at the mid-span of the blade. For real applications k is closer to a value of 0.75. Now the total torque of the old propeller can be written as follows, where F_{old} is the blade drag force acting at the blade's center of pressure:

$$Q_{old} = N_{B_{old}} F_{old} r_{old} = N_{B_{old}} F_{old} kD_{old}/2 \qquad (ii)$$

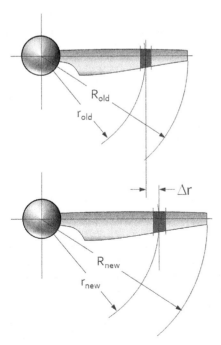

FIGURE 14-33 The geometric difference between two similarly shaped propellers.

Similarly, the total torque of the new prop is given by:

$$Q_{new} = N_{B_{new}} F_{new} r_{new} = N_{B_{new}} F_{new} kD_{new}/2 \qquad (iii)$$

But $Q_{new} = Q_{old}$ because the same engine is swinging the prop; therefore, we can write:

$$Q_{new} = Q_{old} \Rightarrow N_{B_{new}} F_{new} D_{new} = N_{B_{old}} F_{old} D_{old} \qquad (iv)$$

Assume that F is directly related to the area of the blade and the airspeed at the radius of F. This means:

$$F_{old} = \frac{1}{2}\rho V_{old}^2 S_{old} C_{Dold} \quad \text{and} \quad F_{new} = \frac{1}{2}\rho V_{new}^2 S_{new} C_{Dnew}$$
(v)

Assume the drag coefficient of each blade to be constant ($C_{Dold} = C_{Dnew}$) and then divide one equation by the other:

$$\frac{F_{new}}{F_{old}} = \frac{\frac{1}{2}\rho V_{new}^2 S_{new} C_D}{\frac{1}{2}\rho V_{old}^2 S_{old} C_D} \Leftrightarrow F_{new}$$

$$= F_{old}\left(\frac{V_{new}^2}{V_{old}^2}\right)\left(\frac{S_{new}}{S_{old}}\right) \qquad (vi)$$

where $V_{old} = \sqrt{V_0^2 + (\Omega \cdot r)^2} = \sqrt{V_0^2 + (2\pi n r_{old})^2}$ (from Equation (14-9))

$$V_{new} = \sqrt{V_0^2 + (2\pi n r_{new})^2}$$

Inserting this into Equation (vi) and expanding leads to:

$$F_{new} = F_{old}\left(\frac{V_{new}^2}{V_{old}^2}\right)\left(\frac{S_{new}}{S_{old}}\right)$$

$$= F_{old}\left(\frac{V_0^2 + (2\pi n r_{new})^2}{V_0^2 + (2\pi n r_{old})^2}\right)\left(\frac{S_{new}}{S_{old}}\right)$$

Simplifying further yields;

$$F_{new} = F_{old}\left(\frac{V_0^2 + (2\pi nkD_{new}/2)^2}{V_0^2 + (2\pi nkD_{old}/2)^2}\right)\left(\frac{S_{new}}{S_{old}}\right)$$

$$= F_{old}\left(\frac{V_0^2 + (\pi nk)^2 D_{new}^2}{V_0^2 + (\pi nk)^2 D_{old}^2}\right)\left(\frac{S_{new}}{S_{old}}\right)$$

This can now be inserted into Equation (iv) and simplified as follows:

$$N_{B_{new}}\left(\frac{V_0^2 + (\pi nk)^2 D_{new}^2}{V_0^2 + (\pi nk)^2 D_{old}^2}\right)\left(\frac{S_{new}}{S_{old}}\right)D_{new} = N_{B_{old}} D_{old} \qquad (vii)$$

Expansion and subsequent simplification:

$$\left(V_0^2 D_{new} + (\pi nk)^2 D_{new}^3\right)$$
$$= \left(V_0^2 D_{old} + (\pi nk)^2 D_{old}^3\right)\left(\frac{N_{B_{old}}}{N_{B_{new}}}\right)\left(\frac{S_{old}}{S_{new}}\right) \quad \text{(viii)}$$

Consider the special case in which both propellers feature blades of an equal chord. In this case the ratio of the two blade areas can be found from:

$$\left.\begin{array}{l}S_{old} = cD_{old}/2\\ S_{new} = cD_{new}/2\end{array}\right\} \Rightarrow \frac{S_{old}}{S_{new}} = \frac{cD_{old}/2}{cD_{new}/2} = \frac{D_{old}}{D_{new}}$$

Inserting into Equation (viii) leads to:

$$\left(V_0^2 D_{new} + (\pi nk)^2 D_{new}^3\right)$$
$$= \left(V_0^2 D_{old} + (\pi nk)^2 D_{old}^3\right)\left(\frac{N_{B_{old}}}{N_{B_{new}}}\right)\left(\frac{D_{old}}{D_{new}}\right)$$

$$\left(V_0^2 D_{new}^2 + (\pi nk)^2 D_{new}^4\right)$$
$$= \left(V_0^2 D_{old}^2 + (\pi nk)^2 D_{old}^4\right)\left(\frac{N_{B_{old}}}{N_{B_{new}}}\right)$$

QED

14.3.9 Propulsive Efficiency

The power generated by an engine is not completely converted to a thrust force. It is inevitable that some losses occur in the process due to an imperfect conversion process. Consider an aircraft flying at an airspeed V generating thrust T by giving air a backward velocity w. This changes the kinetic energy of the affected air by an amount $\tfrac{1}{2}mw^2$ per unit time[16] and this change must equal the force acting on the mass of air, times its speed, $T \cdot w$. The term $T \cdot w$ is the power consumed (or wasted) in driving the air backwards. The Froude efficiency is defined as the ratio of the power required to propel an object (useful power) to the sum of this power and that which is lost in the conversion:

Froude efficiency:

$$\eta_i = \frac{\text{Thrust Power}}{\text{Thrust Power} + \text{Power lost}} = \frac{TV}{TV + Tw}$$
$$= \frac{V}{V + w} \quad (14\text{-}33)$$

The Froude efficiency represents inviscid losses that result from accelerating the propwash. It is also called the *ideal efficiency* (see Section 14.5.2, *Ideal efficiency*). However, there are also losses due to the viscous drag of the propeller blades. These losses further increase the inefficiency of the power-to-thrust conversion process. This effect is called *viscous profile efficiency* and is denoted by the term η_v. We now define the total propeller efficiency, η_p, as the product of the Froude and profile efficiency as follows:

Propeller efficiency:[17]

$$\eta_p = \eta_i \eta_v \quad (14\text{-}34)$$

A typical maximum value for η_v is about 0.8 to 0.9.

The *propeller efficiency* can also be defined as the fraction of engine power that gets converted into propulsive power (thrust × airspeed) to the total engine output power, P. A common expression of this ratio is:

Propeller efficiency:

$$\eta_p = \frac{TV}{P} = \frac{TV}{550 BHP} = J\frac{C_T}{C_P} \quad (14\text{-}35)$$

Derivation of Equation (14-35)

Using the definitions of the power and thrust coefficients, we write:

$$P = \rho n^3 D^5 C_P \quad \text{and} \quad T = \rho n^2 D^4 C_T$$

Inserting into Equation (14-22) and manipulating leads to Equation (14-23).

$$\eta_p = \frac{TV}{P} = \frac{\rho n^2 D^4 C_T \cdot V}{\rho n^3 D^5 C_P} = \frac{C_T \cdot V}{nDC_P} = \frac{V}{nD}\frac{C_T}{C_P} = J\frac{C_T}{C_P}$$

QED

Propeller Efficiency Map

Sometimes the propeller manufacturer will present the propeller efficiency in a mapped format similar to what is shown in Figure 14-34. In this format, the isobars represent constant values of propeller efficiency. Such maps are defined in terms of a specific propeller blade activity factor and are read as follows. First, determine the advance ratio using Equation (14-23). Then, determine the power coefficient using Equation (14-26). Finally, locate the value of the propeller efficiency using the two previous values. The propeller developer prepares such maps for different prop designs. The designer can review such maps for a specific propeller and decide whether it fits the desired performance of the aircraft.

---

[16] Note that power can be defined as a rate of change of energy, i.e. $P = dE/dt$.

[17] Sometimes casually called prop efficiency.

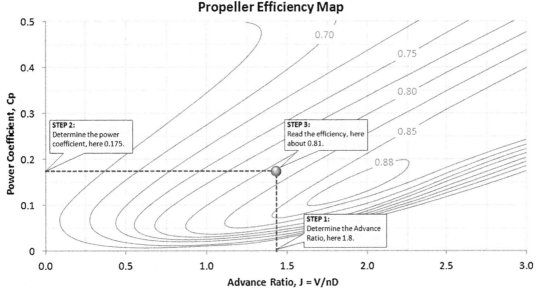

FIGURE 14-34 An idealized propeller efficiency map. Such maps are always drawn for a specific propeller activity factor. Note the island where $\eta_p = 0.88$ is where the propeller should be sized to operate.

It is evident that the propeller efficiency map does not lend itself to modern analysis methods. It is presented here merely for the sake of completeness. It is imperative that the data it represents be converted into a digitized format that allows the efficiency to be extracted using rapid computational approaches, using the modern spreadsheet or other numerical methods.

14.3.10 Moment of Inertia of the Propeller

As has been showed before, the moment of inertia of the propeller is of interest to the aircraft designer. This can be estimated using the fundamental definition for the moment of inertia of a body about a given axis:

$$I = \int_V \rho r^2 dV \qquad (14\text{-}36)$$

The term $\rho \cdot dV$ is the mass of an infinitesimal volume of the propeller blade, located a distance r from the reference axis (i.e. the axis of rotation of the propeller). While the planform of the propeller blade does not always yield an elegant closed-form solution of Equation (14-36), a reasonably accurate estimate of its moment of inertia can be made by idealizing each blade as a thin rod of a uniform cross-sectional area A (see Figure 14-35). If the radius of the prop is R, it weight is W_{prop}, and the number of blades is N_B, then its moment of inertia can be estimated from:

$$I_{prop} = \frac{W_{prop} \cdot R^2}{3g} \qquad (14\text{-}37)$$

where g is the acceleration due to gravity.

Derivation of Equation (14-37)

The mass, m, of the propeller must be close to:

$$m = \rho_{prop} \cdot A \cdot R \cdot N_B \qquad (i)$$

where ρ_{prop} is the density of the material used in the propeller blades. Its mass can also be computed if its weight, W_{prop}, is known, i.e. from $m = W_{prop}/g$. It follows that the constant cross-sectional area is given by:

$$A = \frac{m}{\rho_{prop} \cdot R \cdot N_B} = \frac{W_{prop}}{\rho_{prop} \cdot g \cdot R \cdot N_B} \qquad (ii)$$

Then, using Equation (14-36) we can compute the moment of inertia of a single rod extending from 0 to R:

$$I_{prop} = \int_V \rho r^2 dV = \int_0^R \rho_{prop} r^2 A dr$$

$$= \int_0^R \rho_{prop} r^2 \left(\frac{W_{prop}}{\rho_{prop} \cdot g \cdot R \cdot N_B}\right) dr$$

$$= \int_0^R r^2 \left(\frac{W_{prop}}{g \cdot R \cdot N_B}\right) dr = \left(\frac{W_B}{g \cdot R \cdot N_B}\right)\left[\frac{r^3}{3}\right]_0^R$$

$$= \frac{W_{prop} R^3}{3g \cdot R \cdot N_B} = \frac{W_{prop} R^2}{3g \cdot N_B}$$

This result is for a single blade. If we have N_B blades, this simplifies to:

$$I_{prop} = N_B \times \frac{W_{prop}R^2}{3g \cdot N_B} = \frac{W_{prop}R^2}{3g}$$

QED

EXAMPLE 14-7

To evaluate this result of Equation (14-37), let's compare it to the result using the standard expression for the moment of inertia of a thin rod. To do this determine the moment of inertia of a thin rod of mass $m = 10$ kg and length $2R = 2$ m using each method.

Solution

Using the standard expression of moment of inertia we find that:

$$I = ml^2/12 = (10)(2)^2/12 = 3.333 \text{ kg} \cdot \text{m}^2.$$

To prepare the method above, we note that the weight of the example rod is $10 \times 9.807 = 98.07$ N and its radius is 1 m. Therefore, using Equation (14-37) we get:

$$I_{prop} = W_{prop} \cdot R^2/3g = (98.07) \cdot (1)^2/3(9.807)$$
$$= 3.333 \text{ kg} \cdot \text{m}^2$$

This is precisely what we would expect.

EXAMPLE 14-8

Compute the moment of inertia of a two-bladed aluminum propeller, whose weight is 40 lb$_f$ and diameter is 80 inches.

Solution

$$I_{prop} = \frac{W_{prop} \cdot R^2}{3g} = \frac{40 \cdot (40/12)^2}{3(32.174)} = 4.605 \text{ slugs} \cdot \text{ft}^2$$

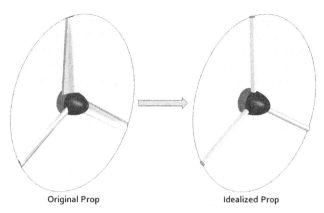

FIGURE 14-35 Idealizing the prop as constant-diameter rods allows for a fast estimation of its moment of inertia.

14.4 DETERMINATION OF PROPELLER THRUST

The conversion of engine power to a propeller thrust force is not nearly as straightforward as that of the jet engine. This is due to the complex interaction between the propeller's rotation speed, forward speed, and geometry. The designer can resort to a number of methods in order to estimate thrust, ranging from the simple to the complex. This section will present useful methods for this task.

14.4.1 Converting Piston BHP to Thrust

The performance analysis of piston engine aircraft requires brake horsepower (BHP) to be converted into thrust.[18] In the UK system, the thrust is calculated using

---

[18]When the crankshaft is connected to a gear reduction drive the BHP is normally referred to as the shaft horsepower, or SHP. SHP is also used to denote the power of turboprops. In the interests of simplicity, the term BHP can mean either; i.e. the power available at the hub of the propeller.

airspeed (in ft/s), engine power (in BHP), P_{BHP}, and propeller efficiency (dimensionless), by solving Equation (14-23) for the thrust as follows:

$$T = \frac{\eta_p P}{V} = \frac{\eta_p \times 550 \times P_{BHP}}{V} \qquad (14\text{-}38)$$

The factor 550 converts the BHP into ft lb$_f$/s. In the SI system the conversion is not necessary as power is given in kW. The expression indicates that as the airspeed V approaches zero the thrust will trend toward infinity. This is physically impossible. What happens in reality can be understood by noting that the magnitude of the velocity through the propeller disc requires a constant flow of air toward it. This, in turn, means V is never zero. Thrust at zero forward airspeed can be estimated using the momentum theory (Section 14.5) or blade element theory (Section 14.6).

Derivation of Equation (14-38)

Using propeller efficiency:

$$\eta_p = \frac{\text{Thrust Power}}{\text{Engine Power}} = \frac{TV}{P} = \frac{TV}{550 \times P_{BHP}}$$

Solving for T yields Equation (14-38).

QED

EXAMPLE 14-9

An airplane is flying at 150 KTAS at a power setting that generates 150 BHP. If the propeller used on this airplane is known to be 80% effective, what is the thrust being generated?

Solution

$$T = \frac{\eta_p \times 550 \times P_{BHP}}{V} = \frac{(0.8) \times 550 \times (150)}{(150 \times 1.688)} = 260.7 \text{ lb}_f$$

14.4.2 Propeller Thrust at Low Airspeeds

This section presents two methods to estimate propeller thrust at low airspeeds, which is essential for the prediction of low-speed operation such as that of T-O performance. Equation (14-38) is extremely useful in the performance analysis of aircraft. However, it presents an important difficulty to the user: it requires the propeller efficiency, η_p, to be known, but, being a function of the airplane's airspeed, propeller geometry, and RPM, it is a hard number to estimate. Therefore, a more accurate way to write Equation (14-38) is shown below:

$$T = \frac{\eta_p(V, H, D_p, RPM, P) \times P(H, RPM, MAP, OAT)}{V} \qquad (14\text{-}39)$$

where H is the altitude, D_p is the propeller diameter, RPM is the engine speed in revolutions per minute, MAP is the manifold pressure, and OAT is the outside air temperature. Naturally, this is far more cumbersome to write, which explains the short form shown in Equation (14-38).

The notion that η_p can be treated as a constant leads to a serious error in thrust calculations at low airspeeds because V is the denominator and it is becoming smaller, so the thrust grows unrealistically large. In performance analyses, low-speed performance cannot be ignored — for instance, it is needed to accurately predict take-off performance.

Method 1: Quadratic Interpolation

The *quadratic interpolation method* is based on the assumption that the propeller thrust varies from the static thrust at zero airspeed to the thrust at the maximum level airspeed (V_{max} or V_H) (whatever it comes out to be). This makes it possible to approximate the thrust using a quadratic polynomial of the form $T = AV^2 + BV + C$.

The method requires the propeller efficiency, η_p, at V_{max} to be used. Figure 14-36 shows how the logic works, as applied to the SR22 (although this method should be limited to fixed-pitch propellers only). Equation (14-38) becomes hugely erroneous at low airspeeds and has a mathematical limit approaching infinity at $V = 0$ (a physical impossibility). The reader must keep in mind that the η_p in the figure is kept constant over the entire speed range, but this is only justifiable for the upper range of airspeeds, thanks to the airplane's constant-speed propeller. In reality, the thrust ranges from T_{STATIC} at rest to T_{max} at V_{max} — a trend much better approximated by the quadratic interpolation formula, shown below:

$$T(V) = \left(\frac{T_{STATIC} - 2T_{max}}{V_{max}^2}\right)V^2 + \left(\frac{3T_{max} - 2T_{STATIC}}{V_{max}}\right)V + T_{STATIC}$$

(14-40)

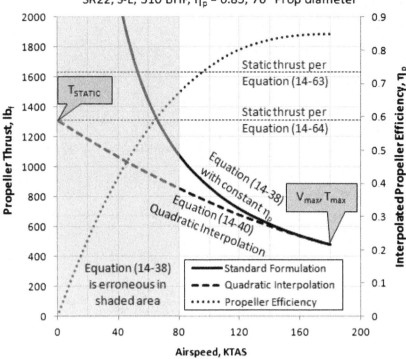

FIGURE 14-36 Application of the quadratic interpolation method to the SR22.

Derivation of Equation (14-40)

The goal of this derivation is to present a polynomial with three unknown variables, A, B, and C. This requires three independent equations:

At the first airspeed ($V = 0$):

$$T_{STATIC} = AV^2 + BV + C = C \quad \text{(i)}$$

At the second airspeed ($V = V_{\max}$):

$$T_{\max} = AV_{\max}^2 + BV_{\max} + C$$
$$= AV_{\max}^2 + BV_{\max} + T_{STATIC} \quad \text{(ii)}$$

where $T_{\max}$ refers to the thrust at $V_{\max}$ and not the maximum thrust possible (the static thrust). The slope at $V_{\max}$ can be used to get the third equation, where the slope is calculated from:

$$\frac{dT}{dV} = \frac{d}{dV}\left(\frac{\eta_p \times 550 \times P_{BHP}}{V}\right) = -\frac{\eta_p \times 550 \times P_{BHP}}{V^2}$$

Therefore, the third equation is:

$$\left.\frac{dT}{dV}\right|_{V_{\max}} = 2AV_{\max} + B = -\frac{\eta_p \times 550 \times P_{BHP}}{V_{\max}^2} \quad \text{(iii)}$$

Equation (iii) gives:

$$B = -2AV_{\max} - \frac{\eta_p \times 550 \times P_{BHP}}{V_{\max}^2} = -2AV_{\max} - \frac{T_{\max}}{V_{\max}}$$

Inserting this into Equation (ii) yields:

$$T_{\max} = AV_{\max}^2 + BV_{\max} + T_{STATIC}$$
$$= AV_{\max}^2 - \left(2AV_{\max} + \frac{T_{\max}}{V_{\max}}\right)V_{\max} + T_{STATIC}$$

Algebraic manipulations yield:

$$A = (T_{STATIC} - 2T_{\max})/V_{\max}^2$$

Thus, the coefficient B can be rewritten as follows:

$$B = \frac{3T_{\max} - 2T_{STATIC}}{V_{\max}}$$

which allows us to combine the coefficients as follows:

$$T(V) = \left(\frac{T_{STATIC} - 2T_{\max}}{V_{\max}^2}\right)V^2$$
$$+ \left(\frac{3T_{\max} - 2T_{STATIC}}{V_{\max}}\right)V + T_{STATIC}$$

QED

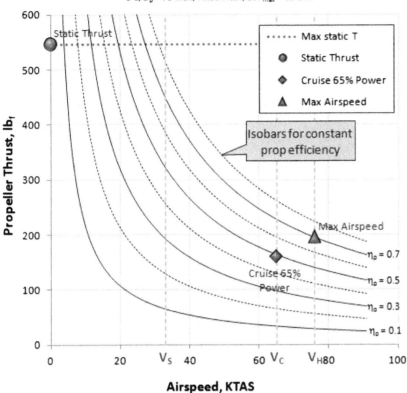

FIGURE 14-37 Thrust isobars for the Piper J-3 Cub. The isobars are calculated for 65 BHP.

Method 2: Cubic Spline Method for Fixed-Pitch Propellers

Consider a performance analysis for an aircraft like the Piper J-3 Cub. The J-3 is powered by a 65 BHP Continental A-65 engine, has a stalling speed (V_S) of about 33 KCAS, a cruising speed (V_C) of about 65 KTAS, and a maximum speed (V_H) of some 76 KTAS. A properly matched propeller should yield a η_p of 0.75 at around 65 KTAS, when the power is about 65% (42 BHP). On the other hand, it may be around 0.70 at the maximum airspeed of 76 KTAS. A typical J-3 propeller diameter is some 70 inches and, using Equation (14-63), the static thrust can be found to equal 546 lb$_f$ (not including the reduction factor 0.85 discussed in Section 14.5.3, *Maximum static thrust*). These values have been identified on the graph shown in Figure 14-37.

Figure 14-37 also shows thrust isobars for constant propeller efficiencies, η_p, at maximum power of 65 BHP. Note that the cruise point is slightly below the $\eta_p = 0.5$ isobar and not at 0.75 (as one might think at first, based on the previous paragraph). This is no mistake — the cruise power is based on 42 BHP and not 65 BHP. The isobars shown are based on full power. If they were based on 42 BHP they would all be sitting much lower and the cruise point would indeed be between 0.7 and 0.8 isobars. The two points to which the isobars pertain are labeled "Static Thrust" and "Max Airspeed." The former is planted on $V = 0$ KTAS and $T = 546$ lb$_f$ and the latter on $V = V_H = 76$ KTAS and the isobar where $\eta_p = 0.70$ (based on the assumption in the previous paragraph). It should be clear that as the airplane accelerates from standstill to maximum airspeed its thrust does not follow a particular isobar, but rather travels across them, showing that the propeller efficiency (and thrust) does indeed change with airspeed. Since propeller theory shows that the thrust is always less than the static thrust (see Section 14.4), whatever path the thrust follows between the two extremes, intuition holds that, generally, the change in thrust would move along a path indicated by the arrow shown in Figure 14-38. The question is only — how? Is this path a straight line or is it curved? Let's try to answer those questions in detail.

Ultimately, the purpose here is to develop a tool that can be used to estimate the thrust for a *fixed-pitch and constant speed propeller at a constant power setting* with reasonable accuracy. This tool can then be used to estimate thrust at airspeeds as low as 0 ft/s,

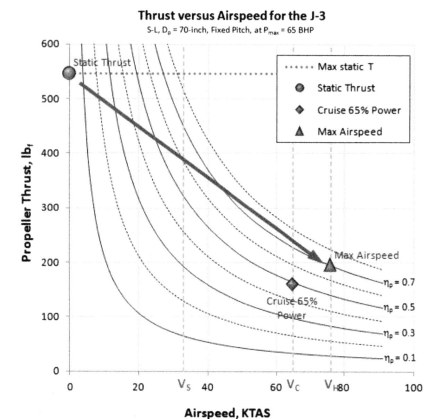

FIGURE 14-38 As the Piper J-3 Cub accelerates from rest to V_H, it moves through the isobars as shown.

sorely needed for T-O analysis. The propeller thrust drops relatively sharply at first, but the drop is reduced as the airspeed increases. In other words, the derivative dT/dV is negative and steep at first; however, eventually it will level off as the airspeed and propeller efficiency increase. Since the maximum airspeed is usually an off-design point, dT/dV will become increasingly negative again. Based on the limited information we have (effectively static thrust and thrust or drag force at cruise and maximum airspeed) it is possible to represent this behavior using a cubic spline. This is the crux of the *cubic spline method*; it allows thrust to be approximated at a given power setting as a function of airspeed, using a cubic spline. A properly developed method allows such off-design conditions to be taken into account.

Such a cubic spline will be of the form:

$$T(V) = A \cdot V^3 + B \cdot V^2 + C \cdot V + D \quad (14\text{-}41)$$

where the constants $A, B, C,$ and D are determined using the following matrix equation:

$$\begin{bmatrix} 0 & 0 & 0 & 1 \\ V_C^3 & V_C^2 & V_C & 1 \\ 3V_C^2 & 2V_C & 1 & 0 \\ V_H^3 & V_H^2 & V_H & 1 \end{bmatrix} \begin{Bmatrix} A \\ B \\ C \\ D \end{Bmatrix} = \begin{Bmatrix} T_{STATIC} \\ T_C \\ -\eta_p \cdot 325.8 \cdot P_{BHP}/V_C^2 \\ T_H \end{Bmatrix} \quad (14\text{-}42)$$

where the airspeeds V_C and V_H are in KTAS. The constants $A, B, C,$ and D can be obtained by inverting the left square matrix and multiplying it by the array on the right-hand side. The use of this method will be demonstrated in Example 14-10 by applying it to the Piper J-3 Cub.

Derivation of Equation (14-42)

The trick is to determine the constants $A, B, C,$ and D. To do this we need four equations. Observation shows that when $V = 0$, thrust is static thrust, T_{STATIC}. Using

Equation (14-63), repeated here for convenience, we can easily write the constant D as follows:

$$D = T_{STATIC} = P^{2/3}(2\rho A_2)^{1/3} \quad (14\text{-}63)$$

Note that it may be more appropriate to use Equation (14-64) rather than (14-63), as it better accounts for the effect of blockage, hub and tip, and the presence of the spinner. However, for this derivation the theoretical maximum static thrust is used. Since $A_2 = \pi \cdot D_P^2/4$ and $P = 550 \cdot P_{BHP}$, the *first equation* of the four can be written as follows (after a simple manipulation):

$$T(0) = A \cdot 0^3 + B \cdot 0^2 + C \cdot 0 + D = T_{STATIC}$$
$$= 78.034 \times (P_{BHP})^{2/3}\left(\rho D_p^2\right)^{1/3} \quad (14\text{-}43)$$

Note that the power is the maximum power at the altitude. The next two equations are based on the airspeed for which the propeller has been optimized — here the cruising speed. Note, for a climb propeller it would be a different airspeed, although this should be denoted by V_C as well. The importance of this airspeed is that this is where the propeller efficiency becomes a maximum. The assumption allows an additional equation to be generated, allowing the problem to be solved. The *second equation* of the four can now be written as follows:

V_C in ft/s

$$T(V_C) = AV_C^3 + BV_C^2 + CV_C + D = T_C$$
$$= \eta_{p\ max} \cdot 550 \cdot P_{BHP}/V_C \quad (14\text{-}44)$$

Note that this equation uses the maximum power and not the cruise power, as this is the maximum thrust the engine would be capable of at V_C. If the intent is to use airspeed in terms of KTAS rather than ft/s, Equation (14-44) must be rewritten as follows (note that in this case all the other equations must be in terms of KTAS as well):

V_C in KTAS

$$T(V_C) = AV_C^3 + BV_C^2 + CV_C + D = T_C$$
$$= \eta_{p\ max} \cdot 325.8 \cdot P_{BHP}/V_C \quad (14\text{-}45)$$

The third equation is based on the fact that the propeller efficiency at V_C, given by $\eta_{p\ max}$, is the highest it will become. This is where in η_p-V space the derivative $d\eta_p/dV = 0$. In the T-V space the slope must be dT/dV at V_C. This derivative can be calculated as shown below:

$$\frac{dT}{dV} = \frac{d}{dV}\left(\frac{\eta_p \cdot 550 \cdot P_{BHP}}{V}\right)$$

$$= \left(\eta_p \cdot 550 \cdot P_{BHP}\right)\frac{V\dfrac{d\eta_p}{dV} - \eta_p \dfrac{dV}{dV}}{V^2}$$

$$= \left(\eta_p \cdot 550 \cdot P_{BHP}\right)\frac{V\dfrac{d\eta_p}{dV} - \eta_p}{V^2}$$

But by definition, at $V = V_C$ the derivative $d\eta_p/dV = 0$; therefore:

V_C in ft/s

$$\left(\frac{dT}{dV}\right)_{V=V_C} = \frac{-\eta_p \cdot 550 \cdot P_{BHP}}{V_C^2} = -\frac{T_C}{V_C} \quad (14\text{-}46)$$

Note that the units for the above derivative is in terms of $\text{lb}_f/(\text{ft/s})$. If used with graphs such as that of Figure 14-37 or Figure 14-38, which are presented in terms of lb_f versus KTAS, the derivative must be divided by 1.688 to convert the units to lb_f/KTAS. This can be seen by noting that the derivative, when represented along the compressed KTAS axis, will be steeper by the factor 1.688. Therefore, to use Equation (14-46) with V_C in terms of KTAS, use the following form:

V_C in KTAS

$$\left(\frac{dT}{dV}\right)_{V=V_C} = \frac{-\eta_p \cdot 325.8 \cdot P_{BHP}}{V_C^2} \quad (14\text{-}47)$$

This form is used with Equation (14-42) above. Thus, the *third equation* is based on the derivative at V_C and is written as follows (here in terms of KTAS):

$$T(V_H) = 3AV_C^2 + 2BV_C + C = -\eta_p \cdot 325.8 \cdot P_{BHP}/V_C^2$$
$$(14\text{-}48)$$

Then, let's consider the fourth and final equation, which describes thrust, T_H, at maximum airspeed, V_H. This expression requires the thrust to be determined by some means. Four ways come immediately to mind. (1) Use η_p provided by the propeller manufacturer to calculate thrust using Equation (14-38). (2) Estimate η_p using a method like the blade element theory (see Section 14.6) to calculate thrust. (3) Estimate η_p using (very good) engineering judgment based on knowledge of other airplanes to calculate thrust. (4) Calculate the drag of the airplane at the condition and use it as the thrust. Note that for a fixed-pitch propeller, η_p is less than $\eta_{p\ max}$. Assuming that η_p is known, we may write the *fourth equation* as follows:

V_H in ft/s

$$T(V_H) = AV_H^3 + BV_H^2 + CV_H + D = T_H$$
$$= \frac{\eta_p \cdot 550 \cdot P_{BHP}}{V_H} \qquad (14\text{-}49)$$

V_H in KTAS

$$T(V_H) = AV_H^3 + BV_H^2 + CV_H + D = T_H$$
$$= \frac{\eta_p \cdot 325.8 \cdot P_{BHP}}{V_H} \qquad (14\text{-}50)$$

These four equations can now be assembled into the following matrix form of Equation (14-42).

QED

EXAMPLE 14-10

Use the *cubic spline method* to determine a function $T(V_{KTAS})$ that describes the full power thrust of the J-3 as a function of airspeed in KTAS. The maximum airspeed, V_H, is 76 KTAS, where the thrust is 195 lb$_f$, and static thrust, T_{STATIC} is 546 lb$_f$. The propeller efficiency at cruise speed (V_C) of 65 KTAS is assumed 0.75 and 0.70 at V_H. Plot the resulting equation. Also plot the propeller efficiency that results from using the cubic spline.

Solution

Start by setting up the matrix of Equation (14-42):

$$\begin{bmatrix} 0 & 0 & 0 & 1 \\ 65^3 & 65^2 & 65 & 1 \\ 3\cdot 65^2 & 2\cdot 65 & 1 & 0 \\ 76^3 & 76^2 & 76 & 1 \end{bmatrix} \begin{Bmatrix} A \\ B \\ C \\ D \end{Bmatrix}$$
$$= \begin{Bmatrix} 546 \\ 0.75 \times 550 \times 65/(65 \times 1.688) \\ -0.75 \times 325.8 \times 65/65^2 \\ 0.70 \times 550 \times 76/(76 \times 1.688) \end{Bmatrix}$$

Evaluating the numbers yields:

$$\begin{bmatrix} 0 & 0 & 0 & 1 \\ 274625 & 4225 & 65 & 1 \\ 12675 & 130 & 1 & 0 \\ 438976 & 5776 & 76 & 1 \end{bmatrix} \begin{Bmatrix} A \\ B \\ C \\ D \end{Bmatrix} = \begin{Bmatrix} 546 \\ 244 \\ -3.759 \\ 195 \end{Bmatrix}$$

Inverting the matrix to solve for A, B, C, and D results in:

$$\begin{Bmatrix} A \\ B \\ C \\ D \end{Bmatrix} = \begin{bmatrix} 0 & 0 & 0 & 1 \\ 274625 & 4225 & 65 & 1 \\ 12675 & 130 & 1 & 0 \\ 438976 & 5776 & 76 & 1 \end{bmatrix}^{-1} \begin{Bmatrix} 546 \\ 244 \\ -3.759 \\ 195 \end{Bmatrix}$$
$$= \begin{Bmatrix} -0.001011 \\ 0.1451 \\ -9.806 \\ 546 \end{Bmatrix}$$

Therefore, the thrust at full power can be expressed as the following function of airspeed:

$$T(V_{KTAS}) = -0.001011 V_{KTAS}^3 + 0.1451 V_{KTAS}^2 - 9.806 V_{KTAS} + 546$$

This is plotted in Figure 14-39. Note that the drag force for the J-3 has also been plotted, but it assumes the drag coefficient can be expressed using the simplified drag model, $C_D = C_{Dmin} + kC_L^2$, where $C_{Dmin} = 0.4976$ and $k = 0.05441$, obtained from an unrelated analysis.

It can be seen the drag curve goes through both the 65% power cruise and maximum airspeed points, as is to be expected. The thrust curve $T(V)$ intersects the max airspeed point only, which is in accordance with performance theory for steady level flight. Note how the cubic spline extends between the isobars of constant $\eta_p = 0.7$ and 0.8, indicating it captures the higher prop efficiency associated with V_C.

To plot the propeller efficiency that results from using the cubic spline we simply back out the propeller efficiency based on Equation (14-38), that is:

$$T = \frac{\eta_p \times 550 \times P_{BHP}}{V_{KTAS} \times 1.688} \Leftrightarrow \eta_p = \frac{T(V_{KTAS} \times 1.688)}{550 \times P_{BHP}}$$

The factor 1.688 is used to convert V, which so far has been in KTAS, to ft/s. This is plotted in Figure 14-40, showing the standard shape for propeller efficiency as a function of airspeed.

EXAMPLE 14-10 (cont'd)

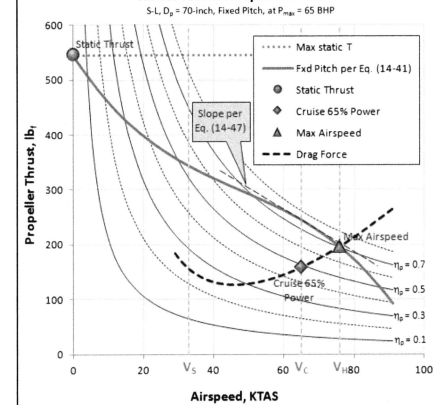

FIGURE 14-39 Plotting $T(V_{KTAS})$ at full power and S-L for the Piper J-3 Cub with a fixed-pitch propeller.

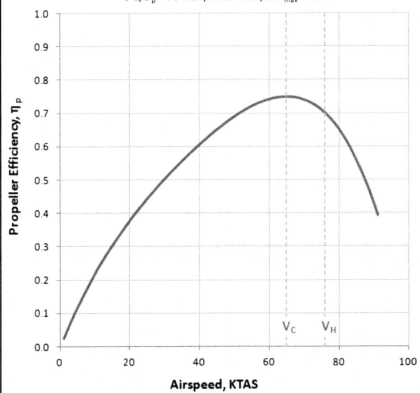

FIGURE 14-40 Propeller efficiency based on the spline.

Method 3: Cubic Spline Method for Constant-Speed Propellers

The preceding treatment of the fixed-pitch propeller raises the question whether it can be extended to include constant-speed propellers. The answer is indeed yes. In particular, consider the unlikely design of a constant-speed propeller for the J-3 Cub. Say it is designed to generate a η_p of 0.85 at V_C. It should be realized that one of the consequences of such a propeller is an increased cruising and maximum speed. Determining these two becomes one of the primary tasks in applying the method to constant-speed propellers.

For the most part, this method is identical to the one for the fixed-pitch propeller. It also uses Equation (14-42) with one major difference; the new V_C and V_H must be determined. This can be done if we assume that the η_p at V_H will be the same as that at V_C. The justification can be seen in Figure 14-48. It shows that the nature of constant-speed propellers is to yield a relatively constant η_p at high airspeeds. The determination of V_H requires the drag properties of the aircraft to be known. If so, the method of Section 19.2.14, *COMPUTER CODE: Determining maximum level airspeed, V_{max}, for a propeller aircraft*, can be used to estimate both V_C and V_H. Then, Equation (14-42) can be populated and the constants A, B, C, and D of the cubic spline of Equation (14-41) determined. The use of this method will now be demonstrated by, again, applying it to the Piper J-3 Cub.

EXAMPLE 14-11

Use the cubic spline method to derive a function, $T(V_{KTAS})$, that describes the thrust of the J-3 at full power using a constant-speed propeller designed to give a propeller efficiency of 0.85 at cruise and maximum airspeed. Also plot the propeller efficiency that results from using the cubic spline.

Solution

By using the computer code of Section 19.2.14 the cruising speed at 65% power and maximum airspeed of the J-3 with the new propeller was estimated to equal 68.7 and 81.6 KTAS, respectively. The corresponding thrust (using $\eta_p = 0.85$) were found to equal 262 and 221 lb$_f$, respectively. The same static thrust, $T_{STATIC} = 546$ lb$_f$, is used. Note that since the method was implemented using Microsoft Excel it is not possible to display all the decimals retained by the program. Then, set up the matrix of Equation (14-42):

$$\begin{bmatrix} 0 & 0 & 0 & 1 \\ 68.7^3 & 68.7^2 & 68.7 & 1 \\ 3\cdot 68.7^2 & 2\cdot 68.7 & 1 & 0 \\ 81.6^3 & 81.6^2 & 81.6 & 1 \end{bmatrix} \begin{Bmatrix} A \\ B \\ C \\ D \end{Bmatrix} = \begin{Bmatrix} 546 \\ 262 \\ -(0.85\cdot 325.8\cdot 65)/68.7^2 \\ 221 \end{Bmatrix}$$

Evaluating the numbers yields:

$$\begin{bmatrix} 0 & 0 & 0 & 1 \\ 324243 & 4720 & 68.7 & 1 \\ 14159 & 137.4 & 1 & 0 \\ 543338 & 6659 & 81.6 & 1 \end{bmatrix} \begin{Bmatrix} A \\ B \\ C \\ D \end{Bmatrix} = \begin{Bmatrix} 546 \\ 262 \\ -3.814 \\ 221 \end{Bmatrix}$$

Inverting the matrix to solve for A, B, C, and D results in:

$$\begin{Bmatrix} A \\ B \\ C \\ D \end{Bmatrix} = \begin{bmatrix} 0 & 0 & 0 & 1 \\ 324243 & 4720 & 68.7 & 1 \\ 14159 & 137.4 & 1 & 0 \\ 543338 & 6659 & 81.6 & 1 \end{bmatrix}^{-1} \begin{Bmatrix} 546 \\ 262 \\ -3.815 \\ 221 \end{Bmatrix}$$

$$= \begin{Bmatrix} 0.0005467 \\ -0.07046 \\ -1.873 \\ 546 \end{Bmatrix}$$

Therefore, the thrust at full power can be expressed as follows:

$$T(V_{KTAS}) = 0.0005467 V_{KTAS}^3 - 0.07046 V_{KTAS}^2 - 1.873 V_{KTAS} + 546$$

This is plotted in Figure 14-41. The drag force and the thrust curve for the fixed-pitch propeller have also been plotted to better realize the difference between the two thrust curves. Again, this assumes the drag coefficient can be expressed using the simplified drag model, $C_D = C_{Dmin} + kC_L^2$, where $C_{Dmin} = 0.4976$ and $k = 0.05441$, obtained from an unrelated analysis.

To plot the propeller efficiency that results from using the cubic spline, the same method is used, as was shown in Example 14-11; we simply back out the propeller efficiency based on Equation (14-38), that is:

$$\eta_p = \frac{T(V_{KTAS} \times 1.688)}{550 \times P_{BHP}}$$

This is plotted in Figure 14-42, showing the standard shape for propeller efficiency for a constant-speed

EXAMPLE 14-11 (cont'd)

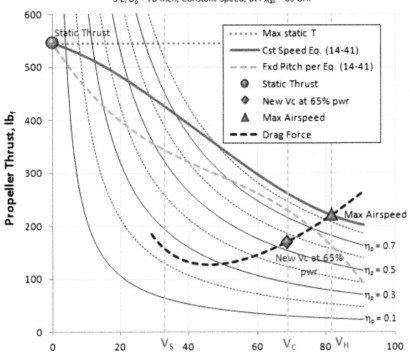

FIGURE 14-41 Plotting $T(V_{KTAS})$ at full power and S-L for the Piper J-3 Cub with a constant-speed propeller.

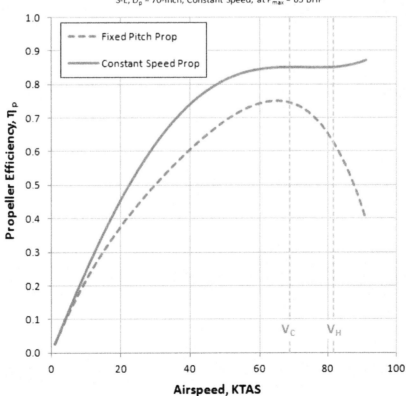

FIGURE 14-42 Propeller efficiency based on the spline.

> **EXAMPLE 14-11** (cont'd)
>
> propeller as a function of airspeed. It can be seen the general shape is maintained up to V_H, as the spline is forced into the appropriate shape, after which an error begins to creep in; therefore, do not extrapolate the thrust.
>
> It is important to understand that while the new propeller efficiency deviates at airspeeds higher than V_H, the method still provides a simple means to perform performance analysis with acceptable accuracy at other airspeeds. It is possible to improve the shape of the spline by fitting a quartic or higher-order spline through the known data, although this becomes progressively more cumbersome. It appears that a cubic spline offers the right amount of complexity and practicality for this kind of analysis.

14.4.3 Step-by-step: Determining Thrust Using a Propeller Efficiency Table

Sometimes the propeller manufacturer provides the designer with a propeller efficiency table like the one shown in Figure 14-43. In it the propeller efficiency can be read based on the computed value of the advance ratio (left-most column) and power coefficient (top row). Using such a table presents a very convenient way to extract the propeller efficiency for use with Equation (14-38). To do this the user must calculate the power coefficient, C_P (defined in Section 14.3.7, *Definition of power- and thrust-related coefficients*), and advance ratio, J (defined in Section 14.3.5, *Advance ratio*), and use these to extract the prop efficiency. The sample table below has the advance ratio in rows, ranging from 0.20 to 1.20, and the C_P in columns, ranging from 0.02 to 0.13. Values of J or C_P that fall between the tabulated values must be extracted by double-interpolation.

Using such a table, the method of extracting thrust is as follows:

Step 1: Determine Advance Ratio

Calculate advance ratio from Equation (14-23):

$$J = \frac{V}{nD} = \frac{60 \cdot V}{RPM \cdot D}$$

J/Cp	0.02	0.03	0.04	0.05	0.06	0.07	0.08	0.09	0.1	0.11	0.12	0.13
0.20	0.50	0.46	0.41	0.36	0.31	0.25	0.20	0.17	0.14	0.12	0.10	0.09
0.25	0.57	0.54	0.50	0.45	0.39	0.33	0.26	0.22	0.18	0.15	0.13	0.11
0.30	0.62	0.61	0.57	0.52	0.47	0.40	0.33	0.27	0.22	0.19	0.16	0.14
0.35	0.67	0.66	0.63	0.59	0.54	0.48	0.40	0.32	0.27	0.23	0.20	0.17
0.40	0.70	0.70	0.68	0.64	0.60	0.55	0.48	0.38	0.32	0.27	0.23	0.20
0.45	0.72	0.74	0.71	0.69	0.65	0.61	0.55	0.46	0.37	0.31	0.27	0.23
0.50	0.74	0.76	0.75	0.72	0.69	0.66	0.61	0.53	0.43	0.36	0.30	0.26
0.55	0.75	0.78	0.77	0.75	0.72	0.69	0.66	0.61	0.50	0.40	0.34	0.29
0.60	0.76	0.79	0.79	0.77	0.75	0.72	0.69	0.65	0.59	0.46	0.39	0.33
0.65	0.77	0.81	0.80	0.78	0.76	0.74	0.71	0.68	0.64	0.55	0.44	0.37
0.70	0.77	0.81	0.81	0.80	0.78	0.76	0.73	0.71	0.67	0.64	0.50	0.42
0.75	0.78	0.82	0.82	0.81	0.79	0.77	0.75	0.72	0.69	0.66	0.61	0.49
0.80	0.78	0.82	0.83	0.82	0.80	0.78	0.76	0.74	0.71	0.68	0.65	0.57
0.85	0.78	0.83	0.83	0.82	0.81	0.79	0.77	0.75	0.73	0.70	0.67	0.64
0.90	0.78	0.83	0.83	0.83	0.81	0.80	0.78	0.76	0.74	0.71	0.68	0.65
0.95	0.78	0.83	0.83	0.83	0.82	0.80	0.79	0.77	0.74	0.72	0.69	0.67
1.00	0.77	0.83	0.84	0.83	0.82	0.80	0.79	0.77	0.75	0.73	0.70	0.68
1.05	0.77	0.82	0.83	0.83	0.82	0.81	0.79	0.77	0.76	0.73	0.71	0.68
1.10	0.76	0.82	0.83	0.83	0.82	0.81	0.79	0.78	0.76	0.74	0.71	0.69
1.15	0.76	0.82	0.83	0.83	0.82	0.81	0.79	0.78	0.76	0.74	0.72	0.69
1.20	0.75	0.81	0.83	0.82	0.82	0.81	0.79	0.78	0.76	0.74	0.72	0.70

FIGURE 14-43 A typical propeller efficiency table for a constant-speed propeller (no specific propeller type).

Step 2: Determine Power Coefficient

Calculate power coefficient from Equation (14-26):

$$C_P = \frac{P}{\rho n^3 D^5}$$

Step 3: Extract Propeller Efficiency

Use a method such as the one in Section E.5.11, *Step-by-step: Rapid interpolation of 2D lookup tables*, to extract the propeller efficiency, using the calculated C_P and J.

Step 4: Calculate Thrust

Calculate thrust from Equation (14-38):

$$T = \frac{\eta_p \times 550 \times P_{BHP}}{V}$$

EXAMPLE 14-12

An airplane is equipped with a 70″ diameter constant-speed propeller driven by a 150 BHP piston engine. If flying at sea-level at 130 KTAS with the prop swinging at 2700 RPM and max power, using the table in Figure 14-43, what is the magnitude of the thrust?

Solution

Step 1: Advance ratio

$$J = \frac{60 \cdot V}{RPM \cdot D} = \frac{60 \cdot (1.688 \times 130)}{2700 \cdot (70/12)} = 0.8360$$

Step 2: Power Coefficient

$$C_P = \frac{P}{\rho n^3 D^5} = \frac{(550 \times 150)}{(0.002378)(2700/60)^3 (70/12)^5} = 0.05637$$

Step 3: Extract prop efficiency – here, by observation $\eta \approx 0.81$.

Step 4: Thrust

$$T = \frac{\eta_p \times 550 \times P_{BHP}}{V} = \frac{0.81 \times 550 \times 150}{1.688 \times 130} = 304 \text{ lb}_f$$

14.4.4 Estimating Thrust From Manufacturer's Data

As said earlier, it is always wise for the aircraft designer to establish a good relationship with propeller manufacturers by committing to using their products. Such relationships may (1) encourage the propeller manufacturer to provide the designer with useful propeller performance information, and (2) if the business case is promising enough, even persuade the manufacturer to design a propeller to match the designer's desired characteristics.

An example of typical data provided by a propeller manufacturer is shown in Figure 14-44. The data is for a two-bladed, 69 inch diameter propeller designed for an LSA aircraft. These propeller characteristics are plotted in Figure 14-45, with the appropriate curve fits. So, now we have the properties of the propeller formulation as a function of advance ratio.

$$\eta_P = 0.096574 + 1.703049J - 0.952281J^2$$
$$C_T = 0.162133 - 0.106480J - 0.038208J^2$$
$$C_P = 0.058005 + 0.084893J - 0.120439J^2$$

EXAMPLE 14-13

An airplane is equipped with a 69″ diameter fixed-pitch propeller driven by a 100 BHP piston engine. If flying at sea-level at 110 KCAS with the prop swinging at 2400 RPM and 75% power, using the above equations, what is the magnitude of the thrust?

Solution

Step 1: Advance ratio

$$J = \frac{60 \cdot V}{RPM \cdot D} = \frac{60 \cdot (1.688 \times 110)}{2400 \cdot (69/12)} = 0.8073$$

EXAMPLE 14-13 (cont'd)

Step 2: Propeller efficiency

$$\eta_P = 0.096574 + 1.703049J - 0.952281J^2 = 0.8508$$

Step 3: Thrust

$$T = \frac{\eta_P \times 550 \times P_{BHP}}{V} = \frac{0.8508 \times 550 \times 75}{1.688 \times 110} = 189 \text{ lb}_f$$

14.4.5 Other Analytical Methods

Other analytical methods that allow one to extract thrust and other properties from propellers are extensive enough to be presented in their own sections. These are the *Rankine-Froude momentum theory* (Section 14.5) and *Blade element theory* (Section 14.6).

J	C_T	C_P	η
0.3142	0.1250	0.0734	0.535
0.4398	0.1077	0.0710	0.667
0.5655	0.0898	0.0674	0.754
0.6912	0.0703	0.0596	0.815
0.8168	0.0497	0.0475	0.854
0.9425	0.0278	0.0306	0.856

Theoretical calculated performance
Includes inflow and spinner/hub drag

FIGURE 14-44 A typical propeller efficiency table.

14.5 RANKINE-FROUDE MOMENTUM THEORY

The *momentum theory* is a mathematical method that models the performance of a propeller. It is also known as the *actuator disc theory*. The theory was originally developed in 1865 by the famed Scottish physicist William John Macquorn Rankine (1820–1872), who had a particularly prolific career as a scientist and to whom the Rankine temperature scale is credited. Important contributions were made by the English hydrodynamicist William Froude (1810–1879). Rankine's contribution is generally dated to 1865, and that of Froude over the period from 1878 to 1889 [9].

The theory can be used to estimate propeller thrust, although the results are generally optimistic. It allows the airspeed inside the propeller streamtube to be estimated (the so-called *propeller-induced velocity*), which is an important result used with the blade element theory,

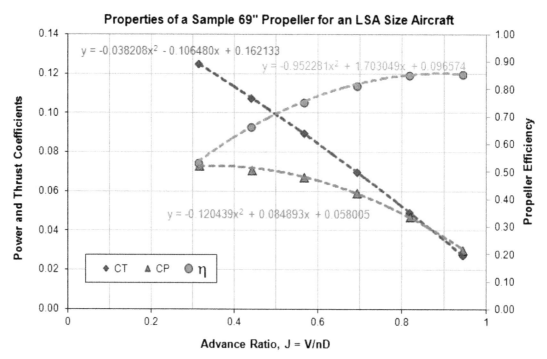

FIGURE 14-45 A typical propeller efficiency graph for a fixed-pitch propeller (no specific propeller type).

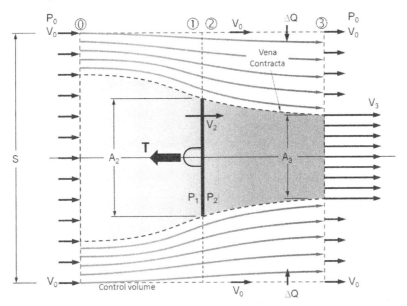

FIGURE 14-46 Idealized flow model for the Froude-Rankine momentum theory.

introduced in Section 14.6. The momentum theory is based on the following assumptions:

(1) The propeller is replaced by an infinitesimally thin actuator disc that offers no resistance to air passing through it.
(2) The disc is uniformly loaded and, therefore, experiences uniform flow passing through the actuator disc. Consequently, it is assumed to impart uniform acceleration to the air passing through it.
(3) A control volume surrounds the streamtube and sharply separates the flow going through it from the surrounding air.
(4) Flow outside the streamtube has constant stagnation pressure (which means no work is imparted on it).
(5) In the far-field, in front and behind the disc, streamlines are parallel and, there, the pressure inside the streamtube is equal to the far-field pressure.
(6) The propeller does not impart rotation to the flow.
(7) The theory assumes inviscid (no drag, no momentum diffusion) and incompressible flow.

14.5.1 Formulation

Consider Figure 14-46, which shows an idealized flow model for the actuator disc. The model consists of the four planes; ⓪, ①, ②, and ③. From Assumption (5), Plane ⓪ is far enough ahead of the actuator disc for the streamlines to be parallel. Plane ③ is far enough behind the disc for the streamlines to have become parallel again. Consequently, the static pressure at either plane is constant and equal to the far-field pressure P_0.

The front face of the control volume has surface area S. The flux of fluid entering through Plane ⓪ is $Q_0 = S \cdot V_0$. The flux exiting through Plane ③ is $Q_3 = (S - A_3) \cdot V_0 + A_3 \cdot V_3$. The change in flux between the front and aft surfaces is thus given by:

$$\Delta Q = Q_3 - Q_0 = (S - A_3) \cdot V_0 + A_3 \cdot V_3 - S \cdot V_0$$
$$= A_3 \cdot (V_3 - V_0)$$

The non-zero value of the flux implies fluid must be entering the control volume from the sides. Now in order to determine the thrust it is necessary to apply the momentum theorem, repeated here in its conventional form;

$$F = \int_A \rho V \left(\vec{V} \cdot \vec{n}\right) dA \qquad \text{(C-19)}$$

where $\mathbf{n}$ = normal vector to the cross-sectional area

V = airspeed
A = area
ρ = density

Note that the momentum flux of the fluid entering Plane ⓪ is given by $-\rho S V_0^2$ (the negative sign indicates it is entering the volume). Similarly, the momentum flux of the fluid entering Plane ③ is given by $\rho[(S - A_3)V_0^2 + A_3 V_3^2]$ (the positive sign indicates it is leaving the volume). Finally, the momentum flux of the fluid entering the sides is given by $-\rho \Delta Q V_0$ (the negative sign indicates it is entering the volume). This generates a force that points in the direction of the incoming flow (here renaming F as T for thrust):

$$T = \int_A \rho V \left(\vec{V} \cdot \vec{n}\right) dA$$
$$= \rho\left[(S - A_3)V_0^2 + A_3V_3^2\right] - \rho S V_0^2 - \rho \Delta Q V_0$$
$$= \rho A_3(V_3^2 - V_0^2) - \rho \Delta Q V_0$$

Inserting our previous result for ΔQ leads to:

$$T = \rho A_3(V_3^2 - V_0^2) - \rho \Delta Q V_0 = \rho A_3 \cdot V_3(V_3 - V_0) \quad (14\text{-}51)$$

Considering Figure 14-46 it can be seen that T can also be computed from:

$$T = A_2(P_2 - P_1) \quad (14\text{-}52)$$

where

P_i = pressure at planes 1 and 2
A = propeller disc area

The changes in pressure and airspeed are depicted in Figure 14-47. It can be seen that the propeller energy is added as a rise in stagnation pressure. Its magnitude depends on the geometry of the propeller and the power delivered by the power plant.

We can calculate P_1 and P_2 using Bernoulli's principle assuming streamlines exist up to the actuator disc and immediately behind it. We must note this principle is not applicable through the actuator itself, although it is reasonable to assume the velocity is continuous through the disc. Thus, we can determine P_1 using the flow conditions at Plane ⓪, and P_2 using the flow conditions at Plane ③ as follows:

Pressure at Plane ①:

$$P_0 + \frac{1}{2}\rho V_0^2 = P_1 + \frac{1}{2}\rho V_1^2 \quad (i)$$

Pressure at Plane ②:

$$P_0 + \frac{1}{2}\rho V_3^2 = P_2 + \frac{1}{2}\rho V_2^2 \quad (ii)$$

From which we find the pressure difference by subtracting Equation (i) from (ii).

$$P_2 - P_1 + \frac{1}{2}\rho(V_2^2 - V_1^2) = \frac{1}{2}\rho(V_3^2 - V_0^2)$$

Note that since $V_1 = V_2$ we can simplify:

$$P_2 - P_1 = \frac{1}{2}\rho(V_3^2 - V_0^2) \quad (14\text{-}53)$$

Now, insert Equation (14-53) into (14-51) to get:

$$T = A_2(P_2 - P_1) = A_2 \times \frac{1}{2}\rho(V_3^2 - V_0^2) \quad (iii)$$

Equate this with Equation (14-51):

$$T = \rho A_3 \cdot V_3(V_3 - V_0) = A_2 \times \frac{1}{2}\rho(V_3^2 - V_0^2) \quad (iv)$$

Simplify to get:

$$2A_3 \cdot V_3(V_3 - V_0) = A_2(V_3^2 - V_0^2)$$
$$= A_2(V_3 - V_0)(V_3 + V_0) \quad (v)$$

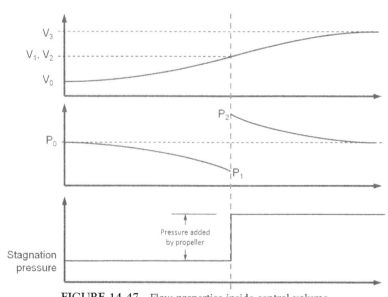

FIGURE 14-47 Flow properties inside control volume.

Deleting like terms yields:

$$2A_3 \cdot V_3 = A_2(V_3 + V_0) \quad \text{(vi)}$$

Mass conservation requires:

$$A_2 V_2 = A_3 V_3 \;\Rightarrow\; A_3 = \frac{A_2 V_2}{V_3}$$

Insert this into (vi);

$$2A_3 \cdot V_3 = 2\frac{A_2 V_2}{V_3} \cdot V_3 = A_2(V_3 + V_0) \quad \text{(vii)}$$

And simplify to get:

$$V_2 = \frac{(V_3 + V_0)}{2} \quad (14\text{-}54)$$

The above equation is recognized as *Froude's Theorem* [10] and is essential in propeller design. It states that the airspeed through the propeller is the average of the far-field airspeed (V_0) far ahead of it and the airspeed in the streamtube far behind (V_3). Let's now define the following differences:

$$w = V_2 - V_0 = V_3 - V_2 \quad \text{(viii)}$$

where w is called the *propeller-induced velocity*. This allows V_2 to be redefined as follows:

$$V_2 = V_0 + w \quad \text{(ix)}$$

and V_3 as:

$$V_3 = V_0 + 2w \quad \text{(x)}$$

Inserting Equations (ix) and (x) into Equation (14-51) results in:

$$T = \rho A_3 \cdot V_3 (V_3 - V_0) = \rho \frac{A_2 V_2}{V_3} \cdot V_3 (V_3 - V_0)$$
$$= \rho A_2 V_2 (V_3 - V_0) = \rho A_2 (V_0 + w)((V_0 + 2w) - V_0)$$

Simplifying yields:

$$T = 2\rho A_2 (V_0 + w) w \quad (14\text{-}55)$$

The power for this system is given by $P = T \cdot V$ or;

$$P = T(V_0 + w) \quad (14\text{-}56)$$

The above result shows the power required by the propeller is a combination of two terms; *useful power* and *induced power*:

Useful power:

$$P_U = TV_0 \quad \text{(xi)}$$

Induced power:

$$P_i = Tw \quad \text{(xii)}$$

We can determine the induced velocity, w, through the propeller disc from Equation (14-55):

$$T = 2\rho A_2 (V_0 + w) w \;\Leftrightarrow\; w^2 + V_0 w - \frac{T}{2\rho A_2} = 0$$

This is a quadratic equation in terms of w and can be solved using Equation (E-12):

$$w^2 + V_0 w - \frac{T}{2\rho A_2} = 0 \;\Leftrightarrow\; w = \frac{-V_0 \pm \sqrt{V_0^2 + \frac{2T}{\rho A_2}}}{2}$$

Simplifying and recognizing the negative sign in front of the radical represents a non-physical condition, we write this as:

$$w = \frac{1}{2}\left[-V_0 + \sqrt{V_0^2 + \frac{2T}{\rho A_2}}\right] \quad (14\text{-}57)$$

Using the above analysis, we can consider two important cases; ideal efficiency and maximum static thrust.

14.5.2 Ideal Efficiency

Ideal efficiency can be defined as the ratio of useful power (Equation (xi)) to the total power per Equation (14-56):

$$\eta_i = \frac{TV_0}{T(V_0 + w)} = \frac{1}{(1 + w/V_0)} \quad (14\text{-}58)$$

We can introduce the fraction w/V_0 to Equation (14-57), yielding:

$$\frac{w}{V_0} = \frac{1}{2}\left[-1 + \sqrt{1 + \frac{2T}{\rho V_0^2 A_2}}\right] = \frac{1}{2}\left[-1 + \sqrt{1 + C_T}\right]$$
(xv)

where C_T is the coefficient of thrust (not to be confused with the thrust coefficient of Equation (14-27)) as follows:

$$C_T = \frac{2T}{\rho V_0^2 A_2} = \frac{T}{q A_2} \quad (14\text{-}59)$$

Inserting this into Equation (xiv) leads to:

$$\eta_i = \frac{1}{(1 + w/V_0)} = \frac{2}{1 + \sqrt{1 + C_T}} \quad (14\text{-}60)$$

14.5.3 Maximum Static Thrust

Consider the special case when $V_0 = 0$. In this case, Equation (14-57) the velocity through the propeller reduces to:

$$w = \sqrt{\frac{T}{2\rho A_2}} \qquad (14\text{-}61)$$

Similarly, the power from Equation (14-56) becomes:

$$P = Tw = \frac{T^{1.5}}{\sqrt{2\rho A_2}} \qquad (14\text{-}62)$$

This yields the following expression for static thrust, T_{STATIC}:

$$P = \frac{T^{1.5}}{\sqrt{2\rho A_2}} \quad \Rightarrow \quad T_{STATIC} = P^{2/3}(2\rho A_2)^{1/3} \qquad (14\text{-}63)$$

As a rule of thumb, Equation (14-63) overestimates static thrust by some 15–20% for various reasons, such as blockage effects, the presence of the hub, which reduces the disk area, reduction of lift distribution near the tip and hub, and so on. Therefore, for design purposes, for typical aircraft, estimate T_{STATIC} by the following empirical correction:

$$T_{STATIC} = 0.85 P^{2/3}(2\rho A_2)^{1/3}\left(1 - \frac{A_{spinner}}{A_2}\right) \qquad (14\text{-}64)$$

EXAMPLE 14-14

A two-bladed propeller whose diameter is 76″ is observed to generate 1500 lb$_f$ of thrust. What is the pressure rise (ΔP) through the propeller disc according to the momentum theory?

Solution

The propeller disc area is:

$$A_2 = \frac{\pi D^2}{4} = \frac{\pi (76)^2}{4} = 4536\ in^2 = 31.50\ ft^2$$

Therefore, from Equation (14-52):

$$T = A_2(P_2 - P_1) = A_2 \Delta P$$
$$\Rightarrow \quad \Delta P = \frac{T}{A_2} = \frac{1500}{31.50} = 47.6\ psf$$

EXAMPLE 14-15

An airplane is powered by a two-bladed propeller whose diameter is 76″ is driven by a 310 BHP engine. What is the maximum static thrust one can expect from it at sea-level? What is the static thrust if the spinner diameter is 18″? What is the speed of air through the propeller disc?

Solution

The propeller disc area is:

$$A_2 = \frac{\pi D^2}{4} = 31.50\ ft^2$$

The spinner area is:

$$A_{spinner} = \frac{\pi (18/12)^2}{4} = 1.78\ ft^2$$

Power is:

$$P = 550 \times P_{BHP} = 170500\ ft\cdot lb_f/sec$$

Using Equation (14-63) we determine the maximum static thrust as follows:

$$T_{STATIC} = P^{2/3}(2\rho A_2)^{1/3}$$
$$= (170500)^{2/3}(2(0.002378)(31.50))^{1/3} = 1633\ lb_f$$

Static thrust corrected for spinner:

$$T_{STATIC} = 0.85 P^{2/3}(2\rho A_2)^{1/3}\left(1 - \frac{A_{spinner}}{A_2}\right)$$
$$= 0.85(1633)\left(1 - \frac{1.78}{31.5}\right) = 1310\ lb_f$$

EXAMPLE 14-15 (cont'd)

The speed of air through the propeller disc is found from Equation (14-61):

$$w = \sqrt{\frac{T}{2\rho A_2}} = \sqrt{\frac{1633}{2(0.002378)(31.50)}} = 104.4 \text{ ft/s}$$

The static thrust in the example is overestimated as it does not account for the fact that the distribution of lift along the blade of the propeller is not constant and necessarily becomes zero at the tip. Thus the value represents a theoretical upper limit on the thrust.

EXAMPLE 14-16

The aircraft of the previous example is cruising at 185 KTAS at 8000 ft at a power setting of 233 BHP and it has been estimated its drag amounts to 470 lb$_f$. What is the propeller efficiency?

Solution

Air density at 8000 ft from Equation (16-18):

$$\rho = 0.002378(1 - 0.0000068756 \times 8000)^{4.2561}$$
$$= 0.001869 \text{ slugs/ft}^3$$

Assuming $T = D$ we get the following value for the coefficient of thrust:

$$C_T = \frac{2T}{\rho V_0^2 A_2} = \frac{2(470)}{(0.001869)(185 \times 1.688)^2(31.50)} = 0.1637$$

Therefore, the prop efficiency is:

$$\eta_i = \frac{2}{1 + \sqrt{1 + C_T}} = \frac{2}{1 + \sqrt{1 + 0.1637}} = 0.9621$$

EXAMPLE 14-17

The propeller of Example 14-15 is found to generate 556 lb$_f$ of thrust at 160 KTAS (270.1 ft/s). Determine the induced velocity and ideal efficiency at that condition.

Solution

Area was determined to be 31.50 ft^2 in Example 14-15. Therefore, the induced velocity is found from Equation (14-57):

$$w = \frac{1}{2}\left[-V_0 + \sqrt{V_0^2 + \frac{2T}{\rho A_2}}\right]$$
$$= \frac{1}{2}\left[-270.1 + \sqrt{270.1^2 + \frac{2(556)}{(0.002378)(31.5)}}\right]$$
$$= 13.1 \text{ ft/s}$$

The ideal propeller efficiency is found from Equation (14-60):

$$\eta_i = \frac{1}{(1 + w/V_0)} = \frac{1}{(1 + 13.1/270.1)} = 0.9537$$

Just as in the case of the static thrust, the propeller efficiency is overestimated as well. The example above is for an airplane in the Cirrus SR22/Cessna Corvalis class, for which a more accurate efficiency is in the 0.86–0.88 range for the cited flight condition.

14.5.4 Computer code: Estimation of Propeller Efficiency Using the Momentum Theorem

Propeller efficiency is required to predict the performance of propeller-powered aircraft. During the development of such aircraft, the designer must be able to predict with a reasonable level of accuracy the propeller efficiency. One way of doing this is to use the *blade element theory* (BET), which is presented in Section 14.6, *Blade element theory*. However, there are times, in particular during the early phases of development, during which it may simply seem too unwieldy. This is due to the fact that detailed information about the geometry of the propeller must be known in advance, but this may be impossible at that stage of development when perhaps the only thing known is that the airplane will be powered by a propeller. The following method, which uses the momentum theorem, is intended to remedy this problem. As such, it is far simpler to implement than the BET, although it is also best implemented using software such as a spreadsheet or mathematical script software like MATLAB. Here, a function written in Visual Basic for Applications (VBA), intended for use with Microsoft Excel, is presented.

The momentum theory can also be used to extract propeller efficiency from flight test data. To do this, details of the flight condition (airspeed, altitude, etc.) and power setting are recorded and reduced in the manner shown below. It must be kept in mind that initially, neither thrust, T, nor propeller efficiency, η_p, is known. The method requires the *viscous profile efficiency*, η_v (see Section 14.3.9, *Propulsive efficiency*) for the propeller to be known. Its value may be provided by the propeller manufacturer and typically ranges from 0.7 to 0.9, depending on propeller type (i.e. fixed or constant speed) and geometry. Sometimes it is given as a function of airspeed. From a certain point of view it can be considered the maximum value or ceiling of the propeller efficiency, η_p.

We begin the process by estimating token thrust using some initial value of the propeller efficiency, denoted by η_p. This is then used to estimate the induced airspeed in the propeller streamtube, which, in turn, is used to calculate the ideal propeller efficiency. This is then used to calculate the new propeller efficiency, η_{pnew}, as the product of the ideal and the viscous profile efficiencies. If the difference between the original and new values is larger than some selected level of accuracy (e.g. 0.0001) then the value of η_p is replaced with that of η_{pnew} and the process is repeated until the difference has reached the desired accuracy.

Step 1: Preliminaries

Establish power, P, airspeed, V, altitude, H, propeller disc area, A, and the estimated viscous profile efficiency.

Step 2: Set Initial Values

Set an initial value for the propeller efficiency, η_p. For instance, start with $\eta_p = 0.5$. Also, estimate the propeller profile efficiency, η_v. In the absence of an analysis use $\eta_v = 0.85$.

Step 3: Determine Thrust

Calculate thrust using P, V, and η_p, using Equation (14-38):

$$T = \frac{\eta_p \times 550 \times P_{BHP}}{V}$$

Step 4: Determine Induced Velocity

Calculate w using Equation (14-57):

$$w = \frac{1}{2}\left[-V + \sqrt{V^2 + \frac{2T}{\rho A}}\right]$$

Step 5: Determine Ideal Efficiency

Calculate ideal propeller efficiency using Equation (14-58):

$$\eta_i = \frac{1}{(1 + w/V)}$$

Step 6: Determine the Next Propeller Efficiency

Recalculate new propeller efficiency:

$$\eta_{p_{new}} = \eta_v \cdot \eta_i$$

Step 7: Determine the Change in Propeller Efficiency

Calculate the difference between the new and original propeller efficiencies:

$$\Delta = \eta_p - \eta_{p_{new}}$$

Step 8: Plan the Next Step

If the difference, Δ, is larger than, say, 0.0001 then set $\eta_p = \eta_{p_{new}}$ and return to Step 3.

If the difference, Δ, is less than 0.0001 then stop.

The following Visual Basic subroutine (or function) implements the above methodology and can be used without modification in Microsoft Excel. The function is written as a VBA (Visual Basic for Applications) function and can then be referenced from within Excel as any other function (e.g. SIN(), COS(), etc.). The arguments (BHP, V, H, etc.) can be cell references (e.g, E4, E5, etc.) or number.

```
Function PropEfficiency(BHP As Single, V As Single, H As Single, Dp As Single, Nv As Single) As Single
'This routine estimates the propeller efficiency for a constant speed propeller,
'by iterating the propulsive disk equations with a user supplied viscous profile efficiency.
'
'Input variables:
'    BHP = Engine horse power at condition in BHP
'    V   = Velocity at condition in ft/s
'    H   = Pressure Altitude in ft
'    Dp  = Propeller diameter in ft
'    Nv  = User entered viscous profile efficiency
'
'Initialize
    Dim A As Single, rho As Single
    Dim Np As Single, Ni As Single, Npnew As Single
    Dim w As Single, T As Single
'Presets
    A = 3.14159265358979 * Dp ^ 2 / 4                        'Prop area (ft!2)
    rho = 0.002378 * (1 - 0.0000068756 * H) ^ 4.2561         'Air density at pressure alt
    Np = 0.5                                                 'Initial propeller efficiency
'Check the value of V to prevent a function crash
    If V = 0 Then
        PropEfficiency = 0
        Exit Function
    End If
'Iterate to get a solution
    Do
        T = Np * BHP * 550 / V                               'Thrust (lbf)
        w = 0.5 * (Sqr(V * V + 2 * T / (rho * A)) - V)       'Induced velocity (ft/s)
        Ni = 1 / (1 + w / V)                                 'Ideal efficiency
        Npnew = Nv * Ni   'New efficiency
        Delta = Abs(Np - Npnew)                              'Difference
        Np = Npnew                                           'Set a new Np before next iteration
    Loop Until Delta < 0.0001
    PropEfficiency = Np
End Function
```

Using this routine, it is possible to estimate the propeller efficiency as a function of airspeed. There are a few caveats though. For instance, the magnitude of the variable η_v (the viscous profile efficiency), denoted by Nv in the VBA function, functions as a limit on the magnitude of the ideal propeller efficiency, η_i, denoted by Ni. This is shown in Figure 14-48 and shows how the total propeller efficiency becomes an asymptote to the value of η_v, indicating this is the maximum value it can acquire. This would be possible only if the ideal efficiency could reach 1. The arguments (W, D_p, and BHP) of the curves in Figure 14-48 correspond to those of the SR22, which has been used throughout this book as an example aircraft.

A more realistic scenario is one in which η_v is not constant, but a function of the airspeed. Such effects can be attributed to viscosity at lower airspeeds and compressibility effects at higher airspeeds. Both affect the shape of the resulting propeller efficiency in profound ways, as can be seen in Figure 14-49. The graph shows how the airspeed changes modify the η_v and lead to a distinct maximum value of the propeller efficiency, rather than a horizontal asymptote. The effect can be extended to fixed-pitch propellers by the proper selection of a curve that describes the changes in the viscous profile efficiency. In this way the propeller efficiency can be forced into a shape that is a reasonable approximation of the real propeller, allowing for more precise performance estimation.

The reader must be mindful that this method is simply a tool intended to make the performance analyses of propeller aircraft easier in the absence of actual manufacturer's data. Of course, in order for it to be practical, it must align well with real data. For instance, we might know that a given propeller is optimized at a given airspeed. Judging from whether the propeller is a fixed or constant-speed propeller, we might guess that its maximum prop efficiency is, say, 0.75 for a fixed pitch and 0.85 for a constant-speed prop. The curve representing η_v should then be selected such that at a low

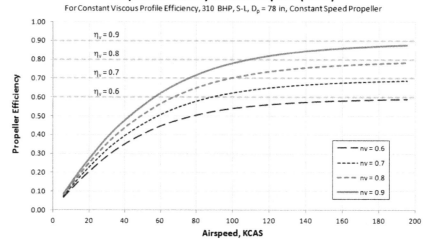

FIGURE 14-48 Propeller efficiencies for selected values of η_v.

airspeed the product $\eta_v \cdot \eta_i$ is about 0.1 and at intended cruising speed (for which the propeller is optimized) the product reaches the maximum values (0.75 or 0.85). At a higher airspeed this would be reduced.

14.6 BLADE ELEMENT THEORY

The *blade element theory* (BET) attempts to estimate the thrust of a propeller by dividing each blade into a number of segments, called *blade elements*. The theory treats each element as an independent two-dimensional airfoil, which allows the aerodynamic forces to be calculated based on the local flow conditions at the element. Then, once the aerodynamic properties have been determined, they are summed up to evaluate the properties of the complete propeller. Propellers usually feature morphing airfoil shapes, as the airfoil changes gradually from a thick airfoil at the hub to at thin one at the tip. The blade element theory can handle any such changes, although solution schemes must be prepared with such changes in mind.

The BET was first devised by the Polish scientist Stefan Drzewiecki (1844−1938), who between 1892 and 1920 almost entirely developed the so-called *primitive blade element theory*, which was published in 1920 in his book *Théorie Générale de L'Hélice* [11] (General Theory of Helixes). This simplified theory ignored the presence of the induced flow inside the propeller streamtube [12]. Consequently, predicted thrust is higher than experimental thrust at the same condition. The method was improved by a contribution from William Froude (1810−1879), himself one of the two people behind the momentum theory (see Section 14.5, *Rankine-Froude momentum theory*). An accurate depiction of the propeller is realized by the BET only through an estimation of the

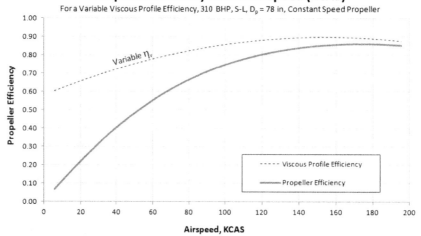

FIGURE 14-49 Propeller efficiencies for a variable η_v.

so-called propeller-induced velocity, which changes the AOA seen by the blade elements. This is caused by the fact that the airspeed inside the propeller streamtube moves faster than the surrounding air. One way of modeling this effect is to use the momentum theory to describe the induced velocity inside the streamtube.

The BET offers a number of advantages when compared to the momentum theory. It can account for varying blade geometry, change in the airfoil's chord, angle-of-pitch, and aerodynamic characteristics. It allows torque to be estimated, which allows the designer to determine the power required to swing it. It also allows important non-linearities, such as that of a standard lift curve, to be modeled. These are the primary reasons for the method's popularity in propeller design.

The BET also suffers from a number of limitations. The theory assumes the flow inside the streamtube is uniform, which does not hold for distorted streamtubes (for instance, blockage effects). The assumption that the forces on each blade element can be treated as two-dimensional neglects spanwise flow. Spanwise flow is important if large pressure gradients exist along the propeller blade. The theory assumes the flow is mostly steady and, therefore, does not model thrust lag caused by change in operating conditions (e.g. change in RPM). The theory also assumes the blade is rigid, ignoring aeroelastic effects. Also, it is a drawback that a special correction has to be made at the hub and tip to account for diminished lift and for a skewed inflow, such as a propeller at a high AOA (e.g. see Figure 14-17).

14.6.1 Formulation

Figure 14-50 shows a propeller of a radius R rotating with an angular velocity of Ω radians/sec. A representative blade element is shown whose chord is denoted by $c(r)$ and width by dr. Its centerline is located at a distance r. A cross-sectional view of this blade element, detailing airspeed components and angles, is shown in Figure 14-51.

The following parameters are identified in Figure 14-51:

$r =$	arbitrary distance from hub to blade element
$V_0 =$	forward airspeed of the airplane
$V_E =$	effective resultant velocity
$V_R =$	resultant velocity
$w =$	elemental induced velocity; due to flow in the streamtube being faster than the far-field airspeed
$\Omega =$	angular velocity of the propeller
$\alpha =$	elemental angle-of-attack
$\alpha_i =$	induced angle-of-attack that results from the induced velocity of air
$\alpha_{ZL} =$	airfoil's zero lift angle
$\beta =$	pitch angle, which is defined as the angle between the rotation plane and the zero lift line of the blade element airfoil
$\phi =$	helix angle

The differential lift and drag forces of the element, dL and dD, can now be written as follows:

$$dL = \frac{1}{2}\rho V_E^2 \cdot c(r) \cdot C_l \cdot dr \qquad (14\text{-}65)$$

$$dD = \frac{1}{2}\rho V_E^2 \cdot c(r) \cdot C_d \cdot dr \qquad (14\text{-}66)$$

where

$c(r) =$ chord at blade station r, in ft or m
$C_d =$ section drag coefficient of the element airfoil
$C_l =$ section lift coefficient of the element airfoil
$\rho =$ air density, in slugs/ft^3 or kg/m^3

The section lift coefficient of the element airfoil can also be written as follows:

$$C_l = C_{l_\alpha}(\beta - \phi - \alpha_i) \qquad (14\text{-}67)$$

where the helix angle is calculated from the relation:

$$\phi = \tan^{-1}\left(\frac{V_0}{V_E}\right) \qquad (14\text{-}68)$$

Observation 1

Note that the C_L-α tables for airfoils for ordinary lifting surfaces always use the chordline as datum, whereas for propeller airfoils the zero-lift line is the datum. This must be accounted for when determining the lift and drag coefficients using such data.

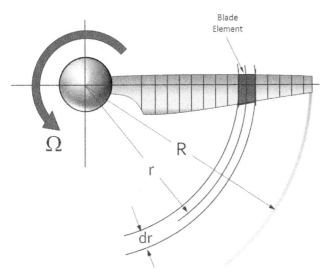

FIGURE 14-50 Definition of terms.

Also note from Figure 14-51 that dL is perpendicular to the vector V_E and dD is normal to it. With that in mind, the differential thrust, torque, and power can be calculated as follows. Recall the definition of power as force × speed. Therefore, the power applied to each blade element equals the amount required to move it through air with force dF_Q at a speed $\Omega \cdot r$:

Differential thrust:

$$dT = dL\cos(\phi + \alpha_i) - dD\sin(\phi + \alpha_i) \tag{14-69}$$

Differential torque:

$$dQ = r \cdot dF_Q = r[dL\sin(\phi + \alpha_i) + dD\cos(\phi + \alpha_i)] \tag{14-70}$$

Differential power:

$$dP = \Omega r \cdot dF_Q = \Omega r[dL\sin(\phi + \alpha_i) + dD\cos(\phi + \alpha_i)] \tag{14-71}$$

This allows us to calculate thrust and torque as follows:

Thrust:

$$T = N_B \int_{R_{hub}}^{R} dT = N_B \int_{R_{hub}}^{R} dL\cos(\phi + \alpha_i)$$
$$- N_B \int_{R_{hub}}^{R} dD\sin(\phi + \alpha_i) \tag{14-72}$$

Torque:

$$Q = N_B \int_{R_{hub}}^{R} r \cdot dF_Q = N_B \int_{R_{hub}}^{R} r \cdot dL\sin(\phi + \alpha_i)$$
$$+ N_B \int_{R_{hub}}^{R} r \cdot dD\cos(\phi + \alpha_i) \tag{14-73}$$

Power:

$$P = N_B \int_{R_{hub}}^{R} \Omega r \cdot dF_Q = N_B \int_{R_{hub}}^{R} \Omega r \cdot dL\sin(\phi + \alpha_i)$$
$$+ N_B \int_{R_{hub}}^{R} \Omega r \cdot dD\cos(\phi + \alpha_i) \tag{14-74}$$

where N_B is the number of propeller blades, R is the tip radius and R_{hub} is the hub radius. Note that in the UK system the unit for torque is ft·lb$_f$ and ft·lb$_f$/s for power. In the SI system the units are N·m (or J) and N·m/s (or W), respectively.

The remaining problem is to determine the induced AOA and its effect on the thrust and torque. It can be seen in Figure 14-51 that the presence of w will reduce the AOA and thus modify the thrust and torque.

The reader should recognize that modern-day methodologies do not evaluate the integrals of Equations (14-72) through (14-74) directly, but rather solve them

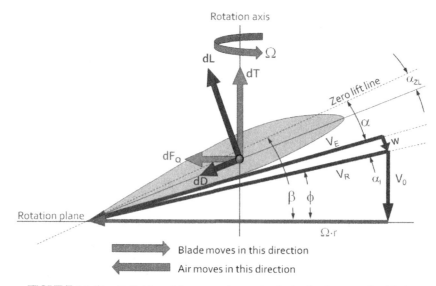

FIGURE 14-51 Definition of forces, angles, and velocity for the propeller blade.

14.6 BLADE ELEMENT THEORY

using numerical integration. This is easy to accomplish using the modern-day spreadsheet or a computer program. Spreadsheet methods increase flexibility and arguably result in improved accuracy, because a number of simplifications would have to be made in order to allow the resulting integrals to be evaluated (for instance when incorporating the helix angle). Additionally, an explicit integral solution would have to be prepared for any individual geometry, whereas a spreadsheet formulating the numerical integration method can be easily modified to incorporate any geometric complexity desired. This is the approach that will be used in this text. The following example displays how such a spreadsheet is constructed and incorporates real complexities such as a tapered blade chord, variable pitch angle, and drag coefficient which is a function of the lift coefficient.

EXAMPLE 14-18

Determine the thrust, torque, and power for a three-bladed propeller whose diameter is 76 inches, hub diameter is 12 inches, a chord whose linear taper is described by the expression $6-2r/R$ inches, and pitch which varies linearly from an angle of 65° at the hub to 20° at the tip. The propeller is rotating at 2500 RPM at a forward speed of 160 KTAS at S-L. The propeller's airfoil is a Clark Y, whose lift and drag coefficients are given by NACA R-502. Initially, assume that the induced velocity, w, is 0, but account for the non-linearity of the lift and drag coefficients.

Note that this example assumes there is no propeller-induced velocity inside the streamtube. As such, it represents the results of Drzewiecki's *primitive blade element theory*. The next example demonstrates the effect of introducing propeller-induced velocity on the derived properties.

Solution

The primary purpose of this problem is to show how to extract propeller properties using the blade element theory, and help the reader understand the importance of accounting for the induced flow field going through the propeller. For this reason it is also important the reader also studies Examples 14-19 and 14-20.

This problem will be solved numerically and is prepared as shown in the table in Figure 14-53. In the interests of space, the table will be broken into three sub-tables, consisting of columns 2–9, 11–17, and 18–25, and each will be given more space so their content is legible. Note that the blue shading in row 10 indicates the values that are calculated as detailed below. However, before we start we must calculate several preliminary values that must be used to complete the table.

The solution as prepared here breaks the propeller blades into 20 elements of equal width. Let's begin the procedure by calculating some preliminaries:

Revolutions per second:

$$n = \frac{RPM}{60} = \frac{2500}{60} = 41.7 \text{ rad/s}$$

Angular velocity:

$$\Omega = 2\pi \left(\frac{RPM}{60}\right) = 2\pi \left(\frac{2500}{60}\right) = 261.8 \text{ rad/s}$$

Hub radius:

$$R_{hub} = \left(\frac{12''}{2}\right)\left(\frac{1\text{ft}}{12''}\right) = 0.500 \text{ ft}$$

Tip radius:

$$R = \left(\frac{76''}{2}\right)\left(\frac{1\text{ft}}{12''}\right) = 3.167 \text{ ft}$$

Element width:

$$\Delta r = \left(\frac{76'' - 12''}{2}\right)\left(\frac{1 \text{ ft}}{12''}\right) \frac{1}{20 \text{ elements}}$$
$$= 0.133 \frac{\text{ft}}{\text{element}}$$

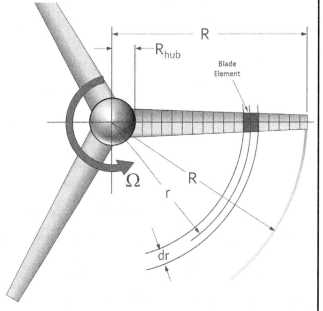

FIGURE 14-52 Definition of propeller geometry.

EXAMPLE 14-18 (cont'd)

Speed of sound at S-L:

$$a = 1116 \text{ ft/s}$$

Thus, the first blade station in Figure 14-54 (see row 1), is the hub radius plus one half of the element width because the radials are positioned at the center of the elements (see Figure 14-52). Keeping in mind that numerical subscripts refer to the row number, we get:

$$r_1 = R_{hub} + \frac{\Delta r}{2} = 0.500 + \frac{0.133}{2} = 0.567 \text{ ft}$$

Consequently, subsequent blade stations can be calculated by simply adding Δr to the previous blade station.

1	2	3	4	5	6	7	8	9	10	11	12	13	14	15	16	17	18	19	20	21	22	23	24	25
	Blade Geometry				Airspeed Components, ft/s				Flow Angles								Lift and Drag Coefficient			Blade Element Differentials				
	r	x = r/R	c(r)	ΔA	V_e	Ω·r	V_R	M	ϕ		α_i		β		$\alpha = \beta-\phi-\alpha_i+\alpha_{\text{ZL}}$		Re	C_l	C_d	dL	dD	dT	dQ	dP
ID	ft		ft	ft²	ft/s	ft/s	ft/s		radians	degrees	radians	degrees	radians	degrees	radians	degrees				lb$_f$	lb$_f$	lb$_f$	ft·lb$_f$	ft·lb$_f$/s
1	0.478	0.1663	0.4723	0.0581	168.8	125.2	210.1	0.188	0.933	53.4	0.000	0.0	0.945	54.1	-0.096	-5.5	635192	0.052	0.010	0.2	0.03	0.1	0.1	18
2	0.601	0.2091	0.4652	0.0572	168.8	157.4	230.8	0.207	0.820	47.0	0.000	0.0	0.914	52.4	-0.015	-0.8	686993	0.382	0.017	1.4	0.06	0.9	0.6	166
3	0.724	0.2518	0.4580	0.0563	168.8	189.5	253.8	0.227	0.728	41.7	0.000	0.0	0.884	50.6	0.048	2.7	743997	0.642	0.032	2.8	0.14	2.0	1.4	368
4	0.847	0.2946	0.4509	0.0554	168.8	221.7	278.7	0.250	0.651	37.3	0.000	0.0	0.853	48.9	0.094	5.4	804146	0.832	0.050	4.3	0.25	3.2	2.4	617
5	0.970	0.3373	0.4438	0.0545	168.8	253.9	304.9	0.273	0.587	33.6	0.000	0.0	0.822	47.1	0.128	7.3	865928	0.960	0.066	5.8	0.40	4.6	3.4	899
6	1.093	0.3801	0.4367	0.0537	168.8	286.1	332.2	0.298	0.533	30.5	0.000	0.0	0.792	45.4	0.151	8.6	928248	1.042	0.080	7.3	0.56	6.0	4.6	1205
7	1.216	0.4228	0.4295	0.0528	168.8	318.2	360.2	0.323	0.488	27.9	0.000	0.0	0.761	43.6	0.166	9.5	990308	1.089	0.090	8.9	0.73	7.5	5.8	1529
8	1.339	0.4656	0.4224	0.0519	168.8	350.4	389.0	0.349	0.449	25.7	0.000	0.0	0.731	41.9	0.174	10.0	1051522	1.114	0.096	10.4	0.89	9.0	7.1	1864
9	1.461	0.5083	0.4153	0.0510	168.8	382.6	418.2	0.375	0.415	23.8	0.000	0.0	0.700	40.1	0.177	10.1	1111457	1.122	0.098	11.9	1.04	10.5	8.4	2202
10	1.584	0.5511	0.4082	0.0502	168.8	414.8	447.8	0.401	0.386	22.1	0.000	0.0	0.670	38.4	0.175	10.0	1169784	1.118	0.096	13.4	1.15	12.0	9.7	2534
11	1.707	0.5938	0.4010	0.0493	168.8	447.0	477.8	0.428	0.361	20.7	0.000	0.0	0.639	36.6	0.170	9.7	1226256	1.103	0.093	14.8	1.24	13.4	10.9	2849
12	1.830	0.6366	0.3939	0.0484	168.8	479.1	508.0	0.455	0.339	19.4	0.000	0.0	0.609	34.9	0.162	9.3	1280680	1.078	0.087	16.0	1.30	14.7	12.0	3135
13	1.953	0.6793	0.3868	0.0475	168.8	511.3	538.5	0.482	0.319	18.3	0.000	0.0	0.578	33.1	0.151	8.7	1332905	1.043	0.080	17.1	1.32	15.8	12.9	3380
14	2.076	0.7221	0.3796	0.0467	168.8	543.5	569.1	0.510	0.301	17.3	0.000	0.0	0.548	31.4	0.138	7.9	1382814	0.999	0.073	18.0	1.30	16.8	13.6	3571
15	2.199	0.7649	0.3725	0.0458	168.8	575.7	599.9	0.538	0.285	16.3	0.000	0.0	0.517	29.6	0.124	7.1	1430309	0.946	0.064	18.5	1.26	17.4	14.1	3699
16	2.322	0.8076	0.3654	0.0449	168.8	607.9	630.9	0.565	0.271	15.5	0.000	0.0	0.487	27.9	0.107	6.2	1475316	0.884	0.056	18.8	1.19	17.8	14.3	3754
17	2.445	0.8504	0.3583	0.0440	168.8	640.0	661.9	0.593	0.258	14.8	0.000	0.0	0.456	26.1	0.090	5.2	1517770	0.815	0.048	18.7	1.10	17.8	14.3	3733
18	2.568	0.8931	0.3511	0.0432	168.8	672.2	693.1	0.621	0.246	14.1	0.000	0.0	0.425	24.4	0.071	4.1	1557620	0.739	0.040	18.2	1.00	17.4	13.9	3631
19	2.691	0.9359	0.3440	0.0423	168.8	704.4	724.3	0.649	0.235	13.5	0.000	0.0	0.395	22.6	0.051	2.9	1594823	0.657	0.033	17.3	0.88	16.7	13.2	3449
20	2.814	0.9786	0.3369	0.0414	168.8	736.6	755.7	0.677	0.225	12.9	0.000	0.0	0.364	20.9	0.031	1.8	1629343	0.571	0.027	16.1	0.77	15.5	12.2	3192
																					218.9	174.9		45796

FIGURE 14-53 Completed spreadsheet analysis.

1	2	3	4	5	6	7	8	9
	Blade Geometry				Airspeed Components, ft/s			
	r	x = r/R	c(r)	ΔA	V	Ω·r	V_R	M
ID	ft		ft	ft²	ft/s	ft/s	ft/s	
1	0.567	0.1789	0.4702	0.0627	270.1	148.4	308.1	0.276
2	0.700	0.2211	0.4632	0.0618	270.1	183.3	326.4	0.292
3	0.833	0.2632	0.4561	0.0608	270.1	218.2	347.2	0.311
4	0.967	0.3053	0.4491	0.0599	270.1	253.1	370.1	0.332
5	1.100	0.3474	0.4421	0.0589	270.1	288.0	394.8	0.354
6	1.233	0.3895	0.4351	0.0580	270.1	322.9	420.9	0.377
7	1.367	0.4316	0.4281	0.0571	270.1	357.8	448.3	0.402
8	1.500	0.4737	0.4211	0.0561	270.1	392.7	476.6	0.427
9	1.633	0.5158	0.4140	0.0552	270.1	427.6	505.8	0.453
10	1.767	0.5579	0.4070	0.0543	270.1	462.5	535.6	0.480
11	1.900	0.6000	0.4000	0.0533	270.1	497.4	566.0	0.507
12	2.033	0.6421	0.3930	0.0524	270.1	532.3	596.9	0.535
13	2.167	0.6842	0.3860	0.0515	270.1	567.2	628.2	0.563
14	2.300	0.7263	0.3789	0.0505	270.1	602.1	659.9	0.591
15	2.433	0.7684	0.3719	0.0496	270.1	637.0	691.9	0.620
16	2.567	0.8105	0.3649	0.0487	270.1	672.0	724.2	0.649
17	2.700	0.8526	0.3579	0.0477	270.1	706.9	756.7	0.678
18	2.833	0.8947	0.3509	0.0468	270.1	741.8	789.4	0.707
19	2.967	0.9368	0.3439	0.0458	270.1	776.7	822.3	0.737
20	3.100	0.9789	0.3368	0.0449	270.1	811.6	855.3	0.766

FIGURE 14-54 Results for columns 1 through 9.

Step 1: Table Columns 2–9

Refer to Figure 14-54. Column 2 is calculated for each row by adding the element width Δr to the blade station (or radial) in the previous row. For example, to calculate the radial in row 10, we look at the value in row 9, which equals 1.633 and add Δr to it, or:

$$r_{10} = r_9 + \Delta r = 1.633 + 0.133 = 1.767 \text{ ft}$$

Column 3 is calculated to allow plotting of results with respect to the fraction of the blade span (see for instance Figure 14-58). The value in row 10 is calculated as follows:

$$x_{10} = \frac{r_{10}}{R} = \frac{1.767}{3.167} = 0.5579$$

Column 4 is the chord at the radial and for row 10 is calculated as follows:

$$c(r_{10}) = 6 - 2\frac{r_{10}}{R} = 6 - 2\frac{1.767}{3.167} = 0.4070 \text{ ft}$$

Column 5 is the area of the blade element in row 10 and is calculated as follows:

$$\Delta A_{10} = c(r) \cdot \Delta r = 0.4070 \times 0.133 = 0.0543 \text{ ft}^2$$

Column 6 is the forward speed (i.e. the airplane's airspeed) converted to ft/s, i.e. 160 KTAS × 1.688 = 270.1 ft/s and is equal for all the rows. Column 7 is the blade's rotational speed and is calculated as follows for row 10:

$$\Omega \cdot r_{10} = 261.8 \times 1.767 = 462.5 \text{ ft/s}$$

V_R is given by Equation (14-9) and is calculated in column 8. The value in row 10 is found from:

$$V_{R_{10}} = \sqrt{V_0^2 + \Omega^2 r_{10}^2} = \sqrt{270.1^2 + 462.5^2} = 535.6 \text{ ft/s}$$

Finally, the Mach number is calculated in column 9 and for row 10 is found to equal:

$$M_{10} = \frac{V_{R_{10}}}{a} = \frac{535.6}{1116} = 0.480$$

Step 2: Table Columns 11–17

The next sub-table contains the flow angles, as shown in Figure 14-55. Columns 10 and 11 contain the helix angle in radians and degrees, respectively. For rows 10 and 11 this is calculated as follows:

$$\phi_{10} = \tan^{-1}\left(\frac{V_0}{\Omega \cdot r_{10}}\right) = \tan^{-1}\left(\frac{270.1}{462.5}\right) = 0.529 \text{ rad}$$
$$= 30.3°$$

Columns 12 and 13 contain the induced flow angle, α_i, ahead of the propeller, which reduces the overall AOA on the blade element. As stated in the introduction to the problem, it is assumed that the induced velocity equals zero. We will study in Example 14-19 the impact of a non-zero induced velocity on the thrust, torque, and power. At any rate, since $w = 0$ the values in columns 12 and 13 are all zero.

1	10	11	12	13	14	15	16	17
	Flow Angles							
	ϕ		α_i		β		$\alpha = \beta - \phi - \alpha_i + \alpha_{ZL}$	
ID	radians	degrees	radians	degrees	radians	degrees	radians	degrees
1	1.068	61.2	0.000	0.0	1.115	63.9	-0.062	-3.545
2	0.975	55.8	0.000	0.0	1.076	61.6	-0.007	-0.417
3	0.891	51.1	0.000	0.0	1.036	59.4	0.037	2.106
4	0.818	46.9	0.000	0.0	0.997	57.1	0.071	4.063
5	0.753	43.2	0.000	0.0	0.958	54.9	0.096	5.512
6	0.697	39.9	0.000	0.0	0.918	52.6	0.114	6.514
7	0.647	37.0	0.000	0.0	0.879	50.4	0.124	7.128
8	0.602	34.5	0.000	0.0	0.840	48.1	0.129	7.407
9	0.563	32.3	0.000	0.0	0.801	45.9	0.129	7.398
10	0.529	30.3	0.000	0.0	0.761	43.6	0.125	7.143
11	0.497	28.5	0.000	0.0	0.722	41.4	0.116	6.675
12	0.470	26.9	0.000	0.0	0.683	39.1	0.105	6.024
13	0.444	25.5	0.000	0.0	0.644	36.9	0.091	5.214
14	0.422	24.2	0.000	0.0	0.604	34.6	0.074	4.267
15	0.401	23.0	0.000	0.0	0.565	32.4	0.056	3.200
16	0.382	21.9	0.000	0.0	0.526	30.1	0.035	2.028
17	0.365	20.9	0.000	0.0	0.487	27.9	0.013	0.764
18	0.349	20.0	0.000	0.0	0.447	25.6	-0.010	-0.582
19	0.335	19.2	0.000	0.0	0.408	23.4	-0.035	-2.000
20	0.321	18.4	0.000	0.0	0.369	21.1	-0.061	-3.482

FIGURE 14-55 Results for columns 10 through 17.

4 BY 24 CLARK Y AIRFOIL CHARACTERISTICS
R.N.: ZERO LIFT=1.55×10⁶, MAX. LIFT=1.48×10⁶

C_L	α	C_D	L/D	c.p.	$C_{m_{c/4}}$	C_{D_0}	α_0
−0.2	−9.0	0.0120	−16.7	−13.1	−0.076	0.0098	−8.4
−.1	−7.7	.0099	−10.1	−50.7	−.076	.0093	−7.3
0	−6.2	.0092	0		−.076	.0088	−6.2
.1	−4.9	.0093	10.8	99.9	−.074	.0087	−5.3
.2	−3.4	.0112	17.9	61.7	−.073	.0090	−4.1
.3	−2.0	.0145	20.7	48.7	−.071	.0095	−3.1
.4	−.6	.0192	20.8	42.0	−.068	.0103	−2.0
.5	.8	.0242	20.6	38.0	−.065	.0103	−1.0
.6	2.2	.0312	19.2	35.6	−.064	.0112	.1
.7	3.6	.0395	17.9	33.8	−.062	.0122	1.1
.8	5.0	.0485	16.5	32.7	−.062	.0129	2.1
.9	6.4	.0582	15.5	31.9	−.062	.0132	3.2
1.0	7.9	.0700	14.3	31.2	−.062	.0143	4.3
1.1	9.6	.0860	12.8	30.7	−.063	.0186	5.7
1.2	11.7	.1093	11.0	30.3	−.064	.0291	7.4
1.255	13.7	.1540	8.1	31.0	−.075	.0664	9.2
1.2	16.4	.2118	5.7	32.8	−.094	.1316	12.1
1.1	19.4	.2660	4.1	34.5	−.107	.1986	15.5
1.0	20.4	.2900	3.5	36.4	−.118	.2343	16.8
.9	20.9	.3039	3.0	38.3	−.128	.2589	17.7

FIGURE 14-56 Original data from Ref. 13 used to model the propeller airfoil.

Columns 14 and 15 contain the geometric pitch angle β which is specified as 65° at the hub (r = 0.5 ft) and 20° at the tip. Using the parametric formulation of Appendix E.5.5 we write:

$$\beta_j = 65 \times \left(1 - \frac{j - 0.5}{20}\right) + 20 \times \left(\frac{j - 0.5}{20}\right)$$

$$\beta_{10} = 65 \times \left(1 - \frac{10 - 0.5}{20}\right) + 20 \times \left(\frac{10 - 0.5}{20}\right) = 43.6°$$

Note that 0.5 is subtracted from the row index because when j = 1 (row 1) the parameter should be zero, but we are at the middle of the first element. Columns 16 and 17 contain the AOA of the airfoil. Referring to Figure 14-55 we see that α is determined from Equation (14-36) as follows:

$$\alpha = \beta - \phi - \alpha_i$$

However, as mentioned in OBSERVATION 1 in the derivation section, the aerodynamic data obtained from NACA R-502 [13], which is used to calculate the section C_l and C_d, uses the chordline for reference datum (see Figure 14-56) rather than the zero-lift AOA line. For this reason the expression must incorporate a correction as shown below:

$$\alpha = \beta - \phi - \alpha_i + \alpha_{ZL}$$

Where α_{ZL} is the zero lift AOA and amounts to −6.2° as can be seen in Figure 14-56 or Figure 14-57. Therefore, for row 10 we get the following AOA:

$$\alpha = \beta - \phi - \alpha_i + \alpha_{ZL}$$
$$= 43.6° - 30.3° - 0° - 6.2° = 7.1°$$

The resulting angles are shown plotted versus the blade station in Figure 14-58.

Step 3: Table Columns 18–25

Column 18 contains the Reynolds number for the blade elements and is calculated from Equation (8-29) (which is only valid for S-L, where the analysis is performed):

$$\text{Re}_{10} \approx 6400 \cdot V_{R_{10}} \cdot c(r_{10}) = 6400 \times 535.6 \times 0.4070$$
$$= 1395175$$

The R_e for the propeller ranges from approximately 930,000 to 1,840,000. For this reason Table III of NACA R-502 was selected (Figure 14-56) as it contains airfoil properties at R_e between these extremes. The C_l and C_d in Table III are plotted in Figure 14-57 and curve-fitted using a best-fit polynomial. These curve-fits are presented below and calculated for row 10 using the previously calculated α = 7.1° for that station:

$$C_{l_{10}} = -0.000004582 \cdot \alpha^4 - 0.00002926 \cdot \alpha^3$$
$$+ 0.000249 \cdot \alpha^2 + 0.07239 \cdot \alpha + 0.4426 = 0.950$$
$$C_{d_{10}} = 0.000006844 \cdot \alpha^3 + 0.0003439 \cdot \alpha^2 + 0.003488 \cdot \alpha$$
$$+ 0.01996 = 0.065$$

14.6 BLADE ELEMENT THEORY

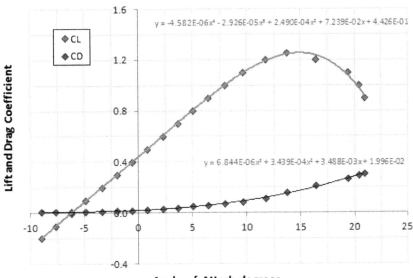

FIGURE 14-57 Original data from Ref. 13 plotted and curve-fitted.

Columns 21 and 22 contain the differential lift and drag acting on the element and are calculated using Equations (14-65) and (14-66). For row 10 this becomes:

$$dL = \frac{1}{2}\rho V_E^2 \cdot c(r) \cdot C_l \cdot \Delta r$$
$$= \frac{1}{2}(0.002378)(535.6)^2 \cdot (0.4070) \cdot (0.950) \cdot (0.133)$$
$$= 17.6 \text{ lb}_f$$

$$dD = \frac{1}{2}\rho V_E^2 \cdot c(r) \cdot C_d \cdot \Delta r$$
$$= \frac{1}{2}(0.002378)(535.6)^2 \cdot (0.4070) \cdot (0.065) \cdot (0.133)$$
$$= 1.20 \text{ lb}_f$$

Column 23 contains the differential thrust calculated using Equation (14-69):

$$dT = dL\cos(\phi + \alpha_i) - dD\sin(\phi + \alpha_i)$$
$$= 17.6 \cdot \cos(30.3 + 0) - 1.20 \cdot \sin(30.3 + 0) = 14.6 \text{ lb}_f$$

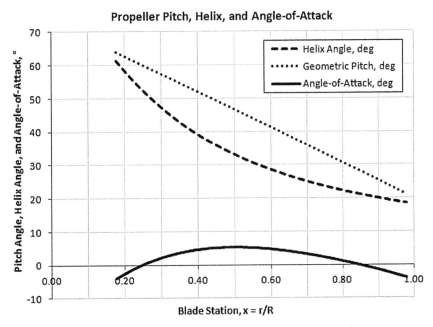

FIGURE 14-58 Results from angular analysis for the condition.

1	18	19	20	21	22	23	24	25
	Lift and Drag Coefficient			Blade Element Differentials				
	Re	C_l	C_d	dL	dD	dT	dQ	dP
ID				lb_f	lb_f	lb_f	$ft{\cdot}lb_f$	$ft{\cdot}lb_f/s$
1	927239	0.190	0.012	1.3	0.08	0.6	0.7	180
2	967474	0.412	0.019	3.2	0.15	1.7	1.9	504
3	1013546	0.596	0.029	5.2	0.25	3.1	3.5	916
4	1063869	0.738	0.040	7.2	0.39	4.6	5.3	1397
5	1117106	0.840	0.051	9.2	0.55	6.3	7.4	1925
6	1172160	0.908	0.059	11.1	0.72	8.1	9.5	2479
7	1228141	0.949	0.065	12.9	0.88	9.8	11.6	3041
8	1284335	0.967	0.067	14.7	1.02	11.5	13.7	3593
9	1340166	0.966	0.067	16.2	1.13	13.1	15.7	4113
10	1395175	0.950	0.065	17.6	1.20	14.6	17.5	4580
11	1448989	0.919	0.061	18.7	1.23	15.8	19.0	4970
12	1501306	0.875	0.055	19.4	1.22	16.8	20.1	5259
13	1551882	0.819	0.048	19.8	1.17	17.4	20.7	5424
14	1600516	0.752	0.042	19.7	1.09	17.5	20.8	5449
15	1647041	0.675	0.035	19.1	0.98	17.2	20.3	5318
16	1691319	0.590	0.029	17.9	0.86	16.3	19.2	5026
17	1733237	0.498	0.023	16.2	0.74	14.8	17.5	4572
18	1772696	0.401	0.018	13.9	0.63	12.8	15.1	3960
19	1809616	0.299	0.014	11.0	0.53	10.2	12.2	3198
20	1843928	0.194	0.012	7.6	0.46	7.1	8.8	2296
						219.2	**260.5**	**68199**

FIGURE 14-59 Results for columns 18 through 25.

Column 24 contains the differential torque calculated using Equation (14-70):

$$\begin{aligned} dQ &= r[dL\sin(\phi+\alpha_i)+dD\cos(\phi+\alpha_i)] \\ &= 1.767\times[17.6\cdot\sin(30.3+0)+1.20\cdot\cos(30.3+0)] \\ &= 17.5 \ \text{ft}\cdot\text{lb}_\text{f} \end{aligned}$$

Column 25 contains the differential thrust calculated using Equation (14-71):

$$\begin{aligned} dP &= \Omega r[dL\sin(\phi+\alpha_i)+dD\cos(\phi+\alpha_i)] \\ &= 261.8\times 1.767 \\ &\quad \times[17.6\cdot\sin(30.3+0)+1.20\cdot\cos(30.3+0)] \\ &= 4580 \ \text{ft}\cdot\text{lb}_\text{f}/\text{s} \end{aligned}$$

Once all the rows have been calculated in this fashion the next step is to sum up the *dT*, *dQ*, and *dP*, but the sum is the thrust, torque, and power generated by a single blade. These values must be multiplied by 3 to account for the three blades on the propeller. The resulting values are:

Thrust:

$$T = 3\times 219.2 = 658 \ \text{lb}_\text{f}$$

Torque:

$$Q = 3\times 260.5 = 782 \ \text{ft}\cdot\text{lb}_\text{f}$$

Power:

$$\begin{aligned} P &= 3\times 68199 = 204597 \ \text{ft}\cdot\text{lb}_\text{f}/\text{s} = \frac{204597}{550} \\ &= 372 \ \text{BHP} \end{aligned}$$

We can calculate additional information as shown below:

Power coefficient:

$$C_P = \frac{P}{\rho n^3 D^5} = \frac{204597}{(0.002378)(41.7)^3(76/12)^5} = 0.1167$$

Thrust coefficient:

$$C_T = \frac{T}{\rho n^2 D^4} = \frac{658}{(0.002378)(41.7)^2(76/12)^4} = 0.0990$$

Torque coefficient:

$$C_Q = \frac{Q}{\rho n^2 D^5} = \frac{782}{(0.002378)(41.7)^2(76/12)^5} = 0.0186$$

Advance ratio:

$$J = \frac{V}{nD} = \frac{270.1}{(41.7)(76/12)} = 1.023$$

Propeller efficiency:

$$\eta_p = J\frac{C_T}{C_P} = 1.023\frac{0.0990}{0.1167} = 0.8681$$

Finally, note that the power can also be calculated using the torque coefficient as follows:

$$C_P = 2\pi C_Q = 2\pi(0.0186) = 0.1167$$

$$P = \rho n^3 D^5 C_P = (0.002378)(41.7)^3(76/12)^5(0.1167)$$
$$= 204597 \text{ ft·lb}_f/s$$

EXAMPLE 14-19

Using the spreadsheet of Example 14-18 determine, tabulate, and plot the variation of T, Q, P, η_p for non-zero values of induced velocity, such that α_i ranges from 0.5° to 5° stepping through a 0.5° interval. The purpose of this example is to evaluate the influence of the induced airspeed on the above parameters. Note that these angles are entered for demonstration only. Example 14-20 estimates these angles using the momentum theory.

Solution

The results are tabulated below and plotted in Figure 14-60. We can see that the induced velocity significantly affects all of the parameters requested.

The aircraft designer should be aware of and concerned with such changes. Even a modest change in the induced flow angle, from 0° to 2°, reduces thrust from 219 lb$_f$ to 166 lb$_f$, by almost 25%! Additionally, the blade element

α_i		T	Q	P	P	η_p
radians	degrees	lb$_f$	ft·lb$_f$	ft·lb$_f$/s	BHP	
0.000	0.000	658	782	204598	372	0.8681
0.009	0.500	618	747	195632	356	0.8527
0.017	1.000	578	712	186324	339	0.8373
0.026	1.500	538	675	176684	321	0.8219
0.035	2.000	498	637	166720	303	0.8066
0.044	2.500	458	598	156441	284	0.7912
0.052	3.000	419	557	145850	265	0.7756
0.061	3.500	380	515	134951	245	0.7598
0.070	4.000	341	473	123746	225	0.7437
0.079	4.500	302	429	112233	204	0.7269
0.087	5.000	264	384	100410	183	0.7093

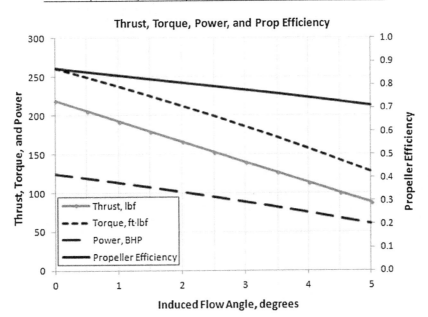

FIGURE 14-60 Variation of thrust, torque, power, and propeller efficiency with α_i.

EXAMPLE 14-19 (cont'd)

method does not include tip effects, which may reduce thrust by additional 10% or so. This shows that ignoring this flow-field phenomenon has a major effect on accuracy and must be taken into account. This effect will be formulated in Section 14.6.4, *Step-by-step: Prandtl's tip and hub loss corrections.*

14.6.2 Determination of α_i Using the Momentum Theory

Having demonstrated the importance of induced *AOA*, α_i, on propeller thrust and power (see Example 14-18), we will now develop a method to help determine the induced velocity, w. One way of accomplishing this is to use the actuator disc or momentum theory (see Section 14.5). Considering Figure 14-51 we see that α_i depends on w, which, in turn, depends on the blade loading. The method shown here is best implemented in a spreadsheet or a computer program. We will incorporate it into the spreadsheet just developed in the preceding examples.

Consider the propeller of Figure 14-61. A circular area (or annulus) of width dr located at blade station r is shown. The total area of the annulus is $dA = 2\pi r \cdot dr$. The total thrust of the propeller using the Rankine-Froude momentum theory of Section 14.5 is given by Equation (14-29), repeated here for convenience:

$$T = 2\rho A_2 (V_0 + w)w \quad (14\text{-}75)$$

where

A_2 = area of the actuator disc = πR^2
V_0 = far-field airspeed
w = propeller induced velocity

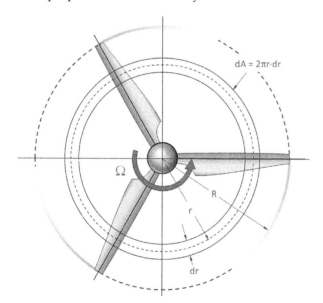

FIGURE 14-61 Annulus of an actuator disc.

Therefore, the thrust generated by the annulus according to the momentum theory is:

$$dT = 2\rho dA(V_0 + w)w = 2\rho(2\pi r \cdot dr)(V_0 + w)w \quad (i)$$

However, dT is also defined by the blade element theory using Equation (14-38). Using that expression and by combining Equations (14-34) for dL and (14-35) for dD and noting that dT is the product of the number of blades, N_B, we get:

$$dT = N_B[dL\cos(\phi + \alpha_i) - dD\sin(\phi + \alpha_i)]$$
$$= N_B\left[\frac{1}{2}\rho V_E^2 \cdot c(r) \cdot C_l \cdot dr\right]\cos(\phi + \alpha_i)$$
$$- N_B\left[\frac{1}{2}\rho V_E^2 \cdot c(r) \cdot C_d \cdot dr\right]\sin(\phi + \alpha_i)$$

Simplified, this becomes:

$$dT = \frac{N_B}{2}\rho V_E^2 \cdot c(r) \cdot [C_l\cos(\phi + \alpha_i) - C_d\sin(\phi + \alpha_i)] \cdot dr$$

(ii)

Considering Figure 14-51 again shows that we can make the following approximations for V_E, w, and the angle $\phi + \alpha_i$:

Effective resultant velocity:

$$V_E \cong \sqrt{(w + V_0)^2 + \Omega^2 r^2} \quad (iii)$$

Local induced velocity:

$$w \cong V_E \cdot \tan\alpha_i \quad (iv)$$

Cosine of $\phi + \alpha_i$:

$$\cos(\phi + \alpha_i) \cong \frac{\Omega r}{V_E} = \frac{\Omega r}{\sqrt{(w + V_0)^2 + \Omega^2 r^2}} \quad (v)$$

Sine of $\phi + \alpha_i$:

$$\sin(\phi + \alpha_i) \cong \frac{w + V_0}{V_E} = \frac{w + V_0}{\sqrt{(w + V_0)^2 + \Omega^2 r^2}} \quad (vi)$$

14.6 BLADE ELEMENT THEORY

Insert this into Equation (ii) to get:

$$dT = \frac{N_B}{2}\rho V_E^2 \cdot c(r) \cdot [C_l \cos(\phi + \alpha_i) - C_d \sin(\phi + \alpha_i)] \cdot dr$$

$$= \frac{N_B}{2}\rho \cdot c(r) \cdot \left((w + V_0)^2 + \Omega^2 r^2\right)$$

$$\cdot \left[C_l \frac{\Omega r}{\sqrt{(w+V_0)^2 + \Omega^2 r^2}} - C_d \frac{w + V_0}{\sqrt{(w+V_0)^2 + \Omega^2 r^2}} \right] \cdot dr$$

Simplifying leads to:

$$dT = \frac{N_B}{2}\rho \cdot c(r) \cdot \sqrt{(w+V_0)^2 + \Omega^2 r^2}$$
$$\cdot [C_l(\Omega r) - C_d(w+V_0)] \cdot dr \qquad \text{(vii)}$$

It must be pointed out that the induced flow angle will reduce the value of the C_l and consequently change the magnitude of C_d. This, in turn, will change the loading on the blades and therefore affect the induced velocity.

Equating Equations (i) and (vii) leads to (rewrite $c(r) = c$ for clarity):

$$2\rho(2\pi r \cdot dr)(V_0 + w)w$$
$$= \frac{N_B}{2}\rho \cdot c \cdot \sqrt{(w+V_0)^2 + \Omega^2 r^2}$$
$$\cdot [C_l(\Omega r) - C_d(w+V_0)] \cdot dr$$

Canceling terms and expanding:

$$8(\pi r)(V_0 + w)w = N_B \cdot c \cdot \sqrt{(w+V_0)^2 + \Omega^2 r^2}$$
$$\cdot [C_l(\Omega r) - C_d(w+V_0)]$$

Simplify by writing:

$$\frac{8\pi r}{N_B \cdot c} w = \frac{\sqrt{(w+V_0)^2 + \Omega^2 r^2}}{(V_0 + w)} \cdot [C_l(\Omega r) - C_d(w+V_0)] \qquad \text{(viii)}$$

The purpose of these manipulations is to ultimately retrieve the value of w at a given radial station r. Equation (viii) is best solved by an iterative scheme, such as Newton-Raphson (see Appendix E.6.19), but this also requires a specific function, call it $f(w)$, and its derivative, i.e. $f'(w)$.

Let's define a function $f(w)$ as follows:

$$f(w) = \frac{8\pi r}{N_B \cdot c} w - \frac{\sqrt{(w+V_0)^2 + \Omega^2 r^2}}{(V_0 + w)} \qquad \text{(ix)}$$
$$\cdot [C_l(\Omega r) - C_d(w+V_0)]$$

Rewriting it using the definition of V_E we get:

$$f(w) = \frac{8\pi r}{N_B \cdot c} w - \frac{V_E}{(V_0 + w)} \cdot [C_l(\Omega r) - C_d(w+V_0)]$$

(14-76)

The derivative of $f(w)$ is determined as follows:

$$f'(w) = \frac{8\pi r}{N_B \cdot c}$$
$$- \frac{d}{dw}\left[\frac{\left((V_0+w)^2 + \Omega^2 r^2\right)^{0.5}}{(V_0 + w)} \cdot [C_l(\Omega r) - C_d(w+V_0)] \right]$$

$$= \frac{8\pi r}{N_B \cdot c} - \frac{d}{dw}\left[\frac{\left((V_0+w)^2 + \Omega^2 r^2\right)^{0.5} \cdot C_l(\Omega r)}{(V_0 + w)} \right.$$
$$\left. - \left((V_0+w)^2 + \Omega^2 r^2\right)^{0.5} \cdot C_d \right]$$

Carrying the differentiation further:

$$f'(w) = \frac{8\pi r}{N_B \cdot c} - C_l(\Omega r) \frac{d}{dw} \frac{\left((V_0+w)^2 + \Omega^2 r^2\right)^{0.5}}{(V_0 + w)}$$
$$+ C_d \frac{d}{dw}\left((V_0+w)^2 + \Omega^2 r^2\right)^{0.5} \qquad \text{(x)}$$

Differentiate the right-most term using the chain rule (see Appendix E.6.2),

$$\frac{d}{dw}\left((V_0+w)^2 + \Omega^2 r^2\right)^{0.5}$$
$$= \frac{d}{dw}(w^2 + 2V_0 w + V_0^2 + \Omega^2 r^2)^{0.5}$$
$$= 0.5(w^2 + 2V_0 w + V_0^2 + \Omega^2 r^2)^{-0.5} \cdot (2w + 2V_0)$$
$$= \frac{(w + V_0)}{\left((w+V_0)^2 + \Omega^2 r^2\right)^{0.5}}$$

Differentiate the center term:

$$\frac{d}{dw}\frac{(w^2 + 2V_0 w + V_0^2 + \Omega^2 r^2)^{0.5}}{(V_0 + w)} = \frac{\frac{d}{dw}(w^2 + 2V_0 w + V_0^2 + \Omega^2 r^2)^{0.5}(V_0+w) - (w^2 + 2V_0 w + V_0^2 + \Omega^2 r^2)^{0.5}\frac{d}{dw}(V_0 + w)}{(V_0 + w)^2}$$

Completing differentiating the center term leads to:

$$\frac{d}{dw}\frac{(w^2 + 2V_0 w + V_0^2 + \Omega^2 r^2)^{0.5}}{(V_0 + w)}$$

$$= \frac{\frac{(V_0+w)^2}{(w^2+2V_0 w+V_0^2+\Omega^2 r^2)^{0.5}} - (w^2 + 2V_0 w + V_0^2 + \Omega^2 r^2)^{0.5}}{(V_0 + w)^2}$$

$$= \frac{1}{\left((w+V_0)^2 + \Omega^2 r^2\right)^{0.5}} - \frac{\left((w+V_0)^2 + \Omega^2 r^2\right)^{0.5}}{(V_0+w)^2}$$

Insert into Equation (x) and manipulate.

$$f' = \frac{8\pi r}{N_B \cdot c} - C_l(\Omega r)\frac{d}{dw}\frac{(w^2 + 2V_0 w + V_0^2 + \Omega^2 r^2)^{0.5}}{(V_0 + w)}$$
$$+ C_d \frac{d}{dw}(w^2 + 2V_0 w + V_0^2 + \Omega^2 r^2)^{0.5}$$

$$f' = \frac{8\pi r}{N_B \cdot c} - C_l(\Omega r)$$
$$\times \left(\frac{1}{\left((w+V_0)^2 + \Omega^2 r^2\right)^{0.5}} - \frac{\left((w+V_0)^2 + \Omega^2 r^2\right)^{0.5}}{(V_0+w)^2}\right)$$
$$+ C_d \frac{(w+V_0)}{\left((w+V_0)^2 + \Omega^2 r^2\right)^{0.5}}$$

Finally, rewrite by using the effective resultant velocity, V_E:

$$f'(w) = \frac{8\pi r}{N_B \cdot c} - C_l(\Omega r)\left(\frac{1}{V_E} - \frac{V_E}{(V_0+w)^2}\right)$$
$$+ C_d \frac{(w+V_0)}{V_E} \quad (14\text{-}77)$$

Having defined $f(w)$ and $f'(w)$, we can now determine the root of f, i.e. the value of w that results in $f(w) = 0$. The value of w is the induced velocity at the blade element at blade station r. Using the Newton-Raphson iterative scheme (mentioned earlier) this would be implemented as follows.

Step 1: Initial Value
Start with an initial value of w, call it w_0.

Step 2: Determine the Next w
Calculate the next value of w, call it w_1, using the following expression;

$$w_1 = w_0 - \frac{f(w_0)}{f'(w_0)}$$

where $f(w)$ is calculated from Equation (14-76) and $f'(w)$ from Equation (14-77).

Step 3: Determine the Difference
Calculate the difference $w_1 - w_0$. If this value is larger than, say, 0.01 then go back to Step 1. Otherwise, the value of w is the current value of w_1, completing the solution.

EXAMPLE 14-20

Using the spreadsheet of Example 14-18 determine and tabulate the induced velocity, w, for each element. Then, calculate the average w for the propeller and compare it to that from the actuator-disc theory. Also, compare the thrust, torque, and power required to that of Example 14-18.

Solution

A considerable amount of iterative computational labor is required to solve this problem. For this reason, in the interests of space, only a sample calculation, representing the first iteration of the 10th element, can be demonstrated. What will not be demonstrated is the final outcome of a number of such iterations that are necessary to yield the final theoretical value of the induced velocity for that element (denoted by w_{10}). The same process must be repeated for all the other elements that make up the propeller. To complicate things, this initial w_{10} changes the induced AOA, α_i, of the 10th element and, thus, the section lift and drag coefficients, denoted by C_l and C_d. This changes the loading on the propeller, which in turn calls for a new iteration until the propeller loading no longer changes. As a consequence, subsequent iterations of the w on each blade element requires the use of the new values of the C_l and C_d. This procedure has to be repeated a few times until the solution converges to a fixed value of w (or α_i) and therefore a final value of the C_l and C_d. This is illustrated in the procedure below. Preliminaries:

EXAMPLE 14-20 (cont'd)

Number of blades	$N_B = 3$
Forward speed	$V_0 = 270.1$ ft/s
Angular velocity	$\Omega = 261.8$ rad/s
Radial station	$r = 1.767$ ft (10th element)
Blade chord	$c = 0.4070$ ft
Initial value for w	$w_{10} = 1.00$ ft/s
Initial AOA	$\alpha_{10} = 7.143°$

Step 1: Start the iterative process to compute w_{10}. Begin by calculating a current value for V_E from Equation (iii) using the current (or initial) value of w_{10};

$$V_E = \sqrt{(w+V_0)^2 + \Omega^2 r^2}$$
$$= \sqrt{(1+270.1)^2 + (261.8)^2(1.767)^2} = 536.2 \text{ ft/s}$$

Step 2: Use Equation (14-76) to calculate the value of the function $f(w)$, noting that $C_l(\alpha_{10}) = 0.950$ and $C_d(\alpha_{10}) = 0.065$ are obtained from the curve-fit expression derived in Example 14-18.

$$f(w) = f(w) = \frac{8\pi r}{N_B \cdot c} w - \frac{V_E}{(V_0 + w)} \cdot [C_l(\Omega r) - C_d(w + V_0)]$$
$$= \frac{8\pi(1.767)}{3 \times 0.4070}(1) - \frac{536.2}{(270.1 + 1)}$$
$$\cdot [0.950(261.8)(1.767) - 0.065(1 + 270.1)] = -798.0$$

Step 3: Use Equation (14-77) to calculate the derivative of the function $f'(w)$;

$$f'(w) = \frac{8\pi r}{N_B \cdot c} - C_l(\Omega r)\left(\frac{1}{V_E} - \frac{V_E}{(V_0 + w)^2}\right) + C_d \frac{(w+V_0)}{V_E}$$
$$= \frac{8\pi(1.767)}{3 \times 0.4070} - 0.950(261.8)(1.767)$$
$$\times \left(\frac{1}{536.2} - \frac{536.2}{(270.1 + 1)^2}\right)$$
$$+ 0.065 \frac{(1 + 270.1)}{536.2} = 38.73$$

Step 4: Calculate the next value of w_{10} using the Newton-Raphson iterative algorithm of Appendix E.6.19.

$$w_{10} = w_{10\,old} - \frac{f(w_{10\,old})}{f'(w_{10\,old})} = 1 - \frac{-798.0}{38.73} = 21.60 \text{ ft/s}$$

Step 5: Comparing the original and new w_{10}, we see the difference is $21.60 - 1.00 = 20.60$, a comparatively large change between iterations. We want the difference to be much smaller, say 0.0001, before considering our solution fully converged. Before starting the next iteration we must proceed to Step 6 and calculate the induced angle-of-attack, $\alpha_{i\,10}$.

Step 6: Calculate the induced angle-of-attack, $\alpha_{i\,10}$, per Equation (iv).

$$\alpha_{10} = \tan^{-1}(w/V_E) = \tan^{-1}(21.60/536.2) = 2.30°$$

This value will be used to recalculate C_l and C_d in the next iteration. The iteration is needed to bypass the problem of circular references in the spreadsheet, and is avoided by copying the induced angles-of-attack and pasting their values (in radians and degrees) into columns 12 and 13, in turn changing the angle of attack in columns 16 and 17 and thus C_l and C_d.

Implementing the methodology in a spreadsheet is relatively easy. Therefore, the next step is to return to Step 1 with the new value of w_{10} and $\alpha_{i\,10}$ and repeat the calculations, until convergence is achieved. Then go to Step 7. The final value of w_{10} will be the value of w at blade element number 10.

Note that it is far faster to use the built-in power of Microsoft Excel and create this as functions as shown in Section 14.6.5, *Computer code: Determination of the propeller induced velocity*, to do these iterations. It returns w without the hassle of having to perform the iteration demonstrated here.

Step 7: Columns 30–36 show how Figure 14-62 shows the values for $\alpha_{i\,10}$ changed in the seven iterations required. The initial value of $\alpha_{i\,10}$ was 0°. After the first iteration $\alpha_{i\,10} = 2.269°$, and 1.936° after the second and so on. The corresponding induced flow velocity is shown in Figure 14-63. It can be seen that after the 7th iteration, $w_{10} = 18.86$ ft/s.

Step 8: The average of w in column 26 in Figure 14-63 is 13.38 ft/s. Let's compare this value to the induced flow velocity from the actuator disc theory, but this is

EXAMPLE 14-20 (cont'd)

considered an average value for the entire disc. Using Equation (14-57) we get:

$$w = \frac{1}{2}\left[-V_0 + \sqrt{V_0^2 + \frac{2T}{\rho A_2}}\right]$$

$$= \frac{1}{2}\left[-270.1 + \sqrt{(270.1)^2 + \frac{2 \times 658}{0.002378 \times 31.50}}\right]$$

$$= 13.09 \text{ ft/s}$$

Step 9: Compare T, Q, and P for the propeller, without and with the induced flow:

T lb$_f$		
w = 0	w > 0	% diff
658	556	15.5

Q ft·lb$_f$		
w = 0	w > 0	% diff
782	693	11.3

P BHP		
w = 0	w > 0	% diff
372	330	11.3

1	10	11	12	13	14	15	16	17	30	31	32	33	34	35	36
	Flow Angles														
	ϕ		α_i		β		$\alpha = \beta - \phi - \alpha_i + \alpha_{ZL}$		$\alpha_i = \tan(w/V_E)$						
ID	radians	degrees	radians	degrees	radians	degrees	radians	degrees	Iter 1	Iter 2	Iter 3	Iter 4	Iter 5	Iter 6	Iter 7
1	1.068	61.2	0.008	0.430	1.115	63.9	-0.069	-4.0	0.000	0.519	0.411	0.434	0.429	0.430	0.430
2	0.975	55.8	0.017	0.962	1.076	61.6	-0.024	-1.4	0.000	1.152	0.924	0.969	0.960	0.962	0.961
3	0.891	51.1	0.024	1.370	1.036	59.4	0.013	0.7	0.000	1.628	1.319	1.379	1.367	1.370	1.369
4	0.818	46.9	0.029	1.668	0.997	57.1	0.042	2.4	0.000	1.966	1.613	1.678	1.666	1.668	1.668
5	0.753	43.2	0.033	1.873	0.958	54.9	0.064	3.6	0.000	2.188	1.817	1.882	1.870	1.873	1.872
6	0.697	39.9	0.035	1.999	0.918	52.6	0.079	4.5	0.000	2.317	1.945	2.007	1.997	1.999	1.998
7	0.647	37.0	0.036	2.061	0.879	50.4	0.088	5.1	0.000	2.375	2.010	2.069	2.060	2.061	2.061
8	0.602	34.5	0.036	2.073	0.840	48.1	0.093	5.3	0.000	2.379	2.024	2.081	2.072	2.073	2.073
9	0.563	32.3	0.036	2.044	0.801	45.9	0.093	5.4	0.000	2.341	1.997	2.051	2.043	2.044	2.044
10	0.529	30.3	0.035	1.981	0.761	43.6	0.090	5.2	0.000	2.269	1.936	1.988	1.980	1.981	1.981
11	0.497	28.5	0.033	1.889	0.722	41.4	0.084	4.8	0.000	2.167	1.846	1.896	1.888	1.889	1.889
12	0.470	26.9	0.031	1.774	0.683	39.1	0.074	4.2	0.000	2.038	1.733	1.780	1.773	1.774	1.774
13	0.444	25.5	0.029	1.639	0.644	36.9	0.062	3.6	0.000	1.885	1.600	1.644	1.638	1.639	1.638
14	0.422	24.2	0.026	1.486	0.604	34.6	0.049	2.8	0.000	1.712	1.450	1.491	1.485	1.486	1.486
15	0.401	23.0	0.023	1.319	0.565	32.4	0.033	1.9	0.000	1.521	1.288	1.324	1.318	1.319	1.319
16	0.382	21.9	0.020	1.141	0.526	30.1	0.015	0.9	0.000	1.316	1.114	1.145	1.140	1.141	1.141
17	0.365	20.9	0.017	0.954	0.487	27.9	-0.003	-0.2	0.000	1.100	0.932	0.958	0.954	0.954	0.954
18	0.349	20.0	0.013	0.761	0.447	25.6	-0.023	-1.3	0.000	0.876	0.743	0.764	0.761	0.761	0.761
19	0.335	19.2	0.010	0.563	0.408	23.4	-0.045	-2.6	0.000	0.647	0.550	0.565	0.563	0.563	0.563
20	0.321	18.4	0.006	0.361	0.369	21.1	-0.067	-3.8	0.000	0.415	0.353	0.362	0.361	0.361	0.361

FIGURE 14-62 Inclusion of induced flow angles.

14.6.3 Compressibility Corrections

The local velocity of the blade elements residing near the outboard part of a normally operating propeller is in the high subsonic range. Consequently, these elements should be corrected for compressibility effects. These effects call for corrections of the drag that differ from those for lift. The corrections are applied to the drag and lift coefficients.

Correction of Lift

The lift can be corrected using Prandtl-Glauert, Kármán-Tsien, or Laitone methods (see Section 8.3.6, *Compressibility modeling*). The latter two must be applied to the pressure coefficients and, thus, require a knowledge of the chordwise distribution of pressure. The first one, Prandtl-Glauert, while simple to apply, becomes gradually inaccurate for airspeeds in excess of M 0.70–0.75.

Prandtl-Glauert:

$$C_{l_\alpha} = \frac{C_{l_{\alpha_0}}}{\sqrt{1-M^2}} \qquad (14\text{-}78)$$

where

M = helical Mach number (dependent on rotation and forward airspeed per Equation (14-9))
C_{l_α} = airfoil lift curve slope at the specific Mach number
$C_{l_{\alpha_0}}$ = incompressible airfoil lift curve slope

14.6 BLADE ELEMENT THEORY

	26	27	28	29
	\multicolumn{4}{c}{Induced Velocity Analysis}			
	w	V_E	\multicolumn{2}{c}{$\alpha_i = \tan(w/V_E)$}	
ID	ft/s	ft/s	radians	degrees
1	2.33	310.2	0.008	0.430
2	5.55	331.0	0.017	0.961
3	8.46	353.8	0.024	1.369
4	11.01	378.2	0.029	1.668
5	13.20	404.0	0.033	1.872
6	15.03	430.7	0.035	1.998
7	16.50	458.4	0.036	2.061
8	17.62	486.8	0.036	2.073
9	18.41	515.8	0.036	2.044
10	18.86	545.3	0.035	1.981
11	18.98	575.3	0.033	1.889
12	18.76	605.6	0.031	1.774
13	18.20	636.3	0.029	1.638
14	17.31	667.2	0.026	1.486
15	16.08	698.4	0.023	1.319
16	14.53	729.7	0.020	1.141
17	12.68	761.3	0.017	0.954
18	10.53	793.1	0.013	0.761
19	8.10	825.0	0.010	0.563
20	5.40	857.1	0.006	0.361
	13.38			

FIGURE 14-63 Resulting induced flow velocity.

Correction of Drag

The propeller drag at high subsonic speeds must be corrected for skin friction and wave drag. Skin friction can be corrected using the Frankl-Voishel formula of Equation (8-59), where the Mach number at each blade element is evaluated using Equation (14-9). Then, if the Mach number is greater than M_{crit} (see Section 8.3.7, *The Critical Mach Number, M_{crit}*) the drag rise at the corresponding Mach number must be determined and applied.

14.6.4 Step-by-step: Prandtl's Tip and Hub Loss Corrections [14]

The section lift coefficient at the tip of any lifting surface must necessarily go to zero. Unmodified BET ignores this effect and effectively assumes the blade element at the tip generates undiminished lift. This error was recognized by early scientists. Most prominent of those was Ludwig Prandtl (1875–1953), who developed a correction that is easily and conveniently incorporated into the BET. This correction method has been expanded to include corrections at the hub as well (which is another radial where the lift goes to zero). Applying the corrections is a three-step process.

Step 1: Tip Correction Parameter

Calculate a tip correction parameter, P_{tip}:

$$P_{tip} = \frac{N_B}{2} \frac{R-r}{r \sin \phi} \quad (14\text{-}79)$$

Step 2: Tip Correction

Calculate the tip correction factor, F_{tip}:

$$F_{tip} = \frac{2}{\pi} \cos^{-1}\left(e^{-P_{tip}}\right) \quad (14\text{-}80)$$

Step 3: Hub Correction Parameter

Calculate a hub correction parameter, P_{hub}:

$$P_{hub} = \frac{N_B}{2} \frac{r - R_{hub}}{r \sin \phi} \quad (14\text{-}81)$$

Step 4: Hub Correction

Calculate a hub correction factor, F_{hub}:

$$F_{hub} = \frac{2}{\pi} \cos^{-1}\left(e^{-P_{hub}}\right) \quad (14\text{-}82)$$

Step 5: Prandtl Correction

Calculate a common correction factor, F_P, for Prandtl correction:

$$F_P = F_{hub} \cdot F_{tip} \quad (14\text{-}83)$$

The factor F_P should vary from 0 at the hub to a value close to 1 over the mid span of the blade, and back to 0 at the tip. This factor is then used by modifying Equations (14-72) through (14-74) as follows:

Thrust:

$$T = N_B \cdot F_P \int_{R_{hub}}^{R} dL \cos(\phi + \alpha_i)$$
$$- N_B \cdot F_P \int_{R_{hub}}^{R} dD \sin(\phi + \alpha_i) \quad (14\text{-}84)$$

Torque:

$$Q = N_B \cdot F_P \int_{R_{hub}}^{R} r \cdot dL \sin(\phi + \alpha_i)$$
$$+ N_B \cdot F_P \int_{R_{hub}}^{R} r \cdot dD \cos(\phi + \alpha_i) \quad (14\text{-}85)$$

Power:

$$P = N_B \cdot F_P \int_{R_{hub}}^{R} \Omega r \cdot dL \sin(\phi + \alpha_i)$$

$$+ N_B \cdot F_P \int_{R_{hub}}^{R} \Omega r \cdot dD \cos(\phi + \alpha_i) \quad (14\text{-}86)$$

14.6.5 Computer code: Determination of the Propeller Induced Velocity

The solution procedure here was implemented in Microsoft Excel. The program allows the iterative determination of w using the Newton-Raphson algorithm to be written in Visual Basic for Applications (VBA) and is shown below. The function **F_of_w** calculates the value of $f(w)$ using a number of input variables, V_0, Ω, r, N_B, $c(r)$, C_d, C_l, and w, using Equation (14-76).

```
Function F_of_w(V0 As Single, Omega As Single, r As Single, NB As Single, CofR As Single,Cd As Single, Cl
As Single, w As Single) As Single
'NOTE: 8 * 3.14159265358979 = 25.1327412287183
    F_of_w = 25.1327412287183 * w * r / (NB * CofR) - Sqr(1 + (Omega * r / (V0 + w)) ^ 2) * (Cl * Omega
* r - Cd * (w + V0))
End Function
```

The function **InducedVelocity** returns the final iterated w. Note that rather than calculating $f'(w)$ using Equation (14-77) it uses a Taylor series finite difference scheme for the derivative (the variable *FprimeofX0*). It also counts the number of iterations and exits the do-loop if it takes more than 10 interations to converge. The code generally converges in fewer than 10 iterations. This function can be accessed from within Excel by a simple cell reference such as '=InducedVelocity(*list of input cells*)'.

```
Function InducedVelocity(V0 As Single, Omega As Single, r As Single, NB As Single, CofR As Single, Cd As
Single, Cl As Single) As Single
'Initialize
    Dim X0 As Single, X1 As Single, Count As Integer
    Dim F1 As Single, F2 As Single, FofX0 As Single, FprimeofX0 As Single
'Presets
    X0 = 1
'Iterate
    Do
        'Advance counter
        Count = Count + 1
        'Calculating the left hand side, at a distance of 0.01
        F1 = F_of_w(V0, Omega, r, NB, CofR, Cd, Cl, X0 - 0.01)
        'Calculating the right hand side, at a distance of 0.01
        F2 = F_of_w(V0, Omega, r, NB, CofR, Cd, Cl, X0 + 0.01)
        'Calculating slope, f'(X0)
        FprimeofX0 = 0.5 * (F2 - F1) / 0.01
        'Calculate the function itself, f(X0)
        FofX0 = F_of_w(V0, Omega, r, NB, CofR, Cd, Cl, X0)
        'Calculate the next value of the root, using Newton-Raphson
        If FprimeofX0 = 0 Then
            Count = 10
        Else
            X1 = X0 - FofX0 / FprimeofX0
            If Abs(X1 - X0) < 0.0001 Then Count = 10
            X0 = X1
        End If
    Loop Until Count = 10
'Return the current value of the induced velocity
    InducedVelocity = X0
End Function
```

VARIABLES

Symbol	Description	Units (UK and SI)
$(I_{XX})_{prop}$	Moment of inertia of propeller about the axis of rotation	slugs ft^2 or kg m^2
$C_{l_{a_0}}$	Incompressible airfoil lift curve slope	/degree or /radian
A	Inflow angle	degrees or radians
a_0	Speed of sound	ft/s or m/s
AF	Activity factor	
AF_{TOT}	Total activity factor	
BHP	Brake horsepower	BHP or HP
C	Propeller circumference	ft or m
c	Blade chord length	ft or m
$C_{D\ new}$	Drag coefficient of new propeller	
$C_{D\ old}$	Drag coefficient of new propeller	
C_l	Section lift coefficient of airfoil	
$C_{l\alpha}$	Airfoil lift curve slope	/degree or /radian
C_P	Power coefficient	
C_Q	Torque coefficient	
C_T	Thrust coefficient	
D	Propeller diameter	ft or m
d_L	Distance from center of thrust of left propeller to the CG along y-axis	ft or m
D_{NAC}	Drag due to nacelle	lb$_f$ or N
d_{NAC}	Distance from nacelle to the CG along y-axis	ft or m
D_{new}	Diameter of new propeller	ft or m
D_{old}	Diameter of old propeller	ft or m
d_R	Distance from center of thrust of right propeller to the CG along y-axis	ft or m
F_{hub}	Hub correction factor	
F_N	Normal force from propeller	lb$_f$ or N
F_{new}	Drag force of new propeller blade	lb$_f$ or N
F_{old}	Drag force of old propeller blade	lb$_f$ or N
F_P	Common correction factor for propeller tip and hub	
F_S	Side force due to propeller	lb$_f$ or N
F_{tip}	Tip correction factor	
f_Ω	Frequency of rotation	Hz
GR	Gear ratio	
H	Altitude	ft or m
h_{SR}	Angular momentum of a spinning about an axis of rotation	slugs·ft^2/s or kg·m^2/s
h_{SRx}	Angular momentum of a spinning about the x-axis	slugs·ft^2/s or kg·m^2/s
h_{SRy}	Angular momentum of a spinning about the y-axis	slugs·ft^2/s or kg·m^2/s
h_{SRz}	Angular momentum of a spinning about the z-axis	slugs·ft^2/s or kg·m^2/s
i	x-component of unit vector	
I_{prop}	Moment of inertia of propeller	slugs·ft^2 or kg·m^2
j	y-component of unit vector	
J	Advance ratio	
k	z-component of unit vector	
k	Fraction spanwise location of blade center of pressure	
K_p	Constant used for required propeller diameter	
L	Lift	lb$_f$ or N
M	Moment about CG along the z-axis due to unbalanced thrust (context dependent)	ft·lb$_f$ or N·m
M	Mach number (context dependent)	
m	Mass of propeller	slugs or kg
M_{tip}	Mach at propeller tip	
M_x	Gyroscopic couple moment about the x-axis	ft·lb$_f$ or N·m
M_y	Gyroscopic couple moment about the y-axis	ft·lb$_f$ or N·m
M_z	Gyroscopic couple moment about the z-axis	ft·lb$_f$ or N·m
n	Revolutions per second	rps
N_B	Number of blades	
$N_{B\ new}$	Number of blades on new propeller	
$N_{B\ old}$	Number of blades on old propeller	
p	Rotational rate about the x-axis (context dependent)	radians/s
P	Power (context dependent)	ft·lb$_f$/s or N·m/s
P_D	Geometric pitch distance	ft or m
$P_{D\ ideal}$	Ideal geometric pitch distance	ft or m
$P_{D\ true}$	True geometric pitch distance	ft or m
P_{hub}	Power at propeller hub	ft·lb$_f$/s or N·m/s
P_i	Induced power	ft·lb$_f$/s or N·m/s
P_R	Required power	ft·lb$_f$/s or N·m/s

(Continued)

Symbol	Description	Units (UK and SI)
P_U	Useful power	ft·lb$_f$/s or N·m/s
P_Ω	Period of rotation	seconds
q	Rotational rate about the y-axis (context dependent)	radians/s
q	Dynamic pressure (context dependent)	lb$_f$/ft^2 or Pa
Q	Propeller torque (context dependent)	ft·lb$_f$ or N·m
Q	Flux of front or aft surface (context dependent)	ft^3/s or m^3/s
Q_{new}	Torque of new propeller	ft·lb$_f$ or N·m
Q_{old}	Torque of old propeller	
R	Blade radius	ft or m
r	Radius of arbitrary blade station (context dependent)	ft or m
r	Rotational rate about the z-axis (context dependent)	radians/s
r_{new}	Arbitrary radius of new propeller	ft or m
ROC	Rate of climb	ft/s or m/s
r_{old}	Arbitrary radius of old propeller	ft or m
RPM	Revolutions per minute	rpm
r_{ref}	Reference radius	ft or m
S_{new}	Planform area of new propeller blade	ft^2 or m^2
S_{old}	Planform area of old propeller blade	ft^2 or m^2
T	Thrust	lb$_f$ or N
T_H	Thrust at maximum velocity	lb$_f$ or N
T_{max}	Maximum thrust	lb$_f$ or N
T_{REQ}	Required thrust	lb$_f$ or N
T_{STATIC}	Thrust at rest	lb$_f$ or N
V_0	Forward velocity	ft/s or m/s
V_C	Cruising velocity	ft/s or m/s
V_H	Maximum velocity	ft/s or m/s
V_{KTAS}	True airspeed	knots
V_{max}	Maximum velocity	ft/s or m/s
V_{MC}	Minimum controllable airspeed	ft/s or m/s
V_{MO}	Maximum airspeed	ft/s or m/s
V_{new}	Forward velocity of new propeller	ft/s or m/s
V_{old}	Forward velocity of new propeller	ft/s or m/s
V_R	Resultant airspeed	ft/s or m/s
V_{ROT}	Propeller rotational speed	ft/s or m/s
V_S	Stall speed	ft/s or m/s
V_{TAS}	True airspeed	ft/s or m/s
V_{tip}	Velocity at propeller tip	ft/s or m/s
V_{TRmin}	Airspeed of minimum required thrust	ft/s or m/s
V_{YSE}	Maximum rate of climb in OEI configuration	ft/s or m/s
W	Weight of aircraft	lb$_f$ or N
w	Propeller induced velocity	ft/s or m/s
W_0	Maximum aircraft weight	lb$_f$ or N
W_{min}	Minimum aircraft weight	lb$_f$ or N
W_{prop}	Propeller weight	lb$_f$ or N
$\Delta C_{D\ OEI}$	Change in drag coefficient due to OEI	
$\Delta D_{windmill}$	Increase in drag due to windmill propeller	lb$_f$ or N
ΔL_{VT}	Change in lift on vertical tail	lb$_f$ or N
ΔQ	Change in flux between front and aft surface	ft^3/s or m^3/s
Δr	Change in radius of old propeller blade to new propeller blade	ft or m
Ω	Rotation rate	degrees/s or radians/s
Ω_x	Angular velocity of propeller about the x-axis	Radians/sec or rad/min
α	AOA	degrees or radians
α_i	Induced angle of attack	degree or radian
α_{ZL}	Angle between zero-lift angle and pitch angle	degrees or radians
β	Geometric pitch angle	degrees or radians
$\delta_{r\ trim}$	Rudder deflection required for trim	degrees or radians
ϕ	Helix angle (context dependent)	degrees or radians
ϕ	Roll angle (context dependent)	degrees or radians
η_i	Froude efficiency	
η_p	Propeller efficiency	
η_v	Viscous profile efficiency	
ρ	Density	slugs/ft^3 or kg/m^3
ρ_{prop}	Density of propeller material	slugs/ft^3 or kg/m^3

References

[1] NACA R-339 Full-Scale Wind-Tunnel Tests with a Series of Propellers of Different Diameters on a Single Fuselage, 1931. Weick, Fred E.

[2] Flight International, November 17, 1921. last paragraph p. 761.

[3] Stinton, D. The Design of the Aeroplane. Collins Professional and Technical Books. p. 304.

[4] Beechcraft C90 King Air, Pilot Training Manual, 2002. FlightSafety International.

[5] http://www.faa.gov/regulations_policies/faa_regulations/

[6] AIR1407 Prediction Procedure for Near-Field and Far-Field Propeller Noise, 1977. Anonymous.

[7] Raymer, D.P. Aircraft Design: A Conceptual Approach, 4th edn. AIAA Eduation Series.

[8] TCDS A00009CH Cirrus Design Corporation. Revision 18, 12/29/2011. FAA.

[9] Stepniewski, W.Z. Keys, C.N. Rotary-Wing Aerodynamics. Dover. p. 46.

[10] Massey, Stanford, B., Ward-Smith, J. Mechanics of Fluids, 7th ed., vol. 1. Nelson Thornes.

[11] Drzewiecki, S. Théorie Générale de L'Hélice. Gauthiers-Villars et Cie Éditeurs.

[12] Stepniewski, W.Z. Keys, C.N. Rotary-Wing Aerodynamics. Dover, 94.

[13] NACA TR-502 Scale Effect on Clark Y Airfoil Characteristics from NACA Full-Scale Wind-Tunnel Tests, 1935. Silverstein, Abe.

[14] Glauert, H. Airplane Propellers. Aerodynamic Theory. In: Durand, W.F. (Ed.). Springer Verlag, Berlin. Div. L, Chapter XI.

CHAPTER 15

Aircraft Drag Analysis

OUTLINE

15.1 Introduction 663
   15.1.1 The Content of this Chapter 665

15.2 The Drag Model 665
   15.2.1 Basic Drag Modeling 666
      Total Drag of Subsonic Aircraft 666
      Minimum Drag, Profile Drag, and Zero-Lift Drag, C_{Dmin} 666
   15.2.2 Quadratic Drag Modeling 666
      Lift-Induced Drag Constant, k 667
      Effective Aspect Ratio, AR_e 668
      C_D Dependency on α and β 668
      Compressibility Effects 668
      Drag Counts 668
      Equivalent Flat Plate Area (EFPA) 668
      Limitations of the Quadratic Drag Model 668
      Drag of Airfoils and Wings 668
   15.2.3 Approximating the Drag Coefficient at High Lift Coefficients 670
   15.2.4 Non-Quadratic Drag Modeling 672
   15.2.5 Lift-Induced Drag Correction Factors 672
      Oswald Efficiency 672
      Span Efficiency 673
   15.2.6 Graphical Determination of L/D_{max} 673
   15.2.7 Comparing the Accuracy of the Simplified and Adjusted Drag Models 673

15.3 Deconstructing the Drag Model: the Drag Coefficients 674
   15.3.1 Basic Drag Coefficient: C_{D0} 674
   15.3.2 The Skin Friction Drag Coefficient: C_{Df} 675
      Standard Formulation to Estimate Skin Friction Coefficient 678
   15.3.3 Step-by-step: Calculating the Skin Friction Drag Coefficient 680
      Step 1: Determine the Viscosity of Air 680
      Step 2: Determine the Reynolds Number 680
      Step 3: Cutoff Reynolds Number 680
      Step 4: Skin Friction Coefficient for Fully Laminar or Fully Turbulent Boundary Layer 680
      Step 5: Mixed Laminar-Turbulent Flow Skin Friction 680
      Step 6: Mixed Laminar-Turbulent Flow Skin Friction 681
      Step 7: Compute Skin Friction Drag Coefficient 681
      Step 10: Compute the Total Skin Friction Drag Force 681
      Step 1: Determine the Viscosity of Air 681
      Step 2: Determine the Reynold's Number for Root Airfoil 681
      Step 3: Determine the Reynold's Number for Tip Airfoil 682
      Step 4: Fictitious Turbulent BL on Root Airfoil – Upper Surface 682
      Step 5: Fictitious Turbulent BL on Root Airfoil – Lower Surface 682
      Step 6: Fictitious Turbulent BL on Tip Airfoil – Upper Surface 682
      Step 7: Fictitious Turbulent BL on Tip Airfoil – Lower Surface 682
      Step 8: Skin Friction for Root Airfoil – Upper Surface 682
      Step 9: Skin Friction for Root Airfoil – Lower Surface 683
      Step 10: Average Skin Friction for Root Airfoil 683
      Step 11: Skin Friction for Tip Airfoil – Upper Surface 683
      Step 12: Skin Friction for Tip Airfoil – Lower Surface 683
      Step 13: Average Skin Friction for Tip Airfoil 683
      Step 14: Average Skin Friction for Complete Wing 683
      Step 15: Wing's Wetted Area 683
      Step 16: Skin Friction Drag Coefficient for Complete Wing 683
      Step 17: Skin Friction Drag Force for Complete Wing 683

General Aviation Aircraft Design   Copyright © 2014 Elsevier Inc. All rights reserved.

Step 18: Skin Friction Coefficient for 100% Laminar Flow — 683
Step 19: Skin Friction Coefficient for 100% Turbulent Flow — 683
Step 20: Comparison — 684

15.3.4 *The Lift-Induced Drag Coefficient: C_{Di}* — 686
Method 1: Lift-Induced Drag from the Momentum Theorem — 686
Method 2: Generic Formulation of the Lift-Induced Drag Coefficient — 687
Derivation of Equation (15-40) — 688
Method 3: Simplified $k \cdot C_L^2$ Method — 689
Method 4: Adjusted $k \cdot (C_L - C_{LminD})^2$ Method — 689
Method 5: Lift-Induced Drag Using the Lifting-Line Method — 689
Method 6: Prandtl-Betz Integration in the Trefftz Plane — 690

15.3.5 *Total Drag Coefficient: C_D* — 691

15.3.6 *Various Means to Reduce Drag* — 691
Reduction of Drag on Wings via Laminar Flow Control (LFC) — 691
Winglets — 692
NLF Airfoils — 692
Reduction of Drag of Fuselages — 692
Reduction of Drag of the Fuselage/Wing Juncture — 693

15.4 The Drag Characteristics of the Airplane as a Whole — 693

15.4.1 *The Effect of Aspect Ratio on a Three-Dimensional Wing* — 693
15.4.2 *The Effect of Mach Number* — 695
15.4.3 *The Effect of Yaw Angle β* — 695
15.4.4 *The Effect of Control Surface Deflection — Trim Drag* — 695
15.4.5 *The Rapid Drag Estimation Method* — 696
15.4.6 *The Component Drag Build-Up Method* — 697
15.4.7 *Component Interference Factors* — 700
15.4.8 *Form Factors for Wing, HT, VT, Struts, Pylons* — 700
15.4.9 *Form Factors for a Fuselage and a Smooth Canopy* — 701
Fuselage as a Body of Revolution — 702
Form Factors at Subcritical Reynolds Numbers — 702
Form Factors at Supercritical Reynolds Numbers — 703
Form Factor for Nacelle and Smooth External Store — 703
Form Factors for Airship Hulls and Similar Geometries — 703

15.5 Miscellaneous or Additive Drag — 708
Derivation of Equation (15-79) — 709

15.5.1 *Cumulative Result of Undesirable Drag (CRUD)* — 709
15.5.2 *Trim Drag* — 710
Trim Drag of a Simple Wing-Horizontal Tail Combination — 711
Derivation of Equation (15-80) — 711
Trim Drag of a Wing-Horizontal Tail-Thrustline Combination — 712
Derivation of Equation (15-81) — 712
15.5.3 *Cooling Drag* — 714
Derivation of Equation (15-82) — 714
15.5.4 *Drag of Simple Wing-Like Surfaces* — 715
15.5.5 *Drag of Streamlined Struts and Landing Gear Pant Fairings* — 715
Thick Fairings — 717
15.5.6 *Drag of Landing Gear* — 718
Drag of Tires Only — 718
Drag of Tires with Wheel Fairings — 718
Drag of Fixed Landing Gear Struts with Tires — 718
Drag of Retractable Landing Gear — 722
Drag of Nose Landing Gear — 723
15.5.7 *Drag of Floats* — 724
15.5.8 *Drag of Deployed Flaps* — 725
Increase of C_{Dmin} Due to Flaps — 725
The Function Δ_1 — 726
The Function Δ_2 — 726
15.5.9 *Drag Correction for Cockpit Windows* — 726
Drag of Conventional Cockpit Windows — 727
Drag of Blunt Ordinary and Blunt Undercut Cockpit Windows — 727
15.5.10 *Drag of Canopies* — 727
15.5.11 *Drag of Blisters* — 728
15.5.12 *Drag Due to Compressibility Effects* — 730
Derivation of Equations (15-106) through (15-107) — 730
15.5.13 *Drag of Windmilling and Stopped Propellers* — 731
Drag Due to Windmilling Propellers — 731
Drag Due to Stopped Propellers — 731
15.5.14 *Drag of Antennas* — 731
15.5.15 *Drag of Various Geometry* — 732
Three-Dimensional Drag of Two-Dimensional Cross-Sections of Given Length — 733
The Cross-Flow Principle — 734
Drag of Three-Dimensional Objects — 734
15.5.16 *Drag of Parachutes* — 735
15.5.17 *Drag of Various External Sources* — 736
Drag of Sanded Walkway on Wing — 736
Drag of Gun Ports in the Nose of an Airplane — 736
Drag of Streamlined External Fuel Tanks — 736
Drag Due to Wing Washout — 736
Drag Due to Ice Accretion — 736

15.5.18	Corrections of the Lift-Induced Drag	737		
	Wingtip Correction	737		
	Correction of Lift-Induced Drag in Ground Effect	738		

- 15.5.18 Corrections of the Lift-Induced Drag — 737
  - Wingtip Correction — 737
  - Correction of Lift-Induced Drag in Ground Effect — 738

15.6 Special Topics Involving Drag — 740

- 15.6.1 Step-by-step: Extracting Drag from L/D$_{max}$ — 740
  - Step 1: Gather Information from the Vehicle's POH — 740
  - Step 2: Convert V$_{LDmax}$ into Units of ft/s — 740
  - Step 3: Calculate the Best Glide Lift Coefficient — 741
  - Step 4: Calculate Span Efficiency — 741
  - Step 5: Compute Minimum Drag — 741
  - Derivation of Equation (15-123) — 741
- 15.6.2 Step-by-step: Extracting Drag from a Flight Polar Using the Quadratic Spline Method — 741
  - Step 1: Select Representative Points from the Flight Polar — 741
  - Step 2: Tabulate — 741
  - Step 3: Fill in the Conversion Matrix and Invert — 742
  - Step 4: Determine the Coefficients to the Quadratic Spline — 742
  - Step 5: Extract Aerodynamic Properties — 742
  - Derivation of Equations (15-124) through (15-127) — 742
- 15.6.3 Step-by-step: Extracting Drag Coefficient for a Piston-Powered Propeller Aircraft — 744
  - Method 1: Extraction C$_{Dmin}$ Using Cruise Performance — 744
  - Derivation of Equation (15-128) — 745
  - Method 2: Extracting C$_{Dmin}$ Using Climb Performance — 746
  - Derivation of Equation (15-129) — 747
- 15.6.4 Computer code 15-1: Extracting Drag Coefficient from Published Data for Piston Aircraft — 748
- 15.6.5 Step-by-step: Extracting Drag Coefficient for a Jet Aircraft — 749
  - Derivation of Equation (15-130) — 750
  - Derivation of Equation (15-131) — 750
- 15.6.6 Determining Drag Characteristics from Wind Tunnel Data — 750
  - Derivation of Equations (15-132) through (15-134) — 751

15.7 Additional Information – Drag of Selected Aircraft — 752

- 15.7.1 General Range of Subsonic Minimum Drag Coefficients — 752
- 15.7.2 Drag of Various Aircraft by Class — 752

Exercises — 756

Variables — 757

References — 759

15.1 INTRODUCTION

Few tasks in aircraft design are as daunting as the estimation of drag. Not only does drag put a lid on what is possible, it can also convert what seems like a promising idea into a terrible one. As far as aircraft design is concerned, the primary objective is usually to minimize drag. However, there is much more to drag than meets the eye. In spite of a desire to keep it as low as possible, ideally we want to control it. Sometimes it is preferable to temporarily increase it and even to shift it around. During climb and cruise, the drag should be kept as low as possible. However, during approach to landing, higher drag helps slow down the airplane and makes it easier to control it during landing. Sailplane pilots can attest to how hard it would be to land a sailplane if such aircraft did not feature spoilers to help them avoid overshooting the runway. For some aircraft, including many sailplanes, it is possible to shift the drag polar around using a cruise flap. This permits the pilot to move the maximum glide ratio to a higher airspeed, which is desirable if one wants to glide it as far as possible. The bottom line is that a good understanding of what the drag consists of, what affects it and what does not, is imperative to the designer. The purpose of this section is to provide the aircraft designer with methods to estimate drag and details of its causes and prevention.

Many areas of the aircraft design process rely on accurate drag estimation. This includes performance analysis, engine selection, and requirements for fuel capacity, to name a few. The subject must be approached with great respect and caution. Drag is extremely hard to predict accurately, but instead easy to over- and underestimate. This does not mean that the calculations themselves are exceedingly difficult, but rather that the sources that contribute to the drag may be hard to identify. The accuracy of the calculations is vitally important as an underestimation will result in an airplane that performs far worse than predicted, risking a costly "drag-cleanup" effort if not the cancellation of an otherwise viable program. By the same token, an overestimation by an overly conservative approach may render the design so bad on paper the project might be cancelled before it even begins.

The aspiring designer should not think that the problem of drag estimation can be solved by simply modeling the aircraft using a Navier-Stokes computational fluid dynamics (NS CFD) solver. This view is often heard uttered by students of aerospace engineering who have yet to be humbled by Mother Nature by comparing predictions to actual wind tunnel or flight testing. While CFD is both a very promising and exciting scientific advance, the technology is not yet robust enough to allow a novice user to estimate drag reliably.[1] In 2001, the AIAA[2] conducted the first of four workshops on drag estimation using NS CFD methods. A clean Airbus-style passenger transport aircraft, consisting of a fuselage and wing only, was modeled by 18 participants using 14 different NS solvers. Then the analysis results were compared to the wind tunnel data, which was not made available to the participants until after the analysis results had been submitted. In short, the lift and drag predictions were all over the map. Some models predicted lift and drag close to the wind tunnel data, while for others the deviation of the drag polar ranged from approximately 50% to 200% of the wind tunnel value [1,2]. These predictions were performed by scientists who were experts in the use of these codes. If they have a hard time performing such predictions, the novice analyst should take caution. The drag analyst is in many ways between a rock and a hard place — a dreaded place shared by the weights engineer.

In this section, classical methods will be used to estimate the magnitude of the drag. In spite of what is stated above, these methods (like the NS methods) can yield good predictions as long as they are applied with caution. As stated in Section 8.1, drag, D, is defined as the component of the resultant force, R, which is parallel to the trajectory of motion (see Figure 15-1). The force of drag differs from the force of lift in that its constituent contributors are both pressure difference and friction. Like lift, drag is a component of the resultant force that results from the pressure differential over a body. Friction, however, is a force that acts parallel to the airspeed, which explains why it contributes only to the drag and not the lift. The purpose of performing a drag analysis is to estimate

FIGURE 15-1 No aircraft generate as little drag as the modern sailplane. A Rolladen-Schneider LS4 sailplane, moments before touchdown with spoilers deployed. *(Photo by Phil Rademacher)*

the magnitude of this force and understand how the geometry, as well as attitude of the aircraft and flight condition, affects its magnitude.

Classical drag estimation methods attempt to predict drag based on geometry and flow properties. Any useful drag estimation method must account for:

- Laminar boundary layer,
- Turbulent boundary layer,
- Location of laminar-to-turbulent transition,
- Flow separation regions, and
- Compressibility.

Non-dimensional coefficients are essential when working with aerodynamic forces and moments. With respect to the drag of aircraft, the coefficient form that represents the total drag is referred to as the *drag model*. A drag model is a mathematical expression that when multiplied by dynamic pressure and a reference area will yield the drag force acting on the aircraft. It can be used in a number of important ways, ranging from plotting the drag polar (a graph in which the drag coefficient is plotted as a function of the lift coefficient) to evaluating important performance characteristics such as the lift coefficient to maintain in flight to achieve the longest glide distance or fly farthest.

Generally, the total drag of an airplane is broken into two classes attributed to *flow separation* (pressure drag) and *skin friction* (skin friction drag). Thus, the drag consists of the following contributions:

(1) **Basic pressure drag** is caused by the pressure differential formed by the airplane that acts parallel to the tangent to the flight path.
(2) **Skin friction drag** is caused by the "rubbing" of molecules along the surfaces of the airplane.
(3) **Lift-induced drag** is caused by the circulation around the wing, which tilts the lift vector

---

[1] Then again, many would argue that novices shouldn't be estimating drag in the first place.
[2] AIAA is the American Institute of Aeronautics and Astronautics.

backwards, creating a force component that adds to the total drag.

(4) **Wave drag** is caused by the rise in pressure around a body due to the formation of a normal shock wave on the airplane. This effect begins at high subsonic Mach numbers.

(5) **Miscellaneous drag** is caused by a number of "small" contributions that are often easily overlooked, such as small inlets and outlets, access panels, fuel caps, to name a few.

In this section we will present methods to determine all of these types of drag, although wave drag will be treated with far less detail than the others since it is a high subsonic phenomenon. To do this, a number of drag modeling methods will be presented and the effects of various flow phenomena will be explained. Then, we will present a method to estimate the skin friction on lifting surfaces that are assumed smooth and continuous. The method assumes a minimum level of flow separation and assumes a mixed boundary layer. A mixed boundary layer is one in which a laminar boundary layer is allowed to transition into a turbulent one. This method is superior to methods that consider the boundary layer to be either laminar or turbulent, as it treats the flow more realistically than either of these. Then, we will introduce the effect that changing the *AOA* or *AOY* has on the drag, in terms of both increased level of separated flow and increase in drag due to lift. Then, special methods will be introduced that allow the total drag to be estimated for an airplane as a whole. This will be followed by a number of specialized methods to estimate the drag of parts of the airplane, such as drag due to engine cooling requirements. Finally, methods will be presented that allow the drag of an existing aircraft to be extracted using specific, publically available data.

Today, the state-of-the-art in low drag aircraft is the modern sailplane (see Figure 15-1). Natural laminar flow (NLF) airfoils, segmented tapered wing planform, tadpole fuselage, sealed control surfaces, carefully tailored wing root fairings, T-tail, and disciplined attention to any detail that reduces drag, combine to make some sailplanes capable of achieving glide ratios in excess of 1:50. No other aircraft are capable of that — flying wing or not. The aircraft designer interested in developing a low-drag aircraft should pay special attention to such aircraft. There is a lot that can be learned.

15.1.1 The Content of this Chapter

- **Section 15.2** discusses the nature of drag, what contributes to it, how it is modeled, and a number ways drag can be evaluated. The section also discusses various limitations to how it is evaluated.
- **Section 15.3** discusses the drag coefficient, including the *basic drag coefficient*, C_{D0}, *skin friction drag coefficient*, C_{Df}, and *lift-induced drag coefficient*, C_{Di}.
- **Section 15.4** discusses the drag characteristics of the airplane as a whole and presents methods to estimate the total drag of aircraft.
- **Section 15.5** presents methods to estimate the drag caused by the addition of necessary imperfections to aircraft, such as antennas, fairings, landing gear, and so on. Additionally, methods to estimate trim drag and cooling drag are presented.
- **Section 15.6** presents methods to estimate the drag of existing aircraft, using data such as published performance data.
- **Section 15.7** presents the drag characteristics of selected aircraft, accumulated from various sources.

15.2 THE DRAG MODEL

A drag model is a mathematical expression of the drag coefficient that describes how the drag of the body changes as a function of its orientation in the flow field. The determination of this model is a challenging task, although good analytical approximations are possible at lower *AOA*s, as we will show shortly. The difficulty in devising good drag models at higher *AOA*s stems from the fact that they are highly dependent on the size of the flow separation regions and these are very difficult to predict accurately. And as stated in the introduction to this section, even state-of-the-art formulations of fluid flow implemented in Navier-Stokes solvers have a hard time doing this well. The most reliable methods remain wind tunnel or flight testing.

Since our purpose here is to build tools to help us predict the drag of the airplane before a wind tunnel testing is conducted, let alone flight testing, we more than anything else seek a realistic model of the drag. First we note that a realistic mathematical presentation for drag, *D*, can be written as shown below:

$$D = f(geometry, \alpha, \beta, \rho, V_\infty, \text{Re}, M) \qquad (15\text{-}1)$$

where *geometry* refers to reference and wetted area.

α = angle-of-attack
β = angle-of-yaw
ρ = air density
V_∞ = far-field airspeed
R_e = Reynolds number
M = Mach number

The standard way to estimate the drag is to represent the dependency on airspeed and density through the dynamic pressure, $\frac{1}{2}\rho V_\infty^2$, the geometry using a reference area, S_ref, and the remaining dependencies are lumped into the drag coefficient, denoted by C_D. Thus, we compute the drag, D, using the expression:

$$D = \frac{1}{2}\rho V_\infty^2 S_{ref} C_D \qquad (15\text{-}2)$$

where

C_D = total drag coefficient
S_ref = reference area (typically wing area)
V_∞ = far-field airspeed
ρ = air density

The drag coefficient in Equation (15-2) is the *drag model*.

15.2.1 Basic Drag Modeling

Basic drag modeling is the mathematical combination of all sources of drag for a vehicle, such that the effect of changing its orientation with respect to its path of motion and fluid velocity is realistically replicated. This modeling culminates in the determination of the total drag coefficient, C_D.

As stated in the introduction, the total drag coefficient comprises the effect of basic pressure drag, skin friction drag, lift-induced drag, wave drag, and contributions from other sources, commonly referred to as miscellaneous drag. Typically, basic pressure drag, skin friction drag, and miscellaneous drag are lumped together and are represented using a single number, called the minimum drag, as is done in the so-called *quadratic drag modeling*. This is accomplished using a special method called the *component drag build-up method*, presented later in the chapter.

In short, the method estimates the total drag by determining the skin friction drag of a particular component (e.g. the wing or the HT). Then the basic pressure drag is accounted for using a booster factor called a *form factor*, presented in detail later in this chapter. Thus, the designer estimates the skin friction coefficient for the surface and then multiplies it by the form factor, yielding a number that, as stated earlier, combines the effect of friction and pressure. Additionally, it is necessary to account for the numerous protrusions and discontinuities in the generally smooth surfaces of the aircraft, as it results in greater than expected pressure drag. This drag is called *miscellaneous drag*.

As the *AOA* or *AOY* changes, the contribution of the pressure drag terms increases, adding non-linearity to the formulation that is usually not accounted for in the modeling. It is acceptable to omit this effect for low values of *AOA* and *AOY*. However, once a certain orientation is reached, the pressure drag grows to a magnitude so large that it can no longer be ignored. Then, standard drag models are no longer acceptable unless a means to capture this effect is included. A method to account for this high AOA effect is presented in Section 15.2.3 of this chapter.

Total Drag of Subsonic Aircraft

In an ideal world, the total drag coefficient can be considered as the combination of a number of constituent components:

$$C_D = C_{Do} + C_{Df} + C_{Di} + C_{Dw} + C_{Dmisc} \qquad (15\text{-}3)$$

where

C_D = total drag coefficient
C_{Do} = basic drag coefficient (pressure drag)
C_{Df} = skin friction drag coefficient
C_{Di} = induced drag coefficient (pressure drag)
C_{Dw} = wave drag
C_{Dmisc} = miscellaneous or additive drag

Each component depends on the aircraft's geometry, as well as its orientation and airspeed. For low subsonic aircraft the norm is to omit wave drag, which simplifies Equation (15-3) as follows:

$$C_D = C_{Do} + C_{Df} + C_{Di} + C_{Dmisc} \qquad (15\text{-}4)$$

Minimum Drag, Profile Drag, and Zero-Lift Drag, C_{Dmin}

As stated above, C_{Do}, C_{Df}, and C_{Dmisc} are lumped in a single number that from now on will be referred to as C_{Dmin} or *minimum drag coefficient*. This is also referred to as a *profile drag* or *parasitic drag* or *zero-lift drag*, although this book will only use the term minimum drag. The minimum drag coefficient represents the lowest drag the vehicle will generate. The advantage of combining the C_{Do}, C_{Df}, and C_{Dmisc} in this fashion is that it allows for a simple evaluation of the total drag coefficient. However, the convenience hides the contribution of the wetted area on the overall airplane drag.

15.2.2 Quadratic Drag Modeling

A standard way to present the drag coefficient is to relate it to the lift coefficient using a quadratic polynomial in accordance with the derivation presented in Section 15.3.4, *The lift-induced drag coefficient: C_{Di}*. The method can provide an accurate prediction over a range of lift coefficients, although the accuracy drops rapidly at the extremes of the drag polar. A standard "simplified" quadratic presentation for the drag coefficient is:

$$C_D = C_{D_{min}} + \frac{C_L^2}{\pi \cdot AR \cdot e} = C_{D_{min}} + k \cdot C_L^2 \qquad (15\text{-}5)$$

where

C_L = lift coefficient
C_D = total drag coefficient
C_{Dmin} = minimum drag coefficient
AR = reference aspect ratio
e = Oswald efficiency
k = lift-induced drag constant

A far more "realistic" presentation for the drag coefficient is the *adjusted drag model*, represented graphically in Figure 15-2.

$$C_D = C_{D_{min}} + k \cdot \left(C_L - C_{L_{minD}}\right)^2 \qquad (15\text{-}6)$$

where $C_{L\ minD}$ = the lift coefficient where drag becomes a minimum.

Two important properties of the adjusted drag polar are shown in Figure 15-2. Changing the C_{LminD} will only shift the polar sideways. If $C_{LminD} < 0$ then the curve will be shifted left, but if $C_{LminD} > 0$ it will be shifted right. Similarly, changing the value of the C_{Dmin} will shift the polar up or down. Of course C_{Dmin} is always larger than zero, whereas C_{LminD} can range from a negative to a positive number depending on the camber of the airfoil, or, in the case of a complete aircraft, the effective camber of the entire geometry.

Lift-Induced Drag Constant, k

This is the constant whose product with the lift coefficient squared yields the lift-induced drag. It is given by:

$$k = \frac{1}{\pi \cdot AR \cdot e} = \frac{1}{\pi \cdot AR_e} \qquad (15\text{-}7)$$

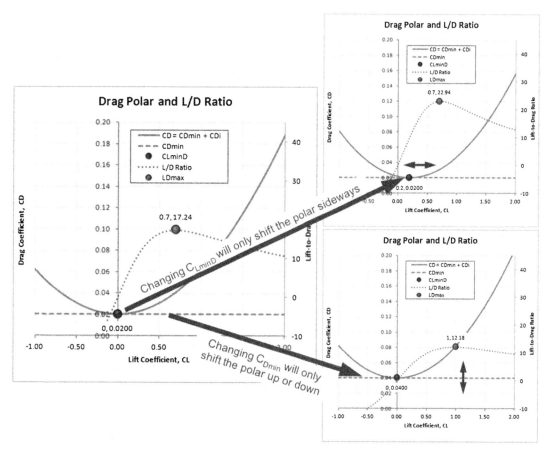

FIGURE 15-2 A schematic showing the effect of changing C_{LminD} and C_{Dmin} on the drag polar.

Effective Aspect Ratio, AR_e

The product $AR \cdot e$ is often referred to as the *effective aspect ratio*, denoted by AR_e. In this respect, the Oswald (or spanwise) efficiency can be considered a factor that renders the AR less effective than the geometric value would indicate. The designer can consider planform and other geometric modifications, such as endplates or winglets, to increase the effective AR, but should be aware that such modifications increase the skin friction drag. Note that for a clean wing, if $AR_e \to \infty$ then $k \to 0$ and therefore $C_D \to C_d$, i.e. the drag coefficient effectively becomes that of the airfoil.

C_D Dependency on α and β

Sometimes, C_{Do} and C_{Df} are treated as if they are constant with respect to α and β. This is not true in real airflow and they are treated this way for convenience. Changes in α and β will move the laminar to turbulent flow transition line and reshape flow separation regions. This changes the pressure drag, modifying the basic drag coefficient.

Consequently, changing α and β *will* change C_{Dmin}, but this change is not to be confused with the simultaneous change in the induced drag, C_{Di}, whose magnitude is related to the lift coefficient, C_L. If the airspeed is mostly unchanged, the change in C_{Dmin} can considered solely due to a change in the size of flow separation areas over the airplane, distinguishing it from pressure changes that directly modify lift generation and, thus, affect C_{Di}.

Compressibility Effects

The effect of compressibility is accounted for by modifying C_{Dmin} at high subsonic airspeeds using a special correction factor (see Section 15.5.12, *Drag due to compressibility effects*). In addition to this effect, C_{Df} should also be corrected using methods such as the Frankl-Voishel one of Section 8.3.6, *Compressibility modeling*. This will also modify C_{Dmin}.

Drag Counts

A "drag count" is the drag coefficient multiplied by a factor of 10000. For instance, 250 drag counts is equivalent to a $C_D = 0.0250$; 363 drag counts is equivalent to a $C_D = 0.0363$, and so on.

Equivalent Flat Plate Area (EFPA)

The *equivalent flat plate area* (denoted by f) is a value that is helpful when comparing the relative drag of different aircraft. It is simply the product of the minimum drag coefficient and the reference area, as shown below. Alternatively, this is nothing but the minimum drag force at the given airspeed, D_{min}, divided by the dynamic pressure, q:

$$f = S_{ref} \times C_{Dmin} = \frac{D_{min}}{q} \qquad (15\text{-}8)$$

The concept assumes the drag of the airplane is equivalent to that of a fictitious plate that has a drag coefficient $C_D = 1.0$. Thus, if the flat plate area of an airplane is 10 ft^2, it means its drag amounts to that of a flat plate of the same area moving normal to the flight path. The concept is bogus in many respects. For instance, it disregards the effect of Reynolds and Mach numbers (the C_D of a flat plate at R_e around 10^5 is actually closer to 1.17; e.g. see Figure 15-66), and no notion is given as to the true geometry of this "plate" (i.e. is it rectangular or circular or any other shape?). In spite of these shortcomings, as stated earlier, it is helpful mostly for comparison purposes. Table 15-18 lists the EFPA for a number of different aircraft.

Limitations of the Quadratic Drag Model

The drag of some aircraft cannot be accurately represented with the quadratic drag model. A sailplane is an example of such an aircraft. Super-clean aerodynamics and natural laminar flow (NLF) result in a drag bucket for the entire vehicle. Consequently, the quadratic approximation will give erroneous values of the max lift-to-drag ratio and where it occurs. The quadratic model works well for airplanes that do not have a noticeable drag bucket, except at very high or very low lift coefficients (see Figure 15-3). The designer must be aware of this limitation as it leads to erroneous prediction of best endurance and range airspeeds, in particular of airplanes with very low wing loading (LSA aircraft). See Section 15.2.7, *Comparing the accuracy of the simplified and adjusted drag models* for a comparison of the quadratic drag models to actual wind tunnel data for further understanding of the limitation of these models. Also see Section 15.2.3, *Approximating the drag coefficient at high lift coefficients* for a method to help approximating the deviation at higher lift coefficients.

Drag of Airfoils and Wings

Figure 15-4 shows the effect of taking the leap from a two-dimensional airfoil to a three-dimensional wing (featuring the airfoil). The two-dimensional drag polar for the airfoil consists of (1) a constant drag component (C_{dmin}), which is attributed to skin friction, and (2) a parasitic increase caused by increase in pressure drag due to the flow separation on the airfoil, which increases with the *AOA*. Typically this depends on the lift coefficient squared as shown in Equations (15-5) and (15-6).

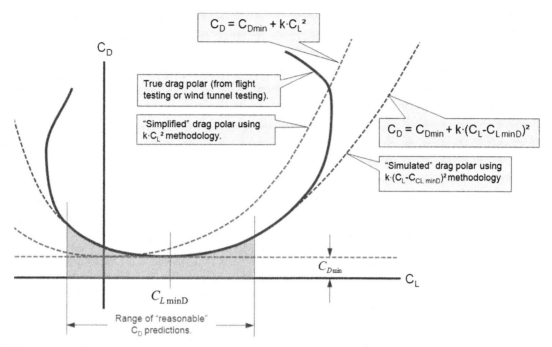

FIGURE 15-3 Curve-fitting the true drag polar. Note that if the airfoil (or vehicle) features a camber, the simplified drag polar is no longer a valid representation of the drag polar.

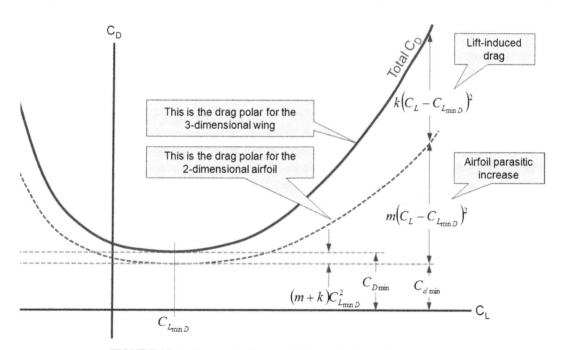

FIGURE 15-4 Drag polar for an airfoil introduced to a finite AR wing.

Introducing this airfoil to a finite-aspect-ratio wing will increase the drag further due to the three-dimensional effects: the *lift-induced drag*. In this case, we express the drag as follows:

$$C_D = C_{d_{min}} + m \cdot (C_L - C_{L_{minD}})^2 + k \cdot (C_L - C_{L_{minD}})^2 \quad (15\text{-}9)$$

where

C_L = lift coefficient
$C_{L\ minD}$ = lift coefficient where drag becomes a minimum
C_D = total three-dimensional drag coefficient
$k = 1/(\pi \cdot AR \cdot e)$ = lift-induced drag constant (see Section 15.3.4)

m = coefficient indicating the parasitic drag increase of the airfoil, obtained from its drag polar
AR = reference aspect ratio
e = span efficiency (note difference from Equations (15-5) and (15-6))

Note that this can be expanded to give the following expression:

$$C_D = \left\{ C_{d_{min}} + (m+k)C_{L_{minD}}^2 \right\} - 2(m+k)C_{L_{minD}}C_L + (m+k)C_L^2 \quad (15\text{-}10)$$

The expansion shows that the minimum drag increases by a small factor, $(m+k)C_{L_{minD}}^2$, which shifts the polar a small amount vertically. This is shown in Figure 15-4.

15.2.3 Approximating the Drag Coefficient at High Lift Coefficients

Figure 15-3 reveals a critical problem in all drag modeling; at higher C_Ls, the drag model deviates drastically from the actual drag and underpredicts it severely. The deviation is caused by a rapid growth of flow separation with AOA and the second-order approximation cannot keep up with the resulting increase in drag. This is a serious problem when estimating aircraft performance at low airspeeds. Therefore, important airspeeds, such as best angle of climb, minimum rate of descent, and others, are shifted to lower airspeeds than observed in practice. This section addresses this shortcoming and develops a method to approximate the rapid rise in drag at higher lift coefficients.

Consider the hypothetical wind tunnel test data in Figure 15-5. Simply stated, the data presented cannot be approximated satisfactorily using a quadratic polynomial. Neither the *simplified* nor the *adjusted* drag models will provide acceptable drag prediction at high (or low) lift coefficients. For this reason, only the data points at lower lift coefficients are used to create a curve-fit (or a trend curve). For the wind tunnel data shown in Figure 15-5, assume that a standard adjusted drag model has been determined and, as shown, it agrees well with the measured drag at lower lift coefficients. This model is given by Equation (15-6), reproduced below for convenience:

$$C_D = C_{Dmin} + k(C_L - C_{LminD})^2 \quad (15\text{-}6)$$

As is very evident from the graph, the test data begins to deviate sharply from the curve starting at $C_L = 1.15$ (and $C_L = -0.45$). Of course we are primarily interested in the positive lift coefficient, as this is needed for low-speed performance predictions. In order to work around this predicament, the author has used the following methodology in the past with very satisfactory results. In this method, once the C_L exceeds a certain value, which here will be called C_{Lm}, a quadratic (or cubic) spline is created to simply replace the values of the adjusted drag model.

The method presented here effectively defines a new quadratic polynomial and splices it to the adjusted drag model at C_{Lm}. Other splines are certainly possible; however, the advantage of the quadratic spline is the simplicity of its determination and the acceptable accuracy it provides. The method allows the spline to blend smoothly with the underlying adjusted drag model. The first step is to define a *modified* drag coefficient to be used for $C_L > C_{Lm}$. It can be represented by:

$$C_{Dmod} = AC_L^2 + BC_L + C \quad (15\text{-}11)$$

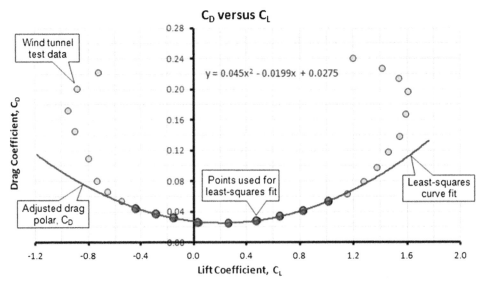

FIGURE 15-5 A hypothetical drag polar for an aircraft, showing the inaccuracy of the quadratic drag model at higher (or lower) lift coefficient.

15.2 THE DRAG MODEL

Ultimately, the task is to define the constants A, B, and C so C_{Dmod} can be evaluated. To do this, two conditions have to be satisfied at C_{Lm}:

(1) Equal drag at C_{Lm}:

$$C_{Dmod}(C_{Lm}) = C_D(C_{Lm}) \quad \text{and}$$

(2) Equal slope at C_{Lm}:

$$\left.\frac{\partial C_{Dmod}}{\partial C_L}\right|_{at\ C_{Lm}} = \left.\frac{\partial C_D}{\partial C_L}\right|_{at\ C_{Lm}}$$

A third condition is needed to finalize the determination of the coefficients; it is that the value of C_{Dmod} at C_{Lmax} must match that of the wind tunnel data. That aside, the function for C_{Dmod} has three constants (A, B, C) so three equations are required to determine them. One of the equations requires the derivative of both C_D and C_{Dmod} to be determined. These are presented below:

Slope of the adjusted drag model:

$$\frac{\partial C_D}{\partial C_L} = \frac{\partial}{\partial C_L}\left[C_{Dmin} + k(C_L - C_{LminD})^2\right]$$
$$= 2k(C_L - C_{LminD})$$

Slope of the modified drag model:

$$\frac{\partial C_{Dmod}}{\partial C_L} = \frac{\partial}{\partial C_L}\left[AC_L^2 + BC_L + C\right] = 2AC_L + B$$

Now the three equations that allow the constants A, B, and C to be determined can be written as follows:

Equation (1):

$$AC_{Lm}^2 + BC_{Lm} + C = C_{Dmin} + k(C_{Lm} - C_{LminD})^2$$

Equation (2):

$$2AC_{Lm} + B = 2k(C_{Lm} - C_{LminD})$$

Equation (3):

$$AC_{Lmax}^2 + BC_{Lmax} + C = C_{Dstall}$$

Rearranging this in a matrix form yields the following expression that allows A, B, and C to be determined using any matrix method, Cramer's rule or matrix inversion methods:

$$\begin{bmatrix} C_{Lm}^2 & C_{Lm} & 1 \\ 2C_{Lm} & 1 & 0 \\ C_{Lmax}^2 & C_{Lmax} & 1 \end{bmatrix} \begin{Bmatrix} A \\ B \\ C \end{Bmatrix} = \begin{Bmatrix} C_{Dmin} + k(C_{Lm} - C_{LminD})^2 \\ 2k(C_{Lm} - C_{LminD}) \\ C_{Dstall} \end{Bmatrix} \quad (15\text{-}12)$$

Then, once A, B, and C have been determined the drag model is further refined as follows:

$$C_D = \begin{cases} C_{Dmin} + k(C_L - C_{LminD})^2 & \text{if } C_L \leq C_{Lm} \\ AC_L^2 + BC_L + C & \text{if } C_L > C_{Lm} \end{cases}$$

This model has been implemented in Figure 15-6. The improvements in the capability of the drag model are demonstrated by how the quadratic spline, represented by C_{Dmod}, smoothly follows the wind tunnel data starting at C_{Lm}. Note that although wind tunnel data is assumed here, the same approach can also be used for analytical estimation of the drag coefficient. For such work, it is reasonable to select C_{Lm} as the value ½$(C_{LminD} + C_{Lmax})$ as a first guess.

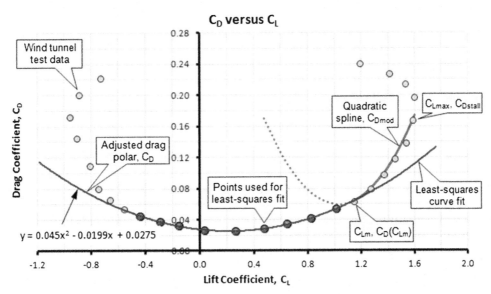

FIGURE 15-6 The same hypothetical drag polar, showing the improvement in prediction accuracy at higher lift coefficients by the introduction of the quadratic spline, C_{Dmod}.

EXAMPLE 15-1

Consider the hypothetical wind tunnel data shown in Figure 15-6. Assume it represents an airplane whose $AR = 9$. Further assume the least-squares quadratic curve-fit for the data points has been determined and is given by:

$$C_D = 0.045 C_L^2 - 0.0199 C_L + 0.0275$$

Determine the coefficients for a quadratic spline, assuming a $C_{Lm} = 1.15$, and write the complete drag coefficient using the spline. Note the resulting graph is shown in Figure 15-7.

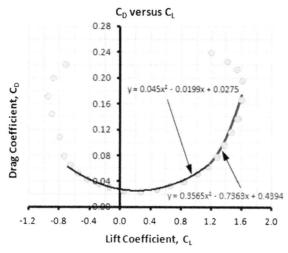

FIGURE 15-7 The modified drag model fits the wind tunnel data better than a regular quadratic model.

Solution

First, using the above drag polar, extract C_{Dmin}, C_{LminD}, and the Oswald span efficiency (e) using the conversion method of Section 15.6.6, *Determining drag characteristics from wind tunnel data*. In particular, see Equations (15-132), (15-133), and (15-134). The section presents an example as well, so only the results based on the above polynomial will be presented.

$$C_{Dmin} = 0.02530$$

$$C_{LminD} = 0.2211$$

$$e = 0.78595$$

Using e, we find the $k = 1/(\pi \cdot AR \cdot e) = 0.045$. Using this data, the matrix of Equation (15-12) becomes:

$$\begin{bmatrix} 1.3225 & 1.15 & 1 \\ 2.3 & 1 & 0 \\ 2.5113 & 1.5847 & 1 \end{bmatrix} \begin{Bmatrix} A \\ B \\ C \end{Bmatrix} = \begin{Bmatrix} 0.06413 \\ 0.08360 \\ 0.16783 \end{Bmatrix}$$

From which we find that $A = 0.3565$, $B = -0.7363$, and $C = 0.4394$. The resulting C_D can thus be represented by:

$$C_D = \begin{cases} 0.0253 + 0.045 \cdot (C_L - 0.2211)^2 & \text{if } C_L \leq 1.15 \\ 0.3565 \cdot C_L^2 - 0.7363 \cdot C_L + 0.4394 & \text{if } C_L > 1.15 \end{cases}$$

15.2.4 Non-Quadratic Drag Modeling

Figure 15-3 shows that the quadratic model may indeed work well for a specific range of lift coefficients provided a proper value of span efficiency, e, is selected. However, this is not always the case. There are two other instances where the quadratic modeling is simply inadequate and must be abandoned. The first one involves very high or very low lift coefficients, when large areas of separated flow have formed on the airplane. This was treated in the previous section. The second one pertains to aerodynamically clean aircraft, such as sailplanes, for which a well-defined drag bucket exists and is not masked by the drag of other sources that cannot be described by the quadratic formulation. These problematic regions are easily visible in Figure 15-8. In this case, the performance formulation in Chapters 13 through 18 must be revised to account for those cases.

Generally, the presence of the drag bucket will prevent the use of polynomials, as these are not capable of following the sharp change in curvature of the drag polar. Even polynomials of very high order (16+) will not be adequate, as they tend to oscillate inside the prediction region. There are primarily two other methods that can be used to represent such drag polars; a spline (e.g. B-spline) or a lookup table.

15.2.5 Lift-Induced Drag Correction Factors

The estimation of drag due to lift of a three-dimensional wing requires a correction of the two-dimensional airfoil data. This is expressed in Equations (15-5) and (15-6) using the factor e with the induced drag components. There are generally two kinds of such correction factors (although some use them interchangeably):

Oswald Efficiency

The *Oswald efficiency* is used when calculating the induced drag coefficient for a wing or an aircraft whose dependence on the lift coefficient can be represented

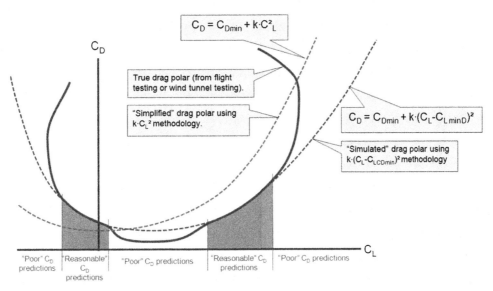

FIGURE 15-8 Curve-fitting the true drag polar featuring a drag bucket.

with the simplified or adjusted drag models. This is the factor e in Equations (15-5) and (15-6).

Span Efficiency

The *span efficiency* is used when calculating the drag increase purely based on the three-dimensional increase as shown in Figure 15-4. It does not account for the parasitic drag increase as shown in the figure, but only the three-dimensional effects. This is the factor e in Equation (15-9).

Lift-induced drag is presented in more detail in Section 15.3.4, *The lift-induced drag coefficient: C_{Di}*.

15.2.6 Graphical Determination of L/D_{max}

In the absence of knowing the exact numbers of a particular drag polar, it is possible to determine the maximum lift-to-drag ratio, as well as the lift coefficient at which it occurs, graphically. This is shown in Figure 15-9 for a "conventional" drag polar and one for a NLF airfoil (or aircraft) featuring a drag bucket. This is done by fairing a line from the origin so it becomes a tangent to the polar. The reason why this yields L/D_{max} can be visualized by recognizing that the tangent aligns with the smallest value of C_D at the greatest value of C_L.

15.2.7 Comparing the Accuracy of the Simplified and Adjusted Drag Models

The accuracy of the simplified and adjusted drag models can best be illustrated by comparing them to wind tunnel data. One such comparison is shown in Figure 15-10, which compares experimental wind tunnel test results to the simplified and adjusted drag models using Equations (15-5) and (15-6), respectively. Of course it should be emphasized that ordinarily one would not use these models with the drag polar of a two-dimensional airfoil, but a three-dimensional wing. However, this particular illustration can be justified based on Equation (15-9) as the purpose is to show the importance of the term C_{lminD}, here using the lower-case subscript for forces.

The comparison was implemented using wind tunnel data for the NACA 2412 airfoil obtained from NACA R-824. This airfoil is used on a number of Cessna aircraft designs, for instance the Model 150, 152, 172, and 182, to

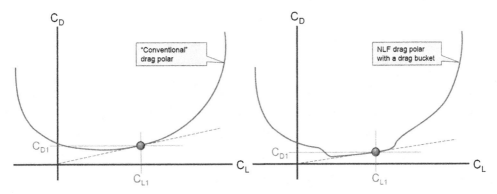

FIGURE 15-9 Graphical determination of the maximum lift-to-drag ratio.

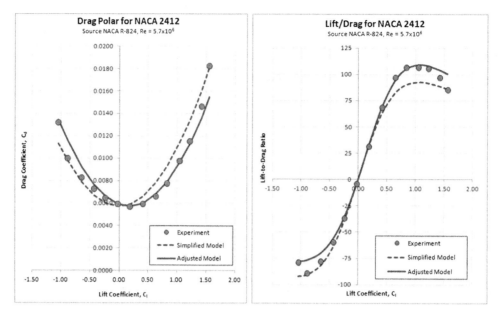

FIGURE 15-10 A comparison of the simplified and adjusted drag models to experimental data.

name a few. First, the drag polar was digitized and the smallest experimental value of the C_d selected for use as the C_{dmin} for each drag model. Second, the corresponding C_{lminD} to use with the adjusted model was determined by trial and error. Third, the induced drag contribution was calculated using the expressions $C_{di} = k \cdot C_l^2$ for the simplified model and $C_{di} = k \cdot (C_l - C_{lminD})^2$ for the adjusted model and added to the C_{dmin}. Then, k was varied to get the best fit of each model (note both models used the same k). The value of k that best fit the experiment was found to be $k \approx 62$.[3] The graphs show that the adjusted drag model provides a substantially better fit than the simplified one, in particular for the positive values of the C_l. Another lesson to be learned is the dire consequences of estimating range for an aircraft using the simplified model. Since range depends so explicitly on the C_L/C_D (see Section 20.2, *Range Analysis*) a prediction using the simplified drag model would yield a range much less than the adjusted (or experiment) indicates. The opposite can also happen, i.e. the C_L/C_D using the simplified model is sometimes higher than experiment. The important lesson is that the adjusted model better matches experiment; the simplified does not and should be avoided.

15.3 DECONSTRUCTING THE DRAG MODEL: THE DRAG COEFFICIENTS

In this section we will discuss the constituent parts of the drag model; the *basic, skin friction,* and *lift-induced* drag coefficients.

15.3.1 Basic Drag Coefficient: C_{D0}

The basic drag is a pressure drag force caused by resultant pressure distribution over the surface of body. It can be thought of as the component of the pressure force parallel to the tangent to the flight path. For instance, consider a cylinder in a moving fluid. Its drag consists of the friction between its surface and the moving fluid, and the difference in pressure along its surface. The latter will be the focus in this section. The force is the product of the pressure acting on a cross-sectional area of the body, normal to the flight path.

$$D_0 = \int_S (P \cdot \mathbf{n}) dA \qquad (15\text{-}13)$$

where

$\mathbf{n}$ = normal to the flight path
S = surface of the body

A simple interpretation of the basic drag is shown in Figure 15-11, which shows a sphere moving through a fluid forming a high-pressure region in front of it (left) and a low-pressure region behind it as the turbulent wake. The pressure differential across its cross-sectional area yields a drag force. Equation (15-13) can be rewritten for simple geometry and uniform pressure distributions as follows:

$$D = \text{Pressure} \times \text{Area} = \Delta P \cdot A \qquad (15\text{-}14)$$

---

[3]Using a value of k higher than 62 (e.g. 200, 400, or higher) will effectively make the polar flatter and flatter.

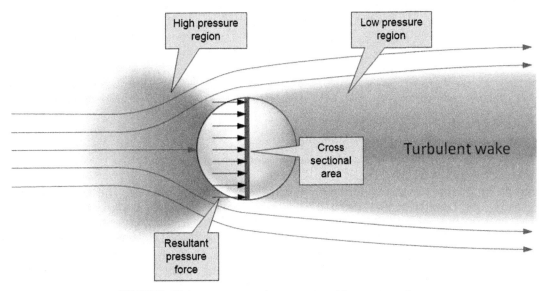

FIGURE 15-11 Important characteristics of flow over a sphere.

For other geometries, this force increases at high (and low) *AOA*s as the separation regions grow. If the basic drag force is known, the basic drag coefficient it is defined as follows:

$$C_{D0} = \frac{2D_0}{\rho V_\infty^2 S_{ref}} \qquad (15\text{-}15)$$

where

D_0 = basic drag force in lb_f (UK system) or N (SI system)
ρ = air density, typically in slugs/ft^3 or kg/m^3
V_∞ = far-field airspeed, typically in ft/s or m/s
S_{ref} = reference area, typically the wing area in ft^2 or m^2
C_{D0} = basic drag coefficient, dimensionless

The basic drag coefficient is typically a function of:

$$C_{D0} = f(geometry, M, R_e, \alpha, \beta) \qquad (15\text{-}16)$$

where

M = Mach number
R_e = Reynolds number
α = angle-of-attack
β = angle-of-yaw

Ultimately, the basic drag force can be thought of as the increase in the skin friction forces due to the applied form factor (FF), i.e. the difference (FF − 1). However, and as will be seen later, this distinction is better suited for explanation than utility, as it is far more practical to combine the two (i.e. the skin friction and pressure drag contributions). This is due to the complex interaction between the two and our desire to maintain a level of simplicity in our analyses.

15.3.2 The Skin Friction Drag Coefficient: C_{Df}

The skin friction drag coefficient is defined as follows:

$$C_{Df} = \frac{2D_f}{\rho V_\infty^2 S_{ref}} = C_f \left(\frac{S_{wet}}{S_{ref}}\right) \qquad (15\text{-}17)$$

where

D_f = skin friction drag force in lb_f (UK system) or N (SI system)
ρ = air density, typically in slugs/ft^3 or kg/m^3
V_∞ = far-field airspeed, typically in ft/s or m/s
S_{wet} = wetted area, typically in ft^2 or m^2
C_{Df} = skin friction drag coefficient, dimensionless
C_f = skin friction coefficient, dimensionless

Note the difference between the two coefficients C_f and C_{Df}.

Skin friction is caused by a fluid's viscosity as it flows over a surface. Its magnitude depends on the viscosity of the fluid and the wetted (or total) surface area in contact with it, as well as the surface roughness. The analysis of skin friction drag is complicated by the process of transition, when the laminar boundary layer becomes turbulent. Consequently, this is called a *mixed boundary layer*. A realistic drag analysis always assumes a mixed boundary layer.

An airfoil that can sustain laminar flow as far aft as 50% of the chord, naturally and on its own merit, is referred to as a natural laminar flow airfoil (NLF). Such airfoils generate substantially less drag than airfoils not capable of this.

Figure 15-12 shows an airfoil immersed in airflow, on which the laminar boundary layer extends from the leading edge to a point on the upper surface denoted by X_{tr_upper} and X_{tr_lower} on the lower surface. At those

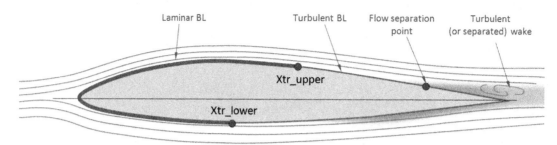

FIGURE 15-12 Important characteristics of flow over an airfoil.

points, enough instability has developed in the boundary layer to change the laminar profile into a thicker turbulent one. This is the consequence of dynamic flow forces becoming larger than the viscous forces (see Figure 15-13) and tends to occur once the Reynolds number (R_e), as measured from the leading edge (LE) along the surface, approaches and exceeds 1 million (on a flat and smooth plate) [3, pp. 2–8]. As the speed of the fluid increases (increasing the R_e) the transition points will move farther and farther forward, although they will never fully get to the LE. Therefore, if smooth, the LE will always develop some laminar BL.

The stability of the boundary layer also depends on the geometry and quality of the surface over which the fluid flows. The presence of rivet-heads, uneven plate joints, insects, and even paint chips can destabilize the laminar boundary layer and initiate an early transition (see Section 8.3.13, *The effect of leading edge roughness and surface smoothness*). Figure 15-14 depicts how the boundary layer changes with *AOA*. At a low *AOA*, the transition location on the upper surface is close to the trailing edge, stabilized by the high pressure along the upper surface. At a high *AOA*, the transition point moves forward, leaving less area covered by the laminar boundary layer. However, and this is important to keep in mind, the laminar boundary layer is more stable on the lower surface because of a net reduction in airspeed.

The boundary layer develops on the surfaces of a "streamlined" three-dimensional shape in a similar fashion as it does on a flat plate and is affected by all the same shortcomings. The velocity distribution in a laminar or turbulent boundary is also similar between the two, albeit being affected by the pressure distribution of the three-dimensional geometry [3, pp. 2–6].

Figure 15-15 shows an important property of flow around an airfoil that ensures the maintenance of the laminar boundary layer (assuming the surface quality is smooth). This property is an extensive region of a favorable pressure gradient. The figure shows the pressure distribution along the chord of two airfoils of 12% thickness-to-chord ratio; a NACA 64-012A and NACA 0012. Both airfoils are at an *AOA* of 0°. The pressure distribution of the former is represented by the solid line and the latter by the dashed line. The figure also shows that the pressure along the surfaces of the NACA 64-012A airfoil is dropping, starting behind the leading edge until 40% of the chord. The pressure on the NACA 0012 airfoil drops sharply until 12% of the chord. The laminar flow of the 64-012A extends to 55% of the chord, versus 50% of the 0012, and, consequently, it generates slightly less drag. In practice, the transition point of the NACA 0012 would be expected to be farther forward than that.

One way to achieve such an extensive laminar boundary layer is to place the maximum thickness of the airfoil as far aft as possible. The goal is to allow the fluid to accelerate as far aft as possible and then decelerate without separation. Ultimately, how far aft the maximum thickness can be placed depends on the distance along which the fluid is allowed to decelerate.

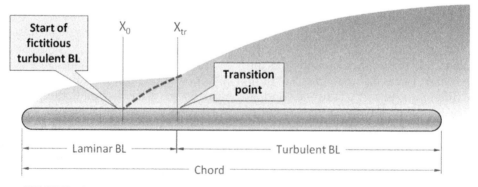

FIGURE 15-13 Transition from laminar to turbulent flow inside a mixed boundary layer.

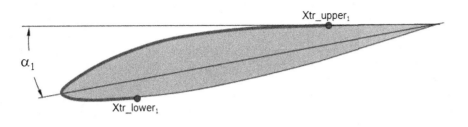

FIGURE 15-14 Movement of transition points with angle-of-attack.

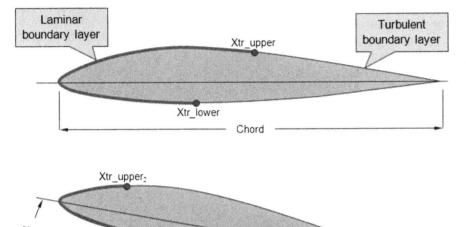

FIGURE 15-15 Difference in the chordwise pressure distribution of "laminar" and "turbulent" BL airfoils.

Too short a distance and the fluid will separate near the trailing edge and increase the drag. Figure 15-15 shows that the negative ("favorable") pressure gradient of the former extends to 40% of the chord, but merely 12% of the latter. A favorable pressure gradient means that fluid molecules are accelerating along the surface and the acceleration stabilizes the boundary layer, allowing the laminar boundary layer to extend farther toward the trailing edge. A direct computational analysis, using the airfoil analysis software Xfoil, predicts the transition point of the NACA 64-012A airfoil to be at 55% of the chord and 50% on the NACA 0012 airfoil at a Reynolds number of 3 million and Mach number of 0.10. Bringing the transition point beyond this will require the maximum thickness to be placed even farther aft. This requires a careful attention to detail and depth of understanding of the aerodynamics of the particular geometry. Some NLF airfoils sustain laminar BL as far aft as 70% of the chord.

So far we have seen how the laminar boundary layer depends on the geometry of the surface over which it flows. The greatest drawback of such flow is its sensitivity to geometry and even uncontrollable factors like the quality of the airflow. Figure 15-16 shows how the flat plate drag coefficient changes with Reynolds number. It is evident how laminar flow reduces the magnitude of the drag coefficient. As an example, the laminar boundary layer drag coefficient at $R_e = 1 \times 10^6$ is approximately 0.00135, while it is approximately 0.00445 for a turbulent boundary layer; the turbulent skin friction is about 3.3 times the laminar one.

Figure 15-16 also shows how susceptible to transitioning the laminar boundary layer is as it flows over a flat plate, operating within Reynolds numbers ranging from 4×10^5 to 1×10^7. A method to estimate this transition is presented below.

Standard Formulation to Estimate Skin Friction Coefficient

The formulation presented below is developed using boundary layer theories, such as those presented by Schlichting [4], Young [5], and Schetz and Bowersox [6]. The scope of the derivation does not lend itself to a convenient fit in this text. For this reason only results important to aircraft drag analysis are presented. The resulting relationships are plotted in Figure 15-16.

(1) *Complete laminar flow:* if it is assumed that laminar flow is fully sustained over a flat plate surface (a theoretical possibility only), the classical Blasius solution for a laminar boundary layer is used to estimate the skin friction:

100% Laminar BL:

$$C_{f_{lam}} = \frac{1.328}{\sqrt{R_e}} \tag{15-18}$$

where R_e = Reynolds number.

(2) *Complete turbulent flow:* if it is assumed that turbulent flow is fully sustained over a flat plate surface, the skin friction coefficient is given by the so-called Schlichting relation [4, p. 438], which is in excellent agreement with experiment:

100% Turbulent BL:

$$C_{f_{turb}} = \frac{0.455}{(\log_{10} R_e)^{2.58}} \tag{15-19}$$

The skin friction coefficient for 100% turbulent boundary layer that accounts for compressibility using a variation of the Schlichting relation is given below:

$$C_{f_{turb}} = \frac{0.455}{(\log_{10} R_e)^{2.58}(1 + 0.144 M^2)^{0.65}} \tag{15-20}$$

where M = Mach number.

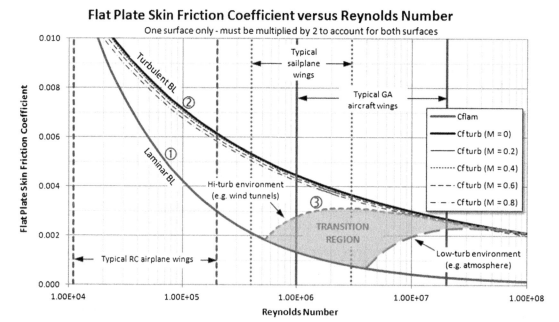

FIGURE 15-16 Change in skin friction coefficient with Reynolds number. Note the transition region, inside which it is challenging to sustain laminar boundary layer. ① is Equation (15-18), ② is Equation (15-19), ③ is Equation (15-21).

TABLE 15-1	Transition Parameters			
Condition	High Turbulence (e.g. Wind Tunnels)		Low-Turbulence (e.g. Atmosphere)	
Critical R_e	0.3×10^6	0.6×10^6	1.0×10^6	3.0×10^6
Constant A	1050	1700	3300	8700

(3) Laminar-to-turbulent transition: the transition of laminar to turbulent boundary layer can be estimated using the so-called *Prandtl-Schlichting skin friction formula for a smooth flat plate at zero incidence* [4, p. 439], presented below. The expression is valid for Reynolds numbers up to 10^9.

$$C_{f_{turb}} = \frac{0.455}{(\log_{10} R_e)^{2.58}} - \frac{A}{R_e} \quad (15\text{-}21)$$

where A is read from the Table 15-1 below and is selected based on the R_e at which transition is expected (critical R_e).

(4) Mixed laminar-turbulent flow skin friction: Young [5, pp. 162–164] presents a two-step method to calculate the skin friction coefficient when the extent of the laminar flow is known. This is the *mixed laminar-turbulent theory*. The first step requires the location of the transition point to be calculated:

$$\left(\frac{X_0}{C}\right) = 36.9 \times \left(\frac{X_{tr}}{C}\right)^{0.625} \left(\frac{1}{\text{Re}}\right)^{0.375} \quad (15\text{-}22)$$

where

X_0 = location of the fictitious turbulent boundary layer
X_{tr} = location of where laminar BL becomes turbulent
C = reference length of the plate (chord)

Then, the skin friction coefficient is determined as follows:

$$C_f = \frac{0.074}{\text{Re}^{0.2}} \left(1 - \left(\frac{X_{tr} - X_0}{C}\right)\right)^{0.8} \quad (15\text{-}23)$$

If the Mach number of the airplane is higher than 0.5, it is prudent to correct this value using the compressibility correction of Frankl-Voishel, presented in Equation (8-59).

It is of interest to compare Equations (15-18), (15-19), and (15-23). A good mixed boundary layer theory should bridge the gap between 100% laminar and 100% turbulent theory. In other words, when $X_{tr}/C = 0$ the result should equal the fully turbulent result and when $X_{tr}/C = 100$ it should equal the fully laminar result. Figure 15-17 shows the three theories compared. As is to be expected, the first two are constant with respect to transition location, whereas only the mixed boundary layer theory is dependent on the location of the transition point. For most R_e applicable to aircraft aerodynamics it deviates only from the turbulent theory when $X_{tr}/C < 0.06$. As such, it is applicable even to turbulent boundary layer airfoils, as these will sustain laminar flow beyond 6% of the chord.

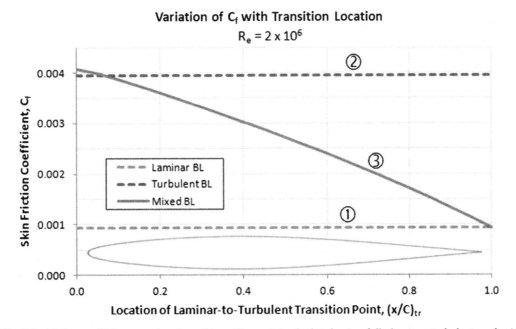

FIGURE 15-17 Skin friction coefficient as a function of transition point calculated using fully laminar, turbulent, and mixed theory. ① is Equation (15-18), ② is Equation (15-19), ③ is Equation (15-23).

15.3.3 Step-by-step: Calculating the Skin Friction Drag Coefficient

Step 1: Determine the Viscosity of Air

In the UK system the temperature is in °R and the viscosity can be found from the following expressions [7]:

$$\mu = 3.170 \times 10^{-11} T^{1.5} \left(\frac{734.7}{T + 216} \right) \text{lb}_f \cdot \text{s}/\text{ft}^2 \quad (16\text{-}20)$$

In the SI system the temperature is in K and the viscosity can be found from:

$$\mu = 1.458 \times 10^{-6} T^{1.5} \left(\frac{1}{T + 110.4} \right) \text{N} \cdot \text{s}/\text{m}^2 \quad (16\text{-}21)$$

where

T = outside air temperature, in °R or K
μ = air viscosity, in $\text{lb}_f \cdot \text{s}/\text{ft}^2$ or $\text{N} \cdot \text{s}/\text{m}^2$

Step 2: Determine the Reynolds Number

Using Equations (8-28) through (8-30), the first one being repeated here for convenience as Equation (15-24):

$$R_e = \frac{\rho V C}{\mu} \quad (15\text{-}24)$$

where

C = reference length (e.g. wing chord being analyzed), in ft or m
V = reference airspeed, in ft/s or m/s

Step 3: Cutoff Reynolds Number [8]

If surface qualities are less than ideal, the actual skin friction will be higher than indicated by Equations (15-18) and (15-19). Accounting for this trend requires a special R_e to be calculated, which is called a *cutoff* R_e. If the actual R_e (calculated above) is larger than this cutoff R_e, then the cutoff R_e should be used instead of the actual R_e.

Subsonic:

$$R_{e\ cutoff} = 38.21 \left(\frac{C}{\kappa} \right)^{1.053} \quad (15\text{-}25)$$

TABLE 15-2 Skin Roughness Values (Based on Ref. [8])

Surface Type	κ (C in ft)
Camouflage paint on aluminum	3.33×10^{-5}
Smooth paint	2.08×10^{-5}
Production sheet metal	1.33×10^{-5}
Polished sheet metal	0.50×10^{-5}
Smooth molded composite	0.17×10^{-5}

Trans and supersonic:

$$R_{e\ cutoff} = 44.62 \left(\frac{C}{\kappa} \right)^{1.053} M^{1.16} \quad (15\text{-}26)$$

where

M = Mach number
κ = skin roughness value (see Table 15-2)

Step 4: Skin Friction Coefficient for Fully Laminar or Fully Turbulent Boundary Layer

First, if the intent is to calculate the skin friction coefficient assuming a *mixed boundary layer* go to Step 5. Otherwise, decide whether to treat the boundary layer as either fully laminar or fully turbulent.

If it is assumed that laminar flow is fully sustained over a flat plate surface (a theoretical possibility only) calculate the skin friction coefficient using Equation (15-18):

100% Laminar BL:

$$C_{f_{lam}} = \frac{1.328}{\sqrt{R_e}} \quad (15\text{-}18)$$

If it is assumed that turbulent flow is fully sustained over a flat plate surface, calculate the skin friction coefficient using Equation (15-19):

100% Turbulent BL:

$$C_{f_{turb}} = \frac{0.455}{\left(\log_{10}(R_e) \right)^{2.58}} \quad (15\text{-}19)$$

If it is assumed that turbulent flow is fully sustained over a flat plate surface and subject to compressibility effects, calculate the skin friction coefficient using Equation (15-20):

$$C_{f_{turb}} = \frac{0.455}{\left(\log_{10}(R_e) \right)^{2.58} (1 + 0.144 M^2)^{0.65}} \quad (15\text{-}20)$$

where

M = Mach number
R_e = Reynolds number

If the skin friction coefficient was calculated in this step, skip Step 5 and 6 and go directly to Step 7.

Step 5: Mixed Laminar-Turbulent Flow Skin Friction

Compute the location of the transition point using Equation (15-22):

$$\left(\frac{X_0}{C} \right) = 36.9 \times \left(\frac{X_{tr}}{C} \right)^{0.625} \left(\frac{1}{R_e} \right)^{0.375} \quad (15\text{-}22)$$

Step 6: Mixed Laminar-Turbulent Flow Skin Friction

Then, the skin friction coefficient is determined using Equation (15-23):

$$C_f = \frac{0.074}{R_e^{0.2}}\left(1 - \left(\frac{X_{tr} - X_0}{C}\right)\right)^{0.8} \quad (15\text{-}23)$$

Step 7: Compute Skin Friction Drag Coefficient

Then, the skin friction coefficient for multiple surfaces is determined as follows:

$$C_f = \frac{\sum_{i=1}^{N} C_{fi} \times S_{wet\ i}}{S_{wet}} \quad (15\text{-}27)$$

$$C_{Df} = \left(\frac{S_{wet}}{S_{ref}}\right)C_f = \left(\frac{S_{wet}}{S_{ref}}\right)\frac{\sum_{i=1}^{N} C_{fi} \times S_{wet\ i}}{S_{wet}}$$

$$= \left(\frac{1}{S_{ref}}\right)\sum_{i=1}^{N} C_{fi} \times S_{wet\ i} \quad (15\text{-}28)$$

where

C_{fi} = skin friction coefficient of surface i
N = number of surfaces
$S_{wet\ i}$ = wetted area of surface i (in ft^2 or m^2)
S_{wet} = wetted area of all surfaces (in ft^2 or m^2)

The skin friction drag force on each surface is determined using the wetted area of the individual surface as follows:

$$D_{fi} = \frac{1}{2}\rho V^2 C_{fi} S_{wet_i} \quad (15\text{-}29)$$

Step 10: Compute the Total Skin Friction Drag Force

The total skin friction drag force on the vehicle can now be determined using the following expressions:

Using the skin friction coefficient:

$$D_f = \frac{1}{2}\rho V^2 C_f S_{wet} \quad (15\text{-}30)$$

Using the skin friction drag coefficient:

$$D_f = \frac{1}{2}\rho V^2 C_{Df} S_{ref} \quad (15\text{-}31)$$

EXAMPLE 15-2: SKIN FRICTION DRAG OF A WING

In this example the skin friction drag of the SR22 wing will be evaluated, neglecting the interference between it and the fuselage (this is done in Example 15-6). The wing's pertinent dimensions are shown in Figure 15-18, but these were obtained by scaling by the reported wing span of 38.3 ft. The actual wings features three distinct NLF airfoils, but for this example assume a single airfoil NACA 65$_2$-415, capable of sustaining a laminar boundary layer on the upper and lower surfaces as indicated in the figure. It is assumed that the wetted area is 7% greater than that of the shaded planform area indicated in Figure 15-18.

If the airplane is cruising at 185 KTAS at S-L ISA, determine the skin friction drag coefficient and force acting on the wing due to the mixed laminar and turbulent BL regions. Compare to a wing with fully laminar or fully turbulent BL.

Solution

Note that the solution below was calculated using the spreadsheet software Microsoft Excel. The reader entering the numbers into a calculator will probably notice slight difference from the numbers shown. This is normal and is due to the fact that Microsoft Excel retains 16 significant digits whereas the typical calculator retains 8 significant digits, not to mentioned user round-off differences.

Step 1: Determine the Viscosity of Air

Use Sutherland's formula to compute the viscosity assuming an OAT of 518.67 °R:

$$\mu = 3.170 \times 10^{-11} T^{1.5}\left(\frac{734.7}{T + 216}\right)$$

$$= 3.170 \times 10^{-11}(518.67)^{1.5}\left(\frac{734.7}{518.67 + 216}\right)$$

$$= 3.745 \times 10^{-7}\ \text{lb}_f\cdot\text{s/ft}^2$$

Step 2: Determine the Reynold's Number for Root Airfoil

Compute R_e for the root airfoil, using ISA density of 0.002378 slugs/ft^3 and an airspeed of 185 KTAS.

$$R_{e1} = \frac{\rho V C_r}{\mu} = \frac{(0.002378)(185 \times 1.688)(4.88)}{3.745 \times 10^{-7}}$$

$$= 9667562$$

EXAMPLE 15-2: SKIN FRICTION DRAG OF A WING (cont'd)

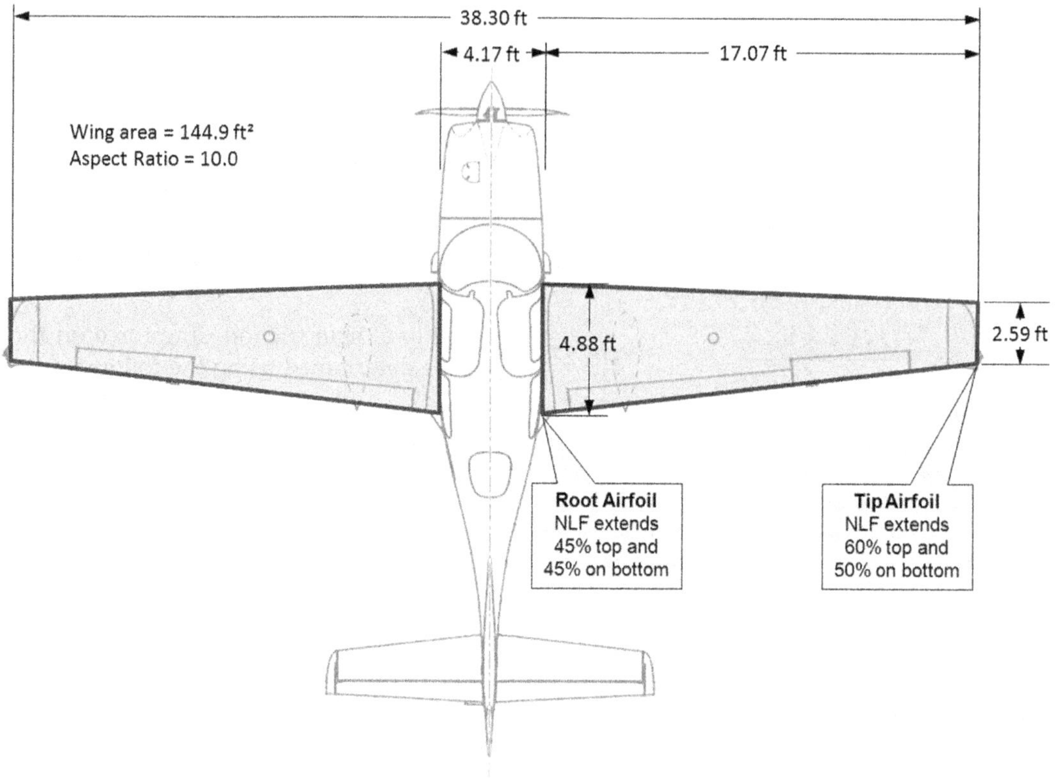

FIGURE 15-18 Scaling the top view based on the wing span yields the following dimensions *(courtesy of Cirrus Aircraft).*

Step 3: Determine the Reynold's Number for Tip Airfoil

Compute R_e for the tip airfoil:

$$R_{e2} = \frac{\rho V C_t}{\mu} = \frac{(0.002378)(185 \times 1.688)(2.59)}{3.745 \times 10^{-7}}$$
$$= 5126287$$

Step 4: Fictitious Turbulent BL on Root Airfoil – Upper Surface

For the upper surface of the root airfoil (45% coverage) we get:

$$\left(\frac{X_0}{C_r}\right) = 36.9 \times \left(\frac{X_{tr}}{C_r}\right)^{0.625}\left(\frac{1}{R_{e1}}\right)^{0.375}$$
$$= 36.9 \times (0.45)^{0.625}\left(\frac{1}{9667562}\right)^{0.375} = 0.05380$$

Step 5: Fictitious Turbulent BL on Root Airfoil – Lower Surface

For the lower surface of the root airfoil (45% coverage) we get the same value as on the upper surface:

$$(X_0/C_r) = 0.05380$$

Step 6: Fictitious Turbulent BL on Tip Airfoil – Upper Surface

For the upper surface of the tip airfoil (60% coverage) we get the same value as on the lower surface:

$$\left(\frac{X_0}{C_t}\right) = 36.9 \times (0.60)^{0.625}\left(\frac{1}{5126287}\right)^{0.375} = 0.08169$$

Step 7: Fictitious Turbulent BL on Tip Airfoil – Lower Surface

For the lower surface of the tip airfoil (50% coverage) we get:

$$\left(\frac{X_0}{C_t}\right) = 36.9 \times (0.50)^{0.625}\left(\frac{1}{5126287}\right)^{0.375} = 0.07290$$

Step 8: Skin Friction for Root Airfoil – Upper Surface

For the upper surface of the root airfoil (45% coverage) we get:

$$\left(C_f\right)_{upper\ 1} = \frac{0.074}{R_{e1}^{0.2}}\left(1 - \left(\frac{X_{tr} - X_0}{C_r}\right)\right)^{0.8}$$
$$= \frac{0.074}{9667562^{0.2}}(1 - (0.45 - 0.05380))^{0.8}$$
$$= 0.001981$$

EXAMPLE 15-2: SKIN FRICTION DRAG OF A WING (cont'd)

Step 9: Skin Friction for Root Airfoil – Lower Surface

For the lower surface of the root airfoil (45% coverage) we get:

$$(C_f)_{lower\ 1} = 0.001981$$

Step 10: Average Skin Friction for Root Airfoil

The average of the upper and lower surfaces of the root airfoil yields:

$$C_{f_1} = \frac{1}{2}(0.001981 + 0.001981) = 0.001981$$

Step 11: Skin Friction for Tip Airfoil – Upper Surface

For the upper surface of the tip airfoil (60% coverage) we get:

$$(C_f)_{upper\ 2} = \frac{0.074}{R_{e_2}^{0.2}}\left(1 - \left(\frac{X_{tr} - X_0}{C_t}\right)\right)^{0.8}$$

$$= \frac{0.074}{5126287^{0.2}}(1 - (0.60 - 0.08169))^{0.8}$$

$$= 0.001877$$

Step 12: Skin Friction for Tip Airfoil – Lower Surface

For the lower surface of the tip airfoil (50% coverage) we get:

$$(C_f)_{lower\ 2} = \frac{0.074}{R_{e_2}^{0.2}}\left(1 - \left(\frac{X_{tr} - X_0}{C_t}\right)\right)^{0.8}$$

$$= \frac{0.074}{5126287^{0.2}}(1 - (0.50 - 0.07290))^{0.8}$$

$$= 0.002156$$

Step 13: Average Skin Friction for Tip Airfoil

The average of the upper and lower surfaces of the tip airfoil yields:

$$C_{f_2} = \frac{1}{2}(0.001877 + 0.002156) = 0.002017$$

Step 14: Average Skin Friction for Complete Wing

The skin friction coefficient for the total wetted surface is simply the average of the average coefficient for both airfoils, i.e.:

$$C_f = \frac{1}{2}(0.001981 + 0.002017) = 0.001999$$

Step 15: Wing's Wetted Area

The wing's total wetted area is:

$$S_{wet} = 2 \times 1.07 \times \left[\frac{1}{2}(4.88 + 2.59) \times (38.3 - 4.17)\right]$$

$$= 272.8 \text{ ft}^2$$

Step 16: Skin Friction Drag Coefficient for Complete Wing

The wing's skin friction drag coefficient is:

$$C_{Df} = \left(\frac{S_{wet}}{S_{ref}}\right)C_f = \left(\frac{272.8}{144.9}\right)(0.001999) = 0.003758$$

Step 17: Skin Friction Drag Force for Complete Wing

Estimate skin friction drag due to the mixed boundary layer.

$$D_f = \frac{1}{2}\rho V^2 \times S_{wet} \times C_f$$

$$= \frac{1}{2}(0.002378)(185 \times 1.688)^2(272.4)(0.001999)$$

$$= 63.1 \text{ lb}_f$$

This means that the total flat plate skin friction drag for the wing only of the SR22 at cruising speed amounts to 63 lb$_f$. The contributions of interference with the fuselage and airfoil shape are not accounted for. It also omits drag due to the presence of control surfaces.

Step 18: Skin Friction Coefficient for 100% Laminar Flow

Laminar flow coefficient for the root:

$$C_{f_{lam}} = \frac{1.328}{\sqrt{R_{e1}}} = \frac{1.328}{\sqrt{9667562}} = 0.0004271$$

Laminar flow coefficient for the tip:

$$C_{f_{lam}} = \frac{1.328}{\sqrt{R_{e2}}} = \frac{1.328}{\sqrt{5126287}} = 0.0005865$$

Average for the wing:

$$C_f = \frac{0.0004271 + 0.0005865}{2} = 0.0005068$$

Step 19: Skin Friction Coefficient for 100% Turbulent Flow

Turbulent flow coefficient for the root:

$$C_{f_{turb}} = \frac{0.455}{(\log_{10}(R_{e1}))^{2.58}} = \frac{0.455}{(\log_{10}(9667562))^{2.58}}$$

$$= 0.003020$$

EXAMPLE 15-2: SKIN FRICTION DRAG OF A WING (cont'd)

Turbulent flow coefficient for the tip:

$$C_{f_{turb}} = \frac{0.455}{(\log_{10}(\text{Re}_2))^{2.58}} = \frac{0.455}{(\log_{10}(5126287))^{2.58}}$$
$$= 0.003350$$

Average for the wing:

$$C_f = \frac{1}{2}(0.003020 + 0.003350) = 0.003185$$

Step 20: Comparison

See Table 15-3 for a comparison of the the analysis techniques.

The most important observation from this comparison is that if it were possible to maintain laminar flow over the entire wing, its drag would be 25% of that predicted by the mixed boundary layer theory. Naturally, this is impossible to achieve, but rather represents an extreme. Alternatively, if the wing sustained turbulent BL only, it would be almost 60% draggier than with the NLF airfoils. This represents a far more realistic comparison and demonstrates the value of employing such airfoils.

TABLE 15-3 Comparing Skin Friction Analysis Methods

Method	C_f	Comparison
Fully Laminar BL	0.0005068	25%
Fully Turbulent BL	0.003185	159%
Mixed BL	0.001999	100%

EXAMPLE 15-3: TOTAL SKIN FRICTION OF A MULTI-PANEL WING

A halfspan of a Schuemann style wing is shown in Figure 15-19. The planform consists of three sections with the dimension and the total wing and wetted areas shown. The skin friction coefficient for each section has been determined and is tabulated with the corresponding planform areas. If this wing (the full wing) is exposed to S-L conditions at 100 KCAS and using the wing area as the reference area and assuming the wetted area is 2× the surface area multiplied by a wetted area booster factor, K_b, of 1.1, estimate the following:

Total skin friction coefficient
Total skin friction drag coefficient
Skin friction drag force for individual surfaces
Skin friction drag for the total wing

Solution

(a) Begin by creating Table 15-4 as follows (where $S_{\text{wet } i} = 2 \times K_b \times S_i$). Thus, the wetted area for panel ① is given by $S_{\text{wet } ①} = 2 \times K_b \times S_① = 2 \times (1.1) \times (35.0) = 77.0$ ft². Subsequent multiplication using $C_{f\,①} = 0.0050$ and $q = \frac{1}{2}\rho V^2 = \frac{1}{2}(0.002378)(100 \times 1.688)^2 = 33.88$ lb$_f$/ft² yield the other two columns. Next, the totals of the three right-most columns are calculated. Note that these are for only one of the two wing halves.

Then the total skin friction coefficient can be found by the weighted contribution from each panel:

$$C_f = \frac{\sum_{i=1}^{N} C_{fi} \times S_{\text{wet } i}}{S_{\text{wet}}}$$
$$= \frac{77.0 \times 0.0050 + 69.3 \times 0.0055 + 46.2 \times 0.0065}{77.0 + 69.3 + 46.2}$$
$$= \frac{1.100}{192.5} = 0.00554$$

This should be considered the representative skin friction coefficient for the wing. Note that the application of form factors is not needed when accounting for each panel as all the panels belong to the same unit.

(b) The total skin friction drag coefficient is always based on the reference area, here chosen as the wing area $S_{\text{ref}} = 160$ ft²:

$$C_{Df} = \frac{S_{\text{wet}}}{S_{\text{ref}}} C_f = \frac{385}{175} \times 0.00554 = 0.01219$$

This amounts to 121.9 drag counts.

(c) And (d) since the skin friction coefficient differs from one panel to the next, the drag force on each surface must be determined using the wetted area of the individual surface as follows:

$$D_{fi} = \frac{1}{2}\rho V^2 C_{f_i} S_{\text{wet}_i} = q C_{f_i} S_{\text{wet}_i}$$

EXAMPLE 15-3: TOTAL SKIN FRICTION OF A MULTI-PANEL WING (cont'd)

Therefore the total skin friction drag force for both wing halves can be found by summing up the individual contributions:

$$D_f = 2 \times q \sum_{i=1}^{3} C_{fi} S_{wet\ i}$$
$$= 2 \times (33.88)[(0.0050)(77.0) + (0.0055)(69.3)$$
$$+ (0.0065)(46.2)]$$
$$= 2 \times (33.88)[0.3850 + 0.3812 + 0.3003] = 72.3\ \text{lb}_f$$

or:

$$D_f = \frac{1}{2} \rho V^2 C_f S_{wet} = (33.88)(0.00554)(385)$$
$$= 72.3\ \text{lb}_f$$

or:

$$D_f = \frac{1}{2} \rho V^2 C_{Df} S_{ref} = (33.88)(0.01219)(175)$$
$$= 72.3\ \text{lb}_f$$

We can also calculate the skin friction drag coefficient, C_{Dfi}, for each panel for each wing half (assuming only half of the reference wing area) as shown below. Note the results are presented in Table 15-5.

$$C_{Df\ 1} = \frac{C_{f1} \cdot S_{wet\ 1}}{S_{ref}} = \frac{0.0050 \cdot 77.0}{(175/2)} = 0.004400$$

$$C_{Df\ 2} = \frac{C_{f2} \cdot S_{wet\ 2}}{S_{ref}} = \frac{0.0055 \cdot 69.3}{(175/2)} = 0.004356$$

$$C_{Df\ 3} = \frac{C_{f3} \cdot S_{wet\ 3}}{S_{ref}} = \frac{0.0065 \cdot 46.2}{(175/2)} = 0.003432$$

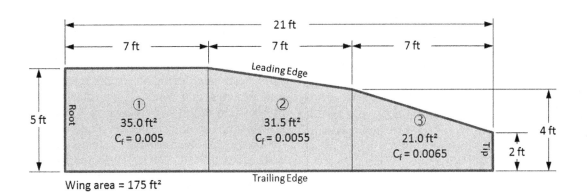

FIGURE 15-19 The multi-panel wing used in this example.

TABLE 15-4 Comparing Skin Friction Analysis Methods

	S_i	C_{fi}	$S_{wet\,i}$	$S_{wet\,i} \cdot C_{fi}$	$q \cdot S_{wet\,i} \cdot C_{fi}$
Part ①	35.0	0.0050	77.0	0.3850	13.04
Part ②	31.5	0.0055	69.3	0.3812	12.91
Part ③	21.0	0.0065	46.2	0.3003	10.17
			TOTALS: 192.5	1.1	36.1

EXAMPLE 15-3: TOTAL SKIN FRICTION OF A MULTI-PANEL WING (cont'd)

TABLE 15-5 Total Skin Friction of the Multi-Panel Wing

	S_i	C_{fi}	$S_{wet\,i}$	$S_{wet\,i} \cdot C_{fi}$	$q \cdot S_{wet\,i} \cdot C_{fi}$	$C_{Df\,i}$
Part ①	35.0	0.0050	77.0	0.3850	13.04	0.004400
Part ②	31.5	0.0055	69.3	0.3812	12.91	0.004356
Part ③	21.0	0.0065	46.2	0.3003	10.17	0.003432
		TOTALS:	192.5	1.1	36.1	0.01219

In calculating the above drag, two important influences were neglected: *form* and *interference* effects. A form effect comes from the fact that more information than just wetted area is required to correctly estimate the drag of the component. This is easy to realize when comparing a sphere or a box of equal wetted area. Intuitively it is obvious to us that the drag of the box will be higher than that of the sphere. This effect is accounted for using the so-called *form factor*, to be dealt with later. Also, when calculating the combined effect of panels comprising a larger surface (such as the wing in the above example) there are no specific interferences that have to be considered. In other words, the proximity of components when mounting a wing to a fuselage or a horizontal tail to a vertical tail, and so on, affects the drag. This effect is called an *interference drag* and it increases the drag beyond what the above example indicates. Interference is accounted for using a so-called *interference factor*. Both the form and interference factors account for the effects of viscous separation (pressure drag) and are evaluated in Section 15.4.6, *The component drag build-up method*. In particular see Example 15-6.

15.3.4 The Lift-Induced Drag Coefficient: C_{Di}

Consider a finite wing featuring a single airfoil mounted in space in airflow at some specific *AOA* and airspeed. Contrast that with a straight wing featuring exactly the same airfoil, extending from wall to wall in a wind tunnel at the same *AOA* and airspeed (ignoring typical wind-tunnel interferences). In this situation, the finite wing will generate less lift per unit span than the one in the wind tunnel. This is caused by the fact that the airflow makes its path around the wingtips of the finite wing through the formation of vortices, one at each wingtip. The short consequence of this flow is that in order for the wing to generate a specific amount of lift, it must always be at a higher *AOA* than the corresponding two-dimensional airfoil in the wind tunnel. This leads to the formation of drag that increases the wing drag beyond that of the airfoil only. This drag is called a *lift-induced*, or *induced*, or *vortex* drag and is denoted by D_i. As usual the standard formulation of this drag involves the determination of a *lift-induced drag coefficient*. A number of methods to determine this drag coefficient are presented below. The lift-induced drag is a pressure drag force.

Method 1: Lift-Induced Drag from the Momentum Theorem

A wing can be considered a device that generates lift by deflecting a stream of fluid downward; an event that changes the fluid's momentum. The downward motion of the fluid is called *downwash*. The magnitude of this lift can be estimated using Newton's second law of motion, which states that force is the rate of change of momentum. In doing so it is assumed the diameter of the stream tube being deflected equals the wingspan,[4] denoted by b. The mass flow rate inside this cylinder is denoted by $\dot{m}$. If the vertical downwash at some distance behind the wing is denoted by w, the rate of change of momentum (lift force) can be estimated from:

$$L = \dot{m}w \qquad (15\text{-}32)$$

The mass flow rate in the stream tube is given by $\dot{m} = \rho A_{tube} V$, where A_{tube} is the cross-sectional area of the stream tube. Thus, the lift can be rewritten as follow:

$$L = \dot{m}w = \rho A_{tube} V w = \rho\left(\frac{\pi}{4}\right)b^2 V w \qquad (15\text{-}33)$$

Equating this with the standard expression for lift (Equation (8-8)) allows the magnitude of the downwash to be estimated:

$$L = \rho\left(\frac{\pi}{4}\right)b^2 V w = \frac{1}{2}\rho V^2 S C_L \quad \Leftrightarrow \quad w = 2V\frac{S}{\pi b^2} C_L \qquad (15\text{-}34)$$

---

[4] Note that some authors, e.g. Stinton, describe diameter as being $(b\sqrt{2})$, but most other texts use the presentation shown.

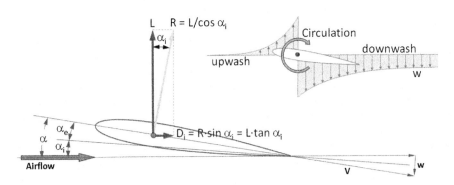

FIGURE 15-20 A schematic showing how the induced drag is a component of the lift force.

Note that since b^2/S is the aspect ratio, AR, we can write:

$$\frac{w}{V} = \frac{2C_L}{\pi AR} \quad (15\text{-}35)$$

This is shown in Figure 15-20. The circulation of fluid around the airfoil will cause a reduction in the *geometric AOA*, denoted by α, which is the angle between the flight path and chordline. The reduction is called *induced AOA* and is denoted by α_i. The circulation introduces an upwash into the airflow (and downwash behind it), reducing the geometric AOA. The difference between the two is called the *effective AOA*, denoted by α_e. It can be seen that the downwash angle will be given by w/V and that α_i is approximately one-half of that angle. For this reason, using Equation (15-35) and small-angle relations, we can estimate the induced AOA as follows:

$$\alpha_i \approx \tan \alpha_i = \frac{1}{2}\frac{w}{V} = \frac{C_L}{\pi AR} \quad (15\text{-}36)$$

Then, again referring to Figure 15-20, it can be seen that the presence of the induced AOA effectively tilts the lift force back by that amount, in the process forming an additional force component D_i, which we call *lift-induced drag*. This drag can be calculated as shown in the figure.

$$D_i = R \sin \alpha_i = \frac{L}{\cos \alpha_i} \sin \alpha_i = L \tan \alpha_i \quad (15\text{-}37)$$

In a coefficient form this becomes:

$$C_{Di} = C_L \tan \alpha_i = C_L \left(\frac{C_L}{\pi AR}\right) = \frac{C_L^2}{\pi AR} \quad (15\text{-}38)$$

The above derivation is based on the assumption that the lift distribution is elliptical or, in other words, the spanwise distribution of section lift coefficients is constant. As has been shown in Chapter 9, *The anatomy of the wing*, this requires an elliptical planform. However, most wing planform shapes are not elliptical and, consequently, the distribution of section lift coefficients is far from being constant. This requires corrections to be made. Such corrections are discussed below.

The following representation of the lift-induced drag is of importance from an aircraft design standpoint.

Using Equation (15-38) to determine the induced drag force yields:

$$D_i = qS\left(\frac{C_L^2}{\pi AR}\right) = qS\left(\frac{(W/qS)^2}{\pi(b^2/S)}\right)$$
$$= qS\left(\frac{W^2 S}{\pi b^2 q^2 S^2}\right) = \frac{W^2}{\pi b^2 q} \quad (15\text{-}39)$$

The result shows that the lift-induced drag force depends on the wing span and not the wing area. This means that only an increase in span will have beneficial effect on the drag. Reducing the chord to increase AR will not help.

Method 2: Generic Formulation of the Lift-Induced Drag Coefficient

Generic formulation of wing characteristics is presented in NACA-TR-572 [9], based on the work of Glauert [10] and Hueber [11]. It allows a number of wing characteristics to be evaluated for tapered wing planform shapes, with and without rounded wingtips (which were popular during the time when it was written). A generic formulation of the lift-induced drag, C_{Di}, is presented as follows:

$$C_{Di} = \frac{2}{S}\int_0^{b/2} \alpha_i \cdot C_l \cdot C \cdot dy = \frac{2}{S}\int_0^{b/2} \left(\alpha - \frac{C_l}{C_{l\alpha}}\right) \cdot C_l \cdot C \cdot dy$$
$$(15\text{-}40)$$

where

b = wing span
C = wing chord (as a function of y)
C_l = section lift coefficient
S = wing area
y = spanwise location along the halfspan
α_i = induced angle-of-attack

A more general version of this expression is possible, which can treat asymmetric wing loading. It is presented below:

$$C_{Di} = \frac{1}{S}\int_{-b/2}^{b/2} \alpha_i \cdot C_l \cdot C \cdot dy \quad (15\text{-}41)$$

Derivation of Equation (15-40)

The induced AOA at any section along the wing can be computed if the section lift coefficient, C_l, and section lift curve slope, $C_{l\alpha}$, are known:

$$\alpha_i = \alpha - \frac{C_l}{C_{l\alpha}} \quad (15\text{-}42)$$

Therefore, using small-angle relations, the induced drag of the airfoil section can be computed from:

$$C_{di} = \alpha_i C_l = \left(\alpha - \frac{C_l}{C_{l\alpha}}\right) C_l = \left(\alpha C_l - \frac{C_l^2}{C_{l\alpha}}\right) \quad (15\text{-}43)$$

The total lift-induced drag for the wing will then be the sum of the weighted contribution of all the sections, extending from tip to tip, or:

$$C_{Di} = \frac{2}{S} \int_0^{b/2} C_{di} \cdot C \cdot dy \quad (15\text{-}44)$$

The weighted form is necessary as the planform shape is may be changing or affected by washout. This way, multiplying it with the chord, C, Inserting the result from Equation (15-43) and manipulate will yield Equation (15-40).

QED

EXAMPLE 15-4

Estimate the induced drag coefficient for a Hershey bar wing with the following characteristics:

Wing area, $S = 145 \text{ ft}^2$
Wing span, $b = 38 \text{ ft}$
Wing chord, constant $C = 4 \text{ ft}$

The wing is operating at an $AOA = 5° = 0.08727$ radians and it has been determined its spanwise distribution of section lift coefficients can be approximated by $C_l = 0.5 \cdot \cos(\pi y/b)$ and is shown in Figure 15-21. The wing features an airfoil whose $C_{l\alpha} = 5.730$ per radian and is untwisted.

Solution

Solution is obtained by evaluating the integral of the distribution of the section lift coefficients along the entire span of the aircraft per Equation (15-40):

$$C_{Di} = \frac{2}{S} \int_0^{b/2} \left(\alpha - \frac{C_l}{C_{l\alpha}}\right) \cdot C_l \cdot C \cdot dy$$

$$= \frac{2}{145} \int_0^{38/2} \left(0.08727 - \frac{0.5 \cos(\pi y/38)}{5.730}\right)$$

$$\cdot (0.5 \cos(\pi y/38)) \cdot (4) \cdot dy$$

$$= \frac{20}{145} \int_0^{19} \left(0.08727 \cos(\pi y/38) - 0.08727 \cos^2(\pi y/38)\right) \cdot dy$$

$$= 0.001204 \int_0^{19} \left(\cos(\pi y/38) - \cos^2(\pi y/38)\right) \cdot dy$$

Note that the solution of the integral is of the form:

$$\int \cos ay - \cos^2 ay \, dy = \frac{1}{4a}(4 \sin ay - \sin 2ay) - \frac{y}{2}$$

This results in:

$$C_{Di} = 0.001204 \int_0^{19} \left(\cos(\pi y/38) - \cos^2(\pi y/38)\right) \cdot dy$$

which becomes:

$$C_{Di} = 0.001204 \left[\frac{1}{4\left(\frac{\pi}{38}\right)}\left(4\sin\left(\frac{\pi}{38}\right)y - \sin 2\left(\frac{\pi}{38}\right)y\right) - \frac{y}{2}\right]_0^{19}$$

$$= 0.001204 \left[\frac{9.5}{\pi}\left(\left(4\sin\frac{\pi}{2} - 4\sin 0\right) - (\sin\pi - \sin 0)\right) - \frac{19}{2}\right]$$

$$= 0.001204 \left[\frac{38}{\pi} - \frac{19}{2}\right] = 0.003125$$

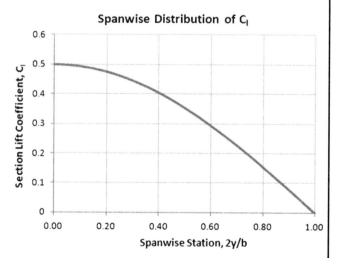

FIGURE 15-21 Spanwise distribution of section lift coefficients.

Method 3: Simplified $k \cdot C_L^2$ Method

This is the simplest representation of lift-induced drag. As has already been demonstrated, it can be derived directly from the momentum theorem, or using the lifting line method presented in Section 9.7, *Numerical analysis of the wing*.

$$C_{Di} = \frac{C_L^2}{\pi \cdot AR \cdot e} = k \cdot C_L^2 \qquad (15\text{-}45)$$

where

C_L = lift coefficient
AR = reference aspect ratio
e = Oswald efficiency
k = lift-induced drag factor

The most difficult parameter to determine is the Oswald efficiency. It can be estimated using Method 5 below. Several methods to estimate it are also provided in Section 9.5.14, Estimation of Oswald's span efficiency. Note an important dependency of the induced drag coefficient on wing area, S, and wing span, b:

$$C_{Di} = \frac{C_L^2}{\pi \cdot AR \cdot e} = \left(\frac{S}{b^2}\right) \frac{C_L^2}{\pi \cdot e} \qquad (15\text{-}46)$$

This result shows that for a given wing area, S, the induced drag is highly dependent on the wing span. It is an important consideration for many applications that feature large area but small wingspan (e.g. a delta wing).

Method 4: Adjusted $k \cdot (C_L - C_{LminD})^2$ Method

As has already been emphasized in the preceding discussion, the adjusted drag model is a far more accurate method of lift-induced drag estimation than the simplified model. This model is presented below:

$$C_{Di} = \frac{(C_L - C_{L_{minD}})^2}{\pi \cdot AR \cdot e} = k \cdot (C_L - C_{L_{minD}})^2 \qquad (15\text{-}47)$$

where $C_{L\ minD}$ = the lift coefficient where drag becomes a minimum.

Method 5: Lift-Induced Drag Using the Lifting-Line Method

In order to remedy the difficulty in determining the lift-induced drag constant, k, shown in Equations (15-7), (15-45), and (15-46), numerical methods, such as the lifting-line or vortex-lattice methods, may be used. Of the two, the lifting-line method, presented in Section 9.7, *Numerical analysis of the wing*, is relatively easy to apply, although it requires a matrix solver to calculate the constants of simultaneous linear equations. These are then used to evaluate a special constant, called the *lift-induced drag factor*, denoted by the Greek letter δ. Once this is known, the lift-induced drag coefficient can be calculated from:

Simplified drag model:

$$C_{Di} = \frac{C_L^2}{\pi \cdot AR}(1+\delta) \qquad (15\text{-}48)$$

Adjusted drag model:

$$C_{Di} = \frac{(C_L - C_{L_{minD}})^2}{\pi \cdot AR}(1+\delta) \qquad (15\text{-}49)$$

Figure 15-22 shows the variation of the lift-induced drag factor, δ, with a range of taper ratios, λ, and aspect ratios, AR, as calculated by the lifting-line method presented in Section 9.7. A code snippet, using Visual Basic for Applications, is also presented in the section, allowing the reader to determine the factor using software such as Microsoft Excel. Note that the Oswald's span efficiency, e, is related to the lift-induced drag factor as shown below:

$$e = 1/(1+\delta) \qquad (15\text{-}50)$$

EXAMPLE 15-5

The SR22 has an $AR = 10$ and a $\lambda = 0.5$. If flying at a condition that generates a $C_L = 0.5$, determine the lift-induced drag coefficient using Figure 15-22. For this example, assume $C_{LminD} = 0$.

Solution

From Figure 15-22 it can be seen that $\delta \approx 0.022$. This means that the lift-induced drag coefficient can be determined as follows:

$$C_{Di} = \frac{C_L^2}{\pi \cdot AR}(1+\delta) = \frac{0.5^2}{\pi \cdot (10)}(1+0.022) = 0.008133$$

This value is 2.2% higher than that for an elliptical wing of the same AR.

Method 6: Prandtl-Betz Integration in the Trefftz Plane

This method was developed by Ludwig Prandtl (1875−1953) and Albert Betz (1885−1968) around the year 1918. The method computes the induced drag of a wing based on disturbance it causes to the fluid flow in the far-field (see Figure 15-23). By evaluating the disturbances on a plane infinitely behind the wing (Trefftz plane, named after Erich Trefftz (1888−1937)), the velocity component in the x-direction (denoted by u) can be eliminated from the integration. In this way, a volumetric integration can be reduced to a surface integration.

$$D_i = \frac{1}{\rho} \iint\limits_{\substack{Trefftz \\ Plane}} (v^2 + w^2) dS \tag{15-51}$$

The method is often applied using computational fluid dynamics (CFD) methods, such as the vortex-lattice method, and is primarily presented here for completeness.

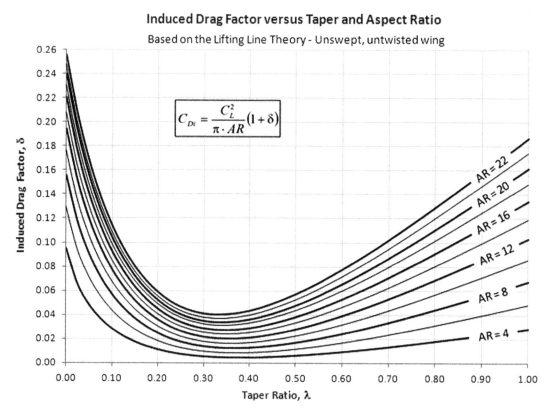

FIGURE 15-22 Induced drag factor for TR ranging from 0 to 1 and AR ranging from 4 to 22.

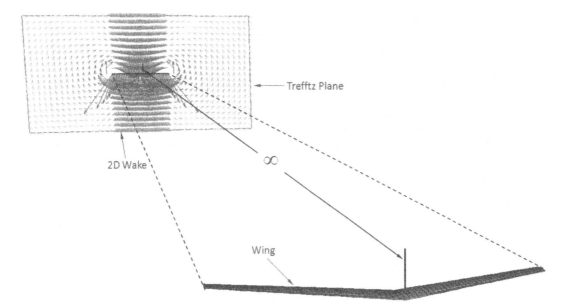

FIGURE 15-23 Calculation of induced drag in the Trefftz plane.

15.3.5 Total Drag Coefficient: C_D

Once the constituent drag contributions have been estimated, the total drag coefficient is simply determined by addition using the following expression:

$$C_D = C_{Do} + C_{Df} + C_{Di} \qquad (15\text{-}52)$$

Note that the combination $C_{Do} + C_{Df}$ is highly internally dependent and, therefore, in this book, they are combined and called the minimum drag, C_{Dmin}. However, in light of the above definition, inserting the explicit forms of the other coefficients yields:

$$C_D = \frac{2D_o}{\rho V^2 S_{ref}} + \frac{2D_f}{\rho V^2 S_{wet}} + \frac{2D_i}{\rho V^2 S_{ref}} \qquad (15\text{-}53)$$

For internal consistency it is better to write:

$$\begin{aligned} C_D &= \frac{2D_o}{\rho V^2 S_{ref}} + \frac{2D_f}{\rho V^2 S_{wet}}\left(\frac{S_{ref}}{S_{ref}}\right) + \frac{2D_i}{\rho V^2 S_{ref}} \\ &= \left(\frac{2}{\rho V^2 S_{ref}}\right)\left(D_o + D_f\left(\frac{S_{ref}}{S_{wet}}\right) + D_i\right) \end{aligned} \qquad (15\text{-}54)$$

15.3.6 Various Means to Reduce Drag

As has already been stated, drag is generally a detrimental force in aircraft design, in particular for operational missions that require fuel efficiency. In addition to "obvious" means to reduce drag, such as retractable landing gear and smooth NLF surfaces, people have been creative in their attempts to reduce drag. Below are a selected number of ideas conceived with this intent, presented here to inspire the reader.

Reduction of Drag on Wings via Laminar Flow Control (LFC)

The history of LFC dates back to 1930s and is well documented by Chambers [12] and in particular Braslow [13]. LFC is also referred to as *artificial laminar flow* (ALF) to contrast with *natural laminar flow* (NLF). LFC is an attempt to maintain the laminar boundary layer over a large part of the wing by effectively "sucking" the turbulent boundary layer through tiny perforations in the wing skin. A variation of LFC, called hybrid LFC or HLFC, uses NLF for a larger portion of the wing to reduce the power required to eliminate the turbulent boundary layer growth. The primary drawback of such methods is that they are *active* rather than *passive*. In other words, additional energy is required to lower the pressure inside the wing and draw in the external boundary layer. Furthermore, it is a serious detriment that the perforations negatively affect the structural integrity of the wing and the operator must deal with nuisances such as cleaning the remains of insects that can clog the perforations, reducing system performance.

Figure 15-24 shows the difference in airflow over an airfoil with and without LFC. The left photo shows the airfoil at an *AOA* of 20° and clearly shows fully separated wake behind it. The right photo shows the same airfoil with the LFC turned on. The change in the nature of the flow is clearly evident, with the separation being eliminated as far aft as 70% of the chord. This "unstalls" the airfoil, giving it a higher C_{lmax}, in addition to a reduction in drag. It demonstrates the promising potential of LFC technology, although being severly hampered by a number of factors, as is discussed in Ref. [12].

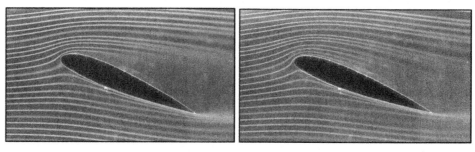

FIGURE 15-24 The effect of laminar flow control. The *AOA* is approximately 20°. *(Photos by Phil Rademacher)*

An early example involves the Northrop X-21A, which was a heavily modified Douglas WB-66D aircraft. Its two original Allison J71 engines, originally on the wing, were replaced with two GE XJ79s that were mounted to the rear fuselage. The bleed air from these engines was then used to drive a compressor that, effectively, sucked the wing boundary layer through slots in the wing. The system, while demonstrating LFC was effective, turned out to be a manufacturing and maintenance nightmare and too costly to be practical.

In 1999, Marshall [14] investigated the effectiveness of a variable-porosity suction glove on the F-16XL-2 aircraft to demonstrate the feasibility of boundary layer suction for supersonic operation. The research showed that at the test condition of Mach 2 at 53000 ft (R_e of the order of 22.7×10^6) the laminar boundary layer was sustained as far aft as 46% of the chord.

Finally, van de Wal [15] showed merely 3.2% reduction in total drag of a small GA aircraft (ENAER Namcu), reporting that installing a boundary layer suction system was not beneficial to its operation.

Winglets

Winglets and the potential for a reduction in lift-induced drag are discussed in Section 10.5.7, *The winglet*.

NLF Airfoils

Natural Laminar Flow airfoils are an obvious choice when trying to reduce drag. Of course they have to be selected before the airplane is built. NLF airfoils are discussed in great detail earlier in this chapter and Chapter 8, *The anatomy of the airfoil*.

Reduction of Drag of Fuselages

Wortman [16] suggests that the installation of relatively large fixed-pitch vortex generators on the bottom near the beginning of the upsweep of the lower fuselage of transport aircraft can reduce the total drag by 1–2%. The idea was validated in extensive wind tunnel tests using fuselage models of the Boeing 747 and Lockheed C-5 Galaxy transport aircraft. The author suggests the vortex generators can be installed on such aircraft for a fraction of the cost of their monthly operational cost.

Such vortex generators are shown mounted to the aft lower fuselage of the B-52 Stratofortress in Figure 15-25.

Kentfield [17] suggests that using a stepped afterbody can significantly reduce the drag of an axis-symmetric fuselage style bodies. The unorthodox idea is to allow an entrapped vortex to form at each step of the conical after-body, which allows the airflow to better follow its geometry, ultimately reducing its drag. The method is unorthodox and results in an unusual afterbody geometry that would be hard to justify from an aesthetics standpoint, not to mention there could be some structural challenges.

A clear way to reduce fuselage drag is to employ tadpole fuselages, like those used for sailplanes. Naturally, such fuselages are not always practical considering the mission of the airplane. Tadpole fuselages are discussed in Appendix C4, *Design of sailplanes*, and Section 12.2.3, *The tadpole fuselage*.

FIGURE 15-25 Vortex generators on the aft fuselage of a B-52 Stratofortress. *(Photo by Phil Rademacher)*

In evaluating the importance of smooth surfaces in maintaining NLF on lifting surfaces (wing, HT, and VT), Quast and Horstmann [18] demonstrate the magnitude of fuselage drag. Using the Airbus 300 as an example, they estimate the drag of the fuselage alone amounts to about 49% of the minimum drag. Studies of this nature are an important reminder that it is easy to spend a tremendous amount of effort getting a few drag counts out of the lifting surfaces, while overlooking the greatest source of drag altogether – the fuselage.

Reduction of Drag of the Fuselage/Wing Juncture

The juncture of the fuselage and wing can be particularly bothersome as it is not always easy to estimate tendency for flow separation. Due to steep adverse pressure gradients that often result in the region such separation can occur even at low AOAs. The implication is increased drag during climb and even cruise and thus reduced performance.

Modern methodologies are being developed that use state-of-the-art Navier-Stokes computational fluid dynamics (N-S CFD) tools to shape the wing fairing using genetic algorithms. Peigin and Epstein [19] suggest the use of such tools and demonstrated a 43 drag count drag reduction in a before-and-after investigation of a business jet.

15.4 THE DRAG CHARACTERISTICS OF THE AIRPLANE AS A WHOLE

For some, the pinnacle of aircraft design is the estimation of the drag of the whole airplane. We have already discussed the shortcomings of such analyses and emphasized enough the care that must be exercised. In this section methods will be presented to estimate the drag of the new design. We will use the Cirrus SR22 as an example by evaluating its drag and comparing it to the drag coefficient that can be extracted from published performance data (i.e. POH values obtained through certification flight testing).

The ideal airplane designed for cruise should be designed to operate near or at its minimum drag coefficient, C_{Dmin}. This is actually hard to accomplish in practice, but is still the goal. If this can be achieved, practically any deviation from this condition will increase the drag of the airplane: changing the airspeed will change the AOA and increase the drag coefficient; moving the CG to a new position (if possible) will also increase the drag coefficient, as will deflecting the control surfaces. We can expand on this imaginary situation by considering changes in surface smoothness, addition of inlets and outlets, antennas, and the like, all which will increase the drag. Such a thought exercise helps the realization that drag is inherently hard to reduce but easy to increase. The aspiring designer must anticipate this effect while attempting to manage and resist requests for features from many directions; typically from management, potential customers, systems group, and similar.

Many effects of some specific sources of drag on the aircraft have already been presented in Chapter 8, *The anatomy of the airfoil*. Here, we will only add ones that are specifically limited to three-dimensional aircraft.

15.4.1 The Effect of Aspect Ratio on a Three-Dimensional Wing

In 1923, Prandtl [20] presented the results of well-known experiment which depict well the effect aspect ratio has on the generation of lift and drag of a wing. A copy of of his actual results is shown in Figure 15-26. The graph to the right shows what has already been discussed in Chapter 9, *The anatomy of the wing*, the impact of AR on lift. The graph to the left shows the effect of AR on drag. It shows that two wings (call them Wing 1 and Wing 2) whose geometry differs only in the AR (wing area and airfoil, and thus the minimum drag of the airfoil, C_{dmin}, are assumed the same) will generate drag coefficient at the same C_L, which can be expressed as follows:

$$C_{D_1} = C_{dmin} + \frac{C_L^2}{\pi \cdot AR_1 \cdot e} \quad \text{and}$$

$$C_{D_2} = C_{dmin} + \frac{C_L^2}{\pi \cdot AR_2 \cdot e} \quad (15\text{-}55)$$

From this it is evident if the drag of one wing, say Wing 1, is known, the drag of the other one can be estimated from the difference of the two drag polars:

$$C_{D_2} = C_{D_1} + \frac{C_L^2}{\pi \cdot e}\left(\frac{1}{AR_2} - \frac{1}{AR_1}\right) \quad (15\text{-}56)$$

Of course, the expression holds as long as the difference in the ARs is not too great. This is because the chord length of the airfoil will affect the minimum drag of the airfoil, C_{dmin}, and the AR will affect the Oswald span efficiency factor, e.

Figure 15-27 shows the effect of changing the AR of a three-dimensional wing on the drag polar using modern notation of coefficients. It shows that as the magnitude of the aspect ratio does not change the minimum drag (not accurate if the change is large), but rather the width of the drag polar is reduced. This is a consequence of a reduction in the magnitude of the C_{Lmax} and the lift curve slope, $C_{L\alpha}$. The reduction is easily estimated through the use of Equations (15-38), (15-45), and (15-47).

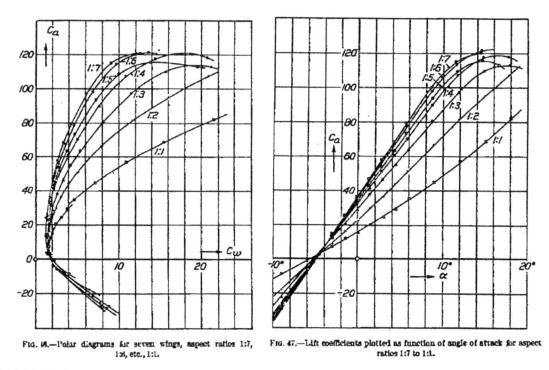

FIGURE 15-26 Ludwig Prandtl's original research demonstrates how lift and drag of a wing change with aspect ratio (from Ref. [20]).

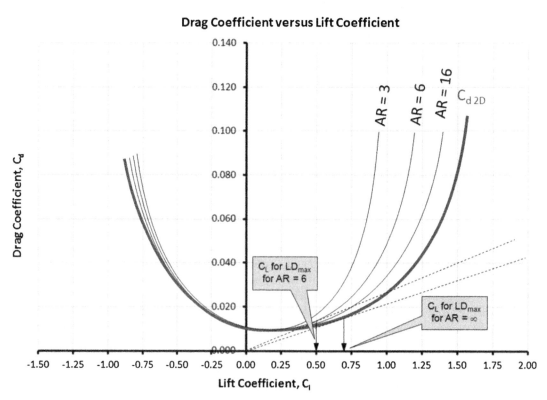

FIGURE 15-27 The effect of different values of the AR on the drag polar assuming $S = $ constant (based on Ref. [20]).

15.4.2 The Effect of Mach Number

Generally, as the airplane approaches the speed of sound, the drag coefficient will begin to rise sharply. The Mach number at which the rise begins depends on the geometry of the airplane. An airplane with thick airfoils will experience it perhaps as early as $M = 0.6$, whereas a sleek high performance jetliner begins to experience it at $M \approx 0.85$. This rise is handled in the drag estimation by the addition of a *compressible drag coefficient*, ΔC_{DC}, which is most accurately estimated in wind tunnel testing or by sophisticated CFD methods. The Mach number at which this happens is called the *drag-divergence Mach number*, denoted by M_{dd}, and it occurs slightly above the *critical Mach number* (see Section 8.3.7, *The critical Mach number, M_{crit}*).

The drag-divergence Mach number is defined either when $\Delta C_{DC} = 0.002$ or when $dC_{DC}/dM = 0.10$.

Figure 15-28 shows a hypothetical scenario in which an airplane flying at a constant C_L experiences a sharp rise in the compressible drag coefficient near $M = 0.7$. As stated earlier, this is the *critical Mach number*, denoted by M_{CRIT}. Its value largely depends on the geometry of the aircraft and details such as the thickness of the wing airfoils and the geometry of the wing/fuselage juncture, to name few. In fact, any part of the airplane which features interferences due to the joining of disparate components is suspect. Such interferences typically result in increased local airspeeds, which promote the formation of normal shocks (and therefore higher drag) in the juncture.

A method to estimate this contribution to the total drag is presented in the Section 15.5.12, *Drag due to compressibility effects*.

15.4.3 The Effect of Yaw Angle β

As discussed in Section 15.3.4, *The Lift-Induced Drag Coefficient: C_{Di}*, drag varies greatly with *AOA*. This effect extends to yaw as well, as shown in Figure 15-29; it too increases the drag and is usually minimum when β = 0°. This fact has been used for a long time by pilots when landing. Pilots "coming in too high" for a landing will yaw the airplane, decreasing its L/D ratio, allowing the airplane to temporarily lose altitude more rapidly. This addition depends entirely on the geometry of the aircraft. Generally, it is an acceptable approximation to assume the flow to stay attached on the aircraft for a β up to ±10°. However, beyond that, flow separation is certain and the associated drag increase must be accounted for.

15.4.4 The Effect of Control Surface Deflection — Trim Drag

Deflecting control surfaces usually increases the drag of the aircraft. For instance, deflecting the elevator shifts

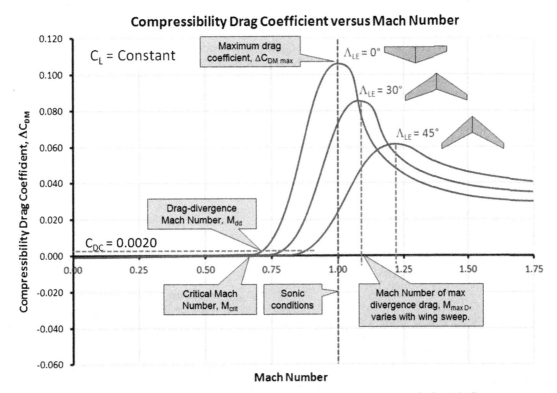

FIGURE 15-28 The effect of compressibility on drag (based on Refs [21] and [22]).

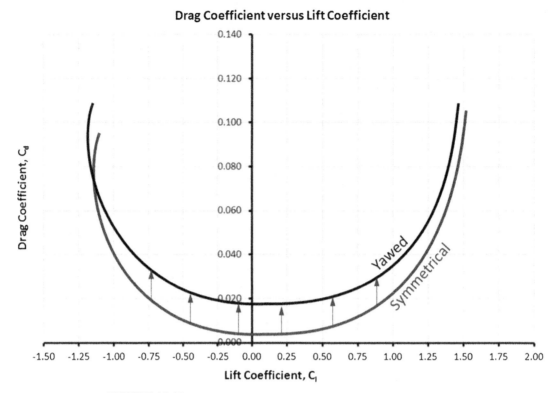

FIGURE 15-29 The effect of yawing the airplane to some yaw angle β.

the drag polar vertically, increasing the C_{Dmin}. This change is referred to as the *trim drag*. If a conventional tail-aft airplane is loaded with its CG far forward, a higher deflection of the elevator will be required to trim it. This implies additional drag, in addition to the extra lift the wing must generate that increases lift-induced drag. This is why it is important to size the tail so the elevator is close to neutral in cruise (usually referred to as an elevator "in trail").

Trim drag is most accurately estimated using precise wind tunnel tests, although it can certainly be estimated using analytical methods too. A graph similar to the one in Figure 15-30 is usually obtained, with several curves showing the drag polar for an elevator deflection of, say, 0°, 2°, 4°, 6°, 10°, and so on. These allow the performance engineer to better estimate the capability of the airplane at different loading and flight conditions. In the absence of wind tunnel testing, it is imperative to be able to assess it with some level of accuracy, in particular if the airplane features a high (or low) thrustline.

A method to estimate trim drag based on CG location and thrust setting is presented in Section 15.5.2, *Trim drag*.

15.4.5 The Rapid Drag Estimation Method

The *rapid drag estimation method* is without question the fastest way to estimate the drag of an airplane. It is based on the assumption that there is a correlation between the averaged skin friction coefficient of an airplane, its total wetted area, and its minimum drag coefficient. Naturally, with this speed comes inaccuracy, so the method should only be used to figure out a "ballpark" value for the drag to compare to other methods. For instance, the method does not account for any peculiarities in the airplane design, such as extent of laminar flow, specific geometric features, and others. It is possible to account for the rudimentary effects of flaps and landing gear, but again, the designer should not use this method for anything but to get an idea of the magnitude of the drag coefficient, and not be surprised if it is vastly different from the value obtained by more accurate methods. The method calculates the minimum drag coefficient using the concept of EFPA (see Section 15.2.2, *Quadratic drag modeling*) and Equation (15-8):

$$C_{D_{min}} = \frac{f}{S_{ref}} \qquad (15\text{-}57)$$

where

C_{Dmin} = minimum drag coefficient
f = equivalent flat plate (parasite) area
S_{ref} = reference area

The method involves reading the value of f for a "clean" aircraft from Figure 15-31, using the approach outlined in the figure. Ideally, the estimate of the aircraft's wetted area should be as precise as possible.

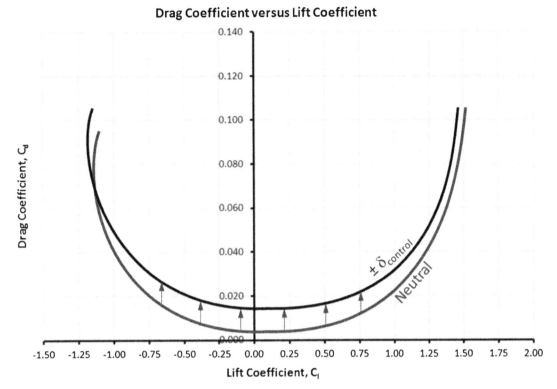

FIGURE 15-30 The effect of deflecting a control surface.

Also note that the *equivalent skin friction coefficient* means a normalized value representative for the entire aircraft. It is inevitable, considering the variety in the surface quality of a normal airplane, that some engineering judgment will be required to assess the value.

The drag increments due to flaps, landing gear, speed brakes, etc., are accounted for using the following relation and then added to the C_{Dmin}:

$$\Delta C_{D_{min}} = \frac{1}{S_{ref}} \sum C_{D_\pi} A_\pi \qquad (15\text{-}58)$$

where

ΔC_{Dmin} = drag increment
$C_{D\pi}$ = component equivalent drag coefficient
A_π = equivalent parasite area

The equivalent parasite areas and coefficients are obtained from the Table 15-6, where A_C stands for maximum cross-sectional area, S_{ref} is the wing reference area and S_{HT} is the HT planform area.

15.4.6 The Component Drag Build-Up Method

The *component drag build-up method* (CDBM) is used to estimate the drag of the complete aircraft. The method is primarily based on the estimation of flat plate skin friction over the surfaces of the airplane that are exposed to the airflow (wetted area). The method accounts for geometric differences between components and changes in drag resulting from bringing one component into the neighborhood of another (interferences).

The CDBM bases the parasitic drag (the combination of $C_{D0} + C_{Df}$) using flat plate skin friction coefficients that are modified using two special factors called a component *form factor* (FF) and *interference factor* (IF). The FF is a measure of the pressure drag due to viscous separation. It accounts for the fact that the drag force generated by a sphere and a box of equal wetted area is different from one another. Interference effects between aircraft parts (wing and fuselage, engine and wing, HT and VT, etc.) are accounted for using a special factor called component *interference factor* (IF). This factor is based on the fact that as two bodies are brought together in fluid flow, the drag of the combination is greater than the drag of the individual bodies on their own. The IF is denoted by Q in Raymer, but by IF by most other authors.

A flow chart showing the procedure is shown in Figure 15-32. In the flow chart, the "+" sign means that contributions are added, whereas the "×" sign means multiplication.

The method requires the skin friction coefficients to be calculated for all components in direct contact with air (or wetted by air) ①. The components are parts like the wings, horizontal tail, vertical tail, and so on. Even though the wing, HT, VT, and fuselage are shown, other components, such as engine nacelles, pylons, winglets,

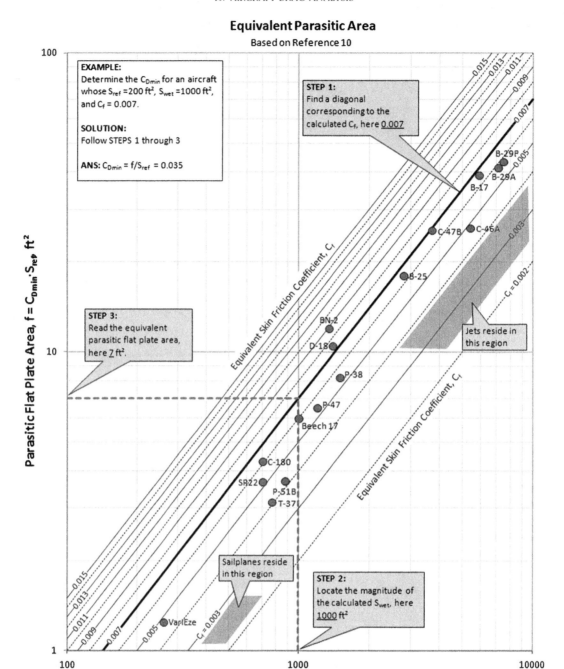

FIGURE 15-31 Determination of equivalent parasite area (based on Refs [23] and [24]).

external fuel tanks, dorsal, or ventral fins, should be added to the list if present. Naturally, they should be excluded only if other means to estimate their drag is selected (e.g. see Section 15.5, *Miscellaneous or additive drag*) and then be included as miscellaneous drag, C_{Dmisc}. These are adjusted with FF ② through a multiplication operation, but this ensures dissimilar components don't contribute equally to the overall drag even if both have identical wetted areas. Then, IFs are applied ③, also through a multiplication operation, to account for the increase in drag when individual components are brought into close proximity.

Once the skin friction drag (with its associated pressure and interference drag boosting) has been determined, the next step is to determine all remaining sources of drag. This is simply called *miscellaneous* or *additive drag* ④. This is drag attributed to miscellaneous sources, such as antennas, fuel caps, air flow through small gaps (such as those between control surfaces and the lifting surface to which they are mounted), as well

15.4 THE DRAG CHARACTERISTICS OF THE AIRPLANE AS A WHOLE

TABLE 15-6 Equivalent Parasite Areas and Coefficients (Based on Refs [23] and [24])

Item	Comment	$C_{D\pi}$	Area on which $C_{D\pi}$ is based
Wing	Standard operational roughness, airfoil t/c = 10%–20%.	0.005–0.009	S_{ref}
HT and VT	Standard operational roughness, airfoil t/c = 8%–12%.	0.006–0.008	S_{HT}
Wing flaps	Plain, 60% span at 30° deflection	0.02–0.03	S_{ref}
Fuselage	Streamlined and very smooth body	0.05	A_C
Fuselage	Small aircraft with engine in nose	0.09–0.13	A_C
Fuselage	Large transport aircraft (DC-4, DC-5)	0.07–0.10	A_C
Fuselage	Bomber (B-17)	0.08–0.12	A_C
Landing gear	Typical light twin, wheel wells closed	0.014	S_{ref}
Landing gear	Typical light twin, wheel wells open	0.017	S_{ref}
Nacelle, propeller	Above wing on a small aircraft (e.g. Cessna 310 type)	0.25	A_C
Nacelle, propeller	Relatively small leading edge nacelle on a large aircraft	0.05–0.09	A_C
Nacelle, turbojet	Mounted on wing (e.g. Me-262)	0.05–0.07	A_C
Wingtip tank	Suspended below wingtip	0.10	A_C
Wingtip tank	Centrally mounted at wingtip	0.06	A_C
Wing tank	Suspended below wing, incl. support	0.19–0.21	A_C
Bomb	Suspended below wing, incl. support	0.22–0.25	A_C
Cooling flaps Speed brakes	Depend very strongly on size, no realistic representative data can be given		

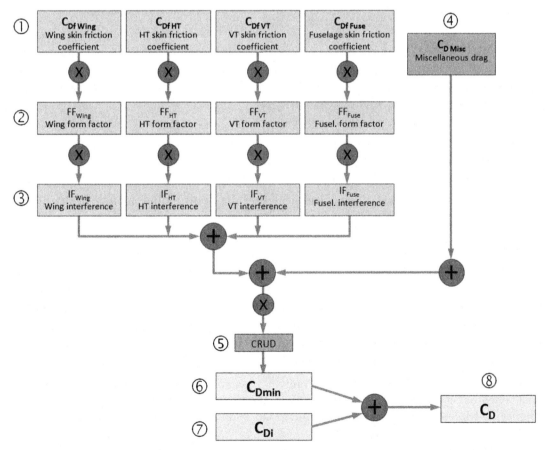

FIGURE 15-32 Flow chart describing the component drag build-up method. Recall that $C_{Df\,i} = C_{f\,i} \cdot (S_{wet\,i}/S_{ref})$.

as control system components (bellcranks, hinges, etc.), inlets, outlets, antennas, and so on. These are then summed up and multiplied by a crud-factor ⑤, which accounts for contributions that are practically impossible to account for otherwise, such as surface panel misalignments, dents, small vents and outlets, and so on. The crud factor is typically 25%, which means that the sum of the above contributions is multiplied by 1.25. This operation returns the minimum drag coefficient ⑥. We add to it the lift-induced drag coefficient at the flight condition ⑦ to obtain the total drag coefficient ⑧.

The total component drag build-up is expressed as follows:

$$C_{Dmin} = \frac{1}{S_{ref}} \sum_{i=1}^{N} C_{f_i} \times FF_i \times IF_i \times S_{wet_i} + C_{D_{misc}}$$

(15-59)

where

C_{Dmin} = minimum, zero-lift, or parasitic drag coefficient
C_f = surface skin friction coefficient
FF = form factor
IF = interference factor
S_{ref} = reference area
S_{wet} = surface wetted area
i = component index number
N = total number of components accounted for
C_{Dmisc} = miscellaneous drag coefficient

15.4.7 Component Interference Factors

As previously stated, interference factors (IF) are used to account for the proximity of one component to another. For instance consider the juncture between the wing and fuselage. The presence of both bodies constrains the airflow compared to that of the individual components, increasing the local airspeeds greatly, which increases the drag. However, this does not account for additional drag that may arise due to early separation due to a poorly designed wing/fuselage juncture.

TABLE 15-7 Typical Interference Factors (in Part Based on Refs [3] and [25])

Component	IF
Nacelle or external store, directly under a fuselage	1.5
Nacelle or external store, under a fuselage, less than about 1 diameter away	1.3
Nacelle or external store, under a fuselage, more than about 1 diameter away	1.0
Object, such as a fuel tank, mounted to a wingtip	1.25
High wing or mid wing with carefully designed fairing	1.0
Un-filleted low wing	1.1–1.4
Whitcomb winglet	1.04
"Airbus" style winglet	1.04
Modern blended winglet	1.00–1.01
Aerodynamic or square leaf-spring main landing gear strut entering wing or fuselage	1.10
Aerodynamic wing strut entering wing on one end and fuselage on the other	1.10
Boundary layer diverter	~1.0
Conventional tail	1.04–1.05
Cruciform tail	1.06
V-tail	1.03
H-tail (B-25 Mitchell or A-10 Warthog style)	1.08
H-tail (Lockheed Model 10 Electra style)	1.13
H-tail (Beech D-18 style)	1.06
Triple-tail (Lockheed Constellation style)	1.10
T-tail	1.04

Table 15-7 lists typical IFs that are partially derived from Refs [3] and [25] and partially using factors that have worked well in drag analyses performed by the authors. No claim is made about their accuracy beyond that. Note that when using the factors for multiple objects, for instance, a triple-tail, the presented IF must be applied to all surfaces. Thus, when summing up the four surfaces of such a tail (one HT and three VTs), their corresponding C_{Df} (= $C_{f\,i} \cdot (S_{wet\,i}/S_{ref})$) must be multiplied each time. The following formulation illustrates how its contribution would be accounted for in the CDBM:

$$\frac{C_{Df_{HT}} \times FF_{HT} \times IF_{HT} + C_{Df_{VT1}} \times FF_{VT1} \times IF_{VT1} + + C_{Df_{VT2}} \times FF_{VT2} \times IF_{VT2} + + C_{Df_{VT3}} \times FF_{VT3} \times IF_{VT3}}{S_{ref}}$$

15.4.8 Form Factors for Wing, HT, VT, Struts, Pylons

As stated earlier, a form factor (FF) reflects the geometric shape of components and, therefore, methods to estimate this value varies greatly with classes of components. Thus there is a set of *FF*s that are only used with geometry capable of generating lift ("wing-like" surfaces). Others are only used with geometry that serves as fuselages, and so on. The following expressions are used to estimate *FF* for lift for wing-like surfaces. Such surfaces include wings, horizontal tail, vertical tail, struts, and pylons, but can also be extended to wing-shaped antennas and landing gear pant fairings. These form factors are typically derived by semi-empirical methods that emphasize the thickness-to-chord ratio of the structure. The following form factors are found in the literature.

The inevitable question that comes up is "which one do I pick?" Unless a specific application is cited (like the ones by Hoerner) the answer is often based on engineering judgment. In that case the unsure designer can take the average of two or three methods.

Hoerner [3, p. 6-6] suggests the following form factors for lifting surfaces featuring airfoils whose $(x/c)_{max} = 30\%$:

$$FF = 1 + 2\left(\frac{t}{c}\right) + 60\left(\frac{t}{c}\right)^4 \qquad (15\text{-}60)$$

Hoerner [3] suggests the following form factors for lifting surfaces featuring airfoils whose $(x/c)_{max} = 40\%$ to 50%, such as NACA 64 and 65 series airfoils:

$$FF = 1 + 1.2\left(\frac{t}{c}\right) + 70\left(\frac{t}{c}\right)^4 \qquad (15\text{-}61)$$

Neither of the above models account for wing sweep or compressibility effects. Torenbeek [26] suggests the following form factors for lifting surfaces featuring airfoils whose $t/c \leq 21\%$:

$$FF = 1 + 2.7\left(\frac{t}{c}\right) + 100\left(\frac{t}{c}\right)^4 \qquad (15\text{-}62)$$

Shevell [27, p. 178] suggests the following form factors for lifting surfaces and introduces compressibility and sweep effects:

$$FF = 1 + \frac{(2 - M^2)\cos \Lambda_{C/4}}{\sqrt{1 - M^2 \cos^2 \Lambda_{C/4}}}\left(\frac{t}{c}\right) + 100\left(\frac{t}{c}\right)^4 \qquad (15\text{-}63)$$

Nicolai [28] and Raymer [25] suggest the following form factor for lifting surfaces that also corrects for compressibility but considers the sweep of the maximum thickness line rather than that of the quarter chord. The equation, as shown, is only valid as long as $M > 0.2$ because the compressibility correction (the bracket on the right-hand side) becomes less than 1 at a lower value. For this reason the compressibility correction term should be set to 1 for airspeeds below $M = 0.2$.

$$FF = \left[1 + \frac{0.6}{(x/c)_{max}}\left(\frac{t}{c}\right) + 100\left(\frac{t}{c}\right)^4\right]$$
$$\times \left[1.34 M^{0.18}(\cos \Lambda_{t max})^{0.28}\right] \qquad (15\text{-}64)$$

Jenkinson [29] suggests two kinds of form factors: one for the wing and another for tail surfaces. The form factor for the wing is given by:

Wing:

$$FF = \left[3.3\left(\frac{t}{c}\right) - 0.008\left(\frac{t}{c}\right)^2 + 27.0\left(\frac{t}{c}\right)^3\right]\cos^2 \Lambda_{C/2} + 1 \qquad (15\text{-}65)$$

Furthermore, the reference recommends the interference factor to use with the expression is $IF = 1.0$ for well filleted low or mid wings, and 1.1–1.4 for small or no fillet. The form factor for tail surfaces is given by:

Tail surfaces:

$$FF = \left[3.52\left(\frac{t}{c}\right)\right]\cos^2 \Lambda_{C/2} + 1 \qquad (15\text{-}66)$$

The reference recommends an $IF = 1.2$ for tail surfaces. In the above equations:

M = Mach number
$\Lambda_{C/4}$ = sweep angle of the quarter chord line
$\Lambda_{C/2}$ = sweep angle of the mid-chord line
$\Lambda_{t\ max}$ = sweep angle of maximum thickness line
$(x/c)_{max}$ = location of maximum airfoil thickness
(t/c) = airfoil thickness ratio.

Note that both Equations (15-63) and (15-64) approach Torenbeek's form, shown in Equation (15-62), for wings whose quarter-chord (Shevell) or maximum thickness (Nicolai, Raymer) sweep angle is $0°$, when $M = 0$ (Shevell) or $M = 0.2$ (Nicolai, Raymer).

15.4.9 Form Factors for a Fuselage and a Smooth Canopy

This section discusses form factors intended for use with geometry that represent fuselages, so-called *streamlined bodies*, and *smooth canopies*. These form factors are typically represented in terms of the fineness ratio, defined as length (l) divided by the average diameter (d), as shown in Figure 15-33 and Figure 15-34. The figure indicates the proper way to evaluate the required parameters. When it comes to fuselages, it is of importance to focus on the body itself. The upper part of the figure shows a fuselage with the tail and engine pods. The lower part shows the fuselage stripped of these, representing the geometry used in this analysis method. Sometimes the distinction is not so

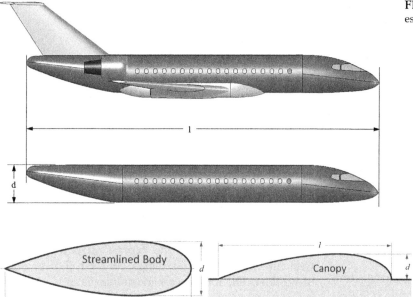

FIGURE 15-33 Definition of terms for use with the estimation of form factors for a fuselage.

FIGURE 15-34 Definition of terms for use with streamlined bodies and canopies.

clear, leaving no option but to depend on engineering judgment.

A streamlined body usually refers to a body-of-revolution, whose cross-section is similar to what is shown in Figure 15-34. A streamlined body is used to represent many types of fuselages, but also the hulls of airships. They can also be used to represent engine nacelles in drag estimation. A canopy refers to the external shape of an airplane's glass helmet and is sized geometrically as shown in Figure 15-34. For more realistic canopies, refer to Section 15.5.10, *Drag of canopies*.

Fuselage as a Body of Revolution

In the following expressions, the *fineness ratio*, appropriately denoted by the letter f, is used to indicate the slenderness of a body of revolution. It is defined as follows:

$$f = \frac{l}{d} \qquad (15\text{-}67)$$

where l = length of the body and d = diameter of the body.

Although devised for bodies of revolution, the expressions for the form factors below are also applicable to bodies that are not truly ones of revolution, but feature cross-sectional shapes other than circular (see Figure 15-35). For instance, a cross-section might have the shape of a silhouette of an egg, or be box-shaped, or rectangular with round corners. These are treated by determining the shape's maximum cross-sectional area, A_{max}, and then relating it to that of a circular cross-section using a "representative" fineness ratio as shown below:

$$f = l/d = l \Big/ \sqrt{\frac{4 A_{max}}{\pi}} \qquad (15\text{-}68)$$

Form Factors at Subcritical Reynolds Numbers

Hoerner [3, p. 6-16] derives and suggests the following form factors for streamlined bodies at subcritical Reynolds numbers ($R_e < 10^5$). The magnitude of the R_e implies it could be the fuselages of small vehicles, such as radio-controlled aircraft, or small unmanned aerial vehicles. Note that since the expression includes the pressure drag component, it is not represented as a stand-alone form factor, but rather the product of the $C_{fi} \times FF_i$ in Equation (15-59). It also requires the skin friction coefficient for laminar boundary layer of Equation (15-18) to be used:

$$C_f \cdot FF = C_{f_{lam}} \left[1 + \frac{1}{f^{1.5}} \right] + \frac{0.11}{f^2} = \frac{1.328}{\sqrt{R_e}} \left[1 + \frac{1}{f^{1.5}} \right] + \frac{0.11}{f^2}$$

(15-69)

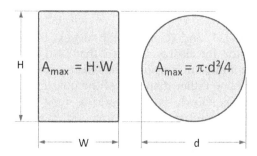

FIGURE 15-35 Two geometric shapes of equal area.

Form Factors at Supercritical Reynolds Numbers

Hoerner [3, p. 6-17] suggests the following form factor for streamlined bodies in airflow whose $R_e > 10^5$:

$$FF = 1 + \frac{1.5}{f^{1.5}} + \frac{7}{f^3} \quad (15\text{-}70)$$

Torenbeek [26] suggests the following form factor for a generic fuselage:

$$FF = 1 + \frac{2.2}{f^{1.5}} + \frac{3.8}{f^3} \quad (15\text{-}71)$$

Likewise, Nicolai [28] and Raymer [25] present the following form factor for a fuselage:

$$FF = 1 + \frac{60}{f^3} + \frac{f}{400} \quad (15\text{-}72)$$

Shevell [27, p. 179] also provides a model for a fuselage given by:

$$FF = 2.939 - 0.7666f + 0.1328f^2$$
$$- 0.01074f^3 + 3.275 \times 10^{-4}f^4 \quad (15\text{-}73)$$

Finally, Jenkinson's model [29] is given by:

$$FF = 1 + \frac{2.2}{f^{1.5}} - \frac{0.9}{f^3} \quad (15\text{-}74)$$

where

FF = form factor
A_{max} = maximum fuselage cross-sectional area
f = fineness ratio and: $f = l/d = l/\sqrt{4A_{max}/\pi}$

Form Factor for Nacelle and Smooth External Store

Raymer [25] suggest the following form factors for nacelles:

$$FF = 1 + \frac{0.35}{f} \quad (15\text{-}75)$$

Jenkinson [29, p. 172] recommends simply using the constant $FF \cdot IF = 1.2$ for wing mounted nacelles and $FF \cdot IF = 1.44$ for nacelles mounted on the rear of the airplane (a Sud-Est Caravelle configuration).

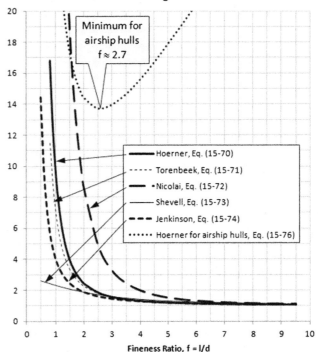

FIGURE 15-36 Comparison of various FFs.

Form Factors for Airship Hulls and Similar Geometries

If the fuselage closely resembles the hull of an airship (which, granted, most fuselages do not), then Equation (15-76), derived by Hoerner [3, p. 6-18] should be used.

$$FF = 3f + \frac{4.5}{\sqrt{f}} + \frac{21}{f^2} \quad (15\text{-}76)$$

Note that it leads to substantially higher value of the form factor than do the other ones. However, it has a very distinct minimum around $f \approx 2.7$.

A minimum fineness ratio of 2.7–3.5 for such bodies is well supported by experimental data and explains why fuselages of transport aircraft typically feature this fineness ratio for the aft end. A number of the above form factors are plotted in Figure 15-36.

EXAMPLE 15-6

Determine the minimum drag coefficient for the SR22, whose top view is shown in Figure 15-37, using the component drag build-up method. Note that the complete minimum drag coefficient is not just worked out here, but in a number of subsequent examples. The complete minimum drag coefficient is presented in Example 15-17.

Assume:

(1) Flight condition at S-L and 185 KTAS.
(2) Wing sustains laminar flow as detailed in Example 15-2.
(3) HT sustains 50% laminar flow on lower and upper surfaces.

EXAMPLE 15-6 (cont'd)

(4) VT sustains 50% laminar flow on left and right surfaces.
(5) Fuselage sustains 5% laminar flow, cut short due to engine cowling assembly.
(6) Wetted area booster coefficient of 1.07 for wing and 1.05 for HT and VT.
(7) Max thickness for wing, HT, and VT is at 50% of the chord for all.
(8) $t/c = 0.15$ for wing, 0.1 for HT and VT, and $4.17/22.42 = 0.186$ for the fuselage.
(9) Skin roughness value from Table (15-2).
(10) Assume the max thickness lines for the wing, HT, and VT to be 0°, 5°, and 18°, respectively.
(11) C_{Dmisc} is 6 drag counts and $C_{DL\&P}$ is 3 drag counts.
(12) CRUD (see Section 15.5.1) is 25% of the sum of the above ($\times 1.25$).

Solution

The solution to this problem is relatively extensive due to the number of components that must be included. However, since many of the calculations are identical they are conveniently implemented in a spreadsheet as shown in Figure 15-38, Figure 15-39 and Figure 15-40. Each row has been numbered for convenience and sample calculations provided below for selected rows. The spreadsheet was designed to allow the user to enter geometry for other simple aircraft and determine the drag polar. Note that green cells are intended for user entry, whereas blue cells contain formulas.

Begin by considering Figure 15-38, which shows cells with the given information (green cells) and five rows which show some calculation results (blue cells). Note the setup of the columns dedicated to the inboard and outboard wing elements, as well as the HT and VT.

Line 1: Airspeed is 185 KTAS × 1.688 ft/s per KTAS = 312.3 ft/s.

Line 2: Viscosity of air:

$$\mu = 3.170 \times 10^{-11} T^{1.5} \left(\frac{734.7}{T+216}\right)$$
$$= 3.170 \times 10^{-11} (518.67)^{1.5} \left(\frac{734.7}{518.67+216}\right)$$
$$= 3.745 \times 10^{-7} \text{ lb}_f \cdot \text{s/ft}^2$$

Line 4: Density of air:

$$\rho = 0.002378(1 - 0.0000068756H)^{4.2561}$$
$$= 0.002378 \text{ slugs/ft}^3$$

Line 5: Dynamic pressure:

$$q = \frac{1}{2}\rho V^2 = \frac{1}{2}(0.002378)(312.3)^2 = 115.95 \text{ lb}_f/\text{ft}^3$$

Lines 6-8, 10-13, and 15-17: Geometric data is obtained from Figure 15-37 and other data is entered based on the problem statement. Note that the exposed halfspan (and planform area) exclude the area of the surface that is inside the fuselage. Here it will be assumed that, of the components shown, only a part of the wing is inside the fuselage and this is reflected as the exposed halfspan and planform area.

Line 9: Exposed planform area.

Wing:

$$S_{WING} = (38.30 - 4.17)\frac{(4.875 + 2.585)}{2} = 127.3 \text{ ft}^2$$

HT:

$$S_{HT} = \frac{1}{2}(2.792 + 1.834) \times 6.442 = 29.79 \text{ ft}^2$$

VT:

$$S_{VT} = \frac{1}{2}(4.083 + 2.050) \times 5.313 = 16.29 \text{ ft}^2$$

Fuselage: not needed since a different method is used to determine its wetted area.

Line 14: Total wetted area of all components (entire exposed wing, HT, VT, and fuselage):

Wing wetted area:

$$S_{WING_{wet}} = 1.07 \times 2 \times S_{WING} = 272.4 \text{ ft}^2$$

HT wetted area:

$$S_{HT_{wet}} = 1.05 \times 2 \times S_{HT} = 62.6 \text{ ft}^2$$

VT wetted area:

$$S_{VT_{wet}} = 1.05 \times 2 \times S_{VT} = 34.2 \text{ ft}^2$$

The total wetted area of the fuselage is estimated using the method of Section 12.4.3, *Surface areas and volumes of a typical tubular fuselage*, and the fuselage geometry of Figure 15-37, using Equation (12-13), repeated here for convenience:

EXAMPLE 15-6 (cont'd)

$$S_{FUSE} = \frac{\pi \cdot D}{4}\left(\frac{1}{3L_1^2}\cdot\left[\left(4L_1^2+\frac{D^2}{4}\right)^{1.5}-\frac{D^3}{8}\right]-D+4L_2 + 2\sqrt{L_3^2+\frac{D^2}{4}}\right)$$

(12-13)

where

D = maximum fuselage diameter = 4.17 ft
L_1 = length of nose section = 5.50 ft
L_2 = length of center section = 8.22 ft
L_3 = length of nose section = 8.70 ft

Plugging and chugging those numbers into this equation yields 203 ft². However, the author's own approximation, which accounts for more details in the fuselage and spinner geometry returned the following value.

Fuselage wetted area: S_{FUSE} = 257.6 ft²

Note that it would be prudent to subtract the cross-sectional area of the wing on the left and right sides, where it enters the fuselage, but in the interests of simplicity it is left out of these calculations. To calculate this cross-sectional area, the reader can for instance use the approximation of Section 9.2.4, *Approximation of airfoil cross-sectional area*. Now, let's consider Figure 15-39, which shows the skin friction analysis for the root and tip chords.

Lines 18-19 and 28-29: Data entered is based on the problem statement.

Line 20: Reynolds number for the root chord (using the wing column as an example).

$$R_{e\ root} = \frac{\rho V L}{\mu} = \frac{(0.002378)(312.3)(4.875)}{3.745\times 10^{-7}} = 9667562$$

Lines 21-22 and 30-31: Here, the calculations for Lines 21-22 are used. The cutoff Reynolds number for the root chord (again using the wing column as an example) using Equation (15-25).

$$R_{e\ cutoff\ root} = 38.21\left(\frac{C_R}{\kappa}\right)^{1.053}$$
$$= 38.21\left(\frac{4.875}{1.7\times 10^{-6}}\right)^{1.053} = 240963686$$

Since the R_e based on C_R is less than this value, it will be used throughout the remainder of these calculations. If the opposite had been the case, then the cutoff R_e would been used.

Line 23-27 and 33-38: Steps 4 through 14 of Example 15-2 detail how the values for the wing column were obtained. All the values in the other three columns are calculated in an identical fashion, yielding the table shown in Line 42. Finally, consider Figure 15-40, which shows how the minimum drag coefficient is calculated.

Line 39: Thickness sweep angles for the wing, HT, and VT are given in the problem statement as 0°, 5°, and 18°, respectively.

Line 40: The form factor for the wing, HT, and VT are calculated using Equation (15-64) and Equation (15-72) for the fuselage. Note that 185 KTAS at sea level corresponds to $M \approx 0.28$.

Wing FF:

$$FF = \left[1+\frac{0.6}{(x/c)_{max}}\left(\frac{t}{c}\right)+100\left(\frac{t}{c}\right)^4\right]$$
$$\times\left[1.34 M^{0.18}(\cos \Lambda_{tmax})^{0.28}\right]$$

$$= \left[1+\frac{0.6}{0.50}(0.15)+100(0.15)^4\right]$$
$$\times\left[1.34(0.28)^{0.18}(\cos 0°)^{0.28}\right]$$
$$= 1.311$$

HT FF:

$$FF = \left[1+\frac{0.6}{0.50}(0.10)+100(0.10)^4\right]$$
$$\times\left[1.34(0.28)^{0.18}(\cos 5°)^{0.28}\right]$$
$$= 1.203$$

VT FF:

$$FF = \left[1+\frac{0.6}{0.50}(0.10)+100(0.10)^4\right]$$
$$\times\left[1.34(0.1513)^{0.18}(\cos 18°)^{0.28}\right]$$
$$= 1.187$$

Fuselage:

$$FF = \left[1+\frac{60}{f^3}+\frac{f}{400}\right] = \left[1+\frac{60}{5.376^3}+\frac{5.376}{400}\right] = 1.399$$

Where f is the fineness ratio, calculated from $f = \frac{l}{d} = \frac{22.42}{4.17} = 5.376$

Line 41: Interference factors are selected from Table 15-7.

EXAMPLE 15-6 (cont'd)

Line 42: Weighted drag factor calculated as shown in table. For instance, for the wing we get;

$$C_{f\ i} \cdot FF \cdot IF \cdot S_{wet\ i} = 0.001999 \times 1.311 \times 1 \times 272.4$$
$$= 0.7140$$

Line 43: Skin friction drag is calculated by summing up the cells in Line 42 and dividing by the reference area of 144.9 ft². The resulting total skin friction coefficient, which includes interference and form drag, is:

$$\frac{1}{S_{ref}} \sum_{i=1}^{N} C_{f_i} \times FF_i \times IF_i \times S_{wet_i}$$

$$= \frac{1}{144.9}(0.7140 + 0.1762 + 0.1120 + 0.7607)$$

$$= 0.01217$$

This only partially completes the drag analysis — drag due to miscellaneous sources remains to be determined and added to the above value, in addition to other corrections must be made. This will be done in the next section.

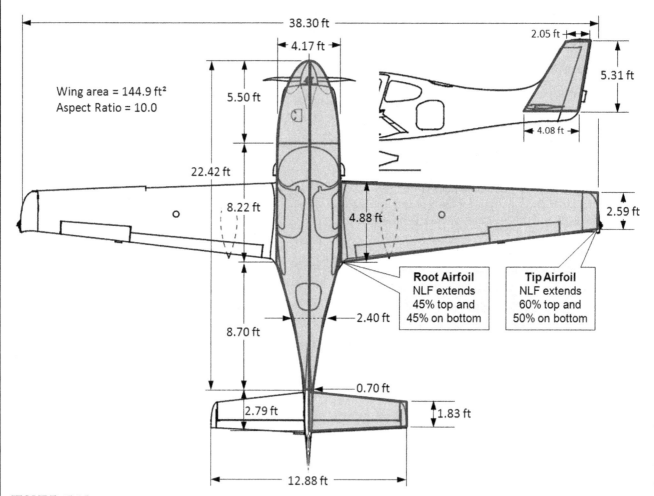

FIGURE 15-37 Geometry of the example aircraft showing the approximation of its wetted area. Note that for clarity some dimensions have been rounded off to two decimals.

EXAMPLE 15-6 (cont'd)

FLIGHT CONDITION

#					
1	Airspeed	$V =$	312.3	ft/s	User entry
2	Viscosity of air	$\mu =$	3.745E-07	$lb_f \cdot s/ft^2$	Formula
3	Altitude	$H =$	0		
4	Density of air	$\rho =$	0.002378	slugs/ft^3	
5	Dynamic pressure	$q =$	115.95	lb_f/ft^2	

#			WING	HT	VT	FUSELAGE	
6	Chord, root	$C_R =$	4.875	2.792	4.083	22.42	ft
7	Chord, tip	$C_T =$	2.585	1.834	2.050	22.42	ft
8	Exposed halfspan	$b/2 =$	17.07	6.442	5.313	2.083	ft
9	Exposed planform area	$S_i =$	127.30	29.79	16.29		ft^2
10	Reference area	$S_{ref} =$	144.9				ft^2
11	Reference wing span	$b =$	38.30				ft
12	Aspect Ratio	$AR = b^2/S_{ref} =$	10.12				
13	Wetted area booster	$k_b =$	1.07	1.05	1.05		
14	Wetted area	$S_{wet\,i} = 2 \cdot S_i \cdot k_b =$	272.4	62.6	34.2	257.6	ft^2
15	Location of max t	$(x/c)_{max} =$	0.50	0.50	0.50	0.43	
16	Thickness-to-chord ratio	$t/c =$	0.150	0.100	0.100	0.186	
17	Skin roughness value	$k =$	1.7E-06	1.7E-06	1.7E-06	1.7E-06	

FIGURE 15-38 Geometric information of the example aircraft.

ROOT CHORD

#						
18	Upper surface transition	$X_{tr\,upper} =$	0.45	0.50	0.30	0.05
19	Lower surface transition	$X_{tr\,lower} =$	0.45	0.50	0.30	0.05
20	Reynolds Number based on C_R	$Re_{CR} =$	9667562	5536125	8097616	44460868
21	Cutoff Reynolds Number	$Re_{cutoff\,root} =$	240963686	133970338	199946103	1201527234
22	Analysis Reynolds Number	$Re_{root} =$	9667562	5536125	8097616	44460868
23	Upper surface fictitious trans	$X_{0\,upper} =$	0.05380	0.07082	0.04463	0.00769
24	Upper surface skin friction	$C_{f\,upper} =$	0.001981	0.002117	0.002427	0.002112
25	Lower surface fictitious trans	$X_{0\,lower} =$	0.05380	0.07082	0.04463	0.00769
26	Lower surface skin friction	$C_{f\,lower} =$	0.001981	0.002117	0.002427	0.002112
27	Average root skin friction	$C_{f\,avg} =$	0.001981	0.002117	0.002427	0.002112

TIP CHORD

#						
28	Upper surface transition	$X_{tr\,upper} =$	0.60	0.50	0.30	0.05
29	Lower surface transition	$X_{tr\,lower} =$	0.50	0.50	0.30	0.05
30	Reynolds Number based on C_T	$Re_{CT} =$	5126287	3635995	4065334	44460868
31	Cutoff Reynolds Number	$Re_{cutoff\,1} =$	240963686	133970338	199946103	1201527234
32	Analysis Reynolds Number	$Re_{tip} =$	5126287	3635995	4065334	44460868
33	Upper surface fictitious trans	$X_{0\,upper} =$	0.08169	0.08292	0.05779	0.00769
34	Upper surface skin friction	$C_{f\,upper} =$	0.001877	0.002342	0.002825	0.002112
35	Lower surface fictitious trans	$X_{0\,lower} =$	0.07290	0.08292	0.05779	0.00769
36	Lower surface skin friction	$C_{f\,lower} =$	0.002156	0.002342	0.002825	0.002112
37	Average root skin friction	$C_{f\,avg} =$	0.002017	0.002342	0.002825	0.002112
38	$C_{f\,avg\,panel}$	$C_f =$	0.001999	0.002230	0.002626	0.002112

FIGURE 15-39 Calculated skin friction coefficients.

EXAMPLE 15-6 (cont'd)

DRAG ANALYSIS

			Wing	HT	VT	Fuselage
39	Sweep angle	$\Lambda_{t\,max} =$	0.00	5.00	18.00	
40	Form Factor	$FF =$	1.311	1.203	1.187	1.399
41	Intereference Factor, IF	$IF =$	1	1.05	1.05	1
42	Weighted drag factor	$C_{fi} \cdot FF \cdot IF \cdot S_{wet\,i} =$	0.7140	0.1762	0.1120	0.7607
43	Skin friction drag	$(1/S_{ref}) \sum C_{fi} \cdot FF \cdot IF \cdot S_{wet\,i} =$	0.01217			

FIGURE 15-40 Minimum drag analysis. See discussion about Line 43 in the text.

15.5 MISCELLANEOUS OR ADDITIVE DRAG

In its most elementary form, the skin friction drag estimated using the *component drag build-up method* (CDBM) excludes drag caused by landing gear, antennas, sharp corners, joints, fasteners, inlets, outlets, fairings, and miscellaneous other protuberances all airplanes feature to some extent. This drag, which is justifiably referred to as *miscellaneous* or *additive drag*, must be accounted for as it increases the total drag well beyond what is predicted by CDBM. It is the purpose of this section to correct this deficiency by presenting a number of methods to account for it. As shown in Equation (15-59) this drag is treated using a special term, C_{Dmisc}. Thus, C_{Dmisc} is estimated as the sum of a number of contributing sources as shown below:

$$C_{D_{misc}} = \Delta C_{D_1} + \Delta C_{D_2} + \cdots + \Delta C_{D_N} \quad (15\text{-}77)$$

where the series ΔC_{D1}, ΔC_{D2}, ... are the contributions of all known components that add to the drag and were not accounted for using the CDBM (e.g. landing gear, antennas, etc.). Note that each contribution becomes an integral part of the minimum drag coefficient, C_{Dmin}. It, on the other hand, is used to calculate the total drag of the airplane using the reference wing area, S_{ref}, as a primary variable. It follows that each contribution must be adjusted in terms of this reference area. Usually, the drag contributions are presented based on the geometry of the component itself, e.g. frontal area or side area or similar. This requires additional conversions to take place, as is explained below, to ensure it references S_{ref}.

The geometric shape of protrusions typical to aircraft is inherently irregular. For instance, consider the dissimilar geometry of antennas, blisters, or the main landing gear typical of ordinary aircraft. None shares common dimensions like those shared by most lifting surfaces (e.g. wing, horizontal, and vertical tails). Each is geometrically complex enough to call for a drag estimation based on empirical formulation. This, in turn, means that the drag of the protrusion must be measured in a wind tunnel and then related to its own geometry and the Reynolds number at the test condition. Only then can it be applied to something as practical as aircraft design. The drag of such protrusions is frequently based on shape parameters, like the thickness-to-length ratio (or fineness ratio) of a particular component, or the location of is maximum thickness, or the frontal area, the lengthwise distribution of thickness, leading edge radius, trailing edge angle, and the camber, to name a few.

It is helpful to use descriptive subscripts when determining the drag coefficient of each such source. Thus, the drag coefficient of a specific antenna might be referred to as $\Delta C_{Dantenna}$ rather than, say, ΔC_{D7}. The following sections assume such naming conventions. Now consider some geometric shape of interest whose drag force has been measured in a wind tunnel at a specific Reynolds number. Furthermore, assume this drag to be denoted by D_S (where the subscript stands for *source*) and that, ultimately, we want the data to be applicable to an unrelated project. In order to accomplish this, the measurement must be converted into a drag coefficient, which here will be called ΔC_{DS}. Furthermore, assume it references some geometry (typically some reference area other than the wing area) given by S_S. Then, the drag coefficient of the component, ΔC_{DS}, would be calculated from:

$$\Delta C_{D_S} = \frac{D_S}{qS_S} \quad (15\text{-}78)$$

However, as stated above, we want to refer this value to the reference wing area S_{ref} and must thus scale the coefficient accordingly. We might refer to this scaled coefficient using the name of the source. For instance, the scaling of the aforementioned antenna, $\Delta C_{Dantenna}$, would then be accomplished using the expression below:

$$\Delta C_{D_{antenna}} = \Delta C_{D_S} \left(\frac{S_S}{S_{ref}} \right) \quad (15\text{-}79)$$

Derivation of Equation (15-79)

This expression is simply derived by noting the same value of the drag must be calculated regardless of whether one uses the source area or the reference area. Mathematically:

$$D_S = qS_S \Delta C_{D_S} = qS_{ref}\Delta C_{D_{antenna}}$$
$$\Rightarrow S_{ref}\Delta C_{D_{antenna}} = S_S\Delta C_{D_S}$$
$$\Rightarrow \Delta C_{D_{antenna}} = \Delta C_{D_S}\left(\frac{S_S}{S_{ref}}\right)$$

QED

15.5.1 Cumulative Result of Undesirable Drag (CRUD)

Of course the above title is just a play on words. The word crud means dirt, filth, or refuse. In aircraft design it stands for the undesirable drag caused by things like exhaust stacks, misaligned sheet-metal panels, antennas, small inlets and outlets, sanded walkways, and so on. These parts are easily overlooked when performing drag analysis, primarily because the evaluation of their contribution is next to impossible. For instance, consider misaligned sheet-metal panels for aluminum aircraft. The problem is not that there isn't a method to estimate the drag of a misaligned panel, because there is, but rather that it is impossible to estimate how poorly or well the panels align until the actual aircraft is built. It is imperative the aspiring designer is aware of this drag and accounts for it appropriately. This section introduces how this is typically done.

In 1940 NACA released the wartime report WR-L-489 in which 11 military aircraft were investigated to determine why they didn't meet predicted performance [30]. The report detailed wind tunnel testing of the aircraft in the NACA Full-Scale Wind Tunnel in Langley, Virginia. The aircraft were stripped to a clean configuration by gradually removing components and the drag coefficient was measured after each removal step. The results give a timeless insight into the cumulative impact of small and easily ignored design details on the total drag.

One of the aircraft investigated was the Seversky P-35 that had been designed for the US Army Air Corps in the early 1930s (see Figure 15-41). At the time, it was an innovative single-engine fighter that offered a number of firsts. It was the Air Corp's first all-metal fighter, the first to feature a retractable landing gear, and the first with an enclosed cockpit. The P-35 had a predicted maximum airspeed in excess of 260 KTAS. However, in practice, it fell way short of that goal as the maximum airspeed was found to be some 20 KTAS less than that. While a part of the problem was that its engine could not develop the advertised power, this did not fully explain the large difference between practice and prediction. This simply implied its drag was higher than

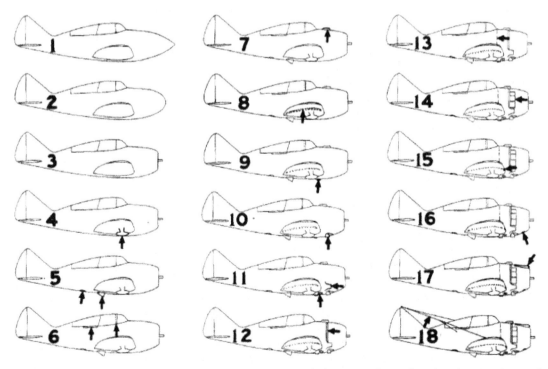

FIGURE 15-41 Order of modifications made to the Seversky P-35 aircraft during wind tunnel testing. Arrow points to the location of changes. *Figure from NACA WR-L-489.*

predicted — something common to other aircraft of the period as well. After all, that is why 11 aircraft, rather than a single one, were studied in the report. This highlights a scenario that anyone estimating the drag of a new airplane may find himself in — to assume the cleanliness of Airplane 1 in Figure 15-41, when in reality it will be closer to Airplane 18.

Such under-predictions are not limited to aircraft of the yesteryear. They still happen. A recently developed small twin-engine business jet was unable to meet the predicted performance advertised on the company website. This led to a costly redevelopment effort and eventually the airplane was shown to meet the promised capabilities. A small single-engine propeller aircraft developed by a prominent car manufacturer also didn't meet the predicted performance, allegedly due to drag higher than predicted. Both aircraft were designed by people who cannot be accused of not knowing what they were doing. It shows that no matter one's proficiency, under-prediction of drag is a likely scenario and the subject must be handled with utmost caution.

Going back to the NACA WR-L-489, the source presents a detailed listing of the changes made to the airplane presented in a graphic and tabulates their effect on the total drag. The graphic is reproduced as Figure 15-41 and the table is recreated as Table 15-8. It should be of great interest to the aircraft designer to inspect the table and consider how the contributions of seemingly insignificant modifications, such as a sanded walkway or an opened cowl flap, affect the overall drag of the airplane. The problem is not the value of individual component, but rather their cumulative effect.

15.5.2 Trim Drag

Trim drag is the penalty paid for providing static stability. Technically, trim drag is the combination of two sources.

(1) The difference in lift-induced drag of the airplane with and without balancing forces and (2) the increase in drag due to the deflection of the elevator (see Section 15.4.4, *The Effect of Control Surface Deflection — Trim Drag*). This is denoted using the variable $\Delta C_{D_{\delta_e}}$. Its value typically ranges from 0.0000 to 0.0005 per degree of elevator deflection. It must be multiplied by the elevator deflection to trim, δ_e, to determine the additive drag coefficient.

If a longitudinally stable airplane could sustain steady level flight without having to generate a balancing force there would be no trim drag. The balancing force of a statically stable conventional tail-aft aircraft is added to the weight. This implies the total lift generated by the wing must be greater than the weight alone. This requires it to operate at a higher *AOA* and, thus, at a higher lift-induced drag than otherwise. It does not matter

TABLE 15-8 Component Contribution to Drag for the Seversky P-35 [30]

ID	Airplane Conditions	C_D	ΔC_D	ΔC_D %
1	Completely faired condition with long nose fairing	0.0166	0	
2	Completely faired condition with blunt nose fairing	0.0169		
3	Completely faired condition with original NACA cowling (no air flowing through cowling)	0.0186	0.0020	12.0%
4	Same as 3 except landing gear seals and fairing removed	0.0188	0.0002	1.2%
5	Same as 4 except original oil cooler installed	0.0205	0.0017	10.2%
6	Same as 5 except canopy fairing removed	0.0203	-0.0002	-1.2%
7	Same as 6 except carburetor scoop added	0.0209	0.0006	3.6%
8	Same as 7 except sanded walkway added	0.0216	0.0007	4.2%
9	Same as 8 except ejector chute added	0.0219	0.0003	1.8%
10	Same as 9 except exhaust stacks added	0.0225	0.0006	3.6%
11	Same as 10 except intercoolers added	0.0236	0.0011	6.6%
12	Same as 11 except cowling exit opened	0.0247	0.0011	6.6%
13	Same as 12 except accessory exit opened	0.0252	0.0005	3.0%
14	Same as 13 except cowling fairing and seals removed	0.0261	0.0009	5.4%
15	Same as 14 except cockpit ventilator opened	0.0262	0.0001	0.6%
16	Same as 15 except cowling venturis installed	0.0264	0.0002	1.2%
17	Same as 16 except blast tubes added	0.0267	0.0003	1.8%
18	Same as 17 except radio aerial installed	0.0275	0.0008	4.8%
	TOTAL DRAG CHANGE		0.0109	65.7%

Percentages are based on ID 1

whether this stability is generated using a horizontal tail, such as that of the statically stable tail-aft configuration, or using airfoils and elevons, such as that of a flying wing; the total drag of the airplane is increased. This section presents a method to estimate the trim drag.

Since trim drag is a consequence of the longitudinal stability, it is easy to derive a complicated and unwieldy formulation that is of limited practical use. Ordinarily, trim drag constitutes a small fraction of the total drag of the airplane; it should range from 1% to 2%. During the design process, it is more practical to estimate the trim drag using a simple formulation and this is what will be demonstrated here. The same methodology can be revised to account for more complex situations. Note that the method assumes the airplane is operating at a low AOA, allowing for simplification.

Trim Drag of a Simple Wing-Horizontal Tail Combination

Consider the simple wing-HT system shown in Figure 15-42. Note the equation is presented in a form that lends itself well when considering changes in the location of the CG. For this system the trim drag is given by:

$$\Delta C_{D_{trim}} = \frac{kW^2}{\left(qS_{ref}\right)^2}\left[\left(1 - \frac{h - h_{AC}}{l_{HT}}\right)^2 - 1\right] + \Delta C_{D_{\delta_e}} \cdot \delta_e$$

(15-80)

where

W = weight at condition
k = lift-induced drag constant
q = dynamic pressure
S_{ref} = reference wing area
C_{MGC} = wing mean geometric chord
l_{HT} = distance between the wing AC ($\approx C_{MGC}/4$) and the C/4 of the HT
h = distance between the wing LE and the CG
h_{AC} = distance between the wing LE and AC ($\approx C_{MGC}/4$)
$\Delta C_{D_{\delta_e}}$ = additive drag due to the elevator deflection

Derivation of Equation (15-80)

Referring to Figure 15-42, statics requires the following to hold in steady level flight:

$$\sum F_z = 0; \quad L_W + L_{HT} - W = 0$$

$$\sum M_{CG} = 0; \quad (h - h_{AC})L_W - (l_{HT} - h + h_{AC})L_{HT} = 0$$

These can be solved for the balancing force the HT must generate:

$$(h - h_{AC})L_W - (l_{HT} - h + h_{AC})L_{HT} = 0$$
$$\Rightarrow (l_{HT} - h + h_{AC})L_{HT} = (h - h_{AC})L_W$$

Therefore, the balancing force of the HT is given by:

$$L_{HT} = \frac{h - h_{AC}}{l_{HT} - h + h_{AC}} L_W$$

Inserting this result into the force equation leads to:

$$L_W + L_{HT} - W = L_W + \left(\frac{h - h_{AC}}{l_{HT} - h + h_{AC}}\right)L_W - W = 0$$

$$\Rightarrow \left(1 + \frac{h - h_{AC}}{l_{HT} - h + h_{AC}}\right)L_W = \left(\frac{l_{HT} - h + h_{AC} + h - h_{AC}}{l_{HT} - h + h_{AC}}\right)L_W$$

$$= \frac{l_{HT}}{l_{HT} - h + h_{AC}} L_W = W$$

Since $L_W = qS_{ref}C_{LW}$, solving for C_{LW} yields:

$$C_{L_W} = \frac{W}{qS_{ref}}\left(1 - \frac{h - h_{AC}}{l_{HT}}\right) \qquad (i)$$

The minimum lift-induced drag is generated when the HT generates no (or negligible) balancing force. This happens when the weight is precisely at the

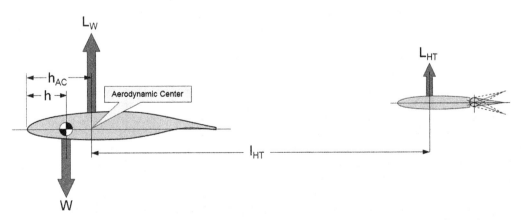

FIGURE 15-42 A simple free-body diagram used to derive the simplest formulation of trim drag.

aerodynamic center. In that case, the lift-induced drag (using the simplified drag model) is given by:

$$C_{D_i} = kC_{L_W}^2 = k\frac{W^2}{\left(qS_{ref}\right)^2} \quad \text{(ii)}$$

On the other hand, when balancing force is required, Equation (i) can be used to calculate the higher lift-induced drag, i.e.:

$$C_{D_i} = kC_{L_W}^2 = \frac{kW^2}{\left(qS_{ref}\right)^2}\left(1 - \frac{h - h_{AC}}{l_{HT}}\right)^2 \quad \text{(iii)}$$

The trim drag is the difference between Equations (iii) and (ii):

$$\Delta C_{D_{trim}} = \frac{kW^2}{\left(qS_{ref}\right)^2}\left(1 - \frac{h - h_{AC}}{l_{HT}}\right)^2 - \frac{kW^2}{\left(qS_{ref}\right)^2}$$

$$= \frac{kW^2}{\left(qS_{ref}\right)^2}\left[\left(1 - \frac{h - h_{AC}}{l_{HT}}\right)^2 - 1\right]$$

QED

Trim Drag of a Wing-Horizontal Tail-Thrustline Combination

Accounting for a high or low thrustline, and wing pitching moment improves the accuracy of the method and is particularly important for long-range aircraft with the thrustline far above or below the CG (see Figure 15-43). Additionally, it is reasonable to account for the wing pitching moment and a possible reduction through the use of a cruise flap. Taking both of these into account yields the following expression to estimate the trim drag:

$$\Delta C_{D_{trim}} = \frac{k}{\left(qS_{ref}\right)^2}\left(1 - \frac{h - h_{AC}}{l_{HT}}\right)^2\left(W - \frac{M_W - Tz_T}{l_{HT} - h + h_{AC}}\right)^2$$

$$- \frac{kW^2}{\left(qS_{ref}\right)^2} + \Delta C_{D_{\delta_e}} \cdot \delta_e$$

(15-81)

where

M_W = wing pitching moment = $qS_{ref}C_{MGC}C_{m_W}$
C_{mW} = wing pitching moment coefficient
z_T = distance between the CG and the thrustline. It is positive if the thrustline is above the CG
T = engine thrust in lb$_f$ or N

Comparing Equation (15-81) to (15-80) shows that the wing pitching moment (whose value is <0) and high thrustline ($z_{HT} > 0$) lead to a higher trim drag.

Derivation of Equation (15-81)

Referring to Figure 15-42, statics requires the following to hold in steady level flight:

$$\sum F_z = 0; \quad L_W + L_{HT} - W = 0$$
$$\sum M_{CG} = 0; \quad M_W + (h - h_{AC})L_W - (l_{HT} - h + h_{AC})L_{HT}$$
$$- Tz_T = 0$$

These can be solved for the balancing force the HT must generate:

$$M_W + (h - h_{AC})L_W - (l_{HT} - h + h_{AC})L_{HT} - Tz_T = 0$$
$$\Rightarrow L_{HT} = \frac{M_W + (h - h_{AC})L_W - Tz_T}{(l_{HT} - h + h_{AC})}$$

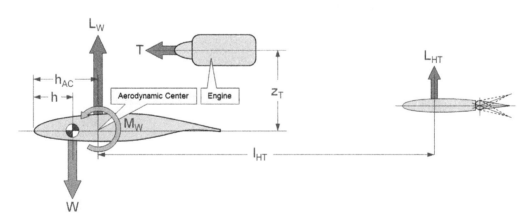

FIGURE 15-43 A simple free-body diagram used to derive trim drag for a wing-HT-thrustline combination.

Inserting this result into the force equation and, as before, using the relation $L_W = qSC_{LW}$, C_{LW} can be determined:

$$L_W + L_{HT} - W = L_W + \frac{M_W + (h - h_{AC})L_W - Tz_T}{(l_{HT} - h + h_{AC})} - W = 0$$

$$\Rightarrow L_W = \left(1 - \frac{h - h_{AC}}{l_{HT}}\right)\left(W - \frac{M_W - Tz_T}{(l_{HT} - h + h_{AC})}\right)$$

Therefore, the lift coefficient is given by:

$$C_{L_W} = \frac{1}{qS_{ref}}\left(1 - \frac{h - h_{AC}}{l_{HT}}\right)\left(W - \frac{M_W - Tz_T}{(l_{HT} - h + h_{AC})}\right)$$

Using the same logic as in the previous derivation, the trim drag can be found to equal:

$$\Delta C_{D_{trim}} = \frac{k}{(qS_{ref})^2}\left(1 - \frac{h - h_{AC}}{l_{HT}}\right)^2\left(W - \frac{M_W - Tz_T}{l_{HT} - h + h_{AC}}\right)^2 - \frac{kW^2}{(qS_{ref})^2}$$

QED

EXAMPLE 15-7

Estimate the trim drag coefficient and drag force for the SR22 at S-L and 185 KTAS assuming the following parameters:

$W = 3400$ lb$_f$	$S_{ref} = 144.9$ ft^2	$T = 450$ lb$_f$
$k = 0.04207$	$C_{mW} = -0.060$	$\Delta C_{D_{\delta_e}} = 0$
$l_{HT} = 14.06$ ft	$z_T = 0.6$ ft	
$h = 0.10\, C_{MGC}$ and $0.20\, C_{MGC}$	$h_{AC} = 0.25\, C_{MGC}$	$C_{MGC} = 3.783$ ft

Solution

Dynamic pressure:

$$q = \frac{1}{2}\rho V^2 = \frac{1}{2}(0.002378)(185 \times 1.688)^2 = 115.9 \text{ lb}_f/\text{ft}^2$$

Wing pitching moment:

$$M_W = qS_{ref}C_{MGC}C_{m_W} = (115.9)(144.9)(3.783)(-0.060)$$
$$= -3812 \text{ ft·lb}_f$$

Determine trim drag in accordance with Equation (15-81). For simplicity, rewrite the equation as shown below:

$$\Delta C_{D_{trim}} = \frac{k}{(qS_{ref})^2}\left(1 - \frac{h - h_{AC}}{l_{HT}}\right)^2\left(W - \frac{M_W - Tz_T}{l_{HT} - h + h_{AC}}\right)^2 - \frac{kW^2}{(qS_{ref})^2} = AB^2C^2 - AW^2$$

Where:

$$A = \frac{k}{(qS_{ref})^2} = \frac{0.04207}{(115.9 \times 144.9)^2}$$
$$= 1.492 \times 10^{-10} \quad (1/\text{lb}_f^2)$$

$$B = 1 - \frac{h - h_{AC}}{l_{HT}} = 1 - \frac{h - 0.25 C_{MGC}}{14.06}$$
$$= \begin{cases} 1.0404 & \text{if } h = 0.10 C_{MGC} \\ 1.0135 & \text{if } h = 0.20 C_{MGC} \end{cases}$$

$$C = W - \frac{M_W - Tz_T}{l_{HT} - h + h_{AC}}$$
$$= 3400 - \frac{-3812 - (450)(0.6)}{14.06 - (h - 0.25)(3.783)}$$
$$= \begin{cases} 3679 \text{ lb}_f & \text{if } h = 0.10 C_{MGC} \\ 3686 \text{ lb}_f & \text{if } h = 0.20 C_{MGC} \end{cases}$$

Therefore, the trim drag coefficient is:

$$\Delta C_{D_{trim}} = AB^2C^2 - AW^2 = \begin{cases} 0.000461 & \text{if } h = 0.10 \\ 0.000358 & \text{if } h = 0.20 \end{cases}$$

This is just about 1.4% to 1.8% of the airplane's minimum drag coefficient, shown to be 0.02541 in Example 15-18. Using, $D_{trim} = qS_{ref} \cdot \Delta C_{D_{trim}}$, the drag is found to equal 7.7 and 6.0 lb$_f$, respectively.

15.5.3 Cooling Drag

The operation of powered aircraft calls for heat transfer using heat exchangers that are exposed to the free stream airflow. Examples of such heat exchangers are the cylinder head fins of a piston engine as well as radiators for oil and water cooling. An important element of the exchange of energy is the restriction to air flow demanded by the radiator. The flow entering the radiator has a given total head of which some is lost as the air flows through it. This results in a drop in the total pressure of the flow, extracting energy from it. Some of this loss in energy is made up by adding heat to the flow. However, if the heat energy added is less than the energy lost due to the pressure drop, the momentum flux will be reduced. This reduction is experienced as a drag force and is referred to as *cooling drag*.

Cooling drag is hard to estimate due to the complexity of the flow field inside the engine compartment (see idealization in Figure 15-44). Typically, this is estimated using empirical methods based on testing performed by the engine manufacturer. The following method for estimating cooling drag is an example of such a methodology; here largely based on McCormick [31]:

Cooling drag coefficient:

$$\Delta C_{D_{cool}} = \frac{\dot{m}(V_0 - V_E)}{qS_{ref}} \qquad (15\text{-}82)$$

where

$\dot{m}$ = mass flow rate through the engine compartment
V_0 = far field airspeed, represents the inlet airspeed
V_E = average airspeed at the exit of the engine cowling

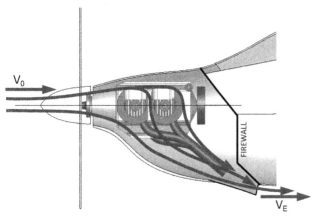

FIGURE 15-44 Idealization of a conventional engine installation.

Derivation of Equation (15-82)

This derivation is in part based on Ref. [31]. First the work extracted from the flow is evaluated. The work-energy theorem says that if an external force acts upon a rigid object, causing its kinetic energy to change from E_0 to E_E, then the mechanical work (W) is given by:

$$W = \Delta E = E_E - E_0 = \frac{1}{2}m(V_E^2 - V_0^2) \qquad (i)$$

The rate at which work is being extracted from the flow is then

$$\Delta W = \frac{dE}{dt} = \frac{d}{dt}\left[\frac{1}{2}m(V_0^2 - V_E^2)\right] = \frac{1}{2}\dot{m}(V_0^2 - V_E^2) \qquad (ii)$$

Note that algebraically this expression can be rewritten as follows:

$$\Delta W = \frac{1}{2}\dot{m}(V_0^2 - V_E^2) = \dot{m}\frac{1}{2}(V_0 - V_E)(V_0 + V_E)$$
$$= \dot{m}(V_0 - V_E)\frac{(V_0 + V_E)}{2} \qquad (iii)$$

Recall that force is the rate of change of momentum:

$$F = \frac{dMomentum}{dTime} = \frac{d(mv)}{dt} = m\frac{dv}{dt} + v\frac{dm}{dt} \qquad (iv)$$

Also, work is the product of force and speed:

$$\Delta W = Force \times Speed = (\dot{m}(V_0 - V_E)) \times \left(\frac{V_0 + V_E}{2}\right) \qquad (v)$$

Therefore;

$$\Delta W = Force \times Speed = D_{cool}\left(\frac{V_0 + V_E}{2}\right) \qquad (vi)$$

where D_{cool} is called the cooling drag, found from:

$$D_{cool} = \dot{m}(V_0 - V_\infty) \qquad (15\text{-}83)$$

In order to convert this to an additive drag coefficient, we write:

$$D_{cool} = qS_{ref}C_{D_{cool}} = \dot{m}(V_0 - V_\infty)$$
$$\Rightarrow \Delta C_{D_{cool}} = \frac{\dot{m}(V_0 - V_\infty)}{qS_{ref}} \qquad (vii)$$

QED

EXAMPLE 15-8

Estimate the cooling drag and cooling drag coefficient of the airplane of Example 7-6, if its wing area is 144.9 ft^2.

Solution

The cooling drag can be found from Equation (15-83) where the input values are taken from Example 7-6:

$$D_{cool} = \dot{m}(V_0 - V_E) = \frac{2.7}{32.174}(312.3 - 199.3) = 9.48 \text{ lb}_f$$

The cooling drag coefficient is estimated from Equation (15-82):

$$\Delta C_{D_{cool}} = \frac{\dot{m}(V_0 - V_\infty)}{qS_{ref}} = \frac{\frac{2.7}{32.174}(312.3 - 199.3)}{(80.60)(144.9)}$$
$$= 0.000812$$

This amounts to about 8.12 drag counts.

15.5.4 Drag of Simple Wing-Like Surfaces

Consider a wing-like surface, such as an aerodynamically shaped antenna or some fin (see Figure 15-45), which features a constant airfoil whose drag characteristics are known. Then, the drag of the entire surface can be determined by estimating the two-dimensional skin friction coefficient at the MGC or the average chord of the surface. The additive drag coefficient of this surface, denoted here by ΔC_{Dfin}, can be estimated from:

$$\Delta C_{D_{fin}} = C_f \left[1 + 2.7\left(\frac{t}{c}\right) + 100\left(\frac{t}{c}\right)^4 \right] \frac{h(C_R + C_T)}{2S_{ref}} \quad (15\text{-}84)$$

This additive drag coefficient is based on the reference area, as can be seen in the equation. Here the form factor is based on Equation (15-62).

EXAMPLE 15-9

Determine the additive drag coefficient for a COM antenna for the SR22 airplane, if its root chord is 4.5 inches, tip is 2 inches, and height is 13.5 inches. Assume a skin friction coefficient of 0.0035 and an average thickness-to-chord ratio of 0.25.

Solution

$$\Delta C_{D_{fin}} = C_f \left[1 + 2.7\left(\frac{t}{c}\right) + 100\left(\frac{t}{c}\right)^4 \right] \frac{h(C_R + C_T)}{2S_{ref}}$$

$$= 0.0035 \left[1 + 2.7(0.25) + 100(0.25)^4 \right] \frac{\left(\frac{13.5}{12}\right)\frac{(4.5+2)}{12}}{2(144.9)}$$

$$= 0.00001520$$

This amounts to about 0.152 drag counts.

15.5.5 Drag of Streamlined Struts and Landing Gear Pant Fairings

The cross-sections in Figure 15-46 are typical of those used for wing struts or to reduce the drag of leaf-spring landing gear legs. Such shapes typically operate in a low R_e region, where the R_e is based on their chord, denoted by c. The drag of such sections is typically related to their thickness-to-chord ratios.

The additional drag coefficient for the strut, ΔC_{DS}, is given by Hoerner [3, p. 6-5] and is based on empirical results:

$$\Delta C_{D_S} = \frac{D_S}{qS_S} = 2C_f \left[1 + \left(\frac{t}{c}\right) \right] + \left(\frac{t}{c}\right)^2 \quad (15\text{-}85)$$

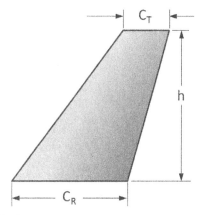

FIGURE 15-45 Geometric definition of a small wing-like surface.

where

c = chord of strut
C_f = skin friction coefficient of strut
S_S = planform area of the strut (i.e. its length × chord)
t = thickness of strut
ΔC_{DS} = drag of strut in terms of the reference area

The miscellaneous drag coefficient for a strut of length L and chord c in terms of S_{ref} would thus be estimated from:

$$\Delta C_{Dstrut} = \left[2C_f\left(1+\frac{t}{c}\right) + \left(\frac{t}{c}\right)^2\right]\left(\frac{L \times c}{S_{ref}}\right) \quad (15\text{-}86)$$

Note that this would be the contribution of the strut to the total miscellaneous drag. Also note that since all the cross-sections have the same form factor, the difference comes in the evaluation of the skin friction coefficient, C_f.

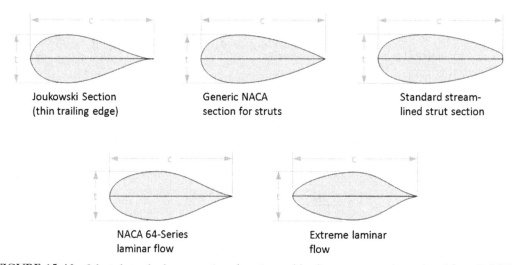

FIGURE 15-46 Selected standard cross sections for wing and landing gear struts (reproduced from Ref. [3]).

EXAMPLE 15-10

Determine the additive drag of a (Cessna-like) wing strut whose length is 5 ft, chord is 4 inches, and t/c is 0.2 at S-L at airspeed of 110 KTAS. The reference area is 160 ft². Assume no interference and $C_f = 0.008$.

Solution

Using Equation (15-86) we get:

$$\Delta C_{Dstrut} = \left[2C_f\left(1+\frac{t}{c}\right) + \left(\frac{t}{c}\right)^2\right]\left(\frac{L \times c}{S_{ref}}\right)$$

$$= \left[2(0.008)(1+0.2) + (0.2)^2\right]\left(\frac{5 \times (4/12)}{160}\right)$$

$$= 0.0006167$$

This amounts to about 6.2 drag counts. Therefore, the resulting drag amounts to:

$$D_{strut} = \frac{1}{2}\rho V^2 S_{ref}\Delta C_{Dstrut}$$

$$= \frac{1}{2}(0.002378)(100 \times 1.688)^2(160)(0.0006167)$$

$$= 3.34 \text{ lb}_f$$

This means that each strut (assuming there are two) adds 3.34 lb$_f$ of drag to the total drag.

EXAMPLE 15-11

Determine the additive drag coefficient of a step onto the wing to help occupants enter the cabin of an SR22 aircraft. There are two such entry steps, whose cross-section is the *standard streamlined strut section* that are approximately 12 inches long, 3-inch chord and 1-inch thickness. Assume no interference, ignore the break in the step where it changes from a step to a strut and $C_f = 0.008$.

Solution

The t/c is about $1/3 = 0.333$. Therefore, using Equation (15-86) we get for each step:

$$\Delta C_{D strut} = \left[2(0.008)(1+0.333) + (0.333)^2\right] \times \left(\frac{(12/12) \times (3/12)}{144.9}\right)$$
$$= 0.0002281$$

This amounts to about 2.28 drag counts.

Thick Fairings

A fairing is a streamlined structure whose purpose is to reduce the drag that would be caused by the underlying geometry. Hoerner [3, p. 6-9] presents the following expression to determine the (two-dimensional) drag of a fairing whose chord is c, height is t, and length is L:

$$\Delta C_{d\ fairing} = C_f \left(4 + \frac{2}{(t/c)} + 120(t/c)^3\right) \quad (15\text{-}87)$$

The term inside the parenthesis is the form factor. It accounts for pressure and friction (see Figure 15-47). This means of course that for a strut of length L and chord c in terms of S_{ref} would thus be estimated from:

$$\Delta C_{Dfairing} = C_f \left(4 + \frac{2}{(t/c)} + 120(t/c)^3\right)\left(\frac{L \times c}{S_{ref}}\right) \quad (15\text{-}88)$$

This allows the optimum thickness-to-chord ratio to be determined by determining the derivative, setting it equal to zero, and solve for the optimum t/c, as shown below:

$$FF = \left(4 + \frac{2}{(t/c)} + 120(t/c)^3\right)$$

$$\Rightarrow \frac{dFF}{d(t/c)} = 360(t/c)^2 - \frac{2}{(t/c)^2} = 0$$

$$\Rightarrow (t/c)_{opt} = \sqrt[4]{\frac{1}{180}} = 0.273$$

Therefore, the optimum thickness-to-chord ratio is 0.273. This corresponds to a fineness ratio of ≈ 3.7.

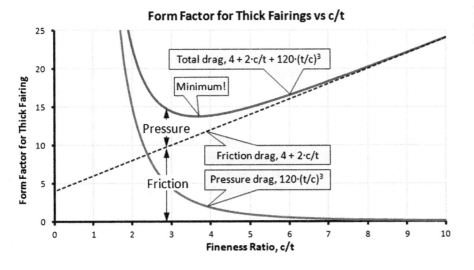

FIGURE 15-47 Form Factor plotted against the fineness ratio for wing and landing gear struts.

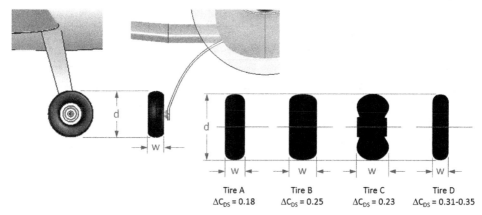

FIGURE 15-48   Most common types of modern tires for aircraft landing gear (based on Ref. [32]).

15.5.6 Drag of Landing Gear

Drag of Tires Only

Figure 15-48 shows several types of aircraft tires, identified as A, B, C, and D. The drag generated by these styles is the topic of NACA R-485 [32]. It is convenient to express the drag coefficient of tires in terms of their frontal area, which is defined as their diameter, d, multiplied by their width, w. This has been done in Figure 15-48 and Table 15-9. The additive drag coefficient of the tire can be estimated from:

$$\Delta C_{D_{tire}} = \frac{(d \times w)}{S_{ref}} \Delta C_{D_S} \qquad (15\text{-}89)$$

TABLE 15-9   Drag of Tires (d = Diameter of Tire, w = Width of Tire)

Tire Type	Corresponds to	Reference Area	ΔC_{DS}	Reference
A	Three Part Type (GA)	$d \times w$	0.18	NACA R-485
B	Type III	$d \times w$	0.25	
C	Type III high floatation (tundra)	$d \times w$	0.23	
D	Old-fashioned disc wheel types	$d \times w$	0.31-0.35	

Drag of Tires with Wheel Fairings

The purpose of wheel fairings is to improve the aerodynamic geometry of the tire and thus reduce its drag. Figure 15-49 shows several fairing styles and Table 15-10 lists the applicable drag coefficients based on (1) the frontal area of the fairing and (2) the frontal area of the tire. It is left to the reader to select which one to use. While the data is based on a Type III tires, for preliminary design purposes it may be assumed the drag coefficients are independent of the type of tire. The additive drag coefficient of the tire with the fairing can be estimated from:

$$\Delta C_{D_{fairing}} = \frac{(H \times W)}{S_{ref}} \Delta C_{D_S} \qquad (15\text{-}90)$$

Note that this coefficient is for a single tire with a fairing. This is emphasized because some texts present the drag coefficient for two wheels (e.g. assuming both main landing gear). However, some aircraft feature fairings on the main wheel only, while others have all three wheels (main and nose landing gear) with fairings.

Drag of Fixed Landing Gear Struts with Tires

Drag coefficients for a number of typical fixed main landing gear with tires are presented in Table 15-10. All the drag coefficients are based on the dimensions of a

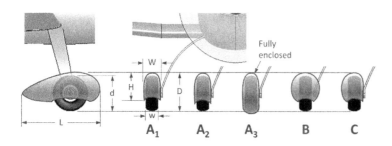

FIGURE 15-49   Selected types of landing gear wheel fairings (based on Ref. [32]). The drag of the landing gear wheel fairing styles denoted by "A" through "C" is presented in Table 15-10.

15.5 MISCELLANEOUS OR ADDITIVE DRAG

TABLE 15-10 Drag of Tires with Fairings (H = Fairing Height, W = Fairing Width, d = Tire Diameter, w = Tire Width)

Fairing Type	Tire Type	ΔC_{DS} Reference Area = $H \times W$	ΔC_{DS} Reference Area = $d \times w$	Reference
A1	Type III (B)	0.130	0.143	NACA R-485
A2	Type III (B)	0.090	0.119	
A3 (tire fully covered)	Type III (B)	0.044	0.070	
B	Type III (B)	0.117	0.217	
C	Type III (B)	0.129	0.188	

single tire per side but apply to the entire installation (both wheels and support structure). The coefficients include interference drag, but exclude the nose landing gear, however. The additive drag coefficient of the fixed landing gear with the fairing can be estimated from:

$$\Delta C_{D_{fixed}} = \frac{(d \times w)}{S_{ref}} \Delta C_{D_S} \qquad (15\text{-}91)$$

Refer to Figure 15-50 for the shape of the landing gear configuration. Each configuration is identified with a letter ranging from A to H. Reference [32] presents results for the said configurations, of which some feature more than one type of tire and even fairings. These differences are presented in Table 15-11 using numbers following the letter. Thus, configurations A and B both feature a Type III tire, whereas configuration C is presented with five different tires, one configuration of which is supported by a streamlined tension wire and the other by a tubular tension support. The remainder of the configurations utilize that same tubular support.

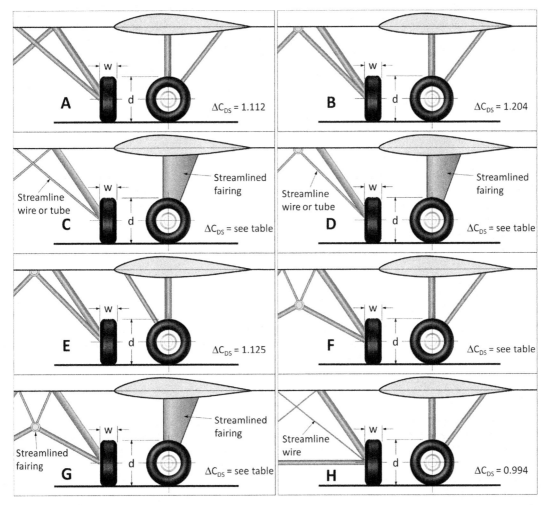

FIGURE 15-50 Drag of selected types of fixed landing gear installations. All struts are streamlined. All drag coefficients are based on the tire geometry and pertain to the entire installation (two main gear) (based on Refs [23] and [32]). The drag of the fixed landing gear installation styles denoted by "A" through "Q" is presented in Table 15-11.

15. AIRCRAFT DRAG ANALYSIS

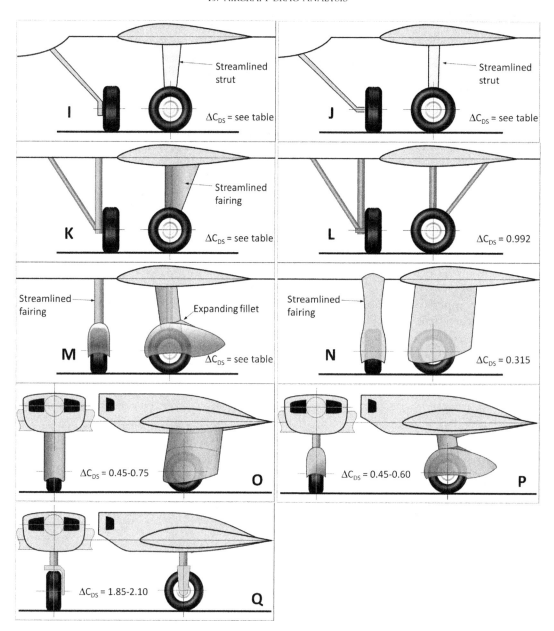

FIGURE 15-50 (*Continued*)

TABLE 15-11 Drag of Landing Gear Struts with and without Fairings (h = Height of Tire, w = Width of Tire)

Strut Type	Tire Type (letter corresponds to Table 15-9)		Reference Area	ΔC_{DS}	Reference
A	8.5-10	B		1.112	
B	8.5-10	B		1.204	
C1	8.5-10 + streamline wire	B		1.151	
C2	8.5-10 + tubular support	B	$d \times w$	1.178	NACA R-485
C3	27 inch streamline + tube	A		1.082	
C4	25x11-4 X-low press + tube	C		0.940	
C5	30x5 disk wheel hi-press + tube	D1		1.779	

TABLE 15-11 Drag of Landing Gear Struts with and without Fairings (h = Height of Tire, w = Width of Tire)—cont'd

Strut Type	Tire Type (letter corresponds to Table 15-9)		Reference Area	ΔC_{DS}	Reference
C6	32x6 disk wheel hi-press + tube	D2		1.373	
D1	8.5-10	B		1.230	
D2	8.5-10	B		1.191	
E	8.5-10	B		1.125	
F1	8.5-10	B		1.138	
F2	8.5-10 + Fairing C	B		0.877	
F3	27 inch streamline + tube	A		1.014	
F4	25x11-4 X-low press + tube	C		0.858	
F5	30x5 disk wheel hi-press + tube	D1		1.628	
G1	8.5-10	B		1.151	
G2	8.5-10+Fairing A2	B		0.733	
H	8.5-10	B		0.994	
I1	8.5-10 + Fairing B	B		0.536	
I2	8.5-10 + Fairing C	B	$d \times w$	0.484	NACA R-485
I4	27 inch streamline + tube	A		0.564	
I5	27 inch streamline + tube	A		0.496	
J1	8.5-10	B		0.615	
J2	8.5-10 + Fairing A1	B		0.458	
J3	27 inch streamline	A		0.485	
K1	8.5-10	B		0.981	
K2	8.5-10 + Fairing C	B		0.641	
L	8.5-10	B		0.992	
M1	8.5-10 + Fairing A1	B		0.484	
M2	8.5-10 + Fairing A1 + Expanding fillet	B		0.315	
N	8.5-10	B		0.315	

EXAMPLE 15-12

Determine the additive drag coefficient for the fixed landing gear of the SR22 airplane. Its wing area is 144.9 ft² and assume the main landing gear tire dimensions are 15 inch diameter and 6 inch width and the nose gear tire is 14 by 5 inches.

Solution

Approximate the main landing gear using Configuration I2 since the landing gear fairing (Style C) is similar in some ways to that of the SR22. The additive drag coefficient for I2 is $\Delta C_{DS} = 0.484$. Similarly, assume the nose landing gear can be represented using Configuration M1, for which $\Delta C_{DS} = 0.484$ as well. In this case note that the drag refers to the entire installation, which consists of two wheels. Since the nose gear is a single wheel, we reduce this number by a factor of two, i.e. $\Delta C_{DS} = 0.242$ for the nose gear. Thus we estimate the drag contribution of the main and nose landing gear to be represented by (note that a factor of 144 is used to convert in² to ft²:

EXAMPLE 15-12 (cont'd)

Main:

$$\Delta C_{D_{main}} = \frac{(d \times w)}{S_{ref}} \Delta C_{D_S} = \frac{(15 \times 6)/144}{144.9}(0.484)$$
$$= 0.00209$$

Nose:

$$\Delta C_{D_{nose}} = \frac{(d \times w)}{S_{ref}} \Delta C_{D_S} = \frac{(14 \times 5)/144}{144.9}(0.242) = 0.00081$$

This amounts to about 20.9 drag counts due to the main gear and 8.1 counts due to the nose landing gear.

Drag of Retractable Landing Gear

Austyn-Mair and Birdsall [33] give the following empirical expressions for the additive drag of landing gear in the absence and presence of flaps. In other words, one expression applies to the landing gear down and flaps retracted, the other to the both the landing gear and flaps extended. The equations are based on historical data and are presented in terms of weight, W (in lb_f) in the UK system, or mass, m (in kg) in the SI system. They are representative of commercial jetliners and business jet, and should not be used with lighter GA aircraft. The drag of landing gear with the flaps stowed is given as follows in the SI and UK systems, respectively:

Flaps retracted:

$$\Delta C_{D_{RG}} = \begin{cases} \frac{5.698 \times 10^{-4} \cdot m^{0.785}}{S_{ref}} & \text{SI system} \\ \frac{0.003294 \cdot W^{0.785}}{S_{ref}} & \text{UK system} \end{cases} \quad (15\text{-}92)$$

If the flaps are fully deflected the expression becomes:

Flaps deployed:

$$\Delta C_{D_{RG}} = \begin{cases} \frac{3.099 \times 10^{-4} \cdot m^{0.785}}{S_{ref}} & \text{SI system} \\ \frac{0.001792 \cdot W^{0.785}}{S_{ref}} & \text{UK system} \end{cases} \quad (15\text{-}93)$$

EXAMPLE 15-13

Determine the additive drag coefficient of the retractable landing gear for an airplane that weighs 22000 lb_f and whose wing area is 300 ft^2, with and without flaps.

Solution

Using Equation (15-92) and (15-93) we get:

Without flaps:

$$\Delta C_{D_{wheel}} = \frac{0.003294 W^{0.785}}{S} = \frac{0.003294(22000)^{0.785}}{300}$$
$$= 0.02814$$

With flaps:

$$\Delta C_{D_{wheel}} = \frac{0.001792 W^{0.785}}{S}$$
$$= \frac{0.001792(22000)^{0.785}}{300}$$
$$= 0.01531$$

This amounts to 281 and 153 drag counts respectively.

Jenkinson et al. [34] give the following expression for the drag of the landing gear of commercial jetliners. Formulation for two classes of jetliners is given; for medium to large jetliners like the Boeing 747, 757, 767, DC-10, L-1011, etc. and for smaller jetliners, such as the F-100, DC-9, and B-737. The expressions are presented in terms of the flat plate area, $\Delta D/q$. However, these have been modified to adhere to the presentation in this book.

Medium to large jetliners:

$$\Delta C_{D_{RG}} = \begin{cases} \dfrac{0.00157 \cdot (m_{NLG}+m_{MLG})^{0.73}}{S_{ref}} & SI-system \\ \dfrac{0.0025 \cdot (W_{NLG}+W_{MLG})^{0.73}}{S_{ref}} & UK-system \end{cases}$$

(15-94)

Small jetliners:

$$\Delta C_{D_{RG}} = \begin{cases} \dfrac{0.00093 \cdot (m_{NLG}+m_{MLG})^{0.73}}{S_{ref}} & SI-system \\ \dfrac{0.006 \cdot (W_{NLG}+W_{MLG})^{0.73}}{S_{ref}} & UK-system \end{cases}$$

(15-95)

where W_{NLG} and W_{MLG} is the weight of the nose and main landing gear, respectively (in lb_f) and m_{NLG} and m_{MLG} is the mass of the nose and main landing gear, respectively (in kg).

Roskam [23] presents a simple method for determining the additive drag of retractable landing gear. It uses the ratio of the actual frontal area of the landing gear to the area of a rectangle enclosing the gear to calculate the additive drag coefficient (see Figure 15-51). The method assumes an open wheel well, but suggests a 7% reduction to correct for closed wells. The resulting formulation is presented below, for configurations with both open and closed wheel wells:

Open wheel wells:

$$\Delta C_{D_{RG}} = 0.05328 \cdot e^{5.615 \cdot S_A/(d \cdot w)} \dfrac{(d \times w)}{S_{ref}} \quad (15\text{-}96)$$

Closed wheel wells:

$$\Delta C_{D_{RG}} = 0.04955 \cdot e^{5.615 \cdot S_A/(d \cdot w)} \dfrac{(d \times w)}{S_{ref}} \quad (15\text{-}97)$$

Note that even though the drag coefficient is based on the aforementioned ratio of the actual frontal to enclosed area of one wheel, the value of $\Delta C_{D_{RG}}$ applies to both legs of the main landing gear (the nose landing gear is not included). Also note a common error is to forget to convert d and w (which often are in inches) to ft, to ensure unit consistency with S_{ref} in ft^2.

Drag of Nose Landing Gear

Drag of nose landing gear is presented in the graph in Figure 15-52. The method requires the distance of the gear from the nose, a, total length of the extended landing gear, e, and tire diameter, d, to be known. The ratio a/d is first determined and used to select the appropriate

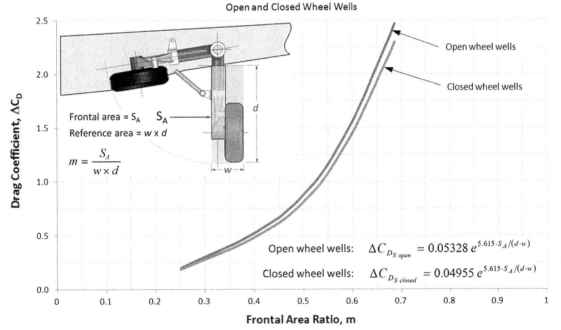

FIGURE 15-51 Estimating the drag contribution of a retractable landing gear (based on Ref. [23]).

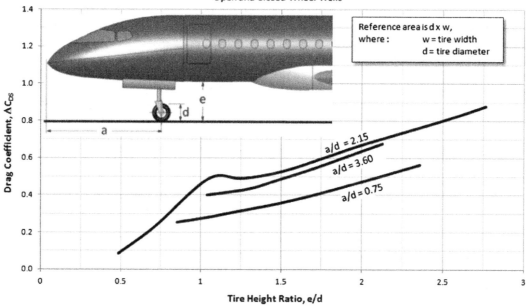

FIGURE 15-52 Estimating the drag contribution of a nose landing gear (based on Ref. [23]).

curve. Then the wheel height ratio, e/d, is calculated and used with the selected curve to read the drag coefficient on the vertical axis. Then, this can be converted into the additive drag coefficient, which is based on the reference wing area as follows:

$$\Delta C_{D_{nose}} = \frac{(d \times w)}{S_{ref}} \Delta C_{D_S} \qquad (15\text{-}98)$$

15.5.7 Drag of Floats

Floats are a popular option for many types of GA aircraft. Their primary drawback is a healthy dose of additive drag, inherent destabilizing moment, and weight. This section presents a method to estimate the drag. Reference [35] investigated the drag of four full-scale floats used as a single float for large single-engine military aircraft, such as the Vought OS2U Kingfisher. The floats varied in length, ranging from about 24.8 to 26.6 ft, with maximum cross-sectional area ranging from 6.63 to 9.25 ft². While this is larger than what is typically used for most single engine GA aircraft, the results are ideal to use for estimating the drag of smaller floats. Among conclusions cited by the reference is that the step adds about 10% of the drag, adding a tail fairing reduced it by some 8%, and there was negligible benefit gained by using counter-sunk rivets versus normal universal-head rivets aft of the step.

The following expressions refer to the configurations A, B, and C, shown in Figure 15-53. The term A_{max} is the maximum cross-sectional area of the float and α is the AOA with respect to the horizontal upper surface. No provision was made to reduce the drag specifically, e.g. by removing hardware. The drag measurements included support fairings, except wire bracings were not present. The Reynolds number is 25 million, based on the length of the float. The drag was measure using an α-sweep from $-6°$ to $+6°$ at an airspeed of 87 KTAS. The drag applies to a single float only, so if using two floats the additive drag coefficient must be doubled:

Reference [36] investigated special NACA-designed floats, referred to as the NACA 57 series. It was found that the dead rise angle (the angle between the horizontal and the "v" of the float) had a small effect on the total drag: the greater the dead rise angle the higher the drag. The drag coefficients increase with the AOA similar to that reflected by Equation (15-99); however, the minimum drag (ΔC_{DS}) is less or about 0.13 to 0.155.

Float A:

$$\Delta C_{D_{float}} = \left(\frac{A_{max}}{S_{ref}}\right)(0.00165\alpha^2 + 0.00413\alpha + 0.2142)$$
(15-99)

Float B:

$$\Delta C_{D_{float}} = \left(\frac{A_{max}}{S_{ref}}\right)(0.00109\alpha^2 + 0.00052\alpha + 0.1771)$$
(15-100)

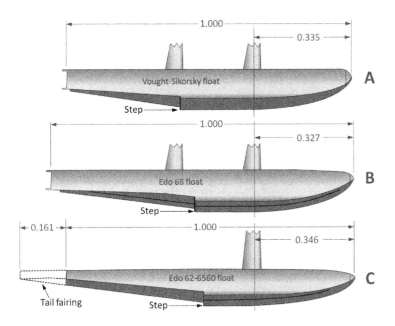

FIGURE 15-53 Float geometries (based on Ref. [35]). The drag of the float styles denoted by "A", "B", and "C" is presented in Equations (15-99), (15-100), and (15-101), respectively.

Float C:

$$\Delta C_{D_{float}} = \left(\frac{A_{max}}{S_{ref}}\right)(0.00176\alpha^2 - 0.00105\alpha + 0.1981)$$

(15-101)

15.5.8 Drag of Deployed Flaps

Deflecting flaps will introduce two modifications to the drag model: (1) the minimum drag will increase and (2) the lift induced drag will increase. This section introduced methods to account for this change.

Increase of C_{Dmin} Due to Flaps

Young [37] presents an empirical method to estimate the drag of a number of types of flaps. The estimation depends on the flap type, flap chord (C_f), deflection angle (δ_f), and its span (b_f). The following expression is used for this estimation and it requires the two functions Δ_1 and Δ_2 to be determined. The former accounts for the contribution of the flap chord to the flap drag and the latter for the contribution of the flap deflection.

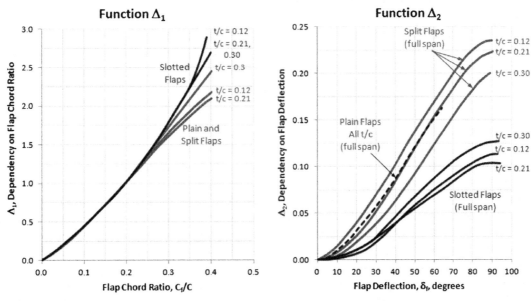

FIGURE 15-54 Estimation of the drag contribution of the flaps calls for the Δ_1 and Δ_2 functions to be determined (based on Ref. [37]).

TABLE 15-12 Polynomial Representations of the Function Δ_1 for Split, Plain, and Slotted Flaps

t/c	Function ($R_f = C_f/C$)
SPLIT AND PLAIN FLAPS	
0.12	$\Delta_1 = -21.090 R_f^3 + 14.091 R_f^2 + 3.165 R_f - 0.00103$
0.21	$\Delta_1 = -19.988 R_f^3 + 12.682 R_f^2 + 3.363 R_f - 0.0050$
0.30	$\Delta_1 = 4.6945 R_f^2 + 4.3721 R_f - 0.031$
SLOTTED FLAPS	
0.12	$\Delta_1 = 179.32 R_f^4 - 111.6 R_f^3 + 28.929 R_f^2 + 2.3705 R_f - 0.0089$
0.21, 0.30	$\Delta_1 = 8.2658 R_f^2 + 3.4564 R_f + 0.0054$

These are detailed in the two graphs of Figure 15-54, and are discussed further below.

$$\Delta C_{D_{flap}} = \Delta_1 \Delta_2 \left(\frac{S_{flap}}{S_{ref}} \right) \qquad (15\text{-}102)$$

The Function Δ_1

As stated earlier, the Δ_1 function represents the contribution of the flap chord to the flap drag. This contribution is shown in the left image of Figure 15-54, for plain, split, and slotted flaps, for airfoils of three thickness-to-chord ratios; 0.12, 0.21, and 0.30. These graphs have been digitized from the original document and can be reconstructed using the polynomial curve-fits shown in Table 15-12. Note that the parameter R_f is the flap chord ratio, i.e. R_f = flap chord/wing chord = C_f/C. This ratio typically ranges from 0.20 to 0.35.

The Function Δ_2

The Δ_2 function represents the contribution of the flap deflection angle to the flap drag, shown in the right image of Figure 15-54, for plain, split, and slotted flaps, for airfoils of three thickness-to-chord ratios; 0.12, 0.21, and 0.30. These are given by the polynomial curve-fits shown in Table 15-13.

Corke [38] presents additive drag for several types of flaps at the 30° and 50° deflection. These are reproduced in Table 15-14. It is a limitation that these apply to the specific flap span and chord of 60% and 25%, respectively. However, these can come in handy for initial estimation of flap drag.

15.5.9 Drag Correction for Cockpit Windows

The fuselage form factor estimation methods presented earlier assume a fuselage with a smooth round forward geometry that provides a smooth contour for air to flow across. Such smoothness is normally not achieved in airplanes; they often feature a sharp discontinuity around the cockpit windows. Cockpit windows are often flat, in particular in pressurized aircraft, which have heated windscreens for improved bird-strike resistance and to minimize optical distortion.

TABLE 15-13 Polynomial Representations of the Function Δ_2 for Split, Plain, and Slotted Flaps

t/c	Function
SPLIT FLAPS	
0.12	$\Delta_2 = -4.161 \times 10^{-7} \delta_f^3 + 5.5496 \times 10^{-5} \delta_f^2 + 1.0110 \times 10^{-3} \delta_f - 2.219 \times 10^{-5}$
0.21	$\Delta_2 = -5.1007 \times 10^{-7} \delta_f^3 + 7.4060 \times 10^{-5} \delta_f^2 - 4.8877 \times 10^{-5} \delta_f + 8.1775 \times 10^{-4}$
0.30	$\Delta_2 = -3.2740 \times 10^{-7} \delta_f^3 + 5.598 \times 10^{-5} \delta_f^2 - 1.2443 \times 10^{-4} \delta_f + 5.1647 \times 10^{-4}$
PLAIN FLAPS	
0.12	$\Delta_2 = -3.795 \times 10^{-7} \delta_f^3 + 5.387 \times 10^{-5} \delta_f^2 + 6.843 \times 10^{-4} \delta_f - 1.4729 \times 10^{-3}$
SLOTTED FLAPS	
0.12	$\Delta_2 = -3.9877 \times 10^{-12} \delta_f^6 + 1.1685 \times 10^{-9} \delta_f^5 - 1.2846 \times 10^{-7} \delta_f^4 + 6.1742 \times 10^{-6} \delta_f^3 +$ $-9.89444 \times 10^{-5} \delta_f^2 + 6.8324 \times 10^{-4} \delta_f - 3.892 \times 10^{-4}$
0.21	$\Delta_2 = -4.6025 \times 10^{-11} \delta_f^5 + 1.0025 \times 10^{-8} \delta_f^4 - 9.8465 \times 10^{-7} \delta_f^3 + 5.6732 \times 10^{-5} \delta_f^2 +$ $-2.64884 \times 10^{-4} \delta_f + 3.3591 \times 10^{-4}$
0.30	$\Delta_2 = -3.6841 \times 10^{-7} \delta_f^3 + 5.3342 \times 10^{-5} \delta_f^2 - 4.1677 \times 10^{-3} \delta_f + 6.749 \times 10^{-4}$

TABLE 15-14 Additive Flap Drag Coefficients (Based on Ref. [38]): Assumes 60% Flap Span and 25% Chord

Flap Type	Reference Deflection, δ_f	$\Delta C_{D_{flaps}}$
Split or plain flap	30°	0.05
	50°	0.10
Slotted flap	30°	0.02
	50°	0.05
Fowler flap	30°	0.032
	50°	0.083

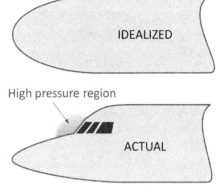

FIGURE 15-55 Idealized and actual forward geometry of typical aircraft.

Drag of Conventional Cockpit Windows

A consequence of this discontinuity is a high-pressure region, caused by the reduction in airspeed over the geometry. This increases the drag of the airplane, which is corrected using an additive drag contribution. Reference [23] presents such a method, assuming the general geometry shown in Figure 15-55. The drag coefficients refer to the maximum frontal area of the fuselage (A_{max}), which can be estimated using a method such as that shown in Figure 15-35. The source drag coefficients are given in Table 15-15.

$$\Delta C_{D_{window}} = \Delta C_{D_s} \frac{A_{max}}{S_{ref}} \quad (15\text{-}103)$$

TABLE 15-15 Additive Drag Coefficients for Cockpit Windows

Description (Drag Refers to the Fuselage Frontal Area)	ΔC_{D_s}
Flat windscreen with a protruding frame	0.016
Flat windscreen with a flush frame	0.011
Curved windscreen with a sharp upper edge	0.005
Curved windscreen with a round upper edge	0.002

EXAMPLE 15-14

Estimate the additive drag of the SR22 due to the cockpit window, assuming a curved windscreen with a round upper edge. The fuselage cross-sectional area is approximately 14 ft².

Solution

$$\Delta C_{D_{window}} = \Delta C_{D_s} \frac{A_{max}}{S_{ref}} = (0.002)\frac{14}{144.9} = 0.0001932$$

This amounts to about 1.93 drag counts.

Drag of Blunt Ordinary and Blunt Undercut Cockpit Windows

A cockpit window installation can be classified as blunt if the windscreen faces the oncoming airflow at angles ranging from normal to around 22° in the horizontal plane and ±20° from the vertical, as shown in the top and side views of Figure 15-56, respectively. The vertical angle is used to define the *ordinary* and *undercut* configurations. Generally, the blunt installation leads to higher drag than conventional curved geometry. In particular, note that the undercut installation, a popular approach in the 1930s to reduce the reflection of instrument lights at night (e.g. on the Boeing 247 [39] and the Vultee V-1), is a very draggy configuration and should be avoided by any means. The source drag coefficients shown in the figure are used with Equation (15-103).

15.5.10 Drag of Canopies

Many single- to four-seat aircraft are designed with canopies rather than roofed cabins to improve the field-of-view. This section presents a simple method to estimate the drag caused by such geometric protrusions. It is based on experimental data presented in Ref. [40],

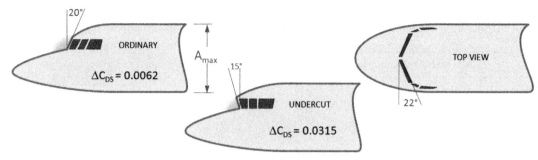

FIGURE 15-56 Drag of blunt and undercut cockpit windows (based on Refs [3] and [50]).

in which a number of dissimilar canopies were investigated at Mach numbers ranging from 0.19 to 0.71 and AOAs up to 6°. The geometric representations of the canopies are shown in Figure 15-57. The reference found that the contribution of a well-designed canopy to the total drag of the airplane was approximately 2% of the total drag, whereas a poorly designed one could easily exceed that 10-fold — and be 20% of the total drag. This shows that awareness of canopy drag is of great importance in the aircraft design. The drag is presented in terms of the maximum cross-sectional area of the window, A_{max}, using the following expression:

$$\Delta C_{D_{canopy}} = \Delta C_{D_S} \frac{A_{max}}{S_{ref}} \quad (15\text{-}104)$$

The values of the ΔC_{D_S} for the various canopy geometries are shown in Figure 15-58, plotted for the cited range of Mach numbers. This data is only provided for a low AOA, although information for higher AOAs (up to 6°) is provided in the reference document. The lower AOAs are likely to be of greater interest to the designer of efficient aircraft featuring canopies. However, if the design mission involves prolonged cruise at or near LD_{max} or high-speed maneuvers (where compressibility effects prevail), the higher AOAs become important. The reference also presents the distribution of pressure coefficients, allowing for the estimation of critical Mach number (see Section 8.3.7, *The critical Mach number, M_{crit}*).

15.5.11 Drag of Blisters

Hoerner [3, p. 8-5] collected information on drag for various shapes of blisters from a number of sources. Blisters are fairings that cover components that extend beyond the original outside mold line. Blisters are ideal when estimating the drag of GPS antennas. The drag of blisters can be calculated from:

$$\Delta C_{D_{blister}} = \Delta C_{D_S} \left(\frac{A_{max}}{S_{ref}} \right) \quad (15\text{-}105)$$

where A_{max} = cross-sectional area of shape as shown in Figure 15-59.

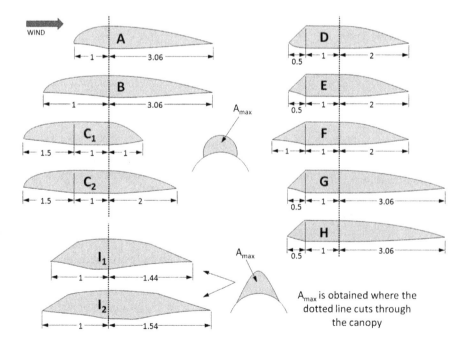

FIGURE 15-57 Canopy styles evaluated by Ref. [40]. The drag of the canopy styles denoted by "A" through "I" is presented in the graph of Figure (15-58).

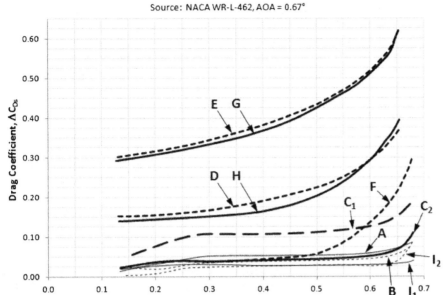

FIGURE 15-58 Drag coefficients for the canopy styles of Figure 15-57 (reproduced from Ref. [40]).

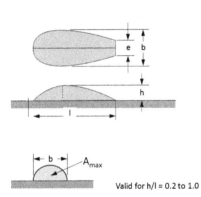
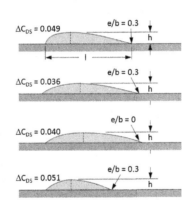

FIGURE 15-59 Drag contribution of typical blisters (bumps) (based on Ref. [3]).

EXAMPLE 15-15

Estimate the additive drag of a GPS blister antenna for the SR22, assuming a length × width × height of 4.7 × 3.0 × 0.78 inches. Assume the base (e) is 30% of the width (b) and that its side view resembles that of the top blister in Figure 15-59. Furthermore, assume a maximum cross-sectional area of 14.1 in².

Solution

The height ratio is about $0.78/4.7 = 0.17$, so it is a tad outside the limits cited in Figure 15-59. However, since it is fairly close, we will assume the drag can indeed be estimated using Equation (15-105).

$$\Delta C_{D_{blister}} = \Delta C_{D_S}\left(\frac{A_{max}}{S_{ref}}\right)$$
$$= 0.049\left(\left(\frac{14.1}{12 \times 12}\right)\bigg/144.9\right)$$
$$= 0.00003311$$

This amounts to approximately 0.33 drag counts. Note this antenna is about two times "draggier" than the wing shaped antenna of Example 15-9.

15.5.12 Drag Due to Compressibility Effects

Drag due to drag divergence was introduced in the Section 15.4.2, *The effect of Mach number*. The following method can be used to account for this drag increase at high Mach numbers, provided the three following parameters are known: (1) the critical Mach number, M_{crit}, (2) the maximum drag due to Mach drag divergence, $\Delta C_{Dmax\,D}$, and (3) the Mach number at which it occurs, $M_{max\,D}$ (see Figure 15-60). If these are known, the additional drag for Mach numbers ranging from 0 to $M_{max\,D}$ can be estimated using the following expression:

$$\Delta C_{D_M} = \frac{\Delta C_{DmaxD}}{2}(1 + \tanh(AM + B)) \quad (15\text{-}106)$$

where M is the Mach number and the constants A and B are determined using the following expressions:

$$A = \frac{\tanh^{-1}\left(\frac{2\Delta C_{DmaxD} - 0.0002}{\Delta C_{DmaxD}} - 1\right) - \tanh^{-1}\left(\frac{0.0002}{\Delta C_{DmaxD}} - 1\right)}{M_{maxD} - M_{crit}} \quad (15\text{-}107)$$

$$B = \tanh^{-1}\left(\frac{0.0002}{\Delta C_{DmaxD}} - 1\right) - AM_{crit} \quad (15\text{-}108)$$

The beauty of this method is that the compressible drag contribution can be used as an additive drag component for any Mach numbers as long as it is less than $M_{max\,D}$. This is convenient for programming or spreadsheet use, as it helps avoiding having to feature *IF* statements to control when this component is added to the incompressible drag. (Note that ΔC_{DM} is added to the sum of miscellaneous drag contributions like any other in this section.)

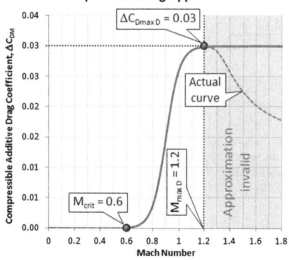

FIGURE 15-60 Method shown in action using $M_{crit} = 0.6$, $M_{maxD} = 1.2$, and $\Delta C_{DmaxD} = 0.03$.

Derivation of Equations (15-106) through (15-107)

Assume a spline function of the type:

$$\Delta C_{D_M} = \frac{\Delta C_{DmaxD}}{2}(1 + \tanh(AM + B))$$

Since the hyperbolic tangent has asymptotes at $y = -1$ and $y = 1$, it is necessary to divide ΔC_{DmaxD} by 2 (because the asymptotes are separated by a value of 2). The value "1" is used to shift tanh vertically, so the lower asymptote will be at $y = 0$, rather than $y = -1$. The function can be used as a spline to approximate the drag divergence by determining the constants A and B such the function goes through some specific points on the drag versus Mach curve (see Figure 15-28). These points are: (1) M_{crit}, where the drag begins to rise and (2) M_{max}, where it reaches its maximum value, ΔC_{DmaxD}. To work around the lower asymptote, we assume a very small increase in ΔC_{DM} at M_{crit}; Here we will assume 1 drag count or 0.0001. To work around the upper asymptote, where ΔC_{DM} reaches its maximum value (ΔC_{DmaxD}) we assume its value is $\Delta C_{DM} = \Delta C_{DmaxD} - 0.0001$. Therefore, we can write:

At $M = M_{crit}$:

$$\Delta C_{D_M} = \frac{\Delta C_{DmaxD}}{2}(1 + \tanh(AM_{crit} + B)) = 0.0001$$

At $M = M_{max\,D}$:

$$\Delta C_{D_M} = \frac{\Delta C_{DmaxD}}{2}(1 + \tanh(AM_{maxD} + B))$$
$$= \Delta C_{DmaxD} - 0.0001$$

We readily see that we can solve for the argument of the hyperbolic tangent as follows:

$$AM + B = \tanh^{-1}\left(\frac{2\Delta C_{D_M}}{\Delta C_{DmaxD}} - 1\right)$$

This allows us to write:

At $M = M_{crit}$:

$$AM_{crit} + B = \tanh^{-1}\left(\frac{2\Delta C_{D_M}}{\Delta C_{DmaxD}} - 1\right)$$
$$= \tanh^{-1}\left(\frac{0.0002}{\Delta C_{DmaxD}} - 1\right)$$

At $M = M_{max\,D}$:

$$AM_{maxD} + B = \tanh^{-1}\left(\frac{2\Delta C_{D_M}}{\Delta C_{DmaxD}} - 1\right)$$
$$= \tanh^{-1}\left(\frac{2\Delta C_{DmaxD} - 0.0002}{\Delta C_{DmaxD}} - 1\right)$$

Solving for A and B, thus, yields Equations (15-107) and (15-108).

QED

15.5.13 Drag of Windmilling and Stopped Propellers

Drag Due to Windmilling Propellers

A windmilling propeller is usually associated with an engine failure in flight. Compounding the loss of thrust is a large amount of drag added to the airplane. This can create a serious problem for its continued operation. For single-engine aircraft, the added drag is a serious detriment to its glide capability. For multi-engine aircraft it severely reduces range and contributes to the asymmetric moment that must be balanced by rudder and aileron deflection.

The rotating propeller acts as a wind turbine that drives the engine and must develop enough torque to overcome the internal friction of the engine. A thorough analysis of the phenomenon is beyond the scope of this book, but some experimental data is provided in Ref. [41]. The following methods are provided for initial estimation only.

Hoerner [3] suggests a method in which the power required to turn the engine is 10% of its rated power and the propeller efficiency expected through the windmilling is of the order of 50%. In other words, 50% of the drag power is converted into the rotational power. This allows us to write:

$$0.1 \times 550 \times BHP_{max} = 0.5 \times D_{windmill} \times V$$

Assuming the windmilling drag to depend on dynamic pressure and propeller disk, we can write:

$$D_{windmill} = \frac{1}{2}\rho V^2 A_{prop} C_{DS} = \frac{1}{2}\rho V^2 S_{ref} \Delta C_{Dwindmill}$$

Inserting this into the previous expression, solving for the drag coefficient and referencing it to the reference area yields:

$$\Delta C_{Dwindmill} = \frac{220 \times BHP_{max}}{\rho V^3 S_{ref}} \quad (15\text{-}109)$$

where V is the glide speed and ρ the density at condition. Comparison to existing aircraft reveals that Equation (15-109) over-predicts the additive drag by a factor of 2 to 4. For this reason, for small engines with low internal friction multiply the value by 0.25 and for larger engines with a high internal friction multiply by 0.33. The author's unpublished study of single-engine aircraft reveals that windmilling propellers increased drag by approximately:

$$\Delta C_{Dwindmill} \approx 0.0150 \quad (15\text{-}110)$$

with several excursions nearing and even exceeding 300 drag counts! The above value will easily reduce the LD_{max} of a sleek airplane by a whopping one-third.

Drag Due to Stopped Propellers

If the propeller of a malfunctioning engine does not windmill, but stops, the drag will indeed be less. It is possible to estimate the drag of such propellers based on the planform area of the blades, assuming they generate drag similar to a flat plate at a specific blade angle. In this context, the blade angle is defined as the angle between the chord of the blade airfoil at 0.7 radius to the rotation plane. The angle is close to the pitch angle of the airfoil (e.g. see Figure 14-9). Thus, when the angle is zero, the blade drag is high and when low, the drag is much lower. Hoerner [3], citing experiments from Ref. [41], provides the following expression to estimate the drag coefficient of such a blade:

$$C_{DS} = 0.1 + \cos^2 \beta$$

where β is the blade angle at the 0.7 radius and which may or may not be equal to the pitch angle referenced in Chapter 14, *The anatomy of the propeller*. This shows the blade drag coefficient varies between 0.1 and 1.1. This allows the drag of the stopped propeller, denoted by D_{sp}, to be written as follows:

$$D_{sp} = \frac{1}{2}\rho V^2 N_{blades} S_{blade} C_{DS} = \frac{1}{2}\rho V^2 S_{ref} \Delta C_{Dsp}$$

where N_{blades} is the number of blades and S_{blade} is the planform area of the blade. Solving for the drag coefficient that uses the reference area (S_{ref}) and is denoted by ΔC_{Dsp}, yields the following expression:

$$\Delta C_{Dsp} = \left(\frac{S_{blade}}{S_{ref}}\right) N_{blades} C_{DS}$$
$$= \left(\frac{S_{blade}}{S_{ref}}\right) N_{blades} (0.1 + \cos^2 \beta) \quad (15\text{-}111)$$

15.5.14 Drag of Antennas

There are typically three kinds of antenna geometries planted on aircraft: (1) blister type, (2) wire type, and (3) wing type. Their drag can be estimated using the methods already presented in here. Typical placement and shapes of such antennas is shown in Figure 15-61. Their number can easily turn your nice, smooth airplane into something resembling a porcupine! If possible, try to mount as many antennas internally as practical, although often this is impossible due to reduced effectiveness of the antenna. Also, ask the manufacturer for an additive drag coefficient associated with their antenna: some have this readily available — it may save you analysis work. Others do not and for those you will have to estimate the drag based on the geometry using the approximations below.

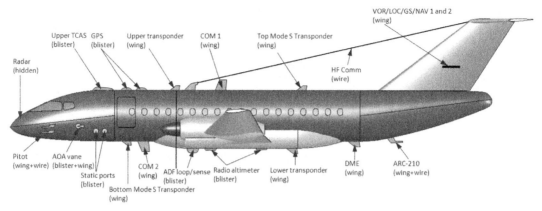

FIGURE 15-61 Typical placement and shape of various antennas and flight control components. Actual size is exaggerated for clarity.

- For wire antennas perpendicular to the airstream use Equation (15-112).
- For wire antennas at an angle θ to the airstream use Equation (15-115).
- For blister antennas use Equation (15-105).
- For wing antennas use Equation (15-84).

15.5.15 Drag of Various Geometry

Figure 15-62 shows the three-dimensional drag coefficient for a sphere and a circular cylinder, based on Schlichting [4]. The drag coefficient is defined using Equation (15-78), where S_S is the cross-sectional area ($\pi D^2/4$ for the sphere and $D \cdot L$, where L is the length of the cylinder). Schematics showing the nature of the separation have been superimposed to demonstrate how the drag coefficient depends on the nature of the flow separation that occurs. The dashed line indicates a specific region where the said flow nature takes place. Of importance is the sharp dip near a R_e of 300,000 (sphere) and 500,000 (cylinder), first discovered by Alexander Gustave Eiffel (1832–1923).[5] This dip is indicative of the formation of a turbulent boundary layer that better follows the geometric shape of the solid, reducing the size of the wake and, thus, the pressure drag generated by the object. The figure also shows the relatively constant drag coefficient of geometry over the range $10^3 < R_e < 10^5$, but this is indicative of the

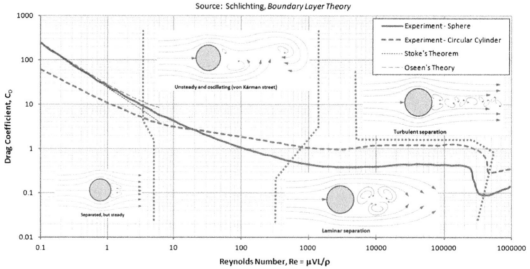

FIGURE 15-62 The drag coefficient of a sphere and a circular cylinder as a function of R_e. Inserts are schematics showing the nature of the separation, whose consequence is the C_D shown in the graph.

---

[5]Eiffel is considered by many to be the father of aerodynamics, while others place the honor on Cayley.

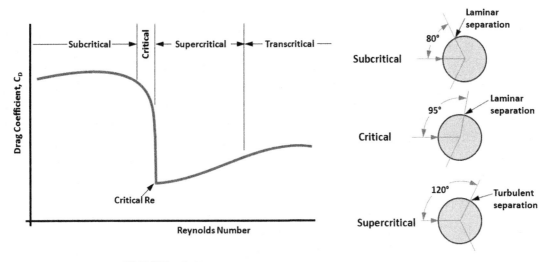

FIGURE 15-63 The classification of flow regions on a sphere.

formation of a laminar boundary layer (as long as the surface is smooth).

The range of R_e starting at 10^3 and above is of great interest to the aircraft designer because this covers most airplanes, even small radio-controlled aircraft. This region is usually broken into four separate sub-regions called *subcritical, critical, supercritical,* and *transcritical* (see Figure 15-63).

In the subcritical region, laminar boundary layer is formed that separates once it flows to latitude of approximately 80° (see the schematic to the right in Figure 15-63). As stated above, the C_D in this range ($10^3 < R_e < 10^5$) is practically constant (i.e. independent of the R_e). In the critical region, the C_D drops sharply over a relatively narrow range of R_e, reaching a minimum value called the *critical Reynolds number*. This drop is caused by a sudden movement of the laminar boundary layer separation point to latitude of almost 95°. At the critical R_e, a separation bubble is formed at this location that forces the laminar boundary layer to separate into a turbulent one. This, in turn, can more easily follow the shape of the geometry, moving the separation point farther downstream to approximately 120°. This dramatically reshapes the separation region, reducing its diameter and consequently the pressure drag, which explains the reduction in the drag coefficient. In the supercritical region, laminar-to-turbulent transition occurs in the attached boundary layer, causing the separation to begin to crawl upstream, slowly increasing the drag coefficient again. In the transcritical region the transition point has moved upstream closer to the stagnation point, eventually causing the drag to become independent of the R_e.

Figure 15-64 and Figure 15-66 show drag coefficients for selected geometry reproduced from Hoerner [3].

Three-Dimensional Drag of Two-Dimensional Cross-Sections of Given Length

Research shows that the drag coefficients for the two-dimensional shapes shown in Figure 15-64 depends on their fineness ratio, here denoted by h/d (shown on the triangular shape in the center, lower row). However,

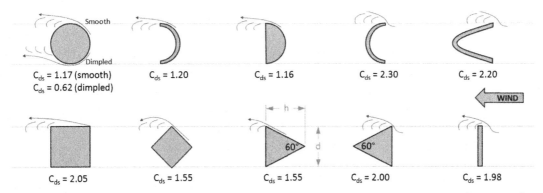

FIGURE 15-64 Two-dimensional drag coefficients of several cross-sections. Valid only for $10^4 < R_e < 10^5$.

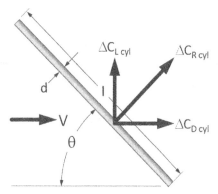

FIGURE 15-65 Dimensions for the applications of the cross-flow principle.

their drag is determined using their frontal areas. Thus, the drag of the shapes is calculated from:

$$\Delta C_{D_{2D-shape}} = \Delta C_{d_S}\left(\frac{d \times l}{S_{ref}}\right) \quad (15\text{-}112)$$

where d = thickness of shape as shown in Figure 15-64 and l = length of shape in the out-of-plane direction.

The Cross-Flow Principle

Hoerner [3] presents a very practical method to calculate the drag of wires that are inclined with respect to the airflow (see Figure 15-65). This is referred to as the *cross-flow principle*. It can be used to estimate the drag and lift of a tube or cylinder of a given length, l, and particular cross-section, whose two-dimensional drag coefficient is known. The formulation below is used to calculate the coefficients in terms of the reference wing area so it (primarily the drag) can be added directly to the miscellaneous drag coefficient.

Note that the absolute sign in Equation (15-115) guarantees the drag is always greater than zero. Also note that the inclination angle of $\theta = 90°$ means the cylinder is perpendicular to airstream.

The cross-flow principle is very helpful in determining the drag of external aircraft components, such as HF radio wire antennas.

Resultant coefficient:

$$\Delta C_{R_{cyl}} = \Delta C_{d_S}\left(\frac{d \times l}{S_{ref}}\right)\sin^2\theta \quad (15\text{-}113)$$

Lift coefficient:

$$\Delta C_{L_{cyl}} = \Delta C_{d_S}\left(\frac{d \times l}{S_{ref}}\right)\sin^2\theta\cos\theta \quad (15\text{-}114)$$

Drag coefficient:

$$\Delta C_{D_{cyl}} = \Delta C_{d_S}\left(\frac{d \times l}{S_{ref}}\right)\cdot\left|\sin^3\theta\right| \quad (15\text{-}115)$$

where θ = the angle of inclination (see Figure 15-65).

Drag of Three-Dimensional Objects

Figure 15-66 shows a number of three-dimensional objects and the corresponding drag coefficients. The drag of these objects is also based on the cross-sectional area normal to the flow direction. Thus, the cross-sectional area normal to the flow direction for the sphere is given by $\pi\cdot d^2/4$, where d is its diameter. In general, if S_N denotes this area (i.e. $S_N = \pi\cdot d^2/4$), the three-dimensional drag coefficient is calculated from:

$$\Delta C_{D_{3D-shape}} = \Delta C_{D_S}\left(\frac{S_N}{S_{ref}}\right) \quad (15\text{-}116)$$

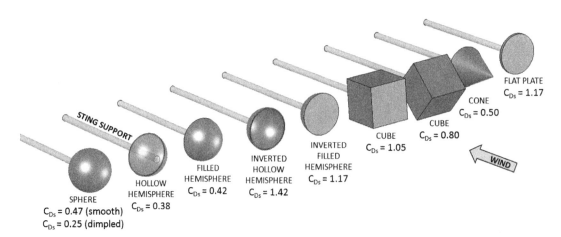

FIGURE 15-66 Three-dimensional drag coefficients of various geometric shapes. Valid only for $10^4 < R_e < 10^5$.

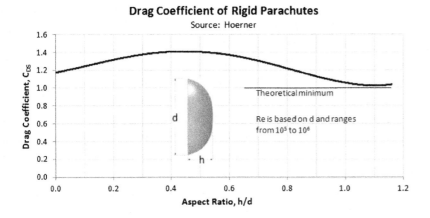

FIGURE 15-67 Drag coefficients of rigid objects shaped like parachutes. Valid only for $10^5 < R_e < 10^6$ and $h/d < 1.1$ (reproduced from Ref. [3]).

15.5.16 Drag of Parachutes

While the drag of parachutes may appear simple, in reality it is surprisingly complicated. Accurate estimation of parachute drag and the time history of the drag generated during deployment is a serious scientific discipline, applicable to a range of applications, including the deployment of re-entry parachutes or ejection seats.

Hoerner [3] provides a practical insight into the drag of parachutes. In general, the drag coefficient of inflated parachutes is based on simple geometric features, such as the height, diameter, and the projected frontal area of the inflated canopy. For initial sizing, the drag coefficient, $C_{D_{parachute}}$, can be estimated based on the aspect ratio (AR) of the parachute, defined as its inflated height, h, by the inflated diameter, d. This is shown in the graph of Figure 15-67, which shows the inflated drag coefficient as a function of h/d. An empirical expression based on the graph is given below. It is valid only for $h/d < 1.1$ and $10^5 < R_e < 10^6$, where R_e is based on the inflated diameter. The maximum drag coefficient is obtained for an AR of 0.5, which represents a hemi-spherical geometry. Further increase of the AR will make the parachute partially "fill in" the flow separation region, which will reduce the drag coefficient until it reaches a theoretical minimum of 1.0.

$$C_{D_{parachute}} = 2.239\left(\frac{h}{d}\right)^4 - 4.202\left(\frac{h}{d}\right)^3 + 1.227\left(\frac{h}{d}\right)^2 + 0.6167\left(\frac{h}{d}\right) + 1.174$$

(15-117)

The coefficient is then used to evaluate the total drag force of the parachute using the following expression, where S is the projected area of the chute and $S = \pi \cdot d^2/4$:

$$D = qSC_{D_{parachute}} = \frac{1}{2}\rho V^2 SC_{D_{parachute}} \qquad (15\text{-}118)$$

EXAMPLE 15-16

The POH for the SR22 states that the "the airplane will descend at less than 1700 feet per minute with a lateral speed equal to the velocity of the surface wind." The company website states the diameter of the canopy is 55 ft. This means that, at the gross weight of 3400 lb$_f$, the airplane will not exceed 1700 fpm (or 28.3 ft/s). Using this information, determine the drag coefficient of the parachute.

Solution

Solve for $C_{Dparachute}$ using Equation (15-118):

$$C_{D_{parachute}} = \frac{2D}{\rho V^2 S} = \frac{2W}{\rho V^2 S}$$
$$= \frac{2(3400)}{(0.002378)\left(\frac{1700}{60}\right)^2\left(\frac{\pi}{4}55^2\right)} = 1.499$$

Note that the Reynolds number is close to 10^7.

15.5.17 Drag of Various External Sources

This section presents the drag of various external sources that are important when present on the design.

Drag of Sanded Walkway on Wing

As shown in Table 15-8, Ref. [30] indicates adding a sanded walkway (both sides of the fuselage) adds 0.0007 drag counts to the total drag. For this reason assume the following drag increase per side:

$$\Delta C_{D_{walkway}} = 0.00035 \quad (15\text{-}119)$$

Drag of Gun Ports in the Nose of an Airplane

NACA WR-L-502 [42] presents the results from drag analysis of the introduction of openings for eight machine gun barrels in a P-38 Lightning style fuselage. It found the drag increase amounted to 5 drag counts or 0.0005 total, based on the wing reference area. Based on this result it is possible to estimate drag increase per opening as follows:

$$\Delta C_{D_{Gun\ Port}} = 0.0000625 \quad (15\text{-}120)$$

Drag of Streamlined External Fuel Tanks

External fuel tanks are a typical supply of additional fuel for military aircraft. However, they are also a plausible solution to a long-range operation of some GA aircraft. A similar shape is often used to house weather radars for GA aircraft. It is for this reason their drag is presented here.

The drag of streamlined tanks is highly dependent on the interference of between the tank and the wing. Careless installation can easily increase drag by a factor of four, as shown in Figure 15-68. The installation should always be improved using streamlined fairings. The drag coefficients shown in the figure are based on the frontal area of the tank, S_{tank}, and are related to the reference area as follows:

$$\Delta C_{D_{i\ washout}} = 0.00004(\phi_{tip} - \phi_{MGC}) = 0.00004 \cdot \Delta\phi \quad (15\text{-}122)$$

Drag Due to Wing Washout

Horner presents the following expression to account for increase in the lift-induced drag of a wing as a consequence of wing washout. It has been shown in Chapter 8, *The anatomy of the wing*, how the lift-distribution is altered as wing twist is introduced. This can lead to an appreciable lift-induced drag, even when the wing is at an *AOA* for which no lift is generated.

$$\Delta C_{D_{i\ washout}} = 0.00004(\phi_{tip} - \phi_{MGC}) = 0.00004 \cdot \Delta\phi \quad (15\text{-}122)$$

where

ϕ_{tip} = angle of the tip with respect to the root of the wing, in degrees

ϕ_{MGC} = angle of the MGC with respect to the root of the wing, in degrees

So the angular difference is between the wingtip and the airfoil at the MGC.

Drag Due to Ice Accretion

Drag due to ice accretion on the aircraft as a whole poses a very serious challenge to safe flight. All aircraft can be classified as those that have been certified for *flight into known ice* (FIKI) and those that have not. Of course, all aircraft, regardless of classification, can accrete ice during operation. The problem of ice formation was studied at least as early as 1938 by Gulick [43], who found that the section drag coefficient of the airfoil studied almost doubled and the maximum lift coefficient reduced from 1.32 to 0.80, without changing the angle of stall. Further research took place in the early

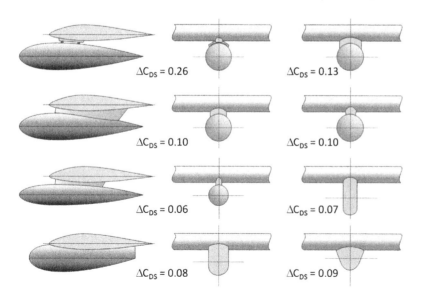

FIGURE 15-68 Drag of external fuel tanks (based on Ref. [3]). Note that the bottom configuration also resembles geometry often used to house wing mounted weather radars for small GA aircraft.

1950s (e.g. see a paper by Gray and Glahn [44]). Since then, tremendous research effort has been dedicated to the subject. In fact, the sheer volume of papers that has been published cannot be adequately presented here. The interested reader is directed toward work done by NASA and AIAA.

15.5.18 Corrections of the Lift-Induced Drag

Wingtip Correction

In addition to the effect of AR and λ, the lift-induced drag is also affected by the wingtip geometry. As soon as the low- and high-pressure regions form on the wing, respectively, a vortex begins to form at each wingtip (see Figure 15-23). High pressure on the lower surface of the wing moves in a spanwise direction outboard and "rolls" over the wingtip toward the low-pressure region on the upper wing. Generally, it is assumed that the distance between the core of the two vortices equals that of the wingspan. However, in reality the wingtip affects how the roll-up of the vortices takes place (see Figure 15-69). Thus, the three-dimensional flow field is modified by the wingtip geometry and this shifts the wingtip vortices inboard or outboard with respect to the wingtip. If the separation of the wingtip vortices is increased in this manner, it is akin to increasing the AR of the wing. Similarly, if the separation of the vortices is decreased, it is as if the AR has been reduced.

Hoerner [3, p. 7-5] presents a number of wingtips and their empirical effects on the AR of the wing, reproduced in Figure 15-70. The terms ΔAR are values that should be added to the geometric AR. The resulting value is then used when calculating lift-induced drag, C_{Di}. For instance, consider a wing whose AR is 7. The selection

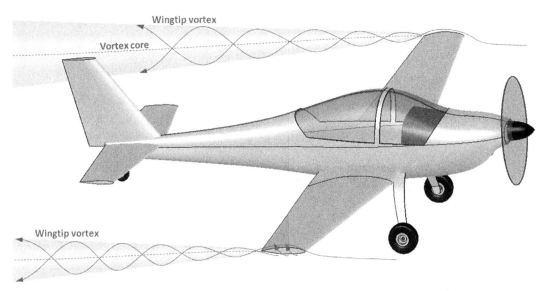

FIGURE 15-69 Pressure difference between the upper and lower surfaces forms the wingtip vortex, as the high-pressure field on the lower surface "rolls" over the wingtip toward the low-pressure field on the upper surface.

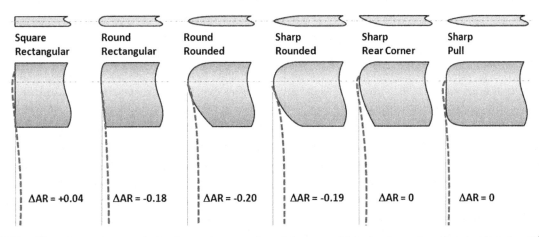

FIGURE 15-70 Effect of several types of wingtips on the separation of the cores of the wingtip vortices. Tested at $R_e = 1 \times 10^6$ and $AR = 3$ (reproduced from Ref. [3]).

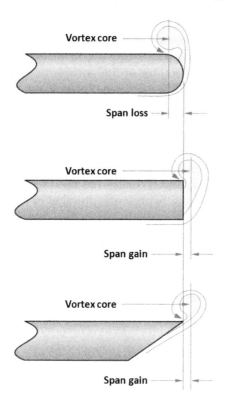

FIGURE 15-71 Location of vortex core for three different wingtip shapes.

of a round frontal view and round top view shows a $\Delta AR = -0.20$. Therefore, the AR to use with Equations (15-45) and (15-47) would be 6.8 and not 7. Generally, the figure shows that rounded tips reduce the effectiveness of the wing — it is simply better to feature a square rectangular tip. It is cheaper too. Other wingtips are presented in Section 10.5, *Wingtip design*.

The effect of how the wingtip shape influences the lift induced drag is based on how air flows around the wingtip. Three examples are shown in Figure 15-71. The top wingtip shape is round. It allows air to flow to the upper surfaces so the vortex core resides inside and on top of the wingtip. This results in a wingtip vortex that is closer to the plane of symmetry than the physical wingtip, effectively reducing the wingspan. The center wingtip is square. It forces the spanwise flow component of the lower surface to make the turn around a very sharp corner. This forces the vortex core to reside farther away from the plane of symmetry than the actual wingspan — it increases the effective wingspan, albeit by a fraction. The bottom wingtip is representative of the so-called Hoerner wingtip. It promotes spanwise flow outboard and upward that is forced to make a turn around a sharp corner.

This forces the vortex core to reside even farther outboard than the square wingtip in the middle. This idea is extended to the so-called upturned or downturned booster wingtip that helps place the vortex such that a small effective wingspan increase is achieved, although this is not always realized in practice.

Correction of Lift-Induced Drag in Ground Effect

As shown in Section 9.5.8, *Ground effect*, the lift-induced drag is reduced when the aircraft is operated close to the ground. This fact should be taken into account during T-O and landing analysis. The effect is favorable during the T-O ground run as the total drag of the airplane is reduced. It is unfavorable during the landing ground roll, for the same reason. Use any of the methods in the section to correct the lift-induced drag based on the height if the MGC above the ground.

EXAMPLE 15-17

(a) Evaluate the complete minimum drag coefficient of the SR22 assuming the results from the component drag build-up method of Example 15-6, and the various additive drag contributions evaluated in this section. How does it compare to the $C_{Dmin} = 0.02541$ calculated from published performance information?

(b) Perform this evaluation at other altitudes as well (only change will be in the skin friction). How will this affect the C_{Dmin}?

Solution

See Table 15-16 for the result and the discussion that follows:

The table shows the estimated drag is in reasonable agreement with the one reverse-engineered from published performance data, as it is $100 \times 0.02495/0.02541 = 98.2\%$.

There are many areas that one can argue could be refined; for instance the airplane's tie-down rings and control surface gaps are not included. These would add to the total. On the other hand, it is debatable whether a 20 drag counts penalty due to the presence of the engine cowling is justifiable (see row 44 in the table). This penalty is attributed to the fact the front part of the airplane features a cowling with an inlet and exit, which generates substantially higher drag than the smooth nose shape assumed by the fuselage skin friction estimation of Example 15-6. The 20 drag counts in row 44 were based on an airplane (the P-35) that has a radial engine. This should be expected to have higher drag than the horizontally opposed piston engine of the SR22. The important point is that while a careful prediction ought to put one in the neighborhood of the drag obtained by experiment, any contribution should be carefully reviewed and justified.

EXAMPLE 15-17 (cont'd)

(b) Since the drag analysis was prepared in a spreadsheet, it is easy to change the altitude and get a new estimate. This has been done for altitudes ranging from S-L to 14,000 ft and this is shown in the left graph of Figure 15-72. The flight test value of $C_{Dmin} = 0.02541$ is shown as the vertical dashed line. Interestingly, this value was obtained using cruise data at 8000 ft (see Example 15-18). The right graph of Figure 15-72 shows the contribution of various sources of drag at two airspeeds. Note the large contribution of the lift-induced drag the lower airspeed.

TABLE 15-16 Drag Analysis

			Wing	HT	VT	Fuselage	
	DRAG ANALYSIS						
39	Sweep angle	$\Lambda_{t\,max} =$	0.00	5.00	18.00		°
40	Form Factor	FF =	1.311	1.203	1.187	1.399	
41	Intereference Factor, IF	IF =	1	1.05	1.05	1	
42	Weighted drag factor	$C_{fi} \cdot FF \cdot IF \cdot S_{wet\,i} =$	0.7140	0.1762	0.1120	0.7607	
43	Skin friction drag	$(1/S_{ref}) \sum C_{fi} \cdot FF \cdot IF \cdot S_{wet\,i} =$	0.01217				
44	Presence of engine cowling	$\Delta C_{Dcowl} =$	0.002000	Same penalty as that of the Severski P-35			
45	Trim drag, 20% Static Margin	$\Delta C_{Dtrim} =$	0.000358	See Example 14-6			
46	Cooling drag	$\Delta C_{Dcool} =$	0.000812	See Example 14-7			
47	Two COM antennas	$2 \times \Delta C_{Dfin} =$	0.000030	See Example 14-8			
48	Four half-sized "COM" antennas	$4 \times 0.5 \times \Delta C_{Dfin} =$	0.000030	See Example 14-8			
49	Two GPS antennas	$2 \times \Delta C_{Dblister} =$	0.000066	See Example 14-14			
50	Two wing entry steps	$\Delta C_{Dstrut} =$	0.000456	See Example 14-10			
51	6 Flap fairings (double "COM" drag)	$6 \times 2 \times \Delta C_{Dfin} =$	0.000182	Each fairing assumed 2x the COM geometry			
52	External wing tip nav lights	$2 \times \Delta C_{Dblister} =$	0.000066	Assumed to cause same drag as GPS ant.			
53	Main landing gear	$\Delta C_{Dmain} =$	0.002090	See Example 14-11			
54	Nose landing gear	$\Delta C_{Dnose} =$	0.000810	See Example 14-11			
55	Cockpit window drag correction	$\Delta C_{Dwindow} =$	0.000193	See Example 14-13			
56	Sanded walkway, both sides	$\Delta C_{Dwalkway} =$	0.000700				
57	Miscellaneous drag coeff.	$C_{D\,misc} =$	0.00780				
58	CRUD	CRUD =	1.25				
59	Minimum drag coefficient	$C_{Dmin} =$	0.02495	$= (0.01217 + 0.00780) \cdot 1.25$			

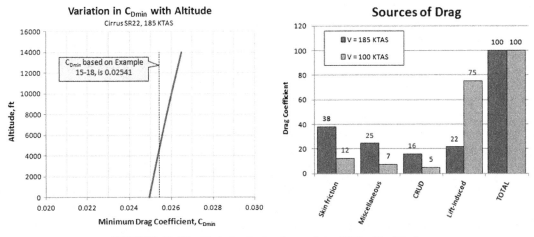

FIGURE 15-72 Variation in the C_{Dmin} of the SR22 with altitude.

15.6 SPECIAL TOPICS INVOLVING DRAG

Sometimes it is necessary to extract drag (reverse-engineer) from existing airplanes to validate the estimates of a new design of a similar geometry. This section presents several methods for this purpose. Of course there are several shortcomings to these methods and some extract less data than others. All the methods assume either the simplified or adjusted drag models of Section 15.2, *The drag model*. Drag characteristics featuring drag buckets require a large number of data and even wind tunnel testing and are not treated by the methods below. For this reason, applying the methods in this section to sailplanes is erroneous (even though one such is presented as an example later in this section).

The reader must use both caution and sound judgment when extracting numbers, as units, such as those for airspeed, are often ambiguous. Thus, the context of the units must be understood. Sometimes numbers come from advertising brochures which have been prepared by the marketing department. Such documents usually feature artistic presentation that is pleasing for the eyes, but sparse on details. They are often prepared by graphic artists who are not pilots themselves, let alone engineers. For instance, airspeed is commonly presented using knots. To such a person a knot is a knot. However, to an engineer (and pilots) a knot is not just a knot. There is a huge difference between an indicated knot (KIAS), a calibrated knot (KCAS), and a true knot (KTAS). The lack of detail in the preparation of advertising documents forces the engineer to apply sound judgment when using such numbers and the type of knot must be inferred from the type of airspeed.

If confronted by such a predicament the reader should be mindful that marketing departments thrive on extremes. They want to report the highest this or the lowest that, so they can separate their product from the competition. For instance, we want our airplane to have a low stalling speed and a high cruising speed. Knowing this, advertising brochures will report the stalling speed using KCAS because it is lower than the KTAS value (assuming the airplane is stalled at some altitude above ground). This is of course justifiable, because pilots want to know the indicated stall speed (KIAS) and by assuming low instrument and position error (approximately ±3 knots or so) one can get an idea of the airplane's low-speed capabilities. Therefore, stalling speeds reported as kts (e.g. 42 kts) in such brochures should be assumed to reflect knots calibrated airspeed (i.e. 42 KCAS). On the other hand, marketing departments want the cruising speed to appear as large as possible and will therefore use KTAS because that number is much larger than the corresponding KCAS (or the KIAS) value, in particular if the airplane cruises at high altitudes. Again, this is justifiable because the pilot or the customer may want to know how long it will take to fly certain distances and the true airspeed is needed for that assessment. Therefore, a cruising speed reported as kts (e.g. 180 kts) in such a brochure should be assumed to reflect knots true airspeed (i.e. 180 KTAS).

15.6.1 Step-by-step: Extracting Drag from L/D_{max}

The simplest method uses the L/D_{max} and the airspeed at which it is achieved. This information is commonly available from aircraft Pilot's Operating Handbooks (POH) and is almost always based on actual flight testing. The information this particular method extracts is the C_{Dmin}, and assumes the UK system of units. It is a limitation of the method that it uses the simplified drag model. However, since the C_{LminD} will only shift the drag polar sideways (see Section 15.2.2, *Quadratic drag modeling*) there really is no error introduced in the extraction of the C_{Dmin}. Another issue is the extraction of drag for propeller powered aircraft. The L/D_{max} reported in the POH will include drag from the windmilling propeller (since the purpose is to present the pilot with potentially life-saving information). This can easily double the minimum drag coefficient when compared to the other methods in this section and renders it much higher than required for accurate performance analyses. Therefore, care must be exercised when using these numbers.

Step 1: Gather Information from the Vehicle's POH

Assuming the user has access to the aircraft's POH, gather the following information: gross weight (W_0) in lb$_f$, best glide airspeed (V_{LDmax}), wing area (S) in ft^2, and wing aspect ratio (AR). If AR is not known, use wing span (b) in ft and compute it from b^2/S.

Step 2: Convert V_{LDmax} into Units of ft/s

Note 1: It is important that consistent units are used. Therefore, if V is read in KTAS is must be converted to ft/s. Similarly, if V_V is given in fpm (ft/min) it must be converted to ft/s. Use the following conversion factors:

Convert KTAS or mph to ft/s:

$$V_{ft/s} = 1.688 \times V_{KTAS} = 1.688 \times 1.15 \times V_{mph}$$

Convert fpm to ft/s:

$$V_{ft/s} = \frac{V_{fpm}}{60}$$

Note 2: Often the POH will report the best glide airspeed in units of KIAS or KCAS. This must be converted to KTAS for this method is to be applied at altitude (see Section 16.3.2, *Airspeeds: instrument, calibrated, equivalent, true, and ground airspeeds*, for methods).

Step 3: Calculate the Best Glide Lift Coefficient

Calculate the lift coefficient at the best glide speed from:

$$C_L = \frac{2W_0}{\rho V_{LDmax}^2 S}$$

Step 4: Calculate Span Efficiency

Estimate the span efficiency, e, from any of the methods of Section 9.5.14, *Estimation of Oswald's span efficiency*.

Step 5: Compute Minimum Drag

Compute the minimum drag from the following expression:

$$C_{Dmin} = \frac{C_L}{LD_{max}} - \frac{C_L^2}{\pi \cdot AR \cdot e} \qquad (15\text{-}123)$$

Derivation of Equation (15-123)

The simplified drag model is given by:

$$C_D = C_{Dmin} + \frac{C_L^2}{\pi \cdot AR \cdot e}$$

Knowledge of the lift-to-drag ratio at a specific condition (here conveniently selected to be the LD_{max}, since aircraft manufacturers so graciously report this for us), and the lift coefficient associated with it can then be used to extract the minimum drag coefficient.

$$LD_{max} = \frac{C_L}{C_D} = \frac{C_L}{C_{Dmin} + \frac{C_L^2}{\pi \cdot AR \cdot e}} \Leftrightarrow$$

$$C_{Dmin} = \frac{C_L}{LD_{max}} - \frac{C_L^2}{\pi \cdot AR \cdot e}$$

QED

15.6.2 Step-by-step: Extracting Drag from a Flight Polar Using the Quadratic Spline Method

This method can be used if the flight polar (or rate-of-sink versus airspeed or V_V versus V plot) is available, as it often is for sailplanes, gliders, and motor gliders. Usually, this polar is not available for powered airplanes. It can be used to extract C_{Dmin}, C_{LminD}, and e. This method will not retrieve a drag model that features a drag bucket, but a quadratic one. Also, it is important the reader reviews Appendix E.5.7, *Quadratic curve-fitting*, for further clarification of the method used here.

Consider the sample flight polar in Figure 15-73. Follow the following the stepwise procedure to extract C_{Dmin}, C_{LminD}, and e.

Step 1: Select Representative Points from the Flight Polar

Select three arbitrary points on the flight polar and record the corresponding V_V and V. For instance, select two points that enclose the minimum (near 50 KTAS in Figure 15-73) and one at a higher speed, for instance near 100 or 120 KTAS.

Step 2: Tabulate

Fill in the table below by entering the V and V_V selected in Step 1 in columns 1 and 2 below. Calculate the values in columns 3 though 5.

	1	2	3	4	5
ID	V	V_V	$x = V^2$	$x^2 = V^4$	$y = V \cdot V_V$
1	V_1	V_{V1}	V_1^2	V_1^4	$V_1 \cdot V_{V1}$
2	V_2	V_{V2}	V_2^2	V_2^4	$V_2 \cdot V_{V2}$
3	V_3	V_{V3}	V_3^2	V_3^4	$V_3 \cdot V_{V3}$

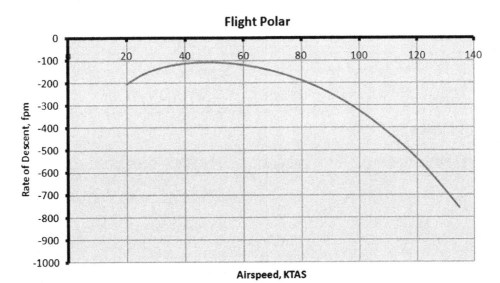

FIGURE 15-73 A typical flight polar.

Note 1: It is important that consistent units be used. Therefore, if V is read in KTAS is must be converted to ft/s. Similarly, if V_V is given in fpm (ft/min) it must be converted to ft/s. Use the conversion factors shown under the Note 1 of Step 2 of Section 15.6.1, *Step-by-step: Extracting drag from L/D$_{max}$*:

Note 2: This formulation is based on the use of the absolute value of the rate-of-sink. If the V_V is reported with a negative sign it must be converted to a positive number.

Step 3: Fill in the Conversion Matrix and Invert

Fill in the matrix below using the values in column 4 for the first row in the matrix and column 3 for the second row. This order is imperative. Then, invert the matrix. This can be done by some software, for instance, any spreadsheet software will offer means to invert matrices:

$$\begin{bmatrix} V_1^4 & V_1^2 & 1 \\ V_2^4 & V_2^2 & 1 \\ V_3^4 & V_3^2 & 1 \end{bmatrix} \rightarrow \begin{bmatrix} V_1^4 & V_1^2 & 1 \\ V_2^4 & V_2^2 & 1 \\ V_3^4 & V_3^2 & 1 \end{bmatrix}^{-1}$$

Step 4: Determine the Coefficients to the Quadratic Spline

Calculate the constants A, B, C by multiplying the inverted matrix in Step 3 with the matrix formed by column 5 in the table above:

$$\begin{Bmatrix} A \\ B \\ C \end{Bmatrix} = \begin{bmatrix} V_1^4 & V_1^2 & 1 \\ V_2^4 & V_2^2 & 1 \\ V_3^4 & V_3^2 & 1 \end{bmatrix}^{-1} \begin{Bmatrix} V_1 \cdot V_{V1} \\ V_2 \cdot V_{V2} \\ V_3 \cdot V_{V3} \end{Bmatrix}$$

Step 5: Extract Aerodynamic Properties

Using the constants A, B, C calculated in the previous step, extract the aerodynamic properties:

Induced drag constant:

$$k = \left(\frac{\rho S}{2W}\right) C \qquad (15\text{-}124)$$

Location of minimum drag:

$$C_{LminD} = -\frac{B}{2k} \qquad (15\text{-}125)$$

Minimum drag coefficient:

$$C_{Dmin} = A\left(\frac{2W}{\rho S}\right) - kC_{LminD}^2 \qquad (15\text{-}126)$$

Oswald efficiency:

$$e = \frac{1}{\pi \cdot AR \cdot k} \qquad (15\text{-}127)$$

Derivation of Equations (15-124) through (15-127)

The adjusted drag model is given by:

$$C_D = C_{Dmin} + k(C_L - C_{LminD})^2$$

This can be expanded as follows:

$$\begin{aligned} C_D &= C_{Dmin} + k(C_L - C_{LminD})^2 \\ &= C_{Dmin} + k(C_L^2 - 2C_L C_{LminD} + C_{LminD}^2) \\ &= (C_{Dmin} + kC_{LminD}^2) - 2kC_{LminD}C_L + kC_L^2 \end{aligned}$$

Also recall that the lift coefficient is given by:

$$C_L = \frac{2W}{\rho V^2 S}$$

And the rate of descent (ROD) is given by:

$$V_V = \frac{DV}{W} = \frac{\rho V^3 S C_D}{2W}$$

Insert the expression for C_D into the ROD and expand:

$$\begin{aligned} V_V &= \frac{\rho V^3 S C_D}{2W} \\ &= \frac{\rho V^3 S\left((C_{Dmin} + kC_{LminD}^2) + (-2kC_{LminD})C_L + (k)C_L^2\right)}{2W} \\ &= \frac{\rho V^3 S\left(kC_L^2 + (-2kC_{LminD})C_L + (C_{Dmin} + kC_{LminD}^2)\right)}{2W} \end{aligned}$$

Insert the expression for the C_L:

$$V_V = \frac{\rho V^3 S\left(kC_L^2 + (-2kC_{LminD})C_L + (C_{Dmin} + kC_{LminD}^2)\right)}{2W}$$

Expand to get:

$$\begin{aligned} V_V &= \frac{\rho V^3 S\left(k\left(\frac{2W}{\rho V^2 S}\right)^2 + (-2kC_{LminD})\left(\frac{2W}{\rho V^2 S}\right) + (C_{Dmin} + kC_{LminD}^2)\right)}{2W} \\ &= \frac{\rho V^3 S\left(k\frac{4W^2}{\rho^2 V^4 S^2} + (-2kC_{LminD})\frac{2W}{\rho V^2 S} + (C_{Dmin} + kC_{LminD}^2)\right)}{2W} \end{aligned}$$

Simplify:

$$\begin{aligned} V_V &= \frac{\rho V^3 S\left(k\frac{4W^2}{\rho^2 V^4 S^2} + (-2kC_{LminD})\frac{2W}{\rho V^2 S} + (C_{Dmin} + kC_{LminD}^2)\right)}{2W} \\ &= \frac{k\frac{4W^2}{\rho^2 V^4 S^2}\rho V^3 S + (-2kC_{LminD})\frac{2W}{\rho V^2 S}\rho V^3 S + (C_{Dmin} + kC_{LminD}^2)\rho V^3 S}{2W} \\ &= \frac{k\frac{4W^2}{\rho V S} + (-2kC_{LminD})2WV + (C_{Dmin} + kC_{LminD}^2)\rho V^3 S}{2W} \\ &= k\frac{2W}{\rho V S} + (-2kC_{LminD})V + (C_{Dmin} + kC_{LminD}^2)\frac{\rho V^3 S}{2W} \end{aligned}$$

Manipulate some more:

$$V_V = k\frac{2W}{\rho VS} + (-2kC_{LminD})V + (C_{Dmin} + kC_{LminD}^2)\frac{\rho V^3 S}{2W}$$

$$= k\left(\frac{2W}{\rho S}\right)\frac{1}{V} + (-2kC_{LminD})V + (C_{Dmin} + kC_{LminD}^2)\left(\frac{\rho S}{2W}\right)V^3$$

Multiply through by V to get:

$$V \cdot V_V = k\left(\frac{2W}{\rho S}\right) + (-2kC_{LminD})V^2$$
$$+ (C_{Dmin} + kC_{LminD}^2)\left(\frac{\rho S}{2W}\right)V^4$$
$$= C + BV^2 + AV^4$$

where

$$A = (C_{Dmin} + kC_{LminD}^2)\left(\frac{\rho S}{2W}\right)$$

$$B = -2kC_{LminD} \qquad C = k\left(\frac{2W}{\rho S}\right)$$

Therefore we can rewrite:

$$V \cdot V_V = C + BV^2 + AV^4 = C + Bx + Ax^2$$

where $x = V^2$

We need three points (i.e. V_V corresponding to a V) to determine these parameters:

$$\left.\begin{matrix} y_1 = Ax_1^2 + Bx_1 + C \\ y_2 = Ax_2^2 + Bx_2 + C \\ y_3 = Ax_3^2 + Bx_3 + C \end{matrix}\right\} \Rightarrow \begin{bmatrix} x_1^2 & x_1 & 1 \\ x_2^2 & x_2 & 1 \\ x_3^2 & x_3 & 1 \end{bmatrix}\begin{Bmatrix} A \\ B \\ C \end{Bmatrix} = \begin{Bmatrix} y_1 \\ y_2 \\ y_3 \end{Bmatrix}$$

Therefore:

$$\begin{Bmatrix} A \\ B \\ C \end{Bmatrix} = \begin{bmatrix} x_1^2 & x_1 & 1 \\ x_2^2 & x_2 & 1 \\ x_3^2 & x_3 & 1 \end{bmatrix}^{-1}\begin{Bmatrix} y_1 \\ y_2 \\ y_3 \end{Bmatrix}$$

Then, the coefficients can be found as follows:

$$C = k\left(\frac{2W}{\rho S}\right) \qquad \Leftrightarrow \qquad k = \left(\frac{\rho S}{2W}\right)C$$

$$B = -2kC_{LminD} \qquad \Leftrightarrow \qquad C_{LminD} = -\frac{B}{2k}$$

$$A = (C_{Dmin} + kC_{LminD}^2)\left(\frac{\rho S}{2W}\right) \qquad \Leftrightarrow$$

$$C_{Dmin} = A\left(\frac{2W}{\rho S}\right) - kC_{LminD}^2$$

Also note that:

$$k = \frac{1}{\pi \cdot AR \cdot e} \qquad \Leftrightarrow \qquad e = \frac{1}{\pi \cdot AR \cdot k}$$

QED

EXAMPLE 15-18

Consider the flight polar for a powered sailplane shown in Figure 15-74. Extract its drag characteristics, ignoring the existence of a drag bucket and assuming it can be described using the adjusted drag model.

Solution

Some properties of the aircraft that are necessary in the following calculations are:

Gross weight:	$W_0 = 1876$ lb$_f$
Wing area:	$S = 202$ ft^2
Wing aspect ratio:	$AR = 29.29$

Step 1 through 3: The three selected points are shown in Figure 15-74 and are tabulated below:

$$\begin{Bmatrix} A \\ B \\ C \end{Bmatrix} = \begin{bmatrix} x_1^2 & x_1 & 1 \\ x_2^2 & x_2 & 1 \\ x_3^2 & x_3 & 1 \end{bmatrix}^{-1}\begin{Bmatrix} y_1 \\ y_2 \\ y_3 \end{Bmatrix}$$

$$= \begin{bmatrix} 28267394 & 5317 & 1 \\ 368383509 & 19193 & 1 \\ 1104195089 & 33229 & 1 \end{bmatrix}^{-1}\begin{Bmatrix} 133.7 \\ 461.8 \\ 1215.3 \end{Bmatrix}$$

Step 4: Calculate the constants A, B, C:

$$\begin{Bmatrix} A \\ B \\ C \end{Bmatrix} = \begin{Bmatrix} 1.07602 \times 10^{-6} \\ -0.00272772 \\ 117.7650518 \end{Bmatrix}$$

	Selected Data						
ID	V km/h	V_V, fpm	V KCAS	V ft/s	$x = V^2$	$x^2 = V^4$	$Y = V \cdot V_V$
1	80	110	43	73	5317	28267394	133.7
2	152	200	82	139	19193	368383509	461.8
3	200	400	108	182	33229	1104195089	1215.3

EXAMPLE 15-18 (cont'd)

Step 5: Extract the aerodynamic properties using the constants A, B, C:

Induced drag constant:

$$k = \left(\frac{\rho S}{2W}\right)C = \left(\frac{(0.002378)(201.3)}{2(1876)}\right)(117.8) = 0.015074$$

Location of minimum drag:

$$C_{LminD} = -\frac{B}{2k} = -\frac{-0.00272772}{2 \times 0.015074} = 0.090478$$

Minimum drag coefficient:

$$C_{Dmin} = A\left(\frac{2W}{\rho S}\right) - kC_{LminD}^2 = 0.008283$$

Oswald efficiency:

$$e = \frac{1}{\pi \cdot AR \cdot k} = \frac{1}{\pi \cdot (29.29) \cdot (0.015074)} = 0.72094$$

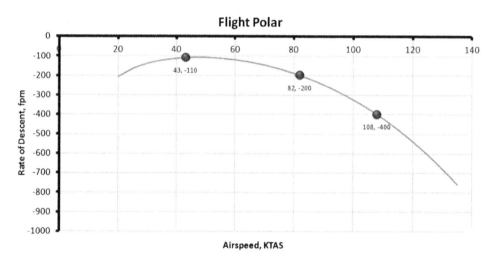

FIGURE 15-74 A flight polar for a powered sailplane.

15.6.3 Step-by-step: Extracting Drag Coefficient for a Piston-Powered Propeller Aircraft

The drag coefficient of piston-powered airplanes can be extracted from typical data provided in the Pilots Operating Handbooks (POH). The accuracy is contingent upon the quality of the data provided by the corresponding manufacturer, but some manufacturers have been guilty of bolstering performance data so the actual aircraft did not meet the values in the POH. Therefore, the results should be taken with a grain of salt. Additionally, as far as piston airplanes are concerned, the hardest parameter to obtain is the propeller efficiency. An educated guess may have to be used. The methods of Chapter 14, *The anatomy of the propeller*, should be helpful.

Method 1: Extraction C_{Dmin} Using Cruise Performance

The following procedure is intended to illustrate how such data extraction would take place using cruise performance. The reader may have to modify the procedure for selected aircraft, although it is thought that most POH feature data of the kind shown. Here, we will illustrate the method using a data sheet for a Cirrus SR22.

Figure 15-75 shows a page from a POH for a Cirrus SR22. The same information can also be obtained from the Type Certificate Data Sheet, also in the public domain and can be downloaded from the FAA official

15.6 SPECIAL TOPICS INVOLVING DRAG

Section 5
Performance Data

Cirrus Design SR22

FIGURE 15-75 Excerpt from a POH for Cirrus SR22 *(courtesy of Cirrus Aircraft)*.

Cruise Performance

8000 Feet Pressure Altitude

RPM	MAP	ISA − 30° C (−31° C)			ISA (−1° C)			ISA + 30° C (29° C)		
		PWR	KTAS	GPH	PWR	KTAS	GPH	PWR	KTAS	GPH
2700	21.7	83%	183	19.7	78%	183	18.6	75%	178	17.7
2600	21.7	79%	180	18.8	75%	180	17.8	71%	175	17.0
2500	21.7	75%	176	17.7	71%	176	16.8	67%	171	16.0
2500	20.7	70%	172	16.7	66%	172	15.8	63%	167	15.0
2500	19.7	66%	168	15.6	62%	168	14.8	59%	163	14.0
2500	18.7	61%	163	14.5	58%	163	13.8	55%	158	13.1
2500	17.7	57%	159	13.5	54%	159	12.8	51%	153	12.1

website (www.faa.gov). The excerpt shown displays the form typically used for General Aviation aircraft.

Step 1: Determine the following parameters for the type: reference area (S) in ft^2, aspect ratio (AR), rated engine power (P_A) in BHP, and gross weight (W_0) in lbf.

Step 2: Using the POH, extract the cruising speed and the associated altitude. Usually, these numbers are normalized to the gross weight. Typically, manufacturers present the cruising speed in KTAS, since this will make the number look larger and thus "better" in the POH. However, be alert in the unlikely event the data is provided as KIAS or KCAS. For this procedure, the airspeed in KTAS must be used.

Step 3: Determine the density of air, ρ, at the cruise altitude using Equation (16-18). Convert the airspeed in KTAS to ft/s by multiplying it by 1.688. This airspeed is V.

Step 4: Determine the Oswald efficiency and the lift-induced drag constant at the condition. For a medium AR aircraft, Equation (9-89) can be used, repeated here for convenience:

Oswald's efficiency:

$$e = 1.78(1 - 0.045 AR^{0.68}) - 0.64$$

Lift-induced drag constant:

$$k = 1/(\pi \cdot AR \cdot e)$$

Step 5: Calculate the minimum drag coefficient from:

$$C_{Dmin} = \frac{\eta_p \times 1100 \times P_{BHP}}{\rho V^3 S} - k\left(\frac{2W}{\rho V^2 S}\right)^2 \quad (15\text{-}128)$$

Note that the selection of η_p can be tricky. If the propeller is a fixed-pitch "climb" propeller use a value ranging from 0.65 to 0.70. If the propeller is a fixed-pitch "cruise" propeller use a value ranging from 0.75 to 0.80. If the airplane is equipped with a constant-speed propeller, use a value ranging from 0.80 to 0.86.

Derivation of Equation (15-128)

The minimum drag can be estimated from:

$$C_D = C_{Dmin} + C_{Di} \quad \Leftrightarrow \quad C_{Dmin} = C_D - C_{Di}$$

Since thrust equals drag at this condition, we can determine the drag coefficient as follows:

$$T = D = \frac{1}{2}\rho V^2 S C_D \quad \Leftrightarrow \quad C_D = \frac{2T}{\rho V^2 S}$$

$$C_D = \frac{2(\eta_p \times 550 \times P_{BHP}/V)}{\rho V^2 S} = \frac{\eta_p \times 1100 \times P_{BHP}}{\rho V^3 S}$$

Furthermore, using the simplified drag model and the lift coefficient the lift-induced drag is given by:

$$C_{Di} = kC_L^2 = k\left(\frac{2W}{\rho V^2 S}\right)^2$$

Inserting this into the expression for C_{Dmin} yields:

$$C_{Dmin} = C_D - C_{Di} = \frac{\eta_p \times 1100 \times P_{BHP}}{\rho V^3 S} - k\left(\frac{2W}{\rho V^2 S}\right)^2$$

QED

EXAMPLE 15-19

Estimate the total, induced, and minimum drag coefficient for the Cirrus SR22, using the data in Figure 15-75 using the published cruising speed of 183 KTAS at 8000 ft and 78% power at ISA condition.

Solution

Step 1–2:

Reference area	$S = 144.9$ ft^2
Aspect ratio	$AR = 10.1$ (using a wing span of 38.3 ft)
Engine power at condition	$P_A = 0.78 \times 310$ BHP $= 241.8$ BHP
Gross weight	$W_0 = 3400$ lb$_f$

Cruising speed is 183 KTAS at 8000 ft.

Step 3:

Density of air at 8000 ft is

$$\rho = 0.002378(1 - 0.0000068756 \times 8000)^{4.2561}$$
$$= 0.001869 \text{ slugs/ft}^3$$

Cruising speed in ft/s is $V = 183 \times 1.688 = 308.9$ ft/s

Step 4: Both the Oswald efficiency factor and lift-induced drag constant have been calculated multiple times throughout this book. Of the two we need the latter one for this problem and it has been shown to be $k = 0.04207$.

Step 5: Since we don't know the propeller efficiency we will have to guess. The propeller is a constant-speed propeller, specifically designed for the aircraft, so it is not unlikely its propeller efficiency is as high as 0.85 at this condition. The reader must remember that this is a guess so the resulting drag coefficient must be regarded with caution. Therefore, we can estimate the minimum drag coefficient for the airplane as follows:

$$C_{Dmin} = \frac{\eta_p \times 1100 \times P_{BHP}}{\rho V^3 S} - k\left(\frac{2W}{\rho V^2 S}\right)^2$$

$$= \frac{0.85 \times 1100 \times 241.8}{(0.001869)(308.9)^3(144.9)} - (0.04207)$$

$$\times \left(\frac{2(3400)}{(0.001869)(308.9)^2(144.9)}\right)^2 = 0.02541$$

The total drag can be estimated using the thrust relation shown in the above derivation:

Total drag coefficient:

$$C_D = \frac{\eta_p \times 1100 \times P_A}{\rho V^3 S} = \frac{0.85 \times 1100 \times 241.8}{(0.001869)(308.9)^3(144.9)}$$

$$= 0.02832$$

And the lift-induced drag is the difference between the total and minimum drag:

$$C_{Di} = C_D - C_{Dmin} = 0.02832 - 0.02541 = 0.00291$$

Method 2: Extracting C_{Dmin} Using Climb Performance

The following procedure is intended to illustrate how such data extraction would take place using climb performance. This requires the rate-of-climb (ROC) and the associated airspeed and altitude to be determined. Ideally, this should be the best ROC, obtained at V_Y (the best rate-of-climb airspeed). The reason is that V_Y occurs at a relatively low *AOA*, which means less flow separation than at, say, V_X (the best angle-of-climb airspeed) and, thus, closer adherence to the quadratic drag model. It is also better to use data at S-L as this will require fewer corrections.

Figure 15-75 shows a page from a POH for a Cirrus SR22. The same information can also be obtained from the Type Certificate Data Sheet, which is in the public domain and can be downloaded from the FAA official website (www.faa.gov). The excerpt shown displays the form typically used for General Aviation aircraft.

Step 1: Same as Step 1 of Method 1.

Step 2: Using the POH, extract the ROC, the associated climb airspeed, and altitude. Ideally this would be V_Y at S-L. Usually, these numbers are normalized to the gross weight. It is common for manufacturers to present the climb speed in KIAS. If this is the case, it must be corrected to KCAS before it is converted to KTAS, which is required for this procedure. This conversion should be implemented using the methods of Section 16, *Performance — introduction*.

Step 3: Determine the density of air, ρ, at the climb altitude using Equation (16-18). Ideally, this should be at S-L, in which case $\rho = 0.002378$ slugs/ft^3. Convert the ROC in fpm to V_V by dividing it by 60 seconds/minute, i.e. $V_V = \text{ROC}/60$. Convert the airspeed in KTAS to ft/s by multiplying it by 1.688. This airspeed is V.

Step 4: Same as Step 4 of Method 1.

Step 5: Calculate the minimum drag coefficient from:

$$C_{Dmin} = \frac{\eta_p 1100 P_{BHP} - 2WV_V}{\rho V^3 S} - k\left(\frac{2W}{\rho V^2 S}\right)^2 \quad (15\text{-}129)$$

Again, note that if the propeller efficiency, η_p, is not known, one must rely on engineering judgment to assess its value, and this can be tricky. Assuming the flight condition to be at V_Y, use a value ranging from 0.65 to 0.70 if the propeller is a fixed-pitch "climb" propeller. If the propeller is a fixed-pitch "cruise" propeller, use a value ranging from 0.55 to 0.65, and if equipped with a constant-speed propeller, use a value ranging from 0.65 to 0.70. Also note that, just like for Method 1, correcting the power to the altitude at which the aircraft is operating is crucial. For this, use the Gagg and Ferrar model of Equation (7-16).

Derivation of Equation (15-129)

The vertical speed, V_V, is determined from:

$$V_V = \frac{ROC}{60} = \frac{TV - DV}{W}$$

Replacing the thrust, T, and drag, D, with the standard expressions yields:

$$V_V = \frac{\left(\frac{\eta_p 550 P_{BHP}}{V}\right)V - \left(\frac{1}{2}\rho V^2 S C_D\right)V}{W}$$

$$= \frac{\eta_p 550 P_{BHP} - \frac{1}{2}\rho V^3 S C_D}{W}$$

Mulitplying through by W and inserting the simplified drag model and, then, the lift coefficient leads to:

$$WV_V = \eta_p 550 P_{BHP} - \frac{1}{2}\rho V^3 S (C_{Dmin} + kC_L^2)$$

$$= \eta_p 550 P_{BHP} - \frac{1}{2}\rho V^3 S \left(C_{Dmin} + k\left(\frac{2W}{\rho V^2 S}\right)^2\right)$$

Solving for the C_{Dmin} returns:

$$C_{Dmin} = \frac{\eta_p 1100 P_{BHP} - 2WV_V}{\rho V^3 S} - k\left(\frac{2W}{\rho V^2 S}\right)^2$$

QED

EXAMPLE 15-20

Estimate the minimum drag coefficient for the Cirrus SR22, using the climb performance extraction method at V_Y and the following data in addition to S, k, and W from the previous example.

Solution

Step 1–2:

Engine power at condition	$P_A = 310$ BHP		
Best rate-of-climb	ROC = 1398 fpm	=>	$V_V = 1398/60 = 23.3$ ft/s
Best rate-of-climb airspeed	$V_Y = 101$ KIAS = 101 KCAS = 101 KTAS = 170.5 ft/s (at S-L)		
Propeller efficiency at V_Y		$\eta_p = 0.7$	(guess)

$$C_{Dmin} = \frac{\eta_p 1100 P_{BHP} - 2WV_V}{\rho V^3 S} - k\left(\frac{2W}{\rho V^2 S}\right)^2$$

$$= \frac{(0.7)1100(310) - 2(3400)(23.3)}{(0.002378)(170.5)^3(144.9)}$$

$$- (0.04207)\left(\frac{2(3400)}{(0.002378)(170.5)^2(144.9)}\right)^2$$

$$= 0.02761$$

Note that if the propeller efficiency is 0.689, rather than 0.700, the C_{Dmin} will be 0.02541, which is the same as that of Method 1. This shows the importance of properly selecting the propeller efficiency.

15.6.4 Computer code 15-1: Extracting Drag Coefficient from Published Data for Piston Aircraft

The following Visual Basic for Applications routine can be used with Microsoft Excel to extract various drag-related parameters using the above method. Note that, as shown, it is only valid for aircraft with straight wings. Also, note that the power is not necessarily the rated power. For instance, the rated power for the SR22 is 310 BHP. However, at the condition used in Example 15-18, the power was only 241.8 BHP. Regardless of whether the number is due to altitude effects (see Equation (7-16)) or power setting or a combination of the two, the power at the condition must be used.

```
Function Extract_CD(Sref As Single, AR As Single, WO As Single, P_BHP As Single, eta As Single, Vktas As
Single, H As Single, Mode As Byte) As Single
'This function uses the method of Section 15.6.3 to extract CD, CDi, CDmin and others
'for a piston powered prop aircraft. Only valid for propeller aircraft with straight
'wings.
'
'Variables: Sref  = Reference wing area (ex. 145 ft²)
'           AR    = Aspect Ratio (ex. 10)
'           WO    = Weight at condition (e.g. altitude and airspeed) (ex. 3100 lbf)
'           P_BHP = Horsepower at condition (e.g. 205 BHP)
'           eta   = Estimated propeller efficiency at condition (ex. 0.85)
'           Vktas = Airspeed in KTAS at condition
'           H     = Altitude in ft at condition
'           Mode  = What to return: =0 for CD, =1 for CDi, =2 for CDmin,
'                                   =3 for CL, =10 for Thrust
'
'Initialize
    Dim rho As Single, V As Single
    Dim T As Single, e As Single
    Dim CL As Single, CD As Single, CDi As Single, CDmin As Single
'Presets
    V = 1.688 * Vktas                                    'Airspeed in ft/s
    If V < 30 Then                                       'Illegitimate airspeed. Exit
        Extract_CD = -1                                  'Indicate error
        Exit Function
    End If
    rho = 0.002378 * (1 - 0.0000068756 * H) ^ 4.2561     'Density
    T = eta * 550 * P_BHP / V                            'Thrust
    CL = 2 * WO / (rho * V ^ 2 * Sref)                   'Lift coefficient
'Calculate CD
    CD = eta * 1100 * P_BHP / (rho * V ^ 3 * Sref)
'Calculate CDi
    e = 1.78 * (1 - 0.045 * AR ^ 0.68) - 0.64
    CDi = CL ^ 2 / (3.14159265 * AR * e)
'Calculate CDmin
    CDmin = CD - CDi
'Return
    Select Case Mode
    Case 0  'CD
        Extract_CD = CD
    Case 1  'CDi
        Extract_CD = CDi
    Case 2  'CDmin
        Extract_CD = CDmin
    Case 3  'CL
        Extract_CD = CL
    Case 10 'T
        Extract_CD = T
    End Select
End Function
```

15.6.5 Step-by-step: Extracting Drag Coefficient for a Jet Aircraft

The drag coefficient of jets can be extracted from typical data published by similar means, e.g. the Pilots Operating Handbooks (POH) or Pilot's Flight Manual (PFM), with all the limitations as before. Sometimes, a surprising amount of information can be learned from document such as Ref. [45]. The primary difficulty is that the maximum level airspeed of many jets is not limited by engine thrust, but by compressibility effects. This is denoted by the M_{MO} or the maximum operating Mach number (maximum and operating are denoted by the subscript). As a matter of fact, the engines often have enough thrust to accelerate the aircraft beyond the M_{MO} value and thrust must therefore be set to prevent this from taking place. Thus, different jets require different fractions of maximum thrust for operation, making the determination of their minimum drag based on cruise information that much harder.

For this reason, we will attempt to extract the minimum drag based on other kind of information – the best rate-of-climb (ROC). This important form of climb is usually presented in performance handbooks. This can be done by at least two means, based on the available information, and the assumption that the reported climb is V_Y and performed at max thrust and with all high-lift devices retracted.

Step 1: Using the PFM or other reliable sources, extract the best rate of climb airspeed, the associated altitude. Now the two following scenarios are possible: (1) The ROC given is not necessarily the best ROC, but has an associated airspeed presented. In this case use Method 1 in Step 7 below. (2) The ROC is not given but the airspeed for best ROC (V_Y). In this case use Method 2.

For jets, these numbers are not necessarily normalized to the gross weight but to some specific operational weights. If so the appropriate weight must be used throughout. Typically, manufacturers present the cruising speed in KTAS or Mach numbers. In either case, the corresponding numbers must be converted to ft/s true airspeed.

Step 2: The following parameters for the type are required based on the discussion in Step 1:

(1) Reference area (S) in ft^2.
(2) Aspect Ratio (AR).
(3) Maximum total rated engine thrust (T_A), in lb$_f$.
(4) Aircraft weight (W) at condition, in lb$_f$.
(5) Altitude at condition, in ft.
(6) Airspeed, in M or KCAS, and ROC in fpm at condition.

Step 3: Determine the density of air, ρ, at the cruise altitude using Equation (16-18).

Step 4: Convert Mach numbers to ft/s by calculating the speed of sound at altitude and multiplying by the Mach number, per Equation (16-30).

Convert the airspeed in KCAS to ft/s by multiplying it by 1.688 and then dividing by the square root of the density ratio, per Equation (16-33).

Step 5: Correct the rated thrust for airspeed and altitude using the Mattingly method for turbojets or turbofans in Sections 7.2.3, *Turbojets*, or Section 7.2.4, *Turbofans*.

Important: It is imperative that the thrust be corrected with respect to the altitude and airspeed, as this is much less than the rated thrust. Otherwise, erroneous results will be returned.

Step 6: Determine the Oswald efficiency at the condition using either Equation (9-90) or (9-91), repeated here for convenience:

$$e = 4.61(1 - 0.045 AR^{0.68})(\cos \Lambda_{LE})^{0.15} - 3.1 \quad (9\text{-}90)$$

$$e = \frac{2}{2 - AR + \sqrt{4 + AR^2(1 + \tan^2 \Lambda_{t\max})}} \quad (9\text{-}91)$$

Of these, the author considers Equation (9-85) more suitable for typical modern transportation jets. Then calculate the lift-induced drag constant:

$$k = \frac{1}{\pi \cdot AR \cdot e}$$

Step 7a – Method 1 (airspeed other than V_Y and the associated ROC are known):

Calculate the lift coefficient from:

$$C_L = \frac{2W}{\rho V^2 S}$$

Calculate the minimum drag coefficient from (remember that (1) V must be in ft/s true airspeed and may or may not be V_Y and (2) the ROC must correspond to V and be in ft/min):

Method 1:

$$C_{Dmin} = \frac{1}{\rho V^2 S}\left(2T - \frac{ROC \cdot W}{30V}\right) - kC_L^2 \quad (15\text{-}130)$$

Step 7b – Method 2 (the best rate-of-climb and associated airspeed are known):

Calculate the minimum drag coefficient from (remember that V_Y must be in ft/s true airspeed):

Method 2:

$$C_{Dmin} = \frac{(T/W)(W/S)}{3\rho V_Y^2 \left(1 - \frac{4k}{\rho V_Y^2}\frac{(W/S)}{(T/W)}\right)} \quad (15\text{-}131)$$

Derivation of Equation (15-130)

The derivation *assumes the simplified drag model* is applicable and begins with the conversion of Equation (18-17):

$$ROC = 60\frac{TV - DV}{W} = \frac{60V}{W}(T - D)$$
$$\Rightarrow D = T - \frac{ROC \cdot W}{60V}$$

Insert Equation (8-8) with the simplified drag model:

$$\frac{1}{2}\rho V^2 S\left(C_{Dmin} + \frac{C_L^2}{\pi \cdot AR \cdot e}\right) = \frac{1}{2}\rho V^2 S(C_{Dmin} + kC_L^2)$$
$$= T - \frac{ROC \cdot W}{60V}$$

Finally, solve for C_{Dmin}:

$$C_{Dmin} = \frac{1}{\rho V^2 S}\left(2T - \frac{ROC \cdot W}{30V}\right) - kC_L^2$$

Note that this expression can also be written as follows:

$$C_{Dmin} = \left(\frac{2T}{\rho V^2 S} - \frac{2W}{\rho V^2 S}\frac{ROC}{60V}\right) - kC_L^2$$
$$= \frac{2T}{\rho V^2 S} - kC_L^2 - \frac{ROC}{60V}C_L$$

QED

Derivation of Equation (15-131)

The derivation *assumes the simplified drag model* is applicable and begins with the conversion of Equation (18-24):

$$V_Y = \sqrt{\frac{(T/W)(W/S)}{3\rho C_{Dmin}}\left[1 + \frac{3}{LD_{max}^2(T/W)^2}\right]} \Leftrightarrow$$

$$V_Y^2 = \frac{(T/W)(W/S)}{3\rho C_{Dmin}}\left[1 + \frac{3}{LD_{max}^2(T/W)^2}\right]$$

Then, replace LD_{max} using Equation (19-18):

$$C_{Dmin} = \frac{(T/W)(W/S)}{3\rho V_Y^2}\left[1 + \frac{3}{LD_{max}^2(T/W)^2}\right]$$

$$= \frac{(T/W)(W/S)}{3\rho V_Y^2}\left[1 + \frac{3}{\frac{1}{4C_{Dmin}k}(T/W)^2}\right]$$

Expand:

$$C_{Dmin} = \frac{(T/W)(W/S)}{3\rho V_Y^2}\left[1 + \frac{12C_{Dmin}k}{(T/W)^2}\right]$$

$$= \left[\frac{(T/W)(W/S)}{3\rho V_Y^2} + \frac{12C_{Dmin}k}{(T/W)^2}\frac{(T/W)(W/S)}{3\rho V_Y^2}\right]$$

And then collect and solve for C_{Dmin}.

$$C_{Dmin} - C_{Dmin}\frac{12k}{3\rho V_Y^2}\frac{(W/S)}{(T/W)} = \frac{(T/W)(W/S)}{3\rho V_Y^2}$$

$$\Rightarrow C_{Dmin} = \frac{(T/W)(W/S)}{3\rho V_Y^2\left(1 - \frac{4k}{\rho V_Y^2}\frac{(W/S)}{(T/W)}\right)}$$

QED

15.6.6 Determining Drag Characteristics from Wind Tunnel Data

Standard wind tunnel testing yields a number of static force and moment coefficients such as C_D, C_L, C_Y, C_l, C_m, and C_n. For this section the focus is on the first two: the lift and drag coefficients. A conventional *AOA* sweep (or alpha-sweep as it is most often called) consists of changing the *AOA* from a given minimum value (e.g. $-5°$) to a maximum value (e.g. $+20°$), perhaps $1°$ at a time. Therefore, the sweep returns a listing of the coefficients as a function of α. For standard aircraft that do not generate a noticeable drag bucket, the relationship between C_L and C_D can then be obtained using a quadratic least-squares curve-fit, which returns a polynomial of the form $C_D = A \cdot C_L^2 + B \cdot C_L + C$. The coefficients of this polynomial can be used to extract the coefficient C_{Dmin}, C_{LminD}, and the Oswald span efficiency factor, e, provided the airplane's *AR* has been established. If so, it can then be shown that these parameters are related to the constants of the curve-fit polynomial as follows:

$$e = \frac{1}{\pi \cdot AR \cdot A} \quad (15\text{-}132)$$

$$C_{LminD} = -\frac{B}{2A} \quad (15\text{-}133)$$

$$C_{Dmin} = C - \frac{B^2}{4A} \quad (15\text{-}134)$$

Derivation of Equations (15-132) through (15-134)

Begin by equating the two forms of the drag coefficients: the quadratic curve-fit and the adjusted drag model of Equation (15-6):

$$C_D = AC_L^2 + BC_L + C$$
$$= C_{Dmin} + \frac{1}{\pi \cdot AR \cdot e}(C_L - C_{LminD})^2$$

Expand and sort coefficients based on their dependency on C_L^2 and C_L:

$$AC_L^2 + BC_L + C = \frac{1}{\pi \cdot AR \cdot e}C_L^2 - \frac{2C_{LminD}}{\pi \cdot AR \cdot e}C_L + C_{Dmin}$$
$$+ \frac{1}{\pi \cdot AR \cdot e}C_{LminD}^2$$

By observation we see that A, B, and C are related to the constants of the adjusted drag polar as follows:

$$A = \frac{1}{\pi \cdot AR \cdot e} \quad \text{(i)}$$

$$B = -\frac{2C_{LminD}}{\pi \cdot AR \cdot e} \quad \text{(ii)}$$

$$C = C_{Dmin} + \frac{1}{\pi \cdot AR \cdot e}C_{LminD}^2 \quad \text{(iii)}$$

From Equation (i) we get:

$$e = \frac{1}{\pi \cdot AR \cdot A}$$

Using this result with Equation (ii) we get:

$$B = -\frac{2C_{LminD}}{\pi \cdot AR \cdot e} = -\frac{2C_{LminD}}{\pi \cdot AR \cdot \frac{1}{\pi \cdot AR \cdot A}}$$
$$= -2C_{LminD}A \quad \Rightarrow \quad C_{LminD} = -\frac{B}{2A}$$

Using the two previous results with Equation (iii) we get:

$$C = C_{Dmin} + \frac{1}{\pi \cdot AR \cdot e}C_{LminD}^2 = C_{Dmin} + \frac{1}{\pi \cdot AR \cdot \frac{1}{\pi \cdot AR \cdot A}}\left(-\frac{B}{2A}\right)^2$$
$$= C_{Dmin} + \frac{B^2}{4A} \quad \Rightarrow \quad C_{Dmin} = C - \frac{B^2}{4A}$$

QED

EXAMPLE 15-21

The data from a wind tunnel test of a complete aircraft configuration is given in the following table:

Cl	-0.4649	-0.3240	-0.1917	-0.0767	0.0240	0.1217	0.2367	0.3518	0.4668	0.5617	0.6537	0.7169	0.7888	0.8492
CD	0.0591	0.0488	0.0400	0.0334	0.0292	0.0280	0.0298	0.0334	0.0405	0.0478	0.0571	0.0669	0.0778	0.0891

Determine the parameters C_{Dmin}, C_{LminD}, and the Oswald span efficiency factor, e, using a quadratic least-squares curve-fit to the data of the form $C_D = A \cdot C_L^2 + B \cdot C_L + C$, if the aspect ratio, AR, = 6.

Solution

One way to obtain the curve-fit is to use commercial spreadsheet software like Microsoft Excel, enter the data and plot using a scatter graph. Then, the user can select the curve and add a trendline with the associated equation and correlation coefficient display. This is shown in Figure 15-76. The resulting curve-fit shows the polynomial constants are given by $A = 0.1056$, $B = -0.0226$, and $C = 0.0292$. Then the drag parameters can be determined as follows:

$$e = \frac{1}{\pi \cdot AR \cdot A} = \frac{1}{\pi \cdot 6 \cdot 0.1056} = 0.5024$$

$$C_{LminD} = -\frac{B}{2A} = \frac{-0.0229}{2 \cdot 0.1056} = 0.1070$$

$$C_{Dmin} = C - \frac{B^2}{4A}$$
$$= 0.0292 - \frac{(-0.0229)^2}{4 \cdot 0.1056} = 0.02799$$

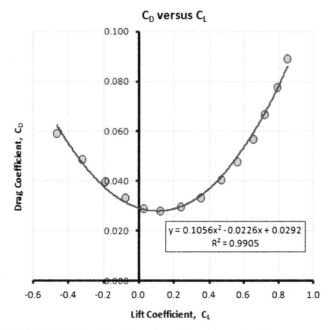

FIGURE 15-76 Hypothetical wind tunnel test data for some complete aircraft plotted with a trendline generated in Microsoft Excel.

15.7 ADDITIONAL INFORMATION — DRAG OF SELECTED AIRCRAFT

15.7.1 General Range of Subsonic Minimum Drag Coefficients

Table 15-17 shows expected ranges of values of the subsonic minimum drag coefficient for several classes of aircraft. These numbers do not bracket all possible aircraft configurations — i.e. there may be specific aircraft that are outside the range shown. However, most aircraft will be inside the lower and upper limits of the range.

15.7.2 Drag of Various Aircraft by Class

Table 15-18 lists selected drag related figures for a number of aircraft of different classes. Some of these are displayed in Figure 15-31. The data was gathered from a number of sources. Note that some of the data was retrieved from graphs using "careful eyeballing" and should be regarded with care.

When estimating the drag of a new aircraft design, it is strongly recommended that the designer compares his own results to that of the aircraft in the tables below that most similarly resembles the new aircraft. This can help flag a possible over- or underestimation.

The numbers come from a variety of sources; Perkins and Hage [24], Stinton [46], Roskam [47], Nicolai [28], NASA SP-468 [48], NASA CR-114494 [49], and the author's own estimates. The author's estimations utilize

TABLE 15-17 Range of Subsonic Minimum Drag Coefficients, C_{Dmin}

Class of Aircraft	Range of C_{Dmin}	
	Lower	Upper
World War I Era Aircraft 1914-1918	0.0317	0.0771
Interwar Era Aircraft 1918-1939	0.0182	0.0585
Multi-Engine WW-II Bombers (Piston)	0.0239	0.0406
Single and Multi-Engine WW-II Fighters (Piston)	0.0157	0.0314
Multi-Engine Commercial Transport Aircraft (Piston)	0.0191	0.0258
Kitplanes (Homebuilt) and LSA	0.0119	0.0447
Single Engine GA Aircraft (Piston and Turboprop)	0.0174	0.0680
GA ag-aircraft, single engine, propeller, clean	0.0550	0.0600
GA ag-aircraft, single engine, propeller, spray-system	0.0700	0.0800
Twin Engine GA Aircraft (Piston and Turboprop)	0.0242	0.0369
Flying Boats	0.0233	0.0899
Selected Jet Fighter/Trainer Aircraft	0.0083	0.0240
Selected Jet Bomber and Attack Aircraft	0.0068	0.0160
Commercial Jetliners	0.0160	0.0219
Various Subsonic Military Aircraft (Props and Jets)	0.0145	0.0250
High performance sailplane	0.0060	0.0100
Tailless aircraft, propeller	0.0150	0.0200
Tailless aircraft, jet	0.0080	0.0140
Low altitude subsonic cruise missile (high W/S)	0.0300	0.0400

Method 1 of Section 15.6.3, *Step-by-step: Extracting drag coefficient for a piston-powered propeller aircraft*, and 15.6.5, *Step-by-step: Extracting drag coefficient for a jet aircraft*, utilizing performance data from the corresponding aircraft's Pilots Operating Handbook (POH) or other reliable sources. The LD_{max} estimated by the author for propeller aircraft assumes no additional drag due to windmilling or stopped propellers. Refer to Section 15.5.13, *Drag of windmilling and stopped propellers*, for methods on how to account for this drag. Windmilling propellers can easily increase the minimum drag coefficient by 150 drag counts or more. Additionally, although expected, there is no guarantee the manufacturer has not bolstered performance values in the POH. For this reason, treat all drag data with caution.

TABLE 15-18 Drag Characteristics of Selected Aircraft

		World War I Era Aircraft 1914-1918							
Make	Model	AR	S_{ref}, ft²	S_{wet}, ft²	f, ft²	C_f	C_{Dmin}	$(L/D)_{max}$	Reference
Albatross	D-III	4.65	231	-	10.74	-	0.0465	7.5	NASA SP-468
B.E.	2c	4.47	371	-	13.66	-	0.0368	8.2	NASA SP-468
Caproni	CA.42	5.43	2223	-	98.70	-	0.0444	8.2	NASA SP-468
De Havilland	DH-2	3.88	249	-	10.71	-	0.0430	7.0	NASA SP-468
	DH-4	4.97	440	-	21.82	-	0.0422	8.1	NASA SP-468
Fokker	D-VII	4.70	221	-	8.93	-	0.0404	8.1	NASA SP-468
	D-VIII	6.58	115	-	6.34	-	0.0552	8.1	NASA SP-468
	E-III	5.70	172	-	12.61	-	0.0771	6.4	NASA SP-468
	Dr.-1 Triplane	4.04	207	-	6.69	-	0.0323	8.0	NASA SP-468
Gotha	G.V	7.61	963	-	68.45	-	0.0711	7.7	NASA SP-468
Handley Page	0/400	7.31	1655	-	70.67	-	0.0427	9.7	NASA SP-468
Junkers	D-I	5.46	159	-	9.75	-	0.0612	7.0	NASA SP-468
	J-I	6.40	522	-	17.50	-	0.0335	10.3	NASA SP-468
Nieuport	Model 17	5.51	159	-	7.81	-	0.0491	7.9	NASA SP-468
Sopwith	5F.1 Dolphin	4.85	263	-	8.35	-	0.0317	9.2	NASA SP-468
	F.1 Camel	4.11	231	-	8.73	-	0.0378	7.7	NASA SP-468
SPAD	XIII C.1	3.69	227	-	8.33	-	0.0367	7.4	NASA SP-468

		Interwar Era Aircraft 1918-1939							
Make	Model	AR	S_{ref}, ft²	S_{wet}, ft²	f, ft²	C_f	C_{Dmin}	$(L/D)_{max}$	Reference
Beechcraft	D17S	4.18	296	-	5.39	-	0.0182	11.7	NASA SP-468
Boeing	P-26A	5.24	149	-	6.68	-	0.0448	8.3	NASA SP-468
Curtiss	Hawk P-6E	4.76	252	-	9.35	-	0.0371	8.7	NASA SP-468
	JN-4H	7.76	352.6	-	17.64	-	0.0500	9.24	NASA SP-468
	R2C-1	4.18	140	-	2.88	-	0.0206	10.9	NASA SP-468
	Robin	7.54	223	-	13.10	-	0.0585	8.7	NASA SP-468
Dayton Wright RB	Hall Scott L-62	4.38	102	-	3.22	-	0.0316	9	NASA SP-468
Fokker	F-2	7.10	452	-	21.10	-	0.0466	9.4	NASA SP-468
Ford	5-AT	7.26	835	-	39.33	-	0.0471	9.5	NASA SP-468
Handley Page	W8F	4.67	1456	-	79.93	-	0.0549	7.1	NASA SP-468
Lockheed	Orion 9D	7.01	262	-	5.50	-	0.0210	14.1	NASA SP-468
	Vega 5C	6.11	275	-	7.65	-	0.0278	11.4	NASA SP-468
Northrop	Alpha	5.93	312	-	8.55	-	0.0274	11.3	NASA SP-468
Ryan	NYP	6.63	319	-	12.10	-	0.0379	10.1	NASA SP-468
Stinson	SR-8B	6.84	256	-	8.91	-	0.0348	10.8	NASA SP-468
Supermarine	S-4	6.84	136	-	3.73	-	0.0274	12.1	NASA SP-468

		Multi-Engine WW-II Bombers (Piston)							
Make	Model	AR	S_{ref}, ft²	S_{wet}, ft²	f, ft²	C_f	C_{Dmin}	$(L/D)_{max}$	Reference
Boeing	B-17	7.58	1420	5900	39.00	0.0067	0.0275	-	Perkins & Hage
	B-17G	7.58	1420	-	42.83	-	0.0302	12.7	NASA SP-468
	B-29	11.50	1736	-	41.16	-	0.0241	16.8	NASA SP-468
	B-29A	11.49	1736	7500	43.50	0.0058	0.0251	-	Roskam
	B-29B	11.49	1736	7100	41.50	0.0058	0.0239	-	Roskam
Consolidated	B-24J	11.55	1048	-	42.54	-	0.0406	12.9	NASA SP-468
North American	B-25	7.49	610	2800	18.00	0.0064	0.0295	-	Perkins & Hage

		Single and Multi-Engine WW-II Fighters (Piston)							
Make	Model	AR	S_{ref}, ft²	S_{wet}, ft²	f, ft²	C_f	C_{Dmin}	$(L/D)_{max}$	Reference
Curtiss	SB2C-1	5.88	422	-	9.52	-	0.0225	12.4	NASA SP-468
Grumman	F6F-3	5.34	334	-	7.05	-	0.0211	12.2	NASA SP-468
Lockheed	P-38	8.26	327.5	1500	8.20	0.0055	0.0250	-	Perkins & Hage
	P-38L	8.26	327.5	-	8.78	-	0.0268	13.5	NASA SP-468
Martin	B-26F	7.66	658	-	20.66	-	0.0314	12.0	NASA SP-468
North American	P-51B	5.83	235	880	3.70	0.0042	0.0157	-	Roskam
	P-51D	5.86	233	-	3.80	-	0.0163	14.6	NASA SP-468
Republic	P-47	5.54	300	1200	6.50	0.0054	0.0217	-	Perkins & Hage
Seversky	P-35	5.89	220	-	5.52	-	0.0251	11.8	NASA SP-468

(Continued)

TABLE 15-18 Drag Characteristics of Selected Aircraft—cont'd

	Selected Jet Fighter/Trainer Aircraft							
Make	Model	AR	S_{ref}, ft²	S_{wet}, ft²	f, ft²	C_f	C_{Dmin}	$(L/D)_{max}$	Reference
Cessna	T-37	6.20	183.9	770	3.15	0.0041	0.0171	-	Roskam
Convair	F-106A	2.10	697.8	-	5.80	-	0.0083	12.1	NASA SP-468
General Dynamics	F-111D	7.56	525	-	9.36	-	0.0186	15.8	NASA SP-468
Lockheed	F-104G	2.45	196.1	-	3.37	-	0.0172	9.2	NASA SP-468
	P-80A	6.37	237.6	-	3.20	-	0.0134	17.7	NASA SP-468
McDonnell-Douglas	F-4C	2.82	530	-	12.72	-	0.0240	8.72	Nicolai
	F-4E	2.77	530	-	11.87	-	0.0224	8.58	NASA SP-468
North American	F-100D	3.76	400.2	-	5.00	-	0.0130	13.9	NASA SP-468
	F-86E	4.78	287.9	-	3.80	-	0.0132	15.1	NASA SP-468
Northrop	F-5E	3.86	186	-	3.40	-	0.0200	10	NASA SP-468
Republic	F-105D	3.16	385	-	6.65	-	0.0173	10.4	NASA SP-468
Vought	F-8H	3.42	375	-	5.00	-	0.0133	12.8	NASA SP-468

	Selected Jet Bomber and Attack Aircraft							
Make	Model	AR	S_{ref}, ft²	S_{wet}, ft²	f, ft²	C_f	C_{Dmin}	$(L/D)_{max}$	Reference
Boeing	B-47E	9.42	1428	-	21.13	-	0.0148	20	NASA SP-468
	B-52H	8.56	4000	-	47.60	-	0.0119	21.5	NASA SP-468
Convair	B-58A	2.09	1542	-	10.49	-	0.0068	11.3	NASA SP-468
Grumman	A-6E	5.31	529	-	7.64	-	0.0144	15.2	NASA SP-468
Martin	B-57B	4.27	960	-	11.45	-	0.0119	15	NASA SP-468
North American	B-45C	6.74	1175	-	18.80	-	0.0160	16.3	NASA SP-468

	Various Subsonic Military Aircraft (Props and Jets)							
Make	Model	AR	S_{ref}, ft²	S_{wet}, ft²	f, ft²	C_f	C_{Dmin}	$(L/D)_{max}$	Reference
Grumman	S2F Tracker	6.89	485	-	15.25	-	0.0315	12.0	Author
Lockheed	U-2	10.61	1000	-	10.95	-	0.0110	23.0	Author
Lockheed	C-141	7.90	3228	-	46.81	-	0.0145	18.9	Nicolai

	Flying Boats							
Make	Model	AR	S_{ref}, ft²	S_{wet}, ft²	f, ft²	C_f	C_{Dmin}	$(L/D)_{max}$	Reference
Boeing	314	8.06	2867	-	78.56	-	0.0274	13.00	NASA SP-468
Consolidated	Commodore	9.00	1110	-	62.38	-	0.0562	9.40	NASA SP-468
	PB2Y-3	7.43	1780	-	50.02	-	0.0281	12.30	NASA SP-468
	PBY-5A	7.73	1400	-	43.26	-	0.0309	11.90	NASA SP-468
Curtiss	Curtiss F-5L	9.33	1397	-	96.95	-	0.0694	8.60	NASA SP-468
	Curtiss H16	9.40	1164	-	85.92	-	0.0768	8.40	NASA SP-468
	Curtiss HS-2L	8.24	803	-	54.24	-	0.0676	8.20	NASA SP-468
Dornier	Do X	5.12	4844	-	228.64	-	0.0472	7.70	NASA SP-468
Douglas	Dolphin	6.08	592	-	25.46	-	0.0430	8.82	NASA SP-468
Fleetwings	F-5	7.25	235	-	8.11	-	0.0345	10.60	NASA SP-468
Grumman	G-21	6.40	375	-	12.19	-	0.0325	10.50	NASA SP-468
Hall	Hall XP2H-1	5.54	2742	-	79.76	-	0.0291	10.20	NASA SP-468
Loening	Loening OA-1C	4.86	504	-	23.08	-	0.0458	7.64	NASA SP-468
Martin	130	7.88	2170	-	65.75	-	0.0303	11.90	NASA SP-468
	JRM-1	10.86	3683	-	84.34	-	0.0233	16.40	NASA SP-468
	Martin PM-1	5.19	1236	-	59.08	-	0.0478	7.70	NASA SP-468
	P5M-2	9.92	1406	-	38.67	-	0.0275	14.40	NASA SP-468
	PBM-3D	9.89	1408	-	46.04	-	0.0327	13.20	NASA SP-468
Navy-Curtiss	Navy-Curtiss NC-4	8.07	2380	-	213.96	-	0.0899	7.00	NASA SP-468
Sikorsky	S-42	9.73	1340	-	47.51	-	0.0362	12.20	NASA SP-468
	Sikorsky S-38B	7.14	720	-	39.10	-	0.0543	8.50	NASA SP-468

TABLE 15-18 Drag Characteristics of Selected Aircraft—cont'd

	Kitplanes (Homebuilt) and LSA							
Make	Model	AR	S_{ref}, ft²	S_{wet}, ft²	f, ft²	C_f	C_{Dmin}	$(L/D)_{max}$	Reference
Bede	BD-5B	9.75	47.4	170	0.80	0.0048	0.0169	-	Stinton
Cozy	Mark IV	8.94	88.3	-	1.78	-	0.0201	16.5	Author
Glassair	III (Standard Wing)	6.68	81.3	-	1.86	-	0.0229	14.0	Author
Flight Design	CTSW	7.29	107.4	-	3.64	-	0.0338	11.9	Author
Icon	A5 (amphib)	9.03	128	-	5.72	-	0.0447	11.1	Author
Osprey	2 (amphib)	5.20	130	-	5.31	-	0.0408	9.5	Author
	GP-4	5.54	104	-	1.23	-	0.0119	18.0	Author
Progressive Aerodyne	Searey (amphib)	6.06	157	-	6.93	-	0.0441	9.7	Author
Pipistrel	Virus 912 LSA	14.10	118	-	1.50	-	0.0127	23.9	Author
Rutan	LongEz	8.31	82.0	-	1.88	-	0.0230	16.2	Author
	VariEze (wheelpants)	9.28	53.6	-	1.11	-	0.0207	16.5	Author
	VariEze (no whlpnts)	9.28	53.6	-	1.21	-	0.0226	15.8	Author
	VariEze	9.20	53.6	260	1.25	0.0047	0.0233	-	Stinton

	Single Engine GA Aircraft (Piston and Turboprop)							
Make	Model	AR	S_{ref}, ft²	S_{wet}, ft²	f, ft²	C_f	C_{Dmin}	$(L/D)_{max}$	Reference
Beechcraft	Beech 17	6.91	296.5	1000	6.00	0.0059	0.0202	-	Stinton
	Bonanza V-35	6.20	181	-	3.48	-	0.0192	13.8	NASA SP-468
Cessna	150	6.93	159.5	-	4.70	-	0.0295	12.5	Author
	152	6.93	159.5	-	5.52	-	0.0346	11.5	Author
	162 Skycatcher	7.67	120	-	3.95	-	0.0329	12.2	Author
	172 Skyhawk	7.45	174	-	5.06	-	0.0291	12.9	Author
	172 Skyhawk	7.32	175	-	5.58	-	0.0319	11.6	NASA SP-468
	177 (flaps dn)	7.40	175	-	8.23	-	0.047	8.3	Roskam
	177 (flaps up)	7.40	175	-	4.55	-	0.026	11.2	Roskam
	177 Cardinal	7.24	174	-	5.42	-	0.0312	12.3	Author
	177 Cardinal RG	7.24	174	-	3.79	-	0.0218	14.7	Author
	177 Cardinal RG II	7.66	174	-	3.88	-	0.0223	14.2	NASA SP-468
	180 Skywagon	7.45	174	700	4.30	0.0085	0.0247	-	Stinton
	180 Skywagon	7.45	174	-	5.16	-	0.0297	12.8	Author
	182 Skylane	7.45	174	-	4.69	-	0.0270	13.4	Author
	208 Caravan (FG-T)	9.71	279.4	-	9.92	-	0.0355	12.8	Author
Cirrus	SR20	9.16	135.2	-	3.45	-	0.0255	14.8	Author
	SR22	10.00	144.9	-	3.68	-	0.0254	15.3	Author
Commander	114B	7.06	152	-	4.37	-	0.0287	12.7	Author
De Havilland of Canada	DHC-2 Beaver (land)	9.22	250	-	12.19	-	0.0488	10.7	Author
	DHC-2 Beaver (floats)	9.22	250	-	17.00	-	0.0680	9.1	Author
Mooney	M20J	7.45	174.8	-	3.04	-	0.0174	16.7	Author
	231	7.45	174.8	-	3.30	-	0.0189	16.0	AIAA-85-8038
Pacific Aerospace	PAC-750XL	5.78	305	-	9.77	-	0.0320	11.1	Author
Piper	J-3 Cub	5.81	178	-	6.64	-	0.0373	9.6	NASA SP-468
	PA-28 Cherokee	6.02	170	-	6.09	-	0.0258	10.0	NASA SP-468
	PA-38 Tomahawk	9.27	124.7	-	4.85	-	0.0389	12.1	Author
	PA-46 Malibu	10.57	175	-	4.14	-	0.0237	16.1	Author
SOCATA	TB10 (with WF)	8.22	128.04	-	5.70	-	0.0445	10.8	Author
	TB10 (without WF)	8.22	128.04	-	7.16	-	0.0559	9.6	Author
	TBM-700 (RG-T)	8.93	193.7	-	4.78	-	0.0247	14.9	Author
	TBM-850 (RG-T)	8.93	193.7	-	4.51	-	0.0233	15.4	Author

(Continued)

TABLE 15-18 Drag Characteristics of Selected Aircraft—cont'd

Twin Engine GA Aircraft (Piston and Turboprop)									
Make	Model	AR	S_{ref}, ft²	S_{wet}, ft²	f, ft²	C_f	C_{Dmin}	$(L/D)_{max}$	Reference
Aero	Ae-45	8.78	184	-	4.45	-	0.0242	15.0	Author
Beechcraft	D-18	6.51	349	1400	10.50	0.0080	0.0301	-	Stinton
	B55 Baron	7.19	199.2	-	4.59	-	0.0230	14.3	Author
	B8 Queen Air	8.59	293.9	-	6.82	-	0.0232	15.2	Author
Britten-Norman	BN-2	7.39	325	1350	12.00	0.0088	0.0369	-	Stinton
Cessna	310 II	7.61	179	-	4.78	-	0.0267	13.0	NASA SP-468
	310 (clean)	7.00	175.00	-	5.16	-	0.0295	12.3	Roskam
	310 (gear dn, flps 15°)	7.00	175.00	-	11.38	-	0.0650	8.6	Roskam
	310 (gear dn, flps 45°)	7.00	175.00	-	20.48	-	0.1170	6.5	Roskam
	337 Skymaster	7.18	201	-	6.08	-	0.0302	12.5	Author
Partenavia	P-68 Victor	7.74	200.2	-	5.89	-	0.0294	13.0	Author
Piper	PA-23-250 Aztec	6.67	207.6	-	6.74	-	0.0325	11.7	Author

Multi-Engine Commercial Transport Aircraft (Piston)									
Make	Model	AR	S_{ref}, ft²	S_{wet}, ft²	f, ft²	C_f	C_{Dmin}	$(L/D)_{max}$	Reference
Boeing	247D	6.55	836	-	17.72	-	0.0212	13.5	NASA SP-468
Curtiss	C-46A	8.59	1358	5400	26.00	0.0047	0.0191	-	Roskam
Douglas	C-47B	9.24	987	3700	25.50	0.0069	0.0258	-	Roskam
	DC-3	9.14	987	-	24.58	-	0.0249	14.7	NASA SP-468
Lockheed	L-1049G	9.17	1650	-	34.82	-	0.0211	16.0	NASA SP-468

Commercial Jetliners									
Make	Model	AR	S_{ref}, ft²	S_{wet}, ft²	f, ft²	C_f	C_{Dmin}	$(L/D)_{max}$	Reference
Boeing	B-747 (M = 0.7)	7.00	5500	-	88.00	-	0.0160	-	NASA-CR-114494
	B-747 (M = 0.8)	7.00	5500	-	96.80	-	0.0176	-	NASA-CR-114494
	B-747 (M = 0.9)	7.00	5500	-	120.45	-	0.0219	-	NASA-CR-114494

RG = retractable gear, FG = fixed gear, T = turboprop, WF = wheel fairings, f = equivalent flat plate area = $C_{Dmin} \times S_{ref}$.

EXERCISES

(1) Estimate the skin friction coefficient for an airfoil whose chord is 5.25 ft at 25,000 ft and 250 KTAS airspeed on a day on which the outside air temperature is 30 °F warmer than a standard day. Do this using the following assumptions:
  (a) fully laminar boundary layer
  (b) fully turbulent boundary layer assuming incompressible flow,
  (c) fully turbulent boundary layer assuming compressible flow,
  (d) mixed boundary layer for which the transition on the upper surface occurs at chord station 1.3 ft and at 65% chord on the lower surface.

Answer: (a) 0.0005037, (b) 0.003186, (c) 0.003139, (d) 0.002119.

(2) An aircraft has a drag polar given by $C_D = 0.035 + 0.052 C_L^2$. Determine the C_L where LD_{max} occurs and the magnitude of the LD_{max}. (Hint: $\frac{d}{dx}\left(\frac{u}{v}\right) = \frac{u' \cdot v - u \cdot v'}{v^2}$)

Answer: C_L where LD_{max} occurs is 0.8549, LD_{max} is 11.71.

(3) Consider the wing shown in Figure 15-77 and for which the representative skin friction coefficients for each of the three wing segments have already been calculated. Note that all data required for geometric evaluation is given in the figure. Assume a wetted area booster factor of 1.07. Determine the total skin friction coefficient and total skin friction drag coefficient for the wing half. Estimate the total skin friction drag force of both wing halves at airspeed of 150 KTAS at S-L on a standard day.

Answer: $C_f = 0.005045$, $C_{Df} = 0.01080$, $D_f = 183$ lb$_f$.

(4) An airplane has two dissimilar airfoils at the root and tip of the wing (see Figure 15-78). Important dimensions to use are:

$$b = 18 \text{ ft}, C_R = 3 \text{ ft}, C_T = 2 \text{ ft} \quad \text{and} \quad S_{ref} = 45 \text{ ft}^2.$$

The root airfoil is a NLF airfoil capable of sustaining 55% laminar flow on the upper surface and 35% on the lower. The tip airfoil is a turbulent flow airfoil that sustains laminar flow to 15% on the upper and lower surfaces.

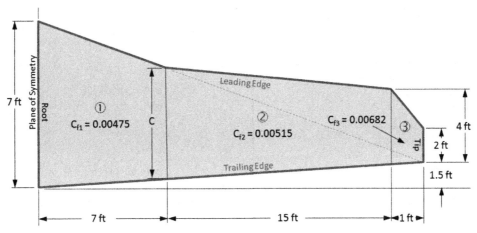

FIGURE 15-77 Wing used in Exercise (3).

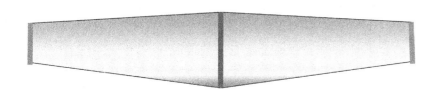

FIGURE 15-78 Wing used in Exercise (4).

If the airplane is cruising at 100 KTAS at S-L ISA, determine the skin friction drag coefficient and force acting on the wing due to the mixed laminar and turbulent BL regions. Compare to a wing with fully laminar or fully turbulent BL.

Answer: $C_f = 0.003115$, $C_{Df} = 0.006230$, $D_f = 9.5$ lb$_f$.

(5) Extract the total, induced, and minimum drag coefficient for the single-piston-engine propeller-powered Cessna 172N Skyhawk II using the data below obtained from its POH. Hint: use Equation (9-89) to estimate the Oswald span efficiency and assume a propeller efficiency of 0.80.

Wing span is 36.08 ft, wing area is 174 ft^2, cruising speed at 8000 ft and 75% power is 122 KTAS. Engine is a Lycoming O-320 rated at 160 BHP at S-L. Gross weight is 2300 lb$_f$.

Answer: $C_D = 0.03714$, $C_{Dmin} = 0.03141$, $C_{Di} = 0.005728$.

(6) Extract the C_{LminD}, C_{Dmin}, and e for the wind tunnel data (points) shown in the table and graph of Figure 15-79. The solid line is a least-squares curve-fit whose constants and correlation coefficients are shown in the legend. The AR for the airplane being tested is 6.

Answer: $C_{LminD} = 0.2574$, $C_{Dmin} = 0.0252$, $e = 0.5584$.

VARIABLES

Symbol	Description	Units (UK and SI)
$\dot{m}$	Mass (or weight) flow rate	lb$_m$/s or kg/s
$\dot{Q}$	Heat transfer	ft·lb$_f$/s or J/s=W
$(x/c)_{max}$	Location of maximum airfoil thickness	
A, B, C	Constants	
A_E	Exit area of a nozzle	ft^2 or m^2
A_{IN}	Inlet area of a diffuser	ft^2 or m^2
A_{max}	Maximum fuselage cross-sectional area	ft^2 or m^2
AOA	Angle-of-attack	Degrees or radians
AR	Wing aspect ratio (context dependent)	
A_π	Equivalent parasite area	ft^2 or m^2
b	Wing span	ft or m
C	Reference length (e.g. chord)	ft or m
C_D	Drag coefficient	
C_{Df}	Skin friction drag coefficient	
C_{Di}	Induced drag coefficient	
$C_{DL\&P}$	Leakage and protuberance drag coefficient	

Symbol	Description	Units (UK and SI)
C_{Dmin}	Minimum drag coefficient	
C_{Dmisc}	Miscellaneous drag coefficient	
C_{Do}	Basic drag coefficient	
C_{Dw}	Wave drag coefficient	
$C_{D\pi}$	Component equivalent drag coefficient	
C_f	Skin friction coefficient	
$C_{f\ lam}$	Skin friction coefficient for laminar boundary layer	
$C_{f\ turb}$	Skin friction coefficient for turbulent boundary layer	
C_{fo}	Reference skin friction coefficient	
C_L	Lift coefficient	
C_{L0}	Basic lift coefficient, i.e. where $\alpha = 0$ (context dependent)	
C_{L0}	Incompressible lift coefficient (context dependent)	
C_{LminD}	Lift coefficient of minimum drag	
C_M	Pitching moment coefficient	
C_{M0}	Incompressible pitching moment coefficient	
C_P	Pressure coefficient	
C_{Po}	Incompressible pressure coefficient	
C_R	Root chord	ft or m
C_T	Tip chord	ft or m
d	Reference width	ft or m
D	Drag force	lb_f or N
D_0	Basic drag force	lb_f or N
D_C	Cooling drag force	lb_f or N
D_f	Skin friction drag force	lb_f or N
D_i	Lift-induced drag force	lb_f or N
e	Span or Oswald efficiency	
E	Kinetic energy	$ft \cdot lb_f$; $N \cdot m$ or J
E_∞	Kinetic energy at some specific condition	$ft \cdot lb_f$; $N \cdot m$ or J
E_0	Kinetic energy at some specific condition	$ft \cdot lb_f$; $N \cdot m$ or J
f	Equivalent flat plate (parasite) area	ft^2 or m^2
f	Fineness ratio	
FF	Form factor	
H_P	Pressure altitude	ft or m
IF	Interference factor	
k	Lift-induced drag coefficient (context dependent)	

Symbol	Description	Units (UK and SI)
k	Pressure recovery factor (context dependent)	
l	Reference length	ft or m
L	Lift force (context dependent)	lb_f or N
L	Reference length (context dependent)	ft or m
m	Mass	lb_m or slugs; kg
M	Mach number (context dependent)	
N	Number of surfaces or components	
P or p	Pressure	psi or psf; N/m^2 or Pa
P_0 or P_∞	Far-field pressure	psi or psf; N/m^2 or Pa
q	Dynamic pressure	psi or psf; N/m^2 or Pa
Q_B	Heat flow into heat exchanger	BTU/s; J/s = W
R	Resultant force (context dependent)	lb_f or N
R	Specific gas constant for air (context dependent)	$ft \cdot lb_f/slug \cdot °R$ or J/kg K
R_e	Reynolds number	
$R_{e\ cutoff}$	Cutoff Reynolds number	
S or S_{ref}	Wing reference area	ft^2 or m^2
S_{wet}	Wetted reference area	ft^2 or m^2
T	Outside air temperature	°F or °R; °C or K
t	Time	seconds
T	Temperature	°F or °R; °C or K
T_∞	Temperature in the streamtube behind the nozzle	°F or °R; °C or K
t/c	Thickness-to-chord ratio	
T_0	Far-field temperature	°F or °R; °C or K
u, v, w	x, y, z components of the velocity of air	ft/s or m/s
V or V_∞	Far-field airspeed	ft/s or m/s
V_∞	Airspeed in the streamtube behind the nozzle	ft/s or m/s
V_0 or V_∞	Far-field airspeed	ft/s or m/s
V_E	Airspeed at the exit	ft/s or m/s
V_{fpm}	Airspeed in ft/min	ft/min
$V_{ft/s}$	Airspeed in ft/s	ft/s
V_{KTAS}	Airspeed in knots, true airspeed	knots
V_V	Vertical airspeed	ft/s or m/s
W	Mechanical work	$ft \cdot lb_f$; $N \cdot m$ or J
x	Generic distance from LE to some specific point	ft or m

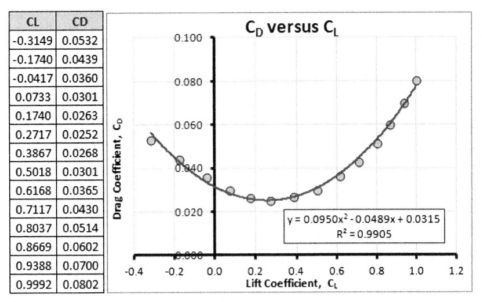

FIGURE 15-79 Information used in Exercise (6).

Symbol	Description	Units (UK and SI)
X_0	Location where fictitious turbulent boundary layer starts	ft or m
X_{tr}	Transition point	ft or m
X_{tr_lower}	Transition point on upper wing surface	ft or m
X_{tr_upper}	Transition point on upper wing surface	ft or m
z	Generic height above some surface	ft or m
ΔC_{Dmin}	Drag increment	
ΔW	Rate at which work is being extracted	ft·lb$_f$/s; N·m/s or J/s
$\Lambda_{t\,max}$	Sweep angle of maximum thickness line	Degrees or radians
α	Angle-of-attack	Degrees or radians
β	Yaw angle (context dependent)	Degrees or radians
β	Prandtl-Glauert compressibility corrector (context dependent)	Degrees or radians
γ	Ratio of specific heats = 1.4 for air	
κ	Skin roughness value	
μ	Air viscosity	lb$_f$·s/ft^2 or N·s/m^2
ρ	Air density	slugs/ft^3 or kg/m^3
ρ_{B1}	Air density at the baffle front face	slugs/ft^3 or kg/m^3
ρ_{B2}	Air density at the baffle aft face	slugs/ft^3 or kg/m^3
ρ_E	Air density at the nozzle exit	slugs/ft^3 or kg/m^3

References

[1] Levy David. AIAA CFD Drag Prediction Workshop, Data Summary and Comparison June 9–10, 2001.
[2] http://aaac.larc.nasa.gov/tsab/cfdlarc/aiaa-dpw/.
[3] Hoerner Sighard F. Fluid-Dynamic Drag. L. Hoerner; 1965.
[4] Schlichting Hermann. Boundary Layer Theory. English ed. Pergamon Press; 1955. p. 16.
[5] Young AD. Boundary Layers. AIAA Education Series; 1989.
[6] Schetz Joseph A, Rodney D W Bowersox. Boundary Layer Analysis. AIAA Education Series; 2011.
[7] Roskam Jan, Chuan-Tau Edward Lan. Airplane Aerodynamics and Performance. DARcorporation; 1997. Equations (2.90) and (2.91).
[8] Hoak DE. USAF Stability and Control DATCOM. Flight Control Division, Air Force Flight Dynamics Laboratory; 1978. Section 4.1.5.1.
[9] NACA-TR-572. Determination of the Characteristics of Tapered Wings. Anderson, Raymond F.; 1936.
[10] R.&M. No. 1226, The Characteristics of a Tapered and Twisted Wing with Sweep-Back. Aeronautical Research Council, Glauert, H., and S. B. Gates; 1929.
[11] Hueber J. Die Aerodynamischen Eigenschaften von Doppel-Trapezförmigen Tragflügeln. Z.F.M.; 13 May 1933. pp. 249–251; 29 May 1933, pp. 269–272.
[12] NASA SP-2005-4539. Innovation In Flight: Research of the Nasa Langley Research Center on Revolutionary Advanced Concepts for Aeronautics. Chambers, Joseph R.; 2005. pp. 123–161.
[13] Braslow Albert L. A History of Suction-Type Laminar-Flow Control with Emphasis on Flight Research. Monographs in Aerospace History Number 13, NASA; 1999.
[14] Marshall Laurie A. Boundary-Layer Transition Results From the F-16XL-2 Supersonic Laminar Flow Control Experiment. NASA TM-1999–209013; 1999.
[15] van de Wal HJB. Design of a Wing with Boundary Layer Suction. M.Sc. Thesis. Delft University of Technology; 2010.
[16] Wortman Andrzej. Reduction of Fuselage Form Drag by Vortex Flows. Journal of Aircraft, May–June 1999;vol. 36(No. 3).
[17] Kentfield JAC. Drag Reduction by Means of Controlled Separated Flows. AIAA-1985-1800-493; 1985.
[18] NASA TM-77785 Profile Design for Wings and Propellers. Quast, A., and K. H. Horstmann; Nov. 1984.

[19] Peigin S, Epstein B. Aerodynamic Optimization of Essentially Three-Dimensional Shapes for Wing-Body Fairing. AIAA Journal, July 2008;vol. 46(No. 7).
[20] NACA R-116 Applications of Modern Hydrodynamics to Aeronautics. Prandtl, Ludwig; 1923.
[21] Whitford Ray. Design for Air Combat. Jane's; 1987.
[22] Torenbeek Egbert. Synthesis of Subsonic Aircraft Design. 3rd ed. Delft University Press; 1986. p. 499.
[23] Roskam Jan. Methods for Estimating Drag Polars of Subsonic Airplanes. 4th printing 1984.
[24] Perkins Courtland D, Robert EHage. Airplane Performance, Stability, and Control. John Wiley & Sons; 1949.
[25] Raymer Daniel. Aircraft Design: A Conceptual Approach. 4th ed. AIAA Education Series; 2006.
[26] Torenbeek Egbert. Synthesis of Subsonic Aircraft Design. 3rd ed. Delft University Press; 1986.
[27] Shevell Richard S. Fundamentals of Flight. Prentice Hall; 1983.
[28] Nicolai Leland. Fundamentals of Aircraft Design. 2nd ed. 1984.
[29] Jenkinson Lloyd R. Civil Jet Aircraft Design. Arnold; 1999.
[30] NACA WR-L-489. Drag Analysis of Single-Engine Military Airplanes Tested in the NACA Full-Scale Wind Tunnel Dearborn. C. H., and Abe Silverstein; 1940.
[31] McCormick Barnes W. Aerodynamics, Aeronautics, and Flight Mechanics. John Wiley & Sons; 1979.
[32] NACA R-485 The Drag of Airplane Wheels, Wheel Fairings, and Landing Gears – I. Herrnstein, William H., and David Biermann; 1935.
[33] Austyn-Mair W, Birdsall David L. Aircraft Performance. Cambridge, England: Cambridge University Press; 1992. p. 124. Method is based on ESDU 79015, *Undercarriage Drag Prediction Methods*.
[34] Jenkinson Lloyd R, Simpkin Paul, Rhodes Darren. Civil Jet Aircraft Design. AIAA Education Series; 1999.
[35] NACA WR-L-238 Wind-Tunnel Tests of Four Full-Scale Seaplane Floats. Conway, Robert N., and Julian D. Maynard; 1943.
[36] NACA TN-716 Hydrodynamic and Aerodynamic Tests of a Family of Models of Seaplane Floats with Varying Angles of Dead Rise. Parkinson, John B., Roland E. Olson, and Rufus O. House; 1939.
[37] Young AD. The Aerodynamic Characteristics of Flaps. R.&M. No. 2622, British A. R. C.; 1947.
[38] Corke Thomas C. Design of Aircraft. Prentice-Hall; 2003.
[39] Linden F, Robert Van Der. The Boeing 247: The First Modern Airliner. University of Washington Press; 1991.
[40] NACA WR-L-462 Investigation of Drag and Pressure Distribution of Windshields at High Speeds. Wright, Ray M.; 1942.
[41] NACA R-464 Negative Thrust and Torque Characteristics of an Adjustable-Pitch Metal Propeller. Hartman, Edwin P.; 1934.
[42] NACA WR-L-502 High-Speed Wind-Tunnel Tests of Gun Openings in the Nose of the Fuselage of a 1/4-Scale Model. Fedziuk, Henry A.; 1942.
[43] NACA WR-L-292 Effects of Simulated Ice Formation on the Aerodynamic Characteristics of an Airfoil. Gulick, Beverly G.; 1938.
[44] NACA TN-2962 Effect of Ice and Frost Formations on Drag of NACA 65(sub 1)-212 Airfoil for Various Modes of Thermal Ice Protection. Gray, V. H., and U. H. Von Glahn; 1953.
[45] *Jane's All the World's Aircraft*. Various editors. Janes Yearbooks, various years.
[46] Stinton Darrol. The Design of the Aeroplane. Collins; 1983.
[47] Roskam Jan. Airplane Design. Part VI. DARcorporation; 2000.
[48] Loftin Jr Laurence K. Quest for Performance; The Evolution of Modern Aircraft. NASA SP-468; 1985.
[49] Hanke CR, Nordwall DR. The Simulation of a Jumbo Jet Transport Aircraft, Volume 2: Modeling Data. NASA-CR-114494; 1970.
[50] NACA TN-525 The Aerodynamic Drag of Flying-Boat Hull Model as Measured in the NACA 20-foot Wind Tunnel I. Hartman, Edwin P.; 1935.

CHAPTER 16

Performance — Introduction

OUTLINE

16.1 Introduction 761
   16.1.1 The Content of this Chapter 762
   16.1.2 Performance Padding Policy 762

16.2 Atmospheric Modeling 763
   16.2.1 Atmospheric Ambient Temperature 763
   16.2.2 Atmospheric Pressure and Density for Altitudes below 36,089 ft (11,000 m) 764
   16.2.3 Atmospheric Property Ratios 764
   16.2.4 Pressure and Density Altitudes below 36,089 ft (11,000 m) 765
   16.2.5 Density of Air Deviations from a Standard Atmosphere 765
      Change in Density Due to Humidity 766
   16.2.6 Frequently Used Formulas for a Standard Atmosphere 767

16.3 The Nature of Airspeed 768
   16.3.1 Airspeed Indication Systems 768
   16.3.2 Airspeeds: Instrument, Calibrated, Equivalent, True, and Ground 769
      Instrument Airspeed 769
      Speed of Sound 770
      Calibrated Airspeed 770
      Equivalent Airspeed 770
      True Airspeed 770
      Ground Speed 770
   16.3.3 Important Airspeeds for Aircraft Design and Operation 771

16.4 The Flight Envelope 774
   16.4.1 Step-by-Step: Maneuvering Loads and Design Airspeeds 775
      Step 1: Establish Load Factors n_+ and n_- 776
      Step 2: Design Cruising Speed, V_C 776
      Step 3: Design Dive Speed, V_D 776
      Step 4: Design Maneuvering Speed, V_A 777
      Step 5: Design Speed for Maximum Gust Intensity, V_B 777
      Step 6: Set Up the Initial Diagram 777
   16.4.2 Step-by-Step: Gust Loads 778
      Step 7: Calculated Gust-related Parameters 779
      Step 8: Calculate Gust Load Factor as a Function of Airspeed 780
      Step 9: Calculate Gust Load Factor as a Function of Airspeed 780
      Step 10: Finalize Gust Diagram 780
      Convenient Relations to Determine Location of Intersections 780
   16.4.3 Step-by-Step: Completing the Flight Envelope 782
      V-n Diagrams with Deployed High-lift Devices 782
   16.4.4 Flight Envelopes for Various GA Aircraft 782

16.5 Sample Aircraft 783
   16.5.1 *Cirrus SR22* 783
   16.5.2 *Learjet 45XR* 785

Exercises 786

Variables 787

References 789

16.1 INTRODUCTION

Any flight can be split into a number of phases that are clearly distinct by their nature. These are the *take-off*, *climb*, *cruise*, *descent*, and *landing*. In addition to these, there are a large number of maneuvers performed while airborne that involve acceleration of one kind or another; for instance, turning flight, rolls, loops, and many others. The purpose of the next six chapters is to present aircraft performance theory in a systematic manner intended to be particularly helpful to the aircraft designer. The titles of these chapters are in an order of occurrence during the flight:

17. Performance − take-off.
18. Performance − climb.
19. Performance − cruise.
20. Performance − range and endurance.
21. Performance − descent.
22. Performance − landing.

A proper prediction of aircraft performance is another extremely important step in the entire design process. Performance and payload are what sell an aircraft more than anything else. The same scenario holds here as for the estimation of weight and drag − an erroneous prediction will manifest itself as soon as the aircraft takes off for the first time and can devastate a development program, if not cancel it altogether.

This section serves as a prologue to the performance methods. Here, topics that apply to all the performance methods, such as types of airspeed and the atmospheric model, will be presented. It will also introduce a couple of example aircraft to be used in the subsequent chapters. The methods presented here are proven and standard in the industry. However, there are important limitations that must be brought up. The quality of the drag model weighs the most. The aspiring aircraft designer is advised to acquire experience with these methods by applying them to existing aircraft for which performance data has been published. This will build experience and an understanding of their accuracy that serves well when assessing the performance of new designs.

As discussed in Chapter 15, *Aircraft Drag Analysis*, there are typically three drag models: the simplified, adjusted, and non-quadratic. The first two models assume the induced drag can be represented in terms of the lift coefficient squared (they are quadratic). Both become inaccurate at high *AOA*s, in particular the simplified model and therefore, generally, it should be avoided. The non-quadratic is typically used to evaluate the performance of sailplanes, as they feature a drag bucket that cannot be described by the quadratic models.

The simplified drag model is only usable for aircraft whose C_{LminD} is zero. This is rare as most airplanes use cambered airfoils. Some fighter aircraft and aerobatic aircraft are designed with airfoils that have a very small camber or are fully symmetrical. Such airplanes may have $C_{LminD} = 0$, although the geometry of their fuselages may shift their drag polars, rendering a non-zero C_{LminD}. The non-quadratic drag model, on the other hand, must usually be represented in the form of a lookup table. The performance methods presented will work equally with all the models, although the accuracy of the predictions will depend on the selected model. The reader should keep in mind that subsequent sections will invariably use the simplified drag model to present a close-form expression of performance concepts. The primary advantage of the simplified drag model is that it allows for clear and concise formulation, which is much harder to accomplish using the adjusted or non-quadratic drag models. Several methods will also derive expressions using the adjusted drag model.

16.1.1 The Content of this Chapter

- **Section 16.2** presents a description of how to model the atmosphere.
- **Section 16.3** presents methods to calculate true airspeed, equivalent airspeed, and other important airspeeds.
- **Section 16.4** presents the V-n diagram and shows how to create it.
- **Section 16.5** presents two sample aircraft that will be used to demonstrate performance calculations: the Cirrus SR22 and the Learjet 45XR.

16.1.2 Performance Padding Policy

Performance information constitutes a very sensitive portion of the aircraft development. It is important for the design team to recognize that performance predictions released internally can be used in both a constructive and a deconstructive manner. One concern is that if the marketing department gets their hands on such predictions, they will use it to sell airplanes that have yet to be built and flown. Of course, selling airplanes is very desirable; however, to the marketing people, a predicted cruising speed of 276 KTAS means a real cruising speed of 276 KTAS. On the other hand, to the performance analyst who understands the shortcomings of the predictions, this cruising speed means a possible 270 to 276 KTAS. All of this is fine, unless of course the airplane turns out to be capable of 270 KTAS, or worse yet, only 265 KTAS. Then, the manufacturer will have to spend considerable effort and loss of revenue pleasing unhappy customers who were promised a new airplane capable of 276 KTAS.

To prevent such annoyances, many businesses require conservatism in prediction through padding factors. The cognizant design lead is urged to consider such factors and ensure the performance prediction team establishes a padding policy that enjoys a *need-to-know-only* status (i.e. no one outside the Aero group knows the actual padding factors). So, if the predicted cruising speed is 276 KTAS, then marketing is told 272 KTAS, or some other reasonable figure. The purpose is not deceit but financial protection.

TABLE 16-1 Recommended Padding Factors

Performance phase	Parameter	Multiply by	
		Normal performance	High performance
Taxi and take-off	Fuel quantity in lb$_f$	1.25	1.50 to 1.75
Take-off	T-O field length, ft	1.10	1.10
Climb	Rate-of-climb, fpm	0.95	0.97
Cruise	Cruising speed, KTAS	KTAS − 4	KTAS − 6
Descent	LD_{max}	0.90	0.95
Landing	Landing field length, ft	1.10	1.10
Range	Total range, nm	0.96	0.98

Additionally, disclosure of performance information should be done with care, because the information can hurt the competitive edge of the company. For instance, an ill-tempered test pilot shouting something along the lines of "the darn thing barely climbs" in a moment of frustration may be echoed elsewhere, perhaps by a careless technician who happened to overhear the pilot. He or she might be unaware that the airplane was being tested with deployed landing gear, at gross weight, and at a high-density altitude. Something as seemingly innocent as that can easily start a rumor that is used by a competitor against the developer.

Table 16-1 shows recommended padding factors for normal and high-performance GA aircraft. As an example of how it is used, consider an aircraft whose T-O field length has been predicted to be 1200 ft. The table suggests the performance team should report 1320 ft.

16.2 ATMOSPHERIC MODELING

Atmospheric modeling is the determination of the properties of air in which an airplane is operated. The properties are outside air temperature, pressure, density, and viscosity. The ability to accurately quantify them is absolutely imperative for the evaluation of the large number of aerodynamic characteristics of an airplane. Our understanding of the atmosphere is extensive, although it is by no means complete. Among important discoveries is that the atmosphere is stratified. The most active layer is the one closest to ground level; the *troposphere*. While most aircraft operate in this layer, flying above it is becoming more common. Atmospheric science has also revealed there can be substantial wind speeds in the layers above the troposphere. The current altitude record set by an aircraft is held by a modified Russian MiG-25 fighter aircraft, which on August 31, 1977, climbed to an estimated altitude of 123,524 ft (37.65 km) [1]. In August 2001, an unmanned experimental solar-powered aircraft, aptly named Helios, designed and built by NASA, climbed to an altitude of 96,863 ft (31.78 km) [2], an official world record altitude for a propeller-powered aircraft. In June 2003, the aircraft broke up in midair after an encounter with atmospheric turbulence (Accident report is available from Ref. [3]).

A detailed atmospheric model, based on document *US Standard Atmosphere 1976*, published by NOAA (National Oceanic and Atmospheric Administration), NASA (National Aeronautics and Space Administration), and the US Air Force, extending up to approximately 85 km (280,000 ft) is provided in Appendix A, *Atmospheric modeling*. The reader is directed toward the appendix for information regarding the higher altitudes, as well as a Visual Basic for Applications code intended for use with Microsoft Excel, which calculates the aforementioned properties with ease. In the interests of space, only fundamental equations needed to obtain temperature, pressure, and density in the troposphere are provided in this section. All derivations are provided in the appendix. The reader is well advised to review the appendix as it contains a large number of very useful equations intended to estimate other properties of the atmosphere, as well as deviations from standard atmosphere.

16.2.1 Atmospheric Ambient Temperature

Let's start by considering the temperature; T. Change in air temperature with altitude can be approximated using a linear function:

$$T = T_0 + a(h - h_0) \qquad (16\text{-}1)$$

An alternative form of Equation (16-1) is:

$$T = T_0(1 + \kappa \cdot h) \qquad (16\text{-}2)$$

where

a = lapse rate
h = altitude in ft or m

h_0 = reference altitude h_0
T = temperature at altitude h
T_0 = temperature at reference altitude h_0
κ = lapse rate constant = a/T_0

In the world of aviation T is usually referred to as the "outside air temperature" or "ambient temperature." Practically all aircraft operate inside the altitude band ranging from S-L to 11,000 m (36,089 ft). That part of the atmosphere is, thus, of considerable interest to us. The variables to be used with the above equations in that band are summarized in Table 16-2.

Note that the author prefers to write the constants explicitly, rather than in the scientific format (e.g. -6.8756×10^{-6}) because it is simply faster to enter using calculators. The author recognizes that this may annoy some readers and empathizes if this is the case. On the flipside, using the scientific format for a number with only five zeros can be annoying and therefore the author uses the scientific format for six or more zeros. The following mnemonic will help you remember the number of zeros when entering the lapse rate constant. Say "one zero, two zeros, three zeros" while typing '0.', '00', '000'. Then enter '68756.'

16.2.2 Atmospheric Pressure and Density for Altitudes below 36,089 ft (11,000 m)

The hydrostatic equilibrium equations allow the pressure, p, and density, ρ, to be calculated as functions of altitude, h, as follows:

Pressure:

$$p = p_0(1 + \kappa \cdot h)^{5.2561} \quad (16\text{-}3)$$

Density:

$$\rho = \rho_0(1 + \kappa \cdot h)^{4.2561} \quad (16\text{-}4)$$

where

p_0 = reference S-L pressure
ρ_0 = reference S-L density
κ = lapse rate constant, which is obtained from Table 16-2

16.2.3 Atmospheric Property Ratios

The pressure, density, and temperature often appear in formulation as fractions of their baseline values. Consequently, they are identified using special characters and are called pressure ratio, density ratio, and temperature ratio.

Temperature ratio:

$$\theta = \frac{T}{T_0} = (1 - 0.0000068756h) \quad (16\text{-}5)$$

Pressure ratio:

$$\delta = \frac{p}{P_0} = (1 - 0.0000068756h)^{5.2561} = \theta^{5.2561} = \sigma^{1.235} \quad (16\text{-}6)$$

Density ratio:

$$\sigma = \frac{\rho}{\rho_0} = (1 - 0.0000068756h)^{4.2561} = \theta^{4.2561} = \frac{\delta}{\theta}$$
$$= \delta^{0.8097} \quad (16\text{-}7)$$

TABLE 16-2 Common Temperature Constants in the Troposphere

	Symbol	UK system	SI system
Reference altitude	H or h	ft	m or km
Reference temperature	T_0	518.67 °R	273.15 K
Lapse rate[a]	a	−0.00356616 °R/ft	−0.0065 K/m −6.5 K/km
Lapse rate constant	κ	$\kappa = -0.003566/518.666$ $= -0.0000068756/\text{ft}$	$\kappa = -0.0065/288.15$ $= -0.000022558/\text{m}$ $\kappa = -6.5/288.15$ $= -0.022558/\text{km}$

[a] "The rate at which air cools or warms depends on the moisture status of the air. If the air is dry, the rate of temperature change is 1 °C/100 meters and is called the dry adiabatic rate (DAR). If the air is saturated, the rate of temperature change is 0.6 °C/100 meters and is called the saturated adiabatic rate (SAR). The DAR is a constant value, that is, it's always 1 °C/100 meters. The SAR varies somewhat with how much moisture is in the air, but we'll assume it to be a constant value here. The reason for the difference in the two rates is due to the liberation of latent heat as a result of condensation. As saturated air rises and cools, condensation takes place. Recall that as water vapor condenses, latent heat is released. This heat is transferred into the other molecules of air inside the parcel causing a reduction in the rate of cooling" [4].

EXAMPLE 16-1

Determine the state of the atmosphere at 8500 ft on a standard day, using the UK system.

Solution

Pressure:
$$p = 2116(1 - 0.0000068756 \times 8500)^{5.2561} = 1542 \text{ psf}$$

Density:
$$\rho = 0.002378(1 - 0.0000068756 \times 8500)^{4.2561}$$
$$= 0.001840 \text{ slugs/ft}^3$$

Temperature:
$$T = 518.67(1 - 0.0000068756 \times 8500) = 488.4 \text{ °R}$$
$$= 28.7 \text{ °F}$$

Temperature ratio:
$$\theta = \frac{T}{T_0} = (1 - 0.0000068756 \times 8500) = 0.9416$$

Pressure ratio:
$$\delta = \frac{p}{P_0} = (1 - 0.0000068756 \times 8500)^{5.2561} = 0.7287$$

Density ratio:
$$\sigma = \frac{\rho}{\rho_0} = (1 - 0.0000068756 \times 8500)^{4.2561} = 0.7739$$

16.2.4 Pressure and Density Altitudes below 36,089 ft (11,000 m)

Sometimes the pressure or density ratios are known for one reason or another. It is then possible to determine the altitudes to which they correspond. For instance, if the pressure ratio is known, we can calculate the altitude to which it corresponds. The altitude is then called *pressure altitude*. Similarly, from the density ratio we can determine the *density altitude*.

Pressure altitude:

$$h_P = 145442 \left[1 - \left(\frac{p}{p_0}\right)^{0.19026} \right] \quad (16\text{-}8)$$

Density altitude:

$$h_\rho = 145442 \left[1 - \left(\frac{\rho}{\rho_0}\right)^{0.234957} \right] \quad (16\text{-}9)$$

16.2.5 Density of Air Deviations from a Standard Atmosphere

Atmospheric conditions often deviate from the models shown above. Often it is because the atmospheric temperature differs from the average temperature, due to meteorological conditions. Such deviations can be handled using the equation of state $\rho = p/RT$, and calculating the pressure using Equation (16-3) and introducing the temperature deviation directly. This is reflected below, using the UK system:

$$\rho = \frac{1.233(1 + \kappa \cdot h)^{5.2561}}{(T + \Delta T_{ISA})} \quad (16\text{-}10)$$

where

h = reference altitude in ft
T = standard day temperature at the given altitude per the International Standard Atmosphere in °R
At S-L it would be 518.67 °R, at 10,000 ft it would be 483 °R, and so on
ΔT_{ISA} = deviation from the International Standard Atmosphere in °F or °R

SI system:

$$\rho = \frac{352.6(1 + \kappa \cdot h)^{5.2561}}{(T + \Delta T_{ISA})} \quad (16\text{-}11)$$

where

h = reference altitude in m
T = standard day temperature at the given altitude per the International Standard Atmosphere, in degrees K. At S-L it would be 288.15 K, at 10,000 ft it would be 483 °R, and so on
ΔT_{ISA} = deviation from the International Standard Atmosphere in K or C

For non-standard atmosphere, use a negative sign for colder and a positive sign for warmer than ISA for ΔT_{ISA}.

EXAMPLE 16-2

Determine the density of air at 8500 ft on a day that is 30 °F colder and then 30 F hotter than a standard day.

Solution

First, determine the ISA temperature at 8500 ft using the value of $\kappa = -0.0000068756/\text{ft}$ (from Table 16-3):

$$T = 518.67(1 - 0.0000068756 \cdot 8500) = 488.4 \text{ °R}$$

Then, "plug and chug" into Equation (16-10):
30 °F colder:

$$\rho = \frac{1.233}{(488.4 - 30)}(1 - 0.0000068756 \cdot 8500)^{5.2561}$$

$$= 0.001960 \text{ slugs/ft}^3$$

30 °F warmer:

$$\rho = \frac{1.233}{(488.4 + 30)}(1 - 0.0000068756 \cdot 8500)^{5.2561}$$

$$= 0.001733 \text{ slugs/ft}^3$$

These are 6.5% greater and 5.8% lighter than standard day density (0.001840 slugs/ft^3), respectively.

TABLE 16-3 Standard Properties of Air at S-L

Property	Symbol	UK system	SI system
Specific gas constant for air	R	1716 ft·lb$_f$/(slug·°R)	286.9 m^2/(K·s^2)
Specific gas constant for water vapor	R_{H_2O}	2760 ft·lb$_f$/(slug·°R)	461.5 m^2/(K·s^2)
Pressure	P	2116.2 lb$_f$/ft^2 14.696 lb$_f$/in^2 29.92 inHg	1.01325 × 10^5 N/m^2 (or Pa) 760 mmHg
Density	ρ	0.002378 slugs/ft^3	1.225 kg/m^3
Temperature	T	518.69 °R 59.0 °F	288.16 K 15.0 C
Absolute viscosity	μ	3.737 × 10^{-7} lb$_f$·s/ft^2	1.789 × 10^{-3} N·s/m^2
Kinematic viscosity	ν	1.572 × 10^{-4} 1/(ft^2·s)	1.460 × 10^{-3} 1/(m^2·s)
Speed of sound	a	1116.4 ft/s	340.3 m/s

Change in Density Due to Humidity

In addition to temperature, humidity also affects density. Under certain circumstances, it may be necessary to account for this phenomenon when estimating aircraft performance — in particular T-O and landing performance. This section presents a method to account for humidity. Humidity is the amount of water vapor present in air. Humidity is typically expressed using any of the following methods:

- *Absolute humidity*, which is the mass of water vapor per unit volume of air. It is presented as a dimensionless number.
- *Relative humidity*, which is the ratio of the water vapor pressure present in air to the vapor pressure that would saturate it [5] (and cause precipitation) — if 1.00 (or 100%), precipitation will occur. This is what is usually reported by weather forecasters on TV.
- *Specific humidity*, which is the mass of water vapor per unit mass of air, including the water vapor — usually expressed as grams of H$_2$O per kilogram of air. Also referred to as *humidity ratio*.

As a rule of thumb, *the density of dry air is higher than that of humid air*. The presence of water molecules in air displaces the oxygen and nitrogen atoms so their amount per unit volume decreases. As a consequence, the mass of a unit volume of the humid air decreases and the density is reduced. A general expression for the density of moist air is given below:

$$\rho = \rho_{std}\left(\frac{1+x}{1+xR_{H_2O}/R}\right) = \rho_{std}\left(\frac{1+x}{1+1.609x}\right) \quad (16\text{-}12)$$

where

ρ_{std} = density at altitude, calculated by standard methods

R = specific gas constant for air, see Table 16-3
R_{H_2O} = specific gas constant for water vapor, see Table 16-3
x = humidity ratio in kg water vapor per kg of air

Humidity is commonly represented using relative humidity (*RH*), presented as a percentage (e.g. 50% humidity). If the ambient temperature is known in °C, this can be converted into a humidity ratio using the following relation:

$$x = \left(\frac{RH}{100}\right) 0.003878 \cdot e^{0.0656 \cdot (T \, °C)} \quad (16\text{-}13)$$

Pressure in Pa and mbar (h is in m):

$$\begin{aligned} p_{Pa} &= 101325 \cdot (1 - 0.000022558 \cdot h)^{5.2561} \\ p_{mbar} &= 1013.25 \cdot (1 - 0.000022558 \cdot h)^{5.2561} \end{aligned} \quad (16\text{-}17)$$

Density in slugs/ft³ (h is in ft):

$$\rho_{slugs/ft^3} = 0.002378 \cdot (1 - 0.0000068756 \cdot h)^{4.2561} \quad (16\text{-}18)$$

Density in kg/m³ (h is in m):

$$\rho_{kg/m^3} = 1.225 \cdot (1 - 0.000022558 \cdot h)^{4.2561} \quad (16\text{-}19)$$

EXAMPLE 16-3

If the outside temperature and relative humidity on a standard day at S-L are reported as 15°C and 50%, respectively, determine the density of air.

Solution

On a standard day, this yields a humidity ratio of $x = \frac{50}{100} 0.003878 \cdot e^{0.0656 \cdot (15)} = 0.00519$. The resulting density at sea level is thus:

$$\rho = \rho_{std}\left(\frac{1+x}{1+1.609x}\right)$$

$$= 0.002378\left(\frac{1+0.00519}{1+1.609 \times 0.00519}\right)$$

$$= 0.002371 \text{ slugs/ft}^3$$

16.2.6 Frequently Used Formulas for a Standard Atmosphere

For convenience, the equations for temperature, pressure, and density are summarized below for a standard atmosphere for both the SI and UK systems of units by inserting the appropriate constants. Note that the subscripts represent the units for each value.

Temperature in degrees Rankine and Fahrenheit (h is in ft):

$$\begin{aligned} T_{°R} &= 518.67 \cdot (1 - 0.0000068756 \cdot h) \\ T_{°F} &= 518.67 \cdot (1 - 0.0000068756 \cdot h) - 459.67 \end{aligned} \quad (16\text{-}14)$$

Temperature in degrees Kelvin and Celsius (h is in m):

$$\begin{aligned} T_K &= 288.15 \cdot (1 - 0.000022558 \cdot h) \\ T_{°C} &= 288.15 \cdot (1 - 0.000022558 \cdot h) - 273.15 \end{aligned} \quad (16\text{-}15)$$

Pressure in psf and psi (h is in ft):

$$\begin{aligned} p_{psf} &= 2116 \cdot (1 - 0.0000068756 \cdot h)^{5.2561} \\ p_{psi} &= 14.694 \cdot (1 - 0.0000068756 \cdot h)^{5.2561} \end{aligned} \quad (16\text{-}16)$$

Pressure altitude:

$$h_P = 145442\left[1 - \left(\frac{P}{P_0}\right)^{0.19026}\right]$$

Density altitude:

$$h_\rho = 145442\left[1 - \left(\frac{\rho}{\rho_0}\right)^{0.234957}\right]$$

Viscosity, UK system, °R:

$$\mu = 3.170 \times 10^{-11} T^{1.5}\left(\frac{734.7}{T+216}\right) \text{lb}_f \cdot \text{s/ft}^2 \quad (16\text{-}20)$$

Viscosity, SI system, K:

$$\mu = 1.458 \times 10^{-6} T^{1.5}\left(\frac{1}{T+110.4}\right) \text{N} \cdot \text{s/m}^2 \quad (16\text{-}21)$$

Reynolds number:

$$\text{Re} = \frac{\rho V L}{\mu} \quad (16\text{-}22)$$

16.3 THE NATURE OF AIRSPEED

Arguably, the airspeed indicator (ASI) is the most important instrument in any aircraft. This is because the pilot operates the airplane based on the airspeed. He or she base their decision to deflect the elevator to lift off the runway when the airplane has accelerated to a specific airspeed, knowing that once airborne they must maintain a specific airspeed in order to maximize either the rate of climb or the angle of climb. The pilot knows that he or she must maintain airspeed higher than the stalling speed and that they must establish and maintain a specific airspeed during cruise, based on intent to maximize range, endurance, or simply comply with a direction set by air traffic controllers. And the pilot knows that he or she must maintain a specific airspeed during descent. The ASI also tells the pilot when it is safe to retract or deploy the landing gear or flaps, and how best to perform approach to landing. Practically all maneuvering is based on the airspeed at which the airplane is flying. No other instrument is used quite like the ASI.

This section focuses on the airspeed. It details how to determine the number of types of airspeed that are of importance to designers and pilots alike, such as true airspeed, calibrated airspeed and others. And it defines specific airspeeds that are important from a regulatory and operational angle — the V-speeds.

16.3.1 Airspeed Indication Systems

Pilots are often overheard talking about the various V-speeds. This is not some jargon that is limited to the pilot community, but rather it originates with engineers in the aviation industry. The term V-speeds denotes nothing but common symbols for types of airspeeds that are important when describing the capabilities of an aircraft. It is imperative to be familiar with these terms.

Figure 16-1 shows the dial of a typical analog airspeed indicator (ASI). The one shown presents the airspeed using units of knots (KIAS). Some airspeed indicators show the airspeed in miles per hour or kilometers per hour. However, units of knots and ft/s are exclusively used in this book. It is imperative to convert all important airspeeds to knots, as this is the unit of airspeed most widely used in the world of aviation. Also, a number of noticeable markings are shown, but these have a specific meaning described in the figure. Pilots are trained to operate the aircraft using these airspeeds. The airspeeds are discussed in greater detail elsewhere in the text.

Modern aircraft feature computer-drawn airspeed indicators that are displayed on special monitors called the primary flight display (PFD). These allow for sophisticated depiction of information for the pilot, all but eliminating the need for guess work when operating the aircraft. The PFD shown in Figure 16-2 is an example of a typical such screen in a modern passenger jet. Among others, the device directly relays to the pilot the calibrated airspeed (V_{CAS}), Mach number (M), and ground speed (GS), which in Figure 16-2 can be seen to amount to 262 KCAS, 0.792, and 381 KGS, respectively. It also displays high and low speed limitations. For instance, the V_{MO}/M_{MO} limitations can be seen as the thick dotted ribbon extending from 285 KCAS and upward. The thin lined ribbon extending from 275 to 285 KCAS indicates the maximum maneuvering speed, which typically provides a 1.3 g margin for maneuvering. This means that the pilot can increase the speed by that amount, which would be required to maintain a level 40° bank angle at that altitude. The thin lined ribbon extending from the bottom to approximately 249 KCAS indicates that the airplane's stick shaker is activated. Some of these values are weight dependent and are determined in real time by the airplane's flight computer system.

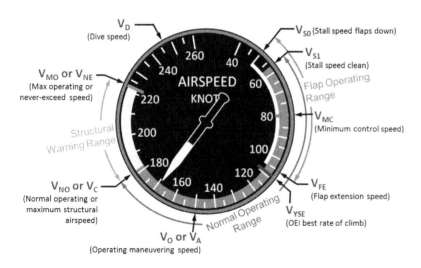

FIGURE 16-1 Markings of a typical analog airspeed indicator.

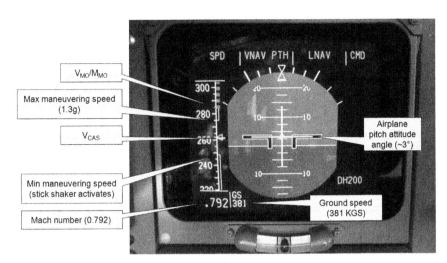

FIGURE 16-2 Markings of a modern airspeed indicator shown on a primary flight display (PFD). *(Photo by Gudbjartur Runarsson)*

An airspeed indicator needs two pressure sources: dynamic pressure and atmospheric. The dynamic pressure is obtained using a pressure probe called a *pitot tube* or simply a *pitot* (pronounced *pee-toh*). The pitot tube has an opening that faces the flow direction and senses stagnation pressure. The atmospheric pressure probe, called a *static source*, is oriented perpendicular to the flow direction. It senses static pressure. Then the ASI displays the airspeed based on the difference between the two pressure sources. Of the two, measuring the static pressure is far more difficult than the dynamic pressure. The static source is typically located on the fuselage, although other locations are certainly possible. If installed on a fuselage, we would ideally like to install the static source in a location where the pressure equals the static pressure in the far-field, at all *AOA*s and airspeeds. However, this task in encumbered by the localized distortion in the flow field around the airplane. The distortion depends on factors such as airspeed, altitude, and *AOA* (also on Mach number and Reynolds number, but these are airspeed dependent), sometimes rendering it impossible to find a location on the fuselage where pressure matches that of the far-field. Figure 16-3 shows a depiction of pressure variation along a fuselage. It can be seen that four locations on the fuselage are suitable for placing the static source.

16.3.2 Airspeeds: Instrument, Calibrated, Equivalent, True, and Ground

Instrument Airspeed

This is the airspeed the pilot reads off the airspeed indicator. The reading can be affected by three kinds of error:

(1) Indication error (due to flaws in the instrument itself).
(2) Position error (due to incorrect location of static or pitot sensors).
(3) Pressure lag error (due to rapid change in pressure, such as when a fighter climbs so rapidly the indication system doesn't keep up with the change in pressure and lags).

When referring to an instrument airspeed using knots as units, it is denoted by the variable V_{IAS}. In the SI system, the units are typically m/s or kmh. In the UK system, the units are ft/s, mph, or knots. It is useful to identify this type of a measurement using the unit KIAS, which stands for *knots, indicated airspeed*.

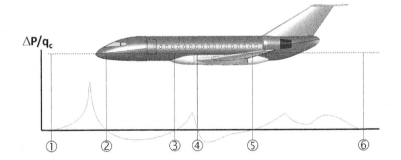

FIGURE 16-3 Pressure variation along a fuselage.

Speed of Sound

This is the speed at which pressure propagates through fluid. For air, it can be estimated in terms of ft/s using the following expression:

$$a = \sqrt{\gamma R T} \qquad (16\text{-}23)$$

For altitudes at which GA aircraft are most frequently operated, the ratio of specific heats is 1.4 and the universal gas constant is 1716 ft·lb$_f$/slug·°R, so Equation (16-23) can be simplified as follows:

$$a = \sqrt{\gamma R T} \approx 49\sqrt{T} \qquad (16\text{-}24)$$

Calibrated Airspeed

If the error in the airspeed indicator is known, and denoted by the term Δerror, the calibrated airspeed is given by:

$$V_{CAS} = V_{IAS} + \Delta error \qquad (16\text{-}25)$$

For GA aircraft, compliance to 14 CFR Part 23, §23.1587(d)(10), *Performance Information*, requires a correlation between IAS and CAS to be determined and presented to the operator of the aircraft.

Equivalent Airspeed

The equivalent airspeed is the airspeed the airplane would have to maintain at sea level in order to generate the same compressible dynamic pressure as that experienced at the specific flight condition (altitude and true airspeed). It relates to the true airspeed as follows:

$$V_{EAS} = V_{TAS}\sqrt{\rho/\rho_{SL}} \qquad (16\text{-}26)$$

The equivalent airspeed can also be calculated from the calibrated airspeed as follows:

$$V_{EAS} = V_{CAS}\sqrt{\frac{P}{P_0}\frac{\left(\frac{q_c}{P}+1\right)^{0.286}-1}{\left(\frac{q_c}{P_0}+1\right)^{0.286}-1}} \qquad (16\text{-}27)$$

Note that the power 0.286 is the ratio 1/3.5. It is sometimes useful to convert equivalent airspeed to calibrated airspeed. The following expression can be used for this:

$$V_{CAS} = V_{EAS}\sqrt{\frac{P_0}{P}\frac{\left(\frac{q_c}{P_0}+1\right)^{0.286}-1}{\left(\frac{q_c}{P}+1\right)^{0.286}-1}} \qquad (16\text{-}28)$$

where the compressible dynamic pressure is given by:

$$q_c = P\left[\left(1+0.2M^2\right)^{3.5}-1\right] \qquad (16\text{-}29)$$

and Mach number:

$$M = V/a \qquad (16\text{-}30)$$

If the calibrated airspeed is known, the Mach number for compressible flow conditions can be determined from:

$$M = \sqrt{\frac{2}{\gamma-1}\left(\left[\frac{1}{\delta}\left[\left\{1+\frac{\gamma-1}{2}\left(\frac{V_{CAS}}{661.2}\right)^2\right\}^{\frac{\gamma}{\gamma-1}}-1\right]+1\right]^{\frac{\gamma-1}{\gamma}}-1\right)} \qquad (16\text{-}31)$$

where 661.2 is the standard day speed of sound at S-L in knots, V_{CAS} is in KCAS, δ is the pressure ratio, and γ is the specific heat ratio (1.4). Inserting the appropriate values and simplifying allows Equation (16-31) to be written in the following form, which is easier to enter when preparing spreadsheet solutions:

$$M = 2.236\sqrt{\left(\frac{\left(1+4.575\times 10^{-7}V_{CAS}^2\right)^{3.5}-1}{\delta}+1\right)^{0.2857}-1} \qquad (16\text{-}32)$$

True Airspeed

True airspeed is the airspeed at which the air molecules in the far-field pass the aircraft (since the local molecules accelerate as they pass the airplane). The following expression is used to convert equivalent airspeed to true airspeed:

$$V_{TAS} = \frac{V_{EAS}}{\sqrt{\rho/\rho_{SL}}} \approx \frac{V_{CAS}}{\sqrt{\rho/\rho_{SL}}} \qquad (16\text{-}33)$$

Ground Speed

Ground speed is the speed at which the aircraft moves along the ground. This speed equals the true airspeed if there is no wind aloft (perfectly calm). However, if windy, the component of the wind parallel to the direction of the aircraft will either add (tailwind) or subtract (headwind) from the true airspeed. If this parallel wind component, denoted by w, is known, then the following expression is used to convert the true airspeed to ground speed:

$$V_{GS} = V_{TAS} + w \qquad (16\text{-}34)$$

EXAMPLE 16-4: DETERMINATION OF EAS

Determining EAS may require a sort of inverse process. We would like to be able to look at an airspeed indicator (ASI), read the KIAS and convert it to KEAS. However, instead we have to figure out the KTAS first and then determine what KIAS it corresponds to. It is best to show this process in an example. Determine KTAS, KEAS, and KCAS for an airplane flying at $M = 0.45$ at 15,000 ft on a standard day.

Solution

Step 1: Compute ambient temperature using Equation (16-14):

$$T = T_0 \cdot (1 - 0.0000068756 \cdot h)$$
$$= 518.67 \times (1 - 0.0000068756 \cdot 15000) = 465.2 \ °R$$

Step 2: Compute speed of sound from Equation (16-23):

$$a = \sqrt{\gamma R T} = \sqrt{(1.4)(1716)(465.2)} = 1057 \ \text{ft/s}$$

Step 3: Compute true airspeed in ft/s from Equation (16-30):

$$M = \frac{V_{TAS}}{a} \ \Rightarrow \ V_{TAS} = Ma = 0.45 \times 1057$$
$$= 475.7 \ \text{ft/s}$$

Step 4: Compute atmospheric pressure at condition using Equation (16-15):

$$p = 2116(1 - 0.0000068756 \times 15000)^{5.2561} = 1194 \ \text{psf}$$

Step 5: Compute density at condition using the equation of state:

$$\rho = \frac{p}{RT} = \frac{1194}{(1716)(465.2)} = 0.001496 \ \text{slugs/ft}^3$$

Step 6: Compute dynamic pressure, q_c, at condition using Equation (16-29):

$$q_c = P\left[(1 + 0.2M^2)^{3.5} - 1\right]$$
$$= 1194 \times \left[(1 + 0.2(0.45)^2)^{3.5} - 1\right] = 178.0 \ \text{psf}$$

Step 7: Compute equivalent airspeed in ft/s at condition using Equation (16-26):

$$V_{EAS} = V_{TAS}\sqrt{\rho/\rho_{SL}} = 475.7\sqrt{\frac{0.001496}{0.002378}} = 377.3 \ \text{ft/s}$$

Step 8: Compute calibrated airspeed in ft/s at condition using Equation (16-28):

$$V_{CAS} = V_{EAS}\sqrt{\frac{P_0}{P}}\sqrt{\frac{\left(\frac{q_c}{P_0}+1\right)^{0.2857}-1}{\left(\frac{q_c}{P}+1\right)^{0.2857}-1}}$$

$$= 377.3\sqrt{\frac{2116}{1194}}\sqrt{\frac{\left(\frac{178}{2116}+1\right)^{0.2857}-1}{\left(\frac{178}{1194}+1\right)^{0.2857}-1}} = 381.3 \ \text{ft/s}$$

Step 9: Convert all airspeeds to knots:

$$\Rightarrow \ V_{TAS} = 475.7/1.688 = 281.8 \ \text{KTAS}$$
$$\Rightarrow \ V_{EAS} = 377.3/1.688 = 223.5 \ \text{KEAS}$$
$$\Rightarrow \ V_{CAS} = 381.3/1.688 = 225.9 \ \text{KCAS}$$

Also, see Figure 16-4 for additional explanations.

16.3.3 Important Airspeeds for Aircraft Design and Operation

The "types" of airspeeds listed in Table 16-4 are of interest to the aircraft designer, from both a performance and certification standpoint. Many can be determined by the analysis methods presented in here. Others are requirements that must be complied with if the aircraft is to be certified.

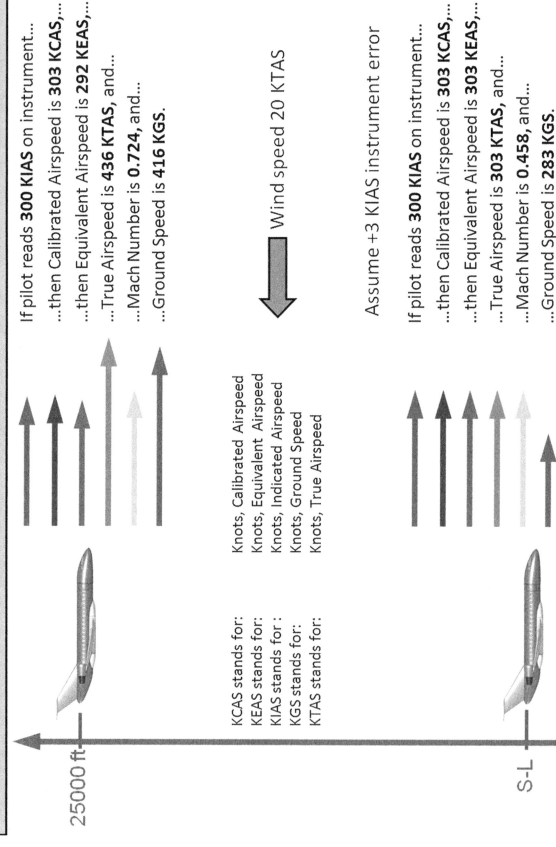

FIGURE 16-4 Graphical depiction of various airspeed types.

TABLE 16-4 Important Airspeeds for Aircraft Design and Operation

V-speed	Description	Article or 14 CFR Part 23
M_C	Cruising speed in terms of Mach number.	§23.335
M_D	Dive Mach number.	§23.335
M_{MO}	Maximum operating Mach number.	§23.1505
V_1	Maximum speed during take-off at which a pilot can either safely stop the aircraft without leaving the runway or safely continue to V_2 take-off even if a critical engine fails (between V_1 and V_2).	17.1.3
V_2	Take-off safety speed. Airspeed the airplane must be capable of reaching 35 ft above the ground.	17.1.2 §23.57
V_{2min}	Minimum take-off safety speed. Minimum value of the V_2 airspeed (see V_2). Defined for commercial aircraft per 14 CFR 25.	§25.107
V_3	Flap retraction speed.	—
V_A	Maneuvering speed. A certification airspeed below which the airplane must be capable of full deflection of aerodynamic controls.	16.4.1 §23.335
V_B	Design speed for maximum gust intensity. Most often used for commuter-class aircraft. See Figure 16-13.	§23.335
V_{BA}	Minimum rate-of-descent airspeed, which yields the least altitude lost in a unit time.	21.3.5
V_{BG}	Best glide speed. Minimum or best angle-of-descent airspeed. This speed will result in the shallowest glide angle and will yield the longest range, should the airplane lose engine power.	19.2.8 21.3.7
V_{BR}	Airspeed when pilot begins to apply brakes after touch-down.	22.2.5
V_{LOF}	Lift-off speed.	17.1.2
V_{max}	See V_H.	19.2.10
V_{MC}	Minimum control speed with the critical engine inoperative.	§23.149
V_{MCA}	Minimum control speed while airborne. See V_{MC}.	—
V_{MCG}	Minimum control speed on the ground. The minimum airspeed required to counteract an asymmetric yawing moment on the ground due to an engine failure on a multiengine aircraft.	§23.149
V_{MO}	Maximum operating speed.	§23.1505
V_{MU}	Minimum unstick speed. The airspeed at which the airplane no longer "sticks" to the ground. It is a function of the ground attitude (or AOA) of the airplane. The minimum is achieved when the ground attitude is a C_{Lmax} or α_{stall}. Defined for commercial aircraft per 14 CFR 25.	§25.107
V_{NE}	Never-exceed speed or maximum structural airspeed.	§23.1505
V_{NO}	Normal operating speed; also called *maximum structural cruising speed*. It is the speed that should not be exceeded except in smooth air, and then only with caution.	§23.1505
V_O	The maximum operating maneuvering speed, V_O, is established by the manufacturer as an operating limitation and is not greater than $V_S \sqrt{n}$ established in §23.335(c).	§23.1507
V_R	Rotation speed. The speed at which the airplane's nosewheel leaves the ground. It is high enough to ensure the aircraft can reach V_2 at 50 ft (GA) or 35 ft (commercial) in the case of an engine failure on a multiengine aircraft.	17.1.2 §23.51
V_{REF}	Landing reference speed or threshold crossing speed, typically $1.2 \cdot V_{S0}$ to $1.3 \cdot V_{S0}$. The factor 1.2 is typically used for military aircraft, but 1.3 for civilian aircraft per §23.73.[a]	22.2.5 §23.73
V_{Rmax}	Best range speed. The airspeed that results in maximum distance flown.	19.2.9
V_S	Stalling speed or minimum steady flight speed for which the aircraft is still controllable.	19.2.6 §23.49, §23.335

TABLE 16-4 Important Airspeeds for Aircraft Design and Operation—Cont'd

V-speed	Description	Article or 14 CFR Part 23
V_{S0}	Stalling speed or minimum steady flight speed for which the aircraft is still controllable in the landing configuration.	19.2.6 §23.49, §23.335
V_{SR}	Reference stalling speed. The stalling speed of the airplane at some condition other than gross weight. Important for heavy aircraft that consume a lot of fuel during a particular mission. The stalling speed at the start of the mission will be higher than at the end.	19.2.6 22.2.5
V_{SR0}	Reference stalling speed in landing configuration at some condition other than gross weight.	19.2.6 22.2.5
V_{SR1}	Reference stalling speed in a specific configuration at some condition other than gross weight.	19.2.6 22.2.5
V_{SW}	Speed at which the stall warning will occur.	§23.207
V_{TD}	Touch-down airspeed.	22.2.5
V_{TR}	Transition airspeed; the average of V_{LOF} and V_2.	17.1.2
V_X	The best angle-of-climb airspeed (max altitude gain per unit distance).	18.3.4 (jet) 18.3.8 (prop)
V_Y	The best rate-of-climb airspeed (max altitude gain per unit time).	18.3.6 (jet) 18.3.9 (prop)
V_{YSE}	The best rate-of-climb airspeed in a multi engine aircraft with one engine inoperative.	18.3.6 (jet) 18.3.9 (prop)

^a Per 14 CFR Part 23, §23.73 Reference Landing Approach Speed.

16.4 THE FLIGHT ENVELOPE

The purpose of this section is to detail the use of the *V-n diagram* or *flight envelope*. The flight envelope shows specific load factors versus airspeed that the airplane has been designed to operate within (see Figure 16-5). It is of primary interest to the structural engineer, but also helps the pilot better understand the limitations of his or her airplane; at what airspeed they can fully deflect control surfaces, what is the dive speed, or the

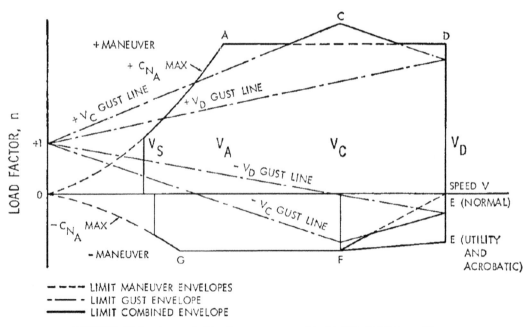

FIGURE 16-5 A generic V-n diagram presented in 14 CFR 23.333 for guidance.

airspeed at which he or she may have to slow down should they encounter turbulent atmospheric conditions, and the list goes on. The V-n diagram is usually prepared in accordance with instructions found in aviation regulations such as 14 CFR Part 23 [6], Part 25 [7], or ASTM F2245 [8], depending on aircraft class. In this section, the construction of a V-n diagram will be shown using 14 CFR Part 23.

Generally, several V-n diagrams are prepared to represent various conditions. Among those are:

(1) Configuration variations (e.g. in T-O, cruise, and landing configuration).
(2) Altitudes (e.g. covering the altitudes from S-L to 20,000 ft and then from 20,000 ft to the design cruise altitude, necessitated by gust loads).
(3) Weight (e.g. empty weight, gross weight, and perhaps some intermediary weights).

An airplane at rest on the ground is acted upon by the force of gravity alone. It is then said to be exerted on by a load factor of 1. If it accelerates for some reason, say upward, because of a force two times larger than its weight, it is said to be subjected to a load factor of 2, and so on. Simply, the load factor is the ratio of the force acting on a body to its unaccelerated weight. A more common expression for this is reacting a 1 g load, a 2 g load, and so on. This is clearly defined in 14 CFR 23.321 General as follows:

> Flight load factors represent the ratio of the aerodynamic force component (acting normal to the assumed longitudinal axis of the airplane) to the weight of the airplane. A positive flight load factor is one in which the aerodynamic force acts upward, with respect to the airplane.

The V-n diagram can be thought of as composed of two separate events superimposed on each other to form a complete diagram: maneuvering and gust loading. The aviation regulations always specify how to construct the effect of each event in applicable paragraphs that can be followed almost like instructions in a cookbook. This can be better seen in a moment. To generate a V-n diagram in accordance with 14 CFR 23, follow paragraphs 23.321 through 23.341. The reader not familiar with the regulations is urged to have those handy when following the discussion below for the first time, although the experienced engineers need not. This will be very helpful in understanding the language being used.

In short, the process is as follows: using 14 CFR Part 23, first determine the category the aircraft is to be certified within; i.e. is it a normal, utility, aerobatic, or commuter aircraft? This is imperative as the magnitude of the maneuvering loads depends on this classification. The second step is to gather important and applicable information about the airplane. This includes weight, wing area, lift curve slope, maximum and minimum lift coefficients, to name a few. Then, prepare the maneuvering diagram, followed by the gust diagram. Finally, trace the outline of the diagram. This process is better shown using an example. Remember that all airspeeds used in the V-n diagram are in terms of equivalent airspeed (e.g. KEAS) or Mach number. Here, we will use the former.

Note that rather than using either of the two sample aircraft to be presented in the next section (the Cirrus SR22 and the Learjet 45XR), a hypothetical aircraft of very light wing loading will be used. This introduces complexity to the generation of the diagram that may perplex even the seasoned aircraft designer. Diagrams for other aircraft are destined to be simpler than the one to be made here and if you can generate the one that follows, you can manage those for the sample aircraft. The sample aircraft for this exercise has the characteristics shown in Table 16-5 and it will be assumed this aircraft is to be certified in the Normal category. Note that it is assumed the aircraft has already been built and flown, as is indicated by the "maximum demonstrated level airspeed" in the table.

16.4.1 Step-by-Step: Maneuvering Loads and Design Airspeeds

This article shows how to step through the regulations to prepare the maneuvering diagram for the above aircraft.

TABLE 16-5 Applicable Properties of the Sample Aircraft

Item	Symbol	Value
Certification category		Normal
Wingspan	b	38 ft
Wing area	S	130 ft²
Mean geometric chord	MGC	3.42 ft
Gross weight	W	1320 lb$_f$
Minimum flying weight	W_{min}	900 lb$_f$
Stalling speed (+ means normal)	$V_{S(+)}$	46.3 KEAS
Stalling speed (− means inverted)	$V_{S(-)}$	54.7 KEAS
Maximum demonstrated level airspeed	V_H	120 KEAS
Three-dimensional lift curve slope	$C_{L\alpha}$	5.25 per radian
Maximum C_L (flaps up)	C_{Lmax}	1.45
Minimum C_L (flaps up)	C_{Lmin}	−1.00
Maximum C_L (flaps down)	C_{Lmax}	2.10
Minimum C_L (flaps down)	C_{Lmin}	−0.75

Step 1: Establish Load Factors n_+ and n_-

Per 14 CFR 23.337(a)(1) estimate the positive load factor that must be used for the aircraft (note that it does not have to be higher than 3.80:

Load factor per 23.337(a)(1):

$$n_+ = 2.1 + \frac{24{,}000}{W_0 + 10{,}000} = 2.1 + \frac{24{,}000}{1320 + 10{,}000} = 4.22 \quad (16\text{-}35)$$

Since $n_+ > 3.80$, we can establish it as 3.80 if we so desire and this is what we will do. Then, 23.337(b)(1) stipulates that the negative load factor, n_-, "may not be less than" 0.4 times n_+. This clumsily phrased sentence actually means the opposite: n_- may not be larger than $-0.4 \cdot n_+$; it may be less on the other hand. Of course a lower value presents a limited benefit to the aircraft manufacturer, as this will almost certainly lead to heavier and more expensive to manufacture aircraft. Therefore, $n_- = -0.4 \cdot n_+ = -1.52$.

Knowing the load factors, we can now calculate the following design airspeeds for the gross weight condition.

Step 2: Design Cruising Speed, V_C

The airframe of the airplane must be designed to react gust loads at this airspeed. The purpose is to design the airplane for operation in turbulent air, let alone the possibility of encountering clear air turbulence (CAT), while in cruising flight. The regulations stipulate this must be done assuming a certain minimum value of the cruising speed. Again, the designer can select any speed above (and including) this value, although one must realize the ramifications of selecting a higher cruising speed. The cruising speed should be carefully picked for the following reasons:

(1) Selecting a "certified" cruising speed lower than the cruising speed the airplane is capable of (and at which it will usually be operated) will require the airplane to be slowed down to that speed every time atmospheric turbulence is present. This will irritate even the most docile pilot and may negatively affect the "reputation" of the type in the long run.
(2) Selecting a "certified" cruising speed higher than the cruising speed the airplane is capable of will push up the dive speed (see Step 3) and result in a heavier airframe that has less useful load.
Therefore, the proper value of V_C should be no higher than the typical and expected cruising speed of the aircraft.

The selection process requires the minimum and maximum cruising speeds to be determined and, then, a representative value between the two to be selected.

Per 23.335(a)(1) the minimum cruising speed may not be less than:

Minimum design cruising speed:

$$V_{Cmin} = 33\sqrt{\frac{W_0}{S}} = 33\sqrt{\frac{1320}{130}} = 105.2 \text{ KEAS} \quad (16\text{-}36)$$

Since the wing loading is $1320/130 = 10.2$ lb$_f$/ft^2 23.335(a)(2) does not apply, but we must check 23.335(a)(3), which says that the cruising speed "need not be more than $0.9 \cdot V_H$."

Maximum design cruising speed:

$$V_{Cmax} = 0.9 V_H = 0.9(120) = 108 \text{ KEAS} \quad (16\text{-}37)$$

For this airplane it is tempting to just go with the upper speed, i.e. $V_C = 108$ KEAS. But first let's consider what dive speeds these render, assuming it will be 40% greater (as per the regulations, as will be shown shortly). The minimum cruising speed leads to a minimum dive speed of 147.3 KEAS, and the maximum cruising speed requires at least 151 KEAS. However, it might be of interest to ask: why not just select a cruising speed that will result in a dive speed of 150 KEAS? After all, such a number will be easy for the engineering team as well as pilots to remember. It should be stressed that basing the selection of this airspeed on "convenience" is not always the right thing to do, although the author argues it makes sense here. After all, the value is easier to remember than either extreme and guarantees the resulting cruising speed falls between the minimum (105.2 KEAS) and maximum (108 KEAS). It turns out that $150/1.4 = 107$ KEAS will result in a $V_D = 150$ KEAS. Therefore, let's select that as the design cruising speed.

Also, since this is a slow-flying aircraft the stipulations of 23.335(a)(4) does not apply either.

Step 3: Design Dive Speed, V_D

This is the maximum airspeed the airplane's airframe is designed to resist. It is very important as the aircraft must also be free of flutter at airspeed no less than $1.2 V_D$ (per 23.629). The dive speed is calculated per 23.335(b)(2) as follows:

Dive speed per 23.335(b)(2):

$$V_D > 1.40 V_{Cmin} = 1.40 \times 105.2 = 147.3 \text{ KEAS} \quad (16\text{-}38)$$

However, we decided earlier to design the airplane for $1.40 \cdot 107 = 150$ KEAS and this will comply with the above minimum. This also complies with the previous requirement of 23.335(b)(1). Note that paragraph 23.335(b)(3) does not apply to this aircraft. And paragraph 23.335(b)(4) can be used to reduce the dive speed. However, compliance requires sophisticated analysis and confirmation by flight testing and at this point we will just ignore it and accept the higher dive speed of 150 KEAS.

Step 4: Design Maneuvering Speed, V_A

The maneuvering speed is the airspeed below which the aircraft must be capable of withstanding the full deflection of control surfaces.

Maneuvering speed per 23.335(c)(1):

$$V_A = V_S\sqrt{n_+} = 46.3\sqrt{3.8} = 90.3 \text{ KEAS} \quad (16\text{-}39)$$

This airspeed is less than V_C (107 KEAS) and, therefore, complies with 23.335(c)(2). Also, calculate the negative or inverted maneuvering speed, denoted by G in Figure 16-5. This can be done per Equation (19-8) of Section 19.2.6, *Stalling speed*, V_S, using $n = |-1.52| = 1.52$, $C_{Lmin} = |-1.00| = 1.00$, and $\rho = 0.002378$ slugs/ft^3.

Negative maneuvering speed:

$$V_G = \sqrt{\frac{2|n_-|W}{\rho S C_{Lmin}}} = 113.9 \text{ ft/s} = 67.5 \text{ KEAS} \quad (16\text{-}40)$$

Step 5: Design Speed for Maximum Gust Intensity, V_B

It is best to consider this airspeed after the gust load diagram has been explained. This airspeed is determined in Section 16.4.2, *Step-BY-Step: Gust loads*.

Step 6: Set Up the Initial Diagram

Having completed determining the above load factors and airspeeds, it is now possible to begin drafting the V-n diagram. This will also help clarify what the above values actually mean. First consider Figure 16-6, which shows a graph with the velocity axis extending from 0 to 160 KEAS and the load factor axis extending from −3 to 5. A number of vertical and horizontal construction lines have been drawn on this graph, but these are used for guidance with the rest of the diagram. The steps taken to accomplish this are labeled as well.

Next perform the following actions illustrated in Figure 16-7. First plot the vertical lines representing the positive and negative stall speeds (see Steps 9 and 10). Next, plot the *positive* and *negative stall lines* given by the following expressions:

Positive stall line:

$$n_+(V) = 0.003388 \frac{V^2 S C_{Lmax}}{W} \quad (16\text{-}41)$$

Negative stall line:

$$n_-(V) = -0.003388 \frac{V^2 S C_{Lmin}}{W} \quad (16\text{-}42)$$

where

ρ = air density = 0.002378 slugs/ft^3
V = aircraft airspeed in KEAS

The constant 0.003388 is simply the product of the S-L density (0.002378 slugs/ft^3) and the knots-to-ft/s

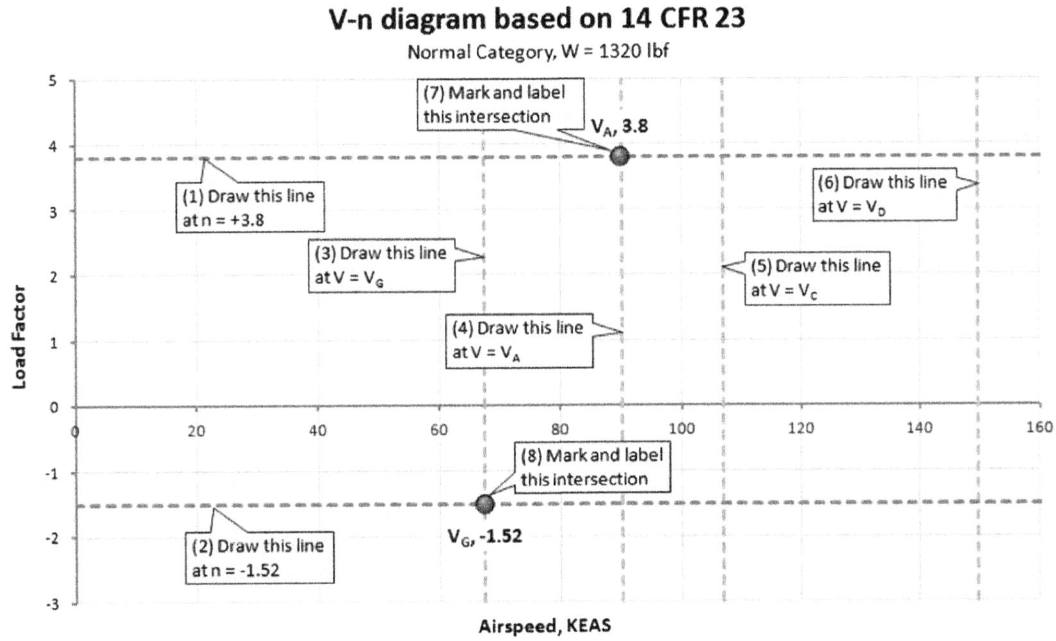

FIGURE 16-6 The first step taken to generate the V-n diagram. The dashed lines are "construction lines" that are removed once the diagram is completed.

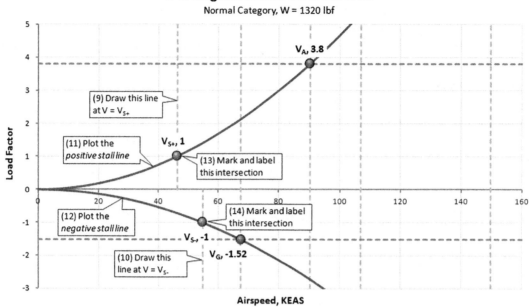

FIGURE 16-7   The second step taken to generate the V-n diagram.

conversion factor (1.688 — which must be squared due to the term V^2), divided by the factor "2" (as in $W \approx L = \frac{1}{2}\rho V^2 \cdot S \cdot C_{Lmax}$). In other words: $0.002378 \times 1.688^2 / 2 = 0.003388$. The resulting expression conveniently allows the airspeed to be entered in terms of KEAS. With the two curves plotted, the V-n diagram now looks as shown in Figure 16-7. Next create solid lines from the construction lines to represent the outlines of the maneuvering envelope, as shown in Figure 16-8. The next step is to add the gust lines and refine the diagram. This is demonstrated in the next section.

16.4.2 Step-by-Step: Gust Loads

All airplanes are subjected to vertical gusts in level flight. These can be caused by thermals, mountain

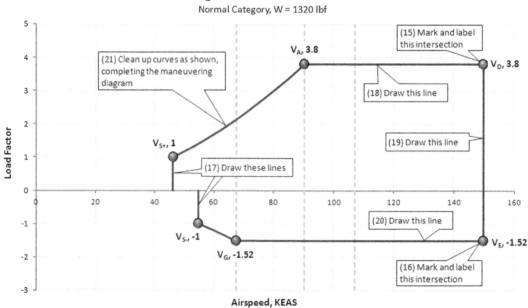

FIGURE 16-8   Completing the maneuvering portion of the V-n diagram.

waves, and other similar atmospheric phenomena. As the airplane penetrates a rising (or sinking) column of air this momentarily changes the angle-of-attack. This either increases (rising column) or decreases (sinking column) the lift of the wing, causing the familiar "bumpiness" to be detected by the occupants as well as the airframe. This is the gust loading. It depends on the forward airspeed of the aircraft and the vertical speed of the rising (or sinking) air penetrated. The aviation authorities specify the gust load requirements in terms of the strength of the vertical speed of the gust.

In magnitude, the gust load ranges from being "annoying" to being so severe it may cause structural failure. For this reason the gust loads must be taken into account when designing the airframe. It must be remembered that the change in *AOA* does not take place instantaneously, but rather is an event that takes a finite (albeit short) time. Consequently, the resulting gust loads are lessened or "alleviated." The aviation authorities allow the applicant to reduce the gust load factors by calculating a special gust alleviation factor (see Step 8).

The gust load factors for GA aircraft are determined in accordance with 14 CFR 23.333(c) in accordance with the following rules (see Figure 16-9):

(1) Positive and negative gust velocity of 50 ft/s must be considered at V_C at altitudes from S-L to 20,000 ft. Above 20,000 ft, the gust velocity may be reduced linearly to 25 ft/s at 50,000 ft.

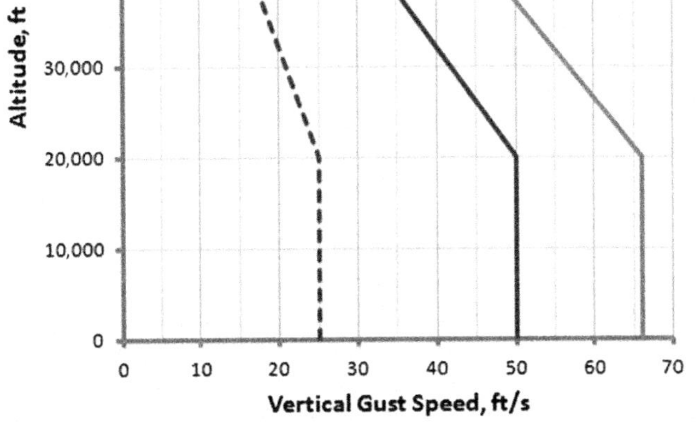

(2) Positive and negative gust velocity of 25 ft/s must be considered at V_D at altitudes from S-L to 20,000 ft. Above 20,000 ft, the gust velocity may be reduced linearly to 12.5 ft/s at 50,000 ft.

(3) Only applicable to commuter category aircraft, a positive and negative gust velocity of 66 ft/s must be considered at V_D at altitudes from S-L to 20,000 ft. Above 20,000 ft, the gust velocity may be reduced linearly to 38 ft/s at 50,000 ft. This gust must be applied at V_B.

As stated earlier, the gust load factor is alleviated as calculated in paragraph 23.341(c). This alleviation is based on the assumption that the shape of the gust follows a sinusoidal shape. In other words, the gust gradually rises to the maximum vertical rate as described using the following formula:

$$U = \frac{U_{de}}{2}\left(1 - \cos\frac{2\pi x}{25 \cdot MGC}\right) \quad (16\text{-}43)$$

where

U_{de} = vertical gust velocity per the above discussion (ft/s)
x = distance penetrated into the gust (ft)

Ordinarily, this formula is only used to demonstrate the nature of the gust. It is not needed for the actual gust load calculations, which are shown below. The gust load portion of the diagram is created using paragraph 23.341(c). This process requires the gust response to be determined for the four above vertical gust velocities per 23.333(c) (that is ±50 ft/s at V_C and ±25 ft/s at V_D). For this part, the designer must estimate the lift curve slope for the entire aircraft, although the wing one will usually be sufficient.

It is imperative that the V-n diagram be constructed for all critical configurations, altitudes, and weights. Thus, separate V-n diagrams must be made for the airplane in the clean, take-off, and landing configurations. Making a diagram for the minimum flying weight is equally important to the one at gross weight. While air loads on primary structures like wing spars and skin are greater at gross weight, inertia loads on components that do not directly react aerodynamic loads (e.g. engines, avionics) are higher at minimum weight than at gross weight.

Step 7: Calculated Gust-related Parameters

Start by computing the aircraft mass ratio per 23.341(c) as follows:

$$\mu_g = \frac{2(W/S)}{\rho \cdot MGC \cdot C_{L\alpha} \cdot g} = \frac{2(1320/130)}{0.002378 \cdot 3.42 \cdot 5.25 \cdot 32.174}$$
$$= 14.77$$

(16-44)

FIGURE 16-9 Regulations allow vertical gusts to be lessened at altitudes higher than 20,000 ft.

where

g = acceleration due to gravity = 32.174 ft/s^2

Then use this to compute the gust alleviation factor as shown below:

Gust alleviation factor:

$$K_g = \frac{0.88\mu_g}{5.3+\mu_g} = \frac{0.88(14.77)}{5.3+14.77} = 0.6476 \quad (16\text{-}45)$$

Step 8: Calculate Gust Load Factor as a Function of Airspeed

With the data in Step 7 available, the load factor for each of the four vertical gust velocities (± 50 ft/s at V_C and ± 25 ft/s at V_D) can be determined from:

$$n_g = 1 + \frac{K_g \cdot U_{de} \cdot V \cdot C_{L\alpha}}{498(W/S)} \quad (16\text{-}46)$$

where

V = aircraft airspeed in KEAS

Equation (16-46) is used to plot the four gust lines seen in Figure 16-10. The next step is to harmonize the two diagrams and form a single one. Note that a potential issue has surfaced, which is that the positive 50 ft/s gust line goes above V_A and not below as shown in Figure 16-5. This will be treated in the next step.

Step 9: Calculate Gust Load Factor as a Function of Airspeed

At this point, it is of interest to determine directly the positive and negative gust load factors. This can be done easily using Equation (16-46):

$$n_g = 1 + \frac{0.6476 \cdot 50 \cdot 107 \cdot 5.25}{498(1320/130)} = 4.60$$

Similarly, the negative gust load factor can be found from:

$$n_g = 1 + \frac{0.6476 \cdot (-50) \cdot 107 \cdot 5.25}{498(1320/130)} = -2.60$$

Step 10: Finalize Gust Diagram

Now, the final presentation of the diagram can be prepared. This is done by identifying all inter sections between the two diagrams, connecting as needed with lines, and resolving any possible trouble areas (see Figure 16-11). It is here that the reason for selecting the particular airplane characteristics becomes apparent. In the author's opinion the diagram in Figure 16-5 is particularly benign in nature. However, the one being created here poses a problem that may seem daunting at first, but it is important to be able to handle it with confidence. The problem is primarily limited to aircraft with low wing loading and is a consequence of the gust lines being too steep.

Convenient Relations to Determine Location of Intersections

When plotting the gust load factors it is often necessary to determine the airspeed at which a specific load factor occurs. This is useful when determining the

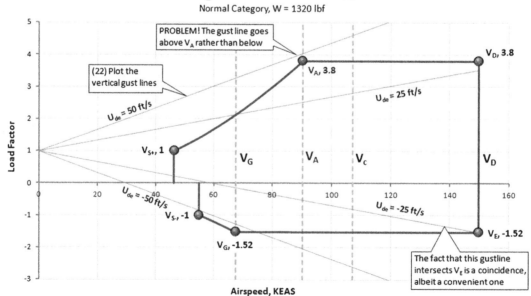

FIGURE 16-10 Initial plotting of the gust lines.

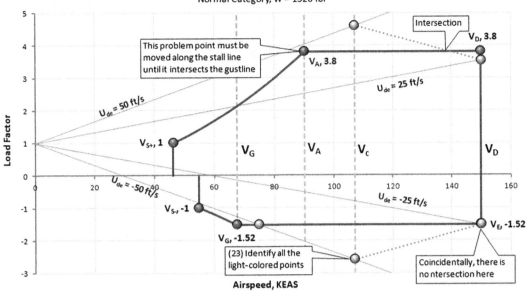

FIGURE 16-11 Refining the influence of the gust lines on the overall V-n diagram.

intersection of maneuvering and gust lines. The following expressions can be used for this:

If $n_g > 0$ use this form:

$$V = \frac{498(W/S)(n_g - 1)}{K_g \cdot U_{de} \cdot C_{L\alpha}} \qquad (16\text{-}47)$$

If $n_g < 0$ use this form:

$$V = \frac{498(W/S)(1 - n_g)}{K_g \cdot U_{de} \cdot C_{L\alpha}} \qquad (16\text{-}48)$$

complicated, but the intersection depends on the solution to the following quadratic equation:

$$\begin{aligned} n_g &= n_+(V) \\ \Rightarrow\ & (709.5 \cdot \rho \cdot C_{Lmax}) \cdot V^2 \\ & - (K_g \cdot U_{de} \cdot C_{L\alpha}) \cdot V - 498(W/S) = 0 \end{aligned} \qquad (16\text{-}49)$$

Thus, the new V_A becomes (ignoring the negative solution):

$$(709.5(0.002378)(1.4)) \cdot V_A^2 - ((0.6476)(50)(5.25)) \cdot V_A - 498(1320/130) = 0$$

$$\Rightarrow\ 2.362 \cdot V_A^2 - 170.0 \cdot V_A - 5057 = 0 \ \Rightarrow\ V_A = 94.6 \text{ KEAS}$$

To determine how to extend the positive stall line to the positive 50 ft/s or 66 ft/s gust lines, the load factor calculated per Equation (16-41) must equal that of Equation (16-46). Unfortunately, the expression is more

Equation (16-49) also conveniently allows the gust penetration speed, V_B, to be determined (noting that the vertical gust velocity for the condition is given by $U_{de} = 66$ ft/s):

$$(709.5 \cdot \rho \cdot C_{Lmax}) \cdot V_B^2 - 66(K_g \cdot C_{L\alpha}) \cdot V_B - 498(W/S) = 0 \ \Leftrightarrow$$

$$V_B = \frac{33 \cdot (K_g \cdot C_{L\alpha}) + \sqrt{1089(K_g \cdot C_{L\alpha})^2 + 498 \cdot 709.5 \cdot (\rho \cdot C_{Lmax})(W/S)}}{709.5 \cdot \rho \cdot C_{Lmax}} \qquad (16\text{-}50)$$

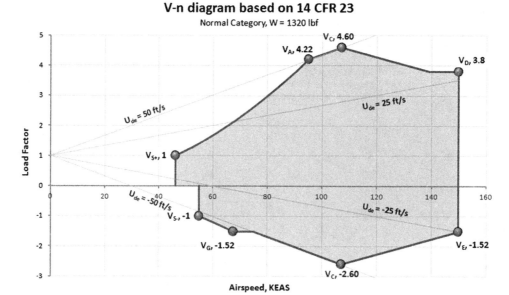

FIGURE 16-12 The completed V-n diagram.

16.4.3 Step-by-Step: Completing the Flight Envelope

The final step in the preparation of the diagram is to combine the maneuvering and gust envelopes. This is shown in Figure 16-12. The shaded region is where the airplane must be demonstrated to operate safely during certification. All the load factors represent <u>limit loads</u> and must be multiplied by a factor of safety of 1.5 to get ultimate loads.

V-n Diagrams with Deployed High-lift Devices

Note that deploying high-lift devices effectively requires a separate envelope to be prepared. This envelope will have its own V_A, V_D (which is called *flap extension speed*, denoted by V_{FE}), maneuvering load factor, and so on. The regulatory requirements to establish these values are presented in the appropriate regulations.

The flap is often superimposed on the V-n diagram for a clean configuration.

16.4.4 Flight Envelopes for Various GA Aircraft

Although the details of creating a V-n diagram are spelled out in the applicable regulations, not all aspects of this guidance are clear. For this reason, it is easy to become uncertain about the shape of the final diagram. The images in Figure 16-13 provide some guidance

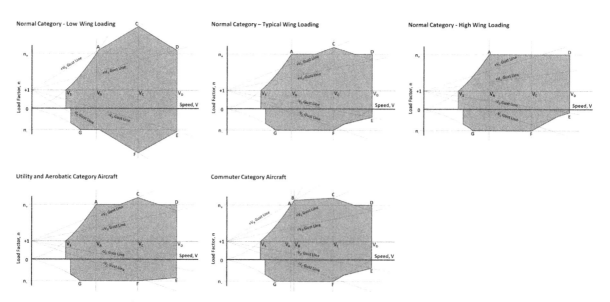

FIGURE 16-13 Typical V-n diagrams for selected certification categories.

as to what shape to expect based on certification category.

IMPORTANT: The V-n diagrams of Figure 16-13 do not reflect the use of high-lift devices.

16.5 SAMPLE AIRCRAFT

The astute aircraft designer is always concerned about the accuracy of calculations. For this reason he or she recognizes the importance of comparing results from the various calculation methods to actual aircraft. A method unable to accurately predict the performance of actual aircraft is limited at best and suspect at worst. The designer must be aware of such limitation, but such knowledge will help him select the right method. This section will introduce two sample aircraft that will be used to demonstrate performance concepts — the Cirrus SR22 and the Learjet 45XR. The former is a piston-powered propeller aircraft and the latter is a twin-engine business jet. Using these aircraft will allow the calculated values to be compared to published data, giving a valuable insight into the accuracy of the methods.

16.5.1 Cirrus SR22

The SR22 (see Figure 16-14) is designed and manufactured by Cirrus Aircraft, of Duluth, Minnesota. The aircraft was conceived as a more powerful derivative of the all-composite SR20, which was designed in the

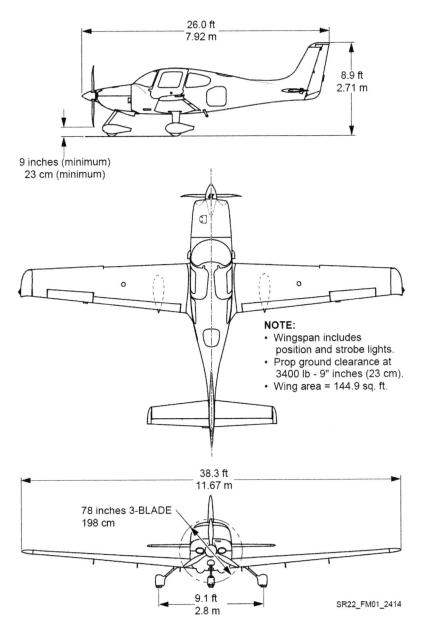

FIGURE 16-14 A three-view of the Cirrus SR22. *(Courtesy of Cirrus Aircraft)*

TABLE 16-6 General Properties of the Cirrus SR22

Item	Symbol	Value
Wingspan	b	38.3 ft
Wing area	S	144.9 ft²
Wing chord equation	$c(y)$	$5.18 - 0.1342 \cdot y$ (ft)
Wing mean geometric chord	C_{MGC}	4.03 ft
Spanwise location of C_{MGC}	y_{MGC}	8.51 ft
Wing AR	AR	10.12
Gross weight	W	3400 lb$_f$
Engine		1xContinental IO-550
Max power at S-L	P	310 BHP
Zero lift drag coefficient	C_{Dmin}	0.02541
C_L of minimum drag	C_{LminD}	0.20
Maximum C_L (flaps up)	C_{Lmax}	1.41
Maximum C_L (flaps down)	C_{Lmax}	1.99
Basic ($\alpha = 0$) C_L	C_{L0}	0.0

mid 1990s. The SR22 is a four-seat, single-engine, high-performance touring aircraft, and has remained the best-selling aircraft in its class. Powered by a 310 BHP Continental IO-550, six-cylinder horizontally opposed piston engine, it is capable of cruising at 185 KTAS at 75% power at 8000 ft. While boasting high performance, it is designed with safety of operation in mind. The SR20 was the first aircraft in the history of aviation to be certified with an emergency parachute capable of lowering the entire airframe in case of an emergency. By 2013, the parachute system, called Cirrus Airframe Parachute System (CAPS), had saved the lives of some 53 people with 32 deployments [9]. Table 16-6 shows properties of this airplane commonly used in examples in this text. All the data is obtained from the manufacturer's website (www.cirrusaircraft.com), using published performance data and the analysis methods provided in this book. All geometric data was obtained using the three-view in a manner shown in Figure 16-15. The reader can easily generate data tables of similar nature for all other aircraft using such basics.

The reader should be mindful that the data extraction to be implemented for the SR22 can just as well be accomplished for any other type of aircraft. One only needs published geometric, inertia, and performance

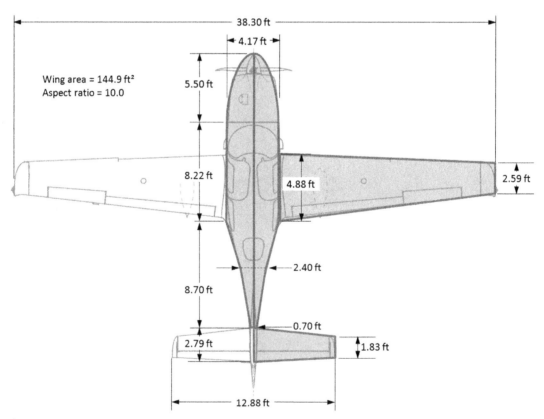

FIGURE 16-15 Scaling the top-view based on the wingspan in the three-view of Figure 16-14 yields the following dimensions. *(Courtesy of Cirrus Aircraft)*

data, and a proportionally correct three-view drawing. Such data can be obtained from both the type certificate data sheet (TCDS [10]) and the Pilots Operation Handbook (POH), both of which are readily available; the former from the FAA website (www.faa.gov) and the latter from pilots. Be careful, however: POH data is copyrighted and cannot be made public in the manner shown here. Cirrus Aircraft has graciously given permission for the presentation of the data extracted in this text and this is fortunate, because the SR22 has far more exciting performance and handling characteristics than most aircraft in its class and, thus, offers great learning potential on how to design fast and efficient single-engine aircraft.

16.5.2 Learjet 45XR

Jet performance concepts will be demonstrated using the Learjet 45XR business jet (see Figure 16-16). Learjet was founded by an American inventor and a very original and influential business man, William P. Lear (1902—1978). Learjet produced a number of well-known business jets, such as the original Learjet 23, the first in a family of high-performance aircraft. In 1969, Learjet merged with Gates Aviation, forming Gates Learjet Corporation, and in 1990 the company was acquired by Bombardier Aerospace. The development of the 45XR was announced by Bombardier in September 1992 and the first flight of the prototype aircraft took place on October 7, 1995. FAA certification was granted in September 1997. The aircraft is powered by two FADEC-controlled Honeywell TFE731-20 engines and is equipped with an internal auxiliary power unit (APU) for ground power. The Learjet 45XR is a special version of the Learjet 45 and was introduced in June 2004. The 45XR offers higher take-off weight, faster cruising speeds and faster rate-of-climb than its predecessor, thanks to a more powerful engine. Simplified and "assumed" values for the 45XR are presented in Table 16-7.

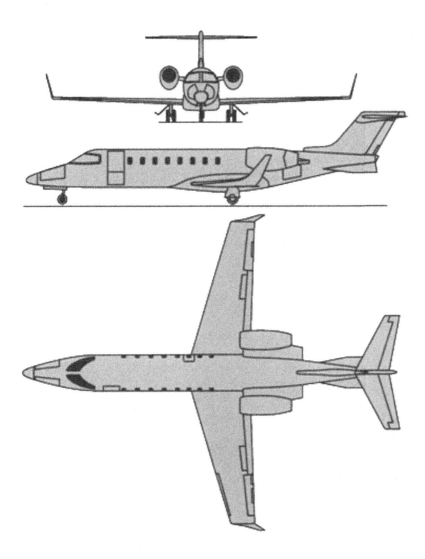

FIGURE 16-16 A three-view of the Learjet 45XR.

TABLE 16-7 General Properties of the Learjet 45XR

Item	Symbol	Value
Wingspan	b	47.78 ft
Wing area	S	311.6 ft^2
Wing AR	AR	7.33
Wing TR	TR	0.391
Gross weight	W	21,500 lb$_f$
Engines		2 × Honeywell TFE731-20
Max thrust at S-L	T	2 × 3500 lb$_f$
Bypass ratio	λ	3.9
Zero lift drag coefficient (clean)	C_{Dmin}	0.020
Zero lift drag coefficient (T-O)	C_{Dmin}	0.035
C_L of minimum drag	C_{LminD}	0.20
Maximum C_L (flaps up)	C_{Lmax}	1.30
Maximum C_L (flaps for T-O)	C_{Lmax}	1.60
Basic ($\alpha = 0$) C_L	C_{L0}	0.0

EXERCISES

(1) Determine the pressure, temperature and density and the corresponding ratios for the following conditions:
  (a) Altitude of 10,000 ft for outside air temperature (OAT) differing some −30 °F, 0 °F, and 30 °F from the ISA temperature (ISA is the 0 °F condition).
  (b) Altitude of 8.4 km for OAT differing some −30 °C, 0 °C, and 30 °C from the ISA temperature.
(2) Convert the following airspeeds into ft/s:
  (a) 189 kmh
  (b) 16.6 in/hr
  (c) 795 mph
  (d) 321 knots
  (e) 56.1 m/s
  (f) 23 furlongs/fortnight
  (g) The average of 121 mph, 65 knots, and 110 kmh
  (h) 19510 nm/week
(3) Convert the following airspeeds into knots:
  (a) 125 kmh
  (b) 65,800 in/hr
  (c) 698 mph
  (d) 452 km/day
  (e) 195.3 m/s
  (f) 22.5 furlongs/fortnight
  (g) The average of 32 mph, 48 m/s, and 62 kmh
  (h) 582 km/week
(4) Determine the KTAS for the following airspeeds and altitudes (all airspeeds are calibrated airspeeds. Ignore compressibility):
  (a) 150 m/s at an altitude of 10,550 m
  (b) 276 kmh at an altitude of 3.3 statute miles
  (c) 432 ft/s at an altitude of 7.6 km
  (d) 299 mph at an altitude of 5.6 nm
(5) Determine the KTAS for the following airspeeds and altitudes and OATs (ignore compressibility):
  (a) 225 KCAS at an altitude of 6000 ft at ISA
  (b) 145 KIAS at an altitude of 15,000 ft at ISA+20 °C. Instrument error is −3 KIAS
  (c) 270 KIAS at an altitude of 35,000 ft at ISA−20 °C. Instrument error is +1.2%
(6) Determine the KIAS for the following airspeeds and altitudes and OATs (ignore compressibility):
  (a) 225 KTAS at an altitude of 6000 ft at ISA
  (b) 145 KTAS at an altitude of 15,000 ft at ISA+20°C. Instrument error is −3 KIAS
  (c) 270 KTAS at an altitude of 35,000 ft at ISA−20°C. Instrument error is +1.2%
(7) Determine the KGS for the following airspeeds and altitudes and OATs and wind speeds (ignore compressibility):
  (a) 225 KTAS at an altitude of 6000 ft at ISA. Headwind is 35 knots
  (b) 227 KIAS at an altitude of 15,000 ft at ISA+15°C. Instrument error is +2.5 KIAS. Headwind is −18 knots
  (c) 270 KIAS at an altitude of 35,000 ft at ISA-30°C. Instrument error is +2.8%. Headwind is −95 knots
(8) Determine KTAS, KEAS, and KCAS for an airplane flying at $M = 0.8$ at 36,000 ft on a standard day. If the instrument error is −3.5 KIAS, determine the indicated airspeed as well.
(9) Create a V-n diagram for the aircraft presented in the table below.

Item	Symbol	Value
Certification category		Normal
Wingspan	b	39 ft
Wing area	S	220 ft^2
Mean geometric chord	MGC	4.5 ft
Gross weight	W	6000 lb$_f$
Minimum flying weight	W_{min}	4000 lb$_f$
Stalling speed (gross weight, flaps down)	$V_{S(+)}$	59 KEAS
Stalling speed (gross weight, flaps down)	$V_{S(-)}$	68 KEAS

(Continued)

Item	Symbol	Value
Maximum level airspeed	V_H	350 KEAS
Three-dimensional lift curve slope	$C_{L\alpha}$	4.6 per radian
Maximum C_L (flaps up)	C_{Lmax}	1.50
Minimum C_L (flaps up)	C_{Lmin}	−1.10
Maximum C_L (flaps down)	C_{Lmax}	2.31
Minimum C_L (flaps down)	C_{Lmin}	−1.74

VARIABLES

Symbol	Description	Units (UK and SI)
a	Lapse rate	
a	Speed of sound	
a_0	Speed of sound at S-L on a standard day	
AR	Aspect ratio	
b	Wingspan	ft or m
$c(y)$	Function for determining wing chord at specified span location	ft or m
C_{L0}	Lift coefficient at zero AOA	
C_{Lmax}	Maximum lift coefficient	
C_{Lmin}	Minimum lift coefficient	
C_{LminD}	Lift coefficient at minimum drag	
$C_{L\alpha}$	3D lift curve slope	/degree or /radian
h	Altitude	ft or m
h_0	Reference altitude	ft or m
h_P	Pressure altitude	ft or m
h_ρ	Density altitude	ft or m
k	Lapse rate constant	
K_g	Gust alleviation factor	
M_C	Cruising Mach number	
M_D	Diving Mach number	
MGC	Mean geometric chord	ft or m
M_{MO}	Maximum operating Mach number	
n_-	Negative load factor	
n_+	Positive load factor	
n_g	Gust load factor	
p	Pressure	lb_f/ft^2 or Pa

Symbol	Description	Units (UK and SI)
P	Maximum power at S-L	$ft \cdot lb_f/s$ or $N \cdot m/s$
p_0	Reference S-L pressure	lb_f/ft^2 or Pa
p_{mbar}	Pressure in mbar	mbar
p_{Pa}	Pressure in Pa	Pa
p_{psf}	Pressure in psf	lb_f/ft^2
p_{psi}	Pressure in psi	lb_f/in^2
q_c	Compressible dynamic pressure	lb_f/ft^2 or Pa
R	Specific gas constant for air	$ft \cdot lbf/(slug \cdot °R)$ or $m^2/(K \cdot s)$
R_e	Reynolds number	
RH	Relative humidity	
R_{H2O}	Specific gas constant for water vapor	$ft \cdot lbf/(slug \cdot °R)$ or $m^2/(K \cdot s)$
S	Wing area	ft^2 or m^2
T	Temperature	°R or K
$T_{°C}$	Temperature in degrees Celsius	°C
$T_{°F}$	Temperature in degrees Fahrenheit	°F
$T_{°R}$	Temperature in degrees Rankine	°R
T_0	Temperature at reference altitude	°R or K
T_K	Temperature in degrees Kelvin	K
TR	Taper ratio	
U	Maximum vertical gust rate	
U_{de}	Vertical gust velocity	ft/s or m/s
V_1	Maximum speed at which a multiengine aircraft can be stopped if critical engine fails during take-off	Knots (typ.) or ft/s or m/s
V_2	Take-off safety speed	Knots (typ.) or ft/s or m/s
V_{2min}	Minimum take-off safety speed	Knots (typ.) or ft/s or m/s
V_3	Flap retraction speed	Knots (typ.) or ft/s or m/s
V_A	Maneuvering speed	Knots (typ.) or ft/s or m/s
V_B	Design speed for maximum gust intensity	Knots (typ.) or ft/s or m/s
V_{BA}	Minimum rate-of-descent airspeed	Knots (typ.) or ft/s or m/s
V_{BG}	Best glide speed	Knots (typ.) or ft/s or m/s
V_{BR}	Airspeed when pilot begins to apply brakes after touch-down	Knots (typ.) or ft/s or m/s
V_C	Design cruising speed or maximum structural speed	Knots (typ.) or ft/s or m/s

Symbol	Description	Units (UK and SI)
$V_{C55\%}$	Cruising speed at 55% power	Knots (typ.) or ft/s or m/s
$V_{C65\%}$	Cruising speed at 65% power	Knots (typ.) or ft/s or m/s
$V_{C75\%}$	Cruising speed at 75% power	Knots (typ.) or ft/s or m/s
V_{CAS}	Calibrated airspeed	Knots (typ.) or ft/s or m/s
V_{Cmax}	Maximum cruising speed	Knots (typ.) or ft/s or m/s
V_{Cmin}	Minimum design cruising speed	Knots (typ.) or ft/s or m/s
V_D	Dive speed	Knots (typ.) or ft/s or m/s
V_{EAS}	Equivalent airspeed	Knots (typ.) or ft/s or m/s
V_{EF}	Speed at which critical engine is assumed to fail during take-off	Knots (typ.) or ft/s or m/s
V_{Emax}	Best endurance speed	Knots (typ.) or ft/s or m/s
V_F	Design cruising speed for negative load factor	Knots (typ.) or ft/s or m/s
V_{FE}	Maximum flap extension speed	Knots (typ.) or ft/s or m/s
V_{FLR}	Airspeed for initiating flare maneuver	Knots (typ.) or ft/s or m/s
V_{FTO}	Final take-off speed	Knots (typ.) or ft/s or m/s
V_G	Negative maneuver speed	Knots (typ.) or ft/s or m/s
V_{GS}	Ground speed	Knots (typ.) or ft/s or m/s
V_{IAS}	Indicated airspeed	Knots (typ.) or ft/s or m/s
V_{LDmax}	Best glide speed	Knots (typ.) or ft/s or m/s
V_{LE}	Maximum landing gear extended speed	Knots (typ.) or ft/s or m/s
V_{LO}	Maximum landing gear operating speed	Knots (typ.) or ft/s or m/s
V_{LOF}	Lift-off speed	Knots (typ.) or ft/s or m/s
V_{max}	Maximum obtainable level airspeed	Knots (typ.) or ft/s or m/s
V_{MC}	Minimum control speed with critical engine inoperative	Knots (typ.) or ft/s or m/s
V_{MCA}	Minimum control speed while airborne	Knots (typ.) or ft/s or m/s
V_{MCG}	Minimum control speed on the ground	Knots (typ.) or ft/s or m/s
V_{MO}	Maximum operating speed	Knots (typ.) or ft/s or m/s
V_{MU}	Minimum unstuck speed	Knots (typ.) or ft/s or m/s
V_{NE}	Never-exceed speed	Knots (typ.) or ft/s or m/s
V_{NO}	Normal operating speed	Knots (typ.) or ft/s or m/s
V_O	Maximum operating maneuvering speed	Knots (typ.) or ft/s or m/s
V_R	Rotation speed	Knots (typ.) or ft/s or m/s
V_{REF}	Landing reference speed	Knots (typ.) or ft/s or m/s
V_{Rmax}	Speed of best range	Knots (typ.) or ft/s or m/s
V_{S0}	Stall speed flaps down	Knots (typ.) or ft/s or m/s
V_{S1}	Stall speed clean	Knots (typ.) or ft/s or m/s
V_{SR}	Reference stalling speed	Knots (typ.) or ft/s or m/s
V_{SR0}	Reference stalling speed in landing configuration	Knots (typ.) or ft/s or m/s
V_{SW}	Speed at which stall warning occurs	Knots (typ.) or ft/s or m/s
V_{TAS}	True airspeed	Knots (typ.) or ft/s or m/s
V_{TD}	Touch-down speed	Knots (typ.) or ft/s or m/s
V_{TR}	Transition airspeed	Knots (typ.) or ft/s or m/s
V_X	Best angle-of-climb airspeed	Knots (typ.) or ft/s or m/s
V_Y	Best rate-of-climb airspeed	Knots (typ.) or ft/s or m/s
V_{YSE}	OEI best rate-of-climb	Knots (typ.) or ft/s or m/s
W	Gross weight	lb_f or N
W_0	Gross weight	lb_f or N
W_{min}	Minimum flying weight	lb_f or N
x	Humidity ratio (context dependent)	ft or m
x	distance penetrated into gust (context dependent)	ft or m
$\Delta error$	Error in airspeed indicator	ft/s or m/s
ΔT_{ISA}	Deviation from International Standard Atmosphere	°R or K
δ	Pressure ratio	
γ	Specific heat ratio	
λ	Bypass ratio	
μ	Viscosity	$lb_f \cdot s/ft^2$ or $N \cdot s/m^2$
μ_g	Aircraft mass ratio	
ν	Kinematic viscosity	$1/(ft^2 \cdot s)$ or $1/(m^2 \cdot s)$
θ	Temperature ratio	
ρ	Density	$slugs/ft^3$ or kg/m^3
ρ_0	Reference S-L density	$slugs/ft^3$ or kg/m^3
ρ_{kg/m^3}	Density in kg/m^3	kg/m^3
ρ_{S-L}	Density at sea level	
ρ_{slugs/ft^3}	Density in $slugs/ft^3$	$slugs/ft^3$
ρ_{std}	Density at altitude, calculated by standard methods	$slugs/ft^3$ or kg/m^3
σ	Density ratio	

References

[1] http://www.fai.org/records/powered-aeroplanes-records.
[2] http://www.nasa.gov/centers/dryden/history/pastprojects/Helios/index.html.
[3] http://www.nasa.gov/pdf/64317main_helios.pdf.
[4] http://www.uwsp.edu/geo/faculty/ritter/geog101/textbook/atmospheric_moisture/lapse_rates_1.html.
[5] http://www.vaisala.com/humiditycalculator/help/index.html#calculating-humidity
[6] Code of Federal Regulations, 14 CFR Part 23.
[7] Code of Federal Regulations, 14 CFR Part 25.
[8] ASTM International, formerly known as the American Society for Testing and Materials.
[9] http://www.cirruspilots.org/Content/CAPSHistory.aspx.
[10] TCDS A00009CH, Cirrus Design Corporation, Revision 18, 12/29/2011, FAA.

CHAPTER 17

Performance – Take-Off

OUTLINE

17.1 Introduction 791
  17.1.1 The Content of this Chapter 792
  17.1.2 Important Segments of the T-O Phase 792
  17.1.3 Definition of a Balanced Field Length 795

17.2 Fundamental Relations for the Take-off Run 797
  17.2.1 General Free-body Diagram of the T-O Ground Run 797
  17.2.2 The Equation of Motion for a T-O Ground Run 798
    Derivation of Equations (17-4) and (17-5) 799
  17.2.3 Review of Kinematics 799
  17.2.4 Formulation of Required Aerodynamic Forces 800
  17.2.5 Ground Roll Friction Coefficients 800
  17.2.6 Determination of the Lift-off Speed 800
    Requirements for T-O Speeds Per 14 CFR Part 23 for GA Aircraft 800
    Requirements for T-O Speeds Per 14 CFR Part 25 for Commercial Aviation Aircraft 801
  17.2.7 Determination of Time to Lift-off 802

17.3 Solving the Equation of Motion of the T-O 802
  17.3.1 Method 1: General Solution of the Equation of Motion 802
    Step 1: Lift-off Speed 803
    Step 2: Lift-induced Drag in Ground Effect 803
    Step 3: Lift at the Reduced Lift-off Speed 803
    Step 4: Drag at the Reduced Lift-off Speed 803
    Step 5: Thrust at the Reduced Lift-off Speed 803
    Step 6: Ground Run 804
  17.3.2 Method 2: Rapid T-O Distance Estimation for a Piston-powered Airplane 805
    Derivation of Equation (17-21) 805
  17.3.3 Method 3: Solution Using Numerical Integration Method 807
    Propeller Thrust at Low Airspeeds 807
  17.3.4 Determination of Distance During Rotation 813
  17.3.5 Determination of Distance for Transition 813
    Derivation of Equations (17-26) and (17-27) 813
  17.3.6 Determination of Distance for Climb Over an Obstacle 814
  17.3.7 Treatment of T-O Run for a Taildragger 815
  17.3.8 Take-off Sensitivity Studies 816

17.4 Database – T-O Performance of Selected Aircraft 817

Exercises 817

Variables 819

References 820

17.1 INTRODUCTION

It is appropriate to start the performance analysis with one of the most important maneuvers performed by any aircraft; the take-off (T-O). Figure 17-1 shows an organizational map displaying the T-O among other parts of the performance theory. It is of utmost importance that the designer not only understands the T-O capabilities of the new design, but also recognizes its limitations and sensitivity. This chapter will present the formulation of and the solution of the equation of motion for the entire T-O maneuver and present practical as well as numerical solution methodologies that can be used both for propeller- and jet-powered aircraft.

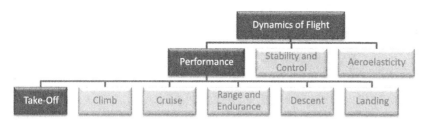

FIGURE 17-1 An organizational map placing performance theory among the disciplines of dynamics of flight, and highlighting the focus of this section; T-O performance analysis.

T-O performance typically refers to the distance required for the aircraft to accelerate from a standstill to lift-off, as well as the distance required to attain an initial and steady climb. Most aircraft are designed to meet specific runway length requirements, and these may dictate the power plant necessary, as most aircraft use far less power for cruise than for the T-O and climb. As an example of the constraints confronting the designer, commercial aircraft must be at least capable of operating from runways used by the competition, and preferably from even shorter runways, as this might expand their marketability and give them a competitive edge.

Aircraft are sometimes required to meet some T-O distance requirements specified in a request for proposal (RFP) or specific design requirements. In order to meet such requirements the designer must not only consider the T-O distance at ISA and S-L, but also runways that present the design with an uphill slope, as well as on a hot day and high-altitude conditions. All can seriously tax the capability of the aircraft. There may even be a combination of high temperatures in locations at high elevation. These are altogether easy to overlook. As an example, the Mariscal Sucre International Airport in Quito, Ecuador, is at an altitude of 9228 ft and presents a serious high-altitude challenge to commercial aircraft. The largest type of aircraft to regularly operate from it is the Airbus A-340.

In order to evaluate the capability of the aircraft the designer should prepare a T-O sensitivity graph, which shows the T-O run as a function of density altitude and shows the impact of a selected parameter, e.g. weight or outside air temperature, on the operation of the aircraft. A typical such graph is shown in Section 17.3.8, *Take-off sensitivity studies*. Preparing the analysis in a spreadsheet is very convenient and a considerable amount of information about the airplane's capabilities can be learned from such a tool.

A capable T-O performance analysis also accounts for the type of landing gear featured on the aircraft. The analysis in this text assumes conventional tricycle or taildragger configurations. Accounting for landing gear is particularly important for taildraggers. A taildragger lifts the tailwheel off the runway as soon as certain airspeed is achieved. The airspeed at which this takes place may range from standstill to half of the airplane's lift-off speed. However, as soon as this happens, the T-O formulation must be modified to account for two rather than three landing gear contact points and the associated reduction in drag of the more horizontal configuration. This is discussed further in Section 17.3.7, *Treatment of T-O run for a taildragger*.

In general, the methods presented here are the industry standard and mirror those presented by a variety of authors, e.g. Perkins and Hage [1], Torenbeek [2], Nicolai [3], Roskam [4], Hale [5], Anderson [6] and many, many others.

17.1.1 The Content of this Chapter

- **Section 17.2** presents the fundamental relations of the T-O ground run, including the equation of motion for a T-O ground run and the kinematics of the T-O run.
- **Section 17.3** presents several methods to solve the equation of motion.
- **Section 17.4** presents the T-O properties of selected aircraft types.

17.1.2 Important Segments of the T-O Phase

Generally, the T-O phase is split into the segments shown in Figure 17-2. The *ground roll* is the distance from brake release to the initiation of the rotation, when the pilot pulls the control wheel (or stick or yoke) backward in order to raise the nose of the aircraft. This maneuver is required to increase the AOA of the airplane to help it become airborne. The aircraft typically remains in this attitude for some 1–3 seconds, depending on aircraft size, before the tires lose contact with the ground. This segment is called the *rotation*. The rotation phase concludes as the aircraft lifts off the ground and begins the *transition* and subsequent *climb phase*.

In short, the purpose of the T-O analysis is to estimate the total T-O distance by breaking it up into the aforementioned segments and analyzing each step using simplified physics. Typically, requirements for aircraft design call for the ground roll and the total T-O distance to be specified. The first step prior to formulating the

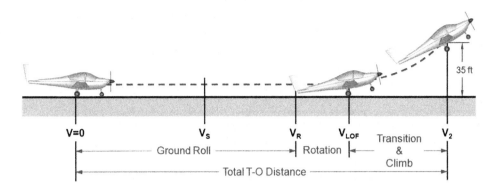

FIGURE 17-2 Important segments of the T-O phase.

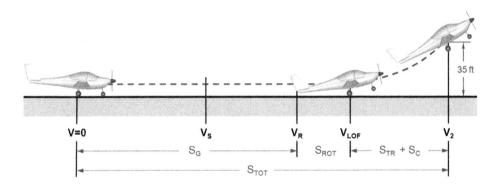

FIGURE 17-3 Nomenclature for important segments of the T-O phase.

problem is to define the segments using variable denotation that will be carried through the remainder of this section. This is shown in Figure 17-3.

The airspeeds referenced in Figure 17-3 used in the T-O analysis are shown in Table 17-1.

The nature of the formulation of the T-O segments depends in part on the airplane's landing gear. To better understand why, consider Figure 17-4, which shows the T-O maneuver for two of the most common landing gear configurations: a tricycle and a taildragger.

In Figure 17-4, V is the instantaneous airspeed of the aircraft and V_S is the stalling speed in the T-O configuration. This means that if the airplane features high-lift devices, V_S refers to the stalling speed with flaps deployed. The important point is that the airplane

TABLE 17-1 Definition of Important Airspeeds for the T-O

Name	Airspeed	GA Aircraft (FAR 23)[a]
Ground run	V_R	$1.1 V_{S1}$
Rotation	V_{LOF}	$1.1 V_{S1}$
Transition	V_{TR}	$1.15 V_{S1}$
Climb	V_2	$1.2 V_{S1}$

[a] *Airspeeds are formally established per 14 CFR Part 23, § 23.51 Takeoff Speeds.*

must accelerate from standstill to a given airspeed, called the lift-off speed, before it can, well, lift off. The tricycle configuration accelerates with the main and nose landing gears in contact with the ground. However, the taildragger, initially, has the main landing gear and tailwheel in contact with the ground until a combination of forward airspeed, thrust, and propwash over the horizontal tail allows it to lift the tailwheel off the ground. For some taildraggers, typically light ones, this happens as soon as the engine generates T-O thrust. For others, primarily larger aircraft, some forward airspeed must be acquired before the tail can be raised off the ground. Sections 13.3.4, *Tricycle landing gear reaction loads* and 13.3.5, *Taildragger landing gear reaction loads*, provide methods to estimate the airspeed at which a tricycle configuration can lift the nose gear off the ground and at which a taildragger can lift the tailwheel off the ground.

A T-O analysis of a tricycle and taildragger aircraft differs primarily in having to account for the tail of the latter being raised off the ground. Initially, the taildragger geometry is one of an airplane at a high *AOA*, whereas once the tailwheel is off the ground it transforms into one at a low *AOA*. A proper representation of the transformation is required for accurate estimation of the T-O for such airplanes.

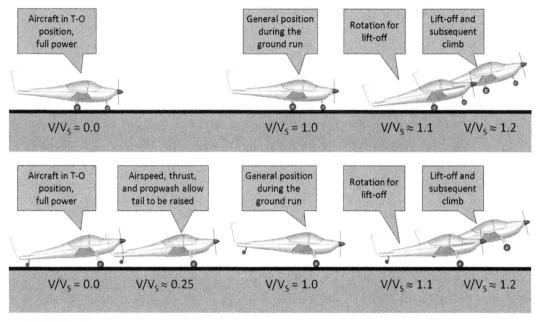

FIGURE 17-4 A sequence of images showing T-O ground run for a tricycle, and taildragger aircraft.

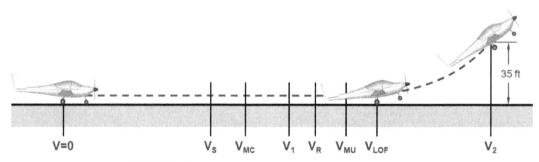

FIGURE 17-5 Important airspeeds during the T-O phase.

The distance covered in the specific segments shown in Figure 17-3 are determined in the sections shown in Table 17-2.

A schematic of the T-O run showing other important airspeeds is shown in Figure 17-5. See Table 16-4 for the definition of the various airspeeds. It should be made clear that the stalling speed (V_S) must be exceeded before the airplane can become airborne, and that it is always based on the configuration of the aircraft during the T-O run. Thus, if the airplane takes off with its flaps deflected (typical deflection is somewhere between 10° and 20°) its stalling speed will be less than in the clean configuration. The minimum control speed with one engine inoperative (V_{MC}) only applies to multiengine aircraft.

The unstick speed is the airspeed at which the airplane no longer 'sticks' to the runway and lifts off whether one wants it to or not. It depends on the attitude (or AOA) of the airplane during the ground run. It is high if the ground run attitude is low and reduces as the attitude is increased (in other words: the higher the nose, the lower the unstick speed). It follows that the minimum unstick speed (V_{MU}) is achieved when the airplane is in a tail-strike position (at its maximum rotation angle). Clearly, V_{MU} is will be lower than the lift-off speed (V_{LOF}), as the airplane is only rotated to its maximum rotation angle due to pilot error (which may be compounded by very aft CG).

TABLE 17-2 Sections Used to Estimate Various Segments of the T-O Run

Segment Name	Symbol	Section
Ground roll	S_G	17.3.1 through 17.3.3
Rotation	S_R	17.3.4 *Determination of distance during rotation*
Transition	S_{TR}	17.3.5 *Determination of distance for transition*
Climb	S_C	17.3.6 *Determination of distance for climb over an obstacle*

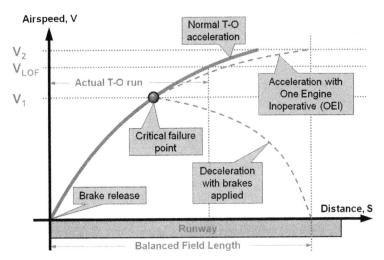

FIGURE 17-6 Definition of a balanced field length.

17.1.3 Definition of a Balanced Field Length

Consider Figure 17-6, which shows how the airspeed of a typical aircraft changes with respect to distance from brake release during a T-O run. Assume this graph reflects the relationship between airspeed and runway distance of a multiengine aircraft. Initially, while at stand-still, the airspeed and runway distance are both zero. However, as soon as the pilot increases thrust and releases the brakes, the aircraft begins to accelerate and the distance from brake release increases.

Now, assume that some distance from brake release one engine becomes inoperative. The airplane will continue to accelerate, only much slower than before. Now two things can happen:

(1) If the airspeed is low (and the distance covered is short), the pilot can simply step on the brakes and bring the aircraft to a complete stop before running out of runway. Problem solved. This is shown in the left-hand graph in Figure 17-7.
(2) On the other hand, if the airspeed is high (assuming the airplane is still on the ground) there may not be enough runway ahead of the plane to fully stop it in time, so braking is not an option. In this case, two new things can happen:

(a) There is insufficient thrust remaining to accelerate to lift-off before running out of runway. Disaster strikes. (The pilot should never have attempted the T-O run.)
(b) There is indeed enough thrust remaining to accelerate to lift-off before running out of runway. Disaster is avoided. This is shown in the right-hand graph in Figure 17-7.

While this logic is sound it requires analysis effort which may not be possible in a time of crisis. Also note that the following can be deduced from the graphs in Figure 17-7. If there is an airspeed at which the airplane can be stopped in time (as shown in the left graph) and there is an airspeed at which it cannot be stopped in time, but accelerated to lift-off (as shown in the right graph), then there must be an airspeed between the two at which it can both be stopped and accelerated to lift-off in time (if not, then the runway is too short for a safe operation of the airplane). There is indeed such an airspeed and it has been given a special name: V_1.

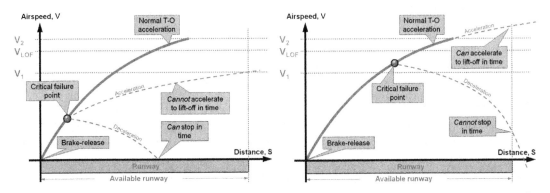

FIGURE 17-7 If the failure occurs early enough it is possible to stop the airplane in time (left). However, if it is moving too fast it may not be able to stop in time, but may still be able to lift-off before running out of runway (right).

In order to avoid exerting analysis effort in a time of crisis and, that way, increase safety, it is simpler to tell the pilot that if below V_1, step on the brakes; and if above, continue the take-off. The distance required to accelerate the aircraft to V_1 and then decelerate it to a complete stop by applying hard braking, is also given a special name; *balanced field length* (sometimes called *accelerate-stop* distance). Strictly speaking it is only applicable to multiengine airplanes, since a single-engine airplane has no choice but to apply the brakes. The importance of the BFL is that it gives the pilot two safe options.

Torenbeek [2, pp. 168–169] developed the following empirical expression to estimate the balanced field length for a multiengine aircraft, presented here using the original symbols:

$$BFL = \frac{0.863}{1 + 2.3\Delta\gamma_2}\left(\frac{W_{to}/S}{\rho g C_{L2}} + h_{to}\right)$$
$$\times \left(2.7 + \frac{1}{\overline{T}/W_{to} - \mu'}\right) + \left(\frac{\Delta S_{to}}{\sqrt{\sigma}}\right) \quad (17\text{-}1)$$

where

C_{L2} = magnitude of the lift coefficient at V_2. If $V_2 = 1.2V_S$, then it can be shown that $C_{L2} = 0.694 C_{Lmax}$
g = acceleration due to gravity = 32.174 ft/s² or 9.807 m/s²
h_{to} = obstacle height = 35 ft (10.7 m) for commercial jetliners and 50 ft (15.2 m) for GA aircraft
S = reference wing area, ft² or m²
W_{to} = take-off weight, lb$_f$ or N
$\overline{T}$ = average thrust during the T-O run, given by Equations (17-2) and (17-3).
ρ = air density, in slugs/ft³ or kg/m³
$\mu' = 0.01 \cdot C_{Lmax} + 0.02$, where C_{Lmax} is that of the T-O configuration
ΔS_{to} = inertia distance = 655 ft (200 m)

$\Delta\gamma_2 = \gamma_2 - \gamma_{2min}$
$\gamma_2 = \sin^{-1}\left(\dfrac{T_{OEI} - D_2}{W_{to}}\right)$ and
$\gamma_{2min} = \begin{cases} 0.024 & \text{for 2 engines} \\ 0.027 & \text{for 3 engines} \\ 0.030 & \text{for 4 engines} \end{cases}$
D_2 = drag at V_2, lb$_f$ or N

Torenbeek's method requires the average thrust during the T-O run to be estimated for jets and propeller-powered aircraft using the following expressions:

For a jet aircraft:

$$\overline{T} = 0.75 T_{to}\left(\frac{5 + \lambda}{4 + \lambda}\right) \quad (17\text{-}2)$$

where

T_{to} = maximum static thrust, in lb$_f$ or N
λ = turbofan bypass ratio

For an aircraft with a constant-speed propeller:

$$\overline{T} = K_p P_{to}\left(\frac{\sigma N_e D_p^2}{P_{to}}\right)^{1/3} \quad (17\text{-}3)$$

where

D_p = propeller diameter, in ft or m
P_{to} = maximum engine power, in BHP or kg·m/s (which is power in watts divided by g)
σ = density ratio

$K_p = \begin{cases} 5.75 & \text{for } P_{to} \text{ in BHP} \\ 0.321 & \text{for } P_{to} \text{ in kg·m/s} \end{cases}$

Note that for a fixed-pitch propeller, the average thrust will be approximately 15–20% lower than that for the constant-speed propeller.

EXAMPLE 17-1

Estimate the balanced field length for the Learjet 45XR assuming a maximum lift coefficient in the T-O configuration to be 1.60 and $V_2 = 1.2V_S$. Also, assume an obstacle height of 50 ft and gross weight of 21,500 lb$_f$. Note that other properties of the airplane are given in Table 16-7. Assume the simplified drag model and that the Oswald span efficiency factor amounts to 0.8294 and $C_{Dmin} = 0.035$ in the T-O configuration. Furthermore, assume the maximum thrust with one engine inoperable (OEI) to equal one half of the average thrust calculated in Step 4 below.

Solution

Step 1: Determine the constant "k":

$$k = \frac{1}{\pi \cdot AR \cdot e} = \frac{1}{\pi \cdot (7.33) \cdot (0.8294)} = 0.05236$$

Step 2: Calculate the stalling speed in the T-O configuration using the given C_{Lmax} and Equation (19-7):

$$V_S = \sqrt{\frac{2W}{\rho S C_{Lmax}}} = \sqrt{\frac{2(21{,}500)}{(0.002378)(311.6)(1.60)}}$$
$$= 190.4 \text{ ft/s} \quad (\approx 113 \text{ KCAS})$$

Step 3: Therefore, V_2 amounts to:

$$V_2 = 1.2 V_S = 1.2 \times 190.4 = 228.5 \text{ ft/s} \quad (\approx 135 \text{ KCAS})$$

EXAMPLE 17-1 (cont'd)

Step 4: Calculate the average thrust during the T-O run, $\overline{T}$, using Equation (17-2):

$$\overline{T} = 0.75T_{to}\left(\frac{5+\lambda}{4+\lambda}\right) = 0.75 \times 7000 \times \left(\frac{5+3.9}{4+3.9}\right)$$
$$= 5915 \text{ lb}_f$$

Step 5: Determine the factor μ': $\mu' = 0.01 \cdot C_{Lmax} + 0.02 = 0.036$.

Step 6: Estimate the drag at the airspeed V_2. Note that this requires the drag polar for the aircraft to be known. Here, assume the following drag polar for the aircraft in the T-O configuration: $C_D = C_{Dmin} + k \cdot C_L^2 = 0.035 + 0.05236\, C_L^2$.

Lift coefficient at V_2:

$$C_{L2} = 0.694 C_{Lmax} = 0.694 \times 1.60 = 1.11$$

Drag coefficient at V_2:

$$C_D = 0.035 + 0.05236(1.11)^2 = 0.09951$$

Drag force at V_2:

$$D_2 = \frac{1}{2}\rho V_2^2 S C_D = \frac{1}{2}(0.002378)(228.5)^2(311.6)(0.09951)$$
$$= 1925 \text{ lb}_f$$

Step 7: Estimate climb angle at V_2 and the difference $\Delta\gamma_2$. As stated in the introduction to this example, it will be assumed that the thrust with OEI amounts to one half that of $\overline{T}$ calculated in Step 4, or 2955 lb$_f$. This yields the following climb angle:

$$\gamma_2 = \sin^{-1}\left(\frac{T_{OEI} - D_2}{W_{to}}\right) = \sin^{-1}\left(\frac{2955 - 1925}{21{,}500}\right)$$
$$= 0.04804 \text{ rad}$$

Therefore:

$$\Delta\gamma_2 = \gamma_2 - \gamma_{2min} = 0.04804 - 0.024 = 0.02404 \text{ rad}$$

Step 8: Finally calculate the BFL using Equation (17-1):

$$BFL = \frac{0.863}{1 + 2.3\Delta\gamma_2}\left(\frac{W_{to}/S}{\rho g C_{L2}} + h_{to}\right)\left(2.7 + \frac{1}{\overline{T}/W_{to} - \mu}\right) + \left(\frac{\Delta S_{to}}{\sqrt{\sigma}}\right)$$

$$= \frac{0.863}{1 + 2.3(0.02404)}\left(\frac{21{,}500/311.6}{(0.002378)(32.174)(1.11)} + 50\right)$$

$$\times \left(2.7 + \frac{1}{5915/21{,}500 - 0.036}\right) + \left(\frac{655}{\sqrt{1}}\right)$$

$$= (0.8178)(862.5)(6.882) + 655 = 5509 \text{ ft}$$

This compares to 5040 ft published by Ref. [7] for the type.

17.2 FUNDAMENTAL RELATIONS FOR THE TAKE-OFF RUN

In this section, the equation of motion for the T-O will be derived, as well as some elementary relationships that can be used to evaluate the ground run segment of the T-O maneuver. Both conventional and taildragger configurations will be considered.

17.2.1 General Free-body Diagram of the T-O Ground Run

Figure 17-8 and Figure 17-9 show the free-body diagrams of tricycle and taildragger configurations during a developed T-O ground run.

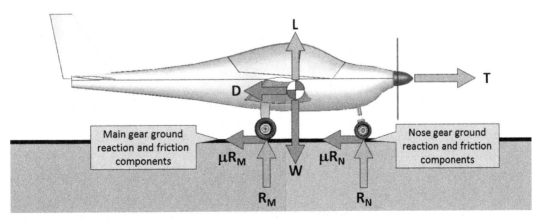

FIGURE 17-8 A balanced 2D free-body (forces only) of the T-O ground run for an aircraft with a tricycle landing gear.

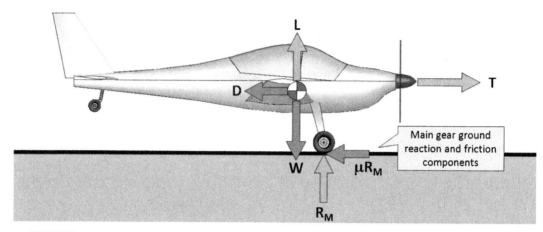

FIGURE 17-9 A balanced 2D free-body (forces only) of the T-O ground run for a taildragger aircraft.

TABLE 17-3 Ground Roll Friction Coefficients

Surface Type	Ground Friction Coefficient, μ	
	Brakes Off	Braking
Dry asphalt or concrete	0.03–0.05	0.3–0.5
Wet asphalt or concrete	0.05	0.15–0.3
Icy asphalt or concrete	0.02	0.06–0.10
Hard turf	0.05	0.4
Firm dirt	0.04	0.3
Soft turf	0.07	0.2
Wet grass	0.08	0.2

17.2.2 The Equation of Motion for a T-O Ground Run

The equation of motion for an aircraft during the ground run on a perfectly horizontal and flat runway can be estimated from:

Acceleration on a flat runway:

$$\frac{dV}{dt} = \frac{g}{W}[T - D - \mu(W - L)] \qquad (17\text{-}4)$$

where

D = drag as a function of V, in lb_f or N
g = acceleration due to gravity, ft/s^2 or m/s^2
L = lift as a function of V, in lb_f or N
T = thrust, in lb_f or N
W = weight, assumed constant in lb_f or N
μ = ground friction coefficient (see Table 17-3)

If the runway is not perfectly horizontal, but has an uphill or downhill slope γ (see Figure 17-10), the acceleration of the aircraft should be estimated from:

Acceleration on an uphill slope γ:

$$\frac{dV}{dt} = \frac{g}{W}[T - D - \mu(W\cos\gamma - L) - W\sin\gamma]$$

$$(17\text{-}5)$$

where γ = slope of runway (if uphill the sign is positive but negative if downhill), in °.

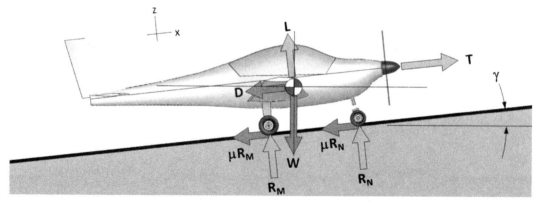

FIGURE 17-10 A balanced 2D free-body (forces only) of the T-O ground run for a tricycle gear aircraft on an uphill slope runway.

Note that Equation (17-5) assumes the reference frame to be aligned along the runway. This is a reasonable assumption because the runway length reported will be along the slope. We can also rewrite this to get the acceleration in terms of the thrust-to-weight ratios:

$$\frac{dV}{dt} = \frac{g}{W}[T - D - \mu(W\cos\gamma - L) - W\sin\gamma]$$
$$= g\left[\frac{T}{W} - \frac{D}{W} - \mu\left(\cos\gamma - \frac{L}{W}\right) - \sin\gamma\right] \quad (17\text{-}6)$$

Derivation of Equations (17-4) and (17-5)

Consider the aircraft in Figure 17-8 and Figure 17-9. Its motion can be completely described using the standard equations of motion as follows. First, the sum of the forces in the x-direction must equal the airplane's horizontal acceleration:

$$\sum F_x = ma_x = \frac{W}{g}\frac{dV}{dt} \Rightarrow T - D - \mu(W - L) = \frac{W}{g}\frac{dV}{dt} \quad \text{(i)}$$

where μ is the coefficient of friction caused by a ground friction. Its value depends on the runway surface. Suitable values can be seen in Table 17-3. Additionally, the forces in the y-direction must yield no net vertical acceleration:

$$\sum F_y = ma_y = 0 \Rightarrow R + L - W = 0$$
$$\Rightarrow R = W - L$$

Solving for the horizontal acceleration dV/dt results in:

$$T - D - \mu(W - L) = \frac{W}{g}\frac{dV}{dt}$$

The resulting expression is shown as Equation (17-4). Note that if the aircraft is accelerating along an uphill or downhill runway, whose slope is given by γ, then Equation (i) must include an additional component caused by its weight:

$$\sum F_x = ma_x$$
$$= \frac{W}{g}\frac{dV}{dt} \Rightarrow T - D - \mu(W - L) - W\sin\gamma$$
$$= \frac{W}{g}\frac{dV}{dt} \quad \text{(ii)}$$

The resulting expression is shown as Equation (17-5). If the runway is uphill, the sign for γ should be positive, as this will cause the resulting acceleration to be less than on a perfectly horizontal runway. Likewise, if the runway is downhill, the sign for γ should be negative, as this will cause the resulting acceleration to be higher than on a perfectly horizontal runway.

QED

17.2.3 Review of Kinematics

Kinematics is the study of the motion of objects that only involves the motion itself (e.g. acceleration, speed, and distance) and not what causes it (forces and moments). The kinematic formulation is essential for the study of the take-off run. The most basic formulation is presented below:

Velocity:

$$V = \int_0^t a\cdot dt = V_0 + a\cdot t \quad (17\text{-}7)$$

Distance:

$$S = \int_0^t V\cdot dt = \int_0^t (V_0 + a\cdot t)\cdot dt = S_0 + V_0 t + \frac{1}{2}a\cdot t^2 \quad (17\text{-}8)$$

Alternative expression for distance:

$$S - S_0 = \frac{V^2 - V_0^2}{2a} \quad (17\text{-}9)$$

EXAMPLE 17-2

The pilot's operating handbook (POH) for the single engine, four-seat, SR22 gives a lift-off speed of 73 KCAS and T-O distance of 1020 ft (ISA at S-L). Estimate the average acceleration and time in seconds from brake release to lift-off.

Solution

Average acceleration:

$$S_G = \frac{V^2 - V_0^2}{2a} \Rightarrow a = \frac{V^2 - V_0^2}{2S_{LOF}} = \frac{(73 \times 1.688)^2 - 0}{2 \times 1020}$$
$$= 7.44 \text{ ft/s}^2$$

Time from brake release to lift-off:

$$V = V_0 + a\cdot t \Rightarrow t = \frac{V - V_0}{a} = \frac{73 \times 1.688 - 0}{7.44}$$
$$= 16.6 \text{ s}$$

17.2.4 Formulation of Required Aerodynamic Forces

In this section, the following formulation of aerodynamic forces is assumed when considering the T-O maneuver.

Lift:

$$L = \frac{1}{2}\rho V^2 S C_L = \frac{1}{2}\rho V^2 S(C_{L0} + C_{L_\alpha}\alpha) \qquad (17\text{-}10)$$

General drag:

$$D = \frac{1}{2}\rho V^2 S C_D = \frac{1}{2}\rho V^2 S(C_{Dmin} + C_{Di}) \qquad (17\text{-}11)$$

Simplified drag:

$$D = \frac{1}{2}\rho V^2 S C_D = \frac{1}{2}\rho V^2 S(C_{Dmin} + kC_L^2) \qquad (17\text{-}12)$$

Adjusted drag:

$$D = \frac{1}{2}\rho V^2 S C_D = \frac{1}{2}\rho V^2 S\left(C_{Dmin} + k(C_L - C_{L_{minD}})^2\right) \qquad (17\text{-}13)$$

C_L during T-O:

$$C_{L\ TO} = C_{L0} + C_{L\alpha} \times \alpha_{TO} \qquad (17\text{-}14)$$

C_D during T-O:

$$C_{D\ TO} = C_{Dmin} + C_{Di}(C_{L\ TO}) \qquad (17\text{-}15)$$

where

$C_{Di}(C_{L\ TO})$ = induced drag coefficient of aircraft during the T-O run
α_{TO} = angle-of-attack of aircraft during the T-O run

Note that the induced drag must be corrected for ground effect, in particular if the airplane uses flaps or if its attitude is such that its ground run *AOA* is high. Refer to Section 9.5.8, *Ground effect*, for a correction method.

17.2.5 Ground Roll Friction Coefficients

The airplane has to overcome aerodynamic drag and ground friction during the ground roll. The ground friction depends on the weight on wheels and the properties of the ground, which are assessed using the ground roll friction coefficients tabulated below.

17.2.6 Determination of the Lift-off Speed

In this document and unless otherwise specified, the lift-off speed, V_{LOF}, is assumed to be 1.1 times the stalling speed in that particular configuration (e.g. with flaps deployed, landing gear extended, etc.), V_{S1}. Also, V_R will be assumed to be about 1.1 times V_{S1}.

If we assume the lift-off speed to be 10% higher than the stalling speed in the T-O configuration, V_{LOF} can be calculated directly using the maximum lift coefficient for the T-O configuration, S for the reference area, and ρ for density as follows:

$$V_{LOF} = 1.1 \times V_{S1}$$
$$= 1.1\sqrt{\frac{2W}{\rho S C_{Lmax}}} \approx 1.556\sqrt{\frac{W}{\rho S C_{Lmax}}} \qquad (17\text{-}16)$$

Most small airplanes lift off as soon as the pilot rotates the airplane at that speed, while larger ones take some 3–5 seconds to lift off after rotation initiates, due to their higher inertia. For the rapid take-off estimation method of Section 17.3.2, *Method 2: Rapid T-O distance estimation for a piston-powered airplane*, it will be assumed that V_R and V_{LOF} are the same value.

Note that determining V_{LOF} and other airspeeds for use in more detailed analysis must comply with regulations. Excerpts that deal with T-O analysis per 14 CFR Part 23 and 25 are provided below.

Requirements for T-O Speeds Per 14 CFR Part 23 for GA Aircraft

---

§23.51 TAKE-OFF SPEEDS

(a) For normal, utility, and acrobatic category airplanes, rotation speed, V_R, is the speed at which the pilot makes a control input, with the intention of lifting the airplane out of contact with the runway or water surface.
  (1) For multiengine landplanes, V_R, must not be less than the greater of 1.05 V_{MC}; or 1.10 V_{S1}.
  (2) For single-engine landplanes, V_R, must not be less than V_{S1}.
  (3) For seaplanes and amphibians taking off from water, V_R, may be any speed that is shown to be safe under all reasonably expected conditions, including turbulence and complete failure of the critical engine.

(b) For normal, utility, and acrobatic category airplanes, the speed at 50 feet above the take-off surface level must not be less than:
  (1) or multiengine airplanes, the highest of—
    (i) A speed that is shown to be safe for continued flight (or emergency landing, if applicable) under all reasonably expected conditions,

including turbulence and complete failure of the critical engine;
  (ii) 1.10 V_{MC}; or
  (iii) 1.20 V_{S1}.
(2) For single-engine airplanes, the higher of—
  (i) A speed that is shown to be safe under all reasonably expected conditions, including turbulence and complete engine failure; or
  (ii) 1.20 V_{S1}.
(c) For commuter category airplanes, the following apply:
  (1) V_1 must be established in relation to V_{EF} as follows:
    (i) V_{EF} is the calibrated airspeed at which the critical engine is assumed to fail. V_{EF} must be selected by the applicant but must not be less than 1.05 V_{MC} determined under §23.149(b) or, at the option of the applicant, not less than V_{MC} determined under §23.149(f).
    (ii) The take-off decision speed, V_1, is the calibrated airspeed on the ground at which, as a result of engine failure or other reasons, the pilot is assumed to have made a decision to continue or discontinue the take-off. The take-off decision speed, V_1, must be selected by the applicant but must not be less than V_{EF} plus the speed gained with the critical engine inoperative during the time interval between the instant at which the critical engine is failed and the instant at which the pilot recognizes and reacts to the engine failure, as indicated by the pilot's application of the first retarding means during the accelerate-stop determination of §23.55.
  (2) The rotation speed, V_R, in terms of calibrated airspeed, must be selected by the applicant and must not be less than the greatest of the following:
    (i) V_1;
    (ii) 1.05 V_{MC} determined under §23.149(b);
    (iii) 1.10 V_{S1}; or
    (iv) The speed that allows attaining the initial climb-out speed, V_2, before reaching a height of 35 feet above the take-off surface in accordance with §23.57(c)(2).
  (3) For any given set of conditions, such as weight, altitude, temperature, and configuration, a single value of V_R must be used to show compliance with both the one-engine-inoperative take-off and all-engines-operating take-off requirements.
  (4) The take-off safety speed, V_2, in terms of calibrated airspeed, must be selected by the applicant so as to allow the gradient of climb required in §23.67 (c)(1) and (c)(2) but must not be less than 1.10 V_{MC} or less than 1.20 V_{S1}.
  (5) The one-engine-inoperative take-off distance, using a normal rotation rate at a speed 5 knots less than V_R, established in accordance with paragraph (c)(2) of this section, must be shown not to exceed the corresponding one-engine-inoperative take-off distance, determined in accordance with §23.57 and §23.59(a)(1), using the established V_R. The take-off, otherwise performed in accordance with §23.57, must be continued safely from the point at which the airplane is 35 feet above the take-off surface and at a speed not less than the established V_2 minus 5 knots.
  (6) The applicant must show, with all engines operating, that marked increases in the scheduled take-off distances, determined in accordance with §23.59(a)(2), do not result from over-rotation of the airplane or out-of-trim conditions.

Requirements for T-O Speeds Per 14 CFR Part 25 for Commercial Aviation Aircraft

§25.107 TAKE-OFF SPEEDS

(a) V_1 must be established in relation to V_{EF} as follows:
  (1) V_{EF} is the calibrated airspeed at which the critical engine is assumed to fail. V_{EF} must be selected by the applicant, but may not be less than V_{MC} determined under Sec. 25.149(e).
  (2) V_1, in terms of calibrated airspeed, is the take-off decision speed selected by the applicant; however, V_1 may not be less than V_{EF} plus the speed gained with the critical engine inoperative during the time interval between the instant at which the critical engine is failed, and the instant at which the pilot recognizes and reacts to the engine failure, as indicated by the pilot's application of the first retarding means during accelerate stop tests.
(b) V_{2MIN}, in terms of calibrated airspeed, may not be less than—
  (1) 1.2 V_S for–
    (i) Two-engine and three-engine turbopropeller and reciprocating engine powered airplanes; and

(ii) Turbojet powered airplanes without provisions for obtaining a significant reduction in the one-engine-inoperative power-on stalling speed;

(2) 1.15 V_S for–
   (i) Turbopropeller and reciprocating engine powered airplanes with more than three engines; and
   (ii) Turbojet powered airplanes with provisions for obtaining a significant reduction in the one-engine-inoperative power-on stalling speed; and

(3) 1.10 times V_{MC} established under Sec. 25.149.

(c) V_2, in terms of calibrated airspeed, must be selected by the applicant to provide at least the gradient of climb required by Sec. 25.121(b) but may not be less than—
   (1) V_{2MIN}, and
   (2) V_R plus the speed increment attained (in accordance with Sec. 25.111 (c)(2)) before reaching a height of 35 feet above the take-off surface.

(d) V_{MU} is the calibrated airspeed at and above which the airplane can safely lift off the ground, and continue the take-off. V_{MU} speeds must be selected by the applicant throughout the range of thrust-to-weight ratios to be certificated. These speeds may be established from free air data if these data are verified by ground take-off tests.

(e) V_R, in terms of calibrated airspeed, must be selected in accordance with the conditions of paragraphs (e) (1) through (4) of this section:
   (1) V_R may not be less than–
      (i) V_1;
      (ii) 105 percent of V_{MC};
      (iii) The speed (determined in accordance with Sec. 25.111(c)(2)) that allows reaching V_2 before reaching a height of 35 feet above the take-off surface; or
      (iv) A speed that, if the airplane is rotated at its maximum practicable rate, will result in a V_{LOF} of not less than 110 percent of V_{MU} in the all-engines-operating condition and not less than 105 percent of V_{MU} determined at the thrust-to-weight ratio corresponding to the one-engine-inoperative condition.

   (2) For any given set of conditions (such as weight, configuration, and temperature), a single value of V_R, obtained in accordance with this paragraph, must be used to show compliance with both the one-engine-inoperative and the all-engines-operating take-off provisions.

   (3) It must be shown that the one-engine-inoperative take-off distance, using a rotation speed of 5 knots less than V_R established in accordance with paragraphs (e)(1) and (2) of this section, does not exceed the corresponding one-engine-inoperative take-off distance using the established V_R. The take-off distances must be determined in accordance with Sec. 25.113(a)(1).

   (4) Reasonably expected variations in service from the established take-off procedures for the operation of the airplane (such as over-rotation of the airplane and out-of-trim conditions) may not result in unsafe flight characteristics or in marked increases in the scheduled take-off distances established in accordance with Sec. 25.113(a).

(f) V_{LOF} is the calibrated airspeed at which the airplane first becomes airborne.

17.2.7 Determination of Time to Lift-off

The time from brake-release to lift-off can be approximated using the simple expression below, which assumes an average acceleration, a, is known:

$$S_G = \frac{1}{2}at^2 \quad \Rightarrow \quad t = \sqrt{\frac{2S_G}{a}} \qquad (17\text{-}17)$$

If the aircraft is "small," add 1 second to account for the rotation. If "large," add 3 seconds.

17.3 SOLVING THE EQUATION OF MOTION OF THE T-O

Now that the equation of motion (EOM) describing the T-O run has been derived, several solution methods will be presented. A solution to the EOM yields information such as the acceleration – average or instantaneous, depending on the solution method – ground run distance, and duration of the ground run. In this section, three methods to estimate the ground run will be introduced to solve the EOM. The first method is a simple solution and is applicable to all aircraft as long as the thrust can be quantified. The second method is intended for piston-powered propeller aircraft only. The third method uses numerical integration to solve the EOM and is, by far, the most powerful of the three.

17.3.1 Method 1: General Solution of the Equation of Motion

This method is applicable to both propeller-powered airplanes as well as jets; the only difference lies in

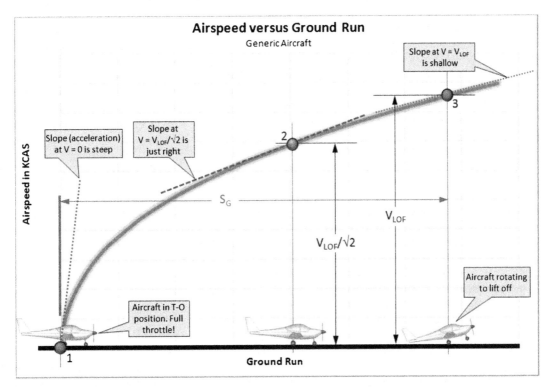

FIGURE 17-11 The reason for the term $V_{LOF}/\sqrt{2}$ explained.

how the thrust is calculated. The method uses Equation (17-4) or (17-5) to calculate the acceleration of the aircraft. With the acceleration in hand, Equation (17-9) is used to estimate the ground run distance. As can be seen, Equations (17-4) and (17-5) both require the estimation of thrust, T, drag, D, and lift, L. The only problem is that all are functions of the airspeed. This therefore begs the question: what airspeed should be used to evaluate them? To answer the question, consider Figure 17-11, which shows how the airspeed of the airplane typically changes during the ground run.

Initially, the acceleration is relatively large, but it gradually diminishes as the airspeed increases. With this in mind, first consider the point labeled '1', but this represents the aircraft in the T-O position, when both $V = 0$ and $S_G = 0$. It is assumed the engines are allowed to develop full thrust before the airplane begins to accelerate. This way, it achieves maximum acceleration upon brake release (maximum thrust and minimum drag). If the rate of change of speed with distance is denoted with the derivative dV/dS, it is evident that, at this point, it has the steepest slope during the entire ground run. Now consider point '3'. This marks the lift-off point and the extent of the ground run. When compared to the start of the ground run, the value of dV/dS has reduced considerably and has reached its lowest value over the entire ground run. It should be clear that if the acceleration at point '1' is used with Equation (17-9), then the estimated ground run will be much less than the actual one, as it is based on a high value of dV/dS. By the same token, if the acceleration at point '3' is used, the ground run estimate will be much larger than experienced, as it is based on a low value of dV/dS. This implies that somewhere between these two extremes exists an airspeed for which the value of dV/dS will give a ground run distance that is in good agreement with experiment. This airspeed is $V_{LOF}/\sqrt{2}$, where V_{LOF} is the lift-off speed.

Step 1: Lift-off Speed

Calculate a lift-off speed (V_{LOF}) per Equation (17-16).

Step 2: Lift-induced Drag in Ground Effect

Calculate $(C_{Di})_{IGE}$ per Equations (9-65) through (9-68).

Step 3: Lift at the Reduced Lift-off Speed

Calculate lift (L) per Equation (17-10) at $V_{LOF}/\sqrt{2}$.

Step 4: Drag at the Reduced Lift-off Speed

Calculate drag (D) per Equation (17-11) at $V_{LOF}/\sqrt{2}$.

Step 5: Thrust at the Reduced Lift-off Speed

Calculate thrust (T) at $V_{LOF}/\sqrt{2}$, depending on engine type as shown below.

Thrust for piston-powered aircraft:

$$T = \frac{\eta_P(550 \times P_{BHP})}{V_{LOF}/\sqrt{2}} \qquad (17\text{-}18)$$

Thrust for jet-powered aircraft:

$$T = T\left(V_{LOF}/\sqrt{2}\right) \qquad (17\text{-}19)$$

where

P_{BHP} = piston engine horsepower
$T()$ = jet engine thrust function[1] using V_{LOF} as an argument
η_P = propeller efficiency at V_{LOF}

Step 6: Ground Run

Calculate the ground run, using Equation (17-4) or (17-5) with Equation (17-9) of Section 17.2.3, *Review of Kinematics*, as shown below. Select an appropriate ground friction coefficient, μ, from Table 17-3.

$$S = \frac{V^2}{2a} \Rightarrow S_G = \frac{V_{LOF}^2 W}{2g[T - D - \mu(W - L)]_{at\ (V_{LOF}/\sqrt{2})}} \qquad (17\text{-}20)$$

EXAMPLE 17-3

During the T-O phase, the Continental IO-550 engine powering the SR22 generates 310 BHP with an assumed propeller efficiency of 0.65. What is the thrust generated at $V = 100$ ft/s? What about 120 ft/s (assuming the same propeller efficiency)? If this airplane weighs 3400 lb$_f$, what is the acceleration for both airspeeds (ignoring drag and ground friction)? Discuss the results.

Solution

Thrust at $V = 100$ ft/s:

$$T = \frac{\eta_P \cdot 550 \cdot P_{BHP}}{V} = \frac{0.65 \cdot 550 \cdot 310}{100} = 1108\ lb_f$$

Acceleration at $V = 100$ ft/s:

$$a = \frac{T}{m} = \frac{Tg}{W} = \frac{1108 \times 32.174}{3400} = 10.48\ ft/s^2$$

Thrust at $V = 120$ ft/s:

$$T = \frac{\eta_P \cdot 550 \cdot P_{BHP}}{V} = \frac{0.65 \cdot 550 \cdot 310}{120} = 924\ lb_f$$

Acceleration at $V = 120$ ft/s:

$$a = \frac{T}{m} = \frac{Tg}{W} = \frac{924 \times 32.174}{3400} = 8.74\ ft/s^2$$

These represent absolutely the greatest acceleration the vehicle could possibly have at that condition (assuming the propeller efficiency is valid). The designer should be aware that even though this kind of analysis is simplistic, it serves superbly as a quick "sanity check" when performing a more sophisticated T-O analysis. In this case, it indicates that if using alternative analysis methods, an answer exceeding 10.48 ft/s^2 for an SR22 class aircraft is highly suspicious.

EXAMPLE 17-4

The airplane of the previous example is observed to have an average acceleration of 8.74 ft/s^2 once it has acquired airspeed of 120 ft/s. If it is assumed this is an "average" acceleration, what distance will the airplane have covered at that airspeed?

Solution

This problem can easily be solved using elementary kinematics:

$$S_G = \frac{V^2}{2a} = \frac{120^2}{2 \times 8.74} = 824\ ft$$

This problem is presented as another kind of "sanity check." The astute aircraft designer will perform such simple and fast checks to evaluate whether errors may have crept into more sophisticated calculations.

---

[1] For jet engines, this is typically obtained using the engine manufacturer's engine deck.

EXAMPLE 17-5

Estimate the ground run for a Learjet 45XR using the following properties:

$W = 21{,}500$ lb$_f$	$C_{Lmax} = 1.65$
$S = 311.6$ ft^2	T at $V_{LOF}/\sqrt{2} = 7000$ lb$_f$
$C_{L\ TO} = 0.90$	$\mu = 0.02$
$C_{D\ TO} = 0.040$	

Solution

Lift-off speed:

$$V_{LOF} \approx 1.556\sqrt{\frac{W}{\rho S C_{Lmax}}} = 1.556\sqrt{\frac{21{,}500}{(0.002378)(311.6)(1.65)}}$$
$$= 206.7 \text{ ft/s (122 KCAS)}$$

The value for the airspeed at $V_{LOF}/\sqrt{2}$ is thus:

$$\frac{V_{LOF}}{\sqrt{2}} = 146.2 \text{ ft/s (86.6 KCAS)}$$

Lift at that airspeed:

$$L = \frac{1}{2}(0.002378)(146.2)^2(311.6)(0.90) = 7127 \text{ lb}_f$$

Drag at that airspeed:

$$D = \frac{1}{2}(0.002378)(146.2)^2(311.6)(0.040) = 316.8 \text{ lb}_f$$

Average acceleration:

$$\frac{dV}{dt} = \frac{g}{W}[T - D - \mu(W - L)]$$
$$= \frac{32.174}{21500}[7000 - 316.8 - (0.02)(21500 - 7127)]$$
$$= 5.700 \text{ ft/s}^2$$

Ground run to lift-off is thus:

$$S_G = \frac{V_{LOF}^2}{2a} = \frac{206.7^2}{2 \times 5.700} = 3748 \text{ ft}$$

17.3.2 Method 2: Rapid T-O Distance Estimation for a Piston-powered Airplane [5]

This is Method 1 but specifically adapted to the typical piston-engine configuration.

$$S_G = \frac{V_{LOF}^2 W}{\frac{50051 \times \eta_P \times P_{BHP}}{V_{LOF}} + 16.09\rho V_{LOF}^2 S(\mu C_{L\ TO} - C_{D\ TO}) - 64.35\mu W} \quad (17\text{-}21)$$

NOTE: If the propeller efficiency, η_P, is not known, the following values can be used as expected approximations:

Fixed pitch climb propeller $\Rightarrow \eta_P = 0.45\text{-}0.50$
Fixed pitch cruise propeller $\Rightarrow \eta_P = 0.35\text{-}0.45$
Constant speed propeller $\Rightarrow \eta_P = 0.45\text{-}0.60$

Derivation of Equation (17-21)

We begin with Equation (17-4) and rewrite the acceleration as follows:

$$a = \frac{g}{W}[T - D - \mu(W - L)] \quad (i)$$

Inserting Equations (17-10), (17-11), and (14-38) yields;

$$a = \frac{g}{W}\left[\frac{\eta_P(550 \times P_{BHP})}{V} - \frac{1}{2}\rho V^2 S(C_{Dmin} + C_{Di})\right.$$
$$\left. - \mu\left(W - \frac{1}{2}\rho V^2 S(C_{L0} + C_{L\alpha}\cdot\alpha)\right)\right]$$

$$= \frac{g}{W}\left[\frac{\eta_P(550 \times P_{BHP})}{V} + \frac{1}{2}\rho V^2 S(\mu(C_{L0} + C_{L\alpha}\cdot\alpha)\right.$$
$$\left. - (C_{Dmin} + C_{Di})) - \mu W\right]$$

Let's denote the lift and drag coefficients during the T-O run with $C_{L\ TO}$ and $C_{D\ TO}$:

$$a = \frac{g}{W}\left[\frac{\eta_P(550 \times P_{BHP})}{V}\right.$$
$$\left. + \frac{1}{2}\rho V^2 S(\mu C_{L\ TO} - C_{D\ TO}) - \mu W\right] \quad (ii)$$

Insert this into the expression for distance to lift-off to get:

$$S_G = \frac{V_{LOF}^2}{2a}$$

$$= \frac{V_{LOF}^2 W}{2g\left[\frac{\eta_P(550 \times P_{BHP})}{V} + \frac{1}{2}\rho V^2 S(\mu C_{L\ TO} - C_{D\ TO}) - \mu W\right]}$$

(iii)

Note that since we evaluate the acceleration at $V = V_{LOF}/\sqrt{2}$, we incorporate this as follows:

$$S_G = \frac{V_{LOF}^2 W}{2g\left[\frac{\eta_P(550 \times P_{BHP})}{V_{LOF}/\sqrt{2}} + \frac{1}{2}\rho(V_{LOF}/\sqrt{2})^2 S(\mu C_{L\ TO} - C_{D\ TO}) - \mu W\right]}$$

$$= \frac{V_{LOF}^2 W}{2g\left[\frac{\eta_P(550 \times P_{BHP})\sqrt{2}}{V_{LOF}} + \frac{1}{4}\rho V_{LOF}^2 S(\mu C_{L\ TO} - C_{D\ TO}) - \mu W\right]}$$

Simplify further:

$$S_G = \frac{V_{LOF}^2 W}{\frac{\eta_P(550 \times P_{BHP})2\sqrt{2}}{V_{LOF}}g + \frac{1}{2}g\rho V_{LOF}^2 S(\mu C_{L\ TO} - C_{D\ TO}) - 2g\mu W}$$

(iv)

Inserting $g = 32.174$ ft/s² and arithmetically evaluating the constants leads to:

$$S_G = \frac{V_{LOF}^2 W}{\frac{50,051 \times \eta_P \times P_{BHP}}{V_{LOF}} + 16.09\rho V_{LOF}^2 S(\mu C_{L\ TO} - C_{D\ TO}) - 64.35\mu W}$$

(v)

The advantage of this method is that it combines several steps in Method 1, and thus, lends itself better to parametric studies.

QED

EXAMPLE 17-6

Determine the lift-off speed, T-O distance, and time from brake release to lift-off at S-L for the Cirrus SR22 if it has the following characteristics:

$W = 3400$ lb$_f$	$C_{L\ TO} = 0.590$ (assumed value)
$P_{BHP} = 310$	$C_{D\ TO} = 0.0414$ (assumed value)
$S = 144.9$ ft²	$\mu = 0.04$ (sample value)
$C_{Lmax} = 1.69$	
(based on POH for T-O)	

Compare the results to published data from the airplane's pilot's operating handbook (POH), which gives a lift-off speed of 73 KCAS, T-O distance of 1020 ft (ISA at S-L) and the calculated time to lift-off of 16.6 seconds from EXAMPLE 17-2.

Solution

Lift-off speed:

$$V_{LOF} \approx 1.556\sqrt{\frac{W}{\rho S C_{Lmax}}}$$

$$= 1.556\sqrt{\frac{1}{(0.002378)(1.69)}\left(\frac{3400}{144.9}\right)}$$

$$= 118.9 \text{ ft/s (70.4 KCAS)}$$

The value for the airspeed at $V_{LOF}/\sqrt{2}$ is thus:

$$\frac{V_{LOF}}{\sqrt{2}} = 84.1 \text{ ft/s (50 KCAS)}$$

Since the propeller efficiency is unknown, let's pick $\eta_P = 0.50$ and calculate lift-off distance:

Thrust at that airspeed:

$$T = \frac{0.5 \times 550 \times 310}{84.1} = 1015 \text{ lb}_f$$

Lift at that airspeed:

$$L = \frac{1}{2}(0.002378)(84.1)^2(144.9)(0.590) = 719 \text{ lb}_f$$

Drag at that airspeed:

$$D = \frac{1}{2}(0.002378)(84.1)^2(145)(0.0414) = 50 \text{ lb}_f$$

Average acceleration:

$$\frac{dV}{dt} = \frac{g}{W}[T - D - \mu(W - L)]$$

$$= \frac{32.174}{3400}[1015 - 50 - (0.04)(3400 - 719)] = 8.12 \text{ ft/s}^2$$

EXAMPLE 17-6 (cont'd)

Ground run to lift-off is thus:

$$S_G = \frac{V_{LOF}^2}{2a} = \frac{118.9^2}{2 \times 8.12} = 871 \text{ ft}$$

Time from brake release to lift-off:

$$t = \sqrt{\frac{2S_{LOF}}{a}} = \sqrt{\frac{2 \times 871}{8.12}} = 14.6 \text{ s}$$

Note that this example does not account for the rotation, which would add 1 second to the time to lift-off and 118.9 ft to the total distance. This would bring the true ground run SG + SROT to 990 ft, which compares favorably to the POH value of 1028 ft.

17.3.3 Method 3: Solution Using Numerical Integration Method

Since a closed-form integration of Equation (17-4), which accurately accounts for various speed dependencies, is prohibitively hard, it is generally solved numerically. Solving the equation of motion using numerical integration provides the designer with a very powerful technique. When properly applied, the technique even allows time-dependent events to be accounted for during the integration. For instance, consider some specialized T-O technique being modeled in which the flaps are dropped just before lift-off to minimize drag during the ground run. Or, consider the T-O run of many radial piston-powered airplanes of the past. During the ground run, the large radials of many of these old airplanes were operated under limited power until the airplane had accelerated to a certain airspeed, say 60 KIAS. Such complexities are relatively easy to account for using this method. It is the recommended method for serious analysis work.

In this section, we will demonstrate how to setup this method using a spreadsheet. Additionally, the method is very robust as it handles discontinuities with ease, something numerical differentiation schemes would not tolerate as well.

The first step in the scheme is to convert the kinematic Equations (17-7) and (17-8) into the following discrete form:

$$V_i = \int_0^t a \cdot dt = \sum_{i=0}^{N}(V_{i-1} + a_i \cdot (t_i - t_{i-1})) \quad (17\text{-}22)$$

$$S_i = \int_0^t V \cdot dt = \int_0^t (V_0 + a \cdot t) \cdot dt$$

$$= \sum_{i=0}^{N}\left(S_{i-1} + V_{i-1} \cdot \Delta t_i + \frac{1}{2} a_i \cdot \Delta t_i^2\right) \quad (17\text{-}23)$$

The variables are depicted in Figure 17-12.

Propeller Thrust at Low Airspeeds

Since the T-O run involves airspeeds ranging from 0 to the lift-off speed, in which the highest acceleration occurs at low speeds, accurate estimation of thrust is imperative. This is problematic for propeller-powered aircraft, where the standard method to extract propeller thrust is that of Equation (14-38), repeated here for convenience:

$$T = \frac{\eta_p \times 550 \times P_{BHP}}{V} \quad (14\text{-}38)$$

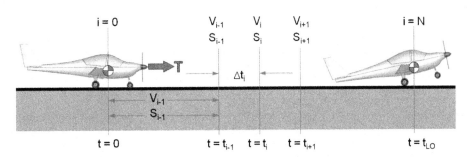

FIGURE 17-12 Nomenclature used for the numerical integration of the T-O ground run.

EXAMPLE 17-7

A vehicle accelerates with a constant $a = 10$ m/s². Determine its (a) speed, and (b) distance after 5 seconds using numerical integration and compare to the exact solution.

Solution

(a) Numerical integration of the speed: see Figure 17-13.

(b) Numerical integration of the distance: see Figure 17-14. Exact solutions:

$$V = at = 10 \times 5 = 50 \text{ ft/s}$$

$$S = \frac{1}{2}at^2 = \frac{1}{2} \times 10 \times 5^2 = 125 \text{ ft}$$

INDEX	t (sec)	Δt (sec)	$V_i = V_{i-1} + a \cdot \Delta t$ (ft/s)	$V_{i-1} \cdot \Delta t$ (ft)	$\frac{1}{2} \cdot a \cdot \Delta t^2$ (ft)	$\Delta S_i = V_{i-1} \cdot \Delta t + \frac{1}{2} \cdot a \cdot \Delta t^2$ (ft)	$S_i = S_{i-1} + \Delta S_i$ (ft)
0	0	0	0				
1	1	1-0 = 1	0 + 10 x 1 = 10				
2	2	2-1 = 1	10 + 10 x 1 = 20				
3	3	3-2 = 1	20 + 10 x 1 = 30				
4	4	4-3 = 1	30 + 10 x 1 = 40				
5	5	5-4 = 1	40 + 10 x 1 = 50				

FIGURE 17-13 Determination of speed using a numerical integration scheme.

INDEX	t (sec)	Δt (sec)	$V_i = V_{i-1} + a \cdot \Delta t$ (ft/s)	$V_{i-1} \cdot \Delta t$ (ft)	$\frac{1}{2} \cdot a \cdot \Delta t^2$ (ft)	$\Delta S_i = V_{i-1} \cdot \Delta t + \frac{1}{2} \cdot a \cdot \Delta t^2$ (ft)	$S_i = S_{i-1} + \Delta S_i$ (ft)
0	0	0	0	0	0	0	0
1	1	1	10	0	5	5	0+5 = 5
2	2	1	20	10	5	15	5+15 = 20
3	3	1	30	20	5	25	20+25 = 45
4	4	1	40	30	5	35	45+35 = 80
5	5	1	50	40	5	45	80+45 = 125

FIGURE 17-14 Determination of distance using a numerical integration scheme.

The inaccuracy is a consequence of a low V and variable and hard-to-predict η_p. This can lead to a thrust estimation that, well, is simply too large. There are two ways to work around this problem. Two methods are presented in Section 14.4.2, *Propeller thrust at low airspeeds*. One of the two methods is demonstrated in Example 17-8 below.

EXAMPLE 17-8

Determine the lift-off speed, T-O distance, and time from brake release to lift-off at S-L for the Cirrus SR22 by solving the equations of motion using the *numerical integration method* with the following parameters:

EXAMPLE 17-8 (cont'd)

$W = 3400$ lb$_f$
$P_{BHP} = 310$ BHP
$S = 144.9$ ft^2
Propeller diameter
$D_p = 76$ inches
$V_{LOF} = 73$ KCAS
(per POH)

$C_{Lmax} = 1.69$ (based on POH for T-O)
$\mu = 0.04$ (sample value)
$C_{L\,TO} = 0.590$
$C_{D\,TO} = 0.0414$
$V_H = 186$ KTAS
(at 2000 ft, per POH)

Use the method in Section 14.4.2, *Propeller thrust at low airspeeds*, to approximate the thrust of the constant-speed propeller at low airspeeds. This method approximates the thrust as a function of airspeed using the following cubic spline, reproduced from the above section:

Such a cubic spline will be of the form:

$$T(V) = A \cdot V^3 + B \cdot V^2 + C \cdot V + D \quad (14\text{-}41)$$

where the constants A, B, C, and D are determined using the following matrix:

$$\begin{bmatrix} 0 & 0 & 0 & 1 \\ V_C^3 & V_C^2 & V_C & 1 \\ 3V_C^2 & 2V_C & 1 & 0 \\ V_H^3 & V_H^2 & V_H & 1 \end{bmatrix} \begin{Bmatrix} A \\ B \\ C \\ D \end{Bmatrix} = \begin{Bmatrix} T_{STATIC} \\ T_C \\ -\eta_p \cdot 325.8 \cdot P_{BHP}/V_C^2 \\ T_H \end{Bmatrix}$$
$$(14\text{-}42)$$

Compare the results to published data from the airplane's pilot's operating handbook (POH), which give a lift-off speed of 73 KCAS, T-O distance of 1020 ft (ISA at S-L) and the calculated time to lift-off of 16.6 seconds from EXAMPLE 17-2.

Solution

The solution is implemented as shown in Table 17-4, which was solved using Microsoft Excel. First, focus on the table itself as the reason behind the analysis performed above will become apparent when discussing column 4. Note that the lift-off speed, which is dependent on the C_{Lmax} stated above, was calculated in Example 17-6 and found to equal 118.9 ft/s or 70.4 KCAS. This serves as a flag in the calculations below — once the airspeed is greater than 118.9 ft/s, the ground run is assumed completed.

In order to help explain steps required to setup the spreadsheet, the columns have been numbered from 1 through 15. The first column (i) lists the indexes used, here ranging from 1 through 30. Effectively, all rows from 2 through 30 contain the same formulas, but row 1 differs as it is required to contain the initial conditions.

Column 1 is the index 'i' shown in Equations (17-22) and (17-23). Note the row with the index '10' has been shaded to draw attention to it as it is used as an example of calculations.

Column 2 is the time. Here, the event starts when $t = 0$ sec and ends when $t = 20$ sec. The initial time (when $i = 1$) is zero by definition. All subsequent times are obtained by simply adding 0.5 sec to the value in the cell above. It is ultimately the user's decision how many rows to use to represent a time interval in which the event (T-O ground run) takes place.

Column 3 is the time step. Note that each row represents a specific time step, Δt, in the analysis. Here all the time steps are equal. Thus the 10th time step (when $i = 10$) takes place when $t = 4.5$ sec. It is recommended that the method is implemented using 0.25 sec or 0.5 sec time steps. The time step is simply calculated by subtracting the previous time from the current time. Thus, the time step in row 10 is obtained by $\Delta t = 4.5 - 4.0 = 0.5$. The same result is obtained for the other rows, except the first one, as it does not have a preceding time. Note that even though each time step in the table is 0.5 second, it does not have to be so; time steps can be of variable size. This allows higher definition around a specific event of interest. For instance consider modeling flaps being deployed sometime during the ground run over a period of time. It is appropriate to relax time steps before and after events that take place during the T-O run and use smaller time steps as deemed appropriate.

Column 4 is the propeller thrust force. It is calculated using the cubic spline method presented in method 3 of Section 14.4.2, *Propeller thrust at low airspeeds*. The method creates a cubic polynomial that is a function of time. Its general form is $T(V) = AV^3 + BV^2 + CV + D$, where A, B, C, and D are determined based on the static thrust, T_{STATIC}, and the airspeeds V_C and V_H, and the thrust at those airspeed, denoted by T_C and T_H. The airspeed, denoted by V, is obtained from column 11. Note that in order to prevent circular error, the value of V is taken from the previous row. Thus, the value of V used in row 10 is that calculated in row 9.

Column 5 is the dynamic pressure, here calculated using density at S-L and the airspeed from column 11. The value of V in Row 10 is taken from Row 9 in order to prevent circular error from occurring in Microsoft Excel. The difference is dynamic pressure is considered acceptable, although it highlights why the time steps should be as small as practical. The dynamic pressure in Row 10 is 2.1 lb$_f$/ft^2.

EXAMPLE 17-8 (cont'd)

Column 6 is the lift force, here calculated using density at S-L, the airspeed from column 11, a wing area of 144.9 ft^2, and the $C_{L\ TO}$ given in the problem statement. Calculated from $L = q \cdot S \cdot C_{L\ TO}$. The lift in row 10 is 179 lb$_f$.

Column 7 is the drag force, using the same parameters as for column 6, with the exception of the drag coefficient given by $C_{D\ TO}$. Calculated from $D = q \cdot S \cdot C_{D\ TO}$. The drag force in row 10 is 13 lb$_f$.

TABLE 17-4 Ground Run Analysis Using the Numerical Integration Method

GROUND RUN ANALYSIS

The calculations below are performed to determine the thrust function, using the Cubic Spline Method

INPUT VALUES

Max rated power	P =	310	BHP
Propeller diameter	D_p =	6.33	ft
Propeller disc area	A_2 =	31.50	ft^2
Static thrust	T_{STATIC} =	1310	lb$_f$

$$\begin{bmatrix} 0 & 0 & 0 & 1 \\ V_C^3 & V_C^2 & V_C & 1 \\ 3V_C^2 & 2V_C & 1 & 0 \\ V_H^3 & V_H^2 & V_H & 1 \end{bmatrix} \begin{bmatrix} A \\ B \\ C \\ D \end{bmatrix} = \begin{bmatrix} T_{STATIC} \\ T_C \\ -\eta_p \cdot 325.8 \cdot BHP/V_C^2 \\ T_H \end{bmatrix}$$

Equation (14-64) $T_{STATIC} = 0.85 P^{2/3}(2\rho A_2)^{1/3}\left(1 - \frac{A_{spinner}}{A_2}\right)$

V_C =	170	KCAS	287.0	ft/s	⇨	η_p =	0.8500	⇨	$T(V_C)$ =	505	lb$_f$
V_H =	186	KCAS	314.0	ft/s	⇨	η_p =	0.8500	⇨	$T(V_H)$ =	462	lb$_f$

4x4 Matrix

0	0	0	1
4913000	28900	170	1
86700	340	1	0
6434856	34596	186	1

4x1

A =	1310
B =	505
C =	-2.9702
D =	462

Inverted 4x4 Matrix

-2E-07	-2.08E-05	-0.0004	2.1E-05
9.8E-05	0.007043	0.13088	-0.0071
-0.0171	-0.589798	-11.625	0.60694
1	0	0	0

Constants

A =	2.986E-05
B =	0.0002455
C =	-5.6427
D =	1310.43

Thrust equation: $T(V) = AV^3 + BV^2 + CV + D =$
$T(V) = 2.98629511500757\text{E-}05 \cdot V^3 + 0.000245459594814701 \cdot V^2 + (-5.6427475231261) \cdot V + 1310.43163455626$

1	2	3	4	5	6	7	8	9	10	11	12	13	14
					Forces, lb$_f$				Airspeed, V				
i	Time, s	Δt	T(V)	q, lb$_f$/ft^2	L	D	μ(W-L)	a = dV/dt	ft/s	KTAS	$V_{i-1} \cdot \Delta t$	½·a·Δt^2	$S_i = S_{i-1} + \Delta S_i$
1	0	-	1310	0.0	0	0	136	11.1	0.0	0.0	-	-	0
2	0.5	0.5	1310	0.0	0	0	136	11.1	5.6	3.3	0.0	1.4	1.4
3	1.0	0.5	1292	0.0	0	3	136	10.9	11.0	6.5	2.8	1.4	5.6
4	1.5	0.5	1274	0.1	12	1	136	10.8	16.4	9.7	5.5	1.4	12.4
5	2.0	0.5	1256	0.3	27	2	135	10.6	21.7	12.9	8.2	1.3	22.0
6	2.5	0.5	1238	0.6	48	3	134	10.4	26.9	15.9	10.8	1.3	34.2
7	3.0	0.5	1221	0.9	74	5	133	10.2	32.0	19.0	13.5	1.3	48.9
8	3.5	0.5	1204	1.2	104	7	132	10.1	37.1	22.0	16.0	1.3	66.2
9	4.0	0.5	1187	1.6	140	10	130	9.9	42.0	24.9	18.5	1.3	86.0
10	4.5	0.5	1171	2.1	179	13	129	9.7	46.9	27.8	21.0	1.2	108.2
11	5.0	0.5	1155	2.6	223	16	127	9.6	51.7	30.6	23.4	1.2	132.9
12	5.5	0.5	1139	3.2	271	19	125	9.4	56.4	33.4	25.8	1.2	159.9
13	6.0	0.5	1123	3.8	323	23	123	9.3	61.0	36.1	28.2	1.2	189.3
14	6.5	0.5	1108	4.4	378	27	121	9.1	65.6	38.8	30.5	1.2	221.0
15	7.0	0.5	1093	5.1	437	31	119	8.9	70.0	41.5	32.8	1.1	254.9
16	7.5	0.5	1079	5.8	498	35	116	8.8	74.4	44.1	35.0	1.1	291.0
17	8.0	0.5	1065	6.6	563	39	113	8.6	78.7	46.6	37.2	1.1	329.3
18	8.5	0.5	1051	7.4	630	44	111	8.5	83.0	49.1	39.4	1.1	369.7
19	9.0	0.5	1037	8.2	699	49	108	8.3	87.1	51.6	41.5	1.1	412.3
20	9.5	0.5	1024	9.0	771	54	105	8.2	91.2	54.0	43.6	1.0	456.9
21	10.0	0.5	1011	9.9	846	59	102	8.0	95.2	56.4	45.6	1.0	503.5
22	10.5	0.5	998	10.8	922	65	99	7.9	99.2	58.8	47.6	1.0	552.1
23	11.0	0.5	986	11.7	1000	70	96	7.8	103.1	61.1	49.6	1.0	602.7
24	11.5	0.5	974	12.6	1079	76	93	7.6	106.9	63.3	51.5	1.0	655.2
25	12.0	0.5	962	13.6	1161	81	90	7.5	110.6	65.5	53.4	1.0	709.6
26	12.5	0.5	950	14.5	1243	87	86	7.3	114.3	67.7	55.3	0.9	765.9
27	13.0	0.5	939	15.5	1327	93	83	7.2	117.9	69.8	57.1	0.9	823.9
28	13.5	0.5	928	16.5	1412	99	80	7.1	121.4	71.9	58.9	0.9	883.8
29	14.0	0.5	917	17.5	1499	105	76	7.0	124.9	74.0	60.7	0.9	945.4
30	14.5	0.5	906	18.6	1586	111	73	6.8	128.3	76.0	62.5	0.9	1008.7

EXAMPLE 17-8 (cont'd)

Column 8 is the ground friction force, using the values from column 6, the weight (3400 lbf), and the ground friction constant μ, given in the problem statement. Calculated from $\mu \cdot (W-L)$. The ground friction in row 10 is 129 lb$_f$.

Column 9 is the acceleration. Calculated by summing the forces and dividing by the mass. The mass is simply the weight divided by the acceleration due to gravity, $g = 32.174$ fts^2. Calculated from:

$$a = dV/dt = (T - D - \mu \cdot (W - L))/(W/g)$$

Thus, the acceleration in row 10 is obtained as follows:

$$a_{10} = (T_{10} - D_{10} - \mu \cdot (W - L_{10}))/(W/g)$$
$$= (1171 - 13 - 0.04 \cdot (3400 - 179))/(3400/32.174)$$
$$= 9.74 \text{ ft/s}^2.$$

Columns 10 and **11** is the airspeed. The value in column 10 is calculated using the simple kinematic expression of Equation (17-22), i.e.: $V_i = V_{i-1} + a_i \cdot \Delta t_i$. Using row 10 as an example, $V_9 = 42.0$ ft/s, $a_{10} = 9.74$ ft/s^2, and $\Delta t_{10} = 0.5$ sec. Therefore, $V_{10} = V_9 + a_{10} \cdot \Delta t_{10} = 43.1 + 9.74 \cdot 0.5 = 46.9$ ft/s. The value in column 11 amounts to $V_i/1.688$, converting ft/s to knots.

Columns 12 through **15** form the distance calculations shown in Equation (17-23). Column 12 is calculated from the expression $V_{i-1} \cdot \Delta t_i$. Thus, row 10 is calculated from $V_9 \cdot \Delta t_{10} = 42.0 \cdot 0.5 = 21.0$ ft. Column 13 is calculated from the expression $\frac{1}{2} \cdot a_{i-1} \cdot (\Delta t_i)^2$, which for row 10 translates to $\frac{1}{2} \cdot a_9 \cdot (\Delta t_{10})^2 = \frac{1}{2} \cdot 9.7 \cdot (0.5)^2 = 1.2$ ft. Column 14 is calculated using the expression $S_i = S_{i-1} + V_{i-1} \cdot \Delta t_i + \frac{1}{2} \cdot a_{i-1} \cdot \Delta t_i^2$, which for row 10 becomes: $S_{10} = S_9 + V_9 \cdot \Delta t_{10} + \frac{1}{2} \cdot a_9 \cdot \Delta t_{10}^2 = 86.0 + 21.0 + 1.2 = 108.2$ ft. This represents the cumulative distance up to that point in time.

With these explanations behind us, it is now possible to explore some of the results. Table 17-4 shows that the airplane accelerates to lift-off speed between rows 27 and 28 (between 13.0 and 13.5 seconds). Interpolating between these two values yields $t = 13.1$ seconds and $S_G = 840$ ft. Adding the rotation segment, the distance increases by $S_{ROT} = |V_{LOF}| = 119$ ft, yields a total ground run of 959 ft, and time to lift-off is 14.1 seconds. The distance is 61 ft shorter than the POH figure. On the other hand, if the airplane is allowed to accelerate to 73 knots, as the POH states, the time to lift-off speed is $t = (13.8 + 1)$ sec and $S_G = 915$ ft, leading to a total distance of 1038 ft (note that 73 knots is about 123 ft). Other numbers are compared in Table 17-5 below.

Considering other results, Figure 17-15 shows the thrust and propeller efficiency models used for this analysis. The shaded region in the graph represents the focus of the above work and shows that the cubic spline method allows the thrust to be determined to a zero airspeed.

The ground run numbers in the POH for the SR22 includes the rotation phase, which here is assumed to be 1 second. Therefore, a 1-second rotation segment is added to the POH value for the ground run analysis. Consequently, the comparable POH ground run should be $1020 - 123 = 897$ ft. It is also possible to adjust the time for lift-off by subtracting the 1 second from the 16.6 sec in EXAMPLE 17-2. From the graph we can see that V_{LOF} calculated (70.4 KCAS) using $1.1V_S$ is a tad slower than the value found in the POH (73 KCAS). Therefore, a comparison assuming that airspeed is included as well. The conclusion is that based on the average acceleration, there are some parameters of the prediction that might be improved with an adjustment. For instance, it is possible that the predicted static thrust is too high, although it included the empirical

TABLE 17-5 Analysis Results

	Parameter	POH Value	Calculated Value	Percent Difference
Numerical Integration: Lift-off at 13.1 sec.	V_{LOF}	73 KCAS	70.4 KCAS	3.6%
	S_G	897 ft	840 ft	6.4%
	$S_G + S_{ROT}$	1028 ft	959 ft	6.7%
	t_{LOF}	16.6 sec	14.1 sec	15.1%
	a_{avg}	7.44 ft/s^2	9.14 ft/s^2	22.8%
Numerical Integration: Lift-off at 14.8 sec.	V_{LOF}	73 KCAS	73 KCAS	-
	S_G	897 ft	915 ft	2.0%
	$S_G + S_{ROT}$	1028 ft	1038 ft	1.0%
	t_{LOF}	16.6 sec	14.8 sec	10.8%

EXAMPLE 17-8 (cont'd)

reduction factor discussed in Section 14.5.3, *Maximum static thrust*. Also, it is possible that drag for the T-O configuration is underestimated, and so on. Regardless, if the lift-off speeds (73 KCAS actual versus the predicted 70.4 KCAS) are comparable, the method and experiment are likely to agree well (see Figure 17-16).

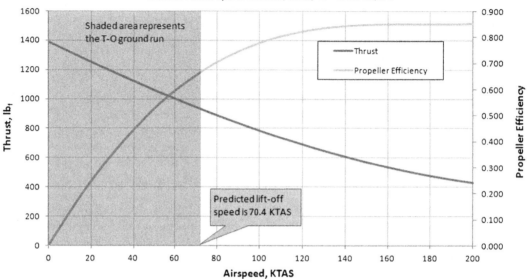

FIGURE 17-15 The thrust and propeller efficiency model for the SR22 used in the T-O ground run analysis.

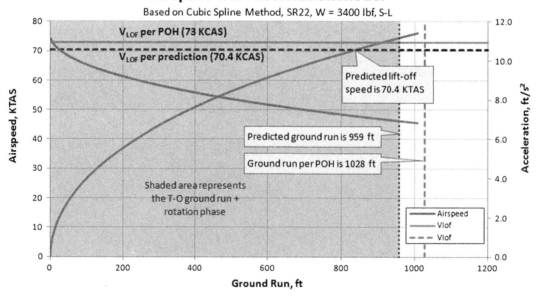

FIGURE 17-16 Resulting acceleration graph for the SR22.

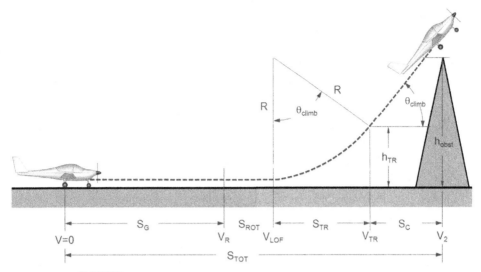

FIGURE 17-17 Determining the segments constituting the T-O run.

17.3.4 Determination of Distance During Rotation

Figure 17-17 shows an aircraft taking off and clearing an obstruction of predetermined height shortly after lift-off. Refer to this figure for the remaining segments of the T-O run, starting with the rotation. Rotation is a very transient event during the T-O. Small airplanes lift off almost as soon as the pilot initiates the rotation (by deflecting the elevator to raise the nose). For large, heavy aircraft the rotation may last anywhere from 2 to 5 seconds. Accounting for the change in drag and lift during the rotation can be implemented using the numerical integration scheme; however, it is far simpler to account for distance traveled during rotation by assuming the distance a small aircraft travels in 1 second, and a large aircraft in 3 seconds. Mathematically:

Small aircraft:

$$S_{ROT} = |V_{LOF}| \qquad (17\text{-}24)$$

Large aircraft:

$$S_{ROT} = 3|V_{LOF}| \qquad (17\text{-}25)$$

It is recognized that the boundary between "small" and "large" aircraft is somewhat subjective, but it is somewhere between a Beech 99 King Air and a British Aerospace BAe 146, with a Fokker F-27 or F-50 in that gray area.

17.3.5 Determination of Distance for Transition

Referring again to Figure 17-17, the transition segment begins with the lift-off and ends with the aircraft achieving a climb angle to be maintained until the obstacle height is achieved. As stated earlier, we want to determine the total distance from the lift-off point to the location where the airplane clears the obstacle, call it S_{obst}. There are typically two issues that one must contend with when determining this distance: is the obstacle cleared before or after the transition segment is completed? A methodology to evaluate the distance denoted by S_{obst} will now be presented. Refer to Figure 17-18 for a more detailed look at this scenario. The following parameters are essential to the analysis of this phase of the T-O maneuver:

Climb angle:

$$\sin \theta_{climb} = \frac{T - D}{W} = \frac{T}{W} - \frac{1}{L/D} \qquad (18\text{-}20)$$

Transition distance:

$$S_{TR} = R \sin \theta_{climb} \approx 0.2156 \times V_{S1}^2 \times \left(\frac{T}{W} - \frac{1}{L/D} \right) \qquad (17\text{-}26)$$

Transition height:

$$h_{TR} = R(1 - \cos \theta_{climb}) \qquad (17\text{-}27)$$

Derivation of Equations (17-26) and (17-27)

Transition generally implies acceleration from $1.1V_{S1}$ to a climb speed of $1.2V_{S1}$. Note that the climb speed may not necessarily be the *best rate of* or *best angle of climb* for the airplane. The average speed for the two limits is of course $1.15V_{S1}$ and this is used in addition to an assumed lift coefficient of about $0.9C_{Lmax}$ to determine distance traveled as follows:

Step 1: Average vertical acceleration in terms of load factor.

$$n = \frac{L}{W} = \frac{\frac{1}{2}\rho(1.15V_{S1})^2 \cdot S \cdot (0.9C_{Lmax})}{\frac{1}{2}\rho V_{S1}^2 \cdot S \cdot C_{Lmax}} = 1.1903 \qquad (17\text{-}28)$$

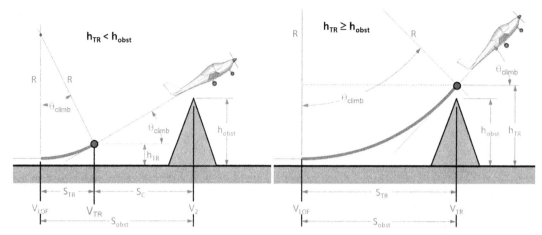

FIGURE 17-18 Evaluation of the T-O over an obstacle. The left image shows the aircraft crossing the obstacle *after* transitioning from the curved into the straight climb. In this case the climb segment must be added to the total. The right image shows the aircraft crossing before it transitions. Only the distance to where it reaches h_{obst} is needed and no climb calculations are necessary.

Step 2: Use the load factor to determine the radius of the curved segment, by using the expression for the centripetal acceleration required. Load factor in terms of centripetal force:

$$n = \frac{L}{W} = \frac{W + m\frac{V^2}{R}}{W} = \frac{W + \frac{W}{g}\frac{V^2}{R}}{W} = 1 + \frac{V^2}{Rg}$$

Solving for the radius yields:

$$n = 1 + \frac{V^2}{Rg} \quad \Leftrightarrow \quad R = \frac{V^2}{g(n-1)} \quad (17\text{-}29)$$

Therefore, using the transition airspeed, we get:

$$R = \frac{V_{TR}^2}{g(n-1)} = \frac{(1.15 V_{S1})^2}{g(1.1903 - 1)} \approx 0.2156 \times V_{S1}^2 \quad (17\text{-}30)$$

Step 3: Determine the angle through which the rotation takes place using Equation (18-20) by solving for θ.

Step 4 — Transition occurs BELOW obstacle height: refer to Figure 17-18, the left part of which shows a schematic of the initial climb in which the aircraft clears the obstacle *after* transitioning from the curved into the straight climb. In this case, the straight climb segment must be added to the total T-O distance. The trick is to determine S_{TR} and S_C. Here, the combination of the two is denoted by S_{obst}. By inspection it can be seen that the horizontal distance (S_{TR}) must equal $R \cdot \sin\theta_{climb}$ and the altitude gained (h_{TR}) during the transition amounts to $R \cdot (1 - \cos\theta_{climb})$. These are expressed in Equations (17-26) and (17-27).

Step 4 — Transition occurs ABOVE obstacle height: the right part of Figure 17-18 shows the aircraft passing the obstacle before completing the transition. Consequently, we only need to determine the distance to where it cleared the obstacle and no climb calculations are needed. This distance, again, is denoted by S_{obst}. It can be seen that the horizontal distance (S_{TR}) is given by the Pythagorean rule:

$$R^2 = S_{TR}^2 + (R - h_{TR})^2 \quad \Leftrightarrow \quad S_{TR} = \sqrt{R^2 - (R - h_{TR})^2} \quad (17\text{-}31)$$

Therefore, the distance required to clear the obstacle can be approximated by considering the ratio between the height and

$$R^2 = S_{obst}^2 + (R - h_{obst})^2 \quad \Leftrightarrow$$

$$S_{obst} = \sqrt{R^2 - (R - h_{obst})^2} \quad (17\text{-}32)$$

QED

17.3.6 Determination of Distance for Climb Over an Obstacle

As stated above, if the value of h_{TR} is less than the obstacle height, the airplane covers additional distance while climbing. The required obstacle clearing height is 50 ft for military and GA aircraft, and 35 ft for commercial aircraft. This distance can be observed from Figure 17-18 as follows:

$$S_C \times \tan\theta_{climb} = h_{obst} - h_{TR} \quad \Leftrightarrow \quad S_C$$

$$= \frac{h_{obst} - h_{TR}}{\tan\theta_{climb}} \quad (17\text{-}33)$$

EXAMPLE 17-9

Determine the transition distance and, if found necessary, the climb distance for the Cirrus SR22. Use data from Example 17-8 as needed. Assume the simplified drag model and the following characteristics:

$W = 3400$ lb$_f$	$C_{Dmin} = 0.0350$
$S = 144.9$ ft^2	(T-O configuration, assumed)
$C_{Lmax} = 1.69$ (based on POH for T-O)	$k = 0.04207$

Compare the results to published data from the airplane's Pilot's Operating Handbook (POH), which gives a lift-off speed of 73 KCAS, T-O distance of 1020 ft (ISA at S-L) and the calculated time to lift-off of 16.6 seconds from Example 17-2.

Solution

Stalling speed:

$$V_{S1} = 108.0 \text{ ft/s} \quad (64 \text{ KCAS}) \quad \text{(from POH)}$$

Transition speed:

$$V_{TR} = 1.15 V_{S1} = 1.15(108.0) = 124.2 \text{ ft/s } (73.6 \text{ KCAS})$$

Lift coefficient at V_{TR}:

$$C_L = \frac{2W}{\rho V^2 S} = \frac{2(3400)}{(0.002378)(124.2)^2(144.9)} = 1.279$$

Drag coefficient at V_{TR}:

$$C_D = C_{Dmin} + kC_L^2 = 0.0350 + 0.04207 \times 1.279^2 = 0.1038$$

Lift-to-drag ratio:

$$L/D = 1.279/0.1038 = 12.3$$

Thrust at V_{TR}:

$$T(73.6) = 908 \text{ lb}_f \quad \text{(using thrust function of EXAMPLE 17-8)}$$

Climb angle at V_{TR}:

$$\theta_{climb} = \sin^{-1}\left(\frac{T}{W} - \frac{1}{L/D}\right) = \sin^{-1}\left(\frac{908}{3400} - \frac{1}{12.3}\right)$$
$$= 10.7°$$

Transition radius:

$$R \approx 0.2156 \times V_{S1}^2 = 0.2156 \times (108)^2 = 2515 \text{ ft}$$

Transition distance:

$$S_{TR} \approx 0.2156 \times V_{S1}^2 \times \left(\frac{T}{W} - \frac{1}{L/D}\right)$$
$$= 0.2156 \times (108)^2 \times \left(\frac{908}{3400} - \frac{1}{12.3}\right) = 467 \text{ ft}$$

Transition height:

$$h_{TR} = R(1 - \cos\theta_{climb}) = 2515(1 - \cos 10.7°) = 43.7 \text{ ft}$$

Since the h_{TR} is less than the obstacle height of 50 ft, the climb segment must be included as well.

Climb distance:

$$S_C = \frac{h_{obst} - h_{TR}}{\tan\theta_{climb}} = \frac{50 - 43.7}{\tan 10.7°} = 33 \text{ ft}$$

Therefore, the total T-O distance over 50 ft is $959 + 467 + 33 = 1459$ ft, if the estimated V_{LOF} of 70.4 KCAS is used, and $1038 + 467 + 33 = 1538$ ft, if the POH V_{LOF} of 73 KCAS is used. The numbers compare to 1594 ft in the POH.

17.3.7 Treatment of T-O Run for a Taildragger

Fundamentally, only the ground run analysis for a taildragger differs from that of a conventional tricycle aircraft. This is because early in the ground run the taildragger can lift its tailwheel off the ground, effectively rendering it a different aircraft configuration. How quickly this happens depends on the aircraft itself. Some taildraggers generate enough thrust to lift the tail as soon as the engine power is increased. Others must accelerate to some airspeed before enough lift is generated by the HT to raise the tailwheel.

For this reason, the taildragger must really be considered as two separate configurations: one has the tailwheel on the ground and the other off the ground. Each configuration is subject to different lift and drag coefficients. It is easiest to treat the T-O run using the numerical method of Section 17.3.3, *Method 3: Solution*

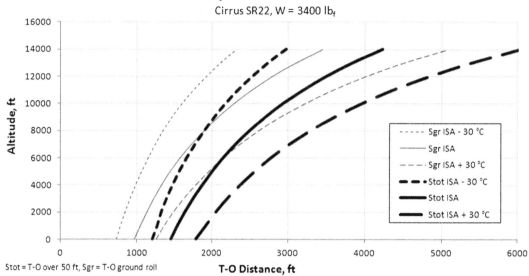

FIGURE 17-19 Sensitivity study showing the effect of altitude and temperature on the T-O distance of the SR22. The study was performed using the analysis method presented in this section.

using numerical integration method. This will even permit a transition to be incorporated, i.e. the lift-and-drag coefficients are functions of the *AOA* of the vehicle, which changes from the tail-on-ground angle to tail-off-ground angle over a period of 1 or 2 seconds. The airspeed at which this takes place can be calculated using the method from Section 13.3.5, *Taildragger landing gear reaction loads.*

17.3.8 Take-off Sensitivity Studies

Once the proper formulation for the T-O run has been prepared, it is helpful to study the impact of variation in atmospheric conditions, weight, runway slope, and other deviations on the operation of the airplane. Three examples of sensitivity studies have been prepared for the SR22 and are shown in Figure 17-19, Figure 17-20, and Figure 17-21. The first one is the sensitivity of the T-O distance to changes in temperature and altitude. Among other things, it shows that a T-O from an airfield at a 10,000 ft elevation on a standard day results in a doubling of the T-O distance over 50 ft. On a day that is 30 °C hotter the distance increases by a factor of 2.7.

The sensitivity being studied in Figure 17-20 shows how weight and altitude affect the T-O distance. A study

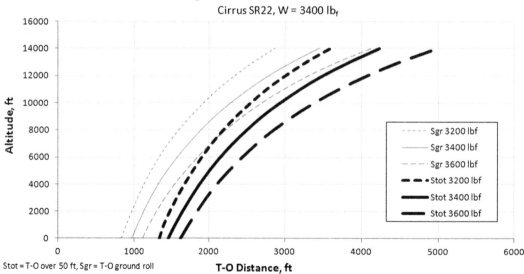

FIGURE 17-20 Sensitivity study showing the effect of weight change on the T-O distance of the SR22.

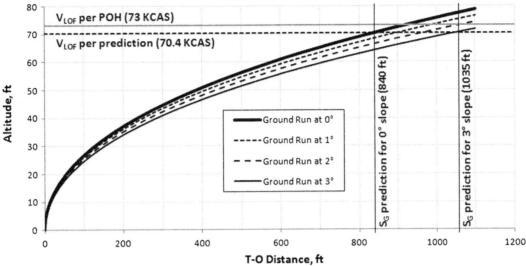

FIGURE 17-21 Sensitivity study showing the effect of runway slope change on the T-O distance of the SR22.

of this nature is helpful when evaluating the impact of a target gross weight not being met. Generally, increasing the gross weight by 200 lb$_f$ will increase the ground run distance by 150 to 675 ft and T-O over 50 ft by 155 to 750 ft, depending on altitude.

The impact of a steep runway slope is presented in Figure 17-21 and is based on Equation (17-5). The figure shows that an uphill runway slope of some 3° will increase the ground run distance by almost 200 ft, from 840 ft to 1035 ft. While this information is helpful for some preliminary design studies, it is vital for the pilot once the aircraft is operational. Most runways feature some degree of uphill or downhill slope; understanding the detrimental effect of uphill slopes, in particular, helps the pilot in the decision-making process.

17.4 DATABASE – T-O PERFORMANCE OF SELECTED AIRCRAFT

Table 17-6 shows the ground run and T-O distance to reach 50 ft altitude above ground level. This data is very helpful when evaluating the accuracy of own calculations.

EXERCISES

(1) The six-seat Beech B58 Baron has two 260 BHP Continental IO-470 engines that swing 78-inch diameter propellers. Its gross weight is 5100 lb$_f$, and it has a 199.2 ft^2 wing area and a 37.83 ft wing span. Estimate its balanced field length at S-L on a standard day, assuming the maximum lift coefficient in the T-O configuration to is 1.60, minimum drag coefficient is 0.035, lift-induced drag constant is 0.05906, and $V_2 = 1.2V_S$ for an obstacle height of 50 ft. Assume the simplified drag model. Assume the maximum thrust with one engine inoperable (OEI) to equal one half of the average T-O thrust. Compare your number to the published value of about 2300 ft.

(2) Estimate the average acceleration and time from break release to lift-off for the following aircraft types, based on the ground run and lift-off speed specified in their pilot's operating handbooks (assume S-L conditions).

(a) Beechcraft B58 Baron, $S_G = 2000$ ft, $V_{LOF} = 84$ KCAS.
(b) SOCATA TBM-850, $S_G = 1017$ ft, $V_{LOF} = 77$ KCAS.
(c) Cessna C-208 Grand Caravan, $S_G = 1405$ ft, $V_{LOF} = 86$ KCAS.
(d) Piper PA-46 Malibu Mirage, $S_G = 1100$ ft, $V_{LOF} = 69$ KCAS.

(3) Estimate the ground run for the three-engine Dassault Falcon 7X business jet, using the following properties:

$W = 70{,}000$ lb$_f$ $C_{Lmax} = 1.5$
$S = 761$ ft^2 T at $V_{LOF}/\sqrt{2} = 3 \times 6000$ lb$_f$
$C_{L\,TO} = 0.85$ $\mu = 0.03$
$C_{D\,TO} = 0.045$

TABLE 17-6 T-O Performance of Selected Aircraft

Aircraft	Gross wgt, lb$_f$	Ground Run, ft	T-O to 50 ft, ft	Reference
SINGLE-ENGINE GA AIRCRAFT				
Aero Boero 115	1697	380	500	Jane's 1976–77
Beech A36 Bonanza	3600	1140	2040	Jane's 1978–79
Beech Sundowner	2030	1130	1955	
Beech V35 Bonanza	3400	1002	1769	
Cessna 152	1670	725	1340	
Cessna 172 Skyhawk	2300	805	1440	
Cessna 177 Cardinal	2800	890	1585	
Cessna 185 Skylane	2950	705	1350	
Cessna 210 Centurion	3800	1250	2030	
Cirrus SR20	3050	1478	2221	POH
Cirrus SR22	3400	1028	1594	
Piper PA-18 Super Cub	1750	200	500	Jane's 1978–79
Piper PA-28 Warrior II	2325	963	1490	
Piper PA-32 Lance II	3600	960	1690	
Piper PA-38 Tomahawk	1670	820	1460	
SIAI-Marchetti SF-260	2425	1837	2543	Jane's 1976–77

Aircraft	Gross wgt, lb$_f$	Ground run, ft	T-O to 50 ft, ft	BFL[a], ft	Reference
MULTIENGINE GA AIRCRAFT					
Beech Duchess 76	3900	1017	2119	-	Jane's 1978–79
Beech Duke B60	6775	2075	2626	-	
Beech King Air C90	9650	-	2261	3498	
Cessna 310	5500	1335	1700	-	
Cessna 337 Skymaster	4630	1000	1675	-	
Cessna 340	5990	1615	2175	-	
DHC-6 Twin Otter	12500	700–860	1200–1500	-	Jane's 1976–77
Partenavia P-68B Victor	4321	912	1539	-	
Piper PA-23 Aztec	5200	945	1695	1985	Jane's 1978–79
Piper PA-31 Cheyenne	9000	-	1980	3140	
Piper PA-31 Chieftain	7000	1360	2490	2100	
Piper PA-34 Seneca II	4570	900	1240	2520	

Aircraft	Gross wgt, lb$_f$	Ground run, ft	T-O to 50 ft, ft	Reference
COMMUTER AIRCRAFT				
Let L-410 Turbolet	12566	1627	1850	Jane's 1976–77
GAF N22B Nomad	8500	600-970	1260-1350	
Embraer EMB-110 Bandeirante	13010	1245	1770	
DHC-5 Buffalo	49200	950-2300	1250-2875	Jane's 1978–79
Fokker F-27 Friendship	43952	3240-3970	-	Jane's 1976–77

[a]BFL = *balanced field length*.

(4) Determine the lift-off speed (V_{LOF}), V_2, and total T-O distance to an obstacle height of 50 ft (includes ground roll, rotation, transition, and climb distances) for the turboprop-powered SOCATA TBM-850, assuming the following characteristics:

$W = 7394$ lb$_f$	$C_{L\ TO} = 0.750$ (assumed value)
$P = 850$ SHP (PT6A)	$C_{D\ TO} = 0.045$ (assumed value)
$S = 193.7$ ft^2	$\mu = 0.03$ (sample value)
$C_{Lmax} = 1.90$ (based on POH for T-O)	$\eta_p = 0.65$

Compare the results to published data from the airplane's pilot's operating handbook (POH), which gives a $V_{LOF} = 90$ KCAS, $V_2 = 99$ KCAS, ground roll distance of 2035 ft (ISA at S-L) and total T-O distance (to 50 ft) of 2840 ft.

VARIABLES

Symbol	Description	Units (UK and SI)
a	Instantaneous acceleration	ft/s^2 or m/s^2
a_{avg}	Average acceleration	ft/s^2 or m/s^2
AR	Aspect ratio	
BFL	Balanced field length	ft or m
P_{BHP}	Piston engine horsepower	BHP
C_D	Drag coefficient	
$C_{D\ TO}$	Take-off drag coefficient	
C_{Di}	Induced drag coefficient	
$(C_{Di})_{IGE}$	Induced drag coefficient, in ground effect	
C_{Dmin}	Minimum drag coefficient	
C_L	Lift coefficient	
$C_{L\ TO}$	Take-off lift coefficient	
C_{L0}	Zero AOA lift coefficient	
C_{L2}	Magnitude of lift coefficient at V_2	
C_{Lmax}	Maximum lift coefficient (of take-off configuration)	
C_{LminD}	Minimum drag lift coefficient	
$C_{L\alpha}$	3D lift curve slope	Per degree or per radian
D	Drag	lb$_f$ or N

Symbol	Description	Units (UK and SI)
D_2	Drag at V_2	lb$_f$ or N
D_p	Propeller diameter	ft or m
e	Oswald span efficiency factor	
g	Gravitational acceleration	ft/s^2 or m/s^2
h_{obst}	Obstacle height	ft or m
h_{to}	Take-off obstacle height	ft or m
h_{TR}	Take-off transition height	ft or m
k	Constant relating aspect ratio and span efficiency	
K_P	Constant for propellers (dependent on units)	
L	Lift	lb$_f$ or N
P_{to}	Maximum engine power	BHP or kg·m/s
R	Radius of take-off transition path	ft or m
R_M	Main gear reaction force	lb$_f$ or N
R_N	Nose gear reaction force	lb$_f$ or N
S	Instantaneous position	ft or m
S	Reference wing area	ft^2 or m^2
S_0	Initial position	ft or m
S_C	(Horizontal) climb distance	ft or m
S_G	Ground run	ft or m
S_{obst}	Obstacle clearance distance	
S_{ROT}	Rotation distance	ft or m
S_{TOT}	Total take-off distance	ft or m
S_{TR}	Transition distance	ft or m
t	Time	seconds
T	Thrust	lb$_f$ or N
$\overline{T}$	Average thrust during the T-O run	lb$_f$ or N
t_{LOF}	Time to lift-off	seconds
T_{to}	Maximum static thrust	lb$_f$ or N
V	Airspeed	ft/s or m/s
V_0	Initial velocity	ft/s or m/s
V_1	Balanced field length velocity	ft/s or m/s
V_2	Obstacle clearance speed	ft/s or m/s
V_{LOF}	Lift-off speed	ft/s or m/s
V_{MC}	Minimum control speed (OEI)	ft/s or m/s
V_{MU}	Minimum unstick speed	ft/s or m/s

Symbol	Description	Units (UK and SI)
V_R	Rotation speed	ft/s or m/s
V_S	Stalling speed with flaps	ft/s or m/s
V_{S1}	Stalling speed without flaps	ft/s or m/s
W	Weight	lb$_f$ or N
W_{to}	Take-off weight	lb$_f$ or N
ΔS_{to}	Inertia distance	ft or m
$\Delta \gamma_2$	Change in climb angle	Degrees or radians
α_{TO}	Take-off angle-of-attack	Degrees or radians
γ	Slope of runway	Degrees
γ_2	Climb angle	Degrees or radians
γ_{2min}	Angle dependent on aircraft configuration	Degrees or radians
η_p	Propeller efficiency (at V_{LOF})	
λ	Turbofan bypass ratio	

Symbol	Description	Units (UK and SI)
μ	Ground friction coefficient	
μ'	$0.01 \cdot C_{Lmax} + 0.02$	
θ_{climb}	Climb angle	Degrees or radians
ρ	Density of air	slugs/ft^3 or N/m^3
σ	Density ratio	

References

[1] Perkins CD, Hage RE. Airplane Performance, Stability, and Control. John Wiley & Sons; 1949.
[2] Torenbeek E. Synthesis of Subsonic Aircraft Design. 3rd ed. Delft University Press; 1986.
[3] Nicolai L. Fundamentals of Aircraft Design. 2nd ed. 1984.
[4] Roskam J, Lan Chuan-Tau Edward. Airplane Aerodynamics and Performance. DARcorporation; 1997.
[5] Hale FJ. Aircraft Performance, Selection, and Design. John Wiley & Sons; 1984. pp. 137−138.
[6] Anderson Jr JD. Aircraft Performance & Design. 1st ed. McGraw-Hill; 1998.
[7] http://www.learjet.com/.

CHAPTER 18

Performance – Climb

OUTLINE

18.1 Introduction 821
   18.1.1 The Content of this Chapter 822

18.2 Fundamental Relations for the Climb Maneuver 822
   18.2.1 General Two-dimensional Free-body Diagram for an Airplane 822
   18.2.2 General Planar Equations of Motion for an Airplane 823
   18.2.3 Equations of Motion for Climbing Flight 823
   18.2.4 Horizontal and Vertical Airspeed 824
   18.2.5 Power Available, Power Required, and Excess Power 824
   18.2.6 Vertical Airspeed in Terms of Thrust or Power 824
      Derivation of Equation (18-15) 824
      Derivation of Equation (18-16) 825
   18.2.7 Rate-of-climb 825
      Climb Gradient 825

18.3 General Climb Analysis Methods 825
   18.3.1 General Rate-of-climb 825
      Derivation of Equation (18-18) 826
   18.3.2 General Climb Angle 827
      Derivation of Equations (18-19) and (18-20) 827
   18.3.3 Max Climb Angle for a Jet 827
      Derivation of Equation (18-21) 828
   18.3.4 Airspeed for θ_{max} for a Jet (Best Angle-of-climb Speed) 828
      Derivation of Equation (18-22) 828
   18.3.5 ROC for θ_{max} for a Jet 829
      Derivation of Equation (18-23) 829
   18.3.6 Airspeed for Best ROC for a Jet 829
      Derivation of Equation (18-24) 830
   18.3.7 Best ROC for a Jet 831
      Derivation of Equation (18-25) 831
   18.3.8 Airspeed for θ_{max} for a Propeller-powered Airplane 832
      Derivation of Equation (18-26) 832
   18.3.9 Airspeed for Best ROC for a Propeller-powered Airplane 833
      Derivation of Equation (18-27) 834
   18.3.10 Best Rate-of-climb for a Propeller-powered Airplane 834
      Derivation of Equations (18-29) and (18-30) 835
   18.3.11 Time to Altitude 836
      Rapid Approximation 836
      Linear Approximation 837
      Derivation of Equation (18-31) 837
      Derivation of Equation (18-33) 837
      Derivation of Equation (18-34) 837
   18.3.12 Absolute/Service Ceiling Altitude 838
   18.3.13 Numerical Analysis of the Climb Maneuver – Sensitivity Studies 840
      Altitude Sensitivity 842
      Weight Sensitivity 842
      Propeller Efficiency Sensitivity 842

18.4 Aircraft Database – Rate-of-climb of Selected Aircraft 842

Variables 844

References 845

18.1 INTRODUCTION

The T-O maneuver is followed by the climb maneuver. It is vital for the aircraft designer to understand how rapidly or how steeply an aircraft can climb. Great climb performance sells aircraft. The climb affects not only how quickly the airplane reaches a desired cruise altitude but also how its noise footprint is perceived. The competent designer will always try to maximize the rate-of-climb and the angle-of-climb. The study of

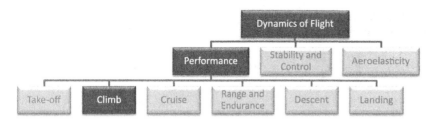

FIGURE 18-1 An organizational map placing performance theory among the disciplines of dynamics of flight, and highlighting the focus of this section; climb performance analysis.

the climb primarily involves the determination of the rate at which the airplane increases its altitude (called the rate-of-climb or *ROC*) and the angle its flight path makes to the ground (called the angle-of-climb or *AOC*). Figure 18-1 shows an organizational map displaying the T-O among other parts of the performance theory.

This chapter will present the formulation of and the solution of the equation of motion for the climb, and present practical as well as numerical solution methodologies that can be used for both propeller- and jet-powered aircraft. The presentation is prepared in terms of independent analysis methods. When appropriate, each method will be accompanied by an illustrative example using the sample aircraft. The primary information we want to extract from this analysis is characteristics like maximum *ROC*, best (highest) *AOC*, the corresponding airspeeds, the *ROC* and *AOC* for a given power setting, and climb range, to name a few.

In general, the methods presented here are the "industry standard" and mirror those presented by a variety of authors, e.g. Perkins and Hage [1]; Torenbeek [2]; Nicolai [3]; Roskam [4]; Hale [5]; Anderson [6]; and many, many others.

18.1.1 The Content of this Chapter

- **Section 18.2** develops fundamental relationships necessary to evaluate climb characteristics, most importantly, the equations of motion for climbing flight.
- **Section 18.3** presents an assortment of methods to predict the various climb characteristics of an airplane.
- **Section 18.4** presents the climb performance of selected aircraft types.

18.2 FUNDAMENTAL RELATIONS FOR THE CLIMB MANEUVER

In this section we will derive the equation of motion for the climb maneuver, as well as all fundamental relationships used to evaluate its most important characteristics. First, a general two-dimensional free-body diagram will be presented to allow the formulation to be developed. Only the two-dimensional version of the equation will be determined as this is sufficient for all aspects of aircraft design.

18.2.1 General Two-dimensional Free-body Diagram for an Airplane

Figure 18-2 shows a free-body diagram of an airplane moving along a trajectory, which we call the *flight path*. The x- and z-axes are attached to the center of gravity of the airplane such that the x-axis is the tangent to the flight path. The z-axis is perpendicular to the flight path. The airspeed is defined as the component of its velocity parallel to the tangent to the flight path. Also, a *datum* has been drawn on the airplane to represent the chord line of the MGC of the wing. The angle between the datum and the tangent to the flight path is called the *angle-of-attack*. The force (or thrust) generated by the airplane's powerplant, T, may be at an angle ε with respect to the x-axis. The figure shows that this coordinate system can change its orientation with respect to the *horizon* depending on the maneuver being performed. Then, the angle between the horizon and the x-axis is called the *climb angle*, denoted by θ. If $\theta > 0$, then the aircraft is said to be *climbing*. If $\theta = 0$, then the aircraft is said to be *flying straight and level* (cruising). If $\theta < 0$, then the aircraft is said to be *descending*. This chapter only considers the first scenario.

The free-body diagram of Figure 18-2 is considered balanced in terms of inertia, mechanical, and aerodynamic forces. The lift is the component of the resultant aerodynamic force generated by the aircraft that is perpendicular to the flight path (along its z-axis). The drag is the component of the aerodynamic force that is parallel (along its x-axis). These are balanced by the weight, W, and the corresponding components of T. The presentation of Figure 18-2 can now be used to derive the *planar equations of motion* for the airplane, so called because the motion is assumed two-dimensional and assumes there is not yaw. This simplification is sufficient to accurately predict the vast majority of climb maneuvers.

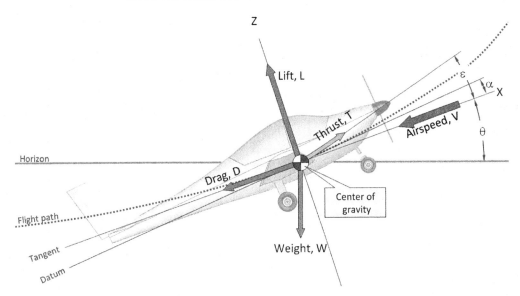

FIGURE 18-2 A two-dimensional free-body of the airplane in climbing flight.

18.2.2 General Planar Equations of Motion for an Airplane

Planar equations of motion (assume no or steady rotation about the *y*-axis) are obtained by summing the forces depicted in Figure 18-1 about the *x*- and *z*-axes as follows:

$$L - W \cos\theta + T \sin\varepsilon = \frac{W}{g}\frac{dV_Z}{dt} \quad (18\text{-}1)$$

$$-D - W \sin\theta + T \cos\varepsilon = \frac{W}{g}\frac{dV_X}{dt} \quad (18\text{-}2)$$

18.2.3 Equations of Motion for Climbing Flight

The equations of motion can be adapted to a steady climbing flight by making the following assumptions:

(1) Steady motion implies $dV/dt = 0$.
(2) The climb angle, θ, is a non-zero quantity.
(3) The angle-of-attack, α, is small.
(4) The thrust angle, ε, is $0°$.

Equations of motion for a steady climb:

$$L - W\cos\theta = 0 \quad \Rightarrow \quad L = W\cos\theta \quad (18\text{-}3)$$

$$-D - W\sin\theta + T = 0 \quad \Rightarrow \quad T - D = W\sin\theta \quad (18\text{-}4)$$

Equation (18-3) shows that lift-in-climb is actually less than the weight (the difference is balanced by the vertical component of the thrust):

$$L = W\cos\theta \quad (18\text{-}5)$$

The lift coefficient at this condition is thus:

$$C_L = \frac{2W\cos\theta}{\rho V^2 S} = \frac{W\cos\theta}{qS} \quad (18\text{-}6)$$

The drag force, using the simplified drag model, is given by:

$$D = qS\left(C_{Dmin} + k \cdot C_L^2\right) \quad (18\text{-}7)$$

Inserting Equation (18-6) into Equation (18-7) yields:

$$D = qS\left(C_{Dmin} + k \cdot \left(\frac{W\cos\theta}{qS}\right)^2\right) \quad (18\text{-}8)$$

Expanding:

$$D = qS\left(C_{Dmin} + k \cdot \frac{W^2\cos^2\theta}{q^2 S^2}\right)$$

$$\Rightarrow \quad D = \left(qSC_{Dmin} + k \cdot \frac{W^2\cos^2\theta}{qS}\right) \quad (18\text{-}9)$$

Note that drag, D, as calculated by Equation (18-9), should be used with Equations (18-15) and (18-17), and would ordinarily require an iterative scheme to solve for *ROC* and θ. However, as demonstrated by Austyn-Mair and Birdsall [7], assuming that $\cos\theta \sim 1$ holds indeed yields an acceptable accuracy for modest climb angles. In particular, the error that results for GA aircraft is small, as their angle-of-climb is usually less than 15°. According to Figure 4.1 of Ref. [7], at the best rate-of-climb airspeed, even an angle-of-climb of 20° will deviate about 0.2° off the exact value. At an angle-of-climb of 40° the deviation is a hair short of 1.0°. The assumption is that cos θ ~ 1 is warranted as it allows for a considerable time saving in analysis work, with low difference from the exact method.

18.2.4 Horizontal and Vertical Airspeed

The primary purpose of the methods in this section is the evaluation of the climb performance of aircraft. We want to determine characteristics like best rate-of-climb (*ROC*), best (largest) angle-of-climb (*AOC*), and the airspeeds at which these materialize. The first important step is to define the *horizontal airspeed*. It is important when estimating the horizontal distance covered during a long climb to altitude:

$$V_H = V \cos \theta \qquad (18\text{-}10)$$

Then it is possible to define the *vertical airspeed*, also called the *rate-of-climb*:

$$V_V = V \sin \theta \qquad (18\text{-}11)$$

Both can be derived by observation from Figure 18-3. Note that in terms of calibrated airspeed, V is the airspeed indicated on the airspeed indicator; V_V is observed on the vertical speed indicator (VSI); and V_H is the ground speed. Note that this V_H should not be confused with the maximum level airspeed to be discussed in Chapter 19.

18.2.5 Power Available, Power Required, and Excess Power

These three concepts are imperative in the climb analysis as they define the climb capability of the aircraft. Note that for aircraft propelled by jet engines, the power available is estimated by multiplying its thrust by the airspeed. For aircraft powered by propellers, the power is obtained by multiplying the engine power by the propeller efficiency. Since the engine power is usually presented in terms of BHP or SHP, this number must be converted from horsepower to ft·lb$_f$/s by multiplying by a factor of 550, if using the UK system. If using the SI system, the horsepower number must be multiplied by a factor of 745.7 to convert to watts.

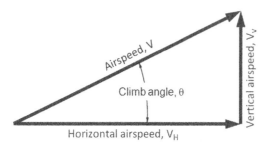

FIGURE 18-3 Airspeed components during climb.

Power available:

$$P_{AV} \equiv Force \times Speed = TV = \eta_p \cdot P_{ENG} \qquad (18\text{-}12)$$

Power required:

$$P_{REQ} \equiv Force \times Speed = DV \qquad (18\text{-}13)$$

Excess power:

$$P_{EX} \equiv P_{AV} - P_{REQ} \qquad (18\text{-}14)$$

18.2.6 Vertical Airspeed in Terms of Thrust or Power

The vertical airspeed can be estimated if thrust or power and drag characteristics of the aircraft are known.

For jets:

$$V_V \equiv \frac{TV - DV}{W} = V \sin \theta \qquad (18\text{-}15)$$

For propellers:

$$V_V \equiv \frac{P_{AV} - P_{REQ}}{W} = \frac{\eta_p \cdot P_{ENG} - P_{REQ}}{W} = V \sin \theta \qquad (18\text{-}16)$$

Note that the above expressions are some of the most important equations in the entire climb analysis methodology. Ultimately, we want to determine some specific values of V_V, for instance the maximum value, or the one that results in the steepest climb possible, and so on. Knowing these are of great importance to the pilot and, as it turns out, also for the certifiability of the aircraft. Additionally, the formulation shows that in order for an airplane to increase its altitude, its thrust power (TV) or available power (P_{AV}) must be larger than the drag power (DV) or required power (P_{REQ}) for level flight.

> ***Derivation of Equation (18-15)***
>
> Multiply Equation (18-4) by V/W to get:
>
> $$\frac{V}{W}(-D - W \sin \theta + T) = 0$$
>
> $$\Rightarrow \quad -\frac{V}{W}D - \frac{V}{W}W \sin \theta + \frac{V}{W}T = 0$$
>
> $$\Rightarrow \quad \frac{TV}{W} - \frac{DV}{W} = V \sin \theta$$
>
> $$\Rightarrow \quad \frac{TV - DV}{W} = V \sin \theta$$
>
> **QED**

Derivation of Equation (18-16)

We note that power is defined as force × speed:

$$\frac{TV - DV}{W} = \frac{\text{Power from engines} - \text{Aerodynamic power}}{W}$$

$$= V \sin \theta$$

$$\Rightarrow \frac{TV - DV}{W} = \frac{P_{AV} - P_{REQ}}{W} = V \sin \theta$$

where $P_{AV} = \eta_p \cdot P_{ENG}$.

QED

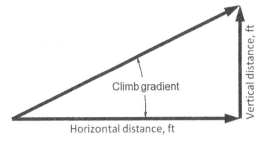

FIGURE 18-4 Distance components during climb.

$$ROC = 0.1 \times \frac{100 \text{ nm/hr}}{60 \text{ min/hr}} \times 6076 \text{ ft/nm} = 1013 \text{ ft/min}$$

18.2.7 Rate-of-climb

The rate-of-climb (ROC) is of great importance to the pilot, as well as a superb indicator of the airplane's capability as it is directly dependent on its thrust and drag characteristics. If the thrust and drag can be quantified at a flight condition, the instantaneous ROC can be calculated as follows:

$$ROC \equiv 60\left(\frac{TV - DV}{W}\right) = 60\left(\frac{P_{AV} - P_{REQ}}{W}\right) \quad (18\text{-}17)$$

NOTE 1: The units for ROC in Equation (18-15) are commonly ft/min or fpm, which is why we multiply it by 60 to convert the ROC in ft/s into fpm.
NOTE 2: In the SI system, ROC is usually in terms of m/s, rendering the factor 60 unnecessary. The reader must be aware of the difference in the representation of time between the UK and SI systems.
NOTE 3: Unless otherwise specified the ROC is in fpm. *It is also possible that the ROC might be given in ft/s.*

Climb Gradient

Climb is sometimes expressed in terms of % climb gradient. For instance, 14 CFR Part 23 refers to climb gradients in this fashion, in lieu of fpm or m/s, in order to present the regulatory requirements in a form applicable to all aircraft. The concept assumes no wind conditions and is defined as follows (see Figure 18-4):

$$\text{Climb Gradient} = \frac{\text{Vertical Distance}}{\text{Horizontal Distance}}$$

$$= \frac{\text{Vertical Distance}/\Delta t}{\text{Horizontal Distance}/\Delta t}$$

Consider an airplane whose climb gradient is 0.1 at 100 KTAS (nm/hr) in no wind conditions. Its rate-of-climb in fpm would be:

18.3 GENERAL CLIMB ANALYSIS METHODS

Armed with the equation of motion derived in the previous section, we can now begin to evaluate the climb characteristics of the new aircraft design. In this section, we will introduce a number of methods to estimate the most important climb properties of the aircraft. Note that the methods presented utilize the simplified drag model. As has been stated before, this can lead to significant inaccuracies for aircraft whose $C_{LminD} > 0$, as is the case for most aircraft that feature cambered airfoils (which most aircraft do). From that standpoint the presentation is somewhat misleading. The reader should regard the methods as an introduction to concepts that are commonly used in the industry. A method that allows for a detailed climb analysis of real aircraft, with adjusted drag polars, and even ones with a drag bucket, will be presented at the end of this section. The concepts that are presented at first will then come in handy.

18.3.1 General Rate-of-climb

This general expression is used to estimate the ROC based on thrust-to-weight ratio and wing loading. It is handy for evaluating climb performance during the design stage, but is also applicable to general climb performance analyses. *It assumes the simplified drag model* and returns the vertical airspeed in terms of ft/s or m/s.

It is computed from:

$$V_V = V \sin \theta$$

$$= V\left(\frac{T}{W} - q\left(\frac{S}{W}\right)C_{Dmin} - k \cdot \left(\frac{W}{S}\right)\frac{\cos^2 \theta}{q}\right)$$

(18-18)

Derivation of Equation (18-18)

Insert Equation (18-9) into (18-15) and manipulate algebraically, as shown:

$$V \sin\theta = \frac{TV - DV}{W} = \frac{TV}{W} - \frac{\left(qSVC_{Dmin} + k \cdot V \frac{W^2 \cos^2\theta}{qS}\right)}{W}$$

Then, simplify to get:

$$V \sin\theta = \frac{TV}{W} - qV\frac{SC_{Dmin}}{W} - k \cdot V \frac{W \cos^2\theta}{qS}$$

Then, complete by rearranging:

$$V \sin\theta = V\left(\frac{T}{W} - q\left(\frac{S}{W}\right)C_{Dmin} - k \cdot \left(\frac{W}{S}\right)\frac{\cos^2\theta}{q}\right)$$

QED

EXAMPLE 18-1

Determine the vertical airspeed for the sample aircraft flying at 250 KCAS at S-L at maximum thrust and at a weight of 20,000 lb$_f$.

Solution

Dynamic pressure:

$$q = \frac{1}{2}\rho V^2 = \frac{1}{2}(0.002378)(250 \times 1.688)^2$$
$$= 211.7 \text{ lb}_f/\text{ft}^2$$

Insert and evaluate the rate-of-climb at the condition:

$$V \sin\theta = V\left(\frac{T}{W} - q\left(\frac{S}{W}\right)C_{Dmin} - k \cdot \left(\frac{W}{S}\right)\frac{\cos^2\theta}{q}\right)$$

$$= (250 \times 1.688)\left(\frac{7000}{20{,}000} - (211.7)\left(\frac{311.6}{20{,}000}\right)(0.020)\right.$$

$$\left. - (0.05236)\left(\frac{20{,}000}{311.6}\right)\frac{\sim 1}{(211.7)}\right)$$

$$= (422)(0.35 - 0.065966 - 0.015875) = 113.2 \text{ ft/s}$$

This amounts to 6792 fpm. Using this method, the excess power and rate-of-climb can be plotted for any altitude. Figure 18-5 depicts such these using the more realistic thrust modeling:

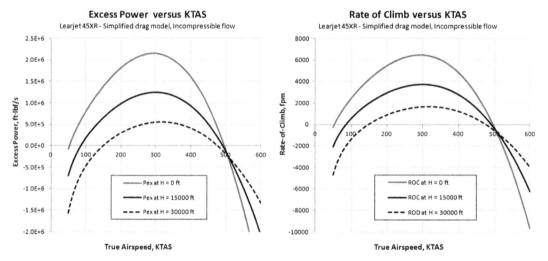

FIGURE 18-5 Excess power and rate-of-climb at three altitudes.

18.3.2 General Climb Angle

The climb angle is of great importance when it comes to obstruction clearance, or when showing compliance with noise regulations (14 CFR Part 36), as well as evaluation of deck angle during climb. *It assumes the simplified drag model.*

Computed from:

$$\sin\theta = \frac{T}{W} - q\left(\frac{S}{W}\right)C_{Dmin} - k\cdot\left(\frac{W}{S}\right)\frac{\cos^2\theta}{q} \qquad (18\text{-}19)$$

General angle-of-climb:

$$\sin\theta = \frac{T}{W} - \frac{1}{L/D} \qquad (18\text{-}20)$$

Derivation of Equations (18-19) and (18-20)

To get Equation (18-19), divide by V on either side of Equation (18-18):

$$V\sin\theta = V\left(\frac{T}{W} - q\left(\frac{S}{W}\right)C_{Dmin} - k\cdot\left(\frac{W}{S}\right)\frac{\cos^2\theta}{q}\right)$$

QED

To get Equation (18-20), divide by V on either side of Equation (18-15):

$$V\sin\theta = \frac{TV - DV}{W}$$

$$\Rightarrow \sin\theta = \frac{T-D}{W} \approx \frac{T}{W} - \frac{D}{L} = \frac{T}{W} - \frac{1}{L/D}$$

QED

EXAMPLE 18-2

Determine the angle of climb for the sample aircraft flying at 250 KCAS at S-L at maximum thrust and at a weight of 20,000 lb$_f$.

Solution

Dynamic pressure:

$$q = \frac{1}{2}\rho V^2 = \frac{1}{2}(0.002378)(250 \times 1.688)^2$$
$$= 211.7 \text{ lb}_f/\text{ft}^2$$

$$\sin\theta = \frac{T}{W} - q\left(\frac{S}{W}\right)C_{Dmin} - k\cdot\left(\frac{W}{S}\right)\frac{\cos^2\theta}{q}$$

$$= \left(\frac{7000}{20,000} - (211.7)\left(\frac{311.6}{20,000}\right)(0.020)\right.$$

$$\left. -(0.05236)\left(\frac{20,000}{311.6}\right)\frac{\sim 1}{(211.7)}\right)$$

$$= (0.35 - 0.065966 - 0.015875)$$

$$= 0.2682 \text{ radians}$$

This amounts to 15.4°.

18.3.3 Max Climb Angle for a Jet

Determining the maximum climb angle is of great importance as this can be used to evaluate the capability of the aircraft to take off from runways in mountainous regions. Typical operational procedures would require an aircraft departing a runway surrounded by high mountains to climb at or near this value, at least until threatening terrain has been cleared. But there is another and very important reason to evaluate the maximum climb angle: noise certification (14 CFR Part 36). The steeper this angle, the farther away from the sound level meter will the airplane be when it is right above it (a regulatory requirement). The maximum climb angle is computed from the following expression and *assumes the simplified drag model*:

$$\theta_{max} \approx \sin^{-1}\left(\frac{T_{max}}{W} - \sqrt{4\cdot C_{Dmin}\cdot k}\right) \qquad (18\text{-}21)$$

Derivation of Equation (18-21)

Rewrite Equation (18-15):

$$V \sin \theta = \frac{TV - DV}{W} \Leftrightarrow \sin \theta = \frac{T}{W} - \frac{D}{W}$$

Insert Equation (18-5):

$$W = \frac{L}{\cos \theta} \Rightarrow \sin \theta = \frac{T}{W} - \frac{D}{W} = \frac{T}{W} - \frac{\cos \theta}{L/D}$$

Assuming $\cos \theta \approx 1$:

$$\sin \theta \approx \frac{T}{W} - \frac{1}{L/D}$$

The climb angle will reach an upper limit when the L/D is maximum, LD_{max}. In other words:

$$\sin \theta_{max} \approx \frac{T}{W} - \frac{1}{LD_{max}}$$

Recalling Equation (19-18), LD_{max} can be rewritten as follows:

$$\sin \theta_{max} \approx \frac{T}{W} - \sqrt{4 \cdot C_{Dmin} \cdot k}$$

QED

EXAMPLE 18-3

Determine the maximum angle-of-climb for the sample aircraft at S-L, maximum thrust and at a weight of 20,000 lb$_f$.

Solution

$$\theta_{max} \approx \sin^{-1}\left(\frac{T_{max}}{W} - \sqrt{4 \cdot C_{Dmin} \cdot k}\right)$$

$$= \sin^{-1}\left(\frac{7000}{20{,}000} - \sqrt{4(0.020)(0.05236)}\right) = 16.6°$$

18.3.4 Airspeed for θ_{max} for a Jet (Best Angle-of-climb Speed)

To continue the discussion of obstruction clearance or compliance with noise regulations from previous analyses, the pilot would establish the best angle-of-climb as soon as possible after take-off by reaching and maintaining the best angle-of-climb airspeed. In the case of a jet aircraft, this airspeed can be calculated from the following expression, which *assumes the simplified drag model*. Note that this result is not valid for propeller-powered aircraft. The airspeed returned is in units of ft/s if the input values are in the UK system, but m/s if the SI system is used.

$$V_X = \sqrt{\frac{2}{\rho}\left(\frac{W}{S}\right)\sqrt{\frac{k}{C_{Dmin}}} \cos \theta_{max}} \quad (18\text{-}22)$$

This airspeed is recognized by pilots and regulation authorities as V_X. In short, it results in the steepest possible climb angle for a jet aircraft, or the largest gain of altitude per unit horizontal distance.

Derivation of Equation (18-22)

Rewrite Equation (18-3):

$$L = W \cos \theta = \frac{1}{2}\rho V^2 S C_L$$

Lift coefficient for LD_{max}:

$$C_L = \sqrt{\frac{C_{Dmin}}{k}}$$

We insert this back into Equation (18-3), where θ and V become θ_{max} and V_X:

$$W \cos \theta_{max} = \frac{1}{2}\rho V_X^2 S C_L = \frac{1}{2}\rho V_X^2 S \sqrt{\frac{C_{Dmin}}{k}}$$

$$\Leftrightarrow V_X = \sqrt{\frac{2}{\rho}\left(\frac{W \cos \theta_{max}}{S}\right)\sqrt{\frac{C_{Dmin}}{k}}}$$

$$= \sqrt{\frac{2}{\rho}\left(\frac{W}{S}\right)\sqrt{\frac{C_{Dmin}}{k}} \cos \theta_{max}}$$

QED

EXAMPLE 18-4

Determine the airspeed for maximum angle-of-climb for the sample aircraft at S-L, maximum thrust and at a weight of 20,000 lb$_f$.

Solution

$$V_X = \sqrt{\frac{2}{\rho}\left(\frac{W}{S}\right)}\sqrt{\frac{k}{C_{Dmin}}\cos\theta_{max}}$$

$$= \sqrt{\frac{2}{0.002378}\left(\frac{20{,}000}{311.6}\right)}\sqrt{\frac{0.05236}{0.020}(\sim 1)} = 295.5 \text{ ft/s}$$

This amounts to 175 KTAS (or KCAS since this is S-L).

18.3.5 ROC for θ_{max} for a Jet

It is also of interest to calculate the *ROC* associated with V_X. Note that this is less than the *ROC* associated with V_Y (the largest gain in altitude per unit time). For a jet climbing while maintaining its best angle-of-climb airspeed, V_X, the *ROC* can be calculated from:

$$ROC_X = 60 \cdot V_X \cdot \sin\theta_{max} \qquad (18\text{-}23)$$

Derivation of Equation (18-23)

Use θ_{max} and V_X with Equation (18-11):

$$V_V = V\sin\theta \;\Rightarrow\; ROC_X = 60\cdot V_X \cdot \sin\theta_{max}$$

QED

18.3.6 Airspeed for Best ROC for a Jet

As the airspeed of an airplane is changed at a given power setting, so is its *ROC*. At a particular airspeed the *ROC* will reach its maximum value. At that airspeed, the airplane will increase its altitude in the least amount of time. This airspeed is particularly important for fuel-thirsty jets and allows them to reach their cruise altitude with the least amount of fuel consumed. This airspeed can be computed from the expression below that *assumes the simplified drag model*:

$$V_Y = \sqrt{\frac{(T/W)(W/S)}{3\rho C_{Dmin}}\left[1+\sqrt{1+\frac{3}{LD_{max}^2(T/W)^2}}\right]}$$

(18-24)

EXAMPLE 18-5

Determine the *ROC* at maximum angle-of-climb for the sample aircraft at S-L, maximum thrust and at a weight of 20,000 lb$_f$.

Solution

$$ROC_X = 60 \cdot V_X \cdot \sin\theta_{max} = 60 \times (295.5) \times \sin(16.6°)$$
$$= 5065 \text{ fpm}$$

Note that the above formulation can also be used to estimate the best rate-of-climb airspeed for a multiengine aircraft suffering a one engine inoperative (OEI) condition. This airspeed is denoted by V_{YSE}. The thrust, of course, must be reduced by the contribution of the failed engine and the minimum drag C_{Dmin} must be increased to account for the asymmetric attitude of the airplane necessary to fly straight and level. Also, LD_{max} must be recalculated as it reduces at this condition.

Derivation of Equation (18-24)

The general rate-of-climb is given by Equation (18-18):

$$V \sin \theta = V\left(\frac{T}{W} - q\left(\frac{S}{W}\right)C_{Dmin} - k\cdot\left(\frac{W}{S}\right)\frac{\cos^2\theta}{q}\right)$$

Assume $\cos\theta \sim 1$ and differentiate with respect to V, as follows:

$$\frac{d(ROC)}{dV} = \frac{d(V)}{dV}\left(\frac{T}{W} - \frac{1}{2}\rho V^2\left(\frac{S}{W}\right)C_{Dmin} - \left(\frac{W}{S}\right)\frac{2k}{\rho V^2}\right)$$
$$+ V\frac{d}{dV}\left(\frac{T}{W} - \frac{1}{2}\rho V^2\left(\frac{S}{W}\right)C_{Dmin} - \left(\frac{W}{S}\right)\frac{2k}{\rho V^2}\right)$$

Manipulating algebraically;

$$\frac{d(ROC)}{dV} = \left(\frac{T}{W} - \frac{1}{2}\rho V^2\left(\frac{S}{W}\right)C_{Dmin} - \left(\frac{W}{S}\right)\frac{2k}{\rho V^2}\right)$$
$$+ V\left(-\rho V\left(\frac{S}{W}\right)C_{Dmin} + 2\left(\frac{W}{S}\right)\frac{2k}{\rho V^3}\right)$$

$$\frac{d(ROC)}{dV} = \frac{T}{W} - \frac{1}{2}\rho V^2\left(\frac{S}{W}\right)C_{Dmin} - \left(\frac{W}{S}\right)\frac{2k}{\rho V^2}$$
$$- \rho V^2\left(\frac{S}{W}\right)C_{Dmin} + 2\left(\frac{W}{S}\right)\frac{2k}{\rho V^2}$$
$$= \frac{T}{W} - \frac{3}{2}\rho V^2\left(\frac{S}{W}\right)C_{Dmin} + \left(\frac{W}{S}\right)\frac{2k}{\rho V^2}$$

As usual, maximum (or minimum) can be found where the derivative equals zero. Setting the result to zero and multiply through by V^2 leads to:

$$\frac{T}{W} - \frac{3}{2}\rho V^2\left(\frac{S}{W}\right)C_{Dmin} + \left(\frac{W}{S}\right)\frac{2k}{\rho V^2} = 0$$

$$\frac{3}{2}\rho V^4\left(\frac{S}{W}\right)C_{Dmin} - \frac{T}{W}V^2 - \left(\frac{W}{S}\right)\frac{2k}{\rho} = 0 \quad (i)$$

Then, divide through by the constant multiplied to V^4:

$$V^4 - \frac{2}{3}\frac{T}{\rho S C_{Dmin}}V^2 - \frac{4}{3}\left(\frac{W}{S}\right)^2\frac{k}{\rho^2 C_{Dmin}} = 0 \quad (ii)$$

Then, note that the last term resembles the expression for max L/D (see Equation (19-18)):

$$LD_{max} = \frac{1}{\sqrt{4\cdot C_{Dmin}\cdot k}} \Leftrightarrow 4\cdot C_{Dmin}\cdot k = \frac{1}{LD_{max}^2}$$

Therefore:

$$V^4 - \frac{2}{3}\frac{T}{\rho S C_{Dmin}}V^2 - \left(\frac{W}{S}\right)^2\frac{1}{3\rho^2 C_{Dmin}^2 LD_{max}^2} = 0 \quad (iii)$$

Let's rewrite Equation (iii), noting that $T/S = (T/W)(W/S)$:

$$V^4 - \frac{2}{3}\frac{(T/W)(W/S)}{\rho C_{Dmin}}V^2 - \left(\frac{W}{S}\right)^2\frac{1}{3\rho^2 C_{Dmin}^2 LD_{max}^2} = 0 \quad (iv)$$

For convenience, define the variables Q and x as follows:

$$Q = \frac{W/S}{3\rho C_{Dmin}} \quad \text{and} \quad x = V^2$$

Let's rewrite Equation (iv) using these definitions:

$$x^2 - 2\frac{T}{W}Qx - \frac{3Q^2}{LD_{max}^2} = 0 \quad (v)$$

This is a quadratic equation in terms of x (or V^2) whose solution is given by:

$$x = \frac{2(T/W)Q \pm \sqrt{4(T/W)^2 Q^2 + 12Q^2/LD_{max}^2}}{2} \quad (vi)$$

Factor $(T/W)Q$ out of the radical to get:

$$x = \frac{T}{W}Q \pm \frac{T}{W}Q\sqrt{1 + 3/\left(LD_{max}^2 (T/W)^2\right)} \quad (vii)$$

or:

$$x = \frac{T}{W}Q\left[1 \pm \sqrt{1 + \frac{3}{LD_{max}^2(T/W)^2}}\right] \quad (viii)$$

Only the positive sign in front of the radical makes physical sense. Writing this in terms of the original definitions of Q and x leads to:

$$V_Y = \sqrt{\frac{(T/W)(W/S)}{3\rho C_{Dmin}}\left[1 + \sqrt{1 + \frac{3}{LD_{max}^2(T/W)^2}}\right]} \quad (ix)$$

QED

EXAMPLE 18-6

Determine the airspeed for maximum rate-of-climb for the sample aircraft at S-L, maximum thrust and at a weight of 20,000 lb$_f$.

Solution

Step 1: Determine LD_{max}:

$$LD_{max} = \frac{1}{\sqrt{4(0.020)(0.05236)}} = 15.45 \Rightarrow LD_{max}^2 = 238.73$$

Step 2: Determine T/W and W/S:

$$\frac{T}{W} = \frac{7000}{20,000} = 0.35 \quad \text{and} \quad \frac{W}{S} = \frac{20,000}{311.6} = 64.2 \text{ lbf/ft}^2$$

Step 3: Best rate-of-climb speed is therefore:

$$V_Y = \sqrt{\frac{(T/W)(W/S)}{3\rho C_{Dmin}}\left[1 + \sqrt{1 + \frac{3}{LD_{max}^2(T/W)^2}}\right]}$$

$$= \sqrt{\frac{(0.35)(64.2)}{3(0.002378)(0.020)}\left[1 + \sqrt{1 + \frac{3}{(238.73)(0.35)^2}}\right]}$$

$$= 568.2 \text{ ft/s} (\sim 337 \text{ KTAS})$$

18.3.7 Best ROC for a Jet

It is clearly evident from Figure 18-5 that the *ROC* varies with airspeed. Its maximum value is called the best rate-of-climb. For a jet, the value of this *ROC* can be computed from the following expression. Note that the expression *assumes the simplified drag model*:

$$ROC_{max} = \sqrt{\frac{(W/S)Z}{3\rho C_{Dmin}}}\left(\frac{T}{W}\right)^{3/2} \times \left(1 - \frac{Z}{6} - \frac{3\cos^2\theta}{2(T/W)^2 LD_{max}^2 Z}\right) \quad (18\text{-}25)$$

Derivation of Equation (18-25)

The equation for the best *ROC* is obtained by substituting V_{ROCmax} from Equation (18-24) in Equation (18-18). In order to simplify the resulting expression, let's define the variable Z such that:

$$Z = 1 + \sqrt{1 + \frac{3}{LD_{max}^2(T/W)^2}} \quad (i)$$

Then, Equation (18-24) becomes:

$$V_Y = \sqrt{\frac{(T/W)(W/S)Z}{3\rho C_{Dmin}}} \quad (ii)$$

Then, substitute Equation (ii) into Equation (18-18):

$$V_Y \sin\theta = V_Y\left(\frac{T}{W} - q\left(\frac{S}{W}\right)C_{Dmin}\right) - k\cdot\left(\frac{W}{S}\right)\frac{\cos^2\theta}{q}$$

This results in:

$$RC_{max} = \sqrt{\frac{(T/W)(W/S)Z}{3\rho C_{Dmin}}}$$

$$\times \left(\frac{T}{W} - \frac{1}{2}\rho\left(\frac{(T/W)(W/S)Z}{3\rho C_{Dmin}}\right)\left(\frac{S}{W}\right)C_{Dmin}\right.$$

$$\left. -2k\cdot\left(\frac{W}{S}\right)\frac{\cos^2\theta}{\rho\frac{(T/W)(W/S)Z}{3\rho C_{Dmin}}}\right)$$

Simplifying yields:

$$RC_{max} = \sqrt{\frac{(T/W)(W/S)Z}{3\rho C_{Dmin}}} \times \left(\frac{T}{W} - \frac{Z}{6}\frac{T}{W} - \frac{6k\cdot C_{Dmin}\cos^2\theta}{(T/W)Z}\right) \quad (iii)$$

Again, resorting to the term for max L/D of Equation (19-18):

$$4 \cdot C_{Dmin} \cdot k = 1/LD_{max}^2$$

We use this to modify the last term of Equation (iii) (and of course noting that the following arithmetic scheme holds: $6 = 12/2 = 4 \cdot 3/2 = (3/2) \cdot 4$):

$$\frac{6k \cdot C_{Dmin} \cos^2 \theta}{(T/W)Z} = \frac{(T/W)}{(T/W)^2} \frac{(3/2)(4 \cdot k \cdot C_{Dmin}) \cos^2 \theta}{Z}$$

$$= \frac{3T/W}{2(T/W)^2} \frac{\cos^2 \theta}{LD_{max}^2 Z} \quad \text{(iv)}$$

Inserting this into Equation (iii) we can now rewrite:

$$RC_{max} = \sqrt{\frac{(T/W)(W/S)Z}{3\rho C_{Dmin}}}$$

$$\times \left(\frac{T}{W} - \frac{Z}{6} \frac{T}{W} - \frac{3T/W}{2(T/W)^2} \frac{\cos^2 \theta}{LD_{max}^2 Z} \right)$$

Rearranging:

$$RC_{max} = \sqrt{\frac{(W/S)Z}{3\rho C_{Dmin}}} \sqrt{\frac{T}{W}}$$

$$\times \left(\frac{T}{W} - \frac{Z}{6} \frac{T}{W} - \frac{3T/W}{2(T/W)^2} \frac{\cos^2 \theta}{LD_{max}^2 Z} \right)$$

And finally:

$$RC_{max} = \sqrt{\frac{(W/S)Z}{3\rho C_{Dmin}}} \left(\frac{T}{W} \right)^{3/2}$$

$$\times \left(1 - \frac{Z}{6} - \frac{3\cos^2 \theta}{2(T/W)^2 LD_{max}^2 Z} \right)$$

QED

18.3.8 Airspeed for θ_{max} for a Propeller-powered Airplane

The best angle-of-climb for a propeller powered airplane is found by solving for V_X using the following expression:

$$V_X^4 + \frac{\eta \cdot 550 \cdot BHP}{\rho S C_{Dmin}} V_X - \left(\frac{W}{S} \right)^2 \frac{4k}{\rho^2 C_{Dmin}} = 0 \quad (18\text{-}26)$$

A closed-form solution of this equation is not known and its solution requires an iterative numerical scheme. Note that in order to obtain the θ_{max} the equation is solved for V_X. Then, this can be used with Equations (18-18) through (18-20) to obtain θ_{max}.

It is important to note that the value of V_X is frequently *less* than the stalling speed — in particular for light aircraft. Theoretically, this means the airplane cannot achieve a maximum θ_{max}. However, it is important to keep in mind the lessons of Chapter 15, *Aircraft drag analysis*, regarding the simplified drag model on which Equation (18-26) is based. As is clearly demonstrated (for instance, see Section 15.2.2, *Quadratic drag modeling*), this drag model is invalid at low AOAs and, consequently, the airplane no longer complies with any formulation based on the model. In practice, real airplanes all have a maximum climb angle higher than their stalling speed. To obtain it analytically, a more sophisticated drag modeling must be employed, for instance using the method presented in Section 15.2.3, *Approximating the drag coefficient at high lift coefficients*.

Derivation of Equation (18-26)

The thrust of a propeller driven airplane is given by Equation (14-38):

$$T = \frac{\eta_p \cdot 550 \cdot BHP}{V}$$

Additionally, the climb angle is given by Equation (18-19):

$$\sin \theta = \frac{T}{W} - q \left(\frac{S}{W} \right) C_{Dmin} - k \cdot \left(\frac{W}{S} \right) \frac{\cos^2 \theta}{q}$$

Inserting the expression for thrust into the above equation, and writing it explicitly in terms of V, results in:

$$\sin \theta = \frac{\eta \cdot 550 \cdot BHP}{WV} - \frac{1}{2} \rho V^2 \left(\frac{S}{W} \right) C_{Dmin}$$

$$- \left(\frac{W}{S} \right) \frac{2k \cdot \cos^2 \theta}{\rho V^2}$$

Then, differentiate with respect to V:

$$\frac{d(\sin\theta)}{dV} = -\frac{\eta \cdot 550 \cdot BHP}{WV^2} - \rho V\left(\frac{S}{W}\right)C_{Dmin}$$
$$+ \left(\frac{W}{S}\right)\frac{4k \cdot \cos^2\theta}{\rho V^3}$$

Set to zero to get the maximum:

$$-\frac{\eta \cdot 550 \cdot BHP}{WV^2} - \rho V\left(\frac{S}{W}\right)C_{Dmin} + \left(\frac{W}{S}\right)\frac{4k \cdot \cos^2\theta}{\rho V^3} = 0$$

Multiply by V^3 for convenience and prepare as a polynomial:

$$-\frac{\eta \cdot 550 \cdot BHP \cdot V}{W} - \rho V^4\left(\frac{S}{W}\right)C_{Dmin} + \left(\frac{W}{S}\right)\frac{4k \cdot \cos^2\theta}{\rho} = 0$$

Simplify further:

$$V^4 + \frac{\eta \cdot 550 \cdot BHP}{\rho S C_{Dmin}}V - \left(\frac{W}{S}\right)^2 \frac{4k \cdot \cos^2\theta}{\rho^2 C_{Dmin}} = 0$$

The solution can be found by an iterative scheme and may include assuming $\cos\theta \sim 1$, but this yields the following expression, for which V is rewritten as V_X:

$$V_X^4 + \frac{\eta \cdot 550 \cdot BHP}{\rho S C_{Dmin}}V_X - \left(\frac{W}{S}\right)^2 \frac{4k}{\rho^2 C_{Dmin}} = 0$$

QED

18.3.9 Airspeed for Best ROC for a Propeller-powered Airplane

The airspeed for best *ROC* for a propeller powered aircraft, assuming the *simplified drag model* is calculated from the following expression:

$$V_Y = V_E = \sqrt{\frac{2}{\rho}\left(\frac{W}{S}\right)}\sqrt{\frac{k}{3 \cdot C_{Dmin}}} \qquad (18\text{-}27)$$

The units are in terms of ft/s or m/s, depending on input values.

The expression identifies the location of the maximum excess power in terms of airspeed. Figure 18-6 plots power available and power required versus true airspeed for a typical small, propeller-powered aircraft and assumes constant engine power with airspeed. The method presented here shows that the best *ROC* airspeed will occur where the difference between the two is the greatest, at around 45 KTAS.

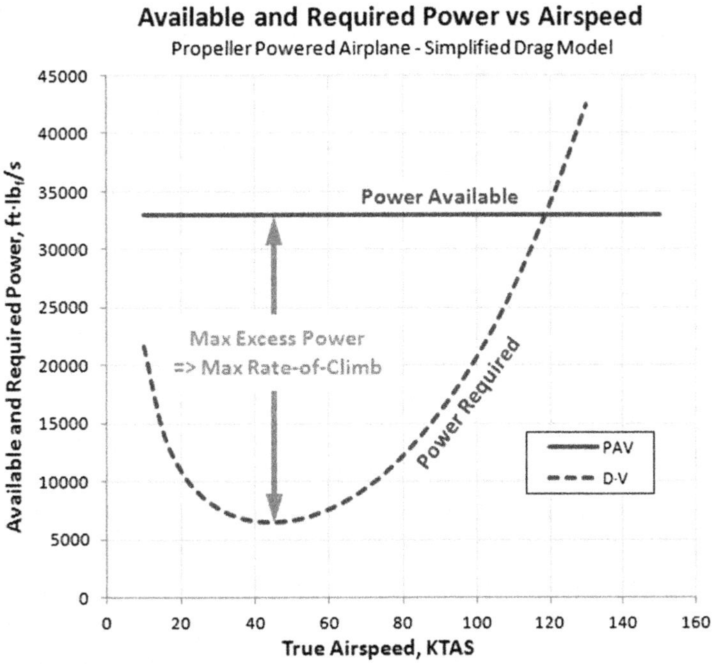

FIGURE 18-6 Power available and power required for a propeller-powered airplane.

As discussed in Section 18.3.6, *Airspeed for best ROC for a jet*, the above formulation can also be used to estimate the best rate-of-climb airspeed for a multiengine aircraft during an OEI condition. This airspeed is denoted by V_{YSE}. This can be done by reducing the contribution of the failed engine to the total power available and the minimum drag, C_{Dmin}, must be increased and LD_{max} must be reduced to account for the asymmetric attitude of the airplane.

Derivation of Equation (18-27)

Equation (18-15) shows that the maximum ROC occurs when (seen graphically in Figures 18-7 and 18-4):

$$ROC_{max} = \frac{\text{Maximum Excess Power}}{W} \qquad (18\text{-}28)$$

From the power studies (see Analyses 3 through 18 in Chapter 19, *Performance − cruise*) we know that the maximum excess power occurs at the airspeed where required power is minimum, but this required the ratio $C_L^{1.5}/C_D$ to be at its maximum.

Therefore, the maximum rate-of-climb occurs at the airspeed of minimum power required, but this is given by Equation (18-27).

QED

EXAMPLE 18-7

Since Equation (18-27) is based on the simplified drag model, it is of interest to evaluate its accuracy for a real airplane. Here, determine the airspeed for best rate-of-climb for the SR22 at a weight of 3400 lb$_f$ at S-L and 10,000 ft. Compare to the POH value (V_Y = 101 KCAS at S-L and 96 KCAS at 10,000 ft). Use the minimum drag coefficient extracted for the airplane in Example 15-18 ($C_{Dmin} = 0.02541$), $S = 144.9$ ft^2, $AR = 10$, and the Oswald efficiency as also calculated in the example, which amounts to 0.7566.

Solution

The density of air at S-L is 0.002378 slugs/ft^3 and at 10,000 ft it is 0.001756 slugs/ft^3. Also, the lift-induced drag constant $k = 1/(\pi \cdot AR \cdot e) = 0.04207$. We can calculate the V_Y by substituting the values given in the table into the Equation (18-27), first at S-L as follows:

$$V_Y = \sqrt{\frac{2}{\rho}\left(\frac{W}{S}\right)}\sqrt{\frac{k}{3 \cdot C_{Dmin}}}$$

$$= \sqrt{\frac{2}{0.002378}\left(\frac{3400}{144.9}\right)}\sqrt{\frac{0.04207}{3 \times 0.02541}} = 121.1 \text{ ft/s}$$

$$= 71.7 \text{ KCAS}$$

And then at 10,000 ft:

$$V_Y = \sqrt{\frac{2}{0.001756}\left(\frac{3400}{144.9}\right)}\sqrt{\frac{0.04207}{3 \times 0.02541}} = 140.9 \text{ ft/s}$$

$$= 83.5 \text{ KTAS}$$

This also amounts to 71.7 KCAS, which contrasts with 101 KCAS at S-L and 96 KCAS at 10,000 ft for the real aircraft. It can be seen that the simplified drag model yields a poor approximation to the real airplane and the reader must be aware of this shortcoming. This is in part due to the absence of the term C_{LminD} in the simplified drag model.

18.3.10 Best Rate-of-climb for a Propeller-powered Airplane

The maximum ROC in ft/s or m/s for a propeller-powered aircraft can be calculated from:

$$V_Y \sin\theta = \frac{\eta_p P}{W} - \sqrt{\frac{2}{\rho}\left(\frac{W}{S}\right)}\sqrt{\frac{k}{3 \cdot C_{Dmin}}}\frac{1.1547}{LD_{max}}$$

$$(18\text{-}29)$$

If the best rate-of-climb airspeed, V_Y, is known and the value is desired in terms of fpm, it can be calculated from:

$$ROC_{max} = 60\left(\frac{\eta_p P}{W} - V_Y\frac{1.1547}{LD_{max}}\right) \qquad (18\text{-}30)$$

Note that the power, P, must be in terms of ft·lb$_f$/s. For this reason, if power is given in BHP, it must be multiplied by the factor 550 to be converted to the proper units.

Derivation of Equations (18-29) and (18-30)

For a propeller-powered airplane the power available is given by: $P_A = TV = \eta_p P$

An expression for the best ROC can be obtained by inserting Equation (18-27) into Equation (18-18):

$$V_{E\,max} = \sqrt{\frac{2}{\rho}\left(\frac{W}{S}\right)\sqrt{\frac{k}{3 \cdot C_{D\,min}}}} \quad \Longrightarrow$$

$$V\sin\theta = V\left(\frac{T}{W} - q\left(\frac{S}{W}\right)C_{D\,min} - k\cdot\left(\frac{W}{S}\right)\frac{\cos^2\theta}{q}\right)$$

This is accomplished in the following manner.

Begin by expanding:

$$V\sin\theta = \frac{TV}{W} - V\left[\frac{1}{2}\rho V^2\left(\frac{S}{W}\right)C_{Dmin} + k\cdot\left(\frac{W}{S}\right)\frac{\cos^2\theta}{\frac{1}{2}\rho V^2}\right] \quad (i)$$

Modify the term TV in Equation (i) by introducing power available for a piston engine and propeller efficiency:

$$V\sin\theta = \frac{\eta_p P}{W} - V\left[\frac{1}{2}\rho V^2\left(\frac{S}{W}\right)C_{Dmin} + k\cdot\left(\frac{W}{S}\right)\frac{2\cos^2\theta}{\rho V^2}\right] \quad (ii)$$

Now, let's insert Equation (18-27) in a specific manner into Equation (ii) and let's use V_Y for the best ROC airspeed:

$$V_Y\sin\theta = \frac{\eta_p P}{W}$$

$$-V_Y\left[\frac{1}{2}\rho\left(\frac{2}{\rho}\left(\frac{W}{S}\right)\sqrt{\frac{k}{3\cdot C_{Dmin}}}\right)\left(\frac{S}{W}\right)C_{Dmin}\right.$$

$$\left. + k\cdot\left(\frac{W}{S}\right)\frac{2\cos^2\theta}{\rho\left(\frac{2}{\rho}\left(\frac{W}{S}\right)\sqrt{\frac{k}{3\cdot C_{Dmin}}}\right)}\right]$$

Further manipulation leads to:

$$V_Y\sin\theta = \frac{\eta_p P}{W} - V_Y\left[\sqrt{\frac{k\cdot C_{Dmin}}{3}} + \sqrt{3k\cdot C_{Dmin}}\cos^2\theta\right]$$

$$= \frac{\eta_p P}{W} - V_Y\left[\frac{1}{\sqrt{3}}\sqrt{k\cdot C_{Dmin}} + \sqrt{3}\sqrt{k\cdot C_{Dmin}}\cos^2\theta\right]$$

$$= \frac{\eta_p P}{W} - V_Y\sqrt{k\cdot C_{Dmin}}\left[\frac{1}{\sqrt{3}} + \sqrt{3}\cos^2\theta\right]$$

Assuming $\cos^2\theta \sim 1$ and using Equation (19-18) for LD_{max} we simplify further:

$$V_Y\sin\theta = \frac{\eta_p P}{W} - V_Y\sqrt{k\cdot C_{Dmin}}\left[\frac{1}{\sqrt{3}} + \sqrt{3}\cos^2\theta\right]$$

$$V_Y\sin\theta = \frac{\eta_p P}{W} - V_Y\frac{\sqrt{4\cdot k\cdot C_{Dmin}}}{2}\left[\frac{1}{\sqrt{3}} + \sqrt{3}\right]$$

$$= \frac{\eta_p P}{W} - V_Y\frac{1}{LD_{max}}\frac{\left[\frac{1}{\sqrt{3}} + \sqrt{3}\right]}{2} = \frac{\eta P}{W} - V_{ROCmax}\frac{1.1547}{LD_{max}}$$

Finally, replacing the explicit expression for V_{ROCmax} into this equation leads to:

$$V_Y\sin\theta = \frac{\eta_p P}{W} - V_Y\frac{1.1547}{LD_{max}}$$

$$= \frac{\eta P}{W} - \sqrt{\frac{2}{\rho}\left(\frac{W}{S}\right)\sqrt{\frac{k}{3\cdot C_{Dmin}}}}\frac{1.1547}{LD_{max}}$$

QED

EXAMPLE 18-8

Let's evaluate the accuracy of Equation (18-30) in the same way that we did in Example 18-7. Here, determine the best rate-of-climb for the SR22 at a weight of 3400 lb$_f$ at S-L and compare to the POH value (ROC_{max} = 1398 fpm at S-L). Use the same constants as used in Example 18-7, but calculate and compare for three propeller efficiencies, η_p = 0.6, 0.7, and 0.8. The maximum engine power at S-L is 310 BHP. For this example, use the value of V_Y determined in Example 18-7, even though it has been shown to be incorrect.

EXAMPLE 18-8 (cont'd)

Solution

The rate-of-climb will only be calculated using the first propeller efficiency, but the results for the other two will be shown. First, calculate the ROC at S-L assuming $\eta_p = 0.6$:

$$ROC_{max} = 60\left(\frac{\eta_p P}{W} - V_Y \frac{1.1547}{LD_{max}}\right)$$

$$= 60\left(\frac{0.6 \times 550 \times 310}{3400} - 121.1\frac{1.1547}{15.6}\right)$$

$$= 1267 \text{ fpm}$$

Results for the other values of the propeller efficiency over a range of hypothetical best rate-of-climb airspeeds are plotted in Figure 18-7. It indicates the propeller is a little over 70% efficient during climb, but also shows there is considerable discrepancy between the predicted and actual V_Y for the airplane. This is the risk of using the simplified drag model — it is convenience versus precision.

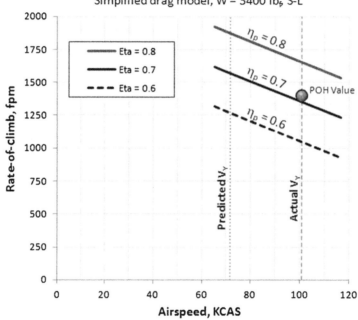

FIGURE 18-7 Best ROC (fpm) as a function of airspeed (KCAS) and propeller efficiency.

18.3.11 Time to Altitude

The time required to increase altitude, h, can be determined from the following expression. If the ROC is in terms of fpm, it will return time in minutes. If ROC is in terms of ft/s or m/s then the time will be in seconds.

$$t = \int_{h_0}^{h_1} \frac{dh}{ROC} \tag{18-31}$$

In this expression, h_1 is the target altitude and h_0 is the initial altitude (e.g. the airplane may begin a climb at 10,000 ft and level out at 30,000 ft). The minimum time to altitude is achieved if the pilot maintains the best ROC airspeed (V_Y) through the entire climb maneuver.

Rapid Approximation

For mathematical simplicity, it may be convenient to assume a constant value of ROC and take it out of the integral sign. The value of the ROC should be a representative value between the initial and final altitudes, denoted by the symbol ROC_a and called the *representative ROC*. In the absence of a better value, the average of the ROC between the initial and final altitudes can be used, although the true value should be biased toward the higher altitude, as the aircraft will spend more time completing the last half of the climb than the first half. Since this approach treats the representative ROC as a constant, it can be taken out of the integral of Equation (18-31), yielding the following expression:

$$t = \int_{h_0}^{h_1} \frac{dh}{ROC} = \frac{1}{ROC_a} \int_{h_0}^{h_1} dh = \frac{(h_1 - h_0)}{ROC_a} \quad (18\text{-}32)$$

If the ROC is known as a function of altitude and given by $ROC(h) = A \cdot h + B$, then, using Equation (18-34) below, the value of ROC_a can be found from the expression:

$$ROC_a = \frac{A(h_1 - h_0)}{\ln(Ah_1 + B) - \ln(Ah_0 + B)} \quad (18\text{-}33)$$

If the initial and final altitudes are close, the ROC_a can be approximated as the average of the ROC at the initial and final altitudes.

Linear Approximation

If a particular airspeed, such as V_Y, is maintained through the climb, the ROC will decrease in a fashion that is close to being linear. In this case, the ROC can be approximated with an equation of a line: $ROC(h) = A \cdot h + B$. In this case, the time to altitude is given by:

$$t = \int_{h_0}^{h_1} \frac{dh}{Ah + B} = \frac{\ln(Ah_1 + B) - \ln(Ah_0 + B)}{A}$$

(18-34)

Derivation of Equation (18-31)

ROC is the rate of change of altitude dh/dt, that is:

$$\frac{dh}{dt} = ROC \quad \Leftrightarrow \quad dt = \frac{dh}{ROC} \quad \Rightarrow$$

$$t = \int \frac{dh}{ROC}$$

QED

Derivation of Equation (18-33)

If ROC as a function of altitude can be approximated using the linear expression $ROC(h) = A \cdot h + B$, then the representative ROC can be determined using:

$$t = \frac{\ln(Ah_1 + B) - \ln(Ah_0 + B)}{A} = \frac{h_1 - h_0}{ROC_a} \quad \Rightarrow \quad ROC_a$$

$$= \frac{A(h_1 - h_0)}{\ln(Ah_1 + B) - \ln(Ah_0 + B)}$$

QED

Derivation of Equation (18-34)

If ROC as a function of altitude can be approximated using the linear expression $ROC(h) = A \cdot h + B$, then the time to altitude can be found from:

$$t = \int_{h_0}^{h_1} \frac{dh}{ROC} = \int_{h_0}^{h_1} \frac{dh}{Ah + B} = \frac{\ln(Ah + B)}{A}\bigg]_{h_0}^{h_1}$$

$$= \frac{\ln(Ah_1 + B) - \ln(Ah_0 + B)}{A}$$

QED

EXAMPLE 18-9

Consider the ROC versus altitude graph for the Cirrus SR22 is shown in Figure 18-8. Assume the aircraft weighs 2900 lb_f and is flying at 4000 ft when a climb at V_Y is initiated. How long will it take to get to 20,000 ft? Determine using the three following approaches: (1) ROC_a is average of the ROC at the initial and final altitude. (2) Calculate ROC_a using Equation (18-33). (3) Use Equation (18-34) directly.

Solution

(1) Calculate ROC at the initial and final altitudes using the trendline for 2900 lb_f shown in Figure 18-8:

At 4000 ft:

$$ROC = 1731 - 0.0674 \times 4000 = 1461 \text{ fpm}$$

EXAMPLE 18-9 (cont'd)

At 20,000 ft:

$$ROC = 1731 - 0.0674 \times 20{,}000 = 383 \text{ fpm}$$

Representative ROC:

$$ROC_a = \frac{1461 + 383}{2} = 922 \text{ fpm}$$

From Equation (18-32):

$$t = \frac{(h_1 - h_0)}{ROC_a} = \frac{(20{,}000 - 4000)}{922} = 17.4 \text{ min}$$

(2) Calculate ROC_a using Equation (18-33) where $A = -0.0674$ and $B = 1731$:

$$ROC_a = \frac{A(h_1 - h_0)}{\ln(Ah_1 + B) - \ln(Ah_0 + B)}$$

$$= \frac{-0.0674 \cdot (20{,}000 - 4000)}{\ln(-0.0674 \cdot 20000 + 1731) - \ln(-0.0674 \cdot 4000 + 1731)}$$

$$= \frac{-0.0674 \cdot (20{,}000 - 4000)}{\ln(383) - \ln(1461)} = 805.3 \text{ fpm}$$

From Equation (18-32):

$$t = \frac{(h_1 - h_0)}{ROC_a} = \frac{(20{,}000 - 4000)}{805.3} = 19.9 \text{ min}$$

(3) Use Equation (18-34) directly:

$$t = \frac{\ln(Ah_1 + B) - \ln(Ah_0 + B)}{A} = \frac{\ln(383) - \ln(1461)}{-0.0674}$$

$$= 19.9 \text{ min}$$

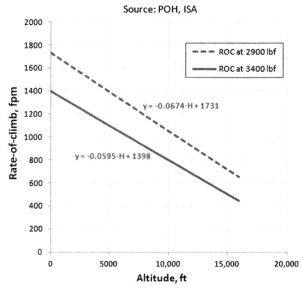

FIGURE 18-8 Rate-of-climb (fpm) as a function of altitude (ft) for the Cirrus SR22 single-engine aircraft for two different weights — gross weight of 3400 lb$_f$, and intermediary weight of 2900 lb$_f$. *(Source: SR22 POH, courtesy of Cirrus Aircraft)*

18.3.12 Absolute/Service Ceiling Altitude

Two frequently referenced performance parameters are the absolute and service ceilings. The *absolute ceiling* is the maximum altitude at which the airplane can maintain level flight. The *service ceiling* is the altitude at which the aircraft is capable of some 100 fpm rate-of-climb. Generally, the designer must keep two things in mind regarding these altitudes and, thus, should treat them with caution.

EXAMPLE 18-10

Determine the absolute and service ceilings for the Cirrus SR22 at 2900 and 3400 lb$_f$ weights using the data from its POH, shown in Figure 18-9.

Solution

If one has the pilot's operating handbook (as we do for this example), it is relatively easy to determine a trendline. If working with a new design, a POH will not be ready, so the max ROC must be calculated for a number of altitudes e.g. S-L, 7000, and 14,000 ft or similar.

With respect to the SR22, using information from its POH we find that the resulting linear fit is given by:

At 2900 lb$_f$:

$$H = 25{,}666 - 14.828 \times ROC_{max}$$

At 3400 lb$_f$:

$$H = 23496 - 16.807 \times ROC_{max}$$

From which we see that at the lighter weight (2900 lb$_f$), the absolute ceiling is 25,666 ft and 23,496 ft at maximum gross weight (3400 lb$_f$). Of course, the airplane would always weigh less than its maximum gross weight, as it would have to burn a considerable amount of fuel to get there in the first place. By the same token, it is easy to evaluate the service ceiling based on a 100 fpm climb rate:

At 2900 lb$_f$:

$$H = 25{,}666 - 14.828 \times 100 = 24{,}183 \text{ ft}$$

At 3400 lb$_f$:

$$H = 23{,}496 - 16.807 \times 100 = 21{,}815 \text{ ft}$$

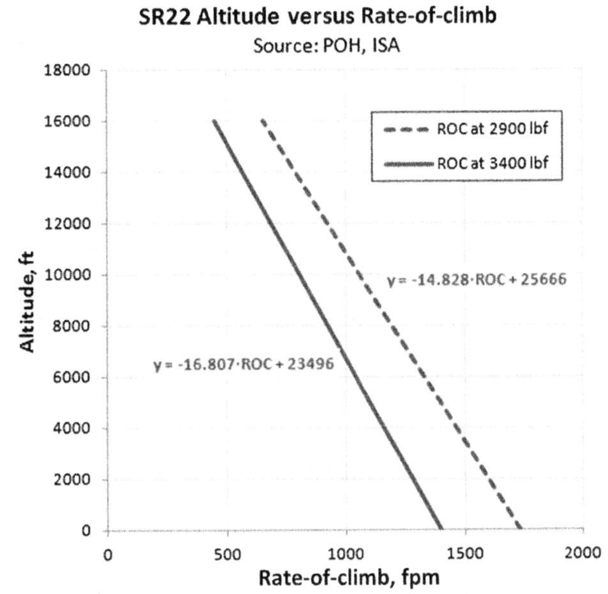

FIGURE 18-9 Determination of service and absolute ceilings.

First, each ceiling is highly dependent on the weight of the aircraft as well as atmospheric conditions and can deviate thousands of feet (or meters) from the calculated values. Second, modern-day regulations are often the arbitrator of altitudes rather than the capability of the aircraft. For instance, CFR 14 Part 23 stipulates requirements for an airplane to fly higher than 25,000 ft or 28,000 ft. These requirements have everything to do with the equipment the aircraft features and not its ability to reach those altitudes. Some business jets have theoretical service ceilings in the neighborhood of 35,000 ft or even higher, but are certified to fly only as high as 28,000 ft.

The theoretical absolute and service ceilings can be computed by the following method:

(1) Compute ROC_{max} at a number of altitudes.
(2) Create a trendline in the form of a line or a polynomial.

(3) Solve the trendline for $ROC = 100$ fpm. This is the *service ceiling*.
(4) Solve the trendline for $ROC = 0$ fpm. This is the *absolute ceiling*.

18.3.13 Numerical Analysis of the Climb Maneuver — Sensitivity Studies

The purpose of this section is to introduce a powerful method to calculate the *ROC* of a generic aircraft using numerical analysis. This will be accomplished through the preparation of an analysis spreadsheet, which here is prepared for a propeller aircraft. The spreadsheet offers far greater analysis power to the aircraft designer because it can handle all the non-linearity the preceding analysis methods cannot. It will allow the designer to estimate the *ROC* of an airplane whose $C_{LminD} > 0$, and even aircraft whose drag polar features a drag bucket or lift-curve that becomes non-linear as a result of an early flow separation. A screen capture of the spreadsheet is shown in Figure 18-10. A description is given below:

The general input values are self-explanatory in light of the preceding discussion and will not be elaborated on. The columns in the main table labeled from 1 through 13, on the other hand, require some explanation.

Column 1 contains a range of calibrated airspeeds (50 KCAS increasing by 5 KCAS to 150 KCAS). This is used to calculate the true airspeed in column 2 using Equation (16-33) as follows (using 100 KCAS as an example):

$$V_{KTAS} = \frac{V_{KCAS}}{\sqrt{\sigma}} = \frac{100}{\sqrt{0.8617}} = 107.7 \text{ KTAS}$$

This is converted to ft/s in column 3 by multiplying by 1.688.

Column 4 contains dynamic pressure, calculated as follows:

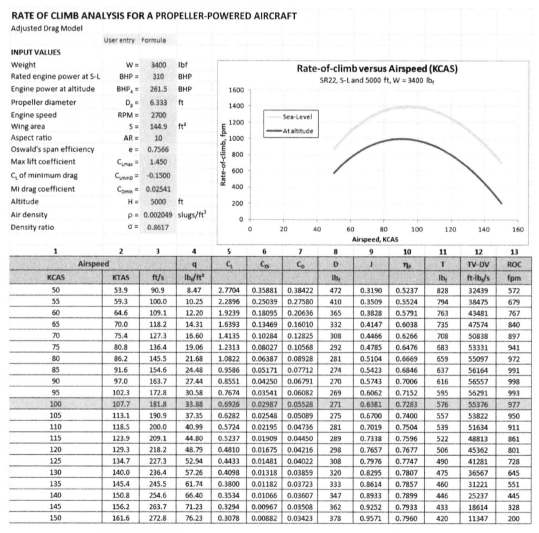

FIGURE 18-10 Spreadsheet designed to estimate climb performance.

$$q = \frac{1}{2}\rho V^2 = \frac{1}{2}(0.002049)(181.8)^2 = 33.88 \text{ slugs/ft}^3$$

Column 5 contains the lift coefficient, calculated as follows:

$$C_L = \frac{W}{qS} = \frac{3400}{(33.88)(144.9)} = 0.6926$$

Column 6 contains the lift-induced drag coefficient, calculated as follows:

$$C_{Di} = \frac{C_L^2}{\pi \cdot AR \cdot e} = \frac{0.6926^2}{\pi \cdot (10)(0.7566)} = 0.02987$$

Column 7 contains the total drag coefficient, calculated as follows:

$$C_D = C_{Dmin} + C_{Di} = 0.02541 + 0.02987 = 0.05528$$

Column 8 contains the total drag, calculated as follows:

$$D = qSC_D = (33.88)(144.9)(0.05528) = 271 \text{ lb}_f$$

Note that, fundamentally, the method 'doesn't care' how the drag is determined. For instance, even though the adjusted drag model is used here, columns 6 through 8 could just as easily contain a non-quadratic drag coefficient. For instance, a lookup table containing a drag model with a drag bucket could simply replace CD (with columns 6 and 7 simply omitted). This gives the approach substantial power, because it is independent of the nature of the drag coefficient.

Column 9 contains the advance ratio, calculated using Equation (14-23) as follows:

$$J = \frac{V}{nD_p} = \frac{(181.8)}{(2700/60)(6.333)} = 0.6381$$

The advance ratio is calculated because it is used in the expression for η_p, which was prepared using information from the propeller manufacturer. Note that the polynomial shown in the next step is prepared for this example and does not pertain to the actual aircraft.

Column 10 contains the propeller efficiency, η_p, calculated using the following hypothetical polynomial, just designed to generate reasonable token values:

$$\eta_p = 0.00000009703704 \cdot J^3 - 0.00005646032 \cdot J^2 + 0.01086138 \cdot J + 0.1096825$$

When using the value of J calculated in column 9, this expression yields 0.7283.

Column 11 contains the propeller thrust, calculated using Equation (14-38), where the engine power at S-L (310 BHP) has been corrected to the altitude (here 5000 ft) using Equation (7-16):

$$T = \eta_p \times 550 \times BHP_a/V$$
$$= (0.7283 \times 550 \times 261.5)/(181.8) = 576 \text{ lb}_f$$

Column 12 contains the excess power, calculated using Equation (18-14):

$$P_{EX} = TV - DV = (576)(181.8) - (271)(181.8)$$
$$= 55376 \text{ ft} \cdot \text{lb}_f/s$$

Column 13 (finally) contains the rate-of-climb, calculated using Equation (18-17):

$$ROC = 60\left(\frac{TV - DV}{W}\right) = 60\left(\frac{55376}{3400}\right) = 977 \text{ ft/min}$$

By performing the same calculations for the other row, it is trivial to extract the maximum ROC using Microsoft Excel's MAX() function, here found to equal 998 fpm at 90 KCAS (at 5000 ft). Note that more accurate calculations should take into account the weight reduction of the aircraft with altitude. For instance, the airplane will burn several gallons of fuel climbing to 10,000 ft and this will improve the ROC. Also note that the graph accompanying the spreadsheet contains a reference curve showing the climb performance at S-L. This is primarily done for convenience to help the designer realize performance degradation with altitude. It is left as an exercise for the reader to figure out how to do this (hint — it is simple). Once the spreadsheet is completed, it can be used to perform various sensitivity studies, of which three are shown below.

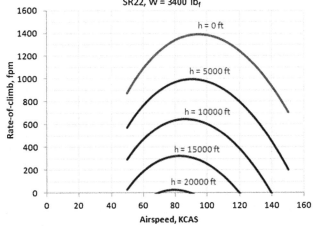

FIGURE 18-11 Altitude sensitivity plot.

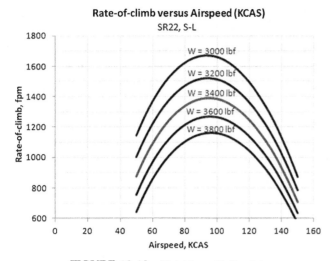

FIGURE 18-12 Weight sensitivity plot.

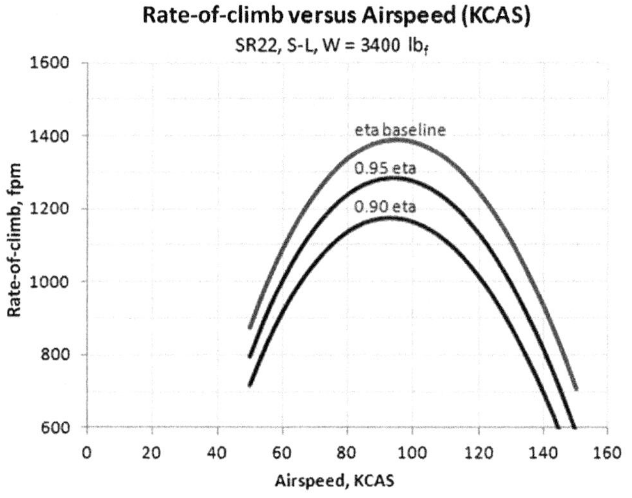

FIGURE 18-13 Propeller efficiency sensitivity plot.

Altitude Sensitivity

Altitude sensitivity reveals how the design will be affected when operated at high altitudes or, worse yet, at high altitude on a hot day (high-density altitude). This is important when considering departure from high-altitude airports in mountainous regions. Such departures can pose serious challenges for pilots, particularly if the aircraft is fully loaded. For instance, the graph in Figure 18-11 shows that at 10,000 ft, the airplane is capable of climbing at about 650 fpm, less than ½ of its S-L capability.

Weight Sensitivity

Weight sensitivity reveals how the design will be affected by deviations from the target design weight. This is important when demonstrating the importance of meeting target design weights. For instance, assume that the target gross weight of the airplane shown in Figure 18-12 is 3400 lb$_f$. If the target is not met and the manufacturer is forced to increase it to, say, 3600 lb$_f$, then the airplane's best ROC is likely to drop from about 1400 fpm to 1260 fpm. This could have a significant impact on the competitiveness of the design project.

Propeller Efficiency Sensitivity

Propeller efficiency sensitivity reveals how the design will be affected if a substandard propeller is purchased, if an "aerodynamically" damaged propeller is used, or if the propeller does not meet the manufacturer's claims of performance. For instance, if the selected propeller for the airplane shown in Figure 18-13 delivers merely 90% of the claimed propeller efficiency, then the best ROC is likely to drop from about 1400 fpm to 1170 fpm.

It can also be helpful to extend the sensitivity evaluation to one that includes propeller and weight sensitivities.

18.4 AIRCRAFT DATABASE — RATE-OF-CLIMB OF SELECTED AIRCRAFT

Table 18-1 shows the rate-of-climb, best angle-of-climb, and best rate-of-climb airspeeds for selected classes of aircraft. This data is very helpful when evaluating the accuracy of one's own calculations. Note that for heavier aircraft, it is practically impossible to specify a single V_X or V_Y as these vary greatly with weight and are usually determined for the pilot on a trip-to-trip basis.

TABLE 18-1 T-O Performance of Selected Aircraft

Name	Gross Weight lb$_f$	Wing Area ft^2	Rate-of-climb fpm	V_X (Best Angle) KCAS	V_Y (Best Rate) KCAS	Ref.
SINGLE-ENGINE						
Aerotec A-122A	1825	145	836	-	-	8
Bede BD-5A Micro	660	47	900	-	-	8
Cessna 162 Skycatcher	1320	160	880	57	62	9
Cessna 172N Skyhawk	2450	174	770	59	73	9
Cessna 182 Skylane	2950	174	1010	54	78	9

TABLE 18-1 T-O Performance of Selected Aircraft—cont'd

Name	Gross Weight lb$_f$	Wing Area ft^2	Rate-of-climb fpm	V_X (Best Angle) KCAS	V_Y (Best Rate) KCAS	Ref.
Cessna 208B Grand Caravan	8750	279	925	87	104	9
Cirrus SR20	3050	145	828	89	96	8, 9
Cirrus SR22	3400	145	1400	91	101	8, 9
Embraer EMB-201	3417	194	1050	-	-	8
Let Z-37A Cmelak (Bumble-Bee)	3855	256	925	-	-	8
Neiva N621A	3306	185	1770	-	-	8
Piper PA-46-350 Malibu	4340	175	1220	81	110	8, 9
Taylor J.T.1 Monoplane	610	76	1000	-	-	8
Transavia PL-12 Airtruk	3800	256	1500	-	-	8
TWIN-ENGINE, PROPELLER						
Beechcraft Duke B60	6775	213	1601	99	120	Unconf
Beechcraft Baron 55	5100	199	1577	91	100	8, 9
Beechcraft Queen Air B80	8800	294	1275	-	-	8
Cessna 421 Golden Eagle	4501	215	1940	-	-	Unconf
Cesssna 337 Skymaster	4630	201	1100	79	96	Unconf
Partenavia P.68	1960	200	1500	76	90	Unconf
Pilatus Britten-Norman BN-2B Islander	2993	325	970	-	-	Unconf
Piper Seminole	3800	184	1200	82	88	9
Rockwell Commander 112A	2650	152	900	72	100	9
Vulcanair P.68 Observer	4594	200	-	-	-	Unconf
COMMUTER TURBOPROPS						
Beechcraft King Air A100	10600	298	2200	-	-	8
Beechcraft King Air C90	9650	294	2000	101	112	8, 9
Casa C-212 Aviocar	13889	431	1800	-	-	8
Frakes Turbo-Mallard	14000	444	1350	-	-	8
Let L-140 Turbolet (L-140)	11905	354	1615	-	-	8
Lockheed Model 85 Orion P-3	135000	1300	1950	-	-	Unconf
Nomad N22	8000	324	1410	-	-	8
Piper PA-31P Pressurized Navajo	7800	229	1740	-	-	8
Rockwell Commander 690A	10250	266	2849	-	-	8
Shorts SD3-30	22000	453	1210	-	-	8

(Continued)

TABLE 18-1 T-O Performance of Selected Aircraft—cont'd

Name	Gross Weight lb$_f$	Wing Area ft^2	Rate-of-climb fpm	V_X (Best Angle) KCAS	V_Y (Best Rate) KCAS	Ref.
BUSINESS JETS						
Beechjet 400a	16100	241	3770	-	-	Unconf
Cessna Citation CJ1	10700	240	3200	-	-	Unconf
Cessna Citation Mustang	8645	210	3010	-	-	Unconf
Dasault Falcon 900	45500	527	3700	-	-	Unconf
Dassault-Breguet Mystère-Falco 900	20640	528	2000	-	-	Unconf
Embraer Phenom 100	10472	202		-	-	Unconf
Falcon 2000X	41000	527	3730	-	-	Unconf
Gates Learjet 24D	13500	232	6800	-	-	8
Gulfstream Aerospace IV	71700	950	4219	-	-	Unconf
Safire S-26	5130	143	2900	-	-	8
COMMERCIAL JETLINERS						
A300-B4	360000	2800	-	-	-	Unconf
A320-200	170000	1320	2400	-	-	Unconf
A330	520000	3892	4000	-	-	9
A340-200	610000	3892	4000	-	-	9
B737-400	150000	1135	3760	-	-	8
B757-200	255000	1994	3000	-	-	Unconf
B777-200A	247200	4605	3800	-	-	Unconf
Boeing 727-200	172000	1700	1800	-	-	9
Fokker 100	95000	1006	-	-	-	Unconf
Lockheed L-1011-1 TriStar	430000	2800	2800	-	-	8

Unconf = unconfirmed. A reliable source has not been located for the specified value. Treat values with caution.

VARIABLES

Symbol	Description	Units (UK and SI)
AOA	Angle-of-attack	Degrees or radians
AOC	Angle-of-climb	Degrees or radians
AR	Reference aspect ratio	
b	Reference (typically wing) span	ft or m
BHP	Brake horsepower	BHP or hp
C_D	Drag coefficient	
C_{Di}	Induced drag coefficient	
C_{Dmin}	Minimum drag coefficient	
C_L	Lift coefficient	
C_{L0}	Basic lift coefficient	
C_{LminD}	Lift coefficient of minimum drag	
$C_{L\alpha}$	Lift curve slope	Per radian or per degree
D	Drag	lb$_f$ or N
D_P	Propeller diameter	ft or m
e	Oswald span efficiency factor	

Symbol	Description	Units (UK and SI)
g	Acceleration due to gravity	ft/s² or m/s²
H, h	Altitude	ft or m
J	Advance ratio	
k	Lift-induced drag constant	
L	Lift	lb$_f$ or N
LD_{max}	Maximum lift-to-drag ratio	
P	Engine power	ft·lb$_f$/s or W = J/s
P_{AV}	Power available	ft·lb$_f$/s or W
P_{ENG}	Engine power	ft·lb$_f$/s or W
P_{EX}	Excess power	ft·lb$_f$/s or W
P_{REQ}	Power required	ft·lb$_f$/s or W
q	Dynamic pressure	lb$_f$/ft² or lb$_f$/in² or N/m²
RC_{max}	Maximum rate-of-climb	ft/s or fpm, m/s
ROC	Rate-of-climb	ft/min or m/s (typical)
ROC_{max}	Best rate-of-climb	ft/min or m/s (typical)
ROC_X	Rate-of-climb at best angle-of-climb speed	ft/min or m/s (typical)
S	Reference wing area	ft² or m²
T	Engine thrust (context dependent)	lb$_f$ or N
t	Time	seconds
T_{max}	Maximum engine thrust	lb$_f$ or N
T_{REQ}	Thrust required	lb$_f$ or N
V	Airspeed	ft/s or m/s
V_{Emax}	Airspeed for max endurance for a propeller airplane	ft/s or knots, m/s or kmh
V_H	Maximum level (horizontal) airspeed	ft/s or m/s
V_{ROCmax}	Airspeed for best rate-of-climb	ft/s or m/s
V_V	Vertical airspeed	ft/s or m/s
V_X	Best angle-of-climb airspeed (context dependent)	ft/s or fpm, m/s
V_X	Airspeed along x-axis (see Figure 18-2) (context dependent)	ft/s or m/s
V_Y	Best rate-of-climb airspeed	ft/s or fpm, m/s
V_{YSE}	Airspeed for best ROC at OEI condition	ft/s or m/s
V_Z	Airspeed along z-axis (see Figure 18-2)	ft/s or m/s
W	Weight	lb$_f$ or N
Z	Simplification relating LD_{max} and T/W	
α	Angle-of-attack	Degrees or radians
ε	Thrust angle	Degrees or radians
η_p	Propeller efficiency	
θ	Aircraft climb angle (relative to horizon)	Degrees or radians
θ_{max}	Maximum climb angle	Degrees or radians
ρ	Air density	slugs/ft³ or kg/m³
ρ_{SL}	Air density at sea level	slugs/ft³ or kg/m³
σ	Density ratio	

References

[1] Perkins CD, Hage RE. Airplane Performance, Stability, and Control. John Wiley & Sons; 1949.
[2] Torenbeek E. Synthesis of Subsonic Aircraft Design. 3rd ed. Delft University Press; 1986.
[3] Nicolai L. Fundamentals of Aircraft Design. 2nd ed. 1984.
[4] Roskam J, Lan Chuan-Tau Edward. Airplane Aerodynamics and Performance. DARcorporation; 1997.
[5] Hale FJ. Aircraft Performance, Selection, and Design. John Wiley & Sons; 1984. pp. 137−138.
[6] Anderson Jr JD. Aircraft Performance & Design. 1st ed. McGraw-Hill; 1998.
[7] Austyn-Mair W, Birdsall DL. Aircraft Performance. Cambridge University Press; 1992. pp. 47−49.
[8] Jane's All the World's Aircraft. Various editors, Jane's Yearbooks, various years.
[9] Type Pilot's Operating Handbook or Airman's Information Manual.

CHAPTER 19

Performance — Cruise

OUTLINE

19.1 Introduction   848
    19.1.1 The Content of this Chapter   849
    19.1.2 General Free-body Diagram for Steady Level Flight   849
    19.1.3 Planar Equations of Motion (Assumes No Rotation About y-axis)   849
    19.1.4 Important Airspeeds for Propeller Aircraft   850
    19.1.5 Important Airspeeds for Subsonic Jet Aircraft   851

19.2 General Cruise Analysis Methods for Steady Flight   851
    19.2.1 Plotting the Drag Polar   852
    19.2.2 Drag Breakdown   852
    19.2.3 Required Versus Available Thrust   855
        Region of Speed Stability   856
        Region of Speed Instability   856
        Introducing the Effect of Thrust and Power   857
    19.2.4 Airspeed in Terms of Thrust   857
        Derivation of Equation (19-3)   858
        Derivation of Equation (19-4)   858
    19.2.5 Minimum Airspeed, V_{min} 860
        Derivation of Equation (19-5)   860
    19.2.6 Stalling Speed, V_S 860
        Level Stalling Speed with a Load Factor n   861
        Stalling Speed During Banking   861
        Level Stalling Speed with Thrust, Flap, and CG Effects   861
        Derivation of Equations (19-7), (19-8), and (19-9)   862
        Derivation of Equation (19-10)   862
    19.2.7 Airspeed of Minimum Power Required, V_{PRmin} 864
        Requirement for Max Endurance for a Propeller-powered Airplane   864
        Derivation of Equation (19-11) and (19-12)   864
        Derivation of Equation (19-13)   865
        Maximum Endurance Airspeed for a Propeller-powered Aircraft, V_{Emax} 866
        Best Rate-of-climb Airspeed for a Propeller-powered Aircraft, V_Y 867
        Derivation of Equation (19-14)   867
        Comparison to the Best Glide Speed   867
    19.2.8 Airspeed of Minimum Thrust Required, V_{TRmin}, or Best Glide Speed, V_{BG}, V_{LDmax} 867
        Derivation of Equation (19-15)   867
        Derivation of Equation (19-16)   868
        Maximum L/D Ratio   869
        Derivation of Equation (19-18)   869
        Derivation of Equation (19-19)   870
        Airspeed for Maximum L/D Ratio   873
        Derivation of Equation (19-20)   873
        Derivation of Equation (19-21)   873
        Various Lift-to-drag Ratios   874
    19.2.9 Best Range Airspeed for a Jet, V_{Rmax} 875
        Requirement for Maximum Range for a Jet-powered Airplane   875
        Derivation of Equation (19-22)   876
        Best Range Airspeed for a Jet   876
        Derivation of Equation (19-23)   877
        Comparison to the Best Glide Speed   877
        Carson's Airspeed   877
        Derivation of Equation (19-24)   877
    19.2.10 Maximum Level Airspeed, V_{max} 877
        Special Case: Propeller Aircraft   878
        Derivation of Equations (19-25) and (19-26)   878
        Derivation of Equation (19-27)   878
    19.2.11 Flight Envelope   880
        Coffin Corner   881
    19.2.12 Power Required   882
        Derivation of Equation (19-30)   883
    19.2.13 Power Available for a Piston-powered Aircraft   883
    19.2.14 Computer code: Determining Maximum Level Airspeed, V_{max}, for a Propeller Aircraft   883

19.2.15 *Computer code: Determining Maximum Level Airspeed, V_{max}, for a Jet* ... 884

19.3 General Analysis Methods for Accelerated Flight ... 885
   19.3.1 *Analysis of a General Level Constant-velocity Turn* ... 885
      Derivation of Equation (19-35) ... 888
      Derivation of Equation (19-36) ... 888
      Derivation of Equation (19-37) ... 888
      Derivation of Equation (19-38) ... 888
      Derivation of Equation (19-39) ... 888
      Derivation of Equation (19-40) ... 888
      Derivation of Equation (19-41) ... 888
   19.3.2 *Extremes of Constant-velocity Turns* ... 889
      Maximum Sustainable Load Factor, n_{max} ... 889
      Maximum Sustainable Turn Rate, $\dot{\psi}_{max}$ ... 889
      Minimum Sustainable Turning Radius ... 889
      Maximum Bank Angle ... 889
      Airspeed for Maximum Bank Angle ... 890
      Derivation of Equation (19-47) ... 890
   19.3.3 *Energy State* ... 891
      Energy Height ... 891
      Specific Energy and Energy Height ... 891
      Specific Excess Power ... 891
      Derivation of Equation (19-52) ... 892
      Constructing a Specific Excess Power Contour Plot ... 893

Variables ... 893

References ... 894

19.1 INTRODUCTION

The cruise maneuver is what the typical aircraft is designed to perform. That should highlight its importance. While fundamentally a simple maneuver, the goal of the designer is to ensure the airplane accomplishes this efficiently. In this sense, the term "efficiently" means the highest possible airspeed for a given fuel consumption. Figure 19-1 shows an organizational map displaying the cruise among other members of the performance theory.

Cruise can be defined as a straight and level flight at constant airspeed. *Straight* means the absence of a roll, which would result in a heading change, and *level* means no change in altitude. In the interests of accuracy, it would be more appropriate to say mostly straight and level flight at a mostly constant airspeed, as there exist cruise methods that require a slow change in both altitude and airspeed, and a change in heading is often required to fly to a destination. Such cruise methods are presented in Chapter 20, *Performance — range analysis* (although heading changes are not accounted for). However, these changes are so slow and gradual that the flight itself can be treated as if these were constant. The cruise maneuver in this chapter entails all variations of this type of flying, ranging from a very slow airspeed to the highest airspeed the airplane can achieve in level flight. As usual, the simplified drag model is used to develop expressions that highlight the various characteristics of cruise, such as best range or endurance, to name a few.

In general, the methods presented in here are the "industry standard" and mirror those presented by a variety of authors, e.g. Perkins and Hage [1], Torenbeek [2], Nicolai [3], Roskam [4], Hale [5], Anderson [6] and many, many others.

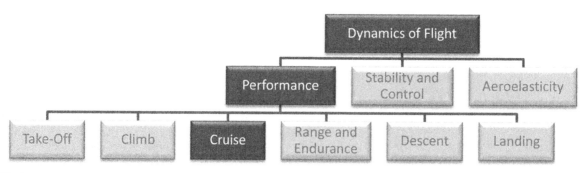

FIGURE 19-1 An organizational map placing performance theory among the disciplines of dynamics of flight, and highlighting the focus of this chapter: cruise performance analysis.

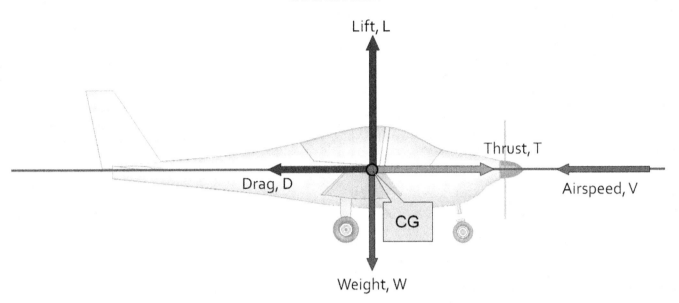

FIGURE 19-2 A two-dimensional free body of the airplane in level flight.

19.1.1 The Content of this Chapter

- **Section 19.2** presents classic analysis methods to determine a number of important steady level cruise performance characteristics. Among those are the most important types of cruising speeds encountered in the operation of aircraft.
- **Section 19.3** presents classic analysis methods to determine a number of important performance parameters for accelerated flight.

19.1.2 General Free-body Diagram for Steady Level Flight

A general free-body diagram for steady level flight is presented in Figure 19-2. It is based on the dynamic diagram of Section 18.2.1, *General two-dimensional free-body diagram for an aircraft*, and Equations (18-1) and (18-2). However, it has been modified to represent level flight, yielding a form familiar to many pilots, in which $L = W$ and $T = D$. However, as shown in Chapters 18 and 21, this is the only instance during the flight of the airplane where this simplification is applicable. The free body assumes all forces are applied at the center of gravity (CG) and all moments are balanced. It assumes a steady motion, that the climb angle, θ, is $0°$, the angle-of-attack, α, is small, and the thrust angle, ε, is $0°$.

19.1.3 Planar Equations of Motion (Assumes No Rotation About y-axis)

Using the assumptions in the image above, Equations (18-1) and (18-2) can be simplified as shown below (steady motion means $dV/dt = 0$):

$$L = W \qquad (19\text{-}1)$$

$$D = T \qquad (19\text{-}2)$$

The solution of the equations of motion for level flight reveals a large number of very important characteristics of the airplane. Among those are the minimum and maximum airspeed the aircraft can (theoretically) achieve, as well as the minimum airspeed the airplane can maintain in level flight (stall). Additionally, there are a number of airspeeds that represent selected optimum conditions, such as best endurance, best range, and the so-called Carson's airspeed, to name a few. Figures 19-3 and 19-4 show a number of important airspeeds for a typical aircraft. The formulas shown, excluding the one for the stalling speed, V_S, are all based on the simplified drag model. The sections in the chapter are ordered from the lowest to the highest with respect to Figures 19-3 and 19-4.

Note that in this chapter, formulation for unsteady flight is also developed.

19.1.4 Important Airspeeds for Propeller Aircraft

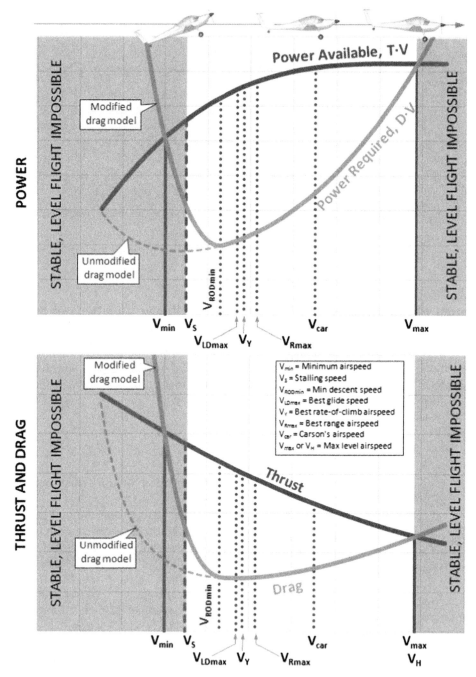

FIGURE 19-3 Important airspeeds for a propeller aircraft in cruising flight. Note that the modified drag model uses the method of Section 15.2.3, *Approximating the drag coefficient at high lift coefficients*.

19.1.5 Important Airspeeds for Subsonic Jet Aircraft

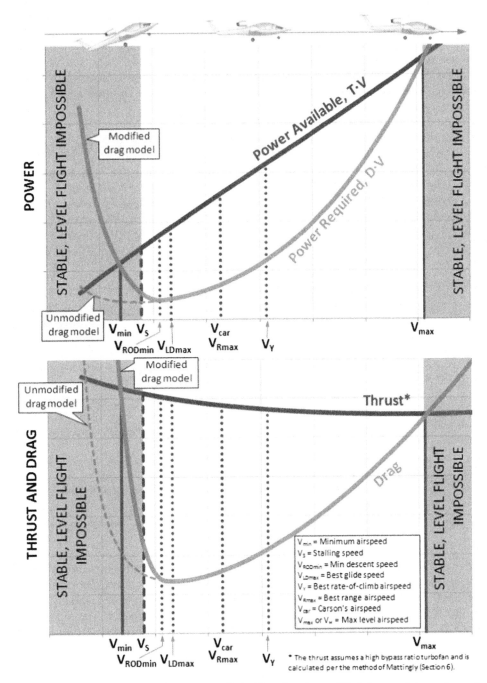

FIGURE 19-4 Important airspeeds for a jet aircraft in cruising flight. Note that the modified drag model uses the method of Section 15.2.3, *Approximating the drag coefficient at high lift coefficients.*

19.2 GENERAL CRUISE ANALYSIS METHODS FOR STEADY FLIGHT

In this chapter we will introduce a number of common cruise analysis methods for un-accelerated flight. All depend on our knowledge of drag as well as thrust. The reader should keep in mind that, generally, three different drag models are used for this purpose: (1) the simplified, (2) adjusted, and (3) all others. The latter could be a situation that involves a sailplane with a well-defined drag bucket that calls for the use of a spline or a lookup table. Performance characteristics can be evaluated using closed-form solutions as long as the drag model is well defined using smooth and continuous mathematical expressions. As such, only the simplified and adjusted drag models allow general closed-form solutions to be developed. In this chapter, analysis methods are developed that utilize those two models, although most are based on the simplified drag model. The reader should recognize that these closed-form solutions can only be used for the said drag models. Using them with other representations will yield erroneous results.

It should be stated that modern-day performance analyses shy away from using the simplified drag model, as its results are suspect for the aircraft that features cambered airfoils. Additionally, it is a good practice to prepare the methods in a spreadsheet. The spreadsheet has become one of the most powerful tools at the disposal of the aerospace engineer. It allows the designer to set up numerical solutions that are highly non-linear and, when properly formulated, allow very complicated problems to be analyzed. Such methodologies are essential when analyzing the performance of very sleek aircraft, such as powered sailplanes and other aircraft that operate outside normal quadratic drag behavior. Regardless, the strength of the simplified model is — well — its simplicity. This allows various hard-to-grasp concepts to be explained and presented in a clear manner using closed-form analytical solutions.

19.2.1 Plotting the Drag Polar

The drag polar presents the drag coefficient, C_D, as a function of the lift coefficient, C_L. The generation of the drag polar is discussed in detail in Chapter 15, *Aircraft drag analysis*. Figure 19-6 shows the subsonic drag polar for the Learjet 45XR, separating the contribution of induced drag, C_{Di}, from the minimum drag, C_{Dmin}.

EXAMPLE 19-1: PLOTTING THE DRAG POLAR

Plot the drag polar for the Learjet 45XR aircraft. Plot C_{Dmin}, C_{Di}, and C_D, assuming Oswald's span efficiency factor amounts to 0.8294.

Solution

Step 1: Determine the constant "k":

$$k = \frac{1}{\pi \cdot AR \cdot e} = \frac{1}{\pi \cdot (7.33) \cdot (0.8294)} = 0.05236$$

Step 2: Calculate C_D using the simplified and adjusted drag models using $C_{LminD} = 0.2$. As an example consider the value for a $C_L = 0.5$:

Simplified drag model:

$$C_D = 0.020 + (0.05236)(0.5)^2 = 0.0331$$

Adjusted drag model:

$$C_D = 0.020 + (0.05236)(0.5 - 0.2)^2 = 0.0247$$

Figure 19-5 compares the constituent components of the drag model. The reader must be mindful that both models, the simplified and adjusted, represent the simplest "reasonable" forms of the drag polar. For instance, in the figure, the effect of increased flow separation will force C_{Dmin} upward at low and high lift coefficients; it would not be a horizontal line. This would make the total drag coefficient rise substantially, well beyond what is represented in the figure and more like that of the modified drag model of Figures 19-3 and 19-4.

Figure 19-6 compares the simplified and adjusted drag models. The offset of the adjusted drag model will have significant effects on very important performance parameters, such as airspeed for minimum power required; best glide speed, and others. Airplanes that feature cambered airfoils almost always have a $C_{LminD} > 0$ and, thus, require the adjusted model.

Another lesson the reader must be mindful of when comparing the two drag models is that it would appear the adjusted drag model leads to much higher L/D ratio on other performance parameters. This is primarily caused by the assumption here that the C_{Dmin} for both models is equal. However, this in not always true in practice.

19.2.2 Drag Breakdown

Chapter 15, *Aircraft drag analysis*, describes the generation of drag and its constituent components. It can be very useful to break the total drag into these components and consider their magnitudes as functions of airspeed. For instance, consider the hypothetical installation of a winglet on an existing airplane. A winglet is a device that reduces the induced drag of the aircraft.

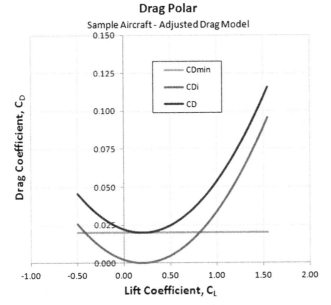

FIGURE 19-5 The drag polar for the Learjet 45XR.

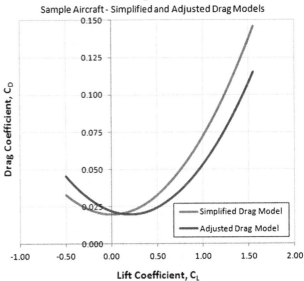

FIGURE 19-6 Comparing the two drag models for the Learjet 45XR.

However, it adds to the wetted area of the aircraft, increasing C_{Dmin} by a magnitude of ΔC_{Dmin}. Its installation will thus only be beneficial at airspeeds where (and if) the decrease in induced drag, ΔC_{Di}, is greater than the increase ΔC_{Dmin}. In order to find out the total drag must be broken down into its constituent components. The following example shows how this is done for the Learjet 45XR.

EXAMPLE 19-2: PLOTTING THE DRAG POLAR

The Learjet 45XR is cruising at 30,000 ft and at a weight of 20,000 lb$_f$. Plot the following parameters:

(1) Lift coefficient and *AOA* versus airspeed assuming $C_{Lo} = 0$.
(2) *Zero-lift drag coefficient* (C_{Dmin}) and *drag coefficient due to lift* (C_{Di}) versus airspeed. On the same graph plot the total drag coefficient (C_D).
(3) *Zero-lift drag force* (D_{min}) and *drag force due to lift* (D_i) versus airspeed. On the same graph plot the total drag force (D).

Solution

The resulting plots can be seen in Figure 19-7, Figure 19-8, and Figure 19-9. Sample calculation for $V = 400$ KTAS using the simplified drag model:

Step 1: Determine density at 30,000 ft.

$$\rho = 0.002378(1 - 0.0000068753 \times 30000)^{4.2561}$$
$$= 0.0008897 \text{ slugs/ft}^3$$

Step 2: Determine lift coefficient.

$$C_L = \frac{2W}{\rho V^2 S} = \frac{2(20000)}{(0.0008897)(1.688 \times 400)^2 (311.6)} = 0.316$$

Step 3: Determine drag coefficient.

$$C_D = 0.020 + (0.05236)(0.316)^2 = 0.02523$$

Step 4: Determine drag force.

$$D = \frac{1}{2}\rho V^2 S C_D$$
$$= \frac{1}{2}(0.0008897)(1.688 \times 400)^2 (311.6)(0.02523)$$
$$= 1594 \text{ lb}_f$$

Step 5: Determine minimum drag and induced drag components, respectively, noting we have already calculated density and the lift coefficient in Steps 1 and 2.

EXAMPLE 19-2: PLOTTING THE DRAG POLAR (cont'd)

$D_{\min} = \frac{1}{2}\rho V^2 S C_{D\min} = 1264 \text{ lb}_f$ and

$D_i = \frac{1}{2}\rho V^2 S C_{Di} = 330 \text{ lb}_f$

Step 3: Determine drag coefficient.

$C_D = 0.020 + (0.05236)(0.316 - 0.200)^2 = 0.02071$

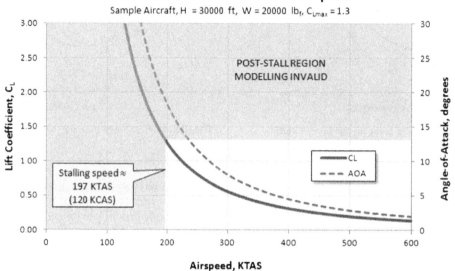

FIGURE 19-7 Variation of the lift coefficient with airspeed at 30,000 ft.

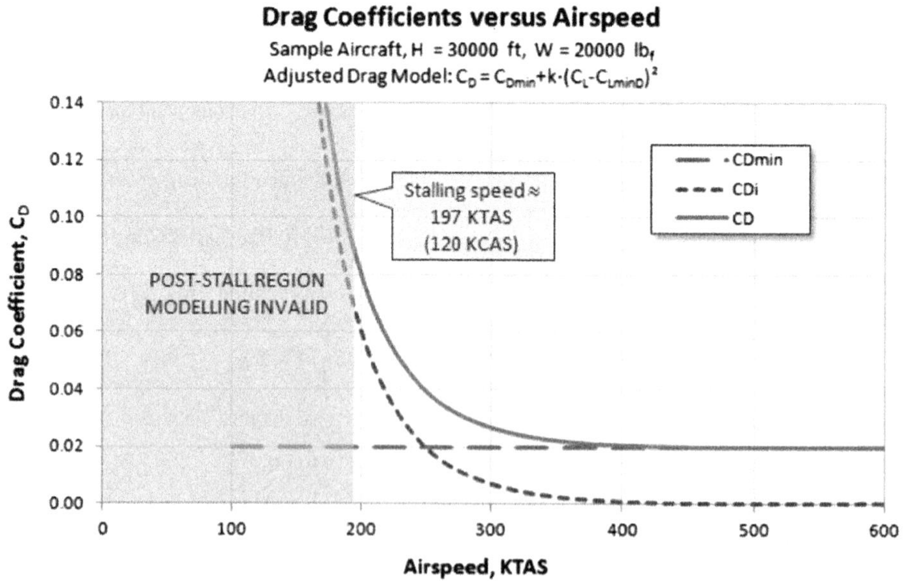

FIGURE 19-8 Breakdown of constituent drag contributions with airspeed at 30,000 ft.

EXAMPLE 19-2: PLOTTING THE DRAG POLAR (cont'd)

Step 4: Determine drag force.

$$D = \frac{1}{2}\rho V^2 S C_D$$

$$= \frac{1}{2}(0.0008897)(1.688 \times 400)^2 (311.6)(0.02071)$$

$$= 1309 \text{ lb}_f$$

Step 5: Determine minimum drag and induced drag components, respectively.

$$D_{min} = \frac{1}{2}\rho V^2 S C_{Dmin} = 1264 \text{ lb}_f \quad \text{and}$$

$$D_i = \frac{1}{2}\rho V^2 S C_{Di} = 45 \text{ lb}_f$$

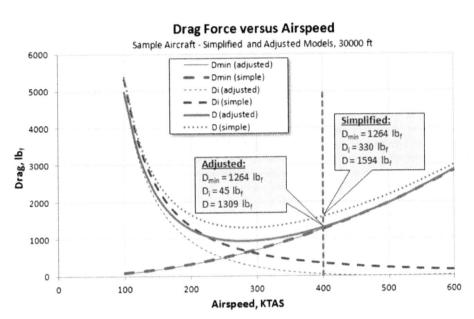

FIGURE 19-9 Breakdown of drag forces with airspeed using the simplified model.

19.2.3 Required Versus Available Thrust

As shown in Section 18.2.6, *Vertical airspeed in terms of thrust or power*, an airplane can only increase its altitude if it is equipped with an engine capable of delivering more power than required to maintain level flight. It was shown at the beginning of this section that the thrust required for level flight equals the drag of the airplane. While the requirement holds for any airplane, the analysis method for airplanes with piston engines differs in the sense that power is used rather than force. However, the general philosophy of analysis remains the same for both.

Consider the Learjet 45XR as it cruises at 30,000 ft and at a weight of 20,000 lb$_f$. The thrust required to cruise at 400 KTAS can be obtained from Equation (17-11) through (17-13).

From Equation (17-12) using the simplified drag model:

$$T_R = D = 1594 \text{ lb}_f$$

From Equation (17-13) using the adjusted drag model:

$$T_R = D = 1309 \text{ lb}_f$$

There are two important points to be made regarding the required thrust, T_R. The first point is the difference in the magnitude of the drag force between the simple and adjusted drag models (see the curves in Figure 19-10). Two regions of importance are shown and of which the designer must be aware. These are referred to as the regions of speed stability and instability. The second point is shown in Figure 19-11; it shows that each power

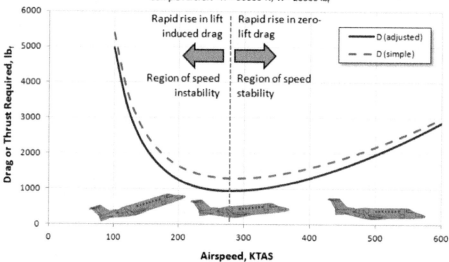

FIGURE 19-10　Regions of speed stability and instability.

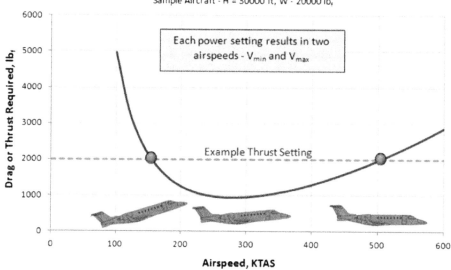

FIGURE 19-11　Each power setting results in two distinct airspeeds.

setting (here displayed as 2000 lb$_f$ of thrust) results in two distinct airspeeds.

Region of Speed Stability

Consider Figure 19-10. Assume the airplane is flying at 400 KTAS when a sudden gust or change in its attitude reduces the speed of the airplane. As a consequence, its drag decreases but this increases the airspeed, eventually bringing it back to 400 KTAS (since the thrust setting has not been changed). By the same token, should the airplane experience a speed increase, its drag now increases, reducing the airspeed and eventually bringing it back to the original airspeed (as the thrust has not been changed). This phenomenon is recognized by pilots as *speed stability*.

Region of Speed Instability

Again consider Figure 19-10. Assume we are flying at 200 KTAS when a sudden gust or change in airplane attitude reduces the speed of the airplane. As this happens, its drag increases, which reduces the speed further. By the same token, should the airplane experience a speed increase, its drag now reduces, increasing the airspeed further. This phenomenon is recognized by pilots as

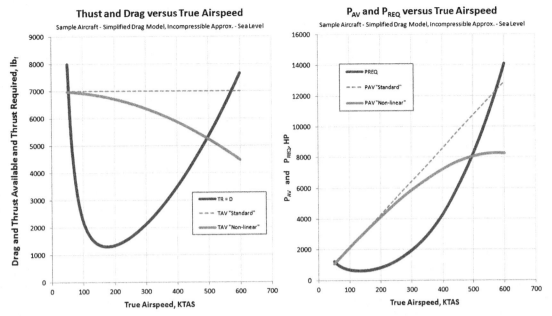

FIGURE 19-12 Required and available thrust and required and available power based on Figure 19-12 The "Standard" is an idealized turbojet, whereas "Non-linear" resembles a more realistic turbofan engine.

speed instability. Speed instability is also known by pilots as flying "on the back side of the power curve." Speed instability can be helpful during landing, as it helps the pilot slow down the airplane during flare.

Figure 19-11 shows an important property when comparing available to required thrust: each power setting results in two possible equilibrium airspeeds, one low and one high. The lower the available thrust, the closer will these speeds be to each other. One and only one airspeed will result in a condition at which the two acquire the same value; this is the *airspeed of minimum thrust required*.

Introducing the Effect of Thrust and Power

The preceding graphs have all demonstrated how the drag of the airplane changes with airspeed. The effects of thrust and power have not yet been presented. It is conventional to superimpose those on the same graph, as this will give important clues to the capability of the aircraft. Figure 19-12 shows a typical such representation, here based on the Learjet 45XR and depicting both the "standard" representation of thrust and power (linear) and one that is more realistic (curved). Both graphs show that where the thrust intersects the drag curve, or the power available intersects the power required curve, it holds that $T = D$. In other words, it is a point of equilibrium. There are two such points on each graph: the minimum and maximum airspeed. Of course there is a serious shortcoming to the lower airspeed — it is usually less than the stalling speed, which is then the true minimum speed of the aircraft. Also, the drag at this point is higher than shown — a limitation of the simplified drag model being used.

19.2.4 Airspeed in Terms of Thrust

Based on Figure 19-11 it is possible to determine two equilibrium airspeeds associated with any arbitrary power setting. Thus, one could predict the airspeeds associated with, say, 50% or 75% thrust settings, or for that matter any other setting of interest. If the thrust (T), weight (W), and drag characteristics of the aircraft are known, the two airspeeds can be estimated as follows, based on the two drag models:

Formulation (simplified drag model):

$$V = \sqrt{\frac{T \pm \sqrt{T^2 - 4C_{Dmin}kW^2}}{\rho S C_{Dmin}}} \qquad (19\text{-}3)$$

Formulation (adjusted drag model):

$$V = \sqrt{\frac{T + 2WkC_{LminD} \pm \sqrt{(T + 2WkC_{LminD})^2 - 4W^2k(C_{Dmin} + kC_{LminD}^2)}}{\rho S (C_{Dmin} + kC_{LminD}^2)}} \qquad (19\text{-}4)$$

Note that if $C_{LminD} = 0$ then, as expected, the ratio under the radical reduces to the simplified C_{Di}:

$$\frac{T + 0 \pm \sqrt{(T+0)^2 - 4W^2k(C_{Dmin} + 0)}}{\rho S(C_{Dmin} + 0)}$$

$$= \frac{T \pm \sqrt{T^2 - 4W^2kC_{Dmin}}}{\rho S C_{Dmin}}$$

Derivation of Equation (19-3)

First, write the thrust in terms of the drag model:

$$T = D = qS(C_{Dmin} + kC_L^2) = qS\left(C_{Dmin} + k\left(\frac{W}{qS}\right)^2\right)$$

$$= qSC_{Dmin} + qS\frac{kW^2}{q^2S^2} = qSC_{Dmin} + \frac{kW^2}{qS}$$

Then, try to eliminate the term $1/q$:

$$qT = q\left(qSC_{Dmin} + \frac{kW^2}{qS}\right) = q^2 SC_{Dmin} + \frac{kW^2}{S}$$

$$\Rightarrow q^2 SC_{Dmin} - qT + \frac{kW^2}{S} = 0$$

This is a quadratic equation in terms of q:

$$a = SC_{D0}, \quad b = -T, \quad c = \frac{kW^2}{S}$$

$$aq^2 + bq + c = 0 \Leftrightarrow q = \frac{-b \pm \sqrt{b^2 - 4ac}}{2a}$$

Solving the quadratic equation:

$$q = \frac{-b \pm \sqrt{b^2 - 4ac}}{2a}$$

$$= \frac{T \pm \sqrt{(-T)^2 - 4(SC_{Dmin})\left(\frac{kW^2}{S}\right)}}{2(SC_{Dmin})}$$

$$= \frac{T \pm \sqrt{T^2 - 4C_{Dmin}kW^2}}{2SC_{Dmin}}$$

Writing q explicitly leads to:

$$q = \frac{T \pm \sqrt{T^2 - 4C_{Dmin}kW^2}}{2SC_{Dmin}} = \frac{1}{2}\rho V^2 \Leftrightarrow$$

$$V = \sqrt{\frac{T \pm \sqrt{T^2 - 4C_{Dmin}kW^2}}{\rho S C_{Dmin}}}$$

QED

Derivation of Equation (19-4)

First, write the thrust in terms of the drag model:

$$T = D = qS\left(C_{Dmin} + k(C_L - C_{LminD})^2\right)$$

$$= qS\left(C_{Dmin} + k\left(\frac{W}{qS} - C_{LminD}\right)^2\right)$$

Therefore;

$$T = qS\left(C_{Dmin} + k\left(\frac{W^2}{q^2S^2} - 2\frac{W}{qS}C_{LminD} + C_{LminD}^2\right)\right)$$

$$= qS\left(C_{Dmin} + \frac{W^2k}{q^2S^2} - \frac{2Wk}{qS}C_{LminD} + kC_{LminD}^2\right)$$

Simplifying further:

$$T = qSC_{Dmin} + qS\frac{W^2k}{q^2S^2} - qS\frac{2Wk}{qS}C_{LminD} + qSkC_{LminD}^2$$

$$= qSC_{Dmin} + \frac{W^2k}{qS} - 2WkC_{LminD} + qSkC_{LminD}^2$$

Then, try to eliminate the term $1/q$:

$$qT = q\left(qSC_{Dmin} + \frac{W^2k}{qS} - 2WkC_{LminD} + qSkC_{LminD}^2\right)$$

$$= \left(q^2SC_{Dmin} + q\frac{W^2k}{qS} - q2WkC_{LminD} + q^2SkC_{LminD}^2\right)$$

$$= q^2SC_{Dmin} + q^2SkC_{LminD}^2 - q2WkC_{LminD} + \frac{W^2k}{S}$$

$$= q^2S(C_{Dmin} + kC_{LminD}^2) - q2WkC_{LminD} + \frac{W^2k}{S}$$

$$\Rightarrow q^2 S(C_{Dmin} + kC_{LminD}^2) - q2WkC_{LminD}$$
$$+ \frac{W^2 k}{S} - qT_R = 0$$
$$\Rightarrow q^2 S(C_{Dmin} + kC_{LminD}^2) - q(T_R + 2WkC_{LminD})$$
$$+ \frac{W^2 k}{S} = 0$$

This is a quadratic equation in terms of q:

$$a = S(C_{Dmin} + kC_{LminD}^2),$$
$$b = -(T + 2WkC_{LminD}), \quad c = \frac{W^2 k}{S}$$
$$aq^2 + bq + c = 0 \iff q = \frac{-b \pm \sqrt{b^2 - 4ac}}{2a}$$

Finally, solve the quadratic equation:

$$q = \frac{-(-(T + 2WkC_{LminD})) \pm \sqrt{(-(T + 2WkC_{LminD}))^2 - 4(S(C_{Dmin} + kC_{LminD}^2))\left(\frac{W^2 k}{S}\right)}}{2(S(C_{Dmin} + kC_{LminD}^2))}$$

$$q = \frac{T + 2WkC_{LminD} \pm \sqrt{(T + 2WkC_{LminD})^2 - 4W^2 k(C_{Dmin} + kC_{LminD}^2)}}{2S(C_{Dmin} + kC_{LminD}^2)} = \frac{1}{2}\rho V^2$$

$$\Rightarrow V = \sqrt{\frac{T + 2WkC_{LminD} \pm \sqrt{(T + 2WkC_{LminD})^2 - 4W^2 k(C_{Dmin} + kC_{LminD}^2)}}{\rho S(C_{Dmin} + kC_{LminD}^2)}}$$

QED

EXAMPLE 19-3

The Learjet 45XR is cruising at 30,000 ft, where it weighs 20,000 lb$_f$. Determine the low and high airspeeds if the thrust amounts to 2000 lb$_f$ total.

Solution

Simplified C_{Di}:

$$V = \sqrt{\frac{T \pm \sqrt{T^2 - 4C_{Dmin} k W^2}}{\rho S C_{Dmin}}}$$

$$V = \sqrt{\frac{2000 \pm \sqrt{2000^2 - 4(0.020)(0.05236)(20000)^2}}{(0.0008897)(311.6)(0.020)}}$$

$$= \sqrt{\frac{2000 \pm \sqrt{2324480}}{0.005545}} = \begin{cases} 797.3 \text{ ft/s} = 472.3 \text{ KTAS} \\ 292.8 \text{ ft/s} = 173.5 \text{ KTAS} \end{cases}$$

Adjusted C_{Di}:

$$V = \sqrt{\frac{T + 2WkC_{LminD} \pm \sqrt{(T + 2WkC_{LminD})^2 - 4W^2 k(C_{Dmin} + kC_{LminD}^2)}}{\rho S(C_{Dmin} + kC_{LminD}^2)}}$$

$$V = \sqrt{\frac{2418.88 \pm \sqrt{4000000}}{0.006125}}$$

$$= \begin{cases} 849.4 \text{ ft/s} = 503.2 \text{ KTAS} \\ 261.5 \text{ ft/s} = 154.9 \text{ KTAS} \end{cases}$$

These points are shown in the graph of Figure 19-13.

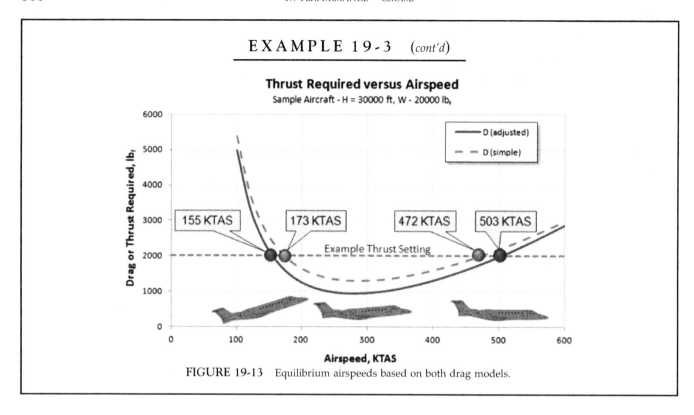

EXAMPLE 19-3 (cont'd)

FIGURE 19-13 Equilibrium airspeeds based on both drag models.

19.2.5 Minimum Airspeed, V_{min}

The *minimum level airspeed* is the lower of the two airspeeds at which the power required and power available are equal (see Section 19.2.4, *Airspeed in terms of thrust*). The absolute theoretical minimum airspeed will be achieved at the maximum thrust. When estimated using the simplified or adjusted drag models, this airspeed is often lower than the stalling speed and, thus, really meaningless. As stated before, this is caused by inaccuracies in the simplified and adjusted drag models at high lift coefficients (refer to Section 15.2.3 for a remedy). Furthermore, the following estimation of the minimum level airspeed assumes (1) the simplified drag model and (2) thrust does not contribute to vertical force (lift).

$$V_{min} = \sqrt{\frac{T_{max} - \sqrt{T_{max}^2 - 4C_{Dmin}kW^2}}{\rho S C_{Dmin}}} \quad (19\text{-}5)$$

Note that if the minimum speed is smaller than the stalling speed of the aircraft, then the stalling speed becomes the V_{min}.

For the adjusted drag model use:

Derivation of Equation (19-5)

Formulation is obtained from Equation (19-3) and (19-4), selecting the negative sign. Thus, for the simplified model, Equation (19-3) simply becomes:

$$V = \sqrt{\frac{T_{max} \pm \sqrt{T_{max}^2 - 4C_{Dmin}kW^2}}{\rho S C_{Dmin}}} \Leftrightarrow$$

$$V_{min} = \sqrt{\frac{T_{max} - \sqrt{T_{max}^2 - 4C_{Dmin}kW^2}}{\rho S C_{Dmin}}}$$

The process is identical for the adjusted drag model.
QED

19.2.6 Stalling Speed, V_S

The stalling speed, V_S, is the minimum speed at which an airplane can maintain altitude. If the airspeed is slowed a hair below the stalling speed this will cause a well-designed airplane to drop its nose uncontrollably, while maintaining wings level. This nose drop, in turn, will increase the airspeed of the airplane so it becomes

$$V_{min} = \sqrt{\frac{T + 2WkC_{LminD} - \sqrt{(T + 2WkC_{LminD})^2 - 4W^2k(C_{Dmin} + kC_{LminD}^2)}}{\rho S(C_{Dmin} + kC_{LminD}^2)}} \quad (19\text{-}6)$$

airborne again. Due to stability and control issues, some aircraft do not achieve a stall break, but rather descend at some AOA below their stall AOA. The resulting airspeed is considered the minimum airspeed by the aviation authorities, although it is higher than the true stalling speed of such aircraft. A determination of the stalling speed during the design phase requires the engineer to estimate the maximum lift coefficient and then calculate it as follows:

Level 1g stalling speed:

$$V_S = \sqrt{\frac{2W}{\rho S C_{L_{max}}}} \quad (19\text{-}7)$$

Figure 19-14 shows how the stalling speed changes with altitude, while the calibrated airspeed remains unchanged for aircraft with a low subsonic stalling speed. On the other hand, the true airspeed increases with altitude, something the designer must keep in mind as it affects impact loads in emergency landings and durability of brakes for airplanes consistently operated from high-altitude airports.

Level Stalling Speed with a Load Factor n

If the airplane is performing some specific wings-level maneuver, for instance a *loop*, it will experience a change in the load factor. If the total load factor is denoted by n (note $n = 1$ for level flight), the stalling speed will change and can be calculated by making the following modification to Equation (19-7):

Level stalling speed at load factor n:

$$V_S = \sqrt{\frac{2nW}{\rho S C_{L_{max}}}} \quad (19\text{-}8)$$

Stalling Speed During Banking

When an airplane banks at an angle ϕ while maintaining altitude (level constant-speed turn), the load factor acting on it increases. This is identical to carrying greater weight; its stalling speed increases. Using the preceding formulation, this stalling speed at any given angle of bank ϕ can be estimated using the following expression:

Stalling speed at angle of bank ϕ:

$$V_S = \sqrt{\frac{2W}{\rho S C_{L_{max}} \cos \phi}} = \frac{V_{S_{level}}}{\sqrt{\cos \phi}} \quad (19\text{-}9)$$

where $V_{S_{level}}$ is the stalling speed with wings level. Note that 14 CFR Part 23 denotes the stalling speed of an airplane in the landing configuration using the term V_{S0}. This implies flaps are fully deflected and retractable landing gear is deployed. Similarly, the term V_{S1} refers to the stalling speed of the airplane in the take-off configuration. This implies flaps are in a take-off position and retractable landing gear is deployed.

Level Stalling Speed with Thrust, Flap, and CG Effects

In practice, the stalling speed is affected by engine thrust, flap deflection, and the CG location. Thrust and deployed flaps will reduce the stalling speed, while a forward location of the CG increases it. The forward CG requires higher download to be generated by the HT for trim. This load must be added to the weight of the aircraft, which means the wing must generate lift greater than the weight of the airplane and therefore, its stalling speed is higher than indicated by the above methods. The same stabilizing effect due to flaps will also render the stall speed reduction smaller than in its

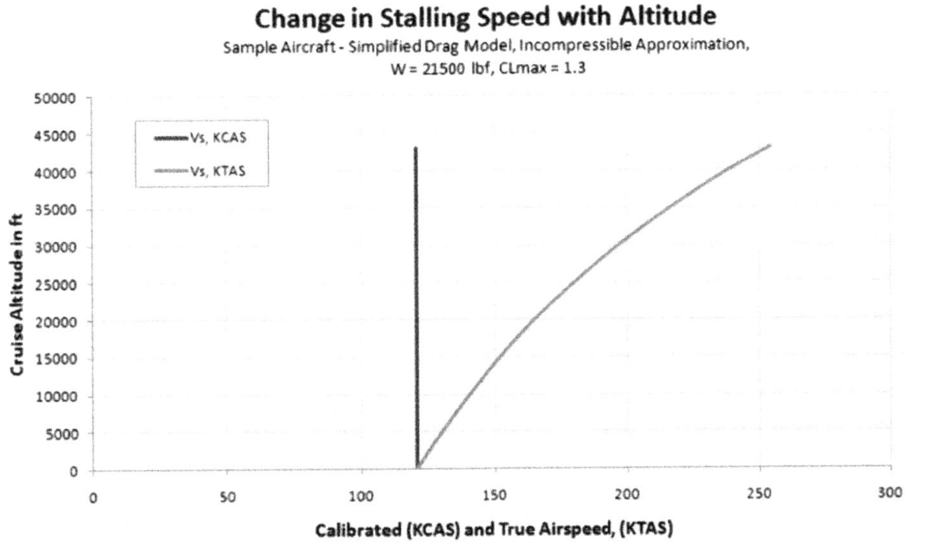

FIGURE 19-14 Stalling speed in terms of calibrated (KCAS) and true (KTAS) airspeed as a function of altitude.

absence. To include these effects, the stalling speed must be calculated as shown below:

$$V_S = \sqrt{\frac{2}{\rho S C_{L_{max}}}\left(1 - \frac{h - h_{AC}}{l_{HT}}\right)\left[W - \frac{M_W}{l_{HT} - h + h_{AC}} - \left(\frac{x_T \cdot \sin\alpha - z_T \cdot \cos\alpha}{l_{HT} - h + h_{AC}} + \sin\alpha\right)T\right]} \qquad (19\text{-}10)$$

where

W = weight at condition
S = wing area
M_W = Pitching moment of wing, landing gear, fuselage, etc. about the aerodynamic center
T = engine thrust

Refer to Figure 19-15 for the dimensions h, h_{AC}, x_T, z_T, and l_{HT}.

Note that this expression is suitable for conventional tail-aft configurations only. Strictly speaking it requires an iterative procedure to solve, because both M_W and T depend on airspeed. If the expected stalling speed is known, it can be used to calculate both M_W and T, which can then be used as constants in the formulation with acceptable accuracy. However, implementing the calculations in a spreadsheet will make it easy to solve iteratively. Note that the effects of flaps are accounted for in the variable M_W. Example 19-4 demonstrates the use of this equation and evaluates its accuracy.

Derivation of Equations (19-7), (19-8), and (19-9)

Formulation is obtained from Equation (9-47), by solving for V and applies to all three equations:

$$L = \frac{1}{2}\rho V^2 S C_L \quad \Leftrightarrow \quad V = \sqrt{\frac{2W}{\rho S C_L}}$$

QED

Derivation of Equation (19-10)

Referring to Figure 19-15, statics requires the following to hold in steady level flight, where at the point of stall, the lift of the wing, L_W, will depend on the maximum lift coefficient of the wing.

Referring to Figure 19-15, statics requires the following to hold in steady level flight:

$\sum F_z = 0; \qquad L_W + L_{HT} + T\sin\alpha - W = 0$

$\sum M_{CG} = 0; \quad M_W + (h - h_{AC})L_W - (l_{HT} - h + h_{AC})L_{HT}$
$\qquad\qquad + x_T \cdot T\sin\alpha - z_T \cdot T\cos\alpha = 0$

Following a similar process as for the derivation of Equation (15-81), the moment equation can be solved for the balancing force the HT must generate:

$$L_{HT} = \frac{M_W + (h - h_{AC})L_W + x_T \cdot T\sin\alpha - z_T \cdot T\cos\alpha}{(l_{HT} - h + h_{AC})}$$

Inserting this result into the force equation and simplifying leads to the expression below:

$$\left(\frac{l_{HT}}{l_{HT} - h + h_{AC}}\right)L_W = W - \frac{M_W}{l_{HT} - h + h_{AC}}$$
$$- \left(\frac{x_T \cdot \sin\alpha - z_T \cdot \cos\alpha}{l_{HT} - h + h_{AC}} + \sin\alpha\right)T$$

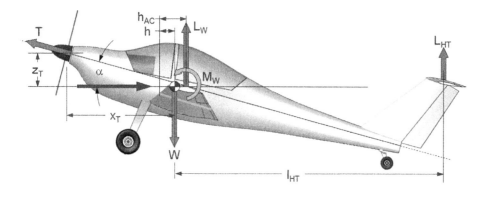

FIGURE 19-15 A simple system used to derive the simplest formulation of trim drag.

19.2 GENERAL CRUISE ANALYSIS METHODS FOR STEADY FLIGHT

Yielding the following relation holds between the wing lift and the other contributors:

$$L_W = \left(1 - \frac{h - h_{AC}}{l_{HT}}\right)\left[W - \frac{M_W}{l_{HT} - h + h_{AC}} - \left(\frac{x_T \cdot \sin\alpha - z_T \cdot \cos\alpha}{l_{HT} - h + h_{AC}} + \sin\alpha\right)T\right]$$

At stall, $L_W = \frac{1}{2}\rho V_S^2 \cdot S \cdot C_{L\max}$, so solving for V_S yields:

$$V_S = \sqrt{\frac{2}{\rho S C_{L\max}}\left(1 - \frac{h - h_{AC}}{l_{HT}}\right)\left[W - \frac{M_W}{l_{HT} - h + h_{AC}} - \left(\frac{x_T \cdot \sin\alpha - z_T \cdot \cos\alpha}{l_{HT} - h + h_{AC}} + \sin\alpha\right)T\right]}$$

QED

EXAMPLE 19-4

Determine the S-L stalling speed (i.e. in KCAS) of the SR22 at its forward and aft CG limits at gross weight, with and without full power, using Equation (19-10). Assume the stall AOA is 16°, maximum lift coefficient of 1.45, and the following parameters:

S = wing area = 144.9 ft²	C_{MGC} = 3.783 ft	W = gross weight = 3400 lb$_f$
C_{mW} = −0.06	l_{HT} = 14.06 ft	
h_{FWD} = 0.192·C_{MGC}	h_{AFT} = 0.192·C_{MGC}	h_{AC} = 0.25·C_{MGC}
x_T = 5 ft	z_T = 1	η_P = propeller efficiency = 0.65

The forward CG limit is at 19.2% C_{MGC}, and the aft at 31.5% C_{MGC}, according to the POH. Compare the power off values to the POH (V_S at forward CG is 70 KCAS and at aft CG is 69 KCAS). Therefore, use 70 KCAS to calculate the value of M_W.

Solution

A solution will be demonstrated for only one case of the four cases being requested: the forward CG power off case. Only the results for the others will be presented.

Pitching moment at 70 KCAS is:

$$M_W = \frac{1}{2}\rho V_S^2 S C_{MGC} C_{m_W}$$
$$= \frac{1}{2}(0.002378)(70 \times 1.688)^2(144.9)(3.783)(-0.06)$$
$$= -546 \text{ ft·lb}_f$$

For the power-on cases use the following thrust at 70 KCAS, given by Equation (14-38):

$$T = \frac{\eta_p \times 550 \times P_{BHP}}{V} = \frac{0.65 \times 550 \times 310}{70 \times 1.688} = 938 \text{ lb}_f$$

Simplify Equation (19-10) by rewriting it as follows:

$$V_S = \sqrt{A \cdot B \cdot C}$$

Here, the power off-forward CG case is being solved. This yields the following values of A, B, and C:

$$A = \frac{2}{\rho S C_{L\max}} = \frac{2}{(0.002378)(144.9)(1.45)} = 4.003$$

$$B = 1 - \frac{h - h_{AC}}{l_{HT}} = \frac{2}{(0.002378)(144.9)(1.45)} = 1.0156$$

$$C = \left[W - \frac{M_W}{l_{HT} - h + h_{AC}} - \left(\frac{x_T \cdot \sin\alpha - z_T \cdot \cos\alpha}{l_{HT} - h + h_{AC}} + \sin\alpha\right)T\right] = 3438 \text{ lb}_f$$

Therefore, the power-off stalling speed at the forward CG limit is:

$$V_S = \sqrt{A \cdot B \cdot C} = \sqrt{(4.003)(1.0156)(3438)} = 118.2 \text{ ft/s}$$
$$= 70.0 \text{ KCAS}$$

> **EXAMPLE 19-4** (cont'd)
>
> Similarly, the power-off stalling speed at the aft CG limit is:
>
> $$V_S = 68.9 \text{ KCAS}$$
>
> Power-on stall at the forward CG limit is:
>
> $$V_S = 67.1 \text{ KCAS}$$
>
> And power-on stall at the aft CG limit is:
>
> $$V_S = 66.0 \text{ KCAS}$$
>
> The analysis shows that the power-off results pretty much reflect the numbers in the POH.

19.2.7 Airspeed of Minimum Power Required, V_{PRmin}

The upper graph in Figure 19-3 shows there is a distinct minimum in the power require curve. For propeller-powered aircraft this implies that the engine can be operated at a minimum power and, thus, with minimum consumption of fuel. For this reason, this airspeed is of great importance to the aircraft designer as it will allow the aircraft to stay aloft the longest. This airspeed is also called the *maximum endurance speed*, V_{Emax}, for the propeller aircraft. Additionally, as shown in 18.3.9, *Airspeed for best ROC for a propeller-powered airplane*, this airspeed is also the *best rate of climb speed*, V_Y, for a propeller-powered aircraft.

Requirement for Max Endurance for a Propeller-powered Airplane

Equation (20-21) of Section 20.4.2, *Endurance profile 2: constant attitude/altitude cruise*, as well as 19.2.12, *Power required*, show that the maximum endurance of a propeller-powered airplane depends on the maximum of the ratio $C_L^{1.5}/C_D$, and this maximum can be found from the following expression:

$$\left(\frac{C_L^{1.5}}{C_D}\right)_{max} = \frac{1}{4}\left(\frac{3}{k \cdot C_{Dmin}^{1/3}}\right)^{3/4} \quad (19\text{-}11)$$

The lift coefficient of this condition can be determined using the relations below:

Using the simplified drag model:

$$C_L = \sqrt{\frac{3C_{Dmin}}{k}} \quad (19\text{-}12)$$

Using the adjusted drag model:

$$C_L = \sqrt{\frac{3C_{Dmin}}{k} + 4C_{LminD}^2} - C_{LminD} \quad (19\text{-}13)$$

Derivation of Equation (19-11) and (19-12)

Start by writing the expression for L/D:

$$\frac{C_L^{3/2}}{C_D} = \frac{C_L^{3/2}}{C_{Dmin} + k \cdot C_L^2} \quad (i)$$

Differentiate with respect to C_L:

$$\frac{d}{dC_L}\left(\frac{C_L^{3/2}}{C_D}\right) = \frac{d}{dC_L}(f \cdot g) = f'g + fg'$$

$$= \frac{d\left(C_L^{3/2}\right)}{dC_L}\left(\frac{1}{C_{Dmin} + k \cdot C_L^2}\right)$$

$$+ \left(C_L^{3/2}\right)\frac{d}{dC_L}\left(\frac{1}{C_{Dmin} + k \cdot C_L^2}\right)$$

Expanding;

$$\frac{d}{dC_L}\left(\frac{C_L^{3/2}}{C_D}\right) = \frac{3}{2}\frac{\sqrt{C_L}}{C_{Dmin} + k \cdot C_L^2}$$

$$+ \left(C_L^{3/2}\right)\frac{d}{dC_L}(C_{Dmin} + k \cdot C_L^2)^{-1}$$

This leads to:

$$\frac{d}{dC_L}\left(\frac{C_L^{3/2}}{C_D}\right) = \frac{3}{2}\frac{\sqrt{C_L}}{C_{Dmin} + k \cdot C_L^2}$$

$$- \left(C_L^{3/2}\right)(C_{Dmin} + k \cdot C_L^2)^{-2}(2k \cdot C_L)$$

Manipulate algebraically;

$$\frac{d}{dC_L}\left(\frac{C_L^{3/2}}{C_D}\right) = \frac{3}{2}\frac{\sqrt{C_L}}{C_{Dmin}+k\cdot C_L^2}$$
$$-\left(C_L^{3/2}\right)\left(C_{Dmin}+k\cdot C_L^2\right)^{-2}(2k\cdot C_L)$$
$$= \frac{\frac{3}{2}\sqrt{C_L}(C_{Dmin}+k\cdot C_L^2)-2k\cdot C_L^{5/2}}{\left(C_{Dmin}+k\cdot C_L^2\right)^2} = 0$$

The maximum can be found when the numerator equals zero, or;

$$\Leftrightarrow \frac{3}{2}\sqrt{C_L}(C_{Dmin}+k\cdot C_L^2)-2k\cdot C_L^{5/2} = 0$$
$$\Rightarrow \frac{3}{2}C_L^{1/2}C_{Dmin}+\frac{3}{2}C_L^{5/2}k-2k\cdot C_L^{5/2} = 0$$
$$\Rightarrow \frac{3}{2}C_L^{1/2}C_{Dmin}-\frac{1}{2}k\cdot C_L^{5/2} = 0$$
$$\Rightarrow 3C_{Dmin} = k\cdot C_L^2 \quad \Leftrightarrow \quad C_L = \sqrt{\frac{3C_{Dmin}}{k}} \quad \text{(ii)}$$

Insert this result into the original expression and manipulate:

$$\left(\frac{C_L^{3/2}}{C_D}\right)_{max} = \left(\frac{C_L^{3/2}}{C_{Dmin}+k\cdot C_L^2}\right)_{max}$$
$$= \frac{\left(\frac{3C_{Dmin}}{k}\right)^{3/4}}{C_{Dmin}+k\cdot \frac{3C_{Dmin}}{k}} \quad \text{(iii)}$$
$$= \frac{1}{4C_{Dmin}}\left(\frac{3C_{Dmin}}{k}\right)^{3/4}$$

QED

Derivation of Equation (19-13)

Start by writing the expression for L/D:

$$\frac{C_L^{3/2}}{C_D} = \frac{C_L^{3/2}}{C_{Dmin}+k\cdot(C_L-C_{LminD})^2}$$
$$= \frac{C_L^{3/2}}{C_{Dmin}+k\cdot C_L^2 - 2\cdot k\cdot C_L\cdot C_{LminD}+k\cdot C_{LminD}^2}$$
$$= \frac{C_L^{3/2}}{k\cdot C_L^2 - 2\cdot k\cdot C_{LminD}\cdot C_L + (C_{Dmin}+k\cdot C_{LminD}^2)}$$
$$= \frac{C_L^{3/2}}{AC_L^2+BC_L+C}$$

Where;

$$A = k$$
$$B = -2\cdot k\cdot C_{LminD}$$
$$C = C_{Dmin}+k\cdot C_{LminD}^2$$

Differentiate with respect to C_L:

$$\frac{d}{dC_L}\left(\frac{C_L^{3/2}}{C_D}\right) = \frac{d}{dC_L}(f\cdot g) = f'g+fg'$$

$$\frac{d}{dC_L}\left(\frac{C_L^{3/2}}{C_D}\right) = \frac{d\left(C_L^{3/2}\right)}{dC_L}\left(\frac{1}{AC_L^2+BC_L+C}\right)$$
$$+\left(C_L^{3/2}\right)\frac{d}{dC_L}\left(\frac{1}{AC_L^2+BC_L+C}\right)$$

Expanding;

$$\frac{d}{dC_L}\left(\frac{C_L^{3/2}}{C_D}\right) = \frac{3}{2}\frac{\sqrt{C_L}}{AC_L^2+BC_L+C}+\left(C_L^{3/2}\right)$$
$$\times \frac{d}{dC_L}(AC_L^2+BC_L+C)^{-1}$$
$$= \frac{3}{2}\frac{\sqrt{C_L}}{AC_L^2+BC_L+C}-\left(C_L^{3/2}\right)$$
$$\times (AC_L^2+BC_L+C)^{-2}(2AC_L+B)$$

Manipulate algebraically;

$$\frac{d}{dC_L}\left(\frac{C_L^{3/2}}{C_D}\right) = \frac{3}{2}\frac{\sqrt{C_L}}{AC_L^2+BC_L+C}$$
$$-\left(C_L^{3/2}\right)(AC_L^2+BC_L+C)^{-2}(2AC_L+B)$$

$$\frac{d}{dC_L}\left(\frac{C_L^{3/2}}{C_D}\right) = \frac{\frac{3}{2}\sqrt{C_L}(AC_L^2 + BC_L + C) - \left(2AC_L^{5/2} + BC_L^{3/2}\right)}{(AC_L^2 + BC_L + C)^2} = 0$$

The maximum can be found when the numerator equals zero, or;

$$\Leftrightarrow \frac{3}{2}\sqrt{C_L}(AC_L^2 + BC_L + C) - \left(2AC_L^{5/2} + BC_L^{3/2}\right) = 0$$

$$\Rightarrow \frac{3}{2}C_L^{1/2}(AC_L^2 + BC_L + C) = 2AC_L^{5/2} + BC_L^{3/2}$$

$$\Rightarrow 3C_L^{1/2}(AC_L^2 + BC_L + C) = C_L^{1/2}\left(4AC_L^{4/2} + 2BC_L^{2/2}\right)$$

$$\Rightarrow 3AC_L^2 + 3BC_L + 3C = 4AC_L^2 + 2BC_L$$

$$\Rightarrow AC_L^2 - BC_L - 3C = 0$$

We have quadratic formulation that can be solved using Equation (E-12), here repeated for convenience:

$$x = \frac{-B \pm \sqrt{B^2 - 4AC}}{2A}$$

Therefore:

$$C_L = \frac{B \pm \sqrt{B^2 + 12AC}}{2A}$$

$$= \frac{-2 \cdot k \cdot C_{LminD} \pm \sqrt{4 \cdot k^2 \cdot C_{LminD}^2 + 12k(C_{Dmin} + k \cdot C_{LminD}^2)}}{2k}$$

$$= \frac{-2 \cdot k \cdot C_{LminD}}{2k} \pm \sqrt{\frac{4 \cdot k^2 \cdot C_{LminD}^2 + 12k(C_{Dmin} + k \cdot C_{LminD}^2)}{4k^2}}$$

Therefore:

$$C_L = -C_{LminD} \pm \sqrt{4C_{LminD}^2 + \frac{3C_{Dmin}}{k}}$$

By observation we also note that only when $C_L > 0$ does the solution make physical sense (because the airplane needs a positive C_L to maintain altitude). Therefore, we must select the plus sign in front of the radical.

$$C_L = \sqrt{4C_{LminD}^2 + \frac{3C_{Dmin}}{k}} - C_{LminD} \qquad (iv)$$

As expected, we can see that if $C_{LminD} = 0$ we get the same expression as that of the simplified drag model. Insert this result into Equation (iii) to determine the ratio.

QED

EXAMPLE 19-5

Determine the maximum endurance ratio for the Learjet 45XR:

Solution

$$\left(\frac{C_L^{3/2}}{C_D}\right)_{max} = \frac{1}{4C_{D_0}}\left(\frac{3C_{Dmin}}{k}\right)^{3/4}$$

$$= \frac{1}{4 \cdot 0.020}\left(\frac{3 \cdot 0.020}{0.05236}\right)^{3/4} = 13.84$$

Maximum Endurance Airspeed for a Propeller-powered Aircraft, V_{Emax}

Since a condition for maximum endurance of a propeller-powered airplane has been identified, it becomes imperative to determine the airspeed at which this occurs. This will allow the Pilot's Operating Handbook (POH) to specify a specific airspeed for loitering or remaining in holding flight due to air traffic control requirements (although airspeeds greater than this can certainly be expected). Also refer to 19.2.8, *Airspeed of minimum thrust required*, for a similar airspeed for jets. *This expression is only valid for the simplified drag model.*

Computed from:

$$V_{PRmin} = V_{Emax} = \sqrt{\frac{2}{\rho}\left(\frac{W}{S}\right)\sqrt{\frac{k}{3 \cdot C_{Dmin}}}} \qquad (19\text{-}14)$$

Best Rate-of-climb Airspeed for a Propeller-powered Aircraft, V_Y

Equation (19-14) is also derived in Section 18.3.9, *Airspeed for best ROC for a propeller-powered airplane*, as Equation (18-27), where it is used to determine the best ROC for a propeller aircraft.

Derivation of Equation (19-14)

We demonstrated that C_L for P_{Rmin} was given by:

$$C_L = \sqrt{\frac{3C_{Dmin}}{k}} \text{ (see Equation (19-12))}$$

Insert this into the lift equation and solve for V:

$$V = \sqrt{\frac{2W}{\rho S C_L}} = \sqrt{\frac{2W}{\rho S \sqrt{\frac{3C_{Dmin}}{k}}}} = \sqrt{\frac{2}{\rho}\left(\frac{W}{S}\right)}\sqrt{\frac{k}{3C_{Dmin}}}$$

QED

Comparison to the Best Glide Speed

Using the specific results based on the simplified drag model, it is of interest to compare Equation (19-14) to Equation (19-20), repeated below for convenience:

$$V_{Emax} = \sqrt{\frac{2}{\rho}\left(\frac{W}{S}\right)}\sqrt{\frac{k}{3 \cdot C_{Dmin}}} \quad \text{and}$$

$$V_{LD_{max}} = \sqrt{\frac{2}{\rho}\left(\frac{W}{S}\right)}\sqrt{\frac{k}{C_{Dmin}}}.$$

Dividing the former by the latter reveals the following difference between the two:

$$\frac{V_{Emax}}{V_{LD_{max}}} = \frac{\sqrt{\frac{2}{\rho}\left(\frac{W}{S}\right)}\sqrt{\frac{k}{3 \cdot C_{Dmin}}}}{\sqrt{\frac{2}{\rho}\left(\frac{W}{S}\right)}\sqrt{\frac{k}{C_{Dmin}}}} = \left(\frac{1}{3}\right)^{1/4} \approx 0.76$$

In other words, the best endurance speed is about 76% of the speed for best L/D.

19.2.8 Airspeed of Minimum Thrust Required, V_{TRmin}, or Best Glide Speed, V_{BG}, V_{LDmax}

This airspeed is also the best glide speed, V_{BG} or V_{LDmax}, for both a jet and propeller-powered aircraft. Furthermore, it is the maximum endurance airspeed, V_{Emax}, for a jet and the best range airspeed, V_{Rmax}, for a propeller-powered aircraft.

It is vitally important for the operation of jets that the designer (and operator) can determine the airspeed that requires the least amount of thrust. It is logical to assume that this airspeed must also be that which requires the least amount of fuel to be consumed. The airspeed of minimum thrust required is particularly important for airplanes whose mission may demand prolonged periods of loitering. Examples of such aircraft are reconnaissance aircraft and even fighter aircraft. Their missions often require them to stay for a prolonged period of time in a specific geographic location. However, the high fuel consumption of jet engines in general renders this airspeed imperative for all jets, regardless of mission. Albeit rare, situations may arise that require the pilot to slow down to the airspeed of minimum thrust required, V_{TRmin}, which is calculated from:

Using the simplified drag model:

$$V_{TRmin} = \sqrt{\frac{2}{\rho}\left(\frac{W}{S}\right)}\sqrt{\frac{k}{C_{Dmin}}} \quad (19\text{-}15)$$

Using the adjusted drag model:

$$V_{TRmin} = \sqrt{\frac{T}{\rho S C_{Dmin}}}$$

$$= \sqrt{\left(\frac{2}{\rho}\right)\left(\frac{W}{S}\right)\frac{-kC_{LminD} \pm \sqrt{k(C_{Dmin} + kC_{LminD}^2)}}{C_{Dmin}}}$$

(19-16)

where

$$T = -2WkC_{LminD} \pm 2W\sqrt{k(C_{Dmin} + kC_{LminD}^2)} \quad (19\text{-}17)$$

When using the adjusted drag model, the primary trick is to first determine the two roots from Equation (19-17) and then select proper thrust (T) from the two (the one with the positive sign). This is then used with Equation (19-16) to calculate the airspeed.

Derivation of Equation (19-15)

Airspeed as a function of thrust was determined from Equation (19-3):

$$V = \sqrt{\frac{T \pm \sqrt{T^2 - 4C_{Dmin}kW^2}}{\rho S C_{Dmin}}}$$

When the quantity under the radical is zero there is only one solution: *the airspeed at which the required thrust is minimum*. The thrust for this condition can be found from:

$$T^2 - 4C_{Dmin}kW^2 = 0 \Leftrightarrow T = 2W\sqrt{C_{Dmin}k}$$

When the quantity under the radical is zero the minimum required thrust can be written as follows:

$$V_{TRmin} = \sqrt{\frac{T_R}{\rho S C_{Dmin}}} = \sqrt{\frac{2W\sqrt{C_{Dmin}k}}{\rho S C_{Dmin}}} = \sqrt{\frac{2W}{\rho S}}\sqrt{\frac{k}{C_{Dmin}}}$$

QED

Derivation of Equation (19-16)

Airspeed as a function of thrust was found from Equation (19-4):

$$V = \sqrt{\frac{T + 2WkC_{LminD} \pm \sqrt{(T + 2WkC_{LminD})^2 - 4W^2k(C_{Dmin} + kC_{LminD}^2)}}{\rho S(C_{Dmin} + kC_{LminD}^2)}}$$

As before, when the quantity under the radical is zero there is only one solution: the airspeed at which required thrust is minimum.

$$(T + 2WkC_{LminD})^2 - 4W^2k(C_{Dmin} + kC_{LminD}^2) = 0$$

$$\Rightarrow T^2 + 4WkC_{LminD}T + 4W^2k^2C_{LminD}^2 - 4W^2k(C_{Dmin} + kC_{LminD}^2) = 0$$

$$\Rightarrow T^2 + 4WkC_{LminD}T + 4W^2k^2C_{LminD}^2 - 4W^2kC_{Dmin} - 4W^2k^2C_{LminD}^2 = 0$$

$$\Rightarrow T^2 + 4WkC_{LminD}T - 4W^2kC_{Dmin} = 0$$

Therefore, the solution to the quadratic formulation is:

$$T = \frac{-4WkC_{LminD} \pm \sqrt{(4WkC_{LminD})^2 + 16W^2kC_{D0}}}{2}$$

$$\Leftrightarrow T = \frac{-4WkC_{LminD} \pm \sqrt{16W^2k^2C_{LminD}^2 + 16W^2kC_{D0}}}{2}$$

$$\Leftrightarrow T = -2WkC_{LminD} \pm 2W\sqrt{k(C_{D0} + kC_{LminD}^2)}$$

Therefore,

$$V_{TRmin} = \sqrt{\frac{T}{\rho S C_{Dmin}}}$$

$$= \frac{\sqrt{-2WkC_{LminD} \pm 2W\sqrt{k(C_{Dmin} + kC_{LminD}^2)}}}{\sqrt{\rho S C_{Dmin}}}$$

$$V_{TRmin} = \sqrt{\left(\frac{2}{\rho}\right)\left(\frac{W}{S}\right)\frac{-kC_{LminD} \pm \sqrt{k(C_{Dmin} + kC_{LminD}^2)}}{C_{Dmin}}}$$

QED

EXAMPLE 19-6

The Learjet 45XR is cruising at 30,000 ft and at a weight of 20,000 lb$_f$. Determine minimum thrust and the corresponding airspeed the pilot should be flying at:

Solution

Simplified C_{Di}:

$$T_R = 2W\sqrt{C_{Dmin}k} = 2(20000)\sqrt{(0.020)(0.05236)}$$
$$= 1294 \text{ lb}_f$$

$$V_{TRmin} = \sqrt{\frac{2(20000)}{(0.0008897)(311.6)}}\sqrt{\frac{0.05236}{0.020}}$$
$$= 483.2 \text{ ft/s} = 286.2 \text{ KTAS}$$

EXAMPLE 19-6 (cont'd)

$$V_{TRmin} = 286.2\sqrt{\frac{0.0008897}{0.002378}} = 175 \text{ KCAS}$$

This answer means that in order to conserve fuel, the pilot must bring the aircraft to an airspeed of 175 KCAS and he will do so by reducing the power setting such that a mere 1294 lb$_f$ of total thrust needs to be generated. This amounts to each engine generating 647 lb$_f$ of thrust, which is about 18.5% of the rated maximum static thrust of 3500 lb$_f$ at S-L.

It is also of interest to solve the problem using the adjusted drag model. This yields the following thrust required:

$$T_R = -2WkC_{LminD} \pm 2W\sqrt{k(C_{Dmin} + kC_{LminD}^2)}$$

$$= -2(20000)(0.05236)(0.20)$$

$$\pm 2(20000)\sqrt{(0.05236)\left(0.02 + (0.05236)(0.2)^2\right)}$$

$$= -418.88 \pm 40000\sqrt{k(C_{Dmin} + kC_{LminD}^2)}$$

$$= \begin{cases} 941.6 \text{ lbf} \\ -1779 \text{ lbf} \end{cases}$$

Clearly, only the positive value of the thrust makes physical sense. Thus, the adjusted drag model indicates that the required thrust will be even less, or 941.6 lb$_f$. Inserting this value into Equation (19-16) yields the airspeed the pilot must maintain:

$$V_{TRmin} = \sqrt{\frac{T_R}{\rho S C_{Dmin}}} = \sqrt{\frac{941.6}{(0.0008897)(311.6)(0.02)}}$$

$$= 412.1 \text{ ft/s} = 244.1 \text{ KTAS}$$

$$\Rightarrow V_{TRmin} = 244.1\sqrt{0.0008897/0.002378}$$

$$= 149 \text{ KCAS}$$

Maximum L/D Ratio

One of the most important performance parameters of an aircraft is its maximum lift-to-drag ratio, LD_{max}. This ratio indicates not only how far an airplane will glide from a given altitude (see Section 21.3.8, *Glide Distance*), but also, in the case of a propeller aircraft, how far it can fly and, in the case of a jet, how long it can stay aloft. Table 15-18 shows typical values for the maximum L/D ratio for several classes of aircraft.

The ratio is also known as the *best glide ratio* and, less commonly, *minimum-thrust-required-to-weight-ratio*. Here, two methods to calculate the best glide ratio are presented: one using the simplified and the other the adjusted drag models. When using the simplified drag model, the LD_{max} can be calculated from:

$$LD_{max} = \left(\frac{C_L}{C_D}\right)_{max} = \frac{1}{\sqrt{4 \cdot C_{Dmin} \cdot k}} \quad (19\text{-}18)$$

The expression shows that the magnitude of the LD_{max} is independent of altitude. Figure 19-17 and Figure 19-18 show that the airspeed of LD_{max} changes with altitude in terms of KTAS, but not KCAS. This is important and helpful to the pilot, who only has to remember one number; the KCAS value, as he operates the airplane using KCAS. For the adjusted drag model, the LD_{max} is determined using the following expression:

$$LD_{max} = \left(\frac{C_L}{C_D}\right)_{max}$$

$$= \frac{1}{\sqrt{4 \cdot C_{Dmin} \cdot k + (2k \cdot C_{LminD})^2} - 2k \cdot C_{LminD}}$$

(19-19)

Derivation of Equation (19-18)

Start by writing the expression for L/D:

$$\frac{L}{D} = \frac{C_L}{C_D} = \frac{C_L}{C_{Dmin} + k \cdot C_L^2}$$

Differentiate with respect to C_L:

$$\frac{d}{dC_L}\left(\frac{C_L}{C_D}\right) = \frac{d}{dC_L}(f \cdot g) = f'g + fg'$$

$$= \frac{d(C_L)}{dC_L}\left(\frac{1}{C_{Dmin} + k \cdot C_L^2}\right) + (C_L)\frac{d}{dC_L}\left(\frac{1}{C_{Dmin} + k \cdot C_L^2}\right)$$

$$= \frac{1}{C_{Dmin} + k \cdot C_L^2} + (C_L)\frac{d}{dC_L}(C_{Dmin} + k \cdot C_L^2)^{-1}$$

$$= \frac{1}{C_{Dmin} + k \cdot C_L^2} - (C_L)(C_{Dmin} + k \cdot C_L^2)^{-2}(2k \cdot C_L)$$

Manipulating algebraically;

$$\frac{d}{dC_L}\left(\frac{C_L}{C_D}\right) = \frac{1}{C_{Dmin} + k \cdot C_L^2} - \frac{2k \cdot C_L^2}{(C_{Dmin} + k \cdot C_L^2)^2}$$

$$= \frac{C_{Dmin} + k \cdot C_L^2 - 2k \cdot C_L^2}{(C_{Dmin} + k \cdot C_L^2)^2} = 0$$

Therefore:

$$\Rightarrow \quad C_{Dmin} - k \cdot C_L^2 = 0 \quad \Leftrightarrow \quad C_L = \sqrt{\frac{C_{Dmin}}{k}}$$

Insert this into the original expression to get:

$$\left(\frac{C_L}{C_D}\right)_{max} = \left(\frac{C_L}{C_{Dmin} + k \cdot C_L^2}\right)_{max} = \frac{\sqrt{C_{Dmin}/k}}{C_{Dmin} + k \cdot C_{D_0}/k}$$

$$= \frac{1}{\sqrt{4 \cdot C_{Dmin} \cdot k}}$$

QED

Derivation of Equation (19-19)

Start by writing the expression for L/D:

$$\frac{L}{D} = \frac{C_L}{C_D} = \frac{C_L}{C_{Dmin} + k \cdot (C_L - C_{LminD})^2}$$

$$= \frac{C_L}{k \cdot C_L^2 - 2k \cdot C_{LminD} \cdot C_L + (C_{Dmin} + k \cdot C_{LminD}^2)}$$

Differentiate with respect to C_L:

$$\frac{d}{dC_L}\left(\frac{C_L}{C_D}\right) = \frac{d}{dC_L}(f \cdot g) = f'g + fg'$$

$$= \frac{d(C_L)}{dC_L}\left(\frac{1}{k \cdot C_L^2 - 2k \cdot C_{LminD} \cdot C_L + (C_{Dmin} + k \cdot C_{LminD}^2)}\right)$$

$$+ (C_L)\frac{d}{dC_L}\left(\frac{1}{k \cdot C_L^2 - 2k \cdot C_{LminD} \cdot C_L + (C_{Dmin} + k \cdot C_{LminD}^2)}\right)$$

Completing the differentiation:

$$\frac{d}{dC_L}\left(\frac{C_L}{C_D}\right) = \frac{1}{k \cdot C_L^2 - 2k \cdot C_{LminD} \cdot C_L + (C_{Dmin} + k \cdot C_{LminD}^2)}$$

$$+ C_L \frac{d}{dC_L}(k \cdot C_L^2 - 2k \cdot C_{LminD} \cdot C_L + (C_{Dmin} + k \cdot C_{LminD}^2))^{-1}$$

Complete differentiation and expand further to get:

$$\frac{d}{dC_L}\left(\frac{C_L}{C_D}\right) = \frac{1}{k \cdot C_L^2 - 2k \cdot C_{LminD} \cdot C_L + (C_{Dmin} + k \cdot C_{LminD}^2)}$$

$$+ -1 \cdot C_L \frac{2k \cdot C_L - 2k \cdot C_{LminD}}{(k \cdot C_L^2 - 2k \cdot C_{LminD} \cdot C_L + (C_{Dmin} + k \cdot C_{LminD}^2))^2}$$

Leading to:

$$\frac{d}{dC_L}\left(\frac{C_L}{C_D}\right) = \frac{1}{C_{Dmin} + k \cdot (C_L - C_{LminD})^2}$$

$$- \frac{2k \cdot C_L^2 - 2k \cdot C_{LminD} \cdot C_L}{\left(C_{Dmin} + k \cdot (C_L - C_{LminD})^2\right)^2}$$

Manipulating further leads to:

$$\frac{d}{dC_L}\left(\frac{C_L}{C_D}\right) = \frac{C_{Dmin} + k \cdot (C_L - C_{LminD})^2 - 2k \cdot C_L^2 + 2k \cdot C_{LminD} \cdot C_L}{\left(C_{Dmin} + k \cdot (C_L - C_{LminD})^2\right)^2}$$

$$= \frac{k \cdot C_L^2 - 2k \cdot C_{LminD} \cdot C_L + C_{Dmin} + k \cdot C_{LminD}^2 - 2k \cdot C_L^2 + 2k \cdot C_{LminD} \cdot C_L}{\left(C_{Dmin} + k \cdot (C_L - C_{LminD})^2\right)^2}$$

Further manipulations then yield:

$$\frac{d}{dC_L}\left(\frac{C_L}{C_D}\right) = \frac{-k \cdot C_L^2 + C_{Dmin} + k \cdot C_{LminD}^2}{\left(C_{Dmin} + k \cdot (C_L - C_{LminD})^2\right)^2}$$

Therefore, the maximum can be found for C_L when the numerator equals zero, or:

$$-k \cdot C_L^2 + C_{Dmin} + k \cdot C_{LminD}^2 = 0$$

Using this expression, the value of C_L reduces to (note that it reduces to the result for the simplified model when $C_{LminD} = 0$):

$$C_L = \sqrt{\frac{C_{Dmin} + k \cdot C_{LminD}^2}{k}} = \sqrt{\frac{C_{Dmin}}{k} + C_{LminD}^2}$$

Insert this into the original expression to get:

$$\left(\frac{C_L}{C_D}\right)_{max} = \left(\frac{C_L}{C_{Dmin} + k \cdot (C_L - C_{LminD})^2}\right)_{max}$$

$$= \frac{\sqrt{C_{Dmin}/k + C_{LminD}^2}}{C_{Dmin} + k \cdot (C_L - C_{LminD})^2}$$

For clarity let's treat the denominator separately:

$$k \cdot C_L^2 - 2k \cdot C_{LminD} \cdot C_L + (C_{Dmin} + k \cdot C_{LminD}^2)$$

$$= k \cdot \left(\frac{C_{Dmin}}{k} + C_{LminD}^2\right) - 2k \cdot C_{LminD} \cdot \sqrt{\frac{C_{Dmin}}{k} + C_{LminD}^2}$$
$$+ (C_{Dmin} + k \cdot C_{LminD}^2)$$

$$= 2(C_{Dmin} + k \cdot C_{LminD}^2) - 2k \cdot C_{LminD} \cdot \sqrt{\frac{C_{Dmin}}{k} + C_{LminD}^2}$$

Noting that $(C_L/C_D)_{max} = (C_D/C_L)_{min}$ we get:

$$\left(\frac{C_D}{C_L}\right)_{min}$$

$$= \frac{2(C_{Dmin} + k \cdot C_{LminD}^2) - 2k \cdot C_{LminD} \cdot \sqrt{\frac{C_{Dmin}}{k} + C_{LminD}^2}}{\sqrt{\frac{C_{Dmin}}{k} + C_{LminD}^2}}$$

$$\left(\frac{C_D}{C_L}\right)_{min} = \frac{2k\left(\frac{C_{Dmin}}{k} + C_{LminD}^2\right)}{\sqrt{\frac{C_{Dmin}}{k} + C_{LminD}^2}} - 2k \cdot C_{LminD}$$

$$= \sqrt{4 \cdot C_{Dmin} \cdot k + (2k \cdot C_{LminD})^2}$$

$$- 2k \cdot C_{LminD} \Leftrightarrow \left(\frac{C_L}{C_D}\right)_{max}$$

$$= \frac{1}{\sqrt{4 \cdot C_{Dmin} \cdot k + (2k \cdot C_{LminD})^2} - 2k \cdot C_{LminD}}$$

QED

EXAMPLE 19-7

Determine the maximum lift-to-drag ratio for the Learjet 45XR using the simplified and adjusted drag models. What is the C_L at which LD_{max} occurs for each? Plot the variation of L/D as a function of lift coefficient and airspeed (KTAS and KCAS).

Solution

Using the simplified model we get:

$$LD_{max} = \frac{1}{\sqrt{4 \cdot C_{Dmin} \cdot k}} = \frac{1}{\sqrt{4 \cdot (0.020) \cdot (0.05236)}}$$

$$= 15.45$$

$$C_L = \sqrt{\frac{C_{Dmin}}{k}} = \sqrt{\frac{0.020}{0.05236}} = 0.6180$$

Using the adjusted model we start by calculating the lift coefficient of LD_{max} as follows:

$$LD_{max} = \frac{1}{\sqrt{4 \cdot C_{Dmin} \cdot k + (2k \cdot C_{LminD})^2} - 2k \cdot C_{LminD}}$$

EXAMPLE 19-7 (cont'd)

$$LD_{max} = \frac{1}{\sqrt{4 \cdot (0.020) \cdot (0.05236) + 4 \cdot (0.20)^2 \cdot (0.05236)^2 - 2(0.05236) \cdot (0.20)}} = 21.24$$

$$C_L = \sqrt{\frac{C_{Dmin}}{k} + C_{LminD}^2} = \sqrt{\frac{0.020}{0.05236} + 0.20^2}$$
$$= 0.6500$$

The L/D as a function of lift coefficient and airspeed is plotted in Figure 19-16, Figure 19-17, and Figure 19-18. Note the difference in presentation of the two latter figures, which show the same ratios plotted versus the airspeed in KTAS and KCAS. The implications are that the magnitude of LD_{max} and the calibrated airspeed at which it occurs are effectively independent of the altitude at which the airplane is operated. This is very convenient, as the pilot does not have to worry about a great variation in these factors with altitude. Nevertheless, it is important to realize that the formulation does not account for a change in the minimum drag coefficient, C_{Dmin}, due to change in Reynolds number with altitude, nor the presence of wave drag at higher airspeeds.

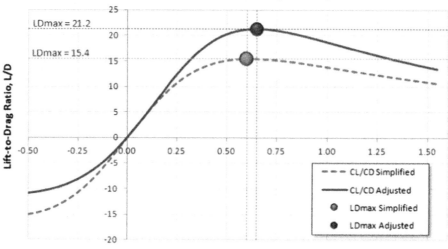

FIGURE 19-16 Maximum lift-to-drag ratio (L/D) for the Learjet 45XR using each drag model.

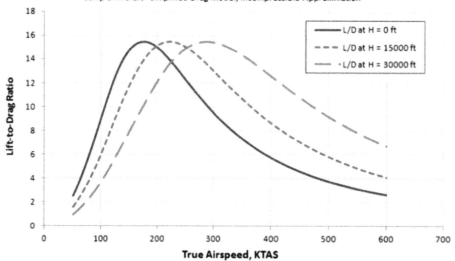

FIGURE 19-17 Change in L/D with true airspeed, using the simplified drag model.

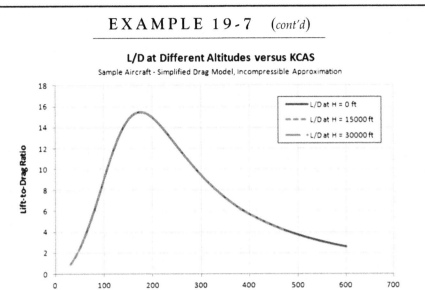

EXAMPLE 19-7 (cont'd)

FIGURE 19-18 Change in L/D with calibrated airspeed, using the simplified drag model. Note that all the curves are superimposed on top of each other and, thus, will appear independent of altitude to the pilot. This is very fortunate for the pilot, who must remember only one number as the best glide speed.

Airspeed for Maximum L/D Ratio

Knowing the airspeed at which the maximum L/D ratio is achieved is imperative, not only from a standpoint of safety but also as it is the airspeed of minimum thrust required (see 19.2.8, *Airspeed of minimum thrust required*). Pilots of single-engine aircraft are trained to establish this airspeed as soon as possible in the case of engine failure as it will result in a maximum glide distance, therefore improving survivability. It is also known as the *airspeed for minimum thrust required*. *Using the simplified drag model*, the airspeed for maximum L/D ratio is:

$$V_{LD_{max}} = \sqrt{\frac{2}{\rho}\left(\frac{W}{S}\right)}\sqrt{\frac{k}{C_{Dmin}}} \qquad (19\text{-}20)$$

Not that this result is the same as Equation (19-15) of 19.2.8, *Airspeed of minimum thrust required*. *Using the adjusted drag model*, it is given by:

$$V_{LD_{max}} = \sqrt{\frac{2}{\rho}\left(\frac{W}{S}\right)}\sqrt{\frac{k}{C_{Dmin} + kC_{LminD}^2}} \qquad (19\text{-}21)$$

Derivation of Equation (19-20)

We showed that C_L for LD_{max} was given by:

$$C_L = \sqrt{\frac{C_{Dmin}}{k}}$$

Insert this into the lift equation and solve for V:

$$V = \sqrt{\frac{2W}{\rho S C_L}} = \sqrt{\frac{2W}{\rho S \sqrt{\frac{C_{Dmin}}{k}}}} = \sqrt{\frac{2}{\rho}\left(\frac{W}{S}\right)}\sqrt{\frac{k}{C_{Dmin}}}$$

QED

Derivation of Equation (19-21)

The C_L for LD_{max} using the adjusted drag model is:

$$C_L = \sqrt{\frac{C_{Dmin}}{k} + C_{LminD}^2}$$

Insert this into the lift equation and solve for V:

$$V = \sqrt{\frac{2W}{\rho S C_L}} = \sqrt{\frac{2W}{\rho S \sqrt{\frac{C_{Dmin}}{k} + C_{LminD}^2}}}$$

$$= \sqrt{\frac{2}{\rho}\left(\frac{W}{S}\right)}\sqrt{\frac{k}{C_{Dmin} + kC_{LminD}^2}}$$

QED

EXAMPLE 19-8

Determine the airspeed the pilot should maintain in order to achieve maximum lift-to-drag ratio for the Learjet 45XR at 30,000 ft and at two weights, 15,000 and 20,000 lb$_f$. Also, plot the lift-to-drag ratio and thrust required at the latter weight as a function of the true airspeed at 30,000 ft on the same plot.

Solution

Determine the airspeed at 30,000 ft and 15,000 lb$_f$:

$$V_{LD_{max}} = \sqrt{\frac{2}{\rho}\left(\frac{W}{S}\right)\sqrt{\frac{k}{C_{Dmin}}}}$$

$$= \sqrt{\frac{2}{0.0008897}\left(\frac{15000}{311.6}\right)\sqrt{\frac{0.05236}{0.020}}}$$

$$= 418.4 \text{ ft/s}$$

$$V_{LD_{max}} = \left(\frac{418.4}{1.688}\right)\sqrt{\frac{0.0008897}{0.002378}} \approx 152 \text{ KCAS}$$

Determine the airspeed at 30,000 ft and 20,000 lb$_f$:

$$V_{LD_{max}} = \sqrt{\frac{2}{\rho}\left(\frac{W}{S}\right)\sqrt{\frac{k}{C_{Dmin}}}}$$

$$= \sqrt{\frac{2}{0.0008897}\left(\frac{20000}{311.6}\right)\sqrt{\frac{0.05236}{0.020}}}$$

$$= 487.5 \text{ ft/s}$$

$$V_{LD_{max}} = \left(\frac{487.5}{1.688}\right)\sqrt{\frac{0.0008897}{0.002378}} \approx 177 \text{ KCAS}$$

Thrust and L/D are plotted as functions of the true airspeed in Figure 19-19.

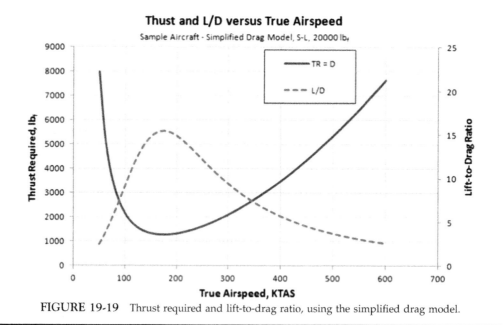

FIGURE 19-19 Thrust required and lift-to-drag ratio, using the simplified drag model.

Various Lift-to-drag Ratios

As is evident from several of the analysis methodologies presented thus far, a number of important optimum properties of the airplane can be extracted once the drag polar has been defined. In particular, three cruising speeds based on the optimization on the various products and ratios of lift, L, drag, D, and airspeed, V, are of interest to the designer. The first is that of optimum range (maximum L/D), endurance (minimum $D \cdot V$), and cruise efficiency (minimum D/V).

Figure 19-20 shows three types of lift-to-drag ratios and their corresponding optimums. Note that the ratio $C_L^{1.5}/C_D$ pertains to propeller aircraft, which renders it moot for the Learjet 45XR. However, it is included for completeness. Figure 19-21 shows the L/D and thrust required for the Learjet 45XR sample aircraft plotted at three altitudes (S-L, 15,000 ft, and 30,000 ft). The graph shows well how the thrust required changes with altitude and how high airspeed is more easily achieved at altitude. The simplified drag model used for this demonstration does not account for compressibility, so the high-speed range (450+ KTAS) is erroneous.

19.2.9 Best Range Airspeed for a Jet, V_{Rmax}

This airspeed is also the so-called Carson's speed, V_{CAR}, for both a jet and a propeller aircraft.

Requirement for Maximum Range for a Jet-powered Airplane

It can be seen from Equation (20-11) that the maximum range for a jet-powered airplane depends on the maximum of the ratio $C_L^{0.5}/C_D$. This maximum can be determined from the following expression, *which is only valid for the simplified drag model*:

$$\left(\frac{C_L^{0.5}}{C_D}\right)_{max} = \frac{3}{4}\left(\frac{1}{3k \cdot C_{Dmin}^3}\right)^{1/4} \quad (19-22)$$

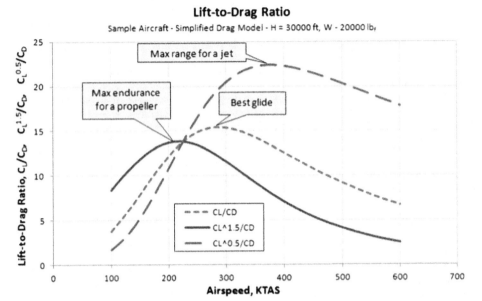

FIGURE 19-20 Various specific lift-to-drag ratios, using the simplified drag model.

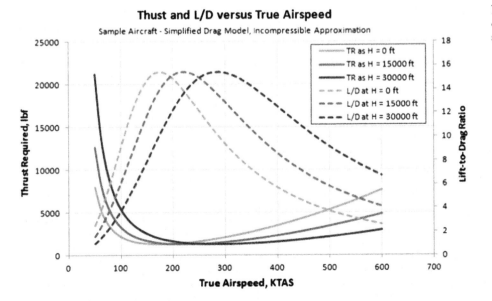

FIGURE 19-21 T_{Rmin} and LD_{max} occur at the same airspeed, using the simplified drag model.

This assumes that the TSFC is constant with power setting, but this ensures the minimum thrust for level flight results in minimum fuel consumption. Note that this may not be the case for real engines equipped with the Full Authority Digital Engine Control system (FADEC).

Derivation of Equation (19-22)

Start by writing the expression for L/D:

$$\frac{C_L^{1/2}}{C_D} = \frac{C_L^{1/2}}{C_{Dmin} + k \cdot C_L^2}$$

Differentiate with respect to C_L:

$$\frac{d}{dC_L}\left(\frac{C_L^{1/2}}{C_D}\right) = \frac{d}{dC_L}(f \cdot g) = f'g + fg'$$

$$\frac{d}{dC_L}\left(\frac{C_L^{1/2}}{C_D}\right) = \frac{d\left(C_L^{1/2}\right)}{dC_L}\left(\frac{1}{C_{Dmin} + k \cdot C_L^2}\right)$$

$$+ \left(C_L^{1/2}\right)\frac{d}{dC_L}\left(\frac{1}{C_{Dmin} + k \cdot C_L^2}\right)$$

$$= \frac{1}{2}\frac{C_L^{-1/2}}{C_{Dmin} + k \cdot C_L^2} + \left(C_L^{1/2}\right)\frac{d}{dC_L}(C_{Dmin} + k \cdot C_L^2)^{-1}$$

$$= \frac{1}{2}\frac{C_L^{-1/2}}{C_{Dmin} + k \cdot C_L^2} - \left(C_L^{1/2}\right)(C_{Dmin} + k \cdot C_L^2)^{-2}(2k \cdot C_L)$$

Further manipulation leads to:

$$\frac{d}{dC_L}\left(\frac{C_L^{3/2}}{C_D}\right) = \frac{1}{2}\frac{C_L^{-1/2}}{C_{Dmin} + k \cdot C_L^2} - \left(C_L^{1/2}\right)(C_{Dmin} + k \cdot C_L^2)^{-2}(2k \cdot C_L)$$

$$= \frac{\frac{1}{2}C_L^{-1/2}(C_{Dmin} + k \cdot C_L^2) - 2k \cdot C_L^{3/2}}{(C_{Dmin} + k \cdot C_L^2)^2} = 0$$

$$\Leftrightarrow \frac{1}{2}C_L^{-1/2}(C_{Dmin} + k \cdot C_L^2) - 2k \cdot C_L^{3/2} = 0$$

$$\Rightarrow \frac{1}{2}C_L^{-1/2}C_{Dmin} + \frac{1}{2}C_L^{-1/2}k \cdot C_L^2 - 2k \cdot C_L^{3/2} = 0$$

$$\Rightarrow \frac{1}{2}C_L^{-1/2}C_{Dmin} + \frac{1}{2}k \cdot C_L^{3/2} - 2k \cdot C_L^{3/2} = 0$$

$$\Rightarrow \frac{1}{2}C_L^{-1/2}C_{Dmin} - \frac{3}{2}k \cdot C_L^{3/2} = 0$$

$$\Rightarrow \frac{1}{2}C_L^{-1/2}C_{Dmin} = \frac{3}{2}k \cdot C_L^{3/2} \Leftrightarrow C_L = \sqrt{\frac{C_{Dmin}}{3k}}$$

Insert this result into the original expression and manipulate:

$$\left(\frac{C_L^{1/2}}{C_D}\right)_{max} = \left(\frac{C_L^{1/2}}{C_{Dmin} + k \cdot C_L^2}\right)_{max} = \frac{\left(\frac{C_{Dmin}}{3k}\right)^{1/4}}{C_{Dmin} + k \cdot \frac{C_{Dmin}}{3k}}$$

$$= \frac{3\left(\frac{C_{Dmin}}{3k}\right)^{1/4}}{4 C_{Dmin}} = \frac{3}{4}\left(\frac{1}{3kC_{Dmin}^3}\right)^{1/4}$$

QED

EXAMPLE 19-9

Determine the maximum range ratio for the Learjet 45XR:

Solution

$$\left(\frac{C_L^{1/2}}{C_D}\right)_{max} = \frac{3}{4}\left(\frac{1}{3kC_{Dmin}^3}\right)^{1/4} = \frac{3}{4}\left(\frac{1}{3 \cdot 0.05236 \cdot 0.020}\right)^{1/4}$$

$$= 22.4$$

Best Range Airspeed for a Jet

The airspeed that yields the maximum range of a jet is clearly an important parameter. The following expression, which is *only valid for the simplified drag model*, can be used to calculate this airspeed. It should be added that this airspeed is almost always too low to be practical for passenger transportation. As usual, the units for the airspeed are ft/s in the UK system and m/s in the SI system.

$$V_{R\max} = \sqrt{\frac{2}{\rho}\left(\frac{W}{S}\right)\sqrt{\frac{3k}{C_{D\min}}}} \qquad (19\text{-}23)$$

Derivation of Equation (19-23)

We showed that C_L for $C_L^{0.5}/C_D$ was given by:

$$C_L = \sqrt{\frac{C_{D\min}}{3k}}$$

Insert this into the lift equation and solve for V:

$$V = \sqrt{\frac{2W}{\rho S C_L}} = \sqrt{\frac{2W}{\rho S \sqrt{\frac{C_{D\min}}{3k}}}} = \sqrt{\frac{2}{\rho}\left(\frac{W}{S}\right)\sqrt{\frac{3k}{C_{D\min}}}}$$

QED

Comparison to the Best Glide Speed

Similar to an earlier comparison that used the specific results based on the simplified drag model, it is of interest to compare Equation (19-23) to Equation (19-20), repeated below for convenience:

$$V_{R\max} = \sqrt{\frac{2}{\rho}\left(\frac{W}{S}\right)\sqrt{\frac{3k}{C_{D\min}}}} \quad \text{and}$$

$$V_{LD_{\max}} = \sqrt{\frac{2}{\rho}\left(\frac{W}{S}\right)\sqrt{\frac{k}{C_{D\min}}}}.$$

We observe the following difference between the two:

$$\frac{V_{R\max}}{V_{LD_{\max}}} = \frac{\sqrt{\frac{2}{\rho}\left(\frac{W}{S}\right)\sqrt{\frac{3k}{C_{D\min}}}}}{\sqrt{\frac{2}{\rho}\left(\frac{W}{S}\right)\sqrt{\frac{k}{C_{D\min}}}}} = (3)^{1/4} \approx 1.32$$

In other words, the best range speed is about 32% greater than the speed for best L/D.

Carson's Airspeed

In a paper titled *Fuel Efficiency of Small Aircraft* [7], B. H. Carson discusses the mismatch between the amounts of power required for climb and cruise in small propeller-driven aircraft. In the paper Carson demonstrates that this excess power can be used more efficiently by bringing it closer to the so-called Gabrielli-Von Kárman [8] limit for vehicle performance. In the paper Carson states that (1) aircraft fuel economy is directly proportional to the L/D ratio, and the optimum is usually realized only at unacceptably low airspeeds. (2) Power required for climb results in aircraft airspeeds well beyond this optimum. (3) This results in greater fuel penalties than otherwise. In short, Carson suggests that flying at a speed faster than the airspeed for $LD_{\max}$ is more advantageous, as the flying public generally value a shorter time en route more than fuel efficiency. From a certain point of view, Carson's airspeed can be considered the "fastest efficient airspeed" to fly.

The most frequently cited measure of efficiency is the so-called *transport efficiency*, defined as WV/P, where W is the vehicle weight, V its speed of travel, and P is the installed power. Expanding on this idea, Carson derives a relation for a specific airspeed that is about 32% higher than the best glide airspeed (see Equation (19-20)). This speed is now recognized as Carson's airspeed.

Carson's airspeed:

$$V_{CAR} = 3^{0.25} V_{LD\max} \approx 1.32 V_{LD\max} \qquad (19\text{-}24)$$

Derivation of Equation (19-24)

The reader is directed to Ref. [7] for derivation and a paper by Smith [9] for additional discussion.

19.2.10 Maximum Level Airspeed, $V_{\max}$

This airspeed is also denoted by the variable V_H in 14 CFR Part 23.

The *maximum level airspeed* is of great interest in the marketing, certification, and operation of the aircraft. High maximum airspeed has a great marketing appeal, especially when compared to slower rival aircraft. From the standpoint of certification it is indicative not only of the magnitude of the maximum loads the airframe must react but also of the required aeroelastic resistance. Its impact on operation is reflected in its efficiency and this directly affects fuel consumption and, therefore, how it will be used. The maximum level airspeed is obtained as the positive value of the radical of Equation (19-3). It leads to the following expression, *which is only valid for the simplified drag model*:

$$V_{\max} = \sqrt{\frac{T_{\max} + \sqrt{T_{\max}^2 - 4C_{D\min}kW^2}}{\rho S C_{D\min}}} \qquad (19\text{-}25)$$

For the adjusted drag model use:

$$V_{max} = \sqrt{\frac{T + 2WkC_{LminD} + \sqrt{(T + 2WkC_{LminD})^2 - 4W^2k(C_{Dmin} + kC_{LminD}^2)}}{\rho S(C_{Dmin} + kC_{LminD}^2)}} \qquad (19\text{-}26)$$

The following observations can be made: (1) V_{max} increases with T_{max}; (2) V_{max} increases with W/S; (3) V_{max} increases if C_{Dmin} and/or k decreases (AR increases).

Special Case: Propeller Aircraft

It is unlikely that the thrust for a propeller-powered aircraft will be known at V_{max}, as it is a function of the airspeed itself. For this reason, the airspeed must be determined by iteratively solving the equation below.

$$\rho S C_{Dmin} V_{max}^3 = 550 \eta_p P_{BHP}$$
$$+ \sqrt{\left(550\eta_p P_{BHP}\right)^2 - 4W^2 V_{max}^2 C_{Dmin} k} \qquad (19\text{-}27)$$

Equation (19-27) can be solved using a multitude of methods, for instance the bisection method, regula falsi, or others. These functions require a single function to be solved, in which case the equation can be rewritten as the function $f(V_{max})$:

$$f(V_{max}) = \rho S C_{Dmin} V_{max}^3 - 550\eta_p P_{BHP}$$
$$- \sqrt{\left(550\eta_p P_{BHP}\right)^2 - 4W^2 V_{max}^2 C_{Dmin} k} \qquad (19\text{-}28)$$

A possible initial condition would then be written for $V_{max} = 0$:

$$f(0) = -1100 \eta_p P_{BHP}$$

Note that for other values of Equation (19-27), terms under the radical require the following to hold:

$$\left(550\eta_p P_{BHP}\right)^2 > 4W^2 V_{max}^2 C_{Dmin} k$$

Both methods are implemented in the computer codes in Sections 19.2.14 and 19.2.15.

Derivation of Equations (19-25) and (19-26)

Formulation is obtained from Equation (19-3) and (19-4), selecting the negative sign. Thus, for the simplified model, Equation (19-3) simply becomes:

$$V = \sqrt{\frac{T_{max} \pm \sqrt{T_{max}^2 - 4C_{Dmin}kW^2}}{\rho S C_{Dmin}}} \Leftrightarrow$$

$$V_{max} = \sqrt{\frac{T_{max} + \sqrt{T_{max}^2 - 4C_{Dmin}kW^2}}{\rho S C_{Dmin}}}$$

The process is identical for the adjusted drag model.

QED

Derivation of Equation (19-27)

Formulation is obtained by inserting Equation (14-38) into Equation (19-25):

$$V_{max} = \sqrt{\frac{T_{max} + \sqrt{T_{max}^2 - 4C_{Dmin}kW^2}}{\rho S C_{Dmin}}}$$

$$= \sqrt{\frac{\frac{550\eta_p P_{BHP}}{V_{max}} + \sqrt{\left(\frac{550\eta_p P_{BHP}}{V_{max}}\right)^2 - 4C_{Dmin}kW^2}}{\rho S C_{Dmin}}}$$

where P_{BHP} is the engine power rating in BHP. Manipulate algebraically:

$$V_{max} = \sqrt{\frac{\frac{550\eta_p P_{BHP}}{V_{max}} + \sqrt{\frac{(550\eta_p P_{BHP})^2}{V_{max}^2} - 4C_{Dmin}kW^2 \frac{V_{max}^2}{V_{max}^2}}}{\rho S C_{Dmin}}}$$

$$= \sqrt{\frac{550\eta_p P_{BHP} + \sqrt{(550\eta_p P_{BHP})^2 - 4W^2 V_{max}^2 C_{Dmin} k}}{V_{max} \rho S C_{Dmin}}}$$

From which we get:

$$V_{max}^2 = \frac{550\eta_p P_{BHP} + \sqrt{\left(550\eta_p P_{BHP}\right)^2 - 4W^2 V_{max}^2 C_{Dmin} k}}{V_{max} \rho S C_{Dmin}}$$

$$\Rightarrow \rho S C_{Dmin} V_{max}^3$$

$$= 550\eta_p P_{BHP} + \sqrt{\left(550\eta_p P_{BHP}\right)^2 - 4W^2 V_{max}^2 C_{Dmin} k}$$

QED

EXAMPLE 19-10

The POH for the SR22 gives a cruising speed of 169 KTAS at 55% power at 14,000 ft. Using the minimum drag coefficient extracted for the airplane in Example 15-18 ($C_{Dmin} = 0.02541$), solve Equation (19-27) iteratively to estimate V_{max} (which here is the cruising speed at 55%). The Oswald efficiency was calculated in Example 15-18 and amounts to 0.7566. Wing area is 144.9 ft² and density at 14,000 ft is 0.001546 slugs/ft³. Assume a propeller efficiency of 0.85 and max rated S-L power of 310 BHP.

Solution

Begin by calculating all coefficient products in Equation (19-27) (omitting to show units):

$$\rho S C_{Dmin} = 0.001546 \times 144.9 \times 0.02541 = 0.005691$$

$$550\eta_p P_{BHP} = 550 \times 0.85 \times (0.55 \times 310) = 79708.75$$

$$4W^2 C_{Dmin} k = 4W^2 C_{Dmin} \frac{1}{\pi \cdot AR \cdot e} = 48274.421$$

Inserting these into Equation (19-27) leads to:

$$\rho S C_{Dmin} V_{max}^3 = 550\eta_p P_{BHP}$$

$$+ \sqrt{\left(550\eta_p P_{BHP}\right)^2 - 4W^2 V_{max}^2 C_{Dmin} k}$$

$$\Rightarrow 0.005691 V_{max}^3$$

$$= 79709 + \sqrt{79709^2 - 48274 V_{max}^2}$$

Using this expression it is now possible to select an initial value of V_{max} and iterate until both sides are equal. It is computationally more convenient to calculate the difference between the two sides. Thus, if the difference is larger than zero, we will have to lower the value of V_{max} and vice versa. To do this, change the expression as follows:

$$\Delta = 0.005691 V_{max}^3$$

$$- \left(79709 + \sqrt{79709^2 - 48274 V_{max}^2}\right)$$

Let's begin by picking a token initial value for V_{max}, say 270 ft/s:

Iteration 1 — $V_{max} = 270$ ft/s:

$$\Delta_1 = 0.005691(270)^3 - \left(79709 + \sqrt{79709^2 - 48274(270)^2}\right)$$

$$= 112020.1698 - 132946.702 = -20926.53226$$

The value is negative so increase V_{max}.

Iteration 2 — $V_{max} = 280$ ft/s:

$$\Delta_2 = 0.005691(280)^3$$

$$- \left(79709 + \sqrt{79709^2 - 48274(280)^2}\right)$$

$$= -5458.251944$$

Since the difference is smaller than before, we are getting closer to a solution. It is still negative so let's increase V_{max} a tad, say 283.5 ft/s.

Iteration 3 — $V_{max} = 283.5$ ft/s:

$$\Delta_3 = 0.005691(283.5)^3$$

$$- \left(79709 + \sqrt{79709^2 - 48274(283.5)^2}\right)$$

$$= 233.6918365$$

Continuing in this manner, the value of V_{max} that yields a zero difference is 283.4 ft/s. This amounts to 167.9 KTAS, which compares very favorably with the POH value. Naturally, a more graceful scheme than this one should be attempted, for instance the so-called Newton-Raphson scheme or the bisection method.

EXAMPLE 19-11

The POH for the SR22 gives maximum cruising speed for altitudes ranging from 2000 to 17,000 ft. Using the routine above with Microsoft Excel, estimate the maximum level airspeed up to 25,000 ft and compare to the POH values. Use the same properties used in the previous examples, but account for the fact that the weight of the aircraft must reduce with altitude. That is, if the airplane takes off at 3400 lb$_f$, it will necessarily consume fuel while climbing and thus reduce in weight. The amount of fuel consumed is also given in the POH and by interpolating this data, the following expression was derived to calculate weight as a function of altitude:

$$W(H) = W_0 - 0.000000062809 \cdot H^2 - 0.0013875 \cdot H - 0.90965$$

where W_0 = gross weight = 3400 lb$_f$.

Solution

The solution was implemented in Microsoft Excel and the results presented in Figure 19-22 were obtained.

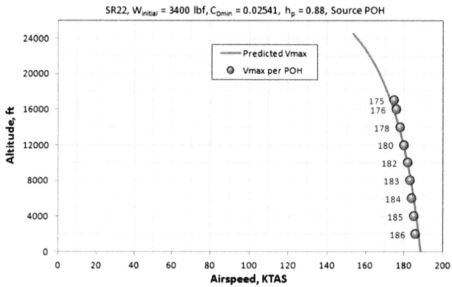

FIGURE 19-22 Comparing maximum airspeed predictions generated using the computer code presented in this chapter to actual flight data from the Pilot's Operating Handbook (POH) shows good agreement.

19.2.11 Flight Envelope

The *flight envelope* is one of several ways to demonstrate the capability of the aircraft in terms of its minimum and maximum airspeed with altitude. The preceding analyses make it possible to begin building such a diagram by plotting the stalling speed and maximum airspeed with altitude. The flight envelope is based on the fact that the performance of the airplane is a function of altitude. An example of this is shown in Figure 19-23 for the Learjet 45XR. It shows how altitude affects the drag and thrust of the airplane and modifies where the two curves intersect, where the maximum difference between the two occurs, and so on. A different presentation of this information is shown in Figure 19-24. It plots the maximum and minimum airspeed for the 45XR, assuming a maximum "clean configuration" lift coefficient of 1.3.

The flight envelope has very specific utility for the aircraft designer. First, it shows the capability of the aircraft over its entire altitude range. The graph is fundamentally simple and can thus be helpful to management and customers and is relatively simple even for laypeople to understand. It is appropriate to also superimpose V_X and V_Y on the graph to further indicate the

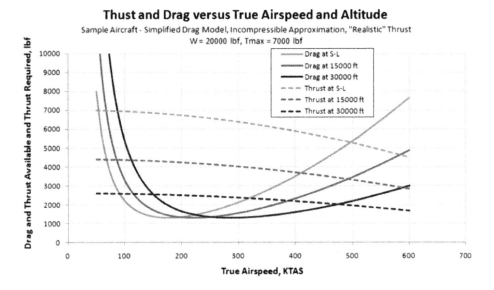

FIGURE 19-23 Required and available thrust at selected altitudes.

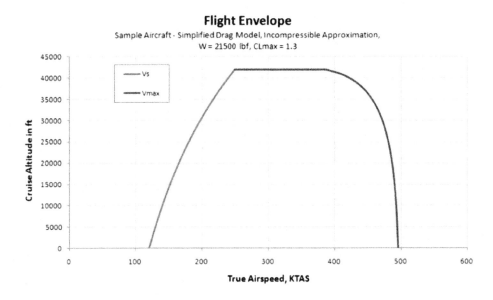

FIGURE 19-24 Flight envelope for the Learjet 45XR at gross design weight.

airplane's performance relative to its low- and high-speed limitations. Second, it should be used to create a set of airspeeds and altitudes at which the dynamic stability characteristics and aeroelastic test matrix for the aircraft can be established. For instance, the designer should map a matrix of points (e.g. minimum, maximum, and the average of the two) at selected altitudes (e.g. 0, 5000, 10,000 ft, etc.) and evaluate phugoid and dutch roll damping at those points. It is possible the flight envelope has to be limited further if points of instability are discovered. Third, a similar matrix should be prepared for aeroelastic evaluation of the structure. This allows the flight test team to plan where structural shakers should be activated during flutter testing.

Coffin Corner

The maximum airspeed of high-flying jet aircraft is called the M_{MO}, or the *maximum operating Mach number*. Soon after exceeding this value, local airspeeds over the airplane begin to exceed Mach 1 and a weak shock begins to form. This will cause several anomalies: first, there is a sharp rise in drag. Second, there is a change in pitching moment as the center of lift on the lifting surfaces begins to move to a different location, eventually moving as far aft as 50% of the MGC. This greatly increases the distance between the CG and the center of lift, leading to the phenomenon called *Mach tuck*. A Mach tuck is essentially a powerful nose pitch-down moment that is uncontrollable for aircraft not designed to handle it. Therefore, the airplane will dive, increasing its speed further, and this may lead to structural failure during recovery, and certain demise if recovery is not possible. The side effects are more complicated than that, as the shock formation may also lead to abnormal shaking and un-commanded wing rocking. Third, compounding the problem, the formation of shockwaves on the lifting surfaces leads to a *shock-stall* (see Figure 8-43). This means that the control surfaces are now operating in separated wake, rendering them far less effective.

As shown above, as the airplane increases its altitude, the stalling speed in terms of true airspeed increases as well, while its calibrated airspeed remains the same. Consider the graph of Figure 19-25, which shows Mach isobars extending from S-L to 80,000 ft plotted against the airspeed in KCAS. The graph allows the altitude at which a given airspeed reaches a set Mach number to be determined. For instance, 200 KCAS becomes Mach 0.7 at approximately 40,000 ft. Now consider a high-flying aircraft whose M_{MO} is 0.80 and stalling speed, V_S, is 150 KCAS. This is represented as the thick solid and dotted curves in the figure.

As the aircraft continues to climb, and assuming a constant airspeed in KCAS, its Mach number increases. Eventually, the aircraft will near its M_{MO} and the indicated airspeed (or KCAS assuming an ideal airspeed indication system) will have to be reduced. For instance, if the indicated airspeed during climb is 200 KCAS it can be seen that once the airplane approaches 45,000 ft, the pilot will have to reduce the airspeed to avoid reaching the M_{MO}. Then, if the climb is continued, at approximately 57,000 ft a new dilemma appears: the required indicated airspeed is now nearing the stalling speed. Thus, if the pilot slows down further, the plane will stall and stalling at this condition may easily result in a dive that takes the airplane beyond M_{MO}. If the airplane accelerates, the plane will hit M_{MO}. This peculiar situation is called the *coffin corner* because any change in airspeed can potentially lead to a very dangerous situation. Flying near the coffin corner requires the pilot to be alert and precise, because it only takes flying into a mass of air whose temperature changes rapidly or encountering a clear air turbulence (CAT) to upset the equilibrium.

Reality is a tad more complicated than reflected here, as a typical airplane would reduce its weight as it climbs and for that reason the vertical stall speed line in Figure 19-25 would be slanted to the left. However, since the stalling speed is really a function of the equivalent airspeed the effect would slant it to the right. The resulting stall speed line would thus depend on the compressibility and the weight. However, in spite of that, the effects are reasonably represented for demonstration purposes in this graph.

In order to create the graph of Figure 19-25 the calibrated airspeed is determined for a given Mach number and altitude. For instance, the true airspeed in knots corresponding to $M = 0.1$ at S-L is $0.1 \times$ (1116 ft/s)/(1.688 ft/s per knot) or 66.1 KTAS. To convert this to KCAS use Equation (16-33). Do this for a range of altitudes and Mach numbers ranging from 0.1 to 0.9 as shown.

19.2.12 Power Required

If we know the lift and drag coefficient associated with a specific flight condition, then it is possible to calculate the power required for level flight at that condition. This can be determined using the expression below:

$$P_{REQ} = \sqrt{\frac{2W^3 C_D^2}{\rho S C_L^3}} \qquad (19\text{-}29)$$

In the UK system, the resulting value can be converted to BHP by a division by 550. In the SI system the units are in watts or joules/second. Note that by inspection it can be seen that the power required to propel

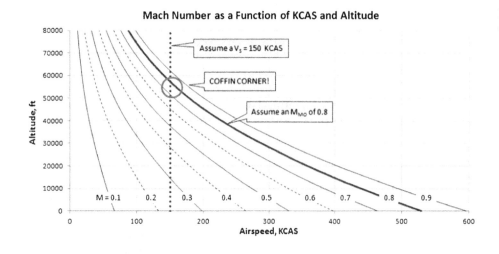

FIGURE 19-25 Determination of the coffin corner.

the aircraft at a given flight condition (specified through C_L and C_D) is:

$$P_{REQ} \propto \frac{C_D}{C_L^{3/2}} \quad (19\text{-}30)$$

So the smaller this ratio, the less is the power required to maintain level flight. When this ratio becomes a minimum the airspeed of minimum power required has been achieved, but this is also the maximum of the ratio $C_L^{3/2}/C_D$. This can be calculated from Equation (19-11) and explains why this represents the condition for minimum power required or max endurance for a propeller-powered airplane.

Derivation of Equation (19-30)

Preliminary definitions:

$$P_{REQ} = T_{REQ} \cdot V \quad \text{and} \quad \frac{C_L}{C_D} = \frac{L}{D} = \frac{W}{T} \quad \text{and}$$

$$V = \sqrt{\frac{2W}{\rho S C_L}}$$

Therefore:

$$P_{REQ} = T_{REQ} \cdot V = \left(\frac{W}{C_L/C_D}\right)_{REQ} \cdot \sqrt{\frac{2W}{\rho S C_L}}$$

$$= \frac{W}{C_L/C_D} \cdot \sqrt{\frac{2W}{\rho S C_L}} = \sqrt{\frac{2W^3 C_D^2}{\rho S C_L^3}}$$

QED

19.2.13 Power Available for a Piston-powered Aircraft

If the power generated by piston engine and propeller efficiency is known, the power available for propulsion is defined using the expression below:

$$P_{AV} = T_{AV} \cdot V = \eta_p \cdot 550 \cdot P_{BHP} \quad (19\text{-}31)$$

Note that η_p is the propeller efficiency and it is a function of the airspeed, RPM, and propeller geometry. Refer to Chapter 14, *The anatomy of the propeller*, for more information.

19.2.14 Computer code: Determining Maximum Level Airspeed, V_{max}, for a Propeller Aircraft

The following Visual Basic for Applications functions are used to determine V_{max} for a propeller-powered aircraft by solving Equation (19-27) using the bisection method (see Article E.6.20 STEP-BY-STEP: *Finding roots using the bisection method*). Both functions must be used, as the first one calls the second one. Note that as long as the value under the radical of Equation (19-27) is greater than zero, a solution is possible. The arguments are explained in the comment section of the code. Also note that the routine calculates the density (rho) by calling the routine AtmosProperty (see Article A.2.10, *Computer code A-1: Atmospheric modeling*) using the argument H. This line can easily be replaced with Equation (16-18).

```
Function PERF_Vmax_Prop(S As Single, k As Single, CDmin As Single, W As Single, H As Single, BHP As
Single, eta As Single) As Single
'This routine determines the maximum speed of a propeller powered aircraft by solving
'Equation (19-27) using the Bisection Method. It returns the airspeed in ft/s.
'
'Input values:   S     = Wing area in ft²
'                k     = Lift-induced drag constant
'                CDmin = Minimum drag coefficient
'                W     = Weight in lbf
'                H     = Altitude in ft
'                BHP   = Piston engine power in BHP
'                eta   = Propeller efficiency
'
'Initialize
    Dim Counter As Single, flag As Byte
    'Variables for the Bisection Method
    Dim V0 As Single, V1 As Single, Vmid As Single
    Dim F0 As Single, Fmid As Single
    Dim rho As Single
```

```
'Presets
    rho = AtmosProperty(H, 12)
    Counter = 0
    V0 = 0
    V1 = 500
    F0 = -1100 * eta * BHP
'Process
    Do
        'Set flag
        flag = 1
        'Advance counter
        Counter = Counter + 1
        'Compute midpoint values
        Vmid = 0.5 * (V0 + V1)
        Fmid = PERF_f_of_V(rho, S, CDmin, k, W, BHP, eta, Vmid)
        'Use logic
        If F0 * Fmid < 0 Then
            V1 = Vmid
        Else
            V0 = Vmid
            F0 = Fmid
        End If
        'Evaluate difference and adjust Vini for next iteration
        If Abs(V1 - V0) < 0.0001 Then flag = 0
    Loop Until flag = 0 Or Counter = 100
'Return results (return -1 if solution was not found)
    PERF_Vmax_Prop = 0.5 * (V0 + V1)
    If Counter = 100 Then PERF_Vmax_Prop = -1
End Function
Function PERF_f_of_V(rho As Single, S As Single, CDmin As Single, k As Single, W As Single, BHP As Single,
eta As Single, V As Single) As Single
'This routine calculates the value of Equation (19-27).
'
'Initialize
    Dim K1 As Single
'Presets
    K1 = (550 * eta * BHP) ^ 2 - (2 * W * V) ^ 2 * CDmin * k
'This is a trick to ensure the routine can solve to higher altitudes
    If K1 < 0 Then K1 = 0
'Calculate the value of Equation (19-27)
    PERF_f_of_V = rho * S * CDmin * V ^ 3 - 550 * eta * BHP
    PERF_f_of_V = PERF_f_of_V - Sqr(K1)
End Function
```

19.2.15 Computer code: Determining Maximum Level Airspeed, V_{max}, for a Jet

The following Visual Basic for Applications routine can be used to determine V_{max} based on Equation (19-25).

```
Function PERF_Vmax_Jet(H As Single, Tmax As Single, S As Single, W As Single, CDmin As Single, CLminD As
Single, k As Single, Mode As Byte) As Single
'This routine calculate the maximum airspeed for an aircraft whose maximum thrust
'is specified. The routine is only valid for the
'
'Input values:   H       = Altitude in ft
'                Tmax    = Maximum thrust in lbf
```

```
•              S     = Wing area in ft²
•              W     = Weight in lbf
•              CDmin = Minimum drag coefficient
•              CLminD = Lift coefficient where drag is minimum
•              k     = Lift-induced drag constant
•              Mode  = 0 to use the simplified drag model, = 1 to use adj. model
•
'Initialize
    Dim Radical As Single
'Process
    If Mode = 0 Then 'Simplified drag model
        Radical = Tmax ^ 2 - 4 * k * CDmin * W ^ 2
        If Radical >= 0 Then
            PERF_Vmax_Jet = Sqr((Tmax + Sqr(Radical)) / (AtmosProperty(H, 12) * S * CDmin))
        Else
            PERF_Vmax_Jet = 0
        End If
    ElseIf Mode = 1 Then 'Adjusted dragmodel
        Radical = (Tmax + 2 * W * k * CLminD) ^ 2 - 4 * k * (CDmin + k * CLminD ^ 2) * W ^ 2
        If Radical >= 0 Then
            PERF_Vmax_Jet = Sqr((Tmax + 2 * W * k * CLminD + Sqr(Radical)) / (AtmosProperty(H, 12) * S *
                (CDmin + k * CLminD ^ 2)))
        Else
            PERF_Vmax_Jet = 0
        End If
    End If
End Function
```

19.3 GENERAL ANALYSIS METHODS FOR ACCELERATED FLIGHT

The previous section developed analysis methods for aircraft in steady level flight. In this section we will introduce a number of common analysis methods intended to evaluate a number of maneuvers that involve accelerated flight. Such maneuvers include turning flight, pull-up (loop), and accelerated rate-of-climb. All utilize the simplified drag model.

19.3.1 Analysis of a General Level Constant-velocity Turn

Consider the aircraft in Figure 19-26, which is banking at an angle ϕ. In order for the airplane to maintain altitude (no slipping or skidding) its lift must balance the weight while generating a centripetal force component that balances the centrifugal force component. The resulting motion renders a steady heading change. This requires the magnitude of the lift to be larger than the weight of the aircraft (otherwise the airplane will lose altitude). Consequently, the airframe is loaded up beyond what would happen in level flight and this additional loading is represented in the load factor, defined as $n = L/W$. The set of equations describing the motion of the airplane in this condition is written as follows:

Fore-aft forces:
$$T - D = 0 \quad (19\text{-}32)$$

Vertical forces:
$$L \cos \phi - W = 0 \quad (19\text{-}33)$$

Lateral forces:
$$L \sin \phi - \frac{W}{g} \frac{V^2}{R_{turn}} = 0 \quad (19\text{-}34)$$

The level constant velocity turn can be analyzed using the following set of equations:

Bank angle:
$$\phi = \cos^{-1}\left(\frac{1}{n}\right) \quad (19\text{-}35)$$

Load factor:
$$n = \frac{1}{\cos \phi} = \left(\frac{T}{W}\right)\left(\frac{L}{D}\right) \quad (19\text{-}36)$$

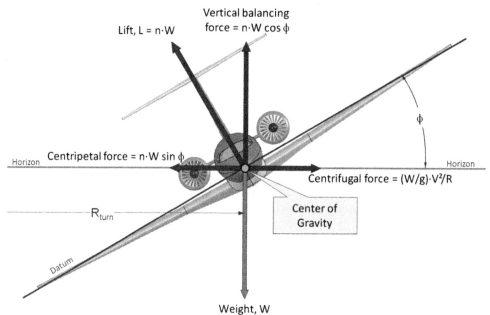

FIGURE 19-26 Forces on an aircraft in a level constant-velocity turn.

Turn radius:

$$R_{turn} = \frac{V^2}{n \cdot g \cdot \sin\phi} = \frac{V^2}{g\sqrt{n^2-1}} \quad (19\text{-}37)$$

Time to turn ψ degrees:

$$t_\psi = \frac{R_{turn}}{V}\left(\psi\frac{\pi}{180}\right) \quad (19\text{-}38)$$

Turn rate in radians/sec:

$$\dot\psi = \frac{g\sqrt{n^2-1}}{V} = \frac{V}{R_{turn}} \quad (19\text{-}39)$$

Thrust required at a load factor n:

$$T_R = qS\left[C_{Dmin} + k\left(\frac{nW}{qS}\right)^2\right] + D_{trim} \quad (19\text{-}40)$$

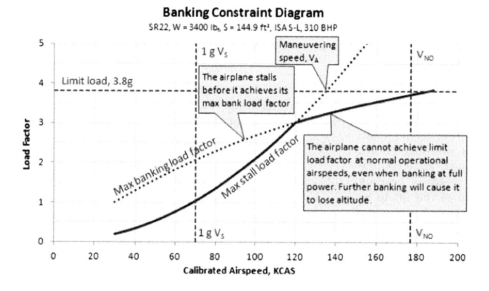

FIGURE 19-27 A banking constraint diagram for the SR22. The maximum stall load factor is calculated by solving Equation (19-8) for the load factor. The maximum banking load factor is calculated using Equation (19-41).

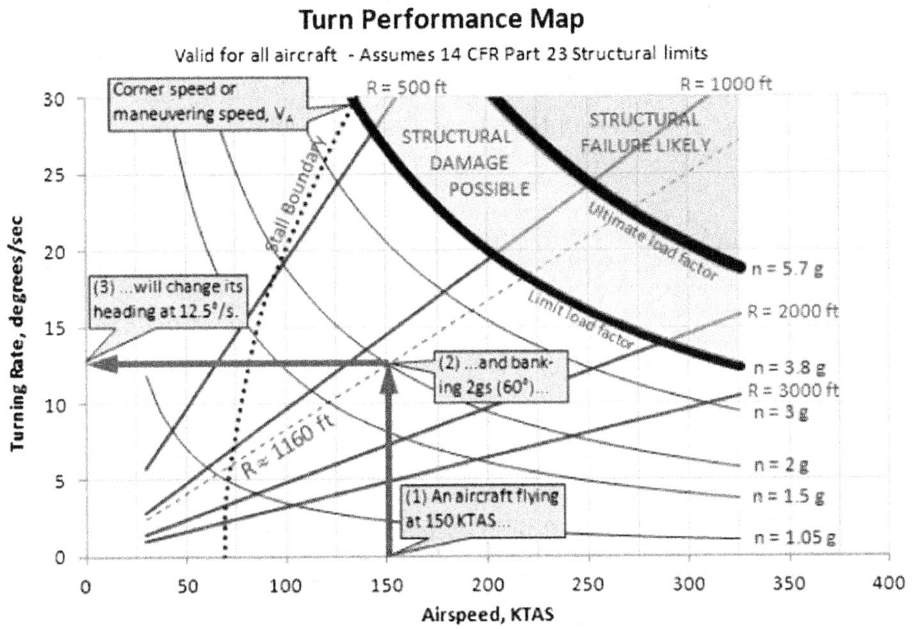

Turn Performance Map

Valid for all aircraft - Assumes 14 CFR Part 23 Structural limits

FIGURE 19-28 A turn performance map is constructed using Equation (19-39).

The load factor that can be sustained at a given thrust, T, and airspeed, V, can be obtained as shown below. Note that inserting the maximum thrust will yield the maximum load factor at a given airspeed:

$$n = \frac{qS}{W}\sqrt{\frac{1}{k}\left(\frac{T}{qS} - C_{Dmin}\right)} \quad (19\text{-}41)$$

This expression is used to plot a part of the banking constraint diagram of Figure 19-27.

Airspeed for a given C_L:

$$V = \sqrt{\frac{2W}{\rho S C_L}\left(\frac{1}{\cos\phi}\right)} \quad (19\text{-}42)$$

A common way to present the turn performance of an aircraft can be seen in the *banking constraint diagram* of Figure 19-27 and the *turn performance map* of Figure 19-28. Both present a convenient way to show the banking capability of an aircraft. First, consider the banking constraint diagram of Figure 19-27, here based on the SR22 sample aircraft. The straight dashed horizontal line shows the limit load factor of $3.8g$. The two vertical dashed lines show the clean stalling speed at normal $1g$ loading (V_S, to the left) and the normal operating speed (V_{NO}, to the right). These two lines effectively enclose the normal speed range of the aircraft. The solid curve, labeled "Max stall load factor," shows the stalling speed of the aircraft at various load factors. Thus, it can be seen that at a load factor of $3g$, the airplane will stall at about 120 KCAS. This curve is a part of the standard V-n diagram. The dotted curve, labeled "Max banking load factor," shows the maximum gs the aircraft can bank at while maintaining altitude. It can be seen that between V_S and 120 KCAS, the airplane will simply stall before achieving its maximum "theoretical" banking load factor. Thus, at 100 KCAS, if the airplane didn't stall first (at $n \approx 2.1g$) it could achieve $n = 2.1g$ before it would begin to lose altitude. At airspeeds beyond 120 KCAS, the airplane cannot achieve $3.8g$ (its limit load factor) while maintaining altitude. This means that when flight testing the aircraft for structural flight tests (e.g. per 14 CFR 23.307, *Proof of structure*), means other than constant-altitude banking may have to be considered. Conversely, for some other aircraft, it might reveal the aircraft can indeed exceed the limit load factor in some airspeed range.

The turn performance map of Figure 19-28 is a cross-plot of Equation (19-39). It is generated by plotting curves for constant turn radius (the straight lines) and then for constant load factors (the curves). Since the equation does not involve any variables dependent on particular aircraft geometry, it is valid for all aircraft, although Figure 19-28 has been drawn up for aircraft that comply with 14 CFR Part 23 (and the stall boundary varies from airplane to airplane). The map shows how rapidly an aircraft can maneuver at specific airspeeds. The maneuvering speed is where the stall boundary intersects the curve for the limit load factor. It is also called the *corner speed* and is the lowest airspeed where the airplane achieves its maximum bank angle, most rapid heading change, and minimum turning radius.

Derivation of Equation (19-35)

Load factor n is defined as $n = L/W$. From Figure 19-26 we readily see that:

$$W = L\cos\phi = nW\cos\phi \iff \cos\phi = \frac{1}{n}$$

QED

Derivation of Equation (19-36)

Divide Equation (19-33) by (19-32):

$$\frac{L\cos\phi - W = 0}{T - D = 0} \Rightarrow \frac{L\cos\phi}{D} = \frac{W}{T}$$

$$\Rightarrow \frac{L}{D}\frac{1}{n} = \frac{W}{T} \iff n = \left(\frac{T}{W}\right)\left(\frac{L}{D}\right)$$

QED

Derivation of Equation (19-37)

Centrifugal force corresponding to the force diagram in Figure 19-26 can be found from the standard curvilinear relation mV^2/R:

$$m\frac{V^2}{R_{turn}} = \left(\frac{W}{g}\right)\frac{V^2}{R_{turn}} = nW\sin\phi$$

Manipulating algebraically leads to:

$$\left(\frac{W}{g}\right)\frac{V^2}{R_{turn}} = nW\sin\phi \Rightarrow \frac{V^2}{R_{turn}}$$

$$= ng\sin\phi \Rightarrow R_{turn} = \frac{V^2}{ng\sin\phi}$$

QED

Derivation of Equation (19-38)

The distance the airplane covers in the turn at an airspeed V in time t_ψ is equal to the standard arc length of a circle of radius R_{turn} through the angle ψ. In other words:

$$\text{Distance} = R_{turn} \times \psi = V \times t_\psi$$

By solving for t_ψ in the above expression and noting the angle to be used must be in radians (note the conversion factor $\pi/180$) we get the expression for the time to turn.

QED

Derivation of Equation (19-39)

Turn rate is the change in heading with respect to time and can be written as follows:

$$\dot{\psi} = \frac{d\psi}{dt} \cong \frac{\psi}{t_{turn}} = \frac{\psi}{\frac{R_{turn} \times \psi}{V}} = \frac{V}{R_{turn}} = \frac{V}{\frac{V^2}{g\sqrt{n^2-1}}}$$

$$= \frac{g\sqrt{n^2-1}}{V}$$

QED

Derivation of Equation (19-40)

The level constant-velocity turn requires thrust to equal the drag of the airplane in the turn, in other words (assuming the simplified drag model):

$$T_R = D = qSC_D \Rightarrow T_R = qS\left[C_{Dmin} + kC_L^2\right]$$

$$= qS\left[C_{Dmin} + k\left(\frac{nW}{qS}\right)^2\right]$$

Since elevator is required to trim the airplane in the turn, the increase in trim (trim drag) should be considered if it is significant. This term, D_{trim}, is shown in Equation (19-40).

QED

Derivation of Equation (19-41)

First note that, typically, D_{trim} is around 1–2% of the total drag of the airplane and, thus, ignoring it will yield acceptable accuracy. Use Equation (19-40), assuming $D_{trim} = 0$ and solve for the load factor, n.

$$T_R = qS\left[C_{Dmin} + k\left(\frac{nW}{qS}\right)^2\right] \Rightarrow \frac{T_R}{qS} - C_{Dmin}$$

$$= k\left(\frac{nW}{qS}\right)^2 \Rightarrow n = \frac{qS}{W}\sqrt{\frac{1}{k}\left(\frac{T_R}{qS} - C_{Dmin}\right)}$$

Then, assuming a given thrust, T, the maximum load factor can be determined.

QED

EXAMPLE 19-12

What bank angle will require a 2g load to be reacted by the aircraft? If it is flying at 200 KTAS at 10,000 ft, what is the radius of the turn? How much time will it take to complete a full circle?

Solution

Angle of bank:

$$\phi = \cos^{-1}\left(\frac{1}{n}\right) = \cos^{-1}\left(\frac{1}{2}\right) = 60°$$

Turn radius:

$$R_{turn} = \frac{V^2}{g\sqrt{n^2-1}} = \frac{(200 \times 1.688)^2}{(32.174)\sqrt{(2)^2-1}} = 2045 \text{ ft}$$

Time to turn 360°:

$$t_\psi = \frac{R_{turn}}{V}\left(\psi\frac{\pi}{180}\right) = t_\psi = \frac{2045}{(200 \times 1.688)}(2\pi)$$
$$= 38 \text{ sec}$$

19.3.2 Extremes of Constant-velocity Turns

Maximum Sustainable Load Factor, n_{max}

The maximum load factor that the aircraft can sustain without stalling is obtained from Equation (19-36) when the thrust-to-weight and lift-to-drag ratios are at their maximum values:

Max sustainable load factor:

$$n_{max} = \left(\frac{T_{max}}{W}\right)LD_{max} \qquad (19\text{-}43)$$

Maximum Sustainable Turn Rate, $\dot{\psi}_{max}$

The maximum turn rate is a very important indicator of an airplane's maneuverability. A large T/W and $AR \cdot e$ combined with a low W/S and altitude yield the smallest turning radius. This is the fastest heading change the airplane can perform and is given by the following relation:

Turn rate in radians/sec:

$$\dot{\psi}_{max} = \frac{g\sqrt{n_{max}^2 - 1}}{V_{max\dot{\psi}}} \qquad (19\text{-}44)$$

where $V_{max\dot{\psi}}$ = fastest turn velocity, given by Equation (19-15), repeated below for convenience.

$$V_{max\dot{\psi}} = V_{TRmin} = \sqrt{\frac{2}{\rho}\left(\frac{W}{S}\right)}\sqrt{\frac{k}{C_{Dmin}}} \qquad (19\text{-}15)$$

A derivation of this result is given by Asselin [10].

Minimum Sustainable Turning Radius

The minimum sustainable turning radius is another important indicator of an airplane's maneuverability. A large T/W and $AR \cdot e$ combined with a low W/S and altitude yield the smallest turning radius. It can be calculated from the following relation:

Turn radius in radians:

$$R_{min} = \frac{V_{R_{min}}^2}{g\sqrt{n_{R_{min}}^2 - 1}} \qquad (19\text{-}45)$$

where $n_{R_{min}} = \sqrt{2 - 1/n_{max}^2}$ = load factor for minimum turning radius

$V_{R_{min}} = 2\sqrt{\frac{(W/S)}{(T/W)}\frac{k}{\rho}}$ = airspeed for minimum turning radius

A derivation of this result is given by Asselin [10].

Maximum Bank Angle

This is the maximum angle the aircraft can bank while maintaining altitude (provided it has enough power or thrust) and sustain the limit load factor it has been designed to. It can simply be determined using Equation (19-35) with n_{lim} being the limit load factor:

Maximum level bank angle:

$$\phi_{max} = \cos^{-1}\left(\frac{1}{n_{max}}\right) \qquad (19\text{-}46)$$

Airspeed for Maximum Bank Angle

Using the simplified drag model, the airspeed required to reach the limit load factor for a given thrust setting is given by the following expression:

$$V_{\lim} = \sqrt{\frac{\left(T \pm \sqrt{T^2 - 4kC_{Dmin}(n_{\lim}W)^2}\right)}{\rho S C_{Dmin}}} \qquad (19\text{-}47)$$

The expression will return two airspeeds, one for each sign. These represent low- and high-speed conditions.

Derivation of Equation (19-47)

Consider Equation (19-41) for turning load factor:

$$n = \frac{qS}{W}\sqrt{\frac{1}{k}\left(\frac{T}{qS} - C_{Dmin}\right)}$$

Begin by solving for the dynamic pressure when banking at the limit load, $n_{\lim}$:

$$n_{\lim}^2 = \left(\frac{q_{\lim}S}{W}\right)^2\left(\frac{1}{k}\left(\frac{T}{q_{\lim}S} - C_{Dmin}\right)\right)$$

$$\Leftrightarrow \frac{kn_{\lim}^2 W^2}{S^2} = (q_{\lim}^2)\left(\frac{T}{q_{\lim}S} - C_{Dmin}\right)$$

$$\Rightarrow k\left(\frac{n_{\lim}W}{S}\right)^2 = \frac{T}{S}q_{\lim} - C_{Dmin}q_{\lim}^2$$

$$\Rightarrow C_{Dmin}q_{\lim}^2 - \frac{T}{S}q_{\lim} + k\left(\frac{n_{\lim}W}{S}\right)^2 = 0$$

This can be solved as a quadratic equation as shown below:

$$q_{\lim} = \frac{1}{2}\rho V_{\lim}^2 = \frac{\frac{T}{S} \pm \sqrt{\left(\frac{T}{S}\right)^2 - 4(C_{Dmin})k\left(\frac{n_{\lim}W}{S}\right)^2}}{2C_{Dmin}}$$

Further manipulations lead to:

$$V_{\lim} = \sqrt{\frac{\frac{T}{S} \pm \sqrt{\left(\frac{T}{S}\right)^2 - 4(C_{Dmin})k\left(\frac{n_{\lim}W}{S}\right)^2}}{\rho C_{Dmin}}}$$

$$= \sqrt{\frac{\left(T \pm \sqrt{T^2 - 4kC_{Dmin}(n_{\lim}W)^2}\right)}{\rho S C_{Dmin}}}$$

QED

EXAMPLE 19-13

What is the maximum bank angle of the Learjet 45XR and the airspeed at which a limit load factor of 3.5g can be achieved at S-L, if it weighs 20,000 lbf with 7000 lb$_f$ of thrust?

Solution

Maximum bank angle:

$$\phi_{\max} = \cos^{-1}\left(\frac{1}{n_{\lim}}\right) = \cos^{-1}\left(\frac{1}{3.5}\right) = 73.4°$$

Airspeed required to achieve the limit load factor:

$$V_{\lim} = \sqrt{\frac{\left(T \pm \sqrt{T^2 - 4kC_{Dmin}(n_{\lim}W)^2}\right)}{\rho S C_{Dmin}}}$$

$$V_{\lim} = \sqrt{\frac{\left(7000 \pm \sqrt{7000^2 - 4(0.05236)(0.020)(3.5 \times 20000)^2}\right)}{(0.002378)(311.6)(0.020)}}$$

$$= \begin{cases} 335.1 \text{ ft/s } (199 \text{ KTAS}) \\ 912.4 \text{ ft/s } (541 \text{ KTAS}) \end{cases}$$

19.3.3 Energy State

Energy Height

The total energy of the airplane whose mass and weight are given by m and W, respectively, flying at an altitude h and airspeed V is a linear combination of its potential and kinetic energy and can be computed from:

$$E_{total} = mgh + \frac{1}{2}mV^2 = Wh + \frac{1}{2}\frac{W}{g}V^2 \qquad (19\text{-}48)$$

Specific Energy and Energy Height

The *specific energy* is defined as the total energy per unit weight and can be computed as follows:

$$H_E \equiv \frac{E_{total}}{W} = h + \frac{V^2}{2g} \qquad (19\text{-}49)$$

Since the units of specific energy are that of height (ft or m) it is also called *energy height*. This concept highlights that the maneuvering of an airplane can be considered an exchange of potential and kinetic energy. To explain what this means, consider an airplane cruising at an altitude of 10,000 ft at airspeed of 236 KTAS (400 ft/s) as shown in Figure 19-29. Its specific energy is then $10,000 + 400^2/(2 \cdot 32.174) = 12,768$ ft. This means that if the pilot exchanged all the kinetic energy into potential energy, by raising the nose of the aircraft and allowing it to climb until the airspeed drains to zero (this is a maneuver called *zooming*), the airplane would reach an altitude of 12,768 ft.

The graph of Figure 19-29 is called a *constant energy height map*. It consists of isobars of constant energy height that extend from the vertical to the horizontal axis. The airspeed at any altitude can be calculated for a given energy height by solving for the airspeed in Equation (19-49) as follows:

$$V = \sqrt{2g(H_E - h)} \qquad (19\text{-}50)$$

Equation (19-50) was used to create the constant energy height map of Figure 19-29. The figure shows isobars for energy heights (H_E) of 5000, 10,000, 15,000, and 20,000 ft, with the one of 12,768 ft shown as a dashed line. Furthermore, the exchange from the initial altitude of 10,000 ft and 250 KTAS to 12768 and 0 KTAS is shown as well. The plot applies to all aircraft, regardless of weight. A more type-dependent representation is obtained by determining and plotting the *specific excess power* contour plots (see below).

Specific Excess Power

Just as the specific energy was defined as the total energy per unit weight, we also define *specific excess power* as the excess power (per Equation (18-15)):

$$P_S \equiv \frac{P_{EX}}{W} = \frac{TV - DV}{W} \qquad (19\text{-}51)$$

The specific excess power can also be written as follows:

$$P_S = \frac{dh}{dt} + \frac{V}{g}\frac{dV}{dt} \qquad (19\text{-}52)$$

The expression shows that the specific excess power of an airplane is the combination of its rate-of-climb

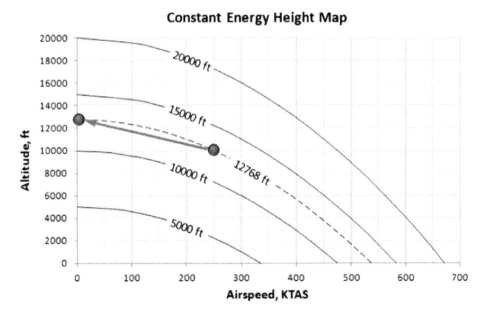

FIGURE 19-29 Constant energy height isobars.

(dh/dt) and forward acceleration ($V/g \cdot dV/dt$). Thus, if $dV/dt = 0$ (steady state), the specific excess power is simply the *ROC* of the airplane. Likewise, if $dh/dt = 0$, the specific excess power is simply its acceleration. An inspection of Equation (19-49) reveals its time derivative equals the specific excess power, that is:

$$\frac{d}{dt}\left(h + \frac{V^2}{2g}\right) = \frac{dH_e}{dt} = \frac{dh}{dt} + \frac{V}{g}\frac{dV}{dt} \quad (19\text{-}53)$$

In short, the specific excess power is the time rate of change of the energy height.

$$P_S = \frac{dH_e}{dt} \quad (19\text{-}54)$$

Derivation of Equation (19-52)

Begin with the dynamic version of the equations of motion, i.e. Equation (18-2), repeated here for convenience (assuming the thrust angle $\varepsilon = 0$):

$$-D - W\sin\theta + T = \frac{W}{g}\frac{dV}{dt} \quad (18\text{-}2)$$

This can be rewritten as follows:

$$T - D = W\sin\theta + \frac{W}{g}\frac{dV}{dt} = W\left(\sin\theta + \frac{1}{g}\frac{dV}{dt}\right) \quad (i)$$

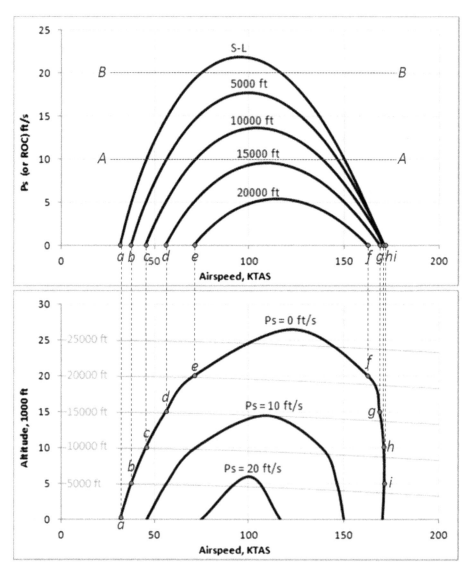

FIGURE 19-30 Constructing a specific excess power contour plot for an SR22 class aircraft. The light-colored curves (labeled 5000 ft through 25000 ft) are the constant energy height isobars plotted in Figure 19-29.

Multiply by V/W:

$$\frac{TV - DV}{W} = \frac{P_{EX}}{W} = P_S = V \sin\theta + \frac{V}{g}\frac{dV}{dt} \quad \text{(ii)}$$

Noticing that $V\sin\theta = $ rate-of-climb $= dh/dt$ we can rewrite Equation (ii) as follows:

$$P_S = \frac{dh}{dt} + \frac{V}{g}\frac{dV}{dt} \quad (19\text{-}52)$$

QED

Constructing a Specific Excess Power Contour Plot

A specific excess power contour plot is used to demonstrate the maneuvering capability of the aircraft throughout its operational airspeed and altitude range. An example of such a contour plot is shown as the lower graph of Figure 19-30. It is typically constructed using the plot of rate-of-climb for a range of altitudes, as shown in the upper graph of Figure 19-30. Thus, the P_S contour for zero ROC ($P_S = 0$ ft/s, which represents the minimum and maximum airspeed) is constructed by transferring the points labeled a through i to the lower graph. The P_S contour for the $P_S = 10$ ft/s (an ROC of 600 fpm) is constructed in a similar manner, by transferring the points along the line A–A to the lower graph. And the P_S contour for the $P_S = 20$ ft/s (an ROC of 1200 fpm) is constructed by transferring the points along the line B–B to the lower graph.

VARIABLES

Symbol	Description	Units (UK and SI)
AR	Aspect ratio	
BHP	Brake horsepower (propeller engines)	HP
C_D	Drag coefficient	
C_{D0}	Zero AOA drag coefficient	
C_{Di}	Induced drag coefficient	
C_{Dmin}	Minimum (zero-lift) drag coefficient	
C_L	Lift coefficient	
C_{L0}	Zero AOA lift coefficient	
C_{LminD}	Lift coefficient at minimum drag	
C_{MGC}	Chord length of the MGC	ft or m
C_{mW}	Moment coefficient of the wing	
D	Drag	lb_f or N
D_i	Induced drag force	lb_f or N
D_{min}	Zero-lift drag force	lb_f or N
D_{trim}	Drag at trim condition	lb_f or N
e	Oswald's span efficiency factor	
E_{total}	Total energy	$lb_f \cdot$ft or J
g	Gravitational acceleration	ft/s^2 or m/s^2
h	Altitude	ft or m
h	x-distance from LE of MGC to aircraft CG	ft or m
h_{AC}	x-distance from LE of MGC to aircraft aerodynamic center	ft or m
H_E	Specific energy/energy height	ft or m
k	Constant relating AR and e	
L	Lift	lb_f or N
LD_{max}	Maximum lift-to-drag ratio	
l_{HT}	Distance from aircraft CG to AC of horizontal tail	ft or m
L_{HT}	Lift of the horizontal tail	lb_f or N
L_W	Lift of the wing	lb_f or N
M	Mach number	
M_{MO}	Maximum operating Mach number	
M_W	Pitching moment of aircraft about the aerodynamic center	ft$\cdot$lb$_f$ or N$\cdot$m
n	Load factor	
n_{max}	Maximum sustainable load factor	
n_{Rmin}	Load factor for minimum turning radius	
P_{AV}	Available power	$lb_f\cdot$ft/s or N$\cdot$m/s
P_{EX}	Excess power	$lb_f\cdot$ft/s or N$\cdot$m/s
P_{REQ}	Required power	$lb_f\cdot$ft/s or N$\cdot$m/s
P_S	Specific excess power	ft/s or m/s
q	Dynamic pressure	$lb_f/$ft^2 or N/m^2
R	Turn radius	ft or m
R_{min}	Minimum sustainable turning radius	ft or m
R_{turn}	Turn radius	ft or m
S	Wing planform area	ft^2 or m^2
T	Thrust	lb_f or N
T_{AV}	Available thrust	lb_f or N
T_R, T_{REQ}	Required thrust	lb_f or N
t_ψ	Time to turn ψ degrees	seconds
V	Airspeed	ft/s or m/s
V_A	Corner speed or maneuvering speed	ft/s or m/s
V_{BG}	Best glide airspeed	ft/s or m/s

Symbol	Description	Units (UK and SI)
V_{car}	Carson's airspeed	ft/s or m/s
V_{Emax}	Maximum endurance airspeed	ft/s or m/s
V_H	Maximum level airspeed	ft/s or m/s
V_{LDmax}	Best glide speed	ft/s or m/s
V_{lim}	Maximum bank angle airspeed	ft/s or m/s
V_{max}	Maximum level airspeed	ft/s or m/s
V_{min}	Minimum airspeed	ft/s or m/s
V_{NO}	Normal operating airspeed	ft/s or m/s
V_{PRmin}	Minimum power required airspeed	ft/s or m/s
V_{Rmax}	Best range airspeed	ft/s or m/s
V_{Rmin}	Minimum turning radius airspeed	ft/s or m/s
V_{RODmin}	Minimum descent speed	ft/s or m/s
V_S	Stalling speed	ft/s or m/s
V_{S0}	Stalling speed in landing configuration	ft/s or m/s
V_{S1}	Stalling speed in takeoff configuration	ft/s or m/s
V_{Slevel}	Stalling speed with wings level	ft/s or m/s
V_{TRmin}	Minimum thrust required airspeed	ft/s or m/s
V_Y	Best rate-of-climb airspeed	ft/s or m/s
W	Weight	lb_f or N
W_0	Gross weight	lb_f or N
x_T	x-distance from thrust line to aircraft CG	ft or m
z_T	z-distance from thrust line to aircraft CG	ft or m
$\dot{\psi}_{max}$	Maximum sustainable turn rate	Radians/second
$\dot{\psi}$	Turn rate	Radians/second
α	Angle-of-attack	Degrees or radians
ε	Thrust angle	Degrees or radians
ϕ	Banking angle	Degrees or radians
ϕ_{max}	Maximum bank angle	Degrees or radians
η_p	Propeller efficiency	
θ	Climb angle	Degrees or radians
ρ	Density of air	$slugs/ft^3$ or kg/m^3
ψ	Desired change in heading angle	Degrees

References

[1] Perkins CD, Hage RE. Airplane Performance, Stability, and Control. John Wiley & Sons; 1949.
[2] Torenbeek E. Synthesis of Subsonic Aircraft Design. 3rd edn. Delft University Press; 1986.
[3] Nicolai L. Fundamentals of Aircraft Design. 2nd edn 1984.
[4] Roskam J, Lan Chuan-Tau Edward. Airplane Aerodynamics and Performance. DARcorporation; 1997.
[5] Hale FJ. Aircraft Performance, Selection, and Design. John Wiley & Sons; 1984. 137–138.
[6] Anderson Jr JD. Aircraft Performance & Design. 1st edn. McGraw-Hill; 1998.
[7] AIAA-80-1847. Fuel Efficiency of Small Aircraft. Carson, B.H., AIAA; 1980.
[8] Gabrielli G, von Kármán T. What price speed? Specific power required for propulsion of vehicles. Mechanical Engineering 1950;72(10):775–81.
[9] Smith, H.C. *An Application of the Carson Cruise Optimum Airspeed – A Compromise Between Speed and Efficiency.* SAE SP-621, Paper 850867, p. 95.
[10] Asselin M. An Introduction to Aircraft Performance. AIAA Education Series; 1997.

CHAPTER 20

Performance — Range Analysis

OUTLINE

20.1 Introduction 896
   20.1.1 The Content of this Chapter 896
   20.1.2 Basic Cruise Segment for Range Analysis 896
   20.1.3 Basic Cruise Segment in Terms of Range Versus Weight 896
   20.1.4 The "Breguet" Range Equation 897
   20.1.5 Basic Cruise Segment for Endurance Analysis 897
   20.1.6 The "Breguet" Endurance Equation 898
   20.1.7 Notes on SFC and TSFC 898
      Thrust Specific Fuel Consumption for a Jet 898
      Thrust Specific Fuel Consumption for a Piston Engine 898
      Derivation of Equation (20-9) 898

20.2 Range Analysis 899
   20.2.1 Mission Profiles 899
   20.2.2 Range Profile 1: Constant Airspeed/Constant Altitude Cruise 899
      Derivation of Equation (20-10) 899
   20.2.3 Range Profile 2: Constant Altitude/Constant Attitude Cruise 901
      Derivation of Equation (20-11) 902
   20.2.4 Range Profile 3: Constant Airspeed/Constant Attitude Cruise 902
      Derivation of Equation (20-12) 902
   20.2.5 Range Profile 4: Cruise Range in the Absence of Weight Change 903
   20.2.6 Determining Fuel Required for a Mission 907
      Range Profile 1: Constant Airspeed/Altitude Cruise 907
      Derivation of Equation (20-14) 907
      Range Profile 3: Constant Airspeed/Attitude Cruise 908
      Derivation of Equation (20-15) 908

   20.2.7 Range Sensitivity Studies 908
      Empty Weight Sensitivity 909
      Drag Sensitivity 909
      Aspect Ratio Sensitivity 909

20.3 Specific Range 909
   20.3.1 Definitions 909
   20.3.2 CAFE Foundation Challenge 910

20.4 Fundamental Relations for Endurance Analysis 911
   20.4.1 Endurance Profile 1: Constant Airspeed/Constant Altitude Cruise 911
      Derivation of Equation (20-19) 911
   20.4.2 Endurance Profile 2: Constant Attitude/Altitude Cruise 912
      Derivation of Equation (20-20) 912
      Derivation of Equation (20-21) 912
   20.4.3 Endurance Profile 3: Constant Airspeed/Attitude Cruise 913
      Derivation of Equation (20-22) 913

20.5 Analysis of Mission Profile 914
   20.5.1 Basics of Mission Profile Analysis 915
      Weight ratios for selected segments 915
   20.5.2 Methodology for Mission Analysis 916
   20.5.3 Special Range Mission 1: IFR Cruise Mission 918
   20.5.4 Special Range Mission 2: NBAA Cruise Mission 919
   20.5.5 Payload-Range Sensitivity Study 919

Exercises 921

Variables 922

References 923

20.1 INTRODUCTION

The majority of aircraft are designed to carry people or freight from a place of origin to some destination. Such airplanes emphasize range or endurance above other characteristics. While aircraft are certainly designed to satisfy other requirements, such as that of speed or maneuverability, range and endurance are almost always included as some of the most important ones. Even the fastest and most maneuverable aircraft would not amount to much if it didn't also offer acceptable range or endurance. In this section we will present methods to estimate range and endurance for aircraft.

Range is the distance an airplane can fly in a given time. This distance is of great importance in aircraft design and is often the parameter used to determine whether a particular design is viable. The range can be broken into a basic segment, called a cruise segment, during which some specific boundary conditions are established. Such boundary conditions allow the amount of energy required (i.e. fuel) to be evaluated. This evaluation impacts the weight requirements for that aircraft, which, in turn, impacts its size, and so on. Figure 20-1 shows how range analysis fits among the other branches of Performance theory.

In general, the methods presented in here are the "industry standard" and mirror those presented by a variety of authors, e.g. Perkins and Hage [1], Torenbeek [2], Nicolai [3], Roskam [4], Hale [5], Anderson [6] and many, many others.

20.1.1 The Content of this Chapter

- **Section 20.1** presents fundamental theory of range analysis, as well as information required to complete such analysis.
- **Section 20.2** presents classic range analysis methods using three different cruise profiles for aircraft powered by fossil fuels, and one aimed at aircraft that use electric power. These profiles are referred to as a *constant airspeed/altitude*, *constant attitude/altitude*, *constant airspeed/attitude*, and *constant weight* profiles, respectively. The end of the section presents methods to evaluate range sensitivity.
- **Section 20.3** presents the concept of specific range.
- **Section 20.4** presents classic endurance analysis methods in a manner similar to that for range analysis.
- **Section 20.5** presents methods to analyze the mission profile of the aircraft, but these are needed to determine realistic fuel requirements for the new aircraft. The section also introduces two important and common mission profiles to which many GA aircraft are designed; the standard *IFR cruise mission* and *NBAA cruise mission*. The section also presents the important *payload-range* analysis.

20.1.2 Basic Cruise Segment for Range Analysis

The basic cruise segment is shown in Figure 20-2. The aircraft begins the cruise at some initial weight, W_{ini}, and after covering some distance, R, will now possess some final weight, W_{fin}. This weight is less than the initial weight if the aircraft is powered by engines that burn fossil fuel, but unchanged if the source of energy is electric power. Both energy sources will be considered in this chapter. Note that the three curves in the figure indicate the aircraft may initially burn more (blue curve) or less fuel (lavender curve) than later in the segment. However, for simplicity, it is often assumed the fuel burn is pretty much linear for the segment and this approximation is in fact accurate (straight line). It is customary to break segments of hugely varying fuel consumption into smaller segments for which the linear assumption holds.

20.1.3 Basic Cruise Segment in Terms of Range Versus Weight

For mathematical convenience it is useful to transpose the axes in Figure 20-2 to what is shown in Figure 20-3. In this figure the weight becomes the horizontal axis and the range the vertical one. It is evident that the range changes from 0 at W_{ini} to the final range, R_{fin}, at W_{fin}.

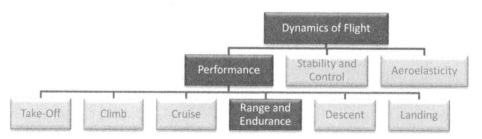

FIGURE 20-1 An organizational map placing performance theory among the disciplines of dynamics of flight, and highlighting the focus of this section: range and endurance.

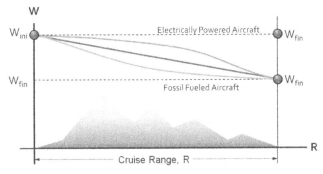

FIGURE 20-2 The basic cruise segment.

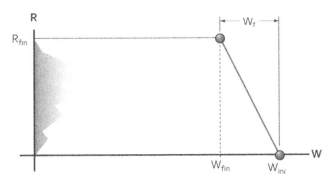

FIGURE 20-3 The basic cruise segment with transposed axes.

From this we can define the *change in range* as follows:

$$\frac{dR}{dW} = \frac{\text{Rate of change of distance}}{\text{Rate of change of weight}} = \frac{V}{-c_t T} \quad (20\text{-}1)$$

where

c_t = thrust specific fuel consumption (in 1/sec)
V = airspeed in ft/s or m/s
T = thrust in lb$_f$ or N

During cruise it is reasonable to assume that $T = D$ and $D = W/(L/D)$. For this reason we can rewrite the change in range as follows:

$$\frac{dR}{dW} = \frac{V}{-c_t T} = \frac{V}{-c_t D} = \frac{V(L/D)}{-c_t W} \quad (20\text{-}2)$$

Note that expressions (20-1) and (20-2) are only valid for aircraft that burn fuel. An expression for electrically powered aircraft does not depend on the change in weight.

20.1.4 The "Breguet" Range Equation

Equation (20-2) is solved for the range by integration, in which the limits are the initial and final weight during that segment. This equation was developed by the French aircraft designer Louis Charles Breguet (1880–1955), who was one of the early pioneers of aviation. Therefore, it is referred to as the "Breguet" range equation:

$$R = \int_{W_{ini}}^{W_{ini}-W_f} \frac{V(L/D)}{-c_t W} dW = \int_{W_{fin}}^{W_{ini}} \frac{V}{c_t} \frac{C_L}{C_D} \frac{1}{W} dW \quad (20\text{-}3)$$

In order to solve Equation (20-3) we must incorporate the dependency of V, L/D, and c_t on W, but these are established in accordance with what "kind" of a cruise we intend to fly. The Breguet equation lends itself well to numerical integration. However, it is common to give closed-form solutions to problems using some simplifying assumptions. All such solutions assume that the specific fuel consumption is constant and an "average" value for the entire range can be determined. Several well known closed-form solutions of the Breguet equation are provided in Section 20.2, *Range analysis*. Section 20.2.5, *Range profile 4: cruise range in the absence of weight change*, provides a method to handle range in the absence of weight change (for electrically powered aircraft).

20.1.5 Basic Cruise Segment for Endurance Analysis

Endurance is the length of time an airplane can stay aloft while consuming a specific amount of fuel. Like range, this length of time is of great importance in aircraft design, particularly for some military aircraft, such as fighters, tankers, and UAVs. As in the case of the range, endurance is considered in terms of a *cruise segment*, during which some specific boundary conditions are established that are then used to evaluate this parameter. Such a segment is shown in Figure 20-4 and is identical to Figure 20-3, except that the vertical axis features time.

Begin by defining the thrust specific fuel consumption as follows:

$$c_t \equiv \frac{\dot{w}_{fuel}}{T} = \frac{dW/dt}{T} \quad \Rightarrow \quad dW/dt = -c_t T \quad (20\text{-}4)$$

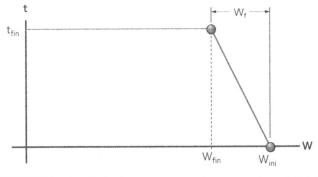

FIGURE 20-4 The basic cruise segment in terms of time of flight.

The inverse of the term dW/dt is simply the rate of change in time with respect to weight. This allows us to write the change in time aloft as follows:

$$\frac{dW}{dt} = -c_t T \quad \Leftrightarrow \quad dt = \frac{dW}{-c_t T} \qquad (20\text{-}5)$$

Expand this expression by introducing the same assumptions as for the range, i.e. noting that $T = D = W/(L/D)$ as done for Equation (20-2), which yields:

$$dt = \frac{1}{-c_t T} dW = \frac{1}{-c_t D} dW = \frac{(L/D)}{-c_t W} dW \qquad (20\text{-}6)$$

In order to solve Equation (20-6) we must incorporate the dependency of V, L/D, and c_t on W.

20.1.6 The "Breguet" Endurance Equation

As for the range, Equation (20-6) is solved for the endurance by integration, in which the limits are the initial and final weight during that segment. It is referred to as the *"Breguet" endurance equation*:

$$E = \int_{W_{ini}}^{W_{ini}-W_f} \frac{(L/D)}{-c_t W} dW = \int_{W_{fin}}^{W_{ini}} \frac{1}{c_t} \frac{C_L}{C_D} \frac{1}{W} dW \qquad (20\text{-}7)$$

The solution of Equation (20-7) requires the dependency of V, L/D, and c_t on W to be established in the same manner as for the range. As before, closed-form solutions exist for the same cases as for the range and are based on similar assumptions. All of the closed-form solutions given below assume the simplified drag model of Chapter 15.

20.1.7 Notes on SFC and TSFC

Thrust Specific Fuel Consumption for a Jet

In the UK system, the specific fuel consumption (c_{jet}) for a jet is given in terms of $lb_f/hr/lb_f$. But since V is in terms of ft/s the units must be made consistent.

$$c_t = c_{jet} \left\langle \frac{lb_f/hr}{lb_f} \right\rangle \left\langle \frac{1}{3600 \text{ sec}/hr} \right\rangle = \left(\frac{c_{jet}}{3600}\right) \left\langle \frac{lb_f/\sec}{lb_f} \right\rangle$$

Therefore, the TSFC for a jet is given by (unit is 1/s or $lb_f/\sec/lb_f$):

$$c_t = \left(\frac{c_{jet}}{3600}\right) \qquad (20\text{-}8)$$

Thrust Specific Fuel Consumption for a Piston Engine

In the UK system, the specific fuel consumption (c_{bhp}) for a piston engine is given in terms of $lb_f/hr/BHP$ (in terms of power). So, it must be converted to reflect *thrust* specific fuel consumption. The resulting expression for the TSFC for a piston engine aircraft is given by (units is 1/s):

$$c_t = \frac{c_{bhp} V}{1980000 \eta_p} \qquad (20\text{-}9)$$

Derivation of Equation (20-9)

By definition for pistons:

$$c_{bhp} \equiv \frac{\dot{w}_{fuel}}{P} \qquad (i)$$

By definition for jets:

$$c_t \equiv \frac{\dot{w}_{fuel}}{T} \qquad (ii)$$

Combining the two to get:

$$\dot{w}_{fuel} = c_{bhp} P = c_t T \quad \Leftrightarrow \quad c_{bhp} P = c_t T \quad \Leftrightarrow \quad c_t = \frac{c_{bhp} P}{T} \qquad (iii)$$

The available power, call it P_A, is related to the available engine power as follows:

$$P_A = \eta_p \times P \qquad (iv)$$

However, P_A can also be written as:

$$P_A = TV \qquad (v)$$

Inserting Equations (iv) and (v) into Equation (iii) yields:

$$c_t = \frac{c_{bhp} P}{T} = \frac{c_{bhp}\left(\frac{P_A}{\eta_p}\right)}{\left(\frac{P_A}{V}\right)} = \frac{c_{bhp} V}{\eta_p} \qquad (vi)$$

Consider the units for Equation (vi):

$$c_t = \frac{c_{bhp} V}{\eta_p} = \frac{\left\langle\frac{lb_f/hr}{bhp}\right\rangle \left\langle\frac{ft}{\sec}\right\rangle}{\langle\,\rangle} = \left\langle\frac{lb_f/hr}{bhp}\right\rangle\left\langle\frac{ft}{\sec}\right\rangle$$

For proper results, the dependency of the SFC on hours and BHP must be eliminated. To do this, divide by $60 \times 60 = 3600$ sec/hr and 550 ft·lb$_f$/sec. Thus, Equation (vi) becomes:

$$c_t = \frac{c_{bhp}V}{\eta_p} = \frac{c_{bhp}V}{\eta_p}\left\langle \frac{lb_f}{hr \times bhp}\right\rangle\left\langle \frac{ft}{sec}\right\rangle\left\langle \frac{1\ hr}{3600\ sec}\right\rangle$$

$$\times \left\langle \frac{1\ bhp}{550\ ft\cdot lb_f/sec}\right\rangle$$

$$= \frac{c_{bhp}V}{\eta_p}\left\langle \frac{1}{sec}\right\rangle\left\langle \frac{1}{3600 \times 550}\right\rangle$$

$$= \frac{c_{bhp}V}{1980000\eta_p}\left\langle \frac{1}{sec}\right\rangle$$

QED

20.2 RANGE ANALYSIS

Range analysis is an investigation of how far an aircraft can fly, how quickly, and at what cost. The analysis does not only consider what airspeed an airplane must maintain in order to obtain optimum range, but also evaluates what impact it has when other airspeeds are chosen. It also allows various sensitivities to be evaluated, such as the effect of fuel weight.

20.2.1 Mission Profiles

Generally, airplanes follow three different types of operation during cruise, based on selected combinations of the following physical and mathematical interpretations. These are shown in Table 20-1.

Closed-form solutions are provided for the combinations of parameters in Table 20-2.

It is helpful to keep Equation (9-47) in mind when considering these combinations, repeated here for convenience:

$$L = \frac{1}{2}\rho V^2 \cdot S \cdot C_L \quad (9\text{-}47)$$

TABLE 20-1 Physical and Mathematical Interpretation of Parameters for Mission Analysis

Physical Interpretation	Mathematical Interpretation
Constant airspeed implies…	… V = Constant
Constant altitude implies…	… ρ = Constant
Constant attitude (i.e. AOA) implies…	… C_L/C_D = Constant

TABLE 20-2 Sample Table for Determining the Best Range for an Electrically Powered Aircraft

Section	Type of Cruise	V	ρ	C_L/C_D
20.2.2	Constant airspeed/altitude	Constant	Constant	
20.2.3	Constant attitude/altitude		Constant	Constant
20.2.4	Constant airspeed/attitude	Constant		Constant
20.2.5	Constant weight	Constant	Constant	Constant

20.2.2 Range Profile 1: Constant Airspeed/Constant Altitude Cruise

This type of cruise requires the airspeed and altitude to be maintained between Points 2 and 3 (the cruise segment) in Figure 20-5. Since the weight of the airplane reduces with time as fuel is consumed, this requires the AOA or the attitude of the aircraft to be reduced accordingly. Reducing the AOA reduces the lift coefficient, which, as can be seen from Equation (9-47), is the only way to reduce the lift if V and ρ are constant. Note that Equation (20-10) below only yields the distance covered between Points 2 and 3.

Cruise Type 1

Airspeed – constant during cruise (V = Constant)
Altitude – constant during cruise (ρ = Constant)
AOA – must be reduced during cruise
(L/D = Variable)

$$R = \frac{V}{c_t\sqrt{kC_{Dmin}}}\left[\tan^{-1}\left(\frac{2\sqrt{k}}{\rho V^2 S\sqrt{C_{Dmin}}}W_{ini}\right)\right.$$
$$\left. - \tan^{-1}\left(\frac{2\sqrt{k}}{\rho V^2 S\sqrt{C_{Dmin}}}W_{fin}\right)\right] \quad (20\text{-}10)$$

Derivation of Equation (20-10)

In all three cases, the thrust specific fuel consumption, c_t, is constant. Additionally for this case, the airspeed, V, is also constant. However, since C_L/C_D is dependent on the change in weight this must remain inside the integral. Therefore, we write Equation (20-3) as follows:

$$R = \int_{W_{fin}}^{W_{ini}} \frac{V}{c_t}\frac{C_L}{C_D}\frac{1}{W}dW = \frac{V}{c_t}\int_{W_{fin}}^{W_{ini}}\frac{C_L}{C_D}\frac{1}{W}dW$$

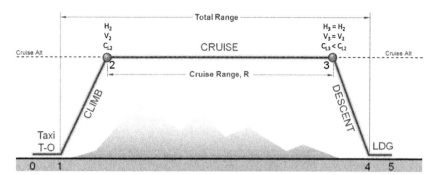

FIGURE 20-5 Constant airspeed/constant altitude cruise mission

Insert the expression for drag (here using the simplified drag model of Equation (15-5) to get:

$$R = \frac{V}{c_t} \int_{W_{fin}}^{W_{ini}} \frac{C_L}{(C_{Dmin} + kC_L^2)} \frac{1}{W} dW$$

$$= \frac{V}{c_t} \int_{W_{fin}}^{W_{ini}} \frac{\frac{2W}{\rho V^2 S}}{\left(C_{Dmin} + k\left(\frac{2W}{\rho V^2 S}\right)^2\right)} \frac{1}{W} dW$$

$$R = \frac{V}{c_t} \frac{2}{\rho V^2 S} \int_{W_{fin}}^{W_{ini}} \frac{W}{\left(C_{Dmin} + k\frac{4W^2}{\rho^2 V^4 S^2}\right)} \frac{1}{W} dW$$

Expanding further leads to:

$$R = \frac{V}{c_t} \frac{2}{\rho V^2 S} \int_{W_{fin}}^{W_{ini}} \frac{1}{\left(C_{Dmin}\frac{4k}{\rho^2 V^4 S^2}\frac{\rho^2 V^4 S^2}{4k} + \frac{4k}{\rho^2 V^4 S^2}W^2\right)} dW$$

$$= \frac{V}{c_t} \frac{2}{\rho V^2 S} \frac{\rho^2 V^4 S^2}{4k} \int_{W_{fin}}^{W_{ini}} \frac{1}{\left(C_{Dmin}\frac{\rho^2 V^4 S^2}{4k} + W^2\right)} dW$$

$$= \frac{1}{c_t} \frac{\rho V^3 S}{2k} \int_{W_{fin}}^{W_{ini}} \frac{1}{\left(C_{Dmin}\frac{\rho^2 V^4 S^2}{4k} + W^2\right)} dW$$

Define a such that:

$$a = \frac{\sqrt{C_{Dmin}}}{\sqrt{k}} \frac{\rho V^2 S}{2} \Leftrightarrow a^2 = C_{Dmin} \frac{\rho^2 V^4 S^2}{4k}$$

Also note that:

$$\frac{1}{a} = \frac{2\sqrt{k}}{\rho V^2 S \sqrt{C_{Dmin}}}$$

And that:

$$\int \frac{du}{a^2 + u^2} = \frac{1}{a}\tan^{-1}\frac{u}{a}$$

Therefore:

$$R = \frac{1}{c_t} \frac{\rho V^3 S}{2k} \int_{W_{fin}}^{W_{ini}} \frac{1}{(a^2 + W^2)} dW$$

$$= \frac{1}{c_t} \frac{\rho V^3 S}{4k} \frac{1}{a} \left[\tan^{-1}\frac{W}{a}\right]_{W_{fin}}^{W_{ini}}$$

$$R = \frac{1}{c_t} \frac{\rho V^3 S}{2k} \frac{1}{a} \left[\tan^{-1}\frac{W_{ini}}{a} - \tan^{-1}\frac{W_{fin}}{a}\right]$$

$$R = \frac{1}{c_t} \frac{\rho V^3 S}{2k} \frac{2\sqrt{k}}{\rho V^2 S \sqrt{C_{Dmin}}}$$

$$\times \left[\tan^{-1}\left(\frac{2\sqrt{k}}{\rho V^2 S \sqrt{C_{Dmin}}}W_{ini}\right) - \tan^{-1}\left(\frac{2\sqrt{k}}{\rho V^2 S \sqrt{C_{Dmin}}}W_{fin}\right)\right]$$

$$= \frac{V}{c_t \sqrt{kC_{Dmin}}}$$

$$\times \left[\tan^{-1}\left(\frac{2\sqrt{k}}{\rho V^2 S \sqrt{C_{Dmin}}}W_{ini}\right) - \tan^{-1}\left(\frac{2\sqrt{k}}{\rho V^2 S \sqrt{C_{Dmin}}}W_{fin}\right)\right]$$

QED

EXAMPLE 20-1

A light airplane starts its cruise segment at 150 KTAS at 10,000 ft when it weighs 3200 lbs. After cruising for a period of time at that altitude it is noticed it now weighs 2800 lbs.

Using a wing area of $S = 145$ ft^2 compute the initial and final lift coefficients if the pilot maintains constant airspeed.

Solution

Density at altitude:

$$\rho = 0.002378(1 - 0.0000068753 \times H)^{4.2561}$$
$$= 0.002378(1 - 0.0000068753 \times 10000)^{4.2561}$$

$$\rho = 0.001756 \text{ slugs/ft}^3$$

Using the constant airspeed/altitude cruise:

Initial lift coefficient:

$$C_L = \frac{2W}{\rho V^2 S} = \frac{2(3200)}{(0.001756)(1.688 \times 150)^2(145)}$$
$$= 0.3921$$

Final lift coefficient:

$$C_L = \frac{2W}{\rho V^2 S} = \frac{2(2800)}{(0.001756)(1.688 \times 150)^2(145)}$$
$$= 0.3431$$

20.2.3 Range Profile 2: Constant Altitude/Constant Attitude Cruise

This type of cruise requires the altitude and attitude (*AOA*) to be maintained between Points 2 and 3 (the cruise segment) in Figure 20-6. Since the weight of the airplane reduces with time as fuel is consumed, this requires the airspeed of the aircraft to be reduced and this, as can be seen from Equation (9-47), is the only way to reduce the lift (since C_L and ρ will not change). Note that Equation (20-11) below is the distance covered between Points 2 and 3.

Cruise Type 2

Airspeed — must be reduced during cruise (V = Variable)
Altitude — constant during cruise (ρ = Constant)
AOA — constant during cruise (L/D = Constant)

$$R = \frac{1}{c_t} \frac{\sqrt{C_L}}{C_D} \frac{2\sqrt{2}}{\sqrt{\rho S}} \left(\sqrt{W_{ini}} - \sqrt{W_{fin}} \right)$$
$$= \frac{1}{c_t C_D} \sqrt{\frac{8 C_L}{\rho S}} \left(\sqrt{W_{ini}} - \sqrt{W_{fin}} \right) \quad (20\text{-}11)$$

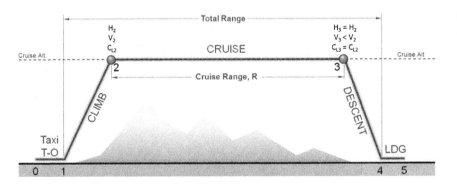

FIGURE 20-6 Constant altitude/constant attitude cruise mission

Derivation of Equation (20-11)

For this case, the airspeed, V, is a variable and C_L/C_D is a constant and, thus, can come outside the integral.

$$R = \int_{W_{fin}}^{W_{ini}} \frac{V}{c_t} \frac{C_L}{C_D} \frac{1}{W} dW = \frac{1}{c_t} \frac{C_L}{C_D} \int_{W_{fin}}^{W_{ini}} \frac{V}{W} dW$$

$$= \frac{1}{c_t} \frac{C_L}{C_D} \int_{W_{fin}}^{W_{ini}} \frac{\sqrt{\frac{2W}{\rho S C_L}}}{W} dW = \frac{1}{c_t} \frac{C_L}{C_D} \frac{\sqrt{2}}{\sqrt{\rho S C_L}} \int_{W_{fin}}^{W_{ini}} \frac{\sqrt{W}}{W} dW$$

$$= \frac{1}{c_t} \frac{C_L}{C_D} \frac{\sqrt{2}}{\sqrt{\rho S C_L}} \int_{W_{fin}}^{W_{ini}} (W)^{-1/2} dW = \frac{1}{c_t} \frac{C_L}{C_D} \frac{\sqrt{2}}{\sqrt{\rho S C_L}} \left[\frac{\sqrt{W}}{1/2} \right]_{W_{fin}}^{W_{ini}}$$

Therefore:

$$R = \frac{1}{c_t} \frac{\sqrt{C_L}}{C_D} \frac{2\sqrt{2}}{\sqrt{\rho S}} \left(\sqrt{W_{ini}} - \sqrt{W_{fin}} \right)$$

$$= \frac{1}{c_t C_D} \sqrt{\frac{8 C_L}{\rho S}} \left(\sqrt{W_{ini}} - \sqrt{W_{fin}} \right)$$

QED

EXAMPLE 20-2

The light airplane of EXAMPLE 20-1 again starts its cruise segment at 150 KTAS at 10,000 ft when it weighs 3200 lbs. After cruising for a period of time at that altitude it is noticed it now weighs 2800 lbs.

Using a wing area of $S = 145$ ft^2 compute the final airspeed if the pilot maintains a constant lift coefficient.

Solution

Density at altitude:

$$\rho = 0.001756 \text{ slugs/ft}^3 \quad \text{(see EXAMPLE 20-1)}$$

Using constant attitude/altitude cruise:

Initial and final lift coefficient:

$$C_L = \frac{2W}{\rho V^2 S} = \frac{2(3200)}{(0.001756)(1.688 \times 150)^2 (145)} = 0.3921$$

Final airspeed:

$$V = \sqrt{\frac{2W}{\rho S C_L}} = \sqrt{\frac{2(2800)}{(0.001756)(145)(0.3921)}} = 237 \text{ ft/s}$$

$$= 140.3 \text{ KTAS}$$

20.2.4 Range Profile 3: Constant Airspeed/Constant Attitude Cruise

The third cruise profile is the constant airspeed/constant attitude cruise. In this cruise mode, the airplane must climb as the fuel is consumed to ensure less lift is generated (per Equation (9-47)) at the constant airspeed between Points 2 and 3 (the cruise segment) in Figure 20-7. Note that Equation (20-12) below is the distance covered between Points 2 and 3.

Cruise Type 3

Airspeed — constant during cruise ($V =$ Constant)
Altitude — increases during cruise ($\rho =$ Variable)
AOA — constant during cruise ($L/D =$ Constant)

$$R = \frac{V}{c_t} \frac{C_L}{C_D} \ln \left(\frac{W_{ini}}{W_{fin}} \right) \qquad (20\text{-}12)$$

Derivation of Equation (20-12)

For this case, the airspeed, V, and C_L/C_D are constant and, thus, can come outside of the integral.

$$R = \int_{W_{fin}}^{W_{ini}} \frac{V}{c_t} \frac{C_L}{C_D} \frac{1}{W} dW = \frac{V}{c_t} \frac{C_L}{C_D} \int_{W_{fin}}^{W_{ini}} \frac{1}{W} dW$$

$$= \frac{V}{c_t} \frac{C_L}{C_D} [\ln(W)]_{W_{fin}}^{W_{ini}} = \frac{V}{c_t} \frac{C_L}{C_D} \ln \left(\frac{W_{ini}}{W_{fin}} \right)$$

QED

EXAMPLE 20-3

The light airplane of Example 20-1 again starts its cruise segment at 150 KTAS at 10,000 ft when it weighs 3200 lbs and cruises for a period of time until it weighs 2800 lbs.

Using a wing area of $S = 145$ ft^2 compute the final density altitude if the pilot maintains a constant airspeed and AOA. Compute dynamic pressure at the beginning and end of the cruise.

Solution

Density at altitude:

$$\rho = 0.001756 \text{ slugs/ft}^3 \quad \text{(see Example 20-1)}$$

Initial lift coefficient:

$$C_L = \frac{2W}{\rho V^2 S} = \frac{2(3200)}{(0.001756)(1.688 \times 150)^2 (145)}$$
$$= 0.3921$$

Density at end of cruise:

$$\rho = \frac{2W}{V^2 S C_L} = \frac{2(2800)}{(1.688 \times 150)^2 (145)(0.3921)}$$
$$= 0.001536 \text{ slugs/ft}^3$$

Initial dynamic pressure:

$$q = \frac{1}{2}\rho V^2 = \frac{1}{2}(0.001756)(1.688 \times 150)^2 = 56.3 \text{ lb}_f/\text{ft}^2$$

Final dynamic pressure:

$$q = \frac{1}{2}\rho V^2 = \frac{1}{2}(0.001536)(1.688 \times 150)^2 = 49.2 \text{ lb}_f/\text{ft}^2$$

Density altitude:

$$H_\rho = 145448 \left(1 - \left(\frac{\rho}{\rho_0}\right)^{0.234957}\right)$$
$$= 145448 \left(1 - \left(\frac{0.001536}{0.002378}\right)^{0.234957}\right) = 14195 \text{ ft}$$

20.2.5 Range Profile 4: Cruise Range in the Absence of Weight Change

At the time of writing, electric airplanes capable of carrying people are steadily gaining popularity. The advent of batteries with high enough energy density to be practical for use in such aircraft is likely to change the face of aviation in the future. Airplanes powered with solar energy are being developed and a number of such airplanes have already flown. Electric airplanes differ from the aircraft of previous sections in that their weight does not change with range or, in the case of fuel cells, changes very slightly. For this reason the Breguet formulation is not applicable. This section will focus on how to compute the range of airplanes powered by electric motors driven by batteries. It is assumed the

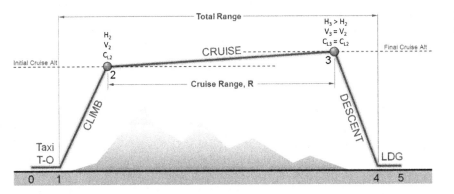

FIGURE 20-7 Constant airspeed/attitude cruise mission.

TABLE 20-3 Sample Table for Determining the Best Range for an Electrically Powered Aircraft

V	C_L	C_D	D	$P_{REQ} = D \cdot V$	t_{TOT}	R
KTAS	-	-	-	Watts	hrs	nm
V_1	C_{L1}	C_{D1}	D_1	$D_1 \cdot V_1$	$E/(D_1 \cdot V_1)$	$t_{TOT1} \cdot V_1$
V_2	C_{L2}	C_{D2}	D_2	$D_2 \cdot V_2$	$E/(D_2 \cdot V_2)$	$t_{TOT2} \cdot V_2$
...	...	...	...	...	...	...
V_N	C_{LN}	C_{DN}	D_N	$D_N \cdot V_N$	$E/(D_N \cdot V_N)$	$t_{TOTN} \cdot V_N$

airplane flies at a constant altitude. For this particular section it is assumed the profile of Figure 20-5 applies.

Let m_b be the mass (or weight) of the battery system and let E_{BATT} be the energy density of the battery. Then the total energy stored in the battery system will be $E = m_b \times E_{BATT}$. Suppose the power of the motor is $P_{ELECTRIC}$, then the time to run the motor is:

$$t_{TOT} = \frac{E}{P_{ELECTRIC}} = \frac{m_b \times E_{BATT}}{P_{ELECTRIC}} \qquad (20\text{-}13)$$

In order to determine maximum range, the designer should tabulate the power required for a range of airspeeds as set up in Table 20-3 below and then extract the best range. Note that P_{REQ} is the power the electric motor must generate and is thus a direct indication of the power setting required. *Also, although not directly shown in the table, it is assumed the proper conversion factors are employed to ensure the units displayed.*

EXAMPLE 20-4

An electric airplane is being designed to feature 200 kg of lithium polymer (LiPo) batteries. Its electric motor is rated at 60 kW at full power (or 60,000 W). If the energy density is 130 Wh/kg and the airplane's minimum power required is 10,000 ft·lb$_f$/s at the intended cruise altitude, determine the percentage power required and endurance. If the minimum power airspeed is 83 KTAS, determine the range. Note that disparate units are being used for demonstration purposes only.

Solution

Total energy stored in the batteries:

$$E = m_b \times E_{BATT} = (200 \text{ kg}) \times (130 \text{ Wh/kg})$$
$$= 26000 \text{ Wh}$$

Power required:

$$P_{ELECTRIC} = (0.746 \text{ kW/hp}) \left(\frac{10000 \text{ ft} \cdot \text{lb}_f/\text{s}}{550} \right)$$
$$= (0.746 \text{ kW/hp})(18.2 \text{ hp}) = 13564 \text{ W}$$

Power setting:

$$P = 100 \left(\frac{13564}{60000} \right) = 22.6\%$$

Endurance:

$$t_{TOT} = \left(\frac{26000}{13564} \right) = 1.92 \text{ hrs}$$

Range at 83 KTAS:

$$R = V \times t_{TOT} = (83)(1.92) = 159 \text{ nm}$$

EXAMPLE 20-5

Consider a Cirrus SR22 as it begins its cruise segment at 135 KCAS at 10,000 ft, weighing 3200 lb$_f$, and cruises for a period of time at that altitude until it weighs 2800 lb$_f$. The following additional data is given:

Wing area	$S = 145$ ft^2
Aspect ratio	$AR = 10$
Minimum drag coefficient	$C_{Dmin} = 0.025$
Average fuel burn	15 gallons/hr
Average BHP	150 BHP
Propeller efficiency	$\eta_p = 0.85$

Compute the range of this airplane using each of the three cruise profiles.

Solution

Density at altitude:

$$\rho = 0.001756 \text{ slugs/ft}^3 \quad \text{(see Example 20-1)}$$

Airspeed, KTAS:

$$V_{TAS} \approx \frac{V_{CAS}}{\sqrt{\rho/\rho_0}} = \frac{(135)}{\sqrt{\frac{(0.001756)}{(0.002378)}}} = 157 \text{ KTAS}$$

Airspeed, ft/s:

$$V = 1.688 V_{TAS} = 1.688 \times (157) = 265.2 \text{ ft/s}$$

Span efficiency:

$$e = 1.78(1 - 0.045 AR^{0.68}) - 0.64$$
$$= 1.78(1 - 0.045(10)^{0.68}) - 0.64 = 0.7566$$

Coefficient "k":

$$k = \frac{1}{\pi \cdot AR \cdot e} = \frac{1}{\pi \cdot (10) \cdot (0.7566)} = 0.04207$$

SFC:

$$c_{bhp} = \frac{15 \text{ gal/hr}}{150 \text{ BHP}} = \frac{(15 \times 6) \text{ lb}_f/\text{hr}}{150 \text{ BHP}}$$
$$= 0.600 \text{ lb}_f/\text{hr/BHP}$$

TSFC:

$$c_t = \frac{c_{bhp} V}{1980000 \eta_p} = \frac{(0.600)(265.2)}{1980000(0.85)} = 0.00009455 \frac{1}{\text{sec}}$$

Lift coefficient:

$$C_L = \frac{2W}{\rho V^2 S} = \frac{2(3200)}{(0.001756)(265.2)^2(145)} = 0.3574$$

Drag coefficient:

$$C_D = C_{Dmin} + kC_L^2 = (0.025) + (0.04207)(0.3574)^2$$
$$= 0.03037$$

Glide ratio:

$$\left(\frac{L}{D}\right) = \frac{C_L}{C_D} = \frac{0.3574}{0.03037} = 11.77$$

As a "sanity check" consider the following:

Total fuel burned:

$$W_f = 3200 - 2800 = 400 \text{ lb}_f$$

Time to burn W_f:

$$T_f = \frac{W_f}{15 \text{ gals/hr} \times 6 \text{ lb}_f/\text{gal}} = \frac{400 \text{ lb}_f}{90 \text{ lb}_f/\text{hr}} = 4.444 \text{ hrs}$$

Distance covered:

$$R = V \times T_f = (157 \text{ nm/hr})(4.444 \text{ hrs}) = 698 \text{ nm}$$

Range profile 1: constant airspeed/altitude cruise

$$R = \frac{V}{c_t \sqrt{kC_{Dmin}}} \left[\tan^{-1}\left(\frac{2\sqrt{k}}{\rho V^2 S \sqrt{C_{Dmin}}} W_{ini}\right) \right.$$
$$\left. - \tan^{-1}\left(\frac{2\sqrt{k}}{\rho V^2 S \sqrt{C_{Dmin}}} W_{fin}\right) \right]$$

$$A = \frac{2\sqrt{k}}{\rho V^2 S \sqrt{C_{Dmin}}} = \frac{2\sqrt{0.04207}}{(0.001756)(265.2)^2(145)\sqrt{0.025}}$$
$$= 0.0001449$$

$$R = \frac{(265.2)}{(0.00009455)\sqrt{(0.04207)(0.025)}}$$
$$\times \left[\tan^{-1}(A(3200)) - \tan^{-1}(A(2800)) \right]$$
$$= (86487974.5)\left[\tan^{-1}(0.46361) - \tan^{-1}(0.40566)\right]$$
$$= (86487974.5)[0.43412 - 0.38538]$$
$$= 4215887 \text{ ft} \stackrel{\div 6076 \text{ ft/nm}}{=} 694 \text{ nm}$$

Range profile 2: constant attitude/altitude cruise

$$R = \frac{1}{c_t} \frac{\sqrt{C_L}}{C_D} \frac{2\sqrt{2}}{\sqrt{\rho S}} \left(\sqrt{W_{ini}} - \sqrt{W_{fin}}\right)$$

EXAMPLE 20-5 (cont'd)

$$R = \frac{1}{(0.00009455)} \frac{\sqrt{0.3574}}{(0.03037)} \frac{2\sqrt{2}}{\sqrt{(0.001756)(145)}}$$
$$\times \left(\sqrt{3200} - \sqrt{2800} \right)$$
$$= 4263642.0 \text{ ft} \overset{\div 6076 \text{ ft/nm}}{=} 702 \text{ nm}$$

Range profile 3: constant airspeed/attitude cruise

$$R = \frac{V}{c_t} \frac{C_L}{C_D} \ln\left(\frac{W_{ini}}{W_{fin}}\right) = \frac{(265.2)}{(0.00009455)} \frac{(0.3574)}{(0.03037)} \ln\left(\frac{3200}{2800}\right)$$
$$= 4407629.9 \text{ ft} \overset{\div 6076 \text{ ft/nm}}{=} 725.4 \text{ nm}$$

EXAMPLE 20-6

The light airplane of EXAMPLE 20-5 starts its cruise segment at 135 KCAS at 10,000 ft when it weighs 3200 lbs and cruises for a period of time at that altitude until it weighs 2800 lbs. Use the same additional data as given in that example to estimate the range at $(L/D)_{max}$ for this airplane. This is reflected in Figure 20-8.

Solution

Air density:	$\rho = 0.001756 \text{ slugs/ft}^3$
Span efficiency:	$e = 0.7566$
Coefficient "k":	$k = 0.04207$

$(L/D)_{max}$:

$$\left(\frac{L}{D}\right)_{max} = \left(\frac{C_L}{C_D}\right)_{max} = \sqrt{\frac{1}{4C_{Dmin}k}}$$
$$= \sqrt{1/(4(0.025)(0.04207))} = 15.42$$

Airspeed of $(L/D)_{max}$:

$$V_{(L/D)_{max}} = \sqrt{\frac{2}{\rho}\sqrt{\frac{k}{C_{Dmin}}}\left(\frac{W}{S}\right)}$$
$$= \sqrt{\frac{2}{(0.001756)}\sqrt{\frac{(0.04207)}{(0.025)}}\left(\frac{3200}{145}\right)} = 180.6 \text{ ft/s}$$
$$= 107 \text{ KTAS}$$

SFC:

$$c_{bhp} = (15 \text{ gal/hr})/(150 \text{ BHP})$$
$$= \left((15 \times 6) \text{ lb}_f/\text{hr}\right)/(150 \text{ BHP})$$
$$= 0.600 \text{ lb}_f/\text{hr/BHP}$$

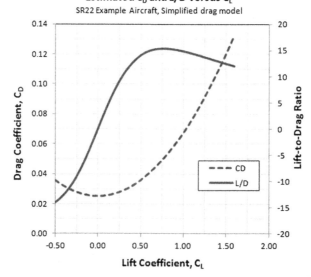

FIGURE 20-8 Drag polar and lift-to-drag ratio estimations for the SR22.

EXAMPLE 20-6 (cont'd)

TSFC:

$$c_t = (c_{bhp}V)/(1980000\eta_p)$$

$$= ((0.600 \times 180.6))/(1980000 \times 0.85)$$

$$= 0.00006439 \ \frac{1}{\sec}$$

Range at $(L/D)_{max}$:

$$R = \frac{V}{c_t}\frac{C_L}{C_D}\ln\left(\frac{W_{ini}}{W_{fin}}\right)$$

$$= \frac{(180.6)}{(0.00006439)}(15.42)\ln\left(\frac{3200}{2800}\right)$$

$$= 5775646.7 \text{ ft} \stackrel{\div 6076 \text{ ft/nm}}{=} 951 \text{ nm}$$

Figure 20-8 shows the drag polar and L/D graph for this airplane. A summary of results is provided in Table 20-4.

TABLE 20-4 Summary of Results

Method	Range, nm
"Dumbed down" method	698
Method 1: constant airspeed/altitude cruise	694
Method 2: constant attitude/altitude cruise	702
Method 3: constant airspeed/attitude cruise	725
Method 4: range at max L/D airspeed (Example 20-6)	951

20.2.6 Determining Fuel Required for a Mission

Sometimes it is necessary to determine the amount of fuel required for a mission. This section determines the fuel required to fly a known distance, assuming the airplane's weight at the beginning of the cruise segment is known.

Range Profile 1: Constant Airspeed/Altitude Cruise

This condition is the norm for commercial and freight aircraft, as they are directed by air traffic controllers (ATC) to maintain uniform airspeed and altitude to simplify management of air traffic. Unfortunately, the resulting expression is transcendental and must thus be solved iteratively.

$$\tan^{-1}\left(\frac{2\sqrt{k}}{\rho V^2 S\sqrt{C_{Dmin}}}W_{ini}\right)$$

$$- \tan^{-1}\left(\frac{2\sqrt{k}}{\rho V^2 S\sqrt{C_{Dmin}}}(W_{ini} - W_f)\right)$$

$$= \frac{R \cdot c_t \sqrt{kC_{Dmin}}}{V} \quad (20\text{-}14)$$

Derivation of Equation (20-14)

The Breguet range equation for this condition is given by Equation (20-10):

$$R = \frac{V}{c_t\sqrt{kC_{Dmin}}}\left[\tan^{-1}\left(\frac{2\sqrt{k}}{\rho V^2 S\sqrt{C_{Dmin}}}W_{ini}\right)\right.$$

$$\left.- \tan^{-1}\left(\frac{2\sqrt{k}}{\rho V^2 S\sqrt{C_{Dmin}}}W_{fin}\right)\right]$$

$$\Rightarrow \tan^{-1}\left(\frac{2\sqrt{k}}{\rho V^2 S\sqrt{C_{Dmin}}}W_{ini}\right) - \tan^{-1}\left(\frac{2\sqrt{k}}{\rho V^2 S\sqrt{C_{Dmin}}}W_{fin}\right)$$

$$= \frac{R \cdot c_t \sqrt{kC_{Dmin}}}{V}$$

Since the fuel weight is given by:

$$W_f = W_{fin} - W_{ini}$$

we can write:

$$\tan^{-1}\left(\frac{2\sqrt{k}}{\rho V^2 S\sqrt{C_{Dmin}}}W_{ini}\right)$$

$$- \tan^{-1}\left(\frac{2\sqrt{k}}{\rho V^2 S\sqrt{C_{Dmin}}}(W_{ini} - W_f)\right)$$

$$= \frac{R \cdot c_t \sqrt{kC_{Dmin}}}{V}$$

QED

Range Profile 3: Constant Airspeed/Attitude Cruise

This profile is far less common in normal operation of aircraft, but would be used to optimize long range for specialized aircraft.

$$W_f = W_{ini}\left(e^{-\left(\frac{c_t R}{V}\right)\left(\frac{1}{C_L/C_D}\right)} - 1\right) \quad (20\text{-}15)$$

Derivation of Equation (20-15)

The Breguet range equation for this condition is given by Equation (20-12):

$$R = \frac{V}{c_t}\frac{C_L}{C_D}\ln\left(\frac{W_{ini}}{W_{fin}}\right) \Rightarrow \left(\frac{c_t \cdot R}{V}\right)\left(\frac{1}{C_L/C_D}\right)$$

$$= \ln\left(\frac{W_{ini}}{W_{fin}}\right) \Rightarrow \frac{W_{ini}}{W_{fin}} = e^{\left(\frac{c_t R}{V}\right)\left(\frac{1}{C_L/C_D}\right)}$$

The weight of the aircraft at the end of the segment is:

$$\frac{W_{ini}}{W_{fin}} = e^{\left(\frac{c_t R}{V}\right)\left(\frac{1}{C_L/C_D}\right)} \Rightarrow W_{ini} = W_{fin} \cdot e^{\left(\frac{c_t R}{V}\right)\left(\frac{1}{C_L/C_D}\right)}$$

$$\Rightarrow W_{fin} = W_{ini} \cdot e^{-\left(\frac{c_t R}{V}\right)\left(\frac{1}{C_L/C_D}\right)}$$

Since the fuel weight is given by:

$$W_f = W_{fin} - W_{ini}$$

we can write:

$$W_f = W_{fin} - W_{ini} = W_{ini} \cdot e^{-\left(\frac{c_t R}{V}\right)\left(\frac{1}{C_L/C_D}\right)}$$

$$- W_{ini} = W_{ini}\left(e^{-\left(\frac{c_t R}{V}\right)\left(\frac{1}{C_L/C_D}\right)} - 1\right)$$

QED

20.2.7 Range Sensitivity Studies

As stated in the introduction to this section, range and endurance are of paramount importance to most aircraft design projects. Many types of aircraft are sold only because of their range or endurance. If the predicted range is not met, the entire development project may be compromised. For this reason, it is imperative to assess how sensitive the design will be to deviations from standard operational parameters. For instance, consider a project in which the empty weight ends up being higher than initially expected. This, inevitably, cuts into the fuel quantity that can be carried for the design mission. A sensitivity study can help expose a possible weakness in the design early enough to justify modifications that remedy the situation.

The following examples use the methods presented in this section. For simplicity they do not account for the range accrued during climb or descent. As usual, a sensitivity study usually implies that specific variables are fixed while allowing a target parameter of interest to vary.

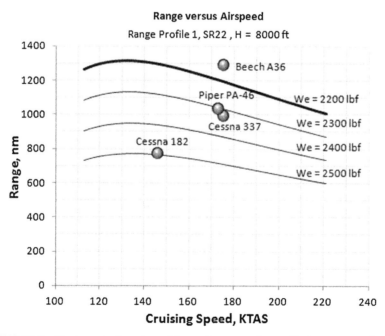

FIGURE 20-9 The impact of empty weight on the range of a small piston-powered aircraft.

Empty Weight Sensitivity

The empty weight sensitivity shows the effect of deviations in empty weight on the range of the airplane (see Figure 20-9). Here, assume the target empty weight to amount to 2200 lb$_f$. If the production aircraft ends up being 300 lb$_f$ heavier than anticipated, the maximum range drops from about 1315 nm to about 771 nm for the particular loading (which here consists of two 200 lb$_f$ individuals and 50 lb$_f$ of baggage).

It is helpful to plot the capability of the competition on the weight sensitivity graph. This will help relay the comparative capability of the new design.

Drag Sensitivity

The drag sensitivity shows the effect of deviations in drag coefficient on the range of the airplane (see Figure 20-10). Here, the target empty weight is assumed 2200 lb$_f$. If the production aircraft generates 20 drag counts above expected value, the baseline range of 1315 nm drops to 1240nm and 1180 nm if it generates 40 additional drag counts. The additional drag can stem from a number of sources. For instance, larger than expected flow separation regions during cruise (poor wing fairing design, poor geometric quality), non-materialization of NLF (turbulent boundary layer in regions where NLF is expected), protrusions such as antennas and entry steps, cooling drag, CRUD, and so on.

Aspect Ratio Sensitivity

The aspect ratio sensitivity reveals the effect of the wing AR on the range of the airplane (see Figure 20-11). Here, the target AR of 10 yields a baseline range of 1315 nm. Reducing this to, say, 6 results in a drop of 115 nm to 1200 nm.

The primary advantage of AR is reduction in lift induced drag. Most of this is realized at lower dynamic pressures, when the airplane must fly at higher AOAs.

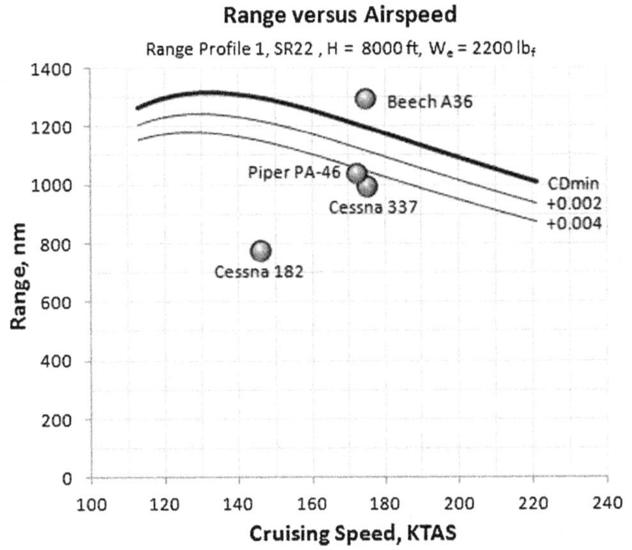

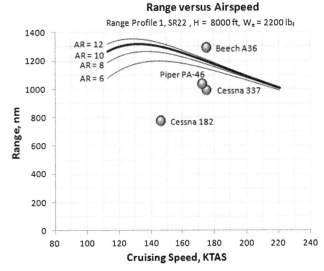

FIGURE 20-10 The impact of minimum drag coefficient on the range of a small piston-powered aircraft.

FIGURE 20-11 The impact of AR on the range of a small piston-powered aircraft.

20.3 SPECIFIC RANGE

20.3.1 Definitions

Specific range (SR) is the distance an airplane can fly on a given amount of fuel. This quantity is important when comparing the efficiency of different aircraft types or different airspeeds for an individual aircraft, for instance, when determining at which airspeed a particular airplane is the most efficient.

$$SR \equiv \frac{\text{Distance}}{\text{Quantity of Fuel}} \quad (20\text{-}16)$$

Specific range is analogous to the term gas mileage as used for cars; the primary difference is that when used for airplanes one usually specifies fuel quantity in terms of lb$_f$ rather than gallons. Knowing the range flown and weight of the fuel consumed during that segment, we can calculate the *average SR* from the following expression:

$$SR = \frac{R}{W_f} \quad (20\text{-}17)$$

We can also compute the *instantaneous SR* as follows:

$$SR = \frac{\Delta R/\Delta t}{W_f/\Delta t} = \frac{V_{TAS}}{\dot{w}_{fuel}} = \frac{\text{True Airspeed}}{\text{Fuel Weight Flow}} \quad (20\text{-}18)$$

EXAMPLE 20-7

A light airplane is cruising at a true airspeed of 157 KTAS when it is noticed it is burning 15 gals/hr. What is its *instantaneous specific range*? If it travels 702 nm using 400 lb$_f$ of fuel, what is its *average specific range*?

Solution

Instantaneous SR:

$$SR = \frac{V_{TAS}}{\dot{w}_{fuel}} = \frac{157 \text{ nm/hr}}{(15 \text{ gals/hr})(6 \text{ lb}_f/\text{gal})}$$
$$= 1.744 \text{ nm/lb}_f$$

Average SR:

$$SR = \frac{R}{W_f} = \frac{702 \text{ nm}}{400 \text{ lb}_f} = 1.754 \text{ nm/lb}_f$$

EXAMPLE 20-8

Calculate the average SR for the airplane of Examples 20-5 and 20-6.

Solution

A modified summary of results is provided in Table 20-5, where the SR is simply calculated using Equation (20-17).

TABLE 20-5 Summary of Results

Method	Range, nm	SR
"Dumbed down" method	698	1.746
Method 1: constant airspeed/altitude cruise	694	1.735
Method 2: constant attitude/altitude cruise	702	1.754
Method 3: constant airspeed/attitude cruise	725	1.813
Method 4: range at max L/D Airspeed (Example 20-6)	951	2.378

20.3.2 CAFE Foundation Challenge

In 2010 the CAFE Foundation announced a competition aimed at encouraging the aviation industry to go "green". To lead the way, the $1.65M NASA-funded CAFE Green Flight Challenge is a flight competition for quiet, practical, "green" aircraft that took place July 11–17, 2011, at the CAFE Foundation Flight Test Center at Charles M. Schulz Sonoma County Airport in Santa Rosa, California. The winning aircraft (Team Pipistrel-USA.com), a four-seat electric-powered version of the Taurus G4, flew 192 statute-miles non-stop, achieving an astounding mileage of 403.5 passenger·mi/gal. This amounts to one person being able to travel 403.5 mi (351 nm) on one gallon of fuel (or two persons travelling 202 mi per gallon of fuel). It demonstrated the possibilities offered by electric aircraft of the future.

The goal of the challenge was to achieve 200 pax·mi/gal. In essence, this requires the following efficiency:

$$\text{Efficiency} = 200 \frac{\text{pax} \cdot \text{mi}}{\text{gal}} = \frac{200}{6 \text{ lb}_f/\text{gal}} \frac{\text{pax} \cdot \text{mi}}{\text{gal}}$$
$$= 33.3 \frac{\text{pax} \cdot \text{mi}}{\text{lb}_f}$$

For the two people on board this corresponds to a SR of:

$$SR = \frac{33.3}{2 \text{ pax}} \frac{\text{pax} \cdot \text{mi}}{\text{lb}_f} = \left(\frac{5280}{6076}\right) 16.67 \frac{\text{mi}}{\text{lb}_f} = 14.48 \frac{\text{nm}}{\text{lb}_f}$$

This value is called "specific range". In terms of the Brequet flight profile 3, the range per pound of fuel consumed amounts to:

$$R = \frac{V}{c_t} \frac{C_L}{C_D} \ln\left(\frac{W_{ini}}{W_{fin}}\right) = \frac{V}{c_t} \frac{C_L}{C_D} \ln\left(\frac{W_{ini}}{W_{ini} - 1}\right) = 14.48 \text{ nm}$$

EXAMPLE 20-9

Estimate the efficiency of the SR22 (see Figure 20-12) if, per its POH, the specific range at 75% power at 8000 ft is 10.3 nm/gal.

Solution

Convert the specific range to nm/lb$_f$ of fuel:

$$SR = 10.3 \frac{nm}{gal} = 1.72 \frac{nm}{lb_f}$$

Therefore, the efficiency can be calculated as follows:

$$Efficiency = 4 \text{ pax} \times \frac{1.72 \text{ nm}}{lb_f} \times \frac{6076}{5280}$$

$$= 7.91 \frac{pax \cdot mi}{lb_f}$$

FIGURE 20-12 The Cirrus SR22 General Aviation aircraft. (Photo by Phil Rademacher)

20.4 FUNDAMENTAL RELATIONS FOR ENDURANCE ANALYSIS

The following mathematical expressions can be used to estimate the endurance with the cited restrictions. An inspection of the resulting equations demonstrates a certain commonality between all. In other words, in order to achieve a long endurance capability the airplane must be designed with the following features in mind:

(1) It must have a high operational L/D ratio. In other words, it must feature as low drag as possible.
(2) It must have a low specific fuel consumption (or low thrust specific fuel consumption).
(3) If driven by a propeller it must feature the highest possible propeller efficiency and operate at a low altitude (high ρ).
(4) It must carry as much fuel as possible.

Formulation for the estimation of endurance will now be developed assuming the same profiles used for the development of range.

20.4.1 Endurance Profile 1: Constant Airspeed/Constant Altitude Cruise

Refer to Section 20.2.2, *Range profile 1: constant airspeed/constant altitude cruise* and Figure 20-5 for more details on this cruise profile. *This result is valid only for the simplified drag model.*

$$E = \frac{1}{c_t \sqrt{kC_{Dmin}}} \left[\tan^{-1}\left(\frac{2\sqrt{k}}{\rho V^2 S \sqrt{C_{Dmin}}} W_{ini} \right) - \tan^{-1}\left(\frac{2\sqrt{k}}{\rho V^2 S \sqrt{C_{Dmin}}} W_{fin} \right) \right]$$

(20-19)

Derivation of Equation (20-19)

In all three cases, the thrust specific fuel consumption, c_t, is constant. Additionally for this case, the airspeed, V, is also constant. However, C_L/C_D is not. Therefore, we write Equation (20-7), introducing the simplified drag model as follows:

$$E = \int_{W_{fin}}^{W_{ini}} \frac{1}{c_t} \frac{C_L}{C_D} \frac{1}{W} dW = \frac{1}{c_t} \int_{W_{fin}}^{W_{ini}} \frac{C_L}{C_D} \frac{1}{W} dW$$

$$= \frac{1}{c_t} \int_{W_{fin}}^{W_{ini}} \frac{C_L}{(C_{Dmin} + kC_L^2)} \frac{1}{W} dW$$

Expanding by inserting the expressions for the lift coefficient we get:

$$E = \frac{1}{c_t}\int_{W_{fin}}^{W_{ini}}\frac{C_L}{(C_{Dmin}+kC_L^2)}\frac{1}{W}dW$$

$$= \frac{1}{c_t}\int_{W_{fin}}^{W_{ini}}\frac{\frac{2W}{\rho V^2 S}}{\left(C_{Dmin}+k\left(\frac{2W}{\rho V^2 S}\right)^2\right)}\frac{1}{W}dW$$

Manipulate algebraically to get:

$$E = \frac{1}{c_t}\frac{2}{\rho V^2 S}\int_{W_{fin}}^{W_{ini}}\frac{1}{\left(C_{Dmin}+k\frac{4W^2}{\rho^2 V^4 S^2}\right)}dW$$

$$= \frac{1}{c_t}\frac{2}{\rho V^2 S}\int_{W_{fin}}^{W_{ini}}\frac{1}{\left(C_{Dmin}\frac{4k}{\rho^2 V^4 S^2}\frac{\rho^2 V^4 S^2}{4k}+\frac{4k}{\rho^2 V^4 S^2}W^2\right)}dW$$

$$E = \frac{1}{c_t}\frac{2}{\rho V^2 S}\frac{\rho^2 V^4 S^2}{4k}\int_{W_{fin}}^{W_{ini}}\frac{1}{\left(C_{Dmin}\frac{\rho^2 V^4 S^2}{4k}+W^2\right)}dW$$

$$= \frac{1}{c_t}\frac{\rho V^2 S}{2k}\int_{W_{fin}}^{W_{ini}}\frac{1}{\left(C_{Dmin}\frac{\rho^2 V^4 S^2}{4k}+W^2\right)}dW$$

Let's define the constant a as follows:

$$a = \frac{\sqrt{C_{Dmin}}}{\sqrt{k}}\frac{\rho V^2 S}{2} \Rightarrow a^2 = \frac{C_{Dmin}\rho^2 V^4 S^2}{4k}$$

$$\Rightarrow \frac{1}{a} = \frac{\sqrt{k}}{\sqrt{C_{Dmin}}}\frac{2}{\rho V^2 S}$$

Then insert and solve:

$$E = \frac{1}{c_t}\frac{\rho V^2 S}{2k}\int_{W_{fin}}^{W_{ini}}\frac{1}{(a^2+W^2)}dW = \frac{1}{c_t}\frac{\rho V^2 S}{2k}\frac{1}{a}\left[\tan^{-1}\frac{W}{a}\right]_{W_{fin}}^{W_{ini}}$$

$$= \frac{1}{c_t\sqrt{kC_{Dmin}}}\left[\tan^{-1}\frac{\sqrt{k}}{\sqrt{C_{Dmin}}}\frac{2}{\rho V^2 S}W\right]_{W_{fin}}^{W_{ini}}$$

$$= \frac{1}{c_t\sqrt{kC_{Dmin}}}$$
$$\times\left[\tan^{-1}\left(\frac{2\sqrt{k}}{\rho V^2 S\sqrt{C_{Dmin}}}W_{ini}\right)-\tan^{-1}\left(\frac{2\sqrt{k}}{\rho V^2 S\sqrt{C_{Dmin}}}W_{fin}\right)\right]$$

QED

20.4.2 Endurance Profile 2: Constant Attitude/Altitude Cruise

Refer to Section 20.2.3, *Range profile 2: constant altitude/constant attitude cruise* and Figure 20-5 for more details on this cruise profile. Both following results are valid regardless of drag model choice.

For a jet:

$$E = \frac{1}{c_t}\frac{C_L}{C_D}\ln\left(\frac{W_{ini}}{W_{fin}}\right) \quad (20\text{-}20)$$

Note that a special version of this expression extends to propeller aircraft due to the fact that the thrust specific fuel consumption is dependent on the airspeed (see Section *20.1.7, Notes on SFC and TSFC*).

For a propeller:

$$E = \left(\frac{1980000\eta_P}{c_{bhp}}\right)\sqrt{2\rho S}\left(\frac{C_L^{1.5}}{C_D}\right)\left[\frac{1}{\sqrt{W_{fin}}}-\frac{1}{\sqrt{W_{ini}}}\right]$$
(20-21)

The constant 1980000 allows the SFC, represented by the variable c, to be entered in terms of lb_f of fuel per hour per BHP (i.e. lb_f/hr/BHP), but this is conveniently the most common presentation of this important parameter the designer will come across.

Derivation of Equation (20-20)

The derivation is straightforward. Since the attitude is constant, move C_L/C_D outside the integral and integrate:

$$E = \frac{1}{c_t}\int_{W_{fin}}^{W_{ini}}\frac{C_L}{C_D}\frac{1}{W}dW = \frac{1}{c_t}\frac{C_L}{C_D}\int_{W_{fin}}^{W_{ini}}\frac{1}{W}dW$$

$$= \frac{1}{c_t}\frac{C_L}{C_D}[\ln W]_{W_{fin}}^{W_{ini}} = \frac{1}{c_t}\frac{C_L}{C_D}\ln\left(\frac{W_{ini}}{W_{fin}}\right)$$

QED

Derivation of Equation (20-21)

Since the TSFC is airspeed dependent, it cannot be taken out of the integral of Equation (20-7). Instead we insert Equation (20-9) and write:

$$E = \int_{W_{fin}}^{W_{ini}}\frac{1}{c_t}\frac{C_L}{C_D}\frac{1}{W}dW = \int_{W_{fin}}^{W_{ini}}\left(\frac{1980000\eta_P}{c_{bhp}V}\right)\frac{C_L}{C_D}\frac{1}{W}dW$$

Rewrite the lift equation, Equation (9-47), in terms of V and insert:

$$E = \int_{W_{fin}}^{W_{ini}} \left(\frac{1980000\eta_P}{c_{bhp}V}\right) \frac{C_L}{C_D} \frac{1}{W} dW = \int_{W_{fin}}^{W_{ini}} \left(\frac{1980000\eta_P}{c_{bhp}}\right) \sqrt{\frac{\rho S C_L}{2W}} \frac{C_L}{C_D} \frac{1}{W} dW$$

$$= \left(\frac{1980000\eta_P}{c_{bhp}}\right) \int_{W_{fin}}^{W_{ini}} \sqrt{\frac{\rho S}{2}} \frac{C_L^{1.5}}{C_D} \frac{1}{W^{1.5}} dW = \left(\frac{1980000\eta_P}{c_{bhp}}\right) \sqrt{\frac{\rho S}{2}} \left(\frac{C_L^{1.5}}{C_D}\right) \int_{W_{fin}}^{W_{ini}} \frac{dW}{W^{1.5}}$$

$$= \left(\frac{1980000\eta_P}{c_{bhp}}\right) \sqrt{\frac{\rho S}{2}} \left(\frac{C_L^{1.5}}{C_D}\right) \left[\frac{W^{-1.5+1}}{-1.5+1}\right]_{W_{fin}}^{W_{ini}} = \left(\frac{1980000\eta_P}{c_{bhp}}\right) \sqrt{\frac{\rho S}{2}} \left(\frac{C_L^{1.5}}{C_D}\right) \left[\frac{W^{-0.5}}{-0.5}\right]_{W_{fin}}^{W_{ini}}$$

$$= 2\left(\frac{1980000\eta_P}{c_{bhp}}\right) \sqrt{\frac{\rho S}{2}} \left(\frac{C_L^{1.5}}{C_D}\right) \left[\frac{1}{\sqrt{W_{fin}}} - \frac{1}{\sqrt{W_{ini}}}\right] = \left(\frac{1980000\eta_P}{c_{bhp}}\right) \sqrt{2\rho S} \left(\frac{C_L^{1.5}}{C_D}\right) \left[\frac{1}{\sqrt{W_{fin}}} - \frac{1}{\sqrt{W_{ini}}}\right]$$

QED

20.4.3 Endurance Profile 3: Constant Airspeed/Attitude Cruise

Refer to Section 20.2.4, *Range profile 3: constant airspeed/constant attitude cruise* and Figure 20-7 for more details on this cruise profile. Valid for both jets and propeller-driven aircraft and independent of the drag model:

$$E = \frac{1}{c_t} \frac{C_L}{C_D} \ln\left(\frac{W_{ini}}{W_{fin}}\right) \qquad (20\text{-}22)$$

Derivation of Equation (20-22)

Again, the derivation is straight forward. Since the attitude is constant, move C_L/C_D outside the integral and integrate:

$$E = \frac{1}{c_t} \int_{W_{fin}}^{W_{ini}} \frac{C_L}{C_D} \frac{1}{W} dW = \frac{1}{c_t} \frac{C_L}{C_D} \int_{W_{fin}}^{W_{ini}} \frac{1}{W} dW$$

$$= \frac{1}{c_t} \frac{C_L}{C_D} [\ln W]_{W_{fin}}^{W_{ini}} = \frac{1}{c_t} \frac{C_L}{C_D} \ln\left(\frac{W_{ini}}{W_{fin}}\right)$$

QED

EXAMPLE 20-10

The light airplane of EXAMPLE 20-1 starts its cruise segment at 135 KCAS at 10,000 ft when it weighs 3200 lbs. After cruising for a period of time at that altitude it is noticed it now weighs 2800 lbs. The following additional data is given:

Wing area	$S = 145$ ft^2
Aspect ratio	$AR = 10$
Minimum drag coefficient	$C_{Dmin} = 0.025$
Average fuel burn	15 gallons/hr
Average BHP	150 BHP
Propeller efficiency	$\eta_P = 0.85$

Compute the endurance of this airplane using the first profile.

Note that this is EXAMPLE 20-5 from the range portion solved for endurance.

Solution

Air density:

$$\rho = 0.001756 \text{ slugs/ft}^3 \quad \text{(see Example 20-1)}$$

Airspeed, KTAS:

$$V_{TAS} = 157 \text{ KTAS}$$

EXAMPLE 20-10 (cont'd)

Airspeed, ft/s:

$$V = 1.688 V_{TAS} = 265.2 \text{ ft/s}$$

Span efficiency:

$$e = 0.7566$$

Coefficient "k":

$$k = \frac{1}{\pi \cdot AR \cdot e} = 0.04207$$

SFC:

$$c_{bhp} = \frac{15 \text{ gal/hr}}{150 \text{ BHP}} = \frac{(15 \times 6) \text{ lb}_f/\text{hr}}{150 \text{ BHP}}$$
$$= 0.600 \text{ lb}_f/\text{hr/BHP}$$

TSFC:

$$c_t = \frac{c_{bhp} V}{1980000 \eta_p} = \frac{(0.600)(265.2)}{1980000(0.85)} = 0.00009455 \frac{1}{\text{sec}}$$

Lift coefficient:

$$C_L = \frac{2W}{\rho V^2 S} = \frac{2(3200)}{(0.001756)(265.2)^2(145)} = 0.3574$$

Drag coefficient:

$$C_D = C_{Dmin} + kC_L^2 = (0.025) + (0.04207)(0.3574)^2$$
$$= 0.03037$$

Glide ratio:

$$\left(\frac{L}{D}\right) = \frac{C_L}{C_D} = \frac{0.3574}{0.03037} = 11.77$$

As a "sanity check" consider the following:

Total fuel burned:

$$W_f = 3200 - 2800 = 400 \text{ lb}_f$$

Time to burn W_f:

$$T_f = \frac{W_f}{15 \text{ gals/hr} \times 6 \text{ lb}_f/\text{gal}} = \frac{400 \text{ lb}_f}{90 \text{ lb}_f/\text{hr}} = 4.444 \text{ hrs}$$

So we are expecting a result in the neighborhood of 4.4 hrs.
Endurance profile 1: constant airspeed/altitude cruise

$$E = \frac{1}{c_t \sqrt{kC_{Dmin}}} \left[\tan^{-1}\left(\frac{2\sqrt{k}}{\rho V^2 S \sqrt{C_{Dmin}}} W_{ini}\right) \right.$$
$$\left. - \tan^{-1}\left(\frac{2\sqrt{k}}{\rho V^2 S \sqrt{C_{Dmin}}} W_{fin}\right) \right]$$

$$A = \frac{2\sqrt{k}}{\rho V^2 S \sqrt{C_{Dmin}}} = \frac{2\sqrt{0.04207}}{(0.001756)(265.2)^2(145)\sqrt{0.025}}$$
$$= 0.0001449$$

$$E = \frac{1}{(0.00009455)\sqrt{(0.04207)(0.025)}}$$
$$\times \left[\tan^{-1}(A(3200)) - \tan^{-1}(A(2800)) \right]$$
$$= (326123.6) \left[\tan^{-1}(0.46361) - \tan^{-1}(0.40566) \right]$$
$$= (326123.6)[0.43412 - 0.38538]$$
$$= 15895.3 \text{ ft} \stackrel{\div 3600 \text{ set/hr}}{=} 4.42 \text{ hrs}$$

20.5 ANALYSIS OF MISSION PROFILE

Now that the formulation to estimate range and endurance has been developed, it is appropriate to introduce the fundamentals of mission analysis. A *mission analysis* is the investigation of an entire flight of an airplane, from engine start to shut-down. Generally, such analysis breaks the mission into several distinct segments, which allows for a simpler analysis of each and important properties of the airplane, such as weight, airspeed, and others to be determined more easily.

Any serious mission planning must account for (a) fuel required to complete the mission and (b) additional fuel to allow the airplane to be diverted from the destination airport to some alternative airport should inclement weather compromise a safe landing.

The additional fuel is called *reserve fuel*. The preceding sections have not accounted for this fuel, but rather assumed this as a part of the empty weight (which it is not) or simply ignored it. Any range or endurance analysis should accommodate reserve fuel in the final weight (W_{fin}) – naturally, reducing the total fuel available for the design mission. The reserve fuel must be sufficient to allow the airplane to climb to a new cruise altitude after an attempted and missed approach for landing, and continue flying to an alternative airport that may be as far as 100 or 200 nm from the original destination. There are generally two operational cruise missions that designers of civilian aircraft should be aware of: IFR and NBAA missions. These will be discussed later in this section. First, fundamentals of *mission analysis* will be presented.

20.5.1 Basics of Mission Profile Analysis

Clearly, the weight of the fuel required to complete the design mission is of paramount importance and accurate range estimation must account for all the fuel that will be consumed. Consider a scenario in which the mission, from engine start until engine shut-down, consists of the segments shown in Figure 20-13. Each segment is denoted as a line drawn between two points, or nodes. Each node is numbered from 0 to 7. Of course the mission may be more complicated than shown here; however, it can also be far simpler. At any rate, the important element is to recognize that as the airplane covers each segment, assuming it uses fossil fuels, its weight is reduced. This way, the weight of the airplane at Node 1 is less than at Node 0. Similarly, after climbing to its cruise altitude, denoted by Node 2, its weight is less than after T-O at Node 1, and so on. Clearly, the difference between each subsequent node is the amount of fuel consumed during the segment, although it may also include the weight of parachutists or external stores.

In order to analyze the entire mission, we observe that (note the difference W_O {"O" for gross weight} versus W_0 {"zero" for Node 0}, which indicates that the airplane does not always take off at gross weight):

Segment 0-1:

$$\frac{W_1}{W_0} < 1 \qquad (20\text{-}23)$$

Segment 1-2:

$$\frac{W_2}{W_1} < 1 \qquad (20\text{-}24)$$

And so forth. If we know the average fuel consumption in fuel weight/unit time during the segment and the time it takes to complete the segment we can always estimate the ratio as follows. Note that if the fuel consumption (or "fuel flow") is in terms of lb_f/hr, the time would have to be in terms of hours:

$$\frac{W_i}{W_{i-1}} = \frac{W_{i-1} - \dot{w}_j(t_i - t_{i-1})}{W_{i-1}} = 1 - \frac{\dot{w}_j(t_i - t_{i-1})}{W_{i-1}} \qquad (20\text{-}25)$$

where

i = node index
j = segment index. Segment 1 is between Nodes 0 and 1, and so on.
t_i = time at Node i
$\dot{w}_j$ = average fuel flow in weight/unit time during segment j

Naturally, if the airplane features items that can be jettisoned, such as military ordnance or fuel tanks, or fertilizer from agricultural aircraft, or even fuel during an emergency fuel dump operation, a different set of relations would have to be established.

Weight ratios for selected segments

Since the cruise segment often remains the unknown variable in mission analyses (see EXAMPLE 20-12), the weight ratios associated with the other segments must be known. As an example, some of the segments shown

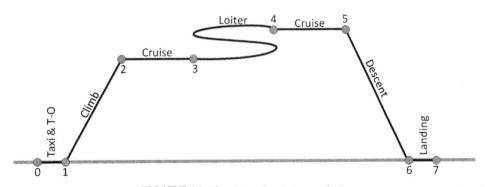

FIGURE 20-13 Aircraft mission analysis.

TABLE 20-6 Weight Ratios for General Aviation Aircraft (W_i/W_{i-1})

	Single-Engine Piston	Twin-Engine Piston	Gas Turbine[a]
Warm-up and takeoff	0.995	0.994	0.970
Climb	0.988	0.985	0.985
Descent	0.997	0.996	-
Landing	0.995	0.995	0.995

[a]From Ref. [7].

in Figure 20-13 lend themselves to a generalized approach: for instance the engine start, warm-up, takeoff, climb, and landing. It is a reasonable assumption that the ratios for these phases remain constant for any particular class of aircraft. In this context, Table 20-6 it may be helpful for filling in some of the gaps in the mission analysis.

20.5.2 Methodology for Mission Analysis

An example of a simple mission is shown in Figure 20-14. It consists of segments that entail engine start, warm-up, run-up, taxi into T-O position, the T-O, climb to cruise altitude, cruise, descent, landing, and finally taxi and parking. In the interests of clarity, the segments have been combined and are identified at each intersection by the six nodes. Each node has a weight associated with it, denoted by $W_0, W_1, ..., W_5$. Then, the mission is analyzed by determining the weight (and therefore fuel consumed) at each node.

As an example of the analysis methodology, consider a scenario in which we have to determine the fuel available for cruise (cruise range); this is the fraction W_3/W_2 in Figure 20-14. In most cases, the engine start-up weight, W_0, and mission fuel weight, W_{fm}, are known. Assume it has been estimated (for instance using historical data or knowledge of engine fuel consumption) that the engine start-up, warm-up, and taxi to T-O position consumes fuel so at Node 1 the airplane weighs 99% of W_0. In this case it is possible to write the weight at Node 1 as follows:

$$W_1 = \left(\frac{W_1}{W_0}\right) W_0 = 0.99 W_0$$

Similarly, assume that during the climb to the cruise altitude the airplane consumes fuel such it weighs 93% of the weight at Node 1. This can be based off of the weight at Node 1 as follows:

$$W_2 = \left(\frac{W_2}{W_1}\right) W_1 = 0.93 W_1$$

In effect, this would allow the weight at Node 2 to be determined based on the initial weight, W_0, as shown below:

$$W_2 = \left(\frac{W_2}{W_1}\right) \times W_1 = \left(\frac{W_2}{W_1}\right) \times \left(\frac{W_1}{W_0}\right) W_0$$
$$= (0.93) \times (0.99) W_0 = 0.9207 W_0$$

This simply means that when the airplane reaches its cruise altitude, it weighs 92.07% of its engine start-up weight. In fact, the entire mission can be represented as the product of the segment weight ratios:

$$\left(\frac{W_1}{W_0}\right)\left(\frac{W_2}{W_1}\right)\left(\frac{W_3}{W_2}\right)\left(\frac{W_4}{W_3}\right)\left(\frac{W_5}{W_4}\right) = \frac{W_5}{W_0}$$

This representation boils the problem of determining the amount of fuel available for cruise (represented by W_3/W_2) down to determining the other weight ratios and using them to estimate the desired ratio.

To see how this is accomplished, assume that we have determined that the weight at the point of touch-down

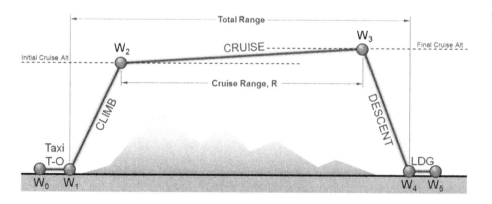

FIGURE 20-14 The break-down of a mission into a number of segments.

(Node 4) is 96% of the weight at cruise let-down (Node 3). Also, assume that at the time of engine shut-down (Node 5), the weight is 99% of the weight at Node 4. If we assume we run out of the mission fuel at Node 5, we can write the weight at Node 5 as follows:

$$W_5 = W_0 - W_{fm}$$

where W_{fm} = total mission fuel weight (known).

This allows the mission weight ratio to be rewritten as follows:

$$\frac{W_5}{W_0} = \frac{W_0 - W_{fm}}{W_0} = \frac{W_1}{W_0}\frac{W_2}{W_1}\frac{W_3}{W_2}\frac{W_4}{W_3}\frac{W_5}{W_4}$$

Inserting known ratios leads to:

$$\frac{W_5}{W_0} = (0.99)(0.93)\frac{W_3}{W_2}(0.96)(0.99) = 0.8750\frac{W_3}{W_2}$$

This allows the weight ratio at the start and end of cruise to be written as show below:

$$\frac{W_3}{W_2} = \frac{1}{0.8750}\frac{W_5}{W_0} = 1.143\frac{W_5}{W_0}$$

Using the value of W_2 already determined above and simple algebra, the amount of fuel weight available for the cruise portion amounts to:

$$W_2 - W_3 = 1.0524 W_{fm} - 0.1317 W_0$$

One important observation is that the fuel required to just get the airplane to cruising altitude and return to base (no actual cruise range) is determined by setting the difference $W_2 - W_3 = 0$ and solving for W_{fm}. The resulting minimum fuel amounts to $0.1251 W_0$, or 12.5% of gross weight. If this represented a real aircraft, it would be placed at a significant competitive disadvantage. Thus, if the gross weight of the airplane is 3400 lb$_f$ and it has 500 lb$_f$ of fuel on board intended for the mission, it will have $1.0524 \times 500 - 0.1317 \times 3400 = 78.4$ lb$_f$ of fuel available for the cruise range. The fuel for the entire mission can be broken down as follows:

$W_0 = 3400$ lb$_f$
$W_1 = 0.99 W_0 = 0.99 \times 3400 = 3366$ lb$_f$
$W_2 = 0.93 W_1 = 0.93 \times 3366 = 3130$ lb$_f$
$W_3 = W_2 - 1.0524 W_{fm} + 0.1317 W_0 = 3130 - 1.0524(500) + 0.1317(3400) = 3052$ lb$_f$
$W_4 = 0.96 W_3 = 0.96 \times 3052 = 2929$ lb$_f$
$W_5 = 0.99 W_3 = 0.99 \times 2929 = 2900$ lb$_f$

EXAMPLE 20-11

A small single-engine aircraft is being designed for the mission of Figure 20-13 and it has been established that it consumes 10 lb$_f$ of fuel during Segment 0-1. During climb, which starts at sea level, the average fuel burn is 15 gallons per hour and the average rate of climb is 1000 fpm. Determine the weight ratios for the first two segments (taxi & T-O and climb) if its mission cruise altitude is 15,000ft. A gallon of fuel weighs 6 lb$_f$. If its start-up weight is 3400 lb$_f$, what is its weight at Node 2?

Solution

Segment 0-1:

$$\frac{W_1}{W_0} = 1 - \frac{\dot{w}(t_1 - t_0)}{W_0} = 1 - \frac{10}{3400} = 0.9971$$

For Segment 1-2 we see that if the average rate of climb is 1000 fpm it will take 15 min or 0.25 hrs to climb to 15000 ft. Also note that W_1 is obtained from the previous step using $W_1 = 0.9971 \cdot W_0$. Therefore:

Segment 1-2:

$$\frac{W_2}{W_1} = 1 - \frac{\dot{w}(t_2 - t_1)}{W_1} = 1 - \frac{(15 \times 6)(0.25)}{0.9971 \times 3400} = 0.9933$$

Therefore, the weight of the aircraft at Node 2 can be found from:

$$\frac{W_2}{W_0} = \left(\frac{W_1}{W_0}\right)\left(\frac{W_2}{W_1}\right) \Leftrightarrow$$
$$W_2 = W_0\left(\frac{W_1}{W_0}\right)\left(\frac{W_2}{W_1}\right) \Leftrightarrow$$
$$W_2 = (3400)(0.9971)(0.9933) = 3367 \text{ lb}_f$$

EXAMPLE 20-12

The aircraft of EXAMPLE 20-11 is being evaluated based on the mission of Figure 20-13 for fuel requirements between Nodes 2 and 5 assuming all fuel is exhausted at Node 7 and the total fuel weight ratio is 0.15. The following weight ratios have been estimated:

Segment 0-1: $\frac{W_1}{W_0} = 0.9971$ Segment 1-2: $\frac{W_2}{W_1} = 0.9933$
Segment 5-6: $\frac{W_6}{W_5} = 0.998$ Segment 6-7: $\frac{W_7}{W_6} = 0.999$

Determine the fuel remaining to complete segments 2-5.

Solution

If we assume all the fuel to be consumed at Node 7, it follows that the total weight ratio between Nodes 0 and 7 will be (noting that a fuel weight ratio of 0.15 means $0.15 \times W_0$):

$$\frac{W_7}{W_0} = \frac{W_0 - W_f}{W_0} = \frac{W_0 - 0.15 \cdot W_0}{W_0} = 0.85$$

We can write the change in weight along the mission shown in Figure 20-13 as follows:

$$\frac{W_7}{W_0} = \frac{W_0 - W_f}{W_0} = \frac{W_1}{W_0} \times \frac{W_2}{W_1} \times \frac{W_5}{W_2} \times \frac{W_6}{W_5} \times \frac{W_7}{W_6} = 0.85$$

The expression reveals that only the ratio W_5/W_2 is not yet known. We will now solve the above expression for this ratio:

$$(0.9971) \times (0.9933) \times \frac{W_5}{W_2} \times (0.998) \times (0.999) = 0.85$$

$$\Leftrightarrow \frac{W_5}{W_2} = \frac{0.85}{0.9875} = 0.8608$$

We have already calculated the weight of the airplane at Node 2 in EXAMPLE 20-11. Using that result we can easily estimate the weight of the aircraft at Node 5:

$$\frac{W_5}{W_2} = 0.8608$$

$$\Leftrightarrow W_5 = 0.8608 \cdot W_2 = 0.8608 \cdot (3367) = 2898 \text{ lb}_f$$

Therefore, the fuel required between Nodes 2 and 5 is $3367 - 2898 = 469 \text{ lb}_f$ or just over 78 gallons.

20.5.3 Special Range Mission 1: IFR Cruise Mission

GA aircraft intended for operation under Instrument Flight Rules (IFR) must carry fuel in compliance with 14 CFR Part 91, § 91.167, Fuel Requirements for Flight in IFR Conditions. The regulation is intended to prevent accidents due to fuel starvation or fuel mismanagement, but such accidents, in spite of the regulation, remain fairly common, with 75 such accidents in 2008 and 74 in 2009 in the US alone [8]. GA aircraft operated in IFR conditions must carry fuel as stipulated below:

§91.167 FUEL REQUIREMENTS FOR FLIGHT IN IFR CONDITIONS

(a) No person may operate a civil aircraft in IFR conditions unless it carries enough fuel (considering weather reports and forecasts and weather conditions) to—
   (1) Complete the flight to the first airport of intended landing;
   (2) Except as provided in paragraph (b) of this section, fly from that airport to the alternate airport; and
   (3) Fly after that for 45 minutes at normal cruising speed or, for helicopters, fly after that for 30 minutes at normal cruising speed.
(b) Paragraph (a)(2) of this section does not apply if:
   (1) Part 97 of this chapter prescribes a standard instrument approach procedure to, or a special instrument approach procedure has been issued by the Administrator to the operator for, the first airport of intended landing; and

(2) Appropriate weather reports or weather forecasts, or a combination of them, indicate the following:
  (i) For aircraft other than helicopters. For at least 1 hour before and for 1 hour after the estimated time of arrival, the ceiling will be at least 2,000 feet above the airport elevation and the visibility will be at least 3 statute miles.
  (ii) For helicopters. At the estimated time of arrival and for 1 hour after the estimated time of arrival, the ceiling will be at least 1,000 feet above the airport elevation, or at least 400 feet above the lowest applicable approach minima, whichever is higher, and the visibility will be at least 2 statute miles.

The paragraph requires the operator of the aircraft to carry enough fuel to complete the flight to the initial airport, then fly from that airport to the alternative airport (if one is required) and, then fly for 45 minutes (30 minutes for helicopters) at normal cruising speed [9]. In short, it is fuel to the destination plus fuel to the alternative plus fuel for 45 minutes of additional flying. It is important for the designer to be mindful of such regulatory "curveballs" so that the range (or size of fuel tanks) won't be incorrectly estimated.

A schematic of an IFR flight profile is shown in Figure 20-15. The term STD IFR APP stands for Standard IFR Approach. Even though such flight planning is the responsibility of the operator, it is only prudent the designer should prepare range estimation for comparison with other aircraft by assuming a 100 or 200 nm diverted flight to an alternative, depending on the size of the aircraft. This will shed a much clearer light on the utility of the new design.

20.5.4 Special Range Mission 2: NBAA Cruise Mission

The National Business Aviation Association (NBAA) is an organization of people who try to promote and expand the use of General Aviation aircraft for business purposes globally [10]. It was founded in 1947 and has its headquarters in Washington DC. According to its Statement of Purpose, the organization represents more than 8000 companies and organizes the world's largest aviation trades show; the NBAA Annual Meeting & Convention. When it comes to the concept of range, NBAA has promoted a special range profile that carries its name. A schematic of this profile is provided in Figure 20-16. This profile is typically used in the marketing of business aircraft.

20.5.5 Payload-Range Sensitivity Study

It is helpful, even when not mandated, to conduct a payload-range sensitivity study to help evaluate how payload affects the range of the aircraft. Most small aircraft have issues with the combination of fuel and occupant loading. This manifests itself as an inability to take off simultaneously with the fuel tanks and seats filled without busting the max gross weight limit. Thus, most small aircraft have to limit fuel weight when flying with full occupancy and this can reduce their range significantly.

Figure 20-17 illustrates the problem using a hypothetical single-pilot-operated ten-seat turboprop commuter design of a maximum design gross weight of 7500 lb$_f$. Assume the airplane is designed to carry as much as 2000 lb$_f$ of Jet A-1, which gives it a range of 1100 nm. This means that no matter the combination of fuel, passengers, and baggage, the airplane shall not exceed 7500 lb$_f$ at the start of the T-O run. Furthermore, assume that based on historical data it is expected an empty

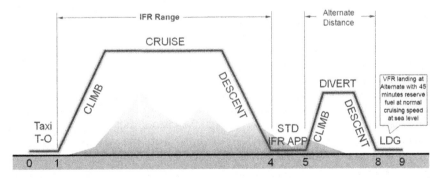

FIGURE 20-15 A schematic for an *Instrument Flight Regulation* mission for a GA aircraft.

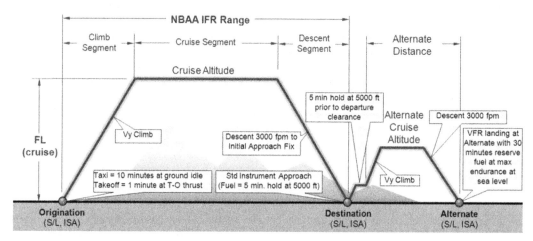

FIGURE 20-16 A schematic for an NBAA mission for a Part 23 Business aircraft.

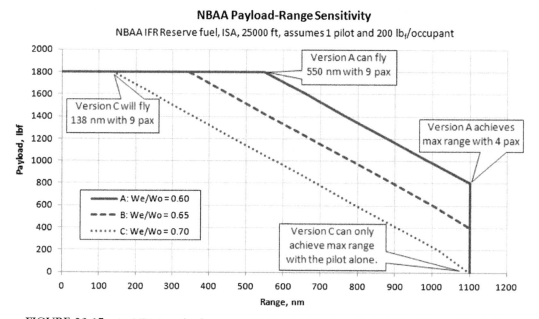

FIGURE 20-17 An NBAA payload-range sensitivity plot for a hypothetical 10-seat commuter design.

weight ratio of 0.6 can be achieved. This corresponds to an empty weight of 4500 lb$_f$ and useful load of 7500 − 4500 = 3000 lb$_f$. If each occupant can weigh as much as 200 lb$_f$, including baggage, it is easy to see that, with 10 people onboard, the airplane will weigh 4500 + 10 × 200 = 6500 lb$_f$; this corresponds to a payload of nine people (the pilot is not a part of the payload) or 1800 lb$_f$. As a consequence the pilot can only add 1000 lb$_f$ of fuel and, thus, will not achieve the full tank range of 1100 nm. Here, for simplicity, it is assumed the aircraft can achieve the hypothetical range of 0.55 nm per lb$_f$ of fuel.

Results for other payloads are presented in Table 20-7 (and graphically in Figure 20-17). It is intended to help with the preparation of such sensitivity graphs. It displays not only how range is affected by payload, but also how detrimental it may be if the desired empty weight ratio is not met. Payload-range sensitivity studies are very helpful when evaluating new aircraft designs and clearly highlights the importance of achieving the design empty-weight ratio.

Table 20-7 presents the payload-range sensitivity for three empty weight ratios; 0.60, 0.65, and 0.70, designated as Versions A, B, and C, respectively. Thus, Version A can carry four passengers and pilot 1100 nm, while it can only achieve 550 nm with nine passengers onboard. However, if the empty weight ratio of 0.60 is not met and, say, it ends up being 0.65 (Version B), the airplane will be much less competitive as it can only fly 1100 nm with two passengers onboard. If the empty weight ratio ends up being 0.70 (Version C), the market prospects of the resulting aircraft may in fact

TABLE 20-7 Example of Payload-Range Analysis

Gross weight, lb_f	$W_o =$	7500								
Max fuel weight, lb_f	$W_f =$	2000								
Weight of occupant, lb_f	$W_{occupant} =$	200								
			VERSION A			VERSION B			VERSION C		
Empty weight ratio	$W_e/W_o =$		0.6			0.65			0.7		
Empty weight, lb_f	$W_e =$		4500			4875			5250		
Passengers	Pilot	Payload, lb_f	Fuel, lb_f	Gross, lb_f	Range, nm	Fuel, lb_f	Gross, lb_f	Range, nm	Fuel, lb_f	Gross, lb_f	Range, nm
0	1	0	2000	6700	1100	2000	7075	1100	2000	7450	1100
1	1	200	2000	6900	1100	2000	7275	1100	1850	7500	1018
2	1	400	2000	7100	1100	2000	7475	1100	1650	7500	908
3	1	600	2000	7300	1100	1825	7500	1004	1450	7500	798
4	1	800	2000	7500	1100	1625	7500	894	1250	7500	688
5	1	1000	1800	7500	990	1425	7500	784	1050	7500	578
6	1	1200	1600	7500	880	1225	7500	674	850	7500	468
7	1	1400	1400	7500	770	1025	7500	564	650	7500	358
8	1	1600	1200	7500	660	825	7500	454	450	7500	248
9	1	1800	1000	7500	550	625	7500	344	250	7500	138

be catastrophic. This highlights its sensitivity to deviations from the target empty weight ratio.

EXERCISES

(1) (a) A twin-piston-engine aircraft is being designed for the mission of Figure 20-18 and it has been established that it consumes 30 lb_f of fuel during Segment 0-1. During climb, which starts at sea level, the average fuel burn is 44 gallons per hour and the average rate of climb is 1600 fpm. Determine the weight ratios for the taxi & T-O and climb segments if its design mission cruise altitude is 20,000 ft. A gallon of fuel weighs 6 lb_f. If its start-up weight is 5500 lb_f, what is its weight at Node 2?

(b) Using the mission of Figure 20-18 estimate the fuel requirements between Nodes 2 and 3 assuming all fuel is exhausted at Node 5 and the total fuel weight ratio is 0.18. Use the weight ratios of Table 20-6 for descent and landing.

(2) You have been asked to design a single-engine, turbofan-powered research aircraft that, on a no-wind standard day, is capable of an 800 nm loiter range. The loitering segment will begin at an initial altitude of 20,000 ft and requires a constant airspeed of 250 KTAS to be maintained. The research equipment on board must be calibrated at the beginning of the segment, after which the autopilot is programmed to maintain constant attitude of the aircraft with respect to the ground. The SFC of the turbofan is 0.6 lb_f/hr/lb_f, wing area is 350 ft^2, and the weight at the start of cruise is 15,000 lb_f. The average fuel consumption of the engine during the loitering segment is 115 gallons/hr of Jet-A. If you are considering a configuration featuring a straight tapered wing with an AR of 10, estimate (a) the highest value the minimum drag coefficient, C_{Dmin}, can take and still meet the range requirement. (b) Also, estimate thrust required at the start of cruise. Assume the simplified drag model applies.

Answer: (a) $C_{Dmin} = 0.0295$, (b) $T = 1403$ lb_f.

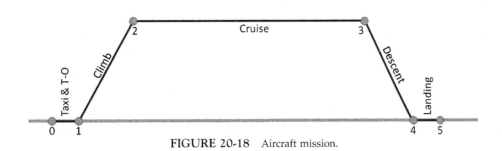

FIGURE 20-18 Aircraft mission.

VARIABLES

Symbol	Description	Units (UK and SI)
A	Constant in constant airspeed/constant cruise range equation	
a	Constant in constant airspeed/constant altitude endurance equation	
AR	Aspect ratio	
c_{bhp}	Thrust specific fuel consumption for a piston	$lb_f/hr/BHP$ or $N/hr/kW$
C_D	Coefficient of drag	
C_{Dmin}	Minimum drag coefficient	
c_{jet}	Thrust specific fuel consumption for a jet	$lb_f/hr/lb_f$ or $N/hr/N$
C_L	Coefficient of lift	
C_{L2}	Lift coefficient at beginning of cruise segment for a given mission	
C_{L3}	Lift coefficient at end of cruise segment for a given mission	
c_t	Thrust specific fuel consumption	$1/sec$
D	Drag	lb_f or N
dR	Rate of change of distance	ft/s or nm/s; m/s or km/s
dW	Rate of change of weight	lb_f/s or N/s
E	Total energy saved in battery system	Btu or Joules
e	Oswald efficiency	
E	Endurance	hours
E_{BATT}	Battery energy density	Jouls/Liter
gal	Fuel quantity in gallons	gal
H_2	Altitude at beginning of cruise segment for a given mission	ft or m
H_3	Altitude at end of cruise segment for a given mission	ft or m
i	Mission node index	
j	Mission segment index	
k	Coefficient for lift induced drag	
L	Lift	lb_f or N
lb_f	Fuel quantity in pound force	lb_f
m_b	Mass of battery	lb_m or kg
mi	Range in miles	ft or m
P	Power	$ft \cdot lb_f/s$ or $N \cdot m/s$
P_A	Available power	$ft \cdot lb_f/s$ or $N \cdot m/s$
pax	Number of passengers	
$P_{ELECTRIC}$	Power of motor	$ft \cdot lb_f/s$ or $N \cdot m/s$
P_{REQ}	Required power	$ft \cdot lb_f/s$ or $N \cdot m/s$
q	Dynamic pressure	lb_f/ft^2 or Pa
R	Cruise range	Watts
R_{fin}	Final range	Watts
S	Wing area	ft^2 or m^2
SR	Specific range	nm/lb_f or km/kg
T	Thrust	lb_f or N
T_f	Time to burn fuel weight	sec, min, or hrs
t_i	Time at node index	sec, min, or hrs
t_{TOT}	Total time to run motor	sec, min, or hrs
V	Airspeed	ft/s or m/s
$V_{(L/D)max}$	Velocity at maximum lift to drag ratio	ft/s or m/s
V_2	Velocity at beginning of cruise segment for a given mission	ft/s or knots; m/s or kmh
V_3	Velocity at end of cruise segment for a given mission	ft/s or knots; m/s or kmh
V_{CAS}	Calibrated airspeed	ft/s or knots; m/s or kmh
V_{TAS}	True airspeed	ft/s or knots; m/s or kmh
W	Weight	lb_f or N
W_0	Gross weight	lb_f or N
W_e	Empty weight	lb_f or N
W_f	Fuel weight	lb_f or N
W_{fin}	Final weight	lb_f or N
$\dot{w}_{fuel}$	Fuel flow rate	lb_f/s or N/s
W_i	Weight of aircraft at segment being analyzed	lb_f or N
W_{i-1}	Weight of aircraft at node before the current node being analyzed	lb_f or N
W_{ini}	Initial weight	lb_f or N
$\dot{w}_j$	Fuel flow rate during mission segment	lb_f/s or N/s
ΔR	Change in distance	ft or m
Δt	Change in time	sec, min, or hrs
η_p	Propeller efficiency	
ρ	Density	$slugs/ft^3$ or kg/m^3
ρ_o	Sea level density	$slugs/ft^3$ or kg/m^3

References

[1] Perkins CD, Hage RE. Airplane Performance, Stability, and Control. John Wiley & Sons; 1949.
[2] Torenbeek E. Synthesis of Subsonic Aircraft Design. 3rd edn. Delft University Press; 1986.
[3] Nicolai L. Fundamentals of Aircraft Design. 2nd edn 1984.
[4] Roskam J, Lan Chuan-Tau Edward. Airplane Aerodynamics and Performance. DARcorporation 1997.
[5] Hale FJ. Aircraft Performance, Selection, and Design. John Wiley & Sons; 1984. 137–138.
[6] Anderson Jr JD. Aircraft Performance & Design. 1st edn. McGraw-Hill; 1998.
[7] Raymer D. Aircraft Design: A Conceptual Approach. AIAA Education Series; 1996.
[8] The Nall Report 2010, Air Safety Foundation, p. 17.
[9] InFO 08004, Federal Aviation Administration, 02/07/2008.
[10] http://www.nbaa.org/about/.

CHAPTER 21

Performance − Descent

OUTLINE

21.1 Introduction 925
   21.1.1 The Content of this Chapter 926

21.2 Fundamental Relations for the Descent Maneuver 926
   21.2.1 General Two-dimensional Free-body Diagram for an Aircraft 926
   21.2.2 Planar Equations of Motion (Assumes No Rotation about Y-axis) 927

21.3 General Descent Analysis Methods 927
   21.3.1 General Angle-of-descent 927
      Derivation of Equation (21-9) 927
   21.3.2 General Rate-of-descent 927
      Derivation of Equation (21-10) 928
   21.3.3 Equilibrium Glide Speed 929

      Derivation of Equation (21-11) 930
   21.3.4 Sink Rate 930
      Derivation of Equations (21-12) and (21-13) 930
   21.3.5 Airspeed of Minimum Sink Rate, V_{BA} 931
      Derivation of Equation (21-14) 931
   21.3.6 Minimum Angle-of-descent 931
      Derivation of Equation (21-15) 931
   21.3.7 Best Glide Speed, V_{BG} 931
      Derivation of Equation (21-16) 932
   21.3.8 Glide Distance 932
      Derivation of Equation (21-17) 933

Variables 933

References 934

21.1 INTRODUCTION

For powered aircraft, the analysis of gliding flight is essential from the standpoint of safety. For unpowered flight, glide performance is what sets one sailplane apart from another. Analysis of descent provides a very important insight into how efficient an airplane is. It can even expose possible handling problems. For instance, a powered airplane with a high glide ratio (L/D ratio) will find this feature very favorable if it experiences engine failure and must rely on this property to get to the nearest emergency airport. However, if the L/D ratio is very high, this will actually make it harder to land as it would have to fly at a very shallow approach path angle to keep the airspeed low. The shallow angle would not only make it more difficult to assess where it will touch-down, but would also tend to make the airplane float once it enters ground effect, compounding the difficulty. Figure 21-1 shows an organizational map displaying the descent among other members of the performance theory.

This section will present the formulation of and the solution of the equation of motion for the descent, and present practical, as well as numerical solution methodologies, that can be used both for propeller and jet-powered aircraft. When appropriate, each method will be accompanied by an illustrative example using the sample aircraft. The primary information we want to extract from this analysis is characteristics like minimum rate-of-descent (ROD), best (lowest) angle-of-descent (AOD), the corresponding airspeeds, the AOD for a given power setting, and unpowered glide distance.

In general, the methods presented here are the "industry standard" and mirror those presented by a variety of authors, e.g. Perkins and Hage [1], Torenbeek [2], Nicolai [3], Roskam [4], Hale [5], Anderson [6] and many, many others. Also note that sailplane design

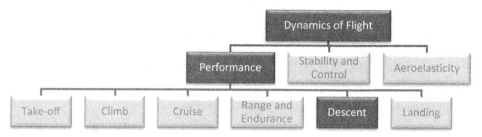

FIGURE 21-1 An organizational map placing performance theory among the disciplines of dynamics of flight, and highlighting the focus of this section: descent performance analysis.

and glide analyses methods are detailed in Appendix C4, Design of Sailplanes.

21.1.1 The Content of this Chapter

- **Section 21.2** develops fundamental relationships necessary to evaluate the characteristics of gliding flight, most importantly the equations of motion for descending flight.
- **Section 21.3** presents an assortment of methods to predict the various descent characteristics of an airplane.

21.2 FUNDAMENTAL RELATIONS FOR THE DESCENT MANEUVER

In this section, we will derive the equation of motion for the descent maneuver, as well as all fundamental relationships used to evaluate its most important characteristics. First, a general two-dimensional free-body diagram will be presented to allow the formulation to be developed. Only the two-dimensional version of the equation will be determined as this is sufficient for all aspects of aircraft design.

21.2.1 General Two-dimensional Free-body Diagram for an Aircraft

The free body for the descending flight is developed in a similar manner to that for the climbing flight. Figure 21-2 shows a free-body diagram of an airplane moving along a flight path. The x- and z-axes are attached to the center of gravity (CG) of the airplane, just as for the climb and have identical orientation with respect to the *datum* of the airplane. The angle between the datum and the tangent to the flight path is the *angle-of-attack* and the force (or thrust) generated by the airplane's power plant, T, may be at some angle ε with respect to the x-axis. The figure shows that this coordinate system can change its orientation with respect to the *horizon* depending on the maneuver being performed. Then, the angle between the horizon and the x-axis, the *descent angle*, is denoted by θ, where these three simple rules apply: If $\theta > 0$, then the aircraft is said to be *climbing*. If $\theta = 0$ then the aircraft is said to be *flying straight and level* (cruising). If $\theta < 0$, then the aircraft is said to be *descending*. This chapter only considers the final rule.

The free-body diagram of Figure 21-2 is balanced in terms of inertia, mechanical, and aerodynamic forces. The lift is the component of the resultant aerodynamic force generated by the aircraft that is perpendicular to the flight path (along its z-axis). The drag is the component

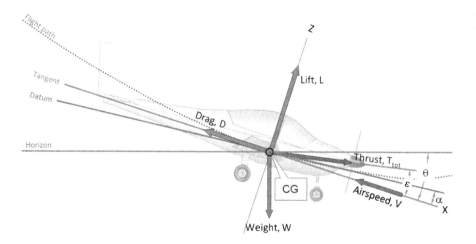

FIGURE 21-2 A two-dimensional free-body of the airplane in a powered gliding flight.

of the aerodynamic force that is parallel (along its *x*-axis). These are balanced by the weight, *W*, and the corresponding components of *T*. The presentation of Figure 21-2 can now be used to derive the *planar equations of motion* for the descent maneuver, which are sufficient to accurately predict the vast majority of descent maneuvers.

21.2.2 Planar Equations of Motion (Assumes No Rotation about Y-axis)

The equations of motion for gliding flight can be derived using the free-body diagram in Figure 21-2:

$$L - W\cos\theta + T\sin\varepsilon = \frac{W}{g}\frac{dV_Z}{dt} \quad (21\text{-}1)$$

$$-D + W\sin\theta + T\cos\varepsilon = \frac{W}{g}\frac{dV_X}{dt} \quad (21\text{-}2)$$

The equations of motion can be adapted for descending flight by making the following assumptions:

(1) Steady motion implies $dV/dt = 0$.
(2) The descent angle, θ, is a non-zero quantity.
(3) The angle-of-attack, α, is small.
(4) The thrust angle, ε, is $0°$.

Equations of motion for a steady unpowered ($T = 0$) descent:

$$L - W\cos\theta = 0 \quad \Rightarrow \quad L = W\cos\theta \quad (21\text{-}3)$$

$$-D - W\sin\theta = 0 \quad \Rightarrow \quad D = W\sin\theta \quad (21\text{-}4)$$

Equations of motion for a steady powered ($T > 0$) descent:

$$L - W\cos\theta = 0 \quad \Rightarrow \quad L = W\cos\theta \quad (21\text{-}5)$$

$$-D + W\sin\theta + T = 0 \quad \Rightarrow \quad D = T + W\sin\theta \quad (21\text{-}6)$$

Vertical airspeed:

$$V_V = V\sin\theta \quad (21\text{-}7)$$

AOD is also known as *angle-of-glide* (AOG) or *glide angle*.

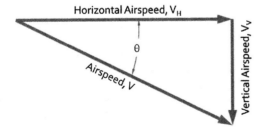

FIGURE 21-3 Airspeed components during climb.

21.3 GENERAL DESCENT ANALYSIS METHODS

21.3.1 General Angle-of-descent

The angle-of-descent is the flight path angle to the horizontal and is computed from:

Unpowered descent:

$$\tan\theta = \frac{D}{L} = \frac{1}{L/D} \approx \frac{D}{W} \quad (21\text{-}8)$$

Powered descent:

$$\sin\theta = \frac{D}{W} - \frac{T}{W} \approx \frac{1}{L/D} - \frac{T}{W} \quad (21\text{-}9)$$

The right approximations ($\approx$) are valid for low descent angles, θ, and when the CG is not too far forward, as this can put a high load on the stabilizing surface and invalidate the approximation $L \approx W$. Many airplanes, in particular sailplanes, have such high glide ratios that landing becomes difficult. For this reason, they are equipped with speed brakes or spoilers, which are panels that deflect from the wing surface and cause flow separation, increasing drag and reducing lift. The same holds for high-speed jets.

Derivation of Equation (21-9)

We get Equation (21-8) by dividing Equation (21-4) by (21-3):

$$\frac{D}{L} = \frac{W\sin\theta}{W\cos\theta} = \tan\theta$$

We get Equation (21-9) from Equation (21-6):

$$D = T + W\sin\theta \quad \Leftrightarrow \quad \sin\theta = \frac{D-T}{W} = \frac{D}{W} - \frac{T}{W}$$

QED

21.3.2 General Rate-of-descent

The rate at which an aircraft reduces altitude is given below:

$$V_V = \frac{DV}{W} = \frac{V}{(C_L/C_D)} \quad (21\text{-}10)$$

The above expression has units of ft/s or m/s. Generally, the units preferred by pilots are in terms of feet per minute or fpm for general aviation, commercial aviation, and military, but m/s for sailplanes and some nations that use the metric system. To convert Equation (21-10) into units of fpm multiply by 60.

Derivation of Equation (21-10)

Begin by multiplying Equation (21-4) by V, and then rewrite $V \sin \theta$ using Equation (21-7):

$$D = W \sin\theta \quad \text{and} \quad V_V = V \sin\theta \quad \Rightarrow \quad DV = WV\sin\theta$$
$$= WV_V \quad \Leftrightarrow \quad V_V = \frac{DV}{W}$$

QED

EXAMPLE 21-1

Plot the rate-of-descent for the Learjet 45XR at S-L, 15,000 ft, and 30,000 ft at a weight of 20,000 lb$_f$ (assuming no thrust). Plot the descent rate as a function of true airspeed in knots (KTAS).

Solution

Sample calculation for the sample aircraft gliding at 175 KCAS (LD_{max}) at S-L and at 20,000 lb$_f$ (no thrust). Note that LD_{max} is calculated in EXAMPLE 19-5.

$$D = \frac{W}{LD_{max}} = \frac{20{,}000}{15.45} = 1294 \text{ lb}_f$$

At S-L the density is 0.002378 slugs/ft^3. Therefore:

$$V_V = \frac{DV}{W} = \frac{(1294)(175 \cdot 1.688)}{20000} = 19.1 \text{ ft/s}$$

This amounts to 1147 fpm. The rate-of-descent for other airspeeds is plotted in Figure 21-4. Figure 21-5 shows the corresponding glide angle and L/D for the aircraft at S-L. Figure 21-6 shows how important performance characteristics, such as the airspeed for minimum power required and best glide ratio, can be extracted from the rate-of-descent plot.

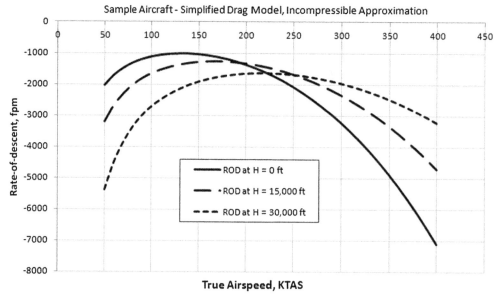

FIGURE 21-4 A flight polar, also known as ROD vs airspeed graph.

EXAMPLE 21-1 (cont'd)

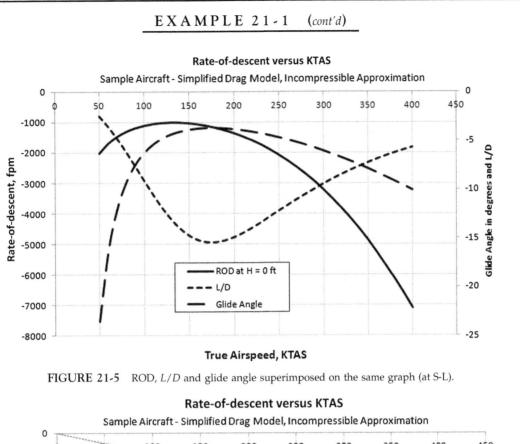

FIGURE 21-5 ROD, L/D and glide angle superimposed on the same graph (at S-L).

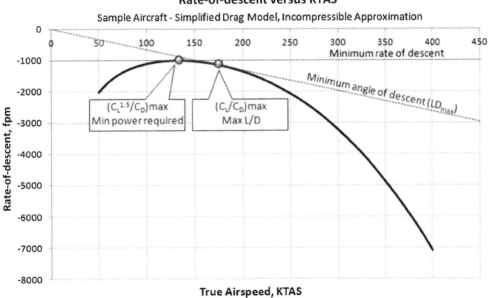

FIGURE 21-6 Important characteristics extracted from the flight polar (at S-L).

21.3.3 Equilibrium Glide Speed

Equilibrium glide speed is the airspeed that must be maintained to achieve a specific glide angle for a specific AOA. One common use is to determine the airspeed required to maintain a specific flight path angle, θ.

$$V = \sqrt{\frac{2\cos\theta}{\rho C_L} \frac{W}{S}} \qquad (21\text{-}11)$$

The lift coefficient can be determined based on the AOA required for the airspeed using $C_L = C_{Lo} + C_{L\alpha} \cdot \alpha$

Derivation of Equation (21-11)

From Equation (21-3) we get:

$$L = W\cos\theta \Leftrightarrow \frac{1}{2}\rho V^2 S C_L = W\cos\theta \Leftrightarrow V = \sqrt{\frac{2\cos\theta}{\rho C_L}\frac{W}{S}}$$

QED

EXAMPLE 21-2

During flight testing, the pilot of an SR22 wants to maintain a 3° glide path angle at an *AOA* of 5°. What airspeed must be maintained? Assume a test weight of 3250 lb$_f$, ISA at S-L conditions, $C_{L_o} = 0.4$ and $C_{L\alpha} = 5.5$ /rad.

Solution

Lift coefficient:

$$C_L = C_{L_0} + C_{L_\alpha} \cdot \alpha = 0.4 + 5.5 \times \left(\frac{5 \times \pi}{180°}\right) = 0.8800$$

Knowing that the wing area is 144.9 ft², we can now compute the airspeed necessary to maintain the said glide path angle:

$$V = \sqrt{\frac{2\cos\theta}{\rho C_L}\frac{W}{S}}$$

$$= \sqrt{\frac{2\cos(5°)}{(0.002378)(0.8800)}\frac{3250}{144.9}}$$

$$= 146 \text{ ft/s} = 86.6 \text{ KTAS}$$

21.3.4 Sink Rate

Sink rate is the rate at which an aircraft loses altitude. This is most commonly expressed in terms of feet per minute or meters per second. If the lift and drag coefficients can be determined for a specific glide condition (e.g. from knowing the *AOA*), the sink rate can be computed from:

Straight and level sink:

$$V_V = \sqrt{\frac{2}{\rho(C_L^3/C_D^2)}\frac{W}{S}} = \frac{C_D}{C_L^{3/2}}\sqrt{\frac{2}{\rho}\frac{W}{S}} \qquad (21\text{-}12)$$

The above expression will return the sink rate in terms of ft/s or m/s. To convert to fpm multiply by 60. If the airplane is turning and the bank angle is given by the bank angle φ, then the sink rate increases and amounts to:

Sink rate while banking at φ:

$$V_V = \sqrt{\frac{2}{\rho\left((C_L\cos\phi)^3/C_D^2\right)}\frac{W}{S}} = \frac{C_D}{C_L^{3/2}\cos^{3/2}\phi}\sqrt{\frac{2}{\rho}\frac{W}{S}}$$

$$(21\text{-}13)$$

Derivation of Equations (21-12) and (21-13)

Substitute Equation (21-11) into (21-7) to get:

$$V_V = V\sin\theta = \sqrt{\frac{2\cos\theta}{\rho C_L}\frac{W}{S}}\sin\theta \qquad (i)$$

Divide Equation (21-4) by (21-3) to get:

$$\frac{\sin\theta}{\cos\theta} = \frac{D}{L} \Leftrightarrow \sin\theta = \frac{D}{L}\cos\theta = \frac{C_D}{C_L}\cos\theta \qquad (ii)$$

Substitute Equation (ii) into (i) to get:

$$V_V = \sqrt{\frac{2\cos\theta}{\rho C_L}\frac{W}{S}}\sin\theta = \sqrt{\frac{2\cos\theta}{\rho C_L}\frac{W}{S}}\left(\frac{C_D}{C_L}\cos\theta\right)$$

$$= \sqrt{\frac{2\cos^3\theta}{\rho(C_L^3/C_D^2)}\frac{W}{S}}$$

If we assume $\cos\theta \sim 1$ we get Equation (21-12).

21.3.5 Airspeed of Minimum Sink Rate, V_{BA}

Just like the rate-of-climb, the magnitude of the sink rate varies with airspeed. This implies it has a minimum value that would be of interest to the operator of the vehicle as the kinetic energy of the vertical speed is then also at a minimum and, thus, may have an impact on survivability in an unpowered glide (as impact energy is a function of the square of the speed). It turns out that if *the simplified drag model applies*, the minimum sink speed can be calculated directly as follows. Note that this expression only holds in air mass that is neither rising nor sinking (see Appendix C4, Design of Sailplanes).

$$V_{BA} = V_{Emax} = \sqrt{\frac{2}{\rho}\left(\frac{W}{S}\right)\sqrt{\frac{k}{3 \cdot C_{Dmin}}}} \quad (21\text{-}14)$$

To get Equation (21-13) we refer to Figure 19-26 and see that $W = L \cdot \cos \phi = qS \cdot C_L \cdot \cos \phi$. When level ($\phi = 0°$) the same relationship is $W = L = qS \cdot C_L$. This shows that the lift really depends on the product $C_L \cdot \cos \phi$. Therefore, it is more accurate to replace the lift coefficient in the above formulation with the product.

QED

$$\tan\theta_{min} = \frac{1}{LD_{max}} = \sqrt{4 \cdot k \cdot C_{Dmin}} \quad (21\text{-}15)$$

Note that the drag model yields a θ_{min} which is independent of altitude.

Derivation of Equation (21-15)

Equation (19-18) gives the maximum lift-to-drag ratio for the simplified drag model (repeated below for convenience) and is inserted into Equation (21-8):

$$LD_{max} = \frac{1}{\sqrt{4 \cdot C_{Dmin} \cdot k}}$$

QED

EXAMPLE 21-3

Determine the minimum angle-of-descent for the sample aircraft flying at S-L at a weight of 20,000 lb$_f$.

Solution

$$\tan\theta_{min} = \frac{1}{LD_{max}} = \sqrt{4 \cdot k \cdot C_{Dmin}} = \sqrt{4 \cdot 0.05236 \cdot 0.020}$$
$$= 0.0647 \text{ rad}$$

This amounts to 3.7°.

Derivation of Equation (21-14)

Inspection of Equation (21-12) reveals that when $C_L^{1.5}/C_D$ is maximum, V_V is minimum. The airspeed at which this takes place has already been derived as Equation (19-14).

QED

21.3.6 Minimum Angle-of-descent

This angle results in a maximum glide distance from a given altitude and is of great importance to both glider pilots and pilots of powered aircraft.

21.3.7 Best Glide Speed, V_{BG}

The best glide speed is the airspeed at which the airplane will achieve maximum range in glide. It is a matter of life and death for the occupants of an aircraft, as is evident from its inclusion in 14 CFR 23.1587(c)(6), *Performance Information*. A part of pilot training requires this airspeed to be remembered in case of an engine failure. It can be calculated using Equation (21-16) below. Note that this expression only holds in air mass that is neither rising nor sinking (see Appendix C4, Design of Sailplanes).

$$V_{BG} = V_{LDmax} = \sqrt{\frac{2}{\rho}\sqrt{\frac{k}{C_{Dmin}}}\frac{W}{S}} \quad (21\text{-}16)$$

Derivation of Equation (21-16)

Using Equation (21-11) and the assumption that at the best glide angle $\cos\theta \approx 1$, we get:

$$V = \sqrt{\frac{2\cos\theta}{\rho C_L}\frac{W}{S}} \Rightarrow V = \sqrt{\frac{2}{\rho C_L}\frac{W}{S}}$$

It was demonstrated in the derivation for Equation (19-18) that at LD_{max} the lift coefficient $C_L = \sqrt{C_{Dmin}/k}$.

Inserting this into the above expression and manipulating leads to:

$$V = \sqrt{\frac{2}{\rho C_L}\frac{W}{S}} = \sqrt{\frac{2}{\rho\sqrt{\frac{C_{Dmin}}{k}}}\frac{W}{S}} = \sqrt{\frac{2}{\rho}\sqrt{\frac{k}{C_{Dmin}}}\frac{W}{S}}$$

QED

EXAMPLE 21-4

Determine the airspeed of minimum angle-of-descent for the sample aircraft flying at 30,000 ft and S-L at a weight of 20,000 lb$_f$.

Solution

At 30,000 ft the density is 0.0008897 slugs/ft^3. Therefore:

$$V_{LDmax} = \sqrt{\frac{2}{\rho}\sqrt{\frac{k}{C_{Dmin}}}\frac{W}{S}}$$

$$= \sqrt{\frac{2}{0.0008897}\sqrt{\frac{0.05236}{0.020}}\frac{20000}{311.6}}$$

$$= 483.2 \text{ ft/s (286 KTAS)}$$

At S-L the density is 0.002378 slugs/ft^3. Therefore:

$$V_{LDmax} = \sqrt{\frac{2}{\rho}\sqrt{\frac{k}{C_{Dmin}}}\frac{W}{S}}$$

$$= \sqrt{\frac{2}{0.002378}\sqrt{\frac{0.05236}{0.020}}\frac{20000}{311.6}}$$

$$= 295.9 \text{ ft/s (175 KTAS)}$$

This amounts to 175 KCAS in both cases.

21.3.8 Glide Distance

For a powered airplane, knowing how far one can glide in case of an emergency is not just a matter of safety, but of survivability. Such information is required information for the operation of GA aircraft by 14 CFR Part 23, §23.1587(d)(10), *Performance Information*, and must be determined per §23.71, *Glide: Single-engine Airplanes* (see below) and presented to the operator of the aircraft. This is typically done in the form of a glide chart, which shows clearly how far the airplane will glide for every 1000 ft lost in altitude.

---

§23.71 GLIDE: SINGLE-ENGINE AIRPLANES

The maximum horizontal distance traveled in still air, in nautical miles, per 1000 feet of altitude lost in a glide, and the speed necessary to achieve this must be determined with the engine inoperative, its propeller in the minimum drag position, and landing gear and wing flaps in the most favorable available position.

During the design phase, this distance can be calculated from the following expression (which is shown schematically in Figure 21-7). Note that this expression only holds in air mass that is neither rising nor sinking (see Appendix C4, Design of Sailplanes).

$$R_{glide} = h\cdot\left(\frac{L}{D}\right) = h\cdot\left(\frac{C_L}{C_D}\right) \qquad (21\text{-}17)$$

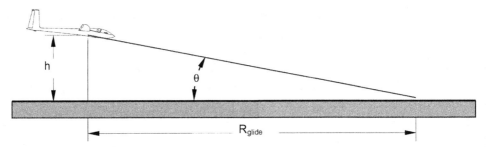

FIGURE 21-7 Distance covered during glide can be estimated using the L/D ratio.

Derivation of Equation (21-17)

First we note the following relation between the speed and distance:

$$\frac{V_V}{V_H} = \frac{h/\Delta t}{R_{glide}/\Delta t} = \frac{h}{R_{glide}}$$

Assuming that θ is small, we can say that $V_H \approx V$. Therefore, using Equation (21-5) we get:

$$V_V = \frac{DV}{W} \Rightarrow \frac{V_V}{V} = \frac{h}{R_{glide}} = \frac{D}{W}$$

Using Equation (21-3) and the assumption that for small angles $\cos\theta \approx 1$, we get:

$$L = W\cos\theta \Rightarrow \frac{h}{R_{glide}} = \frac{D}{W} = \frac{D}{L/\cos\theta} \Rightarrow \frac{h}{R_{glide}}$$

$$= \left(\frac{D}{L}\right)\cos\theta \approx \left(\frac{D}{L}\right)$$

QED

EXAMPLE 21-5

Determine the maximum glide distance for the sample aircraft flying at 30,000 ft.

Solution

Using the maximum LD calculated in EXAMPLE 19-7 we get ($LD_{max} = 15.45$):

$$R_{glide} = h \cdot \left(\frac{L}{D}\right)_{max}$$
$$= 30,000 \times (15.45)$$
$$= 463,500 \text{ ft } (76.3 \text{ nm})$$

VARIABLES

Symbol	Description	Units (UK and SI)
C_{L_α}	3D lift curve slope	/deg or /rad
C_D	Drag coefficient	
C_{Dmin}	Minimum drag coefficient	
C_L	Lift coefficient	
C_{L0}	Lift coefficient at zero AOA	
D	Drag	lb_f or N
g	Acceleration due to gravity	ft/s^2 or m/s^2
h	Altitude	ft or m
k	Coefficient for lift-induced drag	
L	Lift	lb_f or N
LD_{max}	Maximum lift-to-drag ratio	
R_{glide}	Glide distance	ft or m

Symbol	Description	Units (UK and SI)
S	Wing area	ft^2 or m^2
T	Thrust	lb_f or N
V	Airspeed	ft/s or m/s
V_{BA}	Airspeed of minimum sink rate	ft/s or m/s
V_{BG}	Best glide airspeed	ft/s or m/s
V_{Emax}	Airspeed of maximum endurance	ft/s or m/s
V_H	Horizontal airspeed	ft/s or m/s
V_{LDmax}	Velocity of maximum lift-to-drag ratio	ft/s or m/s
V_V	Rate-of-descent	ft/s or m/s
V_X	Horizontal velocity	ft/s or m/s
V_Z	Vertical velocity	ft/s or m/s
W	Weight	lb_f or N
Δt	Change in time	sec
α	Angle-of-attack	deg or rad
ε	Thrust angle	deg or rad
ϕ	Banking angle	deg or rad
θ	Descent angle	deg or rad
θ_{min}	Minimum angle-of-descent	deg or rad
ρ	Density	$slugs/ft^3$ or kg/m^3

References

[1] Perkins CD, Hage RE. Airplane Performance, Stability, and Control. John Wiley & Sons; 1949.
[2] Torenbeek E. Synthesis of Subsonic Aircraft Design. 3rd ed. Delft University Press; 1986.
[3] Nicolai L. Fundamentals of Aircraft Design. 2nd ed. 1984.
[4] Roskam J, Lan Chuan-Tau Edward. Airplane Aerodynamics and Performance. DARcorporation; 1997.
[5] Hale FJ. Aircraft Performance, Selection, and Design. John Wiley & Sons; 1984. pp. 137–138.
[6] Anderson Jr JD. Aircraft Performance & Design. 1st ed. McGraw-Hill; 1998.
[7] Raymer D. Aircraft Design: A Conceptual Approach. AIAA Education Series; 1996.

CHAPTER 22

Performance − Landing

OUTLINE

22.1 Introduction 935
  22.1.1 The Content of this Chapter 936
  22.1.2 Important Segments of the Landing Phase 936
22.2 Fundamental Relations for the Landing Phase 938
  22.2.1 General Free-body Diagram of the Landing Roll 938
  22.2.2 The Equation of Motion for the Landing Roll 938
  22.2.3 Formulation of Required Aerodynamic Forces 938
  22.2.4 Ground Roll Friction Coefficients 939
  22.2.5 Determination of the Approach Distance, S_A 939
    Derivation of Equations (22-5) and (22-6) 940
  22.2.6 Determination of the Flare Distance, S_F 940
  22.2.7 Determination of the Free-roll Distance, S_{FR} 940
  22.2.8 Determination of the Braking Distance, S_{BR} 940
  22.2.9 Landing Distance Sensitivity Studies 942
  22.2.10 Computer code: Estimation of Landing Performance 942
22.3 Database − Landing Performance of Selected Aircraft 944
Variables 945
References 946

22.1 INTRODUCTION

Just as the sections about performance began with the T-O, it is appropriate to end with the landing phase. Figure 22-1 shows an organizational map displaying the landing phase among other members of the performance theory. It is equally important that the designer understands the limitations and sensitivity of this important maneuver as that of the T-O. This section will present the formulation of and the solution of the equation of motion specifically for the landing, and present practical as well as numerical solution methodologies that can be used both for propeller and jet-powered aircraft.

The landing phase is in many ways the inverse of the T-O phase. It begins with the approach to landing in the form of a steady descent. This is followed by a flare maneuver, in which the pilot raises the nose of the aircraft in order to slow it down for a soft touch-down on the runway. The phase terminates with a deceleration from the touch-down airspeed to a complete standstill. The purpose of the analysis methods presented in this section is to determine the total distance the entire maneuver takes.

The landing phase can inflict a serious challenge on aircraft, in particular if its approach speed is high. As the airplane nears the runway it succumbs to ground effect, in which induced drag decreases and lift increases. This may cause the airplane to 'float' before it eventually settles on the runway. If its airspeed is high, a significant portion of the runway may be consumed before the airplane can even begin to decelerate to a full stop. Understanding the types and length of runways the airplane is likely to operate from will help size the wing area, choose the landing gear system, and select high-lift and speed-control systems to make it possible to meet the prescribed requirements.

In general, the methods presented here are the industry standard and mirror those presented by a variety of authors, e.g. Perkins and Hage [1], Torenbeek [2], Nicolai [3], Roskam [4], Hale [5], Anderson [6] and many, many others.

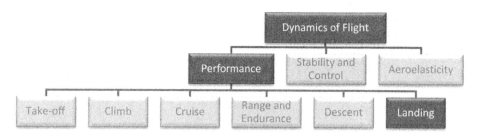

FIGURE 22-1 An organizational map placing performance theory among the disciplines of dynamics of flight, and highlighting the focus of this section: landing performance analysis.

22.1.1 The Content of this Chapter

- **Section 22.2** presents the fundamental relations of the landing phase, including the equation of motion for a landing ground run and its kinematics.
- **Section 22.3** presents several methods to solve the equation of motion.
- **Section 22.4** presents the landing properties of selected aircraft types.

22.1.2 Important Segments of the Landing Phase

Generally, the landing phase is split into the segments shown in Figure 22-2. The *approach distance* is measured from the point at which the airplane is 50 ft above ground; the point where the pilot initiates the next maneuver — the flare — by pulling the control wheel (or stick or yoke) in order to raise the nose of the aircraft. This maneuver is required to slow the descent of the airplane in an attempt to help it gently touch down on the runway. The next distance is the *flare distance*, which extends to the point where the airplane touches down. Then, the airplane typically rolls briefly before the pilot applies the brake system. This distance is referred to as the *free-roll distance*. Finally, the *braking distance* accounts for the length of runway required to apply brakes (thrust reversers, drag chute, wheel brakes, etc.) and bring the aircraft to a complete stop.

In short, the purpose of the landing analysis is to estimate the total landing distance by breaking it up into the aforementioned segments and analyzing each of those using simplified physics that pertain primarily to those specific segments. For analysis, the segments are denoted by the nomenclature indicated in Figure 22-3.

Certification requirements for GA aircraft are largely stipulated in 14 CFR Part 23, paragraphs 23.73 through 23.77. Paragraph 23.75 details requirements for the landing distance, shown below:

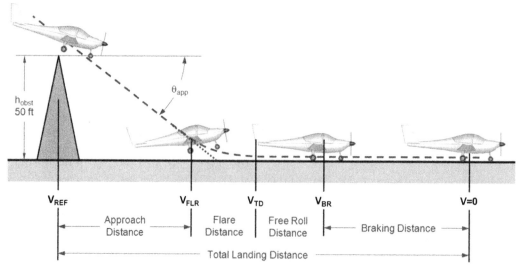

FIGURE 22-2 Important segments of the landing phase.

§23.75 LANDING DISTANCE

The horizontal distance necessary to land and come to a complete stop from a point 50-feet above the landing surface must be determined, for standard temperatures at each weight and altitude within the operational limits established for landing, as follows:

(a) A steady approach at not less than V_{REF}, determined in accordance with §23.73 (a), (b), or (c), as appropriate, must be maintained down to the 50-foot height and—
  (1) The steady approach must be at a gradient of descent not greater than 5.2% (3 degrees) down to the 50-foot height.
  (2) In addition, an applicant may demonstrate by tests that a maximum steady approach gradient steeper than 5.2%, down to the 50-foot height, is safe. The gradient must be established as an operating limitation and the information necessary to display the gradient must be available to the pilot by an appropriate instrument.
(b) A constant configuration must be maintained throughout the maneuver.
(c) The landing must be made without excessive vertical acceleration or tendency to bounce, nose over, ground loop, porpoise, or water loop.
(d) It must be shown that a safe transition to the balked landing conditions of §23.77 can be made from the conditions that exist at the 50-foot height, at maximum landing weight, or at the maximum landing weight for altitude and temperature of §23.63 (c)(2) or (d)(2), as appropriate.
(e) The brakes must be used so as to not cause excessive wear of brakes or tires.
(f) Retardation means, other than wheel brakes may be used if that means—
  (1) It is safe and reliable; and
  (2) It is used so that consistent results can be expected in service.
(g) If any device is used that depends on the operation of any engine, and the landing distance would be increased when a landing is made with that engine inoperative, the landing distance must be determined with that engine inoperative unless the use of other compensating means will result in a landing distance not more than that with each engine operating.

[Amdt. 23—21, 43 FR 2318, Jan. 16, 1978, as amended by Amdt. 23—34, 52 FR 1828, Jan. 15, 1987; Amdt. 23—42, 56 FR 351, Jan. 3, 1991; Amdt. 23—50, 61 FR 5187, Feb. 9, 1996]

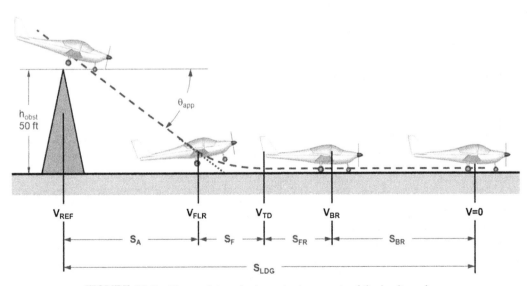

FIGURE 22-3 Nomenclature for important segments of the landing phase.

TABLE 22-1 Sections Used to Estimate Various Segments of the Landing Run

Segment Name	Symbol	Section
Approach distance	S_A	22.2.5 *Determination of the approach distance*
Flare distance	S_F	22.2.6 *Determination of the flare distance*
Free-roll distance	S_{FR}	22.2.7 *Determination of the free-roll distance*
Braking distance	S_{BR}	22.2.8 *Determination of the braking distance*

The designer should be particularly concerned with paragraph 14 CFR 23.77, *Balked Landing*, as this may inflict serious demands for control surface authority on the airplane.

The specific segments shown in Figure 22-3 are determined in the sections listed in Table 22-1.

22.2 FUNDAMENTAL RELATIONS FOR THE LANDING PHASE

In this section, we will derive the equation of motion for the landing, as well as all fundamental relationships used to evaluate the ground run segment of the landing maneuver. We will consider both conventional and taildragger configurations.

22.2.1 General Free-body Diagram of the Landing Roll

For a free-body diagram of the aircraft during descent refer to Figure 21-2 and for a free-body diagram of the aircraft after touch-down, refer to Figures 17-8 and 17-9, which apply for tricycle and taildragger configurations, respectively. Note that the drag, D, should be modified to reflect the aircraft in its landing configuration and the application of braking devices.

22.2.2 The Equation of Motion for the Landing Roll

The equation of motion for an aircraft during ground roll after touch-down on a perfectly horizontal and flat runway can be estimated from Equation (15-1) with slight modifications. This is simply the inclusion of the effect of braking devices, such as drag chutes, deployed spoilers or speed brakes. The application of mechanical brakes is treated using the ground friction coefficient, μ. Note that the drag coefficient must be that of the aircraft in the landing configuration. Deployed flaps and slats will greatly increase the drag of the aircraft. Additionally, the magnitude of the "acceleration" should always be less than 0 for a deceleration:

Deceleration on a flat runway:

$$\frac{dV}{dt} = \frac{g}{W}\left[T - D_{ldg} - \mu(W - L)\right] \quad (22\text{-}1)$$

where

D_{ldg} = drag in the landing configuration as a function of V, in lb$_f$ or N
g = acceleration due to gravity, ft/s^2 or m/s^2
L = lift of the airplane in the landing configuration as a function of V, in lb$_f$ or N
T = thrust (small during landing, but not necessarily negligible), in lb$_f$
W = weight, assumed constant, in lb$_f$
μ = ground friction coefficient (see Table 22-2)

We can also formulate the deceleration of the airplane on a downhill slope, which will increase the total landing distance. This formulation is based on Equation (15-2), but again features minor modifications.

Deceleration on a downhill slope γ:

$$\frac{dV}{dt} = \frac{g}{W}\left[T - D_{ldg} - \mu(W - L) + W\sin\gamma\right] \quad (22\text{-}2)$$

22.2.3 Formulation of Required Aerodynamic Forces

Refer to Section 17.2.4, *Formulation of required aerodynamic forces*, with the exception of the following:

C_L after touchdown:

$$C_{L\,LDG} = C_{Lo} + C_{L\alpha} \times \alpha_{LDG} \quad (22\text{-}3)$$

TABLE 22-2 Ground Roll Friction Coefficients

Surface Type	Ground Friction Coefficient, μ	
	Brakes Off	Braking
Dry asphalt or concrete	0.03–0.05	0.3–0.5
Wet asphalt or concrete	0.05	0.15–0.3
Icy asphalt or concrete	0.02	0.06–0.10
Hard turf	0.05	0.4
Firm dirt	0.04	0.3
Soft turf	0.07	0.2
Wet grass	0.08	0.2

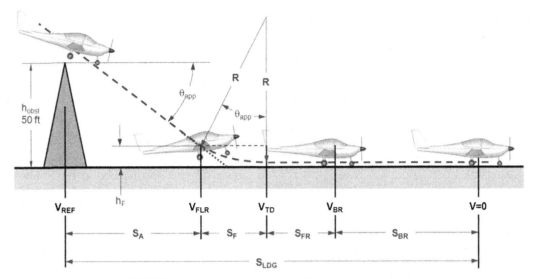

FIGURE 22-4 Evaluation of the landing over an obstacle.

C_D after touchdown:

$$C_{D\ LDG} = C_{Dmin} + C_{Di}(C_{L\ LDG}) + \Delta C_{D_{flaps}} \quad (22\text{-}4)$$

where

$C_{Di}(C_{L\ LDG})$ = induced drag coefficient of aircraft during the landing run after touchdown
α_{LDG} = angle-of-attack of aircraft during the landing run after touchdown
$\Delta C_{D_{flaps}}$ = added drag due to flaps (see below)

Use the methods of Section 15.5.8, *Drag of deployed flaps*, to estimate $\Delta C_{D_{flaps}}$. Also note that the induced drag must be corrected for ground effect, in particular if the airplane uses flaps or if its attitude is such its ground run *AOA* is high. Refer to Section 9.5.8, *Ground effect* for a correction method.

22.2.4 Ground Roll Friction Coefficients

The airplane has to overcome aerodynamic drag and ground friction during the ground roll. The ground friction depends on the weight on wheels and the properties of the ground, which are assessed using the ground roll friction coefficients Table 22-2. This is the same table as Table 17-3, and is merely repeated for convenience.

22.2.5 Determination of the Approach Distance, S_A

The geometry of the approach for landing is shown in Figure 22-4. In order to determine the distance from the obstacle to the point where the pilot initiates flare, the approach angle, θ_{app}, and flare height, h_F, must be computed. The target approach angle for most airplanes is 3°. For instance, airport approach lights (VASIS or visual approach slope indicator system) give the correct light signal when the airplane is approaching at that descent angle. For other applications, the approach angle can also be calculated using Equations (21-8) or (21-9) for unpowered or powered descents, repeated here for convenience:

Unpowered descent:

$$\tan\theta = \frac{D}{L} = \frac{1}{L/D} \approx \frac{D}{W} \quad (21\text{-}8)$$

Powered descent:

$$\sin\theta = \frac{D}{W} - \frac{T}{W} \approx \frac{1}{L/D} - \frac{T}{W} \quad (21\text{-}9)$$

However, as stated above, simply assuming $\theta_{app} = 3°$ is a good initial assumption. Knowing θ_{app} allows the approach distance, S_A, to be estimated by first

TABLE 22-3 Definition of Important Airspeeds for the Landing Run

Name	Airspeed	GA Aircraft (FAR 23)[a]	Commercial (FAR 25)[b]	Military
Reference	V_{REF}	$1.3V_{S0}$	$<1.23V_{S0}$	$1.2V_{S0}$
Flare	V_{FLR}	$1.3V_{S0}$	$<1.23V_{S0}$	$1.2V_{S0}$
Touch-down	V_{TD}	$1.1V_{S0}$	$1.1V_{S0}$	$1.1V_{S0}$
Braking	V_{BR}	$1.1V_{S0}$	$1.1V_{S0}$	$1.1V_{S0}$

[a] *Per 14 CFR Part 23, § 23.73 Reference Landing Approach Speed.*
[b] *Actual V_{REF} is established per 14 CFR Part 25, § 25.125 Landing.*

calculating the flare height, h_F, from the following expression:

$$h_F = R(1 - \cos \theta_{app}) \approx 0.1512 \times V_{S0}^2 \times (1 - \cos \theta_{app}) \quad (22\text{-}5)$$

Then, the flare distance, S_A, can be determined from:

$$S_A = \frac{(h_{obst} - h_F)}{\tan \theta_{app}} \quad (22\text{-}6)$$

The airspeeds used in the analysis are listed in Table 22-3 below:

Derivation of Equations (22-5) and (22-6)

Assume the flare involves slowing down from V_{REF} ($= 1.3V_{S0}$) to airspeed just above stall, where V_{S0} is the stalling speed in the landing configuration. Many pilots of light aircraft slow down enough to initiate the audible stall warning of the airplane, reminding us that pilot technique can have great influence on the total landing distance. The aural warning often begins to sound some 5 knots above the stalling speed. This may be less than 10% of the stalling speed. This is further influenced by the 'floating' of the airplane in ground effect. For analysis purposes, we must rigorously define the characteristics of the landing maneuver, while at the same time being mindful of the importance of pilot technique. Larger GA aircraft slow down to perhaps 10–15% above stall. Here the airspeed immediately before touchdown will be assumed to be 10% of the stalling speed or $V_{TD} = 1.1V_{S0}$. This implies an average airspeed of $1.2V_{S0}$ during the maneuver and this is used in addition to an assumed lift coefficient of about $0.9C_{Lmax}$ to determine distance traveled as follows:

STEP 1: Average vertical acceleration in terms of load factor.

$$n = \frac{L}{W} = \frac{\frac{1}{2}\rho(1.2V_{S0})^2 \cdot S \cdot (0.9C_{Lmax})}{\frac{1}{2}\rho V_{S0}^2 \cdot S \cdot C_{Lmax}} = 1.296 \quad (22\text{-}7)$$

STEP 2: Using Equation (19-37) (derived in Chapter 19) we calculate the transition radius, R:

$$R = \frac{V_{TR}^2}{g(n-1)} = \frac{(1.20V_{S1})^2}{g(1.296-1)} \approx 0.1512 \times V_{S0}^2 \quad (22\text{-}8)$$

where the resulting constant of 0.1512 is applicable to GA aircraft only.

STEP 3: Determine the flare height:

$$h_F = R(1 - \cos \theta_{app}) \approx 0.1512 \times V_{S0}^2 \times (1 - \cos \theta_{app}) \quad (22\text{-}9)$$

QED

22.2.6 Determination of the Flare Distance, S_F

By observation using Figure 22-4, the flare distance, S_F, can be determined from:

$$S_F = R \sin \theta_{app} \approx 0.1512 \times V_{S0}^2 \times \sin \theta_{app} \quad (22\text{-}10)$$

22.2.7 Determination of the Free-roll Distance, S_{FR}

The free-roll distance is determined in a similar manner to rotation distance in the analysis of the T-O. The airplane is assumed to roll freely for a few seconds only. Thus, the airspeed is assumed constant at V_{TD} (see Figure 22-4), which of course implies that $V_{TD} = V_{BR}$. Here it is appropriate to assume that small aircraft travel freely for 1 second and large aircraft for 3 seconds. This results in similar equations to those of Section 17.3.4, *Determination of Distance during rotation*, or:

Small aircraft:

$$S_{FR} = |V_{TD}| \quad (22\text{-}11)$$

Large aircraft:

$$S_{FR} = 3|V_{TD}| \quad (22\text{-}12)$$

22.2.8 Determination of the Braking Distance, S_{BR}

The final segment involves the determination of the distance when the pilot begins to apply any braking device until the airplane comes to a complete stop. Braking devices can be conventional mechanical brakes, deployed spoiler, drogue (or drag) chutes, thrust reversers, and so on. For analytical convenience, the

TABLE 22-4 Definition of Important Airspeeds for the Landing Run

Type of Engine	Sign of Thrust	Comment
Fixed-pitch propeller	$T > 0$	Assume $T \approx$ 5–7% of static thrust; 5% for cruise propellers and 7% for climb propellers
Constant-speed propeller	$T > 0$	Assume $T \approx$ 7% of static thrust
Reverse-thrust props – piston	$T < 0$	Assume $T \approx -40\%$ of static thrust
Reverse-thrust props – turboprop	$T < 0$	Assume $T \approx -60\%$ of static thrust
Jet – no thrust reverser	$T > 0$	Assume $T = T_{idle}$
Jet – thrust reverser	$T < 0$	Assume $T \approx -40\%$ to -60% of static thrust. Can typically only be operated at airspeeds higher than 40–50 KCAS

deceleration can be considered to consist of the simultaneous contribution of all such braking devices. The scenario in which a drogue chute is deployed, followed by the application of mechanical brakes, or similar complex application of braking devices, requires a numerical integration with respect to time, similar to that of Section 17.3.3, *Method 3: Solution using numerical integration*. A simpler approach can be based on the method presented in Section 17.3.1, *Method 1: General solution of the equation of motion*. The approach is applicable to both piston-engine airplanes as well as jets; the only difference lies in how the thrust is calculated. The ratios inside the brackets are evaluated when $V = V_{BR}/\sqrt{2}$, where V_{BR} is the airspeed when the braking is first applied:

$$S = \frac{V^2}{2a} \Rightarrow$$
$$S_{BR} = -\frac{V_{BR}^2 W}{2g\left[T - D_{ldg} - \mu(W-L)\right]_{at\ (V_{BR}/\sqrt{2})}}$$
(22-13)

Note that the thrust, T, can be either positive or negative (see Table 22-4). It is positive for fixed-pitch or constant-speed propellers, but is negative if reverse thrust is applied. Also note that the negative sign in front of the expression ensures that the distance will be a positive value, since the denominator (the deceleration) will have a negative sign.

The term D_{lgd} is the drag of the airplane in the landing configuration with braking. This usually implies deployed high-lift devices (flaps and slats), but also spoilers or speed brakes. Typically, speed brakes are deployed (often automatically) as soon as the aircraft touches down and then later are retracted before the airplane comes to a complete stop. A similar scenario holds for the thrust reversers. Equation (22-13) is not applicable to such intermittent use of braking devices. These must be solved using numerical integration as mentioned above.

If the propeller efficiency and engine power at $V = V_{BR}/\sqrt{2}$ is known, Equation (22-13) can be rewritten as follows:

$$S = \frac{V^2}{2a} \Rightarrow$$
$$S_{BR} = -\frac{V_{BR}^2 W}{2g\left[\frac{\sqrt{2}\cdot \eta_p \cdot 550 \cdot P_{BHP}}{V_{BR}} - D_{ldg} - \mu(W-L)\right]_{at\ (V_{BR}/\sqrt{2})}}$$
(22-14)

EXAMPLE 22-1

Determine the landing distance over a 50-ft obstacle and landing run for the SR22 on a standard day at S-L. Consider a runway whose braking conditions are estimated using the ground friction coefficient of $\mu = 0.3$. Assume a 3° glide path (θ_{app}), a landing weight of 3400 lb$_f$, propeller efficiency of 0.45 and idle engine power of 50 BHP at the airspeed $V = V_{BR}/\sqrt{2}$. Furthermore, assume a $C_{D\ LDG} = 0.040$ and $C_{L\ LDG} = 1.0$. Compare to the POH values of 1141-ft ground roll and 2344 ft over 50 ft.

Solution

Begin by determining the four airspeeds of interest, V_{REF}, V_{FLR}, V_{TD}, and V_{BR}, based on GA aircraft and noting that the stalling speed in the landing configuration is $V_{S0} = 59$ KCAS.

$$V_{REF} = 1.3 \cdot V_{SO} = 76.7 \text{ KCAS} = 129.5 \text{ ft/s}$$
$$V_{FLR} = 1.3 \cdot V_{SO} = 76.7 \text{ KCAS} = 129.5 \text{ ft/s}$$
$$V_{TD} = 1.1 \cdot V_{SO} = 64.9 \text{ KCAS} = 109.6 \text{ ft/s}$$
$$V_{BR} = 1.1 \cdot V_{SO} = 64.9 \text{ KCAS} = 109.6 \text{ ft/s}$$

Then, the approach distance must be determined. The flare height, h_F, is estimated using Equation (22-5):

$$h_F \approx 0.1512 \cdot V_{SO}^2 \cdot (1 - \cos\theta_{app})$$
$$= 0.1512 \cdot (59 \times 1.688)^2 \cdot (1 - \cos 3°) = 2.1 \text{ ft}$$

The obstacle height, h_{obst}, for GA aircraft is 50 ft. The distance covered from an altitude of 50 ft to the beginning of the flare is determined using Equation (22-6):

$$S_A = (h_{obst} - h_F)/\tan\theta_{app} = (50 - 2.1)/\tan 3°$$
$$= 914 \text{ ft}$$

The distance covered during the flare is given by Equation (22-10):

$$S_F \approx 0.1512 \times V_{SO}^2 \times \sin\theta_{app}$$
$$= 0.1512(59 \times 1.688)^2 \sin 3° = 78.5 \text{ ft}$$

The airplane will roll for a brief moment after touch-down before the pilot applies brakes. The distance covered during this time, which for a GA aircraft is considered 1

> **EXAMPLE 22-1** (*cont'd*)
>
> second, is called the free-roll distance, S_{FR}. It is obtained using Equation (22-11):
>
> $$S_{FR} = V_{TD} = 109.6 \text{ ft}$$
>
> Finally, the braking distance is calculated using Equation (22-14):
>
> $$S_{BR} = \frac{V_{BR}^2 W}{2g\left[\sqrt{2}\cdot\eta_p\cdot 550\cdot P_{BHP}/V_{BR} - D_{ldg} - \mu(W-L)\right]_{at\ (V_{BR}/\sqrt{2})}}$$
>
> To solve this, it is easiest to calculate the lift and drag separately. Note that the value of $C_{L\ LDG}$, given as 0.9, is simply the value at an *AOA* of 0° (or whatever landing attitude is considered realistic). Note that here this means that with the flaps deflected in the landing setting (for the SR22 this is 32°) it is assumed to equal 0.9.
>
> Lift:
>
> $$L = \frac{1}{2}\rho\left(\frac{V_{BR}}{\sqrt{2}}\right)^2 SC_{L\ LDG}$$
> $$= \frac{1}{2}(0.002378)\left(\frac{109.6}{\sqrt{2}}\right)^2 (144.9)(1.0) = 1035 \text{ lb}_f$$
>
> Drag:
>
> $$D = \frac{1}{2}\rho\left(\frac{V_{BR}}{\sqrt{2}}\right)^2 SC_{D\ LDG}$$
> $$= \frac{1}{2}(0.002378)\left(\frac{109.6}{\sqrt{2}}\right)^2 (144.9)(0.04) = 41.4 \text{ lb}_f$$
>
> Inserting these numbers into Equation (22-14), yields:
>
> $$S_{BR} = \frac{(109.6)^2(3400)}{2(32.174)\left[\frac{\sqrt{2}\cdot(0.45)\cdot 550\cdot(50)}{109.6} - 41.4 - (0.3)(3400-1035)\right]}$$
> $$= 1074 \text{ ft}$$
>
> The total landing distance from 50 ft is thus:
>
> $$S_{LDG} = S_A + S_F + S_{FR} + S_{BR}$$
> $$= 914 + 78.5 + 109.6 + 1074 = 2176 \text{ ft}$$
>
> The landing ground roll is:
>
> $$S_{FR} + S_{BR} = 109.6 + 1074 = 1184 \text{ ft}$$
>
> These numbers deviate as follows from the POH values:
>
> Total landing distance:
>
> $$100\left(\frac{2176-2344}{2344}\right) = -7.2\%$$
>
> Ground roll distance:
>
> $$100\left(\frac{1184-1141}{1141}\right) = +3.8\%$$
>
> This method is in good agreement with the book values.

22.2.9 Landing Distance Sensitivity Studies

Landing at high elevations and on hot days, when density is much lower than on a standard day at S-L, poses serious challenges to airplane operations. Since the airplane necessarily moves faster with respect to the ground than at lower altitudes, the landing distance increases and the brake energy required to slow down from that higher airspeed is much higher (it depends on the kinetic energy). These effects are corrected through the density and the use of true airspeed.

Using this method, the sensitivity of the landing distance to altitude and temperature is presented in Figure 22-5. The graph plots the estimated landing distance for the SR22 from S-L to 14,000 ft for ±30°C deviations from ISA. It can be seen that at S-L the temperature deviation results in about ±120-ft change in landing over a 50-ft obstacle, when compared to standard day conditions.

22.2.10 Computer code: Estimation of Landing Performance

The following Visual Basic for Applications routine can be used to determine the landing distance for an

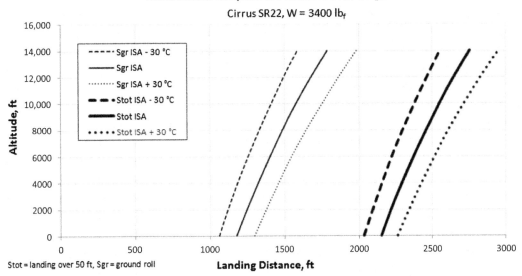

FIGURE 22-5 Sensitivity study showing the effect of altitude and temperature on the landing distance of the SR22. The study was performed using the analysis method presented in this section.

aircraft. The routine assumes that as long as the obstacle height is larger than zero the entire landing distance will be determined. The simple trick in order to extract the landing ground roll only is to set the obstacle height to zero. This way, the routine works for both evaluations. Note that the routine makes calls to the routine AtmosProperty (see Section A.2.10, *COMPUTER CODE A-1: Atmospheric modeling*) in order to allow the user to evaluate landing distances at hot and high airports. This is done using the arguments H and deltaISA, where the temperature deviation is that from the standard temperature at that altitude. This code was used with Microsoft Excel to generate the graph of Figure 22-5.

```
Function PERF_Landing(W As Single, S As Single, CL_ldg As Single, CD_ldg As Single, T As Single,
theta_app As Single, mu As Single, Vso As Single, H As Single, deltaISA As Single, Hobst As Byte)
As Single
  'This routine will calculate the landing distance for an aircraft using the method of
  'Section 22, Performance - Landing.
  '
  'Variables:    W         = Weight in lbf
  '              S         = Reference wing area in ft²
  '              CL_ldg    = Lift coefficient during ground run
  '              CD_ldg    = Drag coefficient during ground run
  '              T         = Thrust at VBR/sqr(2)
  '              theta_app = Approach angle in degrees (typ. 3°)
  '              mu        = Ground friction coefficient
  '              Vso       = Stalling speed in the landing configuration in ft/s
  '              H         = Airport elevation in ft
  '              deltaISA  = Deviation from ISA in °R
  '              Hobst     = Obstacle height in ft. Set Hobst=0 to get ground roll only.
  '
  'Note that this routine calls the function AtmosProperty, so it must be present.
  '
  'Initialize
    Dim Vref As Single, Vflr As Single, Vtd As Single, Vbr As Single
    Dim SA As Single, SF As Single, SFR As Single, SBR As Single
    Dim Hf As Single, theta As Single
    Dim OAT As Single, P As Single, rho As Single, sigma As Single
    Dim L As Single, Dldg As Single
```

```
'Presets
    theta = theta_app * 3.14159265 / 180    'Convert approach angle to radians
    OAT = AtmosProperty(H, 10) + deltaISA   'Corrected temperature at altitude
    P = AtmosProperty(H, 11)                'Pressure at altitude
    rho = P / (1716 * OAT)                  'Corrected density at altitude
    sigma = rho / 0.002378                  'Density ratio at altitude
    Vref = 1.3 * Vso * 1.688 / Sqr(sigma)   'Reference airspeed
    Vflr = 1.3 * Vso * 1.688 / Sqr(sigma)   'Flare airspeed
    Vtd = 1.1 * Vso * 1.688 / Sqr(sigma)    'Touchdown airspeed
    Vbr = 1.1 * Vso * 1.688 / Sqr(sigma)    'Braking airspeed
'Determine approach distance, SA
    Hf = 0.1512 * Vso ^ 2 * (1 - Cos(theta))
    SA = (Hobst - Hf) / Tan(theta)
'Determine flare distance, SF
    SF = 0.1512 * Vso ^ 2 * Sin(theta)
'Determine free roll distance, SFR
    SFR = Vtd
'Determine braking distance, SBR (note that 0.5*rho*(Vbr/Sqr(2))^2 = 0.25*rho*Vbr!2
    Dldg = 0.25 * rho * Vbr ^ 2 * S * CD_ldg
    L = 0.25 * rho * Vbr!2 * S * CL_ldg
    SBR = -Vbr ^ 2 * W / (2 * 32.174 * (T - Dldg - mu * (W - L)))
'Determine distance
    If Hobst = 0 Then 'Ground run only is being requested
    PERF_Landing = SFR + SBR
    Else
    PERF_Landing = SA + SF + SFR + SBR
    End If
End Function
```

22.3 Database – Landing Performance of Selected Aircraft

Table 22-5 shows the landing run and distance from a 50-ft altitude above ground level. This data is very helpful when evaluating the accuracy of one's own calculations.

TABLE 22-5 T-O Performance of Selected Aircraft

Aircraft	Gross wgt, lb$_f$	Ldg from 50 ft, ft	Landing Run, ft	Reference
Single-engine GA Aircraft				
Aero Boero 115	1697	500	150	Jane's 1976–77
Beech A36 Bonanza	3600	1450	840	Jane's 1978–79
Beech Sundowner	2030	1484	703	
Beech V35 Bonanza	3400	1324	763	
Cessna 152	1670	1200	475	
Cessna 172 Skyhawk	2300	1250	520	
Cessna 177 Cardinal	2800	1350	730	
Cessna 185 Skylane	2950	1350	590	
Cessna 210 Centurion	3800	1500	765	
Cirrus SR20	3050	2636	853	POH
Cirrus SR22	3400	2344	1141	
Piper PA-18 Super Cub	1750	885	350	Jane's 1978–79

TABLE 22-5 T-O Performance of Selected Aircraft—cont'd

Aircraft	Gross wgt, lb$_f$	Ldg from 50 ft, ft	Landing Run, ft	Reference
Piper PA-28 Warrior II	2325	1115	595	
Piper PA-32 Lance II	3600	1710	880	
Piper PA-38 Tomahawk	1670	1465	625	
SIAI-Marchetti SF-260	2425	2264	1132	Jane's 1976–77
Multiengine GA Aircraft				
Beech Duchess 76	3900	1881	1000	Jane's 1978–79
Beech Duke B60	6775	3065	1318	
Beech King Air C90	9650	2010	1075	
Cessna 310	5500	1790	640	
Cessna 337 Skymaster	4630	1650	700	
Cessna 340	5990	1615	2175	
DHC-6 Twin Otter	12500	1050-1940	515-950	Jane's 1976–77
Partenavia P-68B Victor	4321	1295	820	
Piper PA-23 Aztec	5200	1310	-	Jane's 1978–79
Piper PA-31 Cheyenne	9000	1860	-	
Piper PA-31 Chieftain	7000	1880	1045	
Piper PA-34 Seneca II	4570	2090	1380	

VARIABLES

Symbol	Description	Units (UK and SI)
C_{L_α}	3D lift curve slope	/deg or /rad
$C_{D\ LDG}$	Drag coefficient after touch-down	
$C_{Di}(C_{L\ LDG})$	Induced drag coefficient of aircraft	
$C_{L\ LDG}$	Lift coefficient after touchdown	
C_{L0}	Lift coefficient at zero AOA	
D_{ldg}	Drag during landing configuration	lb$_f$ or N
dt	Change in time	sec
dV	Change in velocity	ft/s or m/s
g	Acceleration due to gravity	ft/s^2 or m/s^2
h_F	Flare height	ft or m
h_{obst}	Height of obstacle	ft or m
L	Lift	lb$_f$ or N
P_{BHP}	Brake horsepower	BHP
R	Transition radius	ft or m
S_A	Approach distance	ft or m
S_{BR}	Braking distance	ft or m
S_F	Distance traveled during flare	ft or m

Symbol	Description	Units (UK and SI)
S_{FR}	Free rolling distance	ft or m
S_{LDG}	Total distance for landing	ft or m
T	Thrust	lb$_f$ or N
V_{BR}	Velocity when initiating braking	ft/s or m/s
V_{FLR}	Flare velocity	ft/s or m/s
V_{REF}	Velocity at obstacle	ft/s or m/s
V_{S0}	Stall speed in landing configuration	ft/s or m/s
V_{S1}	Stall speed in regular flight configuration	ft/s or m/s
V_{TD}	Velocity at touchdown	ft/s or m/s
W	Weight	lb$_f$ or N
$\Delta C_{D\ flaps}$	Added drag due to flaps	
α_{LDG}	AOA during landing after touchdown	deg or rad
γ	Angle of uphill slope	deg or rad
η_P	Propeller efficiency	
μ	Ground friction coefficient	
θ_{app}	Approach angle	deg or rad

References

[1] Perkins CD, Hage RE. Airplane Performance, Stability, and Control. John Wiley & Sons; 1949.
[2] Torenbeek E. Synthesis of Subsonic Aircraft Design. 3rd edn. Delft University Press; 1986.
[3] Nicolai L. Fundamentals of Aircraft Design. 2nd edn. 1984.
[4] Roskam J, Lan Chuan-Tau Edward. Airplane Aerodynamics and Performance. DARcorporation; 1997.
[5] Hale FJ. Aircraft Performance, Selection, and Design. John Wiley & Sons; 1984. 137–138.
[6] Anderson Jr JD. Aircraft Performance & Design. 1st edn. McGraw-Hill; 1998.

CHAPTER 23

Miscellaneous Design Notes

OUTLINE

23.1 Introduction 948
   23.1.1 The Content of this Chapter 948
23.2 Control Surface Sizing 948
   23.2.1 Introduction to Control Surface Hinge Moments 948
   23.2.2 Fundamentals of Roll Control 949
      Plain Flap Ailerons 951
      Frise Ailerons 951
      Spoiler-Flap Ailerons 951
      Slot-Lip Ailerons 951
      Flaperon 952
      Elevon 952
      Differential Ailerons 952
      Aileron Design Requirements 952
      Aileron Authority Derivative $C_{l_{\delta a}}$ 953
      Derivation of Equation (23-3) 953
      Derivation of Equations (23-4) and (23-5) 954
      Roll Damping Derivative C_{l_p} 955
      Derivation of Equations (23-6) and (23-8) 956
      Derivation of Equations (23-9) and (23-10) 959
   23.2.3 Aileron sizing 960
      Estimating Steady-State Roll Rate 960
      STEP 1: Initial Dimensions 960
      STEP 2: Estimate Roll Damping 960
      STEP 3: Estimate Roll Authority 960
      STEP 4: Estimate Roll Helix Angle 960
      Special Case Aileron Sizing: Constant-Chord Wing 960
      Derivation of Equation (23-11) 961
      Derivation of Equation (23-13) 961
      Maximizing responsiveness 962
   23.2.4 Fundamentals of Pitch Control 962
   23.2.5 Fundamentals of Yaw Control 964
23.3 General Aviation Aircraft Design Checklist 964
   23.3.1 Crosswind Capability at Touch-Down 965
      How to Assess 965
   23.3.2 Balked Landing Capability 965
      How to Assess 966
   23.3.3 Take-Off Rotation Capability 966
      How to Assess 966
   23.3.4 Trim at Stall and Flare at Landing Capability 966
      How to Assess 966
   23.3.5 Stall Handling Capability 966
      How to Assess 967
   23.3.6 Stall Margin for Horizontal Tail 967
      How to Assess 967
   23.3.7 Roll Authority 967
      Take-Off Requirement 967
      Approach Requirement 968
      How to Assess 968
   23.3.8 Control System Harmony 968
   23.3.9 Climb Capability 968
      How to Assess 968
   23.3.10 One-Engine-Inoperative Trim and Climb Capability 969
      How to Assess 969
   23.3.11 Natural Damping Capability 969
      How to Assess 969
   23.3.12 Fuel Tank Selector 969
      How to Assess 969
   23.3.13 Control System Stretching 969
      How to Assess 970
   23.3.14 Control System Jamming 970
      How to Assess 970
   23.3.15 Ground Impact Resistance 970
      How to Assess 970
   23.3.16 Reliance Upon Analysis Technology 971
   23.3.17 Weight Estimation Pitfalls 972
   23.3.18 Drag Estimation Pitfalls 972
   23.3.19 Center of Gravity Travel During Flight 972
   23.3.20 Wing/Fuselage Juncture Flow Separation 972
23.4 Faults and Fixes 972
   23.4.1 Stability and Control – Dorsal Fin and Rudder Locking 973

23.4.2	Stability and Control – Ventral Fin and Deep Stall	973	23.4.11 Stall Handling – Wing Droop (Cuffs, Leading Edge Droop)	978
23.4.3	Stability and Control – Ventral Fin and Dutch Roll	973	23.4.12 Flow Improvement – Vortex Generators	978
23.4.4	Stability and Control – Forebody Strakes	973	23.4.13 Trailing Edge Tabs for Multi-Element Airfoils	980
23.4.5	Stability and Control – Taillets and Stabilons	974	23.4.14 Flow Improvement – Nacelle Strakes	981
23.4.6	Stability and Control – Control Horns	975	23.4.15 Flow Improvement – Bubble-Drag, Turbulators, and Transition Ramps	981
23.4.7	Stall Handling – Stall Strips	975	**Variables**	**982**
23.4.8	Stall Handling – Wing Fence	976	**References**	**983**
23.4.9	Stall Handling – Wing Pylons	977		
23.4.10	Stall Handling – Vortilons	978		

23.1 INTRODUCTION

The purpose of this final section is to gather in one place information considered important by the author, but hard to place in any other specific section.

23.1.1 The Content of this Chapter

- **Section 23.2** presents requirements for control surface design. It also details a method to size ailerons based on roll rate responsiveness and fundamentals of pitch and yaw controls.
- **Section 23.3** presents a design checklist intended to help the project engineer identify important capabilities and how to design them into the airplane. There are many areas where one can go wrong in the design of a new aircraft. The checklist helps to steer away from some of them.
- **Section 23.4** presents examples of possible fixes to faulty aerodynamic characteristics. Since the airplane is a compromise of many focus areas, it is always possible the design will suffer in one or more areas. Aircraft designers have been imaginative in the past and have come up with various means to improve this or fix that, without it having been a part of the original design. It is helpful to the aspiring designer to know there are possible solutions to an imperfect design.

23.2 CONTROL SURFACE SIZING

The proper sizing of control surfaces is of crucial importance in the development of a new aircraft. The handling of aircraft is often described using two adjectives; *sluggish* or *responsive*. For the designer, the former is to be avoided and the latter is the goal. However, the aircraft should be designed to be *just* responsive enough for its mission. Surprisingly, like insufficient responsiveness, too much is detrimental as well, as it inevitably results in greater structural loads that would call for a heavier airframe. It can also lead to *pilot-induced oscillation* (PIO). For instance, there is no need to design a cargo transport aircraft to roll 360° per second – it is a maneuver not needed to complete its mission. Instead, the responsiveness only needs to amount to what is needed for handling during landing in turbulent wind conditions.

There are two aspects to control surface sizing: (1) responsiveness and (2) hinge moment tailoring. The former is imperative during the conceptual design phase, while the latter is important during the preliminary one. In this section, simple methods to size the three primary control surfaces, aileron, elevator, and rudder, will be presented. Only a rudimentary introduction will be given on hinge moments.

23.2.1 Introduction to Control Surface Hinge Moments

In a conventional, human-operated control system, the pilot must react the aerodynamic hinge moments that results from deflecting a given control surface. Consider the control surface in image A of Figure 23-1, which consists of an airfoil and a flap. Hinge moment analysis effectively returns the magnitude of the actuation force F, which multiplied by the bellcrank dimension d is the hinge moment. The force required to deflect the surface depends on parameters like the geometry of the airfoil and flap, deflection angle (δ), and the *AOA* of the combination. The force is applied using a stick, a yoke, a "steering wheel," foot pedals, and, sometimes, a hydraulic or electrical actuation system. There are a number of ways to affect the magnitude

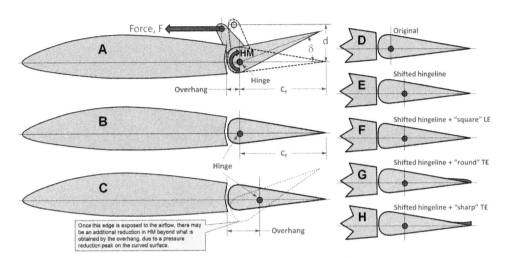

FIGURE 23-1 The hinge moment (HM) is reacted as a force, either directly by the pilot or by a control system (typically hydraulic or electric). It can be modified by various means.

of the aerodynamic hinge moment for any given geometry:

(1) One of the most effective means is to modify the *overhang* by shifting the hingeline (see B and C in Figure 23-1). Shifting it aft reduces the hinge moment. Shifting it forward increases it.

(2) The geometry of the leading edge can also be used to modify the moments, although it must be used with an aft-shifted hingeline, because this will expose it to the airflow (see bottom image of Figure 23-1). For instance, a more "square"-shaped leading edge (see E and F) will decrease the hinge moment as it generates a strong suction peak on the flap ahead of the hinge.

(3) The geometry of the trailing edge can also be modified (see G and H). For instance, a rounded TE (G) will generate a low-pressure region that is located far aft of the hinge —reducing the hinge moment. On the other hand, a sharp protrusion (H) will induce a high-pressure region and increase it.

Hinge moments are highly affected by the geometry of the controls, including the hinge location and shape of the control surface. A general expression for the hinge moment is given below:

$$HM = \frac{1}{2}\rho V^2 C_f S_f C_h \qquad (23\text{-}1)$$

where

C_f = flap chord (aft of hingeline)
C_h = hinge moment coefficient
S_f = flap area (aft of hingeline)
V = airspeed,
ρ = density of air

The hinge moment coefficient is then given by:

$$C_h = C_{h_0} + C_{h_\alpha} \cdot \alpha + C_{h_\delta} \cdot \delta + C_{h_{\delta_t}} \cdot \delta_t \qquad (23\text{-}2)$$

The tailoring of hinge moments is a process used to fine tune the hinge moments of a particular control surface so that some desirable stick forces will be sensed by the pilot. This is an essential task for GA aircraft, especially smaller ones, in which the pilot generates the actuation forces. It is also important for the development of boosted control actuation and autopilot integration.

Table 23-1 shows typical flap chord sizes by control surface type and their usual deflection limits. Of course, there are exceptions from these numbers, but typical values are between the numbers presented.

23.2.2 Fundamentals of Roll Control

The purpose of the ailerons is to provide control about the airplane's roll axis, by modifying the rolling

TABLE 23-1 Typical Ranges of Control Surface Flap Chords and Deflection Angles

Control Surface	Chord Size, C_f	Deflection Angles, δ
Aileron	0.20 to 0.35	15° to 25° TEU 15° to 20° TED
Elevator	0.20 to 0.35	20° to 27° TEU 15° to 20° TED
Rudder	0.20 to 0.40	20° to 25° TEL 20° to 25° TER
Flaps	0.25 to 0.40	0° to 10° TEU (cruise flaps) 25° to 50° TED

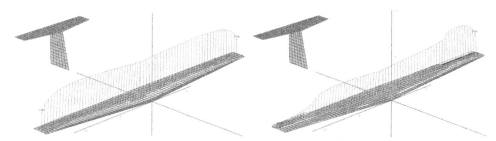

FIGURE 23-2 Typical impact of deflecting ailerons on the section lift coefficients.[1]

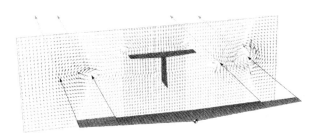

FIGURE 23-3 Impact of aileron deflection on the flow field behind the aircraft predicted by the potential flow theory. Note the difference in the size and source of the various wingtip vortices.[2]

The effect of deflecting ailerons on the distribution of section lift coefficients predicted by potential flow theory can be seen in Figure 23-2. The ailerons are deflected some 15° and the wing's *AOA* amounts to 8° at 100 KCAS. Also, Figure 23-3 shows the resulting impact on the flow field behind the wing. It highlights how the wing that travels up generates higher lift-induced drag than the opposite wing, causing a phenomenon known as *adverse yaw*.

There are three common types of ailerons used in modern airplanes; *plain flap ailerons, Frise ailerons*, and *spoiler-flap ailerons*, also called *spoileron* (a combination of spoilers and ailerons). Schematics of these are shown in Figure 23-4. Other aileron types include the *flaperon* (a combination of flaps and ailerons) and *elevon* (a combination of elevators and ailerons). There also exist some highly specialized types of ailerons such as the *slotted-lip aileron* and some research project ailerons such as *microjet ailerons*.

moment coefficient, C_l (do not confuse with the section lift coefficient, which shares the same variable name). The most effective way to accomplish this is to modify the airfoil in the outboard region of the wing by changing its camber by deflecting a control surface. This modifies the distribution of lift along the wing, un-symmetrically, which generates a rolling moment.

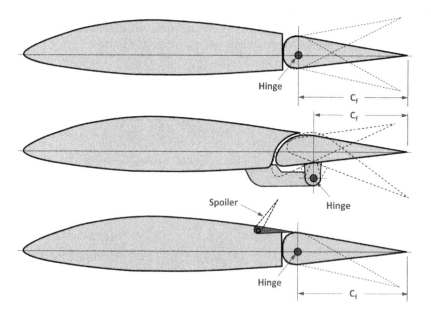

FIGURE 23-4 Three common types of ailerons: plain flap (top), Frise (middle), and spoiler-flap (bottom).

---

[1]Generated with the vortex-lattice code SURFACES.

[2]Generated with the vortex-lattice code SURFACES.

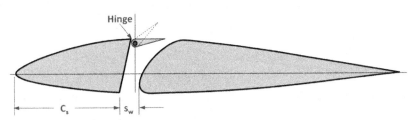

FIGURE 23-5 A schematic of the slot-lip aileron.

Plain Flap Ailerons

The plain flap is the most common type of aileron configuration. They are very effective and inexpensive to manufacture. For this reason, they can be found on a wide range of aircraft, ranging from primary trainers to commercial aircraft. As can be seen in Figure 23-4 the aileron on the up-going wing is deflected trailing edge down (TED) and the down-going wing is deflected trailing edge up (TEU).

Frise Ailerons

The Frise aileron (see Figure 23-4) was invented by the famed designer Leslie George Frise BSc FRAeS (1897–1979).[3] Frise ailerons were first developed during the 1930s [1]. Their purpose is twofold: (1) to reduce stick-forces at high airspeeds and (2) to eliminate adverse yaw. This is accomplished by offsetting the hingeline below the wing surface, introducing a large radius motion to the control surface. The LE of the one deflected TEU will be exposed to the airstream, which, in turn, generates a suction peak which helps reduce the hinge moment. The one deflected TED has its aerodynamic center offset from the hingeline, also helping to reduce the hinge moment.

The aileron reduces or eliminates adverse yaw by forcing the leading edge of the aileron deflected TEU, downward and outside the regular outside mold line. This exposes it to the airstream and increases the drag on that side of the wing (the down-going side). The drag generates a yawing moment and reduces the tendency of the wing to yaw "out of the turn" or opposite the bank. If the leading edge of the aileron is round like the one shown in the figure, a powerful low-pressure region is generated that lowers the hinge moment. This explains its use in both fast and large aircraft before the advent of hydraulically boosted control systems. If the nose is too sharp the lower surface may stall, which can cause severe buffeting [1].

Frise ailerons have seen use on many different aircraft types, among them the B-17, Bell P-39, Grumman F6F-3 and TBF, the Spitfire, Hurricane, Focke-Wulf 190, Curtiss Wright C-46 and DC-4, and many Cessna models.

Spoiler-Flap Ailerons

Several airplanes feature this type of aileron (e.g. Mitsubishi MU-2, Boeing B-52, and others). This aileron type usually features a flap that is deflected TED on the up-going wing and a small spoiler on the down-going side. As the spoiler is deployed it reduces lift on the down-going side, but also increases drag, therefore counteracting adverse yaw tendency. The aileron on that side may or may not deflect TEU at the same time. However, a common complaint is that such aileron systems tend to be sluggish at low airspeeds, as separated flow creeps forward toward the leading edge of the wing, and reduces the effectiveness of the spoiler. This control system may display peculiar side-effects on swept wings. As an example, it is well known that B-52 pilots complain about a significant nose pitch-up moment associated with aileron deflection. It turns out that as the spoilers are deployed the center of lift moves forward, destabilizing the aircraft so it pitches nose-up. An assertive nose pitch-down correction is required by the pilot.

Slot-Lip Ailerons [2,3]

The slot-lip aileron is a lateral control device that regulates the flow of air through a slot made into the wing using a small flap (see Figure 23-5). The flap conforms to the shape of the upper surface of the wing when not in use and is designed reduce the flow of air through the slot. The slot must not be completely closed in the neutral deflection. However, when deployed, air will flow freely through the slot and modify the wing flowfield such that the aircraft can be rolled.

Slot-lip ailerons were investigated in the 1930s in the reports NACA TN-5475 and NACA R-6026. Among results were excessive lag in the control response, which was found to depend on the distance of the device from leading edge (denoted by C_s in Figure 23-5). This lag was found to be excessive unless the device is located some 80% of the chord measured from the leading edge. In general, the aileron was found to result in sluggish response, which explains its rare use in aircraft design.

---

[3] Among well-known aircraft whose design he contributed to are the Bristol Fighter (1916), Bristol Bulldog (1927), Bristol Beaufighter (1941), and the Hunting Percival Jet Provost.

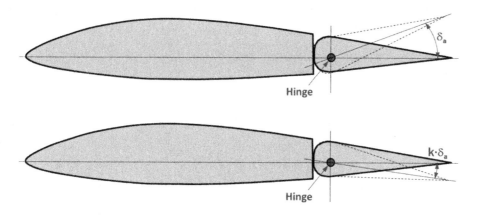

FIGURE 23-6 Differential ailerons.

It was also found that the slot must be open, albeit less open when not in use for roll control. The device delays the stall of the wing tip, while maintaining roll damping, although this is only true if the device is located farther aft than 50% of chord from LE (which is where the resulting lag is high). The aileron was found to increase the drag of the wing by about 10% in cruise and 35% in climb. If simplicity and safety are of higher importance than cruise and climb performance the slot-lip aileron might be useful. It is featured on the Fairchild F-22 test airplane and the GAF N-22 Nomad.

Flaperon

A flaperon is a control surface that serves the role of a flap and aileron. It is typically used with full span flaps. Flaperons are presented in Section 10.4.1, *Lift distribution on wings with flaps deflected*.

Elevon

An elevon is a control surface for which the elevator serves as an aileron in addition to its normal function. It is used for flying wings and tailless delta-wing aircraft. It can also be found on some variable-sweep winged supersonic fighters.

Differential Ailerons

Figure 8-60 of Section 8.3.10, *The effect of deflecting a flap*, shows that as a flap is deflected the drag of the airfoil will increase. On a typical aileron configuration, consisting of two plain flap ailerons, the aircraft is banked by deflecting one aileron up and the other one down. The drag of the down-deflected aileron increases over that of the opposite wing, introducing a yawing moment that tends to point the nose of the aircraft in a heading opposite to the banked one.

As stated earlier, this phenomenon is called *adverse yaw*. Adverse yaw is more pronounced for aircraft with large aspect ratios ($AR > 8$). In order to help remedy this tendency, *differential ailerons* are employed. For such ailerons, the deflection angle for the down-deflected aileron is smaller than that of the up-deflected one (see Figure 23-6). This results in less increase in drag and, thus, less adverse yaw. A typical ratio between the up and down traveling ailerons is 1:0.75. This means that if one aileron is rotated 10° TEU, the opposite one will rotate 7.5° TED.

Aileron Design Requirements

Among critical design points for ailerons are the following:

- Responsiveness at slow speeds.
- Responsiveness at high speeds with low deflections.
- Comfortable stick forces throughout flight envelope.

Another term for responsiveness is *roll authority*. Although responsiveness at slow speeds is imperative (low dynamic pressure requires greater deflection or control surface area, or a combination of the two), high-speed functionality is of great importance as well. It has been known for a long time that a pilot's conception of adequate roll control is tied to the helix angle made by the wing as the airplane rolls at a given airspeed [4, p. 352], denoted by $pb/2V$, where p is the roll rate in radians per second for full aileron deflection, b is the wing span (in ft or m), and V is the airspeed (in ft/s or m/s). Thus, it is recommended that for specific types of aircraft the ratios in Table 23-2 are met or exceeded:

For the aircraft designer this means that the physical dimensions of the ailerons can be determined based on the desired roll rate. A step-by-step design approach is presented below. However, first we must define two important stability derivatives: the *aileron authority derivative* and *roll damping*.

TABLE 23-2 Typical Roll Helix Angles (Radians)

Cargo or Heavy-Lift Aircraft:	Fighter Aircraft:
$\frac{pb}{2V} > 0.07$	$\frac{pb}{2V} > 0.09$

Aileron Authority Derivative $C_{l_{\delta_a}}$

The aileron authority is used to evaluate the roll capability of an airplane. A method, called the *strip integration method*, can be used to estimate it. However, it requires the change in the airfoil's lift coefficient with aileron deflection (denoted by $c_{l\delta a}$) to be known. Note that capitalization is used to separate $C_{l\delta a}$ (aileron authority) from $c_{l\delta a}$ (change in lift coefficient with aileron deflection).

$$C_{l_{\delta_a}} = \frac{dC_l}{d\delta_a} = \frac{2c_{l_{\delta_a}}}{Sb} \int_{b_1}^{b_2} c \cdot y \cdot dy \qquad (23\text{-}3)$$

where

b = wing span, in ft or m
S = wing area, in ft² or m²
c = wing chord, in ft or m
y = wing station, in ft or m.

Units: per radian or per degree.

The dimensions b_1 and b_2 can be seen in Figure 23-7. Note that the expression overestimates the roll authority by not accounting for end effects on either side of each aileron. Example 23-1 compares the analytical $C_{l\delta a}$ to one calculated using potential flow theory, which does account for the tip effects.

Let's simplify the solution of Equation (23-3) for two common wing planform shapes.

CASE 1: Straight tapered wing with taper ratio λ:

$$C_{l_{\delta_a}} = \frac{c_{l_{\delta_a}} C_R}{Sb} \left[\left(b_2^2 - b_1^2 \right) + \frac{4(\lambda - 1)}{3b} \left(b_2^3 - b_1^3 \right) \right] \qquad (23\text{-}4)$$

CASE 2: Rectangular wing ($\lambda = 1$):

$$C_{l_{\delta_a}} = \frac{c_{l_{\delta_a}} \left(b_2^2 - b_1^2 \right)}{b^2} \qquad (23\text{-}5)$$

Derivation of Equation (23-3)

Consider Figure 23-7, which shows the top view of a wing planform. For this derivation it is assumed the aileron on the wing that rotates up is deflected trailing edge down (TED), while the opposite one is deflected trailing edge up (TEU). The lift increases on the former and decreases on the latter, resulting in asymmetry in the lift which generates a rolling moment.

When the aileron is deflected TED, the section lift coefficient in the area of the ailerons will increase by a magnitude $c_{l\delta a} \cdot \delta_a$ and decrease by that same amount on the opposite wing (assuming there is no aileron differential). Thus, the section lift coefficient at a specific spanwise station y becomes:

Up-going wing:

$$c_{l_{up}} = c_{l_y} + c_{l_{\delta_a}} \cdot \delta_a$$

Down-going wing:

$$c_{l_{down}} = c_{l_y} - c_{l_{\delta_a}} \cdot \delta_a$$

where c_{ly} is the value of the section lift coefficient at station y with the aileron in the neutral position. The general definition of the rolling moment coefficient is:

$$C_l = \frac{L}{qSb} = \frac{L_a + L_p}{qSb} \qquad \text{(iii)}$$

where L is rolling moment in ft·lb$_f$ or Nm and is typically a combination of moment due to the aileron deflection, L_a, and restoring moment due to roll rate, L_p. Here we are only concerned with the former. We start by writing the section lift for the up- and down-moving sides at an arbitrary spanwise station y as follows:

$$dF_{up} = q \cdot c_{l_{up}} \cdot dS = q \cdot \left(c_{l_y} + c_{l_{\delta_a}} \cdot \delta_a \right) \cdot c \cdot dy$$

$$dF_{down} = q \cdot c_{l_{down}} \cdot dS = q \cdot \left(c_{l_y} - c_{l_{\delta_a}} \cdot \delta_a \right) \cdot c \cdot dy$$

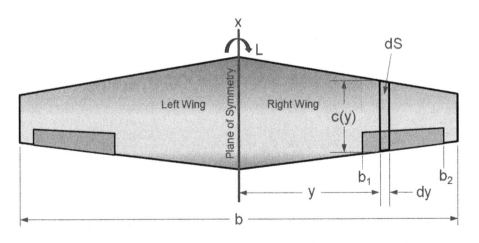

FIGURE 23-7 Definition of the aileron geometry.

The infinitesimal moment attributed to the area at spanwise station y will thus be the difference between the two forces acting at a distance y from the plane of symmetry:

$$dL_a = y \cdot dF_{up} - y \cdot dF_{down}$$
$$= y \cdot q \cdot \left(c_{l_y} + c_{l_{\delta_a}} \cdot \delta_a\right) \cdot c \cdot dy - y \cdot q \cdot \left(c_{l_y} - c_{l_{\delta_a}} \cdot \delta_a\right) \cdot c \cdot dy$$
$$= y \cdot q \cdot \left[2 c_{l_{\delta_a}} \cdot \delta_a\right] \cdot c \cdot dy$$

This allows us to write the infinitesimal rolling moment due to δ_a as follows:

$$dC_l = \frac{dL_a}{qSb} = \frac{y \cdot q \cdot \left[2 c_{l_{\delta_a}} \cdot \delta_a\right] \cdot c \cdot dy}{qSb} = \frac{2 c_{l_{\delta_a}} \cdot \delta_a \cdot c}{Sb} y \cdot dy \quad \text{(iv)}$$

The total moment over the entire span of the ailerons is thus (here the term C_l is used to avoid complicating things with subscripts, even though we are only considering the influence of the aileron deflection):

$$C_l = \int_{b_1}^{b_2} dC_l = \int_{b_1}^{b_2} \frac{2 c_{l_{\delta_a}} \cdot \delta_a \cdot c}{Sb} y \cdot dy = \frac{2 c_{l_{\delta_a}} \cdot \delta_a}{Sb} \int_{b_1}^{b_2} c \cdot y \cdot dy$$

(v)

from which we get the following dependency on the aileron deflection (notice the difference between C_l (rolling moment coefficient) and c_l (the section lift coefficient)):

$$C_{l_{\delta_a}} = \frac{dC_l}{d\delta_a} = \frac{2 c_{l_{\delta_a}}}{Sb} \int_{b_1}^{b_2} c \cdot y \cdot dy \quad \text{(vi)}$$

QED

Derivation of Equations (23-4) and (23-5)

Assume a straight tapered wing whose taper ratio is λ, root chord C_R, and span is b. A parametric representation of the chord as a function of y is given by Equation (9-21):

$$c(y) = C_R \left(1 + \frac{2(\lambda - 1)}{b} y\right) \quad (9\text{-}21)$$

If the wing features a single airfoil, the lift curve slope with may be assumed constant. Therefore, we can write the aileron authority as follows:

$$C_{l_{\delta_a}} = \frac{2 c_{l_{\delta_a}}}{Sb} \int_{b_1}^{b_2} c \cdot y \cdot dy = \frac{2 c_{l_{\delta_a}}}{Sb} \int_{b_1}^{b_2} C_R \left(1 + \frac{2(\lambda - 1)}{b} y\right) \cdot y \cdot dy$$

$$= \frac{2 c_{l_{\delta_a}} C_R}{Sb} \int_{b_1}^{b_2} \left(y + \frac{2(\lambda - 1)}{b} y^2\right) \cdot dy = \frac{2 c_{l_{\delta_a}} C_R}{Sb} \left[\frac{y^2}{2} + \frac{2(\lambda - 1)}{3b} y^3\right]_{b_1}^{b_2}$$

Inserting the limits and manipulating algebraically yields:

$$C_{l_{\delta_a}} = \frac{2 c_{l_{\delta_a}} C_R}{Sb} \left[\frac{1}{2}\left(b_2^2 - b_1^2\right) + \frac{2(\lambda - 1)}{3b}\left(b_2^3 - b_1^3\right)\right]$$

$$= \frac{c_{l_{\delta_a}} C_R}{Sb} \left[\left(b_2^2 - b_1^2\right) + \frac{4(\lambda - 1)}{3b}\left(b_2^3 - b_1^3\right)\right]$$

Consider the special case for a rectangular wing, when $\lambda = 1$, for which $S = C_R \cdot b$:

$$C_{l_{\delta_a}} = \frac{c_{l_{\delta_a}}\left(b_2^2 - b_1^2\right) C_R}{C_R b \cdot b} = \frac{c_{l_{\delta_a}}\left(b_2^2 - b_1^2\right)}{b^2}$$

QED

EXAMPLE 23-1

A UAV is being designed with a Hershey bar wing whose dimensions are shown in Figure 23-8. Determine the aileron authority if change in lift with aileron deflection, $C_{l\delta a}$, has been found to equal 0.05524 per degree (i.e. 10° deflection increases C_l by 0.5524).

Solution

From the figure we see that $b_1 = 3$ ft and $b_2 = 6$ ft. We also see that the wing area S amounts to $12 \times 1 = 12$ ft², the wing span is 12 ft, and the chord is a constant 1 ft.

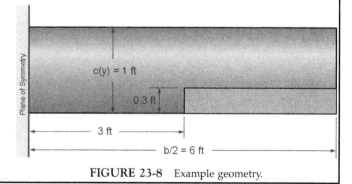

FIGURE 23-8 Example geometry.

EXAMPLE 23-1 (cont'd)

Therefore we can evaluate the aileron authority from Equation (23-3), noting that the chord is constant:

$$C_{l_{\delta_a}} = \frac{dC_l}{d\delta_a} = \frac{2c_{l_{\delta_a}}}{Sb} c \cdot \int_{b_1}^{b_2} y \cdot dy = \frac{2(0.05524)}{12 \times 12}(1)\left[\frac{y^2}{2}\right]_3^6$$
$$= 0.01036 \ /° = 0.5936 \ /rad$$

Compare this to Equation (23-5), which results in:

$$C_{l_{\delta_a}} = \frac{c_{l_{\delta_a}}\left(b_2^2 - b_1^2\right)}{b^2} = \frac{0.05524(6^2 - 3^2)}{12^2}$$
$$= 0.01036 \ /°$$

Additional information: a potential flow model of this wing is shown Figure 23-9, and assumes a symmetrical airfoil. In fact, the value of $c_{l\delta a} = 0.05524 \ /°$ came from this model and was obtained by deflecting the entire trailing edge (as C_L with flap neutral at $2° = 0.51742$. C_L with flap deflected $1°$ at $2° = 0.57266$; therefore, $c_{l\delta a} = (0.57266 - 0.51742)/1° = 0.05524 \ /°$). Therefore, it already includes some tip effects, although this will be compounded when a smaller span aileron is deflected.

The VL solution accounts for end effects and yields a $C_{l\delta a}$ of 0.4134 /rad. The distribution of section lift coefficients along the span is shown in the figure. It clearly shows how the section lift coefficients rise gradually rather than instantly, as the strip integration method assumes. Therefore, the roll authority is less than indicated by Equation (23-3).

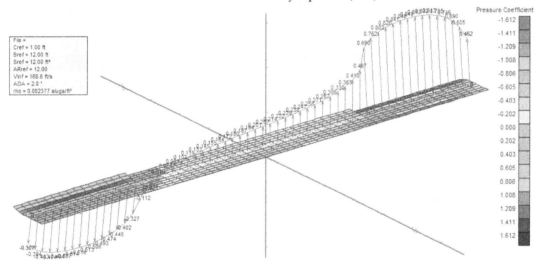

FIGURE 23-9 Potential flow results for the wing.

Roll Damping Derivative C_{l_p}

As the wing rolls it becomes asymmetrically loaded as one half rotates up while the opposite one rotates down (see Figure 23-10). This rotation changes the local *AOA* along the wing. The wing rotating downward is subject to increased local *AOA*s, while the opposite holds for the wing that rotates up. This asymmetry in the spanwise *AOA*s creates a moment that resists the roll of the wing. It is called *roll damping* and is a function of the roll rate, p (in units of degrees/second or, preferably, in radians/second). The derivative is evaluated with respect to the ratio $(pb/2V)$, where b is the wing span, and V is the airspeed. The product $pb/2$ represents the rotation speed of the wingtip (in units of m/s or ft/s, provided p is in rad/s). When divided by the airspeed V the ratio represents the helix angle (in radians) formed by the motion.

If S and b are the reference wing area and span, $c(y)$ is the wing chord as a function of the spanwise station y, and q is the dynamic pressure, the rolling moment can be calculated using the following expression:

$$L_p = \frac{2qp(c_{l_\alpha} + c_{do})}{V} \int_0^{b/2} y^2 \cdot c(y) dy \qquad (23\text{-}6)$$

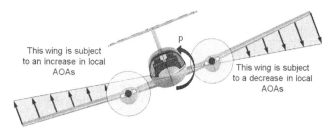

FIGURE 23-10 Arrows show relative airspeed due to roll to the right. This increases the AOA on the right wing (which is to the left in the figure) and decreases it on the left one.

The rolling moment coefficient due to roll damping in terms of $(pb/2V)$ is then calculated as shown below:

$$C_l = \frac{L_p}{qSb} = -\left(\frac{pb}{2V}\right)\frac{4(c_{l_\alpha} + c_{do})}{Sb^2}\int_0^{b/2} y^2 \cdot c(y)\,dy \quad (23\text{-}7)$$

The change in rolling moment coefficient due to roll rate p is called the *roll damping coefficient* and can be found from:

$$C_{l_p} = -\frac{4(c_{l_\alpha} + c_{do})}{Sb^2}\int_0^{b/2} y^2 \cdot c(y)\,dy \quad (23\text{-}8)$$

Units: per radian or per degree. The closed-form solution for two common wing planform shapes is given below.

Let's simplify the solution of Equation (23-8) for two special but common wing planform shapes.

CASE 1: Straight tapered wing with taper ratio λ:

$$C_{l_p} = -\frac{(c_{l_\alpha} + c_{do}) \cdot C_R b}{24S}[1 + 3\lambda] \quad (23\text{-}9)$$

CASE 2: Rectangular wing ($\lambda = 1$):

$$C_{l_p} = -\frac{c_{l_\alpha} + c_{do}}{6} \quad (23\text{-}10)$$

Note that the lift curve slope is the predominant factor in Equation (23-10) as its magnitude is generally much larger than that of the drag coefficient. With this in mind, consider Figure 9-47, which shows how the lift curve changes as a function of AOA. Although the figure is for an entire airplane, airfoils behave in a similar manner. Below stall, at low AOA, the value of c_{l_α} is positive, so the value of $C_{lp} < 0$ and this will slow down rotation. However, in the immediate post-stall region c_{l_α} is negative and the value of $C_{lp} > 0$, which is a pro-rotation effect. Consequently, the roll damping derivative plays an essential role in the nature of autorotation during spinning.

Derivation of Equations (23-6) and (23-8)

Consider Figure 23-10, noting that the roll rate p (in deg/s or rad/s) changes the local angle of attack by an amount $\Delta\alpha$ along the entire wing, where $\Delta\alpha$ is a function of the spanwise station y and airspeed V and is given by:

$$\Delta\alpha = \frac{py}{V} \quad \text{(i)}$$

Figure 23-11 shows a cross-section of the airfoil at an arbitrary wing station y on the up- and down-going

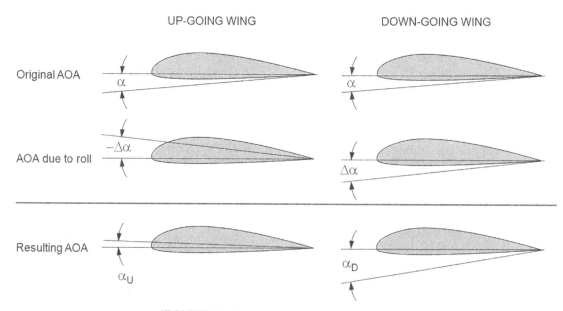

FIGURE 23-11 Changes in local AOA due to roll.

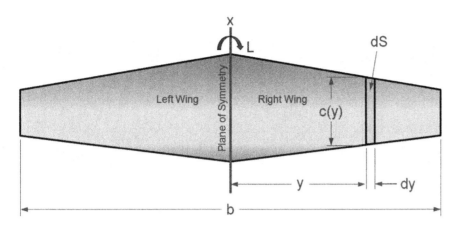

FIGURE 23-12   Wing geometry.

wings. The first row shows the original angle-of-attack (*AOA*). The second row shows the change in *AOA* due to the roll, and the bottom row shows the combination of the two. Figure 23-12 shows pertinent wing dimensions for this derivation.

The change in the section lift coefficient due to the roll rate, Δc_{ly}, can now be expressed as shown below:

$$\Delta c_{l_y} = c_{l_\alpha}(y) \cdot \Delta\alpha = c_{l_\alpha}(y) \cdot \frac{py}{V} \quad \text{(ii)}$$

The lift coefficient is thus:

$$c_{l_y} = c_{l_y}(\alpha) + c_{l_\alpha}(y) \cdot \Delta\alpha = c_{l_y}(\alpha) + c_{l_\alpha}(y) \cdot \frac{py}{V} \quad \text{(iii)}$$

Where $c_{l\alpha}(y)$ and c_{do} are the airfoil's 2D lift curve slope and drag coefficient at the specific flight condition at wing station y. Now, consider the drag coefficient of the airfoil at that station:

$$c_{d_y} = c_{do} + c_{d\alpha} \cdot \Delta\alpha = c_{do} + c_{d\alpha}(\alpha) \cdot \frac{py}{V} \quad \text{(iv)}$$

We can now write the infinitesimal lift and drag as follows:

Lift:
$$dL = q \cdot dS \cdot c_{l_y} = q \cdot c_{l_y}(y) \cdot c(y) \cdot dy \quad \text{(v)}$$

Drag:
$$dD = q \cdot dS \cdot c_{d_y} = q \cdot c_{d_y}(y) \cdot c(y) \cdot dy \quad \text{(vi)}$$

where q is the dynamic pressure and $c(y)$ is the chord at station y.

Now we must estimate the projections of the lift and drag onto the z-axis, which is normal to the axis of rotation (the x-axis). Refer to Figure 23-13, which shows the alignment of the x-z axes with respect to the lift and drag. The rotation takes place about the x-axis and the retarding moment (roll damping) is generated through the force that acts perpendicular to it, the z-axis. Therefore, we can determine the roll damping force:

$$dZ = -dL \cdot \cos(\Delta\alpha) - dD \cdot \sin(\Delta\alpha) \quad \text{(vii)}$$

Ultimately, the moment due to roll is caused by the asymmetric loading on the wing, or, more precisely, its projection onto the z-axis. The image shows we can define the projections of the lift and drag coefficient on the z-axis as follows (note that positive z is down):

Wing moving up:

$$c_{LU} = \left[c_{l_y}(\alpha) - c_{l_\alpha}(y) \cdot \frac{py}{V}\right] \cdot \cos\left(\frac{py}{V}\right) \quad \text{(viii)}$$

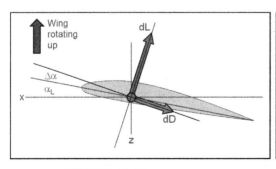 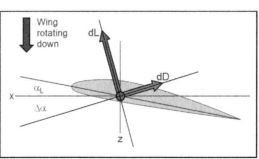

FIGURE 23-13   Changes in *AOA* due to roll on either side of the plane of symmetry.

$$c_{DU} = \left[c_{do} - c_{d\alpha}(\alpha) \cdot \frac{py}{V}\right] \cdot \sin\left(\frac{py}{V}\right) \quad \text{(ix)}$$

The elemental force:
$$dZ_U = -q \cdot [c_{LU} - c_{DU}] \cdot c(y) dy$$
$$= q \cdot c_{DU} \cdot c(y) dy - q \cdot c_{LU} \cdot c(y) dy \quad \text{(x)}$$

Wing moving down:
$$c_{LD} = \left[c_{l_y}(\alpha) + c_{l_\alpha}(y) \cdot \frac{py}{V}\right] \cdot \cos\left(\frac{py}{V}\right) \quad \text{(xi)}$$

$$c_{DD} = \left[c_{do} + c_{d\alpha}(\alpha) \cdot \frac{py}{V}\right] \cdot \sin\left(\frac{py}{V}\right) \quad \text{(xii)}$$

The elemental force:
$$dZ_D = -q \cdot [c_{LD} + c_{DD}] \cdot c(y) dy$$
$$= -q \cdot c_{LD} \cdot c(y) dy - q \cdot c_{DD} \cdot c(y) dy \quad \text{(xiii)}$$

The roll damping is caused by the difference between dZ_U and dZ_D, integrated across the wing (note that the limits of integration are 0 to $b/2$). The difference between the two is given by:

$$dZ_U - dZ_D = q \cdot c_{LD} \cdot c(y) dy - q \cdot c_{LU} \cdot c(y) dy$$
$$+ q \cdot c_{DD} \cdot c(y) dy + q \cdot c_{DU} \cdot c(y) dy$$
$$= q \cdot (c_{LD} - c_{LU} + c_{DD} + c_{DU}) \cdot c(y) dy$$

Consider the summation of the lift and drag coefficients. Inserting the coefficients as defined above results in:

$$c_{LD} - c_{LU} + c_{DD} + c_{DU}$$
$$= \left[c_{l_y}(\alpha) + c_{l_\alpha}(y) \cdot \frac{py}{V}\right] \cdot \cos\left(\frac{py}{V}\right)$$
$$- \left[c_{l_y}(\alpha) - c_{l_\alpha}(y) \cdot \frac{py}{V}\right] \cdot \cos\left(\frac{py}{V}\right)$$
$$+ \left[c_{do} + c_{d\alpha}(\alpha) \cdot \frac{py}{V}\right] \cdot \sin\left(\frac{py}{V}\right)$$
$$+ \left[c_{do} - c_{d\alpha}(\alpha) \cdot \frac{py}{V}\right] \cdot \sin\left(\frac{py}{V}\right)$$

Simplifying and assuming the 2D $c_{l\alpha}$ to be constant along the span:

$$c_{LD} - c_{LU} + c_{DD} + c_{DU}$$
$$= \left[c_{l_y}(\alpha) + c_{l_\alpha} \cdot \frac{py}{V} - c_{l_y}(\alpha) + c_{l_\alpha} \cdot \frac{py}{V}\right] \cdot \cos\left(\frac{py}{V}\right)$$
$$+ \left[c_{do} + c_{d\alpha}(\alpha) \cdot \frac{py}{V} + c_{do} - c_{d\alpha}(\alpha) \cdot \frac{py}{V}\right] \cdot \sin\left(\frac{py}{V}\right)$$

Therefore, the summation is:
$$c_{LD} - c_{LU} + c_{DD} + c_{DU} = 2c_{l_\alpha} \cdot \frac{py}{V} \cdot \cos\left(\frac{py}{V}\right)$$
$$+ 2c_{do} \cdot \sin\left(\frac{py}{V}\right) \quad \text{(xiv)}$$

If the roll rate is low enough to assume small angle relations ($\cos \Delta\alpha \approx 1$ and $\sin \Delta\alpha \approx \Delta\alpha$), we can rewrite Equation (xiv) as follows:

$$c_{LD} - c_{LU} + c_{DD} + c_{DU} = 2c_{l_\alpha} \cdot \frac{py}{V} + 2c_{do} \cdot \frac{py}{V}$$
$$= 2(c_{l_\alpha} + c_{do}) \cdot \frac{py}{V} \quad \text{(xv)}$$

The total restoring moment due to roll, L_p, is thus:
$$L_p = \int_0^{b/2} y \cdot (dZ_U - dZ_D)$$
$$= \int_0^{b/2} y \cdot q \cdot (c_{LD} - c_{LU} + c_{DD} + c_{DU}) \cdot c(y) dy \quad \text{(xvi)}$$

Or:
$$L_p = \int_0^{b/2} y \cdot q \cdot \left(2(c_{l_\alpha} + c_{do}) \cdot \frac{py}{V}\right) \cdot c(y) dy$$
$$= \frac{2qp(c_{l_\alpha} + c_{do})}{V} \int_0^{b/2} y^2 \cdot c(y) dy \quad \text{(xvii)}$$

From the definition of the rolling moment coefficient we get:
$$C_l = \frac{L}{qSb} = \frac{2qp(c_{l_\alpha} + c_{do})}{qSbV} \int_0^{b/2} y^2 \cdot c(y) dy$$
$$= \frac{2p(c_{l_\alpha} + c_{do})}{SbV} \int_0^{b/2} y^2 \cdot c(y) dy \quad \text{(xviii)}$$

Preparing this in terms of $(pb/2V)$ (since that's our intended to differential) leads to:

$$C_l = \frac{2p(c_{l_\alpha} + c_{do})}{SbV} \int_0^{b/2} y^2 \cdot c(y) dy$$
$$= \left(\frac{pb}{2V}\right) \frac{4(c_{l_\alpha} + c_{do})}{Sb^2} \int_0^{b/2} y^2 \cdot c(y) dy \quad \text{(xix)}$$

Finally, we can determine the roll damping derivative by differentiating with respect to $(pb/2V)$ and then evaluate the geometric integral. Note that since the effect retards the motion it is always negative:

$$C_{l_p} = \frac{\partial C_l}{\partial \left(\frac{pb}{2V}\right)}$$

$$= \frac{\partial}{\partial \left(\frac{pb}{2V}\right)} \left[\left(\frac{pb}{2V}\right) \frac{4(c_{l_\alpha} + c_{do})}{Sb^2} \int_0^{b/2} y^2 \cdot c(y) dy \right]$$

$$= -\frac{4(c_{l_\alpha} + c_{do})}{Sb^2} \int_0^{b/2} y^2 \cdot c(y) dy \qquad \text{(xx)}$$

QED

Derivation of Equations (23-9) and (23-10)

Assume a straight tapered wing whose taper ratio is λ, root chord C_R, and span is b. A parametric representation of the chord as a function of y is given by Equation (9-21):

$$c(y) = C_R \left(1 + \frac{2(\lambda - 1)}{b} y\right) \qquad (9\text{-}21)$$

Inserting this equation into Equation (xx) and expanding we get Equation (23-9);

$$C_{l_p} = -\frac{4(c_{l_\alpha} + c_{do})}{Sb^2} \int_0^{b/2} y^2 \cdot c(y) dy$$

$$= -\frac{4(c_{l_\alpha} + c_{do})}{Sb^2} \int_0^{b/2} y^2 \cdot C_R \left(1 + \frac{2(\lambda - 1)}{b} y\right) dy$$

$$= -\frac{4(c_{l_\alpha} + c_{do}) \cdot C_R}{Sb^2} \int_0^{b/2} \left(y^2 + \frac{2(\lambda - 1)}{b} y^3\right) dy$$

$$= -\frac{4(c_{l_\alpha} + c_{do}) \cdot C_R}{Sb^2} \left[\frac{y^3}{3} + \frac{2(\lambda - 1)}{b} \frac{y^4}{4}\right]_0^{b/2}$$

$$= -\frac{(c_{l_\alpha} + c_{do}) \cdot C_R b}{24S} [1 + 3\lambda]$$

We get Equation (23-10) for a rectangular wing, when $\lambda = 1$, for which $S = C_R \cdot b$:

$$C_{l_p} = -\frac{(c_{l_\alpha} + c_{do}) \cdot C_R b}{24S} [1 + 3\lambda]$$

$$= -\frac{(c_{l_\alpha} + c_{do}) \cdot C_R b}{24(C_R b)} [1 + 3] = -\frac{c_{l_\alpha} + c_{do}}{6}$$

QED

EXAMPLE 23-2

Determine the roll damping of the wing of EXAMPLE 23-1 and compare to that of the vortex-lattice method. Assume $c_{do} = 0.010$.

Solution

Begin by calculating the lift curve slope for the wing, $C_{l\alpha}$, here using Equation (9-57), assuming $M \approx 0$, and the airfoil's 2D lift curve is 2π. Also assume the following wing characteristics:

AR = wing aspect ratio = 12
β = Mach number parameter (Prandtl-Glauert) = $(1 - M^2)^{0.5} \approx 1$
κ = Ratio of 2D lift curve slope to $2\pi \approx 1$
$\Lambda_{c/2}$ = Sweepback of mid-chord = $0°$

Therefore:

$$C_{l_\alpha} = \frac{2\pi \cdot 12}{2 + \sqrt{\frac{144 \times 1^2}{1^2}\left(1 + \frac{\tan^2(0°)}{1^2}\right) + 4}}$$

$$= 5.322 \text{ per rad} = 0.09290 \text{ per deg}$$

So we find that the roll damping using Equation (23-10) amounts to:

$$C_{l_p} = -\frac{C_{l_\alpha} + c_{do}}{6} = -\frac{5.322 + 0.010}{6} = -0.8887 \text{ rad}$$

In comparison, using the same vortex-lattice model as shown in Figure 23-9, the roll damping was found to equal -0.6366 /rad with ailerons deflected $15°$ and -0.6388 /rad with neutral ailerons. The model analysis accounts for tip effects, which is not the case for the analytical method of Equation (23-10).

23.2.3 Aileron sizing

With the preliminaries out of the way in the preceding section, it is now possible to present the following method to help with the sizing of the ailerons. The procedure involves estimating the steady-state roll rate using the aileron authority and roll damping derivatives.

Estimating Steady-State Roll Rate

When the ailerons are deflected the airplane begins to roll (see Figure 23-14). The motion consists of a transient acceleration that disappears once the roll rate increases and after which there is a steady roll rate. This the steady-state roll rate, which is helpful in determining a pilot-desired responsiveness.

The steady-state roll helix angle is determined from:

$$\frac{pb}{2V} = -\frac{C_{l_{\delta_a}}}{C_{l_p}} \delta_a \qquad (23\text{-}11)$$

Note that the steady-state roll rate (in rad/sec) can be determined from:

$$p = -\frac{C_{l_{\delta_a}}}{C_{l_p}} \delta_a \left(\frac{2V}{b}\right) \qquad (23\text{-}12)$$

FIGURE 23-14 Change in lift due to aileron deflection.

STEP 1: Initial Dimensions

Establish initial dimensions based on Figure 23-15. Also determine the likely aileron deflection angle, δ_a. Note that the control system will stretch in flight, reducing the maximum ground deflection. This means that a control system designed for a maximum deflection of, say, 15° on the ground may only deflect as much as 75% of that in flight. Some control systems are so poorly designed[4] that they may only achieve 25% of the maximum deflection. At any rate, 75% is a reasonable "first stab" estimate for an average control system. This would mean that a maximum deflection of 15° is closer to 11.3° in flight.

STEP 2: Estimate Roll Damping

Using the geometry from STEP 1, estimate roll damping, C_{l_p}. Use Equation (23-8) for an arbitrary wing planform shape, Equation (23-9) for a straight tapered wing, and Equation (23-10) for a "Hershey bar" wing.

STEP 3: Estimate Roll Authority

Using the geometry from STEP 1, estimate the roll authority, $C_{l_{\delta_a}}$. Use Equation (23-3) for an arbitrary wing planform shape. Use Equation (23-4) for a straight tapered wing, and Equation (23-5) for a "Hershey bar" wing.

STEP 4: Estimate Roll Helix Angle

Determine a desired "target" roll helix angle per Section 26.4.2 using Equation (23-5). If the calculated value is less than the selected target enlarge b_1 or b_2, or both. Note that b_2 can never be larger than $b/2$ and $0 < b_1 < b_2$.

Special Case Aileron Sizing: Constant-Chord Wing

The following expression can be used to determine the spanwise location of the inboard edge of the aileron, for a given outboard edge. *It only applies to constant-chord ("Hershey bar") wings.*

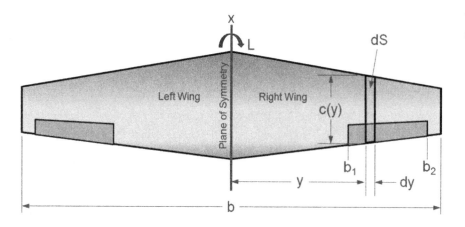

FIGURE 23-15 Definition of the aileron geometry.

---

[4] Poorly designed here mean that it results in excessive stretching.

Physical dimension:

$$b_1 = \sqrt{\left(\frac{pb}{2V}\right)\frac{C_{l_p}}{c_{l_{\delta_a}} \cdot \delta_a}b^2 + b_2^2} \qquad (23\text{-}13)$$

Fractional dimension:

$$\frac{b_1}{b} = \sqrt{\left(\frac{pb}{2V}\right)\frac{C_{l_p}}{c_{l_{\delta_a}} \cdot \delta_a} + \left(\frac{b_1}{b}\right)^2} \qquad (23\text{-}14)$$

where

b_1 = spanwise station for the inboard edge of the aileron, in ft or m
b_2 = spanwise station for the outboard edge of the aileron, in ft or m

Derivation of Equation (23-11)

Assume the wing is rigid and the rolling motion is caused by deflecting the ailerons to an angle δ_a. Further assume the roll rate p is impeded by the roll damping due to a local change in *AOA* along the wing (with a minor contribution from the vertical tail). Therefore we can write the equation of rolling motion for the aircraft as follows:

$$I_{XX}\dot{p} = \frac{\partial L}{\partial p}\left(\frac{pb}{2V}\right) + \frac{\partial L}{\partial \delta_a}\delta_a \qquad (i)$$

where I_{XX} is the aircraft's moment of inertia (in slugs·ft² or kg·m²) $\dot{p}$ is the roll acceleration in rad/s², L is the rolling moment in ft·lb_f or N·m, and δ_a is the aileron deflection in degrees.

$$0 = \frac{\partial L}{\partial p}\left(\frac{pb}{2V}\right) + \frac{\partial L}{\partial \delta_a}\delta_a \Leftrightarrow \frac{\partial L}{\partial p}\left(\frac{pb}{2V}\right) = -\frac{\partial L}{\partial \delta_a}\delta_a$$

$$\Rightarrow \left(\frac{pb}{2V}\right) = -\frac{\partial L/\partial \delta_a}{\partial L/\partial p}\delta_a$$

In terms of stability derivatives we can write:

$$\left(\frac{pb}{2V}\right) = -\frac{C_{l_{\delta_a}}}{C_{l_p}}\delta_a \qquad (ii)$$

QED

Derivation of Equation (23-13)

Begin with Equation (23-11) and solve for $C_{l\delta a}$:

$$\left(\frac{pb}{2V}\right) = -\frac{C_{l_{\delta_a}}}{C_{l_p}}\delta_a \Rightarrow C_{l_{\delta_a}} = -\left(\frac{pb}{2V}\right)\frac{C_{l_p}}{\delta_a} \qquad (i)$$

Insert Equation (23-5):

$$C_{l_{\delta_a}} = -\left(\frac{pb}{2V}\right)\frac{C_{l_p}}{\delta_a} = \frac{c_{l_{\delta_a}}\left(b_2^2 - b_1^2\right)}{b^2} \qquad (ii)$$

Since our target is to determine the inboard station, b_1, for the aileron we solve for it:

$$b_1 = \sqrt{\left(\frac{pb}{2V}\right)\frac{C_{l_p}}{\delta_a}\frac{b^2}{c_{l_{\delta_a}}} + b_2^2} \qquad (iii)$$

QED

EXAMPLE 23-3

A UAV is being designed with a Hershey bar wing whose dimensions are shown in Figure 23-16. The maximum aileron ground deflection is 20°. Assuming that 75% of that will be achievable in flight, determine the roll rate for maximum aileron deflection at $V = 100$ KTAS if change in lift with aileron deflection, $c_{l\delta a}$, has been found to equal 0.05524 per degree. NOTE that this geometry is analyzed in EXAMPLES 22-1 and 22-3. Compare the results to that obtained from the *Vortex-Lat*tice code SURFACES presented in the same examples.

Solution

STEP 1: Determine the Derivative $C_{l\delta a}$

This was done in EXAMPLE 23-1 and was shown to amount to;

$$C_{l_{\delta_a}} = \frac{dC_l}{d\delta_a} = \frac{2c_{l_{\delta_a}}}{Sb}c \cdot \int_{b_1}^{b_2} y \cdot dy = \frac{2(0.05524)}{12 \times 12}(1)\left[\frac{y^2}{2}\right]_3^6$$

$$= 0.01036 \ /° = 0.5936 \ /\text{rad}$$

EXAMPLE 23-3 (cont'd)

STEP 2: Determine the Derivative C_{l_p}

This was done in Example 23-2 and was shown to equal;

$$C_{l_p} = -\frac{c_{l_\alpha} + c_{do}}{6} = -\frac{5.322 + 0.010}{6} = -0.8887 \text{ /rad}$$

STEP 3: Determine the Roll Helix Angle

Based on this the roll rate at 100 KTAS can be found from Equation (23-5), where the maximum achievable deflection amounts to $20° \times 0.75 = 15°$:

$$\frac{pb}{2V} = -\frac{C_{l_{\delta_a}}}{C_{l_p}} \delta_a = -\frac{0.5936}{-0.8887}\left(15° \frac{\pi}{180}\right) = 10.02 \text{ deg}$$

STEP 4: Determine the Roll Rate p

$$p = -\frac{C_{l_{\delta_a}}}{C_{l_p}} \delta_a \left(\frac{2V}{b}\right) = -\frac{0.5936}{-0.8887}(15°)\left(\frac{2 \times 168.8}{12}\right)$$
$$= 282.9 \text{ °/s}$$

Determining the roll rate using the vortex-lattice solution from SURFACES (see Examples 23-1 and 23-2):

$$p = -\frac{C_{l_{\delta_a}}}{C_{l_p}} \delta_a \left(\frac{2V}{b}\right) = -\frac{0.4134}{-0.6336}(15°)\left(\frac{2 \times 168.8}{12}\right)$$
$$= 275.3 \text{ °/s}$$

The VL solution accounts for tip effects.

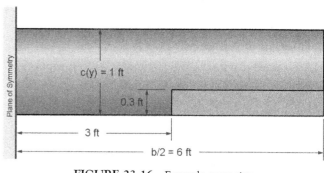

FIGURE 23-16 Example geometry.

Maximizing responsiveness

Some airplanes require flap span to be maximized to meet requirements for stall speed. This means that the aileron span is less than ideal. The designer can attempt to improve the effectiveness of the remaining ailerons by positioning the centroid of the aileron planform as close as possible to the location where their spanwise section moment reaches maximum.

Consider the tapered wing planform (halfspan) in Figure 23-17, which shows the spanwise section moment (analogous to section lift or section lift coefficient) van be defined as follows:

$$\Delta C_m = y \cdot C_l \quad \Leftrightarrow \quad \Delta M_X = y \cdot (q \cdot C_l \cdot \Delta S)$$

where

ΔC_m = spanwise moment coefficient
C_l = section lift coefficient
ΔM_X = elemental rolling moment
y = spanwise station
q = dynamic pressure
ΔS = area of elemental strip

Figure 23-18 shows the variation of the section rolling moment coefficient along the span of the wing shown in Figure 23-2. It shows that the peak of the section moment coefficients on the wing with the ailerons neutral occurs near the 85% spanwise station. This is where one should try to place the aileron centroid of small ailerons to help achieve maximum responsiveness. The location of this peak is highly dependent on the wing geometry.

23.2.4 Fundamentals of Pitch Control

The purpose of the elevator is to provide control about the airplane's pitch axis, by modifying the pitching moment coefficient, C_m. Figure 23-19 shows the pitching moment coefficient, C_m, versus angle-of-attack (AOA) for an airplane with a conventional tail. This graph does not belong to any specific aircraft type, but

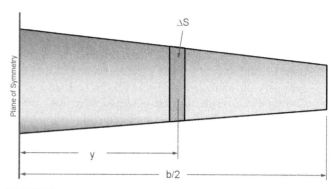

FIGURE 23-17 Definition of an infinitesimal segment ΔS on the wing.

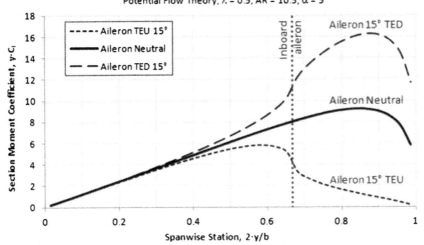

FIGURE 23-18 Distribution of section moments along the wing (in Figure 23-2).

rather shows typical results from a wind tunnel test for such an aircraft. The three curves represent the C_m for the complete aircraft with different elevator deflections and show how deflecting the elevator will allow the pitching moment curve to be moved around. The top curve shows the C_m with the elevator deflected 20° TEU ($\delta_e = -20°$) as would happen for a nose pitch up (climb or stall). The center curve depicts the elevator in a neutral position ($\delta_e = 0°$), and the bottom depicts the elevator deflected 15° TED ($\delta_e = +15°$), as would happen if the airplane's nose were pitched down (e.g. a dive). The separation between the curves represents the effective *elevator authority* for the aircraft. The curves show the classic linear downward slope required for a naturally statically stable aircraft and then a transition into nonlinear curves associated with the post-stall AOAs. When the airplane is operated at low AOAs, the airflow is mostly attached and the C_m changes linearly. However, with higher and higher AOAs larger areas of the airplane become covered with separated flow and this is the main cause for the non-linearity.

Now, consider the linear section of the curves as they intersect the horizontal axis at different AOAs. For instance, the top curve (labeled with $\delta_e = -20°$)

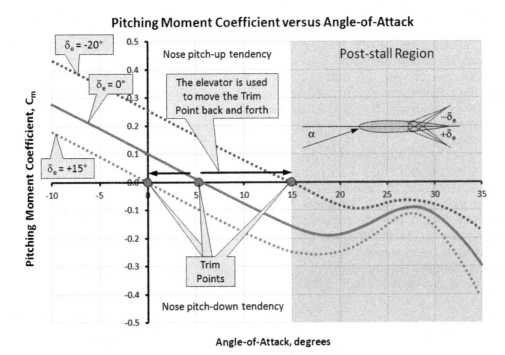

FIGURE 23-19 Pitch control explained.

intersects the AOA axis at a $\alpha \approx 15°$, the center curve at a $\alpha \approx 5.5°$, and the bottom curve at a $\alpha \approx 0°$. Each of the three points indicates that the airplane's C_m has reached zero — in other words, the aircraft is trimmed at that particular AOA. For that reason, they are called *trim points*. Trimmed means that as long as the slope of the curve is negative, the airplane will "want" to stay at that attitude (AOA) until disturbed again — the trim point is an equilibrium point. It is important to keep the following in mind. In order for the airplane to remain trimmed in this fashion, two things have to happen: (1) The slope of the C_m must be negative, and (2) the value of C_m must be zero. If the slope is positive, the airplane will continue to pitch and will not stop at $C_m = 0$. The whole purpose of putting a tail on an airplane is to force the pitching moment curve for the airplane to have a negative slope. And the whole purpose of featuring an elevator is to allow the pilot to move the trim point back and forth. This way, there exists one and only one trim point for any airspeed and the airplane must also have enough thrust to maintain flight at that airspeed, whether it is generated mechanically or by gravity (as for sailplanes). Note that for real airplanes there is only one curve, and not three as shown in the graph, and the pilot can move it back and forth at will. The three curves are only shown to indicate the trim range.

The importance of the three curves is that they also indicate the controllability or the control authority of the aircraft. For instance, if the pilot moves the elevator to the neutral position (the center curve), the aircraft would be trimmed at approximately $\alpha \approx 5.5°$ as stated earlier. If he deflects it to $-20°$ it would trim at $\alpha \approx 15°$ and so on. The airplane might have a maximum elevator deflection of $-25°$, allowing the pilot to bring it to an even higher AOA and to stall if necessary. The curves show that the pilot can control the trim AOA by simply deflecting the elevator and that way move the trim point to a higher or a lower AOA. This, of course, is elementary information that any student of aerospace engineering is exposed to.

The required size of the elevator should be based on the investigation of the capability of the aircraft at the extreme conditions stipulated in Sections 11.2, 23.3, and the additional tools provided in Appendix C1.6. Such an investigation should plot the required AOA and elevator deflection over the entire speed range of the aircraft and include the extremes of the CG envelope. The sizing should ensure the required elevator deflection is less than what is available, per Section 23.3.

23.2.5 Fundamentals of Yaw Control

The rudder sizing depends generally on two conditions: crosswind requirements and, if the aircraft is multiengine, on one-engine-inoperable (OEI) asymmetric thrust requirements. Additional requirements would be spin recovery; however, this is particularly challenging to analyze because of the separated flow field the rudder operates in. For this reason, the rudder is generally sized for the crosswind and asymmetric thrust and there is a good chance it will be adequately sized for spin recovery, provided the fin is not blanketed by a horizontal tail or fuselage.

Figure 23-20 shows the yawing moment coefficient, C_n, versus angle-of-yaw (AOY) for an airplane with a conventional tail. Like Figure 23-19, this graph does not belong to any specific aircraft type either, but rather shows what results one would expect from a wind tunnel test. The three curves represent the C_n for a complete aircraft with the rudder deflected left, neutral, and right. As with the elevator, it shows how deflecting the rudder will shift the yawing moment curve up or down, so the airplane is trimmed with the nose pointed to the left or right. The left-most curve shows the C_n with the rudder deflected 20° trailing edge left (TEL) ($\delta_r = +20°$) as would happen for a nose-left slip. The center curve depicts the C_n with the rudder in a neutral position ($\delta_r = 0°$), and the bottom one depicts the rudder deflected 20° trailing edge right (TER) ($\delta_r = -20°$), as would happen for a nose-right slip. The separation between the curves represents the effective *rudder authority* for the aircraft. The curves show the classic linear upward slope required for a naturally statically and directionally stable aircraft. It also shows that below a certain AOY (typically around 12–15°) the C_n is linear, but beyond becomes nonlinear curves associated with the post-stall AOYs.

As discussed in Section 11.2.8, *Requirements for static directional stability*, the slope of the C_n curve must be positive as shown in Figure 23-20 for the aircraft to be directionally stable. This is indicated by the magnitude of the stability derivative $C_{n\beta}$, i.e. the *directional stability*. The required size of the rudder should be based on the investigation of the capability of the tail at the extreme yaw conditions stipulated in Sections 23.3. Such an investigation should evaluate the AOY the airplane can be trimmed to with a given rudder deflection over the entire speed range of the aircraft and take into account the extremes of the CG envelope. The sizing should ensure the required rudder deflection is less than what is available, per Section 23.3.

23.3 GENERAL AVIATION AIRCRAFT DESIGN CHECKLIST

The purpose of the following checklist, which is compiled based on the author's own experience, is to

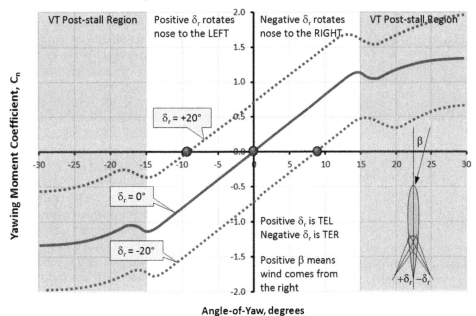

FIGURE 23-20 Yaw control explained. VT stands for vertical tail, TEL for trailing edge left, and TER for trailing edge right.

help the designer of new aircraft steer away from likely pitfalls during the design process. The designer should try and meet all the capabilities listed below, most of which are intended to increase the safety of the airplane. The list is based on aircraft that boast a long and safe operational history. Any airplane that meets the requirements below will provide ample controllability and good handling. Note that many of the concepts are introduced in various chapterss of the book, but are collected here for convenience.

23.3.1 Crosswind Capability at Touch-Down

The airplane should be capable of maintaining straight and level[5] flight while yawed 12–15° (β) at an airspeed amounting to $1.2 \times V_{S0}$, where V_{S0} is the stalling speed in the landing configuration, with sufficient power to maintain the airspeed. This corresponds to a crosswind component of approximately:

$$V_W \approx 1.2 \cdot V_{S0} \cdot \tan(15°) = 0.322 \cdot V_{S0} \quad (23\text{-}15)$$

How to Assess

This capability should be evaluated by a trim analysis at $\beta = 15°$ and $V = 1.2 V_{S0}$, with the CG in the most adverse location and with power required to establish a 3° approach angle. The control surface deflection required to trim the airplane should comply with the following:

(1) Aileron deflection, $\delta_a < 0.25\, \delta_{a\ max}$.
(2) Elevator deflection, $\delta_e < 0.50\, \delta_{e\ max}$.
(3) Rudder deflection, $\delta_r < 0.75\, \delta_{r\ max}$.

A factor like 0.75 is intended to leave 25% deflection capability remaining to allow the pilot to respond to gusts. If met, this requirement prevents the control surfaces from being undersized.

23.3.2 Balked Landing Capability

The airplane should be capable of climbing in the landing configuration with full power at the reference airspeed and the CG at the most adverse location (forward for conventional aircraft). At this condition the dynamic pressure is low, which will require large control surface deflection. If met, this requirement prevents the HT from being undersized (also see Section 23.3.3, *Take-off rotation capability*). Paragraph 14 CFR 23.73 defines reference airspeed as follows:

Normal, Utility, and Aerobatic piston-powered aircraft with $W_0 \leq 6000$ lb$_f$:

$$V_{REF} = \text{the larger of } \begin{cases} 1.3 \cdot V_{S1} \\ V_{MC} \end{cases} \quad (23\text{-}16)$$

---

[5]This means constant heading and altitude.

Normal, Utility, and Aerobatic piston- and turbine-powered aircraft with $W_0 > 6000$ lb$_f$:

$$V_{REF} = \text{the larger of} \begin{cases} 1.3 \cdot V_{S0} \\ V_{MC} \end{cases} \quad (23\text{-}17)$$

Commuter category aircraft with $W_0 > 6000$ lb$_f$:

$$V_{REF} = \text{the larger of} \begin{cases} 1.3 \cdot V_{S0} \\ 1.05 \cdot V_{MC} \end{cases} \quad (23\text{-}18)$$

where

V_{S0} = stalling speed in the landing configuration.
V_{S1} = stalling speed in the take-off configuration (max T-O flap setting).

How to Assess

The capability should be evaluated by a trim analysis. It must be possible to trim the aircraft at V_{REF} and it must be capable of a 3% climb gradient while at:

(1) maximum power,
(2) at the forward end of the proposed CG envelope,
(3) in the most adverse configuration (typically the landing configuration).

Control system stretch must be taken into account (see Section 23.3.13, *Control system stretching*).

23.3.3 Take-Off Rotation Capability

The aircraft must offer a large enough elevator authority to allow the airplane to be rotated at airspeeds below the liftoff speed, V_{LOF}, with the CG in the most adverse location (forward for conventional aircraft). If met, this requirement (also see Section 23.3.2, *Balked landing capability*) will prevent the HT from being undersized for elevator authority.

How to Assess

This capability can be evaluated using the formulation required to comply with Section 13.3.4, *Tricycle landing gear reaction loads*. Also, see Example 13-1. Strictly speaking, the analysis should demonstrate that the load on the nose landing gear can indeed be brought to zero before the airplane reaches its lift-off airspeed, V_{LOF}.

23.3.4 Trim at Stall and Flare at Landing Capability

The aircraft must be designed with enough elevator authority to drive it to the desired stalling speed in the most adverse configuration as detailed below. The stalling speed is the lowest airspeed at which the airplane can maintain level flight. If the airplane has insufficient elevator authority it will not experience a normal "stall-break" (i.e. the sudden nose-drop that indicates the stall event). Instead it will descend rapidly at airspeed higher than the stalling speed. As far as aviation authorities are concerned, this becomes the airplane's minimum airspeed. This can result in a serious design predicament, especially if the aircraft is designed to comply with regulations such as 14 CFR Part 23, which has a stall speed limit of 61 KCAS for single-engine aircraft, or Light-Sport Aircraft regulations, which have a stall speed limit of 45 KCAS (14 CFR 1.1).

Another easily overlooked characteristic is the ability for the pilot to flare the aircraft in the most adverse configuration during touch-down. There are examples of aircraft that have been certified with insufficient elevator authority. Such airplanes pose a serious risk to unsuspecting pilots. It is the purpose of regulatory authorities to ensure such aircraft don't slip through the cracks. Any certified aircraft discovered to suffer from insufficient elevator authority should have a forward CG limitation imposed on it.

Strictly speaking, the stall speed is the airspeed below which the airplane's nose pitches down uncontrollably. When we say trim at stall (see below), we mean the airspeed just above the stall speed.

How to Assess

It must be possible to trim the aircraft at V_{S0} while at:

(1) minimum power,
(2) at the forward end of the proposed CG envelope,
(3) in the most adverse configuration (typically the landing configuration).

Control system stretch must be taken into account (see Section 23.3.13, *Control system stretching*).

23.3.5 Stall Handling Capability

The aircraft should be designed with forgiving stall characteristics. This means it should offer natural resistance to wing roll at stall (which often leads to inadvertent spins) and be devoid of deep stall tendencies. The designer can tailor the wing to help promote resistance to roll off (see Section 9.6.4, *Tailoring the stall progression*). This can be accomplished with geometric wing twist, or washout, but also by selecting a high-lift airfoil as a tip airfoil, even at the cost of cruise performance. While such tailoring is debatable for aircraft such as high-speed passenger airplanes equipped with stick-pushers or shakers, it should be the norm for General Aviation aircraft. Deep stall tendency is prevented by a proper location of the horizontal tail. T-tails are prone to this condition; however, they are often desirable due to stylish appearance or some other reasons, such as to introduce end-plate effects for directional stability or lengthing the tail arm of short fuselages. Such airplanes should be tested to validate that the condition does not exist and, if it does, they should be fitted with adequate ventral fins to remedy it.

How to Assess

Roll-off at stall — provide adequate wing washout, even if it means a slight detriment to cruise efficiency. Consider an airfoil at the wingtip that has higher C_{lmax} than the inboard airfoils, even if this calls for an airfoil that has a turbulent boundary layer. Refer to Section 9.6.4, Tailoring the stall progression for more details.

Deep-stall condition — demonstrate compliance in a wind tunnel (or by other reliable and acceptable means) that no trim points exist in the post-stall region, in particular where elevator authority is diminished due to flow separation.

23.3.6 Stall Margin for Horizontal Tail

Airplanes of conventional tail configuration may suffer from a horizontal tail stall if subjected to extreme wing downwash conditions, such as those that result from the deployment of high-lift devices. This can increase the angle-of-attack on the stabilizing surface so it stalls. This may cause nuisance such as a broken nose gear and ground-struck propeller, if the HT stalls during flare, to dangerous in-flight handling problems that may cause the airplane to pitch over, nose-first, when flaps are deployed.

How to Assess

During the design phase perform an analytical stability check on the airplane in its most adverse condition, with full flaps and most adverse forward CG, and estimate the spanwise distribution of section lift coefficients along the HT. This check should be accomplished using a vortex-lattice or doublet-lattice solver as the minimum and, ideally, a Navier-Stokes solver or a wind tunnel test. It is not enough to calculate the total lift coefficient required by the HT as this will not fully indicate how far from this value the section lift coefficients deviate. For instance, the total lift coefficient generated by the HT in Figure 23-21 amounts to about 0.95, while the section lift coefficients peak at almost 1.1. The maximum section lift coefficient on the control surface should be no more than about 0.2 from the airfoil's maximum lift coefficient in the worst-case scenario.

23.3.7 Roll Authority

The airplane must meet the desired or the most appropriate roll capability that suits the class the airplane is being designed for. This will prevent the ailerons from being undersized. An aerobatic airplane will have a steady-state roll rate exceeding 130°/sec at its cruising speed and sometimes in excess of 360°/sec. A responsive GA aircraft can roll as fast as 90°/sec and a sluggish one at perhaps around 30°/sec. Roll requirements are stipulated in 14 CFR 23.157, *Rate of Roll*. The paragraph is split into two parts: take-off and approach for landing, as follows.

Take-Off Requirement

The airplane must have flaps in the T-O position and retractable landing gear should be in the stowed position. Single-engine aircraft must be at maximum T-O power. Multiengine aircraft must have the critical engine inoperative (see Chapter 14, *The anatomy of the propeller*) and the others at the maximum T-O power. The airplane shall be trimmed at airspeed equal to the greater of

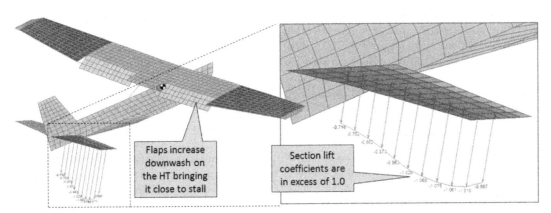

FIGURE 23-21 A vortex-lattice model of a high-wing aircraft with an all-movable stabilator reveals the section lift coefficients on the stabilator are close to their maximum values. This corresponds to a condition in which the airplane is being flared for landing. Airspeed is 55 KCAS, AOA is 10°, flaps are deflected 30° and the stabilator 15°.

$1.2V_{S1}$ or $1.1V_{MC}$ or as close as possible in trim for straight flight. In this condition, the airplane must be capable of rolling from $-30°$ to $+30°$ (or the other way around) within the following time constraints:

$$t = \begin{cases} 5 \text{ sec} & \text{if } W \leq 6000 \text{ lb}_f \\ \left(\dfrac{W+500}{1300}\right) \text{ up to 10 sec} & \text{if } W > 6000 \text{ lb}_f \end{cases}$$

(23-19)

Approach Requirement

The airplane must have flaps in the landing position and retractable landing gear must be extended. All engines must be operating a power setting for a $3°$ approach angle. The airplane shall be trimmed at V_{REF}. In this condition, the airplane must be capable of rolling from $-30°$ to $+30°$ (or the other way around) within the following time constraints:

$$t = \begin{cases} 4 \text{ sec} & \text{if } W \leq 6000 \text{ lb}_f \\ \left(\dfrac{W+2800}{2200}\right) \text{ up to 7 sec} & \text{if } W > 6000 \text{ lb}_f \end{cases}$$

(23-20)

How to Assess

During the conceptual or preliminary design phase, the designer should evaluate the aileron authority derivative (see Section 23.2.2, *Fundamentals of roll control*) and calculate the steady-state roll rate at the high- and low-speed ends of the flight envelope (see Section 19.2.11, *Flight envelope*). This should then be compared to the requirements of 14 CFR 23.157 to evaluate whether the design will comply. It is imperative not to overlook the fact that the aileron span cuts into the span of the flaps. Therefore, if the solution to a potentially sluggish design is to add aileron span, the impact on the stalling speed of the vehicle should be evaluated as well. However, the evaluation should not stop there. The designer should calculate the aileron authority derivative for other aircraft as well for comparison; in particular, aircraft that are known to be considered sluggish by its pilots. Sluggishness should be designed out of the aircraft during the conceptual design phase — failing to do so may require difficult design fixes once the proof-of-concept aircraft demonstrates lack of responsiveness.

23.3.8 Control System Harmony

Design the control system so the harmony in control forces will be aileron:elevator:rudder = 1:2:4. This means that the effort required to actuate the elevator will be about 2 times that of the aileron and the rudder about 4 times that of the aileron. This calls for awareness of how to estimate hinge moments of control surfaces and control system gains.

23.3.9 Climb Capability

With all engines operating, the airplane must have enough power to maintain the following steady climb gradients (also see Section 18.3, *General climb analysis methods*, on how to calculate V_Y) at sea level:

Normal, Utility, and Aerobatic piston-powered aircraft with $W_0 \leq 6000$ lb$_f$:

Landplanes:

$$ROC > 0.083 \dfrac{V_Y}{60} \times 6076 = 8.405 V_Y$$

(23-21)

Seaplanes and amphibians:

$$ROC > 0.067 \dfrac{V_Y}{60} \times 6076 = 6.785 V_Y$$

(23-22)

where V_Y = best rate of climb airspeed in KTAS.

These are legal minimums[6] at maximum continuous power (MCP) conditions at sea level. The airplane may have landing gear retracted and flaps in the take-off position. The value of V_Y for multiengine airplanes shall be greater or equal to the greater of $1.1V_{MC}$ and $1.2V_{S1}$. For single-engine aircraft V_Y shall be greater or equal to $1.2V_{S1}$.

Normal, Utility, and Aerobatic pistons and turbine powered aircraft with $W_0 > 6000$ lb$_f$:

Landplanes:

$$ROC > 0.040 \dfrac{V_Y}{60} \times 6076 = 4.051 V_Y$$

(23-23)

These are legal minimums at take-off power at sea level. The airplane shall have landing gear extracted unless it can be retracted in fewer than 7 seconds. Flaps shall be in the take-off position.

How to Assess

Demonstrate through analysis, assuming conservative conditions, by evaluating the capability on a hot day (ISA+20°C) at S-L and diminished propeller efficiency or jet engine thrust. Although not a direct requirement, multiengine aircraft should be evaluated in the take-off condition. Consider redesign if the condition is marginally met, including more powerful engines.

---

[6]Per 14 CFR, Part 23.

23.3.10 One-Engine-Inoperative Trim and Climb Capability

Multiengine aircraft, in particular twins, must offer safe trim capability and enough power to climb with one engine inoperative (see Chapter 14, *The anatomy of the propeller*, for important details). Such aircraft must be capable of straight and level flight (albeit with control surfaces deflected).

How to Assess

During the development phase of the aircraft, this capability can be demonstrated through analysis using the analyses methods presented in this book; for instance those of Chapter 18, *Performance – climb*, using reduced engine power, increased drag coefficient (due to asymmetry), at gross weight, and on a hot day (ISA+20°C). Consider redesign if the condition is marginally met, including more powerful engines.

23.3.11 Natural Damping Capability

The airplane should be designed such that dynamic stability modes are naturally damped. Diverging non-oscillatory modes may be acceptable, albeit undesirable, provided the time to double amplitude is long enough to allow the pilot to effortlessly correct them. However, oscillatory modes should always be convergent.

How to Assess

Perform a detailed dynamic stability analysis of the airplane in all operational configurations (take-off, cruise, and landing) over the operational flight envelope of aircraft. All oscillatory modes should be convergent, as required by applicable airworthiness regulations. Identify regions of divergent dynamic modes, if these exist, and consider the introduction of aerodynamic fixes or limitations of the flight envelope. Ultimately, these can be lifted if findings are rejected by flight test work.

23.3.12 Fuel Tank Selector

Airplanes equipped with multiple fuel tanks and a fuel tank selector should always place it inside the cockpit in the field of view of the pilot. Never place it so the pilot has to look away or "feel" to figure out the currently selected fuel tank. A large number of aircraft crash every year because of fuel starvation attributed to a pilot running out of fuel after "thinking" he had selected the correct tank. Also, when flying by Instrument Flight Rules (IFR), vertigo can result if the pilot is forced to turn away from the instrument panel to view the fuel selector. The convenience of the pilot takes precedence over the convenience of the engineer or the technician routing or installing the fuel selector valve.

How to Assess

When designing the cockpit ensure the fuel tank selector is inside the pilot's field of view and range of reach. This important control should be treated as if it was between the primary (ailerons, elevator, rudder, throttles, etc.) and secondary controls (flaps, trim controls, speed brakes, etc.) in importance. The detail may have to be developed using a cockpit mockup.

23.3.13 Control System Stretching

As the primary controls (ailerons, elevator, and rudder) are deflected in flight, the aerodynamic loading will stretch cables and loads will be reacted by pulleys and other parts of the control system. This can result in a substantially smaller deflection than what is measured on the ground (see Figure 23-22). The consequence is a serious reduction in control responsiveness.

Some control systems are so poorly designed that perhaps some 25% of the ground deflection is available in flight. The result is sluggish responsiveness, if not outright dangerous handling characteristics. Consider an airplane initiating a flare maneuver just before landing at a forward CG. This maneuver may require, say, 10° of elevator deflection. A flexible control system could only result in, say, 5°, with the stick fully aft. An unsuspecting pilot might discover this at the time of touch down and the result may be a broken nose landing gear if not worse. This would most probably be discovered in the flying prototype, where a mishap could jeopardize the success of an otherwise promising design.

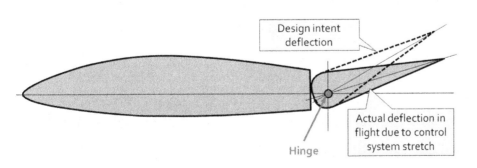

FIGURE 23-22 The impact of control system stretch.

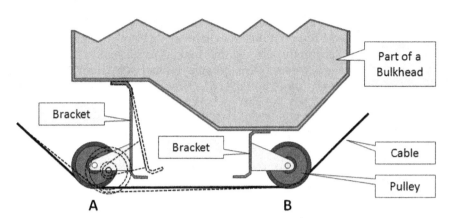

FIGURE 23-23 Control system loads cause the cable-pulley system to flex excessively, rendering it ineffective.

A typical design mistake is displayed in Figure 23-23. A couple of pulleys, called A and B, have been mounted to brackets (or intercostals) that are attached to a bulkhead. The bracket for pulley A is considerably longer than for pulley B. As a consequence, when the control system is loaded up, pulley A will be displaced considerably, which will cause cable slack, which will results in a less control surface deflection. It is imperative that the control system designer is aware of such pitfalls so they can be avoided.

How to Assess

Always perform a load analysis of the control system. In particular, evaluate reaction forces in hardpoint areas, where pulleys are mounted to the airframe. Knowing the reaction forces at each hardpoint, determine the deflection of the structure to which the pulleys are attached. With this information estimate the total deformation of the control system. For instance, consider an elevator control system designed to deflect to a maximum deflection of 20° on the ground. The pilot should be capable of deflecting it to at least 15° with maximum aerodynamic loads applied. If unable to, the control system is improperly anchored to the airframe.

23.3.14 Control System Jamming

Control system jamming is a serious threat in brand new airplane designs. It is easy to overlook how control surfaces could get stuck (or jam) as they flex due to air loads, or how the deformation of the control system might get cables and pushrods into positions that might get things stuck.

How to Assess

Conduct a design review with cognizant engineers fully dedicated to looking at possible control system jamming scenarios. Do not dismiss possible scenarios that seem "outlandish." It is those scenarios that seem unlikely that may manifest themselves in practice. Get fresh eyes to look at the design as well. Be critical.

23.3.15 Ground Impact Resistance

Airplanes with forward-facing firewalls should feature a canted firewall (see Figure 23-24). This may prevent the aircraft from "digging in" during a possible nose-down-attitude crash and improve the vulnerability of the occupants that might result from the sudden deceleration from an otherwise survivable mishap.

How to Assess

Design a forward-facing firewall so its lower edge is canted aft as shown in Figure 23-24. Awareness is the key and will render the topic a non-issue if this is considered during the conceptual/preliminary design phase.

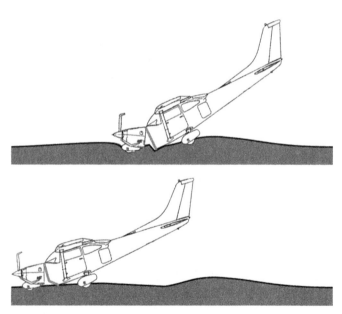

FIGURE 23-24 The difference between canted and uncanted firewalls is improved survivability.

23.3.16 Reliance Upon Analysis Technology

The current advances in computer analysis methods present a particularly serious pitfall for the aircraft designer. Extremely compelling images generated by such software make it easy for even the most seasoned designer to forget to second-guess the results. However, such results may be way off. Figure 23-25 shows an example of two computational methods as compared to an experiment for the NACA 4415 airfoil as reported in the report NACA R-824 [5]. One uses the widely used code Xfoil and the other is the vortex-lattice code SURFACES. Two important observations can be made by studying the figure.

First, inviscid computational methods such as the vortex-lattice method (VLM) do not predict flow separation. While the VLM results are in good agreement with the experiment, this is only true in the linear range. The speed and flexibility of the VLM make it a great tool for the aircraft designer, but only as long as the airflow is mostly attached. For instance, in this particular example, the VLM results are acceptable over the AOA range of $-12° < \alpha < 10°$.

Second, while Xfoil predicts a "gentle" stall behavior as shown in the experiment, its predictions are off in slope, and both C_{Lmax} and α_{CLmax} are too large, just to name a few. The important point is that it is easy for the aircraft designer to take such predictions at face value and, in this case, greatly underestimate wing area. The best advice is to check results for similar airfoils first and fully understand the limitations of the computational method. For this particular example, it is possible the input data was inadequate or some other explanations apply. Perhaps the user did not select the proper options in the program. These words are not intended to judge the software itself, but rather the operator. Just because a computational method offers viscous approximation does not mean the method is bulletproof.

Figure 23-26 shows stall progression predictions by two Navier-Stokes solvers. The left wing half was meshed using a structured grid, and the right one using an unstructured grid. The image shows flow separation patterns at an AOA of $18°$. The disparity in predicted pattern is clearly evident. How do you know which prediction better resembles reality? Neither? Obviously it is impossible to say without experiment; however, an aerodynamicist armed with only one solver has no choice but to rely on it — often erroneously. Be careful — your development program may be at stake.

Always check the results of any computer analysis software by conducting validation tests. When dealing with aerodynamics design, *validation is only acceptable when it consists of comparison to actual wind tunnel tests.* When possible, select wind tunnel test results for geometry that most closely resembles the one being analyzed. If such wind tunnel test results cannot be found, select a different geometry as long as it is trustworthy. When dealing with structural tests, validation is only acceptable when compared to actual load tests. Do not compare one analysis method to another one,

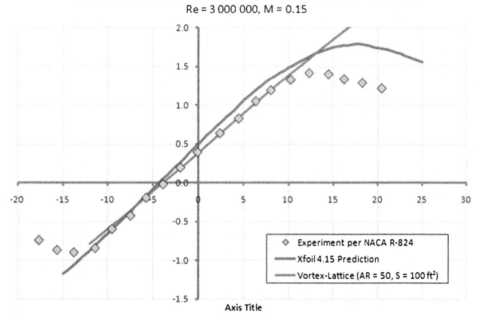

FIGURE 23-25 Experimental lift curve compared to two computational techniques.

FIGURE 23-26 Two dissimilar predictions of stall progression on a wing design. The left side was predicted using a structured mesh and the right using an unstructured mesh. The left and right sides are geometric mirror images. The left side ended up better resembling experiment and had favorable stall characteristics. AOA is 18°.

unless it is validated this way. Such comparison will build trust in software and expose the limitations of the methods. Only an understanding of the limitations of a particular computational method will lead to a proper use.

23.3.17 Weight Estimation Pitfalls

The most common pitfall in the estimation of weight is underestimation. This usually stems from the omission of systems whose weight is poorly defined, but sometimes simply from a human trait — optimism. This is a peculiar human condition that, if left unsupervised, can lead to trouble. Inevitably, an inexperienced designer asked to guess the weight of a component inside some weight range is likely to pick the lower end of the range if the component is destined for use in his own design. Thus, if a component is estimated to be within a weight range of, say, 80 to 120 lb_f, the designer is likely to assume it will weigh 80–100 lb_f in his aircraft. Of course, this is less of a problem among experienced designers, but a possible pitfall it is nevertheless. The remedy? Realism enforced by thorough checks by people other than the weights engineer.

23.3.18 Drag Estimation Pitfalls

There are two common pitfalls when estimating drag: over- and underestimation. Drag is notoriously hard to estimate correctly. Overestimation may make an otherwise promising idea appear as a slug on paper and lead to the termination of a project before it even begins. Underestimation can lead to a devastating disappointment once the airplane flies, if not program cancellation and possibly a financial catastrophe. It is often caused by significant protrusions being "left out" of the calculations during the conceptual design phase (for instance see Section 15.5.1, *Cumulative result of undesirable drag (CRUD)*). The remedy? Ensure the drag estimation is thoroughly checked, if not performed, by experienced people, although this does not always suffice. Estimate the drag of a number of airplanes that appear similar to the one being designed. If the drag estimate of the design deviates greatly from those this should raise a flag. While this discussion pertains to the state of the project before a wind tunnel model is even considered, it is possible the right answer won't emerge until it has been conducted.

23.3.19 Center of Gravity Travel During Flight

Most airplanes experience a movement of the center of gravity (CG) during flight. This is most often caused by the fuel being consumed, but sometimes by mass being purposely dropped as a part of a mission; for instance a military aircraft dropping ordnance, or an aerial firefighter dropping water, or parachutists exiting. The designer should aggressively plot the CG travel as a function of such weight reduction and ensure that at no time will it travel beyond the prescribed CG limits. An airplane with a wide CG limit will be far more forgiving than one with a narrow one. This is discussed in more detail in Section 6.6.13, *In-flight movement of the CG*.

23.3.20 Wing/Fuselage Juncture Flow Separation

The juncture where the wing joins the fuselage can often bring a surprising increase in drag. It is one of those areas often left to flight testing to evaluate, but by then it might be too late and require expensive design changes. Established companies will spend resources on developing proper wing root fairings; however, the less established ones will often wait until flight tests to assess the severity of the flow separation in the juncture. The initial geometry of the fairing can be designed keeping some straightforward rules of thumb in mind, although ultimately a Navier-Stokes solver or wind tunnel testing is required to complete the shape, as the flow field in this region is simply too complicated to assess by other means.

23.4 FAULTS AND FIXES

While aircraft design is clearly more Science than Art, the combination of the two sometimes plays a profound role in the final product. This can be seen in many decisions that are made during the design process that are biased by contributing influences that range from what is perceived to constitute good looks to improved ergonomics of operation. As a consequence, once operational, the complete geometry often displays characteristics that are undesirable. Usually, such undesirables are not severe enough to terminate the design project, but nevertheless require aerodynamic fixes that help eliminate the undesirable characteristic. This section will address common faults and fixes for various detrimental characteristics and is intended to help the practicing engineer find solutions to possible challenges.

Naturally, some flaws that lead to detrimental characteristics have nothing to do with the interplay of art and science, but are rather a direct consequence of the nature of air, which separates at high AOAs or forms shockwaves when airspeeds are high. We have already seen that some airplanes are designed to operate over a wide range of airspeeds, from mid 100 knots to a couple of times the speed of sound. Such airplanes are presented with serious design challenges that call for the introduction of complex geometry to allows them to operate safely and effectively at either end of the airspeed spectrum. Other aircraft are larger and heavier derivatives of previous aircraft and in order to keep manufacturing costs down use the same external parts (e.g. vertical or horizontal tail). This may lead to the discovery of undesirable flight or handling characteristics at some speed or altitude, which are fixed with the addition of some external geometry.

23.4.1 Stability and Control — Dorsal Fin and Rudder Locking

As stated in Section 11.2.11, *The dorsal fin*, the primary reason for the installation of a dorsal fin is to prevent *rudder locking*, a phenomenon that may occur if the yaw angle of the aircraft becomes too great. The phenomenon is discussed in great detail in the section.

23.4.2 Stability and Control — Ventral Fin and Deep Stall

As stated in Section 11.2.12, *The ventral fin*, there are two kinds of ventral fins: for deep stall or for Dutch roll. The T-tail configuration of the Lockheed F-104 Starfighter was intended to increase its effectiveness through an endplate effect (Whitford [6]) but other airplanes feature T-tails for other reasons, for instance aesthetics. Some of those require a ventral fin to fix a deep stall tendency as a consequence of the design decision. Reynolds [7] presents a good discussion of the development of ventral fins on the Learjet 55. Figure 23-27 shows the ventral fins on a Learjet 60.

FIGURE 23-27 The ventral fin on this Learjet 60 adds longitudinal stability that prevents it from entering deep stall. *(Photo by Phil Rademacher)*

23.4.3 Stability and Control — Ventral Fin and Dutch Roll

A ventral fin can also be introduced to increase the vertical or side silhouette of an airplane for the purpose of improving its Dutch roll damping or increase the directional stability of a multiengine aircraft when operating with one engine inoperative (OEI). Ventral fins intended for this purpose are installed in a far more vertical position than those intended to fix deep stall.

23.4.4 Stability and Control — Forebody Strakes

As was discussed in Section 23.3, *Preliminary aircraft design checklist*, conventional sizing of the vertical tail is contingent upon critical asymmetric flight conditions: cross-wind landings, balked landing, one-engine-inoperative trim capability, and many others. The resulting fin size can be large and increases both drag and airframe weight. The large tail can, therefore, be costly for transport aircraft in terms of fuel costs. Consequently, alternative methods that do not increase drag are attractive and of great interest to the aircraft designer. Additional benefits of such methods are realized when time comes to "stretching" or lengthening the fuselage of an existing design in order to develop a variant aircraft that can carry more passengers or payload. Such changes are frequent in the commercial aircraft industry. Extending the fuselage will destabilize the aircraft, longitudinally and directionally, and can easily call for enlarged stabilizing surfaces.

Research on forebody strakes has been active for a long time. One of the best-known efforts was run by NASA's Dryden Flight Research Center, Edwards, CA, in which an originally retired F-18 Hornet fighter was restored and equipped as a *high angle-of-attack research vehicle* (HARV) in a flight research program lasting from April 1987 until September 1996 [8]. The program focused on AOA well above 30° and identified the shape of the nose as a key player in the lateral stability of the test vehicle at high AOAs [9]. The resulting cross-flow on the forebody is considerable and the resulting asymmetry in pressure on the nose can cause a significant side force. This force acts across a large distance between the forebody and the CG, creating a large yawing moment. The effort successfully demonstrated that directional control could be exerted using deployable strakes installed in the nose. This research led to a cooperation between NASA Langley Research Center and McDonnell Douglas Corporation (now Boeing) to develop strake technology for use with transport aircraft. Of course, such aircraft operate at much lower AOAs, something closer to 8° during approach for landing, so a different kind of strake had to be developed.

Shah and Granda [10] studied the use of forebody strakes to improve the directional stability of a wind tunnel model of a generic commercial jetliner. This was done as part of a cooperative research between NASA and Boeing referred to as Strake Technology Research Application to Transport Aircraft (STRATA). The strake is a flat plate that is mounted on the forward part of the fuselage, of which an example can be seen in Figure 23-28 and Figure 23-29. These were mounted to the 50-series of the McDonnell-Douglas DC-9 when it was stretched from the 30-series aircraft. These were also installed on the DC-9-80, also known as the MD-80. Their findings can be summarized as follows:

1. A set of baseline strakes similar in planform and location to those on the MD-80 aircraft improved the model's static directional stability. As *AOA* increased, the *AOY* at which the strakes became effective decreased. This indicates that a critical level of fuselage cross-flow is required for the strakes to become effective.
2. It was found that it is the leeward strake that alters the flow-field and improves stability; the windward strake has negligible effect.
3. The improvement in directional stability is approximately proportional to the span of the strake.
4. The effect of strake chord on stability is non-linear; most of the effectiveness above stall can be achieved with a strake with a small chord.
5. A nose-up incidence of the strake has minimal effect on longitudinal stability; negative incidence has a non-linear and degrading effect.
6. The effectiveness of single or differentially deflected strakes as a directional control device was small and very nonlinear; however, only a limited investigation was conducted in this area.

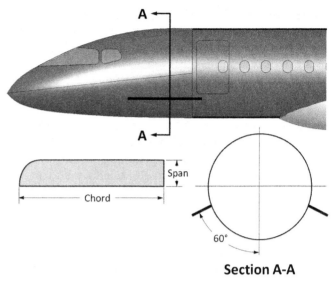

FIGURE 23-29 Forebody strake geometry and positioning.

7. The forebody strakes are de-coupled control effectors — very little impact on static longitudinal and lateral characteristics was seen.

23.4.5 Stability and Control – Taillets and Stabilons

Taillets and *stabilons* are terms coined for the modification made to the twin-engine Beech 1900 commuter aircraft (see Figure 23-30). The design dates back to the late 1970s and early 1980s, developed from the Beech Super King Air to compete with the Swearingen Metro and the British Aerospace Jetstream. It accommodates 19 passengers in a pressurized cabin and is powered by two 1100 SHP Pratt & Whitney PT6 turboprops driving four-blade propellers, capable of 235 KTAS at 25,000 ft. The Beech 1900D is a variant of the 1900 with an enlarged fuselage that offers a stand-up cabin. This was accomplished without changing the dimensions of the original tail.

The increase in drag required more powerful engines, which, combined with the reduction in directional stability due to the larger fuselage, called for the addition of ventral fins and taillets. Each taillet adds about 1.67 ft^2 of area to the total vertical stabilizing area and improves Dutch roll damping. This prevents the airplane's yaw damper from being a non-dispatch critical component [11].

The stabilon refers to a couple of 7.75 ft^2 horizontal ventral fins that are mounted to the lower aft part of the fuselage (again see Figure 23-30). These increase longitudinal stability and allow the center of gravity to be farther aft than otherwise, making the airplane less sensitive to passenger and baggage loading. According to the source [11] (which cites Beech sources) the stabilons

FIGURE 23-28 A forebody strake is mounted to the nose section of this DC-9 commercial jetliner. *(Photo by Phil Rademacher)*

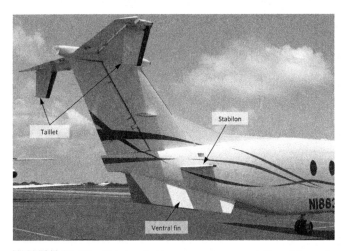

FIGURE 23-30 This is what happens when an existing airframe is adapted to modified requirements. A Beech 1900 commuter boasts an assortment of aerodynamic fixes: taillets (small vertical surfaces on each stabilizer), ventral fins (bottom of fuselage), and stabilons (above the lower ventral fins). *(Photo by Phil Rademacher)*

increase the aft CG limit from 32% MGC to 40%. They also improve "deep" stall characteristics (see discussion in Section 11.2.12, *The ventral fin*).

23.4.6 Stability and Control − Control Horns

Control horns are extensions that are placed on control surfaces, such as ailerons, elevators, and rudders, for reducing aerodynamic hinge moments. This is accomplished by adding area of the surface ahead of the hingeline. Control horns come in many sizes and shapes. There are two common types: *unshielded* and *shielded*. Shielded control horns are recommended for airplanes that are certified to fly into known icing (FIKI − flying into known icing) as they do not accrete ice like control horns that are more directly exposed to the elements.

23.4.7 Stall Handling − Stall Strips

A stall strip is a small triangular strip made out of metal or rubber that is bonded to the leading edge of a wing to help control the stall characteristics of an airplane (see Figure 23-31). The strip can be seen on the leading edges of most manufactured aircraft, as they must comply with regulations such as 14 CFR Part 23.201, Wings level stall.

The difficulty in maintaining production tolerances effectively renders all serial numbers of a given aircraft type similar and not precisely the same. Yes, all Cessna 172 aircraft look so similar to the untrained eye that they may appear identical. However, in fact, when compared to the intended outside mold line (OML) they all deviate in one way or another, albeit in minute ways. The same holds for all other mass-produced aircraft. For this reason, when airplanes come off the production line, they are test flown before delivery by the production flight test team and stalled to confirm stall behavior. Then, based on the observation of the test pilot, stall strips are bonded in a specific location on the leading edge.

Stall strips are detrimental in that they may cause an early separation even at climb *AOA*. This will be manifested as a reduction in climb performance. They can also be detrimental on NLF wings as they may destabilize the boundary layer aft of their location, causing an earlier transition than otherwise.

There are primarily two positions the aerodynamicist needs to be aware of and are important to a properly located stall strip. *spanwise station* and *clocking* (see Figure 23-32). During the development phase of a new aircraft, the location of stall strips on the wing becomes a process of trial and error. The experienced aerodynamicist will create a test matrix with potential locations and clocking positions and have the test pilot stall the aircraft and note change using terms like "good," "bad,"

FIGURE 23-31 Stall strips on a Cessna Citation X and Diamond DA-42 Twinstar aircraft. *(Photo by Phil Rademacher)*

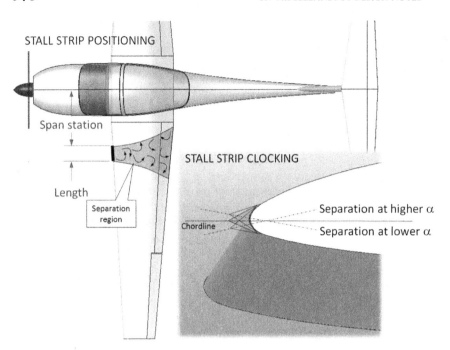

FIGURE 23-32 Placing a stall strip on the wing is a trial and error endeavor. Proper positioning is essential.

"improved," or similar. The essential task is to use information from each stall to its utmost as each flight is costly. Using CFD may help, provided there is evidence to support it correctly predicts flow separation. Such evidence should comprise pictures of the flow field and separation area as seen using a tufted wing and subsequent comparison to CFD predictions. In the absence of such evidence, CFD should be used with care great or not at all — incorrect predictions may simply muddy the waters.

23.4.8 Stall Handling — Wing Fence

A *fence* is a vertical panel mounted to the upper surface of swept-back wings (see Figure 23-33). The invention is attributed to research from 1937 by the German aerodynamicist Wolfgang Liebe at the Deutsche Versuchsanstalt für Luftfahrt [12, p. 224] (or the DVL — The German Aeronautical Test Establishment). This work was carried out in order to help remedy dangerous wing roll-offs during stall of aircraft such as the Heinkel He 70 and Messerschmitt Me 109.

Early versions of the fence extended from the leading to the trailing edge, while the more modern configuration extends a relatively short distance back. The wing fence is a less common solution today than before and is mostly found on some business jets (e.g. Cessna Citation III, Beechcraft 400 Beechjet, Learjet 35, and Gulfstream II). Earlier commercial types featuring wing fences include the Boeing 727 (see Figure 23-33), Comet 1, Sud-Est Caravelle, and Tupolev Tu-154, to name a few. Wing fences can also be found on many fighter aircraft, such as the MiG-15, North American F-86 Sabre, and even relatively recent aircraft such as the Hawker Hawk (US designation T-45 Goshawk), Fiat G91 and SIAI-Marchetti S211. It can even be found on GA aircraft such as the German Akaflieg Braunschweig SB-13 sailplane and the Australian Eagle 150 tandem wing aircraft.

Fences are sometimes referred to as "boundary-layer fences." However, a fence can serve in a number of important ways:

(1) It modifies the distribution of section lift coefficients, delaying wingtip stall. This can be seen as the reduction in the magnitude of the distribution of section lift coefficients in Figure 23-34.

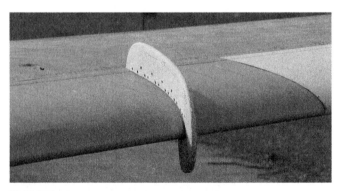

FIGURE 23-33 The wing fence of a Boeing 727 commercial jetliner. *(Photo by Phil Rademacher)*

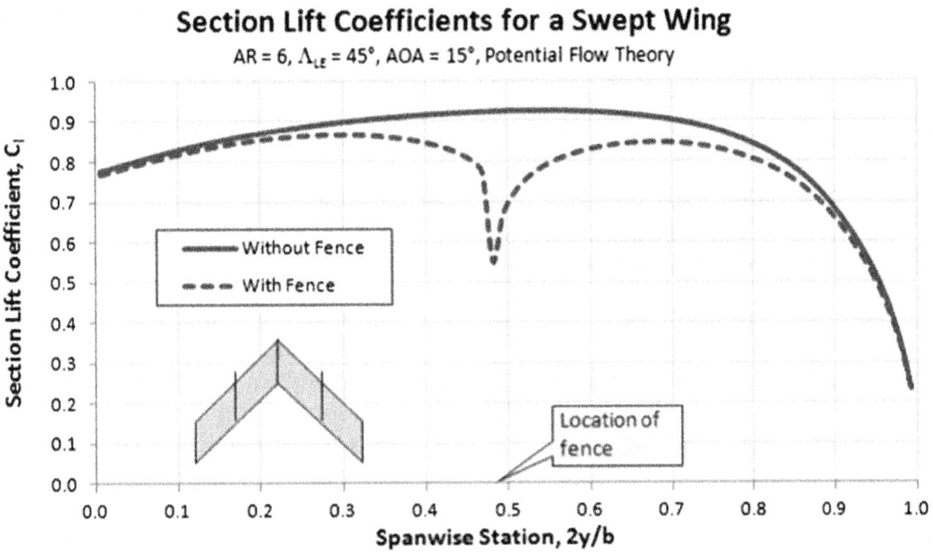

Section Lift Coefficients for a Swept Wing

AR = 6, Λ_{LE} = 45°, AOA = 15°, Potential Flow Theory

FIGURE 23-34 While the effect of the wing fence is highly viscous, it also affects the flow field in a manner that can be approximated using inviscid potential flow theory. Here, the reduction in section lift coefficients mid-span and outboard is clearly visible, helping to delay tip stall.

(2) The fence blocks the outboard spanwise flow of the low-energy boundary layer and reduces the possibility of wingtip stall on swept wings.

(3) The spanwise flow effectively places the fence at a small *AOA*, forming a vortex on its outboard side. When the *AOA* of the wing is low, a single vortex is formed on the outboard side. However, at higher *AOA*s this develops into a complex two-core vortex with opposite rotation [12, p. 107]. The fence is thus a type of leading edge discontinuity that segments the wing into two lower AR parts whose stall *AOA*s are higher.

23.4.9 Stall Handling — Wing Pylons

Jet engine pylons on the wing have been shown to improve the stall characteristics of jet aircraft. Abzug and Larrabee [1, p. 174] explain it by pointing out that the bound vortex on the wing induces side-wash on the engine pylon, as shown in Figure 23-35. Consequently, an outboard side force is generated on the nacelle and pylon. The pylon generates a vortex on the upper surface of the wing, whose flow direction can be seen to oppose the outboard spanwise flow. This suppresses early wingtip flow separation.

It is alleged in Ref. [1] that the effect was discovered by Boeing, most likely when developing the B-47 Stratojet in the late 1940s. Both the Boeing 707 and B-52 Stratofortress feature two pylons per wing and neither had wing fences, so common on swept-wing aircraft of the era. This spurred the development of another aerodynamic fix — a small "pylon" without the nacelle. The device was conceived by Douglas Aircraft aerodynamicists and was mounted to the DC-9, whose engines are mounted at the back of the fuselage. The device was patented under the name *vortilon*. It was developed as a part of a three-part solution to a possible deep-stall scenario on the DC-9 [13] and mounted relatively inboard, where the vortex it generated at high *AOA*s favorably improved flow over the horizontal tail. The other two were an enlarged horizontal tail and hydraulically boosted nose pitch-down control.

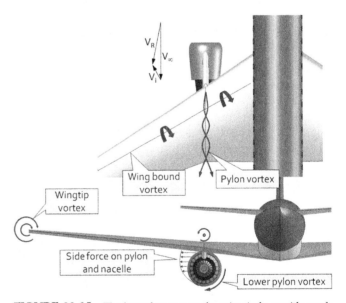

FIGURE 23-35 The bound vortex on the wing induces side-wash on the engine pylon. This forms a vortex on the upper surface of the wing that opposes the wingtip vortex and, thus, causes the normal outflow of air in the spanwise direction. This, in turn creates a vortex on the upper surface whose effect suppresses early wingtip flow separation *(Based on Ref. [1])*.

FIGURE 23-36 Vortilons on the outboard wing of an Embraer ERJ 145. *(Photo by Phil Rademacher)*

23.4.10 Stall Handling – Vortilons

As explained above, a *vortilon* is a small pylon-shaped fence mounted to the lower surface of a wing, usually a swept-back wing. It can be mounted alone (as it is on the inboard wing of the DC-9 commercial jetliner) or in a row (as it is on the outboard wing of the ERJ-145, as shown in Figure 23-36). Its purpose is to shed a vortex at high AOAs that aerodynamically partitions the wing into several low-AR segments, reducing section lift coefficients and delaying stall (see Figure 23-37). It is not as effective as the fence, but adds less drag to the airplane [7].

23.4.11 Stall Handling – Wing Droop (Cuffs, Leading Edge Droop)

The term *wing droop* (also called a *cuff* or a *leading edge droop*) is used to describe the enlarging of wing airfoils in the outboard wing region (see Figure 23-38). The cuff presents a discontinuity to the leading edge of the wing. This means that as the AOA increases, the wing is effectively partitioned into two smaller segments. This can be seen in the vortex that begins to form at the discontinuity. The wing partioning results in a reduced AR of the outboard (and inboard) segments, not unlike the one obtained using the wing fence. As we have seen before, a reduction in the AR of a lifting surface means a more shallow lift curve slope and delay of the stall to a higher AOA. Consequently, the cuff provides improved roll stability at stall.

In 1982, NASA conducted an investigation on the effectiveness of leading edge devices for stall departure and spin resistance on the Piper PA-32 Lance with a T-tail [14]. The investigation revealed that the partial span leading edges eliminated the abrupt stall tendency of the 1/6th scale wind tunnel model and a tendency for autorotation, improving spin resistance. While some of the results are dependent on the geometry of the airplane and Reynolds number, they are still important indicators of the potential benefits of cuffs. Among results were:

(1) The inboard wing of the model stalled at an AOA of 12°, whereas the tip with the wing droop stalled at 32°, improving roll stability and providing spin resistance.
(2) A wing droop starting at span station 0.55 to 0.60 was found to provide the most favorable outboard aerodynamics.
(3) Partitioning of the wing into a lower-AR wing, evident from item (1), lowers the section lift coefficients as the lift curve slope becomes shallower and stalls at a higher AOA.
(4) The wing chord extension means that the lift is generated along a longer chord and this brings down the peak C_p and delays flow separation.
(5) The airfoil of the LEX is modified as well and with the combination of a larger LE radius and camber, C_{lmax} is pushed up as well.

The most serious drawback of wing droops is that they may make spin recovery harder. The feature that improves roll stability at stall – higher lift at the wingtip – is also larger during spin, requiring greater moment to stop. Among other drawbacks is higher drag and added manufacturing complexity. Cuffs for composite aircraft are typically manufactured by two means: as an integrated structure or as an add-on. Integrating the cuff into the wing skin is desirable, as it reduces part count and weight. However, the quality of the laminate layup may suffer due to fiber dryness often associated with sharp corners.

23.4.12 Flow Improvement – Vortex Generators

Strictly speaking, a vortex generator (also known as a VG) is any device that starts and maintains vortex motion in fluid flow. However, generally, the term applies to devices that generate localized vortices on lifting surfaces to improve the airflow over them. There are typically two reasons for their installation; to alleviate buffeting and loss of control effectiveness due to shock interaction at transonic airspeeds [15] or to suppress flow separation at high AOAs at low subsonic airspeeds. The latter will be the focus here. Figure 23-39 shows a common application of VGs: on the lower surface of the HT where it increases the effectiveness of the elevator by delaying separation at high deflection angles and low dynamic pressures.

There is a large amount of information available on VGs in technical journals, like NACA, NASA, AIAA,

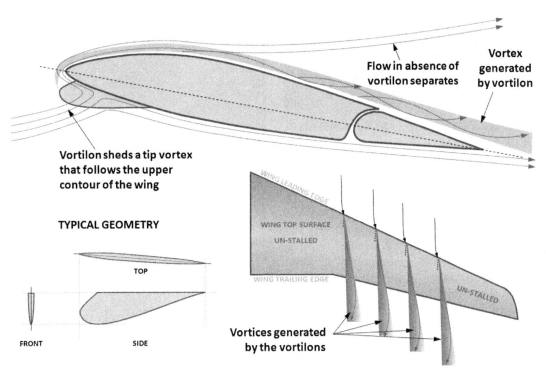

FIGURE 23-37 The vortilon in action *(Based on Ref. [7])*.

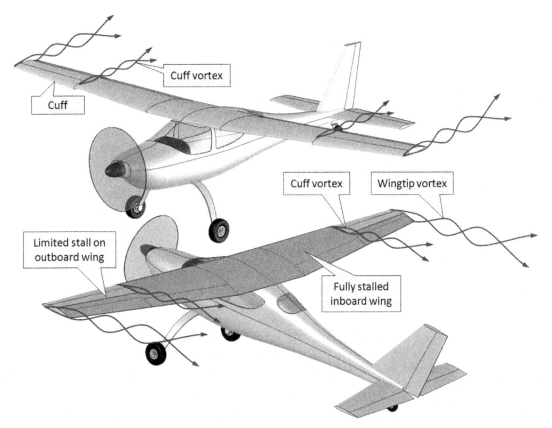

FIGURE 23-38 The workings of the leading edge extention (cuff). It partitions the higher-*AR* wing into three segments. The outboard segments, effectively, are low-*AR* wings that stall at a higher *AOA*. This helps the aircraft retain roll stability.

FIGURE 23-39 A row of vortex generators on the lower HT surface of the Aero L-39 Albatros trainer aircraft delays flow separation on the elevator at low speeds and large deflection (i.e. during flare), improving elevator authority. *(Photo by Phil Rademacher)*

FIGURE 23-40 A row of micro-vortex generators on a flap. *(Photo by S. Gudmundsson)*

ESDU, and others, so only basic information will be given here. A good insight is given by Wentz and Seetharam [16], who investigated the effect of VGs on a GA(W)-1 airfoil. They found the VGs to be very effective in suppressing flow separation. The VGs increased the $C_{l\alpha}$ and C_{lmax}, but lowered the α_{stall}. There was a substantial increase in the section drag coefficient, of about 25 drag counts. However, the drag at higher AOAs was lower due to the delay in separation, although the cross-over point was at a C_L of about 1.1. Guidelines for the sizing of VGs are given below and are obtained from various sources, including Refs 15, 16 and 17.

Height:	1 to 2 times the boundary layer thickness at their position
Aspect ratio:	0.25 to 1.00 (AR = height/chord)
Taper ratio:	0.6 to 0.8 (recommended to help "load up" the VG and make it more effective)
Angle-of-incidence:	10° to 20°
Spacing:	5 to 10 times the height
Chordwise location:	approximately 20 boundary layer thicknesses ahead of the separation point
Orientation:	for unswept wing use counter-rotating VGs (possibly even in pairs). For swept wings use co-rotating VGs.

Very small VGs, called micro-VGs, are used in applications where size, or lack thereof, is of importance; for instance to improve flow over control surfaces and flaps. An example is shown in Figure 23-40. Their purpose is to delay flow separation over a specific control surface. In the case of flaps, they have the potential to increase C_{Lmax} while having no impact on cruise drag, as they are only exposed to the airflow when the flap is deployed. It is a drawback that they may cause an interference with the free motion and stowage of the flap. Refer to Lin [18] for recommendations regarding sizing, spacing, and location.

23.4.13 Trailing Edge Tabs for Multi-Element Airfoils

Ross et al. [19,20] investigated the installation of a special lift-enhancing tab (or a *cove-tab*), similar to a Gurney flap on the pressure side of a two-element airfoil (NACA 63_2-215) with 30% chord single-slotted flap and on a multi-element airfoil. The idea has been patented under US Patent 5,249,080.

A cove-tab with typical dimensions is shown in Figure 23-41. The research indicates the tab must be sized and located based on the geometry of the flap and cove. For this reason, the dimensions shown in the figure do not necessarily hold for all applications. The tab also presents an issue during flap retraction and deployment and may, therefore, have to be mechanically actuated, potentially adding complexity to the installation. It is an advantage that the cove-tab is passive — in other words, it does not require a separate power source to function, like for instance a jet-flap.

The installation and wind tunnel testing of the cove-tab on the two-element airfoil was the focus of Ref. [19]. It was found that it increased the maximum lift coefficient by 10.3% with the flap deflected at 42°, attributed to a reduction in flow separation on the upper surface of the flap. The tab was found to be less effective (even detrimental) for lower deflections. However, a combination of a cove-tab with a Gurney flap and vortex generators on the flap increased the maximum lift at a high flap deflection by 17% over the optimum flap position of the baseline.

Reference [20] investigated various flow mechanics of the cove-tab installation using a Navier-Stokes solver. Among notable conclusions is that as the flap element is moved away from its optimum position the flow separation increases (which explains the reduction in lift).

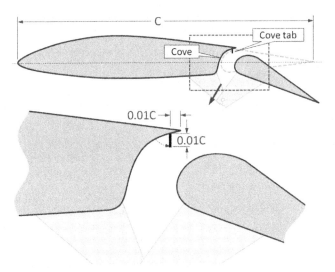

FIGURE 23-41 Typical positioning of a cove-tab.

FIGURE 23-42 A nacelle strake on an Airbus A319 commercial jetliner. *(Photo by Phil Rademacher)*

Such a movement can be the consequence of tolerance stack-up or, simply, incorrect positioning during design. Adding the cove-tap will reduce the separation and restore the flap lift to within 1% of the optimum flap lift. This way, it can serve as a "fix" for incorrectly positioned flaps.

The investigation revealed the tab is detrimental to the lift at low flap deflection (T-O configuration) as it appears to reduce the suction peak on the leading edge of the flap. However, at large flap deflections (landing configuration) it delayed flow separation on the flap, greatly improving its capability.

23.4.14 Flow Improvement – Nacelle Strakes

Nacelle strakes are vortex generators commonly found on the engines of modern jet transport aircraft, civilian and military. At high *AOA*s the strake generates a powerful vortex that makes up for the flow separation and loss of lift due to the presence of the nacelle. The strake typically has a LE sweep of approximately 70° and it is aligned with the airstream at cruise to minimize its interference drag. It can improve the maximum lift coefficient by as much as 0.05–0.1. An example of a nacelle strake on an Airbus A319 commercial jetliner is shown in Figure 23-42.

23.4.15 Flow Improvement – Bubble-Drag, Turbulators, and Transition Ramps

The wings of small radio-controlled aircraft and sailplanes that operate at very low Reynolds numbers ($60,000 < R_e < 500,000$) are often subject to the formation of a laminar boundary layer separation bubble along the leading edge. An example of such a separation bubble is shown in Figure 8-18. Such bubbles can be very detrimental to the flight characteristics of the corresponding aircraft, detrimentally affecting the lift and drag (forming so-called *bubble-drag*). They can even lead to unpredictable and sudden changes in the magnitude of these forces. Generally, two methods can be used to reduce the bubble-drag: (1) *turbulators* and (2) tailoring of the *transition curve*.

Turbulators (also called *trip strips* or *boundary layer trips*) are often installed near the leading edge of such aircraft. They are intended to force the laminar boundary layer to transition immediately to a turbulent one to prevent the formation of or reduce the size of the more detrimental separation bubble. The presence of a turbulator has important effects on the lift and drag of the airfoil. It is difficult to accurately predict its influence, calling for trial-and-error approaches in wind tunnel or flight tests. This is further compounded by the fact that a trip strip configuration found suitable for a specific airfoil at a given *AOA* may be detrimental to it at another operating condition [21].

Tailoring the *transition curve* refers to the proactive design of airfoils to encourage a more rapid movement of the laminar-to-turbulent BL transition point on the upper surface toward the leading edge. This is shown in Figure 23-43, which depicts how the derivative $\partial C_l/\partial(X_{tr}/C)$ is steeper for airfoil A than B. This is referred to as the shape of the *transition ramp*. In other words, in the lift coefficient range $0 < C_l < 1.0$, the transition point, X_{tr}/C for airfoil A "moves faster" toward the LE than airfoil B. Gopalarathnam et al. [21] show that this results in a lower bubble-drag for airfoil A. This observation makes it possible to design airfoils for operation in low Reynolds number flow that are less likely to suffer this undesirable type of pressure drag.

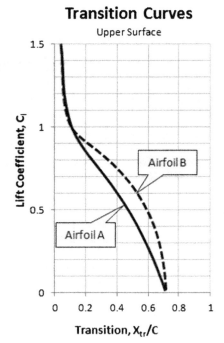

FIGURE 23-43 The shape of the transition curve on the upper surface is a good indicator of a tendency for bubble-drag formation *(Based on Ref. [21])*.

VARIABLES

Symbol	Description	Units (UK and SI)
AR	Aspect ratio	
b	Wing span	ft or m
b_1	Spanwise station for the inboard edge of the aileron	ft or m
b_2	Spanwise station for the outboard edge of the aileron	ft or m
c	Wing chord	ft or m
c_{d0}	Zero AOA drag coefficient	
C_f	Flap chord (aft of hingeline)	ft or m
C_h	Hinge moment coefficient	
C_{h0}	Zero AOA hinge moment coefficient	
$C_{h\alpha}$	Hinge moment coefficient curve slope	Per degree or per radian
$C_{h\delta}$	Hinge moment coefficient caused by flap deflection	Per degree or per radian
$C_{h\delta t}$	Hinge moment coefficient caused by tab deflection	Per degree or per radian
C_l	Rolling moment coefficient	
C_{lmax}	Maximum 2D lift coefficient	
C_{lp}	Roll damping derivative	Per degree or per radian
$c_{l\alpha}$	Lift curve slope	Per degree or per radian

Symbol	Description	Units (UK and SI)
$C_{l\delta a}$	Aileron authority derivative	Per degree or per radian
$c_{l\delta a}$	Change in lift coefficient with aileron deflection	
C_m	Pitching moment coefficient	
C_n	Yawing moment coefficient	
$C_{N\beta}$	Directional stability	
C_R	Root chord length	ft or m
F	Actuation force	lb_f or N
HM	Hinge moment	$ft \cdot lb_f$ or $N \cdot m$
I_{XX}	Moment of inertia of the aircraft	$slugs \cdot ft^2$ or $kg \cdot m^2$
L	Lift (context dependent)	lb_f or N
L	Rolling moment (context dependent)	$ft \cdot lb_f$ or $N \cdot m$
L_p	Rolling moment (due to change in roll rate)	$ft \cdot lb_f$ or $N \cdot m$
p	Roll rate	Deg/s or rad/s
$\dot{p}$	Roll acceleration	Deg/s^2 or rad/s^2
q	Dynamic pressure	lb_f/ft^2 or N/m^2
ROC	Rate-of-climb	ft/s or m/s
S	Wing area	ft^2 or m^2
S_f	Flap area (aft of hingeline)	ft^2 or m^2
t	Time	seconds
V	Airspeed	ft/s or m/s
V_{LOF}	Liftoff airspeed	ft/s or m/s
V_{MC}	Minimum control airspeed	ft/s or m/s
V_{REF}	Reference airspeed	ft/s or m/s
V_{S0}	Stalling speed in landing configuration	ft/s or m/s
V_{S1}	Stalling speed in takeoff configuration	ft/s or m/s
V_Y	Best rate-of-climb airspeed	KTAS
W	Weight	lb_f or N
W_0	Aircraft gross weight	lb_f or N
y	Wing station	ft or m
ΔC_m	Spanwise moment coefficient	
ΔM_x	Elemental rolling moment	$ft \cdot lb_f$ or $N \cdot m$
ΔS	Area of elemental strip	ft^2 or m^2
$\Lambda_{C/2}$	Sweep of mid-chord	Degrees or radians
α	Angle-of-attack	Degrees or radians
β	Prandtl-Glauert Mach number parameter	
β	Sideslip angle	Degrees or radians

Symbol	Description	Units (UK and SI)
δ	Deflection angle (of flap)	Degrees or radians
δ_a	Deflection angle of aileron	Degrees or radians
δ_e	Deflection angle of elevator	Degrees or radians
δ_r	Deflection angle of rudder	Degrees or radians
δ_t	Deflection angle of tab	Degrees or radians
κ	Ratio of 2D lift curve slope to 2π	
λ	Taper ratio	
ρ	Density of air	slugs/ft^3 or kg/m^3

References

[1] Abzug MJ, Larrabee EE. Airplane Stability and Control — A History of the Technologies that Made Aviation Possible. 2nd ed. Cambridge Aerospace Series; 2002.

[2] NACA TN-547. Development of the N.A.C.A. Slot-Lip Aileron. Weick, Fred E., and Joseph A. Shortal; 1935.

[3] NACA R-602. Wind Tunnel and Flight Test of Slot-Lip Ailerons. Shortal, Joseph A; 1937.

[4] Perkins CD, Hage RE. Airplane Performance, Stability, and Control. John Wiley & Sons; 1949.

[5] NACA-TR-824. Summary of Airfoil Data. Abbott, Ira H., Albert E. von Doenhoff and Louis S. Stivers Jr.; 1945.

[6] Whitford R. Design for Air Combat. Jane's Publishing Company Limited; 1987.

[7] Reynolds P. Ten Years of Stall Testing. AIAA-1990-1268; 1990.

[8] http://www.nasa.gov/centers/dryden/news/FactSheets/FS-002-DFRC.html.

[9] NASA-TM-112360. Overview of HATP Experimental Aerodynamics Data for the Baseline F/A-18 Configuration. Hall, Robert M., et al; 1996.

[10] Shah GH, Nijel Granda J. Application of Forebody Strakes for Directional Stability and Control of Transport Aircraft. AIAA-98-4448; 1998.

[11] Flight International November 20th, 1982.

[12] Meier Hans-Ulrich. German Development of the Swept Wing 1935-1945. AIAA; 2006.

[13] Anonymous. The DC-9 and the Deep Stall. Flight International March 25th, 1965.

[14] NASA CR-3636. Wind-Tunnel Investigation of Effects of Wing-Leading-Edge Modifications on the High Angle-of-Attack Characteristics of a T-tail Low-Wing General-Aviation Aircraft. White, E. R.; 1982.

[15] Edwards JBW. Free-Flight Tests of Vortex Generator Configurations at Transonic Speeds. C.P. No. 729, British A. R. C.; 1966.

[16] NASA CR-2443. Development of a Fowler Flap System for a High Performance General Aviation Airfoil. Wentz, W. H., Jr., and H. C. Seetharam; 1974.

[17] Tanner LH, Pearcey HH, Tracy CM. Vortex Generators, Their Design and Their Effects on Turbulent Boundary-Layers. A.R.C. 16.487; January, 1954.

[18] Lin JC. Control of Turbulent Boundary-Layer Separation Using Micro-Vortex Generators. AIAA 99–3404; 1999.

[19] Ross JC, Storms BL. An Experimental Study of Lift-Enhancing Tabs on a Two-Element Airfoil. AIAA-1994-1868; 1994.

[20] Ross JC, Carrannanto PG, Stroms BL, Cummings RM. Navier-Stokes Anaysis of Lift-Enhancing Tabs on Multi-Element Airfoils. AIAA-1994-50; 1994.

[21] Gopalarathnam A, et al. Design of Low Reynolds Number Airfoils with Trips. Journal of Aircraft July-August 2003; 40(4).

APPENDIX A
Atmospheric Modeling

A.1 INTRODUCTION

A number of organizations and scientists have developed sophisticated models of the atmosphere that allow atmospheric properties at different altitudes to be determined. As an example, the National Oceanic and Atmospheric Administration (NOAA) has developed one of the best known of these, the U.S. Standard Atmosphere 1976 [1]. However, far more sophisticated models than that have been developed. One such model is the NRLMSISE-00 (Naval Research Laboratory Mass Spectrometer and Incoherent Scatter, where E means from surface of the Earth to the Exosphere). This model requests input data in the form of year, day, time of day, altitude, geodetic latitude and longitude, and many others. It returns information such as temperature, mass density, and molecular densities of oxygen (O_2), nitrogen (N_2), mono-atomic oxygen (O) and nitrogen (N), argon (Ar), and hydrogen (H). These can be used to estimate other properties, such as specific gas constant (typically denoted by R), pressure, and the ratio of specific heats (typically denoted by γ). Among numerous applications, this model is used to predict the orbital decay of satellites due to atmospheric drag and to study the effect of atmospheric gravity waves. An example of output of temperature and density from this atmospheric model is shown in Figure A-1. The figure shows the two gas states up to an altitude of 500 km, well beyond the so-called *von Kármán line*, which is considered the edge of the atmosphere, as the point where a vehicle would have to fly faster than its orbital escape speed to generate a dynamic pressure large enough to provide aerodynamic lift.

In this text, all atmospheric data is based on the US Standard Atmosphere 1976, unless otherwise specified. This is done because it can be conveniently represented using simple formulation. Additionally, examples in this book take place in the troposphere, below 36,089 ft (see Figure A-2).

A.1.1 General Information About the Atmosphere

An atmosphere is the mixture of gases surrounding a celestial object (i.e. planet) whose gravitational field is strong enough to prevent its molecules from escaping. In particular, the atmosphere refers to the gaseous envelope of the Earth.

Formation of

The current mixture of gases in the air is thought to have taken some 4.5 billion years to evolve. The early atmosphere is believed to have consisted of volcanic gases alone. Since gases from erupting volcanoes today are mostly a mixture of water vapor (H_2O), carbon dioxide (CO_2), sulfur dioxide (SO_2), and nitrogen (N) it is postulated that this was probably the composition of the early atmosphere as well. It follows that a number of chemical processes must have preceded the mixture making the atmosphere of our time.

One of those processes is thought to have been condensation, which was a natural consequence of the cooling of the earth's crust and early atmosphere. This condensation is thought to have slowly but surely filled valleys in the barren landscape, forming the earliest oceans. Some CO_2 would have reacted with the rocks of the earth's crust to form carbonate minerals, while some would have dissolved in the new rising oceans. Later, as primitive life capable of photosynthesis evolved in the oceans, new marine organisms began producing oxygen. Almost all the free oxygen in the air today is attributed to this process; by photosynthetic combination of CO_2 with water.

About 570 million years ago, the oxygen content of the atmosphere and oceans became high enough to permit marine life capable of respiration; 170 million years later, the atmosphere would have contained enough oxygen for air-breathing animals to emerge from the seas.

A.1.2 Chemical Composition of Standard Air

Research shows the chemical composition of the atmosphere is practically independent of altitude from ground level to at least 88 km (55 mi). The continuous stirring produced by atmospheric currents counteracts the tendency of the heavier gases to settle below the lighter ones. In the lower atmosphere ozone is present in

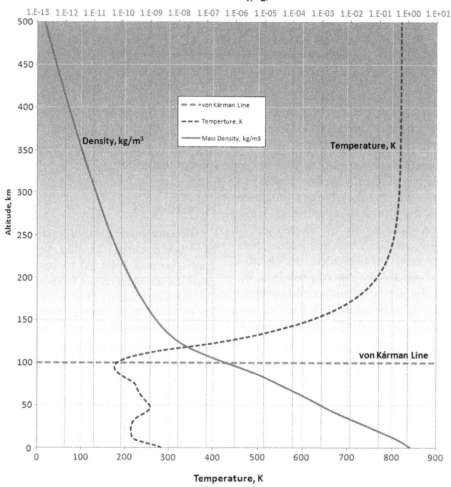

FIGURE A-1 An output from the NRLMSISE-00 atmospheric model showing the variation of temperature and mass density as a function of altitude ranging from S-L to 500 km. The von Kármán Line is considered the "boundary between Earth's atmosphere and outer space." It is the altitude where aerodynamic forces can no longer provide support to maintain altitude, so vehicles must be in orbit in order to do so.

extremely low concentrations. The layer of atmosphere from 19 to 48 km (12 to 30 mi) up contains more ozone, produced by the action of ultraviolet radiation from the sun. Even in this layer, however, the percentage of ozone is only 0.001 by volume. Atmospheric disturbances and downdrafts carry varying amounts of this ozone to the surface of the earth. Human activity adds to ozone in the lower atmosphere, where it becomes a pollutant that can cause extensive crop damage. Table A-1 lists the chemical composition of the atmosphere.

A.1.3 Layer Classification of the Atmosphere

The atmosphere is generally divided into several layers based on some specific characteristics (see Table A-2). The *troposphere* extends from the ground to some 11–16 km (6.8–10 mi) and this is where most clouds occur and weather (winds and precipitation) are most active. It transitions into the next layer; the *stratosphere*, through a thin region called the *tropopause*. The bulk of the atmosphere is found within these two lowest layers. Above the stratosphere is the *mesosphere*, which is characterized by a decrease in temperature with altitude. Research of propagation and reflection of radio waves starting at an altitude of 80 km (50 mi) to some 640 km (400 mi) indicates that ultraviolet radiation, X-rays, and showers of electrons from the sun ionize this layer of the atmosphere, causing it to conduct electricity and reflect radio waves of certain frequencies back to earth. For this reason, it is called the *ionosphere*. It is also termed the *thermosphere*, because of the relatively higher temperatures in this layer. Above it is the *exosphere*, which extends to the outer limit of the atmosphere, at about 9600 km (about 6000 mi). Figure A-2 shows how temperature changes through the lowest layers of the atmosphere. The classification of the atmosphere is based on an average height of the layers.

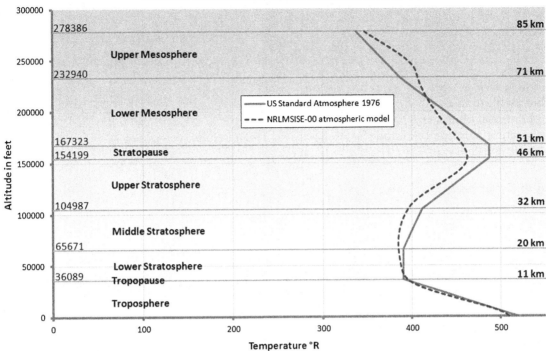

FIGURE A-2 A comparison of temperature changes with altitude up to 85 km, using the US Standard Atmosphere 1976 and NRLMSISE-00 atmospheric models. The former represents standard conditions, whereas the latter is at a geodesic location N45° W80° on January 1st, 2012.

A.2 MODELING ATMOSPHERIC PROPERTIES

A.2.1 Atmospheric Ambient Temperature

Let's start by considering the temperature, T. Change in air temperature with altitude can be approximated using a linear function:

$$T = T_0 + a(h - h_0) \tag{A-1}$$

TABLE A-1 Chemical Composition of Standard Air per Ref. [1], p. 3

Name of Chemical	Chemical Formula	Molecular Weight (kg/kmol)	Fractional Volume (Dimensionless)
Nitrogen	N_2	28.0134	0.78084
Oxygen	O_2	31.9988	0.209476
Argon	Ar	39.948	0.00934
Carbon dioxide	CO_2	44.00995	0.000314
Neon	Ne	20.183	0.00001818
Helium	He	4.0026	0.00000524
Krypton	Kr	83.80	0.00000114
Xenon	Xe	131.30	0.000000087
Methane	CH_4	16.04303	0.000002
Hydrogen	H_2	2.01594	0.0000005
Water vapor (varying)	H_2O	Varying	Varying
Ozone	O_3	-	-
Carbon monoxide	CO	-	-

TABLE A-2 Layer Classification of the Atmosphere

Name of Layer	Altitude in km	Altitude in Statute Miles
Troposphere[a]	0–11 km	0–6.8 sm
Tropopause	11–11.5 km	6.8–1 sm
Stratosphere	11.5–46 km	11.5–29 sm
Stratopause	46–51 km	29–32 sm
Mesosphere	51–85 km	32–53 sm
Ionosphere (Thermosphere)	85–640 km	53–400 sm
Exosphere	640–9600 km	400–6000 sm

[a]*In temperate latitudes this is approximately 0–9.7 km (6 mi). The troposphere can extend to 15 km in the tropics.*

An alternative form of Equation (A-1) is:

$$T = T_0(1 + \kappa \cdot h) \quad \text{(A-2)}$$

where

a = Lapse rate
h = altitude in ft or m
h_0 = reference altitude h_0
T = temperature at altitude h
T_0 = temperature at reference altitude h_0
κ = lapse rate constant = a/T_0

Derivation of Equation (A-2)

This can simply be derived from Equation (A-1), by setting $h_0 = 0$:

$$T = T_0 + ah = T_0 \cdot \left(1 + \frac{a}{T_0} \cdot h\right) = T_0 \cdot (1 + \kappa \cdot h)$$

QED

A.2.2 Atmospheric Pressure and Density for Altitudes below 36,089 ft (11,000 m)

The hydrostatic equilibrium equations allow the pressure, p, and density, ρ, to be calculated as functions of altitude, h, as follows:

Pressure:

$$p = p_0(1 + \kappa \cdot h)^{5.2561} \quad \text{(A-3)}$$

Density:

$$\rho = \rho_0(1 + \kappa \cdot h)^{4.2561} \quad \text{(A-4)}$$

Derivation of Equations (A-3) and (A-4)

Begin with the hydrostatic equilibrium equation and divide by the ideal gas relation as shown below:

$$dp = -\rho \cdot g \cdot dh \quad \text{and} \quad p = \rho \cdot g \cdot R_0 T$$

$$\Rightarrow \frac{dp}{p} = \frac{-\rho \cdot g \cdot dh}{\rho \cdot g \cdot R_0 T} = -\frac{dh}{R_0 T} \quad \text{(i)}$$

Differentiate Equation (A-1) (which is $T(h)$):

$$\frac{dT}{dh} = \frac{d}{dh}(T_0 + a(h - h_0)) = a \quad \Rightarrow \quad dT = a \cdot dh$$

Use this to replace dh in Equation (i):

$$\frac{dp}{p} = -\frac{dh}{R_0 T} = -\frac{dT/a}{R_0 T} = -\frac{1}{aR_0}\frac{dT}{T} \quad \text{(ii)}$$

Integrate:

$$\int \frac{dp}{p} = -\frac{1}{aR_0}\int \frac{dT}{T} \quad \Leftrightarrow \quad [\ln p]_{p_0}^{p} = -\frac{1}{aR_0}[\ln T]_{T_0}^{T}$$

Manipulate algebraically:

$$\ln p - \ln p_0 = -\frac{1}{aR_0}(\ln T - \ln T_0)$$

Therefore:

$$\ln\left(\frac{p}{p_0}\right) = \ln\left(\frac{T}{T_0}\right)^{-\frac{1}{aR_0}} \quad \Leftrightarrow \quad \frac{p}{p_0} = \left(\frac{T}{T_0}\right)^{-\frac{1}{aR_0}} \quad \text{(iii)}$$

Insert our expression for the temperature:

$$\frac{p}{p_0} = \left(\frac{T}{T_0}\right)^{-\frac{1}{aR_0}} = \left(\frac{T_0 + a(h - h_0)}{T_0}\right)^{-\frac{1}{aR_0}}$$

$$= \left(1 + \frac{a(h - h_0)}{T_0}\right)^{-\frac{1}{aR_0}}$$

Insert standard day coefficients for troposphere:

a = lapse rate = -0.00356616 °F/ft = -0.0065 K/m
h = altitude in ft or m
$h_0 = 0$ ft
$T_0 = 59 + 459.67 = 518.67$ °R
$R_0 = 53.35$ ft/°R = 29.26 m/K

$$\frac{p}{p_0} = \left(1 + \frac{a(h - h_0)}{T_0}\right)^{-\frac{1}{aR_0}} = \left(1 + \frac{a(h - 0)}{T_0}\right)^{\beta}$$

$$= \left(1 + \frac{ah}{T_0}\right)^{\beta}$$

where b is given by:

$$\beta = -\frac{1}{aR_0} = -\frac{1}{(-0.00356616)(53.35)}$$

$$= -\frac{1}{(-0.0065)(29.26)} = 5.2561$$

Simplify to get:

$$p = p_0(1 + \kappa \cdot h)^{5.2561} \quad \text{(A-3)}$$

We'd also like to derive an expression for density as a function of altitude. To do this, we start by rewriting Equation (C-6) in terms of density:

$$p = \rho g R_0 T \quad \Rightarrow \quad \rho = \frac{p}{gR_0 T} = \frac{p}{RT}$$

Then, we insert the expressions for temperature and density and expand as follows:

$$\rho = \frac{p}{gR_0 T} = \frac{p_0(1+\kappa \cdot h)^{5.2561}}{gR_0 \cdot T_0 \cdot (1+\kappa \cdot h)}$$

$$= \frac{p_0}{gR_0 \cdot T_0}(1+\kappa \cdot h)^{4.2561}$$

Simplifying yields:

$$\rho = \rho_0(1+\kappa \cdot h)^{4.2561} \qquad (A\text{-}4)$$

QED

A.2.3 Density of Air Deviations From a Standard Atmosphere

Atmospheric conditions often deviate from models shown above. Such deviations can be handled as reflected below, using the UK system:

UK system:

$$\rho = \frac{1.233(1+\kappa \cdot h)^{4.2561}}{(T + \Delta T_{ISA})} \qquad (A\text{-}5)$$

where
h = reference altitude in ft
T = standard day temperature at the given altitude per the International Standard Atmosphere. in °R; at S-L it would be 518.67 °R, at 10,000 ft it would be 483 °R, and so on
ΔT_{ISA} = deviation from International Standard Atmosphere in °F or °R

SI system:

$$\rho = \frac{352.6(1+\kappa \cdot h)^{4.2561}}{(T + \Delta T_{ISA})} \qquad (A\text{-}6)$$

where
h = reference altitude in m
T = standard day temperature at the given altitude per the International Standard Atmosphere. in degrees K; at S-L it would be 288.15 K, at 10,000 ft it would be 483 °R, and so on
ΔT_{ISA} = deviation from International Standard Atmosphere in °F or °R

For non-standard atmosphere, use a negative sign for colder and a positive sign for warmer than ISA for ΔT_{ISA}.

Derivation of Equations (A-5) and (A-6)

Starting with the equation of state we get:

$$\rho = \frac{P}{RT} = \frac{P}{R(T+\Delta T_{ISA})} = \frac{P_0 \cdot (1+\kappa \cdot h)^{5.2561}}{R \cdot (T + \Delta T_{ISA})}$$

where T is the standard day temperature in °R at the altitude h, and P_0 is the S-L pressure. If working with the UK system, this can be written in a simpler form as follows:

$$\rho = \frac{P_0 \cdot (1+\kappa \cdot h)^{5.2561}}{R \cdot (T + \Delta T_{ISA})} = \frac{2116.2 \cdot (1+\kappa \cdot h)^{5.2561}}{1716 \cdot (T + \Delta T_{ISA})}$$

$$\approx \frac{1.233 \cdot (1+\kappa \cdot h)^{5.2561}}{(T + \Delta T_{ISA})}$$

Conversely, if working with the SI system, this can be written as:

$$\rho = \frac{P_0 \cdot (1+\kappa \cdot h)^{5.2561}}{R \cdot (T + \Delta T_{ISA})} = \frac{1.012 \times 10^5 \cdot (1+\kappa \cdot h)^{5.2561}}{287 \cdot (T + \Delta T_{ISA})}$$

$$\approx \frac{352.6 \cdot (1+\kappa \cdot h)^{5.2561}}{(T + \Delta T_{ISA})}$$

QED

A.2.4 Atmospheric Property Ratios

The pressure, density, and temperature often appear in formulation as fractions of their baseline values. As a consequence, they are identified using special characters and are called pressure ratio, density ratio, and temperature ratio.

Temperature ratio:

$$\theta = \frac{T}{T_0} = (1 - 0.0000068756 h) \qquad (A\text{-}7)$$

Pressure ratio:

$$\delta = \frac{p}{P_0} = (1 - 0.0000068756 h)^{5.2561}$$
$$= \theta^{5.2561} \qquad (A\text{-}8)$$

Density ratio:

$$\sigma = \frac{\rho}{\rho_0} = (1 - 0.0000068756 h)^{4.2561}$$
$$= \theta^{4.2561} = \frac{\delta}{\theta} \qquad (A\text{-}9)$$

EXAMPLE A-1

Determine the state in the atmosphere at 8500 ft on a standard day, using the UK system.

Solution

Pressure:

$$p = 2116(1 - 0.0000068756 \times 8500)^{5.2561} = 1542 \text{ psf}$$

Density:

$$\rho = 0.002378(1 - 0.0000068756 \times 8500)^{4.2561}$$
$$= 0.001840 \text{ slugs/ft}^3$$

Temperature:

$$T = 518.67(1 - 0.0000068756 \times 8500) = 488.4 \text{ }°R$$
$$= 28.7 \text{ }°F$$

Temperature ratio:

$$\theta = \frac{T}{T_0} = (1 - 0.0000068756 \times 8500) = 0.9416$$

Pressure ratio:

$$\delta = \frac{p}{p_0} = (1 - 0.0000068756 \times 8500)^{5.2561} = 0.7287$$

Density ratio:

$$\sigma = \frac{\rho}{\rho_0} = (1 - 0.0000068756 \times 8500)^{4.2561} = 0.7739$$

A.2.5 Pressure and Density Altitudes Below 36,089 ft (11,000 m)

Sometimes the pressure or density ratios are known for one reason or another. It is then possible to determine the altitudes to which they correspond. For instance, if the pressure ratio is known, we can calculate the altitude to which it corresponds. The altitude is then called *pressure altitude*. Similarly, from the density ratio we can determine the *density altitude*.

Pressure altitude in ft:

$$h_P = 145442\left[1 - \left(\frac{p}{p_0}\right)^{0.19026}\right] \quad \text{(A-10)}$$

Density altitude in ft:

$$h_\rho = 145442\left[1 - \left(\frac{\rho}{\rho_0}\right)^{0.234957}\right] \quad \text{(A-11)}$$

Derivation of Equations (A-10) and (A-11)

Begin with Equation (A-3) and solve for the altitude h, through the following algebraic maneuvers:

$$p = p_0(1 + \kappa \cdot h)^{5.2561}$$

$$\Rightarrow \frac{p}{p_0} = (1 + \kappa \cdot h)^{5.2561} \Rightarrow \left(\frac{p}{p_0}\right)^{1/5.2561} = 1 + \kappa \cdot h$$

$$\Rightarrow h = -\frac{1}{\kappa}\left[1 - \left(\frac{p}{p_0}\right)^{1/5.2561}\right]$$

Inserting the coefficient $\kappa = -0.0000068756/\text{ft}$ (from Table 16-2) and carrying out the arithmetic yields Equation (A-10). The derivation of Equation (A-11) is identical.

QED

A.2.6 Viscosity

Dynamic or Absolute Viscosity

Viscosity is a measure of a fluid's internal resistance to deformation and is generally defined as the ratio of the shearing stress to the velocity gradient in the fluid as it flows over a surface. Mathematically this is expressed using the following expression:

$$\tau = \mu \frac{\partial u}{\partial y} \quad \text{(A-12)}$$

where $\partial u / \partial y$ = velocity gradient in a fluid as it moves over a surface

u = velocity
y = height above the surface
τ = shear stress in the fluid
μ = viscosity coefficient

The viscosity coefficient we are primarily interested in is that of air. It can be determined using the following empirical expression, which assumes the UK system of units and, therefore, the temperature in °R (Ref. [2], Equation (2.90)):

$$\mu = 3.170 \times 10^{-11} \, T^{1.5} \left(\frac{734.7}{T + 216} \right) \quad \text{lb}_f \cdot \text{s/ft}^2 \quad \text{(A-13)}$$

In the SI system the temperature is given in K and the viscosity can be found from (Ref. [2], Equation (2.91)):

$$\mu = 1.458 \times 10^{-6} \, T^{1.5} \left(\frac{1}{T + 110.4} \right) \quad \text{N} \cdot \text{s/m}^2 \quad \text{(A-14)}$$

where

T = outside air temperature, in °R or K
μ = air viscosity, in $\text{lb}_f \cdot \text{s/ft}^2$ or $\text{N} \cdot \text{s/m}^2$

Kinematic Viscosity

This is defined as the dynamic viscosity divided by the fluid density:

$$\nu = \frac{\mu}{\rho} \quad \text{(A-15)}$$

The units for kinematic viscosity are $1/(\text{ft}^2 \cdot \text{s})$ in the UK system and $1/(\text{m}^2 \cdot \text{s})$ in the SI system.

A.2.7 Reynolds Number

The Reynolds number of a fluid is determined from the relationship below:

$$Re = \frac{\rho V L}{\mu} = \frac{V L}{\nu} \quad \text{(A-16)}$$

where

L = reference length (for instance MAC), in ft or m
V = reference airspeed, in ft/s or m/s
ρ = fluid density, in slugs/ft^3 or kg/m^3
μ = fluid (dynamic) viscosity, in $\text{lb}_f \cdot \text{s/ft}^2$ or $\text{N} \cdot \text{s/m}^2$
$\nu = \mu/\rho$ = fluid kinematic viscosity, in $\text{lb}_f \cdot \text{s/ft}$ or $\text{N} \cdot \text{s/m}^2$
ρ = slugs/ft^3 or kg/m^3
μ = $\text{lb}_f \cdot \text{s/ft}^2$ or $\text{N} \cdot \text{s/m}^2$

A simple expression, valid for the UK system at sea-level conditions only is (V and L are in ft/s and ft, respectively):

$$R_e \approx 6400 V L \quad \text{(A-17)}$$

A simple expression, valid for the SI system at sea-level conditions only is (V and L are in m/s and m, respectively):

$$R_e \approx 68500 V L \quad \text{(A-18)}$$

A.2.8 Speed of Sound and Mach Number

The speed of sound is retrieved from the expression below:

Speed of sound:

$$a_0 = \sqrt{\gamma R T} \approx 1116 \sqrt{1 - 0.0000068756 H} \quad \text{(A-19)}$$

Mach number:

$$M = \frac{V}{a_0} \quad \text{(A-20)}$$

where

V = airspeed
R = universal gas constant (1716 ft·lb_f/slug·°R)
T = ambient temperature (in °R)
γ = ratio of specific heats = 1.4 for air

EXAMPLE A-2

Determine the speed of sound at 8500 ft on a standard day using both forms of Equation (C-14).

Solution

From Example A-1 we get

$$\theta = \frac{T}{T_0} = 0.9416 \Rightarrow T = T_0 \theta = (518.67)(0.9416)$$
$$= 488.4 \, °R$$

Form 1:

$$a_0 = \sqrt{\gamma R T} = \sqrt{(1.4)(1716)(488.4)} = 1083 \text{ ft/s}$$

Form 2:

$$a_0 = 1116\sqrt{1 - 0.0000068756 H}$$
$$= 1116\sqrt{1 - 0.0000068756(8500)} = 1083 \text{ ft/s}$$

A.2.9 Atmospheric Modeling

As stated earlier, the properties of the atmosphere above the troposphere are detailed in the document *US Standard Atmosphere 1976*, published by NOAA,[1] NASA,[2] and the US Air Force. The formulation in Table A-3 is based on a summary from http://www.atmosculator.com/, which is based on the US Standard Atmosphere 1976. The temperature (in °R), pressure (in lb_f/ft^2), and density (in $slugs/ft^3$) are plotted in Figure A-3 from S-L to 278,386.

TABLE A-3 Formulation for the US Standard Atmosphere 1976

$0 \leq h \leq 36089$ ft	$h \leq 6.8$ mi	Troposphere
Temperature ratio:	$\theta = (1 - 0.0000068756 h) = (1 - h/145442)$	
Pressure ratio:	$\delta = (1 - 0.0000068756 h)^{5.2561} = (1 - h/145442)^{5.2561}$	
Density ratio:	$\sigma = (1 - 0.0000068756 h)^{4.2561} = (1 - h/145442)^{4.2561}$	
$36089 \leq h \leq 65617$ ft	$6.8 \leq h \leq 12.4$ mi	Lower stratosphere Isothermal Segment
Temperature ratio:	$\theta = 0.751865$	
Pressure ratio:	$\delta = 0.223361 \cdot e^{-(h-36089)/20806}$	
Density ratio:	$\sigma = 0.297076 \cdot e^{-(h-36089)/20806}$	
$65617 \leq h \leq 104987$ ft	$12.4 \leq h \leq 19.9$ mi	Middle Stratosphere Temperature Inversion Segment
Temperature ratio:	$\theta = 0.682457 + h/945374$	
Pressure ratio:	$\delta = (0.988626 + h/652600)^{-34.16320}$	
Density ratio:	$\sigma = (0.978261 + h/659515)^{-35.16320}$	
$104987 \leq h \leq 154199$ ft	$19.9 \leq h \leq 29.2$ mi	Upper Stratosphere
Temperature ratio:	$\theta = 0.482561 + h/337634$	
Pressure ratio:	$\delta = (0.898309 + h/181373)^{-12.20114}$	
Density ratio:	$\sigma = (0.857003 + h/190115)^{-13.20114}$	
$154199 \leq h \leq 167323$ ft	$29.2 \leq h \leq 31.7$ mi	Isothermal
Temperature ratio:	$\theta = 0.939268$	
Pressure ratio:	$\delta = 0.00109456 \cdot e^{-(h-154199)/25992}$	
Density ratio:	$\sigma = 0.00116533 \cdot e^{-(h-154199)/25992}$	
$167323 \leq h \leq 232940$ ft	$31.7 \leq h \leq 44.1$ mi	Lower Mesosphere
Temperature ratio:	$\theta = 1.434843 - h/337634$	
Pressure ratio:	$\delta = (0.838263 - h/577922)^{12.20114}$	
Density ratio:	$\sigma = (0.798990 - h/606330)^{11.20114}$	
$232940 \leq h \leq 278386$ ft	$44.1 \leq h \leq 52.7$ mi	Upper Mesosphere
Temperature ratio:	$\theta = 1.237723 - h/472687$	
Pressure ratio:	$\delta = (0.917131 - h/637919)^{17.08160}$	
Density ratio:	$\sigma = (0.900194 - h/649922)^{16.08160}$	

[1] NOAA = National Oceanic and Atmospheric Administration.
[2] NASA = National Aeronautics and Space Administration.

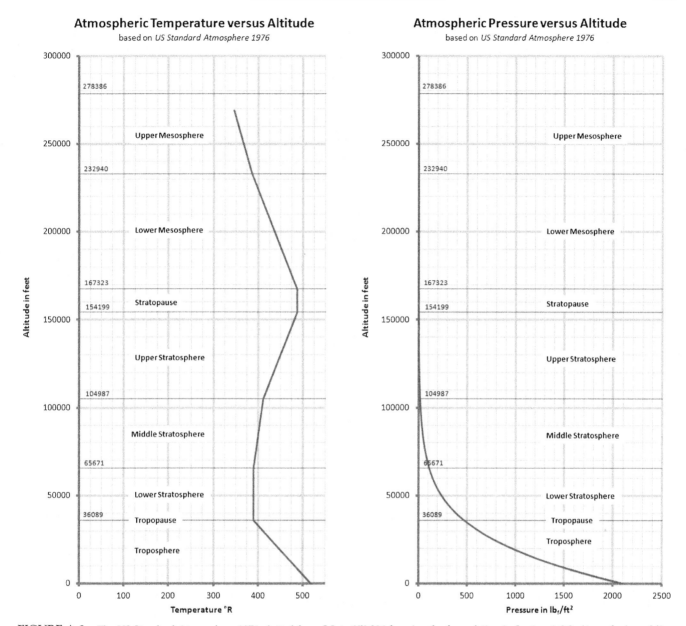

FIGURE A-3 The US Standard Atmosphere 1976 plotted from S-L to 278,386 ft, using the formulation in Section A.2.9, *Atmospheric modeling*.

A.2.10 Computer code A-1: Atmospheric Modeling

The following function, written in Visual Basic for Applications, can be used in Microsoft Excel to determine temperature, pressure, or density at any altitude up to 278,386 ft. To use, insert a VBA module into the spreadsheet and enter the function. Assume we have entered an altitude in cell A1. Then calls are made to it from any other cell by entering a statement like "=AtmosProperty(A1,0)." The rightmost argument (i.e. the ", 0") , called the PropertyID, would cause the function to return the Temperature ratio. If the PropertyID were 10 (i.e. the ", 10") it would return the temperature, and so on. The allowable values for PropertyID are shown in the comment lines below.

Atmospheric Density versus Altitude
based on *US Standard Atmosphere 1976*

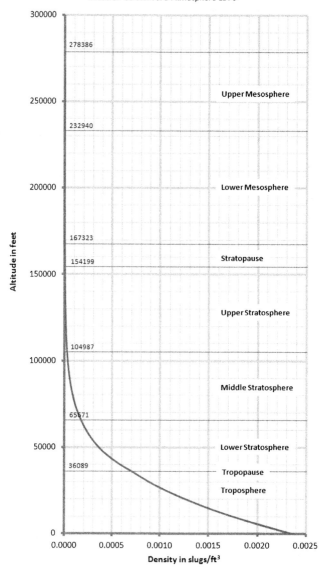

FIGURE A-3 *(Continued).*

```
Function AtmosProperty(H As Single, PropertyID As Byte) As Single
'This function calculates an atmospheric property based on the variable
'PropertyID at the given altitude H in ft, where:
'
'    If PropertyID = 0 then return Temperature ratio
'    If PropertyID = 1 then return Pressure ratio
'    If PropertyID = 2 then return Density ratio
'    If PropertyID = 10 then return Temperature
'    If PropertyID = 11 then return Pressure
'    If PropertyID = 12 then return Density
'
'Initialize
    Dim TempRatio As Single, R As Single
    Dim PressRatio As Single
    Dim DensRatio As Single
```

```
'Select altitude
    If H < 36089 Then
        R = 1 - 0.0000068756 * H
        TempRatio = R
        PressRatio = R ↥ 5.2561
        DensRatio = R ↥ 4.2561
    ElseIf H >= 36089 And H < 65671 Then
        R = -(H - 36089) / 20806
        TempRatio = 0.751865
        PressRatio = 0.223361 * Exp(R)
        DensRatio = 0.297176 * Exp(R)
    ElseIf H >= 65671 And H < 104987 Then
        TempRatio = 0.682457 + H / 945374
        PressRatio = (0.988626 + H / 652600) ↥ -34.1632
        DensRatio = (0.978261 + H / 659515) ↥ -35.1632
    ElseIf H >= 104987 And H < 154199 Then
        TempRatio = 0.482561 + H / 337634
        PressRatio = (0.898309 + H / 181373) ↥ -12.20114
        DensRatio = (0.857003 + H / 190115) ↥ -13.20114
    ElseIf H >= 154199 And H < 167323 Then
        R = -(H - 154199) / 25992
        TempRatio = 0.939268
        PressRatio = 0.00109456 * Exp(R)
        DensRatio = 0.00116533 * Exp(R)
    ElseIf H >= 167323 And H < 232940 Then
        TempRatio = 1.434843 - H / 337634
        PressRatio = (0.838263 - H / 577922) ↥ 12.20114
        DensRatio = (0.79899 - H / 606330) ↥ 11.20114
    ElseIf H >= 232940 And H < 278386 Then
        TempRatio = 1.237723 - H / 472687
        PressRatio = (0.917131 - H / 637919) ↥ 17.0816
        DensRatio = (0.900194 - H / 649922) ↥ 16.0816
    End If
'Output
    Select Case PropertyID
    Case 0
        AtmosProperty = TempRatio
    Case 1
        AtmosProperty = PressRatio
    Case 2
        AtmosProperty = DensRatio
    Case 10
        AtmosProperty = TempRatio * 518.67
    Case 11
        AtmosProperty = PressRatio * 2116
    Case 12
        AtmosProperty = DensRatio * 0.002378
    End Select
End Function
```

References

[1] U.S. Standard Atmosphere. National Oceanic and Atmospheric Administration; 1976. 1976.

[2] Roskam J, Lan Chuan-Tau Edward. Airplane Aerodynamics and Performance. DARcorporation; 1997.

APPENDIX B

The Aerospace Engineer's Formula Sheet

B.1 COST ANALYSIS

Quantity discount factor:

$$QDF = (F_{EXP})^{1.4427 \cdot \ln N}$$

Number of engineering manhours:

$$H_{ENG} = 0.0396 \cdot W_{airframe}^{0.791} \cdot V_H^{1.526} \cdot N^{0.183} \cdot F_{CERT} \cdot F_{CF}$$
$$\cdot F_{COMP} \cdot F_{PRESS}$$

Number of tooling manhours:

$$H_{TOOL} = 1.0032 \cdot W_{airframe}^{0.764} \cdot V_H^{0.899} \cdot N^{0.178} \cdot Q_m^{0.066} \cdot F_{TAPER}$$
$$\cdot F_{CF} \cdot F_{COMP} \cdot F_{PRESS}$$

Number of manufacturing labor manhours:

$$H_{MFG} = 9.6613 \cdot W_{airframe}^{0.74} \cdot V_H^{0.543} \cdot N^{0.524} \cdot F_{CERT} \cdot F_{CF} \cdot F_{COMP}$$

Break-even analysis:

$$N_{BE} = \frac{\text{Total Fixed Cost}}{\text{Unit Sales Price} - \text{Unit Variable Cost}}$$

B.2 CONSTRAINT ANALYSIS

T/W for a level constant velocity turn:

$$\frac{T}{W} = q \left[\frac{C_{Dmin}}{(W/S)} + k \left(\frac{n}{q} \right)^2 \left(\frac{W}{S} \right) \right]$$

T/W for a desired specific energy level:

$$\frac{T}{W} = q \left[\frac{C_{Dmin}}{(W/S)} + k \left(\frac{n}{q} \right)^2 \left(\frac{W}{S} \right) \right] + \frac{P_S}{V}$$

T/W for a desired rate-of-climb:

$$\frac{T}{W} = \frac{V_V}{V} + \frac{q}{(W/S)} C_{Dmin} + \frac{k}{q} \cdot \left(\frac{W}{S} \right)$$

T/W for a desired T-O distance:

$$\frac{T}{W} = \frac{V_{LOF}^2}{2g \cdot S_G} + \frac{q \cdot C_{D\ TO}}{W/S} + \mu \left(1 - \frac{q \cdot C_{L\ TO}}{W/S} \right)$$

T/W for a desired cruise airspeed:

$$\frac{T}{W} = qC_{Dmin} \left(\frac{1}{W/S} \right) + k \left(\frac{1}{q} \right) \left(\frac{W}{S} \right)$$

T/W for a service ceiling:

$$\frac{T}{W} = \frac{V_V}{\sqrt{\frac{2}{\rho} \left(\frac{W}{S} \right)} \sqrt{\frac{k}{3 \cdot C_{Dmin}}}} + 4\sqrt{\frac{k \cdot C_{Dmin}}{3}}$$

C_{Lmax} for a desired stalling speed:

$$C_{Lmax} = \frac{1}{q_{stall}} \left(\frac{W}{S} \right)$$

B.3 WEIGHT ANALYSIS

Design gross weight:

$$W_0 = W_e + W_u$$

Useful load:

$$W_u = W_c + W_f + W_p$$

W_0 consists of:

$$W_0 = W_e + W_c + W_f + W_p$$

Fuel weight ratio:

$$W_f = \left(\frac{W_f}{W_0} \right) W_0$$

Empty weight ratio:

$$W_e = \left(\frac{W_e}{W_0} \right) W_0$$

This can be solved for W_0:

$$W_0 = \frac{W_c + W_p}{\left[1 - \left(\frac{W_e}{W_0}\right) - \left(\frac{W_f}{W_0}\right)\right]}$$

Sailplanes:

$$\frac{W_e}{W_0} = \begin{cases} 0.2950 + 0.0386 \cdot \ln W_0 & \text{if } W_0 \text{ is in lb}_f \\ 0.3255 + 0.0386 \cdot \ln W_0 & \text{if } W_0 \text{ is in kg} \end{cases}$$

Powered sailplanes:

$$\frac{W_e}{W_0} = \begin{cases} 0.3068 + 0.0510 \cdot \ln W_0 & \text{if } W_0 \text{ is in lb}_f \\ 0.3471 + 0.0510 \cdot \ln W_0 & \text{if } W_0 \text{ is in kg} \end{cases}$$

Light Sport Aircraft (land):

$$\frac{W_e}{W_0} = \begin{cases} 1.5451 - 0.1402 \cdot \ln W_0 & \text{if } W_0 \text{ is in lb}_f \\ 1.4343 - 0.1402 \cdot \ln W_0 & \text{if } W_0 \text{ is in kg} \end{cases}$$

Light Sport Aircraft (amphib):

$$\frac{W_e}{W_0} = \begin{cases} 1.6351 - 0.1402 \cdot \ln W_0 & \text{if } W_0 \text{ is in lb}_f \\ 1.5243 - 0.1402 \cdot \ln W_0 & \text{if } W_0 \text{ is in kg} \end{cases}$$

GA single engine:

$$\frac{W_e}{W_0} = \begin{cases} 0.8841 - 0.0333 \cdot \ln W_0 & \text{if } W_0 \text{ is in lb}_f \\ 0.8578 - 0.0333 \cdot \ln W_0 & \text{if } W_0 \text{ is in kg} \end{cases}$$

GA twin piston:

$$\frac{W_e}{W_0} = \begin{cases} 0.4074 + 0.0253 \cdot \ln W_0 & \text{if } W_0 \text{ is in lb}_f \\ 0.4274 + 0.0253 \cdot \ln W_0 & \text{if } W_0 \text{ is in kg} \end{cases}$$

GA twin turboprop:

$$\frac{W_e}{W_0} = \begin{cases} 0.5319 + 0.0066 \cdot \ln W_0 & \text{if } W_0 \text{ is in lb}_f \\ 0.5371 + 0.0066 \cdot \ln W_0 & \text{if } W_0 \text{ is in kg} \end{cases}$$

Agricultural:

$$\frac{W_e}{W_0} = \begin{cases} 1.4029 - 0.0995 \cdot \ln W_0 & \text{if } W_0 \text{ is in lb}_f \\ 1.3242 - 0.0995 \cdot \ln W_0 & \text{if } W_0 \text{ is in kg} \end{cases}$$

Wing Weight

Raymer:

$$W_W = 0.036 \cdot S_W^{0.758} W_{FW}^{0.0035} \left(\frac{AR_W}{\cos^2 \Lambda_{C/4}}\right)^{0.6} q^{0.006} \lambda^{0.04}$$
$$\times \left(\frac{100 \cdot t/c}{\cos \Lambda_{C/4}}\right)^{-0.3} (n_z W_O)^{0.49}$$

Nicolai:

$$W_W = 96.948 \cdot \left[\left(\frac{n_z W_O}{10^5}\right)^{0.65} \left(\frac{AR_W}{\cos^2 \Lambda_{C/4}}\right)^{0.57} \left(\frac{S_W}{100}\right)^{0.61} \right.$$
$$\left. \times \left(\frac{1+\lambda}{2(t/c)}\right)^{0.36} \sqrt{1 + \frac{V_H}{500}}\right]^{0.993}$$

Horizontal Tail (HT) Weight

Raymer:

$$W_{HT} = 0.016 (n_z W_O)^{0.414} q^{0.168} S_{HT}^{0.896} \left(\frac{100 \cdot t/c}{\cos \Lambda_{HT}}\right)^{-0.12}$$
$$\cdot \left(\frac{AR_{HT}}{\cos^2 \Lambda_{HT}}\right)^{0.043} \lambda_{HT}^{-0.02}$$

Nicolai:

$$W_{HT} = 127 \left[\left(\frac{n_z W_O}{10^5}\right)^{0.87} \left(\frac{S_{HT}}{100}\right)^{1.2} \left(\frac{l_{HT}}{10}\right)^{0.483} \sqrt{\frac{b_{HT}}{t_{HT\max}}}\right]^{0.458}$$

Vertical Tail (VT) Weight

Raymer:

$$W_{VT} = 0.073 (1 + 0.2 F_{tail})(n_z W_O)^{0.376} q^{0.122} S_{VT}^{0.873}$$
$$\times \left(\frac{100 \cdot t/c}{\cos \Lambda_{VT}}\right)^{-0.49} \cdot \left(\frac{AR_{VT}}{\cos^2 \Lambda_{VT}}\right)^{0.357} \lambda_{VT}^{0.039}$$

Nicolai:

$$W_{VT} = 98.5 \left[\left(\frac{n_z W_O}{10^5}\right)^{0.87} \left(\frac{S_{VT}}{100}\right)^{1.2} \sqrt{\frac{b_{VT}}{t_{VT\max}}}\right]$$

Fuselage Weight

Raymer:

$$W_{FUS} = 0.052 \cdot S_{FUS}^{1.086} (n_z W_O)^{0.177} l_{HT}^{-0.051} \left(\frac{l_{FS}}{d_{FS}}\right)^{-0.072} q^{0.241}$$
$$+ 11.9 (V_P \Delta P)^{0.271}$$

Nicolai:

$$W_{FUS} = 200\left[\left(\frac{n_z W_O}{10^5}\right)^{0.286}\left(\frac{l_F}{10}\right)^{0.857}\left(\frac{w_F + d_F}{10}\right)\right.$$
$$\left.\times \left(\frac{V_H}{100}\right)^{0.338}\right]^{1.1}$$

Main Landing Gear Weight

Raymer:

$$W_{MLG} = 0.095(n_l W_l)^{0.768}(L_m/12)^{0.409}$$

Nicolai:

$$W_{MNLG} = 0.054(n_l W_l)^{0.684}(L_m/12)^{0.601}$$

Nose Landing Gear Weight

Raymer:

$$W_{NLG} = 0.125(n_l W_l)^{0.566}(L_n/12)^{0.845}$$

Installed Engine Weight

Raymer:

$$W_{EI} = 2.575 W_{ENG}^{0.922} N_{ENG}$$

Nicolai:

$$W_{EI} = 2.575 W_{ENG}^{0.922} N_{ENG}$$

Fuel System Weight

Raymer:

$$W_{FS} = 2.49 Q_{tot}^{0.726}\left(\frac{Q_{tot}}{Q_{tot} + Q_{int}}\right)^{0.363} N_{TANK}^{0.242} N_{ENG}^{0.157}$$

Nicolai:

$$W_{FS} = 2.49\left[Q_{tot}^{0.6}\left(\frac{Q_{tot}}{Q_{tot} + Q_{int}}\right)^{0.3} N_{TANK}^{0.2} N_{ENG}^{0.13}\right]^{1.21}$$

Flight Control System Weight

Raymer:

$$W_{CTRL} = 0.053 l_{FS}^{1.536} b^{0.371}\left(n_z W_O \times 10^{-4}\right)^{0.80}$$

Nicolai:

$$W_{CTRL} = 1.08 W_O^{0.7} \quad \text{(Powered control system)}$$

$$W_{CTRL} = 1.066 W_O^{0.626} \quad \text{(Manual control system)}$$

Hydraulic System Weight

Raymer:

$$W_{HYD} = 0.001 W_O$$

Avionics Systems Weight

Raymer:

$$W_{AV} = 2.117 W_{UAV}^{0.933}$$

Nicolai:

$$W_{AV} = 2.117 W_{UAV}^{0.933}$$

Electrical System

Raymer:

$$W_{EL} = 12.57(W_{FS} + W_{AV})^{0.51}$$

Nicolai:

$$W_{EL} = 12.57(W_{FS} + W_{AV})^{0.51}$$

Air-conditioning and Anti-icing

Raymer:

$$W_{AC} = 0.265 W_O^{0.52} N_{OCC}^{0.68} W_{AV}^{0.17} M^{0.08}$$

Nicolai:

$$W_{AC} = 0.265 W_O^{0.52} N_{OCC}^{0.68} W_{AV}^{0.17} M^{0.08}$$

Furnishings

Raymer:

$$W_{FURN} = 0.0582 W_O - 65$$

Nicolai:

$$W_{FURN} = 34.5 N_{CREW} q_H^{0.25}$$

Center of gravity:

$$X_{CG} = \frac{M_X}{W_{tot}} = \frac{\sum_{i=1}^{N} W_i \times X_i}{W_{tot}}$$

$$Y_{CG} = \frac{M_Y}{W_{tot}} = \frac{\sum_{i=1}^{N} W_i \times Y_i}{W_{tot}}$$

$$Z_{CG} = \frac{M_Z}{W_{tot}} = \frac{\sum_{i=1}^{N} W_i \times Z_i}{W_{tot}}$$

CG in terms of %MGC:

$$X_{CG_{MGC}} = 100 \times \left(\frac{X_{CG} - X_{MGC}}{MGC}\right)$$

To convert torque and RPM to SHP:

$$SHP = \frac{torque \times 2\pi \times RPM}{33000} = \frac{torque \times RPM}{5252}$$

B.4 POWER PLANT

The Basics of Energy, Work, and Power

Concept	Formulation	Units SI System	UK System
Energy			
The conservation of mass-energy is one of the fundamental conservation laws of physics. It basically says that energy can neither be created nor destroyed, but it changes form. The form of energy refers to potential, kinetic, electrical, nuclear, chemical, and other forms of energy.	Kinetic energy: $KE = \frac{1}{2}mV^2$ Potential energy: $PE = mgh$	Joules (J) kWh $1 \text{ kWh} = 3.6 \times 10^6 \text{ J}$	BTU
Work			
Work is defined as the product of force applied to move an object a given distance. Work is also the same as *torque*.	Work $\equiv$ Force $\times$ Distance	Joules N·m	ft·lb$_f$
Power			
Power is defined as the amount of work done in a given time. It is also possible to define it as shown.	Power $\equiv \frac{\text{Work}}{\text{Time}}$ $\equiv \frac{\text{Force} \times \text{Distance}}{\text{Time}}$ $\equiv$ Force $\times$ Speed $\equiv \frac{\text{Torque}}{\text{Time}}$	W J/sec N·m/s	hp ft·lb$_f$/sec
One "horsepower"			
		746 W 0.746 kW	33,000 ft·lb$_f$/min 550 ft·lb$_f$/sec

UK system (T in lb$_f$, V in ft/s):

$$THP = \frac{TV}{550}$$

SI system (T in N, V in m/s):

$$THP = \frac{TV}{746}$$

Gagg and Ferrar model:

$$P = P_{SL}\left(\sigma - \frac{(1-\sigma)}{7.55}\right) = P_{SL}(1.132\sigma - 0.132)$$
$$= P_{SL}\frac{(\sigma - 0.117)}{0.883}$$

Use the Mattingly method to estimate the effect of airspeed and altitude on the thrust of gas turbine engines.

Electric Power

Voltage: $V = \begin{cases} I \times R \\ P/I \\ \sqrt{P \times R} \end{cases}$ Volts   Current: $I = \begin{cases} \sqrt{P/R} \\ P/V \\ V/R \end{cases}$ Amps

Resistance: $R = \begin{cases} V/I \\ V^2/P \\ P/I^2 \end{cases}$ Ohms   Power: $P = \begin{cases} V^2/R \\ R \times I^2 \\ V \times I \end{cases}$ Watts

B.5 WING PLANFORM

Aspect ratio – general:
$AR = b^2/S$

Wing span from AR:
$b = \sqrt{AR \cdot S}$

Aspect ratio – constant chord:
$AR = b/\bar{c}$

Average chord:
$C_{avg} = \left(\dfrac{C_r + C_t}{2}\right) = \dfrac{b}{AR}$

Taper ratio: $\lambda = \dfrac{C_t}{C_r}$

Root chord: $C_r = \dfrac{2S}{b(1+\lambda)}$

Wing loading: $w = \dfrac{W}{S}$

Mean geometric chord:
$MGC = \left(\dfrac{2}{3}\right)C_r\left(\dfrac{1+\lambda+\lambda^2}{1+\lambda}\right)$

y-location of MGC_{LE}:
$y_{MGC} = \left(\dfrac{b}{6}\right)\left(\dfrac{1+2\lambda}{1+\lambda}\right)$

Mean aerodynamic chord:
$MAC \approx MGC$

x-location of MGC_{LE}:
$x_{MGC} = y_{MGC} \tan \Lambda_{LE}$

Wing area: $S = b\left(\dfrac{C_r + C_t}{2}\right)$

Angle of quarter-chord line:

$\tan \Lambda_{C/4} = \tan \Lambda_{LE} + \dfrac{C_r}{2b}(\lambda - 1)$ Average chord:
$C_{avg} = \dfrac{C_R + C_T}{2} = \dfrac{C_R}{2}(1+\lambda)$

General lift curve slope:

$$C_{L\alpha} = \dfrac{2\pi \cdot AR}{2 + \sqrt{\left(\dfrac{AR \cdot \beta}{\kappa}\right)^2 \left(1 + \dfrac{\tan^2 \Lambda_{C/2}}{\beta^2}\right) + 4}}$$

The Law of Effectiveness:

$$P_{MGC} = P_{root} + \dfrac{2y_{MGC}}{b}\left(P_{tip} - P_{root}\right)$$

METHOD 1: Empirical estimation for straight wings:

$$e = 1.78(1 - 0.045 AR^{0.68}) - 0.64$$

METHOD 2: Empirical estimation for swept wings:

Raymer:
$$e = 4.61(1 - 0.045 AR^{0.68})(\cos \Lambda_{LE})^{0.15} - 3.1$$

Brandt:
$$e = \dfrac{2}{2 - AR + \sqrt{4 + AR^2(1 + \tan^2 \Lambda_{tmax})}}$$

METHOD 3: Douglas method:

$$e = \dfrac{1}{\pi \cdot AR \cdot r \cdot C_{Dmin} + 1/((1 + 0.03t - 2t^2)u)}$$

B.6 TAIL SIZING

Horizontal tail volume:
$$V_{HT} = \dfrac{L_{HT} \cdot S_{HT}}{\overline{C}_{REF} \cdot S_{REF}}$$

Vertical tail volume:
$$V_{VT} = \dfrac{L_{VT} \cdot S_{VT}}{b_{REF} \cdot S_{REF}}$$

HT dependent tail arm:
$$L_{HT} = \sqrt{\dfrac{2 \cdot V_{HT} \cdot S_{REF} \cdot C_{REF}}{\pi(R_1 + R_2)}}$$

VT dependent tail arm:
$$L_{VT} = \sqrt{\dfrac{2 \cdot V_{VT} \cdot S_{REF} \cdot b_{REF}}{\pi(R_1 + R_2)}}$$

HT and VT dependent tail arm:
$$L_T = \sqrt{\dfrac{2 \cdot S_{REF}(V_{HT} \cdot C_{REF} + V_{VT} \cdot b_{REF})}{\pi(R_1 + R_2)}}$$

B.7 LIFT AND DRAG

Dynamic pressure: $q = \dfrac{1}{2}\rho V^2$

Induced drag constant k: $k = \dfrac{1}{\pi \cdot AR \cdot e}$

Drag: $D = \dfrac{1}{2}\rho V^2 S C_D = D_{min} + D_i$

Simplified drag model: $C_D = \dfrac{2D}{\rho V^2 S} = \dfrac{2T}{\rho V^2 S} = C_{Dmin} + k C_L^2$

Lift: $L = \dfrac{1}{2}\rho V^2 S C_L$

Lift coefficient: $C_L = \dfrac{2L}{\rho V^2 S} \approx \dfrac{2W}{\rho V^2 S}$

Span efficiency:
$e = 1.78(1 - 0.045 AR^{0.68}) - 0.64$ Straight wing
$e = 4.61(1 - 0.045 AR^{0.68})(\cos \Lambda_{LE})^{0.15} - 3.1$ Swept back

Adjusted drag model: $C_D = C_{Dmin} + k(C_L - C_{LminD})^2$

B.8 THE PROPELLER

Power coefficient:
$$C_P = \frac{P}{\rho n^3 D^5} = \frac{550 \times P_{BHP}}{\rho \left(\frac{RPM}{60}\right)^3 D^5} = \frac{118\,800\,000 \times P_{BHP}}{\rho \cdot RPM^3 \cdot D^5}$$

Thrust coefficient:
$$C_T = \frac{T}{\rho n^2 D^4} = \frac{3600 \cdot T}{\rho \cdot RPM^2 D^4}$$

Torque coefficient:
$$C_Q = \frac{Q}{\rho n^2 D^5} = \frac{3600 \cdot Q}{\rho \cdot RPM^2 \cdot D^5} = \frac{C_P}{2\pi}$$

Power-torque relation:
$$C_Q = \frac{Q}{\rho n^2 D^5} = \frac{C_P}{2\pi} = \frac{P/\rho n^3 D^5}{2\pi} \;\Rightarrow\; P = 2\pi n Q$$

Propeller efficiency:
$$\eta_p = \frac{TV}{P} = \frac{TV}{550\,BHP} = J\frac{C_T}{C_P}$$

Propeller thrust:
$$T = \frac{\eta_p P}{V} = \frac{\eta_p \times 550 \times P_{BHP}}{V}$$

Thrust quadratic spline:
$$T(V) = \left(\frac{T_{STATIC} - 2T_{max}}{V_{max}^2}\right)V^2 + \left(\frac{3T_{max} - 2T_{STATIC}}{V_{max}}\right)V + T_{STATIC}$$

Ideal efficiency:
$$\eta_i = \frac{1}{(1 + w/V_0)} = \frac{2}{1 + \sqrt{1 + C_T}}$$

Static thrust, T_{STATIC}:
$$T_{STATIC} = P^{2/3}(2\rho A_2)^{1/3}$$

Corrected static thrust:
$$T_{STATIC} = 0.85 P^{2/3}(2\rho A_2)^{1/3}\left(1 - \frac{A_{spinner}}{A_2}\right)$$

B.9 THE ATMOSPHERE

Pressure: $p = p_0(1 - 0.0000068756h)^{5.2561}$
Density: $\rho = \rho_0(1 - 0.0000068756h)^{4.2561}$
Temperature: $T = T_0(1 - 0.0000068756h)$
Air pressure:
$P = 2216(1 - 0.0000068753 \times H_{REF})^{5.2561}$
Air density:
$\rho = 0.002378(1 - 0.0000068753 \times H_{REF})^{4.2561}$
Pressure ratio:
$\delta = \frac{p}{p_0} = (1 - 0.0000068756h)^{5.2561} = \theta^{5.2561}$
Density ratio:
$\sigma = \frac{\rho}{\rho_0} = (1 - 0.0000068756h)^{4.2561} = \theta^{4.2561} = \frac{\delta}{\theta}$
Temperature ratio: $\theta = \frac{T}{T_0} = (1 - 0.0000068756h)$
Pressure altitude: $H_P = 145442\left(1 - \left(\frac{p}{p_0}\right)^{0.19026}\right)$
Density altitude: $H_\rho = 145442\left(1 - \left(\frac{\rho}{\rho_0}\right)^{0.234957}\right)$

B.10 AIRSPEEDS

Calibrated airspeed:
$$KCAS = KIAS + \Delta error$$

Equivalent airspeed
$$KEAS = KCAS\sqrt{\frac{P}{P_0}}\sqrt{\frac{\left(\frac{q_c}{P}+1\right)^{0.286}-1}{\left(\frac{q_c}{P_0}+1\right)^{0.286}-1}}$$

True airspeed:
$$KTAS = \frac{KEAS}{\sqrt{\rho/\rho_{SL}}} \approx \frac{KCAS}{\sqrt{\rho/\rho_{SL}}} = \frac{KCAS}{\sqrt{\sigma}}$$

Mach number:
$$M = \frac{V}{a_0} = \frac{V}{\sqrt{\gamma RT}} = \frac{V}{1116\sqrt{1 - 0.0000068756H}}$$

Speed of sound:
$$a_0 = \sqrt{\gamma RT} \approx 49.01\sqrt{OAT_{°R}}$$

B.11 TAKE-OFF

Torenbeek's balanced field length:

$$BFL = \frac{0.863}{1 + 2.3\Delta\gamma_2}\left(\frac{W_{to}/S}{\rho g C_{L2}} + h_{to}\right)\left(2.7 + \frac{1}{\overline{T}/W_{to} - \mu}\right) + \left(\frac{\Delta S_{to}}{\sqrt{\sigma}}\right)$$

Acceleration on a flat runway:

$$\frac{dV}{dt} = \frac{g}{W}[T - D - \mu(W - L)]$$

Acceleration on an uphill slope γ:

$$\frac{dV}{dt} = \frac{g}{W}[T - D - \mu(W\cos\gamma - L) - W\sin\gamma]$$

Lift-off speed:

$$V_{LOF} = 1.1 \times V_{S1} = 1.1\sqrt{\frac{2W}{\rho S C_{Lmax}}} \approx 1.556\sqrt{\frac{W}{\rho S C_{Lmax}}}$$

Time to lift-off:

$$t = \sqrt{\frac{2S_G}{a}} + \begin{cases} 1 & \text{for small aircraft} \\ 3 & \text{for large aircraft} \end{cases}$$

Ground run:

$$S_G = \frac{V_{LOF}^2 W}{2g[T - D - \mu(W - L)]_{at\ (V_{LOF}/\sqrt{2})}}$$

Ground run for piston props:

$$S_G = \frac{V_{LOF}^2 W}{\frac{50051 \times \eta_P \times P_{BHP}}{V_{LOF}} + 16.09\rho V_{LOF}^2 S(\mu C_{L\ TO} - C_{D\ TO}) - 64.35\mu W}$$

Rotation distance:

$$S_{ROT} = \begin{cases} |V_{LOF}| & \text{for small aircraft} \\ 3|V_{LOF}| & \text{for large aircraft} \end{cases}$$

Climb angle:

$$\sin\theta_{climb} = \frac{T - D}{W} = \frac{T}{W} - \frac{1}{L/D}$$

Transition distance:

$$S_{TR} = R\sin\theta_{climb} \approx 0.2156 \times V_{S1}^2 \times \left(\frac{T}{W} - \frac{1}{L/D}\right)$$

Transition height:

$$h_{TR} = R(1 - \cos\theta_{climb})$$

B.12 CLIMB, CRUISE, AND MANEUVERING FLIGHT

Jets in climb:

$$V_V \equiv \frac{TV - DV}{W} = V\sin\theta$$

Propellers in climb:

$$V_V \equiv \frac{P_{AV} - P_{REQ}}{W} = \frac{\eta_p \cdot P_{ENG} - P_{REQ}}{W} = V\sin\theta$$

Rate-of-climb:

$$ROC \equiv 60\left(\frac{TV - DV}{W}\right) = 60\left(\frac{P_{AV} - P_{REQ}}{W}\right)$$

General angle-of-climb (AOC):

$$\sin\theta = \frac{T}{W} - \frac{1}{L/D}$$

Min/max airspeed:

$$V = \sqrt{\frac{T \pm \sqrt{T^2 - 4C_{Dmin}kW^2}}{\rho S C_{Dmin}}}$$

Stalling speed:

$$V_S = \sqrt{\frac{2W}{\rho S C_{L_{max}}}}$$

Level stalling speed at load factor n:

$$V_S = \sqrt{\frac{2nW}{\rho S C_{L_{max}}}}$$

Stalling speed at angle of bank ϕ:

$$V_S = \sqrt{\frac{2W}{\rho S C_{L_{max}} \cos\phi}} = \frac{V_{S_{level}}}{\sqrt{\cos\phi}}$$

Minimum thrust required, V_{TRmin}, or best glide speed, V_{BG}, V_{LDmax}:

$$V_{TRmin} = \sqrt{\frac{2}{\rho}\left(\frac{W}{S}\right)}\sqrt{\frac{k}{C_{Dmin}}}$$

Max L/D:

$$LD_{max} = \left(\frac{C_L}{C_D}\right)_{max} = \frac{1}{\sqrt{4 \cdot C_{Dmin} \cdot k}}$$

Carson's airspeed:

$$V_{CAR} = 3^{0.25} V_{LDmax} \approx 1.32 V_{LDmax}$$

Power required:

$$P_{REQ} = \sqrt{\frac{2W^3 C_D^2}{\rho S C_L^3}}$$

Bank angle:

$$\phi = \cos^{-1}\left(\frac{1}{n}\right)$$

Load factor:

$$n = \frac{1}{\cos \phi} = \left(\frac{T}{W}\right)\left(\frac{L}{D}\right)$$

Turn radius:

$$R_{turn} = \frac{V^2}{n \cdot g \cdot \sin \phi} = \frac{V^2}{g \sqrt{n^2 - 1}}$$

Time to turn ψ degrees:

$$t_\psi = \frac{R_{turn}}{V}\left(\psi \frac{\pi}{180}\right)$$

Turn rate in radians/sec:

$$\dot{\psi} = \frac{g\sqrt{n^2 - 1}}{V} = \frac{V}{R_{turn}}$$

Thrust required at a load factor n:

$$T_R = qS\left[C_{Dmin} + k\left(\frac{nW}{qS}\right)^2\right] + D_{trim}$$

Load factor that can be sustained at a given thrust T and airspeed V:

$$n = \frac{qS}{W}\sqrt{\frac{1}{k}\left(\frac{T}{qS} - C_{Dmin}\right)}$$

Airspeed for a given C_L:

$$V = \sqrt{\frac{2W}{\rho S C_L}\left(\frac{1}{\cos \phi}\right)}$$

Max sustainable load factor:

$$n_{max} = \left(\frac{T_{max}}{W}\right) LD_{max}$$

Maximum sustainable turn rate, $\dot{\psi}_{max}$:

$$\dot{\psi}_{max} = \frac{g\sqrt{n_{max}^2 - 1}}{V_{max\dot{\psi}}}$$

Minimum sustainable turning radius:

$$R_{min} = \frac{V_{R_{min}}^2}{g\sqrt{n_{R_{min}}^2 - 1}}$$

where: $n_{R_{min}} = \sqrt{2 - 1/n_{max}^2}$ = load factor and $V_{R_{min}} = 2\sqrt{\frac{(W/S)}{(T/W)}\frac{k}{\rho}}$ = airspeed for minimum turning radius

Maximum level bank angle:

$$\phi_{max} = \cos^{-1}\left(\frac{1}{n_{max}}\right)$$

Specific excess power:

$$P_S \equiv \frac{P_{EX}}{W} = \frac{TV - DV}{W} = \frac{dh}{dt} + \frac{V}{g}\frac{dV}{dt}$$

Specific energy and height:

$$E_{total} = mgh + \frac{1}{2}mV^2 = Wh + \frac{1}{2}\frac{W}{g}V^2 \quad \text{and} \quad H_E \equiv \frac{E_{total}}{W}$$
$$= h + \frac{V^2}{2g}$$

B.13 RANGE AND ENDURANCE

Range profile 1:

$$R = \frac{V}{c_t\sqrt{kC_{D0}}}\left[\tan^{-1}\left(\frac{2\sqrt{k}}{\rho V^2 S \sqrt{C_{D0}}} W_{ini}\right) - \tan^{-1}\left(\frac{2\sqrt{k}}{\rho V^2 S \sqrt{C_{D0}}} W_{fin}\right)\right]$$

Range profile 2:
$$R = \frac{1}{c_t}\frac{\sqrt{C_L}}{C_D}\frac{2\sqrt{2}}{\sqrt{\rho S}}\left(\sqrt{W_{ini}} - \sqrt{W_{fin}}\right)$$
$$= \frac{1}{c_t C_D}\sqrt{\frac{8C_L}{\rho S}}\left(\sqrt{W_{ini}} - \sqrt{W_{fin}}\right)$$

Range profile 3:
$$R = \frac{V}{c_t}\frac{C_L}{C_D}\ln\left(\frac{W_{ini}}{W_{fin}}\right)$$

Endurance profile 1:
$$E = \frac{1}{c_t\sqrt{kC_{D0}}}\left[\tan^{-1}\left(\frac{2\sqrt{k}}{\rho V^2 S\sqrt{C_{D0}}}W_{ini}\right) - \tan^{-1}\left(\frac{2\sqrt{k}}{\rho V^2 S\sqrt{C_{D0}}}W_{fin}\right)\right]$$

Endurance profile 2:
$$E = \frac{1}{c_t}\frac{C_L}{C_D}\ln\left(\frac{W_{ini}}{W_{fin}}\right)$$

Endurance profile 3:
$$E = \frac{1}{c_t}\frac{C_L}{C_D}\ln\left(\frac{W_{ini}}{W_{fin}}\right)$$

TSFC for a jet:
$$c_t = \left(\frac{c_{jet}}{3600}\right)\left\langle\frac{lb_f/\sec}{lb_f}\right\rangle$$

where c_{jet} = SFC for a jet

TSFC for a piston engine:
$$c_t = \frac{c_{bhp}V}{1980000\eta_p}\left\langle\frac{1}{\sec}\right\rangle$$

where c_{bhp} = SFC for a piston

Specific range:
$$SR \equiv \frac{\text{Distance}}{\text{Quantity of Fuel}} = \frac{R}{W_f} = \frac{V_{TAS}}{\dot{w}_{fuel}}$$
$$= \frac{\text{True Airspeed}}{\text{Fuel Weight Flow}}$$

Index

Note: Page numbers with "*f*" denote figures; "*t*" tables; and "*b*" boxes.

A

A-tail, 495–496
Absolute ceiling, 840–842, 841f
Absolute viscosity. *See* Dynamic viscosity
AC. *See* Advisory circular; Standard airworthiness certificate
Acceleration
  on flat runway, 48, 798
  fuselage reducing, 602–603
  graph for SR22, 812f
  ground run distance estimation with, 802–803
  to higher airspeed, 289
  thrust estimation, 807
  thrust-to-weight ratios, 799
  on uphill slope, 48, 798
AD. *See* Airworthiness Directive
Additive drag, 698–700, 708
  of antennas, 731
    placement and shape, 732f
  of blisters, 728
    drag contribution, 729f
    example, 729b
  of canopies, 727
    canopy styles, 728f
    drag coefficients for canopy styles, 728, 729f
  coefficient, 708, 715
  due to compressibility effects, 730
  cooling drag, 714
    cooling drag coefficient, 714
    idealization of engine installation, 714f
  CRUD, 709
    component contribution, 710t
    Seversky P-35, 709, 709f
    twin-engine business jet, 710
    WR-L-489, 709–710
  of deployed flaps, 725
    Δ_1 function, 726
    Δ_2 function, 726, 727t
    increase of C_{Dmin} due to flaps, 725–726
  drag correction for cockpit windows, 726
    drag of blunt ordinary and undercut, 727, 728f
    drag of conventional cockpit windows, 727
  drag of various geometry
    critical Reynolds number, 733
    cross-flow principle, 734, 734f
    drag of 3D objects, 734–735
    flow regions on sphere, 733, 733f
    3D drag coefficient, 732, 732f, 734f
    2D cross-section, 3D drag of, 733–734
    2D drag coefficient, 733f
  of floats, 724
    float geometries, 725f
    special NACA-designed floats, 724
  geometric shape of protrusions, 708
  of gun ports in airplane nose, 736
  due to ice accretion, 736–737
  landing gear drag
    of fixed landing gear struts with tires, 718–722
    of nose landing gear, 723–724
    of retractable landing gear, 722–723
    of tires only, 718
    of tires with wheel fairings, 718
  landing gear pant fairings, 715–718
    thick fairings, 717–718
  lift-induced drag corrections
    example, 738b–739b
    in ground effect, 738–739
    pressure difference, 737f
    vortex core location, 738f
    wingtip correction, 737–738
    wingtips effect, 737f
  of parachutes, 735
    drag coefficient, 735, 735f
    example, 735b
    total drag force evaluation, 735
  of sanded walkway on wing, 736
  scaling, 708
  stopped propellers, 731
  of streamlined external fuel tanks, 736, 736f
  streamlined struts drag, 715–718
  trim drag, 710
    balancing force, 710
    consequence of longitudinal stability, 711
    wing-horizontal tail combination, 711
    wing-horizontal tail-thrustline combination, 712
    windmilling propellers, 711
  due to wing washout, 736
  wing-like surface drag, 715
Adiabatic compression, 192, 217
Adiabatic expansion, 217, 224
Advance ratio, 611–613, 630, 841
Advisory circular (AC), 15
Aerobatic category, 482
Aerofoil software, 254–255
Aesthetics, 7–8, 10, 77–78, 461
Aft spar, 122
Aft swept planform, 336, 374–375
Aileron
  design requirements, 952
  differential, 952

Aileron (*Continued*)
    frise, 951
    hinges, 123
    plain flap, 951
    sizing, 960
        maximizing responsiveness, 962
        steady-state roll rate estimation, 960
    slot-lip, 951–952
    spoiler-flap, 951
Aileron authority, 953, 954b–955b, 968
Aileron deflection, 596
    angle, 960
    change in lift coefficient due to, 953, 954–955b, 960f
    impact on flow field, 950f
Aircraft conceptual design algorithm, 15
    for GA aircraft, 16, 17t
    implementation of, 15–16
        modern spreadsheet software, 16
        organizational hierarchy of spreadsheet, 18f
        tail sizing worksheet, 16
    modern spreadsheet, 15
Aircraft design process, 11
    elementary outline, 11, 12f
    for GA aircraft, 12f, 13
    per Torenbeek, 12–13, 12f
    regulatory concepts, 13–15
        advisory circular, 15
        airworthiness directives, 14
        maintenance requirements, 14
        parts manufacturer approval, 15
        service bulletin, 15
        special airworthiness certificate, 14
        standard airworthiness certificate, 14
        supplemental type certificate, 14
        technical standard order, 15
        technical standard order authorization, 15
        type certificate, 13–14
Aircraft development cost, 36
    of business aircraft, 44
        avionics cost, 46
        certify, total cost to, 46
        development support, total cost of, 45
        engineering, total cost of, 45
        engineering man-hours, 44–45
        flight test operations, total cost of, 45
        manufacturing, total cost of, 45
        manufacturing labor man-hours, 45
        materials, total cost of, 45–46
        power plant cost, 46
        quality control, total cost of, 45
        retractable landing gear cost, 46
        tooling, total cost of, 45
        tooling man-hours, 45
    of GA aircraft, 37
        avionics cost, 41
        break-even analysis, 43, 44b
        certify, total cost of, 40–41, 40b–41b
        cost analysis, 39
        development support, total cost of, 39–40
        engineering, total cost of, 39
        engineering man-hours, 37–38
        example, 38b
        flight test operations, total cost of, 40
        manufacturing, total cost of, 40
        manufacturing labor man-hours, 38–39
        materials, total cost of, 40
        power plant cost, 41–43, 42b–43b
        product liability costs, 37
        quality Control, total cost of, 40
        retractable landing gear cost, 41
        tooling, total cost of, 40
        tooling man-hours, 38
    project cost analysis, 37t
    QDF, 36
        depends on experience effectiveness, 36–37, 36f
        experience effectiveness adjustment factor, 36
Airfoil cross-sectional area, 308–309
Airfoil design, 254
    AeroFoil software, 255
    design process, 256
    JavaFoil software, 256
    PROFILE software, 255
    types, 254–255
    XFLR5, 255, 255f
    Xfoil, 255
Airfoil direct design, 256
Airfoil inverse design, 256
Airfoil selection matrix, 289
    critical Mach number, 289
    guidelines, 290–293
    impact on drag, 289
    impact on flow separation, 289
    impact on longitudinal trim, 289
    impact on maximum lift and stall handling, 289
    impact on structural depth, 289–290
    impact on wing-fuselage juncture, 289
    NACA recommended criteria, 294
    target zero-lift AOA evaluation, 290
Airload actuated leading edge slat, 413–414, 413f
    on McDonnell-Douglas A-4 Skyhawk, 414f
Airships, 3t
    form factors for airship hulls, 703–708
Airspeed effect on turbofan thrust, 201–202
Airspeed effect on turbojet thrust, 200
Airspeed effect on turboprop thrust, 198–199
Airspeed indicator (ASI), 768
    markings of analog, 768f
    markings of modern, 769f
    pressure sources, 769
Airworthiness Directive (AD), 14
Altitude
    absolute/service ceiling, 16
    sensitivity, 29
    sensitivity plot, 20f
    thrust as function of, 28
    time to, 15
Altitude effect on piston engine power, 192–193
    air compression, 194
    altitude impact on engine, 194
    altitude-dependency model, 193
    Gagg and Ferrar model, 193
    Ideal manifold pressure, 194
    initial pressure in cylinders, 193
    model comparison, 193, 193f
    power estimation, 194
    power settings, 194
    supercharger, 194
    turbo-normalization, 194
    turbocharger, 194

Altitude effect on turbofan thrust, 201–202
Altitude effect on turbojet thrust, 200
Altitude effect on turboprop thrust, 198–199
Aluminum alloy, 103
    applications in GA aircraft, 106t
    designation of, 104
    extrusion for, 101
    flaws, 104
        endurance limit, 104–105
        galvanic corrosion, 105–106
        stress corrosion, 105
    identification, 104
    pre-cure, 114
    properties of, 103–104, 106t
    sheet metal thicknesses for, 106t
Amphibian, 87, 968
Analysis technology, 971–972
Angle-of-attack (AOA), 238, 822, 926
    airspeed, lift coefficient and, 344
    change in induced, 387
    for design lift coefficient, 343
    for maximum lift coefficient, 343
    movement of transition points with, 677f
    non-linear lift curve, 343
    stall, 358
    wing stall, 359–360
    at zero lift, 239, 240f, 343
Angle-of-climb (AOC), 821–824
Angle-of-descent, 927
Angle-of-glide (AOG), 927
Angle-of-incidence (AOI), 467
    lift coefficient value effect, 241
    recommended HT, 327
    symmetrical airfoil at, 127
    wing, 325–329
Angle-of-yaw (AOY), 964
Angular momentum, 596
Anhedral, 318–319
AOA. See Angle-of-attack
AOC. See Angle-of-climb
AOG. See Angle-of-glide
AOI. See Angle-of-incidence
AOY. See Angle-of-yaw
Apophenia, 301
Approach distance, 936, 939–940
Approach for landing, 939
ASI. See Airspeed indicator
Aspect ratio. See wing aspect ratio
Asphalt, 798t, 938t
ASTM standards, 3–4
Asymmetric aircraft, 378–379
Atmosphere, 985
Atmosphere, chemical composition, 985–986
Atmosphere, formation of, 985
Atmosphere, layer classification, 986
Atmospheric modeling, 763, 985, 992
    air deviation density, 765–767, 766b, 989
        change due to humidity, 766–767, 767b
        formulas for standard atmosphere, 767
        standard properties, 766t
    atmospheric ambient temperature, 763–764, 987–988
    atmospheric pressure and density for altitudes, 764, 989
    atmospheric property ratios, 764–765, 989
    computer code A-1, 993
    density altitude, 765, 990

    example, 764b
    Mach number, 991
    pressure altitude, 767, 990
    Reynolds number, 991
    sound speed, 991, 991b
    temperature constants in troposphere, 764t
    viscosity
        dynamic or absolute, 990–991
        kinematic viscosity, 991
Atmospheric pressure, density, temperature, 763–764
Atmospheric property ratio, 764–765, 989
Automated Handley-Page leading edge slat. See Airload actuated
Average chord, 304, 307
    for HT, 507, 514
    for VT, 511
Avgas, 187

B
Balanced field length, 795–796, 795f
    analysis effort, 795
    empirical expression, 796
    example, 796b
    Torenbeek's method, 796–797
Basic drag, 666
Basic drag coefficient, 674–675
Best angle-of-climb, 832
Best angle-of-climb airspeed, 746, 828
Best endurance airspeed, 668, 849, 867
Best glide
    lift coefficient, 741
    ration, 869
Best glide airspeed, 867, 931–932
    comparison to, 867, 877
Best range airspeed, 867
    for Jet, 875–877
        best glide speed comparison, 877
        Carson's airspeed, 877
        requirement for maximum range, 875–876
Best rate-of-climb, 59, 749, 831
    for propeller-powered airplane, 834
Best rate-of-climb airspeed, 746
    example, 835b
    for propeller-powered aircraft, 867
    for selected classes of aircraft, 842, 842t
BET. See Blade element theory
Beta range, 586
BHP. See Brake horsepower
Biot-Savart law, 351, 380
Biplane, 77, 86
Blade element, 640
    chord, 641
    induced velocity at, 652
    local velocity of, 654
    Reynolds number for, 646
Blade element theory (BET), 437–438, 583, 638, 640
    advantages, 641
    compressibility corrections, 654
        drag correction, 655
        lift correction, 654
    formulation
        differential lift and drag forces, 641
        lift coefficient of element airfoil, 641
        observation 1, 641–643, 643b–644b
        parameters, 641

Blade element theory (BET) (Continued)
 propeller, 641, 641f
  table columns 11–17, 645–646
  table columns 18–25, 646–650
  table columns 2–9, 645
 hub loss corrections, 655–656
 induced AOA, 650–654
  difference calculation, 652–654
  example, 652b–654b
  initial value, 652
  next value calculation, 652
 limitations, 641
 Prandtl's tip, 655–656
 primitive, 640–641
 propeller induced velocity, 656
Blended wing-body aircraft (BWB), 342
Blind-rivet, 102
Boundary layer
 fences, 976–977
 laminar, 242–243, 250
  6-series airfoils, 264
  airfoil selection effect, 373
  nature of fluid flow, 248f
  skin friction coefficient, 540
  theoretical extent of, 374f
 mixed, 665, 675, 676f
 Reynolds number, 247
 thickness estimation, 250, 250f
 transition, 248–249
 turbulent, 248–249
  local airspeeds, 526
  nature of fluid flow, 248f
  skin friction coefficient, 542
Brake caliper, 563–564
Brake horsepower (BHP), 185, 620–621
Brake-release, 802
Brakes, 559–560
Braking, 549
 devices, 940–941
Braking, landing, 549
Braking distance, 936
Break-even analysis, 43, 44f
Breguet endurance equation, 898
Breguet range equation, 897
Bubble-drag, 981
Buckingham's Pi-theorem, 237–238
Bulkhead, 117
 firewalls, 209
 pressure, 130
Bungee landing gear, 560–562, 563f
Butterfly-tail. See V-tail
BWB. See Blended wing-body aircraft

C

Cabin dimensions, 533, 543t–544t, 545
CAFÉ, 182, 910–911
Calibrated airspeed, 47, 770, 861
Camber, 256
 effect of, 276
 line, 266
 polyhedral, 454f
 relative magnitude of, 262
Cambered-span wing. See Polyhedral wing
Canard, 496–497
Canonical pressure coefficient, 241–242

Canopy, 88
 acrylic, 88–89
 smooth, 701
 styles evaluation, 728f
Cantilever wing, 587–588
Carbon, 106–107
Carpet plot, 69f
 creation, 68–69
 cruise speed, 67–69
 stalling speed, 353f
Carson's airspeed, 8, 849, 877
Cast alloy, 103–104
Castering, 553
Casting, 549
Center of gravity (CG), 46, 84, 164
 travel during flight, 972
Center of mass (CM), 164
Center of pressure, 243–244, 244f
Certification
 basis for classes of aircraft, 3t
 in LSA category, 21
 requirements for GA aircraft, 936
Certification Standard (CS), 6, 13
CFD. See Computational fluid dynamics
CG. See Center of gravity
CG envelope, 168
Chord, 307
Chordline, 256
Circular Advisory. See Advisory circular (AC)
Climb, 825
 angle, 802–804, 813
 capability, 968
 gradient, 825
 T/W for desired rate of, 58–59
Climb airspeed, 749
Climb angle, 49, 813, 822, 827
Climb gradient, 800–801
Climb performance, 594
 evaluation of, 824
 extracting C_{Dmin} using, 746
 spreadsheet to estimating, 840f
Climb propeller, 594
CM. See Center of mass
Cockpit dimensions, 533f
Cockpit layout, 532–535
 in large business jet, 536f
 typical seating for, 535f
Coefficient of drag
 3D, 732, 732f, 734f
 2D, 733f
Coefficient of lift. See Lift coefficient
Commercial aircraft
 passenger door in, 130
 pitch of seats in, 536f
 turboprops and turbofans for, 183
 wing flex in, 349
Commercial aviation regulations, 7, 78, 168
Commuter aircraft, 124, 818t
Commuter category, 779
Composite images, 5
Composite material, 108
 aircraft construction methodologies, 114
 fabrication methods, 114–115
 fibers, 111–112
 glass transition temperature, 113

pre-cure, 114
pros and cons of, 110–111
resin, 112–113
    thermoplastics, 113
    thermosets, 112–113
sandwich core materials, 113
structural analysis of, 109–110
types of, 108–109
Composites, 6, 8, 99, 108, 111
Compound surface flex, 100–101, 100f
Compound taper, 335, 336f
Compressibility, 278
    corrections, 281
        of drag, 655
        of lift, 654
        method, 281
    modeling, 280–281
Compressibility effect, 278, 668
    on drag, 278–279
    drag due to, 730
    for F-104 Starfighter, 468
    on lift, 278
    on lift and drag exemplified, 279
    on pitching moment, 279–280
Compressible Bernoulli equation, 219, 246
Computational fluid dynamics (CFD), 27–28
    advances in, 430
    Kutta-Joukowski theorem use in, 247
    Prandtl-Betz Integration, 690
    software use, 246
Conceptual design, 10, 15
    algorithm for GA aircraft, 16
    implementation of, 16
    non-planar wings, 453
    of propeller-powered aircraft, 608
Configuration layout
    cabin configurations, 88–89, 89f
    configuration selection matrix, 92–93
    engine placement, 89–91
    fundamentals, 82–93
    landing gear configurations, 91–92
    propeller configuration, 89, 89f
    tail configurations, 92
    vertical wing location, 82–86
        Boeing 737–800 in landing configuration, 83f
        Consolidated PBY-5 Catalina, 85f
        field-of-view, 84
        impact on airframe design, 84–85
        impact on flight, 85
        nomenclature, 84f
        operational characteristics, 85
        parasol wings, 85–86
        properties of aircraft, 83t
    wing configuration, 86, 86f
    wing dihedral, 86–87
        effect, 87f
        nomenclature, 87f
    wing structural configuration, 87–88
        cantilever or braced with struts, 88f
        shear and moment diagrams, 88f
Constant airspeed/constant altitude, 899–900, 907
Constant altitude/constant attitude, 901–902
Constant chord planform (Hershey bar), 303
Constant speed propeller, 41–43, 586, 594

cubic spline method for, 628–630
fixed vs., 594–595, 595f
propeller efficiency table for, 630f
section view of inside, 595f
Constraint analysis, 56–57
    general methodology, 58–63
        additional notes, 59–61
        cruise airspeed, T/W for desired, 59
        example, 64b
        level constant-velocity turn, T/W for, 58
        rate of climb, T/W for desired, 58–59
        service ceiling, T/W for, 59
        specific energy level, T/W for desired, 58
        T-O distance, T/W for desired, 59
    optimum design points, 58
    typical design space, 57f
Constraint diagram, 60f
    banking, 887
    C_{Lmax} for desired stalling speed, 66
    stall speed limits, 65–66, 66f
Consumer Price Index (CPI), 37
Control horn, 975
Control surface
    deflection effect, 109f
    fabrication and installation of, 126–127
    sizing, 948–964
        aileron sizing, 960–962
        control surface hinge moments, 948–949
        pitch control fundamentals, 962–964
        roll control fundamentals, 949–960
        yaw control fundamentals, 964
Control system
    flap, 285, 462
    HM reaction, 949f
    mixer, 491
    side-effects on swept wings, 951
Control system harmony, 968
Control system jamming, 970
Control system stretching, 969–970
Conventional-tail, 92, 483–485, 484f
Cooling, 213
    air, 213
    drag, 714, 715
    of pusher configurations, 213
    quenching, 107
Cost analysis, 1, 39
    methods, 6
    project, 37t
Cost function, 72, 72f, 72b
Cost-effectiveness, 56
Cowl flap, 216–217
Cowling, 587, 603–604
CPI. See Consumer Price Index
Cranked, 338–340
Cranked dihedral, 87
Crew weight, 135
Critical Mach number, 273, 278–279, 281–282
    for airfoil selection, 289
    correction to, 317
    limitation with, 375–376
    NACA 6-series airfoils, 294
    sweep angle impact on, 317
Crosswind
    capability at touch-down, 78
    snow bank collision effect, 85

Cruciform-tail, 92, 486
Cruise airspeed, 59
Cruise flap, 284–285, 286f
Cruise lift coefficient, 313
    airfoils drag at, 294
    critical Mach numbers, 294
Cruise performance, 744
Cruise propeller, 594, 745, 747
Cruise segment, 896, 897f
    for endurance analysis, 897–898
    for range analysis, 896
    range *vs.* weight, 896–897
    in terms of time of flight, 897f
    with transposed axes, 897f
Cruising speed, 119
    AOA, 367
    design, 776
    streamtube, 225
CS. *See* Certification Standard
Cuffs, 978
Cutaway drawings, 30

D
DAPCA-IV, 6, 34, 36
Davis wing, 272–273
Dead rise angle, 724
Decalage angle, 329f
Decalage for biplane, 328–329
Delta wing, 340–341, 355
Delta wing planform, 340f
Density
    of air deviations, 765–767
    of aviation gasoline, 187
    change due to humidity, 766–767
    energy, 204
Density altitude, 765, 767
Descent
    positions, 415f
    at specific condition, 161–162
    weight ratios for, 916t, 921
Descent performance
    descent analysis methods
        airspeed of minimum sink rate, 931
        best glide speed, 931–932
        equilibrium glide speed, 929
        general angle-of-descent, 927
        general rate-of-descent, 927, 928b
        glide distance, 932, 933b
        minimum angle-of-descent, 931, 932b
        sink rate, 930
    descent maneuver, fundamental relations, 926
        general 2D free-body diagram for aircraft, 926–927
        planar equations of motion, 927
Design airspeed for maximum gust intensity, 777
Design algorithm, 779
    for GA aircraft, 766t, 783–785
    implementation, 785
    modern spreadsheet, 778–779
Design gross weight, 38, 137–138
Design lift coefficient, 343
    angle-of-attack for, 343
    in NACA 6-series airfoils, 265–266
Design of experiments (DOE), 56, 69–72
Design process. *See* Aircraft design process; Airfoil design
Design space, 57–58, 57f

Detail design, 10–11, 13
Detailed weight analysis, 134–135, 141–142
Development cost. *See* Aircraft development cost
Development program phase, 11
Differential ailerons, 952, 952f
Diffuser, 223
    inlet, 224–225
Diffuser length, 224–225
Dihedral, 318
    configurations, 87
    wing dihedral, 86–87
        effect, 87f
        nomenclature, 87f
Dihedral angle, 86–87, 318
Dihedral effect, 86, 87f, 477
Directional stability, 71, 476, 964
    formulation of, 500–501
    improvement to, 494
    requirement for, 476f
    trends of, 479f
Disc brake, 559
Dive airspeed, 776–777
DOE. *See* Design of experiments
Dorsal fin, 477–480, 479f, 973
    on Douglas DC-4, 480f
    prevents rudder lock, 481f
Double slotted flap, 427–430
    effect of, 378f
    articulating-vane, 428, 429f
    fixed vane, 428, 428f
    main/aft, 428, 429f
    reference geometry schematics for, 435f
Double-delta, 341
Down-selection matrix, 309t
Drag, 663–664
    of aircraft by class, 752
    airfoil impact on, 289
    of airfoils and wings, 668–670
    analysis, 739t
    of antennas, 731
    basic drag coefficient, 674–675
    basic drag modeling, 666
    of blisters, 728
    breakdown, 852–853
    of canopies, 727
    CDBM, 697–700
    characteristics of Gurney flap, 433f
    compressibility effect on, 278–279, 730
    cooling, 714
    correction for cockpit windows, 726
    correction of, 444
    CRUD, 709
    of deployed flaps, 725
    estimation pitfalls, 972
    of external sources, 736
    of floats, 724
    of landing gear, 718
    of landing gear pant fairings, 715
    lift-Induced drag coefficient, 686–690
    means to reducing, 691–693
    models for airfoils, 287–289
    of parachutes, 735
    polar
        for airfoil, 406f
        of NACA series airfoils, 268f

quadratic drag modeling, 666–670
sensitivity, 909
simple wing-like surfaces, 715
skin friction drag coefficient, 675–679
of stopped propellers, 731
of streamlined struts, 715
total drag coefficient, 691
trim, 444, 710
of various geometry, 732
of windmilling propellers, 731
Droop nose leading edge. *See* Hinged
Drum brake, 559
Dynamic pressure, 591, 769
Dynamic stability, 127–128, 462
Dynamic viscosity, 991

E
EASA. *See* European Aviation Safety Agency
EASA regulations, 13
Eastlake model
  business aircraft, development cost, 44
    avionics cost, 46
    certify, total cost to, 46
    development support, total cost of, 45
    engineering, total cost of, 45
    engineering man-hours, 44–45
    flight test operations, total cost of, 45
    manufacturing, total cost of, 45
    manufacturing labor man-hours, 45
    materials, total cost of, 45–46
    power plant cost, 46
    quality control, total cost of, 45
    retractable landing gear cost, 46
    tooling, total cost of, 45
    tooling man-hours, 45
  GA aircraft, development cost, 37
    avionics cost, 41
    break-even analysis, 43, 44b
    certify, total cost of, 40–41, 40b
    cost analysis, 39
    development support, total cost of, 39–40
    engineering, total cost of, 39
    engineering man-hours, 37–38
    example, 38b
    flight test operations, total cost of, 40
    manufacturing, total cost of, 40
    manufacturing labor man-hours, 38–39
    materials, total cost of, 40
    power plant cost, 41–43, 42b
    product liability costs, 37
    quality Control, total cost of, 40
    retractable landing gear cost, 41
    tooling, total cost of, 40
    tooling man-hours, 38
EHP. *See* Equivalent horsepower
Electric
  airplanes, 206, 903–904
  propulsion, 206
Electric motor, 182, 190, 203–206
Elevator, 495
  limit isobar, 170–171
  stall limit, 170
Elevator authority, 484, 488, 493
Elevator deflection, 460
Elevon, 460, 950, 952
Elliptic wing planform, 331
Empty weight, 134, 161f
  of aircraft ranges, 138
  fractions, 139–141, 140f
  impact, 148f
  ratio, 137
  sensitivity, 909
Endplate wingtip, 176–177
Endurance performance, 897
Endurance Profile 1, 9, 911–912
Endurance Profile 2, 9, 912
Endurance Profile 3, 9, 913
Engine cooling, 665
Engine power, 186–187
  airspeed effect on, 192
  altitude effect on, 192–195
  manifold pressure and RPM effect, 195–196
  piston, 187
  temperature effect on, 195
Engineering
  lean, 8–9
  man-hours, 37–38, 44–45
Engineering cost, 39, 45
Engineering drawings, 32
Engineering reports, 30–32
EOM. *See* Equation of motion
Epoxy, 111
Eppler, 256
Eppler-code, 255
Equation of motion (EOM), 802
  for climb maneuver, 822–825
  for descent maneuver, 822
  general solution of, 802–804
  for landing roll, 938
  for T-O ground run, 798–799
Equilibrium glide speed, 929
Equivalent airspeed, 47, 770
Equivalent horsepower (EHP), 185
Exit
  area, 216–217
  sizing, 213–219
Experience effectiveness, 36, 36f
External flap. *See* Junkers flap
Extrusion, 39

F
FAA. *See* Federal Aviation Administration
FAA regulations, 46, 111
Famous airfoils
  Clark Y airfoil, 267–268, 270f
  Davis wing airfoil, 272–273
  GA(W)-1 airfoil, 271–272, 272f
  Joukowski Airfoils, 274–275, 275f
  Liebeck Airfoils, 275, 275f
  NACA 23012 airfoil, 268–271, 271f, 272f
  "peaky" airfoil, 273–274, 273f
  supercritical airfoils, 274
  USA-35B airfoil, 268, 270f
FAR. *See* Federal Aviation Regulations
FAR 14 CFR Part 23, 825, 861
  business aircraft certification, 44
  fire extinguishing systems, 523
  GA aircraft certification, 4

FAR 14 CFR Part 23 (*Continued*)
  requirements for T-O speeds, 800–801
  restrictions for aircraft classes certification, 4t
  Subpart E—Powerplant, 209
Fasteners, 103, 212
FEA. *See* Finite element analysis
Feathering propeller, 586
Features and Upgradability, 7
Federal Aviation Administration (FAA), 3, 13, 99
Federal Aviation Regulations (FAR), 13
Fiberglass, 108–109
  boat glass, 111
  R-glass, 112
Fibers, 111–112
  aramid, 111
  boron, 111
  carbon, 111
  graphite, 112
  unidirectional, 319
Field-of-view, 84
Final weight, 896, 914–915
Finite element analysis (FEA), 27–28
Firewall, 210, 523
Fishbone diagram, 19
  categories, 20
  during design process, 21f
  for preliminary airplane design, 19
  typical, 20f
Fixed landing gear, 17t
Fixed slot leading edge, 407–408, 412
Fixed-pitch propeller, 42, 586, 594
  cubic spline method for, 623–624
  desired pitch for, 592
  propeller efficiency graph for, 632f
Flap
  deflecting effect of, 284
  Gurney flap, 432–434
  Junkers flap, 423–425
  Krüger flap, 408–411
  plain flap, 417–420
  single-slotted flap, 425–426
  split flap, 420
  Zap flap, 420–422
Flap Extension airspeed, 782
Flaperon, 437, 952
Flare, landing, 966
Flare distance, 936, 940
Flare maneuver, 935
Flexible wings, 349–350
Flight envelope, 774–775
  completion, 782
  design airspeeds, 775–778
  for GA aircraft, 782–783
  gust loads, 778–781
  maneuvering loads, 775–778
Floatplane, 925
Floats, 724
Flow separation, 247, 249–250
  effect of, 247–248
  effect of early, 282, 283f
  growth on aircraft, 367–369
  impact on, 289
  trends for simple diffusers, 227f
Flying boat, 117–118

Flying wing, 460
  wing twist of, 125–126
Folding bull-nose Krüger flap leading edge, 411
Forebody strakes, 973–974
Forging, 100–101
Forward swept planform, 337–338
Fowler flap, 430
  aerodynamic properties, 432
  feature, 430
  general design guidelines, 431–432
  single-slotted, 430–431
Free roll
  determination, 940
  distance, 938t
Free-body
  reaction forces on, 566f
  T-O ground run, 797
  two-dimensional, 822
Free-roll distance, 938t, 940
Frise aileron, 951
Frustum
  frustum-shaped fuselage, 523
  geometry, 507f
Frustum fuselage, 523–524
Fuel, 84
  FF, 188–189
  fuel cell, 205
  fuel grades for jet, 188
  for mission, 907–908
  operational cost, 51
  piston engine installation, 210–213
  system weight, 144
  wing area sizing, 56
Fuel consumption
  air-to-fuel ratio, 192
  aspirated piston engines, 192t
  comparable turbofan aircraft, 495
  SFC, 188–189
    for jets, 189
    for pistons, 189
  typical, 196
Fuel system, 212–213
  gravity-fed, 84
  weight, 144
Fuel tank selector, 969
Fuel weight
  Justification for maximum zero, 136f
  range analysis, 899
  ratio, 38, 137
Fuel-cell, 205, 205f
Fuselage design, 522, 526–527
Fuselage geometry
  large aircraft, 534
  surface areas and volumes, 544
  tadpole fuselage, 524–526
Fuselage internal dimensions, 531–532
Fuselage sizing, 526
  cockpit layout, 532–535
  external shape
    initial design, 526–529
    refining, 529–531
  internal dimensions of fuselage, 531–532
Fuselage volume, 26
Fuselage width, height, 364

G

Gagg-Ferrar, 59–61
Galvanic corrosion, 105–106
Galvanic corrosion, 105–106, 125–126
GAMA. *See* General Aviation Manufacturers Association
Gantt diagram, 19, 20f
Gap
   effect on magnitude, 420f
   jet airspeeds, 428–429
   NLF airfoil, 414
Gearbox, 227–228
General Aviation aircraft
   AR values for, 309t
   CG envelope for light, 169f
   Cirrus SR22, 911f
   design checklist, 964–965
      balked landing capability, 965–966
      center of gravity travel during flight, 972
      climb capability, 968
      control system harmony, 968
      control system jamming, 970
      control system stretching, 969–970
      crosswind capability at touch-down, 965
      drag estimation pitfalls, 972
      fuel tank selector, 969
      ground impact resistance, 970
      natural damping capability, 969
      one-engine-inoperative trim and climb capability, 969
      reliance upon analysis technology, 971–972
      roll authority, 967–968
      stall handling capability, 966–967
      stall margin for horizontal tail, 967
      take-off rotation capability, 966
      trim at stall and flare at landing capability, 966
      weight estimation pitfalls, 972
      wing/fuselage juncture flow separation, 972
   weight ratios for, 916t
General Aviation Manufacturers Association (GAMA), 34, 103
General process of aircraft design
   design process, 11
      elementary outline, 11–12
      for GA aircraft, 13
      per torenbeek, 12–13
   regulatory concepts
      advisory circular, 15
      airworthiness directives, 14
      authorization, 15
      maintenance requirements, 14
      parts manufacturer approval, 15
      service bulletin, 15
      special airworthiness certificate, 14
      standard airworthiness certificate, 14
      supplemental type certificate, 14
      technical standard order, 15
      type certificate, 13–14
Glass-transition temperature, 113
Glide, 3–4
   comparison, 867
   equilibrium glide speed, 929
   glide distance, 932
   lift coefficient, 664
   POH, 740–741
   transport efficiency, 877
Glide distance, 873, 932
Goettinger, 443

Graphite, 112
Gross weight
   using historical relations, 138–140
   maximum lift coefficients, 361t–362t
   preliminary data, 67t
   properties, 83t
Ground adjustable propeller, 586
Ground effect, 350
   airplane in, 350f
   high-wing aircraft, 85
   lift-induced drag correction in, 738–739
Ground fine, propeller, 586
Ground friction
   ground roll friction coefficients, 554t
   main gear tire, 565–566
   weight on wheels, 800
Ground impact resistance, 970
Ground roll, Take-Off, 800
Ground speed (GS), 555, 770
Ground-loop, 550t
Growth
   features and upgradability, 7
   flow separation on aircraft, 367–369
   intermittent periods, 35f
   NASA's Glenn research center, 252
GS. *See* Ground speed
Gull-wing dihedral, 87
Gurney flap, 432–434, 433f
Gust load, 156
   for airframe loads, 347
   design cruising speed, 776
   step-by-step, 778–781
Gyroscopic effects, 595–607

H

H-tail, 127–128, 494–495
Handling requirements, 5
Helix
   angle, 590–591
   geometric pitch angle, 588–589
   propeller, 587f
Helmholz's vortex theorem, 381
Hershey bar. *See* Constant chord planform
Hershey Bar wing. *See* Wing planform
High wing location, 92
High-wing
   airplanes, 82
   configuration, 84
   ground effect, 85
   vortex-lattice model, 967f
   wing-struts, 84
Hinge moment (HM), 949f
   coefficient, 949
   control surfaces, 243, 948–949
   high pressure, 127
   pressure distribution, 242
   zap flap, 422
Hinged leading edge, 403–406
HM. *See* Hinge moment
Hoerner wingtip, 446
Hoop frame, 85
   aluminum semi-monocoque fuselage, 128
   landing gear loads, 522
   underlying fuselage structure, 129f
Horizontal airspeed, 824

Horizontal tail (HT), 89, 460
    aspect ratio for, 305
    downwash angle, 467
    impact on longitudinal trim, 289
    initial tail sizing optimization, 503–509
    for KC-135, 500f
    magnitude effect, 448
    stall margin for, 967
    total weight, 162
    trim drag, 711–712
    weight, 143
    weight data for, 176t
Horizontal tail volume, 501
    initial tail sizing optimization, 503–509
    on location of stick-fixed neutral point, 502f
    tail sizing, 47
House of Quality (HQ), 21–27, 25f
HT. *See* Horizontal tail
Hub, 16
    blades, 210
    constant-pitch propeller, 590
    correction parameter calculation, 655
    and tip effects, 604–605
Hull speed, 703–708
Humidity, 768–769
Hybrid electric aircraft, 206
Hydrostatic stability, 5, 764
Ice-accretion, 252–253

I

IFR range, 914–923
Important elements of new aircraft design, 237–238
    aesthetics, 240
    aircraft design process phases, 241
    certifiability, 239
    development program phase, 242–243
    ease of manufacturing, 239
    features and upgradability, 239–240
    handling requirements, 239–241
    integrated product teams, 241
    lean engineering and lean manufacturing, 241
    maintainability, 240
    mission, 238–239
    performance requirements and sensitivity, 239
    post-development programs, 243
Incidence angle
    decalage angle, 328–329
    determination, 328f
    wing, 255f
Incompressible Bernoulli equation, 246
Indicated airspeed, 769, 882
Inflation pressure, 555
    selection of tire sizes, 555
    tires and tire, 137
    for typical aircraft, 557t
Initial weight
    analysis methods, 138–141
    cruise segment, 896
Initial weight estimation, 134
Inlet
    diffuser Inlet, 224–225
    inlet types for jet engines, 223–224
    inlet-radiator-exit method, 217–219
    jet engine inlet sizing, 223–227

    NACA duct for, 79–80
    piston engine inlet, 213–219
Inlet lip radius, 226
Inlet-Radiator-Exit method, 217–219
Installation
    aircraft power plant, 209
        danger zones around propeller aircraft, 210
        fireproofing, 209–210
        firewall, 210
        fuel system, 212–213
        gas turbines, 222
        jet engine inlet sizing, 223–227
        piston engine inlet and exit sizing, 213–219
        piston engine installation, 210–213
    braking system, 554
    cockpit window, 727
    control surfaces, 46–51
    cost of avionics, 46
    STC, 14
    turboprop on agricultural aircraft, 196f
    ventral fin, 480
Insurance
    annual insurance cost, 47
    cost, 47
    form of liability, 35
Integrated Product Teams (IPT), 9
Inverted gull-wing dihedral, 87
Inverted V-tail, 493, 493f
Inverted Y-tail, 494, 494f
IPT. *See* Integrated Product Teams
Ishikawa diagram. *See* Fishbone diagram

J

Javafoil software, 256
Jet A, Jet A-1, Jet B, TS-1, 188t
Jet engine inlet, 223–227
Jodel wing, 87, 117–118
Joining, 102–103
Joukowski airfoil, 274–275, 275f
Junkers flap, 423–425

K

KCAS. *See* Knots calibrated airspeed
KEAS. *See* Knots equivalent airspeed
KGS. *See* Knots ground speed
KIAS. *See* Knots indicated airspeed
Kinematic viscosity
    dynamic viscosity, 7
    properties, 766t
    Reynolds number, 252
Kinetic energy (KE), 560
    derivation of equation, 560
    energy height, 891
    propulsive efficiency, 618–619
    rate-of-climb, 931
Knots
    using airspeeds, 740
    preliminary data, 67
Knots calibrated airspeed (KCAS), 740
Knots equivalent airspeed (KEAS), 775
    airspeeds, 47
    design maneuvering speed, 777
    properties, 775t
Knots ground speed (KGS), 557t, 768
Knots indicated airspeed (KIAS), 769

Knots true airspeed (KTAS), 740
Krüger flap leading edge, 408–411
KTAS. *See* Knots true airspeed

L

Labor cost, 38–39, 45
Laminar boundary layer, 242–243
   boundary layer transition, 248
   leading edge to force, 250
   transition, 379
   wing layout, 374f
   Xfoil, 255
Landing gear forces, 565
Landing over 50 ft, 942
Landing performance, 942–944
Landing roll, 938
Landing wire bracing, 724
Lapse rate
   atmospheric ambient temperature, 763–764
   temperature constants, 764t
Lateral directional stability, 462–483
Lateral stability
   directional and roll stability derivatives, 478t
   overturn angle, 569
   requirement for, 477, 477f
   roll or bank, 475
   slipping or sideslip, 475
   winglets, 448–449
Law of effectiveness, 46–47, 349
Leading edge extension, 302
   AOA, 379
   lift distribution, 356f
Leading edge radius
   flow over an object with, 249f
   LE radius, 256–257
Leading edge slat, 412–416
Leaf spring landing gear, 563–564
   landing gear legs, 715
   for small aircraft, 564f
Leaf-spring, 560–562, 563f
Lean Engineering, 8–9
Lean Manufacturing, 8–9
Learjet 45XR, 785
   drag polar, 852, 853f
   example, 853b
   flight envelope for, 881f
   properties, 786t
   three-view, 785f
Learning curve, 36
Liability cost, 37
Liebeck airfoil, 275, 275f
Lift
   airfoil stall characteristics, 251
   angle-of-attack at zero lift, 240
   compound tapered planform, 335
   compressibility effect on, 278
   and drag, 5
   engine placement, 89
   force, 149
   forward-swept planform, 337–338
   fuselage, 81
   generation, 245
      Bernoulli theorem, 246
      Kutta-Joukowski circulation theorem, 246–247
      momentum theorem, 245–246
   lift coefficient of minimum drag, 241
   lift curve slope, 239–240, 343
   maximum lift coefficient, 66
   performance efficiency, 56
   positive relation, 26
   Reynolds number effect, 276f
   SAS, 5
   section lift coefficient, 239
   smeaton lift equation, 239
   spanwise location, 155
   turbulators, 981
   USA-35B, 268
Lift Bernoulli theorem, 246
Lift coefficient, 344
   AOA, 86
   camber effect, 276
   comparison of section, 331f
   compressibility, 279
   linear range, 240
   maximum
      for desired stalling speed, 66
      gap effect on magnitude, 420f
      max lift ratio, 357
      maximum theoretical lift coefficient, 276–277
      nacelle strakes, 981
      for selected aircraft, 361t–362t
      standard lift curve graphs, 276
      impact of sweep angle on, 318
      of thin airfoils, 252
      USA-35B, 268
   minimum drag, 241
   numbering system, 262
   relationship, 344
   section, 86
   wide-range lift curve, 344
Lift curve
   airfoil stall characteristics, 251
   airfoil's, 283–284
   derivation of equation, 346
   flap effect to airfoil on, 285f
   NACA series airfoils, 268f
   Reynolds number effect on, 276f
   three-dimensional wing, 343
   two-dimensional, 240f, 276
Lift curve slope, 239–240, 343
   complete aircraft, 347–348
   determination, 345–347
   using equation, 344
   property, 349
Lift distribution
   conventional, 242–243
   drag due to wing washout, 736
   Fourier series, 386
   ideal, actual and wasted, 329f
   methods to present spanwise, 331–332
   optimum, 256
   un-flexed and flexed wings, 350f
Lift distribution with flaps deflected, 336
Lift Kutta-Joukowski circulation theorem, 606–607
Lift momentum theorem, 245–246, 638–640
Lift-induced drag
   correction factors, 672
      Oswald efficiency, 672–673
      span efficiency, 673

Lift-induced drag (*Continued*)
  corrections
    in ground effect, 738
    wingtip correction, 737
  using lifting-line method, 689
  magnitude, 309
  Oswald span efficiency, 363
  Prandtl-Betz integration, 690
  pros, 334
Lift-induced drag coefficient, 314–315, 383–384
  AOA and airspeed, 686
  generic formulation, 687
  using lifting-line method, 689
  from momentum theorem, 686–687
  monoplane equation, 386
  Prandtl-Betz integration, 690
Lift-to-drag ratio
  Liebeck Airfoils, 275
  maximum
    calculation, 351
    determination, 326f
    graphical determination, 673f
    for modern sailplanes and powered sailplanes, 316f
  performance efficiency, 56
Light sport aircraft (LSA), 3
  aircraft classes, 3t
  establishing weight ratios for, 139t
  KCAS for, 58
  stalling speed, 353
Linear range, 240, 343
Loading cloud, 171–173, 529f
Longeron, 101, 117
Longitudinal static stability, 463–466
Low wing location, 82–86
Low-wing, 82
  aircraft, 85
  configuration, 84
  fuel system, 212
LSA. *See* Light sport aircraft

M

Mach number, 249
  compressibility, 278–279
  computer code, 207–209
  critical, 282f, 289
  effect on lift and drag, 278f
  effect on pitching moment, 280f
  throttle ratio, 196–197
  thrust ratio, 199f, 200f
Machining, 101, 124
Macro-and micromechanics, 109
Main landing gear
  comparison, 174t
  main wheel tires, 556–557
  structural capabilities, 170
  structural limits, 170f
  weight, 37–38, 111–112
Main spar, 120–121
  cross sections, 121, 121f
  leading edge rib, 123
  main ribs, 122
Maintainability, 8
Maintenance, 14
  cost, 46, 49
  to flight hour ratio, 46–47, 49

laminar boundary layer, 676
protocols, 585–586
repair stations, 487–488
requirements, 14
Maintenance requirements, 14
Maneuvering airspeed, 775–778
Maneuvering load, 210–211, 775–778
Manhours
  for engineering development, 6
  number of engineering, 37–38
  number of manufacturing labor, 38–39
  number of tooling, 38, 45
Manufacturing labor
  cost analysis, 39
  number of man-hours, 38–39
  total cost of, 40
Materials
  aircraft cost, 56
  aircraft fabrication and, 98–115
  CET and TET, 196–197
  composite materials, 108
    aircraft construction methodologies, 114
    fabrication methods, 114–115
    fibers, 111–112
    gelcoat, 113–114
    glass transition temperature, 113
    pre-cure, 114
    pros and cons, 110–111
    resin, 112–113
    sandwich core materials, 113
    structural analysis, 109–110
    types of composite, 108–109
  technical standard order, 15
  total cost of, 40
Mattingly, 4, 197–198
Max zero fuel weight, 135
Maximum landing weight, 135
Maximum lift coefficient
  for desired stalling speed, 66
  gap effect on magnitude, 420f
  max lift ratio, 357
  maximum theoretical lift coefficient, 276–277
  nacelle strakes, 981
  for selected aircraft, 361t–362t
  standard lift curve graphs, 276
  impact of sweep angle on, 318
  of thin airfoils, 252
  USA-35B, 268
Maximum operating airspeed, 787–789
Maximum zero fuel weight, 134–135, 137
Maxwell leading edge slot, 414–415
MDO. *See* Multi-disciplinary optimization
Mean Geometric Chord (MGC), 84
  Cirrus SR22 properties, 784t
  comparison, 158t
  effectiveness law, 349
  MAC, 304
  spanwise location, 149
  trapezoidal wing, 155
Mean-line, 256
  NACA four-digit airfoils, 257–258
  slope calculation, 259–260
  y-value computation for, 259
MGC. *See* Mean Geometric Chord

Mid-wing
  aircraft design, 85
  configurations, 84
Minimum angle-of-descent, 931
Minimum control airspeed, 601
Minimum descent airspeed, 893–894
Minimum lift coefficient, 239, 343
Minimum power required airspeed, 864–867
Minimum sink rate, 931
Minimum unstick airspeed, 183
Minimum wetted area, 503, 510
Miscellaneous drag. See Additive drag
Mission definition, 5, 78–79
Mission definition, 5
Mission profile, 899, 914–921
Mission range, 899–909
Molding, 100, 111
Moment of inertia
  comparison, 158t
  inertia properties, 161t
  mass, 166f
  parallel-axis theorem, 156
  propeller, 619–620
  system of discrete point loads, 167–168
  trapezoidal wing, 155
Mono-wheel
  landing gear with outriggers, 571
  with outriggers, 92
  structural weight, 92
Monocoque, 117
Monoplane, 77
  aspect ratio, 310
  decalage angle for, 328, 329
  equation, 386
Multi-disciplinary optimization (MDO), 56
  estimation, 145
  software for, 57
Munk-Multhopp method, 473

N

NACA 1-series airfoil, 263–264
NACA 4-digit airfoil, 257–258
NACA 5-digit airfoil, 261f, 263
NACA 6-series airfoil, 264–266, 265f
NACA 7-series airfoil, 266–267, 266f
NACA 8-series airfoil, 267
NACA airfoil
  with Highest C_{lmax}, 270t
  lift and drag properties, 269–271
  with Lowest C_{dmin}, 270t
  properties of selected, 267, 269t
  pros and cons, 267, 267t
Nacelle, 81
  configurations, 585t
  propeller configurations, 584
  stabilizing effects, 473
  strake on Airbus A319 commercial jetliner, 981f
  twin-engine turboprop aircraft, 90
Nacelle strakes, 981
Natural Laminar Flow (NLF), 665
  airfoil, 675, 692
  LFC, 691
  quadratic drag model, 668
National Oceanic and Atmospheric Administration (NOAA), 1, 763
NBAA range, 919

Never-exceed airspeed, 773t–774t
95th percentile human, 533f, 544–545
NLF. See Natural Laminar Flow
NLF airfoil, 692
  advantage, 289
  chordwise distribution for, 243f
  composite sandwich construction, 119
  pressure-recovery region, 526
  square trailing edge, 257
  for stabilizing surfaces, 126–127
NOAA. See National Oceanic and Atmospheric Administration
Noise
  advisory circular, 15
  in cabin, 485
  climb angle, 827
  propeller, 606–607
Non-Conventional aircraft, 3t
Non-planar wing. See Polyhedral wing
Normal Category
  aluminum alloys, 106t
  applicable properties, 775t
  40-ft wingspan, 125
Normal force, 238–239
  moment equation, 469–470
  propeller normal and side force, 598–599
  workaround for, 470
Nose landing gear
  CG location, 165
  comparison, 174t
  drag of, 723–724
  geometric definitions, 552f
  nosewheel tires, 557
  weight, 38, 143
NRLMSISE-00, 2
Numerical integration method
  equation of take-off motion
    closed-form integration, 807
    nomenclature, 807f
    using spreadsheet, 807
  ground run analysis, 810t
  propeller thrust at low airspeeds, 807–808

O

Oleo-strut landing gear, 564f
Operational cost
  aircraft estimation
    aircraft ownership, 46
    manufacturing and selling airplanes, 46
  business aircraft, 49
    annual fuel cost, 49
    annual insurance cost, 49
    engine overhaul fund, 49
    hourly crew, 49
    maintenance cost, 49
    maintenance to flight hour ratio, 49
    storage cost, 49
    total yearly cost, 51
  GA aircraft, 46
    annual fuel cost, 47
    annual inspection cost, 47
    annual insurance cost, 47
    annual loan payment, 47
    cost per flight hour, 48
    engine overhaul fund, 47
    maintenance cost, 46

Operational cost (*Continued*)
  maintenance to flight hour ratio, 46–47
  monthly loan payment, 47
  storage cost, 47
Optimum glide in headwind or tailwind, 770–771
Optimum glide in rising air, 778–779
Optimum glide in sinking air, 778–779
Optimum lift, 329–330
Oswald's span efficiency
  aerodynamic properties calculation, 67
  determination, 67t
  estimation, 363
    definition, 363
    Douglas method, 364
    lifting line theory, 364
    straight wings, empirical estimation for, 363
    swept wings, empirical estimation for, 363–364
    USAF DATCOM method, 364–365
  lift-induced drag coefficient, 314–315
Outrigger
  design guidelines, 572f
  monowheel landing gear with, 571
  monowheel with, 92
  tailwheel and, 92
Overhaul
  annual insurance cost, 49
  engine overhaul bank, 46
  engine overhaul fund, 47
  TBO, 47

P
P-factor, 595–596, 598
Parasitic drag, 666
  CDBM, 697
  endplates, 448
  lift-induced drag, 442
  span efficiency, 673
Parasol wing location, 82–86
Parasol-wing, 85–86
  configurations, 84
  consolidated PBY-5 Catalina, 85f
  dihedral effect, 86
  lower lift-induced drag, 86
  wing configurations, 86f
Parts manufacturer approval (PMA), 15
Payload, 135
  aircraft performance, 762
  fuselage, 526
  payload-range sensitivity study, 919–921
Payload-range
  analysis, 920–921, 921t
  NBAA payload-range sensitivity plot, 920f
  sensitivity study, 919–921
PE. *See* Potential energy
Performance chart
  engine, 195–196
    extracting piston power, 228–229
    using petty equation, 229–230
    RPM, 230f
  piston-engine performance chart, 229f
Performance padding, 762–763
Personal jet, 200
Petty-equation, 229–230
Philosophy of design, 3
Piston engine, 182, 190

air-to-fuel ratio, 192
airspeed effect on engine power, 192
altitude effect on engine power, 192–195
BHP, 185
common fuel grades for, 188t
compression and pressure ratios, 192
cost of power plant, 41
displacement, 192
energy content of fuel for, 187
example, 189b
four-stroke engine operation, 191
fuel consumption, 189
in GA aircraft, 183
inertia loads, 98
installation, 210
  application, 211f
  danger zones, 211f
  fuel system, 212–213
  loads generated by, 210–211
  systems integration, 212
  torque, 211
  types of engine mounts, 212
manifold pressure and RPM effect on, 195–196
manufacturers, 228
performance analysis, 620–621
power plant thermodynamics, 183
specific fuel consumption for, 192
STC, 14
temperature effect on engine power, 195
turboprop engines, 222
two-stroke versus four-stroke engines, 190–191
weight of, 145, 145f
Piston engine exit
  exit area and cowl flaps, 216–217
  inlet-exit-dependent heat transfer, 215–216
  inlet-radiator-exit method, 217–219
Piston engine inlet, 213
  adequate cooling, 213
  airflow, 215f, 216f
  exit area and cowl flaps, 216–217
  fuel system, 214f
  inlet-exit-dependent heat transfer, 215–216
  inlet-radiator-exit method, 217–219
  proper sizing, 214
  pusher configurations, 213
  for selected aircraft, 215f
  tractor and pusher aircraft configurations, 213
  updraft or a downdraft methodology, 213
Pitch control, 607–608, 962–964, 963f
Pitching
  airfoils, 244
  law of effectiveness, 349
  Mach number effect, 280f
  NACA airfoils, 267
  sweep angle
    on stall characteristics, 318
    on structural loads, 318
  wind-tunnel model, 451
Pitching moment
  aerodynamic center, 104
  airfoil at angles-of-attack, 256
  airfoil's pitching moments, 121–122
  airfoils, 104
  combinations, 378f
  impact of ground effect on, 352f

Mach number effect on, 280f
modeling for simple wing-HT system, 466
NACA series airfoils, 268f
short-bubble leading edge stall, 252
stall effect types on, 123f
swept-back and straight wing configurations, 498f
wing partition method, 437–438

Pitching moment coefficient
compressibility effects, 278–280
estimation, 439–441
impact on longitudinal trim, 289
magnitude, 256
NACA five-digit airfoils, 261
three-dimensional objects, 239

Plain flap, 417–418, 418f
aerodynamic properties, 419–420
ailerons, 951
differential ailerons, 952
drawbacks, 418
flap area, 418
general design guidelines, 418–419
polynomial representations, 726t
single-slotted flap, 425
streamlines, 418f

Planar
elliptic planar wing, 450
motion equations for airplane, 823
solution, 849
steady motion, 849

PMA. *See* Parts manufacturer approval

Polyhedral wing, 452
comparison, 454f
Glaser-Dirks DG-1000 sailplane features, 453f
non-planar wings, 453
using potential flow theory, 453
pros and cons, 453t
sailplanes, 453
straight-wing design, 453

Polyhedral wingtip. *See* Polyhedral wing

Post-development program phase, 11

Potential energy (PE), 184t
power plant, 40–41
zooming, 891

Power
airspeed effect on engine, 192
basics, 184t
in BHP, 61f
BHP, 185
correlation, 145f
cost of power plant, 41–43, 46
EHP, 185
GA and experimental aircraft, 191t
modern computer, 57
normalization, 59–61
noticeable power effects, 89–90
number of blades effect on, 615–618
optimum design points, 58
piston engine, 190
plant expert, 9
power-related coefficients, 614–615
propulsive or thrust, 595
for same-displacement engine, 190
SHP, 185
temperature effect on Engine, 195
THP, 185–186

three halves power law, 524

Prandtl's lifting line theory, 379–380
Biot-Savart law, 380–381
Helmholz's vortex theorems, 381
lifting line formulation, 381–384
vortex filament law, 380–381

Preliminary design, 10
detail design phase, 10
development program phase, 11
hypothetical conceptual design, 20f
selection of tire sizes, 555
Torenbeek's diagram, 13
weight tolerancing, 174

Pressure
aerodynamic loads, 98
center of, 243–244
contemporary types, 557t
in cylinders, 193
dynamic
actual distribution, 330
conversion, 213
example, 826b, 827b
pitot tube or pitot, 769
stalling speed expression, 66
force, 237
geometry, 444
gradient, 368
high compressor, 197
inflation, 555
lag error, 769
MAP, 228, 229f
molding, 100
numbering system, 267
piston engines, 183
slipstream effects, 597–598
static port, 79–80
static source, 769
tires and tire inflation, 549
tube fuselage, 523–524, 524f
vessels, 129–130

Pressure altitude, 6
example, 219b
in ft, 990
hydrostatic equilibrium equations, 764
pressure or density ratios, 765

Pressure coefficient, 241
canonical, 241–242
compressibility modeling, 280–281
correction of lift, 654

Pressure distribution
chordwise, 242
conventional lift distribution, 242–243
using pressure coefficient, 242
stratford distribution, 243
difference in chordwise, 677f
drag coefficient, 674–675
equation, 674–675
properties, 242f

Pressure drag
AOA or *AOY* changes, 666
C_{Do} and C_{Df}, 668
drag modeling, 666
drag of subsonic aircraft, 666
flow separation, 367
form factor, 666

Pressure drag (*Continued*)
  generated by object, 732
  Reynolds number, 981
Pressure recovery, 226
  in airplane piston engine cowling inlets, 217
  airspeed effect on engine power, 192
  AOA and AOY, 223–224
  diffuser inlet, 224
  at front face of compressor, 91
  scoop-type inlets for, 213
Pressure-tube
  fuselage, 523–524, 524f
  reasonable accuracy, 536
Pressurization
  for an aircraft, 9
  cabin, 129–130, 130f
  characteristics, 10
  correction factors, 36
  special considerations, 129–131
Pressurized fuselage, 97–98
Product of inertia
  inertia properties, 161t
  parallel-axis theorem, 167
  symmetry of object, 166
  system of discrete point loads, 168
PROFILE software, 255–256
Project cost
  analysis, 37t
  comparison, 43t
Project management, 19
  communication skills, 19
  engineering project, 19
  fishbone diagram, 19–21, 20f
  Gantt diagrams, 19
  house of quality, 21–27
  managing compliance, 21
  project plan and task management, 21
  quality function deployment, 21–27
  scheduled deadlines, 19
  time management, 19
Proof-of-concept aircraft, 11
Propeller, 13, 582
  blade element theory, 640
    compressibility corrections, 654–655
    computer code, 656
    determination, 650–654
    formulation, 641–650
    Prandtl's tip and hub loss corrections, 655–656
  blade-element theory, 583
  configurations, 584–586
  constant-speed, 42
  cost of power plant, 41–43
  determination, 620
    analytical methods, 632
    constant-speed propellers, cubic spline method for, 628–630
    converting piston BHP to thrust, 620–621
    estimating thrust from manufacturer's data, 631–632
    fixed-pitch propellers, cubic spline method for, 623–624
    propeller thrust at low airspeeds, 621–630
    quadratic interpolation method, 621
    step-by-step, 630–631
  fixed *vs.* constant-speed propellers, 594–595
  and gas turbines, 185
  geometric propeller pitch, 588–589
    constant-pitch propeller, 590
    determination, 592
    fundamental formulation, 589–590
    pitch angle or geometric pitch, 590
    propeller rotation relationships, 591
    variable-pitch propeller, 590–591
  geometry, 587–588
  ground clearance, 549
  McDonnell XF-88B, 582–583
  nomenclature, 586–587
  normal force, 469–470
  propeller effects, 595–596
    angular momentum and gyroscopic effects, 596–597
    asymmetric yaw effect, 599
    blockage effects, 602–604
    constant-speed propeller, 595f
    effects of high tip speed, 605
    hub and tip effects, 604–605
    propeller noise, 606–607
    propeller normal and side force, 598–599
    skewed wake effects, 605–606
    slipstream effects, 597–598
    twin-engine aircraft, 599–602
  properties and selection, 607
    activity factor, 613–614
    advance ratio, 611–613
    effect of number of blades on power, 615–618
    moment of inertia of the propeller, 619–620
    power-and thrust-related coefficients, 614–615
    prop diameter estimation, 608–609
    propeller efficiency estimation, 610–611
    propeller pitch estimation, 610
    propulsive efficiency, 618–619
    tips for selecting suitable propeller, 607–608
  propulsive or thrust power, 595
  pros and cons of, 585t
  Rankine-Froude momentum theory
    computer code, 638–640
    flow properties inside control volume, 634f
    formulation, 633–635
    ideal efficiency, 635
    idealized flow model for, 633f
    maximum static thrust, 636–638
    propeller-induced velocity, 632–633
  rotating, 90
  series hybrid, 206
  single piston-engine propeller airplane, 64b
  thrust for, 572
  tractor configuration, 89
  Tupolev Tu-114 passenger aircraft, 583f
  turboprop, 196
  windmilling propellers, 593–594
Propeller $A \cdot q$ loads, 605–606
Propeller activity factor, 613–614, 619f
Propeller asymmetric thrust effect
  airplane operating with, 476
  characteristics, 480
  moment, 600
  OEI, 643
  rotation axis, 599–600
Propeller blade, 582
  forces, angles and velocity for, 642f
  geometry of metal, 588f
  rotating at static conditions, 592f
  section lift coefficients for, 604f
  spanwise flow, 641

Propeller blockage effect, 602–604, 604f
Propeller diameter
  cubic spline method, 623
  in ft or meters, 609
  number of blades, 615–616
Propeller disc
  blockage effects, 602–603
  designer, 590–591
  normal force, 598
  preliminaries, 638
Propeller efficiency, 47
  airspeed-power map, 613f
  computer code, 638–640
  estimation, 610–611
  example, 626b
  fixed-pitch climb propeller, 594
  function, 64b, 67–68
  graph for fixed-pitch propeller, 632f
  high tip speed, 605
  sensitivity plot, 842, 842f
  on spline, 629f
  step-by-step, 626b
  variation, 649f
Propeller hub effect, 210, 587
Propeller induced airspeed, 638
Propeller normal force effect, 598
Propeller number of blades
  effect on power, 615–618
  engine characteristics, 607–608
Propeller pitch angle, 589–590, 590f
Propeller propulsive efficiency
  conversion process, 618
  Froude efficiency, 618
  propeller efficiency, 618
  propeller efficiency map, 618–619
Propeller side force effect, 598–599
Propeller thrust, 47
  CG of engine, 211
  determination, 620
    analytical methods, 632
    constant-speed propellers, cubic spline method for, 628–630
    converting piston BHP to thrust, 620–621
    estimating thrust from manufacturer's data, 631–632
    fixed-pitch propellers, cubic spline method for, 623–624
    propeller thrust at low airspeeds, 621–630
    quadratic interpolation method, 621
    step-by-step, 630–631
  at low airspeeds, 605
  momentum theory, 632
  tractor propeller configuration, 91
Propeller tip effect, 583
  Ground Clearance, 549
  SHP, 185
Propeller tip speed, 612b
Propwash
  advantage, 483
  blocked and unblocked, 603f
  Froude efficiency, 618
  HT, 473, 485
  tractor configuration, 89
  VT surfaces, 494
Pulsejet engine, 182
  electric motors, 182
  THP, 185–186

Pure electric aircraft, 206
Pusher propeller, 89
  drawbacks, 91
  propeller structure, 585–586
  pros and cons, 585t

Q
Quality control
  flight test operations, 45
  total cost of, 40
Quantity discount, 36
Quantity discount factor (QDF), 36
  depends on experience effectiveness, 36–37, 36f
  experience effectiveness adjustment factor, 36

R
Rake angle, 227, 446–448
Raked wingtip, 446
  Boeing 777 commercial transport aircraft, 447f
  effectiveness, 446–448
  geometry, 447f
  Hoerner wing tip, 447f
  lift-induced drag, 447f
  positive and negative sweep rakes, 446
  Rakelet, 448
Ramp weight, 135
Range, 896
  airspeed for Jet, 875
    best range airspeed for, 876–877
    Carson's airspeed, 877
    comparison to best glide speed, 877
    requirement for maximum range, 875–876
  AR values, 309t
  Breguet endurance equation, 898
  Breguet range equation, 897
  cruise segment for, 896
  determining fuel required for mission, 907–908
  endurance analysis, 897–898
  inflation pressures for aircraft, 557t
  linear, 240, 343
  range analysis, 899
    mission profiles, 899
    physical and mathematical interpretation, 899t
    range profile 1, 899–900
    range profile 2, 901–902
    range profile 3, 902
    range profile 4, 903–904
  sensitivity studies, 908
    aspect ratio sensitivity, 909
    drag sensitivity, 909
    empty weight sensitivity, 909
  SFC and TSFC, 898–899
  specific range
    Brequet flight profile, 910–911
    CAFE foundation challenge, 910
    Cirrus SR22 general aviation aircraft, 911f
    efficiency, 910
    fuel quantity, 909
    instantaneous, 909–910
    quadratic model, 672
    range and endurance, 9
  subsonic minimum drag coefficients, 752
  vs. weight, 896–897
Range efficiency, 349–350
Range performance, 343

Range Profile 1
    constant airspeed/attitude cruise, 907
    constant airspeed/constant attitude cruise, 899–900
Range Profile 2, 901–902
Range Profile 3
    constant airspeed/attitude cruise, 908
    constant airspeed/constant attitude cruise, 902
Range requirements, 921
Range sensitivity, 908
    aspect ratio sensitivity, 909
    drag sensitivity, 909
    empty weight sensitivity, 909
    payload-range sensitivity study, 919–921
Rankine-Froude momentum theory, 632
    computer code
        ideal and viscous profile, 638
        Plan Next Step, 638–640
        Preliminaries, 638
        propeller efficiency, 638
        Set Initial Values, 638
    flow properties, 634f
    formulation, 633–635
    ideal efficiency, 635
    idealized flow model for, 633f
    maximum static thrust, 636–638
    propeller-induced velocity, 632–633
Rapid pattern recognition, 2
Rate-of-climb (ROC), 824–825
    airplane, 59
    airspeeds, 773t–774t
    climb gradient, 825
    in OEI configuration, 600–601
    performance handbooks, 749
    for propeller-powered airplane, 834
Rate-of-descent
    aircraft reduces altitude, 927
    in airplanes, 562
    derivation of equation, 927
Reference altitude
    atmospheric conditions, 765
    temperature constants, 764t
Reference area, 302
    airplane's design, 302
    for Boeing KC-135 Stratotanker, 302f
    drag of tires, 718t
    EFPA, 668
    reference geometry, 499
    T-O configuration, 800
Reference speed, 773t–774t
Reference temperature, 764t
Regulations, 7
    aircraft classes certification, 3t
    center of gravity envelope, 168
    GA aircraft, 4
    Klegecell®, 111
    modern-day, 838–839
    safety in commercial aviation, 6–7
    spin, 482
    standards, 13
    V-n diagram, 775
    vertical gusts, 779f
Regulatory concepts, 13
    advisory circular, 15
    airworthiness directives, 14
    CAR, CAA, FAA and EASA, 13

harmonization, 13
    maintenance requirements, 14
    parts manufacturer approval, 15
    service bulletin, 15
    special airworthiness certificate, 14
    standard airworthiness certificate, 14
    supplemental type certificate, 14
    technical standard order, 15
        authorization, 15
    type certificate, 13–14
Requirements for static stability, 464
Requirements phase, 10
Resin
    composites, 6
    fibers, 111
    fibrous composites, 108
    pre-preg, 114
    properties, 115t
    purpose, 112
    RTM, 111
    thermoplastics, 113
    thermosets, 112–113
Responsiveness
    aileron design requirements, 952
    control surface sizing, 948
    control systems, 969
    maximizing, 962
    stubby planform, 309–310
Retractable
    approach requirement, 968
    landing gear, 41, 553–555
    nacelles, 91
    take-off requirement, 967–968
    wing structure, 120
Retractable landing gear
    advent of, 553–554
    aluminum Mooney Ovation, 119
    approach requirement, 968
    benefit of, 554
    cost per airplane, 41
    drag of, 722, 723f
    fads in aircraft design, 79t
    internal volume, 554–555
    kinematics, 554
    retraction and extension, 554
    stick diagrams, 555
    take-off requirement, 967–968
Reversing propeller, 586
Reynolds number, 247, 985–995
    aerodynamic properties, 406
    change in skin friction coefficient with, 678f
    critical, 733
    determination, 680
    effects, 276f, 277–278
    form factors at subcritical, 702
    form factors at supercritical, 703
    linear range, 240
    using Owen's criterion, 252
    SI system, 247
    turbulent boundary layer, 248
Ribs
    joining, 102
    leading edge rib, 123
    parallel, 123
    plywood, 118

structural member, 121–122
stub ribs, 123
swept-back wings, 122–123
in trailing edge, 264–265
Rivet
blind-riveting, 102
bucking-bar, 102
head types, 102–103
stability, 676
standard procedure, 102f
ROC. *See* Rate-of-climb
Rocket engine, 182
Roll control, 437
aileron authority derivative, 953
aileron design requirements, 952
ailerons purpose, 949–950
differential ailerons, 952
elevon, 952
flaperon, 437
flaperon, 952
frise ailerons, 951
plain flap ailerons, 951
roll damping derivative, 955–956
slot-lip ailerons, 951–952
spoiler-flap ailerons, 951
types, 950, 950f
Roll damping, 468
aileron sizing, 960
derivative, 475–476
Dutch, 494
estimation, 960
Roll stability at stall
aerodynamic washout, 319
configuration, 333
forward-swept planform, 337
Rolling
airplanes, 413–414
approach requirement, 968
cold, 107
Rolling moment, 87
calculation, 955
requirements, 477, 477f
Rolling moment coefficient, 953
calculation, 956
roll damping coefficient, 956
Roofed-cabin, 88
Root chord, 303
quarter-chord line, 303–304
straight tapered wing planform, 334
wing, 308
wing planform, 41
Rotation, 813
distance, 49
double-subscripts, 166
mass moment of inertia, 165
plain flap, 417–418
propeller geometry, 587
propeller relationships, 591
single-slotted fowler flap, 430–431
zap flap, 420–422
Rotation, Take-Off, 170, 462
inverted V-tail configuration, 493
operation, 303
pros and cons, 585t
take-off rotation capability, 495

Rotation airspeed, 575f
Round wingtip, 443, 443f
Rubber doughnut landing gear, 560–562, 563f
Rudder, 475
airplanes feature plain flaps for, 418
endplates, 448
functionality, 127–128
rudder-lock, 477–479, 973
severity, 480
stability and control theory, 460
standard practice, 601f
vertical ventral fin, 493
Rudder authority
for aircraft, 964
low directional stability, 964
V-tail, 493
Y-tail, 493
Rudder deflection, 460
clockwise propeller rotation, 599–600
RPM, 599
standard practice, 601f
Rudder pedal
control system interaction, 492–493
pilots, 600
side-slip airplane on, 491
steering, 553
Rutan VariEze, 114
canard configuration, 496
Whitcomb winglet, 450–451
wing layout properties, 320t–322t

S
SAC. *See* Special airworthiness certificate
Sailplane
AR values for, 309t
certification, 3t
cruise flaps, 284
dihedral configurations, 87
empirical formulation, 315
high-aspect-ratio wings, 349
monowheel with outriggers, 92
properties, 83t
spoilers, 285–286
T/W for, 185t
wing layout properties, 320t–322t
Sailplane operation
GA aircraft, 539–540
maximum lift-to-drag ratio, 316f
V-tails, 489, 503t
Sailplanes regulations
culver twist formula, 324
using quadratic spline method, 741
Schuemann wing, 339–340
Sandwich construction, 119–120
SB. *See* Service bulletin
Schuemann planform, 339–340
Schuemann wing, 339–340, 340f
Seaplane
and amphibians, 968
noticeable power effects, 89–90
Seaplane hull, 549
Seaplane operation, 91
Section lift coefficient, 239
aerodynamic washout, 319
angle-of-attack at zero lift, 240

Section lift coefficient (*Continued*)
   comparison of fractional, 332f
   compound tapered wing planform, 335
   compressibility, 279
   endplate wingtip, 448
   hub and tip effects, 604–605
   lift coefficient of minimum drag, 241
   lift curve slope, 239–240
   linear range, 240
   maximum and minimum lift coefficients, 239
   minimum drag coefficient, 241
   for propeller blade, 604f
   vortex-lattice model, 967f
Section lift coefficient distribution
   aerodynamic washout, 319
   with and without endplates, 449f
   CFD methods, 369
   constant-chord wing, 315–316
   elliptical planform, 333f
   geometry and lifting characteristics, 341f
   pros, 334
   spanwise, 688f
     cause roll instability, 437
     for full span flaps, 438f
     for partial span flaps, 437f
   taper ratio effect on, 316f
Semi-tapered planform, 339
Sensitivity
   climb maneuver, 840
     altitude sensitivity, 841f, 842
     calibrated airspeeds, 840
     propeller efficiency, 841
     spreadsheet, 840f
   landing distance sensitivity studies, 942
   low-power, 602
   NLF wings, 379
   payload-range sensitivity study, 919–921
   performance, 5
   propeller efficiency, 842
   range sensitivity studies, 908
     aspect ratio sensitivity, 909
     drag sensitivity, 909
     empty weight sensitivity, 909
   take-off sensitivity studies, 816–817
   weight, 842
Separated boundary layer, 665
Separated flow, 223
   configurations, 482
   flow field, 497
   rudder sizing, 964
   streamlines, 249
Service bulletin (SB), 15
Service Bulletin, 15
Service ceiling, 838–839
   business jets, 839–840
   T/W for, 59
Sesquiplane, 8, 86
SFC. *See* Specific fuel consumption
Shaft horsepower (SHP), 185
Shear web, 84
   aft, 121–122
   main spar, 120–121
   structural member, 122
   stub rib, 123
Sheet metal forming, 100–101

Shimmy, 553, 564–565
Shimmy damper, 553, 554t
SHP. *See* Shaft horsepower
Side force, 598–599
   inflation pressure, 555
   jet engine pylons, 977
   span efficiency, 363
   for tractor configuration, 599f
   V-tail slope, 493
   winglet, 448–449
Simple Krüger flap leading edge, 409
   in action, 410f
   aerodynamic properties, 411
   folding, bull-nose Krüger flap, 411
Simple surface flex, 100–101
Single slotted flap, 425
   aerodynamic properties, 426
   general design guidelines, 426
   leading edge devices, 425
   special lift-enhancing tab, 980
   trailing-edge high-lift devices, 425–426
   translation, 430–431
   versions, 425
Sink rate, 930
Skin
   curvatures, 100
   friction, 655
     analysis methods, 685t
     coefficients calculation, 707f
     multi-panel wing, 686t
   fuselage, 128
   joining, 102
   plywood, 118
   sheet metal for, 374
   stringers, 124
   weight, 123
   wing, 978
Skin friction drag, 664
   CDBM, 708
   computation, 681
   force for complete wing, 683
   Frankl-Voishel correction for, 280–281
   miscellaneous or additive drag, 698–700
   separation and flat roof, 243
   using surface wetted area, 681
Skin friction drag coefficient, 675
   boundary layer stability, 676
   calculation, 680–686
   change in, 678, 678f
   characteristics, 675–676, 676f
   by fluid's viscosity, 675
   laminar boundary layer, 676
   maximum thickness, 676–678
   natural laminar flow airfoil, 675
   standard formulation to estimate, 678–679
   streamlined three-dimensional shape, 676
   transition, 676f
Slipstream, 597–598
Slot-lip
   aileron, 951, 951f
   cruise and climb performance, 952
   NACA TN-5475 and NACA R-6026, 951
Slotted flap
   cruise flaps, 285

double-slotted flaps, 427–428
   articulating-vane double-slotted flap, 428, 429f
   difference, 428
   fixed-vane double-slotted flap, 428, 428f
   general design guidelines, 430
   main/aft double-slotted flap, 428, 429f
   triple-slotted flap, 428–430, 430f
on pitching moment, 378f
polynomial representations, 726t
reference geometry schematics, 434f
single-element, 425, 425f
   aerodynamic properties, 426
   general design guidelines, 426
   plain flaps, 425
   trailing-edge high-lift devices, 425–426
   versions, 425
Smeaton lift equation, 239
Solid modeling
   components, 147
   standard three-view drawing, 29f
Spar
   carry-through for small GA aircraft, 30f
   composite aircraft, 119–120
   cross sections for GA aircraft, 121f
   light aircraft, 126
   low wing structure, 84
   low-wing aircraft, 523
   main, 120–121
   spar-rib-stringers-skin, 301
Spar cap
   extrusion, 101
   main spar cross sections, 121
   weight, 157, 158t
Special airworthiness certificate (SAC), 14
Special Airworthiness Type Certificate, 14
Specific fuel consumption (SFC), 188
   aspirated piston engines for aircraft, 192t
   conventional piston and jet engines, 188
   engines, 188
   fuel flow, 188–189
   for Jets, 189
   for Pistons, 189
   propeller aircraft, 912
   T-O power and, 197t
   thrust specific, 898
   in UK system, 898
Specific range (SR), 909
   CAFE foundation challenge, 910
      Brequet flight profile, 910–911
      Cirrus SR22 general aviation aircraft, 911f
      efficiency, 910
   fuel quantity, 909
   instantaneous, 909–910
   quadratic model, 672
   range and endurance, 9
Speed of sound, 770
   on airfoil, 273
   airplanes, 682
   airspeeds, 48
   drag coefficient, 695
   and mach number, 991
Speed Stability, 855–856
Speed-to-Fly, 877
Spherical wingtip, 443–444, 444f

Spin recovery
   OEI asymmetric thrust, 964
   potential solution to, 483f
   stability and control analysis, 461
   T-tail aircraft post-stall at, 487f
   tail design and, 482–483
   wing droops, 978
Spinner, 587
   empirical correction, 636–638
   piston-engine aircraft, 603–604
Split flap, 420
   aerodynamic properties, 422–423
   design guidelines, 406
   dive brake, 422
   high-pressure region on, 420
   polynomial representations, 726t
   zap flap, 420–422
Spoiler-flap, 950
   ailerons, 951
   types, 950f
Square TE, 257
Square wingtip, 441, 444
SR. *See* Specific range
SR22
   application, 622f
   banking constraint diagram, 886f
   Cirrus, 194, 783–785
   composite sandwich construction, 119
   drag model for, 610
   drag polar and lift-to-drag ratio estimations, 906f
   flat plate skin friction, 683
   maximum lift coefficients, 361t–362t
   sensitivity studies, 816
   T-O performance of selected aircraft, 944t–945t
   variation, 739f
Stability and Control
   airfoils pitching moment, 244
   AOA and *AOY*, 460
   control horns, 975
   dorsal fin and rudder locking, 973
   forebody strakes, 973–974
   handling requirements, 951
   reference area, 302
   static, 462
      airplane, 462
      $C_{m\alpha}$ historical values, 468
      coordinate system, 462
      dorsal fin, 477–480
      longitudinal equilibrium, 468–472
      pitching moment modeling, 466
      requirements for lateral stability, 477
      requirements for static directional stability, 476–477
      static directional and lateral stability, 475–476
      static longitudinal stability, 463–466
      stick-fixed and stick-free neutral points, 472–473
      tail design and spin recovery, 482–483
      ventral fin, 480–482
   tail sizing worksheet, 16
   taillets and stabilons, 974–975
   ventral fin
      and deep stall, 973
      and dutch roll, 973
Stabilons, 974
Stagger, 227
Stagger angle, 227

Stall, 251
- angle-of-attack, 358
- deep stall tendency, 966–967
- long-bubble leading edge, 252
- margin for horizontal tail, 967
- progression on selected wing planforms, 371f
- recovery phase, 366, 367f
- short-bubble leading edge, 252
- speed limits, 65–66
- strips, 975
- TE, 75
- types, 253f

Stall, leading edge, 251–252
Stall, trailing edge, 251
Stall AOA, 277–278
- aerodynamic washout on, 322f
- C_{lmax}, 323
- cons, 334–335
- flow separation effect, 283f
- Krüger flap, 410
- planform shapes, 310f
- tail leaves, 487

Stall characteristics
- airfoil, 251
  - LE stall, 251–252
  - TE stall, 251
- impact on flow separation, 289
- jet aircraft, 977
- NACA 23012 airfoil, 261
- sharp drop, 291f
- stall strips, 975
- impact of sweep angle, 257
- wing, 366
  - deviation from generic stall patterns, 369
  - flow separation growth, 367–369
  - influence of manufacturing tolerances, 378–379
  - pitch-up stall boundary for, 375–378
  - swept-back wing planform, 374–375
  - tailoring stall progression, 369–374

Stall handling
- aerodynamic effectiveness, 442
- capability, 966–967
- impact on maximum lift and, 289
- stall strips, 975–976
- vortilons, 978
- wing droop, 978
- wing fence, 976–977
- wing pylons, 977

Stall margin, 967
Stall progression
- deviation from generic stall patterns, 369
- dissimilar predictions, 972f
- Krüger flap, 411
- on selected wing planforms, 370f
- on straight tapered wing planforms, 372f
- tailoring
  - CFD methods, 369
  - design guidelines, 369–371
  - multi-airfoil wings, 373–374
  - stall characteristics, 369
  - wings with multiple airfoils, 371–373

Stall speed
- C_{Lmax} for desired, 66
- constraint diagram with, 66f
- cruise speed carpet plot, 67–69

KIAS, 740
- limits into constraint diagram, 65–66

Stall strips
- stall handling, 975–976
- stall progression, 330

Stall tailoring, 303
- flow visualization, 369–370
- washout effect on, 373f

Stalling airspeed
- C_{Lmax} for desired, 66
- level stalling speed with load factor, 62
- sizing of wing area, 56
- stalling speed during banking, 62
- with thrust, flap and CG effects, 63
- wing area function, 68t

Standard airworthiness certificate (AC), 14
Standard Airworthiness Type Certificate, 14
Standard atmosphere. *See* US standard
Standing mountain wave, 778–779
State of industry, 34–35
Statistical weight analysis, 142
- aircraft, 142
- statistical aircraft component methods
  - air conditioning and anti-icing, 144
  - avionics systems weight, 144
  - electrical system, 144
  - equations, 142
  - flight control-system weight, 144
  - fuel system weight, 144
  - furnishings, 144–145
  - fuselage weight, 143
  - guidance, 142
  - HT weight, 143
  - hydraulic system weight, 144
  - installed engine weight, 144
  - main landing gear weight, 143
  - nose landing gear weight, 143
  - VT weight, 143
  - wing weight, 142
- statistical methods to engine weight estimation, 145
  - weight of piston engines, 145
  - weight of turbofan engines, 146
  - weight of turboprop engines, 146

STC. *See* Supplemental type certificate
Steady climb, 792
- climb capability, 968
- motion equations, 823

Steel
- alloy, 106–107
- endurance limit, 104
- extrusion process, 101
- forging metals, 101
- low-carbon-grade, 101
- properties, 107t
- truss, 119
- types, 564f

Steel alloy, 106–107
Steel truss, 119
Step
- cruise speed carpet plot
  - aerodynamic properties calculation, 67
  - carpet plot creation, 67
  - carpet plot creation, 68–69
  - decide plot limits, 67

preliminary data, 67
preparation for plotting, 68
stall speed, 67
tabulate maximum airspeeds, 67–68
tabulate stall speeds, 67
engine performance charts, 228–229
fuselage external shape initial design, 526–529
geometric layout
   taildragger landing gear, 569–570
   tricycle landing gear layout, 567–569
maneuvering loads and design airspeeds, 775–778
NACA four-digit airfoils
   airfoil ordinates computation, 258–259
   airfoil resolution, 259
   applications, 258
   generation of NACA 4415, 260–261
   numbering system, 258
   ordinate rotation angle calculation, 260
   preliminary values, 259
   prepare ordinate table, 259
   slope of mean-line calculation, 259–260
   thickness calculation, 259
   upper and lower ordinates calculation, 260
   y-value for mean-line, 259
production steps, 8–9
propeller efficiency table, 630–631
quality function deployment
   comparison matrix, 27
   customer requirements, 22–24
   GA airplane, 22
   HQ preparation, 22
   interrelationship matrix, 26
   QFD, 22
   roof, 25–26
   survey responses, 21–22
   targets, 27
   technical requirements, 24
skin friction drag coefficient calculation, 680–686
turboprop engine thrust, 198–200
weight of ribs, 151t
weight of wing skin, 149t–150t
weight of wing shear web, 150t
weight of wing spar caps, 151t
Stick-fixed neutral point, 170
   conventional aircraft, 472
   design guidelines, 501–502
   determination, 473
   distinction, 472
   impact of horizontal tail volume, 502f
Stick-free neutral point, 472
   conventional aircraft, 472
   distinction, 472
   on hinge moments, 501
Storage
   cost, 47, 49
   delta planform shape, 340–341
   energy on board an aircraft, 206
   wing available for fuel, 308
Straight-tapered, 439
Stratford distribution, 243, 243f
Stress corrosion, 105
Stressed-skin construction, 117, 119
Strict liability, 34

Structural layout, 97–98
   airframe, 116
      fuselage structure layout, 126–127
      horizontal vertical tail structure layout, 126–128
      structural concepts, 116–120
      wing structure layout, 120–126
Strut-braced wing
   cantilever configuration, 88
   maximum shear and bending loads, 87–88
Supercritical airfoil, 274
Supplemental type certificate (STC), 14
Supplemental Type Certificate, 14
Surface area
   body of revolution, 536
   cone, 537–538
   elliptic cylinder, 537
   frustum, 503, 513
   paraboloid, 537–538
   pod-style fuselage, 544
   uniform cylinder, 537
   wingtip device, 442
SURFACES software, 971
Swept
   empirical estimation, 5, 41–43
   forward-swept wing, 318
   for high-speed aircraft, 317
   planforms, 336
      aft-swept planform, 336
      cons, 336–337
      forward-swept, 337–338
      pros, 336
      variable swept, 338
   swept-back wings, 122–123
      aircraft inspection with, 123
      rib layouts for, 122f
   USAF DATCOM method for, 364–365

T

T-O rotation
   inverted V-tail configuration, 493
   limit, 170
   operation, 303
T-O weight
   maximum, 204
   mission airplane design, 135
   restrictions, 4t
Tadpole fuselage, 524
   advantages, 524
   approximation, 538f
   modern sailplane, 665
   properties, 524–525
   reduction, 526, 526f
   Rolladen-Schneider LS4 sailplane boasts, 525f
   surface areas and volumes, 539–542
   transition and total fuselage drag, 525f
Tail
   conventional, 483–485
   cruciform, 486
   design and spin recovery, 482–483
   HT weight, 36–37, 143
   inverted U-Tail, 496
   types, 554t
   VT weight, 37–43, 143

Tail (*Continued*)
  weight data for, 176t
  wheel reaction, 576
Tail arm, 69
  effect of changes in, 71–72
  determination, 503–507
  directional moment, 69–71
  horizontal, 143
  KC-135, 500f
  tail sizing, 502–503
  weight data, 176t
Tail configuration, 92
  A-tail configuration, 495f
  conventional, 484f
  cruciform tail, 486
    aft podded engine configuration, 486
    drawbacks, 486
  H-tail configuration, 494f
  inverted V-tail configuration, 493f
  rudder during spin on, 483f
  U-Tail configuration, 496
  V-tail or butterfly tail, 489–493
  Y-tail configuration, 493f
Tail landing gear, 576
Tail surface area, 17t–18t
Tail upsweep, 529
Tail wheel
  castering nose and, 553
  positioning, 549
  reaction, 576
Taildragger
  advantages, 91
  aircraft, 92
  castering-wheel configurations, 553f
  configurations, 92
  free-body diagram, 938
  geometric layout, 569–570
  ground characteristics, 565–567
  T-O analysis, 793
  treatment of T-O run for, 815–816
Tailless aircraft
  culver twist formula, 324
  designer, 324
  Panknin and Culver formulas, 325f
  range of subsonic minimum drag coefficients, 752t
Taillet, 974–975, 975f
Take-off, 44–45, 791
  aircraft with swept wings, 337
  dry, 186
  padding factors, 763t
  performance, 818t
  requirement, 967–968
  rotation capability, 966
  three-position slat, 415
  wet, 186
Take-Off over 50 ft, 816
  sensitivity, 816–817
  steep runway slope impact, 817
Take-Off performance, 621, 792f
Takeoff safety airspeed, 773t–774t
Tandem
  configuration, 92
  fixed landing gear, 92
  monowheel with outriggers, 92

Tandem wheel
  configuration, 92
  fixed landing gear, 92
  pros and cons, 550t
Tandem wing, 976
Taper ratio, 303
  during design phase, 307
  mathematical expressions, 304–305
  original and reduced wing, 394t
  wing, 309, 315–317
  wing planform, 41
Tapered planform, 307f
  compound, 335
  semi-tapered planform, 339
  straight, 334
    airplane types, 334
    cons, 334–335
    drawback, 334
    geometry and lifting characteristics, 334f
    pros, 334
TC. *See* Type certificate
Technical Standard, Authorization Order, 15
Technical standard order (TSO), 15
Technical Standard Order, 15
Technical standard order authorization (TSOA), 15
Temperature
  in adiabatic compression, 217
  atmospheric ambient, 763–764
  on engine power, 195
  equations for, 767
  glass transition, 113
  high cabin, 88
  TET, 186
Thermal
  cruise flaps, 284
  Klegecell®, 111
  properties, 110–111
Thermodynamics
  gas turbines, 183–184
  piston engines, 183
  power plant, 183
THP. *See* Thrust horsepower
3-dimensional lift coefficient, 237
Three position leading edge slat, 415, 415f
Three-surface configuration, 495, 495f
Three-view drawings, 28
Throttle ratio
  aircraft engine design, 197–198
  ambient air, 196
  CET and TET, 196–197
  function, 207
  High compressor pressure ratio, 197
  prediction of engine thrust, 198
  theta-break, 198
Thrust, 41
  airspeed in, 857–858
  analytical methods, 632
  asymmetric, 476
  coefficients, 614–615
  computer code, 207–209
  constraint analysis, 56
  efficiency model for, 812f
  elevator authority, 89

flat rating, 187
generation, 184–185
for jet-powered aircraft, 804
level stalling speed with, 861–862
maximum
   climb power, 187
   cruise power, 187
optimum design points, 58
Prandtl correction, 655
for propeller, 572
propulsive power, 595
ratio, 202f
using simplified drag model, 874f
T-O and *SFC*, 199t, 202t
THP, 185–186
thrust-to-weight ratio, 185
Thrust coefficient, 614
   calculation, 648–649
   fraction, 635
   propeller, 47
Thrust effects, 91
   mechanical energy, 184
   on stability and control, 224
Thrust generation, 184
   net force, 184–185
   propeller and gas turbines, 185
   rockets, 185
   take-off, 186
   theoretical representation, 184f
   thrust-to-weight ratio, 185
Thrust horsepower (THP), 185–186
Thrust specific fuel consumption, 197
   for jet, 898
   for piston engine, 898
   special version, 912
   variables, 922–923
Thrustline
   free-body diagram, 712f
   effect of high or low, 90f
   trim drag, 712
Time to altitude, 836–837
Tip chord, 303
   leading edge, 303–304
   spherical wingtip, 443
   straight tapered wing planform, 334
Tire
   drag of tires, 718t
   footprint, 552–553, 553f
   geometry, 555, 667
   inflation pressure, 549
   sizes, 555
   types, 555, 558f
Tire footprint, 552–553, 553f
Tire geometry, 555, 667
Tire inflation pressure, 549
Tire sizes, 555
Titanium, 107
   firewall, 681
   properties, 107–108, 108t
Titanium alloy, 107–108, 108t
Tooling, 10
   control surfaces, 271–272
   cost analysis, 34
   man-hours number, 6
   total cost of, 40

Tooling cost, 40, 52–53
Torque, 183
   AOA, 594
   calculation, 642
   conversion, 186
   power-torque relation, 44–46
   RPM, 186
   turboprop aircraft, 186
   variation, 649f
Tort reform, 35
Total distance, 192
   determination, 813
   landing phase, 935
Touch-down
   airspeeds, 939t
   landing phase, 935
   motion equation, 938
   pilot to flare aircraft, 966
   weight at point of, 916–917
Touch-down airspeed, 935
Tractor propeller
   configuration, 91
   designing team, 531
Trailing edge tab, 980–981
Trailing link landing gear, 564
Transition, 248
   boundary layer, 248–249
   distance determination, 813–814
   FRPs and GRPs, 113
   from laminar to turbulent flow, 676f
   laminar-to-turbulent, 679
   movement, 677f
   parameters, 679t
   pressure distribution, 242
   Surface roughness, 249
   T-O run segments, 794t
Transition after Take-Off, 44–45, 793t
Transition ramp
   separation bubbles, 250
   transition curve, 981
Trapezoidal planform, 303
   leading edge, 303
   MGC for, 307f
   trapezoidal wing planform, 304f
Tricycle
   configuration, 91
   fixed landing gear, 92
   ground stability, 567f
   landing gear, 565–566
   landing gear reaction loads, 571–572
      aerodynamic loads, 572–573
      design guidelines, 572f
      static loads, 572
   location, 568f
   pros and cons, 550t
   stable on, 92
   taildragger configuration, 92
Triplane
   aspect ratio, 305
   drag characteristics, 753t
   primary advantage, 86
Triple slotted flap, 428–429, 430f
   for commercial jetliners, 430
   heavy mechanical system, 429

True airspeed, 770
  airspeeds, 48
  equivalent airspeed, 770
  ground speed, 770–771
  landing distance, 942
  using simplified drag mode, 872f
TSO. *See* Technical standard order
TSOA. *See* Technical standard order authorization
Tubular, 102
  fuselage, 523
  landing gear struts drag, 720t–721t
  streamlined tension wire, 719
Turbo-normalizing, 193–194
Turbocharger, 194
Turbofan engine, 41
  altitude and airspeed effect, 201
  GA aircraft, 200
  generic-low-bypass ratio thrust, 202
  mounted on pylons, 5
Turbojet engine, 41
  altitude and airspeed effect, 200
  fuel consumption, 199
  T-O thrust and *SFC*, 199t, 202t
  thrust of generic, 200
Turboprop engine, 41
  altitude and airspeed effect, 198–199
  firewall, 210
  installation, 222
  reversing propeller, 586
  T-O Power and *SFC*, 197t
Turbulator, 981
Turbulent boundary layer, 248
  drag sensitivity, 909
  fluid flow inside laminar, 248f
  laminar-to-turbulent transition, 679
  skin friction coefficient for, 680
Turbulent flow
  dependency, 668
  skin friction coefficient, 678–679, 683–684
  tip airfoil, 756
  transition, 676f
Turf, 798t, 938t
Turning radius
  aircraft, 552
  distance to turning center, 552
  geometric definitions for, 552f
  minimum sustainable, 889
2-dimensional lift coefficient. *See* Section lift coefficient
Two position leading edge slat, 412
  geometric parameters, 413f
  mechanical aspect, 413f
Two position propeller, 586
Type certificate, 13–14
  propeller, 210
  STC, 14
  TCDS, 744–745, 784–785
Type III tire, 719
Type metric tire, 557t
Type radial tire, 558t
Type Three-part tire, 977
Type VII tire, 549

U
U-tail
  inverted U-Tail, 496, 496f
  propeller configurations, 496
  twin tail-boom configuration, 496
US standard atmosphere 1976, 1, 8t
USAF DATCOM, 303–304
  arbitrary chord line angle, 305
  C_{Lmax} estimation, 355–360
  method for swept wings, 364–365
Useful load, weight, 919–920, 997–1000
Utility Category, 125f

V
V-tail
  advantages, 490
  configurations, 92, 490f
  difference in yaw response, 492f
  GA aircraft, 489
  inverted, 493
  pitch-up moment, 490
  Rudlicki V-tail, 489
  simplified theory, 493
  unconventional tails, 127–128
Variable camber Krüger flap leading edge, 411
Variable camber leading edge, 406–407
Ventral fin, 480
  AOA condition, 480
  installation, 482
  on Learjet 60, 973f
  pitching moment curve, 482f
  solid curve, 480–482
  stability and control, 973
Vertical airspeed, 824
  general rate-of-climb, 825–826
  in thrust or power, 824–825
Vertical tail, 69, 126
  aircraft components, 80–81
  control surfaces, 126–127
  empennage, 81
  heavy aircraft, 126
  larger airplanes feature, 126
  spar of light aircraft, 126
  unconventional tails, 127–128
  volumes, 500–501
  weight, 37–43
  yaw control, 965f
Vertical tail volume, 500–501, 503t
Viscous profile efficiency, 618
  fixed-pitch propellers, 639
  magnitude, 639
  momentum theory, 638
Volume
  absolute humidity, 766
  break-even analysis, 43
  compression ratio, 192
  fuselage, 26
  infinitesimal, 619
  mid-wing configuration, 85
  passenger, 84
  structural standpoint, 130
Vortex
  flow improvement, 978–980
  generators on aft fuselage, 692f
  pylon, 977
  stall handling, 978
  wingtip correction, 737

Vortex filament, 380
  Biot-Savart law, 380
  constant-strength straight, 381f
  Helmholz's vortex theorems, 381
  lifting line formulation, 381–384
Vortex generator, 432
  on aft fuselage, 692f
  installation and wind tunnel testing, 980
  large fixed-pitch, 692
  on lower HT surface, 980f
  nacelle strakes, 981
Vortilons, 978, 978f

W

Washin, 240, 319–325
Washout
  aerodynamic, 319–323
  AR values for, 309t
  combination, 323
  drag due to wing, 736
  geometric, 319, 322f
  on probable stall progression, 373f
  wing twist, 319–325
Water spray, 89–90
Weight budget, 173
  acceptable for test vehicle, 174
  actual weights, 174t
  aircraft into categories, 174
  weight analysis, 141
  weight reduction, 173–174
Weight ratio
  cruise, 917
  mission, 917
  payloads, 920
Weight tolerances, 174–176
Welding, 101–102
Wet grass, 798t, 938t
Wheel
  castering-wheel configurations, 553f
  configurations, 92
  fairings, 718
  forward compressor, 200–201
  gearbox, 228f
  modern and aluminum wheels, 558–559
  open and closed, 723
  positioning, 549
  T-O phase, 792
Wheel track, 551
  ground instability, 566f
  overturn angle, 569
Whistling, 565
Wind-milling propeller, 593–594, 731
Wing area, 15
  comparing results, 394t
  constraint analysis, 56
  maximum airspeeds, 67–68
  stall speeds, 67, 68t
  trade study, 66–67
  zap flap, 420–422
Wing aspect ratio
  aircraft properties, 83t
  impact of aspect ratio on, 310t
  lift-induced drag magnitude, 309
Wing attachment, 29
  aft, 126
  extrusions, 101
  fastener orientation, 124–125
  transfer wing torsion, 122
Wing droop, 978
Wing fence, 976–977
Wing planform
  aerospace engineer's formula sheet, 997
  arbitrary, 960
  constant-chord sweptback, 337f
  crescent, 339
  delta, 340f, 341f
  elliptical, 333f
    formidable fighters, 333
    stall progression on, 372f
  for generic airplane, 329
  ideal, actual and wasted lift distributions, 329f
  Reynolds numbers for, 330–331
  sweep angle, 318
  trapezoidal, 303–309
Wing pylons, 977
Wing skin, 118
  Cuffs for composite aircraft, 978
  multi-airfoil wings, 373
  translation, 430–431
  turbulent boundary layer, 691
Wing span, 307
  DATCOM, 355
  physical and angular stations relationship, 391f
  physical dimensions, 331–332
  roll damping derivative, 955
  scaling top view based on, 682f
  taper ratio, 315–316
Wing taper ratio, 315–316
  general rule-of-thumb, 317
  passenger-carrying aircraft, 317
  spanwise distribution, 316f
Winglet, 274
  blended winglet, 451–452
    design and patent, 452
    modern airliners, 452
  comparison, 449f
  Dutch roll damping, 448–449
  familiar airbus, 448f
  generation of lift, 450
  hypothetical installation, 852–853
  interference factors, 700t
  lift-induced drag to distribution, 450
  skin friction and interference drag, 449–450
  Whitcomb winglet
    development, 450–451
    flight test evaluation, 451
    on McDonnell-Douglas, 451f
    wind-tunnel model, 451
Wingtip, 80–81
  aerodynamic effectiveness, 442
  booster, 444–446
  cons, 333
  correction, 737–738
  design, 441
  on drag polar, 442f
  endplate, 448
  hoerner, 446
  lifting characteristics, 340
  parasite areas and coefficients, 699t
  raked, 446–448

Wingtip (*Continued*)
   round, 443
   spherical, 443–444
   square, 444
   stalls, 319
   tip-loading, 317
Wood construction, 117–119
Work, 184t
   engineering reports, 30
   fabrication, 8–9
   hardening, 101
   IPTs, 9
   Lachmann's original, 412
   torque, 186
   ventral fins, 480
   wing fence, 976
Wrought alloy, 103, 104t

X

XFLR software, 255
   airfoils, 254–255
   user interface, 255f
Xfoil software, 255
   airfoils, 255
   capabilities, 255f
   Reynolds number, 272–273
   vortex-lattice code, 971

Y

Y-tail, 92
   configuration, 493f
   inverted, 494, 494f
   shorter-span V-tail, 494
   V-tail variation, 493

Yaw, 431–432, 460
   adverse, 492, 949–950
   aerodynamic properties, 310t
   airplane, 475
   angle, 71
   dorsal fin, 479f
   frise ailerons, 951
   stability and control theory, 460
Yaw control, 460
   fundamentals, 964
   VT and TEL, 965f
Yawing, 491
   airplane effect, 696f
   stabilizing moment, 566
Yawing moment, 69–71
   H-Tail, 494
   magnitude, 597–598
   nacelles, 91
Yawing moment coefficient, 964, 982–983
Yehudi flap, 79t

Z

Zap flap, 420–422
   data for, 423
   on full-scale aircraft, 422
   hinge moment, 422
Zero lift angle-of-attack
   airfoil at, 240f
   design lift coefficient, 343
   midpoint cruise value, 326
   variables, 397–398

CPSIA information can be obtained
at www.ICGtesting.com
Printed in the USA
BVOW09s1225061216
469937BV00006B/12/P

9 780128 099988